the biology of CANCER

THIRD EDITION

Robert A. Weinberg,
Massachusetts Institute
of Technology

With chapters by:

Richard Goldsby, Amherst College

Michael Hemann, Massachusetts Institute of Technology

Tyler Jacks, Massachusetts Institute of Technology

W. W. NORTON & COMPANY
Celebrating a Century of Independent Publishing

W. W. Norton & Company has been independent since its founding in 1923, when William Warder Norton and Mary D. Herter Norton first published lectures delivered at the People's Institute, the adult education division of New York City's Cooper Union. The firm soon expanded its program beyond the Institute, publishing books by celebrated academics from America and abroad. By midcentury, the two major pillars of Norton's publishing program—trade books and college texts—were firmly established. In the 1950s, the Norton family transferred control of the company to its employees, and today—with a staff of five hundred and hundreds of trade, college, and professional titles published each year—W. W. Norton & Company stands as the largest and oldest publishing house owned wholly by its employees.

Editors: Cailin Barrett-Bressack and Betsy Twitchell
Editorial Advisor: Denise Schanck
Senior Project Editor: Jennifer Barnhardt
Assistant Editor: Maggie Stephens
Managing Editor, College: Marian Johnson
Developmental Editor: Judith Kromm
Copyeditors: Richard Mickey and Marjorie Anderson
Senior Production Managers: Richard Bretan and Eric Pier-Hocking
Media Editor: Gina Forsythe
Associate Media Editor: Jasmine Ribeaux
Media Project Editor: Jesse Newkirk
Media Editorial Assistant: Mandy Lin
Managing Editor, College Digital Media: Kim Yi
Ebook Producer: Kate Barnes
Marketing Manager, Biology: Holly Plank
Design Director: Rubina Yeh
Design: Matthew McClements, Blink Studio, Ltd.
Director of College Permissions: Megan Schindel
Permissions Associate: Patricia Wong
Photo Editors: Stacey Stambaugh and Thomas Persano
Composition: Graphic World
Illustrations: Nigel Orme
Manufacturing: Transcontinental

Permission to use copyrighted material appears alongside the appropriate content.

ISBN: 978-0-393-69021-7

W. W. Norton & Company, Inc., 500 Fifth Avenue, New York, N.Y. 10110
  www.wwnorton.com
W. W. Norton & Company Ltd., 15 Carlisle Street, London W1D 3BS

1 2 3 4 5 6 7 8 9 0

About the Authors

Robert A. Weinberg is a Founding Member of the Whitehead Institute for Biomedical Research. He is the Daniel K. Ludwig Professor for Cancer Research and Professor in the Massachusetts Institute of Technology (MIT) Department of Biology. His research is focused on the molecular and cellular mechanisms of cancer pathogenesis. He was awarded the U.S. National Medal of Science in 1997.

Contributing authors

Richard Goldsby is a Professor of Biology emeritus, Amherst College, Amherst, with a longstanding involvement and expertise in immunology.

Michael Hemann is a Member, Koch Institute for Integrative Cancer Research and Professor of Biology, MIT with an interest in how current cancer therapies succeed or fail.

Tyler Jacks is a Member, Koch Institute for Integrative Cancer Research and Professor of Biology, MIT with an interest in how mouse models of human cancer shed critical light on mechanisms of disease pathogenesis.

Front Cover

When propagated in a three-dimensional gel matrix *in vitro*, mammary epithelial cells can be induced to form the alveoli and ducts closely resembling comparable structures developed naturally in the mammary gland. The fluorescent stains reveal cytokeratins associated with the basal and with the luminal epithelial cells in this experimental model. (From E. S. Sokol, D.H. Miller, A. Breggia et al., *Breast Cancer. Res.* 18:19–32, 2016).

Dedication

Robert A. Weinberg dedicates this third edition, as the first and second ones, to his dear wife, Amy Shulman Weinberg, who endured long hours of inattention, hearing from him repeatedly the claim that the writing of this edition was almost complete, when in fact years of work lay ahead. She deserved much better! With much love.

Preface

"No one can trust you scientists. You're always changing your minds." So said a neighbor of mine in the green woods of New Hampshire, to which I responded "If it makes any difference to you, 99% of what I learned as an undergraduate is still considered to be accurate and true." This brief exchange highlighted the two worlds that scientists live in, surrounded in one world by the fixed certainties of accepted wisdom and, when moving through the other, trying to maintain balance on constantly shifting ground, never knowing how much the landscape will change from one year to the next.

This dichotomy afflicted the textbook authors in major ways, in this edition myself (R.A.W.) and the three co-authors—Tyler Jacks, Michael Hemann, and Richard A. Goldsby—who have wrestled repeatedly with this question. What should we portray with a voice of certainty and conviction and what to describe in a very tentative and uncertain voice? If there is a measure of uncertainty, should we present something nevertheless, doing so with the conviction that several years down the road, it will be seen as the precursor of a major, highly topical breakthrough? Which recent discoveries will emerge as truly important and which will fade into insignificance, being nothing more than trivial distractions?

Then, there is a second major problem: How much material to present? This issue is, if anything, an even greater challenge. The problem is dictated by the subject matter, in this case the vast and ever-expanding area of contemporary cancer research. Some readers are entranced by the wealth of precisely documented detail presented in a textbook, but far more will, instead, be driven away by too much. Independent of the limitations of the human mind to absorb and process the details is the far more pressing problem, specifically, that of losing would-be readers who, having confronted a forbidding mass of complex information, will simply run away and search for more readily accessible, simple prose and logic. Cancer research, as the authors of this textbook truly believe, should be truly accessible with relatively little advanced preparation, unlike theoretical physics or complex organic chemistry. Turning away would-be readers will thwart our achieving the major goal of this book—to recruit a new generation of young people to the ranks of the cancer research community.

These dynamics also influence the detailed contents of each chapter in another way: We consciously aspired to present concepts and the logical development of various lines of cancer research rather than providing encyclopedic assemblies of all known facts. Hence, this book is decidedly not a reference book but rather one that hopes to explain scientific logic and paths of discovery. This also touches on a related agenda of this book: to provide by example instruction to students who are learning precisely how biomedical research actually proceeds.

We embrace the existence of a logical backbone that threads its way through each of the chapters and the research subspecialties presented here. Cancer research is an area that does indeed have a conceptual coherence rather than being a grab bag of unrelated and arbitrarily encountered phenomena. All four of us feel a need to move away from compilations of facts and focus instead on underlying principles. In my own case, this drive comes from a chance encounter in the early 1980s with an older, by-then cynical researcher in England, who was disillusioned by what he perceived was the mindless phenomenology of cancer research at the time and warned me sternly that "you should never ever confuse cancer research with science." This textbook is written with the intent of proving him wrong.

Reassuringly, the initial chapters of this book, which reflect the foundations of this field of modern cancer research, have not changed much from the earlier editions. We present these chapters in their largely unchanged form without apology, because they continue to represent the sturdy logical foundations on which all that follows rests. A need to revise them substantially would indicate that many of our assumptions of what were solid scientific truths were nothing more than illusions. The later chapters of this book, in contrast, address the more dynamic, rapidly changing areas of cancer research whose ultimate outcomes are not yet totally clear.

Compared with other areas of biological research, the science of molecular oncology is a recent arrival; its beginning can be traced with precision to a milestone discovery in 1975. In that year, the laboratory of Harold Varmus and J. Michael Bishop in San Francisco, California demonstrated that normal cell genomes carry a gene—they called it a proto-oncogene—that has the potential, following alteration, to incite cancer. Before that time, we knew essentially nothing about the molecular mechanisms underlying cancer formation; since that time an abundance of information has accumulated that now reveals how normal cells become transformed into tumor cells and how these neoplastic cells collaborate to form life-threatening tumors.

In spite of this abundance of findings, there are major gaps in our knowledge that remain to be filled. Most critically, the lessons about cancer's origins, laid out extensively in this book, have not yet been successfully applied to make major inroads into the prevention and cure of many common types of neoplastic disease. This represents the major frustration of contemporary cancer research: the lessons of disease causation have rarely been followed, as day follows night, by the development of definitive cures.

And yes, there are still major questions that remain murky and poorly resolved. We still do not understand precisely how many types of human cancers are triggered. We still possess a woefully incomplete understanding of the role of the immune system in preventing cancer development. And while we know much about the individual signaling molecules operating inside individual human cells, we lack a clear understanding of how the complex signaling circuitry assembled by these molecules makes the life-and-death decisions that govern the fate of individual cells within our body. Those decisions ultimately determine whether or not one or another of our cells begins the journey down the long road leading to cancerous proliferation and, finally, to a life-threatening tumor.

Contemporary cancer research has enriched numerous other areas of modern biomedical research. Consequently, much of what you will learn from this book will be useful in understanding many aspects of cell biology, immunology, neurobiology, developmental biology, and a dozen other biomedical research fields. Enjoy the ride!

On behalf of my three co-authors, Richard Goldsby, Michael Hemann, and Tyler Jacks,

Robert A. Weinberg
Cambridge, Massachusetts
March 2023

A Note to the Reader on the Third Edition

The third edition of this book is organized into 17 chapters of differing lengths. The conceptual structure that was established in the first two editions still seemed to be highly appropriate for this one, and so it was retained. The chapters are meant to be read in the order that they appear, in that each builds on the ideas that have been presented in the chapters before it. The first chapter is a condensed refresher course for undergraduate biology majors and pre-doctoral students; it lays out many of the foundational concepts that are assumed in the subsequent chapters. The driving force of these three editions has been a belief that modern cancer research represents a conceptually coherent field of science that can be presented as a clear, logical progression. Embedded in these discussions is an anticipation that much of this information will one day prove useful in devising novel diagnostic and therapeutic strategies that can be deployed in oncology clinics. Each text chapter has pedagogical elements that will help you navigate these concepts. Also, certain experiments are described in detail to indicate the logic supporting many of these concepts.

- You will find numerous **schematic drawings**, often coupled with **micrographs**, that will help you to appreciate how experimental results have been assembled, piece-by-piece, generating the syntheses that underlie molecular oncology.

- Scattered about the text are "**Sidebars**," which consist of commentaries that represent detours from the main thrust of the discussion. Often these Sidebars contain anecdotes or elaborate on ideas presented in the main text. Read them if you are interested, or skip over them if you find them too distracting. They are presented to provide additional interest—a bit of extra seasoning in the rich stew of ideas that constitutes contemporary research in this area.

- The same can be said about the "**Supplementary Sidebars**," which are available in the ebook. These also elaborate upon topics that are laid out in the main text and are cross-referenced throughout the book, and they appear in a section at the end of each ebook chapter. Space constraints dictated that the Supplementary Sidebars could not be included in the hardcopy versions of the textbook.

- Throughout the main text you will find extensive **cross-references** whenever topics under discussion have been introduced or described elsewhere. Many of these have been inserted in the event that you read the chapters in an order different from their presentation here. These cross-references should not provoke you to continually leaf through other chapters in order to track down cited sections or figures. If you feel that you will benefit from earlier introductions to a topic, use these cross-references; otherwise, ignore them.

- Each chapter except the first one ends with a forward-looking summary titled "**Synopsis and Prospects**." This section synthesizes the main concepts of the chapter and often addresses ideas that remain matters of contention. It also considers where research might go in the future. This overview is extended by a list of key concepts and a set of questions. Some of the questions are deliberately challenging and we hope they will provoke you to think more deeply about many of the issues and concepts developed.

- Each of the chapters has at its end an **Additional Reading** section, which presents a list of especially useful articles from research journals. Most of these are reviews of specific, circumscribed areas of the scientific literature. These will be useful if you wish to explore a particular topic in more detail.

- Perhaps the most important goal of this book is to enable you to move beyond the textbook and jump directly into the primary research literature. This explains why some of the text is directed toward teaching the elaborate, specialized vocabulary of the cancer research literature, and many of its terms are defined in the **Glossary**. Boldface type has been used throughout to introduce and highlight key terms that you should understand.

- Cancer research, like most areas of contemporary biomedical research, is plagued by numerous abbreviations and acronyms that pepper the text of many published reports; the book provides a key to deciphering this alphabet soup by defining these acronyms. You will find a list of such **abbreviations** in the back.

Because this book describes an area of research in which new and exciting findings are being announced all the time, some of the details and interpretations presented here will become outdated (or, equally likely, proven to be wrong) once this book is in print. Still, the primary concepts presented here will remain, as they rest on solid foundations of experimental results.

The author and the publisher would greatly appreciate your feedback. Every effort has been made to minimize errors. Nonetheless, you may find them here and there, and it would be of great benefit if you took the trouble to communicate them. Even more importantly, much of the science described herein will require reinterpretation in coming years as new discoveries are made. Please email us at biology@wwnorton.com with your suggestions, which will be considered for incorporation into future editions.

Digital Resources for Instructors and Students

The third edition has a number of exciting digital resources for instructors and students to use that are available at wwnorton.com.

Instructor Resources

Image files

Much of the art from the book, sized for classroom display, is available in PowerPoint with alt text and in JPEG format.

Test Bank

Norton uses evidence-based assessment practices to deliver high quality and pedagogically effective quizzes and testing materials. The framework to develop our test banks, quizzes, and support materials is the result of a collaboration with leading academic researchers and advisers. Questions are classified by section and difficulty, making it easy to construct tests and quizzes that are meaningful and diagnostic. Norton Testmaker brings Norton's high-quality testing materials online. Create assessments for your course from anywhere with an Internet connection, without downloading files or installing specialized software. Search and filter test bank questions by chapter, type, difficulty, and other criteria. You can also customize test bank questions to fit your course. Easily export your tests to Microsoft Word or Common Cartridge files for your LMS.

Primary Research Suggestions with Discussion Questions

Primary literature suggestions for every chapter are available to instructors for download on wwnorton.com/instructors along with questions to prompt discussion.

Student Resources

Ebook

Norton Ebooks offer an enhanced reading experience at a fraction of the cost of a print textbook. They provide an active reading experience, enabling students to highlight, take notes, search, and read offline. Instructors can add their own content and notes that students can see as they are reading the text. Supplementary Sidebars that elaborate on and further illustrate concepts and figures in the main text are available for the first time within the ebook in a section at the end of each chapter. Norton Ebooks can be viewed on all devices and are born accessible, with content and features designed from the start for all learners. Available at digital.wwnorton.com/cancer3.

Animations

Narrated animations cover selected topics. These animations for *The Biology of Cancer* are available online and in the ebook.

MP3 mini-lectures

Audio files of mini-lectures by the author are available at digital.wwnorton.com/cancer3.

Acknowledgments

The science described in this book is the opus of a large, highly interactive research community stretching across the globe. Its members have moved forward our understanding of cancer immeasurably over the past generation. The colleagues listed below have helped the authors in countless ways, large and small, by providing sound advice, referring us to critical scientific literature, analyzing complex and occasionally contentious scientific issues, reviewing individual chapters, and providing much-appreciated critiques over the past three editions. Their scientific expertise and their insights into pedagogical clarity have proven to be invaluable. Their help extends and complements the help of an equally large roster of colleagues who helped with the preparation of the first two editions. These individuals are representatives of a community, whose members are, virtually without exception, prepared and pleased to provide a helping hand to those who request it. We are most grateful to them. Not listed below are the many colleagues who generously provided high-quality versions of their published images; they are acknowledged through the literature citations in the figure legends.

Daniel Abankwa
Mohammed Abba
Michael Abler
Rudolf Aebersold
Andrew Aguirre
Donna Albertson
Adriana Albini
Sarah Allinson
C. David Allis
Mohammad Alzrigat
L. Mario Amzel
Dimitris Anastassiou
Kristin Ardlie
Cheryl Arrowsmith
Steven Artandi
Laura Attardi
David E. Axelrod
Gustavo Ayala
Yael Aylon
Melanie Badtke
Chris Bailey
Frances Balkwill
Allan Balmain
Xiaomin Bao
Mariano Barbacid
David Bartel
Stephen Baylin
Gregory Beatty
Robert Benezra
A. Elif Erson Bensan
Anton Berns
Bradley Bernstein
Rameen Beroukhim

Brian Bierie
J. Michael Bishop
Stacy Blain
Paul Blainey
Jennifer Blasé
Gerd Blobel
AnneMarie Block
C. Richard Boland
Jean-Paul Borg
Mariana Brait
Greg N. Brooke
Joan Brugge
Thijn Brummelkamp
Christopher Burge
Tony Burgess
Mike Calderwood
Carlos Carmona-Fontaine
Peter Campbell
Judith Campisi
Clare-Anne Edwards Canfield
Carmen Cantemir-Stone
Lewis Cantley
Leah Caplan
Lisa Carey
Scott Carter
Tony Cesare
Iain Cheeseman
Andrew Cherniack
Paola Chiarugi
Shin-Heng Chiou
Karen Cichowski
Hans Clevers
David Cobrinik

Robert Coffey
Ofir Cohen
Jonathan Coloff
Carla Concepcion
John Condeelis
Amanda Coutts
Pau Creixell
Edwin Cuppen
Christina Curtis
Chi Van Dang
Karin de Visser
João Pedro de Magalhães
Titia de Lange
Karin de Visser
Ralph DeBerardinis
James DeCaprio
James DeGregori
George Demetri
Gina DeNicola
Channing Der
Rik Derynck
Mark Dewhirst
John Dick
Deborah Dillon
Elie Dolgin
Anushka Dongre
Steven Dowdy
Julian Downward
Glenn Dranoff
Denis Duboule
Andrew Duncan
Michel Dupage
Nicholas Dyson

Gail Eckhardt

Mikala Egeblad

Martin Eilers

Robert Eisenman

Kevin Elias

Eran Elinav

Stephen Elledge

Neta Erez

Manel Esteller

Andrew Ewald

Yi Fan

Douglas Fearon

Napoleone Ferrara

David Fisher

Garrett Frampton

Christopher French

Peter Friedl

Sylvia Fromherz

Julia Fröse

Leslie Gaffney

Giulio Genovese

Marios Georgiou

Jacqueline Gerritsen

Frank Gertler

Gad Getz

Filippo Giancotti

Charles Giardina

Luke Gilbert

Richard Gilbertson

Christine Gilles

Thomas Gilmore

Richard Goldsby

Richard Gomer

Farhad Islami Gomeshtapeh

Vera Gorbunova

Anna Grabowska

Charles Graham

Kenneth Gray

Jack Griffith

Frank Grosveld

Amy Groth

Kun-Liang Guan

Kishore Guda

Lorraine Gudas

Vincent Guen

Malini Guha

Theresa Guise

Barry Gumbiner

Piyush Gupta

Silvio Gutkind

Daniel Haber

William Hahn

Kevin Haigis

Bashar Hamza

Alfizah Hanafiah

Douglas Hanahan

Edward Harhaj

Masanori Hatakeyama

James Haughian

Daniel Hayes

Whitney Henry

Yusuf Hannun

Mary Hitt

Aaron Hobbs

David A. Hoekstra II

Matan Hofree

Eric Holland

Andrew Hoyt

Ralph Hruban

Karen Hubbard

Chi-Chung Hui

Tony Hunter

Bonnie Hylander

Richard Hynes

Antonio Iavarone

Mitsuhiko Ikura

Sonia Iyer

Emily K. Jackson

Rudolf Jaenisch

Siddhartha Jaiswal

Stefanie Jeffrey

Ahmedin Jemal

Nicolas Jonckheere

Johanna Joyce

FuiBoon Kai

William Kaelin

Eric Kalkhoven

Yibin Kang

Madhuri Kango-Singh

Loukia Karacosta

Jan Karlseder

Edward Kastenhuber

Zuzana Keckesova

Alan Kelly

Trey Kidd

Carla Kim

Alec Kimmelman

Nicole King

Kenneth Kinzler

Jesse Kirkpatrick

Keith Knutson

Angela Koehler

Eugene Koonin

Carl Koschmann

Adrian Krainer

Jordan Krall

Barry Kramer

Merideth Krevosky

Cornelia Kröger

Heather Kuruvilla

Lindsay LaFave

Arthur Lambert

Laurens Lambert

David Lane

Ju-Seog Lee

Greg Lemke

Anthony Letai

Erez Levanon

Arnold Levine

Daniel Levy

Xin Li

Daniel Link

David Livingston

Josep Llovet

Gregory Longmore

Jari Louhelainen

Scott Lowe

Kiven Erique Lukong

Salvador Macip

Ilaria Malanchi

Carlo Maley

Scott Manalis

Mathias Mann

Sanford Markowitz

Ronen Marmorstein

Sarah-Anne Martin

Stuart Martin

Yosef Maruvka

Joan Massagué

Roberto Mayor

Sandra McAllister

Andrea McClatchey

Frank McCormick

Donald McDonald

Matthew Meyerson

Thomas Michniacki

Kim Miller

Mari Mino-Kenudson

Arindam Mitra

Saili Moghe

Shabaz Mohammed

Aristidis Moustakas

Paul Mueller

Karl Munger

James Murphy

Cornelis Murre

Hyman Muss

Mandar Muzumdar

Honami Naora

Nicholas Navin

Jens Nielsen

M. Angela Nieto Toledano

Martin Noble

Roeland Nusse

Ruth Nussinov

Robert O'Donnell

Fergal O'Farrell

Esther Oliva

Moshe Oren

Terry Orr-Weaver

Barbara Osborne

Timothy Padera

Raymond Pagliarini

Duojia Pan

Klaus Pantel

Luis Parada

David Pellman

Eran Perlson
Charles Perou
Julian Peto
Stefano Piccolo
Eli Pikarsky
Mikael Pittet
Leonid Pobezinsky
Jeffrey Pollard
Supriya Prasanth
David Price
Justin Pritchard
Carol Prives
Sidharth Puram
Bruce Pyenson
Jayaraj Rajagopal
Joe Ramos, Jr.
Kodimangalam Ravichandran
Roger Reddel
E. Premkumar Reddy
Elizabeth Repasky
Lyndsay Rhodes
David Richardson
Anne Ridley
Gregory Riely
Samantha Riesenfeld
Douglas Robertson
Rodrigo Romero
Amy Romesburg
Jatin Roper
Jeffrey Rosen
Richard Rothenberg
David Rowley
Orit Rozenblatt-Rosen
David Sabatini
Masoud Sadaghiani
Julien Sage
Yardena Samuels
Francisco Sanchez-Rivera
Leona Samson
Ioannis Sanidas
Pierre Savagner
Antje Schaefer
Alexandra Schambony
William Schiemann
John Schiller
Joseph Schlessinger
Ralph Scully
Darren Seals
Sendurai Mani

Javier Leon Serrano
Jeffrey Settleman
Lawrence Shapiro
Norman Sharpless
Jerry Shay
Kuang Shen
Dean Sheppard
Darryl Shibata
Tsukasa Shibue
Peter Sicinski
Rebecca Siegel
Daniel Silver
Dhirendra Simanshu
Robert Singer
Michael Sixt
Pierre Sonveaux
Steven Stacker
Ben Stanger
Phoebe Stavride
David Stern
Sheila Stewart
Michael Stratton
Mario Suva
Charles Swanton
Clifford Tabin
Maria Letizia Taddei
Makoto Taketo
Wai Leong Tam
Dan Theodorescu
Steven Theroux
Marc Therrien
Jean Paul Thiery
Itay Tirosh
Marc Tollis
Cristian Tomasetti
Nick Tonks
Jeffrey Townsend
Samra Turajlic
David Tuveson
Ruben van Boxtel
Jan van Deursen
Michael van Dyke
Jacco van Rheenen
Frans van Roy
Bert van de Kooij
Sander van den Heuvel
Louise van der Weyden
Matthew Vander Heiden
Vinay Varadan

Ajit Varki
Harold Varmus
Victor Velculescu
Subramanian Venkatesan
Andrea Ventura
Marc Vidal
Bert Vogelstein
Peter Vogt
Ulrich von Andrian
Karen Vousden
Johanna Wagner
Graham Walker
Timothy C. Wang
Xiao-Fan Wang
Yuan Wang
Fiona Watt
Valerie Weaver
Wenyi Wei
David Weinstock
Matthew Weirauch
William Weis
Kipp Weiskopf
Robin Weiss
Zena Werb
Jukka Westermarck
Amanda Whipple
Forest White
Daniel Wicklein
Traci Wilgus
Monte Winslow
Jedd Wolchok
Brian Wong
Victoria Wu
Shujuan Xia
Michael Yaffe
Jing Yang
Yosef Yarden
Xin Ye
Kam Chi Yeung
Ömer Yilmaz
Richard Young
Stuart Yuspa
Jinghui Zhang
Xiang-Dong Zhang
Yun Zhang
Martin Zornig
Lee Zou

Special thanks to Prof. Makoto Mark Taketo for truly extraordinary help in critiquing the previous editions of this textbook.

W. W. Norton & Company: The authors would like to thank the extraordinary staffs at W.W. Norton, and before them, at Garland Science, who made the assembling of this book possible. While the prose of the authors is the most visible component of this book, supporting it and lying behind it are hundreds of hours of detailed editing, production, permissions, and sales efforts on the part of these staffs, including most prominently the leading editorial roles of Denise Schanck at Garland Science and Cailin Barrett-Bressack at W.W. Norton, as well as the work of our extraordinary illustrator, Nigel Orme. While their respective contributions are critical to publication of a book like this one, their work often passes unrecognized, and so we, the authors, would like to thank them for their greatly appreciated work!

Media contributors: A sincere thank you to all of those involved in writing, curating, or reviewing the digital resources for the third edition. Your hard work, creative vision, and thoughtful contributions to the test bank, animations, audio lectures, ebook, and primary research discussion questions will provide support to many instructors and students.

Table of Contents

Detailed Contents

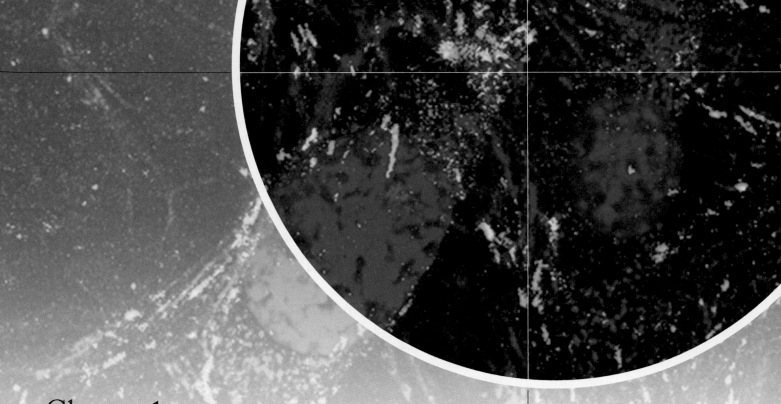

Chapter 1

The Biology and Genetics of Cells and Organisms

> Protoplasm, simple or nucleated, is the formal basis of all life... Thus it becomes clear that all living powers are cognate, and that all living forms are fundamentally of one character. The researches of the chemist have revealed a no less striking uniformity of material composition in living matter.
>
> Thomas Henry Huxley, evolutionary biologist, 1868

> Anything found to be true of *E. coli* must also be true of elephants.
>
> Jacques Monod, pioneer molecular biologist, 1954

The biological revolution of the twentieth century totally reshaped all fields of biomedical study, cancer research being only one of them. The fruits of this revolution were revelations of both the outlines and the minute details of genetics and heredity, of how cells grow and divide, how they assemble to form tissues, and how the tissues develop under the control of specific genes. Everything that follows in this text draws directly or indirectly on this new knowledge.

This revolution, which began in mid-century and was triggered by Watson and Crick's discovery of the DNA double helix, continues to this day. Indeed, we are still too close to this breakthrough to properly understand its true importance and its long-term ramifications. The discipline of molecular biology, which grew from this discovery, delivered solutions to the most profound problem of twentieth-century biology—how does the genetic constitution of a cell or organism determine its appearance and function?

Without this molecular foundation, modern cancer research, like many other biological disciplines, would have remained a descriptive science that cataloged diverse biological phenomena without being able to explain the mechanics of how they occur.

Today, our understanding of how cancers arise is being continually enriched by discoveries in diverse fields of biological research, most of which draw on the sciences of molecular biology and genetics. Perhaps unexpectedly, many of our insights into the origins of malignant disease are not coming from the laboratory benches of cancer researchers. Instead, the study of diverse organisms, ranging from yeast to worms to flies, provides us with much of the intellectual capital that fuels the forward thrust of the rapidly moving field of cancer research.

Those who fired up this biological revolution stood on the shoulders of nineteenth-century giants, specifically, Darwin and Mendel. Without the concepts established by these two, which influence all aspects of modern biological thinking, molecular biology and contemporary cancer research would be inconceivable. So, throughout this chapter, we frequently make reference to evolutionary processes as proposed by Charles Darwin and genetic systems as conceived by Gregor Mendel.

1.1 Mendel establishes the basic rules of genetics

Many of the basic rules of genetics that govern how genes are passed from one complex organism to the next were discovered in the 1860s by Gregor Mendel and have come to us basically unchanged. Mendel's work, which tracked the breeding of pea plants, was soon forgotten, only to be rediscovered independently by three researchers in 1900. During the decade that followed, it became clear that these rules—we now call them Mendelian genetics—apply to virtually all sexual organisms, including **metazoa** (multicellular animals), as well as **metaphyta** (multicellular plants).

Mendel's most fundamental insight came from his realization that genetic information is passed in particulate form from an organism to its offspring. This implied that the entire repertoire of an organism's genetic information—its genome, in today's terminology—is organized as a collection of discrete, separable information packets, now called genes. Only in recent years have we begun to know with any precision how many distinct genes are present in the genomes of mammals; many current analyses of the human genome—the best studied of these—place the number in the range of 20,000 protein-coding genes, somewhat more than the ~14,500 genes identified in the genome of the fruit fly, *Drosophila melanogaster*. In addition, a still-unresolved number of genes—numbering in the thousands—that specify functionally important noncoding RNAs reside in the mammalian genome.

Mendel's research implied that the genetic constitution of an organism (its **genotype**) could be divided into hundreds, perhaps thousands of discrete information packets, paralleling the observable appearance of the organism (its **phenotype**), which also could be divided into a large number of discrete physical traits (**Figure 1.1**).

Mendel's thinking launched a century-long research project among geneticists, who applied his principles to studying thousands of traits in a variety of experimental animals, including flies (*Drosophila melanogaster*), worms (*Caenorhabditis elegans*), and mice (*Mus musculus*). In the mid-twentieth century, geneticists also began to apply Mendelian principles to the study of the genetic behavior of single-celled organisms, such as the bacterium *Escherichia coli* and baker's yeast, *Saccharomyces cerevisiae*. The principle of genotype governing phenotype was directly transferable to these simpler organisms and their genetic systems.

While Mendelian genetics represents the foundation of contemporary genetics, it has been adapted and extended in myriad ways since its embodiments of 1865 and 1900. For example, the notion that each attribute of an organism can be traced to instructions carried in a single gene was realized to be simplistic. The great majority of observable traits of an organism are traceable to the cooperative interactions of multiple genes. Conversely, almost all the genes carried in the genome of a complex organism play roles in the development and maintenance of multiple organs, tissues, and physiologic processes.

Mendelian genetics revealed for the first time that genetic information is carried redundantly in the genomes of complex plants and animals. Mendel deduced that

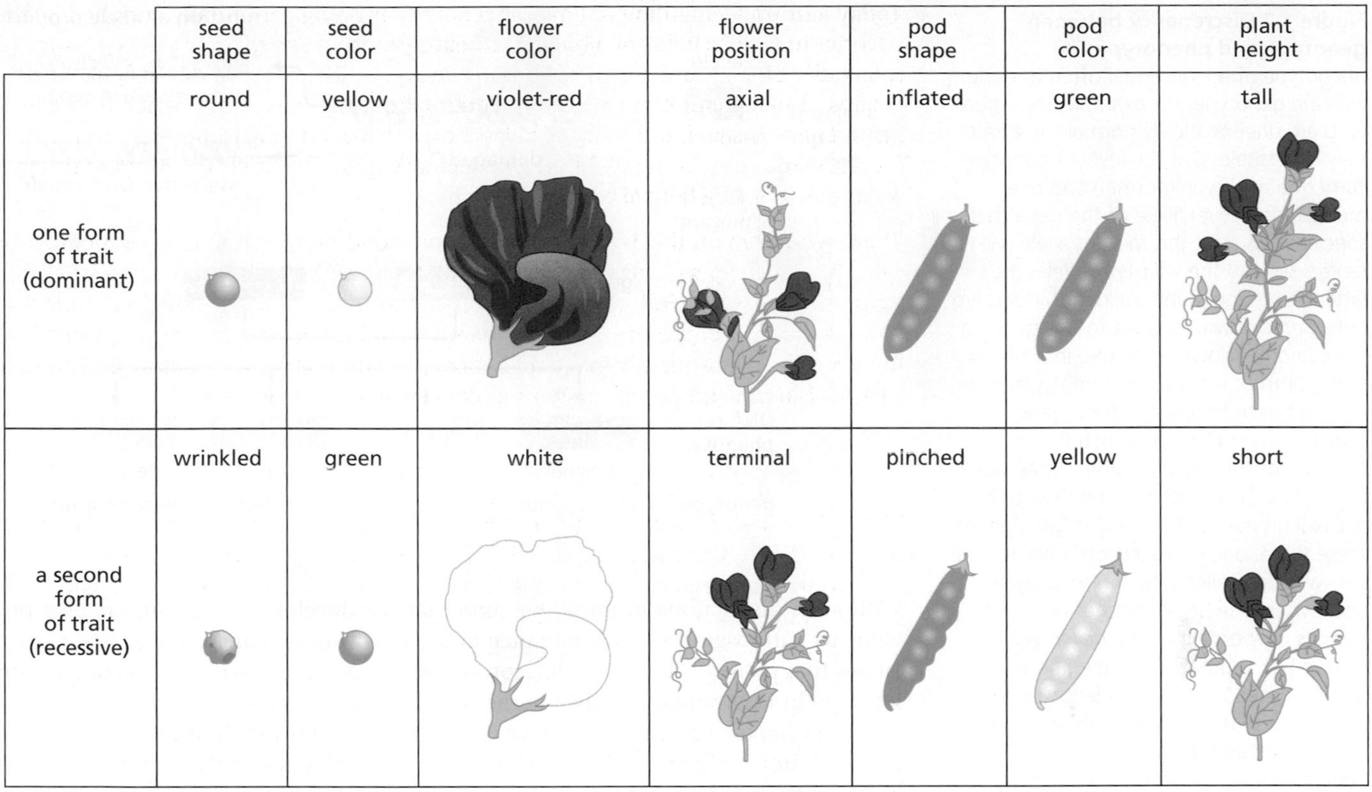

seed shape	seed color	flower color	flower position	pod shape	pod color	plant height
one form of trait (dominant) round	yellow	violet-red	axial	inflated	green	tall
a second form of trait (recessive) wrinkled	green	white	terminal	pinched	yellow	short

there were two copies of a gene for flower color and two for pea shape. Today we know that this twofold redundancy applies to the entire genome with the exception of the genes carried in the sex chromosomes. Hence, the genomes of higher organisms are termed **diploid**.

Mendel's observations also indicated that the two copies of a gene could convey different, possibly conflicting information. Thus, one gene copy might specify rough-surfaced and the other smooth-surfaced peas. In the twentieth century, these different versions of a gene came to be called **alleles**. An organism may carry two identical alleles of a gene, in which case, with respect to this gene, it is said to be **homozygous**. Conversely, the presence of two different alleles of a gene in an organism's genome renders this organism **heterozygous** with respect to this gene.

Because the two alleles of a gene may carry conflicting instructions, our views of how genotype determines phenotype become more complicated. Mendel found that in many instances, the action of one allele may dominate over that of the other in deciding the ultimate appearance of a trait. For example, a pea genome may be heterozygous for the gene that determines the shape of peas, carrying one round and one wrinkled allele. However, the pea plant carrying this pair of alleles will invariably produce round peas. This indicates that the round allele is **dominant**, and that it will invariably overrule its **recessive** counterpart allele (wrinkled) in determining phenotype (see Figure 1.1). (Strictly speaking, using proper genetic parlance, we would say that the phenotype encoded by one allele of a gene is dominant with respect to the phenotype encoded by another allele, the latter phenotype being recessive.)

In fact, classifying alleles as being either dominant or recessive oversimplifies biological realities. The alleles of some genes may be **co-dominant**, in that an expressed phenotype may represent a blend of the actions of the two alleles. Equally common are examples of **incomplete penetrance**, in which case a dominant allele may be present but its phenotype is not manifested because of the actions of other genes within the organism's genome. Therefore, the dominance of an allele is gauged by its interactions with other allelic versions of its gene, rather than its ability to dictate phenotype.

Figure 1.1 A particulate theory of inheritance One of Gregor Mendel's principal insights was that the genetic content of an organism consists of discrete parcels of information, each responsible for a distinct observable trait. Shown are the seven pea-plant traits that Mendel studied through breeding experiments. Each trait had two observable (phenotypic) manifestations, which we now know to be specified by the alternative versions of genes that we call alleles. When the two alternative alleles coexisted within a single plant, the "dominant" trait (*above*) was always observed while the "recessive" trait (*below*) was never observed. (Courtesy of J. Postlethwait and J. Hopson.)

Figure 1.2 Discrepancy between genotype and phenotype The phenotype of an individual often does not indicate genotype. For example, individuals who are phenotypically normal for a trait may nevertheless, at the level of genotype, carry one wild-type (normal) and one mutant (defective) allele of the gene that specifies this trait; this mutant allele will be recessive to the wild-type allele, the latter being dominant. Such individuals are heterozygous with respect to this gene. In the example shown here, two individuals mate, both of whom are phenotypically normal but heterozygous for a gene specifying a DNA repair function. On average, of their four children, three will be phenotypically normal and their cells will exhibit normal DNA repair function; of these three, one of the children will receive two wild-type alleles (be a homozygote) and two will be heterozygotes like their parents. A fourth child, however, will receive two mutant alleles (i.e., be a homozygote) and will be phenotypically mutant, in that this child's cells will lack the DNA repair function specified by this gene. Individuals whose cells lack proper DNA repair function are often cancer-prone, as described in Chapter 12.

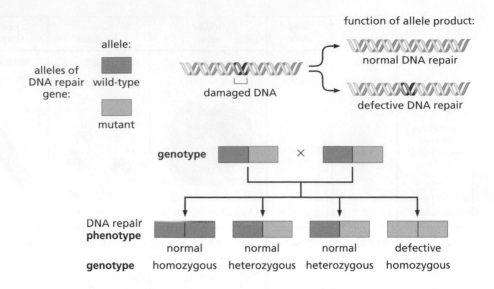

With such distinctions in mind, we note that the development of tumors also provides us with examples of dominance and recessiveness. For instance, one class of alleles that predispose cells to develop cancer encode defective versions of enzymes involved in DNA repair and thus in the maintenance of genomic integrity (discussed again in Chapter 12). These defective alleles are relatively rare in the general population and function recessively. Consequently, their presence in the genomes of many **heterozygotes** (of a normal/mutant genotype) is not apparent. However, two heterozygotes carrying recessive defective alleles of the same DNA repair gene may mate. One-fourth of the offspring of such mating pairs, on average, will inherit two defective alleles, exhibit a specific DNA repair defect in their cells, and develop certain types of cancer at greatly increased rates (**Figure 1.2**).

1.2 Mendelian genetics helps to explain Darwinian evolution

In the early twentieth century, it was not apparent how the distinct allelic versions of a gene arise. At first, this variability in information content seemed to have been present in the collective gene pool of a species from its earliest evolutionary beginnings. This perception changed only later, beginning in the 1920s and 1930s, when it became apparent that genetic information is corruptible. The information content in genetic texts, like that in all texts, can be altered. **Mutations** were found to be responsible for changing the information content of a gene, thereby converting one allele into another or creating a new allele from one previously widespread within a species. An allele that is present in the great majority of individuals within a species is usually termed **wild type**, the term implying that such an allele, being naturally present in large numbers of apparently healthy organisms, is compatible with normal structure and function.

Mutations alter genomes continually throughout the evolutionary life span of a species, which usually extends over millions of years. They strike the genome and its constituent genes randomly. Mutations provide a species with a method for continually tinkering with its genome, for trying out new versions of genes that offer the prospect of novel, possibly improved phenotypes. The lifetime of a species begins with a small founding, ancestral population. During the subsequent evolutionary history of the species, a continuing rain of mutations on the genomes carried by its individual members creates a progressive increase in the genetic diversity within this species. Thus, the collection of alleles present in the genomes of all members of a species—the **gene pool** of this species—becomes progressively more heterogeneous as the species grows older.

The continuing diversification of alleles in a species' genome, occurring over millions of years, is countered to some extent by the forces of natural selection that Charles

Darwin first described. Some alleles of a gene may confer more advantageous phenotypes than others, so individuals carrying these alleles have a greater probability of leaving numerous descendants than do those members of the same species that lack them. Consequently, natural selection results in a continual discarding of many of the alleles that have been generated by random mutations. In the long run, all things being equal, disadvantageous alleles are lost from the pool of alleles carried by the members of a species, advantageous alleles increase in number, and the overall fitness of the species improves incrementally.

Now, more than a century after Mendel was rediscovered and Mendelian genetics revived, we have come to realize that most of the genetic information in our own genome—indeed, in the genomes of all mammals—does not seem to specify phenotype and is often not associated with specific genes. Reflecting the discovery in 1944 that genetic information is encoded in DNA molecules, these "noncoding" stretches in the genome have been called, for want of a better term, "**junk DNA**" (**Figure 1.3**). Only about 1.5% of a mammal's genomic DNA carries sequence information that encodes the structures of proteins. Recent sequence comparisons of human, mouse, and dog genomes suggest that another ~2% encodes important information regulating gene expression and mediating other, still-poorly understood functions. (While some have proposed that as much as 80% of the human genome has important functions, a 2017 estimate of how much of the genome is likely to be truly important for organismic development and viability suggested an upper limit of 25%. So, the debate about the appropriateness of the term "junk DNA" continues unabated, with as much as one-quarter of it having possible biological function.)

Because mutations act randomly on a genome, altering true genes and junk DNA indiscriminately, the great majority of mutations alter genetic information—nucleotide sequences in the DNA—that have no effect on cellular or organismic phenotype. These mutations remain silent phenotypically and are said, from the point of view of natural selection, to be **neutral mutations**, being neither advantageous nor disadvantageous (**Figure 1.4**). Since the alleles created by these mutations are functionally silent, their existence could not be discerned by early geneticists whose work depended on gauging phenotypes. However, with the advent of DNA sequencing techniques, it became apparent that upward of a million functionally silent mutations can be found scattered throughout the genomes of organisms such as humans. The genome of each human carries its own unique array of these functionally silent genetic alterations.

The term *polymorphism* was originally used to describe variations in shape and form that distinguish normal individuals within a species from each other. These days,

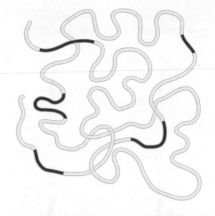

Figure 1.3 Biologically important sequences in the human genome The human genome can be characterized as a collection of relatively small islands of biologically important sequences (~3.5% of the total genome; *red*) floating amid a sea of "junk DNA" (*yellow*). The proportion of sequences carrying biological information has been greatly exaggerated for the sake of illustration. (With the passage of time, genes that appear to play important roles in cell and organismic physiology and specify certain noncoding RNA species have been localized to these intergenic regions, as have short DNA segments involved in regulating gene expression; hence the depiction of all genomic sequences localized between a human cell's ~20,000 protein-coding genes as useless junk is simplistic and incorrect.)

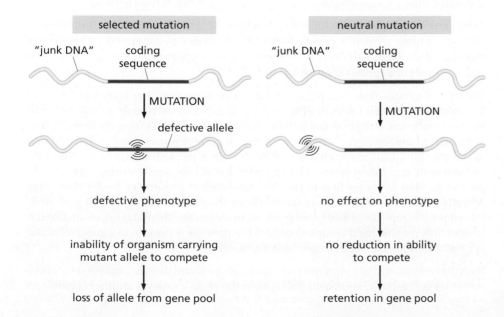

Figure 1.4 Neutral mutations and evolution The coding sequences (*red*) of most genes that encode protein structure were optimized in the distant evolutionary past. Hence, many mutations affecting amino acid sequence and thus protein structure (*left*) often create alleles that compromise the organism's ability to survive. For this reason, these mutant alleles are likely to be eliminated from the species' gene pool by the process of natural selection. In contrast, mutations striking "junk DNA" (*yellow*) have no effect on phenotype and are therefore often preserved in the species' gene pool (*right*). This explains why, over extended periods of evolutionary time, coding DNA sequences change slowly, while noncoding DNA sequences change far more rapidly.

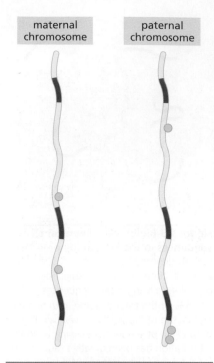

maternal chromosome   **paternal chromosome**

Figure 1.5 Polymorphic diversity in the human gene pool Because the great majority of human genomic DNA does not encode biologically important information (*yellow*), it has evolved relatively rapidly and has accumulated many subtle differences in sequences—polymorphisms—that are phenotypically silent (see Figure 1.4). Such polymorphisms are transmitted like Mendelian alleles, but their presence in a genome can be ascertained only by molecular techniques such as DNA sequencing. The dots (*green*) indicate where the sequence on this chromosome differs from the sequence that is most common in the human gene pool. For example, the prevalent sequence in one stretch may be TAACTGG, while the variant sequence TTACTGG may be carried by a minority of humans and constitute a polymorphism. The presence of a polymorphism in one chromosome but not the other represents a region of heterozygosity, even though a nearby gene (*red*) may be present in the identical allelic version on both chromosomes and therefore exist in a homozygous configuration.

geneticists use the term **genetic polymorphisms** to describe the inter-individual differences in DNA sequence at various sites throughout the genome, most of which, as discussed above, are registered as minor variations in sequence that have no effect on biological function (**Figure 1.5**). Recent estimates of the rate of spontaneously arising point mutations (affecting single DNA nucleotides) striking the genomes of proliferating normal human cells suggest that these mutations are accumulated at a rate of less than one nucleotide per genome per cell division, that is, only about one base pair per three billion base pairs per division—a stunning testimonial to the fidelity of DNA replication and repair. (Even lower mutational rates may affect the genes in germ cells, that is, those that give rise to eggs and sperm.) As we will see later (Chapter 12), these rates are increased in the genomes of replicating cancer cells.

During the course of evolution, the portion of the genome that encodes and affects biological function behaves much differently from the junk DNA. Junk DNA sequences suffer mutations that have no effect on the viability of an organism. Consequently, countless mutations in the noncoding sequences of a species' genome survive in its gene pool and accumulate progressively during its evolutionary history. In contrast, mutations affecting, for example, the protein-coding sequences of genes (see Section 1.6) often lead to loss of function and, as a consequence, reduction in organismic viability; hence, these mutations are usually weeded out of the gene pool by the hand of natural selection, explaining why genetic sequences that do indeed specify biological functions generally change very slowly over long evolutionary time periods (**Sidebar 1.1**).

1.3 Mendelian genetics governs how both genes and chromosomes behave

In the first decade of the twentieth century, Mendel's rules of genetics were found to have a striking parallel in the behavior of the chromosomes that were then being visualized under the light microscope: Mendel's genes and the chromosomes were both found to be present in pairs. Soon it became clear that an identical set of chromosomes is present in almost all the cells of a complex organism. This chromosomal array, often termed the **karyotype**, was found to be duplicated each time a cell went through a cycle of growth and division.

The parallels between the behaviors of genes and chromosomes led to the speculation, soon validated in hundreds of different ways, that the mysterious information packets called genes were carried by the chromosomes. Each chromosome was realized to carry its own unique set of genes in a linear array. Today we know that as many as several thousand genes may be arrayed along a mammalian chromosome. (Human Chromosome 1—the largest of the set—carries ~8% of the human genome.)

Each gene was found to be localized to a specific site on a specific chromosome. This site is often termed a genetic **locus**. Much effort was expended by geneticists throughout the twentieth century to map the sites of genes—genetic loci—along the chromosomes of a species (**Figure 1.7**).

The diploid genetic state that reigns in most cells throughout the body was found to be violated in the *germ cells*, sperm and egg. These cells carry only a single copy of each chromosome and gene and thus are said to be **haploid**. During the formation of germ cells in the testes and ovaries, each pair of chromosomes is separated and one of the pair (and thus associated genes) is chosen at random for incorporation into the sperm or egg. When sperm and egg combine subsequently during fertilization, the two haploid genomes fuse to yield the new diploid genome of the fertilized egg. All cells in the organism descend directly from this diploid cell and, if all goes well, inherit precise replicas of its diploid genome. In a large multicellular organism like the human, this means that a complete copy of the genome is present in almost all of the approximately 3×10^{13} cells throughout the body!

With the realization that genes reside in chromosomes, and that a complete set of chromosomes is present in almost all cell types in the body, came yet another conclusion

Sidebar 1.1 Evolutionary forces dictate that certain genes are highly conserved Many genes encode cellular traits that are essential for the continued viability of the cell. These genes, like all others in the genome, are susceptible to the ever-tinkering hand of mutation, which is continually creating new gene sequences by altering existing ones. Natural selection tests these novel sequences and determines whether they specify phenotypes that are more advantageous than the preexisting ones.

The actual agents that create phenotypes are the proteins encoded by the genes. The amino acid sequences of the proteins required for cell and therefore organismic viability were already optimized hundreds of millions of years ago. Consequently, almost all subsequent changes in the amino acid sequences of such critical proteins (resulting from mutations in the genes encoding them) would have been deleterious and would have compromised the viability of the cell and, in turn, the organism. These mutant alleles were soon lost from the gene pool of a species, because the mutant organisms carrying them failed to leave descendants. This dynamic explains why the sequences of many proteins have been highly conserved over vast evolutionary time periods. This dynamic can also be viewed in a different light: once certain genes and proteins developed, dozens and then hundreds of subsequent evolutionary inventions depended on the continued functioning of these earlier inventions. Any changes in the latter would cause the entire biological edifice to crumble, much like the removal of a playing card from a house of cards or a key foundation stone from a building. As we will read later in this book (Chapter 11), similar evolutionary dynamics influence the development of tumors within the human body.

In fact, the almost half of the proteins that are present in our own cells and are required for cell viability developed during the evolution of single-cell **eukaryotes**. This is indicated by numerous observations showing that many of our proteins have clearly recognizable counterparts in single-cell eukaryotes, such as baker's yeast. Another large repertoire of highly conserved genes and proteins is traceable to the appearance of the first multicellular animals (metazoa); these genes enabled the development of distinct organs and of organismic physiology. Hence, another large group of our own genes and proteins is present in counterpart form in worms and flies (**Figure 1.6**).

By the time the ancestor of all mammals first appeared more than 200 million years ago, virtually all the biochemical and molecular features present in contemporary mammals had already been developed. The fact that they have changed little in the intervening time points to their optimization and essentiality long before the appearance of the various mammalian orders. This explains why the embryogenesis, physiology, and biochemistry of all mammals is very similar, indeed, so similar that lessons learned through the study of laboratory mice are almost always transferable to an understanding of human biology.

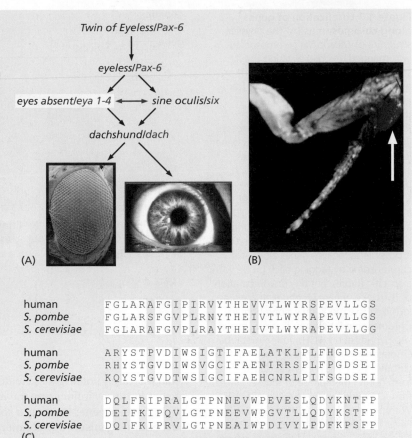

Twin of Eyeless/Pax-6

eyeless/Pax-6

eyes absent/eya 1-4 ⟷ *sine oculis/six*

dachshund/dach

(A)　　　　　　　　　　(B)

human	FGLARAFGIPIRVYTHEVVTLWYRSPEVLLGS
S. pombe	FGLARSFGVPLRNYTHEIVTLWYRAPEVLLGS
S. cerevisiae	FGLARAFGVPLRAYTHEIVTLWYRAPEVLLGG

human	ARYSTPVDIWSIGTIFAELATKLPLFHGDSEI
S. pombe	RHYSTGVDIWSVGCIFAENIRRSPLFPGDSEI
S. cerevisiae	KQYSTGVDTWSIGCIFAEHCNRLPIFSGDSEI

human	DQLFRIPRALGTPNNEVWPEVESLQDYKNTFP
S. pombe	DEIFKIPQVLGTPNEEVWPGVTLLQDYKSTFP
S. cerevisiae	DQIFKIPRVLGTPNEAIWPDIVYLPDFKPSFP

(C)

Figure 1.6 Extraordinary conservation of gene function
The last common ancestor of flies and mammals lived more than 600 million years ago. Moreover, fly (i.e., arthropod) eyes and mammalian eyes show totally different architectures. Nevertheless, the genes that orchestrate their development (*eyeless* in the fly, *Pax-6/small eye* in the mouse) are interchangeable—the gene from one organism can replace the corresponding mutant gene from the other and restore wild-type function. (A) Thus, the genes encoding components of the signal transduction cascades that operate downstream of these master regulators to trigger eye development (*black* for flies, *pink* for mice) are also highly conserved and interchangeable. (B) The expression of the mouse *Pax-6/small eye* gene, like the *Drosophila eyeless* gene, in an ectopic (inappropriate) location in a fly embryo results in the fly developing a fly eye on its leg (*arrow*), demonstrating the interchangeability of the two genes. (C) The conservation of genetic function over vast evolutionary distances is often manifested in the amino acid sequences of homologous proteins. Here, the amino acid sequence of a human protein is given together with the sequences of the corresponding proteins from two yeast species, *Schizosaccharomyces pombe* and *S. cerevisiae*. The amino acid residues present identically in the three proteins are indicated in a *white* background whereas those that have diverged from one another are indicated in a *light blue* background. The last common ancestor of the corresponding organisms is estimated to have lived more than 1 billion years ago. (A, left from I. Rebay et al., *Genetics 154*: 695-712, 2000. With permission from Oxford University Press; right, courtesy of Helga Kolb. B, courtesy of Walter Gehring, University of Basel. C, adapted from B. Alberts et al., Essential Cell Biology, 5th ed. New York: W. W. Norton & Company, 2019.)

Figure 1.7 Localization of genes along chromosomes (A) The physical structure of *Drosophila* chromosomes was mapped (*middle*) by using the fly's salivary gland chromosomes, which exhibit banding patterns resulting from alternating light (sparse) and dark (condensed) chromosomal regions (*bottom*). Independently, genetic crosses yielded linear maps (*top*) of various genetic loci arrayed along the chromosomes. These loci could then be aligned with physical banding maps, like the one shown here for the beginning of the left arm of *Drosophila* chromosome 1. (B) The availability of DNA probes that hybridize specifically to various genes now makes it possible to localize genes along a chromosome by tagging each probe with a specific fluorescent dye or combination of dyes. Shown are six genes that were localized to various sites along human Chromosome 5 by using fluorescence *in situ* hybridization (FISH) during metaphase. (There are two dots for each gene because chromosomes are present in duplicate form during metaphase of mitosis.) (A, top from C.B. Bridges, *J. Hered.* 26:60, 1935. With permission from Oxford University Press; bottom from M. Singer and P. Berg, *Genes and Genomes*. University Science Books, 1991. With permission from University Science Books. B, courtesy of David C. Ward.)

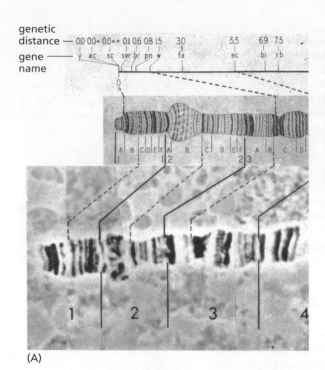

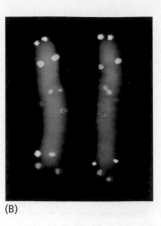

(B)

(A)

that was rarely noted: genes create the phenotypes of an organism through their ability to act *locally* by influencing the behavior of its individual cells. The alternative—that a single set of genes residing at some unique anatomical site in the organism controls the entire organism's development and physiology—was now discredited.

The rule of paired, similarly appearing chromosomes was found to be violated by some of the sex chromosomes. In the cells of female placental mammals, there are two similarly appearing X chromosomes, and these behave like the **autosomes** (the nonsex chromosomes). But in males, an X chromosome is paired with a Y chromosome, which is smaller and carries a much smaller repertoire of genes. In humans, the X chromosome is thought to carry about 900 genes, compared with the 78 distinct genes on the Y chromosome, which, because of redundancy, specify only 27 distinct proteins (see Supplementary Sidebar 1.1).

This asymmetry in the configuration of the sex chromosomes puts males at a biological disadvantage. Many of the 900 or so genes on the X chromosome are vital to normal organismic development and function. The twofold redundancy created by the paired X chromosomes in female cells guarantees more robust biology. If a gene copy on one of the X chromosomes is defective (that is, represents a nonfunctional mutant allele), chances are that the second copy of the gene on the other X chromosome can continue to carry out the task of the gene, ensuring normal biological function. Males lack this genetic fail-safe system in their sex chromosomes. One of the more benign consequences of this is color blindness, which strikes males frequently and females infrequently, due to the localization on the X chromosome of the genes encoding the color-sensing proteins of the retina.

Color blindness reveals the virtues of having two redundant gene copies around to ensure that biological function is maintained. If one copy is lost through mutational inactivation, the surviving gene copy is often capable of specifying a wild-type phenotype. Such functional redundancy operates for the great majority of genes carried by the autosomes. As we will see later, this dynamic plays an important role in cancer development, since virtually all of the genes that operate to prevent runaway proliferation of cells are present in two redundant copies, both of which must be inactivated in a cell before their growth-suppressing functions are lost and malignant cell proliferation can occur. Of additional interest, the asymmetry of X-chromosome complements between males and females is compensated in part by the mechanism of X-chromosome inactivation (see Supplementary Sidebar 1.1).

1.4 Chromosomes are altered in most types of cancer cells

Individual genes are far too small to be seen with a light microscope, and subtle mutations within a gene are smaller still. Consequently, the great majority of the mutations that play a part in cancer cannot be visualized through microscopy. However, the examination of chromosomes through the light microscope can give evidence of large-scale alterations of the cell genome. Indeed, such alterations were noted as early as 1892, specifically in cancer cells.

Today, we know that cancer cells often exhibit aberrantly structured chromosomes of various sorts, the loss of entire chromosomes, the presence of extra copies of others, and the fusion of the arm of one chromosome with part of another (**Sidebar 1.2**; see Figure 1.8). These changes in overall chromosomal configuration expand our conception of how mutations can affect the genome: since alterations of overall chromosomal structure and number also constitute types of genetic change, these changes must be considered to be the consequences of mutations. And importantly, the abnormal chromosomes seen initially in cancer cells provided the first clue that these cells might be genetically aberrant, that is, that they were mutants.

The normal configuration of chromosomes is often termed the **euploid** karyotypic state. Euploidy implies that each of the autosomes is present in normally structured pairs and that the X and Y chromosomes are present in the numbers appropriate for the sex of the individual carrying them. Deviation from the euploid karyotype—the state termed **aneuploidy**—is seen, as mentioned above, in many cancer cells. Indeed, the cells of perhaps 90% of human solid tumors carry aneuploid karyotypes (see Figure 1.9); often this aneuploidy is merely a consequence of the general chaos that reigns within a cancer cell. However, this connection between aneuploidy and malignant cell proliferation also hints at a theme that we will return to repeatedly in this book: the acquisition of extra copies of one chromosome or the loss of another can create a genetic configuration that somehow benefits the cancer cell and its agenda of runaway proliferation.

1.5 Mutations causing cancer occur in both the germ line and the soma

Mutations alter the information content of genes, and the resulting mutant alleles of a gene can be passed from parent to offspring. This transmission from one generation to the next, made possible by the germ cells (sperm and egg), is said to occur via the **germ line** (**Figure 1.10**). Importantly, the germ-line transmission of a recently created mutant allele from one organism to its offspring can occur only if a precondition has been met: the responsible mutation must strike a gene carried in the genome of sperm or egg (**gametes**) or in the genome of one of the cell types that are immediate precursors of the sperm or egg within the gonads. Mutations affecting the genomes of cells everywhere else in the body—which constitute the **soma**—create **somatic mutations**. The resulting mutant alleles have no prospect of being transmitted to the offspring of an organism, because they are not incorporated into the vehicles of generation-to-generation genetic transmission—the chromosomes of sperm or eggs.

Somatic mutations are of central importance to the process of cancer formation. As described repeatedly throughout this book, a somatic mutation can affect the behavior of the cell in which it occurs and, through repeated rounds of cell growth and division, can be passed on to all descendant cells within a tissue. These direct descendants of a single progenitor cell, which may ultimately number in the millions or even billions, are said to constitute a cell **clone**, in that all members of this group of cells trace their lineage directly back to the single ancestral cell in which the mutation originally occurred.

An elaborate repair apparatus within each cell continuously monitors the cell's genome and, with great efficiency, eradicates mutant sequences, replacing them with appropriate wild-type sequences and eliminating cells with inappropriate karyotype. (We will examine this repair apparatus in depth in Chapter 12.) This apparatus

Sidebar 1.2 Cancer cells are often aneuploid The presence of abnormally structured chromosomes and changes in chromosome number provided the first clue, early in the twentieth century, that changes in cell genotype often accompany and perhaps cause the uncontrolled proliferation of malignant cells. These deviations from the normal euploid karyotype can be placed into a number of categories. Chromosomes that seem to be structurally normal may accumulate in extra copies, leading to three, four, or even more copies of these chromosomes per cancer cell nucleus (**Figure 1.8**); such deviations from normal chromosome number are manifestations of *aneuploidy*.

Alternatively, chromosomes may undergo changes in their structure. A segment may be broken off one chromosomal arm and become fused to the arm of another chromosome, resulting in a chromosomal **translocation** (see Figure 1.8C). Moreover, chromosomal segments may be exchanged between

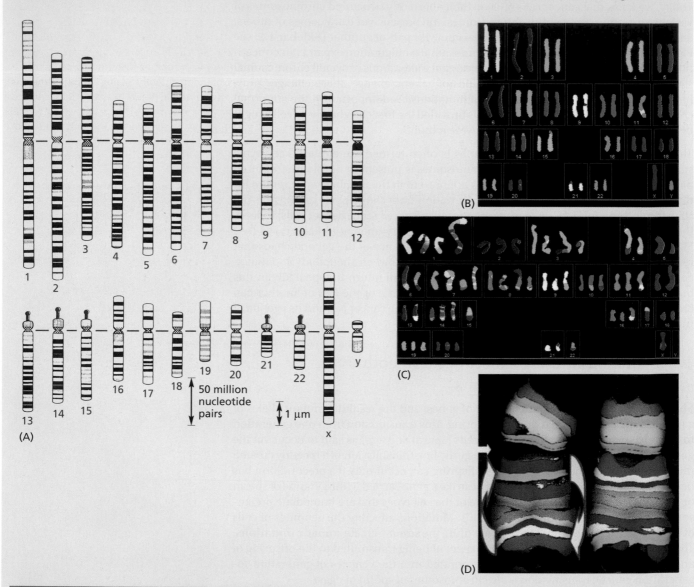

Figure 1.8 Normal and abnormal chromosomal complements (A) Staining of metaphase chromosomes reveals a characteristic light-and-dark banding pattern for each. The full array of human chromosomes is depicted; their centromeres are aligned (*pink line*). (B) The techniques of spectral karyotype (SKY) analysis and multicolor fluorescence *in situ* hybridization (mFISH) allow an experimenter to "paint" each metaphase chromosome with a distinct color (by hybridizing chromosome-specific DNA probes labeled with various fluorescing dyes to the chromosomes). The actual colors in images such as these are generated by computer. The diploid karyotype of a normal human male cell is presented. (The small regions in certain chromosomes that differ from the bulk of these chromosomes represent hybridization artifacts.) (C) The aneuploid karyotype of a human pancreatic cancer cell, in which some chromosomes are present in inappropriate numbers and in which numerous translocations (exchanges of segments between chromosomes) are apparent. (D) Here, mFISH (multicolor FISH) was used to label intrachromosomal subregions with specific fluorescent dyes, revealing that a large portion of an arm of normal human Chromosome 5 (*right*) has been inverted (*left*) in cells of a worker who had been exposed to plutonium in the nuclear weapons industry of the former Soviet Union. The site where the normal chromosome was broken and then rejoined with the inverted remainder of the chromosome is indicated by a *white arrow*. (A, adapted from U. Francke, *Cytogenet. Cell Genet.* 31:24–32, 1981. B and C, courtesy of M. Grigorova, J.M. Staines, and P.A.W. Edwards. D, from M.P. Hande et al., *Am. J. Hum. Genet.* 72:1162–1170, 2003. With permission from Elsevier.)

chromosomes from different chromosome pairs, resulting in **reciprocal translocations**. A chromosomal segment may also become inverted, which may affect the regulation of genes that are located near the breakage and fusion points (see Figure 1.8D).

A segment of a chromosome may be copied many times over, and the resulting extra copies may be fused head-to-tail in long arrays within a chromosomal segment that is termed an HSR (**homogeneously staining region**; Figure 1.9A). A segment may also be cleaved out of a chromosome, replicate as an autonomous, extrachromosomal entity, and increase to many copies per nucleus, resulting in the appearance of subchromosomal fragments termed DMs (**double minutes**; Figure 1.9B). These latter two changes cause increases in the copy number of genes carried in such segments, resulting in **gene amplification**. Sometimes, both types of gene amplification coexist in the same cell. In general, gene amplification can favor the growth of cancer cells by increasing the copy number of growth-promoting genes.

On some occasions, certain growth-inhibiting genes may be discarded or inactivated by cancer cells during their development. For example, when a segment in the middle of a chromosomal arm is discarded and the flanking chromosomal regions are joined, this can result in an **interstitial deletion** (Figure 1.9C), which liberates the incipient cancer cell from the inhibitory actions of the now-deleted gene or genes.

These descriptions of copy-number changes in genes, involving both amplifications and deletions, might suggest widespread chaos in the genomes of cancer cells, with gene amplifications and deletions occurring randomly. However, as the karyotypes and genomes of human tumors have been examined more intensively, it has become clear that certain regions of the genome tend to be lost characteristically in certain tumor types but not in others; this non-random activation or inactivation of a particular gene in a number of independently arising human tumors is termed a **recurrent mutation** and points to a critical role played by the altered gene and encoded protein in specific steps of tumor development. This suggests a theme that we will pursue in great detail throughout this book—that the gains and losses of particular genes favor the proliferation of specific types of tumors.

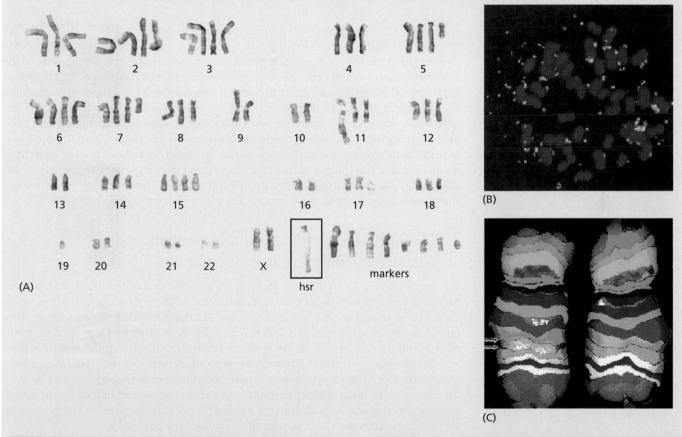

Figure 1.9 Increases and decreases in copy number of chromosomal segments (A) The amplification in the gene copy number of the *myc* oncogene (that is, a cancer-causing gene; see Section 4.3) in a human neuroendocrinal tumor, arising via end-to-end assembly of multiple copies of the oncogene, has caused an entire stretch of chromosome to stain *white* (*rectangle*), creating a homogeneously staining region (HSR). Almost all other chromosomes are present in multiple copies, many with structural abnormalities. (B) Double-minute chromosomes (DMs) derive from chromosomal segments that have broken loose from their original sites and have been replicated repeatedly as extrachromosomal genetic elements; like normal chromatids, these structures are doubled during metaphase of mitosis. FISH reveals the presence of amplified copies of the *HER2/neu* oncogene borne on DMs (*yellow dots*) in a mouse breast cancer cell. (C) The use of multicolor FISH (mFISH) revealed that a segment within normal human Chromosome 5 (*paired arrows, light purple segment, left*) has been deleted, yielding an interstitial deletion in which this segment is absent (*right chromosome*) following extensive exposure to radiation from plutonium. (A, from J.-M. Wen et al., *Cancer Genet. Cytogenet.* 135:91–95, 2002. With permission from Elsevier. B, from C. Montagna et al., *Oncogene* 21:890–898, 2002. With permission from Nature. C, from M.P. Hande et al., *Am. J. Hum. Genet.* 72:1162–1170, 2003. With permission from Elsevier.)

Figure 1.10 Germ-line versus somatic mutations The fate of the novel allele created by a mutation depends upon the particular cell in which the mutation strikes. Mutation A, which occurs in the genome of a germ-line cell in the gonads, can be passed from parent (*above left*) to offspring via gametes—sperm or egg (*green half circles*). Once incorporated into the fertilized egg (zygote), the mutant alleles can then be transmitted to all of the cells in the body of the offspring (2nd generation) outside of the gonads, i.e., its soma, as well as being transmitted via germ-line cells and gametes to a third generation (*not shown*). However, mutation B (*left*), which strikes the genome of a somatic cell in the parent, can be passed (*light brown half circles*) only to the lineal descendants of that mutant cell within the body of the parent and cannot be transmitted to offspring. (Adapted from B. Alberts et al., Essential Cell Biology, 5th ed. New York: W. W. Norton & Company, 2019.)

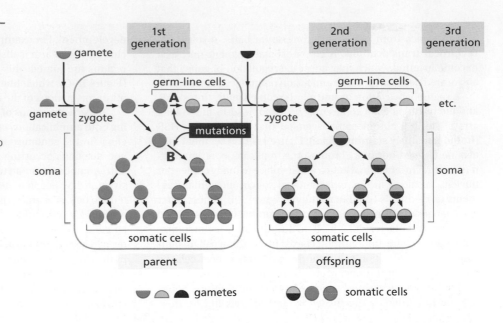

maintains genomic integrity by minimizing the number of mutations that strike the genome and, in the absence of proper repair, are then perpetuated by transmission to descendant cells. The challenge that the repair apparatus confronts is suggested by the fact that as many as 10^{16} cells are formed in the human body during a single lifetime, each created following the replication of cell genomes and the proper segregation of chromosomes; the occasional errors made by these processes must be detected and resolved, lest mutant genomes be perpetuated and transmitted to descendant cells. One stunning indication of the efficiency of genome repair comes from the successes of organismic cloning: the ability to generate an entire organism from the nucleus of a differentiated cell (prepared from an adult tissue) indicates that the genome of this adult cell is essentially a faithful replica of the genome of a fertilized egg, which existed many years and many cell generations earlier (Supplementary Sidebar 1.2).

1.6 Genotype embodied in DNA sequences creates phenotype through proteins

The genes studied in Mendelian genetics are essentially mathematical abstractions. Mendelian genetics explains their transmission, but it sheds no light on how genes create cellular and organismic phenotypes. Phenotypic attributes can range from complex, genetically templated behavioral traits to the **morphology** (shape, form) of cells and subcellular organelles to the biochemistry of cell metabolism. This mystery of how genotype creates phenotype represented the major problem of twentieth-century biology. Indeed, attempts at forging a connection between these two became the obsession of many molecular biologists during the second half of the twentieth century and continue as such into the twenty-first, if only because we still possess an incomplete understanding of how genotype influences phenotype.

Molecular biology has provided the basic conceptual scaffold for understanding the connection between genotype and phenotype. In 1944, DNA was proven to be the chemical entity in which the genetic information of cells is carried. Nine years later, Watson and Crick elucidated the double-helical structure of DNA. Eleven years after that, in 1964, it became clear that the sequences in the bases of the DNA double-helix determine precisely the sequence of amino acids in proteins. The unique structure and function of each type of protein in the cell is determined by its sequence of amino acids. Therefore, the specification of amino acid sequence, which is accomplished by base sequences in the DNA, provides almost all the information that is required to construct a protein.

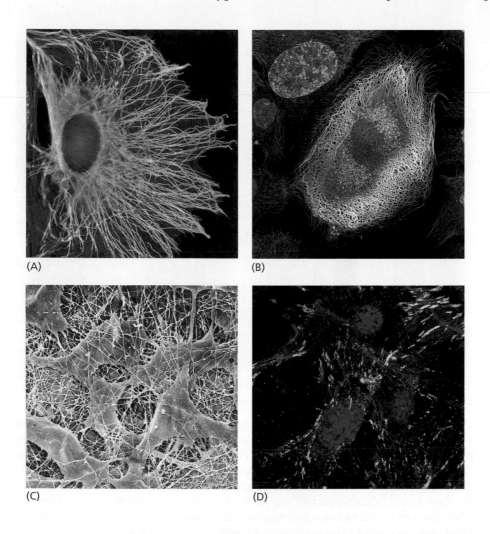

(A)   (B)   (C)   (D)

Figure 1.11 Intracellular and extracellular scaffolding The cytoskeleton is assembled from complex networks of intermediate filaments, actin microfilaments, and microtubules. Together, they generate the shape of a cell and enable its motion. (A) In this cultured cell, microfilaments composed of actin (*orange*) form bundles that lie parallel to the cell surface while microtubules composed of tubulin (*green*) radiate outward from the nucleus (*blue*). Both types of fibers are involved in the formation of protrusions from the cell surface. (B) Here, important intermediate filaments of epithelial cells—cytokeratins—are detected through use of hybrid proteins in which a fluorescent reporter protein (eYFP, *light yellow*) has been fused to both cytokeratins 8 and 18. (This is achieved through the creation of hybrid mRNAs in which the corresponding reading frames have been fused to one another using recombinant DNA techniques that insert eYFP-encoding DNA segments into the two cytokeratin genes.) (C) Cells secrete a diverse array of proteins that are assembled into the extracellular matrix (ECM). A scanning electron micrograph reveals the complex meshwork of collagen fibers, glycoproteins, hyaluronan, and proteoglycans, in which fibroblasts (connective tissue cells) are embedded in the rat cornea. (D) Mouse breast cancer cells propagated in a monolayer on the bottom surface of a Petri dish are attached to the dish via fibronectin molecules that they have secreted (*red*). Points of physical contact with the ECM formed by the fibronectin are achieved by integrin molecules (*green*; see Section 5.9) displayed on the surface of these cells that form *focal adhesions* that physically tether the cells to the fibronectin of the ECM. Nuclei are stained in *blue*. (A, courtesy of A. Tousson, High-Resolution Imaging Facility, University of Alabama at Birmingham. B, courtesy of Bram van den Broek and Kees Jalink. C, courtesy of T. Nishida. D, courtesy of Tsukasa Shibue.)

Once synthesized within cells, proteins proceed to create phenotype, doing so in a variety of ways. Some proteins assemble within the cell to create intracellular structure—the components of the **cytoarchitecture**, or, more specifically, the **cytoskeleton** (**Figure 1.11A and B**). When secreted into the space between cells, other proteins form the **extracellular matrix** (ECM), which ties cells together, enabling them to form complex, physically robust tissues (Figure 1.11C and D). As we will see later, the structure of the ECM is often disturbed by malignant cancer cells, enabling them to migrate to sites within a tissue and organism that are usually forbidden to them. Examples of how individual protein molecules can assemble to form the cytoskeleton and the ECM can be found in **Figure 1.12**.

Many proteins function as enzymes that catalyze the thousands of biochemical reactions that together are termed **intermediary metabolism;** without the active intervention of enzymes, few of these reactions would occur spontaneously. In recent years, it has become apparent that the metabolism of cancer cells is often quite different from that of corresponding normal cells. Proteins can also contract and create cellular movement as well as muscle contraction. Cellular **motility**, described in more detail in Chapter 14, plays an important role in cancer development by allowing cancer cells to spread through tissues and migrate to distant organs.

And most important for the process of cancer formation, proteins can convey signals between cells, thereby enabling complex tissues to maintain the appropriate numbers of constituent cell types. Within individual cells, certain proteins receive signals from an extracellular source, process these signals, and pass them on to other proteins within the cell; such signal-processing functions, often termed signal **transduction**, are also central to the creation of cancers, since many of the

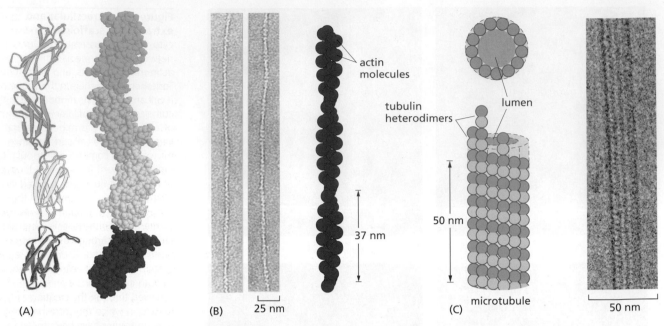

Figure 1.12 Structures of proteins and multiprotein assemblies (A) The three-dimensional structure of part of fibronectin, an important extracellular matrix protein (see Figure 1.11C and D), is depicted as a ribbon diagram (*left*), which illustrates the path taken by its amino acid chain; alternatively, the space-filling model (*right*) shows the positions of the individual atoms. One portion of fibronectin is composed of four distinct, similarly structured domains, which are shown here with different colors. (B) The actin fibers (*left*), which constitute an important component of the cytoskeleton (see Figure 1.11A), are composed of assemblies of individual protein molecules, each of which is illustrated here as a distinct two-lobed body (*right*). (C) Microtubules are assembled as hollow tubes made from heterodimers of two types of tubulin (*light green, dark green*); in addition to helping form the microskeleton, they form the fibers of the mitotic apparatus and the spines of flagella and cilia. (A, adapted from D.J. Leahy, I. Aukhil and H.P. Erickson, *Cell* 84:155–164, 1996. With permission from Elsevier. B, left, courtesy of Roger Craig; right, from B. Alberts et al., Molecular Biology of the Cell, 6th ed. New York: W. W. Norton & Company, 2015. C, left, from B. Alberts et al., Molecular Biology of the Cell, 6th ed. New York: W. W. Norton & Company, 2015; right, courtesy of R. Wade.)

abnormal-growth phenotypes of cancer cells are the result of aberrantly functioning intracellular signal-transducing molecules.

The functional versatility of the countless proteins within a cell (**Sidebar 1.3**) makes it apparent that almost all aspects of cell and organismic phenotype can be created by their actions. Once we realize this, we can depict genotype and phenotype in the simplest of molecular terms: genotype resides in the sequences of bases in DNA, while phenotype derives from the actions of proteins. (In fact, this depiction is simplistic, because it ignores the important role of RNA molecules as intermediaries between DNA sequences and protein structure and the abilities of some RNA molecules to function as enzymes and others to act as regulators of the expression of certain genes.)

In the complex **eukaryotic** (nucleated) cells of animals, as in the simpler **prokaryotic** cells of bacteria, DNA sequences are copied into RNA molecules in the process of **transcription**; a gene that is being transcribed is said to be actively **expressed**, while a gene that is not being transcribed is often considered to be **repressed**. In its simplest version, the transcription of a gene yields an RNA molecule of length comparable to the gene itself. Once synthesized, the base sequences in the RNA molecule are **translated** by the protein-synthesizing factories in the cell, its **ribosomes**, into a sequence of amino acids. The resulting macromolecule, which may be hundreds, even thousands of amino acids long, folds up into a unique three-dimensional configuration and becomes a functional protein.

In eukaryotic cells—the main subject of this book—the synthesis of RNA is itself a complex process. An RNA destined to become a messenger RNA (mRNA) molecule (often termed a pre-mRNA), when transcribed from its parent gene in the nucleus, may initially be almost as long as that gene. However, while it is being elongated, segments of the pre-mRNA molecule, some very small and others enormous, may be cleaved out of the growing RNA molecule. These segments, termed **introns**, are soon discarded and consequently have no impact on the subsequent coding ability of the RNA molecule (**Figure 1.13**). The retained portions of the initial RNA transcript, termed **exons**, once fused together to form a mature mRNA, are exported to the cytoplasm (see Sidebar 1.3). The entire process of editing pre-mRNA molecules via deletion of introns and fusion of exons is termed **splicing**.

Of note, an initially transcribed pre-mRNA may be processed through **alternative splicing** into a series of distinct mRNA molecules that retain different combinations of exons (see Figure 1.13B). Indeed, the pre-mRNAs arising from ~93% of the genes in our genome are subject to alternative splicing. The resulting alternatively spliced mRNAs

(A)

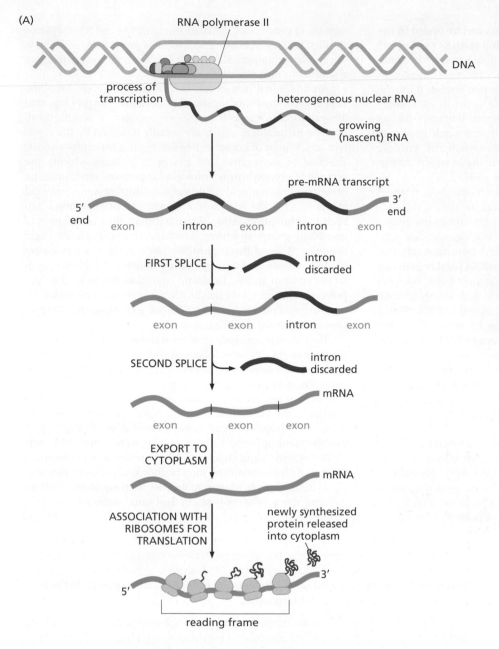

Figure 1.13 Processing of pre-mRNA
(A) By synthesizing a complementary RNA copy of one of the two DNA strands of a gene, RNA polymerase II creates a molecule of heterogeneous nuclear RNA (hnRNA; *red and blue*). The subset of hnRNA molecules that are processed into mRNAs are termed pre-mRNA. The progressive removal of the introns (*red*) leads to a processed mRNA containing only exons (*blue*). (B) A given pre-mRNA molecule may be spliced in a number of alternative ways, yielding distinct mRNAs that may encode distinct protein molecules. Illustrated here are the tissue-specific alternative splicing patterns of the α-tropomyosin pre-mRNA molecule, whose mRNA products specify important components of cell (and thus muscle) contractility. In this case, the introns are indicated as *black carets* while the exons are indicated as *blue rectangles*. (B, adapted from B. Alberts et al., Molecular Biology of the Cell, 6th ed. New York: W. W. Norton & Company, 2015.)

(B)

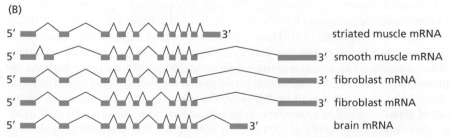

may carry altered reading frames, explaining, for example, the distinctive **isoforms** of certain proteins that are found in cancer cells but not in their normal counterparts. Alternatively, these splicing events may affect untranslated regions of mRNAs, such as those targeted by microRNAs (miRNAs; Section 1.9); these interactions with miRNAs can alter the function of an mRNA, by regulating either its translation or its stability. Alternative splicing events are currently estimated to generate mRNAs encoding ~70,000 distinct proteins arising from ~20,000 human genes. Importantly, the order

Sidebar 1.3 How many distinct proteins can be found in the human body, and how are the encoding mRNAs produced? It is difficult to extrapolate from the number of human genes (~20,000) to the total number of distinct proteins encoded in the human genome. The simplest estimate comes from the assumption that each gene encodes the structure of a single protein. But this assumption is naive, because it ignores the fact that the pre-mRNA transcript deriving from a single gene may be subjected to several *alternative splicing* patterns, yielding multiple, distinctly structured mRNAs, many of which may in turn encode distinct proteins (see Figure 1.13).

In more detail, flanking each intron are two retained sequences, the exons, which are fused together during this process of splicing. The initially synthesized RNA molecule and its derivatives found at various stages of splicing, together with nuclear RNA transcripts being processed from thousands of other genes, collectively constitute the **hnRNA (heterogeneous nuclear RNA)**. The end product of these post-transcriptional modifications may be an RNA molecule that is only a small fraction of the length of its initially synthesized hnRNA precursor. This final, mature RNA molecule may be exported into the cytoplasm, where, as an **mRNA (messenger RNA)** molecule, it serves as the template on which ribosomes assemble the amino acids that form the proteins. (The term **pre-mRNAs** is often used to designate those hnRNAs that are precursors of cytoplasmic mRNAs.) The complexity of post-transcriptional modifications of RNA yields an enormous array of distinct protein species within the cell.

In some cells, splicing may include certain exons in the final mRNA molecule made from a pre-mRNA, while in other cells, these exons may be absent (Figure 1.13B). Such **alternative splicing** patterns can generate mRNAs having greatly differing structures and protein-encoding sequences. In one admittedly extreme case, a single *Drosophila* gene has been found to be capable of generating 38,016 distinct mRNAs and thus proteins through various alternative splices of its pre-mRNA; genes having similarly complex alternative splicing patterns are likely to reside in our own genome.

An additional dimension of complexity derives from the post-translational modifications of proteins. The proteins that are exported to the cell surface or secreted in soluble form into the extracellular space are usually modified by the covalent attachment of complex trees of sugar molecules—usually attached to asparagine side chains of proteins—during the process of **glycosylation**. Intracellular proteins often undergo other types of chemical modifications. Thus, proteins involved in transducing the signals that govern cell proliferation often undergo **phosphorylation** through the covalent attachment of phosphate groups to serine, threonine, or tyrosine amino acid residues. Many of these phosphorylations affect some aspect of the functioning of these proteins. Similarly, the histone proteins that wrap around DNA and control its access by the RNA polymerases that synthesize hnRNA are subject to methylation, acetylation, and phosphorylation, as well as more complex post-translational modifications.

The polypeptide chains that form proteins may also undergo cleavage at specific sites following their initial assembly, often yielding small proteins showing functions that were not apparent in the uncleaved precursor proteins. Later, we will describe how certain signals may be transmitted through the cell via a cascade of the protein-cleaving enzymes termed **proteases**. In these cases, protein A may cleave protein B, activating its previously latent protease activity; thus activated, protein B may cleave protein C, and so forth. Taken together, alternative splicing and post-translational modifications of proteins generate vastly more distinct protein molecules than are apparent from counting the number of genes in the human genome.

---

of exons in the mature, spliced mRNA reflects the order of these segments in the initial unprocessed primary pre-mRNA transcript.

Interestingly, a protein that orchestrates an alternative splicing pattern of certain pre-mRNAs has been reported, when expressed in excessively high levels in cells, to favor their **transformation** (conversion) from a normal to a cancerous growth state. In certain **hematopoietic** malignancies described in the next chapter, the genes encoding several other **splicing factors (SFs)** have been found to have suffered **missense** mutations in specific codons, suggesting that these tumors gained proliferative advantage from the resulting subtle changes in protein structure (**Sidebar 1.4**). These findings are surprising, since one might imagine that proteins that regulate splicing would mediate the processing of many or all pre-mRNAs within the cell rather than affecting only a subset of genes involved in a specific cell-biological function, such as cell transformation. Moreover, a 2008 survey of alternatively spliced mRNAs found 41 that showed a distinct pattern of alternative splicing in human breast cancer cells compared with normal mammary cells; indeed, these alternatively spliced mRNAs could be used as diagnostic markers of the cancerous state of these cells. Even more dramatic, in 2010 as many as 1000 pre-mRNAs were found to undergo alternative splicing as cells passed through an **epithelial–mesenchymal transition (EMT)**, an important transdifferentiation step that carcinoma cells utilize in order to acquire traits of high-grade malignancy, as will be discussed in Chapters 13 and 14. And a year later, the genes encoding three splicing factors were found to be targets of recurrent mutations in a number of independently arising cases of myelodysplastic syndrome (MDS)—a **pre-neoplastic** condition of the bone marrow that is sometimes termed a "pre-leukemia."

Post-translational modification of the initially synthesized protein may result in the covalent attachment of certain chemical groups to the side chains of amino acid residues in a protein chain; included among these modifications are, notably, phosphates, complex sugar chains, and methyl, acetyl, succinyl, and lipid groups, as well as **ubiquitin** and **SUMO** groups. In the case of certain critical cellular proteins, this can lead to an extraordinary complexity: thus, histone H3, which is associated with chromosomal DNA and is involved in the regulation of transcription, has been estimated to exist in perhaps 1000 distinct isoforms, each defined by a different combination of post-translational modifications (PTMs). Some PTMs are almost universal: cell-surface proteins and those secreted in soluble form into the extracellular space almost invariably display carbohydrate side chains that are attached via the process of **glycosylation**, as cited in Sidebar 1.3; proteins of the Ras family, which are located in the cytoplasm and play important roles in cancer development, contain lipid groups attached to their carboxy termini. An equally important post-translational modification involves the cleavage of one protein by a second protein termed a **protease**, which has the ability to cut amino acid chains at certain sites, also mentioned in Sidebar 1.3. Accordingly, the final, mature form of a protein chain may include far fewer amino acid residues than were present in the initially synthesized protein. Following their synthesis, many proteins are dispatched to specific sites within the cell or are exported from the cell through the process of secretion; these alternative destinations are specified in the newly synthesized proteins by short amino acid (oligopeptide) sequences that function, much like postal addresses, to ensure the diversion of these proteins to specific intracellular sites.

With the passage of time, subtleties in the molecular biology of cells continue to be discovered. One of these discoveries involves long noncoding RNAs, discussed in more detail in Section 1.9. Yet another involves ribosomes, which have traditionally been depicted as workshops that are highly flexible in being able to translate essentially all mRNAs in the cytoplasm. Of late, ribosomes have been found to have physically associated proteins that specialize them to translate certain mRNAs but not others. Such specialization could not have been imagined by those who first articulated the DNA → RNA → protein paradigm more than half a century ago. The discovery of these specialized ribosomes leaves open the possibility that the formation of some of them is favored in certain types of cancer cells.

1.7 Gene expression patterns also control phenotype

Approximately 20,000 genes in the mammalian genome, acting combinatorially within individual cells, are able to create the extraordinarily complex organismic phenotypes of the mammalian body. A central goal of twenty-first-century biology is to relate the functioning of this large repertoire of genes to organismic physiology, developmental biology, and disease development. The complexity of this problem is illustrated by the fact that there are at least several hundred distinct cell types within the mammalian body, each with its own behavior, its own distinct metabolism, and its own physiology. Indeed, as the collections of RNA transcripts are surveyed in individual cells throughout the body, more and more cell types are uncovered, so that this number may eventually grow into many hundreds, possibly even several thousand.

This complexity is acquired during the process of organismic development, and its study is the purview of developmental biologists. They wrestle with a problem that is inherent in the organization of all multicellular organisms. All of the cells in the body of an animal are the lineal descendants of a fertilized egg. Moreover, almost all of these cells carry genomes that are reasonably accurate copies of the genome that was initially present in this fertilized egg (see Supplementary Sidebar 1.2). The fact that cells throughout the body are phenotypically quite distinct from one another (for example, a skin cell versus a brain cell) while being genetically identical creates a central problem of developmental biology: how do these various cell types acquire distinct phenotypes if they all carry identical genetic templates? The answer, documented in thousands of ways over the past four decades, lies in the selective reading of the genome by different cell types. Newer technologies, notably the recently gained ability to analyze the transcriptional programs (often termed the **transcriptomes**) of

Sidebar 1.4 Splicing factors and cancer Splicing, (including alternative splicing) is regulated by a large cohort of **splicing factors (SFs)** that function by recognizing and binding specific oligonucleotide segments in pre-mRNA, thereby guiding the general splicing machinery to the appropriate sites of cleavage and fusion of segments within the individual pre-mRNA molecules. When portrayed most simply, all splicing factors would seem to be dedicated to monitoring the processing of all the pre-mRNAs produced within a cell. However, the truth is more interesting: some appear to be specialized to process specific pre-mRNAs, including some SFs ostensibly involved in the control of cell proliferation. In addition to the observation that overexpression of at least one of these has been found to **transform** normal cells in culture into tumor cells, three other SFs have been found to appear repeatedly in mutant form in the genomes of various types of cancer cells, indicating that their alteration favors tumor development. Hence, certain SFs may be specialized to control the processing of genes with highly specialized cell-biological functions, including in this case the control of cell growth and proliferation.

Figure 1.14 Global surveys of gene expression arrays
Multiple experimental strategies are available to survey the gene expression programs of cells. (A) One procedure makes it possible to survey the expression levels of thousands of genes within a given type of cell (i.e., its transcriptome) using sequence-specific probes immobilized in a *microarray* to register the expression of cognate mRNAs. In this image, the most highly expressed genes and the most weakly expressed genes in each of the five major subtypes of breast cancer were used to choose DNA probes for stratification (*ordinate*). Hierarchical clustering was used to generate the *dendrogram* (*above*), which plots the degrees of similarity of the transcriptomes of these five groups of cells, doing so by clustering similar transcriptomes together on individual branches of the larger tree. In this image, higher-than-average levels of RNA expression are indicated as *red* pixels, while lower-than-average levels are indicated by *green* pixels. Average-level expression is indicated by *black* pixels. This analysis reveals that breast cancers, which at one level appear to be superficially similar, can actually be stratified into at least five distinct subtypes indicated by the five vertical colored bars on the right. Subsequent work has used such stratification to develop specific therapies for each subtype, in part based on information in this microarray. (B) In 2015, the newer RNA-seq technology overtook microarray analysis as the most common method for analyzing transcriptomes. It depends on the reverse transcription of a large collection of mRNAs, followed by sequencing of resulting cDNAs to measure the presence and abundance of the reverse transcripts. In this figure, the mRNAs from six different human breast cancer cell lines residing in different phenotypic states along the epithelial-to-mesenchymal spectrum (see Chapter 14) have been analyzed by RNA-seq. The 525 genes analyzed (*vertical axis*) were chosen because of their variability between two of these cell lines. Genes that are expressed at average levels are indicated in *yellow*, and overexpressed genes in *red*; while relatively underexpressed genes are in *blue*. Hierarchical clustering (*above*) has been used to organize these six samples, placing those with greater similarity closer to one another in the dendrogram tree. (C) RNA-seq was used to quantify gene expression across a variety of human tissues (*left panel*). The gene expression profiles (transcriptomes) obtained from each tissue sample were projected onto a two-dimensional space, using a statistical approach that groups samples with similar transcriptomes close to one another. This approach, termed the T-distributed Stochastic Neighbor Embedding (t-SNE) machine learning algorithm, takes a high-dimensional dataset and reduces it to a two-dimensional projection of the data (indicated here by the t-SNE 1 and t-SNE 2 axes of the two scatter plots presented here). A similar approach was applied to the transcriptomes of individual cells (*right panel*), measured using the more recently developed technology of single-cell sequencing (scRNA-seq). In this case, two of the tissues examined by conventional RNA-seq—hippocampus and frontal cortex—were subjected to scRNA-seq analysis, revealing that these two tissues, when mixed, actually together harbor ten distinct subtypes of cells (indicated by the acronyms). These images provide a direct visual validation of the fact that various differentiated cell types and the tissues that they form express distinct sets of transcripts, which presumably underlie their distinct tissue-specific functions. (Such single-cell RNA-seq reveals that the number of distinct subtypes of cells within many tissues is far larger than deduced from previous analyses of the transcriptomes of pooled cells prepared from various tissues.) (A, Charles M. Perou. B, Courtesy of Yun Zhang. C, left from F. Aguet and K.G. Ardlie, *Curr. Genet. Med. Rep.* 4:163–169, 2016. With permission from Springer. Right, from N. Habib, I. Avraham-Davidi, A. Basu et al., *Nat. Methods* 14:955–958, 2017. With permission from Nature.)

individual cells, have been confirming existing classifications of distinct differentiated cell types, while revealing that certain of these cell types, previously thought to represent a single homogeneous population, actually include a variety of distinct subtypes as mentioned above (**Figure 1.14**).

As cells in the early embryo pass through repeated cycles of growth and division, the cells located in different parts of the embryo begin to assume distinct phenotypes, this being the process of **differentiation**. Differentiating cells become committed to form one type of tissue rather than another, for example, gut as opposed to nervous system. All the while, they retain the same set of genes. This discrepancy leads to a simplifying conclusion: sooner or later, differentiation must be understood in terms of the sets of genes that are expressed (that is, transcribed) in some cells but not in others.

By being expressed in a particular cell type, a suite of genes dictates the synthesis of a cohort of proteins and RNA molecules that collaborate to create a specific cell phenotype. Accordingly, the phenotype of each kind of differentiated cell in the body should, in principle, be understandable in terms of the specific subset of genes that is expressed in that cell type.

The genes within mammalian cells have been grouped into two broad functional classes—the **housekeeping genes** and the **tissue-specific genes**. Many genes encode proteins that are universally required among all cell types throughout the body and maintain cell viability by carrying out certain biological functions common to all of them. These commonly expressed genes are classified as housekeeping genes. The size of the repertoire of human housekeeping genes is of more than passing interest to cancer biologists (Supplementary Sidebar 1.3).

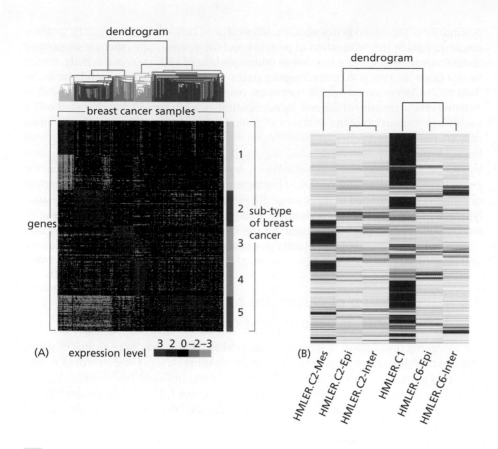

(A)   expression level   3 2 0 –2 –3

dendrogram

breast cancer samples

genes

1
2   sub-type of breast cancer
3
4
5

(B)   dendrogram

HMLER.C2-Mes
HMLER.C2-Epi
HMLER.C2-Inter
HMLER.C1
HMLER.C6-Epi
HMLER.C6-Inter

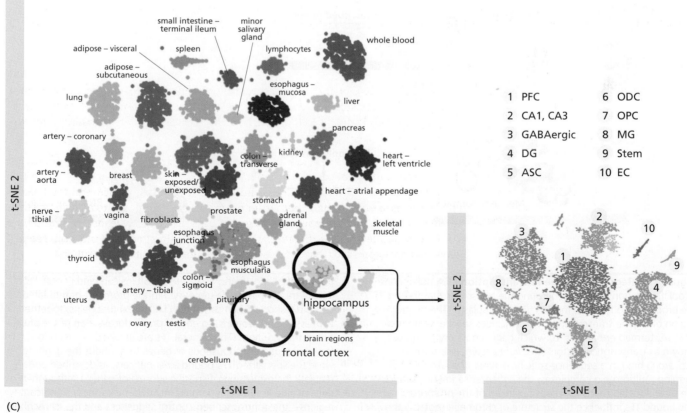

(C)

t-SNE 2

t-SNE 1

small intestine – terminal ileum
minor salivary gland
adipose – visceral
spleen
lymphocytes
whole blood
adipose – subcutaneous
esophagus – mucosa
liver
lung
pancreas
artery – coronary
kidney
heart – left ventricle
colon – transverse
artery – aorta
breast
skin – exposed/unexposed
stomach
heart – atrial appendage
nerve – tibial
vagina
fibroblasts
prostate
adrenal gland
skeletal muscle
esophagus junction
thyroid
esophagus muscularia
colon – sigmoid
artery – tibial
hippocampus
uterus
pituitary
frontal cortex
ovary
testis
brain regions
cerebellum

1 PFC   6 ODC
2 CA1, CA3   7 OPC
3 GABAergic 8 MG
4 DG   9 Stem
5 ASC   10 EC

t-SNE 2

t-SNE 1

A minority of expressed genes within a differentiated cell—the tissue-specific genes—are dedicated to the production of proteins and thus phenotypes that are associated specifically with this and a number of other specialized cell types in the body. It may be, for example, that 6000 housekeeping genes are expressed by a cell while far fewer than 3000 differentiation-specific genes are responsible for the distinguishing, differentiated characteristics of the cell. By implication, in each type of differentiated cell, a significant proportion of the 20,000 or so genes in the genome are unexpressed, since they are not required either for the cell's specific differentiation program or for general housekeeping purposes. As detailed sequencing of the mRNAs expressed by various tissues proceeds, the distinctions between housekeeping genes and tissue-specific genes have blurred, rendering these classifications more difficult to justify. The census of proteins co-expressed in the average differentiated mammalian cell using the analytical tools of **proteomics** reveals about 10,000 concomitantly expressed proteins; on its own, this census does not account for the diversification of protein structures that results from post-translational modifications such as phosphorylation, glycosylation, and proteolysis (see Section 1.6).

1.8 Modification of chromatin proteins and DNA controls gene expression

The foregoing description of differentiation makes it clear that large groups of genes must be coordinately expressed while other genes must be repressed in order for cells to display complex, tissue-specific phenotypes. Such coordination of expression is the job of proteins called **transcription factors** (TFs; **Figure 1.15**). Multiple transcription factors associated with a gene each bind to a specific DNA sequence in the control

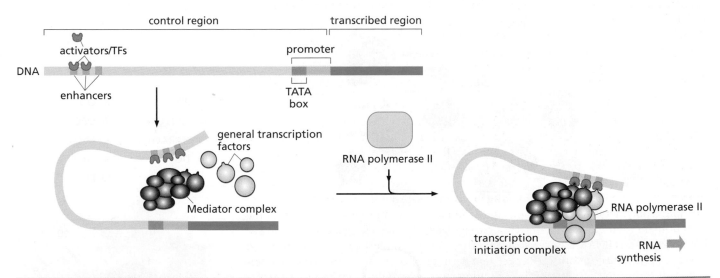

Figure 1.15 Regulation of gene transcription RNA polymerase II (pol II) is responsible for generating the transcripts that in turn are processed to become mRNAs in the cytoplasm. The control region of a gene transcribed by pol II includes specific segments of DNA, termed *enhancers*, to which gene regulatory proteins, known as transcription factors (TFs, *light brown*; also termed *activators*) bind in a sequence-specific manner, the latter often forming multiprotein complexes; the *enhancer* segments (*orange*) are often located many kilobases upstream of the *promoter*. The bound TFs influence the structure of chromatin (notably the histone proteins that package DNA; see Figure 1.16), creating a localized chromatin environment that enables pol II to produce an RNA transcript (*orange-red arrow, right*). In addition, the bound TFs interact with the transcription initiation complex via the Mediator complex (of ~30 protein subunits). The promoter of the gene (*dark, light green*) contains sequences to which RNA polymerase II (pol II) and associated proteins can bind in order to initiate transcription of a gene; within the promoter of regulated genes is a short AT-rich sequence (TATA Box) that helps, together with associated proteins, to ensure proper localization of the start site of transcription. (The general TFs are involved in initiating the transcription of thousands of genes throughout the genome, while the specialized TFs bound to enhancers, as described above, regulate the expression of specific subsets of genes.) Although in general a gene can be separated into two functionally significant regions—the nontranscribed control sequences and the transcribed sequences represented in pre-mRNA and mRNA molecules—some of the regulatory sequences (enhancers) may be located within the transcribed region of a gene, often in introns. (From B. Alberts et al., Molecular Biology of the Cell, 6th ed. New York: W.W. Norton & Company, 2015.)

region of the gene and determine whether or not the gene will be transcribed. The specific stretch of nucleotide sequence to which each TF binds, often called a **sequence motif**, is usually quite short, typically 5–10 nucleotides long. In ways that are still incompletely understood at the molecular level, some TFs operate to provide the RNA polymerase enzyme (RNA polymerase II in the case of pre-mRNAs) with access to a gene. Yet other TFs may block such access and thereby ensure that a gene is transcriptionally repressed.

Transcription factors can exercise great power, since a single type of TF can simultaneously affect the expression of a large cohort of downstream responder genes, each of which carries the recognition sequence that allows this TF to bind its promoter (see Figure 1.15). This ability of a single master regulator, such as a TF, to elicit multiple changes within a cell or organism is often termed **pleiotropy**. In the case of cancer cells, a single malfunctioning, pleiotropically acting TF may simultaneously orchestrate the expression of a large cohort of responder genes that together proceed to create major components of the cancer cell phenotype. A 2018 enumeration of the genes in the human genome that are likely to encode TFs has listed 1639 distinct genes (about 8% of the total protein-coding genes). Not included in this list were variant versions of these proteins arising through alternative splicing of pre-mRNAs.

The transcription of most genes is dependent upon the actions of several distinct TFs that must sit down together, each at its appropriate sequence site (that is, **enhancer**) in or near the gene promoter, and collaborate to activate gene expression (see Figure 1.15). This means that the expression of a gene is most often the result of the combinatorial actions of several TFs. Therefore, the coordinated expression of multiple genes within a cell, often called its gene **expression program**, is dependent on the actions of multiple TFs acting in combination on large numbers of gene promoters. In a variety of differentiated human cells, large clusters of multiple distinct TFs (each binding to its own enhancer sequence) assemble together with **general transcription factors** to drive high-level transcription of nearby non-housekeeping genes; these clusters of TFs, which may number in the hundreds within a given cell type, are termed **super-enhancers** (even though, more appropriately, the term should refer to the DNA sequences to which the component TFs bind). Some have proposed that the DNA sequences in the human genome responsible for regulating transcription, such as the enhancers, exceed those sequences encoding protein structure.

Figure 1.15 implies that modulation of gene expression is achieved by controlling initiation of transcription by RNA polymerase II (pol II) and that transcription proceeds in one direction. In fact, for many genes, possibly the majority, pol II molecules initially sit down on the promoter of a gene and proceed to transcribe the DNA in both directions. After extending nascent RNA transcripts for 20–60 nucleotides, pol II halts—the process termed **transcriptional pausing** or **promoter-proximal pausing**. A subset of the stalled polymerase complexes that have initiated in the appropriate transcriptional direction are then induced by physiologic signals to resume elongation, resulting in full-length pre-mRNA transcripts, while other pol II complexes remain stalled and never resume transcription. The factors that permit stalled pol II to proceed with elongation of transcripts are incompletely understood but would seem to be as important as the conventionally defined TFs in regulating gene expression. One important cancer-causing protein, termed Myc, has been found to act as an anti-pausing protein whose actions permit thousands of cellular genes to be fully transcribed.

Figure 1.15 also implies that both TFs and RNA polymerase interact only with DNA. In fact, in eukaryotic cells, DNA is packaged in a complex mixture of proteins that, together with the DNA, form the **chromatin** (**Figure 1.16**). The chromatin proteins are responsible for controlling the interactions of TFs and RNA polymerases with DNA and therefore play critical roles in governing gene expression. The core of chromatin is formed by DNA bound to nucleosomes, the latter being octamers consisting of two copies of each of four standard histone species (H2A, H2B, H3, and H4) with a fifth histone species—H1—bound to some but not all nucleosome octamers. This basic organization of chromatin structure, which resembles beads on a string, is found throughout the chromosomes. Altogether, about 30 million nucleosomes package the genomic DNA in the nucleus of each human cell.

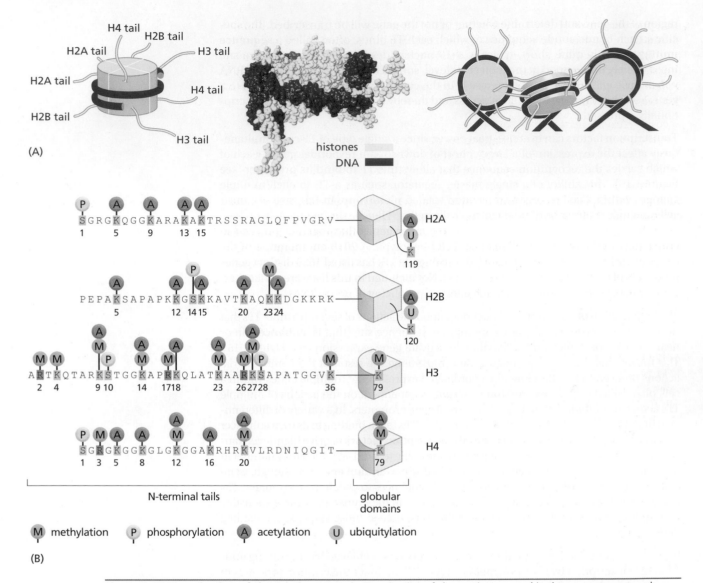

Figure 1.16 Organization of chromatin structure Examination of chromatin prepared in the proper way reveals that it appears like beads on a string (*not shown*). We now understand this structure in molecular detail. (A) The DNA double-helix (*red*) is wrapped ~1.67 times around each nucleosome, which consists of a core (*yellow*) formed as an octamer of four different histone molecules, each present in two copies (*left*); the N-terminal tail of each of the histones protrudes from the core nucleosome. X-ray crystallography has revealed (*middle*) that the core of the nucleosome (*yellow*) is disc-shaped; the N-terminal histone tails are not clearly revealed in this space-filling molecular model. The N-terminal tails of the histones (*right, yellow*) regulate the degree of compaction of chromatin, in part through their role in mediating interactions between nucleosomes bound along the DNA. (B) Each of these N-terminal histone tails can be modified by the covalent attachment of a variety of chemical groups, most commonly methyl, acetyl, phosphate, and ubiquitin groups. These post-translational modifications (PTMs) are attached by histone "writers," enzymes that alter the structure and the function of the chromatin proteins and determine whether or not a given segment of chromatin will be transcribed or not. (A, Adapted from Alberts et al., Molecular Biology of the Cell, 6th ed. New York: W. W. Norton & Company, 2015. PDB code: 1KX5. B, from H. Santos-Rosa and C. Caldas, *Eur. J. Cancer* 41:2381–2402, 2005. With permission from Elsevier.)

The globular core of the nucleosome represents the basic scaffold of chromatin that is modified in two ways. First, some of the standard histones (sometimes termed **canonical** histones), such as histones H2A and H3, may be replaced in a minority of nucleosomes by variant forms that are specified by genes distinct from those encoding the standard histones. Indeed, there are several dozen variant histones scattered here and there throughout the chromatin; the precise contributions of most of these to the regulation of chromatin structure and function remain poorly understood. One exception to this rule is the much-studied H2AX, which represents 10–15% of the H2A molecules in the nucleosomes. As discussed in Chapter 12, the phosphorylation of H2AX occurs in the nucleosomes that flank sites of double-strand DNA breaks, facilitating subsequent DNA repair. Yet another variant of histone H2A, termed H2A.Z, operates specifically to activate genes encoding epithelial functions and to repress those specifying mesenchymal functions; the resulting shifts in transcription are responsible for the programming of malignant cell traits that we will encounter in detail in Chapter 14.

Second, chromatin structure and transcription are strongly affected by covalent modifications of the standard four histones that occur after these histones assemble into nucleosomes. Most of these modifications affect lysine and arginine residues in the N-terminal tails of the core histones (see Figure 1.16B), which extend outward from the globular core of each nucleosome. Included here are modifications such as methylation, acetylation, phosphorylation, and ubiquitylation; other, less common modifications have also been observed, including biotinylation and sumoylation. A minority of this bewildering array of modifications have been associated with specific cell-biological programs, such as meiosis, apoptosis, and X-chromosome inactivation. During the cell growth-and-division cycle, one type of histone phosphorylation is associated with the condensation of chromatin that occurs during mitosis and the related global shutdown of gene expression. At other times in the cell cycle, acetylation of core histones is generally associated with active gene expression, while methylation is generally correlated with gene repression. Certain histone-modifying enzymes are specialized to elicit specific phenotypic changes in cells, doing so by altering transcriptional programs in ways that favor cancer cell growth and survival and, in turn, the outgrowth of tumors. [Recent measurements have suggested that various combinations of post-translational modifications (PTMs) generate as many as 1000 distinct isoforms of the H3 proteins present within individual human cells.]

Decades of research have revealed that the phenotypes and thus transcriptional programs of cells are transmitted stably over dozens, even hundreds of successive cycles of cell growth and division. For example, the descendants of a fibroblast are highly likely to preserve and exhibit a fibroblastic phenotype. This implies that specific states of histone modification are also transmitted faithfully from mother cells to their daughters in order to ensure the perpetuation of phenotypic states over many cell generations. However, the biochemical mechanisms responsible for the transmission of histone modification states are incompletely understood. This fidelity and associated heritability of cell-biological phenotypes is also critical to cancer biology, since many of the specific phenotypes of cancer cells are inherited from normal ancestral cells (often termed **cells-of-origin**) and resist disruption by the actions of activated oncogenes and inactivated tumor suppressor genes. Since the perpetuation of specific cellular phenotypes over multiple cell generations does not depend on specific changes in nucleotide sequence, this transmission of phenotype is considered a product of **epigenetic inheritance**.

Importantly, histone modifications, acting on their own, represent only one mechanism to ensure the stable intergenerational transmission of gene expression programs. In fact, a second key mechanism that enables epigenetic inheritance depends on covalent modification of DNA, specifically by DNA methyltransferases—enzymes that attach methyl groups covalently to cytosine bases of CpG dinucleotides in the DNA double helix. (The designation CpG indicates that the sequence is a cytidine positioned 5′ immediately before a guanosine.) The affected CpG dinucleotides are often located near transcriptional promoters, and the resulting methylation generally causes repression of nearby genes. The biochemical mechanism of maintenance methylation is well understood: maintenance DNA methyltransferase enzymes recognize **hemi-methylated** segments of recently replicated DNA and proceed to methylate any unmethylated CpG dinucleotides that are complementary to already-methylated CpGs in the other DNA strand (**Figure 1.17**). Extensive CpG methylation in the vicinity of a gene promoter often results in its repression and, in turn, in the continuing repression of this gene in descendant cells. As might be concluded from the previous paragraph, the mechanisms of transcriptional repression operate through close coordination between DNA methylation and enzymes responsible for histone

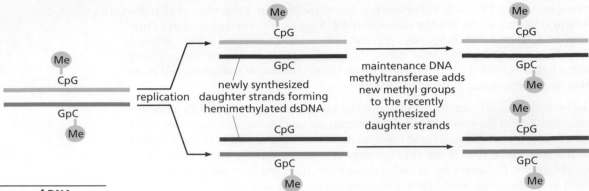

Figure 1.17 Maintenance of DNA methylation following replication
When a DNA double helix that is methylated (*green* Me groups, *left*) at CpG sites undergoes replication, the newly formed daughter helices will initially lack methyl groups attached to CpGs in the recently synthesized daughter strands (*purple, red*) and will therefore be *hemi-methylated*. Shortly after their synthesis, however, a maintenance DNA methyltransferase will detect the hemi-methylated DNA and attach methyl groups (*green*) to these CpGs, thereby regenerating the same configuration of methyl groups that existed in the parental helix prior to replication. CpG sites that are unmethylated in the parental helix (*not shown*) will be ignored by the maintenance methyltransferase and will therefore remain so in the newly synthesized strands.

modification. The mechanisms of CpG methylation assume a major role in ensuring that certain tumor suppressor genes are transcriptionally repressed, as will be discussed in more detail in Chapter 7.

1.9 Unconventional RNA molecules also affect the expression of genes

The Central Dogma of molecular biology, developed in the decade after the 1953 discovery of the DNA double helix, proposed that information flows in cells from DNA via mRNA to proteins. In addition, non-informational RNA molecules—ribosomal and transfer RNAs—were implicated as components of the translational machinery, and small nuclear RNAs were found to play key roles in the splicing and maturation of pre-mRNAs. In the 1980s, the view of RNA's functions was expanded through the discovery that certain RNA species can act as enzymes, thereby taking their place alongside proteins as catalysts of certain biochemical reactions.

The 1990s revealed an entirely new type of RNA molecule that functions to control either the levels of certain mRNAs in the cytoplasm, the efficiency of translating these mRNAs, or both. These **microRNAs** (miRNAs) are only ~22 nucleotides long and are generated as cleavage products of far larger nuclear RNA precursors. As outlined in **Figure 1.18**, the post-transcriptional processing of a primary miRNA transcript results in the formation in the cytoplasm of a miRNA that is part of a RISC (RNA-induced silencing complex) nucleoprotein. This complex associates with a spectrum of mRNA targets that contain, usually in their untranslated region, a sequence that is partially or completely complementary to the miRNA in the complex. Such association can result in either the inhibition of translation of the mRNA or its degradation, or both.

More than 500 distinct miRNA species have been found and characterized in human cells. Although it remains unclear precisely how many of these miRNAs are actually involved in regulating the translation and stability of mRNAs, those that do affect mRNA function are thought to regulate expression of more than 60% of all genes in the human genome. Moreover, a single miRNA species can target and thus regulate the functioning of dozens of distinct mRNA species, enabling it to act pleiotropically on a variety of cellular processes.

The potential importance of miRNAs in regulating gene expression is suggested by one survey of mRNAs and corresponding proteins in a group of 76 lung cancers. Only about 20% of the genes studied showed a close correlation between mRNA expression and protein expression levels. Hence, in the remaining 80%, the rate of protein synthesis (which can be strongly influenced by miRNAs) and the post-translational lifetime of proteins strongly influenced actual protein levels. Since proteins, rather than mRNA, are responsible for creating cell phenotypes, this also reveals the limitations of interpreting mRNA levels as reliable indicators of gene activity.

Let-7, a miRNA expressed by the *C. elegans* worm, was one of the first characterized miRNAs. It was found to suppress expression of the *ras* gene in worms and later in

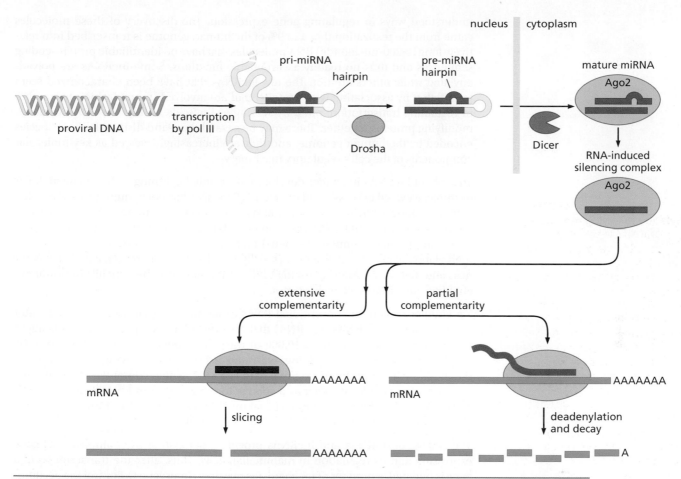

Figure 1.18 MicroRNAs and gene regulation A primary microRNA (pri-miRNA) is transcribed from a gene, and an enzyme complex involving the Drosha protein excises a small segment of the pri-miRNA that has formed a double-stranded RNA hairpin because of the self-complementarity of nucleotide sequences. The resulting pre-miRNA is exported to the cytoplasm, where it is further processed by the Dicer enzyme to generate a mature miRNA of 21 to 25 nucleotides. The resulting double-stranded miRNA is then bound to Argonaute (Ago2), which discards one of the two strands to generate the nucleoprotein complex termed RISC (RNA-induced silencing complex) and associates with mRNAs in the cytoplasm with which it has either precise or partial sequence complementarity, resulting in either degradation of the mRNA or inhibition of its translation. Several dozen miRNAs have been found to regulate various steps of tumor formation, either favoring or blocking critical steps of this process. Mutations of the gene encoding the Dicer enzyme (as well as those encoding other proteins involved in miRNA processing) have been associated with cancer progression, and analyses of miRNA expression patterns, much like expression array analyses of mRNAs (see Figure 1.14), have proved useful in classifying various types of cancer. (Courtesy of P.A. Sharp and D. Bartel.)

mammals. As we will read later (Chapters 4 through 6), the Ras proteins play critical roles in the development of many types of common human cancers. Since this pioneering work, the overexpression or loss of more than a dozen miRNA species has been associated with the formation of a variety of human cancers and the acquisition by tumors of malignant traits. The list of these miRNAs, which have garnered the term "oncoMiRs," continues to lengthen. In addition, loss of the Dicer processing enzyme (see Figure 1.18) and of other miRNA-processing enzymes involved in creating mature miRNAs has been found to facilitate the formation of tumors in mice and humans, doing so through still-unknown mechanisms. Interestingly, inheritance of a variant of the *K-ras* gene, which causes a single nucleotide change in the 3′ untranslated region (3′ UTR) of its mRNA, prevents recognition by *Let-7* and is associated with higher levels of the growth-promoting K-Ras protein and as much as a twofold increased risk of certain forms of lung and ovarian cancers.

A decade after the discovery of microRNAs, yet another unusual class of RNAs appeared on the scene: a diverse assay of lncRNA molecules (long noncoding RNAs) were found in the nucleus and cytoplasm and were shown to be involved in still-poorly understood ways in regulating gene expression. The discovery of these molecules came from the realization that 4 to 9% of the human genome is transcribed into relatively long (>200-nucleotide) RNA molecules that have no identifiable protein-coding sequences and thus no readily ascertainable functions. Some lncRNAs are polyadenylated while others are not. The few lncRNAs that have been characterized seem to function by associating with proteins that are involved in one fashion or another in regulating transcription, often by serving as scaffolds to hold certain chromatin-modifying proteins together. There may be several thousand distinct lncRNA species encoded by the human genome, and they are increasingly viewed as key molecular components of the cell's regulatory machinery.

The role of lncRNAs in cancer development is only beginning to be uncovered. For example, elevated expression of the *HOTAIR* lncRNA has been found to be correlated with metastatic behavior of human breast and colorectal carcinomas. More important, forced expression of *HOTAIR* in carcinoma cells causes localization of a transcription-repressing protein complex, termed PRC2, to certain chromosomal sites, altered methylation of histone H3 lysine 27 (see Figure 1.16B), and increased cancer invasiveness and metastasis. Another lncRNA has been reported to be critically important in enabling melanoma cell invasiveness.

Yet another class of unconventional RNAs came into clear view in the decade after 2010: small circular RNAs (circRNA) that are created as side products of splicing in the nucleus. While more than 10,000 of these have been identified, most may be incidental by-products of splicing, with only a tiny subset playing validated functional roles. One prominent example here is ciRS-7, which contains ~70 copies of a sequence that enables it to act as a sponge to attract and bind molecules of miR-7, thereby diverting the latter from its usual mRNA targets. miR-7, in turn, is an important tumor-suppressing microRNA for reasons described above.

The actions of miRNAs and lncRNAs provide a glimpse of the complexity of gene expression and its regulation in mammalian cells. Thus, after the transcription of a gene is initiated, a number of mechanisms may then intervene to control the accumulation of its ultimate product—a protein that does the actual work of the gene. Among these mechanisms are (1) promoter-proximal pausing; (2) post-transcriptional processing of pre-mRNA transcripts, including alternative splicing patterns; (3) stabilization or degradation in the cytoplasm of the mRNA product; (4) regulation of mRNA translation; and (5) post-translational modification, stabilization, or degradation of the protein product. These mechanisms reinforce the notion, cited above, that the rate of transcription of a gene often provides little insight into the levels of its protein product within a cell. Hence, as we will see, distinct patterns of mRNA expression may help us to distinguish various neoplastic cells from one another but, on their own, tell us rather little about how these cells are likely to behave.

1.10 Metazoa are formed from components conserved over vast evolutionary time periods

The following descriptions of cell biology, genetics, and evolution are informed in part by our knowledge of the history of life on Earth. Metazoa probably arose only once during the evolution of life on this planet, likely more than 650 million years ago. Once the principal mechanisms governing their genetics, biochemistry, and embryonic development were developed, these mechanisms remained largely unchanged in the descendant organisms up to the present (see Figure 1.6). This sharing of conserved traits among various animal phyla and orders has profound consequences for cancer research, since many lessons learned from the study of more primitive but genetically tractable organisms, such as flies and worms, have proven to be directly transferable to our understanding of how mammalian tissues, including those of humans, develop and function.

A survey of the diverse organisms grouped within the mammalian class reveals that the differences in biochemistry and cell biology are minimal (**Figure 1.19**). For this reason, throughout this book we will move effortlessly back and forth between mouse

tortoise   chick   pig   calf   rabbit   human

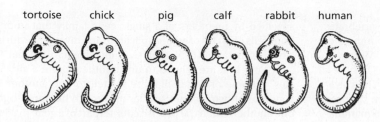

Figure 1.19 Visual evidence of the conservation of vertebrate biological traits A stunning visual demonstration that contemporary metazoa develop through embryonic pathways that have changed little since mammals and reptiles diverged from one another 200+ million years ago comes from an examination of mammalian embryos and one reptile—a tortoise. These morphological similarities do not begin to reveal the underlying biochemical pathways of cells and organisms that are equally well conserved. These 1874 drawings have been criticized for oversimplifying the similarities between these embryos. (In G.J. Romanes, Darwin and after Darwin, London, Vol. 1, 1892, adapted in turn from Haeckel, E., Anthropogenie; oder, Entwickelungsgeschichte des menschen. Keimes- und stammesgeschichte.)

biology and human biology, treating them as if they are essentially identical. On occasion, where species-specific differences are important, these will be pointed out.

The complex signaling circuits operating within cells seem to be organized in virtually identical fashion in all types of mammals. Even more stunning is the interchangeability of the component parts. It is rare that a human protein cannot function in place of its counterpart *orthologous* protein (**Sidebar 1.5**) in mouse cells. In the case of many types of proteins, this conservation of both function and structure is so profound that proteins can be swapped between organisms that are separated by far greater evolutionary distances. A striking example, noted earlier (see Figure 1.6), is provided by the gene and thus the protein that specifies eye formation in mammals and in flies. Extending even further back in our evolutionary history are the histones and the mechanisms of chromatin remodeling discussed earlier. In fact, the counterparts of many molecules and biochemical mechanisms that operate in mammalian cells are already apparent in protozoa.

1.11 Gene cloning techniques revolutionized the study of normal and malignant cells

Until the mid-1970s, the molecular analysis of mammalian genes was confined largely to the genomes of DNA tumor viruses, indeed the viruses described later in Chapter 3. These viruses have relatively simple genomes that accumulate to a high copy number (that is, number of molecules) per cell. This made it possible for biologists to readily purify and study the detailed structure and functioning of viral genes that operate much like the genes of the host cells in which these viruses multiplied. In contrast, molecular analysis of cellular genes was essentially impossible, since there are so many of them (tens of thousands per haploid genome) and they are embedded in a genome of daunting complexity (~3.2 billion base pairs of DNA per haploid cellular genome).

All this changed with the advent of gene cloning (Supplementary Sidebar 1.4). Thereafter, cellular genomes could be fragmented and used to create the collections of DNA fragments known as genomic **libraries**. Various DNA hybridization techniques could then be used to identify the genomic fragments within these libraries that were of special interest to the experimenter, in particular the DNA fragment that carried part or all of a gene under study. The retrieval of such a fragment from the library and the amplification of this retrieved fragment into millions of identical copies yielded a purified, **cloned** fragment of DNA and thus a cloned gene. Yet other techniques were used to generate DNA copies of the mRNAs that are synthesized in the nucleus and exported to the cytoplasm, where they serve as the templates for protein synthesis. Discovery of the enzyme **reverse transcriptase** (RT; see Figure 3.12) was of central importance here. Use of this enzyme made it possible to synthesize *in vitro* (that is, in the test tube) **complementary DNA** copies of mRNA molecules. These DNA molecules, termed cDNAs, carry the sequence information that is present in an mRNA molecule after the process of splicing has removed all introns. While we will refer frequently throughout this book to DNA clones of the genomic (that is, chromosomal) versions of genes and to cDNAs generated from the mRNA transcripts of such genes, space limitations preclude any detailed descriptions of the cloning procedures *per se*. With segments of cloned gene segments in hand, cancer biologists could begin to suppress or eliminate the function of specific targeted genes through several alternative experimental strategies (Supplementary Sidebar 1.5).

Sidebar 1.5 Orthologs and homologs All higher vertebrates (birds and mammals) seem to have comparable numbers of protein-encoding genes—in the range of 20,000. Moreover, almost every gene present in the bird genome seems to have a closely related counterpart in the human genome. The correspondence between mouse and human genes is even stronger, given the closer evolutionary relatedness of these two mammalian species.

Within the genome of any single species, there are genes that are clearly related to one another in their information content and in the related structures of the proteins they specify. Such genes form a **gene family**. For example, the group of genes in the human genome encoding globins constitutes such a group. It is clear that these related genes arose at some point in the evolutionary past through repeated cycles of the process in which an existing gene is duplicated followed by the progressive divergence of the two duplicated nucleotide sequences from one another (**Figure 1.20**). More directly related to cancer development are the more than 500 protein **kinases** encoded by the human genome. The kinase enzymes attach phosphate groups to their protein substrates, and almost all of these enzymes are specified by members of a single gene family that underwent hundreds of cycles of gene duplication and divergence during the course of metazoan evolution (see Supplementary Sidebar 17.7).

Genes that are related to one another within a single species' genome or genes that are related to one another in the genomes of two distinct species are said to be **homologous** to one another. Often the precise counterpart of a gene in a human can be found in the genome of another species. These two closely related genes are said to be **orthologs** of one another. Thus, the precise counterpart—the ortholog—of the c-*myc* gene in humans is the c-*myc* gene in chickens. To the extent that there are other *myc*-like genes harbored by the human genome (that is, N-*myc* and L-*myc*), they are members of the same gene family as c-*myc* but are not orthologs of one another or of the c-*myc* gene in chickens; instead, the c-*myc*, N-*myc*, and L-*myc* genes within a mammalian genome are termed **paralogs** due to their descent from a common ancestral gene.

Throughout this book we will often refer to genes without making reference to the species from which they were isolated. This is done consciously, since in the great majority of cases, the functioning of a mouse gene (and encoded protein) is indistinguishable from that of its human or chicken ortholog.

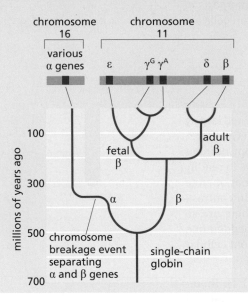

Figure 1.20 Evolutionary development of gene families The evolution of organismic complexity has been enabled, in part, by the development of increasingly specialized proteins. New proteins are "invented" largely through a process of gene duplication followed by diverging evolution of the two resulting genes via incremental changes that ultimately lead to structurally and functionally distinct proteins. Repeated cycles of such gene duplications followed by divergence have led to the development of large numbers of multi-gene families. During vertebrate evolution, an ancestral globin gene, shown here as an example, which encoded the protein component of hemoglobin, was duplicated repeatedly, leading to the α and multiple distinct β-like globin genes in the modern mammalian genome that are present on two human chromosomes. Because these globins have distinct amino acid sequences, each can serve a specific physiologic function. (From B. Alberts et al., Molecular Biology of the Cell, 6th ed. New York: W. W. Norton & Company, 2015.)

For cancer researchers, gene cloning arrived just at the right time. As we will see in the next chapters, research in the 1970s diminished the candidacy of tumor viruses as the cause of most human cancers. As these viruses moved off center stage, cellular genes took their place as the most important agents responsible for the formation of human tumors. Study of these genes would have been impossible without the newly developed gene cloning technology, which became widely available in the late 1970s, just when it was needed by the community of scientists intent on finding the root causes of cancer.

Additional reading

Allis CD & Jenuwein T (2016) The molecular hallmarks of epigenetic control. *Nat. Rev. Genet.* 17, 487–500.

Bannister AJ & Kouzarides T (2011) Regulation of chromatin by histone modifications. *Cell Res.* 21, 381–395.

Bartel DP (2018) Metazoan microRNAs. *Cell* 173, 20–51.

Bartonicek N, Maag JLV & Dinger ME (2016) Long noncoding RNAs in cancer: mechanisms of action and technological advancements. *Mol. Cancer* 15, 43–52.

Eifler K & Vertegaal ACO (2015) SUMOylation-mediated regulation of cell cycle progression and cancer. *Trends Biochem. Sci.* 40, 779–793.

Freeley KP & Edmonds MD (2018) Hiding in plain sight: rediscovering the importance of noncoding RNA in human malignancy. *Cancer Res.* 78, 2149–2158.

Hnisz D, Day DS & Young RA (2016) Insulated neighborhoods: structural and functional units of mammalian gene control. *Cell* 167, 1188–1200.

Jones PA & Liang G (2009) Rethinking how DNA methylation patterns are maintained. *Nat. Rev. Genet.* 10, 805–811.

Lambert SA, Jolma A, Campitelli LF et al. (2018) The human transcription factors. *Cell* 172, 650–665.

Martire S & Banaszynski LA (2020) The roles of histone variants in fine-tuning chromatin organization and structure. *Nat. Rev. Mol. Cell Biol.* 21, 522–541.

Nacev BA, Jones KB, Intlekofer AM et al. (2020) The epigenomics of sarcoma. *Nat. Rev. Cancer.* 20, 608–623.

Pennisi E (2012) ENCODE Project writes eulogy for junk DNA. *Science* 337, 1159–1161.

Rape M (2018) Ubiquitylation at the crossroads of development and disease. *Nat. Rev. Mol. Cell Biol.* 19, 59–70.

Rasmussen KD & Helin K (2016) Role of TET enzymes in DNA methylation, development and cancer. *Genes Dev.* 30, 733–750.

Salzman J (2016) Circular RNA expression: its potential regulation and function. *Trends Genet.* 32, 309–316.

Seeler JS & Dejean A (2017) SUMO and the robustness of cancer. *Nat. Rev. Cancer* 17, 184–197.

Soshnev AA, Josefowicz SZ & Allis CD (2016) Greater than the sum of parts: complexity of the dynamic epigenome. *Mol. Cell* 62:681–694.

Tessarz P & Kouzarides T (2014) Histone core modifications regulating nucleosomal structure and dynamics. *Nat. Rev. Mol. Cell Biol.* 15, 703–708.

Venkatesh S & Workman JL (2015) Histone exchange, chromatin structure and the regulation of transcription. *Nat. Rev. Mol. Cell Biol.* 16, 178–189.

Zhang R, Xia LQ, Lu WW, Zhang J & Zhu JS (2016) LncRNAs and cancer. *Oncol. Lett.* 12, 1233–1239.

Zhou S, Treloard AE & Lupien M (2016) Emergence of the noncoding cancer genome: a target of genetic and epigenetic alterations. *Cancer Discov.* 6, 1215–1229.

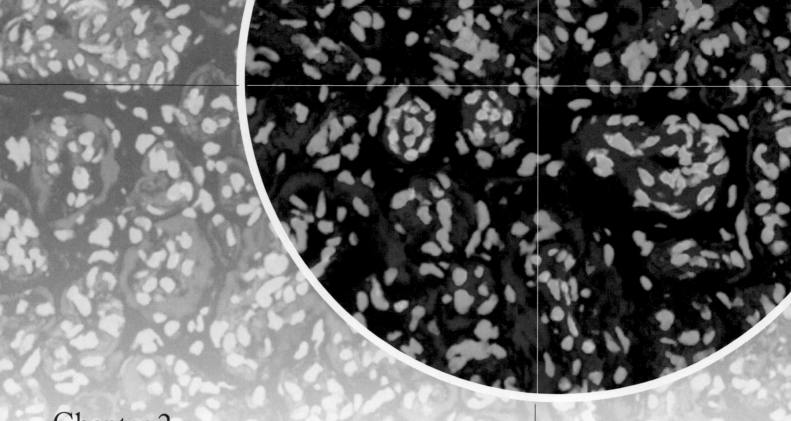

Chapter 2

The Nature of Cancer

When I published the results of my experiments on the development of double-fertilized sea-urchin eggs in 1902, I added the suggestion that malignant tumors might be the result of a certain abnormal condition of the chromosomes, which may arise from multipolar mitosis. . . . So I have carried on for a long time the kind of experiments I suggested, which are so far without success, but my conviction remains unshaken.

Theodor Boveri, pathologist, 1914

Tumors destroy man in a unique and appalling way, as flesh of his own flesh which has somehow been rendered proliferative, rampant, predatory and ungovernable. They are the most concrete and formidable of human maladies, yet despite more than 70 years of experimental study they remain the least understood.

Francis Peyton Rous, tumor virologist, Nobel lecture, 1966

The cellular organization of metazoan tissues has made possible the evolution of an extraordinary diversity of anatomical designs. Much of this plasticity in design can be traced to the fact that the building blocks of tissue and organ construction—individual cells—are endowed with great autonomy and versatility. Most types of cells in the metazoan body carry a complete organismic genome—far more information than any one of these cells will ever require. And many cells retain the ability to grow and divide long after organismic development has been completed. This retained ability to proliferate and to participate in tissue **morphogenesis** (the creation of shape) makes possible the maintenance of adult tissues throughout the life span of an organism. Such maintenance may involve the repair of wounds and the replacement of cells that have suffered attrition after extended periods of service.

At the same time, this versatility and autonomy pose a grave danger, in that individual cells within the organism may gain access to information in their genomes that is normally denied to them and assume roles that are inappropriate for normal tissue

31

maintenance and function. Moreover, their genomic sequences are subject to corruption by various mechanisms that alter the structure and hence information content of the genome. The resulting mutated genes may divert cells into acquiring novel, often highly abnormal phenotypes. Such changes may be incompatible with the normally assigned roles of these cells in organismic structure and physiology. Among these inappropriate changes may be alterations in cellular proliferation programs, and these in turn can lead to the appearance of large populations of cells that no longer obey the rules governing normal tissue construction and maintenance.

When portrayed in this way, the renegade cells that form a tumor are the result of normal development gone awry. In spite of extraordinary safeguards taken by the organism to prevent their appearance, cancer cells somehow learn to thrive. Normal cells are carefully programmed to collaborate with one another in constructing the diverse tissues that make possible organismic survival. Cancer cells have a quite different and more focused agenda. They appear to be motivated by only one consideration: making more copies of themselves.

2.1 Tumors arise from normal tissues

A confluence of discoveries in the mid- and late nineteenth century led to our current understanding of how tissues and complex organisms arise from fertilized eggs. The most fundamental of these was the discovery that all tissues are composed of cells and cell products, and that all cells arise through the division of preexisting cells. Taken together, these two revelations led to the deduction, so obvious to us now, that all the cells in the body of a complex organism are members of cell lineages that can be traced back to the fertilized egg. Conversely, the fertilized egg is able to spawn all the cells in the body, doing so through repeated cycles of cell growth and division.

These realizations had a profound impact on how tumors were perceived. Previously, many had portrayed tumors as foreign bodies that had somehow entered and taken root in an afflicted person. Now, tumors, like normal tissues, could be examined under the microscope by researchers in the then-new science of **histology**. These examinations of tissue **sections** (slices) revealed that tumors, like normal tissues, were composed of masses of cells (**Figure 2.1**). Contemporary cancer research makes frequent use of a variety of histological techniques; the most frequently used of these are illustrated in Supplementary Sidebar 2.1.

Evidence accumulated that tumors of various types, rather than invading the body from the outside world, often derive directly from the normal tissues in which they

Figure 2.1 Normal versus neoplastic tissue (A) This histological section of the lining of the ileum in the small intestine, viewed at low magnification, reveals the continuity between normal and cancerous tissue. At the far left is the normal epithelial lining, the mucosa. In the middle is mucosal tissue that has become highly abnormal, or "dysplastic." To the right is an obvious tumor—an adenocarcinoma—which has begun to invade underlying tissues. (B) This pair of sections, viewed at high magnification, shows how normal tissue architecture becomes deranged in tumors. In the normal human mammary gland (*upper panel*), a milk duct is lined by epithelial cells (*dark purple nuclei*). The ducts are surrounded by mesenchymal tissue (termed "stroma"), which consists of connective tissue cells, such as fibroblasts and adipocytes, and collagen matrix (*pink*). In an invasive ductal breast carcinoma (*lower panel*), the cancer cells, which arise from the epithelial cells lining the normal ducts, exhibit abnormally large nuclei (*purple*), no longer form well-structured ducts, and have invaded the stroma (*pink*). (A, from A.T. Skarin, Atlas of Diagnostic Oncology, 4th ed. Philadelphia: Elsevier Science Ltd., 2010. With permission from Elsevier. B, Courtesy of A. Orimo.)

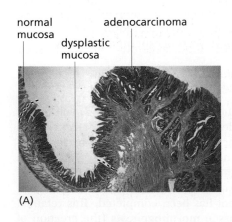

normal mucosa   dysplastic mucosa   adenocarcinoma

(A)

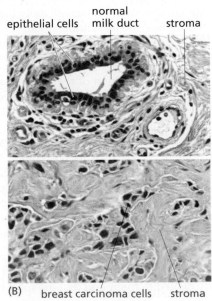

epithelial cells   normal milk duct   stroma

(B)   breast carcinoma cells   stroma

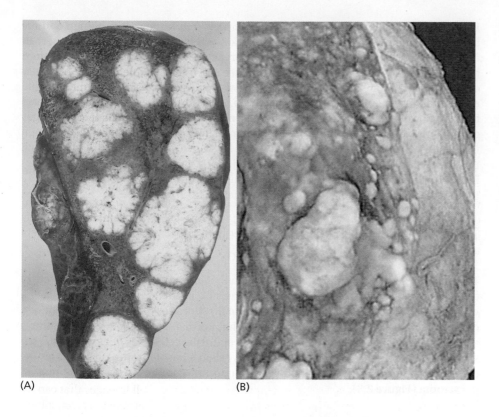

(A)   (B)

Figure 2.2 Metastasis of cancer cells to distant sites Many types of tumors eventually release cancer cells that migrate to distant sites in the body, where they form the secondary tumors known as metastases. (A) Metastases (*white*) in the liver often arise in patients with advanced colon carcinomas. The portal vein, which drains blood from the colon into the liver, provides a route for metastasizing colon cancer cells to migrate directly into the liver, where they are trapped in large numbers in microvessels. (B) Breast cancer often metastasizes to the brain. Here, large metastases are revealed post mortem in the right side of a brain where the dura (membrane covering; shown intact at *right*) of the brain has been removed. (A, courtesy of Peter Isaacson. B, from H. Okazaki and B.W. Scheithauer, Atlas of Neuropathology. New York: Gower Medical Publishing, 1988. With permission of Mayo Foundation for Medical Education and Research.)

are first discovered (see Figure 2.1A). However, certain tumors did seem to be capable of moving within the confines of the human body: in many patients, multiple tumors were discovered at anatomical sites quite distant from where their disease first began, a consequence of the tendency of cancers to spread throughout the body and to establish new colonies of cancer cells (**Figure 2.2**). These new settlements, termed **metastases**, were often traceable directly back to the site where the disease of cancer had begun—the founding or **primary tumor**.

Invariably, detailed examination of the organization of cells within tumor masses gave evidence of a tissue architecture that was less organized than the architecture of nearby normal tissues (see Figure 2.1). These **histopathological** comparisons provided the first seeds of an idea that would take the greater part of the twentieth century to prove: tumors are composed of cells that have lost the ability to assemble and create tissues of normal form and function. Stated more simply, cancer came to be viewed as a disease of malfunctioning cells, and all of the properties of individual tumors must derive, directly or indirectly, from the behavior of the cancerous cells within such tumors.

While the microarchitecture of tumors differed in obvious ways from that of normal tissue, tumors nevertheless exhibited certain histological features that resembled those of normal tissue. This suggested that all tumors should, in principle, be traceable back to the specific tissue or organ site in which they first arose, using the histopathological analyses of tumor sections to provide critical clues. This simple idea led to a new way of classifying these growths, which depended on their presumed tissues of origin. The resulting classifications often united under one roof cancers that arise in tissues and organs that have radically different functions in the body but share common types of tissue organization.

The science of histopathology also made it possible to understand the relationship between the clinical behavior of a tumor (that is, the effects that the tumor had on the patient's body) and its microscopic features. Most important here were the criteria that segregated tumors into two broad categories depending on their degree of aggressive growth. Those that grew locally without invading adjacent tissues were classified as **benign**, while others that invaded nearby tissues and possibly spawned metastases were termed **malignant**.

In fact, the great majority of primary tumors arising in humans are benign and are harmless to their hosts, except in the rare cases where the expansion of these localized masses causes them to press on vital organs or tissues. Some benign tumors, however, may cause clinical problems because they release dangerously high levels of hormones that create physiologic imbalances in the body. For example, thyroid **adenomas** (premalignant epithelial growths) may cause excessive release of thyroid hormone into the circulation, leading to **hyperthyroidism**; pituitary adenomas may release growth hormone into the circulation, causing excessive growth of certain tissues—a condition known as **acromegaly**. Nonetheless, deaths caused by benign tumors are relatively uncommon. The vast majority of cancer-related mortality derives from malignant tumors. More specifically, it is the metastases spawned by these tumors that are responsible for some 90% of deaths from cancer.

2.2 Tumors arise from many specialized cell types throughout the body

The majority of human tumors arise from epithelial tissues. **Epithelia** are sheets of cells that line the walls of cavities and channels or, in the case of skin, serve as the outside protective covering of the body. By the first decades of the twentieth century, detailed histological analyses had revealed that normal tissues containing epithelia are all structured similarly. Thus, beneath the epithelial cell layers in each of these tissues lies a **basement membrane** (sometimes called a **basal lamina**); it separates the epithelial cells from the underlying layer of supporting connective tissue cells, termed the **stroma** (**Figure 2.3**).

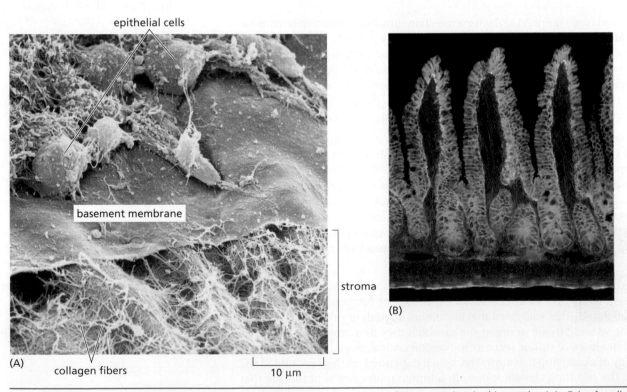

Figure 2.3 Basement membranes (A) This scanning electron micrograph of a chick corneal epithelium illustrates the basic plan of epithelial tissues, in which epithelial cells are tethered to one side of the basement membrane, sometimes termed the "basal lamina." Seen here as a continuous sheet, it is formed as a meshwork of extracellular matrix proteins. A network of collagen fibers anchors the underside of the basement membrane to the extracellular matrix (ECM) of the stroma. (B) While basement membranes cannot be detected using conventional staining techniques, use of immunofluorescence with an antibody against a basement membrane protein—in this case laminin 5 (*red*)—allows its visualization. Epithelial cells coating the villi of mouse small intestine have been stained with an antibody against E-cadherin (*green*), while all cell nuclei are stained *blue*. Here the convoluted basement membrane separates the outer villus layer of epithelial cells, termed enterocytes, from the mesenchymal cells forming the core of each villus (*not stained*). [A, courtesy of Robert Trelstad. B, from Z.X. Mahoney et al., *J. Cell Sci.* 121:2493–2502, 2008 (cover image). With permission from the authors.]

The basement membrane is a specialized type of **extracellular matrix (ECM)** and is assembled from proteins secreted largely by the epithelial cells. Another type of basement membrane separates **endothelial** cells, which form the inner linings of capillaries and larger vessels, from an outer layer of specialized smooth muscle cells. In all cases, these basement membranes serve as a structural scaffolding of the tissue. In addition, as we will learn later, cells attach a variety of biologically active signaling molecules to both the ECM and basement membrane; such molecules, including the **growth factors** to be described in Chapter 5, are then liberated later to stimulate epithelial cell proliferation.

Epithelia are of special interest here, because they spawn the most common human cancers—the **carcinomas**. These tumors are responsible for more than 80% of the cancer-related deaths in the Western world. Included among the carcinomas are tumors arising from the epithelial cell layers of the gastrointestinal tract—which includes mouth, esophagus, stomach, and small and large intestines—as well as the skin, mammary gland, pancreas, lung, liver, ovary, uterus, prostate, gallbladder, and urinary bladder. Examples of normal epithelial tissues are presented in **Figure 2.4**.

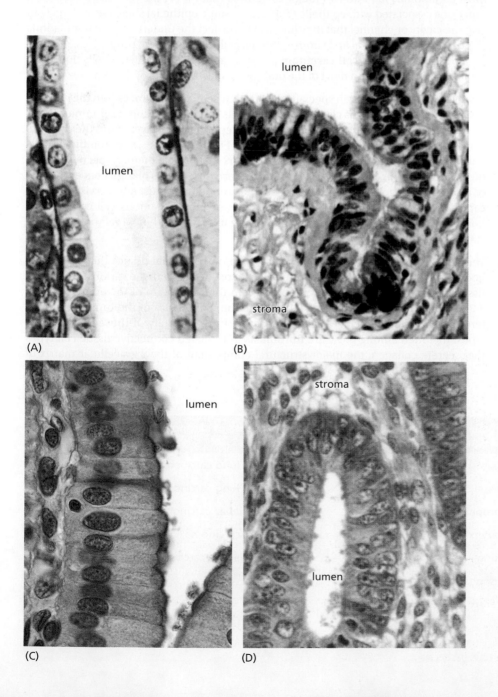

(A)

(B)

(C)

(D)

Figure 2.4 Architecture of epithelial tissues A common organizational plan describes most of the epithelial tissues in the body: The mature, differentiated epithelial cells are at the exposed surface of an epithelium. In many tissues, underlying these epithelia are less differentiated epithelial cells, not seen in this figure. Beneath the epithelial cell layer lies a basement membrane (see Figure 2.3), which is usually difficult to visualize in the light microscope. Shown here are epithelia of (A) a collecting tubule of the kidney, (B) a bronchiole of the lung with its thickened basement membrane, (C) the columnar epithelium of the gallbladder, and (D) the endometrium of the uterus. In each case, the epithelial cells protect the underlying tissue from the contents of the lumen (cavity) that they are lining. Panel C illustrates another property that is characteristic of the epithelial cells forming an epithelium: the state of apico-basal polarity, in which individual epithelial cells are organized to present their apical surface toward the lumen (*right*) and their basal surface toward the underlying basement membrane. (The side-by-side lateral interactions between adjacent epithelial cells are often included together with the basal surface to constitute the actions of the basolateral domain.) This polarization involves the asymmetric localization of the nuclei, which are more basally located, along with hundreds of cell surface (and associated cytoskeletal) proteins (*not shown*) that are specifically localized either to the apical, basal, or lateral surface of these cells. Moreover, the lateral surfaces of the epithelial cells establish several distinct types of junctions with their adjacent epithelial neighbors that bind together the epithelial cell sheet. (A, B, and D, from B. Young et al., Wheater's Functional Histology, 5th ed. Edinburgh: Churchill Livingstone, 2006. With permission from Elsevier. C, Jose Luis Calvo/Shutterstock.)

This group of tissues encompasses cell types that arise from all three of the primitive cell layers in the early vertebrate embryo. Thus, the epithelia of the lungs, liver, gallbladder, pancreas, esophagus, stomach, and intestines all derive from the inner cell layer, the **endoderm**. Skin arises from the outer embryonic cell layer, termed the **ectoderm**, while the ovaries originate embryologically from the middle layer, the **mesoderm** (Supplementary Sidebar 2.2). Therefore, in the case of carcinomas, histopathological classification is not informed by the early developmental history of the tissue of origin.

The epithelial and stromal cells of these various tissues collaborate in forming and maintaining the epithelial sheets. When viewed from the perspective of evolution, it now seems that the embryologic mechanisms for organizing and structuring epithelial tissues were invented early in metazoan evolution, likely more than 600 million years ago, and that these structural principles have been applied time and again during metazoan evolution to construct tissues and organs having a wide array of physiologic functions.

Most carcinomas fall into two major categories that reflect the two major biological functions associated with epithelia (**Table 2.1**). Some epithelial sheets serve largely to seal the cavity or channel that they line and to protect the underlying cell populations (**Figure 2.5**). Tumors that arise from epithelial cells forming these protective cell layers are termed **squamous cell carcinomas**. For example, the epithelial cells lining the skin (**keratinocytes**) and most of the oral cavity spawn tumors of this type.

Many epithelia also contain specialized cells that secrete substances into the ducts or cavities that they line. This class of epithelial cells generates **adenocarcinomas**. Often these secreted products are used to protect the epithelial cell layers from the contents of the cavities (**lumina**) that they surround (see Figure 2.5). Thus, some epithelial cells lining the lung and stomach secrete mucus layers that protect them, respectively, from the air (and airborne particles) and from the corrosive effects of high concentrations of acid. The epithelia in some organs, such as the lung, uterus, and cervix, have the capacity to give rise to pure adenocarcinomas or pure squamous cell carcinomas; quite frequently, however, tumors in these organs are found in which both types of carcinoma cells coexist.

The remainder of malignant tumors arise from nonepithelial tissues throughout the body. The first major class of nonepithelial cancers derive from the various connective tissues, all of which share a common origin in the mesoderm of the embryo (**Table 2.2**). These tumors, the **sarcomas**, constitute only about 1% of the tumors encountered in the oncology clinic. Sarcomas derive from a variety of **mesenchymal** cell types. Included among these are **fibroblasts** and related connective tissue cell types that secrete collagen, the major structural component of the extracellular matrix of

Table 2.1 Carcinomas

Tissue sites of more common types of adenocarcinoma	Tissue sites of more common types of squamous cell carcinoma	Other types of carcinoma
lung	skin	small-cell lung carcinoma
colon	nasal cavity	large-cell lung carcinoma
breast	oropharynx	hepatocellular carcinoma
pancreas	larynx	renal cell carcinoma
stomach	lung	transitional-cell carcinoma (of urinary bladder)
esophagus	esophagus	
prostate	cervix	
endometrium		
ovary		

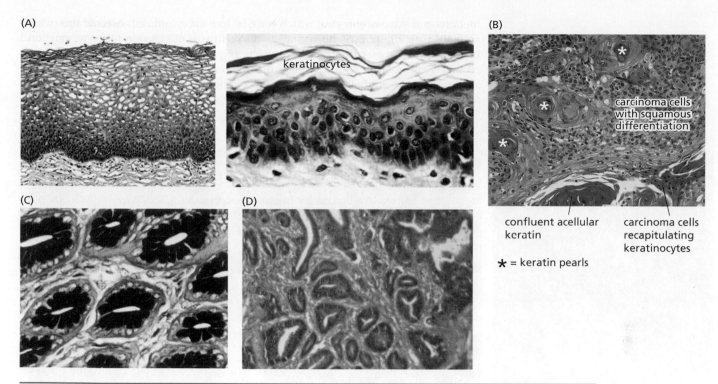

Figure 2.5 Epithelia and derived carcinomas Epithelia can be classified into subtypes depending on their normal shape and function and on the carcinomas arising from them. Two major subtypes and their respective derived carcinomas—squamous cell carcinomas and adenocarcinomas—are seen here. Squamous cell carcinomas derive from normal squamous epithelial cells whereas adenocarcinomas derive from epithelial cells with secretory function. (A) Normal squamous cells are often flattened and function to protect the epithelium and underlying tissue from the contents of the lumen or, in the case of skin, from the outside world. The squamous epithelia of the cervix of the uterus (*left*) and the skin (*right*) are organized quite similarly, with mature flattened cells at the surface being continually shed (for example, the dead keratinocytes of the skin) and replaced by less differentiated cells that move upward and proceed to differentiate. (B) Keratin pearl formation is a characteristic feature among a malignant tumor cell population that shows squamous differentiation. Confluent areas of accumulation of acellular keratin (*white asterisk*) and tumor cells recapitulating the appearance of normal keratinocytes are common in addition to large numbers of carcinoma cells exhibiting a morphology characteristic of squamous cell carcinomas. (C) In some tissues, the glandular cells within epithelia secrete mucopolysaccharides to protect the epithelium; in other tissues, they secrete proteins that function within the lumina (cavities) of ducts or are distributed to distant sites in the body. Pits in the stomach wall are lined by mucus-secreting cells (*dark red*). (D) This adenocarcinoma of the stomach shows multiple ductal elements, which are clear indications of its derivation from secretory epithelia such as those in panel C. (A and C, from B. Young et al., Wheater's Functional Histology, 5th ed. Edinburgh: Churchill Livingstone, 2006. With permission from Elsevier. B, Dr. Vickle Young Jo. D, from A.T. Skarin, Atlas of Diagnostic Oncology, 3rd ed. Philadelphia: Elsevier Science Ltd., 2003. With permission from Elsevier.)

Table 2.2 Various types of more common sarcomas

Type of tumor	Presumed cell lineage of founding cell
Osteosarcoma	osteoblast (bone-forming cell)
Liposarcoma	adipocyte (fat cell)
Leiomyosarcoma	smooth muscle cell (e.g., in gut)
Rhabdomyosarcoma	striated/skeletal muscle cell
Malignant fibrous histiocytoma	adipocyte/muscle cell
Fibrosarcoma	fibroblast (connective tissue cell)
Angiosarcoma	endothelial cells (lining of blood vessels)
Chondrosarcoma	chondrocyte (cartilage-forming cell)

tendons and skin; **adipocytes**, which store fat in their cytoplasm; **osteoblasts**, which assemble calcium phosphate crystals within matrices of collagen to form bone; and **myocytes**, which assemble to form muscle (**Figure 2.6**). **Hemangiomas**, which are relatively common in children, arise from precursors of the **endothelial** cells. The stromal layers of epithelial tissues include a variety of these mesenchymal cell types, notably fibroblasts and adipocytes.

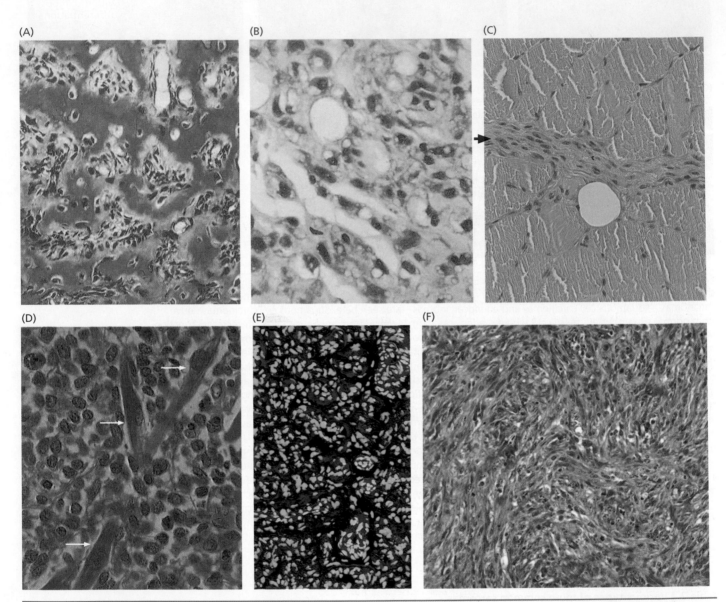

Figure 2.6 Mesenchymal tumors (A) Seen in this osteosarcoma are malignant bone-forming cells—osteoblasts (*dark purple nuclei*)—amid the mineralized bone (*pink*) they have assembled in the surrounding extracellular matrix. (B) A liposarcoma arises from cells closely related to adipocytes, which store lipid globules in various tissues. The presence of these globules throughout this tumor gives it a foamy appearance. (C) This leiomyosarcoma (*arrow, dark purple nuclei*), which arises in cells that form smooth muscle, is dispatching individual tumor cells to grow among adjacent normal muscle fibers (*light purple*). (D) Rhabdomyosarcomas arise from the cells forming striated skeletal muscles; the cancer cells (*dark red nuclei*) are seen here amid several normal muscle cells (*arrows*). (E) Hemangiomas—common benign tumors in infants that regress spontaneously—derive from the endothelial cells that form the lining of the lumina of small and large blood vessels. The densely packed capillaries in this particular tumor are formed from endothelial cells with cell nuclei stained *green* and cytoplasms stained *red*. Like epithelial cells, endothelial cells form basement membranes to which they attach, seen here in *blue*. (F) A dedifferentiated, high-grade spindle cell sarcoma. The dedifferentiated component may show an osteosarcoma, a fibrosarcoma, a malignant fibrous histiocytoma, or rarer subtypes including angiosarcoma and rhabdomyosarcoma, and therefore may mimic the common mesenchymal precursor of these various mesenchymal cell types. (A–C, from A.T. Skarin, Atlas of Diagnostic Oncology, 3rd ed. Philadelphia: Elsevier Science Ltd., 2003. With permission from Elsevier. D, from H. Okazaki and B.W. Scheithauer, Atlas of Neuropathology. New York: Gower Medical Publishing, 1988. With permission of Mayo Foundation for Medical Education and Research. E, from M.R Ritter et al., *Proc. Natl. Acad. Sci. USA* 99:7455–7460, 2002. With permission of National Academy of Sciences, U.S.A. F, Courtesy of Dharam M. Ramnani.)

Table 2.3 Various types of more common hematopoietic malignancies

Acute lymphocytic leukemia (ALL)
Acute myelogenous leukemia (AML)
Chronic myelogenous leukemia (CML)
Chronic lymphocytic leukemia (CLL)
Multiple myeloma (MM)
Non-Hodgkin's lymphoma[a] (NHL)
Hodgkin's lymphoma (HL)

[a]The non-Hodgkin's lymphoma types, also known as lymphocytic lymphomas, can be placed in as many as 15–20 distinct subcategories, depending upon classification system.

The second group of nonepithelial cancers arise from the various cell types that constitute the blood-forming (**hematopoietic**) tissues, including the cells of the immune system (**Table 2.3** and **Figure 2.7**); these cells also derive from the embryonic mesoderm. Among them are cells destined to form **erythrocytes** (red blood cells), antibody-secreting (**plasma**) cells, and T and B **lymphocytes**. The term **leukemia** (literally "white blood") refers to malignant derivatives of several of these hematopoietic cell lineages that move freely through the circulation and, unlike the red blood cells, are nonpigmented. **Lymphomas** include tumors of the **lymphoid** lineages (B and T lymphocytes) that aggregate to form solid tumor masses, most frequently found in lymph nodes, rather than the dispersed, single-cell populations of tumor cells seen in leukemias. This class of tumors is responsible for ~7% of cancer-associated mortality in the United States.

The third and last major grouping of nonepithelial tumors arises from cells that form various components of the central and peripheral nervous systems (**Table 2.4**). These are often termed **neuroectodermal** tumors to reflect their origins in the outer cell layer of the early embryo. Included in this category are **gliomas, glioblastomas, neuroblastomas, schwannomas,** and **medulloblastomas** (**Figure 2.8**). While comprising

Table 2.4 Various types of neuroectodermal malignancies

Name of tumor	Lineage of founding cell
Glioblastoma multiforme	highly progressed astrocytoma
Astrocytoma	astrocyte (type of glial cell)[a]
Meningioma	arachnoidal cells of meninges[b]
Schwannoma	Schwann cell around axons[c]
Retinoblastoma	cone cell in retina[d]
Neuroblastoma[e]	cells of peripheral nervous system
Ependymoma	glial cells lining ventricles of brain[f]
Oligodendroglioma	oligodendrocyte covering axons[g]
Medulloblastoma	granular cells of cerebellum[h]

[a]Nonneuronal cell of central nervous system that supports neurons.
[b]Membranous covering of brain.
[c]Constructs insulating myelin sheath around axons in peripheral nervous system.
[d]Photosensor for color vision during daylight.
[e]These tumors arise from cells of the sympathetic nervous system.
[f]Fluid-filled cavities in brain.
[g]Similar to Schwann cells but in brain.
[h]Cells of the lower level of cerebellar cortex (as a related example, see Figure 2.8B).

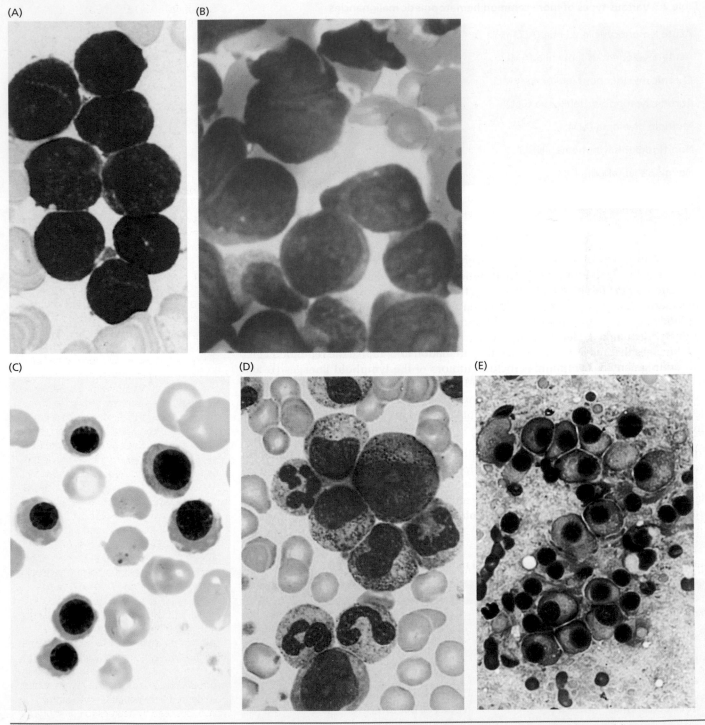

Figure 2.7 Hematologic malignancies (A) Acute lymphocytic leukemias (ALLs) arise from both the B-cell (80%) and T-cell (20%) lineages of lymphocytes. The cells forming this particular tumor (*red-purple*) exhibited the antigenic markers indicating origin from pre-B cells. (B) As in many hematopoietic malignancies, these acute myelogenous leukemia (AML) cells (*blue*) have only a small rim of cytoplasm around their large nuclei. They derive from precursor cells of the lineage that forms various types of granulocytes as well as monocytes, the latter developing, in turn, into macrophages, dendritic cells, osteoclasts, and other tissue-specific phagocytic cells. (C) The large erythroblasts in this erythroleukemia (*dark purple*) closely resemble the precursors of differentiated red blood cells—erythrocytes. (D) In chronic myelogenous leukemia (CML), a variety of leukemic cells of the myeloid (marrow) lineage are apparent (*red nuclei*), suggesting the differentiation of myeloid leukemia stem cells into several distinct cell types. (E) Multiple myeloma (MM) is a malignancy of the plasma cells of the B-cell lineage, which secrete antibody molecules, explaining their relatively large cytoplasms in which proteins destined for secretion are processed and matured. Seen here are plasma cells of MM at various stages of differentiation (*purple nuclei*). In some of these micrographs, numerous lightly staining erythrocytes are seen in the background. (A, B, and D, from A.T. Skarin, Atlas of Diagnostic Oncology, 3rd ed. Philadelphia: Elsevier Science Ltd., 2003. With permission from Elsevier. C, Richard J. Green / Science Source. E, David A. Litman / Shutterstock.)

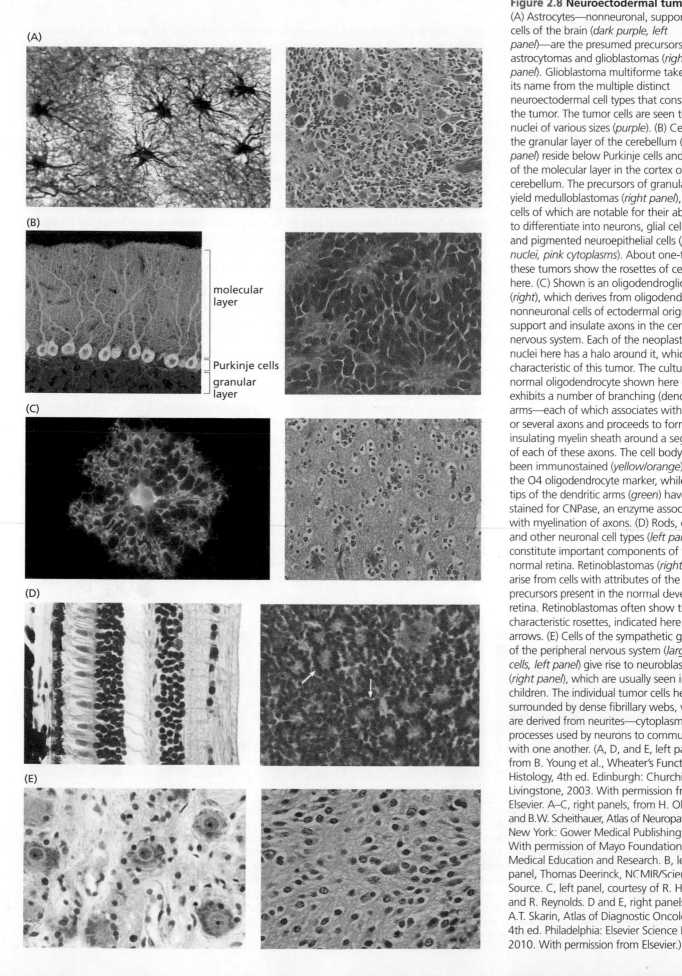

Figure 2.8 Neuroectodermal tumors
(A) Astrocytes—nonneuronal, supporting cells of the brain (*dark purple, left panel*)—are the presumed precursors of astrocytomas and glioblastomas (*right panel*). Glioblastoma multiforme takes its name from the multiple distinct neuroectodermal cell types that constitute the tumor. The tumor cells are seen to have nuclei of various sizes (*purple*). (B) Cells of the granular layer of the cerebellum (*left panel*) reside below Purkinje cells and cells of the molecular layer in the cortex of the cerebellum. The precursors of granular cells yield medulloblastomas (*right panel*), the cells of which are notable for their ability to differentiate into neurons, glial cells, and pigmented neuroepithelial cells (*purple nuclei, pink cytoplasms*). About one-third of these tumors show the rosettes of cells seen here. (C) Shown is an oligodendroglioma (*right*), which derives from oligodendrocytes, nonneuronal cells of ectodermal origin that support and insulate axons in the central nervous system. Each of the neoplastic cell nuclei here has a halo around it, which is characteristic of this tumor. The cultured normal oligodendrocyte shown here (*left*) exhibits a number of branching (dendritic) arms—each of which associates with one or several axons and proceeds to form an insulating myelin sheath around a segment of each of these axons. The cell body has been immunostained (*yellow/orange*) for the O4 oligodendrocyte marker, while the tips of the dendritic arms (*green*) have been stained for CNPase, an enzyme associated with myelination of axons. (D) Rods, cones, and other neuronal cell types (*left panel*) constitute important components of the normal retina. Retinoblastomas (*right panel*) arise from cells with attributes of the cone precursors present in the normal developing retina. Retinoblastomas often show the characteristic rosettes, indicated here with arrows. (E) Cells of the sympathetic ganglia of the peripheral nervous system (*larger cells, left panel*) give rise to neuroblastomas (*right panel*), which are usually seen in children. The individual tumor cells here are surrounded by dense fibrillary webs, which are derived from neurites—cytoplasmic processes used by neurons to communicate with one another. (A, D, and E, left panels, from B. Young et al., Wheater's Functional Histology, 4th ed. Edinburgh: Churchill Livingstone, 2003. With permission from Elsevier. A–C, right panels, from H. Okazaki and B.W. Scheithauer, Atlas of Neuropathology. New York: Gower Medical Publishing, 1988. With permission of Mayo Foundation for Medical Education and Research. B, left panel, Thomas Deerinck, NCMIR/Science Source. C, left panel, courtesy of R. Hardy and R. Reynolds. D and E, right panels, from A.T. Skarin, Atlas of Diagnostic Oncology, 4th ed. Philadelphia: Elsevier Science Ltd., 2010. With permission from Elsevier.)

only 1.3% of all diagnosed cancers, these are responsible for about 2.5% of cancer-related deaths. Of note, the developmental origin of much of the central nervous system (CNS) from a primitive neuroepithelial cell layer seems to explain some of the striking similarities between CNS tumors and the carcinomas listed earlier.

2.3 Some types of tumors do not fit into the major classifications

Not all tumors fall neatly into one of these four major groups. For example, **melanomas** derive from melanocytes, the pigmented cells of the skin and the retina. The melanocytes, in turn, arise from a primitive embryonic structure termed the **neural crest**. While having an embryonic origin close to that of the neuroectodermal cells, the precursors of melanocytes end up during development as wanderers that settle in the skin and the eye and provide pigment to these tissues, but establish no direct connections with the nervous system (**Figure 2.9**). Moreover, melanoma cells share many features in common with certain types of carcinoma cells.

Small-cell lung carcinomas (SCLCs) contain cells having many attributes of **neurosecretory** cells, such as those of neural crest origin in the **adrenal** glands that sit above the kidneys. Such cells, often in response to neuronal signaling, secrete biologically active peptides. It appears that these tumors originate in endodermal stem cell populations of the lung that have shed some of their epithelial characteristics and acquired others that are characteristic of a neuroectodermal lineage.

This switching of tissue lineage and resulting acquisition of an entirely new set of differentiated characteristics is often termed **transdifferentiation**. The term implies that the commitments that cells have made by individual cells during embryogenesis to enter into one or another tissue and cell lineage are not irreversible, and that under certain conditions, descendant cells can move from one differentiation lineage to another. Such a change in phenotype may affect both normal and cancer cells. For example, at the borders of many carcinomas, epithelial cancer cells often change shape and gene expression programs and take on certain attributes that are usually associated with cells of mesenchymal origin. This dramatic shift in cell phenotype, termed the **epithelial–mesenchymal transition**, or simply EMT, implies great phenotypic plasticity on the part of cells that normally seem to be fully committed to behaving like epithelial cells. As described later (Chapters 13 and 14), this transition may often accompany and enable the invasion by carcinoma cells into adjacent

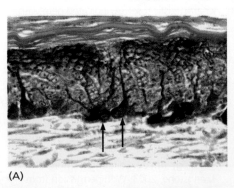
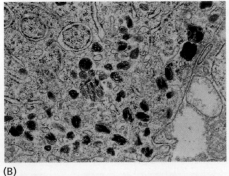
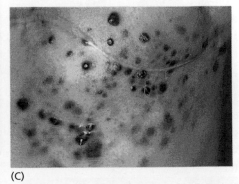

(A)   (B)   (C)

Figure 2.9 Melanocytes and melanomas (A) Melanocytes (*arrows*), which form melanin pigment granules, are normally scattered among the basal keratinocytes of the skin. They extend long, thin cytoplasmic processes through which they deposit these granules in the cytoplasm of keratinocytes, which form the bulk of the epithelium (see Figure 2.5A). Layers of dead keratinocytes at the surface of the skin (*above*) and stroma cells (*below*) are also apparent. (B) The pigment granules, visualized here by transmission electron microscopy, have made melanomas favored objects of research because the metastases that they form are easily visualized. Once melanomas have begun to invade vertically from the superficial layers of the skin into the underlying stroma, they have a high tendency to metastasize to distant tissues. (C) This case of cutaneous melanoma dramatizes the metastatic nature of the disease and the readily observed, pigmented metastases. (A, from W.J. Bacha Jr. et al., *Color Atlas of Veterinary Histology*, 2nd ed. Ames, IA: Wiley–Blackwell, 2012. With permission from John Wiley & Sons. B and C, from A.T. Skarin, *Atlas of Diagnostic Oncology*, 4th ed. Philadelphia: Elsevier Science Ltd., 2010. With permission from Elsevier.)

normal tissues. (This carcinoma-associated EMT exploits a cell-biological program operating in normal tissues during the healing of damaged epithelial cell sheets.) Cells activating an EMT program usually do so reversibly, in contrast to most types of transdifferentiation in which cells shift their tissue-specific phenotypes irreversibly. Importantly, these interconversions involve changes in gene expression programs rather than mutations in the genomes of carcinoma cells.

Of the atypical tumor types, **teratomas** are arguably the most bizarre of all, in part because they defy all attempts at classification. While only ~10,000 cases are diagnosed worldwide annually, teratomas deserve mention because they are unique and shed light on the biology of **embryonic stem (ES) cells**, which have become so important to biologists; ES cells enable genetic manipulation of the mouse germ line and are central to certain types of stem cell therapies currently under development. Teratomas seem to arise from **germ cell** (egg and sperm) diploid precursors (see Section 1.3) that have failed to migrate to their proper destinations during embryonic development and persist at **ectopic** (inappropriate) sites in the developing fetus. They retain the **pluripotency** of early embryonic cells—the ability to generate most and possibly all of the differentiated cell types present in the fully developed fetus. The cells in different sectors of common "mature" teratomas—which are largely benign, localized growths—differentiate to create tissues that are very similar at the histopathological level to those found in a variety of adult tissues (**Figure 2.10**). Typically, representatives of the three cell layers of the embryo—endoderm, mesoderm, and ectoderm (see Supplementary Sidebar 2.2)—coexist within a single tumor and often develop into recognizable tissue structures, such as those of teeth, hair, and bones. Occasionally these tumors progress to become highly malignant and thus life-threatening.

Of special interest is the fact that careful karyotypic and molecular analyses of benign, mature teratomas have suggested that the associated tumor cells are genetically wild type. This suggests that such teratoma cells are unique, being the only type of tumorigenic cell whose genomes are truly wild type, in contrast to the cells of essentially all other tumor types described in this book, which almost always carry identifiable genetic aberrations, often in large number.

The occasional rule-breaking exceptions, such as those represented by teratomas and the products of the EMT, do not detract from one major biological principle that seems to govern the vast majority of cancers: while cancer cells deviate substantially in behavior from their normal cellular precursors, they almost always retain many of the distinctive attributes of the normal cell types from which they descend. These vestigial, differentiation-associated traits provide critical clues about the origins of most tumors; they enable pathologists to examine tumor biopsies under the microscope and assign a tissue-of-origin and tumor classification, even without prior knowledge of the anatomical sites from which these biopsies were prepared.

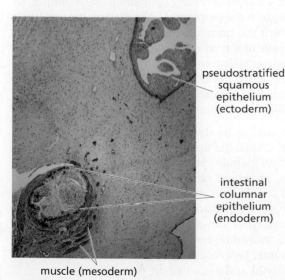

pseudostratified squamous epithelium (ectoderm)

intestinal columnar epithelium (endoderm)

muscle (mesoderm)

Figure 2.10 Teratomas This teratoma was created by implanting human embryonic stem (ES) cells into a mouse, yielding a tumor that is a phenocopy of the spontaneous teratomas found in children; such "mature" teratomas contain fully differentiated cells and are localized, noninvasive tumors. (Courtesy of Sumita Gokhale.)

In a small minority of cases (2–5%), the tumor biopsies given to pathologists for analysis have shed virtually all of the characteristic tissue-specific, differentiated traits of the corresponding normal precursor tissues. The cells in such tumors are said to have **dedifferentiated**, and the tumors as a whole are **anaplastic**, in that it is no longer possible to use histopathological criteria to identify the tissues from which they have arisen (Supplementary Sidebar 2.3). A tumor of this type, by definition detected at a site of metastatic dissemination, is often classified as a **cancer of unknown primary** (CUP), reflecting the difficulty of identifying the original site of primary tumor formation in the patient. In recent years, however, analysis of molecular markers, such as the methylation patterns of DNA and gene expression patterns, has enabled the identification of the tissues-of-origin of almost 90% of CUPs that could not be classified using only histopathological criteria. It is unclear what forces have driven the shedding of tissue-specific differentiation markers; their elimination may reflect the potentially highly mutable genomes of CUPs (as discussed in Chapter 12), or immunological attacks may have weeded out cancer cell subpopulations within growing tumors that express differentiation-specific proteins (as described in Chapter 15).

2.4 Cancers seem to develop progressively

Between the two extremes of fully normal and highly malignant tissue architectures lies a broad spectrum of tissues of intermediate histopathological appearance. The different gradations of abnormality may well reflect cell populations that are evolving progressively toward greater degrees of aggressive and invasive behavior. Thus, each type of abnormal growth detected within a tissue may represent a distinct step along this evolutionary pathway. If so, these architectures suggest, but hardly prove, that the development of tumors is a complex, multi-step process, a subject that is discussed in great detail in Chapter 11.

Some growths contain cells that deviate only minimally from those of normal tissues but may nevertheless be abnormal in that they contain excessive *numbers* of cells. Such growths are termed **hyperplastic** (**Figure 2.11**). In spite of their apparently deregulated proliferation, the cells forming hyperplastic growths have retained the ability to assemble into tissues that appear reasonably normal.

An equally minimal deviation from normal is seen in **metaplasia**, where one type of normal cell layer within a tissue is displaced by cells of another type that are not normally encountered there. These invaders, although present in the wrong location, often appear completely normal under the microscope. Metaplasia is most frequent in epithelial transition zones where one type of epithelium meets another. Transition zones like these are found at the junction of the cervix with the uterus and the junction of the esophagus and the stomach. In both locations, a squamous epithelium normally undergoes an abrupt transition into a mucus-secreting epithelium. For example, an early indication of premalignant change in the esophagus is the metaplastic condition termed **Barrett's esophagus**, in which the normally present squamous epithelium is replaced by secretory epithelial cells of a type usually found within the stomach (**Figure 2.12**). Even though these gastric cells have a quite normal appearance, this metaplasia is considered an early step in the development of esophageal adenocarcinomas. Indeed, patients suffering from Barrett's esophagus have a thirtyfold increased risk of eventually developing these highly malignant tumors.

A slightly more abnormal tissue is said to be **dysplastic**. Cells within a dysplasia are usually abnormal **cytologically**, that is, the appearance of individual cells is no longer normal. The cytological changes include variability in nuclear size and shape, increased nuclear staining by dyes, increased ratio of nuclear versus cytoplasmic size, increased mitotic activity, and lack of the cytoplasmic features associated with the normal differentiated cells of the tissue (**Figure 2.13**). In dysplastic growths, the relative proportions of the various cell types seen in the normal tissue are no longer observed. Together, these changes in individual cells and in cell numbers have major effects on the overall tissue architecture. Dysplasia is considered to be a transitional state between completely benign growths and those that are premalignant.

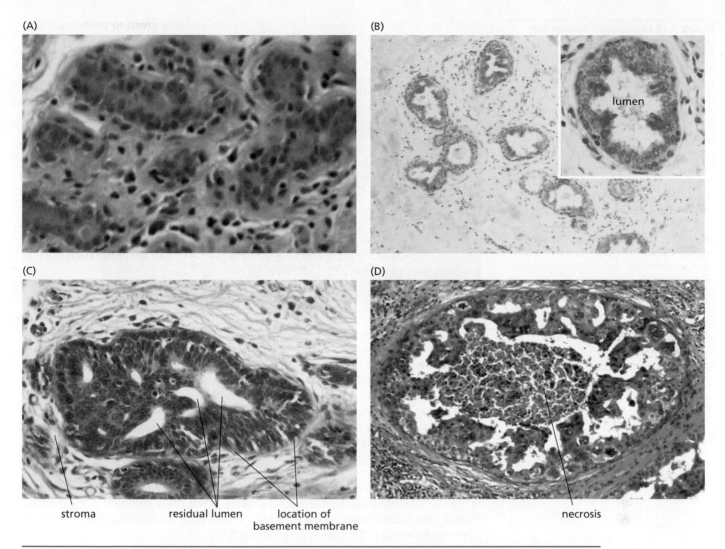

stroma   residual lumen   location of   necrosis
                                basement membrane

Figure 2.11 Progression of normal mammary tissue to ductal carcinoma *in situ* The morphology of the normal ductal epithelium of the mammary gland, termed a normal terminal ductal lobular unit (A), can be compared with mildly hyperplastic milk ducts (B), shown at low magnification and high magnification (*inset*), in which mammary epithelial cells have begun to form piles that protrude into the lumina. (C) A more advanced hyperplastic mammary duct shows epithelial cells that are crowded together and almost completely fill the lumen. (D) In this ductal carcinoma *in situ* (DCIS) the pre-neoplastic cells have proliferated to generate a mass that fills the lumen of a duct and has then suffered necrosis, ostensibly because they lacked access to an adequate blood supply. However, the involved cells have not penetrated the basement membrane (*not visible*) and invaded the surrounding stroma. (A and D, Courtesy of Deborah Dillon. B and C, from A.T. Skarin, Atlas of Diagnostic Oncology, 4th ed. Philadelphia: Elsevier Science Ltd., 2010. With permission from Elsevier.)

Even more abnormal are the growths that are seen in epithelial tissues and termed variously **adenomas, polyps,** adenomatous polyps, **papillomas,** and, in skin, warts (**Figure 2.14**). These are often large growths that can be readily detected with the naked eye. They contain all the cell types found in the normal epithelial tissue, but this assemblage of cells has launched a program of substantial expansion, creating a macroscopic mass. Under the microscope, the tissue within these adenomatous growths is seen to be dysplastic. These tumors usually grow to a certain size and then stop growing, and they respect the boundary created by the basement membrane, which continues to separate them from the underlying stroma. Since adenomatous growths do not penetrate the basement membrane and invade underlying tissues, they are considered to be benign.

A further degree of abnormality is represented by growths that do indeed invade underlying tissues. In the case of carcinomas, this incursion is signaled the moment carcinoma cells break through a basement membrane and invade into the adjacent

Figure 2.12 Metaplastic conversion of epithelia In certain precancerous conditions, the normally present epithelium is replaced by an epithelium from a nearby tissue—the process of metaplasia. For example, in Barrett's esophagus (sometimes termed Barrett's esophagitis), the squamous cells that normally line the wall of the esophagus (residual squamous mucosa) are replaced by secretory cells that migrate from the lining of the stomach (metaplastic Barrett's epithelium). This particular metaplasia, which is provoked by chronic acid reflux from the stomach, can progress to an esophageal carcinoma, which occurs in ~0.5% of Barrett's esophagus patients per year. As seen here, the *left three-quarters* of this image shows the invading columnar epithelium, which is typical of the gastric mucosa, while the *right quarter* shows the remaining squamous epithelium that is typical of the normal esophagus. (From New England Journal of Medicine, Stuart J. Spechler, M.D., Rhonda F. Souza, M.D., Barrett's Esophagus, 371, 836-45. Copyright © 2014 Massachusetts Medical Society. Reprinted with permission from Massachusetts Medical Society. Image: Dr. Robert Genta.)

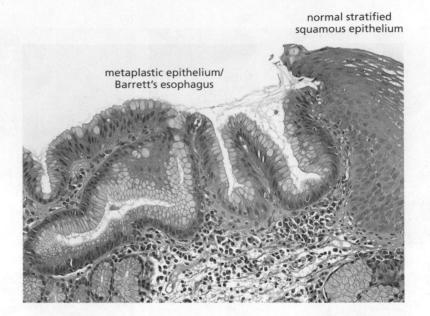

normal stratified squamous epithelium

metaplastic epithelium/ Barrett's esophagus

stroma (**Figure 2.15**). Here, for the first time, we encounter malignant cells that have a substantial potential of threatening the life of the individual who carries them. Clinical oncologists and surgeons often reserve the word **cancer** for these and even more abnormal growths. However, in this book, as in much of contemporary cancer research, the word cancer is used more loosely to include all types of abnormal growths. (In the case of epithelial tissues, the term "carcinoma" is usually applied to growths that have acquired this degree of invasiveness.) This disparate collection of growths—both benign and malignant—are called collectively **neoplasms**, that is, new types of tissue. (Some reserve the term "neoplasm" for malignant tumors.) A summary of the overall pathological classification scheme of tumors is provided in **Figure 2.16**. A short discussion of some of the principles and techniques enabling these classifications can be found in Supplementary Sidebar 2.4.

Figure 2.13 Formation of dysplastic epithelium In this intraepithelial squamous neoplasia of the cervix (*to right of white dotted line, large white arrow*), the epithelial cells have not broken through the basement membrane (*not visible, indicated by white dashed line*) and invaded the underlying stroma. The cells in this dysplasia continue to be densely packed all the way to the luminal surface (*above*), in contrast to the more diffuse distribution of cells in the normal epithelium (*left*), whose cytoplasms (*light pink*) increase in size as the cells differentiate. Numerous mitotic figures are also apparent in the dysplasia (*small white arrows*), indicating extensive cell proliferation. (Courtesy of T.A. Ince.)

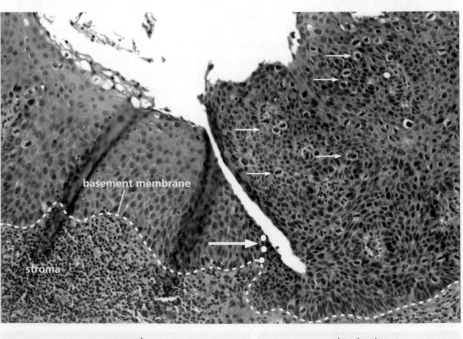

basement membrane

stroma

normal   dysplastic

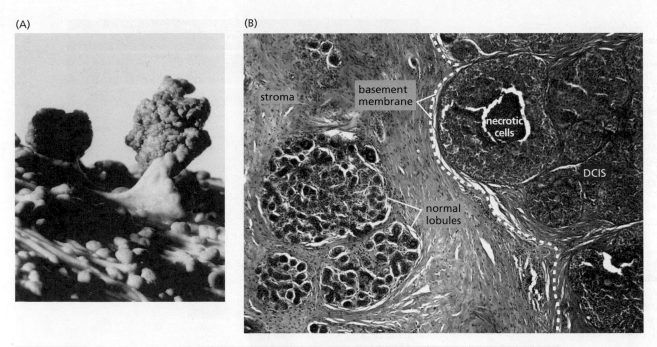

Figure 2.14 Pre-invasive adenomas and carcinomas Adenomatous growths, termed polyps in certain organs, have a morphology that sets them clearly apart from normal and dysplastic epithelium. (A) In the colon, pre-invasive growths appear as either flat thickenings of the colonic wall (sessile polyps; *not shown*) or as the stalk-like growths (pedunculated polyps) shown here in a photograph. These growths, also termed "adenomas," have not penetrated the basement membrane and invaded the underlying stroma. (B) The lobules of the normal human breast (*purple islands, left half of figure*), each containing numerous small alveoli in which milk is produced, are surrounded by extensive fibrous stroma (*pink*). The cells of an intraductal carcinoma, often called a ductal carcinoma *in situ* (DCIS; *purple, to right of dashed line*), fill and distend ducts but have not invaded through the basement membrane surrounding the ducts into the stroma. In the middle of one of these ducts is an island of necrotic carcinoma cells (*dark red*) that have died ostensibly because of inadequate access to the circulation. (A, courtesy of J. Northover and Cancer Research, UK. B, courtesy of T.A. Ince.)

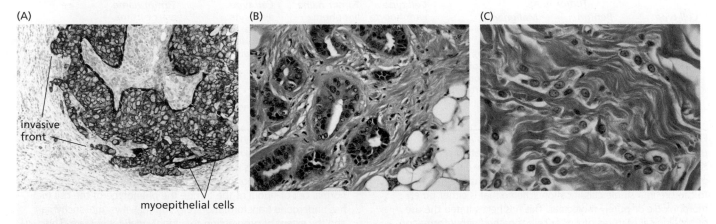

Figure 2.15 Invasive carcinomas Tumors are considered malignant only after they have breached the basement membrane and invaded the surrounding stroma. (A) These breast cancer cells (*dark red*), which previously constituted a ductal carcinoma *in situ* (DCIS; see Figure 2.14B), have now broken through on a broad front the layer of myoepithelial cells (*dark brown, lower right*) and underlying attached basement membrane (*not visible*) into the stroma; this indicates that they have acquired a new trait: invasiveness. (B) After breaching the basement membrane, invasive cancer cells can appear in various configurations amid the stroma. In this invasive ductal carcinoma of the breast, islands of epithelial cancer cells (*dark purple*) are interspersed amid the stroma (*dark pink*). The ductal nature of this carcinoma is revealed by the numerous rudimentary ducts formed by the breast cancer cells. (C) In this invasive lobular carcinoma of the breast, individual carcinoma cells (*dark purple nuclei*) have ventured into the stroma (*red-orange*), often doing so in single-file formation. (A, from F. Koerner, Diagnostic Problems in Breast Pathology. Philadelphia: Saunders/Elsevier, 2008. With permission from Elsevier. B and C, courtesy of T.A. Ince.)

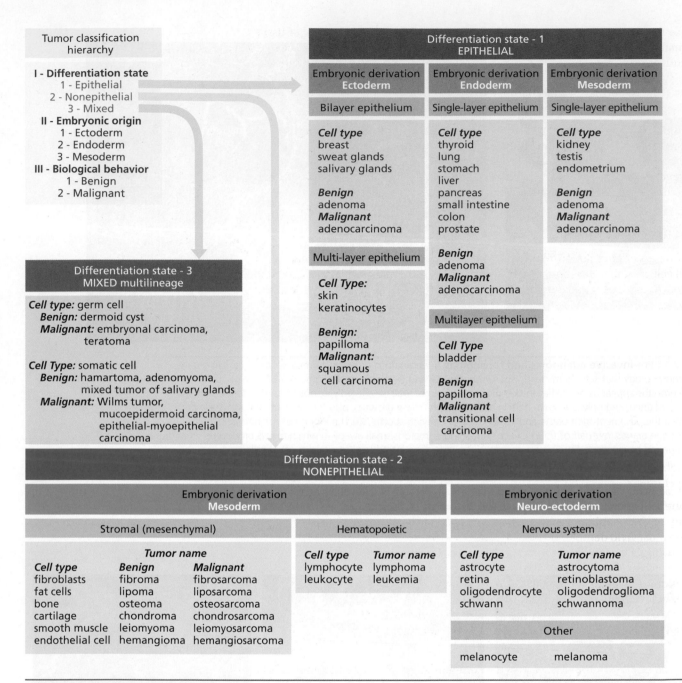

Figure 2.16 Classification scheme of tumors A clear understanding of the histopathological classification of tumors is essential for the study of cancer. However, the entire spectrum of tumors arising in various organs and tissues has been difficult to capture in a single classification scheme that is either purely morphologic or purely molecular. This has necessitated the use of histological features of tumor cells together with information about their respective tissues-of-origin, differentiation states, and biological behaviors; together these make it possible to develop a taxonomy of human tumors that has proven useful for the diagnosis and clinical management of most tumors. The scheme for classifying tumors presented here responds to three critical determinants of tumor biology: the embryonic tissue-of-origin and normal cell-of-origin of the tumor, the phenotype of the cell that has undergone transformation (e.g., epithelial vs. mesenchymal), and the extent of progression to a highly malignant state. This scheme allows classification of the great majority of, but not all, human tumors. (Courtesy of T.A. Ince.)

As mentioned earlier, cells in an initially formed primary tumor may seed new tumor colonies at distant sites in the body through the process of metastasis. This process is itself extraordinarily complex, and it depends upon the ability of cancer cells to invade adjacent tissues, to enter into blood and lymph vessels, to migrate through these vessels to distant anatomical sites, to leave the vessels and invade adjacent tissue, and to found new tumor cell colonies at the distant sites. These steps are the subject of detailed discussion in Chapter 14.

Because the various growths cataloged here represent increasing degrees of tissue abnormality, it would seem likely that they are distinct stopping points along the road of **tumor progression**, in which a normal tissue evolves progressively into one that is highly malignant. However, the precursor–product relationships of these various growths (that is, normal → hyperplastic → dysplastic → neoplastic → metastatic) are only suggested by the above descriptions but by no means proven.

We will postpone until Chapter 11 an in-depth discussion of multi-step tumor progression. Not discussed here are another set of changes that occur in cancer cells as they progress toward high-grade malignancy: their profound changes in metabolism, largely involving energy metabolism. The observed shifts in metabolic pathways appear to be critical for tumor growth, but for reasons that are not yet clear.

2.5 Tumors are monoclonal growths

Even if we accept the notion that tumors arise through the progressive alteration of normal cells, another question remains unanswered: How many normal cells are the ancestors of those cells that congregate to form a tumor (**Figure 2.17A**)? Do the tumor cells descend from a single ancestral cell that crossed over the boundary from normal to abnormal growth? Or did a large cohort of normal cells undergo this change, each becoming the ancestor of a distinct subpopulation of cells within a tumor mass?

The most effective way of addressing this issue is to determine whether all the cells in a tumor share a common, highly unique genetic or biochemical marker. For example, a randomly occurring somatic mutation might mark a cell in a very unusual way. If this particular genetic marker were present in all cells within a tumor, this would indicate that they all descend from an initially mutated cell. Such a population of cells, all of which derive from a common ancestral cell, is said to be **monoclonal**. Alternatively, if the tumor mass were composed of a series of genetically distinct subpopulations of cells that give no indication of a common origin, it can considered to be **polyclonal**.

A number of experimental strategies have been pursued to demonstrate the monoclonality of the neoplastic cells within a tumor (Supplementary Sidebar 2.5). Perhaps the most vivid demonstrations of tumor monoclonality have come from cancer cells sporting a variety of chromosomal aberrations that can be visualized microscopically when chromosomes condense during metaphase of mitosis. Often, a very peculiar chromosomal abnormality—the clear result of a rare random genetic accident—is seen in all the cancer cells within a tumor mass (Figure 2.17B). This observation makes it obvious that all the malignant cells within this tumor descend from the single ancestral cell in which this unusual chromosomal restructuring originally occurred.

While such observations seem to provide compelling proof that tumor populations are monoclonal, tumorigenesis may actually be more complex. Let us imagine, as a counterexample, that 10 normal cells in a tissue simultaneously crossed over the border from being normal to being malignant (or at least premalignant) and that each of these cells, and its descendants in turn, proliferated uncontrollably. Each of these founding cells would spawn a large monoclonal population, and the tumor mass, as a whole, consisting of a mixture of these 10 cell populations, would be polyclonal.

It is highly likely that each of these 10 clonal populations varies subtly from the other 9 in a number of characteristics, among them the time required for their cells to double. Simple mathematics indicates that a cell population that exhibits a slightly shorter doubling time will, sooner or later, outgrow all the others, and that the descendants of these cells will dominate in the tumor mass, creating what will appear to be a monoclonal tumor. In fact, many tumors seem to develop over decades, which is plenty of time for one clonal subpopulation to dominate in the overall tumor cell population. Hence, the monoclonality of the cells in a large tumor mass hardly proves that this tumor was strictly monoclonal during the early stages of its development.

Figure 2.17 Monoclonality versus polyclonality of tumors (A) In theory, tumors may be polyclonal or monoclonal in origin. In a polyclonal tumor (*right*), multiple cells cross over the border from normalcy to malignancy to become the ancestors of several genetically distinct subpopulations of cells within a tumor mass. In a monoclonal tumor (*left*), only a single cell is transformed from normal to cancerous behavior to become the ancestor of the cells in a tumor mass. (B) Illustrated is an unusual translocation (*arrow*) that involves exchange of segments between two different (nonhomologous) chromosomes—a *red* and a *yellow* chromosome. (Only one of the two chromosomal products of the translocation is shown here.) The translocation creates a characteristic "signature" that distinguishes the affected cell from the surrounding population of karyotypically normal cells (*top row*). Since all of the cancer cells within a subsequently arising tumor carry the identical, rare translocation (*bottom row*), this indicates their descent from a common progenitor in which this translocation initially occurred. (C) As tumors progress, an initial monoclonal cell population descended from a founder cell becomes increasingly heterogeneous due to progressive accumulation of random genetic errors affecting labeled genes; only some of these mutations may create functionally important alleles that drive further tumor progression. Each mutation creates a new genetically distinct subpopulation, and thus each subpopulation is characterized by the unique combination of mutant alleles that it carries. The resulting formation of multiple genetically distinct cancer cell subpopulations, however, disguises the fact that all of these subpopulations of cells trace their origins to a common genetic founding population that already carried its own set of mutations (e.g., *VHL*); these commonly shared (ubiquitous) mutant alleles are often termed truncal mutations to reflect their presence in the trunk (*blue*) of the genealogic tree that depicts the genetic relationships between the multiple subpopulations of cells coexisting within an advanced kidney carcinoma under study. The length of the branches is calculated by the numbers of random genetic errors accumulated in intergenic DNA. R, specific sub-regions of the primary tumor; M, metastasis; PreP, primary tumor sample isolated before chemotherapy; PreM, metastasis isolated before chemotherapy; private, found uniquely in this sample. (R4a and R4b are the subclones detected in R4. A question mark indicates that the detected *SETD2* splice-site mutation probably resides in R4a, whereas R4b most likely shares the *SETD2* frameshift mutation also found in other primary-tumor regions.) The fact that three subpopulations independently acquired a mutation in the *SETD2* gene indicates that each acquired significant proliferative advantage from mutation of this particular gene beyond the advantage conferred by the existing truncal mutations. (C, from M. Gerlinger et al., *New Engl. J. Med.* 366:883–892, 2012. With permission from Massachusetts Medical Society.)

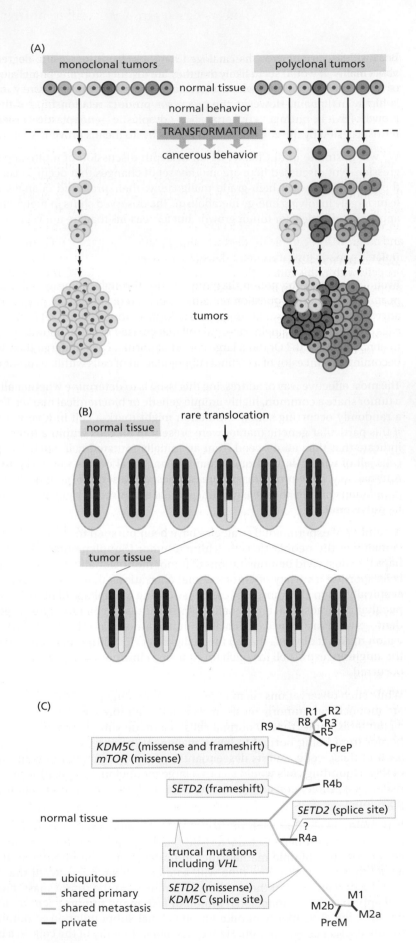

As we will discuss in great detail in Chapters 11 and 12, the population of cells within a tumor may begin as a relatively homogeneous collection of cells (thereby constituting a true monoclonal growth), but soon may become quite heterogeneous because of the continual acquisition of new mutant alleles by some of its cells, this being the consequence of the elevated genetic instability observed in most neoplastic cells (described in Chapter 12). The great majority of acquired variant alleles may have minimal if any effects on the phenotype of the evolving cancer cells belonging to a subpopulation of cancer cells, neither aiding nor hindering their proliferation and thus their representation among the larger populations of cancer cells within a tumor. However, each of these mutant alleles demarcates a genetically distinct subclone and the tumor, as a whole, becomes polyclonal. In fact, analyses of the genomes of the cells within such a tumor generates the analog of a genealogical tree, with repeated cycles of branching and outgrowth of genetically distinct subpopulations (Figure 2.17C). Indeed, the accumulated genetic intra-tumoral heterogeneity may mask the true monoclonal origin of the cells within this tumor. However, as the tree of Figure 2.17C reveals, all the variant subpopulations within a tumor will still share in common a mutation or set of mutant alleles that can be ascribed to a common ancestral population at the trunk of the tree, leading to the increasingly used term of **truncal** mutations. As we will see in Chapters 16 and 17, this genetic diversification of tumor cells holds important implications for cancer therapy and the successful eradication of populations of cancer cells.

2.6 Cancers occur with vastly different frequencies in different human populations

The nature of cancer suggests that it is a disease of chaos, a breakdown of existing biological order within the body. More specifically, the disorder seen in cancer appears to derive directly from malfunctioning of the controls that are normally responsible for determining when and where cells throughout the body will multiply. In fact, there is ample opportunity for the disorder of cancer to strike a human body. Most of the more than the $\sim 3 \times 10^{13}$ cells in the adult body continue to carry the genetic information that previously allowed them to come into existence and might, in the future, allow them to multiply once again. This explains why the risk of uncontrolled cell proliferation in countless sites throughout the body is substantial throughout the lives of mammals like ourselves.

To be more precise, the risk of cancer is far greater than the $\sim 3 \times 10^{13}$ cell population size would suggest, since this number only represents the average, **steady-state** population of cells in the body at any point in time during adulthood. The aggregate number of cells that are formed during an average human lifetime is about 10^{16}, a number that testifies to the enormous amount of cell turnover—involving cell death and replacement (almost 10^7 events per second)—that occurs continuously in many tissues in the adult body. As discussed in various ways in Chapters 10 and 12, each time a new cell is formed by the complex process of cell growth and division, there are many opportunities for things to go awry. Hence, the chance for disaster to strike, including the inadvertent formation of cancer cells, is great.

Since a normal biological process (incessant cell division) is likely to create a substantial risk of cancer, it would seem logical that human populations throughout the world would experience similar frequencies of cancer. However, when cancer **incidence** rates (that is, the rates with which the disease is diagnosed) are examined in various countries, we learn that the risks of many types of cancer seem to vary dramatically (**Table 2.5**), while other cancers (not indicated in this table) do indeed show comparable incidence rates across the globe. So, our speculation that all cancers should strike different human populations at comparable rates is simply wrong. Some do and some don't. But we need to be cautious here, simply because incidence rates for certain cancers may be highly misleading—the consequence of **diagnostic bias** (Sidebar 2.1).

Table 2.5 indicates that, in spite of these reservations about cancer incidence and mortality rates, some cancers vary dramatically from country to country. Yet other tumor

Table 2.5 Geographic variation in cancer incidence and death rates

Countries showing highest and lowest incidence of specific types of cancer[a]			
Cancer site	Country of highest risk	Country of lowest risk	Relative risk H/L[b]
Skin (melanoma)	Australia (Queensland)	Japan	155
Lip	Canada (Newfoundland)	Japan	151
Nasopharynx	Hong Kong	United Kingdom	100
Prostate	U.S. (African American)	China	70
Liver	China (Shanghai)	Canada (Nova Scotia)	49
Penis	Brazil	Israel (Ashkenazic)	42
Cervix (uterus)	Brazil	Israel (non-Jews)	28
Stomach	Japan	Kuwait	22
Lung	U.S. (Louisiana, African American)	India (Madras)	19
Pancreas	U.S. (Los Angeles, Korean American)	India	11
Ovary	New Zealand (Polynesian)	Kuwait	8
Geographic areas showing highest and lowest death rates from specific types of cancer[c]			
Cancer site	Area of highest risk	Area of lowest risk	Relative risk H/L[b]
Lung, male	Eastern Europe	West Africa	33
Esophagus	Southern Africa	West Africa	16
Colon, male	Australia, New Zealand	Middle Africa	15
Breast, female	Northern Europe	China	6

[a]See C. Muir, J. Waterhouse, T. Mack et al., eds., Cancer Incidence in Five Continents, vol. 5. Lyon: International Agency for Research on Cancer, 1987. Excerpted by V.T. DeVita, S. Hellman and S.A. Rosenberg, Cancer: Principles and Practice of Oncology. Philadelphia: Lippincott, 1993.
[b]Relative risk: age-adjusted incidence or death rate in highest country or area (H) divided by age-adjusted incidence or death rate in lowest country or area (L). These numbers refer to age-adjusted rates, for example, the relative risk of a 60-year-old dying from a specific type of tumor in one country compared with a 60-year-old in another country.
[c]See P. Pisani, D.M. Parkin, F. Bray and J. Ferlay, Int. J. Cancer 83:18–29, 1999. This survey divided the human population into 23 geographic areas and surveyed the relative mortality rates of various cancer types in each area.

Sidebar 2.1 Cancer incidence and mortality rates can be highly misleading Cancer incidence rates, (reflecting the number of cases diagnosed) and mortality rates are usually presented in an **age-adjusted** form to compensate for the facts that (1) different cancers tend to arise at different times during the human lifetime and (2) different human populations have different age distributions. These numbers are then corrected for population sizes and often presented in rates per 100,000 of the general population. These corrections, on their own, cannot compensate for the difficulties in gauging the frequencies with which different tumors arise, since the resulting incidence rates are strongly influenced by diagnostic practices and conventions, reflecting the complex problem of **diagnostic bias**. Thus, as a prominent example, it is unclear how much of the sixfold difference in breast cancer incidence between China and the Western world derives from the relative occurrence of potentially life-threatening cancers in these two populations and how much derives from population-wide screening techniques, in this case by the highly sensitive technique of **mammography**, which reveals many small growths in the breast that would have remained indolent during the lifetime of the women bearing them and, because of their small size, remained

undetected. (In the case of the United States, it is plausible that at least two-thirds of currently diagnosed "breast cancers" fall into this class.)

This explains why incidence rates often reflect the frequency with which population-wide, highly sensitive screening techniques are in force, indicating that for many types of cancer, the alternative metric—age-adjusted mortality rates—provides a more reliable index of the frequency with which life-threatening cancers strike in different populations (simply because deaths represent an unambiguous and easily measured outcome). Even here, we need to be cautious when attempting to assess the true rates at which cancers strike, since different types of tumor are treated with more or less success in different countries. For example, cancer mortality rates in the United States have declined by almost 35% over the past generation, a reflection of successes in treating them; this decline on its own, would cause us to increase by about 35% our estimate of the rates with which true life-threatening breast cancers are formed each year in the United States. In addition, in many parts of the world, assessments about the causes of death, especially for diseases of internal organs, may not properly reflect the true disease state that led to death.

types, not listed in this table, seem to strike with comparable frequencies. Such tumors seem to be caused by random, essentially unavoidable accidents of nature. This seems to be especially true for a number of pediatric tumors that affect children with comparable frequencies across the globe.

Still, as Table 2.5 makes clear, other factors and processes appear to intervene in certain populations to increase dramatically the total number of cancer cases. Some of these differences are so dramatic as to essentially rule out important contributions from diagnostic bias and effectiveness of anti-cancer therapies. The two obvious factors contributing to these stark differences are heredity and environment. Which of these two alternatives is the dominant determinant of the country-to-country variability of cancer incidence and mortality rates? While many types of inherited disease-causing alleles are distributed unequally in the gene pools of different human populations, these inherited genetic factors do not seem to explain the dramatically different incidence and mortality rates of various cancers throughout the world. This point is demonstrated most dramatically by measuring cancer rates in migrant populations. For example, Japanese experience rates of stomach cancer that are 6 to 8 times higher than those of Americans. However, when Japanese settle in the United States, within a generation their offspring exhibit a stomach cancer rate that is comparable to that of the surrounding population (**Figure 2.18**). (Since the majority of those diagnosed with stomach cancer eventually die from the disease, the incidence rates track closely with mortality rates and therefore reflect how often life-threatening disease strikes.)

As indicated in Table 2.5, the incidence (that is, the risk of developing diagnosed disease) of breast cancer in China was, at one point in time, about one-sixth as great as in the United States or Northern Europe (see Sidebar 2.1). Having excluded genetic contributions to this difference, we might conclude, perhaps simplistically, that as many as 85% of the breast cancers in the United States might in theory be avoidable if only American women were to experience an environment and lifestyle comparable to those of their Chinese counterparts; however, as we read above, these numbers need to be taken with great reservation, given the profound influence of diagnostic bias.

Even within the American population, there are vast differences in cancer mortality: the Seventh-day Adventists, whose religion discourages smoking, heavy drinking, and the consumption of animal protein, die from cancer at rates that are 40% lower for men and 24% lower for women than the general population. These numbers, which are robustly documented, reveal dramatically the key roles of environment within a

Figure 2.18 Country-to-country comparisons of cancer incidence Public health records reveal dramatic differences in the incidence of certain cancers in different countries. Here, the relative incidence rates of a group of cancers in Japan and in the American state of Hawaii are presented. Invariably, after Japanese have immigrated to Hawaii, within a generation their cancer rates approach those of the population that settled there before them. Cancer incidence numbers are strongly influenced by diagnostic bias (see Sidebar 2.1), especially in the cases of prostate and breast cancer. In contrast, the incidence rates of colon and stomach cancer are not strongly influenced by diagnostic bias and are thus more reflective of the frequency of developing life-threatening disease. This indicates that the differing cancer rates are not due to genetic differences between the Japanese and the American populations. M, male; F, female. (From J. Peto, *Nature* 411:390–395, 2001. With permission from Nature.)

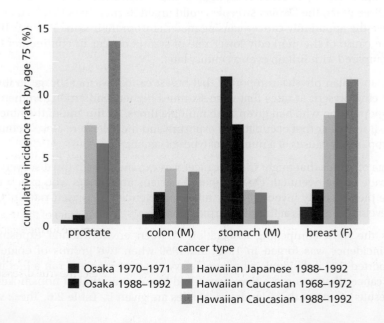

Osaka 1970–1971
Osaka 1988–1992
Hawaiian Japanese 1988–1992
Hawaiian Caucasian 1968–1972
Hawaiian Caucasian 1988–1992

human population, where "environment" encompasses both the surrounding physical environment and, more importantly in this case, lifestyle practices including prominently diet.

For those who wish to understand the **etiologic** (causative) mechanisms of cancer, these often-dramatic differences in incidence and mortality lead to an inescapable conclusion: the great majority of the commonly occurring cancers are caused by factors that can vary dramatically between one population and another, and between individuals within each of these populations. Most of these might be traced to "lifestyle factors," including smoking, diet and obesity, and physical activity, and to various types of infections; in addition, substantial variations in cancer risk may be attributable to differences in reproductive behavior and the resulting dramatic effects on the hormonal environment within the human female body. When comparing cancer risks across the globe, only a small proportion of the differences in cancer incidence and mortality can be ascribed to differences in genetic predisposition and, it would seem, environmental pollution.

Let us imagine, for the sake of argument, that avoidance of certain cancer-causing factors in diet and overall lifestyle, some discussed below, resulted in a 50% reduction in the risk of dying from cancer in the West, leaving the disease of cancer as the cause of only about 10% of overall mortality in this population. Under these conditions, given the approximately 10^{16} mitoses occurring in each human body during a normal life span, we calculate that only 1 in 10^{17} cell divisions—the total number in aggregate of cell divisions occurring in the bodies of 10 individuals during their lifetimes—would lead directly or indirectly to a clinically detectable cancer. Now, we become persuaded that in spite of the enormous intrinsic risk of developing cancer, the body must be able to mount highly effective defenses that usually succeed in holding off the disease for the 70 or 80 years that most of us spend on this planet. These built-in defenses are the subject of many discussions throughout this book.

2.7 The risks of cancers often seem to be increased by assignable influences, including lifestyle

Evidence that certain kinds of cancers are associated with specific exposures or lifestyles is actually quite old, predating modern epidemiology by more than a century. The first known report comes from the observations of the English physician John Hill, who in 1761 noted the connection between the development of nasal cancer and the excessive use of tobacco snuff. Fourteen years later, Percivall Pott, a surgeon in London, reported that he had encountered a substantial number of skin cancers of the scrotum in adolescent men who, in their youth, had worked as chimney sweeps. Within three years, the Danish sweepers guild urged its members to take daily baths to remove the apparently cancer-causing material from their skin. This practice was likely the cause of the markedly lower rate of scrotal cancer in continental Europe when compared with Britain even a century later.

In 1839, an Italian physician reported that breast cancer was a scourge in the nunneries, being present at rates that were six times higher than among women in the general population who had given birth multiple times. By the end of the nineteenth century, it was clear that occupational exposure and lifestyle were closely connected to and apparently causes of a number of types of cancer.

The range of agents that might trigger cancer was expanded with the discovery in the first decade of the twentieth century that physicians and others who experimented with the then-recently invented X-ray tubes experienced increased rates of cancer, often developing tumors at the site of irradiation.

Perhaps the most compelling association between environmental exposure and cancer incidence was forged in 1949 and 1950 when two groups of epidemiologists reported that individuals who were heavy cigarette smokers ran a lifetime risk of lung cancer that was more than twentyfold higher than that of nonsmokers. The initial results of one of these landmark studies are given in **Table 2.6**. These various

Table 2.6 Relative risk of lung cancer as a function of the number of cigarettes smoked per day[a]

	Lifelong non-smoker	Smokers			
Most recent number of cigarettes smoked (by subjects) per day before onset of disease	—	≥1, <5	≥5, <15	≥15, <25	≥25
Relative risk	1	8	12	14	27

[a]The relative risk indicates the risk of contracting lung cancer compared with that of a nonsmoker, which is set at 1.
Adapted from R. Doll and A.B. Hill, *BMJ* 2:739–748, 1950.

epidemiologic correlations proved to be critical for subsequent cancer research, since they suggested that cancers often had specific, assignable causes, and that a chain of causality might one day be traced between these ultimate causes and the cancerous changes observed in certain human tissues. Indeed, in the half century that followed the 1949–1950 reports, epidemiologists identified a variety of lifestyle factors, infections (Table 2.7), and exposures to specific agents (see Table 2.8) that were strongly

Table 2.7 Known or suspected causes of human cancers

Environmental and lifestyle factors known or suspected to be etiologic for human cancers in the United States[a]	
Type	% of total cases[b]
Cancers due to occupational exposures	1–2
Lifestyle cancers	
Tobacco-related (sites: e.g., lung, bladder, kidney)	34
Diet (low in vegetables, high in nitrates, salt) (sites: e.g., stomach, esophagus)	5
Diet (high fat, low fiber, broiled/fried foods) (sites: e.g., bowel, pancreas, prostate, breast)	37
Tobacco plus alcohol (sites: mouth, throat)	2

Specific carcinogenic agents implicated in the causation of certain cancers[c]	
Cancer	Exposure
Scrotal carcinomas	chimney smoke condensates
Liver angiosarcoma	vinyl chloride
Acute leukemias	benzene
Nasal adenocarcinoma	hardwood dust
Osteosarcoma	radium
Skin carcinoma	arsenic
Mesothelioma	asbestos
Vaginal carcinoma	diethylstilbestrol
Oral carcinoma	snuff
ER+ breast cancer[d]	hormone replacement therapy (E + P)[e]

[a]Adapted from American Cancer Society. Cancer Facts & Figures 1990. Atlanta: American Cancer Society, Inc.
[b]A large number of cancers are thought to be provoked by a diet high in calories acting in combination with many of these lifestyle factors.
[c]Adapted from S. Wilson, L. Jones, C. Coussens and K. Hanna, eds., Cancer and the Environment: Gene–Environment Interaction. Washington, DC: National Academy Press, 2002.
[d]ER+, estrogen receptor–positive.
[e]E + P, therapy containing both estrogen and progesterone.

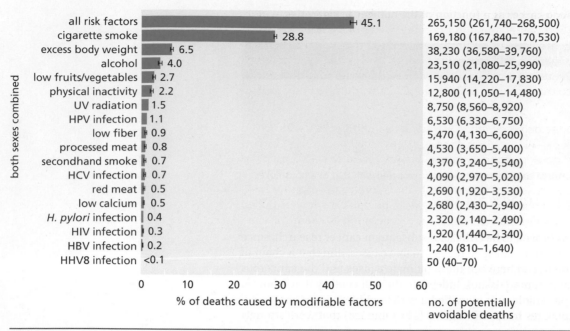

Figure 2.19 Percentage of avoidable cancers Epidemiologists have undertaken long-term studies in which the incidence and mortality of various cancers have been correlated with various lifestyle factors and infections. These studies have employed large population cohorts and attempted to factor out extraneous influences that might have otherwise confounded the calculations of the associations of specific cancers with certain assignable risk factors. A 2018 analysis undertook to determine what effects changes in existing, potentially modifiable lifestyle factors and infections would have had on overall cancer incidence and mortality in the United States in 2014, which totaled ~587,000 deaths. The numbers on the right are the average calculated avoidable deaths with a lower to upper range of variation that could be stated with 95% confidence. The overall reduction in mortality rate was calculated to be 45% for both sexes; less conservative predictions of avoidable cancer deaths have ranged from 50 to 60%. Smoking, obesity, and alcohol intake together represent the lion's share (39.3%) of this potential reduction. HPV, Human papillomavirus; HIV, human immunodeficiency virus; *H. pylori*, *Helicobacter pylori*; HCV, hepatitis C virus; HBV, hepatitis B virus; HHV8, human herpesvirus 8. (From F. Islami et al., *CA Cancer J. Clin.* 68:31–54, 2018. With permission from John Wiley & Sons.)

correlated with the incidence of certain cancers (**Figure 2.19**); in a subset of these cases, researchers have been able to discover the specific biological mechanisms through which these etiologic factors act.

2.8 Specific chemical agents can induce cancer

Coal tar condensates, much like those implicated in cancer causation by Percivall Pott's work, were used in Japan at the beginning of the twentieth century to induce skin cancers in rabbits. Repeated painting of localized areas of the skin of their ears resulted, after many months, in the outgrowth of carcinomas. This work, first reported by Katsusaburo Yamagiwa in 1915, was little noticed in the international scientific community of the time; instead, the 1926 Nobel Prize in Medicine was given to a Dane whose research on the experimental induction of cancer was soon discredited (Supplementary Sidebar 2.6). In retrospect, Yamagiwa's work represented a landmark advance, because it directly implicated chemicals (those in coal tar) in cancer causation. Equally important, Yamagiwa's work, together with that of Peyton Rous (to be described in Chapter 3), demonstrated that cancer could be induced at will in laboratory animals. Before these breakthroughs, researchers had been forced to wait for tumors to appear spontaneously in wild or domesticated animals. Now, cancers could be produced according to a predictable schedule, often involving many months of experimental treatment of animals.

By 1940, British chemists had purified several of the components of coal tar that were particularly **carcinogenic** (that is, cancer-causing), as demonstrated by the ability of these compounds to induce cancers on the skin of laboratory mice. Compounds such as 3-methylcholanthrene, benzo[*a*]pyrene, and 1,2,4,5-dibenz[*a,h*]anthracene were common products of combustion, and some of these hydrocarbons, notably benzo[*a*]pyrene, were subsequently found in the condensates of cigarette smoke as well (Supplementary Sidebar 2.7). These findings suggested that certain chemical species that entered into the human body could perturb tissues and cells and ultimately provoke the emergence of a tumor. The same could be said of X-rays, which were also able to produce cancers, ostensibly through a quite different mechanism of action.

While these discoveries were being reported, an independent line of research developed that portrayed cancer as an infectious disease. As described in detail in Chapter 3, researchers in the first decade of the twentieth century found that viruses could cause leukemias and sarcomas in infected chickens. By mid-century, a wide variety of viruses had been found able to induce cancer in rabbits, chickens, mice, and rats. As a consequence, those intent on uncovering the origins of human cancer were pulled in three different directions, since the evidence of cancer causation by chemical, viral, and radioactive agents had become compelling.

2.9 Both physical and chemical carcinogens act as mutagens

The confusion caused by the three competing theories of carcinogenesis was reduced significantly by discoveries made in the field of fruit fly genetics. In 1927, Hermann Muller discovered that he could induce mutations in the genome of *Drosophila melanogaster* by exposing these flies to X-rays. Most important, this discovery revealed that the genome of an animal was **mutable**, that is, that its information content could be changed through specific treatments, notably irradiation. At the same time, it suggested at least one mechanism by which X-rays could induce cancer: perhaps radiation was able to mutate the genes of normal cells, thereby creating mutant cells that grew in a malignant fashion.

By the late 1940s, a series of chemicals, many of them alkylating agents of the type that had been used in World War I mustard gas warfare, were also found to be **mutagenic** for fruit flies. Soon thereafter, some of these same compounds were shown to be carcinogenic in laboratory animals. These findings caused several geneticists to speculate that cancer was a disease of mutant genes, and that carcinogenic agents, such as X-rays and certain chemicals, succeeded in inducing cancer through their ability to mutate genes.

These speculations were hardly the first ones of this sort. As early as 1914, the German biologist Theodor Boveri, drawing on yet older observations of others, suggested that chromosomes, which by then had been implicated as carriers of genetic information, were aberrant within cancer cells, and that cancer cells might therefore be mutants. Boveri's notion, along with many other speculations on the origin of cancer, gained few adherents, however, until the discovery in 1960 of an abnormally configured chromosome in a large proportion of cases of chronic myelogenous leukemia (CML). This chromosome, soon called the Philadelphia chromosome after the place of its discovery, was clearly a distinctive characteristic of this type of cancer (**Figure 2.20**). Its reproducible association with this class of tumor cells suggested, but hardly proved, that this genetic alteration played a causal role in tumorigenesis.

In 1975 Bruce Ames, a bacterial geneticist working at the University of California in Berkeley, reported experimental results that lent great weight to the theory that carcinogens can function as mutagens. By then, decades of experiments with laboratory mice and rats had demonstrated that chemical carcinogens acted with vastly different potencies, differing by as much as 1 million-fold (per weight unit) in their ability to induce cancers. Such experiments showed, for example, that one microgram of aflatoxin, a compound produced by molds growing on peanuts and wheat, was as potent a carcinogen as was a 10,000 times greater weight of the synthetic compound benzidine. Ames posed the question whether these various compounds were also mutagenic, more specifically, whether compounds that were potent carcinogens also happened to be potent mutagens.

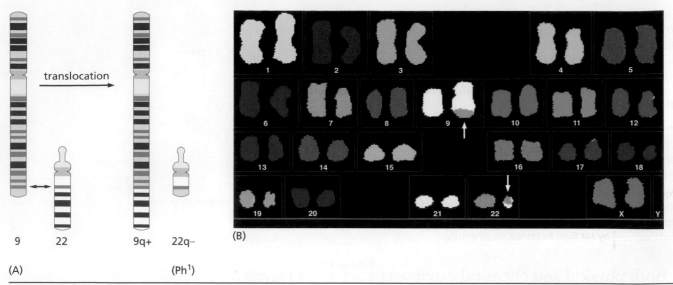

Figure 2.20 Structure of the Philadelphia chromosome
Analyses of the banding patterns of stained metaphase chromosomes of chronic myelogenous leukemia (CML) cells first revealed the characteristic tiny chromosome (called the "Philadelphia chromosome" or Ph[1]) that is present in the leukemia cells of most CML patients. (A) This banding pattern, determined through light-microscopic surveys, is illustrated here schematically. While the chromosomal translocation generating the two altered chromosomes (9q+ and 22q−) is reciprocal (that is, involving a loss and a gain by both), the sizes of the exchanged chromosomal arms are unequal, leading to the greatly truncated Chromosome 22 (22q−). The small arrow indicates the point of crossing over, known as the translocation breakpoint. Colors are added only to clarify banding patterns of chromosomes. (B) The relatively minor change to the tumor cell karyotype that is created by the CML translocation is apparent in this multicolor spectral karyotype (SKY) analysis, in which chromosome-specific probes are used, together with fluorescent dyes and computer-generated coloring, to visualize the entire chromosomal complement of CML cells. As is apparent, one of the two Chromosomes 9 has acquired a *light purple* segment (a color normally assigned to Chromosome 22) at the end of its long arm. Reciprocally, one of the two Chromosomes 22 has acquired a *white* region (normally characteristic of Chromosome 9) at the end of its long arm (*arrows*). (A, from B. Alberts et al., Molecular Biology of the Cell, 6th ed. New York: W.W. Norton, 2015. B, courtesy of T. Ried and N. McNeil.)

At the time, there were no good ways of measuring the relative mutagenic potencies of various chemical species. So Ames devised his own method. It consisted of applying various carcinogenic chemicals to a population of *Salmonella* bacteria growing in Petri dishes and then scoring for the abilities of these carcinogens to mutate the bacteria. The readout here was the number of colonies of *Salmonella* that grew following exposure to one or another chemical. (Implicit in the design of his experiments was the notion that chemicals that were mutagenic for bacterial genomes, and thus for bacterial DNA, would also act as mutagens for mammalian genomes and DNA.)

In detail, Ames used a mutant strain of *Salmonella* that was unable to grow in medium lacking the amino acid histidine. The mutant allele that caused this phenotype was susceptible to back-mutation to a wild-type allele. Once the wild-type allele had formed in response to exposure to a mutagen, a bacterium carrying this allele became capable of growing in Ames's selective medium, multiplying until it formed a colony that could be scored by eye (**Figure 2.21A**).

In principle, Ames needed only to introduce a test compound into a Petri dish containing his special *Salmonella* strain and count the bacterial colonies that later appeared. There remained, however, one substantial obstacle to the success of this mutagenesis assay. Detailed studies by other researchers had shown that after carcinogenic molecules entered the tissues of laboratory animals, they were metabolized into yet other chemical species. In many cases, the resulting products of metabolism, rather than the initially introduced chemicals, seemed to be the agents that were directly responsible for the observed cancer induction. These metabolized compounds were found to be highly reactive chemically and able to form covalent bonds with the various macromolecules known to be present in cells—DNA, RNA, and protein.

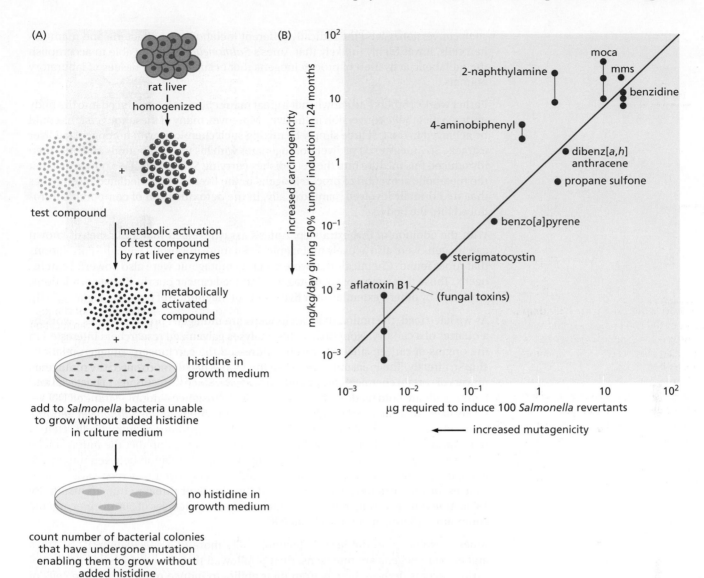

Figure 2.21 The Ames test for gauging mutagenicity The Ames test makes it possible to quantitatively assess the mutagenic potency of a test compound. (A) To begin, the liver of a rat (or other species) is homogenized to release its enzymes. The liver homogenate (*brown dots*) is then mixed with the test compound (*orange*), which often results in the liver enzymes metabolically converting the test compound to a chemically activated, mutagenic state (*red*). This mixture (still containing the liver homogenate, *not shown*) is applied to a dish of mutant *Salmonella* bacteria (*small green dots*) that require the amino acid histidine in their culture medium in order to grow. Since histidine is left out of the medium, only those bacteria that are mutated to a histidine-independent genotype (and phenotype) will be able to grow, and each of these will yield a large colony (*green*) that can be counted with the naked eye, indicating how many mutant bacteria (and thus mutant alleles) were generated by the brief exposure to the activated compound. (B) On this log–log plot, the relative carcinogenic potencies of a group of chemicals (*ordinate*) that have been used to treat laboratory animals (rats and mice) are plotted as a function of their mutagenic potencies (*abscissa*), as gauged by the Ames test (panel A). Since both the ordinate and abscissa are plotted as the amount of compound required to elicit an observable effect (yielding tumors in 50% of treated animals or 100 colonies of mutant *Salmonella* bacteria, termed here "revertants"), the compounds that are the most potent mutagens and most potent carcinogens appear in the lower left of this graph. Note that both parameters vary by 5 orders of magnitude. moca, 4,4'-methylenebis(2-chloroaniline), used in manufacture of polyurethane; mms, methyl methanesulfonate, an alkylating mutagen. (Adapted from M. Meselson and K. Russell, in H.H. Hiatt, J.D. Watson, and J.A. Winsten (eds.), Origins of Human Cancer, Book C: Human Risk Assessment. Cold Spring Harbor, NY: Cold Spring Harbor Laboratory Press, 1977. With permission from Cold Spring Harbor Laboratory Press.)

The original, unmodified compounds that were introduced into laboratory animals came to be called **procarcinogens** to indicate their ability to become converted into actively carcinogenic compounds, which were labeled **ultimate carcinogens**. This chemical conversion complicated the design of Ames's mutagenesis assay. If many compounds required metabolic activation before their carcinogenicity was apparent, it seemed plausible that their mutagenic powers would also be evident only after such conversion. Given the radically different metabolisms of bacteria and mammalian cells, it was highly unlikely that Ames's *Salmonella* would be able to accomplish the metabolic activation of procarcinogens that occurred in the tissues of laboratory animals.

Earlier work of others had shown that a great many chemicals introduced into the body undergo metabolic conversion in the liver. Moreover, many of these conversions could be achieved in the test tube simply by mixing such chemicals with homogenized liver extracts. So Ames mixed rat liver homogenates with his test compounds and only then introduced this mixture into the Petri dishes carrying *Salmonella*. (We now know that the metabolic activation of procarcinogens in the liver is often mediated by enzymes that are normally involved, paradoxically, in the **detoxification** of compounds introduced into the body.)

With the addition of this extra step, Ames's assay revealed that a number of known carcinogens were also actively mutagenic. Even more important were the correlations that Ames found. Chemicals that were potently mutagenic were also powerful carcinogens. Those that were weakly mutagenic induced cancer poorly. These correlations, as plotted by others, extended over five orders of magnitude of potency (Figure 2.21B).

As we have read, the notion that carcinogens are mutagens predated Ames's work by a quarter of a century. Nonetheless, his analyses galvanized researchers interested in the origins of cancer, since the results addressed the carcinogen–mutagen relationship so directly. Their reasoning went like this: Ames had demonstrated the mutagenic powers of certain chemical compounds in bacteria. Since the genomes of bacterial and animal cells are both made of the same chemical substance—double-stranded DNA—it was likely that the compounds that induced mutations in the *Salmonella* genome were similarly capable of inducing mutations in the genomes of animal cells. Hence, the "**Ames test**," as it came to be known, should be able to predict the mutagenicity of these compounds in mammals. And in light of the correlation between mutagenic and carcinogenic potency, the Ames test could be employed to screen various substances for their carcinogenic powers, and thus for their threat to human health. By 1976, Ames and his group reported on the mutagenic potencies of 300 distinct organic compounds (Supplementary Sidebar 2.8).

Ames's results led to the next deduction, really more of a speculation: if, as Ames argued, carcinogens are mutagens, then it followed that the carcinogenic powers of various agents derived directly from their ability to induce mutations in the cells of target tissues. As a further deduction, it seemed inescapable that the cancer cells created by chemical carcinogens carried mutated genes. These mutated genes, whatever their identity, must in some way be responsible for the aberrant growth phenotypes of such cancer cells.

This logic was transferable to X-ray carcinogenesis as well. Since X-rays were mutagens and carcinogens, it followed that they also induced cancer through their ability to mutate genes. This convergence of cancer research with genetics had a profound effect on researchers intent on puzzling out the origins of cancer. Though still unproven, it appeared likely that the disease of cancer could be understood in terms of the mutant genes carried by cancer cells.

2.10 Mutagens may be responsible for some human cancers

The connection between carcinogenesis and mutagenesis seemed to shed light on how human tumors arise. Perhaps many of these neoplasms were the direct consequence of the mutagenic actions of chemical and physical carcinogens. The mutagenic chemicals, specifically, procarcinogens, need not derive exclusively from the combustion of carbon compounds and the resulting formation of coal tars. It seemed plausible that chemical species present naturally in foodstuffs (see Table S2.1) or generated during cooking could also induce cancer. Even if many foods did not contain ultimate carcinogens, chemical conversions carried out by liver cells or by the abundant bacteria in the colon might well succeed in creating actively mutagenic and thus carcinogenic chemical species. In the years after Ames's work, a variety of other tests of mutagenic potency have been developed (Supplementary Sidebar 2.9).

As this research on the causes of human cancer proceeded, the understanding of carcinogenic agents became more nuanced. Thus, it became apparent that virtually all compounds that are mutagenic in human cells are likely to be carcinogenic as well. However, the converse does not seem to hold: chemical compounds that are carcinogenic are not necessarily mutagenic. By the 1990s, extensive use of the Ames test showed that as many as 40% of the compounds that were known to be carcinogenic in rodents showed no obvious mutagenicity in the *Salmonella* mutation assay. So, the conclusions drawn from the initial applications of Ames's test required major revision: some carcinogens act through their ability to mutate DNA, while others promote the appearance of tumors through nongenetic mechanisms. We will encounter these nonmutagenic carcinogens, often called tumor **promoters**, again in Chapter 11. Such promoters, including inflammation-inducing factors, may ultimately prove to be far more important human carcinogens than the much-studied mutagenic agents identified by the Ames test.

Ames and others eventually used his test to catalog the mutagenic powers of a diverse group of chemicals and natural foodstuffs, including many of the plants that are common and abundant in the Western diet. As he argued, the presence of such compounds in foodstuffs derived from plants was hardly surprising, since plants have evolved thousands, possibly millions of distinct toxic chemical compounds in order to defend themselves from predation by insects and larger animals. Some of these naturally toxic compounds, initially developed as anti-predator defenses, might also, as an unintended side effect, be mutagenic (see Table S2.1).

A diverse set of discoveries led to the model that a significant proportion of human cancer is attributable directly to the consumption of foodstuffs that can generate mutagenic compounds and, for this reason, are carcinogenic. This model remains unproven in many of its aspects to this day.

The difficulties in proving this model derive from several sources. Each of the plant and animal foodstuffs in our diet is composed of thousands of diverse chemical species present in vastly differing concentrations. Almost all of these compounds undergo metabolic conversions once ingested, first in the gastrointestinal tract and often thereafter in the liver. Moreover, the hundreds of distinct bacterial species inhabiting our gastrointestinal tract, which together constitute much of our **microbiome**, have diverse metabolisms and thus powers to generate potentially carcinogenic compounds. Accordingly, the number of distinct chemical species that are introduced into our bodies is incalculable. Each of these introduced compounds may then be concentrated in some cells or quickly metabolized and excreted, creating a further dimension of complexity.

Moreover, the actual mutagenicity of various compounds in different cell types may vary enormously because of metabolic differences in these cells. For example, some cells, such as **hepatocytes** in the liver, express high levels of biochemical species designed to scavenge and inactivate mutagenic compounds, while others, such as fibroblasts, express far lower levels. In sum, the ability to relate the biological activities of foodstuffs to actual rates of mutagenesis and carcinogenesis in the human body is far beyond our reach at present—a problem of seemingly intractable complexity. In addition, the limitations of using animal models to detect human carcinogens indicate the profound difficulties of determining in a biological relevant model—a living mammal—how carcinogenic various substances actually are (**Sidebar 2.2**).

2.11 Synopsis and prospects

The descriptions of cancer and cancer cells developed during the second half of the nineteenth century and the first half of the twentieth indicated that tumors were nothing more than normal cell populations that had run amok. Moreover, many tumors seemed to be composed largely of the descendants of a single cell that had

Sidebar 2.2 The frustrating search for elusive human carcinogens Ideally, the identification of important human carcinogens should have been aided by the use of *in vitro* assays, such as the Ames test (see Section 2.9), and *in vivo* tests—exposure of laboratory animals to agents suspected of causing cancer. In truth, however, these various types of laboratory tests have failed to register important human carcinogens. Instead, we have learned about their carcinogenicity in humans because of various epidemiologic studies. For example, the most important known human carcinogen—tobacco smoke—would likely have escaped detection because it is a relatively weak carcinogen in laboratory rodents; and another known human carcinogen—asbestos—would have eluded detection by both *in vitro* and *in vivo* laboratory tests. Conversely, some frequently used drugs, such as phenobarbital and isoniazid, register positive in the Ames test, and saccharin registers as a carcinogen in male laboratory rats, but epidemiologic evidence indicates conclusively that none of these is actually associated with increased cancer risk in humans who have been exposed to these compounds over long periods of time. Moreover, as mentioned, many human carcinogens act through non-mutational mechanisms and therefore do not register in the Ames test or, indeed, in a variety of other laboratory tests. Hence, the development of tests to accurately identify a wide range of human carcinogenic agents still lies in the future.

crossed over the border from normalcy to malignancy and proceeded to spawn the billions of descendant cells constituting these neoplastic masses. This model drew attention to the nature of the cells that founded tumors and to the mechanisms that led to their transformation into cancer cells. If one could understand why a cell multiplied uncontrollably, somehow other pieces of the cancer puzzle were likely to fall into place.

Still, existing observations and experimental techniques offered little prospect of revealing precisely why a cell altered its behavior, transforming itself from a normal into a malignant cell. The carcinogen = mutagen theory seemed to offer some clarification, since it implicated mutant cellular genes as the agents responsible for disease development and, therefore, for the aberrant behavior of cancer cells. Perhaps there were mutant genes operating inside cancer cells that programmed the runaway proliferation of these cells, but the prospects for discovering such genes and understanding their actions seemed remote. No one knew how many genes were present in the human genome and how to analyze them. If mutant genes really did play a major part in cancer causation, they were likely to be small in number and dwarfed by the apparently vast number of genes present in the genome as a whole. They seemed to be the proverbial needles in the haystack, in this case a vast haystack of unknown size.

This theorizing about cancer's origins was further complicated by two other important considerations. First, many apparent carcinogens failed the Ames test, suggesting that they were nonmutagenic. Second, certain viral infections seemed to be closely connected to the incidence of a small but significant subset of human cancer types. Somehow, their carcinogenic powers had to be reconciled with the actions of mutagenic carcinogens and mutant cellular genes.

By the mid-1970s, recombinant DNA technology, including gene cloning, began to influence a wide variety of biomedical research areas. While appreciating the powers of this new technology to isolate and characterize genes, cancer researchers were unable, at least initially, to exploit it to track down the elusive mutant genes that were responsible for cancer. One thing was clear, however. Sooner or later, the process of cancer **pathogenesis** (disease development) needed to be explained and understood in molecular terms. Somehow, the paradigm of DNA, RNA, and proteins, so powerful in elucidating a vast range of biological processes, would need to be brought to bear on the cancer problem.

In the end, the breakthrough came from study of the tumor viruses, which by most accounts were minor players in human cancer development. Tumor viruses were genetically simple, and yet they possessed potent carcinogenic powers. To understand these viruses and their import, we need to move back, once again, to the beginning of the twentieth century and confront another of the ancient roots of modern cancer research. This will be the subject of Chapter 3.

Moving back to the present, we confront an ongoing major challenge for cancer research: how various biological and environmental factors, the latter including lifestyle, contribute to the incidence of cancers, many of them quite common ones. For example, as indicated in part in Table 2.5, the incidence of cancers, such as those of the colon, breast, and prostate, shows enormous geographic variation—dramatic differences that cannot be ascribed to differing genetic susceptibilities. In fact, epidemiologists have uncovered many correlations between the frequencies of these and other cancer types and exposure to various agents. However, with rare exception, our understanding of the biological and biochemical mechanisms by which these factors increase (or reduce) disease incidence is either incomplete or nonexistent (**Table 2.8**). Indeed, these correlations represent one of the major unsolved mysteries confronting contemporary cancer researchers.

Until we understand how various biological and lifestyle factors succeed in triggering or preventing tumor development, our ability to prevent new cancers (which is usually far more effective than trying to cure them after they have been diagnosed) will be limited. Many of the chapters that follow provide critical information that may ultimately help to unravel these mysteries of cancer etiology.

Table 2.8 Examples of etiologic mysteries: epidemiologic correlations between environmental/lifestyle factors and cancer incidence that lack a clear explanation of causal mechanism[a]

Lifestyle, dietary factor, or medical condition	Altered cancer risk
High birth weight	premenopausal breast cancer ↑
	infant acute leukemia ↑
Processed red meat[b]	ER+ breast cancer ↑
	squamous cell and adenocarcinoma of lung ↑
Childhood soy consumption	breast cancer ↓
Well-done red meat	prostate cancer ↑
Western diet—high in fat, high in red meat	colorectal, esophageal, liver, and lung cancer ↑
Exercise	hormone-responsive breast cancer ↓
Diet with cruciferous vegetables	prostate cancer ↓
High body-mass index (BMI)	multiple cancer types ↑
Higher ratio of number of daughters to number of sons born to a woman	ovarian carcinoma ↑
Parkinson's disease	melanoma ↑
Low circulating vitamin D	breast cancer incidence, CRC mortality ↑
Periodontal disease	esophageal carcinoma ↑
Coffee consumption	hepatocellular carcinoma ↓

[a]Relative risk (RR) is not given, because not all studies used the same criteria to gauge RR. ↑ = increased risk; ↓ = decreased risk.
[b]Processed red meat generally refers to meat that has been preserved by smoking, curing, salting, or adding chemical preservatives.
Abbreviations: ER+ = estrogen receptor–positive; CRC = colorectal cancer.

Key concepts

- The nineteenth-century discovery that all cells of an organism descend from the fertilized egg led to the realization that tumors are not foreign bodies but growths derived from normal tissues. The comparatively disorganized tissue architecture of tumors pointed toward cancer as being a disease of malfunctioning cells.

- Tumors can be either benign (localized, noninvasive) or malignant (invasive, metastatic). The metastases spawned by malignant tumors are responsible for almost all deaths from cancer.

- With some exceptions, most tumors are classified into four major groups according to their origin (epithelial, mesenchymal, hematopoietic, and neuroectodermal).

- Virtually all cell types can give rise to cancer, but the most common human cancers are of epithelial origin—the carcinomas. Most carcinomas fall into two categories: squamous cell carcinomas arise from epithelia that form protective cell layers, while adenocarcinomas arise from secretory epithelia.

- Nonepithelial malignant tumors include (1) sarcomas, which originate from mesenchymal cells; (2) hematopoietic cancers, which arise from the precursors of blood cells; and (3) neuroectodermal tumors, which originate from components of the nervous system.

- If a tumor's cells have dedifferentiated (lost all tissue-specific traits), its origin cannot be readily identified; such tumors are said to be anaplastic.

- Cancers seem to develop progressively, with tumors demonstrating different gradations of abnormality along the way from benign to metastatic.

- Benign tumors may be hyperplastic or metaplastic. Hyperplastic tissues appear normal except for an excessive number of cells, whereas metaplastic tissues show displacement of normal cells by normal cell types not usually encountered in those tissues. Metaplasias are most frequent in the transition zones between two types of epithelial populations.

- Dysplastic tumors contain cells that are cytologically abnormal. Dysplasia is a transitional state between completely benign and premalignant. Adenomatous growths (adenomas, polyps, papillomas, and warts) are dysplastic epithelial tumors that are considered to be benign because they respect the boundary created by the basement membrane.

- Tumors that breach the basement membrane and invade underlying tissue are malignant. An even further degree of abnormality is metastasis, the seeding of tumor colonies to other sites in the body. Metastasis requires not only invasiveness but also such newly acquired traits as motility and adaptation to foreign tissue environments.

- Biochemical and genetic markers seem to indicate that human tumors are monoclonal (descended from one ancestral cell) rather than polyclonal (descended from multiple ancestral cells, each of which independently spawned a population of cancer cells).

- The incidence of many (but not all) cancers varies dramatically by country, an indication that they cannot be due simply to a normal biologic process gone awry by chance. While differences in either heredity or environment could explain these variations, epidemiologic studies show that environment (including lifestyle factors) is the dominant determinant of the country-by-country variations in cancer incidence.

- Laboratory research supported the epidemiologic studies by directly implicating chemical and physical agents (tobacco, coal dust, X-rays) as causes of cancers. However, the possibility of cancer as an infectious disease arose when viruses were found to cause leukemias and sarcomas in chickens.

- A possible mechanism that supported carcinogenesis by physical and chemical agents surfaced when mutations were induced in fruit flies by exposing them to either X-rays or chemicals, indicating that they were mutagenic. Since these agents were also known to be carcinogenic in laboratory animals, this led to the speculation that cancer was a disease of mutant genes and that carcinogenic agents induced cancer through their ability to mutate genes.

- In 1975 the Ames test provided support for this idea by showing that many carcinogens can act as mutagens. Additional research showed that although almost all compounds that are mutagenic are likely to be carcinogens, the converse does not hold true. So, some carcinogens act through their ability to mutate DNA, while others promote tumorigenesis through nongenetic mechanisms. Such nonmutagenic carcinogens are called tumor promoters.

Thought questions

1. What types of observation allow a pathologist to identify the tissue of origin of a tumor? And why are certain tumors extremely difficult to assign to a specific tissue of origin?

2. Under certain circumstances, all tumors of a class can be traced to a specific embryonic cell layer, while in other classes of tumors, no such association can be made. What tumors would fit into each of these two groupings?

3. What evidence persuades us that a cancer arises from the native tissues of an individual rather than invading the body from outside and thus being of foreign origin?

4. How compelling are the arguments for the monoclonality of tumor cell populations, and what logic and observations undermine the conclusion of monoclonality?

5. How can we estimate what percentage of cancers in a population are avoidable (through virtuous lifestyles) and what percentage occur independently of lifestyle?

6. What limitations does the Ames test have in predicting the carcinogenicity of various agents?

7. In the absence of being able to directly detect mutant genes within cancer cells, what types of observation allow one to infer that cancer is a disease of mutant cells?

Additional reading

Ames BN (1983) Dietary carcinogens and anticarcinogens. *Science* 231, 1256–1264.

Bickers DR & Lowy DR (1989) Carcinogenesis: a fifty-year historical perspective. *J. Invest. Dermatol.* 92, 121S–131S.

Greenwald P & Dunn BK (2009) Landmarks in the history of cancer epidemiology. *Cancer Res.* 69, 2151–2162.

Jemal A, Center MM, DeSantis C & Ward EM (2010) Global patterns of cancer incidence and mortality rates and trends. *Cancer Epidemiol. Biomarkers Prev.* 19, 1893–1907.

Loeb LA & Harris CC (2008) Advances in chemical carcinogenesis: a historical review and prospective. *Cancer Res.* 68, 6863–6872.

Mukherjee, S (2010) The Emperor of All Maladies: A Biography of Cancer. New York: Scribner.

Peto J (2001) Cancer epidemiology in the last century and the next decade. *Nature* 411, 390–395.

Preston-Martin S, Pike MC, Ross RK et al. (1990) Increased cell division as a cause of human cancer. *Cancer Res.* 50, 7415–7421.

Rassy E & Pavlidis N (2020) Progress in refining the clinical management of cancer of unknown primary in the molecular era. *Nat. Rev. Clin. Oncol.* 17:541–554.

Siegel RL, Miller KD & Jemal A (2020) Cancer statistics, 2020. *CA Cancer J. Clin.* 70, 7–30.

Vander Heiden MG, Cantley LC & Thompson CB (2009) Understanding the Warburg effect: the metabolic requirements of cell proliferation. *Science* 324, 1029–1033.

Wilson S, Jones L, Coussens C & Hanna K (eds) (2002) Cancer and the Environment: Gene-Environment Interaction. Washington, DC: National Academy Press.

Xu L, Wang SS, Healey MA et al. (2011) The Ninth Annual AACR International Conference on the Frontiers of Cancer Prevention Research. *Cancer Prev. Res. (Phila.)* 4, 616–621.

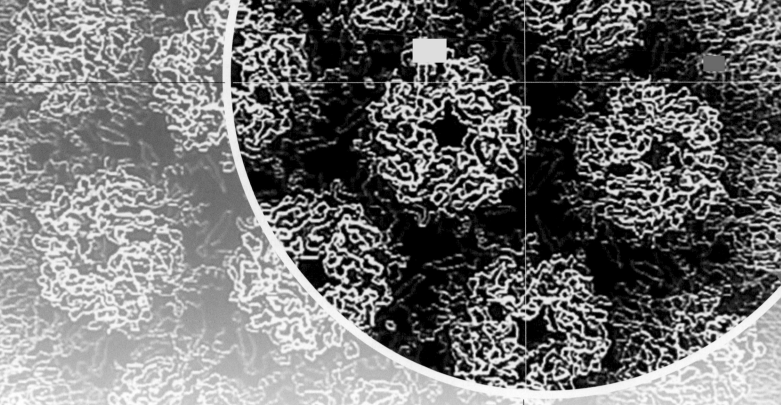

Chapter 3

Cancer as an Infectious Disease

> A tumor of the chicken ... has been propagated in this laboratory since October, 1909. The behavior of this new growth has been throughout that of a true neoplasm, for which reason the fact of its transmission by means of a cell-free filtrate assumes exceptional importance.
>
> Francis Peyton Rous, cancer biologist, 1911

Viruses are capable of causing a wide variety of human diseases, ranging from rabies to smallpox to the common cold and the acute respiratory syndrome triggered by the COVID-19 virus. The great majority of these infectious agents do harm through their ability to multiply inside infected host cells, to kill these cells, and to release progeny virus particles that proceed to infect other hosts nearby. The **cytopathic** (cell-killing) effects of viruses, together with their ability to spread rapidly throughout a tissue, enable these agents to leave wide swaths of destruction in their wake.

But the peculiarities of certain viral **replication cycles** may on occasion yield quite another outcome. Rather than killing infected cells, some viruses may, quite paradoxically, force their cellular hosts to thrive, indeed, to proliferate uncontrollably. In so doing, such viruses—often called tumor viruses—can create cancer.

At one time, beginning in the early 1970s, tumor viruses were studied intensively because they were suspected to be the cause of many common human cancers. This notion was eventually rejected based on the evidence subsequently gathered during that decade, which indicated that virus-induced cancers represent only a minority of the cancer types afflicting humans. Nonetheless, this line of research proved to be invaluable for cancer biologists: study of various tumor viruses provided the key for opening many of the long-hidden secrets of human cancers, including the great majority of cancers that have no connection with infectious agents.

As we will see, tumor virus research had a highly variable history over the course of the last century. These infectious agents were discovered in the first decade of the twentieth century and then retreated from the center stage of science. Half a century later, interest in these agents revived, culminating in the frenetic pace of tumor virus research during the 1970s.

The cancer-causing powers of tumor viruses drove many researchers to ask precisely how they succeed in creating disease. Most of these viruses possess relatively simple genomes containing only a few genes, yet some were found able to overwhelm an infected cell and its vastly more complex genome and to redirect cell growth. Such behavior indicated that tumor viruses have developed extremely potent genes to perturb the complex regulatory circuitry of the host cells that they infect.

By studying tumor viruses and their mechanisms of action, researchers changed the entire mindset of cancer research. Cancer became a disease of genes and thus a condition that was susceptible to analysis by the newly invented tools of molecular biology and genetics. When this story began, no one anticipated how obscure tumor viruses would one day revolutionize the study of human cancer pathogenesis.

3.1 Peyton Rous discovers a chicken sarcoma virus

In the last two decades of the nineteenth century, the research of Louis Pasteur and Robert Koch uncovered the infectious agents that were responsible for dysentery, cholera, rabies, and a number of other diseases. By the end of the century, these agents had been placed into two distinct categories, depending on their behavior upon filtration. Solutions of infectious agents that were trapped in the pores of filters were considered to contain bacteria. The other agents, which were small enough to pass through the filters, were classified as viruses. On the basis of this criterion, the agents responsible for rabies, foot-and-mouth disease, and smallpox were categorized as viruses.

Cancer, too, was considered a candidate infectious disease. As early as 1876, a researcher in Russia reported the transmission of a tumor from one dog to another: chunks of tumor tissue from the first dog were experimentally implanted into the second, whereupon a tumor appeared several weeks later. This success was followed by many others using rat and mouse tumors.

The significance of these early experiments remained controversial. Some researchers interpreted these outcomes as proof that cancer was a transmissible disease. Yet others dismissed these transplantation experiments, since in their eyes, such work showed only that tumors, like normal tissues, could be excised from one animal and forced to grow as a graft in the body of a second animal.

In 1908, two researchers in Copenhagen reported extracting a filterable agent from chicken leukemia cells and transmitting this agent to other birds, which then contracted the disease. The two Danes did not follow up on their initial discovery, and it remained for Peyton Rous, working at the Rockefeller Institute in New York, to found the discipline of tumor virology (Supplementary Sidebar 3.1).

A year later, Rous began his study of a sarcoma that had appeared in the breast muscle of a hen. In initial experiments, Rous succeeded in transmitting the tumor by implanting small fragments of it into other birds of the same breed. Later, as a variation of this experiment, he ground up a sarcoma fragment in sand and filtered the resulting homogenate under conditions in which only subcellular particles could pass through the filter (**Figure 3.1**). When he injected the resulting **filtrate** into young birds, they too developed tumors, sometimes within several weeks. He subsequently found that these induced tumors could also be homogenized to yield, once again, an infectious agent that could be transmitted, following filtration, to yet other birds, which also developed sarcomas at the sites of injection.

These serial passages of the sarcoma-inducing agent from one animal to another yielded a number of conclusions that are obvious to us now but at the time were nothing less than revolutionary. The carcinogenic agent, whatever its nature, was clearly very small, since it could pass through a filter. Hence, it was a virus (**Sidebar 3.1**). This virus could cause the development of a sarcoma in an injected chicken, doing so on a

chicken with sarcoma in breast muscle → remove sarcoma and break up into small chunks of tissue → grind up sarcoma with sand → collect filtrate that has passed through fine-pore filter → inject filtrate into young chicken → observe sarcoma in injected chicken → remove sarcoma and break up into small chunks of tissue → repeat cycle of virus preparation and infection etc.

Figure 3.1 Rous's protocol for inducing sarcomas in chickens Rous removed a sarcoma from the breast muscle of a chicken, ground it with sand, and passed the resulting homogenate through a fine-pore filter. He then injected the filtrate (the liquid that passed through the filter) into the wing web of a young chicken and observed the development of a sarcoma many weeks later. He then ground up this new sarcoma and repeated the cycle of homogenization, filtration, and injection, once again observing a tumor in another young chicken. These cycles could be repeated indefinitely; after repeated serial passaging, the virus produced sarcomas far more rapidly than the original viral isolate.

predictable timetable. Such an infectious agent offered researchers the unique opportunity to induce cancers at will rather than relying on the spontaneous and unpredictable appearance of tumors in animals or humans. In addition to its ability to induce cancer, this agent, which came to be called Rous sarcoma virus (RSV), was capable of multiplying within the tissues of the chicken; far more virus could be recovered from an infected tumor tissue than was originally injected.

3.2 Rous sarcoma virus is discovered to transform infected cells in culture

The rebirth of Rous sarcoma virus research began largely at the California Institute of Technology in Pasadena, in the virology laboratory of Renato Dulbecco. This group found that when stocks of RSV were introduced into Petri dishes carrying cultures of chicken embryo fibroblasts, the RSV-infected cells survived, apparently indefinitely. (These experiments depended on the then-recently developed procedures of **tissue culture**, that is, the ability to propagate cells in culture dishes outside of living tissues; Supplementary Sidebar 3.2.) It seemed that RSV parasitized these cells, forcing them to produce a steady stream of progeny virus particles for many days, weeks, even months (**Figure 3.3A, B**). Most other viruses, in contrast, were known to enter into host cells, multiply, and quickly kill their hosts; the multitude of progeny virus particles released from dying cells could then proceed to infect yet other susceptible cells in the vicinity, repeating the cycle of infection, multiplication, and cell destruction (see Sidebar 3.1).

Most important, the RSV-infected cells in these cultures displayed many of the traits associated with cancer cells. Thus, foci (clusters) of cells appeared after infection. Under the microscope, these cells strongly resembled the cells isolated from chicken sarcomas, exhibiting the characteristic rounded morphology of cancer cells (Figure 3.3C). They also had a metabolism reminiscent of that seen in cells isolated from tumors. This resemblance led Howard Temin (a student in Dulbecco's laboratory; Supplementary Sidebar 3.3) and Harry Rubin (a post-doctoral fellow) to conclude that the process of cell **transformation**—conversion of a normal cell into a tumor cell—could be accomplished within the confines of a Petri dish, not just in the complex and difficult-to-study environment of a living tissue.

These simple observations radically changed the course of twentieth-century cancer research, because they clearly demonstrated that cancer formation could be studied at the level of individual cells whose behavior could be tracked closely under the microscope. This insight suggested the further possibility that the entire complex biology of tumors could one day be understood by studying the transformed cells forming tumor masses. So, an increasing number of biologists began to view cancer as a disease of malfunctioning cells rather than abnormally developing tissues.

Temin and Rubin, soon followed by many others, used this experimental model to learn some basic principles about cell transformation. They were interested in the fate of a cell that was initially infected by RSV. How did such a cell proliferate when compared with uninfected neighboring cells? After exposure of cells to a solution of virus particles (often called a **virus stock**), the two researchers would place a layer of agar above the cell layer growing at the bottom of the Petri dish, thereby preventing free virus particles from spreading from initially infected cells to uninfected cells in other

Sidebar 3.1 Viruses have simple life cycles The term "virus" refers to a diverse array of particles that infect and multiply within a wide variety of cells, ranging from bacteria to the cells of plants and metazoa. Relative to the cells that they infect, individual virus particles, often termed **virions**, are tiny. This explains the original experimental definition of a virus: an infectious particle that could pass through a filter that was designed to trap larger infectious particles, namely bacteria. Virions are generally simple in structure, with a nucleic acid (DNA or RNA) genome wrapped in a protein coat (a **capsid**) and, in some cases, a lipid membrane surrounding the capsid. In isolation, viruses are metabolically inert. They can multiply only by infecting and parasitizing a suitable host cell. The viral genome, once introduced into the cell, provides instructions for the synthesis of progeny virus particles. The host cell, for its part, provides the low–molecular-weight precursors needed for the synthesis of viral proteins and nucleic acids, the protein-synthetic machinery, and, in many cases, the polymerases required for replicating and transcribing the viral genome.

The endpoint of the resulting infectious cycle is the production of hundreds, even thousands of progeny virus particles that can then leave the infected cell and proceed to infect other susceptible cells. The interaction of the virus with the host cell can be either a **virulent** one, in which the host cell is destroyed during the infectious cycle, or a persistent one, in which the host cell survives for extended periods, all the while harboring the viral genome and releasing progeny virus particles.

Many viruses carrying double-stranded DNA (dsDNA) genomes replicate in a fashion that closely parallels the macromolecular metabolism of their host cells (**Figure 3.2**). These viruses use host-cell DNA polymerases to replicate their DNA, host-cell RNA polymerases to transcribe the viral mRNAs from viral dsDNA templates, and host ribosomes to translate the resulting viral mRNAs. Once synthesized, viral proteins are used to coat (**encapsidate**) the newly synthesized viral genomes, resulting in the assembly of complete progeny virions, which then are released from the infected cell.

Since cells do not express enzymes that can replicate RNA molecules, the genomes of most RNA-containing virus particles encode their own RNA-dependent RNA polymerases to replicate their genomes. Poliovirus, as an example, makes such an enzyme, as does rabies virus. RNA tumor viruses like Rous sarcoma virus, as we will learn later in this chapter, follow a much more circuitous route for replicating their viral RNA.

In 1911, when Rous finally published his work, yet another report appeared on a transmissible virus of rabbit tumors, called myxomas. Soon thereafter, Rous and his collaborators found two other chicken viruses, and yet another chicken sarcoma virus was reported by others in Japan. Four years later, Rous himself gave up his research on the virus he had discovered, since human cancer did not seem to behave like an infectious disease. Then, research on tumor viruses languished for

two decades until other novel tumor viruses were discovered. The molecular nature of viruses and the means by which they multiplied would remain mysteries for more than half a century after Rous's initial discovery.

Still, his finding of a sarcoma virus reinforced the convictions of those who believed that virtually all human diseases were provoked by infectious agents. In their eyes, cancer could be added to the lengthening list of diseases, such as cholera, tuberculosis, rabies, and sepsis, whose causes could be associated with a specific microbial agent. Like many others, however, Rous focused his research in the decades that followed on chemicals as causes of human cancer. As described in Section 2.8, the field of chemical carcinogenesis to which Rous turned was established in the early twentieth century following the discovery of chemicals that were clearly carcinogenic. The research community turned its back on Rous sarcoma virus and the other tumor viruses for decades after his initial discoveries.

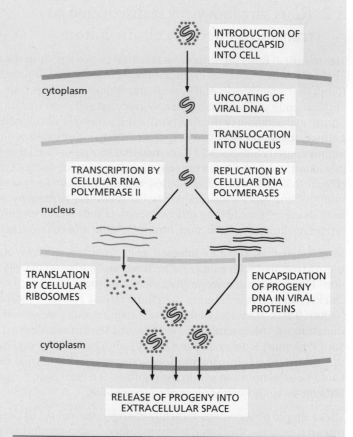

Figure 3.2 Life cycle of viruses with dsDNA genomes The life cycle of viruses with double-stranded DNA (dsDNA) genomes closely parallels that of the host cell. Almost all of the steps leading to the synthesis of viral DNA, RNA, and proteins can be achieved by using the synthetic machinery provided by the infected host cell. (Adapted from B. Alberts et al., Molecular Biology of the Cell, 6th ed. New York: W. W. Norton & Company, 2015.)

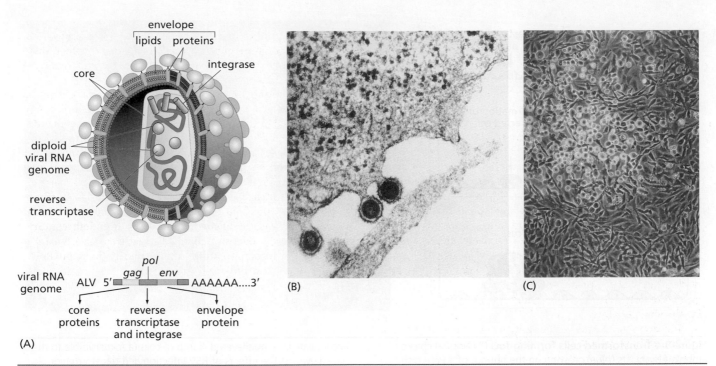

Figure 3.3 The virion of RSV and transformation of infected cells (A) An artist's reconstruction of the structure of a retrovirus virion, such as that of Rous sarcoma virus, which has four major types of viral proteins. The glycoprotein spikes (encoded by the viral *env* gene) that protrude from the lipid bilayer membrane enable the virion to *adsorb* to the surface of a cell and to introduce its contents into the cytoplasm. Beneath this lipid membrane lies a protein shell formed by the several core proteins encoded by the viral *gag* gene. Within this protein shell are two identical copies of the viral genomic RNA and a number of reverse transcriptase and integrase molecules specified by the viral *pol* gene. (B) Transmission electron micrograph showing murine leukemia virus (MLV) particles budding from the surface of a chronically infected cell; avian leukosis virus (ALV) particles behave identically. As the nucleocapsid cores (containing the viral gag proteins, the virion RNA, and the reverse transcriptase and integrase enzymes) push through the plasma membrane of the infected cell, they wrap themselves with a patch of lipid bilayer taken from the plasma membrane of the infected cell. (C) This phase-contrast micrograph reveals a focus of Rous sarcoma virus–transformed chicken embryo fibroblasts surrounded by a monolayer of uninfected cells. The focus stands out because it is composed of rounded, refractile cells that are beginning to pile up on one another, in contrast to the flattened morphology of the surrounding normal cells, which stop proliferating after they have produced a layer of cells one cell thick. (A, adapted from H. Fan et al., The Biology of AIDS. Boston: Jones and Bartlett Publishers, 1989. With permission of Jones and Bartlett. B, reprinted from *Cell Press*, Vol. 107, 55–65, Jennifer E. Garrus, Uta K. von Schwedler, Owen W. Pornillos, Scott G. Morham, Kenton H. Zavitz, Hubert E. Wang, Daniel A. Wettstein, Kirsten M. Stray, Mélanie Côté, Rebecca L. Rich, David G. Myszka, Wesley I. Sundquist, Tsg101 and the Vacuolar Protein Sorting Pathway Are Essential for HIV-1 Budding, 55-65, Copyright (2001), with permission from Elsevier. C, courtesy of P.K. Vogt.)

parts of the dish. Hence, any changes in cell behavior were the direct result of the initial infection by virus particles of chicken embryo fibroblasts.

The foci that Temin and Rubin studied revealed dramatic differences in the behavior of normal versus transformed cells. When first introduced into a Petri dish, normal cells formed islands scattered across the bottom of the dish. They then proliferated and eventually filled up all the space in the bottom of the dish, thereby creating **confluent** cultures. Once they reached confluence, however, these normal cells stopped proliferating, resulting in a one-cell-thick (or slightly thicker) layer of cells, often called a cell **monolayer** (Figure 3.4A).

The cessation of growth of these normal cells after forming confluent monolayers was the result of a process that came to be called **contact inhibition**, **density inhibition**, or **topoinhibition**. Somehow, high cell density or contact with neighbors caused these cells to stop dividing. This behavior of normal cells contrasted starkly with that of the **transformants** within the RSV-induced foci (Figure 3.4B). The latter clearly had lost contact (density) inhibition and consequently continued to proliferate, piling up on top of one another and creating multilayered cell foci so thick that they could often be seen with the naked eye.

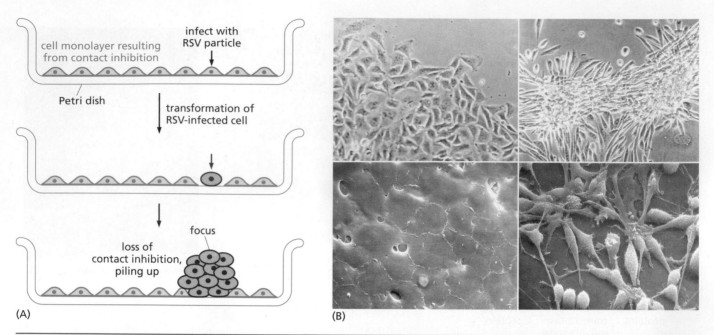

Figure 3.4 Transformed cells forming foci (A) Normal chicken embryo fibroblasts (*blue*) growing on the bottom of a Petri dish form a layer one cell thick—a monolayer. This is formed because these cells cease proliferating when they touch one another— the behavior termed *contact inhibition*. However, if one of these cells is infected by RSV prior to reaching confluence, this cell (*red*) and its descendants acquire a rounded morphology and lose contact inhibition. As a consequence, they continue to proliferate in spite of touching one another and eventually accumulate in a multilayered clump of cells (a *focus*) visible to the naked eye. (B) The effects of RSV infection and transformation can be observed through both phase-contrast microscopy (*upper panels*) and scanning electron microscopy (*lower panels*). Normal chicken embryo fibroblasts are spread out and form a continuous monolayer (*left panels*). However, upon transformation by RSV, they round up, become refractile (*white halos, upper right panel*), and pile up on one another (*right panels*). (B, courtesy of L.B. Chen.)

Under certain experimental conditions, it could be shown that all the cells within a given focus were the descendants of a single progenitor cell that had been infected and presumably transformed by an RSV particle. Today, we would term such a flock of descendant cells a cell **clone**, and the focus as a whole, a clonal outgrowth.

The behavior of these foci gave support to one speculation about the possible similarities between the cell transformation triggered by RSV in the Petri dish and the processes that led to the appearance of tumors in living animals, including humans: maybe all the cells within a spontaneously arising human tumor mass also constitute a clonal outgrowth and therefore are the descendants of a single common progenitor cell that somehow underwent transformation and then launched a program of replication that led eventually to the millions, even billions, of descendant cells that together formed the mass. As discussed earlier (see Section 2.5), detailed genetic analyses of human tumor cells were required, in the end, to test this notion in a truly definitive way.

3.3 The continued presence of RSV is needed to maintain transformation

The behavior of the cells within an RSV-induced focus indicated that the transformation phenotype was transmitted from an initially infected, transformed chicken cell to its direct descendants. This transmission provoked another set of questions: Did an RSV particle infect and transform the progenitor cell of the focus and, later on, continue to influence the behavior of all of its direct descendants, ensuring that they also remained transformed? Or, as an alternative, did RSV act in a "hit-and-run" fashion by striking the initially infected progenitor cell, altering its behavior, and then leaving the scene of the crime? According to this second scenario, the progenitor cell could somehow transmit the phenotype of cancerous growth to its descendants without the continued presence of RSV.

Temin and Rubin's work made it clear that the descendants of an RSV-infected cell continued to harbor copies of the RSV genome, but that evidence, on its own, settled little. The real question was: Did the transformed state of the descendant cells actually depend on some continuing influence exerted by the RSV genomes that they carried?

An experiment performed in 1970 at the University of California, Berkeley, settled this issue unambiguously. A mutant of RSV was developed that was capable of transforming chicken cells when these cells were cultured at 37°C but not at 41°C (the latter being the normal temperature at which chicken cells grow). **Temperature-sensitive** (ts) mutants like this one were known to encode partially defective proteins, which retain their normal structure and function at one temperature and lose their function at another temperature, presumably through thermal **denaturation** of the structure of the mutant protein.

After the chicken embryo fibroblasts were infected with the ts mutant of RSV, these cells became transformed if they were subsequently cultured at the lower (**permissive**) temperature of 37°C, as anticipated. Indeed, these cells could be propagated for many cell generations at this lower temperature and continued to grow and divide just like cancer cells, showing their characteristic transformed morphology. But weeks later, if the temperature of these infected cultures was raised to 41°C (the nonpermissive temperature), these cells lost their transformed shape and quickly reverted to the shape and growth pattern of cells that had never experienced an RSV infection. If these RSV-infected cells were again returned to the permissive temperature, they regained their transformed appearance (**Figure 3.5**).

The Berkeley experiments led to simple and yet profoundly important conclusions. Since the cells that descended from a ts RSV–infected cell continued to show the temperature-sensitive growth trait, it was obvious that copies of the genome of the infecting virus persisted in these cells for weeks after the initial infection. These copies of the RSV genome in the descendant cells continued to make some temperature-sensitive protein (whose precise identity was not known). Most important, the continuing actions of this protein were required in order to maintain the transformed growth phenotype of the RSV-infected cells.

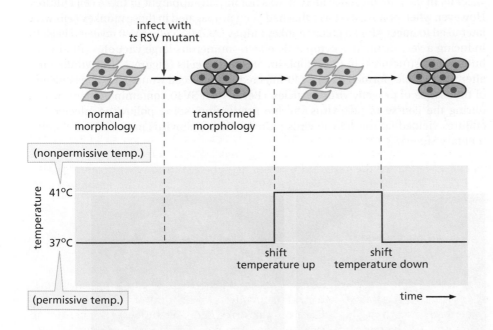

Figure 3.5 A temperature-sensitive mutant and the maintenance of transformation by RSV When chicken embryo fibroblasts were infected with a temperature-sensitive mutant of RSV and cultured at a *permissive* temperature of 37°C, the viral transforming function could be expressed and the cells became transformed. However, when the cultures containing these infected cells were shifted to a *nonpermissive* temperature of 41°C, the cells reverted to a normal, nontransformed morphology. Many cell generations later, with a shift of the cultures back to the permissive temperature, the cells once again exhibited a transformed morphology. The loss of the transformed phenotype upon temperature shift-up demonstrated that the continuous action of some temperature-sensitive viral protein was required in order to maintain this phenotype. The reacquisition of the transformed phenotype after temperature shift-down to the permissive temperature many cell generations later indicated that the viral genome continued to be present in these cells at the high temperature in spite of their normal appearance.

This work showed that cell transformation, at least that induced by RSV, was not a hit-and-run affair. In the language of the tumor virologists, the viral transforming gene was required to both initiate and maintain the transformed phenotype of virus-infected cells.

3.4 Viruses containing DNA molecules are also able to induce cancer

RSV was only one of a disparate group of viruses that were found able to induce tumors in infected animals. By 1960, four other classes of tumor viruses had become equally attractive agents for study by cancer biologists. One of them was a tumor virus that was discovered almost a quarter century after Rous's pioneering work. This virus, discovered by Richard Shope in rabbits, caused **papillomas** (warts) on the skin. These were actually benign lesions, which on rare occasions progressed to true cancers—squamous cell carcinomas of the skin.

By the late 1950s, it became clear that the Shope papillomavirus was constructed very differently from RSV. The papillomavirus particles carried DNA genomes, whereas RSV particles were known to carry RNA molecules. Also, the Shope virus particles were sheathed in a protein coat (**Figure 3.6**), whereas RSV clearly had, in addition, a lipid membrane coating on the outside (see Figure 3.3A and B). In the decades that followed, more than 100 distinct human papillomavirus (HPV) types, all related to the Shope virus, would be discovered.

In the 1950s and 1960s, various other DNA tumor viruses were isolated (**Table 3.1**). The murine polyomavirus (MPyV), named for its ability to induce a variety of distinct tumor types in newborn mice, was discovered in 1953. Closely related to MPyV in its size and chemical makeup was SV40 virus (the 40th simian virus in a series of isolates). This monkey virus had originally been discovered as a contaminant of the poliovirus vaccine stocks prepared in the mid- and late 1950s (**Figure 3.7A**). Clever virological sleuthing revealed that SV40 particles often hid out in cultures of the rhesus and cynomolgus monkey kidney cells used to propagate poliovirus during the preparation of vaccine. In fact, the presence of SV40 was not initially apparent in these cell cultures. However, when poliovirus stocks that had been propagated in these monkey cells were later used to infect African green monkey kidney (AGMK) cells, SV40 revealed itself by inducing a very distinctive **cytopathic** effect—numerous large **vacuoles** (fluid-filled bubble-like structures) in the cytoplasm of infected cells (Figure 3.7B). Within a day after the vacuoles formed in an SV40-infected cell, this cell would lyse, releasing tens of thousands of progeny virus particles. (Because of SV40 contamination, occurring during the course of poliovirus vaccine production, some poliovirus-infected cell cultures yielded far more SV40 virus particles than poliovirus particles! See Supplementary Sidebar 3.4.)

Figure 3.6 Papillomavirus virion
(A) A cryogenic electron microscopy 3D reconstruction and (B) a cryoelectron micrograph, together with image enhancement, reveal the structure of papillomavirus particles. (A, from B.L. Trus et al., *Nat. Struct. Biol.* 4:413–420, 1997. With permission from Springer Nature. B, Hyunwook Lee, Sarah A. Brendle, Stephanie M. Bywaters, Jian Guan, Robert E. Ashley, Joshua D. Yoder, Alexander M. Makhov, James F. Conway, Neil D. Christensen, Susan Hafensteinl. *Journal of Virology*, 2015 Jan 15; 89(2): 1428–1438. doi: 10.1128/JVI.02898-14. Reproduced with permission from American Society for Microbiology.)

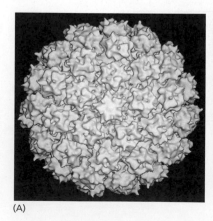
(A)

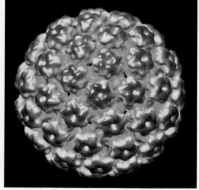
(B)

Table 3.1 Tumor virus genomes

	Virus family	Approximate size of genome (kb)
	DNA viruses	
Hepatitis B virus (HBV)	hepadna	3
SV40/murine polyomavirus (MPyV)	polyoma	5
Human papillomavirus 16 (HPV) / Shope papillomavirus (SPV)	papilloma	8
Human adenovirus 5	adenovirus	35
Human herpesvirus 8 (HHV-8; KSHV) / Human herpesvirus 4 (HHV-4; EBV)	herpesviruses	165
Shope fibroma virus (SFV)	poxviruses	160
	RNA viruses	
Human T-cell leukemia virus (HTLV-I)	retrovirus	8.2
Rous sarcoma virus (RSV)	retrovirus	9.3
Hepatitis C virus (HCV)	flavivirus	9.6

Adapted in part from G.M. Cooper, Oncogenes, 2nd ed. Boston: Jones and Bartlett Publishers, 1995.

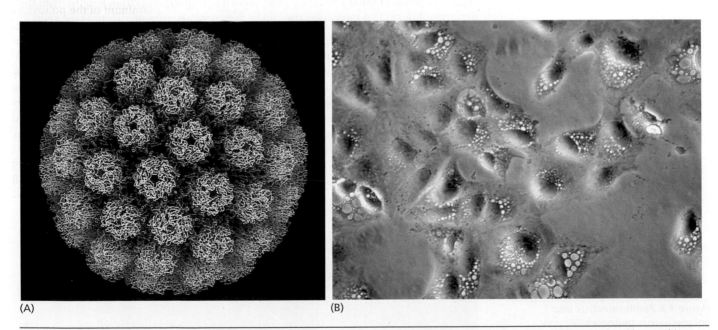

(A) (B)

Figure 3.7 SV40 virus (A) As determined by X-ray diffraction, the protein capsid of the SV40 DNA tumor virus consists of three virus-encoded proteins that are assembled into pentamers and hexamers that assemble to create a virion organized via icosahedral symmetry. The circular dsDNA genome of SV40 is carried within this capsid. (B) SV40 launches a lytic cycle in permissive host cells, such as the kidney cells of several monkey species. The resulting cytopathic effect, seen here, involves the formation of large cytoplasmic vacuoles prior to the death of the cell and the release of tens of thousands of progeny virus particles. (A, courtesy of R. Grant, S. Crainic, and J.M. Hogle. B, courtesy of A. Oppenheim.)

This **lytic cycle** of SV40 in monkey cells contrasted starkly with its behavior in cells prepared from mouse, rat, or hamster embryos. SV40 was unable to replicate in these rodent cells, which were therefore considered to be **nonpermissive** hosts. But on occasion, in one cell out of thousands in a nonpermissive infected cell population, a transformant grew out that shared many characteristics with RSV-transformed cells, that is, the cell had undergone changes in morphology and loss of contact inhibition, and had acquired the ability to seed tumors *in vivo*. On the basis of this, SV40 was classified as a tumor virus.

By some estimates, between one-third and two-thirds of the polio vaccines—the oral, live Sabin vaccine and the inactivated, injected Salk vaccine—administered between 1955 and 1963 contained SV40 virus as a contaminant, and between 10 and 30 million people were exposed to this virus through vaccination. In 1960, the fear was first voiced that the SV40 contaminant might trigger cancer in many of those who were vaccinated. Reassuringly, epidemiologic analyses conducted over the succeeding four decades indicated little, if any, increased risk of cancer among those exposed to these two vaccines.

Shope's papillomavirus, the mouse polyomavirus, and SV40 were initially grouped together as the papovavirus class of DNA tumor viruses, the term signifying papilloma, polyoma, and the vacuoles induced by SV40 during its lytic infection. In 2000, these viruses were recategorized as two distinct families, the first including HPV and the Shope papillomavirus as part of the papillomavirus family, and the second including SV40 and the murine polyomavirus as part of the polyomavirus family. By the mid-1960s, it was apparent that the genomes of the papillomavirus and polyomavirus were all formed from circular double-stranded DNA molecules (**Figure 3.8**). This represented a great convenience for experimenters studying the viral genomes, since there were several techniques in use at the time that made it possible to separate these relatively small, circular DNA molecules [about 5–8 kilobases (kb) in length] from the far larger, linear DNA molecules present in the chromosomes of infected host cells.

The group of DNA tumor viruses grew further with the discovery that human adenovirus, known to be responsible for upper respiratory infections in humans, was able to induce tumors in infected hamsters. Here was a striking parallel with the behavior of SV40. The two viruses could multiply freely in their natural host cells, which were therefore considered to be **permissive**. During the resulting lytic cycles of the virus, permissive host cells were rapidly killed in concert with the release of progeny virus particles. But when introduced into nonpermissive rodent cells, both adenovirus and SV40 failed to replicate and instead left behind, albeit at very low frequency, clones of transformants.

Figure 3.8 Papillomavirus and polyomavirus genomes Electron-microscopic analyses revealed the structure of DNA genomes of both the papillomavirus and the polyomavirus to be double-stranded and circular. Shown here are the closed-circular, supercoiled form of SV40 DNA (form I; kinked DNAs) and the relaxed, nicked circular form (form II; open, spread circles). Form III, which is a linear form of the viral DNA resulting from a double-strand DNA break, is not shown. (Courtesy of J.D. Griffith, University of North Carolina.)

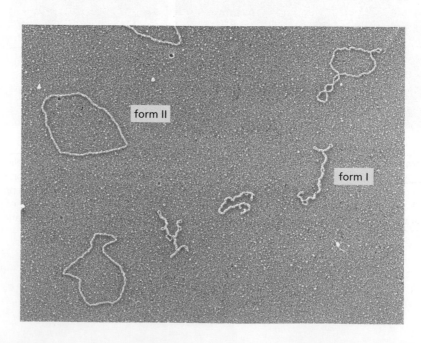

Other entrants into the class of DNA tumor viruses were members of the herpesvirus group. While human herpesvirus types 1 and 2 were apparently not **tumorigenic** (capable of inducing tumors), a distantly related herpesvirus of Saimiri monkeys provoked rapid and fatal lymphomas when injected into monkeys from several other species. Another distantly related member of the herpesvirus family—Epstein–Barr virus (EBV)—was discovered to play a causal role in provoking the development of Burkitt's lymphomas in young children in Equatorial Africa and New Guinea as well as nasopharyngeal carcinomas in Southeast Asia. Finally, at least two members of the poxvirus class, which includes smallpox virus, were found to be tumorigenic: Shope fibroma virus (SFV) and Yaba monkey virus cause benign skin lesions in rabbits and rhesus monkeys, respectively. The tumorigenic powers of poxviruses, which have very large genomes (135–160 kb), remain poorly understood to this day.

Researchers found that adenovirus and herpesvirus particles contain long, linear double-stranded DNA (dsDNA) molecules, which, like the genomes of RSV, carry the information required for both viral replication and virus-induced cell transformation. Compared with the papillomavirus and polyomavirus, the herpesviruses had genomes of enormous size (see Table 3.1), suggesting that they carried a proportionately larger number of genes. In the end, it was the relatively small sizes of papillomavirus and polyomavirus genomes that made them attractive objects of study.

Most of the small group of genes in a papillomavirus and polyomavirus genome were apparently required to program viral replication; included among these were several genes specifying the proteins that form the capsid coat of the virus particle. This dictated that papillomaviruses and polyomaviruses could devote only a small number of their genes to the process of cell transformation.

This realization offered the prospect of greatly simplifying the cancer problem by reducing the array of responsible genes and causal mechanisms down to a very small number. Without such simplification, cancer biologists were forced to study the cancer-causing genes that were thought to be present in the genomes of cells transformed by nonviral mechanisms. Cellular genomes clearly harbored large arrays of genes, possibly more than a hundred thousand. At the time, the ability to analyze genomes of such vast complexity and to isolate individual genes from these genomes was a distant prospect.

3.5 Tumor viruses induce multiple changes in cell phenotype including acquisition of tumorigenicity

Like RSV-transformed chicken cells, the rodent cells transformed by SV40 showed profoundly altered shape and piled up on one another. This loss of contact inhibition was only one of a number of changes exhibited by virus-transformed cells (**Table 3.2**). As discussed later in this book (in Chapter 5), normal cells in culture will not proliferate unless they are provided with serum and serum-associated growth-stimulating factors. Cells transformed by a variety of tumor viruses were often found to have substantially reduced requirements for these factors in their culture medium.

Yet another hallmark of the transformed state is an ability to proliferate in culture for unusually long time periods. Normal cells have a limited proliferative potential in culture and ultimately stop multiplying after a certain, apparently predetermined number of cell divisions. Cancer cells seemed to be able to proliferate indefinitely in culture, and thus to undergo **immortalization** (as discussed in Chapter 10).

When transformed cells were suspended in an agar gel, they were able to proliferate into spherical colonies containing dozens, even hundreds of cells (**Figure 3.9A**). This ability to multiply without attachment to the solid substrate at the bottom of the Petri dish was termed the trait of **anchorage independence**. Normal epithelial cells and fibroblasts, in contrast, demonstrated an absolute requirement for tethering to a solid substrate before they would grow and were therefore considered to be **anchorage-dependent**. The ability of cells to grow in an anchorage-independent fashion *in vitro* usually served as a good predictor of their ability to form tumors *in vivo* following injection into appropriate host animals (Figure 3.9B).

Table 3.2 Properties of transformed cells

Altered morphology (rounded shape, refractile in phase-contrast microscope)
Loss of contact inhibition (ability to grow over one another)
Ability to grow without attachment to solid substrate (anchorage independence)
Ability to proliferate indefinitely (immortalization)
Reduced requirement for exposure to mitogenic growth factors
High saturation density (ability to accumulate large numbers of cells in culture dish)
Inability to halt proliferation in response to deprivation of growth factors
Increased import of glucose
Glycolysis in presence of oxygen
Tumorigenicity

Adapted in part from S.J. Flint, L.W. Enquist, R.M. Krug et al., Principles of Virology. Washington, DC: ASM Press, 2000.

Figure 3.9 Anchorage-independent growth and tumorigenesis (A) A photomicrograph of colonies of cells growing in an anchorage-independent fashion. (Each of the larger colonies seen here may contain several hundred cells, all the descendants of a single transformed ancestor.) This test is usually performed by suspending cells in a semi-solid medium, such as agarose or methylcellulose, to prevent their attachment to a solid substrate, specifically, the bottom of the Petri dish. The ability of cells to proliferate while held in suspension—the phenotype of anchorage independence—is usually a good (but hardly infallible) predictor of their ability to form tumors *in vivo*. (B) Mice of the Nude strain offer two advantages in tests of tumorigenicity. Lacking a thymus, they are highly immunocompromised and therefore relatively receptive to engrafted cells from genetically unrelated sources (i.e., xenografts), including those from foreign species. In addition, because these mice are hairless, it is easy to monitor closely the progress of tumor formation (*arrow*) after transformed cells have been injected under their skin. (A, courtesy of A. Orimo. B, courtesy of X.G. Wang and E. Dadachova; see also X.G. Wang et al., *PLoS ONE* 2(10): e1114, 2007.)

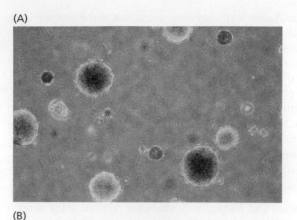
(A)

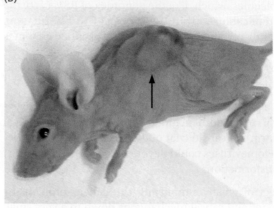
(B)

This tumor-forming ability—the phenotype of tumorigenicity—represented the acid test of whether cells were fully transformed, that is, had acquired the full repertoire of neoplastic traits. One test for tumorigenicity could be performed by injecting mouse cells prepared from one strain of mice and then transformed in an *in vitro* transformation experiment into host mice of the same strain (**Sidebar 3.2**). Since the host and injected cells came from the same genetic strain, the immune systems of such **syngeneic** host mice would not recognize the transformed cells as being foreign bodies and therefore would not attempt to eliminate them—the process of **tumor rejection** (a process to which we will return in Chapter 15). This allowed injected cells to survive in their animal hosts, enabling them to multiply into large tumors if, indeed, they had acquired the tumorigenic phenotype. (Populations of injected non-tumorigenic cells, in contrast, might survive in a syngeneic host as small clumps for weeks, even months, without increasing in size.)

Often, it was impossible to test the tumorigenicity of tumor virus–infected cells in a syngeneic host animal, simply because the cells being studied came from a species in which inbred syngeneic hosts were not available. This forced the use of **immunocompromised** hosts whose immune systems were defective and thus tolerant of a wide variety of foreign cell types, including those from other species (see Sidebar 3.2). Mice of the Nude strain soon became the most commonly used hosts to test the tumorigenicity of a wide variety of cells, including those of human origin. Quite frequently, candidate tumor cells are injected **subcutaneously**, that is, directly under the skin of these animals. Since they also happen to lack the ability to grow hair, Nude mice provide the additional advantage of allowing the experimenter to closely monitor the behavior of implanted tumor cells, specifically the growth of any tumors that they may spawn (see Figure 3.9B).

As mentioned, the small size of the genomes of RNA tumor viruses (for example, RSV) and the papillomavirus and polyomavirus dictated that each of these employ only a small number of genes (perhaps as few as one) to elicit multiple changes in the cells that they infect and transform. Recall that the ability of a gene to concomitantly induce a number of distinct alterations in a cell is termed pleiotropy. Accordingly, though little direct evidence about gene number was yet in hand, it seemed highly likely that the genes used by tumor viruses to induce cell transformation were acting pleiotropically on a variety of molecular targets within cells.

3.6 Tumor virus genomes persist in virus-transformed cells by becoming part of host-cell DNA

The Berkeley experiment (see Section 3.3) provided strong evidence that the continued actions of the RSV genome were required to maintain the transformed state of cells, including those that were many cell generations removed from an initially infected progenitor cell. This meant that some or all of the viral genetic information needed to be perpetuated in some form, being passed from a transformed mother cell to its two daughters and, further on, to descendant cells through many cycles of cell growth and division. Conversely, a failure to faithfully transmit these viral genes to descendant cells would result in the reversion of such cells to those showing normal growth behavior.

Paralleling the behavior of RSV, cell transformation achieved by two intensively studied DNA tumor viruses—SV40 and the murine polyomavirus—also seemed to depend on the continued presence of viral genomes in the descendants of an initially transformed cell. The evidence proving this came in a roundabout way, largely from the discovery of tumor-associated proteins (**T antigens**) that were found in cancers induced by these two viruses. For example, sera prepared from mice carrying an SV40-induced tumor showed strong reactivity with a nuclear protein that was present characteristically in tumors triggered by SV40 and absent in tumors induced by other carcinogenic agents. The implication was that the viral genome residing in tumor cells encoded a protein (in this case, the SV40 T antigen) that induced a strong immunological response in the tumor-bearing mouse or rat host (**Figure 3.10**).

The display of the virus-induced T antigen correlated directly with the transformed state of these cells. Therefore, cells that lost the T antigen would also lose the transformation phenotype induced by the virus. This correlation suggested, but hardly proved, that the viral gene sequences responsible for transformation were associated with or closely linked to viral sequences encoding the T antigen.

The cell-to-cell transmission of viral genomes over many cell generations represented a major conceptual problem. Cellular genes were clearly transmitted with almost total

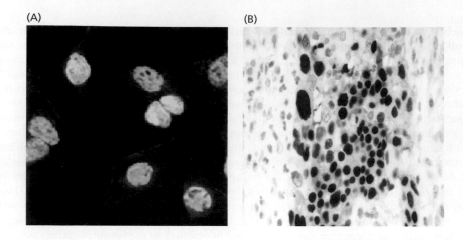

(A)   (B)

Figure 3.10 Expression of the SV40 T antigen When SV40-transformed cells are introduced into a syngeneic host, they provoke a strong immune response against the SV40-encoded T antigen (tumor antigen), which is usually termed "large T" (LT). Anti-LT serum, together with appropriate immunofluorescent tags, can be used to stain virus-transformed cells. (A) In the light micrograph shown here, an anti-LT monoclonal antibody (rather than anti-tumor serum) has been used to stain mouse cells that have been transformed by SV40, revealing the nuclear localization of LT (*green*); the cytoplasm of each cell has been stained with phalloidin (*red*), which reveals the F-actin fibers of the cytoskeleton. (B) While SV40 itself rarely if ever causes human tumors (see Section 3.4), immunosuppression can allow an otherwise latent close cousin of SV40, termed BK virus, to become active and occasionally oncogenic. Seen here is a urothelial (bladder) carcinoma arising in a kidney transplant patient. The tumor cells show strong staining for T antigen (*brown*), while the nearby normal urothelial cells are unstained. (The tumor cell nuclei show dramatic variations in size.) (A, courtesy of S.S. Tevethia. B, from I.S.D. Roberts et al., *Brit. J. Cancer* 99:1383–1386, 2008. With permission from Springer Nature.)

fidelity from mother cells to daughter cells through the carefully programmed processes of chromosomal DNA replication and mitosis that occur during each cellular growth-and-division cycle. How could viral genomes succeed in being replicated and transmitted efficiently through an unlimited number of cell generations? This was especially puzzling, since viral genomes seemed to lack the genetic elements, specifically those forming **centromeres**, that were thought to be required for proper allocation of chromosomes to daughter cells during mitosis.

Adding to this problem was the fact that the DNA metabolism of papovaviruses, such as papillomaviruses and polyomaviruses, was very different from that of the host cells that they preyed upon. For example, when SV40 and the murine polyomavirus infected permissive host cells, the viral DNAs were replicated as autonomous, extra-chromosomal molecules. Both viruses could form many tens of thousands of circular, double-stranded DNA genomes of about 5 kb in size, all derived from a single viral DNA genome initially introduced by an infecting virus particle (see Figure 3.8). While the viral DNA replication exploited a number of host-cell DNA replication enzymes (see Figure 3.2), it proceeded independently of the infected cells' chromosomal DNA replication. This nonchromosomal replication occurring during the lytic cycles of SV40 and the murine polyomavirus shed no light on how these viral genomes were perpetuated in populations of virus-transformed cells. The latter were, after all, nonpermissive hosts and therefore prevented these viruses from replicating their DNA, both in the initially infected cell and, presumably, in the descendants of this cell.

A solution to this puzzle came in 1968, when it was discovered that the viral DNA in SV40-transformed mouse cells was tightly associated with their chromosomal DNA. Using centrifugation techniques to gauge the molecular weights of DNA molecules, it became clear that the SV40 DNA in these cells no longer sedimented like a small (~5 kb) viral DNA genome. Instead, the SV40 DNA sequences in virus-transformed cells co-sedimented with the high–molecular-weight (>50 kb) chromosomal DNA of the host cells (**Figure 3.11A**). In fact, the viral DNA in these transformed cells could not be separated from the cells' chromosomal DNA by even the most stringent methods of dissociation, including the harsh treatment of exposure to alkaline pH.

These results indicated that SV40 DNA in virus-transformed cells had been inserted into the cells' chromosomes, becoming covalently linked to the chromosomal DNA. Such **integration** of the viral genome into a host-cell chromosome solved an important problem in viral transformation: transmission of viral DNA sequences from a mother cell to its offspring could be guaranteed, since the viral DNA would be co-replicated together with the cell's chromosomal DNA during the S (DNA synthesis) phase of each cell cycle. In effect, by integrating into the chromosome, the viral DNA sequences became as much a part of a cell's genome as the cell's own native genes (Figure 3.11B).

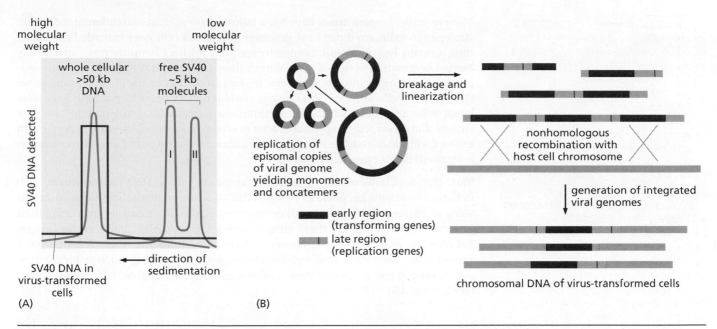

Figure 3.11 Integration of SV40 DNA (A) DNA molecules from SV40-transformed cells were isolated and sedimented by centrifugation through an alkaline solution stabilized by a sucrose gradient (composed of more concentrated sucrose at the bottom and more dilute sucrose at the top and used to prevent mixing of different fluid strata within the centrifuge tube). Under these conditions, the high–molecular-weight cellular DNA (*blue*) sedimented a substantial distance down the sucrose gradient (*left side of graph*). In contrast, the SV40 DNA isolated from virus particles (*green*) sedimented more slowly, indicative of its lower molecular weight. Forms I and II viral DNA refer to the closed circular and nicked circular DNAs of SV40, respectively (see Figure 3.8). Use of an SV40-specific radioactive probe revealed that the SV40 DNA sequences in SV40 virus–transformed cells co-sedimented with the high–molecular-weight chromosomal DNA of the virus-transformed cells. Because high pH conditions disrupt all noncovalent interactions, this co-sedimentation indicated the covalent association of the viral genomic DNA (*red*). with that of the host cell (*blue*). (B) When nonchromosomal (that is, episomal) SV40 genomes (*top leftmost circles*) replicate, they often generate large, tandemly repeated, circular genomes termed concatemers. Monomeric and concatemerized SV40 DNA genomes can undergo linearization and thereafter recombine at low efficiency with the host-cell chromosomal DNA (*gray*), doing so at random sites in the host genome; this leads to the covalent linkage of viral and cellular DNA sequences. Since the viral and host-cell genomes lack significant sequence identity, this type of recombination is termed nonhomologous or illegitimate recombination. Use of restriction enzyme cleavage-site mapping revealed the configurations of SV40 genomes integrated into the chromosomal DNA of virus-transformed cells. In some cells, only a portion of the viral genome was present; in others, full, multiple, head-to-tail tandem arrays were present. Cell transformation by SV40 requires only the genes in the "early region" of the viral genome (*dark red*). Consequently, all virus-transformed cells contained at least one uninterrupted copy of this early region but often lacked segments of the viral late region (*pink*), which is functionally unimportant for cell transformation. (A, adapted from J. Sambrook et al., *Proc. Natl. Acad. Sci. USA* 60:1288–1295, 1968. B, adapted from M. Botchan, W. Topp and J. Sambrook, *Cell* 9:269–287, 1976.)

Some years later, this ability of the papillomavirus and polyomavirus genomes to integrate into host-cell genomes became highly relevant to the pathogenesis of one common form of human cancer—cervical carcinoma. Almost all (>99.7%) of these tumors have been found to carry fragments of human papillomavirus (HPV) genomes integrated into their chromosomal DNA. Provocatively, intact viral genomes are rarely discovered to be present in integrated form in cancer cell genomes. Instead, only the portion of the viral genome that contains **oncogenic** (cancer-causing) information has been found in the chromosomal DNA of these cancer cells, while the portion that enables these viruses to replicate and construct progeny virus particles is almost always absent or present in only fragmentary form (see Figure 3.11B). Interestingly, among the HPV genes discarded or disrupted during the chromosomal integration process is the viral E2 gene, whose product serves to repress transcription of viral **oncogenes** (genes contributing to cell transformation); in the absence of E2, the viral E6 and E7 oncogenes present in the integrated viral genome segments (discussed in Chapters 8 and 9) are actively transcribed, thereby driving cervical epithelial cells toward a neoplastic phenotype. (An analogous situation is seen in SV40-transformed rodent cells in which the viral genes responsible for viral replication are also often absent.)

More recently, herpesviruses have been found to use at least two different molecular strategies to maintain intact viral genomes in infected cells over extended periods of time. Certain **lymphotropic** herpesviruses (which infect lymphocytes), specifically human herpesvirus 6 (HHV-6) and Marek's disease virus (MDV) of chickens, can integrate their genomes into the telomeric repeats (see Chapter 10) of the chromosomal DNA of their cellular hosts. This strategy enables these viruses to establish long-term latent infections, in which the viral genomes are maintained indefinitely while the viruses shut down active replication; after great delay, however, the viral genomes can excise themselves and launch active replication of the resulting non-chromosome-integrated (that is **episomal**) genomes.

More common, however, are molecular strategies that allow DNA tumor viruses, such as Kaposi's sarcoma herpesvirus (KSHV/HHV-8), to maintain their genomes continuously as **episomes** over many cell generations. Rather than covalently linking their genomes to cellular DNA, these viruses deploy viral proteins that serve as bridges between the episomal viral genomes and cellular chromatin. This allows the viral genomes to "hitchhike" with cellular chromosomes, ensuring nuclear localization during mitosis and proper allocation of viral genomes to daughter cells (Supplementary Sidebar 3.5).

3.7 Retroviral genomes become integrated into the chromosomes of infected cells

The ability of SV40 and the other polyomaviruses to integrate copies of their genomes into host-cell chromosomal DNA solved one problem but created another that seemed much less solvable: how could RSV succeed in transmitting its genome through many generations within a cell lineage? The genome of RSV is made of single-stranded RNA, which clearly could not be integrated directly into the chromosomal DNA of an infected cell. Still, RSV succeeded in transmitting its genetic information through many successive cycles of cell growth and division.

This puzzle consumed Temin in the mid- and late 1960s and caused him to propose a solution so unorthodox that it was ridiculed by many, landing him in the scientific wilderness. Temin argued that after RSV particles (and those of related viruses) infected a cell, they made double-stranded DNA (dsDNA) copies of their RNA genomes. It was these dsDNA versions of the viral genome, he said, that became established in the chromosomal DNA of the host cell. Once established, the DNA version of the viral genome—which he called a **provirus**—then assumed the molecular configuration of a cellular gene and would be replicated each time the cell replicated its chromosomal DNA. In addition, the proviral DNA could then serve as a template for transcription by cellular RNA polymerase, thereby yielding RNA molecules that could be incorporated into progeny virus particles or, alternatively, could function as messenger RNA (mRNA) that was used for the synthesis of viral proteins (**Figure 3.12**). (The provirus concept was inspired, in part, by the 1952 discovery of prophages—latent forms of bacteriophages that reside in bacterial genomes.)

The process of **reverse transcription** that Temin proposed—making DNA copies of RNA—was without precedent in the molecular biology of the time, which recognized information flow only in a single direction, specifically, DNA → RNA → proteins. But the idea prevailed, receiving strong support from Temin's and David Baltimore's simultaneous discoveries in 1970 that RSV and related virus particles carry the enzyme **reverse transcriptase**. As both research groups discovered, this enzyme has the capacity to execute the key copying step that Temin had predicted—the step required in order for RSV to transmit its genome through many cycles of cell growth and division.

It soon became apparent that RSV was only one of a large group of similarly constructed viruses (including HIV; Supplementary Sidebar 3.6), which together came

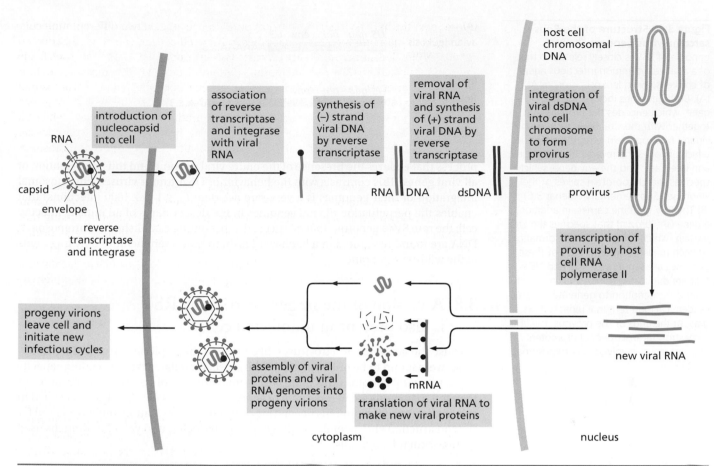

Figure 3.12 The life cycle of an RNA tumor virus like RSV An infecting virion introduces its single-stranded (ss) RNA genome (*blue*) into the cytoplasm of a cell together with the reverse transcriptase enzyme (RT, *purple*). The RT makes a single-stranded DNA using the viral RNA as template complementary to the viral RNA (*red*), and then a second DNA strand (*red*). The reverse-transcribed dsDNA then moves into the nucleus, where it becomes integrated through the actions of the virus-encoded integrase into the cellular chromosomal DNA (*orange*). The resulting integrated provirus is then transcribed by host-cell RNA polymerase II into progeny viral RNA molecules (blue). The progeny RNA molecules are exported to the cytoplasm, where they either serve as mRNAs to make viral proteins or are packaged into progeny virus particles that leave the cell and initiate a new round of infection. (Not explained by this scheme is the mechanistic rationale for the incorporation of RT molecules within the virions. Thus, the indicated replication strategy could in principle be achieved if the RT were not carried within virions but instead was synthesized *de novo* on viral RNA templates only after the virions had entered the cell.) (Adapted from B. Alberts et al., Essential Cell Biology, 3rd ed. New York: Garland Science, 2010.)

to be called **retroviruses** to reflect the fact that their cycle of replication depends on information flowing "backward" from RNA to DNA. Within a year of the discovery of reverse transcriptase, the presence of proviral DNA was detected in the chromosomal DNA of RSV-infected cells. Hence, like SV40 and polyomavirus, retroviruses rely on integration of their genomes into the chromosome to ensure the stable retention and transmission of their genomes.

There is, however, an important distinction between the integration mechanisms used by retroviruses like RSV (see Figure 3.12) and the DNA tumor viruses such as SV40 and the murine polyomaviruses (see Figure 3.11B). Integration is a normal, essential part of the replication cycle of retroviruses. Indeed, retrovirus genomes carry the gene specifying an **integrase**. This enzyme is encoded by the viral *pol* gene (**Figure 3.13**) and is synthesized together with the reverse transcriptase as part of a larger **polyprotein**. Integrase is dedicated to mediating the process of chromosomal integration. Accordingly, during each cycle of infection and replication, the retroviral genome—its provirus—is integrated into the host chromosomal DNA through a precise, ordered

Figure 3.13 Structure of the Rous sarcoma virus genome The RNA genome of RSV is closely related to that of a relatively common infectious agent of chickens, avian leukosis virus, or ALV (A). Both genomes include the *gag* gene, which encodes the proteins that, together with the viral RNA, form each virion's nucleoprotein core; the *pol* gene, which encodes the reverse transcriptase and integrase; and the *env* gene, which specifies the glycoprotein spikes of the virion (see Supplementary Sidebar 3.6). (B) The RSV genome carries, in addition, a gene (*src, brown*) that specifies the Src protein, which causes cell transformation. At both the left and right ends of these genomes are sequences—*U3* and *U5*— that are duplicated during the course of reverse transcription to generate the pair of identical long terminal repeats (LTRs, *gray*) at the ends of the provirus; the LTRs carry strong transcriptional promoters and sequences involved in chromosomal integration.

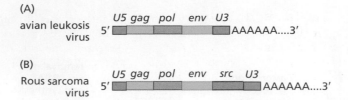

process that ensures the presence of the entire viral genome and thus the retention of all viral genes. This contrasts with the behavior of DNA tumor viruses: chromosomal integration of their genomes is a very rare accident (<<1 per 1000 infections) that enables the perpetuation of viral genomes in the descendants of an initially infected cell; the rare SV40 genomes that do succeed in becoming established in chromosomal DNA are found integrated in a haphazard fashion that often includes only fragments of the wild-type genome

3.8 A version of the *src* gene carried by RSV is also present in uninfected cells

Because the genomes of retroviruses, like those of the papillomavirus and polyomavirus, were found to be quite small (<10 kb), it seemed likely that the coding capacity of the retroviral genomes was limited to a small number of genes, probably far fewer than ten. Using this small repertoire of genes, retroviruses nevertheless succeeded in specifying some viral proteins needed for viral genome replication, others required for the construction and assembly of progeny virus particles, and yet other proteins used to transform infected cells.

In the case of RSV, the use of mutant viruses revealed that the functions of viral replication (including reverse transcription and the construction of progeny virions) required one set of genes, while the function of viral transformation required another. Thus, some mutant versions of RSV could replicate perfectly well in infected cells, producing large numbers of progeny virus particles, yet such mutants lacked transforming function. Conversely, other mutant derivatives of RSV could transform cells but had lost the ability to replicate and make progeny virions in these transformed cells.

At least three retroviral genes were implicated in viral replication. Two of these encode structural proteins that are required for assembly of virus particles; a third specifies both the reverse transcriptase (RT) enzyme, which copies viral RNA into double-stranded DNA shortly after retrovirus particles enter into host cells, and the integrase enzyme, which is responsible for integrating the newly synthesized viral DNA into the host-cell genome (see Figures 3.12 and 3.13). A comparison of the RNA genome of RSV with the genomes of related retroviruses lacking transforming ability suggested that there was rather little information in the RSV genome devoted to encoding the remaining known viral function—transformation. Consequently, geneticists speculated that all of the viral transforming functions of RSV resided in a single gene, which they termed *src* (pronounced "sark"), to indicate its role in triggering the formation of sarcomas in infected chickens (see Figure 3.13B).

In 1974, the laboratory run jointly by J. Michael Bishop and Harold Varmus at the University of California, San Francisco, undertook to make a DNA **probe** that specifically recognized the transformation-associated (that is, *src*) sequences of the RSV genome in order to understand its origins and functions (**Sidebar 3.3**). This *src*-specific probe was then used to follow the fate of the *src* gene after cells were infected with RSV. The notion here was that uninfected chicken cells would carry no *src*-related DNA sequences in their genomes, which *src* sequences would become readily detectable in cells following RSV infection, having been introduced by the infecting viral genome.

Sidebar 3.3 The making of a *src*-specific DNA probe In order to make a *src*-specific DNA probe, a researcher in Bishop and Varmus's laboratory exploited two types of RSV strains: a wild-type RSV genome that carried all of the sequences needed for viral replication and transformation, and a mutant RSV genome that was able to replicate but had lost, by deletion, the *src* sequences required for transformation (**Figure 3.14**). Using reverse transcriptase, he made a DNA copy of the wild-type sequences, yielding single-stranded DNA molecules complementary to the viral RNA genome. He fragmented the complementary DNA (often termed cDNA) and hybridized it to the RNA genome of the RSV deletion mutant that lacked *src* sequences, creating DNA–RNA hybrid molecules. The bulk of the cDNA fragments annealed to the RNA genome of the deletion mutant, with the exception of cDNA fragments derived from the region encoding the *src* sequences. He then retrieved the ssDNA molecules that failed to form DNA–RNA hybrids (discarding the hybrids). The result was ssDNA fragments that specifically recognized sequences contained within the deleted portion of the mutant RSV genome, that is, those lying within the *src* gene.

Because the initial reverse transcription of the wild-type RSV RNA was carried out in the presence of radiolabeled deoxyribonucleoside triphosphates, the *src*-specific DNA fragments (which constituted the *src* "probe") were also labeled with radioisotope. This made it possible to discover whether DNAs

of interest (such as the DNAs prepared from virus-infected or uninfected cells) also carried *src* sequences by determining whether or not the *src* probe (with its associated radioactivity) was able to hybridize to these cellular DNAs.

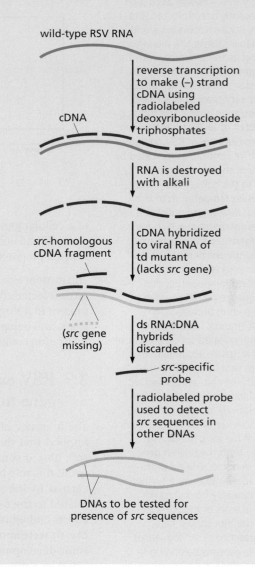

Figure 3.14 The construction of a *src*-specific DNA probe Wild-type (wt) RSV RNA (*blue*) was reverse-transcribed under conditions where only a single-stranded (ss) complementary DNA molecule (a cDNA; *red*) was synthesized. The wt single-stranded viral DNA was then annealed (hybridized) to viral RNA (green) of a transformation-defective (td) RSV mutant, which had lost its transforming function and apparently deleted its *src* gene. The resulting RNA:DNA hybrids were discarded, leaving behind the ssDNA fragment of the wtDNA that failed to hybridize to the RNA of the td mutant. This surviving DNA fragment, if radiolabeled, could then be used as a *src*-specific probe to detect *src*-related sequences in various cellular DNAs (*orange*) via DNA:DNA hybridization.

The actual outcome of this experiment was, however, totally different from expectation. In 1975, this research group, using their *src*-specific probe, found that *src* sequences were clearly present among the DNA sequences of uninfected chicken cells. These *src* sequences were present as single-copy cellular genes; that is, two copies of the *src*-related DNA sequences were present per diploid chicken cell genome—precisely the representation of the great majority of genes in the cellular genome.

The presence of *src* sequences in the chicken cell genome could not be dismissed as some artifact of the hybridization procedure used to detect them. Moreover, careful characterization of these *src* sequences made it unlikely that they had been inserted into the chicken genome by some retrovirus. For example, *src*-related DNA sequences were readily detectable in the genomes of several related bird species, and, more distantly on the evolutionary tree, in the DNAs of several mammals (**Figure 3.15**). The more distant the evolutionary relatedness of a species was to chickens, the weaker was the reactivity of the *src* probe with its DNA. This was precisely the behavior expected

Figure 3.15 Evolutionary tree of the *src* gene The *src*-specific cDNA probe (see Figure 3.14) was annealed with the DNAs of a variety of vertebrate species—whose relatedness to the chicken is indicated on this evolutionary tree—as well as to a rat cell that had recently been infected and transformed by Rous sarcoma virus and by definition contained the entire sequence of RSV including its *src* gene (RSV-rat). The abscissa indicates the time when the last common ancestor shared by the chicken and each other vertebrate species lived. The best hybridization of the RSV *src* probe was with the DNA of an RSV-transformed rat cell (RSV-rat) carrying an RSV provirus (used as a positive control), as indicated by the percentage of the probe radioactivity that annealed to this cellular DNA and the higher melting temperature of the resulting hybrid. (Melting temperature refers to the temperature at which two nucleic acid strands separate from one another, that is, the temperature at which a double-helix is denatured. The higher the melting temperature, the greater the sequence complementarity and thus base-pairing between two annealing DNA strands; hence, genes borne by distantly related species, because of sequence divergence accumulated as they evolved, should hybridize only imperfectly with the chicken-derived *src* probe, and resulting hybrids should be disrupted at relatively low temperature.) The results for chicken DNA indicate that the sequence of the *src* gene of RSV deviates slightly from the related gene in the chicken genome. The *src* proto-oncogenes of species that were further removed evolutionarily from chickens showed lesser extents of hybridization. Subsequent work, not shown here, revealed the presence of *src* sequences in the DNA of organisms belonging to other metazoan phyla, such as arthropods (e.g., *Drosophila melanogaster*) and even a sponge, whose ancestors diverged from those of vertebrates and other metazoa ~550 million years ago. The extensive evolutionary conservation of the *src* gene indicates that it played a vital role in the lives of all the organisms that carried it in their genomes. (Courtesy of H.E. Varmus and adapted from D.H. Spector, H.E. Varmus and J.M. Bishop, *Proc. Natl. Acad. Sci. USA* 75:4102–4106, 1978.)

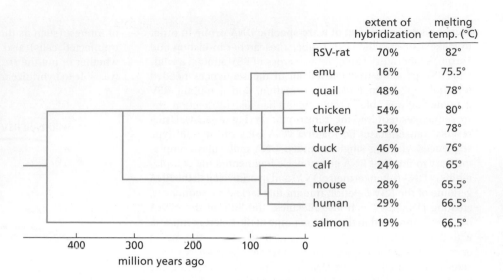

	extent of hybridization	melting temp. (°C)
RSV-rat	70%	82°
emu	16%	75.5°
quail	48%	78°
chicken	54%	80°
turkey	53%	78°
duck	46%	76°
calf	24%	65°
mouse	28%	65.5°
human	29%	66.5°
salmon	19%	66.5°

of a cellular gene that had been present in the genome of a common ancestral species and had acquired increasingly divergent DNA sequences as descendant species evolved progressively away from one another over the course of millions of years.

The evidence converged on the idea that the *src* sequences present in the genome of an uninfected chicken cell possessed all the properties of a normal cellular gene, being present in a single copy per haploid genome, evolving slowly over tens of millions of years, and being present in species that were ancestral to all modern vertebrates. This realization created a revolution in thinking about the origins of cancer.

3.9 RSV exploits a kidnapped cellular gene to transform cells

The presence of a highly conserved *src* gene in the genome of a normal organism implied that this cellular version of *src*, sometimes termed c-*src* (that is, cellular *src*), played some role in the life of this organism (the chicken) and its cells. How could this role be reconciled with the presence of a powerful transforming *src* gene carried in the genome of RSV? This viral transforming gene (v-*src*) was closely related to the c-*src* gene of the chicken, yet the two genes had drastically different effects and apparent functions. When ensconced in the cellular genome, the actions of c-*src* were apparently compatible with normal cellular behavior and normal organismic development. In contrast, the very similar v-*src* gene borne by the RSV genome acted as a potent **oncogene**—a gene capable, following its introduction into a cell by an infecting virus particle, of transforming the previously normal cell into a tumor cell.

One solution to this puzzle came from considering the possibility that perhaps the *src* gene of RSV was not naturally present in the retrovirus ancestral to RSV. This hypothetical viral ancestor, while lacking *src* sequences, would be perfectly capable of replicating in chicken cells. In fact, such a *src*-negative retrovirus—avian leukosis virus (ALV)—was common in chickens and was capable of infectious spread from one chicken to another. This suggested that during the course of infecting a chicken cell, an ancestral virus, similar to this common chicken virus, somehow acquired sequences from the genome of a cell that it had infected (**Figure 3.16**), doing so through some genetic trick. The acquired cellular sequences (that is, the *src* sequences) were then incorporated into the viral genome, thereby adding a fourth gene to the existing three genes that this retrovirus used for its replication in infected cells (see Figure 3.13). Once present in the genome of RSV, the kidnapped *src* gene could then be altered and exploited by this virus to transform subsequently infected cells.

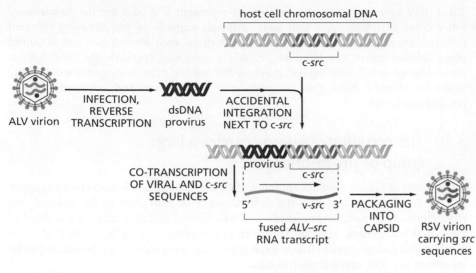

host cell chromosomal DNA

c-src

ALV virion → INFECTION, REVERSE TRANSCRIPTION → dsDNA provirus → ACCIDENTAL INTEGRATION NEXT TO c-src

provirus   c-src

CO-TRANSCRIPTION OF VIRAL AND c-src SEQUENCES

5'   v-src   3' → PACKAGING INTO CAPSID → RSV virion carrying src sequences

fused ALV–src RNA transcript

Figure 3.16 Capture of *src* by avian leukosis virus The precise mechanism by which an avian leukosis virus (ALV) captured a cellular *src* gene (c-*src*) is not known. According to the most plausible model, ALV proviral DNA (*red*) became integrated (by chance) next to a c-*src* proto-oncogene (*green*) residing in the chromosomal DNA of an infected chicken cell. The ALV provirus and adjacent c-*src* gene were then co-transcribed into a single hybrid RNA transcript (*blue and brown*). After splicing out c-*src* introns (not shown), this hybrid viral RNA was packaged into a virus particle that became the ancestor of Rous sarcoma virus (RSV). (Not shown is the acquisition during a subsequent infectious cycle of ALV sequences at the 3' side of v-*src*, with the result that this viral oncogene became flanked on both sides by ALV sequences.)

This scheme attributed great cleverness to retroviruses by implying that they had the ability to pick up and exploit preexisting cellular genes for their own purposes. Such behavior is most unusual for a virus, since virtually all other types of viruses carry genes that have little if any relatedness to DNA sequences native to the cells that they infect (Supplementary Sidebar 3.7).

But there was an even more important lesson to be learned here, this one concerning the c-*src* gene. This cellular gene, one among tens of thousands in the chicken cell genome, could be converted into a potent viral oncogene following some slight remodeling by a retrovirus such as RSV. Because it was a precursor to an active **oncogene**, c-*src* was called a **proto-oncogene**. The very concept of a proto-oncogene was revolutionary: it implied that the genomes of normal vertebrate cells carry a gene that has the potential, under certain circumstances, to induce cell transformation and thus cancer.

The structures of the c-*src* proto-oncogene and the v-*src* oncogene were worked out rapidly in the years that followed these discoveries in 1975 and 1976. Just as the viral geneticists had speculated, all of the viral transforming sequences resided in a single viral oncogene. Within the RSV RNA genome, the v-*src* gene was found at the downstream, 3' end of the genome, added to the three preexisting retroviral genes that were involved in viral replication (see Figure 3.13). Amusingly, the c-*src* gene may have been acquired by ALV on only one occasion documented in the history of modern cancer research (see Supplementary Sidebar 3.7).

This scenario of acquisition and activation of c-*src* by a retrovirus led to three further ideas. First, if retroviruses could activate this proto-oncogene into a potent oncogene, perhaps other types of mutational mechanisms might reshape the normal c-*src* gene and yield a similar outcome. Maybe such mechanisms could activate a cellular proto-oncogene without removing the normal gene from its normal roosting site on the cellular chromosome. Maybe the information for inducing cancer was already present in the normal cell genome, waiting to be unmasked.

Second, it became clear that all of the transforming powers of RSV derived from the presence of a single gene—v-*src*—in its genome. This was of great importance, because it implied that a single oncogene could, as long suspected, elicit a large number of changes in the shape, metabolism, and growth behavior of a cell. More generally, this suggested that other cancer-causing genes could also act pleiotropically. Accordingly, if a transformed, tumorigenic cell differed from a normal cell in 20 or 30 distinct traits, perhaps these multiple changes were not dependent on the alteration of 20 or 30 different genes; instead, maybe a small number of master control genes would suffice to transform a normal cell into a tumorigenic cell.

Third, RSV and its v-*src* oncogene might represent a model for the behavior of other types of retroviruses that were similarly capable of transforming infected cells *in vitro* and inducing tumors *in vivo*. Perhaps such retroviruses had acquired other cellular genes unrelated to *src*. While c-*src* was certainly the first cellular proto-oncogene to be discovered, maybe other cellular proto-oncogenes were hiding in the vertebrate cellular genome, waiting to be picked up and activated by some passing retrovirus.

3.10 The vertebrate genome carries a large group of proto-oncogenes

An accident of history—an encounter in 1909 between a Long Island chicken farmer and Peyton Rous—made RSV the first tumorigenic retrovirus to be isolated and characterized in detail. Consequently, RSV was favored initially with the most detailed molecular and genetic analysis. However, in the 1950s and 1960s, a group of other chicken and rodent tumor viruses were found that were subsequently realized to be members, like RSV, of the retrovirus class.

The diversity of these other transforming retroviruses and the diseases that they caused suggested that they, like RSV, might be carrying kidnapped cellular proto-oncogenes and might be using these acquired genes to transform infected cells. So, within a year of the discovery of v-*src* and c-*src*, the race began to find additional viruses that had traveled down a similar genetic path and picked up other, potentially interesting proto-oncogenes.

Another chicken retrovirus—the MC29 myelocytomatosis virus—which was known to be capable of inducing a malignancy of blood-forming cells in chickens, was one of this class. MC29 was also found to carry an acquired cellular gene in its genome, termed the v-*myc* oncogene, which this virus exploited to induce rapidly growing tumors in infected chickens. As was the case with v-*src*, the origin of the v-*myc* gene could be traced to a corresponding proto-oncogene residing in the normal chicken genome. Like *src*, *myc* underwent some remodeling after being incorporated into the retroviral genome. This remodeling imparted potent oncogenic powers to a gene that previously had played a normal physiologic, apparently essential role in the life of normal chicken cells.

Of additional interest was the discovery that MC29, like RSV, descended from avian leukosis virus (ALV). This reinforced the notion that retroviruses like ALV were adept at acquiring random pieces of a cell's genome. The biological powers of the resulting hybrid viruses would presumably depend on which particular cellular genes had been picked up. In the case of the large majority of randomly acquired cellular genes, hybrid viruses carrying such genes would show no obvious phenotypes such as tumor-inducing potential. Only on rare occasion, however, when a growth-promoting cellular gene—a proto-oncogene—happens to be acquired, the hybrid virus might exhibit a cancer-inducing phenotype that could lead to its discovery and eventual isolation by a virologist.

Mammals were also found to harbor retroviruses that are distantly related to ALV and, like ALV, are capable of acquiring cellular proto-oncogenes and converting them into potent oncogenes. Among these is the feline leukemia virus, which acquired the *fes* oncogene in its genome, yielding feline sarcoma virus, and a hybrid rat–mouse leukemia virus, which on separate occasions acquired two distinct proto-oncogenes: the resulting transforming retroviruses, Harvey and Kirsten sarcoma viruses, carry the H-*ras* and K-*ras* oncogenes, respectively, in their genomes. Within a decade, the repertoire of retrovirus-associated oncogenes had increased to more than two dozen, many named after the viruses in which they were originally discovered (**Table 3.3**). By now, more than fifty distinct vertebrate proto-oncogenes have been discovered through various experimental routes, thirty of which were initially identified through their association with a transforming retrovirus. [Many more genes can be classified as proto-oncogenes based on observations that their copy number is increased by gene amplification (see Section 4.2) in many human tumors.]

Table 3.3 Examples of acutely transforming retroviruses and the oncogenes that they have acquired[a]

Name of virus	Viral oncogene	Species	Major disease	Nature of oncoprotein
Hardy–Zuckerman feline sarcoma	kit	cat	sarcoma	steel factor RTK
Simian sarcoma	sis	woolly monkey	sarcoma	PDGF
AKT8	akt	mouse	lymphoma	ser/thr kinase
Avian virus S13	sea	chicken	erythroblastic leukemia[b]	RTK; unknown ligand
Myeloproliferative leukemia	mpl	mouse	myeloproliferation	TPO receptor
Regional Poultry Lab v. 30	eyk	chicken	sarcoma	RTK; unknown ligand
Avian sarcoma virus CT10	crk	chicken	sarcoma	SH2/SH3 adaptor
Avian sarcoma virus 17	jun	chicken	sarcoma	transcription factor
Avian sarcoma virus 31	qin	chicken	sarcoma	transcription factor[c]
AS42 sarcoma virus	maf	chicken	sarcoma	transcription factor
Cas NS-1 virus	cbl	mouse	lymphoma	SH2-dependent ubiquitylation factor
ASV-16	p3k	chicken	hemangiosarcoma	lipid kinase
Abelson leukemia virus	abl	mouse	lymphosarcoma	tyr kinase

[a]Not all viruses that have yielded these oncogenes are indicated here. "Species" denotes the animal species from which the virus was initially isolated.
[b]Also causes granulocytic leukemias and sarcomas.
[c]Functions as a transcriptional repressor.
Abbreviations: PDGF, platelet-derived growth factor; RTK, receptor tyrosine kinase; ser/thr, serine/threonine; SH, src-homology segment; TK, tyrosine kinase; TPO, thrombopoietin.
Adapted in part from S.J. Flint, L.W. Enquist, R.M. Krug et al., Principles of Virology. Washington, DC: ASM Press, 2000. Also in part from G.M. Cooper, Oncogenes. Boston: Jones and Bartlett Publishers, 1995.

In each case, a proto-oncogene found in the DNA of a mammalian or avian species was readily detectable in the genomes of all other vertebrates. There were, for example, chicken, mouse, and human versions of c-*myc*, and these genes seemed to function identically in their respective hosts. The same could be said of all the other proto-oncogenes that were uncovered. Soon it became clear that this large repertoire of proto-oncogenes must have been present in the genome of the vertebrate that was the common ancestor of all mammals and birds, and that this group of genes, like most others in the vertebrate genome, was inherited by all of the modern descendant species.

Now, almost half a century later, we realize that these transforming retroviruses provided cancer researchers with a convenient window through which to view the cellular genome and its cohort of proto-oncogenes. Without these retroviruses, the discovery of many proto-oncogenes would have been exceedingly difficult. By fishing these genes out of the cellular genome and revealing their latent powers, these viruses catapulted cancer research forward by decades.

3.11 Slowly transforming retroviruses activate proto-oncogenes by inserting their genomes adjacent to these cellular genes

As described above, each of the various tumorigenic retroviruses arose when a nontransforming retrovirus, such as avian leukosis virus (ALV) or murine leukemia virus (MLV), acquired a proto-oncogene from the genome of an infected host cell. The result, in each case was an "acutely transforming" retrovirus whose transforming powers could usually be observed by its effects on monolayer cultures of infected cells (see Figure 3.4) and on virus-infected animal hosts, which developed tumors rapidly after infection. In fact, the "nontransforming" precursor viruses could also induce cancers, but they were able to do so only on a much more extended timetable; often months passed before these viruses succeeded in producing cancers. The oncogene-bearing acutely transforming retroviruses, in contrast, often induced tumors within days or weeks after they were injected into host animals.

When the rapidly transforming retroviruses were used to infect cells in culture, the cells usually responded by undergoing changes in morphology and growth behavior that typify the behavior of cancer cells (see Table 3.2). In contrast, when the slowly tumorigenic, nontransforming viruses, such as ALV and MLV, infected cells, these cells released progeny virus particles but did not show any apparent changes in shape or growth behavior. This lack of change in cell phenotypes was consistent with the fact that these viruses appeared to lack oncogenes in their genomes.

These facts, when taken together, represented a major puzzle. How could viruses like MLV or ALV induce a malignancy if their genomes carried no oncogenes?

The solution came in 1981 from study of the leukemias that ALV induced in chickens, more specifically, from detailed analysis of the genomic DNAs of the leukemic cells. These cells invariably carried copies of the ALV provirus integrated into their genomes. By the time these experiments began, a decade after Temin and Baltimore's discovery, it had become clear that the integration of retroviral proviruses occurs at myriad sites throughout the chromosomal DNA of infected host cells. Given the size of the chicken genome, there might be many millions of distinct chromosomal sites used by ALV to integrate its provirus; moreover, each infected leukemia cell might carry one or several proviruses in its genome. Knowing all this, researchers nevertheless undertook to map the genomic sites in ALV-induced leukemias where ALV proviruses had integrated.

This molecular analysis of a series of ALV-induced leukemias, each arising independently in a separate chicken, revealed a picture very different from the expected collection of random integration sites. In a majority (>80%) of these leukemia cell genomes, the ALV provirus was found to be integrated into the chromosomal DNA immediately adjacent to or within the c-*myc* proto-oncogene (**Figure 3.17A**)! This observation, on its own, was difficult to reconcile with the notion that provirus integration occurs randomly at millions of sites throughout the genomes of infected cells.

It soon became clear that the close physical association of an integrated proviral genome and the cellular *myc* proto-oncogene led to a functional link between these two genetic elements. The viral transcriptional promoter and enhancer (see Section 1.8), nested within the ALV provirus, overrode and disrupted the control mechanisms that normally govern expression of the c-*myc* gene (see Figure 3.17A). Now, instead of being regulated by its own native gene promoter, the cellular *myc* gene was placed directly under viral transcriptional control. As a consequence, rather than being regulated up and down by the finely tuned control circuitry of the cell, c-*myc* expression was taken over by a foreign usurper that drove its expression unceasingly and at a high rate. In essence, this hybrid viral–cellular gene residing in the chromosomes of leukemic cells now functioned much like the v-*myc* oncogene carried by avian myelocytomatosis virus.

Suddenly, all the clues needed to solve the puzzle of **leukemogenesis** (leukemia formation) by ALV fell into place. The solution went like this. During the course of infecting a chicken, ALV spread to thousands, then millions of cells in the hematopoietic system of this bird. Soon, the infection was so successful that the bird would become **viremic**, that is, its bloodstream carried high concentrations of virus particles. Each of the tens of millions of infections occurring within this bird resulted in the insertion of an ALV provirus at some random location in the genome of an infected cell. In the vast majority of cases, this provirus integration had no effect on the infected host cell, aside from forcing the host to produce large numbers of progeny virus

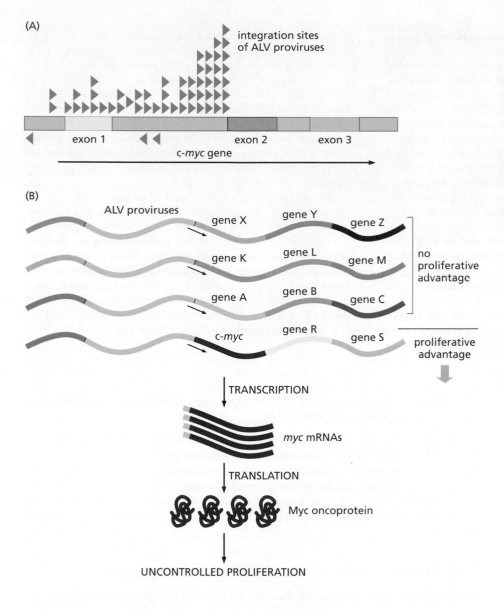

Figure 3.17 Insertional mutagenesis The oncogenic actions of viruses, such as avian leukosis virus (ALV), that lack acquired oncogenes could be explained by the integration of their proviral DNA adjacent to a cellular proto-oncogene. (A) Analysis of numerous B-cell lymphomas that were induced in chickens by ALV revealed that a large proportion of the ALV proviruses were integrated (*filled blue triangles*) into the chromosomal DNA segment carrying the c-*myc* proto-oncogene, usually in the same transcriptional orientation as that of the c-*myc* gene; the majority were integrated between the first noncoding exon of c-*myc* and the second exon, in which the *myc* reading frame begins. (B) ALV's oncogenic behavior could be rationalized as follows. In the course of ALV infection of chicken lymphocytes, ALV proviruses (*green*) become integrated randomly at millions of different sites in the chromosomal DNA of these cells. (Chromosomal DNA maps of only four are illustrated schematically here.) On rare occasions, an ALV provirus becomes integrated (by chance) within the c-*myc* proto-oncogene (*red*). This may then cause transcription of the c-*myc* gene to be driven by the strong, constitutively acting ALV promoter, notably the one found at the 3′ (*right*) end of the provirus (see Figure 3.13). Because high levels of the Myc protein are potent in driving cell proliferation, the cell carrying this particular integrated provirus and activated c-*myc* gene will now multiply uncontrollably, eventually spawning a large host of descendants that will constitute a lymphoma. [A, adapted from J.R. Nevins and P.K. Vogt, "Cell Transformation by Viruses" in *Fields Virology*, Third Ed. (Lippincott-Raven, 1996). With permission from Wolters Kluwer.]

particles. But on rare occasions, perhaps in 1 out of 10 million infections, a provirus became integrated by chance next to the c-*myc* gene (Figure 3.17B). This jackpot event led to an explosive outcome—conversion of the resident c-*myc* gene into a potent oncogene whose unceasing expression was now driven by the adjacently integrated provirus and its transcriptional promoter. The rare cell carrying this deregulated *myc* gene then initiated a program of uncontrolled proliferation, and within weeks, some of the progeny cells evolved further into more aggressive cancer cells that constituted a leukemia. It was this neoplasia that alerted experimenters to the likelihood of an infection by a retrovirus.

This scenario explains the slow kinetics with which these leukemias arise after initial viral infection of a bird. Since activation of the c-*myc* gene through provirus integration is a low-probability event, many weeks and many millions of infectious events are required before these malignancies are triggered. This particular mechanism of proto-oncogene activation came to be called **insertional mutagenesis**; it explains, as well, the leukemogenic powers of other slowly acting retroviruses, such as MLV. By now, study of avian and murine retrovirus-induced infections has demonstrated integration events occurring next to more than 25 distinct cellular proto-oncogenes. Indeed, insertional mutagenesis can be used as a powerful strategy to find new proto-oncogenes (**Sidebar 3.4**; see Table 3.4). Moreover, insertional mutagenesis has had

Sidebar 3.4 Insertional mutagenesis uncovers novel proto-oncogenes As described earlier (see Section 3.10), the analysis of the genomes of rapidly transforming retroviruses enabled investigators to identify a large cohort of proto-oncogenes. The phenomenon of insertional mutagenesis, first discovered through the insertion of an ALV genome adjacent to the c-*myc* proto-oncogene, offered an alternative strategy for discovering these cellular genes. Thus, a researcher could study a series of independently arising tumors, all of which had been induced by a retrovirus, such as ALV or MLV, that was known to lack its own oncogene. More specifically, this researcher could analyze the host-cell sequences that lay immediately adjacent to the integrated proviruses in the chromosomal DNA of the virus-infected, transformed cells. The hope was that the proviruses might be found to be integrated repeatedly next to a (possibly still-unknown) cellular gene whose activation was triggered by the transcriptional promoter of the provirus. The adjacent gene could be readily cloned, since it was effectively tagged by the closely linked proviral DNA.

The initial fruits of this strategy came from studying the breast cancers induced by mouse mammary tumor virus (MMTV), another retrovirus that lacked an oncogene in its genome. Researchers mapped the integration sites of MMTV proviruses in the genomes of mouse breast cancers that had been induced by this virus. Most of the proviruses were found to be integrated in one of three chromosomal locations, clustering next to cellular genes that were then called *int-1*, *int-2*, and *int-3* (Table 3.4). Each of these genes was later discovered to encode a protein involved in stimulating cell proliferation in one way or another. The deregulated expression of each of these genes, resulting from nearby MMTV provirus integration, seemed to be responsible for triggering the cell proliferation that led to the appearance of mammary tumors.

The *int-1* gene, which was found to be homologous to the *wingless* gene of *Drosophila*, was renamed *Wnt-1*, and was the forerunner of a whole series of *Wnt* genes that have proven to be important vertebrate mitogens and **morphogens**, that is, factors controlling morphogenesis. This set of genes and the associated Wnt pathway is upregulated in many types of human cancer (see Chapters 5 and 6). More recently, this search strategy has uncovered a large group of other cellular genes, each of which, when activated by insertional mutagenesis mediated by MLV, triggers leukemias in mice (see Table 3.4).

implications for the development of new approaches to curing disease using gene therapy (see Supplementary Sidebar 3.8).

Unvoiced by those who uncovered insertional mutagenesis was another provocative idea: maybe it was possible that nonviral carcinogens could achieve the same end result as ALV did. Perhaps these other carcinogens, including X-rays and mutagenic chemicals, could alter cellular proto-oncogenes while these genes resided in their normal sites in cellular chromosomes. The result might be a disruption of cellular growth control that was just as destabilizing as the events that led to ALV-induced leukemias.

3.12 Some retroviruses naturally carry oncogenes

The descriptions of retroviruses in this chapter indicate that they essentially fall into two classes. Some, such as ALV and MLV, carry no oncogenes but can induce tumors that erupt only after a long latent period (that is, many weeks) following the initial infection of a host animal. Other viruses, such as RSV, can induce cancer rapidly (that is, in days or several weeks), having acquired an oncogene from a cellular proto-oncogene precursor.

In reality, there is a third class of retroviruses that conforms to neither of these patterns. Most prominent among them is human T-lymphotropic leukemia virus type 1 (HTLV-I), which chronically infects about 1% of the inhabitants of Kyushu, the south island of Japan; in Japan as a whole about one thousand people die each year from HTLV-I–induced leukemias. An endemic infection is also present, albeit at a lower rate, in some islands of the Caribbean. Lifelong HTLV-I infection carries a 3–4% risk of developing adult T-cell leukemia, and the virus seems to be maintained in the population largely via milk-borne, mother-to-infant transmission.

There are no indications, in spite of extensive molecular surveys, that HTLV-I provirus integration sites are clustered in certain chromosomal regions. Accordingly, it appears highly unlikely that HTLV-I uses insertional mutagenesis to incite leukemias. Instead, its leukemogenic powers seem to be traceable to one or more viral proteins that are naturally encoded by the viral genome. The best understood of these is the viral Tax protein, whose product is responsible for activating transcription of proviral DNA

Table 3.4 Examples of cellular genes found to be activated by insertional mutagenesis

Gene	Insertional mutagen	Tumor type	Species	Type of oncoprotein
Myc	ALV	B-cell lymphoma	chicken	transcription factor
Myc	ALV, FeLV	T-cell lymphoma	chicken, cat	transcription factor
Nov	ALV	nephroblastoma	chicken	growth factor
erbB	ALV	erythroblastosis	chicken	receptor TK
mos	IAP	plasmacytoma	mouse	ser/thr kinase
int-1[a]	MMTV	mammary carcinoma	mouse	growth factor
int-2[b]	MMTV	mammary carcinoma	mouse	growth factor
int-3	MMTV	mammary carcinoma	mouse	receptor[c]
int-H/int-5	MMTV	mammary carcinoma	mouse	enzyme[d]
pim-1	Mo-MLV	T-cell lymphoma	mouse	ser/thr kinase
pim-2	Mo-MLV	B-cell lymphoma	mouse	ser/thr kinase
bmi-1	Mo-MLV	T-cell lymphoma	mouse	transcription repressor
tpl-2	Mo-MLV	T-cell lymphoma	mouse	ser/thr kinase
lck	Mo-MLV	T-cell lymphoma	mouse	non-receptor TK
p53	Mo-MLV	T-cell lymphoma	mouse	transcription factor
GM-CSF	IAP	myelomonocytic leukemia	mouse	growth factor
IL2	GaLV	T-cell lymphoma	gibbon ape	cytokine[e]
IL3	IAP	T-cell lymphoma	mouse	cytokine
K-ras	F-MLV	T-cell lymphoma	mouse	small G protein
CycD1	F-MLV	T-cell lymphoma	mouse	G1 cyclin
CycD2	Mo-MLV	T-cell lymphoma	mouse	G1 cyclin

Abbreviations: ALV, avian leukosis virus; FeLV, feline leukemia virus; F-MLV, Friend murine leukemia virus; GaLV, gibbon ape leukemia virus; IAP, intracisternal A particle (a retrovirus-like genome that is endogenous to cells); Mo-MLV, Moloney murine leukemia virus; MMTV, mouse mammary tumor virus; ser/thr, serine/threonine; TK, tyrosine kinase.
[a]Subsequently renamed *Wnt-1*.
[b]Subsequently identified as a gene encoding a fibroblast growth factor (FGF).
[c]Related to Notch receptors.
[d]Enzyme that converts androgens to estrogens.
[e]Cytokines are growth factors that largely regulate various types of hematopoietic cells.
Adapted in part from J. Butel, *Carcinogenesis* 21:405–426, 2000; and from N. Rosenberg and P. Jolicoeur, in J.M. Coffin, S.H. Hughes and H.E. Varmus (eds.), Retroviruses. Cold Spring Harbor, NY: Cold Spring Harbor Laboratory Press, 1997. Also in part from G.M. Cooper, Oncogenes, 2nd ed. Boston: Jones and Bartlett Publishers, 1995.

sequences, thereby enabling production of progeny RNA genomes. At the same time, the *tax* gene product—the Tax **oncoprotein**—appears to activate two cellular signaling pathways involving NF-κB and PI3 kinase (Chapter 6). These pathways proceed to stimulate the proliferation of T lymphocytes (Chapter 15) and to suppress apoptosis (Chapter 9). While such induced proliferation, on its own, does not directly create a leukemia, it seems that populations of these HTLV-I–stimulated cells may progress at a low but predictable frequency to spawn variants that are indeed neoplastic. In this instance, the expression of certain viral oncogenes, notably tax, appears to be an intrinsic and essential component of a retroviral replication cycle within host animals, rather than the consequence of rare genetic accidents that yield unusual hybrid genomes, such as the genome of RSV. Jaagsiekte sheep retrovirus represents a second example of an acutely transforming retrovirus that carries a viral oncogene exhibiting

Figure 3.18 Perturbation of intracellular signaling pathways by *H. pylori* CagA strains of *Helicobacter pylori* carry various versions of the *CagA* pathogenicity island (PAI) in their genomes; those with *CagA*-associated alleles that are especially virulent are associated with greatly increased incidence of gastric adenocarcinomas. (A) A bacterial micro-syringe functioning as a type IV secretion system (T4SS) forms a channel in the plasma membrane of host gastric epithelial cells through which it delivers the *H. pylori* CagA protein into the cytoplasm of these cells. PAR1 (partitioning-defective protein 1) is an important regulator of the polarity of an epithelial cell. The inhibition of PAR1 leads to loss of epithelial cell polarity, whereas the activation of Erk signaling (see Section 6.5) stimulates cell proliferation. These *H. pylori*–induced changes, together with other factors, such as those of dietary origin, lead to gastritis (gastric epithelial inflammation) and, with some frequency, eventual formation of gastric adenocarcinomas. SHP2 is a phosphotyrosine phosphatase involved in regulating mitogenic signals downstream of growth factor receptors, as described in Chapter 6. (B) The biochemical changes elicited by CagA within cultured gastric carcinoma cells create the characteristic "hummingbird" morphological phenotype in these cells, in this case forced to express a virulent form of CagA (detected by immunofluorescence; *green*); F-actin cytoskeletal filaments are labeled with phalloidin (*red*). Indeed, this unique cell morphology is, on its own, a signature of the actions of the CagA protein. (A, courtesy of M. Hatakeyama. B, Republished with permission of American Society for Biochemistry & Molecular Biology, from Role of Partitioning defective 1/Microtubule Affinity-regulating Kinases in the Morphogenetic Activity of Helicobacterpylori CagA, Huaisheng Lu, Naoko Murata-Kamiya, Yasuhiro Saito, Masanori Hatakeyama, 2009 Aug 21; 284(34):23024-36. doi: 10.1074/jbc .M109.001008; permission conveyed through Copyright Clearance Center, Inc.)

no apparent homology with a normal cellular gene; this poorly studied virus deploys the glycoprotein product of its *env* gene (see Figure 3.13) to transform infected cells, leading to contagious pulmonary carcinomas that spread through flocks of sheep.

3.13 Bacterial cancers

This chapter has focused on tumor viruses for two reasons: They opened the window on discovering the genes in our genomes that, in mutant form, can function to drive cancer cell proliferation. In addition, they are responsible for almost 20% of lethal cancers worldwide. [Most virus-induced cancers derive from infections by hepatitis B and C viruses (~600,000 deaths), which produce liver cancers after years of inducing chronic inflammation in this organ; as well as human papillomaviruses, whose genomes encode viral oncoproteins (~610,000 deaths.)] However, another class of infectious agents—bacteria—seem to drive the formation of a considerable proportion of a common cancer: gastric carcinomas. Worldwide, almost 1 million cases were diagnosed in 2012, three-quarters of which were likely to prove lethal. Indeed, while stomach carcinoma mortality rates are declining, these tumors still represent the third leading cause of cancer-associated death. It is estimated that at least 75% of these deaths are due to chronic infection by pathogenic strains of the common bacterium *Helicobacter pylori*. In Japan, >99% of cases of gastric carcinomas are associated with infection by some specific strains of *H. pylori*.

H. pylori actually causes gastric carcinomas through two mechanisms. Like hepatitis B and C viruses, it establishes chronic, long-term infections and associated inflammation; the latter condition serves to incubate and drive cancer formation, a topic discussed in more detail in Chapter 11. In addition, certain strains of *H. pylori* carry a 40-kb stretch of genes, termed the *Cag* pathogenicity island (PAI), which encodes ~30 protein products that enable the insertion of the CagA bacterial protein—also encoded by this PAI—into cells lining the stomach (**Figure 3.18**; Supplementary Sidebar 3.9). Once inside the cell cytoplasm, the introduced CagA protein participates in the deregulation of a series of cellular signals and thereby confers multiple cell-biological changes on these cells, including changes in cell shape, disruption of junctions formed between affected cells and their neighbors, invasiveness, enhanced proliferation, inhibition of apoptosis,

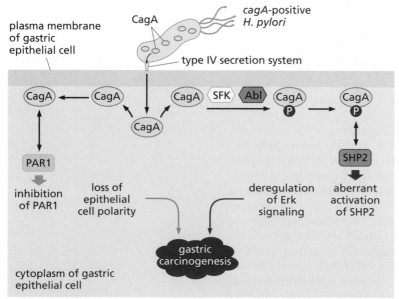

(A)

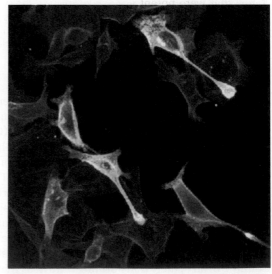

(B)

and activation of gastric stem cells. Each of these changes in cellular phenotype will be addressed in later chapters of this book. For the moment, we note that these multiple changes in cell phenotype are, strikingly, precisely the ones that are associated with shifts occurring during the cell transformation orchestrated by well-characterized oncoproteins, such as those cited in this chapter. Hence, by these criteria, the bacterial CagA protein can be considered a bona fide pleiotropically acting oncoprotein.

Unanswered by these descriptions is the rationale of why certain *H. pylori* strains carry highly pathogenic versions of the *CagA* gene, that is, what proliferative advantage do these bacteria gain by inducing various changes in the cells lining the stomach? The answer seems to derive from the fact that CagA-activated gastric epithelium appears to offer a tissue microenvironment that is more hospitable to ongoing bacterial survival, shielding these cells from the contents of the hostile, highly acidic gastric lumen.

The incubation time of gastric adenocarcinomas is generally more than five decades, since infection generally occurs very early in life whereas tumors usually appear in individuals over the age of 50. While *H. pylori* infections are common worldwide, only those strains of the bacterium harboring the *Cag* PAI are thought to be highly effective in triggering gastric cancer. Of additional interest is the fact that CagA function, while being critical to the formation of many gastric carcinomas, is no longer required once these tumors erupt. Hence, it operates a "hit-and-run" oncoprotein, unlike Src, whose continued presence is required to maintain the transformed cell phenotype (see Section 3.3).

Finally, bacteria harboring the most virulent alleles of the *Cag* PAI are commonly present in East Asia while being relatively uncommon in the Western Hemisphere and Europe; these differences seem to explain the extraordinarily high rate of gastric cancer in certain parts of East Asia compared with the rest of the world. Importantly, the pathogenesis of most gastric cancers appears to depend on the collaborative effects of *H. pylori* infection with certain co-factors, such as dietary components. The bacterium seems to be spread largely through **vertical transmission** (from parent to offspring). Of additional interest, a version of the *H. pylori*–associated CagA protein that was present in the Western Hemisphere prior to European colonization shows genetic affinity with CagA proteins of East Asia, suggesting a chain of transmission extending over the >12,000 years since the trans-Siberian migration of the ancestors of Amerindians.

3.14 Synopsis and prospects

By studying tumors in laboratory and domesticated animals, cancer biologists discovered a wide array of cancer-causing viruses during the twentieth century. Many of these viruses, having either DNA or RNA genomes, were found able to infect cultured cells and transform them into tumorigenic cells. These transforming powers pointed to the presence of powerful oncogenes in the genomes of the viruses, indeed, oncogenes that were potent enough to induce many of the phenotypes associated with cancer cells (see Table 3.2). Moreover, the ability of these viruses to create transformed cells in the culture dish shed light on the mechanisms by which such viruses could induce cancers in the tissues of infected host animals.

A major conceptual revolution came from the detailed study of RNA tumor viruses, specifically, Rous sarcoma virus (RSV). Its oncogene, termed v-*src*, was found to have originated in a normal cellular gene, c-*src*. This discovery revealed the ability of nontransforming, slowly tumorigenic retroviruses, such as ALV (avian leukosis virus), to acquire certain normal cellular genes and convert these captured genes into potently transforming oncogenes. The hybrid viruses that arose following these genetic acquisitions were now able to rapidly induce tumors in infected hosts.

Even more important were the implications of finding the c-*src* gene. Its presence in uninfected cells demonstrated that the normal cellular genome carries a proto-oncogene that can be converted into an oncogene by altering its sequences. (The details of these alterations will be described in the next several chapters.) Soon a number of retroviruses of both avian and mammalian origin were discovered to carry other oncogenes that had been acquired in similar fashion from the genomes of infected cells. While each of these proto-oncogenes was found initially in the genome of one or another vertebrate species, we now know that all of these genes are represented in the genomes of all mammals and likely all vertebrates. Consequently, the generic vertebrate genome carries dozens of such normal genes, each of which has the potential to become converted into an active oncogene.

Yet other proto-oncogenes were discovered by studying the integration sites of proviruses in the genomes of tumors that had been induced by nontransforming retroviruses, such as murine leukemia virus (MLV) and ALV. The random integration of these proviruses into chromosomal DNA occasionally yielded, through the process of insertional mutagenesis, the conversion of a proto-oncogene into an activated oncogene that could readily be isolated because of its close physical linkage in the cell genome to the provirus. On many occasions, insertional mutagenesis led to rediscovery of a proto-oncogene that was already known because of its presence in an acutely transforming retrovirus; *myc* and avian myelocytomatosis virus (AMV) exemplify this situation. On other occasions, entirely novel proto-oncogenes were discovered through study of provirus integration sites; the *int-1* gene activated by mouse mammary tumor virus (MMTV) provides a striking example of this route of discovery. This gene (the prototype of a class of genes now termed *Wnts*) is now known to encode a key component of a cellular signaling pathway frequently corrupted in human cancers. In fact, the process of insertional mutagenesis remained little more than a laboratory curiosity, of interest to only a small cadre of cancer biologists, until it was reported to lead to leukemias in several children being treated by gene therapy (Supplementary Sidebar 3.9).

The discoveries of proto-oncogenes and oncogenes, as profound as they were, provoked as many questions as they answered. It remained unclear how the retrovirus-encoded oncogene proteins (that is, oncoproteins) differed functionally from the closely related proteins encoded by corresponding proto-oncogenes. The biochemical mechanisms used by these oncoproteins to transform cells were also obscure.

The biochemical mechanisms used by DNA tumor viruses to transform infected cells were even more elusive, since these viruses seemed to specify oncoproteins that were very different from the proteins normally made by their host cells. Such differences suggested that these viral oncoproteins could not insinuate themselves into the cellular growth-regulating machinery in any easy, obvious way. Only in the mid-1980s, ten years after this research began, did their transforming mechanisms become apparent, as we will see in Chapters 8 and 9.

For many cancer researchers, and for the public that supported this research, there was a single overriding issue that had motivated much of this work in the first place: did any of these viruses and the proto-oncogenes that they activated play key roles in causing human cancers? In fact, about one-fifth of the human cancer burden worldwide is associated with infectious agents. Hepatitis B and C viruses (HBV, HCV), Epstein–Barr virus (EBV), human papillomaviruses (HPVs), and *H. pylori* play key roles in triggering commonly occurring cancers. Indeed, even infrequently occurring human tumors that seem to be familial have been traced in recent years to viral infections (see Supplementary Sidebar 3.5). So, this recognized role of viruses in cancer pathogenesis is substantial and growing. Still, these various infectious agents—now understood to cause many cancers throughout the world—do not include the retroviruses and the DNA tumor viruses, such as SV40, that were intensively studied in the 1970s. Hence, from the perspective of improving public health, was the research described in this chapter truly justified?

The response to this question is a resounding yes. Even if the tumor viruses described were not responsible for inciting a single case of human cancer, the contributions of this research to understanding the mechanisms of human cancer development were clear: this research opened the curtain on the genes in our genome that play central roles in all types of human cancer. It accelerated by decades our understanding of cancer pathogenesis at the level of genes and molecules. It catapulted cancer research from a descriptive science into one where complex phenomena could finally be understood and explained in precise, mechanistic terms.

Key concepts

- Decades after Peyton Rous's 1910 discovery that a virus could induce tumors in chickens, the transformation of cultured cells into tumor cells upon infection with Rous sarcoma virus (RSV) resurrected tumor virus research and led to the realization that cancer could be studied at the level of the cell.

- Transformed cells in culture have numerous unusual characteristics, including altered morphology, lack of contact inhibition, anchorage independence, proliferation in the absence of growth factors, immortalization, and tumorigenicity, the latter being the acid test for full cellular transformation.

- The transformation phenotype induced by RSV infection was found to be transmitted to progeny cells and to depend on the continued activity of an RSV gene product.

- In addition to RNA viruses like RSV, several classes of DNA viruses—including the papillomavirus, polyomavirus, human adenovirus, herpesvirus, and poxvirus—were found to induce cancers in laboratory animals or humans.

- While the genomes of RNA tumor viruses consist of single-stranded RNA, the genomes of DNA tumor viruses consist of double-stranded DNA (dsDNA).

- Since replication of viral DNA genomes normally occurs independently of the host cells' DNA and since viral genomes lack the DNA segments (centromeres) to properly segregate during mitosis, the transmission of DNA tumor virus genomes from one cell generation to the next posed a conceptual problem, until it was discovered that DNA tumor virus genomes integrate into host-cell chromosomal DNA.

- Since the genomes of RNA tumor viruses consist of single-stranded RNA that cannot be incorporated into host DNA, and since re-infection does not explain the persistence of the transformed state in descendant cells, Howard Temin postulated that RNA viruses make double-stranded DNA copies of their genomes—the process of reverse transcription—and that these DNA copies are integrated into the host's chromosomal DNA as a part of the normal viral replication cycle. This represents a major distinction from most DNA tumor viruses, for which integration of their genomes into chromosomal DNA is a very rare, haphazard event and not an integral part of viral replication.

- Because their replication cycle depends on information flowing backward (that is, from RNA to DNA), RNA viruses came to be called retroviruses and the DNA version of their viral genomes was called a provirus.

- Working with RSV, researchers found that viral replication and cell transformation were specified by separate genes, with the transforming function residing in a single gene called *src*.

- Use of a DNA probe that specifically recognized the transformation-associated (that is, *src*) sequences of the RSV genome led to the unexpected discovery that *src*-related sequences were present in the DNA of uninfected chicken cells. Further research indicated that the *src* gene was a normal, highly conserved gene of all vertebrate species (as later proved true of many other such genes).

- The difference between the actions of the cellular version of *src* (c-*src*), which supports normal cell function, and the viral version (v-*src*), which acts as an oncogene, could be explained if v-*src* were an altered version of a c-*src* that had originally been kidnapped from a cellular genome by an ancestor of RSV and thereafter reshaped into an active oncogene.

- Because it can serve as a precursor to an oncogene, c-*src* was called a proto-oncogene, a term implying that normal vertebrate cells contain genes that have the intrinsic potential to become converted into oncogenes that can induce cancer.

- The actions of v-*src* indicated that a single oncogene could act pleiotropically to evoke a multiplicity of changes in cellular traits. In addition, the discovery of c-*src* in the normal cellular genome suggested the possibility that other mutational mechanisms, notably those not involving viruses, might activate proto-oncogenes that continued to reside in their normal sites in cellular chromosomes.

- In contrast to RSV, nontransforming, slowly tumorigenic retroviruses lack their own oncogenes and occasionally cause tumors by integrating their genomes adjacent to proto-oncogenes in the chromosomal DNA of infected host cells, a process called insertional mutagenesis. This chance occurrence places the proto-oncogene under the control of the viral transcriptional promoter, which deregulates the gene's expression and leads to uncontrolled cell proliferation. Insertional mutagenesis can also be exploited to find new proto-oncogenes.

- The formation of most oncogene-bearing, rapidly transforming retroviruses involves the replacement of viral replicative genes by acquired cellular oncogenic sequences. This results in the formation of a potently transforming virus that cannot proliferate on its own due to the loss of viral genes that are essential for replication.

- In addition to the nontransforming retroviruses (which work via insertional mutagenesis) and the acutely transforming ones (which work via acquired oncogenes), retroviruses exist whose carcinogenic powers are traceable to their own viral gene products. Needed for viral replication, these viral proteins have the side effect of also activating the expression of cellular genes involved in cell growth and proliferation.

- Stomach cancer, which is responsible for almost a million deaths internationally each year, is often provoked by strains of *H. pylori* that inhabit the gastric mucosa and inject the CagA bacterial protein into the cytoplasm of gastric epithelial cells, evoking a series of transformation-related phenotypes in these cells that set the stage for the formation of full-blown stomach cancers.

Thought questions

1. What observations favor or argue against the notion that cancer is an infectious disease?

2. How can one prove that tumor virus genomes must be present in order to maintain the transformed state of a virus-induced tumor? What genetic mechanisms, do you imagine, might enable this process to become "hit-and-run," in which the continued presence of a tumor virus is not required to maintain the tumorigenic phenotype after a certain time?

3. Why are oncogene-bearing viruses like Rous sarcoma virus so rarely encountered in wild populations of chickens?

4. What evidence suggests that the phenotypes of cells transformed by tumor viruses *in vitro* reflect comparable phenotypes of tumor cells *in vivo*?

5. What logic suggests that the chromosomal integration of tumor virus genomes is an intrinsic, obligatory part of the replication cycle of RNA tumor viruses but an inadvertent side product of DNA tumor virus replication?

6. What evidence suggests that a proto-oncogene like c-*src* is actually a normal cellular gene rather than a gene that has been inserted into the germ line by an infecting retrovirus?

7. How do you imagine that DNA tumor viruses and retroviruses like avian leukosis virus arose in the distant evolutionary past?

8. Why do retroviruses like avian leukosis virus take so long to induce cancer?

Additional reading

Aguayo F, Muñoz JP, Perez-Dominguez F et al. (2020) High-risk human papillomavirus and tobacco smoke interactions in epithelial carcinogenesis. *Cancers (Basel)* 12, E2201. doi: 10.3390/cancers12082201.

Brower V (2004) Connecting viruses to cancer: how research moves from association to causation. *J. Natl. Cancer Inst.* 96, 256–257.

Butel J & Fan H (2012) The diversity of human cancer viruses. *Curr. Opin. Virol.* 3, 449–452.

Carbone M, Pass HI, Miele L & Bocchetta M (2003) New developments about the association of SV40 with human mesothelioma. *Oncogene* 22, 5173–5180.

Coffin JM, Hughes SH & Varmus HE (eds) (1997) Retroviruses. Cold Spring Harbor, NY: Cold Spring Harbor Laboratory Press.

Cover TL (2016) *Helicobacter pylori* diversity and gastric cancer risk. *mBio* 7:e01869-15.

Coffin JM & Fan H (2016) The discovery of reverse transcriptase. *Annu. Rev. Virol.* 3, 29–51.

de Martel C, Ferlay J, Franceschi S et al. (2012) Global burden of cancers attributable to infections in 2008: a review and synthetic analysis. *Lancer Oncol.* 13, 607–615.

DiMaio D & Liao JB (2006) Human papillomaviruses and cervical cancer. *Adv. Virus Res.* 66, 125–159.

Drost J and Clevers H (2018) Organoids in cancer research. *Nat. Rev. Cancer* 18, 407–418.

Flint SJ, Enquist LW, Krug RM et al. (2000) Principles of Virology. Washington, DC: ASM Press.

Gaglia MM & Munger K (2018) More than just oncogenes: mechanisms of tumorigenesis by human cancer viruses. *Curr. Opin. Virol.* 32, 48–59.

Howley PM & Livingston DM (2009) Small DNA tumor viruses: large contributors to biomedical sciences. *Virology* 384, 256–259.

Hsu JL & Glaser SL (2000) Epstein-Barr virus-associated malignancies: epidemiologic patterns and etiologic implications. *Crit. Rev. Oncol. Hematol.* 34, 27–53.

Javier RT & Butel JS (2008) The history of tumor virology. *Cancer Res.* 68, 7693–7706.

Martin D & Gutkind JS (2008) Human tumor-associated viruses and new insights into the molecular mechanisms of cancer. *Oncogene* 27, S31–S42.

Mesri E, Feitelson MA & Munger K (2014) Human viral oncogenesis: a cancer hallmark analysis. *Cell Host & Microbe* 15, 266–282.

Moore PS & Chang Y (2010) Why do viruses cause cancer? Highlights of the first century of human tumor virology. *Nat. Rev. Cancer* 10, 878–889.

Parsonnet J (ed) (1999) Microbes and Malignancy: Infection as a Cause of Human Cancers. Oxford, UK: Oxford University Press.

Phillips AC & Vousden KH (1999) Human papillomavirus and cancer: the viral transforming genes. *Cancer Surv.* 33, 55–74.

Pisani P, Parkin DM, Muñoz N & Ferlay J (1997) Cancer and infection: estimates of the attributable fraction in 1990. *Cancer Epidemiol. Biomarkers Prevent.* 6, 387–400.

Taylor GS, Long HM, Brooks JM et al. (2015) The immunology of Epstein-Barr virus-induced disease. *Annu. Rev. Immunol.* 33, 787–821.

Wallace MC, Preen D, Jeffrey GP & Adams LA (2015) The evolving epidemiology of hepatocellular carcinoma: a global perspective. *Expert Rev. Gastroenterol. Hepatol.* 9, 765–779.

Weiss R, Teich N, Varmus H & Coffin J (eds) (1985) Molecular Biology of Tumor Viruses: RNA Tumor Viruses, 2nd ed. Cold Spring Harbor, NY: Cold Spring Harbor Laboratory Press.

zur Hausen H (2009) Papillomaviruses in the causation of human cancers: a brief historical account. *Virology* 384, 260–265.

zur Hausen H (2019) Cancers in humans: a lifelong search for contributions of infectious agents, autobiographic notes. *Annu. Rev. Virol.* 6, 1–28.

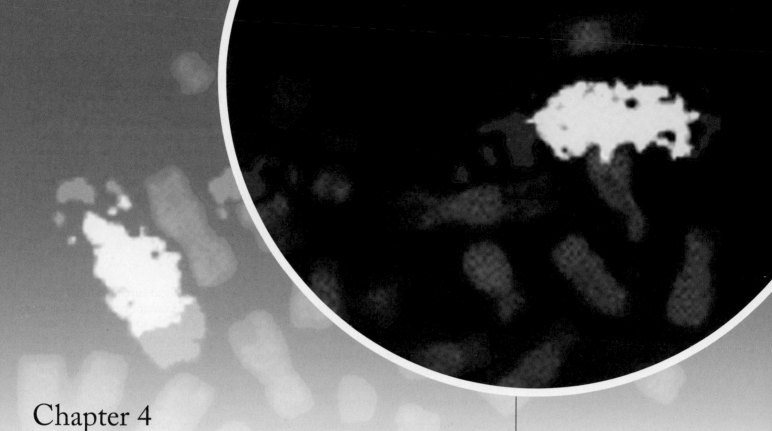

Chapter 4

Cellular Oncogenes

The viral origin of the majority of all malignant tumors ... has now been documented beyond any reasonable doubt. It ... would be rather difficult to assume a fundamentally different etiology for human tumors.

Ludwik Gross, tumor virologist, 1970

The DNA and RNA tumor viruses characterized in the 1970s provided cancer biologists with a simple and powerful theory of how human tumors could arise. Viruses that occurred commonly in the human population might, with some frequency, infect susceptible tissues and cause the transformation of infected cells. These cells, in turn, would begin to multiply and, sooner or later, form the large cell masses that were encountered frequently in the oncology clinic. Since tumor viruses succeeded in transforming normal rodent and chicken cells into tumor cells with only a small number of introduced genes, these viruses might have similar powers in transforming human cells as well.

With the passage of time, this scenario, attractive as it was, became increasingly difficult to reconcile with the biology and epidemiology of human cancer. Most types of human cancer clearly did not spread from one individual to another as an infectious disease. Significant clusters of cancer cases—mini-epidemics of disease—were hard to find. Even more important, attempts undertaken during the 1970s to isolate viruses from most types of human tumors were unsuccessful. Of the hundred and more tumor types encountered in the oncology clinic, only two commonly occurring tumor types in the Western world—**hepatomas** (liver carcinomas) and cervical carcinomas—were known or suspected to be tied to specific viral causative agents.

These realizations evoked two responses. Those who hung tenaciously to tumor viruses as causative agents of all human cancers argued that chemical and physical carcinogens interacted with viruses that normally hid within the body's cells, activating their latent cancer-causing powers. Other researchers responded by jettisoning viruses entirely and began looking at another potential source of the genes responsible for human cancers—the cellular genome with its tens of thousands of genes. This second tack eventually triumphed, and by the late 1980s, the normal human genome was recognized to be a rich source of the genes that drive human cancer cell proliferation.

So, tumor viruses, once viewed as the key agents triggering all human cancers, failed to live up to these high expectations. Ironically, however, tumor virus research proved to be critical in uncovering many of the cellular genes that are indeed responsible for the neoplastic cell phenotype. The large catalog of cellular cancer-causing genes assembled over the ensuing decades—oncogenes and tumor suppressor genes— derives directly from these early efforts to find infectious cancer-causing agents in human populations.

4.1 Transfection of DNA provides a strategy for detecting nonviral oncogenes

In the 1970s some researchers attempted to connect chemical carcinogenesis of the type that Yamagiwa first developed (Section 2.8) with viral carcinogenesis. In particular, they sought an explanation that came from the fact that latent retroviral genomes could be found in abundance in the germ lines of various mammalian species. Such **endogenous retroviral genomes**, they argued, while normally transcriptionally silent, could be activated by various carcinogens, provoking virus production, viremia, and eventually cancer induction (Supplementary Sidebar 4.1). As attractive as this mechanism was, it failed to attract experimental support.

The demise of the endogenous retrovirus theory left one viable theory on the table. According to this theory, carcinogens function as mutagens (see Section 2.9). Whether physical (for example, X-rays) or chemical (for example, tobacco tars), these agents could induce cancer through their ability to mutate critical growth-controlling genes in the genomes of susceptible cells. Such growth-controlling genes might possibly be normal cellular genes, such as the proto-oncogenes discovered by the retrovirologists. Once these genes were mutated, the resulting mutant alleles might thereafter function as active oncogenes, driving the cancerous growth of the cells that carried them. All this could happen without the involvement of tumor viruses and their genomes.

Stated differently, this model—really a speculation—predicted that chemically transformed cells carried mutated cellular genes and that these genes were responsible for programming the aberrant growth of these cells. In fact, the notion that cancer cells were mutant cells traced its origins back to the beginning of the twentieth century and the speculations of the pathologist David von Hansemann and the embryologist Theodor Boveri (Supplementary Sidebar 4.2). It was impossible to predict the number of such mutated genes present in the genomes of these cells. More important, experimental proofs of the existence of these cancer-causing genes represented a daunting challenge. If they were really present in the genomes of chemically transformed cells, including perhaps human tumor cells, how could they possibly be found? If these genes were mutant versions of normal cellular genes, then they were embedded in cancer cell genomes together with tens of thousands, perhaps even a hundred thousand other genes, each present in at least one copy per cell genome. These cancer genes, if they existed, were clearly tiny needles buried in very large haystacks.

To determine whether nonviral oncogenes existed in chemically transformed cells, a novel experimental strategy was devised. It involved introducing DNA (and thus the genes) of cancer cells into normal cells, and then determining whether the recipient cells became transformed in response to the introduced tumor cell DNA. Such transformation could be scored by the appearance of foci in the cultures of recipient cells several weeks after their exposure to tumor cell DNA—essentially the assay that Howard Temin had used to score for the presence of infectious transforming Rous sarcoma virus particles in monolayers of chick embryo fibroblasts (Section 3.2). This strategy depended on several experimental advances, including the development of an effective gene transfer procedure. Additionally, appropriate cancer cells from which to extract DNA and suitable recipient cells needed to be identified.

In 1972, a highly effective gene transfer procedure, soon termed **transfection**, was developed. This procedure, which made possible the introduction of naked DNA

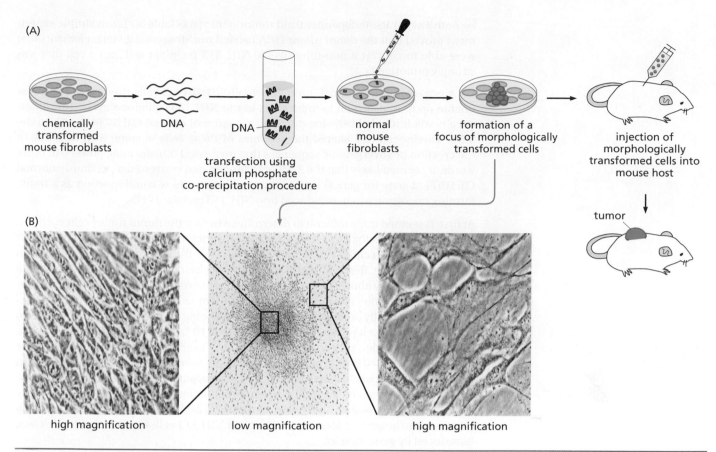

Figure 4.1 Transfection of DNA of transformed cells (A) The procedure of transfection can be used to detect oncogenes in the DNA of cancer cells. DNA is extracted from cancer cells (*pink*) growing in a Petri dish. (For simplicity, the double-stranded DNA is depicted as single lines.) DNA is then introduced into a phosphate buffer. When calcium ions are added, a co-precipitate of DNA and calcium phosphate crystals is formed (*pink and purple*). These crystals are added to a monolayer culture of normal cells (*green*). The calcium phosphate crystals facilitate the uptake of DNA fragments by cells. If a transforming gene (oncogene) is present in the donor DNA, it may become incorporated into the genome of one of the recipient cells and transform the latter. This transformed cell will now proliferate, and its descendants will form a clump (focus, *blue*) of cells that is visible to the naked eye. Injection of these cells into a host mouse and resulting tumor formation can be used to confirm the transformed state of these cells. (B) In addition to the initially used mouse tumor DNAs (panel A), human tumor DNAs could also yield foci of transformants following transfection. Seen at low magnification (*middle*) is a focus generated by transfection of DNA from the T24 human bladder carcinoma cell line. At high magnification the transformed cells within this focus (*left*), like many transformed fibroblasts, are spindle-shaped, refractile, and piled up densely on one another (see also Figure 3.4A). At the same magnification, the cells in the surrounding monolayer of untransformed NIH 3T3 cells (*right*), like normal fibroblasts, have wide, extended cytoplasms and are not piled on one another. (B, from M. Perucho et al., *Cell* 27:467–476, 1981. With permission from Elsevier.)

molecules from donor cells directly into mammalian cells serving as recipients (**Figure 4.1**), was based on co-precipitation of DNA with calcium phosphate crystals. Cells of the NIH 3T3 cell line, derived originally from mouse embryo fibroblasts, turned out to be especially adept at taking up and integrating into their genomes the foreign DNA introduced in this fashion.

Researchers chose donor tumor cells derived from mouse fibroblasts of the C3H10T1/2 mouse cell line; these cells had been treated repeatedly with the potent carcinogen and mutagen 3-methylcholanthrene (3-MC), a known component of coal tars (see Section 2.8), which resulted in their transformation. Importantly, these transformed cells bore no traces of either tumor virus infection or activated endogenous retroviral genomes. Hence, any transforming oncogenes detected in the genome of these donor cells would, with great likelihood, be of cellular origin, that is, mutant versions of normal cellular genes.

In 1978–1979, DNAs extracted from several such 3-MC–transformed mouse cell lines were transfected into cultures of NIH 3T3 recipient cells, yielding large numbers of foci after several weeks. The cells plucked from the resulting foci were later found to be both anchorage-independent and tumorigenic (see Table 3.2). This simple experiment proved that the donor tumor DNA carried one or several genetic elements that were able to convert a non-tumorigenic NIH 3T3 recipient cell into a cell that was strongly tumorigenic.

DNA extracted from normal, untransformed C3H10T1/2 cells used as controls was unable upon transfection to induce foci in the NIH 3T3 cell monolayers. This difference made it highly likely that previous exposure of normal C3H10T1/2 cells to the 3-MC carcinogen had altered the genomes of these cells in some way, resulting in the creation of novel genetic sequences that possessed transforming powers. In other words, it seemed likely that the 3-MC carcinogen had converted a previously normal C3H10T1/2 gene (or genes) into a mutant allele that now could function as a transforming oncogene when introduced into NIH 3T3 recipient cells.

At first, it seemed quite difficult to determine whether the donor tumor cells carried a single oncogene in their genomes or several distinct oncogenes that acted in concert to transform the recipient cells. Careful analysis of the transfection procedure soon resolved this issue. Researchers discovered that when cellular DNA was applied to a recipient cell, only about 0.1% of a cell genome's worth of donor DNA became established in the genome of each transfected recipient cell. The probability of two independent, genetically unlinked donor genes both being concomitantly introduced into a single recipient cell was therefore $10^{-3} \times 10^{-3} = 10^{-6}$, that is, a highly unlikely event. From this calculation they could infer that only a single gene was responsible for the transformation of NIH 3T3 cells following transfection of donor tumor cell DNA. This led, in turn, to the conclusion that, years earlier, exposure of normal C3H10T1/2 mouse cells to the 3-MC carcinogen had caused the formation of a single mutant oncogenic allele; this allele was able, on its own, to transform both the C3H10T1/2 cells in which this mutation arose and, later on, the recipient NIH 3T3 cells into which this allele was introduced by gene transfer.

These transfection experiments provided strong indication that oncogenes can arise in the genomes of cells through mechanisms that have no apparent connection with viral infection. Perhaps human tumor cells, which likewise appeared to arise via nonviral mechanisms, also carried transfectable oncogenes. Would human oncogenes, if present in the genomes of these cells, also be able to alter the behavior of mouse cells?

Both of these questions were soon answered in the affirmative. DNAs extracted from cell lines derived from human bladder, lung, and colon carcinomas, as well as DNA from a human promyelocytic leukemia, were all found capable of transforming recipient NIH 3T3 mouse fibroblasts. This meant that the oncogenes in these cell lines, whatever their nature, were capable of acting across species and tissue boundaries to induce cell transformation. Most important, this work demonstrated directly that chemically transformed cells as well as certain human tumor cells contained mutant genes that were capable of driving such cells into a neoplastic growth state.

4.2 Oncogenes discovered in human tumor cell lines are related to those carried by transforming retroviruses

The oncogenes detected by transfection in the genomes of various human tumor cells were ostensibly derived from preexisting normal cellular genes that lacked oncogenic function. This seemed to parallel the process that led to the appearance of transforming retroviruses (see Section 3.9). Recall that during the formation of these viruses, preexisting normal cellular genes—proto-oncogenes—became associated with viral genomes and activated into potent oncogenes, albeit through an entirely different genetic mechanism.

These apparent parallels led to an obvious question: Could the same group of cellular proto-oncogenes become activated into oncogenes by marauding retroviruses in one context and by nonviral mutagens in another? Or, alternatively, did the retrovirus-associated oncogenes and those activated by nonviral mechanisms arise from two very distinct groups of cellular proto-oncogenes?

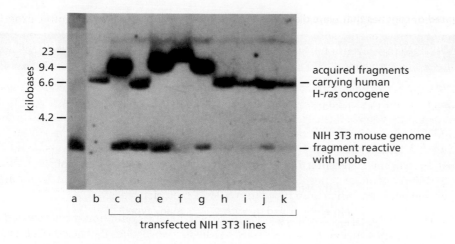

acquired fragments
carrying human
H-*ras* oncogene

NIH 3T3 mouse genome
fragment reactive
with probe

a b c d e f g h i j k

transfected NIH 3T3 lines

Figure 4.2 Homology between transfected oncogenes and retroviral oncogenes The Southern blot procedure (see Supplementary Sidebar 4.3) was used to determine whether there was any relatedness between retrovirus-associated oncogenes and those discovered by transfection of tumor cell DNA. Cloned retroviral oncogene DNAs were used to make radiolabeled probes, while the restriction enzyme–cleaved genomic DNAs from transfected cells were analyzed by the Southern blot procedure. Shown here is the annealing between a radiolabeled H-*ras* oncogene probe (cloned from the genome of Harvey murine sarcoma virus; see Section 3.10) and the genomic DNAs from a series of nine lines of NIH 3T3 cells (*lanes b–j*) that had been transformed by transfection of DNA extracted from the EJ/T24 human bladder carcinoma cell line; the DNA of untransfected NIH 3T3 cells was analyzed in channel a and the genomic DNA of T24 human bladder carcinoma cells was analyzed in lane k. In many of the lines of transfected NIH 3T3 cells (*lanes b–j*), in addition to the acquired DNA fragments carrying the transfected oncogene, a lower–molecular-weight fragment representing the endogenous mouse H-*ras* proto-oncogene was also detected. In many of these transfected cell lines, the size of the DNA fragment carrying the transfected H-*ras* oncogene was affected by random DNA breakage events that had occurred in prior rounds of transfection and separated previously linked restriction enzyme cleavage sites from the DNA segment carrying the oncogene. The human tumor DNA in channel a indicates the size of the restriction fragment carrying the human H-*ras* proto-oncogene. The variable intensity of this band indicates unequal loading of DNA samples in various gel channels. (From L.F. Parada et al., *Nature* 297:474–478, 1982. With permission from Nature.)

Use of DNA probes specific for the retrovirus-associated oncogenes provided the answers in short order. Using the Southern blot procedure (Supplementary Sidebar 4.3), a DNA probe derived from the H-*ras* oncogene present in Harvey rat sarcoma virus was able to recognize and form hybrids with the oncogene detected by transfection in the DNA of a human bladder carcinoma cell (**Figure 4.2**). A related oncogene, termed K-*ras* from its presence in the genome of Kirsten sarcoma virus, was able to anneal with the oncogene detected by transfection of DNA from a human colon carcinoma cell line.

The list of connections between the retrovirus-associated oncogenes and onco-genes present in non-virus-induced human tumors soon grew by leaps and bounds (**Table 4.1**). In fact, many of these connections were forged following the discovery that the retrovirus-associated oncogenes were present in increased copy number in human tumor cell genomes. The *myc* oncogene, originally known from its pres-ence in avian myelocytomatosis virus (MC29; see Section 3.10), was found to be present in multiple copies in the DNA of the HL-60 human promyelocytic leukemia cell line. These extra copies of the *myc* gene (about 10–20 per diploid genome) were the result of the process of **gene amplification** and the apparent cause of the proportionately increased levels of its protein product, the Myc oncoprotein; the excess Myc protein somehow favored the proliferation of the cancer cells. The *erbB* gene, first discovered through its presence in the genome of avian erythroblastosis virus (AEV; refer to Table 3.3), was discovered to be present in increased copy number in the DNAs of human stomach, breast, and brain tumor cells. (**Erythro-blastosis** is a malignancy of red blood cell precursors.) Elevated expression of the homolog of the *erbB* gene is now thought to be present in the majority of human carcinomas.

In 1987, amplification of the *erbB*-related gene known variously as *erbB2*, *neu*, or *HER2* was reported in many breast cancers (**Figure 4.3A**). Increases in gene copy number of more than five copies per cancer cell were found to correlate with a decrease in the number of patients who, following initial treatment, survived without recur-rence of disease (Figure 4.3B). (**Kaplan–Meier plots**, like the one in the figure, will be used repeatedly throughout this book. In each case, the status of patients—either disease-free survival, progression-free survival, overall survival, or another clinical parameter—is plotted as a function of the time elapsed following initial diagnosis or treatment.) Significantly, the observed amplification of the *erbB2/HER2* gene was often but not always correlated with an increased expression of its encoded protein (see Figure 4.3A). Among a large group of breast cancer patients, those whose tumors expressed normal levels of this protein showed a median survival of 6 to 7 years after diagnosis, while those patients whose tumors expressed elevated levels had a median survival of only 3 years. This inverse correlation between *erbB2/HER2* expression levels and long-term patient survival provided a strong indication that this gene, in amplified form, was causally involved in driving the malignant growth of the breast cancer cells (but see Supplementary Sidebar 4.4).

Table 4.1 Initial examples of retrovirus-associated oncogenes that were discovered in altered form in human cancers

Name of virus	Species	Oncogene	Type of oncoprotein	Homologous oncogene found in human tumors
Abelson leukemia	mouse	abl	non-receptor TK	CML
Avian erythroblastosis	chicken	erbB	receptor TK	gastric, lung, breast[a]
McDonough feline sarcoma	cat	fms	receptor TK	AML[b]
H-Z feline	cat	kit	receptor TK[c]	gastrointestinal stromal
Murine sarcoma 3611	mouse	raf	ser/thr kinase[d]	bladder carcinoma
Simian sarcoma	monkey	sis	platelet-derived growth factor (PDGF)	many types[e]
Harvey sarcoma	mouse/rat	H-ras[f]	small G protein	bladder carcinoma
Kirsten sarcoma	mouse/rat	K-ras[f]	small G protein	many types
Avian erythroblastosis	chicken	erbA	nuclear receptor[g]	liver, kidney, pituitary
Avian myeloblastosis E26	chicken	ets	transcription factor	leukemia[h]
Avian myelocytoma	chicken	Myc[i]	transcription factor	many types
Reticuloendotheliosis	turkey	Rel[j]	transcription factor	lymphoma

[a]Receptor for EGF; the related erbB2/HER2/Neu protein is overexpressed in 30% of breast cancers.
[b]Fms, the receptor for colony-stimulating factor (CSF-1), is found in mutant form in a small number of AMLs; the related Flt3 (Fms-like tyrosine kinase-3) protein is frequently found in mutant form in these leukemias.
[c]Receptor for stem cell factor.
[d]The closely related B-Raf protein is mutant in the majority of melanomas.
[e]Protein is overexpressed in many types of tumors.
[f]The related N-ras gene is found in mutant form in a variety of human tumors.
[g]Receptor for thyroid hormone.
[h]27 distinct members of the Ets family of transcription factors are encoded in the human genome. Ets-1 is overexpressed in many types of tumors; others are involved in chromosomal translocations in AML, Ewing's sarcoma, and prostate carcinoma.
[i]The related N-myc gene is overexpressed in pediatric neuroblastomas and small-cell lung carcinomas.
[j]Rel is a member of a family of proteins that constitute the NF-κB transcription factor, which is constitutively activated in a wide range of human tumors.
Abbreviations: AML, acute myelogenous leukemia; CML, chronic myelogenous leukemia; TK, tyrosine kinase.
Adapted in part from J. Butel, *Carcinogenesis* 21:405–426, 2000; and G.M. Cooper, Oncogenes, 2nd ed. Boston and London: Jones and Bartlett, 1995.

In the years that followed, yet other techniques, including fluorescence *in situ* hybridization (FISH), have been used to determine gene amplification in these and other tumor types. By 2010, a variety of techniques had been brought to bear for validating the oncogenic roles of genes that are amplified in a variety of human cancers; included among these techniques were tests of the transforming powers of these genes. This work resulted in a roster of 77 genes that are, with great likelihood, contributing to tumor development when they are present in amplified form in human cancer cell genomes. (Since the amplification of a gene is often accompanied by the co-amplification of other genes that happen to be located nearby on a chromosome, the gene whose amplification is thought to be causal in fostering tumor development is often termed a **driver**, while the nearby flanking genes that "may go along for the ride" but are mechanistically unimportant in tumor development are termed **passengers**.)

As genetic technology has improved, the ability to survey entire tumor cell genomes for chromosomal regions that have suffered changes in copy number has improved immeasurably, and genome-wide surveys of both gene amplifications and deletions have become almost routine. In the case of cancer cells, the amplifications presumably result in the overexpression of growth-promoting genes (that is, proto-oncogenes), while the deletions involve the loss of putative growth-retarding **tumor suppressor genes**, which are discussed in Chapter 7. As **Figure 4.4** makes clear, different cell types undergo distinct sets of genomic alterations during their transformation into full-fledged tumor cells. The analyses of cancer cell genomes by a variety of methods will be discussed in detail in Chapter 12.

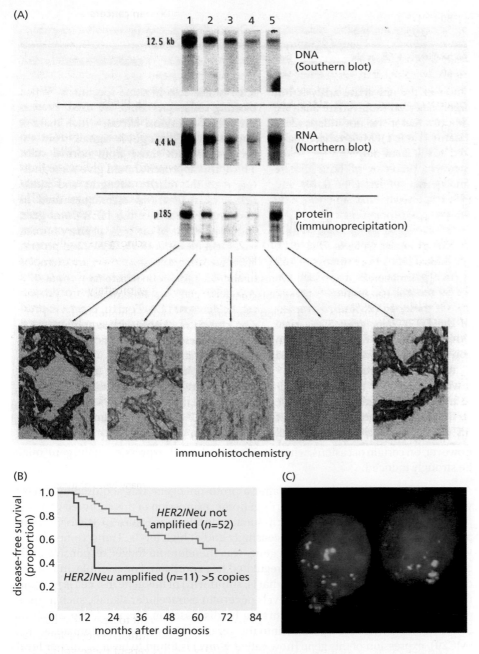

immunohistochemistry

Figure 4.3 Amplification of the *erbB2/ HER2/neu* oncogene in breast cancers
(A) The Southern blotting procedure (see Supplementary Sidebar 4.3) was used to determine whether the DNA of five human breast carcinomas carried extra (that is, amplified) copies of the *erbB2/ neu* oncogene (also termed *HER2*), a close relative of the *erbB* oncogene. As indicated by the dark bands representing restriction enzyme fragments, some human breast carcinomas carried extra copies of this gene. Subsequent work indicated that while the *erbB2/HER2* oncogene was amplified in some tumors ("DNA," lanes 1, 2, and 3), others overexpressed the mRNA without gene amplification ("RNA," lane 5). Levels of the encoded ErbB2/HER2 protein could also be gauged via immunoprecipitation and immunohistochemistry (see Supplementary Sidebar 2.1) by using an antibody that reacts with this protein; these levels generally correlated with relative levels of the encoding mRNA. (B) This gene amplification is correlated with a poor prognosis for the breast cancer patient, as indicated by this *Kaplan–Meier* plot. Those patients whose tumors carried more than five copies of the *erbB2/neu* gene were far more likely to experience a relapse in the first 18 months after diagnosis and treatment than were those patients whose tumors lacked this amplification. (All patients in this study had breast cancer cells that had spread to the lymph nodes draining the involved breast. *n*, number of patients.) (C) Use of fluorescence *in situ* hybridization (FISH) has shown that in some human breast cancers, there are multiple spots in each cell nucleus that react with a fluorescence-labeled DNA probe that recognizes the *erbB2/HER2* gene (*light green*) and yet other spots that react with a DNA probe specific for the *CCND1/cyclin D1* gene (*orange/ pink*). Each of these probes generates only two spots when used to analyze nuclei of normal diploid cells. (As described in Chapter 8, cyclin D1 promotes progression of cells through their growth-and-division cycles.) (A, from D.J. Slamon et al., *Science* 244:707–712, 1989. B, from D.J. Slamon et al., *Science* 235:177–182, 1987. Both with permission from AAAS. C, from J. Hicks et al., *Genome Res.* 16:1465–1479, 2006. With permission from Cold Spring Harbor Laboratory Press.)

Ironically, mutant alleles of the c-*src* proto-oncogene (see Section 3.9), the first cellular oncogene to be discovered, proved to be elusive in human tumor cell genomes. More specifically, repeated attempts to find mutant alleles of c-*src* that are present in human cancer cell genomes and affect the structure of the encoded Src protein have consistently failed.

The lesson taught by these numerous cross connections was simple and clear: many (but not all) of the oncogenes originally discovered through their association with avian and mammalian retroviruses could be found in a mutated, activated state in human tumor cell genomes. This meant that a common set of cellular proto-oncogenes might be activated either by retroviruses (in animals) or, alternatively, by nonviral mutational mechanisms operating during the formation of human cancers.

4.3 Proto-oncogenes can be activated by genetic changes affecting either protein expression level or structure

While a number of proto-oncogenes were found in activated, oncogenic form in human tumor genomes, the precise genetic alterations that led to many of these activations remained unclear. In the case of retrovirus-associated oncogenes, one

Figure 4.4 Nonrandom amplifications and deletions of chromosomal regions A survey of nine different types of pediatric cancers indicates that each cancer type has characteristic gene amplifications and deletions affecting a specific set of chromosomal regions. This survey examined the transcripts deriving from genes carried by the 22 non-sex chromosomes. For example, neuroblastomas (*pink*) often have changes in the copy numbers of genes on Chromosomes 1 and 17 and corresponding changes in the levels of the transcripts expressed by these genes. In this analysis, measurable changes in the copy number of 493 genes were found to he paralleled by changes in mRNA expression levels—that is, when DNA copy number of a chromosomal region was increased (that is, amplified) or reduced, there was a parallel change in the level of the corresponding RNA, this being observed in multiple tumors of a given type. (Of these 493 chromosomal sites, which were scattered across the human genome, 440 could be associated with genes of known biological function.) This parallel behavior made it possible to assess changes in gene copy number by measuring levels of RNA transcripts in these tumors rather than by directly measuring DNA copy number of the corresponding genes. (The "altered transcripts" indicated in the ordinate denotes RNAs whose concentrations were significantly increased above or below normal in the tumors being examined.) As is apparent, Chromosome 1 carries regions that are most commonly found to be amplified or deleted in a variety of these tumors, which might be expected from its relatively large size. This does not explain, however, the relatively infrequent amplification or deletion of regions of Chromosomes 2, 3, 4, and 5, which are almost as large. Some tumors, such as lymphomas (*darker green*), show changes at diverse chromosomal sites, while the aforementioned neuroblastomas (*pink*) usually show changes mapping to one or two chromosomes. (From G. Neale et al., *Clin. Cancer Res.* 14:4572–4583, 2008. With permission from the American Association for Cancer Research.)

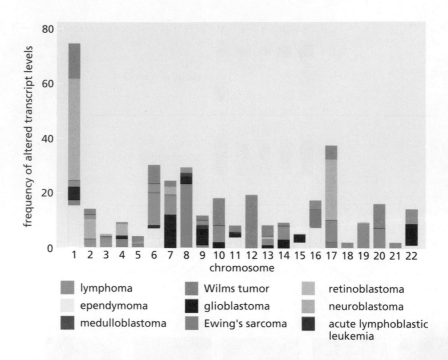

mechanism became obvious once the organization of the transforming retrovirus genomes was known. In the normal cell, the expression of each proto-oncogene was regulated by its own transcriptional promoter—the DNA sequence that controls the level of its transcription. The promoter of each proto-oncogene enabled the gene to respond to a variety of physiologic signals. Often the needs of the cell, communicated through these signals, caused a proto-oncogene to be expressed at very low levels; however, on certain occasions, when required by the cell, expression of the gene might be strongly induced.

A quite different situation resulted after a proto-oncogene was acquired by a retrovirus. After insertion into the retrovirus genome, expression of this captured gene was now controlled by a retroviral transcriptional promoter (see Figure 3.13), which invariably drove the gene's expression unceasingly and at high levels. Transcription of this virus-associated gene, now an oncogene, was therefore no longer responsive to the cellular signals that had previously regulated its expression. For example, in the case of c-*myc*, expression or **repression** (that is, shutdown) of this gene in a cell is normally tightly controlled by the changing levels of certain extracellular signals, such as those conveyed by mitogenic growth factors (to be discussed in Chapter 5), to which the cell is exposed. Once incorporated into the genome of avian myelocytomatosis virus (MC29), expression of this gene (now called v-*myc*) is found to be at far higher levels than are seen normally in cells, and this expression occurs at a constant (sometimes termed **constitutive**) level.

But how did a normal human H-*ras* proto-oncogene become converted into the potent oncogene that was detected by transfection of human bladder carcinoma DNA (see Sections 4.1 and 4.2)? Gene amplification could not be invoked to explain its activation, since this oncogene seemed to be present in bladder carcinoma DNA as a single-copy gene. The puzzle grew when this H-*ras* bladder carcinoma oncogene was isolated by molecular cloning (Supplementary Sidebar 4.5). It was localized to a genomic DNA fragment of 6.6 kilobases in length. Provocatively, an identically sized DNA fragment was found in normal human DNAs. The latter fragment clearly represented the human H-*ras* proto-oncogene—the normal gene that suffered some type of mutation that converted it into an oncogene during the formation of the bladder carcinoma.

While their overall DNA structures were very similar, these two versions of the H-*ras* gene performed in dramatically different ways. The oncogene that had been cloned from human bladder carcinoma cells caused transformation of NIH 3T3 cells, while its normal proto-oncogene counterpart lacked this ability. The mystery deepened when more detailed mapping of the physical structures of these two DNA segments—achieved by making maps of the cleavage sites of various restriction enzymes—revealed that the two versions of the gene had overall physical structures that were indistinguishable from one another.

Yet clearly, the two versions of the H-*ras* gene had some significant difference in their sequences, because they functioned so differently. The critical sequence difference was initially localized by recombining segments of the cloned proto-oncogene with other segments deriving from the oncogene (Supplementary Sidebar 4.6). This made it possible to narrow down the critical difference to a segment only 350 base pairs long.

The puzzle was finally solved when the corresponding 350-bp segments from the proto-oncogene and oncogene were subjected to DNA sequence analysis. The critical difference was extraordinarily subtle—a single base substitution in which a G (guanosine) residue in the proto-oncogene was replaced by a T (thymidine) in the oncogene. This single base-pair replacement—a **point mutation**—appeared to be all that was required to convert the normal gene into a potent oncogene (**Figure 4.5**)! This important discovery was made simultaneously in three laboratories, eliminating all doubt about its correctness.

The discovery of this point mutation represented a significant milestone in cancer research. It was the first time that a specific mutation was discovered in a gene that contributed causally to the neoplastic growth of a human cancer. Equally important, it seemed that this genetic change arose as a somatic mutation. And finally, this work revealed the powers of the then-recently developed recombinant DNA technology, which could trace a critical oncogene-activating mutation to a change in a single nucleotide out of the ~3.2 billion nucleotides that comprise the haploid human genome.

With this information in hand, researchers could devise a likely explanation for the origin of the bladder carcinoma and, by extension, other similar tumors. The particular bladder carcinoma from which the H-*ras* oncogene had been cloned was said to have arisen in a middle-aged man who had been smoking for four decades. During this time, carcinogens present in cigarette smoke were introduced in large amounts into his lungs and passed from there through the bloodstream to his kidneys, which excreted these chemical species with the urine. While in the bladder, some of the carcinogen molecules present in the urine had entered **urothelial** cells lining the bladder and attacked their DNA. On one occasion, a mutagenic carcinogen introduced a point mutation in the H-*ras* proto-oncogene of a urothelial cell. Thereafter, this mutant cell and its descendants proliferated, being driven by the potent transforming action of the

```
                                    CCCGGG CCGCAGGCCC TTCAGGAGCG
                                            ↓
                                        ┌───┐ ⌐proto-oncogene
                                        │gly│
met thr glu tyr lys leu val val val gly ala │GGC│ gly val gly lys ser ala leu thr
ATG ACG GAA TAT AAG CTG GTG GTG GTG GGC GCC │GTC│ GGT GTG GGC AAG AGT GCG CTG ACC
                                        │val│ ⌐oncogene
                                        └───┘              splice
                                                            ↓
ile gln leu ile gln asn his phe val asp glu tyr asp pro thr ile glu
ATC CAG CTG ATC CAG AAC CAT TTT GTG GAC GAA TAC GAC CCC ACT ATA GAG GTGAGCCTGC

GCCGCCGTCC AGGTGCCAGC AGCTGCTGCG GGCGAGCCCA GGACACAGCC AGGATAGGGC TGGCTGCAGC

CCCTGGTCCC CTGCATGGTG CTGTGGCCCT GTCTCCTGCT TCCTCTAGAG GAGGGGAGTC CCTCGTCTCA

GCACCCCAGG AGAGGAGGGG GCATGAGGGG CATGAGAGGT ACC
```

Figure 4.5 Mutation responsible for H-*ras* oncogene activation As described in Supplementary Sidebar 4.6, the critical difference between the human EJ/T24 bladder carcinoma oncogene and its corresponding proto-oncogene could be localized to an intragenic restriction enzyme–generated fragment of 350 base pairs. The sequences of the two 350-nucleotide-long DNA fragments from the oncogene and proto-oncogene were then determined. The two differed at a single nucleotide, which affected the 12th codon of the H-*ras* reading frame (*arrow*), converting the normally present glycine-encoding codon to one specifying valine. (From C.J. Tabin et al., *Nature* 300:143–149, 1982. With permission from Nature.)

H-*ras* oncogene that they carried. The result, years later, was the large tumor mass that was eventually diagnosed in this patient. [As we will learn later (Chapter 11), acquisition of this H-*ras* mutation must have been followed by additional, subsequently acquired mutations in other growth-controlling genes before these cells could behave in a fully neoplastic manner.]

Importantly, this base-pair substitution occurred in the **reading frame** of the H-*ras* gene—the portion of the gene dedicated to encoding amino acid sequence (see Figure 4.5). In particular, this point mutation caused the substitution of a glycine residue present at residue position 12 in the normal H-*ras*–encoded protein by a valine residue. The effects of this amino acid substitution on the function of the Ras oncoprotein will be discussed later in Chapters 5 and 6.

The discovery of this point mutation established a mechanism for oncogene activation that was quite different from that responsible for the creation of *myc* oncogenes. In the case of H-*ras*, a change in the structure of the encoded protein appeared to be critical. In the contrasting case of *myc*, deregulation of its expression level seemed to be important for imparting oncogenic powers to this gene.

Within a decade, a large number of human tumors were found that carried point mutations in one of the three *ras* genes present in the mammalian genome: H-*ras*, K-*ras*, and N-*ras*. Significantly, in each of these tumors, the point mutation that was uncovered was present in one of three specific codons in the reading frame of a *ras* gene. Consequently, all Ras oncoproteins (whether made by the H-, K-, or N-*ras* gene) were found to carry amino acid substitutions in residue 12 or, less frequently, residues 13 and 61 (**Figure 4.6**); yet other sites of mutated codons have been reported on occasion. (This striking concentration of mutations affecting only a small number of residues in the Ras protein occurs in spite of the fact that all of the coding sequences in the *ras* proto-oncogenes presumably have similar vulnerability to mutation. The resolution of this puzzle comes from studying the structure and function of the Ras protein, topics that we will address in the next chapter.) A survey of 40,000 human tumor genomes conducted many years later revealed that 22%, 8.2%, and 3.7% of these tumors had activating mutations in either the K-, N-, or H-*ras* genes, respectively.

For unknown reasons, the frequency with which these mutant oncogenes are found varies dramatically from one tissue site to another (**Table 4.2**). One clue, however, has come from genetically engineered mice in which the reading frame of the K-*ras* proto-oncogene has been replaced ("**knocked-in**") with the reading frame of an H-*ras* oncogene. Wild-type mice usually develop lung tumors carrying K-*ras* oncogenes in response to treatment by a chemical carcinogen; however, the vast majority of the knock-in mice develop lung tumors carrying H-*ras* oncogenes following this treatment. This indicates that the *regulatory sequences* controlling the expression of the K-*ras* proto-oncogene, rather than the *structure* of the encoded protein, determine which *ras* proto-oncogene, following mutation by a carcinogen, appears in mutant form in the genome of a particular type of carcinoma.

Figure 4.6 Concentration of point mutations leading to activation of the K-*ras* oncogene The sequencing of almost 10,000 K-*ras* oncogenes in human tumors has revealed that, rather than occurring randomly throughout the reading frame of the gene, which encompasses 189 codons, mutations affect almost exclusively the nucleotides forming codon 12; mutations affect codon 13 only about 12% as frequently and codon 61 only on rare occasions. Since all the codons of the K-*ras* reading frame are likely to suffer mutations at comparable rates during tumor development, this suggests that an extraordinarily strong selective pressure favors the outgrowth of cells having mutations affecting codons 12, 13, and 61, while mutations altering other codons in the reading frame either have no effect on cell phenotype or are actively disadvantageous for the cells bearing them. (Data from Catalogue of Somatic Mutations in Cancer (COSMIC database), http://www.sanger.ac.uk.)

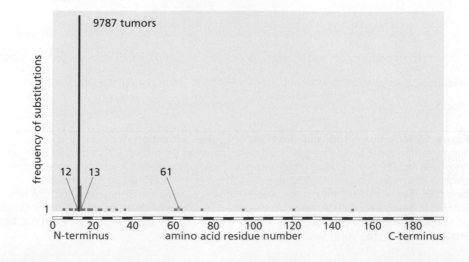

Table 4.2 A list of point-mutated *ras* oncogenes carried by a variety of human tumor cells

Tumor type	Proportion (%) of tumors carrying a point-mutated *ras* gene[a]
Pancreas	98 (K)
Thyroid (papillary)	60 (H, K, N)
Thyroid (follicular)	55 (H, K, N)
Colorectal	52 (K, N)
Seminoma	45 (K, N)
Myeloma	43 (N, K)
Lung (non-small-cell)	32 (K)
Melanoma	30 (N)
Uterine endometrioid	25 (H, K,N)
Acute myelogenous leukemia	11 (N)
Bladder	10 (H, K, N)

[a]H, K, and N refer to the human *H-RAS, K-RAS*, and *N-RAS*, respectively.
Adapted from J. Downward, *Nat. Rev. Cancer* 3:11–22, 2003; N. Vasan et al., *Clin. Cancer Res.* 20:3921–3930; and A.D. Cox et al., *Nat. Rev. Drug Discov.* 13, 828–851.

Both activation mechanisms—regulatory and structural—might collaborate to create an active oncogene. In the case of the *myc* oncogene carried by avian myelocytomatosis virus, for example, expression of this gene was found to be strongly deregulated by the viral transcription promoter as mentioned above. At the same time, some subtle alterations in the reading frame of the *myc* oncogene (and thus changes in the structure of its encoded oncoprotein, Myc) further enhanced the already-potent transforming powers of this oncogene. Similarly, the H-*ras* oncogene carried by Harvey sarcoma virus was found to carry a point mutation in its reading frame (like that discovered in the bladder carcinoma oncogene); at the same time, this viral oncogene was greatly overexpressed relative to its expression levels in normal cells, being driven by the retroviral transcriptional promoter. Indeed, in many animal and human tumors carrying a mutant, activated *ras* oncogene, the initially arising oncogene is found to subsequently undergo the process of gene amplification (see Section 4.2), enabling the tumor cells to acquire increased signaling from the mutant oncoprotein.

4.4 Variations on a theme: the *myc* oncogene can arise via at least three additional distinct mechanisms

The observation that the v-*myc* oncogene of avian myelocytomatosis virus (MC29) arose largely through deregulation of its expression only hints at the diverse mechanisms that are capable of creating this oncogene. As mentioned in Section 4.2, in some human tumors, expression of the *myc* gene continues to be driven by its own natural transcriptional promoter but the copy number of this gene is found to be elevated to levels many times higher than the two copies present in the normal human genome. In 30% of childhood neuroblastomas, a close relative of the c-*myc* gene, termed N-*myc*, has also been found to be amplified, specifically in the more aggressive tumors of this type (**Sidebar 4.1**). In both instances, the increased gene copy numbers result in corresponding increases in the level of gene products—the Myc and N-Myc proteins. As we will discuss later in Chapter 8, proteins of the Myc family possess potent growth-promoting powers. Consequently, when present at excessive levels, these proteins seem to drive uncontrolled cell proliferation.

Sidebar 4.1 N-*myc* amplification and childhood neuroblastomas Amplification of the N-*myc* gene occurs in about 30% of advanced pediatric neuroblastomas, which are tumors of the peripheral nervous system. This amplification, which is associated with the formation of either double minutes (DMs) or homogeneously staining regions (HSRs; see Figure 1.9A and B), represents a bad prognosis for the patient (**Figure 4.7**). The HSRs, which contain multiple copies of the genomic region encompassing the N-*myc* gene, are often found to have broken away from the normal chromosomal mapping site of N-*myc* and, in one study, to have become associated with at least 18 other different chromosomal regions. While N-*myc* amplification was originally thought to be a peculiarity of neuroblastomas (and thus a specific diagnostic marker for this particular disease), it has now been found in a variety of neuroectodermal tumors, including astrocytomas and retinoblastomas; in addition, small-cell lung carcinomas, which have neuroendocrine traits, often exhibit amplified N-*myc* genes.

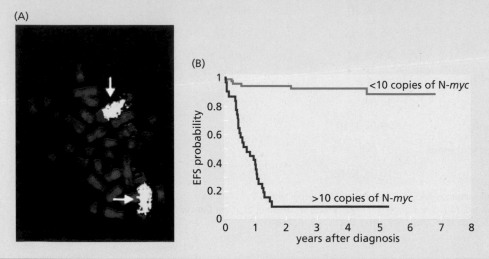

Figure 4.7 N-*myc* amplification and neuroblastoma prognosis
(A) The N-*myc* gene is often amplified in human childhood neuroblastomas. Here, use of fluorescence *in situ* hybridization (FISH) has revealed multiple copies of this gene (*yellow*). The fact that these N-*myc* gene copies are present as tandem arrays within chromosomes means that they constitute homogeneous staining regions (HSRs) rather than extrachromosomal particles—double minutes (DMs)—which are also frequently seen in these tumors (see also Figure 1.10). (B) This Kaplan–Meier plot illustrates the event-free survival (EFS) of children suffering from neuroblastoma, that is, no clinically significant cancer-related symptoms in the indicated years following initial diagnosis and treatment. Those whose tumor cells have minimal or no N-*myc* amplification have a very good prognosis and minimal clinical events, while those whose tumors exhibit extensive N-*myc* amplification have a dramatically poorer prognosis and therefore short survival times after diagnosis. (A, from C. Lengauer et al., *Nature* 396:643–649, 1998. With permission from Nature. B, from M.L. Schmidt et al., *J. Clin. Oncol.* 18:1260–1268, 2000. With permission from Wolters Kluwer.)

Note, by the way, the notations that are used here and throughout this text. Non-human oncogenes are usually written as uncapitalized three-letter words in italics (for example, *myc*), while their protein products are written in roman font with an initial capital (for example, Myc). The *myc* proto-oncogene itself is often termed c-*myc* to distinguish it from its two cousin genes, N-*myc* and L-*myc*. To make matters more confusing, human genes follow a different nomenclature, so that the human *myc* gene is denoted as *MYC* and its protein product is written as MYC. We will generally use the nonhuman acronym conventions throughout this book.

The gene amplification process, which is responsible for increases in *myc* copy number, occurs through the preferential replication of a limited region of chromosomal DNA, leaving the more distantly located chromosomal regions unaffected (see Supplementary Sidebar 4.4). Since the region of chromosomal DNA that undergoes amplification—the **amplicon**—usually includes a stretch of DNA far longer than the c-*myc* or N-*myc* gene (for example, typically including 0.5 to 10 megabases of DNA), the amplified chromosomal regions are often large enough to be observed at the metaphase of mitosis through a light microscope. Often gene amplification yields large, repeating end-to-end linear arrays of the chromosomal region, which appear as homogeneously staining regions (HSRs) in the microscope (see Figure 1.9A). Alternatively, the chromosomal region carrying a *myc* or N-*myc* gene may break away from the chromosome and can be seen as small, independently replicating, extrachromosomal particles (double minutes; see Figure 1.9B). Indeed, we now know that a number of proto-oncogenes can be found in amplified gene copy number in various types of human tumors (**Table 4.3**). A 2010 survey found that altogether 77 genes in

Table 4.3 Some frequently amplified chromosomal regions and the genes they are known to carry

Name of oncogene[a]	Human chromosomal location	Human cancers	Nature of protein[b]
MDM4/MDMX	1q32	breast, colon, lung, pre-B leukemias	p53 inhibitor
PIK3CA	3q26.3	lung SCC, ovarian, breast carcinomas	PI kinase
erbB1/EGFR	7q12–13	glioblastomas (~30%), squamous cell carcinomas (10–20%)	RTK
cab1–erbB2–grb7	17q12	gastric, ovarian, breast carcinomas (10–25%)	RTK, adaptor protein
k-sam	7q26	gastric, breast carcinomas (10–20%)	RTK
FGF-R1	8p12	breast carcinomas (10%)	RTK
Met	7q31	gastric carcinomas (20%)	RTK
k-ras	12p12.1	lung, ovarian, colorectal, bladder, gastric carcinomas (5–20%)	small G protein
n-ras	1p13	head-and-neck cancers (30%)	small G protein
h-ras	11p15	colorectal carcinomas (30%)	small G protein
c-myc	8q24	various leukemias, carcinomas (10–50%)	TF
l-myc	1p32	lung carcinomas (10%)	TF
n-myc–DDX1	2p24–25	neuroblastomas, lung carcinomas (30%)	TF
akt-1	14q32–33	gastric cancers (20%)	ser/thr kinase
akt-2	19q13	ovarian carcinomas	ser/thr kinase
cyclin D1–exp1–hst1–ems1	11q13	breast, liver and squamous cell carcinomas (25–50%)	G1 cyclin
cdk4–mdm2–sas–gli	12q13	sarcomas (10–30%), HNSCC (40%), B-cell lymphomas (25%)	CDK, p53 antagonist
cyclin E	19q12	gastric cancers (15%)	cyclin
akt2	19q13	pancreatic, ovarian cancers (10-30%)	ser/thr kinase
AIB1, BTAK	20q12–13	breast cancers (15%)	receptor co-activator
cdk6	19q21–22	gliomas (5%)	CDK
myb	6q23–24	colon carcinoma (5–20%), leukemias	TF
ets-1	11q23	lymphoma	TF
gli	12q13	glioblastomas	TF

[a]The listing of several genes indicates the frequent co-amplification of a number of closely linked genes; only the products of the most frequently amplified genes are described in the right column.
[b]Abbreviations: TF, transcription factor; RTK, receptor tyrosine kinase; CDK, cyclin-dependent kinase; G protein, guanine nucleotide-binding protein; HNSCC, head-and-neck squamous cell carcinomas; SCC, squamous cell carcinomas.
Courtesy of M. Terada, Tokyo, and adapted from G.M. Cooper, Oncogenes, 2nd ed., Boston and London: Jones and Bartlett, 1995; and C.M Croce, *N. Engl. J. Med.* 358:502–511, 2008.

the genomes of human cancer cells are found to be recurrently amplified, indicating that increased copy number of each of these confers an advantageous phenotype on the cells that develop them.

As Figure 4.4 indicates, certain types of tumor, such as Wilms tumor, show multiple regions of chromosomal amplification, while others, such as Ewing's sarcoma, usually exhibit amplification of only a single chromosomal region. In most cases, the identities of the critical growth-promoting genes in these chromosomal regions remain elusive. Even within a single tumor, such as the breast cancers described above, multiple chromosomal regions that have undergone gene amplification are often found (see

Table 4.3). As discussed later (Chapter 11), these multiple changes are thought to collaborate with one another to create the fully neoplastic phenotypes of cancer cells. The molecular mechanisms responsible for generating gene amplifications remain poorly understood; however, in the case of tandem duplications that yield HSRs, detailed structural analyses of amplified genes have provided good clues as to how these may occur (see Supplementary Sidebar 4.4).

An even more unusual way of deregulating *myc* expression levels has already been cited (see Section 3.11). Recall that the insertional mutagenesis mechanism generated by integrated retrovirus proviruses causes the expression of the c-*myc* proto-oncogene to be placed under the transcriptional control of an ALV provirus that has integrated nearby in the chromosomal DNA. The resulting constitutive overexpression of c-*myc* RNA and thus Myc protein results, once again, in flooding of the cell with excessive growth-promoting signals.

Such activation by provirus integration hinted at a more general mode of activation of the c-*myc* proto-oncogene: even while continuing to reside at its normal chromosomal site, c-*myc* can become involved in cancer development if it happens to come under the control of foreign transcriptional promoters. In the disease of Burkitt's lymphoma (BL), this principle was validated in a most dramatic way. This tumor occurs with significant frequency among young children in East and Central Africa (Supplementary Sidebar 4.7). The etiologic agents of this disease include chronic infections both by Epstein–Barr virus (EBV, a distant relative of human herpesviruses; Section 3.4) and by malarial parasites, which appear to contribute to tumor pathogenesis by suppressing the immune response to EBV.

Neither of these etiologic factors sheds light on the nature of a critical genetic change occurring inside Burkitt's lymphoma cells that is responsible for their runaway proliferation. However, careful examination of metaphase chromosome spreads of these tumor cells uncovered a striking clue: the tumor cells almost invariably carried chromosomal **translocations** (see also Figure 2.20). Such alterations fuse a region from one chromosome with a region from a second, unrelated chromosome (**Figure 4.8**). Translocations were usually found to be *reciprocal*, in the sense that a region from

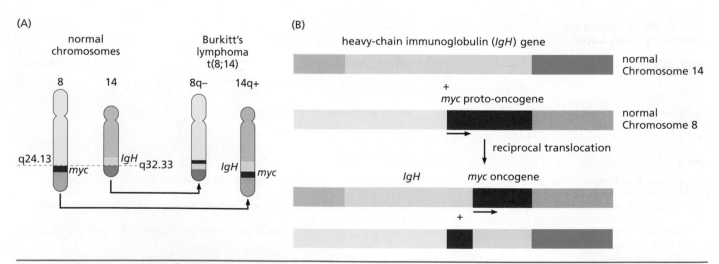

Figure 4.8 Chromosomal translocations in Burkitt's lymphomas (A) In the genomes of Burkitt's lymphoma (BL) cells, the expression of the c-*myc* gene is placed under control of the transcription-controlling enhancer sequences (see Figure 1.15) of an immunoglobulin gene as a direct consequence of reciprocal chromosomal translocations. These translocations juxtapose immunoglobulin genes on Chromosomes 2, 14, or 22 with the *myc* gene on Chromosome 8. (Translocations involving either the κ, λ, or heavy immunoglobulin chain occur in 9%, 16%, and 75% of BLs, respectively.) The most common translocation, t(8;14), is shown here. (B) Depicted is a molecular map of the translocation event that places the c-*myc* gene (*red rectangle*) on Chromosome 8 under the control of the immunoglobulin heavy-chain sequences (*IgH; gray rectangle*) on human Chromosome 14. Because the immunoglobulin enhancer sequences (see Figure 1.15) direct high, constitutive expression, the normal modulation of *myc* expression in response to physiologic signals is abrogated. The resulting *myc* oncogene initially makes structurally normal Myc protein but in abnormally high amounts. (Subsequently occurring point mutations in the *myc* reading frame may further potentiate the function of the Myc oncoprotein.) (Data from P. Leder et al., *Science* 222:765–771, 1983.)

chromosome A lands on chromosome B, while the displaced segment of chromosome B ends up being linked to chromosome A. In the case of Burkitt's lymphomas, three distinct, alternative chromosomal translocations were found, involving human Chromosomes 2, 14, or 22. The three translocations were united by the fact that in each case, a region from one of these three chromosomes was fused to a section of Chromosome 8.

In 1982 researchers realized that the *myc* (that is, c-*myc*) proto-oncogene could be found in the region on human Chromosome 8 that is involved in all three of these translocations. On the other side of the fusion site (often termed the chromosomal **breakpoint**) were found the transcription-promoting sequences from any one of three distinct **immunoglobulin** (antibody) genes. Thus, the immunoglobulin heavy-chain gene cluster is found on Chromosome 14, the κ antibody light-chain gene is located on Chromosome 2, and the λ antibody light-chain gene is found on Chromosome 22. There is clear evidence that the enzymes responsible for rearranging the sequences of antibody genes during the development of the immune system (described in Chapter 15) occasionally lose specificity and, instead of creating a rearranged antibody gene, inadvertently fuse part of an antibody gene with the *myc* proto-oncogene. Parenthetically, none of this explains the role of EBV in the pathogenesis of Burkitt's lymphoma (Supplementary Sidebar 4.8).

Suddenly, the grand design underlying these complex chromosomal changes became clear, and it was a simple one: these translocations separate the *myc* gene from its normal transcriptional promoter and place it, instead, under the control of one of three highly active transcriptional regulators, each part of an immunoglobulin gene (see Figure 4.8). Once its expression is subjugated by the antibody gene promoters, *myc* becomes a potent oncogene and drives the relentless proliferation of lymphoid cells in which these transcriptional promoters are highly active. Hence, the proliferation of the rare cell that happens to acquire such a deregulated *myc* gene will be strongly favored.

Since these discoveries, several dozen distinct chromosomal translocations have been found to cause deregulated expression of known proto-oncogenes; many of these genes remain poorly characterized (**Table 4.4**). Altogether, more than 300 distinct

Table 4.4 Translocations in human tumors that deregulate proto-oncogene expression and thereby create oncogenes

Oncogene	Neoplasm
myc	Burkitt's lymphoma; other B- and T-cell malignancies
bcl-2	follicular B-cell lymphoma
bcl-3	chronic B-cell lymphoma
bcl-6	diffuse B-cell lymphoma
hox1	acute T-cell leukemia
lyl	acute T-cell leukemia
rhom-1	acute T-cell leukemia
rhom-2	acute T-cell leukemia
tal-1	acute T-cell leukemia
tal-2	acute T-cell leukemia
tan-1	acute T-cell leukemia
etv-1, *etv-4*	prostate carcinoma
erg	prostate carcinoma

Adapted from G.M. Cooper, Oncogenes, 2nd ed. Boston and London: Jones and Bartlett, 1995. With permission from Jones and Bartlett.

recurring translocations (that is, those that have been encountered in multiple, independently arising human tumors) have been cataloged, and more than 100 of the novel hybrid genes created by these translocations have been isolated by molecular cloning. More often than not, each of these translocations can be found in a small set of hematopoietic tumors. However, solid tumors are increasingly found to harbor specific translocations, some quite common. For example, in 2005, ~70% of prostate carcinomas were found to harbor chromosomal translocations that lead to deregulated expression of one of two transcription factors (termed ERG and ETV1); given the high incidence of prostate cancer in Western populations, these may represent the most frequent translocations in human cancer. (Indeed, the dearth of translocations discovered to date in solid tumors may simply be a consequence of the difficulties of dissociating solid tumors and studying the karyotypes of their constituent cells.) Yet other chromosomal translocations function in an entirely different fashion to foster cancer outgrowth (Sidebar 4.2).

Returning to *myc*, we see that there are three alternative ways of activating the c-*myc* proto-oncogene—through provirus integration, gene amplification, or chromosomal translocation—all converging on a common mechanistic theme. Invariably, the gene is deprived of its normal physiologic regulation and is forced instead to be expressed at high, constitutive levels.

In general, the mechanisms leading to the overexpression of genes in cancer cells remain poorly understood. Some overexpression, as indicated here, is achieved through gene amplification and chromosomal translocation. But even more frequently, genes that are present in normal configuration and at normal copy number are transcribed at excessively high levels in cancer cells through the actions of deregulated transcription factors; the latter are largely uncharacterized. To complicate matters, gene amplification does not always lead to overexpression. Instead, only 40 to 60% of the genes that are found to be amplified in cancer cell genomes show corresponding increases in their RNA transcripts (and thus proteins). Such observations indicate that the expression levels of many genes are regulated by complex **negative-feedback** mechanisms that ensure physiologically appropriate levels of expression even in the presence of excess copies of these genes.

Sidebar 4.2 Chromosomal translocations can act indirectly to perturb levels of gene expression The discovery of microRNAs and the roles they play in gene regulation (see Section 1.9) have made it possible to understand certain chromosomal translocations whose precise molecular mechanisms-of-action were previously obscure. In one case, a small non-histone nuclear protein, termed HMGA2, is known to be overexpressed in a variety of benign and malignant tumors and to be capable of functioning as an oncoprotein. Its gene is frequently found to undergo translocations that sever the sequences specifying the 3′ untranslated region (3′UTR) from the reading frame of the *HMGA2* mRNA while joining the reading frame of the *HMGA2* gene with that of a second gene (Figure 4.9). It was long assumed that the resulting fusion protein enhanced the oncogenic functions of HMGA2 (similar to situations that we will encounter shortly in Section 4.5). However, deletion of the protein sequences that became fused to HMGA2 did not affect the transforming powers of the hybrid protein. This led to the realization that the translocation served another purpose: it caused deletion of seven recognition sites for the *Let-7* microRNA (see Section 1.9) that are normally present in the 3′UTR of

the *HMGA2* mRNA. This loss, in turn, enables the resulting mRNA to escape *Let-7*-mediated translational inhibition and eventual degradation. The end result is greatly increased levels of the *HMGA2* mRNA and protein, which proceeds, via still poorly understood mechanisms, to alter chromatin configuration (see Section 1.8) and thereby to facilitate cell transformation. In this instance, translocations serve to liberate a proto-oncogene from negative regulation rather than fusing it with a positive regulator.

In another case, chromosomal translocations, like those involving *myc*, have been found to drive enhanced expression of a microRNA by fusing the miRNA-encoding gene with a foreign transcriptional promoter. In tumors with such translocations—**myelodysplasia** and associated acute myelogenous leukemias—the resulting overexpressed miRNA blocks differentiation of cells, trapping them in a phenotypic state where they can evolve into highly aggressive tumors. This discovery, along with others that followed, resolves the long-standing mystery of how it is possible for chromosomal translocations that do not involve protein-encoding genes to affect cell behavior.

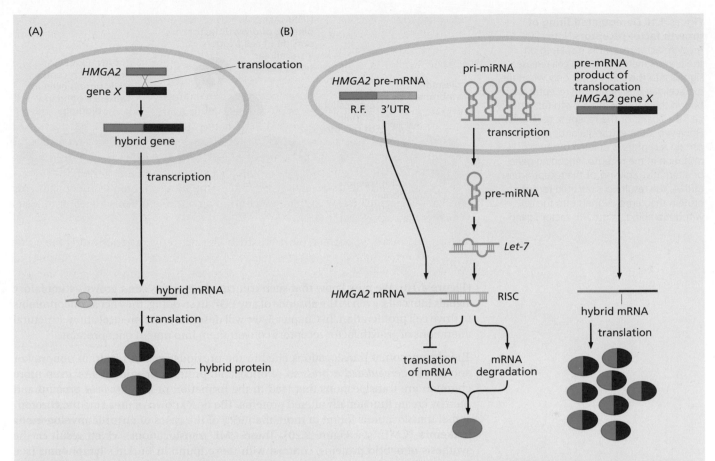

Figure 4.9 Translocations liberating an mRNA from miRNA inhibition The *HMGA2* gene encodes a chromatin-binding, proliferation-promoting protein whose biochemical function is still poorly understood; this protein (often present as a hybrid protein in which HMGA2 is fused to the protein encoded by a second gene) is expressed at elevated levels in a variety of cancer cell types in the absence of detectable gene amplification. (A) A chromosomal translocation that fuses the *HMGA2* (*aquamarine*) gene with a second gene (gene *X, red*) was initially assumed to create a hybrid fusion protein whose two domains collaborated to form a functional oncoprotein. (B) The true functional consequence of such a translocation was later realized to derive from severing the reading frame (R.F.) of the *HMGA2* mRNA (*blue*) from its 3′ untranslated region (3′UTR, *orange*). In the 3′UTR of this mRNA there are normally seven recognition sites that are bound by the *Let-7* microRNA (*green*); this binding reduces translation of the mRNA and drives its degradation, resulting in suppression of HMGA2 protein synthesis. (Only a single binding site is illustrated here.) These recognition sites are absent in the novel hybrid mRNA, allowing this mRNA to escape inhibition by *Let-7*; this results in synthesis of greatly increased amounts of the hybrid protein (right). Importantly, the domain of the novel protein that is encoded by the fusion partner gene (gene *X, red*) does not contribute to the oncogenic functions of the resulting protein. (Adapted from A.R. Young and M. Narita, *Genes Dev.* 21:1005–1009, 2007. With permission from Cold Spring Harbor Laboratory Press.)

4.5 A diverse array of structural changes in proteins can also lead to oncogene activation

The point mutation discovered in *ras* genes was the first of many mutations that were found to affect the structures of proto-oncogene–encoded proteins (rather than their expression levels), converting them into active oncoproteins. As an important example, the formation of certain human tumors, such as gastric and mammary carcinomas and glioblastoma brain tumors, involves the protein that serves as the cell surface **receptor** for epidermal growth factor (EGF). As we will discuss in detail in the next chapter, this receptor protein extends from the extracellular space through the plasma membrane of cells into their cytoplasm. Normally, the EGF receptor, like almost 60 similarly structured receptors, recognizes the presence of one of its cognate **ligands** (notably EGF) in the extracellular space and, having bound this ligand, informs the cell interior of this encounter. In about one-third of glioblastomas, however, the EGF receptor has been found to be decapitated, lacking most of its extracellular domain

Figure 4.10 Deregulated firing of growth factor receptors Normally, growth factor receptors displayed on the plasma membrane of a cell release signals into the cell interior only when the extracellular domain of the receptor has bound the appropriate growth factor (GF; i.e., the *ligand* of the receptor, *dark green*). However, if the extracellular domain of certain receptors is deleted because of a mutation in the receptor-encoding gene or alternative splicing of the receptor pre-mRNA, the resulting truncated receptor protein then emits signals into the cell without binding its growth factor ligand.

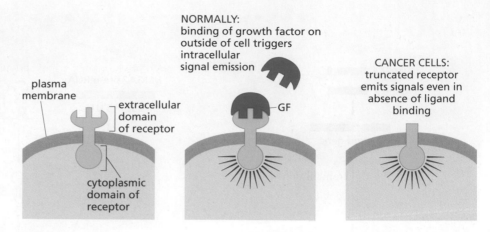

(**Figure 4.10**). We now know that such truncated receptors send growth-stimulatory signals into cells, even in the absence of any EGF. In so doing, they act as oncoproteins to drive cell proliferation. In Chapter 5, we will describe more precisely how structural alterations of growth factor receptors convert them into potent oncoproteins.

The chromosomal translocations cited in the previous section result, in one way or another, in *deregulated expression* of oncogenic proteins or microRNAs; even more common are translocations that lead to the formation of *novel hybrid proteins* and thereby create functionally altered proteins. The best known of these are the chromosomal translocations found in more than 95% of the cases of **chronic myelogenous leukemia** (CML; see Figure 2.20). These CML translocations, which result in the synthesis of hybrid proteins, contrast with those found in Burkitt's lymphomas (see Figure 4.8), in which the *transcription-regulating* sequences of one gene are juxtaposed with the reading frame of a second gene.

To the right of the CML translocation breakpoint illustrated in **Figure 4.11** are sequences that encode the protein made by the *abl* proto-oncogene. The *abl* gene was originally discovered by virtue of its presence as an acquired oncogene in **Abel**son murine leukemia virus, a rapidly tumorigenic retrovirus (refer to Table 3.3). The *abl* gene, which maps to Chromosome 9q34 (that is, the fourth band of the third region of the long arm of human Chromosome 9), was found, upon examination of various CMLs, to be fused with sequences that are clustered in a narrow region in Chromosome 22q11. [The standard notation used here is t(9;22)(q34;q11), where t signifies a translocation, q the long arm, and p the short arm of a chromosome.] This area on Chromosome 22 was called the **b**reakpoint **c**luster **r**egion. Subsequently, all of these breakpoints were found to lie within the gene that came to be called *bcr*. The resulting fusion of Abl with Bcr amino acid sequences deregulates the normally well-controlled Abl protein, causing it to emit growth-promoting signals in a strong, deregulated fashion. (Two other alternative breakpoints in the *bcr* gene result in the formation of the slightly different Bcr–Abl fusion proteins found in other hematopoietic malignancies; these alternative Bcr–Abl proteins are not illustrated in Figure 4.11.)

Since the discovery of the *bcr–abl* translocation, dozens of other, quite distinct translocations have been documented that also result in the formation of hybrid proteins, most but not all of which are encountered in small subsets of hematopoietic tumors (for examples, see **Table 4.5**). Possibly the greatest diversity derives from the *MLL1* (mixed-lineage leukemia) gene, occasionally termed *ALL1*, which encodes a histone methylase (see Section 1.8); it participates in translocations with at least 50 distinct fusion-partner genes, resulting in a diverse array of hybrid proteins, all of which appear to affect chromatin structure and function. Because each type of translocation is found only rarely, the functions of many of the resulting hybrid proteins remain obscure. In the case of solid tumors, an example of such fusion proteins comes from

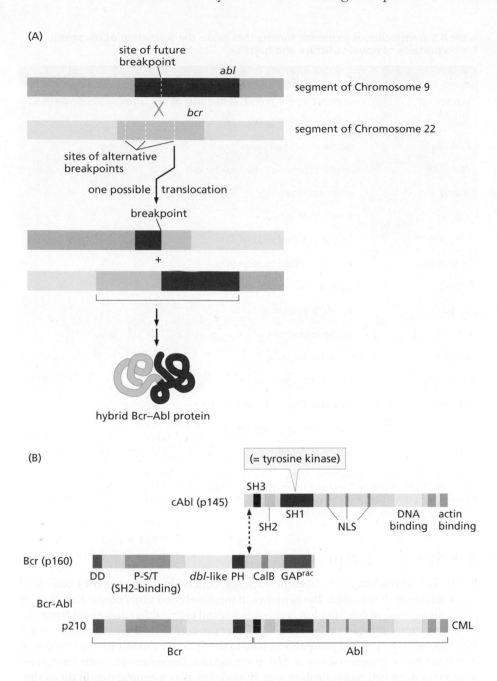

Figure 4.11 Formation of the *bcr–abl* oncogene (A) Reciprocal chromosomal translocations between human Chromosomes 9 and 22, which carry the *abl* and *bcr* genes, respectively, result in the formation of fused, hybrid genes that encode hybrid Bcr–Abl proteins commonly found in chronic myelogenous leukemias (CML). Two other alternative breakpoint sites in the *bcr* gene are involved in *bcr–abl* translocations (*not shown here*) arising in certain other hematopoietic malignancies. (B) The domain structures of the proteins encoded by the normal *bcr* and *abl* genes are shown here (*above and middle*). Each is a multidomain, multifunctional protein, as indicated by the labels attached to the colored areas. The chromosomal translocation found in almost all cases of CML results in the fusion of most of the reading frame of the *Bcr* gene with most of the reading frame of the *Abl* gene (*below*). The approximate locations of the translocation breakpoint are indicated by the double-headed arrow. The fusion causes deregulated firing of the tyrosine kinase of the fusion protein, which is equivalent to the SH1 domain labeled here; this firing (to be discussed in Section 6.3) is responsible for most of the transforming effects of the newly formed oncoprotein. (B, adapted from A.M. Pendergast, in A.M. Carella, *Chronic Myeloid Leukaemia: Biology and Treatment*. London: Martin Dunitz, 2001. With permission from Taylor and Francis.)

a class of pediatric tumors—alveolar rhabdomyosarcomas—80% of which are found to harbor a translocation that fuses the domains of two transcription factors (PAX3 or PAX7 and FKHR), thereby yielding a protein with far more potent transcription-activating powers than either of its two normal precursors possessed. (This novel transcription factor seems to expand the population of skeletal muscle progenitor cells, thereby setting the stage for acquisition of additional mutations that then lead to the eruption of highly aggressive tumors.) Yet other less common chromosomal rearrangements, such as inversions of discrete segments of DNA, may also serve to generate novel oncogenes and often novel oncoproteins.

Every chromosomal rearrangement, even those occurring repeatedly in many tumors of a given type, appears to represent the product of an exceedingly rare genetic accident. Some of these accidents generate deregulated gene or protein expression while others encode novel hybrid proteins. However, once formed, these translocations clearly favor the outgrowth of the cells that carry them, this being indicated by the presence of these chromosomal translocations in all of the neoplastic cells within resulting tumors.

Table 4.5 Translocations in human tumors that cause the formation of oncogenic fusion proteins of novel structure and function

Oncogene	Neoplasm
bcr/abl	chronic myelogenous leukemia; acute lymphocytic leukemia
dek/can	acute myeloid leukemia
E2A/pbx1	acute pre-B-cell leukemia
PML/RAR	acute promyelocytic leukemia
tls/erg	myeloid leukemia
irel/urg	B-cell lymphoma
CBFβ/MYH11	acute myeloid leukemia
aml1/mtg8	acute myeloid leukemia
ews/fli	Ewing's sarcoma
lyt-10/Cα1	B-cell lymphoma
hrx/enl	acute leukemias
hrx/af4	acute leukemias
NPM/ALK	large-cell lymphomas
PAX3/FKHR	alveolar rhabdomyosarcoma
EML4/ALK[a]	non-small-cell lung cancer
MLL/various	acute leukemias

[a]While often termed a translocation, this gene fusion arises from inversion of a short segment of the short arm of Chromosome 2 without the involvement of other chromosomal arms.
Adapted from G.M. Cooper, Oncogenes, 2nd ed. Boston and London: Jones and Bartlett, 1995. With permission from Jones and Bartlett.

4.6 Synopsis and prospects

By the late 1970s, disparate lines of evidence concerning cancer genes coalesced into a relatively simple idea. The genomes of mammals and birds contain a cohort of proto-oncogenes, which function to regulate normal cell proliferation and differentiation. Alterations of these genes that affect either the control of their expression or the structure of their encoded proteins can lead to overly active growth-promoting genes, which appear in cancer cells as activated oncogenes. Once formed, such oncogenes proceed to drive cell multiplication and, in so doing, play a central role in the pathogenesis of cancer.

Many of these cellular genes were originally identified because of their presence in the genomes of rapidly transforming retroviruses, such as Rous sarcoma virus, avian erythroblastosis virus, and Harvey sarcoma virus. Subsequently, transfection experiments revealed the presence of potent transforming genes in the genomes of cells that had been transformed by exposure to chemical carcinogens and cells derived from spontaneously arising human tumors. These tumor cells had no associations with retrovirus infections. Nonetheless, the oncogenes that they carried were found to be related to those carried by transforming retroviruses. This meant that a common repertoire of proto-oncogenes could be activated by two alternative routes: retrovirus acquisition or somatic mutation.

The somatic mutations that caused proto-oncogene activation could be divided into two categories—those that caused changes in the structure of encoded proteins and those that led to elevated, deregulated expression of these proteins. Mutations affecting structure included the point mutations affecting *ras* proto-oncogenes and the chromosomal translocations that yielded hybrid genes such as *bcr–abl*. Elevated expression could be achieved in human tumors through gene amplification or chromosomal translocations, such as those that place the *myc* gene under the control of immunoglobulin enhancer sequences.

These revelations about the role of mutant cellular genes in cancer pathogenesis eclipsed for some years the observations that certain human cancers, some of them quite common, are associated with and likely caused by infectious agents, notably viruses and bacteria. With the passage of time, however, the importance of infections in human cancer pathogenesis became clear. We now know that about one-fifth of human deaths from cancer worldwide are associated with tumors triggered by infectious agents. Thus, most of the 9% of worldwide cancer mortality caused by stomach cancer is associated with long-term infections by *Helicobacter pylori* bacteria. Six percent of cancer mortality is caused by carcinomas of the liver (hepatomas), almost all of which are associated with chronic hepatitis B and C virus infections. And the 5% of cancer mortality from cervical carcinomas is almost entirely attributable to human papillomavirus (HPV) infections (Table 4.6).

Somehow, we must integrate the carcinogenic mechanisms activated by these various infectious agents into the larger scheme of how human cancers arise—the scheme that rests largely on the discoveries, described in this chapter, of cellular oncogenes. In the case of most of the cancer-causing viruses, it is clear that they introduce viral oncogenes into cells that contribute to the transformation phenotype. Included here are HPV, MCV, Epstein–Barr virus (EBV) and human herpesvirus type 8 (HHV-8).

Most of these viruses carry at least one gene that alters the physiology of the cells that it infects. In each case, it seems that these virus-induced changes are driven by the need of the virus to create an intracellular environment that is more hospitable for its own replication. Of note, the viral genes that induce these changes, which can fairly be termed viral oncogenes, have resided in viral genomes for millions—if not hundreds of millions—of years rather than being rare aberrations of the sort that have led to the formation of transforming retroviruses like Rous sarcoma virus (RSV). Hence, cancer, to the extent that it ensues from infections by these viruses, is an unintended side-effect of the changes favoring viral replication. In Chapters 8 and 9 we shall see precisely how several

Table 4.6 **Viruses implicated in human cancer causation**

Virus[a]	Virus family	Cells infected	Human malignancy	Transmission route
EBV[b]	Herpesviridae	B cells	Burkitt's lymphoma	saliva
		oropharyngeal epithelial cells	nasopharyngeal carcinoma	saliva
		lymphoid	Hodgkin's disease[c]	?
HTLV-I	Retroviridae	T cells	non-Hodgkin's lymphoma	parenteral, venereal[d]
HHV-8[e]	Herpesviridae	endothelial cells	Kaposi's sarcoma, body cavity lymphoma	venereal, vertical[d]
HBV	Hepadnaviridae	hepatocytes	hepatocellular carcinoma	parenteral, venereal
HCV	Flaviviridae	hepatocytes	hepatocellular carcinoma	parenteral
HPV	Papillomaviridae	cervical epithelial	cervical carcinoma	venereal
JCV[f]	Polyomaviridae	central nervous system	astrocytoma, glioblastoma	?
MCV	Polyomaviridae	Merkel cells in skin	Merkel cell carcinoma	?

[a]Most of these viruses carry one or more growth-promoting genes/oncogenes in their genomes.
[b]EBV is also involved in transplant-related lymphoproliferative disorder, gastric cancer, and cutaneous leiomyosarcoma (in patients with AIDS).
[c]These lymphomas, which bear copies of EBV genomes, appear in immunosuppressed patients.
[d]Parenteral, blood-borne; venereal, via sexual intercourse; vertical, parent to child.
[e]Also known as KSHV, Kaposi's sarcoma herpesvirus.
[f]JCV (JC virus, a close relative of SV40) infects more than 75% of the human population by age 15, but the listed virus-containing tumors are not common. Much correlative evidence supports the role of JCV in the transformation of human central nervous system cells, but evidence of a causal role in tumor formation is lacking.
Adapted in part from J. Butel, *Carcinogenesis* 21:405–426, 2000.

of these viral oncogenes function to deregulate cell proliferation and protect cells from apoptosis (programmed cell death), two critical changes that contribute to the formation of human cancers. Only in Chapter 11 do we return to HBV and HCV and the mechanisms by which they induce cancer without deploying viral oncogenes and oncoproteins.

In the next two chapters we focus on the cellular genes that trigger cancer as a consequence of somatic mutations and the various biochemical mechanisms-of-action of the encoded oncoproteins. At first, the discoveries of cellular oncogenes seemed, on their own, to provide the definitive answers about the origins and growth of human tumors. But soon it became clear that cellular oncogenes and their actions could explain only a part of human cancer development. Yet other types of genetic elements were clearly involved in programming the runaway proliferation of cancer cells. Moreover, a definitive understanding of cancer development depended on understanding the complex machinery inside cells that enables them to respond to the growth-promoting signals released by oncoproteins. So, the discovery of cellular oncogenes was a beginning, a very good one, but still only a beginning. Two decades of further work were required before a more complete understanding of cancer pathogenesis could be assembled.

Key concepts

- Unable to find tumor viruses in the majority of human cancers, researchers in the mid-1970s were left with one main theory of how most human cancers arise: the mutation by carcinogens of normal growth-controlling genes, converting them into oncogenes.

- To verify this model's prediction that transformed cells carry mutated genes functioning as oncogenes, a novel experimental strategy was devised: DNA from chemically transformed cells was introduced into normal cells—the procedure of transfection—and the recipient cells were then monitored to determine whether they too had become transformed.

- Cultures of NIH 3T3 cells that had been transfected with DNA from chemically transformed mouse cells yielded numerous transformants, which proved to be both anchorage-independent and tumorigenic; this indicated that the chemically transformed cells carried genes that could function as oncogenes and that oncogenes could arise in the genomes of cells independently of viral infections.

- Further experiments using human tumor donor cells transfected into murine cells showed that the oncogenes could act across species and tissue boundaries to induce cell transformation.

- The oncogenes detected in human tumor cells by transfection experiments and the oncogenes of transforming retroviruses were both found to derive from the same group of preexisting normal cellular genes. This meant that normal proto-oncogenes could be activated either by retrovirus-induced modifications or by somatic mutations.

- Proto-oncogenes were often found to be present in increased copy number in human tumor cell genomes, which suggested that gene amplification resulted in increased levels of protein products that favored the proliferation of cancer cells, an example being the amplification of the *myc* gene in a variety of human cancer types.

- The activation of proto-oncogenes via deregulation by retro viral transcriptional promoters was well documented. However, the mechanism(s) by which normal human proto-oncogenes became converted into oncogenes in the absence of viruses was not evident until the discovery of a point mutation in the H-*ras* gene that yielded a structurally altered protein with aberrant behavior. This established a new mechanism for oncogene activation based on a change in the structure of an oncogene-encoded protein rather than the expression levels of such a protein.

- Both activation mechanisms—regulatory and structural—might collaborate to create an active oncogene.

- The *myc* oncogene was initially discovered in the genome of an avian retrovirus. This oncogene was found to be activated through three alternative mechanisms in cancer cells: provirus integration, gene amplification, and chromosomal translocation.

- Gene amplification occurs through preferential replication of a segment (an amplicon) of chromosomal DNA. The result may be repeating end-to-end linear arrays of the segment, which appear under the light microscope as homogeneously staining regions (HSRs) of a chromosome. Alternatively, the region carrying the amplified segment may break away from the chromosome and be seen as small, independently replicating extrachromosomal particles (double minutes, DMs).

- Translocation involves the fusion of a region from one chromosome to a nonhomologous chromosome. Translocation can place a gene under the control of a foreign transcriptional promoter and lead to its overexpression, as is the case with the *myc* oncogenes in Burkitt's lymphomas. Translocation may also free an mRNA from inhibition by a microRNA by removing mRNA sequences normally recognized by the miRNA. Alternatively, a translocation can result in the fusion of two protein-coding sequences, leading to a hybrid protein that functions differently than the two normal proteins from which it arose, as is seen in the Bcr–Abl protein encountered in chronic myelogenous leukemias.

- Besides the amino acid substitutions that activate signaling by the Ras oncoprotein and the fusion of protein domains, as is seen in the Bcr–Abl protein, yet other changes in protein structure can lead to oncogene activation. For example, truncation of the EGF receptor leads this protein to emit growth-promoting signals in an unremitting fashion.

Thought questions

1. What evidence do we have that suggests that endogenous retrovirus genomes play little if any role in the development of human cancers?

2. Why might an assay like the transfection-focus assay fail to detect certain types of human tumor-associated oncogenes?

3. What molecular mechanisms might cause a certain region of chromosomal DNA to accidentally undergo amplification?

4. How many distinct molecular mechanisms might be responsible for converting a single proto-oncogene into a potent oncogene?

5. How many distinct molecular mechanisms might allow chromosomal translocations to activate proto-oncogenes into oncogenes?

6. What experimental search strategies would you propose if you wished to launch a systematic screening of a vertebrate genome in order to enumerate all of the proto-oncogenes that it harbors?

7. Since proto-oncogenes represent distinct liabilities for an organism, in that they can incite cancer, why have these genes not been eliminated from the genomes of chordates?

Additional reading

Albertson DG (2006) Gene amplification in cancer. *Trends Genet.* 22, 447–455.

Bos JL (1989) *Ras* oncogenes in human cancer: a review. *Cancer Res.* 49, 4682–4689.

Bos JL, Rehmann H & Wittinghofer A (2007) GEFs and GAPs: critical elements in the control of small G proteins. *Cell* 129, 865–877.

Boxer LM & Dang CV (2001) Translocations involving c-*myc* and c-*myc* function. *Oncogene* 20, 5595–5610.

Brodeur G (2003) Neuroblastoma: biological insights into a clinical enigma. *Nat. Rev. Cancer* 3, 203–216.

Butel J (2000) Viral carcinogenesis: revelation of molecular mechanisms and etiology of human disease. *Carcinogenesis* 21, 405–426.

Edwards PAW (2010) Fusion genes and chromosome translocations in the common epithelial cancers. *J. Pathol.* 220, 244–254.

Martin GS (2001) The hunting of the Src. *Nat. Rev. Mol. Cell Biol.* 2, 467–475.

Mitelman F, Johansson B & Mertens F (eds) (2004) Mitelman Database of Chromosome Aberrations in Cancer. http://cgap.nci.nih.gov/Chromosomes/Mitelman.

Mitelman F, Johansson B & Mertens F (2007) The impact of translocations and gene fusions on cancer causation. *Nat. Rev. Cancer* 7, 233–245.

Moasser M (2007) The oncogene HER2: its signaling and transforming functions and its role in human cancer pathogenesis. *Oncogene* 26, 6469–6487.

Moore PS & Chang Y (2010) Why do viruses cause cancer? Highlights of the first century of human tumour virology. *Nat. Rev. Cancer* 10, 879–889.

Morris DS, Tomlins SA, Montie JE & Chinnaiyan AM (2008) The discovery and application of gene fusions in prostate cancer. *BJU Int.* 102, 276–281.

Nesbit CE, Tersak JM & Prochownik EV (1999) *Myc* oncogenes and human neoplastic disease. *Oncogene* 18, 3004–3016.

Pylayeva-Gupta Y, Grabocka E & Bar-Sagi D (2011) RAS oncogenes: weaving a tumorigenic web. *Nat. Rev. Cancer* 11, 761–774.

Santarius T, Shipley J, Brewers D et al. (2010) A census of amplified and overexpressed human cancer genes. *Nat. Rev. Cancer* 10, 59–64.

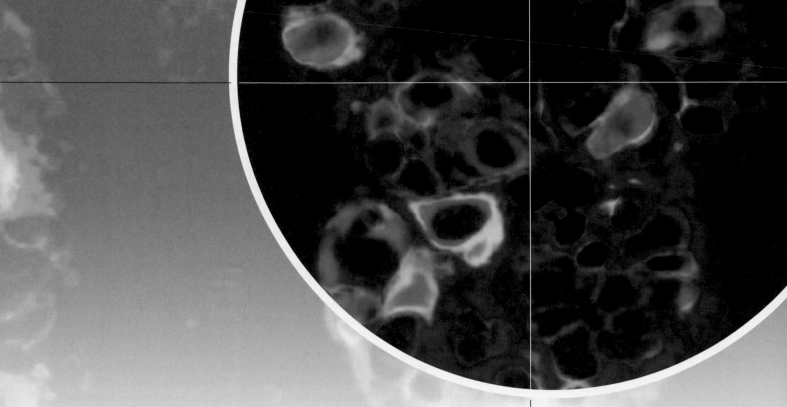

Chapter 5

Growth Factors, Receptors, and Cancer

Chapter opener photo: From J.S. de Jong et al., *J. Pathol.* 184:44-52, 1998. With permission from John Wiley & Sons.

The ability of certain fetal serums to stimulate cell growth and the decreased requirement for such factors by transformed cells may be due to the fact that these serum factors are the same or similar to the transforming factors synthesized by some embryonic or neoplastic cells.

David E. Comings, geneticist, 1973

The discovery of oncogenes and their precursors, the proto-oncogenes, stimulated a variety of questions. The most pointed of these centered on the issue of how the oncogenes, acting through their encoded protein products, succeed in perturbing cell behavior so profoundly. A variety of cell phenotypes were concomitantly altered by the actions of oncoproteins such as Src and Ras, the products of the *src* and *ras* oncogenes. How could a single protein species succeed in changing so many different cellular regulatory pathways at the same time?

The vital clues about oncoprotein functioning came from detailed studies of how normal cells regulate their growth and division. Normal cells receive growth-stimulatory signals from their surroundings. These signals are processed and integrated by complex circuits within the cell, which decide whether or not cell growth and division are appropriate.

This need to receive extracellular signals at the cell surface and to transfer them into the cytoplasm creates a challenging biochemical problem. The extra- and intracellular spaces are separated by the lipid bilayer that is the plasma membrane. This membrane is a barrier that effectively blocks the movement of virtually all but the smallest molecules through it, resulting in dramatically differing concentrations of many types of molecules (including ions) on its two sides. How have cells managed to solve the problem of passing (**transducing**) signals through a membrane that is almost

Movies in this chapter

5.1 Cellular Effects of EGF vs. HGF

5.2 Activation of Kit Receptor Signaling by SCF

5.3 Ras

5.4 IGF Receptors and Monoclonal Antibodies

impervious? And given this barrier, how can the inside of the cell possibly know what is going on in the surrounding extracellular space?

These signaling processes are part of the larger problem of cell-to-cell communication. Indeed, the evolution of the first multicellular animals (metazoa) more than 600 million years ago depended on the development of biochemical mechanisms that allow cells to receive and process signals arising from their neighbors within tissues. Without effective intercellular communication, the behavior of individual cells could not be coordinated, and the formation of architecturally complex tissues and organisms was inconceivable. Obviously, such communication depended on the ability of some cells to emit signals and of others to receive them and respond in specific ways. By necessity, these signals must pass through extracellular spaces en route from signal-emitting cells to those cells that are destined to receive these signals.

In very large part, the signals passed between cells are conveyed by proteins. This fact also required evolutionary innovations: signal emission requires an ability by some cells to release proteins into the nearby extracellular space. Such release—the process of protein secretion—is also complicated by the imperviousness of the plasma membrane. Once released into the extracellular space, the designated recipient cells must be able to *sense* the presence of these proteins in their surroundings. Much of this chapter is focused on this second problem—how normal cells receive signals from the environment that surrounds them. As we will see, the deregulation of this signaling is central to the formation of cancer cells.

5.1 Normal metazoan cells control each other's lives

As already implied, the normal versions of oncogene-encoded proteins often serve as components of the machinery that enables cells to receive and process biochemical signals regulating cell proliferation. Therefore, to truly appreciate the complexities of oncogene and oncoprotein function, we need to understand the details of how normal cell proliferation is governed.

Our entrée into this discussion comes from a basic and far-reaching principle. Proper tissue architecture depends absolutely on maintaining appropriate proportions of different constituent cell types within a tissue, on the replacement of missing cells, and on discarding extra, unneeded cells (**Figure 5.1**). Wounds must be repaired, and attacks by foreign infectious agents must be warded off through the concerted actions of many cells within the tissue.

Decisions about growth versus no-growth must be made for the welfare of the entire tissue and whole organism, not for the benefit of its individual component cells. For this reason, no single cell within the condominium of cells that is a living tissue can be granted the autonomy to decide on its own whether it should proliferate or remain in a nongrowing, quiescent state. Instead, this weighty decision can be undertaken only after consultation with other cells within the tissue. These neighbors may provide a particular cell with growth-stimulating signals that stimulate its proliferation or release growth-inhibitory signals that discourage it. In the end, all the decisions made by an individual cell about its proliferation must, by necessity, represent a consensus decision shared with the cells that reside in its neighborhood. Much of this incessant back-and-forth chatter is conveyed by **growth factors (GFs)**. These are relatively small proteins that are released by some cells, make their way through intercellular space, and eventually impinge on yet other cells, carrying with them specific biological messages. Growth factors convey many of the signals that tie the cells within a tissue together into a single community, all members of which are in continuous communication with their respective neighbors.

The dependence of individual cells on their surroundings is illustrated nicely by the behavior of normal cells when they are removed from living tissue and propagated in a Petri dish—the procedures that are termed tissue culture (see Supplementary Sidebar 3.2). Even though the liquid medium placed above the cells contains all the

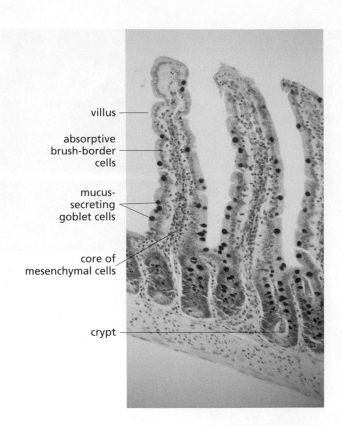

villus

absorptive
brush-border
cells

mucus-
secreting
goblet cells

core of
mesenchymal cells

crypt

Figure 5.1 Maintenance of tissue architecture The epithelial lining of the small intestine shown here illustrates the fact that a number of distinct cell types coexist within a tissue. The relative numbers and positions of each cell type must be tightly controlled in order to ensure proper tissue structure and function. This control is largely achieved via the exchange of signals between neighboring cells within the tissue. In this particular epithelium, cell-to-cell signaling also ensures that new epithelial cells—termed *enterocytes* in the intestine—are continually being generated in the crypts (at the bases of fingerlike villi) in order to replace others that have migrated up the sides of the villi to the tips, where they are sloughed off. Included among the epithelial cells are mucus-secreting goblet cells (*dark red*) and absorptive cells, as well as enteroendocrine cells and Paneth cells (*not shown*). In addition, in the middle of each villus is a core of mesenchymal cell types, which together constitute the stroma and include fibroblasts, endothelial cells, pericytes, and macrophages. (From B. Alberts et al., Molecular Biology of the Cell, 6th ed. New York: W.W. Norton, 2015.)

nutrients required to sustain their growth and division, including amino acids, vitamins, glucose, and salts, such medium, on its own, does not suffice to induce these cells to proliferate. Instead, this decision depends upon the addition to the medium of serum, usually prepared from the blood of calves or fetal calves. Serum contains the growth factors that persuade cells to multiply.

Serum is produced when blood is allowed to clot. The blood platelets adhere to one another and form a matrix that gradually contracts and traps most of the cellular components of the blood, including both the white and red cells. In the context of a wounded tissue, this clot formation is designed to stanch further bleeding. The clear fluid that remains after clot formation and retraction constitutes the serum.

While the platelets in a wound site aggregate as part of clot formation, they also initiate the wound healing process. This process is mediated via the release into the extracellular space of granules carrying a complex mixture of GFs (see Figure S5.1), notably platelet-derived growth factor (PDGF), epidermal growth factor (EGF), transforming growth factor-β1 (TGF-β1), vascular endothelial growth factor (VEGF), insulin-like growth factor (IGF), and fibroblast growth factor 2 (FGF-2) (**Figure 5.2**), which act on diverse cell types in the damaged tissue, all of which must be enlisted in the task of reconstructing the damaged tissue. Portrayed differently, this complex mixture of released GFs hints at the complexity of the wound-healing process, in which the proliferation of these multiple distinct cell types must be coordinated.

Secreted proliferation-stimulating factors, such as most of these GFs, are often termed **mitogens**, to indicate their ability to induce cells to proliferate. PDGF, for its part, is a potent stimulator of fibroblasts, which form much of the connective tissue including the cell layers beneath epithelia (see Figure 2.4). Its release during clot formation attracts fibroblasts into the wound site and then stimulates their proliferation (see, for example, Figure 5.2A). Other released GFs, such as epidermal growth factor (EGF), can cause dramatic changes in cell shape in addition to stimulating the proliferation of epithelial cells (see Figure 5.2B). Without stimulation by serum-derived PDGF, cultured fibroblasts will remain viable and metabolically active for weeks in a Petri dish, but they will not grow and divide.

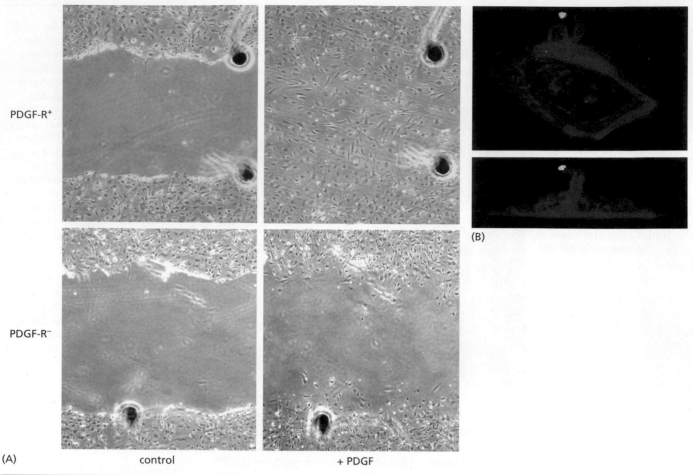

(A)   control   + PDGF

(B)

PDGF-R⁺

PDGF-R⁻

Figure 5.2 Effects of growth factors on cells (A) The PDGF released by platelets upon clotting is a potent attractant and mitogen for fibroblasts, which swarm into a wound site in order to reconstruct the tissue. In this *in vitro* model of wound healing, two monolayers of fibroblasts in Petri dishes have been wounded by the tip of a pipette (*left two panels*) and one form of PDGF (PDGF-BB) has been added to their growth medium. While large numbers of wild-type fibroblasts (*top left*) proliferate and migrate into the wound site within 48 hours (*top right*), mutant fibroblasts that have been rendered unresponsive to PDGF (through the loss of a PDGF cell-surface receptor, PDGF-R, that enables them to sense and respond to the presence of PDGF; *lower left*) fail to do so (*lower right*). (B) Five minutes after epidermal growth factor (EGF) that has been immobilized on a bead (*green dot*) is applied to a mouse mammary carcinoma cell, the cell has reorganized its actin cytoskeleton (*red*) and extended an arm of cytoplasm toward the growth factor (*above*, top view; *below*, side view). (A, from Z. Gao et al., *J. Biol. Chem.* 280:9375–9389, 2005. Copyright © 2005 ASBMB. This article is distributed under a Creative Commons Attribution 4.0 International license. B, from S.J. Kempiak et al., *J. Cell Biol.* 162:781–787, 2003. With permission from Rockefeller University Press.)

Sidebar 5.1 Cell-to-cell communication involves far more than secreted growth factors In fact, soluble growth factors represent only a small subset of the proteins that are secreted by cells and potentially contribute to the intercommunication between them. Most important is the network of secreted proteins that assemble to form the extracellular matrix (ECM) as discussed in Chapter 1 (see Figure 1.11C). More than 250 distinct proteins form the structural backbone of the ECM with up to three times as many additional secreted proteins associating in various ways with the ECM and modifying its structure. Some of the latter, largely proteases, continuously remodel the ECM and the GFs that are present in the extracellular space. We focus in this chapter largely on the GFs. Nevertheless, our narrative overlooks many of the still-poorly understood functions of almost all ECM components, ECM-associated proteins, and secreted proteases. Together, they form the **secretomes** of cells.

As developed in the rest of this chapter, many oncogene-encoded proteins usurp the signaling molecules that enable cells to sense the presence of growth factors in their surroundings, to convey this information into the cell interior, and to process this information. By taking charge of the natural growth-stimulating machinery of the cell, oncoproteins are able to delude a cell into believing that it has encountered growth factors in its surroundings. Once this deception has taken place, the cell responds slavishly by beginning to proliferate, just as it would have done if abundant growth factors were actually present in the medium around it.

This dependence of fibroblasts on growth-stimulating signals released by a second cell type, in this case, blood platelets, mirrors hundreds of similar cell-to-cell communication routes that operate within living tissues to encourage or discourage cell proliferation. Moreover, as discussed in greater detail below, PDGF and EGF are only two of a large and disparate group of growth factors that help to convey important growth-controlling messages from one cell to another (**Sidebar 5.1**).

5.2 The *Src* protein functions as a tyrosine kinase

The first clues to how cell-to-cell signaling via growth factors operates came from biochemical analysis of the v-*src* oncogene and the protein that it specifies. The trail of clues led, step-by-step, from this protein to the **receptors** used by cells to detect growth factors and, in turn, to the intracellular signaling pathways that control normal and malignant cell proliferation.

The initial characterization of the v-*src*–encoded oncoprotein attracted enormous interest. This was the first cellular oncoprotein to be studied and therefore had the prospect of giving cancer researchers their first view of the biochemical mechanisms of cell transformation. After the molecular cloning of the *src* oncogene, the amino acid sequence of its protein product was deduced directly from the nucleotide sequence of the cloned gene. This sequence revealed that the encoded protein was synthesized as a polypeptide chain of 533 amino acid residues and had a mass of almost precisely 60 kilodaltons (kD).

The amino acid sequence of Src provided few clues as to how this oncoprotein functions to promote cell proliferation and transformation. Avian and mammalian cells transformed by the v-*src* oncogene were known to exhibit a radically altered shape, to pump in glucose from the surrounding medium more rapidly than normal cells, to grow in an anchorage-independent fashion, to lose contact inhibition, and to form tumors. Whatever the precise mechanisms of action, it was obvious that Src affected, directly or indirectly, a wide variety of intracellular targets.

Clever biochemical sleuthing in 1977–1978 solved the puzzle of how Src operates. Antibodies were produced that specifically recognized and bound Src (Supplementary Sidebar 5.1). The antibody molecules were found to become phosphorylated when they were incubated in solution with both Src and adenosine triphosphate (ATP), the universal phosphate donor of the cell. Phosphorylation was known to involve the covalent attachment of phosphate groups to the side chains of specific amino acid residues. Hence, it was clear that Src operates as a protein **kinase**—an enzyme that removes a high-energy phosphate group from ATP and transfers it to a suitable protein substrate, in this instance, an antibody molecule (**Figure 5.3**). While Src does not normally phosphorylate antibody molecules, its ability to do so in these early experiments suggested that its usual mode of action also involved the phosphorylation of certain protein substrates within cells.

Independent of this kinase activity, Src was itself a **phosphoprotein**, that is, it carried phosphate groups attached covalently to one or more of its amino acid side chains. This indicated that Src also served as a substrate for phosphorylation by a protein kinase—either phosphorylating itself (**autophosphorylation**) or serving as the substrate of yet another kinase. Its molecular weight and phosphoprotein status caused it initially to be called pp60^{src}, although hereafter we will refer to it simply as Src.

The fact that Src functions as a kinase was a major revelation. In principle, a protein kinase can phosphorylate multiple, distinct substrate proteins within a cell. In the case of Src, more than 50 distinct substrates have since been enumerated (**Figure 5.4A**). Once phosphorylated, each of these substrate proteins may be functionally altered and proceed, in turn, to alter the functions of its own set of downstream targets. This mode of action (Figure 5.4B and C) seemed to explain how a protein kinase could act pleiotropically to perturb multiple cell phenotypes. At the time of this discovery, other protein kinases had already been found able to regulate complex circuits, notably those involved in carbohydrate metabolism.

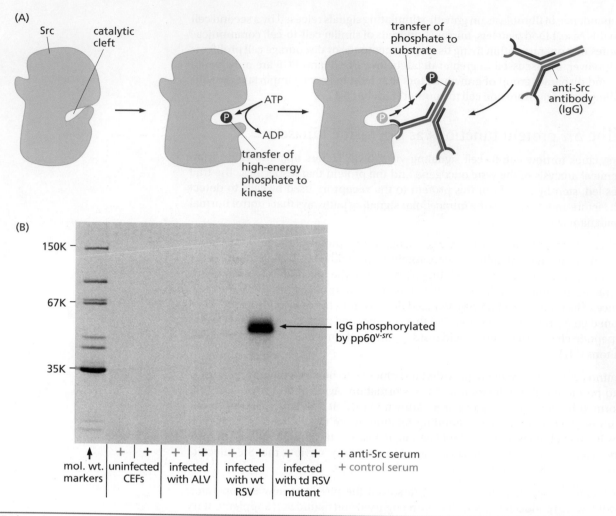

Figure 5.3 Phosphorylation of a precipitating antibody molecule by Src Protein kinases operate by removing the high-energy γ phosphate group from ATP and attaching it to the hydroxyl groups in the side chains of serine, threonine, or tyrosine residues of substrate proteins. (A) The antibody molecule that was used to immunoprecipitate Src molecules (see Supplementary Sidebar 5.1) also could serve as a substrate for phosphorylation by this kinase. In cells, Src phosphorylates a wide range of protein substrates. (B) This experiment revealed, for the first time, the biochemical activity associated with an oncoprotein. Normal rabbit serum (*blue "+" signs*) or serum from a rabbit bearing an RSV-induced, Src-expressing tumor (which contained antibodies against Src; *red "+" signs*) was used to immunoprecipitate cell lysates that were incubated with $^{32}PO_4$-radiolabeled ATP. Lysates were prepared from uninfected chicken embryo fibroblasts (CEFs); CEFs infected with avian leukosis virus (ALV), which lacks a *src* oncogene; CEFs infected with wild-type (wt) RSV; or CEFs infected with a transformation-defective (td) mutant of RSV, which lacks a functional *src* gene. Only the combination of tumor-bearing rabbit serum and a lysate from wt RSV–infected CEFs yielded a strongly ^{32}P-labeled protein that co-migrated with the precipitating antibody, indicating that the antibody molecule had become phosphorylated by the Src oncoprotein. (B, from M.S. Collett and R.L. Erikson, *Proc. Natl. Acad. Sci. USA* 75:2021–2024, 1978. With permission from the authors.)

Figure 5.4 Actions of protein kinases (A) This autoradiograph of proteins resolved by gel electrophoresis illustrates the fact that the phosphorylation state of a number of distinct protein species within a cell can be altered by the actions of the Src kinase expressed in cells by Rous sarcoma virus. This immunoblot analysis, using an anti-phosphotyrosine antibody, specifically detects proteins that carry phosphotyrosine residues, the known products of Src kinase action (see Supplementary Sidebar 5.2). The left channel contains proteins from a lysate of untransformed mouse NIH 3T3 fibroblasts and is essentially blank, indicating the relative absence of phosphotyrosine-containing proteins in cells that lack Src expression. The right channel shows the phosphorylated proteins in cells expressing an active Src kinase protein. (B) The pleiotropic actions of a protein kinase usually derive from its ability to phosphorylate and thereby modify the functional state of a number of distinct substrate proteins. Illustrated here are the actions of the Akt/PKB kinase, a serine/threonine kinase that can influence a wide variety of biological processes through phosphorylation of the indicated major control proteins. (A number of other known substrates of this kinase are not shown here.) Thus, Akt/PKB can inactivate the antiproliferative actions of GSK-3β, the pro-apoptotic powers of Bad, and the translation-inhibiting actions of TSC2, and it can activate the angiogenic (blood vessel–inducing) powers of HIF-1α. In this diagram and throughout this book, arrowheads denote stimulatory signals, while lines at right angles (crossbars) denote an inhibitory signal. (C) A more typical behavior of a protein kinase is illustrated here. The ability of a serine/threonine kinase termed ATM, which evokes a highly complex set of biochemical changes in a cell in response to detected DNA damage (discussed in Chapter 12), has been associated with its phosphorylation of many hundreds of distinct protein substrates, most validated by direct biochemical analyses. (A, from S.M. Ulrich et al., *Biochemistry* 42:7915–7921, 2003. With permission from American Chemical Society. C, courtesy of Y. Shiloh.)

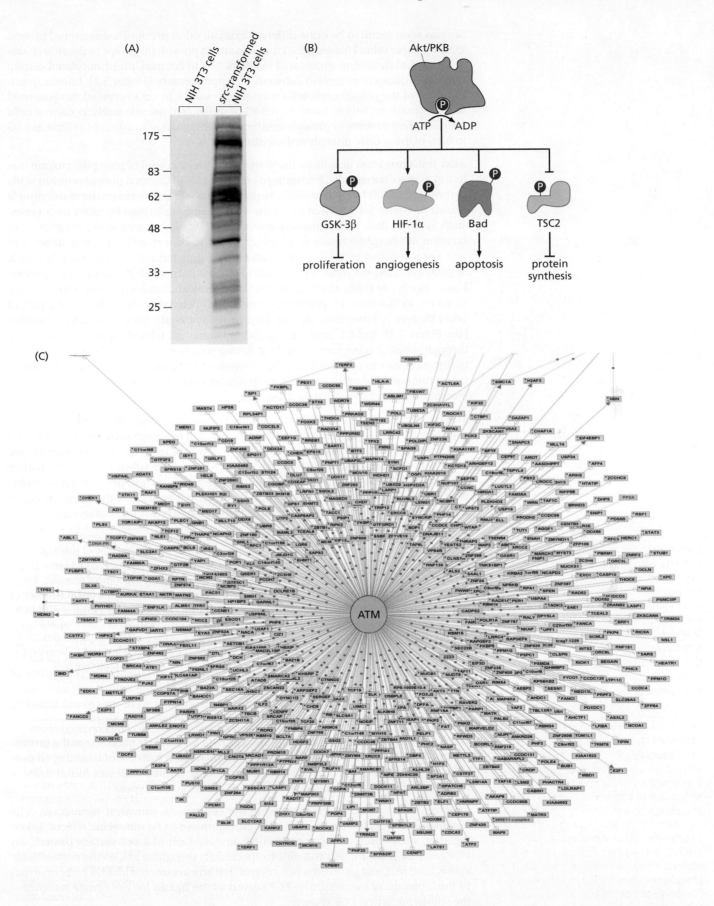

Src was soon found to be quite different from all other previously discovered protein kinases. These other kinases were known to attach phosphate groups to the side chains of serine and threonine amino acid residues. Src, in contrast, phosphorylated certain tyrosine residues of its protein substrates (Supplementary Sidebar 5.2). Careful quantification of the phosphorylated amino acid residues in cells revealed how unusual this enzymatic activity is. More than 99% of the phosphoamino acids in normal cells are phosphothreonine or phosphoserine; phosphotyrosine constitutes as little as 0.05 to 0.1% of these cells' total phosphoamino acids.

After transformation of cells by the v-*src* oncogene, the level of phosphotyrosine was found to rise dramatically, becoming as much as 1% of the total phosphoamino acids in these cells. (This helps to explain the dramatic differences between the two channels shown in Figure 5.4A.) When the same cells were transformed by other oncogenes, such as H-*ras*, their phosphoamino acid content hardly changed at all. Therefore, the creation of phosphotyrosine residues appeared to be a specific attribute of Src and was not associated with all mechanisms of cell transformation. The multiple bands of proteins phosphorylated by Src provided another insight: the ability of a protein kinase like Src to evoke multiple distinct biochemical changes in a cell (representing its ability to function in a **pleiotropic** fashion) is derived from its ability, and that of other kinases, to phosphorylate a multitude of functionally distinct protein substrates (see Figure 5.4B and C). Later studies yielded a second, related conclusion: signaling through tyrosine phosphorylation is a device that is used largely by mitogenic signaling pathways in mammalian cells, whereas the kinases that are involved in regulating thousands of other processes rely almost exclusively on serine and threonine phosphorylation to convey their messages.

Mutant forms of Src that had lost the ability to phosphorylate substrate proteins also lost their transforming powers. So the accumulating evidence converged on the idea that Src succeeds in transforming cells through its ability to act as a tyrosine kinase (TK) and phosphorylate a still-uncharacterized group of substrate proteins within these cells. While an important initial clue, these advances only allowed a reformulation of the major question surrounding this research area. Now the problem could be restated as follows: How did the phosphorylation of tyrosine residues result in cell transformation?

5.3 The EGF receptor functions as a tyrosine kinase

After the cloning and sequencing of the v-*src* oncogene, the oncogenes of a number of other acutely transforming retroviruses were isolated by molecular cloning and subjected to DNA sequencing. Unfortunately, in the great majority of cases, the amino acid sequences of their protein products provided few clues about biochemical function. The next major insights came instead from another area of research. Cell biologists interested in how cells exchange signals with one another isolated a variety of proteins involved in cell-to-cell signaling and determined the amino acid sequences of these proteins. Quite unexpectedly, close connections were uncovered between these signaling proteins and the protein products of certain oncogenes.

This line of research began with epidermal growth factor (EGF), the first of the growth factors to be discovered. EGF was initially characterized because of its ability to provoke premature eye opening in newborn mice. Soon after, EGF was found to have mitogenic effects when applied to a variety of epithelial cell types.

EGF was able to bind to the surfaces of the cells whose growth it stimulated; cells to which EGF was unable to bind were unresponsive to its mitogenic effects. Taken together, such observations suggested the involvement of a cell surface protein, an EGF *receptor* (EGF-R), that was able to specifically recognize EGF in the extracellular space, bind to it, and inform the cell interior that an encounter with EGF had occurred. In the language of biochemistry, EGF served as the **ligand** for its cognate receptor—the still-hypothetical EGF receptor.

Isolation of the EGF-R protein proved challenging because this receptor, like many others, is usually expressed at very low levels in cells. Researchers circumvented

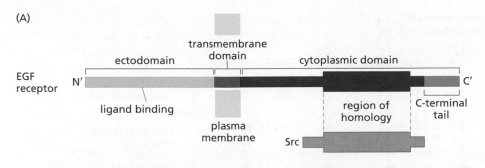

(A)

EGF receptor

Figure 5.5 Structure of the EGF receptor (A) The receptor for epidermal growth factor (EGF-R) is a complex protein with an extracellular domain (ectodomain, *green*), a transmembrane domain that threads its way through the plasma membrane (*brown*), and a cytoplasmic domain (*red, blue*). The ligand-binding domain is responsible for binding EGF. Similarity in amino acid sequences (see panel B) demonstrates that one region of the cytoplasmic domain (*wide red rectangle*) is related to a region in the Src oncoprotein (*gray*). (B) Comparison of the amino acid sequences (using the single-letter code) of the cytoplasmic domains of the EGF-R (labeled "ErbB" here) and Src revealed areas of sequence identity (*green*), suggesting that the EGF-R, like Src, emits signals by functioning as a tyrosine kinase. While the sequence identities seem to be quite scattered, the shared residues nonetheless indicate clear evolutionary relatedness (homology) between Src and ErbB; such homology of structure usually indicates analogous functions. Yet other viral oncoproteins, such as Abl and Fes, some of whose sequences are also shown here, were found to share some sequence similarity with these two. The viral oncoproteins specified by the *raf* and *mos* oncogenes, which function as serine/threonine kinases, are seen to be more distantly related, sharing even fewer sequences with Src and the EGF-R. The dashes indicate amino acid residues that are missing in one protein but are present in one or more of its homologs; these have been introduced to maximize the sequence alignment between homologs. (Adapted from G.M. Cooper, Oncogenes, 2nd ed. Sudbury, MA: Jones and Bartlett, 1995. www.jblearning.com. Reprinted with permission.)

(B)

```
Src   RESLRLEAKLGGGCFGEVVWMGTWNDTTR-----VAIKTLKPGTMSPE--AFLQEAQVMKK
Abl   RTDITMKHKLGGGQYGEVYEGVWKKYSLT----VAVKTLKEDTMEVE--EFLKEAAVMKE
Fes   HEDLVLGEQIGRGNFGEVFSGRLRADNTL----VAVKSCRETLPPDIKAKFLQEAKILKQ
ErbB  ETEFKKVKVLGSGAFGTIYKGLWIPEGEKVKIPVAIKELREATSPKANKEILDEAYVMAS
Raf   ASEVMLSTRIGSGSFGIVYKGKWHGD-------VAVKILKVVDPTPEQLQAFRNEVAVLR
Mos   WEQVCLMHRLGSGGFGSVYKATYHGVP------VAIKQVNKCTEDLRASQRSFWAELNIA

Src   LR-HEKLVQLYAVVSEEP-IYIVIEYMSK-----GSLLDFLKGEMGKY--------LRLP
Abl   IK-HPNLVQLLQVCTREPPFYIITEFMTY-----GNLLDYLRECNRQE--------VSAV
Fes   YS-HPNIVRLIGVCTQKQPIYIVMELVQG-----GDFLTFLRTEGAR---------LRMK
ErbB  VD-NPHVCRLLGICLTST-VQLITQLMPY-----GCLLDYIREHKDN---------IGSQ
Raf   KTRHVNI--LLFMGYMTKDNLAIVTQWCEGSSLYKHLHVQETK------------FQMF
Mos   GLRHDNIVRVVAASTRTPEDSNSLGTIIMEFGGNVTLHQVIYDATRSPEPLSCRKQLSLG

Src   QLVDMAAQIASGMAYVERMNYVHRDLRAANILVGENLVCKVADFGLARLIEDNEYTARQG
Abl   VLLYMATQISSAMEYLEKKNFIHRDLAARNCLVGENHLVKVADFGLSRLMTGDTYTAHAG
Fes   TLLQMVGDAAAGMEYLESKCCIHRDLAARNCLVTEKNVLKISDFGMSREEADGVYAASGG
ErbB  YLLNWCVQIAKGMNYLEERRLVHRDLAARNVLVKTPQHVKITDFGLAKLLGADEKEYHAE
Raf   QLIDIARQTAQGMDYLHAKNIIHRDMKSNNLFLHEGLTVKIGDFGLATVKSRWSGSQQVE
Mos   KCLKYSLDVVNGLLFLHSQSILHLDLKPANILISEQDVCKISDFGCSQKLQDLRGRQASP

Src   AKF--PIKWTAPEAALYGR---FTIKSDVWSFGILLTELTTKGRVPYPGMVNREVLDQVE
Abl   AKF--PIKWTAPESLAYNK---FSIKSDVWAFGVLLWEIATYGMSPYPGIDLSQVYELLE
Fes   LRLV-PVKWTAPEALNYGR---YSSESDVWSFGILLWETFSLGASPYPNLSNQQTREFVE
ErbB  GGKV-PIKWMALESILHRI---YTHQSDVWSYGVTVWELMTFGSKPYDGIPASEISSVLE
Raf   QPTG-SVLWMAPEVIRMQDDNPFSFQSDVYSYGIVLYELMA-GELPYAHINNRDQIIFMV
Mos   PHIGGTYTHQAPEILKGEI---ATPKADIYSFGITLWQMTT-REVPYSGEPQYVQYAVVA

Src   R--GYRMPCPPECPESLHD----LMCQCWRKDPEERPTFKYLGAQLLPA
Abl   K--DYRMERPEGCPEKVYE----LMRACWQWNPSDRPSFAEIHQAFETM
Fes   K--GGRLPCPELCPDAVFR----LMEQCWAYEPGQRPSFSAFYQELQSI
ErbB  K--GERLPQPPICTIDVYM----IMVKCWMIDADSRPKFRELIAEFSKM
Raf   GR-GYASPDLSRLYKNCPKAIKRLVADCVKKVKEERPLFPQILSSIELL
Mos   YNLRPSLAGAVFTASLTGKALQNIIQSCWEARGLQRPSAELLGRDLKAF
```

this problem by taking advantage of a human tumor cell line (an epidermoid carcinoma of the uterus) that expresses the EGF-R at elevated levels—as much as 100-fold higher than normal. These cells yielded an abundance of the receptor protein, which could then be purified biochemically and subjected to amino acid sequence analysis.

The sequence of the EGF-R protein provided important insights into its overall structural features and how this structure enables it to function (**Figure 5.5A**). Its large N-terminal domain of 621 amino acid residues, which protrudes into the extracellular space (and is therefore termed its **ectodomain**), was clearly involved in recognizing and binding the EGF ligand. The EGF receptor possessed a second distinctive domain that is common to many cell surface glycoproteins; this **transmembrane** domain of 23 amino acid residues threads its way from outside the cell through the lipid bilayer of the plasma membrane into the cytoplasm. The existence of this transmembrane domain could be deduced from the presence of a continuous stretch of **hydrophobic** amino acid residues located in the middle of the protein's sequence; this domain of EGF-R protein resides comfortably in the highly hydrophobic environment of the lipid bilayer that forms the plasma membrane. Finally, a third domain of 542 residues at the C-terminus (that is, the end of the chain synthesized last) of the EGF receptor was found to extend into the cytoplasm.

The overall structure of the EGF receptor suggested in outline how it functions. After its ectodomain binds EGF, a signal is transmitted through the plasma membrane to activate the cytoplasmic domain of the receptor; once activated, the latter emits signals

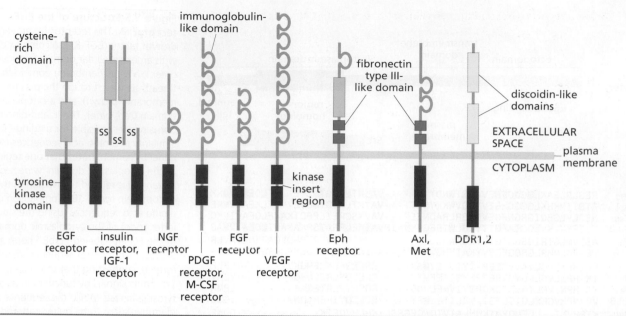

Figure 5.6 Structure of tyrosine kinase receptors The EGF receptor (see Figure 5.5A) is only one of many similarly structured receptors that are encoded by the human genome. These tyrosine kinase receptors (RTKs) can be placed into distinct families, depending on the details of their structure. Representatives of most of these families are shown here. All have in common quite similar cytoplasmic tyrosine kinase domains (*red*), although in some cases (e.g., the VEGF receptor) these domains are interrupted by small "insert" regions. The RTK ectodomains (which protrude into the extracellular space; *green, gray, yellow*) have highly variable structures, reflecting the fact that they recognize and bind a wide variety of extracellular ligands. (From B. Alberts et al., Molecular Biology of the Cell, 6th ed. New York: W.W. Norton, 2015.)

that induce a cell to grow and divide. Provocatively, examination of this cytoplasmic domain revealed a clear sequence similarity with the already-known sequence of the Src protein (Figure 5.5B). (Such shared structure indicates similar functions and a shared descent from a common precursor in the distant evolutionary past.)

Suddenly, it became clear how the EGF receptor emits signals in the cell interior: once its ectodomain binds EGF, somehow the Src-like kinase in its cytoplasmic domain becomes activated, proceeds to phosphorylate tyrosines on certain cytoplasmic proteins, and thereby causes a cell to proliferate. Subsequent sequencing efforts revealed overall sequence similarities among a variety of tyrosine kinases, many of which can function as oncoproteins (see Figure 5.5B).

Since these initial analyses of the EGF receptor structure, a large number of other, similarly structured receptors have been characterized (**Figure 5.6**). As discussed later, each of these receptors has its own growth factor ligand or set of ligands (**Table 5.1**). Depending on the particular growth factor–receptor pair, the binding of a ligand to its receptor can trigger multiple biological responses in the cell in addition to stimulating growth and division. Important among these are changes in cell shape, cell survival, cell metabolism, and cell **motility**.

Interestingly, these receptors and their ligands represent relatively recent evolutionary inventions, and their antecedents cannot be found among the proteins made by almost all single-cell eukaryotes. Many were invented just before metazoan life first arose and have been retained in clearly recognizable form in all modern metazoa. This explains why research on the cell-to-cell communication systems of simpler metazoan organisms, such as worms and flies, has so greatly enriched our understanding of growth factors and how they signal. Unexpectedly, recent research has revealed the presence of a tyrosine kinase receptor gene in the genome of at least one single-cell eukaryote; this organism, which may occasionally form multicellular aggregates, may be closely related to the common protozoan ancestor of all metazoa, and its receptor may be the prototype of many dozens of tyrosine kinase receptors present in modern metazoa including ourselves (Supplementary Sidebar 5.3).

Table 5.1 Growth factors (GFs) and tyrosine kinase receptors that are often involved in tumor pathogenesis

Name of GF	Name of receptor	Cells responding to GF
PDGF[a]	PDGF-R	endothelial, VSMCs, fibroblasts, other mesenchymal cells, glial cells
EGF[b]	EGF-R[c]	many types of epithelial cells, some mesenchymal cells
NGF	Trk	neurons
FGF[d]	FGF-R[e]	endothelial, fibroblasts, other mesenchymal cells, VSMCs, neuroectodermal cells
HGF/SF	Met	various epithelial cells
VEGF[f]	VEGF-R[g]	endothelial cells in capillaries, lymph ducts
IGF[h]	IGF-R1	wide variety of cell types
GDNF	Ret	neuroectodermal cells
SCF	Kit	hematopoietic, mesenchymal cells

[a]PDGF is represented by four distinct polypeptides, PDGF-A, -B, -C, and -D. The PDGF-Rs consist of at least two distinct species, α and β, that can homodimerize or heterodimerize and associate with these ligands in different ways.

[b]The EGF family of ligands, all of which bind to the EGF-R (ErbB1) and/or heterodimers of erbB1 and one of its related receptors (footnote c), includes—in addition to EGF—TGF-α, HB-EGF, amphiregulin, betacellulin, and epiregulin. In addition, other related ligands bind to heterodimers of ErbB2 and ErbB3 or ErbB4; these include epigen and a variety of proteins generated by alternatively spliced neuregulin (NRG) mRNAs, including heregulin (HRG), glial growth factor (GGF), and less well-studied factors such as sensory and motor neuron–derived factor (SMDF).

[c]The EGF-R family of receptors consists of four distinct proteins, ErbB1 (EGF-R), ErbB2 (HER2, Neu), ErbB3 (HER3), and ErbB4 (HER4). They often bind ligands as heterodimeric receptors, for example, ErbB1 + ErbB3, ErbB1 + ErbB2, or ErbB2 + ErbB4; ErbB3 is devoid of kinase activity and is phosphorylated by ErbB2 when the two form heterodimers. ErbB2 has no ligand of its own but does have strong tyrosine kinase activity. ErbB3 and ErbB4 bind neuregulins, a family of more than 15 ligands that are generated by alternative splicing.

[d]FGFs constitute a large family of GFs. The prototypes are acidic FGF (aFGF) and basic FGF (bFGF); in addition there are other known members of this family.

[e]There are four well-characterized FGF-Rs.

[f]There are four known VEGFs. VEGF-A and -B are involved in angiogenesis, while VEGF-C and -D are involved predominantly in lymphangiogenesis.

[g]There are three known VEGF-Rs: VEGF-R1 (also known as Flt-1) and VEGF-R2 (also known as Flk-1/KDR), involved in angiogenesis; and VEGF-R3, involved in lymphangiogenesis.

[h]The two known IGFs, IGF-1 and IGF-2, both related in structure to insulin, stimulate cell growth (i.e., increase in size) and survival; they also appear to be weakly mitogenic.

Abbreviation: VSMC, vascular smooth muscle cell.

Adapted in part from B. Alberts et al., Molecular Biology of the Cell, 6th ed. New York: W.W. Norton, 2015.

5.4 An altered growth factor receptor can function as an oncoprotein

A real bombshell fell in 1984 when the sequence of the EGF receptor was recognized to be closely related to the sequence of a known oncoprotein specified by the *erbB* oncogene. This oncogene had been discovered originally in the genome of avian erythroblastosis virus (AEV), a transforming retrovirus that rapidly induces a leukemia of the red blood cell precursors (an **erythroleukemia**). In one stroke, two areas of cell biology were united. A protein used by the cell to sense the presence of a growth factor in its surroundings had been appropriated (in its avian form) and converted into a potent retrovirus-encoded oncoprotein.

Once examined in detail, the oncoprotein specified by the *erbB* oncogene of avian erythroblastosis virus was found to lack sequences present in the N-terminal ectodomain of the EGF receptor (**Figure 5.7**). Without these N-terminal sequences, the ErbB oncoprotein clearly cannot recognize and bind EGF, and yet it functions as a potent stimulator of cell proliferation. This realization led to an interesting speculation that was soon validated: somehow, deletion of the ectodomain enables the resulting truncated EGF receptor protein to send growth-stimulating signals into cells in a **constitutive** fashion, fully independent of EGF. Years later, such truncated EGF receptors were discovered to be present in one-third of human glioblastomas and in smaller proportions of a variety of other human tumors. Still later, the presence in breast cancers of similarly truncated versions of ErbB/EGF-R's close cousin,

Figure 5.7 The EGF receptor and v-ErbB The EGF receptor and the v-ErbB oncoprotein of avian erythroblastosis virus are closely related. More specifically, the v-ErbB protein is specified by an altered version of the gene encoding the chicken EGF receptor, which encodes a truncated form of EGF-R that lacks most of the normally present ectodomain (*green*). In the absence of ligand, signaling by the intact receptor molecule is silent, whereas the truncated receptor can emit mitogenic signals constitutively, i.e., without stimulation by EGF ligand.

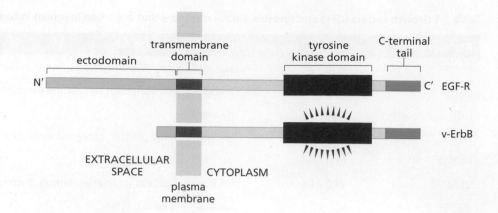

termed variously ErbB2, HER2, or Neu, were found to be associated with particularly poor prognosis.

As indicated schematically in **Figure 5.8A**, mutations in the genes encoding growth factor receptors, including those specifying truncated receptors, may trigger ligand-independent firing by these receptors. Indeed, a variety of mutations, including those creating amino acid substitutions in the transmembrane domain (shown), or in the ectodomain and cytoplasmic domains (not shown), of some receptors can provoke ligand-independent firing. (In fact, Figure 5.8 introduces yet another idea that we will explore in detail shortly: growth factors often activate receptors by triggering receptor dimerization.) In Chapter 17, we will see how structurally altered growth factor receptors influence the responsiveness of human tumors to anti-cancer therapeutic drugs.

These insights into receptor function provided one solution to a long-standing problem in cancer cell biology. As mentioned earlier, normal cells had long been known to require growth factors in their culture medium in order to grow. In contrast, cancer cells propagated in tissue culture were known to have a greatly reduced dependence on growth factors for their proliferation and survival. The discovery of the ErbB–EGF-R connection yielded a simple, neat explanation of this particular trait of cancer cells: the ErbB oncoprotein releases signals very similar to those emitted by a ligand-activated, wild-type EGF receptor. However, unlike the EGF receptor, the ErbB oncoprotein can send a constant, unrelenting stream of growth-stimulating signals into the cell, thereby persuading the cell that substantial amounts of EGF are present in its surroundings when none might be there at all. Stated differently, in the presence of such an oncoprotein, a cell is liberated from dependence on GFs present in its surrounding. More generally, a variety of growth factor receptors that are configured much like the EGF-R have been found to be overexpressed in human tumors or synthesized in structurally altered form (**Table 5.2**).

Figure 5.8 Regulation of growth factor signaling (A) Normally functioning growth factor receptors emit cytoplasmic signals (*red spikes*) in response to binding soluble ligand (*blue, left*). However, mutations in the genes encoding the receptor molecules (*upper right*) can cause subtle alterations in protein structure, such as amino acid substitutions (*red dots*), that cause ligand-independent firing. More drastic alterations in receptor structure, including truncation of the ectodomain (see, for example, Figure 5.7) may also yield such deregulated signaling. In many human tumors, receptor proteins are overexpressed (*lower right*). Excessive numbers of normally structured receptor molecules can also drive ligand-independent receptor firing, apparently by causing these molecules to frequently collide and thereby spontaneously dimerize and release signals (*red spikes*). (B) In general, in the context of paracrine signaling, normal cells do not synthesize and release a growth factor ligand whose cognate receptor they also display. Shown here is the tendency of epithelial cells to make PDGF (*blue*), the ligand for the PDGF-R displayed by mesenchymal cells; conversely, mesenchymal cells often make TGF-α, an EGF-related ligand (*red*), which binds the EGF-R displayed by epithelial cells. However, the PDGF made by an epithelial cell fails to find its cognate receptor on the surface of that cell (*horizontal bar*), preventing inadvertent activation of autocrine signaling. Similarly, the mesenchymal cells releasing TGF-α fail to express the cognate EGF-R that could bind and respond to this ligand. (C) In contrast, in many types of cancer, tumor cells acquire the ability to make a ligand for a growth factor receptor that they also naturally display (*right*). This creates an auto-stimulatory or autocrine signaling loop. (*continued*)

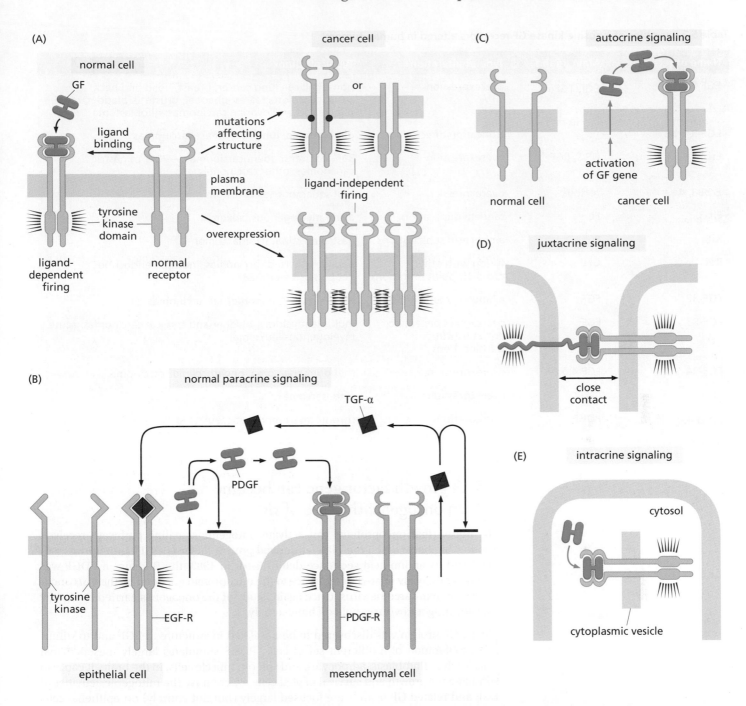

Figure 5.8 Regulation of growth factor signaling (continued) (D) Certain ligand–receptor interactions can occur only when two cells are in direct physical contact with one another. Such juxtracrine signaling involves expression of ligand that remains tethered to the ligand-presenting cells; in the case of one set of juxtracrine ligand:receptor interactions depicted here schematically (involving the Ephrin ligands and Eph receptors), the ligand-presenting cell also receives signals following receptor binding. (E) Intracrine signaling involves ligand–receptor interactions that occur entirely *within* a cell. The ligand and receptor may encounter and bind one another after their synthesis without ever moving to the cell surface. Alternatively, ligand and receptor may encounter and bind one another in a conventional fashion in the extracellular space; thereafter they may be internalized together and this multiprotein complex may continue to release signals within the cytoplasm.

Table 5.2 Examples of tyrosine kinase GF receptors altered in human tumors[a]

Name of receptor	Main ligand	Type of alteration	Types of tumor
EGF-R/ErbB1	EGF, TGFα	overexpression	non-small cell lung cancer; breast, head and neck, stomach, colorectal, esophageal, prostate, bladder, renal, pancreatic, and ovarian carcinomas; glioblastoma
EGF-R/ErbB1		truncation of ectodomain	glioblastoma; lung and breast carcinomas
ErbB2/HER2/Neu	NRG, EGF	overexpression	30% of breast adenocarcinomas, ~20% of gastric carcinomas, other carcinomas
ErbB3, 4	various	overexpression	oral squamous cell carcinoma
Flt-3	FL	tandem duplication	acute myelogenous leukemia
Kit	SCF	amino acid substitutions	gastrointestinal stromal tumor
Ret	GFL	fusion with other proteins, point mutations	papillary thyroid carcinomas; multiple endocrine neoplasias 2A and 2B
FGF-R2	FGF	amino acid substitutions	breast, gastric, endometrial carcinomas
FGF-R3	FGF	overexpression; amino acid substitutions; translocations	multiple myeloma; bladder and cervical carcinomas; acute myelogenous leukemia
FGF-R4	FGF-4,5,6	overexpression	rhabdomyosarcoma, hepatocellular carcinoma
IR	IGF-2	overexpression	osteosarcoma
PDGF-Rβ	PDGF	translocations	chronic myelomonocytic leukemia

[a]See also Figure 5.12.

5.5 A growth factor gene can become an oncogene: the case of *sis*

The notion that oncoproteins can activate mitogenic signaling pathways received another big boost when the platelet-derived growth factor (PDGF) protein was isolated and its amino acid sequence determined. In 1983, the B chain of PDGF was found to be closely related in sequence to the oncoprotein encoded by the v-*sis* oncogene of simian sarcoma virus. Once again, study of the oncogenes carried by rapidly transforming retroviruses paid off handsomely.

The PDGF protein was discovered to be unrelated in structure to EGF and to stimulate proliferation of a different set of cells. PDGF stimulates largely mesenchymal cells, such as fibroblasts, adipocytes, and smooth muscle cells; in the brain, it can also stimulate the growth and survival of glial cells. By contrast, the mitogenic activities of EGF and related GF ligands are focused largely (but not entirely) on epithelial cells (for example, see Figure 5.8B). This specificity of action could be understood once the PDGF receptor was isolated: the PDGF-R was found to be expressed on the surfaces of mesenchymal cells and is not usually displayed by epithelial cells, while the EGF-R largely shows the opposite pattern of expression. (Like the EGF receptor, the PDGF-R uses a tyrosine kinase in its cytoplasmic domain to broadcast signals into the cell.)

The connection between PDGF and the *sis*-encoded oncoprotein suggested another important mechanism by which oncoproteins could transform cells: when simian sarcoma virus infects a cell, its *sis* oncogene causes the infected cell to release copious amounts of the PDGF-like Sis protein into the surrounding extracellular space. Once there, the PDGF-like molecules can attach to the PDGF-R displayed by the same cell that just synthesized and released them (Figure 5.8C). The result is strong activation of this cell's PDGF receptors and, in turn, a flooding of the cell with an unrelenting flux of the growth-stimulating signals released by the ligand-activated PDGF-R.

Table 5.3 Examples of human tumors making autocrine growth factors

Ligand	Receptor	Tumor type(s)
CSF1	CSF1R	breast cancer
HGF	Met	miscellaneous endocrinal tumors, invasive breast and lung cancers, osteosarcoma
IGF-2	IGF-1R	colorectal
IL-6	IL-6R	myeloma, HNSCC
IL-8	IL-8R A	bladder cancer
NRG	ErbB2[a]/ErbB3	ovarian carcinoma
PDGF-BB	PDGF-Rα/β	osteosarcoma, glioma
PDGF-C	PDGF-Rα/β	Ewing's sarcoma
PRL	PRL-R	breast carcinoma
SCF	Kit	Ewing's sarcoma, SCLC
VEGF-A	VEGF-R 1(Flt-1)	neuroblastoma, prostate cancer, Kaposi's sarcoma, osteosarcoma
TGFα	EGF-R	squamous cell lung, breast and prostate adenocarcinoma, pancreatic, mesothelioma
GRP	GRP-R	small-cell lung cancer

[a]Also known as HER2 or Neu receptor.
Abbreviations: HNSCC, head and neck squamous cell carcinoma; SCLC, small cell lung cancer.

These discoveries also resolved a long-standing puzzle. Most types of acutely transforming retroviruses are able to transform a variety of infected cell types. Simian sarcoma virus, however, was known to be able to transform fibroblasts, but it failed to transform epithelial cells. The cell type–specific display of the PDGF-R explained precisely why this virus could transform some cells but not others.

Once again, a close connection had been forged between a protein involved in mitogenic signaling and a viral oncoprotein. In this instance, a virus-infected cell was forced to make and release a growth factor to which it could also respond. Rather than a growth factor signal sent from one type of cell to another cell type located nearby (often termed **paracrine** signaling; Figure 5.8B), or sent via the circulation from cells in one tissue in the body to others in a distant tissue (**endocrine** signaling), this represented an auto-stimulatory, or **autocrine**, signaling loop in which a cell manufactured its own mitogens (see Figure 5.8C and Supplementary Sidebar 5.4).

In fact, a variety of human tumor cells are known to produce and release substantial amounts of growth factors to which these cells can also respond (**Table 5.3**). Some human cancers, such as certain lung cancers, produce at least three distinct growth factors, specifically, transforming growth factor-α (TGF-α), stem cell factor (SCF), and insulin-like growth factor-1 (IGF-1); at the same time, the tumors express the receptors for these three ligands, thereby establishing three autocrine signaling loops simultaneously. Such autocrine signaling loops seem to be functionally important for the clinical behavior of tumors. Melanoma cells that develop resistance to therapeutic drugs designed to inhibit signaling pathways downstream of growth factor receptors often express elevated levels of the EGF-R, PDGF-R, IGF1-R, and AXL receptors (**Sidebar 5.2**) as well as their respective ligands.

Possibly the champion autocrine tumor is Kaposi's sarcoma, a tumor of cells closely related to the endothelial cells that form lymph ducts (see also Supplementary Sidebar 3.5). To date, Kaposi's tumors have been documented to produce VEGFs, Ang-1,

Sidebar 5.2 Receptor overexpression results from a variety of molecular mechanisms Receptor overexpression may cause a cell to become hyper-responsive to low concentrations of growth factor in the extracellular space; alternatively, overexpression may cause ligand-independent receptor firing, simply because **mass-action** effects may drive spontaneous receptor dimerization (see Figure 5.8A), much like the ligand-stimulated receptor dimerization described in the next section (see Figure 5.11).

In some cancer cells, receptors are overexpressed at the cell surface because the encoding gene is transcribed at an elevated rate, resulting in correspondingly increased levels of mRNA and protein. Alternatively, in a number of types of human cancers, these receptor genes are often amplified, leading to increased copy number of the involved genes and, once again, to proportionately increased levels of the corresponding mRNA transcript and receptor protein.

Yet other, more subtle mechanisms may also be responsible for increased numbers of receptor molecules being displayed at the cell surface. Some of these derive from the mechanisms that govern the *lifetime* of receptor molecules at the cell surface. Many receptor molecules are displayed at the surface for only a brief period of time before they are internalized via **endocytosis**, in which a patch of plasma membrane together with associated proteins is pulled into the cytoplasm, where it forms a vesicle. Thereafter, the contents of this vesicle may be dispatched to **lysosomes**, in which they are degraded, or alternatively they may be recycled back to the cell surface.

One protein, termed huntingtin-interacting protein-1 (HIP1), is part of a complex of proteins that facilitate endocytosis. HIP1 has been found to be overexpressed in a variety of human carcinomas. For reasons that are poorly understood, overexpression of HIP1 prevents the normal endocytosis of a number of cell surface proteins, including the EGF receptor (EGF-R). As a consequence, cells overexpressing HIP1 accumulate excessive numbers of EGF-R molecules at the cell surface and, in the case of NIH 3T3 mouse fibroblasts, become hyper-responsive to EGF. This explains why such cells can grow vigorously in the presence of very low levels of growth factor–containing serum (0.1%) that would otherwise cause normal NIH 3T3 cells to retreat into a nongrowing, quiescent state. Moreover, these cells behave like transformed cells in the presence of 10% serum in the tissue culture medium—the serum concentration that is routinely used to propagate untransformed NIH 3T3 cells; both behaviors demonstrate the hyper-responsiveness to GFs resulting from receptor overexpression.

Another protein, called cyclin G–associated kinase (GAK), is a strong promoter of EGF-R endocytosis. When its expression is suppressed, levels of cell surface EGF-R increase by as much as fiftyfold. A third protein, called c-Cbl, is responsible for tagging ligand-activated EGF-R, doing so via a covalent modification termed mono-ubiquitylation (which contrasts with the poly-ubiquitylation that causes degradation of a protein in proteasomes; see Supplementary Sidebar 7.5). This tagging causes the endocytosis of the EGF-R and its subsequent degradation in lysosomes. The v-Cbl viral oncoprotein, as well as several cellular proteins (Sts-1, Sts-2), can block this endocytosis, resulting, once again, in accumulation of elevated levels of EGF-R at the cell surface. In a more general sense, the spontaneous receptor dimerization or hyper-responsiveness to growth factors resulting from receptor overexpression is likely to drive the proliferation of a variety of human cancer cell types *in vivo*.

Ang-2, bFGF, IL-6, IL-8, GRO-α, TNF-β, and ephrin B2—all ligands of cellular origin—as well as the receptors for these ligands. At the same time, the causal agent of this disease, the human herpesvirus-8 (HHV-8) genome that is present in Kaposi's tumor cells, encodes and thus produces two additional ligands—vIL6 and vMIP—whose cognate receptors are also expressed by the endothelial cell precursors that generate these tumors.

Such autocrine signaling loops seem to represent potential perils for tissues and organisms. In normal tissues, the proliferation of individual cells almost always depends on signals received from other cells; such interdependence ensures the stability of cell populations and the constancy of tissue architecture. A cell that has gained the ability to control its own proliferation (by making its own mitogens) therefore creates an imminent danger, since self-reinforcing, positive-feedback loops often lead to gross physiologic imbalances. (Nonetheless, autocrine signaling loops are clearly used in certain normal biological situations, such as those that operate to stably maintain the unique properties of stem cells including cancer stem cells, as we will see in Chapter 11. Accordingly, it seems that some autocrine loops operate normally to maintain specific states of differentiation, whereas others conveying mitogenic signals are confined largely to neoplastic cells.)

We should note that the activation of autocrine mitogenic signaling loops yields an outcome very similar to that occurring when a structurally altered receptor protein such as ErbB/EGF-R is expressed by a cell. In both cases, the cell generates its own mitogenic signals and its dependence upon exogenous mitogens is greatly reduced. Interestingly, these dynamics also have important implications for the development of anti-cancer drugs (Sidebar 5.3).

5.6 Transphosphorylation underlies the operations of many receptor tyrosine kinases

The actions of oncogenes such as *sis* and *erbB* provide a satisfying biological explanation of how a cell can become transformed. By supplying cells with a continuous flux of growth-stimulatory signals, oncoproteins are able to drive the repeated rounds of cell growth and division that are needed in order for large populations of cancer cells to accumulate and for tumors to form. Still, this biological explanation sidesteps an important biochemical question lying at the heart of mitogenic signaling: How do growth factor receptors containing tyrosine kinases (often called receptor tyrosine kinases or RTKs) succeed in transducing signals from the extracellular space into the cytoplasm of cells?

Knowing the presence of the tyrosine kinase enzymatic activity borne by the cytoplasmic domains of these proteins allows us to rephrase this question: How do growth factor receptors use their tyrosine kinase domains to emit intracellular signals in response to ligand binding?

The solution to this problem came from a detailed examination of the proteins that become phosphorylated within seconds after a growth factor such as EGF is applied to cells expressing its cognate receptor, in this case the EGF-R (**Figure 5.9**). It would seem reasonable that a variety of cytoplasmic proteins become phosphorylated on their tyrosine residues following ligand binding to a growth factor receptor. Indeed, there are a number of such proteins. But the most prominent among these phosphotyrosine-bearing proteins is often the receptor molecule itself (**Figure 5.10**)! Hence, these receptors seem to be capable of autophosphorylation.

Another clue came from the structure of many growth factor ligands; they were often found to be **dimeric**, being composed either of two identical protein subunits (**homodimers**) or of very similar but nonidentical subunits (**heterodimers**) (see also Supplementary Sidebar 5.5). A third clue derived from observations that many transmembrane proteins constructed like the EGF and PDGF receptors have lateral mobility in the plane of the plasma membrane. That is, as long as the hydrophobic transmembrane domains of these receptor proteins remain embedded within the

Figure 5.9 Formation of phosphotyrosine on the EGF-R following ligand addition The use of a fluorescent reagent that binds specifically to a phosphotyrosine residue on the EGF-R enables the visualization of receptor activation following ligand binding. Here, receptor activation is measured on a monkey kidney cell at a basal level (0 sec), as well as 30 and 60 sec after EGF addition. In addition, following a 2-minute stimulation by EGF, the effects of a 60-sec treatment by a chemical inhibitor of the EGF-R kinase (AG1478) are shown (*right*), indicating that the EGF-induced activation of the EGF-R can be rapidly reversed. Receptor activity above the basal level is indicated in *blue*, while activity below the basal level is indicated in *red*. The response to AG1478 treatment indicates that a significant basal level of EGF-R activity was present (at 0 sec) even before EGF addition. (From M. Offterdinger et al., *J. Biol. Chem.* 279:36972–36981, 2004. Copyright © 2004 ASBMB. This article is distributed under a Creative Commons Attribution 4.0 International license.)

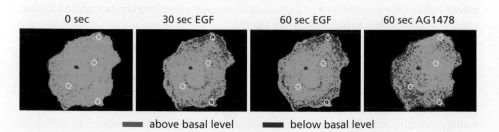

| 0 sec | 30 sec EGF | 60 sec EGF | 60 sec AG1478 |

■ above basal level   ■ below basal level

Figure 5.10 Apparent autophosphorylation of the EGF receptor (A) When human A431 epidermoid carcinoma cells, which greatly overexpress the EGF-R, are incubated in $^{32}PO_4$-containing medium and then exposed to EGF, a radiolabeled protein can be immunoprecipitated with an anti-phosphotyrosine antiserum (*lane 2*); this protein co-migrates upon gel electrophoresis with the EGF-R and is not detectable in the absence of prior EGF treatment (*lane 1*). (B) The $^{32}PO_4$-labeled phosphoamino acids borne by the proteins in such cells in the absence of EGF (as resolved by electrophoresis; see Supplementary Sidebar 5.2) are seen here. (C) Following the addition of EGF to A431 cells, a spot in the lower right becomes darker; internal markers indicate that this is phosphotyrosine. (A, from A.B. Sorokin et al., *FEBS Lett.* 268:121–124, 1990. With permission from John Wiley and Sons. B and C, courtesy of A.R. Frackelton Jr.)

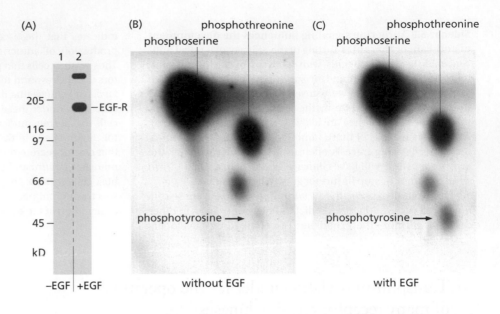

lipid bilayer of this membrane, they are relatively free to wander back and forth across the surface of the cell.

Taken together, these facts led to a simple model (**Figure 5.11A**). In the absence of ligand, a growth factor receptor exists in a **monomeric** (single subunit) form, embedded as always in the plasma membrane. It may wander laterally in the plasma membrane and encounter another identical RTK, with which it forms an unstable homodimeric complex. However, when presented with the appropriate growth factor ligand, the homodimeric receptor complex will be stabilized. Alternatively, a receptor molecule will bind to one of the two subunits of its dimeric ligand. Thereafter, the ligand–receptor complex will wander around the plasma membrane until it encounters another receptor molecule, to which the second, still-unengaged subunit of the ligand can bind. The result is, once again, the noncovalent cross-linking of the two receptor molecules achieved by the dimeric ligand that forms a bridge between them. (In all of these cases, the identity of the particular RTKs that are recruited into these complexes will be dictated by the binding specificities of the GF ligand.)

X-ray crystallographic studies of growth factors bound to the ectodomains of their cognate receptors have revealed a variety of mechanisms by which these ligands are able to induce receptor dimerization (Supplementary Sidebar 5.5). In fact, certain GF ligands don't naturally form dimers. For example, in the case of the EGF-R–related receptors, individual monomeric ligand molecules bind separately to the ectodomains of monomeric receptors. These GF-bound ectodomains respond by undergoing a steric shift that allows them to dimerize with one another via sites quite distant from the sites of ligand binding (see Supplementary Sidebar 5.5).

Once dimerization of the ectodomains of two receptor molecules has been achieved by ligand binding, the cytoplasmic portions are also pulled together. Now each kinase domain phosphorylates tyrosine residues present in the cytoplasmic domain of the other receptor. So the term "autophosphorylation" is actually a misnomer. This bidirectional, reciprocal phosphorylation is best described as a process of **transphosphorylation** (see Figure 5.11A). (Indeed, it is unclear whether a single isolated kinase molecule is ever able to phosphorylate itself.)

The phosphorylation of these tyrosine residues can have at least two consequences. The catalytic cleft of a kinase—the region of the protein where its enzymatic function operates—may normally be partially obstructed by a loop of the protein, preventing the kinase from interacting effectively with its substrates. In many RTKs, transphosphorylation of a critical tyrosine in the obstructing "activation loop" causes the loop to swing out of the way, thereby providing the catalytic cleft with direct access to substrate

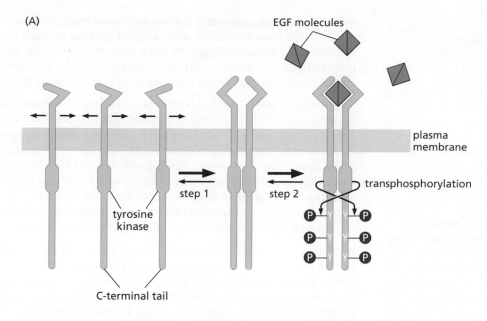

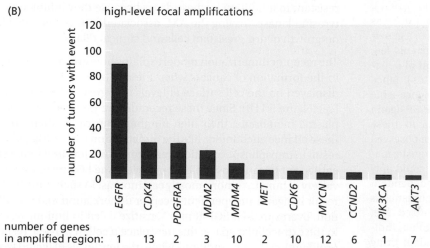

Figure 5.11 Receptor dimerization, ligand binding, transphosphorylation, and glioblastoma pathogenesis (A) In the absence of ligand, many growth factor receptor molecules (*left, green*) are free to move laterally (*diverging arrows*) in the plane of the plasma membrane. The consequences of these lateral migrations have been studied in the greatest detail in the case of the EGF receptor depicted here. When two receptor molecules encounter one another, they form a homodimer that persists for a brief period; at the same time, the ectodomains of these receptors change conformation and acquire increased affinity for binding growth factor ligands (step 1). Subsequent binding of EGF ligand molecules results in further stabilization of the dimer and activation of the linked tyrosine kinase (TK) domains (step 2). Thus, the TK domain of each receptor subunit phosphorylates the C'-terminal cytoplasmic tail of the other subunit—the process of transphosphorylation (*arrows*). Other GF receptors, not yet studied in such detail, seem to behave in a similar fashion. (While the initial receptor–ligand interactions may occur while receptors such as the EGF-R are displayed on the cell surface, the bulk of subsequent signaling events may operate after the receptor–ligand complexes are internalized by endocytosis and are associated, at least transiently, with endosomes. In addition, as shown in Supplementary Sidebar 5.5, the EGF-related ligand TGF-α binds as monomers to opposite-facing domains of the EGF-R ectodomains.) (B) The dynamics of receptor function depicted in panel A shed light on how amplification of *EGFR* genes and resulting EGF-R protein overexpression confers proliferative advantage on glioblastoma cells. Sequence analysis of 206 human glioblastomas has revealed chromosomal regions that are repeatedly amplified (yielding focal amplifications), resulting in copy number alterations (CNAs) of the genes embedded in these chromosomal regions. As seen here, the chromosomal region harboring the *EGFR* gene is amplified in ~90 of these tumors out of the total cohort of 206 tumors analyzed; in contrast, chromosomal regions harboring other growth factor receptors (*PDGFRA, MET*) and oncogenes are found to be amplified far less frequently. (The majority of the amplified *EGFR* genes encode wild-type protein, indicating that any effects of receptor amplification must derive from the increased number of receptor molecules expressed at the cell surface rather than the deregulated firing of structurally altered receptor molecules.) The excessive numbers of EGF-R molecules that are displayed on the surfaces of many of these brain tumor cells appear to drive an increased frequency of random collisions between these molecules and thus dimerization (step 1 of panel A). The number given below each receptor indicates the number of genes that are co-amplified with the receptor gene in the particular amplicon in which this gene resides. (B, from The Cancer Genome Atlas Research Network, *Nature* 455:1061–1068, 2008. With permission from Nature.)

molecules (see Supplementary Sidebar 5.6). In addition, receptor transphosphorylation results in the phosphorylation of an array of tyrosine residues present in C-terminal, cytoplasmic portions of the growth factor receptor outside the kinase domain, as indicated in Figure 5.11A. As we will learn in the next chapter, these phosphorylation events enable the receptor to activate a diverse array of downstream signaling pathways.

As mentioned above, some receptor molecules may heterodimerize with others. For example, the HER2/ErbB2/Neu receptor, which is overexpressed in human breast cancers (see Figure 4.3), often forms heterodimers with its cousins, either the HER1/EGF-R or the HER3 receptor. Since HER2 lacks its own identified ligand, it may be that a single ligand molecule binding either HER1 (= EGF-R) or HER3 suffices to trigger ligand-activated heterodimerization with HER2 and activation of downstream signaling. HER3 actually has a defective kinase domain, but its C-terminal tail can nonetheless serve as a substrate for phosphorylation by the HER2-borne kinase. The resulting phosphotyrosines then enable HER3 to release signals into the cell.

Receptor heterodimerization is increasingly seen to play an important role in human cancer development. For example, the AXL receptor, which is a member of an RTK family distinct from that of HER1/2/3/4 (see Figure 5.6), can be expressed at significant levels and heterodimerize with the EGF-R in lung cancers being treated with drug inhibitors of EGF-R signaling. Signaling by the resulting heterodimers is relatively resistant to inhibition by these drugs because they inhibit only the EGF-R–associated tyrosine kinase, leaving the AXL-associated kinase unaffected, thereby enabling the outgrowth of drug-resistant cells and clinical relapse.

The receptor dimerization model explains how growth factor receptors can participate in the formation of cancers when the receptor molecules are overexpressed, that is, displayed on the cell surface at levels that greatly exceed those seen in normal cells (see Figure 5.11B). Since these receptors are free to move laterally in the plane of the plasma membrane, their high numbers can cause them to collide frequently, and these chance encounters, like the dimerization events triggered by ligand binding, can result in transphosphorylation, receptor activation, and signal emission, as mentioned earlier. For instance, as Table 5.2 indicates, the EGF receptor is overexpressed in a wide variety of human tumors, mostly carcinomas. In such tumors, this overexpression may result in ligand-independent receptor dimerization and firing (Sidebar 5.4). In addition, overexpressed EGF-R may sensitize a cell to limiting amounts of ligand present in the extracellular space; this may place a cell in a hyper-responsive state, driving its proliferation under conditions where the low levels of ligand would fail to activate proliferation. [This touches on an issue that is almost unexplored in contemporary cancer research: how often is the frequently observed elevated receptor firing due to elevated levels of available ligands (Sidebar 5.5)?]

Sidebar 5.4 Clear evidence of ligand-independent receptor firing Experiments with the Met receptor provide clear evidence of ligand-independent receptor dimerization and firing. In these experiments, the human Met receptor [whose ligand is hepatocyte growth factor (HGF), also known as scatter factor (SF)] was overexpressed in liver cells of a genetically altered mouse strain; this led, in turn, to the development of hepatocellular carcinomas (HCCs). Since the human Met receptor is incapable of binding the mouse HGF/SF ligand (the only ligand available in these mice), this dictated that the observed receptor activation, cell transformation, and tumor formation could only be ascribed to a ligand-independent process, specifically, spontaneous dimerization of overexpressed Met receptor molecules. Yet other indications of ligand-independent firing come from studies of structurally altered RTKs present in human tumors (Supplementary Sidebar 5.7).

Sidebar 5.5 Regulation of ligand availability A major unsolved problem in cell biology and therefore in cancer pathogenesis derives from the complex biochemical mechanisms that govern the availability of GF ligands and thus binding to their cognate receptors. As one example, in the case of the EGF-R, nine alternative ligand proteins have been identified that can all bind, via characteristic EGF-like domains, to this receptor; all of these EGF-like ligands are synthesized as transmembrane proteins that are initially displayed on the surface of ligand-producing cells and can thereafter be mobilized by certain proteases that liberate them from their plasma-membrane anchorage, thereby making them available for receptor binding. These proteolytic cleavage events are highly regulated and are clearly rate-limiting steps in ligand availability and thus receptor activation. In the case of the EGF-R, proteolytic cleavage of EpCAM, a cell-surface epithelial adhesion molecule that is unrelated in structure to EGF itself, has been reported to generate a soluble ectodomain of this protein that is capable of activating EGF-R signaling, much like EGF itself.

Members of the TGF-β family of GFs—there are 33 in all—are initially synthesized in inactive complexes that require regulated proteolytic activation, which liberates them from both inactivating subunits and sequestration in the extracellular matrix (ECM). Fibroblast growth factors (FGFs), of which there are 23 distinct species, operating with their cognate receptors (FGF-Rs), play a critical role in various types of morphogenesis and cancer pathogenesis. Their availability as ligands is governed by their binding to heparan proteoglycans in the ECM to which they bind with various affinities. A final example is provided by the process of lymphangiogenesis, which results in the formation of lymph ducts draining lymph from tissues throughout the body. In this case, a critical growth factor, VEGF-C, is required. Once secreted into the extracellular space, the precursor of mature VEGF-C must be cleaved by two distinct proteases that also regulate one another; the resulting cleavage is essential for the proper formation of lymphatic drainage of all tissues. Current research has not even begun to grapple with the complexity of signaling in the extracellular space.

Arguably, the most direct and compelling demonstration of the key role of receptor homodimerization comes from the mechanisms by which fusion proteins formed in many human cancers drive constitutive receptor signaling (**Figure 5.12**; see also Supplementary Sidebar 5.7). Nonetheless, it is important to know that the scheme of receptor dimerization oversimplifies the responses of certain receptors to binding of their cognate ligands: certain receptors are constitutively dimerized even in the absence of ligand binding, and some evidence suggests that higher-order clusters involving large numbers of receptor molecules may form on the surfaces of tumor cells in response to ligand binding. Conversely, some published reports indicate that certain ligand-bound monomeric receptors may also fire, ostensibly by associating with tyrosine kinases that are not covalently associated, like most RTKs, with ligand-binding ectodomains.

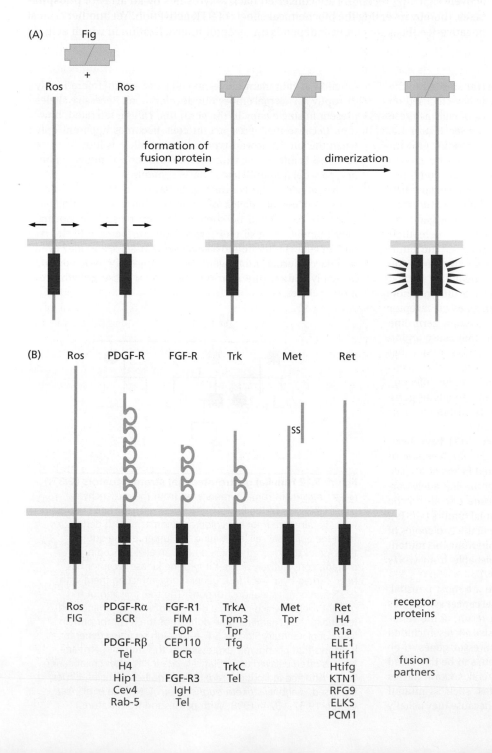

Figure 5.12 Gene fusion causing constitutively dimerized receptors (A) The Ros receptor (*lower left*), like other RTKs, normally is free to move laterally in the plane of the plasma membrane and requires ligand binding for dimerization. However, in a glioblastoma, the reading frame specifying the ectodomain of the *ROS* gene has become fused to the reading frame of a second, unrelated neighboring gene, termed *FIG*, which encodes a protein that normally dimerizes spontaneously (*green, upper left*). In the resulting fusion protein (*right*), the dimerized Fig domain pulls two Ros receptors together, resulting in constitutive, ligand-independent dimerization and signaling. (B) Similarly, in a diverse array of malignant tumors, receptor dimerization occurs when the genes encoding growth factor receptors (*black type*) become fused to unrelated genes that happen to specify proteins that normally dimerize or form higher-order oligomers (*red type*), dragging together pairs of receptor monomers. Listed here are a number of such mutant, fused receptors found in human tumors; in certain cases, a given receptor may be found fused to multiple alternative oligomerizing proteins. (B, courtesy of A. Charest.)

A further source of complexity comes from the fact that not all ligands of RTKs are soluble GFs that are transported through intercellular space. Thus, the DDR receptors depicted in Figure 5.6 and Supplementary Sidebar 5.5 bind various insoluble collagens that form key components of the extracellular matrix and basement membrane; loss of the DDR2 receptor function can compromise the ability of aggressive cancer cells to exhibit motility and invasiveness, key components of the cells of high-grade malignancies. Moreover, the Eph RTKs (see Figures 5.6 and 5.8D) bind ephrin ligands that are stably displayed on the surface of closely apposed neighboring cells, resulting in signals funneled into the cytoplasms of *both* the receptor- and ligand-displaying cells.

Finally, it is increasingly apparent that the strong mitogenic signals released by ligand-activated RTKs must be limited by opposing mechanisms that operate to constrain such signaling. As detailed in the next chapter, the phosphotyrosines created by actively signaling receptors are converted back to tyrosines by an array of **phosphatases**, thereby reversing the biochemical effects of RTK activation. Yet another critical negative-feedback mechanism depends on receptor internalization in which actively

Sidebar 5.6 Mutant receptor genes can be transmitted in the human germ line Our discussions until now have focused on somatic mutations sustained in the genomes of cells as these cells advance toward neoplasia; for example, see Figure 4.5. However, certain mutant alleles of the genes encoding growth-promoting proteins, notably receptors, may also be passed through the germ line, resulting in familial cancer syndromes afflicting multiple members of a family; these represent our first exposure in this book to multi-generational cancer syndromes. The *ret* gene encodes an RTK; mutant alleles of this gene can be transmitted through the human germ line, where their transmission from one generation to the next causes the familial cancer syndromes known as multiple endocrine neoplasia (MEN) types 2A and 2B, as well as familial medullary thyroid carcinoma. Hereditary papillary renal cancer is due to inherited mutant alleles in the *met* gene, which encodes the receptor for hepatocyte growth factor (HGF). These mutant germ-line alleles of *met* usually carry point mutations that cause amino acid substitutions in the tyrosine kinase domain of Met that result in constitutive, ligand-independent firing by the receptor. (In the Rottweiler breed of dogs, a point mutation affecting the Met juxtamembrane region is transmitted through the germ line of ~70% of these pets and appears to contribute to their high risk of developing several types of cancer.)

Members of a Japanese family (**Figure 5.13**) have been reported to transmit a mutant allele of the Kit receptor in their germ line. Like the somatically mutated forms of *kit* (see Supplementary Sidebar 5.8), the mutant germ-line allele carries a mutation affecting the **juxtamembrane** domain of the Kit receptor and yields gastrointestinal stromal tumors (GISTs). When this mutation is introduced into one of the two copies of the *kit* gene in the mouse germ line, mice inheriting this mutant allele develop tumors that are indistinguishable from GISTs seen in humans.

The familial syndromes described here are most unusual, because the great majority of human familial cancer syndromes involve germ-line mutations of genes that result, at the cellular level, in the formation of inactive, recessive alleles. Included among these are alleles of the tumor suppressor genes to be discussed in Chapter 7 and DNA repair genes to be described in Chapter 12. It seems likely that, in general, mutant alleles encoding constitutively active oncoproteins, such as mutant RTKs, cannot be tolerated in the germ line, because they usually

function as dominant alleles in cells and are therefore highly disruptive of normal embryonic development. Recessive alleles present in single copy in the germ line can be tolerated, however, because their presence in cells becomes apparent only when the surviving wild-type dominant allele is lost. Because such loss is infrequent, expression of the cancer-inducing phenotype is delayed until late in development.

Nonetheless, the contrasting behaviors of the dominantly acting, oncogenic alleles of tyrosine kinase receptor genes listed above seem to be compatible with reasonably normal development because their expression (in the case of the *ret* and *kit* genes) is limited to a small set of tissues, and may even be delayed during embryogenesis, allowing embryos to develop normally without the disruptive effects of these growth-promoting alleles.

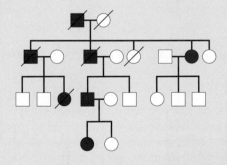

Figure 5.13 Familial gastrointestinal stromal tumors (GISTs)
Gastrointestinal stromal tumors arise from the mesenchymal pacemaker cells of the lower gastrointestinal tract and contrast with the far more frequent colon carcinomas, which derive from the epithelial cell layer lining the lumen of the gut. In the case of the kindred depicted here, an allele encoding a mutant, constitutively active Kit tyrosine kinase receptor has been carried by at least four generations, afflicting them with GISTs and/or intestinal obstructions. The transmission of this mutant allele—which carries a deletion of a single amino acid residue in the juxtamembrane domain of the receptor protein (see Supplementary Sidebar 5.6)—through so many generations indicates that phenotypic expression of this mutant germ-line allele is often delayed until relatively late in life after childbearing years. Affected individuals, *red*; males, *squares*; females, *circles*; deceased, *diagonal strikethroughs*. (From T. Nishida et al., *Nat. Genet.* 19:323–324, 1998. With permission from Nature.)

signaling RTKs are shunted via the process of **endocytosis** into the cytoplasmic vesicles termed **endosomes**, from which their signals continue to be emitted, and soon thereafter into cytoplasmic lysosomes, in which their GF ligands are released and the receptor molecules themselves are degraded. Interestingly, certain ligand-activated RTKs, such as the EGF-R, seem to release the bulk of their signals not from their normal homes on the plasma membrane but instead while residing in endosomes prior to their eventual degradation in lysosomes.

We can also note that of the approximately 20,000 protein-encoding genes in our genome, at least 58 specify protein products having the general structures of the EGF and PDGF receptors, that is, a ligand-binding ectodomain, a transmembrane domain, and a tyrosine kinase cytoplasmic domain. These receptors are further grouped into subclasses according to their structural similarities and differences (see Figure 5.6). Some of these receptors play seemingly unrelated roles in diverse tissues (Supplementary Sidebar 5.8). To date, only a minority of these receptors have been implicated as agents contributing to human cancer formation. It is likely that the involvement of other receptors of this class in human cancer will be uncovered in the future. Significantly, mutant alleles of certain tyrosine kinase receptor genes may also be transmitted in the human germ line. This explains the origins of a number of familial cancer syndromes, in which affected family members show greatly increased risks of contracting certain types of cancer (**Sidebar 5.6**).

5.7 Yet other types of receptors enable mammalian cells to communicate with their environment

The **cytokine** receptors are especially important in receiving signals controlling the development of various hematopoietic cell types. This class of receptors also uses tyrosine kinases. However, in this case, the responsible kinases, termed JAKs (Janus kinases), are separate polypeptides that associate with the cytoplasmic domains of the cytokine receptors through noncovalent links. Indeed, as early as the 1980s, the discovery of a number of other structurally unrelated types of metazoan receptors made it clear that the tyrosine kinase receptors (RTKs) described above represent only one class of receptors among a far larger group of cell-surface molecules belonging to various structural and functional classes. As is the case with RTKs, each of these other receptor types is specialized to detect the presence of specific extracellular ligands or groups of related ligands and to respond to this encounter by transducing signals through the plasma membrane into the cell.

One class of receptors that is especially important in receiving signals controlling the development of various hematopoietic cell types also uses tyrosine kinases. In this case, however, the responsible kinases, termed JAKs (Janus kinases), are separate polypeptides that associate with the cytoplasmic domains of these receptors through noncovalent links. Included here are the receptor for erythropoietin (EPO), which regulates the development of erythrocytes (red blood cells), and the **thrombopoietin** (TPO) receptor, which controls the development of the precursors of blood platelets, called **megakaryocytes**. The receptors for the important immune regulator interferon are also members of this class, as is the large group of interleukin receptors that regulate diverse immune responses. When these various cytokine receptor molecules dimerize in response to ligand binding, the associated JAKs phosphorylate and activate each other (**Figure 5.14**); the activated JAKs then proceed to phosphorylate the C-terminal tails of the receptor molecules, thereby creating receptors that are activated to emit signals, much like the RTKs described in the previous section. We will return to the details of JAK intracellular signaling in the next chapter.

Transforming growth factor-β (TGF-β) and related ligands have receptors that are superficially similar to the tyrosine kinase receptors, in that they have an extracellular ligand-binding domain, a transmembrane domain, and a cytoplasmic kinase domain. However, these receptors invariably function as heterodimers, such as the complex of the type I and type II TGF-β receptors (**Figure 5.15**). Importantly, their kinase domains phosphorylate serine and threonine rather than tyrosine residues. TGF-β ligands play centrally important roles in cancer pathogenesis, since they suppress the proliferation

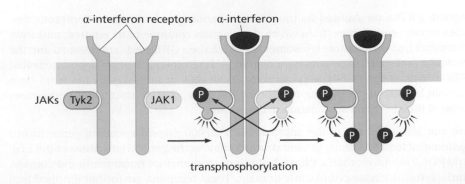

transphosphorylation

Figure 5.14 Structure of cytokine receptors The interferon receptor (IFN-R, *darker green*), like the receptors for erythropoietin and thrombopoietin, carries noncovalently attached tyrosine kinases (TKs) of the JAK family (in this case, Tyk2 and Jak1). Upon ligand binding (in this case α-interferon), the two TKs transphosphorylate and thereby activate one another. Subsequently, they phosphorylate the C-terminal tails of the receptor subunits, placing the receptor in an actively signaling configuration, much like the receptor tyrosine kinases described in Section 5.6.

of normal epithelial cells while promoting the acquisition of malignant properties by already-transformed cells. Upon ligand binding, the type II TGF-β receptor subunit, which carries a constitutively active serine/threonine kinase, is brought into close proximity with the type I TGF-β receptor subunit, which it then phosphorylates. The kinase belonging to the type I TGF-β receptor subunit becomes activated as a consequence, and proceeds to phosphorylate cytosolic proteins that subsequently migrate to the nucleus, where they trigger expression of certain target genes. (A series of TGF-β-related factors, including activin and bone morphogenetic proteins, or BMPs, use similar receptors for signaling; they are not discussed further here, as their role in cancer development has not been as extensively documented.) We postpone more detailed discussion of the effects of the TGF-β signaling system on proliferation (until Chapter 8) and on carcinoma cell aggressiveness (until Chapter 14).

A far more primitive form of a transmembrane signaling system is embodied in the Notch receptor (termed simply Notch) and its multiple alternative ligands (NotchL, Delta, Jagged; **Figure 5.16**). Importantly, these ligands are immobilized cell-surface proteins that are displayed by the signal-emitting cell. Hence, ligand–receptor interactions require close physical association between the signal-emitting and signal-receiving cells, similar to the Ephrin-Eph receptors depicted in Figure 5.8D and unlike other examples of receptor–ligand interactions, where the ligands are secreted as soluble proteins into the extracellular space by a cell that may be located at some distance from the cell destined to respond to this ligand. Hence, the interactions of Notch with its ligands conform to the signaling pattern between closely apposed cells that we have termed **juxtacrine** signaling (Figure 5.8D).

Interestingly, considerable evidence points to a mechanism in which the Notch ligand–presenting cell must attach its ligand to the Notch receptor and then use mechanical force to tear the Notch ectodomain away from the signal-receiving cell. (This mechanical force is generated by the endocytosis of the ligand, which pulls it into the cytoplasm of the ligand-presenting cell!) The mechanical tension initially created by ligand binding seems to stretch in some fashion the Notch (receptor) ectodomain, making it vulnerable to cleavage by a protease; this cleavage is followed by a

Figure 5.15 Structure of the TGF-β receptor The structure of the TGF-β receptor is superficially similar to that of tyrosine kinase receptors (RTKs), in that both types of receptors signal through cytoplasmic kinase domains. However, the kinase domains of TGF-β receptors specifically phosphorylate serine and threonine, rather than tyrosine residues. After the type II receptor (TGF-βRII) is brought into contact with the type I receptor (TGF-βRI) through the binding of TGF-β ligand, it phosphorylates and activates the kinase carried by the type I receptor. The activated type I receptor kinase then emits signals by phosphorylating certain cytoplasmic substrates.

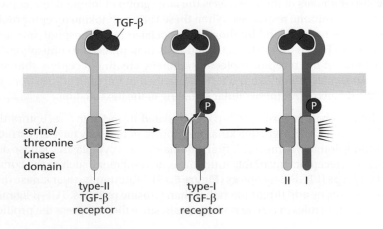

second proteolytic cleavage that releases a fragment from the Notch receptor cytoplasmic domain. The released fragment then migrates to the nucleus, where it functions as part of a complex of transcription factors that activate expression of a cohort of responder genes. Mutant, constitutively active forms of Notch, which fire in a ligand-independent fashion, have been found in half of adult T-cell leukemias. Of note, Notch signaling seems often to direct **cell fate**, that is, the choice of which specific differentiation pathway a stem cell chooses to enter, rather than providing mitogenic signals.

Yet another receptor, termed Patched, is constructed from multiple transmembrane domains that weave their way back and forth through the plasma membrane (**Figure 5.17**); receptors configured in this way are often termed **serpentine** receptors. Curiously, the Patched–Smoothened signaling system depends on a cellular structure that was once dismissed as nothing more than an evolutionary relict, a vestige of our early origins as ciliated or flagellated single-cell eukaryotes. A **primary cilium** protrudes from the surfaces of cells belonging to almost all mammalian cell types, often being deployed by these cells for various forms of sensory input (see Figure 5.17B). Normally, Smoothened (Smo), which is associated with cytoplasmic membranous vesicles, and the Gli2 transcription factor cycle back and forth in and out of the primary cilium (Figure 5.17A and C). However, when a Hedgehog (Hh) ligand binds its Patched receptor in the plasma membrane, Smoothened and Gli2 are permitted to accumulate in the primary cilium, where Gli2 is converted into an inducer of transcription and is dispatched to the nucleus. (The related Gli1 and Gli3 behave in slightly different ways.)

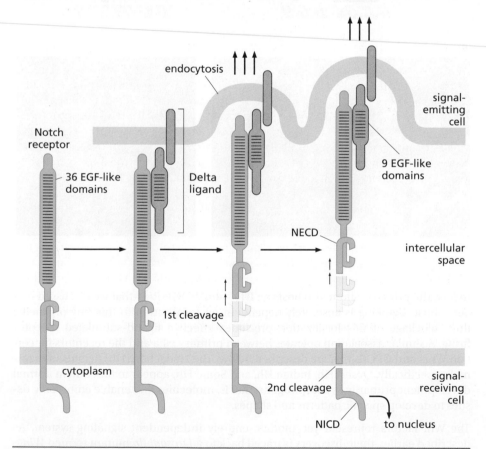

Figure 5.16 Structure of the Notch receptor The Notch receptor (*green*) appears to embody a very primitive type of signaling. The extracellular domain (NECD) of Notch displayed by the recipient cell is composed of two polypeptides that are associated via noncovalent interactions. After it has bound a ligand (for example, NotchL or, as shown here, Delta, *pink*) that is being displayed on the surface of an adjacent signal-emitting cell, the bound ligand (here, Delta) undergoes endocytosis into the ligand-presenting cell. This ligand binding also triggers two successive proteolytic cleavages of the Notch receptor. The resulting C-terminal cytoplasmic fragment (NICD) of the Notch receptor is thereby liberated, allowing it to migrate to the cell nucleus, where it alters the expression of certain genes. Mammals have five distinct Notch ligands and four distinct Notch receptors.

Figure 5.17 The Patched–Smoothened signaling system The Hedgehog pathway utilizes its own, unique signaling system. (A) Normally (*left*), Smoothened (Smo), a seven-membrane-spanning protein structurally related to the Frizzled receptors of the Wnt signaling family (*dark green; see Figure 5.18*), is functionally inert, being confined in cytoplasmic vesicles by the actions of Patched (Ptc; *light green*), which contains 12 membrane-spanning domains (*top left*). This sequestration seems to depend on the catalytic actions of Patched on Smoothened, which prevents Smoothened from moving into primary cilia. Recent research suggests that Ptc acts to control the access of Smo to cholesterol molecules (*not shown*). However, when a Hedgehog (Hh) ligand (*right, blue*) binds to Patched, the latter releases Smoothened from its site of sequestration in the cytoplasm, and Smoothened moves onto the surface of a primary cilium and proceeds, in some unknown fashion, to protect Gli2 from cleavage, enabling Gli2 to move into the nucleus, where it can act with yet other protein partners as an inducer of transcription. (According to an alternative version, Smoothened routinely cycles in and out of the primary cilium. In the absence of Hh binding, Patched, which is associated closely with the primary cilium, prevents Smoothened accumulation in the cilium; in the presence of Hh, however, Patched is inactivated, thereby allowing the accumulation of Smoothened in the cilium.) Two other Gli proteins—not pictured here—operate in slightly different ways. (B) While most discussions of cell surface receptors imply that such receptors are dispersed randomly across the surface of the plasma membrane, in fact many receptors are displayed only in localized, specialized domains of the cell surface. The Smoothened receptor (like the PDGF-α receptor) is displayed on the surface of primary cilia (*white arrow*), which protrude from the plasma membrane of many types of mammalian cells, as is seen here in a scanning electron micrograph of the surface of an epithelial cell. (C) The cycling of Gli back and forth between the cytosol and the tip of the primary cilium can be blocked experimentally, trapping Gli (*green*) at the tip of the primary cilium (*magenta stalk*) and revealing dramatically one phase of its biochemical cycle. (A, courtesy of D. Thomas and adapted from S. Goetz and K.V. Anderson, *Nat. Rev. Genet.* 11:331–344, 2010. B, from C.J. Haycraft et al., *PLoS Genet.* 1(4):e53, 2005. This article is distributed under the terms of the Creative Commons Public Domain declaration. C, from J. Kim, M. Kato and P.A. Beachy, *Proc. Natl. Acad. Sci. USA* 106:21666–21671, 2009. With permission from National Academy of Sciences, U.S.A.)

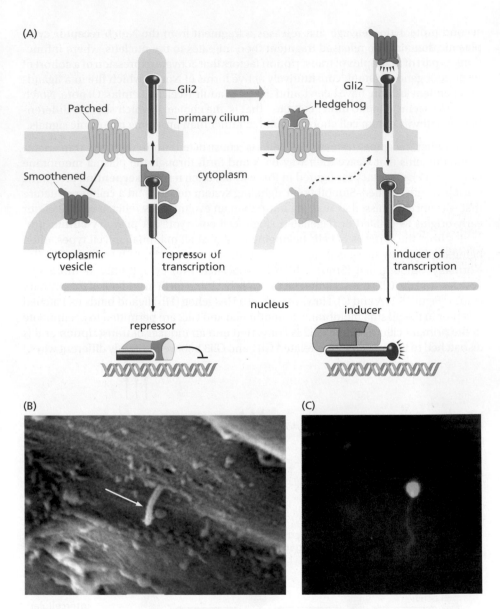

In fact, the primary cilium also hosts an RTK—the PDGF-Rα—that we discussed earlier, whose signaling is absolutely dependent on localization to this tiny organelle; thus, blockage of this localization precludes effective ligand-stimulated receptor firing. A similar association operates between primary cilia and the receptors for certain Wnts and GPCRs that are described below. The Hedgehog (Hh) ligands of mammals, specifically, Desert Hh, Indian Hh, and Sonic Hh, appear to function in normal development primarily as **morphogens**, that is, molecules that enable embryonic tissues to develop specific patterns and shapes.

The Wnt factors represent yet another, entirely independent signaling system. As described earlier, their discovery is traced back to a *Drosophila* mutant termed *Wingless* and a related gene termed *int-1* in mice; the latter was discovered because of insertional mutagenesis by mouse mammary tumor virus (see Sidebar 3.4). These Wnt proteins—humans make at least 19 distinct Wnt species—are tethered tightly to the extracellular matrix (ECM) and, through a lipid tail, to cell membranes, and thus do not appear to be freely diffusible like several of the growth factors that we encountered earlier. Growth factors of the Wnt class activate receptors of the Frizzled (Fzd/Frz) family, which, like Patched, are complex receptors that weave back and forth through the plasma membrane multiple times (**Figure 5.18**). (Their name comes from the behavior of a *Drosophila* mutant allele encoding one of these receptors that creates disordered hairs growing in disorganized orientations from the cuticle of this fruit fly; mammals express ten distinct receptors belonging to this family.)

In the absence of Wnt signaling, the cytoplasmic enzyme glycogen synthase kinase-3 β (GSK-3β) phosphorylates several key proliferation-promoting proteins, including β-catenin, tagging them for destruction. However, the binding of Wnts to Frizzled receptors triggers a cascade of steps that shut down GSK-3β firing, allowing these proteins to escape degradation and promote cell proliferation (see Figure 5.18). This pathway also plays a critical role in cancer pathogenesis, and we will discuss it in great detail later in this book.

The control of β -catenin levels by Wnts, often termed "canonical Wnt signaling," is only one of multiple ways by which Wnts and their Frizzled receptors affect cell behavior. Certain Wnts, acting together with a subset of the Frizzled receptors, signal through entirely different biochemical pathways that together constitute "non-canonical Wnt signaling." Two of these pathways, which play especially important roles in cancer development—termed Wnt-PCP and Wnt-Ca²+—contribute in important ways to the formation of certain cancers and will be described in some detail in the next chapter. Recent research has revealed another four non-canonical Wnt signaling pathways whose respective contributions to cancer cell phenotypes are still being described.

The. structure of Patched and the Frizzled receptors echoes that of a far larger family of 7 membrane-spanning, serpentine receptors that fall into the class of G-protein–coupled receptors (GPCRs). The latter are named because of their ability to activate guanosine nucleotide–binding proteins (**G-proteins**, for short), which function, much like Ras (see Section 5.10), as binary on–off switches that alternate between

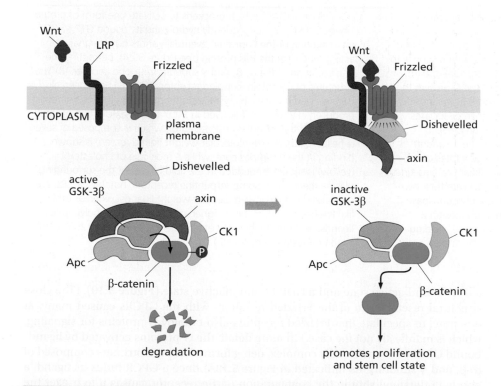

Figure 5.18 Canonical Wnt signaling via Frizzled receptors Receptors of the Wnt proteins are members of the Frizzled (Frz) family of transmembrane proteins. In the canonical Wnt signaling pathway, and in the absence of Wnt ligand binding (*left*), a complex of axin (*brown*) and Apc (*light blue-green*) allows glycogen synthase kinase-3β (GSK-3β, *pink*) to phosphorylate β-catenin (*blue*). This marks β-catenin for rapid destruction by proteolysis. However, when a Wnt ligand binds to certain Frizzled receptors (*right, dark green*) and the LRP co-receptor (*purple*), the resulting activated Frizzled receptor, acting via the Dishevelled protein (*light green*) and axin (*dark brown*), causes inhibition of GSK-3β. This spares β-catenin, which may accumulate in the nucleus and promote cell proliferation. Two other Wnt-activated pathways, termed "non-canonical Wnt pathways," are not shown here.

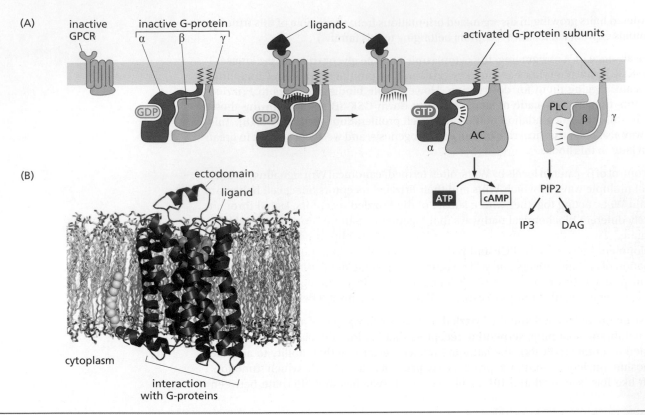

Figure 5.19 Signaling via the G-protein-coupled receptors (GPCRs) (A) In the absence of ligand binding to the GPCR, both the GPCR and its cognate heterotrimeric G-protein (which is composed of α, β, and γ subunits) are in inactive configurations. In response to ligand binding (usually a low molecular weight molecule), the receptor binds to and stimulates the α subunit of the G protein to release its GDP (guanosine diphosphate) and bind GTP (guanosine triphosphate) instead (*light blue*). The α subunit then dissociates from the β + γ pair. Now, the α subunit (*brown*) can proceed to regulate key enzymes: in the case shown here, the α subunit activates an adenylate cyclase (AC) (*pink*), which generates cAMP, which in turn modulates the activity of several cellular proteins kinase A (PKA). In some cases the β + γ complexes can activate their own constituencies of downstream effectors, such as a phospholipase C (PLC) (*orange*) shown here. PLC cleaves phosphatidylinositol 2 (PIP2) embedded in the plasma membrane (*not shown*), yielding two products, that is IP$_3$ which increases intracellular calcium and diacylglycerol (DAG) which proceeds to activate one form of protein kinase C (PKC). By subsequently hydrolyzing its bound GTP to GDP, the α subunit of the G-protein eventually shuts down signaling, acting much like the Ras protein (see Figure 5.24), permitting the re-association of the α, β, and γ subunits with one another. (B) The structures of G-protein–coupled receptors (GPCRs) have been elusive because of difficulties in crystallizing them. A structure of rhodopsin produced in 2000 was followed in 2007 by the elucidation of the structure of the β2-adrenergic receptor, which is composed of seven α-helices (*red*). A low–molecular-weight ligand (*green*) is shown binding in the receptor's core, while a molecule of cholesterol (*yellow*) is shown embedded in the lipid bilayer of the surrounding plasma membrane. Some serpentine receptors, notably GPCRs, are specialized to bind low–molecular-weight ligands, whereas others, like Frizzled receptors, bind higher–molecular-weight polypeptide ligands. (B, from V. Cherezov et al., *Science* 318:1258–1265, 2007. With permission from AAAS.)

a GTP-bound active state and a GDP-bound inactive state (**Figure 5.19**). (The close structural resemblance of the Frizzled receptors with the GPCRs caused many, at one time, to speculate that Frizzled receptors also employ G-proteins for signaling, which is manifestly not the case.) In more detail, the G-proteins activated by ligand-bound GPCRs are structurally complex, being formed as heterotrimers composed of α, β, and γ subunits. As indicated in Figure 5.19A, once a GPCR binds its ligand, a poorly understood shift in the configuration of the receptor causes it to trigger the release of GDP by the α subunit of the G-protein, permitting the binding of GTP instead. The α subunit then dissociates from its β and γ partners, and both the α sub-unit and the complexed β + γ subunits may then proceed to interact with their own respective **effector** proteins, that is, proteins that do the work of the the G-proteins by transmitting signals further into the cell interior. In the case of the GPCR depicted in Figure 5.19A, the effector of the activated α subunit is the adenylate cyclase (AC) enzyme, while the β + γ complex activates a distinct effector—phospholipase C-β (PLC-β). Eventually, an actively signaling α subunit turns itself off by hydrolyzing its bound GTP (converting it to GDP) and reassociates with its β and γ partners. These heterotrimeric complexes then retreat into their inactive, nonsignaling state, awaiting signals once again from a ligand-activated GPCR.

A recent survey of the human genome sequence indicates that it encodes, in addition to the tyrosine kinase receptors, eleven TGFβ-like receptors, four Notch receptors, two Patched receptors, and ten Frizzled receptors. The GPCRs vastly outnumber these. Altogether, there are more than 800 *GPCR* genes in the mammalian genome, constituting ~4% of the coding genes in the genome; of these, 400 or so GPCRs function as olfactory receptors and most of the remainder appear to be nonfunctional **pseudo-genes**. The functional versatility of these receptors is illustrated by their diverse functions beyond olfaction, as some underlie mechanisms of taste in the mouth, detection of photons in the retina, and signaling by dozens of distinct neurotransmitters in the brain. While vast in number, these receptors have to date been found to contribute directly to the formation of only a relatively small number of human cancers (described in Chapter 6). However, a subset of the GPCRs, termed **chemokine** receptors, play an important role in recruiting various immune cells into the tumor-associated stroma, which interact in various complex ways with cancer cells, as described in Chapters 13 and 15. Moreover, hyperactive signaling by certain chemokine receptors also contributes to mitogenic signals in a small subset of human tumors.

5.8 Nuclear receptors sense the presence of low–molecular-weight lipophilic ligands

The ability of cells to sense various extracellular signals is dictated by the chemical properties of signaling ligands and the impermeability of the plasma membrane to polar compounds, including almost all polypeptides. These conditions necessitated the development of two of the major classes of receptors that we have already encountered: because neither polypeptide GFs nor the low–molecular-weight polar ligands of GPCRs can penetrate the plasma membrane, their corresponding receptors display elaborate ligand-binding ectodomains to the extracellular environment.

Another class of signaling ligands, however, consists of molecules that are both of low molecular weight and relatively hydrophobic; included among these are steroid sex hormones, retinoids, and vitamin D. Their hydrophobicity enables these ligands to readily penetrate the plasma membrane and advance further into the nucleus, where they may bind and activate a series of nuclear DNA-binding proteins that function as transcription factors. The complex cytoplasmic signaling cascades that have evolved to function as intermediaries between a diverse array of cell surface receptors and the nuclear transcription machinery are not needed here, since these "**nuclear receptors**" can bind their ligands and directly regulate the expression of target genes by binding to the promoters of these genes. Prominent among the 48 or so human nuclear receptors are those that bind the steroid hormones estrogen, progesterone, and androgens. These three receptors play key roles in the development of common human malignancies, in particular, breast, ovarian, and prostate carcinomas.

Nuclear receptor molecules contain a DNA-binding domain, a hinge region, and a conserved ligand-binding domain, these three being flanked by variable N'- and C'-terminal domains. Some nuclear receptors remain in the cytoplasm until ligand binding triggers their nuclear import, while others are constitutively bound to chromatin. Once associated with the chromatin, these receptors bind as homo- or heterodimers (**Figure 5.20A**) to a pair of recognition sequences in the DNA located within or near the promoters of genes whose expression they control; these recognition sequences are often termed **hormone response elements** (HREs) and are composed of a pair of hexanucleotides separated by a variable number of spacer sequences (Figure 5.20B).

In the case of the estrogen receptor (ER), binding of its natural ligand—17β-estradiol, commonly called estrogen or simply E2 (see Figure 5.20C)—causes a conformational shift of α-helices near the ER ligand-binding pocket; these shifts allow the receptor to attract one of several receptor **co-activators** and release **co-repressors**, that is, proteins that relay signals from the ER to the general transcriptional machinery (see Figure 1.15). These intermediaries often act by modifying

Figure 5.20 Nuclear receptors (A) While the estrogen receptor (ER) often functions as a homodimer, many other nuclear receptors function as heterodimers, with each monomer binding its own ligand and its own recognition site in the DNA. Seen here is the structure of a nuclear receptor complex that has been elucidated in its entirety—the heterodimeric complex of PPAR-γ (peroxisome proliferator-activated receptor-γ, *light purple*) and the RXR (retinoid X receptor, *blue*). The LXLL domain (where L = leucine and X = variable amino acid residue) of the co-activator of each receptor (*red*) enables the co-activator to bind directly to the receptor; the bound co-activators then recruit chromatin-modifying enzymes to the nearby chromatin. The ligands (*green*) of the two receptors are rosiglitazone (bound by PPAR) and 9-*cis*-retinoic acid (bound by RXR). Zinc ions are seen as *red dots*. α-helices are depicted as cylinders. (B) The DNA-binding domains of the two subunits of the receptor shown in (A) are seen here in greater detail, in this case binding to two copies of the sequence AGGTCA (separated by a single base-pair spacer sequence, *light green lines*). Other nuclear receptors have paired DNA recognition sites with spacers of various lengths, and one class of these receptors bind DNA as monomers. Zinc ions are seen as *green dots*. (C) Like the ligands of other nuclear receptors, 17β-estradiol, otherwise called estrogen or simply E2, causes a conformation shift of the estrogen receptor (ER) that enables the receptor to associate physically with a co-activator, resulting in activation of transcription. Binding of pseudoligands, such as the synthetic drug 4-hydroxytamoxifen (4-OHT, *below*), causes an alternative stereochemical shift in the ER that results in the loss of contact with co-activators and the acquisition of contact with co-repressors; this explains why tamoxifen is often used therapeutically to antagonize estrogen signaling. (D) Analysis of the effects of the E2 mimic diethylstilbestrol (DES; *white, left*) versus 4-OHT (*white, right*) reveals a shift in the location of helix 12 of the ER (*red*). Following DES binding (*left*), the ER (*dark purple/blue*) is able to associate with a domain of the GRIP-1 (glucocorticoid receptor-interacting protein-1) transcriptional co-activator (*green*). However, if the E2 antagonist 4-OHT is bound (*right*), the shift in helix 12 of the ER (*red*) precludes the binding of the co-activator, blocking transcriptional activation. (A and B, from V. Chandra et al., *Nature* 456:350–356, 2008. With permission from Nature. D, adapted from A.K. Shiau et al., *Cell* 95:927–937, 1998. With permission from Elsevier.)

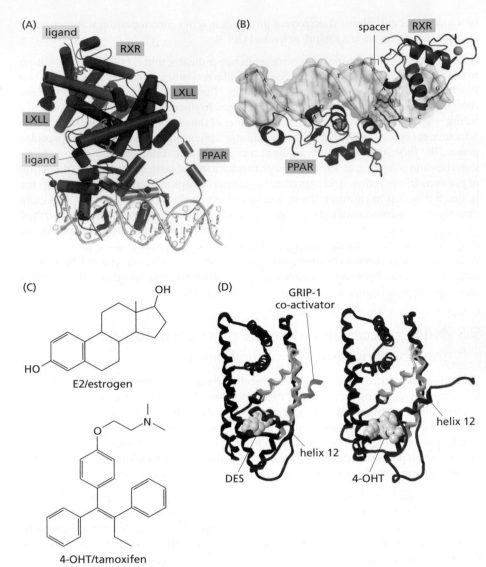

nearby histones or, alternatively, by recruiting histone-modifying enzymes (see Section 1.8). The resulting modified chromatin then either allows or prevents transcription of associated genes by RNA polymerase II. A **selective estrogen receptor modulator** (SERM), as represented here by **tamoxifen**, causes the ER to release co-activators (see Figure 5.20D) and to recruit co-repressors (not shown), resulting in effective blocking of E2 signaling; this shutdown, in turn, can confer great clinical benefit when this SERM is used to treat patients having ER-positive human breast carcinomas. For example, in one clinical trial, blocking E2 signaling with tamoxifen treatment resulted in an almost twofold reduction in the breast cancer relapse rate of postmenopausal women whose ER-positive primary tumors had been surgically removed. Yet other nuclear receptors play key roles in cancer pathogenesis. For example, the vitamin D ligand molecule, acting via its cognate nuclear receptor, has strong effects in normalizing the otherwise highly abnormal stroma of pancreatic adenocarcinomas.

5.9 Integrin receptors sense association between the cell and the extracellular matrix

The biology of normal and transformed cells provides hints of yet another type of cell surface receptor, one dedicated to sensing a quite different class of molecules in the extracellular space. Recall that an important attribute of transformed cells is their ability to grow in an anchorage-independent fashion, that is, to proliferate without attachment to a solid substrate such as the one at the bottom of a Petri dish (see Section 3.5). This behavior contrasts with that of normal cells, which require attachment in order to proliferate. Indeed, in the absence of such attachment, many types of normal cells will activate a version of their death program (**apoptosis**) that is often termed **anoikis**. These cell death processes will be explored in detail in Chapter 9.

Biochemical analyses of the solid substrate to which cells adhere at the bottom of Petri dishes have revealed that, in very large part, cells are not anchored directly to the glass or plastic surface of these dishes. Instead, they attach to a complex network of molecules that closely resembles the extracellular matrix (ECM) usually found in the spaces between cells within most tissues (see Figure 1.11C). The ECM is composed of a series of glycoproteins, including collagens, laminins, proteoglycans, and fibronectin. After cells are introduced into a Petri dish, they secrete ECM components that adhere to the glass or plastic; once a provisional ECM is laid down in this way, the cells attach themselves to this matrix; this process of ECM assembly mirrors in some fashion the formation of ECM in living tissue. Consequently, the anchorage dependence observed *in vitro* reflects the need of normal cells to be tethered within living tissues to ECM components in order to survive and proliferate.

The trait of anchorage dependence makes it obvious that cells are able to sense whether or not they have successfully attached themselves to the ECM. As was learned in the mid-1980s, such sensing depends on specialized receptors that inform cells about the extent of tethering to the ECM and about the identities of specific molecular components of the ECM (for example, collagens, laminins, fibronectin) to which tethering has occurred. Sensing of the collagen fibers in the ECM can be accomplished by an unusual pair of RTKs, termed **discoidin domain receptors** (DDRs; see Figure 5.6). However, most sensing of ECM components is carried out by a specialized class of cell surface receptors termed **integrins**. In effect, the molecular components of the ECM serve as the ligands of the integrin receptors. At the same time, the integrins create mechanical stability in tissues by tethering cells to the scaffolding formed by the ECM.

Integrins constitute a large family of heterodimeric transmembrane cell surface receptors composed of α and β subunits (**Figure 5.21**). At least eighteen α and eight β subunits have been enumerated; together, 24 distinct heterodimers are known. **Table 5.4** indicates that the ectodomain of each integrin heterodimer shows specificity for binding a specific ECM molecule or a small subset of ECM components. The much studied α5β1 integrin, for example, is the main receptor for fibronectin, an important glycoprotein component of the ECM found in vertebrate tissues. Laminins, which are large, multidomain ECM molecules, have been reported to be bound by as many as 12 distinct integrin heterodimers.

Table 5.4 Examples of integrins and their extracellular matrix ligands

Integrin	ECM ligand
α1β1	collagens, laminin
αvβ3	vitronectin, fibrinogen, thrombospondin
α5β1	fibronectin
α6β1	laminin
α7β1	laminin
α2β3	fibrinogen
α6β4	laminin—epithelial hemidesmosomes

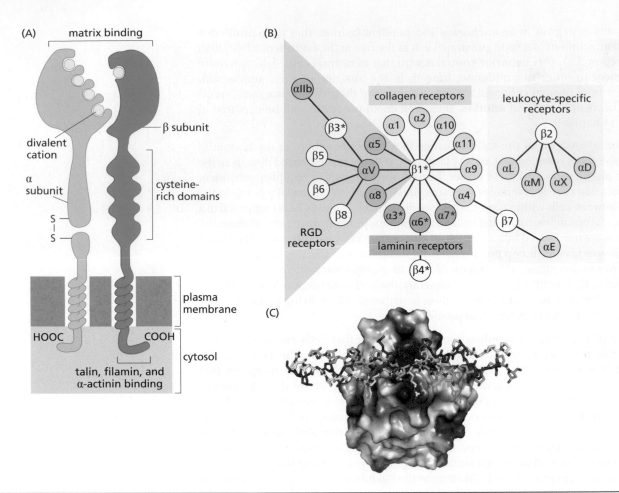

Figure 5.21 Structure of integrins Integrin molecules function as heterodimeric cell surface receptors, each composed of an α plus a β subunit (*green, blue*). (A) The ectodomains of these receptors bind specific components of the extracellular matrix (ECM). Divalent cations (*yellow dots*), notably magnesium, manganese, and calcium, modulate this binding in various ways. At the same time, the cytoplasmic domain (largely that of the β subunit) is linked, via intermediary proteins, to the cytoskeleton (largely that constructed by actin fibers); in addition, the cytoplasmic domains may attract a variety of signal-transducing proteins that become activated when the ectodomain binds an ECM ligand. (B) The α and β integrin subunits associate with one another as heterodimers in various specific combinations to enable the binding of a variety of ECM ligands and to transduce multiple signals into the cell. As is apparent, the set of heterodimers actually found in cells is far less (24) than might be theoretically predicted through combinatorial interactions of each α subunit with each β subunit (144 in all). [The most promiscuous of the β subunits is β1 (*center*), which can associate with 12 alternative α subunits, resulting in many distinct integrin heterodimers.] RGD receptors recognize the arginine–glycine–aspartic acid tripeptide motif that is found in a number of distinct ECM proteins and represents the core sequences recognized by these integrins to initiate such binding. Laminin is a critical component of the basement membrane (see Figure 2.3). Asterisks denote alternatively spliced cytoplasmic domains. (C) One depiction of integrin–ECM binding comes from the crystallographic analysis of the ectodomain of the α2β1 integrin (*space-filling structure*) interacting with collagen I, a triple helix (*stick figures*). The *red* surface patches on the integrin ectodomain contain the more acidic side chains, while the *blue* surface patches contain basic amino acids. (B, from R.O. Hynes, *Cell* 110:673–687, 2002. C, from J. Emsley et al., *Cell* 101:47–56, 2000. Both with permission from Elsevier.)

Having bound their ECM ligands through their ectodomains, integrins cluster to form *focal adhesions* (**Figure 5.22A**). This clustering affects the organization of the cytoskeleton underlying the plasma membrane, since some integrins are linked directly or indirectly via their cytoplasmic domains to important components of the cytoskeleton, such as actin, vinculin, talin, and paxillin (Figure 5.22B and C). The formation of focal adhesions may also cause the cytoplasmic domains of integrins to activate signaling. As an example, by triggering the release of anti-apoptotic signals, integrins reduce the likelihood of anoikis. The functions of integrins continue to be critical during the development of many tumors, indicating that tumor cells usually do not evolve to a stage where they have completely outgrown their dependence on integrins and thus tethering to the ECM in order to acquire survival and proliferation signals (**Sidebar 5.7**).

(A)

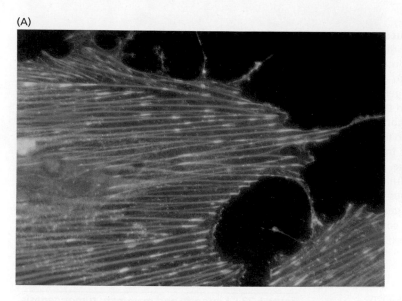

(C)

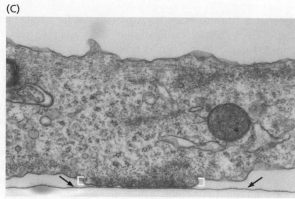

(B)

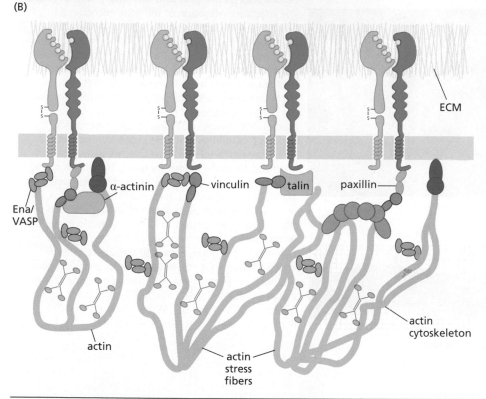

ECM

Ena/
VASP

α-actinin

vinculin

talin

paxillin

actin
cytoskeleton

actin

actin
stress
fibers

Figure 5.22 Integrin tethering to the ECM and cytoskeleton
(A) This fluorescence micrograph illustrates the discrete elongated foci (*yellow/orange*) on the cell surface, termed focal adhesions, at which integrins tether cells to the extracellular matrix (ECM). Most of these foci seen here have been formed on the *ventral* sides of these cells, i.e., the undersides of these cells that are directly apposed to the ECM that is, in turn, adsorbed to the surface of the Petri dish. Formation of these focal adhesions also controls the organization of the actin fibers (*green*) forming the cytoskeleton inside the cell. (B) This schematic figure of the organization of integrins (*green, blue*) indicates their association with the ECM (*green fibers, above*) through their ectodomains and their association with the actin cytoskeleton (*pink chains*) through their cytoplasmic domains largely via the β subunit of each heterodimer. A series of intermediary proteins, such as actinin, vinculin, and talin, allows these linkages to be formed. (C) This

transmission electron micrograph section, which is oriented in a plane perpendicular to the bottom surface of a Petri dish, reveals a single focal adhesion formed between the ventral surface of a chicken lens cell and the underlying solid substrate; in this case, ECM molecules adsorbed to the surface of the Petri dish are seen as a dark line (*arrows*). The electron-dense, cytoplasmic material (*white brackets*) that is immediately above the region of contact with the substrate contains large numbers of integrin-associated proteins and actin fibers that in aggregate constitute this focal adhesion, which may be as much as several microns in width when viewed from this perspective. (The resolution of this image does not afford visualization of individual actin fibers or integrin molecules.) (A, courtesy of Keith Burridge. B, adapted from C. Miranti and J. Brugge, *Nat. Cell Biol.* 4:E83–E90, 2002. With permission from Nature. C, from O. Medalia and B. Geiger, *Curr. Opin. Cell Biol.* 22:659–668, 2010. With permission from Elsevier.)

Sidebar 5.7 Some integrins are essential for tumorigenesis The ability to selectively inactivate ("knock out") genes within targeted tissues has made it possible to evaluate the contributions of various signaling proteins to tumorigenesis (to be described in Sidebar 7.5). For example, inactivation of the gene encoding the β1 integrin in the mouse germ line leads to the absence of this integrin in all cells of the embryo and to embryonic lethality; however, introduction of a modified gene into the mouse germ line makes it possible to inactivate this gene in a targeted fashion in one or another specific organ or tissue. In the experiment shown in **Figure 5.23**, both copies of the β1 integrin allele have been selectively deleted from the genomes of epithelial cells of the mouse mammary gland, leaving the gene unperturbed in all other tissues. Such inactivation of both copies of the β1 integrin gene has minimal effects on the normal development of the mammary gland.

In these mice, a second germ-line alteration involved the insertion of the middle T oncogene of polyomavirus that had been placed under the control of a promoter that allows this **transgene** to be expressed only in the mice's breast tissue. In mice with an intact β1 integrin gene, this oncogenic transgene generally triggers the formation of premalignant hyperplastic nodules that progress into aggressive mammary carcinomas. However, when β1 integrin is absent from these oncogene-expressing mammary epithelial cells (because of selective gene inactivation), the number of normally observed hyperplastic nodules is reduced by more than 75% and the β1 integrin–negative cells in these nodules are fully unable to develop further to form mammary carcinomas (see Figure 5.23C). Detailed analyses indicate that this loss of β1 integrin expression permits the oncogene-expressing mammary epithelial cells to survive but precludes their active proliferation and thus their ability to spawn mammary carcinomas.

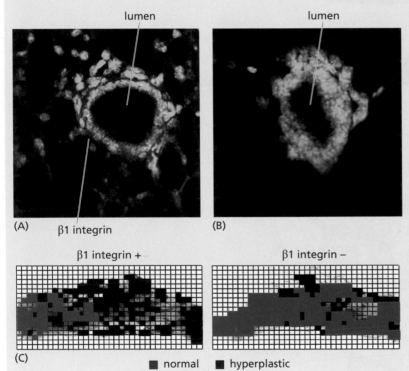

(A) β1 integrin

(B)

(C) β1 integrin +   β1 integrin –

■ normal   ■ hyperplastic

Figure 5.23 β1 integrin and mammary tumorigenesis The selective inactivation of the gene encoding β1 integrin in the mouse mammary gland allows a test of its importance in tumorigenesis. (A) The immunostaining of β1 integrin (*red*) on the surface of the epithelial cells forming a normal mouse mammary duct (*blue*), seen here in cross section. (B) In the absence of β1 integrin, ductal morphogenesis still proceeds normally. Moreover, the overall development of the mammary gland is normal (*not shown*). (C) Premalignant hyperplastic nodules in a transgenic mouse prone to mammary tumorigenesis (because of the selective expression of an oncogene in the mammary gland) have been scored visually by microscopic examination of whole mammary glands (*left*); the nodules are present in abundance and, wherever they have been detected, are represented by *black squares*, in contrast to areas in which only normal epithelium is observed in these glands (*red squares*). However, in mammary glands of transgenic mice in which β1 integrin is absent (because its encoding gene has been inactivated), the transgenic oncogene fails to induce the formation of large numbers of premalignant nodules (*right*). (From D.E. White et al., *Cancer Cell* 6:159–170, 2004. With permission from Elsevier.)

Integrins are most unusual in one other respect. Normally we think of receptors as passing information from outside the cell into the cytoplasm. Integrins surely do this. But in addition, it is clear that signals originating in the cytoplasm are used to control the binding affinities of integrins for their ECM ligands. Such "inside-out" signaling enables cells to modulate their associations with various types of ECM or with various points of contact with an ECM, breaking existing contacts and forging new ones in their place. Rapid modulation of extracellular contacts enables cells to free themselves from one microenvironment within a tissue and move into another, and to invade through a sheet of ECM *in vitro*. Cultured fibroblasts lacking focal adhesion kinase (FAK), one of the signaling molecules that associate with the cytoplasmic domains of integrins, are unable to remodel their focal contacts and lack motility, indicating that the signals transduced throughout the cytoplasm by FAK are important for reconfiguring the cytoskeleton—the structure that enables cells to change shape and to move. Later, in Chapter 14, we will see how cell motility is critical to the ability of cancer cells to invade and metastasize.

These various descriptions of tyrosine kinase receptors and integrins reveal that mammalian cells use specialized cell surface receptors to sense two very distinct classes of extracellular proteins. Some receptors, such as the EGF, PDGF, and Frizzled receptors, sense soluble (or solubilized) growth factors, while others, notably the integrins, sense attachment to the essentially insoluble scaffolding of the ECM. (Sometimes two types of receptors, such as integrins and RTKs, associate laterally with one another and regulate each other's signaling. As we learn more about these cell surface proteins, we increasingly realize the importance of these lateral interactions; see Supplementary Sidebar 5.9.) Together, these various receptors enable a normal cell to determine whether two preconditions have been satisfied before this cell undertakes growth and division: the cell must sense the presence of adequate levels of mitogenic and **trophic** growth factors in its surroundings and the existence of adequate anchoring to specific components of the ECM.

Both of these requirements for cell proliferation are known to be abrogated if a cell carries an activated *ras* oncogene. Thus, *ras*-transformed cells can grow in the presence of relatively low concentrations of serum and therefore serum-associated mitogenic growth factors; in addition, many types of *ras*-transformed cells can proliferate in an anchorage-independent fashion. These behaviors suggest that, in some way, a Ras oncoprotein is able to mimic and thereby supplant the signals normally introduced into the cell by ligand-activated growth factor receptors and by integrins that have engaged ECM components. An understanding of the biochemical basis of these various signals demands some insight into the structure and function of the normal and oncogenic Ras proteins. So, we will move back in history to the early 1980s to see how the puzzle of Ras function was solved.

5.10 The Ras protein, an apparent component of the downstream signaling cascade, functions as a G-protein

The discoveries that two oncogenes—*erbB* and *sis*—encode components of the mitogenic growth factor signaling machinery (Sections 5.4 and 5.5) sparked an intensive effort to relate other known oncoproteins, notably the Ras oncoprotein, to this signaling machinery. The *ras* oncogene clearly triggered many of the same changes in cells that were seen when cells were transformed by either *src*, *erbB*, or *sis*. Was there some type of signaling cascade—a molecular bucket brigade—operating in the cell, in which protein A transferred a signal to protein B, which in turn signaled protein C? And if such a cascade did exist, could the Ras oncoprotein be found somewhere downstream of *erbB* and *sis*? Did the signals emitted by these various proteins all converge on some common target at the bottom of this hypothesized signaling cascade?

At the biochemical level, it was clear that growth factor ligands activated tyrosine kinase receptors, and that these receptors responded by activating their cytoplasmic tyrosine kinase domains. But which proteins were affected thereafter by the resulting receptor phosphorylation events? And how did this phosphorylation lead further to a mitogenic response by the cell—its entrance into a phase of active growth and division? In the early 1980s, progress on these key issues ground to a halt because biochemical experiments offered no obvious way to move beyond the tyrosine kinase receptors to the hypothesized downstream signaling cascades.

All the while, substantial progress was being made in understanding the biochemistry of the Ras protein. In fact, the three distinct *ras* genes in mammalian cells encode four distinct Ras proteins (since K-*ras* specifies a second protein via alternative splicing of a pre-mRNA). Because all four Ras proteins have almost identical structures and function similarly, we refer to them simply as "Ras" in the discussions that follow even though subtle differences in their functions have emerged in recent years. At their C-termini, they all carry covalently attached lipid tails, composed of farnesyl, palmitoyl, or geranylgeranyl groups (or combinations of several of these groups). These lipid moieties enable the Ras proteins, all of about 21 kD mass, to become anchored to cytoplasmic membranes, largely to the cytoplasmic face of the plasma membrane. There are, however, clear indications that isoforms of Ras are also tethered to the membranes of the endoplasmic reticulum and the Golgi apparatus, where they may also emit signals.

Like the heterotrimeric G-proteins (Section 5.7), Ras was found to bind and hydrolyze (that is, cleave) guanosine nucleotides. This activity as a GTPase indicated a mode of action quite different from that of the Src and erbB tyrosine kinases, yet intriguingly, all three oncoproteins had very similar effects on cell behavior.

In further analogy to the G-proteins, Ras was found (1) to bind a GDP molecule when in its inactive state; (2) to jettison its bound GDP after receiving some stimulatory signal from upstream in a signaling cascade; (3) to acquire a GTP molecule in place of the recently evicted GDP; (4) to shift into an activated, signal-emitting configuration while binding this GTP; and (5) to cleave this GTP after a short period, using its own intrinsic GTPase function to do so, thereby placing itself, once again, in its non-signal-emitting configuration (**Figure 5.24**). In effect, the Ras molecule seemed to behave like a light switch that automatically turns itself off after a certain predetermined time.

In more detail, the hypothesized chain of events went like this. Mitogenic signals, perhaps transduced in some way by tyrosine kinase receptors, activated a **guanine nucleotide exchange factor** (GEF) for Ras. This GEF induced an inactive Ras protein to shed its GDP and bind GTP instead. The resulting activated Ras would emit signals to a (still-unknown) downstream target or group of targets. This period of signaling would be terminated, sooner or later, when Ras decided to hydrolyze its bound GTP. By doing so, Ras would force itself to revert to its nonsignaling state. As was learned later, a group of other proteins known as GAPs (GTPase-activating proteins) could actively intervene and encourage Ras to undertake this hydrolysis (see Sidebar 5.8). Implied in this model (see Figure 5.24) is the notion that Ras acts as a signal-transducing protein, in that it receives signals from upstream in a signaling cascade and subsequently passes these signals on to a downstream target, doing so for a circumscribed period of time.

A striking discovery was made in the course of detailed biochemical analysis of a Ras oncoprotein made by a point-mutated *ras* oncogene, specifically, the Ras oncoprotein encoded by Harvey sarcoma virus. Like the normal Ras protein, the Ras oncoprotein could bind GTP. However, the oncoprotein was found to have lost virtually all GTPase activity (**Figure 5.25A**). In such condition, it could be pushed into its active, signal-emitting configuration (Figure 5.25B) by some upstream stimulatory signal

Figure 5.24 The Ras signaling cycle
Detailed study of the biochemistry of Ras proteins revealed that they, like the heterotrimeric G-proteins (see Figure 5.19A), operate as binary switches, binding GDP in their inactive state (*top*) and GTP in their active, signal-emitting state (*bottom*). Thus, an inactive, GDP-binding Ras protein is stimulated by a GEF (guanine nucleotide exchange factor, *orange*) to release its GDP and acquire a GTP in its stead, placing Ras in its active, signaling configuration. This period of signaling is soon halted by the actions of a GTPase activity intrinsic to Ras, which hydrolyzes GTP to GDP. This GTPase activity is strongly stimulated by GAPs (GTPase-activating proteins, *light red*) that Ras might encounter while in its activated state. Amino acid substitutions caused by oncogenic point mutations (see Figures 4.5 and 4.6) block this cycle by preventing the stimulatory GAPs from interacting with Ras, effectively inactivating the intrinsic GTPase activity of Ras; this traps Ras in its activated, signal-emitting state.

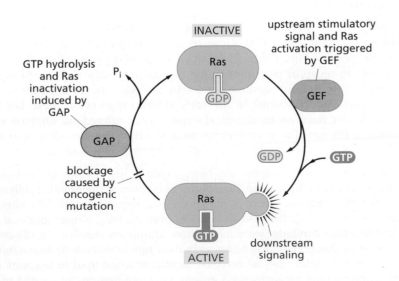

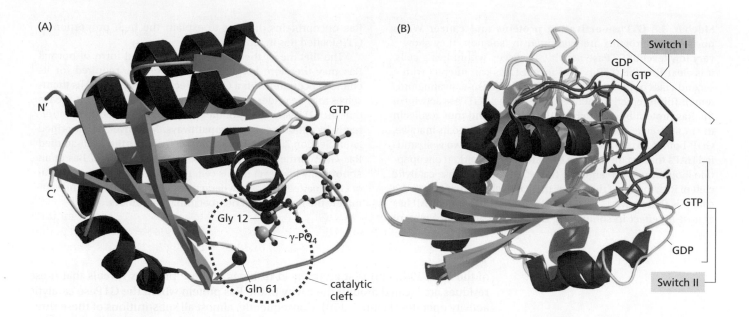

Figure 5.25 The structure of the Ras protein and its response to GTP binding (A) This ribbon diagram of the structure of a Ras protein, as determined by X-ray crystallography, depicts the polypeptide backbone of Ras and its α-helical (*red*) and β-pleated sheet (*green*) domains. GTP is indicated as a stick figure, and two of the amino acid residues often found altered in the Ras oncoproteins present in human tumors —glycine 12 and glutamine 61—are shown as *dark blue balls*. Their close association with the γ-phosphate of GTP (*gray ball*) helps to explain why substitutions of these residues reduce or eliminate the Ras protein's GTPase activity, which normally converts GTP into GDP by removing the γ-phosphate. The catalytic cleft of the GTPase is outlined (*red dotted oval*). This provides an explanation of why the codons specifying these particular residues are commonly mutated in human tumor cell genomes; conversely, amino acid substitutions striking elsewhere in the Ras protein may compromise its intrinsic signal-emitting functions and therefore are unlikely to potentiate Ras function. (B) The replacement of GDP by GTP, which results in the activation of Ras signaling, also causes a shift in two regions of the Ras protein (depicted here from a slightly different perspective as in panel A). Both Switch I (*magenta*) and Switch II (*aquamarine*) regions undergo the GDP-to-GTP conformational shift (as indicated by the positions labeled GDP and GTP). The conformational shift for a domain of Switch I induced by GTP binding is also indicated by the *magenta arrows*. These shifts allow the Ras protein to interact physically with its downstream effectors (see Figures 6.9, 6.10, and 6.15). The guanine nucleotide is indicated by a stick figure. (A and B, courtesy of I.R. Vetter and A. Wittinghofer.)

and guanine nucleotide exchange factors (GEFs). But once in this activated state, the Ras oncoprotein was unable to turn itself off! The negative-feedback loop that is so essential to the operations of the normal Ras protein had been sabotaged by this point mutation and resulting amino acid substitution.

This provided a clear explanation of how Ras can operate as an oncoprotein: rather than sending out short, carefully rationed pulses of growth-stimulating signals (the behavior of a normal RAS protein), the oncoprotein emits signals for a long, possibly indefinite period of time, thereby flooding the cell with these signals. Subsequent analysis of Ras protein biochemistry has revealed additional levels of control that may go awry in cancer cells (**Sidebar 5.8**).

These findings provided a solution to a puzzle created by the sequencing of the mutant *ras* oncogenes present in many human tumors. As we learned (see Section 4.3), invariably the point mutations creating these oncoproteins were **missense** mutations (which caused an amino acid substitution) rather than **nonsense** mutations (which caused premature termination of the growing polypeptide chain). Invariably these mutations struck either the 12th, 13th, or 61st codon of the reading frame of the *ras* genes (with the great majority altering codon 12; see Figure 4.6).

We might imagine that these particular nucleotides represent sites in the genomic DNA that are particularly susceptible to attack and alteration by mutagenic carcinogens. However, detailed study of the structure of the Ras proteins, as generated by X-ray crystallography, has led to a far more compelling model. Examination

Sidebar 5.8 GTPase-activating proteins and cancer When purified Ras proteins are examined in isolation, they show a very low level of GTPase activity. However, within living cells, it is clear that the GTP-bound form of Ras can interact with a second class of proteins termed GAPs (GTPase-activating proteins). This interaction increases the intrinsic GTPase activity of the Ras protein by as much as 100,000-fold and thus results in the rapid conversion of active, GTP-bound Ras into its inactive, GDP-bound form. (The Ras-GAP protein, one of two well-studied GAPs that act on Ras, has been found to insert an oligopeptide loop termed an "arginine finger" into the GTPase catalytic cleft of Ras; the tip of this finger then actively participates in the catalysis that converts GTP to GDP. Interestingly, Ras-GAP has almost no effect in stimulating the GTPase activity of mutant Ras oncoproteins, helping to explain the high proportion of GTP-loaded Ras in mutant cells.)

The lifetime of the activated, GTP-bound form of normal Ras may therefore be governed by the time required for its chance encounter with a GAP molecule. Since the GAPs themselves may be under the control of yet other signals, this means that the GTPase activity of the Ras proteins is modulated indirectly by yet other signaling pathways. Moreover, as described later (Section 7.10), the loss from a cell of the other well-studied Ras-GAP, termed NF1, leaves the GTPase activity of Ras at an abnormally low level in this cell, resulting in the accumulation of GTP-loaded Ras, the hyperactivity of Ras signaling, and ultimately the formation of neuroectodermal tumors.

of the 12th, 13th, and 61st amino acid residues of these proteins reveals that these residues are located around the cavity in the Ras protein where the GTPase catalytic activity operates (Figure 5.25B). Consequently, almost all substitutions of these three amino acids, such as the glycine-to-valine substitution that we encountered earlier, compromise GTPase function. [To be more precise, they compromise the ability of GTPase-activating proteins (GAPs; see Figure 5.24) to trigger hydrolysis of the GTP bound by Ras.]

Knowing this, we can immediately understand why point mutations affecting only a limited number of amino acid residues are found in the *ras* oncogenes carried by human tumor cell genomes. Large-scale alterations of the *ras* proto-oncogenes, such as deletions, are clearly not productive for cancer, since they would result in the elimination of Ras protein function rather than an enhancement of it. Similarly, the vast majority of point mutations striking *ras* proto-oncogenes will yield mutant Ras proteins that have lost rather than gained the ability to emit growth-stimulatory signals. Only when the signal-emitting powers of Ras are left intact and its GTPase negative-feedback mechanism is inactivated selectively (by amino acid substitutions at one of these three amino acid residue sites in the protein) does the Ras protein gain *increased* power to drive cell proliferation and transform the cell.

This explains why only those rare cells that happen to have acquired mutations affecting one of the three amino acid residues of Ras (residues 12, 13, or, rarely, 61; see Figure 4.6) gain proliferative advantage over their wild-type counterparts and thereby stand the chance of becoming the ancestors of the cells in a tumor mass. Stated differently, even though point mutations affecting all the amino acid residues of a Ras protein are likely to occur with comparable frequencies, only those few conferring substantial proliferative advantage will actually be found in tumor cells. Other cells that happen to acquire point mutations affecting other residues in the Ras protein will retain a normal phenotype or may even lose proliferative ability.

5.11 Synopsis and prospects

In the early 1980s, the discovery of oncogenes such as *erbB* and *sis* revealed the intimate connections between growth factor signaling and the mechanisms of cell transformation. More specifically, these connections suggested that cell transformation results from hyperactivation of the mitogenic signaling pathways. This notion galvanized researchers interested in the biochemical mechanisms of carcinogenesis. Many began to study the growth factor receptors and their biochemical mechanisms of action.

A major problem surrounding growth factor receptors was ignorance about their mechanism for transferring signals from outside a cell into its interior. A simple model was soon developed to explain how this occurs. The growth factor ligands of receptors bind to their ectodomains. By doing so, the ligands encourage receptor molecules to come together to form dimers. This dimerization of the ectodomains then encourages dimerization of the receptor's cytoplasmic domains as well. Somehow, this enables the cytoplasmic domains to emit signals.

This model sheds no light on how, at the biochemical level, receptors are able to emit signals. Here, research on the biochemistry of the Src protein proved to be telling. It was found to be a tyrosine kinase (TK), an enzyme that transfers phosphate groups to the side chains of tyrosine residues borne by various substrate proteins. Once the amino acid sequences of several growth factor receptors were analyzed, the cytoplasmic domains of these receptors were found to contain sequences encoding kinases that were structurally (and thus functionally) related to Src. This indicated that tyrosine kinase domains create the means by which growth factor receptors emit biochemical signals.

This information was then integrated into the receptor dimerization model: when growth factor receptors are induced by ligand binding to dimerize, the tyrosine kinase of each receptor monomer transphosphorylates the cytoplasmic domain of the other monomer. This yields receptor molecules with phosphorylated cytoplasmic tails that allow signaling to proceed. Much of the next chapter is focused on these downstream signaling mechanisms. An overview and preview of how one subfamily of RTKs operates is provided in Supplementary Sidebar 5.10. This image, which involves only this group of receptors (all related to the EGF-R), indicates the complexity of these signaling processes and, by implication, the challenges at present encountered by those who would like to portray these interactions with various types of mathematical models.

While highly instructive, the RTK dimerization model has proven to be simplistic: we now realize that in the absence of ligand, certain growth factor receptors are found in large clusters, that ligand binding exerts subtle effects on the stereochemistry of individual receptor molecules, and that the activation of many types of RTKs cannot be understood simply by ligand-driven dimerization.

Hyperactive signaling by these receptors is encountered in many types of human cancer cells. Often, the receptors are overexpressed, resulting in ligand-independent firing via molecular mechanisms that remain unclear. Even more potent ligand-independent firing is achieved by various types of structural alterations of receptors. Most of these occur as the consequences of somatic mutations. However, some mutant alleles of receptor genes have been found in the human gene pool, and inheritance of such alleles is associated with a variety of inborn cancer susceptibility syndromes. In the case of commonly occurring lung cancers, some commonly observed mutations affect the structure of the kinase domain of the EGF receptor and result in ligand-independent firing. Yet other mutations affect the internalization of the receptor, which usually operates following ligand binding; by compromising this negative-feedback mechanism, this second class of mutations results in the accumulation of inappropriately high concentrations of receptor at the cell surface. Both classes of change result in driving the proliferation of these cancer cells.

In our discussions of receptor tyrosine kinases (RTKs) we have focused exclusively on the signaling powers of their cytoplasmic domains, specifically the TK catalytic domains and the multiple tyrosine residues phosphorylated as substrates of these TKs. In doing so, we have overlooked important understudied functions of RTKs: after ligand binding, entire receptors or fragments thereof migrate to the nucleus, where they operate as components of multiprotein complexes involved in the regulation of transcription. As an example, the entire PDGF-Rβ protein is conveyed to the nucleus following binding of its PDGF-BB ligand, where it participates in chromatin remodeling. In another case, in response to ligand binding by HER4/ErbB4 (cousin of the EGF-R), the intracellular domain of this receptor associates with several other transcription-regulating proteins in neural precursor cells, after which the resulting complexes migrate to the nucleus, where they repress genes involved in the formation of glial cells. These cases represent only the tip of the iceberg, since multiple RTKs have been found to move, either entirely or as proteolytic fragments, to the nucleus, where they exert various functions regulating transcription.

Such RTKs represent only one of multiple ways by which cells sense their surroundings. The TGF-β receptors, for example, are superficially similar to the RTKs, in that they have a ligand-binding ectodomain and a signal-emitting kinase domain in their cytoplasmic portion. However, the kinases of the TGF-β receptors are serine/threonine kinases and, as we will learn later, signal through entirely different mechanisms.

A variety of other signal-transducing receptors have been uncovered, including those of the Notch, Patched–Smoothened, and Frizzled classes. These receptors use a diversity of signal-transducing mechanisms to release signals into the cytoplasm. As we will see throughout this book, aberrant signaling by these receptors plays a key role in the pathogenesis of many types of human cancers. Moreover, we will encounter yet another class of receptors in Chapter 9, where their role in programmed cell death—apoptosis—is described. Descriptions of these diverse cell surface receptors do not address yet another class of receptors that operate in the nucleus, bind hydrophobic ligands, and in response function as transcription factors. The nuclear receptors play prominent roles in a group of common human carcinomas—those arising in the breast, ovary, and prostate. In these tumors, the ligands are steroid hormones—estrogen, progesterone, and androgens—and the roles of their corresponding nuclear receptors in tumor growth have been described in tens of thousands of research reports. (To provide an indication of their importance, note that—as of 2021—55,000 studies of the estrogen receptor and 41,000 studies of the progesterone receptor had been published in research journals worldwide.)

Cells must also sense their contacts with the extracellular matrix (ECM)—the cage of secreted glycoproteins and proteoglycans that surrounds every cell in living solid tissues. One class of RTKs binds directly to collagens as their signal-activating ligands. But another class of receptors, the far more common integrins, play a more central and multifaceted role. Having bound components of the ECM, the heterodimeric integrin receptors transduce signals into cells that stimulate proliferation, suppress cells' apoptotic suicide program, and participate in various ways in cell motility. (Altogether, almost 15% of the human proteome of ~20,000 proteins has been predicted to involve cell surface proteins mediating various types of communication of cells with their extracellular environment.)

In recent years an entirely different channel of cell-to-cell communication has been proposed following the discovery of membranous vesicles—termed **exosomes**—that are released into the extracellular space by one cell, travel through the interstitial space between cells, and become fused with the plasma membranes of other cells that become the recipients of signals borne by these vesicles. The precise roles of these vesicles in the cell-to-cell signaling operating in normal and neoplastic tissues are still being uncovered.

A distinct line of research, pursued in parallel with the studies of RTKs, elucidated the biochemistry of the Ras protein, another key player in cancer pathogenesis. Ras operates like a binary switch, continually flipping between the active, signal-emitting state and the quiescent state. The mutant alleles found in cancer cells cause amino acid substitutions in the GTPase pockets of Ras proteins, thereby disabling the mechanism that these proteins normally use to shut themselves off. This traps the Ras proteins in their active, signal-emitting configuration for extended periods of time, causing cells to be flooded with an unrelenting stream of mitogenic signals.

While these investigations revealed how isolated components of the cellular signaling machinery (for example, RTKs, Ras) function, they provided no insights into how these proteins communicate with one another. That is, the *organization* of the intracellular signaling circuitry remained a mystery. Cell physiology hinted strongly that ErbB (that is, the EGF-R) and other tyrosine kinase receptors operated in a common signaling pathway with the Ras oncoprotein. Src fit in somewhere as well. These various oncoproteins, while having diverse biochemical functions, were found to exert very similar effects on cells. They all caused the rounding up of cells in monolayer culture, the loss of contact inhibition, the acquisition of anchorage independence, and reduction of a cell's requirement for mitogenic growth factors in its culture medium. This commonality of function suggested participation in a common signaling cascade.

One possible connection between Ras and the tyrosine kinase receptors came from the discovery that an activated *ras* oncogene causes many types of cells to produce and release growth factors. Prominent among these was transforming growth factor-α (TGF-α), an EGF-like growth factor. Like EGF, TGF-α binds to and activates the EGF receptor. This suggested the following scenario. Once released by a *ras*-transformed cell, TGF-α might function in an autocrine manner (see Figure 5.8C) to activate EGF receptors displayed on the surface of that cell. This, in turn, would evoke a series of responses quite similar to those created by a mutant, constitutively activated EGF receptor. When viewed from this perspective, Ras appeared to operate *upstream* of a growth factor receptor rather than being an important component of the signaling cascade lying *downstream* of the receptor (**Figure 5.26A**).

In the end, this autocrine scheme was able to explain only a small part of *ras* function, since *ras* was found to be able to transform cells that lacked receptors for the growth factors that it induced. In addition, ample biochemical evidence accumulated that ligand binding of growth factor receptors led rapidly to activation of Ras (Figure 5.26B). This left open only one alternative: direct intracellular signaling between ligand-activated growth factor receptors and Ras proteins. So the search was on to find the elusive biochemical links between tyrosine kinase receptors and the proteins like Ras that appeared to form components of the hypothesized downstream signaling cascade.

The use of biochemistry to discover the connections between RTKs and their downstream effectors had serious limitations. As this chapter has made clear, biochemistry provided researchers with powerful tools to elucidate the functioning of individual,

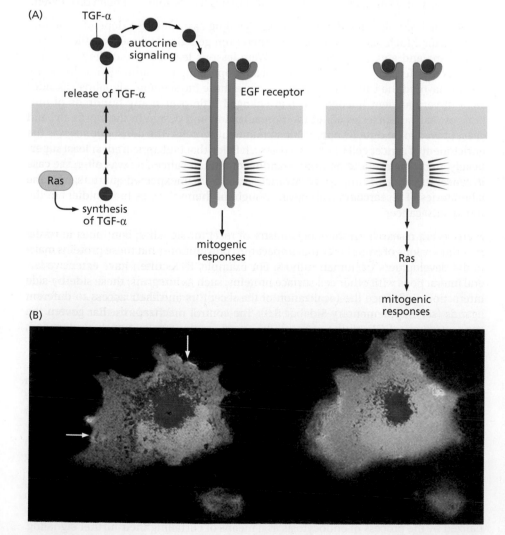

Figure 5.26 Alternative mechanisms of transformation by Ras (A) One hypothetical depiction (*left*) of how Ras (*green*) normally operates to promote cell proliferation was inspired by the observation that *ras*-transformed cells release a number of growth factors into their surroundings, such as TGF-α (*red circles*). Once secreted, the growth factors might act in an autocrine fashion to activate a receptor, such as the EGF receptor shown here, thereby promoting cell proliferation. In this depiction, Ras operates "upstream" of the EGF receptor. An alternative scheme (*right*) placed Ras in the midst of a signaling cascade that operates "downstream" of such growth factor receptors. (B) Extensive evidence favoring the second scheme accumulated. This graphic example, produced years later, shows that 5 minutes after EGF is added to cells, previously inactive Ras protein (*blue shading*) becomes activated near the cell surface (*red–orange shading*). This response is so rapid that it indicates close biochemical communication between the EGF receptor and Ras. [Interestingly, this activation occurred only on the two sides of a cell (*white arrows*) that were not in direct contact with nearby cells.] (B, from N. Mochizuki et al., *Nature* 411:1065–1068, 2001. With permission from Nature.)

isolated signaling proteins. However, it had limited utility in revealing how different proteins might talk to one another within a cell. If there really were linear signaling cascades of the form A → B → C, biochemistry might well reveal how A or C functioned in isolation (for example, as a kinase or GTPase) but would shed little light on the *organization* of the signaling pathway that lay between them.

Detailed biological characterization of human tumor cells overexpressing certain growth factor receptors also suggested another dimension of complexity. In the case of human breast cancers, overexpression of the HER2/Neu receptor (see Section 4.2) was found to be correlated with a bewildering array of phenotypes displayed by the associated cancer cells. Those breast cancer cells expressing elevated levels of this protein showed increased rates of DNA synthesis, better anchorage-independent growth, greater efficiency in forming tumors when implanted into host mice (that is, tumorigenicity), greater tendency to metastasize, and less dependence on estrogen for their growth. Hence, these receptor proteins were acting pleiotropically to confer a number of distinct changes on cancer cells. Such action seemed to be incompatible with a simple linear signaling cascade operating downstream of activated receptors. Instead, it appeared more likely that a number of distinct downstream signaling pathways were radiating out from these receptors, each involved in evoking a distinct cancer cell phenotype. But once again, there seemed to be little prospect of mapping these pathways and their signaling components.

This stalemate could be broken only by exploiting new types of experimental tools and a new way of thinking about the organization of complex signaling circuits. The science of genetics provided the new tool kit and mindset. However, genetics could be practiced with facility only on relatively simple experimental organisms, specifically, flies, worms, and yeast, for which powerful genetics techniques had been developed.

As it turned out, the elucidation of these signaling cascades also depended on one more factor. Much had already been learned from genetically dissecting the simpler metazoans and single-cell eukaryotes. Were lessons learned about their signaling systems transferable to the signaling cascades operating in mammalian cells? Here, fortune favored the cancer biologists. Because these intracellular signaling cascades were of very ancient lineage and had changed little over many hundreds of millions of years, discoveries about their organization and design in the worm, fly, and yeast proved to be directly applicable to mammalian cells. The consequence was the enrichment of cancer cell biology by research fields that had appeared, at least superficially, to be irrelevant to an understanding of human cancer. As was often the case in cancer research, the important advances came from unexpected quarters, from the laboratories of researchers who never thought of themselves as foot soldiers in the war against cancer.

More recent research on the biochemistry of receptor signaling continues to reveal new dimensions of complexity that impact the contributions that these proteins make to the development of human tumors. For example, RTKs often have extensive lateral interactions with other cell surface proteins, such as integrins; these side-by-side interactions influence the localization of the receptors and their access to different ligands (see Supplementary Sidebar 5.9). The control mechanisms that govern the concentrations of GFs in the extracellular space continue to be poorly understood; included among these mechanisms are those that allow proteases to activate latent GF precursors or to liberate GFs from sequestration in the extracellular matrix.

Moreover, a variety of inhibitory proteins are released by cells that prevent GFs from binding to their cognate receptors, doing so by intercepting these ligands in the extracellular space; some of these interceptors may bind and sequester GFs in an inactive state, while others may function as "decoy receptors" that display the ligand-binding domains of bona fide receptors but are incapable of releasing downstream signals. Such inhibitory proteins seem to play especially important roles in regulating Wnt signaling.

Yet other negative-feedback controls operate to constrain receptor signaling, often confining it to brief periods of time. Many ligand-bound receptors are quickly internalized by the process of endocytosis as a means of curtailing their further signaling.

In the case of RTKs, a variety of tyrosine **phosphatases** are often mobilized to quickly reverse the actions of the TK domains of these receptors, doing so by removing the phosphate groups from recently formed phosphotyrosine residues. This represents only one strategy to constrain signaling by RTKs. (In the next chapter, we will gain an appreciation of the fact that the release of stimulatory signals by various regulatory proteins must be counterbalanced by compensating negative-feedback controls in order to ensure proper regulation of the overall signal-processing circuitry operating within individual cells.)

We have mentioned, only in passing, members of another group of RTKs that actually constitute the largest subfamily—the Eph receptors and their ligands, termed ephrins. As is the case with Notch receptors and their cognate ligands, their signaling occurs between two cells that are in direct physical contact with one another. Thus, the ephrins are displayed as proteins tethered to the surface of the ligand-presenting cells; when an ephrin ligand binds and activates an Eph receptor displayed by an adjacent cell, signals are dispatched bidirectionally, that is, transduced into *both cells* that are engaged in these juxtacrine interactions. The roles of these ligand–receptor pairs in cancer pathogenesis are only beginning to emerge.

Yet other subtleties of receptor function remain underexplored. As one example, members of the family of 33 mammalian **tetraspanins** can be found embedded in various membranes throughout the cell. These abundantly expressed proteins are named after the four transmembrane domains carried by each of them. While not directly involved in transduction of ligand-triggered signal emission, they can move laterally in the plane of the plasma membrane and associate physically with other receptors, such as the RTKs and integrins, and influence signaling by the latter. We understand almost nothing about these membrane proteins and their contributions to the runaway signaling associated with the neoplastic cell state.

As we learn more about cell surface receptors, our current depictions of how they function will increasingly be seen as crude approximations of how they really operate within living tissues including tumors. Moreover, our focus on GFs and RTKs as the prime means of regulating communication between cells will be viewed as overly narrow. One hundred and three human phosphotyrosine phosphatases (PTPs), which reverse the actions of TKs, are likely to play key roles in regulating the levels of specific phosphotyrosines within cells, including those acquired by RTKs. Secreted proteases having narrow substrate specificities are also likely to sit astride numerous cell-to-cell signaling channels, controlling, for example, growth factor ligand availability as described above; at present we know virtually nothing about the 300 or so of these human enzymes and how they influence intercellular communication routes. Many of these normal signaling mechanisms are likely to be subverted by cancer cells. All of this suggests that we have explored only the most visible corners of the complex circuitry that drives neoplastic growth.

Implicit in this entire chapter is another idea that we only begin to confront in subsequent chapters: the fate of individual cells, including cancer cells, is heavily influenced by their surroundings. Hence, while we often focus on the intracellular signals governing the life of an individual cell, equally important determinants of a cell's fate come from outside, from its immediate neighboring cells and from others far away in the body.

Key concepts

- Because a multicellular organism can exist only if its individual cells work in a coordinated fashion, the problems of cell-to-cell communication were solved by the time the first metazoa arose. Deregulation of such cell signaling is central to the formation of cancer.

- Src provided the first insight into how oncoproteins function when it was found to operate as a protein kinase—an enzyme that transfers phosphates from ATP to other proteins in the cell. Since multiple distinct protein substrates could be phosphorylated and since each of these could affect its own set of downstream targets, this explained how Src could act pleiotropically to yield the numerous biological changes observed in RSV-transformed cells.

- Most protein kinases phosphorylate threonine or serine residues, but Src functions as a tyrosine kinase: it attaches phosphates to tyrosine residues of its protein substrate. Tyrosine phosphorylation is favored by mitogenic signaling pathways.

- Study of tyrosine kinase receptors (RTKs), such as the epidermal growth factor receptor (EGF-R), elucidated how they transduce signals from a cell's exterior into its cytoplasm. Binding of growth factor ligand to the N-terminal ectodomain of a cognate receptor induces dimerization of the receptor. This activates each monomer's kinase to phosphorylate its partner's tyrosines on the partner's C-terminal cytoplasmic domain. The resulting phosphotyrosines then enable emission of mitogenic signals to downstream target proteins within the cell.

- The dimerization model explains how overexpression of growth factor receptors favors cancer formation: when overexpressed, the high numbers of receptor molecules collide frequently as they move around the plane of the plasma membrane, resulting in dimerization, transphosphorylation, receptor activation, and emission of mitogenic signals.

- Mutations affecting any of the three RTK domains may create ligand-independent firing. An EGF-R with a truncated ectodomain cannot recognize its ligand but nonetheless emits growth-stimulatory signals in a constitutive fashion. Gene fusion events may create receptors that signal unremittingly by virtue of their ectodomains becoming fused with proteins that are naturally prone to dimerize or oligomerize. Amino acid substitutions or deletions in the transmembrane and cytoplasmic domains of receptors are also found in some cancers.

- Another mechanism for cell transformation is exemplified by the Sis oncoprotein of simian sarcoma virus. The virus forces an infected cell to release copious amounts of a PDGF-like protein, which attaches to the PDGF receptors of the same cell. This creates an autocrine signaling loop in which a cell manufactures a mitogen to which it can also respond.

- Other classes of receptors besides RTKs are important in cancer:

  - Cytokine receptors lack tyrosine kinase domains and rely on noncovalently associated JAK tyrosine kinases for signaling instead of covalently associated TK domains.

  - Receptors for TGF-β have cytoplasmic kinase domains that phosphorylate serine and threonine rather than tyrosine residues.

  - Notch receptor forgoes activation by phosphorylation and instead relies on proteases to liberate a cytoplasmic domain fragment that migrates to the nucleus, where it regulates gene expression.

  - The Patched–Smoothened system relies on one transmembrane protein controlling another, which in turn controls a transcriptional regulator.

  - Binding of canonical Wnt factors to a Frizzled receptor triggers a cascade of steps that prevents a cytoplasmic kinase from tagging β-catenin for destruction.

  - G-protein–coupled receptors (GPCRs) induce heterotrimeric G-proteins to flip from an inactive GDP-bound state to a signaling GTP-bound state.

- Integrins constitute a large family of heterodimeric transmembrane cell surface receptors, almost all of which have ECM components as their ligands. Upon ligand binding, integrins form focal adhesions that link their cytoplasmic domains to the actin cytoskeleton. Integrins pass information both into and out of the cell.

- While Ras proteins seemed to be part of a signaling cascade lying downstream of RTKs, the biochemical mechanisms connecting them were obscure, since the tyrosine kinase activity of RTKs had no apparent connection with the GTPase-bearing Ras protein.

- The difference between a normal Ras protein and its oncogenic counterpart is a missense mutation that alters a residue in the GTPase domain of Ras, preventing proper GTP hydrolysis; this in turn causes the mutant Ras protein to remain in its actively signaling state for extended periods of time rather than behaving like the normal protein and shutting itself off.

Thought questions

1. Why is autocrine signaling an intrinsically destabilizing force for a normal tissue?

2. The responsiveness of a cell to exposure to a growth factor is usually attenuated after a period of time (for example, half an hour), after which time it loses this responsiveness. Given what you have already learned about growth factor receptors, what mechanisms might be employed by a cell to reduce its responsiveness to a growth factor?

3. What lines of evidence can you cite to support the notion that growth factor receptor firing following ligand binding is often dependent on the dimerization of a receptor (rather than some other molecular change in the receptor)?

4. In what ways are the heterotrimeric G-proteins similar to the low-molecular-weight G-proteins (for example, Ras), and in what way do they differ fundamentally?

5. In what ways has the study of lower organisms, notably yeast, flies, and worms, proven to be highly revealing in characterizing the intracellular signaling cascades and the receptors operating at the cell surface?

6. Integrins exhibit a novel type of control, termed "inside-out signaling," in which intracellular signals dictate the affinity of these receptors for their various extracellular ligands (that is, components of the extracellular matrix). Why is such signaling essential to the process of cell motility?

Additional reading

Andrae J, Gallini R & Betsholtz C (2008) Role of platelet-derived growth factors in physiology and medicine. *Genes Dev.* 22, 1276–1312.

Angers S & Moon RT (2009) Proximal events in Wnt signal transduction. *Nat. Rev. Mol. Cell Biol.* 20, 468–477.

Arnal JF, Lenfant F, Metivier R et al. (2017) Membrane and nuclear estrogen receptor alpha actions: from tissue specificity to medical implications. *Physiol. Rev.* 97, 1045–1087.

Arnaout MA, Mahalingam B & Xiong JP (2005) Integrin structure, allostery, and bidirectional signaling. *Annu. Rev. Cell Dev. Biol.* 21, 381–410.

Bethani I, Skånland SS, Dikic I & Acker-Palmer A (2010) Spatial organization of transmembrane receptor signaling. *EMBO J.* 29, 2677–2688.

Borza CM & Pozzi A (2014) Discoidin receptors in disease. *Matrix Biol.* 34, 185–192.

Briscoe J & Thérond PP (2013) The mechanisms of Hedgehog signaling and its roles in development and disease. *Nat. Rev. Mol. Cell Biol.* 14, 418–431.

Charrin S, Jouannet S, Boucheix C & Rubinstein E (2014) Tetraspanins at a glance. *J. Cell Sci.* 127, 3641–3648.

Chow MT & Luster AD (2014) Chemokines in cancer. *Cancer Immunol. Res.* 2, 1125–1131.

Cooper J & Giancotti FG (2019) Integrin signaling in cancer: mechanotransduction, stemness, epithelial plasticity, and therapeutic resistance. *Cancer Cell* 35, 347–367.

Desgrosellier JS & Cheresh DA (2010) Integrins in cancer: biological implications and therapeutic opportunities. *Nat. Rev. Cancer* 10, 9–22.

Dorsam RT & Gutkind JS (2007) G-protein-coupled receptors and cancer. *Nat. Rev. Cancer* 7, 79–94.

D'Souza B, Miyamoto A & Weinmaster G (2008) The many facets of Notch ligands. *Oncogene* 27, 5148–5167.

Evangelista M, Tian H & de Sauvage FJ (2006) The Hedgehog signaling pathway in cancer. *Clin. Cancer Res.* 12, 5924–5928.

Fodde R & Brabletz T (2007) Wnt/beta-catenin signaling in cancer stemness and malignant behavior. *Curr. Opin. Cell Biol.* 19, 150–158.

Geiger B & Yamada KM (2011) Molecular architecture and function of matrix adhesions. *Cold Spring Harb. Perspect. Biol.* 3, a005033; doi: 10.1101/cshperspect.a005033.

Gilbert B & Mehlen P (2015) Dependence receptors and cancer: addiction to trophic ligands. *Cancer Res.* 72, 5171–5175.

Goel HL & Mercurio AM (2013) VEGF targets the tumour cell. *Nat. Rev. Cancer* 13, 871–882.

Gómez-Orte E, Saenz-Narciso B, Moreno S & Cabello J (2013) Multiple functions of the noncanonical Wnt pathway. *Trends Genet.* 29, 545–553.

Hanna A & Shevde LA (2016) Hedgehog signaling: modulation of cancer properties and tumor microenvironment. *Mol. Cancer* 15, 24; doi: 10.1186/s12943-016-0509-3.

Hassounah NB, Bunch TA & McDermott KM (2012) Molecular pathways: the role of primary cilia in cancer progression and therapeutics with a focus on Hedgehog signaling. *Clin. Cancer Res.* 18, 2429–2435.

Hynes RO (2002) Integrins: bidirectional, allosteric signaling machines. *Cell* 110, 673–687.

Hynes RO (2009) The extracellular matrix: not just pretty fibrils. *Science* 326, 1216–1219.

Jura N, Zhang X, Endres NF et al. (2011) Catalytic control in the EGF receptor and its connection to general kinase regulatory mechanisms. *Mol. Cell* 42, 9–22.

Klein R (2012) Eph/ephrin signaling during development. *Development* 139, 4105–4109.

Kopan R & Ilagan MX (2009) The canonical Notch signaling pathway: unfolding the activation mechanism. *Cell* 137, 216–233.

Kowatsch C, Woolley RE, Kinnebrew M et al. (2019) Structures of vertebrate Patched and Smoothened reveal intimate links between cholesterol and Hedgehog signalling. *Curr. Opin. Struct. Biol.* 57, 204–214.

Lee JS, Kim KI & Baek SH (2008) Nuclear receptors and coregulators in inflammation and cancer. *Cancer Lett.* 267, 189–196.

Lemmon MA & Schlessinger J (2010) Cell signaling by receptor tyrosine kinases. *Cell* 141, 117–134.

Massagué J (2012) TGF-β signaling in context. *Nat. Rev. Mol. Cell Biol.* 13, 616–630.

Mikels AJ & Nusse R (2006) Wnts as ligands: processing, secretion and reception. *Oncogene* 25, 7461–7468.

Mosesson Y, Mills GB & Yarden Y (2008) Derailed endocytosis: an emerging feature of cancer. *Nat. Rev. Cancer* 8, 835–850.

O'Malley BW & Kumar R (2009) Nuclear receptor coregulators in cancer biology. *Cancer Res.* 69, 8217–8222.

Pencik J, Pham HT, Schmoeller J et al. (2016) JAK-STAT signaling in cancer: from cytokines to non-coding genome. *Cytokine* 87, 26–36.

Schlaepfer DD & Mitra SK (2004) Multiple contacts link FAK to cell motility and invasion. *Curr. Opin. Genet. Dev.* 14, 92–101.

Schou KB, Pedersen LB & Christensen ST (2015) Ins and outs of GPCR signaling in primary cilia. *EMBO Rep.* 16, 1099–1113.

Seguin L, Desgrosellier JS, Weis SM & Cheresh DA (2015) Integrins and cancer: regulators of cancer stemness, metastasis and drug resistance. *Trends Cell Biol.* 25, 234–240.

Sever R & Glass CK (2013) Signaling by nuclear receptors. *Cold Spring Harb. Perspect. Biol.* 5, a016709; doi: 10.1101/cshperspect.a016709.

Sigismund S, Avanzato D & Lanzetti L (2017) Emerging functions of the EGFR in cancer. *Mol. Oncol.* 12, 3–20.

Singh AB & Harris RC (2005) Autocrine, paracrine and juxtacrine signaling by EGFR ligands. *Cell Signal.* 17, 1183–1193.

Singh B & Coffey RJ (2014) Trafficking of epidermal growth factor receptor ligands in polarized epithelial cells. *Annu. Rev. Physiol.* 76, 275–300.

Song S, Rosen KM & Corfas (2013) Biological function of nuclear receptor tyrosine kinase action. *Cold Spring Harb. Persp. Biol.* 5, a009001; doi: 10.1101/cshperspect.a009001.

Streuli CH & Akhtar N (2009) Signal co-operation between integrins and other receptor systems. *Biochem. J.* 418, 491–506.

Tian A-C, Rajan A & Bellen HJ (2009) A Notch updated. *J. Cell Biol.* 184, 621–629.

Valiathan RR, Marco M, Leitinger B et al. (2013) Discoidin domain receptor tyrosine kinases: new players in cancer progression. *Cancer Met. Rev.* 31, 295–321.

Varjosolo M & Taipale J (2010) Hedgehog: functions and mechanisms. *Genes Dev.* 22, 2454–2472.

Wieduwilt MJ & Moasser MM (2008) The epidermal growth factor receptor family: biology driving targeted therapeutics. *Cell. Mol. Life Sci.* 65, 1566–1584.

Witsch E, Sela M & Yarden Y (2010) Roles for growth factors in cancer progression. *Physiology* 25, 85–101.

Yamamoto S, Schulze KL & Bellen HJ (2014) *Methods Mol. Biol.* 1187, 1–14.

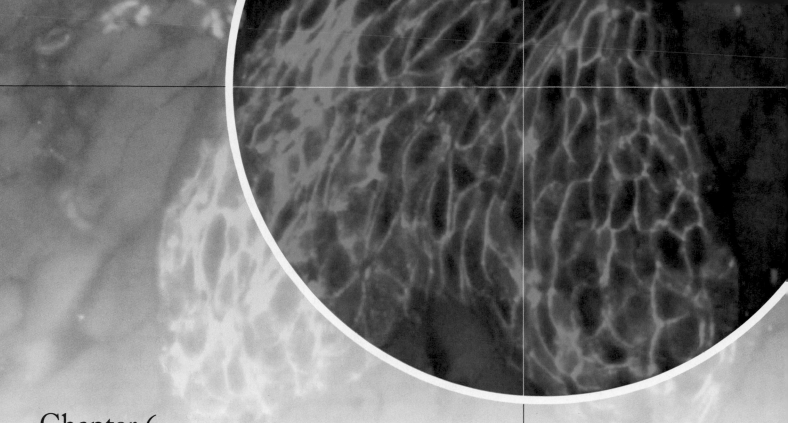

Chapter 6

Cytoplasmic Signaling Circuitry Programs Many of the Traits of Cancer

Chapter opener photo: From R.L. Tuttle et al., Nat. Med. 7:1133-1137, 2001. With permission from Nature.

Considering the wonderful frame of the human body, this infinitely complicated engine, in which, to the due performance of the several functions and offices of life, so many strings and springs, so many receptacles and channels are necessary, and all to be in their right frame and order; and in which, besides the infinite, imperceptible and secret ways of mortality, there are so many sluices and flood-gates to let death in, and life out, it is next to a miracle we survived the day we were born.

James Puckle, inventor, 1711

Any living cell carries with it the experiences of a billion years of experimentation by its ancestors. You cannot expect to explain so wise an old bird in a few simple words.

Max Delbrück, geneticist, 1966

Cancer is a disease of uncontrolled cell proliferation. Because the proliferative behavior of cancer cells is so aberrant, we might imagine that cancer cells invent entirely new ways of programming their growth and division—that the control circuitry within cancer cells is organized quite differently from that of normal, healthy cells. Such thinking greatly exaggerates the actual differences between normal and neoplastic cells. In truth, the two types of cells utilize control circuitry that is almost identical. Cancer cells discover ways of making relatively minor modifications to the control machinery operating inside cells. They tweak existing controls rather than demolishing the entire machinery and assembling a new version from the remnants of the original.

Movies in this chapter

6.1 Regulation of Signaling by the Src Protein

6.2 EGF Receptors and Signaling

The present chapter is about this control machinery—really about the signal-processing circuitry that operates within the cell cytoplasm and governs cell growth and proliferation. An electronic circuit board is assembled from complex arrays of hard-wired components that function as resistors, capacitors, diodes, and transistors. The cell also uses circuits assembled from arrays of intercommunicating components, but these are, almost without exception, proteins. While individual proteins and their functions are relatively simple, the operating systems that can be assembled from these components are often extraordinarily complex.

In Chapter 5, we read about the receptors displayed at the cell surface that gather a wide variety of signals and funnel them into the cytoplasm. Here, we study how these signals, largely those emitted by growth factor receptors, are processed and integrated in the cytoplasm. Many of the outputs of this signal-processing circuitry are then transmitted to the nucleus, where they provide critical inputs to the central machinery that governs cell proliferation. Our discussion of this nuclear governor—the *cell cycle clock*—will follow in Chapter 8.

A single cell may express 20,000 or more distinct proteins, many of which are actively involved in the cytoplasmic regulatory circuits that are described here. These regulatory proteins are found in various concentrations and in different locations throughout the cytoplasm; often intercommunicating proteins congregate in specific organelles or focal areas within the cytoplasm. Moreover, rather than floating around in dilute concentrations in the intracellular water, they form a thick soup. Indeed, as much as 30% of the volume of a cell is taken up by proteins rather than aqueous solvent.

These proteins must be able to talk to one another and to do so with great specificity and precision. The problem of specificity is complicated by the fact that important components of the signaling circuitry, such as kinases, have many **paralogs**—proteins that share a common evolutionary origin and retain common structural features. For example, there are more than 400 similarly structured serine/threonine kinases in mammalian cells. In spite of this complexity, a signaling protein operating in a linear signaling cascade must recognize only those signals that come from its upstream partner protein(s) and pass them on to its intended downstream partner(s). In so doing, it will largely ignore the thousands of other proteins within the cell.

This means that each protein component of a signaling circuit must actually solve two problems. The first is one of specificity: how can it exchange signals only with the small subset of cellular proteins that are its intended signaling partners in the circuit? Second, how can this protein acquire rapid, almost instantaneous access to these signaling partners, doing so while operating in the viscous soup that is present in the cytoplasm and nucleus?

We study this circuitry because its design and operations provide key insights into how cancer cells arise. Thus, many of the oncoproteins described in previous chapters create cancer through their ability to generate signaling imbalances in this normally well regulated system. At one level, cancer is surely a disease of inappropriate cell proliferation, but as we bore deep into the cancer cell, we will come to understand cancer at a very different level: cancer is really a disease of aberrant signal processing.

The present chapter will perhaps be the most challenging of all chapters in this book. The difficulty comes from the sheer complexity of signal transduction biochemistry, a field that is afflicted with many facts and blessed with only a small number of unifying principles. So, absorb this material in pieces; the whole is far too much for one reading.

6.1 A signaling pathway reaches from the cell surface into the nucleus

The growth and division that cells undertake following exposure to mitogens clearly represent complex regulatory programs involving the coordinated actions of hundreds, even thousands of distinct cellular proteins. How can growth factors succeed in evoking all of these changes simply by binding and activating their receptors? And how can the deregulation of these proliferative programs lead to cell transformation? And where does Ras fit into this circuitry, if indeed it does?

A major insight into these issues came in 1981 from examining the behavior of cultured normal cells, specifically, fibroblasts that had been deprived of serum-associated growth factors for several days. During this period of "serum starvation," the cells within a culture were known to enter reversibly into the nongrowing, quiescent state termed G_0 ("G zero"). After remaining in this G_0 state for several days, serum-starved cells were then exposed to fresh serum and thus to abundant amounts of mitogenic growth factors. The intent of this experiment was to induce all the cells in such cultures to enter into a state of active growth and division **synchronously** (coordinately). Indeed, within a period of 9 to 12 hours, the great majority of these previously quiescent cells would begin to replicate their DNA and initiate cell division a number of hours later.

These readily observed changes in cell behavior provided no hint of the large number of less obvious but nonetheless important molecular changes that occurred in these cells, often within minutes of exposure to fresh growth factors. For example, within less than an hour, transcription of an ensemble of more than 100 cellular genes was induced (**Figure 6.1**). Expression of these genes, which we now call **immediate early genes** (IEGs), increased rapidly in the half hour after growth factor stimulation. As is suggested by the contents of **Table 6.1**, the products of some of these genes help cells in various ways to emerge from the quiescent G_0 state into their active growth-and-division cycle. In fact, experiments involving several of these genes showed that blocking their expression prevents the emergence of cells from the G_0 state.

Table 6.1 A sampling of immediate early genes[a]

Name of gene or gene product	Location of gene product	Function of gene product
fos[b]	nucleus	component of AP-1 TF
junB	nucleus	component of AP-1 TF
egr-1	nucleus	zinc finger TF
nur77	nucleus	related to steroid receptors
Srf-1[c]	nucleus	TF
myc	nucleus	bHLH TF
β-actin	cytoplasm	cytoskeleton
γ-actin	cytoplasm	cytoskeleton
tropomyosin	cytoplasm	cytoskeleton
fibronectin	extracellular	extracellular matrix
glucose transporter	plasma membrane	glucose import
JE	extracellular	secreted cytokine
KC	extracellular	secreted cytokine

[a]The genes listed here represent only a small portion of the immediate early genes (IEGs; see Figure 6.1A).

[b]Expression of a group of *fos*-related genes is also induced as IEGs. These include *fosB*, *fra-1*, and *fra-2*.

[c]Srf is a TF that binds to the promoters of other immediate early genes such as *fos*, *fosB*, *junB*, *egr-1*, *egr-2*, and *nur77*; and cytoskeletal genes such as actins and myosins.

Adapted in part from H.R. Herschman, *Annu. Rev. Biochem.* 60:281–319, 1991; and from B.H. Cochran, in R. Grzanna and R. Brown (eds.), Activation of Immediate Early Genes by Drugs of Abuse. Rockville, MD: National Institutes of Health, 1993, pp. 3–24.

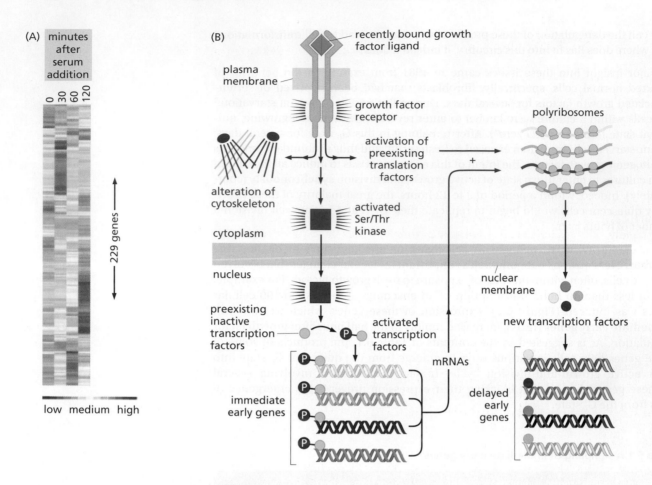

Figure 6.1 Expression of immediate and delayed early genes
(A) Expression of a large array of "immediate early genes" (IEGs) is induced within the first two hours after serum-starved cells are exposed to fresh serum. The expression of IEGs, by definition, occurs even when protein synthesis is blocked during this period, indicating that the transcription factors regulating expression of these genes are already present in the serum-starved cells. Analyses of the expression patterns of the 14,824 genes surveyed here at 30, 60, and 120 minutes following serum addition led to the identification of 229 genes (*unlabeled, arrayed top to bottom*) that behave like IEGs. Complex signaling cascades explain why some IEGs are only induced with some delay. *Red* indicates a high level of RNA expression, while *blue* indicates low expression.

Note that some IEGs are expressed for only a brief period of time before being repressed even though the serum and associated mitogens continue to be present in the culture medium. (B) The profound changes in cell physiology following the application of serum-associated mitogenic growth factors to a cell indicate that growth factor receptors can release a diverse array of biochemical signals once they are activated by ligand binding. The transcription of some IEGs can already be observed within minutes of growth factor stimulation (*lower left*). These contrast with the expression of delayed early genes that requires the synthesis of new transcription factors (*lower right*). (A, from A. Selvaraj and R. Prywes, *BMC Mol. Biol.* 5:13, 2004. With permission from Springer.)

A variation of this 1981 experiment was undertaken in which fresh serum was added together with **cycloheximide**, a drug that shuts down all cellular protein synthesis. In spite of the inhibition of protein synthesis, induction of the immediate early genes proceeded quite normally. This indicated that all of the proteins required to activate transcription of these genes were already in place at the moment when the serum was added to the cell. Stated differently, the induction of these genes did not require ***de novo*** (new) protein synthesis.

These results also demonstrated that, in addition to the growth factor receptors at the cell surface, an array of other proteins was present within the cell that could rapidly convey mitogenic signals from the receptors located at the cell surface to transcription factors (TFs) operating in the nucleus. Clearly, the functional activation of such cytoplasmic signal-transducing proteins did not depend upon increases in their concentration (since cycloheximide had no apparent effect on this signaling). Instead, changes in the proteins' structure, configuration, and intracellular localization appeared to play a dominant role in their functional activation soon after the stimulation of tyrosine kinase receptors by their growth factor ligands.

The immediate early genes encode a number of interesting proteins (Table 6.1). Some of them specify transcription factors that, once synthesized, help to induce a second wave of gene expression. Included here are the *myc*, *fos*, and *jun* genes, which were originally identified through their incorporation as oncogenes into the genomes of transforming retroviruses (see Table 3.3). Yet other immediate early genes encode proteins that are secreted growth factors (**cytokines**) or help to organize the cellular cytoskeleton.

The levels of *myc* mRNA were found to rise steeply following mitogen addition and to collapse quickly following mitogen removal. Moreover, the Myc protein itself turns over rapidly, having a **half-life** ($T_{1/2}$) of only 25 minutes. Such observations indicated that the levels of the Myc protein serve as a sensitive and responsive intracellular indicator of the amount of mitogens present in the nearby extracellular space.

As an aside, we might note that the activity of Myc operating as a signaling protein derives in large part from major changes in its *concentration* within the cell nucleus. This contrasts starkly with the behavior of cytoplasmic signal-transducing proteins such as Ras and Src. These and many other cytoplasmic signaling proteins respond to mitogenic signals by undergoing *relocalization* within the cell as well as alterations in their *structure* rather than exhibiting significant increases in concentration. This difference is mirrored by the changes occurring when the normal *myc*, *ras*, and *src* genes are converted into oncogenes: the *levels* of Myc protein become deregulated (being expressed constitutively rather than being modulated in response to physiologic signals), whereas the *structures* of the Ras and Src proteins undergo alteration but the amounts of these proteins do not necessarily increase.

Within an hour after the immediate early mRNAs appear, a second wave of gene induction was found to occur. Significantly, the induction of these **delayed early genes** was largely blocked by the presence of cycloheximide, indicating that their expression did indeed depend on *de novo* protein synthesis (see Figure 6.1B). (In fact, expression of these delayed early genes seems to depend on transcription factors cited above that are synthesized in the initial wave of immediate early gene expression.)

Addition of growth factors to quiescent cells was found to provoke yet other changes in cell physiology besides the rapid induction of nuclear genes. Following exposure of cells to serum, their rate of protein synthesis increases significantly, this being achieved through functional activation of the proteins that enable ribosomes to initiate translation of mRNAs. Some growth factors were discovered to induce motility in cells, as indicated by the movement across the bottom surfaces of Petri dishes. Yet others can provoke a reorganization of actin fibers that help to construct the cell's cytoskeleton—the intracellular scaffolding that defines cell shape (Supplementary Sidebar 6.1). Subsequently, many growth factors were discovered to provide survival signals to cells, thereby protecting them from inadvertent activation of the cell suicide program known as **apoptosis**, which is described later in Chapter 9.

These diverse responses indicated that a variety of distinct biochemical signals radiate from ligand-activated growth factor receptors, and that these signals impinge on a diverse array of cellular targets. Some of these signals seemed to be channeled directly into the nucleus, where they altered gene expression programs, while others were clearly directed toward cytoplasmic targets, including the protein-synthesizing machinery and the proteins that organize the structure of the cytoskeleton.

Clearly, an understanding of how these signaling cascades operate was very relevant to the cancer problem: if it were true that some oncoproteins flood cells with a continuous stream of mitogenic signals (see Chapter 5), then the transformed state of cancer cells might well represent an exaggerated version of the responses that normal cells exhibit following exposure to growth factors. In fact, many of the traits of cancer cells can indeed be traced to responses evoked by growth factors, while others cannot be predicted by the initial responses of cells to these factors (**Sidebar 6.1**).

Sidebar 6.1 Short-term transcriptional responses to mitogens provide no indication of the gene expression in continuously growing cells Simple logic would dictate that the transcriptional responses observed shortly after growth factor stimulation of a serum-starved cell ought to predict the transcriptional state of cells in which mitogenic signaling occurs continuously, whether in response to ongoing exposure to growth factors or the activity of certain oncoproteins. Accordingly, the spectrum of immediate and delayed early genes induced within the first hour of growth factor stimulation should also be expressed continuously and at a high rate in transformed cells.

In fact, some genes that are members of the immediate and delayed early gene expression groups are found to be expressed at high levels in cancer cells, while others are not. This discordance derives from negative-feedback controls that force the shutdown of certain genes shortly after they have been expressed. For example, expression of the *fos* gene, itself a proto-oncogene, increases rapidly in response to serum stimulation, peaks, and then declines dramatically, all within an hour or so (see, for example, Figure 6.1A). This shutdown is due to the ability of the Fos protein, once synthesized, to act as a transcriptional repressor of the synthesis of more *fos* mRNA. Numerous other negative-feedback mechanisms operate in cells to ensure that many of the initial biochemical responses evoked by growth factors function only transiently. More generally, negative-feedback loops are installed in virtually all signaling pathways to ensure that the flow of signals through these pathways is tightly controlled in both intensity and duration (see Section 6.14).

6.2 The Ras protein stands in the middle of a complex signaling cascade

The disparate responses of cells to growth factors represented a challenge to those interested in intracellular signal transduction, since almost nothing was known about the organization and function of the communication channels operating within the cell. Over a period of a decade (the 1980s), these circuits were slowly pieced together, much like a jigsaw puzzle. The clues came from many sources. The story started with Ras and then moved up and down the signaling cascades until the links in the signaling chains were finally connected.

The advances in Ras biochemistry that we discussed in the last chapter (Section 5.10) seemed to explain much about how it operates as a binary switch, but provided no insight into its context—how this protein is connected with the overall signaling circuitry in which it is embedded. Indeed, it remained possible that the Ras protein functioned in a signaling pathway that was independent of the one controlled by growth factors and their receptors.

In the end, the solution to the Ras problem came from a totally unexpected quarter. The genetics of eye development in the fruit fly *Drosophila melanogaster* revealed a series of genes whose products were essential for the normal development of the ommatidia, the light-sensing units forming the compound eye. One important gene here came to be called *Sevenless*; in its absence, the seventh cell in each ommatidium failed to form (**Figure 6.2A**). Provocatively, after cloning and sequencing, the *Sevenless* gene was found to encode a protein with the general structural features of a tyrosine kinase receptor, specifically a mammalian FGF receptor.

Yet other mutations mimicked the effects of *Sevenless* mutation. Genetic **complementation** tests revealed that these mutations affected genes whose products operate downstream of *Sevenless*, ostensibly in a linear signal transduction cascade. One of these downstream proteins was encoded by a gene that was named *Son of sevenless* or simply *Sos*. Close examination of the Sos amino acid sequence showed it to be related to proteins known from yeast biochemistry to be involved in provoking nucleotide exchange by G (guanine nucleotide–binding) proteins such as Ras. The yeast proteins, often called guanine nucleotide exchange factors (GEFs), were known to induce G proteins to release their bound GDPs, thereby making room for GTPs to jump aboard (see Figure 5.24). The consequence was an activation of these G proteins from their inactive to their active, signal-emitting configuration. This was precisely the effect that Sos exerted on Ras proteins (Figure 6.2B). Therefore, Sos appeared to be the long-sought upstream stimulator that kicked Ras into its active configuration.

Soon other intermediates in the signaling cascade were uncovered. Two of these, Shc (pronounced "shick") and Grb2 (pronounced "grab two"), were discovered through

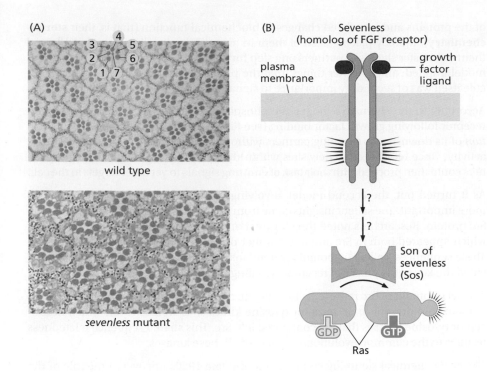

(A) wild type

selenless mutant

(B)

Sevenless
(homolog of FGF receptor)

plasma
membrane

growth
factor
ligand

Son of
sevenless
(Sos)

GDP   GTP

Ras

Figure 6.2 Son of Sevenless and Ras
(A) Study of the light-sensing units in
the Drosophila melanogaster eye—
the ommatidia—revealed that each
ommatidium is normally formed from a
series of seven cells, with six outer ones
surrounding a seventh in the center, as
seen in this electron micrograph section
(*upper panel*). A fly carrying the mutation
in the gene that came to be called
sevenless produced ommatidia lacking the
seventh, central cell (*lower panel*). Later,
sevenless was found to encode a homolog
of the mammalian fibroblast growth
factor (FGF) receptor. (B) Genetic studies
of *Drosophila* eye development revealed
another gene whose product appeared
to lie downstream of *Sevenless* in a
signaling cascade and specified a guanine
nucleotide exchange factor (GEF; *orange*),
capable of activating the Ras protein.
This gene was named *Son of sevenless* or
simply *Sos*. (A, courtesy of E. Hafen.)

genetic and biochemical screens for proteins that interacted with phosphorylated receptors or derived peptides. It became apparent that these two proteins function as adaptors, forming physical bridges between growth factor receptors and Sos; such bridging proteins will be discussed in greater detail later. A third such adaptor protein, called Crk (pronounced "crack"), was identified as the oncoprotein encoded by the CT10 avian sarcoma virus (see Table 3.3).

These discoveries demonstrated the very ancient origins of this signaling pathway, which was already well developed more than 600 million years ago in the common ancestor of all contemporary metazoan phyla. Once in place, the essential components of this cascade remained relatively unchanged in the cells of descendant organisms. Indeed, the individual protein components of these cascades have been so conserved that in many cases the protein components prepared from the cells of members of one phylum (for example, chordates) can be exchanged with those of another (for example, arthropods) to reconstitute a functional signaling pathway.

Together, the genetic and biochemical data from these distantly related animal phyla could be combined in a scheme that suggested a linear signaling cascade organized like this: tyrosine kinase receptor → Shc → Grb → Sos → Ras. While this cascade provided the outline of a signaling channel, it gave little insight into the biochemical interactions that enabled these various proteins to exchange signals with one another.

6.3 Tyrosine phosphorylation controls the location and thereby the actions of many cytoplasmic signaling proteins

Among the major biochemical issues left unanswered, perhaps the most critical were the still-mysterious actions of the kinases carried by many growth factor receptors. Was the phosphorylation of these receptors on tyrosine residues crucial to their ability to signal, or was it a distraction? And if this phosphorylation was important, how could it activate the complex signaling circuitry that lay downstream?

Two alternative mechanisms seemed plausible. The first predicted that a receptor-associated tyrosine kinase would phosphorylate a series of target proteins in the cytoplasm. Such covalent modification would alter the three-dimensional *conformation* of the proteins and associated changes in biochemical function (that is, their **stereochemistry**), thereby placing each of them in an actively signaling state that allowed them to transfer signals to partners one step further down in a signaling cascade. This model implied, among other things, that the phosphorylation of the receptor molecule itself was of secondary importance to signaling.

According to the alternative model, the phosphorylation of the cytoplasmic tail of a receptor following growth factor binding (see Figure 5.11A) affected the *physical location* of its downstream signaling partners without necessarily changing their intrinsic activity. Once relocalized to new sites within the cytoplasm, these downstream partners could then proceed with their task of emitting signals to yet other targets in the cell.

As it turned out, the second model, involving protein relocalization, proved to be more important. The salient insights came from analyzing the detailed structure of the Src protein. Researchers noted three distinct amino acid sequence domains, each of which appeared both in Src and in a number of other, otherwise unrelated proteins. These sequences, called Src homology domains 1, 2, and 3 (SH1, SH2, and SH3), provided the key to the puzzle of receptor signaling (**Figure 6.3**).

The SH1 domain of Src represents its catalytic domain, and indeed similar sequence domains are present in all receptor tyrosine kinases, as well as in the other, nonreceptor tyrosine kinases that are configured like Src. This shared sequence relatedness testifies to the common evolutionary origin of all these kinases.

Clever biochemical sleuthing conducted in the late 1980s uncovered the role of the SH2 domain present in Src and other signaling proteins. This relatively small structural domain of about 100 amino acid residues (**Figure 6.4A and B**) acts as an intracellular "receptor." The "ligand" for this SH2 receptor is a short oligopeptide sequence that contains both a phosphorylated tyrosine and a specific oligopeptide sequence 3 to 6 residues long that flanks the phosphotyrosine on its C-terminal side.

Soon, it became apparent that there are dozens of distinct SH2 domains, each carried by a different protein and each having an affinity for a specific phosphotyrosine-containing

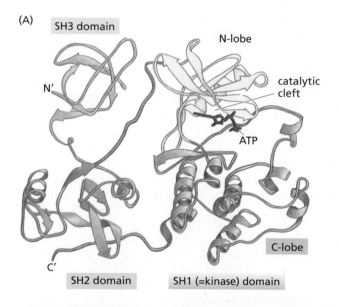

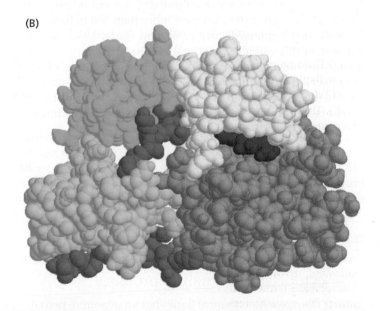

Figure 6.3 Domain structure of the Src protein Sequence analyses of the Src protein revealed three distinct structural domains that are also found in a number of otherwise-unrelated proteins. (A) In this ribbon diagram of Src—the product of X-ray crystallography—coils indicate α-helices, while flattened ribbons indicate β-pleated sheets. Like the great majority of other subsequently analyzed protein kinases, the catalytic SH1 (Src homology) domain is seen to be composed of two subdomains (*right*), termed the N- and C-lobes of the kinase (*yellow, orange; right*). Between these lobes is the ATP-binding site (ATP—*red stick figure*) where catalysis takes place. The SH2 and SH3 domains (*blue, light green; left*) are involved in substrate recognition and regulation of catalytic activity. (B) A space-filling model of Src is presented in the same color scheme. Connecting segments are shown in *dark green*. (From B. Alberts et al., Molecular Biology of the Cell, 6th ed. New York: W.W. Norton & Company, 2015.)

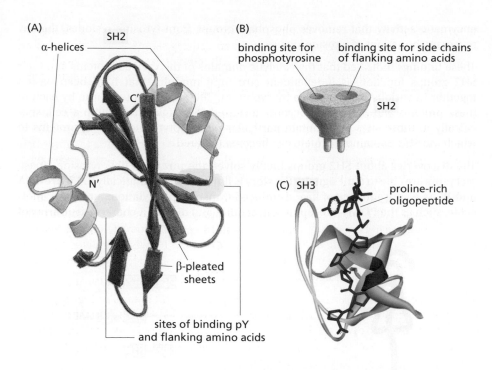

Figure 6.4 Structure and function of SH2 and SH3 domains (A) This SH2 domain is seen in a different orientation from that in Figure 6.3. As revealed by X-ray crystallography, a typical SH2 domain is composed of about 100 amino acid residues and is assembled from a pair of anti-parallel β-pleated sheets (*red*) surrounded by a pair of α-helices (*green*). (The N'- to C'-terminal orientation is indicated by the arrowheads.) The sites of interaction by the SH2 domain with its "ligand"—a phosphotyrosine (pY)-containing peptide (composed of a pY and flanking amino acids)—are indicated by the *yellow* spots. (B) As indicated, the SH2 domain can be thought to function like a modular plug: it recognizes both a phospholyrosine and the amino acids that flank this phosphotyrosine on its C-terminal side. These flanking amino acid residues determine the specificity of binding, i.e., the identities of the particular phosphotyrosine-containing oligopeptides that are recognized and bound by a particular SH2 domain. (C) The SH3 domain (*ribbon diagram*), shown in a different orientation from that of Figure 6.3, recognizes as its ligands proline-rich oligopeptide (*blue stick figure*) sequences in partner proteins. (A, from G. Waksman et al., *Cell* 72:779–790, 1993. With permission from Elsevier. B, from B. Alberts et al., Molecular Biology of the *Cell*, 6th ed. New York: W.W. Norton & Company, 2015. C, adapted from S. Casares et al., *Protein* 67:531–547, 2007.)

oligopeptide sequence that functions as its ligand. By now, the sequence specificities of a substantial number of SH2 groups have been cataloged. The human genome is estimated to encode at least 120 distinct SH2 groups, each constituting a domain of a larger protein and each apparently having an affinity for binding a particular phosphotyrosine together with a flanking oligopeptide sequence. (Ten proteins carry two SH2 domains, yielding a total census of 110 distinct SH2-containing proteins in our cells.)

Most commonly, an SH2 domain enables the protein carrying it to associate with a partner protein that is displaying a specific phosphotyrosine plus flanking amino acid sequence, thereby forming a physical complex between these two proteins. Some SH2-bearing proteins carry no enzymatic activity at all. Other such proteins, in addition to their SH2 domains, carry catalytic sites that are quite different from the tyrosine kinase (TK) activity present in Src itself. For example, one form of phospholipase C carries an SH2 domain, as does the p85 subunit of the enzyme phosphatidylinositol 3-kinase (PI3K). The SHP-1 protein, discussed in **Sidebar 6.2**, carries two SH2 domains linked to a protein tyrosine **phosphatase** (PTP) catalytic domain, that is, a domain with an

---

Sidebar 6.2 A negative-feedback control mechanism leads to an SH2-containing molecule that can determine outcomes of Olympic competitions One of the SH2-containing partners of tyrosine kinase receptors is a phosphotyrosine phosphatase (PTP), an enzyme that removes the phosphate group from phosphotyrosine. Once tethered via its SH2 group to a ligand-activated, tyrosine-phosphorylated receptor, this particular phosphatase, termed SHP-1, begins to chew away the phosphate groups on tyrosine residues that have enabled the receptor to attract other downstream signaling partners. In this fashion, an enzyme such as SHP-1 creates a negative-feedback loop to shut down further receptor firing. In fact, the SHP-1 phosphatase is responsible for shutting down signaling by a variety of receptors, among them the receptor for the erythropoietin (EPO) growth factor, which normally responds to EPO by triggering **erythropoiesis**—red blood cell formation (see Section

6.8 and Figure 6.16). (The catalytic activity of the phosphatase domain of the related SHP-2 protein is increased by >100-fold following binding of its two SH2 groups to phosphotyrosines.)

Some individuals inherit a mutant gene encoding an EPO receptor that lacks the usual docking site for the SHP-1 phosphatase. In individuals whose cells express this mutant receptor, the resulting defective negative-feedback control leads to a hyperactive EPO receptor, and, in turn, to higher-than-normal levels of red cells and thus oxygen-carrying capacity in their blood. In the case of several members of a Finnish family with this inherited genetic defect, this condition seems to have played a key role in their winning Olympic medals in cross-country skiing! Provocatively, expression of SHP-1 is shut down in many leukemias and non-Hodgkin's lymphomas; this permits multiple mitogen receptor molecules to fire excessively in the tumor cells.

enzymatic activity that removes phosphate groups from tyrosine residues, thereby reversing the actions of tyrosine kinases.

These findings indicated that the catalytic domains of the various proteins and their SH2 groups function as independent structural modules that have been pasted together in various combinations by evolution. The SH2 group(s) borne by each of these proteins allows them to become localized to certain sites within a cell, specifically to those sites that contain particular phosphotyrosine-bearing proteins to which the SH2-containing protein can become tethered.

The discoveries about SH2 groups finally solved the puzzle of how tyrosine kinase receptors are able to emit signals. The story is illustrated schematically in **Figure 6.5A and B**. As a consequence of ligand-induced transphosphorylation, a receptor molecule, such as the PDGF-β receptor, will acquire and display a characteristic array of

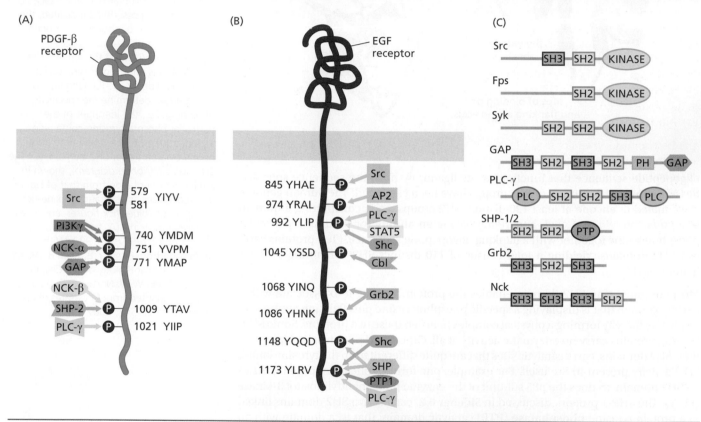

Figure 6.5 Attraction of signal-transducing proteins by phosphorylated receptors (A) This schematic diagram of a molecule of the platelet-derived growth factor-β (PDGF-β) receptor, which omits its tyrosine kinase domain, reveals a large number of tyrosine residues in its C-terminal domain that undergo phosphorylation following ligand binding and receptor activation. (The positions of these tyrosine residues in the receptor polypeptide chain are indicated by the numbers.) Listed to the left are seven distinct cytoplasmic proteins, each of which can bind via its SH2 domain(s) to one or more of the phosphotyrosines of the PDGF-β receptor. The three amino acid residues (denoted by the single-letter code) that flank each tyrosine (Y) residue on its C-terminal side define the unique binding site recognized by the various SH2 domains of these seven associated proteins. (The "kinase insert region" and associated phosphotyrosine in the middle of the PDGF-β receptor tyrosine kinase domain is not shown.) (B) Similarly, a constellation of phosphotyrosines can be formed after the EGF receptor binds its ligand, often forming heterodimers with the HER2 receptor protein. However, the spectrum of SH2-containing proteins that associates with the EGF-R is quite different, allowing this receptor to activate a set of downstream signaling pathways that are different from those associated with the PDGF-β receptor. Not shown here is the binding of Grb2 to Y1045, Y1101, Y1148, and Y1173, as well as the binding of Shc to Y1045, Y1068, Y1086, and Y1101. (C) Specific multiprotein complexes are formed using domains such as the SH2, SH3, and PH domains, which bind phosphotyrosines, proline-rich sequences, and PIP3, respectively (discussed in Section 6.6). As is indicated here, evolution has shuffled these domains in a modular fashion, attaching them to tyrosine kinase catalytic domains in the case of the first three tyrosine kinases (Src, Fps, and Syk), to a GAP (Ras GTPase-activating protein) domain, to a PLC (phospholipase C) domain, or to a protein tyrosine phosphatase domain (SHP-1, SHP-2). At the bottom are shown two adaptor or "bridging" proteins that contain only SH2 and SH3 domains and lack catalytic domains; each of these adaptors serves as a linker between pairs of other proteins, one of which carries a certain phosphotyrosine (to which an SH2 domain can bind) and the other of which carries a proline-rich domain (to which an SH3 domain can bind; see Figure 6.7). (A and C, adapted from T. Pawson, *Adv. Cancer Res.* 64:87–110, 1994. B, adapted from R. Sordella et al., *Science* 305:1163–1167, 2004.)

phosphotyrosine residues on its cytoplasmic tail; likewise, the EGF receptor will display its own spectrum of phosphotyrosines on its cytoplasmic tail. The unique identity of each of the phosphotyrosine residues borne by these tails is dictated by the sequence of amino acid residues flanking the phosphotyrosine on its C-terminal side. These phosphotyrosines become attractive homing sites for various SH2-containing cytoplasmic proteins, specifically, proteins that normally reside in the soluble portion of the cytoplasm (the **cytosol**) and are therefore free to move from one location to another in the cytoplasm. Consequently, shortly after becoming activated by ligand binding, a growth factor receptor becomes decorated with a specific set of these SH2-containing partner proteins that are attracted to its various phosphotyrosines.

For example, one ligand-activated form of the PDGF receptor attracts Src, PI3K, Ras-GAP (which stimulates Ras GTPase activity; refer to Sidebar 5.8), SHP-2 (SH2-containing tyrosine phosphatase 2), and PLC-γ (phospholipase C-γ). Each of these recruited proteins carries at least one SH2 domain (Figure 6.5C), and all of them flock to the PDGF-R after its cytoplasmic tail and a short segment in the middle of its kinase domain (called a "kinase insert region") become tyrosine-phosphorylated. Once tethered to the PDGF-R, some of these SH2-containing proteins may then become substrates for tyrosine phosphorylation by the PDGF-R kinase. Yet other proteins serve as bridges by attracting additional proteins to these multiprotein complexes (see Figure 6.7 in the next section). (It remains unclear whether all of these SH2-containing proteins can bind simultaneously to a single ligand-activated PDGF receptor molecule.)

The biochemical diversity of these associated proteins provides us with at least one solution to the question of how a growth factor receptor is able to elicit a diversity of biochemical and biological responses from a cell: each of these associated proteins controls its own downstream signaling cascade, and each of these cascades, in turn, influences a different aspect of cellular behavior. By becoming tethered to a receptor molecule, these SH2-containing proteins are poised to activate their respective cascades. Some of these SH2-bearing proteins function to activate negative-feedback loops that ultimately shut down receptor signaling, as discussed in Sidebar 6.2 and again later in this chapter.

Because the cytoplasmic domains of tyrosine kinase receptors are found near the inner surface of the plasma membrane, the SH2-containing partner proteins attracted to the receptor are brought into close proximity to the plasma membrane and thus close to other molecules that are, for their own reasons, already tethered to this membrane. This close juxtaposition enables various SH2-containing proteins to interact directly with the membrane-associated proteins and phospholipids, thereby generating a variety of biochemical signals that can thereafter be transmitted to various downstream signal-transducing cascades.

For example, the Ras GTPase-activating protein (Ras-GAP), having become attached to a phosphorylated receptor, is tethered in close proximity to Ras proteins, many of which are permanently anchored to the plasma membrane via their C-terminal lipid tails. This allows Ras-GAP to interact with nearby Ras molecules, stimulating them to hydrolyze their bound GTPs and thereby convert from an active to an inactive signaling state (see Figure 5.24).

Similarly, when phosphatidylinositol 3-kinase (PI3K) becomes tethered via its SH2 group near the plasma membrane, it is able to reach over and phosphorylate inositol lipids embedded in the membrane; we return to this enzyme and its mechanism of action in Section 6.6. Yet another example is provided by phospholipase C-γ (PLC-γ), which, when juxtaposed to the plasma membrane via receptor tethering, is able to cleave a membrane-associated phospholipid [phosphatidylinositol (4,5)-diphosphate, or PIP2] into two products (IP3 and DAG; see Figure 6.11), each of which, on its own, has potent signaling powers. The point here is that many reactions become possible (or can proceed at much higher rates) when enzymes and their substrates are brought into close proximity through tethering of these enzymes to phosphorylated receptors.

The discovery of SH2 domains and their presence in a diverse array of signaling proteins solved a long-standing paradox. The dozens of distinct RTKs expressed by human cells were known to carry essentially identical tyrosine kinase signal-emitting domains, yet these receptors, when activated by ligand binding, clearly induced very different cellular responses. This puzzle was resolved when it was realized that, following ligand binding, each RTK acquires its own particular array of phosphotyrosine-containing peptides (see Figure 6.5A and B), enabling it to attract its own combination of downstream signaling partners. In recent years, systematic surveys have been undertaken to reveal the full range of interactions between RTKs, such as the EGF receptor, and the various SH2 domain–containing proteins in the cytoplasm (Supplementary Sidebar 6.2). In the longer term, such surveys will provide the databases that will make it possible to construct mechanistic models that help to predict how individual receptors, once activated by their ligands, operate to change the biochemistry and physiology of a cell. Nonetheless, even with this information in hand, a significant number of other variables complicate attempts to understand precisely how individual ligand-activated RTKs function (Supplementary Sidebar 6.2).

The last of the sequence motifs present in Src (see Figure 6.3)—SH3 domains— were initially found to bind specifically to certain proline-rich sequence domains in partner proteins; these proline-rich sequences thus serve as the ligands of the SH3 domains (see Figure 6.4C). (A rapidly growing minority of SH3 groups has been found to recognize ligands of a quite different structure.) These SH3 domains are a relatively ancient invention (apparently originating in an array of protozoa more than one billion years ago), since there are at least 28 of them in various proteins of the yeast *S. cerevisiae*. The SH2 domain, in contrast, seems to be of more recent vintage, having been invented and subsequently diversified possibly at the same time as tyrosine phosphorylation came into widespread use—perhaps 900 million years ago when the first metazoa arose.

In the specific case of the SH3 domain of Src, there is evidence that this domain is used to recognize and bind certain proline-rich substrates, which can then be phosphorylated by Src's kinase activity. Yet other evidence points to a proline-rich linker domain within Src itself, to which Src's SH3 domain associates, creating intramolecular binding (**Sidebar 6.3**). Through analyses of the human genome sequence, almost 300 distinct human SH3 domains have been documented, each part of a larger protein.

The SH2 and SH3 domains were the first of a series of specialized protein domains to be recognized and cataloged (see Supplementary Sidebar 6.3). Each of these specialized domains is able to recognize and bind a specific sequence or structure present in a partner molecule. Some of these specialized domains recognize and bind phosphoserine or phosphothreonine on partner proteins, while others bind to phosphorylated forms of certain membrane lipids (**Table 6.2**). These partner-binding domains are critical to ensure the specific, focused intercommunication between appropriate pairs of signaling molecules operating throughout the intracellular signaling circuitry.

Taken together, these various domains illustrate how the lines of communication within a cell are restricted, since the ability of signaling molecules to physically associate in highly specific ways with target molecules ensures that signals are passed only to these intended targets and not to other proteins within the cell. In addition, examination of an array of signaling proteins has revealed that these various domains have been used as modules, being assembled in different combinations through evolution to ensure the specificity of a variety of intermolecular interactions (see Figure 6.5C).

The discovery of SH2 groups also explained how Src, the original oncoprotein, operates within the normal cell, and how Rous sarcoma virus has reconfigured the structure of Src, making it into an oncoprotein (see Sidebar 6.3).

6.4 SH2 and SH3 groups explain how growth factor receptors activate Ras and acquire signaling specificity

The discovery of SH2 and SH3 groups and their mechanisms of action made it possible to solve several important problems. The biochemical clues were now in place to explain precisely how the receptor → Sos → Ras cascade operates.

Sidebar 6.3 The SH2 and SH3 domains of Src both have alternative functions We have portrayed SH2 as a protein domain that enables its carrier to recognize and bind to other phosphotyrosine-containing partner proteins, thereby forming bimolecular complexes. In tyrosine kinases such as Src, however, the SH2 group can also play a quite different role. Like most signaling molecules, the Src kinase is usually held in a functionally silent configuration. It is kept in this inactive state by a negative regulatory domain of Src—a stretch of polypeptide that lies draped across the catalytic site of Src and blocks access to the substrates of this kinase. This obstructing domain (sometimes called an "activation loop" (*dark green*); see Supplementary Sidebar 5.6) is held in place, in part, through the actions of the Src SH2 group, which recognizes and binds a phosphotyrosine in position 527 of Src itself. Hence, this SH2 group is forming an *intra*molecular rather than an *inter*molecular bridge (**Figure 6.6A**).

Some receptors, such as the PDGF receptor, may become activated by ligand binding and display phosphotyrosines in their cytoplasmic domains that are also attractive binding sites for the SH2 group of Src. As a consequence, the fickle Src SH2 group breaks its intramolecular bond and forms an intermolecular bond with the PDGF receptor instead; this results

in the tethering of Src to the PDGF receptor (Figure 6.6B). Concomitantly, the SH3 group of Src breaks its intramolecular association with the linker segment of Src and binds to an oligopeptide in the PDGF receptor. Now Src's catalytic cleft shifts into an active configuration and, following phosphorylation of its activation loop (*red*), begins to phosphorylate its clientele of substrates, thereby amplifying the signal broadcast by the PDGF receptor (Figure 6.6C).

The Src oncoprotein specified by Rous sarcoma virus is a mutant form of the normal c-Src protein, in which the C-terminal oligopeptide domain containing the tyrosine 527 residue has been replaced by another, unrelated amino acid sequence. As a consequence, a phosphotyrosine can never be formed at this site and the intramolecular inhibitory loop cannot form and, in turn, obstruct the Src catalytic site. The viral oncoprotein (v-Src) is therefore given a free hand to fire constitutively. A similar mutant Src, in which the six C-terminal amino acids are deleted (including tyrosine 527), has been reported in liver metastases of human colon carcinomas. While the Src kinase is hyperactive in many colon carcinomas, this is one of the very rare examples of a mutant, structurally altered Src in human cancers.

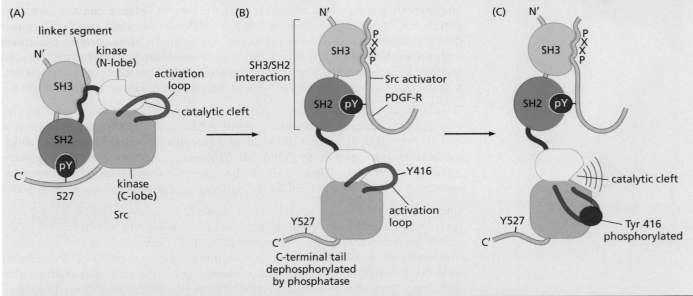

Figure 6.6 Structure and function of Src (A) The SH2 group of Src (*blue*) normally binds in an intramolecular fashion to a phosphotyrosine residue (pY; *red*) at position 527 near the C-terminus of Src. This binding causes the catalytic cleft of Src, which is located between the N-lobe and the C-lobe of the kinase domain (*yellow, light orange*), to be obstructed. At the same time, the SH3 (*light green*) domain binds a proline-rich portion in the linker segment (*dark red*) between the N-lobe of the kinase domain and the SH2 domain. (B) However, when a domain of the PDGF receptor (indicated here as a "Src activator") becomes phosphorylated, one of its phosphotyrosines (*red*) serves as a ligand for this SH2 group of Src, causing the SH2 to switch from intramolecular binding to intermolecular binding. Concomitantly, the SH3 group of Src detaches from its intramolecular binding to bind a proline-rich domain (PXXP) on the receptor. These changes open up the catalytic cleft of Src, placing it in a configuration that poises it to fire. (C) The full tyrosine kinase activity of Src only occurs following a final phosphorylation of tyrosine 416, which moves an obstructing oligopeptide activation loop (*dark green*) out of the way of the catalytic cleft; the formation of Y416 may occur in dimerized Src molecules that transphosphorylate one another. (Adapted from W. Xu et al., *Mol. Cell* 3:629–638, 1999. With permission from Elsevier.)

Table 6.2 Binding domains that are carried by various proteins[a]

Name of domain	Ligand	Examples of proteins carrying this domain
SH2	phosphotyrosine	Src (tyrosine kinase), Grb2 (adaptor protein), Shc (scaffolding protein), SHP-2 (phosphatase), Cbl (ubiquitylation)
PTB	phosphotyrosine	Shc (adaptor protein), IRS-1 (adaptor for insulin RTK signaling), X11 (neuronal protein)
SH3	proline-rich	Src (tyrosine kinase), Crk (adaptor protein), Grb2 (adaptor protein)
14-3-3	phosphoserine	Cdc25 (CDK phosphatase), Bad (apoptosis regulator), Raf (Ser/Thr kinase), PKC (protein kinase C Ser/Thr kinase)
Bromo	acetylated lysine	P/CAF (transcription co-factor), chromatin proteins
PH[b]	phosphorylated inositides	PLC-δ (phospholipase C-δ), Akt/PKB (Ser/Thr kinase), BTK

[a]At least 32 distinct types of binding domains have been identified (see Supplementary Sidebar 6.3). This table presents six of these that are often associated with transduction of mitogenic signals.

[b]The phosphoinositide-binding groups include, in addition to the PH domain, the Fab1, YOTB, Vac1, EEA1 (FYVE), PX, ENTH, and FERM domains.

The final clues came from analyzing the structure of Grb2. It contains two SH3 groups and one SH2 group (Figure 6.5C). Its SH3 domains have an affinity for two distinct proline-rich sequences present in Sos, while its SH2 sequence associates with a phosphotyrosine present on the C-terminus of many growth factor–activated receptors. Consequently, Sos, which seems usually to float freely in the cytoplasm, now becomes tethered via the Grb2 linker molecule to the receptor (**Figure 6.7A**).

Detailed analysis of the structures of Grb2 and similarly configured proteins (for example, Crk) indicates that they carry no other functional domains beyond their SH2 and SH3 domains. Apparently, these adaptor proteins are nothing more than bridge builders designed specifically to link other proteins to one another. (An alternative bridging, also shown in Figure 6.7B, can be achieved when a growth factor receptor associates with Sos via Grb2 and Shc.)

Once Sos becomes anchored via Grb2 (or via Grb2 + Shc) to a receptor, it is brought into close proximity with Ras proteins, most of which seem to be permanently tethered to the inner face of the plasma membrane. Sos is then physically well positioned to interact directly with these Ras molecules, inducing them to release GDPs and bind instead GTPs. This guanine nucleotide exchange causes the activation of Ras protein signaling. Hence, the biochemically defined pathway could now be drawn like this:

Receptor → Grb2 → Sos → Ras   or   Receptor → Shc → Grb2 → Sos → Ras

Because a diverse group of signaling proteins become attracted to ligand-activated receptors (see, for example, Figure 6.5A and B), the Sos–Ras pathway represents only one of a number of signaling cascades radiating from growth factor receptors. Moreover, the SH2 and SH3 groups, as described here, represent only a small subset of the domains within intracellular proteins that are responsible for protein:protein interactions (Supplementary Sidebar 6.3).

6.5 Ras-regulated signaling pathways: A cascade of kinases forms one of three important signaling pathways downstream of Ras

The multiple cellular responses that a growth factor elicits could now be explained by the fact that its *cognate* receptor (that is, the receptor that specifically binds it) is able to activate a specific combination of downstream signaling pathways. Each

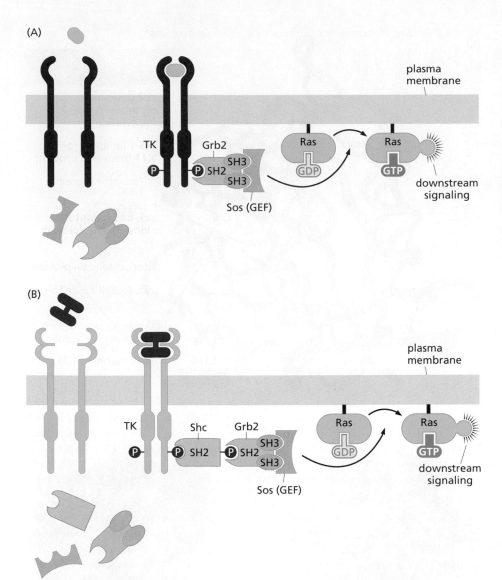

Figure 6.7 Intermolecular links forged by the Grb2 and Shc bridging proteins (A) The association of Sos, the Ras guanine nucleotide exchange factor (GEF), with ligand-activated growth factor receptors is achieved by bridging proteins, such as Grb2, shown here. The two SH3 domains of Grb2 (*pink*) bind to proline-rich domains of Sos (*orange*), while its SH2 domain (*light blue*) binds to a phosphotyrosine located on the C-terminal tail of a receptor. (B) An alternative mode of association can be achieved through a collaboration between two bridging proteins, Grb2 and Shc. Grb2 uses its SH3 domains to bind Sos and its SH2 domain to bind a phosphotyrosine on Shc (*beige*); the latter uses its own SH2 domain to bind a phosphotyrosine on a ligand-activated receptor. (Phosphorylation of Shc is achieved by Src-like tyrosine kinases.) (Adapted from G.M. Cooper, Oncogenes, 2nd ed. Sudbury, MA: Jones and Bartlett, 1995. www.jblearning.com. Reprinted with permission from Jones and Bartlett.)

of these pathways might be responsible for inducing, in turn, one or another of the multiple biological changes occurring after cells are stimulated by this particular growth factor. Moreover, exaggerated forms of this signaling might well be operating in cancer cells that experience continuous growth factor stimulation because of a mutant, constitutively activated receptor or an autocrine signaling loop (see Figure 5.8A and C).

This branching of downstream pathways did not, however, explain precisely how the mutant Ras oncoprotein (which itself represents only one of the multiple pathways downstream of growth factor receptors) is able to evoke a number of distinct changes in cells. Here, once again, signal transduction biochemistry provided critical insights. We now know that at least three major downstream signaling cascades radiate from the Ras protein. (Precisely where within a cell these signals are transmitted remains unclear; see Supplementary Sidebar 6.4.). Their diverse actions help to explain how the Ras oncoprotein causes so many distinct changes in cell behavior.

The layout of these downstream signaling pathways was mapped through a combination of yeast genetics and biochemical analyses of mammalian cells. These approaches made it clear that when Ras binds GTP, two of its "switch domains" shift (see Figure 5.25B), enabling its **effector loop** (**Figure 6.8**) to interact physically with several alternative downstream signaling partners that are known collectively as Ras

Figure 6.8 The Ras effector loop (A) The crystallographic structure of the Ras protein, viewed from a different angle than those illustrated in Figure 5.25, is depicted here in a ribbon diagram (*largely aquamarine*). The guanine nucleotide (*light purple stick figure*) is shown together with the nearby glycine residue, which, upon replacement by a valine in oncogenic Ras, causes inactivation of the GTPase activity (G12V). The effector loop (*yellow*) interacts with at least three important downstream effectors of Ras. Experimentally introduced amino acid substitutions (*arrows*) affecting three residues in the effector loop create three alternative forms of the Ras protein that interact preferentially with either PI3 kinase, Raf/B-Raf, or Ral-GEF. (B) In these space-filling models, the stereochemical shift of the effector domain of Ras (*green, purple*) enables Ras to interact physically with at least three of its downstream effectors listed in panel A. (A, Courtesy of R. Latek and A. Rangarajan, adapted from data in U. Krengel et al., *Cell* 62:539–548, 1990. B, Courtesy of P. Marinec and F. McCormick.)

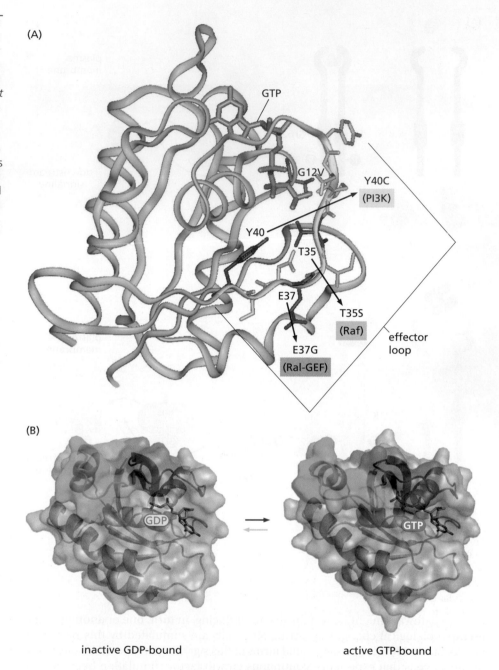

(A)

(B)

inactive GDP-bound   active GTP-bound

effectors, that is, the proteins that carry out the actual work of Ras. Each of these effectors binds quite tightly to the effector loop of the GTP-bound form of Ras protein while having little affinity for the loop presented by its GDP-bound form. The first of these Ras effectors to be discovered was the Raf kinase (**Figure 6.9A**). Like the great majority of protein kinases in the cell, Raf phosphorylates substrate proteins on their serine and threonine residues. Long before its association with Ras was uncovered, the Raf kinase had already been encountered as the oncoprotein specified by both a rapidly transforming murine retrovirus and, in homologous form, a chicken retrovirus.

The activation of Raf by Ras depends upon the relocalization of Raf within the cytoplasm. Recall that Ras proteins are always anchored to a membrane, usually the inner surface of the plasma membrane, through their C-terminal hydrophobic tails. Once it has bound GTP (see Figure 6.9A), the affinity of Ras for Raf increases by three orders of magnitude, enabling Ras to bind Raf relatively tightly, doing so via the Ras effector loop. Before this association, Raf is found in the cytosol; thereafter,

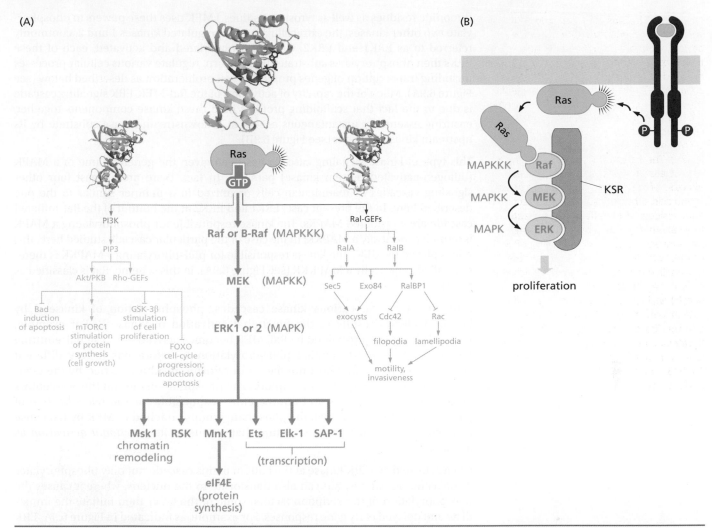

Figure 6.9 The Ras → Raf → MAP kinase pathway (A) This multi-kinase signaling cascade (*aquamarine, center*) is one of a series of similarly organized pathways in mammalian cells that have the overall plan MAPKKK → MAPKK → MAPK. In this particular case, Ras activates the Raf kinase (a MAPKKK), the latter proceeds to phosphorylate and activate MEK (a MAPKK), and MEK then phosphorylates and activates ERK1 and ERK2 (MAPKs). (The two other key effector pathways of Ras shown here will be analyzed further in Figures 6.10 and 6.15.) ERK1/2 can then phosphorylate kinases in the cytoplasm that regulate translation as well as transcription factors in the nucleus. The bimolecular complex, portrayed as a ribbon structure (*above*), illustrates the interaction of the effector domain of Ras (*red, purple*) with the Ras-binding domain of Raf (*aquamarine*). (B) Multi-kinase signaling cascades in eukaryotic cells are often assembled by scaffolding proteins that ensure the direct physical juxtaposition of one kinase to its downstream kinase substrate. The KSR scaffold shown here organizes the highly efficient signal transduction of Raf to ERK. (A, adapted from M.E. Pacold et al., *Cell* 103:931–943, 2000; N. Nassar et al., *Nature* 375:554–560, 1995; and L. Huang et al., *Nat. Struct. Biol.* 5:422–426, 1998. B, adapted from M.C. Good, J.G. Zalatan and W.A. Lim, *Science* 332, 680–686, 2011.)

Raf becomes tethered via Ras to the plasma membrane (or, in certain cases, to another cytoplasmic membrane). Moreover, as has recently been determined, both Ras and Raf function as dimeric proteins, and the dimerization of Raf may even depend on the prior dimerization of Ras molecules (Supplementary Sidebar 6.5). These interactions are further influenced by the fact that the intracellular localization of Ras is affected by whether it has bound GDP or GTP (see Supplementary Sidebar 6.4).

The physical association of Raf with Ras appears to be facilitated by additional proteins that act as scaffolds to hold them together (Figure 6.9B); yet others trigger the phosphorylation and dephosphorylation of key amino acid residues on Raf, altering its three-dimensional configuration. Raf now acquires active signaling powers and proceeds to phosphorylate and thereby activate a second kinase known as MEK. (MEK is actually a "dual-specificity kinase," that is, one that can phosphorylate serine/

threonine residues as well as tyrosine residues.) MEK uses these powers to phosphorylate two other kinases, the *e*xtracellular signal–*r*egulated *k*inases 1 and 2, commonly referred to as ERK1 and ERK2. Once phosphorylated and activated, each of these ERKs then phosphorylates substrates that, in turn, regulate various cellular processes including transcription of genes promoting cell proliferation as described below (see Figure 6.9A). Much of the rapidity of activation of the Raf-MEK-ERK signaling cascade is due to the fact that scaffolding proteins hold these kinase components together, ensuring essentially instantaneous access to a downstream kinase substrate by its upstream kinase activator (see Figure 6.9B).

This type of kinase signaling cascade has been given the generic name of a MAPK (mitogen-activated protein kinase) pathway. In fact, there are at least four other signaling cascades in mammalian cells organized in a manner similar to the one described here. In the present case, ERK1 and ERK2 at the bottom of the Raf-initiated cascade are considered MAPKs. The kinase responsible for phosphorylating a MAPK is termed generically a MAPKK; in the case of the particular cascade studied here, this role is played by MEK. The kinase responsible for phosphorylating a MAPKK is therefore called generically a MAPKKK (see Figure 6.9A); in this scheme, Raf is classified as a MAPKKK.

Note that in these various kinase cascades, phosphorylation of kinase B by kinase A always results in the functional activation of kinase B. For example, when MEK is phosphorylated by Raf, MEK becomes activated as a signal-emitting kinase. Such serine/threonine phosphorylation therefore has a very different consequence from the phosphorylation of tyrosine residues carried by the cytoplasmic tails of growth factor receptors. The phosphotyrosines on these receptors attract signaling partners and are therefore responsible for the *relocalization* of partner proteins; in contrast, the phosphate groups attached to MEK by Raf cause a stereochemical shift in its structure that results in its *functional activation* as a kinase.

Once activated, an ERK kinase at the bottom of this cascade not only phosphorylates cytoplasmic substrates, but can also translocate to the nucleus, where it causes the phosphorylation of transcription factors; some of the latter then initiate the immediate and delayed early gene responses. For example, as indicated in Figure 6.9A, ERK can phosphorylate several transcription factors (as examples, Ets, Elk-1, and SAP-1) directly and, in addition, phosphorylate and thereby activate other kinases, which proceed to phosphorylate and activate yet other transcription factors. These downstream kinases phosphorylate two chromatin-associated proteins (HMG-14 and histone H3), creating modifications that place the chromatin in a configuration that is more hospitable to transcription (see Section 1.8). In the cytoplasm, the Mnk1 kinase, a cytoplasmic substrate of ERK1 and ERK2, phosphorylates the translation initiation factor eIF4E, thereby helping to activate the cellular machinery responsible for protein synthesis.

Once the ERKs phosphorylate the Ets transcription factor, the latter can then proceed to stimulate the expression of important growth-regulating genes, such as those specifying heparin-binding EGF (HB-EGF), cyclin D1, Fos, and p21^{Waf1}. (The functions of the last three are described in detail in Chapter 8.) The actual number of distinct ERK substrates is vastly larger: a 2017 study undertook to enumerate those for which there was reasonably compelling biochemical evidence and came up with 2507 protein targets!

Especially important among the genes induced by the Raf → MEK → ERK pathway are the two immediate early genes encoding the Fos and Jun transcription factors. Once synthesized, these two proteins can associate with one another to form AP-1, a widely acting heterodimeric transcription factor that is often found in hyperactivated form in cancer cells. The special importance and influence of Fos and Jun is indicated by the fact that each was originally identified as the product of a retrovirus-associated oncogene (see Table 3.3).

The Ras → Raf → MEK → ERK pathway is only one of several downstream signaling cascades that are activated by the GTP-bound Ras proteins (see Figures 6.9, 6.10, and 6.15).

The central importance of this kinase cascade is indicated by the fact that in a number of cell types, the Raf protein kinase, when introduced into cells in mutant, oncogenic form, can evoke most of the transformation phenotypes that are induced by the Ras oncoprotein itself. Hence, in such cells, the pathway downstream of Raf is responsible for the lion's share of the transforming powers of Ras oncoproteins. In addition to activating a number of growth-promoting genes, as described above, this pathway confers anchorage independence and loss of contact inhibition. It also contributes to the profound change in cell shape that is associated with transformation by a *ras* oncogene.

In certain cancers, the signaling pathway lying downstream of Raf may be strongly activated without any prior stimulation by Ras. Thus in ~50% of human melanomas a mutant form of B-Raf, a close cousin and functional analog of Raf, is found to carry an amino acid substitution causing a ~500-fold increase in kinase activity. [The mutation commonly occurring in melanomas causes B-Raf's activation loop, which usually obstructs access to the catalytic cleft of this enzyme (as seen in other kinases, see Supplementary Sidebar 5.6), to swing out of the way.] In about one-third of the remaining melanomas, mutant alleles encoding activated N-Ras oncoproteins can be found. In human tumors overall, a *ras* oncogene is rarely found to coexist with mutant alleles of *B-raf*, providing evidence that these two oncogenes have overlapping functions and thus are partially redundant with one another.

Various explanations have been proposed to rationalize why multi-kinase cascades like this one are used in a variety of signaling circuits throughout the cell. In fact, versions of these cascades are already apparent in single-cell eukaryotes, such as yeast, in which they are employed to regulate a variety of cell-biological processes. These kinase cascades—all exhibiting the MAPKKK–MAPKK–MAPK organization—share in common another striking feature that ensures their rapid and efficient firing: the participating kinases are physically held together by scaffolding proteins (see Figure 6.9B). These proteins, such as the KSR featured in this figure, ensure that the signals flowing through the kinase cascades do so rapidly, and that the participating kinases are able to find their substrates instantaneously rather than depending on diffusion-dependent processes to do so. (A second aspect of kinase firing involving dimerization is discussed in Supplementary Sidebar 6.5.)

Of note, versions of ERK that are unable to dimerize and associate with the cytoplasmic scaffolding proteins in response to Ras or Raf signaling cripple the cell transformation usually induced by these oncoproteins; similarly, ERK proteins that are unable to translocate as monomers to the nucleus following their activation in the cytoplasm also preclude transformation by these oncogene-encoded proteins

6.6 Ras-regulated signaling pathways: a second downstream pathway controls inositol lipids and the Akt/PKB kinase

A second important downstream effector of the Ras protein enables Ras to evoke other, quite distinct cellular responses (**Figure 6.10**). In the context of cancer, the most important of these is a suppression of apoptosis. This anti-apoptotic effect is especially critical for cancer cells, since many of them are poised on the brink of activating this cell suicide program, a subject covered in detail in Chapter 9.

Early studies of the biochemistry of the plasma membrane suggested that its phospholipids serve simply as a structural barrier between the aqueous environments in the cell exterior and interior. The biochemistry of these phospholipid molecules showed that they are **amphipathic**—that is, each possesses a hydrophilic head, which likes to be immersed in water, and a hydrophobic tail, which prefers nonaqueous environments (**Figure 6.11A**). This polarity explains the structure of lipid bilayers throughout the cell, in which the hydrophilic groups face and protrude into the extracellular or cytosolic aqueous environments while the hydrophobic tails are buried in the middle of the membrane, from which water is excluded. (The transmembrane domains of growth factor receptors discussed in Chapter 5 consist of hydrophobic acids that permit these domains to be buried stably in this hydrophobic environment.)

Figure 6.10 The PI3 kinase pathway
A second effector pathway of Ras (*dark orange, left*) derives from its ability to associate with and activate phosphatidylinositol 3-kinase (PI3K), with the resulting formation of phosphatidylinositol (3,4,5)-triphosphate (PIP3). This leads in turn to the tethering of PH-containing molecules, such as Akt (also called PKB) and Rho guanine nucleotide exchange proteins (Rho-GEFs), to the plasma membrane. Akt/PKB is able to inactivate Bad (a pro-apoptotic protein); inactivate GSK-3β, an antagonist of growth-promoting proteins such as β-catenin, cyclin D1, and Myc; and activate mTOR. (The latter is a kinase that stimulates protein synthesis and cell growth and also acts reciprocally to phosphorylate and activate Akt/PKB.) The bimolecular complex, seen as a ribbon structure (*above*), illustrates the interaction of the effector domain of Ras (*red, purple*) with the Ras-binding domain of PI3K (*orange*). (Adapted from M.E. Pacold et al., *Cell* 103:931–943, 2000; N. Nassar et al., *Nature* 375:554–560, 1995; and L. Huang et al., *Nat. Struct. Biol.* 5:422–426, 1998.)

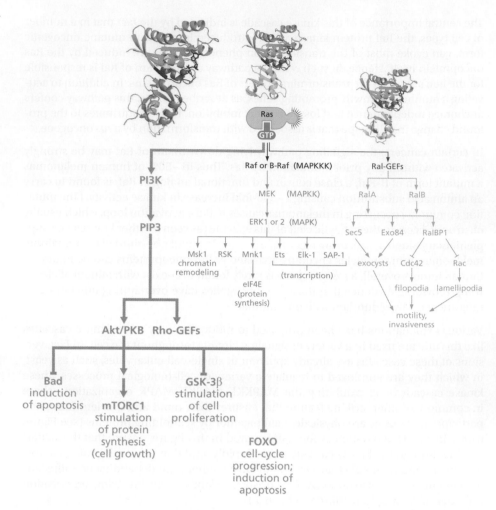

In the 1970s, it became apparent that eukaryotic cells exploit some of the membrane-associated phospholipids for purposes that are fully unrelated to the maintenance of membrane structure. Some phospholipids contain, at their hydrophilic heads, an inositol group. Inositol is a water-soluble carbohydrate molecule (termed more properly a polyalcohol). The inositol moiety of such phospholipids can be modified by the addition of phosphate groups. In some situations, the resulting phosphoinositol may then be cleaved by the phospholipase C (PLC) from the remaining, largely hydrophobic portions of a phospholipid molecule. Since it is purely hydrophilic, this liberated phosphoinositol, termed IP3 or IP_3, can then diffuse away from the membrane (Figure 6.11B), thereby serving as an *intracellular hormone* to dispatch signals from the plasma membrane to distant parts of the cell; such intracellular hormones are often called **second messengers**. The second product of this cleavage, termed diacylglycerol (DAG), which remains tethered to the plasma membrane, can serve to activate key signaling kinases in the cell—certain serine/threonine kinases belonging to the class known generically as protein kinase C (PKC). In other situations, a phosphorylated inositol avoids cleavage, can remain intact, and therefore can remain attached to the remainder of the phospholipid and thus can remain tethered to the plasma membrane; there, it can serve as an anchoring point to which certain cytosolic proteins can become attached, as discussed below.

We now know that the inositol moiety can be modified by several distinct kinases, each of which shows specificity for phosphorylating a particular hydroxyl of inositol. Phosphatidylinositol 3-kinase (PI3K), for example, is responsible for attaching a phosphate group to the 3′ hydroxyl of the inositol moiety of membrane-embedded phosphatidylinositol (PI). While several distinct PI3 kinases have been discovered, the most important of these may well be the PI3K that phosphorylates PI(4,5)P_2 (or, more simply,

(A)

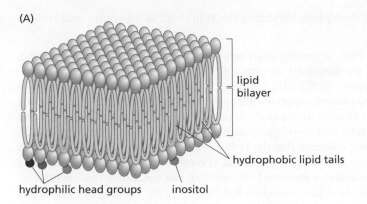

(B)

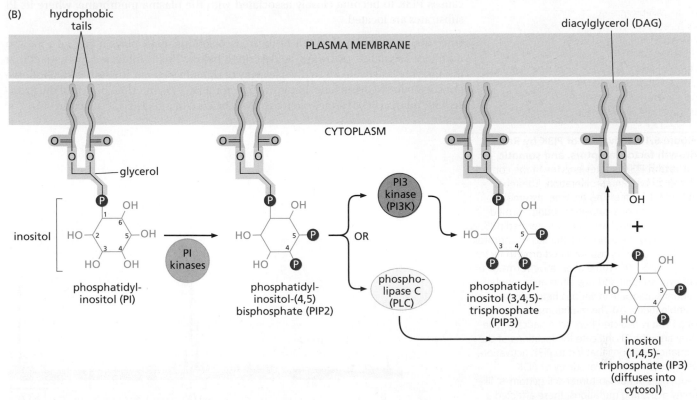

Figure 6.11 Biochemistry of lipid bilayers (A) All cellular membranes are assembled as phospholipid bilayers, in which the hydrophilic head groups (*gray ovals*) protrude into the aqueous solvent above and below the membrane, while the hydrophobic tails (*green, gray*) are confined to the nonaqueous space within the lipid bilayer. A small minority of the various types of hydrophilic head groups (*red, green, blue*) contain inositol sugars (*blue*). (B) Phosphatidylinositol (PI; *left*) is composed of three parts—two fatty acids with long hydrocarbon tails (*green*) inserted into the plasma membrane, glycerol (*gray*), and inositol (*light blue*) attached to the glycerol via a phosphodiester linkage. PI kinases can add phosphate groups to the various hydroxyls of inositol, yielding, for example, the PI (4,5)-diphosphate (PIP2) shown here. Cleavage of PIP2 by phospholipase C yields diacylglycerol (DAG), which can activate protein kinase C (PKC); and inositol (1,4,5)-triphosphate (IP3), which induces release of calcium ions from intracellular stores. Alternatively, PIP2 can be further phosphorylated by PI3 kinase (PI3K) to yield phosphatidylinositol (3,4,5)-triphosphate (PIP3). Once formed, the phosphorylated head group of PIP3 serves to attract proteins carrying PH domains, which thereby become tethered to the inner surface of the plasma membrane.

PIP2). Prior to modification by PI3K, PIP2 already has phosphates attached to the 4′ and 5′ hydroxyl groups of inositol (in addition to the phosphodiester bond that links its 1′ hydroxyl to the membrane-embedded diacyglycerol-lipid moiety and gives these various forms their "PI" name); after this modification, this inositol acquires yet another phosphate group, and PIP2 is converted into PI(3,4,5)P$_3$, that is, PIP3 (see Figure 6.11B).

Note an important difference between the organization of this pathway and the Ras → Raf → MEK → ERK cascade that was described in the previous section. In the present case, one of the critical kinases (that is, PI3K) attaches phosphates to a phospholipid, rather than phosphorylating the side chain of an amino acid residue within a protein.

We first encountered PI3K, in passing, at an earlier point in this chapter (see Section 6.3, Figure 6.5A). There, we described the attraction of the PI3Kγ enzyme to a ligand-activated PDGF receptor via the SH2 groups of its regulatory subunit (resulting, as we now realize, in the phosphorylation of inositol-containing phospholipids already present in the nearby plasma membrane). It turns out that GTP-activated Ras is also able to bind PI3K directly and thereby enhance its functional activation (**Figure 6.12**). (Indeed, there is strong evidence that the H-Ras protein can effectively activate PI3K only when the p85 regulatory subunit of PI3K is bound to a phosphotyrosine on a ligand-activated growth factor receptor.) Hence, PI3K can function as a direct downstream effector of Ras. As is the case when Raf binds Ras, the binding of PI3K to Ras causes PI3K to become closely associated with the plasma membrane, where its PI substrates are located.

Once activated by one or both of these associations, PI3K plays a central role in a number of signaling pathways, as described below. This is indicated by the fact that this enzyme is activated by a diverse array of signaling agents, including PDGF, nerve growth factor (NGF), insulin-like growth factor-1 (IGF-1), interleukin-3, and the extracellular matrix (ECM) attachment achieved by certain integrins (see Section 6.9).

Figure 6.12 Activation of PI3K by Ras, growth factor receptors, and somatic mutation PI3K is a key enzyme in the control of cell growth and proliferation. The most important form of this enzyme, assembled as a complex of a catalytic subunit, termed p110, and a regulatory subunit, termed p85, is activated by the actions of the Ras oncoprotein and by tyrosine-phosphorylated growth factor receptors (RTKs). Three alternative forms of the catalytic subunit and five alternative regulatory subunits associate in various heterodimeric combinations. (A) The multidomain structure of p110α is depicted here schematically. The *gray* arrowheads indicate the locations of the somatic mutations that led to PI3K activation, as reported in an initial survey of PI3K sequences in human tumor cell genomes; like many activating mutations, these affected a limited number of nucleotides in the encoding gene. (B) A ribbon model of p110α with the inter-SH2 domain of p85α is shown here. The Ras-binding domain of p110α (*dark purple*) enables its tethering via Ras to the plasma membrane (*not shown*). The p85α subunit uses its two SH2 groups, which are located at the opposite ends of the inter-SH2 domain of p85 (*dark red*), to bind to phosphotyrosine residues in the cytoplasmic tails of ligand-activated growth factor receptors. The catalytic domain of p110α is shown in *light purple*. It remains unclear whether simultaneous interactions between p85, p110, Ras, and a ligand-activated RTK are required for activation of p110 or, alternatively, whether interactions of p85 with either a ligand-activated RTK or with Ras suffice for this activation. The inter-SH2 domain seems to have an affinity for binding lipid membranes. (A, from Y. Samuels et al., *Science* 304:554, 2004. B, from C.-H. Huang et al., *Science* 318:1744–1748, 2007. Both with permission from AAAS.)

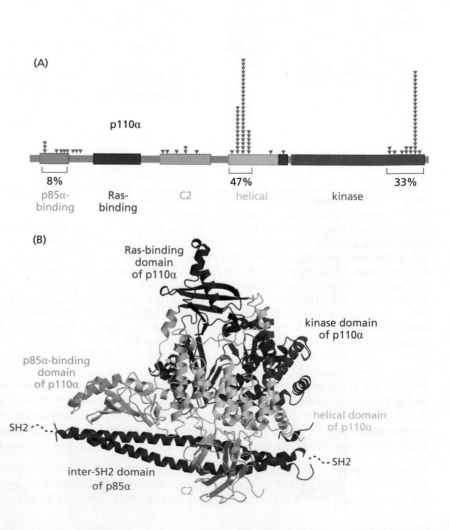

The formation of the phosphorylated inositol head groups of phosphatidylinositol has little if any effect on the overall structure and function of the plasma membrane. (The inositol phospholipids are minor constituents of the plasma membrane.) However, a phosphorylated inositol head group protruding from the plasma membrane may be recognized and bound by certain proteins that are usually floating freely in the cytosol. Once anchored via a phosphoinositol to the plasma membrane, these proteins are then well positioned to release certain types of signals.

The most important of the phosphorylated inositol head groups appears to be PIP3, whose inositol, as mentioned above, carries phosphates at its 3', 4', and 5' positions (see Figure 6.11B). Many of the cytosolic proteins that are attracted to PIP3 carry pleckstrin homology (PH) domains (see Table 6.2) that have strong affinity for this triply phosphorylated inositol head group (**Figures 6.13 and 6.14**). Arguably the most important PH domain–containing protein is the serine/threonine kinase named Akt/ PKB, hereafter simply Akt. Thus, once PIP3 is formed by PI3K, an Akt kinase molecule can become tethered via its PH domain to the inositol head group of PIP3 that protrudes from the plasma membrane into the cytosol (Figure 6.14A). This association of Akt with the plasma membrane (together with phosphorylations of Akt that soon follow) results in the functional activation of Akt as a kinase. (In the absence of PIP3 binding, the PH domain of an Akt molecule obstructs its kinase domain, explaining the inactivity of the cytosolic pool of Akt; see Figure 6.14B.) Once activated, Akt proceeds to phosphorylate a series of protein substrates that have multiple effects on the cell (see Figure 5.4B), including (1) aiding in cell survival by reducing the possibility that the cell apoptotic suicide program will become activated; (2) stimulating cell proliferation; and (3) stimulating cell growth, in the most literal sense of the term, that is, stimulating increases in cell size by increasing protein synthesis (Supplementary Sidebar 6.6). In addition, it also exerts an influence, still poorly understood, on cell motility and on **angiogenesis**—the production of new blood vessels (to be discussed in Chapter 13).

Interestingly, functionally active Akt seems to elicit many of its effects while still anchored via its PH domain to the plasma membrane. In addition, once released from this tethering into the cytosol, Akt persists for some time in its activated state, ostensibly sustained by its pair of phosphorylated residues (see Figure 6.14A);

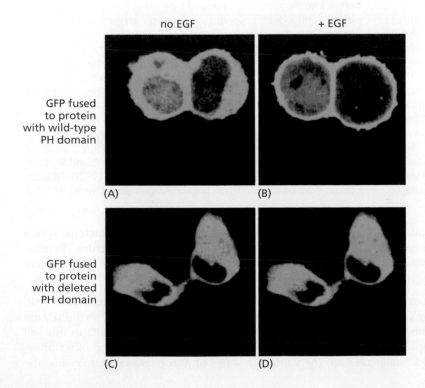

no EGF   + EGF

GFP fused
to protein
with wild-type
PH domain

(A)   (B)

GFP fused
to protein
with deleted
PH domain

(C)   (D)

Figure 6.13 Migration of PH-containing proteins to PIP3 in the plasma membrane After PIP3 head groups are formed by PI3 kinase (see Figure 6.11B), they become attractive docking sites for PH (pleckstrin homology) domain–containing proteins. This phosphorylation and homing occurs within seconds of the activation of PI3 kinase, as revealed by these micrographs, in which green fluorescent protein (GFP) has been fused to a protein carrying a PH domain. This hybrid protein moves from a diffuse distribution throughout the cytoplasm (A) to the inner surface of the plasma membrane (B) within a minute after EGF (epidermal growth factor) has been added to cells. However, if the PH domain of the fusion protein is deleted (C), addition of EGF has no effect on its intracellular distribution (D) and it remains diffusely distributed throughout the cytoplasm. (From K. Venkateswarlu et al., *J. Cell Sci.* 112:1957–1965, 1999. With permission from The Company of Biologists.)

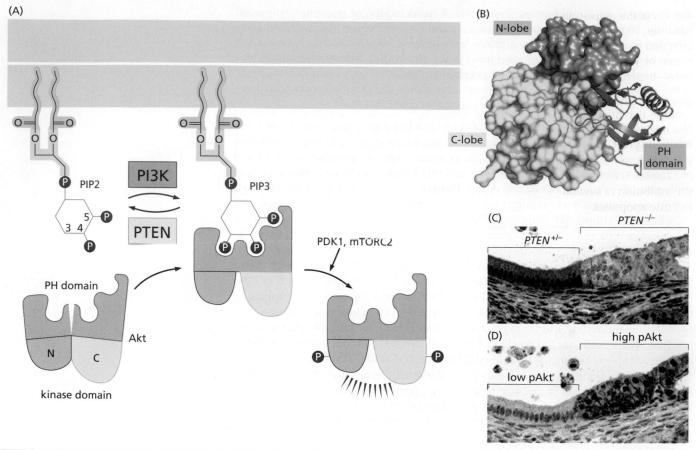

Figure 6.14 Docking of PH domains to PIP3 By forming the triply phosphorylated inositol "head group" (*blue lines*) of PIP3 (see Figure 6.11B), PI3K creates a docking site for cytoplasmic proteins that carry PH domains. The most important of these is the Akt kinase. Once docked to the inner surface of the plasma membrane via its PH domain, the Akt kinase (sometimes termed PKB) becomes doubly phosphorylated at its two lobes (N, C) by two other kinases (PDK1 and mTORC2, respectively) and thereby activated (*lower right*). (Recent evidence indicates that PDK1, which has its own PH domain, can transduce oncogenic signals that do not depend on Akt activation.) Akt then proceeds to phosphorylate a variety of substrates throughout the cell regulating cell proliferation, survival, and size. However, if PTEN is also present and active, it dephosphorylates PIP3 to PIP2, thereby depriving Akt/PKB of a docking site at the plasma membrane. (The size of the PH domain of Akt/PKB relative to the size of its kinase domain has been greatly exaggerated in this drawing.) (B) The crystal structure of Akt reveals that the protein is formed from a kinase domain (*left*), shown here in a space-filling structure with the N- and C-lobes (*pink, yellow*) typical of most kinases (see Figure 6.6), which are connected via a flexible linker to the PH domain, shown as a ribbon diagram (*right,* *orange brown*). The catalytic cleft of the kinase domain sandwiched between its N- and C-lobes is usually inhibited by the PH domain. Upon binding to PIP3, the PH domain shifts away, disinhibiting the kinase domain. A drug molecule that inhibits the kinase and is bound near the catalytic cleft of the kinase is shown in *green*. (C) The dynamics of this interaction are seen in genetically altered mice having a *Pten*[+/−] genotype. Uterine epithelial cells in these mice often lose the remaining *Pten* wild-type allele, yielding *Pten*[−/−] cells, which form localized hyperplasias and cysts. Immunostaining with anti-PTEN antiserum shows that the epithelial cells on the right have lost PTEN expression and become larger, indicating the effects of the PIP3 pathway on cell size. (D) When the same lesion is stained with an anti-phospho-Akt antibody (which detects activated Akt), the cells on the right show high pAkt activity, while those on the left show very low activity, illustrating how even half the normal dose of PTEN present in the heterozygous cells suffices to shut down Akt function. (B, from W.-I. Wu et al., *PLoS One* 5:e12913, 2010. © 2010 Wu et al. This is an open-access article distributed under the terms of the Creative Commons Attribution License. C and D, from K. Podsypanina et al., *Proc. Natl. Acad. Sci.* USA 98:10320–10325, 2001. With permission from National Academy of Sciences, U.S.A.)

indeed, a portion of activated Akt succeeds in migrating into the nucleus, where it phosphorylates and thereby activates certain transcription factors. Eventually, however, Akt signaling is reversed by phosphatases that remove its activating phosphate groups.

The activation of PI3K, and thus of Akt, is under very tight control. In a quiescent, nongrowing cell lacking growth factor stimulation, the intracellular levels of PIP3 are extremely low. Once such a cell encounters mitogens, the levels of PIP3 in this cell increase rapidly (see Figure 6.13). The normally low levels of PIP3 and other phosphorylated PI molecules are the work of a series of phosphatases that reverse the actions of the activating kinases such as PI3K. The best characterized of these phosphatases, PTEN, removes the 3′ phosphate group from PIP3 that has previously been attached by PI3K (Figure 6.14A and C). This removal suggests two distinct mechanisms by which the Akt/PKB signaling pathway can become deregulated in cancer cells—hyperactivity of PI3K or inactivity of PTEN.

The portfolio of diverse biological functions assigned to Akt and described above requires that this kinase be able to phosphorylate a variety of substrates, each involved in a distinct downstream signaling pathway (Table 6.3). The first of these functions—facilitating cell survival—depends on the ability of Akt to suppress any tendencies the cell may have to activate its own built-in suicide program of apoptosis, as discussed in Chapter 9. For the moment, suffice it to say that phosphorylation by Akt results in the inhibition of several proteins that play prominent roles favoring the entrance of a cell into apoptosis.

Table 6.3 Effects of Akt/PKB on survival, proliferation, and cell growth[a]

Biological effect	Substrate of Akt/PKB	Description	Functional consequence
		Anti-apoptotic	
	Bad (pro-apoptotic)	Bcl-X antagonist; like Bad, belongs to Bcl-2 protein family controlling mitochondrial membrane pores (Section 9.13).	inhibition
	caspase-9 (pro-apoptotic)	Component of the protease cascade that affects the apoptotic program (Section 9.13).	inhibition
	IκB kinase, abbreviated IKK (anti-apoptotic)	Activated by Akt/PKB phosphorylation (Section 6.12).	activation
	FOXO1 TF, formerly called FKHR TF (pro-apoptotic)	Phosphorylation prevents its nuclear translocation and activation of pro-apoptotic genes.	inhibition
	Mdm2 (anti-apoptotic)	Activated via phosphorylation by Akt/PKB; it triggers destruction of p53 (Section 9.8).	activation
		Proliferative	
	GSK-3β (anti-proliferative)	Phosphorylates β-catenin, cyclin D1, and Myc (Sections 7.10, 8.3, 8.9), causing their degradation; inactivated via phosphorylation by Akt/PKB.	inhibition
	FOXO4, formerly called AFX (anti-proliferative)	Induces expression of the CDK inhibitor p27^{Kip1} (Section 8.4) gene and some pro-apoptotic genes; exported from the nucleus when phosphorylated by Akt/PKB.	inhibition
	p21^{Cip1} (anti-proliferative)	CDK inhibitor, like p27^{Kip1} (Section 8.4). Exits the nucleus upon phosphorylation by Akt/PKB.	inhibition
		Growth	
	Tsc2 (anti-growth)	Phosphorylation by Akt/PKB causes Tsc1/Tsc2 complex to dissociate, allowing activation of mTOR, which then up-regulates protein synthesis.	inhibition
		Metabolism	
	GSK-3β	glycogen formation	activation
	ATP citrate lyase	fatty acid synthesis	activation
	mTOR	activator of glycolysis	activation
	GLUT4	glucose transporter	activation

[a]Additional effects of Akt on extracellular matrix formation, motility, and yet other cell-biological processes are not indicated here.

The proliferative functions of Akt depend on its ability to perturb proteins that are important for regulating the advance of a cell through its cycle of growth and division, often termed the "cell cycle." We will learn more about this complex program in Chapter 8.

Independent of these effects on proliferation, Akt is able to induce dramatic changes in the proteins that control the rate of protein synthesis in the cell. Akt acts through intermediaries to trigger activation of mTOR kinase; the latter, when acting as a subunit of the mTORC1 complex, proceeds to phosphorylate and inactivate 4E-BP, a potent inhibitor of translation and, at the same time, phosphorylates and thereby activates p70S6 kinase, an activator of translation. These changes allow Akt to increase the efficiency with which the translation of a class of mRNAs is initiated; the resulting elevated rate of protein synthesis favors the accumulation of many cellular proteins and is manifested in the growth (rather than the proliferation) of cells. An illustration of the growth-inducing powers of Akt is provided by transgenic mice that express a mutant, constitutively active Akt in the β cells of the pancreatic islets (see Supplementary Sidebar 6.6); in such mice, these particular cells grow to a size more than twice the cross-sectional area (and almost 4 times the volume) of normal β cells!

There is evidence that yet other cytoplasmic proteins use their PH domains to associate with PIP3, and that this association also favors their functional activation. Among the PH-bearing cellular proteins are a group of guanine nucleotide exchange factors (GEFs) that function analogously to Sos (see Figure 5.24), being responsible for activating various small GTPases that are distant relatives of the Ras proteins. These other GTPases belong to the Rho family of signaling proteins, which includes Rho proper and its two cousins, Rac and Cdc42. Like Ras, these Rho proteins operate as binary switches, cycling between a GTP-bound actively signaling state and a GDP-bound inactive state (Sidebar 6.4). The Rho-GEFs, once activated by binding to PIP3, act on Rho proteins in the same way that Sos acts on Ras. In particular, the Rho-GEFs induce Rho proteins to jettison their bound GDPs, allowing GTP to jump aboard and activate the Rho's.

Activated Rho proteins have functions that differ strongly from those of Ras: they participate in reconfiguring the structure of the cytoskeleton and the attachments that the cell makes with its physical surroundings. (While Ras and Rho proteins retain many commonalities in structure and guanine nucleotide binding, they seem to have diverged from one another at least 1.5 billion years ago during the early evolution of eukaryotic cells.) The Rho-like proteins control cell shape, motility, and, in the case of cancer cells, invasiveness. For example, Cdc42 is involved in reorganizing the actin cytoskeleton of the cell as well as controlling **filopodia**, small, fingerlike extensions from the plasma membrane that the cell uses to explore its environment and form

Sidebar 6.4 Ras is the prototype of a large family of similar proteins Ras is only one of a large superfamily of at least 151 similarly structured mammalian proteins that together are called "small G proteins" to distinguish them from the other class of larger, heterotrimeric guanine nucleotide–binding G proteins that are activated by association with seven-membrane-spanning (serpentine) cell surface receptors (see Figure 5.19A). Most of these small G proteins have been given three-letter names, each derived in some way from the initially named and discovered Ras proteins—for example, Ral, Rac, Ran, Rho, Rheb, and so forth. (The Cdc42 protein is a member of the Rho family that has kept its own unique name because of its initial discovery through yeast genetics.)

Much of the sequence similarity shared among these otherwise diverse proteins is found in the cavities in each that bind and hydrolyze guanine nucleotides. Virtually all of the small G proteins operate like a binary switch, using a GTP–GDP–GTP cycle to flip back and forth between an on and an off state (see Figure 5.24). Like Ras, each of these proteins is thought to have its own specialized guanine nucleotide exchange factors (GEFs) to activate it (by promoting replacement of GDP by GTP) and its own GTPase-activating proteins (GAPs) to trigger its GTPase activity, thereby inactivating it.

Each of these small G proteins is specialized to associate with its own group of effector proteins and control a distinct cell-physiologic or biochemical process, including regulation of the structure of the cytoskeleton, trafficking of intracellular vesicles, and apoptosis. One is even used to regulate the transport of proteins through pores in the nuclear membrane. The only trait shared in common by these diverse cell-biological processes is a need to be regulated by some type of binary switching mechanism—an operation for which the small G proteins are ideally suited.

erythropoiesis (red blood cell formation) and **thrombopoiesis** (platelet formation), respectively.

Following ligand-mediated receptor dimerization, the JAK enzyme associated with each receptor molecule phosphorylates tyrosine residues on the cytoplasmic tails of the partner receptor molecule, much like the transphosphorylation occurring after ligand activation of receptors like the EGF-R and PDGF-R (see Figure 5.11A). The resulting phosphotyrosines attract and are bound by SH2-containing transcription factors termed STATs (signal transducers and activators of transcription), which then become phosphorylated by the JAKs. This creates individual STAT molecules that possess both SH2 groups and newly acquired phosphotyrosines. Importantly, the SH2 groups displayed by the STATs have a specificity for binding the phosphotyrosine residues that have just been created on the STATs. Consequently, STAT–STAT dimers form, in which each STAT uses its SH2 group to bind the phosphotyrosine of its partner. The resulting dimerized STATs migrate to the nucleus, where they function as transcription factors (see Figure 6.16A).

The STATs—seven of which have been documented—activate target genes that are important for cell proliferation and cell survival. Included among these genes are *myc*, the genes specifying cyclins D2 and D3 (which enable cells to advance through their growth-and-division cycles, as described in Chapter 8), and the gene encoding the strongly anti-apoptotic protein Bcl-X$_L$ (described in Chapter 9). In addition to phosphorylating STATs, the JAKs can phosphorylate substrates that activate other mitogenic pathways, including the Ras–MAPK pathway described earlier.

Perhaps the most dramatic demonstration of the contribution of STATs to cancer development has come from a re-engineering of the STAT3 protein through the introduction of a pair of cysteine residues, which causes the resulting mutant STAT3 to dimerize spontaneously, forming stable covalent disulfide cross-linking bonds (Figure 6.16B). These stabilized STAT3 dimers are structural and functional mimics of the dimers that are formed physiologically when STAT3 is phosphorylated by JAKs. This mutant STAT3 protein is now constitutively active as a nuclear transcription factor and can function as an oncoprotein that is capable of transforming NIH 3T3 and other immortalized mouse cells to a tumorigenic state.

STAT3 is also known to be constitutively activated in a number of human cancers. For example, in many melanomas, its activation is apparently attributable to Src, which is also constitutively active in these cancer cells; indeed, at least four other cousins of Src, all nonreceptor tyrosine kinases, can also phosphorylate and thereby activate STAT3. This highlights the fact that STATs can be activated in the cytoplasm by tyrosine kinases other than the receptor-associated JAKs. In the case of the melanoma cells, inhibition of Src, which leads to deactivation of STAT3, triggers apoptosis, pointing to the important contribution of STAT3 in ensuring the survival of these cancer cells.

In the majority of breast cancer cells, STAT3 has been found to be constitutively activated; in some of these, STAT3 is phosphorylated by Src and JAKs acting collaboratively. As is the case with melanoma cells, reversal of STAT3 activation in these breast cancer cells leads to growth inhibition and apoptosis. In almost all head-and-neck cancers, STAT3 is also constitutively activated, possibly through the actions of the EGF receptor. Alternatively, in many glioblastomas, activation of STAT3 is favored by the loss of a phosphatase responsible for removing its phosphotyrosine. STAT3 activation in lung and gastric cancers also appears to be critical to their neoplastic phenotype.

After reaching the nucleus, dimerized activated STAT proteins induce the expression of several hundred genes, many of which favor cell proliferation and protect against apoptosis. Taken together, these various threads of evidence converge on the notion that STATs represent important mediators of transformation in a variety of human cancer cell types. This depiction of the STATs as being invariably tumor-promoting must be tempered by the fact that certain STATs, such as STAT1, can antagonize cell proliferation and survival and therefore act as tumor-suppressing proteins. For example, this particular STAT can act to promote cell cycle arrest and even induce cell death via apoptosis, which we will encounter in Chapter 9.

6.9 Cell adhesion receptors emit signals that converge with those released by growth factor receptors

In the last chapter, we read that cells continuously monitor their attachment to components of the extracellular matrix (ECM; Section 5.9). Successful tethering to the molecules forming the ECM, achieved via integrins, causes survival signals to be released into the cytoplasm that decrease the likelihood that a cell will enter into the form of apoptosis termed anoikis. At the same time, integrin activation, provoked by signals originating inside the cell, can promote cell motility by stimulating integrin molecules located at specific sites on the plasma membrane to forge new linkages with the ECM. Consequently, integrins serve the three functions of (1) physically linking cells to the ECM, (2) informing cells whether or not tethering to certain ECM components has been achieved, and (3) facilitating motility by making and breaking contacts with the ECM. (In fact, there is a fourth but incompletely understood function: integrins and their cytoplasmic signaling partners serve as sensors of physical tension that arises between the ECM and the actin cytoskeleton. Cells respond in complex ways to this sensed tension.)

As described in the last chapter, integrins may cluster and form multiple links to the ECM in small, localized areas termed *focal adhesions* (see Figure 5.22A). Such clustering provokes strong activation of focal adhesion kinase (FAK), a nonreceptor tyrosine kinase like Src. FAK is associated with the cytoplasmic tails of the β subunits of certain integrin molecules and becomes phosphorylated, apparently by transphosphorylation, once integrins congregate in these focal adhesions. One of the resulting phosphotyrosine residues on FAK provides a docking site for various SH2-containing signaling molecules including notably Src, Shc, PI3K, and PLC-γ (**Figure 6.17A**). The binding of Src results in the phosphorylation of additional tyrosines on FAK, and the resulting phosphotyrosines serve as docking sites for yet other SH2-containing signaling molecules, including Grb2, Shc, and PLC-γ. There is also evidence that Grb2, once bound to tyrosine-phosphorylated FAK, can recruit Sos into this complex; Sos then proceeds to activate its normal downstream Ras target. In fact, the full complexity of signaling downstream of integrins still awaits resolution (Figure 6.17B).

The accumulated evidence suggests a pathway that has, minimally, the following components:

$$\text{ECM} \rightarrow \text{integrins} \rightarrow \text{Sos} \rightarrow \text{Ras} \rightarrow \text{Raf, PI3K, and Ral-GEF}$$

The parallels between integrin and receptor tyrosine kinase signaling are striking: tyrosine kinase receptors bind growth factor ligands in the extracellular space, while integrins bind extracellular matrix components as ligands. Once activated, these two classes of sensors release signals that converge on many of the same downstream signal transduction cascades. Moreover, some integrins co-localize on the plasma membrane with certain tyrosine kinase receptors; this lateral physical association appears to enhance the signaling by both types of cell surface proteins.

This convergence of signaling pathways appears to explain one of the important cell-biological effects of the Ras oncoprotein—its ability to enable cells to grow in an anchorage-independent fashion. It seems that cells normally depend on integrin signaling to provide a measure of activation of the normal Ras and Src proteins. In the absence of this signaling, cells come to believe that they have failed to associate with the extracellular matrix. This may block further advance of cells through their cell growth-and-division cycle or, more drastically, may cause cells to enter into anoikis as mentioned above. Avoidance of anoikis triggered by loss of normal attachment to the basement membrane seems to represent one of the first steps in the initiation of breast cancers (**Figure 6.18**).

Taken together, these diverse observations suggest that oncogenic Ras promotes anchorage-independent proliferation by mimicking one of the critical downstream signals that result following tethering by integrins to the extracellular matrix. In effect, oncogenic Ras may delude a cell into believing that its integrins have successfully bound certain ECM components, thereby permitting a *ras*-transformed cell to proliferate, even when no such attachment has actually occurred.

(A)

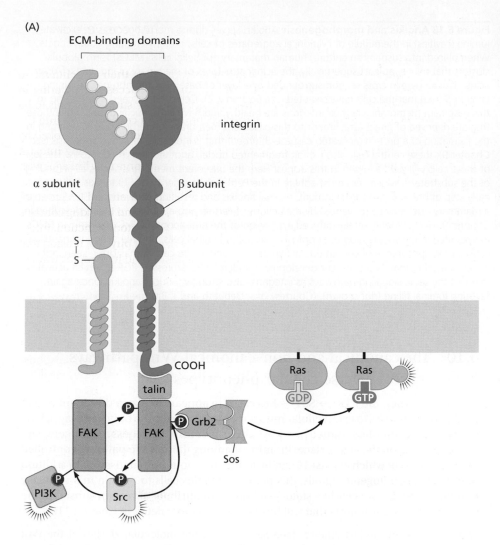

(B)

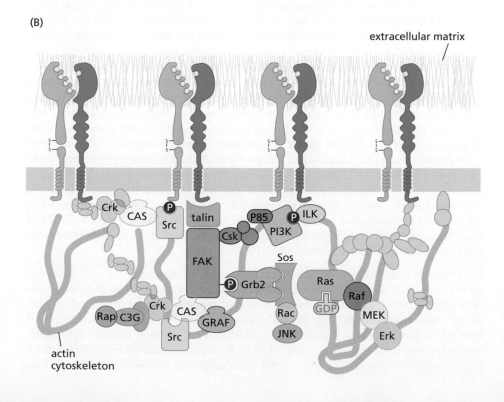

Figure 6.17 Integrin signaling (A) The integrins are assembled as α + β heterodimers. In addition to physically linking the cytoskeleton to the extracellular matrix (ECM; see Figure 5.22B), the binding by the ectodomain of the heterodimer to components of the ECM (*not shown*) triggers the association of a series of cytoplasmic signal-transducing proteins, such as focal adhesion kinase (FAK), with the β subunit. Resulting transphosphorylation events and the attraction of SH2-containing signaling molecules release signals that activate many of the same pathways that are turned on by ligand-activated growth factor receptors. This signaling is amplified by clustering of multiple heterodimers following ECM binding (*not shown*). (B) As indicated here, the molecules shown in panel A represent only part of a complex, still-poorly understood collection of signal-transducing molecules that become physically associated with the cytoplasmic domains of various integrin β subunits following ECM binding. In total, almost 180 distinct proteins have been found to be physically associated with focal adhesions. (Adapted from C. Miranti and J. Brugge, *Nat. Cell Biol.* 4:E83–E90, 2002. With permission from Nature.)

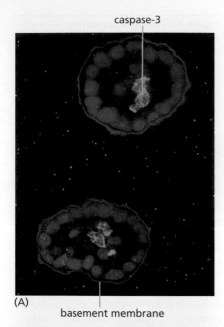

caspase-3

(A)

basement membrane

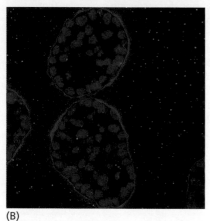

(B)

Figure 6.18 Anoikis and morphogenesis Anoikis serves during morphogenesis to excavate lumina (cavities) in the middle of cylindrical aggregates of cells, thereby creating hollow ducts. When placed into suspension culture, human mammary epithelial cells (MECs) form globular clusters that mimic, at least superficially, the acinar structures of mammary gland ducts. (A) An acinus is seen here in cross section surrounded by a layer of basement membrane containing laminin-5 (*red*) that the cells have secreted (see Section 2.2). Cells that succeed in tethering to this basement membrane via their integrins are healthy and have large nuclei (*blue*), while those that are deprived of direct attachment to this basement membrane undergo anoikis, resulting in the formation of a protein (activated caspase-3; *green*) that is indicative of active apoptosis (see Chapter 9); these central cells also exhibit fragmented nuclei, another sign of apoptosis. The loss of these cells will yield a lumen in this acinus. Here the basement membrane serves as an analog of the substrate to which cells must adhere in the Petri dish in order to avoid anoikis. (B) In an early step of breast cancer development, normal anoikis and resulting excavation of the lumen of a mammary duct may be prevented by the actions of certain anti-apoptotic proteins described in Chapter 9. Here, experimentally induced expression of the anti-apoptotic Bcl-xL protein in MECs has resulted in the suppression of apoptosis, survival of luminal cells, and resulting luminal filling, echoing the appearance of certain ductal hyperplasias and *in situ* carcinomas often seen in the initial steps of human breast cancer development. Indeed, overexpressed EGF-R, which activates many of the same signaling pathways as integrins, also strongly reduces anoikis, once again favoring luminal filling (*not shown*). (Courtesy of J. Debnath and J. Brugge.)

6.10 The canonical and non-canonical Wnt pathways control diverse cellular phenotypes

In most cell types, the RTK → Sos → Ras cascade appears to play a dominant role in mediating responses to extracellular mitogens, but it is hardly the only pathway having this result. Yet other, less studied pathways also confer responsiveness to mitogenic signals impinging on the cell surface. Prominent among these is the pathway controlled by Wnt factors, of which at least 19 can be found in various human tissues. In addition to transducing mitogenic signals, this pathway enables cells to remain in a relatively undifferentiated, stem cell-like state—an important attribute of certain subsets of the cancer cells in many tumors that will be discussed in more detail in Chapter 11.

As we learned in the last chapter (see Section 5.7), the molecular design of the Wnt pathway is totally different from that of the RTK → Sos → Ras → Raf → ERK pathway. Recall that canonical Wnt factors, acting through Frizzled receptors, suppress the activity of glycogen synthase kinase-3β (GSK-3β). In the absence of Wnts, GSK-3β phosphorylates several key substrate proteins, which are thereby tagged for destruction. The most important of these substrates is β-catenin, normally a cytoplasmic protein that exists in three states. It may be bound to the cytoplasmic domain of cell–cell adhesion receptors, notably the E-cadherin that is expressed on the surface of all epithelial cells and serves to knit together epithelial cell sheets (**Figure 6.19A**); E-cadherin molecules from adjacent cells may associate with one another to form **adherens junctions** involving cell-to-cell contacts together with associated cytoplasmic proteins that help to ensure the structural integrity of epithelial sheets and impede metastasis (as discussed in Chapter 14). Moreover, the cytoplasmic association of adherens complexes with the actin cytoskeleton, which depends on the involvement of β-catenin, creates physical coupling between the cytoskeletons of adjacent epithelial cells. Alternatively, in a fully unrelated role, β-catenin exists in a soluble pool in the cytosol, where it turns over very rapidly, having a lifetime of less than 20 minutes. And finally, β-catenin operates in the nucleus as an important component of a transcription factor complex (Figure 6.19B).

The relatively large number of E-cadherin molecules on the surface of epithelial cells serves as a major sink to absorb and stabilize a significant proportion of a cell's pools of β-catenin molecules; the latter are liberated when adherens junctions are dissolved, often achieved through the endocytosis of E-cadherin molecules. We focus here on the pools of these β-catenin molecules that are released into the cytosol following E-cadherin endocytosis. Such cytosolic β-catenin molecules may initially form multiprotein complexes with three other cellular proteins—Apc (the adenomatous polyposis coli protein), Wtx, and axin. These proteins help to bring β-catenin together with its GSK-3β executioner (see Figure 6.19B). By phosphorylating β-catenin, GSK-3β ensures that β-catenin will be tagged by **ubiquitylation**, a process that drives the rapid destruction of β-catenin; this explains the low steady-state concentrations of β-catenin normally present in the cytosol. We will encounter this degradation system later in more detail (see Supplementary Sidebar 7.5).

However, when canonical Wnt signaling is activated by the binding of canonical Wnt ligands to their cognate Frizzled (Fzd) receptors, GSK-3β firing is blocked and β-catenin is saved from rapid destruction; its half-life increases from less than 20 minutes to 1–2 hours, and its steady-state concentrations increase proportionately. Following appropriate changes in their phosphorylation, many of the accumulated β-catenin molecules move into the nucleus and bind to Tcf/Lef proteins (see Figure 6.19B). The resulting multi-subunit transcription factor complexes proceed to activate expression of a number of important genes, including those encoding critical proteins involved in cell growth and proliferation, such as cyclin D1 and Myc, that we will study later in Chapter 8.

In addition, GSK-3β can phosphorylate other crucial growth-regulating proteins besides cytoplasmic β-catenin. By phosphorylating cyclin D1 in the nucleus, GSK-3β also tags this growth-promoting protein for rapid degradation. Hence, the Wnt pathway actually modulates cyclin D1 expression at both the transcriptional and the post-translational levels. These various actions of Wnts indicate that, in addition to their originally discovered powers as **morphogens**, they can also act as potent mitogens, much like the ligands of many tyrosine kinase receptors. Moreover, as we will learn in Chapter 7, high levels of β-catenin in intestinal stem cells ensure that these cells remain in an undifferentiated, stem cell-like state, rather than developing into the specialized cells that line the wall of the intestine (Figure 6.19C). Indeed, there are multiple tissues throughout the body in which the stem cells seem to depend on Wnt signaling to enable their survival and drive their proliferation.

Then there is a major puzzle: Why has evolution invested two such unrelated functions in a single protein? (In worms, these two functions of β-catenin are carried out by distinct proteins encoded by separate genes.) According to one rationalization, the adherens junctions act as sensors of cell–cell contact in epithelial cell sheets, and once this contact is disrupted by various biological mechanisms, β-catenin can rapidly move into the nucleus to re-align the cell's transcriptional program to respond to this new biological state. This may explain how epithelial cells at the edge of a damaged cell sheet (which have lost adherens junctions with neighbors previously on one side) can activate the β-catenin–induced EMT (epithelial–mesenchymal transition) program, doing so in order to gain the motility that enables them to cover over and thereby heal the wound that generated this broken sheet, as described later in Chapter 14.

In many human breast cancers, expression of certain Wnts is increased four- to ten-fold above normal, suggesting the operations of autocrine and paracrine growth-stimulatory pathways. There is clear evidence of nuclear translocation of β-catenin in approximately 20% of advanced prostate carcinomas. Moreover, in 5 to 7% of prostate carcinomas, mutations of the β-catenin gene have yielded a protein that can no longer be phosphorylated by GSK-3β, allowing it, in turn, to accumulate to high levels in the nucleus, where it can operate as a transcription factor. Such mutations have also been documented in carcinomas of the liver, colon, endometrium, and ovary, as well as in melanomas (Supplementary Sidebar 6.7). And as we will see in the next chapter, in virtually all colon carcinomas, β-catenin degradation is flawed, due to defects in one of the proteins—Apc—that usually facilitate its degradation (see Figure 6.19B). It is likely that this pathway is also deregulated in other human cancers by mechanisms that have not yet been uncovered.

The above descriptions do not begin to address the complexity of Wnt signaling. Thus, there are alternatives to the "canonical" Wnt signaling cascade described above. These alternative pathways tend to be triggered by a distinct subset of the Wnt ligands,

Figure 6.19 Canonical Wnt signaling: multiple roles of β-catenin (A) Like many integrins, the cadherins (*light green*) are transmembrane proteins that form attachments in the extracellular space and become linked, via intermediary proteins, to the actin cytoskeleton (*red*). The ectodomains of an epithelial cell's E-cadherin associate with E-cadherin molecules displayed by an adjacent epithelial cell via *homophilic* interactions that form the core of adherens junctions displayed between adjacent epithelial cells in an epithelial cell sheet. Only the most distal, N-terminal subdomain among the five repeating subdomains of an individual E-cadherin ectodomain participate in associating with one displayed by an adjacent cell. Each of the four inter-subdomain linker regions binds three calcium ions whose presence is critical to preserve the rigidity

and adhesive interactions of the protein. The cytoplasmic tail of E-cadherin, for its part, is known to associate directly with p120 and β-catenin (*dark blue, light blue*). These associate in turn with a monomer of α catenin (*dark green*). The precise molecular mechanisms of linkage of the cytoplasmic components are still poorly resolved. Breakdown of this multi-subunit complex, which includes comparable E-cadherin-associated proteins in the adjacent cell (*above, not shown*) often accompanies the acquisition of highly malignant properties by carcinoma cells. With the exception of actin (*red, below*), all of the proteins shown in this figure together with corresponding proteins in the adjacent cell (*not shown*) form, in aggregate, an adherens junction. (*continued*)

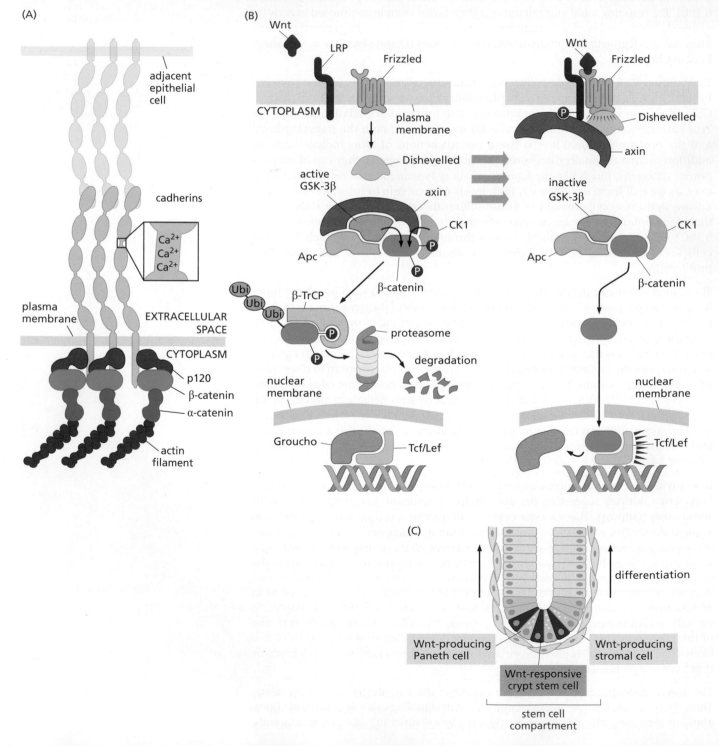

Figure 6.19 Canonical Wnt signaling: multiple roles of β-catenin (*continued*) (B) The Wnt proteins (*red*), by binding to their Frizzled receptors (Fzd; *orange*) and the LRP co-receptor (*purple*), act via Dishevelled to suppress the activity of glycogen synthase kinase-3β (GSK-3β, *pink*); the latter, working together with casein kinase 1 (CK1), would otherwise phosphorylate β-catenin, labeling it for ubiquitylation by the β-TrCP ubiquitin ligase and destruction in proteasomes. Consequently, in the presence of canonical Wnt ligand (*red*), β-catenin escapes degradation and accumulates in the cytoplasm and translocates to the nucleus. Once in the nucleus, β-catenin associates with Tcf/Lef transcription factors, displacing the Groucho repressor protein (*brown*) and thereby activating expression of a variety of genes, including those involved in cell proliferation. Not shown here are additional proteins that associate with β-catenin, notably BCL9-2/BL9 and Pygopus, and enable β-catenin to form productive transcription-promoting complexes with Tcf/Lef in the nucleus. In addition, another component of the β-catenin-containing, multi-protein complex in the cytosol—Wtx1—is not illustrated here. Moreover, β-catenin undergoes complex changes in phosphorylation at different stages of signaling that are not indicated here. The pathway depicted here is often termed the *canonical* Wnt pathway to distinguish it from a group of alternative Wnt pathways that are termed collectively *non-canonical* Wnt pathways and are described in outline in Figure 6.20. (C) Not addressed by most of our discussions until now is the biological state of stemness, that is, the ability of relatively undifferentiated stem cells (SCs) to self-renew by ensuring that one of the daughter cells generated by their cell division remains a SC, whereas the other daughter cell becomes the progenitor of more differentiated cells that assume the role of executing the specialized functions of the tissue in which this SC resides. Wnt ligands play a major role in inducing and maintaining SCs in many adult tissues. Their role in doing so is illustrated most graphically by studies of the crypts of the gastrointestinal tract in which the SCs reside in a specialized SC "niche" that is sustained by the local release of various Wnt factors, both by adjacent Paneth cells and by mesenchymal stromal cells. The lipid tails that are covalently attached to Wnt ligands ensure that signals borne by these ligands travel only short distances. Once intestinal SCs have migrated out of the bottom of the crypts (*upward pointing arrows*), they no longer experience Wnt signals and begin to differentiate into the enterocytes that create the epithelial lining of the gastrointestinal tract. (C, from H. Clevers, K.M. Loh and R. Nusse, *Science* 346:1248012, 2014. With permission from AAAS.)

operate independently of β-catenin signaling, and are grouped together under the rubric of "non-canonical Wnt signaling." Thus, Wnt3a is often used experimentally as a representative of canonical Wnt ligands whereas Wnt5a is used to study non-canonical signaling.

As seen in **Figure 6.20**, non-canonical Wnt signaling encompasses multiple signaling pathways. In the context of cancer, arguably the most important are a group of five tyrosine kinase receptors (RTKs), all of which appear to carry catalytically inactive kinase domains, much like HER3 (see Section 5.6). Among these pseudokinase receptors, Ror1 and Ror2 are best documented as important drivers of cancer-associated cellular phenotypes. These five receptors are clearly distant relatives of active RTKs whose kinase domains have lost catalytic activity through replacement of critical amino acid residues within the catalytic clefts of their respective tyrosine kinase domains. Indeed, like most RTKs, versions of these receptors are evolutionarily highly conserved, being clearly present in the cells of worms and flies.

These RTKs display a structurally diverse group of Wnt-binding ectodomains, some of which are structurally related to the Frizzled receptors cited above; like other "pseudokinase" receptors, these appear to act via dimerization with other actively signaling receptors (see Section 5.6 and Supplementary Sidebar 6.5). In some instances, these RTKs may form heterodimers with the structurally unrelated Frizzled receptors. The downstream signaling mechanisms of these RTKs are complex and still being resolved. However, their roles in cancer pathogenesis are clearly significant, since these receptors have been reported to contribute to a broad spectrum of cell-biological traits relevant to tumor formation and progression, namely, cell growth, proliferation, motility, invasiveness, and inhibition of apoptosis. Some have even suggested that a historically earlier discovery of these Wnt RTKs would have caused them, rather than the much-studied Frizzled proteins, to be portrayed as the "canonical" Wnt receptors.

Yet other non-canonical Wnt pathways have been reported, but their role in cancer pathogenesis is less clear. Some of these distinct subsets of signaling cascades, termed the Wnt-PCP and the Wnt-Ca^{2+} pathways, lie downstream of Frizzled receptors, while others may be activated by the pseudokinase RTKs mentioned above. While the canonical Wnt pathway has been found to play a role in cancer cell proliferation and stemness—and thus the formation of tumor-initiating cancer stem cells—the non-canonical Wnt pathway largely contributes to critical aspects of the malignant cell phenotype, notably cytoskeletal rearrangement, motility, and invasiveness. Canonical Wnt ligands can occasionally activate non-canonical signaling

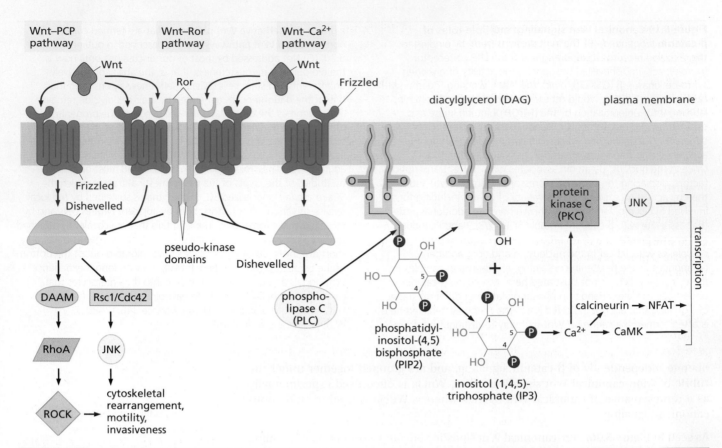

Figure 6.20 Non-canonical Wnt signaling A second set of Wnt signaling pathways operates entirely differently from the canonical Wnt pathway that is described in Figure 6.19 that is involved in regulating cell proliferation and stemness. Some Wnt ligands bind preferentially to the Frizzled (Fzd) receptors that activate these non-canonical Wnt pathways (*dark green*), while other Wnt ligands can bind both canonical and non-canonical Frizzled receptors; in addition, certain Wnts (*purple*) can bind a set of non-canonical receptors [including Ror (*light green*), Ryk, PTK7, and MuSK and represented here by Ror] that are structured like RTKs but have a nonfunctional "pseudokinase" domain. Various canonical and non-canonical Wnt ligands associate with either non-canonical Frizzled receptors, Ror-like receptors, or physical complexes involving both Ror-like receptors and Frizzled receptors and elicit a distinct set of cell-biological processes largely associated, at least in carcinoma cells, with motility, invasiveness, and, ultimately, metastatic dissemination. For example, the *planar cell polarity* (Wnt-PCP) pathway is involved in reorganization of the actin cytoskeleton, motility, and associated invasiveness, while the Wnt-Ca^{2+} pathway also contributes to motility and invasiveness by up-regulating a constituency of nuclear genes. In addition, the Wnt-Ror pathways can also function to suppress canonical Wnt β-catenin signaling. (Not illustrated here are the Wnt ligand-driven heterodimerizations of Ror receptors with certain RTKs.)

and vice versa, and the non-canonical Wnt-PCP pathway has been found in some contexts to inhibit canonical Wnt signaling. For better or worse, as non-canonical Wnt signaling pathways are studied in ever-increasing detail, their complexity grows exponentially.

6.11 G-protein–coupled receptors can also drive normal and neoplastic proliferation

As we read in the last chapter (Section 5.7), G-protein–coupled receptors (GPCRs) are transmembrane proteins that weave their way back and forth through the plasma membrane seven times. Upon binding their extracellular ligands, each of these "serpentine" receptors activates one or more types of cytoplasmic heterotrimeric G protein, so named because of its three distinct subunits (Gα, Gβ, and Gγ), the first of which binds either GDP or GTP. As is the case with the Ras protein, the activated state of Gα is achieved when it binds GTP. Hence, a ligand-activated GPCR can act as a GEF (guanine nucleotide exchange factor) for Gα proteins, much like the GEFs for Ras-like proteins (see Figure 5.24). While the majority of known GPCR ligands are soluble molecules, a small number of these receptors use their ectodomains to sense adhesion to the extracellular matrix (ECM).

Of note, only a relatively small proportion of the ~830 GPCRs encoded in the human genome are involved in cancer pathogenesis. About half of the rest are olfactory receptors, while several dozen others are involved in other types of sensation, such as taste and light, and in signaling via neurotransmitters in the brain. These receptors are of more than passing interest to clinical medicine, since more than 35% of currently approved pharmaceutical agents target various GPCRs. As many as 100 of the GPCRs have no known ligands and, as such, are termed "orphan receptors." The multiplicity of downstream heterotrimeric G proteins derives from the variability of the constituent subunits: there are at least 21 distinct Gα, 6 distinct Gβ, and 12 Gγ subunits, which can associate in various combinations.

Once stimulated by a G-protein–coupled receptor (GPCR), the Gα subunit of a heterotrimeric G protein dissociates from its two partners, Gβ and Gγ, and proceeds to activate a number of distinct cytoplasmic enzymes (**Figure 6.21**). Included among these are adenylyl cyclase (which converts ATP into cyclic AMP) and phospholipase C-β (PLC-β), which cleaves PIP3 to yield diacylglycerol (DAG) and inositol triphosphate (IP3) (see Figure 6.11B). The latter are potent second messengers that can function to stimulate cell proliferation. (The Frizzled receptors enabling canonical Wnt signaling [see Section 6.10] are structurally classified as GPCRs and some have been found to associate with heterodimeric G proteins. However, the importance of Wnt signaling through heterotrimeric G proteins remains unclear.)

There is also evidence that the Src kinase can be activated by certain GTP-bound Gα subunits. Moreover, complexes of the other two subunits of the heterotrimeric G protein—Gβ and Gγ—have been found to stimulate yet other important mitogenic signaling proteins, such as one form of phosphatidylinositol 3-kinase (PI3K). The ability of GPCRs to activate mitogenic pathways suggests that deregulation of signaling by these receptors and associated G proteins may well contribute to cell

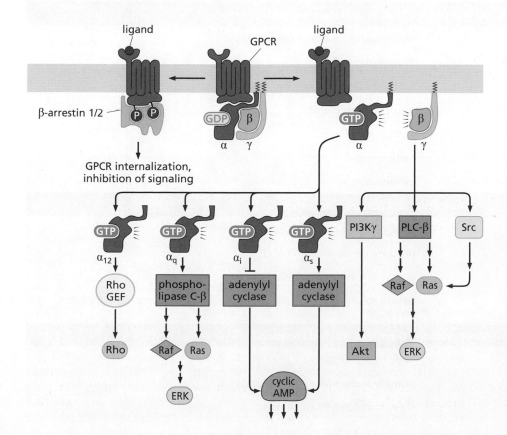

Figure 6.21 G-protein–coupled receptors
The seven-membrane-spanning ("serpentine") receptors (*dark purple*) are able to activate a variety of heterotrimeric G proteins that differ largely in the identity of their Gα subunits (*dark brown*). Once stimulated by a ligand-activated GPCR, a Gα subunit evicts its bound GDP, binding GTP instead, separates from the Gβ + Gγ complex, and proceeds to activate or inhibit a variety of cytoplasmic effectors, only four of which are shown here. In addition to activating enzymes, a guanine nucleotide exchange factor (GEF) for the Rho family of small GTPase proteins (Rho, Rac, and Cdc42) can be activated (*light yellow, lower left*). These enzymes, in turn, can have mitogenic or anti-mitogenic influences, depending upon the cell type. At the same time, the Gβ + Gγ dimer complexes (*right*) can activate their own set of effectors, including phosphatidylinositol 3-kinase γ (PI3Kγ), phospholipase C-β (PLC-β), and Src. Following GPCR activation, these receptors are rapidly phosphorylated and attract β-arrestins 1 and 2 (*light orange, left*), which results in cessation of further GPCR firing and in GPCR internalization. Soon, each Gα subunit will hydrolyze its bound GTP, returning it to the inactive, nonsignaling state (*not shown*). Moreover, this cartoon provides only a hint of the diverse array of G proteins and their effectors that have been discovered. For example, there are nine subtypes of adenylyl cyclase.

transformation and tumorigenesis—a speculation that has been borne out by analyses of a limited number of types of human cancer cells (**Table 6.5**).

The role of GPCRs in human cancer pathogenesis is highlighted by the behavior of small-cell lung carcinomas (SCLCs), a common tumor of cigarette smokers. SCLC cells release a number of distinct peptide factors, some with **neuropeptide**-like properties. In some SCLCs, the tumor cells may simultaneously secrete bombesin (also known as gastrin-releasing peptide, or GRP), bradykinin, cholecystokinin (CCK), gastrin, neurotensin, and vasopressin. At the same time, these cells display the GPCRs that recognize and bind all of these released factors, resulting in the establishment of multiple autocrine signaling loops.

The experimental proof that these autocrine loops are actually responsible for driving the proliferation and/or survival of the SCLC cells is straightforward: SCLC cells can be incubated *in vitro* in the presence of an antibody that binds and neutralizes a secreted autocrine growth factor, such as GRP. In a number of SCLC cell lines, this treatment results in the rapid cessation of growth and even leads to apoptosis. Such a response indicates that these cancer cells depend on a GRP-based autocrine signaling loop for their survival, and it suggested a novel therapy for SCLC patients: in twelve SCLC

Table 6.5 G-protein–coupled receptors and G proteins involved in human cancer pathogenesis

G protein or receptor	Type of tumor
Activating mutations affecting G proteins	
$G\alpha_s$	thyroid adenomas and carcinomas, pituitary adenomas
$G\alpha_{i2}$	ovarian tumors, adrenal cortical tumors
$G\alpha_q$	melanocyte-derived tumors, uveal melanomas
$G\alpha_{11}$	uveal melanomas
Activating mutations affecting G-protein–coupled receptors	
Thyroid-stimulating hormone receptor	thyroid adenomas and carcinomas
Follicle-stimulating hormone receptor	ovarian tumors
Luteinizing hormone receptor	Leydig cell hyperplasias
Cholecystokinin-2 receptor	colorectal carcinomas
Ca^{2+}-sensing receptor	various neoplasms
GPR98	melanomas
GRM3	melanomas
Autocrine and paracrine activation	
Neuromedin B receptor	small cell lung cancer (SCLC)
Neurotensin receptor	prostate carcinomas and SCLC
Gastrin receptor	gastric carcinomas and SCLC
Cholecystokinin receptor	pancreatic hyperplasias and carcinomas, gastrointestinal carcinomas, and SCLC
Virus-encoded G-protein–coupled receptors	
Kaposi's sarcoma herpesvirus (HHV-8)	Kaposi's sarcoma
Herpesvirus saimiri	primate leukemias and lymphomas
Jaagsiekte sheep retrovirus	sheep pulmonary carcinomas

Adapted in part from M.J. Marinissen and J.S. Gutkind, *Trends Pharmacol. Sci.* 22:368–376, 2001. With permission from Elsevier.

patients who were treated with a GRP-neutralizing antibody, one patient showed a complete remission, while four patients exhibited a partial shrinkage of their tumors. (This therapeutic strategy has not, however, been further pursued in recent years.)

In another class of neoplasias—thyroid adenomas and some thyroid carcinomas—the gene encoding the thyroid-stimulating hormone receptor (TSHR), another GPCR, is often found to carry a point mutation. This leads to constitutive, ligand-independent firing of the TSH receptor, which in turn results in the release of strong mitogenic signals into the thyroid epithelial cells (see Table 6.5). In yet other tumors of the thyroid gland, a Gα subunit has suffered a point mutation that functions much like the mutations activating Ras signaling (see Figure 5.24), by depriving this Gα subunit of the ability to shut itself off through its own intrinsic GTPase activity. Altogether, at least 10 of the 17 human genes encoding Gα subunits have been found to function as oncogenes in certain cell types and in various human malignancies.

The most bizarre subversion of these G-protein–coupled receptors occurs during infections by certain herpesviruses, such as human herpesvirus type 8 (HHV-8), also known as the Kaposi's sarcoma herpesvirus (KSHV; refer to Supplementary Sidebar 3.5). This virus is responsible for the vascular tumors that frequently afflict AIDS patients. At one point in its evolutionary past, this virus acquired a cellular gene specifying a G-protein–coupled receptor. The viral form of this gene has been remodeled, so that the encoded receptor releases signals in a ligand-independent manner. Among other consequences, this signaling causes HHV-8–infected endothelial cells (lining the walls of blood vessels) to secrete vascular endothelial growth factor (VEGF); the released VEGF then creates an autocrine signaling loop by binding to its cognate receptors on the surface of these infected endothelial cells and driving their proliferation. Later we will read (Chapter 15) about the role of **chemokines**, which serve as ligands of a special class of GPCRs (altogether 20 in humans), for recruiting various types of immune cells to specific sites within tissues.

6.12 Four additional "dual-address" signaling pathways contribute in various ways to normal and neoplastic proliferation

Both JAK–STAT and Wnt–β-catenin signaling represent "dual-address" pathways, which operate by modifying and dispatching to the nucleus certain signaling proteins that normally reside in the cytoplasm; once in the nucleus, the key proteins in these two pathways (that is, STATs and β-catenin) function as components of specific transcription factor complexes to drive gene expression. In fact, four other dual-address signaling channels play important roles in the pathogenesis of human cancers. Here, we will briefly summarize the downstream actions of these four pathways (**Figure 6.22**) and cite examples of how each is deregulated in one or another type of human tumor (**Figure 6.23**).

Nuclear factor-κB The first indication of the importance of this pathway to cancer pathogenesis came from the discovery of the *rel* oncogene in a rapidly transforming turkey retrovirus responsible for **reticuloendotheliosis**, a hematopoietic tumor of the monocyte/macrophage lineage. Later investigations into the transcription factors responsible for regulating immunoglobulin gene expression revealed Rel to be a member of a family of transcription factors that came to be called collectively NF-κBs. These proteins form homo- and heterodimers in the cytoplasm.

The most common form of NF-κB is a heterodimer composed of a p65 and a p50 subunit. Usually, NF-κB is sequestered in the cytoplasm by a third polypeptide named IκB (inhibitor of NF-κB; see Figure 6.22A); while in this state, this signaling system is kept silent. However, in response to signals originating from a diverse array of sources, IκB becomes phosphorylated and thereby tagged for rapid destruction. (Recall that β-catenin suffers the same fate following its phosphorylation in the cytoplasm.) As a result, NF-κB is liberated from the clutches of IκB, migrates into the nucleus, and proceeds to activate the expression of a cohort of as many as 500 target genes.

Figure 6.22 Four different "dual-address" signaling pathways
(A) The NF-κB family of five transcription factor proteins, which function as homo- and heterodimers represented here by the commonly found p50 and p65 (*light and dark blue*), are sequestered in the cytoplasm by IκB (*red*). A variety of receptors and afferent signals (such as inflammatory cytokines) activate the

IKK heterotetramer, which contains two distinct kinase subunits and phosphorylates IκB, tagging it for proteolytic degradation. Once liberated from IκB, the NF-κB dimer can translocate to the nucleus, where it activates a broad constituency of genes, including anti-apoptotic and mitogenic genes. A prominent gene target of NF-κB transcriptional regulation is the gene encoding IκBα, which (*continued*)

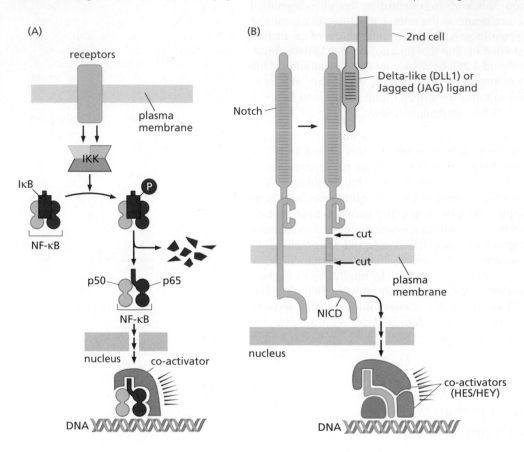

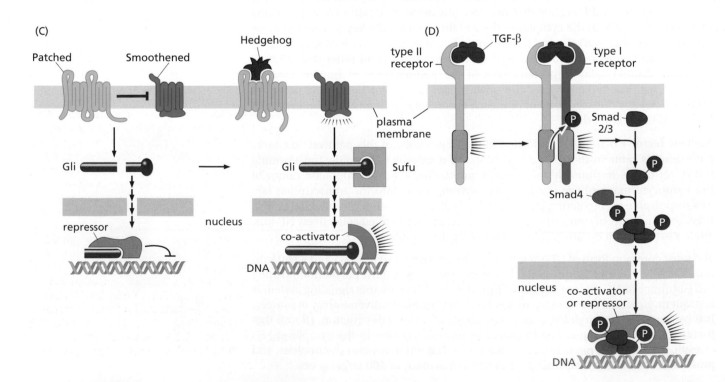

Figure 6.22 Four different "dual-address" signaling pathways (*continued*) generates this inhibitory protein that proceeds to move into the nucleus to shut down further transcription by NF-κB, thereby functioning as a negative-feedback inhibitor (*not shown*). Also not shown are the alternative, non-canonical, and atypical NF-κB pathways. (B) The Notch receptor can bind ligands belonging to the Delta-like and Jagged families that are presented by a closely apposed second cell. Ligand binding causes two proteolytic cuts in Notch (*below*), liberating a cytoplasmic fragment (termed Notch intracellular domain, NICD) that can translocate to the nucleus, where it functions as part of a transcription factor complex. At the same time, the Notch ligand and the ectodomain of the Notch receptor are internalized and then degraded by the ligand-presenting cell (*not shown*; see Figure 5.16). (C) In this simplified depiction of Patched (Ptc)–Smoothened signaling, the binding of a Hedgehog (Hh) ligand releases Smo from Ptc control (see Figure 5.17 for more details). Smo then allows Gli to escape ongoing degradation (where its cleaved form acts as a transcriptional repressor in the nucleus). The resulting full-length Gli protein appears to be sequestered by Sufu in the cytoplasm and in the primary cilium. In response to Hh ligand binding, Sufu releases its grip on the Gli protein, allowing the latter to accumulate in full-length form and translocate to the nucleus, where it functions as a transcriptional inducer or repressor. (In fact, two of the three Gli proteins function as repressors; a third one can act to stimulate transcription under certain conditions.) (D) Binding of the TGF-β ligand to the type II TGF-β receptor brings the type II and type I receptors together and results in the phosphorylation by the type II receptor of the type I receptor. The latter, now activated, phosphorylates cytosolic Smad2 and/or Smad3 proteins; these phospho-Smads then bind to Smad4. The resulting heterotrimeric Smad complexes (composed of Smad 2/2/4, 2/3/4, or 3/3/4 subunits) translocate to the nucleus, where they bind specifically to a tetranucleotide DNA segment and associate with other adjacently bound transcription factors (*light brown*) to induce or repress gene expression.

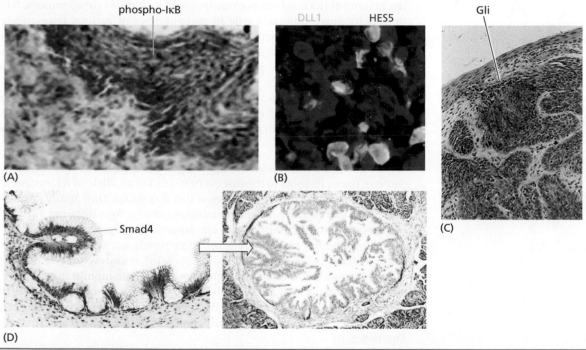

Figure 6.23 Deregulation of dual-address pathways in human cancers (A) NF-κB pathway. Immunostaining for the phosphorylated form of IκB reveals the localized activation of the NF-κB pathway (*brown area*) early in the multi-step development of a human cervical carcinoma, in this case in a low-grade squamous intraepithelial lesion. (B) Notch pathway. This pathway is often deregulated through localized expression of Notch ligands. Characteristic downstream targets of Notch-stimulated transcription are a pair of bHLH transcription factors (see Figure 8.21) termed HES and HEY. In human glioblastomas, the DLL1 Notch ligand is expressed by endothelial cells forming the microvasculature of the tumor (*green*). Close to these endothelial cells are neoplastic glioblastoma cells that have responded to this DLL1-mediated activation of their Notch1 and Notch2 receptors by expressing HES5, a clear marker of activated Notch signaling (*red orange*). This *juxtacrine* signaling also causes the glioblastoma cells to express Nestin (*not shown*), a marker of both normal and neoplastic stem cells in the brain. (C) Hedgehog pathway. This pathway is deregulated in the majority of basal cell carcinomas of the skin (BCCs). Its deregulation is indicated by the elevated concentration of Gli1 protein in BCCs, detected here by immunostaining (*brown*), while being barely evident in normal cells (*light pink*). Northern (RNA) blot analyses (*not shown*) indicate as much as a 40-fold elevation in Gli1 mRNA in BCCs compared with normal skin. (D) TGF-β pathway. This pathway is inactivated in a wide variety of carcinomas through several mechanisms. During pancreatic carcinoma progression, low-grade pancreatic intraepithelial neoplasias (PanIN-1) usually express Smad4 (*brown, left panel*), a critical transcription factor in TGF-β signaling (see Figure 6.22D). These can progress (*arrow*) to high-grade PanIN-3 lesions, which often lose Smad4 expression, typically because of mutation of the encoding gene (*center of right panel*). (A, from A. Nair et al., *Oncogene* 22:50–58, 2003. With permission from Nature. B, from T.S. Zhu et al., *Cancer Res.* 71:6061–6062, 2011. With permission from American Association for Cancer Research. C, from L. Ghali et al., *J. Invest. Dermatol.* 113:595–599, 1999. With permission from Elsevier. D, from R.E. Wilentz et al., *Cancer Res.* 60:2002–2006, 2000. With permission from American Association for Cancer Research.)

The kinase that tags IκB for destruction (named IκB kinase or simply IKK) and thereby activates NF-κB signaling is itself stimulated by signals as diverse as tumor necrosis factor-α and interleukin-1β (extracellular factors involved in the inflammatory response of the immune system), lipopolysaccharide (a component of the outer membrane of certain bacteria and thus an indicator of bacterial infection), reactive oxygen species (ROS), anti-cancer drugs, and gamma irradiation. In the context of cancer, NF-κBs affect cell survival and proliferation. Once they have arrived in the nucleus, the NF-κBs can induce expression of genes specifying a number of key anti-apoptotic proteins, such as Bcl-2 and IAP-1 and -2; we will learn more about these proteins in Chapter 9. At the same time, NF-κBs function in a mitogenic fashion by inducing expression of the *myc* and cyclin D1 genes, components of the cell cycle machinery that we will encounter in Chapter 8. Hence, NF-κBs can protect cancer cells from apoptosis (programmed cell death) and, at the same time, drive their proliferation. An alternative, non-canonical form of NF-κB participates, albeit less frequently, in the development of certain tumors.

While components of the NF-κB signaling cascade are rarely found in mutant form in human cancers, this pathway is frequently found to be constitutively activated. In breast cancers, for example, the pathway is often highly active, ostensibly through the actions of IKKε, which is overexpressed in ~30% of these tumors. NF-κB seems to play its most important role in malignancies of various lymphocyte lineages. The *REL* gene, which encodes one of the subunits of NF-κB, is amplified in about one-fourth of diffuse large B-cell lymphomas, resulting in a 4- to 35-fold increase in expression of its gene product. Translocations affecting the *NFKB2* locus have been found frequently in B- and T-cell lymphomas and in **myelomas** (tumors of antibody-producing cells). And deregulation of this pathway can often be observed early in the development of low-grade malignant growths (see Figure 6.23A). More generally, overexpression of NF-κB is associated with poor prognosis in a variety of common cancers.

Notch Study of another unusual signaling pathway, this one controlled by the Notch protein, traces its roots to the discovery in 1913 of an allele of a *Drosophila* gene that causes notches to form in the edges of this fly's wings. Only many decades later was it realized that Notch is a transmembrane protein; four different varieties of Notch (products of four different genes) are expressed by mammalian cells. As described in the last chapter (see Figure 5.16), after Notch, acting as a cell surface receptor, binds a ligand (a Jagged- or Notch-like protein), it undergoes two proteolytic cleavages, one in its ectodomain, the other within its transmembrane domain. The latter cleavage liberates a largely cytoplasmic protein fragment from its tethering to the plasma membrane. This fragment of Notch then migrates to the nucleus of targeted cells, where it functions, together with partner proteins, as a transcription factor (see Figure 6.22B).

Notch signaling is especially important in governing **cell** fate decisions during development and tumorigenesis, that is, the decision by a less differentiated cell to enter into one or another alternative differentiation pathway. Indeed, the repertoire of genes activated in different cell types by Notch is strongly affected by the differentiation pathways that these cells have entered. In addition, Notch has proliferation-inducing functions independent of its tissue-specific effects; these are most apparent from its ability to activate expression of Myc, which we will see in Chapter 8 is a potent driver of cell proliferation.

This signaling system operates using biochemical mechanisms that are clearly very different from those governing signaling by tyrosine kinase receptors. Each time a Notch receptor binds its ligand, the receptor undergoes irreversible covalent alterations, specifically, the proteolytic cleavages cited above. Hence, receptor firing occurs in direct proportion to the number of ligands encountered in the extracellular space, and each Notch receptor molecule can presumably fire only once after it has bound its ligand and therefore relies on its ability to drive multiple cycles of transcription in the nucleus in order to amplify the signal initiated by Notch ligand binding. Tyrosine kinase receptors, in dramatic contrast, release signals repetitively over an extended period of time following ligand binding, and therefore already amplify the signals initiated by their growth factor ligands in the cytoplasm; this amplification is multiplied further by the ability of their downstream partners, many of which are enumerated in Figure 6.5. In addition, as mentioned in the last chapter, Notch signaling operates in a *juxtacrine* fashion, that is, between two cells that are in direct physical contact with one another. This contrasts with many other signaling interactions that depend on soluble ligands diffusing through the **interstitial** space between cells located near one another within a tissue but not necessarily in direct physical contact, creating the contrasting *paracrine* signaling interactions.

Truncated forms of *Notch* that specify only the cytoplasmic domain of the Notch protein are potent oncogenes for transforming cells, suggesting that altered forms of Notch contribute to human cancer pathogenesis. Indeed, overexpression of one or another of the Notch proteins is seen in the great majority of cervical carcinomas, in a subset of colon and prostate carcinomas, and in squamous cell carcinomas of the lung. This overexpression is often accompanied by nuclear localization of the cytoplasmic cleavage fragment of Notch, indicating that active signaling through this pathway is occurring in tumor cells. Increased expression of two classes of Notch ligands, termed Jagged and Delta, has also been found in some cervical and prostate carcinomas (Figure 6.23B). More dramatically, in about 60% of T-cell acute lymphocytic leukemias (T-ALLs), constitutively active forms of Notch are found; these result from genetic deletions of the portion of the *NOTCH-1* gene encoding the extracellular domain of the protein. Some experiments also suggest that Notch signaling contributes to transformation by *ras* oncogenes. In general, however, variant forms of *Notch* act as oncogenes in hematopoietic malignancies and in an array of other tumors but may function in an opposite fashion as tumor suppressors in certain carcinomas; hence, the contributions of Notch signaling to cancer development seem to be highly dependent on cellular context.

Hedgehog As was discussed previously, binding of the Patched receptor by its ligand, Hedgehog (Hh), somehow alters the interactions between Patched (Ptc) and Smoothened (Smo). As a consequence, Smoothened and Gli then accumulate in the primary cilium, often at its tip (see Figure 5.17). Normally, in the absence of intervention by Smoothened, the Gli precursor protein is cleaved into two fragments, one of which moves into the nucleus, where it functions as a transcriptional repressor. However, following Hedgehog stimulation, Smoothened and Gli interact in the primary cilium, where the Gli precursor is protected from cleavage. The resulting intact Gli protein migrates to the nucleus, where it serves as an activator of transcription (see Figure 6.22C). During the course of embryogenesis, Hh signaling has been associated with development of patterned development in certain tissues as well as commitments to specific cell states (although acting over greater distances than Notch). It can also emit signals promoting cell proliferation.

Gli was first discovered as a highly expressed protein in glioblastomas (whence its name). When overexpressed, Gli and several closely related cousins can function as oncoproteins. Subsequent research into this pathway has revealed other instances where its malfunctioning contributes to tumor development. For example, germline inactivating mutations of the human patched (*PTCH*) gene cause Gorlin syndrome, an inherited cancer susceptibility syndrome, which involves increased risk of multiple basal cell carcinomas of the skin as well as other tumors, notably medulloblastomas—tumors of cells arising from primitive stem cells in the cerebellum. Such inactivating mutations prevent Patched from inhibiting Smoothened, giving the latter a free hand to dispatch an uninterrupted stream of active Gli protein to the cell nucleus.

As many as 50% of sporadic basal cell carcinomas of the skin (which, by definition occur because of somatic rather than germ-line mutations) carry mutant *PTCH1* or *SMO* alleles—the latter encoding Smoothened. These basal cell carcinomas are the most common form of cancer in Western populations and fortunately are usually benign (see Figure 6.23C). Somatically mutated alleles of *PTCH1* have also been found in a variety of other tumors, including medulloblastomas (as noted above) and meningiomas, as well as breast and esophageal carcinomas. These somatic mutations, like those present in mutant germ-line alleles of *PTCH1*, seem to compromise Patched 1 function, once again permitting Smoothened to constitutively activate Gli transcription factors. Recently, germ-line and somatically mutated inactivating alleles of *SUFU*—which otherwise sequesters some full-length Gli proteins (see Figure 6.22C)—have been described in association with human medulloblastoma development.

These various cancers, all involving mutations that affect the structure of various components of the Hedgehog pathway, represent only a small portion of the human tumors in which this signaling cascade is hyperactivated. Thus, a survey of esophageal, gastric, biliary tract, and pancreatic carcinoma cell lines and tumors has revealed that virtually all of these expressed significant levels of the Patched receptor as well as its ligands (either Sonic Hedgehog or Indian Hedgehog). Often the ligands are produced by the epithelial cells within carcinomas while the Patched receptor is expressed in nearby stromal cells, suggesting the operations of a paracrine signaling channel. This notion was confirmed by demonstrating elevated nuclear levels of the Gli factor, the downstream product of the activated pathway, in stromal cells rather than the epithelial carcinoma cells themselves. Moreover, treatment of tumors with anti-Hedgehog antibody caused cessation of proliferation and/or death of tumor cells, suggesting that stromal cells stimulated by Hedgehog ligands reciprocate by sending mitogenic and survival signals back to nearby carcinoma cells. Similar findings have been reported for small-cell lung carcinomas (SCLCs). Equally plausible is an alternative mechanism, in which both the Hh ligands and the Patched receptor are expressed in the carcinoma cells, causing cell proliferation to be driven by autocrine signaling loops. Quite often, Hedgehog-responsive tumor cells have been identified as neoplastic cells with stem cell-like properties. Together, these reports point to an important role of this signaling pathway in the tumors including those that arise in the multiple tissues deriving developmentally from the embryonic gut.

TGF-β A fourth signaling pathway that involves the dispatch of cytoplasmic proteins to the nucleus is represented by the pathway leading from TGF-β receptors (Section 5.7). TGF-β and the signaling pathway that it controls appear to play major roles in the pathogenesis of many if not all carcinomas, both in their early stages, when TGF-β acts to arrest the growth of many cell types, and later in cancer progression, when it contributes, paradoxically, to the phenotype of tumor invasiveness. We will defer detailed discussions of the cell-biological functions of this important pathway until Section 8.10 and Chapter 14. For the moment, suffice it to say that activation of this pathway leads to dispatch of Smad transcription factors to the nucleus, where they can bind a specific tetranucleotide sequence in the DNA and associate with adjacently bound transcription factors to induce the expression of a large constituency of genes and repress many others (see Figure 6.22D). In the absence of critical Smads, epithelial cancer cells can escape the growth-inhibitory actions of TGF-β and thrive—a state that is often observed in the precursors of invasive human pancreatic carcinomas (see Figure 6.23D).

The Smad transcription factors (TFs) are most unusual, because they recognize and bind only a tetranucleotide sequence in chromosomal DNA, as mentioned above. This DNA binding, on its own, is too weak to maintain the stable association of Smads with the DNA. Instead, the Smads often and perhaps invariably bind DNA segments immediately adjacent to the binding sites of other more stably bound TFs with which they establish non-covalent associations. The resulting lateral protein–protein interactions between the Smad and its diverse TF partners ensure that binding by Smads and its partners to DNA is cooperative and quite strong. To date, Smads have been found to form such partnerships with more than 130 distinct TFs; the functions of most of these have not been explored. As one example among many, the complex of Smad2 and Smad4 can associate with the YAP/TAZ +

TEAD complex described in the next section. This versatility explains why TGF-β and its Smad effectors can elicit diverse responses from cells, depending on the signaling context. TGF-β can induce apoptosis in certain cells but in the presence of active PI3K signaling may simply be cytostatic (see Chapter 8). When signaling in the presence of a Ras oncoprotein, TGF-β may induce an epithelial–mesenchymal transition (Chapters 13 and 14).

6.13 The Hippo signaling circuit integrates diverse inputs to govern diverse cell phenotypes

Virtually all of the discussions of signaling in this chapter have been motivated by a simple paradigm: an extracellular signal, conveyed by a soluble or insoluble protein, activates a cell surface receptor, which then proceeds through a dedicated signaling cascade to evoke multiple downstream cell-biological responses in a cell. While fruitful and accessible, such discussions ignore other types of signal-processing circuits that are not dedicated to serving specific cell surface receptors. Instead, these other types of circuits receive incoming (**afferent**) signals from diverse extracellular sources, integrate these signals with others of intracellular origin (for example, monitors of ATP levels), and only then proceed to effect key decisions in the life of a cell. One prominent example of such an intracellular control circuit is the cell-cycle clock, which is explored in detail in Chapter 8. Others involve the circuits that monitor genomic integrity, energy metabolism, and nutrient availability; some of these will be visited in future chapters.

The Hippo signaling circuit, involving prominently the YAP and TAZ transcription factors, represents one of the key intracellular circuit that responds to diverse afferent signals and merits our attention because of the roles that it plays in cancer pathogenesis. It was originally studied for its control of organ size, that is, the mechanisms governing how large individual organs become within a metazoan body, be they of a fly or a human. More recently, its role in governing the proliferation of normal and neoplastic mammalian cells has become apparent. Thus, the Hippo signaling circuit is a central element in the circuitry that processes a diverse set of inputs including the cell's localization within a tissue, its attachment to other cells, and the state of the surrounding extracellular matrix. Hence, unlike other extracellular signals described above, the YAP and TAZ paralogs monitor the overall tissue context of a cell, only some of which is defined by the presence in the surrounding space of discrete, soluble signaling molecules (**Figure 6.24A**). In response to the proper combinations of signals, YAP/TAZ will move into the nucleus (Figure 6.24B) and there encourage gene expression that favors cell proliferation, survival, and stemness. Of additional interest, YAP/TAZ is also a "dual-address" circuit like the six others described earlier, but one in which the modifying signal (protein phosphorylation) results in functional inactivation via sequestration or degradation in the cytoplasm and thus blockage of signals being dispatched to the transcriptional apparatus in the nucleus.

The fact that nuclear YAP/TAZ favors cell proliferation already suggests that its actions are, with great likelihood, deregulated within various human tumors, and indeed this is the case. High-level expression of YAP or TAZ, especially when present in the nuclei of the carcinoma cells in a variety of human tumors, is, on its own, correlated with aggressive clinical behavior, including a tendency to spawn metastases and, in the case of certain breast cancers, a reduced tendency of involved tumors to respond to otherwise-effective therapies. Among carcinomas, ranging from liver and pancreatic tumors to those of lung and colon, these correlations have been documented repeatedly. Conversely, shutting down YAP/TAZ nuclear localization by overexpressing some of their negative regulators results in loss of proliferation and, in certain instances, tumor shrinkage. Unlike most of the signaling circuits discussed in this chapter, this one is still poorly understood because of the daunting complexity of its inputs and the myriad cross-connections between this circuit and other regulatory circuits operating within cancer cells.

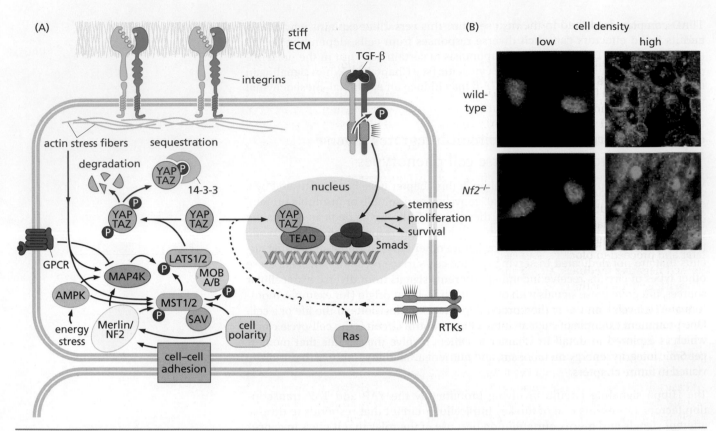

Figure 6.24 YAP-TAZ signaling The YAP/TAZ signaling circuit is designed to integrate diverse inputs and, under the proper conditions, respond by fostering cell proliferation, acquisition of stemness properties, and suppression of apoptosis. (A) The key components are the closely related YAP and TAZ transcriptional co-activators which, when permitted to enter the nucleus, associate with a TEAD transcription factor (there are four in all) to evoke these various downstream responses. A key input is the Hippo pathway, which drives YAP/TAZ phosphorylation, ensuring that these two transcription factors are sequestered in the cytoplasm and even destroyed in proteasomes (*not shown*); this inhibitory phosphorylation is driven by the actions of the LATS1/2 kinases. Upstream of LATS1/2 kinases are yet other regulators. MST1/2 kinases (termed Hippo in *Drosophila* genetics) phosphorylate and thereby activate the LATS1/2 kinases, a step that requires the tumor suppressor NF2, the formation of cell–cell contacts at adherens junctions (depicted in Figure 6.19A), and apico-basal polarity in epithelia (see Figure 2.4C). An overarching input controlling YAP/TAZ activity is cellular *mechanotransduction*, i.e., the ability of cells to sense the rigidity of the ECM around them as well as their own shape. These cues are regulated by the cell's attachment via focal adhesions to the ECM, the resulting tension transferred to the actin cytoskeleton, and the surrounding tissue architecture in which it is embedded. Both canonical and non-canonical Wnt signaling (see Figures 6.19 and 6.20) as well as the TGF-β pathway (see Figure 6.22D) also govern the outputs of this pathway. Finally, cellular metabolism influences this pathway through AMPK (AMP kinase), which monitors the level of AMP vs. ATP in cells. (B) Wild-type and *Nf2* mutant mouse Schwann cells were immunostained with anti-YAP antibody (*green*). In wild-type cells grown in monolayer culture, YAP, which acts as a transcription factor to promote cell proliferation, is enriched in the nucleus at low cell density but is preferentially localized in the cytosol at high cell density and thus effectively excluded from the nucleus; this may explain much of the phenomenon of contact inhibition described in Figure 3.4. However, in cells lacking the Merlin/NF2 tumor suppressor protein, which normally transduces signals from the cell surface to YAP/TAZ, the nuclear localization of YAP is unaffected by the density of cells in monolayer culture, ostensibly allowing cells to continue proliferating long after they have sensed contact with their neighbors. (A, adapted from S. Piccolo, S. Dupont and M. Cordenonsi, *Physiol. Rev.* 94:1287–1312. B, Courtesy of Duojia (DJ) Pan.)

6.14   Well-designed signaling circuits require both negative and positive feedback controls

All well-designed regulatory circuits are fine-tuned by feedback controls that ensure that the flux of signaling through one or another pathway is at appropriate levels for optimal performance. Such well-regulated signal-processing circuits contain inhibitory components that counterbalance and modulate the behavior of other components that emit positive signals. Without these negative controls, positive signals, such as those that promote cell proliferation, may run rampant; in the case of mitogenic signals, this can lead to the runaway proliferation that we associate with neoplastic cells.

In principle, one effective strategy for constraining the signals flowing through the mitogenic signaling pathways that we have encountered could rely on limiting the amounts of growth factor molecules to which a cell is exposed. That way, even if this cell displayed abundant growth factor receptors on its surface, the limited number of available GF ligand molecules in the extracellular space would ensure only minimal firing by these receptors and therefore limited activation of downstream signaling pathways.

However, a more precisely targeted and effective strategy for regulating the flow of such mitogenic signals depends on modulating signal processing *within* the responding cell. Often, the inhibitory components operating in an intracellular signaling circuit are functionally inactive until the circuit becomes active; however, once signaling begins flowing through the circuit, the inhibitory proteins become activated sooner or later and proceed to block further signaling. This type of inducible inhibition is often termed **negative feedback** and is an important feature of many, possibly all, signal-processing circuits within cells.

Actually, without noting this feature explicitly in the last two chapters, we have encountered a number of proteins that are involved in negative-feedback loops. For example, when a GF receptor becomes activated by ligand binding and autophosphorylation, the SH2 domains of Ras-GAP (see Figure 5.24) allow this protein to bind to phosphotyrosines of the receptor. This places Ras-GAP in a position where it can trigger the GTPase function and resulting inactivation of Ras (**Figure 6.25A**); the latter, which had recently been activated into its active, GTP-bound state, is now forced to shut down its signaling. This negative-feedback loop ensures the release of only a small burst of mitogenic signals by Ras. Yet other mechanisms involved in shutting down receptor signaling are depicted in Figure 6.25A, B, and C; in fact, dozens of such negative-feedback loops have been uncovered in various regulatory circuits operating within the cell, and far more are likely to be found in the future. As predicted from this scenario, the loss of various proteins that function as Ras-GAPs can contribute to runaway Ras signaling. For example, mutant *Ras* oncogenes are rarely found in human breast cancers; however, some especially aggressive breast tumors nonetheless acquire strong Ras signaling by inactivating key Ras-GAP proteins.

An alternative mechanism of signal regulation is embodied in **positive-feedback loops**, which function to further *amplify* initial signaling. For example, the neurofibromin (NF1) protein normally functions like Ras-GAP to induce GTP hydrolysis by Ras proteins (Figure 6.25D). Thus, in cells that are not experiencing mitogenic signaling, NF1 ensures that the level of GTP-loaded Ras is held very low. However, soon after mitogenic signaling is initiated by a ligand-bound GF receptor, a cell's pool of NF1 protein is quickly degraded, clearing the way for Ras to rapidly become activated. Here, this **feedforward** loop is exploited to accelerate the initial activation of the signaling event, allowing it to reach its maximum more quickly. (Later on, NF1 levels are restored by yet other regulators, ensuring eventual shutdown of Ras signaling.)

Alternatively, feedforward mechanisms can be used to *stabilize* a decision, such as a decision to *maintain* a certain state of differentiation. Thus, a cell in a certain differentiated state may produce a growth factor or cytokine that functions in an autocrine manner to induce and thereby ensure continued maintenance of this state of differentiation. In Chapter 8, we will encounter yet other examples of positive-feedback loops exploited by cells that have executed a decision and use these loops to guarantee the *irreversibility* of this decision; for example, once a cell decides to advance into the next phase of the cell cycle, positive-feedback signaling loops reinforce and amplify this decision, preventing the cell from inadvertently slipping backward into an earlier phase of the cell cycle. Still, we need to remember that feedforward loops also represent a liability, at least in the case where mitogenic signaling is being regulated: since they allow signaling to become self-perpetuating, feedforward loops may lead to loss of the finely tuned control of cell proliferation that is central to the normal cell phenotype.

Figure 6.25 Negative and positive feedback controls on signaling An elaborate array of negative-feedback loops ensures that excessive signaling in normal cells is reduced to levels that are compatible with maintenance of normal cell behavior. (A) When

RTKs bind their cognate ligands, the resulting activation of the Ras → Raf → MEK → ERK signaling cascade causes Sprouty (*brown, lower left*) to block binding of Grb2 to Sos (*light orange*), thereby shutting down further Ras activation. Sprouty also binds (*continued*)

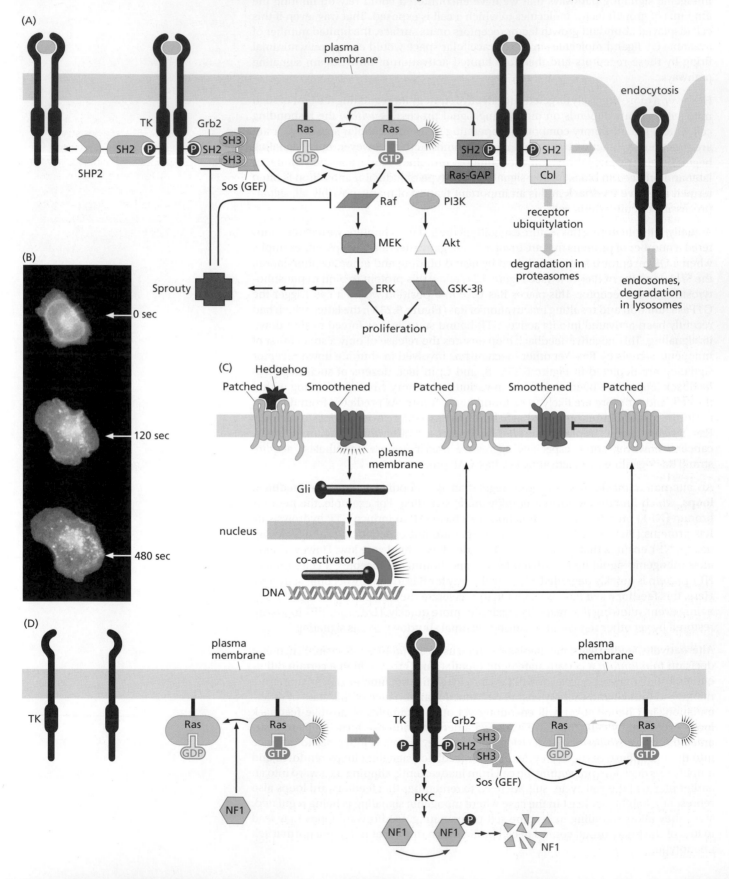

Figure 6.25 Negative and positive feedback controls on signaling (*continued*) and inhibits the Raf kinase as well as other targets (*not shown*). Functioning concomitantly with this negative-feedback loop, the binding of Ras-GAP (*purple*; see also Figure 5.24) via one or both of its SH2 groups to a ligand-activated receptor drives GTP hydrolysis by Ras. Moreover, the binding of Cbl (*light beige*) to a receptor via its SH2 group results in ubiquitylation of the receptor and its destruction within proteasomes. Two more negative-feedback loops operate via the ligand-induced endocytosis of an RTK, such as the EGF-R, and lead to its introduction either into endosomes at low ligand-binding levels (from which it may continue to fire for a period) or into lysosomes (at high ligand-binding levels), in which it is degraded (*right*). Finally, the binding of the SHP-2 phosphatase to the receptor removes the phosphates previously attached to tyrosine residues (*left*). (Not shown is a second SH2 group carried by SHP-2.) See also Sidebar 5.2. (B) As illustrated in part A, an important mechanism of shutting down RTK signaling involves the physical internalization of receptor molecules via endocytosis. When a cell is exposed to epidermal growth factor (EGF), introduced into the medium at the location shown (*white arrows*), the EGF initially causes the cell surface EGF receptor molecules, which were previously localized to the plasma membrane and in this case have been linked to a fluorescent dye, to release mitogenic signals into the cell (*not shown*). However, within 120 seconds, the fluorescing EGF-R molecules have been internalized into the cytosol on the side of the cell facing the EGF source; and by 480 seconds, they are found in internalized clusters throughout the cell, which are either endosomes or lysosomes. (C) Binding of Hedgehog to Patched (*top left*) permits signaling by Smoothened, one of the consequences of which is the activation, via Gli, of transcription of the Patched-encoding gene. The resulting increased numbers of Patched molecules then function to inhibit further signaling by Smoothened. (D) A positive-feedback mechanism is illustrated by the actions of NF1 (neurofibromin), which in resting cells ensures the hydrolysis of Ras-bound GTP and associated inactivation of signaling by Ras. However, when a GF-activated receptor begins to release signals, it activates a form of PKC (protein kinase C) that phosphorylates NF1, causing its ubiquitylation and degradation (see Supplementary Sidebar 7.5), thereby enabling a rapid increase in firing by active, GTP-bound Ras molecules (*right*). (B, from M. Bailly et al., *Mol. Biol. Cell* 11:3873–3883, 2000. With permission from American Society for Cell Biology.)

It turns out that some of the negative-feedback controls can have major effects on the successes or failures of targeted pharmacologic therapies, such as the one that is directed toward shutting down the B-Raf kinase, which is mutated in a majority of human melanomas and in a subset of colorectal carcinoma (CRC) cells. The inhibitory drug termed vemurafinib (discussed in greater detail in Chapter 17) is effective in shutting down the proliferation of the *B-Raf*-mutant melanoma cells but is unable to elicit the same response in the CRC cells. As it happens, a powerful negative-feedback loop operates in CRC cells: B-Raf signaling activates ERK, which in turn acts via a negative-feedback loop to suppress signaling by the EGF-R, which lies upstream in the signaling pathway (Figure 6.25A). However, when MEK is shut down (through B-Raf inhibition), the negative-feedback loop no longer operates and EGF-R quickly begins highly effective firing, now signaling via an alternative downstream pathway—the Ras–PI3K–Akt–GSK-3β cascade—to drive CRC proliferation; this circumvents the anti-mitogenic actions of the B-Raf inhibitor. Melanoma cells do not express significant levels of the EGF-R and therefore cannot compensate for B-Raf inhibition, explaining their sensitivity to B-Raf inhibition. This represents only one of myriad feedback pathways that can thwart the most carefully designed anti-cancer therapies.

6.15 Synopsis and prospects

The signaling pathways that we have described in this chapter represent only the bare bones of the complexity of signal processing that actually operates within mammalian cells. In describing these, we have overlooked many of the endless complexities, some of which are enumerated below. Stated simply, the signaling networks created by the interactions between hundreds of proteins acting in various combinations create a complexity that is vastly beyond our current abilities to assimilate, dooming our attempts at present to understand the behavior of an individual cell as a whole. Moreover, this complexity grows progressively—driven by increasingly powerful analytical tools, so that the overall cellular signaling circuitry can no longer be grasped by a single human mind. Instead, as the years go on, each of the pathways described here has attracted its own coterie of highly specialized researchers.

Why has evolution cobbled together so many different solutions to processing signals within cells and tissues? The answer can only come from appreciating the extraordinary complexity of the metazoan body and the mechanisms that make its normal development possible. Long after the fact, we are trying to tease apart the biochemical solutions that were developed, step-by-step, during the ~1.4 billion years between the emergence of eukaryotes (~2.1×10^9 years ago) and the development of complex metazoan bodies (~0.7×10^9 years ago).

The resulting complexity greatly complicates most current attempts at understanding this circuitry for the purpose of treating human tumors in the oncology clinic. Those researchers who have chosen to develop new anti-cancer therapies are forced to flee into corners of this now-vast field of research and embrace extreme reductionism, focusing as they do on a small number of signal-processing components while purposively ignoring the surrounding complexity; their reductionism is the only rational and pragmatic approach to making solid progress.

Importantly, this complexity has caused cancer researchers to develop new ways to navigate through these thickets. The advent of genome-wide screens (to be discussed in Chapter 17) has allowed the discovery of the vulnerabilities of cancer cells—the choke points in specific circuits that can be revealed even if the detailed operations of the surrounding signaling cascades remain poorly understood. These powerful and incisive approaches to developing clinical treatments of cancer have begun to yield impressive fruit.

Signal transduction in cancer cells As described in this chapter, one of the hallmarks of cancer cells is their ability to generate their own mitogenic signals endogenously. These signals liberate cancer cells from dependence on external mitogenic signals, specifically those conveyed by growth factors. It is now clear that the mitogenic pathway of greatest importance to human cancer pathogenesis is the one that was initially discovered and has been discussed here in detail, that is:

$$GFs \rightarrow RTKs \rightarrow Grb2 \rightarrow Sos \rightarrow Ras \rightarrow Raf \rightarrow MEK \rightarrow ERK$$

As described in Chapter 5 and in this one, a variety of molecular mechanisms cause the upstream portion of this signaling pathway to become continuously activated in cancer cells. Neoplastic cells may acquire the ability to make and release growth factors (GFs) that initiate autocrine signaling loops. Alternatively, the growth factor receptors (RTKs) may suffer significant structural alterations. Even more common are the human tumors whose cells display elevated levels of wild-type versions of receptors such as the EGF receptor or its close relative, ErbB2/HER2/Neu. It is clear that excessively high levels of these receptor molecules, often expressed by breast, brain, and stomach cancer cells, favor increased cell proliferation. Such receptor overexpression is often achieved through either deregulated transcription or amplification of the gene encoding the receptor. Reductions in the elimination via endocytosis from the cell surface of ligand-activated RTKs also seem to contribute frequently.

As we learn more about the regulation of cell signaling, it is becoming apparent that negative-feedback controls like those modulating the signaling in this particular pathway (see Figure 6.25A) operate in essentially all the intracellular signaling pathways, ensuring that signals stimulating cellular growth and division are released in carefully titered amounts to ensure proper **homeostasis**—the proper physiologic balance of cell birth and death that is so critical to tissue and thus organismic viability. We will revisit some of these negative-feedback controls in Chapter 17, since they play an important role in thwarting recently developed therapeutic agents that have been designed to shut down certain key signaling molecules.

Moving down the mitogenic pathway from the receptors, we note that few, if any, alterations in signal-transducing proteins have been documented in human tumors until we reach the Ras proteins. Recall that point mutations in the 12th, 13th, or 61st codon of the reading frames of encoding *ras* genes result in constitutively activated Ras oncoproteins. The power of these Ras oncoproteins comes directly from their ability to activate two major and critical mitogenic pathways—those involving Raf and PI3K. Moving further downstream in the RTK $\rightarrow$ Sos $\rightarrow$ Ras signaling cascade, we once again find a relative dearth of mutant proteins in human cancer cells. The B-Raf protein, a close cousin of the Raf kinase, is one exception. It is found in mutant form in about 8% of a large panel of human tumor lines. Most importantly, almost half of all human melanomas possess point-mutated versions of the *BRAF* gene, whose normal product, like Raf, requires Ras stimulation before it can fire. The resulting mutant B-Raf proteins have a greatly reduced dependence on interaction with Ras.

Another exception is provided by phosphatidylinositol 3-kinase (PI3K), which also operates immediately downstream of Ras. Recall that the PI3K enzyme is activated by Ras, by growth factor receptors, or by the two acting in concert (see Figures 6.12 to 6.14). Its product is phosphatidylinositol triphosphate (PIP3). By 2020, activating mutations in the PI3K gene (termed *PIK3CA*) had been documented in the genomes of about 10% of all human cancers, with almost one-third of certain cancer types, notably those of the breast and endometrium, harboring mutant, oncogenic *PIK3CA* alleles. Far more commonly, however, *inhibition* of this pathway is defective in cancer cells: levels of PIP3, the product of PI3K, are normally held down by the actions of the PTEN phosphatase, which is lost in a wide variety of cancers. Almost 30% of all human tumors suffer loss of *PTEN* gene function, caused by either mutational inactivation or transcriptional repression. The resulting loss of PTEN function may be as effective for increasing PIP3 levels as the hyperactivity of the PI3K that is induced by Ras oncoproteins. In both cases, the increased PIP3 levels drive activation of Akt/PKB, which in turn has widespread effects on cell survival, growth, and proliferation.

In theory, future research may reveal that other pathways, some mentioned in this chapter (involving GPCRs, Patched, Notch, NF-κB, and Hippo), rival the importance of the RTK → Ras pathway in providing the mitogenic signals that drive cell proliferation in certain types of human cancer cells. In reality, however, every year's worth of newly published genomic studies seems to reinforce the primacy of the canonical RTK pathway in human cancer pathogenesis.

Organization of mitogenic signaling pathways Evolution might have constructed complex metazoa like ourselves by using a small number of cellular signaling pathways (or even a single one) to regulate the proliferation of individual cells in various tissues. As it happened, a quite different design plan was cobbled together: each of our cells uses a number of distinct signaling pathways to control its proliferation (with one of these, involving tyrosine kinase receptors, playing an especially prominent role). In each type of cell, the signaling pathways work in a *combinatorial fashion* to ensure that proliferation occurs in the right place and at the right time during development and, in the adult, during tissue maintenance and repair. Moreover, different cell types use different combinations of pathways to regulate their growth and division. This helps to explain why the biochemistry of cancer, as described in this chapter, is so complex.

The signaling circuitry that was described in this chapter operates largely in the cytoplasm and receives inputs from cell surface receptors—those that bind either soluble extracellular ligands, notably growth factors, or insoluble matrix components (the integrins). Having received a complex mixture of afferent (incoming) signals from these receptors, this cytoplasmic circuitry emits signals to molecular targets in both the cytoplasm and the nucleus.

Included among the cytoplasmic targets are the regulators of cell shape and motility, energy metabolism, protein synthesis, and the apoptotic machinery, which decides on the life or death of the cell. The remaining signals are transmitted to the nucleus, where they modulate the transcription of thousands of genes involved in the programming of cell growth and differentiation as well as cell cycle progression. We will return to some of these themes later. Thus, the regulation of the cell cycle and differentiation is described in Chapter 8, while the regulation of programmed cell death—apoptosis—will be presented in Chapter 9. The regulators of cell motility, which is critical to high-grade malignancy, will be described in outline in Chapter 14.

Certain major themes have recurred throughout this chapter. Most apparent are the three ways in which signals are transduced through the signaling pathways of normal and neoplastic cells. First, the *intrinsic activity* of signaling molecules may be modulated. This can be achieved by noncovalent modifications (for example, the binding of GTP by Ras) or receptor dimerization. Alternatively, a signaling molecule may be covalently modified; the phosphorylation of MEK by Raf, the phosphorylation of PIP2 by PI3K, and the proteolytic cleavage of Notch are three examples of this.

Second, the *concentration* of a signaling molecule may be modulated, often by orders of magnitude. Thus, the concentration of β-catenin is strongly reduced by GSK-3β phosphorylation, and the concentration of PIP3 is regulated by PI3K (positively) as well as PTEN (negatively).

Third, the intracellular *localization* of signaling molecules can be regulated, which results in moving them from a site where they are inactive to a new site where they can do their work. The most dramatic examples of these derive from the phosphorylation of the C-terminal tails of receptor tyrosine kinases (RTKs) and the subsequent recruitment of multiple, distinct SH2-containing molecules to the resulting phosphotyrosines (see Figure 6.5). Once anchored to the receptors, many of these signal-transducing molecules are brought in close proximity with other molecules that are associated with the plasma membrane; examples of the latter are the Ras molecule (the target of Sos action) and PIP3 (the target of PI3K action).

The translocation of cytoplasmic molecules to the nucleus is another manifestation of this third class of regulatory mechanisms. For example, the "dual-address" transcription factors dwell in inactive form in the cytoplasm and may be dispatched as active transcription factors (or components thereof) to the nucleus. Examples of these include β-catenin, NF-κB, the Smads, Notch, Gli, YAP/TAZ, and the STATs.

The kinetics with which signals are transduced from one site to another within a cell may also vary enormously. One very rapid mechanism involves the actions of kinase cascades. Thus, the cytoplasmic MAP kinases (MAPKs), including ERKs lying at the bottom of these cascades, are activated almost immediately (in much less than a minute) following mitogen treatment of cells, and, following activation, the ERKs move into the nucleus essentially instantaneously. Once there, they proceed to phosphorylate and functionally activate a number of key transcription factors.

A slightly slower but nonetheless highly effective signaling route derives from the strategy, cited above, of activating dormant transcription factors in the cytoplasm and dispatching them in activated form to the nucleus. By far the slowest signaling mechanisms depend on modulating the *concentrations* of signaling proteins. This mode of regulation depends on changes in the rates of transcription of genes, the translation of their mRNAs, and the post-translational stabilization of their protein products.

The intracellular signaling circuitry encountered in this chapter appears to be organized similarly in various cell types throughout the body. This fact holds important implications for our understanding of cancer and its development. More specifically, it suggests that the biochemical lessons learned from studying the neoplastic transformation of one cell type often prove to be applicable to a number of other cell types throughout the body. The ever-increasing body of information on the genetic aberrations of human cancer cells provides strong support for this notion.

The similar design of the intracellular growth-regulating circuitry in diverse cell types forces us to ask why different kinds of cells behave so differently in response to various external signals. Many of the answers will eventually come from understanding the spectrum of receptors that each cell type displays on its surface. In principle, knowledge of the cell type–specific display of cell surface receptors and of the largely invariant organization of the intracellular circuitry should enable us to predict the behaviors of various cell types following exposure to various mitogens and to growth-inhibitory factors like TGF-β. Similarly, by understanding the array of mutant alleles coexisting within the genome of a cancer cell, we should be able to predict how these alleles perturb behavior and how they conspire to program the neoplastic cell phenotype. In practice, we are still far from being able to do so for at least nine reasons:

1. We possess only a rudimentary knowledge of the layout of the circuit diagram—the map of intermolecular interactions that describe how different proteins and signaling molecules intercommunicate within a cell. One approach to describing this map involves determining all the binary protein–protein physical interactions operating within a single human cell—the sum of which represents the human **interactome**. Given the coding capacity of the human genome, approximately 200 million such binary interactions are theoretically possible. Of these, only

about 64,000 binary physical interactions were validated experimentally by 2016. Nonetheless, even the most rigorous attempts at mapping the binary interactome are unlikely to reveal all of the signaling interactions between pairs of proteins, which often involve only transient, weak physical associations between interacting partner proteins, such as those between an enzyme and its substrate. In addition, physical interactions resulting in the formation of multiprotein complexes with more than two participating subunits are commonly overlooked in existing protein–protein interaction analyses. Such complexes are found everywhere throughout the signaling circuits operating in our cells.

2. In this chapter, we have described only the bare outlines of how these circuits operate and have consciously avoided many details. But such details are critical. For example, individual steps in these pathways are often controlled by multiple, similarly acting paralogs, each of which transduces signals slightly differently from the others. Examples of this are the four structurally distinct Ras proteins; the three forms of Akt/PKB; Raf and its close cousins, A-Raf and B-Raf; and the nine similarly structured Src-family kinases (SFKs), which have partially overlapping functions. In most cases, we remain ignorant of the functional differences between the superficially similar proteins operating within a pathway, and therefore have only an incomplete understanding of how the pathway as a whole operates.

3. Complex negative-feedback and positive-feedback loops serve to dampen or amplify the signals fluxing through each of these pathways. While this chapter touched on these only briefly, negative-feedback loops are an indispensable feature of all signaling pathways, (Section 6.14). The feedback loops discovered to date surely represent only the tip of a vastly larger iceberg. For example, more than 100 phosphotyrosine phosphatase (PTP)–encoding genes have been found in the human genome. PTPs remove the phosphate groups attached by the 90 tyrosine kinases (TKs) present in our cells. We know almost nothing about the substrates of the phosphatase enzymes and how they are regulated in response to tyrosine phosphorylation events.

4. Each component in a signaling circuit acts as a complex signal-processing device that can amplify, attenuate, and/or integrate the upstream signals that it receives before passing them on to downstream targets. We have not begun to understand the quantitative parameters governing the efficiency of signal processing by individual signal-transducing proteins. As one example, the GTP-bound normal Ras protein exhibits different binding affinities to its three major downstream effectors, and these affinities change when amino acid substitutions block Ras-GTPase function and create Ras oncoproteins.

5. Calculations of diffusion times in cells indicate that efficient, rapid signal transduction is likely to occur only when interacting proteins are co-localized within microspaces within a cell. Only very fragmentary information is available about the intracellular localizations of most signaling proteins and, in some cases, the scaffolding proteins that bring them together with their interacting partners (see, for example, Figure 6.9B). The ability of proteins to transduce signals to one another is also affected by their localized intracellular concentrations; here, we have only begun to plumb the depths of this complexity. As one example, the four Ras proteins localize to various cytoplasmic membranes following their activation, where they have more or less access to their downstream effectors.

6. The phosphorylation of proteins has been discussed repeatedly in this chapter. These phosphorylation events, on their own, represent extraordinary complexity: More than 50,000 distinct phosphopeptides have been cataloged in human cells; each of these is associated with one or another cellular protein and the effects of these modifications on protein function are generally not known. In fact, post-translational covalent modifications (PTMs) of proteins are vastly more varied and complex than these phosphorylation events. Proteins may become methylated, acetylated, lipidated, or ubiquitylated (see Supplementary Sidebar 7.5). There are also a variety of ubiquitin-like PTMs that are bound covalently to the amino acids constituting the primary structures of proteins. In almost all cases, the effects on protein function of these bulky PTMs are poorly understood; included among

these modifying groups are NEDD8, ISG15, FAT10, MNSFβ, Ufm1, six Atg8 paralogs, Atg12, Urm1, Ubl5, and four SUMO paralogs. Beyond these at least 40 other types of distinct covalent modifications of proteins have been cataloged. In virtually all cases, the consequences on protein function of these PTMs are entirely obscure. In most cases, the PTMs are presumed to alter the functions of a modified protein, such as its catalytic activity or its interactions with other proteins or its levels within a cell. As one specific example, the highly modified histone H4 has been found to carry prominent acetylation, methylation, and phosphorylation PTMs on its various amino acid residues, yielding combinatorially a calculated 98,304 theoretical distinct **proteoforms** of this single cellular protein.

7. None of these intracellular pathways operates in isolation. Instead, each is influenced by cross-connections with other pathways. Understandably, the search for these cross-connections has been largely postponed until the operations of each primary pathway are elucidated. Once these cross-connections are added to the maps of these primary pathways, the drawings of signal transduction circuitry will likely resemble a weblike structure (see Supplementary Sidebar 6.8) rather than a series of parallel pathways that begin at the plasma membrane and extend linearly from there into the nucleus.

8. Distinct from these extensive cross-connections (such as those operating in the cytoplasm) there exists another dimension of combinatorial complexity: the most apparent endpoints of signaling—specific changes in cell phenotype achieved by transcriptional regulation of certain genes—are the results of interactions between multiple signaling pathways that converge in various combinations on specific gene promoters and their associated enhancers. Once again, this dimension of complexity continues to baffle us.

9. Our depiction of how signals are transmitted through a cell is likely to be fundamentally flawed. We have spoken here, time and again, about potent signaling pulses speeding along signal-transducing pathways and evoking, at endpoints, strong and clearly defined responses within a cell. In reality, each signaling cascade is likely to operate in a finely tuned, dynamic equilibrium, where positive and negative regulators continuously counterbalance one another. Accordingly, a signaling input (for example, a mitogenic stimulus) may operate like the plucking of a fiber in one part of a spider web, which results in small reverberations at distant sites throughout the web. Here, neither our language nor our existing mathematical representations of signaling suffice to clarify our understanding.

Undaunted by these obstructions to understanding, some cancer researchers would like to be able to draw a complete and accurate wiring diagram of the cell—the scheme that depicts how these pathways are interconnected. This project will only be accomplished many years from now. One measure of the difficulties associated with this task comes from censuses of the various classes of genes in the human genome. According to one enumeration, there are 518 distinct genes specifying various types of protein kinases; 40% of these genes make alternatively spliced mRNAs encoding slightly different protein structures, leading to a collection of more than 1000 distinct kinase proteins that may be present in human cells. Of the 518 kinase genes, 90 encode tyrosine kinases, the remainder being serine/threonine kinases. Among the 90 tyrosine kinases, 58 function as the signaling domains of growth factor receptors. These numbers provide some measure of the complexity of the signaling circuitry that underlies one component of cancer pathogenesis, since many of these kinases are involved in regulating the proliferation and survival of cells.

Another dimension of complexity has come from more detailed burrowing into the details of individual signaling circuits. While we have portrayed Ras signaling as focused on three downstream effectors (see Figure 6.9A), directed biochemical analyses have indicated that the Ras proteins command additional effectors to do their work (Supplementary Sidebar 6.9). It remains unclear whether these additional pathways contribute in critical ways to the transforming powers of the corresponding oncoproteins.

The difficulties of explaining cell behavior in terms of the design of signal transduction circuitry underlie our current inability to understand another set of problems, these being related specifically to the genetics of cancer cell genomes: Why do various types of human tumors display highly specific types of genetic alterations (**Figure 6.26**)? Why do 90% of pancreatic carcinomas carry a mutant K-*ras* oncogene, while this mutation is seen in only a small proportion (~10%) of breast cancers? Why are amplifications of the EGF receptor and its cousin, HER2/erbB2, encountered in more than a third of human breast cancers but seen uncommonly in certain other epithelial cancers, even though these receptors are widely displayed in many kinds of normal epithelia?

Why do approximately 50% of colon carcinomas carry a mutant K-*ras* gene, and how do the remaining 50% of these tumors acquire a comparable mitogenic signal? Some of these other tumors have mutant B-Raf or PI3K proteins. Still, this leaves many cancers with no apparent alterations affecting either the Ras → Raf → MAPK signaling cascade or the parallel PI3K → Akt/PKB pathway (a question that we revisit in Chapter 11). Do these tumors harbor still-undiscovered genetic lesions that activate one or the other pathway? Or, alternatively, are the cells in these tumors able to proliferate uncontrollably because of mechanisms that have nothing to do with the operations of these particular pathways?

Finally, there remains a major problem in clinical oncology that remains far from being solved: Why do certain combinations of mutant alleles and protein expression patterns in cancer cells portend good or bad prognoses for patients? Many of the responses to this question will surely derive from our detailed precise understanding of the elusive signal-processing circuitry operating in the cytoplasm. At present, we are largely limited to noting correlations between these patterns and clinical outcomes. But correlations hardly reveal the chains of causality connecting the genomes

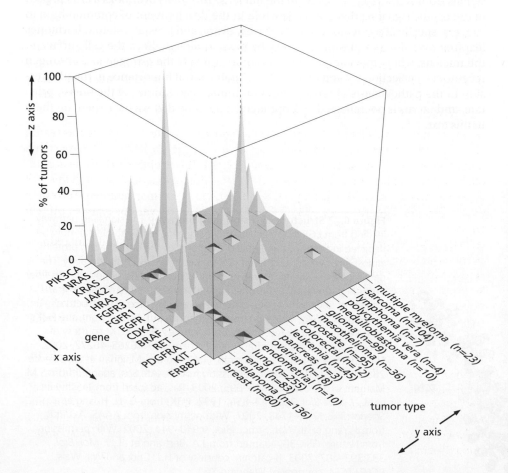

Figure 6.26 Frequencies of oncogene activation in various human tumors This three-dimensional histogram illustrates the great variability with which various commonly studied human oncogenes are found in mutant, activated forms in the genomes of human tumors. The percentage (*z axis*) of tumors of a given type (*arrayed along y axis*) that harbor mutant forms of a given oncogene (*arrayed along x axis*) reveals great differences in the involvement of each of these oncogenes in human tumor pathogenesis; the underlying mechanisms that explain these strong variations between tumor types are largely obscure. (From R.S. Thomas et al., *Nat. Genet.* 39:347–351, 2007. With permission from Nature.)

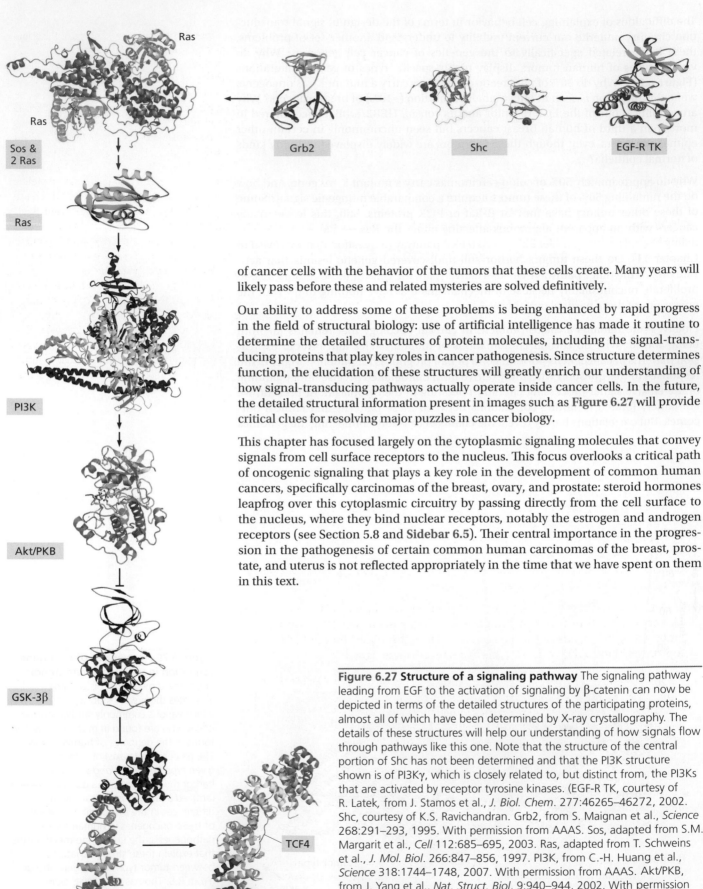

of cancer cells with the behavior of the tumors that these cells create. Many years will likely pass before these and related mysteries are solved definitively.

Our ability to address some of these problems is being enhanced by rapid progress in the field of structural biology: use of artificial intelligence has made it routine to determine the detailed structures of protein molecules, including the signal-transducing proteins that play key roles in cancer pathogenesis. Since structure determines function, the elucidation of these structures will greatly enrich our understanding of how signal-transducing pathways actually operate inside cancer cells. In the future, the detailed structural information present in images such as **Figure 6.27** will provide critical clues for resolving major puzzles in cancer biology.

This chapter has focused largely on the cytoplasmic signaling molecules that convey signals from cell surface receptors to the nucleus. This focus overlooks a critical path of oncogenic signaling that plays a key role in the development of common human cancers, specifically carcinomas of the breast, ovary, and prostate: steroid hormones leapfrog over this cytoplasmic circuitry by passing directly from the cell surface to the nucleus, where they bind nuclear receptors, notably the estrogen and androgen receptors (see Section 5.8 and **Sidebar 6.5**). Their central importance in the progression in the pathogenesis of certain common human carcinomas of the breast, prostate, and uterus is not reflected appropriately in the time that we have spent on them in this text.

Figure 6.27 Structure of a signaling pathway The signaling pathway leading from EGF to the activation of signaling by β-catenin can now be depicted in terms of the detailed structures of the participating proteins, almost all of which have been determined by X-ray crystallography. The details of these structures will help our understanding of how signals flow through pathways like this one. Note that the structure of the central portion of Shc has not been determined and that the PI3K structure shown is of PI3Kγ, which is closely related to, but distinct from, the PI3Ks that are activated by receptor tyrosine kinases. (EGF-R TK, courtesy of R. Latek, from J. Stamos et al., *J. Biol. Chem.* 277:46265–46272, 2002. Shc, courtesy of K.S. Ravichandran. Grb2, from S. Maignan et al., *Science* 268:291–293, 1995. With permission from AAAS. Sos, adapted from S.M. Margarit et al., *Cell* 112:685–695, 2003. Ras, adapted from T. Schweins et al., *J. Mol. Biol.* 266:847–856, 1997. PI3K, from C.-H. Huang et al., *Science* 318:1744–1748, 2007. With permission from AAAS. Akt/PKB, from J. Yang et al., *Nat. Struct. Biol.* 9:940–944, 2002. With permission from Nature. GSK-3β, adapted from J.A. Bertrand et al., *J. Mol. Biol.* 333:393–407, 2003. β-catenin, courtesy of H.J. Choi and W.I. Weis; BCL9+TCF4, courtesy of Wenqing Xu.)

Sidebar 6.5 The mysteries of breast and prostate cancer mitogenesis Breast and prostate cancers are commonly occurring diseases in the West that have been the subjects of extensive research, yet we still have only a vague understanding at the molecular level of how the growth of the majority of these tumors is controlled. A majority of breast cancers express the estrogen receptor (ER), while almost all prostate cancers express the androgen receptor (AR); both of these nuclear receptors act through mechanisms that are very different from those of the cell surface receptors we have studied (see Section 5.8). Like other "nuclear receptors," the ER binds its ligand, estrogen (more precisely, estradiol), and then proceeds to bind to specific DNA sequences in the promoters of certain target genes; this binding, followed by association with transcriptional co-regulators, leads to the transcription of these genes.

Estrogen acts as a critically important mitogen for ER-positive breast cancer cells. This is indicated by the fact that a number of "anti-estrogens" such as tamoxifen, which bind to the ER and block some of its transcription-activating powers, prevent the growth of ER-positive cells and tumors and can actually cause tumor regression. However, none of this seems to explain precisely how, at the biochemical level, estrogen and the ER drive the proliferation of ER-positive breast cancer cells. Some reports indicate that within minutes after estrogen is added to these cells, the Ras → MAPK signaling cascade is activated. This rapid response cannot be explained by the fact that ER acts as a nuclear transcription factor, since changes in nuclear gene transcription are not felt in the cytoplasm (in the form of proteins synthesized on newly formed mRNA templates) for at least half an hour.

One critical clue may come from the observation that some of the ligand-activated ER molecules are found to be tethered to cytoplasmic membranes rather than being localized to their usual sites of action in the nucleus; this association depends on the covalent modification of the ER via palmitoylation, much like the lipid tails attached to the C′-termini of Ras proteins. Following ligand binding, some of this cytoplasmic ER has been reported to be bound to Shc, the important SH2- and SH3-containing bridging protein that participates in Sos activation and Ras signaling (see Figure 6.7B); alternatively, some research reports physical association between the ER and either Src or RTKs in the cytoplasm depending on the epithelial versus mesenchymal phenotype of mammary cells. Moreover, in some cell types, nuclear ER via transcriptional regulation is key to estrogen-induced mitogenesis, whereas in others, the cytoplasmic palmitoylated form of ER appears to play a dominant role. This fragmentary description of a major mitogenic pathway is most unexpected, in light of the fact that breast cancer has, by now, been studied far more intensively than any other type of human malignancy.

Key concepts

- Discovery of Src's three homology domains (SH1, 2, and 3) led to the following model of tyrosine kinase receptor (RTK) signaling cascades: ligand-induced trans-phosphorylation results in an RTK displaying on its cytoplasmic tail an array of phosphotyrosines, each of which attracts and binds a specific SH2-containing cytoplasmic protein. Such relocalization permits the SH2-containing proteins to interact with plasma membrane–associated proteins that can activate their own downstream targets or with other signal-transducing molecules that pass on activating signals to yet other targets.

- One such signal-transducing protein tethered to the plasma membrane is Ras. Linker proteins form a physical bridge between ligand-activated RTKs and Sos, the latter acting as a guanine nucleotide exchange factor (GEF) that induces Ras to replace its bound GDP with GTP, thereby activating Ras signaling.

- Three major downstream signaling cascades emanate from activated Ras via binding of its effector loop with its main downstream signaling partners—Raf kinase, phosphatidylinositol 3-kinase (PI3K), and Ral-GEF. Yet other downstream effectors that bind to activated Ras appear to play secondary roles in Ras signaling.

- Raf phosphorylates residues on MEK, activating the latter. MEK then phosphorylates and activates the extracellular signal–regulated kinases 1 and 2 (ERK 1 and 2), which are classified as MAPKs (mitogen-activated protein kinases). The Raf signaling cascade, often called a MAPK pathway, is the foremost mitogenic pathway in mammalian cells, both normal and neoplastic.

- The phosphatidylinositol 3-kinase (PI3K) pathway depends on kinases phosphorylating a membrane-embedded phosphatidylinositol, converting it to PIP3. Once formed, the latter attracts molecules (such as Akt) that carry a pleckstrin homology (PH) domain. The resulting activation of Akt activates a series of downstream proteins. PIP3 levels are normally kept low in cells by phosphatases, notably PTEN.

- In the third Ras-regulated pathway, Ras binds Ral-GEFs, causing localization of Ral-GEF near the inner surface of the plasma membrane. Ral-GEF then induces nearby Ral to exchange its GDP for GTP, enabling Ral to activate downstream targets. This has multiple effects on the cell, including changes in the cytoskeleton and cellular motility.

- Cytokine receptors use noncovalently bonded TKs of the JAK class to phosphorylate STATs. The STATs form dimers that migrate to the nucleus, where they serve as transcription factors.

- Integrins bind components of the ECM, leading to the formation of focal adhesions and activation of focal adhesion kinase (FAK), which is associated with the integrin cytoplasmic tail. Transphosphorylation of FAK and its recruitment of SH2-containing molecules activates many of the same signaling pathways as RTKs.

- The pathway controlled by Wnt factors enables cells to remain in a relatively undifferentiated state—an important attribute of certain cancer cells. Wnt, acting via Frizzled receptors, prevents glycogen synthase kinase-3β (GSK-3β) from tagging its substrates, including β-catenin, for destruction. The spared β-catenin moves into the nucleus, where it activates transcription of key mitogenic genes and contributes to stemness.

- Ligand-bound G-protein–coupled receptors (GPCRs) activate cytoplasmic heterotrimeric G proteins whose α subunit responds by exchanging its GDP for GTP. The Gα subunit then dissociates from its partners (Gβ + Gγ) and affects a number of cytoplasmic enzymes, while the Gβ + Gγ dimer activates its own effectors, including PI3Kγ, Src, and Rho-GEFs.

- Among the pathways relying on dual-address proteins is the nuclear factor-κB (NF-κB) signaling system. It depends on liberation of cytoplasmic NF-κB homo- and heterodimers from sequestration by IκBs. Following migration of NF-κB subunits into the nucleus, these proteins function as transcription factors that activate expression of at least 150 genes, including cell proliferation and antiapoptotic genes.

- The cleavage of Notch at two sites following ligand binding liberates a cytoplasmic fragment that migrates to the nucleus and functions as part of a transcription factor complex. Notch aberrations are seen in the majority of cervical carcinomas.

- When Hedgehog binds a Patched receptor, Smoothened is released from inhibition and emits signals that protect cytoplasmic Gli protein from cleavage. The resulting intact Gli migrates to the nucleus, where it activates transcription, whereas cleaved Gli acts as a transcriptional repressor.

- The binding of TGF-β to its receptors causes phosphorylation of cytoplasmic Smad proteins and their dispatch to the nucleus, where they collaborate with other transcription factors to activate a large contingent of genes. This pathway plays a major role in the pathogenesis of many carcinomas, both in the early stages when TGF-β fails to arrest cell proliferation and later, when it contributes to tumor cell invasiveness.

- YAP/TAZ signaling represents a major intracellular hub for integrating a diverse array of signals, many of them affected by the physical surroundings of a cell. These transcription factors are normally phosphorylated and sequestered in the cytoplasm and when dephosphorylated migrate to the nucleus, where they drive, among other things, cell proliferation.

Thought questions

1. What molecular mechanisms have evolved to ensure that the signals coursing down a signaling cascade reach the proper endpoint targets rather than being broadcast nonspecifically to "unintended" targets in the cytoplasm?

2. How many distinct molecular mechanisms can you cite that lead to the conversion of a proto-oncogene into an active oncogene?

3. Why are point mutations in *ras* oncogenes confined so narrowly to a small number of nucleotides, while point mutations in genes encoding other cancer-related proteins are generally distributed far more broadly throughout the reading frames?

4. What factors determine the lifetime of the activated state of a Ras oncoprotein?

5. Treatment of cells with proteases that cleave the ectodomain of E-cadherin often result in rapid changes in gene expression. How might you rationalize this response, given what we know about this cell surface protein?

6. What mechanisms have we encountered that ensure that a signal initiated by a growth factor receptor can be greatly amplified as the signal is transduced down a signaling cascade in the cytoplasm? Conversely, what signaling cascade(s) strongly limit the possible amplification of a signal initiated at the cell surface?

7. What quantitative parameters describing individual signal-transducing proteins will need to be determined before the behavior of the signaling cascade formed by these proteins can be predicted through mathematical modeling?

Additional reading

Aebersold R, Agar JN, Amster IJ et al. (2018) How many human proteoforms are there? *Nature Chem. Biol.* 14, 206–214.

Alvarez-Garcia V, Tawil Y, Wise HM & Leslie NR (2019) Mechanisms of PTEN loss in cancer: it's all about diversity. *Semin. Cancer Biol.* 59, 66–79.

Angers S & Moon RT (2009) Proximal events in Wnt signal transduction. *Nat. Rev. Mol. Cell Biol.* 10, 468–477.

Aster JC, Pear WS & Blacklow SC (2017) The varied roles of Notch in cancer. *Annu. Rev. Pathol.* 12, 245–275.

Avraham R & Yarden Y (2011) Feedback regulation of EGFR signalling: decision making by early and delayed loops. *Nat. Rev. Mol. Cell Biol.* 12, 104–117.

Bangs F & Anderson KV (2017) Primary cilia and mammalian Hedgehog signaling. *Cold Spring Harb. Perspect. Biol.* 9, a028175; doi: 10.1101 /cshperspect.a028175.

Bos JL, Rehmann H & Wittinghofer A (2007) GEFs and GAPs: critical elements in the control of small G proteins. *Cell* 129, 865–877.

Cadigan KM & Waterman ML (2012) TCF/LEFs and Wnt signaling in the nucleus. *Cold Spring Harb. Perspect. Biol.* 4, a007906.

Castellano E & Downward J (2011) RAS interaction with PI3K: more than just another effector pathway. *Genes Cancer* 2, 261–274.

Castellano E & Santos E (2011) Functional specificity of Ras isoforms: so similar but so different. *Genes Cancer* 2, 216–231.

Chalhoub N & Baker SJ (2009) PTEN and the PI3-kinase pathway in cancer. *Annu. Rev. Pathol.* 4, 127–150.

Clevers H, Loh KM & Nusse R (2014) Stem cell signaling. An integral program for tissue renewal and regeneration: Wnt signaling and stem cell control. *Science* 346, 1248012.

Clevers H & Nusse R (2012) Wnt/β-catenin signaling and disease. *Cell* 149, 1192–1205.

Colicelli J (2004) Human Ras superfamily and related GTPases. *Sci. STKE* 250, re 13; doi: 10.1126/stke.2502004re13.

Cooper J & Giancotti FG (2019) Integrin signaling in cancer: mechanotransduction, stemness, epithelial plasticity, and therapeutic resistance. *Cancer Cell* 35, 347–367.

Daulat AM & Borg JP (2017) Wnt/planar cell polarity signaling: new opportunities for cancer treatment. *Trends Cancer* 3, 113–125.

DiDonato JA, Mercurio F & Karin M (2012) NF-κB and the link between inflammation and cancer. *Immunol. Rev.* 246, 379–400.

Dobrokhotov O, Samsonov M, Sokabe M & Hirata H (2018) Mechanoregulation and pathology of YAP/TAZ via Hippo and non-Hippo mechanisms. *Clin. Transl. Med.* 7, 23–36.

Endicott JA, Noble MEM & Johnson LN (2012) The structural basis for control of eukaryotic protein kinases. *Annu. Rev. Biochem.* 81, 587–613.

Fey D, Matallanas D, Rauch J et al. (2016) The complexities and versatility of the RAS-to-ERK signalling system in normal and cancer cells. *Semin. Cell Dev. Biol.* 58, 96–107.

Fleuren EDG, Zhang L, Wu J & Daly RJ (2016) The kinome 'at large' in cancer. *Nat. Rev. Cancer* 16, 83–98.

Fortini ME (2009) Notch signaling: the core pathway and its posttranslational regulation. *Dev. Cell* 26, 633–647.

Frodyma D, Neilsen B, Costanzo-Garvey D et al. (2017) Coordinating ERK signaling via the molecular scaffold kinase suppressor of Ras. *F1000Res.* 6, 11895; doi: 10.12688/f1000research. 11895.

Gentry LR, Martin TD, Reiner DJ & Der CJ (2014) Ral small GTPase signaling and oncogenesis: more than just 15 minutes of fame. *Biochim. Biophys. Acta* 1843, 2976–2988.

Good MC, Zalatan JG & Lim WA (2011) Scaffold proteins: hubs for controlling the flow of cellular information. *Science* 332, 680–686.

Green J, Nusse R & van Amerongen R (2014) The role of Ryk and Ror receptor tyrosine kinases in Wnt signal transduction. *Cold Spring Harb. Perspect. Biol.* 6, a009175.

Guo X & Wang XF (2009) Signaling cross-talk between TGF-β/BMP and other pathways. *Cell Res.* 19, 71–88.

Hatakeyama M (2017) Structure and function of *Helicobacter pylori* CagA, the first identified bacterial protein involved in human cancer. *Proc. Jpn. Acad. Ser.* B 93, 196–219.

Hayden MS & Ghosh S (2012) NF-κB, the first quarter-century: remarkable progress and outstanding questions. *Genes Dev.* 26, 203–234.

Hayden MS & Ghosh S (2014) Regulation of NF-κB by TNF cytokines. *Semin. Immunol.* 26, 253–266.

Hennig A, Markwart R, Esparza-Franco M et al. (2015) Ras activation revisited: role of GEF and GAP systems. *Biol. Chem.* 396, 831–848.

Hill CS (2016) Transcriptional control by the SMADs. *Cold Spring Harb. Perspect. Biol.* 8, a22079.

Hori K, Sen A & Artavanis-Tsakonas S (2013) Notch signaling at a glance. *J. Cell Sci.* 126, 2135–2140.

Hun HD, Kim DH, Jeong H-S & Park HW (2019) Regulation of TEAD transcription factors in cancer biology. *Cells* 8, 600–621.

Iqbal J, Sun L & Zaidi M (2010) Complexity in signal transduction. *Ann. N.Y. Acad. Sci.* 1192, 238–244.

Kar S, Deb M, Sengupta D et al. (2012) Intricacies of hedgehog signaling pathways: a perspective in tumorigenesis. *Exp. Cell Res.* 318, 1959–1972.

Kashatus DF (2013) Ral GTPases in tumorigenesis: emerging from the shadows. *Exp. Cell Res.* 319, 2337–2342.

Katoh M (2017) Canonical and non-canonical WNT signaling in cancer stem cells and their niches: cellular heterogeneity, omics reprogramming, targeted therapy and tumor plasticity. *Int. J. Oncol.* 51, 1357–1369.

Lau AW, Fukushima H & Wei W (2012) The Fbw7 and beta-TRCP E3 ubiquitin ligases and their roles in tumorigenesis. *Front. Biosci.* 17, 2197–2212.

Lavoie H & Therrien M (2015) Regulation of RAF protein kinases in ERK signalling. *Nat. Rev. Mol. Cell Biol.* 16, 281–298.

Lawson CD & Ridley AJ (2018) Rho GTPase signaling complexes in cell migration and invasion. *J. Cell Biol.* 217, 447–457.

Lazzara MJ & Lauffenburger DA (2009) Quantitative modeling perspectives on the ErbB system of cell regulatory processes. *Exp. Cell Res.* 315, 717–725.

Lee M & Yaffe M (2016) Protein regulation in signal transduction. *Cold Spring Harb. Perspect. Biol.* 7, a005918.

Lie EC, Dibble CC & Toker A (2017) PI3K signaling in cancer: beyond AKT. *Curr. Opin. Cell Biol.* 45, 62–71.

Liu T, Zhang L, Joo D & Sun S-C (2017) NF-κB signaling in inflammation. *Signal Transduct. Target. Ther.* 2, 17023.

Lloyd AC (2013) The regulation of cell size. *Cell* 154, 1194–1205.

Misra JR & Irvine KD (2018) The Hippo signaling network and its biological functions. *Annu. Rev. Genet.* 52, 65–87.

Mattila PK & Lappalainen P (2008) Filopodia: molecular architecture and cellular functions. *Nat. Rev. Mol. Cell Biol.* 9, 446–454.

Neben CL, Lo M, Jura N & Klein OD (2017) Feedback regulation of RTK signaling in development. *Dev. Biol.* 447, 71–89.

Neilsen BK, Frodyma DE, Roberts RE & Fisher KW (2017) KSR as a therapeutic target for Ras-dependent cancers. *Expert Opin. Ther. Targets* 21, 499–509.

Nieto Gutierrez A & McDonald PH (2018) GPCRs: emerging anti-cancer drug targets. *Cell Signal.* 41, 65–74.

Nusse R & Clevers H (2017) Wnt/β-catenin signaling, disease and emerging therapeutic modalities. *Cell* 169, 985–999.

O'Hayre M, Degese MS & Gutkind JS (2014) Novel insights into G protein and G protein-couple receptor signaling in cancer. *Curr. Opin. Cell Biol.* 27, 126–135.

O'Hayre M, Vázquez-Prado J, Kufareva I et al. (2013) The emerging mutational landscape of G proteins and G-protein-coupled receptors in cancer. *Nat. Rev. Cancer* 13, 412–424.

Ranganathan P, Weaver KL & Capobianco AJ (2011) Notch signalling in solid tumors: a little bit of everything but not all the time. *Nat. Rev. Cancer* 11, 338–351.

Roberts PJ & Der CJ (2007) Targeting the Raf-MEK-ERK mitogen-activated protein kinase cascade for the treatment of cancers. *Oncogene* 26, 3291–3310.

Roy JR, Halford MM & Stacker SA (2018) The biochemistry, signaling and disease relevance of RYK and other WNT-binding receptor tyrosine kinases. *Growth Factors* 36, 15–40.

Saxton RA & Sabatini DM (2017) mTOR signaling in growth, metabolism and disease. *Cell* 168, 960–976.

Sever R & Brugge JS (2014) Signaling in cancer. *Cold Spring Harb. Perspect. Med.* 5: A006098.

Shi Y & Massagué J (2003) Mechanisms of TGF-β signaling from cell membrane to the nucleus. *Cell* 113, 685–700.

Shirakawa R & Horiuchi H (2015) Ral GTPases: crucial mediators of exocytosis and tumourigenesis. *J. Biochem.* 157, 285–299.

Shostak K & Chariot A (2015) EGFR and NF-κB: partners in cancer. *Trends Mol. Med.* 21, 385–393.

Simanshu DK, Nissley DV & McCormick F (2017) Ras proteins and their regulators in human disease. *Cell* 170, 17–33.

Stark GR, Cheon H & Wang Y (2018) Responses to cytokines and interferons that depend upon JAKs and STATs. *Cold Spring Harb. Perspect. Biol.* 10, a028555.

Stephen AG, Esposito D, Bagni RK & McCormick F (2014) Dragging Ras back in the ring. *Cancer Cell* 25, 272–281.

Stricker S, Rauschenberger V & Schambony A (2017) ROR-family receptor tyrosine kinases. *Curr. Top. Dev. Biol.* 123, 105–142.

Thorpe LM, Yuzugullu H & Zhao JJ (2015) PI3K in cancer: divergent roles of isoforms, modes of activation and therapeutic targeting. *Nat. Rev. Cancer* 15, 7–24.

Vidal M (2016) How much of the human protein interactome remains to be mapped? *Sci. Signal.* 9, eg7.

Volinsky N & Kholodenko BN (2013) Complexity of receptor tyrosine kinase signal processing. *Cold Spring Harb. Perspect. Biol.* A009043.

Wagner MJ, Stacey MM, Liu BA & Pawson T (2013) Molecular mechanisms of SH2- and PTB-domain-containing proteins in receptor tyrosine kinase signaling. *Cold Spring Harb. Perspect. Biol.* A008987.

Wennerberg K, Rossman KL & Der CJ (2005) The Ras superfamily at a glance. *J. Cell Sci.* 118, 843–846.

Worby CA & Dixon JE (2014) PTEN. *Annu. Rev. Biochem.* 83, 641–649.

Xia Y, Shen S & Verma IM (2014) NF-κB, an active player in human cancers. *Cancer Immunol. Res.* 2, 823–830.

Yan K, Kuti M & Zhou MM (2002) PTB or not PTB—that is the question. *FEBS Lett.* 513, 67–70.

Yu F-X, Zhao B & Guan K-L (2015) Hippo pathway in organ size control, tissue homeostasis, and cancer. *Cell* 163, 811–828.

Yu H, Lee H, Herrmann A et al. (2014) Revisiting STAT3 signalling in cancer: new and unexpected biological functions. *Nat. Rev. Cancer* 14, 736–746.

Zanconato F, Cordenonsi M & Piccolo S (2016) YAP/TAZ at the roots of cancer. *Cancer Cell* 29, 783–803.

Zhang Q, Lenardo MJ & Baltimore D (2017) 30 years of NF-κB: a blossoming of relevance to human pathobiology. *Cell* 168, 37–57.

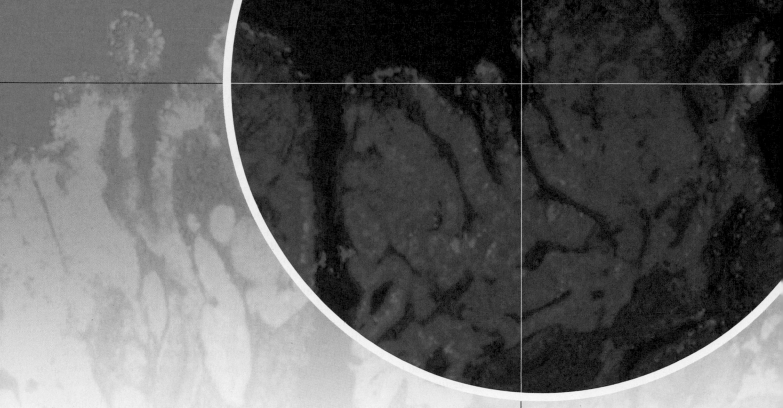

Chapter 7

Tumor Suppressor Genes

Let me add . . . a consideration of the inheritance of tumors. . . . In order
that a tumor may arise in such cases, the homologous elements in both
series of chromosomes must be weakened in the same way.

Theodor Boveri, pathologist, 1914

The discovery of proto-oncogenes and oncogenes, as described in Chapter 6, provided a simple and powerful explanation of how the proliferation of cells is driven. The proteins encoded by proto-oncogenes participate in various ways in receiving and processing growth-stimulating signals that originate in the extracellular environment. When these genes suffer mutation, the flow of these growth-promoting signals becomes deregulated. Instead of emitting them in carefully controlled bursts, oncoproteins release a steady stream of growth-stimulating signals, resulting in the unrelenting proliferation associated with cancer cells.

The logic underlying well-designed control systems dictates, however, that the components promoting a process must be counterbalanced by others that oppose this process. Biological systems seem to follow this logic as well, which leads us to conclude that the growth-promoting genes we have discussed until now provide only part of the story of cellular growth control.

In the 1970s and early 1980s, certain pieces of experimental evidence about cancer cell genetics began to accumulate that were hard to reconcile with the known properties of oncogenes. This evidence hinted at the existence of a second, fundamentally different type of growth-controlling gene—one that operates to constrain or suppress cell proliferation. These antigrowth genes came to be called **tumor suppressor genes** (TSGs). Their involvement in tumor formation seemed to happen when the genes were inactivated or lost, thereby liberating affected cells from their growth-suppressing effects.

Now, more than four decades later, we have come to realize that the inactivation of TSGs plays a role in cancer pathogenesis that is as important to cancer as the activation

Movies in this chapter

7.1 Intestinal Crypt and
APC Loss

233

of oncogenes. Indeed, the inactivation of these growth-suppressing genes may be even more important than oncogene activation for the formation of many kinds of human cancer.

7.1 Cell fusion experiments indicate that the cancer phenotype is recessive

The study of tumor viruses in the 1970s revealed that these infectious agents carried a number of cancer-inducing genes, specifically viral oncogenes, which acted in a dominant fashion when viral genomes were introduced into previously normal cells (see Chapter 3). In particular, the introduction of a tumor virus genome into a normal cell would result in transformation of that cell. This response meant that viral oncogenes could dictate cell behavior in spite of the continued presence and expression of opposing cellular genes within the virus-infected cell that usually functioned to ensure controlled, appropriate cell proliferation. Because the viral oncogenes could overrule these cellular genes, by definition the viral genes displayed a *dominant* phenotype—they were bringing about cellular transformation. This observation suggested, by extension, that cancerous cell growth was a dominant phenotype when compared to normal (wild-type) cell growth, which was therefore considered to be *recessive* (see Section 1.1).

However, as was suspected at the time, and reinforced by research in the 1980s, most human cancers do not seem to arise as consequences of tumor virus infections. Thus, the lessons learned from studying such viruses might well be irrelevant to understanding how human tumors arise. Hence, this issue needed to be revisited, making it possible that human cancer cells might even exhibit a recessive phenotype.

An initial hint of the importance of such recessive cancer-inducing alleles came from experiments undertaken initially at Oxford University in Great Britain. Based on the principles of Mendelian genetics, when an allele that specifies a dominant phenotype is juxtaposed with an allele specifying a recessive phenotype, the dominant allele, by definition, wins out and the cell will display the phenotype associated with the dominant allele. Such battles for phenotypic domination, it turns out, can be engineered experimentally by a technique called *cell fusion*. In this procedure, cells of two different phenotypes—for example, normal cells and cancer cells—are cultured together in a Petri dish. An agent is then used to induce fusion of the plasma membranes of cells that happen to be growing near one another. The result of such treatment is a large cell with a single cytoplasm and multiple nuclei, often termed a **syncytium**. Because these fusing agents act indiscriminately, they could join together cells of the same type—a pair of normal cells, for example—or cells of different types. If the participating cells happen to be of different origins, then the resulting hybrid cell is termed a **heterokaryon** (**Figure 7.1A**), to indicate that it carries genetically distinct nuclei. Genetic techniques can then be used to select for a fused cell that has acquired two genetically distinct nuclei and, at the same time, to eliminate cells that have failed to fuse or happen to carry two identical nuclei. For example, each type of cell may carry a genetic marker that allows it to resist being killed by a particular antibiotic agent to which it would normally be sensitive; heterokaryons would therefore be resistant to killing by a mixture of the two antibiotics.

Cells having large numbers of nuclei are generally inviable, whereas cells having only two nuclei are often viable and will proceed to grow. When these cells divide, the two nuclear membranes will break down, the two sets of chromosomes will intermingle, and each resulting daughter cell will receive a single nucleus with chromosome complements originating from both parental cell types.

Using this cell fusion technique, hybrid cells of disparate origins, including even chicken–human hybrids, could be made. Such hybrids were used to address many interesting biological questions. For our purposes, however, the results of only one type of experiment are of special interest: How did normal + cancer cell hybrids grow in culture and were they capable of forming tumors (Figure 7.1B)?

The smart money at the time bet that cancer, a potent, dominating phenotype, would come out on top when forced into a competition with the normal cell growth phenotype.

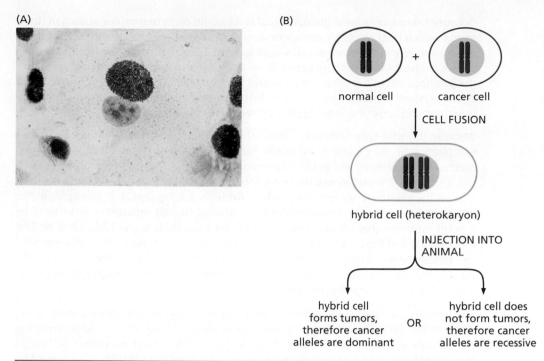

Figure 7.1 Experimental fusion of cells (A) When cells growing adjacent to one another in culture are exposed to a *fusogenic* agent, such as the chemical polyethylene glycol (PEG) or a viral protein that triggers membrane fusion, they initially form a heterokaryon with multiple nuclei. In this experiment, radiolabeled mouse NIH 3T3 cells have been fused (using an inactivated Sendai virus) with monkey kidney cells. The resulting hybrid cell contains two distinct nuclei. The larger mouse nucleus is identified by the silver grains that were formed during subsequent autoradiography. The membrane surrounding this heterokaryon cannot be seen. (B) The use of cell fusion to test the dominance or recessiveness of the cancer cell phenotype can, in theory, yield two different outcomes: tumorigenic or non-tumorigenic hybrid cells. In fact, when cancer cells derived from most kinds of non-virus-induced human tumors (or from chemically induced rodent tumors) were used in these fusions, the hybrid cells were non-tumorigenic. In contrast, when the cancer cell derived from a virus-induced tumor, the hybrids were usually tumorigenic. (A, courtesy of S. Rozenblatt.)

However, in a number of experiments in which tumor cells were fused with normal cells, the hybrid cells that initially formed were found to have lost the ability to form tumors when injected into an appropriate host animal. This unexpected observation was interpreted to mean that the malignant, cancer cell phenotype was recessive to the phenotype of normal, wild-type growth. Alternative explanations could be proposed to explain this result: perhaps the dominant oncogene from the tumor cell failed to be expressed in the hybrid cell, or the hybrid cell might have been eliminated by the immune system of the host animal. Nevertheless, these early experiments were very influential in introducing the concept that the malignant phenotype might behave as a recessive trait.

There was, however, one notable exception to these observations: when the parental cancer cell in the two-cell hybrid had been transformed by tumor virus infection, the malignant phenotype would dominate (see Figure 7.1B). This behavior makes sense, given that tumor viruses are able to deploy the virus-associated oncogenes to transform normal cells. Such viral oncogenes are able, in a variety of ways, to dominate the behavior of the cells that they have infected and thereby act as dominant alleles (with respect to normal host cell genes that regulate related cell functions).

7.2 The recessive nature of the cancer cell phenotype requires a genetic explanation

The results of the cell-fusion experiments led to the notion that normal cells carry functional alleles of genes that inhibit malignant growth and that these growth-inhibitory genes are somehow inactivated or lost in cancer cells.

Because human cells are diploid, normal cells would carry two copies of each of these genes, which help constrain or suppress their proliferation. During the development of a tumor, an evolving cancer cell would have to discard or otherwise inactivate both copies of one or more of these genes. Once these growth-suppressing genes were lost, the proliferation of the cancer cells would likely accelerate. However, re-introducing intact versions of these genes by cell fusion would restore their growth-inhibitory functions and cause the hybrid cells to slow or halt proliferation.

Because the wild-type versions of these (hypothetical) genes antagonize the cancer cell phenotype, they came to be called "tumor suppressor genes." There were arguments both in favor of and against the existence of TSGs. The argument in their favor was that it is far easier to inactivate a gene by various mutational mechanisms than it is to hyperactivate its protein product through a mutation. For example, a *Ras* proto-oncogene can be converted into its hyperactivated oncogenic form only by a point mutation that affects certain specific codons, such as the 12th, 13th, or 61st (Section 4.3 and Figure 4.6). In contrast, a TSG—or, for that matter, any other gene—can readily be inactivated by point mutations that strike at many different sites in its protein-coding sequences, by insertions or deletions that alter its reading frame, or by large-scale chromosomal deletions.

The logical case against the existence of TSGs derived from the diploid state of the mammalian cell genome. If inactivation of TSGs really did play a role in enabling the growth of cancer cells, and if these inactive alleles were recessive, an incipient tumor cell would seemingly reap no benefit from inactivating only one of its two copies of a TSG because a dominant, wild-type copy would remain. Hence, it seemed that both copies of a given TSG would need to be eliminated to produce a malignant phenotype.

This requirement for two separate genetic alterations to the same gene seemed too complex and unwieldy and too improbable to occur during the lifetime of a cell. Because the likelihood of two mutations occurring is the square of the probability of a single mutation, this made it seem highly unlikely that TSGs could be fully inactivated in the time required for tumors to form. For example, if the probability of inactivating a single gene copy by mutation is on the order of 10^{-6} per cell generation, the probability of silencing both copies would be on the order of 10^{-12} per cell generation—exceedingly unlikely to occur in the early phases of tumor development.

7.3 The retinoblastoma tumor provides a solution to the genetic puzzle of TSGs

The arguments for and against the involvement of TSGs in tumor development could never be settled by the cell fusion technique. For one thing, this experimental strategy, on its own, offered little prospect of finding and isolating specific genes whose properties could be studied and used to support the TSG hypothesis. In the end, the important insights came from studying a rare childhood eye tumor, **retinoblastoma**.

Observed in about 1 in 20,000 children, this tumor of the retina arises in the precursors of photoreceptor cells (**Figure 7.2**). It is diagnosed anytime from birth to ages 6 to 8 years, after which the disease is rarely encountered. Retinoblastoma appears in two forms. Children who are born into families with no history of the disease present in the clinic with a single tumor in one eye, a condition termed unilateral retinoblastoma. If this tumor is eliminated—by radiation, **cryoablation**, or by removal of the affected eye—then the child usually has no further risk of retinoblastoma and no elevated risk of tumors elsewhere in the body. This form of retinoblastoma is considered **sporadic**.

A **familial** form of retinoblastoma occurs in children with a parent who also suffered from and was cured of the disease early in life. These children usually experience tumors arising in multiple sites in both eyes, a condition called *bilateral* retinoblastoma. Removal of these tumors does not protect these children from a greatly increased risk (more than 500 times above normal) of bone cancers (osteosarcomas) during adolescence and an elevated susceptibility to developing yet other,

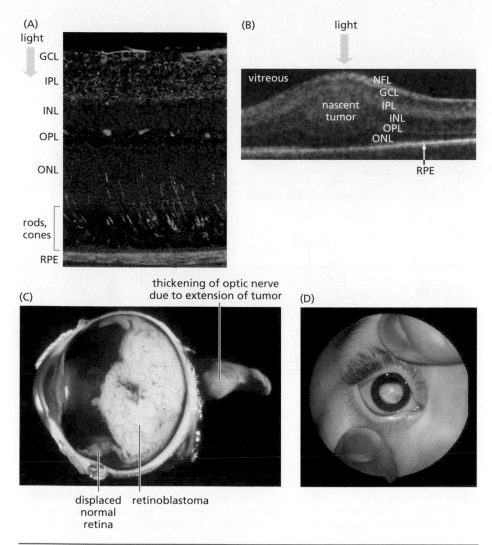

Figure 7.2 Pediatric retinoblastomas Although the *Rb* gene regulates cell proliferation in most cell types throughout the body, inactivation of this TSG leads preferentially to the formation of pediatric tumors in the retina—a small, highly specialized tissue (A). The tumors arise from a stem cell precursor of the cone cells. In this micrograph, which shows a cross section of the retina, the nuclei of the rods and cones (*rcd*) are located in the outer nuclear layer (ONL), and their cell bodies (*dark green*) extend downward toward the retinal pigmented epithelium (RPE). Closer to the front of the eye are the outer plexiform layer (OPL), inner nuclear layer (INL), inner plexiform layer (IPL), and ganglion cell layer (GCL). (B) Created using Fourier domain optical coherence tomography, this image shows a nascent retinoblastoma in the eye of an *Rb*[+/−] infant. The developing tumor appears to be emerging from the INL of the retina. NFL, nerve fiber layer. (C) A section of an eye in which a large retinoblastoma has developed and begun to invade the optic nerve. (D) A retinoblastoma may often initially present in the clinic as an opacity that obscures the retina. (A, from S.W. Wang et al., *Development* 129:467–477, 2002. With permission from the Company of Biologists. B, courtesy of A. Mallipatna, C. Vandenhoven, B.L. Gallie and E. Heon. C and D, courtesy of T.P. Dryja.)

nonretinal tumors later in life (**Figure 7.3A**). Those who survive these tumors and grow to adulthood are usually able to reproduce; in half of their offspring, the familial form of retinoblastoma again rears its head (Figure 7.3B).

Upon studying the kinetics with which retinal tumors appeared in children affected by both forms of the disease, Alfred Knudson concluded in 1971 that the rate of appearance of familial tumors was consistent with a single random event, whereas the sporadic tumors behaved as if two random events were required for their formation. These postulates have become known as Knudson's "one-hit" and "two-hit" hypotheses (**Figure 7.4**).

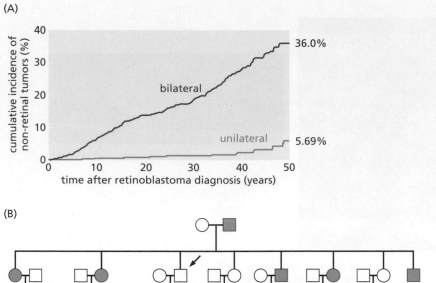

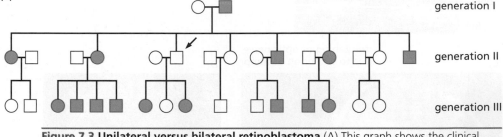

Figure 7.3 Unilateral versus bilateral retinoblastoma (A) This graph shows the clinical courses of 1601 retinoblastoma patients who had been diagnosed between 1914 and 1984. Children diagnosed with bilateral tumors—those with familial retinoblastoma (*red line*)—have a dramatically higher risk of subsequently developing tumors in a variety of organ sites than those with unilateral tumors, who most often have the sporadic form of the disease (*blue line*). (A portion of this elevated risk is attributable to tumors that arose in the vicinity of the eyes because of the *radiotherapy* that was used to eliminate the retinoblastomas when these individuals were young.) (B) This pedigree shows multiple generations of a family with familial retinoblastoma. Males (*squares*), females (*circles*), affected individuals (*green*), unaffected individuals (*hollow shapes*). In one lineage, tumor development skipped a generation (*arrow*); this represents an example of *incomplete penetrance*, in which an individual carrying the mutant allele fails, for complex biological reasons, to develop retinoblastoma. (We can conclude that the indicated man carried the mutant allele because two of his daughters developed the disease.) Multi-generation pedigrees like this one were rarely observed before the advent of modern medicine, which allowed affected children to be cured of the disease, reach reproductive age, and transmit mutant alleles to the next generation. (A, from R.A. Kleinerman et al., *J. Clin. Oncol.* 23:2272–2279, 2005. With permission from the American Society of Clinical Oncology. B, courtesy of T.P. Dryja.)

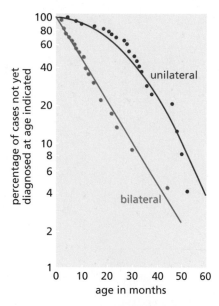

Figure 7.4 Kinetics of appearance of unilateral and bilateral retino-blastomas in children Knudson calculated that the bilateral tumors arose with one-hit kinetics (*blue*), whereas the unilateral tumors arose with two-hit kinetics (*red*). Each of these hits was presumed to represent a somatic mutation. It was realized only later that these two "hits" inactivated two copies of the same gene—*Rb*. (From A.G. Knudson Jr., *Proc. Natl. Acad. Sci. USA* 68:820–823, 1971. With permission from the author.)

The kinetics of tumor formation led to the deduction (or speculation, to be honest) that the development of retinoblastoma depended on the full loss of function of a TSG, which became known as the retinoblastoma gene or *Rb*. For children who inherit a defective allele of the *Rb* gene from one parent, a mutant version of the *Rb* gene will be present in all of the cells of the developing embryo, including the cells of the retina (**Figure 7.5**). Any one of these retinal cells needs to sustain only a single (somatic) mutation to inactivate the remaining wild-type allele; such cells, fully lacking this TSG function, could then spawn a retinoblastoma. Given the need for only one somatic mutational event and the large number of vulnerable retinal cells, children with the familial form of the disease typically develop multiple tumors usually affecting both eyes, that is, the bilateral form of the disease.

Of note, some children with no family history of the disease also develop bilateral retinoblastoma and other cancers characteristic of the familial syndrome. In these cases, a mutation must have arisen during the formation of a parental germ cell. The child would thus inherit a defective allele that would be present in all cells of the body and, like the children with the familial form of the disease, have a higher risk of tumor development.

The dynamics differ for children with the sporadic form of the disease. In these cases, both copies of the *Rb* gene must be inactivated independently, one after the other, before a retinal cell can reach the uncontrolled proliferation that eventually results

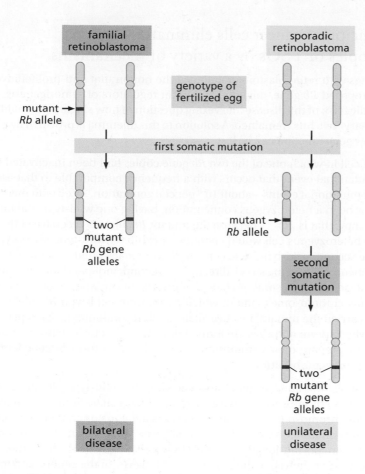

familial retinoblastoma

sporadic retinoblastoma

genotype of fertilized egg

mutant *Rb* allele

first somatic mutation

two mutant *Rb* gene alleles

mutant *Rb* allele

second somatic mutation

two mutant *Rb* gene alleles

bilateral disease

unilateral disease

Figure 7.5 Dynamics of retinoblastoma formation In familial retinoblastoma (*left*), the fertilized egg from which a child develops carries one defective copy of the *Rb* gene; all of this child's retinal cells will therefore carry only a single functional *Rb* gene copy. If this surviving copy of the *Rb* gene is eliminated in a retinal cell by a somatic mutation, that particular cell will lack all *Rb* gene function and will be poised to proliferate into a mass of tumor cells. In sporadic retinoblastoma (*right*), the fertilized egg is genetically wild type at the *Rb* locus, and retinoblastoma development therefore will require two successive somatic mutations striking the two copies of the *Rb* gene carried by a lineage of retinal precursor cells. Because only a single somatic mutation is needed to eliminate *Rb* function in familial cases, multiple cells in both eyes are typically affected. However, the convergence of two somatic mutations—each a rare event—on a single cell, thus its descendants, is mathematically unlikely, explaining why children with sporadic retinoblastoma generally develop only a single tumor, that is, unilateral disease.

in a tumor (see Figure 7.5). Because this chain of events is expected to be rare, affected children tend to have just a solitary tumor affecting one eye (unilateral retinoblastoma).

Strikingly, this depiction of *Rb* gene behavior corresponded exactly to the recessive attributes of TSGs whose existence had been postulated from results of the cell fusion studies (see Figure 7.1). Still, all this remained no more than an attractive hypothesis, because the *Rb* gene had not yet been cloned and no molecular evidence was available demonstrating that *Rb* gene copies had been rendered inactive by mutation. This behavior seemed to create a puzzle: although the inherited, TSG-inactivating mutations associated with retinoblastoma are *recessive* when it comes to the phenotype of a cell, the familial form of the disorder is passed from one generation to the next in a fashion that conforms to the behavior of a Mendelian *dominant* allele. This apparent conundrum is discussed in detail in **Sidebar 7.1**.

Sidebar 7.1 Mutant *Rb* alleles are both dominant and recessive Although we have not yet presented the molecular underpinnings of retinoblastoma, our discussion has already raised an apparent contradiction. At the level of the whole organism, an individual who inherits a mutant, defective allele of *Rb* is almost certain to develop retinoblastoma at some point in childhood, usually multiple tumors affecting both eyes. Hence, the disease-causing *Rb* allele is considered to act dominantly at the level of the *whole organism*. However, if we were to study the behavior of the mutant allele *within a cell* from this individual, we would conclude that the mutant allele acts recessively, because a cell carrying a mutant and a wild-type *Rb* gene copy would behave normally. How can we explain this genetic paradox?

The dominant behavior of the mutant *Rb* allele at the organismic level is determined by the likelihood that one or more cells in the retina of a person who inherits a mutant allele of *Rb* (hence has the genotype $Rb^{+/-}$) will lose the remaining wild-type *Rb* allele and become $Rb^{-/-}$. The likelihood that such a loss will take place depends on the size of the target cell population in the developing retina. For example, if the average retina contains 5×10^6 cells that are susceptible to transformation, and if the rate of loss of the surviving wild-type allele in the cells of an $Rb^{+/-}$ heterozygous individual is 1 per 10^6 cells, then on average 5 cells in each of this individual's eyes will lose the surviving *Rb* allele and therefore all *Rb* gene function. Each of these $Rb^{-/-}$ cells will then be primed to spawn a retinoblastoma, producing multiple (~5) tumors in each eye. Hence the inheritance of a defective *Rb* allele, which produces heterozygous cells in both retinas, virtually preordains disease development.

7.4   Incipient cancer cells eliminate wild-type copies of TSGs by a variety of mechanisms

The findings with retinoblastoma reinforced the notion that anti-proliferative genes, like the proposed *Rb* gene, might be important regulators of tumorigenesis. Still, for the sporadic form of the disease, the vexing question of how single cells could acquire the necessary "two hits" remained. A solution to this dilemma is provided by a broader look at how gene function can be lost.

Imagine a cell in which one of the two *Rb* gene copies had been inactivated by some type of mutational event that occurs with a frequency comparable to that associated with most mutational events—about 10^{-6} per cell generation. A cell with this mutation would now be in a heterozygous configuration, having one wild-type and one defective gene copy, that is, $Rb^{+/-}$. Because the mutant *Rb* allele is recessive at the cellular level, this heterozygous cell would continue to exhibit a wild-type phenotype. Now, what if the second, wild-type gene copy of *Rb* were inactivated by a mechanism that did not depend on its being struck directly by a second, independent, rare mutational event? Instead, perhaps some exchange of genetic information took place between homologous chromosomes, one of which carried the wild-type *Rb* allele, whereas the other carried the mutant, inactive allele. Such recombination takes place during meiosis, when sperm and eggs are formed. It can also occur in actively proliferating cells as they prepare their chromosomes for division—a form of genetic exchange called **mitotic recombination**.

In this fashion, the chromosomal arm carrying the wild-type *Rb* allele might be replaced with a chromosomal arm carrying the mutant allele derived from the paired homologous chromosome. Because this process depends upon reciprocal exchange of genetic information between chromosomal arms, the participating chromosomes would both remain full-length and therefore indistinguishable from their normal counterparts when viewed in the microscope. However, at the genetic and molecular level, one of the cells emerging from this recombination event would have replaced its remaining wild-type *Rb* allele and would therefore have become $Rb^{-/-}$. Importantly, such mitotic recombination is found to occur at a frequency of 10^{-5} to 10^{-4} per cell generation; it therefore provides a far more probable way for a cell to rid itself of the remaining wild-type copy of the *Rb* gene than direct mutational inactivation.

Prior to undergoing mitotic recombination, two homologous chromosomes (in the case of the *Rb* gene, the two human Chromosomes 13) will contain many different alleles, leading them to be heterozygous at many genetic loci (Supplementary Sidebar 7.1). However, following the mitotic recombination that leads to homozygosity at the *Rb* locus (with the resulting homologous chromosomes within a daughter cell possessing either an $Rb^{-/-}$ or an $Rb^{+/+}$ genotype), a number of the alleles of genes that are physically close to *Rb* on Chromosome 13 will be carried along during the exchange, and thus will also become homozygous. This genetic alteration of a gene or an extended chromosomal region is usually termed **loss of heterozygosity**, or simply LOH (**Figure 7.6**).

LOH and the elimination of TSGs such as *Rb* can be achieved in many other ways. One of these derives from the process of **gene conversion** (**Figure 7.7A**). In this mechanism, the enzyme replicating a DNA strand on one chromosome temporarily switches templates and begins to replicate the DNA strand belonging to the homologous chromosome. After progressing some distance along this strand, it disentangles and reverts to using the template strand on which it started (a mechanism sometimes termed **copy choice**). In this manner, this newly synthesized strand of DNA will now contain DNA sequences identical to a stretch on the homologous chromosome, again resulting in LOH. Such gene conversion is known to occur even more frequently per cell generation than mitotic recombination.

LOH can also be achieved by large chromosomal deletions or by simply breaking off and discarding an entire chromosomal region without replacing it with a copy derived from the other homologous chromosome. These events result in **hemizygosity** of the chromosomal region, in which now all the genes on this chromosomal arm or

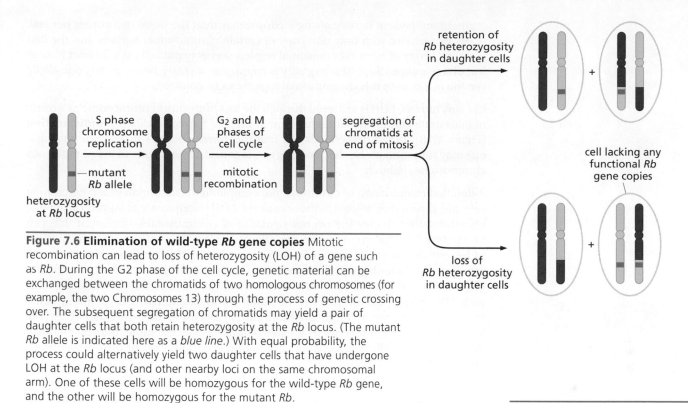

Figure 7.6 Elimination of wild-type *Rb* gene copies Mitotic recombination can lead to loss of heterozygosity (LOH) of a gene such as *Rb*. During the G2 phase of the cell cycle, genetic material can be exchanged between the chromatids of two homologous chromosomes (for example, the two Chromosomes 13) through the process of genetic crossing over. The subsequent segregation of chromatids may yield a pair of daughter cells that both retain heterozygosity at the *Rb* locus. (The mutant *Rb* allele is indicated here as a *blue line*.) With equal probability, the process could alternatively yield two daughter cells that have undergone LOH at the *Rb* locus (and other nearby loci on the same chromosomal arm). One of these cells will be homozygous for the wild-type *Rb* gene, and the other will be homozygous for the mutant *Rb*.

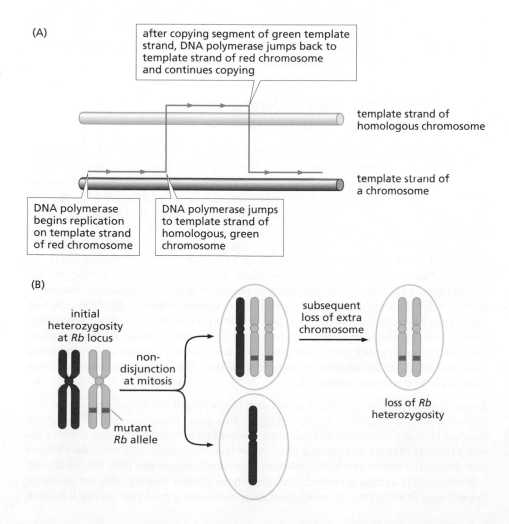

Figure 7.7 Loss of heterozygosity via gene conversion and chromosomal nondisjunction (A) During the process of gene conversion, a DNA polymerase initially begins to use a DNA strand on one chromosome (*red*) as a template for the synthesis of a new daughter strand (*blue*). After advancing some distance down this template strand, the polymerase jumps to the other (homologous) chromosome (*green*) and uses one of its strands as a template for further elongation of the nascent daughter strand. After progressing some distance down this strand, the polymerase may jump back to the originally used template strand and continue replication. In this manner, a mutant TSG allele, such as a mutant allele of *Rb*, may be transmitted from one chromosome (*green*) to its homolog (*red*), replacing the wild-type TSG allele that was previously residing there. (B) LOH can occur through the mis-segregation of chromosomes during mitosis. When such chromosomal nondisjunction takes place, one daughter cell will retain both chromatids of a chromosome rather than the usual allotment of a single chromatid. The resulting triploidy of this chromosome in one daughter cell (*top*) may be disadvantageous, so descendants of this cell may shed an extra chromosome and revert to a diploid state. Because the chromosome to be lost will likely be selected at random, this reversion to diploidy may result in homozygosity for the *Rb⁻* chromosome and loss of the wild-type allele (carried by the *red* chromosome).

segment are present in only a single copy rather than the usual two copies per cell. (Cells can survive with only one copy of certain chromosomal regions, but the loss of a single copy of other chromosomal regions seems to put cells at a distinct biological disadvantage.) Such hemizygosity is counted as an LOH, because only one allelic version of a gene in this chromosomal region can be detected.

In many tumors, LOH is achieved through the loss of an entire chromosome as a result of inaccurate chromosomal segregation at mitosis, a process called **nondisjunction** (Figure 7.7B). In the descendant cell that incorrectly receives three chromosomes, one may subsequently be shed, leaving two identical copies of one of the homologous chromosomes behind.

A detailed examination of chromosomal complements in human colon carcinoma cells has shown that, at least in these cancers, LOH is frequently achieved via genetic alterations that change the overall structure of chromosomes—an event that can be observed in the karyotype of the affected cells. Many of these events appear to be translocations, in that they involve recombination between chromosomal arms on nonhomologous chromosomes. Such recombination events might be triggered by double-stranded DNA breaks followed by fusion of the resulting ends with DNA sequences originating in other chromosomes. In this fashion, some genetic regions may be duplicated and others lost altogether. In Chapter 10 (see Section 10.4) we will discuss in greater detail one specific molecular mechanism in which such nonhomologous recombination events may occur during the development of a tumor.

Because LOH is thought to occur at a far higher frequency than mutational alteration of genes in cells that are $Rb^{+/-}$, the second, surviving wild-type copy of the Rb gene would be far more likely to be lost through LOH than through a mutation directly striking this gene copy (Supplementary Sidebar 7.2). Consequently, most retinoblastoma tumor cells would be predicted to show LOH at the Rb locus and at genetic markers that reside nearby on Chromosome 13; conversely, the cells in only a small minority of tumors would be predicted to carry two distinct mutant alleles of Rb, each inactivated by an independent mutational event.

7.5 The Rb gene often undergoes loss of heterozygosity in tumors

In 1978, the chromosomal localization of the Rb gene was determined by examining the chromosomes present in retinoblastomas. In a small number of retinal tumors, careful analysis of chromosomal karyotypes (see Sidebar 1.2) revealed **interstitial deletions** within the long arm (q) of Chromosome 13. Even though each of these deletions began and ended at different sites within this chromosomal arm, they shared in common the loss of chromosomal material in the 4th band within the 1st region of this chromosomal arm, that is, 13q14 (**Figure 7.8**). These interstitial deletions eliminated many hundreds, often thousands, of kilobases of DNA, indicating that a number of genes in this region had been jettisoned simultaneously by the developing retinal tumor cells. These changes involved the *loss* of genetic information, providing evidence that the Rb gene, which was imagined to lie somewhere in this chromosomal region, had been discarded—precisely the outcome predicted by the tumor suppressor gene theory. (Such large-scale deletions are actually rare. In most retinoblastomas, the mutations that knock out the Rb gene affect far smaller segments of chromosomal DNA and are therefore **submicroscopic**, that is, invisible upon microscopic analysis of metaphase chromosome spreads.)

Through good fortune, a second gene lying in the 13q14 chromosomal region had already been reasonably well characterized. This gene, which encodes the enzyme esterase D, is represented in the human gene pool by two distinct alleles whose protein products migrate at different rates in gel electrophoresis. The esterase D locus thus presented geneticists with a golden opportunity to test the LOH theory. Recall that when LOH occurs, an entire chromosomal region is usually affected. Because the esterase D locus had been mapped close to the Rb gene locus on the long arm

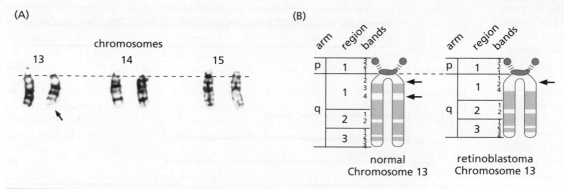

Figure 7.8 Chromosomal localization of the *Rb* locus (A) The characteristic darkly staining bands of each human chromosome delineate that chromosome's specific subregions. In this case, careful study of the karyotype of retinoblastoma cells, performed on the condensed mitotic chromosomes, revealed a deletion in the long arm (q) of Chromosome 13 in a 6-year-old child with retinoblastoma (*arrow*). Note that while the loss of the darkly staining band at 13q13 (*green*) is easily observed, a significant portion of 13q14 (where *Rb* is present) is also lost. For comparison, Chromosomes 14 and 15 in the same cell appeared normal. (B) Careful cytogenetic analysis revealed that the deletion was localized between bands 13q12 and 13q14, that is, between the 2nd and 4th bands of the 1st region of the long arm of Chromosome 13. (A, from U. Francke, *Cytogenet. Cell Genet.* 16:131–134, 1976. With permission from Karger.)

of Chromosome 13, if the *Rb* locus suffered LOH during tumor development, then the esterase D locus should frequently suffer the same fate. In effect, the esterase D locus could act as a **surrogate marker** for the still-mysterious, yet-to-be-cloned *Rb* gene.

Indeed, when researchers scrutinized the cells of several children afflicted with retinoblastoma who were born heterozygous at the esterase D locus (having inherited two distinct esterase D alleles from their parents), they found that their tumor cells had lost the protein product of one esterase D allele and thus must have undergone LOH. This finding suggested that the closely linked *Rb* gene had also undergone LOH. Such a change conformed to the theorized behavior ascribed to a growth-suppressing gene, both of whose wild-type alleles needed to be jettisoned before a cell could proliferate uncontrollably.

In a more general sense, the swapping of information between two chromosomes was likely to involve not only the *Rb* gene but also, as argued above, a large number of genes and genetic markers flanking the *Rb* locus on both sides. Such genetic markers might be short nucleotide base sequences that are not themselves components of any gene. For example, a genetic marker sequence in a neighboring region of Chromosome 13 that was initially AAGCC/AAGTC (hence heterozygous) might now become either AAGCC/AAGCC or AAGTC/AAGTC (and therefore homozygous; **Figure 7.9**). Like the esterase D gene, this neighboring oligonucleotide segment has been "carried along for the ride" with the nearby *Rb* gene during the swapping of sequences between the two homologous chromosomes. Indeed, precisely this type of analysis was performed in 1982, and the results reinforced the conclusion reached from analysis of the esterase D locus. Note, by the way, that such genetic marker sequences could lie many kilobases distant from the *Rb* locus in an intergenic region of Chromosome 13q14. Because they are not known to be associated with any biological function, these marker sequences are termed **anonymous**, and their conversion from a heterozygous to a homozygous configuration needs to

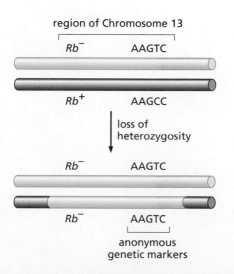

region of Chromosome 13

Figure 7.9 Demonstrations of loss of heterozygosity at the *Rb* locus A genetic marker sequence, which is linked to the *Rb* locus but is not part of any known gene, can also be used to follow changes in this chromosomal region. The pentanucleotide shown here may initially be in a heterozygous configuration (*top*) and during the process that leads to LOH at the *Rb* locus may also undergo LOH.

Figure 7.10 Mutations of the _Rb_ gene
A cloned _Rb_ cDNA was used as a probe in a Southern blot analyses of genomic DNAs from a variety of retinoblastomas and an osteosarcoma. Shown are the 10 protein-coding segments (exons) of the _Rb_ gene (_blue cylinders_). The normal version of the gene encompasses a chromosomal region of approximately 190 kb. However, analysis of a subset of retinoblastoma DNAs indicated that significant portions of this gene had undergone homozygous deletion (the segments within each pair of brackets). Thus, tumors 41 and 9 lost the entire _Rb_ gene, apparently together with flanking chromosomal DNA segments on both sides. Tumors 44, 28, and 3 lost, to differing extents, the right half of the gene together with rightward-lying chromosomal segments. The fact that an osteosarcoma (OS-15) lost a portion of this gene that was contained entirely within the left and right boundaries of the gene argued strongly that this 190-kb chromosomal DNA segment, and not leftward- or rightward-lying DNA segments, was the repeated target of mutational inactivation occurring during the development of these tumors. (From S.H. Friend et al., _Nature_ 323: 643–646, 1986. With permission from Nature.)

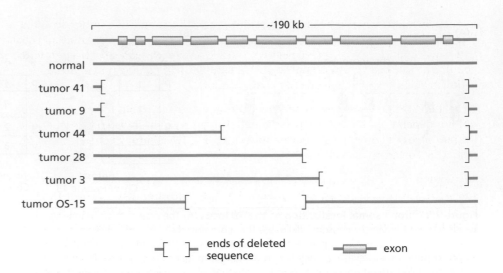

In 1986 the _Rb_ gene was cloned with the aid of a DNA probe that recognized an anonymous genomic sequence that turned out to be located within the _Rb_ gene itself. Additional DNA probes derived from different portions of the cloned gene revealed, as was previously speculated, that the _Rb_ gene in retinoblastomas suffered mutations that resulted in its inactivation (**Figure 7.10**). In some retinoblastomas, these mutations involve large deletions within the _Rb_ gene and its flanking DNA sequences. Analysis of the DNA from these tumors by Southern blotting (Supplementary Sidebar 4.3) revealed that the mutant _Rb_ allele resulting from these deletions was usually present in homozygous configuration. This meant that after the creation of an inactive or **null allele** of _Rb_ on one chromosome, the corresponding region on the other, homologous chromosome was discarded or replaced, leading to hemizygosity or LOH at this locus. These data directly validated many of the predictions of the theoretical models that had been proposed to explain how TSGs behave during the development of cancers.

As mentioned earlier, children who inherit a defective _Rb_ gene copy are also predisposed to osteosarcomas (bone tumors) as adolescents. With an _Rb_ gene probe in hand, it became possible to demonstrate that certain osteosarcomas also carried structurally altered _Rb_ genes (see Figure 7.10). At the same time, these findings highlighted a puzzle that remains largely unsolved to this day: Why does inheritance of an inactive mutant gene such as _Rb_, which encodes a protein that operates in a wide variety of cell types and tissues throughout the body (as we will learn in Chapter 8), predispose individuals predominantly to retinal and bone tumors? Why are not all tissues at equal risk? One answer to this question is that mutation of _Rb_ appears to be sufficient to initiate the development of tumors in the cells that give rise to retinoblastoma. In other tissues, multiple mutational events are required to form tumors (as we discuss in Chapter 11). Interestingly, we now know that _Rb_ is widely inactivated in human cancers, although in each tumor it is typically mutated along with other unrelated genes located elsewhere in the genome. Thus, in such settings, _Rb_ mutation is not sufficient, on its own, to trigger tumor development.

7.6 Loss-of-heterozygosity events can be used to find TSGs

Numerous TSGs that operate like the _Rb_ gene were presumed to lie scattered around the human genome and to play a role in the pathogenesis of many types of human tumors. In the late 1980s, researchers interested in finding these genes were confronted with an experimental quandary: How could one find genes whose existence is most apparent when they are missing from a cell's genome? Dominantly acting oncogenes could be detected readily through their ability to transform normal cells in a transfection-focus assay (see Figure 4.1) or through their presence in a retrovirus genome or within a chromosomal segment that repeatedly underwent gene amplification in a number of independently arising tumors.

To locate TSGs, however, researchers needed a more generalizable strategy—one that did not depend on the chance observation of interstitial chromosomal deletions or the presence of a known gene (for example, esterase D) that, through good fortune, lay near a TSG on a chromosome.

The tendency of TSGs to undergo LOH during tumor development provided cancer researchers with a novel genetic strategy for tracking these genes down. Because the chromosomal region flanking a TSG often experienced LOH along with the TSG itself, investigators should be able to detect the existence of a still-unknown TSG by searching for anonymous genetic markers that repeatedly undergo LOH during the development of a specific type of human tumor. This strategy, which was used to follow the behavior of the *Rb* gene (see Figure 7.9), could be generalized by exploiting DNA segments that were mapped to sites throughout the human genome. Once again, these DNA segments had no obvious affiliations with specific genes. In some cases, one version of the segment contained a nucleotide sequence that could be recognized by a restriction enzyme, whereas another variant possessed a single base-pair substitution that eliminated this sequence, rendering the surrounding DNA segment resistant to cleavage (**Figure 7.11A**). Because such sequence variability occurs as a consequence of normal genetic variability in the human gene pool, such sites were said to represent **polymorphic** genetic markers (see Section 1.2). When the allelic versions of such polymorphic sequences either permit or disallow cleavage by a restriction enzyme, the marker is called a **restriction fragment length polymorphism** (RFLP).

Cancer geneticists used RFLP markers to determine whether various chromosomal regions frequently underwent LOH during the development of certain types of tumors. Figure 7.11B shows the results of using RFLPs to search for regions of LOH arising in a group of human **colorectal tumors**, that is, carcinomas of the colon and rectum. In this case, the long and short arms of most chromosomes were represented by at least one RFLP marker, allowing investigators to track their fate. The results indicate that the short (p) arm of Chromosome 17 and the long (q) arm of Chromosome 18 suffered unusually high rates of LOH in this series of tumors—rising far above the background level of the 15–20% LOH that befell most of the remaining chromosomal arms in the cells of these tumors. (This background level of LOH reveals that all chromosomal regions within developing tumors have some tendency to undergo LOH at a certain rate. However, if the LOH happens to occur in a region harboring a TSG, the proliferation or survival of the cells in which this has occurred may be favored, leading to high numbers of tumors carrying this particular LOH.) The fact that specific arms of Chromosomes 17 and 18 frequently underwent LOH in tumors provided strong evidence that both arms harbored still-unknown TSGs. Thus, this genetic localization provided the investigators with a clear indication of where in the genome they should search for the culprit TSGs.

Another strategy to identify regions of LOH and suspected TSGs depends on the detection of single-nucleotide polymorphisms (SNPs, pronounced "snips") in normal and cancer cell genomes. These SNPs are detected by DNA sequencing or by other methods, such as using polymerase chain reaction (PCR) primers that are complementary to one or the other polymorphic allele (Supplementary Sidebar 7.3). Based on large-scale genome sequencing of the human population, we now know of more that 100 million SNPs in the human genome, which represent a significant fraction of the genetic variation that exists in our species.

Of course, these SNP markers are useful for LOH analyses only if a significant proportion of individuals are heterozygous at these marker loci in their normal tissues. Because about 1 in 10 known SNPs has an allele frequency of greater than 5% across human populations, the average person's genome is likely to harbor a SNP marker in heterozygous configuration every few kilobases of DNA. Such greatly increased density, compared with that of the earlier RFLP markers, allows regions of LOH to

Figure 7.11 Localization of TSGs (A) Genetic markers that are defined by the presence or absence of a restriction endonuclease cleavage site can be used to analyze chromosomal regions. In this illustration, the DNA segment in the chromosome of maternal origin (*red*) can be cleaved by the enzyme EcoRI, while the homologous, paternally derived segment (*green*), contains a single base-pair substitution that renders it resistant to cleavage. This variation represents a restriction fragment length polymorphism (RFLP), and a radioactive probe that recognizes the right end of both DNA fragments can be used in Southern blot analysis to determine whether cleavage has occurred (*below*). The presence or absence of a cleavage fragment makes it possible, in turn, to track the inheritance of DNA sequences that behave genetically like Mendelian alleles. In this example, the DNAs from the normal tissues of two cancer patients (1 and 2) both show heterozygosity in this chromosomal region. However, loss of heterozygosity (LOH) has occurred in their tumor DNAs, with patient 1 showing loss of the maternal allele and patient 2 shows loss of the paternal allele. (*continued*)

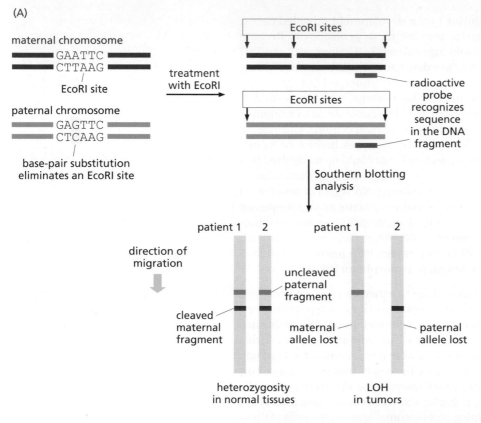

(A)

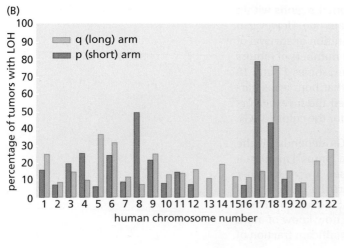

(B)

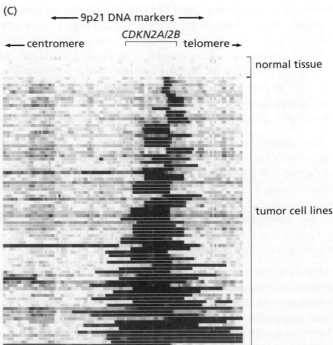

(C)

Figure 7.11 Localization of TSGs (*continued*)
(B) RFLP analysis was used to survey the LOH in entire chromosomal arms in a series of human colon cancers. The long (q) and short (p) arms of the chromosomes are indicated by *green bars* and *orange bars*, respectively. The "allelic deletions" indicated on the ordinate are equivalent to LOH. Although these analyses were imprecise, because very few probes were available at the time to gauge the extent of LOH, Chromosomes 17 and 18 were nevertheless obvious targets of LOH and thus likely to harbor TSGs. Chromosome 8p is also a target of LOH, but its TSG has never been identified. Another TSG inactivated in almost all colorectal cancers, *Apc*, is located on Chromosome 5q. Because this gene is so large (creating a large target of random mutations), both copies are often lost through independent, random mutation, rather than mutation followed by LOH; thus, its importance in the development of colorectal cancer is underestimated in this analysis. The short arms of Chromosomes 13, 14, 15, 21, and 22 contain only ribosomal genes and thus were not included in this analysis. (C) The locations of deleted chromosomal segments reveal the locations of TSGs. Thus, another strategy for localizing and identifying TSGs depends on mapping overlapping regions of relatively small deletions that affect a chromosomal arm. In this analysis, DNAs of 80 tumor cell lines (arrayed top to bottom, names deleted) were analyzed for the presence or absence of sequences recognized by 250,000 probes, present at an average density of one per 12 kb across the genome. Short homozygous deletions (*dark blue bars*) frequently eliminate one or both copies of *CDKN2A* and *CDKN2B*, an adjacent pair of TSGs that map to Chromosome 9p21. In contrast, nearby chromosomal sequences to the right and left are rarely affected by either significant amplification (*light red squares*) or deletions (*light blue squares*). (B, from B. Vogelstein et al., *Science* 244:207–211, 1989. With permission from AAAS. C, from S.M. Rothenberg et al., *Cancer Res.* 70:2158–2164, 2010. With permission from American Association for Cancer Research.)

be localized far more precisely to chromosomal segments that may contain a relatively small number of genes, greatly facilitating the identification of candidate TSGs within these regions. Nonetheless, even prior to the introduction of these and other more powerful mapping techniques, LOH analyses allowed the identification and subsequent cloning of more than 30 TSGs.

SNP-based analysis of collections of tumors has identified many regions of frequent deletion across cancer genomes, indicative of TSG loss. The fruits of one such study can be seen in Figure 7.11C, in which relatively small deletions of a region of human Chromosome 9 were frequently detected in the genomes of 80 cancer cell lines. The relatively short chromosomal DNA segment that was targeted, time after time, by these deletions is known from other work to encode two well-studied TSGs, *CDKN2A*, and *CDKN2B*, which will be discussed in detail in Chapters 8 and 9.

In the current era of genomics, in which cancer cell genomes can be rapidly sequenced and compared with the genome sequences of normal tissues, the identification of candidate TSGs is more straightforward. Such sequencing studies not only reveal regions of chromosomal deletion or LOH, but they can be used to tentatively identify TSGs based on the observation of frequent loss-of-function mutations in these genes across multiple patients with a given cancer type (as discussed in Chapter 12). As an example, the connection to cancer of *Keap1*, a TSG that we will discuss shortly, was revealed by direct DNA sequencing.

Together, these various techniques have uncovered at least two hundred TSGs in the human genome. A representative handful of these TSGs, which have been found on almost every chromosome, is presented in **Table 7.1**. As shown, several of these TSGs are also responsible for important familial cancer syndromes through mutation in the germ line of affected individuals.

7.7 Promoter methylation represents an important mechanism for inactivating TSGs

DNA molecules can be altered covalently by the attachment of methyl groups to cytosine bases (as described in Section 1.8). In mammalian cells, this methylation is found only when cytosine bases are located in a position that is 5′ to guanosines, that is, in the sequence CpG. This modification is often termed "methylated CpG" (meCpG), even though only the cytosine is methylated. When CpG methylation occurs in the vicinity of a gene promoter, it can cause repression of transcription of the associated gene, and conversely, when methyl groups are removed, transcription of the gene is often de-repressed. The means by which this modification influences gene expression is an active area of research, but it appears that methylation attracts proteins that control the packaging of DNA (as discussed in Section 1.8). In any case, extensive research indicates that the methylation of CpGs and the accessibility of promoter DNA are as important as are mutations in shutting down TSGs.

Table 7.1 Examples of human tumor suppressor genes

Name of gene	Chromosomal location	Familial cancer syndrome	Sporadic cancer	Function of gene product
CHD5	1p36.31	cutaneous melanoma	many types	histone reader, transcriptional inducer
FH	1q42.3	familial leiomyomatosis[a]	—	fumarate hydratase
BAP1	3p21.1	mesothelioma, melanoma	mesothelioma, uveal melanoma	ubiquitin hydrolase
RASSF1A	3p21.3	—	many types	multiple functions
TGFB2	3p2.2	HNPCC	colon, gastric, pancreatic carcinomas	TGF-β receptor
VHL	3p25–26	von Hippel–Lindau syndrome	renal cell carcinoma	ubiquitylation of HIF
hCDC4	4q32	—	endometrial carcinoma	ubiquitin ligase
APC	5q21–22	familial adenomatous polyposis coli	colorectal, pancreatic, and stomach carcinomas; prostate carcinoma	β-catenin degradation
miR-124a[b]	8p23.1	—	many types	suppresses CDK6
p16^{INK4A} [c]	9p21	familial melanoma	many types	CDK inhibitor
p14^{ARF} [d]	9p21	—	all types	p53 stabilizer
PTC	9q22.3	nevoid basal cell carcinoma syndrome	medulloblastomas	receptor for hedgehog GF
TSC1	9q34	tuberous sclerosis	—	inhibitor of mTOR[e]
PTEN[f]	10q23.3	Cowden's disease, breast, and gastrointestinal carcinomas	glioblastoma; prostate, breast, and thyroid carcinomas	PIP$_3$ phosphatase
WT1	11p13.5–6	Wilms tumor	Wilms tumor	TF
MEN1	11p13	multiple endocrine neoplasia	—	histone modification, transcriptional repressor
BWS/CDKN1C	11p15.5	Beckwith–Wiedemann syndrome	—	p57^{Kip2} CDK inhibitor
CBL	11q23.3	juvenile myelomonocytic leukemia	adult myelomonocytic leukemia	SH2-containing ubiquitin ligase
RB	13q14.2	retinoblastoma, osteosarcoma	retinoblastoma; sarcomas; bladder, breast, esophageal, and lung carcinomas	transcriptional repression; control of E2Fs
miR-15a/16-1	13q14.3	—	B-cell lymphoma	suppresses Bcl-2, Mcl-1, cyclin D1, Wnt3a
miR-127	14q32.31	—	many types	suppresses Bcl-6
TSC2	16p13.3	tuberous sclerosis	—	inhibitor of mTOR[e]
CBP	16p13.3	Rubinstein–Taybi syndrome	AML[g]	TF co-activator
CYLD	16q12–13	cylindromatosis	—	deubiquitinating enzyme
CDH1	16q22.1	familial gastric carcinoma	invasive cancers	cell–cell adhesion

Table 7.1 Examples of human tumor suppressor genes (*continued*)

Name of gene	Chromosomal location	Familial cancer syndrome	Sporadic cancer	Function of gene product
BHD/FLCN	17p11.2	Birt–Hogg–Dube syndrome	kidney carcinomas, hamartomas	regulator of mTOR[e]
TP53	17p13.1	Li–Fraumeni syndrome	many types	TF
NF1	17q11.2	neurofibromatosis type 1	colon carcinoma, astrocytoma, acute myelogenous leukemia	Ras-GAP
DPC4[h]	18q21.1	juvenile polyposis	pancreatic and colon carcinomas	TGF-β TF
KEAP1	19p13.2	—	many types	suppresses anti-oxidant response
LKB1/STK11	19p13.3	Peutz–Jegher syndrome	hamartomatous colonic polyps	serine/threonine kinase
RUNX1	21q22.12	familial platelet disorder	AML	TF
SNF5[i]	22q11.2	rhabdoid predisposition syndrome	malignant rhabdoid tumors	chromosome remodeling
NF2	22q12.2	neurofibroma-predisposition syndrome	schwannoma, meningioma; ependymoma	cytoskeleton–membrane linkage
WTX	Xq11.1	—	Wilms tumor	β-catenin degradation

[a]Familial leiomyomatosis includes multiple fibroids, cutaneous leiomyomas, and renal cell carcinoma. The gene product is a component of the tricarboxylic cycle.
[b]*miR124a-1* genes are also located at 8q12.3 and 20q13.33l.
[c]Also known as *MTS1*, *CDKN2*, and *p16*.
[d]The human homolog of the murine *p19^ARF* gene.
[e]mTOR is a serine/threonine kinase that controls, among other processes, the rate of translation and activation of Akt/PKB, TSC1 (hamartin), and TSC2 (tuberin), thereby controlling both cell size and cell proliferation.
[f]Also called *MMAC* or *TEP1*.
[g]The *CBP* gene is involved in chromosomal translocations associated with AML. These translocations may reveal a role of a segment of CBP as an oncogene rather than a tumor suppressor gene.
[h]Encodes the Smad4 TF associated with TGF-β signaling; also known as *MADH4* and *SMAD4*.
[i]The human SNF5 protein is a component of the large Swi/Snf complex that is responsible for remodeling chromatin in a way that leads to transcriptional repression through the actions of histone deacetylases. The rhabdoid predisposition syndrome involves susceptibility to atypical teratoid/rhabdoid tumors, choroid plexus carcinomas, medulloblastomas, and extra-renal rhabdoid tumors.
Adapted in part from E.R. Fearon, *Science* 278:1043–1050, 1997; and in part from D.J. Marsh and R.T. Zori, *Cancer Lett.* 181:125–164, 2002.

During the development of all animal species, decisions concerning the transcriptional state of many genes are made early in embryogenesis. Once made, these decisions must be passed on to the descendant cells in various parts of the growing embryo. Because CpG methylation can be transmitted from cell to cell heritably (see Section 1.8), it represents a highly effective way to ensure that descendant cells, many cell generations removed from an early embryo, continue to respect and enforce the methylation decisions made by their embryonic ancestors.

The enzymes that are responsible for methylating CpGs—called DNA **methyltransferases**—are themselves altered in cancer, as are the enzymes involved in regulating the removal of this 5′ methyl group. The end result is that the methylation state of cancer genomes differs from that of the corresponding normal cell. Indeed, in developing cancers, DNA methylation is altered in two opposing directions. In the early stages of tumor development, methylation throughout the cancer cell genome is often found to decrease progressively (as determined by the technique described in Supplementary Sidebar 7.4). This decline could result from the failure of DNA methyltransferases to maintain methylation levels or to an active process of demethylation. Much of this "global hypomethylation" takes place within highly repeated DNA sequences that do

Figure 7.12 Methylation of the RASSF1A promoter (A) A map of the structure of the *RASSF1A* promoter region shows the location of the CpG dinucleotides (ticks) in the CpG island in which the promoter is embedded. (B) A diagram depicts the state of methylation of each CpG dinucleotide (circle) in this island. Filled circles (*blue*) indicate that a CpG has been found to be methylated, whereas open circles indicate that it is unmethylated. Analyses of five DNA samples from a prostate tumor indicate methylation at almost all CpG sites in the *RASSF1A* CpG island; this CpG island is unmethylated in most, but not all, samples of adjacent, ostensibly normal tissue. Analyses of control DNA from a normal individual indicate the absence of any methylation of the CpGs in this CpG island. (These data suggest the presence of some abnormal cells with methylated DNA in the ostensibly normal tissue adjacent to tumor 232.) Analysis was performed by bisulfite sequencing (see Supplementary Sidebar 7.4). (Courtesy of W.A. Schulz and A.R. Florl.)

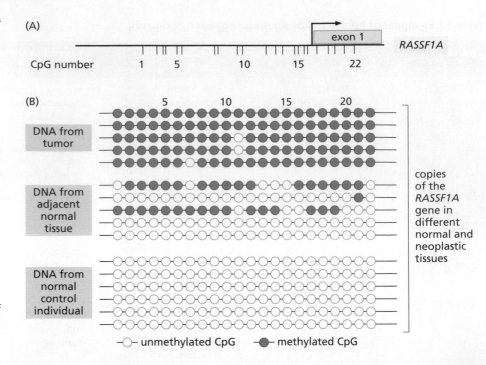

not specify biological functions; although such hypomethylation is correlated with chromosomal instability, it remains unclear whether it directly causes this instability.

Apart from this global hypomethylation, there are also localized regions of DNA—segments with a high density of CpGs called "CpG islands"—that become excessively methylated in the genomes of cancer cells. These CpG islands are often associated with the promoters of genes (**Figure 7.12**), and their methylation by DNA methyltransferases silences transcription, resulting in the shutdown of gene expression. (A recent survey indicated that 60–70% of all genes in the human genome have CpG islands affiliated with their promoters, yielding a total of 12,000 to 14,000 genes that are, in principle, vulnerable to inappropriate shutdown by methylation.)

CpG methylation is effective in shutting down the expression of a gene only if it occurs within the promoter or other regulatory sequences of the gene; conversely, methylation of DNA sequences in the body of the gene, including exonic sequences, seems to have little if any effect on the level of transcription. Because promoter methylation can silence a gene as effectively as a mutation of its nucleotide sequences, we might predict that methylation plays a role in the silencing of TSGs that occurs during tumor progression. In fact, this has proven to be the case. More than half of the TSGs that are inactivated by germ-line mutations in familial cancer syndromes have been found to be silenced in sporadic cancers by promoter methylation. For example, when the TSG *Rb* is mutated in the germ line, it leads to familial retinoblastoma. In sporadic retinoblastomas, however, *Rb* is inactivated either by somatic mutations or by promoter methylation. In addition, the promoters of a variety of other genes that are known or thought to inhibit tumor formation have been found in a methylated state (**Table 7.2**).

The extent of CpG island methylation varies greatly from one tumor genome to the next. In some tumors, the hypermethylation is so extensive that such tumors are said to exhibit the "CpG island methylator phenotype" (CIMP). In the cells of these tumors, as many as 5% of the genes (that is, about 1000 genes) may possess hypermethylated CpG islands. The molecular mechanisms driving this runaway CpG methylation remain under active investigation.

Some candidate TSGs are rarely inactivated by somatic mutations, and their involvement in creating cancer is suggested only by the finding that their promoters are methylated in large numbers of independently arising tumors. The absence of defined mutations makes it challenging to verify that inactivation of a candidate TSG actually

Table 7.2 Examples of hypermethylated genes found in human tumor cell genomes

Name of gene	Nature of protein function	Type of tumor
AR	androgen receptor signaling	prostate
RARβ2	nuclear receptor for differentiation	breast, lung
p57^{Kip2}	CDK inhibitor	gastric, pancreatic, hepatic; AML
IGFBP	sequesters IGF-1 factor	diverse tumors
BRCA1	DNA damage response	breast, ovarian
CDKN2A/p16^{INK4A}	inhibitor of CDK4/6	diverse tumors
CDKN2B/p15^{INK4B}	inhibitor of CDK4/6	diverse tumors
APC	inducer of β-catenin degradation	colon carcinomas
miR-200s	inhibitor of epithelial-mesenchymal transition	colon, bladder, squamous cell carcinoma
ESR1	estrogen receptor signaling	breast
p73	aids p53 to trigger apoptosis	diverse tumors
GSTP1	mutagen inactivator	breast, liver, prostate
MGMT	DNA repair enzyme	colorectal
CDH1	cell–cell adhesion receptor	bladder, breast, colon, gastric
DKK1	Wnt inhibitor	colon
MLH1	DNA mismatch repair enzyme	colon, endometrial, gastric
PTEN	degrades PIP$_3$	diverse tumors
TGFBR2	TGF-β receptor	colon, gastric, small-cell lung
VHL	ubiquitin ligase	kidney, hemangioblastoma
RB	cell cycle regulator	retinoblastoma
CASP8	apoptotic caspase	neuroblastoma, SCLC
APAF1	pro-apoptotic cascade	melanoma

Adapted from H. Heyn and M. Esteller, *Nat. Rev. Genet.* 13:679–692, 2012.

contributes causally to cancer formation. Nevertheless, even in the absence of direct evidence of mutational inactivation, the recurring functional silencing of these genes by CpG methylation is now well accepted as proof of their roles as TSGs.

The elimination of TSG function by promoter methylation might, in principle, occur through either of two routes. Both copies of a TSG might be methylated independently of one another. Alternatively, one copy might be methylated and the second might then suffer an LOH accompanied by a duplication of the already-methylated TSG copy. Actually, this second mechanism seems to operate quite frequently. For example, one study found that a TSG called *p16^{INK4A}* (which will be described in Section 8.4) was methylated in 44% of the (ostensibly) normal lung epithelial cells cultured from current and former smokers and not at all in the comparable cells prepared from those who had never smoked. LOH in this chromosomal region was found in 71 to 73% of the two smoking populations and in 1.5 to 1.7% of nonsmokers. This study also teaches us a second lesson: methylation of critical growth-controlling genes often occurs early in the complex, multi-step process of tumor formation—long before histological changes are apparent in a tissue. These populations of outwardly normal

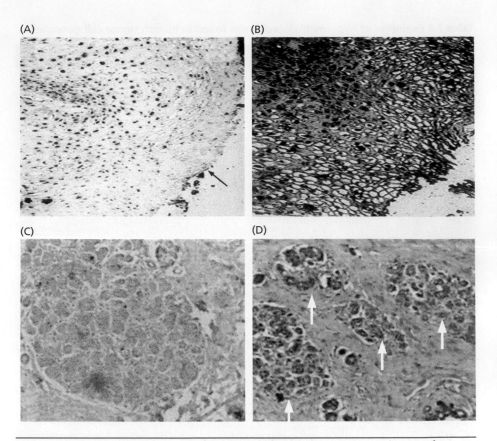

Figure 7.13 *In situ* measurements of DNA methylation The presence or absence of methylation at specific sites in the genomes of cells can be determined in tissue sections that are fixed to a microscope slide using *in situ* hybridization. In these images, the methylation status of the promoter of the TSG *p16^INK4A* is analyzed with a methylation-specific probe, which yields dark staining in areas where this gene promoter is methylated. (A) In a low-grade squamous intraepithelial lesion of the cervix (*left*), nuclei of cells located some distance from the uterine surface show promoter methylation, whereas those near the surface (*arrow*) do not. (B) However, in an adjacent high-grade lesion, which is poised to progress to a cervical carcinoma (*right*), all the cells show promoter methylation. (C, D) In these micrographs of normal breast tissue, some histologically normal lobules show no promoter methylation (C), whereas others (*arrows*) show uniform promoter methylation (D). This suggests that the mammary epithelial cells in such outwardly normal lobules have already undergone a critical initiating step in cancer progression. (A and B, from G.J. Nuovo et al., *Proc. Natl. Acad. Sci. USA* 96:12754–12759, 1999. With permission from National Academy of Sciences, U.S.A. C and D, from C.R. Holst et al., *Cancer Res.* 63:1596–1601, 2003. With permission from American Association for Cancer Research.)

cells presumably provide a fertile soil for the eventual eruption of premalignant and malignant growths. This point is borne out in other tissues as well, such as the histologically normal breast tissues analyzed in **Figure 7.13**.

Genome-wide methylation analysis (Supplementary Sidebar 7.4) has allowed for comprehensive characterization of the methylation status of nearly all known TSGs. As illustrated in **Figure 7.14**, the frequency of methylation of a specific TSG varies dramatically from one type of human tumor to the next. However, the data clearly reinforce the pervasiveness of promoter methylation during the development of a wide variety of human cancers.

Beyond the biological significance of hypermethylation as a mechanism for the silencing of cancer-related genes, the overall state of DNA methylation in neoplastic tissue compared to normal tissue can have implications for disease diagnosis, prognosis, and therapy. For several cancer types, the overall changes in DNA methylation patterns correlate with the severity of the disease and the response to therapy, both of which affect patient survival after diagnosis. In addition, there is a growing list of cancer therapies that are directed at the enzymes that are responsible for DNA methylation and histone modifications.

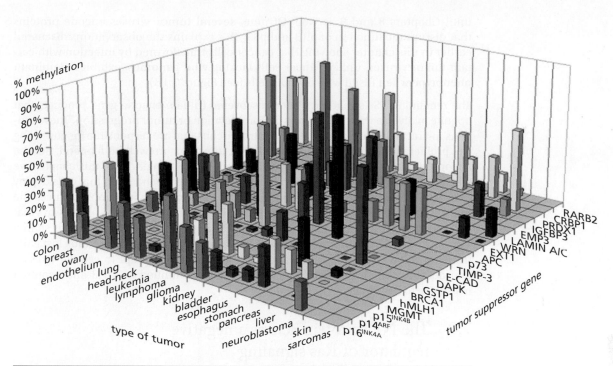

Figure 7.14 Methylation of multiple genes within tumor cell genomes This three-dimensional bar graph summarizes measurements of the methylation state of the CpG islands associated with 20 different genes that are known or presumed to play an important role in suppressing the development of human tumors. Methylation status of each was determined for 18 different tumor types. The height of each bar indicates the proportion of tumors of a given type in which a specific promoter has undergone methylation. All of these promoters are unmethylated (or methylated to an insignificant extent) in normal tissues. The data indicate that the hypermethylation profiles vary by tumor type: several genes are hypermethylated in cancers of the colon, for example, whereas less methylation of these genes takes place in sarcomas. (Adapted from M. Esteller, *Hum. Mol. Genet.* 16:R50–R59, 2001. With permission from Oxford University Press.)

7.8   TSGs and their encoded proteins function in diverse ways

Even a cursory examination of Table 7.1 makes it clear that TSGs specify a diverse array of proteins that operate in many different intracellular sites to reduce the risk of cancer. Indeed, an anti-cancer function is the only property that is shared in common by these otherwise unrelated genes.

Some function to directly suppress the proliferation of cells in response to a variety of signals that inhibit growth and induce differentiation. Others are components of the cellular control circuitry that inhibits proliferation in response to metabolic imbalances and genomic damage.

The first two TSGs that were studied intensively—the *Rb* gene discussed earlier and another called *p53* (or sometimes *Trp53* or *TP53*)—also happen to be two of the more frequently altered TSGs in human cancers. The protein encoded by the *Rb* gene, pRb, governs progress through the growth-and-division cycles of a wide variety of cells, and the growth control imposed by the *Rb* circuit appears to be disrupted in most and perhaps all human tumors. Because of its central importance, we will devote an entire chapter to *Rb* and its function (Chapter 8). The TSG *p53* and its product, p53, play an equally central role in the development of cancers and, consequently, we devote an additional chapter to an extended description of p53 and its cellular functions (Chapter 9). Several of the proteins encoded by the TSGs listed in Table 7.1 happen to be components of the *Rb* and *p53* pathways; included in these circuits are the $p16^{INK4A}$, $p15^{INK4B}$, and $p19^{ARF}$ genes, and so discussion of these will be postponed until Chapters 8 and 9 as well. Of note, several tumor viruses encode proteins that disrupt the function of pRB and p53. This explains the observations discussed earlier in this chapter showing that cancer cells transformed by infection with certain tumor viruses, unlike cancer cells lacking viral genomes, exhibit a dominant phenotype when fused with normal cells (see Figure 7.1B).

The remaining TSGs that have been discovered to date encode gene products that operate in diverse ways to suppress cell proliferation or tumor progression. These proteins interact with virtually all of the control circuits that are responsible for governing cell proliferation and survival. Some of their functions are well understood, but the actions of others remain to be fully elucidated. In the remainder of this chapter, we will focus on three TSGs that reflect the diversity of the mechanisms that cells deploy to control proliferation—and reveal how disruption of their function enhances the fitness of cancer cells. Although these mechanisms do not encompass all the ways that TSGs function, each represents an approach to regulating cell growth that is shared by many other TSGs—from those that constrain proliferation and are thus lost early in tumor initiation to those that allow malignant cells to adapt to cellular stresses and detoxify chemicals, thereby promoting their continued survival and acquired resistance to treatment.

7.9   The NF1 protein acts as a negative regulator of Ras signaling

The disease neurofibromatosis was first described by Friedrich von Recklinghausen in 1862. We now know that neurofibromatosis type 1 (NF1)—sometimes called von Recklinghausen's neurofibromatosis—is a relatively common familial cancer syndrome, with 1 in 3500 individuals affected on average worldwide. The primary feature of this disease is the development of benign tumors, called **neurofibromas**, in the sheaths surrounding nerves of the peripheral nervous system. On occasion, a subclass of these tumors, termed *plexiform neurofibromas*, progress to malignant tumors called **neurofibrosarcomas** or **malignant peripheral nerve sheath tumors** (MPNSTs) (**Figure 7.15A**). Patients suffering from NF1 also have greatly increased risk of gliomas (tumors of the astrocyte lineage in the brain; see Figure 2.8A), **pheochromocytomas** (arising from the adrenal glands), and myelogenous leukemias (see Figure 2.7B). This collection of tumors involves cell types arising from diverse embryonic lineages. Neurofibromatosis type 2, which we do not discuss in detail, has distinct clinical features and arises from mutations in a completely unrelated TSG (see Table 7.1).

Patients with NF1 often suffer from additional abnormalities that involve still other cell types. Among these are **café au lait spots** (Figure 7.15B), which are areas of hyperpigmentation in the skin; subtle alterations in the morphology of the cells in the skin and long bones; cognitive deficits; and benign lesions of the iris called "Lisch nodules" (Figure 7.15C). The risk of developing these additional manifestations is strongly influenced by the patient's **genetic background** (that is, the array of all other genetic alleles in an individual's genome), because siblings inheriting the same mutant allele of the responsible TSG, called *Nf1*, often exhibit dramatically different disease phenotypes. Such variation in the clinical presentation of a genetic disease is referred to as variable **expressivity**.

The *Nf1* gene was cloned in 1990. The genetics of the *Nf1* gene parallels closely that of the *Rb* gene. Thus, mutant, inactivated alleles of the *Nf1* gene transmitted through the germ line act in a dominant fashion to create disease phenotypes. At the cellular level, the originally heterozygous configuration of the gene ($Nf1^{+/-}$) is converted to a homozygous state ($Nf1^{-/-}$) in tumor cells through LOH. Finally, as many as half of the patients with NF1 lack a family history of the disease, indicating that the mutant allele that they carry is the consequence of a *de novo* mutation in the germ line. More than 80% of such *de novo* mutations usually occur during spermatogenesis in the fathers of afflicted patients, a pattern also seen with *de novo* germ-line mutations of the *Rb* gene (see Sidebar 7.2).

Once the cloned *Nf1* gene was sequenced, researchers were able to assign a function to its encoded protein neurofibromin: the amino acid sequence of neurofibromin

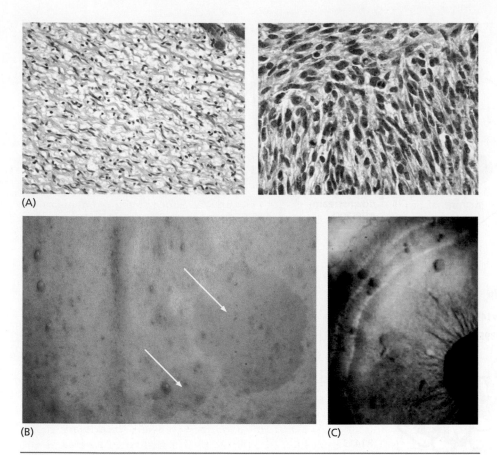

(A)

(B)   (C)

Figure 7.15 Neurofibromatosis Neurofibromatosis type 1 is considered a *syndrome* because a number of distinct conditions and phenotypes are associated with a single genetic state. (A) Histological observation of afflicted tissues indicates the benign neurofibromas (*left*), which occasionally progress to malignant peripheral nerve sheath tumors (MPNSTs, *right*). Whereas benign neurofibromas contain multiple distinct cell types and abundant extracellular matrix material—giving them a more diffuse appearance—malignant MPNSTs consist of a more uniform collection of densely packed cancer cells. The histological complexity of the neurofibromas greatly complicated the identification of the progenitors of these growths, which are now thought to be the precursors of the Schwann cells that wrap around peripheral axons. (B) Notable among the conditions associated with this syndrome are the numerous small, subcutaneous nodules (that is, neurofibromas) and the light brown *café au lait spots* (*arrows*) seen here on the back of a patient. (C) In addition, Lisch nodules are often observed in the iris of the eyes. (A, courtesy of S. Carroll. B, courtesy of B.R. Korf. C, from B.R. Korf, *Postgrad. Res.* 61:3225–3229, 1988 and courtesy of J. Kivlin.)

shows extensive relatedness to that of proteins that function as GTPase-activating proteins for Ras (Ras-GAPs; see Figure 5.24). Eukaryotic cells use Ras proteins to regulate important aspects of their metabolism and proliferation, and Ras-GAPs negatively control the state of Ras activation. These regulatory proteins do so by inducing Ras to activate its intrinsic GTPase activity, thereby forcing Ras to convert itself from its activated GTP-bound form to its inactive, GDP-bound form (**Figure 7.16A**). Indeed, a Ras-GAP may even ambush activated Ras before the latter has had a chance to stimulate its coterie of downstream effectors (see Section 5.10).

This initial insight into *Nf1* function inspired a simple scheme of how defective forms of neurofibromin create disease phenotypes. *Nf1* is expressed widely throughout the body, with especially high concentrations of neurofibromin found in the adult peripheral and central nervous systems. In normal cells, stimulation by growth factors would initially trigger a degradation of neurofibromin, enabling Ras signaling to continue without interference by this negative regulator. However, after 60 to 90 minutes, neurofibromin levels would return to normal, allowing the accumulation of this inhibitory protein to help shut down further Ras signaling—a form of negative-feedback

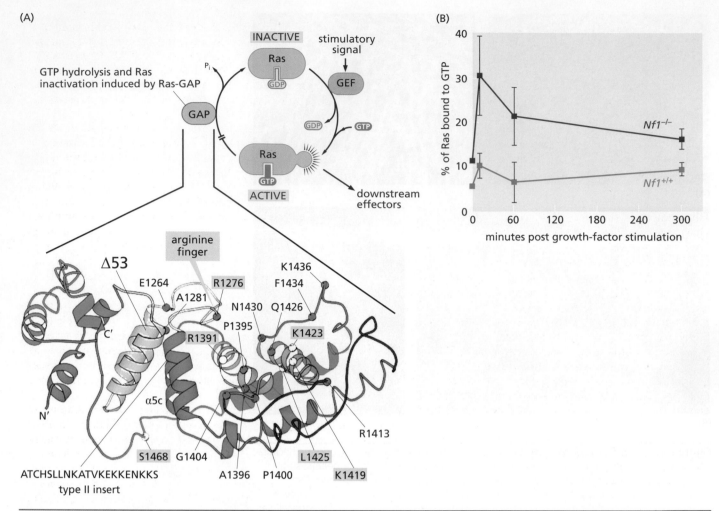

Figure 7.16 Neurofibromin and the Ras signaling cycle (A) As shown at the top, the Ras protein passes through a cycle in which it is activated by a guanine nucleotide exchange factor (GEF), promoting the release of GDP and binding of GTP, and inactivated by a GTPase-activating protein (Ras-GAP). One of the main Ras-GAPs is neurofibromin, which can stimulate the GTPase activity of Ras more than 1000-fold. The structure of the neurofibromin domain that interacts with Ras is illustrated here. A subdomain of neurofibromin, termed the "arginine finger" (*blue box*), carries a critical arginine (R1276) that is inserted into the GTPase cleft of Ras and actively contributes to the hydrolysis of GTP to GDP by Ras. Mutations that cause replacement of this critical arginine residue with another amino acid decrease the GTPase-stimulating activity of neurofibromin by 1000-fold. In this diagram, mutant forms of neurofibromin, which carry amino acid substitutions observed in neurofibromatosis patients, are indicated by *white spheres* that are labeled by *gray boxes*; experimentally generated mutations that produce amino acid substitutions (*gray spheres without boxed labels*) also compromise neurofibromin's GTPase activity. The *white sphere with boxed label* (R1391) indicates a mutation that arose through both routes. In addition, a large-scale deletion (Δ53) and an insertion ("type II insert") found in patients are shown; both are also disease-associated. (B) When myeloid cells of wild-type *Nf1*^{+/+} mice are stimulated with the growth factor GM-CSF, the ligand-activated receptor causes GTP-bound Ras levels to rise rapidly and then decrease slightly over the next five hours (*blue line*). In contrast, in *Nf1*^{−/−} myeloid cells, the basal level of GTP-bound Ras is already elevated, and in response to GM-CSF stimulation, levels of this active form of Ras rise threefold further and remain elevated over this time period (*red line*). (A, from K. Scheffzek et al., *EMBO J.* 17:4313–4327, 1998. With permission from John Wiley and Sons. B, from D. Largaespada et al., *Nat. Genet.* 12:137–143, 1996. With permission from Nature.)

control (see Figure 6.25). In cells that lack *Nf1* function, Ras proteins are predicted to persist in their activated, GTP-bound state for longer-than-normal periods of time. In fact, in **myeloid** cells (a class of cells in bone marrow) that have been genetically engineered to be *Nf1*^{−/−}, the basal (that is, unstimulated) levels of GTP-bound activated Ras are higher than normal and respond to growth factor stimulation by increasing rapidly to far higher levels before declining gradually (Figure 7.16B). Consequently, the loss of *Nf1* function in a cell can mimic functionally the hyperactivation of Ras proteins created by mutant *Ras* oncogenes (see Section 5.10).

The histological complexity of neurofibromas has made it very difficult to identify the normal cell type that yields the bulk of the cells in these growths. Microscopic

Sidebar 7.2 Targeted gene deletion can produce mouse models of cancer The natural process of homologous recombination can be exploited to introduce well-defined genetic changes into the mouse germ line in order to mutate an endogenous gene. This technology has been instrumental in the study of TSG function in the whole organism and for the development of mouse models of human cancer involving this class of genes. The process begins by manipulating the genome of a mouse embryonic stem (ES) cell. A cloned gene fragment that is introduced by electroporation into an ES cell is able to recombine, with a low but significant frequency, with homologous DNA sequences residing in the chromosomal DNA of this cell. Selectable markers—for example, neo^r (a gene that confers resistance to neomycin)—can be exploited to identify those cells that have integrated the exogenous DNA into the ES cell genome. Additional screening methods can then be used to identify the rare cells in which the cloned gene fragment, with its selectable marker, has disrupted one copy of the endogenous TSG by homologous recombination, thus creating an inactivating (or "knockout") mutation. These ES cells are thus heterozygous for the mutation in the targeted TSG (or $TSG^{+/-}$) (Figure 7.17A). A similar approach can be used to replace endogenous sequences with a gene whose activity has been altered more subtly, thereby creating a "knock-in" mutation.

$TSG^{+/-}$ ES cells can then be introduced via micro-injection into the **blastocoel** cavity of a mouse **blastocyst** (an early embryo), and the embryo transferred into a **pseudopregnant** female for further development (Figure 7.17C). Because ES cells are **pluripotent** (able to differentiate into all cell types in the body), the descendants of the injected cells can insert themselves (**chimerize**) into any tissues within the developing embryo, including the germ line. Such **germ-line chimeras** can be bred with wild-type animals to transmit the inactivated TSG allele to their offspring, creating mice that are constitutionally $TSG^{+/-}$ in all of their cells. These animals are genetic analogs of humans with a familial cancer syndrome caused by an inherited mutation in the TSG, and they can be studied for tumor predisposition and other aspects of tumor biology.

In addition, two animals that are $TSG^{+/-}$ can be bred and their $TSG^{-/-}$ offspring can be used to assess the effects of homozygosity for a given TSG mutation (Figure 7.17D). For most TSGs, this condition results in embryonic lethality, indicating the requirement for the TSG in one or more aspects of normal development. Genes required for embryonic development are often referred to as **essential genes**.

To examine the effects of inactivating an essential TSG in adult tissues, researchers have developed strategies to knockout the gene *conditionally*—in a subset of tissues or at a particular time. In this approach, homologous recombination is used in ES cells to flank a functionally important region of the target gene with a pair of short DNA sequences that serve as the recognition sequences for a site-specific recombinase (Figure 7.17B). The most frequently used system involves the use of *LoxP* sites and the recombinase Cre, which were initially derived from bacteriophage P1. Because the *LoxP* sites are inserted in such a way that they do not disrupt the function of the target gene, the alleles carrying flanking *LoxP* sites (or "floxed" alleles) remain functional and can be carried in a homozygous configuration. However, in cells in which Cre is expressed—for example, through the use of a cell type- or tissue-specific transcriptional promoter—the recombinase can catalyze the deletion of the sequences between the *LoxP* sites in both floxed alleles, leading to complete disruption of gene function (see Figure 7.17B).

These approaches can greatly facilitate the development of mouse models involving loss-of-function mutations in TSGs in different tissues in the body. The system can be further engineered so that the elimination of TSG activity can be induced by treating the animal with a drug. For example, the Cre recombinase molecule can be fused to a mutant form of estrogen receptor (ER) that can be activated by the drug tamoxifen, a ligand of this receptor (Figure 7.17E). When this hybrid Cre-ER is initially synthesized, it will be confined to the cytoplasm and will only gain access to the nucleus—where the floxed TSG alleles reside—when tamoxifen is provided. This allows for temporal control of gene deletion.

---

analyses (see Figure 7.15A) reveal that these tumors are composed of a mixture of cell types, including Schwann cells (which wrap around and insulate nerve axons), neurons, perineurial cells (which seem to surround and protect the axon–Schwann cell complexes), fibroblasts, and mast cells, the latter being a type of granulocyte and originating in the bone marrow. In some of these benign tumors, Schwann cells are clearly the prevailing cell type, whereas in others it seems that fibroblasts or perineurial cells may dominate.

The weight of evidence indicates that (1) primitive, undifferentiated **neural crest**-derived cells related to Schwann cell precursors are the primary targets of *Nf1* LOH leading to neurofibromas, and (2) once these cells have lost all *Nf1* function, they initiate the development of these histologically complex growths by inducing the co-proliferation of a variety of other cell types via paracrine signaling (see Section 5.5). This co-proliferation raises an interesting and still-unanswered question, however: Because these other cell types, including those in neighboring tissues and in the bone marrow are heterozygous for *Nf1*—and carry only half the normal dose of functional neurofibromin—might they be hyper-responsive to any growth-stimulatory, paracrine signals released by the neoplastic $Nf1^{-/-}$ cells? This hypothesis is borne out by detailed studies of neurofibromatosis formation in a genetically engineered mouse model of the disease. This approach to producing mouse models of cancer via targeted gene deletion is outlined in **Sidebar 7.2** and in **Figure 7.17**.

Figure 7.17 Homologous recombination can be used to produce TSG knockout or conditional knockout mice (A) Using recombinant DNA techniques, a targeting vector is constructed that contains a cloned mouse TSG in which the *neo^r* gene disrupts or replaces a functionally important segment of the gene. This targeting vector is then introduced via *electroporation* into mouse embryonic stem (ES) cells, and drug selection and genetic screening are performed to identify those ES cells that have undergone homologous recombination between the targeting vector and one allele of the TSG; this recombination creates a "knockout" allele. (B) It is also possible to conditionally knock out a TSG. This begins with a targeting vector carries *LoxP* sites, which function as target sequences for the site-specific recombinase Cre; pairs of *LoxP* sites are used to create a "floxed" allele of the TSG following their insertion into the genome via homologous recombination. Here, the *LoxP* sites do not, on their own, disrupt TSG function and will only do so following Cre-driven recombination (see E below). (C) ES cells carrying properly integrated knockout or floxed alleles (either *TSG^{+/-}* or *TSG^{+/Fl}*) are injected into wild-type mouse blastocysts, whereupon their progeny may become incorporated into the tissues of the resulting embryo. The embryos are implanted into pseudo-pregnant females and, following their development, some of the newborn chimeric mice may carry the altered allele in their germ cells, that is, sperm or eggs; these germ-line chimeras, when bred, enable the germ-line transmission of this DNA fragment to their descendants. (D) *TSG^{+/-}* mice can be studied for tumor predisposition. They can also be bred with one another to determine if homozygosity for the TSG mutation leads to embryonic lethality. If *TSG^{-/-}* animals survive to adulthood, they too can be studied for tumor predisposition. (E) Mice that are bred to be homozygous for a floxed allele of a TSG (*TSG^{Fl/Fl}*) can be made to carry an allele of *Cre* expressed from a tissue-specific promoter (TSP); these mice are used to determine the effects of conditional deletion of the TSG in a given tissue or cell type. Alternatively, a mutant estrogen-receptor fusion of *Cre* (*CreER*) that is activatable by a ligand of the ER can be used to control the timing of Cre activation through the addition of the drug tamoxifen (an estrogen analog and thus ER ligand). Adapted from H. Lodish et al., *Molecular Cell Biology*, 5th ed. New York: Freeman, 2004 and from B. Alberts et al., *Molecular Biology of the Cell*, 7th ed. New York: W.W. Norton & Company, 2022.

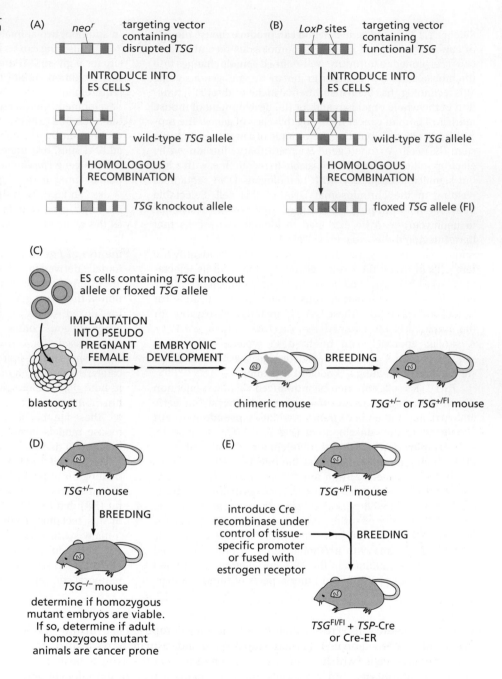

Studies of such mice reveal that recruitment of mast cells into incipient neurofibromas is critical to the formation of these histopathologically complex tumors. In mouse models of NF1 in which Schwann cells are engineered to become *Nf^{-/-}* and the remaining cells—including the cells of the hematopoietic system—are *Nf1^{+/-}*, tumors termed *plexiform* neurofibromas form with high frequency at the roots of nerves emerging from the spinal cord. However, if the hematopoietic system (including the mast cells) of such mice is replaced with cells that are wild type (that is, *Nf1^{+/+}*), these cells are not as responsive to signals released by the *Nf1^{-/-}* Schwann cells and are no longer recruited in large numbers to developing neurofibromas. The result is a drastic decrease in the numbers of these tumors (**Figure 7.18**). Such collaborations between multiple cell types in forming a tumor will be discussed in greater detail in Chapter 13.

The observation that a half-normal dose of neurofibromin in the mast cells (or their immediate precursors) results in a clear and aberrant cellular phenotype raises a more general point about TSGs: although the full phenotypic changes associated with inactivation of these genes may only be felt when both gene copies are lost, a half-dosage

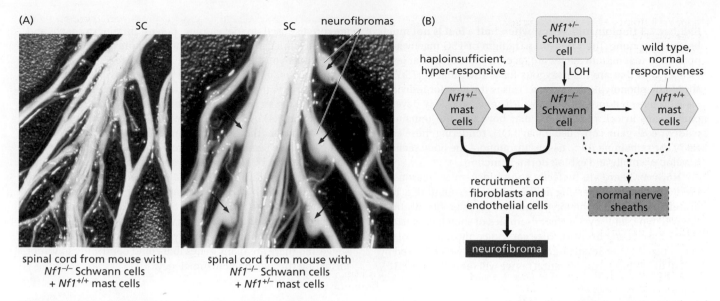

Figure 7.18 Contribution of genotype of surrounding tissue to neurofibroma development (A) In *Nf1*⁺/⁺ mice in which the Schwann cells have specifically been rendered *Nf1*⁻/⁻, the nerve roots emerging from the spinal cord (SC) exhibit a relatively normal morphology (*left*); hence, loss of all *Nf1* function in the target cells—the Schwann cells—does not suffice to allow abnormal neurofibromas to form. However, if, in addition, mast cells originating in the bone marrow are rendered *Nf1*⁺/⁻, both a thickened spinal cord and plexiform neurofibromas (*arrows*) are readily apparent (*right*). Hence, a partial loss of *Nf1* function in mast cells must cooperate with total loss of *Nf1* function in the Schwann cells to generate these growths. (B) The observations in panel A suggest that mast cells originating in the bone marrow are normally recruited to the developing spinal cord, where they contribute to normal morphogenesis. However, when these mast cells are *Nf1*⁺/⁻—and thus are deprived of half of their neurofibromin protein—they and their progeny that are recruited to the developing spinal cord behave abnormally and become hyper-responsive to certain signals released by homozygous mutant *Nf1*⁻/⁻ Schwann cells. This in turn causes the mast cells to trigger aberrant morphogenesis by collaborating with the Schwann cells to recruit yet other cell types, such as fibroblasts and endothelial cells, thereby generating the histopathologically complex neurofibromas. In patients who are born with an *Nf1*⁺/⁻ genotype, some *Nf1*⁺/⁻ Schwann cells may undergo LOH, generating *Nf1*⁻/⁻ derivatives that recruit heterozygous (*Nf1*⁺/⁻) mast cells from the bone marrow and resulting in the formation of neurofibromas. In fact, other work suggests that the *Nf1*⁻/⁻ Schwann cells release elevated levels of stem cell factor (SCF), to which the *Nf1*⁺/⁻ mast cells are hyper-responsive because the intracellular signaling pathway downstream of their SCF receptor is hyperactive as a consequence of hyperactive Ras protein. (A, from F.C. Yang et al., *Cell* 135: 437–448, 2008. With permission from Elsevier.)

of their encoded proteins (which is often observed in cells having a $TSG^{+/-}$ genotype) can in some instances yield subtle but functionally important changes in cell phenotype. Such behavior is a manifestation of **haploinsufficiency**, a state in which the presence of only a single functional copy of a gene yields a mutant or partially mutant phenotype (**Sidebar 7.3**).

Given their role in regulating signaling through the Ras pathway, inactivation of neurofibromin or other Ras-GAPs might also be expected to occur sporadically in human cancers. Indeed, large-scale genome sequence analyses of human cancer specimens (as discussed in Section 12.12) have revealed somatic mutations in *Nf1* in a wide variety of human cancers, most notably in glioblastoma (a malignant tumor of the brain), lung cancer, and melanoma. Moreover, several other of the 13 documented Ras-GAPs encoded in the human genome have been shown to undergo mutation or epigenetic silencing in cancers (**Table 7.3**).

7.10 APC facilitates egress of cells from colonic crypts

Although the great majority (>95%) of colon cancers appear to be sporadic, a small group arises as a consequence of inherited alleles that create substantial lifelong risk for this disease. One form of familial colon cancer, referred to as hereditary nonpolyposis colon cancer (HNPCC) or Lynch syndrome, is caused by inherited mutations in a class of DNA repair genes. We will discuss HNPCC further in Chapter 12. Here we focus on the other major familial colon cancer syndrome, termed adenomatous polyposis coli (APC) or, more commonly, familial adenomatous polyposis (FAP). At relatively young ages, patients with FAP develop hundreds or thousands of growths or polyps in the colon (**Figure 7.19**). Although these polyps are themselves nonmalignant, they are prone to develop into carcinomas at a low but predictable frequency. FAP is responsible for a bit less than 1% of all colon cancers in the West.

The *Apc* gene, which is inherited in mutant form in FAP patients, was cloned in 1991 following a series of genetic mapping studies that pinpointed its location on Chromosome 5 (**Sidebar 7.4**). As in the case of the *Rb* gene, *Apc* follows a classic TSG cancer-associated mutation pattern. In FAP, patients inherit one mutant allele of the gene and accumulate mutations or LOH to inactivate the remaining wild-type allele. In sporadic colon cancer, in which *Apc* gene mutation occurs in ~90% of cases, the complete loss of *Apc* function occurs through two sequential steps, another example of the Knudson "two hit" mechanism (see Figure 7.4). Characterization of the wild-type *Apc* gene product (also called APC) has led to new insights into the control of intestinal stem cells and the normal maintenance of epithelial tissues. As we discuss in greater detail in later chapters, the epithelia in the colon and duodenum are organized in a fashion that is typical of a number of epithelia throughout the body. In these tissues, groups of relatively undifferentiated stem

Table 7.3 Examples of Ras-GAPs mutated or silenced in human cancers

Gene	Ras-GAP protein	Cancers	Mechanism of inactivation
NF1	neurofibromin	neurofibromas, neurofibrosarcomas, glioblastomas, pheochromocytomas, lung cancer, melanoma, myelogenous leukemias	mutation
RASA1	p120 Ras-GAP	lung cancer, melanoma	mutation
RASA2	Ras p21 protein activator 2	range of cancers	mutation
DAB2IP	Disabled homolog 2-interacting protein	malignant progression in breast and prostate cancers	epigenetic silencing
RASAL2	Ras protein activator like 2	breast cancer	mutation and epigenetic silencing

Adapted from D. Simanshu et al., *Cell* 170:17–33, 2017.

Figure 7.19 Familial adenomatous polyposis The wall of a normal colon is convoluted but smooth (*left*); in an individual afflicted with familial adenomatous polyposis (FAP), the colon wall is often carpeted with hundreds of small polyps (*right*). The structure of one type of colonic polyp is shown in Figure 2.14A. (Courtesy of A. Wyllie and M. Arends.)

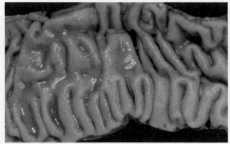

cells yield two distinct daughter cells when they divide: one daughter remains a stem cell, thereby ensuring that the number of stem cells in the tissue remains constant; however, the other daughter cell and its descendants become committed to differentiate.

In the small intestine, some of these differentiated epithelial cells participate in absorbing nutrients from the lumen and transferring these nutrients into the circulation; in the colon, many of them absorb water from the lumen. Yet other specialized epithelial cells secrete mucin (a mucus-like material) that helps to protect the colonic epithelium from the contents of the lumen. In the gastrointestinal tract, these various epithelial cells are termed **enterocytes**. In both parts of the intestine, enterocytes are born in deep, mucin-filled cavities termed **crypts**, which initially shield them from the contents of the intestine (**Figure 7.20A**).

The locations of the stem cells and more differentiated cells in a colon crypt are illustrated in (Figure 7.20B). Although some of the progeny of these stem cells stay behind to maintain a constant number of stem cells, others are dispatched upward and out of the crypts toward the luminal surface of the epithelium, where they will function briefly to form the epithelial lining of the gut; they subsequently die by apoptosis and are shed into the colonic lumen. In the human colon, this entire process of out-migration and death takes only 3 to 4 days.

The scheme depicted in Figure 7.20B represents a highly effective defense mechanism against the development of colon cancer, because almost all cells that may have sustained mutations while on duty protecting the colonic wall are doomed to die within days after they have been formed. By this logic, the only type of mutations that could lead to the development of a cancer would be those mutations (and resulting mutant alleles) that block both the out-migration of colonic epithelial cells from the crypts and the cell death that follows soon thereafter. Should a colonic enterocyte acquire such a mutation, this cell and its descendants could accumulate in the crypt, and any additional mutant alleles acquired subsequently by their progeny would be similarly retained in the crypt, rather than being rapidly lost through out-migration and apoptosis. Such additional mutations might include, for example, genes that push the progeny toward a neoplastic growth state.

Sidebar 7.4 Special human populations facilitate the detection of heritable cancer syndromes and the isolation of responsible genes The Mormons in the state of Utah represent a population that provides a golden opportunity to understand the genetics of various types of heritable human diseases. A substantial proportion of these individuals can trace their ancestry back to a relatively small group of founding settlers who arrived in Utah in the mid-nineteenth century. Mormon couples have traditionally had large numbers of children and have reproduced relatively early in life, leading to large, multigenerational families—ideal subjects of genetic research.

In addition, the Mormon church maintains the world's largest genealogical archive in Salt Lake City, which facilitates the assembly of multigenerational pedigrees. Such extensive pedigrees enabled geneticists to trace with precision how the mutant allele of a gene predisposing a person to familial adenomatous polyposis is transmitted through multiple generations of a single large family and how this allele operates in a dominant fashion to increase the susceptibility to colonic polyps and risk of colon cancer.

The use of linkage analysis, in which the genetic transmission of this predisposing allele was connected to anonymous genetic markers on various human chromosomes, revealed that this allele was repeatedly co-transmitted from one generation to the next together with genetic markers on the long arm of human Chromosome 5. This localization, together with more detailed LOH analyses, eventually made possible the molecular cloning of the *Apc* gene in 1991.

Figure 7.20 β-Catenin and the biology of colonic crypts (A) As seen in these scanning electron micrographs, the linings of the small (*left*) and large (*right*) intestines are organized similarly, with deep crypts in which stem cells reside; new enterocytes produced in these crypts migrate up toward the lumen of the gut, differentiate, and emerge, via small openings (*narrow black arrows*). In the duodenum, however, these newly born cells continue their upward migration along the sides of the fingerlike villi that protrude into the lumen (*broad white arrow*). (B) At the bottom of the colonic crypt, replicating stem cells (*gray*) receive paracrine Wnt signals (*red arrows*) from neighboring stromal cells (*red*), which keeps their levels of β-catenin high. Within individual enterocytes at the bottoms of the crypts, the β-catenin molecules migrate to the nucleus and associate with Tcf/Lef transcription factors; this complex activates genes that enhance proliferation of these cells and prevents their differentiation. In the normal intestine (*left side of crypt*), many of the progeny of these stem cells migrate upward toward the lumen (*above*). As they do so, stimulation by Wnts decreases and intracellular levels of APC increase; these two changes together lead to increased degradation of β-catenin, which results in cessation of proliferation and increased differentiation as these cells approach the lumen; ultimately, the cells undergo apoptosis and are shed into the lumen (*small green arrow*). In contrast, when the APC protein is defective (*right side of crypt*), β-catenin levels remain high, even in the absence of intense Wnt signaling, and proliferating, still-undifferentiated cells (*purple*) fail to migrate upward, accumulate within crypts, and may ultimately generate an adenomatous polyp. (A, from M. Bjerknes and H. Cheng, *Methods Enzymol.* 419:337–383, 2006. B, from M. van de Wetering et al., *Cell* 111: 251–263, 2002. Both with permission from Elsevier.)

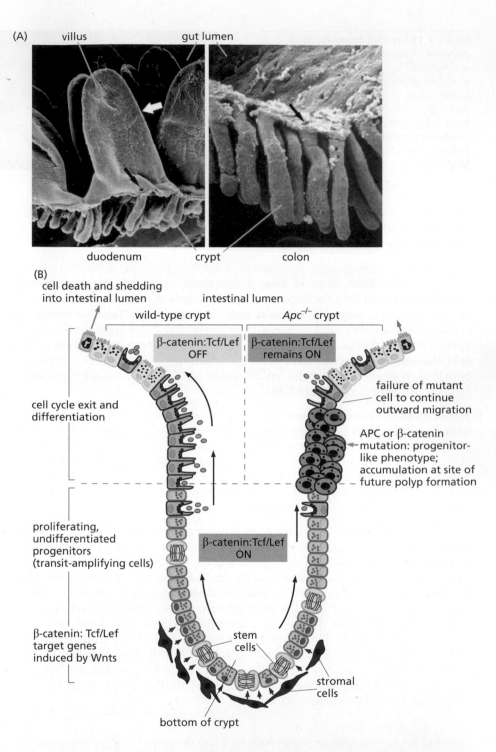

The protein called β-catenin governs much of this out-migration behavior. As described in Section 6.10, levels of soluble β-catenin in the cytoplasm are controlled by the Wnt family of growth factors and other accessory proteins. When Wnts bind cell surface receptors, β-catenin is saved from destruction; the protein then accumulates and migrates to the nucleus, where it associates with a group of DNA-binding proteins termed Tcf/Lef (see Figure 6.19B). The resulting heterodimeric transcription factor complexes ultimately activate expression of a series of target genes that program (in the case of enterocytes) the stem cell phenotype.

In the context of the intestinal crypts, enterocyte stem cells encounter Wnt factors that are released by neighboring Paneth cells in the small intestine and mesechymal stromal cells in the colon, located near the bottom of the crypt, leading to high β-catenin levels in the stem cells (see Figure 7.20B). Indeed, the β-catenin:Tcf4/

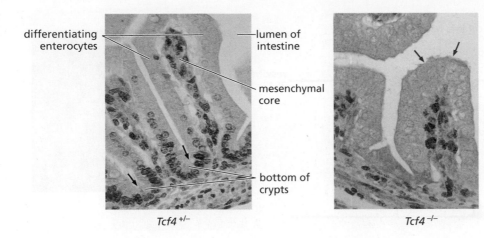

differentiating enterocytes

lumen of intestine

mesenchymal core

bottom of crypts

Tcf4 $^{+/-}$

Tcf4 $^{-/-}$

Figure 7.21 The Tcf4 transcription factor and crypt development The formation of the stem cell compartment near the bottoms of small intestinal crypts depends critically on high levels of β-catenin and on its ability to associate with the TCF4 protein in the nucleus to form an active transcription factor complex. As seen in this figure, epithelial cells in the bottoms of the crypts of the small intestine of *Tcf4*$^{+/-}$ mice (*black arrows, left panel*) are highly proliferative, as evidenced by the staining for Ki67, a marker of cell proliferation (*dark brown nuclei*). As they move out of the crypt up the sides of the villi, they lose proliferative potential. (Although the mesenchymal cells in the core of each villus are also highly proliferative, their behavior is unrelated to that of the epithelial enterocytes.) In *Tcf4*$^{-/-}$ mice (*right panel*), in contrast, a layer of enterocytes is initially formed (*red arrows*), but there are no proliferating stem cells at the bottom of the crypts; therefore, new enterocytes are not continually formed to replace those lost by apoptosis and sloughing into the intestinal lumen. (From V. Korinek et al., *Nat. Genet.* 19:379–383, 1998. With permission from Nature.)

Lef transcription factor complex is required for the formation of these stem cells (**Figure 7.21**). However, as the progeny of these stem cells begin their upward migration, they leave the Wnt signaling zone and their intracellular β-catenin levels fall precipitously. As a consequence, these cells lose their stem cell phenotype, exit the cell cycle, and differentiate into functional enterocytes.

The APC protein is normally responsible for negatively controlling the levels of β-catenin in the cytosol. In stem cells, the action of APC is inhibited through the effects of Wnt signaling, allowing β-catenin to accumulate. However, as cells begin to move out of the crypts, the declining concentration of Wnts and other factors, plus the increased levels of *Apc* expression in these cells, leads to a significant decline in intracellular levels of β-catenin cited above and thus the levels of nuclear β-catenin:Tcf/Lef transcription factor complexes.

The molecular mechanism of action of APC helps to explain these declining levels of intracellular β-catenin: APC is a large protein of 2843 amino acid residues that can associate with β-catenin (**Figure 7.22A**). Together with scaffolding proteins, APC forms a multiprotein complex that brings together glycogen synthase kinase-3β (GSK-3β) and β-catenin (see Figure 6.19B). This association enables GSK-3β to phosphorylate four amino-terminal residues of β-catenin, an event that leads to the degradation of β-catenin via the ubiquitin–proteasome pathway (Supplementary Sidebar 7.5). APC is thus essential for triggering the degradation of β-catenin; in its absence, β-catenin levels accumulate to high levels within cells.

With this mechanism in mind, we can see a clear role for the mutation of *Apc* in colon cancer, whether it is inherited, as in FAP patients, or occurs via two independent mutational events, as in the high percentage of cases of the sporadic form of the disease. Cataloging the spectrum of *Apc* mutations found in human colon cancers reveals many mutations that cause premature termination of translation of the APC protein, thereby removing domains that are important for its ability to associate with β-catenin and the scaffolding protein axin and for the resulting degradation of β-catenin (Figure 7.22B).

The accumulation of β-catenin (Figure 7.22C) is clearly the most important consequence of *Apc* inactivation. One can draw this conclusion from studying the minority of sporadic colon carcinomas (approximately 10%) that carry wild-type *Apc* alleles. In some of these, the *Apc* gene promoter is hypermethylated and rendered inactive. In others, the gene encoding β-catenin carries point mutations, and the resulting mutant β-catenin molecules lose the amino acid residues that are normally phosphorylated by GSK-3β. Because they cannot be properly phosphorylated, these mutant β-catenin molecules escape proteasomal degradation and accumulate—precisely the outcome that is seen when APC is missing.

When β-catenin accumulates in enterocyte precursors—as a result of APC inactivation or alternative mechanisms—the affected cells retain a progenitor cell-like phenotype, which prevents them from migrating out of the crypts. This retention leads, in turn, to the accumulation of large numbers of relatively undifferentiated cells

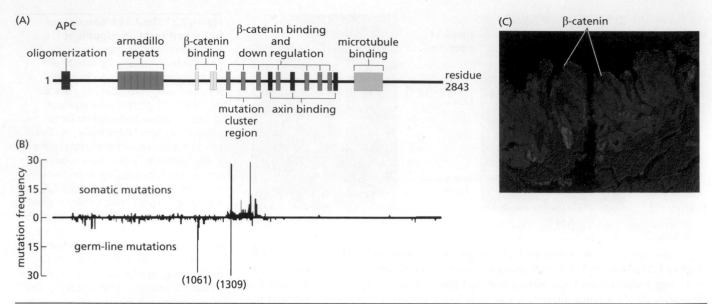

Figure 7.22 *Apc* mutations and β-catenin (A) This diagram outlines the multidomain structure of APC. The interactions of the microtubule-binding domain (*light green*) are important for proper regulation of cell motility and mitosis (see Supplementary Sidebar 7.6). (B) This plot shows the spectrum of *Apc* mutations and where they occur with relation to the domains outlined in (A). As can be seen, the mutations that occur with high frequency tend to affect APC's ability to bind to β-catenin. (C) In the Min mouse model of familial polyposis, an inactivating point mutation was introduced into the *Apc* gene by the actions of a mutagenic chemical. The chromosomal region carrying this mutant allele undergoes LOH in some cells within intestinal crypts. In these cells, which lack all APC activity, β-catenin is no longer degraded, accumulates to high levels, and enters the cell nuclei, where it collaborates with Tcf/Lef transcription factors to drive expression of growth-promoting genes and prevent the usual exit of cells from the crypt. The result is a clonal outgrowth within the crypt, such as the adenoma seen here, in which the β-catenin is visualized by immunostaining (*pink, above*). The nuclei in all cells, including those of the normal crypts (*below*), are seen in *blue*. (A and B, from P. Polakis, *Biochim. Biophys. Acta* 1332:F127–F147, 1997, and adapted from R. Fodde et al., *Nat. Rev. Cancer* 1:55–67, 2001. With permission from Elsevier. C, courtesy of K.M. Haigis and T. Jacks.)

in the colonic crypt (see Figure 7.20B), which eventually form adenomatous polyps. Equally important, these accumulating cells can later sustain additional mutations that enable them to form more advanced polyps and, following even more mutations, carcinomas.

This model explains the sequence of mutations that, over time, leads to the formation of human colon carcinoma cells. The first of these mutations invariably involves inactivation of *Apc* function (or the functionally equivalent changes mentioned above). The resulting cells, now trapped in the crypts, may then suffer mutations in a number of other genes—such as K-*ras* (see Chapter 4) and *p53* (see Chapter 9)—which cause the cells to proliferate more aggressively. Importantly, alterations of the APC–β-catenin pathway nearly always come first, whereas the order of the subsequent genetic changes is more variable.

Many important insights into the function of this pathway in cancer came from the study of mouse models of colorectal cancer. In the early days, such models were generated by random, chemically induced mutagenesis followed by selection for the phenotype of interest—in this case, the appearance of intestinal adenomas in the mutant animals. One of the earliest models of colon cancer, dubbed the *Min* mouse (for its multiple intestinal neoplasias), was developed in this manner. These animals, which can develop 100 polyps in the small intestine in addition to colon tumors, were later found to contain heterozygous point mutations in the *Apc* gene. Tumor cells in these mice were shown to undergo LOH to eliminate the remaining wild-type *Apc* allele.

As discussed above, today biologists have access to more sophisticated genetic tools for inactivating or "knocking out" specific genes in a more targeted fashion—including genes suspected of being TSGs (see Sidebar 7.2 and Figure 7.17). Like *Min* mice, mice engineered to have one inactive *Apc* allele (*Apc*$^{+/-}$) are predisposed to develop intestinal tumors. Using a conditional knockout strategy (see Figure 7.17) to mutate the *Apc* gene specifically in intestinal stem cells, researchers discovered that this targeted inactivation fueled the growth of microscopic tumors at the base of the crypts. These microadenomas then expanded unimpeded, forming adenomas within a matter of weeks. In contrast, elimination of *Apc* function in the short-lived transit-amplifying cells triggered the formation of microadenomas whose growth rapidly fizzled out; full-blown adenomas were rarely found in these mutant mice. These observations reinforce the notion that loss of *Apc* in stem cells is the event that drives colon cancer.

Another form of conditional inhibition of *Apc* function relies on the use of inducible expression of short hairpin RNAs (shRNAs), which target the gene for inactivation through the RNA interference (RNAi) pathway, as will be discussed in Chapter 12. In these experiments, suppressing *Apc* function with an *Apc*-targeted shRNA generates tumors in mice, as would be expected. What is surprising, however, is that shutting off the shRNA—which restores the expression of *Apc*—leads to rapid tumor regression, even in highly advanced tumors. Hence, as discussed later in Chapter 11, the mutant allele that initiated the multi-step development of a tumor continues to be critical to its ongoing growth long after additional mutant alleles have been acquired by the cancer cells. Conditional knockouts and conditional RNAi approaches are now being used widely to interrogate TSG function.

7.11 KEAP1 regulates cellular response to oxidative stress

Thus far, we have focused mainly on TSGs whose primary function is related to constraining cell proliferation. Mutation of these genes would be expected to liberate a developing cancer cell from a variety of growth-inhibitory mechanisms and allow increased, inappropriate proliferation. Consistent with this role in proliferation control, mutations in such genes typically occur in the early stages of tumorigenesis and can serve as the initiating events for certain cancer types. However, TSGs can also control processes linked to disease progression—that is, the acquisition of more aggressive phenotypes that occurs later in the multi-step process of tumorigenesis. Mutant alleles of these types of genes would be selected for only at later stages of tumorigenesis. One prominent example of such a TSG is *Keap1*, a gene that encodes the protein KEAP1 (Kelch-like ECH-associated protein 1), which is mutated in a significant percentage of lung and other cancers.

In normal cells, KEAP1 acts as a sensor for a variety of chemical and environmental stresses—from the carcinogens in cigarette smoke and the phytochemicals produced by the plants we eat to the reactive oxygen species (ROS) generated by cells as a by-product of oxidative metabolism and intracellular signaling pathways. When detected by KEAP1, such chemically reactive, toxic compounds activate a cellular response program that directs the expression of a suite of genes encoding various cellular antioxidants and detoxification enzymes. This defensive response relies on the interaction of KEAP1 and the transcription factor called NRF2.

In the absence of oxidative stress, KEAP1 sequesters NRF2 in the cytoplasm and targets this protein for **ubiquitylation** and degradation; as a result, this protein has a lifetime of only 20 minutes (see Supplementary Sidebar 7.5). This ongoing targeted destruction of NRF2 minimizes the activity of the antioxidant response pathway under normal conditions (**Figure 7.23**). However, when cells are exposed to ROS or other toxic molecules, these chemicals react with key cysteine residues in KEAP1, which causes KEAP1 to release NRF2. Once liberated from KEAP1, NRF2 is free to translocate to the nucleus, where it joins with members of the MAF transcription factor family as well as other transcription factors and binds to genes whose promoters include antioxidant responsive element (ARE) sequences. In this way, NRF2 orchestrates the activation of a series of genes involved in cellular detoxification (see Figure 7.23). Once the reactive compounds have been cleared, the chemical modifications to the KEAP1 cysteines are reversed, allowing the protein to again sequester NRF2 and trigger its degradation in the cytoplasm.

By eliminating carcinogens and disarming ROS that could otherwise damage DNA and other cellular constituents, this cellular detoxification program can prevent cancer initiation and progression. However, in cells that are already malignant, activation of

Figure 7.23 KEAP1, NRF2, and the cellular antioxidant response In normal cells, the KEAP1–NRF2 pathway monitors the cellular environment and, upon exposure to ROS, carcinogens, or other chemically reactive compounds, activates an elaborate transcriptional program that boosts the production of cellular antioxidants and detoxifying enzymes. Under normal conditions, in the absence of these reactive chemicals, KEAP1 sequesters NRF2 in the cytoplasm (*left*). This interaction brings NRF2 in contact with a protein complex containing an E3 ligase—an enzyme that recognizes NRF2 and tags it with a polyubiquitin (Ub) chain. Polyubiquitylated NRF2 is then degraded in a proteasome (as described in Supplementary Sidebar 7.5). This destruction of NRF2 keeps the activity of the antioxidant response pathway low. Exposure to chemically reactive compounds disrupts the binding of KEAP1 and NRF2 (*right*). Once released from KEAP1, NRF2 accumulates and enters the nucleus, where it couples with a MAF transcription factor and activates expression of genes containing an antioxidant-reactive element (ARE) in their promoters. These genes encode a variety of proteins that act as antioxidants and detoxifying enzymes.

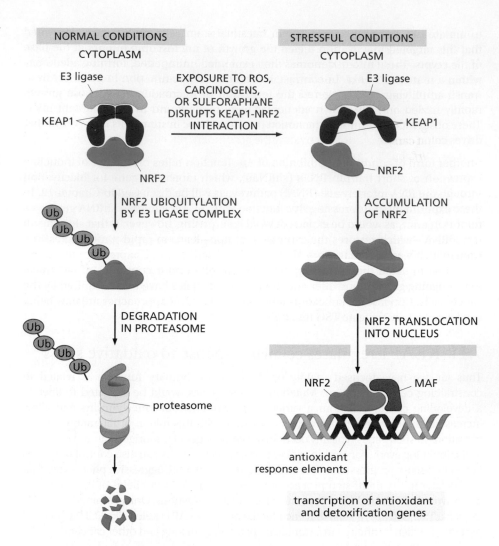

this pathway can provide a selective advantage, both by thwarting chemotherapies and allowing cells to survive and grow under conditions of higher levels of oxidative stress. Indeed, changes in their metabolism often cause cancer cells to produce high concentrations of ROS. Under these conditions, cancer cells that boost their detoxification ability may have a marked selective advantage.

Some cancers achieve this detoxification boost by mutating *Keap1*. Loss-of-function mutations in *Keap1* can chronically upregulate the NRF2-dependent detoxification pathway. In one study, inactivating mutations in *Keap1* were found in the genomes of half of the lung cancer cell lines examined. These mutations included missense mutations (which cause an amino acid substitution in the protein), deletions, frameshift mutations, and nonsense mutations (which cause premature truncation of the protein)—all of which are characteristic of the types of loss-of-function mutations associated with other TSGs. Subsequent large-scale cancer genome sequencing studies (as described in Chapter 12) extended these findings to large numbers of primary cancer specimens. Among the numerous human cancers profiled, adenocarcinomas and squamous cell cancers of the lung were found to have the highest frequency of *Keap1* mutations (17 and 12%, respectively), although *Keap1* mutations were also observed in other cancer types at significant frequency. The same studies also found that the chromosomal region around *Keap1* is a frequent target of LOH in lung cancers and other cancer types. In addition, many cancers—particularly those of lung and prostate—show hypermethylation of the *Keap1* promoter, another common means of functionally inactivating TSGs, as discussed earlier in this chapter.

As an alternative to inactivating mutations in *Keap1*, the KEAP1–NRF2 pathway can also be made constitutively active by mutations in the *Nrf2* gene. In contrast to those seen in *Keap1*—which affect a variety of amino acids throughout the KEAP1 sequence—the cancer-associated mutations in *Nrf2* are narrowly targeted to the sequences encoding the amino acid residues that mediate NRF2 binding to KEAP1. These gain-of-function mutations prevent KEAP1-mediated degradation of NRF2 while preserving its ability to activate antioxidant-responsive genes. Hence, both the inactivating *Keap1* mutations and the gain-of-function *Nrf2* mutations lead to constitutive activation of the pathway. Analogous to the cancer-causing mutations that affect either the APC or the β-catenin genes within a given colorectal carcinoma genome (see Section 7.10), those cancers that carry mutations activating the antioxidant response pathway affect either *Keap1* or *Nrf2* but not both genes.

Functional studies in mice have confirmed the importance of the KEAP1–NRF2 pathway in tumor progression. For example, targeted inactivation of *Keap1* in a mouse model of lung adenocarcinoma using CRISPR-Cas9—a powerful gene editing technique described in Sidebar 12.11—led to accelerated growth and a significant expansion in the volume of primary tumors (**Figure 7.24A**). These tumors exhibited increased levels of nuclear NRF2 as well as increased expression of NRF2 target genes. Interestingly, *Keap1*-deficient tumors and cell lines also show altered metabolic requirements, which may suggest novel targets for cancer therapy. In other mouse models of lung and pancreatic cancer, inactivation of *Nrf2* significantly inhibited tumor development (Figure 7.24B). These animal studies provide strong evidence for the pro-tumorigenic effect of constitutive activation of the antioxidant pathway, as driven by inactivation of KEAP1. The loss of KEAP1 thus creates an intracellular environment that enables evolving neoplastic cells to tolerate a variety of tumor-promoting alterations.

Finally, because the KEAP1–NRF2 pathway is important for detoxifying mutagenic carcinogens, mutations in *Keap1* and *Nrf2* can affect the potency of certain chemical carcinogens, at least in animal studies. Mutations in *Keap1* that lead to pathway activation in the mouse can inhibit tumorigenesis induced experimentally by mutagenic carcinogens, presumably because these chemicals are more effectively detoxified. Interestingly, mutations that inactivate *Nrf2* can *enhance* cancer initiation in response to such mutagenic agents but *reduce* the subsequent progression of these cancers to full-fledged malignancy (because they are less capable of coping with the oxidative stress conditions found in highly malignant cells). This observation reflects the differing roles that can be played by the antioxidant response at different stages of multi-step tumor progression.

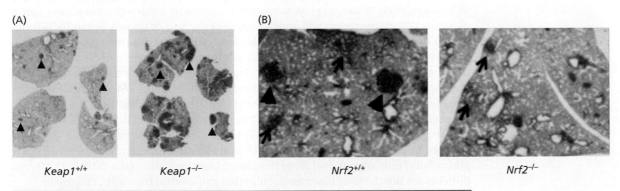

(A)   (B)

Keap1$^{+/+}$ *Keap1*$^{-/-}$ *Nrf2*$^{+/+}$ *Nrf2*$^{-/-}$

Figure 7.24 Effects of inactivating mutations in the *Keap1* and *Nrf2* genes on lung tumor development (A) Inactivation of the *Keap1* gene in a mouse model of lung adenocarcinoma leads to larger and more advanced tumors. The left panel shows hematoxylin plus eosin (H&E) staining of lungs containing adenomas (*black arrowheads*) formed in a genetically engineered mouse model of lung cancer in which *Keap1* is wild type. CRISPR-Cas9–mediated deletion of *Keap1* in this model (*right panel*) results in larger, more advanced tumors (*black arrowheads*). (B) In a related model of lung cancer, germ-line deletion of *Nrf2* (*right panel*) causes a reduction in the size of hyperplasias (*black arrows*) and adenomas (*black arrowheads*) compared to those seen in mice with wild-type *Nrf2* (left panel). Note that images in (B) are taken at higher magnification than those in (A). A, from R. Romero et al. *Nature Medicine* 23(11):1362-1368, 2017. Reprinted with permission from Springer Nature. B, from G. DeNicola et al., *Nature* 475:106-109, 2011. Reprinted with permission from Springer Nature.)

7.12 Not all familial cancers can be explained by inheritance of mutant TSGs

Many of the cloned TSGs listed in Table 7.1—and indeed, many of the TSGs we have discussed in this chapter—are involved in both familial and sporadic cancers. Generally speaking, this means that inheritance of defective copies of these genes creates a greatly increased risk of developing one or another specific type of cancer, often a type of tumor that is otherwise relatively rare in the human population.

However, not all familial cancer syndromes can be traced back to an inherited TSG allele. As we will discuss in Chapter 12, mutant germ-line alleles of a second class of genes also cause cancer predisposition. These other genes are normally responsible for maintaining the cellular genome and thus act to reduce the likelihood of mutations and chromosomal abnormalities will accumulate. Because cancer pathogenesis depends on the accumulation by individual cells of multiple somatic mutations, genes that reduce the mutation rate, such as these genome maintenance genes, are highly effective in suppressing cancer onset. Conversely, defects in genome maintenance often lead to a significant increase in cancer risk because they increase the overall mutation rate of a cell.

Thus, there are really two distinct classes of familial cancer genes: the TSGs described in this chapter and the genome maintenance genes described in Chapter 12. We can rationalize the distinction between the two classes of genes as follows. The TSGs function to directly control the biology of cells by affecting how they proliferate, differentiate, or die; genes functioning in this way are sometimes called **gatekeepers** to indicate their role in allowing or disallowing cells to progress through cycles of growth and division. The DNA maintenance genes affect cancer cell biology only indirectly by controlling the rate at which cells accumulate mutant genes; these genes have been termed **caretakers** to reflect their role in the maintenance of cellular genomes. Unlike mutant gatekeeper and caretaker alleles, mutant, functionally activated versions of proto-oncogenes are rarely transmitted through the germ line and are thus not commonly associated with familial cancer syndromes (**Sidebar 7.5**).

7.13 Synopsis and prospects

TSGs constitute a large group of genes having only one shared attribute: in one way or another, each of these genes normally functions to reduce the likelihood that a clinically detectable tumor will appear in one of the body's tissues. Some TSGs function to decrease the likelihood that early steps of multi-step tumor progression will occur, whereas others operate specifically to suppress completion of the later steps of tumor progression. The dynamics governing these complex processes are the subject of Chapter 11. We therefore should expand the definition of TSGs to include genes that operate to suppress initial benign tumor formation, as well as those that constrain subsequent malignant progression. A further refinement of definitions has come from the demonstration that some TSGs encode microRNAs rather than proteins; the pleiotropic powers of many microRNAs makes them potent regulators of complex cellular phenotypes, including neoplastic transformation (see Table 7.1). Hence, proteins are not the only macromolecules that can control runaway cell growth.

The discovery of TSGs helped to explain one of the major mysteries of human cancer biology—that of familial cancer syndromes. As we discussed in this chapter, inheritance of a defective allele of one of these genes is often compatible with normal embryonic development. The phenotypic effects of this genetic defect may only become apparent with great delay, sometimes in midlife, when its presence is revealed by the loss of the surviving wild-type allele and the outgrowth of a particular type of tumor. Elimination of these wild-type alleles often involves LOH events, and the repeatedly observed LOH in a certain chromosomal region among a group of tumors can serve, on its own, as a clear indication of the presence of a still-unidentified TSG lurking in this region.

The diverse behaviors of these genes highlight an ongoing difficulty in this area of cancer research: What criteria can be used to define a TSG? For our purposes, we

Sidebar 7.5 Why are mutant tumor suppressor genes transmitted through the germ line, whereas mutant proto-oncogenes are usually not? A number of familial cancer syndromes have been associated with the inheritance of mutant alleles of TSGs (see Table 7.1). Very few cancer syndromes, however, are associated with inherited mutant alleles of proto-oncogenes (that is, activated oncogenes). These include Costello syndrome (H-*ras*); familial gastrointestinal stromal tumors (*Kit* and *PDGFR-A*); hereditary papillary renal cell cancer (*Met*); multiple endocrine neoplasia (*Ret*); Noonan syndrome (*PTPN11*, K-*ras*, and *SOS*); and familial neuroblastoma (*ALK*). How can we explain this dramatic difference in the heritability of mutant TSGs and mutant proto-oncogenes?

Mutations that yield activated oncogenes, like those that eliminate the activity of TSGs, are likely to arise with some frequency during **gametogenesis**—the processes that form sperm and egg. Thus, they are likely to be transmitted to fertilized eggs. However, because oncogenes act at the cellular level as dominant alleles, these mutant alleles are likely to perturb the behavior of individual cells in the developing embryo and therefore to disrupt normal tissue development. Consequently, embryos carrying these mutant oncogenic alleles are unlikely to develop to term, and these mutant alleles will disappear from the germ line of a family and thus from the gene pool of the species. (For example, experiments reported in 2004 indicate that mouse embryos arising from sperm carrying a mutant, activated K-*ras* oncogene develop only to mid-gestation; the mutant germ-line alleles of the H-*ras* and K-*ras* genes that cause Costello and Noonan syndromes in humans carry mutations that are much weaker in their transforming powers than those somatically mutated alleles responsible for creating the *Ras* oncogenes that underlie many types of cancer.)

Mutant germ-line alleles of TSGs behave much differently, however. Because these alleles are generally recessive at the cellular level, their presence in most cells of an embryo will not be apparent. For this reason, the presence of inherited mutant TSGs will often be compatible with normal embryonic development, and the cancer phenotypes that they create will become apparent only in a small number of cells and after great delay, allowing an individual carrying these alleles to develop normally and, as is often the case, to survive through much of adulthood.

Still, if mutant TSGs undergo LOH in 1 out of 10^4 or 10^5 cells, and if an adult human has many more than 10^{13} cells, why doesn't a person who inherits a mutant TSG develop tens of thousands, even millions of tumors? The answer stems from the fact that, for most types of cancer, tumorigenesis is a multi-step process (as we discuss in Chapter 11). This means that a mutant gene (whether it be an oncogene or a TSG) may be necessary for tumor formation but will not, on its own, be sufficient. Hence, many cells in an individual heterozygous for a mutant TSG may well undergo LOH of this gene, but only a tiny minority of these cells will ever acquire the additional genetic changes needed to make a clinically detectable tumor. For other, rare cancers—including retinoblastoma—the inactivation of a single gene (*Rb*) may be sufficient to initiate tumor formation, at least in fetal retinal cells. This explains why individuals with the Rb^+/Rb^- genotype are predisposed to this tumor type.

have included in the family of TSGs only those genes whose products operate in some dynamic fashion to constrain cell proliferation, survival, or malignant progression. Other genes that function indirectly to prevent cancer through their abilities to maintain the genome and suppress mutations are described in Chapter 12. From the perspective of a geneticist, this division—the dichotomy between the "gatekeepers" and the "caretakers"—is an artificial one, because the wild-type versions of both classes of genes are often found to be eliminated or inactivated in the genomes of cancer cells. Moreover, the patterns of inheritance of the cancer syndromes associated with defective caretaker genes are formally identical to the mechanisms described in this chapter. Still, for those who would like to understand the biological mechanisms of cancer formation, the distinction between TSGs (the gatekeepers) and genome maintenance genes (the caretakers) is a highly useful one and is therefore widely embraced.

In the years since the first TSGs (*p53* and *Rb*) were cloned, numerous other genes have been touted as "candidate" TSGs because their expression was depressed or absent in cancer cells, while being readily detectable in corresponding normal cells. This criterion for membership in the TSG family was soon realized to be flawed, in no small part because it is often impossible to identify the normal precursor of a cancer cell under study. Certain kinds of tumor cells may not express a particular gene because of a gene expression program that is played out during normal differentiation of the tissue in which these tumor cells have arisen. Hence, the absence of expression of this gene in a cancer cell may reflect only the actions of a normal differentiation program rather than a pathological loss. So, this criterion—absence of gene expression—is not definitive.

In certain instances, expression of a candidate TSG may be present in the clearly identified normal precursors of a group of tumor cells and absent in the tumor cells themselves. This would appear to provide slightly stronger support for the candidacy of this gene. But even this type of evidence is not conclusive, because the absence of gene expression in a cancer cell may often be one of the myriad *consequences* of the transformation process rather than one of its root *causes*. Thus, inactivation of this gene may have played no role whatsoever in the formation of the tumorigenic cell.

Responding to these criticisms, researchers have undertaken to functionally test candidate TSGs. The process involves introducing cloned wild-type versions of candidate genes into cancer cells that lack any detectable expression of these genes. The goal here has been to show that once the wild-type TSG function is restored in these cancer cells, they revert partially or completely to a normal growth phenotype, or they may even enter into apoptosis (as discussed in Chapter 9). However, the results of such experiments are often difficult to interpret, because the expression of genes at unnaturally elevated levels—and in cells or locations where they do not normally belong (termed **ectopic** expression)—often makes cells quite unhappy and causes them to stop growing or even to die. Such responses are often observed following introduction of a variety of genes that would never be considered TSGs.

The ambiguities of these functional tests have necessitated the introduction of genetic criteria to validate the candidacy of many putative TSGs. If a gene repeatedly undergoes LOH in tumor cell genomes, for example, then surely its candidacy becomes far more credible. But here too ambiguity reigns. After all, genes that repeatedly suffer LOH may only be closely linked on a chromosome to a bona fide TSG that is the true target of elimination during tumor development.

These various considerations have led to an even stricter genetic definition of a TSG: a gene can be safely called a TSG only if it undergoes LOH in many tumor cell genomes *and* if the resulting homozygous alleles bear clear and obvious inactivating mutations. (This latter criterion should allow an investigator to discount any bystander genes that happen to be closely linked to bona fide TSGs on human chromosomes.) Recent advances in large-scale genomic sequencing, which we will discuss in Chapter 12, have now revealed more than 200 human genes that meet these criteria.

Not surprisingly, even these quite rigid genetic and molecular criteria have proven to be flawed, because they exclude from consideration certain genes that are likely to be genuine TSGs. The activity of many TSGs can be eliminated by promoter methylation (see Section 7.7), which explains why mutant alleles of a bona fide TSG might rarely be encountered in tumor cell genomes, even though effective silencing of the gene has occurred and contributed, in turn, to tumor outgrowth.

The ability to knock out or otherwise modify the functions of candidate TSGs through manipulation of the mouse genome line (see Sidebar 7.5) provides yet another powerful tool for validating their role in tumorigenesis. Although the biology of mice and humans differs in many respects, the shared, fundamental features of mammalian biology make these genetically altered mice extraordinarily useful for accurately modeling many aspects of human tumor biology.

Many of the genes that are listed in Table 7.1 have been knocked out in the germ line of mice. For the most part, resulting heterozygotes have been found to exhibit increased susceptibility to one or another type of cancer. In many instances, the particular tissue that is affected differs from that observed in humans. For example, *Rb*-heterozygous mice (that is, $Rb^{+/-}$) tend to develop pituitary and thyroid tumors rather than the retinoblastomas observed in humans. Still, the development of any type of tumor at an elevated frequency in such genetically altered mice adds persuasive evidence to support the candidacy of a gene as a TSG.

All this explains why multiple criteria are now brought to bear when evaluating nominees for membership in this exclusive gene club. For us, perhaps the most compelling criterion is a functional one: Can the tumor-suppressing ability of a candidate TSG be rationalized in terms of the biochemical activities of its encoded protein and the known position of this protein in a cell's regulatory circuitry?

In the end, these many complications in validating candidate TSGs have one attribute in common: the very existence of a TSG becomes apparent only when it is absent. This fact sets the stage for all the difficulties that have bedeviled TSG research, and it underscores the difficulties that will continue to impede the validation of new TSGs in the future.

With all these reservations in mind, we can nevertheless distill some generalizations about TSGs that will likely stand the test of time. To begin, the name that we apply to these genes is, in one sense, a misnomer. The normal roles of many are to suppress inappropriate increases in cell number, either by suppressing proliferation or by triggering apoptosis in normal tissues. The participation of these genes in cancer formation only becomes apparent in their absence. Thus, in the absence of these genes, cells survive and proliferate at times and in places where their survival and proliferation are inappropriate. Beyond the resulting effects on cell number are the important roles of some of these genes, as highlighted here, in constraining malignant progression of already-initiated tumors.

Another generalization is also obvious, even without knowing the identities of all TSGs: their protein products do not form any single, integrated signaling network. Instead, these proteins crop up here and there in the wiring diagrams of various regulatory sub-circuits operating in different parts of the cell. This diversity is explained by the simple and obvious rationale with which we began this chapter: all well-designed control systems have both positive and negative regulatory components that counterbalance one another. Thus, for every type of positive signal, such as those signals that flow through mitogenic signaling pathways, there must be negative controllers ensuring that these signaling fluxes are kept within proper limits. Perhaps cancer biologists should have deduced this from first principles, long before they launched their studies of TSGs.

Even after the full repertoire of TSGs are cataloged and characterized, the pRb and p53 proteins will continue to be recognized as the products of TSGs that are of preeminent importance in human tumor pathogenesis. The reasons for this will become apparent in the next two chapters. Although all the pieces of evidence are not yet in hand, it seems increasingly likely that the two signaling pathways controlled by pRb and p53 are deregulated in the great majority of human cancers. Almost all of the remaining TSGs (see Table 7.1) are involved in the development of more circumscribed sets of human tumors.

The fact that certain TSGs are missing from the expressed genes within cancer cells has prompted many to propose the obvious: if only we could replace or repair the missing TSGs in cancer cells, these cells would revert partially or totally to a normal cell phenotype and the problem of cancer would be largely solved. Such "gene therapy" strategies have the additional attraction that the occasional, inadvertent introduction of a TSG into a normal cell should have little if any negative effect if the gene is expressed at an appropriate, physiological level; this reduces the risk of undesired side-effect toxicity to normal tissues.

As attractive as such gene therapy strategies are in concept, they have been extremely difficult to implement in practice. The viral vectors that form the core of most of these procedures are inefficient in delivering intact, wild-type copies of TSGs to the neoplastic cells within tumor masses. Unfortunately, efficient transfer is essential for curative anti-tumor therapies: if significant numbers of cancer cells within a tumor under treatment fail to acquire a vector-borne wild-type TSG, these cells will serve as the progenitors of a reborn tumor mass. Similar concerns about the efficiency of gene delivery lessen enthusiasm for using CRISPR-based genome editing techniques to repair defective TSGs in cancer cell genomes. For these reasons, although TSGs are profoundly important to our understanding of cancer formation, in most cases, transformation of this knowledge into therapeutic practice is still largely beyond our reach.

Given the challenges associated with restoring TSG function as a therapeutic strategy, an alternative approach is to target signaling pathways that are activated following TSG loss. For example, loss of the *PTEN* TSG leads to hyperactivity of the kinase Akt/PKB, on which some types of cancer cells come to depend for their continued viability. This dependence explains why drugs that shut down an essential upstream co-activator of Akt/PKB, termed mTOR, hold promise as therapeutic agents against tumors, such as glioblastomas and prostate carcinomas, in which PTEN activity has been lost.

The loss of a TSG may render a cell more susceptible to other types of intervention. For example, using CRISPR-based methods (see Sidebar 12.11), it is now possible to perform genetic screens to systematically identify genes and pathways that cancer cells lacking the function of a TSG have come to depend on for their continued growth or survival. This strategy is often termed a **"synthetic lethal screen,"** simply because elimination of a gene identified in such a screen is only injurious to a cell when accompanied by the presence of a mutant TSG allele located elsewhere in the genome of a cancer cell. This approach holds the promise of identifying drug targets for the treatment of cancers with particular TSG mutations. At present, we confront a rapidly expanding roster of TSGs and a rudimentary understanding of how most of them work to forestall cancer, which explains why we cannot begin to forecast the clinical ramifications of this work that will be realized only over the coming decades.

Key concepts

- At the cellular level, the neoplastic phenotype of human cancer cells is usually recessive, because expression of this phenotype depends, in part, on the inactivation of tumor suppressor genes (TSGs). This indicates that the loss of genetic information is critical for the development of most if not all human tumors.

- Much of the loss of functionally important genetic information during the formation of tumors is demonstrated by **TSGs**, which are often present in the genomes of cancer cells as inactive, null alleles.

- TSG loss usually affects cell phenotype only when both copies of such a gene are lost in a cell. This means that null alleles of TSGs may be present in a heterozygous state in many cells throughout the body without affecting cell and tissue phenotype.

- The loss of TSG function can occur either through genetic mutation or the epigenetic silencing of genes via promoter methylation.

- Inactivation (by mutation or methylation) of one copy of a TSG may be followed by other mechanisms that facilitate loss of the other gene copy; these mechanisms often depend on loss of heterozygosity (LOH) at the TSG locus and may involve mitotic recombination, loss of a chromosomal region that harbors the gene, inappropriate chromosomal segregation (nondisjunction), or gene conversion stemming from a switch in template strand during DNA replication.

- LOH events usually occur more frequently per cell generation than mutations or promoter methylation, and they occur with different frequencies in different genes.

- Repeated LOH occurring in a given chromosomal region in a number of independently arising tumors often indicates the presence of a TSG in that region.

- In certain instances, the absence of a single functional copy of a TSG leads to changes in cell or tissue behavior—the state of haploinsufficiency.

- When mutant, defective copies of a TSG are inherited in the germ line, the result is often greatly increased susceptibility to one or another specific type of cancer.

- The loss of TSGs occurs far more frequently during the development of a tumor than the activation of proto-oncogenes into oncogenes, and individual tumor cell genomes usually harbor multiple inactivated TSGs.

- TSGs regulate cell proliferation through many biochemical mechanisms. The only theme that unites them is that the loss of any one of them increases the likelihood that a cell will undergo neoplastic transformation or progression to a highly malignant state.

- TSGs are often called "gatekeepers" to signify their involvement in governing the dynamics of cell proliferation and to distinguish them from a second class of genes, the "caretakers," which also increase cancer risk when inherited in defective form but function entirely differently, working to maintain the integrity of the cell genome.

- Replacing or repairing a TSG as a therapeutic strategy for cancer is challenging because of limitations in the efficiency of gene delivery technologies; targeting pathways that become activated upon TSG loss is a promising alternative strategy.

Thought questions

1. Why is the inheritance of mutant, activated oncogenes responsible for only a small proportion of familial cancer syndromes, whereas the inheritance of defective tumor suppressor genes (TSGs) is responsible for the lion's share of these diseases?

2. What factors may determine whether the inactivation of a TSG occurs at a frequency per cell generation higher than the activation of an oncogene?

3. How might the loss of TSG function yield an outcome that is, at the cell-biological level, indistinguishable from the acquisition of an active oncogene?

4. Some TSGs undergo LOH in fewer than 20% of tumors of a given type. Why and how do such low rates of LOH complicate the identification and molecular isolation of such genes?

5. What criteria need to be satisfied before you would be comfortable in categorizing a gene as a TSG?

6. What factors might influence the identities of the tissues affected by an inherited, defective allele of a TSG?

7. What methods could be used to identify genes or pathways that cancer cells harboring a particular TSG mutation become dependent on for growth or survival?

8. What evidence might be produced to demonstrate that a group of candidate TSGs function in the same regulatory pathway within normal cells?

Additional reading

Baylin SB & Jones PA (2016) Epigenetic determinants of cancer. *Cold Spring Harb. Perspect. Biol.* 8:a019505.

Berdasco M & Esteller M (2010) Aberrant epigenetic landscape of cancer: how cellular identity goes awry. *Dev. Cell* 19, 698–711.

Brahimi-Horn MC, Chiche J & Pouysségur J (2007) Hypoxia and cancer. *J. Mol. Med. (Berl.)* 85, 1301–1307.

Comings DE (1973) A general theory of carcinogenesis. *Proc. Natl. Acad. Sci. USA* 70, 3324–3328.

Esteller M (2007) Cancer epigenomics: DNA methylomes and histone-modification maps. *Nat. Rev. Genet.* 8, 286–298.

Esteller M (2009) Epigenetics in cancer. *N. Engl. J. Med.* 358, 1148–1159.

Foulkes WD (2008) Inherited susceptibility to common cancers. *N. Engl. J. Med.* 359, 2143–2153.

Gregorieff A & Clevers H (2005) Wnt signaling in the intestinal epithelium: from endoderm to cancer. *Genes Dev.* 19, 877–890.

Herman JG & Baylin SB (2003) Gene silencing in cancer in association with promoter hypermethylation. *N. Engl. J. Med.* 349, 2042–2054.

Hershko A, Ciechanover A & Varshavsky A (2000) The ubiquitin system. *Nat. Med.* 6, 1073–1081.

Humphries A & Wright NA (2008) Colonic crypt organization and tumorigenesis. *Nat. Rev. Cancer* 8, 415–424.

Jiang F & Doudna JA (2017) CRISPR-Cas9 structures and mechanisms. *Annu. Rev. Biophys.* 45, 505–529.

Jones PA, Issa JP & Baylin S (2016) Targeting the cancer epigenome for therapy. *Nat. Rev. Genet.* 17:630–641.

Kaelin WG Jr (2008) The von Hippel–Lindau tumour suppressor protein: O2 sensing and cancer. *Nat. Rev. Cancer* 8, 865–873.

Krausova M & Korinek V (2014) Wnt signaling in adult intestinal stem cells and cancer. *Cell Signal.* 2014 Mar;26(3):570–579.

Lock R &. Cichowski K (2015) Loss of negative regulators amplifies RAS signaling. *Nat. Genet.* 47:426–427.

Maertens O & Cichowski K (2014) An expanding role for RAS GTPase activating proteins (RAS GAPs) in cancer. *Adv. Biol. Regul.* 55:1–14.

Radtke F & Clevers H (2005) Self-renewal and cancer of the gut: two sides of a coin. *Science* 307, 1904–1907.

Sharma S, Kelly TK & Jones PA (2010) Epigenetics in cancer. *Carcinogenesis* 31, 27–36.

Sherr CJ (2004) Principles of tumor suppression. *Cell* 116, 235–246.

Staser K, Yang FC & Clapp DW (2010) Mast cells and the neurofibroma microenvironment. *Blood* 116, 157–164.

Taguchi K & Yamamoto M (2017) The KEAP1-NRF2 system in cancer. *Front. Oncol.* 7:85.

Ting A, McGarvey KM & Baylin SB (2006) The cancer epigenome: components and functional correlates. *Genes Dev.* 20, 3215–3231.

Varshavsky A (2005) Regulated protein degradation. *Trends Biochem. Sci.* 30, 283–286.

Wang LH, Wu CF, Rajasekaran N & Shin YK (2018) Loss of tumor suppressor gene function in human cancer: an overview. *Cell Physiol. Biochem.* 51(6):2647–2693.

Zhu H, Wang G & Qian J (2016) Transcription factors as readers and effectors of DNA methylation. *Nat. Rev. Genet.* 17:551–565.

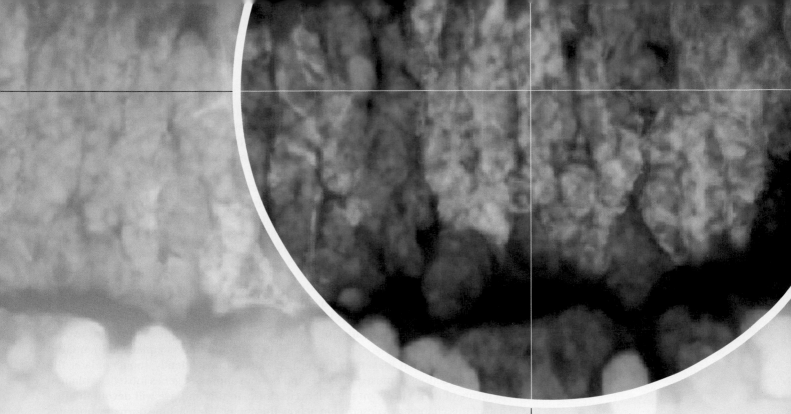

Chapter 8

pRb and Control of the Cell Cycle Clock

The dream of every cell is to become two cells.
> François Jacob, molecular biologist, quoted in 1971

This immediately leads one to ask, if the [hypothetical cellular transformation] loci can get into so much mischief, why keep them around? The logical answer is that they have some necessary function during some stage of the cell cycle, or some stage of embryogenesis.
> David E. Comings, geneticist, 1973

The fate of individual cells throughout the body is dictated by the signals that each receives from its surroundings—a point made repeatedly in earlier chapters. Thus, almost all types of normal cells will not proliferate unless prompted to do so by mitogenic growth factors released largely by their neighbors within a tissue. Yet other signaling proteins, notably transforming growth factor-β (TGF-β), may overrule the messages conveyed by mitogenic factors and force a halt to proliferation. In addition, extracellular signals may persuade a cell to enter into a *post-mitotic*, differentiated state from which it will never re-emerge to resume proliferation.

These disparate signals are collected by dozens of distinct cell surface receptors and then funneled into the complex signal-processing circuitry that operates largely in the cell cytoplasm. In some way, this mixture of signals must be processed, integrated, and ultimately distilled down to some simple, binary decisions made by the cell as to whether it should proliferate or become quiescent, and whether, as a quiescent cell, it will or will not differentiate. These behaviors suggest the existence of some centrally acting governor that operates inside the cell—a master clearinghouse that receives a wide variety of incoming signals and makes major decisions concerning the fate of the cell.

This master governor has been identified. It is the **cell cycle clock**, which operates in the cell nucleus. Its name is really a misnomer, because this clock is hardly a device for counting the passage of time. Nonetheless, we will use this term here for want of a better one. Rather than counting elapsed time, the cell cycle clock is a network of interacting proteins—a signal-processing circuit—that receives signals from various sources both outside and inside the cell, integrates them, and then decides the cell's fate. Should the cell cycle clock decide in favor of proliferation, it proceeds to orchestrate the complex transitions that together constitute the cell's cycle of growth and division; during this cycle, the expression of at least 1000 genes needs to be modulated in highly regulated ways. Should it decide in favor of quiescence, it will use its agents to impose this nonproliferative state on the cell (**Figure 8.1**).

The proliferative behavior of cancer cells indicates that the master governor of the cell's fate is influenced not only by normal proteins but also by oncogene-encoded proteins that insert themselves into various signaling pathways and disrupt normal control mechanisms. Similarly, the deletion of key tumor suppressor proteins evokes equally profound changes in the control circuitry and thus is equally influential in perturbing decision-making by the cell cycle clock. Consequently, sooner or later, the molecular actions of most oncogenes and tumor suppressor genes must be explained in terms of their effects on the cell cycle clock. To this end, we will devote the first half of this chapter to a description of how this molecular machine normally operates and then proceed to study how it is perturbed in human cancer cells.

8.1 Cell growth and division is coordinated by a complex array of regulators

When placed into culture under conditions that encourage exponential multiplication, mammalian cells exhibit a complex cycle of growth and division that is usually referred to as the **cell cycle**. A cell that has recently been formed by the processes of cell division—**mitosis** and **cytokinesis**—must decide soon thereafter whether it will once again initiate a new round of active growth and division or retreat into the nongrowing state that we previously termed G_0. As described earlier (Sections 5.1 and 6.1), this decision is strongly influenced by mitogenic growth factors in the cell's surroundings. Their presence in sufficient concentration will encourage a cell recently formed by mitosis to remain in the active growth-and-division cycle; their

Figure 8.1 The central governor of growth and proliferation The term "cell cycle clock" denotes a molecular circuitry operating in the cell nucleus that processes and integrates a variety of afferent (incoming) signals originating from outside and inside the cell and decides whether or not the cell should enter into the active cell cycle or retreat into a nonproliferating state. In the event that active proliferation is decided upon, this circuitry proceeds to program the complex sequence of biochemical changes in a cell that enable it to double its contents and to divide into two daughter cells. The alternative decision to withdraw from the active cell cycle may allow a cell to retain the option to re-enter into the growth-and-division cycle later or enter into post-mitotic states involving acquisition of differentiated characteristics (*not shown*).

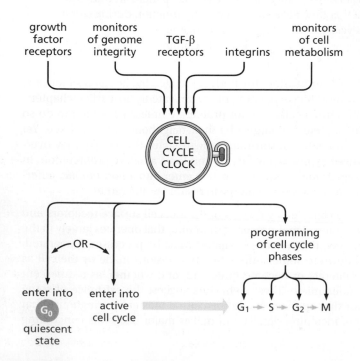

absence will trigger the default decision to proceed from mitosis into the G_0, quiescent state.

Withdrawal from the cell cycle may be actively encouraged by the presence of growth-inhibitory factors in the medium. Prominent among these anti-mitogenic factors is transforming growth factor-β (TGF-β). Withdrawal from the cell cycle into the G_0, quiescent state, whether caused by the absence of mitogenic growth factors or the presence of anti-mitogens such as TGF-β, is often reversible, in that an encounter by a quiescent cell with mitogenic growth factors on some later occasion may induce this cell to re-enter into active growth and division. However, some cells leaving the active cell cycle may do so irreversibly, thereby giving up all option of ever re-initiating active growth and division, in which case they are said to have become **post-mitotic**. Many types of neurons in the brain, for example, are widely assumed to fall into this category.

The decision by a cell recently formed by cell division to remain in the active growth-and-division cycle requires that this cell immediately begin to prepare for the next division. Such preparations entail, among other things, the doubling of the cell's macromolecular constituents and organelles to ensure that the two daughter cells resulting from the next round of cell division will each receive an adequate endowment of them. This accumulation of cellular constituents, which drives an increase in cell size, is termed cell *growth* to distinguish it from the process of cell *division*, which yields, via mitosis and cytokinesis, two daughter cells from a mother cell (**Figure 8.2**). However, in the more common usage and throughout this book, the term "cell growth" implies both the accumulation of cell constituents and the subsequent cell division, that is, the two processes that together yield cell proliferation.

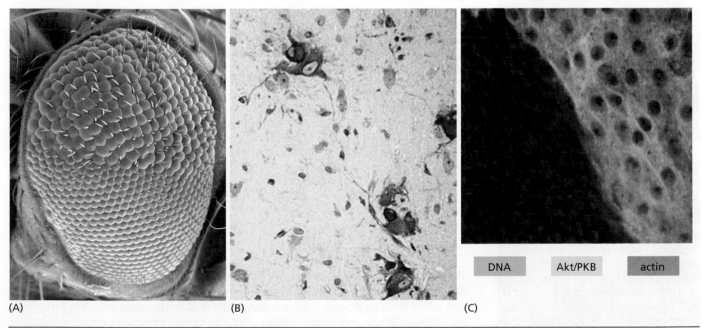

(A)   (B)   (C)

DNA   Akt/PKB   actin

Figure 8.2 Growth vs. proliferation Alterations of certain signaling proteins, such as the one encoded by the *TSC1* tumor suppressor gene, allow the processes of cell growth and division to be uncoupled from one another, enabling cells to grow in size without concomitant proliferation. (A) In this scanning electron micrograph of a *Drosophila* eye, the ommatidial cells in the upper portion of the eye have been deprived of the fly ortholog of *TSC1*; these cells are physically larger than the wild-type cells shown (*below*), because they have grown more during the cell cycles that led to their formation. (B) The same behavior can be seen in the brains of patients suffering from tuberous sclerosis, in which *TSC1* function has been lost through a germ-line mutation and subsequent somatic loss of heterozygosity. Seen here are the giant cells present in a benign growth (a "tuber"). The giant cells (*brown*) are labeled with an antibody against phosphorylated S6, a ribosomal protein important in regulating protein synthesis and thus cell growth; S6 phosphorylation and functional activation is deregulated in cells lacking *TSC1* function. (C) The Akt/PKB protein is frequently hyperactivated in human cancers. In an imaginal disc of a developing *Drosophila* larva, Akt/PKB (*green*) that has been hyperactivated in a portion of cells (*upper right*) causes a great increase in nuclear and overall cell size relative to cells with normal Akt/PKB (*lower left*), as revealed by the spacing of the nuclei (*blue*) in these two sectors; similar increases in mammalian cells are also observed in response to hyperactive Akt/PKB. (A, from X. Gao and D. Pan, *Genes Dev.* 15:1383–1392, 2001. With permission of Cold Spring Harbor Laboratory Press. B, courtesy of J.A. Chan and D.J. Kwiatkowski. C, from J. Verdu et al., *Nat. Cell Biol.* 1:500–506, 1999. With permission from Nature.)

(A)

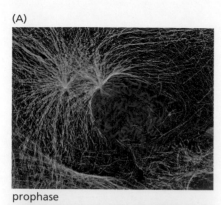

prophase

prometaphase

metaphase

anaphase

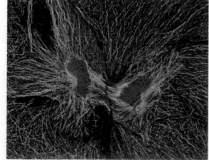

telophase

(B)

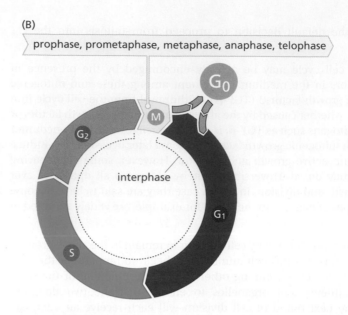

Figure 8.3 The vertebrate cell cycle (A) Immunofluorescence is used here to illustrate the five distinct subphases of mitosis (M phase) in newt lung cells (*top to bottom*). During the *prophase* of mitosis, the chromosomes (*blue*), which were invisible microscopically during *interphase* (the period encompassing S and G_2), begin to condense and become visible, while the centrosomes (*light green radiating bodies*) at the poles of the cell begin to assemble (*top image*). During *prometaphase*, the nuclear membrane disappears (*second image*). During *metaphase*, the chromosomes align along a plane that bisects the cell and become attached to the microtubule fibers of the mitotic spindle (*light green, third image*). During *anaphase*, the two halves of each chromosome—the chromatids—are pulled apart by the mitotic spindle (i.e., they segregate) to the two opposite poles of the cell (*fourth image*). During *telophase*, shortly after the chromatids cluster into the two sets seen here (*bottom image*), the chromatids de-condense and a new nuclear membrane forms around each set of chromatids (now called chromosomes). At the same time, during the process of *cytokinesis*, the cytoplasm of the mother cell divides, yielding two daughter cells (*not shown*). Actin fibers are labeled in *red*. (B) The mammalian growth-and-division cycle is divided into four phases—G_1, S (during which DNA is replicated), G_2, and M (mitosis). A fifth state, G_0 (G zero), denotes the resting, nonproliferating state of cells that have withdrawn from the active cell cycle. Although exit from the active cell cycle into G_0 is depicted here as occurring in early G_1, it is unclear when during G_1 this actually occurs. (Some evidence suggests that cells exiting mitosis decide whether or not to enter G_0 on the basis of signals experienced during the preceding G_2.) (A, photomicrographs by Dr. Conly L. Rieder, Wadsworth Center, Albany, New York 12201-0509.)

The accumulation of a cell's macromolecules involves, among many other molecules, the duplication of the cell's genome. In many prokaryotic cells, this duplication—the process of DNA replication—begins immediately after daughter cells are formed by cell division. But in most mammalian cells, the overall program of macromolecular synthesis is organized quite differently. Although the accumulation of RNA and proteins is initiated immediately after cell division and proceeds continuously until the next cell division, the task of replicating the DNA is deferred for a number of hours (often as many as 12 to 15) after emergence of new daughter cells from mitosis and cytokinesis. During this period between the birth of a daughter cell and the subsequent onset of DNA synthesis, which is termed the G_1 (first gap) phase of the cell cycle (**Figure 8.3**), cells make critical decisions about growth versus quiescence, and whether, as quiescent cells, they will differentiate.

In many types of cultured mammalian cells, the DNA synthesis that follows G_1 often requires 6 to 8 hours to reach completion. This period of DNA synthesis is termed the S (synthetic) phase, and its length is determined in part by the enormous amount of cellular DNA ($\sim$6.4 $\times$ 10^9 bp per diploid genome) that must be replicated with fidelity during this time. The actual length of S phase varies greatly among different kinds of cells, being much shorter in certain cell types, such as rapidly dividing embryonic cells and lymphocytes.

Having passed through S phase, a cell might be thought fit to enter directly into mitosis (M phase). However, most mammalian cells spend 3 to 5 hours in a second gap phase, termed G_2, preparing themselves, in some still-poorly understood fashion, for entrance into M phase and cell division. M phase itself usually encompasses an hour or so, and includes five distinct subphases—**prophase**, **prometaphase**, **metaphase**, **anaphase**, and **telophase**; this culminates in cytokinesis, the division of the cytoplasm that allows the formation of two new cells (see Figure 8.3A). Although these times are commonly observed when studying mammalian cells in culture, they do not reflect the behavior of all cell types under all conditions. For example, actively proliferating lymphocytes may double in 5 to 6 hours, and some cells in the early embryos of certain organisms may do so in as little as 15 minutes, where the G_1 and G_2 phases are essentially absent and all the materials needed to generate the newly formed daughter cells are already present in the fertilized egg.

As is the case with S phase, M phase must proceed with great precision. M phase begins with the two recently duplicated DNA helices within each chromosome; these are carried in the sister **chromatids** of the chromosome, which are aligned adjacent to one another in the nucleus. The allocation during mitosis of the duplicated chromatids to the two future daughter cells must occur flawlessly to ensure that each daughter receives exactly one diploid complement of chromatids—no more and no less. Once present within the nuclei of recently separated daughter cells, these chromatids become the chromosomes of the newly born cells.

Consequently, the endowment of one genome's worth of genetic material to each daughter cell depends on the precise execution of two processes—the faithful replication of a cell's genome during S phase and the proper allocation of the resulting duplicated DNAs to daughter cells during M phase. As will be discussed later in Chapter 12, defects in either of these processes can have disastrous consequences for the cell and the organism, one of which is the disease of cancer.

Like virtually all machinery, the machine that executes the various steps of the cell cycle is subject to malfunction. This fallibility contrasts with the stringent requirement of the cell that the various phases of the cell cycle are completed flawlessly. For this reason, the cell deploys a series of surveillance mechanisms that monitor each step in cell cycle progression and permit the cell to proceed to the next step in the cycle only if a preceding step has been completed successfully. Thus, if specific steps in the execution of a preceding process go awry, these monitors rapidly call a halt to further advance through the cell cycle until these problems have been successfully addressed. Yet other monitors ensure that once a particular step of the cell cycle has been completed, it is not repeated until the cell passes through the next cell cycle. These monitoring mechanisms are termed variously **checkpoints** or **checkpoint controls** (**Figure 8.4**).

One checkpoint ensures that a cell cannot advance from G_1 into S if the genome is in need of repair. Another, operating in S phase, will slow or pause DNA replication in response to DNA damage. (In mammalian cells, this may cause a doubling of the time required to complete DNA synthesis.) A third will not permit the cell to proceed through G_2 to M until the DNA replication of S phase has been completed. DNA damage will trigger another checkpoint control that blocks entrance into M phase. During M phase, highly efficient checkpoint controls block entrance into anaphase; these blocks are only removed once all the chromosomes have been properly attached to the mitotic spindle. Yet other checkpoint controls, not cited here, have been reported. For example, a **decatenation** checkpoint in late G_2 prevents entrance into M until the pair of DNA helices replicated in the previous S phase have been disentangled from one another, a step that requires the active involvement of type II topoisomerases. Defects in some of these checkpoint controls can be observed because of their effects on cells' chromosomes (see Figure 1.9B and Supplementary Sidebar 8.1).

Figure 8.4 Examples of checkpoints in the cell cycle Checkpoints impose quality control to ensure that a cell has properly completed all the requisite steps of one phase of the cell cycle before it is allowed to advance into the next phase. A cell will not be permitted to enter into S phase until all the steps of G_1 have been completed. It will be blocked from entering G_2 until all of its chromosomal DNA has been properly replicated during S phase. Similarly, a cell is not permitted to enter into anaphase (when the paired chromatids are pulled apart) until all of its chromosomes are properly assembled on the mitotic spindle during metaphase. In addition, a cell is not allowed to advance into S or M if its DNA has been damaged and not yet repaired. Other controls (*not shown*) ensure that once a specific step in the cell cycle has been completed, it is not repeated until the next cell cycle.

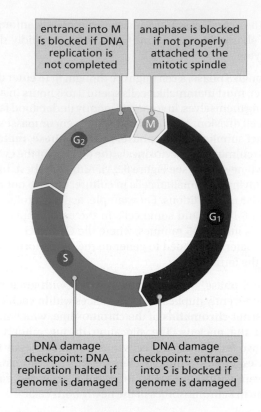

entrance into M is blocked if DNA replication is not completed

anaphase is blocked if not properly attached to the mitotic spindle

DNA damage checkpoint: DNA replication halted if genome is damaged

DNA damage checkpoint: entrance into S is blocked if genome is damaged

The operations of these checkpoints also influence the formation of cancers. As tumor development (often called tumor **progression**, discussed in Chapter 11) proceeds, incipient cancer cells benefit from experimenting with various combinations and permutations of mutant alleles, retaining those that will afford them the greatest proliferative advantage. An increased mutability of their genomes accelerates the rate at which these cells can acquire advantageous combinations of alleles and thus hastens the overall pace of tumor progression. Such mutability and resulting genomic instability are incompatible with normal cell cycle progression, because checkpoint controls usually block advance of the cell through its cycle if its DNA has been damaged or its chromosomes are in disarray. So, in addition to acquiring altered growth-controlling genes (that is, activated oncogenes and inactivated tumor suppressor genes), many types of cancer cells have inactivated one or more of their checkpoint controls. With these controls relaxed, incipient cancer cells can more rapidly accumulate the mutant genes and altered karyotypes that propel their neoplastic growth. The breakdown of the controls responsible for maintaining the cellular genome in an intact state is one of the main subjects of Chapter 12.

8.2 Cells make decisions about growth and quiescence during a specific period in the G_1 phase

As mentioned earlier, virtually all normal cell types in the body require external signals, such as those conveyed by mitogenic growth factors, before they undertake growth and division. The only exceptions to this rule appear to be very early embryonic cells, which seem able to proliferate without receiving growth-stimulating signals from elsewhere (**Sidebar 8.1**). The rationale for the contrasting behavior of the normal differentiated cells throughout our tissues is a simple one: because these cells participate in the formation of precisely structured tissues, their proliferation must, by necessity, be coordinated with neighboring cells in those tissues. Put differently, the body cannot give each of its almost 10^{14} component cells the license to decide on its own whether to grow and divide. To do so would invite chaos.

Evidence accumulated over the past three decades indicates that cells consult their extracellular environment and its growth-regulating signals during a discrete window of time in the active cell cycle, namely from the onset of G_1 phase until an hour or two before the G₁-to-S transition (**Figure 8.5**). The operations of the G₁ decision-making machinery are indicated by the responses of cultured cells to extracellular signals. If we were to remove serum and thus growth factors from cells before they had completed 80 to 90% of G₁, they would fail to proceed further into the cell cycle and would, with great likelihood, revert to the G₀ state. However, once these cells had moved through this G₁ decision-making period and advanced into the final hours of G₁ (the remaining 10 to 20% of G₁), the removal of serum would no longer affect their progress and they would proceed through the remainder of G₁ and onward through the S, G₂, and M phases. Similarly, anti-mitogenic factors, such as TGF-β, are able to impose their growth-inhibitory effects only during this period in the early and mid-G₁ phase. Once a cell has entered into late G₁, it seems to be oblivious to the presence of this negative factor in its surroundings. (There are isolated reports of the effects of TGF-β at the G₁/S boundary and the G₂/M boundary, but their role in regulating overall cell cycle progression remains unclear.)

This schedule of total dependence on extracellular signals followed by entrance in late G₁ into a state of relative independence indicates that a weighty decision must be made toward the end of G₁. Precisely at this point, a cell must make up its mind whether it will remain in G₁, retreat from the active cycle into G₀, or advance into late G₁ and thereafter into the remaining phases of the cell cycle. This critical decision is made at a transition that has been called the **restriction point** or **R point** (see Figure 8.5). In most mammalian cells studied to date, the R point occurs several hours before the G₁/S phase transition.

If a cell should decide at the R point to continue advancing through its growth-and-division cycle, it commits itself to proceed beyond G₁ into S phase and then to complete a rigidly programmed series of events (the entire S, G₂, and M phases) that enable it to divide into two daughter cells. This decision will be respected even if growth factors are no longer present in the extracellular space during these remaining phases of the cell cycle. We know that this series of later steps (S, G₂, and M phases) proceeds on a fixed schedule, because a cell that enters S will, in the absence of a major disaster, invariably complete S and, having done so, enter into G₂ and then advance into M phase.

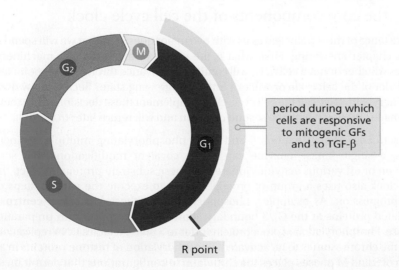

period during which cells are responsive to mitogenic GFs and to TGF-β

R point

Figure 8.5 Responsiveness to extracellular signals during the cell cycle Cells respond to extracellular mitogens and inhibitory factors (such as TGF-β) only in a discrete window of time that begins at the onset of G₁ and ends just before the end of G₁. The end of this time window is often designated as the restriction (R) point, which denotes the point in time when the cell must make the commitment to advance through the remainder of the cell cycle through M phase, to remain in G₁, or to retreat from the active cell cycle into G₀.

For those interested specifically in understanding the deregulated proliferation of cancer cells, this fixed program holds relatively little interest, because the late $G_1 \rightarrow S \rightarrow G_2 \rightarrow M$ progression proceeds similarly in normal cells and cancer cells. Students of cancer therefore focus largely on the G_0/G_1 transition and on the one period in the life of an actively growing cell—the time window encompassing most of G_1—when a cell is given the license to make decisions about its fate.

The commitment to advance through the R point and continue all the way to M phase is, as hinted, not an absolute one. Metabolic, genetic, or physical disasters may intervene during S, G_2, or M and force the cell to call a halt, often temporary, to its further advance through the cell cycle until these conditions have been addressed. Still, in the great majority of cases, cells in living tissues seem to succeed in avoiding these various disasters. This leaves the R-point decision as the critical determinant of whether cells will grow or not. An increasingly large body of evidence indicates that deregulation of the R-point decision-making machinery accompanies the formation of most if not all types of cancer cells.

If a cell does not enjoy the proper attachment to the extracellular matrix (ECM), much of it mediated by integrins (see Section 5.9), it will halt further advance through the cell cycle until proper tethering has been achieved. Alternatively, if it has lost all attachment to the ECM, it may activate **anoikis**, a form of the apoptotic cell suicide program described in the next chapter.

When grown in monolayer culture in Petri dishes, normal cells achieve the required attachment by adhering to ECM that they have deposited on the glass or plastic surfaces of these dishes. This requirement for attachment reflects the phenotype of anchorage dependence (see Section 3.5). Tumorigenic cells, which almost always have lost this dependence and have therefore become **anchorage-independent**, have inactivated or lost this late G_1 checkpoint. In some fashion, still poorly understood, oncoproteins like Ras and Src are able to mislead a cell into thinking that it has achieved extensive anchorage (see, for example, Section 6.9) when, in fact, none may exist at all.

Yet other, still-fragmentary evidence suggests additional environment-sensing checkpoints that operate before and after the R point. For example, one important checkpoint control operating between the R point and the onset of S phase gauges whether a cell has access to adequate levels of nutrients and halts cell cycle progression until such nutrients become available; another appears to determine whether **reactive oxygen species** (which in other contexts are thought to be toxic for the cell) are present in adequately high levels before progression is permitted. Such findings suggest that successful advance through late G_1 depends on passing through a succession of checkpoints in addition to the R point described here, not all of which have been well characterized.

8.3 Cyclins and cyclin-dependent kinases constitute the core components of the cell cycle clock

The existence of the R point leaves us with two major questions that we will spend much of this chapter answering. First, what is the nature of the molecular machinery that decides whether or not a cell in G_1 will continue to advance through the R point and the remainder of the cell cycle or will exit into a nongrowing state? Second, how does this machinery, which we call the cell cycle clock, implement these decisions once they have been made? We begin with the second question and will return later to the first.

As described earlier (Chapters 5 and 6) by phosphorylating multiple distinct substrates, a single kinase enzyme can create covalent modifications that serve to switch on or off various activities inherent in these substrate proteins. In fact, the cell cycle clock also uses a group of protein kinases to execute the various steps of cell cycle progression. As examples, phosphorylation by these kinases of **centrosome**-associated proteins at the G_1/S boundary allows their duplication in preparation for M phase. Phosphorylation of other proteins prior to S phase enables DNA replication sites along the chromosomes to be activated. Phosphorylation of histone proteins in anticipation of S and M phases places the chromatin in configurations that permit these two phases to proceed normally. And the phosphorylation of proteins forming the nuclear membrane (sometimes called the *nuclear envelope*), such as lamin and nucleoporins, causes their dissociation and the dissolution of this membrane early in M phase.

The kinases deployed by the cell cycle machinery are called collectively **cyclin-dependent kinases (CDKs)** to indicate that these enzymes never act on their own; instead, they depend on associated regulatory subunits, the **cyclin** proteins, for proper functioning. Bimolecular complexes of CDKs and their cyclin partners are responsible for dispatching signals from the cell cycle clock to hundreds of effector molecules that carry out the actual work of moving the cell through its growth-and-division cycle. (A number of structurally related CDKs involved in other cellular processes, such as control of transcription, are not discussed further here.)

The CDKs are serine/threonine kinases, in contrast to the tyrosine kinase activities that are associated with growth factor receptors and with nonreceptor kinase molecules like Src. The CDKs show about 40% amino acid sequence identity with one another and therefore are considered to form a distinct subfamily within the large (approximately 430) throng of Ser/Thr kinases encoded by the human genome (see Supplementary Sidebar 17.7). The cyclins associated reversibly with their CDK partners activate the catalytic activity of these partners (Supplementary Sidebar 8.2). (In the well-studied example of the binding of cyclin A to CDK2, the association of the two proteins increases the enzymatic activity of CDK2 a staggering 400,000-fold!) At the same time, the cyclins serve as guide dogs for the CDKs by helping the cyclin–CDK complexes recognize appropriate protein substrates in the cell. The cyclins, for their part, also constitute a distinct family of cellular proteins that share in common a domain of approximately 100 amino acid residues that is involved in the binding and functional activation of CDKs.

It is actually cyclin–CDK complexes that constitute the engine of the cell cycle clock machinery. During much of the G_1 phase of the cell cycle, two similarly acting CDKs—CDK4 and CDK6—are guided by and depend upon their association with one of a trio of related cyclins (D1, D2, and D3) that collectively are called the D-type cyclins (**Figure 8.6**). After the R point in late G_1, the E-type cyclins (E1 and E2) associate with CDK2 to enable the phosphorylation of appropriate substrates required for entry into S phase. As cells enter into S phase, the A-type cyclins (A1 and A2) replace E cyclins as the partners of CDK2 and thereby enable S phase to progress (see Figure 8.6). Later in S phase, the A-type cyclins switch partners, leaving CDK2 and associating instead with another CDK called either CDC2 or CDK1. (We will use CDK1 here.) As the cell moves further into G_2 phase, the A-type cyclins are replaced as CDK1 partners by the B-type cyclins (B1 and B2). Finally, at the onset of M phase, the complexes

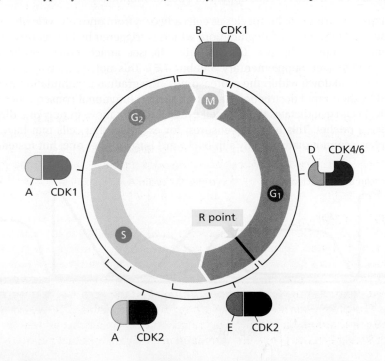

Figure 8.6 Pairing of cyclins with cyclin-dependent kinases Each type of cyclin pairs with a specific cyclin-dependent kinase (CDK) or set of CDKs. The D-type cyclins (D1, D2, and D3) bind CDK4 or CDK6, the E-type cyclins (E1 and E2) bind CDK2, the A-type cyclins (A1 and A2) bind CDK2 or CDK1, and the B-type cyclins (B1 and B2) bind CDK1. The brackets indicate the periods during the cell cycle when these various cyclin–CDK complexes are active.

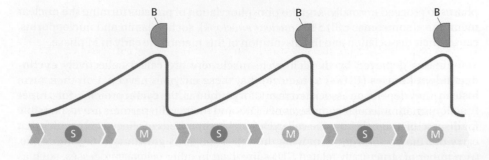

B   B   B

Figure 8.7 Cell cycle–dependent fluctuations in cyclin B levels The cyclic fluctuations in the levels of cyclin B in early frog and sea urchin embryos gave cyclins their name. These fluctuations were noticeable because the cell cycles in these early embryos are synchronous, i.e., all cells in an embryo enter into M phase simultaneously. In these early embryos, the G_1 and G_2 phases of the cell cycle (*orange–pink chevrons*) are virtually absent and the cells, in effect, alternate between M and S phases. (Although cyclin B levels are already substantial prior to the onset of M phase, cyclin B molecules are unable to form catalytically active B–CDK1 (also termed CDC2) complexes until the G_2-to-M transition.)

of CDK1 with the B-type cyclins trigger many of the events of the prophase, metaphase, anaphase, and telophase that together constitute the complex program of mitosis.

As is the case with all well-regulated systems, the activities of the various cyclin–CDK complexes must be modulated in order to impose control on specific steps in the cell cycle. The most important way of achieving this regulation depends upon changing the levels and availability of cyclins during various phases of the cell cycle. In contrast, the levels of most CDKs vary only minimally.

The first insights into cyclin-CDK control came from studies of the governors of mitosis in early frog and sea urchin embryos. As these experiments showed, levels of B-type cyclins increase strongly in anticipation of mitosis, allowing B cyclins and CDK1 to form complexes that initiate entrance into M phase. At the end of M phase, cyclin B levels plummet because of the scheduled degradation of this protein. Early in the next cell cycle, cyclin B is virtually undetectable in cells, and it accumulates gradually later in this cycle in preparation of the next M phase. Because the growth-and-division cycles of all of the cells in these early embryos are **synchronous** (that is, take place coordinately), all cells in the embryo go through S and M at the same time. This behavior results in repeated rounds of cycling of the levels of these cyclin proteins—the behavior that inspired their name (**Figure 8.7**).

This theme of dramatic cell cycle phase–dependent changes in the levels of cyclin B is repeated by other cyclins as well. Cyclin E levels are low throughout most of G_1, rise abruptly after a cell has progressed through the R point, and collapse as the cell enters S phase (**Figure 8.8**), while cyclin A increases in concert with the cell's entrance into S phase. (Although there are at least two subtypes of cyclin A, cyclin B, and cyclin E, we will refer to these simply as cyclins A, B, and E, respectively, because the two subtypes of each of these cyclins appear to operate identically.)

The collapse of various cyclin species as cells advance from one cell cycle phase to the next results from their rapid degradation, which is triggered by the actions of highly coordinated ubiquitin ligases. The ubiquitin ligases attach polyubiquitin chains to these cyclins (see Supplementary Sidebar 7.5). This polyubiquitylation leads to proteolytic breakdown within the proteasomes. The gradual accumulation of cyclins followed by their rapid destruction has an important functional consequence for the cell cycle, because it dictates that the cell cycle clock can move in only one direction, much like a ratchet. This behavior ensures, for example, that cells that have exited one M phase cannot inadvertently slip backward into another one, but instead must

Figure 8.8 Fluctuation of cyclin levels during the cell cycle The levels of most of the mammalian cyclins fluctuate dramatically as cells progress through the cell cycle. For most of these cyclins, these fluctuations are tightly coordinated with the schedule of advances through the various cell cycle phases. However, in the case of the D-type cyclins, extracellular signals, notably those conveyed by growth factors, strongly influence their levels. (Although cyclin D1—and possibly other D-type cyclins—is present in other cell cycle phases besides G_1, following the G_1/S transition it is exported from the nucleus into the cytoplasm, where it can no longer influence cell cycle progression. The precise time point at which cyclin D–CDK4/6 complexes lose activity is not well defined.)

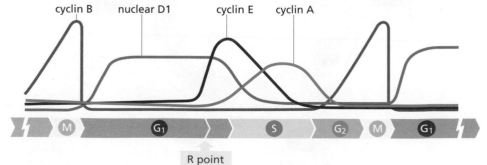

cyclin B   nuclear D1   cyclin E   cyclin A

M G_1 S G_2 M G_1

R point

advance through G_1, S, and G_2 until they have once again accumulated the B cyclins required for entrance into M phase.

The D-type cyclins present the sole exception to these well-programmed fluctuations in the levels of cyclins. The levels of these three, similarly structured cyclins are not found to vary dramatically as a cell advances through the various phases of its growth-and-division cycle. Instead, the levels of D-type cyclins are controlled largely by extra-cellular signals, specifically those conveyed by a variety of mitogenic growth factors. In the case of cyclin D1—the best-studied of the three D-type cyclins—growth factor activation of tyrosine kinase receptors and the resulting stimulation of several down-stream signaling cascades results in rapid accumulation of cyclin D1 (**Figure 8.9**).

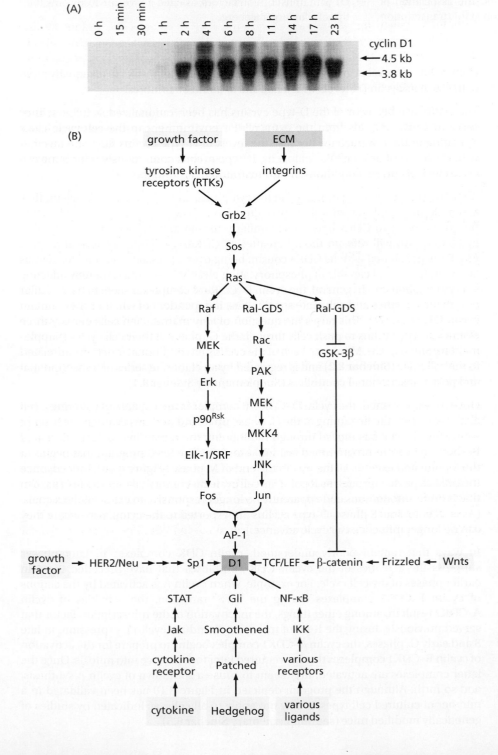

Figure 8.9 Control of cyclin D1 levels
(A) Cyclin D1 was discovered as a protein whose levels are strongly induced by exposure of macrophages to the mitogen CSF-1 (colony-stimulating factor-1). Here, macrophages that had been starved of CSF-1 were exposed to fresh CSF-1 and the amounts of cyclin D1 mRNA at subsequent times thereafter were determined by RNA (Northern) blotting. (B) The control of cyclin D1 levels by extracellular mitogens can be explained, in part, by a signal transduction cascade that leads from growth factor receptors (RTKs) to the AP-1 transcription factor (TF), one of a number of factors that modulate the transcription of the *cyclin D1* gene. (AP-1 is a heterodimeric TF formed from Fos and Jun subunits, each encoded by a proto-oncogene.) In addition, a number of other signaling cascades converge on the promoter of this gene, not all of which are illustrated in this figure. (A, from H. Matsushime et al., *Cell* 65: 701–713, 1991. With permission from Elsevier.)

Sidebar 8.2 Cyclins have other jobs besides cell cycle control Over three decades of cell cycle research have created the impression that cell cycle control is the sole function of cyclins. In fact, cyclin D1 has been shown to associate with both the estrogen receptor (ER) and the transcription factor C/EBPβ. By binding the ER, cyclin D1 may mimic the normal ligand of this receptor—estrogen—in stimulating the receptor's transcriptional activities. The great majority (>70%) of breast cancers express the ER, and its expression explains the mitogenic effects that estrogen has on the cells in these tumors. Because cyclin D1 is overexpressed in most of these tumors and can also activate this receptor, the D1–ER complexes may also play an important role in driving the proliferation of cells in these tumors. The association of cyclin D1 with C/EBPβ results in activation of this transcription factor, which is thought to play a key role in programming the differentiation of a variety of cell types. Knockout of cyclin D1 in the mouse germ line results in viable mice that have severely underdeveloped mammary glands. This effect can be reversed by providing the mammary glands with a mutant cyclin D1 that can bind C/EBPβ but cannot activate CDK4/6. Hence, in this case, the non-CDK-associated functions of this cyclin are far more important than its effects on CDK4/6 activation. More globally, the use of chromatin immuno-precipitation (ChIP) has revealed at least 700 sites in the genome, most affiliated with gene promoters, to which cyclin D1 is bound, indicating that the two associations of cyclin D1 with transcription factors, as cited above, are likely only the tips of a far larger iceberg.

Conversely, removal of growth factors from a cell's medium results in an equally rapid collapse of its cyclin D1 levels with a half-life of about 30 minutes.

The distinctive behavior of the D-type cyclins has been rationalized as follows: They serve to convey signals from the extracellular environment to the cell cycle clock operating in the cell nucleus. Because the levels of D-type cyclins fluctuate together with the levels of extracellular mitogens, D-type cyclins continuously inform the cell cycle clock of current conditions in the environment around the cell.

After D-type cyclins are synthesized in the cytoplasm and migrate to the nucleus, they assemble in complexes with their two alternative CDK partners, CDK4 and CDK6. Because these two CDKs operate so similarly to one another within the cell cycle machinery, we will refer to them hereafter as CDK4/6. (CDK6 has several distinct functions not shared with its CDK4 cousin, being overexpressed in certain leukemias and lymphoma and capable of phosphorylating glycolytic enzymes, thereby affecting energy metabolism.) In general, the cyclin D–CDK4/6 complexes seem to have similar enzymatic activities and substrate specificities, independent of whether they contain cyclin D1, D2, or D3. This raises the question of why mammalian cells employ three distinct D-type cyclins to push cells through the G_1 phase of their cell cycles (Supplementary Sidebar 8.3). Moreover, each of the cyclins has additional functions unrelated to the cell cycle (**Sidebar 8.2**) and is regulated by an elaborate series of transcriptional and post-transcriptional controllers (Supplementary Sidebar 8.4).

Once they are formed, the cyclin D–CDK4/6 complexes are capable of ushering a cell all the way from the beginning of the G_1 phase up to and perhaps through the R-point gate. After the cell has moved through the R point, the remaining cyclins—E, A, and B—behave in a pre-programmed fashion, executing the fixed program that begins at the R point and extends all the way to the end of M phase (**Figure 8.10**). Indeed, once the cell has passed through its R point, its cell cycle machinery takes on a life of its own that is quite autonomous and apparently no longer responsive to extracellular signals. (As cells enter into S phase, D-type cyclins are exported to the cytoplasm, where they can no longer influence cell cycle advance.)

In ways that remain poorly understood, cyclin–CDK complexes in later phases suppress the activities of the cyclin–CDK complexes that have preceded them in earlier phases of the cell cycle. For example, when cyclin A is activated by the actions of cyclin E–CDK2 complexes during the G_1/S transition, the activities of cyclin A–CDK2 result in, among other things, the inactivation of the transcription factor that served previously during the R-point transition to induce cyclin E expression. In late S and early G_2 phases, the cyclin A–CDK1 complex begins to prepare for the activation of cyclin B–CDK1 complexes that are required later for entrance into mitosis. Once the latter complexes are activated, they seem to cause a shutdown of cyclin A synthesis, and so forth. Although the program depicted in Figure 8.10 has been validated in a number of cultured cell types, it may not apply to all cells, as indicated by studies of genetically modified mice (see Supplementary Sidebar 8.5).

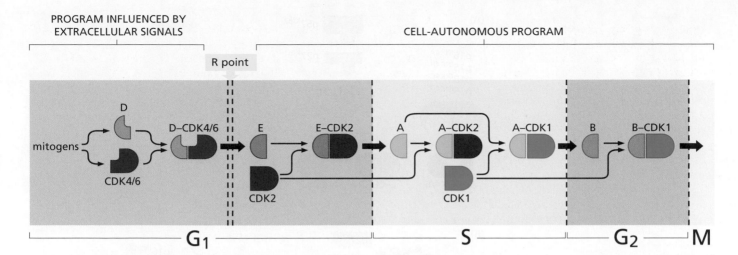

Figure 8.10 Control of cyclin levels during the cell cycle Although extracellular signals strongly influence the levels of D-type cyclins during most of the G_1 phase of the cell cycle, the levels of the remaining cyclins are controlled by intracellular signaling and precisely coordinated with cell cycle advance. Thus, after cells pass through the R point and cyclin E–CDK2 complexes are activated, the activation of the remaining cyclin–CDK complexes occurs on a predictable schedule that appears to be independent of extracellular physiologic signals. In part, this coordination is achieved because the cyclin–CDK complexes in one phase of the cell cycle are responsible for activating those in the subsequent phase (*indicated here*) and for shutting down those that were active in the previous phase (*not shown*).

8.4 Cyclin–CDK complexes are also regulated by CDK inhibitors

The scheme of cell cycle progression laid out above implies that physiological signals are able to influence the activity of the cell cycle clock only through their modulation of cyclin levels. In truth, there are several other layers of control that modulate the activity of the cyclin–CDK complexes and thereby regulate advance through the cell cycle.

The most important of these additional controls is imposed by a class of proteins referred to generically as CDK inhibitors or simply CKIs. To date, seven of these proteins have been found that are able to antagonize the activities of the cyclin–CDK complexes. A group of four of these, the INK4 proteins (named originally as **in**hibitors of CD**K4**), target specifically the CDK4 and CDK6 complexes; they have no effect on CDK1 and CDK2. These inhibitors are p16^{INK4A}, p15^{INK4B}, p18^{INK4C}, and p19^{INK4D}. The three remaining CDK inhibitors, p21^{Cip1} (sometimes termed p21^{Waf1}), p27^{Kip1}, and p57^{Kip2} (encoded by the human *CDKN1A*, *CDKN1B*, and *CDKN1C* genes, respectively) are more widely acting, being able to inhibit the cyclin–CDK complexes that function in G1/S (E–CDK2) or S (A–CDK2) phases of the cell cycle (**Figure 8.11**). (Each one of this trio of CKIs has been found to affect other specific processes unrelated to cell cycle progression, including transcriptional regulation, apoptosis, cell fate determination, cell migration, and cytoskeletal organization; however, because these other functions appear to play relatively minor roles in cancer development, we will not discuss them further.) p57^{Kip2} plays a key role in certain tissues by imposing quiescence on stem cells but a minor role in cancer development; for this reason, our discussion will focus on its two cousins, p21^{Cip1} and p27^{Kip1}.

The actions of these two classes of CDK inhibitors are nicely illustrated by p15^{INK4B} and the pair p21^{Cip1} and p27^{Kip1}. When TGF-β is applied to epithelial cells, it elicits a number of downstream responses that antagonize cell proliferation. Among these are substantial increases in the levels of p15^{INK4B}, which proceeds to block the formation of cyclin D–CDK4/6 complexes (**Figure 8.12**) and to inhibit those that have already formed. Without active D–CDK4/6 complexes, the cell is unable to advance through early and mid-G_1 and reach the R point. Once a cell has passed through the R point, the actions of the D–CDK4/6 complexes seem to become unnecessary. This behavior may explain why TGF-β inhibits growth during early and mid-G_1 and loses most (perhaps all) of its growth-inhibitory powers after a cell has passed through the R point.

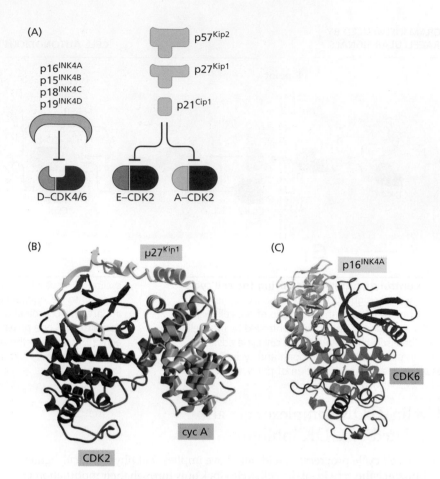

Figure 8.11 Actions of CDK inhibitors (A) The CDK inhibitors block the actions of CDKs at various points in the cell cycle. The four INK4 proteins (p16^{INK4A}, p15^{INK4B}, p18^{INK4C}, and p19^{INK4D}) are specialized to inhibit the D–CDK4 and D–CDK6 complexes that are active in early and mid-G₁. Conversely, the three Cip/Kip CKIs (p21^{Cip1}, p27^{Kip1}, and p57^{Kip2}) can inhibit E–CDK2 and A–CDK2, cyclin–CDK complexes that are active in the G1/S and S phases of the cell cycle. Under certain conditions, signals originating within the cell (e.g., signals indicating damage to the cell's DNA) can halt advance through the cell cycle by blocking the actions of cyclin/CDK complexes that are active in the S and G₂ phases of the cell cycle. Paradoxically, p21^{Cip1} and p27^{Kip1} are known to *promote* formation and catalytic activity of D–CDK4/6 complexes during the G₁ phase (see Figure 8.14). (B) This depiction illustrates how one domain of p27^{Kip1} (*light green*) blocks cyclin A–CDK2 function by obstructing the ATP-binding site in the catalytic cleft of the CDK. (C) Inhibitors of the INK4 class, such as p16^{INK4A} shown here (*brown*), bind to CDK6 (*reddish*) and to CDK4 (*not shown*). These CDK inhibitors distort the cyclin-binding site of CDK6, reducing its affinity for D-type cyclins. At the same time, they distort the ATP-binding site and thereby compromise catalytic activity. Identical interactions likely characterize the responses of CDK4 to p16^{INK4A} binding. (B, from A.A. Russo et al., *Nature* 382:325–331, 1996. C, from A.A. Russo et al., *Nature* 395:237–243, 1998. Both with permission from Nature.)

p21^{Cip1}, a more widely acting CDK inhibitor, is also induced by TGF-β, albeit weakly. Far more important are increases in the levels of p21^{Cip1} that occur in response to various physiological stresses (see Figure 8.12A); once present at significant levels, p21^{Cip1} can act throughout much of the cell cycle to stop a cell in its tracks. Prominent among these stresses is damage to the cell's genome. As long as the genomic DNA remains in an unrepaired state, the p21^{Cip1} that has been induced in response will shut down the activity of already-formed cyclin–CDK complexes—such as E–CDK2 and A–CDK2—that happened to be active when this damage was first incurred; once the damage is repaired, the block imposed by p21^{Cip1} may then be relieved. Such a strategy makes special sense in G₁: if a cell's genome becomes damaged during this period through the actions of various mutagenic agents, p21^{Cip1} will block advance through the R point (by inhibiting E–CDK2 complexes) until the damage has been

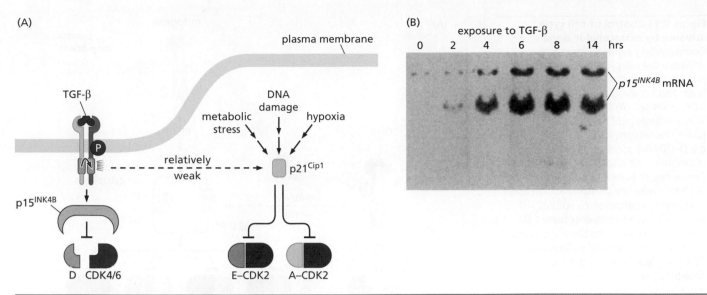

Figure 8.12 Control of cell cycle progression by TGF-β
(A) TGF-β (*top left*) controls the cell cycle machinery in part through its ability to modulate the levels of CDK inhibitors. It acts to strongly induce increased expression of p15^{INK4B} and, weakly, of p21^{Cip1}. The former can block the actions of cyclin D–CDK4/6 complexes, whereas the latter can block the actions of the E–CDK2 and A–CDK2 complexes. Independent of this, damage to cellular DNA causes strong, rapid increases in p21^{Cip1}, which in turn can shut down the cyclin–CDK complexes that are active in the phases of the cell cycle after the cell has passed through the R point in the late G$_1$ phase. (B) When TGF-β is applied to human keratinocytes, it evokes a dramatic 30-fold induction of *p15^{INK4B}* mRNA synthesis as demonstrated here by RNA (Northern) blotting analysis. These cells were exposed to TGF-β for the time periods (in hours) indicated. (B, from G.J. Hannon and D. Beach, *Nature* 371:257–261, 1994. With permission from Nature.)

repaired, ensuring that the cell does not stumble into S phase and inadvertently copy still-damaged DNA sequences. In addition, p21^{Cip1} can inhibit the functions of a key component of the cell's DNA replication apparatus, termed PCNA (proliferating cell nuclear antigen), ensuring that already-initiated DNA synthesis is halted until DNA repair has been completed. We will return to the mechanisms controlling p21^{Cip1} expression in the next chapter.

Although DNA damage and, to a much lesser extent, TGF-β can elicit increases in the levels of p21^{Cip1} (thereby blocking cell cycle advance), mitogens act in an opposing fashion to mute the actions of this CDK inhibitor and in this way favor cell cycle advance. One mechanism by which they do so depends on the phosphatidylinositol 3-kinase (PI3K) pathway (see Section 6.6), which is activated directly or indirectly by the mitogens that stimulate many receptor tyrosine kinases (**Figure 8.13A**). Akt/PKB, the important kinase activated downstream of mitogen-activated PI3K, phosphorylates p21^{Cip1} molecules located in the nucleus, thereby causing them to be exported into the cytoplasm, where they can no longer engage and inhibit cyclin–CDK complexes (Figure 8.13B). Similarly, Akt/PKB phosphorylates p27^{Kip1} (which functions much like p21^{Cip1}) and prevents its export from its cytoplasmic site of synthesis into the nucleus, where it normally does its critical work (Figure 8.13C and Supplementary Sidebar 8.6). Taken together, these signaling responses illustrate how extracellular growth-inhibitory signals (conveyed by TGF-β) impede the advance of the cell cycle clock, while growth-promoting signals promote its forward progress.

These effects on intracellular localization appear to have clinical consequences. For example, in low-grade (that is, less advanced) human mammary carcinomas, levels of activated Akt/PKB are low and p27^{Kip1} is able to carry out its anti-proliferative functions in the cell nucleus. In high-grade tumors, however, activated Akt/PKB is abundant and much of p27^{Kip1} is now found in the cell cytoplasm, where it does not have access to the cell cycle machinery. This intracellular localization is correlated with, and likely causally linked to, the progression of these cancers to fatal endpoints (see Supplementary Sidebar 8.6).

Figure 8.13 Control of cell cycle advance by extracellular signals
Countervailing extracellular signals influence the cell cycle machinery, in part through their ability to control the levels and intracellular localization of CDK inhibitors. (A) During the G_1 phase, TGF-β induces p15^{INK4B} and (weakly) p21^{Cip1} expression, negatively affecting the D–CDK4/6 and E–CDK2 complexes, respectively (see also Figure 8.12). Conversely, mitogens, acting through Akt/PKB, cause the phosphorylation and cytoplasmic localization of both p21^{Cip1} and p27^{Kip1}, which prevents these CDK inhibitors from entering the nucleus and blocking the activities of various cyclin–CDK complexes operating there. (The growth factor receptors can also reduce the levels of the two CKIs depicted here, p21^{Cip1} and p27^{Kip1}; these CKIs can inhibit, in turn, cyclin–CDK complexes active in late G_1 and S. It is unclear, however, whether these cyclin–CDK complexes are actually affected by mitogens in the later phases of the cell cycle.) (B) Ectopic expression of a constitutively active Akt/PKB (*upper row*) causes p21^{Cip1} (*orange, left*) to be localized largely in the cytoplasm. Compare its localization with that of the cell nuclei (*blue, middle*); the overlap of these two images is also seen (*right*). Conversely, expression of a dominant-negative Akt/PKB (*lower row*), which interferes with ongoing Akt/PKB function, allows p21^{Cip1} to localize to the cell nucleus, where it can interact with cyclin–CDK complexes and inhibit cell cycle progression. (C) In normal cells (*first panel*), ectopically expressed wild-type p27^{Kip1} (*green*) is found to be exclusively nuclear. This localization is not changed if a p27^{Kip1} mutant is expressed that lacks the threonine residue normally phosphorylated by Akt/PKB (*second panel*). (The T157A mutant carries an alanine in place of this threonine.) If a mutant, constitutively active Akt/PKB is expressed, however, much of wild-type p27^{Kip1} is now seen in the cytoplasm (*third panel*). But if the p27^{Kip1} mutant that cannot be phosphorylated by Akt/PKB is expressed, it resists the actions of constitutively activated Akt/PKB and remains in the nucleus (*fourth panel*), where it can continue to influence cell cycle advance. (B, from B. Zhou et al., *Nat. Cell Biol.* 3:245–252, 2001. C, from G. Viglietto et al., *Nat. Med.* 8:1136–1144, 2002. Both with permission from Nature.)

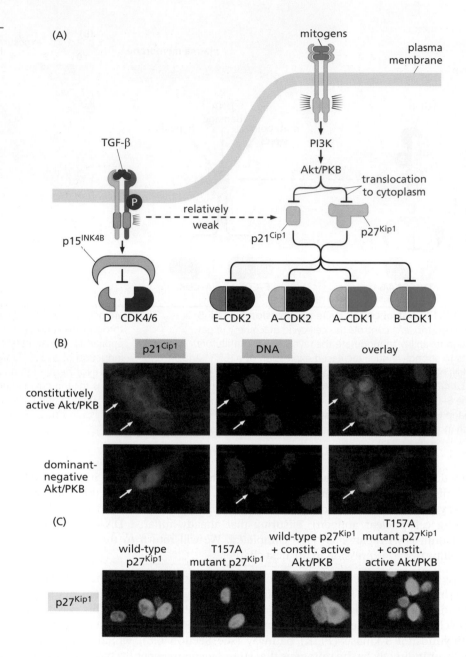

One aspect of the behavior of p21^{Cip1} and p27^{Kip1} seems quite paradoxical. Although they inhibit the actions of cyclin E–CDK2 and cyclin A–CDK2, they actually *stimulate* the formation of active cyclin D–CDK4/6 complexes (**Figure 8.14A**). Moreover, once a ternary (three-part) complex has formed between D–CDK4/6 and either of these CDK inhibitors, this cyclin–CDK complex can still phosphorylate its normal substrates. So the term "CDK inhibitor" is in fact a misnomer: the two proteins (p21^{Cip1} and p27^{Kip1}) act on most cyclin–CDK complexes in an inhibitory manner and on the cyclin D–CDK4/6 complexes in a stimulatory fashion.

Although p27^{Kip1} functions similarly to p21^{Cip1} in many ways, it has its own, highly interesting behavior. When cells are in the G_0 quiescent state, p27^{Kip1} is present in high concentrations, and therefore binds to and suppresses the activity of the few cyclin E–CDK2 complexes that happen to be present in the cell. When cells are exposed to growth factors, cyclin D–CDK4/6 complexes accumulate, because these mitogens cause an increase in the levels of the D-type cyclins. Each time a new cyclin D–CDK4/6 complex forms, it captures and binds another molecule of free p27^{Kip1}; consequently, the pool of free p27^{Kip1} molecules in the cell dwindles as they become progressively sequestered by cyclin D–CDK4/6 complexes. Once a small portion of cyclin E–CDK2:p27^{Kip1} complexes are liberated from the inhibitory effects of p27^{Kip1},

(A)

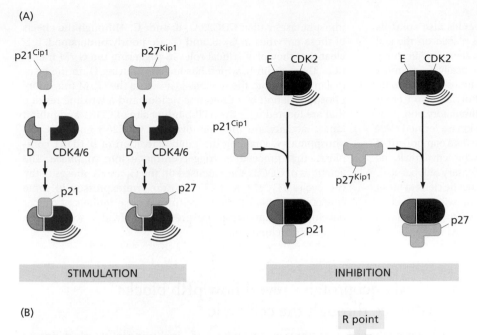

STIMULATION   INHIBITION

(B)

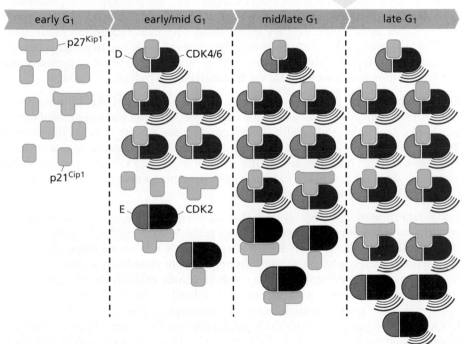

Figure 8.14 Interactions of CDK inhibitors with cyclin–CDK complexes (A) The various CDK inhibitors affect cyclin–CDK complexes in different ways. The CDK inhibitors of the p21^{Cip1} and p27^{Kip1} family, while inhibiting the cyclin–CDK complexes active in late G$_1$ and S phases (involving CDK2; only E–CDK2 is shown here, *right*), actually stimulate the formation of the D–CDK4/6 complexes that are active in the early and mid-G$_1$ phase and nevertheless permit these complexes to be catalytically active (*left*; see also Figure 8.11). (B) Cells in early G$_1$ often have substantial concentrations of p21^{Cip1} and p27^{Kip1} molecules (*green; left*). As cyclin D–CDK4/6 complexes (*pink/red*) accumulate throughout early to mid-G$_1$ (being induced by extracellular mitogens), they bind and thereby sequester an ever-increasing proportion of the p21^{Cip1} and p27^{Kip1} molecules, both those molecules that are present in soluble pools (unbound to any cyclin–CDK complexes) as well as those that are associated with the small number of cyclin E–CDK2 complexes that are present in G$_1$ (*light, dark purple*). The binding by D–CDK4/6 complexes of these two CDK inhibitors does not negatively affect their catalytic activity (see panel A). However, as this process continues, the D–CDK4/6 complexes succeed in abstracting most of the p21^{Cip1} and p27^{Kip1} molecules away from E–CDK2 complexes (*purple*) in late G$_1$ (*right*). This finally liberates the E–CDK2 complexes from control by p27^{Kip1} (*bottom right*), permitting these complexes to hyperphosphorylate pRb, thereby triggering the R-point transition. At the same time, the rapid increase in active E–CDK2 complexes drives the phosphorylation and subsequent degradation of p27^{Kip1} molecules (see Figure 8.20), ensuring that cells advance irreversibly through the R point and into S phase.

these complexes are free to begin triggering passage of the cell through the R point (Figure 8.14B). This means that the *timing* of the R point during G$_1$ is dictated by the rates at which cyclin D–CDK4/6 complexes accumulate in the cell during G$_1$; stated differently, by encouraging the formation of these complexes, mitogens govern the length of time during G$_1$ required for a cell to activate its R point transition.

In the context of cancer development, the overexpression of D-type cyclins, which occurs in a variety of tumors, leads to premature sequestration of p27^{Kip1} and attendant R-point activation, whereas the inactivation of CDK4/6 by pharmacological inhibitors, some of which are employed clinically (see Chapter 17), prevents p27^{Kip1} sequestration and thereby obstructs activation of the R-point transition. In addition to these dynamic changes in CKIs, they also act in a static fashion to impose post-mitotic states on differentiated normal cells (Supplementary Sidebar 8.7). In fact, yet another level of control of cyclin–CDK complexes is imposed by covalent modifications of the CDK molecules themselves (**Sidebar 8.3**).

Sidebar 8.3 Phosphorylation of CDK molecules also controls their activity An additional level of control placed on the cell cycle clock machinery can be traced to the covalent modifications of the CDK molecules themselves. More specifically, CDKs must be phosphorylated at specific amino acid residues in order to be active. At the same time, inhibitory phosphorylations of other amino acid residues must be removed. The stimulatory phosphorylations are imparted by a serine/threonine kinase termed CAK (for CDK-activating kinase). CAK itself operates as a component of a cyclin–CDK complex, which phosphorylates the activation loops (often called T-loops) of CDKs (see Supplementary Sidebar 8.2), causing them to swing out of the way of the catalytic clefts of these serine/threonine kinases which they would otherwise obstruct.

Equally important are the effects of *inhibitory* phosphorylations affecting CDK1 and CDK2, which are removed by a class of phosphatases called CDC25A, -B, and -C. Although the effects of these enzymes in G_1, S, and G_2 are poorly understood, it is clear that they play critical roles in triggering the G_2/M transition. Thus, cyclin B, which has lingered during G_2 in the cytoplasm, moves into the nucleus just before the G_2/M transition; phosphorylation of a threonine residue and a tyrosine residue that are located near the ATP-binding site of CDK1 and inhibit kinase activity are then removed by CDC25B and CDC25C phosphatases, enabling the rapid activation of B–CDK1 complexes that proceed to trigger entrance into M phase. The activities of CDC25A are focused on the G_1 and S phases of the cell cycle. CDC25A and CDC25B are overexpressed in some cancers, and there is evidence from mouse models of cancer development that their overexpression can lead to acceleration of tumor development.

8.5 Viral oncoproteins reveal how pRb blocks advance through the cell cycle

We have viewed the cell cycle from several angles here, including its clearly delineated G_1, S, G_2, and M phases and the cyclins and CDKs that orchestrate advances through these phases. In addition, we have traced these advances back to the extracellular signals that persuade the cell to enter into the active cell cycle and to progress through the restriction (R) point gate. These descriptions, however, fail to reveal precisely how the R-point transition is executed at the molecular level.

The solution to this puzzle was provided by the isolation of the *Rb* tumor suppressor gene, defective versions of which are involved in the pathogenesis of retinoblastomas, sarcomas, and small-cell lung carcinomas as well as other tumors (see Table 7.1). Soon after the *Rb* gene was isolated in 1986, it became apparent that it encodes a nuclear phosphoprotein of a mass of about 105 kD. This protein, called variously pRb or RB, was found to be absent, or present in a defective form, in the cells of many of the above-named tumor types.

Initial experiments with pRb revealed that it undergoes phosphorylation in concert with the advance of cells through their cell cycle. More specifically, pRb is essentially unphosphorylated immediately after cells exit mitosis or when they are in G_0; becomes weakly phosphorylated (**hypophosphorylated**) on serine and threonine residues after entrance into G_1; and becomes extensively phosphorylated (**hyperphosphorylated**) on a much larger number (~14 in all) of serine and threonine residues in concert with advance of cells through the R point (**Figure 8.15**). Once cells have passed through the R point, pRb usually remains hyperphosphorylated throughout the remainder of the cell cycle. After cells exit mitosis, the phosphate groups on pRb are stripped off by the enzyme termed protein phosphatase type 1 (PP1). This removal of phosphate groups, in turn, sets the stage for the next cell cycle and thus for a new cycle of pRb phosphorylation.

The fact that pRb hyperphosphorylation occurs in concert with passage through the R point provided the first hint that this protein is the molecular governor of the R-point transition. An additional, critical clue came from discoveries in 1988 of physical interactions between DNA tumor virus–encoded oncoproteins and pRb. Recall that DNA tumor viruses are able to transform cells and do so through the use of virus-encoded oncoproteins. Unlike the oncoproteins of RNA tumor viruses such as RSV, the DNA tumor viruses employ oncoproteins that have little if any resemblance to proteins that exist naturally within uninfected cells (see Sections 3.4 and 3.6).

The 1988 work revealed one important mechanism whereby a DNA tumor virus–encoded oncoprotein can disrupt the cellular growth-regulating circuitry: biochemical characterization of the oncoprotein made by the E1A oncogene of human adenovirus type 5 showed that in adenovirus-transformed cells, this oncoprotein is tightly

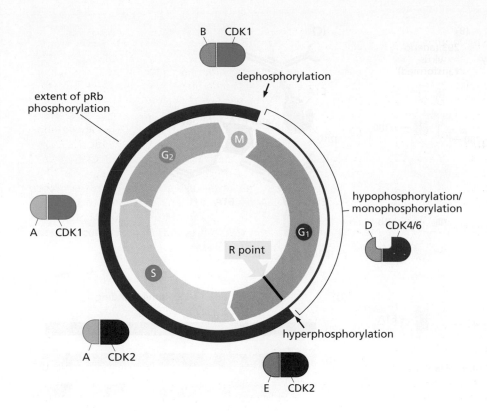

Figure 8.15 Cell cycle–dependent phosphorylation of pRb The phosphorylation state of pRb (*red circle*) is closely coordinated with cell cycle advance. As cells pass through the M/G$_1$ transition, virtually all of the existing phosphate groups are stripped off pRb by phosphatase PP1, leaving it in an *unphosphorylated* configuration. As cells progress through G$_1$, a single phosphate group is attached as any one of ~14 distinct phosphorylation sites (by cyclin D-CDK4/6 complexes), yielding *hypophosphorylated* (more precisely *monophosphorylated*) pRb. Later, when cells pass through the restriction (R) point (*yellow arrow*), cyclin E–CDK2 complexes phosphorylate pRb on at least 12 more sites in this group of 14, placing it in a *hyperphosphorylated* state. Throughout the remainder of the cell cycle, the extent of pRb phosphorylation remains constant or increases slightly until cells enter into M phase.

bound to a number of cellular proteins, among them pRb (**Figure 8.16**; **Sidebar 8.4**). Soon thereafter, the SV40 virus–encoded large T oncoprotein and the E7 oncoprotein of certain strains of human papillomavirus (HPV) implicated in causing human cervical carcinoma were also found capable of forming physical complexes with pRb in virus-transformed cells.

These three different DNA tumor virus oncoproteins (E1A, large T antigen, and E7) are structurally unrelated to one another, yet all of them target a common cellular protein—pRb. Moreover, the molecular biology of adenovirus replication cycles otherwise shows no resemblance to those of SV40 and HPV. This suggests that **convergent evolution** led to the development of viral oncoproteins capable of perturbing cellular pRb. Stated differently, because evolution selects for viruses that can multiply more effectively, these discoveries suggested that sequestration or functional inactivation of pRb by the three viral oncoproteins is necessary for optimal viral replication in infected cells. Finally, to state the obvious, DNA tumor viruses use an entirely different genetic strategy to transform cells than do tumorigenic retroviruses: rather than acquiring and exploiting host-cell proto-oncogenes, the DNA tumor viruses have evolved proteins that interfere with the actions of host-cell tumor suppressor genes, that is, the proteins encoded by these genes. As it turned out, these viral oncoproteins could also bind and ostensibly inactivate two paralogs of pRB—p107 and p130—that exert cell cycle controls complementary to that of pRB (see Figure 8.16B).

The discoveries of these viral–host protein complexes had profound effects on our understanding of how DNA tumor viruses succeed in transforming the cells that they infect. The speculation went like this: Within a normal cell, the *Rb* gene was known to function as a tumor suppressor gene. Hence, the *Rb* gene, acting through its protein product, pRb, must inhibit cell proliferation in some fashion. DNA tumor viruses can sabotage the activities of pRb by dispatching viral oncoproteins that seek out and bind pRb, sequestering and apparently inactivating it. This sequestration removes pRb from the regulatory circuitry of the cell, yielding the same outcome as when a cell loses the two copies of its chromosomal Rb gene through mutations (see Figure 7.5). This suggested, furthermore, that the DNA tumor viruses succeed in transforming cells through their ability to disable this key cellular growth-inhibitory protein, thereby liberating the virus-infected cell from the growth suppression that pRb

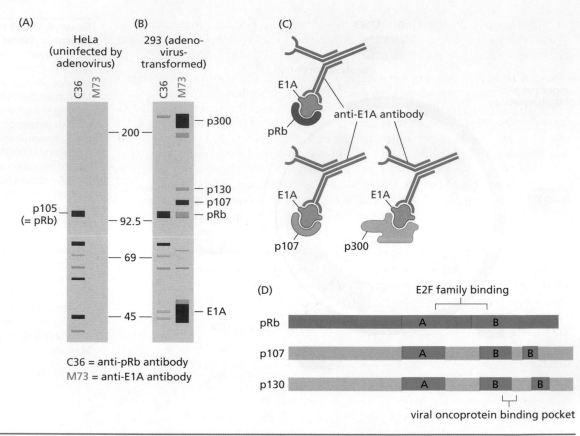

C36 = anti-pRb antibody
M73 = anti-E1A antibody

Figure 8.16 Co-precipitation of cellular proteins with adenovirus E1A (A) Incubation of the C36 anti-pRb monoclonal antibody with a lysate of HeLa cells revealed that pRb migrated with the mobility of a protein of ~105 kD, with more rapidly migrating cleavage fragments evident below it (*left channel*). Use of the M73 antibody reactive with the adenovirus E1A oncoprotein revealed no reactive proteins in these cells, which lack adenovirus genomes. This also indicated that this antibody did not have undesired cross-reactions with cellular proteins. (B) Incubation of the anti-pRb antibody with lysates of adenovirus-transformed 293 cells revealed once again the presence of pRb and some lower–molecular-weight fragments. However, when the M73 anti-E1A antibody was used, it precipitated not only the adenovirus E1A protein but also a series of host-cell proteins, including pRb, p107, p130, and p300. (C) Because the anti-E1A antibody had no reactivity with host-cell proteins (see panel A), E1A must have formed a bridge between the antibody molecules and a number of host-cell proteins with which it was physically complexed in virus-transformed cells. (D) pRb is one of three structurally related proteins that together are often called "pocket proteins." All three are bound by viral oncoproteins, such as adenovirus E1A (see panel B), SV40 large T, and human papillomavirus E7. All three pocket proteins are phosphorylated by cyclin D–CDK4/6 complexes, and each binds its own subset of E2F transcription factors, as discussed later. The "pocket" that interacts with both viral oncoproteins and cellular E2F transcription factors is composed of A and B domains (*blue*). (A, B, Adapted from P. Whyte et al., *Nature* 334:124–129, 1988.)

Sidebar 8.4 Adenovirus deploys a single viral oncoprotein to bind and sequester multiple cellular proteins As seen in Figure 8.16, when a lysate of adenovirus-transformed cells is incubated with an anti-E1A monoclonal antibody, the resulting immunoprecipitate contains the E1A oncoprotein plus a number of other proteins. These other proteins are not bound directly by the antibody; instead, in each case, the E1A oncoprotein acts as a bridge to link the antibody molecule to a second protein, indicating that within adenovirus-transformed cells the E1A protein had previously formed physical complexes with each of these other cellular proteins. (Before the immunoprecipitates were analyzed by gel electrophoresis and autoradiography, these multiprotein complexes were treated with a strong detergent to dissociate the component proteins from one another.)

The first of these associated proteins to be identified was pRb, with two other cellular proteins, termed p107 and p130, being characterized soon thereafter; these two are paralogous cousins of pRb. These observations indicated that during the course of its evolution, adenovirus has configured the E1A protein in a way that enables it to bind these three cellular proteins, ostensibly in order to sequester or inactivate them. pRb and its two cousins are often called generically **pocket proteins** (Figure 8.16D), because each carries a cavity into which viral oncoproteins such as E1A can insert. Later, a higher–molecular-weight cellular partner protein, termed p300, was also characterized, and a number of other cellular targets remain to be characterized.

The rationale for why adenovirus deploys its oncoprotein to sequester these four proteins must be understood in terms of the viral replication cycle. A virus needs to reconfigure the regulatory circuitry of an infected cell in order to create an intracellular environment that supports various steps of viral replication. Most important is its requirement for viral DNA replication, which can apparently occur efficiently only if the virus is able to liberate an infected cell from its quiescent pre-restriction point state by functionally inactivating the cellular proteins that are holding it in that state; this liberation permits the infected cell, in turn, to advance into S phase, where the cellular DNA replication machinery is now active and can be exploited by the virus.

normally imposes. (In fact, the HPV E7 protein goes beyond simple sequestration of pRb, because it also tags pRb for ubiquitylation and thus proteolytic degradation; see Supplementary Sidebar 7.5.)

Another clue to pRb function was provided by the discovery that the oncoproteins of the DNA tumor viruses bind preferentially to the hypophosphorylated forms of pRb, that is, the forms of pRb that are present in the cell through most of the G_1 phase (see Figure 8.15). Conversely, the hyperphosphorylated pRb forms that are present in late G_1 and in the subsequent phases of the cell cycle are ignored by the viral oncoproteins. This observation led to yet another speculation: tumor virus oncoproteins focus their attention on sequestering—and thereby inactivating—the only forms of pRb in the cell that are worthy of their attention, these being the growth-inhibitory forms of pRb found in early and mid-G_1. Conversely, the viral oncoproteins ignore those forms of pRb that are already inactivated by another mechanism, phosphorylation.

Such logic indicated that the hypophosphorylated forms of pRb present in early and mid-G_1 cells are actively growth-inhibitory, whereas the hyperphosphorylated forms found after the R point have lost their growth-inhibitory powers, having been functionally inactivated by this phosphorylation. This meant that the R-point transition, which is defined by a physiological criterion (acquisition of growth factor independence), is accompanied by a biochemical alteration of pRb that converts it from a growth-inhibitory protein to one that is functionally inert. These speculations were soon validated by various types of experiments.

In addition, further research has revealed that the control of pRb phosphorylation is more reversible than is indicated by the diagram in Figure 8.15. Thus, if a cell should experience serious physiological stresses while in S phase or G_2, pRb phosphorylation can be reversed by phosphatases such as PP1, thereby returning pRb to its actively growth-inhibitory state. These stresses include, for example, hypoxia, DNA damage, and disruption of the mitotic spindle. Such reactivation of pRb-mediated growth inhibition is presumably only transient and is reversed once the physiological stresses or damage have been resolved.

8.6   pRb is deployed by the cell cycle clock to serve as a guardian of the restriction-point gate

As described above, various lines of evidence converge on the conclusion that pRb, which inhibits growth through early to mid-G_1 phase, becomes inactivated by extensive phosphorylation when the cell passes through the R-point gate and thereafter is largely or completely inert as a growth suppressor. In effect, pRb serves as a guardian at this gate, holding it shut unless and until it becomes hyperphosphorylated, in which case pRb loses its growth-inhibitory powers, opens the gate, and permits the cell to enter into late G_1 and advance thereafter into the remaining phases of the cell cycle.

Because pRb is the ultimate arbiter of growth versus nongrowth, its phosphorylation must be carefully controlled. Not unexpectedly, its phosphorylation is governed by components of the cell cycle clock. In early and mid-G_1, D-type cyclins together with their CDK4/6 kinase partners are responsible for initiating pRb phosphorylation, leading to its hypophosphorylation. In fact, the "hypophosphorylated pRb" cited above is actually a collection of various *mono*phosphorylated pRb species, each of which carries a single phosphate on one or another of its serine or threonine residues. (Why and how cyclin D—CDK4/6 complexes attach only a single phosphate to each pRb molecule but not more than one remains a mystery.) Because the levels of D-type cyclins appear to be controlled largely by extracellular signals, notably mitogenic growth factors, we can now plot out a direct line of signaling: growth factors induce the expression of D-type cyclins; D-type cyclins, collaborating with their CDK4/6 partners, initiate pRb phosphorylation (**Figure 8.17**).

During a normal cell cycle, this initial hypophosphorylation of pRb on single amino acid residues is not sufficient for the subsequent functional inactivation of pRb at the

early and mid-G_1 late G_1

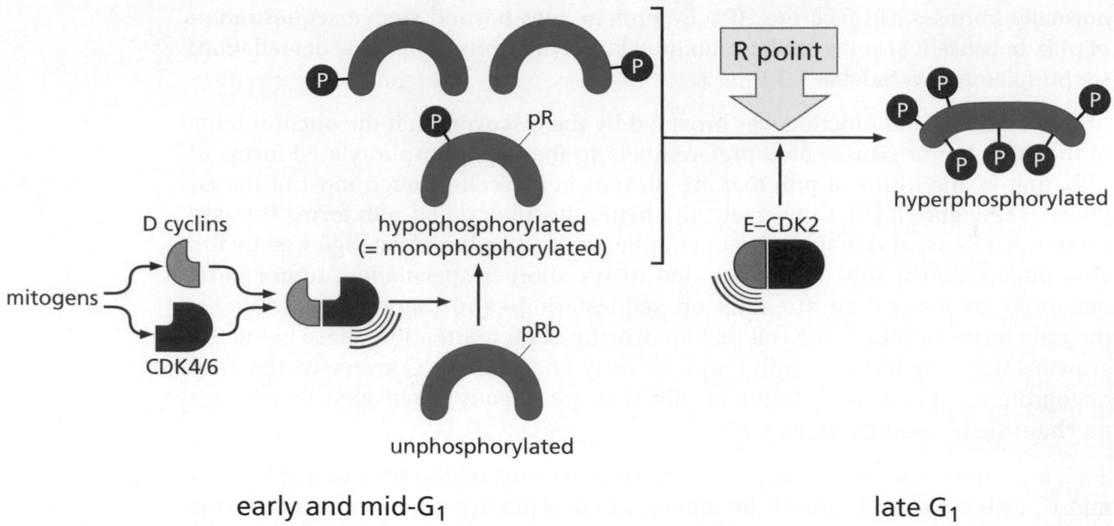

Figure 8.17 Control of the restriction-point transition by mitogens The levels of D-type cyclins are controlled largely by extracellular signals (mitogens, *left*). The D-type cyclins, acting with their CDK4 and CDK6 partners, are able to drive monophosphorylation at multiple alternative serine and threonine residues of pRb. Eventually, accumulation of D-CDK4/6 complexes leads to the abstraction of p27^{Kip1} away from its association with E-CDK2 complexes (see Figure 8.14B) and the resulting activation of E-CDK2 complexes. Cyclin E–CDK2 complexes then assume the work of decommissioning pRb by placing it in a hyperphosphorylated state, causing a >10-fold increase in its phosphorylation that results in the complete functional inactivation of pRb.

R point, which requires its hyperphosphorylation on multiple residues. In fact, cyclin E levels increase dramatically at the R point. The cyclin E then associates with its CDK2 partner, and this complex drives pRb phosphorylation to completion, leaving pRb in its *hyperphosphorylated*, functionally inactive state (see Figure 8.17). The phosphorylation and functional inactivation of pRb's cousins, p107 and p130, are also under the control of cyclin D–CDK4/6 and cyclin E–CDK2 complexes. The roles of these pRb cousins in cancer pathogenesis is unclear and, for this reason, they will not be discussed further here.

Once the cell advances through the R point, the continued hyperphosphorylation and functional inactivation of pRb is apparently maintained and further increased by CDK complexes carrying cyclin E, then cyclin A, and finally cyclin B, none of which is responsive to extracellular signals—precisely the properties of the cell cycle clock machinery that, we imagine, operates after the R-point transition and guarantees execution of the rigidly programmed series of transitions in S, G_2, and M phases.

This scheme reveals why pRb is such a critical player in the regulation of cell proliferation. If its services are lost from the cell (through mutation of the chromosomal *Rb* gene copies, methylation of the *Rb* gene promoter, or the actions of DNA tumor virus oncoproteins), then this protein can no longer stand as the guardian of the R-point gate. Moreover, as we will see, in some cancer cells, pRb phosphorylation is deregulated, resulting in inappropriately phosphorylated and thus functionally inactivated pRb. In certain other cancer cells, there is evidence that the dephosphorylation (and attendant activation) of pRb, which normally happens at the M/G_1 transition through the action of the PP1 phosphatase, never occurs, leaving pRb in its hyperphosphorylated, inactivated state throughout the entire growth-and-division cycle of these cells. Without pRb on watch, cells move through G_1 into S phase without being subjected to the usual controls that are designed to ensure that cell cycle advance can proceed only when certain preconditions have been satisfied. Indeed, as we will see, the deregulation of pRb phosphorylation is so widespread that one begins to think that this important signaling pathway is perturbed in virtually all human tumors.

The activities of pRb seem to overlap with those of its cousins, p107 and p130, raising the question why pRb is the only one of these three pocket proteins to play a clear and obvious role in cancer pathogenesis. A definitive answer to this puzzle is still elusive. It may come from the specific times in the cell cycle when these two pRb-related proteins are active. Whereas p130 is specialized to suppress cell proliferation while a cell resides in G_0, p107 appears to be most active in the late G_1 and S phases. Only pRb is well positioned to control the transition at the R point from mitogen-dependent to mitogen-independent growth. (Yet another physical complex of important regulatory proteins operates at the late G_2/M transition to enable the complicated molecular events associated with mitosis; we focus in this chapter on late G_1 and the R point, simply because it is here that the decisions of cell proliferation versus quiescence are made that are so critical to cancer pathogenesis.)

8.7 E2F transcription factors enable pRb to implement growth-versus-quiescence decisions

As described above, a number of lines of evidence indicate that hypophosphorylated pRb is active in suppressing G_1 advance and loses this ability when it becomes hyperphosphorylated. Still, this behavior provides no clue as to precisely how pRb succeeds in *imposing* this control. Research conducted in the early 1990s indicated that pRb exercises much of this control through its effects on a group of transcription factors termed E2Fs.

When pRb (and its two cousins, p107 and p130) are in their unphosphorylated or hypophosphorylated state, they bind E2Fs, including E2Fs that are already bound to DNA; however, when hyperphosphorylated (see Figure 8.15), pRb and its cousins dissociate from the E2Fs (**Figure 8.18A**). This suggested a simple model of how pRb is able to control cell cycle advance. In the early and middle parts of G_1, E2Fs are associated with the promoters of a number of genes under their control. At the same time, these transcription factors are bound by pocket proteins (pRb, p107, p130). This pocket protein association prevents the E2Fs from acting as stimulators of transcription. Accordingly, during much of the G_1 phase of the cell cycle, genes that depend on E2Fs for their expression remain repressed. However, when the pocket proteins (most importantly, pRb) undergo hyperphosphorylation at the R point in late G_1, they release their grip on the E2Fs, permitting the E2Fs to stimulate transcription of their clientele of genes. The products of these genes, in turn, usher the cell from late G_1 into S phase. Similarly, when viral oncoproteins are present, they mimic pRb hyperphosphorylation by preventing pRb from binding E2Fs, doing so throughout the cell cycle (Figure 8.18B).

Although the above scheme is correct in outline, the details are more interesting and complex. The term "E2F" is now known to include a family of proteins composed of eight members—E2F1 through E2F8; the first six of these can bind to either a DP1 or a DP2 subunit to form heterodimeric transcription factors (Figure 8.18C and D); the two others—E2F7 and E2F8—have two DNA-binding domains (DBDs), allowing each to bind DNA without an associated DP1/2 subunit. (In the discussions that follow, we focus on the behavior of the various E2F subunits, with association to their DP1 or DP2 partner subunits being assumed.) Once assembled, E2F–DP complexes recognize and bind to a sequence motif in the promoters of various genes that seems to be TTTCCCGC or slight variations of this sequence.

Throughout this book, we have portrayed transcription factors as proteins that bind to the promoters of genes and then proceed to activate transcription. In fact, transcription factors operating like E2Fs can exert two opposing effects on transcriptional control. When bound to the promoters of genes in the absence of any associated pocket proteins (that is, pRb, p107, or p130), E2Fs such as E2Fs 1, 2, and 3 can indeed trigger gene expression by attracting other proteins—such as histone acetylases—that function to remodel the nearby chromatin and recruit RNA polymerase to initiate transcription.

However, when pRb is hypophosphorylated, it physically associates with its E2F partners, which themselves are already sitting on the promoters of various genes and remain so after pRb binds to them. Once bound to these E2Fs, hypophosphorylated pRb molecules block the **transactivation domains** of the E2Fs that they employ to activate transcription (see Figure 8.18C). At the same time, pRb recruits other proteins that actively repress transcription. One important means of doing so, but hardly the only one, is to recruit a histone deacetylase (HDAC) to this complex (see Section 1.8); by removing acetyl groups from nearby histone molecules, an HDAC remodels the

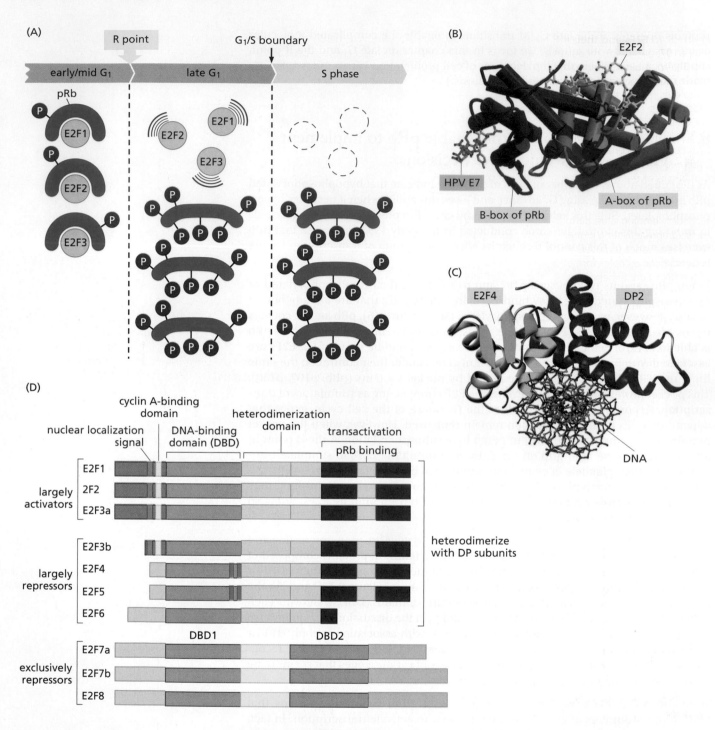

nearby chromatin into a configuration that is incompatible with active transcription (**Figure 8.19A**).

This means that, in addition to occluding (physically blocking) the transactivation domains of E2Fs, pRb actively functions as a transcriptional repressor. E2Fs 4 and 5 seem to be involved primarily in the repression of genes and act largely through associated p107 and p130 proteins to attract repressors of transcription to promoters and thereby shut down gene expression. E2Fs 6, 7, and 8, which do not associate with pocket proteins, appear to act exclusively as transcriptional repressors.

In quiescent G_0 cells, for example, E2F4 and E2F5 are present in abundance (and are associated largely with p130), whereas E2Fs 1, 2, and 3 are hardly present at all, being expressed largely in proliferating cells. Moreover, cells that have been genetically deprived of E2F4 and E2F5 have lost responsiveness to the growth-inhibitory effects of the p16[INK4A] CDK inhibitor. This implies that the lion's share of the downstream

Figure 8.18 E2Fs and their interactions (A) By binding various DNA-bound E2F transcription factors, the three related pocket proteins, pRb, p107, and p130, modulate the expression of E2F-responsive genes in various phases of the cell cycle. When pRb binds to E2Fs 1, 2, and 3 (*left*), it blocks their transcription-activating domains; other proteins that serve to modify chromatin and repress transcription (*not shown*) are attracted to the pRb–E2F complexes. Because they are bound to the promoters of certain genes via the E2Fs, these multiprotein complexes repress the expression of such genes, doing so by attracting chromatin-modifying proteins such as histone methyltransferases and deacetylases. (The behavior of p107 and p130, *not shown*, is similar to that of pRb.) After being hyperphosphorylated at the R point, the pocket proteins release their grip on the E2Fs, allowing the latter to activate transcription of those genes (late G_1, *middle*). However, because E2Fs 1, 2, and 3 induce expression of E2F7 and E2F8, which function as transcriptional repressors (see panel D), the ability of these three E2Fs to activate signaling is curtailed toward the end of G_1. Moreover, as cells enter into S phase (*right*), E2Fs 1, 2, and 3 are inactivated or degraded. Hence, E2Fs 1, 2, and 3 appear to function as active inducers of transcription only in the narrow window of time in late G_1 phase that begins at the R point (*yellow arrow*) and ends as cells enter S phase. (B) The binding of E2Fs, such as E2F2 (*yellow stick figure*) by pRb (*red and blue*), is prevented by several DNA tumor virus oncoproteins. Here we see in a crystallographic structure how a segment of the human papillomavirus (HPV) E7 oncoprotein (*green stick figure, left*) binds to a shallow groove of pRb; this binding, together with other secondary binding sites, perturbs the pocket cavity formed by the B- and A-boxes of pRb, thereby precluding the binding of the C-terminal, transcription-activating domain of E2F2 (*yellow*) to this pocket. This loss of E2F binding effectively neutralizes the ability of pRb to inhibit cell cycle advance (*cylinders* = α-helices). (C) With the exception of E2F7 and E2F8, each of the E2F transcription factors binds DNA as a heterodimeric complex formed by an E2F subunit together with a DP partner. Shown is the binding of an E2F4–DP2 complex (*ribbon diagram*) to the DNA double helix (*stick figure*), as revealed by X-ray crystallography. (D) Because of the existence of variant forms of several E2Fs, they are actually represented by at least 10 distinct protein species. E2Fs 1, 2, and 3a represent transcription-activating E2Fs, whereas E2F3b, E2F4, E2F5, and E2F6 are largely or exclusively involved in transcriptional repression. E2Fs 7a, 7b, and 8 are atypical E2Fs, in that each has a duplicated DNA-binding domain (DBD) that enables DNA binding in the absence of a DP subunit as partner (see Panel C); these particular E2Fs are involved exclusively in repression. E2F1, 2, and 3 induce their expression, suggesting the operation of a negative-feedback loop designed to shut down E2F1, 2, and 3 signaling. Overall, E2Fs 4 and 5 are involved in repression in early/mid-G_1, whereas E2Fs 7 and 8 are involved in repression in late G_1 at the G_1/S transition. (B, courtesy of Y. Cho, from C. Lee et al., *Genes Dev.* 16:3199–3212, 2002. C, courtesy of R. Latek, from N. Zheng et al., *Genes Dev.* 13:666–674, 1999. Both with permission from Cold Spring Harbor Laboratory Press. D, adapted from H.-Z. Chen et al., *Nat. Rev. Cancer* 9:785–797, 2009. With permission from Nature.)

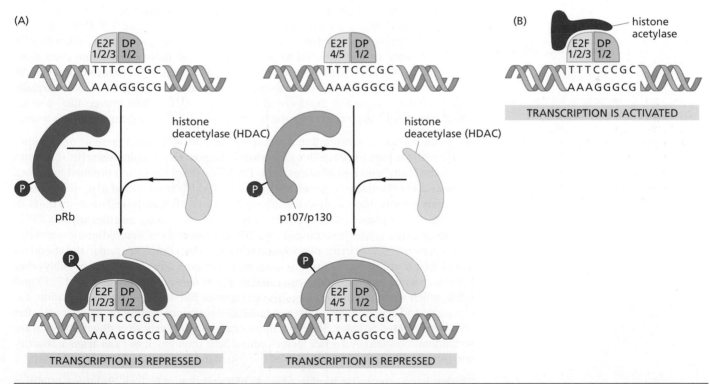

Figure 8.19 Modification of chromatin by pocket proteins (A) Hypophosphorylated pRb (*dark green*) discourages transcription by attracting histone deacetylase (HDAC) enzymes (*gray*) to pRb–E2F complexes that are bound, in turn, to gene promoters (*left*); these enzymes prevent transcription by removing acetyl groups from nearby histones, causing the latter to shift into a configuration that is incompatible with the progression of RNA polymerase through the chromatin. A similar repressive role is exercised by the pRb-related p107 and p130 (*right*), which bind E2F4/5; the latter appear to never be involved in induction of active transcription. (B) When the E2Fs 1/2/3 are not complexed with pRb after the latter undergoes hyperphosphorylation, they attract histone acetylases (*red*), which place chromatin in a configuration that is conducive for transcription, which occurs largely in late G_1.

effects that result from inhibiting the cyclin D–CDK4/6 complexes and thus blocking the hyperphosphorylation of pocket proteins is mediated via the binding of p107 and p130 to these two E2Fs. Ultimately, this binding results in the repression of a constituency of genes that carry TTTCCCGC sequences in their promoters (see Figure 8.19A).

Once hyperphosphorylated, pRb releases its grip on E2F1, 2, and 3 complexes, and these E2Fs can then attract transcription-activating proteins, including histone acetylases, which convert the chromatin into a configuration that encourages transcription (Figure 8.19B). A series of genes that are expressed specifically in late G_1 and are known or assumed to be important for S-phase entry has been found to contain binding sites in their promoters for E2Fs. Included among these are some genes encoding proteins involved in synthesizing DNA precursor nucleotides (such as dihydrofolate synthetase and thymidine kinase), as well other genes involved directly in DNA replication, such as that encoding DNA polymerase α. Yet other target genes are involved in activating certain aspects of energy metabolism. (One survey indicates that at least 500 genes are direct targets of transcriptional activation by E2Fs.)

Prominent among the E2F-activated genes are the genes encoding cyclin E and E2F1. Consequently, the levels of cyclin E and E2F1 mRNAs rise quickly after passage through the R point and the levels of their protein products increase shortly thereafter. Recall that cyclin E, for its part, is responsible, together with CDK2, for driving the hyperphosphorylation of pRb. Hence initial pRb inactivation leads to increases in cyclin E levels, and cyclin E, once formed, drives further pRb inactivation (**Figure 8.20**). These relationships enable the operations of a powerful, self-amplifying positive-feedback loop that is triggered as cells pass through the R point. (Such regulation can also be termed a "feed-forward loop.")

Yet another positive-feedback loop is activated simultaneously at the R point through the following mechanism: cyclin E–CDK2 complexes phosphorylate p27$^{\text{Kip1}}$, and the latter, once phosphorylated, is ubiquitylated and rapidly degraded (Figure 8.20B). The destruction of p27$^{\text{Kip1}}$ molecules, which normally inhibit E–CDK2 firing, liberates additional cyclin E–CDK2 complexes that proceed to phosphorylate and thereby inactivate even more p27$^{\text{Kip1}}$ molecules. Such positive-feedback mechanisms operate in many control circuits to guarantee the rapid execution of a decision once that decision has been made. Equally important, they ensure that the decision, once made, is essentially irreversible—a situation that corresponds precisely with the behavior of cells advancing through the R point.

The period of active transcriptional promotion by E2Fs 1, 2, and 3 appears to be rather short-lived (see Figure 8.18A). This time window begins at the R point when pRb becomes hyperphosphorylated and liberates these E2Fs. The three E2Fs then proceed to induce expression of critical late G_1 genes, the products of which are required to prepare the cell for entrance into S phase, as described above. Soon thereafter, as the cell traverses the G_1/S transition into S phase, cyclin A becomes activated and, acting together with its CDK2 partner, phosphorylates both the E2F and DP subunits of these heterodimeric transcription factors; this results in the dissociation of the E2F–DP complexes and in the attendant loss of their transcription-activating abilities. At the same time, E2F1 (and possibly other E2Fs) are tagged for proteasome-mediated degradation by ubiquitylation, and E2F7 and E2F8, which seem to function largely to antagonize E2F-mediated gene activation, are expressed and appear to block any residual E2F-dependent expression that persists after cells have entered S phase. Together, these changes effectively shut down further transcriptional activation by the E2F factors, whose brief moment of glory as active transcription factors will come again only during the end of the G_1 phase of the next cell cycle.

pRb has been reported to bind to a variety of transcription factors, and these might, in principle, be controlled by pRb and its state of phosphorylation. In the great majority of cases, however, detailed functional analyses of their interactions with pRb have not been undertaken. There is a compelling reason to believe that, among all of the transcription factor clients of pRb, the E2Fs are of preeminent importance in controlling cell cycle advance: when E2F1 protein molecules are micro-injected into certain types of serum-starved, quiescent cells (which are therefore in G_0), these transcription factor molecules, acting on their own, are able to induce these cells to enter G_1 and advance all the way into S phase; E2F2 and E2F3 have similar powers.

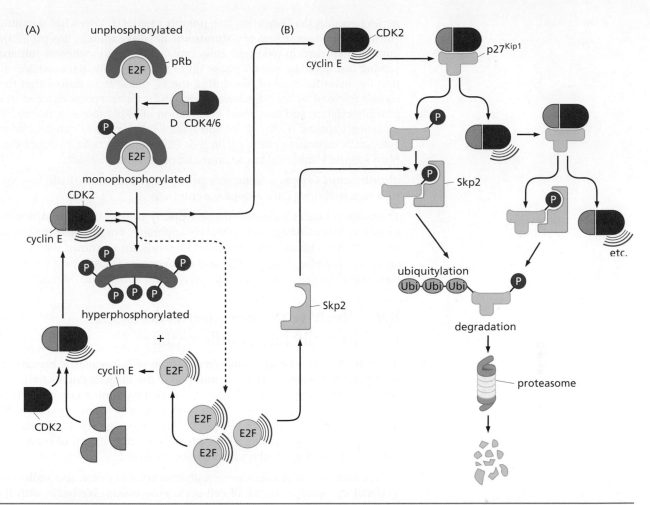

Figure 8.20 Positive-feedback loops and the irreversibility of cell cycle advance The irreversibility of certain key steps in cell cycle progression and the rapidity of their execution is ensured, in part, by the activation of certain interconnected, self-amplifying positive-feedback loops, such as those illustrated here that operate at the R point. This cascade of events is triggered when small numbers of E–CDK2 molecules are liberated from inhibition by p27^{Kip1}, as shown in Figure 8.14B. (A) When the liberated E–CDK2 complexes drive pRb hyperphosphorylation, this liberates small numbers of E2F transcription factors (specifically E2F1/2/3, *below*) from pRb control, enabling the E2Fs to trigger increased transcription of the cyclin E gene, leading in turn to the synthesis of more cyclin E protein and the formation of more E–CDK2 complexes, which then function to drive more pRb hyperphosphorylation. At the same time, the newly liberated E2F proteins drive expression of E2F1 itself, further amplifying E2F activity (*not shown*). Also not shown here is the interaction of monophosphorylated pRb with Skp2, which blocks the ability of the latter to drive p27^{Kip1} degradation, leading to p27^{Kip1} accumulation through much of the G$_1$ phase of the cell cycle; hence, monophosphorylated pRb can also halt cell proliferation via an E2F-independent mechanism. (The active E2F transcription factors shown are actually heterodimers of E2F subunits plus DP subunits, as shown in Figure 8.19B). (B) The initial activation of a small number of active E–CDK2 complexes enables the latter to phosphorylate p27^{Kip1}, marking it for recognition and binding by Skp2, ubiquitylation and destruction in proteasomes. Yet another positive-feedback loop leads to E2F-mediated increased transcription of the *Skp2* gene (as seen in panel A), whose protein product drives ubiquitylation and thus degradation of p27^{Kip1}; this liberates even more E-CDK2 complexes from inhibition by the p27^{Kip1} CKI.

8.8 A variety of mitogenic signaling pathways control the phosphorylation state of pRb

It is worthwhile at this point to step back and summarize briefly the ways in which these various mitogenic signals regulate distinct components of the cell cycle clock. Prominent among the signals impinging on this nuclear regulatory machinery are those conveyed by the mitogenic signaling pathways, as we saw in Figure 8.9B. Because excessive signals flow through these pathways in virtually all types of cancer cells, the connections between these pathways and the cell cycle clock are critical to our understanding of cell transformation and tumor pathogenesis.

Mitogenic signals transduced via Ras result in the mono- and eventual hyperphosphorylation of pRb. This phosphorylation is initiated by inducing the expression of genes encoding the D-type cyclins, notably cyclin D1. When Ras signaling is blocked (through the introduction of a **dominant-negative** mutant Ras protein), cell proliferation is blocked in wild-type cells exposed to serum, whereas mutant $Rb^{-/-}$ cells lacking Ras function enter S phase quite normally. This demonstrates dramatically that the main function of Ras during the G_1 phase is to ensure that the mitogenic signals released by ligand-activated growth factor receptors succeed in causing the phosphorylation and functional inactivation of pRb. Moreover, the block to S-phase entrance imposed by the dominant-negative Ras protein can be circumvented by ectopically expressing either cyclin E or E2F1 genes in cells. Together, these observations suggest a linear chain-of-command of the following sort:

growth factors → growth factor receptors → Ras → cyclins D1 and E → inactivation of pRb → activation of E2Fs → S-phase entrance.

There are yet other mechanisms, still poorly characterized, that enable mitogenic growth factors to energize the cell cycle apparatus. For instance, the physical association of D-type cyclins with CDK4 also depends, for unknown reasons, on mitogenic signaling. The elucidation of this and other connections between growth-promoting signals and the cell cycle clock will surely come from future research.

8.9   The Myc protein governs decisions to proliferate or differentiate

As we have read, pRb function is compromised in neoplastic human cells through multiple mechanisms, including mutation of the *Rb* gene and its functional inactivation through hyperphosphorylation. A functionally similar outcome can be observed in cells that have been infected and transformed by human papillomavirus (HPV); thus, pRb binding by the viral E7 oncoprotein (see Figures 8.16D and 8.18B) plays a key role in the development of the great majority (>99.7%) of cervical carcinomas, which occur in large numbers throughout the world.

In fact, human cancer cells devise a diverse array of alternative molecular strategies to derail the normal control of cell cycle progression. We begin with the Myc protein which, when expressed in a deregulated fashion, operates as an oncoprotein. More than 70% of human tumors overexpress either Myc (often termed c-Myc) or one of its two close cousins—N-Myc and L-Myc. The Myc family oncoproteins function very differently from most of the oncoproteins that we encountered in Chapters 4, 5, and 6. These others, notably Ras and Src, operate in close proximity to cytoplasmic membranes and trigger complex signaling cascades that activate cytoplasmic

Sidebar 8.5 Many oncoproteins function as transcription factors Myc is only one of a large group of transcription factors that can function as oncoproteins (Table 8.1). We first encountered oncoproteins of this type in our discussion of the immediate early genes *myc*, *fos*, and *jun* (see Section 6.1), each of which encodes a transcription factor that can function as an oncoprotein. Later, we read about STAT proteins (Section 6.8), NF-κB, Notch, and Gli (Section 6.12), which also can play these dual roles. These various transcription factors act on a variety of target genes to induce a subset of the phenotypes that we associate with transformed cells.

The constituency of target genes that are acted upon by each of these transcription factors (TFs) remains poorly defined. This lack of information is explained largely by the fact that at the experimental level, it has been challenging to determine the identities of the genes that are directly activated or repressed by various TFs. Even with the determination that various TFs bind to certain chromosomal segments does not, on its own, guarantee that the TF actually regulates the expression of a gene associated with this segment. Indeed, for many TFs, only a relatively small proportion (for example, 20% or 30%) of

the binding sites can actually be validated as functionally important sites of TF action.

Table 8.1 Representatives of the classes of genes encoding TFs that act as oncoproteins[a]

Transcription factor	Representative genes
leucine zipper + basic DNA-binding domain	*fos* and *jun*
helix + loop + helix (bHLH)	*myc*, *N-myc*, *L-myc*, *tal*, and *sci*
zinc finger	*myl/RARa*, *erbA*, *evi-1*, and *gli-1*
homeobox	*pbx* and *Hoxb8*
others	*myb*, *rel*, *ets-1*, *ets-2*, *fli-1*, *spi-1*, *ski*

[a]Genes are grouped according to the structural features of the encoded proteins.
Adapted from T. Hunter, *Cell* 64:249–270, 1991. With permission from Elsevier.

signal-transducing proteins and, ultimately, nuclear transcription factors. Myc and its cousins are found, by contrast, in the nucleus, where they function directly as growth-promoting transcription factors (**Sidebar 8.5**).

These three Myc family proteins are only three of a very large (>100 members) family of bHLH transcription factors. They were named after their shared three-dimensional structures, which include a **b**asic DNA-binding domain followed by amino acid sequences forming an α-**h**elix, a **l**oop, and a second α-**h**elix. Members of this transcription factor family form homo- and heterodimers with themselves and with other members of the family. Such dimeric complexes then associate with specific regulatory sequences, termed E-boxes (composed minimally of the consensus sequence CANNTG, where N can be any nucleotide), which are found in the promoters of target genes that they regulate (**Figure 8.21A**).

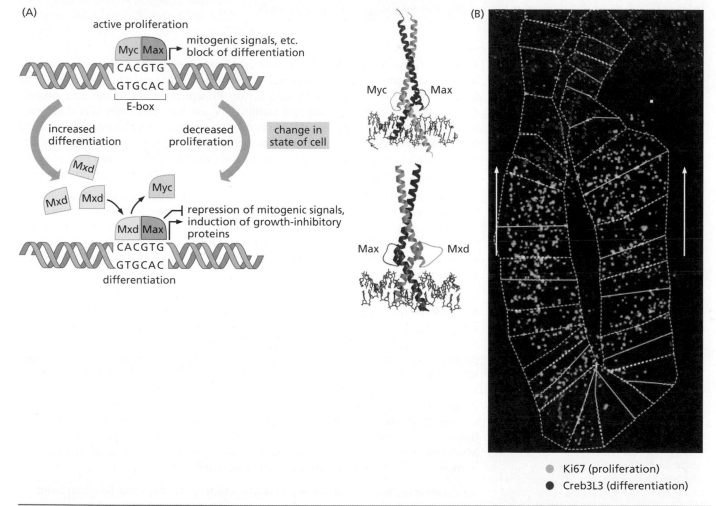

● Ki67 (proliferation)
● Creb3L3 (differentiation)

Figure 8.21 Proliferation vs. differentiation (A) Myc belongs to a family of bHLH (basic helix–loop–helix) transcription factors that act as heterodimers to modulate the transcription of a large cohort of target genes possessing E-box sequences (*left*). Myc–Max complexes act at enhancers like many other transcription factors and, in addition, promote elongation of nascent transcripts by releasing RNA polymerase II complexes from pause sites located immediately downstream of transcription start sites, allowing polymerase II to proceed with transcription and thus to the synthesis of full-length pre-mRNA transcripts; Mxd–Max complexes prevent this release. As cells differentiate, Mad levels increase progressively and Myc is displaced by Mxd, resulting in the disappearance of Myc–Max complexes, which otherwise would block differentiation (*left*). The actual molecular structures of these complexes sitting astride DNA (*right*) reveal why these proteins are termed basic helix–loop–helix transcription factors. (B) The opposition of proliferation vs. differentiation can be observed in various tissues. Here, a highly sensitive fluorescence *in situ* hybridization (FISH) technique has been used to detect individual mRNA molecules in the individual cells (*outlined in dashed lines*) lining a crypt in the mouse small intestine (see Figure 7.20. The *green dots* reveal mRNA molecules encoding Ki67 (indicating active proliferation), and the *red dots* reveal mRNA molecules encoding Creb3L3 (indicating enterocyte differentiation). The expression of these two mRNAs is mutually exclusive and the implied transition from one cell state to another occurs abruptly as cells migrate upward (*white arrows*) from one cell position to the next, apparently within an hour. (The involvement of Myc and its partners in this particular switch has not yet been demonstrated.) (A, right, from S.K. Nair and S.K. Burley, *Cell* 112:193–205, 2003. With permission from Elsevier. B, from S. Itzkovitz et al., *Nat. Cell Biol.* 14:106–114, 2011. With permission from Nature.)

Sidebar 8.6 Myc can operate in a way that differs from that of most transcription factors Recent research indicates that Myc can function quite differently than the other transcription factors discussed throughout this book. The others seem to be involved specifically in regulating transcriptional *initiation* (see Figure 1.18), that is, the formation of multiprotein complexes including RNA polymerase II at sites of initiation. In addition to initiating transcription, Myc has another role: in myriad genes, after RNA polymerase II (pol II) has initiated transcription, it halts at a pausing site ~50 bp downstream of the transcription start site. A Myc–Max heterodimer, bound to DNA, then interacts with the paused pol II complex and releases it from its pause site, allowing it to continue transcribing the bulk of the gene, that is, to *elongate* the RNA transcript; without this intervention by Myc–Max, the pol II that has begun the transcription of many genes is likely to sit at the pause site indefinitely. In one survey, 70% of actively transcribed genes were found to have Myc–Max heterodimers bound to these pause sites. (This number raises an interesting and unresolved paradox: in an actively growing cell, the number of genes found to bind Myc exceeds the absolute number of Myc molecules measured in such cells!) Stated differently, when acting as an anti-pausing factor, Myc serves to *amplify* the expression of genes that have already been activated, by other TFs. This, on its own, has important implications for Myc function: rather than having its own fixed constituency of responder genes that it regulates, many of Myc's targets are determined on a cell-type-specific basis, being governed by the actions of other TFs.

In the particular case of Myc, its actions are determined by its own levels as well as by its associations with partner bHLH proteins that either enhance or suppress its function as a transcription-activating factor. Phosphorylation of the Myc protein also modulates its functioning and stability. When Myc associates with Max, one of its bHLH partners that enhance its transcription-activating powers, the resulting Myc–Max heterodimeric transcription factor drives the expression of a large cohort of target genes, possibly more than one thousand (**Sidebar 8.6**). The products of many of these genes, in turn, have potent effects on the cell cycle, favoring cell proliferation.

Whereas Myc levels are strongly influenced by mitogenic signals, the levels of Max are kept relatively constant within cells. Accordingly, when normal cells are cultured in the presence of serum mitogens, Myc accumulates to substantial levels; conversely, Myc levels collapse when serum mitogens are withdrawn. This means that the levels of the Myc–Max heterodimer are continuously controlled by the flux of mitogenic signals that normal cells are receiving.

A diverse array of transcription factors belonging to the same bHLH family as Myc are involved in orchestrating tissue-specific differentiation programs. Acting through intermediaries, Myc can prevent these other bHLH transcription factors from executing various differentiation programs. Consequently, Myc can simultaneously promote cell proliferation and block cell differentiation; this is precisely the biological behavior that is associated with actively growing cells that have not yet entered into post-mitotic, differentiated states (Figure 8.21B). However, as cells slow their proliferation and become differentiated, the Myc–Max complexes disappear, because increasing levels of Mxd protein (another bHLH partner formerly known as Mad) displace Myc from existing Myc–Max complexes (see Figure 8.21A). The resulting Mxd–Max complexes then lose the ability to stimulate transcription, permitting cells in many human tissues to enter into post-mitotic differentiated states.

Myc interacts closely with the cell cycle machinery. Perhaps the first indication of this came from observations in which pairs of oncogenes were found to collaborate with one another in transforming rodent cells to a tumorigenic state. We will revisit oncogene collaboration later in Chapter 11. For the moment, suffice it to say that the *ras* and *myc* oncogenes were found to be highly effective collaborators in cell transformation, implying that each made its own unique contribution to this process. Similarly, a *ras* oncogene was found to collaborate with the adenovirus E1A oncogene in cell transformation. Together, these observations indicated that the *myc* and E1A oncogenes functioned analogously in this particular experimental setting. Precisely how they did so was a mystery.

The fact that the E1A oncoprotein binds and inactivates pRb and its p107 and p130 cousins (see Sidebar 8.4) suggested that the Myc protein could have similar effects on these vital cellular proteins, especially pRb. However, a direct association between Myc and pRb could not be established. Instead, we have come to realize that Myc induces the expression of other critical components of the cell cycle clock that operate

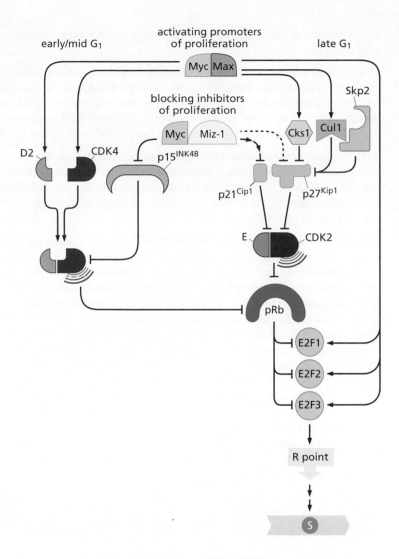

Figure 8.22 Actions of Myc on the cell cycle clock The Myc protein, acting with its Max partner (*top*), induces expression of a number of positive regulators of cell cycle advance, thereby driving cells through the R-point transition. For example, the Myc–Max heterodimer is able to induce expression of the growth-promoting proteins cyclin D2 and CDK4 (*left*), which can promote advance through early G_1; it also contributes to the expression of cyclin E (*not shown*). At the same time, by increasing the expression of Cks1, Cul1, and Skp2 (which are responsible for degrading the p27^{Kip1} CDK inhibitor, *upper right*; see Figure 8.20B) as well as E2Fs 1, 2, and 3 (*below right*), the Myc-Max complex favors advance into S phase. In addition, Myc, acting with its Miz-1 partner, represses directly or indirectly expression of the p15^{INK4B}, p21^{Cip1}, and p27^{Kip1} CDK inhibitors; once again, effects are felt both in early/mid and in late G_1. Also not shown are the actions of active E-CDK2 complexes in phosphorylating p27^{Kip1}, leading to its degradation (see Figure 8.20B).

to favor progression through the cell cycle (**Figure 8.22**). When expression of these components is driven by the abnormally high levels of Myc present in cells carrying *myc* oncogenes, the result is a physiological state similar to that seen in cells lacking pRb function. Both changes deprive a cell of the normal control of progression through the G_1 phase of its cell cycle and deregulate passage through the R point. This helps to explain the ostensibly analogous mechanisms-of-action of the adenovirus E1A oncoprotein and Myc, as mentioned above: both converge on incapacitating pRb, doing so in very different ways.

Among the many targets of Myc is the cyclin D2 gene (see Figure 8.22), whose elevated expression leads, in turn, to the monophosphorylation of pRb (see Figure 8.17). Myc also drives expression of the CDK4 gene; the resulting increased levels of CDK4 drive the formation of the cyclin D–CDK4 complexes that sequester the p27^{Kip1} CDK inhibitor, thereby liberating cyclin E–CDK2 complexes from inhibition (see Figure 8.14B). Myc also drives expression of the Cul1 protein, which plays a central role in driving the degradation of the p27^{Kip1} CDK inhibitor through ubiquitylation. In one way or another, these Myc-driven gene expression changes serve to push the cell through the G_1 phase of its cell cycle.

By associating with a second transcription factor named Miz-1, Myc can also function as a transcriptional repressor (see Figure 8.22). In this role, Myc can repress expression of the genes encoding the p15^{INK4B} and p21^{Cip1} CDK inhibitors, which shut down the actions of CDK4/6 and CDK2, respectively. In fact, as we will read later, TGF-β uses these two CDK inhibitors to block progression through the G_1 phase of the cell cycle. Hence, by preventing the expression of these two CDK inhibitors, Myc confers resistance to the growth-inhibitory actions of TGF-β. This represents an important means by which cancer cells can continue to multiply under conditions (for example, the presence of TGF-β in their surroundings) that would normally preclude their proliferation.

Finally, Myc is able to induce the expression of the genes encoding the E2F1, E2F2, and E2F3 transcription factor proteins (see Figure 8.22). As we read earlier, pRb and its two cousins inhibit the transcription-activating powers of these transcription factors. By causing these growth-promoting transcription factors to accumulate to higher levels within a cell, Myc once again tips the balance in favor of cell proliferation.

Now, we begin to understand how the Myc oncoprotein can function so effectively to deregulate cell proliferation. It reaches in and pulls multiple regulatory levers within the cell cycle clock. The resulting changes in levels of key proteins shift the balance of proliferation-promoting versus -inhibiting mechanisms, strongly favoring pRb inactivation and cell growth.

Perhaps the most graphic demonstration of the potent mitogenic powers of Myc has come from an experiment in which Myc is engineered to be expressed at high, constant levels but as a protein that is fused with the estrogen receptor (ER). In the absence of an estrogen receptor ligand, such as estrogen or tamoxifen, the Myc–ER fusion protein is sequestered in a functionally inactive state in the cytoplasm (see Section 5.8). However, upon addition of ligand, this fusion protein is liberated from sequestration, rushes into the nucleus, and operates like the normal Myc transcription factor. When quiescent, serum-starved cells (in the G_0 state) expressing the Myc–ER fusion protein are treated with either estrogen or tamoxifen (but left in serum-deprived medium), these cells are induced to enter into the G_1 phase and progress to S phase (**Figure 8.23**). This illustrates that Myc, acting on its own and in the absence of serum-associated mitogens, is able to relieve all of the constraints on proliferation that have held these cells in G_0 and is able to shepherd them all the way through G_1—an advance that usually requires substantial, extended stimulation by growth factors. In addition to Myc, a number of other transcription factors display the wide-ranging powers of oncoproteins (see Table 8.1).

In addition to promoting cell cycle progression, Myc is also able to positively regulate fundamental biosynthetic processes, such as ribosome biogenesis, general transcription, protein synthesis and, more generally, **anabolism** (the synthesis of complex

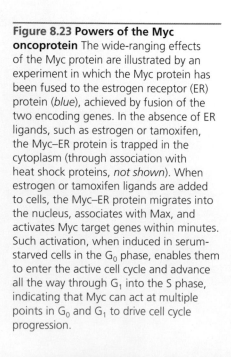

Figure 8.23 Powers of the Myc oncoprotein The wide-ranging effects of the Myc protein are illustrated by an experiment in which the Myc protein has been fused to the estrogen receptor (ER) protein (*blue*), achieved by fusion of the two encoding genes. In the absence of ER ligands, such as estrogen or tamoxifen, the Myc–ER protein is trapped in the cytoplasm (through association with heat shock proteins, *not shown*). When estrogen or tamoxifen ligands are added to cells, the Myc–ER protein migrates into the nucleus, associates with Max, and activates Myc target genes within minutes. Such activation, when induced in serum-starved cells in the G_0 phase, enables them to enter the active cell cycle and advance all the way through G_1 into the S phase, indicating that Myc can act at multiple points in G_0 and G_1 to drive cell cycle progression.

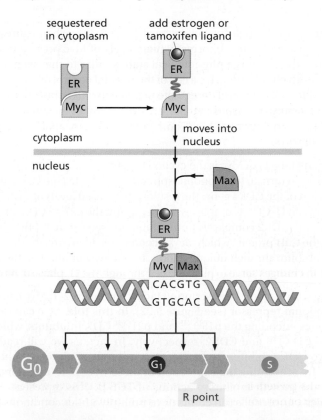

biomolecules). Therefore, we can conclude that Myc acts as an important master regulator that operates to coordinate overall biosynthesis with a cell's advance through the cell cycle, thereby ensuring that a cell that is committed to pass through M phase has acquired the requisite biomass that is needed to generate two healthy daughter cells. Moreover, as might be expected from these multiple effects on the cell around it, a Myc oncoprotein can also perturb the interactions of this cell with others in its surroundings, that is, the neighboring cells that form its tissue microenvironment. Included here are, among other processes, neovascularization, the process of generating new blood vessels, a topic described in some detail in Chapter 13. Finally, elevated Myc expression can induce high levels of ID proteins (see Figure 8.85C), which are direct antagonists of many of the bHLH transcription factors that orchestrate differentiation programs in diverse tissues; indeed, one of these (ID2) is able to bind pRb directly and antagonize the proliferation-inhibiting powers of the latter (Supplementary Sidebar 8.8.)

8.10 TGF-β prevents phosphorylation of pRb and thereby blocks cell cycle progression

TGF-β represents a major growth-inhibitory signal that normal cells, especially epithelial cells, must learn to evade in order to become cancer cells. As we will learn later in Chapters 13 and 14, TGF-β also exerts other, quite distinct effects on cells, by forcing them to change their differentiation programs. Often the resulting changes in cell phenotype actually favor tumor progression, as some of these phenotypic changes enable cancer cells to become anchorage-independent, **angiogenic**, and even invasive.

These two major effects of TGF-β—one antagonizing, the other favoring tumor progression—are clearly in direct conflict with one another. Cancer cells often resolve this dilemma by learning how to evade the cytostatic effects of TGF-β while leaving intact other responses, such as those favoring tumor cell invasiveness. Not surprisingly, the cytostatic actions of TGF-β derive from its direct and indirect effects on pRb, the central controller that determines whether or not a cell will proliferate.

We previously learned that TGF-β has its own, quite unique cell surface receptor, which uses serine/threonine kinases in its cytoplasmic domains (rather than tyrosine kinases) to emit signals. The primary targets of these signals documented to date are proteins of the Smad transcription factor family. Once phosphorylated in the cytoplasm by a TGF-β receptor, a pair of Smad2 (or Smad3) protein molecules associates with a Smad4 protein (which is not a substrate for phosphorylation by the receptor), and the resulting heterotrimeric protein complexes migrate to the nucleus, where they function as transcription factors (see Figure 6.22D).

As we read earlier (Section 8.4), in the context of cell cycle control, the most important targets of action by TGF-β are the genes encoding the two CDK inhibitors $p15^{INK4B}$ and, to a lesser extent, $p21^{Cip1}$ (see Figures 8.12 and 8.13). Each has a CAGAC sequence in its promoter that attracts a heterotrimeric Smad3–Smad4 transcription factor complex. On its own, such a Smad complex cannot activate transcription of these genes because of its intrinsically weak binding to DNA. Instead, it collaborates with yet another transcription factor—the Miz-1 factor mentioned in the last section—that binds to adjacent DNA sequences, allowing side-to-side association of these two transcription factors, cooperative stable binding of this complex to the promoters of target genes, and activation of transcription of such genes, in this case those encoding $p15^{INK4B}$ and $p21^{Cip1}$ (**Figure 8.24**, *left*).

We also noted earlier that the Myc oncoprotein, acting in the opposite direction, can block the induction of these two CDK inhibitors, doing so through its ability to associate with Miz-1. By repressing expression of the genes encoding $p15^{INK4B}$ and $p21^{Cip1}$, Myc removes two major obstacles standing in the way of cell cycle progression (see Figure 8.22). Stated differently, the constitutively high levels of Myc protein made by *myc* oncogenes ensure that expression of these two CDK inhibitors is strongly repressed, thereby paving the way for vigorous cell cycle advance.

In contrast to its powers in cancer cells, TGF-β needs to have the last word in determining whether or not normal cells proliferate. In order to do so, TGF-β must overrule

Figure 8.24 Antagonistic effects of TGF-β and Myc on cell cycle progression As shown here, TGF-β and Myc oppose one another in regulating advance through the G₁ phase of the cell cycle. TGF-β induces expression of the p15^INK4B CKI and, to a lesser extent, the p21^Cip1 CKI, by causing TGF-β-activated phospho-Smads to bind to the CAGAC canonical Smad-binding sequence in promoters adjacent to the binding of Miz-1 transcription factor to its own, quite extended sequence binding site, termed here a Miz box (*left*). Together this complex of physically associated TFs causes expression of the genes encoding the two CKIs, thereby slowing or preventing advance through the G₁ phase up to the R point. At the same time, TGF-β-activated phospho-Smad3 associates physically with E2F4 or E2F5. (The *double-headed arrows* indicate physical interaction.) The E2F box represents the sequence TTTCCCGC. The resulting complexes of p107, E2F4/5, and Smad3 repress transcription of the *myc* gene. The resulting decrease of Myc-Miz-1 complexes (*lower right*) leads to less repression of the genes encoding the p15^INK4B and p21^Cip1 CKIs and a slowdown of advance through G₁.

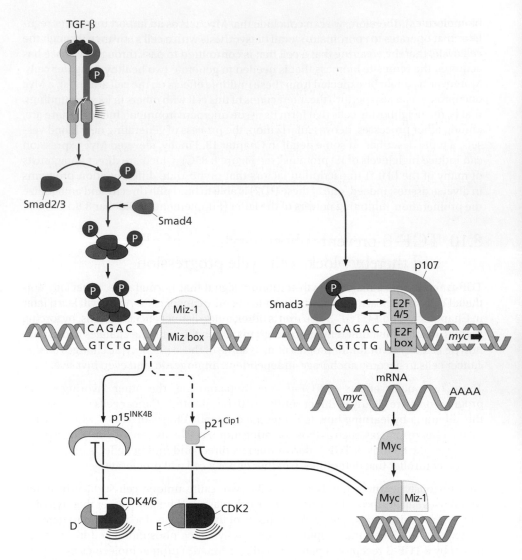

any conflicting signals that Myc might be releasing. To guarantee success in this endeavor, TGF-β must ensure that Myc does not thwart its scheme to activate expression of the p15^INK4B and p21^Cip1 CDK inhibitors. So, in normal cells, TGF-β keeps Myc away from the promoters of these genes by shutting down expression of the *myc* proto-oncogene (see Figure 8.24, *right*). By forming a tripartite complex with E2F4/5 plus p107, Smad3 ensures the shutdown of *myc* transcription, thereby eliminating the growth-promoting effects of the Myc protein from the cell's regulatory circuitry. Once TGF-β has succeeded in removing Myc from the scene—from the promoters of the p15^INK4B and p21^Cip1 CDK inhibitors—TGF-β can then deploy its Smad3–Smad4 complexes, working together with Miz-1, to activate the promoters of these two CDK inhibitors, thereby erecting major obstacles to the advance of a cell through the G₁ phase of the cell cycle (see Figure 8.24, *left*).

Many types of incipient cancer cells must evade TGF-β–imposed growth inhibition if they are to thrive. More specifically, such cancer cells depend on high levels of the Myc transcription factor to drive their proliferation and therefore must liberate *myc* transcription from the repressive actions of TGF-β. This explains why, for example, in a series of 12 human breast carcinomas, *myc* expression in all 12 of these tumors was no longer responsive to TGF-β–imposed shutdown even though many other TGF-β–induced responses were found to be intact in 11 of these tumors. (Because the *myc* gene itself was apparently in a wild-type configuration in these tumors, deregulation of some of the transcription factors controlling its transcription appeared to be responsible for the observed vigorous expression of this gene.) Once protected from the threat of this TGF-β–mediated repression, the *myc* oncogene, more specifically, its

Myc product, proceeds to hold expression of the two CDK inhibitors (that is, p15^{INK4B} and p21^{Cip1}) at a very low level and does so indefinitely. This helps, in turn, to create a condition in the cell permitting rapid, unimpeded proliferation.

In a more general sense, cancer cells place a high premium on escaping the growth-inhibitory influences of TGF-β while leaving other, potentially useful TGF-β responses intact. Often this evasion of TGF-β–mediated growth inhibition is achieved through inactivation of the pRb signaling pathway. In effect, in normal cells pRb operates as the brake lining that calls a halt to cell proliferation in response to TGF-β–initiated signals. Accordingly, if the pRb protein is eliminated from the regulatory circuitry through one of the various mechanisms described earlier, then the ability of TGF-β to impose growth arrest is greatly compromised because the p15^{INK4B} induced by TGF-β now fails to effectively block cell cycle advance.

8.11 pRb function and the controls of differentiation are closely linked

Cell differentiation is a process of crucial importance to cancer pathogenesis and, as such, is mentioned repeatedly in this text. For the moment, we can imagine—a bit simplistically— that cells throughout the body can exist in either of two alternative growth states. They may be found in a relatively undifferentiated state in which they retain the option to divide in the event that mitogenic signals call for their proliferation; this is essentially the behavior of stem cells. Alternatively, cells leave this state and enter into a more differentiated state, whereupon they give up the option of ever proliferating again and thus become post-mitotic. The decisions that govern these fundamental changes in cell biology must be explainable in terms of the molecular controls that determine whether a cell remains in the active growth-and-division cycle, exits reversibly into G$_0$, or exits irreversibly from this cycle into a post-mitotic, differentiated state.

The opposition between cell proliferation and differentiation is seen most clearly in the formation of most types of cancer cells in which differentiation is partially or, more rarely, completely blocked. Knowing this, we might ask whether two independent changes in the cellular control circuitry must be made during the formation of a cancer cell—one that deregulates control of proliferation and another that blocks differentiation. The alternative is simpler: might certain alterations, in a single stroke, deregulate proliferation and prevent differentiation?

Evidence supporting the latter idea, which implies a close coupling between the mechanisms controlling these two processes, has been forthcoming from studies of retinal development *in vivo* and muscle cell differentiation *in vitro*. In the case of the retina, the absence of pRb function in retinal precursor cells prevents the proper differentiation of the rod photoreceptor cells (**Figure 8.25A**). In this micrograph, the *Rb* gene was inactivated sporadically in retinal precursor cells, resulting in patches of cells where subsequent rod development was defective (in which negative regulation of cell proliferation was also presumably lost).

Yet another line of evidence indicating a coupling between differentiation and the cell cycle machinery comes from research on the biochemistry of the Myc protein. As mentioned earlier (see Section 8.9), the Myc protein shifts the balance between proliferation and post-mitotic differentiation in favor of proliferation. A dramatic demonstration of the opposition between the Myc oncoprotein and cell differentiation has come from a mouse model of liver cancer pathogenesis, which depends on the targeted expression of a *myc* transgene in hepatocytes. Large hepatocellular carcinomas form, and these will regress whenever the *myc* transgene is shut down subsequently. At the same time, many of the carcinoma cells, which previously lacked most of the traits of normal hepatocytes, rapidly differentiate into liver cells that appear normal and assemble to reconstruct many of the histological features of the normal liver. Similar dynamics operate during muscle differentiation (see Figure 8.25B and C). Moreover, the differentiation of a variety of cell types is favored by the G$_1$ CDK inhibitors, notably p27^{Kip1}. Finally, tissue-specific, differentiation-inducing transcription factors

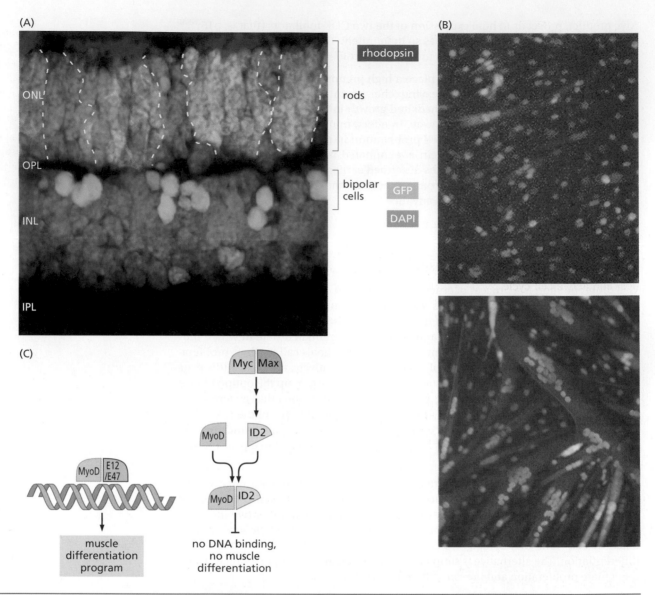

Figure 8.25 Differentiation of retinal cells, and muscle cells (A) The proper differentiation of retinal cells (see Figure 7.2A) depends on the presence of pRb. During normal development, retinal progenitor cells move upward and differentiate into various specialized cell types. In the eyes of 2-week-old mice in an engineered transgenic mouse model, rod cells that successfully differentiated express the rhodopsin photoreceptor protein (*red*), whereas those that failed to do so (*between dashed white lines*) are represented only by the DAPI stain that marks their DNA (*blue*). The latter cells sit directly above GFP-labeled bipolar cells (*green*) that mark the locations where progenitor cells had previously been forced to excise their only copy of their *Rb* gene (through the actions of a tissue-specific Cre-Lox system, see Sidebar 7.2) prior to migrating upward, showing directly that in the absence of pRb, progenitor cells differentiate incompletely and spawn rhodopsin-negative rod cells. (B) Myoblasts, the less differentiated precursors of muscle cells (myocytes), can be cultured *in vitro* and remain in the undifferentiated state (*upper panel*). However, various types of physiologic signals can induce them to differentiate into myocytes, whereupon they fuse to form muscle fibers (*red, lower panel*). For example, removal of serum (and its associated growth factors) from the culture medium leads to such differentiation. Numerous experiments have shown that differentiation can be prevented by forcing pRb phosphorylation and inactivation. Conversely, signals that prevent pRb phosphorylation favor and often induce differentiation. (C) MyoD is a bHLH transcription factor that serves as the master regulator, together with its E12/E47 partner, of the muscle differentiation program. Myc, for its part, acts to antagonize differentiation, in part through its ability to induce synthesis of ID2, an ID protein that has the bHLH structure but lacks DNA-binding ability; by forming heterodimers with MyoD, ID2 can act as a dominant-negative inhibitor of MyoD, because the heterodimer between MyoD and ID2 cannot bind DNA properly. (A, from S.L. Donovan and M.A. Dyer, *Vision Res.* 44:3323–3333, 2004. With permission from Elsevier. B, from E.M. Wilson et al., *Mol. Biol. Cell* 15:497–505, 2004. With permission from American Society for Cell Biology.)

often antagonize cyclin:CDK function and, conversely, cyclin:CDK complexes can antagonize these transcription factors; the resulting double-negative feedback loops create bi-stable controls, in which cells either proliferate or enter into nonproliferative, differentiated states. These various controls on cell differentiation, involving pRb, Myc, and other regulatory proteins, clearly have effects on the formation and development

Sidebar 8.7 Blocked differentiation can accompany the advance of tumor progression Some hematopoietic malignancies, specifically, leukemias, come in two forms—chronic and acute. The chronic diseases, such as chronic myelogenous leukemia (CML), are typified by more mature, differentiated cells and can proceed for years without becoming life-threatening. In the case of CML, after a period of 3 to 5 years, suddenly there is an eruption of a more malignant, aggressive form of the disease that is termed "**blast crisis.**" The cells present in blast crisis clearly originate from the same cell clone that initiated the chronic phase of the disease. However, they are less differentiated and multiply ceaselessly, creating a life-threatening disease that is difficult to treat. It seems that the cells in the chronic phase are derived from a mutant self-renewing stem cell, but this disease remains relatively benign, because the differentiation of these mutant cells into post-mitotic neutrophils proceeds quite normally. The moment that such differentiation is blocked (by some still-poorly understood genetic changes), cells in this population are trapped in a less differentiated stem cell compartment (of either erythroid or myeloid type) and proliferate without a compensating exit of their progeny into more differentiated, post-mitotic compartments.

of various types of cancer, because tumors formed by more differentiated cells are usually less aggressive, whereas those composed of poorly differentiated cells tend to be far more aggressive and carry a worse prognosis for the patient. These observations are only a small part of a growing body of work demonstrating the opposition between active cell cycle progression and entrance into differentiated growth states (**Sidebar 8.7** and Supplementary Sidebar 8.9).

8.12 Control of pRb function is perturbed in most if not all human cancers

Deregulation of the pRb pathway yields an outcome that is an integral part of the cancer cell phenotype—unconstrained proliferation. This explains why normal regulation of the R-point transition, as embodied in pRb phosphorylation, is likely to be disrupted in most if not all types of human tumor cells (**Table 8.2** and **Table 8.3**). These disruptive mechanisms are summarized in **Figure 8.26**. We are already familiar with the most direct mechanism for deregulating advance through the R point— inactivation of the *Rb* gene through mutation. In some tumors, an equivalent outcome is achieved through methylation of the *Rb* gene promoter. In others, pRb, though synthesized in normal amounts, may be functionally inactivated by viral oncoproteins, such as the HPV E7 protein, which prevent pRb from binding and regulating E2Fs.

Yet another strategy used by cancer cells to inactivate pRb function is indicated by the presence of very high levels of cyclin D1 in a variety of human tumor cells; the resulting D1–CDK4/6 complexes sequester $p27^{Kip1}$ molecules (see Figure 8.14B), freeing E-CDK2 complexes to drive hyperphosphorylation and inactivation of pRb. This dynamic is most widely documented in breast cancer cells, in which as many as half of the tumors have been reported to show elevated levels of this G_1 cyclin. In these and other carcinomas, the overexpression is sometimes achieved by increases in the copy number of the cyclin D1 genes (that is, gene amplification; **Figure 8.27**). More frequently, however, breast cancer cells acquire excessive cyclin D1 by altering the upstream signaling pathways (see Section 8.8) that are normally responsible for controlling expression of the cyclin D1 gene.

A more devious ploy is frequently exploited by cancer cells to disable the pRb machinery: they shut down expression of their $p16^{INK4A}$ tumor suppressor protein. Recall that the $p16^{INK4A}$ protein, like its $p15^{INK4B}$ cousin, inhibits the cyclin D–CDK4/6 responsible for initiating pRb phosphorylation. In the absence of the $p16^{INK4A}$ protein, pRb phosphorylation operates without an important braking mechanism, resulting in excessive cyclin D–CDK4/6 kinase activity, $p27^{Kip1}$ sequestration, pRb hyperphosphorylation, and associated inappropriate inactivation of pRb (see Figure 8.26). Individuals suffering from one form of familial melanoma inherit defective versions of the $p16^{INK4A}$ gene. It is unclear why loss of this particular CDK inhibitor, which seems to operate in all cell types throughout the body, should affect specifically the melanocytes in the skin that are the normal precursors of melanoma cells. In sporadic (that is, nonfamilial) tumors arising in various tissues, cancer cells resort far more frequently to another strategy to shed $p16^{INK4A}$ function—they methylate the CpG sequences

Table 8.2 Molecular changes in human cancers leading to deregulation of the cell cycle clock

Specific alteration	Clinical result
Alterations of pRb	
Inactivation of the *Rb* gene by mutation	retinoblastoma, osteosarcoma, small-cell lung carcinoma
Methylation of *Rb* gene promoter	brain tumors, diverse others
Sequestration of pRb by ID1, ID2	diverse carcinomas, neuroblastoma, melanoma
Sequestration of pRb by the HPV E7 viral oncoprotein	cervical carcinoma
Alteration of cyclins	
Cyclin D1 overexpression through amplification of *cyclin D1* gene	breast carcinoma, leukemias
Cyclin D1 overexpression caused by hyperactivity of *cyclin D1* gene promoter driven by upstream mitogenic pathways	diverse tumors
Cyclin D1 overexpression caused by reduced degradation of cyclin D1 because of depressed activity of GSK-3β	diverse tumors
Cyclin D3 overexpression caused by hyperactivity of *cyclin D3* gene	hematopoietic malignancies
Cyclin E overexpression	breast carcinoma
Defective degradation of cyclin E protein from loss of hCDC4	endometrial, breast, and ovarian carcinomas
Alteration of cyclin-dependent kinases	
CDC25A overexpression	breast cancers
CDK4 structural mutation	melanoma
Alteration of CDK inhibitors	
Deletion of *p15^{INK4B}* gene	diverse tumors
Deletion of *p16^{INK4A}* gene	diverse tumors
Methylation of *p16^{INK4A}* gene promoter	melanoma, diverse tumors
Decreased transcription of *p27^{Kip1}* gene because of action of Akt/PKB on Forkhead transcription factor	diverse tumors
Increased degradation of p27^{Kip1} protein because of Skp2 overexpression	breast, colorectal, and lung carcinomas, and lymphomas
Cytoplasmic localization of p27^{Kip1} protein due to Akt/PKB action	breast, esophagus, colon, thyroid carcinomas
Cytoplasmic localization of p21^{Cip1} protein due to Akt/PKB action	diverse tumors
Multiple concomitant alterations by Myc, N-myc, or L-myc	
Increased expression of ID1, ID2 leading to pRb sequestration	diverse tumors
Increased expression of cyclin D2 leading to pRb phosphorylation	diverse tumors
Increased expression of E2F1, E2F2, E2F3 leading to expression of cyclin E	diverse tumors
Increased expression of CDK4 leading to pRb phosphorylation	diverse tumors
Increased expression of Cul1 leading to p27^{Kip1} degradation	diverse tumors
Repression of p15^{INK4B} and p21^{Cip1} expression allowing pRb phosphorylation	diverse tumors

Table 8.3 Alteration of the cell cycle clock in human tumors A plus sign indicates that this gene or gene product is altered in at least 10% of tumors analyzed. Alteration of gene product can include abnormal absence or overexpression. Alteration of gene can include mutation and promoter methylation. More than one of the indicated alterations may be found in a given tumor.

Tumor type	Rb	Cyclin E1	Cyclin D1	p16[INK4A]	p27[Kip1]	CDK4/6	% of tumors with 1 or more changes
glioblastoma	+	+		+	+	+/+	>80
mammary carcinoma	+	+	+	+	+	+/	>80
lung carcinoma	+	+	+	+	+	+/	>90
pancreatic carcinoma			a		+		>80
gastrointestinal carcinoma	+	+	+[b]	+	+	+/[c]	>80
endometrial carcinoma	+	+	+	+	+	+/	>80
bladder carcinoma	+	+	+	+	+		>70
leukemia	+	+	+	+[d]	+	+/	>90
head-and-neck carcinomas	+	+	+	+	+	+/	>90
lymphoma	+	+	+[e]	+[d]	+	/+	>90
melanoma		+	+	+	+	+/	>20
hepatoma	+	+	+	+[d]	+	+/[c]	>90
prostate carcinoma	+	+	+	+	+		>70
testis/ovary carcinomas	+	+	+[b]	+	+	+/	>90
osteosarcoma		+		+		+/	>80
other sarcomas		+	+	+	+	/+	>90

[a]Cyclin D3 (not cyclin D1) is present and is up-regulated in some tumors.
[b]Cyclin D2 also is up-regulated in some tumors.
[c]CDK2 is also found to be up-regulated in some tumors.
[d]p15[INK4B] is also found to be absent in some tumors.
[e]Cyclin D2 and D3 are also found up-regulated in some lymphomas.
Adapted from M. Malumbres and M. Barbacid, *Nat. Rev. Cancer* 1:222–231, 2001.

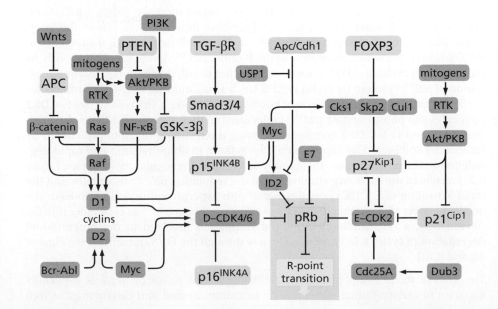

Figure 8.26 Perturbation of the R-point transition in human tumors The decision to advance through the R-point transition (*yellow, middle, bottom*) can be perturbed in a variety of ways in human tumors. Elements that favor advance through the R point are drawn in *red*, whereas those that block this advance are shown in *blue*. Almost all human tumors show either a hyperactivity of one or more of the agents favoring this advance (*red*) or an inactivation of the agents blocking this advance (*blue*). Proteins whose expression levels or activities are not known to change during the process of tumor formation are shown in *gray*.

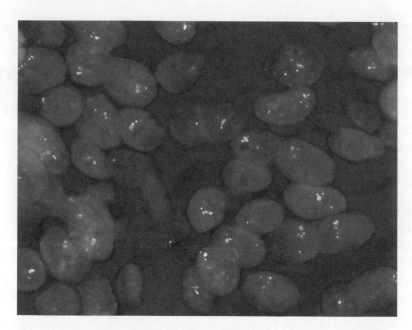

present in the promoter of the p16^{INK4A} gene (see Section 7.7), thereby shutting down transcription of this gene

Evidence of an even more cunning strategy for destabilizing this control circuit has been found in the genomes of a small number of both sporadic and familial melanomas. In these cancers, a point mutation in the CDK4 gene (the R24C mutation) creates CDK4 molecules that are no longer susceptible to inhibition by the family of INK4 molecules (that is, p15, p16, p18, and p19). Although these various CDK inhibitors may be perfectly intact and functional in such tumor cells, their normally responsive CDK4 target now eludes them. Once again, this permits CDK4, together with its cyclin D partners, to drive the initial steps of pRb phosphorylation and sequestration of p27^{Kip1} (see Figure 8.14B) in a deregulated fashion, driving cells through the R point in the late G$_1$ phase of the cell cycle. [Because the R24C mutation creates a dominant allele of CDK4 (at the cellular level), only one of the two copies of the gene encoding this CDK needs to be mutated in order for a cancer cell to derive proliferative benefit. Mice carrying one or two copies of this R24C allele develop a diverse variety of tumors, including those affecting mesenchymal, epithelial, and hematopoietic cell types.]

The most critical CDK inhibitor involved in cancer pathogenesis may well be p27^{Kip1}. As mentioned earlier, p27^{Kip1} is involved largely in inhibiting the activity of cyclin E–CDK2 complexes (for example, see Figure 8.14B). As cells exit the cell cycle into the G$_0$ quiescent state, p27^{Kip1} levels rise (see Supplementary Sidebar 8.8). Conversely, as cells re-enter the cell cycle and advance through its G$_1$ phase, the levels of free p27^{Kip1} are reduced progressively throughout early and mid-G$_1$ through cyclin D:CDK4/6 sequestration. Beginning at the R point, p27^{Kip1} levels fall precipitously by the actions of cyclin E–CDK2 complexes. In more detail, during progressive sequestration of p27^{Kip1} protein by cyclin D:CDK4/6, a small number of E–CDK2 molecules are freed from the inhibitory effects of bound p27^{Kip1} protein. The now-active E–CDK2 molecules can phosphorylate p27^{Kip1} molecules (on their threonine 187 residue) that are still bound to E–CDK2 complexes, causing the resulting phosphoproteins to be recognized and bound by Skp2, which directs them to ubiquitylation and rapid degradation in the proteasomes. This rapidly self-amplifying regulatory loop (see Figure 8.20) results in the explosive elimination of the remaining p27^{Kip1} molecules and the rapid activation of E–CDK2 complexes and pRb hyperphosphorylation. Indeed, the declining levels of Skp2 explain the observed increases in p27^{Kip1} as cells enter into G$_0$. Interestingly, a very similarly structured complex is involved in the programmed degradation of cyclin E later on as cells pass through the G$_1$/S transition (see Figures 8.8 and 8.10).

The inverse relationship between p27^{Kip1} levels and those of Skp2 is especially apparent in various human cancers such as mammary and oral carcinomas, as well as lymphomas (see, for example, Supplementary Sidebar 8.8), with higher levels of Skp2 portending shorter patient survival. Moreover, when the levels of p27^{Kip1} are measured in the cells of human esophageal, breast, colorectal, and lung carcinomas, poor patient survival is correlated with low levels of this CDK inhibitor. In all of these tumors, Skp2 seems to act as a proliferation-promoting oncoprotein by forcing the degradation of the critical p27^{Kip1} CDK inhibitor. One clue to the ultimate source of elevated Skp2 levels comes from research indicating that the Notch protein (see Section 6.12), which is hyperactive in many types of human cancer, increases transcription of the gene encoding Skp2.

We have read repeatedly about the hyperactive state of the Akt/PKB kinase in many human tumors, which is caused by a variety of molecular defects, including hyperactive growth factor receptors, loss of the PTEN tumor suppressor protein, and mutations in the PI3K gene (see Section 6.6). It, too, has effects on the cell cycle clock in cancer cells by reducing the effective levels of important CDK inhibitors (see Figure 8.13). By phosphorylating p21^{Cip1} and p27^{Kip1}, Akt/PKB ensures the cytoplasmic localization of these two critical antagonists of cell cycle advance, thereby marginalizing them. At the same time, Akt/PKB can suppress expression of the gene encoding p27^{Kip1} by phosphorylating a transcription factor of the Forkhead family, which serves to further reduce the overall concentrations of p27^{Kip1} in the cell. As if this were not enough, Akt/PKB ambushes p27^{Kip1} in a third way: by phosphorylating Skp2, Akt/PKB activates the latter, enhancing its ability to bind p27^{Kip1}, resulting thereafter in the increased ubiquitylation and degradation of p27^{Kip1} (see Supplementary Sidebar 8.8).

The diverse genetic and biochemical strategies shown in Figure 8.26 are all focused on one common goal—that of overwhelming and deregulating pRb function, thereby destroying the tight control that it normally imposes on the R-point transition. Darwinian selection, occurring in the microcosm of living tissues, favors the outgrowth of cells that, by hook or by crook, have succeeded in inactivating the critical pRb braking system and thus deregulating the R-point transition.

8.13 Synopsis and prospects

All of the physiological signals and signaling pathways affecting cell proliferation must, sooner or later, be connected in some fashion to the operations of the cell cycle clock. It represents the brain of the cell—the central signal processor that receives afferent signals from diverse sources, integrates them, and makes the final decisions concerning growth versus quiescence, and in the case of the growth-versus-quiescence decision, whether or not the exit from the active cell cycle will be reversible. Connected with the latter decision are the mechanisms governing entrance into tissue-specific differentiation programs.

The core components of the cell cycle clock are already present in clearly recognizable form in single-cell eukaryotes, including the much-studied baker's yeast, *Saccharomyces cerevisiae*. Single-cell organisms such as this one respond to a far smaller range of external signals than do metazoan cells residing within complex tissues. These simple organisms lack the hundred and more distinct types of growth factor receptors that vertebrate cells display on their surfaces, as well as other receptors, such as integrins, that our cells use to sense and control attachment to the extracellular matrix. Yeast cells also lack the growth-inhibitory receptors, such as the TGF-β receptors, that play such a critical role in the economy of mammalian tissues.

All this explains why the peripheral wiring that regulates the core cell cycle machinery of animal cells has been added relatively recently in the history of life on this planet—perhaps 600 million years ago when metazoa may first have appeared. The need to respond to a wide variety of afferent signals explains why so many distinct layers of regulation have been imposed on the core machinery. Without these additional regulators, notably the CDK inhibitors, the core machinery could not be made responsive to the diverse array of signals that impinge on individual metazoan cells and modulate their proliferation within complex tissues.

While these connections between the cell exterior and the cell cycle clock were being forged, other critical regulators became integrated into this complex

Massari ME & Murre C (2000) Helix-loop-helix proteins: regulators of transcription in eucaryotic organisms. *Mol. Cell Biol.* 20, 429–440.

Morgan DO (2007) The Cell Cycle: Principles of Control. Sunderland, MA: Sinauer Associates.

Perk J, Iavarone A & Benezra R (2005) Id family of helix-loop-helix proteins in cancer. *Nat. Rev. Cancer* 5, 603–614.

Reed S (2003) Ratchets and clocks: the cell cycle, ubiquitylation and protein turnover. *Nat. Rev. Mol. Cell Biol.* 4, 855–864.

Sherr CJ (2001) The INK4a/ARF network in tumour suppression. *Nat. Rev. Mol. Cell Biol.* 2, 731–737.

Sherr CJ & McCormick F (2002) The RB and p53 pathways in cancer. *Cancer Cell* 2, 103–112.

Sherr CJ & Roberts JM (2004) Living with or without cyclins and cyclin-dependent kinases. *Genes Dev.* 18, 2699–2711.

Sherr CH (2019) Surprising regulation of cell cycle entry. *Science* 366, 1315–1316.

Siegel PM & Massagué J (2003) Cytostatic and apoptotic actions of TGF-β in homeostasis and cancer. *Nat. Rev. Cancer* 3, 807–820.

Varlakhanova NV & Knoepfler PS (2009) Acting locally and globally: Myc's ever-expanding roles on chromatin. *Cancer Res.* 69, 7487–7490.

Whitfield ML, George LK, Grant GD & Perou CM (2006) Common markers of proliferation. *Nat. Rev. Cancer* 6, 99–106.

Yamasaki L & Pagano M (2004) Cell cycle, proteolysis and cancer. *Curr. Opin. Cell Biol.* 16, 623–628.

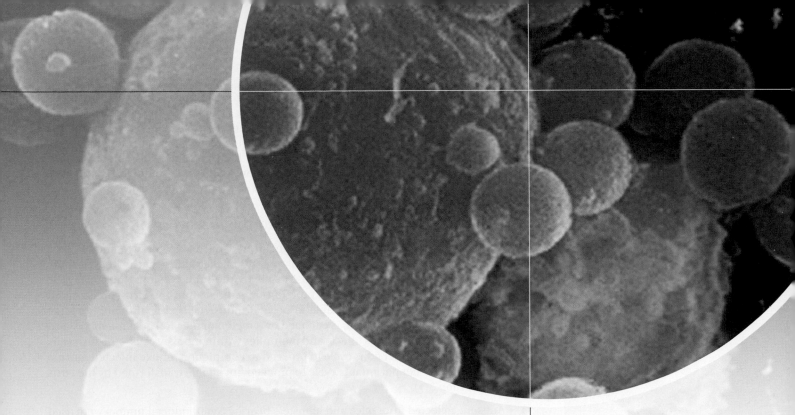

Chapter 9

p53: Master Guardian and Executioner

There cannot however be the least doubt, that the higher organisms, as they are now constructed, contain within themselves the germs of death.

August Weissmann, philosopher of biology, 1889

To examine the causes of life, we must first have recourse to death.

Mary Shelley, Frankenstein, 1831

Multicellular animals have a vital interest in limiting the effects of defective or malfunctioning cells in their tissues. To accomplish this goal, organisms have within their cells the p53 protein. This loyal sentinel serves as a local representative of the organism's interests and is present on-site to ensure that every cell keeps its own household in order.

If p53 receives information indicating a cell is experiencing genetic corruption or metabolic disarray, the protein may arrest the progress through the cell's growth-and-division cycle and, at the same time, orchestrate localized responses to facilitate the repair of damage in that cell. In other circumstances, including situations in which damage is too extensive to be repaired, p53 can emit signals that awaken the cell's normally latent suicide program—**apoptosis**. The consequence of this summons is the rapid death of the cell. Such drastic measures eliminate cells whose continued growth and division might otherwise pose a threat to the overall health and viability of the organism.

The apoptotic program that can be activated by p53 is built into the control circuitry of most cells throughout the body. Apoptosis consists of a series of distinctive molecular changes that function to ensure the disappearance of all traces of a cell, often

Movies in this chapter

9.1   p53 and Tumor Growth
9.2   p53 DNA Complex
9.3   Apoptosis

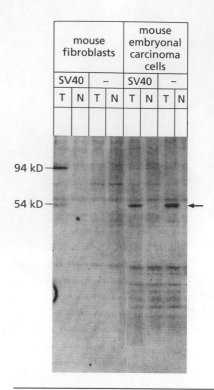

Figure 9.1 The discovery of p53 and its association with SV40 large T
Normal mouse fibroblasts (3T3 cells) and mouse embryonal carcinoma cells (F9 cells) were infected by SV40. Cell lysates prepared from both infected (SV40) and uninfected (−) cells were labeled with ³⁵S-methionine and incubated with either normal hamster serum (N) or with antiserum prepared against SV40-transformed hamster cells (T). In the virus-infected fibroblasts, the anti-tumor serum immunoprecipitated a protein of 94 kD (large T), as well as a second protein that runs slightly ahead of the 54-kD marker. This same smaller protein is immunoprecipitated from the embryonal carcinoma cells, whether or not they had been exposed to SV40 (*arrow*). Although these data, on their own, did not prove a physical association between large T and p53, they did show that p53 was present in elevated amounts in two types of transformed cells, one of which had no association with a tumor virus. (From D.I. Linzer and A.J. Levine, *Cell* 17:43–52, 1979. With permission from Elsevier.)

within an hour of its initial activation. The continued presence of this latent but intact suicide machinery represents an ongoing threat to incipient cancer cells, because this program is poised to eliminate cells that are en route to becoming neoplastic. The role of p53 in orchestrating cell cycle arrest, apoptosis, and a range of other functions, explains why this gene becomes disabled in a high percentage of human tumors. Without a clear description of the many functions of p53, we cannot fully appreciate the processes that limit the creation of virtually all types of human cancer.

9.1   DNA tumor viruses lead to the discovery of p53

When murine cells that have been transformed by the SV40 DNA tumor virus are injected into a mouse of identical genetic background (that is, a **syngeneic** host), the immune system of the host reacts by mounting a strong response. The antibodies produced react with a protein that is present in the virus-transformed cells but is otherwise undetectable in normal mouse cells. This protein, the large tumor (large T or LT) antigen, is encoded by a region of the viral genome that is also expressed when this virus infects and multiplies within monkey cells—the host cells that permit a full infectious cycle to proceed to completion, unlike infected murine cells, which do not permit this cycle to proceed but can be transformed by the virus (see Section 3.4).

Antisera collected from mice and hamsters bearing SV40-induced tumors were used in 1979 to analyze the proteins in SV40-transformed cells. The samples contained antibodies that recognized both large T and an associated protein that exhibited an apparent molecular weight of 53 to 54 kD (**Figure 9.1**). Antisera reactive with the latter, smaller protein could also detect that protein in mouse embryonal carcinoma cells and, later on, in a variety of human and rodent tumor cells—including those that had never been infected by SV40. Monoclonal antibodies directed specifically against large T could also immunoprecipitate the 53- to 54-kD protein—but only in SV40-infected cells.

Taken together, these observations indicated that the large T protein expressed in SV40-transformed cells was tightly bound to a novel protein, which came to be called p53. The observation that p53 could also be detected in certain uninfected cells, such as the mouse embryonal carcinoma cells analyzed in Figure 9.1, indicated that p53 was of cellular rather than viral origin. This conclusion was reinforced by a report in the same year that mouse cells transformed by exposure to a chemical carcinogen also harbored detectable p53.

These various lines of evidence suggested that the viral large T oncoprotein functions, at least in part, by targeting host-cell proteins. (The discovery that large T antigen is also able to bind pRb would come seven years later.) In the years since these discoveries, a number of other DNA viruses and at least one RNA virus have been found to specify oncoproteins that associate with p53 or perturb its function (**Table 9.1**). As is apparent from this table, these viruses also target pRb and function to block apoptosis, an observation that we return to later in this chapter.

9.2   *p53* is discovered to be a tumor suppressor gene

The path to our understanding of p53's function began with a substantial scientific detour: early studies in which a cDNA encoding p53 was transfected into rat embryo fibroblasts revealed that this cDNA, in collaboration with a co-introduced *ras* oncogene, could transform the rodent cells. Such activity suggested that the *p53* gene (which is sometimes termed *Trp53* in mice and *TP53* in humans) could operate as an oncogene—much like the *myc* oncogene, which had previously been shown to collaborate with the *ras* oncogene in rodent cell transformation (see Section 11.8). Like *myc*, the introduced *p53* cDNA seemed to contribute certain growth-inducing signals that resulted in cell transformation in the presence of a co-expressed *ras* oncogene.

But appearances turned out to be deceiving. As later became apparent, the *p53* cDNA used in these experiments had been synthesized from a template mRNA extracted

Table 9.1 Tumor viruses that perturb pRb, p53, and/or apoptotic function

Virus	Viral protein targeting pRb	Viral protein targeting p53	Viral protein targeting apoptosis
SV40	large T[a]	large T[a]	
Adenovirus	E1A	E1B55K	E1B19K[b]
HPV	E7	E6	
Polyomavirus	large T	large T?	middle T (MT)[c]
Herpesvirus saimiri	V cyclin[d]		v-Bcl-2[e]
HHV-8 (KSHV)	K cyclin[d]	LANA-2	v-Bcl-2,[e] v-FLIP[f]
Human cytomegalovirus (HCMV)	IE72[g]	IE86	vICA,[h] pUL37[i]
HTLV-I	Tax[j]	Tax	
Epstein–Barr	EBNA3C	EBNA-1[k]	LMP1[k]

[a]SV40 large T also binds a number of other cellular proteins, including p300, CBP, Cul7, IRS1, Bub1, Nbs1, and Fbw7, thereby perturbing a variety of other regulatory pathways.
[b]Functions like Bcl-2 to block apoptosis.
[c]Activates PI3K and thus Akt/PKB.
[d]Related to D-type cyclins.
[e]Related to cellular Bcl-2 anti-apoptotic protein.
[f]Viral caspase 8 (FLICE) inhibitory protein; blocks an early step in the extrinsic apoptotic cascade.
[g]Interacts with and inhibits p107 and possibly p130; may also target pRb for degradation in proteasomes.
[h]Binds and inhibits procaspase 8.
[i]Inhibits the apoptotic pathway below caspase 8 and before cytochrome c release.
[j]Induces synthesis of cyclin D2 and binds and inactivates p16[INK4A].
[k]LMP1 facilitates p52 NF-κB activation and thereby induces expression of Bcl-2; EBNA-1 acts via a cellular protein, USP7/HAUSP, to reduce p53 levels. EBNA3C interferes with p53 function.

from tumor cells (rather than normal cells). This mRNA was later shown to carry a mutant form of *p53*. Subsequent use of a *p53* cDNA cloned instead from the mRNA of normal cells revealed that this wild-type *p53* cDNA clone, rather than favoring cell transformation, actually *suppressed* it (**Figure 9.2**). Comparison of the sequences of the two cDNAs revealed that they differed by a single base substitution—a point mutation—that caused an amino acid substitution in the p53 protein.

These results suggested that, in normal cells, the wild-type allele of *p53* functions to suppress cell proliferation, and that *p53* only acquires powers that promote growth when it sustains a mutation. Because of this discovery, the *p53* gene was eventually categorized as a tumor suppressor gene (TSG).

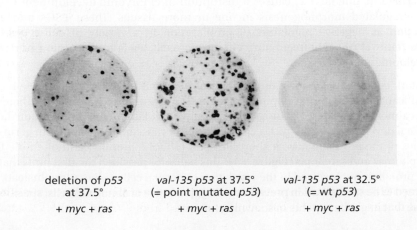

deletion of *p53* at 37.5°	*val-135 p53* at 37.5° (= point mutated *p53*)	*val-135 p53* at 32.5° (= wt *p53*)
+ *myc* + *ras*	+ *myc* + *ras*	+ *myc* + *ras*

Figure 9.2 Effects of p53 on cell transformation A cDNA encoding a *ras* oncogene was co-transfected with several alternative forms of a *p53* cDNA into rat embryo fibroblasts. In the presence of a mutant vector that contains an almost complete deletion of the *p53* reading frame (*left*), a small number of transformed foci appeared. In the presence of a *p53* gene that carries a point mutation (*middle*), a large number of robust foci were formed, which is explained by the dominant-negative effects of the point mutant p53 protein (see Section 9.4). However, in the presence of a *p53* wild-type clone (*right*), almost no foci were formed. The experiment demonstrated that wild-type *p53* is effective at suppressing *ras*-induced transformation, whereas this mutant version appeared to facilitate transformation. (Courtesy of M. Oren; from D. Michalovitz et al., *Cell* 62: 671–680, 1990. With permission from Elsevier.)

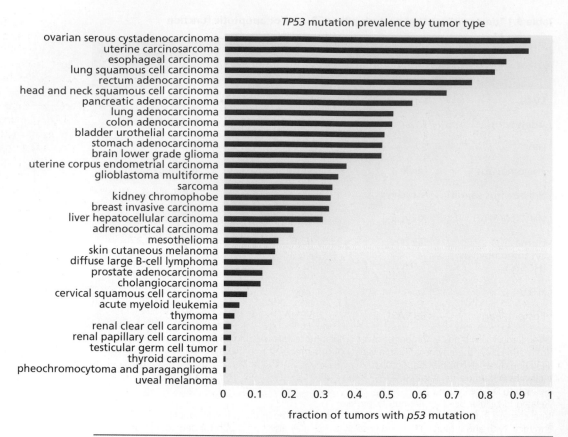

Figure 9.3 Frequency of mutant *p53* alleles in human tumor cell genomes As indicated in this bar graph, mutant alleles of *p53* are found frequently in human tumors. Sequences were obtained from more than 10,000 patients with 32 different cancer types. The bars indicate the percentage of each tumor type found to carry a mutant *p53* allele. Note that greater than 90 percent of ovarian serous carcinomas and uterine carcinosarcomas carry mutant *p53* alleles. (Adapted from L. Donehower et al., *Cell Rep.* 28:1370–1384, 2019.)

By 1987 it became apparent that *p53* alleles carrying a point mutation are common in the genomes of a wide variety of human tumor cells. Data accumulated from diverse studies indicate that the *p53* gene is mutated to varying degrees in human cancers, and may cause more than 90% of some types of cancer (**Figure 9.3**). Indeed, among all the genes examined to date in human cancer cell genomes, *p53* is the gene found to be most frequently mutated.

Further functional analyses of *p53*, conducted much later, made it clear that *p53* is not a typical TSG. For most members of this class of cancer genes, homozygous inactivation in the mouse germ line (using the strategy of targeted gene "knockouts" described in Sidebar 7.2) causes a disruption of embryonic development caused by deregulated morphogenesis in one or more tissues. These TSGs appear to function as negative regulators of proliferation in a variety of cell types, and their removal from the regulatory circuitry of cells leads to disruption of normal development.

In stark contrast, deletion of both copies of *p53* from the mouse germ line has no significant effect on the development of the great majority of *p53*$^{-/-}$ embryos. Therefore, *p53* cannot be considered to be a simple negative regulator of cell proliferation during normal development. Still, *p53* is clearly a TSG, because mice lacking both germ-line copies of the *p53* gene have a short life span (about 5 months) and often die from lymphomas and sarcomas (**Figure 9.4**). This behavior provided the first hints that the p53 protein is not involved in the regulation of normal cell proliferation. Instead, p53 seemed to be specialized in preventing the appearance of abnormal cells, specifically, those that might be capable of spawning tumors.

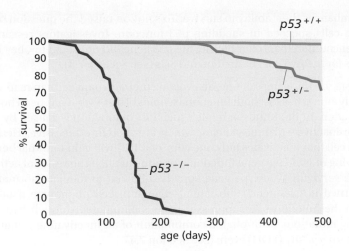

Figure 9.4 Effects of mutant *p53* alleles in the mouse germ line This plot shows the survival of mice with or without functional *p53* alleles. Although the absence of p53 function in the mice carrying two *p53* null alleles (*p53*$^{-/-}$) had relatively little effect on their embryonic development and viability at birth, it resulted in a greatly increased mortality relatively early in life. This effect was caused by the development of cancer, particularly lymphoma. All *p53*$^{-/-}$ homozygotes succumbed to malignancies by about 250 days of age (*red line*), and even *p53*$^{+/-}$ heterozygotes (*blue line*) began to develop tumors at this time; wild-type (*p53*$^{+/+}$) mice (*green line*) showed virtually no mortality until almost 500 days of age. (Adapted from T. Jacks et al., *Curr. Biol.* 4:1–7, 1994. With permission from Elsevier.)

9.3 Inherited mutations affecting *p53* predispose individuals to a variety of tumors

Inherited mutations in tumor suppressor genes underlie a number of human familial cancer syndromes (as we read in Chapter 7). And germ-line mutations in *p53* are likewise associated with significantly increased cancer risk. In 1982, a group of families was identified that displayed a greatly increased susceptibility to a variety of different tumors, including glioblastoma; leukemias; carcinomas of the breast, lung and pancreas; Wilms tumor; and soft-tissue sarcomas. In some families, as many as half of the members were afflicted with one or another of these cancers, and two-thirds developed some type of cancer by the time they reached age 22. Some family members were even afflicted with several types of cancer concurrently. This familial cancer syndrome, termed Li–Fraumeni after the two human geneticists who first identified and characterized it, was notable because of the wide range of tumor types involved (see Table 7.1).

Eight years after the Li–Fraumeni syndrome was first described, researchers discovered that many of the cases arose from a mutant allele at a locus on human Chromosome 17p13—precisely where the *p53* gene is located. In about 70% of these families, mutant alleles of *p53* were found to be transmitted in a Mendelian fashion. Individuals who inherited a single mutant *p53* allele had a high probability of developing some form of malignancy, often early in life, although the age of onset for these various malignancies was found to be quite variable: about 5 years of age for adrenocortical carcinomas, 16 years for sarcomas, 25 years for brain tumors, 37 years for breast cancer, and almost 50 years for lung cancer. In light of the multiple roles that *p53* would be discovered to play in suppressing cancer throughout the body, it seems reasonable that mutant germ-line alleles of this gene should predispose a person to such a diverse group of malignancies.

We now know that the mutant *p53* alleles that are transmitted through the germ lines of Li–Fraumeni families carry a variety of point mutations that are scattered across the *p53* coding sequence. These germ-line mutations, which have been characterized in nearly 900 families, display a distribution reminiscent of that shown by the somatic mutations that have been documented in the genomes of more than 30,000 sporadic human tumors.

9.4 Mutant versions of *p53* interfere with normal p53 function

The observations of frequent mutations in the *p53* gene in familial cancer syndromes and in a large percentage of the genomes present in **sporadic** (nonfamilial) tumors suggested that, for many incipient cancer cells, perturbation or elimination of p53 function enhances their ability to survive. This notion raised the question of precisely how these cells succeed in shedding p53 function. Investigators researching this question encountered another puzzle: the *p53* gene did not seem to obey Knudson's hypothesis for the two-hit elimination of TSGs (see Section 7.3).

According to the Knudson scheme, an evolving premalignant cell can reap substantial benefit only after it has lost both functional copies of a TSG that has been holding back its proliferation. In the Knudson model, such loss of function is caused by mutations that create inactive—or null—alleles, which are therefore recessive. Hence, a premalignant cell that inactivates only one copy of a TSG will gain minimal benefit—due to the halving of effective gene function—or none at all, if the residual activity specified by the surviving wild-type gene copy is sufficient to mediate normal function. As we learned in Chapter 7, substantial change in cell phenotype usually occurs only when the function of a suppressor gene is eliminated through two successive inactivating mutations or through a combination of an inactivating mutation plus a loss-of-heterozygosity (LOH) event (see Section 7.4).

Knudson's model was hard to reconcile with the observed behavior of the mutant *p53* cDNA introduced into rat embryo fibroblasts (see Figure 9.2). The mutant *p53* cDNAs clearly altered cell phenotype, even though these embryo fibroblast cells continued to harbor their own pair of wild-type *p53* gene copies. This meant that the introduced mutant *p53* cDNA could not be functioning as an inactive, recessive allele. It seemed, instead, that the point-mutated *p53* allele was actively exerting some type of dominant function when introduced into these rat embryo cells.

A solution to the genetic puzzle of how mutant *p53* alleles might foster tumor cell formation arose from three lines of research. First, studies in the area of yeast genetics indicated that, for certain genes, mutations can do more than compromise the normal functioning of the encoded gene product; they can confer on the mutant allele (and its protein product) the ability to interfere with or obstruct the ongoing activities of the surviving wild-type copy of the gene. Alleles that show this behavior are termed **dominant-interfering** or **dominant-negative** alleles.

A second telling clue came from analysis of the mutant alleles of the *p53* gene in cancer cells (**Figure 9.5A**). These analyses indicated that most tumor-associated, mutant *p53* alleles carry point mutations in their reading frames that create **missense** codons (resulting in amino acid substitutions) rather than **nonsense** codons (which cause premature termination of the growing polypeptide chain and, quite often, degradation of the resulting truncated protein). Furthermore, larger deletions within the coding sequence of the *p53* gene were found to be relatively uncommon. Consequently, researchers came to the inescapable conclusion that tumor cells can derive more benefit from the presence of a slightly altered p53 protein than from its complete elimination through the creation of **null alleles** by nonsense mutations or by the outright deletion of significant portions of the *p53* gene. These slightly altered p53 proteins—so the thinking went—could somehow function in a dominant-interfering manner to compromise the function of any wild-type p53 protein that might be present in a cell.

The third and final clue to the p53 puzzle came from biochemical and structural analyses of its encoded protein, which revealed that p53 is a DNA-binding transcription factor that normally exists in the cell as a **homotetramer**, that is, an assembly of four identical polypeptide subunits. This observed tetrameric state suggested a plausible mechanism through which a dominant-negative mutant allele of *p53* could actively interfere with the continued functioning of a wild-type *p53* allele being expressed in the same cell.

Assume that a mutant *p53* allele found in a human cancer cell encodes a form of the p53 protein that has lost most of its normal function but has retained the ability to participate in tetramer formation (Figure 9.5B). If one such mutant allele were to coexist with a wild-type allele in this cell, the p53 tetramers assembled in such a cell would contain mixtures of mutant and wild-type proteins in various proportions. In fact, statistically speaking, if the mutant and wild-type proteins were expressed at equal levels, 15 out of the 16 equally possible combinations of mutant and wild-type

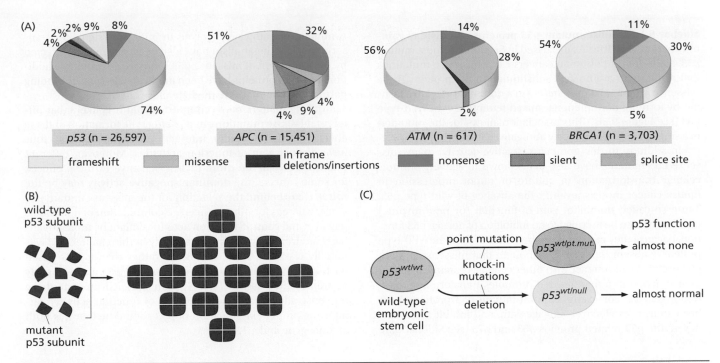

Figure 9.5 *p53* functions as a dominant-negative allele
(A) As indicated in these pie charts, missense mutations that lead to amino acid substitutions (*green*) affect the great majority of the mutant *p53* alleles found in the 30,000 human tumor genomes analyzed here; other types of mutations are seen relatively infrequently in *p53*. In contrast, the mutations striking other well-studied tumor suppressor genes (such as *APC*) or "caretaker" genes involved in maintenance of the genome (*ATM, BRCA1*) and described in Chapter 12 are more often frameshift mutations (*yellow*) or nonsense codons (*blue*); both of these types of mutation disrupt protein structure, usually by creating truncated versions of proteins that are often degraded rapidly in cells, and thus create effectively null alleles. (B) In cells bearing a single mutant *p53* and a single wild-type allele, the mutant protein usually retains its ability to form tetramers but loses its ability to function normally because of a defective DNA binding domain.

Consequently, mixed tetramers composed of differing proportions of wild-type (*blue*) and mutant (*red*) p53 subunits may form, and the presence of even a single mutant protein subunit may compromise the functioning of the entire tetramer. Therefore, in a cell that is heterozygous at the *p53* locus, fifteen-sixteenths of the p53 tetramers may lack fully normal function. (C) Perhaps the most direct demonstration of the dominant-negative mode of *p53* action has come from "knocking in" (see Sidebar 7.5) mutant *p53* alleles into the genome of mouse embryonal stem (ES) cells. In cells in which a point mutation in the DNA-binding domain was knocked into one *p53* gene copy, almost all p53 function was lost. In contrast, when one *p53* gene copy was completely inactivated (yielding a null allele), p53 function was almost normal. (A, from International Agency for Research on Cancer, TP53 genetic variations in human cancer, IARC release R14, 2009; and A.I. Robles et al., *Oncogene* 21:6898–6907, 2002. With permission from Nature.)

p53 monomers would contain at least one mutant p53 subunit (Figure 9.5B). If the presence of even a single mutant p53 protein in a tetramer could interfere with the functioning of the complex as a whole, these mixed tetramers might therefore lack some or all of the activity associated with their fully wild-type counterpart. Consequently, only one-sixteenth of the p53 tetramers assembled in a heterozygous cell (which carries one mutant and one wild-type *p53* gene copy) would be formed purely from wild-type p53 subunits and thus retain full wild-type function. This scheme would also explain why slightly altered versions of p53 were far more effective in driving cancer progression than the absence of p53 protein, that is, the state in which mutant null alleles of *p53* arose in the genome of an evolving, pre-neoplastic cell. Experiments in genetically altered embryonic stem (ES) cells have since provided direct evidence for p53's dominant-negative mode of action (Figure 9.5C).

In an experimental situation in which a mutant *p53* cDNA clone is introduced via an engineered expression vector into cells carrying a pair of wild-type *p53* alleles (see Figure 9.2), the expression of this introduced allele is usually greater than that of the native *p53* gene copies. As a consequence, in such transfected cells, the amount of mutant p53 protein present will be vastly higher than the amount of wild-type protein. Hence, far fewer than one-sixteenth of the p53 tetramers in such cells will be formed purely from wild-type p53 subunits. The relative abundance of the mutant p53 protein explains how introducing a mutant *p53* allele might compromise virtually all p53 function in such cells.

Sidebar 9.1 Missense mutant p53 proteins can display gain-of-function activities As discussed in Section 9.4, most mutant *p53* alleles found in human cancers carry missense mutations that result in an amino acid substitution in the protein. These missense mutant p53 proteins act in a dominant-negative fashion by forming nonfunctional mixed tetramers with wild-type p53 (see Figure 9.5B). This association allows for the mutation of a single *p53* allele to largely abrogate cellular p53 function.

In addition to this dominant-negative activity, many missense p53 mutants have also been shown to actively *promote* cellular transformation in culture or tumor progression in mouse cancer models, even in the absence of wild-type p53. These so-called dominant gain-of-function (or **neomorphic**) activities cannot be ascribed to an inhibition of normal p53 function, because they can manifest their effects when the wild-type protein is absent. Instead, it is thought that missense mutant p53 proteins can interact with other cellular proteins and alter their function in a manner that promotes transformation and tumorigenesis. For example, point-mutant p53 proteins have been shown to selectively engage with, and inhibit the activities of, the p53-related proteins p63 and p73 (see Sidebar 9.5).

A number of pathways involved in mitogenic signaling, transcription, and metabolism have also been shown to be affected by the expression of missense p53 proteins, and multiple additional cellular proteins have been implicated as specific binding partners of mutant p53 to mediate these effects.

Making matters more complex is that it is likely that different missense mutations in p53 affect the formation of these aberrant protein–protein interactions differently and thus have distinct gain-of-function activities. Moreover, different tumor types may be more or less sensitive to these effects: for some cancers, the dominant-negative activity may be the *raison d'etre* behind the selection for the missense mutant *p53* allele, whereas for other cancers, a combination of dominant-negative and gain-of-function activities might be more advantageous. Ironically, as discussed early in this chapter, *p53* was initially classified as an oncogene rather than a TSG based on the inadvertent use of a point-mutant *p53* allele in early co-transfection assays (see Section 9.2). Given that these missense alleles can also have gain-of-function activity, perhaps the proper classification of *p53* should be that it is both an oncogene and a TSG!

The above logic suggests that many incipient human tumor cells would benefit from carrying one wild-type and one co-existing mutant *p53* allele. However, in most human tumor cells that are mutant at the *p53* locus, a subsequent LOH event eliminates the wild-type allele, producing a cell with either two mutant *p53* alleles or only one allele that is mutant. Again, this observation at first seems puzzling. If a mutation that creates a dominant-negative allele can eliminate much of a cell's p53 function (see Figure 9.5B), why should the loss of the surviving wild-type *p53* allele be necessary? The answer, it seems, is that even a fraction of fully normal *p53* gene function is disadvantageous to a cancer cell. So, the most opportunistic cancers eventually evolve to shed the remaining wild-type *p53* allele.

In addition, many experiments demonstrate that point-mutant p53 proteins can also display a "gain-of-function" activity that contributes to tumorigenesis, doing so in the absence of wild-type p53 (**Sidebar 9.1**). This behavior, on its own, indicates that mutant p53 interacts with a variety of other nuclear proteins beyond wild-type p53 subunits and in various ways perturbs their function. Of additional interest are the sites of *p53*-associated point mutations. As seen in **Figure 9.6A**, almost all of them affect the DNA-binding site of p53, preventing this protein from participating in its normal function as a transcriptional regulator (Figure 9.6B), which in turn holds strong implications for how p53 functions in more than half of human cancers.

9.5   p53 protein molecules usually have short lifetimes

Long before p53 was discovered to bind to DNA, the nuclear localization of this protein in many normal and neoplastic cells suggested that it might function as a transcription factor. In general, there are at least three mechanisms that can regulate the activity of transcription factors: (1) modulation of their nuclear concentrations; (2) alteration of their intrinsic activity by covalent modification; (3) variation in the concentrations of transcription factors with which they collaborate. In some instances, all three mechanisms cooperate. For *p53*, the first mechanism—changes in the concentration of the p53 protein—was initially implicated. Measurement of p53 protein levels indicated that they could vary drastically from one cell type to another and, provocatively, would increase rapidly when cells were exposed to certain types of physiological stress.

These observations raised the question of how cells control the concentration of p53. Many protein molecules, once synthesized, persist for tens or even hundreds of hours. Yet other proteins are highly unstable and are degraded almost as soon as they are assembled. One way to measure the half-life of proteins is to treat cells with

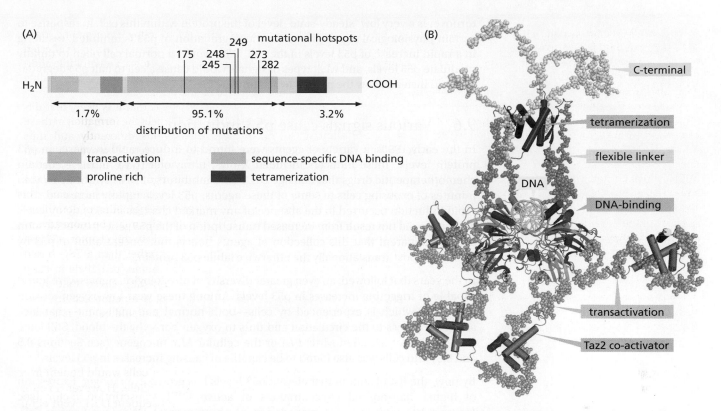

Figure 9.6 Nature of *p53* mutations (A) The locations of the point mutations that cause amino acid substitutions in the p53 protein are shown. Most *p53* mutations (95.1%) affect the DNA-binding domain of the p53 protein. The numbers above the structure indicate the residue numbers of the amino acids that are subject to frequent substitution in human tumors. The transactivation domain enables p53 to interact physically with a number of alternative nuclear partner proteins, including the transcriptional co-activator p300/CBP and the p53 antagonist Mdm2. (B) The overall structure of the DNA-bound p53 tetramer. The four DNA-binding domains are shown in *green* and *blue*, and the four tetramerization domains are seen as *red* and *dark red* α-helices. The DNA double helix (*center*) is shown in *yellow*.

Each of the four DNA-binding domains associates with half of a binding site in the DNA; two copies of the binding site are present in the DNA with a small number of base pairs separating them (see Figure 9.15). Each of the four transactivating domains (*dark pink*) is shown interacting with the Taz2 domain of the p300 co-activator (*light purple*), which functions to stimulate transcription through its ability to acetylate histones and p53 itself. The C-terminal domain (*yellow*) plays important roles in regulating transcription. (A, from K.H. Vousden and X. Lu, *Nat. Rev. Cancer* 2:594–604, 2002; and A.C. Joerger and A.R. Fersht, *Annu. Rev. Biochem.* 77:557–582, 2008. With permission from Nature. B, from A.C. Joerger and A.R. Fersht, *Cold Spring Harb. Perspect. Biol.* 2:000919, 2010. With permission from Cold Spring Harbor Laboratory Press.)

cycloheximide, a drug that blocks ongoing protein synthesis. When such an experiment was performed in cells with wild-type *p53* alleles, the p53 protein disappeared with a half-life of only 20 minutes, leading to the conclusion that p53 is usually a highly unstable protein and one that is broken down by proteolysis soon after it is synthesized.

Such rapid degradation might seem highly wasteful. Why should a cell invest substantial energy and synthetic capacity in making a protein molecule, only to destroy it almost as soon as it has been created? Similar behaviors have been associated with other cellular proteins such as Myc (see Section 6.1).

The rationale underlying this ostensibly "futile cycle" of rapid protein turnover is a simple one: a cell may need to rapidly increase or decrease the concentration of a protein in response to certain physiological signals. In principle, such modulation could be achieved by regulating the level of its encoding mRNA or the rate with which this mRNA is translated. However, far more rapid changes in the levels of a critical protein can be achieved simply by stabilizing or destabilizing the protein itself. In the case of p53, a cell can double the concentration of p53 protein in 20 minutes simply by blocking its degradation.

Under normal conditions, a cell will continuously synthesize p53 molecules at a high rate and subsequently degrade them at an equally high rate. The net result of this rapid turnover is a very low "steady-state" level of the protein within this cell. In response to certain physiological signals, however, the degradation of p53 is inhibited, resulting in a rapid increase of p53 levels in the cell. Why would a normal cell need to rapidly modulate p53 levels, and what types of signals would cause a cell to halt p53 degradation and instead allow the protein to accumulate?

9.6   Various signals cause p53 induction

In the early 1990s, a variety of agents were found to induce rapid increases in p53 protein levels. These inducers included X-rays, ultraviolet (UV) radiation, certain chemotherapeutic drugs that damage DNA, and inhibitors of DNA synthesis. Within minutes of exposing cells to some of these agents, p53 levels rapidly increased. This rapid induction occurred in the absence of any marked changes in *p53* mRNA levels and hence did not result from increased transcription of the *p53* gene. Instead, it soon became apparent that this collection of agents boosts the concentration of p53 by stabilizing post-translationally the otherwise labile p53 protein.

In the years that followed, an even greater diversity of physiological signals were found capable of triggering increases in p53 levels. Among these were low oxygen tension (hypoxia), which is experienced by cells—both normal and malignant—that lack adequate access to the circulation and thus to oxygen borne by the blood. Still later, introduction of the adenovirus *E1A* or the cellular *Myc* oncogene (see Sections 8.5 and 8.9) into cells was also found to be capable of causing increases in p53 levels.

By now, the list of stimuli that elevate p53 levels has grown even longer. Expression of higher-than-normal concentrations of active E2F1 transcription factor (see Figure 8.18), widespread demethylation of chromosomal DNA (see Section 7.7), deficits in the nucleotide precursors of DNA, and increases in intracellular reactive oxygen species (ROS) all elicit p53 accumulation. Exposure of cells to nitrous oxide or to an acidified growth medium, depletion of the intracellular pool of ribonucleotides, and blockage of either RNA or DNA synthesis drive increases of p53 levels as well.

This extensive list of triggers made it clear that a diverse array of sensors is responsible for monitoring the integrity and functioning of various cellular systems. When these sensors detect damage or aberrant functioning, they send signals to p53 and its regulators, resulting in a rapid increase in p53 levels within a cell and the subsequent induction of a variety of downstream responses.

The same physiological signals and DNA-damaging—or **genotoxic**—agents that provoke p53 increases were already known from other work to trigger cell cycle arrest (sometimes referred to as "growth arrest"). In some cases, these stresses could even induce a more permanent form of arrest known as cellular senescence (see Section 9.11). In yet other situations, some of these stressful signals might trigger activation of the apoptotic cell suicide program (see Section 9.13). These observations, when taken together, showed a striking parallel: toxic agents that induce cell cycle arrest, senescence, or apoptosis are also capable of inducing increases in p53 levels. Because such observations were initially only correlations, they did not prove that p53 somehow *caused* cells to arrest their growth or trigger apoptosis following exposure to toxic or stressful stimuli.

The definitive demonstrations of causality came from more detailed examinations of the effects of elevating p53. For example, the rise in cellular p53 levels that is triggered by treatment with genotoxic agents, such as X-rays, is followed by a subsequent increase in the levels of the mRNA encoding the CDK inhibitor p21^{Cip1} (see Section 8.4); this induction is absent in cells expressing mutant p53 protein. The observation suggests that in some cell types, such as fibroblasts, p53 can halt cell cycle progression by inducing expression of this widely acting CDK inhibitor (**Figure 9.7A**, see also Figure 8.12). Indeed, the long-term biological responses to irradiation are often affected by the state of a cell's *p53* gene. For example, mouse embryo fibroblasts carrying two defective *p53* alleles show a greatly decreased tendency to enter into cell cycle arrest following irradiation. Likewise, in cell types such as T lymphocytes, which undergo apoptosis when exposed to radiation, a functional p53 is required to launch this cell's suicide response (Figure 9.7B).

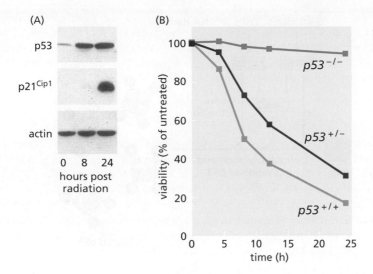

Figure 9.7 p53 and the radiation response Exposure of cells to X-rays strongly increases p53 levels. (A) Once p53 is present in higher concentrations and is functionally activated via various covalent modifications (*not measured here*), it induces expression of the p21^{Cip1} protein (see Section 8.4). p21^{Cip1} acts as a potent CDK inhibitor of the cyclin–CDK complexes that are active in late G_1, S, G_2, and M phases and can thereby halt further cell proliferation at any of these phases of the cell cycle. The actin protein is included in all three samples as a "loading control" to ensure that equal amounts of protein were added to each lane of the gel prior to electrophoresis. (B) Thymocytes (leukocytes derived from the thymus) of wild-type (*p53$^{+/+}$*) mice show an 80% loss of viability relative to untreated control cells during the 25 hours following X-irradiation (*green*), and thymocytes from *p53$^{+/-}$* heterozygous mice (with one wild-type and one null allele) show almost as much loss of viability (*red*). In contrast, thymocytes prepared from *p53$^{-/-}$* homozygous mutant mice exhibit less than a 5% loss of viability during this time period (*blue*). In all cases, the loss of viability was attributable to apoptosis (*not shown*). (A, courtesy of K.H. Vousden. B, from S.W. Lowe et al., *Nature* 362:847–849, 1993. With permission from Nature.)

These various observations can be incorporated into a simple, unifying mechanistic model: a diverse array of cellular surveillance systems relay information to p53, which then triggers appropriate physiological responses: halting cell proliferation, activating DNA repair systems, adjusting metabolism, blocking angiogenesis (the growth of new blood vessels), or engaging the apoptotic suicide program described later (**Figure 9.8**). In fact, a number of stresses—including hypoxia, genomic damage, and imbalances in the signaling pathways governing cell proliferation—are commonly experienced by cancer cells during various stages of tumor development. Consequently, p53 activity must be blunted or even fully eliminated in these cells if they are to survive and prosper. In the same manner, normal cells must also avoid excessive p53 activity, because it threatens to cause their arrest or death, thereby affecting normal bodily functions (Supplementary Sidebar 9.1).

9.7 DNA damage and deregulated growth signals cause p53 stabilization

Three well-studied systems send alarm signals to p53 in the event that they detect damage or signaling imbalances. The first monitors chromosomal DNA for double-strand breaks (DSBs), including those that are created by ionizing radiation such as X-rays. Indeed, in some cell types, even a single DSB anywhere in the genome appears to be sufficient to induce a measurable increase in p53 levels. Many of the proteins that detect such breaks have been discovered; they act through a kinase called ATM.

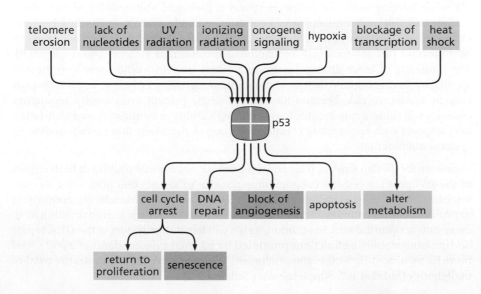

Figure 9.8 p53-activating signals and p53's downstream effects Studies of p53 function have revealed that a variety of cell-physiological stresses can cause a rapid rise in p53 level. The resulting accumulated p53 protein then undergoes post-translational modifications and proceeds to induce a number of responses. A cytostatic response ("cell cycle arrest," often called "growth arrest") can be either irreversible ("senescence") or reversible ("return to proliferation"). DNA repair proteins may be mobilized as well as proteins that alter metabolism or antagonize blood vessel formation ("block of angiogenesis"). As an alternative, in certain circumstances, p53 may trigger apoptosis.

Figure 9.9 Control of p53 levels by various kinases The cycle of p53 synthesis and destruction is modulated by a series of regulators. Extensive ssDNA regions and double-strand DNA breaks (DSBs) activate two kinases, ATR and ATM, respectively. These kinases can phosphorylate p53 directly, or indirectly via Chk1 and Chk2, in all cases in its N-terminal domain (see Figure 9.11B). This phosphorylation prevents the binding of the p53 antagonist Mdm2 (see Figure 9.10).

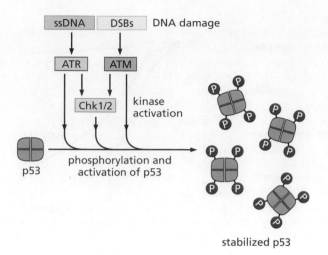

Not only does ATM phosphorylate p53 directly, but it also activates another kinase, Chk2, that proceeds to phosphorylate p53 as well (**Figure 9.9**). This phosphorylation stabilizes the p53 protein and protects it from destruction orchestrated by the protein Mdm2, as discussed in the next section. A deficiency of ATM—the cause of the disease ataxia telangiectasia—leads to a hypersensitivity of cells to X-irradiation (as described in Chapter 12). Thus, without ATM on the job, p53 fails to be properly stabilized and activated by DNA damage; in the absence of p53, cells will lose the ability to rapidly mount their DNA repair program, leading to DNA breaks that can impair their survival.

A second signaling pathway is activated by the presence of single-strand DNA. Single-strand DNA tends to persist at replication forks that have stalled, often because DNA polymerases have encountered nucleotides that have been altered by any one of a wide variety of DNA-damaging agents, including UV radiation and certain chemo-therapeutic drugs. Proteins that detect single-strand DNA activate a kinase called ATR. Like ATM, this kinase both directly phosphorylates p53 and activates an additional kinase—in this case Chk1—which also phosphorylates p53 (see Figure 9.9).

A third pathway leading to p53 activation is triggered by aberrant growth signals, notably those that perturb the pRb–E2F cell cycle control pathway (see Section 8.7). As we will see shortly, this pathway does not depend on activation of intermediary kinases to induce increases in p53 levels and signaling. The mechanisms by which other physiological stresses or imbalances, such as hypoxia, trigger increases in p53 levels remain an active area of investigation.

These converging signaling pathways reveal a profound vulnerability of the mammalian cell. Through the course of evolution, a single protein—p53—has become entrusted with the task of receiving signals from lookouts that monitor a wide variety of important physiological and biochemical intracellular systems (see Figure 9.8). The funneling of these diverse signals to a single protein would seem to represent an elegant and economic design of the cellular signaling circuitry. But it also puts cells at significant risk, because loss of this single protein from a cell's regulatory circuitry will cause a catastrophic loss of the cell's ability to monitor its own well-being and respond with appropriate countermeasures in the event that certain operating systems malfunction.

In one stroke (or two strokes, if we consider the successive inactivation of both copies of the *p53* gene), a cell can become unresponsive to signals that arise from its own internal defects. Thus, a cell lacking intact p53 will continue to blunder on, continuing to proliferate under circumstances that would normally cause it to stop dividing or to enter into apoptotic death. In addition, as we will learn shortly, loss of the DNA repair and genome-stabilizing functions promoted by p53 make descendants of a *p53*$^{-/-}$ cell more likely to acquire further mutations and advance more rapidly down the road of malignancy (**Sidebar 9.2**, Supplementary Sidebar 9.2).

Sidebar 9.2 Sunlight, p53, and skin cancer The p53 protein stands as an important guardian against skin cancer induced by sunlight. When the genome of a **keratinocyte** in the skin suffers extensive damage from ultraviolet-B (UV-B) radiation, p53 can rapidly trigger its apoptotic death. One manifestation of this reaction is the extensive scaling of skin several days after a sunburn. At the same time, UV-B exposure can induce the mutation and functional inactivation of *p53*. Indeed, *p53* alleles in human squamous cell carcinomas of the skin often harbor mutations at dipyrimidine sites—precisely the sites at which UV-B rays induce the formation of pyrimidine–pyrimidine cross-links (see Section 12.6). Such mutant *p53* alleles can also be found in outwardly normal skin that has suffered chronic sun damage. Once p53 function is compromised by these mutations, keratinocytes may be able to survive subsequent extensive exposures to UV-B irradiation, because apoptosis will no longer be triggered by their p53 protein. Moreover, loss of p53 results in a diminished ability to repair subsequent UV-B–induced DNA lesions. Hence, *p53*-mutant cells may subsequently acquire additional mutant alleles that enable them, together with the mutant *p53* alleles, to form a squamous cell carcinoma.

Human papillomaviruses (HPVs) are increasingly implicated as co-factors in many of these squamous cell carcinomas; a key function of the virus-encoded E6 oncoprotein may explain the synergistic actions of UV-B radiation and HPV in these relatively common tumors: E6 tags p53 for destruction by ubiquitylation and degradation in proteasomes, thereby phenocopying mutational inactivation of the *p53* gene (Supplementary Sidebar 9.2). Interestingly, mice that lack functional *p53* gene copies respond to UV-B exposure by developing uveal melanomas—tumors of pigmented cells in the front of the eye; similar tumors are suspected to be caused in humans by UV exposure.

Of additional interest, p53 in keratinocytes has another totally unrelated effect that illustrates its diverse functions: in response to the DNA damage created by UV radiation, p53 causes these cells to release melanocyte-stimulating hormone (αMSH); this hormone stimulates nearby skin melanocytes to produce the melanin pigment to protect against further sunlight-induced damage and, more productively, generate a suntan!

9.8 Mdm2 destroys its own creator

The diverse alarm signals that converge on p53 produce a common result: a rapid increase in the levels of the p53 protein. This response is achieved by a mechanism that we have discussed in previous chapters, specifically the regulation of protein stability by the ubiquitin–proteasome system (see Supplementary Sidebar 7.5). Proteins that are destined to be degraded by this system are initially tagged by the covalent attachment of polyubiquitin side chains, which direct the proteins to proteasomes, where they are digested into oligopeptides. The critical control point in this process is the initial tagging process.

In normal, unperturbed cells, the degradation of p53 is regulated by a protein cited earlier, the ubiquitin ligase termed Mdm2 in mouse cells and MDM2 or HDM2 in human cells. This protein recognizes p53 as a target for ubiquitylation shortly after its synthesis and thereby marks p53 for rapid destruction (**Figure 9.10**). As a consequence, the steady-state concentration of p53 is kept at a low level.

Surprisingly, an opposing control also operates: the expression of Mdm2 in cells is regulated by p53. As mentioned earlier, p53 functions as a transcription factor and the *Mdm2* gene is one of its targets. This relationship between p53 and Mdm2 creates an autoregulatory loop in which p53 induces the synthesis of the protein that drives its own destruction (see Figure 9.10).

The *Mdm2* gene was initially identified because in murine sarcoma cells it is present in multiple copies on extrachromosomal fragments called **double minute** chromosomes (hence, **m**ouse **d**ouble **m**inutes). These derive from segments that break free from their original chromosomes and are replicated repeatedly (see Figure 1.9B). Subsequently, the human homolog of the *Mdm2* gene was discovered to be frequently amplified in sarcomas as well as in other tumor types. In certain other tumor types (e.g., lung carcinomas), the gene is overexpressed through mechanisms that remain unclear. As is the case with other oncogenes, it seemed at first that amplification of the *Mdm2* gene in mouse sarcoma cells afforded the cells some direct, immediate proliferative advantage. Only long after the Mdm2 protein was first identified did its role as the agent of p53 destruction become apparent.

In fact, Mdm2 has an additional effect on the activity of p53. By binding to p53, Mdm2 immediately blocks p53's ability to act as a transcription factor by preventing the protein's interaction with transcriptional co-factors, such as acetyltransferases and histone methyltransferases (see Section 1.8). After it exerts these direct effects on p53's function as a transcription factor, Mdm2 orchestrates the attachment of a ubiquitin

Figure 9.10 Control of p53 levels by Mdm2 After p53 concentrations increase in response to certain physiological signals (*not shown*), the p53 tetramers bind to the promoters of a large constituency of target genes whose transcription they induce (*above*); these target genes include the gene encoding the p53 antagonist Mdm2. Mdm2 molecules then proceed to bind the p53 protein subunits and initiate their ubiquitylation, resulting in their export to the cytoplasm, further ubiquitylation, and, ultimately, degradation in proteasomes. This negative-feedback loop ensures that p53 concentrations eventually fall after danger signals, such as those shown in Figure 9.8, cease; such regulation ensures that p53 concentration will thereafter return to its normally low level.

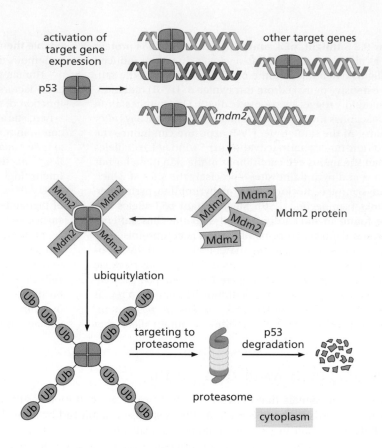

moiety to p53 as well as its export from the nucleus to the cytoplasm. There, polyubiquitylation of p53 ensures its rapid degradation in cytoplasmic proteasomes (see Figure 9.10). The continuous, highly efficient actions of Mdm2 ensure the short, 20-minute half-life of p53 in normal, unstressed cells.

Although the present discussion and Figure 9.10 represent Mdm2 as a monomeric protein, it actually often forms heterodimeric complexes with its close cousin, MdmX (also called Mdm4). This complex may be responsible for much of the ubiquitylation activity that drives p53 degradation. Indeed, there is evidence that without the presence of MdmX, Mdm2 loses the ability to drive p53 degradation. When expressed on its own, however, MdmX seems to be limited to blocking p53-mediated transcriptional activation. Like *Mdm2*, the *MdmX* gene is a transcriptional target of p53, and it thus participates in the elaborate **negative-feedback** loop that controls p53 levels.

When cells suffer certain types of stress or damage, p53 protein molecules must be protected from their Mdm2/MdmX executioners (hereafter simply Mdm2) so that they can accumulate to functionally significant levels in the cell. This protection is often achieved by the phosphorylation of p53 described above (see Figure 9.9). The phosphorylation sites—recognized by kinases such as ATM, ATR, Chk1, and Chk2—lie in the N-terminal region of p53 in a domain that is normally recognized and bound by Mdm2 (**Figure 9.11A**). Thus, phosphorylation of p53 blocks the binding of Mdm2 and subsequent ubiquitylation (Figure 9.11B). At the same time, ATM kinase also phosphorylates Mdm2, causing its functional inactivation and destabilization (**Figure 9.12**). As a consequence of the phosphorylation of both p53 and Mdm2, p53 escapes destruction and its concentrations in the cell increase rapidly. Once p53 has accumulated in substantial amounts, the protein is then poised to evoke a series of downstream responses (see Figure 9.8). These responses will be discussed in detail below.

Note that Mdm2 operates here as an oncoprotein, but one whose mechanism of action is very different from those of the various oncoproteins that we encountered in Chapters 4, 5, and 6. The latter function as components of mitogenic signal cascades and thereby induce cell proliferation by mimicking the signals normally triggered by the binding of growth factors to their receptors. Mdm2, in contrast, operates by

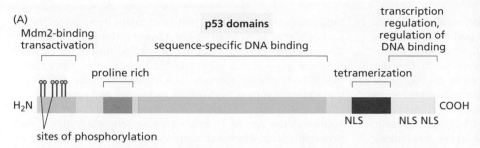

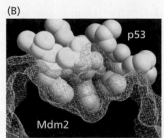

Figure 9.11 Specialized domains of p53 (A) The interaction of p53 with Mdm2 occurs in a small domain near its N-terminus, where the transactivation domain of p53 is also located. The phosphorylation of p53 amino acid residues in this region (*red lollipops; not all are indicated*) blocks Mdm2 binding and thus saves p53 from ubiquitylation and degradation. The nearby proline-rich domain (*salmon*) contributes to p53's pro-apoptotic functions. Its tetramerization domain is located toward its C-terminus. Nearby are nuclear localization signals (NLS), which allow import of recently synthesized p53 into the nucleus, as well as amino acid sequences that regulate its DNA binding. (B) The structure of the interface where p53 and Mdm2 interact has been revealed by X-ray crystallography. The interacting domain of p53 is shown as a yellow space-filling model that includes p53 residues 18 through 27, and the surface of the complementary pocket of Mdm2 is shown as a blue wire mesh. (A, from M. Ashcroft and K.H. Vousden in D.E. Fisher, ed., *Tumor Suppressor Genes in Human Cancer*. Totowa, NJ, Humana Press, 2000. With permission from Springer. B, from P.H. Kussie et al., *Science* 274:948–953, 1996. With permission from AAAS.)

antagonizing p53 and thereby prevents entrance of a cell into cell cycle arrest, senescence, or into the apoptotic suicide program. The final outcome is, however, the same: the actions of traditional oncoproteins and Mdm2 all favor increases in cell number. Of note, inherited genetic polymorphisms that in humans lead to increased Mdm2 expression—or otherwise disable the p53 alarm system—are also associated with familial cancer susceptibility (**Sidebar 9.3**).

The activity and levels of the Mdm2 protein are affected by yet other positive and negative signals. The signaling pathway that favors cell survival through activation of the PI3 kinase (PI3K) pathway (see Section 6.6) leads, via the Akt/PKB kinase, to Mdm2 phosphorylation (at a site different from that altered by the ATM kinase described above). This phosphorylation results in the translocation of Mdm2 from the cytoplasm to the nucleus, where it is well-positioned to bind to and modify p53 (see Figure 9.12). Because PI3K itself is activated by Ras and growth factor receptors, it becomes clear that this mitogenic signaling pathway influences Mdm2 and, indirectly, p53.

In normal, unstressed cells, Mdm2 must be allowed to perform its normal role of keeping p53 levels very low, as is highlighted by the results of inactivating both *Mdm2*

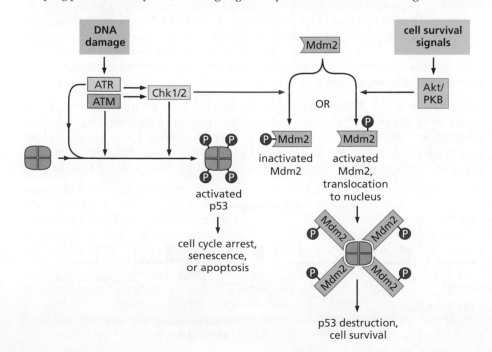

Figure 9.12 The activity of Mdm2 is also regulated by phosphorylation The same DNA damage that induces phosphorylation of p53 by ATM and other kinases (resulting in "activated p53" in this figure) also stimulates phosphorylation of Mdm2. This phosphorylation inactivates and destabilizes Mdm2, preventing its binding to p53, resulting in accumulation of the latter. Mdm2 can also be phosphorylated at a different site by kinases, such as Akt/PKB, which are, in turn, activated by cell growth and survival signals. This phosphorylation promotes translocation of Mdm2 to the nucleus, where it can bind to p53, marking it for destruction, thereby reducing the likelihood of p53-triggered apoptosis.

Figure 9.13 The gene encoding p16^INK4A and p14/p19^ARF Analysis of the *p16^INK4A* gene (*red*) has revealed that it shares its second exon with a gene encoding a 19-kD protein in mice and a 14-kD protein in humans. The *p14/p19* gene uses an alternative transcriptional start site (*blue arrow, left*) located more than 13 kb upstream of the one used by *p16^INK4A* (*red arrow, center*). Because translation of its mRNA uses an alternative reading frame (*green bracket*) in exon 2 (*red, blue*), the resulting protein and thus gene came to be called p19^ARF (or in humans p14^ARF). The patterns of RNA splicing are indicated by the carets connecting the various exons of the two intertwined genes. The boxes indicate exons, and the filled areas within each exon indicate reading frames. Other research indicates the presence of four E2F-binding sites (*not shown here*) in the promoter region of the gene encoding p14^ARF, three upstream of the transcriptional start site and one downstream. (From C. Sherr, *Genes Dev.* 12:2984–2991, 1998. With permission from Cold Spring Harbor Laboratory Press.)

gene copies in the genomes of mouse embryos: these embryos die very early in development, because p53 levels increase to physiologically intolerable levels, preventing the normal proliferation of embryonic cells or causing them to die. These profoundly disruptive effects of *Mdm2* gene inactivation are attributed specifically to runaway p53 activity, as is made clear by studies in which both of these genes in a mouse embryo harbor homozygous inactivating mutations, yielding the *Mdm2^-/-*; *p53^-/-* genotype. Once p53 is eliminated from the embryonic cells, the loss of Mdm2 becomes tolerable and embryonic development proceeds normally!

Yet another important mechanism that controls Mdm2 has been revealed through the discovery of a protein termed p19^ARF in mouse cells and p14^ARF in human cells. Astute sequence analysis led to the discovery of ARF, as we call it hereafter. Its gene was originally uncovered in mouse cells as one whose coding sequence overlaps the gene that encodes p16^INK4A, an important CDK inhibitor involved in regulating the activation of pRb (see Section 8.4).

Transcription of the ARF-encoding gene is directed by a promoter located 13 kb upstream of the *p16^INK4A* promoter; alternative splicing of the resulting transcript produces an mRNA that encodes—in an **a**lternative **r**eading **f**rame relative to that used to encode p16^INK4A—the ARF protein (**Figure 9.13**). Forced expression of an ARF-encoding cDNA in wild-type rodent cells was found to cause a strong inhibition of

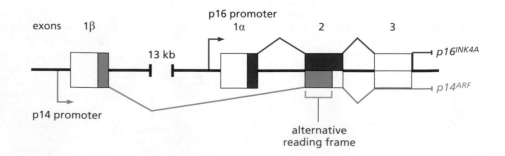

proliferation. However, this inhibition was not observed when the ARF cDNA was introduced into cells that lacked wild-type p53 function. This result indicates that the growth-inhibitory powers of ARF depend absolutely on the presence of functional p53.

Further investigation revealed that in wild-type cells, the expression of ARF causes a rapid increase in p53 levels. We now understand the molecular mechanisms that underlie this response: ARF binds to Mdm2 and prevents its further action by driving its localization from the **nucleoplasm** to the nucleolus where it is sequestered; recall that the nucleolus is the nuclear organelle that is largely devoted to manufacturing ribosomal subunits (**Figure 9.14A**). Once Mdm2 is diverted from interacting with p53, the latter escapes Mdm2-mediated ubiquitylation and therefore accumulates rapidly to high levels in the nucleoplasm, where it is poised to

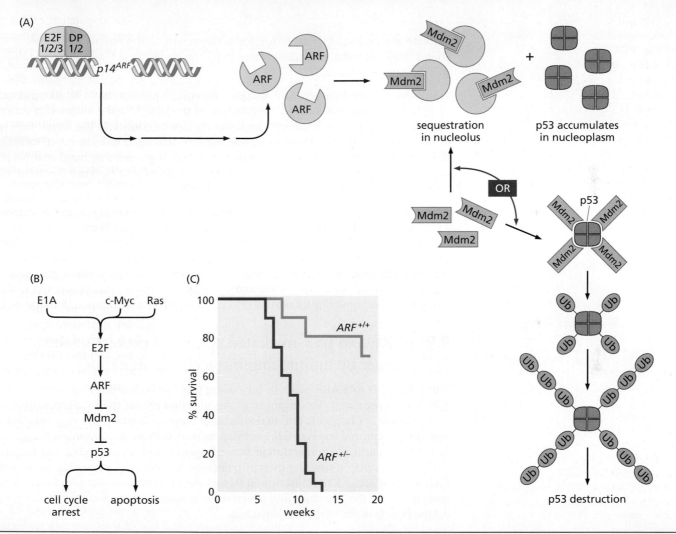

Figure 9.14 Control of apoptosis by ARF (A) The human p14^ARF protein (p19^ARF in mice), termed ARF in the figure (*light blue*), associates with and drags Mdm2 (*gold*) into the nucleolus, where it is sequestered. Once it is neutralized, Mdm2 can no longer tag p53 for destruction. Hence, elevated ARF levels cause increases in the levels of p53. [ARF can also add the ubiquitin-related SUMO groups to Mdm2 in the nucleolus (*not shown*).] (B) Transcription of the *p14^ARF* gene is driven by E2F transcription factors composed of E2F1/2/3 and their DP1/2 partners (see also Figure 8.18). Various oncogenic signals favor apoptosis through their ability to induce E2F activity (see Sections 8.7 and 8.8), which leads, in turn, to increased ARF expression. Among these are the adenovirus E1A, and the Myc (c-Myc) and Ras oncoproteins. Thus, this signaling pathway seems to have evolved to eliminate cells with hyperactive E2F signaling driven by a variety of overly active mitogenic signals, specifically those released by various

oncoproteins. (C) Mice that were engineered to be heterozygous for the *p19* gene (*ARF^+/−*) and wild-type mice (*ARF^+/+*) were mated with mice carrying an Eμ-*myc* transgene known to cause B-cell lymphomas (also described in Figure 9.20). The *ARF^+/−* Eμ-*myc* mice (*red line*) developed fatal tumors far more rapidly than mice carrying only the Eμ-*myc* transgene (*green line*), and cells in these tumors shed their remaining wild-type *ARF* allele. Hence, in the absence of ARF function, the pro-apoptotic effects of a *Myc* oncogene (indicated in B) are largely lost, allowing survival of lymphoma cells and the proliferative effects of *Myc* to dominate and drive tumor formation. (B, from P.J. Iaquinta, A. Alanian, and J.A. Lees, *Cold Spring Harb. Symp. Quant. Biol.* 70:309-316, 2005. With permission from Cold Spring Harbor Laboratory Press. C, from F. Zindy et al., *Proc. Natl. Acad. Sci. USA* 100:15930–15935, 2003. With permission from National Academy of Sciences, U.S.A.)

Sidebar 9.4 Have mammalian cells placed too many eggs in one basket? The discovery of the $p16^{INK4A}/p14^{ARF}$ genetic locus, which is inactivated through one mechanism or another in about half of all human tumors, raises a provocative question: Why have mammalian cells invested a single chromosomal locus with the power to encode two proteins regulating the two most important tumor suppressor pathways, those of pRb and p53? Deletion of this single locus results in the simultaneous loss of normal regulation of both pathways. As was the case with *p53* itself, enormous power has been concentrated in a single genetic locus.

Placing two such vital eggs in a single genetic basket seems foolhardy for the mammalian cell, as it causes the cell to be vulnerable to two types of deregulation through loss of a single gene. To make matters even worse, the gene encoding $p15^{INK4B}$, another important regulator of pRb phosphorylation (see Section 8.4), is closely linked to the $p16^{INK4A}/p14^{ARF}$ locus, indeed so close that all of these genetic elements are often lost through the deletion of only about 40 kb of chromosomal DNA. We have yet to discern the underlying rationale of this genetic arrangement. Maybe there is none, and perhaps mammalian evolution has produced a less-than-optimal design of its control circuitry.

activate expression of its constituency of target genes. The enemy of an enemy is a friend: ARF can induce rapid increases in p53 levels because it kidnaps and inhibits p53's destroyer, Mdm2.

Given the alliance between ARF and p53, one might predict that ARF, like p53, acts as a tumor suppressor. Indeed, inactivation of the $p16^{INK4A}/p14^{ARF}$ locus by genetic mutation or epigenetic promoter methylation (see Section 7.7) has been detected in many human tumors. Once a cell has lost ARF activity, it loses much of its ability to block Mdm2 function. As a consequence, Mdm2 is given a free hand to drive p53 degradation, and the cell is deprived of the services of p53 because the latter can never accumulate to functionally significant levels.

Because ARF has a central role in increasing p53 levels in many cellular contexts, the $p14^{ARF}$ gene is an extremely important TSG. Moreover, it seems likely that many of the human cancer cells that retain wild-type *p53* gene copies have eliminated p53 function by inactivating their two copies of the gene encoding ARF. We should note that the co-localization of the $p16^{INK4A}$ and $p14^{ARF}$ genes in the mammalian genome (see Figure 9.13) would seem to represent another dangerous concentration of power leading to significant vulnerability for normal cells to oncogenic transformation (**Sidebar 9.4**).

9.9 ARF and p53-mediated apoptosis protect against cancer by monitoring intracellular signaling

The influential role ARF plays in increasing p53 levels raises the question of how ARF itself is regulated. The response to this question relates to our discussion of the pRb pathway in Chapter 8, and to the fact that mammalian cells are very sensitive to higher-than-normal levels of E2F1 activity. Indeed, cells seem to monitor the activity level of this particular transcription factor (together with that of E2F2 and E2F3) as a means of detecting whether their pRb circuitry is functioning properly: excessively high levels of active E2F transcription factors are a telltale sign that pRb function has gone awry and that entrance into senescence or apoptosis must quickly be triggered for the benefit of the tissue and organism.

It turns out that the $p14^{ARF}$ gene carries at least four E2F-binding sequences in its promoter (see Figure 9.13), and high levels of E2F1, E2F2, or E2F3A activity indeed induce production of $p14^{ARF}$ mRNA. When the ARF protein appears on the scene, it blocks Mdm2 action and p53 accumulates as described earlier. The elevated levels of p53 then trigger apoptosis. The signaling cascade is configured like this:

$$\text{pRB} \dashv \text{E2F} \rightarrow \text{ARF} \dashv \text{Mdm2} \dashv \text{p53} \rightarrow \text{apoptosis}$$

In fact, evolution has devised yet other ways to eliminate cells that carry too much E2F activity and, by implication, have lost proper pRb control. Runaway E2F1 activity drives expression of a number of genes that encode proteins that participate directly in the apoptotic program (to be described below). Included among these genes are those encoding several **caspases** (types 3, 7, 8, and 9), pro-apoptotic Bcl-2–related proteins (Bim, Noxa, PUMA), Apaf-1, and p53's cousin, p73; these proteins collaborate to drive cells into apoptosis. We will learn more about them later.

These various mechanisms conspire to ensure triggering of the p53-dependent apoptotic program in response to excessive E2F activity. As an example, such E2F-initiated apoptosis contributes to lethality in mouse embryos that have been deprived of both copies of their *Rb* gene. These mice die in mid-gestation from the excessive proliferation and concomitant apoptosis of certain critical cell types, including those involved in **erythropoiesis** (formation of red blood cells) and in placental function.

Some of the effects of the *E1A* and *Myc* oncogenes are related to this mechanism as well. Both deregulate pRb control and are highly effective in inducing apoptosis. The adenovirus E1A oncoprotein binds and effectively sequesters pRb and its cousins, whereas Myc, for its part, induces changes in cell cycle control machinery that ensure that pRb is inactivated through phosphorylation (see Sections 8.5 and 8.8).

Many studies of *Myc* oncogene function indicate that this gene has both potent mitogenic and pro-apoptotic functions. Indeed, the pro-apoptotic effects of *Myc* are so strong that it is highly likely that most cells that happen to acquire an activating *Myc* mutation are rapidly eliminated through apoptosis. Only when the pro-apoptotic program of Myc is blunted or inactivated do its mitogenic actions become apparent.

As an example, when a *Myc* oncogene becomes activated in the lymphoid tissues of a mouse, it prompts a substantial increase in cell proliferation. However, there is no net increase in cell number because the newly formed cells are rapidly lost through apoptosis. If one of the *Myc* oncogene–bearing cells happens subsequently to inactivate its *p53* gene copies, then *Myc*-induced apoptosis is diminished and the cell proliferation driven by the oncogene leads to a net increase in the pool of mutant lymphoid cells. As might be expected from the organization of the pathway shown in Figure 9.14B, a similar effect occurs when one allele of the gene encoding ARF has been knocked out (Figure 9.14C). Loss of the remaining, wild-type *ARF* allele disrupts the pathway linking *Myc* over-expression to p53-dependent apoptosis, resulting in the development of lymphoma.

Although we have focused here on the role of the pRb–E2F axis in activating ARF, at least one other major oncogenic signaling pathway can also contribute significantly to increased ARF expression and thus to p53 activation: the signaling pathway downstream of Ras (see Figure 9.14B). Hyperactive signaling by Ras or its immediate downstream partners, Raf or B-Raf (see Figure 6.9A), drives increased ARF expression and p53 activation. The induction of p53 imposes great selective pressure on incipient cancer cells that experience intense Ras oncoprotein signaling, forcing them to either neutralize p53 function or to confront rapid elimination by p53-dependent apoptosis or senescence.

9.10 p53 functions as a transcription factor that halts cell cycle advance in response to DNA damage and attempts to aid in the repair process

The p53 protein functions by activating the expression of a diverse set of target genes. Its DNA-binding domain consists of a recognition sequence composed of 10 nucleotides repeated twice in tandem, separated by up to 20 nucleotides of random sequence (**Figure 9.15A**). This recognition sequence is present in the promoters or initial introns of dozens of p53 target genes, which collectively control the various downstream functions of p53 (Figure 9.15B). The great majority (>90%) of the mutant *p53* alleles found in human tumor cell genomes contain amino acid substitutions in the DNA-binding domain of p53 (see Figure 9.6). These mutations eliminate p53's ability to bind to its recognition sequences, thereby rendering the mutant protein unable to exert most of its multiple functions. Furthermore, as discussed above, p53 tetramers that contain even one mutant p53 subunit appear to have a reduced ability to bind DNA.

In addition to binding to its recognition sequence, p53 must interact with other factors that modulate its transcriptional functions. These interactions are governed by a complex array of covalent modifications to the p53 protein, including acetylation, glycosylation, methylation, phosphorylation, ribosylation, and sumoylation—involving the

(A)

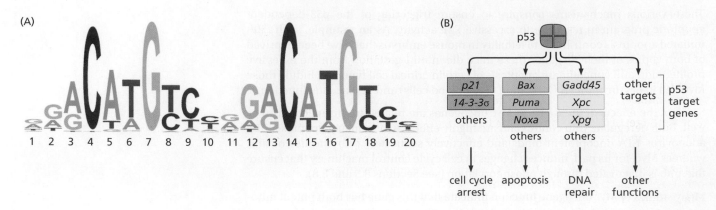

(B)

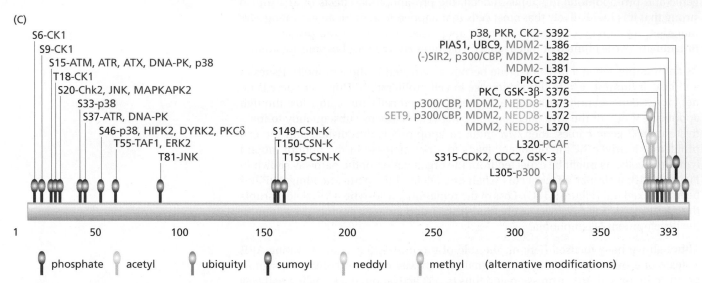

(C)

| phosphate | acetyl | ubiquityl | sumoyl | neddyl | methyl | (alternative modifications) |

Figure 9.15 p53 functions as a transcription factor (A) The consensus oligonucleotide sequence recognized by p53. The sequence was determined by immunoprecipitation of p53–DNA complexes in human cells stressed by exposure to the drug actinomycin, a potent inhibitor of transcription and thus inducer of increased p53 levels. In more detail, the chromatin immunoprecipitation (ChIP) procedure employed anti-p53 antibodies to isolate p53–DNA complexes present in the chromatin of these actinomycin-treated cells. The DNA fragments in the immunoprecipitates were subsequently sequenced, which led to the identification of 1546 sites in the human genome to which p53 is bound; nucleotides that recurred frequently at a particular position in the immunoprecipitated DNA fragments are indicated in *large font* letters, and those that recurred less frequently are indicated by *smaller font* letters. As is apparent, p53 tetramers appear to bind to two similar octanucleotide sequences that are separated from one another by (in this case) two intervening nucleotides; far more intervening nucleotides may separate the two octanucleotide binding sites in other promoters. Interestingly, the majority of these 1546 sites were also recognized and bound by p53's cousins, p63 and p73, also discussed in this chapter. (B) p53 induces expression

of a variety of genes involved in growth arrest, apoptosis, and DNA repair. Examples of genes involved in each response are shown, although many more are known. (C) The phosphorylation of p53 by various kinases (see Figure 9.9) represents only one type of post-translational modification (PTM) that is important to its functioning. In addition to the attachment of phosphate groups (*red*), other enzymes modify the protein by attaching acetyl (*green*), ubiquityl (*blue*), sumoyl (*purple*), neddyl (*yellow*), and methyl (*gray*) groups at the indicated amino acid residues and residue numbers (*black type*). The identities of the modifying enzymes are indicated (*various font colors*) next to the amino acid residue numbers. Only a small portion of these PTMs have been linked to specific p53 functions. Moreover and as an example, one of these PTMs, phosphorylation of serine 46, is accomplished by any of four distinct kinases and activates the pro-apoptotic functions of p53. Both neddyl and sumoyl groups are similar to ubiquityl groups but have very different effects on the proteins to which they become attached. A 2014 proteomics report describes as many as 150 distinct PTMs of p53. The functional consequences of these PTMs are largely unknown. (A, from L. Smeenk et al., *Nucleic Acids Res.* 36:3639–3654, 2008. With permission from Oxford University Press. C, Courtesy of K.P. Olive and T. Jacks.)

attachment of, respectively, acetyl, sugar, methyl, phosphate, ribose, and sumo groups (the latter being a ubiquitin-like peptide that appears to target proteins for localization to specific intracellular sites, often in the nucleus; Figure 9.15C). Phosphorylation of p53's N-terminal transactivation domain (see Figure 9.6B), for example, can increase the protein's ability to bind the p300 co-activator as well as other transcription co-activators. Recruitment of p300 contributes to transcriptional activation by acetylating nearby histones H3 and H4 as well as p53 itself. Indeed, it is highly likely that combinatorial interactions of p53 with other transcription factors determine the identities of the specific target genes that are activated in various circumstances by p53.

In addition to functioning as a transcriptional activator, it is now clear that p53 can repress the expression of certain genes as well. However, the mechanisms underlying p53's transcriptional repression functions are less well understood. We therefore focus our attention on p53's ability to activate the expression of target genes.

One key target of the p53 transcription factor, as discussed earlier, is the *Mdm2* gene, the induction of which leads to a negative-feedback loop that functions to ensure the very low steady-state levels of p53 protein observed in normal, unperturbed cells. The operation of this p53–Mdm2 feedback loop explains a curious aspect of p53 behavior: in human cancer cells that carry mutant, defective *p53* alleles, the p53 protein is almost invariably present in high concentrations (**Figure 9.16**). At first glance, this might appear paradoxical, because high levels of a growth-suppressing protein like p53 would seem to be incompatible with malignant cell proliferation. However, because mutant p53 is unable to induce the expression of Mdm2 (among its other downstream target genes), this mutant, functionally inactive protein escapes Mdm2-driven degradation and accumulates to very high levels. This logic explains why the presence of readily detectable p53 in a population of tumor cells, usually revealed by **immunostaining** (see Figure 9.16), is a telltale sign of the presence of a mutant *p53* allele in the genome of these cells. (Such a conclusion cannot be drawn, however, from analyzing human tissue that has recently been irradiated because radiation can also evoke the widespread expression of p53 throughout a tissue for days, even weeks after radiotherapy.) The large amounts of p53 protein in SV40-infected or SV40-transformed cells can be explained analogously: in this case, sequestration of p53 by the viral large T antigen prevents p53-induced expression of the *Mdm2* gene and resulting p53 degradation. According to one measurement, when large T is expressed in previously normal cells, the half-life of p53 increases from 20 minutes to 24 hours. Alternatively, in cancer cells carrying wild-type *p53* alleles, the presence of chronic DNA damage signals as well as oncogene-induced stress signals may also contribute to the high levels of the p53 protein.

Mdm2 is only one of a large cohort of genes whose expression is induced by p53. As mentioned earlier, another highly important target gene is *p21^{Cip1}* (sometimes called *Cdkn1a* in the mouse), which is key to p53's ability to arrest progression through the cell cycle. In fact, the gene encoding p21^{Cip1} was originally discovered in an effort to uncover genes whose expression is increased by p53. Soon after its discovery, it became apparent that p21^{Cip1} functions as an important inhibitor of a number of the cyclin-dependent kinases (CDKs). The ability of p21^{Cip1} to inhibit two CDKs—CDK2 and CDK1/CDC2—that are active in the late G_1, S, G_2, and M phases of the cell cycle

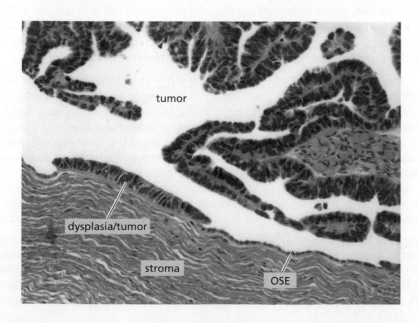

Figure 9.16 Accumulation of p53 in *p53*-mutant cells This microscope section of ovarian tissue has been stained with an anti-p53 antibody coupled to the peroxidase enzyme, resulting in the blackened nuclei seen here. Large patches of epithelial cells in an ovarian carcinoma (*above*) are composed of cells that have high levels of p53; a patch of dysplasia (*left middle*) is also p53-positive. This staining pattern is an indication that relatively early in the course of tumor progression, the dysplastic precursors to this carcinoma already sustained a *p53* somatic mutation. Stromal cells (*small black nuclei, pink matrix, below*) are unstained, as is a patch of normal ovarian surface epithelium (OSE, *below, right*). (Courtesy of R. Drapkin and D.M. Livingston.)

(see Figures 8.6 and 8.11A) explains how p53 is able to block forward progress at multiple points in this cycle.

The consequences of p53 activation can differ at different points in the cell cycle (see Figures 8.6 and 8.11A). For example, if a cell suffers DNA damage during the G_1 phase of the cell cycle, p53 will become activated and will induce p21^{Cip1} synthesis; p21^{Cip1}, in turn, will halt further cell proliferation. At the same time, components of the cellular DNA repair machinery will be mobilized to repair the damage. The genes encoding some of these repair proteins are directly induced by p53. As such, cells lacking functional p53 are unable to efficiently repair the DNA lesions caused by agents such as benzo[a]pyrene (a potent carcinogen present in tars) or the cyclobutane pyrimidine dimers caused by ultraviolet (UV) radiation (described in Chapter 12). In addition, DNA polymerase β, which plays a critical role in reconstructing DNA strands after chemically altered bases have been excised by DNA repair proteins, is much less active in p53-negative cells than in their wild-type counterparts. We will return to these DNA repair proteins and their mechanisms of action in Chapter 12. In the event that the DNA is successfully repaired, the signals that have protected p53 from destruction (see Figure 9.12) will disappear and the concentrations of p53 and, in turn, of p21^{Cip1}, will fall. In the absence of p21^{Cip1}, cell cycle progression will resume, enabling cells to enter S phase, where replication of the now-repaired DNA can proceed.

The rationale for this series of steps is a simple one: by halting cell cycle progression in G_1, p53 prevents a cell from entering S phase and inadvertently copying still-unrepaired DNA, which could be passed on to daughter cells. The importance of these cytostatic actions of p21^{Cip1} can be seen from the phenotype of genetically altered mice in which both germ-line copies of the *p21*^{Cip1} gene have been inactivated. Although not as tumor-prone as p53-null mice, they show an increased incidence of tumors late in life.

If, alternatively, a cell suffering DNA damage has already advanced into S phase and is therefore in the midst of actively replicating its DNA, the p21^{Cip1} induced by p53 can engage the DNA polymerase machinery at the replication forks and halt its further advance along the DNA template molecules. It does so by binding PCNA (proliferating-cell nuclear antigen), which interacts, in turn, with the key DNA polymerases δ and ε, thereby blocking further advance of replication forks along the DNA. Once again, the effect here is to hold DNA replication in abeyance until DNA damage has been successfully repaired.

In addition, the p53 protein uses yet other genes and proteins to halt further cell cycle advance. For example, p53 induces the expression of 14-3-3σ (see Figure 9.15B), which helps to govern the G_2/M transition. The 14-3-3σ protein, for its part, sequesters the cyclin B–CDC2 complex in the cytoplasm, thereby preventing it from moving into the nucleus, where its actions are required to drive the cell into mitosis; a similar mechanism allows 14-3-3σ to bind and sequester the Cdc25A, Cdc25B, and Cdc25C phosphatases, whose actions are also required for entry into mitosis. This mechanism delays mitosis until the chromosomal DNA is determined to be in good repair.

These extensive actions of p53 have caused some to portray this protein as the "guardian of the genome." By preventing cell cycle advance and DNA replication while chromosomal DNA is damaged and by inducing expression of DNA repair enzymes, p53 can reduce the rate at which mutations accumulate in cellular genomes. Moreover, in the event that severe DNA damage has been sustained (for example, damage that exceeds a cell's ability to repair DNA), p53 may trigger apoptosis, thereby eliminating mutant cells and their damaged genomes. This outcome contrasts with the behavior of cells that have lost p53 function: they may replicate their damaged, still-unrepaired DNA, and this can cause them to exhibit relatively **mutable** genomes—that is, genomes that accumulate mutations at an abnormally high rate per cell generation.

In one particularly illustrative experiment, pregnant *p53*$^{+/-}$ mice that had been bred with *p53*$^{+/-}$ males were treated with the highly mutagenic carcinogen ethylnitrosourea (ENU). Of the pups that were subsequently born with the *p53*$^{-/-}$ genotype, 70% developed brain tumors; only 3.6% of the *p53*$^{+/-}$ pups developed similar tumors, whereas none of those that were *p53*$^{+/+}$ did. Hence, in the absence of p53 function, fetal cells that had been mutated by ENU could survive and spawn the progeny forming these lethal tumors.

9.11 Prolonged DNA damage and oncogene activation can induce p53-dependent senescence

As described above, p53 can orchestrate entrance into apoptosis or impose a temporary cell cycle arrest. In fact, as part of its anti-neoplastic functions, p53 can also induce a more permanent form of cell cycle arrest cited above and termed senescence. First observed in experiments performed in the mid-1960s using cells prepared from rodent or human embryos, one form of senescence, "replicative senescence" was associated with limiting the life span of cells growing in tissue culture. Thus, after a certain number of population doublings, the cells would undergo an irreversible cell cycle arrest, accompanied by an enlarged surface area and a "fried egg" appearance (**Figure 9.17A**). Although metabolically active (sometimes for months), these senescent cells would no longer divide, even when presented with adequate nutrients and growth factors.

We now know that senescence can be caused by a number of stimuli, including telomere shortening (which can induce a DNA damage response; see Section 10.4), exposure to high levels of reactive oxygen species (ROS), or other forms of chronic DNA damage. Experiments that showed delayed induction of senescence following expression of the SV40 large T antigen—which inactivates both p53 and the pRB family of proteins (see Sections 8.5 and 9.1)—helped to establish that both of these tumor suppressor pathways are required for the induction of senescence.

Cells that are approaching senescence show increased levels of p53 as well as $p21^{Cip1}$. The cyclin kinase inhibitor $p16^{Ink4a}$ is also strongly induced in senescent cells, which causes pRB to be maintained in a hypo-phosphorylated state (see Section 8.5), contributing to the prolonged cell cycle arrest (Figure 9.17B). Under some conditions, senescent cells undergo pronounced changes in their chromatin, including trimethylation of histone 3 at lysine 9. This modification resembles that present in **heterochromatin** and likely contributes to the irreversibility of the senescent state. Indeed, nuclear regions of **senescence-associated heterochromatin foci** (SAHF) are used as a marker for cells that have undergone senescence rather than temporary cell cycle arrest (Figure 9.17C; see also Supplementary Sidebar 10.3).

Senescent cells can also be characterized by their secretion of a variety of pro-inflammatory cytokines, including several interleukins, IGFBPs, and TGF-β. The cytokines that are released as part of this **senescence-associated secretory phenotype** (SASP) include those that act in an autocrine fashion to promote senescence and those that act in a paracrine fashion to recruit pro-inflammatory cells of the innate immune system. Another commonly used marker of senescence is a senescence-associated isoform of β-galactosidase (SAβ-gal), which we already encountered in Figure 9.17A. Although this enzyme is not mechanistically required for senescence, it is frequently found to be upregulated in senescent cells and can be easily detected by incubating cells with the substrate X-gal at pH 6 (see Figure 9.17A).

If cellular senescence can limit the reproductive life span of cells in the body, as it does in tissue culture, this mechanism could represent a powerful form of tumor suppression. Indeed, as we discuss in greater detail in Chapter 10, overcoming senescence—and thereby achieving cellular immortality—is one of the commonly observed hallmark changes that transformed cells undergo on the path leading to the formation of advanced tumors. Cancer cells frequently mutate one or more of the tumor suppressor genes implicated in triggering senescence, suggesting that diversion of cells into this state may act as an effective barrier to tumorigenesis; still, the same genes and encoded proteins are also involved in regulating reversible growth arrest and apoptosis (see Figure 9.17B), which suggests alternative explanations for their inactivation that is observed in cells en route to high-grade malignancy.

There is accumulating evidence that cellular senescence does indeed occur in living tissues, including the detection of increasing numbers of cells that stain for

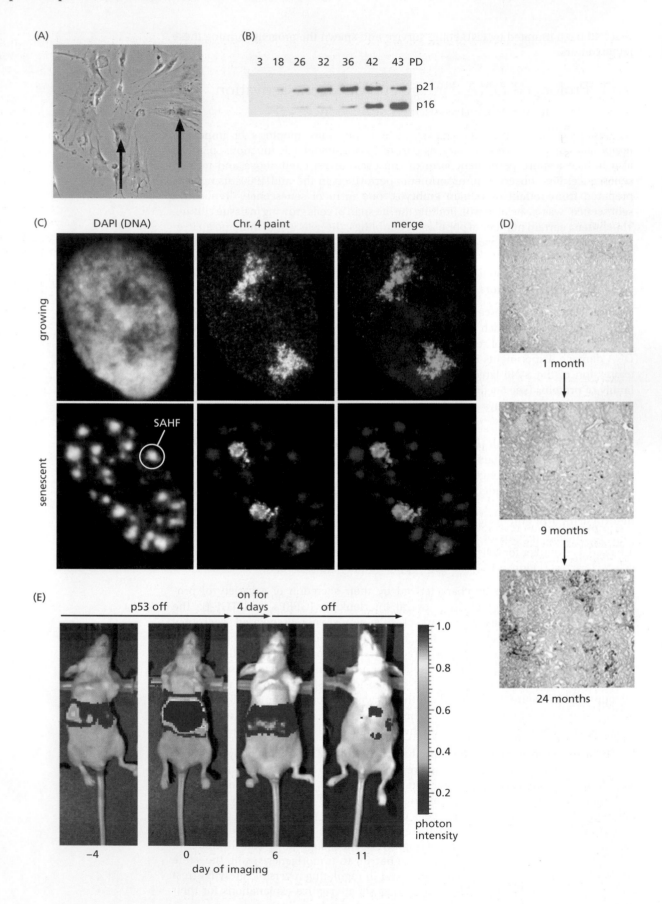

Figure 9.17 Senescent cells arise *in vitro* and *in vivo*
(A) Cells that enter into senescence cease proliferating but remain viable. Many develop extremely large cytoplasms, giving them a "fried egg" appearance. Senescent cells also characteristically express acidic β-galactosidase enzyme, which can be detected by supplying them with a substrate that turns *blue* upon cleavage by this enzyme (*arrows*). (B) When adult human endometrial fibroblasts are propagated *in vitro*, expression of both the p16^{INK4A} and p21^{Cip1} CDK inhibitors (see Section 8.4) is induced, albeit with different kinetics. These cells have an expected life span of 40 to 43 population doublings (PD) in culture, whereupon they enter into senescence. Much of this senescence may reflect a response to the stresses suffered by cells being propagated under sub-optimal conditions of culture. (C) Senescence-associated heterochromatin foci (SAHF) arise during senescence and derive from covalent modifications of the N- and C-terminal tails of histones within the nucleosomes (see Section 1.8). Human fibroblasts were induced to enter senescence through introduction of an activated *ras* oncogene (*lower row*). In the nuclei of control cells that continued active growth (*upper row*), the chromosomal DNA was dispersed throughout the nucleus, as indicated by the DAPI stain (*white*). In contrast, the DNA of the senescent cells was concentrated in a number of discrete foci—SAHFs. The use of *in situ* hybridization with a DNA probe specific for Chromosome 4 (Chr. 4 paint) revealed that the Chromosome 4 DNA was far more dispersed in the nuclei of growing cells than in the senescent cells. The overlap between these two imaging procedures (merge, *green*) shows that the DNA of each Chromosome 4 has coalesced into a single SAHF within a senescent cell. (The behavior of Chromosome 4 is presumed to be representative of the behavior of all other chromosomes, which were not analyzed here.)

Other studies indicate that SAHFs represent chromosomal regions in which transcription is irreversibly repressed. (D) Kidney tissue from rats of the indicated ages was stained for the presence of senescence-associated β-galactosidase (SA-β-gal; *blue*). Observations like these indicate that normal aging is accompanied by increasing numbers of senescent cells within tissues. (E) In a mouse model of liver cancer pathogenesis, an activated H-*ras* oncogene was introduced into cultured hepatoblasts in which p53 expression was suppressed by an inducible siRNA vector construct (see Supplementary Sidebar 1.5). The cells were injected into the spleens of host mice, leading to their subsequent implantation in the liver. The resulting tumors were allowed to grow to a significant size (day 0), at which point expression of the siRNA construct was shut down for 4 days, permitting p53 function to return; p53 expression was then shut down once again for the remaining time of the experiment. By day 6, tumors were markedly smaller, and by day 11, they were almost undetectable. Because the tumors continued to shrink long after p53 expression was suppressed, the brief (4-day-long) period of p53 expression seems to have forced tumor cells to enter irreversibly into the post-mitotic senescence state, and the senescent cells were ostensibly cleared by cells of the innate immune system, such as the macrophages described in Chapter 15. Tumors were imaged *in vivo* through light released by a luciferase gene borne by the tumor cells. (A, courtesy of C. Scheel. B, from S. Brookes et al., *Exp. Cell Res.* 298:549–559, 2004. With permission from Elsevier. C, from R. Zhang, W. Chen, and P.D. Adams, *Mol. Cell Biol.* 27:2343–2358, 2007. With permission from American Society for Microbiology. D, from A. Melk et al., *Kidney Int.* 63:2134–2143, 2003. With permission from International Society of Nephrology. E, from W. Xue et al., *Nature* 445:656–660, 2007. With permission from Nature.)

SAβ-gal activity in aged mice and humans (Figure 9.17D). However, it remains to be determined whether senescence associated with aging, telomere shortening, or chronic DNA damage is an effective anti-tumor mechanism in humans. In fact, there is a growing body of research literature demonstrating that an increase in inflammation, such as that caused by the SASP response, can actually create a pro-tumorigenic microenvironment (see Section 11.15). As such, in some settings, senescent cells in aging animals may actually *promote* cancer development.

Cellular senescence can also be induced in response to inappropriate oncogene expression, as first demonstrated by experiments examining the effects of introducing multiple copies of a *ras* oncogene into wild-type fibroblasts. Rather than driving unrestrained proliferation, the expression of this potent oncogene was found to cause the cells to undergo irreversible cell cycle arrest, accompanied by a rise in SA-β-gal expression and other markers of senescence. The central role of p53 in this response was shown by the finding that oncogenic Ras expression in p53-deficient fibroblasts failed to induce senescence and led instead to cellular transformation. Hence, a *ras* oncogene could collaborate with a mutant *p53* gene in generating transformed cells, illustrating the ability of multiple mutant genes within a cell genome to create neoplastic cells, a theme that we will explore in detail in Chapter 11.

Examples of **o**ncogene-**i**nduced **s**enescence (OIS) have been shown to limit the further development of the pre-neoplastic lesions in animal models as well as in human tissues. For example, in humans, melanomas can arise at low frequency from benign tumors called **nevi** (or moles), which can otherwise remain in a growth-arrested state for decades. Nevi frequently harbor mutations in the B-Raf oncogene but lack the additional mutations that are required for full melanoma development. These pre-neoplastic lesions have been shown to upregulate expression of the *p16^Ink4a^* gene and to induce SA-β-gal activity; expression of the oncogenic B-Raf protein in normal human or mouse melanocytes in culture likewise causes cell cycle arrest and multiple markers of senescence. Finally, using genetically engineered mouse models of cancer in which *p53* alleles can be switched on and off, researchers have observed that restoring p53 function can induce cellular senescence and profound tumor regression in certain mouse models of tumor development, the latter resulting from the actions of immune cells in clearing the senescent cancer cells created by the restored p53 function (Figure 9.17E). These data strongly support the notion that OIS is a hardwired cellular response mechanism designed to eliminate cells that have activated certain oncogenes. Accordingly, OIS can operate as an important tumor-suppressive mechanism and the accumulation of inactivating mutations in *p53* and in genes specifying other mediators of the senescence program may be required in order for incipient cancer cells to circumvent this fate.

9.12 The apoptosis program participates in normal tissue development and maintenance

We have introduced the role of p53 in inducing apoptosis as a protective mechanism for eliminating damaged or pre-cancerous cells. However, this cell suicide program has a larger role in the organism beyond the removal of cells that teeter on the brink of malignancy. Apoptosis is used routinely during normal morphogenesis to chisel away unneeded cell populations during the sculpting of well-formed, functional tissues and organs (**Figure 9.18**). Mice that have been genetically deprived of various key components of the apoptotic machinery show a characteristic set of developmental defects, including excess neurons in the brain, facial abnormalities, delayed removal of the webbing between digits, and abnormalities in the palate and lens. In the case of the developing brain, more than half of all initially formed neurons are normally eliminated by apoptosis during formation of various neural circuits.

Apoptosis also plays an important role in normal adult tissue physiology. In the small intestine, for example, epithelial cells are continually being eliminated by apoptosis after a four- or five-day journey from the bottom of intestinal crypts to the tips of the villi that protrude into the lumen (see Section 7.10). During the development of red blood cells (the process of erythropoiesis), more than 95% of the erythroblasts—the precursors of mature red cells—are destroyed as part of the routine operations of the bone marrow. However, in the event that the rate of oxygen transport by the blood falls below a certain threshold level—as a result of hemorrhage, various types of anemia, or low oxygen tension in the surrounding air—a rapid rise in the concentration of the blood-forming hormone erythropoietin (EPO; see Section 5.7) blocks apoptosis of these erythrocyte precursors, enabling their maturation into functional

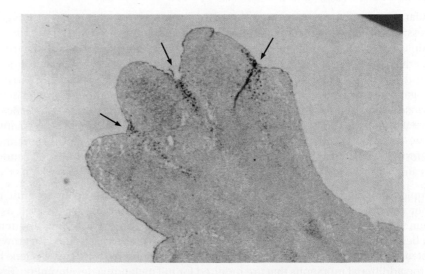

Figure 9.18 Apoptosis and normal morphogenesis The webs of tissue between the future toes of a mouse paw are still in evidence in this embryonic paw. They were preferentially labeled, as indicated by the numerous dark dots (*arrows*), using the TUNEL assay (see Supplementary Sidebar 9.3). As in humans, the apoptosis of cells forming the webs ultimately results in the formation of digits that are joined by webs only near the palm. (Courtesy of Z. Zakeri.)

red blood cells. This mechanism yields a rapid increase in the levels of these cells in the blood.

The regression of the cells in the mammary gland following the weaning of offspring is a particularly dramatic example of the contribution of apoptosis to normal physiology. As many as 90% of the epithelial cells in the mammary gland, which have accumulated in large numbers during pregnancy to produce milk for the newborn, die via apoptosis during this regression, which is usually termed **involution**.

In a more general sense, apoptosis is used to maintain appropriate numbers of different cell types in a wide variety of human tissues. The importance of this process is indicated by the turnover of cells—the number that are newly formed and the equal number that are eliminated in each year of our lives—which exceeds by a factor of ~4 the total number of cells ($\sim 3 \times 10^{13}$) present in the adult body at any one time. Most of these discarded cells appear to be eliminated by apoptosis.

As observed through the microscope, the cellular changes that constitute the apoptotic program proceed according to a precisely coordinated schedule. Within minutes, patches of the plasma membrane herniate to form structures known as **blebs**; indeed, in time-lapse movies, the cell surface appears to be boiling. The nucleus collapses into a dense structure—a state termed **pyknosis**—and fragments, as the chromosomal DNA is cleaved into small segments. Ultimately, usually within an hour, the apoptotic cell breaks up into small pieces, sometimes called *apoptotic bodies*, which are rapidly ingested by neighboring cells in the tissue or by itinerant macrophages, thereby removing all traces of what had recently been a living cell. These changes are illustrated in **Figure 9.19**.

The phagocytosis by neighboring cells and macrophages depends on the display of a specialized "eat-me" signal on the surfaces of cells entering apoptosis. Normal cells contain phosphatidylserine (PS) molecules only in the inner leaflet (facing the cytoplasm) of the lipid bilayer forming the plasma membrane. This asymmetric distribution is lost, however, during the fragmentation of apoptosis. As a result, PS is displayed on the exterior face of an apoptotic cell, where it is recognized by a specialized **annexin** receptor on the surface of phagocytic cells. The resulting binding of neighboring cells or specialized phagocytes (that is, macrophages) to the apoptotic cell surface triggers phagocytosis. A related annexin protein, when coupled to a fluorescent dye, can be used to stain cells that have initiated apoptosis and flipped many of their PS molecules to the outer cell surface. This approach—and other procedures that can be used to detect apoptotic cells within a tissue or in culture—are outlined in Supplementary Sidebar 9.3.

9.13 Apoptosis is a complex biochemical program that often depends on mitochondria

Although the apoptotic program had been recognized as a normal biological phenomenon in animal tissues by nineteenth-century histologists, it was rediscovered and described with far greater precision in 1972. Prior to this rediscovery, cells in metazoan tissues were thought to be eliminated solely by **necrosis**. As indicated in **Table 9.2**, necrosis and apoptosis are actually quite different. By the late 1980s, genetic research on the worm *Caenorhabditis elegans* revealed that apoptosis is exploited to eliminate various cell types as part of the normal developmental program of these tiny animals (see also Figure 10.1). This research led to the recognition that apoptosis is a basic, genetically driven biological process that is common to all metazoa.

The first indication that cancer-associated proteins contribute to the regulation of apoptosis came from an exploration of the functioning of the *Bcl-2* (B-cell lymphoma gene-2) oncogene. The oncogenic version of the *Bcl-2* gene is formed through reciprocal chromosomal translocations occurring during the formation of human follicular B-cell lymphomas. These events—which result in the exchange of portions of the arms of human Chromosomes 14 and 18 observed in these tumors—place the reading frame of the *Bcl-2* gene under the control of a promoter that drives its high, constitutive expression.

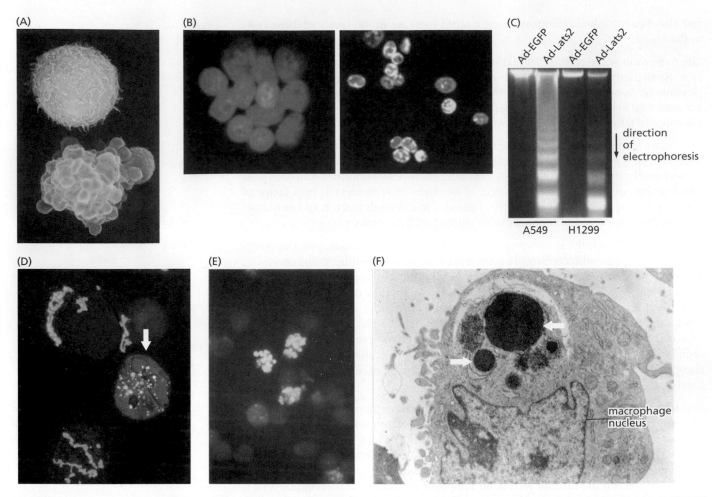

Figure 9.19 Diverse manifestations of the apoptotic program
(A) The upper of these two lymphocytes, visualized by scanning electron microscopy (SEM), is healthy, whereas the lower one has entered into apoptosis, resulting in the numerous blebs protruding from its surface. (B) HeLa cells—a line of human cervical carcinoma cells—have been fused, forming a syncytium (*left*). Prior to fusion, some of these cells were treated with staurosporine, an apoptosis-inducing drug. As a consequence, the nuclei of these cells, which are normally quite large (*left*), have undergone *pyknosis* (*right*), a process that accompanies apoptosis and involves condensation of chromatin (*white*) and collapse of nuclear structure. Fragmentation of nuclei (*not shown*) typically follows soon thereafter. (C) Apoptosis causes the cleavage of a cell's chromosomal DNA into low–molecular-weight fragments that appear as a "DNA ladder" on gel electrophoresis. Cell lines A549 and H1299 were treated with the pro-apoptotic protein Lats2 (*2nd and 4th lanes*), which was introduced via an adenovirus (Ad) vector; cells infected with the vector alone (Ad-EGFP) were included as controls (*1st and 3rd lanes*). (D) The Golgi bodies (*green*) are usually found in perinuclear locations in a normal cell (*upper left*); in an apoptotic cell (*arrow*),

the Golgi bodies have become fragmented. Chromatin is stained *blue*, and an antibody specific for cleaved PARP [poly(ADP-ribose) polymerase], which is degraded during apoptosis, reveals this form of the protein (*red*) in the nucleus of an apoptotic cell. (E) Here, an antibody that specifically reacts with histone 2B molecules (in the chromatin) that are phosphorylated on their serine 14 residues during apoptosis was used to stain apoptotic nuclei (which are already undergoing fragmentation). (F) The end result of apoptosis is the phagocytosis of apoptotic bodies—the fragmented remains of apoptotic cells—by neighboring cells or by macrophages. In this image, the pyknotic nuclear fragments of a phagocytosed apoptotic cell are seen above (*white arrows*) and contrast with the normal nucleus of the phagocytosing macrophage (*below*). (A, courtesy of K.G. Murti. B, from K. Andreau et al., *J. Cell Sci.* 117:5643–5653, 2004. With permission of The Company of Biologists. C, from H. Ke et al., *Exp. Cell Res.* 298:329–338, 2004. With permission from Elsevier. D, from J.D. Lane et al., *J. Cell Biol.* 156:495–509, 2002. With permission from Rockefeller University Press. E, from W.L. Cheung et al., *Cell* 113:507–517, 2003. With permission from Elsevier. F, courtesy of G.I. Evan.)

When such a *Bcl-2* oncogene was inserted into the germ line of mice under conditions that ensured its expression in lymphocyte precursor cells, there was no observable effect on the animals' long-term survival (**Figure 9.20A and B**). In contrast, expression of an oncogenic *Myc* transgene in these cells led to lymphomas and to the death of half the mice within two months of birth. However, when both oncogenes were expressed simultaneously (by breeding *Bcl-2* transgenic mice with *Myc* transgenic mice), the off-spring fared even worse: virtually all of the mice died less than two months after birth.

The inability of *Bcl-2*, on its own, to trigger tumor formation argued against its acting as a typical oncogene, such as *Myc*, which emits potent growth-promoting signals.

Table 9.2 Apoptosis versus necrosis

	Apoptosis	Necrosis
Provoking stimuli		
	programmed tissue remodeling	metabolic stresses
	maintenance of cell pool size	absence of nutrients
	genomic damage	changes in pH, temperature
	metabolic derangement	hypoxia, anoxia
	hypoxia	
	imbalances in signaling pathways	
Morphological changes		
Affected cells	individual cells	groups of cells
Cell volume	decreased	increased
Chromatin	condensed	fragmented
Lysosomes	unaffected	abnormal
Mitochondria	morphologically normal initially	morphologically aberrant
Inflammatory response	none	marked
Cell fate	apoptotic bodies consumed by neighboring cells	lysis
Molecular changes		
Gene activity	required for program	not needed
Chromosomal DNA	cleaved at specific sites	random cleavage
Intracellular calcium	increased	unaffected
Ion pumps	continue to function	lost

Adapted from R.J.B. King, Cancer Biology, 2nd ed. Harlow, UK: Pearson Education, 2000.

Careful study of the lymphocytes in mice carrying only the *Bcl-2* transgene indicated that the effects of this gene on cells were actually quite different from those of *Myc* or *Ras*: *Bcl-2 prolonged the lives* of lymphocytes that were otherwise destined to die rapidly. Indeed, when B lymphoid cells from these transgenic mice were cultured *in vitro*, they showed a remarkable extension of life span. *In vivo*, the lymphocytes in which *Bcl-2* was being expressed accumulated in amounts several-fold above normal; however, these cells were not actively proliferating, which explains the absence of hematopoietic tumors in mice carrying only the *Bcl-2* transgene.

The *Myc* oncogene, when present on its own, acted as a potent mitogen, but its growth-stimulatory powers were attenuated by its apoptosis-inducing effects (see Section 9.9). However, when *Myc* and *Bcl-2* acted collaboratively, they created an aggressive malignancy of the B-lymphocyte lineage; *Myc* would drive rapid cell proliferation, and its accompanying death-inducing effects were neutralized by the life-prolonging actions of *Bcl-2*. Because these were early experiments, even more dramatic examples of synergy between *Myc* and *Bcl-2*-like oncogenes, such as that shown in Figure 9.20C, have been reported. These dynamics parallel the situation we discussed in Section 9.9, in which the death-inducing effects of *Myc* were blunted by inactivation of the *p53* gene.

Once the details of the apoptotic program were clearly understood, the role of Bcl-2 in cancer pathogenesis became apparent. This oncoprotein operates as an anti-apoptotic agent rather than as a mitogen. Hence, Bcl-2 acts in a fashion precisely opposite to

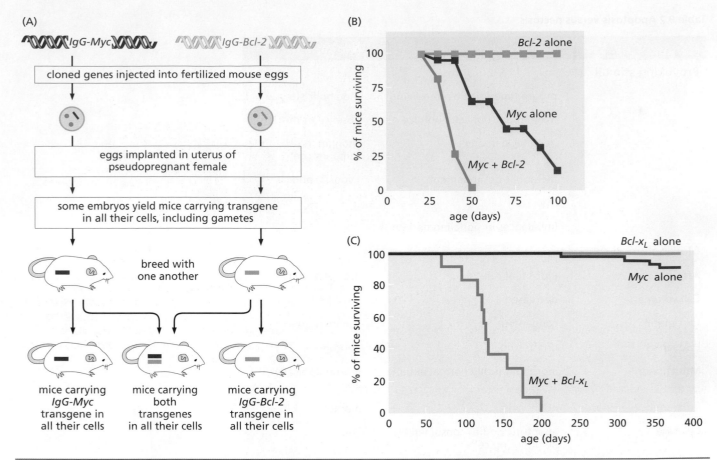

Figure 9.20 Synergy of *Bcl-2*-like genes with *Myc* (A) Clones of the *Myc* and the *Bcl-2* genes were engineered to be expressed under the control of an antibody gene transcriptional promoter (*IgG*). This promoter ensures that both genes will be expressed specifically in cells of the B lymphocyte lineage. The engineered clones are then injected into eggs and, with a certain frequency, the DNA becomes integrated via nonhomologous recombination into a chromosomal site. Southern blotting was subsequently used to verify the presence of the integrated sequences in the cells of the resulting mice and in their progeny. Such a heritable inserted gene is termed a *transgene*. (B) Mice bearing the *IgG–Bcl-2* transgene experienced no increased mortality (*green squares*). In contrast, the *IgG-Myc* transgene led to a greatly increased mortality from lymphomas, with almost all of the mice dying from this disease by 100 days of age (*red squares*). However, when both *Bcl-2* and *Myc* transgenes are present, lymphomagenesis was greatly accelerated and virtually all mice succumbed to lymphomas by 50 days of age (*blue squares*). (C) An even more dramatic outcome was observed when two strains of transgenic mice were developed that expressed the *Myc* or the *Bcl-x*$_L$ oncogenes specifically in the plasma cell (antibody-secreting) lymphocytes of the immune system. (*Bcl-x*$_L$ is a cousin of *Bcl-2* and functions in a very similar fashion.) Normal mice and transgenic mice carrying only the *Bcl-x*$_L$ transgene all survived for more than a year without any tumor development (*green line*); similarly, mice expressing a *Myc* transgene were largely healthy until about 10 months of age, when a small number of them developed plasma cell tumors. However, when the two strains of mice were bred, the double-transgenic progeny developed plasma cell tumors rapidly, beginning at about 75 days of age; all died from these tumors by 200 days of age. (B, adapted from A. Strasser et al., *Nature* 348:331–333, 1990. C, from W.C. Cheung et al., *J. Clin. Invest.* 113:1763–1773, 2004. With permission from American Society for Clinical Investigation.)

that of p53. Whereas p53 promotes apoptosis, Bcl-2 blocks it. Stated in genetic terms, activation of the *Bcl-2* oncogene and inactivation of the *p53* tumor suppressor gene confer similar (but not identical) benefits on cancer cells, in that both reduce the likelihood of inducing the apoptotic program.

Careful biochemical and cell-biological sleuthing eventually revealed the surprising site of Bcl-2 action: the outer membrane of the mitochondrion (**Figure 9.21**). At first, this discovery made little sense. Mitochondria were thought to be specialized for metabolic tasks, namely, generating energy in the form of ATP for the cell. As a by-product of this energy production, mitochondria were also known to make metabolites of glucose that are used in the biosynthesis of some of the cell's diverse biochemical species.

Soon, however, the role of mitochondria in the apoptotic program was clarified. Cytochrome *c*, the central actor in this process, normally resides in the space between the inner and outer mitochondrial membranes (see Figure 9.21B), where it functions to transfer electrons as part of **oxidative phosphorylation**, the main energy-generating

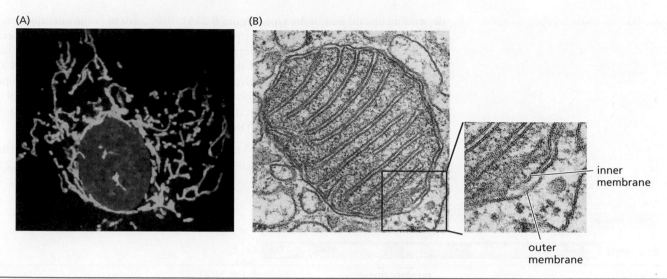

Figure 9.21 Mitochondria (A) Mitochondria (*green*), stained here in the cytoplasm of a human liver cell, were once regarded only as the sites of oxidative phosphorylation and biosynthesis of certain key metabolites. The discovery of their ability to regulate apoptosis through cytochrome *c* release forced a major rethinking of their role in the life of a cell. (B) Cytochrome *c* is stored in the space between the outer mitochondrial membrane and the inner membrane, where it normally plays a key role in electron transport. When this outer membrane is breached (a condition called mitochondrial outer membrane permeabilization, or MOMP), cytochrome *c* escapes into the cytosol. (A, courtesy of Althea Bock-Hughes and Kay Macleod. B, courtesy of D.S. Friend.)

reaction in our cells. However, when certain signals trigger the initiation of apoptosis, pores form in the mitochondrial outer membrane creating a condition known as MOMP (**m**itochondrial **o**uter **m**embrane **p**ermeabilization). Cytochrome *c* then spills out of the organelle and into the surrounding cytosol (**Figure 9.22**). Once present in the cytosol, cytochrome *c* associates with other proteins to trigger a cascade of events that together yield apoptotic death, as discussed below. Smac/DIABLO, another mitochondrial protein that escapes during MOMP, also has a direct role in the induction of apoptosis. Yet other proteins are released from mitochondria during apoptosis, but the evidence supporting their role in regulating the process is less clear.

We now know that the Bcl-2 protein is a member of a large and complex family of proteins, which consists of at least 24 members expressed in different patterns

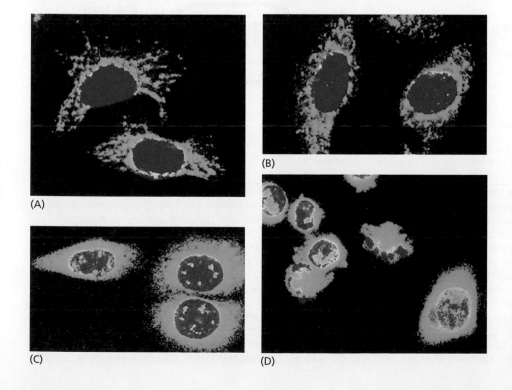

Figure 9.22 Release of cytochrome *c* from mitochondria into the cytosol The presence of cytochrome *c* can be detected by staining with a specific fluorescence-labeled antibody (*green*), which contrasts with the dye-labeled nuclei (*red orange*). (A) Initially, the distribution of cytochrome *c* coincides with the distribution of mitochondria in the cytoplasm (see Figure 9.21A). As apoptosis proceeds, however (B, C, D), cytochrome *c* staining becomes increasingly uniform in the cytoplasm as this protein is released from the mitochondria into the cytosol. At the same time, some nuclei give evidence of the fragmentation that is associated with apoptosis (C, D). (From E. Bossy-Wetzel et al., *EMBO J.* 17:37–49, 1998. With permission from John Wiley & Sons.)

depending on cell and tissue type (**Figure 9.23A**). They have in common one or more protein domains, which are referred to as **B**cl-2 **h**omology (or BH) domains. By acting through a series of physical associations, the Bcl-2 family of proteins collectively controls the formation and function of the mitochondrial outer membrane pores and therefore MOMP. The pores themselves contain homo-oligomers of two of the pro-apoptotic members of the Bcl-2 family: Bax and Bak. Anti-apoptotic members of the family—such as Bcl-2, Bcl-X$_L$, and Mcl1—form heterodimers with these pore-forming proteins, resulting in inhibition of MOMP (Figure 9.23B).

By contrast, the so-called "BH3-only" members of the family (so named because they carry just one of the four conserved BH domains present in Bcl-2; see Figure 9.23A)

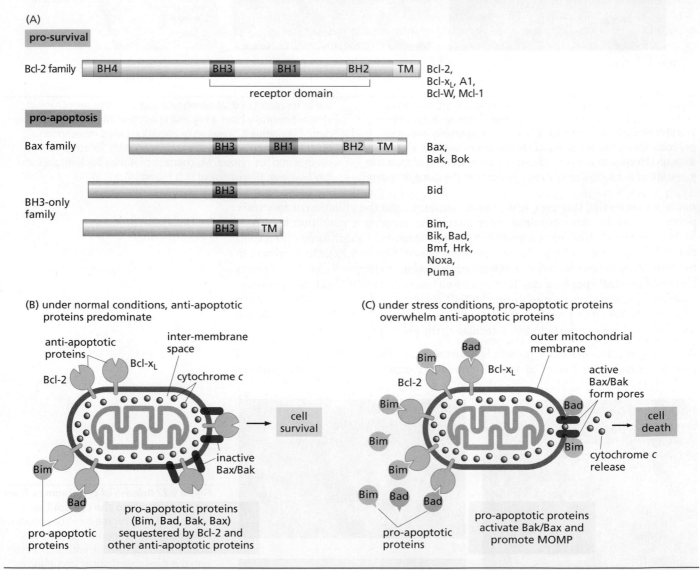

Figure 9.23 Bcl-2 and related proteins and the control of apoptosis. (A) Bcl-2 is a member of a large family of proteins involved both in preventing (*top*) and promoting (*bottom*) apoptosis. Each of these proteins possesses one or more Bcl-2-homology (BH) domains. All have a BH3 domain (*purple*), and some, like Bcl-2 itself, also have BH1 and BH2 domains (*red, aquamarine*). Several also have transmembrane (TM) domains (*light green*), used for membrane anchorage, and BH4 domains (*darker green*), which inhibit apoptosis, in part by blocking release of calcium ions from the endoplasmic reticulum. The pro-apoptotic proteins are subgrouped into the "Bax" family and have several domains that are homologous to domains of Bcl-2; in contrast, the "BH3-only" proteins have only the BH3 domain in common with Bcl-2. Members of the Bax family normally reside in inactive form in the outer mitochondrial membrane or in the cytosol, whereas inactive members of the BH3-only family normally are limited to the cytosol; both classes of proteins are activated and/or translocated to the mitochondria by pro-apoptotic signals. (B) When anti-apoptotic proteins, such as Bcl-2, are present in sufficient concentrations, they can sequester pro-apoptotic BH3-only proteins and thereby inhibit apoptosis. (C) When pro-apoptotic BH3-only proteins are in relative abundance, they stimulate the homo-oligomerization of the pore-forming proteins Bax and Bak. This association promotes MOMP and thereby triggers apoptosis. (A, from S. Cory and J.M. Adams, *Nat. Rev. Cancer* 2:647–656, 2002. With permission from Nature. B and C, adapted from R. Singh, A. Letai, and K. Sarosiek, *Nat. Rev. Mol. Cell Biol.* 20: 175–193, 2019.)

promote MOMP, and thus apoptosis. The BH-3 only proteins—which include Bid, Bim, Noxa, Bad, and PUMA—act either by binding to the pore-forming proteins directly and stimulating homo-oligomerization or by binding to and inhibiting anti-apoptotic Bcl-2 family members (Figure 9.23C). Interestingly, several pro- and anti-apoptotic members of the Bcl-2 family of proteins also associate with the endoplasmic reticulum, where they modulate release of calcium ions from this organelle, which also affects regulation of apoptosis.

The relative levels of the pro- and anti-apoptotic proteins ultimately determine whether MOMP will occur, cytochrome *c* will spill out, and apoptosis will ensue (see Figure 9.23B, C). In this fashion, MOMP determines the life and death of a cell. For example, mice that have been deprived of both germ-line copies of the anti-apoptotic *Bcl-2* gene suffer from kidney failure and immune collapse as a result of widespread apoptotic death of cells in these tissues; these lethal phenotypes can be avoided through the additional deletion from their germ line of just one copy of *Bim*, one of the pro-apoptotic cousins of *Bcl-2* (**Figure 9.24**). These results illustrate how, in tissues like these, the delicate balance between pro- and anti-apoptotic Bcl-2-like proteins governs the fate of individual cells.

A large number of cell-physiological stimuli and signaling pathways converge on the cell death decision, and this complexity is reflected in the plethora of Bcl-2-related proteins that control this process. As an example, the form of apoptosis called **anoikis**, which occurs when epithelial cells undergo detachment from a substratum (also termed loss of anchorage), is dependent on the BH3-only protein Bim. By contrast, p53 induces the expression of the BH3-only proteins PUMA and Noxa, whereas other stimuli act by regulating distinct members of the family (**Figure 9.25A**).

In addition, the various pro- and anti-apoptotic members of the Bcl-2 family of proteins each has its own opposing partner or partners (Figure 9.25C). Each of these proteins also functions to trigger apoptosis in specific cell types; for example, Figure 9.25D shows the key role that Bim-dependent anoikis plays in eliminating cells in the developing mammary gland, allowing lumina to form in the ducts. Finally, these Bcl-2–related proteins can be regulated in different, complex ways: the actions of the BH3-only proteins of the Bcl-2 family (see Figure 9.23A), for example, can be controlled by modulating their expression, their intracellular localization, proteolytic cleavage, or phosphorylation (Figure 9.25B).

Once individual cytochrome *c* molecules have leaked out of mitochondria into the cytoplasm, they associate with seven copies of the Apaf-1 protein with which they form the structure called the **apoptosome** (**Figure 9.26** and **Figure 9.27**). The resulting apoptosome complexes then proceed to activate a normally latent cytoplasmic protease termed procaspase 9, converting it into active caspase 9, which has catalytic activity that is at least one thousand-fold higher than the procaspase (see Figure 9.27).

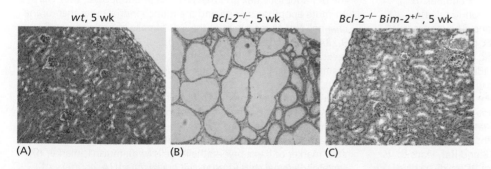

wt, 5 wk	*Bcl-2$^{-/-}$*, 5 wk	*Bcl-2$^{-/-}$ Bim-2$^{+/-}$*, 5 wk

(A)   (B)   (C)

Figure 9.24 The delicate balance between pro- and anti-apoptotic proteins (A) The wild-type mouse kidney is a well-consolidated tissue at 5 weeks of age. (B) However, deletion of both germ-line copies of the anti-apoptotic *Bcl-2* gene (resulting in the *Bcl-2$^{-/-}$* genotype) results in widespread apoptosis and polycystic kidney disease at this age. (C) These effects can be largely reversed if one copy of the pro-apoptotic *Bim* gene is also deleted from the germ line (yielding the *Bcl-2$^{-/-}$ Bim$^{+/-}$* genotype). (From P. Bouillet et al., *Dev. Cell* 1:645–654, 2001. With permission from Elsevier.)

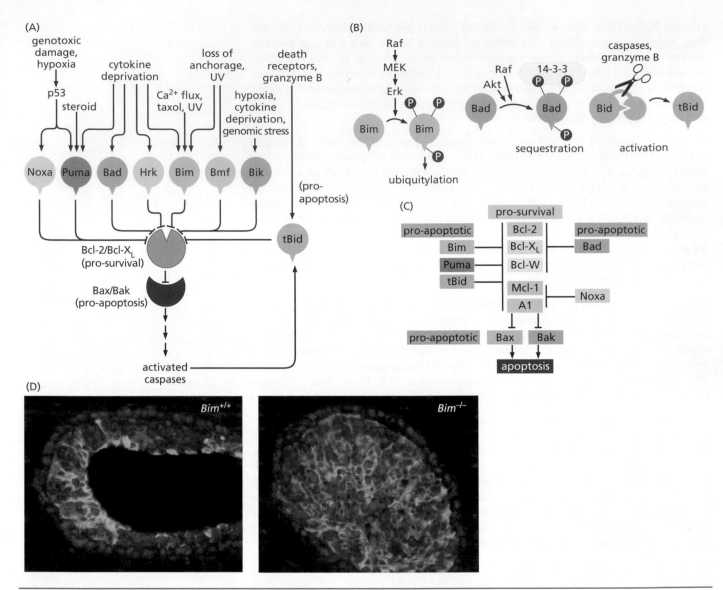

Figure 9.25 Pro-apoptotic signals acting through various Bcl-2–related proteins (A) The rationale for the multiplicity of Bcl-2–like proteins is that various cell-physiological stresses operate through different pro-apoptotic proteins to antagonize anti-apoptotic proteins such as Bcl-2 and Bcl-X$_L$ (*green*). The latter are thereby prevented from neutralizing Bax and Bak (*red*), the dominant pro-apoptotic proteins that function to release cytochrome *c* from the mitochondrial membrane. The caspases that are activated downstream in the apoptotic pathway (see Figure 9.27) can also activate the pro-apoptotic function of Bid by proteolysis (cleaving it to truncated Bid, or tBid). This action amplifies the apoptotic response, as indicated in panel B. (B) The activities of pro-apoptotic BH3-only proteins are controlled in multiple ways. Bim is regulated at the transcriptional level (*not shown*) and post-translationally via phosphorylation by Erk, which results in its ubiquitylation and degradation. Phosphorylation of Bad by anti-apoptotic kinases, such as Akt/PKB and Raf, leads to its sequestration by the 14-3-3 protein and loss of its ability to bind the anti-apoptotic Bcl-2 and Bcl-X$_L$ proteins. The pro-apoptotic Bid protein is liberated from its inhibitory domain by the actions of proteases, such as caspases that are activated downstream in the apoptotic cascade and the granzymes released by cytotoxic cells, the latter described in Chapter 15. (C) Each member of the Bcl-2 family of proteins has its own set of opposing partners.

At the center of this network lie the anti-apoptotic (i.e. pro-survival) proteins Bcl-2, Bcl-X$_L$, Bcl-W, Mcl-1, and A1 (*green rectangles*). Each has its own set of antagonistic, pro-apoptotic partners (*pink rectangles*). For example, Bcl-2, Bcl-X$_L$, and Bcl-W are opposed by the pro-apoptotic Bad protein. At the bottom of this network lie the pro-apoptotic Bax and Bak proteins; if unopposed, they will promote apoptosis by opening the mitochondrial outer membrane channel. Not all members of the Bcl-2 family of proteins (see Figure 9.23A) are indicated here. (D) In wild-type (*Bim*$^{+/+}$) mice, a cavity is excavated in the solid tip (terminal end bud) of an elongating mammary duct in order to form the lumen of the future duct (*left*). However, in *Bim*$^{-/-}$ mice, mammary epithelial cells persist, preventing normal lumen formation (*right*) and suggesting that lumen formation normally depends on anoikis of cells that failed to establish anchorage to the basement membrane that lies immediately outside of the myoepithelial cells (*not visualized*). p63, marker of basal myoepithelial cells (*red*); mucin1, marker of secretory luminal cells (*green*); cell nuclei (*blue*). (A, courtesy of J. Adams; adapted from S. Cory et al., *Oncogene* 22:8590–8607, 2003. With permission from Nature. B, adapted from S. Cory and J.M. Adams, *Nat. Rev. Cancer* 2:647–656, 2002. C, courtesy of D.C.S. Huang. D, from A.A. Mailleux et al., *Dev. Cell* 12:221–234, 2007. With permission from Elsevier.)

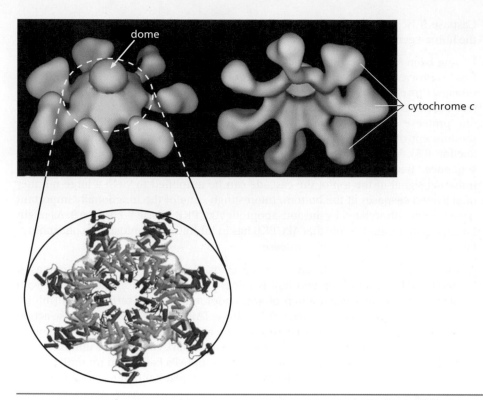

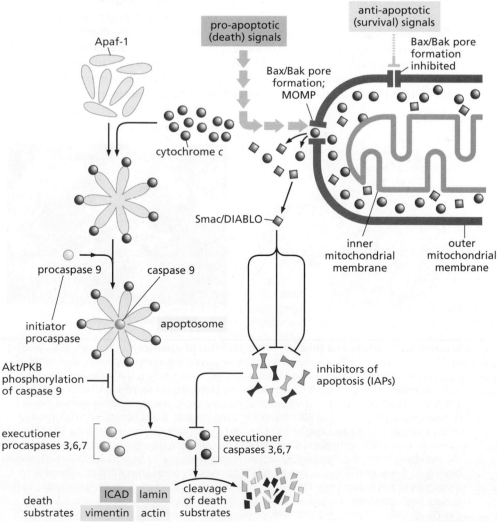

Figure 9.26 The wheel of death The apoptosome—essentially a wheel of death—is assembled in the cytosol when cytochrome *c* molecules are released from mitochondria and associate with Apaf-1 monomers. Cytochrome *c* molecules and Apaf-1 monomers assemble to form the assembly of a seven-spoked wheel, in which Apaf-1 monomers form the spokes and cytochrome *c* molecules form the tips of the spokes (*above*), visualized here by electron cryomicroscopy. Procaspase 9 is attracted to the hub of the wheel ("dome"), where it is converted into active caspase 9. The more detailed structure of the inner circle of the apoptosome (*below*), formed from seven Apaf-1 subunits, is illustrated in the molecular model determined by X-ray crystallography. The *blue* helices in the center interact with procaspase 9 to convert it into caspase 9, which then triggers the apoptotic cascade. (Top, from D. Acehan et al., *Mol. Cell* 9:423–432, 2002. With permission from Elsevier. Bottom, from S.J. Riedl et al., *Nature* 434:926–933, 2005. With permission from Nature.)

Figure 9.27 The apoptotic caspase cascade Various anti-apoptotic signals promote MOMP (see Figure 9.23). Important mitochondrial molecules that may be released into the cytosol by MOMP are cytochrome *c* (*red circles*) and Smac/DIABLO (*brown squares*). Cytochrome *c* molecules proceed to aggregate with Apaf-1 to form the seven-spoked apoptosome (*left*; see Figure 9.26). Once assembled, the apoptosome attracts and converts procaspase 9 into active caspase 9 (*green circle*), which in turn cleaves and activates procaspases 3, 6, and 7 (*bottom left*), converting them into executioner caspases, which then cleave various "death substrates" whose products create the apoptotic cell phenotype. Normally, a number of inhibitors of apoptosis (IAPs; *lower right*) attach to and inactivate caspases. However, the Smac/DIABLO that is also released from mitochondria during the triggering of apoptosis antagonizes these IAPs, thereby protecting caspases from IAP inhibition and setting the stage for full activation of the apoptosis program. Acting in the opposite direction, as part of its anti-apoptotic actions, the Akt/PKB kinase can phosphorylate and thereby inactivate already-activated caspase 9 (*lower left*).

Caspase 9 is just one member of a family of **c**ysteine **asp**artyl–specific prote**ases**; the human genome carries 12 genes that encode members of this protease family.

Having been converted into an active protease by the apoptosome, caspase 9 then cleaves procaspase 3, thereby activating yet another related protease; caspase 3 then proceeds to cleave and activate yet another procaspase, and so on down the line (see Figure 9.27). This sequence of cleavages constitutes a signaling cascade in which one protease activates the next one in the series by cleaving it. Such behavior is reminiscent of the organization of kinase cascades that we read about earlier (see Section 6.5), in which a group of Ser/Thr kinases activate one another in a linear sequence. Because each of these caspases acts catalytically, a relatively minor initiating signal at the top of the cascade can be amplified to yield a large number of activated caspases at the bottom. Interestingly, one of the functionally important proteins phosphorylated by the anti-apoptotic Akt/PKB kinase is caspase 9. As might be anticipated from the role that Akt/PKB has in inhibiting apoptosis, the phosphorylation of caspase 9 inhibits this protease.

At the same time as cytochrome *c* in the cytosol serves to activate the caspases, Smac/DIABLO—another protein that is released from mitochondria together with cytochrome *c*—inactivates a group of anti-apoptotic proteins termed IAPs (inhibitors of apoptosis proteins; see Figure 9.27). These IAPs normally block caspase action in two ways: first, they bind to caspases directly and thereby inhibit their proteolytic activity (**Figure 9.28**); in certain cases, they can also mark caspases for ubiquitylation and degradation. Once the restraining influences of IAPs have been minimized, caspases are free to initiate the proteolytic cleavages that result ultimately in apoptosis.

The cascade of caspase activations, once initiated, proceeds until the final caspases in the cascade cleave "death substrates," that is, proteins whose degradation creates the diverse cellular changes that one associates with the apoptotic death program (see Figure 9.27). The specialized roles of these various caspases have led to their classification into two functional groups: the **initiator caspases**, which trigger the onset of apoptosis by activating the caspase cascade, and the downstream **executioner caspases**, which undertake the actual work of destroying critical components of the cell.

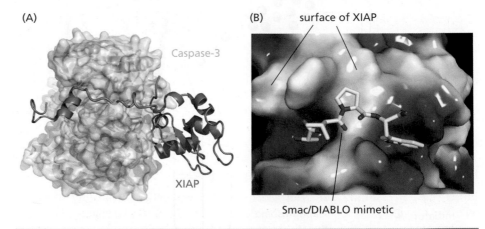

Figure 9.28 Structure and function of an inhibitor of apoptosis IAPs function by binding and inhibiting various caspase molecules, thereby preventing inadvertent activation of the caspase cascade and the triggering of apoptosis. (A) Shown here is the molecular structure of the complex formed by the BIR3 domain of the IAP molecule termed XIAP with caspase 3. The peptide backbone of XIAP is shown (*blue*) in contrast to the space-filling surface model of caspase 3 (*light green*). The BIR3 domain of XIAP binds to the substrate-binding groove of the caspase, thereby blocking its function. (B) Like the naturally occurring Smac/DIABLO molecule, chemically synthesized Smac/DIABLO mimetics can promote apoptosis by keeping IAP molecules from inhibiting caspase action. Here a tetrapeptide Smac/DIABLO mimic is shown binding to the groove in the surface of XIAP that normally binds Smac/DIABLO. Chemical derivatives of this tetrapeptide function as potent apoptosis-inducing drugs. Negatively charged surface (*red*); positively charged surface (*blue*). (A, from P.D. Mace, S. Shirley and C.L. Day, *Cell Death Differ.* 17:46–53, 2010. With permission from Nature. B, from K. Zobel et al., *ACS Chem. Biol.* 1:525–533, 2006. With permission from American Chemical Society.)

A number of key cellular components are cleaved by caspases 3, 6, and 7, which function as the executioner caspases. The degradation of these components causes the profound morphological transformations that accompany the death throes of apoptotic cells. Cleavage of lamins on the inner surface of the nuclear membrane is involved in some fashion in the observed chromatin condensation and nuclear shrinkage, or pyknosis, that is characteristic of the apoptotic program. Cleavage of the inhibitor of caspase-activated DNase (ICAD) liberates this nuclease, which then fragments the chromosomal DNA. Caspase-mediated cleavage of cytoskeletal proteins such as actin, plectin, vimentin, and gelsolin leads to collapse of the cytoskeleton, the formation of blebs protruding from the plasma membrane, and the production of apoptotic bodies—the condensed hulks of cells that remain after all this destruction has occurred (see Figure 9.19).

The executioner caspases even extend their reach back to the mitochondria, where the apoptotic program is initiated: one of their substrates is a protein that is part of the mitochondrial electron transport chain. Its cleavage by caspases leads to disruption of electron transport, loss of ATP production, release of reactive oxygen species (ROS) from the mitochondria, and the loss of mitochondrial structural integrity, thereby amplifying an already-initiated apoptotic program. In addition, as cytochrome c and other components of the mitochondrial intermembrane space leak out into the cytoplasm, a variety of mitochondrial functions are severely compromised.

The efficient execution of many of the steps of the apoptotic program is also enabled by changes in signaling that further amplify the signals initiating apoptosis. For example, cleavage of the Rel subunit of NF-κB—a transcription factor driving expression of anti-apoptotic genes such as *Bcl-2* (see Figure 6.22A) —shifts the regulatory balance further in favor of apoptosis. Some researchers estimate that 400 to 1000 distinct cellular proteins undergo cleavage during apoptosis, but it remains unclear how many of these are active participants in the apoptotic program and how many are victims of the collateral damage that occurs as cells are disassembled from within.

9.14 Both intrinsic and extrinsic apoptotic programs can lead to cell death

This series of events described above is sometimes termed the *intrinsic apoptotic program*, because the signals that trigger it originate within the cell; it is also known as the *stress-activated* apoptotic pathway. While p53 often plays a prominent role in triggering the intrinsic apoptosis program, other types of intracellular signals can also cause the mitochondria to release cytochrome c without involving p53 (see Figure 9.25A). Included in this category are stresses such as a buildup of calcium within the cell, excessive oxidants, certain types of DNA-damaging agents, and agents that disrupt the microtubules—key components of the cytoskeleton and mitotic spindle. Infections by various tumor viruses may also trigger apoptosis. Interestingly, these viruses generally take great pains to ensure that the intrinsic apoptotic program is not initiated, lest their cellular hosts die before their infectious cycles reach completion (**Sidebar 9.5**).

In fact, these diverse intracellular stressors, when taken together, still do not encompass all of the triggers of apoptosis, since this death program can also be triggered through an alternative route—one that is initiated *outside* the cell. This pathway operates in part via the activation of pro-apoptotic cell surface receptors. These transmembrane proteins are often termed **death receptors** to reflect their ability to activate the apoptotic program. Thus, after binding their cognate ligands in the extracellular space, the death receptors activate a cytoplasmic caspase cascade that converges rapidly on the intrinsic apoptotic pathway described above, thereby triggering an apoptotic response similar to the one seen following activation of the intrinsic apoptotic program (see Figure 9.27). Because the signals that activate the death receptors originate from outside the cell, the apoptotic program initiated by these receptors has been called either the *extrinsic apoptotic* program or the *receptor-activated* apoptotic pathway.

The ligands of the death receptors are members of the tumor necrosis factor (TNF) family of proteins. TNF-α was first identified and studied because of its ability to cause the death of cancer cells. Subsequently, this group of protein ligands was found

Table 9.3 Death receptors and their ligands

Alternative names of receptors	Alternative names of ligands
FAS/APO-1/CD95	FasL/CD95L
TNFR1	TNF-α
DR3/APO-3/SWL-1/TRAMP	APO3L
DR4/TRAIL-R1	APO2L/TRAIL
DR5/TRAIL-R2/KILLER	APO2L/TRAIL

capable of causing the death of a wide variety of normal cell types that display appropriate receptors on their surfaces. Similar to growth factors and their receptors, each of these TNF-like proteins binds to its own cognate death receptor (**Table 9.3**). Six members of a superfamily of TNF-like receptors are involved in triggering apoptosis.

Once activated by ligand binding, members of the death receptor family assemble a death-inducing signaling complex (DISC) that triggers the cleavage and activation of caspases 8 and 10 (**Figure 9.29A**). These initiator caspases then cleave and activate the

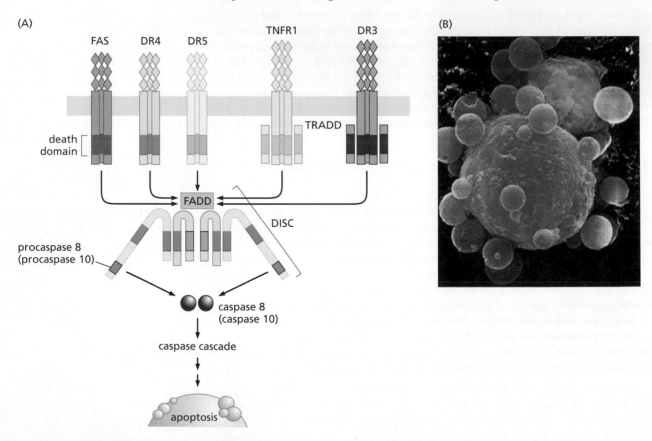

Figure 9.29 Death receptors (A) Five families of "death receptor" proteins are displayed on the surfaces of various types of mammalian cells. For example, the FasL and APO2L/TRAIL ligands bind to the FAS, DR4, and DR5 death receptors (*top left*). The cytoplasmic tails of these receptors act via the FAS-associated death domain (FADD) protein to assemble a death-inducing signaling complex (DISC), and the latter proceeds to convert procaspases 8 and 10 into their respective active caspases. These then converge on the caspase cascade that is activated through the intrinsic apoptotic program. A similar sequence of events occurs following the activation of the TNFR1 and DR3 receptors (*top right*). These death receptors differ from receptor tyrosine kinases (and their ligand-induced dimerization) in that the ligand complexes, sometimes called "death ligands," are homotrimers and appear to work by causing receptor trimerization. [Compare these ligands with the ligands of growth factor receptors (see Figure 5.11A), which drive receptor dimerization.] (B) Death receptor ligands, notably FasL, are frequently used by cytotoxic T lymphocytes (CTLs) to kill cancer cells, which display the FAS receptor on their surfaces. (In addition, granzymes may be injected by CTLs into target cells; see Figure 9.31.) In this colorized scanning electron micrograph, a CTL (*orange*) is attacking a cancer cell, which is in the throes of apoptosis, as evidenced by the numerous blebs on its surface. (A, adapted from A. Ashkenazi, *Nat. Rev. Cancer* 2:420–430, 2002. With permission from Nature. B, from Dr. Andrejs Liepins / Science Source.)

Sidebar 9.5 Why do DNA tumor viruses seek to inactivate p53? As described previously (see Section 8.5), a diverse group of DNA tumor viruses specify oncoproteins that are designed to inactivate pRb. In addition, they undertake to inactivate p53 activity (see Table 9.1), raising the question why both host-cell proteins must be inactivated by these viruses. Because these agents are called "tumor viruses," we might conclude that they must create tumors in order to propagate themselves and that pRb and p53 inactivation is central to this goal. However, careful examination of the life cycles of these viruses indicates that cell transformation is only an inadvertent side effect of their replication. Like all other viruses, these viruses have evolved to optimize only one outcome—their efficient multiplication in tissues of infected hosts. This realization forces a restatement of the question: Why is inactivation of both tumor suppressor proteins critical to viral replication?

DNA tumor viruses parasitize the host-cell DNA replication machinery to replicate their own genomes; this machinery is available only in the late G_1 and S phases of the cell cycle. Consequently, these viruses must inactivate pRb, as well as the two pRb-related proteins p107 and p130, to allow initially quiescent, infected host cells to advance into S phase. (Certain herpesviruses achieve the same end by expressing potent virus-encoded D-type cyclins.) The situation is slightly different in the case of human papillomaviruses (HPVs): they infect replicating cells in the cervical epithelium and block the normally occurring exit from the active cell cycle into a post-mitotic state that takes place as these cells differentiate (**Figure 9.30**).

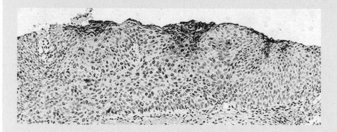

The cells infected by these various viruses respond to pRb inactivation by activating their p53 alarm systems. This response occurs due to the excessive activity of E2Fs that results from the functional inactivation of pRb (see Figure 9.14B); in addition, other, poorly understood virus-induced changes in cells may trigger p53 activation. Such p53 activation threatens to kill the infected host cells through apoptosis long before the viruses have had a chance to multiply and produce progeny. So these viruses proceed to eliminate p53 function. The SV40 large T antigen binds and sequesters p53 (in addition to binding pRb); HPV uses its E6 protein, which functions like Mdm2 to mark p53 for destruction via ubiquitylation (see Figure 9.14A). Adenovirus uses one of its E1B proteins to bind to p53 and pRb, and it deploys a second E1B-encoded protein, which is related to Bcl-2, to antagonize two important pro-apoptotic proteins, Bax and Bak (see Figure 9.25A).

Interestingly, at least one human RNA tumor virus—the HTLV-I retrovirus—takes the trouble to inactivate many of the same host-cell pathways. Perhaps the active transcriptional machinery of rapidly dividing cells confers some advantage on this virus, whose multiplication does not depend on the DNA replication of its host (see Figure 3.12).

Figure 9.30 Human papillomavirus and the pathogenesis of cervical carcinomas This histological section of cervical epithelium infected with a human papillomavirus (HPV) reveals, through immuno staining with an anti-HPV antibody stain (*brown*), clusters of HPV-infected cells in dysplastic areas of the epithelium, termed high-grade squamous intraepithelial lesions (HSILs). The dysplastic cells arise, in large part, from the ability of the HPV E7 oncoprotein to inactivate pRb and thereby block their entrance into a post-mitotic state and the ability of the viral E6 oncoprotein, which targets the cells' p53 protein for destruction, to suppress apoptosis. The resulting lesions progress with a low but significant frequency to cervical carcinomas. Unlike epithelial cells in normal cervical epithelium, these virus-infected cell populations harbor a significant proportion of proliferating, Ki67-positive cells (*not shown*). (From K. Middleton et al., *J. Virol.* 77:10186–10201, 2003. With permission from American Society for Microbiology.)

executioner caspases 3, 6, and 7, thereby converging on the signaling pathway through which the intrinsic apoptotic program operates (see Figure 9.27). Caspase 3 can also cleave and activate the Bcl-2–related protein Bid (see Figure 9.23A), which then initiates MOMP, further amplifying the pro-apoptotic signals of the extrinsic apoptotic program by recruiting elements of the intrinsic program into the process (**Figure 9.31**). In addition, as this figure illustrates, the attack and killing of cancer cells by cytotoxic cells of the immune system, such as the CD8 and NK lymphocytes described in Chapter 15, can be provoked through their ability to inject the granzyme B protease into the cytoplasm; once in the cytosol, the injected protease then proceeds to cleave and activate executioner caspases.

Mapping the intrinsic and extrinsic apoptotic cascades has enriched our understanding of how p53 succeeds in evoking apoptosis (**Figure 9.32**). To begin, p53 induces expression of the genes encoding Bax and PUMA proteins (see Figure 9.25A) that are directly involved in MOMP and the initiation of apoptosis at the mitochondrial membrane. In addition, p53 activates expression of the death receptor FAS and a protein called IGFBP-3. The FAS receptor, once expressed at the cell surface, increases a cell's responsiveness to extracellular death ligands, specifically the Fas Ligand (FasL). IGFBP-3, an IGF-binding protein, when released from the cell, sequesters IGF-1/2 (insulin-like growth factors-1/2), growth factors that provide cells with **trophic** (survival) signals. Because many cells depend on continuous exposure to trophic signals to remain viable, reducing the survival signals induced by IGF-1 places them in grave danger of succumbing to

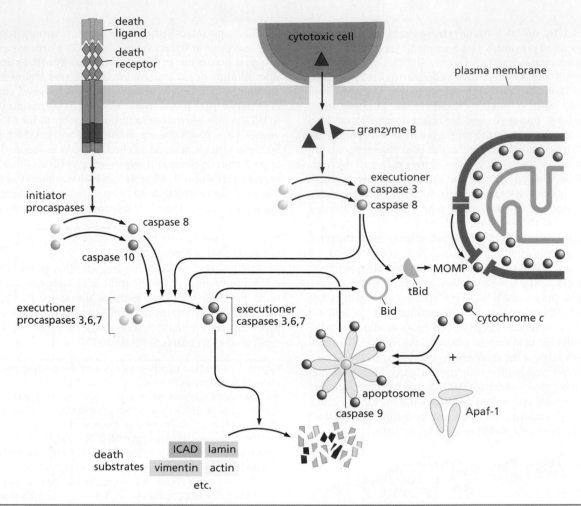

Figure 9.31 Convergence of intrinsic and extrinsic apoptotic pathways After a death receptor activates procaspase 8 or 10, or both, into the corresponding active caspases (*top left*), the latter converge on the intrinsic apoptotic cascade by cleaving and activating the executioner caspases (*middle left*). In addition, after procaspases 3 and 8 have been activated into caspases 3 and 8, the latter cleave and activate Bid, a Bcl-2–related protein (*center, pink circle*). The resulting activated tBid moves from its cytosolic location to the mitochondrion, where it initiates MOMP, further amplifying the apoptotic cascade by promoting the formation of new, activated apoptosomes (*lower right center*). A second, alternative form of the extrinsic apoptotic pathway is initiated by cytotoxic cells that are dispatched by the immune system (*top middle*) and attach to the surfaces of targeted cells, into which they introduce granzyme B molecules; these cleave and thereby activate procaspases 3 and 8 (see also Figure 9.25).

apoptosis, a process that has been called "death by neglect." This explains why many cancer cells find ways of maintaining high concentrations of IGFs in the microenvironments that surround them. Apart from simply secreting abnormally high amounts of IGF-1 or the related IGF-2, cancer cells may reduce the levels of IGFBPs—either by methylating *IGFBP* gene promoters or by increasing the synthesis and secretion of matrix metalloproteinases (MMPs) that cleave and thereby inactivate IGFBP molecules.

We will return to the extrinsic apoptotic program in Chapters 15 and 16, because it can play two important roles in the complex interactions between cancer cells and the immune system. On the immune front, cells of the immune system may release death ligands (such as FasL) in an attempt to trigger apoptosis in cancer cells that they have targeted for elimination (Figure 9.29B). Working in the opposite direction, cancer cells may display or release death ligands that they deploy to kill immune cells that have approached too closely and threaten their survival.

The complexity of the signals flowing through apoptotic pathways underscores the fact that the decision for or against apoptosis is achieved by altering the *balance* of complex arrays of pro- and anti-apoptotic proteins that influence the ultimate decision whether or not to flip the "apoptotic master switch."

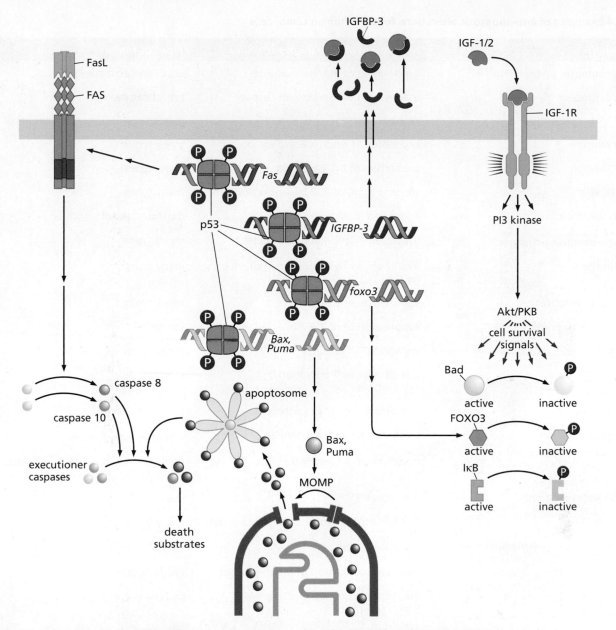

Figure 9.32 Activation of apoptosis by p53 p53 employs multiple signaling pathways in order to activate the apoptotic program. By inducing expression of the gene encoding the FAS receptor, p53 promotes display of this death receptor at the cell surface (*brown, top left*), thereby sensitizing the cell to any Fas ligand (FasL) that may be present in the extracellular space. Activated p53 also drives expression of the FOXO3 pro-apoptotic transcription factor, as well as Bax (*center*), the pro-apoptotic, Bcl-2–related protein; Bax homo-oligomerization triggers MOMP, promoting the release of cytochrome c and other pro-apoptotic proteins from the mitochondria (*bottom center*). In addition, p53 induces expression of IGF-binding protein-3 (IGFBP-3, *red*); when this protein is released into the extracellular space, it binds and sequesters IGF-1 and IGF-2 (*blue, top center*), the pro-survival, anti-apoptotic ligands of the IGF-1 receptor (IGF-1R). In the absence of IGFBP-3, IGF-1 would bind to its receptor (*top right*), causing the activation of PI3 kinase and the subsequent generation of anti-apoptotic signals in the cell, including those leading to the inactivation of the pro-apoptotic proteins Bad (related to Bcl-2), FOXO3, and IκB (antagonist of NF-κB). However, the sequestration of IGF-1/2 by IGFBP-3 prevents it from exerting its critical anti-apoptotic functions (*right*). p53 also inactivates several other anti-apoptotic agents (*not shown*).

9.15 Cancer cells deploy numerous ways to inactivate their apoptotic machinery

As emphasized earlier, cells advancing down the long road toward neoplasia encounter a number of physiological stressors that threaten their existence. Each of these stressors has the ability to trigger apoptosis, which explains why most, if not all, types of cancer cells inactivate key components of their apoptosis-inducing machinery (**Table 9.4**).

Table 9.4 Examples of anti-apoptotic alterations found in human tumor cells

Alteration	Mechanism of anti-apoptotic action	Types of tumors
CASP8 promoter methylation	inactivation of extrinsic cascade	SCLC, pediatric tumors
CASP3 repression	inactivation of executioner caspase	breast carcinomas
Survivin overexpression[a]	caspase inhibitor	mesotheliomas, many carcinomas
ERK activation	repression of caspase 8 expression	many types
ERK activation	protection of Bcl-2 from degradation	many types
Raf activation	sequestration of Bad by 14-3-3 proteins	many types
PI3K mutation/activation	activation of Akt/PKB	gastrointestinal
NF-κB constitutive activation[b]	induction of anti-apoptotic genes	many types
p53 mutation	loss of ability to induce pro-apoptotic genes	many types
*p14*ARF gene inactivation	suppression of p53 levels	many types
Mdm2 overexpression	suppression of p53 levels	sarcomas
IAP-1 gene amplification	antagonist of caspases 3 and 7	esophageal, cervical
APAF1 methylation	loss of caspase 9 activation by cytochrome *c*	melanomas
BAX mutation	loss of pro-apoptotic protein	colon carcinomas
Bcl-2 overexpression	closes mitochondrial channel	~½ of human tumors
PTEN inactivation	hyperactivity of Akt/PKB kinase	glioblastoma, prostate carcinoma, endometrial carcinoma
IGF-1/2 overexpression	activates PI3K	many types
IGFBP repression	loss of anti-apoptotic IGF-1/2 antagonist	many types
Casein kinase II overexpression	activation of NF-κB	many types
TNFR1 methylation	repressed expression of death receptor	Wilms tumor
FLIP overexpression	inhibition of caspase 8 activation by death receptors	melanomas, many others
Akt/PKB activation	phosphorylation and inactivation of pro-apoptotic Bcl-2-like proteins	many types
USP9X overexpression	deubiquitylates Mcl-1	lymphomas
STAT3 activation	induces expression of Bcl-X_L	several types
TRAIL-R1 repression	loss of responsiveness to death ligand	small-cell lung carcinoma
FAP-1 overexpression	inhibition of FAS receptor signaling	pancreatic carcinoma
XAF1 methylation[c]	loss of inhibition of anti-apoptotic XIAP	gastric carcinoma
Wip1 overexpression[d]	suppression of p53 activation	breast and ovarian carcinomas, neuroblastoma

[a]Survivin is an inhibitor of apoptosis (IAP) in gastric, lung, and bladder cancer and melanoma, in addition to the mesotheliomas indicated here. The expression of a number of IAP genes is directly induced by the NF-κB TFs.
[b]Induces synthesis of c-IAPs, XIAP, Bcl-X_L, and other anti-apoptotic proteins.
[c]XAF1 (XIAP-associated factor 1) normally binds and blocks the anti-apoptotic actions of XIAP, the most potent of the IAPs.
[d]Wip1 is a phosphatase that inactivates p38 MAPK, which otherwise would phosphorylate and stimulate the pro-apoptotic actions of p53.

We can also view these changes from the perspective of the organism as a whole: virtually all human cancer cells employ anti-apoptotic strategies, which suggests that the apoptotic machinery integrated into all normal cells is held in active reserve as an anti-neoplastic mechanism that can be rapidly deployed, when needed, to obstruct tumor progression.

Arguably the most important and widely exploited strategy has already been noted: inactivation of the p53 pathway. The *p53* gene is altered in almost half of all human cancer cell genomes, and in a substantial proportion of others, ARF is no longer expressed. This repression is often achieved through outright deletion of the ARF-encoding gene or by promoter methylation, the latter resulting in the silencing of transcription (see Section 1.8). A small percentage of human tumors (mostly sarcomas) will overexpress Mdm2. All of these mechanisms affecting ARF or Mdm2 expression serve to drive down levels of p53 protein in cells. Finally, in some human tumor cells, p53 appears to be inappropriately localized, that is, it is sequestered in the cytoplasm, where it cannot carry out its main task, transcriptional activation.

Still, this elimination of p53 function does not seem to suffice for many types of human tumors, which strive to subvert additional components of the apoptotic machinery. For example, many melanoma cells exhibit methylation and thus functional inactivation of the promoter of the *APAF1* gene, which encodes the cytosolic protein that assembles with released cytochrome *c* to form the apoptosome and activate caspase 9 (see Figure 9.27). The pro-apoptotic *Bax* gene (see Table 9.4) is inactivated by mutation in more than half the colon cancers that show microsatellite instability as described in Chapter 12. Deregulated overexpression of *BCL2* (the human version of *Bcl-2*) is found in follicular B-cell lymphomas, and *BCL2* expression is elevated in large numbers of human tumors of diverse tissue origins; by some estimates, at least half of all human tumors show elevated *BCL2* expression.

Yet another highly effective strategy for acquiring resistance to apoptosis derives from hyperactivation of the PI3K → Akt/PKB pathway (see Section 6.6). Activation of this pathway can be achieved through the collaborative actions of tyrosine kinase receptors and the Ras oncoprotein, both of which activate PI3K; the result is an increased level of PIP3 and activation of Akt/PKB (see Figure 6.14A). Inactivation of the gene encoding PTEN, the phosphatase that breaks down PIP3, has a similar effect. Once Akt/PKB has been activated by the accumulated PIP3, it can phosphorylate and thereby inhibit pro-apoptotic proteins such as Bad and caspase 9, and at the same time phosphorylate and activate Mdm2, the key antagonist of p53.

A diverse collection of cancer cell types, ranging from leukemias to colon carcinomas, have been found to resort to yet another mechanism to protect themselves against apoptosis: they activate expression of NF-κB. This transcription factor is usually sequestered in an inactive form in the cytoplasm and, in response to certain physiological signals, is liberated and translocates to the nucleus, where it activates a large constituency of target genes (see Section 6.12). A number of these genes are involved in pro-inflammatory and angiogenic functions, and a significant cohort function as apoptosis antagonists, blocking both the intrinsic and extrinsic apoptotic programs. The list of NF-κB targets is expanding rapidly, and it is likely that the overexpression of some of these anti-apoptotic genes (and encoded proteins) seen in many types of human cancer cells will be traced directly to activation of NF-κB.

Yet another strategy for avoiding apoptosis is evident in childhood neuroblastomas, in which amplification of the N-*myc* gene normally indicates aggressive tumor growth and short life expectancy. On its own, this amplification actually *threatens* the proliferation of neuroblastoma cells because, as mentioned earlier, the actions of the *Myc* oncogene (and those of its N-*myc* cousin) release potent mitogenic signals, but at the same time are strongly pro-apoptotic. Many neuroblastoma cells that have overexpressed N-*myc* through gene amplification seem to solve this problem, at least in part, by inactivating the gene encoding caspase 8 (**Figure 9.33**), doing so through methylation or deletion.

The involvement of caspase 8, as cited above, reveals that the actions of the extrinsic apoptosis-inducing pathway are also relevant for the evolution of certain types of cancer cells. Others deploy a protein termed FLIP (FLICE-inhibitory protein) that

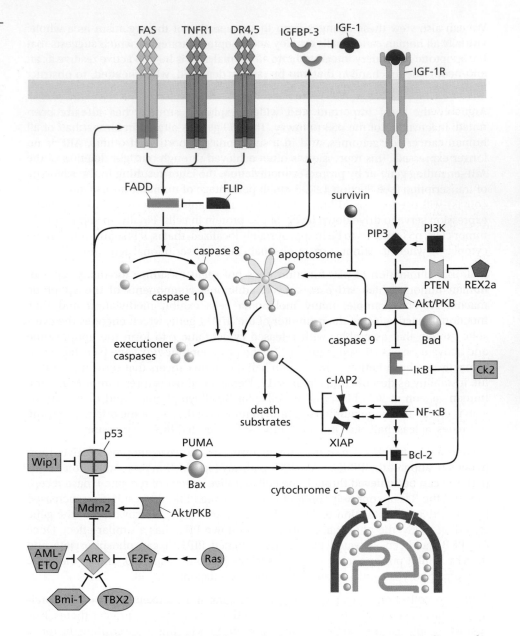

Figure 9.33 Anti-apoptotic strategies used by cancer cells Cancer cells resort to numerous strategies in order to decrease the likelihood of apoptosis. This diagram indicates that in various cancer cell types, the levels or activity of important pro-apoptotic proteins are decreased (*blue*). Conversely, the levels or activity of certain anti-apoptotic proteins may be increased (*red*). (Some of these, because they are downstream of master regulators that are known to be altered in cancer cells, are affected by known changes in their upstream regulators.)

can interfere with the activation of the extrinsic apoptotic cascade by binding to the death domain proteins that associate with death receptors, thereby blocking the activation of the initiator caspase 8. Yet other tumor cells suppress expression of their TNF receptors by methylating the promoter of the encoding *TNFR1* gene. A selection of the mechanisms that disable apoptosis is outlined in Figure 9.33.

These maneuvers provide clear evidence that cancer cells can gain survival advantage by inactivating the extrinsic pathway. This observation leads, in turn, to the speculation that cells of the immune system often besiege cancer cells and attempt to kill them through the release of TNF-like factors that can initiate the extrinsic apoptotic program. Given the complex wiring diagram of the circuitry controlling apoptosis, it is virtually certain that many additional anti-apoptotic strategies will be uncovered in various types of human cancer cells.

9.16 p53 inactivation provides an advantage to incipient cancer cells at a number of steps in tumor progression

As we will learn in Chapter 11, the formation of a malignant human cell involves more than half a dozen distinct steps that usually occur over many years' time. An early step in the formation of a cancer cell often involves activation of an oncogene through some type of mutation. Such oncogene activation can put a cell at great risk of p53-induced senescence or apoptosis. Recall, for example, that a *Myc* oncogene can, on its own, trigger p53-dependent apoptosis. Hence, cells that have acquired such an oncogene can accrue additional growth advantage by shedding p53 function.

Later in tumor development, a growing population of tumor cells may experience hypoxia because cells toward the center of a developing tumor may lack an adequate network of vessels to provide access to blood-borne oxygen. Although normal cells would die from oxygen deprivation, tumor cells may survive because their ancestors managed to inactivate the *p53* gene during an earlier stage of tumor development (having done so initially for a quite different reason). Hypoxia itself could provide a sufficient stress to select for p53-mutant cells.

During much of this long, multi-step process of tumor progression, the absence of p53-triggered responses to genetic damage will permit the survival of cells that are accumulating mutations at a greater-than-normal rate. Such increased mutability increases the rate at which oncogenes become activated and tumor suppressor genes become inactivated; the overall rate of evolution of premalignant cells to a malignant state is thereby accelerated. Telomere collapse, another danger faced by evolving, premalignant cells (see Chapter 10), also selects for the outgrowth of those cells that have lost their p53-dependent DNA damage response.

The advantages to the incipient tumor cell of shedding p53 function do not stop there. One of the important target genes whose expression is increased by p53 is the *TSP-1* gene, which specifies thrombospondin-1. As we will see in Chapter 13, Tsp-1 is a secreted protein that functions in the extracellular space to block the development of new blood vessels. Consequently, a reduction of Tsp-1 expression following p53 loss removes an obstacle that otherwise would prevent clusters of cancer cells from developing an adequate blood supply during the early stages of tumor development. Loss of p53 can also affect major metabolic pathways in cells, stimulate changes in the tumor microenvironment, and promote cell motility and invasion—all of which can favor tumor progression.

Together, these diverse consequences of p53 pathway inactivation illustrate, once again, how the malfunctioning of a small number of key components of the alarm response circuitry permits cancer cells to acquire multiple alterations and survive under conditions that usually lead to the death of normal cells. These multiple benefits accruing to cancer cells explain why the p53 pathway is disrupted in most if not all types of human tumors.

In almost half of these tumors (see Figure 9.3), functioning of the p53 protein itself is compromised by mutations in the *p53* gene on numerous occasions as cited earlier. In many of the remaining tumors, the ARF protein is missing or the Mdm2 or MdmX proteins are overexpressed. In addition, p53 function may be compromised by defects in the complex signaling network in which p53 and its antagonist, Mdm2, are embedded. This network includes the *p53*-related genes *p63* and *p73*, which carry distinct but overlapping functions with p53 (**Sidebar 9.6**). There are reasons to suspect that other, still-undiscovered genetic mechanisms subvert p53 function. Although the organism as a whole benefits greatly from the p53 being stationed in its myriad cells, it suffers grievously once p53 function is lost in some of them, because the resulting p53-negative cells are now free to begin the long, multi-step march toward malignancy.

9.17 Additional forms of cell death may limit the survival of cancer cells

Although much of this chapter has focused on apoptosis as a major obstacle to cancer cell survival, alternative fates may befall a cell en route to full-fledged neoplastic growth. One such fate has already been mentioned, albeit in passing: necrosis. In complex metazoa like ourselves, the cells within our tissues are usually guaranteed access to reasonably constant levels of essential nutrients, such as glucose and amino acids. Cancer cells, however, often lack such access. In the early stages of tumor growth, tumor-directed angiogenesis—the formation of new blood vessels—is not

Sidebar 9.6 p53 has interesting cousins As is often the case with important regulatory proteins, both p53 and Mdm2 have relatives in the cell; we have already encountered Mdm2's cousin, Mdm4 (also known as MdmX; see Section 9.8). In many situations, the ability of Mdm2 to drive down p53 concentrations may depend on its ability to do so as part of multi-protein complexes formed with Mdm4.

p53, for its part, actually has two cousins, p63 and p73. These two have quite different functions in cells. For example, although *p53*-null mice (created by germ-line gene inactivation; see Supplementary Sidebar 7.2) are largely normal at birth, *p63*-null mice have severe developmental defects involving epithelial cell differentiation and die shortly after birth. *p73*-null mice show yet other developmental abnormalities that affect primarily the nervous and secretory tissues. Moreover, like $p53^{+/-}$ mice, $p63^{+/-}$ and $p73^{+/-}$ heterozygous mice show predispositions to cancer, although these phenotypes are not as strongly penetrant as those of *p53* mutants.

These two cousins are important for p53-dependent apoptosis, because mouse embryo fibroblasts that have been deprived of both copies of both genes (that is, are of a $p63^{-/-}$ $p73^{-/-}$ genotype) are no longer able to induce expression of many pro-apoptotic genes and fail to execute p53-dependent apoptosis. The tendency of excessive E2F1 to trigger apoptosis, as described in Section 9.9, may result in part from its ability to induce expression of p73, which then functions as a pro-apoptotic partner of p53. Loss of the *p63* and *p73* genes does not, however, affect the ability of p53 to inhibit proliferation (mediated by $p21^{Cip1}$ induction) or p53's ability to induce Mdm2 expression.

A final dimension of complexity comes from naturally occurring variant forms of these proteins. For example, full-length p63 and p73 function with p53 to promote apoptosis, whereas shorter, N-terminally truncated versions of p63 and p73—both products of alternative splicing—inhibit p53-induced apoptosis. Nonetheless, inactivation of either the *p63* or the *p73* gene in the mouse germ line leads to increased tumor incidence and places these two genes squarely in the camp of TSGs.

fully activated, as we discuss in Chapter 13. Later in tumor progression, many vessels are malformed. In both cases, cancer cells may become nutrient- and oxygen-starved. This **ischemia** can trigger bioenergetic collapse and necrosis. The necrotic cores of many tumors are testimonials to the important role of necrosis in limiting tumor growth. Although the overall size of a tumor often yields the impression of a large and actively growing mass, the bulk of most tumors is actually inert, necrotic tissue that has never been eliminated.

As indicated in Table 9.2, necrotic cell death differs in many respects from apoptosis and results in the lysis of the cells rather than the production of apoptotic bodies. The resulting cell lysis can trigger localized inflammation, as the surrounding tissues become exposed to the intracellular contents of the necrotic cell. Interestingly, if the apoptotic response is blocked through one of the mechanisms described earlier, exposing the cell to a death receptor ligand may instead trigger necrosis.

Before nutrient starvation drives a cancer cell into necrosis, the cell may resort to an alternative strategy to stave off the irreversible collapse of metabolism that triggers necrotic cell death: it may activate a program called **autophagy**, which allows it to cannibalize its own organelles as a means of recycling their contents (**Figure 9.34A**). The recycled molecules can then be used to generate the energy and intermediary metabolites that are required to maintain cell viability, permitting the nutrient-starved cell to survive for an extended period of time.

In the case of cancer cells, this survival may allow them to avoid necrosis altogether if they subsequently succeed in obtaining access to nutrients (Figure 9.34B–D). Under conditions of nutrient deprivation, some cancer cells have been seen to consume two-thirds of their own mass and still remain viable, retaining the ability to return to full size and to a state of active proliferation once nutrients are eventually procured. Importantly, excessive, deregulated autophagy may lead a cell to cannibalize too many of its own constituents, leading to its death by specialized death program termed **autophagic cell death**.

In recent years, a number of additional forms of regulated cell death have been described. They can be distinguished by a combination of the morphological changes that accompany the death process and by the genetic and biochemical pathways that are implicated in their execution. One of these, **necroptosis**, exhibits certain morphological features of necrosis (for example, cell lysis), but unlike necrosis, this form of cell death is orchestrated by specific biochemical pathways. In particular, death receptors, such as the TNF receptor, and the kinases RIPK3 and RIPK1 are implicated in necroptosis. Following activation of this pathway by upstream signals, the protein MLKL is

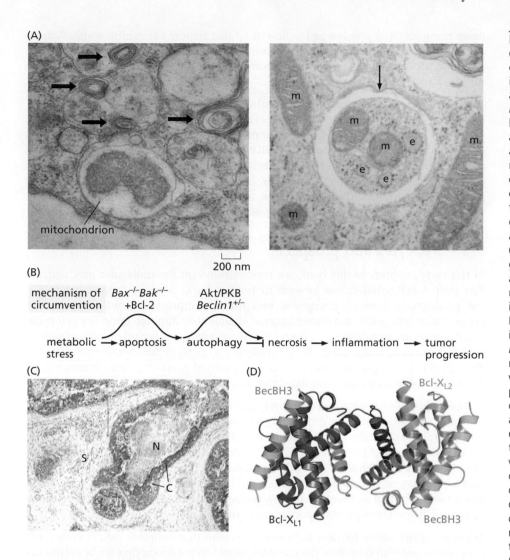

(A)

mitochondrion

m   m   m   e   m
e   e
m

200 nm

(B)

| mechanism of circumvention | $Bax^{-/-}Bak^{-/-}$ +Bcl-2 | Akt/PKB $Beclin1^{+/-}$ | | | |

metabolic stress → apoptosis ⌢ autophagy ⊣ necrosis → inflammation → tumor progression

(C)

S   N   C

(D)

BecBH3   Bcl-X$_{L2}$

Bcl-X$_{L1}$ BecBH3

Figure 9.34 Autophagy and necrosis (A) In a cell undergoing autophagy, an entire mitochondrion has been engulfed in a digestive compartment called an autophagosome (electron micrograph, *left*); the *arrows* indicate structures that have the multilayered appearance of autophagosomes. The *right* panel shows an autophagosome (*arrow*) surrounding mitochondria (m) and fragments of endoplasmic reticulum (e). (B) Under some circumstances, cancer cells may respond to metabolic stress, such as nutrient deprivation, by activating apoptosis. If apoptosis is circumvented experimentally (*curved arrow*) by inactivating pro-apoptotic genes (*Bax* and *Bak*) and by expressing an anti-apoptotic protein (Bcl-2), these cells may activate autophagy. If autophagy is then inhibited by expressing the Akt/PKB kinase (which blocks autophagy) plus inactivating one of the two copies of the *Beclin* gene (whose protein product is a master regulator of autophagy), such cells will enter into necrosis. Necrosis may in turn promote localized inflammation, which often serves to promote tumor progression, as discussed in Chapter 11. (C) A subtype of ductal carcinoma *in situ* (DCIS) of the breast, termed comedo carcinoma, shows a rim of viable cells (C) that are expressing high levels of Beclin-1 (*brown*), the master regulator of autophagy, surrounding a central core of necrotic cells (N, *light blue*). The *Beclin*$^+$ cells have activated autophagy, apparently in response to the limited nutrients available to them from the nearby surrounding stroma (S). Of note, this subtype of DCIS has a worse prognosis than the more common subtypes, in which intraductal necrosis is not seen; indeed, the *Beclin*$^+$ cancer cells seen here are close to becoming actively invasive. (D) The complex interactions between the apoptosis and autophagy programs are highlighted by the fact that Beclin can be bound directly by either Bcl-2 or Bcl-X$_L$, two potent anti-apoptotic proteins; this binding prevents Beclin from causing autophagy. Seen here is a crystallographic structure of a complex of two molecules of the BH3 domain of Beclin bound to two molecules of the BH3 domain of Bcl-X$_L$. (A, left, courtesy of E.M. Haley, B.A. Lund, M.E. Bisher and H.A. Coller; right, from N. Mizushima and B. Levine, *Nat. Cell Biol.* 12:823–830, 2010. With permission from Nature. B, from K. Degenhardt et al., *Cancer Cell* 10:51–64, 2006. With permission from Elsevier. C, from V. Espina and L.A. Liotta, *Nat. Rev. Cancer* 11:68–75, 2011. With permission from Nature. D, from A. Oberstein, P.D. Jeffrey and Y. Shi, *J. Biol. Chem.* 282: 13123–13132, 2007. © 2007 ASBMB. This article is available under the terms of the Creative Commons Attribution 4.0 license.)

phosphorylated by RIPK3 and undergoes oligomerization and translocation to the plasma membrane, leading to membrane rupture and cell lysis. The importance of the MLKL executioner in this pathway has been confirmed via mouse knockout experiments. Thus, although $Mlkl^{-/-}$ mice survive and show no obvious defects, their cells are resistant to the induction of necroptosis.

Pyroptosis derives its name from a form of cell death that is associated with inflammation. Indeed, pyroptosis operates in the context of infection of cells by intracellular pathogens, including viruses and bacteria. In the context of cancer, this program can be induced by exposure of cancer cells to a variety of cytotoxic therapeutic drugs. Once activated, the programmed release of a variety of pro-inflammatory cytokines together with the rupture of the plasma membrane results in the recruitment of several types of cells of the innate immune system, amplifying the host response to these various noxious agents. It is thought that induction of pyroptosis involves several caspases, including caspase 1, which activates the pro-inflammatory cytokine IL-1β, as well as the pore-forming protein gasdermin D (GSDMD). Cleavage of the latter by caspase 1 or other inflammatory caspases leads to the release of its N-terminal domain, which forms pores in the cell membrane and induces cell lysis. These processes explain why cells lacking GSDMD are protected against pyroptosis provoked by certain agents.

Finally, a form of regulated cell death known as **ferroptosis** is associated with iron availability and intracellular lipid peroxidation. Ferroptosis is inhibited by the lipid peroxidase GPX4, and inhibition of this enzyme can lead to induction of this death program. Because ferroptosis depends on iron, various iron chelators also inhibit ferroptosis. This program may be important in the cellular response to certain anti-cancer agents. In addition, carcinoma cells that have moved from an epithelial to a more mesenchymal phenotype (as described in Chapter 14) exhibit a significantly increased susceptibility to induced ferroptosis. Interestingly, recent evidence points to p53 as a positive regulator of ferroptosis, adding to the list of its potential tumor-suppressing mechanisms.

As is true for apoptosis, the genes and pathways that control these other forms of cell death may be the target of mutation or silencing during tumor development, enabling cancer cells to survive in the face of multiple insults and challenges. As an example, expression levels of the genes encoding RIPK and MLKL are lower in certain cancers compared with corresponding normal tissues, implicating suppression of necroptosis as a pro-tumorigenic strategy. Moreover, investigations of these various cell death pathways, undertaken in order to reveal new strategies for eliminating neoplastic cells, represents an active area of investigation.

9.18 Synopsis and prospects

In the early chapters of this book, we concentrated on the molecular mechanisms that propel cell proliferation forward or hold it back. Most of these regulators— the products of (proto-) oncogenes and tumor suppressor genes—manage the extracellular mitogenic and anti-mitogenic signals that impinge upon the cell cycle machinery from outside the cell. These signals are processed within individual cells to enable them to converge on the final decisions that determine whether they advance through the active growth-and-division cycle, retreat reversibly into quiescence, or enter irreversibly into a post-mitotic, differentiated state.

In this chapter, we have focused on entirely different cell-physiological processes, most of which are responsible for monitoring the internal well-being of cells. The monitoring apparatus conducts continuous surveillance of vital cell systems, including access to oxygen and nutrients, the physical state of the genome, and the balance of signals flowing through a cell's growth-regulating circuits. If such monitors determine that a vital system is damaged or malfunctioning within a cell, an alarm is sounded and growth is halted or, more drastically, the cell is induced to enter into senescence or activate its previously latent apoptotic program.

These monitors create barriers to tumor formation that incipient cancer cells must override in order to complete the complex, multi-step agenda that leads eventually to the formation of malignant cancers. Thus, as cells evolve from normalcy to a state of high-grade neoplasia, many of their subcellular systems malfunction in one way or another, leading to the risks of cytostasis, senescence, apoptosis, necroptosis, ferroptosis, pyroptosis, or other fates that threaten to derail tumor development.

Any one of these traps can halt further progress down the long road to malignancy. Indeed, attempts by most would-be cancer cells to evade these pitfalls usually fail, as reflected in the high attrition of the cells in premalignant cell populations. The multiple sources of cell attrition also help to explain one of the mysteries of human cancer pathogenesis: even though each of us experiences ~10^{16} cell divisions in a lifetime, each one of which threatens to create a seed of cancer, most of us will not develop life-threatening malignancies, a testament to the effectiveness of the multiple mechanisms that trigger cell death.

p53 is a central player in many of these responses because elimination of its function is one frequent means of avoiding apoptosis, growth arrest, and senescence, the last representing a permanent exit from the cell cycle (see Section 10.1). The *p53* gene and encoded protein have been the subject of more than 100,000 research reports since the discovery of the p53 protein in 1979. This extraordinary focus on a single gene and protein reflects both the central role that this protein plays in cancer pathogenesis and the frequent presence of mutant *p53* alleles in the genomes of human cancer cells. Still unanswered, however, is the most perplexing puzzle surrounding this protein: Why would cells entrust so many vital alarm functions to a single protein? Once a cell has lost p53 function, it becomes oblivious, in a single stroke, to a range of conditions that would normally call a halt to its proliferation or trigger entrance into the apoptotic program.

As we will see in Chapter 11, the complex, multi-step process that drives tumor development involves changes in multiple cellular control circuits. Because p53 receives signals from so many distinct sensors of damage or cell-physiological distress, the loss of p53 function plays a role in many of these steps. For example, early in the development of bladder, breast, lung, and colon carcinomas, when premalignant growths are present, DNA damage signaling is already active, as indicated by the presence of phosphorylated, functionally active ATM and Chk2 kinases. In addition, the p53 substrate of these kinases is also phosphorylated and apparently active, because the *p53* gene is often still in its wild-type configuration at this stage of tumor progression. This indicates that DNA lesions are widespread in the cells of these growths, even at this early stage of tumorigenesis, and suggests that p53 is striving to constrain the proliferation of the cells in these growths and may even induce with some frequency their elimination by apoptosis. Consequently, these premalignant cells stand to benefit greatly by shedding p53 function, thereby liberating themselves from its cytostatic and pro-apoptotic effects.

At these early stages of tumor development, cells that succeed in eliminating p53 function may also gain special advantage because they have a better chance of avoiding the apoptosis that is triggered by p53 in response to severe hypoxia. Thus, early premalignant growths generally have not yet developed adequate access to the circulation and the oxygen that it brings (via the process of angiogenesis to be described in Chapter 13). In addition, because p53 induces expression of the potently anti-angiogenic molecule thrombospondin (Tsp-1), p53 inactivation can accelerate the development of nearby capillary networks that can cure the hypoxia.

Later in tumor progression, when malignant carcinomas develop, the absence of p53 function will compromise some of the DNA repair systems of the cell and, at the same time, allow cells carrying damaged genomes greater survival advantage; the result is an increased rate of genomic alteration and an associated acceleration in the rate of multi-step tumor development as described in Chapter 11. Additionally, the absence of p53 function may enable these tumor cells to survive the pro-apoptotic side effects of oncogenes such as *Myc*. This short list reveals how truly catastrophic the loss of p53 function can be for a tissue and, ultimately, for the organism as a whole.

The effects of p53 loss may also be manifested even later, in the event that the *p53*-mutant tumor cells become clinically apparent and targeted for chemotherapy and radiation. The success of most existing anti-cancer therapies is predicated on their ability to damage the genomes of tumor cells, thereby provoking the death of these cells by apoptosis. In some cases, the loss of p53 function renders these cancer cells far less responsive to many of these therapeutic strategies (see Figure 9.7B as an example). In other cases, the absence of the p53-dependent cell cycle arrest response allows cells treated with chemotherapeutic agents to progress into the cell cycle with damaged genomes, which can actually *sensitize* them to p53-independent forms of cell death. Hence, the effects of p53 mutation of cancer therapy are highly context dependent.

Our understanding of the p53 protein remains incomplete, in spite of the vast research literature that describes it. One indication of how much more needs to be learned has come from attempts at systematically cataloging the myriad post-translational modifications of p53. As we have seen, some of these modifications affect p53's stability and thus intracellular concentration. In addition, they also affect the interactions of p53 with other nuclear transcription regulators that control the genes targeted by p53 action. The number of enzymes that modify p53—as evidenced by the variety of modifications shown in Figure 9.15C—is likely to reflect the number of distinct signaling pathways that impinge on p53 and regulate its functioning. Thus, each of the post-translational modifications illustrated in this figure is likely to affect the physical interaction of p53 with another partner protein in the cell, often a signaling protein.

In the case of the apoptotic response, loss of p53 function is only one of many strategies that cancer cells employ to protect themselves against this form of cell death. Alternative strategies include perturbing the expression of death receptors, their ability to activate caspases, and the activities of the caspases themselves. Yet other defensive maneuvers undertaken by cancer cells help to prevent mitochondrial outer membrane permeabilization (MOMP). These diverse strategies (see Figure 9.33) illustrate that the apoptotic machinery, like all well-regulated machines, operates through the balanced actions of mutually antagonistic elements. Cancer cells exploit preexisting anti-apoptotic components of this machinery by increasing their levels or activities, thereby nudging the balance away from apoptosis. These changes in specific components of the apoptotic machinery may compound the effects of p53 loss, further increasing the resistance of cancer cells to a variety of signals that usually succeed in provoking cell death.

Later, in Chapter 17, we will learn about novel forms of therapy that are designed to mobilize components of the pro-apoptotic machinery that remain intact in cancer cells. The hope is that at some time in the not-too-distant future, the circuitry governing apoptosis will be understood with sufficient precision so that we will be able to predict its behavior with accuracy and manipulate it at will within cancer cells.

These discussions of the mechanisms governing apoptotic cell death, which seem to be far removed from the real world, may soon be brought to bear on a major puzzle about human cancer: Why does obesity, as gauged by body-mass index (BMI), represent a major factor determining cancer incidence? In the United States, obesity represents the second greatest risk factor for developing cancer after tobacco consumption and will soon outpace tobacco as the leading risk factor for developing cancer. An attractive but still-unproven model proposes that obesity increases cancer incidence by protecting incipient cancer cells from death by apoptosis. If this thinking is substantiated, then apoptosis research will soon assume a major role in explaining the onset of cancer as well as the successes that we have in preventing and treating this disease.

Key concepts

- Organisms attempt to block the development of cancer through the actions of the p53 protein, which can cause cells to enter cell cycle arrest, undergo senescence, or experience apoptosis in the event that the machinery regulating cell proliferation is malfunctioning or the cells are exposed to various types of physiological stress.

- p53 is a nuclear protein that normally exists as a tetramer and functions as a transcription factor.

- The mutant p53 proteins found in many human tumors often carry amino acid substitutions in their DNA-binding domain. When a mutant p53 subunit forms tetrameric complexes with wild-type p53 subunits, it compromises the normal functions of the wild-type subunits and thus with functions of the tetramer as a whole.

- p53 can impose cell cycle arrest through its ability to induce expression of $p21^{Cip1}$ and apoptosis through its ability to promote the expression of a variety of pro-apoptotic proteins.

- p53 normally turns over rapidly. This turnover is blocked when various signals indicate cell-physiological stress, including anoxia, damage to the genome, and signaling imbalances in the intracellular growth-regulating machinery.

- p53 becomes functionally activated when its normally rapid degradation is blocked. In addition, covalent modifications of the resulting accumulated p53 protein modulate its activity as a transcription factor, directing it to activate the expression of the large constituency of genes involved in various cellular responses, notably apoptosis, cytostasis, and senescence.

- p53 levels are controlled by two critical upstream regulators, Mdm2 and $p14^{ARF}$. Mdm2 works to destroy p53, and ARF prevents Mdm2 from driving p53 degradation.

- The apoptotic caspase cascade can be triggered through the induction of mitochondrial outer membrane permeabilization (MOMP), which releases several pro-apoptotic proteins, most importantly cytochrome c.

- MOMP is determined by the relative levels of Bcl-2–related anti-apoptotic and pro-apoptotic proteins.

- Apoptosis involves the activation of a cascade of caspases that results in the destruction of a cell, usually within an hour. Apoptosis can be activated by p53 as well as by signals impinging on the cell from the outside, notably those transduced by cell surface death receptors.

- Loss of apoptotic functions allows cancer cells to survive a variety of cell-physiological stresses, including anoxia, signaling imbalances, DNA damage, and loss of anchorage.

- Cancer cells invent numerous ways to inactivate the apoptotic machinery in order to survive and thrive. Included among these are activation of Akt/PKB firing, increase in the levels of anti-apoptotic Bcl-2–related proteins, inactivation of p53 through changes in the *p53* gene or the upstream regulators of p53, methylation of the promoters of a variety of pro-apoptotic genes, interference with cytochrome *c* release from mitochondria, and inhibition of caspases.

- Apoptosis is only one of several forms of regulated cell death that threaten the survival of pre-neoplastic cells. Others include necroptosis, pyroptosis, and ferroptosis.

Thought questions

1. Considering that DNA tumor viruses must suppress the apoptosis of infected cells in order to multiply, what molecular strategies are available for them to do so?

2. What types of factors influence the decision of p53 to act in a cytostatic versus a pro-apoptotic fashion?

3. How can anti-cancer therapeutics be successful in treating cancer cells that have inactivated components of their apoptotic machinery? Given the physiological stresses that are known to activate p53-induced apoptosis, what types of anti-cancer therapeutic drugs might be created to treat cancer cells?

4. What side effects would you predict could result from a general inhibition of apoptosis in all tissues of the body?

5. How might the loss of components of the apoptosis machinery render cancer cells more susceptible than normal cells to certain types of cell death?

6. Enumerate the range of physiological stresses that induce apoptosis that cancer cells must confront and circumvent during the course of tumor development.

Additional reading

Ashkenazi A (2002) Targeting death and decoy receptors of the tumour-necrosis factor superfamily. *Nat. Rev. Cancer* 2, 410–430.

Baehrecke EH (2002) How death shapes life during development. *Nat. Rev. Mol. Cell Biol.* 3, 779–787.

Bargonetti J and Prives C (2019) Gain-of-function mutant p53: history and speculation. *J. Mol. Cell Biol.* 11, 605–609.

Bartek J, Bartkova J & Lukas J (2007) DNA damage signalling guards against activated oncogenes and tumor progression. *Oncogene* 26, 7773–7779.

Baserga R, Peruzzi F & Reiss K (2003) The IGF-1 receptor in cancer biology. *Int. J. Cancer* 107, 873–877.

Belyi VA, Ak P, Markert E et al. (2010) The origins and evolution of the p53 family of genes. *Cold Spring Harb. Perspect. Biol.* 2, a001198.

Blattner C (2008) Regulation of p53. *Cell Cycle* 7, 3149–3153.

Bouaoun L, Sonkin D, Ardin M et al. (2016) TP53 Variations in Human Cancers: New Lessons from the IARC TP53 Database and Genomics Data. *Hum. Mutat.* 9, 865–876.

Brosh R & Rotter V (2010) When mutants gain new powers: news from the mutant p53 field. *Nat. Rev. Cancer* 9, 701–713.

Bursch W (2001) The autophagosomal-lysosomal compartment in programmed cell death. *Cell Death Differ.* 8, 569–581.

Calle EE & Kakas R (2004) Overweight, obesity, and cancer: epidemiological evidence and proposed mechanisms. *Nat. Rev. Cancer* 4, 579–591.

Chandeck C & Mooi WJ (2010) Oncogene-induced senescence. *Adv. Anat. Pathol.* 17, 42–48.

Cichowski K & Hahn WC (2008) Unexpected pieces to the senescence puzzle. *Cell* 133, 958–961.

Collado M & Serrano M (2010) Senescence in tumours: evidence from mice and humans. *Nat. Rev. Cancer* 10, 51–57.

Coutts AS & La Thangue NB (2007) Mdm2 widens its repertoire. *Cell Cycle* 6, 827–829.

Eisenberg-Lerner A, Bialik S, Simon H-U & Kimchi A (2009) Life and death partners: apoptosis, autophagy and the cross-talk between them. *Cell Death Differ.* 16, 966–975.

El-Deiry WS (2003) The role of p53 in chemosensitivity and radiosensitivity. *Oncogene* 22, 7486–7495.

Faget DV, Ren O & Stewart SA (2019) Unmasking senescence: context-dependent effects of SASP in cancer. *Nat. Rev. Cancer* 19, 439–453.

Fulda S and Kögel D (2015) Cell death by autophagy: emerging mechanisms and implications for cancer therapy. *Oncogene* 34, 5105-5113.

Fürstenberger G & Senn H-G (2002) Insulin-like growth factors and cancer. *Lancet Oncol.* 3, 298–302.

Gencel-Augusto J & Lozano G (2020) p53 tetramerization: at the center of the dominant-negative effect of mutant p53. *Genes Dev.* 34, 1128–1146.

Gil J & Peters G (2006) Regulation of the INK4b–ARF–INK4a tumour suppressor locus: all for one or one for all. *Nat. Rev. Mol. Cell Biol.* 7, 667–677.

Green DR (2019) The coming decade of cell death research: five riddles. *Cell* 177, 1094–1107.

Gyrd-Hansen M & Meier P (2010) IAPs: from caspase inhibitors to modulators of NF-κB, inflammation and cancer. *Nat. Rev. Cancer* 10, 561–574.

Hampton TJ & Vousden KH (2016) Regulation of cellular metabolism and hypoxia by p53. *Cold Spring Harb. Perspect. Med.* 6, a026146.

Hippert MM, O'Toole PS & Thorburn A (2006) Autophagy in cancer: good, bad, or both? *Cancer Res.* 66, 9349–9351.

Hursting SD, DiGiovanni J, Dannenberg AJ et al. (2012) Obesity, energy balance, and cancer: new opportunities for prevention. *Cancer Prev. Res. (Phila.)* 5, 1260–1272.

Joerger AC & Fersht AR (2008) Structural biology of the tumor suppressor p53. *Annu. Rev. Biochem.* 77, 557–582.

Kaiser AM & Attardi LD (2018) Deconstructing networks of p53-mediated tumor suppression in vivo. *Cell Death Differ.* 25, 93–103.

Kale J, Osterlund EJ & Andrews DW (2018) BCL-2 family proteins: changing partners in the dance towards death *Cell Death Differ.* 25, 65–80.

Kalkavan H & Green DR (2018) MOMP, cell suicide as a BCL-2 family business. *Cell Death Differ.* 25, 46–55.

Karin M & Lin A (2002) NF-κB at the crossroads of life and death. *Nat. Immunol.* 3, 221–227.

Kastenhuber ER & Lowe SW (2017) Putting p53 in Context. *Cell* 170, 1062–1078.

Kroemer G & White E (2010) Autophagy for the avoidance of degenerative, inflammatory, infectious and neoplastic disease. *Curr. Opin. Cell Biol.* 22, 121–123.

Kuilman T, Michaloglou C, Mooi WJ & Peeper DS (2010) The essence of senescence. *Genes Dev.* 24, 2463–2479.

Larsson O, Girnita A & Girnita L (2005) Role of insulin-like growth factor I receptor in signaling in cancer. *Brit. J. Cancer* 92, 2097–2101.

Levine AJ (2009) The common mechanisms of transformation by small DNA tumor viruses: the inactivation of tumor suppressor gene products: p53. *Virology* 384, 285–293.

Levine AJ, Feng Z, Mak TW et al. (2006) Coordination and communication between the p53 and IGF1–AKT–TOR signal transduction pathways. *Genes. Dev.* 20, 267–275.

Levine AJ, Hu W & Feng Z (2006) The p53 pathway: what questions remain to be explored? *Cell Death Differ.* 13, 1027–1036.

Levine B & Kroemer G (2008) Autophagy in the pathogenesis of disease. *Cell* 132, 27–42.

Lim YP, Lim TT, Chan YL et al. (2007) The p53 knowledgebase: an integrated information resource for p53 research. *Oncogene* 26, 1517–1521.

Lomonosova E & Chinnadurai G (2009) BH3-only proteins in apoptosis and beyond: an overview. *Oncogene* Suppl. 1, S2–S19.

Maiuri MC, Zaickvar E, Kimchi A & Kroemer G (2007) Self-eating and self-killing: crosstalk between autophagy and apoptosis. *Nat. Rev. Mol. Cell Biol.* 8, 741–752.

Matheu A, Maraver A & Serrano M (2008) The Arf/p53 pathway in cancer and aging. *Cancer Res.* 68, 6031–6034.

Mathew R & White E (2011) Autophagy in tumorigenesis and energy metabolism: friend by day, foe by night. *Curr. Opin. Genet. Dev.* 21, 113–119.

Meek DW (2015) Regulation of the p53 response and its relationship to cancer. *Biochem. J.* 469, 325–346.

Meek DW & Anderson CW (2009) Posttranslational modification of p53: Cooperative integrators of function. *Cold Spring Harb. Perspect. Biol.* a000950.

Meulmeester E, Pereg Y, Shiloh Y & Jochemsen AG (2005) ATM-mediated phosphorylations inhibit Mdmx/Mdm2 stabilization by HAUSP in favor of p53 activation. *Cell Cycle* 4, 1166–1170.

Mooi WJ & Peeper DS (2006) Oncogene-induced cell senescence: halting on the road to cancer. *New Engl. J. Med.* 355, 1037–1045

Oren M & Rotter V (2010) Mutant p53 gain-of-function in cancer. *Cold Spring Harb. Perspect. Biol.* a001107.

Ow Y-LP, Green DR, Hao Z & Mak T (2008) Cytochrome: functions beyond respiration. *Nat. Rev. Mol. Cell Biol.* 9, 532–542.

Paluvai H, Di Giorgio E & Brancolini C (2020) The histone code of senescence. *Cells* 9, 466.

Perry ME (2010) The regulation of the p53-mediated stress response by MDM2 and MDM4. *Cold Spring Harb. Perspect. Biol.* 2, 968.

Pflaum J, Schlosser S & Müller M (2014) p53 family and cellular stress responses in cancer. *Front. Oncol.* 4, 1–15.

Plati J, Bucur O & Khosravi-Far R (2008) Dysregulation of apoptotic signaling in cancer: molecular mechanisms and therapeutic opportunities. *J. Cell. Biochem.* 104, 1124–1149.

Pollak M (2008) Insulin and insulin-like growth factor signaling in neoplasia. *Nat. Rev. Cancer* 8, 915–928.

Proskuryakov SY & Gabai VL (2010) Mechanisms of tumor cell necrosis. *Curr. Pharm. Des.* 16, 56–68.

Rodier F & Campisi J (2011) Four faces of cellular senescence. *J. Cell Biol.* 192, 547–556.

Sharpless NE & Sherr CJ (2015) Forging a signature of in vivo senescence. *Nat. Rev. Cancer* 15, 397–408.

Siegmund D, Lang I & Wajant H (2017) Cell death-independent activities of the death receptors CD95, TRAILR1, and TRAILR2. *FEBS J.* 284:1131–1159.

Singh R, Letai A & Sarosiek K (2019) Regulation of apoptosis in health and disease: the balancing act of BCL-2 family proteins. *Nat. Rev. Mol. Cell Biol.* 20, 175–193.

Sinha S & Levine B (2009) The autophagy effector Beclin 1: a novel BH3-only protein. *Oncogene* 27, S137–S148.

Tang D, Kang R, Berghe TV et al. (2019) The molecular machinery of regulated cell death. *Cell Res.* 29, 347–364.

Varley JM (2003) Germline TP53 mutations and Li-Fraumeni syndrome. *Hum. Mutat.* 21, 313–320.

Wang, Y-Y, Liu X-L & Zhao R (2019) Induction of pyroptosis and its implications for cancer management. *Front. Oncol.* 9, 1–10.

Williams AB & Schumacher B (2016) p53 in the DNA-damage-repair process. *Cold Spring Harb. Perspect. Med.* 6, 1–15

White E & DiPaola RS (2009) The double-edged sword of autophagy modulation in cancer. *Clin. Cancer Res.* 15, 5308–5316.

Xie Z & Klionsky DJ (2007) Autophagosome formation: core machinery and adaptations. *Nat. Cell Biol.* 9, 1102–1109.

Zong WX & Thompson CB (2006) Necrotic death as a cell fate. *Genes Dev.* 20, 1–15.

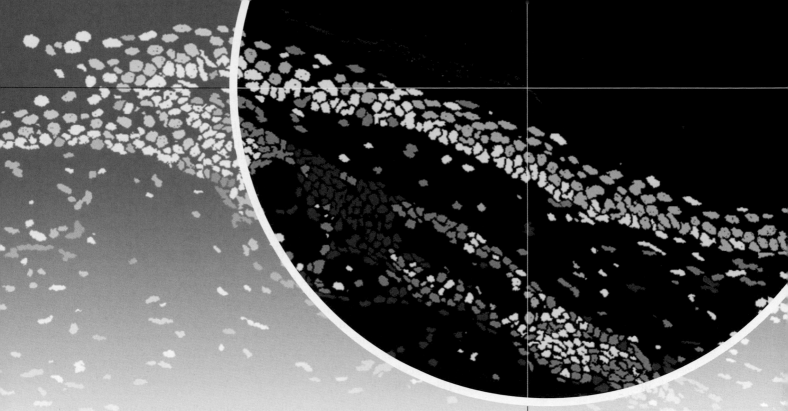

Chapter 10

Eternal Life: Cell Immortalization and Tumorigenesis

Chapter opener photo: From I. Flores et al., *Genes Dev.* 22:654-667, 2008. With permission from Cold Spring Harbor Laboratory Press.

> Death takes place because a worn-out tissue cannot forever renew itself, and because a capacity for increase by means of cell division is not everlasting but finite.
>
> August Weissmann, biologist, 1881

In previous chapters, we read about a number of distinctive traits displayed by cancer cells. In some instances, these traits are acquired through the actions of activated oncogenes; in others, cancer cell–specific traits can be traced back to the loss of tumor suppressor genes. As we will discuss in Chapter 11, the acquisition by human cells of these neoplastic traits (and thus the development of clinically apparent human tumors) usually requires several decades' time.

During this extended period of development, populations of human cells pass through a long succession of growth-and-division cycles as they evolve toward the neoplastic growth state. Such extensive proliferation, however, conflicts with a fundamental property of normal human cells: they are endowed with an ability to replicate only a far smaller number of times. Once normal human cell populations have exhausted their allotment of allowed doublings, the cells in these populations cease proliferating and may even enter senescence or apoptosis.

These facts lead us to a simple, inescapable conclusion: in order to form tumors, incipient cancer cells must breach the barrier that normally limits their proliferative potential. Somehow, they must acquire the ability to multiply for an abnormally large number of growth-and-division cycles, so that they can successfully complete the multiple steps of tumor development.

Movies in this chapter

10.1 Telomere Replication

In this chapter, we explore the nature of the regulatory machinery that limits cell proliferation and will see how it must be neutralized in order for cells to become fully neoplastic and form clinically detectable tumors. By neutralizing this machinery, cells gain the ability to proliferate indefinitely—the phenotype of cell immortality. This immortality is a critical component of the neoplastic growth program of many if not all types of tumor cells.

10.1 Normal cell populations appear to register the number of cell generations separating them from their ancestors in the early embryo

In multicellular (metazoan) animals such as ourselves, the origin of each cell can, in principle, be traced through multiple cell generations back to a single ancestor—the fertilized egg. Looking in the other direction, the sequence of cell divisions that stretches from an ancestral cell that existed in the embryo to a descendant cell that exists many cell generations later is often termed a cell lineage. Indeed, in a relatively simple metazoan—the worm *Caenorhabditis elegans*—the lineages of all 959 somatic cells in the adult body have been traced and can be depicted as a pedigree (**Figure 10.1**).

In large, complex mammals, however, the assembling of a comparable pedigree will never happen, because the total number of cell divisions is astronomical: the adult human body, for example, comprises almost 10^{14} cells, and the organism as a whole

Figure 10.1 The pedigree of cells in the body of a worm (A) The adult worm *Caenorhabditis elegans* is composed of 959 somatic cells, and the lines of descent of all cells in this worm have been traced with precision back to the fertilized egg. (B) As is apparent, many of the cell lineages end abruptly because cells stop proliferating or are discarded (usually by apoptosis). The cells in each of the major branches are destined to form a specific tissue. For example, cells in the major, multigenerational branch on the right are destined to form the gonads, whereas those in a small branch at the far left will form the nervous system. (A, from J.E. Sulston and H.R. Horvitz, *Dev. Biol.* 56:110–156, 1977. With permission from Elsevier. B, from A. Chisholm and Y. Jin, eLS, 2003. With permission from John Wiley & Sons.)

(A)

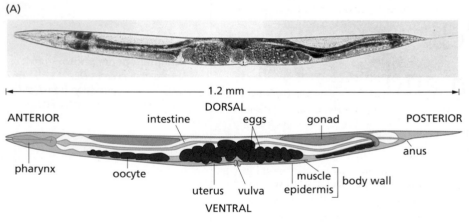

(B)

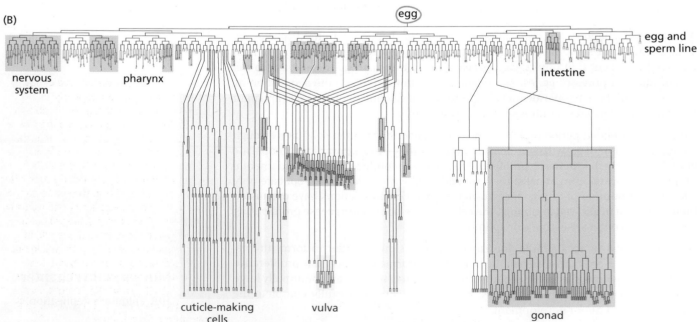

undergoes as many as 10^{16} cell divisions in a lifetime. Still, we can imagine that for each human body such a cell pedigree must exist, if only as a theoretical construct.

As is the case with *C. elegans*, during the course of human development, early embryonic cells become the founders of specific cell lineages that are committed to assuming various tissue-specific cellular phenotypes, that is, to differentiating (see Section 8.11). Indeed, the science of developmental biology focuses much of its attention on how individual cells in various cell lineages acquire the information that causes them to enter into one or another program of differentiation. However, developmental biologists do not address a question of great relevance to tumor development: Are there specific controls that determine the number of cell generations through which the cells belonging to a particular cell lineage can pass during the lifetime of an organism? Can each branch and twig of the cell pedigree grow indefinitely, or is the number of replicative generations in each cell lineage predetermined and limited?

To be sure, some cell populations proliferate often in normal tissues, whereas others proliferate hardly at all, further complicating calculations of cell generations (Supplementary Sidebar 10.1). Currently available techniques do not allow us to determine with any precision how many times the cells within specific normal cellular lineages within the human or mouse body pass through successive growth-and-division cycles. However, a crude measure of the replicative capacity of a cell lineage can be undertaken by culturing cells of interest *in vitro*. For example, one can prepare fibroblasts from living tissue, introduce them into a Petri dish, and determine how many times these cells will double. (In practice, such experiments require **serial passaging**, in which a portion of the cells that have filled one dish are removed and introduced into a second dish and allowed to proliferate, after which some of their number are introduced into a third dish, and so forth.)

As first demonstrated in the mid-1960s, **primary cells** (that is, cells prepared from living tissues and introduced directly into tissue culture) from rodent or human embryos exhibit a limited number of successive replicative cycles in culture. The work of Leonard Hayflick showed that cells would stop growing after a certain, apparently predetermined number of divisions and enter into the state that came to be called **replicative senescence** (**Figure 10.2**), which closely resembles the senescent state discussed in Section 9.11. As previously described, senescent cells remain metabolically active but seem to have lost irreversibly the ability to re-enter the active cell cycle. Such cells will spread out in monolayer culture, acquire a large cytoplasm, and persist for weeks if not months, as long as they are given adequate nutrients and growth factors; such cells are often described as taking on the appearance of a fried egg (see Figure 9.17). The supplied growth factors help to sustain the viability of the senescent cells, but they are unable to elicit the usual proliferative response observed when the same factors are applied to healthy, nonsenescent cells. Correlated with the unusual appearance of senescent cells are a number of biochemical alterations within these cells.

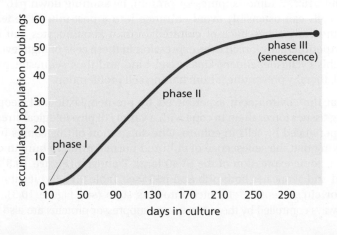

Figure 10.2 The proliferative capacity of cells passaged extensively in culture The ability of human fibroblasts to proliferate in culture was gauged in Leonard Hayflick's work by counting the number of times that the population of cells had doubled (*ordinate*). As is apparent, these cells, beginning soon after explantation from living tissue into culture (phase I), were able to proliferate robustly for about 60 doublings (phase II) before entering into senescence (phase III), in which state they could remain viable but nonproliferating; such senescent cells could remain viable for as long as a year. (From J.W. Shay and W.E. Wright, *Nat. Rev. Mol. Cell Biol.* 1:72–76, 2000. With permission from Nature.)

The precise number of replicative doublings exhibited by cultured cells before they reach senescence depends on the species from which the cells were prepared, on the tissue of origin, and on the age of the donor organism. Some experiments with human cells indicate that cells prepared from newborns are able to double in culture a greater number of times than comparable cells taken from middle-aged or elderly adults (Supplementary Sidebar 10.2). Such behavior suggests that cells from older individuals have already used up part of their allotment of replicative doublings prior to being introduced into tissue culture.

Although this is the simplest explanation, other explanations are equally plausible. For example, lineages of nondividing cells within a living tissue may sustain damage from, say, long-term exposure to reactive oxygen species (ROS) leaking from their mitochondria. In this case, the subsequently observed loss of proliferative capacity *in vitro* may be proportional to the *elapsed time* since these cells were first formed and may be unconnected directly with the *number of successive cell generations* their lineage has passed through. In fact, certain experiments have raised the question of whether replicative senescence, as defined by Hayflick, is a physiological process operating in living tissues or a reflection of the shortcomings of *in vitro* tissue culture techniques (Supplementary Sidebar 10.3).

10.2 Cells need to become immortalized in order to form a cell line

Whether or not replicative senescence is a tissue culture artifact, it stands in the way of propagating cell populations over long periods of time using standard tissue culture techniques. Thus, following their extraction from living tissues and introduction into tissue culture, cells almost invariably enter senescence and halt proliferation after an almost-predictable number of cell doublings. This represents a major obstacle for those investigators who would like to propagate cell populations indefinitely *in vitro*, which requires that the cells in these populations have undergone the process of **immortalization**.

The onset of replicative senescence is accompanied by increasing expression of the two key CDK inhibitors, $p16^{INK4A}$ and $p21^{Cip1}$. As discussed earlier (see Section 8.4), these CDK inhibitors are capable of halting cell cycle advance and driving a cell into a nongrowing state. Taken together, such evidence suggests that these two proteins may actually be responsible for imposing the senescent state on cultured cells. This notion is supported by forcing expression of $p16^{INK4A}$ in cells, which on its own suffices to create a cell-biological state that is similar if not identical to senescence. To complete this logic, various types of cell-physiological stress (including stresses experienced in tissue culture) activate expression of $p16^{INK4A}$ and $p21^{Cip1}$, which proceed to shut down cell proliferation and induce the cell phenotypes associated with senescence.

The prominent role of $p16^{INK4A}$ in triggering senescence is supported by other kinds of experiments: variants of human keratinocytes that succeed in escaping the fate of early senescence *in vitro* are often found to carry inactivated copies of the gene encoding the $p16^{INK4A}$ tumor suppressor protein; by shutting down $p16^{INK4A}$ expression, these cells can ostensibly avoid entrance into a post-mitotic senescent state. Moreover, the early senescence of cultured human keratinocytes can be circumvented by experimentally forced over-expression in these cells of high levels of CDK4; recall that this cyclin-dependent kinase can bind and thus sequester $p16^{INK4A}$ (see Figure 8.11), thereby preventing it from halting cell proliferation.

Imagine that the environment experienced by pre-neoplastic and neoplastic cells within living tissues forces them to cope with a variety of physiological stresses similar to those experienced by cells in culture. One suggestion of this comes from experiments showing that the senescence of cultured normal primary human cells can be avoided by ectopic expression of the SV40 large T antigen (**Figure 10.3**). Large T is able to bind and sequester both pRb and p53 (see Table 9.1) and its ability to do so is critical for circumventing entrance into this state (see Figure 10.3). Recall that (1) the pathways controlled by these two tumor suppressor proteins are also found to be

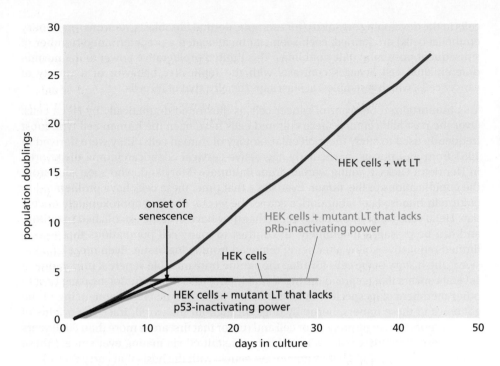

Figure 10.3 **Role of large T antigen in circumventing senescence** Experiments with a variety of human cell types indicate that both pRb and p53 must be functionally inactivated to ensure that these cells do not senesce in culture. This can be achieved through the expression of the SV40 large T antigen (LT) in these cells. As seen here, human embryonic kidney (HEK) cells senesce after 10 to 12 population doublings in culture (*gray line*). However, if they express the wild-type LT protein (*blue line*), they circumvent senescence and continue to proliferate for an extended period of time. Mutants of LT that have lost the ability to sequester either p53 (*brown line*) or pRb (*light green line*) fail to prevent senescence. Importantly, such cells that have bypassed senescence will eventually enter into crisis, which the expression of LT does not prevent. (Adapted from W.C. Hahn et al., *Mol. Cell. Biol.* 22:2111–2123, 2002.)

inactivated in the great majority of human tumors, as we read in Chapters 8 and 9, and (2) pRb and p53 serve as the key downstream effectors of the two proteins—p16^{INK4A} and p21^{Cip1}—that are induced as cells enter replicative senescence. Accordingly, when cancer cells are extracted from human tumors and placed into tissue culture, the previous inactivation of their pRb and p53 pathways that occurred *in vivo* (mimicking the actions of the SV40 large T antigen introduced experimentally *in vitro*) is likely to help these cells resist many of the stresses imposed by *in vitro* culture conditions and thereby to avoid senescence.

This reasoning suggests that avoidance of senescence is critical to the progression and survival of neoplastic cells *in vivo* and that, as a consequence, senescence is indeed a physiological process (rather than being simply an artifact of *in vitro* cell culture). Indeed, diverse observations indicate that a variety of other cell-physiological stresses besides extended *in vitro* propagation can also induce a cellular state that is indistinguishable from replicative senescence. Included among these are hyperoxia (that is, **supranormal** oxygen tensions leading to intracellular oxidative stress), DNA damage created by endogenously generated reactive oxygen species, X-rays, chemotherapeutic drugs, dysfunctional telomeres (see below), and aberrant signaling by certain oncoproteins, as discussed in Section 9.11. Nonetheless, even when tissue culture conditions are optimized and the SV40 large T oncoprotein is deployed to circumvent replicative senescence, as seen in Figure 10.3, cells will still not possess unlimited proliferative ability and instead will eventually be forced to halt proliferation by a process that is discussed in some detail below and is termed **crisis**. Hence, a second obstacle to unlimited proliferation (that is, entrance into an immortalized growth state) operates even after the initial obstacle of senescence has been surmounted. Unlike senescence, the obstacle erected by crisis does indeed seem to count the number of successive replicative doublings through which lineages of cells have passed since their origins in the early embryo.

There are, in fact, naturally occurring cells with unlimited proliferative potential *in vitro*: embryonic stem (ES) cells, which are prepared from the inner cell mass of very early mammalian embryos, retain the ability, under the proper conditions, to seed all the differentiated lineages in the body. When provided with the proper culture media, the ES cells show unlimited replicative potential in culture and are thus said to be **immortal**. (The term is a bit misleading, because it is really a *lineage* of ES cells in culture that is immortal rather than individual ES cells.)

These various observations convey the notion that very early in embryogenesis, cells have an unlimited replicative capacity. However, as specific lineages of differentiated cells in the developing organism (for example, dermal fibroblasts, neurons, mammary epithelial cells) are formed, each seems to be allocated a predetermined number of subsequent postembryonic doublings. This limited proliferative power of the normal differentiated cell lineages contrasts with the replicative behavior of a variety of cancer cells, which resembles, at least superficially, that of ES cells.

This immortalized behavior of cancer cells is illustrated dramatically by HeLa cells. Over the past half century, these cultured cells have been the human cell type most frequently used to study the molecular biology of human cells. They were derived in 1951 from an unusual, particularly aggressive cervical adenocarcinoma discovered in Henrietta Lacks, a young woman from Baltimore, Maryland, who soon died from the complications of this tumor. Ever since that time, these cells have proliferated in culture in hundreds of laboratories across the world, dividing approximately once a day. HeLa cells constitute a **cell line**, in that they have become established in culture and can be passaged indefinitely, in contrast to many cell populations that have a limited replicative ability after being removed from living tissue. Even more impressive is the history of the cells forming the canine transmissible venereal tumor, one of several cancers that is spread during sexual intercourse from a tumor-bearing host to other members of its species via transmission of neoplastic cells. Sequencing of the genomes of these tumors, borne by dogs throughout the world, trace the origins of these tumors to a single progenitor cell and tumor that first arose more than 6000 years ago, whose direct descendants have been transmitted via mating ever since. (These tumors thrive even though they are not syngeneic with the hosts that carry them.)

10.3 Cancer cells need to become immortal in order to form tumors

The observation that cancer cells, once adapted to growth in tissue culture, are often found to be immortal strongly suggests that immortalization is an integral component of the cancer cells' transformation to a neoplastic growth state. That is, cancer cells are immortal because they need to be in order to form a tumor. Why, conversely, are lineages of normal cells deprived of immortalized growth properties? Perhaps the body endows its normal cells with only a limited number of replicative generations as an anti-cancer defense mechanism. For example, if one or another cell in the body were to accidentally acquire certain oncogenes and shed critical tumor suppressor genes, its descendants would begin to proliferate uncontrollably, and the population of these tumor cells might well increase exponentially. However, if endowed with only a limited replicative potential, these cells might exhaust their allotment of cell doublings long before they succeeded in forming a life-threatening tumor mass; as a consequence, tumor development would grind to a halt.

The credibility of this model depends on some critical numbers. Specifically, we need to know how many successive cell generations are required to make a clinically detectable human tumor (**Figure 10.4A**) and how many generations are granted to normal cell lineages throughout the body. We know that human tumors are clonal, in the sense that all the neoplastic cells in the tumor mass descend from a common ancestral cell that underwent transformation at one point in time (see Section 2.5). With this fact in mind, we can ask how many cell generations separate the neoplastic cells forming a large human tumor from their common progenitor.

The arithmetic works out like this. The volume of a cubic centimeter (cm^3) within a tumor cell mass contains about 10^9 cells, and a life-threatening human tumor has a size, say, of 10^3 cm^3. We can calculate that these 10^{12} cells seem to have arisen following 40 cycles of exponential growth and division (Figure 10.4B)—that is, 40 cell generations separate the founding cell from its descendants in the end-stage, highly aggressive tumor ($10^3 \approx 2^{10}$; hence $10^{12} \approx 2^{40}$).

In fact, some types of normal human cells are known to pass through 50 or 60 cycles of growth and division in culture before they stop growing. According to the arithmetic above, and assuming that these numbers shed light on the proliferative potential of normal cells *in vivo*, these 50 or 60 cell generations of exponential growth

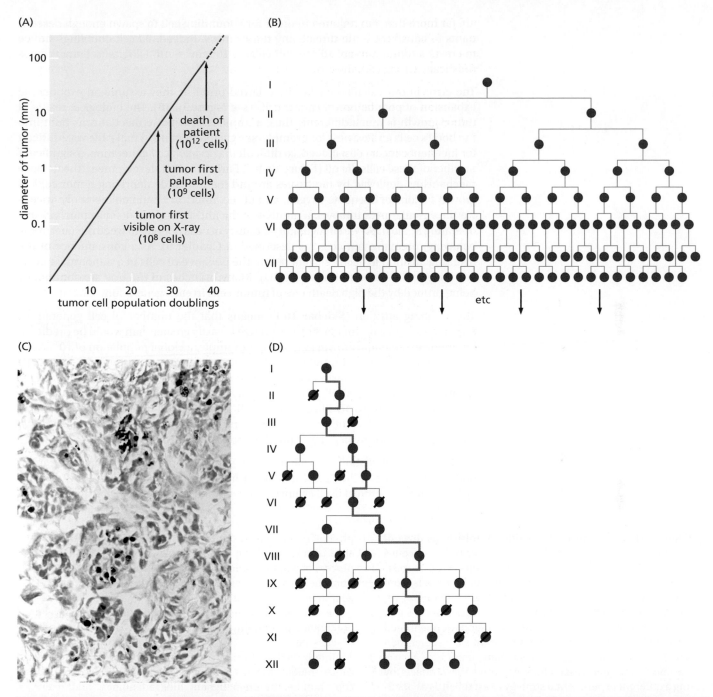

Figure 10.4 Generations of cells forming a tumor mass
(A) The growth of a human tumor—from detectable to palpable
to life-threatening size—can be related to the minimum number
of cell population doublings of the tumor-associated cancer cells.
(B) The plot in panel A assumes that each time the cancer cells in
a tumor divide, this cell population as a whole will double in size.
(Only six doublings are shown here.) Consequently, the number
of successive population doublings theoretically required to
generate a lethal tumor from a single founding ancestral cell can be
calculated from the graph in panel A to be about 40. (C) In reality,
cell populations that are evolving toward the neoplastic state or
are already neoplastic experience substantial attrition during each
cell generation. This attrition is evident from the apoptotic cells—
detected here by the *dark brown* staining of the TUNEL assay (see
Figure 9.12)—in a mammary tumor that arose in a mouse carrying
an oncogenic *Wnt-1* transgene in its germ line. In fact, this assay

greatly underestimates the rate of apoptosis in each cell generation,
because apoptotic cells persist only for an hour before they are
consumed by neighboring cells and by macrophages. Moreover,
cancer cells within a tumor may become senescent and thereby
enter irreversibly into a post-mitotic state. (D) The high rate of
attrition leads to loss of many cells in each cell generation (*diagonal
slashes*). Accordingly, an ancestral founder cell that should have
generated 2^{11} descendants leaves instead only five descendants
in the 12th generation. The number of successive cell generations
required to generate a tumor of life-threatening size is therefore
far larger and, in the absence of precise knowledge of attrition
rates, incalculable. Also shown is the *lineage* of a cell in the 12th
generation (*thick blue line*). (A, from B. Alberts et al., Molecular
Biology of the Cell, 6th ed. New York: W.W. Norton, 2015. C,
courtesy of L.D. Attardi and T. Jacks.)

are far more than are required in order for a founding cell to spawn enough descendants to constitute a life-threatening tumor mass. Indeed, 60 cell doublings suffice to create a tumor mass of 10^{15} to 10^{18} cells $\approx 10^9$ cm^3 $\approx 10^6$ kilograms. Something is drastically wrong with these numbers!

The error in our calculations lies in a flawed premise: they assume an exponential expansion of populations of cancer cells (see Figure 10.4B). The biological reality of tumor growth is much different. Thus, a number of defense mechanisms built into the body's cells and tissues (for example, see Chapters 7 and 8) make life very difficult for incipient cancer cells, indeed, so difficult that in each cell generation, a significant number of these cells die off (Figure 10.4C). Early in tumor development, the defense mechanisms deployed by the tissues around them include depriving tumor cells of growth factors, of adequate oxygen, and of the ability to eliminate metabolic wastes via the vasculature. Moreover, a number of the anti-tumor cell defense mechanisms operating in the hard-wired regulatory circuitry of cells operate to weed out aberrantly behaving, premalignant cells (as described in Chapter 9). As a consequence of the resulting attrition in each cell generation, the pedigree of cells in a tumor mass actually looks quite different (Figure 10.4D). Many branches of the tree are continually being pruned by the high death rate of tumor cells in each generation.

This ongoing attrition (Sidebar 10.1) means that the number of cell generations required to form a tumor mass of a given size is vastly greater than would be predicted by simple exponential growth kinetics. For example, a clonal population of 10^3 tumor cells might be thought, on the basis of its size, to have gone through 10 successive cycles of exponential growth and division since its founding by an ancestral cell; in reality, 20 or 30 or more cell generations may have been required to accumulate this many cells. (Figure 10.4D illustrates as well the *lineage* of a cell in a population of descendant cells.) Yet other calculations provide strong indication of extensive attrition of cells during neoplastic progression (Sidebar 10.2).

These revised calculations provide credible support for the notion that the human body endows its cell lineages with only a limited number of growth-and-division cycles in order to protect itself against the development of tumors. For example, if cell

Sidebar 10.1 A key cause of attrition: a poisonous dose of Ras A *ras* oncogene has been found to induce cell senescence when introduced into cultured cells by a viral vector, suggesting paradoxically that this potent transforming gene acts more to halt cell proliferation than to foster it. This result on its own created a puzzle, in that it suggested that virtually all cells that acquire a *ras* oncogene by somatic mutation were shunted immediately into the post-mitotic senescent state rather than becoming precursors of vigorously growing tumors.

Some years later, researchers explored these effects by using a different experimental strategy to establish a *ras* oncogene in cells: this strategy relied on activating an endogenous, chromosomal *ras* proto-oncogene in a genetically modified strain of mice (see Sidebar 7.5). This activation could be targeted to a specific tissue and induced by an experimental stimulus. Under these conditions, *ras* activation in the lungs and in the gastrointestinal tract resulted in hyperplasias (rather than senescence). Similar results could be observed in the pancreas and in the myeloid compartment of the bone marrow.

The critical difference here could be traced to how the *ras* oncogene was being expressed: in the earlier *in vitro* experiments, its expression was driven by a highly active transcriptional promoter present within a recombinant viral vector (see Supplementary Sidebar 3.9), whereas in the later mouse experiments, the oncogene was present in single copy number per cell and was being driven by its own native transcriptional promoter. This later experimental strategy recapitulated the state of cells that have acquired a *ras* oncogene through the process of point mutation. Hence, the critical difference came from the *level* of expression, which suggested that *ras* oncogenes created by somatic mutations can "fly under the radar" of the surveillance system that is designed to immediately weed out mutant, potentially neoplastic cells through the induction of senescence.

Interestingly, as tumor progression proceeded in the lung cancer model, the *ras* oncogene, initially present in single copy number in premalignant lung adenomas, underwent amplification. Prior to this time, the presence or absence of a functional *p53* gene had little if any effect on these early-stage lung adenomas; however, once the *ras* oncogene underwent amplification, elimination of p53 function became critical to permit the survival and continued expansion of the cells in the resulting, more aggressive adenocarcinomas. (Such p53 inactivation ostensibly allowed the evolving tumor cells to avoid the senescence that would otherwise be provoked by higher levels of Ras signaling; the elevated levels of Ras oncoprotein, for its part, appeared to provide these cells with increased mitogenic stimulation and thus enhanced proliferation.) Of note, this mouse model of lung cancer development reflects the behavior of many human tumors, in which mutant *ras* oncogenes undergo amplification as tumor progression proceeds. As the sixteenth-century alchemist Paracelsus put it, "the dose makes the poison."

Sidebar 10.2 Ongoing attrition makes the numbers add up
The profound effects of continual cell attrition are highlighted by observations of the genomes of the carcinoma cells forming human colorectal primary carcinomas and metastases. By measuring the frequency of randomly acquired point mutations present throughout these genomes, which accumulate at a predictable rate per cell generation together with other metrics of cumulative successive cell generations, investigators have estimated that in some of these tumors, many hundreds of cell generations intervene between the founding neoplastic cell and those cells present in a surgically resected tumor. In the absence of frequent, ongoing attrition, the founding cell would have spawned, as an example, $\sim 2^{500} = \sim 10^{150}$ cells, and such cells would weigh, in aggregate, ~ 1012 kilograms = ~ 10 billion times the weight of a (large) human body.

populations must pass through a succession of 100 replicative generations in order to form a clinically detectable human tumor, it is plausible that most incipient tumor cell populations will exhaust their normal allotment of 50 or 60 cell divisions long before they succeed in creating such a mass.

Having accepted, at least for the moment, this argument and its conclusions, we are now left with some major puzzles: How can normal cells throughout the body possibly remember their replicative history? And how can aspiring cancer cells erase the memory of this history and acquire the ability to proliferate indefinitely, or at least as long as is required to form a macroscopic, clinically detectable tumor?

A solution to the problem of replicative history must, sooner or later, be spelled out in terms of the actions of specific molecules within cells. Whatever its nature, the solution is hardly an obvious one. Organisms as complex as humans possess no innate biological clock for counting the number of (organismic) generations that separate each of us from ancestors who lived 100 or 1000 years ago. When we humans wish to learn the number of these generations over extended periods of time, we usually hire genealogists to chart them for us. No biological counting device inside our bodies can provide us with such answers. How, then, can far simpler biological entities—individual human cells—keep track of their generations?

In addition, this generational counting mechanism is likely to be, in the language of developmental biologists, **cell-autonomous**; that is, it must be intrinsic to a cell and not influenced by the ongoing interactions of the cell with its neighbors and with the body as a whole. Recall the contrasting situation of oncogenes and tumor suppressor genes. The products of most of these genes perturb the pathways responsible for processing the signals received by cells from their surroundings—*non-cell-autonomous* processes. Finally, the hypothesized generation-counting device must be relatively stable biochemically, because it needs to store a record of the past history of a cell lineage in a form that survives over extended periods of time, often many decades.

In principle, the counting processes of this "generational clock" (which measures elapsed cell generations rather than elapsed time) might depend on the concentrations of some intracellular molecule that (1) is synthesized early in development and not thereafter, (2) is present in high concentrations in the embryonic cells that are the ancestors to various lineages, and (3) undergoes progressive dilution by a factor of 2 each time a cell in the lineage divides. Senescence might then be triggered after the levels of this compound fall below some threshold level. Though plausible to mathematicians, this arrangement cannot work in the real world of biology. Biochemistry dictates that no molecule can be present in a cell over a concentration range of 2^{50} or 2^{60}.

This realization forced researchers to look elsewhere for the molecular embodiments of the generational clock. They came upon two regulatory mechanisms that govern the replicative capacity of cells growing *in vitro* and possibly *in vivo*. The first of these appears to measure the cumulative physiological stress that lineages of cells experience over extended periods of time and halts further proliferation once that damage exceeds a certain threshold; this yields the senescence discussed above (see Section 10.1).

The second regulator measures how many replicative generations a cell lineage has passed through and sounds an even more drastic alarm once the allowed quota has been used up; this alarm leads to the state termed crisis, which results in the apoptotic death of almost all cells in a population in culture. Crisis, as discussed below, represents a process and mechanism that can operate to constrain cell proliferation and abort the formation of tumors and, in other settings, paradoxically fosters tumor progression.

10.4 The proliferation of cultured cells is also limited by the telomeres of their chromosomes

Primary cells, as mentioned above, confront two obstacles to their continued long-term proliferation in culture. The first of these is entrance into a state of replicative senescence and the second, subsequent one involves crisis. To contrast the two, senescence represents a halt in cell proliferation with retention of cell viability over extended periods of time, whereas crisis involves death by apoptosis (see Section 9.13; **Figure 10.5**). Senescent cells seem to have a reasonably (but not totally) stable karyotype, whereas cells in crisis show widespread karyotypic disarray. Moreover, experimental evidence indicates clearly that senescence is more than a tissue culture artifact, but instead operates in a variety of biological states *in vivo* to limit or circumscribe further cell proliferation (see Section 9.6). In the case of human embryonic kidney (HEK) cells propagated *in vitro*, expression of the SV40 large T oncoprotein enables bypass of the first hurdle—senescence (see Figure 10.3)—but it will fail, on its own, to immortalize these cells. Thus, after an additional number of cell generations—sometimes 10 to 20—beyond the time when they would usually senesce, cell populations enter into crisis and exhibit widespread apoptosis. The SV40 large T antigen, even with its potent p53-inactivating ability, clearly does not protect against the apoptotic death associated with crisis.

The timing of crisis and the appearance of cells in crisis suggest the involvement of a second mechanism, one operating independently of the mechanisms triggering senescence described above. The workings of this second mechanism have been traced back to the chromosomal DNA of cells. Unlike the mechanisms leading to senescence, the molecular apparatus that initiates crisis is truly a functional counting device that tallies how many successive growth-and-division cycles cell lineages have passed through since their founding in the early embryo.

At first glance, the chromosomal DNA would seem to be an unlikely site for mammalian cells to construct such a "generational clock." We know that the structure of chromosomal DNA is highly stable and therefore should be unchanged from one cell generation to the next. A generational clock, in contrast, must depend on some progressive molecular change that is noted and recorded during each cell generation. How can chromosomal DNA molecules, which seem to be so immutable, register an additional cell generation each time a cell passes through a growth-and-division cycle?

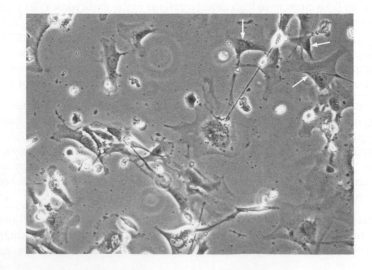

Figure 10.5 Apoptosis associated with cell populations in crisis Cell populations in crisis show widespread apoptosis, with a high percentage of cells in the midst of disintegrating and numerous cell fragments (*white refractile spots*). Because populations of cells enter into crisis asynchronously, some of the cells in such populations still appear relatively normal (*arrows, top right*). (Courtesy of S.A. Stewart.)

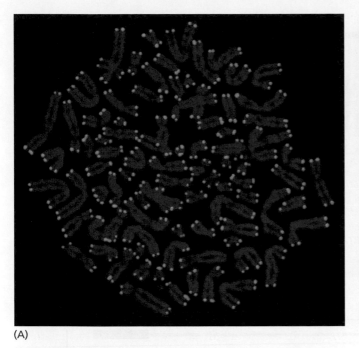

(A)

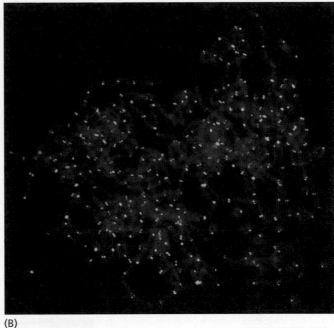

(B)

Figure 10.6 Telomeres detected by fluorescence *in situ* hybridization Telomeres can be detected by the technique of fluorescence *in situ* hybridization (FISH), in which a DNA probe that recognizes the repeated sequence present in telomeric DNA is coupled with a fluorescent label. (A) In this micrograph, human cells have been trapped in metaphase using a microtubule antagonist. The telomeres of the resulting condensed chromosomes (*red*) are then visualized through a probe that labels the telomeric DNA (*yellow green dots*). Each chromatid carries telomeres at both its ends. Because these cells are in metaphase, the paired, recently replicated chromatids have not yet separated. (B) The karyotype seen in panel A can be compared with that of cells that have been deprived of TRF2, a key protein in maintaining normal telomere structure (see Sidebar 10.4). In these cells, the telomeres lose their protective function (although telomeric DNA remains), resulting in dozens of end-to-end fusions involving all of the chromosomes in a cell and yielding one huge chromosome. (A, courtesy of J. Karlseder and T. de Lange. B, from G.B. Celli and T. de Lange, *Nat. Cell Biol.* 7:712–718, 2005. With permission from Nature.)

The key lies in the structure of chromosomal DNA: each mammalian chromosome carries a single, extremely long DNA molecule (ranging from ~247 mb for Chromosome 1 to ~47 mb for Chromosome 21). As it turns out, the two ends of this linear DNA molecule create a serious problem for the cell—a problem dramatically revealed by the experimental technique of DNA transfection (see Section 4.1). After entering cells, transfected linear DNA molecules are rapidly fused end-to-end through the actions of a variety of nucleases and DNA **ligases** that are active in most if not all mammalian cell types. Hence, linear DNA molecules are intrinsically unstable in our cells, yet the linear DNA molecules forming the backbones of chromosomes clearly persist.

The **telomeres** located at the ends of the chromosomes (**Figure 10.6**) explain how these linear DNA molecules can stably coexist with the cell's various DNA-modifying enzymes. Telomeres act to prevent the end-to-end fusion of chromosomal DNA molecules and, hence, the fusion of chromosomes with one another. In effect, telomeres serve as protective shields for the chromosomal ends, much like the aglets safeguarding the tips of shoelaces. As we will see, the catastrophic events of crisis are triggered when cells lose functional telomeres from their chromosomes.

Discoveries reported in 1941 by Barbara McClintock (**Figure 10.7**) concerning her studies of the chromosomes of corn first revealed that chromosomes that have lost functional telomeres at their ends soon fuse, end-to-end, with one another. The resulting megachromosomes possess two or more **centromeres**, the specialized

Figure 10.7 Barbara McClintock Barbara McClintock's detailed studies of the chromosomes of corn (maize) revealed specialized structures at the ends of chromosomes—the telomeres—that protected them from end-to-end fusions. This work, and her demonstration of movable genetic elements in the corn genome, later called transposons, earned her the Nobel Prize in Physiology or Medicine in 1983. (Courtesy of American Philosophical Society.)

Figure 10.8 Shortening of telomeric DNA in concert with cell proliferation (A) Here, a culture of human lymphocytes that had circumvented culture-induced senescence following infection by Epstein–Barr virus was passaged, and a portion of each culture was set aside for analysis of the cells' telomeric DNA. Each passage represented three to six population doublings. The length of telomeres was measured by treating genomic DNA with several restriction enzymes that do not cleave within the repeating TTAGGG sequence that constitutes telomeric DNA but do cleave frequently throughout the rest of the chromosomal DNA. The only high–molecular-weight genomic fragments such treatment leaves behind are the telomeric restriction fragments (TRFs). Their lengths can then be determined by gel electrophoresis followed by Southern blotting analysis using a probe that recognizes the TTAGGG sequence. Because the telomeres within a given cell vary in length and the extent of telomeric DNA shortening differs from one cell to another, the size distribution of telomeric DNA in each sample is heterogeneous. In the analysis shown, the TRFs of mortal (i.e., non-immortalized) human lymphocytes grew shorter with every successive passaging by 50 to 100 bp per cell generation. When the TRFs became less than about 3 kbp long, cells entered crisis. Later, when cells emerged spontaneously from crisis (and became immortalized), they maintained their telomeric DNA at slightly longer sizes than those seen in crisis. (B) Such observations have led to the model shown, in which telomeric DNA (*red*) shortens progressively each time a cell passes through a growth-and-division cycle until the telomeres are so eroded that they can no longer protect the ends of chromosomal DNA. This telomere collapse provokes a cell to enter into crisis. [In fact, the TRFs of panel A include both pure TTAGGG repeats at the terminus of each telomere and long stretches of TTAGGG-like sequences located at the internal (centromeric) end of each telomere; the latter may not be functional in protecting the chromosomal ends. Telomeres may lose their protective function when their regions of pure TTAGGG repeats have eroded down to a length of ~12 repeats.] In most tumors, crisis is ended when cells acquire the expression of hTERT, resulting in slightly longer telomeres that succeed in protecting the ends of the telomeres and thus of the chromosomal DNA. Note that this drawing is not to scale: telomeric DNA is usually 5 to 10 kbp long, whereas the body of chromosomal DNA (*blue*) is tens to hundreds of megabase pairs (Mbp) long. (A, from C.M. Counter et al., *J. Virol.* 68:3410–3414, 1994. With permission of American Society for Microbiology.)

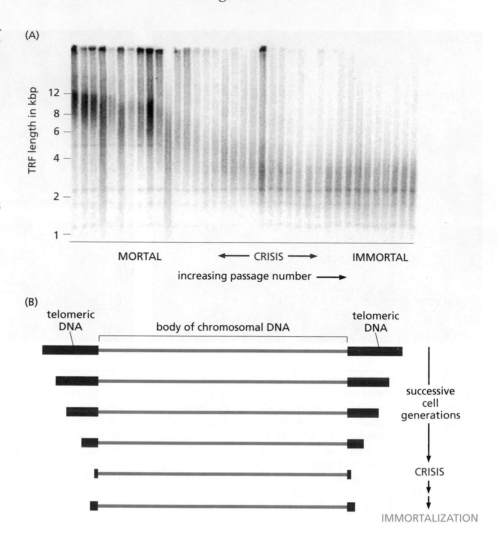

chromosomal structures that become attached during mitosis to the fibers of the mitotic spindle. An extreme form of this fusion is shown in Figure 10.6B, in which virtually all the chromosomes of a cell have fused into one giant chromosome after they all have been deprived of functional telomeres.

As discussed in greater detail below, the DNA component of each telomere in our cells is composed of the 5′-TTAGGG-3′ hexanucleotide sequence, which is tandemly repeated thousands of times; these repeated sequences, together with associated proteins, form the functional telomere. The telomeric DNA (and thus the telomeres) of normal human cells proliferating in culture shorten progressively during each growth-and-division cycle, doing so until they become so short that they can no longer effectively protect the ends of chromosomes (**Figure 10.8**). At this point, crisis occurs, chromosomal DNA ends fuse with one another, and widespread apoptotic death is observable. Hence, in such cells, it is *telomere shortening* that functions to register the number of cell generations through which cell populations have passed since their origin in the early embryo.

In human cells entering crisis, the initial end-to-end fusion events often occur between the eroded telomeric ends of the two sister chromatids that form the two halves of the same chromosome (**Figure 10.9A**). Recall that paired chromatids exist during the G_2 phase of the cell cycle—a period after S phase has created two chromatids from a parental chromosome and before M phase, when these two sister chromatids are destined to be separated from one another (see Figure 8.3). Such fusions between the ends of sister chromatids (rather than between the ends of two unrelated chromosomes) are favored for at least two reasons. First, the two chromatids, and thus their ends, are held in close proximity through their joined centromeres. Second, for unknown reasons, telomeres shorten at different rates on different chromosomal

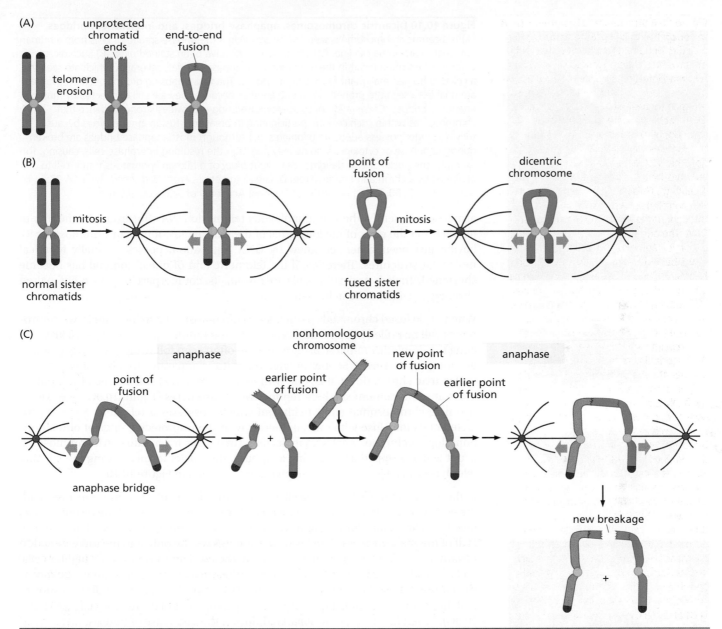

Figure 10.9 Mechanisms of breakage–fusion–bridge cycles (A) Chromosomal breakage, fusion, and bridge formation occur when telomeres (*red*) have become too short to protect the ends of chromosomal DNA. Pictured here is the configuration of a chromosome in the G_2 phase of the cell cycle. During this phase of the cell cycle, its two chromatids remain attached to one another by their shared centromere (*green*). (In fact, it is the multi-subunit centromere-associated protein complex known as the kinetochore, *not shown*, that associates with the centromeric DNA and is responsible during the G_2 phase for holding the two chromatids together and during mitosis for binding the chromosome to microtubules of the mitotic spindle.) As shown, the telomeres on the long arms of this *blue* chromosome have eroded before those of the short arms. Both ends of the long arms are equally eroded since they have recently been duplicated during the previous S phase. The ends of these two arms, held in close proximity by the shared centromere, now fuse with one another, leading to a *dicentric* chromatid, which contains two telomeres and two half-centromeres. (B) During a normal mitosis (*left*), the two sister chromatids will be pulled apart to opposite centrosomes located at the two polar ends of the mitotic spindle (*left*). Likewise, the mitotic spindle will pull the centromeres (*green*) of a dicentric chromosome in opposite directions, unaware that these two chromatids are joined via their long arms. (C) Unlike normally configured chromatid pairs, which will separate cleanly and segregate in groups near the two centrosomes, the dicentric chromosome described in (B) will be unable to do so and will instead create a temporary bridge between the two poles of the mitotic spindle during the anaphase of mitosis, explaining the term anaphase bridge. Eventually, the dicentric chromosome will be ripped apart at some weak point. During the next cell cycle, the larger fragment lacking a telomere at one end (*blue*) may fuse with another atelomeric chromosome (*beige*), creating a new dicentric chromosome, which itself will be pulled apart during a subsequent mitosis, resulting once again in a breakage–fusion–bridge cycle. (The fate of the shorter chromatid fragment generated by the first cycle of breakage is not shown.)

(A)

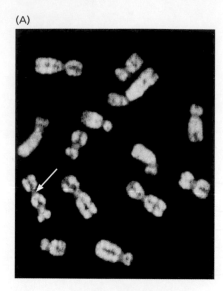

(B)

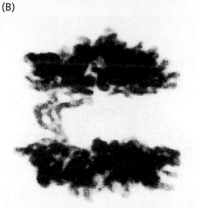

(C)

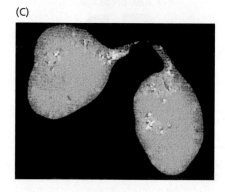

Figure 10.10 Dicentric chromosomes, anaphase bridges, and internuclear bridges
(A) A dicentric chromosome is seen (*white arrow*) in a metaphase spread of cells from a human pancreatic carcinoma cell line. Centromeric regions have been stained *pink*. (B) Such dicentric chromosomes often result in the formation of anaphase bridges, several of which are seen in a cell of a human malignant fibrous histiocytoma. The two opposing groups of chromosomes are unable to separate properly from one another because they are still connected by multiple anaphase bridges. Observation of such anaphase bridges provides strong evidence that chromosomes within such cells are participating in breakage–fusion–bridge cycles because they no longer possess adequate telomeres. (C) Although most anaphase bridges are broken during telophase or cytokinesis, some may persist in the resulting interphase cells, causing, for example, the nuclei of two daughter cells (*light blue*) of a human lipomatous tumor cell to be linked by a chromatin bridge. (From D. Gisselsson et al., *Proc. Natl. Acad. Sci. USA* 97: 5357–5362, 2000. With permission of National Academy of Sciences, U.S.A.)

arms. Consequently, the two homologous telomeric DNAs (for example, the pair of telomeres at the ends of the long arms of the chromatids of human Chromosome 9), having just been generated anew by DNA replication, possess virtually identical molecular structures. Therefore, if the telomeric end of one chromatid has become shortened, frayed, and vulnerable to fusion, its counterpart on the other paired chromatid is likely to be in the same state.

When these fused chromatids participate in the mitosis that follows, their two centromeres will be pulled in opposite directions by the mitotic spindle (Figure 10.9B), creating the dramatic *anaphase bridges* that are often seen following extensive telomere erosion (**Figure 10.10**). Sooner or later, the mitotic apparatus that is pulling on such a **dicentric** chromatid (that is, one with two centromeres) will succeed in ripping it apart at some random site between the two centromeres (Figure 10.9C). This yields two new chromosomal ends, neither of which possesses a telomere. Such unprotected ends may fuse with one another or with the unprotected ends of other, nonhomologous chromosomal DNA molecules. (In the event that a dicentric chromosome is not actually ripped apart during anaphase, the nuclei formed during the ensuing telophase will be joined by the anaphase bridges seen in Figure 10.10.)

In the event that the resulting defective, nontelomeric end of a chromosome fuses with the end of an unrelated (nonhomologous) chromosome, the fate of the resulting new, dicentric chromatid during the next mitosis is more ambiguous (see Figure 10.9C). Half of the time, on average, the two centromeres will be pulled in the same direction toward the centrosome that nucleates one of the two soon-to-be-born daughter cells, and no breakage will occur. The other half of the time, the two centromeres of a dicentric chromatid will become attached to the two opposing spindle bodies, as shown, and thus to the two centrosomes located at opposite sides of the mitotic cell (see Figure 10.9C). When chromatids are separated during the anaphase of mitosis, this second configuration (like the earlier one between sister chromatids) will once again prove to be disastrous because it will involve the tearing apart of a chromatid and the generation of ends that are, as before, unprotected by telomeres.

These newly created chromosomal ends will once again attempt to fuse with yet other chromosomes, yielding more dicentric chromosomes and a new cycle of chromosome breakage. This sequence of events is termed the *breakage-fusion-bridge* (BFB) *cycle* because it involves the initial *breakage* of dicentric chromatids in anaphase, the subsequent *fusion* of the resulting **atelomeric** DNA ends with yet other chromatids, and the formation once again of anaphase bridges by the new dicentric chromosomes resulting from these fusions (see Figure 10.10). While they are occurring, these BFB cycles create karyotypic chaos that has the potential to affect many chromosomes within a cell, as proposed in 1941 by McClintock.

Cells that have entered crisis provide a clear indication that the molecular machinery triggering crisis resides in the telomeres because those cells show precisely the type of karyotypic disarray that is observed when chromosomes lose their telomeres. Moreover, the chromosomes that are fusing within these cells possess especially short telomeres, or none at all. Note that the karyotypic instability occurring during crisis, which has been termed "genetic catastrophe," is so severe that apoptosis often occurs even in the absence of p53 function.

10.5 Telomeres are complex molecular structures that are not easily replicated

Research conducted since the mid-1980s has revealed the molecular structure of telomeres and their DNA. To begin, and as cited earlier, the telomeric DNA of mammalian cells (as well as the cells of many other metazoa) is formed from the repeating hexanucleotide sequence 5'-TTAGGG-3' in one strand (the "G-rich" strand) and the complementary 5'-CCCTAA-3' in the other (the "C-rich" strand). In normal human cells, telomeric DNA is formed from several thousand of these hexanucleotide sequences, resulting in 5- to 15-kbp-long stretches of such repeating sequences at the ends of all human chromosomes.

The telomeric DNA of mammalian cells, and possibly the cells of all higher eukaryotes (metazoa and metaphyta), possesses an additional, distinctive feature: its G-rich strand is longer than its C-rich strand by one hundred to several hundred nucleotides, resulting in a long 3' single-strand overhang (**Figure 10.11A**). This overhanging strand is often found in a most unusual molecular configuration termed the **t-loop**. It was discovered in the late 1990s, when electron microscopy showed the ends of telomeric DNA to be configured in a loop, in effect a lariat (Figure 10.11B). This structure has been interpreted to depend on the formation of a three-stranded complex of DNA (Figure 10.11C). The t-loop may be present at the ends of all telomeres, although it has been observed in only a subset of those viewed, probably because of the technical difficulties associated with preserving and visualizing

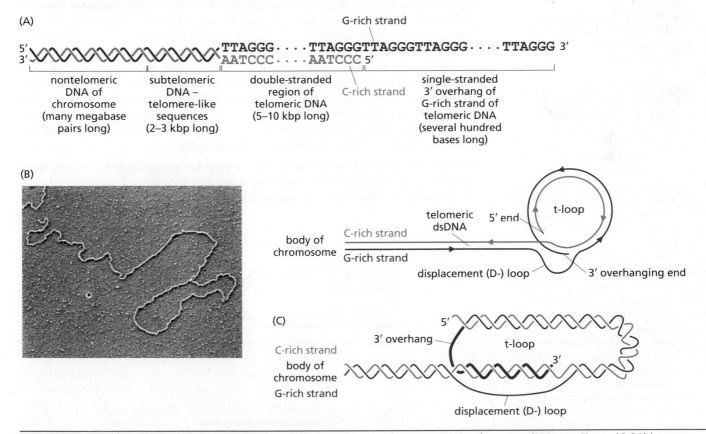

Figure 10.11 The 3' overhang of telomeric DNA and the t-loop
(A) The G-rich strand of telomeric DNA (*pink*) extends beyond the C-rich strand (*blue*). This creates a 3' overhang (*right*) that is often several hundred nucleotides long. This overhang is far shorter than the double-stranded portion of telomeric DNA (*middle*), which is often 5 to 10 kbp long and is not drawn to scale. In addition, a subtelomeric region of DNA lies between the functional telomeric DNA (*middle*) and the nontelomeric chromosomal DNA (*far left*). This subtelomeric DNA contains imperfect copies of the TTAGGG sequence and does not function to protect the end of the chromosome. However, it is usually included in the telomere restriction fragments (TRFs; see Figure 10.8A) because of its telomere-like sequence content. (B) Purification of telomeric DNA from mouse cells (following stabilization of double-stranded structures through chemical cross-linking) reveals, upon electron microscopy, lariats, called *t-loops*, at the ends of chromosomal DNA. (C) The t-loop is illustrated with the double helices drawn and the 3' overhanging end emphasized in a *heavier pink line*. Although this figure depicts the t-loop as being formed by the overhanging end of one of the two strands, there is evidence that the other strand can also occasionally form t-loops. (B, from J.D. Griffith et al., *Cell* 97:503–514, 1999. With permission from Elsevier.)

this structure in the electron microscope. The t-loop clearly helps to protect the ends of linear DNA molecules, because the end of the single-stranded overhanging region is tucked into a double-stranded region, out of harm's way.

Both the relatively long double-stranded telomeric DNA and the short overhanging end are bound by specific proteins. As might be expected, some of the proteins possess domains that specifically recognize and bind to the hexanucleotide sequence present in the double- and the single-stranded regions of telomeric DNA (Supplementary Sidebar 10.4). Together, these telomere-binding proteins, yet other associated proteins that remain to be identified, and the telomeric DNA form the nucleoprotein complexes that function as telomeres.

Although the replication machinery that operates during the S phase of the cell cycle is highly effective at copying the sequences in the middle of linear DNA molecules, such as those in the body of each of our chromosomes, this machinery has great difficulty faithfully copying sequences at the very ends of these molecules. The difficulty can be traced to the requirement that the synthesis of all DNA strands during DNA replication must be initiated at the 3′-hydroxyl end of an existing DNA or RNA strand, which serves as a **primer** to nucleate DNA strand elongation; alternatively, in the absence of an available DNA primer, the 3′ end of a short RNA molecule, can serve as a primer for DNA synthesis (**Figure 10.12**).

If the **primase** enzyme, which is responsible for laying down short RNA primers, happens to deposit a primer at some distance from the 3′ terminus of a template strand on which "lagging-strand synthesis" is occurring (see Figure 10.12), a DNA polymerase will initiate synthesis of a new DNA strand that lacks a substantial number of the bases complementary to the very 3′ end of this template strand. (Even if the primase sits down and constructs an RNA primer at the very end of the template strand, the sequences corresponding to the approximately 10 nucleotides of the RNA primer will not be present in the newly synthesized daughter strand, because the RNA primer will be degraded after it has served its purpose of initiating DNA strand elongation.) The end result is that the nucleotide sequences at the very tip of one of the two template strands of DNA will not be properly replicated.

This **end-replication problem** provides a molecular explanation for the observed shortening of telomeric DNA each time a normal human cell passes through a cell cycle. In addition to the under-replication of telomeric DNA ends, there are

Figure 10.12 Primers and the initiation of DNA synthesis (A) During normal DNA replication, the parental DNA double-helix (*right*) is unwound by a helicase enzyme (*not shown*), creating two template strands (*left*) and allowing the replication process as a whole to progress in a rightward direction (*light brown arrow*). Elongation of the *leading* daughter strand (*thin blue line*) can occur continuously, because its 5′-to-3′ synthesis proceeds in the same direction as the replication fork. In contrast, synthesis of the new *lagging* daughter strand (*thin pink line*) proceeds in a direction opposite to that of the progress of the replication fork, because DNA strand synthesis must occur in a 5′-to-3′ direction. This necessitates the involvement of short RNA primer segments (*green segments*) that are laid down at intervals of several hundred nucleotides by the *primase* enzyme. The 3′-hydroxyl ends of these short RNA primers serve as sites of initiation of newly made daughter-strand segments, which are termed Okazaki fragments prior to the removal of the RNA primers, the filling in by DNA polymerases of the resulting gaps, and the joining of these fragments by a DNA ligase (*not shown*). (B) For these reasons, the leading-strand synthesis (*thin blue line, above*) will proceed all the way to the end of the telomere (*right*) and thereby generate a faithful copy of the telomeric DNA sequences in one of the two new daughter helices. However, the lagging-strand synthesis (*thin red line, below*) will require an RNA primer near the end of the telomeric DNA (*circled*). (C) Because this RNA primer will subsequently be removed and because it may not be deposited precisely at the end of the parental template strand (*thick blue line*), sequences complementary to the 3′ end of the parental strand will not be present in the ultimately synthesized DNA molecule. Consequently, there is under-replication of the parental strand of DNA and loss of a limited stretch of telomeric DNA sequences from the chromosome of a daughter cell. (In addition, certain exonucleases degrade this specific stretch of DNA, compounding the effects of the under-replication.) The 5′-to-3′ orientation of each strand is indicated by the arrowheads. The parental DNA strands are drawn more thickly than the newly synthesized daughter strands.

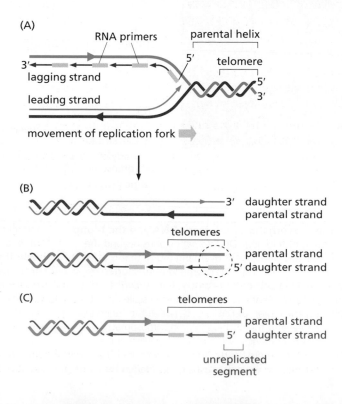

exonucleases within cells that chew on the ends of telomeric DNA in order to maintain the t-loops described above and are likely to contribute significantly to telomere erosion. For whatever reason, in many types of normal human cells, telomeres lose 50 to 100 bp of DNA during each cell generation (see Figure 10.8). This progressive erosion of telomeric DNA represents a simple molecular device that limits how many generations of descendant progeny a cell can spawn and solves the problems, enumerated in Section 10.3, of how to construct a stable, long-lived generation-counting device.

We can imagine, for example, that in human embryonic cells, the telomeric DNA begins rather long, perhaps 8 to 10 kb in length. As various lineages of descendant cells throughout the developing body proceed through their repeated cycles of growth and division, the telomeres in these cells grow progressively shorter. Ultimately, in some cells, the telomeric DNA erodes down to a size that is so short that it can no longer perform its intended function of protecting the ends of the chromosomal DNA, the result being the breakage–fusion–bridge cycles and chromosomal translocations that are illustrated in Figure 10.9. Indeed, it is plausible that the aging of certain tissues derives from the loss by individual cells of replicative potential, and that this loss is attributable, in turn, to telomere erosion.

Although the supporting evidence is still indirect, it is highly likely that telomere shortening, by limiting the replicative potential of cell lineages, creates an obstacle to the accumulation of large populations of cancer cells. Thus, as argued earlier, should a clone of cells acquire oncogenic mutations (involving oncogene activation and tumor suppressor gene inactivation), the ability of this clone to expand to a large, clinically detectable size is likely to be constrained by its eroded telomeres.

10.6 Incipient cancer cells can escape crisis by expressing telomerase

The detailed behavior of pre-neoplastic cell clones *in vivo* is difficult to study, and so we are largely forced to extrapolate from observations of cultured cells, such as human fibroblasts. As mentioned earlier, if these cells are allowed to circumvent senescence through the expression of the SV40 large T oncoprotein, they will continue to replicate another 10 to 20 cell generations and then enter crisis. On rare occasion, a small group of cells will emerge spontaneously from the vast throng of cells in the midst of crisis. These variant cell clones—arising perhaps from a single cell among 10 million cells in crisis—proceed to proliferate and continue to do so indefinitely (for example, see Figure 10.8A). They have become immortalized, seemingly as the consequence of some random event.

This transition, observed with cultured cells *in vitro*, seems to recapitulate the behavior of premalignant cell populations *in vivo*. Thus, both cell populations enter crisis sooner or later, having passed through large numbers of replicative generations and suffered extensive erosion of their telomeres. Both are capable of spawning rare immortalized variants that have apparently solved the problem of telomere collapse, but how?

At first glance, telomere collapse and crisis would appear to be irreversible processes from which cells can never escape. As it turns out, the route to immortality is simple and, in retrospect, obvious: cells can emerge from crisis by regenerating their telomeres, thereby erasing the molecular record (their shortened telomeres) that previously blocked their proliferation and drove them into crisis.

Telomere regeneration can be accomplished through the actions of the **telomerase** enzyme, which functions specifically to elongate telomeric DNA. A striking finding is that telomerase activity is clearly detectable in 85 to 90% of human tumor cell samples, and is present at very low levels in the lysates of most types of normal human cells, as measured by the TRAP assay (Supplementary Sidebar 10.5). Although these low levels of telomerase activity may enable some type of minimal maintenance of the ends of telomeric DNA (such as repair or regeneration of the t-loop), they are clearly unable to prevent the progressive erosion of the double-stranded region of telomeres that accompanies passage of normal human cells through each cell cycle.

In adult humans, there are actually several known exceptions to the generally observed low levels of telomerase activity in normal cells. For example, substantial enzyme activity is present in the germ cells of the testes, and lymphocytes express a burst of telomerase activity when they become functionally activated.

The available evidence indicates that strong expression of the telomerase enzyme is present early in embryogenesis and is largely lost during the cellular differentiation that produces most of the body's tissues. Thus, during the course of early mouse and cow embryogenesis, telomerase is strongly expressed in embryonic cells in the time between the morula and blastocyst stages and seems to disappear thereafter; a similar control of telomerase expression is likely to operate during human embryonic development. Accordingly, most normal human somatic cells, which carry the full complement of genes specifying the multi-subunit telomerase holoenzyme, are denied the services of this enzyme because they do not *express* all of the encoding genes at significant levels.

Among the large populations of cultured cells in crisis, however, rare variants find a way to de-repress the gene or genes encoding telomerase and thereby acquire high levels of constitutively expressed enzyme. This enables them to extend their telomeric DNA to a length that permits further proliferation. Indeed, as long as their descendants continue to express this enzyme, such cells will continue to proliferate and thus will be considered to be immortalized. A similar sequence of events is presumed to occur *in vivo* when populations of pre-neoplastic cells enter into crisis and, on rare occasion, generate immortalized variants that become the progenitors of large neoplastic cell populations.

The telomerase enzyme was developed early in eukaryotic cell evolution and is present in the cells of almost all descendant organisms, both protozoan and metazoan. Its presence in single-cell eukaryotes afforded its initial detection and characterization, first in baker's yeast using genetic analyses and in a ciliate through biochemical purification. It is a complex enzyme composed of a number of distinct subunits, not all of which have been characterized. At the core of the mammalian telomerase **holoenzyme** are two subunits. One subunit is a DNA polymerase, more specifically a reverse transcriptase, which functions, like the related enzymes made by retroviruses (see Section 3.7) and a variety of other viruses and transposable elements, to synthesize DNA from an RNA template. (**Figure 10.13**). However, unlike these other reverse transcriptases (Supplementary Sidebar 10.6), the telomerase holoenzyme cleverly packs its own RNA template—the second essential subunit. A short segment of this 451-nucleotide-long RNA molecule serves as the template that instructs the reverse transcriptase activity of the holoenzyme.

Figure 10.13 Core subunits of the human telomerase holoenzyme
The human telomerase holoenzyme is composed of two core subunits, the hTERT catalytic subunit (*light brown*) and the associated *hTR* RNA subunit (*blue*); its mouse counterparts are termed mTERT and *mTR*. (Five other proteins, some present in two copies, are required to form a functional holoenzyme.) The holoenzyme attaches to the 3′ end of the G-rich strand overhang (*pink*), doing so in part through the hydrogen bonding of *hTR* to the last five nucleotides of the G-rich strand. Subsequently, by reverse transcription of sequences in the *hTR* subunit, hTERT is able to extend the G-rich strand by six nucleotides (*black*). By repeating this process in hexanucleotide increments (*ticks*), the enzyme can extend the G-rich strand by hundreds, even thousands of nucleotides, doing so in a *processive* fashion. Arrowheads indicate the 5′-to-3′ orientation of the nucleic acids. Upon completion of this elongation, a conventional DNA polymerase can fill in the complementary (*blue*) strand after RNA primer molecules have been laid down by a primase enzyme (*not shown*).

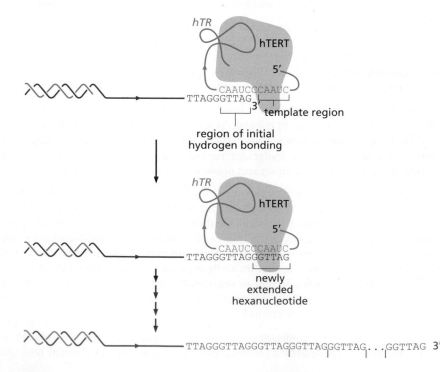

Isolation of substantial quantities of telomerase protein from the ciliate *Euplotes aediculatus* (see Supplementary Sidebar 10.6) allowed determination of the amino acid sequence of its catalytic subunit and, in turn, cloning of the gene encoding the ciliate enzyme as well as the homologous human gene. In human cells, this catalytic subunit, termed hTERT (for **h**uman **te**lomerase **r**everse **t**ranscriptase), synthesizes a DNA molecule that is complementary to six nucleotides present in the telomerase-associated RNA molecule (*hTR*; its encoding gene is sometimes called *TERC*), attaching these nucleotides to the G-rich 3′ overhanging end of the preexisting telomeric DNA (see Figure 10.13). The complementary strand of the telomeric DNA is then presumably synthesized by conventional DNA polymerases.

As would be predicted from the scenario described earlier, before entering crisis, human cells are essentially telomerase-negative and do not express appreciable levels of the *hTERT* mRNA. However, should a rare immortalized variant emerge from a population of cells in crisis, its descendants usually express significant levels of *hTERT* mRNA and exhibit substantial levels of telomerase enzyme activity (**Figure 10.14A**).

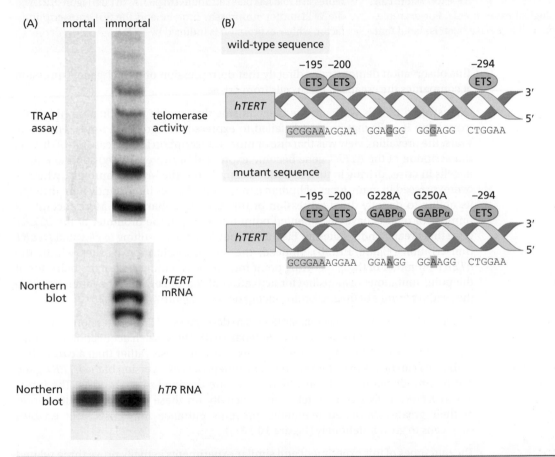

Figure 10.14 Spontaneous activation of telomerase
(A) Individual B lymphocytes can escape at low frequency from a cell population in crisis and begin to grow robustly, thereby becoming immortalized (see Figure 10.8A). The pre-crisis "mortal" cells lack telomerase activity, as measured by the TRAP assay (see Sidebar 10.5), whereas the spontaneously immortalized ("immortal") cells show abundant activity (*top*). The *hTERT* mRNA, which encodes the catalytic subunit of the telomerase holoenzyme, is essentially absent in the pre-crisis cells but clearly present in the immortalized cells (*middle*). However, levels of the *hTR* RNA, which functions as a template for hTERT during telomeric DNA elongation (see Figure 10.13), are increased only slightly (*bottom*), suggesting that levels of hTERT expression determine the levels of enzyme activity in the cell. (B) Point mutations in the *hTERT* promoter have been documented in the genomes of numerous human tumors. The strong localization of the recurring point mutations indicates the significant survival advantage that evolving pre-neoplastic cells acquire by significantly elevating hTERT expression levels. Each of the two mutually exclusively occurring point mutations (at bases −228 and −250 from the start site of transcription) adds a binding site for the GABPα ETS-type transcription factor, substantially increasing *hTERT* transcription. These two mutations have been observed in the genomes of 83% of primary glioblastomas, 66% of bladder carcinomas, and 71% of human melanomas. [Indeed, the presence of these recurring mutations at a small number of sites in the *hTERT* promoter echoes the highly localized, recurring mutations affecting the reading frame of the promoters of the *ras* oncogenes in human tumors (see Figure 4.6).] (A, from M. Meyerson et al., *Cell* 90:785–795, 1997. With permission from Elsevier. B, adapted from G.M. Sizemore et al. *Nature Rev. Cancer* 17:337–351, 2017. With permission from Nature.)

Sidebar 10.3 Oncoproteins and tumor suppressor proteins play critical roles in governing *hTERT* expression The mechanisms that lead to the de-repression of *hTERT* transcription during tumor progression in humans are complex and still quite obscure. Multiple transcription factors appear to collaborate to activate the *hTERT* promoter. Once expressed, the resulting *hTERT* transcripts allow the synthesis of substantial amounts of hTERT protein; the latter, by complexing with other subunits of the telomerase holoenzyme, suffices to generate high levels of telomerase activity. The Myc protein, which acts as a transcription factor that controls the cell cycle clock and thousands of genes involved in other aspects of cell physiology (see Section 8.9), also contributes to *hTERT* transcription. There are at least two recognition sequences for bHLH transcription factors like Myc in the promoters of the mouse and human *TERT* genes.

The connection between Myc and *hTERT* expression can also be demonstrated by biological experiments. For example, cells from a human melanoma cell line have been isolated that express high levels of Myc and are tumorigenic. Three clonal variant subpopulations of this cell line have been isolated that exhibit as much as an eightfold reduction in Myc expression and have lost tumorigenicity. Their tumorigenic growth could be restored if they were supplied with an hTERT-expressing gene construct. Hence, in these cells, the loss of tumorigenicity following reduction of Myc expression levels could be attributed largely, if not exclusively, to an associated loss of *hTERT* gene expression.

In addition to Myc, the E6 oncoprotein of HPV (see Sidebar 9.6) and the ER81 transcription factor, which is activated by the Ras–MAPK pathway, stimulate *hTERT* transcription. Working in the opposite direction, Menin—the product of the *MEN1* tumor suppressor gene (see Table 7.1)—associates with the *hTERT* promoter and represses *hTERT* transcription. A similar repressive role has been attributed to p53, Mxd (see Figure 8.21A), the WT1 tumor suppressor protein, and the Sp1 transcription factor, whose expression is induced by TGF-β.

This observation demonstrates directly that de-repression of *hTERT* gene expression accompanies the escape of these cells from crisis.

Unanswered by this work are the mechanisms that enable the induction of hTERT expression in cells that previously failed to express it at significant levels. For many years, the prevailing view was that one or more transcription factors capable of driving transcription of the *hTERT* gene become expressed or overexpressed spontaneously in cells in crisis, driving in turn its sudden expression. The Myc oncoprotein, which is overexpressed in many types of human cancer cells, figures importantly here through its ability to directly drive expression of this gene (**Sidebar 10.3**). More recently, a number of reports have documented point mutations in the promoter of the *hTERT* gene that create binding sites for Ets transcription factors, resulting in elevated *hTERT* transcription (Figure 10.14B). [Indeed, the strong localization of sites within the *hTERT* promoter of these recurring point mutations is reminiscent of the behavior of the point mutations responsible for activation of *Ras* oncogenes, the latter affecting the reading frames of the *Ras* proto-oncogenes (see Figure 4.6)!]

Cancer cells use other molecular strategies to de-repress *hTERT* expression to acquire high levels of telomerase activity (see Sidebar 10.3). The sudden acquisition of telomerase activity might be only a *correlate* of escape from crisis rather than a *cause*. This ambiguity can be resolved by a simple experiment: a cDNA version of the *hTERT* gene can be introduced into cells just before they are destined to enter crisis. The introduced *hTERT* cDNA confers telomerase activity on these cells, causes elongation of their greatly shortened telomeres, prevents entrance into crisis, and enables such cells to grow indefinitely (**Figure 10.15**).

The outcomes of this experiment and similar experiments actually prove three related points. First, although the telomerase holoenzyme may be composed of multiple, distinct subunits, the only subunit missing from normal, pre-crisis human cells is the catalytic subunit encoded by the *hTERT* gene; the remaining subunits of the telomerase holoenzyme, including the *hTR* telomerase-associated RNA molecule, seem to be present in adequate amounts in pre-crisis cells. Second, because expression of telomerase (rather than another enzyme) allows cells to avoid crisis, and because telomerase acts specifically on telomeric DNA, these observations demonstrate that one cause of crisis must indeed be critically shortened telomeres. Third, this experiment shows that the acquisition of telomerase activity, which can be achieved spontaneously by rare variant cells scattered amid a population of cells in crisis, is sufficient to enable cells to escape crisis and generate descendants that can grow in an immortalized fashion.

We can conclude further that telomerase succeeds in allowing cells to circumvent crisis because it subverts the operations of the generational clock. When telomerase is

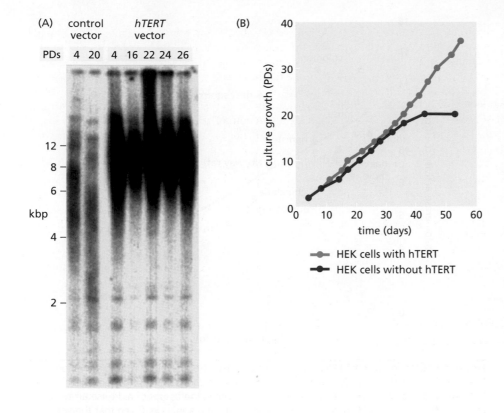

Figure 10.15 Prevention of crisis by expression of telomerase (A) This Southern blot reveals the lengths of telomeric DNAs that were generated by cleaving genomic DNA with a restriction enzyme that does not cut within the repeating hexanucleotide sequence of telomeres, followed by hybridization to a DNA probe that recognizes this same sequence. As is seen, telomerase-negative human embryonic kidney (HEK) cells infected with a control retroviral vector that specifies no protein carry telomeres that decrease in size between the 4th and 20th population doublings (PDs; *left two channels*). However, infection of such cells with a retroviral vector specifying hTERT causes the telomeres to substantially increase in length, maintaining their sizes through the 26th population doubling (*right channels*). (Because these telomeres are much longer, they generate a stronger signal upon Southern blotting.) The size markers are given in kilobase pairs. (B) This population of human embryonic kidney (HEK) cells was destined to enter crisis by 40 days, which is the equivalent of roughly 20 population doublings (PDs; *red line*). However, when hTERT was ectopically expressed in these cells (*blue line*), they gained the ability to proliferate indefinitely. (From C.M. Counter, W.C. Hahn, W. Wei et al., *Proc. Natl. Acad. Sci. USA* 95: 14723-14728, 1998. With permission from National Academy of Sciences, U.S.A.)

expressed at significant levels, telomeres are maintained at lengths that are compatible with unlimited further replication. In effect, the generational clock that depends on progressive telomere shortening is rendered inoperative.

With these insights in mind, the normal biological roles of telomerase now become clearer. In single-cell eukaryotic species, such as ciliates and yeast, the exponential growth of cells requires the continuous presence of high levels of telomerase activity to ensure that telomeres are maintained indefinitely at a length that is compatible with chromosomal stability. By elongating telomeric DNA, the telomerase in these protozoan cells is able to compensate for the continued erosion of telomeric DNA due to the under-replication of DNA termini by the bulk DNA replication machinery as well as degradation by certain exonucleases (see Figure 10.12). Being single-celled, these eukaryotes have no need to fear cancer.

Expression of telomerase activity is programmed differently in complex metazoans, notably humans. Because telomerase expression is largely repressed in postembryonic cell lineages, these lineages are granted only limited postembryonic replicative potential before they enter crisis. This limitation seems to represent one component of the human body's anti-cancer defenses. Thus, the repression of TERT activity has been found to be regulated as a function of the number of cell divisions experienced in an organism's lifetime, supporting the notions that (1) the risk of developing tumors is correlated with the number of cell divisions that an organism experiences in an average life span and (2) one mechanism for reducing cancer risk is to limit the access of normal somatic cells to TERT activity (**Figure 10.16**).

Another perspective on this problem comes from attempting to compare the relative numbers of cell divisions in the lives of laboratory mice relative to humans: mice may experience ~10^{11} cell divisions in a lifetime, whereas humans are thought to undergo about 10^{16}, suggesting that there is a roughly 100,000-fold difference in the risk of acquiring cancer-causing somatic mutations deriving from errors in DNA replication occurring during cell division. Indeed, studies of telomerase activity in a variety of rodents indicates an inverse relationship between body mass and basal telomerase activity (see Figure 10.16). This finding provides further support for the notion that limiting telomerase activity represents one mechanism evolved to compensate for the large number of cell divisions and associated risk of somatic mutations in a variety of mammals.

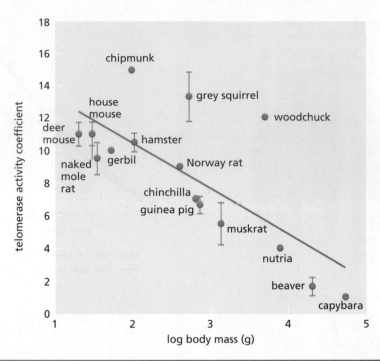

Figure 10.16 Repression of TERT expression as a function of body mass The notion that repression of TERT activity in mammals represents a tumor-suppressing mechanism has been tested by examining the activity of the enzyme in the tissues of a diverse array of rodents whose body sizes vary over almost three orders of magnitude. Given that the sizes of individual proliferating cells appear to be equivalent among these species, body mass measurements are likely to serve as one metric of the cumulative number of cell divisions that an organism experiences in its lifetime. Thus, more cell divisions are required to generate and maintain various tissues in larger organisms, an effect that is compounded because larger mammals generally have longer lifespans. Also, as described in Chapter 12, cell divisions represent a significant source of the somatic mutations that contribute to tumor pathogenesis. Consequently, during the course of their evolution, large, long-lived mammals must have developed increased anti-neoplastic mechanisms to compensate for their intrinsically increased risk of developing cancers. One such mechanism might be to limit TERT activity in somatic tissues, a notion that is supported by this study of TERT activity and body mass among a group of 15 species, all of which are members of a single mammalian order—Rodentia. Rodent body mass varies from 20 g to 55 kg. In this case, the TRAP enzyme assay (see Sidebar 10.5) was used to gauge TERT activity in various differentiated cells of rodents. (From A. Seluanov et al., *Aging Cell* 6:45–52, 2007 and V. Gorbunova et al., *Nat. Rev. Genet.* 15:531–540, 2014. With permission from John Wiley & Sons.)

This mechanistic model also provides a compelling explanation of how human cells can escape from crisis and become immortalized. It fails, however, to address another long-standing mystery: How does experimentally induced expression of telomerase enable certain types of cultured cells to avoid senescence (**Sidebar 10.4, Figure 10.17**)?

10.7 Telomerase plays a key role in the proliferation of human cancer cells

The experiments described above using pre-crisis human cells in culture suggest that the telomerase activity detectable in the great majority of human cancer cells plays a causal role in their immortalization and that such immortalization is a key component of the neoplastic growth state. This mechanistic model has been tested experimentally by suppressing the telomerase activity in cancer cells and following their subsequent responses. Telomerase activity can be reduced through the expression of antisense RNA in the telomerase-positive cells. For example, an RNA that is complementary in sequence to the *hTR* RNA subunit of the telomerase holoenzyme (see Figure 10.13) can be introduced experimentally into telomerase-positive human cancer cells. Such an antisense molecule is presumed to anneal to the *hTR* molecule, forming an RNA–RNA

Sidebar 10.4 The puzzle of senescence and telomeres As described here, a diverse array of experimental data have converged on the conclusion that cells enter senescence in response to a variety of cell-physiological stresses, some sustained over extended periods of time. These include various types of genetic damage, oncogene signaling, oxidative stress, and even metabolic stress. Replicative senescence—manifested by many cell types in response to being introduced into tissue culture—is largely and possibly entirely explained by the stresses that cells suffer from sub-optimal conditions of culture; this is underscored by observations indicating that replicative senescence can be delayed by improving the conditions of culture (see Supplementary Sidebar 10.3).

Once cells have entered into senescence, which is generally an irreversible step, they manifest DNA lesions that seem, for one reason or another, to be irreparable. Indeed, many of the stressors that induce senescence appear to function, directly or indirectly, to induce DNA damage, and the resulting unrepaired lesions seem to be responsible for emitting signals that induce senescence, thereby stably maintaining the cells around them in the senescent state. In human (but not in laboratory mouse) cells, the senescence-associated heterochromatic foci (SAHFs; see Figure 9.17) indicate a massive shutdown of the expression of thousands of genes, many of which are apparently essential for cell proliferation.

This scheme is difficult to reconcile with other solid observations demonstrating that in certain human cell types,

senescence can be circumvented by forced expression of hTERT. Because such hTERT expression also allows a cell subsequently to circumvent crisis, this treatment effectively immortalizes cells. These observations have been incorporated into a scheme suggesting that when telomeres are eroded after short-term culture, some of the shortest telomeres have already eroded enough to lose their ability to protect the ends of telomeric DNA; these shortened telomeres release signals that induce senescence. In the event that senescence is circumvented, cells will then proceed to crisis, where far greater telomere erosion triggers the breakage–fusion–bridge cycles and the widespread apoptosis associated with crisis. Accordingly, telomere shortening acts as a clocking device to trigger both senescence and, later on, crisis.

These two mechanistic models are incompatible with one another: either the timing of senescence is triggered by telomere erosion or it is triggered by cell-physiological stress. A scheme to reconcile these two alternative mechanistic models is not obvious. One possible mechanism is the following (see Figure 10.17): when cells experience certain types of physiological stress, they actively degrade the ends of some or all of their telomeres; the resulting dysfunctional telomeres release a DNA damage signal that in turn triggers senescence. In the event that hTERT is present in significant amounts, such telomere degradation and resulting entrance into senescence is avoided. Yet other mechanisms by which hTERT acts independently of telomeres to prevent entrance into senescence have been proposed.

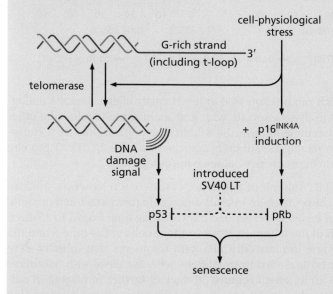

Figure 10.17 Replicative senescence and the actions of telomerase This diagram presents a speculative mechanistic model of how and why telomerase expression (achieved by ectopic expression of the hTERT catalytic subunit) can prevent human cells from entering into replicative senescence. The t-loop of the telomeric DNA (see Figure 10.11) is not illustrated here. The G-rich overhanging strand of telomeric DNA (*pink, top*), which forms the t-loop, has been found in some studies to be largely degraded in cells that have entered replicative senescence. This loss may be triggered by certain cell-physiological stresses. The resulting blunted telomeric DNA may then emit a DNA damage signal that activates p53. Together with p16INK4A expression (see Section 8.4), which is activated by various cell-physiological stresses, this may impose a senescent state. By reversing this loss of the G-rich overhang (and a possible partial degradation of the dsDNA portion of telomeric DNA), the experimentally introduced telomerase prevents certain cell types, such as human fibroblasts, from entering into senescence. Of note, the demonstrated ability of SV40 large T antigen to circumvent senescence depends on its ability to sequester both pRB and p53, thereby neutralizing the cytostatic effects of increased expression of p16INK4A and p53 (see Figure 10.3).

double helix, thereby blocking the ability of the *hTR* subunit to participate in the synthesis of telomeric DNA. In fact, it is difficult to achieve total inhibition of hTR function using this strategy. Nonetheless, such an antisense experiment performed with HeLa cells (from the line of human cervical carcinoma cells cited earlier) caused them to stop growing 23 to 26 days after they were initially exposed to the antisense RNA.

An alternative experimental strategy that is more effective derives from the use of a mutant hTERT enzyme carrying amino acid substitutions in its catalytic cleft; these alterations yield a catalytically inactive enzyme. When overexpressed in telomerase-positive cells, the mutant hTERT protein can act in a dominant-negative fashion (dn; see Section 9.4) to interfere with the existing endogenous telomerase activity of these

Figure 10.18 Suppression of telomerase activity and resulting loss of the neoplastic growth program Expression vectors specifying a mutant dominant-negative (dn) hTERT protein (*red lines*), a wild-type hTERT enzyme (*green lines*), or an empty control vector (*blue lines*) have dramatically different effects on four different human cancer cell lines. The respective lengths of their telomeric DNAs, as determined by TRF Southern blots (see Figure 10.8A) at the onset of the experiment, are indicated in kilobases (kb). The empty control vector and the hTERT-expressing vector permit the continued proliferation of these cells, as measured in population doublings (PD). However, introduction of the dn hTERT enzyme, which differs from the wild-type enzyme by a single amino acid substitution in its catalytic cleft and is presumed to shut down endogenous hTERT function, causes these cells to stop growing, doing so with various lag times (in days). Microscopic examination revealed that the cells that ceased proliferating entered into crisis and showed widespread apoptosis. Cells of the LoVo cell line (*top left*) entered crisis immediately after expressing the dn hTERT enzyme, and therefore no cells were available to count. (Because the TRFs also contain nonfunctional subtelomeric DNA that carries imperfect hexanucleotide repeats, the lengths of functional telomeric sequences are much shorter than the observed lengths of the TRFs.) (From W.C. Hahn et al., *Nat. Med.* 5:1164–1170, 1999. With permission from Nature.)

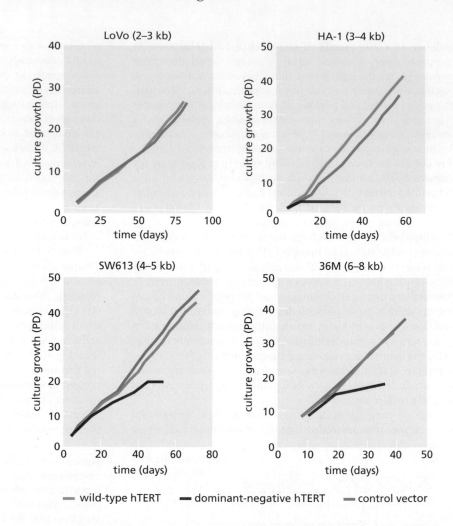

— wild-type hTERT   — dominant-negative hTERT   — control vector

cells. The dn hTERT, which can be expressed at levels vastly higher than the endogenous hTERT protein, is likely to associate with and thereby monopolize the other subunits that normally assemble to form the telomerase holoenzyme. As a consequence, when these other telomerase subunits associate with the dn hTERT protein, they become recruited into unproductive holoenzyme complexes.

Expression of the dn hTERT subunit in a number of different telomerase-positive human tumor cell lines causes them to lose all detectable telomerase activity and, with some delay, to enter crisis. The crisis occurs with a lag time from 5 to 25 days, depending on the length of the telomeric DNA in these cells at the time when the dn hTERT subunit was first introduced. Cells with telomeres that initially were 2 to 3 kb in length enter crisis almost immediately, whereas those with telomeres that were initially 4 to 5 kb in length require 30 days of further passaging in culture before entering crisis (**Figure 10.18**). These delayed reactions suggest that after the dn hTERT is expressed in cancer cells, telomeres shorten progressively, and that crisis ensues after the initially present telomeric DNA erodes down to some threshold length that is incompatible with protecting the ends of chromosomal DNA.

Importantly, the dn hTERT enzyme has no observable effect on the growth of telomerase-positive cells up to the point when they reach crisis. This rules out an alternative explanation: that the mutant enzyme is intrinsically cytotoxic and that its effects on the cell are attributable to some nonspecific toxicity. Taken together, such experiments lead to the conclusion that the continued activities of the telomerase enzyme are as important to the proliferation of these cancer cells as the actions of oncogenes and the inaction of tumor suppressor genes. This conclusion is strengthened by the dramatic outcomes of studies of certain human pediatric tumors (**Sidebar 10.5, Figure 10.19**).

Sidebar 10.5 Telomerase fuels the growth of some pediatric tumors Neuroblastomas are tumors of cells in the peripheral (sympathetic) nervous system and are usually encountered in very young children. These tumors have highly variable outcomes, with some regressing spontaneously and others progressing into invasive, metastatic tumors that ultimately prove fatal. As described in Section 4.4, the N-*myc* gene often undergoes amplification in these tumors, and greater copy numbers of N-*myc* indicate a worse prognosis for the patient. In fact, telomerase activity is an even more useful indicator of the eventual outcome of the disease. As shown in Figure 10.19A, children whose neuroblastomas are telomerase-negative (as judged by the TRAP assay; see Supplementary Sidebar 10.5) do very well in response to therapy, whereas those whose tumors are telomerase-positive do poorly. Expression of these two genes—N-*myc* and the telomerase-encoding *hTERT* gene—is likely to be functionally linked because N-*myc*'s cousin, *myc*, is already known to be a strong inducer of *hTERT* transcription (see Sidebar 10.3). Similar dynamics operate in another childhood tumor—Ewing's sarcoma—which may arise from mesenchymal stem cells (Figure 10.19B).

Some pathologists argue that neuroblastomas are extremely common in children under the age of one year, and that almost all of these regress spontaneously and never become clinically apparent. The causes of this regression and the benign outcomes of many clinically detected neuroblastomas that respond well to treatment may now have found a molecular explanation: without significant hTERT expression, neuroblastoma cells lose telomeres and progress into crisis, from which they fail to emerge.

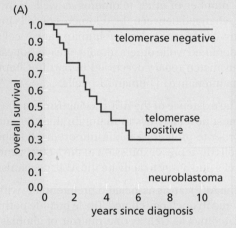

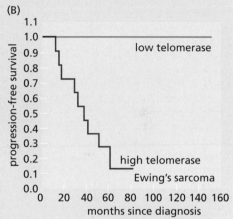

Figure 10.19 Telomerase activity and the prognosis of pediatric tumors (A) This Kaplan–Meier plot illustrates the survival of a group of pediatric neuroblastoma patients who had received no cytotoxic chemotherapy treatment and who were tracked for the time spans indicated (*abscissa*) following diagnosis. These patients were segregated into two groups—those whose tumors were telomerase activity–positive (*red*), and those whose tumors appeared to lack this activity (*blue*). The ordinate indicates overall survival, i.e., the proportion of patients who remained alive at various times after initial diagnosis. (B) A similar study of children with Ewing's sarcoma was undertaken, in which the patients were segregated into two groups, depending on whether their tumors had high (*red*) or low (*blue*) telomerase activity. In this case, the ordinate indicates the proportion of children who experienced *progression-free survival,* i.e., those whose tumors had not advanced beyond the stage initially encountered in the clinic. (A, from C. Poremba et al., *J. Clin. Oncol.* 18:2582–2592, 2000. B, from A. Ohali et al., *J. Clin. Oncol.* 21:3836–3843, 2003. Both with permission from the American Society of Clinical Oncology.)

The key role of hTERT in human cell transformation is revealed by other data as well. Thus, as described in the next chapter, the experimental transformation of primary human cells into tumorigenic cells requires, among other introduced genes, the introduction of a vector expressing hTERT constitutively. Responding to these various discoveries, some researchers have used an automotive analogy to illustrate the growth of human cancer cells: activated oncogenes are said to be akin to accelerator pedals that are stuck to the floor; inactivated tumor suppressor genes are compared to defective braking systems; and telomerase is likened to an agent that ensures that the runaway car has an endless, self-replenishing supply of gasoline in its tank.

10.8 Some immortalized cells can maintain telomeres without telomerase

As noted earlier, 85 to 90% of human tumors have been found to be telomerase-positive. The remaining 10 to 15% lack readily detectable enzyme activity, yet the cells within this second group of tumors are presumably faced with the need to maintain their telomeres above some minimum length in order to proliferate indefinitely. In fact, many of these cells have learned to maintain their telomeric DNA using a mechanism that does not depend on the actions of telomerase.

This non-telomerase-based mechanism was first discovered in the yeast *Saccharomyces cerevisiae* following inactivation of one of the several yeast genes encoding subunits of its telomerase holoenzyme. Resulting mutant yeast cells soon entered into a state that is analogous to crisis in mammalian cells. The vast majority of the cells died, but rare variants emerged from these populations of dying cells that used the ALT (alternative lengthening of telomeres) mechanism, which is telomerase-independent, to construct and maintain their telomeres. The minority of human tumor cells that lack significant telomerase activity also use the ALT mechanism.

Details of the molecular mechanisms that ALT-positive human cells use to maintain their telomeres are poorly understood. An important clue, however, comes from an experiment in which a traceable molecular marker—a neomycin resistance gene—was introduced into the midst of the telomere repeat sequence of one chromosome carried by a mammalian cell in the ALT state. When the telomeric DNAs were examined in the descendants of this cell, copies of this molecular marker were found in a number of other telomeres as well. Hence, sequence information appears to be exchanged between the telomeres of cells in the ALT state. Some of the enzymes that are used by ALT-positive human cancer cells to maintain their telomeres have been identified while others remain elusive. For example, certain enzymes involved in DNA mismatch repair, described in part in Chapter 12, appear to contribute to telomere maintenance in human ALT cells.

The existence of the ALT mechanism emphasizes that all types of human cancer cells must develop ways to maintain their telomeres above a certain threshold length in order to proliferate in an immortalized fashion. To summarize, most human cancer cells de-repress expression of the *hTERT* gene, but a small minority of incipient cancer cells find a way to activate the ALT mechanism.

The ALT state is associated preferentially with a specific subset of the tumors encountered in oncology clinics. These include perhaps half of osteosarcomas and soft-tissue sarcomas as well as one-quarter of glioblastomas, but only rarely carcinomas. Perhaps these correlations may one day be explained by the characteristic expression of a group of genes in mesenchymal cells that are also associated with repression of the *hTERT* gene, forcing the cancer cells that arise from these cell types to rely on ALT as a more accessible route to immortalization. Additional clues to understanding the origin of this mechanism may also come from observations that the ALT state operates in certain subclasses of normal lymphocytes.

At the moment, the precise reasons why one mechanism (involving hTERT) is usually favored over the other (ALT) remain obscure. Equally unclear are the mechanisms that cause the telomeric DNA of human cells in the ALT state to grow to heterogeneous lengths (often >30 kb) far greater than those usually seen in telomerase-positive cells (5–10 kb). Importantly, the dominant-negative hTERT enzyme (see Figure 10.18), which induces crisis in a number of telomerase-positive human cancer cell lines, fails to do so in a human ALT-positive tumor cell line, providing further support of the notions that (1) ALT-positive cells do not depend on telomerase for their growth, and (2) the dn hTERT enzyme is not intrinsically cytotoxic.

Of note, most of the telomeres in the ALT-positive cells are able to form the t-loops/lariats (see Figure 10.11) in the absence of telomerase function. These structures explain how these telomeres are protected from end-to-end fusion and BFB cycles. At the same time, it illustrates that these t-loops form spontaneously and are directed to do so by the DNA sequences and associated proteins (see Supplementary Sidebar 10.4) found in human cells lacking hTERT function.

hTERT is a very attractive target for researchers who are intent on developing novel types of anti-cancer therapeutics. As we will learn in Chapter 17, the catalytic clefts of enzymes such as hTERT can often be blocked by highly specific therapeutic drug molecules. Significantly, hTERT is a distant relative of the reverse transcriptase of human immunodeficiency virus (HIV; see Supplementary Sidebar 10.6), against which effective drug inhibitors have been successfully developed. These existing successes increase the likelihood that anti-hTERT compounds will one day be produced as well. In addition, this enzyme is expressed in the great majority of human cancers, but is not present at significant levels in normal tissue. This offers the prospect of drug selectivity—being able to affect cancer cells, while leaving cells in normal tissues untouched. However, for reasons that remain unclear, the hTERT enzyme has proven time and again to be a formidable enemy, in that repeated screens for low–molecular-weight inhibitors of this enzyme have failed to yield attractive drug candidates.

The ALT state holds important implications for such drug development. Imagine that an inhibitor of hTERT were indeed developed. It is already clear that as many as 10% of human tumors will never respond to such an anti-telomerase drug, because they are ALT-positive and therefore do not depend on this enzyme for their continued proliferation. The remaining 90% of human tumors, which do indeed exhibit robust telomerase activity, may initially be susceptible to killing by this drug, but sooner or later may spawn variants that have activated the ALT mechanism of telomere maintenance. This might allow the newly arising ALT variants to evade the killing actions of the anti-telomerase drug. In this case, a back-up therapeutic strategy might come from the discovery that inhibitors of the ATR enzyme (see Section 9.7), which participates in DNA replication, recombination and repair, are highly effective in killing ALT cancer cells but not those that express high levels of hTERT.

10.9 Telomeres play different roles in the cells of laboratory mice and in human cells

Our descriptions of telomerase expression repeatedly refer to its actions in human cells, with good reason. Rodent cells, specifically those of laboratory mouse strains, control telomerase expression in a totally different way. These mice do not strongly repress mTERT expression, which seems to explain why the double-stranded region of telomeric DNA in laboratory mice is as much as 30 to 40 kb long—about five times longer than corresponding human telomeric DNA. In fact, laboratory mouse telomeres are so long that they are never in danger of eroding down to critically short lengths during the lifetime of a mouse, even when expression of mouse telomerase is suppressed experimentally.

The lengths of mouse telomeres permit cell lineages to pass through a far larger number of replicative generations than would seem to be required for tumor formation. This finding suggests that laboratory mice do not rely on telomere length to limit the replicative capacity of their normal cell lineages and that telomere erosion cannot serve as a mechanism for constraining tumor development in these rodents (see also Figure 10.16). Moreover, telomerase activity is detectable in the stem cells of certain tissues in the mouse, providing further indication of the absence of global repression of TERT activity in these animals.

The long telomeres of laboratory mice hold additional implications: to the extent that cells derived from these mice show limited replicative ability *in vitro*, that ability is never determined by telomere shortening. In fact, mouse cells can be immortalized relatively easily following propagation in culture (which selects for cells that have avoided senescence and are spontaneously immortalized *in vitro*). Human cells, in contrast to their mouse counterparts, require the introduction of both the SV40 *large T* oncogene (to avoid senescence; see Figure 10.3) and the *hTERT* gene (to avoid crisis). As an aside, one reason for the reduced tendency of mouse cells to become senescent during *in vitro* culture may be that they fail to develop senescence-associated heterochromatic foci (SAHFs; see Section 9.11) in culture, whereas their human counterparts do so readily.

The role of telomeres in the lives of mice has been revealed dramatically by researchers who have inactivated the gene encoding the *mTR* RNA subunit of telomerase holoenzyme in the mouse germ line. As expected, the homozygous mutant offspring of initially created heterozygous mice carrying the mutant *mTR* allele exhibit no telomerase activity in any of their cells. During their lifetimes, the telomeres of these telomerase-negative mice may shorten from approximately 30 kb down to about 25 kb; the latter length is still far longer than is required to protect chromosomal ends.

These *mTR*-negative mice are superficially indistinguishable phenotypically from wild-type mice. This absence of any readily observable mutant phenotype suggests that the telomerase holoenzyme plays no essential role in the tissues of these mammals beyond its task of maintaining telomeres. (However, detailed characterization of genetically altered mice that continue to express *mTR* but have been deprived of *mTERT* expression reveals a subtle morphogenetic defect—the conversion of a thoracic to a lumbar vertebra. This change may be caused by the association of the mTERT protein with other cellular proteins whose functions are totally unrelated to its actions on telomeres; one of these is, as an example, the complex of transcription factors that allow β-catenin to activate gene expression in response to Wnt signaling; see Figure 6.19B. The same association presumably holds true for hTERT.) Beyond this, there are several reports describing the actions of hTERT in protecting neurons against oxidative stress.

The homozygous telomerase-negative (that is, *mTR*$^{-/-}$) mice can be bred with one another through at least three more organismic generations without showing any discernible phenotype. Finally, however, in the fifth generation, telomerase-negative mice begin to show distinctive phenotypes—an indication that after five organismic generations without telomerase, their telomeres have eroded down to dangerously short lengths (**Figure 10.20**). These fifth-generation *mTR*$^{-/-}$ mice and, even more so, their progeny in the sixth generation are sickly and show a diminished capacity to heal wounds, indicative of the inability of their cells to respond properly to mitogenic signals. The sixth-generation *mTR*$^{-/-}$ mice suffer also from substantially reduced fertility.

The *mTR*$^{-/-}$ mice in this sixth generation already exhibit very short telomeres at birth, and their inability to subsequently maintain these telomeres, particularly in mitotically active tissues, results in widespread cell death and loss of tissue function. For example, highly proliferative tissues, such as the gastrointestinal epithelium, the hematopoietic system, and the testes, show substantial **atrophy** (loss of cells). These observations are striking and without parallel in the field of mouse genetics, because they demonstrate organismic phenotypes that only become manifest five or six organismic generations after a mutation has been introduced into the germ line of these animals.

Moreover, such observations dramatically demonstrate the differences between the telomeres of human cells and those of these mutant laboratory mice. Human

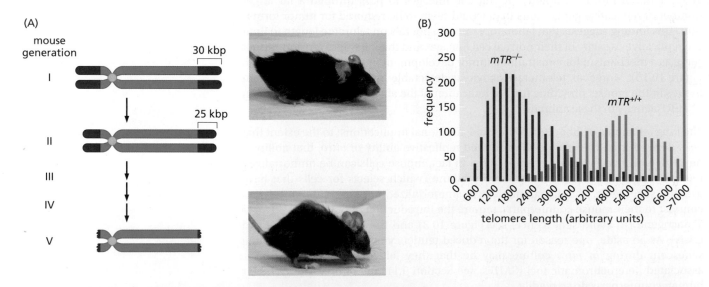

Figure 10.20 Erosion of telomeres over multiple generations in populations of *mTR*$^{-/-}$ mice (A) During the lifetimes of the first *mTR*$^{-/-}$ mouse generation, their telomeres, which are initially 30 kbp long (*red*), are reduced in size by about 5 kbp (*left*). These mice (*top, right*) are phenotypically normal, as are their descendants in the second and third generations. However, by the fifth organismic generation, the telomeres in their cells have eroded to the extent that they can no longer protect the ends of chromosomes; the tissues of these mice begin to lose their ability to renew themselves and heal wounds, and these mice begin to show symptoms of premature aging, including wasting away of muscle tissues and a hunched back (*bottom, right*). The relative sizes of the telomeres in these images have been exaggerated

1000-fold for the sake of illustration. (B) Telomere length can be gauged by fluorescent *in situ* hybridization (FISH) using a probe that is specific for the telomere repeat sequence. Here the *mTR*$^{-}$ allele has been bred into a strain of wild mice that normally have very short telomeres. Already in the first generation, the resulting homozygous *mTR*$^{-/-}$ mice (*red*) have far shorter telomeric DNA than their wild-type (*mTR*$^{+/+}$) counterparts (*blue*). These measurements also reveal the heterogeneous lengths of telomeric DNA within each population of mice, mirroring comparable heterogeneity within the individual cells of a mouse. Lengths are in arbitrary units. All telomeres longer than 7000 units are represented by a single bar (*far right*). (A, courtesy of R.S. Maser and R.A. DePinho. B, from L.-Y. Hao et al., *Cell* 123:1121–1131, 2005. With permission from Elsevier.)

telomeres begin relatively short, and loss of telomerase activity can lead, already in the *first* organismic generation, to severe organismic phenotypes (Supplementary Sidebar 10.7).

10.10 Telomerase-negative mice show both decreased and increased cancer susceptibility

Laboratory mice are susceptible to spontaneous cancers, largely lymphomas and leukemias. This background of susceptibility can be increased by experimentally introducing specific mutations into the mouse germ line that cause activation of proto-oncogenes or inactivation of tumor suppressor genes. These facts, when taken together, raise an interesting question: To what extent do the critically short telomeres in fifth- and sixth-generation telomerase-negative mice also affect the cancer rate in these animals?

The response to this question would seem to be straightforward. We can imagine that the deterioration of cells in these $mTR^{-/-}$ mice leaves a barely adequate number of healthy cells to sustain tissue and organismic viability. Stem cells charged with the task of replenishing the pools of differentiated cells in these tissues will likely have great difficulty doing so.

Because the normal cells in these tissues are barely able to maintain themselves, incipient cancer cells should, by all rights, have an even more difficult time. The process of tumor development requires that clones of premalignant cells evolving toward malignancy must pass through a large number of growth-and-division cycles—a number substantially greater than those experienced by nearby normal cells (**Figure 10.21A**). Consequently, any tumor development that is successfully initiated by these cell clones is likely to be aborted long before it has reached completion (Figure 10.21B and C).

In order to test these predictions, mice were first made cancer-prone by inactivating tumor suppressor genes in their germ line, in this case the locus encoding both the p16^{INK4A} and p19^{ARF} tumor suppressor proteins (see Sections 8.4 and 9.8). Such mice, as might be expected, are highly susceptible to cancer; indeed, they frequently develop lymphomas and fibrosarcomas relatively early in life. This susceptibility is manifested even more dramatically when they are exposed to carcinogens. In one set of experiments, sequential exposure to dimethylbenz[*a*]anthracene (DMBA), a potent carcinogen, followed by repeated exposures to ultraviolet-B (UV-B) radiation, led after 20 weeks to a 90% tumor incidence in the (p16^{INK4A}/p19^{ARF}-negative) mutant mice, whereas a control group of similarly treated wild-type mice had no tumors.

This experiment was extended by introducing the p16^{INK4A}/p19^{ARF} germ-line inactivation into mice that also lacked *mTR*, the gene that is the counterpart of human *hTR* and therefore encodes the RNA subunit of the mouse telomerase holoenzyme (see Figure 10.13). The results were that 64% of control, telomerase-positive, p16^{INK4A}/p19^{ARF}-negative mice contracted tumors, whereas only 31% of the generation 4 and 5 $mTR^{-/-}$ p16^{INK4A}/p19^{ARF}-negative mice developed tumors. These results were even more dramatic when survival of mice after 16 weeks was measured: 88% of the telomerase-positive mice had been lost to cancer, whereas only 46% of the fifth-generation telomerase-negative mice had been lost to this disease.

This reduced rate of cancer in telomerase-negative mice fulfills the prediction that the normal tissues of $mTR^{-/-}$ mice have exhausted most of their endowment of replicative generations even before tumorigenesis has begun (see Figure 10.20B). Once tumors are initiated in these mice, incipient tumor cell populations must pass through many additional doublings before they can create macroscopic tumors. However, relatively early during the course of tumor formation, these aspiring cancer cells will be driven into crisis by telomere collapse and their agenda of forming tumors will be aborted. Indeed, in human tissues, premalignant cells that are poised to become fully neoplastic already show drastically truncated telomeres; this truncation ostensibly arises because these cell populations have passed through many more successive cell cycles than nearby normal cells (see Figure 10.21B and C).

Figure 10.21 Normal and neoplastic cell lineages in *mTR*$^{-/-}$ mice and in humans (A) This diagram indicates the pedigree of normal cells (*open circles*) in a small patch of tissue. A founder cell in the embryo becomes the ancestor of all the cells in this tissue, e.g., all the epithelial cells. At one point, one of these epithelial cells suffers changes that make it the ancestor of the cells (*closed red circles*) that develop into a precancerous cell clone. This clone of cells must pass through many more replicative generations in order to eventually form a tumor mass. Slashes indicate the substantial attrition of cells through apoptosis or necrosis that occurs in almost every cell generation. (B) The large number of cell generations through which premalignant cells must pass en route to neoplasia results in considerable telomere erosion. In these images, telomeres are revealed by fluorescence *in situ* hybridization (FISH, *pink dots*). In normal human mammary ducts (*not shown*), telomeres are apparent in both the luminal epithelial cells and the myoepithelial cells. However, in ductal carcinoma *in situ* (DCIS), the luminal epithelial cells, which constitute the neoplastic cell population (*above dotted line*), have lost all detectable telomeric DNA, whereas the nearby normal myoepithelial cells (*below dotted line*), which have been stained for α-smooth muscle actin (*green*), still carry strongly staining telomeres. Similar behavior is seen in the normal epithelial cells (*lower left*). (In all cases, cell nuclei are stained *blue*.) (C) Similar dynamics apply in the case of prostatic intraepithelial neoplasia (PIN)—a precursor lesion to prostatic carcinoma—in which many of the luminal epithelial cells lining ducts (*below dotted line*) have lost telomeric DNA staining, while the keratin-positive (*green, above dotted line*) basal epithelial cells continue to display a strong telomeric DNA signal. (B and C, courtesy of A.K. Meeker; see also A.K. Meeker et al., *Cancer Res.* 62:6405–6409, 2002.)

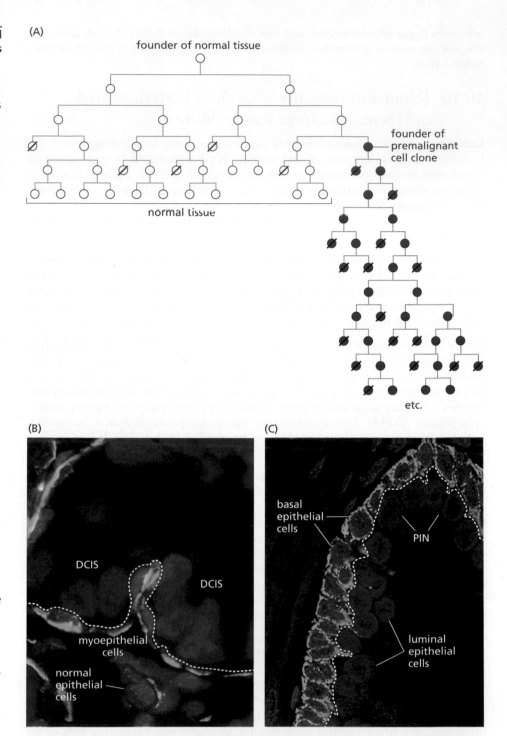

A very different and ultimately far more interesting outcome was seen when mice were used that instead had been deprived of *p53* gene copies in their germ line (that is, rather than having had their p16^{INK4A}/p19^{ARF} locus inactivated). Ordinarily, germ-line inactivation of the *p53* gene in mice leads, on its own, to an increased tumor incidence and resulting mortality, mirroring aspects of the Li–Fraumeni syndrome of humans (see Section 9.3). The mutant *mTR* locus was then introduced into the *p53*$^{+/-}$ genetic background of these mice. [The *p53*$^{+/-}$ genotype was used to forestall the early-onset sarcomas and leukemias that are known to arise in *p53*$^{-/-}$ mice; hence, loss-of-heterozygosity (LOH) and resulting conversion into the *p53*$^{-/-}$ genotype could be postponed for multiple organismic generations.]

When the *p53*$^{+/-}$ genotype was present in the genomes of fifth- and sixth-generation telomerase-negative *mTR*$^{-/-}$ mice, something totally unexpected was observed:

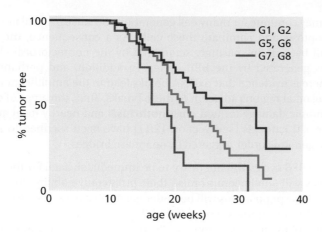

Figure 10.22 Rate of tumor formation in cancer-prone *mTR*^{−/−} *p53*^{−/−} mice These Kaplan–Meier plots reveal that tumor incidence actually increases with increasing organismic generation (G) in mice that lack both the *p53*- and *mTR*-encoding genes in their germ line. In generations 1 and 2 (*red*), about 50% of the mice had readily detectable tumors by the age of 24 weeks, but in generations 5 and 6 (*blue*), the mice developed tumors more rapidly. In generation 7 or 8 (*brown*), about 50% of the mice already exhibited tumors by the age of 17 weeks. (Adapted from L. Chin et al., *Cell* 97:527–538, 1999.)

the rate of cancer formation was significantly *increased* above that created by the *p53*^{−/−} genotype alone. In addition, the spectrum of tumors, namely, lymphomas and angiosarcomas, which are commonly seen in *p53*^{−/−} mice, was shifted in favor of carcinomas—just the types of tumors commonly seen in humans. These trends became even more apparent in the seventh and eighth generation of telomerase-negative mice (**Figure 10.22**).

How can these intriguing results be explained? We can imagine that as fifth- and sixth-generation mice exhaust their telomeres, their cells will begin to experience chromosomal breakage–fusion–bridge (BFB) cycles (see Figures 10.9 and 10.10) and thus crisis. We already know that double-strand DNA breaks, which are formed in BFB cycles, can provoke apoptosis through a p53-dependent pathway (see Section 9.10). Hence, cells that carry functional *p53* genes and experience BFB cycles are likely to be rapidly eliminated from tissues.

As cells of fifth- and sixth-generation *mTR*^{−/−} *p53*^{−/−} mice experience BFB cycles, they may well struggle to stay alive because of the repeated breakage of their chromosomal DNA. However, many of these cells may manage nevertheless to survive, because a key component of their pro-apoptotic response machinery—p53—is missing. These cells will now limp through a number of growth-and-division cycles in spite of the ongoing karyotypic chaos that afflicts their genomes. All the while, their chromosomes will participate in multiple successive BFB cycles (**Figure 10.23**).

Figure 10.23 A mechanistic model of how BFB cycles promote human carcinoma formation During the early steps of tumor progression, inactivation of the *p53* tumor suppressor gene is often favored, because it allows incipient cancer cells to escape apoptotic death from oncogene activation, insufficient trophic signals, anoxia, and poisoning by metabolic wastes caused by inadequate vascularization. As tumor progression proceeds, the telomeric DNA (*red*) of evolving premalignant cell populations eventually erodes below the level needed to protect chromosomal ends. Breakage–fusion–bridge (BFB) cycles (see Figure 10.9) ensue, leading to increasing chromosomal rearrangements and the amplification and deletion of chromosomal segments adjacent to breakpoints (*not shown*); only one BFB cycle is shown here. Centromeres are *dark green* and *dark brown*. Cells undergoing BFB cycles would normally be eliminated by p53-induced apoptosis. In the absence of p53 function, however, such cells may survive, albeit with scrambled karyotypes that slow their proliferation. Emerging from such cell populations will be variants that have learned how to regenerate telomeres and thereby stabilize their karyotype; such cells can once again grow rapidly. In cells of the mutant *mTR*^{−/−} mice, this stabilization may be achieved through the activation of the ALT telomere maintenance system, whereas in wild-type (e.g., *mTR*^{+/+}) organisms (including human cancer patients), de-repression of hTERT expression allows reconstruction of telomeres to proper length. Once telomeres have been regenerated in all chromosomes, further karyotypic disorder resulting from BFB cycles will be halted. However, any karyotypic aberrations generated previously will be perpetuated in descendant cell populations.

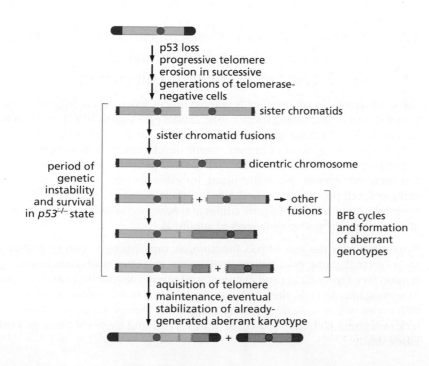

p53 loss
progressive telomere erosion in successive generations of telomerase-negative cells

sister chromatids

sister chromatid fusions

dicentric chromosome

+ → other fusions

period of genetic instability and survival in *p53*^{−/−} state

BFB cycles and formation of aberrant genotypes

+

acquisition of telomere maintenance, eventual stabilization of already-generated aberrant karyotype

+

These cycles may continue for many cell generations because p53 will not be on watch to trigger apoptosis and eliminate these cells. As a consequence, the genomes of such cells will become increasingly scrambled by the nonreciprocal chromosomal translocations generated by the BFB cycles. In addition, and perhaps even more important, there is evidence that dsDNA breaks lead to the amplification or deletion of the chromosomal regions adjacent to these breaks. Thus, the areas of densely concentrated genomic damage termed **chromothripsis** and nearby focal areas of point mutations termed **kataegis** (see Section 12.11) have been ascribed to the localized breakdown of genomic order triggered by anaphase bridges.

Although these BFB cycles will not prove to be immediately fatal for the p53-negative cells, the cycles will surely compromise their proliferative ability. As a result, individual cells in these populations will be under great selective pressure to escape these BFB cycles and to reacquire karyotypic stability; once they do so, they will be able to grow more rapidly. In these mouse cells, however, the acquisition of telomerase activity is not an option (because of the germ-line knockout of both *mTR* gene copies). For this reason, they will be forced to resort to activating the ALT telomere maintenance system (see Section 10.8). Having done so, these cells will stabilize whatever aberrant karyotypes arose since the time when their ancestors first lost functional telomeres and will begin to grow robustly again.

Clearly, the scrambled karyotypes that these cells have acquired generate novel combinations of genes (including many translocated, deleted, and amplified genes), and among the novel genotypes that are created, there will likely be some that fortuitously *favor* neoplastic proliferation. Stated differently, the period of genetic instability appears to *increase* the probability that cancer-promoting genetic configurations will be produced. Such reasoning explains how collapsing telomeres in a p53-negative background can actually promote the development of tumors (see Figure 10.23).

These $mTR^{-/-}$ mice have human counterparts: as described in Supplementary Sidebar 10.7, patients suffering the dyskeratosis congenita syndrome involving germ-line mutations in one or another gene involved in telomere maintenance often go on to develop cancer once their collapsing hematopoietic systems have been successfully reconstructed by bone marrow transplantation. Some develop hematopoietic malignancies (myelodysplasia and acute myelogenous leukemia) or carcinomas of the gastrointestinal tract. These neoplasias arise in those tissues that undergo constant, intense proliferation—precisely the sites where telomeres might become rapidly eroded in the absence of telomerase activity, resulting in repeated breakage–fusion–bridge cycles.

10.11 The mechanisms underlying cancer pathogenesis in telomerase-negative mice may also operate during the development of human tumors

These observations of genetically altered mice, when taken together, lead us to consider a most interesting mechanistic model (see Figure 10.23) that explains why aneuploidy is seen in the great majority of human carcinomas. Imagine that relatively early in the long process of human tumor development, cells in premalignant cell populations are able to jettison their *p53* gene copies. The selective pressure favoring this particular genetic loss might occur, for example, because cells in the early premalignant cell populations experience anoxia and, in turn, suffer from p53-induced apoptosis. Alternatively, *p53* gene copies may have been shed by incipient cancer cells in order to escape oncogene-triggered apoptosis.

Some time after the loss of p53 function, as tumor progression proceeds, cells in these premalignant populations will begin to experience substantial telomere erosion (see Figure 10.21) and eventually telomere collapse because, like most types of normal human cells, they lack significant levels of telomerase activity. Repeated BFB cycles will ensue and result in aneuploidy. However, because these cells also lack functional p53, they will be able to survive and even continue to proliferate, albeit slowly.

Sooner or later, in this population of cells suffering BFB cycles, a cell will emerge that has acquired the ability to de-repress *hTERT* gene expression and, having done so, has gained the ability to prevent further BFB cycles, thereby stabilizing whatever karyotype it acquired during the BFB cycles. With telomerase on hand, we can imagine that this cell will repair the frayed ends of its chromosomal DNA molecules and once again enjoy robust growth. Of course, the belated appearance of telomerase on the scene cannot unscramble the considerable aneuploidy that accumulated during the time window that began with telomere collapse and ended with the acquisition of *hTERT* expression. (The same arguments will apply in the event that the ALT mechanism, rather than telomerase activation, is exploited to protect the genomes of these cells from further BFB cycles.)

Observations of human pancreatic cancer progression provide support for this model. Relatively early in the long, multi-step process of tumorigenesis in the pancreas (see Supplementary Sidebar 11.3), the mitotic cells in low-grade adenomas have few of the anaphase bridges that are characteristic of BFB cycles. However, more advanced, highly dysplastic adenomas show large numbers of these bridges. Still later, the even more advanced, *in situ* carcinomas once again exhibit lower levels of anaphase bridges. This behavior reinforces the model, described above, that proposes that BFB cycles and associated chromosomal instability are found only in a defined window of time during the course of multi-step tumor progression.

Additional support for this model comes from a study of human carcinomas of the esophagus, colon, and breast. In each case, tumors possessing relatively short telomeres are strongly associated with poor long-term prognosis, whereas tumors carrying long telomeres are associated with a far better prognosis, including greater long-term survival. Here, once again, we can imagine that the chromosomes of cancer cells with sub-optimal telomere lengths undergo relatively frequent breakage–fusion–bridge cycles, perhaps because they express hTERT at levels that are barely adequate to maintain telomeric DNA. These BFB cycles will create an ongoing chromosomal instability that continues throughout the life of the tumor to generate novel, scrambled genotypes, some of which may happen to favor advantageous phenotypes, such as more rapid proliferation and increasing aggressiveness. Certain human chronic inflammatory conditions, which are associated with increased risks of cancer, may also be attributable to eroded telomeres and BFB cycles (**Sidebar 10.6**).

This model provides an attractive mechanism that explains how many types of human tumor cells acquire highly aneuploid karyotypes. Although not directly demonstrated, it is widely assumed that the resulting aneuploid genomes confer growth advantages on these cells by creating novel oncogenes through translocations, by increasing the dosage of growth-promoting proto-oncogenes, and by eliminating tumor suppressor genes that have been holding back cell proliferation. Hence, these BFB cycles may be instrumental in accelerating tumor progression because they increase genomic mutability and allow evolving, premalignant cells to explore a multitude of novel, potentially advantageous genotypes.

Yet other observations are consistent with this thinking but hardly prove it. For example, the leukemia cells in about two-thirds of acute myelogenous leukemia (AML) patients exhibit a normal karyotype, whereas cells from the remainder exhibit various types of karyotypic disarray. Those AML tumors with normal karyotype exhibit significantly longer telomeres than do those with a scrambled karyotype, consistent with the notion that eroded telomeres are associated with and possibly responsible for the derangement of normal karyotype. Of additional interest, a **hypomorphic** allele of *hTERT*—one that specifies reduced enzyme function—is carried by as many as 1% of individuals in the general population. Significantly, individuals inheriting this allele—which confers only 60% of normal hTERT function—are overrepresented by a factor of 3 among AML patients. This finding suggests that normal cells throughout the bodies of these patients have a reduced ability to maintain normal telomere lengths, which in turn may predispose these people to the development of this form of leukemia. Indeed, those afflicted with dyskeratosis congenita (see Supplementary Sidebar 10.7) have a greatly increased risk of developing myelodysplastic syndrome, which in turn leads, with significant frequency, to AML.

Sidebar 10.6 Telomere collapse may contribute to cancer in organs affected by chronic inflammation In Chapter 11, we will read about a number of human tumors arising in tissues that suffer continuous loss of cells resulting from chronic infections or inflammation. Such ongoing attrition of cells, which may occur over a period of decades, occurs in tissues affected by, for example, ulcerative colitis, Barrett's esophagus, and hepatitis B or C virus infection. In response to losses of differentiated cells, the stem cells in these tissues are continually producing replacements to ensure maintenance of tissue functions. These stem cells are therefore forced to pass through many more cycles of growth and division than are the corresponding stem cells of normal tissues. The resulting repeated cell cycles may tax the regenerative powers of the stem cell pools and lead eventually to telomere collapse, to the triggering of breakage–

fusion–bridge cycles, and to the generation of karyotypes that favor neoplastic cell proliferation.

Such a mechanism may well account for the observed high rates of cancer in individuals suffering from these inflammatory and infectious diseases. In fact, in the intestinal epithelial cells of individuals suffering from ulcerative colitis, one can often find the anaphase bridges that are telltale signs of BFB cycles and hence of telomere collapse—precisely what we would expect if telomere collapse were contributing to the formation of the colon carcinomas seen in these patients (**Figure 10.24**). Similarly, anaphase bridges have been documented in the cells of Barrett's esophagus (see Figure 2.12), a condition in which the reflux of stomach acid causes a high turnover of esophageal epithelial cells that leads, with significant frequency, to esophageal carcinomas.

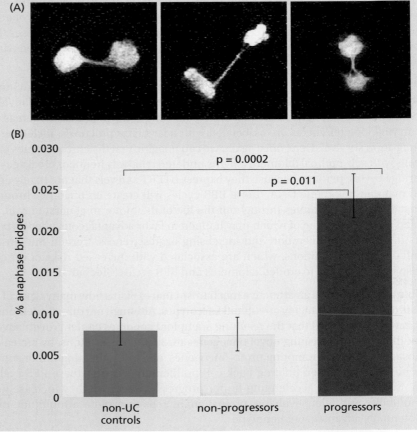

Figure 10.24 Ulcerative colitis and anaphase bridges Ulcerative colitis (UC) involves chronic irritation and inflammation of the colonic epithelium. These provoke ongoing turnover of colonic epithelial cells that can continue for several decades. UC leads, with significant frequency, to the development of colon carcinomas. (A) UC patients can be separated into "progressors," whose colitis has led to the appearance of one or more carcinomas, and "non-progressors," who show no evidence of intestinal tumors. The colonic epithelial cells of progressors frequently display the anaphase bridges characteristic of BFB cycles seen here (see also Figure 10.10), as well as extensive shortening of telomeres and numerous karyotypic aberrations (*not shown*). (B) Colon tissue samples from normal control individuals and UC non-progressors show relatively few anaphase bridges, but much greater numbers are found in the inflamed epithelium of the progressors. This indicates a correlation between the occurrence of anaphase bridges and the onset of tumor development. The p values indicate the probability of these differences arising by chance. (From J.N. O'Sullivan et al., *Nat. Genet.* 32:280–284, 2002. With permission from Nature.)

These observations of premature, pathological shortening of telomeres lead to another question: Is the significantly increased cancer-associated incidence and mortality among the elderly (see Figure 11.1) attributable, at least in part, to *normal* telomere shortening? Exhaustive measurements of telomere lengths in normal humans do indeed indicate marked telomere erosion with increasing age. The data from measurements, such as those provided in **Figure 10.25A**, yield three conclusions: (1) In some (and perhaps most) human tissues, telomeres shorten progressively with increasing age. (2) The rate of shortening differs in different tissues, ostensibly because of differing mitotic activity in the cell lineages and differing expression of telomerase function in various stem cell compartments. (3) The scatter on each curve indicates substantial inter-individual variability in the lengths of telomeric DNA early in life or rates of telomere shortening during a lifetime. (In addition, the telomeres within a single cell exhibit differing, chromosome-specific lengths.)

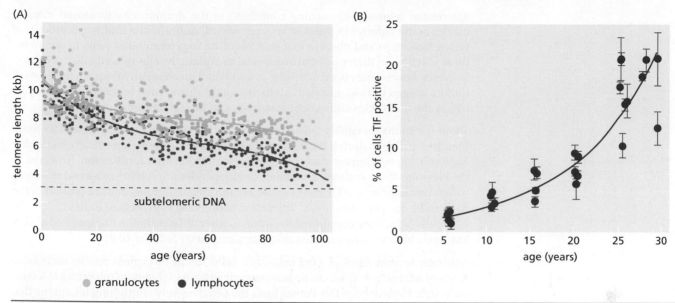

Figure 10.25 Telomere shortening, aging, and cancer
(A) Telomeric DNA lengths were measured in the lymphocytes and granulocytes of 400 individuals. As is apparent, the length of telomeres decreased progressively with increasing age in both human cell types. Inaccuracy in measuring telomere lengths is ±0.5 kb. The horizontal line (*dashed*) represents the threshold length below which telomeric DNA can no longer protect chromosomal ends from end-to-end fusions. (B) TIFs (telomere dysfunction–induced foci) were scored in dermal fibroblasts from baboons of various ages. These TIFs were scored wherever there was a congruence within a cell between γ-H2AX, a marker of double-strand DNA breaks (see Chapter 12), and telomeric DNA; they represent unrepaired and ostensibly irreparable dsDNA breaks. Their accumulation suggests that telomere erosion (and resulting cell senescence) represents one force in driving the aging process in certain tissues and, quite possibly, an important source of the age-related increase in cancer incidence in older humans. (A, courtesy of G.M. Baerlocher and P. Lansdorp. B, from U. Herbig et al., *Science* 311:1257, 2006. With permission from AAAS.)

Other research, not described here in detail, yields a highly relevant observation: individuals whose normal circulating lymphocytes exhibit relatively short telomeres are overrepresented in cohorts of patients suffering from head-and-neck, bladder, lung, and renal carcinomas. In this case, the lymphocytes are presumed to be representative of normal cells throughout the body, suggesting that short telomeres in one's normal tissues represent a lifelong increased risk for developing various types of carcinomas. This also raises the question of whether the behavior of $mTR^{-/-}$ mice described above is applicable to that of aged humans, or more specifically, whether it contributes to the greatly elevated cancer incidence observed in older humans (Figure 10.25B).

These speculations still do not address an issue raised by some observations cited earlier: the repeated BFB cycles in $mTR^{-/-} p53^{-/-}$ mice affect the *types* of tumors that these mice exhibit, causing them to develop carcinomas, which are common in humans, rather than hematopoietic and mesenchymal malignancies, which are frequently seen in mice. Precisely how does telomere biology possibly help to explain why cancers tend to arise in some tissues and not others?

10.12 Synopsis and prospects

Telomeres are now known to be major determinants of the ability of cells to multiply for a limited number of growth-and-division cycles before halting proliferation and entering into crisis. The resulting circumscribed proliferative potential of normal human cells appears to operate as an important barrier to the development of cancers by limiting the proliferation of pre-neoplastic cell clones. This mechanistic model of cell immortalization and cancer pathogenesis is elegant, if only because it is so simple. Still, the simplicity of this model should not obscure the fact that much of telomere structure and function remains poorly understood, and some of the conclusions described in this chapter may one day require substantial revision.

Among the many unresolved issues are the connections between the findings described in this chapter and a major problem of biomedical research: Does age-related telomere shortening contribute to the dramatically increased rates of cancer in the elderly? In the next chapter, we will examine the multi-step process of tumor formation and observe that each requisite step often takes years to complete; these kinetics, on their own, can be invoked to explain the age-related onset of cancer. However, now we may need to invoke an additional mechanism to explain the onset of tumors in aged individuals: perhaps their significantly eroded telomeres occasionally trigger the BFB cycles and genetic instability that fuels the formation of tumors.

These questions inevitably return to a focus on the telomerase enzyme and its function. Interestingly, during the earliest stages of **development**, specifically soon after egg fertilization, telomeres may be relatively short soon after fertilization. However, in the cleavage-stage embryo, a telomerase-independent, interchromosomal recombination mechanism similar to ALT generates the long telomeres that are present in the soon-to-be-formed blastocysts. Thereafter, telomerase is expressed in tissue-specific stem cells (which are committed to one or another differentiation lineage) and at very low levels in more differentiated cells (Supplementary Sidebar 10.8).

Although several types of stem cells have been found to express readily detectable levels of telomerase, this activity is apparently unable to maintain telomeres at a constant size. Evidence for this comes from the progressively shortening telomeres that are seen with increased life span. (Because the differentiated cells that make up the great majority of cells in most tissues are generally a small number of cell generations removed from the stem cells within their respective tissues, the progressive telomere shortening seen in these differentiated cells must reflect the status of telomeres in the corresponding stem cell precursor.)

The experiments with the telomerase-negative mice described here underscore the consequences associated with the absence of telomerase and critically shortened telomeric DNA: in the presence of functional p53, such cells will enter crisis and die in large numbers—a mechanism that serves to obstruct the further expansion of clones of pre-neoplastic cells. However, without p53 on the job, critically shortened telomeres fail to trigger crisis and instead may lead the cell into multiple breakage–fusion–bridge cycles that create widespread genetic instability (**Figure 10.26**); this increased mutability can accelerate the progression of incipient tumors.

Figure 10.26 Karyotypic chaos Here, use of the SKY technique to "paint" each chromosome its own characteristic color revealed that numerous chromosomes belonging to two subclones (A and B) of a human bladder carcinoma cell line have participated in translocations. The translocations are indicated by the many chromosomes that carry segments originating from two or more normal human chromosomes. Each translocation is identified by a "t" followed by the number of each parental chromosome from which these arms derive. An "i" indicates an isochromosome, in which a single arm of a chromosome has been duplicated. The notation "p" and "q" indicates the short and long arms of normal human chromosomes, respectively; "del" indicates that a chromosomal segment has been deleted. The label "clone A and B" indicates aberrant chromosomes shared by the two subclones, while "clone A" and "clone B" indicate chromosomes that are present uniquely in one or another of these subclones. The translocations shared in common by the two subclones are likely to have occurred before these cells were introduced into culture, whereas those present in either one subclone or the other indicate ongoing chromosomal instability in culture. Comparable degrees of aneuploidy can be observed in cancer cells that have recently been introduced into culture. (From H.M. Padilla-Nash et al., *Genes Chromosomes Cancer* 25:53–59, 1999.)

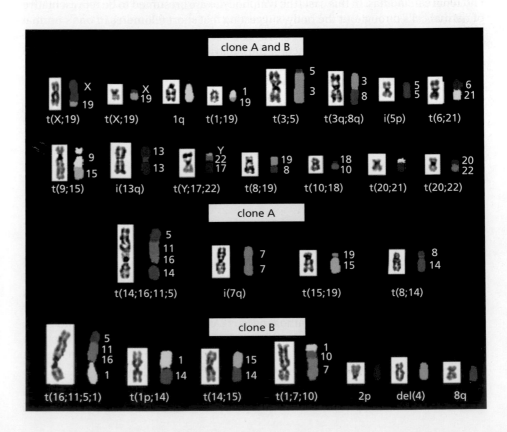

There are still many puzzles associated with hTERT and telomeres that require clarification. Ectopic hTERT expression can operate to protect certain types of cells against p53-induced apoptosis. Provocatively, this protection is also afforded by a mutant of hTERT that lacks catalytic activity. Similarly, in zebrafish, knockdown of the *zTERT* gene results in compromised hematopoiesis (formation of various blood cells); this effect can be reversed by a mutant of zTERT that lacks the TERC-binding domain and therefore lacks bona fide telomerase activity. In addition, overexpression of a TERT mutant lacking reverse transcriptase activity stimulates hair growth in mice. Multiple reports indicate that hTERT lacking its *hTR* subunit can be found in mitochondria, where it appears to have a role in conferring resistance to death triggered by reactive oxygen species (ROS). Yet another mystery is created by the discovery of TERRA (telomere repeat-containing RNA), a long noncoding RNA (lncRNA) of heterogeneous size that is transcribed from the C-rich strand of telomeric DNA and participates in some important fashion in the maintenance of the telomere-associated chromatin.

Certain non-telomerase-associated functions of hTERT may one day be associated with another biochemical activity of hTERT: it associates in the nucleus with the Brg-1 protein, a chromatin-modifying enzyme, and thereby amplifies Wnt-β-catenin–stimulated transcription. This function may explain why the use of immunofluorescence to localize hTERT protein demonstrates that it is distributed along the lengths of chromosomes rather than being concentrated at telomeres.

Cells that have been deprived of hTERT activity by siRNA-mediated knockdown have a markedly decreased ability to repair their DNA following exposure to several types of mutagens. This effect also cannot be easily related to any effects of hTERT on telomeric DNA. Recall as well that certain cells maintaining their telomeres via the ALT mechanism (which generates very long telomeres) cannot be readily transformed by an introduced *ras* oncogene but can readily be transformed if they are supplied with an ectopically expressed hTERT. Diverse observations like these indicate that maintenance of telomere length is only one facet of the function of the apparently multi-talented hTERT protein.

Replicative senescence, as controlled by telomere erosion, has been portrayed as one mechanism whereby the unlimited replicative power of cell lineages is constrained. In recent years, this thinking has been challenged by two types of findings. First, the onset of this process can be affected, sometimes dramatically, by altering conditions of *in vitro* cell cultures. From this perspective, the timing of replicative senescence can no longer be viewed as a cell-autonomous process that counts the number of cell generations separating cells from their ancestors in the early embryo. Second, as cited above, various lines of research indicate that hTERT has multiple functions that are clearly distinct from telomere maintenance. Taken together, these findings raise the question of whether the ability of the hTERT protein to allow cells to circumvent senescence is tied to one of its functions that is unrelated to elongating telomeric DNA.

Importantly, these findings do not jeopardize another key concept: that severe telomere erosion triggers crisis in human cells and that ectopic expression of hTERT can prevent or allow escape from the ensuing, almost-certain death. Moreover, whether or not replicative senescence, as originally defined with cultured cells, is a real phenomenon, it is clear that certain forms of senescence (for example, oncogene-induced senescence) operate in living tissues, where they seem to serve as an impediment to the forward march of pre-neoplastic cells, as discussed in Chapter 9.

Arguably the most direct and compelling proof of the importance of telomeres and telomerase in cancer pathogenesis comes from experiments that will be described in detail in the next chapter: a clone of the *hTERT* gene has been found to be an essential ingredient in the cocktail of introduced genes that are used experimentally to transform normal human cells into tumorigenic derivatives. Without the presence of *hTERT*, experimental human cell transformation fails.

The central role of telomerase in fueling the growth of cancer cells has long suggested a novel, highly effective strategy of anti-cancer therapy. Telomerase inhibitors should prove to be potent in killing cancer cells that carry relatively short telomeres—the situation observed in many telomerase-positive human tumor cells. Such drugs, if developed, should at the same time have little if any effect on normal cells throughout the body, which express relatively low levels of telomerase and then only transiently during S phase. Although attractive in concept, this plan may be derailed by certain realities. First, attempts by a number of pharmaceutical companies to develop novel low–molecular-weight inhibitors of hTERT catalytic function have repeatedly failed to yield potent, highly specific drug molecules. The reasons for these failures are still obscure. Second, no one knows whether the relatively low levels of telomerase enzyme detectable in certain types of normal cells throughout the human body play a significant role in the proliferative capacity and viability of these cells. Consequently, the side effects of anti-telomerase drugs, should these agents be developed, are unpredictable.

The failures to generate drug-like inhibitors of hTERT function have led some to embrace an alternative strategy—the construction of a chemically modified 13-nucleotide-long RNA molecule that is complementary to the template region of the *hTR* RNA subunit of the holoenzyme (see Figure 10.13). This potent inhibitor of the telomerase holoenzyme yields rapid responses in telomerase-positive cancer cells propagated in culture, reducing detectable telomerase activity by 90 to 95%; the testing of its clinical efficacy is under way. Nonetheless, the therapeutic utility of this very promising agent, specifically its ability to elicit durable clinical responses, may be frustrated by one of the realities of telomere biology cited above: short-term decreases in tumor burden may be followed by the emergence of cancer cells and thus tumors that have switched to the ALT state of telomere maintenance, thereby becoming hTERT-independent. Accordingly, the goal of exploiting the widespread expression of hTERT in human tumors to generate novel, cancer-specific therapeutics may continue to prove elusive.

Key concepts

- Two barriers prevent cultured cells from replicating indefinitely in culture—senescence and crisis.

- Senescence involves the long-term residence of cells in a nongrowing but viable state; crisis involves the apoptotic death of cells.

- Senescence is provoked by a variety of physiological stresses that cells experience *in vitro* and also *in vivo*, where it seems to play a significant role in constraining neoplastic progression.

- Crisis is provoked by the erosion of telomeres, which results in widespread end-to-end chromosomal fusions, karyotypic chaos, and cell death.

- Most premalignant cells escape from crisis by activating expression of hTERT, the telomerase enzyme, which is specialized to elongate telomeric DNA by extending it in hexanucleotide increments.

- A minority of incipient cancer cells escape crisis by regenerating their telomeric DNA through the ALT mechanism.

- Cells that have stabilized their telomeres through the actions of telomerase or the ALT mechanism can then proliferate indefinitely and are therefore said to be immortalized.

- Cell immortalization is a step that appears to be a prerequisite to the development of all human cancers.

- The end-to-end chromosomal fusions that accompany crisis lead to repeated breakage–fusion–bridge (BFB) cycles, which appear to be responsible for much of the aneuploidy associated with the karyotypes of many kinds of solid human tumors.

- These BFB cycles may prove to be an important means by which incipient cancer cells acquire mutant alleles, thereby expediting the formation of fully neoplastic cells.

Thought questions

1. Why is the acquisition of an immortalized proliferative potential so important for human tumors?

2. What types of evidence connect telomeres and telomerase to enter into the senescent state, and what alternative mechanisms are responsible for entrance into this state?

3. What complications and side effects might result from the shutdown of telomerase activity by anti-cancer drugs that may be developed in the future and function as specific inhibitors of this enzyme?

4. How does the molecular configuration of the t-loop protect the ends of telomeric DNA?

5. How does the telomerase-associated *hTR* RNA molecule facilitate the maintenance of telomeric DNA by the hTERT enzyme?

6. Precisely how do breakage–fusion–bridge cycles confer an advantage on populations of premalignant cells?

Additional reading

Artandi SE & DePinho RA (2010) Telomeres and telomerase in cancer. *Carcinogenesis* 31:9–18.

Bizard AH & Hickson ID (2018) Anaphase: a fortune-teller of genomic instability. *Curr. Opin. Cell Biol.* 52:112–119.

Boehm JS & Hahn WC (2004) Immortalized cells as experimental models to study cancer. *Cytotechnology* 45,47–59.

Blackburn EH & Collins K (2010) Telomerase: an RNP enzyme synthesizes DNA. *Cold Spring Harb. Perspect. Biol.* a003558.

Campisi J (2013) Aging, cellular senescence, and cancer. *Annu. Rev. Physiol.* 75:685–705.

Campisi J & d'Adda di Fagagna F (2007) Cellular senescence: when bad things happen to good cells. *Nat. Rev. Mol. Cell Biol.* 8:729–740.

Cesare AJ & Reddel RR (2010) Alternative lengthening of telomeres: models, mechanisms and implications. *Nature Rev. Genet.* 11:319–330.

Coppé JP, Desprez PY, Krtolica A et al. (2010) The senescence-associated secretory phenotype: the dark side of tumor suppression. *Annu. Rev. Pathol. Mech. Dis.* 5:99–118.

de Lange T (2018) Shelterin-mediated telomere protection. *Annu. Rev. Genet.* 52:223–247.

Diman A & Decottignies A (2018) Genomic origin and nuclear localization of TERRA telomeric repeat-containing RNA: from darkness to dawn. *FEBS J.* 285:1389–1398.

Guterres AN & Villanueva J (2020) Targeting telomerase for cancer therapy. *Oncogene* 39:5811–5824.

Huang FW, Hodis E, Xu MJ et al. (2013) Highly recurrent TERT promoter mutations in human melanoma. *Science* 339:957–959.

Jäger K & Walter M (2016) Therapeutic targeting of telomerase. *Genes (Basel)* 7:39–53.

Maciejowski J & de Lange T (2017) Telomeres in cancer: tumour suppression and genome instability. *Nat. Rev. Mol. Cell Biol.* 18:175–186.

Min J & Shay, JW (2016) *TERT* promoter mutations enhance telomerase activation by long-range chromatin interactions. *Cancer Discov.* 6:1212–1214.

Martinez P & Blasco MA (2011) Telomeric and extra-telomeric roles for telomerase and the telomere-binding proteins. *Nat. Rev. Cancer* 11, 161–176.

Martinez P & Blasco MA (2017) Telomere-driven diseases and telomere-targeting therapies. *J Cell Biol.* 216:875–887.

Murnane JP (2010) Telomere loss as a mechanism of chromosome instability in human cancer. *Cancer Res.* 70:4255–4259.

Pickett HA & Reddel RR (2015) Molecular mechanisms of activity and derepression of alternative lengthening of telomeres. *Nat. Struct. Mol. Biol.* 22:875–880.

Podlevsky JD & Chen JJ-L (2016) Evolutionary perspectives of telomerase RNA structure and function. *RNA Biol.* 13:720–732.

Roake CM & Artandi SE (2017) Control of cellular aging, tissue function, and cancer by p53 downstream of telomeres. *Cold Spring Harb. Perspect. Med.* 7:a026088.

Roake CM & Artandi SE (2017) Approaching TERRA firma: Genomic functions of telomeric noncoding RNA. *Cell* 170:8–9.

Schmutz I & de Lange T (2016) Shelterin. *Curr. Biol.* 26:R397–R399.

Shay JW & Wright WE (2011) Role of telomeres and telomerase in cancer. *Semin. Cancer Biol.* 21:349–353.

Zhou J, Ding D, Wang M & Cong Y-S (2014) Telomerase reverse transcriptase in the regulation of gene expression. *BMB Rep.* 47:8–14.

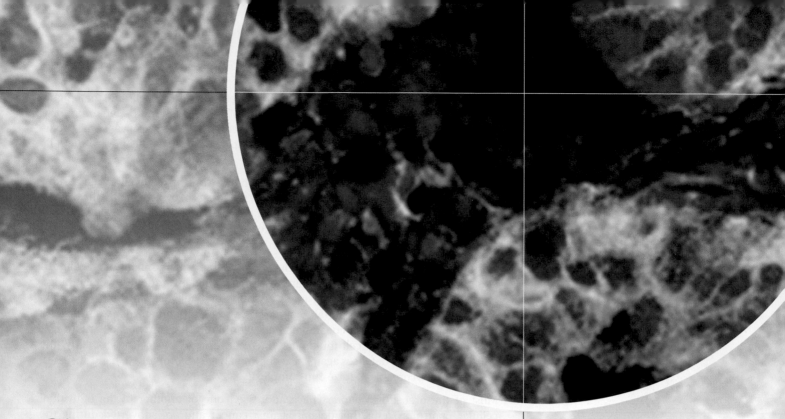

Chapter 11

Multi-Step Tumorigenesis

In the survival of favoured individuals and races, during the constantly-recurring struggle for existence, we see a powerful and ever-acting form of selection.

Charles Darwin, biologist, 1859

Nothing in biology makes sense except in the light of evolution.
Theodosius Dobzhansky, geneticist, 1973

In the evolutionary diversification of mammals, enormous changes in cellular susceptibility to oncogenes have developed.
Richard Peto, epidemiologist, 1977

The chapter that follows is the longest of this book because it attempts to draw together diverse lines of evidence in order to create an overarching conceptual framework for understanding of human cancer formation. It draws on the contents of all preceding chapters and is worthwhile digesting in segments rather than swallowing the contents whole.

The formation of a human tumor is a complex process that usually proceeds over a period of decades. Normal cells evolve into cells with increasingly neoplastic phenotypes through a process termed **tumor progression**. This process takes place at myriad sites throughout the normal human body, advancing further and further as we get older. Rarely does it proceed far enough at any single site to make us aware of its end product, a clinically detectable tumor mass.

Tumor progression is driven by a sequence of randomly occurring mutations, epigenetic alterations of DNA, and epigenetically controlled cell-biological programs that affect the genes governing cell proliferation, survival, and other traits associated with the malignant cell phenotype. The complexity of this process reflects the work of evolution, which has erected a series of barriers between normal cells and their highly neoplastic derivatives. Accordingly, completion of each step of multi-step tumor progression within a tissue can be viewed as the successful breaching of yet another barrier that has been impeding the progress of a clone of normal cells toward the fully malignant state.

One might think that these barriers are the handiwork of relatively recent evolutionary processes. Perhaps, we might imagine, the forces of evolution initially worked to design the architecture and physiology of complex metazoan bodies and, having completed this task, then proceeded to tinker with these plans in order to reduce the risk of cancer.

An alternative scenario is far more likely, however: the risk of uncontrolled cell proliferation has been a constant companion of metazoans from their very beginnings, roughly 600 million years ago. By granting individual cells within their tissues the license to proliferate under certain conditions, even simple metazoans ran the risk that one or another of their constituent cells would turn into a renegade and trigger the disruptive, runaway cell multiplication that we call cancer. Consequently, the erection of multiple defenses against cancer must have accompanied, hand-in-hand, the evolution of organismic size and complexity.

Most of the preceding chapters have addressed one or another of the individual cellular control systems that defend against cancer and are subverted during the process of tumorigenesis. Now, we will begin tying these individual threads together and examine how alterations in these systems contribute to the end product—the formation of a primary tumor. We start by attempting to gauge the scope of the problem at hand: How many different sequential changes are actually required in cells and tissues in order to create a human cancer and how long do these processes take to reach completion?

11.1 Most human cancers develop over many decades of time

Epidemiologic studies have shown that age is a surprisingly large factor in the incidence of cancer. In the United States, the risk of dying from colon cancer is as much as 1000 times greater in a 70-year-old man than in a 10-year-old boy. This fact, on its own, suggests that this type of cancer and, by extension, many other cancers common in adults (**Figure 11.1**), require years if not decades to develop (**Sidebar 11.1**).

The various frequencies of cancers in diverse populations have led to a number of calculations that shed light on the complex processes of cancer formation (Supplementary Sidebar 11.1). In many cases, simple algebraic formulas, a^5 or a^6, may roughly predict cancer incidence as a function of elapsed lifetime (where a = years

Figure 11.1 Cancer incidence at various ages These graphs of diagnoses of various types of epithelial cancers show a steeply rising incidence with increasing age, indicating that the process of tumor formation generally requires decades to reach completion. (Courtesy of W.K. Hong, compiled from *SEER Cancer Statistics Review*.)

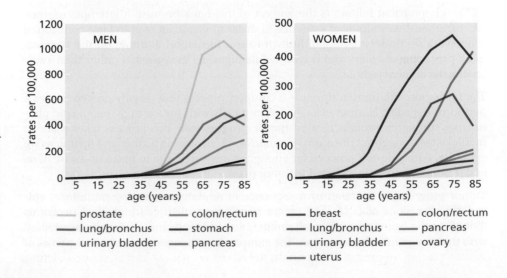

Sidebar 11.1 Estimates of time required for tumor development in humans Attempts to gauge the latency of cancer development—the time window between initial exposure to a carcinogen and the eventual diagnosis of disease—are complicated by our inability in the case of most cancers to identify the timing of exposure to a known carcinogen and the time of subsequent diagnosis of disease. One compelling measurement comes from measures of the incidence of lung cancers among males in the United States. Cigarette smoking was relatively uncommon among this group until World War II, when large numbers of men first acquired the habit, encouraged in part by the cigarettes they received as part of their rations while serving in the U.S. armed forces. Thirty years later, in the mid-1970s, the rate of lung cancer began to climb steeply with kinetics that closely paralleled the increase in size of the smoking population in the U.S. (**Figure 11.2**). Another comparable measurement comes from the development of mesothelioma among British shipyard workers who were exposed to asbestos used in ship construction; in this case, the lag in cancer development could be gauged to be about 40 years after first exposure. In general, however, to the extent that many human cancers derive from chronic exposure to low levels of carcinogen over extended periods of time, it is impossible to estimate the lag period between exposure and disease development.

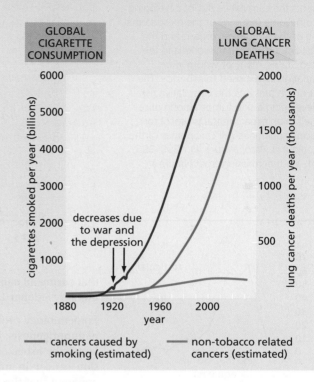

Figure 11.2 Cigarette consumption and lung cancer These curves compare the annual global consumption of cigarettes (*red curve*) with the recorded and predicted annual worldwide mortality from lung cancer (*green curve*). Annual mortality from tobacco-induced lung cancer (in thousands) is estimated to peak sometime in the fourth or fifth decade of the twenty-first century. During the twentieth century, there was an increase in what is judged to be "non-tobacco-related" lung cancer mortality (*blue curve*). The precise number of these cases is unclear, and there is some debate about how many of them are attributable to secondhand tobacco smoke. (From R.N. Proctor, *Nat. Rev. Cancer* 1:82–86, 2001. With permission from Nature.)

lived). Such dynamics imply a sequence of random, independent events that occur at comparable frequencies over extended periods of time. Such calculations can only be made when cancer incidence and age-of-onset of disease are averaged over a large human population. These independent events are likely to be **rate-limiting**, and their probability of occurring per unit of time may vary dramatically from one individual to another, being affected by inherited predisposition, diet, and lifestyle, among other variables. Independent of the measurements of event probability is the conclusion that cancer development is a multi-step process for most kinds of human cancer. (These calculations have limited utility for pediatric cancers, whose dynamics do not conform to this model of disease pathogenesis, as discussed later in this chapter.)

The conclusion that tumorigenesis is a multi-step process hints at another interesting idea. Assume that (1) a sequence of unlikely events is required in order for a tumor to appear and (2) many of these events happen at comparable frequencies in all of us. Together, these assumptions indicate that as we grow older, virtually all of us will carry populations of cells in many locations throughout the body that have completed some but not all of the steps of tumor progression. Because most of us will not live long enough for the full schedule of requisite events to be completed (because we will succeed in dying from accidents or other diseases), we will never realize that any of these tumor progressions had been initiated at various sites within our bodies. Indeed, autopsies reveal that by life's end, 60 to 70% of individuals carry undiagnosed tumors, independent of the cause of death; these autopsies provide minimum estimates of the actual number of minor pre-neoplastic growths in the body. Viewed from this perspective, cancer is an inevitability; if we succeeded in avoiding the death traps set by all the other usual diseases and conditions, sooner or later most of us would become victims of cancer. Interestingly, the late onset of most human cancers dictates

Figure 11.3 Cancer incidence and duration of carcinogen exposure These graphs indicate that cumulative exposure to a carcinogenic stimulus, rather than the age at which this exposure began, determines the likelihood of developing a detectable tumor. The left graph presents the cumulative risk of developing mesothelioma (a tumor of the mesodermal lining of the abdominal organs and the lungs) among insulation workers in the United States, many of whom were occupationally exposed to asbestos. The right graph presents the cumulative risk of developing a skin tumor among mice treated in an experimental protocol for inducing squamous cell carcinomas of the skin. (From J. Peto, *Nature* 411:390–395, 2001. With permission from Nature.)

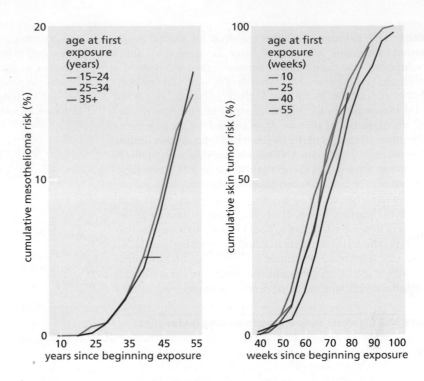

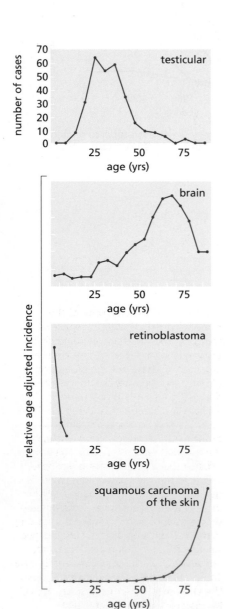

that curing all cancers will have relatively minor effects on overall life span (Supplementary Sidebar 11.2).

Epidemiology provides us with another important insight into the multi-step nature of tumorigenesis. If we examine the frequencies of mesothelioma in humans (caused largely by asbestos exposure plus increased incidence from cigarette smoking) and skin cancer in mice (induced by repeated benzo[*a*]pyrene painting), it becomes apparent that the formation of each of these tumors requires an extended period of repeated exposure to carcinogens and that it is the *duration* of this exposure (rather than the absolute *age* of exposed individuals or the age when exposure began) that determines the timing of the onset of detectable disease (**Figure 11.3**). In these particular cases, tumors are created by the actions of exogenous carcinogens rather than by processes operating entirely within the body's various tissues; these carcinogens increase the rate of tumor development, often by many orders-of-magnitude above the spontaneous background rate. The lion's share of the background rates of human cancers that seem to be formed spontaneously appear to result from the actions of *endogenous* carcinogenic processes, notably (1) errors in DNA replication and repair, (2) accidents occurring during various steps of cell division, and (3) actions of mutagenic molecular species that are by-products of metabolism, notably energy metabolism in the mitochondria.

These generalizations about cancer formation tend to overlook cancers in a minority of adults that occur at ages that do not conform to a simplified algebraic formula of multi-step tumorigenesis as described above. As we see in **Figure 11.4**, a pediatric cancer strikes only early in life, whereas testicular cancer strikes in early adult life. Kinetics like these demonstrate that biological forces operating in certain differentiated tissues may ultimately determine cancer onset, a point made as well in Figure 11.1 by examining the kinetics of breast cancer onset.

Figure 11.4 Cancer incidence as a function of age These graphs show age-dependent incidence rates in a subset of cancers. The rates vary dramatically from the kinetics of appearance of most common cancers. With the exception of testicular cancer, the ordinates in all cases represent incidence relative to the highest incidence rate. Squamous cell skin carcinoma is included to illustrate the incidence kinetics of a cancer that is more typical of adult-onset neoplasias. (Courtesy of B. Westermark; from J. Li et al., *Pediatrics* 121: e1470–e1477, 2008.)

11.2 Histopathology provides evidence of multi-step tumor formation

The notion of human tumor development as a multi-step process has been histologically documented most clearly in the epithelium of the intestine. The intestinal epithelial cells, which face the interior cavity (lumen) of the gastrointestinal tract, form a layer that is only one cell thick in many places (**Figure 11.5**). These epithelial cell populations are in constant flux. Each minute, 20 to 50 million cells in the human duodenum and a tenth as many in the colon die and an equal number of newly minted cells replace them! These epithelial cells, rather than underlying mesenchymal cells, are the sources of cancers arising in the intestine.

Analyses of human colonic biopsies have revealed a variety of tissue states, with degrees of abnormality that range from mildly deviant tissue, which is barely distinguishable from the structure of the normal intestinal **mucosa** (the epithelial lining of the colonic lumen), to the chaotic jumble of cells that form highly malignant tissue (**Figure 11.6**). Like the normal intestinal lining, these growths are composed of various distinct cell types, indeed almost all of the cell types found in the normal tissue.

Some growths that are classified as **hyperplastic** exhibit an almost-normal histology, in that the individual cells within these growths have a normal appearance. However, it is clear that in these areas of **hyperplasia**, the rate of epithelial cell division is unusually high, yielding thicker-than-normal epithelia, often composed of multiple cell layers in an epithelium that is normally only one-cell-layer thick. Yet other growths show abnormal histology, with the individual epithelial cells no longer forming the

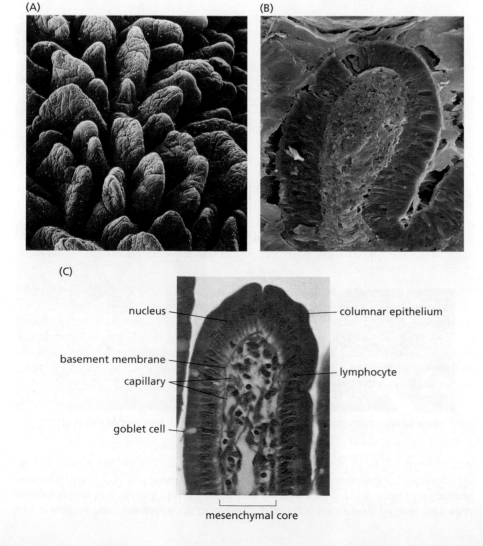

(A)

(B)

(C)

nucleus — columnar epithelium

basement membrane

capillary

lymphocyte

goblet cell

mesenchymal core

Figure 11.5 Microanatomy of the normal intestinal wall The epithelial cells lining the small intestine and those lining the large intestine (colon) have a similar organization. However, the protruding villi visualized here are present only in the small intestine. (A) This scanning electron micrograph of epithelium of the small intestine shows villi (fingers) of the intestinal mucosa extending into the lumen at regular intervals. (B) Each villus is covered with a layer of epithelial cells. (C) The core of each villus, which is separated from the overlying epithelium by a basement membrane, is composed of various types of mesenchymal cells, including fibroblasts, endothelial cells, pericytes, and various cells of the immune system (*not indicated*). (A, © Susumu Nishinaga/Science Source. B, © SPL/Science Source. C, from University of Iowa Virtual Hospital, Atlas of Microscopic Anatomy, Plate 194.)

Figure 11.6 Histopathological alterations of the human colon
The various types of abnormal tissues revealed by histopathological analyses of the human colon can be arrayed in a succession of ever-increasing abnormality. Normal colonic crypts are seen here in longitudinal section (*top left*). (Very similar crypts are found in the small intestine around the base of each villus.) A small adenomatous crypt (*arrow, circled*) is shown, together with normal crypts, in cross section (*top right*). The cells in this abnormal crypt are presumed to have the ability to develop into at least two distinct types of adenomas—tubular and villous. A small tubular adenoma is shown (*left, middle*). The larger tubular adenoma (*below*) is sometimes termed *pedunculated,* indicating its attachment via a stalk to the colonic wall. These two types of adenomas have the ability to evolve into a locally invasive carcinoma, whose presence here is indicated by small islands of carcinoma cells (*circled*) surrounded by extensive stroma (*right, lower middle*). Dissemination of these cells, usually to the liver (*bottom right*), can lead to metastases (*circled*), which are surrounded here by layers of recruited stromal cells. The histopathological progression indicated here would seem to be the most logical way by which normal tissue, in this case the colonic epithelium, is transformed through a series of intermediate steps into carcinomas and ultimately spawns metastatic growths. However, the evidence for most of these precursor–product relationships is actually quite fragmentary. (Courtesy of C. Iacobuzio-Donahue and B. Vogelstein.)

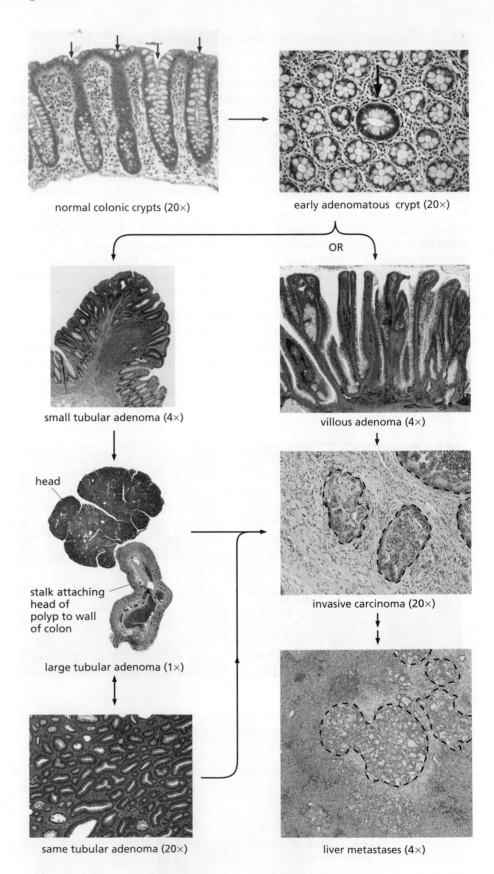

normal colonic crypts (20×)

early adenomatous crypt (20×)

OR

small tubular adenoma (4×)

villous adenoma (4×)

head

stalk attaching head of polyp to wall of colon

large tubular adenoma (1×)

invasive carcinoma (20×)

same tubular adenoma (20×)

liver metastases (4×)

well-ordered cell layer of the normal colonic mucosa and exhibiting a morphology that deviates in subtle ways from that of normal cells; these growths are said to be **dysplastic** (see Figure 2.13). A much larger and more deviant growth that has dysplastic cells and marked thickening is termed a **polyp** or an **adenoma** (see Figures 2.14A and 11.6). In the colon, several distinct types of polyps are encountered; some grow along the wall of the colon, whereas others are tethered to the colonic wall by a stalk. Importantly, these growths are considered benign, in that none has broken through the basement membrane (see Figure 2.15A) and invaded underlying stromal tissue.

The more abnormal growths that have indeed invaded through the basement membrane and beyond have progressed from a previously **benign** to a **malignant** state. There are distinctions among these more aggressive colon carcinomas and associated cancer cells, depending on whether they have penetrated deeply into the stromal layers and smooth muscle and whether cells from these growths have migrated— **metastasized**—to nearby lymph nodes and anatomically distant sites in the body, where they may have succeeded in founding new tumor cell colonies.

Having arrayed these growths in a succession of tissue phenotypes that advance from the normal to the aggressively malignant (see Figure 11.6), we might imagine that this succession depicts with some accuracy the course of tumor development as it actually occurs in the human colon. In truth, the evidence supporting this scheme is quite indirect. Some tumors may well develop through a succession of intermediate growths, such as those arrayed here. Alternatively, it is possible that some of the tissue types depicted as intermediates in this sequence represent dead ends rather than stepping stones to more advanced tumors. In certain cases of colon cancer, it is also possible that the development of the tumor depends on the ability of early growths to leapfrog over intermediate steps, allowing them to arrive at highly malignant endpoints far more quickly than is suggested by this succession of steps. Similar successions have been proposed for various other epithelial cancers (**Figure 11.7A**). On rare occasions, one can actually find a tissue in which multiple stages of cancer progression coexist (Figure 11.7B); however, observations like these provide no direct evidence of precursor–product relationships between the various types of histologically abnormal growths.

In the case of colorectal tumors, at least three types of evidence strongly support the precursor–product relationship between adenomas and carcinomas. First, on rare occasions, one can actually observe a carcinoma growing directly out from an adenomatous polyp (**Figure 11.8A**). We can surmise that outgrowths like these occur routinely during the development of virtually all colon carcinomas and that, more often than not, the rapid expansion of the carcinoma soon overgrows and obliterates the adenoma from which it arose.

Second, clinical studies have been performed on large cohorts of patients who have undergone **colonoscopy**, which is performed routinely to survey the colon for occasional adenomatous polyps and to remove any that are detected (see also Figure 7.19). At present, it is estimated that those patients whose polyps are removed experience, in subsequent years, more than a 60% reduction in the incidence of colon carcinomas (Figure 11.8B). This finding indicates that in this patient population, conservatively, at least 60% of colon carcinomas derive from preexisting, readily detectable adenomas; because colonoscopy often misses polyps, especially those in the right side of the colon, the actual proportion may be higher. (These observations still do not prove that every single colon carcinoma arising in humans must arise from a preexisting adenoma.)

A third type of evidence supporting precursor–product relationships between premalignant and frankly malignant growths comes from **longitudinal** studies of individual patients over a period of time. This is occasionally possible in patients developing sporadic tumors in organs that are accessible to repeated surveillance, such as the skin, colon, and lungs (Figure 11.8C). Familial cancer syndromes also offer a demonstration of such progression in individual patients. Recall, for example, the disease of familial adenomatous polyposis (FAP; see Section 7.10), in which an individual inheriting a mutant form of the *APC* tumor suppressor gene is prone to develop anywhere from dozens to more than a thousand polyps in the intestine (see Figure 7.19). With a certain low but measurable frequency, one or another of these polyps will progress spontaneously into a carcinoma. (The multiplicity of polyps in these patients and the low conversion rate of individual polyps to carcinomas—estimated to be ~2.5 events per

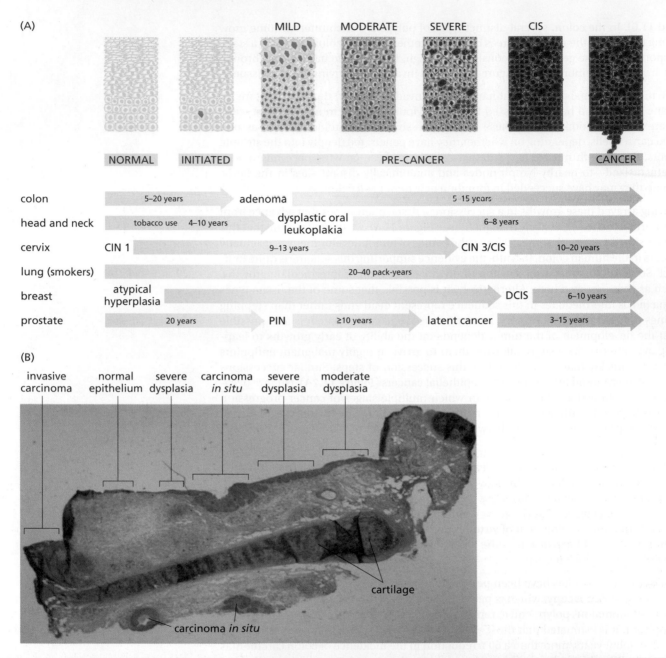

Figure 11.7 Multi-step tumorigenesis in a variety of organ sites (A) The pathogenesis of carcinomas is thought to be governed by very similar biological mechanisms operating in a variety of epithelial tissues. Accordingly, multi-step tumorigenesis involving similar histological entities has been proposed to progress along parallel paths in these various organ sites. These similarities are obscured because the nomenclature is quite variable from one tissue to another. CIS, carcinoma *in situ*; CIN, cervical intraepithelial neoplasia; DCIS, ductal carcinoma *in situ*; PIN, prostatic intraepithelial neoplasia. (B) Because more aggressive growths often overgrow their more benign precursors, it is rare to see the multiple states of tumor progression coexisting in close proximity, as is the case in this lung carcinoma. Importantly, although the close juxtaposition of the aberrant growths suggests some relationship among them, an image like this provides no definitive evidence of precursor–product relationships between these various abnormal tissues. (A, courtesy of W.K. Hong and adapted from J.A. O'Shaughnessy et al., *Clin. Cancer Res.* 8:314–316, 2002. With permission from the American Association for Cancer Research. B, courtesy of A. Gonzalez and P.P. Massion.)

1000 polyps per year—effectively preclude association of a carcinoma with a particular precursor polyp.). This evidence has been garnered through studies of the colon and its proclivity to generate the relatively common intestinal tumors. Indeed, these studies appear to indicate comparable processes operating in a variety of epithelial tissues throughout the body (**Sidebar 11.2**).

Importantly, and as we will discuss again later in this book, the early stages of progression in a tissue do not inevitably progress to later stages, and certain subtypes of

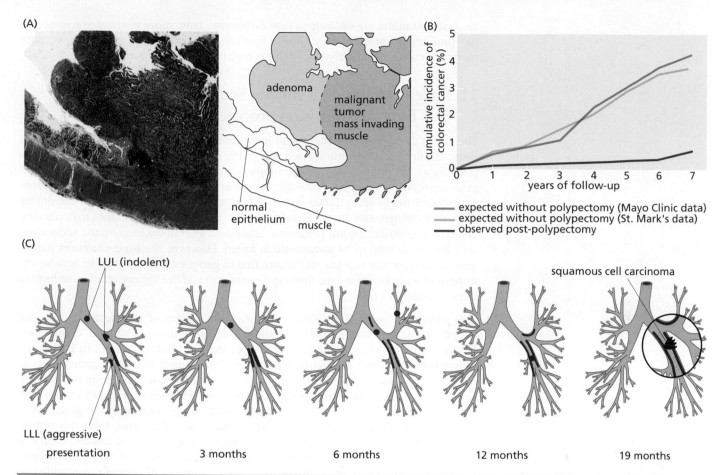

Figure 11.8 Evidence for adenoma-to-carcinoma progression The evidence for precursor–product relationships between premalignant growths (adenomas and *in situ* carcinomas) and frank carcinomas comes from at least three types of clinical observations. (A) On occasion, carcinomas are observed to be growing directly out of adenomas. As seen here, the demarcation between the two can sometimes be drawn with some precision. (B) A second line of evidence derives from clinical studies in which colonoscopy is used to screen large cohorts of patients. Any polyps that are discovered are removed by *polypectomy* (i.e., surgically). As seen in the graph, two independent studies (*yellow-orange, blue lines*) predict a certain expected number of carcinomas in such cohorts on the basis of historical experience. However, in patients who have undergone polypectomy, the number of colorectal cancers diagnosed in subsequent years has been reduced by more than 80% (*red line*). (A more conservative estimate of reduced risks gives the extent of reduction at 60%.) (C) As diagrammed here, three apparently significant premalignant carcinomas *in situ* were detected by *bronchoscopy* in the lungs of a patient upon initial clinical presentation. By 19 months, the lesion in the left lower lobe (LLL; the left side of the lung is depicted here as might be seen from the front of the patient and is located from this perspective in the right side of the lung) had developed into a frank squamous cell carcinoma (*circle*), whereas the two indolent lesions in the left upper lobe (LUL) were no longer apparent. Interestingly, the three initially characterized lesions bore the same relatively rare *p53* mutation, indicating a common origin in spite of their physical separation; this mutation was also present in the subsequently arising carcinoma. (A, courtesy of Paul Edwards. B, courtesy of W.K. Hong, data from S. Winawer et al., *N. Engl. J. Med.* 329:1977–1981, 1993. C, from N.A. Foster et al., *Genes Chromosomes Cancer* 44:65–75, 2005. With permission from John Wiley & Sons.)

Sidebar 11.2 Parallel multi-step pathogenesis in multiple tissues The development of carcinomas in other organ sites throughout the body is thought to resemble, at least in outline, the multi-step progression observed in the colon (see, for example, Figure 11.7A). Many of these other tissues, such as the breast, stomach, lungs, prostate, and pancreas, also exhibit growths that can be called hyperplastic, dysplastic, and adenomatous, and these growths would seem to be the benign precursors of the carcinomas that arise in these organs. However, the histopathological evidence supporting multi-step tumor progression in these tissues is, for the most part, less well developed than for the colon—a consequence of the accessibility of the colon through colonoscopy and the relative inaccessibility of these other tissues. (An exception to this rule is tumor formation in the pancreas, discussed below.) In the case of nonepithelial tissues, including components of the nervous system, the connective tissues, and the hematopoietic system, the histopathological evidence supporting multi-step tumor progression is even more fragmentary.

early-stage tumors are unlikely to ever evolve into later, more aggressive stages. This has profound implications for clinical medicine and diagnostic screening, because many patients bearing early-stage tumors may be treated unnecessarily.

11.3 Cells accumulate genetic and epigenetic alterations as tumor progression proceeds

The forces driving tumor progression in various tissues are hardly obvious from the histopathological descriptions cited above. The progressive changes in cell and tissue phenotypes observed as a tumor develops may arise entirely from changes in various gene expression programs. For example, some or possibly all of the steps in tumor progression might recapitulate specific changes in cell behavior that occur during normal embryogenesis. Because such cell-biological changes are known to depend on changes in gene expression programs rather than alterations in genomic sequences, they are considered to be **epigenetic** in origin. However, the logic of cancer pathogenesis dictates that somatic mutations, that is, genetic changes, must be involved in important ways as major rate-limiting determinants of the formation of most human tumors (**Sidebar 11.3**).

This parallel between the histopathological progression of neoplastic tissues and genetic evolution was initially documented for human colorectal carcinomas (CRCs). Thus, in 1989 researchers at Johns Hopkins Medical School in Baltimore, Maryland recognized the existing evidence that mutant alleles of genes such as *ras* and *p53* can contribute to cell transformation under experimental conditions *in vitro*. They therefore sought to determine whether they could find *in vivo* correlates of this genetically driven transformation by examining the genomes of sizable groups of small colonic adenomas, mid-sized adenomas, large adenomas, and frank colorectal carcinomas. It was plausible that as colonic tissues advanced progressively from normalcy to high-grade malignancy, the epithelial cells in these various tissues would accumulate increasing numbers of somatic mutations in various cancer-causing genes.

Sidebar 11.3 Our large size and long lives has forced the erection in our cells of barriers to cancer formation Given the huge number of cell growth-and-division cycles in an average human lifetime ($\sim 10^{16}$) and the opportunity during each of these cycles for disaster, our own species and indeed all large, long-lived animal species have needed to evolve robust defense mechanisms in order compensate for the countless genetic and epigenetic accidents that inevitably occur in tissues during a normal life span. These randomly occurring changes inevitably lead to progressive degradation of the genomes and **epigenomes** of the cells in these tissues, the latter referring to the entire set of the epigenetic mechanisms that govern the lives of individual cells. Among the consequences of increasing disorder is, inevitably, the runaway proliferation of neoplastic cells.

As argued earlier, we might propose that each of the discrete steps of multi-step tumor pathogenesis represents the successful breaching of one or another functional barrier that has been installed in our cells and designed specifically to impede the forward march of cancer progression. In fact, ever-increasing numbers of such barriers seem to have been needed as our metazoan ancestors evolved from small, short-lived organisms to contemporary long-lived, large animals like ourselves.

This logic dictates that the successful breaching of each functional barrier must have been rendered as improbable as possible, that is, such barriers, once erected, are as resistant to change as possible. In fact, the barriers that are most capable of resisting changes inflicted by random accidents are present in the nucleotide sequences of the genomes of our cells. Thus, the somatic mutations that corrupt cellular genomes occur with extraordinarily low probability per cell generation. This very low rate of accumulated mutations ($\sim 10^{-9}$ per base pair per cell cycle) testifies to the high fidelity of DNA replication and to the efficiency of the DNA repair apparatus that subsequently erases any mistakes that are made during initial replication (both of which we will encounter in Chapter 12).

Stated differently, the strategy of entrusting anti-neoplastic defense mechanisms to highly stable DNA sequences would seem to reduce the likelihood of frequent, spontaneous breakdowns of these defenses and resulting neoplastic cell transformation. This explains why mutations seem to be required to liberate incipient cancer cells from the constraints imposed by these genome-based defense mechanisms. (In contrast, the breaching of anti-neoplastic barriers established by various *epigenetic* defense mechanisms would seem to occur far more often per cell cycle, given the phenotypic plasticity that many cells exhibit, as discussed below.) Taken together, this logic helps to explain why successful tumorigenesis in various adult human tissues generally requires, at a minimum, a significant number of somatic mutations and resulting mutant alleles (perhaps four to six), and why epigenetic changes, on their own, rarely succeed in creating robustly growing adult tumors.

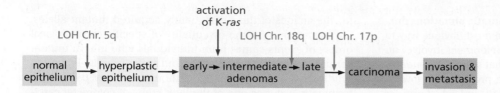

This is just what these scientists discovered (**Figure 11.9**). The genes that they examined included the K-*ras* oncogene and a number of tumor suppressor genes (TSGs). In fact, the identities of the TSGs that participate in colon cancer pathogenesis were not known when they began their work, so they searched instead for chromosomal regions that suffered loss of heterozygosity (LOH) during tumor progression. Recall that a high rate of LOH in a particular chromosomal region provides strong indication that this region harbors a TSG, and that developing cancer cells exploit the LOH mechanism as a means of shedding the still-functional (that is, actively expressed) wild-type alleles of such a TSG (see Section 7.4).

This research demonstrated that early-stage adenomas often showed loss of heterozygosity in the long arm of Chromosome 5 (that is, 5q). Almost half of slightly larger adenomas carried, in addition, a mutant K-*ras* oncogene. Even larger adenomas also tended to have high rates of LOH on the long arm of Chromosome 18 (that is, 18q); about half of all carcinomas showed, in addition, an LOH on the short arm of Chromosome 17 (that is, 17p); (see also Figure 7.11B and Table 7.1).

These observations provided strong support for the idea that as epithelial cells acquire increasingly neoplastic phenotypes during the course of tumor progression, their genomes show a corresponding increase in the number of altered genetic loci. Equally important, these changes involve both the activation of a proto-oncogene into an oncogene (for example, K-*ras*) and the apparent inactivation of at least three distinct TSGs. This was the first suggestion of a phenomenon that is now recognized to be quite common: the number of inactivated TSGs present in a human cancer cell genome often greatly exceeds the number of activated oncogenes within this genome.

The identities of the genetic loci on Chromosomes 5 and 17 were revealed soon after this genetic progression was laid out (**Figure 11.10**). The Chromosome 5q21 gene that is often the target of LOH was found to be the *APC* (adenomatous polyposis coli) tumor suppressor gene (see Section 7.11), and the Chromosome 17p13 gene was identified as the *p53* tumor suppressor gene (see Chapter 9). The identity of the gene or genes on Chromosome 18q that are inactivated during colon carcinoma pathogenesis remains unclear to this day. Of additional importance, tumors that carry a mutant *ras* allele rarely acquire in addition a mutant *PI3K* allele. This suggests that an evolving premalignant cell will acquire an advantageous set of phenotypes from one of these mutations (for example, *ras*) and that descendants of this cell acquire no additional benefit from mutating a second gene conferring a redundant cell-biological function. This sequence of changes, as laid out in Figure 11.10, raises the question of whether these mutations are acquired in a specific or at least preferred sequence during the course of tumor pathogenesis (**Sidebar 11.4**). A more detailed accounting of these genetic changes and those occurring during pancreatic carcinoma development is given in Supplementary Sidebar 11.3.

As discussed in the previous chapters and in this one, incipient cancer cells must undergo a set of fundamental cell-biological changes in order to progress to the state of cells found in high-grade malignancies. These changes in cell-biological behavior,

Figure 11.9 Colon tumor progression and loss of heterozygosity in various chromosomal arms DNA from tissue samples representing various stages of colon cancer progression was analyzed for loss of heterozygosity (LOH) by examining the behavior of chromosomal markers present on the long and short arms of most chromosomes, as described in Figure 7.11B. In general, each chromosomal arm was represented by one or more genetic loci that existed in heterozygous form in the normal tissue of a patient. As tumor progression proceeded, the involved colonic epithelial cells exhibited increasing numbers of chromosomal arms that had lost heterozygosity. Certain chromosomal regions (*blue lettering*) suffered especially high rates of LOH, suggesting that they carried tumor suppressor genes that were being inactivated, in part, through the LOH mechanism. Analyses of K-*ras* oncogene activation (*pink lettering*) indicated additional changes in the genomes of evolving premalignant cells. (Adapted from B. Vogelstein et al., *Science* 244:207–211, 1989.)

Figure 11.10 Tumor suppressor genes and colon carcinoma progression Each of the inactivated TSG(s) on Chromosome 18q remains unclear. About half of colon carcinomas also had mutant, activated alleles of the K-*ras* gene (*pink lettering*), and the genomes of most evolving, pre-neoplastic growths were found to suffer widespread hypomethylation (loss of methylated CpGs; *green lettering*). The precise contribution of DNA hypomethylation to tumor progression remains unclear; some evidence suggests that it creates chromosomal instability.

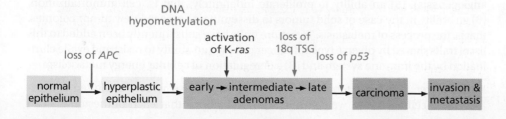

Sidebar 11.4 Is there a fixed order of somatic mutations during multi-step tumor progression? The most obvious way to rationalize the steps in colon cancer development involves an ordered succession of genetic changes that strike the genomes of colonic epithelial cells as they evolve progressively toward malignancy, as illustrated in Figure 11.10. Indeed, the order of acquisition of mutant alleles affects in a major way the efficiency and speed with which tumors are formed. It turns out the most efficient route to primary carcinoma formation involves mutations of the *APC* gene, then in K-*ras*, then *Smad4*, and finally *p53*.

To be more precise, mutation of each of the genes indicates the deregulation of a specific signaling pathway. Hence, the order of disrupting these *pathways* is the ultimate determinant of the rapidity of tumor progression, and the nature of each pathway (rather than the specific identity of a specifically mutated gene) is the critical determinant of ultimate biological effects (**Figure 11.11**). Thus, mutations in the gene encoding β-catenin can substitute for mutant *Apc* alleles, whereas mutations of the *B-Raf* or the gene specifying PI3 kinase can replace a mutant *Ras* allele. Moreover, promoter methylation and shut down of the gene specifying the Type I TGF-β receptor can replace *Smad 4* mutant alleles, and so forth. This theme of preferred order is echoed in other tumor types as well.

As an example, in mice bearing germ-line mutant alleles of the *p53* gene, mutations of *PTEN* are the favored initial somatic mutations during the pathogenesis of thymic lymphomas, followed later by amplification of the cyclin D1-encoding gene and then mutations in the *Ikaros* gene, which encodes a **zinc finger** transcription factor. The well-documented order of these acquired somatic mutations suggests that, in the case of each developing tumor, an earlier mutation sets the stage for the actions of the subsequently acquired mutant alleles. Further support for the essentiality of specific chronological orders of events comes from individuals who inherit mutant germ-line alleles of the *p53* gene and suffer from Li–Fraumeni syndrome (see Section 9.3). Although mutant *p53* alleles frequently occur as one of the later steps of colorectal carcinoma pathogenesis, these individuals rarely develop colorectal tumors and instead develop them elsewhere throughout the body, revealing the unique role of germ-line mutant *Apc* alleles in initiating multi-step colon cancer pathogenesis.

Figure 11.11 Alternative paths during cancer progression
The genetic alterations shown in Figure 11.10 do not accurately represent an invariant program of genetic alterations acquired by pre-neoplastic cells during the progression of all colorectal carcinomas. In actual human tumors, both the *combination* of mutated genes and the *temporal order* in which the responsible mutations are acquired vary greatly. (Only the combinations of mutant genes are depicted in this figure.) Loss of APC function (or functionally equivalent alterations, *left column*) represents a starting point that is common to almost all human colon carcinomas, ostensibly because the resulting mutant cells are trapped in the colonic crypts rather than migrating up and out; all of these changes seem to potentiate β-catenin function and to generate the clusters of aberrant crypts (see Figures 7.20 and 11.24B) encountered in colonoscopy. However, the identities of the genes altered in subsequent steps are variable, as is the precise order of these changes. (For the sake of clarity, the gene names given here reflect the names of their respective protein products.) Alterations in the Ras–Raf–PI3K pathway is shown as the second step, as these changes are often found in adenomas prior to the formation of frank carcinomas. These changes have been grouped functionally (so that the *second column* shows the Ras pathway; the *third column*, the TGF-β pathway; and the *fourth column*, the p53 pathway). Mutations in the *p53* gene are usually seen in carcinomas but not adenomas, indicating that its loss represents a relatively late step. Importantly, the precise number of genetic steps and epigenetic steps (notably promoter methylation) occurring during human colon cancer pathogenesis is not known. Moreover, there is only indirect evidence that all four of these pathways must be altered in order for a human colorectal carcinoma to form. The percentages indicate the proportion of colorectal carcinomas that exhibit a change in the gene in question, including activating mutations in proto-oncogenes and inactivating changes in tumor suppressor genes. (Courtesy of Y. Niitsu.)

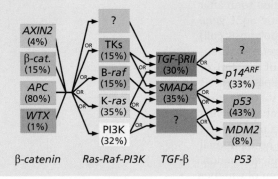

that is, cancer cell phenotypes, have been enumerated and proposed to be shared in common by a wide range of malignant cells and are termed "**Hallmarks of Cancer**." As initially proposed, they include, minimally, the acquired traits of (1) autonomously driven proliferation, (2) resistance to the anti-proliferative effects of certain signals received from the tissue microenvironment, (3) an elevated resistance to several forms of cell death, (4) an ability to induce the in-growth of new blood vessels (neo-angiogenesis), (5) an ability to proliferate indefinitely, that is, cell immortalization, (6) an ability in the case of solid tumors to disseminate and seed new tumor colonies, that is, the process of metastasis. Two more traits have subsequently been added to this list of traits shared by diverse types of cancer cells: (7) an ability to resist and evade elimination by the immune system and (8) a deregulation of cellular energy metabolism.

Having agreed, at least tentatively, upon a set of these widely exhibited cancer-associated cellular traits, we confront the observations that each of these traits can

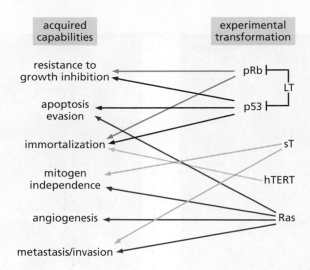

acquired capabilities

experimental transformation

resistance to growth inhibition

apoptosis evasion

immortalization

mitogen independence

angiogenesis

metastasis/invasion

pRb

p53

LT

sT

hTERT

Ras

Figure 11.12 Cancer cell genotypes vs. phenotypes During the course of tumor progression, human cells acquire a number of distinct cancer-associated phenotypes, such as the Hallmarks of Cancer cited here. Independent of this, genetic analyses of human tumors and laboratory experiments reveal that a number of distinct genes must be acquired by or introduced into human cells in order to transform them to a tumorigenic state. How then, does each of these introduced genes contribute to the cell phenotypes associated with tumorigenicity? As indicated here, genes such as that encoding hTERT, the catalytic subunit of telomerase, affect only the phenotype of immortalization, whereas other genes, such as the one encoding SV40 large T (LT), used experimentally to neutralize both p53 and pRb, disrupt two distinct cellular signaling pathways and affect multiple phenotypes. For example, by inactivating p53, LT elicits at least three changes in cell phenotype—resistance to growth inhibition, evasion of apoptosis, and immortalization; by neutralizing pRb, LT also affects two of these phenotypes. The most widely acting protein is likely to be Ras, which affects susceptibility to apoptosis, dependence on exogenous mitogens, angiogenesis, and invasiveness and metastatic ability. Hence, a one-to-one mapping between genes and cancer-associated phenotypes is not possible. (Because hTERT has been found to be physically associated with the β-catenin–TCF/LEF transcription factor complex, its actions may extend beyond cell immortalization.)

be acquired through multiple alternative genetic or biochemical solutions. As an example, signals that drive proliferation can be acquired by structural alteration of growth factor receptors, their overexpression, the acquired ability to produce and con-stitutively secrete the cognate ligand of a receptor (leading to autocrine signaling), or the structural alteration of signal-transducing proteins lying downstream of the recep-tors as components of the complex mitogenic signaling pathway. In addition, there is no one-to-one correspondence between the actions of most somatically mutated genes and acquisition of critical cell-biological phenotypes such as the Hallmarks of Cancer cited above (**Figure 11.12**).

Finally, we should note that these various genetic and epigenetic changes do not represent an upper limit to the number of alterations that contribute in essential ways to colon cancer progression. For example, the type of genetic analysis used to iden-tify this particular series of four chromosomal regions registered only those loss-of-heterozygosity (LOH) events in tumor suppressor genes that occurred at a significant frequency above the general background rate of LOH associated with all chromosomal arms in advanced tumor cells (see Figure 7.11B). Accordingly, LOH events that occur with a relatively low frequency (that is, those present in 20% or fewer of the tumors analyzed) would not be registered in such an analysis, even though these events might lead to the elimination from the genome of an incipient cancer cell of functionally important tumor suppressor genes. Similarly, changes in gene copy number, such as amplification of mutant oncogenes, were also not registered in this early work.

Epigenetic events, including the repression of certain genes through promoter methylation (see Section 7.7) and the de-repression of others through demethylation, may also contribute importantly to tumor progression. Recently, evidence has accu-mulated that hypomethylation (that is, demethylation of normally CpG methylated DNA sequences), such as that observed in early adenomas (see Figure 11.10), has a functional consequence that is independent of its possible effects on gene transcrip-tion: widespread chromosomal instability, which presumably favors tumor progres-sion. How precisely it does so is unclear; a de-repression of usually-silent transposable elements in the cell genome may contribute to this genomic destabilization.

The publication of this "genetic biography" of colon cancer tumorigenesis, as depicted in Figure 11.10 (see also Supplementary Sidebar 11.3), should ideally have been fol-lowed by similar descriptions of a wide variety of other tumor types, with each of such biographies involving alterations of its own particular set of oncogenes and tumor sup-pressor genes as accompaniments of specific histopathological states. Unfortunately, only a handful of such descriptions have actually been reported in detail. At present, therefore, we cannot cite lists of genetic alterations in tumor cell genomes to illustrate the molecular underpinnings of the multi-step nature of cancer progression in most organs. However, with the availability of large-scale, whole-genome sequencing tech-nologies, these biographies will surely be described in the near-term future.

Figure 11.13 Retention of differentiated characteristic in human tumors (A) The tumors arising in a transgenic mouse model of colorectal carcinoma pathogenesis (*left*) would seem to have little if any resemblance to the histology of the normal colon. At higher magnification (*right*), the cells are seen to form ductal structures with clear evidence of their epithelial origin (evidenced by E-cadherin expression, significant levels of cell proliferation, and stemness). (B) The ducts in the normal mammary gland (*middle*) are composed of two major types of epithelial cells: luminal cells that form the lining of ducts (*red-orange*) and underlying basal cells (*green*) that provide various types of physiological support to the luminal cells. These cell types can be distinguished by immunofluorescence staining using antibodies that are specific for the cytokeratin molecules that these cells express: cytokeratins 8 and 18 are expressed characteristically in luminal cells, whereas cytokeratin 5 is expressed in basal cells. These markers can be used to classify breast carcinomas into two major histological classes—luminal (*left*) and basal (*right*)—and the resulting classifications hold strong implication for the prognosis of breast cancer patients. However, some evidence indicates that these classifications may not properly reveal the identities of the respective normal cells of origin, and suggests, for example, that certain basal breast carcinomas may actually originate from luminal progenitors. (A, L. Caplan and J. Roper, courtesy of The Broad Institute, Inc. B, from J.I. Herschkowitz et al., *Genome Biol.* 8:R76, 2007. With permission from Springer Nature.)

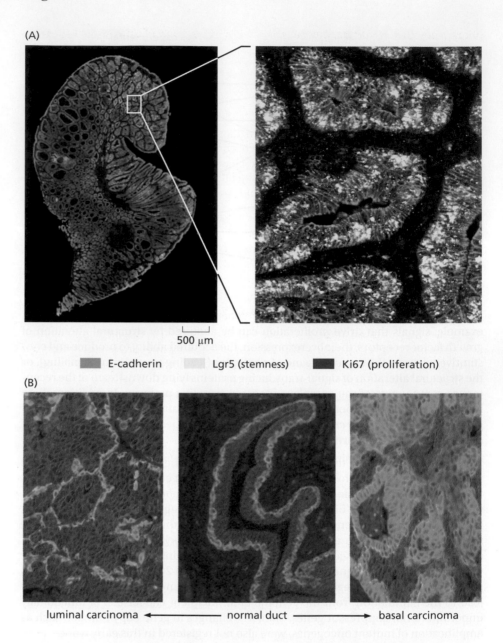

(A)

500 μm

■ E-cadherin   ■ Lgr5 (stemness)   ■ Ki67 (proliferation)

(B)

luminal carcinoma ◄—— normal duct ——► basal carcinoma

The accumulations of multiple genetic and epigenetic changes during tumor formation might suggest that the final cellular end-products of tumorigenesis bear little resemblance to their most recent, fully normal ancestors—the normal **cells-of-origin**. In fact, it seems increasingly likely that many of the differentiation programs of normal cells-of-origin persist in their neoplastic descendants years if not decades later (**Figure 11.13**). One indication of this at the molecular level is the extensive retention of sequence-specific CpG methylation sites present in the genomes of the normal cells-of-origin within the genomes of derived neoplastic cells; these methylated CpG play critical roles in the cell-heritable transmission of differentiation programs. Accordingly, even though certain oncogenic changes in tumor cell genomes (for example, loss of pRB function and gain of Myc function; see Section 8.9) tend to cause loss of differentiated characteristics, this loss is incomplete; that is, these genetic changes fail to eradicate all traces of the differentiation program that pre-exists in the normal tissue. Indeed, the retention of differentiated traits—distinctive morphologies and protein markers—in most (~95%) tumors allows pathologists to classify these growths into different subgroups with reasonable accuracy.

These findings, on their own, raise another issue that complicates our understanding of cancer biology and remains poorly resolved: Precisely how do the differentiation programs of normal cells-of-origin interact with and influence the phenotypes of derived cancer cells bearing various sets of somatically mutated cancer-forming genes?

11.4 Cancer development seems to follow the rules of Darwinian evolution

The observations about colorectal cancer made at Johns Hopkins University demonstrated that the histopathological changes occurring during tumor progression (see Figure 11.10) were *correlated* with genetic changes in cells of the colonic mucosa. More important, however, it became plausible that these genetic changes were actually *causing* the phenotypic progression of these cells and the tissues they form.

Years earlier, in the mid-1970s, researchers had speculated that tumor development could be understood in terms of a biological process that resembles Darwinian evolution. The results of the genetic analyses of human colorectal cancer progression could be mustered to provide further support for this model. (Although Darwin himself knew virtually nothing about genes and genetics, the "modern synthesis" of Darwinian theory introduces Mendelian and population genetics into the evolutionary processes that Darwin first postulated.)

In the case of cancer development, the evolving units are individual cells competing with one another in a population of cells, rather than individual organisms competing with one another within a species. Like the modern depiction of Darwinian evolution, random mutations are presumed to create genetic and thus phenotypic variability in a cell population. Once a genetically heterogeneous cell population has been generated through stochastic genetic and epigenetic events, the forces of selection may then favor the outgrowth of individual cells (and their lineal descendants) that happen to be endowed with mutant alleles and cell-heritable epigenetic alterations conferring advantageous traits, notably traits that favor proliferation and survival in the microenvironment of a living tissue. In the language of evolutionary biologists, such favored cells are said to exhibit increased **fitness**.

Combining Darwinian theory with the assumptions of multi-step tumor progression, researchers could now depict tumorigenesis as a succession of clonal expansions, (where the term "clonal" indicates a cell population all of whose members descend from a single, common ancestor) (**Figure 11.14**). The scheme goes like this: a random mutation creates a cell having particularly advantageous proliferative or survival traits. This cell and its descendants then proliferate more effectively than their neighbors, eventually yielding a large clonal population of descendants that dominates the tissue and crowds out phenotypically less favored neighbors. Sooner or later, this particular cell clone will reach a large enough size (for example, 10^6 cells) that yet another

Figure 11.14 Darwinian evolution and clonal successions Darwinian evolution involves the increase in number of organisms that are endowed with advantageous genotypes and thus phenotypes; a formally similar scheme seems to describe how tumor progression occurs. One cell amid a large cell population sustains an initiating mutation (*red sector, top*) that confers on it a proliferative and/or survival advantage over the other cells. Eventually, the clonal descendants of this mutant cell dominate in a localized area by displacing the cells that lack this mutation, resulting in the first clonal expansion. When this clone expands to a large enough size (e.g., 10^6 cells), a second mutation—one that strikes with a frequency of $\sim 10^{-6}$ per cell generation—may occur (*green sector*), resulting in a doubly mutated cell that has even greater proliferative and/or survival advantage. The process of clonal expansion then repeats itself, and the newly mutated population displaces ("succeeds") the previously formed one, yielding a process that is termed *clonal succession*. This results once again in a large descendant population, in which a third mutation (*blue sector*) occurs, and so forth. Although classical Darwinian evolution is thought to depend on mutations in the genomes of organisms, it is highly likely that other heritable changes in cell populations, notably promoter methylation events (see Section 7.7), can play an equally prominent role in multi-step tumor progression. Importantly, this scheme does not take into account that clonal successions may require greatly different time intervals to reach completion. For example, later successions are likely to proceed far more rapidly than earlier ones because the participating cells, having acquired oncogenic mutations, may proliferate more rapidly and have more mutable genomes.

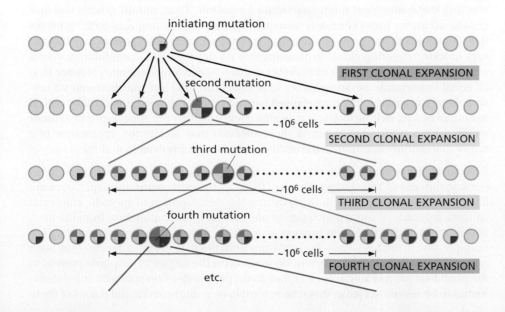

advantageous mutation, which strikes randomly with a probability of about one per 10^6 cell generations, becomes mathematically possible, striking one or another cell within this now-expanded clonal population.

Importantly, this new, secondary mutation alters a gene and encoded protein so that it is functionally *complementary* to the actions of the earlier mutation in the ways that it affects tumor cell proliferation and survival. For example, in the case of colorectal cancer (CRC), if the first mutated gene is *APC,* the subsequently mutated allele, such as K-*ras,* will *collaborate* with one or more mutant *APC* alleles to favor cell outgrowth and survival rather than *duplicating* the actions of the existing mutant *APC* allele (see Figure 11.10). (In the discussion here, advantageous cell-heritable epigenetic alterations are, for the sake of this argument, included under the rubric of genetic mutations.)

Implicit in this scheme is a simple conclusion: in order to operate in a way that is functionally complementary to the actions of previously acquired mutant alleles, a newly acquired mutant allele must provide additional proliferative and/or survival advantage. Conversely, a newly generated allele that is functionally redundant with existing mutant alleles is unlikely to confer such additional advantage, and its retention in the descendants of this cells will accordingly not be favored by the forces of selection.

To be sure, a strict Darwinian model of cancer progression, which involves a series of **clonal successions**, is simplistic. For example, it must be amended to respond to the discoveries that cell-heritable epigenetic alterations of genes, such as promoter methylation and demethylation events (see Section 7.7 and Figure 11.10), play an important role in eliminating the activities of some genes (notably tumor suppressor genes as well as genes involved in the maintenance of the cell genome) and activating others. Here, we encounter a major discordance between tumor progression and Darwinian evolution because heritable epigenetic alterations, such as DNA methylation events, have not been shown to drive the evolution of species.

This scheme is simplistic in other respects as well. Thus, the number of distinct steps in tumor progression, each represented by a clonal expansion, may be underrepresented by counting the number of genetic loci that are altered during this process. As discussed in Chapter 7, the inactivation of a TSG is, almost always, a two-step process. First, one gene copy is mutated (or methylated) to inactivity. Thereafter, the surviving, still-intact gene copy is eliminated, usually through loss of heterozygosity (LOH). Knowing this and the fact that elimination of TSGs often represents the majority of the genomic changes occurring during tumor progression, we conclude that the number of distinct genetic alterations taking place during tumor progression may be almost twice as many as the number of loci that are involved.

In addition, the number of successive clonal expansions is difficult to quantify by studying the *dynamics* of multi-step tumor formation. Thus, certain genetic and epigenetic accidents may occur only infrequently, thereby creating rate-limiting events that slow down the progress of the entire process, whereas other events may happen very quickly, triggering clonal expansions that do not register as rate-limiting events when calculating the multiplicity of events as discussed in Supplementary Sidebar 11.1. Of equal importance, we know rather little about the time intervals between successive clonal expansions. The time elapsed between these expansions is actually critical to human health, as long intervals may effectively preclude the appearance of a tumor within a human life span, whereas short intervals may dictate the appearance of a tumor that requires attention in the oncology clinic, possibly even in mid-life.

Because an infrequently occurring genetic or epigenetic alteration (or a pair of alterations, in the case of TSGs) triggers each of these clonal expansions, the expansions are likely to be spaced far apart in time. During the development of sporadic colorectal cancers, a decade or more may separate one critical genetic alteration from the next, and in many individuals, the process as a whole may stretch over a century. (For example, one of the key steps—the evolution from polyp to invasive cancer—has been estimated to take as long as 17 years, depending on the degree of dysplasia present in the precursor polyp.) Still, it is clear that some people develop sporadic colon carcinomas in far less time, and so this schedule must be compressed in the colons of these individuals. Moreover, there is no reason to think that each clonal expansion requires an interval of time that is comparable to the time required for other such expansions. As an example, for various molecular reasons, the formation of a mutant K-*ras* allele via point mutation may occur more slowly than the elimination of *APC* function through losses of both copies of this gene.

We know almost nothing about the physiological mechanisms that govern the lengths of the time intervals between successive clonal expansions. In the case of the disease of colon cancer, we know that the incidence of this disease varies by as much as twentyfold internationally and that environmental factors, specifically foodstuffs (rather than genetic susceptibility) are responsible for these dramatic differences. (As Figure 2.18 shows, a population migrating from one country to another exhibits a colon and prostate incidence rates typical of its new host country within a generation or two, ruling out genetics as the key determinant of cancer risk in the great majority of these cancer cases.) It is highly likely that certain constituents of diet greatly increase the rate at which clonal successions occur, that is, compress the time intervals between these successions. This might occur through two alternative mechanisms. Mutation rates may increase progressively during the course of multi-step tumorigenesis due to acquired defects in the genes tasked with maintaining genomic integrity, as described in the next chapter. In addition, non-genetic mechanisms, specifically the **tumor promoters** to be described in Section 11.11 may accelerate the rate of clonal expansions, acting exclusively through nongenetic mechanisms. Perhaps in individuals who consume certain foodstuffs, 5 or fewer years intervene between successive clonal expansions rather than the usual 10 or 20. As a consequence, a disease process that usually requires a century to reach completion may reach its neoplastic endpoint in 30 or 40 years — well within a human life span. Finally, individuals who bear mutations in DNA repair genes that are inherited through the germ line may experience increases in the rapidity of clonal expansions. In the context once again of the gut, such people may suffer from greatly accelerated rate of formation of premalignant and malignant growths in the colon—the syndrome of hereditary non-polyposis colon cancer (HNPCC).

In fact, there is another dimension of complexity that we have not yet confronted. Specifically, the tissue microenvironment (TME) in which normal and pre-neoplastic cells find themselves. There is clear evidence that this TME, which is represented by the stroma of normal and neoplastic epithelial cells, changes progressively with age, and that aged tissue microenvironments can be more hospitable to the inception and outgrowth of pre-neoplastic and frankly neoplastic cell clones. Hence, part of the increasing incidence of cancer with age may not derive exclusively from cell-autonomous processes (that is, those operating entirely within the precursors of carcinoma cells) that govern the dynamics of clonal expansion. Instead, tumors may occur more frequently in older people in part because the microenvironments within their aging tissues increasingly favor the eruption of tumors.

11.5 Multi-step tumor progression helps to explain familial polyposis and field cancerization

As described in the previous section, the first step in the development of the great majority of *sporadic* colon carcinomas (that is, those not driven by an inherited mutant gene) involves the inactivation of the *APC* gene—precisely the same gene that is inherited in mutant form by individuals suffering from the *familial* polyposis syndrome (Section 7.10). Now, in the context of multi-step tumorigenesis, we can understand why inheritance of a mutant *APC* allele results in polyposis and colon cancer: the first step in colon cancer progression, which involves the inactivation of an *APC* gene copy, has already occurred in all of the colonic epithelial cells of an individual suffering from familial polyposis. That is, each of their cells, including those in the colon, is $APC^{+/-}$ (rather than $APC^{+/+}$).

Consequently, a common mechanism of inborn susceptibility yielding familial cancers involves an acceleration of a critical rate-limiting step; in the case of the gut, this step involves elimination of the two copies of the *APC* gene in one or another cell.

Thus, one of the two steps in this process no longer depends on infrequently occurring somatic mutations (because it has already occurred and affected the inherited germ-line genome and therefore all of the cells throughout the body of an afflicted individual).

This model of multiple clonal expansions also bears on another important issue in tumorigenesis: tissues affected by sporadic tumors occasionally sprout multiple tumors that arise at similar times in different sites in the tissue and seem to be formed independently of one another. This phenomenon is often called **field cancerization**. Thus, as an example, two or more premalignant or frankly malignant growths may erupt suddenly in an epithelial tissue, separated from one another by many centimeters of apparently normal epithelium. Because the appearance of any single sporadic (that is, nonfamilial) neoplasm is, on its own, a highly improbable event, the simultaneous appearance of two or more seemingly independent sporadic growths would seem to be extremely unlikely. Our understanding of multi-step tumorigenesis provides a key insight into the underlying processes at work here (**Figure 11.15**).

Imagine that a clone of cells undergoes some of the initial genetic alterations that launch these cells partway down the road toward full-fledged neoplasia; these alterations confer a proliferative or survival advantage on these cells but have no outward effect on the histopathological appearance of the tissue in which these cells arise. In the eyes of a pathologist, this tissue will appear entirely normal. Nonetheless, these alterations poise these cells to progress further toward neoplasia.

After many years, this clone of mutant cells, because its constituent cells possess slightly elevated proliferative or survival potential, may have expanded to form a large patch of outwardly normal epithelium. Thereafter, two cells, located far from one another within this patch (that is, "field") and sharing the same set of somatic mutations, may then suffer new oncogenic mutations independently of one another, and each of the resulting multiply-mutated cells may spawn a papilloma or carcinoma (see Figure 11.15A). Indeed, at the experimental level, genetic analyses of apparently normal bladder epithelium, oral mucosa, and skin, as well as of esophageal and lung epithelia, have revealed large patches of cells, some many centimeters wide, that are morphologically normal but already carry a mutant gene, often a mutant allele of *p53* (see Figure 11.15B); alternatively, widely dispersed patches of dysplastic tissue having a common genetic origin can be documented (Figure 11.15C and D).

Strikingly, similar processes operate in the hematopoietic system. Thus, as many individuals age, subpopulations of lymphoid cells in the circulation can be found that share common oncogenic mutations; such subpopulations occur in individuals in which there are no clinical signs of incipient cancers, that is, either lymphomas or leukemias. The common origin of the cells and their association with a specific clonal subpopulation of cells is indicated by the otherwise-rare, highly specific mutation(s) that the cells within one of these subpopulations share, indicating that each of these subpopulations represents a genetically homogeneous cell population descended from a common ancestor in which the mutation first arose. For this reason, the presence of such subpopulations is considered a manifestation of the process of **clonal hematopoiesis**. Of note, individuals harboring such clonal subpopulations in their blood or bone marrow exhibit an elevated risk of eventually developing a hematopoietic malignancy, once again indicating a field effect albeit in a liquid tumor. Hence, the development with increasing age of pre-neoplastic cell populations appears to occur in diverse tissues governed by very different types of regulatory mechanisms.

11.6 Intra-tumor diversification can outrun Darwinian selection

As multi-step tumor progression proceeds, the genomes and epigenomes of cancer cells become increasingly destabilized. Hence, the rates at which genetic and epigenetic changes accumulate increasingly *outpace* the rate at which Darwinian selection can eliminate less-fit variant subpopulations of cells. As a consequence, with each successive step, the populations of cells within a tumor become increasingly diverse

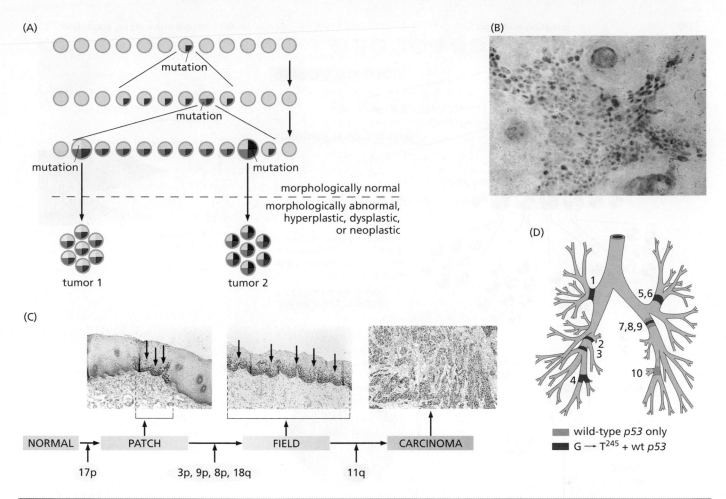

Figure 11.15 Field cancerization A large area of histologically normal but genetically altered epithelium can sprout multiple, apparently independent foci of neoplastic cells, all of which carry the one or more mutations common to this area as well as subsequently acquired mutants that are unique to each focus. (A) According to one model of field cancerization, one cell sustains an initiating mutation (*red sector, top middle*). Following extensive proliferation, some of its clonal descendants may subsequently acquire a second mutant allele (*blue sector*), and the doubly mutated "initiated" cells may then proliferate and eventually occupy a large patch of epithelium. Although these cells have already undergone several of the early steps of multi-step tumor progression, they continue to form an epithelium that is histologically normal. Subsequently, two cells located at different sites within this large patch of initiated cells may independently acquire additional mutations (*yellow sector left; black sector right*); both of these cells will now advance over the boundary (*dashed line*) to acquire a histologically abnormal appearance, spawning hyperplastic, dysplastic, even neoplastic growths. These two tumors will appear to have arisen independently even though they derive from the same clone of initiated cells. (B) Histologically normal patches of p53-mutant keratinocytes, detected here by immunostaining, can be found in sun-exposed human skin and may contain as many as 3000 cells. Such patches appear to represent fertile ground for the inception of basal cell carcinomas. (C) p53-positive (i.e., *p53*-mutant; see Section 9.9) cells, detected by immunostaining, are initially found in small, histologically normal patches of the oral epithelium early in the progression of squamous cell carcinomas in the oral cavity (*left image, brown spots*). As tumor progression proceeds, large fields of *p53*-mutant cells can be found that remain quite normal histologically, although they have often undergone loss of heterozygosity (LOH) of markers on the indicated chromosomal arms (*middle*). Eventually, after additional genetic changes, including LOH of the long arm of Chromosome 11, several of the cells in a field may progress independently to squamous cell carcinomas (*right*). (D) An extreme case of field cancerization can be seen here in lungs of a patient examined upon autopsy. Cells within widely scattered areas in both lobes of the lungs were found to share the same rare *p53* somatic mutation (present either in homozygous or heterozygous configuration, *red*), indicating their common genetic origin (see also Figure 11.8C). Although these areas did indeed exhibit mild squamous metaplasia, no carcinomas could be detected; hence, none of these widely dispersed fields had spawned a carcinoma by the time of the patient's death. (B, courtesy of D.E. Brash. C, from B.J.M. Braakhuis et al., *Cancer Res.* 63:1727–1730, 2003. With permission from American Association for Cancer Research. D, from W.A. Franklin et al., *J. Clin. Invest.* 100:2133–2137, 1997. With permission from American Society for Clinical Investigation.)

at the genetic or epigenetic levels. This increased intra-tumoral heterogeneity can be best documented by isolating multiple single-cell-derived cell clones from a tumor. Such clonal populations originating from the same tumor often differ markedly in their rates of proliferation, susceptibility to apoptosis, tendency to invade and metastasize, and organization of intracellular metabolism (**Figure 11.16 A and B**).

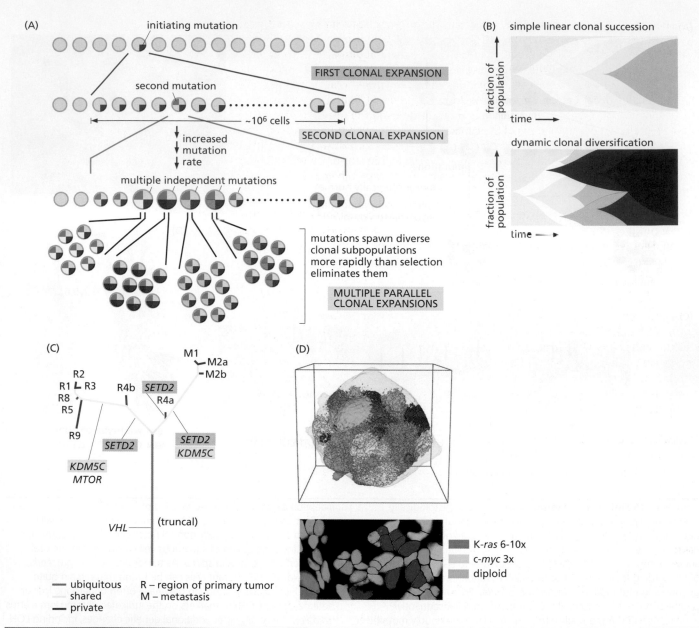

Figure 11.16 Clonal diversification caused by high mutation rates (A) As tumor progression proceeds, the genomes of tumor cells often become increasingly unstable. As this occurs, the rate at which new mutant alleles are generated may exceed the rate at which Darwinian selection can eliminate phenotypically less-fit clones. Consequently, the tumor mass develops an increasing number of sectors, each dominated by a genetically distinct subclone of cells. (B) Recent surveys of both genetic and epigenetic (CpG methylation) heterogeneity within individual tumors have provided experimental support for the extensive diversification of distinct clonal populations within a tumor. Thus, the simple linear model of clonal succession (*above*) should be replaced, at least in late-stage tumors, by the dynamic changes depicted *below* in which multiple distinct subpopulations of cancer cells coexist within a tumor. (C) Detailed sequencing analyses of a primary human renal cell carcinoma and derived metastases provide some indication of the genetic diversification that occurs during tumor progression. As seen here, mutation of the *VHL* gene occurred early and was ultimately transmitted to all resulting neoplastic cells in this tumor; the shape of this tree reveals why this is termed a "truncal" mutation. Regions of the tumor (designated as R) indicate a diversification of an early population into two major branches, one of which spawned metastases (*right*), whereas the other did not (*left*). Multiple mutations of the *SETD2* and *KDM5C* genes occurred independently in distinct

regions of the tumor, suggesting convergent evolution. Additional mutations were accumulated in derived metastases (*top right*). (D) Although the schemes of A and B imply the coexistence of distinct subclones within a tumor, they do not necessarily indicate the topological localization of these clones and their constituent cells. A computationally based model of how a tumor should develop predicts discrete sectors in which genetically (and epigenetically) distinct, individual subclones are localized in a tumor composed of 2 million cells. In contrast, use of fluorescence *in situ* hybridization (FISH) in a section of an actual tumor (*below*), such as the human breast cancer analyzed here, reveals the extensive physical intermingling of the individual cells deriving, in this case, from three major, genetically distinct subclones coexisting within this tumor. The three clones involve those with apparent diploid genomes, those with 3 copies of c-*myc*, and those with 6–10 copies of K-*ras* and a closely linked gene. (It is unclear which of these two topological patterns is more common in actual tumors.) (B, from A. Marusyk and K. Polyak, *Biochim. Biophys. Acta* 1805: 105–117, 2010. With permission from Elsevier. C, from M. Gerlinger, A.J. Rowan, S. Horswell et al. *NEJM* 366:888-892, 2012. With permission from Massachusetts Medical Society. D, from A. Sottoriva et al., *PLoS Comput. Biol.* 7:e1001132, 2011, courtesy of A. Sottoriva; and N. Navin et al., *Genome Res.* 20:68–80, 2010. With permission from Cold Spring Harbor Laboratory Press.)

This heterogeneity present in highly progressed tumors contrasts starkly with the cellular products of selection that operated earlier during tumor progression at a time when less-fit cells and subpopulations were successfully eliminated. Such efficient elimination is often termed **purifying selection**, in which only the most fit cells survive to yield homogeneous, clonal cell populations whose constituents are genetically and epigenetically identical to one another. As described later in Chapter 17, the intensive diversification of cells occurring within certain tumors and the resulting phenotypic heterogeneity of subpopulations of the cancer cells within such tumors greatly complicate anti-cancer therapies. Thus, some applied therapies may be highly effective in eliminating certain subpopulations of neoplastic cells within a tumor, while other co-existing subpopulations that happen to be resistant to therapy can emerge and create clinical relapses.

As a consequence, rather than looking like a linear series of clonal successions, actual tumor progression in many tumor masses is likely to resemble the highly branched scheme shown in Figure 11.16C, in which multiple genetically distinct subclones of cells coexist within a single tumor mass. Indeed, the dynamic nature of the expansion of some clonal subpopulations and the obliteration of others is suggested by Figure 11.16B.

Demonstrations of the genetic diversification of cells within a tumor cell population can be obtained by tracking the state of individual genes of interest within various cells of a primary tumor and derived metastases. For example, in a human pancreatic carcinoma, detailed genome sequence analysis of different sectors of the tumor revealed genetically distinct subclones, each of which was estimated to comprise at least 100 million cells. Interestingly, several of these subclones each spawned its own set of genetically related metastases found at distant sites in the body of the cancer patient—a topic that we will pursue in greater depth in Chapter 14. An illustration of these processes operating during the formation of a renal cell carcinoma is provided by Figure 11.16C, in which an initial, commonly shared mutation (often termed **truncal**) is inherited by derived clonal subpopulations and metastases.

Of note, in Figure 11.16C, the relative *lengths* of the branches were gauged by the number of neutral mutations accumulated throughout the genomes of the cells under analysis. Such functionally silent neutral mutations are often termed **passenger mutations** because they are carried along with **driver mutations**, the latter affecting genes whose alteration is assumed to be functionally important and capable of conferring selective advantage on the cells bearing them. Almost invariably, the passenger mutations are far more numerous than driver mutations and are reflective of widespread genetic instability within tumor cells. The genes affected by driver mutations, as discussed in the next chapter, are often identified because they are repeatedly found in mutant form in the genomes of multiple independently arising tumors, suggesting that mutant alleles of these genes confer a strong phenotypic advantage on those cells that happen to acquire them.

The precise localization of individual subclones and associated cells within a tumor is itself unclear. A computer-based modeling of how such subclones arise (distinguished from one another by heritable differences in DNA sequence or CpG methylation) predicts large sectors occupied by individual subclones, which may indeed represent the configuration of many tumors. However, analysis of individual carcinoma cells within a tumor may well reveal a more complex situation, in which the cells of various subclonal populations become intermingled (Figure 11.16D).

A vivid example of genetic instability and resulting genetic diversification is shown in **Figure 11.17**, in which the FISH (fluorescence *in situ* hybridization) technique was used to determine the copy number of Chromosomes 11 and 17 in individual cells deriving from a human lung carcinoma. As is apparent, the copy number of Chromosome 17 (which carries the gene specifying cyclin D1) varies enormously from one cell to another, whereas the copy number of Chromosome 11 is relatively stable. Such variability indicates that the tumor cell genomes being analyzed here are quite plastic, possibly changing each time the cells pass through a cycle of growth and division.

Figure 11.17 Frequent variations in chromosome number in tumors In high-grade tumors, the numbers of chromosomes often fluctuate wildly from one cell to another, indicative of great cell-to-cell genetic heterogeneity. In this image, the use of FISH revealed the copy numbers of Chromosomes 11 (*green*) and 17 (*pink*) in cells from a *pleural effusion* in a non-small-cell lung carcinoma patient. In addition to fluctuations in chromosome number, highly polyploid giant nuclei are apparent. (From M. Fiegl et al., *J. Clin. Oncol.* 22:474–483, 2004. With permission from American Society of Clinical Oncology.)

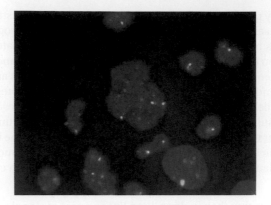

These fluctuations seem to occur almost randomly and at a rate that vastly exceeds the ability of (Darwinian) selection to eliminate less-fit variants.

Yet another source of diversification has not even been addressed here: epigenetic mechanisms may profoundly shift the transcriptomes and thus the phenotypes of cancer cells without any underlying changes in the cancer cell genomes (Supplementary Sidebar 11.4); such epigenetic plasticity would not be registered in the genetic analyses described in Figure 11.16. The resulting phenotypic diversification may be apparent at the histopathological level (**Figure 11.18A**). Alternatively, surveys of protein expression in tumor sections using immunofluorescence provide dramatic testimony to the diversification of small, discrete subpopulations of cancer cells within a tumor (Figure 11.18B). In principle, the variability in the expression of the EGF and VEGF receptors shown in this figure might well result from genetic diversification. Far more likely, however, are epigenetic mechanisms—localized, frequent fluctuations in cell-to-cell signaling in the tumor microenvironment yielding phenotypic diversity that does not directly reflect underlying, slowly occurring genetic diversification. Finally, reversible shifts in cell phenotype make it almost impossible to map with precision the developmental histories of individual tumors (**Sidebar 11.5**).

(A)

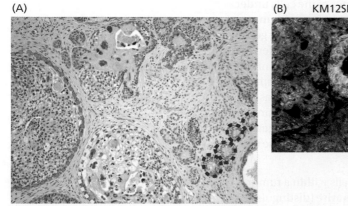

(B)

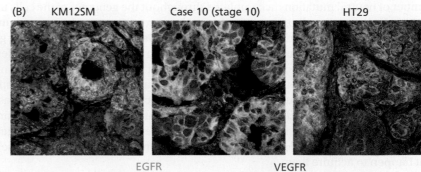

KM12SM   Case 10 (stage 10)   HT29

EGFR   VEGFR

Figure 11.18 Phenotypic diversification within a tumor cell population Independent of the genetic diversification of subpopulations of cells within a tumor (see Figure 11.16) is the ability of tumor cells to enter into new phenotypic states by activating epigenetic programs associated with alternative states of differentiation. (A) An adenoid basal carcinoma (ABC) of the cervix from a 94-year-old woman demonstrates heterogeneous cytologic features. To the far right, a row of small basaloid islands infiltrates the cervical stroma (*red circles*). Above them (*top right*) is an area of columnar cell differentiation within such a nest of basaloid cells (*green circle*). In the center and left are two large nests of neoplastic squamoid cells (*yellow, blue*) exhibiting various degrees of *atypia*. In both cases, the large nests are surrounded by rims of basal cell layers, findings characteristic of ABC. ABC is typically associated with a conventional high-grade squamous intraepithelial lesion (*not shown*) and HPV infection; this case was positive for HPV type 16. The high frequency of these variant morphologies in this particular subtype of carcinoma strongly suggests that these islands of cells represent distinct differentiation programs activated by the various carcinoma cells rather than genetic diversification. (B) Use of immunofluorescence with antibodies that react with the EGF (*green*) and VEGF (*red*) receptors reveals great phenotypic diversity even within small sectors of these tumor xenografts; cells co-expressing both receptors are seen as *yellow*. Tumors analyzed in the *left* and *right* panels derive from cancer cell lines, and the tumor in the *center* panel derives from a surgical specimen. (A, courtesy of M. Hirsch and M. Loda. B, from T. Kuwai et al., *Am. J. Pathol.* 172:358–366, 2008. With permission from American Society for Investigative Pathology.)

11.7 Tumor stem cells further complicate the Darwinian model of clonal succession and tumor progression

As depicted above, the clonal succession model proposes that a mutant cell spawns a large flock of descendants and that among these numerous descendants, a new mutational event (or heritable epigenetic alteration) will confer a novel, advantageous phenotype and thereby trigger yet another wave of clonal expansion. (For the sake of simplicity, we return for the moment to the model of Figure 11.14, which dictates that each clonal expansion generates a genetically homogeneous cell population.) In fact, many experiments have forced a revision of the notion that all of the cells within a genetically homogeneous cell clone are biologically equivalent to one another and therefore equally capable of becoming ancestors of new, more-progressed, successor populations of cells.

These insights came from attempts to resolve and fractionate tumor cell populations that were ostensibly genetically homogeneous into subpopulations that might, in spite of their common genetic identity, differ from one another phenotypically. In initial experiments, the cancer cells within a specific human tumor were separated into distinct subclasses. These separations took advantage of distinct sets of cell-surface proteins displayed by the different subpopulations. In particular, the technique of fluorescence-activated cell sorting (FACS) was used to fractionate living cancer cells after labeling them (via their cell-surface proteins) with monoclonal antibodies linked to fluorescent dyes (Supplementary Sidebar 11.5). Importantly, cells separated by this procedure can be recovered in viable form and used thereafter in biological experiments, including tests of their ability to seed tumors following injection into host mice.

Use of the FACS technique initially enabled researchers to segregate populations of acute myelogenous leukemia (AML) cells into majority and minority populations; in one such experiment, the latter represented less than 1% of the neoplastic cells in the overall leukemia cell population. As few as 5000 cells in the minority subpopulation were able to produce new tumors upon injection into host mice and were therefore deemed to be "**tumorigenic**"; in contrast, as many as 500,000 AML cells from the majority subpopulation were unable to seed a tumor. In fact, the leukemia cells in this majority subpopulation exhibited many of the attributes of differentiated granulocytes or monocytes, and these cells had limited ability to proliferate. These observations provided compelling evidence that the AML tumors were composed of small populations of self-renewing, tumorigenic cells and large populations of more differentiated cells that had little, if any, ability to serve as founders of new cell populations and thus new leukemias.

Subsequent experiments extended these results to human breast cancer cells prepared by culturing cells extracted directly from tumors. In these later experiments, the minority tumorigenic cell population within a tumor represented only about 2% of the overall neoplastic cell population. When implanted in a host mouse, 200 of these minority cells could seed a new tumor, but as many as 20,000 cells from the majority cell population failed to do so (**Figure 11.19**). Of note, the majority and minority breast cancer cell subpopulations both contained comparable proportions of cells in the active growth-and-division cycle, and both subpopulations were purified away

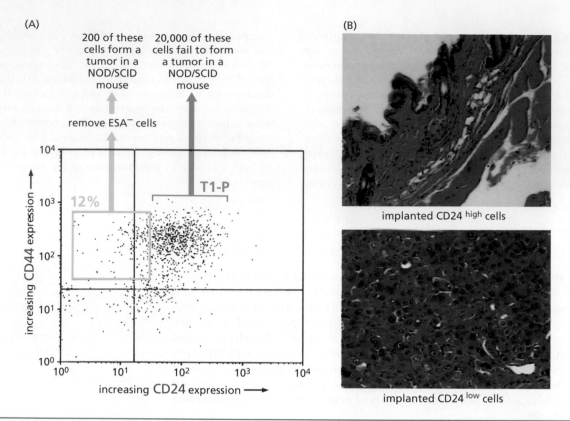

Figure 11.19 Enrichment of breast cancer stem cells
(A) Metastatic human breast carcinoma cells were freed of contaminating noncancerous (stromal) cells and separated from one another (using FACS; see Supplementary Sidebar 11.5). The expression of two distinct cell surface antigens—CD24 and CD44—was gauged simultaneously, each being detected with a specific monoclonal antibody linked to a distinct fluorescent dye. The intensity of staining is plotted logarithmically on each axis. Each black dot in the graph represents the detection of a single cell. In this experiment, a 12% subpopulation of cells that expressed low CD24 and high CD44 antigen staining (*green box*) was separated from cells (T1-P) that showed high CD24 expression and high CD44 expression (*blue bracket*). (B) The cells in the 12% minority population were further enriched by sorting for those that expressed epithelial surface antigen (ESA), which resulted in the elimination of contaminating stromal cells that did not express this antigen. Following injection into a NOD/SCID immunocompromised mouse, 200 of the CD24^{low} CD44^{high} ESA$^+$ cells were able to form a tumor, whereas 20,000 of the CD24^{high} CD44^{high} cells failed to do so. The *upper* micrograph shows a section through the subcutaneous site of implantation of CD24^{high} cells, in which only relatively normal skin and underlying muscle wall can be seen (*above left*); the *lower* micrograph shows a section through the tumor that formed following injection of the CD24^{low} cells. (From M. Al-Hajj et al., *Proc. Natl. Acad. Sci. USA* 100:3983–3988, 2003. With permission from National Academy of Sciences, U.S.A.)

from contaminating nonmalignant cells (such as stromal cells) that were present in the original tumor masses. Also, later genomic analyses confirmed that these two cell populations did not differ from one another genetically.

The tumors that eventually arose from the implanted minority cells were once again composed of minority and majority cell populations that showed, as before, vast differences in their ability to seed a new group of tumors, that is, in their relative **tumorigenicity**. In subsequent years, subpopulations of tumor-initiating cells (TICs) have been demonstrated in a wide spectrum of other solid tumors, such as neuroectodermal, pancreatic, colorectal, and hepatocellular carcinomas, as well as certain hematopoietic malignancies.

Taken together, these experiments indicate that the neoplastic cell populations originating in a variety of tissues are organized much like most normal epithelial tissues (**Figure 11.20**), in which relatively small pools of self-renewing stem cells are able to spawn large majority populations of descendant cells that have only a limited proliferative potential *in vivo*. In each of the experiments shown here, the tumorigenic minority cells expressed a set of antigenic markers on their surfaces that were distinct from those exhibited by cells in the majority populations. These experiments provided further indication that two cancer cell subpopulations can coexist within the same

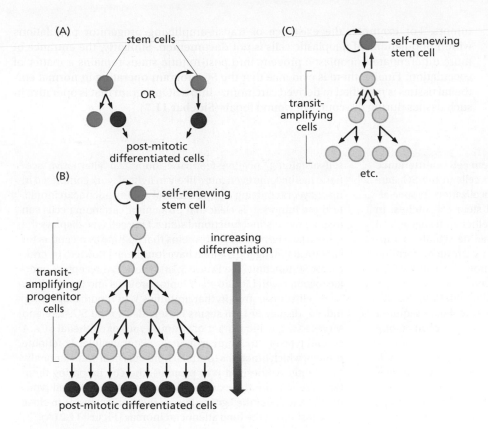

Figure 11.20 Stem cells and their progeny (A) The simplest organization of stem cell behavior involves an asymmetric cell division of a stem cell (*blue*), in which one of the daughter cells becomes a stem cell like its mother, whereas the other daughter (*red*) exits the stem cell state and may initiate a program of differentiation. This can be depicted in two graphic conventions, both of which are shown here. (B) In many tissues, a more complex scheme appears to operate. As in panel A, one of the daughters of a dividing stem cell becomes a stem cell. However, the other becomes a "transit-amplifying cell" (sometimes termed a "progenitor cell"; *gray*), which is committed to enter a differentiation pathway but does not yet participate in end-stage differentiation. Instead, this cell and its progeny undergo a series of symmetric cell divisions before its descendants eventually enter into a fully differentiated, post-mitotic state (*red*). Because a single stem cell can spawn dozens if not hundreds of differentiated descendants that carry out the work of a tissue, this cell does not need to proliferate frequently. Moreover, if the differentiated cells within a tissue cease dividing (and become *post-mitotic*), the great bulk of the cell proliferation within such a tissue is associated with transit-amplifying/progenitor cells. (C) A variation on the scheme shown in panel B recognizes that diverse lines of evidence indicate the reversibility of the processes leading toward increased differentiation (*red arrows*). Such evidence indicates the possibility of dedifferentiation both in normal and neoplastic tissues.

tumor in alternative phenotypic states even though they are genetically identical (**Sidebar 11.6**).

This scheme of normal tissue organization has been appropriated by researchers attempting to understand the organization of the neoplastic cells within a tumor. Thus, the tumor-initiating cell (TIC, that is, one capable of seeding the outgrowth of a tumor), often termed a **cancer stem cell** (CSC), is self-renewing and has the ability to generate the countless neoplastic progeny that together constitute a tumor. Although the CSC and its progeny are genetically identical, the progeny cells have also lost tumor-initiating ability during the course of their initial differentiation. Indeed, as the details of the cellular programs that govern the CSC state are elucidated, it becomes increasingly apparent that these CSC programs are organized in a fashion that is similar to those operating in the corresponding normal tissues, that is, tumors do not invent new stem cell programs but simply appropriate or modify ones operating in their corresponding normal tissues-of-origin.

The hierarchical scheme depicted in Figure 11.20B must surely be amended to reflect what is known and not known about the cell types that co-exist within individual

Sidebar 11.6 Organization of stem cell hierarchies In normal tissues, the cells are organized in a hierarchy, such as that depicted in Figure 11.20B. At the apex of the hierarchy is a stem cell, one of whose daughters remains a stem cell, while the other daughter exits the stem cell state and initiates a program of differentiation. (Because this particular cell division yields two phenotypically distinct daughters, it is often termed *asymmetrical*, in contrast to most cell divisions, which generate two phenotypically identical daughters and are therefore termed *symmetrical*.) The daughter cell that is formed directly by division of the stem cell and has exited the stem cell state is often termed a **transit-amplifying cell**, also termed a **progenitor cell** (see Figure 11.20B). Such cells are present in many normal tissues and represent intermediates between stem cells and their fully differentiated descendants. In addition, in certain tissues, transit-amplifying/progenitor cells may pass through a large succession of symmetrical cell divisions before their descendants eventually become fully differentiated. Hence, the initially formed transit-amplifying/progenitor cell can spawn dozens if not hundreds of differentiated progeny, which means that (1) the stem cell needs to divide only once in order to generate a large flock of fully differentiated descendants; (2) the stem cell may therefore divide only periodically rather than continually, even in a tissue in which differentiated cells are continually being lost and replenished; and (3) the great majority of ongoing cell divisions within a tissue are associated with the mitotically active transit-amplifying cells and not the stem cells.

tumors. For example, the existence of transit-amplifying/progenitor populations within populations of neoplastic cells is not documented. Similarly, the entrance of more differentiated neoplastic progeny into post-mitotic states remains a matter of speculation. Finally, there is evidence that the SC program operating in normal epithelial tissues is modified in derived carcinomas into a SC program that is operative in such tissues during the course of wound repair (**Sidebar 11.7**).

Sidebar 11.7 Conditions required for stem cell maintenance The continued residence of normal stem cells in the SC state generally depends on their continued localization in specialized tissue microcompartments, termed stem cell **niches**, in which the niche-forming cells, acting together with specialized extracellular matrix (ECM), provide paracrine signals to support and sustain the SCs in the SC state; such niche-forming cells represent cell types that are distinct from those of the SCs. In the context of the normal intestinal crypts, as an example (see Figure 7.20B), the SCs reside side by side with **Paneth cells**, which provide them with some of the paracrine signals required for continued residence in the SC state. Most important among these are Wnt3 ligands (see Figure 5.18).

It is less clear how the neoplastic CSCs maintain their residence in the SC state; one possible mechanism derives from the actions of autocrine signaling of the sort first illustrated in Figure 5.8C. Such signaling affords the continuous maintenance of a phenotypic state through the actions of self-sustaining, positive-feedback loops. An alternative scenario has become even more likely in light of work completed in recent years studying a mouse model of lung adenocarcinoma. In these tumors, it is clear that progenitor carcinoma cells can assume two distinct functional states. One cell type displays the cell-surface Frizzled (Fzd) proteins that operate as receptors for canonical Wnt ligands, which have long been known to be critical for SC function (see Figure 5.18). These Fzd receptors operate together with LRP5/6 co-receptor proteins and the ancillary Lgr5 cell-surface protein that amplifies Wnt binding signaling; indeed, display of Lgr5 seems closely tied to the SC state and serves as a specific marker of various types of epithelial SCs. A second type of carcinoma cell displays intracellular Porcupine protein, which functions to attach lipid tails post-translationally to recently synthesized Wnt proteins, thereby preparing them for secretion into the extracellular space. These two cell types (displaying either the Lgr5 or Porcupine marker) reside in close juxtaposition in the lung adenocarcinomas (**Figure 11.21A**).

(A)

(B)

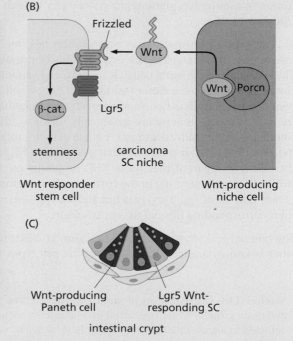

(C)

Porcupine⁺ (= Wnt ligand-producing cells)
Lgr5⁺ (= Wnt ligand-responding cells)

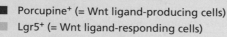

Figure 11.21 Tumors may deploy niche-forming cells to sustain the cancer stem cells (A) The carcinoma cells formed in a transgenic mouse model of lung adenocarcinoma were analyzed for their expression of the Lgr5 epithelial stem-cell marker (*green*) or, alternatively, for expression of the Porcupine enzyme (*red*), which indicates the presence of cells synthesizing and releasing Wnt proteins. Because normal stem cells reside in a niche that is created, in part, by cells secreting canonical Wnt proteins and such stem cells depend on Wnt proteins to sustain their residence in the stem-cell state, such a niche-stem cell interaction is presumed to operate as well in the adenocarcinomas being analyzed here and illustrated in panel B. (C) The close juxtaposition of normal stem cells and niche-forming cells operates as well in the intestinal crypts (see Figure 6.19C). In the latter case, the Wnt-secreting niche-forming cells are Paneth cells, which exert other functions as well in these crypts. β cat.—β-catenin; Porcn— porcupine. (A and B, adapted from T. Tammela, F.J. Sanchez-Rivera, N.M. Cetinbas et al. *Nature* 545:355–359, 2017. With permission from Nature.)

To summarize, Porcupine serves as a good marker for niche-forming, Wnt-secreting cells, whereas Lgr5 functions as a good marker of the SCs that depend on the secreted Wnt protein for their stemness. Moreover, it is clear that **oligopotent** progenitor cells in these tumors differentiate to generate both the SC niche-forming cells as well as the CSCs that reside in these niches (Figure 11.21B). This is strikingly reminiscent of the situation operating in the normal intestinal crypt (Figure 11.21C) where a common progenitor cell generates both the Paneth cells that operate as niche-forming cells as well as the SCs that depend on residing in the Paneth-generated niches (see Figure 7.20B).

An additional dimension of complexity derives from observations made in both normal and neoplastic epithelial cells that the hierarchy depicted in Figure 11.20B does not operate in a rigid fashion. Thus, there is clear evidence in both normal epithelial tissues and in their neoplastic derivatives that non-SCs, such as transit-amplifying cells, can *dedifferentiate* backward into SCs under certain conditions (see Figure 11.20C, *red arrows*). This holds implications for the identities of cellular genomes that are objects of mutation during multi-step tumor development (Supplementary Sidebar 11.6). In normal tissues, this dedifferentiation may be triggered by the depletion of cells from the normal SC compartment, resulting in the upward pointing arrow in this figure and the replenishment of missing SCs. Precisely how such dedifferentiation processes are regulated is poorly understood. These phenomena suggest an additional notion that is yet to be documented: that epithelial cell populations, independent of their state of neoplastic transformation, spontaneously organize themselves into hierarchies of the sort depicted in Figure 11.20B and that phenotypic **plasticity** (that is, an ability to transition between alternative phenotypic states) is essential to create and maintain such hierarchies.

The existence of these hierarchies creates a new dimension of complexity within individual tumors. If virtually all epithelial cell populations organize themselves in this fashion, as speculated here (see Sidebar 11.7), it becomes likely that stem cell hierarchies are established within each of the cell populations that are formed during each of the multiple phases of multi-step progression. Previously, we drew a linear, one-dimensional progression from a normal cell to a tumor cell (**Figure 11.22A**). Now, with this newer information in hand, it seems more appropriate to describe multi-step tumor progression in the form of a two-dimensional map, with the SCs formed at each stage of tumor progression spawning multiple distinct subpopulations of more and less differentiated cells (Figure 11.22B).

The existence of cancer stem cells is of more than academic interest. Because they may reside in a more mesenchymal phenotypic state, they are often more resistant to conventional chemo- and radiotherapies, as described in Chapter 17, and to more

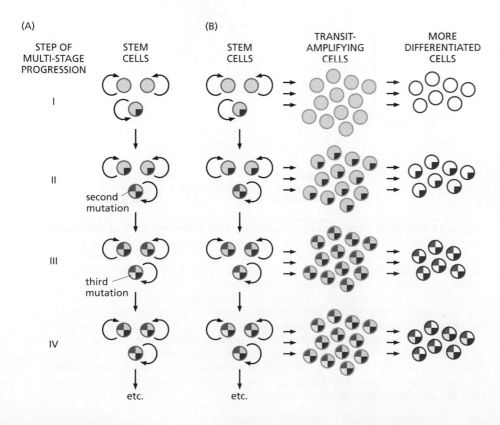

Figure 11.22 Multiple alternative phenotypic states at different stages of tumor progression (A) The scheme of multi-step evolution operating during tumor progression depicted in Figure 11.14 can be simplified graphically to indicate a linear succession of cell populations that succeed one another during this progression. Each of these states is defined by a distinct set of genetic and cell-heritable epigenetic alterations. (B) However, if epithelial cell populations (in the case of carcinomas) tend to spontaneously organize in a hierarchical fashion, as illustrated in Figure 11.20B, then the cell populations constituting *each stage* of tumor progression are also likely to self-organize into a similar hierarchical fashion. (This notion is further supported by the fact that fully normal and fully neoplastic epithelial cell populations both have stem cell subpopulations, providing strong indication that all of the intermediate cell populations formed during multi-step tumor progression will also contain subpopulations of stem cells.)

recently developed checkpoint immunotherapies, as described in Chapters 15 and 16. As a consequence, these cells may survive initial treatment, persist, and ultimately generate the new tumors whose appearance signals clinical relapse. In addition, there is clear evidence indicating that the cells that seed metastases at distant sites in the body must also possess tumor-initiating function, that is, they must operate as cancer stem cells (described in Chapter 14).

In fact, the involvement of CSCs in tumor pathogenesis suggests yet another dimension of complexity that we have not yet mentioned, even in passing: like normal SCs, CSCs may exhibit an ability to launch cells into multiple distinct differentiation lineages, that is, they may function like oligopotent stem cells. Such phenotypic diversification, if it occurs, should operate entirely independently of the genetic diversification that accompanies high-grade tumor progression and illustrated in Figure 11.16.

Recent observations using single-cell RNA sequencing techniques have indeed demonstrated that CSCs within individual tumors can indeed reside in hierarchies that lead, via phenotypic branching to generate multiple, phenotypically distinct subpopulations of cancer cells (**Figure 11.23**). As seen in this image, cells within a pediatric glioma could be resolved using single-cell RNA sequencing (scRNAseq) into three populations—a highly proliferative stem-like cell subpopulation and two distinct subpopulations that were generated by the stem cells and contained cells expressing transcriptomes typical of one or another differentiated normal cell lineage of the central nervous system; these more differentiated cells were nonproliferative and ostensibly post-mitotic. Such depictions of CSC differentiation have been extended by compelling demonstrations of the role of SCs in the formation and growth of skin papillomas (Supplementary Sidebar 11.7).

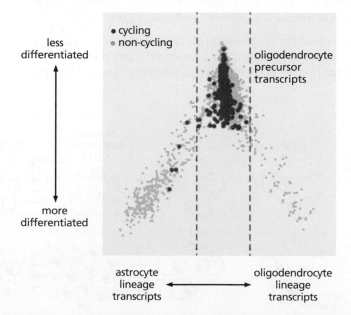

Figure 11.23 Glioma stem-like cells spawn cells that enter into distinct differentiation states The transcripts present in the cells of a subtype of glioma that are formed in the central nervous systems of young children have been analyzed by single-cell RNA sequencing (scRNAseq). These cells fell into three distinct subpopulations based on their expression of a series of cell-specific transcripts. These included stem-like cells that exhibited transcripts characteristic of oligodendrocyte precursors, as well subpopulations that expressed, to differing extents, transcriptomes that were typical of differentiated cells of either the astrocyte or the oligodendrocyte lineage. Increased expression of these lineage-specific transcriptomes is plotted as increasing horizontal distances, *rightward or leftward*, from the less-differentiated precursors. Cells with less-differentiated transcriptomes are localized between the two dashed lines. Actively proliferating cells are indicated as *red dots*, whereas nonproliferating cells are indicated as *gray dots*. (From M.G. Filbin, I. Tirosh, V. Hovestadt et al., *Science* 360:331–335, 2018. With permission from AAAS.)

Moreover, as discussed in great detail in Chapters 13 and 14, the cell-biological program termed the epithelial-mesenchymal transition (EMT) operates epigenetically to drive carcinoma cells into a mixture of cell states that can be arrayed along a spectrum that ranges from highly epithelial to highly mesenchymal phenotypes. Here, once again, we see an example of a nongenetic program that profoundly affects the behavior of cancer cells but is not reflected in the "genetic biographies" discussed in Section 11.3. When taken together, these various considerations tell us that truly accurate depictions of multi-stage tumor pathogenesis will need to be mapped in a multi-dimensional space encompassing both genetically and epigenetically governed traits. In addition, as discussed in Supplementary Sidebar 11.8, the Darwinian model, on which many of the arguments in this section are based, remains unproven in many of its aspects.

A final complication has not even been explored here: recently acquired experimental evidence indicates clearly that within an individual tumor, different clonal subpopulations of tumor cells actively *collaborate* with one another rather than *compete*. Such collaboration flies in the face of the traditional model of Darwinian evolution, which is firmly grounded on mechanisms of competition between individual members of a species which allow more fit subpopulations to outgrow and eventually eliminate closely related, less-fit subpopulations. At present it remains unclear how often such interclonal competition dictates the growth and lethality of various types of human tumors.

11.8 Multiple lines of evidence reveal that normal cells are resistant to transformation by a single mutated gene

The description of multi-step tumorigenesis presented in Figure 11.10 is satisfying because it seems to provide a clear and definitive rationale of how multi-step tumor pathogenesis proceeds. In truth, however, it fails in at least two important ways. First, it describes genetic alterations as *correlates* of histopathological changes without proving *causality*; that is, it fails to demonstrate that the accumulated mutant alleles are actually responsible driving tumorigenesis forward. And second, it provides no explanation of why so many mutant genes are required for tumorigenesis to reach completion.

In the end, these shortcomings could only be addressed by intervening experimentally in the process of tumorigenesis, more specifically, by reconstructing it in detail in the laboratory. To begin, the introduction of well-defined genetic alterations into previously fully normal cells offered the prospect of revealing with precision how specific changes in cell genotype function to create the neoplastic cell phenotype.

In fact, certain early experiments (described in Chapter 4) seemed to indicate that the genetic rules governing the transformation of normal mammalian cells into tumorigenic cells are actually extremely simple. Recall, for example, the experiment in which a mutant, activated H-*ras* oncogene from a human bladder carcinoma was introduced via transfection into previously normal NIH 3T3 mouse fibroblasts (see Section 4.1 and Figure 4.1). Having acquired the mutant *ras* oncogene, these cells became fully transformed, to the point that they were capable of seeding tumors in appropriate host mice.

This behavior of the *ras*-transformed NIH 3T3 cells indicated that the requirements for transforming these cells were quite minimal. A single genetic alteration of these cells—the acquisition of a *ras* oncogene—sufficed to convert them to a transformed, tumorigenic state. Moreover, the mutation that originally created the *ras* oncogene was itself a simple point mutation. Taken together, these findings suggested that a point mutation affecting one of the NIH 3T3 cells' own native H-*ras* proto-oncogenes would yield the identical outcome—full transformation to a neoplastic state. In one stroke, a point mutation should transform a normal cell into a tumor cell.

We know from our earlier discussions of human tumor cell genetics that this conclusion—at least in the case of human cells—cannot be correct. We can verify this conclusion with a simple calculation. Given the rate at which specific point mutations occur randomly in the human genome, the number of point mutations accumulated per cell division, and the number of cell divisions occurring each day in the body ($\sim 3 \times 10^{11}$), some have estimated that several thousand new, point-mutated *ras* oncogenes are created accidentally every day throughout the human body and that the total body burden of cells carrying *ras* oncogenes must number in the millions. Clearly, human beings are not afflicted with a comparable number of new tumors each day.

Something is terribly wrong here, either in these calculations or in the transfection experiments that were relied upon to gauge the genetic complexity of the transformation process. In the end, the flaw in in these calculations could be found in the design of the particular experiments, specifically in the "normal" cells that were used in the transformation assay. The NIH 3T3 cells, as it turns out, are not truly normal, because they constitute a cell line—a population of cells that has been adapted to grow in culture and can be propagated indefinitely (see Section 10.1). This history implies that these cells at some point underwent one or more genetic or epigenetic alterations that enabled them to grow in culture and to proliferate in an immortalized fashion (see Chapter 10).

As a consequence, investigators began in the early 1980s to examine the consequences of introducing a *ras* oncogene into truly normal cells—those from rat, mouse, or hamster embryos that had been recently explanted from living tissues and propagated *in vitro* for only a short period of time before being used in gene transfer experiments. Such cells—sometimes termed **primary cells**—were unlikely to have undergone the alterations that seem to have affected NIH 3T3 cells during their many-months-long adaptation to tissue culture and attendant immortalization.

The results obtained with primary rat and hamster cells differed from those observed previously with NIH 3T3 cells: These primary cells were not susceptible to *ras*-induced transformation. Control experiments left no doubt that these cells had indeed acquired the transfected oncogene and were able to express the encoded Ras oncoprotein, but somehow they did not respond by undergoing transformation and forming foci in the Petri dish. These observations provided the first evidence that the act of adapting rodent cells to culture conditions and selecting for those that have undergone immortalization yields cells that, in a fully unanticipated fashion, also become responsive to transformation by a single introduced gene, in this case an introduced *ras* oncogene.

These *in vitro* experiments with primary cells have been extended by introducing activated *ras* oncogenes into the colonic epithelial cells of mice. The resulting *ras* oncogene–expressing cells create nothing more than hyperplastic epithelia, that is, cells that are present in excessive numbers but are, in other respects, essentially normal (**Figure 11.24A**). Moreover, in a mouse model of skin carcinogenesis described later in Section 11.11, treatment with a mutagenic carcinogen can create H-*ras* oncogene-bearing cells that are detectable within a week, yet the skin remains morphologically normal for another year. In the human colon, clusters of cells bearing both the *ras* oncogene and inactivated copies of the *Apc* tumor suppressor gene form hyperplasias that are, in spite of these two alterations, incapable of forming adenomatous crypts (Figure 11.24B and **Sidebar 11.8**).

The resistance of fully normal, primary rodent cells to *ras*-induced transformation led to an interesting question: Were there yet other oncogenes that could render these cells susceptible to transformation by *ras?* In fact, a line of human promyelocytic leukemia cells had been reported to carry both an activated N-*ras* and an activated *myc* oncogene. This finding suggested the possibility that these two cellular oncogenes were cooperating with one another to create the malignant phenotype of the leukemia cells. Indeed, when a *myc* oncogene was introduced together with an H-*ras* oncogene into primary rat embryo fibroblasts (REFs), the cells responded by becoming morphologically transformed (**Figure 11.25**) and, more important, tumorigenic; neither of these oncogenes, on its own, could create such transformed cells. An adenovirus *E1A* oncogene could be used in place of *myc* in these transformation assays.

This result yielded several interesting conclusions. These two cellular oncogenes clearly affected the phenotype of rodent cells in quite different ways, because they

Sidebar 11.8 Inborn oncogenes may predispose to cancer but do not suffice to create it Certain "experiments of nature" also support the notion that single mutations in human cells are not sufficient for the development of cancers. For example, some individuals are born carrying a germ-line mutation of the gene encoding the Kit growth factor receptor; such mutations create a constitutively active, ligand-independent Kit receptor, which functions as a potent oncoprotein. These individuals are at high risk for developing gastrointestinal stromal tumors (GISTs), but these tumors only become apparent several decades after birth, even though a constitutively active Kit oncoprotein has been functioning in many of their cells since birth (see also Sidebar 5.6). Similarly, some individuals have been documented who carry mutant H-*ras* alleles in their germ lines yet usually develop tumors only after several decades. Early childhood leukemias in monozygotic (that is, identical) twins provide equally dramatic examples of the inability of single mutations, acting on their own, to create clinically apparent tumors (Figure 11.24C).

(A)

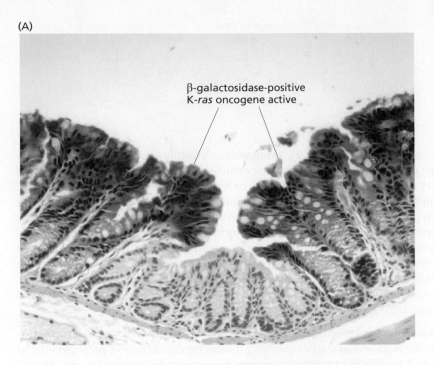

β-galactosidase-positive
K-*ras* oncogene active

(B)

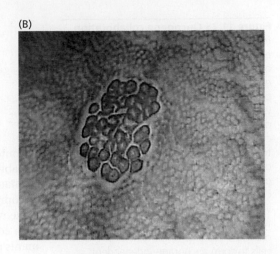

(C)

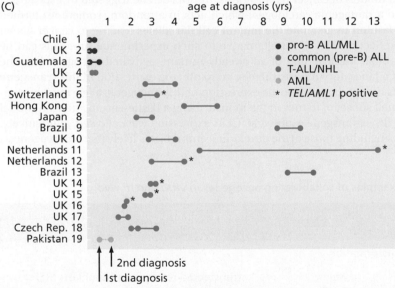

age at diagnosis (yrs)

2nd diagnosis
1st diagnosis

Figure 11.24 One or two oncogenic lesions do not suffice to generate full-blown tumors (A) A mouse germ line was re-engineered to create colonic epithelial cells in which both a mutant K-*ras* oncogene and the β-galactosidase gene could be activated in scattered cells by an infecting adenovirus. As indicated here, colonic epithelial cells in which both β-galactosidase (*blue*) and the K-*ras* oncogene (*not visible*) have become expressed create localized regions of hyperplasia in which the epithelial cells are otherwise normal, indicating that the *ras* oncogene on its own does not suffice to transform these cells into a tumorigenic state. (B) A cluster of aberrant crypt foci (ACFs) in the human colon are clearly revealed by colonoscopy. ACFs like this one frequently carry an inactivated *APC* tumor suppressor gene together with an activated K-*ras* oncogene; they may eventually progress into adenomatous polyps, which themselves are still not full-fledged colon carcinomas. Hence, even two oncogenic genetic lesions, acting together, generate human cells that are still several steps short of being fully transformed. (C) Monozygotic (identical) twin pairs have been documented throughout the world in which both twins develop the same type of leukemia. The leukemias invariably share a common chromosomal marker or mutation, indicating that they derive from the same clone of initiated cells that somehow became shared between the twins. Although many of these leukemias are diagnosed at quite different postnatal ages (*dots*), these initiating somatic mutations, which occurred *in utero*, were not, on their own, sufficient to trigger the formation of a clinically apparent leukemia. The labels (*top right*) and the associated colors denote different subtypes of leukemia identified by distinctive gene markers. (A, courtesy of K.M. Haigis and T. Jacks. B, courtesy of Y. Niitsu. C, from M.G. Greaves et al., *Blood* 102:2321–2333, 2003. With permission from American Society of Hematology.)

were able to complement one another in eliciting cell transformation. Each seemed to be specialized to evoke a subset of the cellular phenotypes associated with the transformed state. For example, *ras* was able to elicit anchorage independence, a rounded, refractile appearance in the phase microscope, and loss of contact inhibition, whereas *myc* helped the cells to become immortalized and reduced somewhat their dependence on growth factors. In subsequent years it became apparent that there were yet other complementary functions of the two. Thus, the Ras oncoprotein could supplant mitogenic signals normally supplied by growth factors, re-program cell metabolism, and suppress apoptosis. Myc, for its part, could cooperate in driving cell proliferation, and promote various phenotypes only apparent *in vivo*, notably angiogenesis, invasiveness and metastasis (described in Chapters 13 and 14).

These oncogene collaboration experiments could be extended by demonstrating other pairs of collaborating oncogenes, some of which operated *in vivo* (**Table 11.1**). Together, such observations suggested a rationale for the complex genetic steps that accompany and cause tumor formation: each of the genetic changes provides the nascent tumor cell with one or more of the cell-biological traits needed in order to

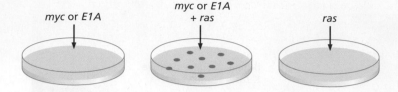

Figure 11.25 Oncogene collaboration in rodent cells *in vitro* Cultures of early-passage rat embryo fibroblasts or baby hamster kidney cells were exposed to cloned DNAs via the calcium phosphate gene transfection procedure (see Figure 4.1). Introduction of a *myc* or adenovirus *E1A* oncogene into these cells (*left*), on its own, did not yield foci of transformants, although the introduced oncogenes did facilitate the establishment of these early-passage cells in long-term culture. Introduction of an H-*ras* oncogene via transfection (*right*) also did not yield foci of transformed cells, although it did allow significant numbers of these cells to form anchorage-independent colonies when they were introduced into a semi-solid medium such as dilute agar. However, the simultaneous introduction of *ras* + *myc* or, alternatively, *ras* + *E1A* did generate foci of transformed cells that were able to form tumors when injected into syngeneic or immunocompromised hosts.

become tumorigenic (**Table 11.2**). These unique contributions seem to derive from the ability of each of these oncogenes to perturb a specific regulatory circuit or set of regulatory circuits within a cell. (Adenovirus E1A oncoprotein could supplant a Myc oncoprotein of entirely different structure and biochemical function, suggesting that these two proteins impinged on a common intracellular signaling circuit rather than on a common target protein within a cell; see Section 8.9.) Moreover, these experiments provided indication that cell proliferation and cell survival are governed by two or more distinct intracellular regulatory circuits, all of which must be perturbed before the cell will become tumorigenic. In the decade that followed, other pairs of collaborating oncogenes became apparent through research in a variety of experimental models (see Table 11.1).

These demonstrations of oncogene collaboration *in vitro* required extension and confirmation through *in vivo* studies involving mouse models of cancer pathogenesis. In many rodent models of cancer pathogenesis, tumors can be triggered by exposure of an animal to a mutagenic carcinogen, which acts in a random (sometimes termed **stochastic**) fashion to generate the mutant cellular alleles leading to cancer (as an example, see Section 11.11). An alternative to such experimental protocols can be achieved through the insertion of an already-mutant, activated oncogene into the germ line of a laboratory mouse, thereby guaranteeing expression of this **transgene** in some of its tissues. In practice, the expression of this oncogenic allele must be confined to a small subset of tissues in the mouse, using a tissue-specific transcriptional promoter to drive transgene expression. (If its expression were allowed, alternatively, in all tissues, including those of the developing embryo, it is likely that embryogenesis

Table 11.1 Examples of collaborating oncogenes *in vitro* and in *vivo*

"*ras*-like" oncogene[a]	"*myc*-like" oncogene[a]	Target cell or organ
***In vitro* transformation**		
Ras	*myc*	transfected rat embryo fibroblasts (REFs)
Ras	*E1A*	transfected rat kidney cells
Ras	SV40 *large T*	transfected REFs
Notch-1	*E1A*	transfected rat kidney cells
***In vivo* tumorigenesis**		
middle T	*large T*	polyomavirus-induced murine tumors
mil (= *raf*)	*myc*	MH2 avian leukemia virus chicken tumors
erbB	*erbA*	avian erythroblastosis virus chicken tumors
pim1	*myc*	mouse leukemia virus tumors
abl	*myc*	mouse leukemia virus tumors
Notch-1/2	*myc*	thymomas in transgenic mice
bcl-2	*myc*	follicular lymphomas in transgenic mice

[a]The terms "*ras*-like" and "*myc*-like" refer to functional classes rather than genes encoding components of a common signaling pathway. "*ras*-like" oncogenes tend to encode components of cytoplasmic signaling cascades, whereas "*myc*-like" oncogenes tend to encode nuclear proteins.

Table 11.2 Physiological mechanisms of oncogene collaboration[a]

Oncogene pair	Cell type	Mechanisms of action
ras + SV40 large T	rat Schwann cells	ras: proliferation + proliferation arrest
		large T: prevents proliferation arrest and reduces mitogen requirement
ras + E1A	mouse embryo fibroblasts	ras: proliferation and senescence
		E1A: prevents senescence
erbB + erbA	chicken erythroblasts	erbB: induces GF-independent proliferation
		erbA: blocks differentiation
TGF-α + myc	mouse mammary epithelial cells	TGF-α: induces proliferation and blocks apoptosis
		myc: induces proliferation and apoptosis
v-sea + v-ski	avian erythroblasts	v-sea: induces proliferation
		v-ski: blocks differentiation
bcl-2 + myc	rat fibroblasts	bcl-2: blocks apoptosis
		myc: induces proliferation and apoptosis
ras + myc	rat fibroblasts	ras: induces anchorage independence
		myc: induces immortalization
raf + myc	chicken macrophages	raf: induces growth factor secretion
		myc: stimulates proliferation
src + myc	rat adrenocortical cells	src: induces anchorage and serum independence
		myc: prolongs proliferation

[a]In each pair, the first oncogene encodes a cytoplasmic oncoprotein, whereas the second oncogene encodes a nuclear oncoprotein.

would be so profoundly disrupted that the developing fetus would die long before the end of gestation.)

An early version of this strategy for creating cancer-prone, **transgenic** mice (see, for example, Section 9.13) involved the insertion of oncogenic alleles of the *ras* and the *myc* genes into the mouse germ line (**Figure 11.26**). In one influential set of experiments, expression of a *ras* oncogene and a *myc* oncogene were each placed under the control of the transcriptional promoter of mouse mammary tumor virus (MMTV), a retrovirus that specifically targets mammary tissues. Such hybrid genes could then be inserted into the mouse germ line, each becoming a transgene whose MMTV promoter ensured expression at significant levels only in mammary glands.

As anticipated, the presence of either one of these oncogenic **transgenes** in the mouse germ line predisposed mice to eventual development of breast cancer. However, in spite of the expression of either a *ras* or a *myc* oncogene in most if not all of the mammary epithelial cells of these mice, their mammary glands showed either minimal morphologic changes (in the case of the *myc* transgene) or at most hyperplasia (in the case of the *ras* transgene). Moreover, breast tumors were observed only beginning at 4 weeks of age (for the MMTV-*ras* mice) and 12 weeks (for the MMTV-*myc* mice)—significantly long **latency** periods (see Figure 11.26). This confirmed that the presence of a single oncogene within a normal cell in living tissue was not, on its own, sufficient to transform this cell into a tumor cell. Instead, the kinetics of breast cancer formation in these mice pointed to the necessary involvement of one or more additional stochastic events before either the *ras* or *myc* oncogene–bearing mammary cells could progress to a tumorigenic state.

Double-transgenic mice that carried both MMTV-*ras* and MMTV-*myc* transgenes were created through mating between the two transgenic strains described above.

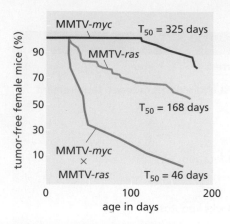

Figure 11.26 Oncogene collaboration in transgenic, cancer-prone mice The ability to create transgenic mice (see Figure 9.20A) made it possible to determine whether the *myc* and *ras* oncogenes are able to collaborate *in vivo* as well as *in vitro* (see Figure 11.25). Use of the MMTV (mouse mammary tumor virus) transcriptional promoter to drive expression of each oncogene ensured expression of both transgenes largely in the mammary glands. Transgenic mice carrying the MMTV-*myc* and the MMTV-*ras* transgene were bred to create double-transgenic mice carrying both transgenes. The incidence of mammary carcinomas in the three strains of mice—those carrying the MMTV-*myc* transgene only (*red curve*), the MMTV-*ras* transgene only (*green curve*), or both transgenes (*blue curve*)—was tracked over many months. The percentage of tumor-free mice (*ordinate*) is plotted vs. the age of the various strains of mice. "T_{50}" indicates the number of days required for one-half of the mice of a particular genotype to develop detectable mammary carcinomas. (From E. Sinn et al., *Cell* 49:465–475, 1987. With permission from Elsevier.)

These double-transgenic mice contracted tumors at a greatly accelerated rate and at high frequency compared with mice inheriting only one of these transgenes (see Figure 11.26). Therefore, the two oncogenic transgenes could collaborate *in vivo* to generate tumors, paralleling and corroborating the conclusions of the *in vitro* experiments described above.

Interestingly, even with two mutant oncogenes expressed in the great majority of mammary cells from early in development, tumors did not appear in these double-transgenic mice soon after birth but instead were seen with significant delay. Hence, the concomitant expression of two powerful oncogenes was still not sufficient to fully transform mouse mammary epithelial cells (MECs); instead, these cells clearly required at least one additional stochastic event, ostensibly a somatic mutation, before they would proliferate like full-blown cancer cells. (A hint about the identity of this third, stochastic event has come from careful analysis of rat cells that have been transformed *in vitro* by the *ras* + *myc* protocol; sooner or later, such cells usually acquire a mutation or a promoter methylation that leads to inactivation of the p53 tumor suppressor pathway; see Chapter 9.)

Note that the collaborative actions of transgenic oncogenes were already mentioned earlier, when we read of the synergistic actions of *myc* and *bcl-2* transgenic oncogenes in promoting lymphomagenesis (see Figure 9.20). In this case, the benefit of *bcl-2* (and its *bcl-x*$_L$ cousin) derives largely from its anti-apoptotic effects. This work illustrates that oncogenes can collaborate through a variety of cell-physiological mechanisms to promote tumor formation, the point made in Table 11.2.

11.9 Human cells are constructed to be highly resistant to immortalization and transformation

The biological lessons derived from studying mice and rats are usually directly transferable to understanding various aspects of human biology. Even though 80 million years may separate us from the most recent common ancestor we share with rodents, the great majority of biological and biochemical attributes of these distant cousins are present in very similar, if not identical, form in humans. The genomes of humans and rodents also seem very similar: essentially all of the roughly 20,000 protein-encoding genes discovered in the human genome have been found to have mouse orthologs. It stands to reason that the cell-biological processes of immortalization and neoplastic transformation should also be essentially identical in rodent and human cells.

The biological reality is, however, quite different. It is easy to immortalize rodent cells simply by propagating them through a relatively small number of passages *in vitro*. Spontaneously immortalized cells arise frequently and become the progenitors of cell lines, such as the NIH 3T3 cells discussed earlier (see Section 11.8). In contrast, human cells rarely, if ever, become immortalized following extended serial passaging in culture (see Chapter 10). Eventually, cultured human cells stop growing and become senescent, and spontaneously immortalized cell clones do not emerge. Moreover, attempts over a period of 15 years to experimentally transform human cells via transfected oncogenes consistently failed. These failures led to the notion that human cells were impregnable, that is, they simply could not be transformed experimentally. In fact, any human cells emerging from such failed transfection experiments were not even immortalized and therefore senesced sooner or later.

These repeated failures at experimental transformation of human cells prevented researchers from addressing a simple yet fundamental problem in cancer biology: How many intracellular regulatory circuits need to be perturbed in order to transform a normal human cell into a cancer cell? In the end, a solution to these failures was inspired by experience with cultured rodent cells, which indicated that once cells were immortalized in culture, they became responsive to transformation by a *ras* oncogene. The stark differences in the telomere biology of rodent and human cells (see Section 10.9) seemed to explain at least part of the difficulty of immortalizing human cells and thus their very different responses to introduced oncogenes. Recall that the cells of laboratory mice usually carry extremely long telomeric DNA (as long as 40 kb), whereas normal human cells have far shorter telomeres. In addition, telomerase activity seems to be regulated differently in cells of the two species, that is, it is constitutively high in murine cells and almost absent in corresponding human cells. Accordingly, the immortalization problem related to human cells might be solved by adding the *hTERT* gene (encoding the catalytic subunit of the human telomerase holoenzyme) to other immortalizing oncogenes introduced into these cells.

As it turned out, introduction of an *hTERT* gene in addition to the SV40 *large T* onco-gene (whose product inactivates both pRb and p53 tumor suppressor proteins) did indeed yield immortalized human cells. (Alternative means of inactivating both pRb and p53, such as introduction of human papillomavirus *E6* and *E7* oncogenes, suc-ceeded as well.) And once immortalization was achieved through these multiple changes, the resulting human cells, so the logic went, could then be transformed mor-phologically in the culture dish by introduction of an activated *ras* oncogene.

However, these morphologically transformed human cells were still not fully trans-formed, as indicated by their inability to form tumors when implanted in immunocom-promised mouse hosts. (Recall that the faulty immune systems of such mice ensure that tissues of foreign origin, such as human tumors, are not eliminated by immunological attack.) In fact, these cells still required one more alteration, this one achieved by intro-duction of the gene encoding the SV40 small t oncoprotein. Small t perturbs a subset of the functions of the abundant cellular enzyme termed protein phosphatase 2A, notably PP2A enzyme complexes containing the B56γ subunit (PP2A; **Sidebar 11.9**).

Sidebar 11.9 The elusive fifth signaling pathway It remains unresolved whether, in addition to the deregulation of the four signaling pathways described here, the fifth alteration—deregulation of the actions of PP2A—commonly occurs during the formation of most human cancers. PP2A is a phosphoserine/phosphothreonine phosphatase composed of multiple alterna-tive regulatory subunits, a scaffolding subunit, and a catalytic subunit; altogether as many as 30 heterotrimeric forms of PP2A are thought to exist, each being defined by its complement of regulatory subunits, which in turn has been implicated in sub-strate recognition by the heterotrimeric holoenzyme complexes (**Figure 11.27**). In a number of cancers, the PR65α regulatory subunit of PP2A has been found in mutant form. And two cellu-lar proteins, termed CIP2A and SET, are over-expressed in many human cancers and operate like the SV40 small T oncoprotein to protect the Myc protein from dephosphorylation by PP2A; the resulting preservation of phosphorylated Myc saves it from degradation. At the same time, CIP2A and SET protect kinases of the Ras–Raf–MEK–ERK pathway (see Section 6.5) from dephosphorylation, which would shut down the activation of this key mitogenic pathway achieved by actively signaling Ras oncoproteins and as well as upstream mitogenic signaling pro-teins such as growth factor receptors. Yet other endogenous cellular inhibitors of PP2A have also been uncovered. Among the effects in transformation mediated by various combinations of PP2A subunits, the most important may well be the dephos-phorylation of Myc, because a nonphosphorylatable mutant of Myc can replace SV40 small t oncoprotein in *in vitro* human cell transformation assays.

alternative PP2A heterotrimers

small T replaces a subset of B-type subunits, preventing their function

Figure 11.27 Protein phosphatase 2A The PP2A holoenzyme is composed of A, B, and C subunits. There are 2 types of both A and C subunits, and 18 types of B subunit. The C subunits carry the catalytic activity of the holoenzyme and the B subunits direct the holoenzyme to specific substrates; the A subunits act as scaffolding to assemble B and C subunits into holoenzyme complexes. The various holoenzymes are responsible for dephosphorylating cellular phosphoproteins bearing either phosphoserine or phosphothreonine residues. The small T (sT) oncoprotein of SV40 can associate with the A and C subunits (*bottom*) and thereby prevent the association of certain B-type subunits with A + C. In so doing, sT prevents PP2A from dephosphorylating a subset of its normal constituency of substrates. This action contributes in an incompletely understood fashion to the experimental transformation of human cells. (Adapted from S. Colella et al., *Int. J. Cancer* 93:798–804, 2001.)

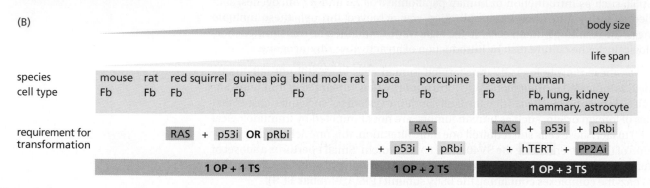

(A) human cell transformation

pathway	Ras	pRb	p53	telomeres	PP2A
alternative genes/ molecules used to deregulate pathway	*ras*, *MEK+ Akt/PKB*, *MEK+IKBKε, PAK1+ Akt/PKB*	**SV40 *LT*,** *CDK4 + D1,* *HPV E7,* *Rb shRNA*	**SV40 *LT*,** *DN p53,* *HPV E6,* *p53 shRNA*	***hTERT,*** *myc* + SV40 *LT*	**SV40 *sT*** in some cells: *myc Akt/PKB+Rac1, PI3K, B56 shRNA*

(B)

body size

life span

species cell type	mouse Fb	rat Fb	red squirrel Fb	guinea pig Fb	blind mole rat Fb	paca Fb	porcupine Fb	beaver Fb	human Fb, lung, kidney mammary, astrocyte

requirement for transformation

RAS + p53i OR pRbi   RAS + p53i + pRbi   RAS + p53i + pRbi + hTERT + PP2Ai

1 OP + 1 TS   1 OP + 2 TS   1 OP + 3 TS

Figure 11.28 Intracellular pathways involved in mammalian cell transformation (A) The experimental transformation of human cells has been achieved through the insertion of various combinations of cloned genes into cells. Initially, a combination of three genes encoding the SV40 early region [which specifies both the large T (LT) and small T (sT) oncoproteins], the hTERT telomerase enzyme, and a *ras* oncogene was found to suffice for the transformation of a variety of normal human cell types to a tumorigenic state (*bolded*). These introduced genes were found to deregulate five distinct regulatory pathways involving (1) Ras mitogenic signaling, (2) pRb-mediated G₁ cell cycle control, (3) p53, (4) telomere maintenance, and (5) protein phosphatase 2A (PP2A). Subsequent work has found that other combinations of cloned genes suffice as well. For example, activation of the Ras signaling pathway can be mimicked by introducing a combination of constitutively active alleles of *MEK* (*MEK^DD*) and *Akt/PKB*. Further dissection of these pathways has also revealed that a combination of *MEK^DD* with *IKBKε* or *PAK1* with *Akt/PKB* also suffices to replace *ras* as an oncogenic activator. The disruption of pRb function can be achieved by a combination of ectopically expressed, CDK inhibitor–resistant CDK4 + cyclin D1 or a short hairpin construct (shRNA) directed against the *Rb* gene; p53 can be disrupted by an introduced dominant-negative *p53* allele or an shRNA directed against the *p53* gene; telomeres can be maintained by activating endogenous *hTERT* expression through a combination of introduced SV40 *LT* + *myc*; and PP2A function can be disrupted by an shRNA that inhibits the synthesis of the B56 subunit of PP2A. It is unknown whether these five pathways are required for the experimental transformation of all human cell types and whether deregulation of all of these five pathways occurs in all spontaneously arising human tumors. A comma in figure indicates "or." (B) The requirements for cell transformation differ in terms of the number of introduced genes needed to have full transformation. Indicated here are the cultured cells of a number of mammalian species (mostly rodents) that have been examined experimentally by introducing oncoproteins targeting or supplying the indicated cellular proteins. The indicated human cell types include fibroblasts, three epithelial cell types and astrocytes. Fb—fibroblast; i—inhibitor; OP —oncoprotein; TS —product of tumor suppressor gene; Rb—pRb. (B, courtesy of J. Westermarck.)

Taken together, these experiments demonstrated that *five* distinct cellular regulatory circuits needed to be altered experimentally before human cells can grow as tumor cells in immunocompromised mice (**Figure 11.28A**). These changes involve: (1) the mitogenic signaling pathway controlled by Ras (see Chapter 6); (2) the cell cycle checkpoint controlled by pRb (see Chapter 8); (3) the alarm pathway controlled by p53 (see Chapter 9); (4) the telomere maintenance pathway controlled by hTERT (see Chapter 10); and (5) the signaling pathways controlled by protein phosphatase 2A (PP2A); the latter include effects modulating the activity of a wide spectrum of ser/thr-phosphorylated proteins, among them those controlling the Ras, mTOR, Myc, β-catenin, and PKB/Akt signaling proteins (see Sidebar 11.9). (It remains possible that perturbation of alternative cellular signaling pathways quite distinct from the five listed here might also permit the experimental transformation of human cells.)

Experiments like these indicate why adult human cells are highly resistant to transformation. At the same time, it remains unclear whether the steps required to experimentally transform human cells *in vitro* accurately reflect the changes that normal cells must undergo within human tissues in order to acquire the attributes of cancer cells. In fact, four of these changes (involving Ras and hTERT activation, and pRb and p53 inactivation) are commonly seen in the cells of human tumors. For example, in a recent genetic survey of glioblastomas, 74% carried changes in the three critical pathways involving the pRb, p53, and Ras–MAPK proteins, and virtually all of these tumors exhibited elevated expression of hTERT as well. Of additional interest, a series of studies involving large-scale sequencing of 2658 cancer cell genomes derived from 38 distinct types of human cancer reported that, on average, these tumors carried four to five driver gene mutations per genome.

Extensions of such work, reported in 2014 and 2015, demonstrated that the targeted alteration of four genes (encoding the Apr, p53, K-ras, and Smad4 proteins) in the genomes of human Lgr5+ colonic crypt stem cells yielded tumorigenic cells with the features of invasive colorectal carcinomas; because these experiments employed cells that developed into colonic **organoids** in three-dimensional culture, they approached far more closely the conditions operating in human tissues *in vivo* (Supplementary Sidebar 11.9). In these experiments, each of the mutations introduced into the genomes of the Lgr5+ stem cells replaced one of four protein factors (Wnt, R-spondin, EGF, Noggin) that was usually present in the culture medium and required to generate proper crypt morphogenesis *in vitro*. This work showed several things: It provided an independent accounting of how many genes needed to undergo mutation in order to create tumorigenic cells, that is, four instead of the five enumerated in Figure 11.28A. In addition, it showed that each of the requisite mutations rendered these colorectal cells independent of one or another specific extracellular signal.

These various lines of investigation seem to be converging on a number, possibly half a dozen, of the critical rate-limiting events involved in the formation of many human solid tumors. Nevertheless, it is likely that some normal human cells require a greater or lesser number of changes in order to undergo neoplastic transformation and the number of required changes differ between species (Figure 11.28B). Each of these requisite changes seems to reflect the need to deregulate or perturb a distinct signaling circuit operating within human cells (see Figure 11.28).

11.10 Mammalian evolution contributed to the complexity of human cell transformation

At this stage, we need to revisit the profound interspecies differences in the behavior of mouse versus human cells with respect to transformation. In particular, these differences require some type of biological rationale. We could begin here with noting that, in principle, the risk of tumor development in an organism is likely to be proportional to the cumulative number of cell divisions experienced by that organism during its lifetime; such cell divisions encompass both those needed to build up tissue mass during development as well as those involved in replacing cells that have been lost through physiologically programmed cell turnover. The complexity of the mammalian cell cycle, as laid out in Chapter 8, offers many opportunities for genetic disaster during the course of DNA replication and mitosis and thus during each of the myriad cell cycles occurring in a human body during a lifetime.

Changes wrought by evolution would seem to be essential to compensate, wherever possible, for the increased risk that, for example, large bodied, long-lived organisms like ourselves confront during an average life span. Thus, the cells in a mouse's body may pass through about 10^{11} mitoses in a mouse lifetime, whereas those in a human body pass through about 10^{16} cell cycles in a human lifetime. These numbers derive from the relative masses of the two organisms (which reflects cell number) times the average organismic life span. Such calculations are in one sense simplistic, because they do not account for other critical parameters, such as differing metabolic rates (which can affect mutation rates, as described in Chapter 12) as well as DNA repair capacity. (It remains to be seen how bowhead whales, which may weigh in at 100 tons and live up to 200 years, cope with the challenge of cancer development, which would seem to be 10,000 times higher than our own lifetime risks of cancer development. Provocatively, a 2021 genomics analysis provided clear evidence of accelerated evolution among whale species of genes involved in holding back cancer development including a number of tumor suppressor genes!)

These numbers, on their own, reveal the enormously increased risk of cancer development that is intrinsic to our biology relative to that of the mouse. Yet a human is not more prone to developing cancer than a mouse, creating a puzzle that is known as **Peto's paradox**. This logic suggests, in turn, that in response to the extreme peril created by our large bodies and long life spans, our bodies and tissues developed compensatory anti-neoplastic mechanisms. In principle, these heightened defenses—progressively erected as our mammalian ancestors evolved to increasingly larger sizes and longer life spans—depended on changes in our systemic physiology, such as defenses mounted by the immune system (as described in Chapters 15 and 16). In practice, however, many of these defenses were hardwired into our individual cells, rendering them far more resistant to transformation.

In the transformation experiments studies described here, these hardwired defenses might be reflected in the number of barriers placed to impede neoplastic cell transformation, each of these barriers being manifested by an additional requirement for transformation, that is, the need for an additional gene mutation. This logic, on its own, provides one rationale for why our cells have been evolved to be far more resistant to neoplastic transformation than have those of mice. Additional support for this thinking comes from the data presented in Figure 11.28B, which shows the positive correlation between the size/life span of mammalian species and number of introduced oncogenes required to transform their cells. Even more dramatic correlations come from measurements of constitutive TERT (telomerase) expression levels in normal somatic cells as they relate to the size of various rodents (Supplementary Sidebar 11.10). In this case the correlations are made entirely within one mammalian order and show inverse correlation—reflecting an apparent strategy of limiting telomerase activity as a way of constraining the outgrowth of neoplastic cells.

The measurement of telomerase levels reveals an anti-neoplastic mechanism that operates quite differently from those depending on requirements for multiple genetic mutations as illustrated in Figure 11.28. In fact, there are likely to be dozens of other still-undiscovered anti-neoplastic mechanisms, whose existence can be deduced directly from the history of mammalian evolution: at least 10 mammalian orders, all descended from a ~25 gram common ancestor, independently invented large body sizes over the past 100 million years. Each of these developments must have required the compensatory co-evolution of highly effective defenses against cancer pathogenesis, many reflected in the hard-wiring of individual cells. This historical zoology yields insights that are of more than passing interest to experimental cancer researchers, simply because they dictate that attempts to develop many animal models of human cancer pathogenesis will often be hobbled because the cells of other mammals are organized differently from our own, at least in the molecular circuitry governing cell proliferation and survival.

Actually, independent of the intrinsic limitations of the cells of other mammalian species, there are cells of many humans that seem to break all of the rules of transformation that were described in Section 11.9. More specifically, the cells of human embryos and fetuses seem to be organized much differently from those present in adult tissues. Thus, a variety of pediatric cancers occur so early in life—some even arising prenatally—that it is difficult to imagine how the cells in these tumors could have had sufficient time to accumulate the cohort of mutations (and epigenetic changes) that seem to be required for the formation of adult malignancies.

This notion, really a speculation, is corroborated by genomic sequencing the genomes of a variety of pediatric tumors. As an example, a report in 2012 described the discovery of dozens to hundreds of mutations in the genomes of individual adult tumors, while the same research team had difficulty finding more than a single recurring mutation in early childhood rhabdoid tumors, similar to the unusually simple mutational spectrum present in the genomes of another important pediatric tumor type—retinoblastoma (see Chapter 8). This suggests the possibility that some pediatric cancers arise directly from certain embryonic cell types, and that such embryonic cells may be more readily transformed (that is, through far fewer alterations) than the cells that serve in various adult tissues as the precursors of adult tumors. In fact, it is hard to imagine how or why this could not be the case.

The extreme case of this simplified genetic route to transformation derives from study of embryonic stem (ES) cells, which can be extracted from very early embryos. ES cells are, by all measurement, genetically wild type, yet are tumorigenic (yielding teratomas) when implanted in syngeneic hosts. They seem to be the only example of genetically wild-type cells that are tumorigenic. This suggests that, in order to become tumorigenic, certain cells within later embryos require a number of genetic changes that are *intermediate* between the number needed by ES cells (zero) and the number required by adult human cells (as an example, five). Here we may be encountering a replay of our ancient, billion-year-old evolutionary history that led single-cell protozoa to evolve into the first metazoa and then into ourselves (Supplementary Sidebar 11.11).

11.11 Nonmutagenic agents, including those favoring cell proliferation, make important contributions to tumorigenesis

The clinical observations and experimental results that we have read about in this chapter provide us with a crude picture of the genetic and epigenetic changes needed to form a human cancer cell. They fail, however, to reveal precisely how these changes are actually acquired during the course of tumor progression. So now, we turn to these issues—the processes occurring *in vivo* that enable cells to accumulate the multiple alterations needed for tumor formation.

As detailed above, a succession of somatic mutations provides much of the force that drives forward tumor progression. Because many of these changes are caused by the actions of mutagens, this implies that cancer progression is fueled largely and perhaps entirely by the genetic hits inflicted by mutagenic carcinogens. This notion was tested by the development and use of the Ames test and led to the conclusion, illustrated in Figure 2.21, that mutagens are carcinogens, and quite possibly, that carcinogens are mutagens.

In fact, attempts to use the Ames test to search for carcinogenic agents that play significant roles in human cancer causation, doing so through their mutagenicity, have repeatedly failed over the past half century. With the exception of tobacco smoke, ultraviolet radiation, aflatoxin-B1, and ethanol (which becomes mutagenic following its rapid conversion in the liver into acetaldehyde), almost all of the myriad other types of human cancer cannot be associated with prior exposure to specific mutagenic agents (Sidebar 11.10).

Sidebar 11.10 The misleading lessons of the Ames test The profound differences in cancer incidence in different human populations (see Table 2.5) provoked extensive searches, many employing the Ames test, to find causative agents in the environment or lifestyle practices. Although Bruce Ames' research that carcinogens act through their mutagenic powers clearly stimulated scientific thinking in the 1970s about the origins of cancer, this thinking has proven to be simplistic in trying to understand human cancer causation. Thus, the failures to demonstrate mutagenic agents that function as significant causes of common human cancers have provided one indication that human cancer formation is driven by mechanisms other than mutagenic agents experienced during our lifetimes.

Perhaps the most graphic demonstration of this discordance between carcinogens and mutagens has come from an extensive series of experiments in which 20 environmental chemicals known or suspected to be human carcinogens were tested for their carcinogenicity in laboratory mice. All of these chemicals were indeed found to be capable of inducing tumors in mice. However, detailed sequence analyses of the genomes of 188 resulting mouse tumors revealed that 3 chemicals caused significantly elevated mutational burden and were therefore clearly mutagenic, whereas the genomes of tumors induced by the remaining 17 carcinogens gave no indication of increased rates of mutation. Hence, 17 of the tested agents are highly likely to have facilitated tumor development through nongenotoxic mechanisms, as described below. This has left us with the generalization that most, if not all, mutagens can be carcinogenic, but the converse is not true: many—possibly most—human carcinogens are not mutagenic.

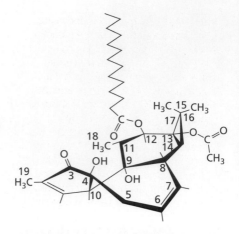

Figure 11.29 TPA, an important promoter of skin tumorigenesis
The stereochemical structure of 12-O-tetradecanoylphorbol-13-acetate (TPA), also known as phorbol-12-myristate-13-acetate (PMA), is shown here. TPA is extracted from the seeds of the croton plant, specifically *Croton tiglium*. Its main target in cells is protein kinase C-α (PKC-α), which it activates.

Indications of the importance of nongenotoxic carcinogens first came from attempts in the early 1940s to develop effective experimental methods for inducing skin cancers in mice. The experimental model used in this research depended on exposing mouse skin to highly carcinogenic tar constituents, such as benzo[*a*]pyrene (BP), 7,12-dimethylbenz[*a*]anthracene (DMBA), or 3-methylcholanthrene (3-MC; see Supplementary Sidebar 2.7), these being coal tar constituents that were already known or suspected to be highly carcinogenic. For example, mice subjected to daily painting of DMBA on a patch of skin would eventually develop skin carcinomas after several months of this treatment. These echoed decades earlier experiments showing that repeated painting of rabbit ears yielded the same outcome.

But another experimental protocol proved to be even more revealing. Following a single painting with an agent like DMBA, the same patch of skin could be treated on a weekly basis with a second agent, termed TPA (12-O-tetradecanoylphorbol-13-acetate; **Figure 11.29**), a potent skin irritant prepared from the seeds of the croton plant. (Another often-used term for TPA is PMA, for phorbol-12-myristate-13-acetate.) Repeated painting of a DMBA-exposed skin area with TPA resulted in the appearance of papillomas after 4 to 8 weeks, depending on the strain of mice being used (**Figure 11.30**). (These papillomas are in many ways analogous to the adenomas observed in early-stage colon cancer progression; see Figure 11.6.)

At first, the survival and growth of these skin papillomas depended upon continued TPA paintings, because cessation of TPA treatments caused the papillomas to regress (see Figure 11.30E). However, if TPA painting was continued for many weeks, TPA-independent papillomas eventually emerged, which would not regress after cessation of TPA painting and instead persisted for extended periods of time (see Figure 11.30F). Some of these TPA-independent papillomas might, with low probability, eventually evolve further into malignant squamous cell carcinomas of the skin after about 6 months.

Importantly, in the absence of prior DMBA treatment, repeated painting with TPA alone failed to provoke either papillomas or carcinomas (see Figure 11.30B). Even more interesting, an area of skin could be treated once with DMBA and then left to rest for a year, giving no outward indication of having been treated. If this particular patch of skin was then treated with a series of TPA paintings (as in Figure 11.30C), it would "remember" that it had been exposed previously to DMBA and respond by forming a papilloma.

These phenomena could be rationalized as follows (**Figure 11.31**). A single treatment by an **initiating agent** (or **initiator**) like DMBA left a stable, long-lived mark on a cell or cluster of cells; this mark was apparently some type of genetic alteration. Subsequent repeated exposures of these "initiated" cells to TPA (termed the **promoting agent** or simply the **promoter**) allowed these cells to proliferate vigorously while having no comparable effect on nearby cells that had not experienced prior DMBA painting and were therefore considered to be uninitiated. (Note that use of the word "promoter" in this context is unconnected with its other meaning—namely, the DNA sequences controlling the transcription of a gene.) The localized proliferation of initiated cells that was encouraged by the promoter would eventually produce a papilloma. However, as mentioned above, if TPA painting was halted after several months, many papillomas would disappear, indicating that the effects of the promoter were reversible and therefore likely to impose a *nongenetic* effect on the cells in the papillomas. Clearly, this nongenetic effect, whatever its nature, could collaborate with the apparent genetic alteration previously created by the initiator to drive the proliferation of cells, causing them to form a papilloma.

As we read above, if the initiated cells were treated with the TPA promoter for many months' time, eventually some papillomas would evolve to become TPA-independent; in this case, even after TPA withdrawal, the papillomas continued to increase in size and some eventually developed into skin carcinomas. This permanent change in cell behavior seemed to reflect, once again, the actions of a second, independent, cell-heritable alteration, that is, an apparent second mutation. Indeed, this evolution to a carcinomatous state could be strongly accelerated by treating a papilloma

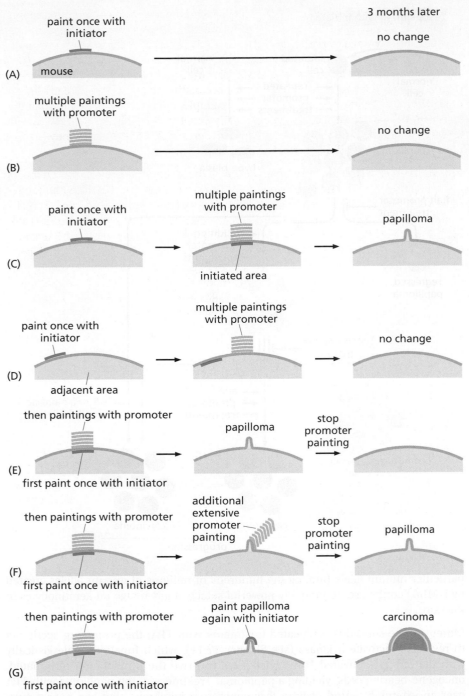

Figure 11.30 Protocols for inducing skin carcinomas in mice The induction of skin carcinomas by painting carcinogens on the backs of mice requires certain combinations of treatments with initiators and promoters. (A) A single treatment *(blue line)* with an initiating carcinogen, such as DMBA (dimethylbenz[a] anthracene), leads to no skin carcinomas 3 months later. (B) Multiple treatments *(orange lines)* with promoting agents, such as TPA (see Figure 11.29), also do not lead to significant numbers of tumors. (C) If an area of skin is painted once with an initiating agent followed by repeated paintings with a promoting agent, a papilloma will often appear several months later. If several months intervene between treatment with the initiating agent and subsequent promoter treatment, papillomas will also appear, indicating that the initiating agent permanently marked an area of the skin. (D) If an area of skin is painted once with an initiating agent and a promoting agent is then used to repeatedly paint another nearby but non-overlapping patch of skin, no papillomas will be seen 3 months later. (E) In a variation of the protocol depicted in panel C, an initiator such as DMBA is applied followed by repeated TPA treatments, which lead to papillomas. However, if repeated painting with the TPA tumor-promoting agent is halted soon after the papillomas appear, they regress, indicating that they are dependent on ongoing promoter stimulation. (F) In another variation of the protocol depicted in panel C, TPA promoter painting is continued for several months after papillomas first appear before being halted. Under these conditions, some of the papillomas will persist, indicating that they have become independent of continued promoter stimulation. (G) If a papilloma is produced by the protocols of panels C or F and the papilloma is then treated with an initiating agent, a carcinoma may appear, even in the absence of further promoter treatment.

with a second dose of the initiating agent, already suspected to be a mutagen (see Figure 11.30G). This third step in tumorigenesis (coming after initiation and promotion) is termed **progression**; this term is used more generally, throughout this chapter and elsewhere, to indicate the stepwise evolution of cells to an increasingly malignant growth state.

Four decades after the mouse skin cancer induction protocol was first developed, the identities of the genes and proteins that are the main actors in this skin tumorigenesis model were discovered (**Figure 11.32**). As long suspected, the DMBA used as the initiating agent is indeed a potent mutagen in the context of skin carcinogenesis. Because it is a randomly acting mutagen, DMBA creates an almost-limitless number of mutations throughout the genomes of exposed cells. However, the skin tumors that emerge invariably bear point-mutated H-*ras* oncogenes (see Section 4.3), indicating that this

Figure 11.31 Scheme of initiation and promotion of epidermal carcinomas in mice The observations of Figure 11.30 can be rationalized as depicted here. The initiating agent converts a normal cell (*gray, top left*) into a mutant, initiated cell (*blue*). Repeated treatment of the initiated cell with the TPA promoter generates a papilloma (*cluster of blue cells*); among these papilloma cells may be ones with a heightened proclivity to progress further (*orange cell*). TPA treatment of normal, adjacent cells (*gray, top right*) may create localized hyperplasia, but papillomas are not formed. Further treatment of the initially formed papilloma can be halted (*middle left*), in which case most papillomas regress (*not shown*) but a subset may persist as promoter-independent papillomas (*bottom left*). Further repeated treatment of this more progressed papilloma with TPA eventually yields, with low frequency, a carcinoma (*bottom middle, red cells*). Alternatively, exposure of the initially formed papilloma to a second treatment by the initiating agent yields, with much higher frequency, cells that form a carcinoma (*bottom right, red cells*).

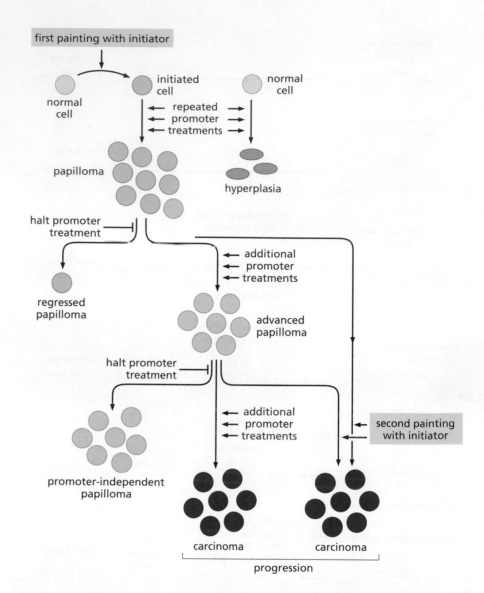

particular mutant allele (among the hundreds of millions of mutant alleles created by DMBA) confers some uniquely powerful selective advantage on keratinocytes in the skin.

Moreover, the subsequent repeated treatments with TPA, the promoting agent, act to reversibly activate a kinase (see Sidebar 11.11), which functions synergistically with the already-activated H-*ras* oncogenes to drive the proliferation of initiated, oncogene-bearing cells, yielding a papilloma. Treatment of the papilloma with TPA over an extended period of time may generate a papilloma that can persist even after TPA treatment is halted. Alternatively, if cells in a papilloma are exposed, once again, to an initiating agent like DMBA, this papilloma may progress into a carcinoma, whose cells now carry, in addition to the H-*ras* oncogene, a mutant *p53* gene. Accordingly, the eventual emergence of a promoter-independent, autonomously growing carcinoma depends on the actions of a *second mutational event* that occurred either spontaneously or through the actions of an experimentally applied mutagen (**Sidebar 11.11**).

Actually, the processes depicted in Figure 11.30 and Figure 11.31 reveal that tumor promoters can act in two distinct mechanistic ways. They may drive the preferential proliferation and clonal expansion of already-initiated, mutant cells, making it possible that—sooner or later—low probability events, such as random mutations of certain genes, may strike the large cell populations of cells produced by this initial clonal expansion. (Because these random events occur with a low probability per cell generation, a large, expanded cell population is required before such a secondary

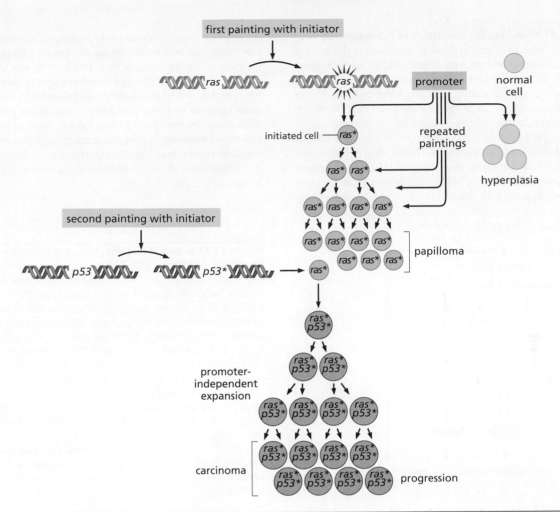

Figure 11.32 Genes and proteins involved in mouse skin carcinogenesis The phenomena of initiation and promotion (see Figures 11.30 and 11.31) can be understood at the genetic level in the manner illustrated here. The initiating agent acts as a mutagen to convert a *ras* proto-oncogene into an active *ras* oncogene (*top left*). This initiation, on its own, has little apparent effect on the behavior of the keratinocyte bearing this mutant allele. However, in the presence of repeated stimulation by a promoting agent (*top right*), the *ras*-bearing cell is induced to pass through repeated growth-and-division cycles, leading to the formation of a papilloma (*blue cells*). Conversely, a cell lacking a *ras* oncogene (*gray, top right*) is not stimulated by the promoting agent and may only generate a localized hyperplasia in response to repeated exposure to this agent. If the papilloma is exposed a second time to a mutagenic initiating agent (*left*), a second genetic lesion, often involving the mutation of the *p53* tumor suppressor gene, is created. This mutant *p53* allele (*p53**) collaborates with the *ras* oncogene to create a population of cells (*red-orange*) that no longer depend on the promoter and are capable of forming a carcinoma. Not shown here is the amplification of the *ras* oncogene, which also occurs as tumor progression proceeds.

mutational event is likely to occur in one or another cell.) Ostensibly, these later mutational events enable descendant cells to acquire a critical second mutant allele that *collaborates* with the initial, primary mutation created by the initiating agent; more specifically, this new mutation can collaborate with the initially sustained mutation to drive neoplastic proliferation, yielding tumor outgrowth.

Independent of this function of driving preferential clonal expansion of already-mutated cells, some promoters can act in an alternative fashion to favor tumorigenesis simply by driving endless rounds of cell proliferation, doing so even in cell populations that lack previously acquired, oncogenic initiation events. Because passage through a cell cycle, on its own, results in mutations at a low but measurable frequency per cell generation (as described in the next chapter), this mitogenic effect is itself indirectly mutagenic and thus carcinogenic.

Finally, it is essential for us to integrate the phenomenon of tumor promotion with the scheme depicted in Figure 11.14. More specifically, it is clear that tumor promoters

Sidebar 11.11 Signaling biochemistry of initiation and promotion If we were to describe these phenomena at the level of signal-transduction biochemistry (**Figure 11.33**), we would say that the TPA promoter functions as a potent stimulator of cell proliferation through its ability to activate the cellular serine/threonine kinase known as protein kinase C-α (PKC-α), which we encountered briefly in the context of cytoplasmic signal transduction (see Section 6.6). More specifically, TPA acts as a functional mimic of diacylglycerol (DAG), the molecule that cells generate endogenously (see Figure 6.11B) as a means of activating protein kinase C (PKC). The downstream effectors of PKC-α collaborate with the H-*ras* oncogene, in still-poorly

understood ways, to drive proliferation of initiated keratinocytes whose descendants form papillomas and, on rare occasion, progress to carcinomas. In fact, TPA is only one example of a diverse spectrum of tumor-promoting agents to be described here. In the case of mouse skin carcinogenesis, these observations leave us with the conclusion that tumor promoters like TPA, which do not directly alter the genomes of cells, are nevertheless important in propelling multi-step tumorigenesis forward. Moreover, they impress upon us the notion that an agent that contributes to cancer formation—a carcinogen—need not be a mutagen.

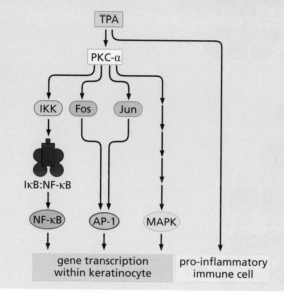

Figure 11.33 Activation of Fos, Jun, and NF-κB by PKC-α The precise mechanism by which an activated *ras* oncogene acts synergistically with TPA-activated protein kinase C-α (PKC-α) to drive the proliferation of keratinocytes is poorly understood. It is clear that once PKC-α is activated by binding TPA, it is able to stimulate transcription via several distinct signaling pathways, including those involving the NF-κB and AP-1 transcription factors and the ERK/MAPK enzyme. This depiction of intracellular, cell-intrinsic signaling within a keratinocyte does not address what are likely to be even more important effects of TPA and similar promoters on the tissue microenvironment around the keratinocyte, notably the cellular components of inflammatory immune cells that congregate in areas of promoter-exposed skin. These various promoter-induced responses, then taken together, have not yet yielded a coherent explanation of why TPA stimulates the proliferation of *ras* oncogene-bearing, initiated keratinocyte, while having minimal effect on nearby wild-type keratinocytes.

can increase the *rate of clonal expansion* of already-mutated cells; the resulting expansion rate may be the critical, rate-limiting determinant that governs the time intervals between successive clonal expansions. Consequently, in the context of human tumor pathogenesis, the incidence of various tumors may be determined by whether or not such clonal expansions occur rapidly enough to generate full-blown tumors within an average human life span. Recent work shows that the accumulation of random mutations occurs at similar rates in various tissues that nevertheless spawn cancers at vastly different frequencies; this suggests, once again, that nongenetic processes often represent rate-limiting determinants of tumor formation, doing so by governing the rapidity of tumor formation and thus, during a lifetime, the incidence of cancer in various human tissues.

These realizations forced a reinterpretation of how diet and lifestyle factors (found by epidemiologists to be important determinants of human cancer incidence) actually work. Rather than exposing the body's tissues to potent mutagens, many if not most of these factors, seem to act through their ability to drive forward certain clonal expansions that elicit nonmutagenic, tumor-promoting mechanisms.

11.12 Mitogenic agents, key governors of human cancer incidence, can act as tumor promoters

The experimental model of mouse skin carcinogenesis, as described above in some detail, is not used today by many researchers to study the dynamics of tumorigenesis. Nevertheless, at the conceptual level, it proved highly useful decades ago in revealing the role of nongenotoxic carcinogens in fostering carcinoma progression, and more specifically for its powers in illustrating the principles of tumor initiation, promotion, and progression. This scheme was a revelation to many, in that it demonstrated that

Sidebar 11.12 Limitations of GEMMs In Sections 7.10 and 11.8 we read about the powers of transgenic mouse models of human cancer pathogenesis (often termed genetically engineered mouse models or GEMMs) to shed light on how tumors may develop in the human body. At the same time, the discussions in the present chapter reveal certain profound limitations of these models that are not readily addressed by experimentally introduced modifications of the germ-line genomes of these laboratory mammals. One obstacle to successful modeling comes from the observations, as described in Section 11.10, that far fewer introduced genes are required to transform an adult mouse cell than to transform a corresponding human cell. This indicates, on its own, that the wiring of the intracellular signaling circuitry of mice is organized in slightly different ways in mice than it is humans.

A second significant difficulty arises from the fact that certain critical determinants of cancer development are often not genotoxic processes, that is, processes creating mutant alleles, but instead nongenotoxic mechanisms, specifically those that serve to promote clonal expansions, such as those described here. In many cases, the nature of the tumor-promoting mechanisms that govern human cancer incidence remain elusive. And in many others, even when their identities are known, such mechanisms cannot be easily recapitulated in mice because of the complex differences between mouse and human physiology. For example, if chronic inflammation within a tissue represents an important tumor-promoting mechanism, it is challenging to re-create an inflamed tissue microenvironment in a mouse that accurately echoes the corresponding state in a human tissue. These considerations represent significant limitations to GEMMs that may or may not be addressed by future research.

cancer formation cannot be attributed simply to mutagens that either enter the body from outside or are generated by endogenous metabolic processes that damage the genomes of human cells (as described in Chapter 12). Still, the experimental model of mouse skin carcinogenesis tells us almost nothing about how analogous tumor-promoting mechanisms actually operate within the human body to create cancer.

In fact, a diverse array of biochemical and biological mechanisms appear to be responsible for the tumor promotion that leads to human cancers, complicating attempts at experimental modeling in laboratory mice. (**Sidebar 11.12**). As we learned in Chapter 2, the incidence rates of various types of human cancer vary dramatically between different human populations globally, indicating that exposure to carcinogenic agents experienced in different environments exerts enormous influence on the development of various types of cancer (see Table 2.5). However, as mentioned above, searches for environmental carcinogens that are mutagenic, including agents present in the human food chain, have been most frustrating, and researchers attempting these identifications have, for the most part, emerged empty-handed. Hence, the variability in cancer incidence (and resulting mortality) must be traced, in large part, to nonmutagenic carcinogens. (The role of nonmutagenic factors is emphasized by the fact that, with the exception of cigarette smoke and UV radiation, most lifestyle factors identified in Figure 2.19 as important factors in cancer causation have no obvious mutagenic function.)

Among known human tumor promoters, arguably the most intensively studied are those involved in breast cancer pathogenesis. Of central importance are the steroid hormones—estrogen, progesterone, and testosterone. In the female body, for example, estrogen and progesterone are involved in programming the proliferation of cells in reproductive tissues. The monthly menstrual cycles of women between **menarche** and menopause result in the proliferation and then regression of the cells forming the epithelia of the ducts in the mammary gland. The endometrial lining of the uterus undergoes similar cycles of proliferation and regression.

Epidemiology makes it clear that the more menstrual cycles a woman experiences in a lifetime, the proportionately higher is her risk of developing breast cancer. By one estimate, lifetime breast cancer risk decreases by 20% for each year that menarche is delayed during adolescence. [The most compelling illustration of the importance of the timing of menarche has come from studies of identical (**monozygotic**) twins, both of whom eventually developed breast cancer; the twin whose menarche occurred earlier had a 5.4-fold greater risk of being the first to be diagnosed with this disease.] Women who stop menstruating before age 45 have only about one-half the risk of breast cancer of those who continue to menstruate to age 55 or beyond.

Removal of the ovaries—the primary source of estrogen in the female body—causes breast cancer risk to plummet. A dramatic example comes from the results of a Dutch study, which has shown that women who enter menopause before the age of 36 due to the side effects of chemotherapy for Hodgkin's lymphoma have a 90% decreased risk of subsequently developing breast cancer. Conversely, postmenopausal women who contract breast cancer have on average a 15% higher level of circulating estrogen than unaffected women.

The effects of estrogen on breast cancer are surely complex, and it appears that this hormone acts on other cells in the mammary gland besides the epithelial cells. Still, it is evident that estrogen, and perhaps other hormones such as progesterone and even prolactin, periodically induce cell proliferation in a way that ultimately enables the progression of initiated mammary epithelial cells (MECs) into the carcinoma cells found in the various subtypes of breast cancers. (Some have argued that metabolites of estrogen are mutagenic, and that these metabolites contribute to breast cancer development; if so, estrogen's mutagenic effects on breast cancer development are surely dwarfed by its power to promote the proliferation of MECs.)

11.13 Chronic inflammation often serves to promote tumor progression in mice and humans

Tumor promoters work in diverse ways to expedite multi-step cancer progression. They share the ability to drive the clonal expansion of initiated cells, positioning these cells to acquire additional cell-heritable changes (including mutations and shifts in epigenetic state). Their diversity derives from the various biochemical and cellular mechanisms of action that enable them to expedite tumor formation. The steroid hormones cited above may act largely as pure mitogens directly driving the proliferation of potentially neoplastic epithelial cells. More common, however, are tumor-promoting mechanisms that do not impinge directly on the progenitors of future neoplastic cells, but instead perturb the tissue microenvironment surrounding such cells.

This other class of tumor-promoting mechanisms often acts by creating various forms of inflammation, thereby assembling a supportive tissue microenvironment for the clonal expansion of initiated tumor cells. And in contrast to the steroid hormones described above, the mechanisms-of-action of these inflammatory states do indeed draw on the lessons taught by the mouse skin carcinogenesis protocols in Section 11.11. More specifically, the TPA promoter that was employed experimentally in mouse skin carcinogenesis was initially chosen because it is a potent irritant of mouse skin and thus an inducer of localized inflammation, including the recruitment of large numbers of neutrophils into promoter-treated skin. Moreover, experimental strategies that inhibit inflammation also block or significantly reduce TPA's tumor-promoting powers in the skin. Some forms of inflammation may operate chronically over extended periods of time, whereas others, being integral components of wound-healing programs, may act only transiently as tissues undertake to heal themselves following various types of damage. (Thus, wounding of the skin is at least as effective as TPA painting in promoting papilloma formation.)

Various types of experiments can be cited to reveal the key role inflammation plays in tumor promotion. As one illustrative example, when cells of a human colonic adenoma cell line were implanted subcutaneously into immunocompromised Nude mice, they were found to be non-tumorigenic. However, when they were introduced into these mice together with an attached fragment of plastic, localized stromal inflammation was induced by the plastic and the adenoma cells grew out to eventually form tumors. The tumorigenic phenotype of these cells persisted even after they were subsequently transplanted (without the plastic) to another host animal, indicating that their neoplastic proliferation was now driven by certain cell-heritable genetic or epigenetic alterations or both. Yet other experimental strategies can illustrate this point, as described in Supplementary Sidebar 11.12.

Chronic localized inflammation plays a clear role in the pathogenesis of a diverse array of human carcinomas (**Table 11.3**). For example, those arising in the gallbladder are usually associated with a decades-long history of gallstones and resulting inflammation

Table 11.3 Inflammatory conditions and tumor development

Human tumor	Inflammatory condition or inflammation-provoking agent
bladder carcinoma	schistosomiasis, chronic cystitis
gastric carcinoma	*H. pylori*–induced gastritis
hepatocellular carcinoma	hepatitis B/C virus
bronchial carcinoma	silica, cigarette smoke
mesothelioma	asbestos
ovarian carcinoma	endometriosis
colorectal carcinoma	inflammatory bowel disease
esophageal carcinoma	chronic acid reflux
papillary thyroid carcinoma	thyroiditis
prostate carcinoma	prostatitis
lung carcinoma	chronic bronchitis
gallbladder carcinoma	chronic cholecystitis

Adapted from F. Balkwill, K.A. Charles and A. Mantovani, *Cancer Cell* 7:211–217, 2005.

of the epithelial lining of the gallbladder (**Figure 11.34A**). Similarly, hepatocellular carcinomas (HCCs), which are common in East Asia, are associated with chronic hepatitis B virus (HBV) infections and accompanying inflammation of the liver (Figure 11.34B, left panel). In many infected individuals, HBV infection is well established in the liver early in life and continues in a chronically active form for decades. The resulting hepatitis may have relatively few outward effects on the individual, because the ongoing HBV-induced killing of hepatocytes (the cells forming the bulk of the liver) is compensated by an equal proliferation of surviving cells. Nevertheless, these chronic infections yield chronic localized inflammation incurred as the immune systems of the infected individuals attempt unsuccessfully to eliminate the virus, and the dying and dead cells killed by the virus release molecules that provoke further inflammation.

Recent research has made it clear that chronic hepatitis C virus (HCV) infections act in a similar way to increase liver cancer rates. Although the two viruses (HBV

Figure 11.34 Chronic inflammation leading to cancer (A) A graphic demonstration of chronic inflammation leading to cancer is provided by carcinomas of the epithelial lining of the gallbladder (*white mass, above*), which are commonly associated with the formation of gallstones arising as precipitates from the bile (*brown masses, below*). (B) Chronic hepatitis B virus (HBV) infection creates continual cell death of hepatocytes together with chronic inflammation, as indicated by the numerous lymphocytes (*small dense nuclei, left panel*). Over a period of decades, these processes can lead to an almost 100-fold increased risk of liver cancer. The inflammation in the liver of someone suffering from chronic hepatitis C virus (HCV) infection is strikingly similar (*right panel*). The inflammatory cells are largely in the right part of this micrograph. These two inflammatory conditions are similar histologically and lead to comparably increased risks of hepatocellular carcinoma (HCC), suggesting that the inflammatory states, rather than some specific aspect of viral function, are responsible for the appearance of HCC in patients infected with either of these viruses. (A, from A.T. Skarin, Atlas of Diagnostic Oncology, 4th ed. Philadelphia: Elsevier Science Ltd., 2010. With permission from Elsevier. B, left, courtesy of A. Perez-Atayde; right, courtesy of A.K. Bhan.)

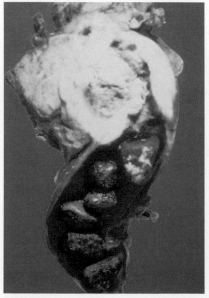

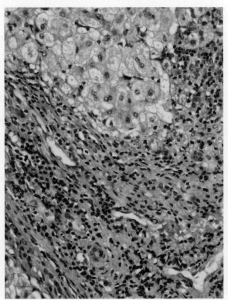

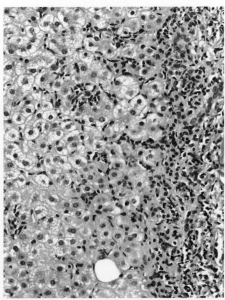

(A)   (B)   HBV   HCV

and HCV) are totally unrelated to one another with respect to genome structure and replication cycles, they evoke very similar biological outcomes through their shared ability to create chronic infections, cytotoxicity, and inflammation in the liver (see Figure 11.34B). (Significantly, a variety of other types of chronic liver injury, including that inflicted by alcoholism, are also associated with increased incidence of hepatocellular carcinoma, although the relative risks are vastly less than those associated with lifelong HBV infection.)

HBV, acting as a tumor promoter, can also function synergistically with aflatoxin-B1, a potently mutagenic compound that is made by *Aspergillus* fungi that proliferate on moldy peanuts, nuts, and corn stored under conditions of high humidity (see Figure 2.21). This combination of HBV infection plus aflatoxin exposure proves to be deadly. In one, relatively small prospective epidemiologic study carried out in Shanghai, infection with HBV increased the risk of hepatocellular carcinoma about 7-fold, whereas exposure to aflatoxin-B1–contaminated food yielded about a 3-fold increased risk. When an individual had experienced both agents, the risk of liver cancer increased about 60-fold. A less dramatic but nonetheless compelling parallel came from a 2021 study of the effects of cigarette smoking on those afflicted with inflammatory bowel disease: those who smoked had an almost 2-fold elevated risk of developing colorectal cancer, revealing once again the synergistic effects of initiators (tobacco-derived mutagens) and promoters (long-term inflammation). The parallels between the pathogenesis of these human carcinomas and the actions of initiators and promoters of mouse skin cancer are striking.

11.14 Inflammation-dependent tumor promotion operates through defined signaling pathways

Table 11.4 lists the diverse array of pro-inflammatory processes that favor tumor formation. The conclusions drawn from this table are reinforced by large-scale studies of populations of adults who took anti-inflammatory drugs, such as aspirin and other non-steroidal anti-inflammatory drugs (NSAIDs) daily over multi-year periods, usually five years. Their rates of cancer development were compared in each case with the rates of cancer development anticipated in the general population as controls. One such study reported such individuals were 46% less likely to develop pancreatic carcinomas, a second reported a ~30% reduced risk of colorectal carcinomas, and a third

Table 11.4 Links between inflammation and cancer pathogenesis

Many inflammatory conditions predispose to cancer
Cancers arise at sites of chronic inflammation
Functional polymorphisms of cytokine genes are associated with cancer susceptibility and severity
Distinct populations of inflammatory cells are detected in many cancers
Extent of tumor-associated macrophage infiltrate correlates with prognosis
Inflammatory cytokines are detected in many cancers; high levels are associated with poor prognosis
Chemokines are detected in many cancers; they are associated with inflammatory infiltrate and cell motility
Deletion of cytokines and chemokines protects against carcinogens, experimental metastases, and lymphoproliferative syndrome
Inflammatory cytokines are implicated in the action of nongenotoxic liver carcinogens
The inflammatory cytokine tumor necrosis factor is directly transforming *in vitro*
Long-term NSAID use decreases mortality from colorectal cancer

From F. Balkwill and A. Mantovani, *Lancet* 357:539–545, 2001. (With permission from Elsevier.)

Indeed, as these cells evolve progressively toward high-grade neoplasia, their internal biochemistry—involving both energy metabolism and the intermediary metabolism of biosynthesis—shifts in concert.

Here, we encounter a fundamental ambiguity that has dogged the field of cancer metabolism research since its inception. Do changes in metabolism represent a fundamental, intrinsic component of the neoplastic cell state that drives cancer cell behavior or, alternatively, is the altered metabolism of cancer cells simply one of the myriad downstream consequences of the neoplastic state? Given the pleiotropic consequences of cell transformation, are these well-documented metabolic shifts, as described below, *causes* or *consequences* of the growth state of cancer cells?

In fact, the peculiarity of tumor cell biochemistry was already appreciated a century ago: the energy metabolism of most cancer cells differs markedly from that of normal cells, as first reported in 1924 by Otto Warburg, who was later honored with a Nobel Prize for discovering the respiratory enzyme now known as cytochrome *c* oxidase. By then, it was already apparent that in the absence of adequate oxygen (**hypoxia**), mammalian cells would resort to glycolysis in contrast to the process that we now term oxidative phosphorylation, which operates during **normoxic** conditions. Glycolysis occurs in the cytosol, whereas oxidative phosphorylation is confined to the mitochondria.

As was more thoroughly documented in the decades that followed, normal cells that experience aerobic, normoxic conditions break down glucose into pyruvate in the cytosol through the process of glycolysis and then dispatch the pyruvate into mitochondria, where it is broken down further into carbon dioxide in the tricarboxylic acid (TCA) cycle (also known as the Krebs cycle and the citric acid cycle; **Figure 11.37**). Under anaerobic or **hypoxic** (low oxygen tension) conditions, however, normal cells are limited to using only glycolysis, generating pyruvate that is reduced to lactate, which is then secreted from cells. Warburg discovered that even when exposed to ample oxygen, many types of cancer cells appeared to rely largely on glycolysis, generating lactate as the breakdown product of glucose (Figure 11.37) and secreting it into the extracellular space.

The use by cancer cells of "**aerobic glycolysis**," as Warburg called it, would seem to make little sense energetically, because the breakdown of one molecule of glucose yields only two molecules of ATP through glycolysis. In contrast, when under aerobic conditions glycolysis is followed by oxidation of pyruvate in the citric acid cycle, which allows 33 to 36 ATPs to be harvested from the breakdown products of a single glucose molecule. Actually, most types of normal cells in the body have continuous access to O_2 conveyed by the blood and therefore metabolize glucose through this energetically far more efficient route. The tendency of cancer cells to apparently limit themselves largely to glycolysis, even when provided with adequate oxygen, stood out as exceedingly unusual behavior.

In the 1950s, Warburg proposed that this altered energy metabolism was the central driving force in the formation of cancer cells, a notion that was discredited in the decades that followed, as described repeatedly in the preceding chapters of this book. Nonetheless, the process of aerobic glycolysis that he uncovered has stood the test of time, being found to operate in a wide variety of human cancer cells. Indeed, it is now thought to represent one of the many characteristics directly associated with cell transformation.

This aerobic glycolysis, sometimes called the **Warburg effect**, remains a subject of much contention, as its rationale in cancer cell biology has never been fully resolved: Why do as many as 80% of cancer cells metabolize most of their glucose via glycolysis when completion of glucose degradation in mitochondria by the citric acid cycle would afford them vastly more ATP to fuel their own growth and proliferation? Is aerobic glycolysis required for maintenance of the cancer cell phenotype, or does it represent nothing more than a side effect of cell transformation that plays no causal role in cell transformation and tumor growth? At least one more of Warburg's claims was refuted in recent decades: he claimed that the Krebs/TCA cycle (which we now know

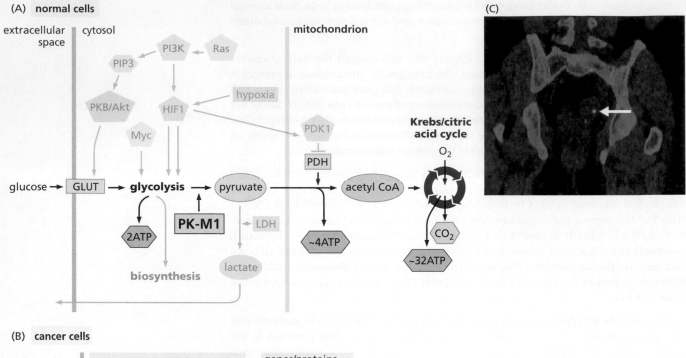

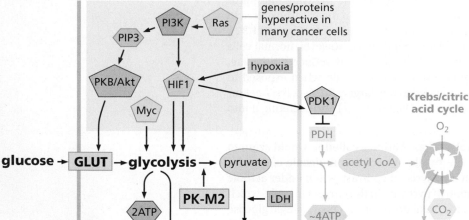

Figure 11.37 Deregulation of energy metabolism during the course of neoplastic transformation (A) In most normal nonproliferating cells having access to adequate oxygen, glucose is imported into the cells by glucose transporters (GLUTs) and then broken down by glycolysis and the citric acid cycle. During the last step of glycolysis, pyruvate kinase form M1 (PK-M1) ensures that its product, pyruvate, is imported into the mitochondria, where it is oxidized by pyruvate dehydrogenase (PDH) into acetyl CoA for processing in the citric acid cycle. Altogether, the mitochondria can generate as much as 36 ATP molecules per glucose molecule. (B) In cancer cells, including those with access to ample oxygen, the GLUT glucose transporters import large amounts of glucose into the cytosol, where it is processed by glycolysis. However, as the last step of glycolysis, pyruvate kinase M2 (PK-M2) causes its pyruvate product to be diverted to lactate dehydrogenase (LDH-A), yielding the lactate that is secreted in abundance by cancer cells. Because relatively little of the initially imported glucose is metabolized by the mitochondria, as few as 2 ATPs are generated per glucose molecule. Moreover, many of the intermediates generated during glycolysis are diverted toward biosynthetic uses. This mode of metabolic regulation resembles the metabolic state of normal, rapidly dividing cells, which also divert a significant portion of their glycolytic intermediates to biosynthetic pathways. Enzymes are in rectangles, glucose metabolites are in ovals, low–molecular-weight compounds are in hexagons, regulatory proteins are in pentagons. (C) 2-Deoxy-2-(^{18}F)fluoro-D-glucose positron-emission tomography (FGD-PET) makes it possible to visualize tumors in the body that have concentrated large amounts of glucose because of the hyperactivity of the GLUT1 transporter in the associated cancer cells. In the case shown here, FDG-PET revealed a small tumor (bright orange; arrow) in the region near an ovary of a woman who was under treatment for breast cancer but was otherwise without symptoms. X-ray-computed tomography (CT) was used at the same time to image the outlines of the tissues of this patient. This highly sensitive technology provided the first indication of an incipient ovarian cancer in this patient. (C, from R.A. Milam, M.R. Milam and R.B. Iyer, *J. Clin. Oncol.* 25:5657–5658, 2007. With permission from Wolters Kluwer.)

operates within the mitochondria) was essentially shut down in cancer cells experiencing normoxia, but this turned out not to be the case. Although glycolysis is clearly a dominant metabolic force in many types of cancer cells, the mitochondria and their Krebs/TCA cycles function quite well in most types of cancer cells.

One explanation of aerobic glycolysis comes from the observation that the cancer cells within a tumor often have inadequate access to oxygen, as we will discuss in detail in Chapter 13. The resulting hypoxic state limits cancer cells to glycolysis and thus to inefficient ATP production—just as normal cells would be limited under these conditions. Because of the Warburg effect, cancer cells would seem to be well adapted to this oxygen starvation, because glycolysis operates normally under hypoxic conditions. Still, this fails to explain why cancer cells, even when provided with abundant oxygen, do not take advantage of this oxygen to generate ATP in far larger quantities but instead continue to funnel much of their glucose through glycolysis without dispatching the end products of glycolysis into the mitochondria, as would occur in normal cells operating under normoxic conditions.

Many cancer cells also exhibit a second, associated abnormality: they import an enormous amount of glucose from the extracellular space. This behavior is observed in many types of cancer cells, including both carcinomas and hematopoietic tumors; they express greatly elevated levels of glucose transporters, particularly GLUT1, which span the plasma membrane and drive the high rates of glucose uptake by these cells. High levels of GLUT1 gene expression have been found to be driven by now-familiar oncogenes, notably *myc*, *ras*, and *src*, as well as the hypoxia-inducible HIF-1α transcription factor. Radiologists take advantage of this elevated glucose uptake by injecting into the circulation radiolabeled glucose [2-deoxy-2-(^{18}F)fluoro-D-glucose, FDG] and observing its rapid concentration in tumors through positron emission tomography (PET; see Figure 11.37). The import of large amounts of glucose by the GLUT1 and other transporter proteins offers cancer cells one clear biological advantage when they are residing within tumors. While glucose is present in abundance in tissue culture media, it is scarce in tissues throughout the body's normal tissues. This scarcity imposes a major constraint on the proliferation of many normal cell types that cancer cells can, however, overcome through their ongoing efficient importation of large amounts of this carbohydrate. Thus, many types of oncogene-transformed cell enjoy only a minimal growth advantage (relative to normal cells) *in vitro*; only when such transformed cells are growing *in vivo* as components of tumors does their growth advantage, specifically the advantage afforded by their expression and utilization of GLUT1, become apparent.

One compelling rationale for the use of aerobic glycolysis derives from the fact that glycolysis actually serves a second role independent of ATP generation: the intermediates in the glycolytic pathway function as precursors of many molecules involved in cell growth, including the biosynthesis of nucleotides and lipids. By blocking the last step of glycolysis, cancer cells ensure the accumulation of earlier intermediates via feedback reactions in this pathway. These glycolytic intermediates can then be diverted into critically important biosynthetic reactions, such as those leading to the formation of macromolecules and lipids whose high-level production is clearly central to successful rapid cell proliferation. This behavior contrasts with that of normal cells, which are generally not actively proliferating, do not require large-scale biosynthetic reactions, and depend largely on ATP to sustain their metabolic activity. (By some estimates, normal cells use more than 30% of their imported glucose to make ATP, whereas cancer cells use only ~1% of their glucose for this purpose—a striking contrast in metabolic organization.)

A complete rationale for why cancer cells use aerobic glycolysis is still not in hand. However, independent of how this question is resolved, there is yet another: how do cancer cells actually manage to avoid mitochondrial processing of glucose metabolites? Pyruvate kinase (PK) catalyzes the last step of glycolysis—the conversion of phosphoenolpyruvate to pyruvate. As noted above, this end product of glycolysis is normally destined for import into the mitochondria, where it is broken down in the citric acid cycle (see Figure 11.37). The M1 isoform of PK typically is expressed in most differentiated adult tissues, whereas the M2 isoform is expressed by early embryonic cells, rapidly growing normal cells, and cancer cells. For reasons that are still poorly understood mechanistically, the commonly expressed M1 isoform of PK ensures that its product, pyruvate, is dispatched from the cytosol directly into the mitochondria, whereas the M2 isoform that is expressed instead in cancer cells causes its pyruvate product to be diverted away from the mitochondria, reduced to lactate in the cytosol, and ultimately secreted. Of interest to cancer cell biology, the M2 isoform of PK that predominates in cancer cells can interact physically with proteins containing phosphotyrosine residues that are abundant in cancer cells (see Sections 5.2 to 5.6). This interaction causes a shutdown of the catalytic activity of this PK, resulting in turn in the accumulation of glycolytic intermediates that are then diverted into various biosynthetic pathways, the latter being termed **anabolic** reactions. Relative to the M1 form of PK, the M2 enzyme has a very slow **turnover number**, which contributes to the backup of glycolytic intermediates and their diversion into biosynthetic pathways. Also, the low catalytic activity of the M2 isoform of PK favors the formation of lactate rather than pyruvate, explaining the high levels of lactate secretion by cancer cells.

Experimental evidence indicates that the growth of tumors actually depends on the expression of the M2 form of PK and on the elevated expression of both the glucose importer GLUT1 and lactate dehydrogenase-A (LDH-A) enzyme, the latter being involved in reducing pyruvate to lactate, which is then secreted (see Figure 11.37B). When any one of these is inhibited, tumor growth slows down, sometimes dramatically. Observations like these provide the first indications that the unusual glucose metabolism of cancer cells creates a physiological state on which cancer cell growth and proliferation depend.

We should also note one other characteristic of cancer cell metabolism—the important role of glutamine. It serves as a critical source of nitrogen for various types of cells and a precursor of the nitrogenous bases that form both RNA and DNA. Moreover, once deamidated, the resulting glutamate molecules are then employed in the biosynthesis of various other amino acids. In fact, glutamine is considered a "nonessential amino acid," because normal cells are perfectly capable of synthesizing it on their own (rather than depending absolutely on the import of this amino acid from various nutrient sources). It turns out, however, that import of large amounts of glutamine from extracellular sources is critically important to the proliferation of normal cells and, even more so, neoplastic cells. This requirement for large amounts of imported glutamine is far more than might be predicted from the use of this amino acid in assembling proteins. This peculiarity already featured in the formulation of tissue culture media more than half a century ago, where high concentrations of glutamine in the media were found to be critical to the robust proliferation of both normal and neoplastic cells *in vitro*.

To conclude, it is important to emphasize that the expression of the Warburg-like, aerobic glycolysis is not an invention of neoplastic cells, as was thought at one point. Instead, it is very clear that highly proliferative normal cells also exhibit the complex metabolic shifts described above that have been associated with cancer cells. From the perspective of signal transduction biochemistry, this similarity makes sense, because exposure of normal cells to various mitogenic growth factors (as described in Chapter 5) triggers dramatic increases in the import of glucose via the GLUT1 transporter as well as a number of the changes in glycolysis that are also associated with cancer cell metabolism.

11.16 Synopsis and prospects

This chapter returns time and again to the theme that the development (that is, the pathogenesis) of a human tumor is a process involving a succession of distinct steps occurring one after the other, each making cells progressively more aggressive. The end product of this succession is a human cell that is altered in multiple distinct ways that together constitute the phenotype of a neoplastic cell. This phenotype is created by deregulation of multiple cellular control systems, including deregulation of cell proliferation, evasion of anti-proliferative signals, the ability to escape detection and elimination by the immune system, a heightened resistance to cell-death programs such as apoptosis, shifts in glucose metabolism, and an ability to multiply without any limits to the number of successive cell generations—the last trait representing the state of immortalization.

The complexity of this process is a direct reflection of the development of multiple anti-neoplastic mechanisms that have been hardwired into the circuitry of our cells and tissues. Many of these mechanisms were developed as our small, short-lived mammalian ancestors evolved progressively over the course of 100 million years into the long-lived, large organisms that we have become. In our own species, such evolution created organisms that experience 10^{16} cell divisions in a lifetime, each of which represents an opportunity for genetic and epigenetic disasters, notably mutations and disruption of normal differentiation programs. This greatly increased risk of biological chaos necessitated the co-evolution of anti-neoplastic mechanisms at both the cellular and organismic level, that is, the erection of multiple concentric lines of defense against cancer. These multiple barriers represent the foundation of the present chapter, because the completion of each of the steps of multi-step tumor pathogenesis reflects the successful breaching of one or another of these multiple anti-neoplastic barriers.

As argued here, each of these steps of multi-step progression involves the clonal expansion of the descendants of a cell that has acquired one or another advantageous phenotype, doing so as a direct consequence of a somatic mutation or a cell-heritable shift in epigenetic state. According to this model, each of these clonal expansions continues until one or another cell in the now-expanded population acquires yet another heritable shift in phenotype that renders this particular cell even more capable of surviving and proliferating than its neighbors in the expanded clonal cell population (see Figure 11.14).

Ideally, this scheme of a succession of multiple clonal expansions should yield a logical framework with which to understand the complexity of human cancer pathogenesis. However, this framework, as compelling as it might be, still does not allow us to truly understand the daunting complexity of human cancer pathogenesis. The scope of this problem is illustrated by the recently expanded powers of genomic sequencing, which have generated a bewildering array of germ-line mutations and somatically mutated human genes. The largest repository of this information, named COSMIC, reported in 2020 that mutant alleles of more than 1% of the genes in the human genome (and therefore more than 200 genes) have been implicated in various ways in cancer causation. More than 90% of these are involved via somatic mutations and 20% are borne as mutant cancer-predisposing genes in the germ line, with half of the latter being involved via both genetic mechanisms of cancer causation. Importantly, these genomic analyses do not even include the considerable contributions of epigenetic alterations to the acquired phenotypes of human cancer cells; recall that such nongenetic controls may even be responsible for display of the *majority* of the distinct phenotypes of cancer cells.

Given the possible combinatorial interactions of these various mutant genes (beyond calculation), the various distinct differentiated cell types involved (far more than 200), and the participation of a multitude of epigenetic programs in driving tumor progression forward (a number entirely unknown), how can we possibly distill these various factors into a simple, accessible, and compelling depiction of how human cancers are formed?

In fact, in this chapter we have come across two approaches to begin to address this seemingly intractable problem. In the first, the experimental transformation of human cells through various introduced genes suggests the operations of a limited number of intracellular signaling circuits that must be perturbed in order for a fully normal human cell to grow as a malignant cell; four to five such pathways appear to be critical, at least in the cells of human solid tumors (see Figure 11.28). An entirely different approach is to define instead a limited number of cell-biological *phenotypes* that are commonly shared by a diverse array of human cancer cell types; such phenotypes have been termed "cancer hallmarks" and help us to understand cancer cell biology in terms of cell physiology (Sidebar 11.13).

Sidebar 11.13 Cancer hallmarks—the distinguishable phenotypes of cancer cells Attempts at enumerating the cell-biological traits shared in common by a diverse array of human cancer cell types has yielded a relatively short list of cellular phenotypes, most having been long-recognized. These are: (1) a reduced dependence on exogenous mitogenic growth factors (see Chapters 5 and 6); (2) an acquired resistance to proliferation-inhibitory signals, such as those conveyed by TGF-β (see Chapter 8); (3) the ability to multiply indefinitely, that is, immortalized cell proliferation (see Chapter 10); (4) a reduced susceptibility to apoptosis (see Chapter 9); (5) the ability to generate new blood vessels—angiogenesis (see Chapter 13); (6) the acquisition of invasiveness and metastatic ability (see Chapter 14); and a seventh whose universality remains less well documented but increasingly seems to be observed among human solid tumors: (7) the ability to evade elimination by the immune system (see Chapters 15 and 16). In addition, (8) the role of aerobic glycolysis, as described in Section 11.15 increasingly seems intrinsic to the pathogenesis of many types of cancer.

Beyond these, two other processes have been cited that, although not intrinsic to the biology of cancer cells themselves, are nevertheless important for the *acquisition* of the eight hallmark phenotypes. Thus, our discussion of clonal evolution and succession concluded that in order for these events to occur at reasonable frequency, mutation rates within evolving preneoplastic cell populations must be abnormally high (as described in the next chapter). In the absence of such increased mutability, clonal populations of these cells may not have sufficient time to accumulate the multiple genetic alterations that are required in aggregate for cancer formation. Yet another enabling process featured prominently in this chapter: inflammation increasingly seems to play a role in creating tissue microenvironments that are favorable for tumor promotion, which accelerates the completion of individual steps multi-step tumorigenesis, allowing the process as a whole to reach completion in a human life span.

In principle, these two very different ways to approach cancer cell biology may one day be interconnected, specifically by assigning distinct cancer cell traits to "subcircuits" of the type that are depicted in **Figure 11.38**. If so, perhaps we will be able to portray tumor progression as the progressive deregulation of a number of distinct subcircuits within the cell. Nonetheless, it is clear that an overarching synthesis of these two analytical approaches is still elusive, indicating that the development of a unifying "theory of cancer" still lies in the distant future.

Among the multiple obstacles to progress is the realization that the Darwinian model of tumor progression is hard to validate in detail. We are hard-pressed, at least at present, in measuring the time intervals between clonal expansions. Some may occur far more rapidly than others and may be influenced by the biology of the individual in which these processes occur, including factors such as inborn genetic susceptibility and various aspects of lifestyle, obesity, and diet. Such time intervals are surely crucial to cancer formation because, by necessity, they govern the time-to-completion of the process as a whole and thereby determine whether or not it creates clinical disease in a human life span.

In Sections 11.13 and 11.14, we read that the great majority of human carcinogens are likely to drive cancer formation through their ability to *accelerate the rates* of multi-step tumor progression rather than through direct effects on the genomes of cells in various target organs. These accelerating mechanisms were portrayed here as acting via their powers to act as non-genotoxic tumor promoters rather than the initiators defined by the mouse skin carcinogenesis model (see Figure 11.30). Indeed, many types of human tumors seem to arise entirely without the contributions of identifiable mutagens originating outside the human body. In such tumors, the genetic damage that is essential to multi-step tumor pathogenesis would, by necessity, be generated entirely by endogenous processes, a topic that we will visit in great depth in the next chapter. Stated differently, in most human tissues, the numbers of endogenously generated mutations (that is, those created entirely by mechanisms operating entirely within an affected cell) are likely to dwarf those of exogenous origin, (that is, those created by genotoxic substances that enter into the human body and its cells from outside).

As described in this chapter, some of the clearly validated tumor-promoting mechanisms operate in a purely mitogenic fashion; the actions of steroid hormones come to mind. Far more numerous and likely influential, however, are the tumor-promoting mechanisms that derive from various types of localized tissue inflammation. Thus, chronic or "smoldering" inflammation in the stromal microenvironment of a tissue seems to provide future neoplastic cell populations with the mitogenic signals needed to drive their clonal expansion.

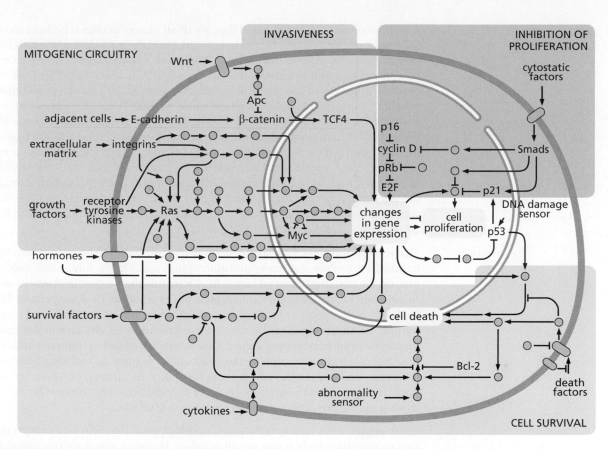

Figure 11.38 The intracellular signaling circuitry and collaboration between cancer-associated genes The design of the signal-transduction circuitry of mammalian cells has been uncovered, piece by piece, over the past four decades. The diagram here indicates only a portion of the proteins (*green circles*) that play critical roles in modulating the flow of signals through the various circuits operating within our cells. As indicated here by the various shadings, growth-promoting, mitogenic circuit (*light red*), the circuit governing growth-inhibitory signals (*light brown*), the circuit governing apoptosis (*light green*), and the circuit governing invasiveness and metastasis (*light blue*) can be assigned to distinct regions in the map of the master circuitry of the mammalian cell. (Note that the circuit governing mitogenesis/proliferation overlaps in part with that governing cancer cell invasiveness, indicating that a common set of proteins mediates both biological responses.) In principle, each of the distinct stages of multi-step tumor progression and each driver gene mutation can be mapped to one of these subcircuits. In practice, this remains challenging because of numerous cross-connections between individual subcircuits, our still-incomplete enumeration of functionally critical signaling proteins, and our fragmentary understanding of how individual signaling proteins operate and interact with one another. Moreover, additional subcircuits still need to be mapped out. By implication, all of these subcircuits need to be perturbed or deregulated in order for a normal human cell to become transformed into a tumorigenic derivative. (Adapted from D. Hanahan and R.A. Weinberg, *Cell* 100:57–70, 2000. With permission from Elsevier.)

In the case of human tumors, the apparently central roles of tumor promoting mechanisms have left us with two major quandaries. First, we do not have assays for detecting human tumor promoters, which may be present in low, limiting amounts in our tissues. Second, we are unable to quantify the concentrations of these agents, which represent critical numbers in judging their contributions to cancer pathogenesis and the importance of eliminating them from our environment. This depiction of the key roles of tumor promotion and localized inflammation within a tissue ultimately bears directly on the agendas of public health campaigns organized to reduce cancer incidence (and resulting cancer-associated mortality). However, nongenotoxic, tumor-promoting mechanisms are, at least at present, very difficult to recapitulate in the research laboratory. Lacking this ability, we still understand relatively little about these nongenetic mechanisms that are so deeply involved in cancer pathogenesis and thus tumor incidence.

Still, in the case of a defined subset of common human cancers pro-inflammatory mechanisms can be traced directly back to clearly identified causal agents. Their identities of these do indeed suggest effective strategies for reducing cancer incidence. On a worldwide scale, it is estimated that 9% of all cancer deaths arise because chronic infections of the stomach by the bacterium *Helicobacter pylori* often lead to gastric carcinomas, and that 6% of all cancer deaths are associated with liver cancers, most of which are caused by chronic infections by hepatitis B and hepatitis C viruses. The chronic infection of the cervical epithelium caused by human papillomaviruses, notably strains 16, 18, and 45, leads in some women to the cervical carcinomas that account for 5% of worldwide cancer mortality. (The chronic inflammatory effects of these papillomavirus infections may be compounded by certain viral oncogenes.) The identification of these etiologic agents presents an enormous opportunity for reducing worldwide cancer incidence and mortality: infectious diseases are ultimately far easier to control and prevent (through immunization, antibiotics, and antiviral drugs) than the effects of foodstuffs that we ingest and the tobacco products that we inhale. By some estimates, 450 million people worldwide are infected chronically with HBV, and 200 million are long-term HCV carriers.

This recognition of the widespread influence of pro-inflammatory processes in cancer pathogenesis offers another opportunity. More specifically, future anti-cancer drugs may derive from the recent advances in elucidating the biochemical details of inflammation and its specific contributions to tumor promotion. Cyclooxygenase-2 (COX-2) is clearly at the center of this process, and a number of anti-inflammatory agents (largely NSAIDs) targeting this enzyme are known to be effective in decreasing the incidence of various types of cancer. Ultimately, successes in reducing inflammation pharmacologically may provide the most effective route to reducing human cancer incidence and thus, in the longer term, mortality. At present, however, the ability to directly target anti-inflammatory agents to specific tissues or organs is limited, leaving us with agents that often carry unacceptable side effects.

There are yet other issues encountered in this chapter that bear on attempts to reduce cancer mortality. Here is one question raised by the research we discussed: Do full-blown, multiply mutated tumors continue to depend on the mutant alleles that were created decades earlier by initiating mutagens? Or do these initially acquired mutant alleles become irrelevant later, when subsequently acquired mutant alleles take over the job of programming cancer cell proliferation?

The logic developed in this chapter seems to provide a clear and direct answer: Mutant alleles acquired late in multi-step progression should never be functionally redundant with those acquired earlier, simply because these late alleles do not confer advantageous phenotypes beyond those that cancer cells have already developed earlier. Accordingly, early mutations and mutant alleles should not be obviated by later ones, and these early mutations remain critical to ongoing robust tumor growth. Transgenic mice shed some light here. In some of these mouse models, the initiating transgenic oncogenes, such as *myc* and *ras* (see Section 11.8), have been inactivated by a variety of experimental tricks long after substantial tumors have formed. As it turns out, the effects on tumor growth that have been observed to date are conflicting. More often than not, these tumors collapse rapidly when deprived of the initiating oncogenes that led originally to their formation. Hence, in these cases, the mutant oncogenic alleles continued to be critically important both for the initiation of the neoplastic cell state as well as the maintenance of this state long afterward. Stated differently, the mutant alleles that were acquired in later steps of multi-step tumorigenesis did not render the earlier ones unnecessary. However, in several research reports, shutdown of the initiating *neu*, *myc*, or *wnt* oncogenes long after tumors had formed led, after short remissions, to the regrowth of these tumors. In these cases, the initiating oncogenes were clearly no longer needed for the expansion of the relapsing tumors.

These observations would seem to hold important implications for the development of new types of anti-cancer therapeutics, a topic that we will return to in Chapter 17. For example, if a mutant *ras* allele is found in a human tumor cell genome, can this tumor be cured by drugs targeted to the Ras oncoprotein, or has this oncoprotein, which almost certainly played a key role in the early steps of tumor formation, become irrelevant, later on, to the continued proliferation and survival of the cells in this tumor?

In spite of our rapidly expanding understanding of cancer pathogenesis, as described in this chapter, we still do not have the means to impose control on most kinds of aggressive human cancers. The images in Figure 11.16 reveal starkly the major difficulty of current anti-cancer therapy—the fact that the extensive diversification of cell populations within individual tumors seems, almost inevitably, to yield therapy-resistant subpopulations of cells that erupt and drive clinical relapses in patients whose tumors previously responded to therapy. We still lack the means to compensate for the rapid pace of this intra-tumoral evolutionary diversification and therefore lack the ability to deliver the definitive cures that so many seek. So, this chapter does not end on a high note, but that is where we are and may be for a long time to come.

Key concepts

- The process of tumor formation is complex, with multiple steps involving multiple alterations of cells and their physiological control mechanisms.

- The complexity of this process is reflected in the long time periods required for most human cancers to develop.

- These changes involve both the activation of oncogenes and the inactivation of tumor suppressor genes.

- The number of steps required to experimentally transform human cells is larger than is needed to transform cells of laboratory mice.

- These alterations affect multiple distinct regulatory circuits within cells and function in a complementary fashion to create the neoplastic cell phenotype.

- Some of these changes occur as the direct result of the actions of exogenous mutagens, and exposure to such mutagens may represent a "rate-limiting" determinant of tumor progression.

- In most instances, however, the rate of tumor progression may be governed by the actions of nonmutagenic tumor-promoting agents, which may determine the rate of expansion of mutant cell clones.

- In many human cancers, these critical nonmutagenic, tumor-promoting stimuli include chronic mitogenic stimulation and inflammation.

- The multiplicity of steps required for human cancers to arise is not known, in part because certain changes may occur rapidly and therefore may not be "rate-limiting," whereas others may require a decade or more to complete.

- Multi-step tumor progression can be depicted as a form of Darwinian evolution occurring within tissues. However, because some of the critical changes occurring during tumorigenesis are epigenetic, and because the rate of genetic diversification can occur very rapidly, the classic depictions of linear Darwinian evolution must be modified.

- In most, but not all, transgenic models of tumorigenesis, the initiating changes continue to be required for continued survival of a tumor, long after the process of tumor progression has reached completion.

- The number of genetic changes found in the genomes of human cancer cells vastly exceeds the number required for tumorigenesis to reach completion, complicating identification of the critical genetic changes that affect driver genes and are causally important in tumor formation.

- The discovery of cancer stem cells greatly changes our concepts about the mechanisms of multi-step tumorigenesis, because these self-renewing cells (or closely related progenitor cells), rather than the bulk populations of cancer cells, may be the objects of genetic alteration and clonal selection.

Thought questions

1. Knowing the various genetic and epigenetic changes that occur during multi-step tumorigenesis, which would be likely to be readily uncovered and which may be difficult to identify? Describe the reasons for these assignments.

2. Some tumor suppressor genes inactivated during multi-step tumorigenesis may be readily identified because of LOH in the chromosomal region carrying them, whereas others may be difficult to identify in this way. Describe the factors that allow or complicate this identification.

3. What arguments favor the notion that all of us carry myriad clones of initiated premalignant cells throughout the body?

4. What arguments can be mustered that favor the notion that most human carcinogens act as promoters rather than initiators of tumorigenesis?

5. What different approaches can be used to estimate the number of steps in multi-step tumor progression, and how is each of these approaches flawed?

6. How does the current available information about multi-step tumor progression provide insights into strategies for the prevention of clinically detectable cancers?

7. What mechanisms enable chronic viral infections to exert a carcinogenic influence on a tissue?

8. Describe the various mechanisms of tumor promotion and the features that they share in common and those that distinguish them from one another.

Additional reading

Adekola, K, Rosen ST & Shanmugam M (2012) Glucose transporters in cancer metabolism. *Curr. Opin. Oncol.* 24, 650–654.

Al-Hajj M & Clarke ME (2004) Self-renewal and solid tumor stem cells. *Oncogene* 23, 7274–7282.

Albuquerque TAF, do Val LD, Doherty A et al. (2018) From humans to hydra: patterns of cancer across the tree of life. *Biol. Rev.* 93, 1715–1734.

Alison MR & Islam S (2009) Attributes of adult stem cells. *J. Pathol.* 217, 144–160.

Altrock PM, Liu LL & Michor F (2015) The mathematics of cancer: integrating quantitative models. *Nature Rev. Cancer* 15, 730–745.

Armitage P & Doll R (1954) The age distribution of cancer and a multistage theory of carcinogenesis. *Br. J. Cancer* 8, 1–12.

Arwert E, Hoste E & Watt FM (2012) Epithelial stem cells, wound healing and cancer. *Nat. Rev. Cancer* 12, 170–180.

Balkwill F, Charles KA & Mantovani A (2005) Smoldering and polarized inflammation in the initiation and promotion of malignant disease. *Cancer Cell* 7, 211–217.

Balkwill F & Mantovani A (2010) Cancer and inflammation: implications for pharmacology and therapeutics. *Clin. Pharmacol. Ther.* 87, 401–406.

Blanpain C & Fuchs E (2006) Epidermal stem cells of the skin. *Annu. Rev. Cell Dev. Biol.* 22, 339–373.

Borovski T, De Sousa e Melo F, Vermeulen L & Medema JP (2011) Cancer stem cell niche: the place to be. *Cancer Res.* 71, 634–639.

Braakhuis BJM, Tabor MP, Kummer JA et al. (2003) A genetic explanation of Slaughter's concept of field cancerization: evidence and clinical implications. *Cancer Res.* 63, 1727–1730.

Brabletz S, Schmalhofer O & Brabletz T (2009) Gastrointestinal stem cells in development and cancer. *J. Pathol.* 217, 307–317.

Cairns J (1975) Mutation, selection and the natural history of cancer. *Nature* 255, 197–200.

Cairns RA, Harris IS & Mak, TW (2011) Regulation of cancer cell metabolism. *Nat. Rev. Cancer* 11, 85–95.

Caldas C (2012) Cancer sequencing unravels clonal evolution. *Nature Biotech.* 30, 408–410.

Cheng J, DeCaprio JA, Fluck MM & Schaffhausen BS (2009) Cellular transformation by simian virus 40 and murine polyoma virus T antigens. *Semin. Cancer Biol.* 19, 218–228.

Cho RW & Clarke MF (2008) Recent advances in cancer stem cells. *Curr. Opin. Genet. Dev.* 18, 1–6.

Crusz S & Balkwill FR (2015) Inflammation and cancer: advances and new agents. *Nat. Rev. Clin. Oncol* 12, 584–596.

Damsky WE & Bosenberg M (2017) Melanocytic nevi and melanoma: unraveling a complex relationship. *Oncogene* 36, 5771–5792.

DeBerardinis RJ & Chandel NS (2016) Fundamentals of cancer metabolism. *Sci. Adv.* 2, e1600200.

de Magalhaes JP (2013) How ageing processes influence cancer. *Nat. Rev. Cancer* 13, 357–365.

Deng T, Lyon CJ, Bergin S et al. (2016) Obesity, inflammation and cancer. *Annu. Rev. Pathol. Mech. Dis.* 11, 421–449.

Dotto GP (2014) Multifocal epithelial tumors and field cancerization: stroma as a primary determinant. *J. Clin. Invest.* 145, 1446–1453.

Eichhorn PJA, Creyghton MP & Bernards R (2009) Protein phosphatase 2A regulatory subunits and cancer. *Biochim. Biophys. Acta* 1795, 1–15.

Garcia SD, Park HS, Novelli M & Wright NA (1999) Field cancerization, clonality, and epithelial stem cells: the spread of mutated clones in epithelial sheets. *J. Pathol.* 187, 61–81.

Gerlinger M, McGranahan N, Dewhurst M et al. (2014) Cancer: evolution within a lifetime. *Annu. Rev. Genet.* 48, 215–236.

Gold LS, Ames BN & Slone TH (2002) Misconceptions about the causes of cancer. In Human and Environmental Risk Assessment: Theory and Practice (D Paustenbach ed), pp 1415–1460. Hoboken, NJ: John Wiley and Sons.

Gonda TA, Tu S & Wang TC (2009) Chronic inflammation, the tumor microenvironment and cancer. *Cell Cycle* 8, 2005–2013.

Gorbunova V, Seluanov A, Zhang Z et al. (2014) Comparative genetics of longevity and cancer: insights from long-lived rodents. *Nat. Rev. Genet.* 15, 531–540.

Greaves M (2015) Evolutionary determinants of cancer. *Cancer Discov.* 5, 806–820.

Griner EM & Kazanietz MG (2007) Protein kinase C and other diacylglycerol effectors in cancer. *Nat. Rev. Cancer* 7, 281–293.

Grivennikov SI, Greten FR & Karin M (2010) Immunity, inflammation, and cancer. *Cell* 140, 883–899.

Haber DA & Settleman J (2007) Cancer: drivers and passengers. *Nature* 446, 145–146.

Hahn WC & Weinberg RA (2002) Modelling the molecular circuitry of cancer. *Nat. Rev. Cancer* 2, 331–341.

Hanahan D & Weinberg RA (2011) Hallmarks of cancer: the next generation. *Cell* 144, 646–674.

Hansel DE, Kern SE & Hruban RH (2003) Molecular pathogenesis of pancreatic cancer. *Annu. Rev. Genomics Hum. Genet.* 4, 237–256.

Henderson BE & Feigelson HS (2000) Hormonal carcinogenesis. *Carcinogenesis* 21, 427–433.

Hirschey MD, DeBerardinis RJ, Diehl AME et al. (2015) Dysregulated metabolism contributes to oncogenesis. *Semin. Cancer Biol.* 35, S129–S150.

Hoxhaj G & Manning BD (2020) The PI3K-AKT network at the interface of oncogenic signalling and cancer metabolism. *Nature Rev. Cancer* 20, 74–88.

Humphries A & Wright NA (2008) Colonic crypt organization and tumorigenesis. *Nat. Rev. Cancer* 8, 415–424.

Karin M & Greten FR (2005) NF-κB: linking inflammation and immunity to cancer development and progression. *Nat. Rev. Immunol.* 5, 749–759.

Kemp CJ (2005) Multistep skin cancer in mice as a model to study the evolution of cancer cells. *Semin. Cancer Biol.* 15, 460–473.

Levine AJ, Jenkins NA & Copeland NG (2019) The roles of initiating truncal mutations in human carcinomas: the order of mutations and tumor cell type matters. *Cancer Cell* 35, 10–15.

Loeb LA & Harris CC (2008) Advances in chemical carcinogenesis: a historical review and prospective. *Cancer Res.* 68, 6863–6872.

Maley CC, Aktipis A, Graham TA et al. (2017) Classifying the evolutionary and ecological features of neoplasms. *Nat. Rev. Cancer* 17, 605–619.

Markowitz SD & Bertagnolli MM (2009) Molecular basis of colorectal cancer. *N. Engl. J. Med.* 361, 2449–2460.

Martincorena I & Campbell PJ (2015) Somatic mutations in cancer and normal cells. *Science* 349, 1483–1489.

Marusyk A, Almendro V & Polyak K (2012) Intra-tumor heterogeneity: a looking glass for cancer? *Nat. Rev. Cancer* 12, 323–334.

McGranahan N & Swanton C (2017) Clonal heterogeneity and tumor evolution: past, present and the future. *Cell* 168, 613–628.

Merlo LMF, Pepper JW, Reid BJ & Maley CC (2006) Cancer as an evolutionary and ecological process. *Nat. Rev. Cancer* 6, 924–935.

Miller EC (1978) Some current perspectives on chemical carcinogenesis in humans and experimental animals: presidential address. *Cancer Res.* 38, 1479–1496.

Navin NE & Hicks J (2010) Tracing the tumor lineage. *Mol. Oncol.* 4, 267–283.

Nowell PC (1976) The clonal evolution of tumor cell populations. *Science* 194, 23–28.

Parsonnet J (ed) (1999) Microbes and Malignancy: Infection as a Cause of Human Cancers. Oxford, UK: Oxford University Press.

Peto R, Roe FJC, Lee PN et al. (1975) Cancer and ageing in mice and men. *Brit. J. Cancer* 32, 411–426.

Potter M, Newport E & Morten KJ (2016) The Warburg effect: 80 years on. *Biochem. Soc. Trans.* 44, 1499–1505.

Reich BJ, Li X, Galipeau PC & Vaughan TL (2010) Barrett's oesophagus and oesophageal adenocarcinoma: time for a new synthesis. *Nat. Rev. Cancer* 10, 87–101.

Renan MJ (1997) How many mutations are required for tumorigenesis? Implications from human cancer data. *Mol. Carcinogen.* 7, 139–146.

Sanchez-Vega F, Mina M, Armenia J et al. (2018) Oncogenic signaling pathways in the Cancer Genome Atlas. *Cell* 173, 321–337.

Schedin P (2006) Pregnancy-associated breast cancer and metastasis. *Nat. Rev. Cancer* 6, 281–291.

Shackleton M, Quintana E, Fearon ER & Morrison SJ (2009) Heterogeneity in cancer: cancer stem cells versus clonal evolution. *Cell* 138, 822–829.

Singh A & Settleman J (2010) EMT, cancer stem cells and drug resistance: an emerging axis of evil in the war on cancer. *Oncogene* 29, 4741–4751.

Stiles CD & Rowitch DH (2008) Glioma stem cells: a midterm exam. *Neuron* 58, 832–846.

Stratton MR, Campbell PJ & Futreal PA (2009) The cancer genome. *Nature* 458, 719–724.

Sullivan LB, Gui DY & Vander Heiden MG (2016) Altered metabolite levels in cancer: implications for tumour biology and cancer therapy. *Nat. Rev. Cancer* 16, 680–693.

Suva ML & Tirosh I (2020) The glioma stem cell model in the era of single-cell genomics. *Cancer Cell* 37, 630–636.

Tabassum DP & Polyak K (2015) Tumorigenesis: it takes a village. *Nat. Rev. Cancer* 15, 473–483.

Thompson PA, Khatami M, Baglole CJ et al. (2015) Environmental immune disruptors, inflammation, and cancer risk. (2015) *Carcinogenesis* 36, S232–S253.

Thun MJ, Henley SJ & Calle EE (2002) Tobacco use and cancer: an epidemiologic perspective for geneticists. *Oncogene* 21, 7307–7325.

Tomatis L, Aitio A, Wilbourn J & Shuker L (1989) Human carcinogens so far identified. *Jpn. J. Cancer Res.* 80, 795–807.

Trichopoulos D, Adama H-O, Ekborn A et al. (2007) Early life events and conditions and breast cancer risk: from epidemiology to etiology. *Int. J. Cancer* 122, 481–485.

Turajlic S, Sottoriva A, Graham T & Swanton C (2019) Resolving genetic heterogeneity in cancer. *Nat. Rev. Genet.* 20:404–416.

Vander Heiden MG, Cantley LC & Thompson CB (2010) Understanding the Warburg effect: The metabolic requirements of cell proliferation. *Science* 324, 1029–1033.

Visvader JE (2011) Cells of origin in cancer. *Nature* 469, 314–322.

Visvader JE & Lindeman GJ (2008) Cancer stem cells in solid tumours: accumulating evidence and unresolved questions. *Nat. Rev. Cancer* 8, 755–768.

Visvader JE & Smith GH (2010) Murine mammary epithelial stem cells: discovery, function and current status. *Cold Spring Harb. Perspect. Biol.* 4, a004879, Oct. 6.

Wang D & DuBois RN (2010) Eicosanoids and cancer. *Nat. Rev. Cancer* 10, 181–193.

Wang D & DuBois RN (2010) The role of COX-2 in intestinal inflammation and colorectal cancer. *Oncogene* 29, 781–788.

Westermarck J & Hahn WC (2008) Multiple pathways regulated by the tumor suppressor PP2A in transformation. *Trends Mol. Med.* 14, 152–160.

Wild CP, Weiderpass E, Stewart BW (eds) (2020). World Cancer Report: Cancer Research for Cancer Prevention. Lyon, France: International Agency for Research on Cancer. Available from: http://publications.iarc.fr/586.

Yap T, Gerlinger M, Futreal PA et al. (2012) Intratumor heterogeneity: seeing the wood for the trees. *Sci. Transl. Med.* 4, 127–130.

Zhao JJ, Roberts TM & Hahn WC (2004) Functional genetics and experimental models of human cancer. *Trends Mol. Med.* 10, 344–350.

Chapter 12

Shaping and Characterizing the Cancer Genome

> The capacity to blunder slightly is the real marvel of DNA. Without this special attribute, we would still be anaerobic bacteria and there would be no music.
>
> Lewis Thomas, biologist, 1979

The fact that human tumor formation is a complex, multi-step process reflects the multiple lines of defense against cancer that have been established within our cells, each maintained by the hard-wiring of a complex regulatory circuit. The human body—actually, its individual cells—entrusts the maintenance of these anti-cancer defenses to its most stable, reliable constituents: DNA molecules. Over extended periods of time, DNA sequences are the most fixed, unchangeable components of cells; most of their other parts are in constant flux, being created and broken down continuously. Despite the inherent stability of DNA, however, the lifelong requirement for DNA replication provides ample opportunity for DNA damage and the accumulation of cancer-associated mutations that drive tumorigenesis.

By current estimates, an adult human has approximately 4×10^{13} cells. In some tissues, such as the brain, there is relatively little cellular turnover and thus relatively little need for cell division. However, in highly regenerative tissues such as the intestine, the skin, or the blood, our bodies are producing cells at a remarkable rate. Indeed, by one estimate we produce approximately 80 billion new cells every hour of every day. Prior to each division, the mother cell is charged with faithfully duplicating its chromosomal DNA in the S phase of the cell cycle and accurately segregating it to the daughter cells in M phase. Given that the total length of the decondensed chromosomes of a single cell is about 2 meters, every hour we make an aggregate of approximately 160 billion meters of DNA! This is enough DNA to wrap around the circumference of the Earth a staggering 4000 times.

479

This prodigious rate of replication is why the mechanisms that govern DNA synthesis and chromosomal segregation must function at high fidelity in order to minimize the likelihood that cells will acquire faulty genomes. Adding to this challenge is that throughout our lifetimes our cells are exposed to a series of endogenously produced mutagenic chemical species as well as mutagens of exogenous origin that can increase the mutation rate significantly. With more mutations, there is an increased chance that a proto-oncogene or tumor suppressor gene will be altered by cancer-promoting mutations. Multiple mutant alleles of these genes must be acquired before adult tumors erupt, and so the rate of formation of these alleles is a critical-rate limiting determinant of cancer formation.

In this chapter, we consider the various mutagenic processes that impinge on normal cells as well as incipient cancer cells and that ultimately shape the cancer genome. Guarding against the effects of these mutagenic events are a battery of DNA repair and genome maintenance pathways that function to restore genomic order. As we discussed in Chapter 7, loss-of-function mutations in many of the genes that function in these pathways (so-called caretaker genes) are frequently themselves mutated in cancer cells, leading to increased mutation frequency and genomic instability.

Over the past decade, the development of high-throughput genomic sequencing technologies has enabled the extensive analysis of cancer genomes from many human cancer types. The emerging data from this new era of cancer genomics underscore the fact that the march from cellular normalcy to malignancy is typically accompanied and indeed driven by widespread genomic alterations. These broad-based cancer genome characterization efforts have provided insights into the mutational landscape of human cancer as well as the mutagenic processes that drive it.

12.1 Tissues are organized to minimize the progressive accumulation of mutations

On several occasions throughout this book, we have described the effects of carcinogens and tumor promoters on target cells throughout the body. However, the specific biological identities of these target cells have never been spelled out. As it turns out, knowledge of the nature of these cells is critical to understanding how genome integrity is maintained and how it breaks down. To explore the topic of genome integrity, we need to delve into the organization of tissues and the types of cells that form them. Their biological behavior furnishes us with insights into the strategies that tissues and cells exploit to minimize the accumulation of genetic lesions.

As described earlier (see Section 11.7), a common scheme seems to explain the construction and maintenance of many tissues throughout the body. Within each tissue, a relatively small number of cells populate its stem cell **compartment**. These self-renewing cells may constitute a minute fraction of the entire cell population within a tissue, sometimes as few as 0.1 to 1% of the total. In truth, in most tissues these numbers represent nothing more than poorly informed guesses. Because stem cells are present in very small numbers, are not particularly distinctive in appearance, and are often scattered among other cell types within tissues, they can be difficult to identify and study. Consequently, much of what is described below rests on inference rather than on direct observation of stem cells and their properties.

As is the case with stem cells of tumors (see Section 11.7), the stem cells in a normal tissue are self-renewing, because at least one of the two daughters of a dividing stem cell will retain the phenotype exhibited by the mother cell prior to cell division (see Figure 11.20). In many tissues, the second daughter cell and its transit-amplifying descendants will pass through a substantial number of cell divisions before entering into a post-mitotic, highly differentiated state (see Sidebar 11.6). These actively dividing cells, which serve as intermediates between a stem cell and its differentiated descendants, generate large flocks of differentiated descendants of the second daughter cell that will be discarded sooner or later, only to be replaced by newly minted differentiated cells (**Figure 12.1**). Implicit in what follows is a simple idea: because the lineage of stem cells represents the only stable repository of genetic information within a tissue, the genomes of stem cells must be protected as much as possible from corruption.

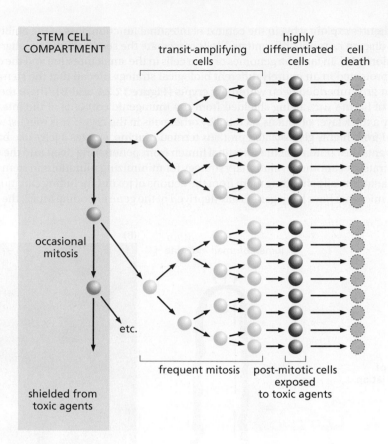

STEM CELL
COMPARTMENT

transit-amplifying
cells

highly
differentiated
cells

cell
death

occasional
mitosis

etc.

frequent mitosis

post-mitotic cells
exposed
to toxic agents

shielded from
toxic agents

Figure 12.1 Tissue organization and protection of the stem cell genome The organization of many epithelial tissues seems to conform to the scheme shown here. Each stem cell (*blue*) divides only occasionally in an asymmetric fashion to generate a new stem cell daughter and a transit-amplifying daughter. These stem cells are often shielded anatomically from toxic agents. The transit-amplifying cells (*green*) undergo repeated rounds of growth and division, expanding their populations exponentially. Eventually, the products of these cell divisions undergo further differentiation into post-mitotic, highly differentiated cells (*red*). The highly differentiated cells, which are often in direct contact with various toxic agents, are shed with some frequency; hence, any mutant alleles that arise in these cells will be lost, sooner or later, from the tissue. This means that the genomes of stem cells are often protected through two mechanisms: infrequent cell division and an anatomical barrier that blocks noxious, potentially mutagenic influences.

The exponential increase in the number of transit-amplifying cells prior to end-stage differentiation means that a stem cell needs to divide only on rare occasion in order to maintain a large pool of end-stage, highly differentiated cells in a tissue. In the colonic crypts, for example, it is estimated that a single stem cell division ultimately results in the formation of as many as 32 differentiated **enterocytes** (see Figure 7.21). Therefore, while one might think that stem cells participate in continual cycles of growth and division, the reality is usually much different: the transit-amplifying cells create the great bulk of mitotic activity in highly proliferative tissues. Because the DNA replication occurring during each cell cycle is inherently prone to making errors, this scheme reduces the risk that mutations will accumulate in the genomes of the long-lived cells within a tissue, that is, its stem cells. Conversely, although the mitotically active transit-amplifying cells may indeed accumulate mutations, their mutant descendants will be discarded sooner or later, reducing the accumulation of mutant cells in the tissue as a whole.

In many epithelial tissues, the differentiated epithelial cells are especially vulnerable to genetic damage of a different origin, because they form cell sheets that line the walls of various ducts and cavities that often contain toxic substances. In the cases of the colon and the bile duct, the epithelial cells confront fecal contents and highly corrosive bile, respectively. The cells lining the alveoli in the lungs cope every day with particulates and pollutants in the air. The keratinocytes in our skin are exposed directly to the outside world and hence are susceptible to sustaining several types of damage, including those inflicted by ultraviolet radiation and various toxic chemicals.

The differentiated end-stage cells (see Figure 12.1) in these and other tissues have, as stated, a finite lifetime and are discarded sooner or later. The epithelial cells in the colon live for 5 to 7 days before they are induced to enter apoptosis and are sloughed off into the lumen of the intestine. The keratinocytes in our skin die within 20 to 30 days of being formed and are shed continually in small flakes of dead skin (see, for example, Figure 2.5A). Once again, any genetic damage they have sustained will disappear with them. (To be sure, some cell types are shed for a different reason: they may simply age and lose their viability, being worn out from carrying out the active business of the tissue. For example, red blood cells have an average lifetime of approximately 120 days, after which they are scavenged by the spleen, broken down, and their contents recycled or excreted.)

Figure 12.2 Stem cells and the organization of gastrointestinal crypts (A) Boxed numbers indicate the position of individual enterocytes counting in cell diameters from the bottom of the crypt. Four to six very slowly cycling stem cells (*red*) are located 4–5 cell diameters from the bottom of the crypts. In addition, more rapidly cycling stem cells are found interspersed among the Paneth cells at the bottom of the crypts (*not shown*). Both groups of stem cells are shielded from the contents of the small intestine by their location and by mucus that prevents fluids in the intestinal lumen from entering the crypt. The stem cells spawn a large number (~150) of highly proliferative transit-amplifying cells (*yellow, green*), which divide every 12 hours or so. Their division eventually yields approximately 3500 enterocytes (*blue*), which cover the finger-like villus. The enterocytes are continuously migrating toward the tip of the villus, where they undergo apoptosis and are shed into the lumen of the small intestine (*gray*). (B) The Lgr5 protein has been found to be a highly useful marker for identifying stem cells in a number of tissues. Located in the bottoms of the crypts of the mouse small intestine (duodenum), these cells are labeled here by a transgene that fuses the coding region of the Lgr5 gene with a sequence encoding green fluorescent protein (GFP, *green*). Other evidence, not shown here, points to the presence of a second stem cell population located at 4 cell diameters (from the crypt bottom) that is normally relatively quiescent but becomes mitotically active in response to tissue injury that requires replacement of the Lgr5$^+$ cells. (C) The emigration of transit-amplifying cells from the crypts of the small intestine can be tracked by injecting a dose of ^{3}H-thymidine into a mouse and following the incorporation of this radiolabel into DNA by autoradiography; radioactive decay is indicated by dark silver grains. Seen here are the cells in the crypts of the duodenum of the mouse at the indicated times after injection of the ^{3}H-thymidine. Cells that multiplied only a small number of times after initial incorporation of ^{3}H-thymidine remain heavily labeled (*broad arrows*), while the great majority underwent multiple additional divisions and therefore exhibit diluted radiolabeling. After four days, virtually all cells carrying genomes that were synthesized at the beginning of the experiment have moved out of the crypts to the tips of the villi. (A, courtesy of C.S. Potten. B, courtesy of N. Barker and H. Clevers. C, from C.S. Potten, *Philos. Trans. R. Soc. Lond. B Biol. Sci.* 353:821–830, 1998. With permission from the Royal Society.)

These schemes explain why, in the context of intestinal function, loss of the ability of a tissue to discard its more differentiated cells can create the seeds of future malignancy (see Section 7.10). In fact, the genomes of stem cells in the small intestine and the colon are also protected in an entirely different biological strategy. Recall that the stem cells in the gut are embedded deep within the crypts (**Figure 12.2A and B**). There they are well out of harm's way, being shielded from the mutagenic contents of the intestinal lumen by a thick layer of mucus secreted by other cells in the crypt. This mucus, which is formed from highly glycosylated proteins termed **mucins**, creates a jelly-like barrier that prevents the contents of the intestinal lumen from penetrating deep into the crypt and illustrates yet another evolutionary strategy for minimizing mutations in stem cells, namely, anatomically shielding them from the actions of toxins, including carcinogens. (Indeed, mice that have been genetically deprived of the gene encoding Muc2, the most

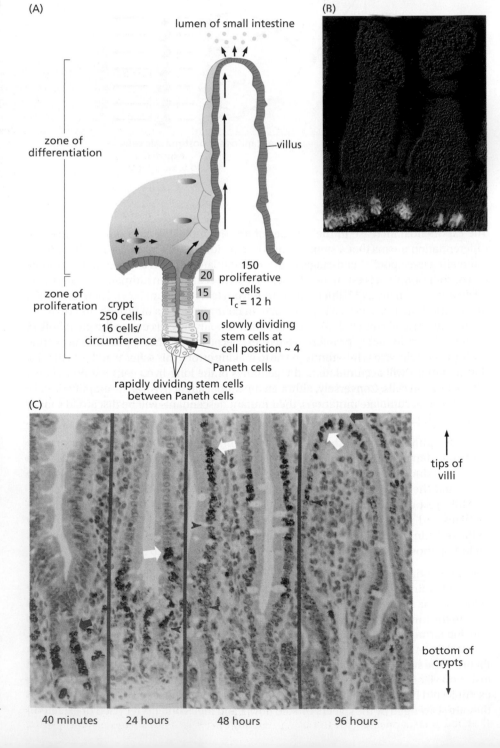

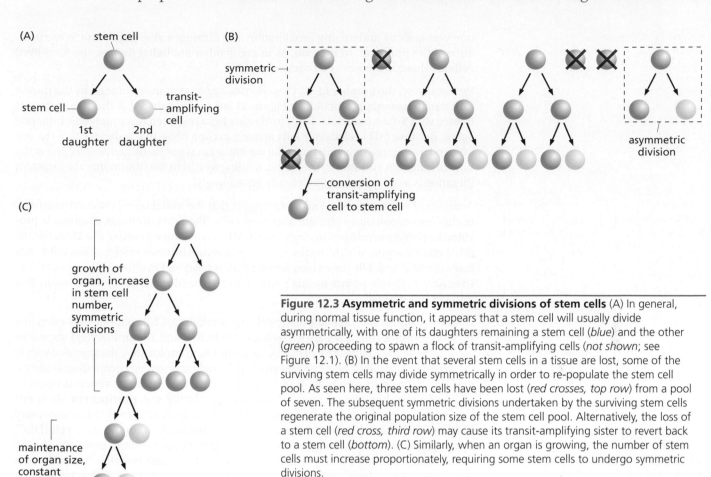

(A) stem cell

stem cell — 1st daughter / 2nd daughter — transit-amplifying cell

(C) growth of organ, increase in stem cell number, symmetric divisions

maintenance of organ size, constant number of stem cells, asymmetric divisions

(B) symmetric division / conversion of transit-amplifying cell to stem cell / asymmetric division

Figure 12.3 Asymmetric and symmetric divisions of stem cells (A) In general, during normal tissue function, it appears that a stem cell will usually divide asymmetrically, with one of its daughters remaining a stem cell (*blue*) and the other (*green*) proceeding to spawn a flock of transit-amplifying cells (*not shown*; see Figure 12.1). (B) In the event that several stem cells in a tissue are lost, some of the surviving stem cells may divide symmetrically in order to re-populate the stem cell pool. As seen here, three stem cells have been lost (*red crosses, top row*) from a pool of seven. The subsequent symmetric divisions undertaken by the surviving stem cells regenerate the original population size of the stem cell pool. Alternatively, the loss of a stem cell (*red cross, third row*) may cause its transit-amplifying sister to revert back to a stem cell (*bottom*). (C) Similarly, when an organ is growing, the number of stem cells must increase proportionately, requiring some stem cells to undergo symmetric divisions.

abundant gastrointestinal mucin, are prone to develop adenomas in the small intestine, many of which progress to adenocarcinomas.) This physical barrier confers further protection on the genomes of stem cells, complementing the dynamics described above, which ensures that enterocytes that may have sustained mutations are continuously flushed out of the crypts and eliminated in less than a week (Figure 12.2C).

In theory, the stem cell compartment within a tissue has an inexhaustible ability to generate differentiated progeny without ever suffering depletion. However, almost inevitably, a stem cell will be lost through one or another mishap. This gap in the ranks must be filled by other stem cells. More specifically, both daughters of a surviving stem cell will need to retain the phenotype of their mother, which therefore undergoes a symmetrical division (**Figure 12.3**). Alternatively, in certain tissues, loss of critical stem cells may trigger the dedifferentiation of transit-amplifying cells, which reverse their earlier exit from the stem cell compartment (see Figure 11.20C).

12.2 The properties of stem cells make them good candidates to be cells-of-origin of cancer

The properties of the intestinal cells (see Figure 12.2) also provide important clues about the identities of the cells that are common targets of carcinogenesis. Actually, we already have come across evidence that points to the central role of long-lived cells in the process of carcinogenesis. In Section 11.11, we read about experimental protocols used to induce skin cancer in mice. One such protocol involved painting a patch of skin with an initiating agent, allowing the patch to remain untouched for some months, and then painting it repeatedly with TPA, a potent skin tumor promoter. Cells that had been exposed to the initiating carcinogen "remembered" the earlier exposure one year later by undergoing proliferation and forming a skin papilloma in the presence of the promoter. In the skin, as in many other epithelial tissues, the long-lived cells are those in the stem cell compartment.

Provocatively, the number of skin papillomas and carcinomas induced by the mouse skin carcinogenesis protocol (see Figure 11.30) is not reduced if the mouse skin is treated with 5-fluorouracil (5-FU) shortly after being exposed to a mutagenic initiating agent. Because 5-FU selectively kills actively cycling cells, this indicates that the cell targeted by carcinogenic mutagens during initiation is not in the active cell cycle at the time of initiation and shortly thereafter, lending weight to the notion that the target for initiation is a cell type that divides only occasionally.

Analyses of several types of leukemia suggest that the initial targets of carcinogenesis in the hematopoietic system are also stem cells. The most dramatic example is provided by chronic myelogenous leukemia (CML). As described earlier, the Philadelphia (Ph[1]) chromosome, which results from a reciprocal chromosomal translocation that fuses the *BCR* and *ABL* genes (see Section 4.5), is observed in almost all cases of CML. Extensive evidence points to this particular translocation as the genetic lesion that initiates this disease.

A number of distinct hematopoietic cell types within a CML patient may carry the Ph[1] chromosome. Included are lymphoid cells (both B and T lymphocytes), as well as cells of the myeloid lineage (including neutrophils, granulocytes, the megakaryocyte precursors of platelets, and erythrocytes). This finding provides compelling evidence that the cell type in which the translocation is harbored is the common progenitor of all of these hematopoietic cell lineages, specifically the **pluripotent** stem cell that serves as the precursor for many types of hematopoietic cells (Supplementary Sidebar 12.1). Like a variety of other stem cells, this hematopoietic stem cell (HSC) is thought to have a very long lifetime in the hematopoietic system, more specifically in the bone marrow. In the particular case of CML, a stem cell that has suffered a critical mutation—formation of the Ph[1] chromosome—retains the option to dispatch its progeny into a number of distinct hematopoietic cell lineages. Yet other indications that stem cells are targets for tumor formation come from other types of hematopoietic disorders (**Sidebar 12.1**). Although stem cells play an important role in the origins of many cancers, current evidence supports the candidacy of other cells within tissues as also being important players in tumor initiation (**Sidebar 12.2**).

Sidebar 12.1 Blocked differentiation is a frequent theme in the development of hematopoietic malignancies Studies of human cancer have revealed dozens of examples of malignancies in which inhibition of differentiation favors the appearance of neoplasia. Possibly the first to be defined genetically involved the avian erythroblastosis virus, a retrovirus that encodes two oncoproteins: its erbB oncogene specifies a constitutively active version of the epidermal growth factor (EGF) receptor (see Section 5.4), which drives the proliferation of erythroblasts (precursors of red blood cells), and its erbA oncogene encodes a nuclear receptor (a homolog of the thyroid hormone receptor), which inhibits differentiation of the hyperproliferating erythroblasts created by erbB. Similarly, in human acute myelogenous leukemia (AML), a large variety of genetic lesions found in the leukemic cells have been assigned to two functional classes: those that are required to drive the proliferation of the myeloid precursor cells, and others that are required in the same cells to block subsequent differentiation.

In the megakaryoblastic leukemias (a malignancy of platelet precursor cells) encountered with some frequency in Down syndrome patients, the gene encoding the GATA1 transcription factor is frequently found to be mutated, preventing the proper maturation and differentiation of these precursors of platelets.

These few examples point to the notion that the exit of cells from stem cell compartments must be impeded in order for tumorigenesis to succeed.

Not addressed by these observations are the precise identities of the cellular targets of transformation. In many cases, the target is not likely to be the pluripotent hematopoietic stem cell, but instead one of its derivatives that is already committed to one or another lineage of differentiation. Such "committed progenitors" (see Supplementary Sidebar 12.1) normally may have significant (but limited) self-renewal capacity and are not yet fully differentiated, and thereby can be considered stem cells. Their transformation from normal to tumor stem cells involves, among other changes, an acquisition of unlimited self-renewal capability.

Compelling observations of the role of stem cells in cancer derive from transgenic mice in which the expression of an activated *ras* oncogene is limited to either the keratinocyte stem cells in the skin (which in this case are located in hair follicles) or the keratinocytes that have begun to enter into a terminally differentiated state. When the transgene directs expression of the *ras* oncogene in the stem cells, the mice develop malignant carcinomas. In contrast, when the same oncogene is expressed in the differentiating keratinocytes, benign papillomas are formed and these tend to regress.

Sidebar 12.2 An alternative target of cancer-driving mutations Most current evidence points clearly to stem cells as being direct targets of the mutations driving tumor initiation and progression. There is, however, an alternative mechanistic model and an alternative set of cells that may often serve as the cellular targets of cancer-forming mutations. This newer thinking comes from extensive evidence that transit-amplifying/progenitor cells can dedifferentiate under certain circumstance and thus re-enter the stem cell pool as mentioned above and as is illustrated in Figure 11.20C. Such dedifferentiation may be triggered, for example, if the pool of stem cells within a stem-cell niche has been depleted for various reasons, requiring replenishment, which can accomplished by dedifferentiation. This notion was explored briefly in Sidebar 11.7. In fact, there are several reasons that transit-amplifying cells are more attractive candidates for sustaining mutations than are stem cells, simply because there are usually far more of them in a tissue, increasing proportionately the chance that a mutation will hit their compartment rather than that of the stem cell. In addition, the transit-amplifying cells, because of their high rate of proliferation, often generate the lion's share of the mitotic activity; if errors in cell proliferation cause mutational events, they are more likely to target these cells, rather than stem cells. Importantly, by dedifferentiating, a mutant transit-amplifying cell can introduce a mutant allele back into the stem cell pool, thereby ensuring the perpetuation of its recently acquired mutation back in the stem cell niche. Indeed, it may be that many of the reported experimental demonstrations of stem cells as direct targets of oncogenic mutations may have ignored the bidirectional interconversions between stem cells and transit-amplifying cells and thus the possibility of this alternative source of cancer-associated mutant genomes. In many tissues, these two alternative mechanisms of cancer-driving mutagenesis remain unresolved experimentally.

12.3 Apoptosis, drug pumps, and DNA replication quality control mechanisms offer tissues a way to minimize the accumulation of mutant preneoplastic cells

The important role played by normal stem cells as targets for transformation requires that the genomes of these cells must be protected by whatever biological and biochemical strategies these cells and the tissues around them can muster. We have already come across two such strategies: the relatively infrequent replication of stem cell DNA and the placement of stem cells in anatomically protected sites. Still, these mechanisms do not seem to suffice, so organisms have developed yet other strategies.

One protective mechanism is suggested by the responses of stem cells in the crypts to massive genetic damage. In the intestinal crypts of the mouse, stem cells that have suffered genetic damage inflicted by X-rays will rapidly initiate apoptosis rather than halt their proliferation and attempt to repair the damage. The motive here seems to be associated with the less-than-perfect nature of DNA repair. As we will learn later, the DNA repair apparatus is highly efficient but hardly perfect, and therefore a residue of unrepaired or incorrectly repaired lesions may remain in the chromosomal DNA. If the DNA replication machinery encounters such lesions, damaged DNA sequences may be copied and passed on as mutations to daughter cells, including those that will themselves become stem cells. So, rather than risk this outcome, stem cells in the mouse crypts are primed to activate apoptosis in response to DNA damage.

Yet another mechanism is suggested by a commonly used technique for separating stem cells from the bulk of cells in a tissue via fluorescence-activated cell sorting (FACS; see Supplementary Sidebar 11.5). Stem cells efficiently pump out certain fluorescent dye molecules, whereas these cells' differentiated derivatives do so much less actively. As a consequence, after exposure of cell populations to such dyes, the stem cells fluoresce much more weakly than all other cells in these populations. The active excretion of these fluorescent dye molecules results from the actions of a plasma membrane protein termed Mdr1 (multi-drug resistance 1), which was discovered because many cancer cells exploit it to pump out, and therefore acquire resistance to, chemotherapeutic drug molecules (see Figure 17.26). The unusually high levels of Mdr1 expressed by many types of stem cells seem to represent a strategy that they use to protect their genomes from potentially mutagenic compounds that may have entered them from outside.

The design of stem cell compartments and the behavior of individual stem cells illustrate several biological strategies used by tissues to reduce the burden of accumulated somatic mutations. Importantly, these strategies represent only the first line of defense against genomic damage. The next line of defense, which is relevant to all cells in the body, is a biochemical one that depends on the ability of their various expressed proteins to recognize and repair damaged DNA molecules.

In fact, DNA molecules are under constant attack by a variety of agents and processes. For the sake of simplicity, we can place these mutagenic processes in three categories. First, the replication of DNA sequences by DNA polymerases during the S phase of the cell cycle is subject to a low but nonetheless significant level of error. Included among these errors are those generated when chemically altered nucleotide precursors are inadvertently incorporated into DNA in place of their normal counterparts. Second, even in the absence of misincorporation of such altered nucleotides, those that have already been incorporated into DNA molecules undergo chemical changes spontaneously; these changes often alter the base sequence and thus the information content of the DNA. Finally, DNA molecules may be attacked by various mutagenic agents, including those molecules generated endogenously by normal cell metabolism as well as agents of exogenous origin—chemical species and physical mutagens (X-rays and UV rays) that enter the body from outside. We will return to the latter two processes in the next sections.

The molecular machinery that is responsible for replicating almost all chromosomal DNA sequences has a remarkably low rate of error. The basic replication machinery in the cell nucleus is powered by the actions of three polymerases, pol-α, pol-δ, and pol-ε. (In all, at least 14 distinct DNA polymerase genes have been cataloged in the human genome; as will be apparent later, most of these are not involved in DNA replication per se but rather in the repair of damaged DNA molecules.)

A cell has two major strategies for detecting and removing the miscopied nucleotides arising during DNA replication. The first strategy rests with the DNA polymerases themselves, which are structurally complex assemblies composed of a number of distinct protein subunits. While they are advancing down single-strand DNA templates and extending nascent DNA strands in a 5′-to-3′ direction, DNA polymerases, such as pol-δ, continuously look backward, scanning the stretch of DNA that they have just polymerized; such monitoring is often called **proofreading**. Should a polymerase detect a copying error, it will use its associated 3′-to-5′ exonuclease activity to move backward and digest the DNA segment that it has just synthesized and then copy this segment once again, with the hope of a better outcome the second time (**Figure 12.4**).

The importance of this proofreading mechanism for the prevention of cancer has been illustrated dramatically by the creation of a mouse strain whose germ-line pol-δ–encoding gene has been subtly altered (by a point mutation encoding a single amino acid substitution). The resulting mutant pol-δ retains its ability to carry out lagging-strand DNA synthesis at the replication fork but has lost its 3′-to-5′ exonuclease activity; this loss eliminates its proofreading function. In a cohort of mice carrying the mutant *PolD* allele in a homozygous configuration, nearly half developed tumors by one year of age, whereas no tumors developed in a group consisting of twice as many heterozygous mice (**Figure 12.5A**). As discussed in Section 12.8, humans with germ-line mutations affecting the proofreading domains of pol-δ or pol-ε (responsible for leading-strand synthesis at the replication fork) are cancer prone (Figure 12.5B). Somatic mutations affecting the proofreading domains of these polymerases are observed in certain human cancers as well (see Figure 12.5B).

These observations demonstrate that the maintenance of wild-type genomic sequences, in this case by two DNA polymerases, represents a critical line of defense against the onset of cancer. Moreover, for us, these observations are the first of many indications that the mutations leading to cancer may arise through endogenous processes associated with normal cellular proliferation rather than being triggered exclusively by exposure to exogenous carcinogenic agents.

Augmenting the DNA polymerases and their proofreading activities is a complex set of **mismatch repair** (MMR) enzymes. These enzymes monitor recently synthesized DNA in order to detect miscopied DNA sequences that have been overlooked by the proofreading mechanisms of the DNA polymerases described above. The actions of

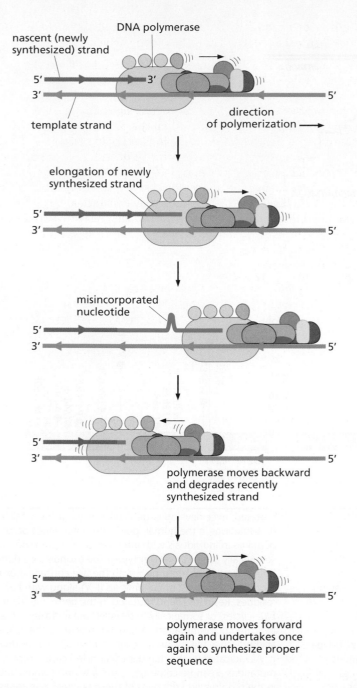

nascent (newly synthesized) strand

DNA polymerase

5′
3′ template strand

direction of polymerization →

elongation of newly synthesized strand

5′
3′

misincorporated nucleotide

5′
3′

5′
3′

polymerase moves backward and degrades recently synthesized strand

5′
3′

polymerase moves forward again and undertakes once again to synthesize proper sequence

Figure 12.4 Proofreading by DNA polymerases A number of DNA polymerases have a proofreading ability that allows them to minimize the number of bases that are misincorporated and retained in the recently synthesized DNA strand. Thus, as a DNA polymerase extends a nascent strand (*dark blue*) in a 5′-to-3′ direction (moving rightward), it will use the existing 3′-OH of the nascent strand as the primer for further elongation (*light blue*). However, if a base has been misincorporated (*third drawing*), the DNA polymerase, which is continuously looking backward to check whether it has incorporated the correct bases in the growing DNA strand, can degrade in a 3′-to-5′ (leftward) direction the recently elongated strand (*fourth drawing*) and undertake once again to synthesize this stretch of nascent strand (*bottom drawing*).

the MMR system become especially critical in regions of the DNA that carry repeated sequences. These sequence blocks include simple mononucleotide repeats (such as AAAAAAA), dinucleotide repeats (such as AGAGAGAG), and repeats of greater sequence complexity. In instances of strand slippage, which occurs in the replication fork when the parental and nascent strands slip out of proper alignment, DNA polymerases appear to occasionally "stutter" while copying these repeats, resulting in incorporation of longer or shorter versions of the repeats into the newly formed daughter strands (**Figure 12.6**). Thus, the sequence AAAAAAA, that is, A_7, might well cause a polymerase to synthesize a T_6 or T_8 sequence in the complementary strand. The resulting insertions or deletions may elude detection by the proofreading components of the DNA polymerases, making them prime targets for recognition and repair by the MMR machinery.

For historical reasons, highly repeated sequences in the genome, often carrying 100 or more nucleotides per repeat unit, have been called "**satellite**" sequences. Because the simple, far shorter sequences discussed here are also found in many places in the

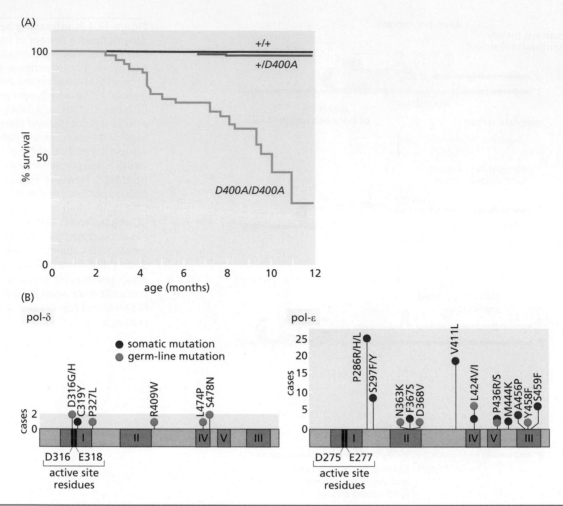

Figure 12.5 Mutations in proofreading domains of DNA polymerases promote cancer (A) A point mutation was introduced into the germ-line copy of the mouse gene encoding DNA polymerase δ (pol-δ), the mammalian DNA polymerase that is responsible for most lagging-strand synthesis. This mutation, termed D400A, alters the amino acid sequence in the proofreading domain of the polymerase by specifying the replacement of an aspartic acid by an alanine at residue position 400 of the polymerase molecule; the synthetic activity of the polymerase is unaffected by this mutation. Shown here is the fate of 53 wild-type mice (+/+), 97 heterozygotes (+/D400A), and 49 homozygous mutants (D400A/D400A). Deaths of the mutant homozygotes were all a result of malignancies; these included lymphomas, squamous cell carcinomas of the skin, lung carcinomas, and several other types of cancer that occurred less frequently. Two of the heterozygotes died from causes that were unrelated to cancer, whereas the homozygous wild-type mice all survived to the age of one year. (The cooperating partner of pol-δ, called pol-ε, carries out leading-strand synthesis and proofreading; when it is similarly mutated, mice develop largely carcinomas of the small intestine.) (B) Mutations in the human pol-δ and pol-ε genes occur both in the germline in individuals with inherited cancer predispositions and somatically in certain cancer types. The exonuclease domains of both polymerases are shown with the five conserved exonuclease motifs marked by Roman numerals (I–V) along with key active site residues. Residues found mutated in the germ line (denoted by *blue circles*) or somatically (*red circles*) are indicated. In each of the indicated point mutations, the letter indicating the wild-type amino acid precedes the codon number and the altered residue follows it. As indicated by the number of reported cases shown, pol-ε mutations are more common. Germ line and somatic mutations in the proofreading domains of these polymerases are associated with colon and endometrial cancers as well as other cancers at lower frequency. Tumors harboring these mutations exhibit "ultra-mutated" genomes with more than 100 mutations detected per megabase of DNA sequenced. (A, from R.E. Goldsby et al., *Nat. Med.* 7:638–639, 2001. B, from E. Rayner et al., *Nat. Rev. Cancer* 16: 71–81, 2016. Both with permission from Nature.)

genome, they have been named **microsatellites**. A defective mismatch repair system that fails to detect and remove stuttering mistakes made by DNA polymerases when copying a microsatellite will result in the expansion or shrinkage of its sequences in progeny cells. This creates the genetic condition known as **microsatellite instability** (MIN; **Figure 12.7A**), which may ultimately involve changes in thousands of microsatellite sequences scattered throughout a cell genome. Importantly, as illustrated in Figure 12.7B, defects in MMR yield base substitutions even more frequently than expansions and contractions of microsatellite sequences; indeed, the genome is

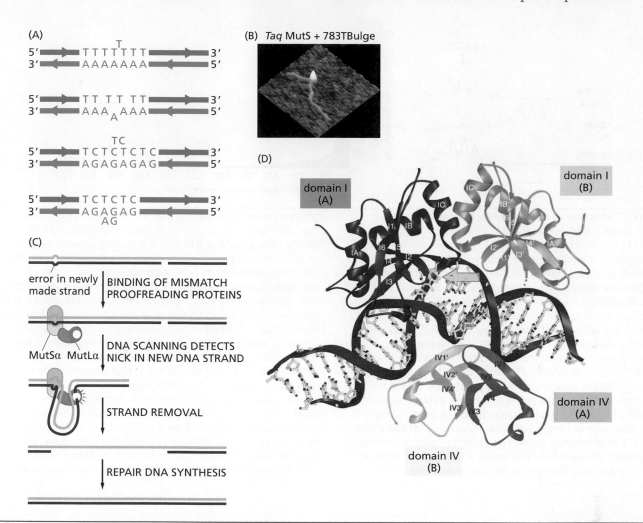

Figure 12.6 DNA polymerase errors and mismatch repair
(A) The DNA polymerases, notably pol-δ, occasionally "stutter," or skip a base when copying a repeating sequence of DNA (e.g., a microsatellite sequence) in the template strand (*blue*). As a consequence, the newly synthesized strand (*green*) either may acquire an extra base that increases the length of the repeating sequence or may lack a base (*top two images*). Identical dynamics may cause similar changes in microsatellite sequences where the repeat unit is a TC dinucleotide segment (*bottom two images*), or a more complex repeating sequence (*not shown*). (B) Mismatch repair (MMR) proteins recognize and repair the mistakes made by DNA polymerases, including misincorporated bases and inaccurate replication of microsatellite sequences. Here, use of atomic-force microscopy reveals the behavior of the MutS MMR protein, a bacterial homolog (from *Thermus aquaticus*) of a number of mammalian MMR proteins. MutS is seen binding to a DNA fragment into which a mismatch has been introduced at a specific nucleotide site. MutS kinks the DNA double helix as it scans for and ultimately finds regions of mismatch (793T Bulge), where it binds in a stable fashion, seen here as a white pyramid. (C) In eukaryotic cells, two components of the MMR apparatus, MutSα and MutLα, collaborate to initiate repair of mismatched DNA. After

MutSα (a heterodimer of MSH6 and MSH2) scans the DNA and locates a mismatch, MutLα (a heterodimer of MLH1 and PMS2) scans the DNA for single-strand nicks, which identify the strand that has recently been synthesized (*red*). In addition, the under-methylation of the recently synthesized strand may aid in this identification (*not shown*). MutLα then triggers degradation of this strand back through the detected mismatch, allowing for repair DNA synthesis to follow. (D) Part of the structure of the *T. aquaticus* MutS homodimeric protein in complex with a mismatched helix (*red*). In addition, is shown. Domains I and IV of subunit A are in *dark blue* and *orange*, and the corresponding domains of subunit B are in *light blue* and *yellow*. An arrow (*yellow*) indicates where phenylalanine residue 39 of domain I of subunit A is associated with an unpaired thymidine in one of the two DNA strands. Defects in the human homolog of this protein play a critical role in triggering hereditary non-polyposis colon cancer (HNPCC), as discussed in Section 12.8. (B, from H. Wang et al., *Proc. Natl. Acad. Sci. USA* 100:14822–14827, 2003. With permission from National Academy of Sciences, U.S.A. C, adapted from B. Alberts et al., Molecular Biology of the Cell, 7th ed. New York: W.W. Norton, 2022, D, from G. Obmolova et al., *Nature* 407:703–710, 2000. With permission from Nature.)

showered with point mutations in an MMR-defective cancer cell. These more subtle alterations may also be detected and erased by mismatch repair proteins, which are highly sensitive to bulges and loops in the double helix caused by inappropriately incorporated nucleotides. Importantly, the mismatch repair machinery must be able to distinguish the recently synthesized DNA strand from the complementary "parental" strand that served as the template during DNA synthesis; in making this

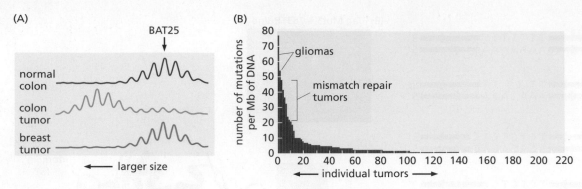

Figure 12.7 Detection of microsatellite instability Defects in mismatch repair (MMR) are responsible for the accumulation of a variety of mutations in the genome. (A) The most obvious type of these mutations is microsatellite instability (MIN), which causes an expansion or contraction of the size of a microsatellite repeat sequence. In the analysis shown here, the size of a mononucleotide repeat is analyzed in a woman suffering from HNPCC (hereditary non-polyposis colon cancer) and presenting in the clinic with both a colorectal and a breast carcinoma; analysis was performed using PCR (polymerase chain reaction). The BAT25 sequence, which is located on human Chromosome 4q12, consists of the sequence TTTTxTxTTTTxT7xxT25, where "x" indicates a nucleotide other than T. As a result of errors made by the polymerase used in the PCR reaction, the products of the PCR reactions show a Gaussian distribution of lengths grouped around the actual length of the genomic DNA segment being amplified. This analysis reveals a clear increase in size of the microsatellite repeat in the colon carcinoma (*leftward shift*), whereas the breast tumor exhibits a microsatellite repeat that is precisely the same as normal, control DNA. (This observation strongly suggests that the breast carcinoma, unlike the colon carcinoma, is unlikely to have been caused by MIN.) (B) The coding exons of 518 kinase-encoding genes in the genomes of 210 human tumors of diverse type were sequenced, revealing 1007 somatic mutations in all. As is apparent, the density of mutations (number per megabase) of all types varied dramatically, with one-third of the tumor genomes showing no mutations at all (*rightward*). Two classes of tumors showed especially high densities of mutations in their genomes. Gliomas/glioblastomas, which had previously been exposed during the course of treatment to high doses of temozolomide, a mutagenic chemotherapeutic agent, exhibited an unusually high density of mutations. In addition, the genomes of a group of five MMR-deficient tumors showed an almost-equivalent density of mutations, consisting largely of base substitutions (14–40 per megabase) plus insertions/deletions of microsatellite repeats (5–12 per megabase). (A, from A. Müller et al., *Cancer Res.* 62:1014–1019, 2002. With permission from American Association for Cancer Research. B, from C. Greenman et al., *Nature* 446:153–158, 2007. With permission from Nature.)

distinction, the MMR apparatus is able to direct its attention to removing and then repairing the recently synthesized and therefore defective DNA strand (see Figure 12.6C). Like the proofreading function described above, mismatch repair involves the excision of the nucleotides that have created the mismatch and a new attempt at synthesis of this strand.

Working together, these various error-correcting mechanisms yield extremely low rates of miscopied bases that survive to become mutant DNA sequences. To begin, DNA polymerases make copying mistakes in only about 1 out of 10^5 polymerized nucleotides. The $3' \rightarrow 5'$ proofreading by the polymerases overlooks only about 1 out of every 10^2 nucleotides initially miscopied by the polymerase, thereby reducing the error rate to about 1 in 10^7 nucleotides. After the DNA polymerase has passed through a stretch of DNA followed by polymerase-associated proofreading, the mismatch repair proteins check the recently synthesized DNA strand a second time. The mismatch repair enzymes fail to correct only about 1 miscopied base out of 100 that have escaped the proofreading actions of the DNA polymerase. Together, this yields a stunningly low mutation rate of only about 1 nucleotide per 10^9 that have been synthesized during DNA replication, yielding only several miscopied bases per haploid genome following each cycle of cell growth and division. As we will see, defects in these error-correcting mechanisms can lead to both familial and sporadic human cancers.

Finally, DNA replication holds yet other dangers for the genome. Some measurements indicate that as many 10 double-strand (ds) DNA breaks occur per cell genome each time a cell passes through S phase. These breaks appear to occur near replication forks, ostensibly because the single-strand DNA at the unwound but not-yet-replicated portion of the parental DNA is susceptible to inadvertent breakage (**Figure 12.8**). Cells have well-developed mechanisms for dealing with such dsDNA breaks, as we will see later. Failure to repair such breaks properly can lead to disastrous consequences, including chromosomal breaks and translocations.

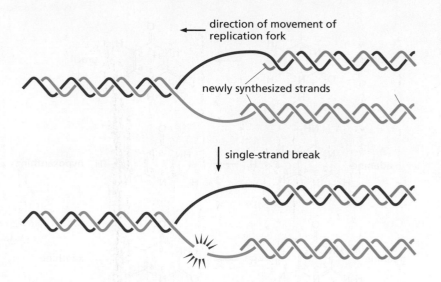

direction of movement of
replication fork

newly synthesized strands

single-strand break

Figure 12.8 Double-strand DNA breaks at replication forks During DNA replication, the DNA molecules are especially vulnerable to breakage in the single-stranded portions near the replication fork that have not yet undergone replication. Such breaks often occur because of the chemical alteration of a base (discussed later in this chapter) that causes the DNA polymerase to stall, being unable to recognize this altered base; polymerase stalling results in a region of single-stranded DNA that may persist for an extended period of time before it is finally protected by successful replication and thus formation of a complementary strand. The breakage of such a single-stranded region (sometimes termed a "collapsed" replication fork) is functionally equivalent to a double-strand break occurring in an already-formed double helix, in that the break leaves two helices unconnected by either strand.

12.4 Cell genomes are under constant attack from endogenous biochemical processes

Most accounts of the origins of contemporary cancer research contain a strong emphasis on the actions of carcinogenic agents that enter the body through various routes, attack DNA molecules within cells, and create mutant cell genomes that occasionally cause the formation of cancer cells (see Section 11.11). Unrecognized by these models are the mutagens and mutagenic mechanisms of endogenous origin. In recent decades, however, analytical techniques of greatly improved sensitivity have allowed researchers to detect altered bases and nucleotides in the DNA of normal cells that have not been exposed to exogenous mutagens. The results of these analyses have caused a profound shift in thinking about the origins of most of the mutant genes present in the genomes of human cells, because they have shown that endogenous biochemical processes usually make far greater contributions to genome mutation than do exogenous mutagens. Because mutagenic events, independent of their origin, are potentially carcinogenic, there has been a rethinking of the cause of many human cancers.

An important contributor of these endogenous mutational processes is deamination, in which the amine groups that protrude from cytosine, adenine, and guanine rings of the bases are lost. This deamination leads respectively to uracil, hypoxanthine, and xanthine (**Figure 12.9**). The uracil, for instance, may be read as a thymine during subsequent DNA replication, thereby causing a C-to-T (C>T) point mutation, known as a **transition** mutation, in which one pyrimidine replaces another. The bases generated by deamination are all foreign to normal DNA, and consequently can be recognized as such and removed by specialized DNA repair enzymes. However, any such altered bases that escape detection and removal represent potential sources of point mutations.

The rate of spontaneous deamination of the 5-methylcytosine base—the methylated form of the cytosine in CpG dinucleotides that we encountered earlier (see Section 7.7)—is even higher, yielding thymine (see Figure 12.9). This creates a significant problem for the DNA repair apparatus, because thymine (unlike the other three products of deamination described above) is a component of normal DNA, and the T:G base pair may therefore escape detection, survive, and ultimately serve as template during a subsequent cycle of DNA replication, leading to a C>T point mutation.

In fact, this deamination of 5-methylcytosine represents a major source of point mutations in human DNA, including in cancer genomes as well as the genomes of otherwise-normal cells as individuals age. As discussed in greater detail in Section 12.13, with the advent of high-throughput DNA sequencing methods, it is now possible to obtain comprehensive DNA sequence information from any number of genomic sources. These analyses have clearly demonstrated that the mutation of

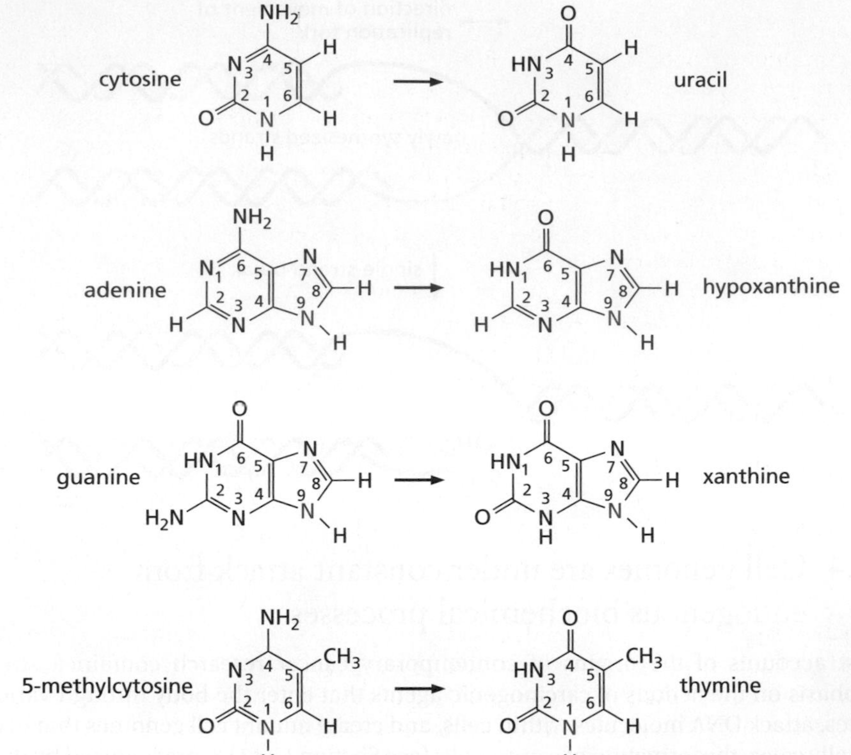

Figure 12.9 Base deamination The deamination reactions affecting purine and pyrimidine bases, which occur spontaneously at various rates at neutral pH, lead to changes in nucleotide sequences unless they are repaired. The deamination of 5-methylcytosine yields thymine (*bottom*); because this base is naturally present in DNA, the repair machinery does not always recognize it as being aberrant, explaining the frequent mutations at sites bearing this methylated base. (In each case, the nitrogen atom in red participates in the formation of a glycosidic bond with the 1-carbon of deoxyribose.) (Adapted from B. Alberts et al., Molecular Biology of the Cell, 7th ed. New York: W.W. Norton, 2022.)

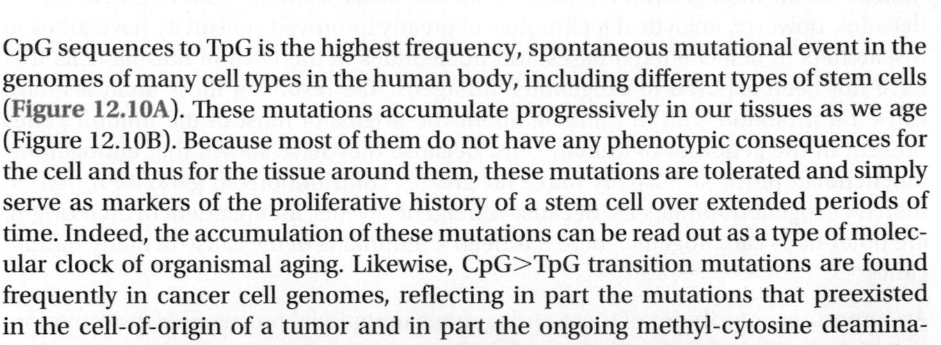

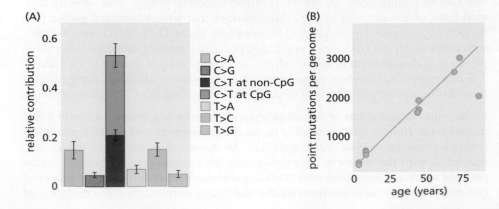

Figure 12.10 Age-associated accumulation of somatic point mutations in human intestinal stem cells Single normal human stem cells were isolated from the small intestines of individuals of different ages, expanded as organoid cultures, and then subjected to whole-genome sequencing (see Section 12.13) to determine the pattern and frequency of mutations as a function of age. Normal blood samples were sequenced from the same individuals for reference. (A) The relative contribution of genome-wide mutations of the pyrimidine bases of base pairs to the three alternative bases is shown for all samples according to the color scheme in the key. Note that C>T mutations occur at the highest frequency, with the majority of these occurring at CpG dinucleotide sequences. This pattern is explained by high frequency of spontaneous deamination of 5-methylcytosine at CpG sites, leading to thymine (see Figure 12.09). (B) Correlation of the number of somatic point mutations in each stem cell sample as a function of the age of the donor. Each data point represents a single small intestinal stem cell sample. The accumulation of CpG to TpG mutations can be considered a type of mutational clock that correlates with organismal age. (From F. Blokzijl et al. *Nature* 538:260–264, 2016. With permission from Nature.)

CpG sequences to TpG is the highest frequency, spontaneous mutational event in the genomes of many cell types in the human body, including different types of stem cells (**Figure 12.10A**). These mutations accumulate progressively in our tissues as we age (Figure 12.10B). Because most of them do not have any phenotypic consequences for the cell and thus for the tissue around them, these mutations are tolerated and simply serve as markers of the proliferative history of a stem cell over extended periods of time. Indeed, the accumulation of these mutations can be read out as a type of molecular clock of organismal aging. Likewise, CpG>TpG transition mutations are found frequently in cancer cell genomes, reflecting in part the mutations that preexisted in the cell-of-origin of a tumor and in part the ongoing methyl-cytosine deamination events occurring during the course of multi-step tumorigenesis. In cancers that are not the product of environmental exposure to highly mutagenic carcinogens (for example, smoking-associated lung cancers or UV-induced melanomas of the skin; see Section 12.13), these CpG>TpG mutations rank among the most common types of genetic change.

The spontaneous breakage of the chemical bond linking a purine or pyrimidine base to its deoxyribose sugar in DNA, termed **depurination** and **depyrimidination**, respectively, also occurs remarkably frequently. By some estimates as many as

10,000 purine bases are lost by depurination each day in a mammalian cell. Depyrimidination occurs at a 20- to 100-fold lower rate, but still results in as many as 500 cytosine and thymine bases lost per cell per day. Estimates of the steady-state level of base-free nucleotides present in a single human genome range from 4000 to 50,000. If not effectively repaired, these events can also lead to mutation.

The intracellular environment holds yet other dangers for the chromosomal DNA. The greatest of these comes from the processes of oxidation, which may inflict far more damage on DNA than the reactions mentioned above. Most important here are the reactions that occur in the mitochondria and generate a variety of intermediates as oxygen is progressively reduced to water:

$$O_2 + e^- \rightarrow \underset{\substack{\text{superoxide} \\ \text{ion}}}{O_2^{\cdot-}} + e^- \rightarrow \underset{\substack{\text{hydrogen} \\ \text{peroxide}}}{H_2O_2} + e^- \rightarrow \underset{\substack{\text{hydroxyl} \\ \text{radical}}}{{}^{\cdot}OH} + e^- \rightarrow H_2O$$

Some of these incompletely reduced intermediates—the so-called reactive oxygen species (ROS)—may leak out of the mitochondria into the cytosol and thence into the rest of the cell. Among them are the superoxide ion, hydrogen peroxide, and the hydroxyl radical—the intermediates in the reactions listed above. (By one estimate, in normal cells 1–2% of oxygen molecules consumed by mitochondria end up as ROS; these levels may increase after mitochondria become more leaky following cell transformation.) Other oxidants arise as by-products of various oxygen-utilizing enzymes, including those in **peroxisomes** (cytoplasmic organelles that are involved in the oxidative breakdown of various cellular constituents, notably lipids); yet others are formed by the spontaneous oxidation of lipids, which results in their peroxidation. Inflammation also provides an important source of the oxidants that favor mutagenesis and therefore carcinogenesis (**Sidebar 12.3**).

The highly reactive molecules produced by these various processes proceed, usually within seconds, to form covalent bonds with many other molecular species in the cell. Among the many targets of ROS attack are the bases within DNA, including both purines and pyrimidines (**Figure 12.11A**). In addition, reactive oxygen species can induce single- and double-stranded DNA breaks, **apurinic** and **apyrimidinic** sites (together known as **abasic** sites), in which bases are cleaved from deoxyribose, as well as DNA–protein cross-links. As described below, many of the resulting altered bases are recognized by specialized components of the DNA repair machinery that proceed to excise them from the DNA. Some of the excised bases, including thymine glycol, which derives from deoxythymidine glycol, and 8-oxoguanine, which derives from 8-oxo-deoxyguanosine (8-oxo-dG), can be detected and quantified in the urine of mammals, providing some indication of the rate at which they are produced throughout the body (Supplementary Sidebar 12.2).

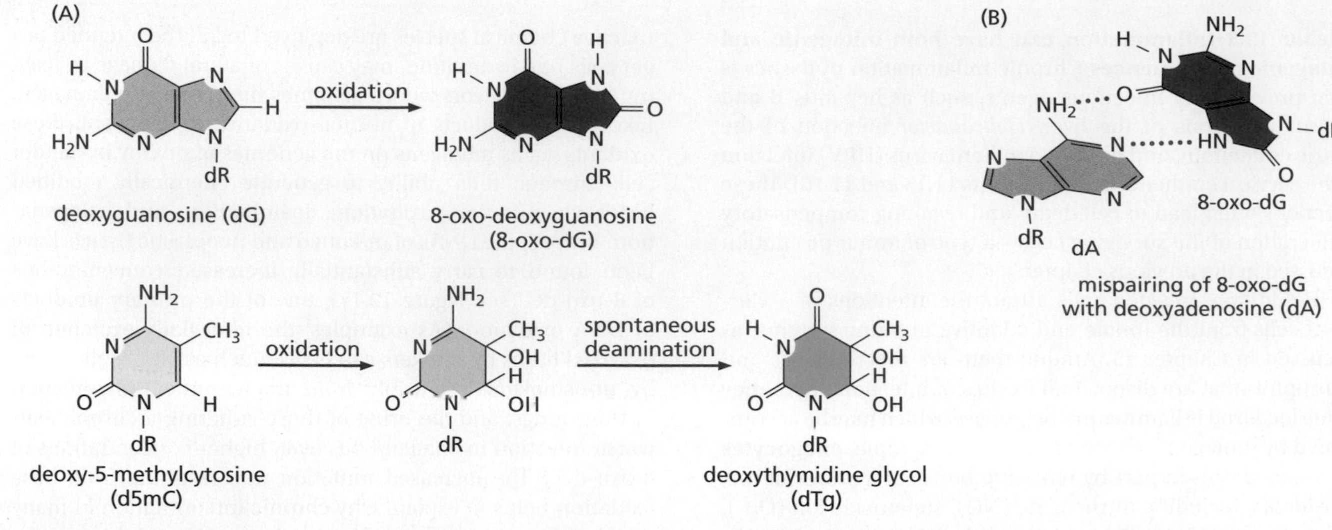

Figure 12.11 **Oxidation of bases in the DNA** The oxidation of DNA bases, which often results from the actions of reactive oxygen species (ROS), can be mutagenic in the absence of subsequent DNA repair reactions. (A) Two frequent oxidation reactions involve deoxyguanosine (dG), which is oxidized to 8-oxo-deoxyguanosine (8-oxo-dG), and deoxy-5-methylcytosine (d5mC), the nucleotide that is present in methylated CpG sequences. Upon oxidation, the latter initially forms an unstable base that rapidly deaminates, yielding deoxythymidine glycol (dTg). (B) The 8-oxo-dG, which arises from the oxidation of dG, can mispair with deoxyadenosine (dA) rather than form a normal base pair with deoxycytosine (dC). Hence, if 8-oxo-dG is not removed from a double helix, the DNA replication machinery may inappropriately incorporate a dA rather than a dC opposite it, resulting in a C>A point mutation. The notation "dR" signifies deoxyribose in all cases. The purines are shown in various shades of red and brown, and the pyrimidines are shown in various shades of green.

Some experiments have shown that the yield of these compounds is directly proportional to the rate of oxidative metabolism in various species. The formation of 8-oxo-dG creates a danger of mutation, as one conformation of this altered base can readily pair with A. This mispairing of bases during DNA replication can lead, in turn, to the replacement of a G:C base pair, via G:A pairing, to a T:A base pair (see Figure 12.11B). Such a G → T replacement of a purine by a pyrimidine (or the opposite) is often termed a **transversion**. Taken together, the continuing hail of damage from oxidation, depurination, deamination, and methylation, which together may alter as many as 100,000 bases per cell genome each day, greatly exceeds the amount of damage created by exogenous mutagenic agents in almost all tissues.

12.5 Cell genomes are under occasional attack from exogenous mutagens and their metabolites

As we have discussed repeatedly in Chapter 11, cellular genomes are also damaged by exogenous carcinogens, including various types of radiation as well as molecules that enter the body via the food we eat and the air we breathe. Among the best studied of the exogenous carcinogens are X-rays, often termed **ionizing radiation** because of the ionized, chemically reactive molecules that this form of electromagnetic energy creates within cells. As much as 80% of the energy deposited in cells by X-rays is thought to be expended in stripping electrons from water molecules. The resulting free radicals proceed to generate ROS that create single- and double-strand breaks in the DNA double helix. As discussed later, these double-strand breaks (DSBs) are often difficult to repair and may, on occasion, generate breaks in a chromosome that are visible microscopically during metaphase.

Ultraviolet (UV) radiation from the sun is a far more common source of environmental radiation than X-rays. Living organisms have had to contend with UV radiation since life first formed on this planet some 3.5 billion years ago. Once oxygen accumulated to high levels in the atmosphere about 0.6 billion years ago, the ozone formed from atmospheric oxygen provided a protective shield that significantly attenuated the flux

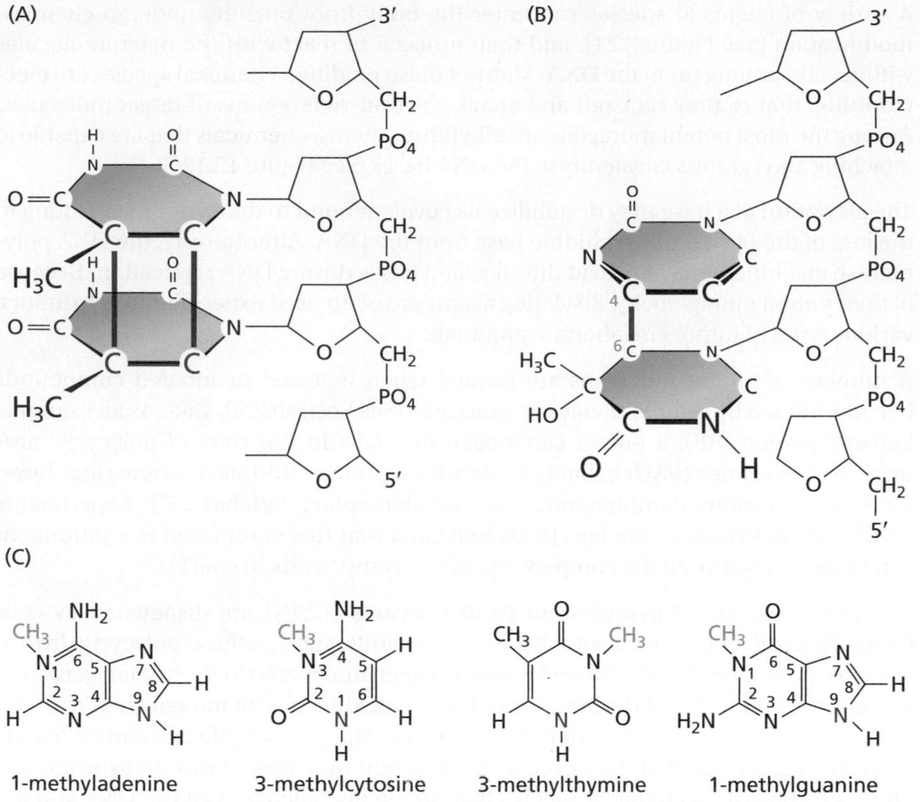

Figure 12.12 Products of UV irradiation and alkylation of DNA Ultraviolet (UV) radiation produces covalent cross-links between adjacent pyrimidine bases in the DNA. These structures are relatively stable chemically and must be removed by transcription-coupled repair and global genomic repair (described in Section 12.7). (A) When purified DNA is irradiated with 254-nm photons, 71% of the photoproducts are cyclobutane pyrimidine dimers (CPD). The cyclobutane ring of a CPD is highlighted in *red*. (B) About 24% of the photoproducts generated from this irradiation are pyrimidine (6–4) pyrimidinone (6–4 PP). The covalent bond linking the 6-position of one pyrimidine to the 4-position of the adjacent pyrimidine is highlighted in *red*. (C) Alkylating agents of exogenous origin can covalently alter DNA bases by the attachment of alkyl groups, such as the methyl groups (*orange*) shown here. Many of these methyl groups may also be generated endogenously by the inadvertent actions of S-adenosyl methionine, which carries a highly reactive methyl that plays a key role in many normal biosynthetic reactions. The nitrogens that form glycosidic linkages with deoxyribose are shown in *pink*.

of UV radiation striking the Earth's surface. Nonetheless, a significant amount of UV still succeeds in penetrating the ozone shield and reaching the biosphere.

Should UV photons strike a DNA molecule in one of our skin cells, a frequent outcome is the formation of pyrimidine dimers—that is, covalent bonds form between two adjacent pyrimidines in the same strand of DNA. In principle, these can form between two adjacent C's, two adjacent T's, or a C and an adjacent T. In mammals, where the percentages of A's, C's, G's, and T's are similar, more than 60% of the pyrimidine dimers are TT and perhaps 30% are CT dimers, with the remaining dipyrimidines being CC dimers. As seen in **Figure 12.12A**, a pair of covalent bonds forms between adjacent pyrimidines, resulting in the creation of a 4-carbon (cyclobutane) ring. Another, less common class of DNA photoproducts, termed pyrimidine (6–4) pyrimidinone, also involves covalent linkage between two adjacent pyrimidines (Figure 12.12B). Once formed, pyrimidine dimers are very stable and can persist for extended periods unless DNA repair enzymes recognize and remove them.

The pattern of mutations observed in genome sequencing studies of melanomas of the skin, as well as basal and squamous skin cancers, demonstrates dramatically that these pyrimidine dimers are mutagenic. In these cancers, the genomes are littered with dipyrimidine substitutions. Although the TT dimer is the one most frequently formed by UV radiation, it is only weakly mutagenic because various DNA repair and replication enzymes, to be discussed in Section 12.7, are able to deal with it effectively. The efficient repair of TT dimers explains why CC>TT substitutions, which arise from CC (rather than TT) dimers, are the most common sources of UV light mutagenesis found in UV-induced cancers. UV photons characteristically cause this mutation, providing further support for the notion that UV rays are directly mutagenic and carcinogenic for the human skin. Interestingly, melanomas that arise in non-sun exposed areas of the body (for example, the soles of the feet) do not show this signature of UV exposure. Other evidence for a direct mutagenic role of UV radiation comes from the observed incidence of squamous cell skin carcinomas in the human population, which doubles with each 10-degree decline in latitude, reaching its peak at the equator where cumulative UV exposure is highest.

A variety of chemical species can enter the body from outside, undergo chemical modification (see Figure 2.21), and then proceed to react with the macromolecules within cells, among them the DNA. Many of these modified chemical species are **electrophilic**, that is, they seek out and attack electron-rich regions of target molecules. Among the most potent mutagens are **alkylating** agents, chemicals that are capable of attaching alkyl groups covalently to the DNA bases (see Figure 12.12C).

The alkylation of a base may destabilize its covalent bond to deoxyribose, resulting in the loss of the purine or pyrimidine base from the DNA. Alternatively, the DNA polymerase machinery may misread the alkylated bases during DNA replication. Because of their potent mutagenicity, alkylating agents are often used experimentally to induce various types of tumors in laboratory animals.

A number of potent mutagens are formed when ingested or inhaled compounds become altered by cellular metabolic processes (see Section 2.9). Take, as an example, benzo[a]pyrene (BP), a potent carcinogen that falls in the class of polycyclic aromatic hydrocarbons (PAHs), that is, molecules carrying multiple benzene rings fused together in various combinations (see Supplementary Sidebar 2.7). Experiments conducted in Britain in the late 1920s indicated that this compound is a prominent carcinogen found amid the complex mixture of compounds in coal tar.

An elaborate array of **cytochrome P450 enzymes** (CYPs) are dispatched by cells largely in the liver to oxidize **xenobiotic** compounds, such as these polycyclic hydrocarbons. (The genes for 57 distinct P450s have been uncovered in the human genome.) The intended function of these enzymes is the detoxification of foreign chemical species, converting them by oxidation into molecules that are soluble and can be readily excreted (**Figure 12.13A**). However, an inadvertent outcome of this detoxification is often the creation of chemical species that are highly reactive with the DNA and are therefore actively mutagenic (Figure 12.13B). As a consequence, chemically inert, unreactive **procarcinogens** are converted into highly reactive **ultimate carcinogens** that can attack DNA molecules directly through their ability to form covalent bonds

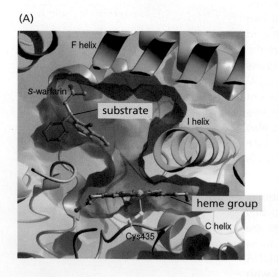

(A)

(B)

benzo[a]pyrene
(procarcinogen)

CYP1A1
+ NADPH + O₂

+ H₂O

CYP1A1
+ NADPH + O₂

benzo[a]pyrenediolepoxide
(BPDE)
(ultimate carcinogen)

Figure 12.13 Actions of cytochromes on procarcinogens (A) Cytochrome P450s (CYPs) are involved in the biosynthesis of a variety of metabolites such as steroid hormones, cholesterol, and bile acids as well as in the degradation of compounds such as fatty acids and steroids. In addition, they aid in the oxidation and associated detoxification of xenobiotics (compounds originating outside the body), such as drugs and carcinogens. The substrate-binding cavity of human CYP2C9, shown here, carries a molecule of a xenobiotic substrate, in this case warfarin (used both as an anticoagulant drug and as a rat poison). The heme ring detoxifies the warfarin by oxidizing it. The large substrate-binding cavities of CYPs allow them to accommodate a wide range of substrates, most of which are quite hydrophobic. (B) Among the xenobiotic compounds entering the body are a variety of polycyclic aromatic hydrocarbons (PAHs) that derive from tobacco smoke, broiled foods, and polluted environments. A common PAH is benzo[a]pyrene (BP, *far left*), which, following two successive oxidation reactions mediated by cytochrome P450 enzymes (largely CYP1A1), is converted to benzo[a] pyrenediolepoxide (BPDE; *far right*). This highly reactive molecule is termed an ultimate carcinogen (Section 2.9) because, unlike its BP precursor, it is able to directly attack and form covalent adducts with DNA bases, which may then generate oncogenic mutations. Importantly, exposure of a cell to PAHs increases synthesis of CYP1A1, thereby accelerating these reactions. (A, from P.A. Williams et al., *Nature* 424:464–468, 2003. With permission from Nature. B, from E.C. Miller, *Cancer Res.* 38:1479–1496, 1978. With permission from American Association for Cancer Research.)

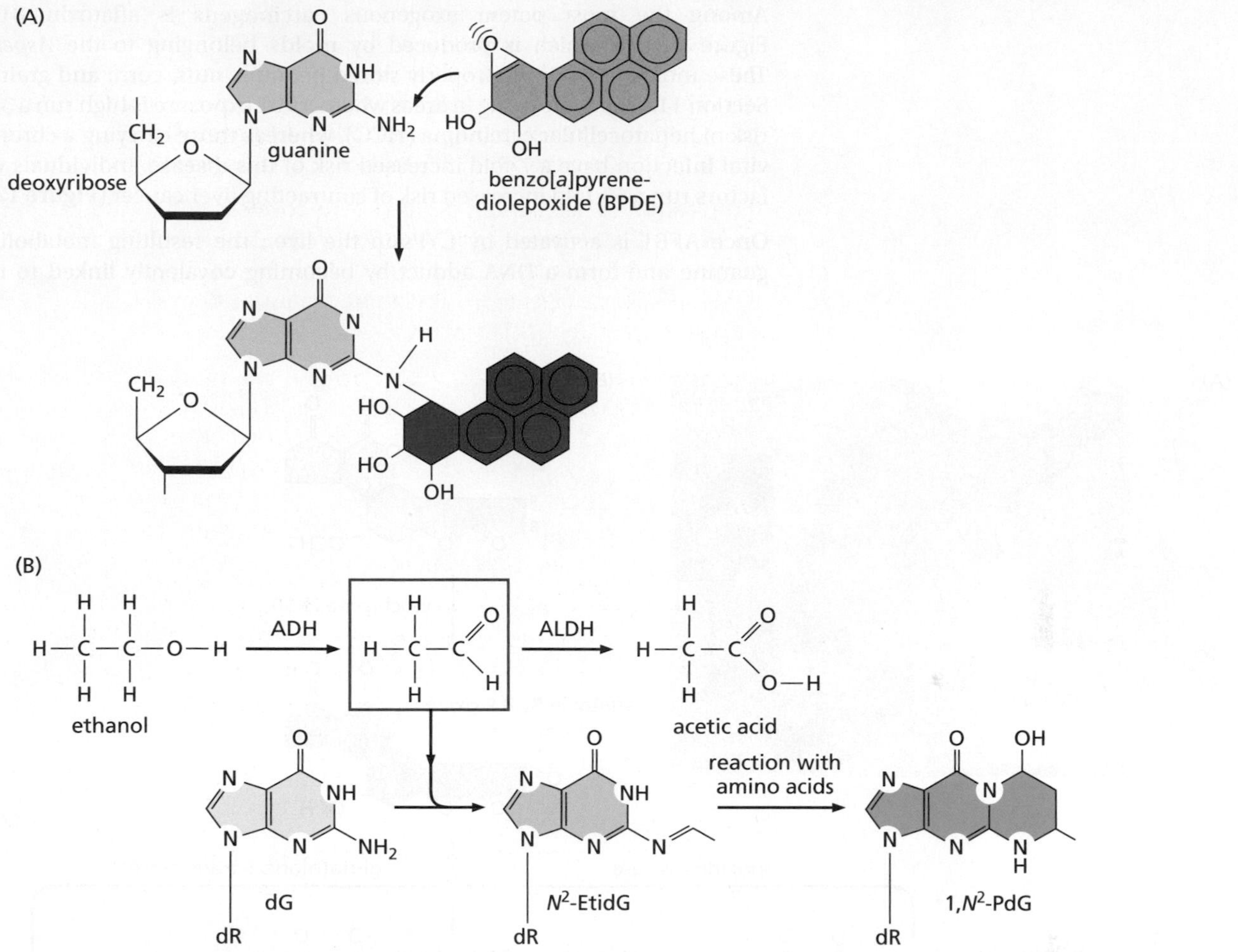

Figure 12.14 DNA adducts (A) The chemically reactive epoxide group of benzo[a]pyrenediolepoxide (BPDE; see Figure 12.13B) can attack a number of chemical sites in DNA, including the extracyclic amine of guanine (shown here), as well as the two ring nitrogens and the O^6 of this base. Because the cell may remove these various adducts with different efficiencies, the O^6 adduct of benzo[a]pyrene (BP) may be more potently mutagenic than the more frequently formed adduct shown here. Even though polycyclic aromatic hydrocarbons and BP have been studied for more than half a century, their precise contributions to human cancer development remain a matter of debate. (B) Ethanol, which is consumed in large quantities by many people, is converted by alcohol dehydrogenase (ADH) to acetaldehyde (*red box*), which has been classified as a mutagen because of its reactivity with DNA. The most abundant DNA adduct resulting from the reaction of acetaldehyde with deoxyguanosine (dG) is *N*^{2}-ethylidene-dG (*N*^{2}-EtidG), a relatively weak mutagen. A more complex adduct that is quite mutagenic is 1,*N*^{2}-propano-2'-deoxyguanosine (1,*N*^{2}-PdG), which forms from the reaction of *N*^{2}-EtidG with amino acids. Heavy drinkers with inborn alleles leading to either highly active ADH or slow acetaldehyde breakdown (via aldehyde dehydrogenase, ALDH) exhibit an especially elevated risk of developing oral and esophageal carcinomas.

with various bases (see Section 2.9). The chemical entity formed after reaction of a carcinogen with a DNA base is often termed a DNA **adduct** (**Figure 12.14**).

In most cases, chemically reactive, ultimate carcinogens attack other molecules almost immediately after being formed. Consequently, they have lifetimes as free molecules that are measured in seconds, and consequently many of the genetic lesions they create arise in the same cells where these molecules underwent initial metabolic activation. For example, BP (benzo[a]pyrene), which is a prominent carcinogenic component of tobacco smoke, is often activated in the first cells that it enters—the epithelial cells (**pneumocytes**) in the lungs of smokers. Once formed, the activated derivative, benzo[a]pyrenediolepoxide (BPDE), proceeds directly to form adducts with the guanosine residues in the DNA of these epithelial cells (see Figure 12.14A). Likewise, the immediate downstream product of ethanol metabolism, acetaldehyde, is highly reactive with deoxyguanosine, forming several distinct DNA adducts that are much smaller than the bulky adduct generated by BP (see Figure 12.14B).

Among the most potent exogenous carcinogens is aflatoxin B1 (AFB1; see Figure 2.21B), which is produced by molds belonging to the *Aspergillus* genus. These molds grow on improperly stored peanuts, nuts, corn, and grains. As cited in Section 11.13, people living in areas where AFB1 exposure is high run a 3-fold elevated risk of hepatocellular carcinoma (HCC), whereas those carrying a chronic hepatitis B viral infection have a 7-fold increased risk of this disease. Individuals with both risk factors run a 60-fold increased risk of contracting liver cancer (**Figure 12.15A**).

Once AFB1 is activated by CYPs in the liver, the resulting metabolite can attack guanine and form a DNA adduct by becoming covalently linked to this base (see

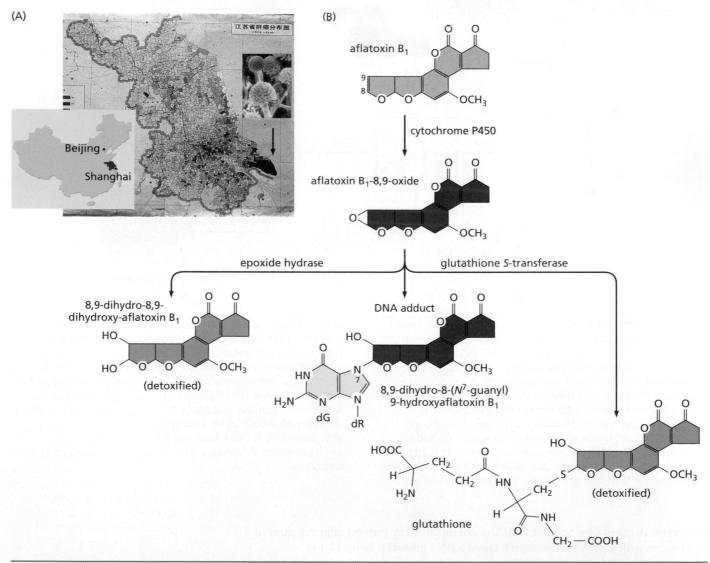

Figure 12.15 Aflatoxin and liver carcinogenesis (A) The fungal toxin aflatoxin B$_1$ (AFB1) is made by molds, largely *Aspergillus flavus* (*inset*), that grow on peanuts and grains that have been stored improperly, notably in areas of high rainfall and humidity; it is one of the most potently mutagenic substances known (see Figure 2.21B). Within the Jiangsu province of eastern China, the incidence of hepatocellular carcinoma (HCC) cases (*brown dots*) is 8-fold higher in the very humid southeastern Qidong peninsula (*arrow*) than in the northwestern parts of the province. The incidence of hepatitis B virus (HBV) infection—a critical co-carcinogen in HCC pathogenesis (see Section 11.13)—is relatively constant across the province. (B) Activation of AFB1 (*pink*) by cytochrome P450s results in formation of the highly reactive 8,9-oxide form (*red*). This form may be detoxified through several side reactions (*bottom left, right*), or it can react directly with DNA, forming a covalent adduct with the N^7 atom of guanosine (*light green*) that is highly mutagenic. Indeed, the liver carcinomas of individuals living in areas of high AFB1 exposure often carry mutant p53 alleles with a characteristic G-to-T transversion in codon 249—precisely the type of mutation that would be expected from the known reactivity of AFB1. (A, from T.W. Kensler et al., *Nat. Rev. Cancer* 3:321–329, 2003. With permission from Nature. Inset, courtesy of CABI.org, UK. B, from J.D. Groopman and L.G. Cain, in C.S. Cooper and P.L. Grover, eds., *Interactions of Fungal and Plant Toxins with DNA in Chemical Carcinogenesis and Mutagenesis*. Berlin: Springer-Verlag, 1990. With permission from Springer.)

Figure 12.15B). AFB1 causes a characteristic G>T mutation in DNA. Such point mutations, where the sequence AGG has been converted to AGT, are found at codon 249 of the *p53* tumor suppressor gene in about half of the hepatocellular carcinomas occurring in individuals exposed to this carcinogen. These characteristic changes in the DNA provide compelling evidence of the direct interaction of this mutagenic carcinogen with bases in the DNA. (These conclusions are important, because in the absence of such evidence, it becomes possible that a carcinogenic agent is actually functioning as a non-mutagenic tumor promoter rather than an "initiating carcinogen"; see Section 11.11).

Another widely studied example of carcinogens of exogenous origin involves the **heterocyclic amines** (HCAs), a class of molecules that are formed in large amounts when meats of various sorts are cooked at high temperatures (**Figure 12.16A**). These

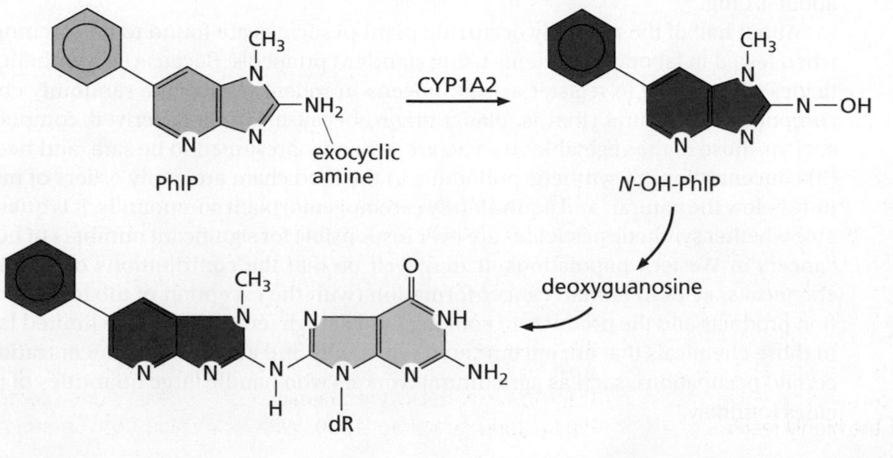

Figure 12.16 Heterocyclic amines
(A) The 10 heterocyclic amines (HCAs) shown here are the most common HCAs to which humans are exposed. This class of compounds derives its name from being composed of multiple fused rings, which are generally formed from both carbon and nitrogen and bear one or more exocyclic amine groups protruding from the ring structure. HCAs arise through cooking foods, notably red meats, at high temperature. PhIP has been estimated to constitute two-thirds of the total dietary intake of HCAs among Americans. (B) The oxidation of the exocyclic amine of PhIP (2-amino-1-methyl-6-phenylimidazo[4,5-b] pyridine; *pink*) by CYP1A2, a cytochrome P450, leads to the highly reactive compound *N*-OH-PhIP (*red*). It can react with the 8-C of deoxyguanosine (*green*) to form a mutagenic adduct (*lower left*). (A, from T. Sugimura, *Carcinogenesis* 21:387–395, 2000. With permission from Oxford University Press. B, adapted from M. Nagao and T. Sugimura, eds., *Food-Borne Carcinogens.* New York: John Wiley & Sons, 2000.)

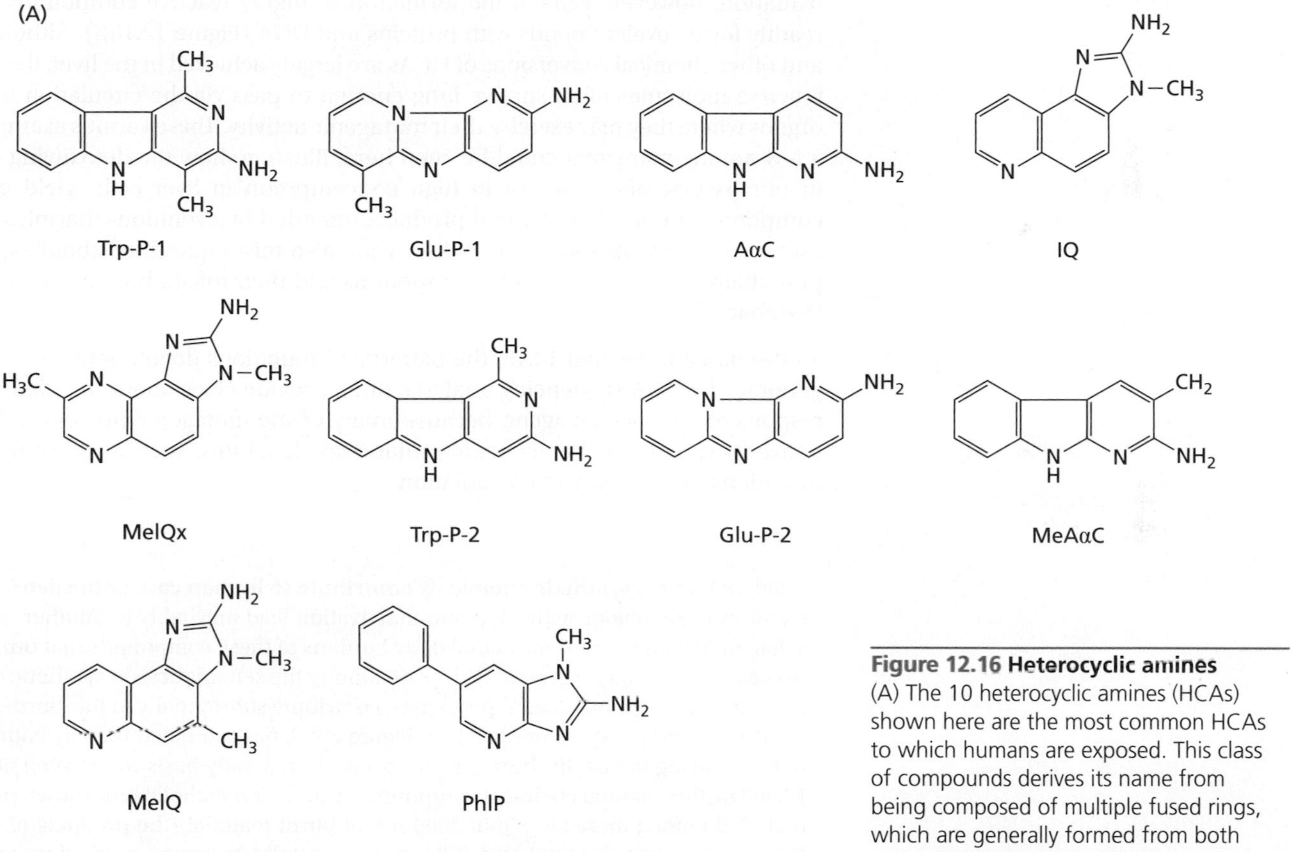

compounds arise through the chemical reactions that take place between naturally occurring molecular species in cells, notably creatine, glucose phosphates, dipeptides, and free amino acids. The HCAs are undoubtedly carcinogenic. For example, the most abundant of these compounds in meats cooked at high temperature—2-amino-1-methyl-6-phenylimidazo[4,5-b]pyridine (PhIP)—is capable of inducing colon and breast carcinomas in rats and lymphomas in mice. PhIP is recognized as being the principal HCA in the human diet. Nonetheless, these and other observations still do not prove that these chemical species are actual causal agents of significant numbers of human tumors, and alternative mechanistic models of meat-induced carcinogenesis have been proposed (Supplementary Sidebar 12.3).

Once heterocyclic amines have entered into cells, CYPs are employed by the cells to oxidize these molecules. Some CYPs will oxidize the rings of heterocyclic amines, whereas others will oxidize the **exocyclic** amine groups, that is, those that protrude from the rings. Ring oxidation by CYPs leads to successful detoxification; amine group oxidation, however, leads to the formation of highly reactive compounds that can readily form covalent bonds with proteins and DNA (Figure 12.16B). Although these and other chemical conversions of HCAs are largely achieved in the liver, the resulting reactive molecules often survive long enough to pass via the circulation into other organs where they may exercise their mutagenic activity. These various examples, only a few among many that could be cited here, illustrate how the detoxifying enzymes in our tissues, often present in high concentration in liver cells, yield genotoxic compounds rather than the end products intended by evolution—harmless, readily excretable chemical species. This discussion also raises questions about exposure to potentially mutagenic synthetic compounds and their role in human carcinogenesis (Sidebar 12.4).

As discussed in Section 12.13, the patterns of mutations documented in cancer cell genomes by DNA sequencing analyses often provide clues about the identity of the responsible mutagenic agent, because many of the mutagenic processes described above leave behind characteristic mutational signatures, such as the dipyrimidine mutations associated with UV radiation.

Sidebar 12.4 Do synthetic chemicals contribute to human cancer burden? The discussions of xenobiotic activation and inactivation lead inevitably to another question: What are the origins and the actual daily burdens of these compounds that our various tissues must routinely contend with? And among the xenobiotics, do synthetic carcinogens, such as the much-feared pesticides, contribute substantially to this burden?

Bruce Ames, of the Ames test (see Figure 2.21), has estimated that, by eating naturally occurring foodstuffs, humans are exposed on a daily basis to between 5000 and 10,000 distinct natural chemical compounds and their metabolic breakdown products. Included among these are about 2000 mg of burnt material (the products of cooking food at high temperatures) and 1500 mg of naturally occurring pesticides (manufactured by plants to protect themselves against insect predators). In contrast, the average daily exposure to all synthetic pesticide residues contaminating the food chain is about 0.1 mg.

About half of the naturally occurring plant pesticides are found to be carcinogenic when tested in laboratory rodents using standard protocols. Because (1) synthetic pesticides are as likely to register as carcinogens in rodent tests as are randomly chosen compounds of natural (that is, plant) origin, because (2) plant-derived compounds, such as those in the vegetables we eat, are generally presumed to be safe, and because (3) concentrations of synthetic pollutants in the food chain are many orders of magnitude below the natural (and equivalently carcinogenic) plant compounds, it is questionable whether synthetic pesticides are ever responsible for significant numbers of human cancers in Western populations. It may well be that the contributions of xenobiotic chemical species to human cancer formation (with the exception of tobacco combustion products and the products of cooking food at high temperature) are limited largely to those chemicals that are encountered repeatedly and at very high concentrations in certain occupations, such as agricultural workers who handle large quantities of pesticides routinely.

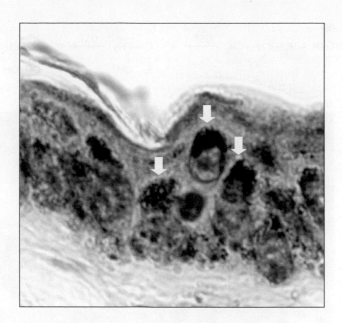

Figure 12.17 Physical shielding of keratinocyte nuclei from ultraviolet radiation The main protection that keratinocytes in the skin have from ultraviolet (UV) radiation, notably UVB photons, derives from the melanosomes—vesicles carrying melanin pigment that have been transferred from melanocytes into keratinocytes in the basal layers of the epidermis. As seen here, once the melanosomes are acquired by the keratinocytes, they are assembled into tiny sun umbrellas (sometimes called supranuclear caps) that sit above keratinocyte nuclei (*arrows*) and shield them from visible and, more importantly, UVB radiation. Keratinocyte nuclei that lack these umbrellas sustain as much as fourfold more UV-induced DNA damage than those that carry them. Moreover, redheads, who produce little or no black eumelanin pigment in their hair and skin (but do make the red/yellow pheomelanin), have an almost fourfold increased risk of developing melanoma relative to those with dark brown or black hair. (Courtesy of D.E. Fisher.)

12.6 Cells deploy a variety of defenses to protect DNA molecules from attack by mutagens

The most effective way for a cell to defend its genome from disruption by mutagenic agents is to *physically* shield its DNA molecules from direct attack. In the case of ultraviolet rays from the sun, these penetrate poorly into the body's tissues, leaving the cells of the skin and pigmented cells in the retina as the only vulnerable targets of mutation. The skin shields itself from UV radiation using the **melanin** pigment that is transferred by **melanocytes** to keratinocytes located in the basal region of the epidermis (the epithelial layer of the skin; **Figure 12.17**).

Skin color in humans is determined by the types and amounts of melanin that are donated by the melanocytes to the keratinocytes. The role of this pigmentation in cancer pathogenesis is highlighted by the oft-cited case of skin cancers in Australia. There, a high flux of UV radiation (because of proximity to the equator) and a lightly pigmented population (deriving until recently largely from the British Isles) combine to create the world's highest incidence of these diseases. In Africa, in contrast, the darkly pigmented human populations living at similar latitudes rarely experience skin cancers. In the case of X-rays and cosmic radiation, by contrast, no effective physical shielding can be erected by the body because these types of radiation can penetrate easily through all biological substances.

These options for protection against physical carcinogens contrast with the large number of mechanisms that cells can deploy to intercept *chemical* carcinogens before they have had the opportunity to damage the cellular genome. The ambushing of ROS and free radicals is the job of a variety of enzymes, including **superoxide dismutase** and **catalase**; they collaborate to detoxify an ROS into unreactive forms of oxygen. The ROS may also be intercepted by a series of free-radical scavengers, including vitamin C, α-tocopherol (vitamin E), bilirubin, and urate. These molecules react chemically with the ROS, thereby detoxifying them.

Yet another important line of defense is erected by enzymes of the class termed **glutathione *S*-transferases** (GSTs), which function to covalently link electrophilic compounds, and thus many carcinogens, with glutathione, thereby detoxifying these compounds and preparing them for further metabolism and excretion (**Figure 12.18A and B**). Significantly, as many as 90% of human prostate adenocarcinomas exhibit a shutdown of glutathione *S*-transferase-π (GST-π) expression as a result of methylation of the promoter of the *GSTP* gene (Figure 12.18C; see Section 7.7). Frequent inactivation of this gene has been reported as well in a number of other human carcinomas. This loss of GST-π expression, which often occurs relatively early in tumor

(A)
glutathione (GSH)

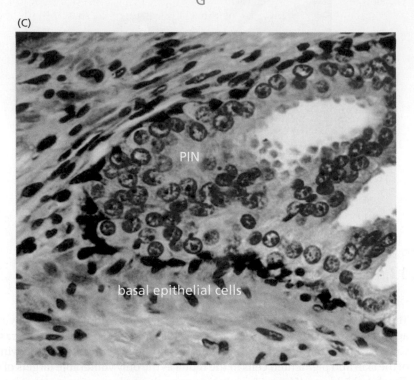

glycine   cysteine   glutamic acid

(B)

(C)

PIN

basal epithelial cells

Figure 12.18 Glutathione *S*-transferase and its loss from tumors (A) Glutathione (GSH) is an unusual tripeptide in which a glutamic acid is attached to the amino group of cysteine via the carboxyl side chain of the glutamic acid. (B) Glutathione *S*-transferase (GST) enzymes use the sulfhydryl (SH; *pink*) group of glutathione's cysteine residue to detoxify a number of reactive compounds before the latter are able to react with cellular target molecules, such as DNA. Shown here is a typical reaction mediated by a GST in which the SH group of glutathione is used to disrupt the highly reactive epoxide group of a compound that has suffered oxidation (where R can be any of a variety of chemical groups). (C) As many as 90% of prostate carcinomas show the loss of expression of an important glutathione *S*-transferase, GST-π; this deprives the tumor cells of the means to detoxify many electrophilic mutagens. In many of these cases, loss of GST-π expression is traceable to methylation of the GSTP1 promoter. In the premalignant PIN (prostatic intraepithelial neoplasia) lesion—a benign precursor of prostate carcinoma—shown here, use of an antibody reactive with GST-π demonstrates the presence of GST-π in the basal epithelial cells (*dark brown, left side*) but the loss of all GST-π expression in the luminal epithelial cells forming the PIN (*light blue, right side*). The luminal epithelial cells are revealed by DAPI, a stain that is specific for DNA (*blue*). (C, from C. Jeronimo et al., *Cancer Epidemiol. Biomarkers Prev.* 11:445–450, 2002. With permission from American Association for Cancer Research.)

progression, suggests that premalignant prostate tumor cells acquire a distinct advantage by inactivating this gene, thereby increasing the mutability of their genomes. Thus, without this enzyme to defuse certain carcinogens, the genomes of these prostate epithelial cells are attacked more often by actively mutagenic carcinogens; the resulting increased rate of mutagenesis likely accelerates the forward march of tumor progression. Another connection between GSTs and other genes involved in carcinogen metabolism and cancer is discussed in Supplementary Sidebar 12.4.

12.7 Repair enzymes fix DNA that has been altered by mutagens

If genotoxic chemicals are not intercepted before they attack DNA, mammalian cells have a backup strategy for minimizing the genetic damage caused by these potential carcinogens. An elaborate DNA repair system exists to continuously monitor the integrity of the genome, to remove inappropriate bases or nucleotides created by chemical or physical attack, and to replace them with the bases/nucleotides that occupied those sites in the DNA prior to the attack. Components of this system also work to stitch together double helices that have been broken by genotoxic agents or by accident during replication. Altogether, mammalian cells depend on more than 160 distinct proteins to ensure that damage to DNA is unlikely to result in this damage being transmitted in the form of a mutation to daughter cells. Some of these DNA repair proteins figure large in the process of human carcinogenesis, because defects in these proteins result in increased rates of mutation, thereby accelerating the overall rate of tumor progression.

Cells deploy a wide variety of enzymes to accomplish the very challenging task of restoring normal DNA structure. Importantly, these functions differ from the mismatch repair (MMR) enzymes described above (see Section 12.3), because the MMR enzymes are largely focused on detecting nucleotides of normal structure that have recently been incorporated by DNA polymerases into the wrong positions, whereas the repair mechanisms discussed in this section detect nucleotides of abnormal chemical structure independent of their origin or location in the genome.

The simplest strategy for restoring the structure of chemically altered DNA involves an enzyme-catalyzed reversal of the chemical reaction that initially created the altered base. For example, one type of DNA alkyltransferase removes methyl and ethyl adducts from the O^6 position of guanine, thereby restoring the structure of the normal base (**Figure 12.19A and B**). The importance of this enzyme [O^6-methylguanine-DNA methyltransferase (MGMT), often referred to simply as DNA alkyltransferase] in the development of certain kinds of human tumors is suggested by observations that the *MGMT* gene is silenced by promoter methylation in as many as 40% of gliomas and

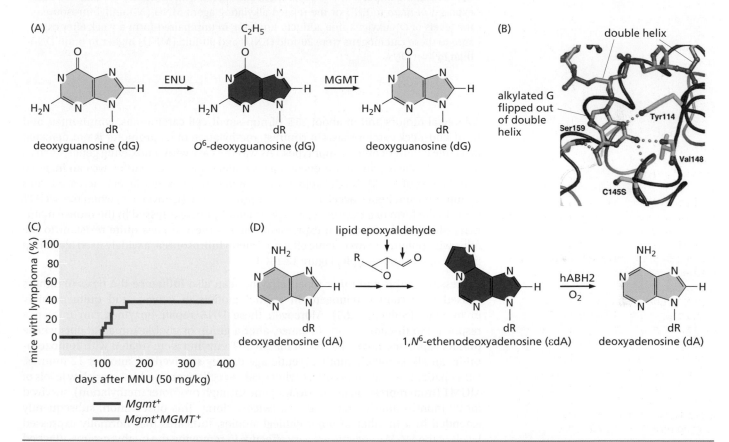

Figure 12.19 Restoration of normal base structure by dealkylating repair enzymes (A) The O^6 position of guanosine is especially vulnerable to alkylation by agents such as ethylnitrosourea (ENU). Unlike many DNA repair enzymes, which respond to altered bases by excising them or the entire nucleotide containing them from the DNA, the enzyme O^6-methylguanine-DNA methyltransferase (MGMT; also known as O^6-alkylguanine-DNA alkyltransferase, AGT) restores an altered deoxyguanosine to its normal structure. It does so by removing the alkyl group from the O^6 atom of guanine. In the absence of such repair, the alkylated deoxyguanosine often leads to a G>A transition mutation. (B) Structural analyses of DNA repair proteins have revealed much about how they function. In the case of the MGMT protein, this enzyme (*below, light and dark blue*) works by flipping the damaged base out of the double helix before removing the alkyl group. Moreover, the reaction between enzyme and substrate is stoichiometric, in that the cysteine 145 (C145S) residue (*below, red and green*) in the enzyme's active site becomes irreversibly alkylated following restoration of normal guanine structure; hence,

each enzyme molecule is able to dealkylate only a single alkylated deoxyguanosine. (C) The effects of MGMT ectopic expression are shown for mice that were exposed to the alkylating carcinogen methylnitrosourea (MNU), which attaches methyl groups to guanosine. Wild-type mice (Mgmt$^+$) were highly susceptible to the induction of thymic lymphomas (*red curve*), whereas transgenic mice that were forced to overexpress the MGMT enzyme because of a transgene in their germ line (Mgmt$^+$ MGMT$^+$, *blue line*) were protected from developing these tumors. (D) Highly reactive lipid epoxyaldehydes (*shown here*) and peroxides (*not shown*), which are common in inflamed tissues, can attack and modify deoxyadenosine (*shown here*) as well as other DNA bases (*not shown*). The AlkB enzyme of bacteria can remove the resulting adducts as well as simpler methyl adducts, such as those shown in Figure 12.12C; a mammalian homolog of AlkB, hABH2 (*indicated here*), acts similarly in human cells. (A, adapted from S.L. Gerson, *Nat. Rev. Cancer* 4: 296–307, 2004. B, from D.S. Daniels et al., *Nat. Struct. Mol. Biol.* 11: 714–720, 2004. With permission from Nature. C, from L.L. Dumenco et al., *Science* 259:219–222, 1993. With permission from AAAS.)

Sidebar 12.5 Expression patterns of repair enzymes explain certain tissue-specific susceptibilities to cancer Mammalian cells seem to express only a single MGMT enzyme. The importance of this activity in influencing carcinogenesis is indicated by experiments in which pregnant rats are exposed to the carcinogen *N*-ethylnitrosourea (ENU) during the fifteenth day of gestation. Virtually all the rat pups that are born succumb to neuroectodermal tumors arising in the central nervous system several months after birth. The MGMT enzyme is expressed at significant levels throughout the bodies of developing embryos and newborns but is only minimally expressed in the central nervous system, which explains the peculiar ability of ENU to preferentially induce these tumors. Consequently, alkylated guanine residues that are formed in the cells of the nervous system persist rather than being quickly removed and are ultimately able to generate the point mutations that are responsible for the creation of the oncogenes in the resulting tumors. In one experiment conducted with newborn rats that had been exposed *in utero* to ENU or the related alkylating agent MNU (*N*-methylnitrosourea), the levels of O^6-alkylguanine adducts surviving in unrepaired form a week after exposure to these carcinogens were 20-fold (ENU) and 90-fold (MNU) higher in brain DNA than in liver DNA.

colorectal tumors and in about 25% of non-small-cell carcinomas, lymphomas, and head-and-neck carcinomas. (In contrast, methylation of its promoter is not detected in a large cohort of other tumor types.) As was the case with detoxifying genes, such as the glutathione *S*-transferase-encoding *GST* alleles discussed earlier, we can imagine that the loss of MGMT's DNA repair function in certain tissues favors increased rates of mutation and hence accelerated tumor progression. (Conversely, when the *MGMT* gene, in the form of a transgene, is experimentally overexpressed in the mouse mammary gland or thymus, such expression renders these tissues quite resistant to the otherwise potently carcinogenic effects of methylnitrosourea, a widely used alkylating mutagen; see, for example, Figure 12.19C).

Expression of this and other repair enzymes can also influence the *types* of tumors caused by certain carcinogens in animal models of cancer, and quite possibly in humans (**Sidebar 12.5**). Moreover, these DNA repair enzymes can influence responses to therapy. In one instance, after a group of glioblastoma patients whose tumors expressed normal levels of the MGMT enzyme were treated with **temozolomide**, an alkylating chemotherapeutic agent, they survived for another 12 months. Other patients in the same cohort, whose cancer cells expressed only very low levels of MGMT (from repression of the *MGMT* gene through promoter methylation), survived for 22 months after therapy—almost twice as long. This observation, subsequently extended in a number of other clinical studies, indicates that normally expressed levels of the MGMT enzyme are very effective in removing the methyl groups attached to DNA by the chemotherapeutic drug, thereby blunting its cytotoxic effects. Cells also use other mechanisms to deal with methylated bases (Supplementary Sidebar 12.5).

Far more important than these dealkylating enzymes, however, are the numerous cellular enzymes that recognize chemically altered bases in the DNA and respond in two other ways, depending on the specific modifications of the DNA. In some cases, specialized enzymes will cleave the bond linking a modified base to the deoxyribose sugar, the process of **base-excision repair** (BER; **Figure 12.20A**). In other cases, the entire nucleotide containing both the base and associated deoxyribose will be cut out, this being the process termed **nucleotide-excision repair** (NER; see Figure 12.20B).

Base-excision repair (BER) tends to repair lesions in the DNA that derive from endogenous sources, such as those attributed to the reactive oxygen species (ROS) and depurination events described earlier (Section 12.4). Nucleotide-excision repair (NER), in contrast, largely repairs lesions created by exogenous agents, such as UV photons and chemical carcinogens (for example, see Figures 12.12 and 12.14). BER seems to concentrate on fixing lesions that do not create structural distortions of the DNA double helix, whereas NER directs its attention to bulky, helix-distorting alterations.

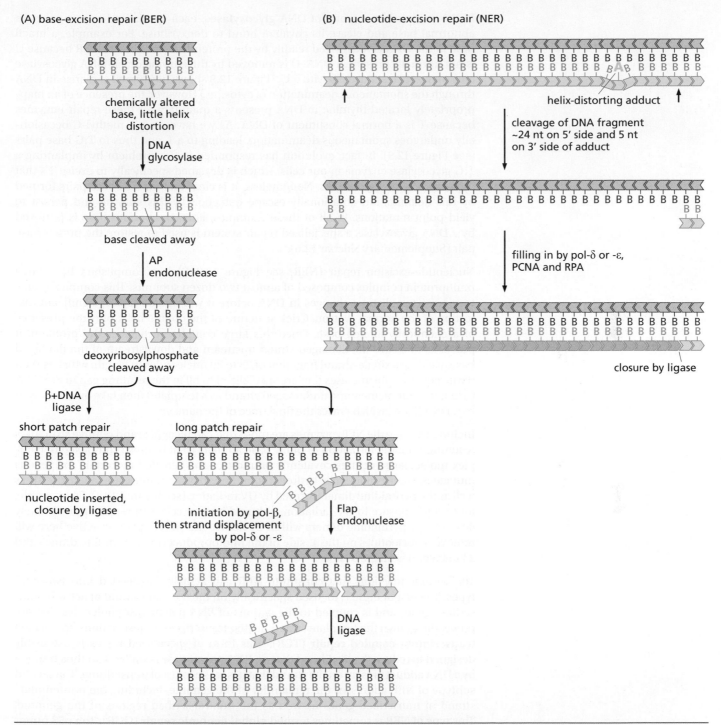

Figure 12.20 Base- and nucleotide-excision repair (A) Base-excision repair (BER) is achieved by enzymes that recognize chemically altered bases having minimal helix-distorting effect. These DNA N-glycosylase enzymes cleave the glycosyl bond linking the altered base (*yellow*) and the deoxyribose. The base-free deoxyribosylphosphate is then excised by an enzyme—apurinic/apyrimidinic endonuclease (APE)—specialized to remove base-free sugars. The resulting single nucleotide gap is filled by DNA polymerase β and sealed by a DNA ligase. In fact, there are two forms of BER. A single nucleotide may be excised and the gap may be filled in by a DNA polymerase (pol-β) and ligated (short patch repair). Alternatively, a strand-displacing DNA polymerase (δ or ε) may, following excision by APE and the addition, as before, of a

single base by pol-β, extend the 3' strand by several nucleotides beyond the original gap. The resulting displaced strand is removed by a Flap endonuclease and the remaining lesion is sealed by a ligase (long patch repair). (B) Nucleotide-excision repair (NER) is accomplished by enzymes that recognize bulky, helix-distorting lesions and cleave the flanking oligonucleotide sequences at sites approximately 24 nucleotides (nt) on the 5' side and about 5 nucleotides on the 3' side. The resulting approximately 29-nt single-strand gap in the DNA is then filled by DNA polymerase δ or ε, acting together with PCNA (proliferating-cell nuclear antigen) and RPA (which binds to single-strand DNA), and is finally sealed by a DNA ligase. The chevrons represent deoxyribose nucleotides, all pointing in a 5'-to-3' direction.

BER is initiated by a group of DNA **glycosylases**, each specialized to recognize an abnormal base and cleave its covalent bond to deoxyribose. For example, a uracil base in the DNA is recognized readily by the proteins responsible for BER because U is not normally present in DNA. U is removed by the enzyme uracil-DNA glycosylase and soon replaced, usually with a C. (Figure 12.9 shows how uracil can arise in DNA through the spontaneous deamination of cytosine.) However, the presence of an inappropriately located thymine in DNA presents a quandary for these repair enzymes because T is a normal constituent of DNA. As we have read, 5-methyl-C occasionally undergoes spontaneous deamination, leading to a T, and thus to T:G base pairs (see Figure 12.9). In fact, evolution has responded to this problem by implanting a T:G glycosylase enzyme in our cells, which is designed specifically to excise T's that happen to arise opposite G's. Nonetheless, it is clear that the T:G base pairs formed by this deamination occasionally escape detection by this enzyme and persist to yield point mutations. In all of these examples, after an aberrant base is removed by a DNA glycosylase, a specialized repair system is used to restore the proper base pair (Supplementary Sidebar 12.6).

Nucleotide-excision repair (NER; see Figure 12.20B) is accomplished by a large multiprotein complex composed of almost two dozen subunits. This complex seems to require two distinct changes in DNA before it will initiate repair: significant distortion of the normal Watson–Crick structure of the double helix plus the presence of a chemically altered base. Once this large complex recognizes the problem, it proceeds to cleave the damaged strand upstream and downstream of the damaged base, yielding a single-strand fragment of 25 to 30 nucleotides in length, which is then removed. DNA polymerases that are specialized to fill in the resulting gap in the DNA (using the complementary, undamaged strand as a template) then take over, followed by a DNA ligase, which erases the final trace of the damage.

Included among the NER enzymes are those that can recognize and remove structures resulting from the formation of bulky base adducts (that is, those composed of complex molecular structures covalently bound to bases) created by certain exogenous mutagens, such as polycyclic hydrocarbons, heterocyclic amines, and aflatoxin B1, as well as the pyrimidine dimers formed by UV radiation (see Figure 12.12). For example, following exposure to UV radiation, cultured human cells can repair approximately 80% of their pyrimidine dimers within 24 hours. The NER apparatus active here will remove 5 nucleotides on the 3′ side of the photoproduct (the pyrimidine dimer) and 24 nucleotides on the 5′ side.

The various reactions that constitute NER can actually be divided into two subtypes. The first of these is focused specifically on the template strand of actively transcribed genes and is coupled to the actions of RNA polymerase molecules that are proceeding down the template strand during transcription; these actions are termed **transcription-coupled repair** (TCR). This form of specialized repair is ostensibly designed to reduce the likelihood that an RNA polymerase is stalled and thus trapped by a DNA adduct present on the strand that it is in the midst of transcribing. The second subtype of NER addresses the remainder of the genome, including the nontemplate strand of transcribed genes as well as the nontranscribed regions of the genome. This type of NER is sometimes termed **global genomic repair** (GGR). The p53 tumor suppressor protein activates expression of several genes encoding NER proteins involved in GGR, explaining the defectiveness of GGR in *p53*-mutant cells; in contrast, transcription-coupled repair is intact in these cells. This defect in GGR holds profound implications for the maintenance of cell genomes in the half of all human tumors in which the *p53* gene is mutant and in many of the remaining tumors in which it is compromised through various alternative mechanisms (see Figure 9.12).

As implied by much of the above discussion, the mechanisms designed to ensure genomic integrity strive to restore DNA molecules prior to their replication in S phase, lest persisting damage in template strands cause the copying of these still-mutant strands and the resulting introduction of mutations into the genomes of daughter cells and their descendants. Beyond these mechanisms, cells have yet another alternative strategy to cope with damaged DNA. This alternative—actually an act of

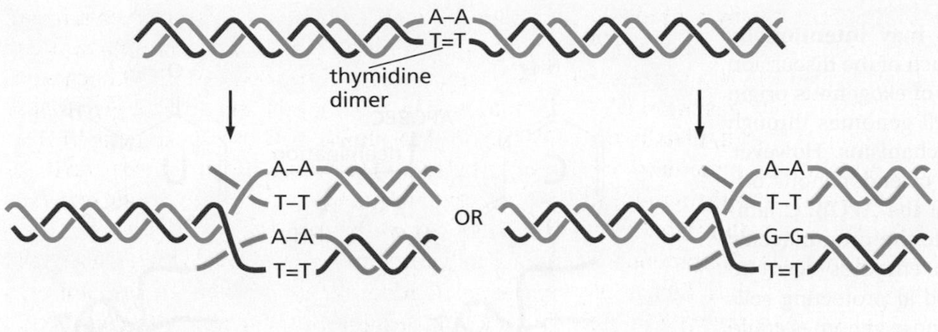

Figure 12.21 Error-prone repair
Error-prone DNA synthesis occurs when an advancing DNA replication fork encounters a still-unrepaired DNA lesion, such as the thymidine dimer shown here. In most cases (left), the error-prone DNA polymerase (sometimes called a bypass polymerase) responds to this lesion in the damaged template strand (*red*) by inserting the appropriate bases (in this instance, an A–A dinucleotide) in the growing DNA strand (*green*). However, in several percent of these encounters (*right*), the error-prone polymerase will fail to properly "guess" the structure of the lesion in the template strand and will instead incorporate a G–G dinucleotide opposite the thymidine dimer.

desperation—involves DNA replication of a still-unrepaired stretch of template-strand DNA occurring in a cell that has already entered S phase. This process is termed *error-prone* DNA replication, because the replication apparatus involved must often "guess" which of the four nucleotides is appropriate for incorporation into the growing DNA strand when it encounters a still-damaged base or set of bases; these guesses are not always correct, leading quite frequently to misincorporated bases and mutant gene sequences (**Figure 12.21**).

To date, at least nine distinct mammalian error-prone human DNA polymerases have been discovered. Some of these can add a nucleotide to a growing strand even when a base in the complementary template strand is missing. Yet others can extend a nascent DNA strand, using as primer a nucleotide that has been misincorporated by another DNA polymerase. A third type can incorporate a base when the corresponding base in the complementary strand carries a bulky, covalently attached DNA adduct that has not yet been removed by nucleotide-excision repair. One of these enzymes, encoded by the *XPV* gene, is highly specialized, being able to recognize the TT thymine dimers created by UV radiation and insert two A's on the opposite strand (see Figure 12.21). Another "bypass polymerase," pol-κ, can also replicate past bulky adducts in the template strand and, in addition, advance through templates containing the much less bulky 8-oxo-deoxyguanosine (see Figure 12.11A); pol-κ incorporates an A more often than a C opposite the 8-oxo-dG, helping to explain the mutagenic effects of this common product of base oxidation. Whereas the DNA polymerases responsible for the bulk of DNA synthesis in a cell have error rates as low as 10^{-5}, these error-prone polymerases generally have error rates as high as 1 misincorporated base per 100 bases replicated.

The mistakes made by the error-prone polymerases would seem to generate unacceptably high rates of mutation in cell genomes. Still, the price paid for accumulating such mutations should be balanced against the alternative: the risk of imminent death confronted by a cell whose DNA replication forks are stalled indefinitely because of difficult-to-copy lesions in its DNA.

DNA polymerase β (pol-β) is error-prone for another reason: This relatively small polymerase molecule lacks the proofreading capabilities of the larger polymerase enzymes (see Section 12.3), which may explain much of its error-prone DNA replication activities. It is usually involved in replacing the nucleotides that have been removed because of BER. In a variety of ovarian carcinoma cell lines, this enzyme has been found to be overexpressed by as much as a factor of 10. The overexpression of the error-prone DNA polymerase β may represent an effective strategy that these cancer cells use to increase the mutability of their genes and hence accelerate the rate of tumor progression. In support of this idea, the forced overexpression of polymerase β in cultured human fibroblasts has been found to encourage microsatellite instability and to increase overall mutation rates as much as threefold.

Although not involved in DNA repair, the APOBEC family of cytosine deaminases is responsible for inserting mutations into the genome of normal and cancer cells. As described in **Sidebar 12.6**, these enzymes are involved in the cellular response to viral infection as well as the hypermutation of immunoglobulin genes during B-cell development. The documentation of APOBEC-mediated mutagenesis is currently increasing rapidly.

Sidebar 12.6 Normal and cancer cells may intentionally inflict damage on their own genomes Much of the discussion in this chapter is centered around agents of exogenous origin that enter cells and inflict damage on cell genomes through a diversity of chemical and physical mechanisms. However, an unexpected source of mutagenesis is of endogenous origin and quite intentional. It derives from the APOBEC family of enzymes, which function to promote the promutagenic process of cytosine deamination. These enzymes seem to have originated in cellular defenses aimed at protecting cells against infecting RNA and DNA viruses, whose genomes could be inactivated by mutation via these enzymes; however, early in vertebrate evolution, at least one of these enzymes became specialized to diversify the genomic sequences that specify the antigen-combining sites of antibodies, a process discussed in detail in Chapter 15.

The best studied of the APOBECs is the **a**ctivation-**i**nduced cytidine **d**eaminase (AID) enzyme, which is normally responsible for deaminating cytidine bases in the 3-kb span downstream of the start site for transcription of the genes specifying antibodies and several other proteins. By stochastically converting cytidine to uridine residues in defined genomic regions (**Figure 12.22**), the APOBEC enzymes effectively insert C-to-T point mutations in the genome; in the case of immunoglobulin (antibody) genes, the resulting "somatic hypermutation" causes diversification of the antigen-binding sites of the encoded antibody molecules, enabling the immune system to develop antibodies of ever-increasing avidity for their antigen ligands. The AID enzyme seems to run amok in some human lymphomas, inserting point mutations in genes that are not its normally intended targets. And when the *AID* gene is ectopically expressed as a transgene in many tissues of the mouse, it causes high rates of T-cell lymphomas, many of which exhibit large numbers of genes, such as *myc*, with point mutations throughout their reading frames. APOBEC3B and other members of the APOBEC family have also been found to be overexpressed in solid cancers, including breast, head and neck, and lung carcinomas. As discussed below in Section 12.13, a number of cancer genome sequencing studies have implicated dysregulated APOBEC gene function in greatly increasing mutation rates in cancer.

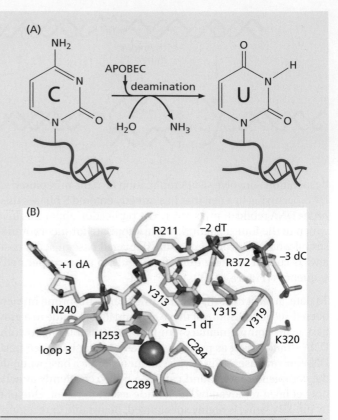

Figure 12.22 APOBEC enzymes induce mutations (A) The APOBEC family of cytosine deaminases catalyzes the deamination of cytosine residues in single-stranded DNA, producing uracil. Failure to repair these uracil residues results in C>T mutations. In addition, DNA repair intermediates such as abasic sites can also lead to C>G mutations and other outcomes. Certain APOBEC enzymes also deaminate cytosine residues in RNA. (B) Model of the active site of APOBEC3B bound to a ssDNA oligonucleotide. The target cytosine base (*red arrow*) can be seen in close proximity to the sidechain of the histidine residue at amino acid 253 (H253). H253 also coordinates a zinc atom (*blue sphere*), which is required for catalysis. (A, from C. Swanton et al., *Cancer Discov.* 5:704–712, 2015. With permission from American Association for Cancer Research. B, from K. Shi et al., *J. Biol. Chem.* 290:28120–28130, 2015. © 2015 ASBMB. This article is available under a Creative Commons Attribution (CC BY 4.0) license.)

12.8 Inherited defects in nucleotide-excision repair, base-excision repair, and mismatch repair lead to specific cancer susceptibility syndromes

In 1874, two Austro–Hungarian physicians, Ferdinand Hebra and Moritz Kaposi, described an unusual syndrome that involved high rates of the development of squamous and basal cell carcinomas of the skin. (Kaposi subsequently described the unusual sarcoma that bears his name; *see* Section 4.6.) As became apparent later, affected individuals have extreme sensitivity to UV radiation, and infants will often suffer severe burning of the skin after only minimal exposure to sunlight (**Sidebar 12.7**). These individuals show dry, parchment-like skin (xeroderma) and many freckles ("pigmentosum"; **Figure 12.23**). In aggregate, individuals suffering from the **xeroderma pigmentosum** (XP) syndrome have a 2000-fold increased risk of skin cancer before the age of 20 compared with the general population and about a 100,000-fold increased risk of squamous cell carcinoma of the tip of the tongue. Skin cancers appear in XP children with a median age of ~10 years, compared with ~60 years in the general population (**Figure 12.24**).

Sidebar 12.7 Degenerate genomes lead to sunlight sensitivity and the need for sunscreen Humans as well as other placental mammals rely totally on the nucleotide-excision repair (NER) system to remove the highly mutagenic UV-induced pyrimidine dimers (see Figure 12.12). This situation contrasts with that seen in bacteria, lower eukaryotes, and plants, where redundant backup systems are in place to deal with these dimers if the first NER line of defense should fail. These simpler organisms, which may exhibit DNA repair capabilities mirroring those of our distant evolutionary ancestors, express (1) enzymes that can monomerize pyrimidine dimers (thereby directly reversing the UV-induced damage), (2) glycosylases to cleave the dimerized bases from deoxyribose residues, and (3) nucleases to incise the DNA strand around the dimers.

It is possible that once our mammalian ancestors learned to grow coats of hair, the evolutionary pressures to retain these important backup DNA repair enzymes receded, and the encoding genes were lost through one or another genetic mechanism. Now, because of a quirk of recent primate evolution, we humans are left without protective coats of hair and, at the same time, without an ability to mobilize the backup repair systems to fix UV-damaged DNA in the event that our only remaining line of defense—the xeroderma pigmentosum genes and encoded proteins—fails to do this task. The lack of these extra layers of protection explains, in part, why the mutant XP phenotypes can be so devastating for those afflicted with this disease and why more light-complexioned people should make lavish use of sunscreen lotion when in the sun.

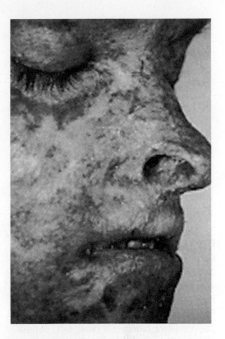

Figure 12.23 A xeroderma pigmentosum patient A patient suffering from xeroderma pigmentosum (XP) has severe and extensive lesions in all areas of sun-exposed skin. These lesions can develop into squamous and basal cell carcinomas as well as melanomas. The tumors develop at rates that are as much as 10,000-fold higher than in the general population. (Courtesy of K.H. Kraemer.)

We now know that inherited defects in any one of eight genes can lead to xeroderma pigmentosum, but this number is arbitrary, as these genes overlap with other genes involved in NER and a variety of other syndromes resulting from defective DNA repair. The genetic complexity of the XP syndrome was first recognized through the use of somatic cell genetics. Thus, cells from two different XP patients were fused in culture in order to determine the repair phenotype of the resulting hybrid cells (**Figure 12.25**; see also Figure 7.1). On many occasions, the hybrids were found to repair DNA normally, indicating that the two parental cells carried defects in DNA repair that were associated with two distinct genes. For example, using nomenclature developed later, cells from an individual carrying a mutant *XPA* gene (and having a wild-type *XPC* gene) were able to repair DNA normally after being fused to cells from an individual carrying a mutant *XPC* gene (and having a wild-type *XPA* gene). Conversely, cells from two individuals, each carrying a mutant *XPA* allele, would continue to exhibit a DNA repair defect. Such collaboration, or "genetic **complementation**," (or lack thereof) led to the classification of XP-associated mutant alleles into eight complementation groups, each ostensibly defined by the identity of a responsible gene. Only years later were the responsible genes isolated by molecular cloning. Almost always, it has been possible to show that an affected individual has inherited two mutant, null alleles of a gene representing one or another XP complementation group. Thus, the pattern of inheritance of XP is autosomal recessive, in contrast to some of the other "caretaker" gene-associated familial cancer syndromes we have discussed in Chapter 7 and will return to below, which exhibit autosomal dominant patterns of inheritance at the organismic level.

Seven of the eight XP-associated genes, termed *XPA* through *XPG*, encode components of the large, multiprotein nucleotide-excision repair (NER) complex. The eighth gene, *XPV*, specifies the error-prone DNA polymerase pol-ε that many cells use when

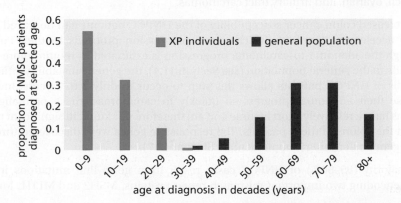

Figure 12.24 Significantly earlier onset skin cancers in individuals with xeroderma pigmentosum (XP) compared to the general population The graph shows the proportion of XP individuals (*blue bars*) and individuals in the general population (*red bars*) who were diagnosed with nonmelanoma skin cancer (NMSC) in different decades of life. The median age of diagnosis was 9 years for XP individuals and 67 years for individuals in the general population. (From P.T. Bradford et al., *J. Med. Genet.* 48:168–176, 2011. With permission from BMJ.)

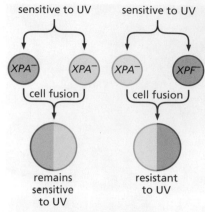

Figure 12.25 Discovery of XP complementation groups Cell fusion experiments using cultured cells from xeroderma pigmentosum (XP) patients have revealed that certain combinations of fibroblasts, each derived from a different patient (*dark blue, light blue, left*), when fused, yielded hybrid tetraploid cells that are as UV-sensitive as the two parental cell populations. On other occasions, however (*right*), fusion of cells from two patients (*light green, light red*) yielded cell hybrids that are as resistant to UV-mediated killing as the cells from normal individuals. Such findings caused the pair of cells that did not complement one another's DNA repair defect (*dark blue, light blue*) to be assigned to the same complementation group, ostensibly because the two populations carried mutations in the same gene. Conversely, successful complementation by other pairs of cells (*light green, light red*) allowed the respective parental cell types to be assigned to two distinct complementation groups, indicating that mutant alleles of at least two distinct genes were involved in predisposing to XP. In this way, eight distinct XP complementation groups were eventually delineated; the responsible genes were isolated by gene cloning years later.

their regular DNA polymerases (for example, pol-δ) are unable to copy over unrepaired DNA lesions in the template strand such as pyrimidine dimers. As mentioned in the last section, error-prone polymerases are able to copy a template strand of DNA containing the still-unrepaired TT dimers created by UV radiation, usually by incorporating two A's in the complementary strand. In general, pol-ε is thought to be so accurate that it incorporates AA nucleotides in the growing DNA strand opposite a TT dimer 95% of the time.

Individuals afflicted with XP also have some increased risk of other diseases, notably neurological problems, which are observed in about 18% of these patients. And mice that have been deprived of one of several XP genes suffer markedly increased susceptibility to tumors following exposure to chemical carcinogens. These two observations provide evidence that components of the nucleotide-excision repair system encoded by some of the XP genes are, not unexpectedly, responsible for repairing genetic damage created by other agents besides UV radiation. This finding raises the following question: Why does a human who lacks the function of one or another XP gene have relatively little increased risk of cancers in internal organs, even though an important component of the NER machinery is missing from all cells throughout this person's body? The simplest and possibly correct explanation is that UV rays are, by far, the most important environmental mutagen to which most humans are exposed, and thus the source of most of the lesions that require repair by the NER machinery. (By one estimate, strong sunlight can inflict as many as 100,000 DNA lesions per skin cell per hour.) Accordingly, skin cancers represent the great majority of cancers in XP patients, with tumors of internal organs being greatly overshadowed and, quite possibly, never registered in these patients.

Several other inherited syndromes are also associated with defects in NER. For example, individuals suffering from Cockayne syndrome (CS) appear to be defective in one of two genes that are involved in transcription-coupled NER. Their cells have increased photosensitivity like those of XP patients. The median age of death of patients from this disease is 12 years of age. This disease is highly unusual, in that the significant defects in DNA repair are not associated with increased rates of cancer. Still unaddressed is the possibility that individuals who are *heterozygous* carriers for one of the XP- or Cockayne syndrome–associated mutant alleles have an elevated risk of developing certain types of cancer. (The fibroblasts of such carriers have been found to exhibit several-fold increased numbers of chromosomal breaks in response to X-irradiation.)

XP was only the first of many human cancer susceptibility syndromes that have been found to be caused by inherited defects in various types of DNA repair (**Table 12.1**). We will explore another one here—hereditary non-polyposis colon cancer (HNPCC; also referred to as **Lynch syndrome**). HNPCC is a familial cancer syndrome that represents a quite common cause of inherited predisposition to colon cancer, being responsible for 2 to 3% of all colon cancer cases. HNPCC is, as its name implies, distinct from the other type of hereditary colon cancer predisposition that we encountered previously—adenomatous polyposis coli (see Section 7.10). A subclass of HNPCC/Lynch syndrome patients, in addition to their 80% lifetime risk of developing colon carcinomas, have increased susceptibility to brain tumors as well as endometrial, stomach, ovarian, and urinary tract carcinomas.

The increased colon cancer susceptibility of the HNPCC patients can be traced back to the accelerated rate with which tumor progression proceeds in their colons: although the adenoma-to-carcinoma progression is estimated to require more than a decade in the general population (see Section 11.4), the genetic instability afflicting the cells of HNPCC patients allows this step to occur in only 2 to 3 years. Indeed, because their adenomas progress so quickly to carcinomas, these premalignant growths have a relatively short lifetime and are therefore not found in significant numbers in the colons of these patients. The responsible genes were discovered through clever genetic sleuthing (Supplementary Sidebar 12.7).

The majority (85–90%) of HNPCC cases result from germ-line mutations in the genes encoding two important mismatch repair proteins, MSH2 and MLH1. Mutant

Table 12.1 Human familial cancer syndromes arising from germ-line defects in DNA repair

Name of syndrome	Name of gene	Cancer phenotype	Enzyme or process affected
HNPCC/Lynch	(4–5 genes)[a]	colonic polyposis	mismatch repair enzymes
XP[b]	(8 genes)[b]	UV-induced skin cancers	nucleotide-excision repair
ataxia telangiectasia (AT)[c]	ATM	leukemia, lymphoma	response to dsDNA breaks
AT-like disorder[c]	MRE11	lung, breast cancers	dsDNA repair by NHEJ
Familial breast, ovarian cancer	*BRCA1, BRCA2,*[d] *BACH1, RAD51C*	breast, ovarian, prostate carcinomas	homology-directed repair of dsDNA breaks
Werner	WRN	sarcomas, other cancers	exonuclease and DNA helicase,[e] replication
Bloom	BLM	leukemias, lymphomas, solid tumors	DNA helicase, replication
Fanconi anemia	(13 genes)[f]	AML, diverse carcinomas	repair of DNA cross-links and ds breaks
Nijmegen breakage[g]	NBS	mostly lymphomas	processing of dsDNA breaks, NHEJ
Li–Fraumeni	TP53	multiple cancers	DNA damage alarm protein
Li–Fraumeni	CHK2	colon, breast carcinomas	kinase signaling DNA damage
Rothmund–Thomson	RECQL4	osteosarcoma	DNA helicase
Familial adenomatosis	MYH	colonic adenomas	base-excision repair
Familial breast cancer	PALB2	breast cancer	dsDNA repair by HR

[a]Five distinct MMR genes are transmitted as mutant alleles in the human germ line. Two MMR genes—*MSH2* and *MLH1*—are commonly involved in HNPCC; two other MMR genes—*MSH6* and *PMS2*—are involved in a small number of cases; in addition, there is elevated risk of developing tumors of the prostate, ureter, ovary, connective tissues, and brain; a fifth gene, *PMS1*, may also be involved in a small number of cases.
[b]Xeroderma pigmentosum; at least eight distinct genes, seven of which are involved in NER. The seven genes are named *XPA* through *XPG*. An eighth gene, *XPV*, encodes DNA polymerase ε.
[c]Ataxia telangiectasia, small number of cases.
[d]Mutant germ-line alleles of *BRCA1* and *BRCA2* together may account for ~5% of all breast cancers and 10–20% of identifiable human familial breast cancers.
[e]An exonuclease digests DNA or RNA from one end inward; a DNA helicase unwinds double-stranded DNA molecules.
[f]Thirteen genes have been cloned and at least thirteen complementation groups have been demonstrated. Complementation group J encodes the BACH1 protein, the partner of BRCA1. A number of the products of the *FANC* genes form a complex that interacts with BRCA1 and its partners; BRCA2 associates with BRCA1 (and *FANCD1* = *BRCA2*). Homozygous absence of either the *RAD51C*, *FANCD1* (= *BRCA2*), *FANJ* (= *BACH1*), or *FANCN* (= *PALB2*) gene leads to Fanconi anemia, whereas lack of only one gene copy leads to breast cancer and/or susceptibility thereto.
[g]The NBS1 protein (termed nibrin) forms a physical complex with the RAD50 and Mre11 proteins; all three are involved in repair of dsDNA breaks. The phenotypes of patients with Nijmegen breakage syndrome are similar but not identical to those suffering from AT.

Adapted in part from B. Alberts et al., Molecular Biology of the Cell, 7th ed. New York: WW Norton & Company, 2022. Also from E.R. Fearon, *Science* 278:1043–1050, 1997.

germ-line alleles of two other MMR genes, *MSH6* and *PMS2*, are involved in a small proportion (~15%) of these cases; however, two other MMR genes (*PMS1*, *MSH3*), which have been found to play equally important roles in DNA repair, are rarely if ever transmitted as mutant alleles in the human germ line. Similar to the genetics of most tumor suppressor genes, patients inherit one defective allele of an MMR gene and the genomes present in tumor cells that arise almost always undergo a loss of heterozygosity (LOH) that results in the discarding of the surviving wild-type gene copy.

The resulting inability to properly detect and repair sequence mismatches leads to, among other consequences, high rates of mutations in genes that have microsatellite repeats nested in their sequences (see Figure 12.7) as well as a very high mutation rate overall. A dramatic and early illustration of the consequences of this repair defect came from study of a group of 11 colorectal cancer cell lines that showed microsatellite instability. In 9 of these cell lines, the gene encoding the type II TGF-β receptor (TGF-βRII) was found to be mutant. More specifically, the wild-type reading frame of this gene carries a stretch of 10 A's in a row (**Figure 12.26**). However, in these 9 tumor cell lines, the TGF-βRII gene was found to have lost 1 or 2 A's of the normally

Figure 12.26 A TGF-β receptor gene affected by microsatellite instability
The type II TGF-β receptor (TGF-βRII) is frequently inactivated in human colon cancers exhibiting microsatellite instability and therefore carrying defects in mismatch repair (MMR) genes. In the particular colon cancer whose DNA was analyzed here, the last 2 of 10 adenines (*boldface*) were deleted from the reading frame of the receptor-encoding gene, ostensibly because of an MMR defect. This particular deletion ($A_{10} \rightarrow A_8$) resulted in a nonsense mutation that caused premature termination of translation of the nascent TGF-βRII protein and hence loss of functionally critical signaling domains in the C-terminus of the receptor. This loss, in turn, allowed the progenitors of the colon carcinoma cells to become resistant to the growth-inhibitory effects of TGF-β. Alterations of this tract of A's in the TGF-βRII gene were subsequently found in 100 of 111 colorectal carcinomas showing defective MMR, in which they caused translation of the TGF-βRII mRNA to yield polypeptides of either 129 or 161 amino acid residues (depending on how many A's were deleted) rather than the 565 amino acid residues in the wild-type receptor. (From C. Lengauer, K.W. Kinzler and B. Vogelstein, *Nature* 396:643–649, 1998. With permission from Nature.)

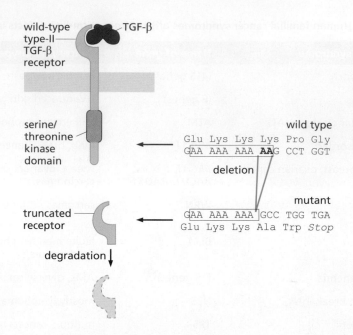

present homopolymeric stretch of 10 A's. These sequence changes forced the coding sequence of the TGF-βRII gene out of its normal reading frame and resulted in nonfunctional TGF-βRII proteins. We can imagine that once tumor cell precursors no longer express functional TGF-βRII, they can escape the inhibiting effects of proliferation of this anti-mitogenic signaling protein (see Section 8.4)—a highly advantageous trait if it is acquired early in tumor progression by epithelial cells. In a subsequent study of a series of 111 colon carcinomas exhibiting microsatellite instability, 100 were found to carry mutant, defective alleles of the TGF-βRII gene, with almost all mutant alleles being present in homozygous configuration. Hence, once one of the receptor-encoding genes suffers an inactivating mutation, the surviving wild-type allele is discarded through loss of heterozygosity.

Later, yet other genes were found to have suffered similar mutations in mismatch repair–defective cancer cells (**Table 12.2**). In most cases, the MMR defect and resulting mutant alleles were discovered in sporadic (rather than familial) cancers. These observations suggested that in nonfamilial tumors, MMR genes, like tumor suppressor genes, can be rendered defective either by somatic mutation or, alternatively, by promoter methylation and resulting transcriptional silencing (see Section 7.7). In fact, the second mechanism is responsible for the lion's share of defective MMR in these tumors: about 15% of sporadic gastric, colorectal, and endometrial tumors show defective MMR, and in almost all of these, the observed microsatellite instability can be traced to the methylation and resulting silencing of the *MLH1* gene. Interestingly, in the histologically normal endometrial tissue directly adjacent to tumors with defective MMR, the *MLH1* gene is often found to be methylated, suggesting that this methylation is one of the earliest events of tumor progression in this tissue (**Figure 12.27**); this alteration would appear to be another manifestation of the field cancerization described in Section 11.5. As we discuss in Section 12.13, genomic analysis of a large number of colon cancers and other cancers carrying MMR gene mutations has revealed an extremely high mutation frequency, involving length changes of microsatellite sequences as well as base-pair substitutions outside of repetitive sequences. Thus, the genes shown in Table 12.2 represent only a small proportion of genes mutated in MMR-deficient tumors.

Although MMR gene mutations are associated with high mutational burden in cancers, an even higher mutation rate is observed in cancers with mutations in the proofreading domains of the replicative DNA polymerases pol-δ and pol-ε. As discussed in Section 12.3, the proofreading mechanism recognizes and removes misincorporated bases during DNA replication. Germ-line mutations in the human *PolD1* and *PolE* genes, which encode the catalytic domains of pol-δ and pol-ε, respectively, lead to predisposition to colon cancer and other malignancies (see Figure 12.5B).

Table 12.2 Genes and proteins that have been inactivated in human cancer cell genomes because of mismatch repair defects

Gene	Function of encoded protein	Wild-type coding sequence	Colon	Stomach	Endometrium
ACTRII	GF receptor	A_8	X		
AIM2	interferon-inducible	A_{10}	X		
APAF1	pro-apoptotic factor	A_8	X	X	
AXIN-2	Wnt signaling	A_6, G_7, C_6	X		
BAX	pro-apoptotic factor	G_8	X	X	X
BCL-10	pro-apoptotic factor	A_8	X	X	X
BLM	DNA damage response	A_9	X	X	X
Caspase-5	pro-apoptotic factor	A_{10}	X	X	X
CDX2	homeobox TF	G_7	X		
CHK1	DNA damage response	A_9	X		X
FAS	pro-apoptotic factor	T_7	X		X
GRB-14	signal transduction	A_9	X	X	
hG4-1	cell cycle	A_8	X		
IGRIIR	decoy GF receptor	G_8	X	X	X
KIAA0977	unknown	T_9	X		
MLH3	MMR	A_9	X		X
MSH3	MMR	A_8	X	X	X
MSH6	MMR	C_8	X	X	X
NADH-UOB	electron transport	T_9	X		
OGT	glycosylation	T_{10}	X		
PTEN	pro-apoptotic	A_6	X		X
RAD50	DNA damage response	A_9	X	X	
RHAMM	cell motility	A_9	X		
RIZ	pro-apoptotic factor	A_8, A_9	X	X	X
SEC63	protein translocation into endoplasmic reticulum	A_9, A_{10}			X
SLC23A1	transporter	C_9	X		
TCF-4	transcription factor	A_{10}	X	X	X
TGF-βRII	TGF-β receptor	A_{10}	X	X	X
WISP-3	growth factor	A_9	X		

From A. Duval and R. Hamelin, *Cancer Res.* 62:2447–2454, 2002. With permission from American Association for Cancer Research.

Abbreviations: GF, growth factor; MMR, mismatch repair; TF, transcription factor.

In functional studies in yeast, several of these mutations have been shown to affect proofreading function and raise the mutation rate. Moreover, somatic mutations in the proofreading-encoding domains of the *PolD1* and *PolE* genes have been observed in endometrial cancer, colon cancer and other tumor types (see Figure 12.5B). Tumors carrying these mutations exhibit what has been characterized as an "ultra-mutated" phenotype, which leads to as many as one million base substitutions in the genomes

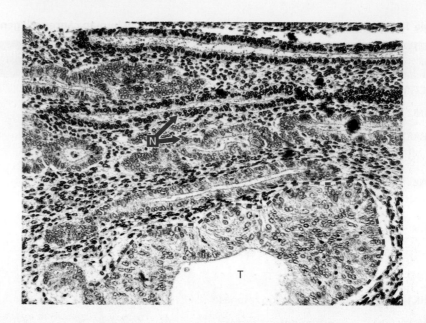

Figure 12.27 Suppression of MLH1 activity in endometrial tissue Anti-hMLH1 antibody coupled to a peroxidase enzyme generates a dark brown spot wherever it binds MLH1 protein. Here, the normal endometrial tissue of an endometrial cancer patient shows areas of intense, dark brown staining, indicating high hMLH1 expression (N, *upper pink arrow*), and areas of weak staining (N, *lower pink arrow*) where cell nuclei have been stained light blue with DAPI (a DNA-specific stain) but little MLH1 staining is seen. The endometrial carcinoma tissue (*below, dashed pink line*, T) is virtually devoid of hMLH1 staining and exhibits only light staining of cell nuclei by DAPI. Molecular analysis (*not shown*) indicated that the promoter of the *hMLH1* gene in the carcinoma cells was strongly methylated. Expression of hMLH1 was reduced or absent in some of the histologically normal tissue adjacent to the tumor, indicating that loss of hMLH1 expression occurred relatively early in tumor progression and preceded the histopathological alterations that led to the formation of this carcinoma. (From T. Kanaya et al., *Oncogene* 22:2352–2360, 2003. With permission from Nature.)

of certain tumors (see Section 12.13). In contrast to MMR-deficient tumors, these tumors do not exhibit microsatellite instability.

12.9 A variety of other DNA repair defects confer increased cancer susceptibility

By far the most notorious genes associated with cancer, at least in the mind of the public, are *BRCA1* and *BRCA2*. Mutant germ-line alleles of either of these genes confer an inborn susceptibility to breast and ovarian carcinomas. More specifically, almost half of all identified familial breast cancers involve germ-line transmission of a mutant *BRCA1* or *BRCA2* allele; by some estimates, 70 to 80% of all familial ovarian cancers result from mutant germ-line alleles of *BRCA1* or *BRCA2*. Stated differently, carriers of mutant germ-line *BRCA1* or *BRCA2* alleles have a 50–70% risk of developing breast cancer before the age of 70. Ovarian carcinoma risk is also high: 40–50% of *BRCA1* mutation carriers and 10–20% of *BRCA2* carriers develop disease before the age of 70. In addition, mutant *BRCA2* germ-line alleles have been associated in males with a >7-fold elevated risk of developing prostate cancer before the age of 65. Germ-line mutations in *BRCA1* and *BRCA2* increase the risk of pancreas cancer as well. Somatic loss of *BRCA1* function also appears to be important: 20–30% of sporadic (non-familial) ovarian carcinomas (that is, arising in women carrying wild-type *BRCA1* and *BRCA2* germ-line alleles) and almost as many sporadic breast cancers exhibit loss of function of one of these genes as a result of promoter methylation (see Section 7.7).

When these two genes were first discovered, it seemed that they should be included among the tumor suppressor genes, which are known to be involved in regulating the dynamics of cell proliferation, survival, and differentiation (see Chapter 7). But diverse lines of evidence built an increasingly persuasive case that these two genes are actually involved in the maintenance of genomic integrity and that their products therefore should be considered "caretakers" (DNA repair proteins) rather than "gatekeepers" (tumor suppressor proteins) (see Section 7.12).

The case for the participation of BRCA1 and BRCA2 in genomic maintenance can be argued using guilt-by-association evidence. The BRCA1 and BRCA2 proteins are found in large multi-subunit complexes with one another and with a large number of other proteins in the cell nucleus. These massive complexes carry, among other components, both the RAD50/Mre11 and the RAD51 proteins—homologs of two proteins in yeast that were initially discovered because of the important roles they play in repairing DNA breaks caused by ionizing radiation (that is, X-rays). Mismatch repair proteins have been found in these complexes as well.

Quite dramatically, treatment of cells with hydroxyurea, which results in a stalling of replication forks during S phase, causes BRCA1 molecules to cluster at discrete sites in the nucleus. Many of these stalled forks are thought to be sites of dsDNA breakage caused by accidental breaks in the still-unreplicated single-stranded DNA at the forks (see Figure 12.8); the breaks are usually fixed by a mechanism that is variously termed homologous recombination–mediated repair or homology-directed repair (abbreviated HR or HDR). As visualized by immunofluorescent microscopy, BRCA1 molecules are normally distributed in a large number of tiny dots throughout the nucleus; hydroxyurea causes the BRCA1 molecules to leave these dots and flock together in a far smaller number of large, discrete spots, in which the proliferating-cell nuclear antigen (PCNA)—known to be localized to replication forks—is also found (**Figure 12.28A**). These spots have also been found to contain a number of other

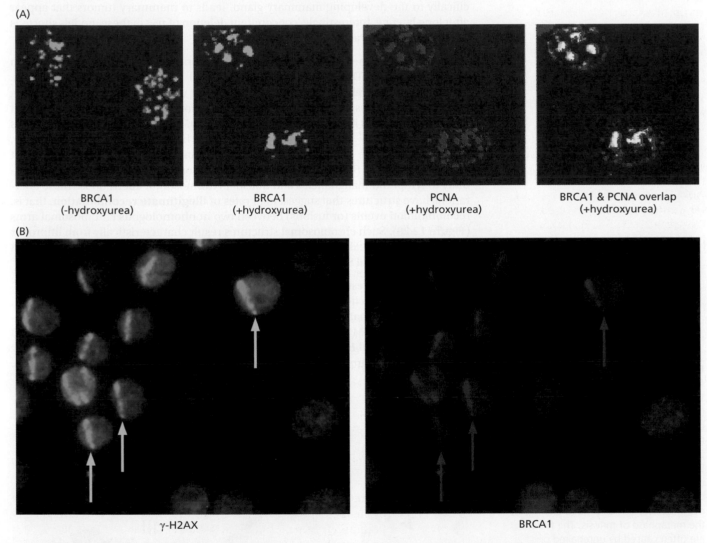

(A)

BRCA1
(-hydroxyurea)

BRCA1
(+hydroxyurea)

PCNA
(+hydroxyurea)

BRCA1 & PCNA overlap
(+hydroxyurea)

(B)

γ-H2AX

BRCA1

Figure 12.28 BRCA1 and the response to DNA damage
(A) The BRCA1 protein, which can be detected by fluorescence-labeled antibodies, is normally found during S phase in numerous discrete, small dots throughout the nucleus (*green, first panel*). However, when cells in S phase are treated with hydroxyurea, which stalls forward progress of the replication forks, the BRCA1 protein molecules leave these dots and congregate in a small number of quite large spots (*green, second panel*). A similar BRCA1 relocalization pattern can be observed with the proliferating-cell nuclear antigen (PCNA), which is known to be associated with replication forks (*red, third panel*). Substantial co-localization of the BRCA1 and PCNA proteins is indicated by the yellow spots (*fourth panel*); i.e., *red + green → yellow*. These stalled replication forks often are sites of double-strand (ds) DNA breaks caused by the accidental breakage of the still-unreplicated (and thus fragile) single-strand DNA (see Figure 12.8); these observations suggest that BRCA1 is recruited to sites of dsDNA breaks. (B) A 355-nm UV laser was used to paint narrow stripes across individual nuclei, which were then analyzed by immunostaining with antibodies reactive with either γ-H2AX (a phosphorylated histone that is known to localize to chromatin flanking dsDNA breaks; *green*) or BRCA1 (*red*). This co-localization indicates that BRCA1 is attracted to areas of dsDNA breaks. (A, from R. Scully et al., *Cell* 90:425–435, 1997. With permission from Elsevier. B, from R.A. Greenberg et al., *Genes Dev.* 20:34–46, 2006. With permission from Cold Spring Harbor Laboratory Press.)

known DNA repair proteins, including RAD50 and RAD51. The BRCA2 protein is also found in these spots, providing additional presumptive evidence of its collaboration in DNA repair processes. Moreover, when dsDNA breaks are intentionally created in discrete areas within cell nuclei using a focused laser beam, the BRCA1 protein co-localizes in these areas together with γ-H2AX, a phosphorylated histone that is present in the chromatin flanking sites of dsDNA damage (Figure 12.28B). Altogether, 20 distinct cellular proteins, most known to be involved in DNA repair, have been found to be recruited to these areas of damage, and the list is growing.

Mouse embryos carrying homozygous *BRCA1* mutations die at various stages of development, depending on the mutant allele. More revealing, however, are genetic strategies that limit the loss of BRCA1 function to specific, already-formed tissues after birth. Thus, **conditional deletion** (see Section 7.9) of segments of *BRCA1*, specifically in the developing mammary gland, leads to mammary tumors that appear after long latency. Interestingly, concomitant deletion of *p53* in these models shortens the latency considerably.

When taken together, these observations are consistent with a role for p53 in responding to the DNA damage caused by failed HDR in BRCA1-deficient cells. This response is likely to explain the long latency of *BRCA1*-mutant mice bearing wild-type *p53* alleles, in which ongoing p53 function appears to lead to significant attrition of incipient breast carcinoma cells. When p53 is also eliminated in such cells (yielding the *BRCA1$^{-/-}$p53$^{-/-}$* genotype), they exhibit high levels of genomic instability, this being indicative of ongoing defects in DNA damage repair. In addition, mutant germ-line alleles of *BRCA2* that cause only partial loss of function result in susceptibility to lymphoid malignancies and unusual chromosomal aberrations. These aberrations have structures that suggest high rates of **illegitimate recombination**, that is, recombination events (or fusions) between two nonhomologous chromosomal arms (**Figure 12.29**). Such chromosomal structures result characteristically from improper repair of dsDNA breaks (DSBs), many of which may arise accidentally at replication forks during a typical S phase of the cell cycle.

BRCA2 deficiency results additionally in deregulation of centrosome number, indicating defective functioning of the mitotic spindle, a condition that generates, in turn, aneuploidy. Cells that have reduced levels of BRCA2 also exhibit prolonged cytokinesis at the end of M phase (see Figure 8.3). Yet other indications of defective DSB repair in *BRCA1* and *BRCA2* mutant cells come from experiments that test the ability of cells to recover from DSBs introduced into their chromosomal DNA by X-rays.

Figure 12.29 Karyotypic alterations from partial loss of BRCA2 function Various karyotypic abnormalities (*arrowheads*) have been observed in cultured fibroblasts prepared from a mouse embryo that was homozygous for an allele encoding a truncated BRCA2 protein. These include fusions between chromosomal arms, resulting in chromosomal translocations that often manifest aberrant chromatid pairings at the metaphase of mitosis. The fusions are often caused by unrepaired or improperly repaired dsDNA breaks. When the hybrid chromatids resulting from such fusions pair with unaffected sister chromatids, structures such as those seen here are observed (*panel I*). Among the aberrations resulting from dsDNA breaks are chromatid breaks (ctb, *panel II*), triradial chromosomes (tr, *panel III*), and quadriradial chromosomes (qr, *panel IV*). Chromatid pairings like these are rarely observed in wild-type cells. (From K.J. Patel et al., *Mol. Cell* 1:347–357, 1998. With permission from Elsevier.)

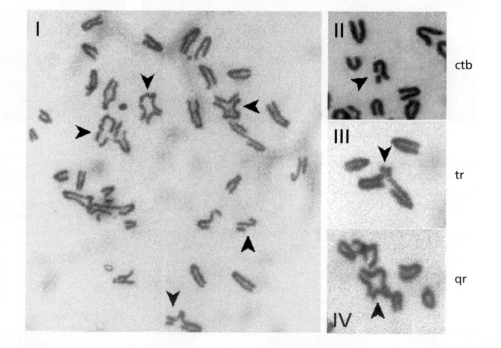

Cultured cells lacking the bulk of BRCA1 function show greatly increased sensitivity to killing by both X-rays and by chemotherapeutic drugs, such as cisplatin, that generate DSBs and inter-strand cross-links, respectively, in the DNA. As cited again in Section 17.3, the increased sensitivity of cells with defective homology-directed repair (such as *BRCA1* and *BRCA2*-mutant cells) to such DNA damaging agents has led to novel treatments for cancers carrying these mutations that take advantage of their reduced DNA repair capacity.

The repair of a DSB formed in one chromatid often depends on the ability of the repair apparatus to consult the undamaged, homologous DNA sequences in a sister chromatid and to use those sequences to instruct the repair apparatus on precisely how to reconstruct the broken double helix (**Figure 12.30**; see Supplementary Sidebar 12.8 for a more detailed molecular description). Thus, such homology-directed repair (HDR) occurs largely during the late S and the G₂ phases of the cell cycle, when the double helix in a sister chromatid can provide the sequence information for repairing the damaged chromatid. Recall that during S phase, DNA replication results in the production of two identical chromatids that remain associated as part of a common chromosome until they are separated during the next mitosis (see Figure 8.3). HDR is also used if inter-strand covalent cross-links within a double helix should arise.

All types of HR are compromised in cells lacking either BRCA1 or BRCA2 function. This compromised function may be explained, in part, by the behavior of the RAD51

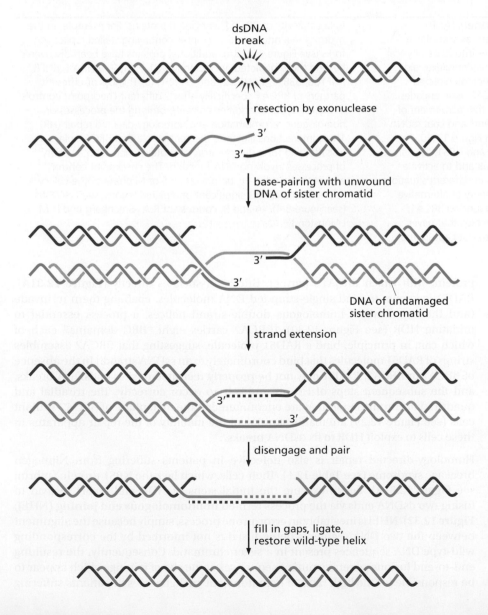

Figure 12.30 Homology-directed repair The repair of dsDNA breaks during the late S phase and G₂ phase of the cell cycle often depends on the ability of the repair apparatus to consult the sequences in the undamaged sister chromatid that was formed, together with the damaged chromatid, during the most recent S phase. Such homology-directed repair (HDR) begins (*top*) with the resection (removal) by an exonuclease of one of the two DNA strands at each of the ends formed by a dsDNA break. Each of the resulting ssDNA strands (*blue, red*) then invades the undamaged sister chromatid, whose double helix (*green, gray*) has been unwound by the repair apparatus in order to accommodate the pairing of the invading ssDNA strands with complementary sequences in the undamaged sister chromatid. The ssDNA strands from the damaged chromatid are then elongated in a 5′-to-3′ direction by a DNA polymerase, using the strands of the sister chromatid's DNA as templates. Thereafter, the extended ssDNA strands are released from the sister chromatid and caused to pair with one another, allowing further elongation by a DNA polymerase and a ligase, which together reconstruct a double helix possessing wild-type DNA sequences. Included among the DNA repair proteins known or thought to facilitate these complex steps of HR are RAD51, BRCA1, and BRCA2.

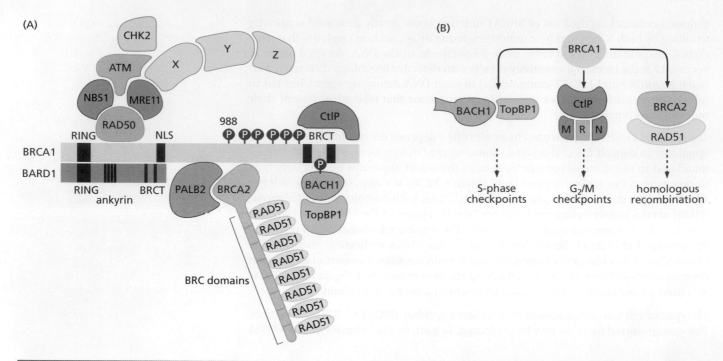

Figure 12.31 BRCA1, BRCA2, and their partners (A) The BRCA1 and BRCA2 proteins act, at least in part, as scaffolds to assemble a cohort of other DNA repair proteins into large physical complexes. Once assembled, these multiprotein complexes aid in the repair of dsDNA breaks, usually via homology-directed repair (HR). For example, one exon of the *BRCA2* gene encodes eight copies of a "BRC domain"; these aid in the recruitment of multiple RAD51 molecules, which form filaments and coat ssDNA strands as part of the HR process illustrated in Figure 12.30. The MRN complex, composed of MRE11, RAD50, and NBS1, appears to recognize the end created by a dsDNA break and to activate ATM kinase function in response. BRCA1 acts in different situations and locales (see panel B) as a scaffold for a variety of alternative partners, not all of which are concomitantly bound to BRCA1; for this reason, this image depicts a single complex that never exists in a living cell. Moreover, this image does not capture the full complexity of BRCA1-associated proteins. For example, in the hydroxyurea-induced nuclear spots containing stalled replication forks (see Figure 12.28A), additional proteins have been discovered beyond those depicted, including MDC1, RPA32, γ-H2AX, RNF8, RNF168, 53BP1, FANCD2, and FANC1. (B) The loss of different partners of BRCA1 specifically affects different checkpoint controls in the cell cycle, in addition to compromising the processes of homologous recombination and homology-directed repair (HR). These interactions illustrate the central role that BRCA1 plays as a scaffolding for a diverse array of proteins mediating a variety of processes involving DNA function. The checkpoint controls ensure that a cell halts before either S or M phase of the cell cycle if its genome carries significant unrepaired lesions, such as DSBs (see Figure 8.4). (A and B, courtesy of R.A. Greenberg and D.M. Livingston.)

protein, with which BRCA1's partner, BRCA2, associates directly (**Figure 12.31A**). RAD51 is known to bind single-stranded DNA molecules, enabling them to invade (and thereby unwind) homologous double-strand helices, a process essential to initiating HDR (see Figure 12.30). [BRCA2 carries eight "BRC domains," each of which can, in principle, bind a RAD51 molecule, suggesting that BRCA2 assembles strings of RAD51 molecules that bind coordinately to an ssDNA strand.] In the absence of BRCA1 or BRCA2, RAD51 may not be properly recruited to sites of dsDNA breaks, and the subsequent steps of HR are unlikely to occur correctly. The triradial and quadriradial chromosomes that are encountered in the metaphase of *BRCA2*-mutant cells (see Figure 12.29) are manifestations of the inability of the repair apparatus in these cells to exploit HDR to fix dsDNA breaks.

Homology-directed repair is also defective in patients suffering from Nijmegen breakage syndrome (see Table 12.1). Their cells, which lack the NBS1 protein (nibrin; see Figure 12.31A), fail to execute the initial steps of HDR and resort instead to fusing two dsDNA ends via the process termed **nonhomologous end joining** (NHEJ; **Figure 12.32**). NHEJ is inevitably an error-prone process, simply because the alignment between the two DNA segments being fused is not informed by the corresponding wild-type DNA sequences present in a sister chromatid. Consequently, the resulting end-to-end fusions generate mutant sequences at the site of joining, which appear to be responsible for the high rates of hematopoietic malignancies in patients suffering

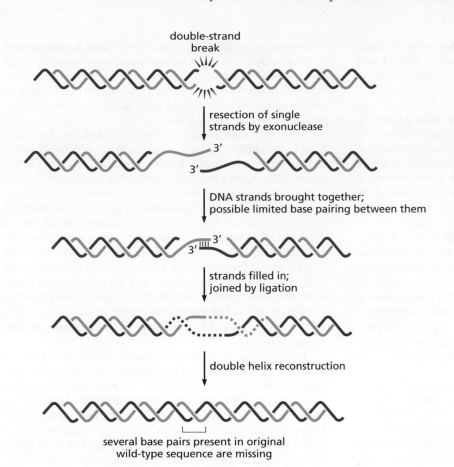

double-strand
break

resection of single
strands by exonuclease

DNA strands brought together;
possible limited base pairing between them

strands filled in;
joined by ligation

double helix reconstruction

several base pairs present in original
wild-type sequence are missing

Figure 12.32 Nonhomologous end joining NHEJ is used to restore a DNA double helix following a double-strand break when the nucleotide sequences from a sister chromatid are not available to instruct the repair apparatus how these ends should be properly joined (see Figure 12.30). In NHEJ, the resection of single strands from both broken ends results in ssDNA overhangs that can then be joined to one another, possibly by a limited degree of base pairing between them; three base pairs are shown here. The subsequent filling in of single-strand gaps and the ligation of any remaining ssDNA breaks results in the reconstruction of a double helix that lacks some of the base pairs that were present in the original undamaged DNA helix. (Adapted from M.R. Lieber et al., *DNA Repair (Amst.)* 3:817–826, 2004.)

from Nijmegen breakage syndrome. Mice that are defective for one or another component of the NHEJ machinery and that also lack p53 function develop lymphomas at extremely high rates.

Interestingly, NHEJ occurs largely in the G_1 phase of the cell cycle, when sister chromatids (formed during S phase and segregated to different daughter cells during mitosis) are not available to support HDR. NHEJ is virtually unique among the DNA repair processes because it plays a role in a normal, physiological function unrelated to repairing DNA damage, namely, the normal process of gene rearrangement that leads to the formation of functional antibodies and T-cell receptors. For example, the formation of the DNA sequences encoding antigen-binding domains depends on the rearrangement and fusion of chromosomal V, D, and J DNA segments (see Section 15.5). In the absence of the full complement of repair proteins needed for NHEJ, such DNA segment fusions cannot occur. This results in the inability to make proper immunoglobulin (antibody) molecules and T-cell receptors, compromising both the humoral and cellular arms of the adaptive immune system described in Chapter 15 and creating the syndrome of severe combined immunodeficiency (SCID).

To return to BRCA1/2, it is still unclear precisely how these two proteins contribute to the maintenance of normal chromosomal structure and thereby ward off cancer. Because BRCA1 and BRCA2 can bind to so many distinct nuclear proteins, many involved in DNA repair (see Figure 12.31A), it appears that these two act, at least in part, as a molecular scaffolding that helps to assemble large complexes of these other proteins and coordinate their actions. Once assembled, these various repair proteins presumably collaborate in fixing lesions, largely double-strand breaks, in the DNA. Altogether, as many as 215 distinct proteins have been found to associate with BRCA1 in multiple combinations and permutations, some of which are implicated in various aspects of DNA repair beyond its much-studied role in HR. For example, as Figure 12.31B illustrates, loss of BRCA1 can cripple certain cell cycle checkpoint controls that normally respond to damaged DNA and halt progress through this cycle until the damage is repaired.

Sidebar 12.8 Ethnic groups with unusual histories are valuable for studying the penetrance of mutant *BRCA1* and *BRCA2* germ-line alleles The precise effects of mutant *BRCA1* and *BRCA2* germ-line alleles on cancer risk have been difficult to gauge because both of these genes are very large and the detection of germ-line mutations in their reading frames is made correspondingly difficult. Moreover, many hundreds of mutant germ-line alleles of these genes have been cataloged in the human population, each having its own penetrance, that is, its relative ability to induce clinical disease in a person carrying it in the germ line. Each of these variant alleles operates in the complex genetic background of an individual, and this background also influences disease development; together these two dimensions of variability (the particular *BRCA1/2* mutation and the genetic background) make it almost impossible to gauge with any precision the disease penetrance of most of these alleles.

Many of these mutant alleles are of ancient vintage, having entered into the human gene pool hundreds, even thousands of years ago. Once present, many have been retained in the gene pool because they usually induce disease long after reproduction has occurred and are therefore not rapidly eliminated by Darwinian selection.

The peculiar histories of certain ethnic groups often simplify these analyses. For example, Ashkenazic Jews, who derive from Central and Eastern Europe, descend from a Jewish population of several million that lived in the late Roman Empire. Genetic analyses suggest, however, that the modern population derives from a founding population of only several hundred or so individuals who lived during the last days of this period. This genetic history seems to explain why this population harbors essentially only three mutant *BRCA1* and *BRCA2* germ-line alleles, all of ancient origin, which in aggregate are carried by some 2.5% of

contemporary Ashkenazim. These three **"founder mutations"** contrast with the situation in the general population, in which more than 1500 distinct inherited alleles of *BRCA1* have been reported (a small portion of which may be functionally neutral polymorphisms). Each of these other alleles has been detected by arduous sequence analysis.

These facts explain why the detection of mutant germ-line alleles of the *BRCA1* and *BRCA2* genes in the Ashkenazic population is relatively simple: polymerase chain reaction (PCR)-based DNA primers can be designed that are focused specifically on the detection of only these three alleles, rather than on the countless others that have been reported to date. The most common Ashkenazic *BRCA1* allele, for example, is the *BRCA1 185delAG* allele, which, as its name indicates, involves the deletion of two nucleotides; this mutant allele was found to be carried by 4.2% of women from this population who presented with invasive breast cancer in a New York City cancer clinic. Further studies revealed that the three mutant alleles conferred comparable risks of invasive breast cancer, which exceeded 80% by the time carriers of one or another of these alleles reached the age of 80.

Provocatively, the risk of developing breast cancer by the age of 50 among carriers of the *185delAG* mutation who were born before 1940 was 24%, whereas this risk was 67% among those carriers born after 1940. This indicates that nongenetic factors play a major role in cancer development, even in those born with a heritable DNA repair defect. (In the case of mutant alleles of *BRCA1*, changes over the past half century in nutrition and reproductive practices may have played major roles.) In addition, this type of epidemiology reveals the power of studying the biological effects of a single mutant allele operating in the context of variable genetic background and lifestyle changes.

We also do not know why inheritance of mutant alleles of the *BRCA1* and *BRCA2* genes leads preferentially to cancers of the breast and ovary, or why somatic mutation of *BRCA2* is occasionally associated with prostate and colon carcinomas. In addition, the **penetrance** of *BRCA1*- and *BRCA2*-mutant germ-line alleles (that is, the degree to which each allele exerts an observable effect on phenotype) has also been difficult to quantify (**Sidebar 12.8**). Beyond their well-defined roles in DNA damage repair, BRCA1 and BRCA2 have additional biochemical functions, the disruption of which contribute to tumor formation as well (Supplementary Sidebar 12.9).

12.10 The karyotype of cancer cells is often changed through alterations in chromosome structure

More than a century intervened between the discovery of aberrant karyotypes in cancer cells and the period when subtleties of DNA damage and repair, such as those described above, came to light. The triradial and quadriradial metaphase chromosomes seen in cells lacking BRCA1 or BRCA2 function are examples of these aberrations (see Figure 12.29). Stepping back for a moment from these particular aberrations, we can recognize that two distinct classes of karyotypic abnormalities can be seen in cancer cells: changes in the *structures* of individual chromosomes and changes in chromosome *number* that have no effect on chromosome structure.

One frequent deviation from the normal diploid karyotype involves an increase or decrease in the number of specific chromosomes. On occasion, through various accidents occurring during mitosis, cancer cells may acquire **polyploid** genomes, where an additional haploid complement of chromosomes is acquired (leading to a **triploid** state) or even an extra diploid complement of chromosomes is acquired (leading to a **tetraploid** state). Alternatively, extra copies of individual chromosomes may be present, or, less commonly, a chromosome copy may be missing.

The term **aneuploidy** is usually reserved for a deviation from a normal (or **euploid**) karyotype that involves changes in chromosome *number*. In recent years, however, use of the term aneuploid has occasionally been broadened to include changes in the *structures* of individual chromosomes, which are prevalent in cells of the great majority (>85%) of solid tumors; a more specific term is "chromosomal aberration," which we will use here. These two major types of karyotypic alteration arise through fundamentally different mechanisms.

Changes in chromosome number are discussed in the next section. Here, we review the mechanisms responsible for changes in chromosome structure, some of which we have already encountered at various points in this text. For example, as we saw in the previous section, unrepaired dsDNA breaks, many of which occur accidentally at DNA replication forks, are thought to be a major source of chromosomal translocations.

In addition, far earlier (see Chapter 4) we learned of a class of cancer-associated chromosomal alterations as part of a discussion of the mechanisms leading to the creation of the *MYC* and *BCR-ABL* oncogenes. Recall the well-studied case of the translocations that fuse the *myc* proto-oncogene to promoter/enhancer sequences deriving from one of three alternative immunoglobulin genes (see Section 4.4). In these and other lymphomas, it is likely that the complex machinery dedicated to rearranging the immunoglobulin and T-cell receptor (TCR) genes misfires. Instead of rearranging the immunoglobulin or TCR gene sequences (see Section 15.5), this machinery inadvertently catalyzes inappropriate interchromosomal recombination events that join the immunoglobulin genes promiscuously with sequences scattered throughout the genome, the *myc* gene being only one of them. Those rare translocations that happen to involve the *myc* proto-oncogene and deregulate its transcription seem to confer special proliferative advantage on cells, resulting in the appearance of cell clones and ultimately lymphomas that carry these very characteristic karyotypic alterations.

The first characterized translocations were documented in a variety of hematopoietic malignancies (see, for example, Tables 4.4 and 4.5), largely a result of the ease of carrying out karyotype analyses of cells derived from these tumors. Although many mechanisms likely contribute to the formation of structurally aberrant chromosomes, one well-studied mechanism was presented in Chapter 10, where we read about telomere collapse resulting in breakage–fusion–bridge (BFB) cycles (see Figure 10.9). These cycles create large-scale aberrations in the structures of individual chromosomes, apparently striking all chromosomes with comparable frequency. On occasion, translocations may generate fusion genes and other rearrangements that provide proliferative advantage to the cells carrying them, which one imagines results in the clonal outgrowth of these cells. Below we consider another disruptive process—**chromothripsis** (or chromosome shattering)—in which a limited stretch of a chromosome undergoes massive fragmentation with subsequent rejoining of the resulting fragments.

With the advent of high-throughput DNA sequencing technologies (see Section 12.12), genome-scale analyses of human cancers have increased dramatically. In addition to the discovery of more subtle mutations (base alterations or small insertions and deletions), many larger-scaled alterations have also been detected across virtually all cancer types. Many of these are termed recurrent because they have been seen on multiple occasions in a series of independently arising human tumors; as described in Section 11.6, these alterations are considered to be "driver mutations" and the affected genes "drivers." By now, hundreds of these recurrent translocations have been cataloged in hematopoietic and solid tumors. Because recurrent translocations map to highly specific chromosomal sites, it would seem that the molecular mechanisms creating them are distinct from the breakage–fusion–bridge cycles cited above, which appear to act totally randomly via BFB cycles in generating fused chromosomal segments.

Chromosomal translocations were thought for many decades to be present uniquely in hematopoietic tumors, as discussed above. However, beginning in 2000, the availability of new DNA sequencing strategies, the resulting sequencing of the human

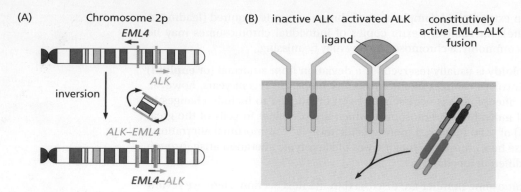

Figure 12.33 Chromosomal inversion produces a fusion oncogene A subset of lung adenocarcinomas carry the EML4–ALK fusion oncogene, which is formed by a chromosomal inversion event. (A) The *EML4* (*purple arrow*) and *ALK* (*green arrow*) genes are oriented in opposite directions on the p arm of Chromosome 2. Following the chromosomal inversion event, the promoter and 5′ exons of EML4 are fused to the 3′ exons of ALK, including exons encoding the kinase domain, forming the EML4-ALK fusion. (B) The wild-type ALK protein is a receptor tyrosine kinase that is normally expressed in cells of the nervous system and not in the adult lung.

Binding to its ligand leads to homodimerization and activation of the kinase domain (*indicated by the dark green bars*), leading to signal transduction (*forked black arrow*). Under control of the *EML4* promoter, which is transcribed in the lung, the EML4-ALK fusion protein is expressed in lung cancer cells. The dimerization domain of the EML4 protein (*dark purple bars*) permits unregulated dimerization of the tyrosine kinase domain of the fusion protein, constitutively activating downstream signaling pathways and contributing to oncogenesis. (From J.N. Rosenbaum et al., *Mod. Pathol.* 31: 791–808, 2018. With permission from Nature.)

genome, and new bioinformatics tools began to reveal chromosomal translocations and other chromosomal aberrations, some quite common, in solid tumors. For example, in 2005, a specific recurrent translocation, involving a gene encoding an androgen-regulated serine protease and an ETS transcription factor, was discovered. This particular translocation (*TMPR32/ERG*) is now known to be present in the tumors of ~50% of the patients bearing localized prostate carcinomas, and translocations involving *TMPR32* and other *ERG*-related transcription factors bring the total to ~70%. These discoveries in a commonly occurring human carcinoma triggered surveys of a variety of solid tumor genomes and a spate of additional discoveries, many of them driven by the powerful new genome sequencing technologies.

In another example, approximately 5% of non-small-cell lung adenocarcinomas express a fusion oncoprotein termed EML4-ALK, which is the product of an intrachromosomal inversion and fusion event. In particular, this inversion event fuses the EML4 protein's N-terminal domain to the tyrosine kinase domain of the ALK RTK, leading to constitutive activation of its kinase function and uncontrolled proliferative signaling (**Figure 12.33**). Like the fusion proteins described in Figure 5.12, this particular one causes constitutive receptor activation, in that the trimerization domain of EML4 drives formation of oligomerized ALK kinase domains. This chromosomal event also places the fusion gene under the control of the *EML4* promoter. In these lung carcinomas, this EML4-ALK fusion is seen to be mutually exclusive with activating mutations of the EGF receptor, indicating that it is likely to function similarly in the transformation process. Interestingly, the kinase function of ALK is also activated in several hematopoietic malignancies via chromosomal translocation events. More generally, comprehensive genomic analyses of many tumor types have revealed that chromosomal translocations and other aberrations occur at greatly differing frequencies among various solid tumors (**Figure 12.34**). The functional rationales that would explain these tissue-specific associations remain elusive.

The enzymes involved in the rearrangement of immunoglobulin genes and T-cell receptor genes (see Section 15.5) are unlikely to be responsible for most of the commonly encountered translocations found in hematopoietic and non-hematopoietic tumors, simply because these enzymes are not expressed in the normal cells of origin of these diverse tumors. Hence, the molecular mechanisms responsible for these translocations are yet to be discovered. Sequence analyses of the DNA flanking translocation breakpoints have revealed duplications, deletions, and inversions of sequence blocs, findings that suggest but hardly prove the involvement of some type

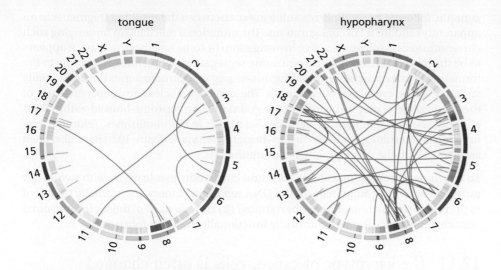

tongue

hypopharynx

Figure 12.34 Chromosomal translocations in head-and-neck carcinomas of smokers This graphic convention of depicting chromosomal translocations places the chromosomes around a circle according to their number or letter (sometimes termed a Circos plot) with fusions between previously unlinked chromosomal segments depicted by purple lines. Short green ticks pointing toward the center of the circle indicate regions of local intrachromosomal rearrangements, including translocations, small amplifications, deletions, and inversions. The genomes of two head-and-neck squamous cell carcinomas (HNSCCs) of smokers—a tongue and a hypopharyngeal carcinoma—are depicted here. None of these translocations were associated with a recurrent translocation documented in other similar tumors. (With permission from N. Stransky et al., *Science* 333:1157-1160, 2011. With permission from AAAS.)

of "error-prone" DNA repair mechanism, such as the nonhomologous end joining (NHEJ) discussed earlier (see Figure 12.32).

Implementation of new DNA sequencing technologies led in 2011 to discovery of a totally novel type of aberration in the genomes of ~25% of bone cancers and a significant fraction of cancers overall—the localized firestorms of chromosomal rearrangements representing the chromothripsis cited earlier. In the example shown in **Figure 12.35**, a single catastrophic event seems to have shattered a limited stretch of the

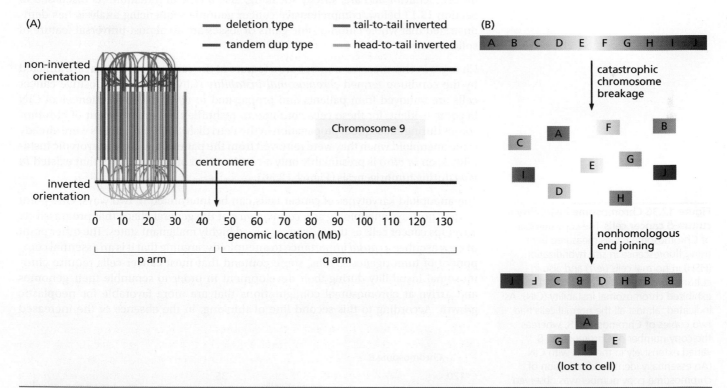

Figure 12.35 Localized firestorms of chromosomal rearrangement The development of high throughput DNA sequencing technologies has made it possible to discover certain types of chromosomal damage that previously eluded detection. (A) In the genome of a thyroid carcinoma, a cluster of 77 distinct chromosomal rearrangements affecting the short arm of Chromosome 9 (i.e., 9p) have been documented. Each of the lines indicates the direct fusion of two DNA segments that are normally located at some distance from one another in this chromosomal arm; these rearrangements involve four distinct types of changes in DNA configuration listed *above right*. (B) The scheme illustrated here provides one explanation of how such a focally localized set of chromosomal rearrangements, which has been termed chromothripsis, may have arisen. According to one attractive model, the formation of single-chromosome-containing micronuclei results in individual chromosomes undergoing extensive intrachromosomal rearrangement prior to becoming reassociated with the larger complement of chromosomes during a subsequent mitosis. (From P.J. Stephens et al., *Cell* 144:27–40, 2011. With permission from Elsevier.)

genome, followed by multiple rejoining events between the resulting fragments in an apparently random set of configurations. The mutational mechanism underlying such chromothripsis is still under active investigation. In some instances, at least, it appears to be the product of improper chromosome segregation during mitosis, leading to the formation of **micronuclei** surrounding mis-segregated chromosomes that are initially stranded in the cytoplasm after mitosis. The defective nuclear envelope structure of these micronuclei may be insufficient to guard the chromosome housed within from endonucleases that initiate the damage that results in chromothripsis. Telomere dysfunction and resulting breakage–fusion–bridge events (see Figure 10.9) have also been associated with the development of chromothripsis.

To conclude, chromosomal aberrations and chromothripsis leave us with two major puzzles: (1) Which components of the DNA repair machinery are normally on guard to prevent the formation of these aberrations? (2) How do most of these chromosomal abnormalities, once formed, contribute functionally to cancer formation?

12.11 The karyotype of cancer cells is often changed through alterations in chromosome number

As stated earlier, some types of genetic instability affect karyotype by altering the *number* of individual chromosomes without affecting their structure. These changes create aneuploidy (in the usual sense of this term). Although the term "mutation" is often reserved for changes in DNA sequence (and thus includes changes in chromosome structure), we should recognize that alterations in chromosome number also represent significant changes in a genome that can have equally profound effects on cell behavior and are, strictly speaking, also a type of mutation. As discussed in Section 12.13 below, comprehensive cancer genome sequencing analysis has demonstrated that whole chromosome gains or losses are an almost-universal feature of solid tumors.

Changes in chromosome number are a feature of the carcinoma cells that are afflicted by the condition termed *chromosomal instability* (CIN). When CIN-positive cancer cells are removed from patients and propagated *in vitro*, the consequences of CIN become evident, for these cells continue to reshuffle their complement of chromosomes during subsequent propagation in the Petri dish. Such cancer cells were already quite aneuploid when they were removed from the patient, and their karyotypic instability seen *in vitro* is presumably only a continuation of the instability that existed *in vivo* during tumorigenesis (**Figure 12.36**).

The aneuploid karyotypes of cancer cells can be interpreted in two ways. One point of view portrays aneuploidy as a *consequence* of the general chaos that progressively envelops cancer cells as they advance toward highly malignant states. The other point of view ascribes a *causal* importance to aneuploidy, arguing that it is an essential component of tumorigenesis. Thus, some contend that most cancer cells require chromosomal instability during their development in order to scramble their genomes and arrive at chromosomal configurations that are more favorable for neoplastic growth. According to this second line of thinking, in the absence of the increased

Figure 12.36 Chromosome instability in cultured cancer cells The copy number of Chromosome 8 was measured here using fluorescence *in situ* hybridization (FISH) in normal cells (*left*) and also in cultured breast cancer cells (*right*) that exhibited chromosomal instability (CIN). As indicated, almost all the normal cells had two copies of Chromosome 8, whereas the copy number of Chromosome 8 varied extensively in the cells with CIN. (An essentially identical distribution of chromosome copy number was observed upon study of a second, arbitrarily chosen chromosome.) This great cell-to-cell variability of chromosome number indicates that fluctuations of this number continue to occur frequently as these cancer cells are propagated in culture. (From G.A. Pihan et al., *Cancer Res.* 58:3974–3985, 1998. With permission from American Association for Cancer Research.)

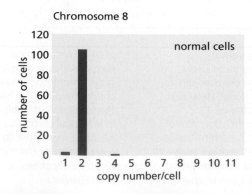

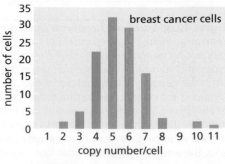

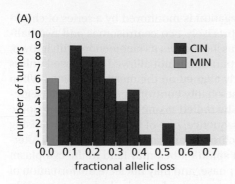

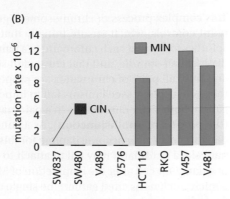

Figure 12.37 Chromosomal instability vs. gene mutation The presence of chromosome instability (CIN) can be gauged by measuring the loss of alleles from chromosomal arms. (A) In the colorectal carcinomas studied here, analyses of a large number of tumors revealed that many lost heterozygosity (LOH; see Section 7.4) at a substantial number of chromosomal loci. On the *abscissa*, 0.3 allelic loss, for example, refers to tumors in which 30% of the loci that were previously heterozygous, as revealed by analyses of chromosomal markers, no longer exhibit heterozygosity (*red bars*). Most of this LOH is attributable to the loss of whole chromosomes. In contrast, among the tumors afflicted with microsatellite instability (MIN; *blue bar*), the loss of alleles and hence the loss of entire chromosomes is negligible. (B) In colorectal tumor cell lines that exhibit CIN, as gauged by the loss of chromosomal markers, the rate of inactivation of the HPRT (hypoxanthine phosphoribosyltransferase) gene is virtually zero (*first four bars, red*). In contrast, in those that exhibit MIN, the rate of mutation of this gene is significant and is occasionally 100-fold higher than in CIN tumor cell lines (*last four bars, blue*). (A, from C. Lengauer, K.W. Kinzler and B. Vogelstein, *Nature* 396:643–649, 1998; and B. Vogelstein et al., *Science* 244:207–211, 1989. With permission from AAAS. B, from C. Lengauer, K.W. Kinzler and B. Vogelstein, *Nature* 396:643–649, 1998; and J.R. Eshleman et al., *Oncogene* 10:33–37, 1995. With permission from Nature.)

mutability associated with aneuploidy, most clones of incipient cancer cells could never succeed in acquiring all the genetic alterations needed to complete multi-step tumorigenesis.

An important observation that will help settle this debate has come from the study of a series of human colon and rectal carcinomas. The few tumors that exhibit microsatellite instability (MIN) show relatively little aneuploidy and virtually no chromosomal instability (CIN). Conversely, the far more numerous tumors that have CIN are not prone to show nucleotide sequence alterations that are characteristic of MIN (**Figure 12.37**). Together, these observations suggest that, at least in the case of colorectal carcinomas, tumor cells must acquire increased mutability of their genomes and that either one or the other of these two mechanisms suffices to provide such mutability. This conclusion begins to persuade one that CIN, like MIN, is an effective mechanism for remodeling the cellular genome in a way that favors evolution toward neoplasia. (Of note, contrary to the depiction in Figure 12.37, CIN and MIN are not always mutually exclusive states, and some tumors exhibit both types of instability concomitantly.)

Whether this logic pertains as well to the genetic mechanisms creating hematopoietic malignancies remains unclear. Unlike carcinoma cells, which almost invariably exhibit widespread karyotypic chaos, hematopoietic tumor cells often exhibit karyotypes that are diploid, with the exception of one or two reciprocal translocations that seem to be responsible for initiating the cancer or triggering a specific step of tumor progression (for example, the one creating the *BCR-ABL* oncogene; see Section 4.5). Still, it is highly unlikely that the small number of observable karyotypic alterations found in most hematopoietic cancer cells suffice on their own to enable full neoplastic proliferation. (In one case—that of chronic myelogenous leukemia—the acquisition of the *BCR-ABL* oncogene is often followed during blast crisis relapse by the loss of p53 function; point mutations cause this loss, and they are, of course, karyotypically invisible.) Moreover, hematopoietic tumor cells have not been reported to suffer from microsatellite instability. It therefore remains unclear which genetic or epigenetic mechanisms enable hematopoietic cells to acquire the entire ensemble of neoplastic functions needed in order for them to proliferate as fully neoplastic cells. Indeed, we do not even know whether the formation of hematopoietic tumors requires as many genetic changes as those required for the formation of solid tumors (see Section 11.9).

The changes in chromosome number that characterize chromosomal instability are usually (and perhaps always) the consequences of **mis-segregation** of chromosomes during mitosis. During the normal M phase of the cell cycle, the chromosomes line up in a plane, the **metaphase plate**, and associate with spindle fibers (see Figure 8.3A). The fibers together form a metaphase spindle, a bipolar structure in which each half spindle is constituted of microtubule fibers, many of which extend from the **kineto-chores** on the chromosomes (the complex nucleoprotein bodies associated with the centromeric DNA segments of the chromosomes) back to the centrosomes; the latter are responsible for organizing the entire metaphase spindle structure. When this apparatus is working properly, the spindle fibers pull sister chromatid pairs apart, so that each chromatid moves toward one of the two centrosomes. This ensures that the two daughter cells that will eventually arise after cell division receive precisely equal allotments of chromosomes.

This complex process of chromosome segregation is monitored by a series of checkpoint controls, which ensure initially that precisely two centrosomes and two half-spindles form; that each chromatid in a pair associates via its kinetochore with its own, distinct half-spindle; and that chromatid separation is not allowed to proceed unless and until all pairs of chromatids are properly aligned on the metaphase plate. When these checkpoint mechanisms fail to impose quality control on chromosomal segregation, both sister chromatids in a pair may be pulled to one or the other centrosome (the process of **nondisjunction**). As a consequence, one of the subsequently arising daughter cells may become haploid for this chromosome and the other triploid. Alternatively, a chromatid may fail to attach to a spindle fiber, may land in the cytoplasm of a daughter cell after the completion of M phase, and may nucleate the formation of a micronucleus as cited earlier; the single chromosome borne in the latter may rejoin the bulk of the chromosomes during the course of the next mitosis or, more frequently, disintegrate and thus fail to contribute to the chromosomal array of descendant cells.

A possibly more important source of aneuploidy derives from malfunction of the complex machinery that normally ensures that each kinetochore is bound appropriately to its own set of 20–25 microtubules that form a spindle fiber; these attachments allow each pair of sister chromatids, initially linked by their paired kinetochores, to be pulled in opposite directions at anaphase. However, in many cancer cells, this control mechanism does not operate properly, and individual kinetochores become associated instead with *too many* spindle fibers. For example, **merotely** occurs when a kinetochore (belonging to a single chromatid) becomes attached simultaneously to two oppositely oriented sets of spindle fibers, which then proceed to engage in a microscopic tug-of-war; this competition is often unresolved by the end of mitosis, leaving the chromatid carrying this kinetochore stranded between the two groups of properly segregating chromatids (now called chromosomes; **Figure 12.38A**). The fate of this orphaned chromosome is unclear; it may be lost entirely or eventually associate with one or another daughter nucleus. A number of studies indicate that merotely represents the major source of CIN in cancer cells. As discussed above, the formation of micronuclei surrounding improperly segregated chromosomes can lead to chromothripsis.

(A)

(B)

(C)

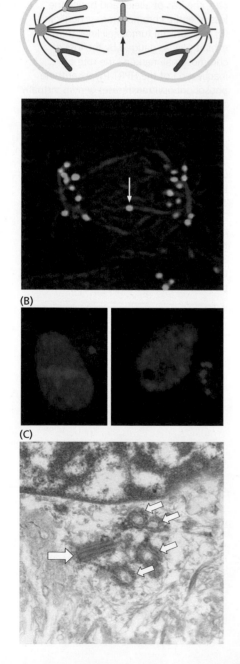

Figure 12.38 Centrosomes and the organization of the mitotic spindle Centrosomes are responsible for organizing the microtubule spindle fibers at mitosis. (A) Above, the kinetochore (*green*) of an individual chromatid (*arrow*) may be connected inappropriately to two opposite spindle fibers (*red*) at the same time—the condition termed merotely—causing the spindle fibers to pull in opposite directions and, quite frequently, leave the chromatid stranded between the two properly segregated populations of chromatids (*left, right*), now called chromosomes. Below, in the micrograph, only the spindle fibers (*red*) and kinetochores (*green*) are visualized by immunofluorescence. The isolated chromosome (*green dot, arrow*) in the middle, revealed only by its kinetochore, may not be incorporated into either of the soon-to-be-formed daughter nuclei (assembled around the kinetochore clusters, *left, right*) and may therefore be lost from descendant cells. More likely, however, it may become associated with one of the daughter nuclei. Association with the inappropriate daughter nucleus will result in aneuploidy. (B) In these immortalized but nonmalignant interphase cells (*left*), the presence of a single centrosome is discernible in the cytoplasm through use of an antibody that detects pericentrin, a centrosome-associated protein (*red*). In normal cells, a centrosome is duplicated at the G_1/S transition to generate the two centrosomes found at the poles of the mitotic spindle. In contrast, during interphase of human breast cancer cells (*right*), multiple centrosomes are often observed. These will frequently create multipolar spindles (see Figure 12.39) when such cells enter mitosis. (C) The pair of centrioles that forms the core of each centrosome can be best seen using transmission electron microscopy (TEM), in this case of a human colon carcinoma cell. Four centrioles are seen here in cross section (*small arrows*), while a side view of a fifth (*large arrow*) is apparent, together indicating a deregulation of centriole number. The nucleus is seen above. (A, above, adapted from D. Cimini, *Biochim. Biophys. Acta* 1786:32–40, 2008; with permission from Elsevier; below, courtesy of D. Cimini and from W.T. Silkworth et al., *PloS ONE* 4:e6564, 2009. © 2009 Silkworth et al. This is an open-access article distributed under the terms of the Creative Commons Attribution License. B, from G.A. Pihan et al., *Cancer Res.* 58:3974–3985, 1998. With permission from American Association for Cancer Research. C, courtesy of M.J. Difilippantonio and T. Ried.)

In normal cells, merotely is usually cured by the time cells advance through metaphase into anaphase (see Figure 8.3A). Such curing is critical to normal cells because merotelic chromosomes fail to trigger the **spindle assembly checkpoint** (SAC), which is designed to halt progress into anaphase in the event that spindle fibers are not properly attached to the kinetochores of all chromatids. (In normal cells, despite these elaborate corrective mechanisms, by some estimates about 1% of mitoses result in some type of mis-segregation; in cancer cells, in stark contrast, as many as one-half of mitoses may result in some form of mis-segregation.) Cancer cells seem to be more tolerant of merotely, which therefore can contribute to their aneuploid karyotypes. Indeed, cancer cells often exhibit defects in one or more of the fourteen distinct proteins that have been implicated in orchestrating the SAC. Of note here, mice that have been genetically engineered to express subnormal levels of certain SAC proteins exhibit elevated levels of spontaneous tumor formation.

More widespread karyotypic chaos may occur if the spindles themselves are not properly assembled. Aberrant mitoses, which result from inappropriate spindle organization, were noticed as early as 1890 and, in retrospect, represented the first clue that cancer cells are genetically abnormal. In normal interphase cells, a single centrosome can be visualized in the cytoplasm (Figure 12.38B); during mitosis, two centrosomes are arrayed at opposite poles within the cell. Cancer cells, however, often show marked defects in this organization, including multiple centrosomes at interphase (see Figure 12.38B and C). The result may be mitotic spindles that have multiple poles rather than the two seen in normal cells (**Figure 12.39A and B**). Often, **supernumerary** (extra) centrosomes coalesce into the normal set of two as cells proceed through mitosis; however, the spindle fibers that they initially generated may end up forming dysfunctional spindle–kinetochore attachments, such as that seen in Figure 12.38A. Alternatively, the extra centrosomes may persist, causing the chromosomal array to be

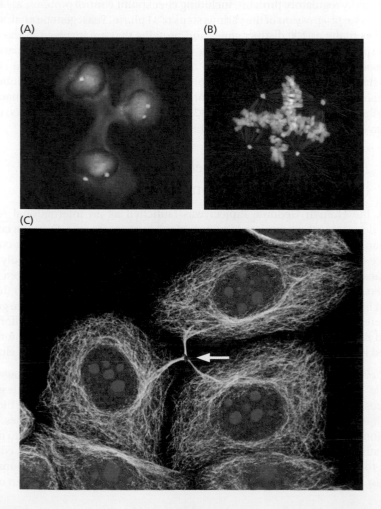

Figure 12.39 Multipolar mitotic apparatuses In a normal mitotic metaphase of cultured cells, the spindle fibers, which are composed of microtubules, reach from the two centrosomes at the mitotic poles to the kinetochores—the multiprotein complexes that are associated with the centromeric DNA of each chromosome. (A) In this micrograph of *p53*−/− mouse embryo fibroblasts, the formation of multiple centrosomes has resulted in a triradial mitotic spindle array. The spindle fibers have been immunostained *red* with an antibody reactive with α- and β-tubulin, and the chromosomal clusters have been stained *blue* with DAPI, a DNA dye. The five centrosomes (immunostained with an antibody reactive with γ-tubulin) are seen as small light yellow or light green spots. Observations like these implicate p53 in the regulation of centrosome number (also see Supplementary Sidebar 12.10). (B) The four centrosomes in this quadriradial mitotic spindle array in a prometaphase human fibroblast have been stained *yellow-green*; the microtubules forming the spindle fibers, *red*; and the chromosomes, *white*. (C) Here, just after cytokinesis, which follows directly after telophase (see Figure 8.3) three daughter cells of a human HeLa cervical carcinoma cell line have formed from a mother cell that ostensibly assembled a triradial spindle during mitosis; these daughters are trying to divide the maternal dowry of chromosomes in three ways rather than the usual two. Microtubules are stained *yellow*; DNA is stained *purple*. The abscission of these cells at the end of cytokinesis clearly did not proceed properly, leaving the cells connected to one another via a chromosomal fragment (*purple dot, arrow*). (A, from P. Tarapore and K. Fukasawa, *Oncogene* 21:6234–6240, 2002. B, from N.J. Ganem, S.A. Godinho and D. Pellman, *Nature* 460: 278–282, 2009. Both with permission from Nature. C, from K.G. Murti, *Biotechniques*, Oct. 2004, cover. Courtesy of K.G. Murti.)

divided among three or more daughter cells (see Figure 12.39C). The resulting mis-segregation of chromosomes into daughter cells may lead to wild fluctuations in chromosome number and overall karyotype. Strong evidence supporting a causal role for supernumerary centrosomes comes from studies in mice genetically engineered to overexpress the Plk4 serine/threonine kinase, which stimulates centrosome duplication. These mice develop aneuploidy and tumor formation in many tissues, and the tumors in these animals exhibit complex, abnormal karyotypes.

It seems that once the complex apparatus designed to ensure proper chromosomal segregation has been damaged, such damage is irreversible. For example, as was seen in Figure 12.36, the enormous cell-to-cell variability in the number of Chromosome 8 copies in certain breast cancer cells indicates that chromosome instability (CIN) persisted in these cells long after tumor progression had reached completion. In this respect, CIN differs from the breakage–fusion–bridge (BFB) cycles (see Section 10.4), which seem to plague the genomes of cancer cells for a limited window of time during tumor progression and then cease once these cells succeed in acquiring telomerase and thus the ability to stabilize their aberrant karyotypes.

In recent years, additional molecular defects that contribute to various types of chromosomal instability have come to light. Not surprisingly, the duplication of centrosomes is closely coordinated with cell cycle advance; it seems to occur at or near the G_1/S transition. Thus, an increasing body of evidence indicates that centrosome duplication is coordinated in some way by the cyclin E– and A–containing cyclin-dependent kinase (CDK) complexes, which are master regulators of entrance into and progression through S phase (see Section 8.3). Defects in the regulation of these processes, including those triggered by cancer-associated viruses such as HBV and EBV, can lead to deregulation of centrosome number (Supplementary Sidebar 12.10).

The great complexity of mitosis causes us to consider how prone this process is to error and how many regulatory proteins, including checkpoint control proteins, are in place to monitor the progression of the various steps of M phase. Yeast genetic analyses have revealed as many as 100 distinct genes and proteins that are involved in the various steps of spindle assembly and dynamics, spindle attachment, and the separation of chromosomes during mitosis; mutation of many of these genes results in chromosome instability in yeast. Many of these proteins are highly conserved evolutionarily, and their homologs are likely to be components of the mammalian mitotic machinery. As described in Supplementary Sidebar 12.11, to date only a small proportion of these genes have been found to be mutated at high frequency in human cancer.

12.12 Advances in genome sequencing technologies have fueled a revolution in cancer genomics

In 1990, the Human Genome Project was launched as an international research effort to determine the complete sequence of the human genome. Over the course of more than a decade, this effort resulted in the publication in 2003 of a near-complete sequence of a "reference" human genome, that is, one that was purportedly typical of humans, although actually being a composite of genome segments from several volunteers. The full extent of the genetic polymorphisms (see Section 7.6) that exist between individual members of our species is not reflected in this reference genome. This reference sequence appears to contain the coding and regulatory information required to construct a human being. (Even now, however, the genetic information harbored in certain difficult-to-sequence segments of the genome remains elusive.)

The importance of this milestone in human cancer biology cannot be exaggerated. Thus, the availability of the human genome sequence together with the genome sequences of a great many other species has greatly accelerated the pace of biological and biomedical research by providing ready access to the sequences of genes, regulatory regions, and other chromosomal elements. Today, it is straightforward to perform comparative genomic analyses between species and highlight differences computationally, to design oligonucleotide primers used to amplify a segment of genomic DNA of interest, and to undertake a diversity of other applications that depend in one way

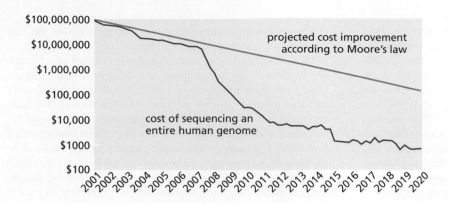

or another on genomic sequence information. Until recently, these procedures were laborious and highly time-consuming.

The opportunities enabled by the availability of the human genome sequence inspired the development of new DNA sequencing technologies that improved the speed and reduced the cost of genome sequence analysis. Within a decade of the publication of the first human genome sequence, a number of alternative technologies to traditional Sanger sequencing—referred to collectively as **next-generation sequencing** (NGS) technologies—were developed, leading to progressively lower sequencing costs (Sidebar 12.9). Whereas the first human genome sequence was estimated to cost nearly $3 billion, the most advanced sequencing technologies available today have allowed the cost of the sequencing of an entire genome to fall well below $1000 (Figure 12.40).

This technological revolution has influenced many fields of biomedical research, but none more than cancer research. Through large-scale cancer genome and whole-exome sequencing efforts like **The Cancer Genome Atlas** (TCGA) and the **International Cancer Genome Consortium** (ICGC), thousands of cancer specimens have been profiled at unprecedented "depths," that is, redundantly, to ensure maximal accuracy. Among the fruits of these labors are measurements of the prevalence of the specific mutations across multiple cancer types; such analyses include surveys of essentially all the cancer genes that we have discussed in this text. In addition, a large number of novel cancer-associated genes have been identified directly through these large-scale cancer genome studies by means of their recurrent involvement in multiple, independently arising tumors and cancer types. By current estimates, nearly 600 genes in the human genome have been found to be recurrently mutated in one or more types of cancer, among them oncogenes and tumor suppressor genes, as well as genes involved in DNA repair and related genome maintenance functions. The mutations that affect these genes range from missense mutations and nonsense mutations to large-scale structural rearrangements. Extensions of these procedures have also enabled analyses of the protein-encoding RNA segments of specific cell types via the process of **whole-exome sequencing**.

12.13 Genomic analysis reveals that human cancers differ with respect to mutational burden, patterns of mutations, and copy number gains and losses

A number of factors can influence the number of mutations present in the genomes of different types of cancer cells. In cancer genomics studies, **tumor mutational burden** (TMB) can be expressed either in terms of mutations detected per megabase of DNA sequenced or as the number of protein-altering (also called **nonsynonymous**) mutations per tumor genome. As shown in Figure 12.42, TMB can range from low in certain pediatric cancers with less than one mutation detected per megabase of DNA to very high in adult cancers associated with carcinogen exposure or in those carrying mutations in genes involved in DNA repair (between 10 and to well more than 100 mutations per megabase). The low TMB in cancers of children is likely explained

Figure 12.40 Falling cost of genome sequencing. The graph illustrates the dramatic reduction in the cost of sequencing a human genome sample beginning in 2001, shortly after the first draft sequence of the human genome was released. With a series of technological improvements in conventional DNA sequencing technology, the costs of sequencing steadily declined over the next several years. With the advent of massively parallel sequencing technologies (described in Figure 12.41), genome sequencing costs dropped precipitously beginning in 2007. Continued improvements have led to even more cost reductions, such that by 2021 the cost of sequencing of an entire human genome sample is well under $600. These cost reductions have fueled multiple large-scale cancer genomic studies as well as the widespread use of numerous other sequencing-based genomic technologies. For comparison, the projected decline in sequencing costs according to Moore's law is shown over this period. Moore's law was initially described in relation to the two-fold improvement in computing power observed in the computer hardware industry every two years, and is considered a mark of outstanding technological improvement. The falling cost of genome sequencing over this two-decade period has vastly outperformed Moore's law (Source: https://www.genome.gov/about-genomics/fact-sheets/DNA-Sequencing-Costs-Data).

Sidebar 12.9 "Next-generation," massively parallel DNA sequencing has enabled large-scale genomic studies The genomics revolution has been fueled by advances in DNA sequencing technologies, which collectively allow for very high-throughput sequencing at a fraction of the cost of what was possible only a decade ago (see Figure 12.40). Today, the most widely adopted next-generation sequencing (NGS) technology has been a method termed "sequencing by synthesis," in which thousands of DNA templates are sequenced simultaneously on a glass slide through the repeated addition of fluorescently labeled **chain-terminating nucleotides**. With each round of nucleotide addition, the identity of the added bases in each reaction is determined by fluorescence imaging. Following removal of the chain-terminating moiety on the inserted nucleotides, the next round of sequencing can be undertaken in the same fashion. With sophisticated automation, hundreds of bases of sequence can be determined per template in only several hours (**Figure 12.41**). Although initially developed for whole-genome sequencing, this method (as well as other NGS methods), has been adapted to allow for sequencing only the exonic portions of the genome (called whole-exome sequencing), which has been used extensively in cancer genomics as described in Section 12.12. In addition, these methods can be used to rapidly sequence cDNA libraries for assessment of the mRNA content of cells or tissues; they can also be used to determine the DNA methylation pattern in genomic DNA as well as other applications.

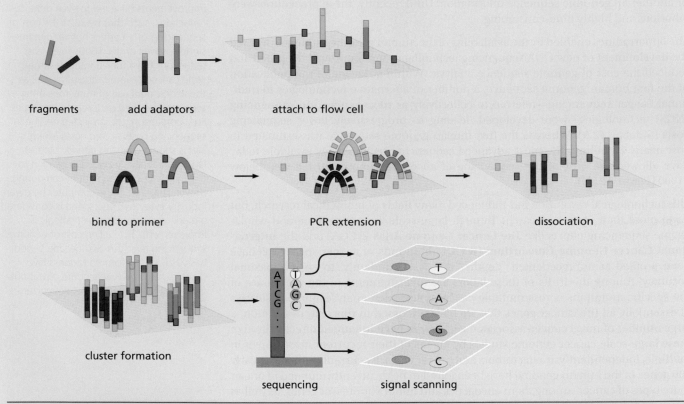

Figure 12.41 Massively parallel DNA sequencing Among several powerful methods for high-throughput DNA sequencing, the one commercialized by the company Illumina has been used most extensively in cancer genomics studies to date. In this method, DNA fragments to be sequenced are modified with short adaptor sequences. These adaptors enable attachment of single strands of the fragments to the surface of a reaction chamber (or flow cell) by hybridization to complementary nucleotides attached to the flow cell. Through a series of local binding, PCR amplification, and dissociation steps, each of the original fragments can be amplified into a cluster of identical molecules ready for sequencing. Sequencing is performed by the addition of a sequencing primer complementary to the adaptor sequence, DNA polymerase, and fluorescent nucleotides (with each nucleotide labeled with a different fluorophore). The addition of only one nucleotide per sequencing round is controlled through the use of reversible chain-termination moieties on the fluorescent nucleotides. At the end of each round of sequencing, the flow cell is imaged to record the identity of the last nucleotide added at each cluster. The next round of sequencing follows the quenching of the fluorescent signal by the previously incorporated nucleotides and the removal of the chain termination moiety. The process is continued in this fashion until hundreds of bases of sequence are read out from each cluster. Computational analysis is then used to align the fragments into larger sequences, such as exomes or whole genomes. (From Y. Lu et al. in J.K. Kulski ed., Next Generation Sequencing - Advances, Applications and Challenges. London: IntechOpen, 2016. © 2016 The Author(s). With permission from the authors.)

by a combination of the relatively low number of cell divisions that the cell-of-origin of the cancer underwent prior to neoplastic transformation as well as the expected lower lifetime exposure to carcinogens in children compared to adults. (At the same time, it leaves unresolved the question of why pediatric tumors, in contrast to almost all adult tumors, require relatively few mutant driver genes to become neoplastic.)

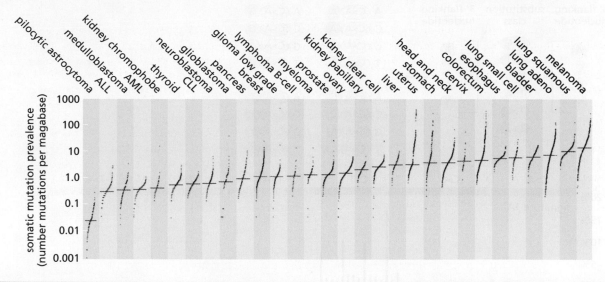

Figure 12.42 Large-scale genome sequencing reveals diverse patterns of mutagenesis The results of sequencing of more than 7000 cancer genomes show vastly differing densities of point mutations. These vary by greater than a factor of 1000 from pediatric tumors (pilocytic astrocytoma, *left*) to some of the solid tumors in adults (*right*) having documented exposure to known carcinogens (tobacco smoke and UV radiation), mutations in DNA repair genes, or overexpression of APOBEC enzymes. Each *black* point in the graph represents the number of point mutations per megabase (Mb) in a single cancer genome. The *red* horizontal lines are the median number of mutations in the respective cancer type. ALL, acute lymphoblastic leukemia; AML, acute myeloid leukemia; CLL, chronic lymphocytic leukemia. (From L.B. Alexandrov et al., *Nature* 500, 415–421, 2013. With permission from Nature.)

Underscoring the importance of carcinogen exposure in driving TMB are the mutations observed in smoking-associated adenocarcinoma of the lung, in which the average number of nonsynonymous mutations is approximately 175; in contrast, the genomes of lung adenocarcinomas in never-smokers bear on average approximately tenfold fewer mutations of this type. At the high end of the TMB spectrum are tumors that carry mutations in genes involved in the DNA repair pathways that we discussed above, including the mismatch repair genes, as well as mutations in the proofreading domains of DNA polymerases genes, *POLE* and *POLD*. For example, as mentioned earlier, colon and endometrial cancers that harbor mutations affecting the fidelity of the pol-δ and pol-ε replicative DNA polymerases have "ultra-mutated" genomes with well over 200 mutations per megabase of DNA sequenced. (As we will learn later in Chapter 15, the nonsynonymous mutations have implications for the immune responses to tumors, because these mutations encode, by definition, structurally altered proteins that may function as "neoantigens" capable of provoking robust immune responses.)

In addition to revealing the TMB in a given cancer or cancer type, in some cases the *patterns* of mutation in cancer genomes—whether they occur in genes or in noncoding regions—as well as the DNA sequence context in which those mutations arose, can be used to infer the nature of the mutagenic processes that were responsible for these mutations (**Figure 12.43**). A clear example of this can be seen in melanomas that arise in areas of sun-exposed areas of skin. Whole-genome or whole-exome sequencing of these cancers clearly shows a "signature" of UV-induced mutations (see Figure 12.43 for examples of mutational signatures), featuring C>T mutations in dipyrimidine sites. As discussed earlier in Section 12.5, these sequences are known to undergo UV-induced intra-strand photo-cross-linking that, if not repaired, can lead to transition mutations at these sites.

The genome-wide analyses have confirmed and extended earlier work that focused on sets of frequently mutated genes such as p53, which also showed a predominance of C>T mutations in cancers associated with UV light exposure. Yet another clear signature is seen from genome sequencing of the smoking-associated lung cancers. Here one finds clear evidence of exposure to the guanine alkylating agents that comprise some of the many carcinogens in tobacco products (see Section 12.5). These tumors have a mutagenic signature (Signature 4 in Figure 12.43B and C) featuring a high frequency of C>A transversion mutations. Other cancer types linked to tobacco use,

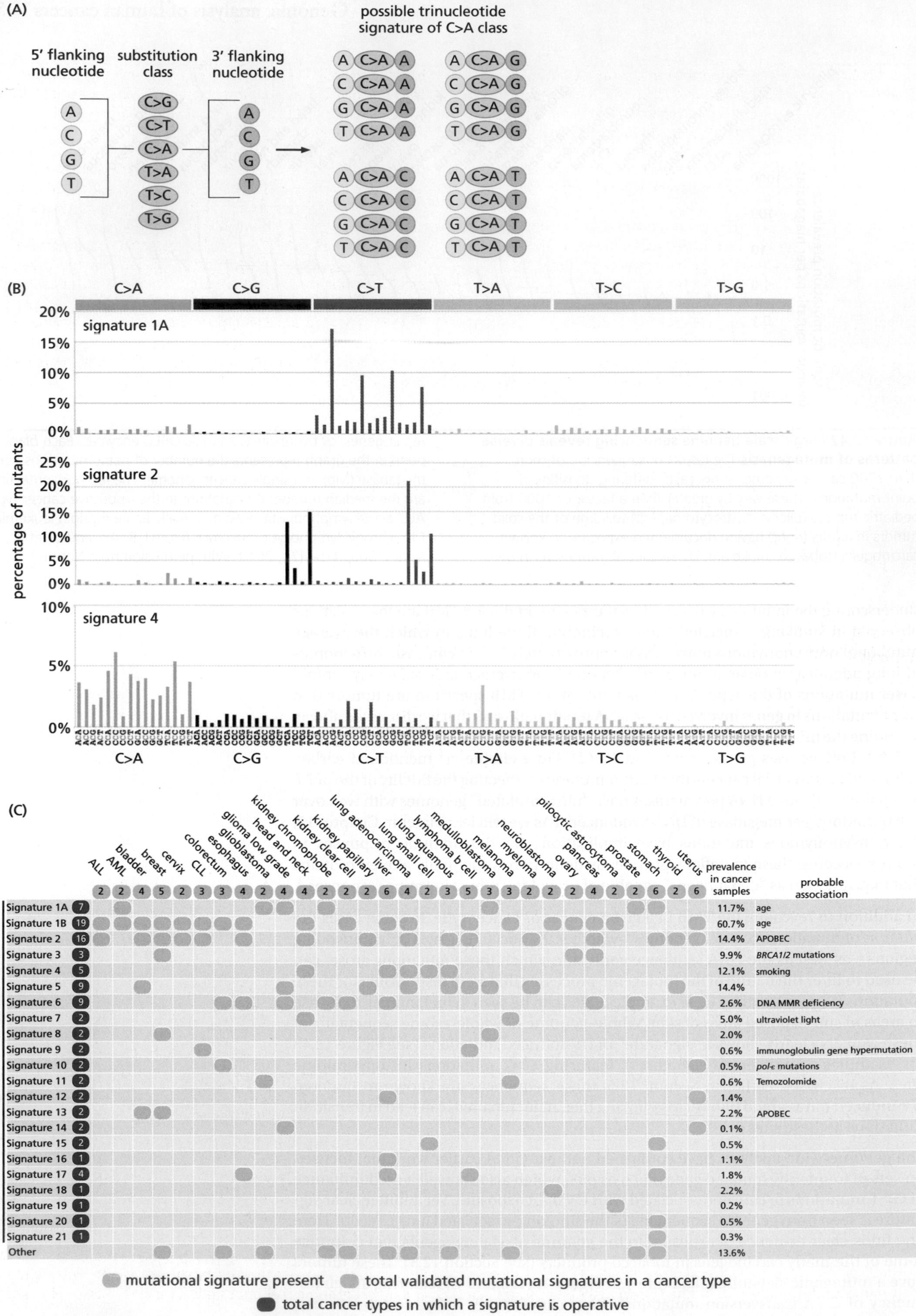
(A)
possible trinucleotide signature of C>A class
5′ flanking nucleotide
substitution class
3′ flanking nucleotide
A C G T
C>G C>T C>A T>A T>C T>G
A C G T
A C>A A
C C>A A
G C>A A
T C>A A
A C>A G
C C>A G
G C>A G
T C>A G
A C>A C
C C>A C
G C>A C
T C>A C
A C>A T
C C>A T
G C>A T
T C>A T
(B)
C>A C>G C>T T>A T>C T>G
percentage of mutants
signature 1A
20% 15% 10% 5% 0%
signature 2
25% 20% 15% 10% 5% 0%
signature 4
10% 5% 0%
C>A C>G C>T T>A T>C T>G
(C)
ALL AML bladder breast cervix CLL colorectum esophagus glioblastoma glioma low grade head and neck kidney chromophobe kidney clear cell kidney papillary liver lung adenocarcinoma lung small cell lung squamous lymphoma b cell medulloblastoma melanoma myeloma neuroblastoma ovary pancreas pilocytic astrocytoma prostate stomach thyroid uterus
2 2 4 5 2 3 3 4 2 4 4 2 2 2 6 4 2 3 5 3 3 2 2 2 4 2 2 6 2 6
prevalence in cancer samples
probable association
validated mutational signatures
Signature 1A 7 11.7% age
Signature 1B 19 60.7% age
Signature 2 16 14.4% APOBEC
Signature 3 3 9.9% *BRCA1/2* mutations
Signature 4 5 12.1% smoking
Signature 5 9 14.4%
Signature 6 9 2.6% DNA MMR deficiency
Signature 7 2 5.0% ultraviolet light
Signature 8 2 2.0%
Signature 9 2 0.6% immunoglobulin gene hypermutation
Signature 10 2 0.5% *pol-ε* mutations
Signature 11 2 0.6% Temozolomide
Signature 12 2 1.4%
Signature 13 2 2.2% APOBEC
Signature 14 2 0.1%
Signature 15 2 0.5%
Signature 16 1 1.1%
Signature 17 4 1.8%
Signature 18 1 2.2%
Signature 19 1 0.2%
Signature 20 1 0.5%
Signature 21 1 0.3%
Other 13.6%
mutational signature present
total validated mutational signatures in a cancer type
total cancer types in which a signature is operative

Figure 12.43 Mutational signatures in cancer genomes To discern patterns of mutations (mutational signatures) in cancer genomes, one can analyze the frequency of base changes in the context of the flanking 5′ and 3′ bases. (A) In the analysis seen here, only the pyrimidine bases (C,T) of the unmutated, wild-type sequence that were initially present were considered; each of these two initial pyrimidine bases has three possible alternative substitutions (C>A, C>G, C>T, T>A, T>C, T>G). Because there are four possible 5′ flanking bases and four possible 3′ flanking bases (that is, 16 combinations) in each case, there are 3 × 16 or 48 possible mutational signatures for each of these two initial pyrimidine nucleotides, making 96 in all. The 16 possible mutation types for the substitution class C>A are shown here. (B) From the analysis of several thousand cancer genomes, up to 30 mutational signatures have now been observed. This panel shows three signatures that are prominently observed in different types of human cancer. Signature 1A (*above*), which features C>T mutations in sequences where the mutated C residue is followed by a G residue are believed to be caused by the spontaneous deamination of 5-methylcytosine residues in CpG dinucleotides (see Section 12.4). Signature 2 (*middle*) is associated with increased expression of members of the APOBEC family of cytosine deaminases and is associated with a high prevalence of C>T and C>G mutations in particular sequence contexts. Signature 4, (*below*) which is marked by C>A mutations, is associated with exposure to cigarette smoke carcinogens. (C) Different human cancer types show contributions of the different mutational signatures (*green ovals*). The total validated number of distinct mutational signatures for each cancer type are shown in *blue ovals* and the number of cancers that show a particular identifiable signature are shown in *red ovals*. The probable or inferred causes of 11 of the mutational signatures are shown at the *right*. Note that for many of the signatures, the mutagenic mechanisms leading to their formation are not known at present. (From L.B. Alexandrov et al., *Nature* 500: 415–421, 2013. With permission from Nature.)

such as squamous cell cancer of the head and neck, also show this mutational signature upon genomic analysis. Hence, mutational signatures in cancer genomes can provide clues to the nature of the etiological agents responsible for tumor development (**Sidebar 12.10**).

Mutational signatures in cancer can also provide tell-tale signs of the failure of DNA repair in repair-deficient tumors or the hyperactivity of mutagenic enzymatic processes. For example, the genome sequences of colon and endometrial cancers harboring mutant mismatch repair genes show numerous short insertion and deletion (**indel**) mutations in homopolymeric or simple repeat sequences. Breast and ovarian cancers with inactivating mutations in the *BRCA1* or *BRCA2* genes, which are involved in homology-directed repair (see Section 12.9), show high incidence of indel mutations throughout the genome, which likely result from these cancer cells needing to rely on error-prone nonhomologous end-joining (NHEJ) mechanisms to repair double-strand breaks in DNA. Overexpression of members of the APOBEC family of cytosine deaminases (see Sidebar 12.6) occurs in many cancers, including those of the lung and bladder. Their conversion of cytosine residues to uracil occurs preferentially in the sequence context TpCpA/T, leading to C>T and C>G mutations at these sites.

In addition to identifying base substitutions and indel mutations, cancer genome sequencing studies also highlight regions of copy number gain and loss, which are the products of amplifications or deletions of segments of a cancer genome. These so-called **copy number variations** (CNVs) are revealed by the number of sequence

Sidebar 12.10 Mutational signatures reveal the likely environmental source of Romanian kidney cancers Searching the genomic sequence data from a particular cancer specimen or cancer type for mutational signatures can provide clues to the endogenous mutational processes or exogenous mutagens that might have contributed to tumor formation. A good example of the value of this approach comes from the genomic analysis of renal cell carcinomas (RCC, a form of kidney cancer) from different countries in Europe. Although the patterns of mutations present in these cancers from most countries were similar, a subset of the RCC specimens from Romania carried a distinctive mutational signature with a preponderance of A>T transversion mutations. Further signature analysis showed that these mutations occurred in the sequence C/TpApA/G. Interestingly, this same mutational signature had previously been found in urothelial cancers in parts of Asia, associated with exposure to aristolochic acid (AA), a known carcinogen present in plants from the genus *Aristolochia* used in herbal medicines.

These observations raised the possibility that AA exposure may have also contributed to RCC development in parts of Romania, where *Aristolochia* species endemic to the area may be inadvertently introduced in the food supply. Subsequent studies have shown that the DNA base adduct caused by a metabolic derivative of AA can be observed in the normal kidney of Romanian RCC patients, strongly supporting a causal role for this environmental mutagen in this form of cancer. Ongoing cancer genomic studies are attempting to use mutational signature analysis to identify environmental carcinogens that are responsible for cancers with unexpectedly high prevalence in certain geographic regions, that is, so-called "cancer clusters."

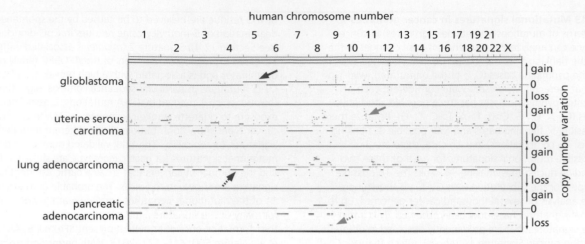

Figure 12.44 Copy number variation in cancer Cancer genomes frequently carry regions of copy number variation (CNV), resulting from whole chromosome gain or loss or regional amplification or deletion within individual chromosomes. These CNVs can be detected using a number of methods, including the whole-genome sequencing (WGS) method employed here. WGS data from four different human cancers are depicted with the read counts plotted according to chromosomal location (*top*). The average read count per sample is considered the copy neutral region for this analysis (denoted 0). Regions of copy number gain have read counts greater than the copy neutral value, and regions of copy number loss have read counts less than the copy neutral value. Note that all four cancers have numerous areas of CNV, including whole chromosome gains (*solid red arrow*), loss of an entire chromosome arm (*dashed red arrow*), focal amplification (*solid blue arrow*), and focal deletion (*dashed blue arrow*). (From A. Cherniack and R. Beroukhim; data from https://portal.gdc.gov/.)

reads from a region of the genome relative to the value expected from a normal diploid (2*n*) genome. CNVs provide clues to the existence of possible oncogenes in regions of gain or tumor suppressor genes in regions of loss. Regions containing CNVs can be limited in length (termed "focal"), in which only one or a small number of genes is affected, or more extensive, in which as much as an entire chromosomal arm can be affected. Whole-chromosome gain or loss can also be recognized by CNV analysis. **Figure 12.44** shows examples of CNVs cancer genomes.

12.14 Cancer genomes contain driver and passenger gene mutations

When familiar genes such as *Kras* or *p53* are found to be mutated in a cancer sample, there is no debate about their causal role in the process of tumor development. However, when a gene that has never been previously associated with cancer is discovered in mutant form in a cancer specimen, it is reasonable to ask whether the mutation was selected for during the development of that cancer (that is, the mutant gene functioned as a driver of tumorigenesis) or whether the mutation might simply have occurred coincidentally and been carried along for the ride together with a selectively advantageous driver mutation (that is, a passenger mutation; see Section 11.6).

Several criteria can be applied to bolster the claim that a gene is a bone fide cancer driver. First, the frequency of mutation of a particular gene in multiple tumors of a given cancer type or across multiple cancer types can be an indication of direct involvement in the tumorigenic process. If such frequencies significantly exceed those expected simply by chance, one can reasonably conclude that the mutations were selectively advantageous. When calculating these expected frequencies, one must bear in mind that large genes will present a larger target for randomly occurring mutations than smaller ones.

In addition, the *pattern* of mutation in a gene might provide a clue. For example, if the gene in question is predicted to encode a kinase and the mutations detected in cancer specimens are analogous to known kinase-activating mutations in well-described oncogenes, then the case in favor of assigning driver mutation status would be strengthened. Likewise, a novel tumor suppressor gene might be identified based in part on a pattern of recurring loss-of-function mutations as well as evidence of mutations affecting both alleles of the gene in a series of cancer specimens. Ultimately, functional tests, either in cell culture transformation assays (see Sections 3.2 and 3.4.) or in whole animal cancer models (see Sidebar 7.2), are the most direct way to distinguish true cancer driver genes from passengers. The use of CRISPR/Cas9-based genome editing methods has greatly accelerated cancer functional genomics (Sidebar 12.11). Although passenger mutations do not contribute to cellular transformation or tumor progression per se, they may still have an impact on tumor development by creating neoantigens, as mentioned earlier and discussed in detail in Chapter 15.

12.15 Cancer genomic studies reveal both inter-tumoral and intra-tumoral heterogeneity

Surveys of the genomes of multiple tumors of a given cancer type reveal considerable inter-tumoral heterogeneity, such that different tumors display different

---

Sidebar 12.11 The CRISPR-Cas9 system has become an important tool in cancer research A powerful set of genome-editing tools has emerged in recent years through the application of CRISPR–Cas (clustered regularly interspaced short palindromic repeats–CRISPR-associated systems). This novel technology has been derived from prokaryotic systems that evolved as an adaptive defense mechanism against bacteriophage infection and have now been redeployed to rapidly mutagenize the genomes of cells from a wide array of species; in addition, the CRISPR-Cas technology can now be used to functionally interrogate cancer genes.

The *Streptococcus pyogenes*-derived type II CRISPR–Cas9 system was the first to be modified for use in mammalian cells. This highly versatile system is composed of two components: an RNA-guided DNA endonuclease (Cas9) and a so-called single-guide RNA (sgRNA). The sgRNA molecule binds to the Cas9 enzyme and directs it to a genomic sequence of interest via base pairing to a predetermined target sequence (**Figure 12.45A**). Combining the expression of Cas9 with an sgRNA that is complementary to a target DNA sequence leads to high-efficiency cleavage of the target site in the genome, resulting in a double-strand DNA break. Subsequent repair by the error-prone NHEJ DNA repair pathway (see Section 12.9) can lead to insertion or deletion mutations at the target site, often causing loss of gene function.

Co-introducing (via transfection) a cloned exogenous DNA template that carries a mutation of interest at the time of the CRISPR-Cas9 cleavage event can enable efficient transfer of the mutation into the genomic site by HDR (see Section 12.9), thereby allowing the precise introduction of gain-of-function or other mutations into a cell's genome (see Figure 12.45A). In both cell-based models and whole-animal models, these methods have been used to functionally test candidate TSGs and oncogenes identified in various cancer genomic studies.

In addition to facilitating the targeting of mutations to individual genes, CRISPR-Cas9-based systems have also been used to carry out large-scale genetic studies in cell lines. These studies use libraries of sgRNAs (typically encoded by pools of recombinant retrovirus vectors), each of which targets a specific gene and, when used together, target hundreds or thousands of genes in populations of cells (Figure 12.45B). The introduction of these libraries into large populations of cultured cells allows the selective outgrowth of individual cells (and thus clonal populations) that have gained increased fitness in the population; this fitness is defined by certain selective conditions imposed on the populations as whole, such as ability to resist killing by a certain cytotoxic drug.

Mutant cell clones that grow out in these experiments can be isolated and their genomes interrogated by DNA sequencing of the sgRNA-encoding portion of the retroviral proviruses that have integrated into the genomes of the cells in these clones, thereby identifying the genes in which loss-of-function mutations provide a selective advantage under the experimental conditions imposed on these cell populations.

One can also use this method to identify those genes in which mutations lead to a selective *disadvantage* to cells. For example, genes required for cell growth in culture or upon transplantation into animals (termed "essential genes") can be identified in CRISPR-Cas9-based genetic screens by comparing the relative abundance of sgRNA coding sequences in the starting retroviral library to their relative abundance after introduction into cells and a period of passage in culture or growth following implantation *in vivo* (Figure 12.45B). These "dropout" screens can also be applied to cancer cell lines in an effort to identify genes that are specifically required in the context of a particular cellular genotype (e.g., in the presence of a mutant oncogene or tumor suppressor gene; see Figure 12.45B). Sometimes termed "synthetic lethal" screens, these experiments can identify genetic vulnerabilities in cancer cells of a certain mutant genotype that have the potential to guide the subsequent development of anti-cancer therapeutic drugs.

The CRISPR-Cas9 system has been further modified to be used to *modulate* gene expression rather than cause genetic mutation. Mutant Cas9 proteins engineered to lack endonuclease activity (thus preventing DNA cleavage) have been fused to protein domains that either stimulate or repress transcription (Figure 12.45C). Here sgRNAs serve to guide the Cas9 fusion proteins to promoters of genes of interest in order to increase the transcription of targeted genes [CRISPR-mediated activation (CRISPRa)] or decrease their expression [CRISPR-mediated inhibition (CRISPRi)]. As with traditional CRISPR-Cas9 strategies described above, CRISPRa and CRISPRi can be used in studies focused on individual genes or in genome-wide screens using sgRNA libraries.

Continued

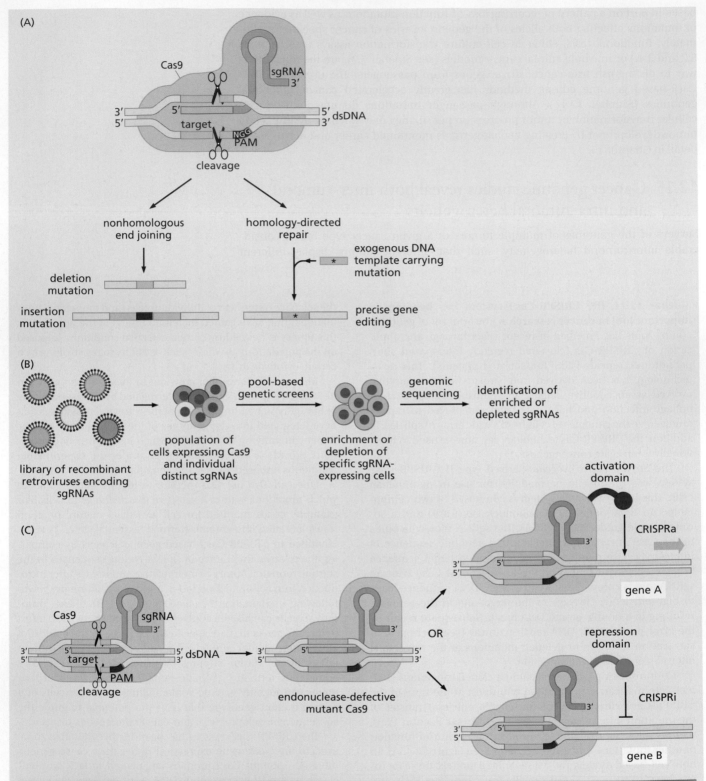

Figure 12.45 CRISPR/Cas9-based genome editing, genetic screens, and modulation of gene expression The versatile CRISPR/Cas9 system can be used to precisely modify genomes and broadly modulate gene expression. (A) The *S. pyogenes*-derived Cas9 DNA endonuclease can be localized to a specific DNA sequence through binding to a single-guide RNA (sgRNA) that also forms base pairs via a complementary sequence (*blue*) to a specific target sequence in the genome (*green*). The target sequence must be adjacent to a protospacer adjacent motif (PAM) sequence, a three-nucleotide sequence composed of any nucleotide (N) followed by two G residues (NGG). Cas9-mediated cleavage (*shown as scissors*) of both strands of the target sequence results in a double-strand DNA break. Repair of the break by the error-prone nonhomologous end joining pathway can produce deletion and insertion mutations, which typically cause loss of gene function. With the addition of an exogenous DNA template carrying a mutation (*asterisk*), homology-directed repair can transfer the mutation into (*continued*)

Figure 12.45 CRISPR/Cas9-based genome editing, genetic screens, and modulation of gene expression (*continued*) the genome. (B) To perform CRISPR-based genetic screens, libraries of recombinant retroviruses are constructed, such that each virus carries a different sgRNA targeting a specific gene (depicted by different colors). These viruses are used to infect Cas9-expressing cells. Subsequent CRISPR/Cas9-mediated mutation leads to the generation of a population of cells with targeted loss-of-function mutations in the corresponding genes. Subjecting these cells to selective conditions causes enrichment or depletion of certain mutant cells. For enrichment-based screens, the identity of selected mutations can be determined following DNA sequencing of the sgRNA portions of the retroviral proviruses present in the remaining mutant cells. For depletion-based screens, one compares the representation of sgRNA sequences in the starting population of cells to that present following growth in the selective conditions; sgRNA sequences that are relatively underrepresented in the final population are indicative of genes in which mutation led to a fitness disadvantage. (C) Cas9 can be rendered endonuclease defective through specific mutations. Such endonuclease-defective mutant Cas9 proteins can still bind to sgRNAs and be targeted to specific sites in the genome. By fusing transcriptional activation or repression domains onto the endonuclease-defective Cas9 proteins and targeting them to the promoter regions of genes of interest via expression of the appropriate sgRNAs, one can affect CRISPR-mediated gene activation (CRISPRa) or inhibition (CRISPRi).

constellations of mutations of known or presumed driver genes. Some of this heterogeneity is explained by the number of alternative cellular signaling pathways that can be activated or inactivated to promote tumor formation and progression. Thus, "many roads lead to Rome" and different tumors may take somewhat different paths en route to full-fledged malignancy (see Figure 11.11).

As an example, recent studies that analyzed larger numbers of lung adenocarcinomas identified almost 40 genes that were mutated to a statistically significantly extent, including some that were found in just a handful of patients. Scanning the data from this and other large collections of tumors, one can conclude that very few tumors of a given type carry precisely the same combination of driver gene mutations and CNVs. At the mechanistic level, we can rationalize this inter-tumoral diversity by recognizing that within a given cellular regulatory pathway, there are multiple alternative genes that can be altered in order to yield the same pro-tumorigenic phenotype, as discussed in Section 11.3. For example, in the lung adenocarcinoma study cited above, genome sequencing revealed frequent mutation of components of the mitogenic signaling pathway that was described in Chapter 6, including those affecting the EGFR growth factor receptor gene, the K-Ras protein, and the GAP-related gene *NF1*. Because mutation of any one of these genes would be expected to activate this particular pathway, one might expect mutation of one, but not multiple, members of this group of genes within a given patient's tumor genome. This type of mutual exclusivity of cancer gene mutations is evident in large-scale cancer genomics datasets (**Figure 12.46**), and indeed, these patterns of mutations can help investigators infer functional relationships between genes ostensibly operating in common signaling pathways.

In addition to this inter-tumoral heterogeneity, genome sequencing studies have revealed extensive intra-tumoral heterogeneity, the product of clonal diversification occurring during the course of the multi-step formation of a single tumor, as we discussed in Section 11.6. By sampling distinct sectors of a tumor, cancer genome studies have revealed that some mutations are shared by most or all cells within the tumor. Such mutations are likely to have arisen early in disease progression and were under continued selective pressure to be maintained throughout the entire subsequent course of multi-stage tumorigenesis. These mutations are sometimes referred to as "truncal mutations" because they can be viewed as arising in the trunk of an evolutionary lineage tree that illustrates the genetic relationships of the cells forming the tumor. In contrast, some mutations can be found in only one sector of the tumor—an indication that they mark separate branches of an evolutionary tree (see Figures 2.17C and 11.16C). Moreover, the fine structure of such trees—the individual twigs—can also be identified through the analysis of individual cancer cells isolated from tumors using single-cell DNA sequencing.

For an individual patient, the sequencing of their tumor DNA (taken from a surgical resection or even a biopsy) can identify mutated genes that may be appropriate for

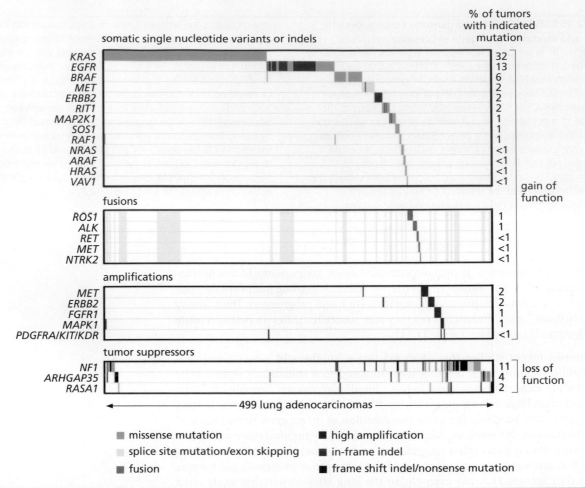

missense mutation

splice site mutation/exon skipping

fusion

high amplification

in-frame indel

frame shift indel/nonsense mutation

Figure 12.46 Patterns of cancer gene mutations Large-scale cancer genomic studies often reveal pattens of cancer gene mutations displaying a high degree of mutual exclusivity. This figure shows data from 499 lung adenocarcinomas, each of which carried an alteration in one or more of 26 genes encoding components of the mitogenic signaling pathway (*listed at left*); the different cancers samples are arrayed *left to right*. The presence of a mutation in a given gene is indicated by a *colored vertical line*. Different classes of mutations are shown in each of the 4 rectangles. The gain-of-function alleles are grouped in the *upper three rectangles*, which affect presumptive oncogenes, while the loss-of-function alleles, which affect presumptive tumor suppressor genes, are grouped in the *lower rectangle*. "Somatic single nucleotide variants" denotes point mutations, whereas "indels" indicates insertions or deletions.

The specific types of mutation are denoted according to the color scheme at bottom. Within each rectangle the tumor samples with the most frequent mutant allele are arrayed to the *left* and the least frequent to the *right*. Note the high degree of mutual exclusivity involving mutations in this class of genes; that is, most tumors have only one mutation that activates oncogenes. Essentially all tumors lacking an identified oncogene-activating mutation carry instead a loss-of-function mutation in a tumor suppressor gene that normally negatively regulates the mitogenic signaling pathway (*lowest rectangle*). The mutual exclusivity is not absolute, however, as can be seen in the subset of tumors that have mutations in both an oncogene (*upper three rectangles*) and a tumor suppressor gene (*lowest rectangle*). (Adapted from J.D. Campbell et al., *Nature Genet.* 48:607–616, 2016. With permission from Nature.)

particular forms of targeted therapy. As discussed in Chapter 17, an increasing number of such therapies, involving both small-molecule drugs and monoclonal antibodies, have been developed to counteract the effects of specific mutant oncogenes. The use of these therapies can be tailored to those patients whose tumor carries a mutant form of a specific gene and sometimes even a particular mutation harbored by that gene. In all cases, drug targeting in response to the discovery of a truncal mutation and derived mutant gene product is most likely to yield a longer clinical response, whereas drug treatments that respond to one or another mutant branch are unlikely to elicit long-term responses, if any at all.

Another important application of cancer genomic studies is tracing the clonal origins of late-stage cancer colonies, that is, metastases, back to the primary tumor. This concept was introduced in Chapter 11, and we will return to it again in Chapters 14 and 17.

For example, by comparing the genome or exome sequence of a series of metastatic lesions from a patient with the lineage information derived from the clonal architecture of their primary tumor, one can often identify a specific subclone within a primary tumor that seeded a given metastasis as well as the clonal relationships between distinct metastatic lesions present in the same patient. Some of the mutant alleles are present in one metastasis but not others and are therefore referred to as "private" (that is, not shared) mutations. Finally, similar cancer genomics-based analyses can be used to determine the clonal origins of a relapsed cancer following initial therapy. By comparing the sequence of a relapsed sample with the lineage information from an individual's initially diagnosed primary tumor, it is often possible to identify the subpopulation of cells within the primary tumor that eventually gave rise to the cells that created the clinical relapse. We return to this subject in Chapter 17.

12.16 Synopsis and prospects

Genome instability has been inherent in life since the first cells appeared 3.5 billion years ago. In the intervening time, living organisms have continually struck a balance between too little and too much instability in their genomes. If they went too far in suppressing the rate at which mutations accumulated, Darwinian evolution, which depends directly on the continued generation of genetic diversity, would have ground to a halt. Conversely, if they allowed mutation rates to increase too much, their ability to reproduce and even their viability would have been seriously compromised. The relatively low level of genomic instability that operates in our cells—specifically those carrying the germ line—represents a compromise between these two conflicting needs.

This balance between too little and too much genetic instability does not need to be struck in the individual cells of our somatic tissues. In these, any genetic instability—mutability, as we have called it—is undesirable because it opens the door to neoplasia. The need to avoid neoplasia would seem to explain why evolution has worked hard to ensure that the genomes of our somatic cells are so stable.

Multiple layers of defense mechanisms operate to hold somatic mutation rates to extremely low levels. At the biological level, they are embodied in the organization of stem cells and their differentiated progeny. At the biochemical level, an array of enzymes and a variety of low–molecular-weight biochemical species are deployed to confront and neutralize mutagens before they succeed in striking the genome. And should damage be inflicted, either because mutagens have slipped through the outer defenses or because of errors in DNA replication, then a large group of DNA repair enzymes—the caretakers—lie in wait, always alert to structural aberrations in the double helix and its nucleotides. More often than not, these enzymes deal very effectively with incurred damage and succeed in restoring the DNA to its pristine state, erasing any sign that damage was ever sustained. In addition, a complex array of proteins ensures that mitosis and meiosis occur only when the chromosomes are aligned properly at the metaphase plate, thereby sustaining the euploid karyotype.

Our perception of DNA and its much-touted stability is changed by an understanding of these caretakers and their multiple roles in maintaining the genome. Previously, we depicted DNA as a rock-solid, unchanging entity within the cell, a unique island of stability sitting amid countless other molecules that are constantly forming and being degraded. Now, we realize that this portrayal was simplistic and an illusion. Like all other molecules in the cell, DNA is vulnerable to many types of damage. Its apparent stability reflects nothing more than a dynamic equilibrium, an ongoing battle between the forces of order and chaos. Any stability that chromosomal DNA does exhibit, and it is considerable, represents a stunning testimonial to the elaborate array of caretakers that are always on watch, ready to fix even the most minor lesion in the double helix.

The implications of this dynamic equilibrium for cancer are simple and clear: if a breakdown of genomic integrity is an essential ingredient in forming human tumors, it can result most readily from weakening of the ever-vigilant repair machinery and its controllers.

Our initial encounter with the breakdown of genomic stability came in Chapter 9, where we learned that the p53 tumor suppressor protein is occasionally called the "guardian of the genome," because cells lacking p53 function acquire a variety of genetic defects at an elevated rate. In large part, this increased mutability, which includes alterations in DNA sequence as well as changes in karyotype, does not reflect p53's role in directly maintaining the genome. Instead, the loss of p53 function creates an environment that is permissive for the survival of mutant cells. In the present chapter, we changed our focus by posing a different question, one that goes beyond p53 inactivation: If p53 loss permits mutant (and highly mutable) cells to survive, how, precisely, do the mutations acquired by these cells arise in the first place?

These mutations occur frequently during tumor progression, and elevated mutability is increasingly accepted as an important element of cancer pathogenesis. As discussed in Section 11.4, a departure from DNA's highly stable state may be essential for the formation of human cancers. This notion derives from calculations of the rates at which mutations accumulate in normal cells and an estimate of the number of genetic alterations that are needed in order for tumor progression to reach completion.

Without such increased mutation rates—so the thinking goes—the time intervals between clonal successions (see Section 11.4) would be far too long. Actually, the readings from the last chapter and this one suggest at least two alternative ways by which clonal successions can be accelerated during multi-step tumor progression. Tumor promoters (including endogenous processes such as inflammation) can compress the time between clonal successions; alternatively, acceleration can be achieved by the destabilization of the genome, as described here. Because the two complex processes often work hand in hand, mathematical modeling of tumor progression becomes very difficult.

We now realize that mathematical analyses of multi-step tumor formation depend on so many quantitative assumptions that their major predictions represent little more than an inspired speculation. As is almost always the case in biology, observations of living systems speak more loudly than theorizing: recent high-throughput sequencing analyses of tumor cell genomes indicate enormous variability in the rates with which various types of tumors accumulate mutations as they pass through multi-step tumor progression. Although increased mutability clearly accelerates the rate of tumor progression (thereby leading to increased tumor incidence), it is clear that certain tumors arise in the absence of large numbers of mutations, as discussed below.

Cancer stem cells (CSCs) are present in many and perhaps all tumors. Their presence complicates our understanding of how mutations accumulate within tumor cell genomes. Because of their ability to self-renew, the lineages of CSCs within tumors are likely to persist long after their more differentiated descendants have entered into post-mitotic states or died; accordingly, CSCs are likely to serve as the long-term repositories of the tumors' genetic information. Hence, advantageous mutations must sooner or later be introduced into the CSC pools in order to ensure transmission to future generations of cancer cells.

This scenario is complicated because SCs in general, and CSCs in particular, are present in relatively small numbers within normal and neoplastic tissues. Because the likelihood that a population of cells will sustain a mutation is directly proportional to the size of this population, pools of CSCs are relatively unlikely (mathematically) to directly acquire the mutations that drive tumor progression forward. This situation creates a dilemma that is at present unresolved: specialized, still-undiscovered mutagenic mechanisms may destabilize the genomes of CSCs. Alternatively, and more likely, most advantageous mutations strike the genomes of transit-amplifying/progenitor cells within tumors, and the latter cells then introduce their advantageous mutations back into the CSC pool by spontaneous dedifferentiation (Supplementary Sidebar 11.6).

Unfortunately, all discussions of cancer cell mutability rest on shaky foundations. Measurements of mutability—number of mutations sustained per cell generation—depend on knowledge of the number of successive cell generations through which a cell lineage has passed since its initiation. At present, we have only very limited information on this critically important parameter (see, for example Figure 10.21A), undermining any attempts at demonstrating heightened mutability definitively.

These discussions of mutation rates generally focus on the submicroscopic changes in genome structure created by defective BER and NER—lesions that are far too small to affect the karyotype of cancer cells. However, as we read in this chapter, cell genomes are also affected by changes occurring on a far larger scale—changes that scramble overall chromosome structure and thereby alter the karyotype of a cell. Such karyotypic alterations are found in the cancer cells from the great majority of solid tumors. In Section 10.4, we read that telomere erosion may be responsible for much of this instability through its ability to trigger breakage–fusion–bridge cycles. Surely, yet other molecular mechanisms will one day be found to contribute to the karyotypic chaos frequently encountered in cancer cells.

Between the minute lesions left behind by imperfect BER and NER and the large-scale rearrangements generated by telomere collapse are genomic changes resulting from the process of **replication stress**, in which the unbalanced mitogenic signals operating in many cancer cells cause uncoordinated firing of replication origins and frequent replication fork collapse of the sort depicted in Figure 12.8. The repair of these collapsed replication forks is often imperfect, providing an explanation for many of the local amplifications and deletions that are present in the genomes of solid tumors (see Figure 12.44). Strikingly, signs of replication stress, in the form of activated DNA repair proteins, are already apparent relatively early in multi-step tumor progression (for example, in dysplastic tissues), long before frankly neoplastic growths have emerged.

Although there seem to be common underlying biochemical principles governing how diverse types of human cells are transformed to a neoplastic state (see Section 11.9), the genetic routes taken by various cells throughout the body to reach this state bear little resemblance to one another, at least as gauged by measurements of copy number change.

Independent of its origins, increased mutability forces us to confront another reality: the resulting disruption of cancer cell genomes is a double-edged sword, because it places these cells at great risk. Frequent errors in DNA replication and in chromosomal segregation make these cells especially vulnerable to death because of lethal, unrepaired defects in their genomes. Indeed, it is possible that much of the attrition of cancer cells during each cell generation (see Figure 10.4D) occurs because of frequent genomic catastrophes that are incompatible with continued cell survival. Hence, cancer cells must weave a fine course between too much and too little mutability, and certain cancer cell types may actually benefit from maintaining their genomes in a relatively stable configuration.

Actually, cancer cells may be exposed to a second danger because of their defective DNA repair apparatus: they may be especially susceptible to certain forms of chemotherapy. We already encountered one example of this in the vulnerability of many gliomas to killing by temozolomide, an alkylating chemotherapeutic. Those tumors that shut down expression of their *MGMT* gene during their initial formation, ostensibly because doing so increases mutability and accelerates tumor progression, pay a price later, because they have lost the ability to remove the toxic methyl groups added to their DNA by this drug (see Section 12.7).

Another example of such increased vulnerability comes from breast and ovarian tumors that have lost BRCA1 or BRCA2 function. The resulting loss of an ability to repair covalent inter-strand cross-links, which depends on homology-directed repair (HDR), makes certain ovarian carcinomas especially sensitive to killing by platinum-based chemotherapeutics (see Section 17.2). Consistent with this, ovarian cancer patients carrying germ-line *BRCA1* or *BRCA2* mutations have longer progression-free survival following platinum therapy than do those identically treated patients with sporadic (that is, nonfamilial) tumors in which BRCA1/BRCA2 function is usually intact.

A more subtly crafted therapeutic strategy derives from examining in great detail the mechanisms of base-excision repair (BER) and homology-directed repair (HDR). As we learned earlier, a relatively small number of double-strand breaks (DSBs) occur in each cell cycle because of the fragility of the replication fork (see Figure 12.8); the repair of these lesions usually depends on HDR. In fact, an alternative and far greater threat comes from the thousands of mutant bases that are created every day by endogenous processes, notably oxidation; the resulting DNA lesions are healed by BER. As described earlier (see Figure 12.20A), BER generates single-nucleotide gaps (through both BER pathways) and thus single-strand breaks (SSBs) that must be filled by a DNA polymerase and then sealed by a DNA ligase.

DNA repair enzymes involved in BER are recruited to SSB sites by the actions of PARP1 [poly-(ADP ribose) polymerase-1]. This enzyme binds to these breaks and proceeds to ADP-ribosylate itself and possibly neighboring proteins as well; **Figure 12.47A**. The resulting poly-ADP tails thereafter serve as docking sites for recruiting the repair enzymes needed to fix the SSBs; hence, when the PARP1 enzyme is inhibited pharmacologically, SSBs persist. These unrepaired SSBs generate a far more serious problem—DSBs—when a replication fork passes through during a subsequent S phase (see Figure 12.47B).

All this suggests an interesting therapy: shut down the PARP enzyme in *BRCA1*- or *BRCA2*-mutant cells with a pharmacological inhibitor. Resulting SSBs will be converted to DSBs during replication, and these DSBs, which are usually repaired by HDR, will not be repairable in cells lacking HDR function (from loss of BRCA1 or BRCA2 function; see Figure 12.47B). Lethal DSBs will then accumulate in large numbers, triggering cell death. Indeed, PARP1 inhibitors have been employed successfully in the clinic for patients with ovarian and breast cancers.

The scenario of acquired genetic instability seems to create a logical quandary: almost all the genetic alterations that occur during tumor progression appear to confer some immediate growth or survival benefit on the cells that acquire them. Many of the earlier chapters in this book documented the specific growth advantages resulting from each of these alterations. As described in a previous chapter, these dynamics cause one to speculate that tumor progression is a process that is analogous to Darwinian evolution (see Section 11.4). However, unlike many traits acquired by cancer cells, the trait of increased genomic instability does not provide an immediate payoff—no marked advantage in proliferation or survival. Instead, acquisition of this trait represents a "long-term investment"; that is, a benefit will only be realized by the distant descendants of this cell.

We can understand this long-term advantage in the following way. Cell clones that have acquired some of the initial genetic alterations leading to cancer seem to arise with great frequency throughout the body's tissues. Nearly all remain in a dormant, premalignant state for an entire human lifetime, unobtrusive and unthreatening, because their stable genomes preclude the acquisition of the additional mutations that would render these cell clones truly dangerous. On rare occasions, however, a cell in one or another of these already-initiated clones acquires an alteration in one of its caretaker genes. Now, for the first time, this cell and its descendants have the opportunity to tinker with their genomes by testing new combinations of genetic elements and new sequences, some of which will allow them to resume their advance down the long road of tumor progression. By opening a floodgate of genetic changes, clones of would-be cancer cells are destined, over the long term, to prosper, while their brethren, lacking this instability, are likely to remain unchanged and indolent for decades.

As we learned in this chapter, a variety of familial cancer syndromes are attributable to the inheritance of mutant forms of genes specifying important components of the DNA repair apparatus. Genomic instability or a tendency toward genetic instability is already implanted in all the cells of such genetically afflicted individuals. In these cases, genetic instability is not an acquired attribute—unlike the situation operating in the most human cancers.

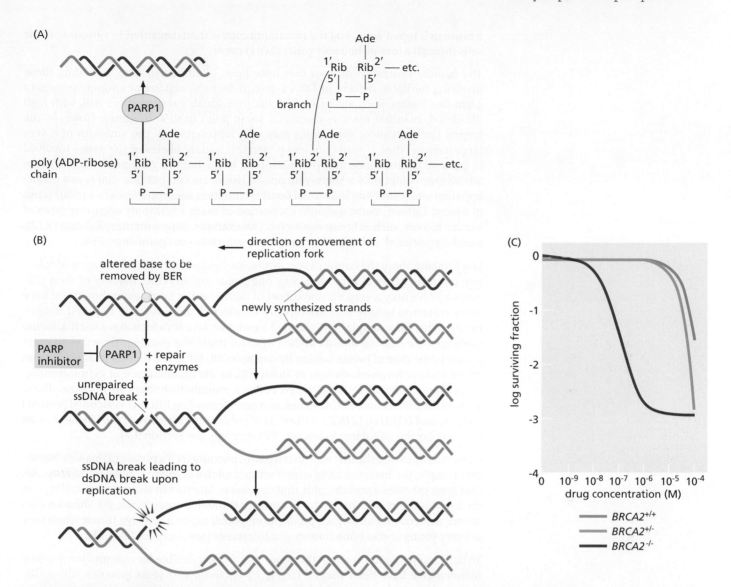

Figure 12.47 Killing of BRCA2-negative cells by an anti-PARP drug The reliance of BRCA1- and BRCA2-negative human breast cancer cells on a redundant HR pathway mediated by poly-(ADP ribose) polymerase-1 (PARP1) makes these cells especially vulnerable to killing by an anti-PARP drug. (A) PARPs constitute a family of related enzymes that act by generating repeating branched polymers of as many as 200 residues (*brackets*) each. These residues are attached both to the PARP enzymes themselves (*green oval*) and to other proteins involved in effecting biochemical responses (*not shown*); these chains may be branched as shown here. In the case of PARP1, it binds to ssDNA breaks (SSBs) and ADP-ribosylates itself; its poly-ADP chains then attract a series of repair enzymes that may be ADP-ribosylated by PARP1 and thereafter proceed to complete various steps of base excision repair (BER). (B) Such SSBs are created routinely by oxidized bases, which are usually removed by BER. Here, an oxidized base (*yellow*) should be excised and the

DNA repaired through the actions of PARP1 and recruited repair enzymes. However, in the presence of a PARP1 inhibitor, these other enzymes are not recruited, BER cannot proceed to completion, and the resulting SSB will be converted to a double-strand DNA break (DSB) when a replication fork moves through during the next S phase. This break can be efficiently restored to a wild-type configuration only through homology-directed repair (HR), which is defective in BRCA1- and BRCA2-mutant cells. Failure to repair a DSB will often lead to cell death. (C) Cultured mouse ES (embryonic stem) cells of three genotypes were treated with KU0058948, a PARP1 inhibitor. The *BRCA2*+/+ (*blue*) and *BRCA2*+/− (*green*) ES cells were killed only by very high drug concentrations, whereas the *BRCA2*−/− ES cells were killed by the drug at concentrations almost 1000-fold lower (*red*). (C, from H. Farmer et al., *Nature* 434:917–921, 2005. With permission from Nature.)

On some occasions, such as those arising in the many subtypes of xeroderma pigmentosum (XP), both copies of a critical DNA repair gene are inherited in defective form. This nullizygous state leaves all the cells in the skin without the means to cope with the UV-induced formation of pyrimidine dimers. In other cases, such as hereditary non-polyposis colon cancer (HNPCC), an individual inherits a defective gene copy of a mismatch repair gene, and the remaining copy is then discarded in various somatic cells through a loss-of-heterozygosity (LOH) event.

The familial cancer syndromes that have been identified to date, including those involving heritable defects in DNA repair, reflect the actions of strongly penetrant germ-line alleles. Such alleles ensure that individuals carrying them will, with high likelihood, manifest obvious disease at some point in their lifetimes. However, the known familial cancer syndromes may well represent only the small tip of a very large iceberg. Thus, a variety of less penetrant, mutant alleles of the genes involved in genome maintenance may be widespread in the human gene pool. Each of these alleles may confer only a slightly increased risk of cancer, but one that is not readily apparent when studying individual families and their susceptibilities to various types of cancer. Consequently, a significant portion of many commonly occurring types of human tumors, such as breast cancer (see, for example, Supplementary Sidebar 12.12), may be associated with inheritance of these still-unknown germ-line alleles.

Our thinking about the family cancer syndromes associated with inheritance of defective caretaker genes is confounded by one major mystery: the majority of these diseases involve only a very narrow subset of tissues in the body, even though we have every reason to believe that the services of affected caretakers are required ubiquitously. Why, for example, do inherited mismatch repair defects have such a strong preference for causing tumors in the intestinal tract? The puzzle is drawn even more starkly in the case of breast cancer. By one account, inherited defects in at least seven genes that are involved, directly or indirectly, in the maintenance of genomic integrity generate substantially increased risks of mammary tumors in humans. These include the *BRCA1* and *BRCA2* genes, two genes encoding BRCA1-associated proteins (*BARD1* and *BACH1*), *CHK2*, *p53*, and *ATM* (which specifies one of the kinase sensors that activate the p53 alarm following DNA damage; see Section 9.7).

Of course, this is not the first time that we have encountered a puzzle of this sort. Recall, for example, the behavior of inherited mutant alleles of the retinoblastoma gene, *Rb*. This gene specifies a protein, pRb, that appears to function in almost every cell type in the body as the critical controller of advance through the cell cycle, yet children who inherit defective *Rb* alleles are predisposed peculiarly to a rare eye tumor when they are very young and to bone tumors as adolescents (see Sections 7.3–7.5).

To date, our perceptions about the genes that are responsible for mammalian genome maintenance have been shaped largely by bacterial and yeast genetics. These disciplines have yielded a wealth of genes that are essential for genome maintenance in these microbes and, by extension, in our own genomes as well. For example, the resonance between yeast genetics and its harvest of mismatch repair genes and the HNPCC syndrome led to the rapid enumeration of six human genes and proteins that are responsible for this type of genome maintenance.

Importantly, microbial genetics may have netted only a portion of the genes in the human genome that are responsible for maintenance of its integrity. How large, then, is the universe of human caretaker genes and proteins? The sequencing of the yeast genome begins to give us a feeling for this number. Recent estimates indicate that this genome consists of approximately 5600 distinct genes. Of these, 153 are classified as being involved in DNA replication and repair, and 88 seem to be involved exclusively with repair. Many of the genes encountered in this chapter are metazoan inventions, which suggests that the size of the cohort of human genes involved in these processes is likely to be much larger. Indeed, by 2001, more than 130 DNA repair genes had already been identified in the human genome; by 2004, the number approached 150, and it reached 176 in 2012. Hence, genes and proteins that play critical roles in the maintenance of the genome and thus in the suppression of cancer are being discovered continually, some in very unexpected places (**Sidebar 12.12**).

This area of cancer research is also having a profound effect on our understanding of another, seemingly unrelated biological process—aging. At least ten distinct premature aging syndromes—each a form of **progeria**—have been traced to inherited defects in one or another component of the DNA repair system. Individuals suffering from these syndromes often show many of the phenotypes of the aged by the time they

Sidebar 12.12 Histones function to reduce cancer risk For many decades, histones and the nucleosomes that they form were depicted as static structural components of chromatin that serve to package and compact the chromosomal DNA. Beginning in the 1990s, it became clear that covalent modification of histones is critical in making regions of chromatin more or less hospitable for transcription (see Section 1.8). Even more recently, histones have been found to play a dynamic role in DNA repair. H2AX is a variant of the abundant H2A histone that is present in virtually all nucleosomes. H2AX constitutes about 15% of the H2A-like histones in cells and substitutes for H2A in some nucleosomes (see also Figure 1.16). When a double-strand break is sustained in chromosomal DNA, H2AX molecules (but not the other histones) become phosphorylated, primarily by the ATM and ATR kinases, on a specific serine residue located four amino acid residues from the carboxyl-terminal end; such phosphorylated H2AX is observed in a large chromosomal region (involving as much as 2 megabases of DNA) flanking the break (**Figure 12.48**). (These two kinases are also responsible for phosphorylating and thereby mobilizing p53; see Figure 9.12.) The phosphorylated H2AX (usually termed γ-H2AX) attracts DNA repair proteins, such as BRCA1 and NBS1, as well as at least four others that aid in the task of rejoining the DNA ends (see also Figures 12.28 and 12.31).

Mice that have lost the H2AX gene (because of germline inactivation) are viable but stunted in growth. Their cells are unable to execute homology-directed repair (HDR; see Section 12.9) and are prone to accumulate structurally abnormal chromosomes. Mice lacking both the H2AX and p53 proteins are highly prone to both hematopoietic and solid tumors. Even mice lacking one copy of the H2AX gene in the context of p53 deficiency show significantly increased rates of lymphomas.

These responses illustrate how highly complex the DNA repair process is, and how defects in any of its individual components—many still unidentified—open the door to the appearance of cancer. Provocatively, the human H2AX gene maps to a chromosomal region (11q23) that frequently undergoes loss of heterozygosity and/or deletion, creating the possibility that this gene and encoded histone are frequently shed by human cells en route to malignancy.

Figure 12.48 γ-H2AX and double-strand DNA breaks The creation of dsDNA breaks by various mechanisms results in the phosphorylation of H2AX histone (a variant form of histone H2A), yielding γ-H2AX. Like H2A, H2AX participates in the formation of nucleosomes. Chromosomal regions as large as two megabases that flank a dsDNA break are found to carry copies of γ-H2AX. In the absence of γ-H2AX , critical DNA repair proteins, including NBS1 and BRCA1 (see Figure 12.31), fail to be recruited to the site of dsDNA breaks. In the image shown here, cells of the Indian muntjac, used because of their small number of chromosomes, have been irradiated with a low dose (0.6 Gy) of X-rays. Late anaphase cells were then stained for γ-H2AX (*yellow-green, arrows*) and for chromosomal DNA (*red*). The outlines of the dividing cell are seen in blue. Note the staining of γ-H2AX at the broken ends of a chromosome. (From E.P. Rogakou et al., *J. Cell Biol*. 146:905–916, 1999. With permission from Rockefeller University Press.)

reach adolescence. These clinical effects of defective DNA repair provide compelling indications that much of the normal aging process will one day be traced to genetic damage that we accumulate in our stem cells throughout life; their progressive attrition seems to lead to the inability of tissues to renew themselves, yielding precisely the changes observed in the aged. Consequently, cancer and aging may be found to share a common root—the progressive deterioration of our genomes as we get older. And both cancer development and aging may one day be forestalled by treatments and lifestyles that protect our genomes from the ongoing attacks that they suffer, decade after decade, deep within our cells.

The nature and pace of cancer research has changed markedly over the past decade as a result of a series of technological developments in the area of cancer genomics. We have discussed the importance of these approaches to developing comprehensive mutational landscapes of most types of human cancer, including single nucleotide mutations, small-scale indel mutations, larger-scale mutations, and copy number variation. Today, whole-exome and whole-genome analysis of cancer specimens from human patients and pre-clinical models are routinely performed, allowing a remarkably complete view of accumulated genomic alterations associated with tumor development. As discussed in this chapter, understanding the patterns of mutations across the genome in different types of cancer can provide important insights into the etiological factors that influence tumor development, including exposure to exogenous carcinogens as well as defects in DNA damage repair pathways. In Chapter 17, we discuss the use of human cancer genomic studies in guiding the use of specific therapies that are targeted against particular mutant genes. Increasingly, oncologists order genome sequence analyses for their patients' cancers precisely for this purpose, and this practice is expected to become routine in the not-too-distant future.

Mutations in protein-coding genes—whether they be in cancer drivers or passengers—can produce neoantigens that distinguish the cancer cells bearing these mutations from normal cells, making them visible to components of the adaptive immune system and, in turn, possible targets for attack by various types of immunotherapy, as described in Chapter 16. This provides the rationale for ongoing, extensive efforts to use cancer genome analyses of a patient's tumor DNA to predict which mutations would be expected to yield "strong antigens" that are highly likely to attract the attention of the immune system and, in turn, might be exploited to produce therapeutic cancer vaccines. We will revisit some of these issues in Chapter 16. At the same time, the vulnerabilities of cancer cells arising from acquired defects in their genome maintenance machinery are being explored with the hope of spawning an entire new generation of anti-cancer therapeutic strategies.

Key concepts

- The structural integrity and thus low mutability of DNA depends on a large and complex set of biological and biochemical mechanisms that work to ensure that somatic mutations accumulate in tissues at very low rates.

- Some of the mechanisms depend on the organization of tissues, in which the long-lived cells (the stem cells) are protected from genetic damage while the short-lived cells (the transit-amplifying and differentiated cells) are vulnerable to such damage but are soon discarded.

- Misincorporated bases generated by errors in DNA replication can contribute to the burden of accumulated mutations. The numbers of these alterations are held down by the low error rates of DNA polymerases together with an array of error-correcting proteins, such as those involved in mismatch repair. Inherited defects in mismatch repair proteins can lead to increased susceptibility to certain types of cancer, notably hereditary non-polyposis colon cancer.

- Cell genomes are under continuous attack by a variety of chemically reactive molecules, many of them deriving from the cellular process of oxidative phosphorylation and the resulting reactive oxygen species that are generated as by-products of this process. Cell genomes may also suffer spontaneous chemical alterations, which affect DNA bases at a low but significant rate.

- In addition, the genomic DNA of cells can be attacked by mutagenic molecules of foreign origin. Such xenobiotics and their chemically reactive derivatives may come from pollutants and, to a far greater extent, from commonly consumed foodstuffs.

- Cells attempt to detoxify many of these compounds before they can attack genomic DNA. However, the side products of these reactions may actually be more reactive and mutagenic than the initially introduced molecular species.

- If the attacking mutagenic agents succeed in damaging DNA, an elaborate array of proteins involved in base-excision and nucleotide-excision repair lies in wait in order to remove most damaged bases. Inherited defects in base-excision repair proteins can lead to various types of cancer susceptibility.

- Other types of DNA damage include double-strand DNA breaks, which can be created by X-rays or, more commonly, by accidental DNA breakage at replication forks.

- Double-strand DNA breaks can be repaired in the G_1 phase of the cell cycle by nonhomologous end joining or in the S and G_2 phases by homology-dependent repair. Inherited defects in double-strand DNA repair can explain the breast and ovarian cancer susceptibility among patients inheriting mutant *BRCA1* or *BRCA2* germ-line alleles.

- Genomes may be scrambled by mechanisms that affect the karyotype of cells. One class of such alterations includes those that affect chromosomal structure, including the translocations that are created by the fusions of unrelated chromosomal arms to one another. Such fusions seem to be commonly triggered by eroded telomeres and collapsed replication forks, both of which generate double-strand DNA breaks.

- Changes in chromosome number are also common in cancer cell genomes and appear to facilitate the accumulation of genes in proportions that favor the proliferation and survival of these neoplastic cells. Many of these changes derive from defects in the mitotic apparatus and its regulators, notably the centrosomes and proteins involved in connecting spindle fibers with chromosomal kinetochores.

- The advent of high-throughput genome sequencing technologies has allowed in-depth analysis of cancer genomes and a comprehensive understanding of the array of mutations that contribute to tumor development.

- The specific nature of mutations in cancer genomes and the sequence context in which they occur can provide clues to possible carcinogen exposure or defects in DNA repair processes that contributed to tumor development.

Thought questions

1. What types of evidence suggest that karyotypic alterations of cell genomes are not absolutely essential for neoplastic transformation?

2. When calculating the rates of mutation required for multi-step tumor progression to reach completion, what parameters must one know in order for such a calculation to accurately describe the actual biological process?

3. How does our understanding of defective DNA repair processes in tumor cells make possible the development of new anti-cancer therapeutic strategies?

4. In which ways do the defectiveness of p53 function and resulting defects in apoptosis and DNA repair facilitate the forward march of tumor progression?

5. What evidence implicates mutagenic chemicals originating outside the body in the pathogenesis of human cancers? How can one gauge their contribution to human carcinogenesis compared with that of mutagens and mutagenic processes of endogenous origin?

6. How do defects in various cell cycle checkpoints allow for accelerated rates of the accumulation of mutations?

7. How do the biological properties of stem cells help to reduce the rate at which tissues accumulate mutant genes?

8. How does the existence of cancer stem cells affect the calculations of the rate at which mutations must be accumulated in order to allow multi-step tumor progression to advance?

9. How does the genetic heterogeneity in the human gene pool affect the functioning of various types of biological defenses that have been erected to prevent the accumulation of mutant alleles in human somatic cells?

10. How does mutual exclusivity in cancer gene mutations in a given cancer type inform the functional relationships between genes in a biological pathway?

11. What does the observation of extensive inter-tumoral heterogeneity in cancer gene mutations for a given cancer type imply with respect to the nature of the pathways required to cause oncogenic transformation?

12. How can cancer genomics allow scientists to investigate intra-tumoral heterogeneity in cancer?

Additional reading

Alexandrov LB, Kim J, Haradhvala NJ et al. (2020) The repertoire of mutational signatures in human cancer. *Nature* 578, 94–101.

Ames BN (1989) Endogenous DNA damage as related to cancer and aging. *Mutat. Res.* 214, 41–46.

Arias EA & Walter JC (2007) Strength in numbers: preventing rereplication via multiple mechanisms in eukaryotic cells. *Genes Dev.* 21, 497–518.

Barnes DE & Lindahl T (2004) Repair and genetic consequences of endogenous DNA base damage in mammalian cells. *Annu. Rev. Genet.* 38, 445–476.

Ben David U & Amon A (2020) Context is everything: aneuploidy in cancer. *Nat. Rev. Genet.* 21(1), 44–62.

Branzel D & Folani M (2008) Regulation of DNA repair throughout the cell cycle. *Nat. Rev. Mol. Cell Biol.* 9, 297–308.

Brenner JC & Chinnaiyan AM (2009) Translocations in epithelial cancers. *Biochim. Biophys. Acta* 1796, 201–215.

Brown JS, O'Carrigan B, Jackson SP & Yap TA (2017) Targeting DNA repair in cancer: Beyond PARP inhibitors. *Cancer Discov.* 7, 20–37.

Campbell BB, Light N, Fabrizio D et al. (2017) Comprehensive analysis of hypermutation in human cancer. *Cell* 171, 1042–1056.

Christmann M & Kaina B (2019) Epigenetic regulation of DNA repair genes and implications for tumor therapy. *Mutat. Res.* 780, 15–28.

Cimini D (2008) Merotelic kinetochore orientation, aneuploidy, and cancer. *Biochim. Biophys. Acta* 1786, 32–40.

Cleaver JE (2005) Cancer in xeroderma pigmentosum and related disorders of DNA repair. *Nat. Rev. Cancer* 5, 564–573.

Couch FJ, Shimelis H, Hu C et al. (2017) Associations between cancer predisposition testing panel genes and breast cancer. *JAMA Oncol.* 3,1190–1196.

D'Assoro AB, Lingle WL & Salisbury JL (2002) Centrosome amplification and the development of cancer. *Oncogene* 21, 6146–6153.

DiGiovanna DJ & Kraemer KH (2012) Shining a light on xeroderma pigmentosum. *J. Invest. Dermatol.* 132, 785–796.

Duensing S & Münger K (2002) Human papillomavirus and centrosome duplication errors: modeling the origins of chromosomal stability. *Oncogene* 21, 6241–6248.

Ferguson LR (2010) Dietary influences on mutagenesis: where is this field going? *Environ. Mol. Mutagen.* 51, 909–918.

Friedberg EC, Aguilera A, Gellert M et al. (2006) DNA repair: from molecular mechanism to human disease. *DNA Repair (Amst.)* 5, 986–996.

Gold LS, Ames BN & Slone TH (2002) Misconceptions about the causes of cancer. In Human and Environmental Risk Assessment: Theory and Practice (D Paustenbach ed), pp 1415–1460. New York: John Wiley & Sons.

Halazonetis T, Gorgoulis VG & Bartek J (2008) An oncogene-induced DNA damage model for cancer development. *Science* 319, 1352–1356.

Harper JW & Elledge SJ (2007) The DNA damage response: ten years after. *Mol. Cell* 28, 739–745.

Harris RS & Dudley JP (2015) APOBECs and viral restriction. *Virology* 479–480, 131–145.

Hartlerode AJ & Scully R (2009) Mechanisms of double-strand break repair in somatic mammalian cells. *Biochem. J.* 423, 157–168.

Hoeijmakers JHJ (2009) DNA damage, aging and cancer. *N. Engl. J. Med.* 361, 1475–1485.

Holland AJ & Cleveland DW (2009) Boveri revisited: chromosomal instability, aneuploidy and tumorigenesis. *Nat. Rev. Mol. Cell Biol.* 10, 478–487.

Hübscher U, Maga G & Spadari S (2002) Eukaryotic DNA polymerases. *Annu. Rev. Biochem.* 71, 133–163.

Huertas P (2010) DNA resection in eukaryotes: deciding how to fix the break. *Nat. Struct. Mol. Biol.* 17, 11–16.

Jachimowicz RD, Goergens J & Reinhardt HC (2019) DNA double-strand break repair pathway choice—from basic biology to clinical exploitation. *Cell Cycle* 18, 1423–1434.

Jiricny J (2006) The multifaceted mismatch-repair system. *Nat. Rev. Mol. Cell Biol.* 7, 335–346.

Kastan MB & Bartek J (2004) Cell cycle checkpoints and cancer. *Nature* 432, 316–323.

Kee Y & D'Andrea AD (2010) Expanded roles of the Fanconi anemia pathway in preserving genomic stability. *Genes Dev.* 24, 1680–1694.

Kim D & Guengerich FP (2005) Cytochrome P450 activation of arylamines and heterocyclic amines. *Annu. Rev. Pharmacol.* 45, 27–49.

Leibowitz ML, Zhang C-Z & Pellman D (2015) Chromothripsis: A new mechanism for rapid karyotype evolution. *Annu. Rev. Genet.* 49,183–211.

Lord CJ & Ashworth A (2017) PARP inhibitors: synthetic lethality in the clinic. *Science* 355, 1152–1158.

Loeb LA, Bielas JH & Beckman RA (2008) Cancer cells exhibit a mutator phenotype: clinical implications. *Cancer Res.* 68, 3551–3557.

Loeb LA, Loeb KR & Anderson JP (2003) Multiple mutations and cancer. *Proc. Natl. Acad. Sci. USA* 100, 776–781.

Lynch HT & de la Chapelle A (2003) Hereditary colorectal cancer. *N. Engl. J. Med.* 348, 919–932.

Marnett LJ & Plastaras JP (2001) Endogenous DNA damage and mutation. *Trends Genet.* 17, 214–221.

Maynard S, Schurman S, Harboe C et al. (2009) Base excision repair of oxidative DNA damage and association with cancer and aging. *Carcinogenesis* 30, 2–10.

Morris DS, Tomlins SA, Montie JE & Chinnaiyan AM (2008) The discovery and application of gene fusions in prostate cancer. *BJU Int.* 102, 276–282.

Nakabeppu Y, Tsuchimoto D, Ichinoe A et al. (2004) Biological significance of the defense mechanisms against oxidative damage in nucleic acids caused by reactive oxygen species: from mitochondria to nuclei. *Ann. NY Acad. Sci.* 1011, 101–111.

Navin NE & Hicks J (2010) Tracing the tumor lineage. *Mol. Oncol.* 4, 267–283.

Negrini S, Gorgoulis VG & Halazonetis TD (2010) Genomic instability: an evolving hallmark of cancer. *Nat. Rev. Mol. Cell Biol.* 11, 220–228.

Nigg EA (2006) Origins and consequences of centrosome aberrations in human cancers. *Int. J. Cancer* 119, 2717–2723.

O'Donovan PJ & Livingston DM (2010) BRCA1 and BRCA2: breast/ovarian cancer susceptibility gene products and participants in double-strand break repair. *Carcinogenesis* 31, 961–967.

Poirier MC (2004) Chemical-induced DNA damage and human cancer risk. *Nat. Rev. Cancer* 4, 630–637.

Rayner E, van Gool IC, Palles C et al. (2016) A panoply of errors: polymerase proofreading domain mutations in cancer. *Nat. Rev. Cancer* 16, 71–81.

Reinhardt HC & Yaffe MB (2009) Kinases that control the cell cycle in response to DNA damage: Chk1, Chk2, and MK2. *Curr. Opin. Cell Biol.* 21, 245–255.

Ricke RM, van Ree JH & van Deursen JM (2008) Whole chromosome instability and cancer: a complex relationship. *Trends Genet.* 24, 457–466.

Sengupta S & Harris CC (2005) p53: traffic cop at the crossroads of DNA repair and recombination. *Nat. Rev. Mol. Cell Biol.* 6, 44–55.

Silver DP & Livingston DM (2012) Mechanisms of BRCA1 tumor suppression. *Cancer Disc.* 2, 679–684.

Singh MS & Michael M (2009) Role of xenobiotic metabolic enzymes in cancer epidemiology. *Methods Mol. Biol.* 472, 243–264.

Smith J, Tho LM, Xu N & Gillespie DA (2010) The ATM-Chk2 and ATR-Chk1 pathways in DNA damage signaling and cancer. *Adv. Cancer Res.* 108, 73–112.

Stolz A, Ertych N & Bastians H (2011) Tumor suppressor *CHK2*, regulator of DNA damage response and mediator of chromosomal stability. *Clin. Cancer Res.* 17, 401–405.

Sugimura T (2000) Nutrition and dietary carcinogens. *Carcinogenesis* 21, 387–395.

Thompson SL & Compton DA (2008) Examining the link between chromosomal instability and aneuploidy in human cells. *J. Cell Biol.* 180, 655–672.

Weinstock DM, Richardson CA, Elliott B & Jasin M (2006) Modeling oncogenic translocations: distinct roles for double-strand break repair pathways in translocation formation in mammalian cells. *DNA Repair (Amst.)* 5, 1065–1074.

Wilch ES & Morton CC (2018) Historical and clinical perspectives on chromosome translocations. *Adv. Exp. Med. Biol.* l044, 1-14.

Wogan GN, Hecht SS, Felton JS et al. (2004) Environmental and chemical carcinogenesis. *Semin. Cancer Biol.* 14, 473–486.

Wood RD, Mitchell M & Lindahl T (2005) Human DNA repair genes, 2005. *Mutat. Res.* 577, 275–283.

Wu X, Zhao H, Suk R & Christiani DC (2004) Genetic susceptibility to tobacco-related cancer. *Oncogene* 23, 6500–6523.

Yang B, Li X, Lei L & Chen J (2017) APOBEC: From mutator to editor. *J. Genetics Genomics* 44, 423–437.

Zhao EY, Jones M & Jones SJM (2019) Whole-genome sequencing in cancer. *Cold Spring Harb. Perspect. Med.* 9/3/a034579.

Zhao W, Wiese C, Kwon Y et al. (2019) The BRCA tumor suppressor network in chromosome damage repair by homologous recombination. *Annu. Rev. Biochem.* 88, 221–245.

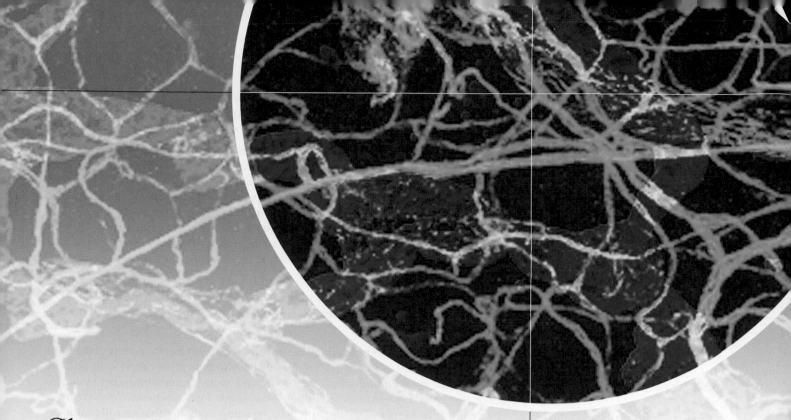

Chapter 13

Dialogue Replaces Monologue: Heterotypic Interactions and the Biology of Angiogenesis

> Simple ideas are wrong. Complicated ideas are unattainable.
>
> Paul Valéry, poet, 1942

> Tumors: wounds that do not heal.
>
> Harold Dvorak, 1986

A simple and powerful conceptual paradigm pervades most of the previous discussions in this book: cancer is a disease of cells, and the phenotypes of cancer cells can be understood by examining the genes and proteins within them. The origins of this idea are clear, being traceable directly back to bacterial and yeast genetics. These two research specialties thrived because they were well served by the postulate that cell genotype determines all aspects of cell phenotype. Indeed, virtually all the attributes of individual bacterial and yeast cells could be shown to derive directly from the genes that these microorganisms carry in their genomes.

Applying this concept to metazoans and their tissues has had obvious advantages for biologists. Metazoan organisms are complex almost beyond measure, and this complexity has frequently prevented researchers from extracting simple and irrefutable truths about organismic function. In response, many researchers, notably molecular biologists, cell biologists, and biochemists, embraced the credo of **reductionist** science: when working with complex systems, the best way to arrive at solid and rigorous conclusions is to take apart these systems into simpler, more tractable components and study each separately.

Movies in this chapter

13.1 Adhesion Junctions Between Cells

13.2 Interactions of Innate Immune Cells with a Mammary Tumor

13.3 Mechanisms Enabling Angiogenesis

Although the lessons learned may relate only to small parts of very large systems, at least these lessons are solid and definitive and will not require substantial revision each time a new generation of researchers revisits these complex systems and their component parts.

Such reductionism has allowed many areas of cancer research to thrive. Witness the progress described in the early chapters of this textbook: when the first proto-oncogene was discovered three-quarters of the way through the twentieth century (see Sections 3.8 and 3.9), almost nothing was known about the genetics and molecular biology of cancer. By the end of the century, information was available in abundance about how cancers begin and progress to highly aggressive, malignant states. Much of this avalanche of information came from taking the reductionist paradigm to its limits—disassembling normal and neoplastic tissues into component cells and the cells into component molecules.

In the case of cancer research, the reductionist pact was extended in another direction as well: all of the traits of a tumor can be traced directly back to the behavior of its constituent cancer cells. Their behavior, if well understood, could be used to deduce that of the tumor as a whole.

By the end of the twentieth century, it became increasingly clear that many of the traits of tumors could not in fact be traced directly back to individual cancer cells and the genes they carry. The grand simplifications agreed upon a generation earlier by cancer researchers had begun to lose their utility. Increasingly, evidence turned up that cancer is actually a disease of tissues, in particular, the complex tissues that we call tumors.

The reductionists' way of thinking had, all along, denied certain realities that clinical oncologists and pathologists confronted on a daily basis, namely, that carcinomas—which constitute more than 80% of the human cancer burden—derive from epithelial tissues of very complex microscopic structure. Histopathological characterizations of these epithelial tumors reveal that they are composed of a number of distinct cell types (**Figure 13.1**).

Figure 13.1 Stromal components of several commonly occurring carcinomas As is shown here, a variety of common carcinomas contain significant proportions of both epithelial cancer cells and recruited stromal cells. (A) A high-grade invasive ductal carcinoma of the breast, in which the membranes (*arrows*) of the epithelial cells are stained *brown*, and the nuclei of the stromal cells are *light blue*. (B) A colon carcinoma (*right side*) and normal colonic mucosa (N, *left side*), in which fibroblast-like cells in the tumor-associated stroma (TAS) are immunostained *dark brown* for PINCH (a protein associated with the cytoplasmic tails of integrins), and the nuclei of epithelial cancer cells lining ducts are *light blue*. (C) A lobular carcinoma *in situ* of the breast showing small clusters of cancer cells (*light blue nuclei*, arrow) surrounded by thin layers of stromal fibroblasts that have been immunostained for PINCH antigen (*dark brown*). (D) An adenocarcinoma of the stomach in which the cancer cells (*purple, lower left*) are adjacent to extensive stroma, which has been stained for collagen (*blue*). (A, from A. Gupta, C.G. Deshpande and S. Badve, *Cancer* 97:2341–2347, 2003. B and C, from J. Wang-Rodriguez et al., *Cancer* 95:1387–1395, 2002. D, from S. Ohno et al., *Int. J. Cancer* 97:770–774, 2002. All with permission from John Wiley & Sons.)

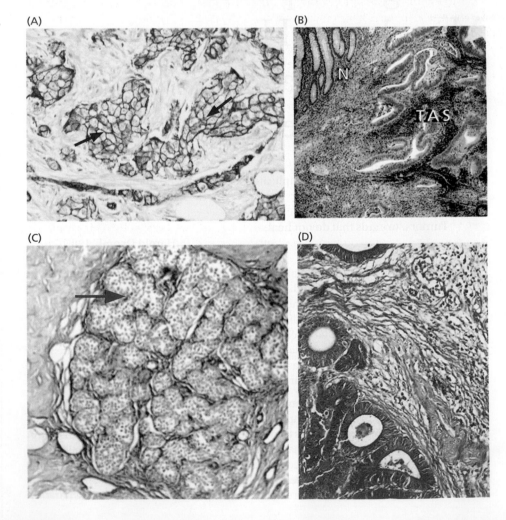

One's first instinct is to dismiss the non-neoplastic cells forming the tumor stroma as distracting contaminants that confound rather than illuminate one's understanding of the biology of tumors. To do so now seems increasingly unwise. Recent research results make it clear that non-neoplastic cells, notably the stromal cells of carcinomas, are active, indeed essential collaborators of the neoplastic epithelial cells within tumor masses, having been recruited and then exploited by the cancer cells. The biology of these recruited cells now appears to be almost as important as that of the neoplastic cells in enabling the growth of tumors.

Consequently, the disease of cancer is far more than cancer cells talking to themselves in endless monologues. We know, instead, that in most tumors, the neoplastic cells are in continuous communication with non-neoplastic host cells. In this chapter, we focus largely on these nearby signaling channels and explain how they affect the fate of incipient tumors and thus the course of clinical cancer development. At the same time, realizing the presence of non-neoplastic stromal cells in the life of a tumor, we can refine our vocabulary: "carcinoma cells" are neoplastic and cancerous, whereas "tumor cells" includes both the neoplastic and non-neoplastic cells within a mass (Sidebar 13.1).

13.1 Normal and neoplastic epithelial tissues are formed from interdependent cell types

In fact, even cursory examinations under the microscope indicate that most carcinomas are as complex histologically as the normal epithelial tissues from which they have arisen. In some cases of commonly occurring carcinomas, such as those of the breast, colon, stomach, and pancreas, the non-neoplastic cells, which together constitute the tumor **stroma**, may account for as many as 90% of the cells within the tumor mass.

The true complexity of the stroma of epithelial cancers becomes apparent only at high magnification under the microscope **Figure 13.2**. In addition to the neoplastic epithelial (that is, carcinoma) cells, a number of stromal cell types populate the tumor. These stromal cells include fibroblasts, myofibroblasts, endothelial cells, pericytes, adipocytes, macrophages, lymphocytes, and mast cells; not shown are the extensions of neurons that thread their way through the bulk of tumor masses. Because most of these cell types constitute only a relatively small proportion of the stroma of many tumors, their presence, when viewed microscopically, becomes apparent only through use of immunostaining with antibodies that bind cell type–specific antigens. Use of immunostaining allows the localization of some of these cell types, as seen in Figure 13.2. Often the stroma of a tumor is depicted as the **tumor microenvironment** (TME, TMEN), with the implication that this collection of cells, having been recruited in various ways to the tumor-associated stroma, thereafter conditions the surroundings of the neoplastic cells and thereby influences growth of the tumor as a whole in various ways.

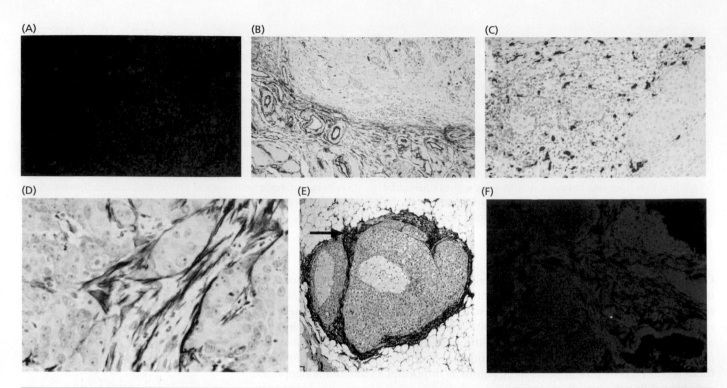

Figure 13.2 A variety of distinct cell types in the stroma of carcinomas Various antibodies can be used to immunostain specific mesenchymal cell types in the stroma of carcinomas. (A) CD4 antigen–positive T lymphocytes, often termed helper T cells (*red*), are scattered among mouse mammary carcinoma cells (*blue-purple nuclei*). The stroma of squamous cell carcinomas of the oral cavity, pharynx, and larynx may contain (B) CD34 antigen–positive fibrocytes (*brown*), (C) CD117 antigen–positive mast cells (*brown*), and (D) α-smooth muscle actin–positive myofibroblasts (*brown*). (E) The stroma of a ductal carcinoma *in situ* of the breast (DCIS) shows PINCH antigen–positive fibroblasts (*dark brown*; see Figure 13.1B) arrayed in a ring immediately around the carcinoma (*arrow*) and surrounded, in turn, by adipocytes (*light blue rims*). (F) Macrophages (*red*), which have been recruited to a mouse mammary carcinoma, play critical roles in tumor invasiveness and angiogenesis. (A and F, from D.G. DeNardo et al., *Cancer Cell* 16:91–102, 2009. With permission from Elsevier. B, C, and D, from P.J. Barth et al., *Virchows Arch.* 444:231–234, 2004. With permission from Springer. E, from J. Wang-Rodriguez et al., *Cancer* 95:1387–1395, 2002. With permission from John Wiley & Sons.)

This impression of cellular complexity provided by Figure 13.2 is reinforced by the techniques of typing multiple antigens expressed in single cells as well as single-cell RNA sequencing (scRNAseq) in which distinct cell types are identified by the antigens or the transcriptomes that they express (**Figure 13.3**).

An additional dimension of complexity comes from the great variations in the proportions of various bone marrow–derived cell types within the stroma of different tumor types. As we will see in this chapter and the next, some of these cells are active participants in processes such as tumor cell invasiveness and metastatic spread; yet others may be dispatched by the immune system to eliminate tumor cells. In light of the important functions that these infiltrating cells can exert, we can imagine that they may contribute in major ways to the distinctive traits exhibited by various types of carcinomas.

[These examples fail to illustrate the great range of variations that are encountered in various types of non-carcinomatous human tumors. At one end of the spectrum is another solid tumor—Hodgkin's disease—a lymphoma in which non-neoplastic cells account for 99% of the cells in a tumor mass (**Figure 13.4A**). At the other end are tumors such as the hemangioma, a relatively common benign tumor of the endothelial cells (Figure 13.4B), in which neoplastic cells form the great bulk of the tumor mass. In what follows, we focus largely on carcinomas, in which the epithelial and stromal cells are often present in comparable numbers.]

Most of the diverse cell types forming the tumor-associated stroma are members of several mesodermal cell lineages that generate both connective tissue and various types of **immunocytes** in the blood and immune tissues, and therefore are biologically

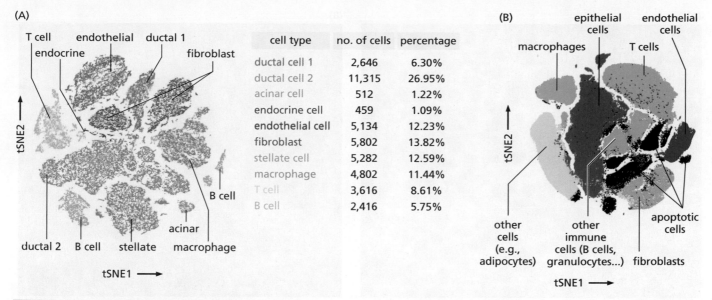

cell type	no. of cells	percentage
ductal cell 1	2,646	6.30%
ductal cell 2	11,315	26.95%
acinar cell	512	1.22%
endocrine cell	459	1.09%
endothelial cell	5,134	12.23%
fibroblast	5,802	13.82%
stellate cell	5,282	12.59%
macrophage	4,802	11.44%
T cell	3,616	8.61%
B cell	2,416	5.75%

Figure 13.3 Analyses of single cells enables surveys of the cellular complexity within tumors Techniques have been developed to analyze both the RNA and protein contents of single cells, making it possible to survey the entire contents of a tumor mass cell by cell. Results in both cases are plotted in 2 dimensions using the t-distributed stochastic neighbor embedding (t-SNE) algorithm, which allows multidimensional data to be plotted in 2 or 3 dimensions. (A) The intra-tumoral cellular heterogeneity has been surveyed here in a sample of a human malignant pancreatic adenocarcinoma, in this case from 57,530 individual cells whose RNA transcripts were analyzed by the procedure of single-cell RNA-seq (scRNAseq). In this analysis, two subtypes of pancreatic epithelial cells lining the ducts (ductal cells) are distinguished, as well as a class of acinar exocrine cells that are responsible for generating and secreting the digestive enzymes made by the pancreas. The endocrine cells include the components of the pancreatic islets that are described in a mouse model of pancreatic islet cells carcinogenesis in Figure 13.35. (B) The expression of 73 distinct protein species in 58,000 human breast tumor cells was examined by using cognate monoclonal antibodies to detect individual protein species followed by mass spectrometry to identify the representation of the individual antigen species that were assigned, in turn, to specific cell types. (A, J. Peng, B.F. Sun, C.Y. Chen et al. *Cell Res.* 29:725–738, 2019. With permission from Nature. B, from J. Wagner, M.A. Rapsomaniki, S. Chevrier et al. *Cell* 177:1330–1345, 2019. With permission from Elsevier.)

very different from the epithelial cells whose transformation drives the growth of carcinomas. (In the case of nonepithelial tumors such as sarcomas, where the cancer cells are themselves of mesodermal origin, the boundaries between the neoplastic cell compartment and the non-neoplastic cells within the tumor mass have never been clearly delineated.)

In the absence of other information, we might explain the presence of these various stromal cell types within a carcinoma in two ways. They might represent the remnants of cells that resided in the stroma of a tissue before tumor development began. During the subsequent expansion of the tumor cell population, groups of cancer cells may have inserted themselves between these preexisting normal stromal cell layers, thereby generating the richly textured tissues that are seen when most carcinomas are examined microscopically.

The alternative explanation for the presence of these numerous stromal cells is more intriguing and is likely to be the correct one for the great majority of tumors as suggested above. The rationale behind this second model depends on our current understanding of how normal, architecturally complex tissues develop and maintain their structure and function. In such tissues, the proper proportions of the various component cells must depend on the continuous intercommunication between them. These interactions involve multiple distinct cell types, each following its own particular differentiation program. Such communication between dissimilar cell types is termed **heterotypic** signaling and is used by each cell type to encourage or limit the proliferation of the other types of cells nearby.

Extending this thinking, we might speculate that many of the heterotypic interactions operating in normal tissues continue to play important roles in the biology of the tumors arising in these tissues. This would suggest that, like normal

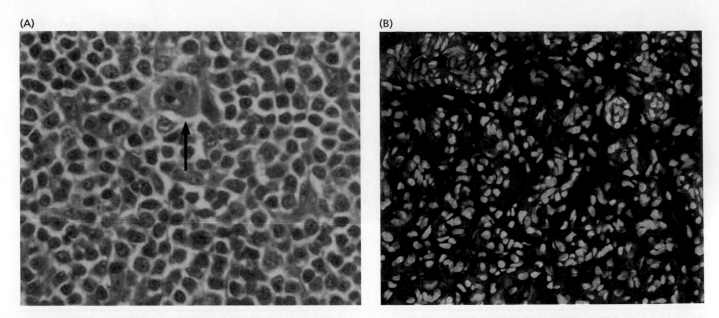

Figure 13.4 Extreme variations in the ratio of neoplastic cells to stroma (A) In certain tumors, normal, non-neoplastic cells may constitute the great majority of living cells. An extreme form of this is illustrated here by Hodgkin's disease, a lymphoma in which as many as 99% of the cells are recruited normal lymphocytes, which surround a rare neoplastic Reed–Sternberg cell (*arrow*). (B) At the other end of the spectrum are tumors such as the benign hemangioma shown here, in which neoplastic endothelial cells (*green nuclei*) form the great bulk of the tumor mass, which is demarcated by basement membranes (*blue-purple*) and occasional capillaries (*red*). (A, from A.T. Skarin, Atlas of Diagnostic Oncology, 4th ed. Philadelphia: Elsevier Science Ltd., 2010. With permission from Elsevier. B, from M.R. Ritter et al., *Proc. Natl. Acad. Sci. USA* 99:7455–7460, 2002. With permission from National Academy of Sciences, U.S.A.)

epithelial cells, carcinoma cells continue to control the populations of stromal cells near them, perhaps by recruiting the latter from nearby normal tissues and distant bone marrow, and then encouraging their proliferation. Operating in the other direction, stromal cells also influence epithelial cell proliferation and survival in the tumor mass.

In normal tissues, these heterotypic signaling channels depend in large part on the exchange of (1) mitogenic growth factors, such as hepatocyte growth factor (HGF), transforming growth factor-α (TGF-α), and platelet-derived growth factor (PDGF); (2) growth-inhibitory signals, such as transforming growth factor-β (TGF-β); and (3) trophic factors, such as insulin-like growth factor-1 and -2 (IGF-1 and -2), which favor cell survival. Although these factors are secreted into the extracellular space, they usually act over short distances between nearby cells (**Figure 13.5**) and their localization within the interstitial spaces between cells is likely to be critically important to their functions (Supplementary Sidebar 13.1).

Within individual tumors the complexity that we will confront exists in at least three dimensions: To begin, there are multiple distinct cell types that are recruited to the stroma (see Figures 13.2 and 13.3). Second, multiple signals are exchanged between these distinct cell types; our focus here will be on bidirectional interactions between cancer cells and individual stromal cell types. Finally, each signal that is exchanged between neoplastic and recruited stromal cells evokes multiple distinct responses in the various cell types recruited to the tumor-associated stroma. Shown in **Figure 13.6** is one example of this complexity—the effects of a single type of released signal, TGF-β , on the various mesoderm-derived cell types present in the stroma of a carcinoma. [Not addressed by this figure is yet another target of TGF-β action—the carcinoma cells themselves. These are often induced by TGF-β to activate their previously latent cell-biological program termed the EMT (epithelial–mesenchymal transition), which confers on them higher degrees of aggressiveness and will be described later in this chapter.] The realities of heterotypic signaling dictate that we will only scratch the surface of its complexity in the present chapter, attempting to understand mechanistic principles rather than to enumerate

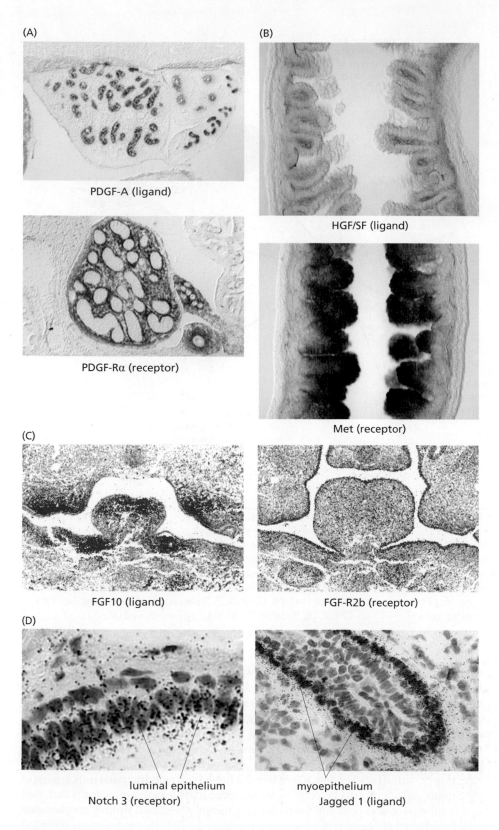

(A)

PDGF-A (ligand)

PDGF-Rα (receptor)

(B)

HGF/SF (ligand)

Met (receptor)

(C)

FGF10 (ligand)

FGF-R2b (receptor)

(D)

luminal epithelium
Notch 3 (receptor)

myoepithelium
Jagged 1 (ligand)

Figure 13.5 Heterotypic ligand–receptor signaling (A) In epithelial tissues, the synthesis of PDGF-A and its receptor (PDGF-Rα) is usually confined to distinct cell types. In the testis, expression of PDGF is confined to the cells of the tubular epithelium (*above*), while expression of its receptor is confined to the surrounding mesenchymal cells (*below*). (B) The reciprocal expression of the Met receptor and its ligand, termed variously hepatocyte growth factor (HGF) or scatter factor (SF), is shown here in the embryonic mouse ileum (the lower section of the small intestine). In each fingerlike villus, seen in longitudinal section (see also Figure 11.5C), HGF/SF mRNA, detected by *in situ* hybridization, is made by the mesenchymal cells in the core of the villus (*above*), whereas Met is expressed by the epithelial cells coating the surface of the villus (*below*). (C) In the developing oral cavity of the mouse, the FGF10 ligand (*red*) is produced by the mesenchymal cells of the stroma of the developing palate (*left*), whereas its cognate receptor, FGF-R2b (*red*), is displayed by the overlying epithelial cells (*right*). Other work not shown here indicates that the epithelial cells respond to resulting ligand-mediated activation of their FGF-R2b by releasing Sonic Hedgehog (Shh), which signals back to Patched (Ptc) receptors displayed by mesenchymal cells in the stroma. (D) In the normal mouse mammary gland, use of antisense probes to detect mRNAs indicates that the Notch 3 receptor (*left*) is expressed (*black dots*) in the luminal epithelial cells, but its ligand, Jagged 1 (*right*), is expressed (*black dots*) in myoepithelial cells. (A, from L. Gnessi et al., *J. Cell Biol.* 149:1019–1026, 2000. With permission from Rockefeller University Press. B, courtesy of S. Britsch and C. Birchmeier. C, from R. Rice et al., *J. Clin. Invest.* 113:1692–1700, 2004. With permission from American Society for Clinical Investigation. D, from M. Reedijk et al., *Cancer Res.* 65:8530–8537, 2005. With permission from American Association for Cancer Research.)

systematically all of the cell–cell signaling pathways that operate within the confines of a single tumor.

Given these complex interactions within tissues, what evidence can we muster to support the notion that such heterotypic interactions are *functionally important* for tumor growth rather than representing vestiges—functionally irrelevant relics—of normal tissue physiology? Until the mid-twentieth century, the role of stromal cells in

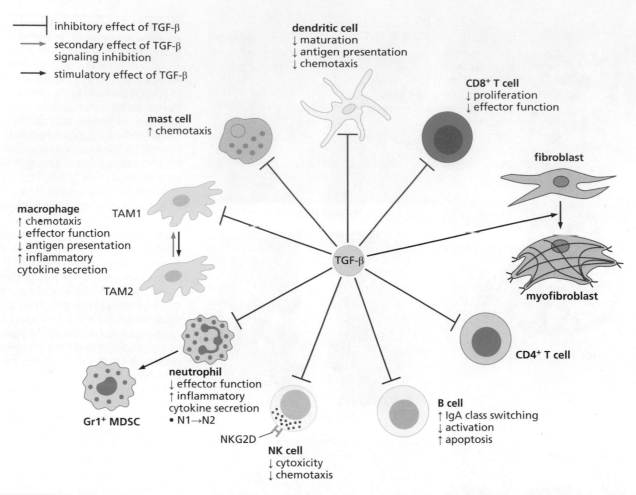

Figure 13.6 Paracrine signaling: the case of TGF-β Various signaling molecules are exchanged between carcinoma cells and the nearby stroma. These molecules often operate similarly in corresponding normal epithelial tissues but their effects may be altered during the course of tumor progression. Among these paracrine signaling factors, the most widely acting is likely to be TGF-β, which perturbs the functions of a wide range of cell types that are recruited to the tumor-associated stroma, often termed the tumor microenvironment (TMEN). Most of the cell types that are illustrated here are components of the innate and adaptive immune systems that are described in detail in Chapter 15. In addition, TFG-β can induce stromal fibroblasts to become myofibroblasts that express α-smooth muscle actin (α-SMA; *right*), which play critical roles in inflammation and wound healing. (Adapted from R. Derynck, S. Turley and R. Akhurst *Nat. Rev. Clin. Oncol.* 18, 9–34, 2021. With permission from Nature.)

supporting the growth of epithelial tumors was not appreciated. Experiments reported in 1961 provided some of the first evidence for the importance of these interactions in tumor biology; these were experiments of a sort that could not be repeated these days, given existing regulations on clinical testing of human subjects. Basal cell carcinomas of the skin were excised from patients and then reimplanted in normal areas of the skin elsewhere in the same patients (**Figure 13.7A**). Because these were autotransplantations, with each patient receiving a graft of his or her own cells, no immune rejection occurred (see Section 3.5). When the implanted cells included only groups of carcinoma cells, the grafts failed to grow; however, if these carcinoma grafts included both epithelial cancer cells and the underlying tumor-associated stroma, they became well established at the sites of implantation (Figure 13.7B). This experiment provided strong indication that the mesenchymal cells of the stroma are indeed essential to supporting the growth of the tumor as a whole.

Now, more than half a century later, the details of many stromal–epithelial heterotypic interactions and their roles in tumor progression have been elucidated in great detail. We now know that they play key roles in virtually all the stages of tumor progression (see Chapter 11), beginning with the initial formation of tumors. For example, carcinoma cells release growth factors, cytokines, and chemokines that recruit

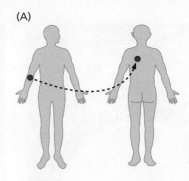

Figure 13.7 Autologous transplantation of skin tumors (A) Basal cell carcinomas of the skin were excised from patients and the carcinoma cells were reimplanted, either together with or without carcinoma-associated stromal cells, in distant anatomical sites on the skin of the back of the patients who had borne the original tumors. (B) The implanted tissues were excised 1 to 5 weeks later. Those that had been implanted together with stroma (*left*) yielded robustly growing carcinoma cells, whereas those that were implanted without concomitantly implanted stroma (*right*) failed to grow and yielded only vacuolated cells and cell debris. (A, courtesy of S. McAllister. B, from E.J. Van Scott and R.P. Reinertson, *J. Invest. Dermatol.* 36, 109–131, 1961. With permission from Elsevier.)

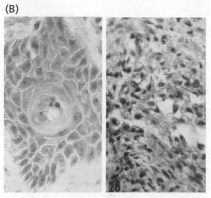

with stroma   without stroma

macrophages, neutrophils, and lymphocytes to the tumor-associated stroma; the recruited cells then orchestrate an inflammatory response that involves the release of TNF-α and prostaglandins, which proceed to stimulate the proliferation of nearby epithelial cells and the process of **neoangiogenesis**—the formation of blood vessels within the tumor stroma.

Many of these heterotypic interactions continue to operate after fully neoplastic tumors have been formed. To cite some examples, epithelial cells within a carcinoma often release PDGF, for which stromal cells—notably fibroblasts, myofibroblasts, and macrophages—possess receptors; the stromal cells reciprocate by releasing IGF-1 (insulin-like growth factor-1), which benefits the growth and survival of the nearby cancer cells. Similarly, the neoplastic cells within melanomas release PDGF, which elicits IGF-2 production from nearby stromal fibroblasts; this IGF-2 helps to maintain the viability of the melanoma cells. Stromal cells in breast cancers release the factor SDF-1/CXCL12 (a **chemokine**) and the HGF/SF growth factor, which stimulate the proliferation and survival of nearby epithelial cancer cells. More generally, by regulating each other's numbers and positions, epithelial and stromal cells ensure the optimal representation and localization of each cell type within both normal and neoplastic tissues. Additional examples are illustrated in Supplementary Sidebar 13.2.

The **endothelial** cells (**Figure 13.8A**), which assemble to form the linings of the walls of capillaries and larger blood vessels (Figure 13.8B) as well as **lymphatic** ducts (Figure 13.8C), represent vital components of the normal and neoplastic stroma. As discussed in greater detail later, proliferation of the endothelial cells forming capillaries is encouraged by other cells in both the epithelium and stroma in order to guarantee access by all of these cells to an adequate blood supply. Once capillaries are successfully assembled and become fully functional, they supply essential nutrients and oxygen to nearby tumor-associated cells; at the same time, they serve the equally important function of evacuating metabolic wastes and carbon dioxide from tissues including tumors.

While capillary formation is proceeding, the endothelial cells secrete growth factors that stimulate the proliferation of nearby nonendothelial cell types. Most important, endothelial cells release PDGF and HB-EGF (heparin-binding EGF), which enables them to attract (see Supplementary Sidebar 13.1) the peri-endothelial cells called **pericytes** and the vascular smooth muscle cells that together create the outer cell layers of capillaries, sometimes called the **mural** cells to distinguish them from the luminal endothelial cells (see Figure 13.8A and B). Once in place, the pericytes (which closely resemble smooth muscle cells) reciprocate by releasing vascular endothelial growth factor (VEGF) and angiopoietin-1 (Ang-1), which provide important survival signals to the endothelial cells that recruited them in the first place. In addition, these mural cells provide the endothelial tubes with structural stability and an ability to resist the forces exerted by the pressure of blood (see Figure 13.8B).

At present we are far from being able to enumerate all the key heterotypic interactions that operate within normal tissues and, by extension, in tumors. Because tumors rarely invent novel signaling mechanisms (rather than exploiting normal ones), a subset of the interactions operating in normal tissues must be operating in tumors, including those that underlie the support of carcinoma cells by nearby recruited stromal cells. Given the multiple distinct cell types present in normal tissues and tumors and the

(A)

(B)

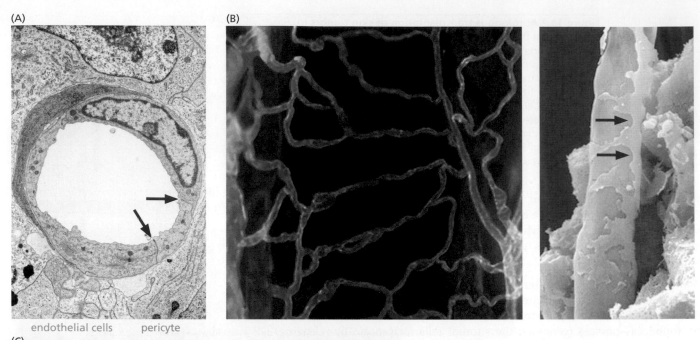

endothelial cells   pericyte

(C)

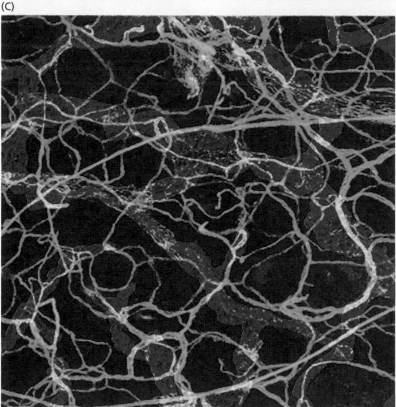

capillaries   lymph ducts

Figure 13.8 Microvessels in normal tissues (A) This transmission electron micrograph presents a cross section of a capillary within the stroma of a tumor formed by mouse Lewis lung carcinoma cells. The endothelial cell and its nucleus have been colorized *green*, and the pericyte, which helps maintain capillary rigidity and intraluminal blood pressure, is colorized *red*. At least one additional endothelial cell has contributed to the formation of this section of the capillary, as indicated by the arrows at the joints between the cytoplasms of the two endothelial cells. (B) Immunofluorescent microscopy (*left*) of the capillaries in a normal mouse trachea reveals (using an antibody that recognizes the endothelial cell–specific CD31 antigen) a well-organized network of microvessels. At the level of scanning electron microscopy (*right*), a capillary is seen to resemble a smooth-walled tube with pericytes (*arrows*) adhering to its outer surface. (C) Most normal tissues are interlaced with networks of capillaries (*green*) and lymphatic vessels (*red-orange*). As is apparent here, the diameters of lymphatic vessels are far larger than those of capillaries. Unlike capillaries, however, the endothelial cells of lymphatic vessels are not supported by underlying mural cells—pericytes and smooth muscle cells (*not shown*). (A, from B. Sennino et al., *Cancer Res.* 67:7358–7367, 2007. With permission from American Association for Cancer Research. B, from T. Inai et al., *Am. J. Pathol.* 165:35–52, 2004. With permission from Elsevier. C, courtesy of T. Tammela and K. Alitalo.)

bidirectional intercommunications between different pairs of these cell types, the total number of these signaling channels, calculated combinatorially, is surely vast.

Stromal cells are often found to be layered between carcinoma cells in most metastases. This arrangement shows that even those cancer cells that are independent enough to disseminate from the primary tumor and to found new tumors at distant sites usually find it necessary to recruit stromal cells and encourage their proliferation (**Figure 13.9A**). On occasion, metastatic cells will insinuate themselves into the existing stroma of the tissue in which they have landed (Figure 13.9B), once again demonstrating their need to secure stromal support by one means or another. Finally,

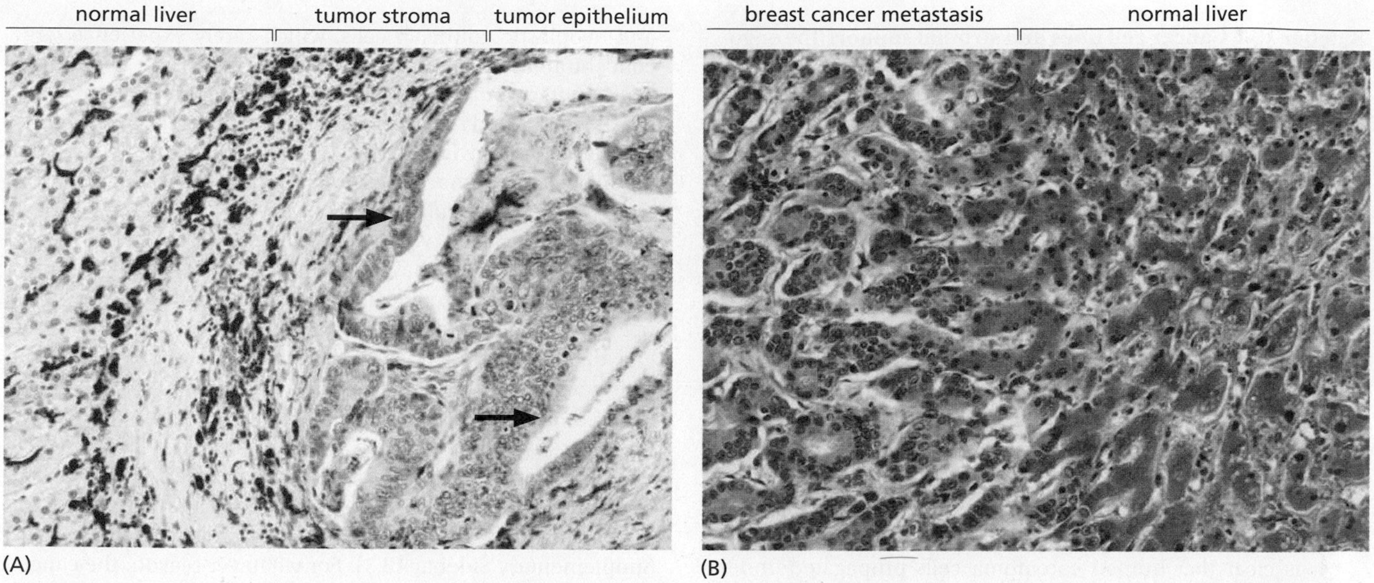

Figure 13.9 **Metastasis and dependence on stroma** (A) This micrograph of a metastasis originating from a primary colon carcinoma reveals that the metastasis found in the liver is as complex histologically as the primary tumors seen in Figure 13.1B. Here, we see that the carcinoma cells that have formed a metastasis (*light purple, right*) have constructed the same ductal structures (*arrows*) as are seen in primary adenocarcinomas of the colon. In addition, the carcinoma cells have recruited substantial stroma (*middle*), which includes macrophages (*dark brown*). (B) Most breast cancer metastases to the liver (*left*) behave quite differently from colon carcinoma metastases to this organ, in that the mammary carcinoma cells infiltrate into normal liver tissue (*right*) and displace resident hepatocytes, thereby taking advantage of the existing stroma, including its vasculature. These metastatic cells appear to be as dependent on stromal support as the colorectal carcinoma cells of (A) but acquire access to it in an entirely different way. (From F. Stessels et al., *Br. J. Cancer* 90:1429–1436, 2004. With permission from Nature.)

we should note that an entire class of stromal cells—immunocytes—plays a critical role in the immunobiology of tumors, indeed a role so critical and far-reaching that we will devote two chapters to these cells and their implications for tumor growth and therapeutic elimination.

Certain classes of metastases teach important overarching lessons about the roles of neoplastic cells in organizing their associated TME. For example, melanomas that have metastasized to the brain recruit abundant T lymphocytes to their stroma, whereas primary brain tumors growing in similar locations—gliomas and glioblastomas—although being rich in macrophages and microglia, attract hardly any of these immune cells. This these vignettes emphasize the role of the neoplastic cells in orchestrating the cellular composition of the TME.

In one extreme version of these scenarios, we might imagine that all heterotypic interactions needed to maintain normal tissue function continue to operate within carcinomas, being essential for the neoplastic cells to thrive and multiply within these tumors. However, such a depiction cannot be literally true, because many of the acquired traits that we associate with cancer cells (see Chapter 3), including a decreased dependence on mitogens, an increased resistance to apoptosis, and the acquisition of anchorage independence, are certain to affect the interactions between epithelial and stromal compartments and to lessen their interdependence. So, we conclude that these acquired traits generally *reduce* the dependence of epithelial cancer cells on stroma but do not seem to *eliminate* it. Finally, there are examples of outliers—cancer cells that appear to be almost totally independent of neighboring stromal cells and their signaling. Certain highly progressed carcinomas will generate free-floating cancer cells that accumulate in various body fluids—generating tumors known as **ascitic** tumors (Supplementary Sidebar 13.3). The ascites formed during ovarian carcinoma progression contain clumps of carcinoma cells—sometimes termed **spheroids**—that often seem to lack identifiable stromal cells.

Human cancer cell lines—there are several dozen in common use—are standard reagents used in many types of cancer research (**Sidebar 13.2**). Cells from virtually all of these can be implanted into immunocompromised host mice, where they proliferate, often vigorously, and form tumors termed **xenografts** (because they result from grafting the tissues of one species into a host animal of another). Mice of the Nude, NOD/SCID, and RAG1/2 strains, lacking fully functional immune systems (see Chapter 15), are used in these experiments because they tolerate the growth of introduced, genetically foreign cells. The resulting xenografts growing in these mice can then be tested for their responsiveness to anti-cancer drugs under development. The outcome of such experiments is not necessarily predictive of eventual clinical responses to these drugs (**Sidebar 13.3**).

Sidebar 13.2 Cancer cell lines and stromal support The ongoing dependence of almost all types of carcinoma cells on stromal support explains the enormous difficulties that researchers have experienced when attempting to adapt tumor cells to *in vitro* culture conditions. Their goal has been to create cancer cell lines—populations of cancer cells that can be propagated in tissue culture indefinitely and can serve as experimental models of the cells encountered in the tumors of human cancer patients. To do so, cells are prepared directly from tumor biopsies and placed into culture, with the hope that colonies of vigorously growing cancer cells will eventually emerge. In most cases, however, such cell colonies do not appear, foiling attempts to generate carcinoma cell lines. More often than not, the cells that do thrive in culture are of stromal origin—specifically the fibroblasts whose growth is favored by the platelet-derived growth factor (PDGF; see Figure 5.8) that is present in abundance in the serum component of standard tissue culture medium.

It is clear that human carcinoma cells propagated under standard conditions of tissue culture are being forced to proliferate in environments that differ from those experienced by their precursors residing in the tumors of cancer patients. For example, the high concentration of serum present in most culture media is growth-inhibitory for many types of normal and neoplastic epithelial cells, which rarely experience large concentrations of the serum-associated factors *in vivo* except acutely after wounding. Even more important, carcinoma cells are being selected *in vitro* for their ability to proliferate without intimate contact with stroma, because the proper balance of epithelial and stromal cells present in tumors cannot be reproduced in the tissue culture dish.

On rare occasion, vigorously proliferating colonies of carcinoma cell populations have indeed emerged following extended culture of tumor fragments in the Petri dish. Because they have been selected *in vitro* for their ability to grow autonomously, and therefore independently of stromal support, such cancer cells have evolved beyond the stage of tumor progression that is reached *in vivo* by the neoplastic cells of most human tumors. In fact, many of the successes in establishing human tumor cell lines have depended on culturing cells from the few tumors that had already progressed *in vivo* to a stage where they no longer depended on stroma for their survival and proliferation (see Supplementary Sidebar 13.3). For whatever reason, the cancer cell lines that are established for long-term growth in culture have, almost invariably, evolved to a biological state that is significantly different from that of the cells in corresponding human tumors.

Sidebar 13.3 The development of anti-cancer therapies has been imperfectly served by the use of existing human cancer cell lines Human cancer cell lines, when grown as tumor xenografts in immunocompromised mice, often respond to the anti-proliferative effects of cytotoxic drugs under development. However, the ability of these xenograft models to predict the clinical responses of patients to these drugs is limited. A 2001 retrospective study of 39 different anti-cancer drugs showed a weak correlation between the responses of xenografted tumors and comparable tumors borne by cancer patients to these drugs. For example, candidate drugs that were effective in halting the growth of more than one-third of a disparate collection of mouse tumor xenografts had only a 50% likelihood of showing any therapeutic activity (including halting tumor growth) in some human tumor types treated in the clinic. And drugs that affected smaller proportions of the mouse xenograft models had essentially no meaningful activity in the clinic.

These disappointing outcomes are not surprising, because the human tumor cell lines commonly used in such drug testing experiments bear little resemblance to cells in the tumors frequently encountered in the oncology clinic. These cell lines have been propagated *in vitro* under conditions that differ dramatically from their sites of origin in patients (see Figure 13.11), including (1) the absence of a supporting stroma, (2) the lack of similarity between existing tissue culture media and the composition of the fluid bathing cells in actual living tissues, (3) the often used two-dimensional culture (on the bottom of Petri dishes) that does not recapitulate the three-dimensional conditions of cells growing *in vivo*, (4) the selection of cancer cell lines that proliferate relatively rapidly in culture and, in that respect, do not correspond to cells growing *in vivo*, which may have vastly slower doubling times, and (5) the absence of components of the adaptive immune system in the mouse hosts in which human tumor xenografts are grown (as described in Chapter 15) that influence stromal composition and behavior. This helps to explain why the tumors implanted that these cell lines form often look quite different histologically from those routinely encountered in a clinical pathology laboratory (see Figure 13.11).

Yet another difficulty arises because cells of various human tumor cell lines are usually implanted in mice subcutaneously, which represents an **ectopic** site within the host that has little resemblance to the tissue in which the tumor originated. For example, pancreatic carcinoma cells may behave differently when growing under the skin than when they are implanted into the host animal's pancreas, their natural, **orthotopic** site, presumably because the stromal microenvironments in these two locations are so different.

One experimental response to these difficulties involves attempts at introducing fragments of tumors directly into immunocompromised mice without intervening propagation in tissue culture. These implanted tumor grafts contain both epithelial and stromal cells, which may co-proliferate to form histologically complex tumors in host mice that resemble the original tumors from which they derive. Unfortunately, like the cancer cells in actual human tumors, the cells in these xenografts usually proliferate very slowly, and their propagation involves labor-intensive transplantation from one host animal to another. These properties preclude their routine use in the testing of anti-cancer drugs.

The inadequacies of currently available human tumor models and the resulting inability to accurately predict the clinical effectiveness of anti-cancer drugs under development cost the pharmaceutical industry hundreds of millions of dollars annually and therefore represent one of the major impediments to the development of new anti-cancer drugs.

13.2 The extracellular matrix represents a critical component of the tumor microenvironment

Our overview of the tumor microenvironment (TME) must include, in addition to the stromal cell types described above, the complex extracellular matrix (ECM) that can be found between individual cells, sometimes termed the **interstitial matrix** (**Figure 13.10A**). We encountered its complexity earlier when we enumerated some of the cell-surface receptors, such as integrins, that epithelial cells deploy in order to physically tether themselves to the ECM and, equally important, to sense the presence of specific ECM components in their immediate surroundings (see Figures 5.21, 5.22, and Supplementary Sidebar 5.9).

In the case of the interface between normal epithelial and stromal cells, recall that it is created by the physical barrier established by the specialized ECM termed the basement membrane (BM) or basal lamina (see Figure 13.10A; see also Figure 2.3). In the case of the epithelial cells of the skin, as an example, the keratinocytes express genes specifying many of the major protein components of the basement membrane, including type IV collagen and laminins; the stromal cells also contribute to its construction, doing so in ways that have yet to be documented in detail. Certain **proteoglycans** in the basement membrane, such as perlecan, provide it with increased hydration and with sites to which growth factors can be attached for long-term storage (see Figure 13.10C). Continuous tethering to the basement membrane, mediated largely by integrins and the hemidesmosomes that they construct (see Section 5.9, Figure 13.10A), is essential for the survival of many kinds of epithelial cells, and loss of this tethering often provokes anoikis, the form of apoptosis resulting from loss of anchorage to a solid substrate (see Section 9.13). In many carcinomas, the ECM constitutes as much as 25% of the total tumor-associated mass. In the extreme seen in pancreatic ductal adenocarcinomas, as much as 90% of the mass is formed by the ECM.

In its most simple depiction, we can imagine that the epithelial and stromal cells of normal and carcinomatous tissues collaborate to lay down the ECM. Once assembled, the ECM provides most of the instructive cues that the epithelial cells employ in order to organize themselves.

These cues are conveyed by the structural scaffolding of the ECM, by the various physical forces that influence epithelial cell biology (such as rigidity and tensile strength), and by the myriad mitogenic and survival factors that are attached to the ECM during its formation, creating a reservoir of signaling molecules that can be mobilized and, in solubilized form, stimulate the various receptors whose signals influence epithelial cell biology, as was described in Chapter 5.

Some have likened the role of the ECM (when acting as a reservoir of signaling proteins) to a "message board" that is consulted with frequency by epithelial cells, guiding various aspects of their behavior. The initial attachment of growth factors (GFs) to the ECM soon after their synthesis and release into the extracellular space obviously complicates our understanding of how these signaling molecules actually operate. Depictions of them floating in soluble form from GF-producing cells to cognate GF receptors displayed by recipient cells overly simplify the multiple steps that make them *available* to their partner receptors.

Careful curation of the genes in the human genomes has led to an enumeration of the proteins forming the "core **matrisome**," that is, the proteins that form the essential structural scaffolding of ECMs. There are about 300 in mammals, and they are distinguished from the numerous proteins that are tethered reversibly to the ECM, such as the GFs cited above, as well as others, such as proteases, that are responsible for continuously modifying the core scaffolding.

The key biological role of the ECM in instructing morphogenesis can be illustrated most dramatically by its essentiality in tissue models constructed *in vitro*. As seen in Figure 13.10D, a **hydrogel** was assembled experimentally that incorporated many of the ECM proteins present in normal human breast tissue, together with a

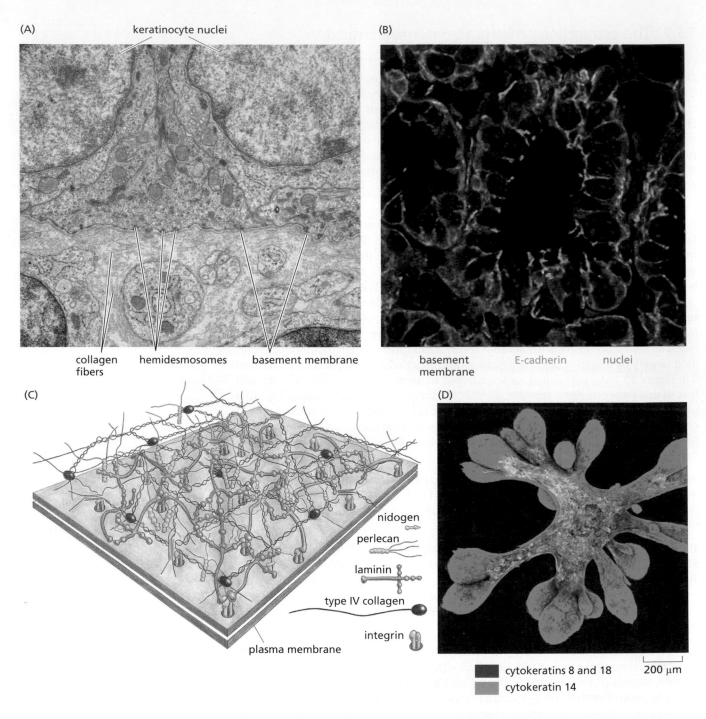

(A)

keratinocyte nuclei

collagen fibers   hemidesmosomes   basement membrane

(B)

basement membrane   E-cadherin   nuclei

(C)

nidogen
perlecan
laminin
type IV collagen
integrin

plasma membrane

(D)

cytokeratins 8 and 18
cytokeratin 14

200 μm

glycosoaminoglycan polysaccharide and three growth factors—insulin, EGF, hydrocortisone. This hydrogel was constructed to have a degree of elasticity comparable to that present in normal mammary ECM. Normal mammary epithelial cells (MECs) were introduced into this gel and their behavior monitored microscopically. This arrangement provided a three-dimensional environment, which contrasted starkly with the highly unnatural two-dimensional environment created by the bottom surfaces of Petri dishes that were used almost universally since the mid-twentieth century to propagate cells in culture.

In this particular tissue culture model, the introduced normal human MECs formed lobules and ducts that closely resembled the ductal and lobular morphologies seen in developing human breast tissues. Gels formed instead exclusively from collagen did not work nor did ECM composed only of basement membrane proteins. Importantly, the **alveoli** that formed developed cytokeratin 8/18-positive luminal cells and cytokeratin 14-positive myoepithelial/basal cells, which reflected the composition of

Figure 13.10 The extracellular matrix and its role in tissue development (A) In epithelial tissues, the specialized extracellular matrix (ECM) termed the basement membrane (BM) separates the epithelial cells (*above*) from the stroma (*below*). As shown in this transmission electron micrograph of the skin of a newborn mouse, the epithelial cells of the skin (termed keratinocytes) are tethered to one side (*above*) of the BM, being anchored in part through structures termed hemidesmosomes, which are composed of clusters of integrins that bind BM proteins, notably laminin. Below the BM are collagen fibers, which represent important components of the interstitial matrix and the mesenchymal cells that constitute the stroma of the skin, i.e., the dermis. (B) This immunofluorescence micrograph of a mammary duct in a lactating mouse reveals the E-cadherin molecules (*green*) that form the lateral junctions tying together the mammary epithelial cells forming the lumen of the duct; the epithelial cells are underlain by a basement membrane, which is revealed here by the presence of laminin (*orange*). Outside of the BM are various types of stromal cells (*not shown*). (C) This schematic drawing indicates that the basement membrane is composed largely of four major ECM proteins, namely laminin, collagen type IV, perlecan, and nidogen. This highly permeable molecular meshwork allows a variety of molecules to pass through in both directions. A specialized BM, not shown here, underlies endothelial cells in capillaries. (D) Mammary epithelial cells were prepared from reduction mammoplasty-derived tissues and propagated in a 3-dimensional gel composed of ECM modeled after the interstitial matrix present in the normal mammary gland plus a series of defined growth factors in the absence of serum. No stromal cells were present in these cultured cells. Cytokeratin 14 is expressed by basal cells in the intact mammary gland, whereas cytokeratins 8 and 18 are expressed by luminal cells. (A, courtesy of H.A. Pasolli and E. Fuchs. B, courtesy of E. Lowe and C. Streuli; from J. Muschler and C. Streuli, *Cold Spring Harb. Perspect. Biol.* 2:a003202, 2010. With permission from Cold Spring Harbor Laboratory Press. C, from B. Alberts et al., Molecular Biology of the Cell, 5th ed. New York: Garland Science, 2008, based on H. Colognato and P.D. Yurchenko, *Dev. Dyn.* 218:213–234, 2000. D, from E. Sokol, D.H. Miller, A. Breggia et al. *Breast Cancer. Res.* 18:19–32, 2016.)

the two cell layers (that is, luminal and basal) forming normal mammary epithelium. About 5% of the mammary epithelial cells developed steroid hormone receptors—ER and PR—and these organoids responded to the steroid hormones by undergoing hollowing of the ducts and lobules. Application of the **prolactin** hormone caused the formation of lipid droplets in the lumina of the alveoli; such droplets are suggestive of the formation of milk components. Moreover, the myoepithelial/basal cells expressed the Slug and Sox9 transcription factors (see Chapter 14) that are known to be responsible for inducing the mammary stem cell state. Strikingly, this essentially normal morphogenesis proceeded in the presence of only ECM and thus in the absence of any admixed stromal cells, indicating that the contextual signals required for this particular morphogenetic program were provided entirely by ECM constituents and thus proceeded without dependence on stromal cells.

The continued dependence of most cancer cell populations on stromal support, including both the cells and the ECM that together form the stroma, has represented a major obstacle to understanding the biology of human tumors, given the inability of existing cancer cell lines to generate tumors that are encountered in the oncology clinic (see Sidebar 13.2). Nevertheless, an ability to grow human cancer cells that are prepared from patient tumor fragments and implanted directly into host mice represents a major step toward generating populations of human cancer cells that more closely resemble the cells living within actual human tumors. Thus, such **patient-derived xenografts** (PDXs), never experience the artificial environment of tissue culture. Presumably, the heterotypic interactions with recruited mouse stromal cells more closely mimic the stroma that was assembled in the original tumors within cancer patients (**Figure 13.11**). However, even the PDXs are associated with certain problems. Some of the heterotypic interactions that operate in human tumors may fail to do so in PDXs because of signaling incompatibilities between mouse growth factors and cytokines and the corresponding human receptors. Moreover, the use of immunocompromised mouse hosts dictates that certain cells responsible for creating an inflammatory stroma (see below) may be missing. Finally, PDX tumors usually grow very slowly, making them impractical for use in many types of experiments.

13.3 Tumors resemble wounded tissues that do not heal

As argued above, heterotypic signaling governs much of the biology of carcinomas, and experimental models of cancer that ignore this important process generate tumors that differ biologically from those found in cancer patients (see Figure 13.11). As mentioned above, the heterotypic interactions operating within tumors are highly complex and involve the exchange of dozens of distinct molecular species that mediate cell-to-cell signaling between the various cell types that together form these growths.

Figure 13.11 Tumors and derived xenografted tumor cell lines
(A) The conditions under which human tumor cell lines are usually propagated (*red font*) differ drastically from those operating in an actual autochthonously arising human tumor, i.e., one that arose spontaneously in a living human tissue (*second column*). Attempts to remedy this by developing culture models that more closely resemble conditions in actual tumors (*third column*) address some but not all the shortcomings of existing means of propagating human tumor cells. (This figure does not indicate the differences in the chemical compositions of the fluids surrounding individual cells or the molecular compositions of the ECMs, which are likely to differ dramatically between the three conditions.) Pa, pascals, unit of pressure here denoting pressure, mechanical stress. (B) The histologies of a primary prostate carcinoma (*left*) and of a colon carcinoma (*right*) that were removed surgically from patients are shown in the top row. Below these (*middle row*) are the histologies arising from implantation of chunks of tumor derived from cancer surgery at subcutaneous sites in immunocompromised SCID mice; these implanted chunks, growing as patient-derived xenografts, contained both epithelial and stromal tumor components. Three to six months after implantation, these engrafted tumor samples, which have grown from a 2-mm to a 1-cm diameter, still retain the architecture of the primary tumors from which they derive. However, the tumors formed by a prostate and a colon cancer cell line (*bottom row*), which have been propagated extensively in tissue culture as pure cancer cell populations, bear little histological resemblance to the tumors typically encountered in the oncology clinic and shown above. (A, from E. Krausz et al., *J. Biomol. Scrren*. 18:54-66, 2013. With permission from Elsevier. B, courtesy of B.L. Hylander and E.A. Repasky.)

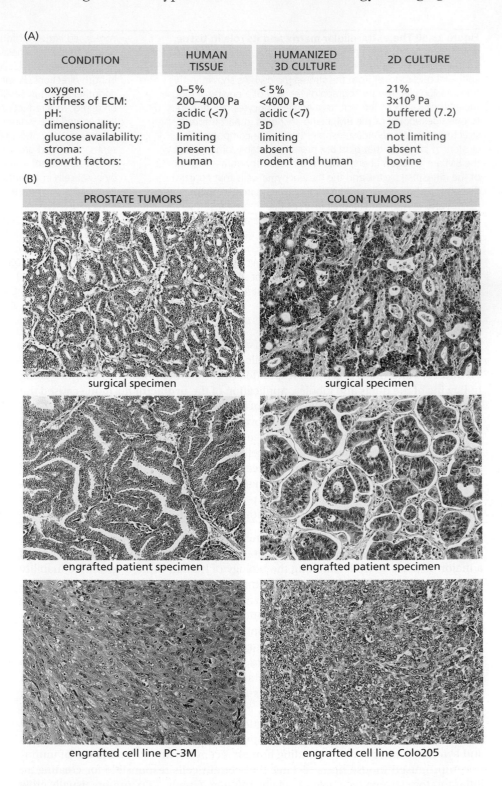

(A)

CONDITION	HUMAN TISSUE	HUMANIZED 3D CULTURE	2D CULTURE
oxygen:	0–5%	< 5%	21%
stiffness of ECM:	200–4000 Pa	<4000 Pa	3×10^9 Pa
pH:	acidic (<7)	acidic (<7)	buffered (7.2)
dimensionality:	3D	3D	2D
glucose availability:	limiting	limiting	not limiting
stroma:	present	absent	absent
growth factors:	human	rodent and human	bovine

(B)

PROSTATE TUMORS	COLON TUMORS
surgical specimen	surgical specimen
engrafted patient specimen	engrafted patient specimen
engrafted cell line PC-3M	engrafted cell line Colo205

This complexity raises some fundamental questions: How do cancer cells develop the capacity to release and respond to an array of diverse heterotypic signals? Are the complex signaling programs and resulting biological responses invented anew, being cobbled together piece by piece, each time normal cells evolve progressively into cancer cells? Or do cancer cells exploit preexisting, multifaceted cell-biological programs that are normally used by tissues for other purposes?

One attractive solution to these questions has come from an insight gained in 1986. A researcher studying the histological appearance of both tumors and wound-healing sites noted the striking resemblance between many of the signaling processes involved in tumor progression and those that occur during wound healing. Indeed,

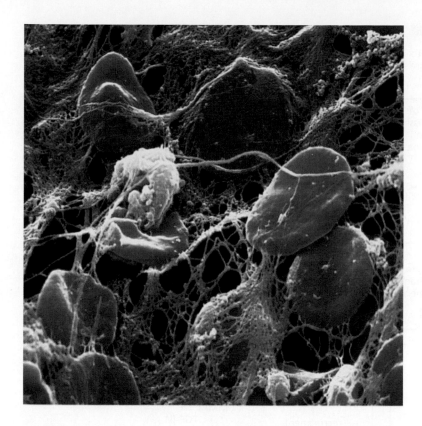

Figure 13.12 Scanning EM of a blood clot This scanning electron micrograph, which has been colorized, shows that clots are composed of dense networks of fibrin fibers *(light brown)* that have trapped erythrocytes/red blood cells *(orange-red)*. Prior to clotting, platelets undergo activation, which leads to the release of the granules seen in Figure S5.1. (Eye of Science/Science Source.)

these similarities have been extensively documented in the years since this notion was first proposed, yielding a clarifying explanation for the complex behavior of neoplastic tissues: they acquire aggressive behavior simply by activating a complex, normally latent cell-physiological program—wound healing—that is encoded in the normal genome. By accessing and exploiting this preexisting biological program, cancer cells are spared the task of re-inventing it anew each time a tumor arises.

Wound healing has been studied most extensively in the context of the skin. Following the formation of a superficial wound to the skin and underlying tissues, blood platelets aggregate and release granules containing, among other factors, platelet-derived growth factor (PDGF) and transforming growth factor-β (TGF-β). Wounding also causes release of **vasoactive** factors, which increase the permeability of blood vessels near the wound. This helps the wound site to acquire fibrinogen molecules from the blood plasma, which, when converted to fibrin, create the scaffolding of the blood clot (**Figure 13.12**). The resulting fibrin bundles, which become entwined around clumps of platelets, help to stanch further bleeding.

The PDGF released by the platelets attracts fibroblasts and stimulates their proliferation (**Figure 13.13**). Thereafter, the platelet-derived TGF-β activates these fibroblasts, converting them into **myofibroblasts**, and induces the latter to release a class of secreted proteases termed **matrix metalloproteinases** (MMPs; **Table 13.1**), which are also produced by macrophages that have been recruited in large numbers into the wound site. Unlike most secreted proteases, which have a serine in their catalytic clefts, MMPs carry zinc ions to aid in catalysis; indeed, these ions inspired their name. Activated fibroblasts also secrete mitogens, such as various fibroblast growth factors (FGFs), that can stimulate the proliferation of certain subtypes of epithelial cells.

Once released, the MMPs begin to degrade specific components of the extracellular matrix (ECM; see Table 13.1). This degradation has two major consequences. On the one hand, it allows for structural remodeling of the ECM, thereby carving out space for newly formed cells. On the other, it results, as cited above, in the release of a variety of growth factors that have been tethered in an inactive form to the proteoglycans of the ECM and now become solubilized and activated. Included among these are basic fibroblast growth factor (bFGF), TGF-β1, PDGF, several EGF-related factors, and

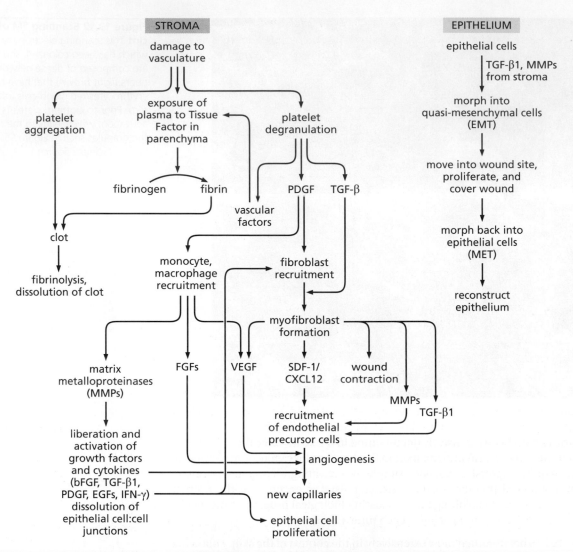

Figure 13.13 Flowcharts of wound healing These flowcharts indicate many of the major events occurring after an epithelial tissue is wounded. Many of the changes occur in the stroma of the wounded tissue (*left*) and thereby provide a foundation for reconstructing the overlying epithelial cell layer (*right*). A short-term effect of wounding is the process of hemostasis—the stanching of further hemorrhage. It involves both the plugging of damaged vessels and the degranulation of platelets, which release several distinct growth factors (GFs), including importantly PDGF and TGF-β; in addition, the exposure of plasma and platelets to Tissue Factor (expressed on the surfaces of a wide variety of nonendothelial cell types) triggers the formation of fibrin and thus clot formation (see Figure 13.12). PDGF and TGF-β recruit and activate both macrophages and fibroblasts. Macrophages, in turn, produce a number of products that are critical to initiating wound healing, including MMPs (to remodel the extracellular matrix and liberate various GFs); this allows recruitment of more macrophages and stimulates the proliferation of epithelial cells. The fibroblasts are activated by the actions of TGF-β and CXCL12/SDF-1 and, functioning thereafter as myofibroblasts, produce more of these factors that function, together with macrophage products, to stimulate angiogenesis. The epithelial cells respond to these signals by activating EMT (epithelial–mesenchymal transition) programs (see Figure 13.15) that provide them with the ability to move into the wound site, proliferate, and eventually revert via an MET (mesenchymal–epithelia transition) program to the epithelial phenotypic state of the undamaged nearby epithelium.

interferon-γ (IFN-γ). (In addition, virtually every MMP has been found to act on a variety of non-ECM proteins in the extracellular space; among these substrates are the latent **pro-enzyme** forms of other proteases, which are converted into active enzymes following cleavage by MMPs.) These wound-healing associated factors appear to provide a compelling explanation for a phenomenon frequently observed in the oncology clinic (**Sidebar 13.4**).

The growth factors released by platelets and those mobilized from the ECM then attract **monocytes** (which soon differentiate into several subtypes of macrophages) and another class of phagocytes, termed **neutrophils**, which infiltrate the wound site. (Yet other types of immune cells, including **eosinophils**, **mast cells**, and **lymphocytes**, are

Table 13.1 Some matrix metalloproteinases and their extracellular matrix substrates

Name of MMP[a]	Alternative name of MMP	Major ECM substrates[b]
MMP-1	collagenase-1	fibrillar collagens (collagen I, II, III)
MMP-2	gelatinase A	collagen IV, denatured collagens/gelatins
MMP-3	stromelysin-1	proteoglycans, glycoproteins (laminin, fibronectin, vitronectin)
MMP-7	matrilysin	similar to MMP-3
MMP-9	gelatinase B	similar to MMP-2
MMP-12	metalloelastase	elastin
MMP-14	MT1-MMP	fibrillar collagens, gelatins

[a]Twenty-three distinct MMPs have been documented in mammalian cells.
[b]In addition to ECM substrates, most of the MMPs also cleave other substrates that are present in the extracellular space and whose activities influence cancer progression. For example, MMP-2 processes the chemokine MCP-3; MMP-3 and MMP-7 cleave E-cadherin; MMP-9 releases VEGF and Kit-L; MMP-14 cleaves CD44. Also, the MMPs are part of a proteolytic cascade involving the conversion of plasminogen to plasmin and activation of latent pro-MMPs.

Courtesy of L. Matrisian and B. Fingleton, adapted in part from information from K. Kessenbrock, V. Plaks and Z. Werb, *Cell* 141: 52–67, 2010.

also recruited to this site.) These recruited cells scavenge and remove foreign matter, bacteria, and tissue debris from the wound site; at the same time, they release and activate mitogenic factors, such as fibroblast growth factors (FGFs) and vascular endothelial growth factor (VEGF). Such factors proceed to stimulate endothelial cells in the vicinity to multiply and to construct new capillaries—the process of angiogenesis (sometimes termed neoangiogenesis). As we will see shortly, VEGF and its cognate receptors play multiple roles in angiogenesis and related hematopoietic processes (Table 13.2).

While all this is occurring, largely in the stromal region of a wound site, the epithelial cells around the edges of the wound are undergoing their own alterations. Their goal is to reconstruct the epithelial sheet that existed prior to wounding. To do so, epithelial cells reduce their adhesion to the ECM, specifically to the basement membrane (see Section 2.2 and Figure 2.3) that separates them from the stromal compartment (Figure 13.14A). By severing these connections, the epithelial cells gain increased mobility.

Table 13.2 Properties of VEGFs

capillary and lymph duct formation

monocyte migration

hematopoiesis

recruitment of hematopoietic progenitor cells from the bone marrow

regulation of the endothelial cell pool during development

capillary permeability

Sidebar 13.4 Breast cancer surgery may lead to the stimulation of tumor growth Epidemiological studies of breast cancer patients undergoing surgery (that is, partial or total mastectomy) for removal of their primary tumors have revealed a peak of recurrence of breast cancers at the primary site as well as in distant anatomical sites about three years after surgery. The timing of these relapses suggests that the surgery is responsible for stimulating their formation. Examination of the fluids draining from the wound sites created by the surgery revealed the presence of potent mitogenic factors that are associated with the wound-healing process. Among these are mitogens for breast cancer cells, especially those cancer cells that overexpress the HER2/Neu receptor protein (see Figure 4.3). In addition, a burst of vascular endothelial growth factor (VEGF) synthesis occurs following surgery; as discussed later in this chapter, VEGF is a potent stimulant of tumor angiogenesis. It is therefore plausible that surgery stimulates the proliferation of residual **micrometastases** (small deposits of metastatic cells)

that are not detected and removed when the primary tumors are excised. Indeed, some have argued that the clinical relapses stimulated by surgery nullify much of the therapeutic benefit that should, by all rights, be achieved by the removal of primary tumors and nearby lymph nodes. The clinical importance of such surgery-stimulated tumor progression remains a matter of great contention.

Another clinical observation may eventually be found to bear on this issue: when Herceptin (a therapeutic monoclonal antibody to be described in Chapter 16 that blocks the action of the HER2/Neu receptor) is applied postoperatively in women who have borne relatively small, low-grade tumors, the rate of clinical relapse is reduced by as much as 50%. This stunning therapeutic success may, quite possibly, be associated with the ability of Herceptin to block stimulation of residual cancer cells by the growth factors elaborated during healing of the surgical wound site.

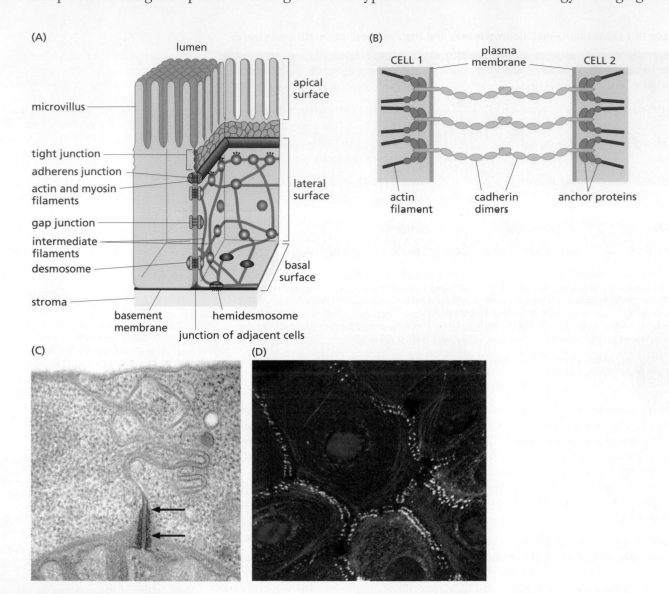

Figure 13.14 Attachments of epithelial cells to their neighbors and the basement membrane (A) Epithelial cells are anchored via hemidesmosomes on their basal surfaces to the underlying basement membrane (*yellow*) and via their lateral surfaces to adjacent epithelial cells (see also Figure 13.10A). These lateral connections include tight junctions (which create a fluid seal between the luminal and basolateral spaces, *light orange*), adherens junctions and desmosomes (which link the cytoskeletons of adjacent epithelial cells, *red, blue*), and gap junctions (which permit low–molecular-weight solutes to pass between adjacent cells, *light orange*). (B) The adherens junctions, which play a central role in regulating epithelial cell behavior, depend on oligomerization between E-cadherin molecules (*light green*) displayed by two adjacent cells. The "anchor proteins" (*blue*), which include prominently α- and β-catenin and p120, form physical bridges between the cytoplasmic tails of E-cadherin proteins and the actin cytoskeleton (*red*; see also Figure 6.19A). As discussed in Section 7.10 and illustrated in Figure 6.19B, β-catenin has a second, quite independent function: when it is not sequestered in adherens junctions, it may move into the nucleus and serve as a

subunit of a transcription factor complex that drives the expression of key genes. (C) This transmission electron micrograph shows the adherens junctions (*arrows*) between two closely apposed plasma membranes of adjacent epithelial cells, in this instance those creating the intestinal lining of the worm *Caenorhabditis elegans*. Virtually identical structures are seen in mammalian cells. (D) This immunofluorescence micrograph illustrates the interactions between the lateral surfaces of neighboring keratinocytes in 2-dimensional monolayer culture. E-cadherin "zippers" (*yellow*) form adherens junctions between neighboring cells and are attached via their cytoplasmic domains and other intermediary proteins (*not shown*) to the actin cytoskeletons of these cells (*red*). Nuclei are stained *blue-purple*. Some of these interactions in actual epithelial tissues can be visualized via immunostaining of E-cadherin and laminin (a basement membrane protein). (A, from H. Lodish et al., Molecular Cell Biology, 6th ed. New York: W.H. Freeman, 2008. With permission from Macmillan. B, from B. Alberts et al., Molecular Biology of the Cell, 4th ed. New York: Garland Science, 2002. C, courtesy of D. Hall. D, from M. Perez-Moreno, C. Jamora and E. Fuchs, *Cell* 112:535–548, 2003. With permission from Elsevier.)

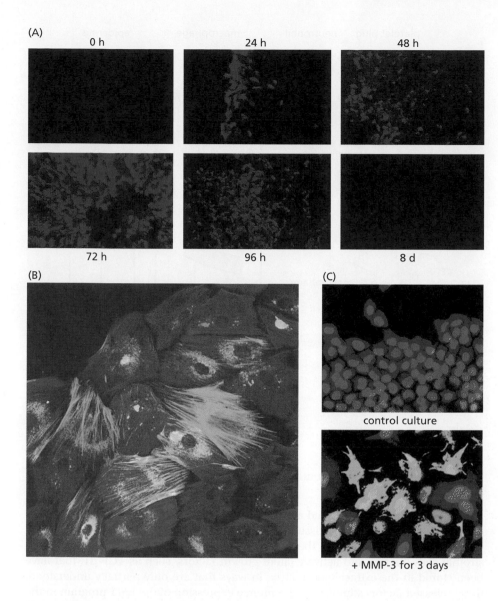

(A)

0 h   24 h   48 h

72 h   96 h   8 d

(B)

(C)

control culture

+ MMP-3 for 3 days

Figure 13.15 The epithelial–mesenchymal transition Multiple signals and conditions trigger activation of EMT programs, including prominently the wounding of epithelial cell sheets studied here. (A) A confluent monolayer of MCF10A cells [a line of nontransformed, immortalized human mammary epithelial cells (MECs)], has been disturbed by removal of a patch of cells (*at left of each image*). Before wounding, no vimentin is expressed by these cells. However, soon after wounding, the epithelial cells at the edge of the wound undergo a partial epithelial–mesenchymal transition (EMT), enter into a quasi-mesenchymal phenotypic state, express vimentin (characteristic of more mesenchymal cells; *red*), migrate into the wound site, proliferate, and fill the wound site. After 8 days, the cell monolayer has become confluent, and these cells revert to a fully epithelial phenotype and shut down vimentin expression [indicating a mesenchymal–epithelial transition (MET)], thereby reconstructing an intact epithelium. (B) In some cell types in culture, individual cells may spontaneously shift from an epithelial phenotype, indicated by cytokeratin expression (*red*), to a more mesenchymal phenotype, indicated by α-smooth muscle actin expression (*green*). This mirrors a phenotypic plasticity that is apparent in certain cells during early embryogenesis and tumorigenesis. (C) An EMT was provoked here by a 3-day-long exposure to matrix metalloproteinase-3 (MMP-3), which may trigger an EMT through its ability to degrade E-cadherin. Thus, some of the epithelial cells have shut down expression of many typical epithelial cell markers, such as cytokeratins (*dark pink*) and E-cadherin (*not shown*), and express instead mesenchymal proteins, such as vimentin (*green*), fibronectin, and N-cadherin (*not shown*). In addition, they change their typically polygonal shape (*top*) to a fibroblastic shape (*bottom*) and concomitantly acquire motility and invasiveness. (A, from C. Gilles et al., *J. Cell Sci.* 112:4615–4625, 1999. With permission from Company of Biologists. B, from O.W. Petersen et al., *Am. J. Pathol.* 162:391–402, 2003. C, from M.D. Sternlicht et al., *Cell* 98:137–146, 1999. Both with permission from Elsevier.)

The epithelial cells also sever their attachments to one another. These side-by-side associations are stabilized in part by **adherens junctions** (Figure 13.14B–D), which are assembled as associations of E-cadherin molecules that are displayed by adjacent epithelial cells and tether the cells' closely **apposed** plasma membranes to one another. Accordingly, E-cadherin expression is suppressed in the epithelial cells that are situated around the edges of a wound site and is often replaced by N-cadherin, another cell–cell adhesion molecule that is normally displayed by mesenchymal cells, notably fibroblasts. (These N-cadherin molecules, though related structurally to E-cadherin, do not reestablish adherens junctions between the epithelial cells, because they bind more weakly to one another.)

While shifting their display of cell-surface cadherins, the epithelial cells at the edge of a wound undergo a major change in phenotype, which causes them to assume, at least superficially, a fibroblastic appearance (**Figure 13.15**). This profound shift is termed an **epithelial–mesenchymal transition** (EMT) and enables the epithelial cells to become motile and invasive. The EMT program that becomes activated in these cells involves profound changes in their biology beyond those apparent under the microscope. Thus, induction of several hundred genes creates a cell-biological *program* that imparts a variety of mesenchymal traits, causing these cells to transit from a fully epithelial state into a mixed epithelial/mesenchymal (**quasi-mesenchymal**) phenotypic state. Having dissolved the adherens junctions that bound them on their lateral surfaces to one another (see Figure 2.4), the resulting quasi-mesenchymal cells are

Figure 13.16 A program of epithelial wound healing This illustration depicts the heterotypic signaling occurring after the skin, a typical epithelial tissue, is wounded. A fibrin clot that carries platelets and red blood cells/erythrocytes (see Figure 13.12) initially fills the wound site. Day 3 of the wound-healing process is sometimes termed its inflammatory phase, because of the involvement of cell types that are associated with inflammation, notably neutrophils and monocytes; they are involved in removing cell debris, the clot, and invading bacteria. In addition, macrophages, which derive from differentiation of extravasated monocytes, release TGF-β1 and TGF-α; these factors stimulate the proliferation of nearby epithelial cells from the epidermis and induce them to undergo an epithelial–mesenchymal transition (EMT). Tongues of the resulting cells, now exhibiting mesenchymal behavior, invade under the clot from both sides in order to cover the wound site (see Figure 13.15A). At the same time, TGF-β1 and PDGF-A and -B, which were released by the platelets during initial clot formation, stimulate the proliferation of fibroblasts (*blue-purple*) in the underlying dermal stroma and induce them to convert into the activated fibroblasts termed myofibroblasts (*red*); because of their contractility, myofibroblasts will pull the sides of the wound together. Myofibroblasts also release angiogenic factors, notably VEGFs and FGF-2 (fibroblast growth factor-2), which stimulate the formation of new blood vessels in the stroma underlying the wound site. Keratinocyte growth factor (KGF), another isoform of FGF, is released by mesenchymal cells and stimulates epithelial cell proliferation. In addition, insulin-like growth factors (IGFs) provide survival signals to various cells in the wound-healing site. Finally, after the wounded stroma is covered, the quasi-mesenchymal cells that had undergone an EMT revert, via a mesenchymal–epithelial transition (MET), to an epithelial state, thereby reconstituting the epithelium (*not shown*). (Redrawn from A.J. Singer and R.A.F. Clark, *N. Engl. J. Med.* 341:738–746, 1999. With permission from Massachusetts Medical Society.)

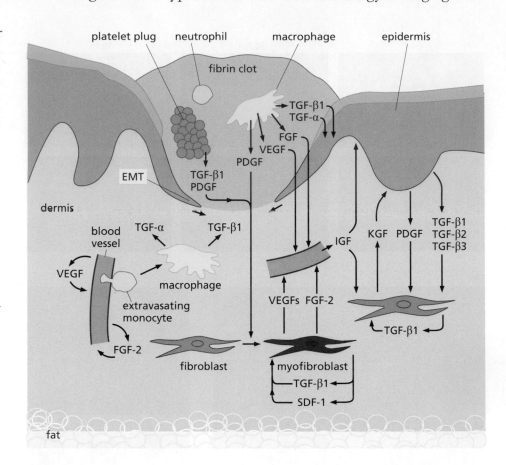

now freed to move into the wound site and fill in the gap in the epithelium previously created by wounding.

Although the EMT, as described above, can be observed in monolayers of cultured epithelial cells, within a living tissue, the triggering of this program is aided by stromal cells that underlie epithelia. The matrix metalloproteinases (MMPs) secreted by stromal cells liberate and activate latent growth factors, such as TGF-β1, that have been stored in the extracellular matrix. In ways that are only partially understood, these released factors stimulate and reinforce expression of the EMT program in the epithelial cells. Importantly, the EMT is only a temporary shift in cell phenotype. After the migrating cells have moved into position, proliferated, and covered the wound site, they reconstruct the epithelium by reverting to an epithelial state via the program termed the **mesenchymal–epithelial transition** (MET; see Figure 13.15A). As a consequence, once wound healing is complete, the cells in the reconstituted epithelium show no trace of having passed transiently through a quasi-mesenchymal state. We will discuss the EMT program in much greater detail in the next chapter because it plays a key role in enabling carcinoma cells to acquire malignant traits. In fact, a complex array of heterotypic signals impinge on epithelial cells to encourage them to activate their previously silent EMT programs, as described in Supplementary Sidebar 13.4.

If we compare these processes (**Figure 13.16**) with the interactions of epithelial cancer cells and their stromal neighbors, we find striking parallels between wound healing and tumorigenesis. Among the similarities between the two is the presence of clumps of fibrin in the tumor-associated stroma. In the case of tumors, this does not derive from traumatic damage to blood vessels. Instead, the capillaries within tumors are constitutively leaky, unlike those in normal tissues (**Figure 13.17A**). Later, we will discuss the causes of this leakiness. For the moment, suffice it to say that the permeability of the capillary walls and **venules** (small veins) allows fibrinogen molecules from the plasma to come in direct contact with cancer cells; this provokes, through a series of intermediate reactions, the conversion of fibrinogen to fibrin and the formation of large bundles of fibrin strands (Figure 13.17B and C).

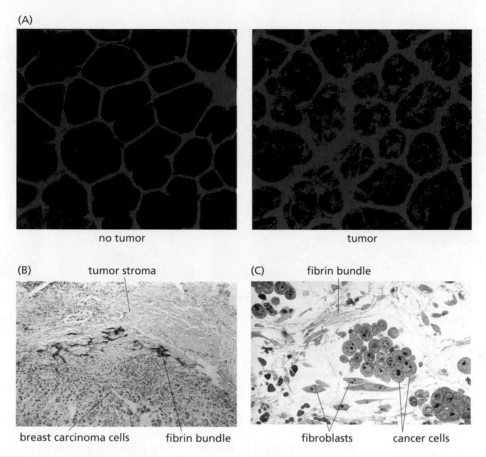

(A)

no tumor

tumor

(B) tumor stroma

breast carcinoma cells   fibrin bundle

(C) fibrin bundle

fibroblasts   cancer cells

Figure 13.17 Capillary leakiness and deposition of fibrin bundles in the tumor-associated stroma (A) Injection of red dextran dye into normal microvasculature reveals that the walls of associated capillaries have relatively low permeability, as indicated by the sharply delineated profiles of these vessels (*left*). However, when injected into tumor-associated microvasculature, the dye leaks out of the capillaries and diffuses into the nearby parenchyma, yielding diffusely staining outlines of vessels (*right*). (B) The leakiness of tumor-associated microvasculature results in the continuous leaking of thrombin and fibrinogen molecules from the plasma into the parenchyma surrounding blood vessels. Tissue Factor, a protein displayed on the surface of cancer cells and many other nonendothelial cell types, activates plasma-derived thrombin upon contact, whereupon thrombin, functioning as a protease, converts fibrinogen

into fibrin (see Figure 13.12). This results in the extensive network of fibrin bundles (*dark red*) seen here at the border between actively growing breast cancer cells (*below*) and the tumor stroma (*above*). (C) The fibrin bundles form an extracellular matrix to which migrating cells, such as fibroblasts (*below cluster of cancer cells*) as well as endothelial cells (*not seen*), can attach, using it as a substrate on which they can move forward. In many tumors, such as this guinea pig liver carcinoma of bile duct origin, the fibrin matrix may eventually be dissolved by the process of *fibrinolysis* and replaced by a collagenous matrix. (A, courtesy of R. Muschel. B, from C.G. Colpaert et al., *Histopathology* 42:530–540, 2003. With permission from John Wiley & Sons. C, from H.F. Dvorak, *Am. J. Pathol.* 162:1747–1757, 2003. With permission from Elsevier.)

Many kinds of cancer cells, including those forming carcinomas of the breast, prostate, colon, and lung, continuously release significant levels of PDGF. This contrasts with the situation in wounds, in which PDGF is released in a brief burst by platelets as they form the initial blood clot (see Figure S5.1). As is the case in wound healing, the targets of the PDGF produced by cancer cells are mesenchymal cells in the stroma that display PDGF receptors, including smooth muscle cells, fibroblasts, and macrophages. PDGF functions here both as an attractant and as a mitogen for these stromal cells and appears to be the most important signaling molecule used by carcinoma cells to recruit and stimulate the proliferation of stromal cells. In breast cancers, for example, the level of PDGF expression generally increases with increased degree of tumor progression, which may explain the high proportions of stroma present in many advanced carcinomas; an extensive stroma, including high levels of collagens, is especially evident in adenocarcinomas of the pancreas. Indeed, in a variety of carcinomas, a high ratio of stromal to epithelial cells represents an unfavorable prognosis for the cancer patient.

(A) (B) (C)

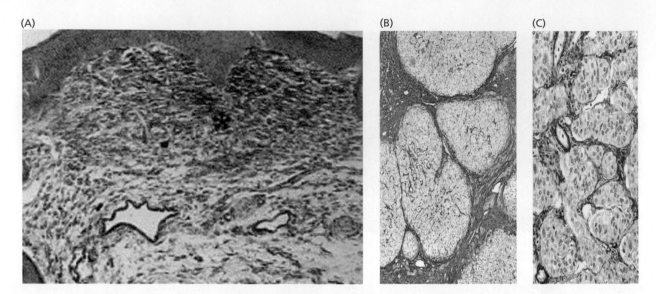

Figure 13.18 Myofibroblasts Myofibroblasts arise in sites of wound healing, during inflammation, and in tumor-associated stroma, in part through differentiation of stromal fibroblasts (see Figure 13.16) and recruitment of myofibroblast precursors, possibly fibrocytes and mesenchymal stem cells (MSCs), from the circulation. Their presence is revealed typically by their expression of α-smooth muscle actin (α-SMA). (A) Staining for α-SMA–positive myofibroblasts (*reddish brown*) shows them present in abundance 3 days after the skin of a mouse has suffered a wound. (B) Chronically inflamed tissues acquire a fibrotic stroma, such as the cirrhotic liver seen here stained with anti-α-SMA antibody, which reveals the presence of masses of myofibroblasts (*brown*). (C) A hepatocellular carcinoma stained with anti-α-SMA antibody (*brown*) reveals the striking resemblance of the stroma of a chronically inflamed tissue (panel B) to the stroma of a carcinoma arising in that tissue. (A, from P. Martin et al., *Curr. Biol.* 13:1122–1128, 2003. With permission from Elsevier. B and C, from A. Desmouliere, A. Guyot and G. Gabbiani, *Int. J. Dev. Biol.* 48:509–517, 2004. With permission from UPV/EHU Press.)

The PDGF released by carcinoma cells initially succeeds in recruiting fibroblasts into the fibrin matrix. At the same time, the fibrin matrix (see Figure 13.17B and C) provides an important scaffolding that helps these recruited mesenchymal cells to attach, via integrins, and migrate. The invading stromal cells remodel this "provisional matrix" by degrading many of the initially formed fibrin molecules and replacing them with a more permanent matrix that is assembled from the collagen secreted by the myofibroblasts (**Figure 13.18**). This sequence of steps loosely parallels those occurring during wound healing.

Recent work has revealed the complexity of carcinoma-associated fibroblasts (CAFs) (**Figure 13.19**). At least in the context of pancreatic adenocarcinoma, those that were near the carcinoma cells expressed α-smooth muscle acting (α-SMA), which is known to be induced by TGF-β released by carcinoma cells. These CAFs produced prodigious amounts of type I collagen, which eventually creates the **desmoplastic** (that is, fibrous, hard) stroma that is typical of this very aggressive tumor. (Quantitative physical measurements of the solid stresses, that is, the hardness of pancreatic carcinomas indicates that these tumors are 100 times harder than, for example, glioblastomas.) Yet other CAFs, situated further away from islands of carcinoma cells lack elevated α-SMA expression and instead released pro-inflammatory molecules, prominently interleukin-6 (IL6).

One of the tasks of the stromal cells in wound healing involves the physical contraction of the wound site in order to close wounds. This contraction is mediated by the specialized fibroblasts cited above to generate the mechanical tension needed for wound closure (see Figure 13.18A). These specialized fibroblasts are termed myofibroblasts (see Figure 13.16); these cells express α-smooth muscle actin (α-SMA) and are capable of using the actin–myosin contractile system (related to the one operating in muscle cells). In addition to their major contribution to the populations of cancer-associated fibroblasts (CAFs), myofibroblasts are also present in sites of chronic wounding, that is, continuously inflamed tissues (see Figure 13.18B). Provocatively, essentially identical α-SMA-positive fibroblasts are prominent components of the stroma present in most advanced carcinomas (for example, see Figure 13.18C). Of note, normal tissues, such as the prostate gland, may contain α-SMA-positive cells that, because they lack vimentin,

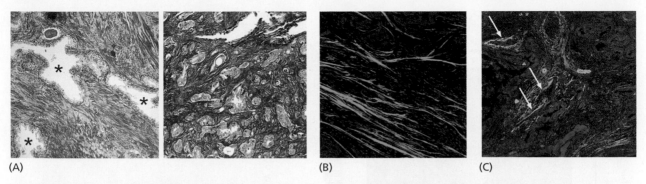

Figure 13.19 Micrograph of normal stroma and desmoplastic stroma The histologically complex stroma of normal tissue may eventually be replaced by the desmoplastic stroma of an advanced carcinoma. Here we see normal human prostate tissue (*left panel*) stained with Masson's trichrome stain, which reveals the extensive smooth muscle cells in the stroma as *pink*; normal ducts (*asterisks*) are scattered throughout this stroma. In contrast, an advanced prostate carcinoma (*right panel*), stained identically and viewed at the same magnification, reveals the extensive desmoplastic stroma, which is rich in extracellular matrix, notably collagen I (*blue-purple*). Islands of prostate carcinoma cells forming small ducts (*pink*) are scattered throughout this desmoplastic stroma, which lacks significant numbers of viable cells such as myofibroblasts and fibroblasts. (B) In this invasive ductal carcinoma of the breast, the deposition of parallel collagen fibers by fibroblasts and myofibroblasts (*green*; visualized here by 2-photon microscopy) creates a stiff extracellular matrix in the desmoplastic stroma that encourages epithelial–mesenchymal transitions (EMTs; described in Chapter 14) in nearby carcinoma cells (*red*). (C) A more complex stroma is seen in a mouse model of pancreatic ductal adenocarcinoma (PDA). At this stage of tumor progression, the glandular carcinoma cells (stained for EpCam, *blue*), are greatly outnumbered by the immune cells that constitute the inflammatory and immunosuppressive components of the stroma (stained for CD45, *red*). The α-SMA-positive cells (*green*) are largely stromal myofibroblasts and, to a small extent, microvessel-associated pericytes (*arrows*). (A, from G. Ayala et al., *Clin. Cancer Res.* 9:4792–4801, 2003. With permission from American Association for Cancer Research. B, courtesy of S. Wei and J. Yang. C, from C. Feig, A. Gopinathan, A. Neesse et al. *Clin. Cancer Res.* 18, 4266-4276, 2012. Reprinted by permission from the American Association for Cancer Research.)

can be identified as smooth muscle cells rather than the myofibroblasts present in sites of wound-healing, inflammation, and tumorigenesis (see Figure 13.19A). Myofibroblasts have been found to release a broad spectrum of growth factors, including HGF, CTGF, EGF, IGF, NGF, bFGF, Wnts, and cytokines (see Table 5.1); this list overlaps partially with key factors released by platelets upon clot formation.

It remains unresolved precisely where most stromal fibroblasts and myofibroblasts originate. During wound healing, fibroblasts may initially be recruited from the stroma of adjacent unaffected tissues by the PDGF released by platelets participating in clot formation; myofibroblasts may then be produced through the TGF-β1–induced transdifferentiation of some of these recruited fibroblasts (see Figure 13.16). In fact, myofibroblasts can be produced *in vitro* simply by exposing normal fibroblasts to TGF-β1. This finding suggests that the TGF-β1 released by many types of carcinoma cells, especially those that have activated EMT programs, is a major factor responsible for the formation of myofibroblasts in the tumor-associated stroma.

Alternative sources of tumor-associated fibroblasts and fibroblasts have also been documented: they may arise from **fibrocytes** and **mesenchymal stem cells** (MSCs) of bone marrow origin. Both types of cells can be found in the circulation and are known to home to areas of tissue damage, where they can differentiate into myofibroblasts and, quite possibly, fibroblasts, thereby facilitating the rapid reconstruction of stroma. Such mechanisms ensure that large numbers of cells can be mobilized rapidly from sources outside a wound site in order to expedite reconstruction of damaged stroma.

Yet another hint of parallels between wound healing and tumorigenesis comes from several studies indicating that agents known to inhibit tumor-associated angiogenesis also antagonize wound healing. Included among these agents are certain soluble inhibitors of the growth factor receptors that play key roles in angiogenesis, as well as an ECM component, termed thrombospondin-1 (Tsp-1), which is a potent anti-angiogenic molecule. The many common mechanisms shared by wound healing and by the epithelial–stromal interactions in tumors also have implications for clinical practice (see Sidebar 13.4).

The importance of various classes of fibroblasts in clinical progression has been illuminated by studies of serum-stimulated fibroblasts propagated *in vitro*, that is,

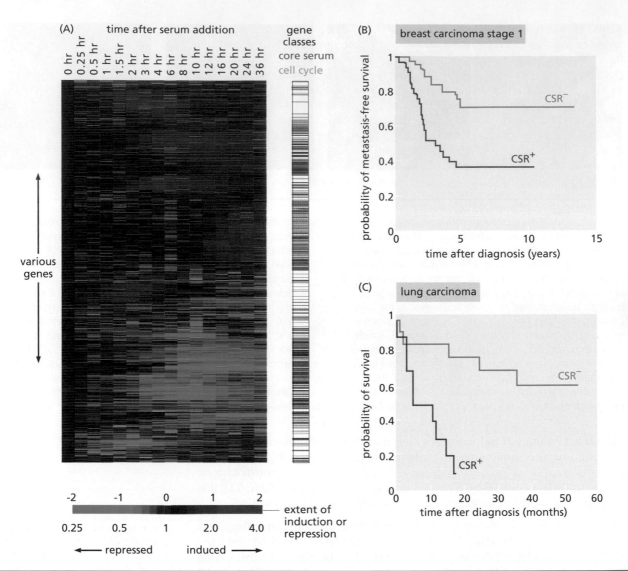

Figure 13.20 Gene expression arrays of tumor-associated fibroblasts and serum-activated fibroblasts (A) In this gene expression analysis, the expression of cellular mRNAs was analyzed following the addition of fresh serum to previously quiescent, serum-starved human fibroblasts; this resulted in these cells' entering into the active cell cycle and the induction or repression of the expression of a large cohort of genes (see also Figure 8.9). Gene expression was measured at various times (*above*) after addition of serum. The cohort of genes analyzed (*not named*) is arrayed vertically from *top to bottom*. Genes that were induced or repressed late after the addition of serum were classified as "cell cycle genes" (*orange lines, right vertical bar*), whereas genes that were induced or repressed early and did not fluctuate with the cell cycle phase were classified as "core serum response" (CSR) genes (*blue lines, right vertical bar*). The horizontal bar (*below*) is a key indicating the degree of induction (*red*) or repression (*green*) following serum addition, this being indicated logarithmically above the bar and in absolute numbers (limited to a 4-fold difference) below the bar. (B) The expression pattern of the CSR genes was analyzed in a series of tumors in women presenting with stage 1 (i.e., early-stage) breast cancer. As indicated here, those whose tumors expressed the CSR gene pattern (*red line*), which presumably reflected the states of stromal fibroblasts and related cell types within these tumors, showed a much greater probability for developing metastases in the years following initial treatment compared with those whose tumors did not show this gene expression pattern (*blue*). (C) Similarly, those patients whose lung adenocarcinomas (including all stages of tumor progression at the time of diagnosis) showed the CSR gene expression signature (*red line*) suffered dramatically higher mortality rates compared with those whose tumors did not show this gene expression signature (*blue line*). (From H.Y. Chang et al., *PLoS Biol.* 2:E7, 2004. © 2004 Chang et al. This is an open-access article distributed under the terms of the Creative Commons Attribution License.)

fibroblasts that had been starved of serum and then exposed to high concentrations of fresh serum. This exposure re-creates the environment of fibroblasts during the initial stages of wound healing (a time when the various serum factors released by activated platelets act on stromal fibroblasts at the wound site, converting them, at least transiently, into myofibroblasts). A relatively small cohort of serum-induced and -repressed genes was extracted from these data and used to represent the characteristic "signature" of serum-stimulated fibroblasts (**Figure 13.20A**).

The RNA expression patterns of a large group of human carcinomas were then analyzed to determine whether the tumors expressed the signature of serum-stimulated fibroblasts. (Because all the genes being analyzed are associated specifically with fibroblasts, these analyses, by necessity, reflected the RNA expression patterns of the CAFs present in each tumor.) Many carcinomas were indeed found to express the signature of serum-stimulated fibroblasts. Significantly, the carcinomas that expressed this signature more intensely were associated with a grimmer clinical prognosis (see Figure 13.18B and C). Similarly, tumors that contain higher proportions of myofibroblasts indicate shorter survival rates of the patients bearing these tumors (Supplementary Sidebar 13.5). This correlation suggests that the activated, myofibroblast-rich stroma represents a potent force driving aggressive tumor progression, that is, invasion and metastasis as discussed in the next chapter.

13.4 Experiments directly demonstrate that stromal cells are active contributors to tumorigenesis

Several biological experiments provide far more direct demonstrations of the profound influence that recruited stromal cells exert on epithelial cell tumorigenesis. In one study, previously non-tumorigenic, immortalized keratinocytes were forced to secrete PDGF at high levels (achieved through the introduction of a PDGF expression vector). The released growth factor had no effect on the proliferation of these epithelial cells *in vitro* because they do not display PDGF receptors on their surface. However, when these cells were implanted into host mice, they acquired the ability to form robustly growing tumors, clearly derived from the ability of the released PDGF to recruit and activate stromal cells. The stromal cells then reciprocated by driving the proliferation of the PDGF-secreting keratinocytes, eventually causing the latter to undergo malignant transformation.

The important role of stromal fibroblasts in supporting tumor growth can be demonstrated by yet another type of experiment: transformed, weakly tumorigenic human mammary epithelial cells (MECs) were found to require more than two months to form a tumor after being introduced into immunocompromised host mice (**Figure 13.21**). However, if these cancer cells were mixed with human mammary stromal fibroblasts (from normal breast tissue) prior to injection into hosts, the cells formed tumors in one-third the time. These admixed fibroblasts clearly obviated the need of the MECs to spend time recruiting fibroblasts from host mice—a process that usually requires many weeks' time; in the absence of these fibroblasts, tumor growth could not take off.

This experiment did not address the issue of whether the stromal cells of normal epithelial tissues can accelerate tumor formation as effectively as the stromal cells of carcinomas. This possibility was addressed by comparing the actions of stromal

Figure 13.21 Admixed fibroblasts and tumor growth When human mammary epithelial cells (HMECs) were transformed through the introduction of the SV40 early region, the *hTERT* gene, and a weakly expressed *ras* oncogene (see Section 11.9), the resulting transformed HMECs (txHMECs) formed tumors, albeit with a long lag time after injection, and then only in about half of the mice into which they had been injected (*orange curve*). When a preparation of extracellular matrix (Matrigel) produced by mouse sarcoma cells was mixed with the txHMECs prior to introduction into host mice (*blue curve*), this slow development was accelerated a bit and tumor-forming efficiency was doubled. However, when mammary stromal fibroblasts from a normal human breast were admixed to the transformed human MECs prior to injection, the cancer cells formed tumors rapidly and with 100% efficiency (*red curve*). This experiment illustrated that the recruitment of stromal fibroblasts (or similarly functioning mesenchymal cells) is an important rate-limiting step in tumor formation. (From B. Elenbaas et al., *Genes Dev.* 15:50–65, 2001. With permission from Cold Spring Harbor Laboratory Press.)

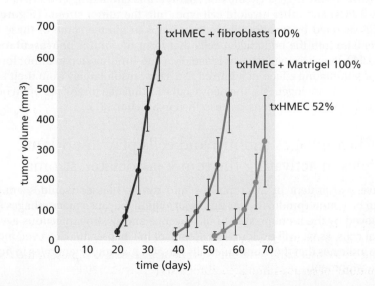

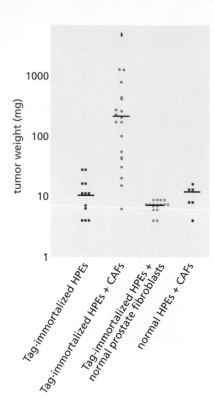

Figure 13.22 Effects of prostatic stromal cells on immortalized prostatic epithelial cells When normal human prostate epithelial cells (HPEs) that had been immortalized by SV40 T antigen (Tag-immortalized HPEs) were introduced into Nude mice, they formed growths of ≉10 mg after many weeks, not significantly larger than the initial inoculum (*red dots*), indicating that these cells were, when implanted on their own, non-tumorigenic. However, if these same cells were mixed with carcinoma-associated fibroblasts (CAFs) prepared from a human prostate carcinoma, tumors arose whose median weight was 10 times larger (Tag-immortalized HPEs + CAFs, *blue dots*). In contrast, when these Tag-immortalized HPEs were mixed with stromal fibroblasts from a normal human prostate, the median weight of the resulting growth was ≉7 mg, possibly even less than the starting inoculum (*orange dots*). And if CAFs were mixed with normal HPEs prior to inoculation, the median weight after many weeks was, once again, only ≉10 mg (*purple dots*). (From A.F. Olumi et al., *Cancer Res.* 59:5002–5011, 1999. With permission from American Association for Cancer Research.)

fibroblasts extracted from a normal epithelial tissue with the carcinoma-associated fibroblasts (CAFs) prepared from carcinomas (in which myofibroblasts are abundant; for example, see Figure 13.18). In one set of experiments, fibroblasts were purified from the stroma of normal human prostate glands, and the CAFs were prepared from the stroma of human prostate carcinomas. Each cell population was then mixed with otherwise non-tumorigenic, SV40 large T antigen–immortalized human prostate epithelial cells and implanted in immunocompromised Nude mice. The results, summarized in **Figure 13.22**, showed dramatic differences in the growth of these mixed tissue grafts. In particular, the grafts containing CAFs plus immortalized prostate epithelial cells formed tumors that were as much as 500 times larger than those containing normal prostate fibroblasts plus immortalized prostate epithelial cells. (When injected alone, the CAFs formed no tumors at all.)

This experiment demonstrated that these CAFs were functionally very different from the stromal fibroblasts present in normal prostatic tissue. Stated differently, during the course of tumor progression, stromal cells become increasingly adept at helping their epithelial neighbors to survive and proliferate. Similar observations have since been made with CAFs extracted from human breast cancers.

Still, these experiments do not reveal precisely *how* the myofibroblast-rich CAF populations accelerate tumor growth. Once established within the tumor-associated stroma, the myofibroblasts, as mentioned earlier, confer multiple benefits on nearby epithelial cancer cells, the most important benefit possibly being angiogenesis. Myofibroblasts expedite angiogenesis through their ability to release stroma-derived factor-1 (SDF-1/CXCL12), a chemokine that recruits circulating endothelial progenitor cells (EPCs) and other myeloid cell types into the tumor stroma (**Figure 13.23**). The VEGF secreted by myofibroblasts then helps to induce some of these recruits to differentiate into the endothelial cells that form the tumor **neovasculature** (see Figure 13.8A). Because angiogenesis is usually a rate-limiting step in tumor formation, the tumor-stimulating effects of admixed CAFs may result largely from their ability to accelerate tumor angiogenesis. The neovascularization of a tumor can often be solved through unconventional means, as described in **Sidebar 13.5**.

13.5 Macrophages and myeloid cells play important roles in activating the tumor-associated stroma

The active recruitment of macrophages into tumor masses would seem, at first glance, to be counterproductive for tumor formation, because macrophages are usually deployed by the immune system to scavenge and destroy infectious agents and abnormal cells, as we will explore in more detail in Chapter 15. However, increasing evidence indicates that these immune cells can also play important roles in *furthering* tumor development.

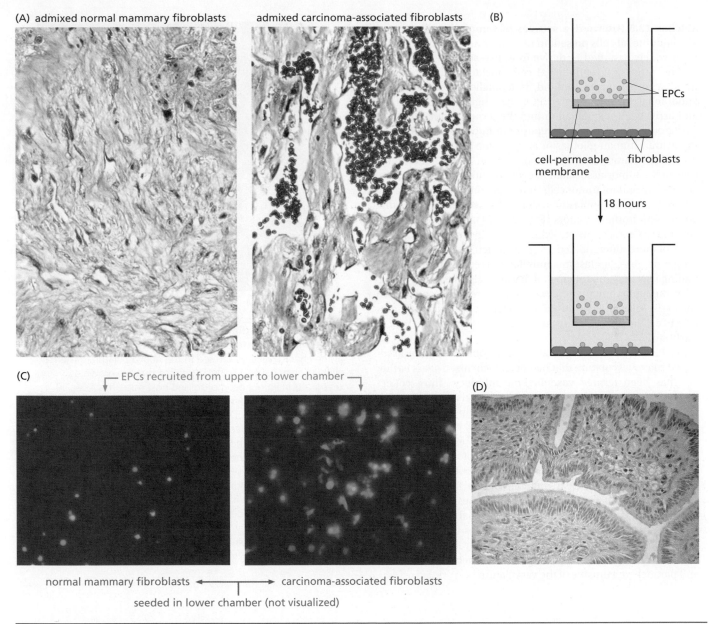

Figure 13.23 Recruitment of endothelial precursor cells and induction of their differentiation by CAFs (A) When normal human mammary stromal fibroblasts are admixed to cells of the human MCF7-ras breast cancer cell line (*left panel*), the resulting tumors show relatively small numbers of blood vessels (*pink-red*) amid the tumor-associated stroma (*blue*). However, admixture of carcinoma-associated fibroblasts (CAFs; *right panel*) results in highly vascularized tumors with large vessels and associated erythrocytes (*red*). The resulting access to the circulation is likely to greatly facilitate tumor growth. (B) The ability of the CAFs, which are largely myofibroblasts, to attract endothelial cells was demonstrated by an *in vitro* experiment, in which green fluorescent protein (GFP)–labeled bone marrow cells expressing surface antigens characteristic of endothelial progenitor cells (EPCs; *green*) were placed above a cell-permeable membrane (*light gray*) in a Boyden chamber. Either normal mammary stromal fibroblasts or CAFs from a breast cancer (*blue*) were placed on the bottom surface of the lower chamber. After 18 hours, the number of EPCs that had migrated through the filter and had been recruited to the bottom of the lower chamber was gauged by fluorescence microscopy. (C) Analysis of the cells attached to the lower chamber indicated that the CAFs (*right panel*) are able to attract far more GFP-labeled EPCs than the normal mammary stromal fibroblasts (*left panel*). This recruitment could be reduced by 60% by placing antiserum that neutralized the SDF-1/CXCL12 chemokine released by the CAFs in these chambers, indicating its important role in mediating this recruitment. (D) In a mouse model of gastric carcinoma development, the myofibroblasts of the tumor-associated stroma are seen, by immunostaining, to express high levels of VEGF-A (*brown*), which induces differentiation of EPCs and the proliferation of resulting endothelial cells. (A and C, from A. Orimo et al., *Cell* 121:335–348, 2005. With permission from Elsevier. D, from X. Guo et al., *J. Biol. Chem.* 283:19864–19871, 2008. © 2008 ASBMB. This is an open access article distributed under the terms of the Creative Commons Attribution license.)

Sidebar 13.5 Alternative sources of tumor vascularization
The endothelial cells present in the tumor neovasculature have been widely assumed to derive from two sources—the proliferation of existing endothelial cells, including those present in adjacent normal tissue and, to a smaller extent, from the endothelial progenitor cells (EPCs) that originate in the marrow and arrive via the circulation into the tumor stroma. In 2010, two groups simultaneously reported a fully unexpected finding: within human glioblastomas (and the xenografts derived from these tumors), a significant portion of the endothelial cells in the tumor-associated neovasculature descend from the genetically mutant tumor cells, more specifically arising via the differentiation of glioblastoma stem cells (see Section 11.7) that are present in these tumors (**Figure 13.24**). Indeed, the glioblastoma CSCs are known to exhibit an ability to spawn a diverse array of more differentiated progeny, which is reflected in their proper name—glioblastoma multiforme (see Figure 2.8A). This finding was most unanticipated, because glioblastomas derive from the embryonic neuroectoderm, whereas the endothelial cells arise ultimately from the mesoderm and thus are members of cell lineages very distant from those forming the bulk of the brain.

In fact, there is a second alternative source of tumor-associated microvasculature that had been recognized years earlier and has been termed **vasculogenic mimicry**. Thus, collections of neoplastic cells can form microvessels through which blood passes. These channels have the superficial appearance of capillaries but clearly are not lined by endothelial cells as usually defined. They may, however, express at least one protein that is a critical component of the endothelial cell-associated, normal microvessels: vascular-endothelial-cadherin (VE-cadherin), a member of the family of adhesion proteins that includes the widely expressed E-cadherin that is so critical to knitting together epithelial cell sheets (see Figure 13.14). Similarly, VE-cadherin enables the side-by-side associations of the endothelial cells in fully formed vessels, thereby creating a continuous tight barrier between the blood and parenchyma outside of the vasculature.

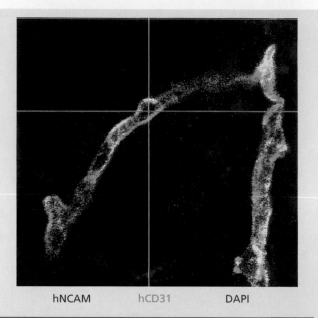

hNCAM   hCD31   DAPI

Figure 13.24 Transdifferentiation of glioblastoma stem cells into tumor-associated endothelial cells The characterization of the endothelial cells formed in the neovasculature of human glioblastomas (GBMs), including those that are grown as xenografts in immunocompromised host mice, has revealed that many of these cells, while expressing a human endothelial cell–specific marker (CD31, *green*), can also be stained with antibodies that are specifically reactive with human NCAM antigen (*red*), which indicates cells of neuroectodermal origin. Overlapping staining between these two colors is seen here as *yellow.* Moreover, these apparent endothelial cells can be shown to carry the same genomic alterations as are present in the bulk of the neoplastic cells in these GBMs and to derive from the differentiation of glioblastoma stem cells. Cell nuclei are stained *blue.* (From R. Wang et al., *Nature* 468:829–833, 2010. With permission from Nature.)

In more detail, monocytes from the myeloid lineage in the bone marrow enter into the general circulation, from which they are recruited by cancer cells into a tumor; once ensconced there, the monocytes are induced to differentiate into macrophages. This recruitment depends on attraction signals that are conveyed by **chemotactic** factors. By definition, these factors provide directional cues to motile cells rather than mitogenic stimulation. In the case of **leukocytes** (white blood cells such as monocytes), the relevant chemotactic factors are often called **chemokines**, that is, chemotactic cytokines.

The chemokine known as monocyte chemotactic protein-1 (sometimes called macrophage chemoattractant protein-1; MCP-1) is expressed in significant quantities by a wide range of neuroectodermal and epithelial cancer cell types. It appears to be a critical signal for attracting monocytes to some tumors and inducing their differentiation into macrophages. In other tumors, vascular endothelial growth factor (VEGF), colony-stimulating factor-1 (CSF-1; often called M-CSF, macrophage colony-stimulating factor), and the PDGF released by tumor cells also seem to help in this recruitment, while CSF-1 aids in stimulating the subsequent monocyte-to-macrophage differentiation (**Figure 13.25**).

Once established in the tumor stroma, macrophages play important roles in stimulating angiogenesis (**Figure 13.26**). Thus, in some cancers, such as breast carcinomas,

(A) *Csf1+/op*

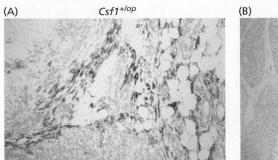

(B) *Csf1op/op*

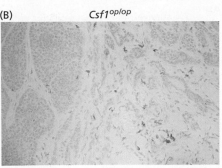

there is a direct correlation between the level of MCP-1 produced, the number of macrophages present, and the level of angiogenesis induced by the various tumors; in other tumors, the density of infiltrating tumor-associated macrophages (TAMs) is correlated with microvessel density (see Figure 13.26D); this density is gauged by counting the number of capillaries per microscopic field in a tumor section. Numerous reports suggest that macrophages play a major role in driving tumor progression by demonstrating a direct correlation between the presence of a high density of macrophages in tumor masses and a poor prognosis for cancer patients. This correlation has been documented for gliomas and for carcinomas of the breast, ovary, prostate, cervix, bladder, and lung.

Such evidence, however, is only *correlative*. More compelling evidence of the *causal* role of macrophages, at least in tumor angiogenesis, comes from experiments in which cancer cells are forced to express higher levels of MCP-1. The expression of this chemokine allows the cancer cells to attract more macrophages, which proceed to secrete important angiogenic factors, notably VEGF and interleukin-8 (IL-8), resulting in marked increases in angiogenic activity and the formation of more extensive tumor-associated neovasculature. In addition, when lung cancer cells are exposed *in vitro* to factors that have been secreted by cultured macrophages, the cancer cells respond by producing IL-8 and several other proteins that promote angiogenesis and cell invasiveness.

Hypoxic areas within tumors attract macrophages, which appear to tolerate hypoxia quite well. Once established in hypoxic regions of tumors, these inflammatory cells begin to secrete significant amounts of VEGF (see Figure 13.26A), which reduces the hypoxia by bringing in endothelial cells, capillaries, and thus oxygen-rich blood. Hypoxic areas of tumors often remain poorly vascularized, which indicates that the recruited macrophages, on their own, are unable to fully cure local defects in angiogenesis. Thus, in rapidly growing tumors, the macrophages and the vasculature that they induce cannot keep pace with the rate of tumor expansion, and large areas within a tumor mass eventually become **necrotic** (that is, filled with dead and dying cells) as a consequence (to be described in Section 13.6).

Like myofibroblasts, macrophages are adept at secreting matrix metalloproteinases (MMPs; see Table 13.1). MMP-9, prominently involved in cancer progression, is produced by tumor-associated macrophages (TAMs; see Figure 13.26E) and, once activated, proceeds to cleave a number of important protein substrates in the extracellular space. In certain invasive carcinomas, including those of the breast, bladder, and ovary, TAMs have proven to be the major source of this enzyme. It contributes to tumor progression by enhancing angiogenesis, by disrupting existing tissue structure, thereby carving out space for expanding tumor masses, and by liberating critical mitogens that have been immobilized through tethering to proteoglycans of the ECM. MMP-9 can also cleave IGFBPs (insulin-like growth factor–binding proteins), which normally sequester IGF molecules in the extracellular space (see Section 9.14). This cleavage liberates the IGFs, notably IGF-1, which then provide survival signals to nearby cells, including cancer cells. In advanced breast cancers, TAMs are also able to help the carcinoma cells directly because they are the major source of the epidermal growth factor (EGF) that drives the proliferation of EGF-R–expressing carcinoma cells.

Figure 13.25 Recruitment of macrophages by CSF-1 In some tumors, colony-stimulating factor-1 (CSF-1) released by tumor cells serves as the key attractant for monocytes and thus macrophages. In the transgenic mice studied here, mammary tumorigenesis was driven by a transgene, which was constructed from an MMTV (mouse mammary tumor virus) transcriptional promoter that drives expression of the polyoma middle T (PyMT) oncogene, doing so almost exclusively in mammary epithelial cells. (A) The tumor analyzed here arose in a transgenic MMTV-PyMT mouse that was heterozygous at the *Csf1* locus, because it carried one null allele of the CSF-1–encoding gene, and was therefore labeled *Csf1+/op*. The presence of macrophages in the stroma of a tumor is revealed by a monoclonal antibody that detects macrophages (*reddish brown*). (B) This tumor arose in an MMTV-PyMT transgenic mouse that was genetically *Csf1op/op* and therefore lacked all CSF-1 activity. Consequently, the tumor cells arising in this mouse were unable to release this chemoattractant, resulting in a mammary tumor in which the macrophages are barely detectable. The absence of macrophages did not affect the growth of the primary tumor but strongly suppressed progression to an invasive and metastatic state, implicating macrophages in such tumor progression. (From E.Y. Lin et al., *J. Exp. Med.* 193: 727–740, 2001. With permission from Rockefeller University Press.)

Figure 13.26 Macrophage involvement in angiogenesis (A) Macrophages play a major role in releasing mitogenic factors for carcinoma cells as well as reorganizing the tumor-associated stroma in order to facilitate angiogenesis and, in some tumors, carcinoma cell invasiveness. MMPs, matrix metalloproteinases. (B) In certain breast tumors (*left panel*), the cancer cells synthesize the potent angiogenic factor VEGF (vascular endothelial growth factor), revealed here (*dark areas*) by staining with an anti-VEGF monoclonal antibody. However, in other breast tumors (*right panel*), VEGF production originates in isolated macrophages within the tumor-associated stroma (*red arrows*). (C) In a variety of tumor types, including the pancreatic islet tumors of Rip-Tag mice (see Figure 13.35) analyzed here, macrophages (*green*) can be found in large numbers next to the tumor-associated neovasculature (*red*). Cell nuclei are stained *blue*. Yet other work (see Figure 13.36B) reveals that such macrophages, working together with recruited mast cells, contribute actively to the process of angiogenesis in the tumor stroma. (D) In a series of human non-small-cell lung carcinoma (NSCLC) specimens, each represented here by a point on the graph, the density of tumor-associated macrophages (TAMs) per microscope field has been plotted vs. the density of microvessels (i.e., capillaries). Such correlations add further weight to the notion that TAMs play a key role in fostering tumor-associated angiogenesis. (E) Macrophages in the stroma of a human colorectal adenocarcinoma produce matrix metalloproteinase-9 (MMP-9; *brown spots*), a key enzyme in angiogenesis that acts by liberating angiogenic factors from the ECM. (A, update courtesy of J. Joyce. B, from R.D. Leek et al., *J. Pathol.* 190:430–436, 2000. With permission from John Wiley & Sons. C, courtesy of V. Gocheva and J. Joyce. D, from J.J.W. Chen et al., *J. Clin. Oncol.* 23:953–964, 2005. With permission from American Society of Clinical Oncology. E, from B.S. Nielsen et al., *Int. J. Cancer* 65:57–62, 1996. With permission from John Wiley & Sons.)

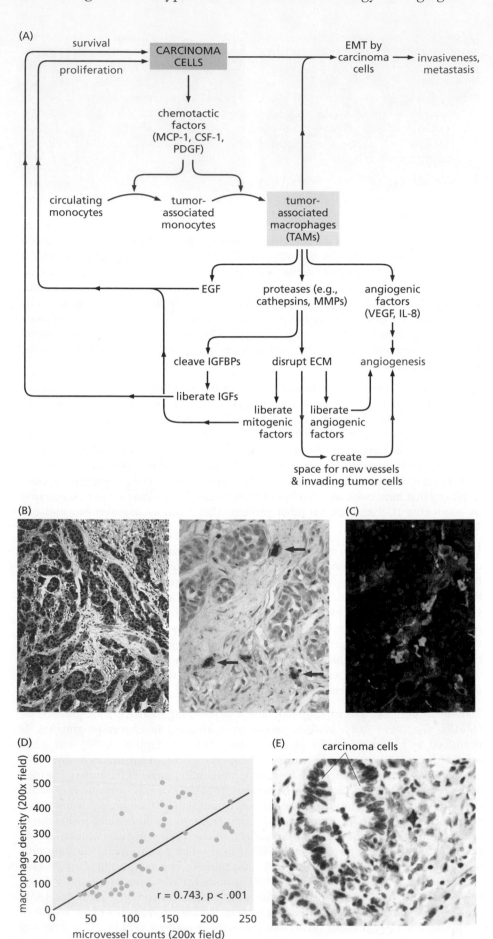

Although there is abundant evidence implicating some macrophages as active collaborators in tumor progression, it is also clear that another, distinct subset of macrophages, by acting as deputies of the immune system, can detect and kill cancer cells. Yet they clearly fail to do so in many types of tumors. Thus, some cancer cells acquire the ability to inactivate or blunt the **tumoricidal** (cancer-killing) actions of this second class of macrophages, while leaving intact the functions of the first type that are involved in aiding tumor progression. As we will learn in Chapters 15 and 16, these two classes of macrophages, termed M1 and M2, can either facilitate or blunt the immune attacks on tumors.

Yet another cell type in the myeloid lineage can also contribute to tumor invasiveness. *Immature myeloid cells* (iMCs) have been best characterized in mouse models of colorectal carcinogenesis. For example, when the genotype of human colorectal carcinomas is modeled in the mouse by inactivating both the *Apc* and *Smad4* genes (see Sections 7.10 and 11.3)—Smad4 being critical to TGF-β signal transduction (see Figure 6.22D)—the resulting adenoma and adenocarcinoma cells release the chemokine CCL9, which recruits iMCs originating in the bone marrow. The iMCs congregate at the invasive front of these tumors, secrete matrix metalloproteinases MMP-2 and MMP-9, and thereby facilitate invasion by the carcinoma cells into underlying tissue layers (**Figure 13.27**). Other research has shown that without MMP-2 and MMP-9, the iMCs cannot foster invasion and metastasis by the colorectal mouse tumors, and that the ability of cancer cells in the primary tumors to metastasize depends on recruitment of iMCs to these tumors.

Interestingly, a certain class of immature myeloid cells have been shown to play an important role in tumor immunology, as discussed in Chapters 15 and 16. In that

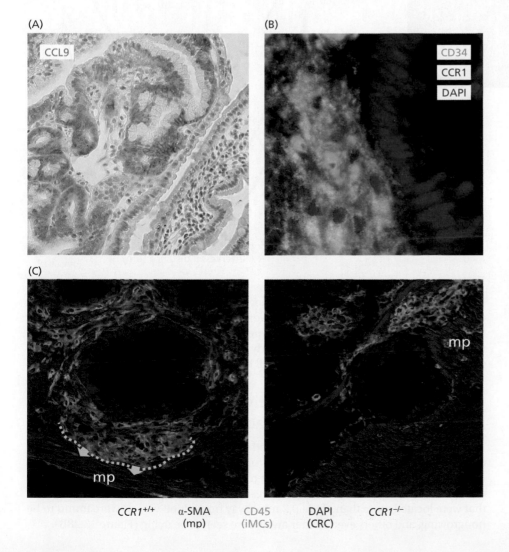

(A) CCL9

(B) CD34 / CCR1 / DAPI

(C)

CCR1+/+   α-SMA   CD45   DAPI   CCR1−/−
(mp)   (iMCs)   (CRC)

Figure 13.27 Immature myeloid cells promote colorectal carcinoma progression In a mouse model of colorectal (CRC) carcinogenesis, epithelial cells that have lost both Apc and Smad4 function—thereby mirroring the state of many human colorectal carcinomas (see Figure 11.11)—recruit a type of immature myeloid cell (iMC) of bone marrow origin into the tumor stroma. (A) The chemokine CCL9 (*brown*) is already released early in tumor progression by the cells in adenomas, which are precursors to frank carcinomas. Normal tissue, stained with DAPI (*light blue nuclei*), is seen in the *lower right*. CCL9 then acts as an attractant to recruit immature myeloid cells (iMCs) into the tumor stroma. (B) The cognate receptor of CCL9, termed CCR1 (*red-orange*), is displayed by many of the CD34+ iMCs that are being recruited into the tumor stroma (*green*). DAPI (*blue*) stains all nuclei. (CD34 is a marker of hematopoietic precursor cells.) (C) When displaying the CCR1 receptor of CCL9 (*left panel*), large numbers of CD45+ iMCs (*green*) congregate (*below*) near colorectal carcinoma cells (*blue nuclei*), and this mixed cell population begins to invade in a broad front (*green dotted line*) the underlying α-smooth muscle actin–expressing *muscularis propria* (mp, *red*). (CD45 is a marker of all types of hematopoietic cells.) However, when the gene encoding the CCR1 receptor is knocked out in the genome of the tumor-bearing host (*right panel*), such CD45+ iMCs are essentially absent and the depth of invasion is reduced 10-fold (*not shown*). The nature of the thin stripe of α-SMA-positive cells above the carcinoma cells is not known. (From T. Kitamura et al., *Nat. Genet.* 39:467–475, 2007. With permission from Nature.)

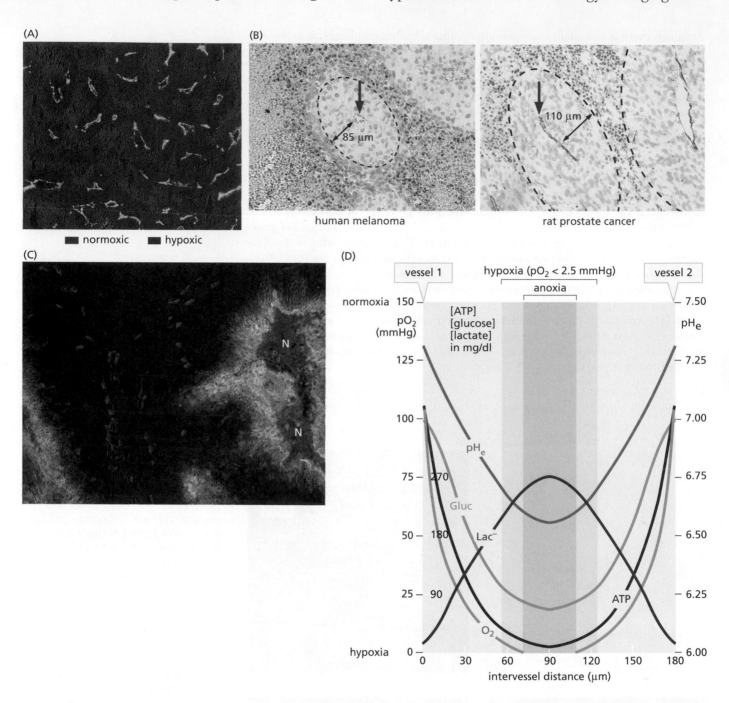

context, they are often termed **myeloid derived suppressor cells** (MDSCs) because of their strongly immunosuppressive powers that can be mustered by tumor cells to help protect themselves against various types of immune attack. As a class, immature myeloid cells represent a diverse and poorly characterized set of cell types whose precise contributions to tumor biology are still being explored.

13.6 Endothelial cells and the vessels that they form ensure tumors adequate access to the circulation

The most obvious stromal support required by tumors has already been cited repeatedly in this chapter: like normal tissues, tumors require access to the circulation in order to grow and survive. As early as the mid-1950s, pathologists noted that cancer cells grew preferentially around blood vessels (**Figure 13.28A**). Those tumor cells that were located more than about 0.2 mm away from blood vessels were found to be nongrowing, and others even farther away were seen to be dying (Figure 13.28B).

Figure 13.28 Hypoxia and necrosis of cells in poorly vascularized sections of tissues Various techniques have been used to demonstrate hypoxia and necrosis in regions of tissues surrounding capillaries. (A) This micrograph reveals capillaries (*green*) and the degrees of oxygenation in the surrounding tumor parenchyma. Immunostaining using an antibody reactive with EF5, a molecule that localizes to hypoxic regions of tissues, reveals hypoxic areas (*red*) at some distance from the capillaries, whereas those cells that are well oxygenated are unstained and therefore appear *dark brown*. (B) These sections of a human melanoma and a rat prostate carcinoma have been immunostained with an anti-CD31 antibody, which reveals capillaries (*red arrows*). Immediately around each capillary is a region of healthy cells (*inside the dashed line*), beyond which necrosis is apparent. The necrotic regions (*granular areas*) begin as close as 85 μm from the melanoma capillary (*left*) and 110 μm from the prostate carcinoma capillary (*right*). This necrosis reveals the limitations of diffusion in conveying oxygen and nutrients from capillaries to cells in the parenchyma of a tumor. (C) The dynamics of oxygenation on a larger scale are apparent in this human head-and-neck squamous cell carcinoma (HNSCC). Where blood vessels (*blue spots*) provide good oxygenation, the tumor appears

black. In contrast, in areas of poor vascularization and moderate hypoxia, a carbonic anhydrase enzyme is expressed (*red-orange*), whereas in areas of extreme hypoxia, the pimonidazole dye is detectable (*green*). Overlap between these two markers appears *orange*. Areas of necrosis, located even further away from the tumor vasculature, are indicated by "N." (D) Measurements of five parameters indicate the dramatic changes that occur with increasing distance from microvessels within tumors. Parameters have been calculated from two arterioles running in parallel, 180 μm apart. pH$_e$ represents the pH of extracellular fluid. These measurements in aggregate explain why cells that are situated at some distance from a microvessel experience a sub-optimal environment that often leads to their death by necrosis. (A, courtesy of B.M. Fenton. B, left, from C.M. Croce et al: Holland Frei Cancer Medicine, 9th ed. Hoboken, N.J.: John Wiley & Sons, 2017. With permission from John Wiley & Sons. B, right, from L. Hlatky, P. Hahnfeldt and J. Folkman, *J. Natl. Cancer Inst.* 94:883–893, 2002. With permission from Oxford University Press. C, from J.H. Kaanders et al., *Cancer Res.* 62: 7066–7074, 2002. With permission from American Association for Cancer Research. D, from P. Vaupel, *Semin. Radiat. Oncol.* 14: 198–206, 2004. With permission from Elsevier.)

We now realize that this threshold of approximately 0.2 mm represents the distance that oxygen can effectively diffuse through living tissues. Cells located within this radius from a blood vessel can rely on diffusion to guarantee them oxygen; those situated further away suffer from moderate or severe hypoxia and low pH (see Figure 13.28B–D). Tissues suffering from hypoxia are in danger of becoming necrotic as illustrated in **Figure 13.29**, an alternative form of cell death that does not depend on apoptosis. Apoptotic death triggered by p53 also threatens hypoxic cells (see Chapter 9), and the inactivation of the p53 signaling system often enables cancer cells to survive beyond the small perimeter surrounding each capillary. Newly arising tumors must negotiate these constraints, doing so in the context of the rules that govern normal angiogenesis (Supplementary Sidebar S13.6).

The existence and powers of angiogenic factors were first revealed by implanting small chunks of tumor on the cornea or the ears of laboratory animals such as

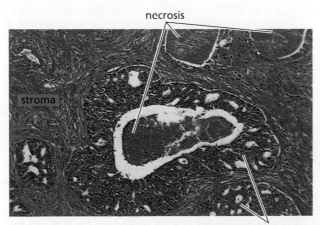

Figure 13.29 Large-scale necrosis within a tumor The inability of large sectors of a tumor to gain access to the vasculature in the stroma is evident in this high-grade adenocarcinoma of the prostate. As seen in the growth in the *center*, a nest of adenocarcinoma cells (*dark red*) is surrounded by extensive stroma (*light red*). In the middle of the nest of adenocarcinoma cells is a large clump of necrotic tissue (*light red*) that has shrunk away from the viable adenocarcinoma cells that surround it; the latter thrive because they have more direct access to the stroma and its associated vasculature. (From M.A. Weiss and S.E. Mills, Atlas of Genitourinary Tract Disorders, London: Gower Medical Publishing, Ltd., 1988. With permission from Elsevier.)

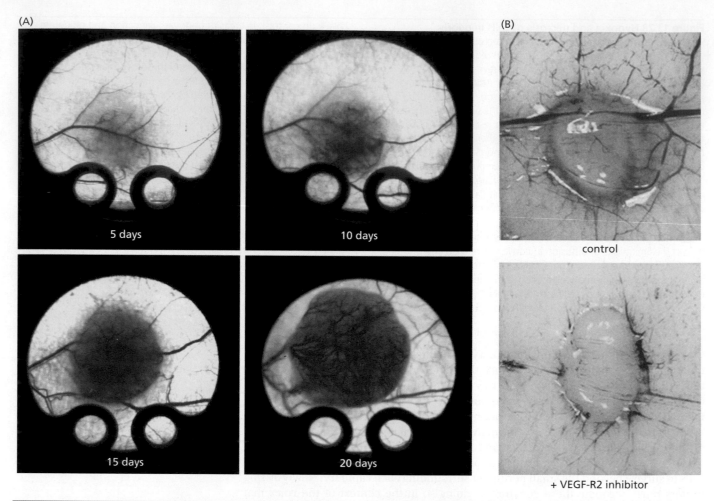

(A)

5 days

10 days

15 days

20 days

(B)

control

+ VEGF-R2 inhibitor

Figure 13.30 Recruitment of capillaries by an implanted tumor
(A) Shown is the growth of a small group of subcutaneously implanted human colorectal adenocarcinoma cells over a period of 20 days. The growth of the tumor-associated vessels (*red*) was observed through a window inserted above the tumor in the skin of the host mouse. (B) Such vascularization can be suppressed by antagonists of the VEGF receptor. Seen here are the effects of ZD6474 (also termed vandetanib), a low–molecular-weight inhibitor of the tyrosine kinase of the VEGF receptor-2, in mice bearing subcutaneously implanted human adenocarcinoma cell xenografts. The widespread *pink* background in the untreated control mouse is indicative of numerous capillaries, and the number of major vessels entering into a tumor mass (*above*) is strongly decreased in the presence of the drug (*below*). (A, from M. Leunig et al., *Cancer Res.* 52:6553–6560, 1992. B, from S.R. Wedge et al., *Cancer Res.* 62:4645–4655, 2002. Both with permission from American Association for Cancer Research.)

rabbits. Within days, dense networks of capillaries and larger vessels were seen to emerge from preexisting capillary beds and to converge on the implanted tumor chunks. Images like these (**Figure 13.30**) have strongly influenced our thinking about the behavior of tumors and their vasculature, because they demonstrate so vividly that tumors actively recruit blood vessels into their midst.

Angiogenesis is actually far more complicated than is suggested by the above discussion. Thus, this complex morphogenetic process involves many other factors in addition to the VEGFs, which clearly play a major role in attracting blood vessels to tumors. These other factors include several forms of TGF-β, basic fibroblast growth factors (notably FGF2), interleukin-8 (IL-8), angiopoietin, angiogenin, and PDGF. Also, several distinct cell types in addition to endothelial cells contribute to the construction of capillaries and larger vessels. Recall that the endothelial cells form the lumen of a capillary; these cells, in turn, are surrounded by the mural pericytes and vascular smooth muscle cells (see Figure 13.8A and B), which are absent from small lymph ducts.

The systematic covering of capillaries by pericytes seen in normal tissues can be contrasted with their chaotic dispersion near tumor-associated capillaries (**Figure 13.31**).

(A)

(B)

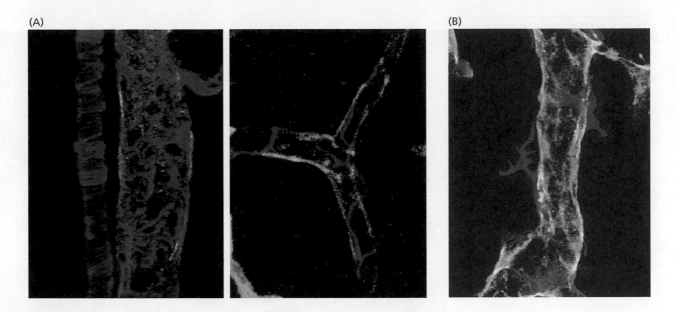

Figure 13.31 Pericytes in normal and tumor-associated vessels (A) This immunofluorescence micrograph shows that pericytes and smooth muscle cells (*both red-orange*) coat the outside of the tubes of endothelial cells (*green*), forming normally structured venules and arterioles (*left panel*). Pericytes and smooth muscle cells often cover these medium-sized vessels so completely that it is difficult to see the underlying endothelial cells. In contrast, in the smaller diameter capillaries (*right panel*), the pericytes are more sparsely disposed, but are nonetheless tightly attached to the endothelial cells. (B) This micrograph reveals the structure of a tumor-associated microvessel, in this case a vessel formed in a mouse by Lewis lung carcinoma (LLC) cells. The endothelial cells forming the lumen of this vessel (*green*) are only partially overlain by pericytes and smooth muscle cells (*red-orange*). The loose attachment of pericytes to the capillaries can be contrasted to their tight attachment seen in Figure 13.8A and B and panel A of this figure. (From S. Morikawa et al., *Am. J. Pathol.* 160:985–1000, 2002. With permission from American Society for Investigative Pathology.)

Capillaries in tumors typically have diameters that are three times greater than their normal counterparts. In addition, the overall layout of blood vessels around and within tumor masses is quite chaotic (**Figure 13.32**). Often vessels stop abruptly in dead-end pouches or circle back and attach to themselves.

The precise reasons that capillaries and larger vessels within tumors are so haphazardly constructed are unclear. One possible factor is likely attributable to the imbalance between two mutually antagonistic growth factors, angiopoietin-1 and -2. Whereas VEGF is responsible for initiating the growth of capillaries by attracting and stimulating endothelial cells, a mix of angiopoietin-1 and -2 induces endothelial cells to recruit the pericytes and smooth muscle cells that enable newly formed capillaries to mature into properly assembled vessels containing appropriate proportions of these three cell types. An imbalance of these two angiopoietins, together with the greatly elevated levels of VEGF within tumors, is suspected to be responsible for much of the defective construction of the vasculature within neoplasms. Indeed, some experiments indicate that most of the morphological abnormalities of tumor-associated microvessels can be mimicked simply by local overexpression of VEGF-A, the main pro-angiogenic factor.

At the submicroscopic level, it is also apparent that the capillaries in tumor masses are assembled haphazardly. This arrangement arises because the plasma membranes of adjacent endothelial cells do not contact one another to form a seamless lining around the capillary lumen but instead leave gaps, often several microns wide (**Figure 13.33A and B**), which allow direct access of the blood plasma to the cells surrounding the capillary. The resulting leakage (Figure 13.33C) seems to be responsible for the deposition of fibrin in the tumor parenchyma described earlier (see Figure 13.17).

Quantitative measures indicate that the walls of capillaries in tumors are about 10 times more permeable than those of normal capillaries. Much of this leakiness is attributable to the defective assembly of capillary walls noted here, which seems, in

Figure 13.32 The chaotic organization of tumor-associated vasculature
Different techniques of imaging vessels in living tissues (sometimes termed *intra-vital* imaging) reveal at different levels of resolution the starkly contrasting organization of normal and neoplastic vasculature. (A) Intra-vital multiphoton microscopy allows the imaging of living tissues at multiple depths, in this case, the microvasculature of an implanted human glioblastoma multiforme growing as a xenograft in the mouse brain (*upper left*). Note the absence of the well-organized hierarchy of major and minor vessels that is apparent in normal brain tissue. (B) Two-photon microscopy has generated this higher-resolution image of the detailed organization of normal vs. neoplastic microvessels. Note the differing overall organization of the microvascular beds in the normal versus tumor tissue as well as the increased diameters of the tumor-associated capillaries. (A, from B.J. Vakoc et al., *Nat. Med.* 15:1219–1223, 2009. B, from R.K. Jain, *Nat. Med.* 9:685–693, 2003. Both with permission from Nature.)

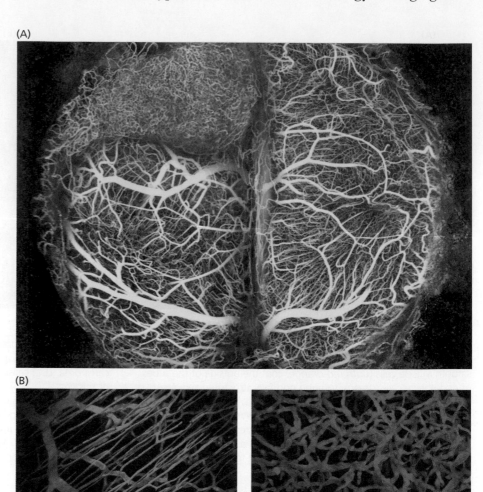

(A)

(B)

normal tissue   tumor

turn, to result from the deregulated production of VEGF by cancer cells and by macrophages and fibroblasts in the tumor-associated stroma. We know this because when an antibody that neutralizes VEGF is introduced into the circulation of tumor-bearing mice, the permeability of the walls of the tumor-associated capillaries is substantially reduced. In fact, a mysterious extracellular factor—vascular permeability factor (VPF)—was first described in 1979, but its biochemical nature only became clear a decade later with the cloning of its gene, which was thereafter termed the VEGF. This permeability of tumor-associated microvessels can have dramatic effects on other processes, such as certain forms of anti-cancer therapy (**Sidebar 13.6**).

Sidebar 13.6 Vessel permeability governs one childhood cancer treatment Medulloblastomas are childhood tumors of the cerebellum and are encountered in four forms in the pediatric oncology clinic, distinguished by various somatically mutated genes. Three forms are treated with various degrees of success, whereas the fourth type, termed WNT-medulloblastomas, is almost invariably curable. The cancer cells of this subtype, which exhibit hyperactive Wnt signaling caused by their expression of mutant β-catenin (see Figure 6.19B), secrete large amounts of Wnt inhibitors into the extracellular space. These inhibitors block normal Wnt signaling, which is critical for the proper formation of capillaries in the brain. Consequently, these tumors develop microvessels that carry many large gaps, termed **fenestrations**, that result in as much as a 300-fold increase in the transport of drugs from the circulation into the tumor parenchyma. This permeability of tumor-associated microvessels renders these tumors highly curable, in that relatively low drug concentrations that have few toxic side effects on the treated children succeed in eliminating these tumors, while having little effect on the other three forms of medulloblastoma. (The four types of medulloblastoma cells are equally susceptible to killing by cytotoxic drugs when propagated in tissue culture.)

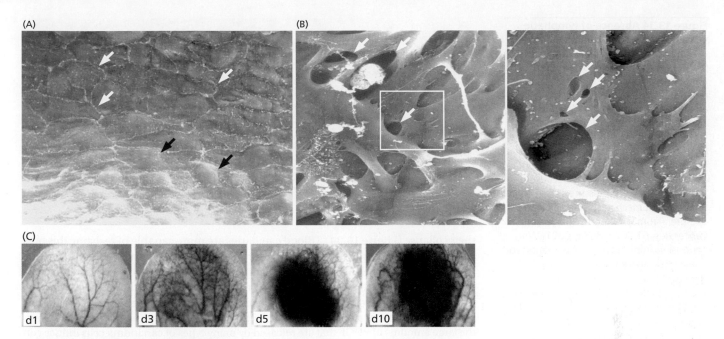

Figure 13.33 Gaps in the tumor microvasculature and resulting leakage (A) This scanning micrograph reveals the tight seals between endothelial cells forming the luminal wall of this venule in a normal mouse mammary gland; these cell–cell junctions, indicated here by *white arrowheads*, do not overlap and block essentially all leakage of fluid from the lumen of the microvessel into the tissue parenchyma. The *black arrows* indicate bulges in the luminal surface deriving from underlying endothelial cell nuclei (see Figure 13.8A). (B) This scanning electron micrograph reveals that the sheets of endothelial cells fail to form a continuous uninterrupted surface in the wall of a capillary within a tumor. Instead, the tumor-associated endothelial cells overlap one another and, as indicated here (*arrows*), show gaps of significant size between them. The box in the *left panel* is shown at higher magnification in the *right panel*. Such gaps permit the seepage of plasma fluids into the interstitial spaces between the cancer cells in the tumor parenchyma, contributing to the high hydrostatic pressure in these spaces. (C) The role of VEGF in inducing capillary permeability is illustrated here in an experiment in which an adenovirus gene expression vector encoding VEGF-A was infected into the ears of mice. At the indicated days thereafter, Evans Blue dye was injected into the general circulation of the mice and the ears were imaged 30 minutes later. With increasing time after initial VEGF-A expression, the microvessels became increasingly permeable, resulting in leakage of the dye molecules from the circulation into the parenchymal tissue of the ear. (A and B, from H. Hashizume et al., *Am. J. Pathol.* 156:1363–1380, 2000. With permission from American Society for Investigative Pathology. C, from J.A. Nagy et al., *Lab. Invest.* 86:767–780, 2006. With permission from Nature.)

As mentioned, the leakiness of tumor-associated capillaries leads to the accumulation of substantial amounts of fluid in the interstitial spaces within a tumor. In normal tissues, these fluids are drained continuously by the lymphatic vessels (see Figure 13.8C), which eventually empty their contents into the venous circulation (**Sidebar 13.7**). However, within solid tumors, the ongoing expansion of cancer cell populations exerts physical pressure on those few lymphatic vessels that do succeed in forming, causing

Sidebar 13.7 Endothelial cells also construct lymph ducts Lymph ducts have two major functions in normal physiology. They drain fluid from the interstices between cells and empty this fluid into the venous circulation. In addition, they allow antigen-presenting cells of the immune system to convey antigens from various tissues to the lymph nodes, where immune responses are often initiated (to be described in Chapter 15).

Interestingly, lymph ducts are assembled from endothelial cells originating in the same embryonic stem cell population that yields the endothelial cells of capillaries and larger blood vessels. During embryonic development, lymphatic vessels can often be observed to bud from developing capillaries before they separate and construct their own parallel network of interconnecting vessels (see Figure 13.8C). In addition, the factors that stimulate **lymphangiogenesis**—vascular endothelial growth factors C and D (VEGF-C and VEGF-D)—are homologous to VEGF-A and -B, which play a major role in stimulating the angiogenesis that creates the blood vasculature.

As might be expected, the receptor for VEGF-C and VEGF-D displayed by lymphatic endothelial cells—VEGF receptor 3 (VEGF-R3)—is structurally related to the dominant VEGF receptor of blood capillaries, VEGF-R2. In addition, there is clear evidence that VEGF-D, which is mainly responsible for driving lymphangiogenesis, may also help stimulate angiogenesis by binding and activating VEGF-R2. So these two systems—the blood and lymphatic networks—derive from common evolutionary roots, develop from common precursors in the embryo, and continue to interact with one another in complex ways within adult tissues. Of note, there are even more ancient evolutionary connections between the mechanisms guiding the growth of capillaries and those guiding nerve axon growth (to be described in Section 13.10.)

Figure 13.34 Absence of lymphatic vessels within solid tumors Analysis of a section of a hepatocellular carcinoma (HCC; liver cancer) reveals (via specific antibody staining) that lymph ducts (*dark red*) are present in the normal tissue above the tumor margin (*dotted line*) but are absent within the tumor mass itself. This absence may be attributed (1) to the lack of formation of these ducts during tumor growth or (2) to the collapse and degeneration of these ducts because of the high hydrostatic pressure within solid tumors; both mechanisms are likely to operate. (From C. Mouta Carreira et al., *Cancer Res.* 61:8079–8084, 2001. With permission from American Association for Cancer Research.)

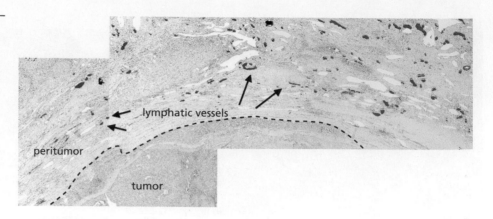

their collapse (**Figure 13.34**). (Blood capillaries are more capable of resisting this pressure because of their own significant internal hydrostatic pressure, which lymphatic vessels lack.) The resulting defective lymphatic drainage within the cores of solid tumors further exacerbates the elevated accumulation of fluid caused by capillary leakage, generating relatively high fluid pressure in the nonvascular parts of tumors.

Yet another mechanism conspires to maintain high hydrostatic pressure within tumors: the large amounts of PDGF that are released by many types of carcinoma cells induce contraction of stromal myofibroblasts, resulting in the squeezing out ("**expression**") of **interstitial** fluid. Hence, at least three mechanisms—lymphatic vessel collapse and resulting loss of fluid drainage, weeping by capillaries, and fibroblast contraction—contribute to high intra-tumoral hydrostatic pressure. This pressure, in turn, greatly complicates the administration of anti-cancer therapeutic drugs, because it prevents the formation of the steep pressure gradient leading from the capillary lumen down to the parenchyma that is needed for the efficient passive transfer of drugs from the circulation into the extravascular spaces of the tumor (Supplementary Sidebar 13.7).

13.7 Tripping the angiogenic switch is essential for tumor expansion

The descriptions of angiogenesis given above would seem to suggest that the release of angiogenic factors is virtually automatic: whenever groups of cells, including cancer cells, suffer hypoxia, they release angiogenic factors and thereby provoke the growth of capillaries into their midst. This sequence of steps cures the hypoxia and results in an appropriate density of capillaries in the tissue harboring these cells.

In fact, the ability to attract blood vessels seems to be a trait that many tumor cell populations initially lack and must acquire as tumor progression proceeds. This idea was first suggested by the observation, cited above, that in certain tumors, cancer cells thrive near capillaries, but those that are located further than 0.2 mm from capillaries stop growing and may enter apoptosis or become necrotic (see Section 13.6). So, even though these cancer cells experience hypoxia, they lack the ability to induce the formation of nearby capillaries.

We know much about these dynamics from detailed study of experimental models of tumorigenesis. The most informative of these has been the Rip-Tag transgenic mouse. It carries a transgene in its germ line, in which the concomitant expression of the SV40 large T antigen (see Sections 8.5 and 9.1) and small T antigen (see Sidebar 11.9) is driven by the promoter of the insulin gene. This promoter ensures expression of these viral oncoproteins in the β cells that form the islets of Langerhans in the pancreas (in which insulin is normally produced).

Tumor progression in the 400 or so islets of the mouse pancreas can be easily followed, because these islets can readily be distinguished from the surrounding tissue of the **exocrine** pancreas, which is involved in manufacturing and secreting digestive enzymes. As many as half of these islets in a Rip-Tag mouse form hyperplastic nodules by 10 weeks of age, and 8 to 12% of the hyperplastic islets eventually progress to become angiogenic, that is, they acquire the ability to recruit new blood vessels. By 12

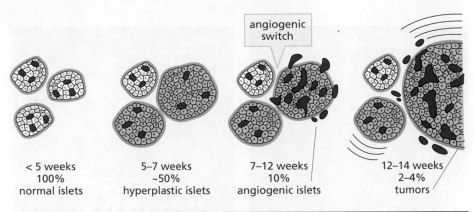

angiogenic
switch

< 5 weeks	5–7 weeks	7–12 weeks	12–14 weeks
100%	~50%	10%	2–4%
normal islets	hyperplastic islets	angiogenic islets	tumors

Figure 13.35 The Rip-Tag model of islet cell tumor progression Multi-step tumorigenesis in the Rip-Tag transgenic mouse model proceeds in distinct stages. Normal pancreatic islets (also known as islets of Langerhans; *left*) carry a small number of capillaries (*red*) to support, among their other cellular constituents, the β cells, which are responsible for synthesizing and secreting insulin into the circulation. By 5–7 weeks of age, about half of these islets become hyperplastic, but the density of blood vessels is not increased. At 7–12 weeks, about 10% of the hyperplastic islets become angiogenic, as indicated by the greatly increased density of capillaries in the surrounding, nearby exocrine pancreas. Finally, by 12–14 weeks of life, 2–4% of the initially formed hyperplastic islets have become invasive carcinomas that grow rapidly and invade the surrounding exocrine pancreas. The exocrine pancreas, in which these islets are embedded, is not illustrated here. (From D. Hanahan and J. Folkman, *Cell* 86:353–364, 1996. With permission from Elsevier.)

to 14 weeks, about 3% of the initially formed hyperplastic islets finally progress to form carcinomas (**Figure 13.35**).

Early in tumor progression in the Rip-Tag mice, the hyperplastic pancreatic islets begin to expand slightly to a small diameter of about 0.1–0.2 mm and then halt their forward march (see Figure 13.35), at least for a while. In these small nests of tumor cells, cell division continues unabated, being driven by the oncogenic SV40 transgene. However, the overall size of the tumor cell population within each islet remains constant because of a compensating attrition of cells occurring through apoptosis. This mouse model suggests that in humans, small tumor nests may also remain in such a dynamic but nongrowing state for many years, unable to break through the barriers that are holding them back.

In principle, a critical barrier constraining the expansion of the tumor cell nests might be a physical one—lack of adequate space within the tissue for these cancer cells to multiply. But detailed histological analysis of these small nests of cancer cells reveals something quite different. Because these cells have not yet become angiogenic, they lack vasculature. The resulting hypoxia that they experience triggers apoptosis, which explains their high rate of attrition. (It is possible that other sub-optimal conditions within these poorly vascularized cell nests, including inadequate nutrient supply, high levels of carbon dioxide and metabolic wastes, and low pH caused by lactic acid accumulation, also contribute to the death of these cells; for example, see Figure 13.28D).

At some point in time, however, small clusters of these pre-neoplastic islet cells suddenly acquire the ability to provoke neoangiogenesis (see Figure 13.35). Once these cells learn how to induce capillaries to form nearby, they and their descendants seem to be liberated from the major constraint that has been holding back their multiplication. This sudden, dramatic change in the behavior of the small tumor masses is said to reflect the operations of the mechanism termed the **angiogenic switch**.

These phenomena suggest an interesting idea, really a speculation: the body purposefully denies its cells the ability to readily induce angiogenesis. By doing so, the body erects yet another impediment to block the development of large neoplasms. According to such thinking, the angiogenic switch—a clearly important step in tumor progression—represents the successful breaching of this defensive barrier and the acquisition by cancer cells of a forbidden fruit: the ability to induce blood vessel growth at will.

Figure 13.36 The angiogenic switch and recruitment of inflammatory cells (A) The normal islet of Langerhans (*oval area, center of left panel*) is poorly vascularized and is sustained largely through diffusion from the microvessels surrounding it. Following tripping of the angiogenic switch, there is an increase in the number of cells within the islet, caused by tumor expansion and, as seen here, a dramatic induction of vessel formation with vessels now threading their way through the islet (*right panel*). Imaging was achieved by whole-mount microscopy of lectin-perfused vessels. (B) According to the scheme presented here, a pre-angiogenic, hyperplastic islet sends signals to the bone marrow or circulation, or both (*not illustrated*), that lead to the recruitment of mast cells and macrophages. Once in the vicinity of the islet, these cells release metalloproteinases, notably MMP-9. MMP-9 then cleaves extracellular matrix (ECM) components (*light orange, orange, lower right*), liberating vascular endothelial growth factor (VEGF) from its sequestered state. The solubilized VEGF proceeds to induce angiogenesis around the islet, thereby tripping the angiogenic switch. (A, courtesy of G. Bergers, D. Hanahan, and L.M. Coussens. See also G. Bergers, D. Hanahan and L.M. Coussens, *Int. J. Dev. Biol.* 42:995–1002, 1998. B, adapted from data in G. Bergers et al., *Nat. Cell Biol.* 2:737–744, 2000. Image of VEGF in B, from Y.A. Muller et al., *Proc. Natl. Acad. Sci. USA* 94:7192–7197, 1997. With permission from National Academy of Sciences, U.S.A.)

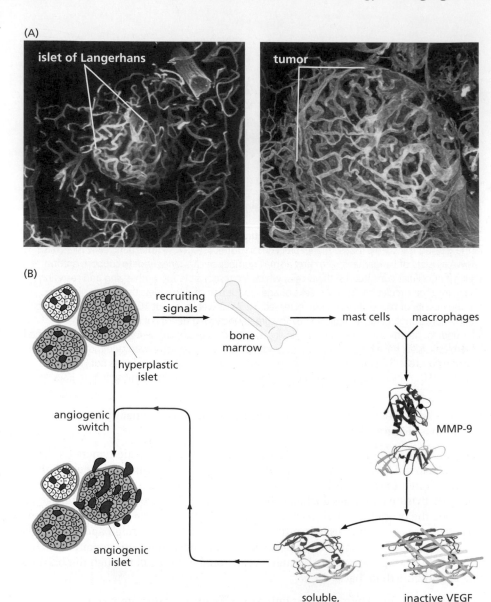

One might conclude that the angiogenic switch in these transgenic mice is driven by the premalignant β cells' suddenly acquiring the ability to express and release VEGFs. Actually, these cells have been making large amounts of VEGFs long before the angiogenic switch occurs, as do fully normal pancreatic islets. However, the VEGF molecules secreted by these β cells are efficiently sequestered by the surrounding extracellular matrix (ECM). As a consequence, the VEGF molecules are functionally inactive and thus unable to stimulate angiogenesis.

This sequestered state of the VEGF explains why the angiogenic switch in the Rip-Tag pancreatic islets is accompanied by the sudden appearance of substantial amounts of matrix metalloproteinase-9 (MMP-9; see Table 13.1). MMP-9 acts in a targeted fashion to cleave specific structural components of the ECM, thereby releasing VEGF for active signaling to nearby endothelial cells (**Figure 13.36**). This MMP-9 is synthesized and released by inflammatory cells—mast cells and macrophages—that have been attracted to the premalignant islets. Hence, in this particular tumor model, tripping of the angiogenic switch ultimately depends on an acquired ability to recruit inflammatory cells.

If the gene encoding VEGF is selectively deleted from the islet cells through genetic engineering, the islets survive and the early steps of tumor progression still proceed normally, but the angiogenic switch is never tripped. This observation reinforces the conclusion that VEGF molecules of islet cell origin play a critical role in triggering the onset of angiogenesis and that recruited stromal cells cannot compensate for this absence of VEGF by bringing in some of their own.

This scenario (see Figure 13.36B) involves heterotypic interactions among three distinct cell types: (1) the release of still-unidentified signals from the premalignant islet cells that recruit mast cells (**Sidebar 13.8**) and, quite possibly, macrophages; (2) the release of MMP-9 by these inflammatory cells to activate previously latent VEGF made by the premalignant islet cells; and (3) the proliferative response of endothelial cells to the activated VEGF. In fact, yet other cell types are likely to be partners in islet cell neoangiogenesis. Thus, the cells creating the walls around microvessels (formed by pericytes and smooth muscle cells) may be formed by progenitor cells originating in the marrow.

Importantly, the Rip-Tag model does not typify the angiogenic mechanisms occurring during the formation of all types of tumors. For example, in some tumors, angiogenesis appears to increase progressively, as if a controlling rheostat is gradually being turned, which contrasts with the behavior described here, in which a binary, on–off switch seems to be tripped.

Other tumors may depend on different angiogenic factors to provoke angiogenesis. For example, when transformed mouse embryonic stem (ES) cells are deprived of both copies of the VEGF-A–encoding gene, they lose almost all their power to make malignant teratomas. In contrast, transformed adult mouse dermal fibroblasts remain highly tumorigenic after they have been deprived of both copies of this gene. And the sarcomas generated by these transformed fibroblasts continue to grow even when the mice bearing them have been treated with an antibody that binds and inactivates VEGF-R2 (the primary endothelial cell receptor that confers responsiveness to both VEGF-A and VEGF-B). This lack of response is likely explained by the fact that transformed dermal fibroblasts can make a complex *mixture* of angiogenic factors—including VEGF-B, acidic, and basic fibroblast growth factors (aFGFs and bFGFs), and transforming growth factor-α (TGF-α). More generally, the deployment of multiple angiogenic factors (**Table 13.3**), often observed in advanced human cancers,

Table 13.3 Important angiogenic factors

Name	Mol. wt. (kD)
vascular endothelial GFs (VEGFs)	40–45
basic fibroblast growth factor (bFGF)	18
acidic fibroblast growth factor (aFGF)	16.4
angiogenin	14.1
transforming growth factor-α (TGF-α)	5.5
transforming growth factor-β1 (TGF-β1)	25
tumor necrosis factor-α (TNF-α)	17
platelet-derived growth factor-B (PDGF-B)	45
granulocyte colony-stimulating factor (G-CSF)	17
placental growth factor	25
interleukin-8 (IL-8)	40
hepatocyte growth factor (HGF)	92
proliferin	35
angiopoietin	70
leptin	16

Sidebar 13.8 A proof that mast cells from the bone marrow play a key role in the angiogenic switch Rip-Tag mice carrying the insulin SV40 transgene in their germ line can be bred with others that lack the Kit growth factor receptor. Kit serves as the receptor for stem cell factor (SCF), an important growth factor that triggers the development of certain subclasses of hematopoietic cells. Among other deficits, $Kit^{-/-}$ mice lack the ability to form mast cells. In Rip-Tag mice lacking Kit receptor, islet cell tumors are initiated at rates routinely observed in standard Rip-Tag mice. However, these tumors never succeed in becoming angiogenic, and the rate of cell proliferation in these small growths is compensated by an equal rate of apoptotic death. Consequently, these tumors remain very small (0.1–0.2 mm diameter).

If these mice are provided with a bone marrow transplant containing wild-type hematopoietic precursor cells, which cures their mast cell deficit, angiogenesis is initiated in the pancreatic islets, tumor cells in the islets are no longer lost through apoptosis, and large, life-threatening neoplasms appear soon thereafter. This outcome demonstrates that the ability to provoke angiogenesis in Rip-Tag tumors depends on the actions of at least one nonendothelial component of the tumor-associated stroma—the mast cells originating in the bone marrow.

Figure 13.37 Signaling through the basement membrane early in tumor progression (A) A human ductal carcinoma *in situ* (DCIS) of the breast contains a collection of carcinoma cells (*purple-blue*) that are noninvasive and therefore have not yet breached the basement membrane (BM, *arrows*) that surrounds them and separates them from the mammary stroma. As is apparent, this DCIS has nonetheless succeeded in transmitting angiogenic signals through the BM into the nearby stroma that have resulted in the growth of a small vessel (*dark brown*) that surrounds the tumor mass but does not penetrate into the tumor itself because of the continued integrity of the BM. (B) In this transgenic mouse model of skin carcinogenesis, the oncoprotein-encoding human papillomavirus (HPV) 16 early region is expressed under the control of a keratin-14 gene promoter, which ensures its expression specifically in keratinocytes of the skin (*upper panel*). In the early hyperplastic stage of tumor progression (*middle panel*), the noninvasive carcinoma cells on the epithelial side of the BM (*above*) are able to transmit signals through the BM (*dashed line*) that provoke angiogenesis on the stromal side (*below*), as evidenced by the increased density of capillaries (*red*), detected in this case through their display of the CD31 antigen. Dysplastic tissue, in which the BM (*dashed line*) has not yet been breached, shows even more intensive angiogenesis in the nearby stroma. (A, courtesy of A.L. Harris. B, from L.M. Coussens et al., *Genes Dev.* 13:1382–1397, 1999. With permission from Cold Spring Harbor Laboratory Press.)

(A)

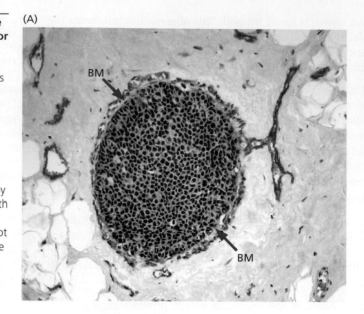

BM

BM

(B)

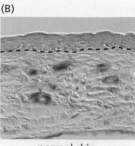

normal skin

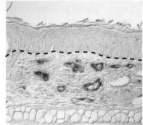

hyperplasia

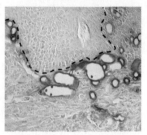

dysplasia

complicates the development of anti-angiogenesis cancer therapies. Moreover, certain classes of tumor-associated microvessels appear to arise independently of the VEGF-driven capillary formation.

13.8 The angiogenic switch initiates a highly complex process

Angiogenesis begins in the stroma surrounding the Rip-Tag tumors long before the basement membrane has been broken down. This behavior typifies that of many tumors in both mice and humans (**Figure 13.37**). Somehow, angiogenic signals are dispatched by benign cancer cells through the porous basement membrane (BM) in order to encourage increased angiogenesis on the stromal side of this membrane. Nevertheless, this early angiogenesis is circumscribed, and it is clear that intense angiogenesis can begin only when cancer cells become invasive, penetrate the basement membrane, and acquire direct, intimate contact with stromal cells (**Figure 13.38**). This suggests that tumor invasiveness and intense angiogenesis are often tightly coupled processes. We will study tumor invasiveness in detail in the next chapter.

In many human tumor types, the density of capillaries per microscope field increases in lockstep with increasing degrees of malignancy. For example, among human breast carcinomas that have already grown to a considerable size, those that have managed to attract dense networks of capillaries into their midst are indicative, on average, of a far worse prognosis than those that are poorly vascularized (**Figure 13.39A**). Moreover, patients whose breast tumors that express large amounts of VEGF (in addition to HER2; see Figure 4.3) fare especially badly following initial diagnosis and treatment (Figure 13.39B). Altogether, these correlations suggest that the angiogenic switch is only the first of many shifts that enable tumors to become progressively more

(A)

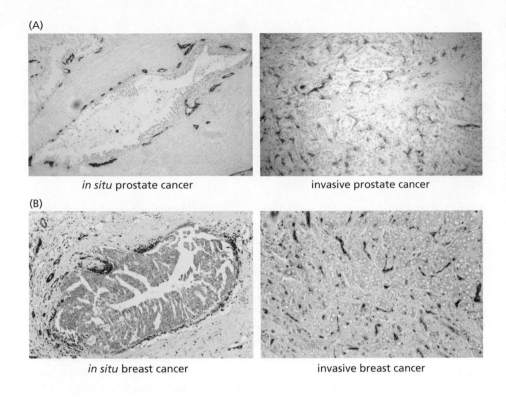

in situ prostate cancer   invasive prostate cancer

in situ breast cancer   invasive breast cancer

Figure 13.38 Angiogenesis before and after acquisition of invasiveness Tumor angiogenesis is circumscribed as long as human carcinomas remain benign. However, once they become invasive, the intensity of angiogenesis increases, leading to a higher density of capillaries (*brown*) threading their way through tumors. (A) The capillaries ringing a benign prostatic intraepithelial neoplasia (PIN) lesion (*left*) are far fewer than those in an invasive prostate carcinoma (*right*). (B) Similarly, those in a benign ductal carcinoma *in situ* (DCIS) of the human breast (*dark red, left panel*) are far fewer than those in an invasive ductal carcinoma (*right panel*). (Courtesy of J. Folkman.)

angiogenic and hence increasingly vascularized, enabling their highly aggressive behavior.

These striking correlations are actually susceptible to two alternative interpretations. It is possible that intense vascularization enables cancer cells to grow more aggressively, thereby leading to poor clinical outcomes. Alternatively, intense angiogenesis might be only a reflection of an underlying aggressive phenotype but not be causally involved in driving high-grade malignancy. Available clinical data do not allow a clear resolution between these alternatives.

We have spoken of the angiogenic switch without knowing which carcinoma cells within a tumor drive this shift; it might represent a concerted shift of all cells within a tumor or, alternatively, the actions of a small number of especially aggressive cells. In fact, analyses of cells isolated from explanted human tumors indicate great heterogeneity in the angiogenic powers of different subpopulations of cancer cells within a given tumor, with some cancer cells being highly angiogenic, whereas others are poorly so. Even within tumor cell lines, individual cancer cell subpopulations show greatly differing angiogenic and tumorigenic powers when engrafted *in vivo* (**Figure 13.40A**). The strong positive correlation between the microvessel density and

Figure 13.39 Clinical outcomes and the intensity of angiogenesis (A) Breast carcinomas were analyzed for the density of capillaries, which was determined as the number of microvessels per microscopic field. This Kaplan–Meier graph demonstrates that those patients whose tumors had a high microvessel count (*red curve*) had a markedly lower probability of disease-free survival in the 20 years following initial diagnosis than those whose tumors had a low microvessel count (*blue curve*). (B) As indicated earlier (see Figure 4.3), breast cancer patients whose tumors overexpress HER2/Neu have a markedly poorer prognosis than those whose tumors do not. In the group analyzed here, all patients showed metastatic cancer cells in one or more of the lymph nodes draining the breast. The differences in survival are even more dramatic when the levels of VEGF produced by their tumors are also considered. In this relatively small clinical study, more than 80% of patients whose tumors expressed low, basal levels of both HER2/Neu and VEGF (HER2⁻ VEGF⁻) were alive 8 years after diagnosis; in contrast, only about 35% of the patients whose tumors expressed elevated levels of both HER2/Neu and VEGF (HER2⁺ VEGF⁺) survived this long. (A, from R. Heimann and S. Hellman, *J. Clin. Oncol.* 16:2686–2692, 1998. With permission from American Society of Clinical Oncology. B, from G.E. Konecny et al., *Clin. Cancer Res.* 10:1706–1716, 2004. With permission from American Association for Cancer Research.)

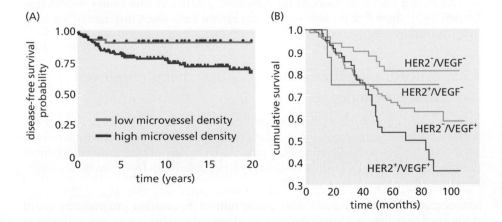

(A)

disease-free survival probability

— low microvessel density
— high microvessel density

time (years)

(B)

cumulative survival

HER2⁻/VEGF⁻
HER2⁺/VEGF⁻
HER2⁻/VEGF⁺
HER2⁺/VEGF⁺

time (months)

Figure 13.40 Heterogeneous degrees of vascularization within a tumor cell population (A) When a population of cells in a human liposarcoma cell line was subjected to single-cell cloning (in which all the cells in each resulting cell clone derive from a single common ancestor), the various cloned cell populations show greatly differing abilities to form tumors (*individual curves*) when implanted into immunocompromised host mice. (B) When the resulting tumors were analyzed for microvessel density (microvessels per microscopic field) and tumor volume, plotted logarithmically here, it appears that they have greatly differing angiogenic capabilities and that angiogenesis behaves as if it were a rate-limiting determinant of overall tumor growth. (From E.G. Achilles et al., *J. Natl. Cancer Inst.* 93:1075–1081, 2001. With permission from Oxford University Press.)

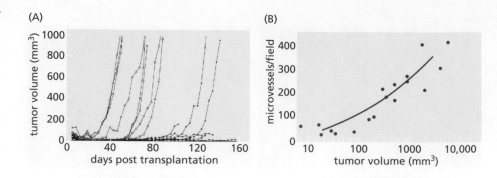

tumor volume (**Figure 13.40B**) provides additional evidence, albeit of a correlative nature, that neoangiogenesis is a critical driving force in tumor growth. Also, these observations introduce yet another idea into our thinking: that within a tumor mass, long after the tripping of the angiogenic switch, the weakly angiogenic cancer cells rely on help from their friends—their strongly angiogenic neighbors—in order to acquire adequate microvasculature.

The angiogenic switch is associated with processes that extend far beyond the immediate vicinity of a tumor. Thus, as mentioned earlier, neovascularization relies in part on the recruitment of endothelial progenitor cells (EPCs) that originate in the marrow and travel via the circulation to the tumor. In fact, VEGF released into the circulation by a tumor stimulates the production of EPCs in the bone marrow and their release into the general circulation; in addition, it helps to attract circulating EPCs to the tumor mass. Detailed analyses have shown that the proportion of endothelial cells derived from circulating EPCs (versus those arising from nearby capillaries) varies greatly from one type of tumor to another.

The importance of EPC recruitment is dramatically illustrated by the behavior of mice that lack one copy of the *Id1* gene and both copies of the *Id3* gene. These *Id* genes encode transcription factors regulating differentiation of a variety of cell types, often inhibiting development of tissue-specific differentiation programs (see Figure 8.25C). Many of these mutant mice survive to adulthood and show a very peculiar phenotype: they are defective in neoangiogenesis and will not permit several types of engrafted tumors to grow (**Figure 13.41A**).

The defect in these $Id1^{+/-}\ Id3^{-/-}$ mice can be cured by transplanting wild-type hematopoietic stem cells into their bone marrow. Alternatively, the subset of bone marrow stem cells whose production is normally spurred by high levels of circulating VEGF can be harvested from a wild-type mouse and transplanted into the bone marrow of $Id1^{+/-}\ Id3^{-/-}$ mice. These mice soon generate large numbers of endothelial precursor cells (EPCs) in their circulation and, following implantation of cancer cells, develop rapidly growing tumors. Taken together, these observations indicate that these tumors rely on VEGF to mobilize EPCs from the bone marrow (Figure 13.41B) and to recruit the resulting circulating cells into the tumor stroma. These findings, as well as others demonstrating the recruitment of myofibroblast precursors into tumor stroma (see Section 13.3), show that primary tumors can extend their reach throughout the body long before they metastasize.

13.9 Anti-angiogenesis therapies have been employed to treat cancer

In finely tuned physiological processes, the actions of positive effectors must be counterbalanced by negative regulators. We have read much about the positive effectors of angiogenesis, such as VEGF and bFGF, as well as a variety of others that conspire to promote the assembly of normal vasculature (see Table 13.3). However, their antagonists have remained offstage until now.

Anti-angiogenic mechanisms clearly play key roles in the creation and maintenance of the normal vasculature. During the process of wound healing, for instance, the burst

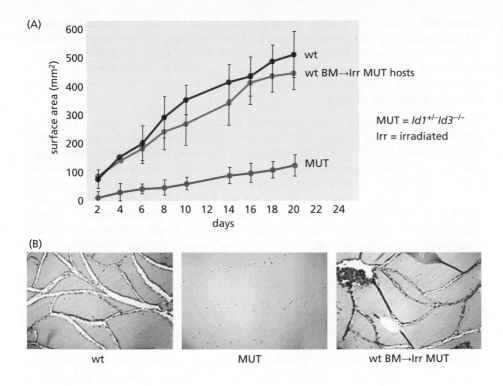

(B)

wt   MUT   wt BM→Irr MUT

Figure 13.41 Defective tumor angiogenesis in *Id1⁺ᐟ⁻ Id3⁻ᐟ⁻* mice
Mutant mice (Id MUT), which lack three of four *Id* gene copies (and have an *Id1⁺ᐟ⁻ Id3⁻ᐟ⁻* genotype), show impaired mobilization of endothelial precursor cells (EPCs) from the bone marrow and thus impaired recruitment of circulating EPCs into tumors that these mice may carry. (A) Wild-type (wt) mice bearing Lewis lung carcinoma (LLC) cells develop rapidly growing tumors (*red curve*), whereas those of the mutant strain (Id MUT) are unable to support vigorous tumor growth (*brown curve*). This defect is essentially reversed if the bone marrow (BM) of the mutant mice is eliminated by irradiation and replaced with transplanted wild-type bone marrow (*blue curve*, wt BM → Irr MUT), indicating that the recruitment of bone marrow–derived cells is defective in the mutant mice and is responsible for the inability of the tumor to grow in them. (B) This defective recruitment can be further localized by the use of plugs of Matrigel (an extracellular matrix material) that have been impregnated with VEGF and are then implanted subcutaneously. These plugs are able to recruit neovasculature when implanted in wild-type mice (*left*) but not in the mutant mice (*center*). However, if the mutant mice receive a graft of wild-type bone marrow cells, the defect is cured and now angiogenesis within these plugs is comparable to that seen in wild-type mice (*right*). (From D. Lyden et al., *Nat. Med.* 7:1194–1201, 2001. With permission from Nature.)

of angiogenesis that is required to repair the wound site must be shut down once the newly formed capillaries have reached a density that suffices to support normal tissue function. This shutdown is achieved, at least in part, by reducing the ongoing hypoxia-induced production of VEGF. In addition to the decline of pro-angiogenic factors, there are also a host of naturally anti-angiogenic proteins that reside within normal tissue and aid in actively shutting down angiogenesis. Their origins and mechanisms of action are described in detail in Supplementary Sidebar 13.8. At least two dozen anti-angiogenic molecular species involved in regulating various steps of vascular morphogenesis have been identified, as described in this Supplementary Sidebar.

The equilibrium created by the counterbalancing pro- and anti-angiogenic factors in each tissue ensures that tissues are supplied with the proper density of microvessels required to sustain their physiological activities. For those researchers intent on developing new types of anti-cancer therapeutics by inhibiting neoangiogenesis, this complexity offers multiple targets for intervention. Thus, highly targeted therapies may be devised to inhibit the several cell types that participate in angiogenesis as well as the multiple signaling channels through which they intercommunicate. Because tumors depend absolutely on angiogenesis to grow above a certain size (≯0.2 mm diameter; see Figure 13.28), any successes in blocking angiogenesis or in undoing the products of angiogenesis should represent a highly effective strategy for treating cancer. Microscopic tumors should be prevented from growing larger, and larger tumors should collapse once their already-established blood supply disintegrates.

In principle, anti-angiogenesis therapies have a major advantage over those directed at the neoplastic cells within tumors. As we will learn in Chapter 17, one of the great frustrations of anti-cancer drug development is that, sooner or later, tumors that initially respond to a drug treatment become **refractory** (resistant) to further treatment by the drug. Almost always, these relapses can be traced to the emergence of drug-resistant genetic variants within tumor cell populations (see Figure 11.16); these variants arise at an almost-predictable frequency and proceed to proliferate and regenerate aggressively growing tumor masses. The emergence of these drug-resistant variants seems to be one of the consequences of the highly unstable genomes of cancer cells (described in Chapter 12) and their resulting ability to spawn mutants at high frequency.

In contrast, many anti-angiogenesis therapies are directed at killing the genetically *normal* cells that have been recruited into tumor masses and co-opted by the cancer cells to do their bidding. There is every reason to believe that the endothelial cells within tumors possess normal, stable genomes and are therefore stable phenotypically. Hence, drug therapies directed against these cells—attractive targets of anti-angiogenic therapies—are unlikely to select for the outgrowth of drug-resistant genetic variants, and tumors should not, in theory, become refractory to anti-angiogenic drug therapy.

This interest in treating tumor-associated endothelial cells is further heightened by the peculiar biology of these cells. They are continually being formed and lost within tumor masses, with lifetimes measured as short as a week, whereas their counterparts that line normal blood vessels elsewhere in the body rarely divide and have lifetimes that are measured in hundreds of days, some being as long as seven years. Cycling cells (that is, cells racing around the active cell cycle) are, almost always, far more sensitive to drug-induced killing than are quiescent cells. For this reason, cytotoxic therapies directed against endothelial cells should have drastic effects on the tumor-associated neovasculature, while leaving blood vessels located elsewhere in the body unscathed.

An experiment reported in 1991 provided one of the earliest indications of the promise of another type of anti-angiogenic therapy. In this work, cancer cells were engineered to release a modified basic fibroblast growth factor, which greatly increased their tumorigenicity in mice, simply because this bFGF strongly enhanced the angiogenic powers of these cells. Use of a monoclonal antibody that specifically bound and neutralized this modified bFGF (but had no effect on the endogenous bFGF of the mouse) blocked the angiogenesis by tumor cells and led to a dramatic reduction in tumor volume. Within two years, a similar experiment was performed with an anti-human–VEGF monoclonal antibody. It succeeded in blocking the proliferation of two human sarcoma cell lines as well as a glioblastoma in Nude (immunocompromised) mouse hosts.

An even more effective version of this therapy came a decade later with the use of a monoclonal antibody that binds and neutralizes VEGF-A, which blocks new vessel formation and destabilizes recently formed ones. This antibody, termed variously Avastin or bevacizumab, showed significant efficacy in certain large-scale clinical trials. For example, patients with metastatic colon carcinoma who were treated with this antibody plus chemotherapy (the drug 5-fluorouracil) survived, on average, four months longer than patients treated with chemotherapy alone, and the addition of Avastin to conventional chemotherapy extended the survival of patients with non-small-cell lung carcinoma (NSCLC) by about two months. Similarly, Avastin could retard the progression of renal cell carcinomas in patients, but in the end had no effect on their long-term survival. It is plausible that the widely observed synergistic effects of Avastin with conventional chemotherapeutic drugs derive from the ability of this VEGF-A inhibitor to normalize tumor-associated vasculature (see Supplementary Sidebar 13.7), thereby greatly facilitating the delivery of drugs to the tumor parenchyma. By 2020, 3.5 million cancer patients worldwide had received this monoclonal antibody as part of their therapy, a stunning testimonial to its widely recognized powers in treating diverse types of human tumors. For many of the cancer types where it is currently being employed, addition of Avastin to existing approved chemotherapeutic regimens leads to a significant slowing of disease progression. It often extends survival, but it does not significantly increase overall survival.

The most informative preclinical studies of angiogenic inhibitors have come from using synthetic receptor inhibitors in the transgenic Rip-Tag model of pancreatic islet tumorigenesis described in Figure 13.35. Two types of low–molecular-weight synthetic compounds have been used in attempts to block various stages of islet tumor progression (**Figure 13.42**). One of these drugs is directed against the VEGF-R2, which is the main receptor driving angiogenesis. By inhibiting the tyrosine kinase of this RTK, the agent termed SU5416 should mimic the effects of Avastin, that is, both should be able to shut down VEGF signaling. Drugs of a second class, such as an agent termed SU6668, are directed primarily against the tyrosine kinase of the PDGF receptor. Recall from Section 13.1 that the role of PDGF in recruiting pericytes and smooth muscle

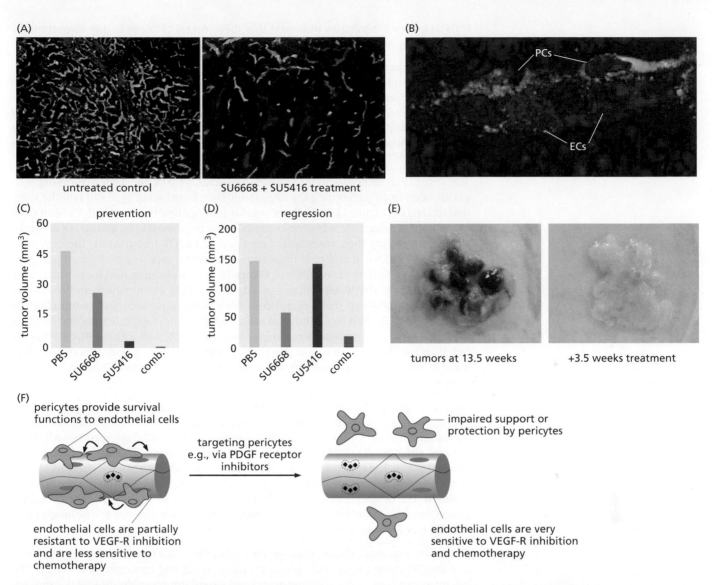

(A) untreated control   SU6668 + SU5416 treatment

(B) PCs   ECs

(C) prevention — tumor volume (mm³): PBS, SU6668, SU5416, comb.

(D) regression — tumor volume (mm³): PBS, SU6668, SU5416, comb.

(E) tumors at 13.5 weeks   +3.5 weeks treatment

(F) pericytes provide survival functions to endothelial cells

targeting pericytes e.g., via PDGF receptor inhibitors

impaired support or protection by pericytes

endothelial cells are partially resistant to VEGF-R inhibition and are less sensitive to chemotherapy

endothelial cells are very sensitive to VEGF-R inhibition and chemotherapy

Figure 13.42 Angiogenesis inhibitors as treatments of islet cell carcinogenesis The Rip-Tag transgenic mouse model of pancreatic islet cell carcinogenesis (see Figure 13.35) makes possible the testing of anti-angiogenic pharmacological inhibitors. (A) The treatment of mice bearing established angiogenic islet tumors (*left*) for 4 weeks with the PDGF receptor inhibitor SU6668 plus the VEGF-R inhibitor SU5416 results in regression of the vasculature (*right*). The lumina of the capillaries are labeled *light green* and the associated pericytes are labeled *red*. (B) The actions of the SU6668 PDGF-R inhibitor could be traced specifically to its effects on the pericytes, which act as support cells for the endothelial cells. In this image, an antibody reactive with the PDGF-R reveals that the only cells within the islets expressing this receptor (*green*) are the pericytes (PCs) that are closely associated with the endothelial cells (ECs, *red*). Nuclei are stained *blue*. (C) Although neither the SU6668 PDGF-R TK inhibitor (*blue bar*) nor the SU5416 VEGF receptor TK inhibitor (*red bar*) was able, on its own, to fully prevent the formation of such tumors (being applied before tumors became apparent), the two agents applied in concert (combined, *brown bar*) succeeded in doing so. PBS, phosphate-buffered saline control (*green*). (D) In an attempt to cause regression of already-formed tumors, SU5416 and SU6668 were introduced either singly or in combination into 12-week-old Rip-Tag mice and the size of tumors was measured 4 weeks later. Although the SU5416 VEGF-R

antagonist on its own (*red bar*) showed only minimal reduction of tumor volume, the SU6668 PDGF-R antagonist showed greater effects (*blue bar*), and a combination of the two applied together (*brown bar*) functioned synergistically to reduce overall tumor volume by approximately 85%. This indicates, once again, that antagonists of the endothelial cells (which depend on the VEGF-R) together with antagonists of the supporting pericytes (which depend on the PDGF-R) can act synergistically in anti-angiogenic therapy. (E) When 13.5-week-old mice with substantial pancreatic islet tumors (*red growths, left*) are treated for another 3.5 weeks with this combination therapy, their tumors largely regress (*right*). (F) A schematic summary of these and other observations indicates that endothelial cells depend on closely apposed pericytes for various types of biological support. Inhibition of PDGF signaling causes dissociation of pericytes from endothelial cells and renders the latter sensitive to various types of subsequent therapy, including inhibition of VEGF-R function. This emphasizes that anti-angiogenesis therapy is most effective when two synergistically acting treatments are applied. (A, B, and E, from G. Bergers et al., *J. Clin. Invest.* 111:1287–1295, 2003. With permission from American Society for Clinical Investigation. C and D, courtesy of D. Hanahan. F, from K. Pietras and D. Hanahan, *J. Clin. Oncol.* 23:939–952, 2005. With permission from American Society for Clinical Oncology.)

cells to growing capillaries indicated that these mural cells are highly important for consolidating and strengthening recently formed capillaries.

SU5416, the anti-VEGF-R agent, was able to block 90% of the early-stage, dysplastic islets from undergoing the angiogenic switch, thereby holding them to a small size and a noninvasive state (see Figure 13.42C). However, it had no effect on late-stage, well-established tumors, which continued to progress despite its presence in equally high concentrations. Hence, in the early stages of angiogenesis, VEGF signaling plays a critical role, whereas later on, this process seems to become increasingly independent of VEGF.

The anti-PDGF receptor agent, SU6668, had a far weaker effect on preventing dysplastic islets from undergoing the angiogenic switch, reducing by about half the islets that did so (see Figure 13.42C). But it was far more potent than the anti-VEGF-R drug in treating the late-stage, advanced tumors, reducing their size by about half and substantially reducing their vascularity (see Figure 13.42D). Importantly, the only cells expressing PDGF-R in and near these tumors were the capillary-associated pericytes and related smooth muscle cells (see Figure 13.42B), indicating that these mural cells were the targets of SU6668 action. Indeed, microscopic examination confirmed that SU6668 had prevented these mural cells from associating with and reinforcing the capillary tubes formed by endothelial cells. Together, these experiments showed that the initial steps of angiogenesis could proceed reasonably well without PDGF receptor function, but that later in tumor progression, PDGF signaling, and thus the involvement of mural cells, became increasingly important to angiogenesis and growth of the tumor masses.

Combination therapy using the two agents proved to be a highly potent way of intervening at various stages of tumor formation (see Figure 13.42C and D). Thus, simultaneous inhibition of both VEGF and PDGF receptors prevented angiogenic switching, holding virtually all (>98%) islets at a pre-angiogenic stage. This drug combination blocked the further expansion of small, already-established angiogenic tumors by 90% and caused an approximately 80% regression of large tumors, as did another PDGF-R inhibitor, the drug Gleevec (discussed in Chapter 17). In some mice, these combination drug therapies held already-formed tumors to a small size for as long as two months (see Figure 13.42E).

These observations indicate that the most effective ways of inhibiting angiogenesis and thus blocking tumor progression are likely to depend on targeting several of the cell types that construct the tumor-associated vasculature (see Figure 13.42F). Indeed, targeting only the endothelial cells yielded only a temporary regression of microvessels: when anti-VEGF therapy is halted, these vessels rapidly regenerate, often growing back in the hollow sleeves formed by the basement membranes and pericytes that are left behind after endothelial cells have succumbed to anti-VEGF therapy.

Of additional interest, neither of the agents described above—neither SU5416 nor SU6668—is currently being used clinically, having been abandoned because of limited efficacy in phase II or phase III clinical trials and showing once again the limited powers of mouse models of cancer pathogenesis to predict human clinical responses. Still, the very real clinical benefits afforded by Avastin treatment demonstrate unequivocally that tumor growth can be constrained by treating the tumor-associated stroma rather than the neoplastic cells themselves. Since these successes, a number of other anti-angiogenic therapies have been approved (Table 13.5). Moreover, the actions of some of the naturally occurring angiogenesis inhibitors described in Supplementary Sidebar 13.8 have been tested in various pre-clinical and clinical research programs, as described in Supplementary Sidebar 13.9. Of note, a number of the therapies listed in Table 13.5 have additional, alternative modes of action in addition to their demonstrated anti-angiogenic effects. In addition, it is worth noting that oncologists have unknowingly employed a treatment strategy—radiotherapy—whose clinical effects may have resulted from anti-angiogenic mechanisms (Supplementary Sidebar 13.10). As attractive as these anti-angiogenic strategies are, it is clear that tumors can develop strategies to resist these therapies and generate clinical relapses in patients whose neoplasms were initially responsive (Sidebar 13.9).

Sidebar 13.9 Tumors may outsmart even the best anti-angiogenic therapies Much of the allure of anti-angiogenic therapies comes from the likelihood that the cells being targeted, notably the endothelial cells, are unlikely to generate drug-resistant variants. The complexity of heterotypic signaling may, however, allow tumors to circumvent even the cleverest anti-angiogenic therapies. Imagine, for example, that we devise a strategy to block the VEGF signaling that is so critical for the formation of new vessels in a tumor and their subsequent maintenance (because VEGF is required for endothelial cell survival). A tumor treated in this way should rapidly collapse because its capillary beds will disintegrate. Indeed, just such a response has been observed in Rip-Tag mice (see Figure 13.42) that developed pancreatic islet tumors: when treated with an anti-VEGF-R2 antibody, their tumors regressed by more than 50%. However, the residual, surviving tumor cells responded by spawning variants that acquired the ability to produce elevated levels of fibroblast growth factors (FGFs), which are also potent angiogenic factors (see Table 13.3). These growth factors then supplanted VEGF as the main conveyor of signals from the islet tumor to the endothelial cells and succeeded in triggering the regeneration of vasculature and, in turn, the rebirth of vigorously growing tumors. Accordingly, truly effective anti-angiogenesis strategies will require the inhibition of multiple angiogenic pathways.

Anti-angiogenesis therapies are also attractive because of their potential selectivity for killing the proliferating endothelial cells within tumors, since many blood vessels in these growths seem to be in a constant state of formation and collapse and are therefore susceptible to cytotoxic drugs that may leave the quiescent endothelial cells in normal tissues untouched. For example, dramatically contrasting growth states of normal and tumor-associated blood vessels can be seen in many mouse models of cancer, in which engrafted tumors expand rapidly. However, it is possible that many of the vessels in slowly growing human tumors may exist for years and therefore may have ample time to consolidate and mature into robust, well-structured vessels in which the endothelial cells turn over slowly. These tumor-associated vessels may then be as resistant to anti-angiogenic drugs as the normal blood vessels elsewhere in the body. Yet another possible source of resistance to existing anti-angiogenic therapies may derive from mechanisms that tumors have to assemble microvessels from cells other those used conventionally by tumors, as described in Sidebar 13.5. These and other mechanisms of acquired resistance to anti-angiogenesis therapies are summarized in Table 13.4.

Table 13.4 Alternative mechanisms of acquired resistance to anti-angiogenesis therapies

EC radioresistance: Hypoxic activation of HIF-1 renders ECs resistant to irradiation.

Vascular mimicry: A fraction of tumor vessels are lined by malignant cells and are thus unresponsive to anti-angiogenic agents. Similarly, certain types of tumor stem cells may differentiate into endothelial-like cells (see Figure 13.23).

Angiogenic switch: The outgrowth of tumor cell clones expressing elevated levels of certain angiogenic factors may be naturally favored at advanced stages of tumor progression or may emerge in response to anti-angiogenic treatment, e.g., upregulation of (1) PlGF, VEGF-C, or FGF-2 in response to VEGF inhibition; (2) VEGF after VEGF-R or EGF-R inhibition; and (3) IL-8 after HIF-1 inhibition.

Vascular independence: Mutant tumor cell clones (e.g., those lacking p53) are able to survive in hypoxic tumors.

Stromal cells: VEGF$^{-/-}$ tumors recruit pro-angiogenic stromal cells such as myofibroblasts via upregulation of PDGF-A and myeloid cells.

Mature, robust vessels: Preexisting vessels are covered by a full complement of supporting pericytes and are not readily pruned by anti-angiogenic treatments.

Bone marrow–derived cells: Tumors or ischemic tissues recruit pro-angiogenic endothelial cells and inflammatory cells independent of VEGF; the recruited cells may produce several pro-angiogenic molecules to rescue vascularization upon VEGF blockage.

RTK inhibitors: These may not synergize with chemotherapy, possibly because they do not block the actions of neuropilin-1.

Abbreviations/definitions: EC, endothelial cell; IL-8, interleukin-8, an angiogenic cytokine; neuropilin, an alternative VEGF receptor; PlGF, placental GF, an angiogenic factor related to VEGFs and a ligand of PlGF-R.

13.10 Nervous tissue contributes to tumor growth

During the second decade of the new millennium a new set of actors emerged on the tumor microenvironment stage: nervous tissue was found to contribute in significant ways to the progression of certain types of solid tumors. Until the arrival of these actors the TMEN was assumed to be formed entirely by cells of mesodermal origin, at least in carcinomas, that is, connective tissues cells of various types and derivatives of the various hematopoietic lineages spawned in the bone marrow.

With the comfort of hindsight, it should have been clear that nervous tissue of various sorts must participate in various ways in the formation of neoplastic tissues. After all, tumors rely on the morphogenetic programs used to construct normal tissues for their own growth, and during the formation of almost all normal tissues, blood vessels and nerves grow in lockstep. Indeed, careful examination of the molecules that govern neurogenesis and angiogenesis reveal a clear common evolutionary origin and the use of homologous signaling systems to guide the growth of both types of tissue.

Table 13.5 Anti-angiogenic drugs approved for clinical use—2018[a]

Treatment	Target/mechanism	Cancer type
Avastin (Bevacizumab)	MoAb against VEGF	OvCa, CRC, RCC, BrCa, prost., NSCLC, GBM
EYLEA (Aflibercept)	VEGFR1/VEGR2 fusion	CRC, prost., NSCLC, SCLC
Erbitux (Cetuximab)	MoAb against EGFR	CRC, gastr., HNSCC
Endostar (endostatin)	endog. inhibitor	NSCLC, mel. NPC, CRC
Nexavar (Sorafenib)	inhib. of VEGFR, PDGFR, Raf	HCC, thyr., OvCa, RCC
Sunitinib/Sutent	inhib. of RTKs	RCC, NSCLC, HCC, prost.
Sprycel (Dasatinib)	inhib. of Src TKs,Bcr/Abl	CML, mel., ACC
Iressa (gefitinib)	inhib. of EGFR	NSCLC, HNSCC, esoph.
Tarceva (Erlotinib)	inhib of EGFR	HCC, panc., NSCLC
Votrient (pazopanib)	inhib. of VEGFR, PDGFR, Kit	RCC, STS, OvCa, thyr.

[a]ACC, adenoid cystic carcinoma; BrCa, breast cancer; CML, chronic myelogenous leukemia; CRC, colorectal carcinoma; EGFR, EGF receptor; esoph., esophageal; GBM, glioblastoma; HCC, hepatocellular carcinoma; HNSCC, head-and-neck squamous cell carcinoma; mel., melanoma; MoAb, monoclonal antibody against; NPC, nasopharyngeal carcinoma; NSCLC, non-small-cell lung carcinoma; OvCa, ovarian carcinoma; panc., pancreatic carcinoma; RCC, renal cell carcinoma; SCLC, small-cell lung carcinoma; STS, soft tissue sarcoma; thyr., thyroid carcinoma; VEGF, vascular endothelial growth factor.

Adapted from T. Li, G. Kang, T. Wang and H. Huang, *Oncol. Lett.* 16, 687–702, 2018. With permission from the authors.

Thus, similar types of "tip cells" serve as the pioneers to guide both capillary growth and axon extension and use homologous proteins as guidance cues. In certain tissues, genetic disruption of genes and proteins involved in axonal guidance has also been found to disrupt the normal patterning of the vasculature morphogenesis. Hence, the deep evolutionary roots that indicate an ancient common origin of these two systems continue to influence a degree of interdependence during the morphogenesis of mammalian tissues.

Diverse observations, some quite recent, have suggested clear associations between innervation, on the one hand, and normal and neoplastic tissue development, on the other, doing so without providing detailed explanations of the underlying mechanisms. In the case of the prostate gland, as an example, this tissue does not develop normally if it is not innervated properly early in its development. Similarly, in a well-studied model of salivary gland development, mesenchymal cells in the mouse glands secrete factors such as growth nerve growth factor (NGF) together with several other **neurotrophic** factors, which recruit nerve axons that proceed to release acetylcholine, a key neuronal signaling molecule, which then induces branching of the developing salivary ducts. In the absence of these and other critical nerve cell-derived factors, salivary gland development fails to proceed normally. Similarly, in the developing pancreas, inability to produce NGF leads to disruption of glandular architecture.

During the course of progression of certain carcinomas, the density of nervous tissue may even double in these tumors. As prostate cancer develops, carcinoma cells often invade adjacent to existing nerves, this representing the process termed **perineural invasion** (PNI), a process first reported by a histopathologist in 1905. Those carcinoma cells located adjacent to nerves exhibit lower rates of apoptosis and higher rates of proliferation. Moreover, surgical severing of the nerves innervating the prostate leads to regression of orthotopically implanted human prostate cancer cells in a mouse xenograft model of the disease. Reinforcing these observations are others demonstrating that severing of innervating nerves in autochthonously arising prostate carcinomas formed in a transgenic mouse model leads to inhibition of tumor outgrowth; similar observations have been made about mouse mammary and oral carcinomas. In a

(A)   (B)

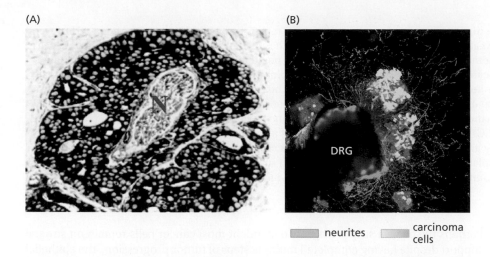

neurites   carcinoma cells

Figure 13.43 Interaction of neuronal tissue with carcinoma cells (A) This section of a human prostate carcinoma has been immunostained with an antibody against NF-κB, which confers survival signals on carcinoma cells. The carcinoma cells (*black*) are seen growing around a cluster of neuronal tissue (*red* N). Cells in closer proximity to nerves show higher levels of NF-κB, higher rates of proliferation, and lower levels of apoptosis (*not shown*). (B) In this *in vitro* cell culture model, a portion of a mouse dorsal root ganglion (DRG), which represents a cluster of neuronal cell bodies that are associated with the spinal cord and relay inputs from sensory receptors (DRG, *black, lower left*) have been cultured together with cells of DU145 prostate carcinoma cell line (*red orange*) that had been transfected with the red fluorescent protein reporter gene. As is apparent, the presence of clusters of prostate cancer cells (*middle*) induces the DRG to extend numerous neurites (*green*), which are immature axon-like extensions of neurons. The prostate carcinoma cells respond reciprocally by growing toward the DRG, and in the presence of co-cultured stromal cells (*not shown*), initiate a model of perineural invasion. In human prostate carcinomas, the carcinoma cells stimulate the formation of new axons. (A, from G.E. Ayala, H. Dai, M. Ittmann et al. *Cancer Res.* 64:6082–6090, 2004. Reprinted by permission from the American Association for Cancer Research. B, Courtesy of Gustavo Ayala, M.D.)

mouse transgenic model of pancreatic carcinoma, the efficacy of chemotherapy was enhanced when **ganglia** that generate the nerves innervating the pancreas were surgically removed.

In vitro assays have demonstrated that cancer cells release substances that stimulate **neurite** outgrowth, that is, the initial provisional protrusion of a neuron that has the potential to be stabilized and mature into a fully functional **axon**. In a number of tumor types, the density of nerves increases as progression proceeds, indicating that the ultimate drivers of such progression, specifically the carcinoma cells, release morphogenetic signals that incite neurite and axon outgrowth (**Figure 13.43**).

When taken together, these various phenomena indicate clearly that in many types of tumors, nervous tissue qualifies as a functionally important component of the stroma that interacts in a bidirectional manner with nearby neoplastic cells. These interactions extend as well to those operating in metastatic colonies. For example, the outgrowth of carcinoma metastases to the brain can be fostered by glutamate released by a subset of neurons that impinge on specific types of glutamate receptors expressed by the carcinoma cells. This research on cancer:neuronal signaling is still in its infancy. Nonetheless, results reported to date already provide clear indications of critical heterotypic interactions appear to represent attractive targets for the development of various types of novel anti-cancer therapeutics.

13.11 Synopsis and prospects

Metazoan tissues are organized as condominiums of various cell types that are continuously communicating with one another. To the developmental biologist, the need for this organizational plan is self-evident: only through such interactions can the proper numbers and locations of each of these cell types be ensured. These heterotypic interactions continue to operate after embryogenesis has been completed in order to support the maintenance and repair of already-formed tissues.

As we have learned in this chapter, this organizational plan confers another, quite distinct benefit on the organism. By making its cells so interdependent, none is easily able to extricate itself from its complex web of interactions and go off on its own. Interdependence imposes the regimentation that wards off the chaos of neoplasia.

Earlier, we viewed multi-step cancer progression as the hurdling of successive barriers placed in the path of developing cancer cells (see Chapter 11). Each successfully completed step, whether achieved by genetic or epigenetic changes operating within incipient cancer cells, removes one of these obstacles and places the developing pre-neoplastic cell incrementally closer to full-fledged malignancy. Now we can revisit these barriers by confronting *non-cell-autonomous* mechanisms: as premalignant cells evolve toward malignancy, they progressively sever some of their ties with their neighbors and their dependence on support by nearby cells, often those residing within the tumor-associated stroma (see however **Sidebar 13.10**).

Perhaps the biggest surprise is how dependent most cancer cells remain on stromal support despite having completed multiple steps of tumor progression. The epithelial cells residing in many carcinomas continue to rely on many of the physiological signals that sustained their precursors in normal tissues; equally elaborate heterotypic signaling may operate in other types of tumors as well. This conservatism is also suggested by microscopic examination of tumors. Thus, a well-trained pathologist can recognize the origins of perhaps 95% of the tumor samples viewed under the microscope, because most of the heterotypic interactions that govern normal morphology appear to be operative in the great majority of cancers.

The remaining approximately 5% of tumors present a challenge, because they are anaplastic and therefore often have minimal losses of the histologic traits that make identification of their tissue of origin possible. The cells in these anaplastic tumors have shed most forms of dependence that tied their precursors to normal neighboring cells. However, most anaplastic tumor cells have not progressed all the way to total independence because they still assemble to form solid tumors. The ultimate independence is achieved only by the cells in those cancers that have advanced so far that they can grow as **pleural effusions** or **ascites** (see Supplementary Sidebar 13.3) and therefore have minimal or no contact with supporting cells and, apparently, with an extracellular matrix (ECM).

Even without detailed knowledge of heterotypic interactions, the dependence of most carcinoma cells on at least one form of stromal support could have been predicted from the known physiology of mammalian tissues: virtually all of them depend on a functional blood supply. Less predictable was the mechanism by which most tumors acquire their vasculature. Rather than invading normal tissue and expropriating existing capillary beds, most tumors actively recruit endothelial cells that proceed to construct capillaries and larger vessels within the tumors, usually by extending the vasculature of nearby normal tissues. They succeed in this endeavor by orchestrating the expression of a complex array of signaling molecules that together ensure the formation of the tumor-associated neovasculature (**Figure 13.44**).

Figure 13.44 Balancing the angiogenic switch This diagram presents the major physiological regulators that work to promote or inhibit angiogenesis within tissues and indicates that it is the balance between these two countervailing groups of regulators that determines whether or not angiogenesis proceeds. Although the angiogenic switch has been described here as an essentially on–off binary decision, in fact diagrams like this one suggest that the rate of neovasculature formation may change continuously within a tumor mass after the switch has been activated and often increases progressively, seemingly governed by a rheostat rather than a binary switch. (From D. Hanahan and J. Folkman, *Cell* 86:353–364, 1996. With permission from Elsevier.)

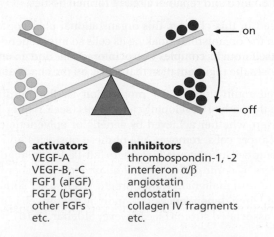

activators	inhibitors
VEGF-A	thrombospondin-1, -2
VEGF-B, -C	interferon α/β
FGF1 (aFGF)	angiostatin
FGF2 (bFGF)	endostatin
other FGFs	collagen IV fragments
etc.	etc.

The sources of these endothelial cells could also not be deduced from first principles. The proliferation of endothelial cells originating in neighboring tissues seems to be the main mechanism for acquiring neovasculature soon after the angiogenic switch has been tripped (see Figure 13.44). However, in some tumors, the initially formed vessels may be constructed largely by endothelial precursor cells (EPCs) that originate in the bone marrow and then differentiate in the tumor stroma into functional endothelial cells. In fact, the relative contributions of EPCs are still unclear in most types of tumors.

Major aspects of angiogenesis are still poorly understood, even paradoxical. For example, it seems apparent that once tumors become angiogenic, they can launch into a prolonged phase of growth and expansion, and that tumors that are more angiogenic can grow even more rapidly than those that are less able to attract new vasculature. Indeed, as we have read, measurements of microvessel density—the number of capillaries per microscope field—correlate quite well with the likelihood that a primary tumor, such as a breast cancer, will progress to a highly malignant endpoint.

The paradox comes from the frequent observations that patients bearing highly hypoxic tumors also confront a poor prognosis. Indeed, hypoxic tumors, including those exposed to anti-angiogenesis therapies, are often more aggressive than their normoxic counterparts. This makes no sense, given that hypoxia should starve tumor cells and lead to their death through apoptosis and, on a larger scale, to extensive necrotic regions within tumors (see Figure 13.29). The paradox may one day be resolved by invoking the actions of the HIF-1 transcription factor, which becomes activated in hypoxic cells and induces production of a large number of other proteins besides VEGF. Indeed, a high level of HIF-1 expression is also an indicator of poor clinical prognosis. Included among the genes activated by HIF-1 are those specifying PDGF, TGF-α, TGF-β, and several matrix metalloproteinases (MMPs), the latter being responsible for remodeling the extracellular matrix (ECM). As we have learned, several of these secreted proteins act as potent mitogens that drive the proliferation of both epithelial cells and their stromal neighbors.

An additional HIF-1–induced gene encodes the Met protein, which functions as the receptor for the hepatocyte growth factor (HGF) ligand, also known as scatter factor (SF). HGF, for its part, seems to be widely available in many human tumors of both epithelial and mesenchymal origin. Consequently, HIF-1–mediated increases in expression of the Met RTK can sensitize tumor cells to HGF molecules that are present in their surroundings, having been released by nearby stromal cells. Once the Met receptor is activated by binding its HGF ligand, it activates a diverse set of responses within epithelial cells, including the epithelial–mesenchymal transition (EMT), a cell-biological program that imparts increased motility and invasiveness (see Section 13.3 and Chapter 14). At the molecular level, the concentrations of two potent EMT-inducing transcription factors—Twist and Snail—are significantly increased in hypoxic cells; as we will see in the next chapter, carcinoma cells that have undergone an EMT acquire many of the phenotypes of the cells forming high-grade malignancies. These diverse products of HIF-1 action and tissue hypoxia help explain the poor prognosis attached to hypoxic tumors, which thrive despite great adversity and actually turn out to be more aggressive than their well-oxygenated counterparts (see also Supplementary Sidebar 13.6).

These paradoxical, indeed counterproductive responses to anti-angiogenesis therapies, together with the evasive maneuvers that tumors may take to circumvent these therapies (see Table 13.4), may explain the modest successes that most of these therapies have achieved in the oncology clinic. Thus, most clinical outcomes have fallen far short of anticipated outcomes. Still, some combinations of anti-angiogenesis therapy and chemotherapy have shown clear utility in terms of increasing progression-free survival (PFS) and even overall survival of cancer patients and thus are finding their way into standard clinical practice. In the longer term, more efficacious anti-angiogenesis therapies may derive from dosing schedules that serve to normalize the tumor-associated microvasculature, enabling more efficient drug delivery, while avoiding the complete vascular collapse that triggers tumor hypoxia and the induction of increased tumor aggressiveness.

Certain recently discovered dynamic interactions between tumors and the bone marrow have been a surprise to most cancer researchers. In the absence of metastasis, most tumors have traditionally been considered to be localized diseases confined to one or another corner of the body. But the more we learn about the cells of the stromal compartment, the more we come to realize that even localized tumors extend their reach far and wide throughout the body in order to recruit the cells that they need in order to support their own survival and proliferation programs. Besides the endothelial precursor cells (EPCs) cited above, carcinomas recruit mast cells and monocytes from the marrow, the latter differentiating on-site into various types of macrophages. Mesenchymal stem cells (MSCs and fibrocytes) of bone marrow origin are also recruited in large numbers to the stroma of many carcinomas; once present within these tumors, they may differentiate into fibroblasts and myofibroblasts, which then release signals favoring the survival and malignant progression of nearby cancer cells.

Throughout most of this chapter, we have focused on epithelial–stromal interactions within tumors and the heterotypic signals that they exchange via *paracrine* signaling channels. The presence of a variety of bone marrow–derived cells in the tumor stroma creates an additional dimension of complexity. Recognizing their existence and their potential contributions to tumor physiology, we now need to include interactions of cancer cells with the faraway bone marrow. Thus, tumors may communicate with the marrow to elicit the production and mobilization into the circulation of certain cell types that can then be recruited from the circulation into the tumor stroma. Such systemic interactions depend, by definition, on information conveyed by *endocrine* signals rather than the paracrine signals that have been a central focus of our discussions here.

Even if we confine our attention to the bidirectional exchanges of paracrine signals between the epithelial and stromal compartments operating within tumors, we confront extraordinary complexity. The tumor stroma is composed of almost a dozen distinct cell types (fibroblasts, myofibroblasts, MSCs, endothelial cells, pericytes, smooth muscle cells, mast cells, monocytes, macrophages, lymphocytes, nerve cells, and, in some tissues, adipocytes), each of which signals to the neoplastic epithelial cells as well as to other cellular components of the stroma. Physicists have struggled, so far unsuccessfully, with general solutions to the three-body problem. Here, we confront a world of a far larger number of distinct cell types, each of which is sending a complex mixture of signals to other types of cells. Examinations of only a single cell type—fibroblasts—reveal the heterogeneity of these cells and their ability to change dynamically in response to the contextual signals that they encounter within the tumor stroma (Supplementary Sidebar 13.11). We have only begun to scratch the surface of the complex epithelial–stromal interactions operating in neoplastic tissues and in their normal counterparts.

For more than half a century, the mindset of oncologists has been focused on eradicating the bulk of the cancer cells within solid tumors, with the hope of achieving the seemingly impossible—cures of common carcinomas that have traditionally been incurable. As we learned, first in Chapter 11 and now in this one, this focus needs to be radically redirected. First, the targets of anti-cancer therapies can no longer be confined to the bulk of neoplastic cells in a tumor, because these cells have limited self-renewal capacity; instead, we are likely to find that truly durable cures can come only from eradicating as well the still-elusive cancer stem cells (CSCs) that hide out, here and there, throughout tumor masses and represent their engines of self-renewal. Second, many of the useful anti-cancer therapies to be developed in the future will not come from targeting the cancer cells themselves. Instead, it may often be far more profitable to attack the cells that provide them with vital physiological support; by undermining the elaborate stromal support network on which most cancer cells depend, truly dramatic regressions of solid tumors may one day be achieved (**Figure 13.45**).

The insight that tumors are "wounds that do not heal" extends and echoes our earlier discussions (see Sections 11.13 and 11.14), in which the critical role of chronic

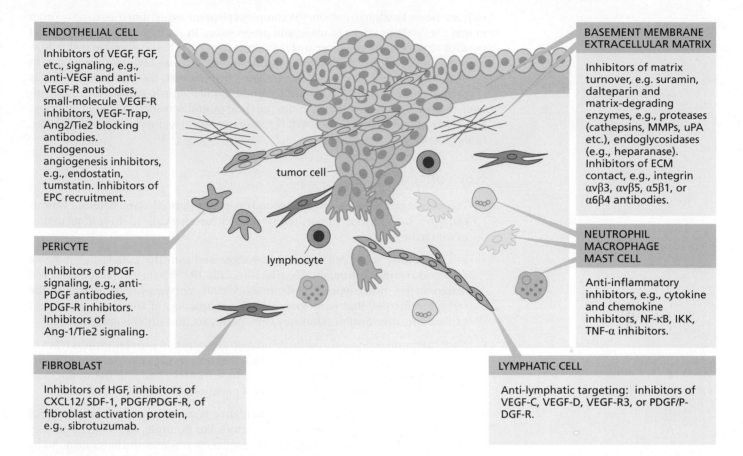

ENDOTHELIAL CELL

Inhibitors of VEGF, FGF, etc., signaling, e.g., anti-VEGF and anti-VEGF-R antibodies, small-molecule VEGF-R inhibitors, VEGF-Trap, Ang2/Tie2 blocking antibodies. Endogenous angiogenesis inhibitors, e.g., endostatin, tumstatin. Inhibitors of EPC recruitment.

PERICYTE

Inhibitors of PDGF signaling, e.g., anti-PDGF antibodies, PDGF-R inhibitors. Inhibitors of Ang-1/Tie2 signaling.

FIBROBLAST

Inhibitors of HGF, inhibitors of CXCL12/ SDF-1, PDGF/PDGF-R, of fibroblast activation protein, e.g., sibrotuzumab.

tumor cell

lymphocyte

BASEMENT MEMBRANE EXTRACELLULAR MATRIX

Inhibitors of matrix turnover, e.g. suramin, dalteparin and matrix-degrading enzymes, e.g., proteases (cathepsins, MMPs, uPA etc.), endoglycosidases (e.g., heparanase). Inhibitors of ECM contact, e.g., integrin $\alpha v\beta 3$, $\alpha v\beta 5$, $\alpha 5\beta 1$, or $\alpha 6\beta 4$ antibodies.

NEUTROPHIL MACROPHAGE MAST CELL

Anti-inflammatory inhibitors, e.g., cytokine and chemokine inhibitors, NF-κB, IKK, TNF-α inhibitors.

LYMPHATIC CELL

Anti-lymphatic targeting: inhibitors of VEGF-C, VEGF-D, VEGF-R3, or PDGF/P-DGF-R.

Figure 13.45 Heterotypic interactions as targets for therapeutic intervention As described in this chapter, cancer cells depend on the nearby stromal microenvironment for a variety of cell-physiological supports. This dependence on heterotypic interactions has inspired development of new types of cancer therapy, some of which have been featured in this chapter. Instead of focusing on the intracellular signaling defects within cancer cells, this new class of therapies is directed toward interrupting heterotypic signaling, thereby depriving cancer cells of essential stromal support. This scheme indicates some of the anti-tumor therapies that are being developed or under consideration (*light blue boxes*). (From J.A. Joyce, *Cancer Cell* 7:513–520, 2005. With permission from Elsevier.)

inflammation in promoting tumor formation was described in great depth. Inflammation and wound healing are intertwined processes, and the mechanisms of inflammation-driven tumor promotion, which lead to the initial formation of cancers, are extended and elaborated by the chronic wound healing that seems to best characterize the biology of the stroma of well-established tumors.

Here, too, there has been another surprise. Inflammatory cells, notably macrophages, have traditionally been depicted as the front-line soldiers of the immune response that deal effectively with infectious agents, such as bacteria, by consuming them and help guide the long-term immune response through antigen presentation, as we will see later in Chapter 15. Now we learn that macrophages can also function as key sources of tumor promotion by producing mitogenic growth factors, liberating angiogenic factors, and remodeling the extracellular matrix (ECM); the latter process is also critical for tumor invasion and metastasis, as we will see in Chapter 14.

So, the traditional job description assignments of various stromal cell types are being extended and blurred. Cells of the immune system, which are purportedly dispatched to protect us from infection and even cancer, are often active collaborators in tumor development. And the deletion of one or another cell type from the immune system, achieved in mice through germ-line re-engineering, often creates a host organism that is, paradoxically, less able to support tumorigenesis.

Normal, multifaceted morphogenetic programs that depend on epithelial–stromal interactions, such as wound healing and the EMT, are likely to explain how carcinoma cells are clever enough to acquire the complex cell phenotypes that they need in order to execute the later stages of malignant progression. In retrospect, this notion is not so surprising, because the more we learn about cancer cells, the more we realize how opportunistic they are in co-opting and exploiting normal biological processes in order to further their own ends. This insight leaves us with the last question of this chapter: Have we begun to truly understand the mechanistic complexity of heterotypic interactions, or is there an entire universe of undiscovered signaling pathways and behavioral programs lurking within tumors waiting, like intergalactic dark matter, to surprise us once again?

Key concepts

- Tumors are complex tissues that depend on intercommunication between various cell types. Indeed, most tumors are as complex histologically as the normal tissues in which they arise.

- In carcinomas, these cell types can be separated into the neoplastic epithelial cells and recruited stromal cells, which include fibroblasts, myofibroblasts, and macrophages, other types of inflammatory cells, nervous tissue, as well as the various cell types that participate in the construction of the tumor-associated vasculature, specifically endothelial cells, pericytes, and smooth muscle cells.

- Most carcinomas depend absolutely on recruited stromal cells for various types of physiological support. This dependence is lost only in the small subset of tumors that progress to an extremely malignant state, notably some of the tumor cells growing in the ascites and pleural fluids of patients suffering advanced cancer.

- At the biochemical level, this interdependence is manifested by the exchange of various types of mitogenic and trophic factors. For example, carcinoma cells may release PDGF to recruit and activate stromal cells, whereas the latter respond by releasing IGFs that sustain the survival of the carcinoma cells.

- The formation of tumor-associated vasculature, formed by the process of neoangiogenesis, is a critical, rate-limiting determinant of the growth of all tumors larger in size than approximately 0.2 mm.

- In the case of carcinomas, the acquisition of tumor-associated stroma closely resembles the process of healing in wounded epithelial tissues. The genesis of stroma therefore relies on the same gene expression programs that are activated during wound healing.

- As tumor progression proceeds, the fibroblast-rich stroma is increasingly replaced by myofibroblasts, which eventually may generate collagen-rich, desmoplastic stroma.

- The recruitment of the cells that participate directly in the construction of the neovasculature of tumors involves the release of factors, such as VEGF, by both the tumor cells and inflammatory cells, notably macrophages.

- Neoangiogenesis represents an attractive target for the development of novel anticancer agents, in that the targeted cells are the various normal, genetically stable stromal cell types participating in angiogenesis rather than the ever-changing cancer cells.

- Anti-angiogenic therapies often provoke paradoxical responses in carcinomas, including progression to higher grade malignancy, tempering the enthusiasm for introducing some of these therapies into clinical practice.

Thought questions

1. What diverse lines of evidence prove directly that most carcinoma cells depend on stromal cell types for various types of physiological support?

2. How might anti-angiogenesis therapies improve (in some cases) or neutralize (in other cases) the efficacy of conventional chemotherapeutic agents?

3. How might macrophages facilitate or antagonize tumorigenesis?

4. Which lines of evidence persuade you that the generation of tumor-associated stroma depends on the same biological programs that are activated during wound healing?

5. Which biological forces cause the tumor-associated vasculature to be defective in so many respects?

6. Which types of anti-cancer therapeutic agents would you deploy in order to encourage the collapse of an established tumor by depriving it of vasculature support?

7. What biochemical strategies can tumor cells use to lessen their dependence on stromal support?

8. How might you determine what proportion of endothelial cells in the vasculature of a tumor derive from the expansion of adjacent vasculature and what proportion of these cells arise from circulating endothelial precursor cells?

Additional reading

Avramides CJ, Garmy-Susini B & Varner JA (2008) Integrins in angiogenesis and lymphangiogenesis. *Nat. Rev. Cancer* 8, 604–617.

Balkwill F, Charles KA & Mantovani A (2005) Smoldering and polarized inflammation in the initiation and promotion of malignant disease. *Cancer Cell* 7, 211–217.

Baluk P, Hashizume H & McDonald DM (2005) Cellular abnormalities of blood vessels as targets in cancer. *Curr. Opin. Genet. Dev.* 15, 102–111.

Battle E & Massagué J (2019) Transforming growth factor-β signaling in immunity and cancer. *Immunity* 50, 924–940.

Baum B & Georgiou M (2011) Dynamics of adherens junctions in epithelial establishment, maintenance, and remodeling. *J. Cell Biol.* 192, 907–917.

Bergers G & Hanahan D (2008) Modes of resistance to anti-angiogenic therapy. *Nat. Rev. Cancer* 8, 592–603.

Bierie B & Moses HL (2006) Tumour microenvironment: TGFβ, the molecular Jekyll and Hyde. *Nat. Rev. Cancer* 6, 506–520.

Brancato SK & Albina JE (2011) Wound macrophages as key regulators of repair. *Am. J. Pathol.* 178, 19–25.

Calon A, Tauriello DVF & Battle E (2014) TGF-beta in CAF-mediated tumor growth and metastasis. *Semin. Cancer Biol.* 24, 15–22.

Carmeliet P & Jain RK (2011) Molecular mechanisms and clinical applications of angiogenesis. *Nature* 473, 298–307.

Carmeliet P & Tessier-Lavigne M (2005) Common mechanisms of nerve and blood vessel wiring. *Nature* 436, 193–200.

Cassetta L, Stamatina F, Sims AH et al. (2019) Human tumor-associated macrophage and monocyte transcriptional landscapes reveal cancer-specific reprogramming, biomarkers and therapeutic targets. *Cancer Cell* 35, 588–602.

De Visser KE, Eichten A & Coussens LM (2006) Paradoxical roles of the immune system during cancer development. *Nat. Rev. Cancer* 6, 24–37.

De Wever O, Demetter P, Mareel M & Bracke M (2008) Stromal myofibroblasts are drivers of invasive cancer growth. *Int. J. Cancer* 123, 2229–2238.

Dvorak HF (1986) Tumors: wounds that do not heal. *N. Engl. J. Med.* 315, 1650–1689.

Dvorak HF (2003) How tumors make bad blood vessels and stroma. *Am. J. Pathol.* 162, 1747–1757.

Ebos JML & Kerbel RS (2011) Antiangiogenic therapy: impact on invasion, disease progression, and metastasis. *Nat. Rev. Clin. Oncol.* 8, 210–221.

Egeblad M, Nakasone ES & Werb Z (2010) Tumors as organs: complex tissues that interface with the entire organism. *Dev. Cell* 18, 884–901.

Ehnman M, Chaabane W, Haglund F & Tsagkozis P (2019) The tumor microenvironment of pediatric sarcoma: mesenchymal mechanisms regulating cell migration and metastasis. *Curr. Oncol. Rep.* 21:90.

Ellis LM & Hicklin DJ (2008) VEGF-targeted therapy: mechanisms of anti-tumor activity. *Nat. Rev. Cancer* 8, 579–591.

Fernandez-Cortes M, Delgado-Bellido D & Oliver FJ (2019) Vasculogenic mimicry: become an endothelial cell "but not so much." *Front. Oncol.* 2019.00803.

Ferrara N, Hillan KJ, Gerber HP & Notovny W (2004) Discovery and development of bevacizumab, an anti-VEGF antibody for treating cancer. *Nat. Rev. Drug Discov.* 3, 391–400.

Franke WW (2009) Discovering the molecular components of intercellular junctions: a historical view. *Cold Spring Harb. Perspect. Biol.* 1, a003061.

Garcia J, Hurwitz HI, Sandler AB et al. (2020) Bevacizumab (Avastin®) in cancer treatment: a review of 15 years of clinical experience and future outlook. *Cancer Treat. Rev.* 86, 102017.

Gerhardt H & Semb H (2008) Pericytes: gatekeepers in tumour cell metastasis? *J. Mol. Med.* 86, 135–144.

Hanahan D & Coussens LM (2012) Accessories to the crime: functions of cells recruited to the tumor microenvironment. *Cancer Cell* 21, 309–322.

Hessmann E, Buchholz SM, Demir IE et al. (2020) Microenvironmental determinants of pancreatic cancer. *Physiol. Rev.* 100, 1707–1751.

Hinz B (2007) Formation and function of the myofibroblast during tissue repair. *J. Invest. Dermatol.* 127, 526–537.

Hynes RO (2012) The evolution of metazoan extracellular matrix. *J. Cell Biol.* 196, 671–679.

Hynes RO & Naba A (2012) Overview of the matrisome—an inventory of extracellular matrix constituents and functions. *Cold Spring Harb. Perspect. Biol.* 4, a004903.

Jain RK (2005) Normalization of tumor vasculature: an emerging concept in antiangiogenic therapy. *Science* 307, 58–62.

Jain RK, Di Tomaso E, Duda DG et al. (2007) Angiogenesis in brain tumors. *Nat. Rev. Neurosci.* 8, 610–622.

Jain RK (2014) Anti-angiogenesis strategies revised: From starving tumors to alleviating hypoxia. *Cancer Cell* 26, 605–622.

Kalluri R & Zeisberg M (2006) Fibroblasts in cancer. *Nat. Rev. Cancer* 6, 392–401.

Kessenbrock K, Plaka V & Werb Z (2010) Matrix metalloproteinases: regulators of the tumor microenvironment. *Cell* 141, 52–67.

Lewis CE & Pollard JW (2006) Distinct role of macrophages in different tumor microenvironments. *Cancer Res.* 124, 263–266.

Li Calzi S, Neu MB, Shaw LC et al. (2010) EPCs and pathological angiogenesis: when good cells go bad. *Microvasc. Res.* 79, 297–216.

Liebig C, Ayala G, Wilks JA et al. (2009) Perineural invasion in cancer. *Cancer* 115, 3379–3391.

Loges S, Mazzone M, Hohensinner P & Carmeliet P (2009) Silencing or fueling metastasis with VEGF inhibitors: antiangiogenesis revisited. *Cancer Cell* 15, 167–170.

Lu P, Weaver VM & Werb Z (2012) The extracellular matrix: a dynamic niche in cancer progression. *J. Cell Biol.* 196, 395–406.

Nagy JA, Chang SH, Dvorak AM & Dvorak HF (2009) Why are tumour blood vessels abnormal and why is it important to know? *Br. J. Cancer* 100, 865–859.

Neel NI, Karsdal M & Willumsen N (2019) Collagens and cancer associated fibroblasts in the reactive stroma and its relation to cancer biology. *J. Exp. Clin. Cancer Res.* 38:115.

Nia H, Munn LL & Jain RK (2020) Physical traits of cancer. *Science* 370, eaaz0868.

Öhlund D, Elyada E & Tuveson D (2014) Fibroblast heterogeneity in the cancer wound. *J. Exp. Med.* 211, 1503–1523.

Page-McCaw A, Ewald AJ & Werb Z (2007) Matrix metalloproteinases and the regulation of tissue remodeling. *Nat. Rev. Mol. Cell Biol.* 8, 221–233.

Quail DF & Joyce JA (2013) Microenvironmental regulation of tumor progression and metastasis. *Nat. Med.* 19, 1423–1437.

Ribatti D, Nico B, Crivellato E et al. (2007) The history of the angiogenic switch concept. *Leukemia* 21, 44–52.

Rolfe D, Fleming JB & Gomer RH (2020) Fibrocytes in the tumor microenvironment. *Adv. Exp. Med. Biol* 1224, 79–85.

Sahai E, Astsaturov I, Cukierman E et al. (2020) A framework for advancing our understanding of cancer-associated fibroblasts. *Nat. Rev. Cancer* 20, 174–186.

Santi A, Kugeratski FG & Zanivan S (2018) Cancer associated fibroblasts: the architects of stroma remodeling. *Proteomics* 18, e1700167.

Servais C & Erez N (2013) From sentinel cells to inflammatory culprits: cancer-associated fibroblasts in tumor-related inflammation. *J. Pathol.* 229, 198–297.

Shibuya M (2011) Vascular endothelial growth factor (VEGF) and its receptor (VEGFR) signaling in angiogenesis. *Genes Cancer* 2, 1097–1105.

Suva ML & Tirosh I (2019) Single-cell RNA sequencing in cancer: lessons learned and emerging challenges. *Mol. Cell* 75, 7–12.

Theunissen J-W & de Sauvage FJ (2009) Paracrine Hedgehog signaling in cancer. *Cancer Res.* 69, 6007–6010.

Thiery JP & Sleeman JP (2006) Complex networks orchestrate epithelial-mesenchymal transitions. *Nat. Rev. Mol. Cell Biol* 7, 131–142.

Thiery JP, Acloque H, Huang RYG & Nieto MA (2009) Epithelial-mesenchymal transitions in development and disease. *Cell* 139, 871–890.

Tlsty TD & Coussens LM (2006) Tumor stroma and regulation of cancer development. *Annu. Rev. Pathol.* 1, 119–150.

Vaahtomeri K, Karaman S, Mäkinen T & Alitalo K (2017) Lymphangiogenesis guidance by paracrine and pericellular factors. *Genes Dev.* 31, 1615–1634.

Wendt MK, Tian M & Schiemann WP (2012) Deconstructing the mechanisms and consequences of TGF-β-induced EMT during cancer progression. *Cell Tissue Res.* 347, 85–101.

Werner S & Grose R (2003) Regulation of wound healing by growth factors and cytokines. *Physiol. Rev.* 83, 835–870.

Zahalka AH & Frenette PS (2020) Nerves in cancer. *Nat. Rev. Cancer* 20, 143–157.

Zecchin A, Kalucka J, Dubois C & Carmeliet P (2017) How endothelial cells adapt their metabolism to form vessels in tumors. *Front Immunol.* 8, 1750.

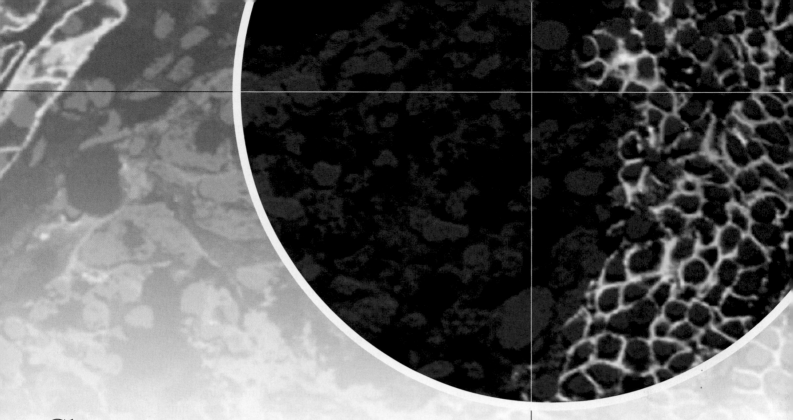

Chapter 14

Moving Out: Invasion and Metastasis

The fact of cells identical with those of the cancer itself being seen in the blood may tend to throw some light upon the mode of origin of multiple tumors existing in the same person.

T.R. Ashworth, physician, 1869

What is it that decides what organs shall suffer in a case of disseminated cancer?

Stephen Paget, surgeon, 1889

It is not birth, marriage, or death, but gastrulation, which is truly the most important time in your life.

Lewis Wolpert, embryologist, 1986

In the early phases of multi-step tumor progression, cancer cells multiply near the site where their ancestors first began uncontrolled proliferation. The result, usually apparent only after many years' time, is a **primary tumor** mass. Given that a cubic centimeter of tissue may contain as many as 10^9 cells, we can easily imagine that tumors may often reach a size of 10^{10} or 10^{11} cells before they become apparent to the individual carrying them or to the clinician in search of them.

Primary tumors in some organ sites—specifically those arising within the peritoneal or pleural space—may well expand without causing any discomfort to the patient, simply because these cavities are expansible and their contents are quite plastic; in other sites, such as the brain, the presence of a tumor is often apparent when it is still relatively small. Sooner or later, however, in all sites throughout the body, tumors of substantial size compromise the functioning of the organs in which they have arisen and begin to evoke symptoms.

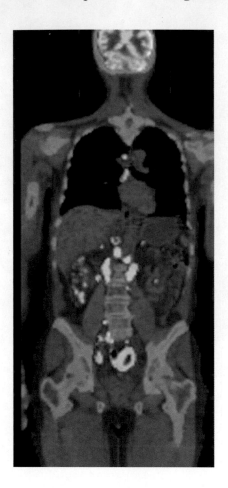

Figure 14.1 Disseminated tumors The diagnosis of metastatic disease often represents a death sentence for a cancer patient, yet the mechanisms by which cancer cells metastasize from a primary tumor to distant sites in the body remain poorly understood. Seen here is a whole-body scan of a patient with metastatic non-Hodgkin's lymphoma (NHL). This is a fusion image of a CT (computed X-ray tomography) scan of the body's tissues (*gray, blue*) and a PET (positron-emission tomography) scan in which the uptake of radioactively labeled fluorodeoxyglucose (FDG) in various tissues (*yellow*) has been detected. FDG uptake indicates regions of high glucose uptake associated with aerobic glycolysis (see Section 11.15). The activity associated with the brain is normal. However, the yellow spots in the abdominal regions indicate multiple NHL metastases. (Courtesy of S.S. Gambhir.)

In many cases, the effects on normal tissue function come from the physical pressure exerted by the expanding tumor masses. In others, cells from the primary tumor mass invade adjacent normal tissues and, in so doing, begin to compromise vital functions. Large tumors in the colon may obstruct passage of digestion products through the lumen, and in tissues such as the liver and pancreas, cancer cells may obstruct the flow of bile through critical ducts. In the lungs, airways may be compromised.

As insidious and corrosive as these primary tumors are, they ultimately are responsible for only about 10% of deaths from cancer. The remaining 90% of patients are struck down by cancerous growths that are discovered at sites often far removed from the locations where their primary tumors first arose (**Figure 14.1**; see also Figure 2.2). These **metastases** are formed by cancer cells that have left the primary tumor mass and traveled by the body's highways—blood and lymphatic vessels—to seek out new sites throughout the body where they may found new colonies (**Figure 14.2**). Breast cancers often spawn metastatic colonies promiscuously in many tissues throughout the body, including the brain, liver, bones, and lungs. Prostate tumors are most often seeded to the bones, whereas colon carcinomas preferentially form new colonies in the liver.

Such wandering cancer cells are the most dangerous manifestations of the cancer process. When they succeed in founding colonies in distant sites, they often wreak great havoc. The female body can dispense with its mammary glands without losing vital physiological functions, and so almost all primary breast carcinomas do not compromise survival while they are confined to the breast. However, the metastatic colonies that breast cancer cells initiate in the bone marrow can cause localized erosion of bone tissue, resulting in agonizing pain and skeletal collapse. Metastases in the brain may rapidly compromise central nervous system function, while those in the lung or liver are similarly threatening to life because of the vital functions of these organs.

One major puzzle concerns the variable tendencies that different tumors have to metastasize. For reasons that remain obscure, primary tumors formed in certain tissues have a high probability of metastasizing, while those arising in other tissues almost never do so. After primary melanomas penetrate a certain distance downward into the tissue underlying the skin, the presence of metastases at distant sites in the body is almost a certainty. In contrast, basal cell carcinomas of the skin and astrocytomas—primary tumors of the glial cells in the brain—rarely spawn metastases. Another major unresolved issue concerns the metastatic *tropism* cited above: Why do tumors originating in a given organ preferentially seed colonies in particular tissues located elsewhere in the body?

In a variety of human tumor types, the dissemination of cancer cells throughout the body has already occurred by the time a primary tumor is first detected; at the time of initial diagnosis, these scattered cells may be clinically inapparent because they form only minute tumor colonies—**micrometastases**. Such behavior provokes questions that we will confront in this chapter and in Chapter 17: Do the properties of a primary tumor reveal whether it has broadcast cancer cells throughout the body that will eventually create life-threatening metastatic disease long after the primary

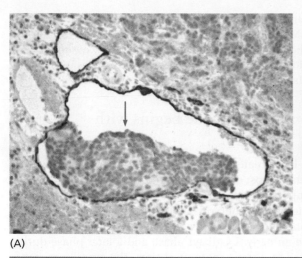
(A)

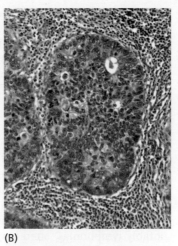
(B)

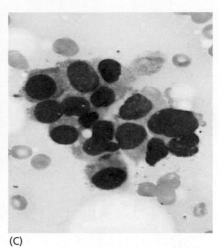
(C)

Figure 14.2 Histology of metastases in various tissues throughout the body
(A) In the Rip-Tag transgenic mouse model of pancreatic islet cell tumorigenesis (see Figure 13.35), metastasis via the lymph nodes can be encouraged through the forced expression of VEGF-C—a lymphangiogenic factor—in the islet tumor cells. Seen here is a small metastasis of islet cells (*arrow*) within a lymphatic vessel that is lined with endothelial cells (*reddish brown*). (B) A small metastasis of a human breast cancer (*center*) is seen growing within a lymph node associated with one of the lymphatic ducts draining the breast. Note that this metastasis exhibits features that are characteristic of many primary tumors of the breast; adjacent to this metastasis are numerous lymphocytes in the surrounding lymph duct (*dark nuclei*). (C) The presence of clumps of metastatic carcinoma cells (*blue*) in the bone marrow can be revealed using specific immunohistochemistry to detect cells displaying epithelial markers, which set them apart from the mesenchymal cells naturally present in the marrow. In this particular case, however, the metastatic deposits were revealed by an alternative technique, use of the Wright–Giemsa stain. (A, from S.J. Mandriota et al., *EMBO J.* 20:672–682, 2001. With permission from John Wiley & Sons. B, courtesy of T.A. Ince. C, Fan FS, Yang C-F and Wang Y-F *Case Rep. Oncol. Med.* Article ID 2946409, 2018.)

tumor has been surgically removed? And equally important, do metastatic growths respond to therapies similarly to the corresponding primary tumors, or have the disseminated tumor cells in these growths evolved to a state of higher resistance to various forms of therapy?

In this chapter, we confront the processes that create these most aggressive products of tumor progression. These processes initially depend within the primary tumor on complex biochemical and biological changes in cancer cells and in the tumor-associated stroma. Most of the steps of cancer formation, as described in earlier chapters, are understood in considerable detail. In contrast, our understanding of invasion and metastasis is still incomplete, explaining why these late steps of tumor progression represent the major unsolved problems of cancer pathogenesis.

Importantly, extensive efforts have been undertaken to document metastasis-specific mutations, that is, mutant alleles that are present recurrently in the cells forming metastases and are apparently absent in the corresponding primary tumors; such mutations would ostensibly be responsible for driving metastatic dissemination or the outgrowth of already-seeded metastases. Such efforts have consistently failed. These failures provide strong suggestion that, although carcinoma cells acquire a variety of somatic mutations in order to form primary tumors, additional metastasis-specific mutations do not seem to be involved in propelling primary tumor cells to distant sites in the body. This explains why the present chapter describes the mechanisms associated with invasion and metastasis as consequences of the actions of non-genetic (that is epigenetic) regulatory programs that are activated during the course of metastatic progression.

As you will see, this chapter is unusually long and takes numerous excursions, doing so for some very good reasons. Most obvious, metastases are responsible for ~90% of all cancer-associated deaths. Then, there is the biological complexity of the process. And finally, the nature of invasion and metastasis has left major puzzles unsolved, if only because critical steps in this process have proven elusive experimentally.

14.1 The invasion–metastasis cascade begins with local invasiveness

The great majority (>80%) of life-threatening cancers occur in epithelial tissues, yielding carcinomas. Consequently, most of our discussions in this chapter, as in the last, will refer to this class of tumors, with the understanding that cancers arising in other tissue types, such as connective and nervous tissues, often follow similar paths when they become invasive and metastatic. Even certain hematopoietic tumors, notably lymphomas, often have an early, localized phase and a later phase during which they become disseminated to distant anatomical sites. The overall process of dissemination, sometimes called the **invasion–metastasis cascade**, involves a complex succession of steps, which are outlined in **Figure 14.3**. As we proceed through this chapter, we will confront each of these steps, one after the other.

Our focus on carcinomas requires us to draw from earlier discussions of these tumors and the epithelial tissues in which they arise (see, for example, Sections 2.2 and 13.1). To recap briefly, the great majority of epithelial tissues are constructed according to a common set of architectural principles; in most cases, relatively thin sheets of epithelial cells sit atop deep, complex layers of various types of mesenchymal cells forming the stroma. Separating the two groups of cells is the specialized type of extracellular matrix (ECM) known as the basement membrane, sometimes termed the basal lamina (see Figure 2.3A and 13.10). This proteinaceous meshwork is constructed collaboratively by proteins secreted by both epithelial and stromal

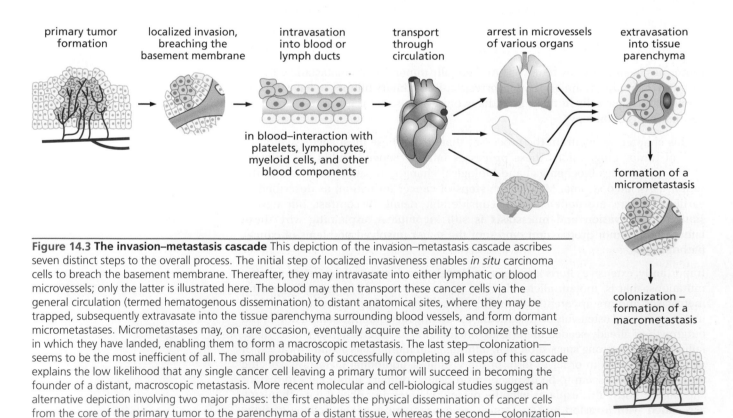

Figure 14.3 The invasion–metastasis cascade This depiction of the invasion–metastasis cascade ascribes seven distinct steps to the overall process. The initial step of localized invasiveness enables *in situ* carcinoma cells to breach the basement membrane. Thereafter, they may intravasate into either lymphatic or blood microvessels; only the latter is illustrated here. The blood may then transport these cancer cells via the general circulation (termed hematogenous dissemination) to distant anatomical sites, where they may be trapped, subsequently extravasate into the tissue parenchyma surrounding blood vessels, and form dormant micrometastases. Micrometastases may, on rare occasion, eventually acquire the ability to colonize the tissue in which they have landed, enabling them to form a macroscopic metastasis. The last step—colonization— seems to be the most inefficient of all. The small probability of successfully completing all steps of this cascade explains the low likelihood that any single cancer cell leaving a primary tumor will succeed in becoming the founder of a distant, macroscopic metastasis. More recent molecular and cell-biological studies suggest an alternative depiction involving two major phases: the first enables the physical dissemination of cancer cells from the core of the primary tumor to the parenchyma of a distant tissue, whereas the second—colonization— depends on the adaptation of disseminated cancer cells to the microenvironment of this tissue. (Adapted from I.J. Fidler, *Nat. Rev. Cancer* 3:453–458, 2003. With permission from Nature.)

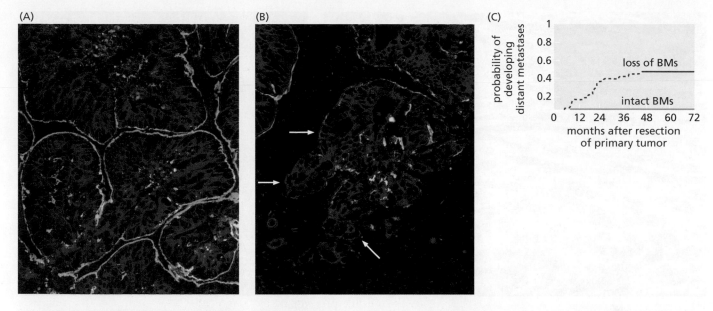

Figure 14.4 Breaching of the basement membrane The basement membrane (BM) can be detected through use of antibodies reactive with laminin, a key component of this membrane (see Figure 13.10C). (A) In the more differentiated, less aggressive portions of a colorectal carcinoma, islands of carcinoma cells labeled with an anti-cytokeratin antibody (*red*) are surrounded by BMs (*green*) that have been detected with an antibody reactive with the α3 chain of laminin. The stroma between islands of carcinoma cells appears *black*. (B) In the portions of the tumor that appear to be less differentiated and more invasive, this staining reveals edges of islands of carcinoma cells (*arrows*) that lack BM, which has been broken down by the invading carcinoma cells and recruited myeloid cells. Now the carcinoma cells are in direct contact with the surrounding stroma. (C) For a group of patients whose high-grade rectal carcinomas (all classified as stage T3) had penetrated at least 5 mm into nearby fatty tissue, the loss of BM around tumor islands directly correlated with the probability of developing distant metastases. Thus, partial or complete loss of BM within their tumors indicated a strong likelihood that metastases would develop during the five years after surgical removal of the primary tumors. (From S. Spaderna et al., *Gastroenterology* 131:830–840, 2006. With permission from Elsevier.)

cells, with most of the basement membrane components being contributed by the epithelial cells.

By definition, carcinomas begin on the epithelial side of the basement membrane (BM) and are considered to be **benign** as long as the cells forming them remain on this side. Sooner or later, however, many carcinomas acquire the ability to breach the basement membrane (**Figure 14.4A and B**). In certain groups of cancer patients, loss of the BM in their tumors predicts the future course of their disease, specifically their subsequent development of metastatic disease (Figure 14.4C); consequently, loss of the BM is often a prelude to eventual dissemination of cells from the primary tumor.

Having breached the BM, cancer cells begin to invade the nearby stroma singly or in coherent groups, the latter process being called **collective invasion** (**Figure 14.5**). The tumors spawning these invasive neoplastic cells are now reclassified as **malignant**. The breakdown of the BM by invading carcinoma cells removes an important physical barrier to the further expansion of tumor cell populations. But in addition, as we learned in the last chapter, by degrading various components of the BM, invasive cells harvest growth and survival factors that have been sequestered by attachment to this specialized extracellular matrix. The invading cohorts of cancer cells that may arise following the breaching of the BM rely on cells at their leading edges that have activated the cell-biological program termed the epithelial–mesenchymal transition (EMT), which we encountered previously in Section 13.3 and will be described in greater detail below.

In the cases of collective invasion, there is evidence that individual leading cells may simultaneously reside in two distinct phenotypic states: the leading halves of each of these cells (facing toward the direction of advance) may exhibit a variety of signs of

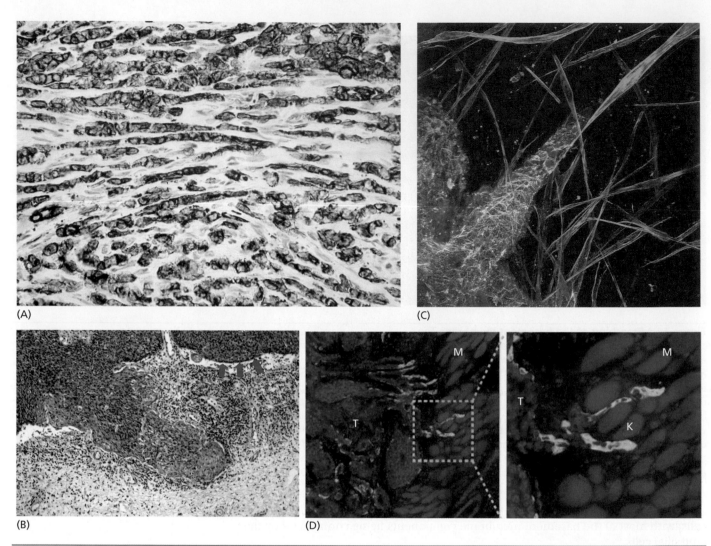

Figure 14.5 Patterns of invasion (A) These invasive lobular mammary carcinoma cells (*brown*) have left the primary tumor (*not shown to the left*) and are proceeding rightward, one by one in single file, through channels they have carved in the adjacent stroma (*white, gray*). (B) Far more typical is the coordinated invasion of a phalanx of carcinoma cells, often termed collective invasion. In this squamous cell carcinoma of the cervix, a tongue of many hundreds of cancer cells (*pink, brown*) has breached the basement membrane and is invading the stroma. The latter is characterized by both fibroblasts and inflammatory cells (*dark green*). The basement membrane is the *dark brown, horizontal line* (*pink arrows*) that separates the bulk of the carcinoma cells (*above*) from the stromal cells (*below*) and is uninterrupted except for a capillary and the tongue of invasive cancer cells. (C) Collective invasion can be modeled experimentally *in vitro*. Here, MCF-7 human breast cancer cells were cultured together with fibroblasts in a 3-dimensional (3D) matrix composed largely of collagen. (Such 3D matrices are thought to recapitulate the *in vivo* tissue environment more closely than 2D monolayer culture conditions.) The breast cancer cells, seen here invading through the collagen matrix, continue to adhere to one another via E-cadherin–containing adherens junctions (*red*). The actin stress fibers of the fibroblasts are stained with phalloidin (*green*); cell nuclei are stained with DAPI (*blue*). (D) In the MMTV-PyMT transgenic mouse model of mammary carcinoma pathogenesis, the bulk of the tumor cells (T, *blue, left*) at the invasive edge of a primary tumor are following highly invasive leader cells (*green*) that have moved from a luminal differentiation state to a basal state, express the basal state-associated cytokeratin 14 marker (K), and have thereby activated a weak form of the EMT program (see Figure 14.12). These leader cells are invading into adjacent muscle tissue (M, *red, right*). (A, courtesy of J. Jonkers. B, courtesy of T.A. Ince. C, from O. Ilina and P. Friedl, *J. Cell Sci.* 122:3203–3208, 2009. With permission from The Company of Biologists. D, from K.J. Cheung, E. Gabrielson, Z. Werb and A.J. Ewald, *Cell* 155:1639–1651, 2013. With permission from Elsevier.)

invasiveness and components of the EMT program, whereas the lagging halves may continue to form adherens junctions with their followers; such junctions ensure the continued attachment of these leaders with the follower cells that form the bulk of the invading cohort (**Figure 14.6**).

Recall that even before carcinoma cells breach the basement membrane, they often succeed in stimulating angiogenesis on the stromal side of the membrane, apparently by dispatching angiogenic factors through this porous barrier to endothelial cells within the stroma (see Figure 13.37B). Signals borne by other soluble factors may also

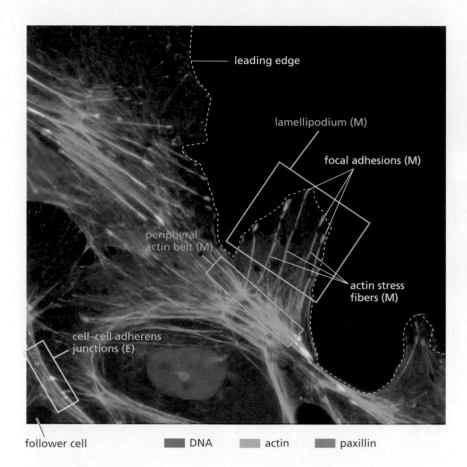

leading edge

lamellipodium (M)

focal adhesions (M)

peripheral
actin belt (M)

actin stress
fibers (M)

cell–cell adherens
junctions (E)

follower cell   ■ DNA   ■ actin   ■ paxillin

Figure 14.6 Cellular polarization at edge of invading epithelial sheet The cells at the edge of a wounded epithelial sheet undertake to fill the wounded area and restore the integrity of the epithelial sheet. To do so, they activate transiently an EMT program (see Figure 13.15A). As they move into the wound site, they retain tight associations with tightly adhering follower cells (*lower left corner*). This continued association is enabled by their undergoing a form of intracellular polarization, with the sector within the cell that faces the leading edge displaying certain mesenchymal (M) traits, while the sector facing the follower cells retaining epithelial (E) traits, as evidenced by the continued assembly of E-cadherin-containing adherens junctions (*lower left*) formed with follower, fully epithelial cells. Included among the mesenchymal traits are the actin stress fibers (*green*) ending with focal adhesions containing paxillin (*orange*). (Courtesy of P. Savagner.)

flow through an intact BM in the other direction: inflammatory cells such as macrophages on the stromal side of this membrane may induce initial EMT on the epithelial side (see Section 13.3). Nevertheless, once the BM is broken down, carcinoma cells are placed in a far better position for executing subsequent steps of the invasion-metastasis cascade.

Carcinoma cells invading the stroma can gain direct access to the blood and lymphatic vessels, which are normally found only on the stromal side of the basement membrane. Closer contact with the capillaries affords cancer cells improved access to the nutrients and oxygen carried by the blood (see also Figure 13.40B). In addition, their invasive properties enable these cancer cells to move through the walls and into the lumina (that is, the bores) of blood and lymphatic vessels. This invasion into vessels is often termed **intravasation** and depends on the ability of individual cancer cells or small clumps of these cells to break away from their neoplastic neighbors and enter into these vessels.

Local invasion seems to depend invariably on the release of secreted proteases, which are required to remodel the extracellular matrix (ECM), thereby carving out space for the advance of cancer cells (**Figure 14.7A**). In some tumors, invading carcinoma cells make their own proteases, such as MMP-2 and MMP-9, while in others a variety of stromal cells, notably fibroblasts and macrophages, are recruited and induced to produce and release these enzymes, allowing flocks of carcinoma cells to follow closely behind them (Figure 14.7B–C).

14.2 Epithelial–mesenchymal transitions profoundly reshape the phenotypes of carcinoma cells

The first of the many steps leading to metastasis—the acquisition of local invasiveness—involves major changes in the phenotype of cancer cells within the primary tumor. As before, we will focus this discussion on epithelial tissues and the carcinomas that they spawn. The organization of the epithelial cell layers in normal tissues is incompatible

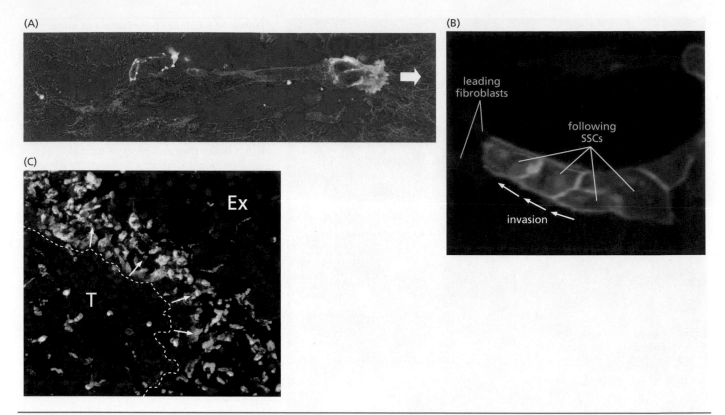

Figure 14.7 Invasion through the extracellular matrix and its control by stromal cells (A) Carcinoma cell invasion requires degradation of the extracellular matrix (ECM), as illustrated by this confocal micrograph of a cohort of 5–10 melanoma cells that are moving together through a collagen matrix (*blue*); like normal melanocytes, they continue to adhere to one another through adherens junctions that are formed by E-cadherin molecules (*red*). Large gaps in the matrix surrounding the melanoma cells (*black*) indicate areas that have been degraded by the advancing cancer cells or associated stromal cells. At the leading invasive edge (*white arrow*), the melanoma cells are displaying β1 integrins (*green*), which enable them to attach to the still-intact ECM lying ahead in their path. (B) In a variety of tumors, cancer cells recruit and co-opt stromal cells, which clear the invasion path by degrading the ECM, with the carcinoma cells following close on their heels. In this case, human oral squamous cell carcinoma cells (SCCs, *green*) are being led by a small group of carcinoma-associated fibroblasts (*red*) in an *in vitro* experimental model of cancer cell invasion. The direction of invasion by these cells (*arrows*) was determined using time-lapse microscopy. (C) In the Rip-Tag transgenic model of pancreatic islet carcinogenesis (see Figure 13.35), the carcinoma cells forming the leading edge [dashed line at edge of the tumor (T)] recruit macrophages (*green*) and, by stimulating them with interleukin-4 (IL-4), cause them to release cathepsins (*red-orange*), a class of secreted proteases distinct from matrix metalloproteinases (MMPs). Release of these cathepsins at the invasive front of the carcinoma cells is essential for invasion by these cells (*white arrows*) into the adjacent exocrine pancreas (Ex). Cathepsin-producing macrophages are seen here in *yellow*. Nuclei are stained *blue* with DAPI (*blue*). (A, from P. Friedl, Y. Hegerfeldt and M. Tusch, *Int. J. Dev. Biol.* 48:441–449, 2004. With permission from UPV/EHU Press. B, from C. Gaggioli et al., *Nat. Cell Biol.* 9:1392–1400, 2007. With permission from Nature. C, courtesy of B.B. Gadea and J.A. Joyce; from V.L. Gocheva et al., *Genes Dev.* 24:241–255, 2010. With permission from Cold Spring Harbor Laboratory Press.)

with the motility and the invasiveness displayed by malignant carcinoma cells (see Figure 2.15), yet elements of this epithelial organization plan continues to be respected in many primary carcinomas. In these tumors, well-organized sheets of epithelial cells are present, although their overall topology may be quite different from that of comparable normal epithelia (see, for example, Figure 2.5).

In order to acquire motility and invasiveness, carcinoma cells must shed many of their epithelial phenotypes, detach from epithelial sheets, and undergo a drastic alteration—the epithelial–mesenchymal transition (EMT), which was mentioned in the context of wound healing (see Section 13.3). Recall that an EMT involves the shedding by epithelial cells of their characteristic morphology and gene expression pattern and the assumption of a shape and transcriptional program more characteristic of mesenchymal cells (**Figure 14.8A**). (The EMT represents a cell-biological "program," in that it involves the coordinated changes of the expression of hundreds of genes and thus proteins, with the implication that this program is templated directly or indirectly by a cell's genome and its developmental history.)

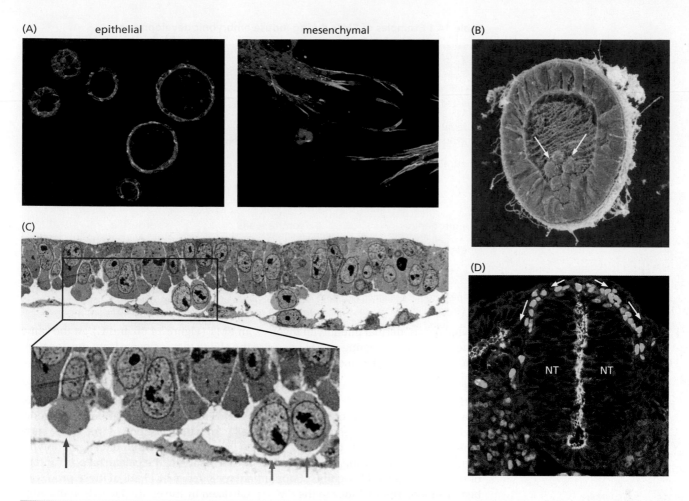

Figure 14.8 The epithelial–mesenchymal transition and embryogenesis (A) Epithelial (MDCK; Madin-Darby canine kidney) cells form ducts when propagated *in vitro* in 3D-matrices designed to model the *in vivo* microenvironment (*left*). However, when grown under identical conditions, MDCK cells forced to express the Prrx1 EMT-TF form networks of mesenchymal cells (*right*). Note that the dramatic morphological changes are accompanied by the acquisition of motility and invasive properties (*not shown*). (Nuclei revealed by DAPI staining, *blue*; actin filaments, *red*; E-cadherin (epithelial marker), *yellow-green,* left panel; vimentin (mesenchymal marker), *green,* right panel.) (B) This scanning electron micrograph shows the delamination of cells from the primitive ectoderm of a sea urchin embryo (*white arrows*) and their migration into the interior of the embryo—the process of gastrulation. These cells have become round and acquired motility in anticipation of their forming the rudiments of the mesoderm—changes associated with an epithelial–mesenchymal transition (EMT). They are migrating along strands of extracellular matrix in the blastocoel of this early embryo. (C) The process of gastrulation in mammals, such as in the early rabbit embryo studied here, is superficially quite different but depends on very similar processes. As revealed by this transmission electron micrograph, one cell of the primitive ectoderm (also termed epiblast; lower panel, *red arrow*) is in the midst of undergoing an EMT, while two other cells (*blue arrows*) have already delaminated from the epiblast and become mesenchymal, thereby initiating the formation of mesodermal tissues. (D) Certain cells derive from the upper region of the neural tube (NT, *center*) and can be immunostained for the Sox 9 transcription factor. These cells (*light blue, top*) arise in the embryonic neural crest of chordates and undergo an EMT, delaminate from this epithelium, acquire motility and invasiveness, and disperse (*white arrows*) throughout the embryo, where they form melanocytes, much of the peripheral nervous systems, and much of the skeleton of the face. The neural epithelium is immunostained for N-cadherin (*yellow*). All cells are stained with phalloidin to reveal their actin cytoskeleton (*red*) and with DAPI to indicate their nuclei (*dark blue*). (A, from M.A. Nieto, *Science* 342:1234850, 2013. With permission from AAAS. B, from G.N. Cherr et al., *Microsc. Res. Tech.* 22:11–22, 1992. With permission from John Wiley & Sons. C, from C. Viebahn, B. Mayer and A. Miething, *Acta Anat. (Basel)* 154:99–110, 1995. With permission from Karger. D, courtesy of J. Briscoe; see also J. Yang and R.A. Weinberg, *Dev. Cell* 14:818–829, 2008.)

In its normal manifestations, the EMT is used in specific morphogenetic steps occurring during embryogenesis, when tissue remodeling depends on EMTs executed by various types of epithelial cells (**Table 14.1**). During one of the steps of **gastrulation**, for example, individual cells peel away from the ectoderm and migrate inward toward the center of the embryo to form the mesoderm, the precursor of mesenchymal tissues, including (in chordates) fibroblasts and hematopoietic cells. This conversion of ectodermal cells, which at this stage are arrayed in an epithelial cell layer, to those

Table 14.1 Examples of EMTs during mouse embryonic development

Process	From	Transition	
		From	**To**
gastrulation	epiblast	mesoderm	
prevalvular mesenchyme in the heart	endothelium	atrial and ventricular septum	
neural crest cells	neural plate	neural crest cells, which can yield bone, muscle, peripheral nervous system	
somitogenesis	somite walls	sclerotome	
palate formation	oral epithelium	mesenchymal cells	
Müllerian duct regression	Müllerian tract	mesenchymal cells	

Adapted from P. Savagner, *BioEssays* 23:912–923, 2001.

having a mesodermal phenotype involves an EMT (Figure 14.8B and C). At the same time, the cells undergoing an EMT acquire the ability to translocate from one location (the outer cell layer) to another (the interior) within the embryo.

The migration of neuroepithelial cells from the neural crest into the mesenchyme of early vertebrate embryos also depends on a transformation of cell phenotype that can best be described as an EMT (Figure 14.8D). Similarly, the migration of myogenic precursor cells (the progenitors of muscle cells) from the dermomyotome of the early embryo to the limb buds depends on an EMT-like transformation of cell phenotype. Even late steps in morphogenesis, such as the formation of the palate, take advantage of the powers of EMT programs (Supplementary Sidebar 14.1). All of these processes bear a striking resemblance to the EMT undertaken by the epithelial cells at the edge of a wound that has damaged an epithelial cell sheet; these cells must undergo a transient EMT in order to migrate into the wound site and close the gap in the epithelial cell sheet that was created by the wounding process (see Figure 13.15A). Hence, EMT programs play important roles in two aspects of normal physiology—developmental morphogenesis and wound healing.

As described below, an EMT can also be seen at the edges of carcinomas that are invading adjacent tissues (**Figure 14.9**; see also Figure 14.5D). This pathological

Figure 14.9 Epithelial–mesenchymal transition at the invasive edge of a tumor (A) In the Rip-Tag transgenic mouse model of pancreatic islet carcinogenesis (see Figure 13.35), neoplastic cells undergo an EMT that enables them to invade the surrounding exocrine pancreas (Ex). Carcinoma cells at the far right continue to reside in an epithelial state (E), exhibit E-cadherin staining (*green*), and in many places exhibit overlapping β-catenin staining (*red*). The overlap of these two proteins, indicating their co-localization in adherens junctions, appears as *yellow*. Those carcinoma cells that have undergone an EMT and become more mesenchymal (M) have lost E-cadherin staining and show greatly increased levels of cytoplasmic β-catenin staining. Nuclei are stained *blue* with DAPI. (B) Colon carcinoma cells at the invasive edge of a primary human tumor undergo changes in gene expression and the localization of certain proteins. Although E-cadherin (*brown*) is strongly expressed on the plasma membranes of cells in the core of a primary tumor, where it forms adherens junctions (*left panel*) that delineate the associations between adjacent cells, its expression decreases substantially in individual invasive cells at the edge of this tumor (*pink arrows, right panel*) and is no longer localized to their plasma membranes, indicating that it no longer participates in forming adherens junctions. (C) Tumor cells in the center of this colon tumor (*dashed lines, left panel*) express β-catenin (*dark red*) under their plasma membranes and diffusely throughout the cytoplasm, whereas tumor cells at the invasive edge (*right panel*) show intense β-catenin staining in their nuclei. (D) The ectodomain of E-cadherin displayed by an epithelial cell dimerizes with the ectodomain of a second E-cadherin molecule displayed by an adjacent epithelial cell (*left*). Calcium ions, which are essential for the normal secondary structure and rigidity of the ectodomains, are shown as *red balls*. At the same time, the cytoplasmic tail of E-cadherin is linked via β-catenin and several other molecules to the actin cytoskeleton. β-Catenin also functions in the cytoplasm as a key intermediary in the Wnt signaling pathway (see Section 6.10). Loss of E-cadherin from the plasma membrane liberates β-catenin molecules, which may then accumulate in the cytoplasm and, under the influence of Wnt-initiated signals, migrate to the nucleus and associate with Tcf/LEF and other transcription factors, thereby inducing expression of genes orchestrating the EMT program. (A, from M. Herzig et al., *Oncogene* 26:2290–2298, 2007. With permission from Nature. B, courtesy of T. Brabletz. C, courtesy of T. Brabletz and T. Kirchner. D, from T.A. Graham et al., *Cell* 103:885–896, 2000, PDB: 2BCT; and from E. Parisini et al., *J. Mol. Biol.* 373:401–411, 2007. Both with permission from Elsevier.)

process is very similar to the EMTs occurring during early embryogenesis and wound healing. It is plausible, though hardly proven, that all types of carcinoma cells must activate components of an EMT program in order to become motile and invasive.

The strong resemblance between the pathological process of tumor invasiveness and normal steps of embryogenesis and wound healing suggests a mechanistic model that is supported by extensive evidence gathered in recent years: the complex program of cellular reorganization exhibited by invasive carcinoma cells depends on the reactivation of a latent behavioral program whose expression is usually confined to early embryogenesis and to damaged adult tissues. Following this thinking, once carcinoma cells acquire access to an expressed EMT program, they can exploit it to profoundly change their own morphology, motility, and ability to invade nearby cell

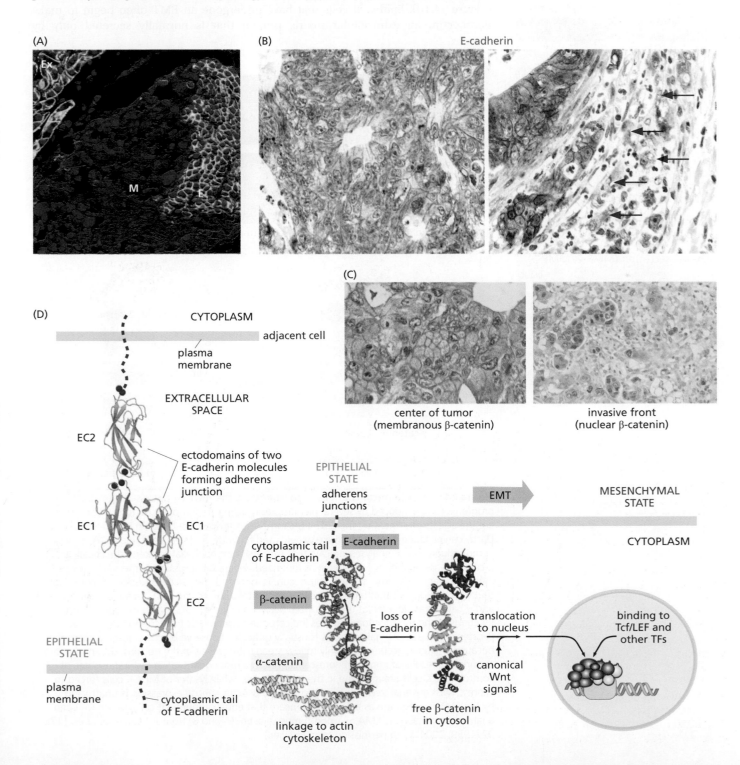

(A)

(B) E-cadherin

(C)

center of tumor
(membranous β-catenin)

invasive front
(nuclear β-catenin)

(D)

CYTOPLASM

adjacent cell

plasma
membrane

EXTRACELLULAR
SPACE

EC2

ectodomains of two
E-cadherin molecules
forming adherens
junction

EPITHELIAL
STATE

adherens
junctions

EMT

MESENCHYMAL
STATE

EC1   EC1

cytoplasmic tail
of E-cadherin

E-cadherin

CYTOPLASM

β-catenin

EC2

loss of
E-cadherin

translocation
to nucleus

binding to
Tcf/LEF and
other TFs

EPITHELIAL
STATE

α-catenin

canonical
Wnt
signals

plasma
membrane

cytoplasmic tail
of E-cadherin

linkage to actin
cytoskeleton

free β-catenin
in cytosol

Table 14.2 Cellular changes associated with an epithelial–mesenchymal transition

Loss of
cytokeratin (intermediate filament) expression
tight junctions and epithelial adherens junctions involving E-cadherin
epithelial cell polarity
epithelial gene expression program

Acquisition of
fibroblast-like shape
motility
invasiveness
increased resistance to apoptosis
increased resistance to radiotherapy
mesenchymal gene expression program including EMT-inducing transcription factors
mesenchymal adherens junction protein (N-cadherin)
protease secretion (MMP-2, MMP-9)
vimentin (intermediate filament) expression
fibronectin secretion
PDGF receptor expression
$\alpha_v\beta_6$ integrin expression
stem cell-like traits

layers. This model implies that the multiple changes in cell phenotype associated with invasiveness, some of which are described below, need not be acquired piecemeal by carcinoma cells. Instead, these cells simply activate a morphogenetic program that is already encoded in their genomes. (This logic echoes our earlier discussion in Section 13.3 of epithelial–stromal interactions, where we argued that cancer cells co-opt entire wound-healing programs in order to acquire an activated stroma.)

The normal and pathological versions of the EMT involve, in addition to changes in shape and the acquisition of motility, fundamental alterations in the gene expression profiles of cells (**Table 14.2**). Expression of E-cadherin and cytokeratins—hallmarks of epithelial cell protein expression—is repressed, while the expression of **vimentin**, an intermediate filament component of the mesenchymal cell cytoskeleton, is induced (**Figure 14.10**). Epithelial cells that have undergone an EMT often begin to make fibronectin, an extracellular matrix protein that is normally secreted only by

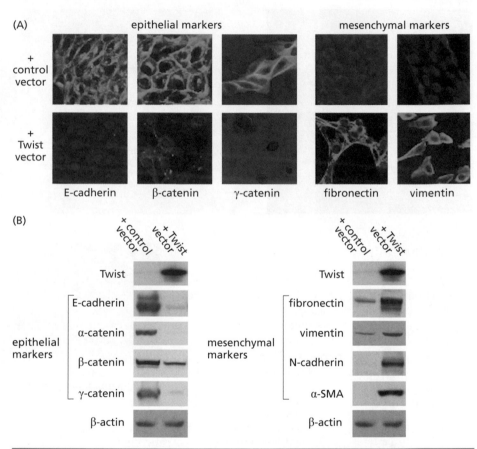

Figure 14.10 Biochemical changes accompanying the EMT As discussed later, EMT programs can be induced by several pleiotropically acting transcription factors. Shown here are the effects of expressing the EMT-inducing Twist transcription factor in MDCK (Madin–Darby canine kidney) cells, which are widely used to study epithelial cell biology. (A) These immunofluorescence analyses indicate that expression of epithelial markers, specifically E-cadherin, β-catenin, and γ-catenin, is repressed, whereas expression of mesenchymal markers, specifically vimentin and fibronectin, is induced by ectopic expression of Twist. Note that in epithelial cells that have not been induced to enter into an EMT, E-cadherin, β-catenin, and γ-catenin are located at the interfaces between adjacent cells, where they form adherens junctions. Although functional studies indicate the activation of transcription by nuclear β-catenin in mesenchymal cells, this is usually difficult to observe by immunofluorescence because β-catenin accumulates only to low steady-state levels in the nucleus. (B) Immunoblots confirm the results of immunofluorescence, but in a more quantitative fashion. Lysates of control MDCK cells are analyzed in the left channels, while lysates of MDCK cells forced to express Twist are analyzed in the right channels. β-Actin, whose expression is unaffected by the EMT, is used here as a control to ensure that equal amounts of cell lysate have been analyzed in all cases. α-SMA, α-smooth muscle actin. (A and B, from J. Yang et al., *Cell* 117: 927–939, 2004. With permission from Elsevier.)

mesenchymal cells such as fibroblasts. At the same time, expression of a typical fibroblastic marker—N-cadherin—is often acquired in place of E-cadherin.

Of all these proteins, the transmembrane E-cadherin molecule plays the dominant role in influencing epithelial versus mesenchymal cell phenotypes. Recall our earlier encounters with E-cadherin and its role in enabling epithelial cells to adhere laterally to one another (see Figures 6.19A and 13.14). In normal epithelia, the ectodomains of E-cadherin molecules extend from the plasma membrane of one epithelial cell to form complexes with other E-cadherin molecules protruding from the surface of an adjacent epithelial cell. This enables homodimeric (and higher-order) bridges to be built between adjacent cells in an epithelial cell layer, resulting in the adherens junctions that are so important to the structural integrity of epithelial cell sheets (see Figure 13.14D).

The cytoplasmic domains of individual E-cadherin molecules are tethered to the actin fibers of the cytoskeleton via a complex of α- and β-catenins (see Figure 14.9D) and other ancillary proteins. The actin cytoskeleton, for its part, provides tensile strength to the cell. Hence, by knitting together the actin cytoskeletons of adjacent cells, E-cadherin molecules help an epithelial cell sheet resist mechanical forces that might otherwise tear it apart. Once E-cadherin expression is suppressed, many of the other cell-physiological changes associated with the EMT seem to follow suit. Some experiments indicate that simply by suppressing the expression of the E-cadherin protein, cells acquire a more mesenchymal morphology and increased motility. Yet other cell–cell physical interactions in both normal and neoplastic tissues are mediated by other members of the large family of cadherin proteins encoded in the human genome. For example, the binding of normal and neoplastic epithelial cells to the endothelial cells forming microvessels depends on **homotypic** interactions mediated by cadherin 5, that is, interactions between two identical molecules displayed by two closely apposed cells.

The pivotal role of E-cadherin in the acquisition of malignant cell phenotypes is further supported by observations indicating that the *CDH1* gene, which specifies E-cadherin, is repressed by promoter methylation in certain invasive human carcinomas (see Table 7.2) and in others by transcriptional repressors; this gene can also be inactivated by reading-frame mutations. For example, an analysis of 26 human breast cancer cell lines indicated that 8 had mutations that led to inactivation of E-cadherin gene expression, 5 had truncating mutations in the E-cadherin reading frame, and 3 had in-frame deletions resulting in the expression of mutant E-cadherin molecules at the cell surface. By now, loss of E-cadherin expression or expression of mutant E-cadherin proteins has been documented in advanced carcinomas of the breast, colon, prostate, stomach, liver, esophagus, skin, kidney, and lung. Finally, mutant germ-line alleles of the *CDH1* gene result in familial gastric cancer (see Table 7.1).

Additionally, in studies of several types of carcinoma cells that had lost E-cadherin expression, re-expression of this protein (achieved experimentally by introduction of an E-cadherin expression vector) strongly suppressed the invasiveness and metastatic dissemination of these cancer cells. Together, these diverse observations indicate that E-cadherin levels are key determinants of the biological behavior of epithelial cancer cells and that the cell-to-cell contacts constructed by E-cadherins impede invasiveness and hence metastasis.

Like E-cadherin, the N-cadherin that is produced in its stead during the course of an EMT participates in homophilic interactions. Consequently, the N-cadherin molecules expressed on the surface of a carcinoma cell that has undergone an EMT increase the affinity of this cancer cell for the stromal cells that normally display N-cadherin, notably the fibroblasts in the stroma underlying the epithelial cell layer. This association seems to help invading carcinoma cells insert themselves amid stromal cell populations. Precisely the same dynamics have been proposed to explain how melanomas develop: normal melanocytes express E-cadherin, which binds them to the skin keratinocytes around them; melanoma cells—the transformed derivatives of melanocytes—express N-cadherin, which facilitates their invasion of the dermal stroma of the skin and their association with its fibroblasts and endothelial cells

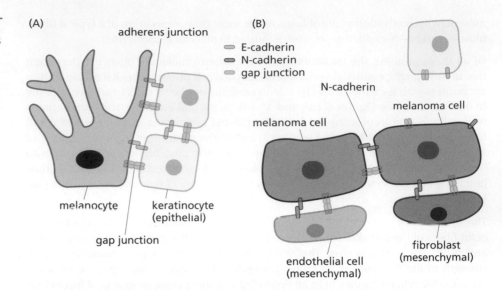

Figure 14.11 Cadherin shifts and melanoma cell invasiveness Melanomas are among the most malignant tumors because of their tendency to metastasize widely once they reach a certain stage of tumor progression. This behavior is attributable, in part, to the reactivation of a cell-biological program that enabled the migratory behavior of their neural crest ancestors (see Figure 14.8D). The shift from E- to N-cadherin expression, which occurs when melanocytes (A) become transformed into melanoma cells (B), has been proposed to facilitate invasion of the stroma, since the shutdown of E-cadherin expression (*green*) enables these cancer cells to extricate themselves from their keratinocyte neighbors in the epidermis, while its replacement by expression of N-cadherin (*reddish pink*) allows these tumor cells to form homotypic interactions with various types of mesenchymal cells, such as fibroblasts and endothelial cells, that reside in the stroma of the skin (i.e., the dermis). Of note, the EMT-inducing Slug transcription factor (see Figure 14.19B) plays a key role in activating aggressive behavior in melanoma cells. (Adapted from N.K. Haass, K.S.M Smalley and M. Herlyn, *J. Mol. Histol.* 35:309–318, 2004. With permission from Springer.)

(**Figure 14.11**). Moreover, E-cadherin:N-cadherin interactions have recently been documented that violate the simple rule limiting cadherins strictly to homophilic interactions. The non-homophilic interactions are less strong and, being more transient, may also help the initial invasion of more epithelial carcinoma cells into the stroma.

The intermolecular homophilic bonds formed between pairs of N-cadherin molecules are weaker than those formed by E-cadherin homodimers. This helps to explain why cell surface N-cadherin molecules actively favor cell motility and thus behave very differently from their E-cadherin cousins, which work to immobilize cells within epithelial cell layers. This point is driven home by experiments in which expression vectors are used to force high levels of N-cadherin expression in otherwise-normal cultured epithelial cells. Such ectopic expression causes the epithelial cells to acquire motility and invasiveness, as indicated by their ability to break through a reconstructed extracellular matrix introduced into a culture dish.

Finally, we should note that the above description of the EMT program is simplistic in at least one very important way: the cancer cells in human carcinomas that have activated an EMT program rarely advance all the way toward a fully mesenchymal state. Instead, they proceed only partway down the road, resulting in a "partial EMT" in which these neoplastic cells co-express certain retained epithelial markers together with newly acquired mesenchymal ones. Hence, the EMT program does not operate as a binary switch that alternates between extreme epithelial and mesenchymal poles. Instead, its actions are manifested by the entrance of cells into a spectrum of phenotypic states arrayed along an epithelial–mesenchymal axis, with the possibility that cells can rapidly interconvert between these intermediate states depending on contextual signals that they may be receiving (**Figure 14.12**). Indeed, it is unclear whether an EMT program, as it operates in spontaneously arising tumors, ever transports a carcinoma cell all the way into a highly mesenchymal state such as that of a typical fibroblast.

14.3 Epithelial–mesenchymal transitions are often induced by contextual signals

Stepping back for a moment, much of the thinking described above could be traced back to pioneering experiments carried out in Austria in the mid-1990s, which extended earlier research on the role of EMT in embryogenesis into the area of cancer pathogenesis. As was demonstrated at the time, exposure of *ras*-transformed EpRas mouse mammary epithelial cells (MECs) to TGF-β resulted in the progressive loss of epithelial morphology and a reduction of epithelial markers, including cytokeratins and E-cadherin (**Figure 14.13**). At the same time, these transformed cells acquired mesenchymal protein markers, such as vimentin, and assumed a morphology resembling that of fibroblasts—all of the hallmarks of the EMT studied earlier by developmental biologists.

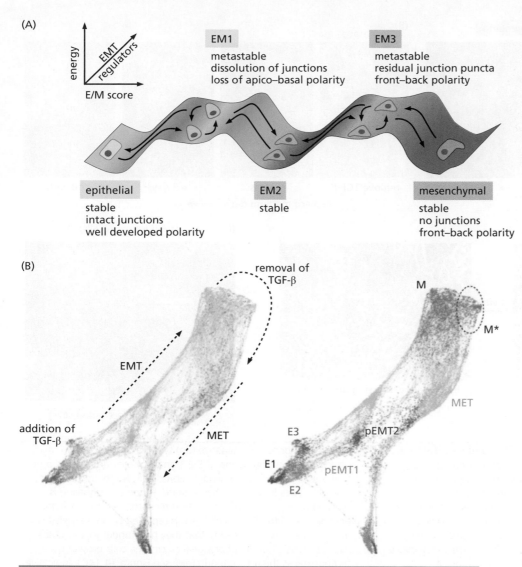

Figure 14.12 Multiple intermediate states between epithelial and mesenchymal phenotypes (A) As depicted here, the EMT does not operate as a binary switch that conveys cells between two alternative states—highly epithelial and highly mesenchymal. Instead, multiple intermediate states with differing proportions of epithelial and mesenchymal traits intervene between the two extreme endpoints of this spectrum of phenotypes. The interconversions between phenotypic states are represented as movements between states plotted on a free energy diagram, with cells residing in a stable or metastable fashion in energy minima. While the mesenchymal state is indicated at the right of this spectrum, in reality spontaneous activation of an EMT program in carcinoma cells during the course of tumor progression rarely if ever generates fully mesenchymal cells. (B) Mass cytometry uses the analytical tools of mass spectrometry to detect isotope-coupled antibody molecules, enabling the simultaneous detection of dozens of different antibody-bound antigens in a biological sample. The resulting multi-dimensional protein expression profiles are presented in a way that collapses multi-dimensional measurements into the 2-dimensional images shown here. The *left* image indicates the course of entrance into successive phenotypic states of cells following application of TGF-β (which induces an EMT program), whereas the reversal of those phenotypes and the return to a more epithelial state following withdrawal of TGF-β is indicated by MET. The *right* image indicates distinct phenotypic states that were assigned to cell populations as they passed between epithelial and mesenchymal cell states. (A, from M.A. Nieto, R.Y.-J. Huang, R.A. Jackson and J.P. Thiery, *Cell*, 166:21–45, 2016. With permission from Elsevier. B, from L.G. Karacosta, B. Anchang, N. Ignatiadis et al., *Nature Commun.* 10:5587, 2019. With permission from Nature.)

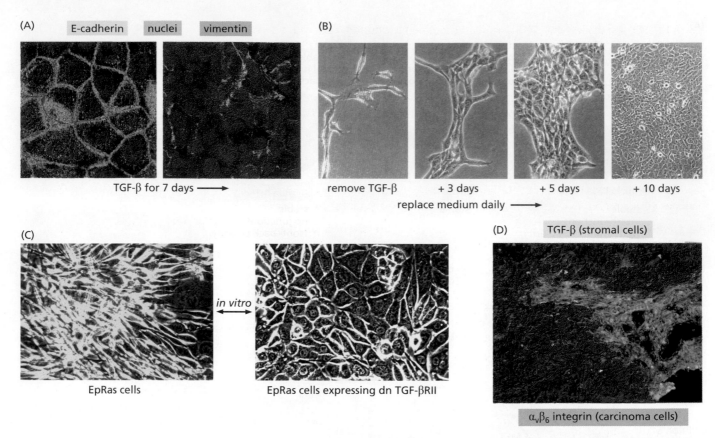

Figure 14.13 Control of the EMT by TGF-β and its effects on tumorigenic cells Immortalized mouse mammary epithelial cells (MECs) of the EpH4 cell line were transformed into EpRas tumor cells by introducing a *ras* oncogene. (A) EpRas cells (*left panel*) usually have an epithelial, cobblestone-like appearance and express E-cadherin (*green*) at their cell–cell junctions. However, when cultured for 7 days in the presence of TGF-β1 (*right panel*), they undergo an EMT and assume an elongated fibroblastic appearance (*not visible here*). In addition, they suppress expression of E-cadherin and express instead vimentin (*red*), thereby shifting from an epithelial to a mesenchymal gene expression program. Nuclei are stained *blue* with DAPI dye. (B) Thereafter, EpRas cells maintain the EMT-induced mesenchymal, fibroblast-like state through their own production of and response to TGF-β1 (i.e., via autocrine signaling; *not shown*). However, as seen here, when these cells are cultured in a medium that lacks added TGF-β1 and their growth medium is changed daily (to remove any TGF-β that they may have secreted into the medium and have accumulated), their appearance gradually reverts to an epithelial cobblestone phenotype, as shown after 3, 5, and 10 days of culture (*left to right*), indicating that they have undergone a mesenchymal–epithelial transition (MET).

(C) Use of a dominant-negative (dn) type II TGF-β receptor (which effectively blocks autocrine TGF-β signaling) provides further proof that autocrine TGF-β signaling by EpRas cells is required to maintain their residence in a mesenchymal state. When this signaling is blocked by expression of this mutant receptor, the mesenchymal appearance of the EpRas cells (*left*) is replaced by an epithelial appearance (*right*), indicating that they have undergone an MET. (D) Detroit 562 human pharyngeal carcinoma cells growing in a tumor express the $\alpha_v\beta_6$ integrin (*red*; see Figure 14.14C), indicative of their having undergone an EMT, whereas TGF-β (*green*) is produced by cells in the nearby tumor-associated stroma. The $\alpha_v\beta_6$ displayed by the tumor cells can be deployed to activate the latent form of TGF-β produced by the stromal cells, thereby creating a self-sustaining, positive-feedback loop (since the activated TGF-β is a strong inducer of the EMT and thus of additional expression of $\alpha_v\beta_6$ integrin by the carcinoma cells, yielding even more activated TGF-β). (A, from E. Janda et al., *J. Cell Biol.* 156:299–314, 2002. With permission from Rockefeller University Press. B, from M. Oft et al., *Genes Dev.* 10:2462–2477, 1996. With permission from Cold Spring Harbor Laboratory Press. C and D, courtesy of D.R. Leone, B.M. Dolinski and S.M. Violette, Biogen MA Inc.)

Provocatively, once these *ras*-transformed cells underwent an EMT, they began to produce their own TGF-β1; this TGF-β1, acting via an autocrine positive-feedback signaling loop (see Figure 5.8C), allowed them to maintain their mesenchymal phenotype for extended periods of time, long after the inciting TGF-β was withdrawn from their culture medium. If they were prevented, however, from accumulating concentrations of TGF-β1 in their immediate surroundings, then they would lapse back to an epithelial state (Figure 14.13B). A similar reversion to an epithelial state was achieved through inhibition of signaling by a key TGF-β1 receptor displayed by these cells (Figure 14.13C; see also Figure 5.15). Moreover, maintenance of TGF-β autocrine signaling through a positive-feedback loop was apparently amplified by the $\alpha_v\beta_6$ integrin, which is displayed characteristically as one component of the EMT program and serves to activate latent TGF-β sequestered in the extracellular space (Figure 14.13D).

Figure 14.13D also reveals that TGF-β is often produced in abundance by the tumor-associated stroma. These experiments, taken together, demonstrated that TGF-β signaling could conspire with oncogenes frequently found in human cancers—members of the *ras* gene family—to force epithelial cancer cells to undergo an EMT.

In retrospect, this work established three principles that were to guide research into the EMT-cancer connection in the quarter century that followed. First, that a contextual signal that is abundant in many human carcinomas could serve as the trigger of EMT and the resulting formation of more-mesenchymal carcinoma cells. Second, that the resulting cells could thereafter retain their more-mesenchymal phenotype through autocrine signaling. Finally, that this residence in a quasi-mesenchymal state was only **metastable**, in that it could be reversed by interrupting certain positive-feedback controls such as the TGF-β-dependent autocrine loop (see Figure 14.13C).

The prominent role of TGF-β in actively promoting the aggressiveness of malignant cancer cells contrasts starkly with our earlier discussions of its anti-proliferative effects (see Section 8.10 and Supplementary Sidebar 14.2). In Chapter 8, we learned that this cytokine could block the advance of cells through the cell cycle, often preventing forward march through the G_1 phase of the cell cycle. In this chapter we encounter TGF-β in an entirely different role in which it actually *fosters* tumor progression by activating EMT programs. The resolution of this apparent paradox comes from an understanding of the effects of somatic mutations acquired during primary tumor formation, such as acquisition of a *ras* oncogene or inactivation of the *Rb* tumor suppressor gene. Both of these changes, as well as other similarly acting mutations, render a carcinoma cell resistant to the cytostatic effects of TGF-β, permitting its EMT-inducing effects to manifest themselves.

Strong support for the notion that TGF-β can favor malignant cell behavior is provided by numerous studies in which the levels of tumor-associated TGF-β (often TGF-β1) have been found to rise in parallel with increasing degrees of tumor invasiveness and general aggressiveness. Indeed, high levels of TGF-β, both within the tumor mass and in the general circulation, augur poorly for the long-term survival of a cancer patient.

In the years that followed the early EMT–TGF-β experiments, studies of the localization within a tumor of the carcinoma cells that underwent an EMT reinforced the notion that contextual signals are often important in triggering the expression of EMT programs in carcinoma cells. As seen in **Figure 14.14**, carcinoma cells in close contact with the surrounding stroma activate their EMT programs, whereas those residing in the interior of islands of carcinoma cells (and thus shielded from direct contact with stroma) fail to do so. The notion that these "edge cells" have indeed activated components of their EMT program is supported by the observations that these cells also exhibit changes in a collection of other proteins expressed following experimental induction of EMT programs in cultured cells.

To summarize, in the primary tumor, a **"reactive stroma"** forms during the long course of primary tumor formation and, once formed, releases EMT-inducing signals (among them TGF-β) back to nearby carcinoma cells, which indeed are the neoplastic cells that previously were responsible for provoking the formation of the reactive stroma in the first place.

These observations and others like them (see Figure 14.14 C–E) clearly indicate the involvement of contextual signals experienced by carcinoma cells at the outer edges of islands of these cells. Actually, these signals may be of several types: In addition to soluble factors released by stromal cells, epithelial cells at the edges of these islands may sense that they no longer enjoy direct interactions on all sides with epithelial neighbors; the absence of a full complement of epithelial–epithelial homotypic interactions (**Figure 14.15**) might, on its own, induce them to activate an EMT program.

The extracellular matrix (ECM) assembled by cells in the stroma also contributes to EMT induction. Some research implicates direct contact of carcinoma cells with collagen type I, which is present in abundance in the ECM of the stroma but is absent in the epithelial compartment of the tumor, as another that favors EMT induction.

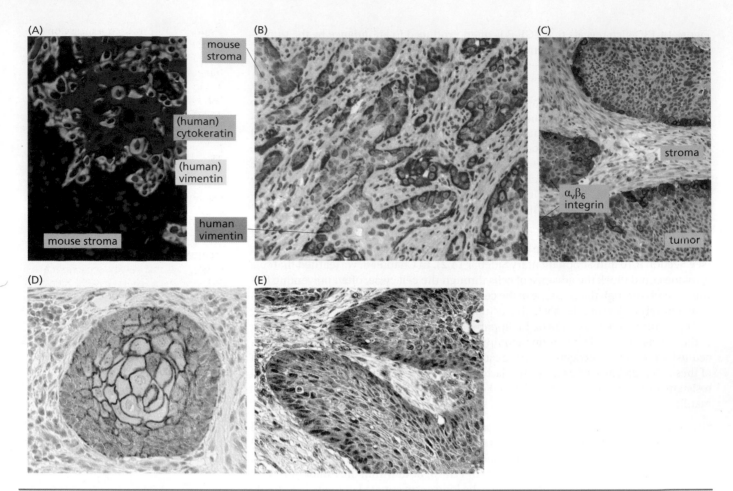

Figure 14.14 Manifestations of the EMT at the interface between tumor epithelium and stroma (A) Experimentally transformed human mammary epithelial cells (MECs) were implanted in an immunocompromised mouse host. The cytokeratin-positive human carcinoma cells (*red*) toward the center of the tumor mass are not in direct contact with the surrounding mouse stromal cells, whose presence is indicated only by their DAPI-stained nuclei (*blue*). However, many of the human MECs that are in contact with the stroma have undergone an EMT, as indicated by their loss of cytokeratin staining (*red*) and their display instead of human-specific vimentin (*green*). [The use of antibody that specifically recognizes human (and not mouse) vimentin ensures that the green cells at the invasive edge derive from the engrafted human cells rather than from the mouse host.] Moreover, some of these cancer cells at the invasive edge have lost the cuboidal shape of the epithelial cancer cells and have assumed, instead, a more elongated, fibroblastic shape. (B) A tumor formed by the same strain of transformed human MECs described in panel B is seen here at lower magnification. The display of human vimentin (*dark brown*), which indicates passage through an EMT, is limited to cells that are in direct contact with the surrounding stroma. Conversely, tumor cells in the interior of these islands do not display human vimentin and have presumably remained in an epithelial state.

(C) The expression of the $\alpha_v\beta_6$ integrin is associated with the EMT program. This integrin is expressed in epithelial tissues that are undergoing wound healing or suffering chronic inflammation; it is also seen at the invasive edge of carcinomas. In a xenografted tumor formed by SCC-14 human pharyngeal carcinoma cells, expression of the $\alpha_v\beta_6$ integrin is exhibited by carcinoma cells at the invasive edge of the tumor (*dark brown*) that are in direct contact with the tumor-associated stroma, suggesting that stromal signals are responsible for its expression in epithelial cells. (D) In this human oral squamous cell carcinoma (OSCC), intense expression of p120 catenin (*light brown*), another component of the adherens junction, is limited to the center of this island of carcinoma cells, whereas cells closer to the surrounding stroma (*light blue nuclei*) have lost p120 catenin and thus the adherens junctions typical of epithelial cells. Cell nuclei are stained *blue* with hematoxylin. (E) In another OSCC, matrix metalloproteinase-2 (MMP-2, *brown*), which is up-regulated during passage through an EMT and deployed by some carcinoma cells to facilitate invasion, is also seen to be expressed preferentially near the stroma. (A, courtesy of K. Hartwell and T.A. Ince. B, courtesy of T.A. Ince. C, courtesy of D.R. Leone, B.M. Dolinski and S.M. Violette, Biogen MA Inc. D and E, from M. Vidal et al., *Am. J. Pathol.* 176:3007–3014, 2010. With permission from Elsevier.)

Equally important is the *physical state* experienced by carcinoma cells: stromal cells may construct a stiff ECM that induces carcinoma cells to activate components of their EMT program (**Figure 14.16**). At the level of signal transduction biochemistry, Twist, a transcription factor and organizer of EMT programs, is liberated from its site of cytoplasmic sequestration and allowed to migrate to the nucleus, where its EMT-promoting powers can be exercised. Similarly, the expression of Snail, another

24 hours

48 hours

72 hours

96 hours

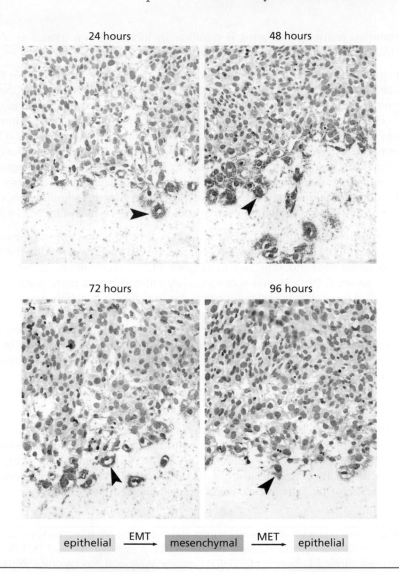

epithelial $\xrightarrow{\text{EMT}}$ mesenchymal $\xrightarrow{\text{MET}}$ epithelial

Figure 14.15 Transient expression of an EMT-inducing transcription factor in wound healing Expression of the Slug transcription factor is induced transiently in a monolayer culture of keratinocytes that has been wounded by scraping away a swath of cells. As seen here, 48 hours after wounding, keratinocytes at the edge of the wound induce expression of the Slug EMT-TF (*dark brown*) as they separate from the contiguous monolayer and begin to make their way as individual cells into the wound site (*bottom of each panel*), doing so in order to reconstruct an intact monolayer. By 96 hours, most of these cells cease expressing Slug and become integrated into a continuous monolayer. Hence, cells that initially underwent an EMT under the influence of the Slug EMT-TF in order to fill in and cover the wound site subsequently undergo an MET (mesenchymal–epithelial transition) in order to revert to the epithelial phenotype of their ancestors and reassemble a contiguous epithelial sheet. (From P. Savagner et al., *J. Cell Physiol.* 202: 858–866, 2004. With permission from John Wiley & Sons.)

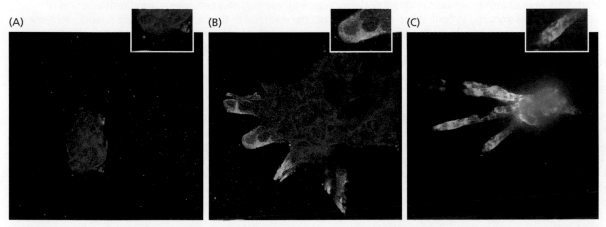

Figure 14.16 TGF-β and stiff extracellular matrix conspire to activate EMT markers Mammary carcinoma cells from the MMTV-PyMT mouse model of adenocarcinoma pathogenesis were propagated *in vitro* in collagen I gels having different degrees of stiffness in the presence or absence of added TGF-β. The gels model one aspect of the extracellular matrix (ECM) encountered by such gels during tumor pathogenesis. The expression of the RNA encoding the Snail EMT-TF was gauged by use of a recombinant construct in which the promoter of the *Snail* gene drove expression of a YFP fluorescent reporter (*green*). In each panel, a higher magnification section of an interface between the introduced carcinoma cells and the surrounding gel is provided in the *top right*. (A) In a soft collagen I, the introduced carcinoma cells formed small noninvasive spheroids. (B) However, if the gel into which the cells were introduced was constructed to provide a relatively stiff surrounding matrix, these colonies began to extend small invasive fingers that expressed the *Snail* RNA at elevated levels. (C) If the introduced cells were propagated in a stiff gel, as in (B), that also contained TGF-β1, then long, highly invasive *Snail*-positive fingers were extended into the gel. (From A.H. Mekhdjian, F. Kai, M.G. Rubashkin et al., *Mol. Biol. Cell* 28:1467–1488, 2017.)

Table 14.3 Contextual signals inducing EMT programs

TGF-β
canonical and non-canonical Wnts
Notch ligands
RTKs such as EGF-R, FGF-R, Met
various cytokines
stiff extracellular matrix
absence of epithelial neighbors
periostin
collagen type1

EMT-inducing transcription factor, is significantly increased in response to stiffness of the ECM as is TGF-β autocrine signaling.

The major class of EMT-inducing signals involves a large cohort of soluble factors that, like the growth factors described in Chapter 5, are released by cells of the reactive stroma and impinge on nearby carcinoma cells located at the outer edges of epithelial cell masses, TGF-β being only one of them. Numerous observations implicate a variety of other factors, including Wnts, TNF-α (tumor necrosis factor-α), EGF (epidermal growth factor), HGF (hepatocyte growth factor), IGF-1 (insulin-like growth factor-1), as well as interleukins 6 and 8. It is likely that these stromal signals act in various combinations to induce epithelial cells to activate their previously latent EMT programs (**Table 14.3**). This raises the question of why this signaling became so complex during the course of metazoan evolution (Supplementary Sidebar 14.3).

This theme of combinations of signals working through paracrine and autocrine signaling channels was supported by experiments in 2011, which extended the notion of collaborating EMT-inducing signals (**Figure 14.17**). Study of the growth factors that were secreted by cells that spontaneously entered into a more mesenchymal state revealed three classes of signaling proteins that appear to collaborate in various ways to actively maintain ongoing residence in this state: canonical Wnt proteins and non-canonical Wnt proteins (see Section 6.10), and the aforementioned TGF-β (see Section 6.12). Ongoing secretion and resulting autocrine stimulation by these factors were essential for continued maintenance of the mesenchymal phenotype. Conversely, if these autocrine signaling loops were interrupted by physiological inhibitors of these signaling loops, cells would revert to and rest in an epithelial state.

In addition, these experiments reinforced the notion that, although an EMT program can be triggered initially by paracrine signals originating in the nearby stroma, once an EMT program is kindled, it can be self-sustaining through autocrine signaling. By revealing the importance of combinations of EMT-inducing signals in triggering activation of this program, studies like this one also suggested the converse: that individual heterotypic signals are usually not capable on their own of triggering such

Figure 14.17 Multiple signals collaborate to induce and then maintain the mesenchymal state Immortalized HMLE human mammary epithelial cells (parental cells) entered into the mesenchymal state either spontaneously or by the forced expression of the Twist EMT-TF (EMT-inducing transcription factor). The parental cells secreted significant amounts of TGF-β1, canonical Wnts, and non-canonical Wnts. (A) However, the secreted Wnts were unable to activate autocrine signaling because these epithelial cells release significant amounts of the DKK1 and SFRP1 inhibitors of Wnt signaling (*green*). When cells moved from the epithelial to the mesenchymal state (*orange, red*), the levels of these two Wnt inhibitors decreased greatly. (B) TGF-β autocrine loops were similarly blocked in the parental epithelial cells by the secretion of an array of BMPs (bone morphogenetic proteins) that function to blunt TGF-β signaling. Entrance into the mesenchymal state resulted in strong decreases in the expression of the mRNAs encoding these TGF-β inhibitors. Ongoing autocrine signaling was required to maintain expression of a variety of mesenchymal phenotypes, because their expression could be reduced in the mesenchymal cells through the addition of the SFRP1 and DKK1 recombinant proteins to the culture medium (*not shown*). (C) Entrance into the mesenchymal state can be triggered in mammary epithelial cells in culture by disrupting adherens junctions with an anti-E-cadherin antibody (which liberates β-catenin), the shutdown of SFRP1 mRNA function (through an shRNA), the addition of a DKK1-neutralizing antibody, the addition of large amounts of TGF-β1, and the addition of Wnt5a (a non-canonical Wnt) to the growth medium. Passage through an EMT and entrance into the mesenchymal state were gauged here by measuring levels of mRNAs encoding N-cadherin and two key EMT-TFs—Zeb1 and Zeb2. "+" denotes the addition of a component to this EMT-inducing cocktail, whereas absence of a component is indicated by a blank square in the *gray* matrix *below*. If the epithelial cells were exposed to this induction cocktail for more than a week, they would maintain their mesenchymal phenotype indefinitely in the absence of further such treatment (*not shown*). (D) These observations among others suggest that residence in the mesenchymal state is maintained following passage through an EMT by the actions of these three autocrine signaling loops (*right*). In the epithelial state (*left*), however, while cells secrete TGF-β and canonical Wnts, the ability of these factors to function in an autocrine manner is blocked by the concomitant secretion of inhibitors—BMPs, DKK1, and SFRP1. Hence, when cells undergo an EMT, they shut down production of BMPs and the Wnt inhibitors, thereby creating a permissive extracellular environment for the firing of EMT-inducing autocrine signaling loops. In addition, upon passing through an EMT, such cells express greatly elevated levels of non-canonical Wnt proteins, which also operate thereafter in an autocrine fashion. The residence of cells in the resulting mesenchymal state is maintained in a metastable fashion, because interrupting the three autocrine loops forces cells to revert to an epithelial state. It is not known whether similar dynamics apply to epithelial cells from other tissues. (E) The stromal cells in inflamed tissues and in many tumors release prostaglandin E$_2$ (PGE$_2$), which can induce an EMT in epithelial cells and thereby may complement or reinforce the stromal signals released by Wnts and TGF-β (see Panels D and F). In this non-small-cell lung carcinoma (NSCLC), the mutually exclusive expression of the E-cadherin epithelial marker (*red*) and COX-2, the enzyme that produces PGE$_2$ (*dark brown*), is apparent at high magnification. COX-2–positive carcinoma cells tended to be discohesive, i.e., in the process of separating from other carcinoma cells. Other work (*not shown*) indicates that expression of E-cadherin and the Zeb1 EMT-inducing transcription factor was mutually exclusive in these tumors. (*continued*)

activation. Moreover, biochemical studies of the intracellular signaling circuitry of mammalian cells (Figure 14.17F) suggest how these diverse signals collaborate with one another to induce and maintain expression of EMT programs.

Compounding these normal processes are genetic and epigenetic alterations acquired by carcinoma cells during primary tumor formation that are likely to increase their *responsiveness* to these various contextual signals and thereby influence the way that EMT programs alter carcinoma cell phenotypes. For example, some experiments indicate that cancer cells that have lost p53 function are more responsive to various EMT-inducing signals, and that the *ras* oncogene activated in various human tumors triggers the production and release of TGF-β, which amplifies the TGF-β of stromal origin to facilitate induction of an EMT. Moreover, the extent of EMT expression exhibited by a variety of human tumor cell lines that are propagated as pure populations *in vitro* (in which heterotypic signals are absent) is echoed by the degree of expression

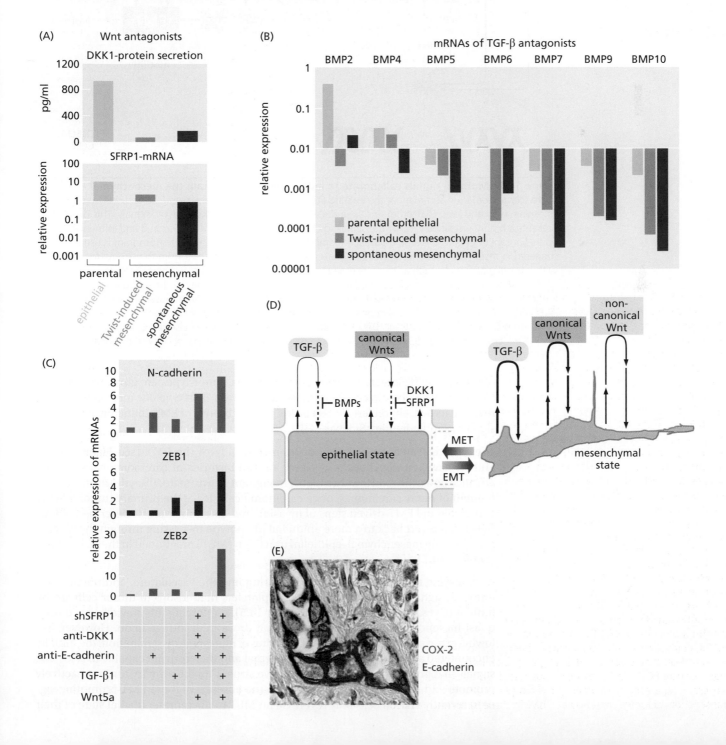

(F)

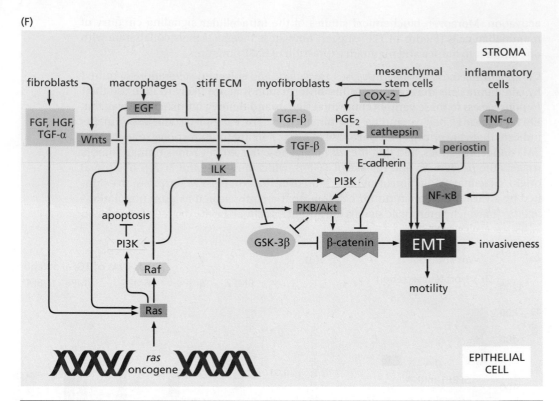

Figure 14.17 Multiple signals collaborate to induce and then maintain the mesenchymal state (*continued*) (F) A summary of the existing literature reveals numerous cross-connections that begin to explain why and how EMT programs are induced by combinations of heterotypic signals. This diagram presents a highly simplified view of the signaling channels that originate in the stroma and influence epithelial cancer cells to undergo a partial or complete EMT. It is likely that an EMT program is usually triggered in response to a convergence of signals that carcinoma cells receive from the stroma together with intracellular signals, such as those released by a *ras* oncogene, as indicated here. The precise identities of these stromal signals and their combinatorial mechanisms of action remain to be elucidated. In the longer term, the interactions depicted in this diagram must be integrated with those indicated in panels A, B, and C, an undertaking that is impossible at present, given the dearth of available information. (A–C, from C. Scheel et al., *Cell* 145:926–940, 2011. With permission from Elsevier. E, from M. Dohadwala et al., *Cancer Res.* 66:5338–5345, 2006. With permission from American Association for Cancer Research.)

of the same EMT programs when corresponding tumor types are growing in human cancer patients. Findings such as these suggest that each type of cancer cell tends to revert to its own characteristic "**set point**" in the absence of EMT-inducing heterotypic signals (Supplementary Sidebar 14.4).

The early research on TGF-β and the reversibility of the EMT (see Figure 14.13B and C) have found echoes in research on the histories of carcinoma cells as they advance down the road toward a highly malignant growth state. Thus, during the development of many carcinomas, once carcinoma cells leave the primary tumor, having completed the EMT-driven steps of invasion and dissemination, as described below, they often revert back to a more epithelial phenotype by passing through the reverse process—the mesenchymal–epithelial transition (MET) mentioned in the last chapter (**Sidebar 14.1**).

After leaving the primary tumor and entering into the vasculature, additional EMT-inducing signals may be released by the platelets that adsorb to tumor cells in the lumina of vessels (Supplementary Sidebar 14.5), reinforcing the previously acquired quasi-mesenchymal state of these cells that drove their intravasation. However, following extravasation of carcinoma cells at sites of metastatic dissemination, the initially encountered stroma is likely to be fully normal and therefore not capable of releasing signals that previously enabled the "reactive stroma" of the primary tumor to actively promote and sustain an EMT program; in the absence of these EMT-promoting signals, recently arrived cells may revert via an MET to the more-epithelial state of their

Sidebar 14.1 Reversal of the EMT at sites of metastasis Although EMT programs may play a central role in enabling the dissemination of primary tumor cells to distant sites in the body, the reversal of these EMT programs that is often observed in subsequently arising metastatic carcinoma colonies may actually be essential to the robust outgrowth of these colonies (**Figure 14.18**). Multiple lines of experimental evidence support this notion. Thus, the histology of primary tumors often indicates a largely epithelial phenotype (Figure 14.18A), whereas the invasive edges of these tumors, which contain the presumed precursors of disseminated tumor cells, show loss of basement membrane, this structure normally serving as a key hallmark of residence in the epithelial state. However, the metastases that ultimately are formed provide clear indication of a basement membrane and a histology that closely resembles that of the primary tumor, indicating a reversion from a more mesenchymal to a more epithelial state. Similarly, in a mouse transgenic model of mammary carcinoma development, experimentally seeded, metastatic carcinoma cells initially show a low representation of cell surface markers indicative of the mesenchymal state. However, the representation of these more mesenchymal cells soon increases more than tenfold as the metastases grow out, but weeks later, as the metastatic colonies expand, the mesenchymal cells once again become a small minority of the cells in these metastases (Figure 14.18B).

The above experiments indicate a strong *correlation* between robust outgrowth of metastatic colonies and residence in an epithelial state. Even more compelling are experiments that demonstrate *causality*, specifically that the continued residence in a more mesenchymal phenotypic state is counterproductive for the formation of robustly growing metastatic colonies (Figure 14.18C). All this can be reconciled with a scheme (Figure 14.18D) in which the EMT program participates in the physical dissemination of carcinoma cells from primary tumors, but its continued expression in the majority of cells within derived metastatic colonies actually hinders the robust outgrowth of metastatic colonies, necessitating a reversal of the EMT program, that is, a mesenchymal–epithelial transition (MET) occurring during the process of metastatic colonization.

The reasons why such METs are needed for optimal metastatic colonization and the eruption of macroscopic metastases remain unclear. In fact, cells that have entered into a more mesenchymal state because of the actions of an EMT program proliferate less rapidly than their epithelial counterparts, but this difference is relatively minor and cannot explain the METs observed in recently seeded metastases. More plausible, however, is a mechanism in which the epithelial subpopulations of carcinoma cells within a metastatic colony actively contribute to the robust outgrowth of the colony as a whole, doing so as part of a hierarchy of cells of the sort depicted in Figure 11.20B.

ancestors in the heart of the primary tumor. Later on, if the initially formed micrometastases succeed in colonizing a distant tissue (thereby forming an actively growing macroscopic tumor), a reactive stroma may eventually be recruited to the resulting metastatic tumor colony, echoing the changes that occurred previously in the primary tumor and inducing, once again, an EMT in the carcinoma cells forming this colony.

Also, as our discussions of the EMT progress, on occasion they will embrace a subtle change in wording: rather than describing a single unitary EMT program, they will refer to EMT *programs* that share a number of core elements but differ from one another in certain details. The existence of multiple alternative EMT programs is dictated by the diverse factors that shape these programs, including the identities of the somatically mutated alleles borne by cancer cells, the differentiation programs of the normal cells-of-origin that spawned various types of neoplastic cells, the nature of the stromal cells that are releasing EMT-inducing heterotypic signals, and the varying extents to which cells have advanced through these programs (see Figure 14.12).

Finally, the phenomena described in this section, when taken together, hold major implications for our thinking about cancer cell biology. Thus, the realization that EMT programs are induced by contextual signals released by other cells indicates that the expression of these programs does not derive directly from mutant alleles carried in the genomes of cancer cells. Hence, the EMT program is an example of an epigenetic rather than a genetically encoded process. In addition, the fact that EMT programs are multifaceted and affect a wide range of cell-biological phenotypes

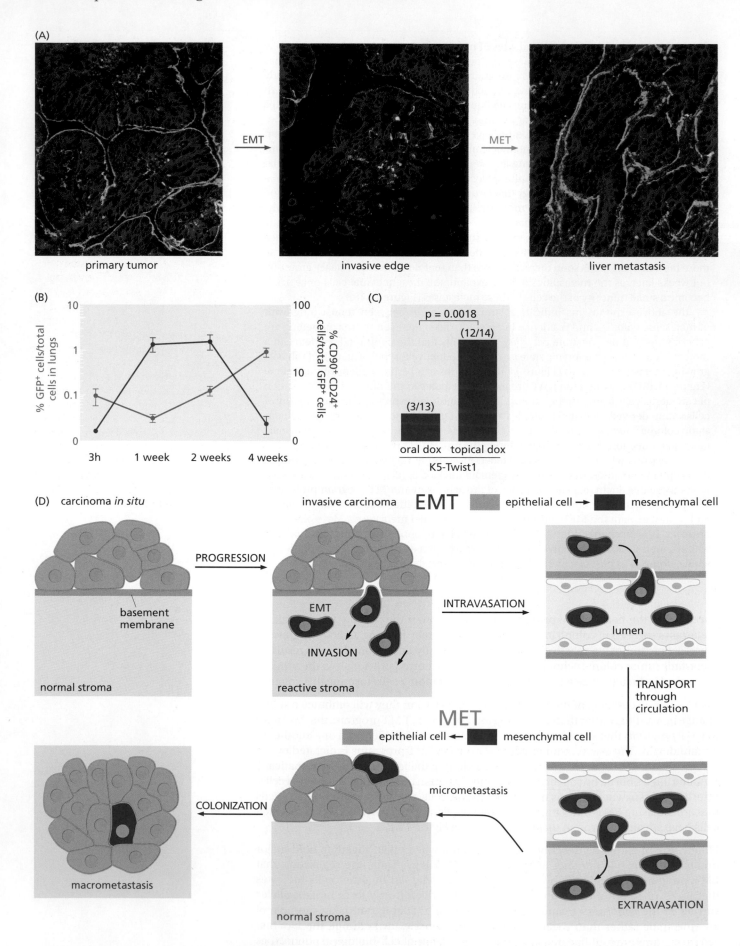

Figure 14.18 Reversibility of EMT Although cells at the invasive edge of a primary carcinoma often give evidence of an EMT, derived metastases may exhibit a histology typical of the center of the primary tumor, suggesting a reversion programmed by a mesenchymal–epithelial transition (MET) program. (A) Release of degradative enzymes, notably matrix metalloproteinases (MMPs), is one of the many manifestations of the EMT. These cells in a primary colorectal carcinoma show expression of both cytokeratin 18 (*red*) and a basement membrane protein (*green*). However, at the invasive edge of this tumor, the cells have undergone a partial EMT, in that they have degraded the adjacent basement membranes while still expressing cytokeratin 18, a key epithelial marker. In a subsequently arising metastasis in this patient, which presumably descended from cells that underwent an EMT and acquired invasiveness en route to metastatic dissemination, the cancer cells form a growth having, once again, the histological appearance of cells in the heart of the primary tumor, suggesting reversibility of the EMT. (B) These dynamics are supported by studies of disseminated cancer cells. Green fluorescent protein (GFP)-expressing cancer cells were prepared from mammary tumors of MMTV-PyMT transgenic mice in which the MMTV promoter drives expression of the polyoma middle T (PyMT) oncogene (established as a transgene in the mouse germline); these mice develop spontaneously metastasizing mammary carcinomas at a high rate. These cell populations were injected into the tail veins of tumor-free host mice and the proportion of GFP$^+$ cells relative to total cells in the lungs (*blue curve*) was followed over the indicated time intervals (*red curve*). Initially, ~0.1% of all cells in the lungs were GFP$^+$ carcinoma cells, which declined to ~0.03% a week later and later increased progressively as metastatic colonization and the growth of metastases proceeded. In contrast, the percentage of GFP$^+$ tumor cells exhibiting a CD90$^+$ CD24$^+$ marker phenotype (representing carcinoma cells with a mesenchymal phenotype; *red curve*) was initially ~1.5%, reflecting the makeup of cells in the primary tumors from which these cells were prepared. This percentage increased almost 20-fold in the first week to over 20%; however, as the individual metastases expanded in size, this percentage decreased back to ~1.8%, almost precisely the level in the initially inoculated cells. Hence, mesenchymal-like carcinoma cells are favored initially to found and establish metastatic colonies, but as colonization proceeds, the representation of mesenchymal/stem-like cells decreases to that seen in the primary tumor. (C) The inducible expression of the Twist1 EMT-TF was used to gauge its effects on metastatic colonization. As indicated, when this EMT-TF was shut down after dissemination (*right bar*), 12 of 14 mice developed metastases, whereas if this EMT-TF continued to be expressed following metastatic dissemination (*left bar*), a far smaller proportion of the mice developed metastatic colonies. This result indicated that the continued high-level expression of this EMT-TF, and thus tumor cell residence in a more mesenchymal state, was counterproductive for metastatic colonization. (D) These observations can be explained by a model in which epithelial cancer cells at the edge of a primary carcinoma (*pink/brown cells*) undergo an EMT as they invade the stroma and become more mesenchymal (*red cells*). This change seems to be triggered by signals that these carcinoma cells receive from the nearby tumor-associated reactive stroma (see Figure 14.17), which is composed of a variety of inflammatory cells that accumulate during the long course of primary tumor formation. The newly acquired mesenchymal state enables the carcinoma cells to invade locally and intravasate, and subsequently to extravasate into the parenchyma of a distant, ostensibly normal tissue. Following extravasation, however, these cancer cells find themselves in a fully normal stroma that lacks the inflammatory cells and therefore does not release EMT-inducing signals. This allows these cells and their descendants to lapse back to an epithelial phenotype (*pink/brown cells*) via a mesenchymal–epithelial transition (MET). (The regeneration of a basement membrane, which occurs as one consequence of an MET, is not illustrated here.) (A, courtesy of T. Brabletz. B, from I. Malanchi et al., *Nature* 481:85–89, 2012. With permission from Nature. C, from J.H. Tsai, J.L. Donaher, D.A. Murphy et al., *Cancer Cell* 22:725–736, 2012. With permission from Elsevier. D, adapted from J.P. Thiery, *Nat. Rev. Cancer* 2:442–454, 2002. With permission from Nature.)

(see Table 14.2), suggests that a significant proportion, possibly the majority, of the distinctive phenotypes of aggressive carcinoma cells are not traceable directly back to somatically mutated alleles of genes (such as the oncogenes and tumor suppressor genes described in earlier chapters) but instead are manifestations of this multifaceted cellular program. This realization moves our thinking away from the notion that cancer cell biology can be understood as a cell-autonomous process, in which study of the genome borne by a neoplastic cell—in its wild-type or mutant configuration—can be used, on its own, to predict the biological phenotypes of this cell (as was implied in Figure 11.12).

14.4 EMTs are programmed by transcription factors that orchestrate key steps of embryogenesis

Execution of an EMT program depends on coordinated changes in the expression of hundreds of distinct genes as well as complex mechanisms of post-transcriptional regulatory controls. These changes affect many aspects of cell biology, not all of which have been enumerated here. The changes include the organization of a cell's intermediate filament cytoskeleton, its motility, its sensitivity to apoptosis, its association with neighboring cells, its release of proteases, and even its display of cell surface integrins and growth factor receptors (see Table 14.2). While extensive evidence, some of it detailed above, implicates stromal signals as key elements in triggering the EMT of carcinoma cells, none of this evidence, on its own, reveals precisely how the complex EMT program is actually coordinated and executed *within* the responding epithelial cells.

The genetics of the early development of a variety of model organisms has provided many of the answers to this question. Like many complex cell-biological programs, EMTs are orchestrated by a small number of pleiotropically acting transcription

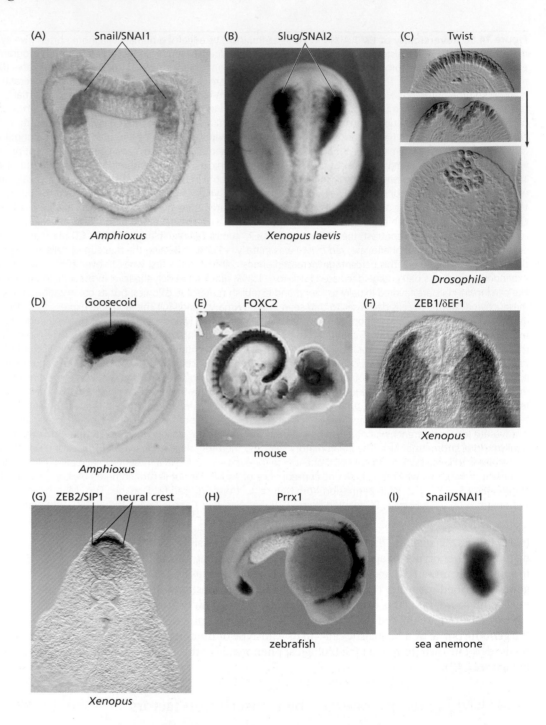

(A) Snail/SNAI1 — Amphioxus

(B) Slug/SNAI2 — Xenopus laevis

(C) Twist — Drosophila

(D) Goosecoid — Amphioxus

(E) FOXC2 — mouse

(F) ZEB1/δEF1 — Xenopus

(G) ZEB2/SIP1 neural crest — Xenopus

(H) Prrx1 — zebrafish

(I) Snail/SNAI1 — sea anemone

factors (TFs). A number of genes specifying these EMT-inducing TFs (EMT-TFs) have been identified (**Figure 14.19**), many initially uncovered in the embryos of the fruit fly, *Drosophila melanogaster*. These genes and the TFs that they encode are conserved in all chordates and in most metazoa and have been found to control key steps in early embryogenesis in various metazoan embryos; these steps involve various versions of the EMT program. (The strong evolutionary conservation of these EMT-TFs indicates that the EMT and key steps of early embryogenesis were developed early in metazoan evolution, long before the radiation of the various metazoan phyla which seems to have occurred before the Cambrian era 550 million years ago.) By activating expression of these TFs, cancer cells gain access to the complex, multifaceted EMT programs that they orchestrate. Implicit in what follows is the following scheme: heterotypic signals from the stroma induce expression of EMT-TFs in carcinoma cells; these TFs, in turn, orchestrate EMT programs within these cells.

Figure 14.19 Embryonic transcription factors programming epithelial–mesenchymal transitions A variety of EMT-inducing transcription factors (EMT-TFs) can induce this program in various stages of embryogenesis, and their expression has been documented in a wide range of organisms. (A) Snail (also known as SNAI1) is shown being expressed in cells of a primitive chordate, specifically in cells (*dark areas*) whose counterparts in higher chordates form the neural crest. (B) The Slug TF (SNAI2), a close relative of Snail, is expressed in *Amphioxus* (*dark blue*) in the embryonic neural crest, here in an embryo of *Xenopus laevis*, the African clawed toad. (C) Twist is shown here in an early *Drosophila* embryo, in which it programs an EMT at the site of gastrulation (*brown*). (D) The Goosecoid TF is expressed at the blastopore lip in gastrulating chordate embryos. Here its expression, which is inducible by the TGF-β signaling pathway, is adjacent to the blastopore 8 hours after fertilization of an *Amphioxus* egg. (E) The FOXC2 TF is expressed in important mesodermal structures in this day 9.5 mouse embryo, including the mesoderm around the developing spinal column as well as in somites, which are precursors to many of the body's muscles. (F) The ZEB1 (δEF1) TF is expressed in the paraxial mesoderm of an early *Xenopus* embryo, where it may play a role in programming the toad's development from axial stem cells. (G) In the same embryo at this stage, expression of the related ZEB2 (SIP1) TF occurs in a complementary pattern, being highest in the neural crest and neural tube, where it appears to be responsible for the cell movements that lead to closure of the neural tube and emigration of cells from the neural crest to other parts of the body. (H) Expression of the PRXX1 EMT-TF at the 22-somite stage of a zebrafish embryo. (I) The antiquity of these TFs is illustrated by the expression of Snail in the blastula stage of the starlet sea anemone embryo, *Nematostella vectensis*. Its ancestors diverged from those of the chordates >600 million years ago, before the evolution of bilateral animals with three cell layers. It contrasts with *Amphioxus*—clearly a chordate with many features of vertebrate ancestors—which diverged ~525 million years ago from our own lineage. (A, from J. Langeland et al., *Dev. Genes Evol.* 208:569-577, 1998. With permission from Springer. B, from C. LaBonne et al., *J. Neurobiol.* 36:175-189, 1999. With permission from John Wiley & Sons. C, courtesy of M. Leptin. D, from A.H. Neidert, G. Panopoulou and J.A. Langeland, *Evol. Dev.* 2:303–310, 2000. With permission from John Wiley & Sons. E, from T. Furumoto et al., *Dev. Biol.* 210:15–29, 1999. With permission from Elsevier. F, from L.A. van Grunsven et al., *Dev. Dyn.* 235:1491–1500, 2006. With permission from John Wiley & Sons. G, courtesy of E.J. Bellefroid. H, from O.H. Ocaña, R. Corcoles, A. Fabra I et al., *Cancer Cell* 22:709–724, 2012. With permission from Elsevier. I, from M.Q. Martindale, K. Pang and J.R. Finnerty, *Development* 131:2463–2474, 2004. With permission from The Company of Biologists.)

More than half a dozen EMT-TFs have been described, each capable of inducing an EMT when ectopically expressed in certain epithelial cells (**Table 14.4**). For example, Snail is an EMT-TF that was first described in *Drosophila* and has since been discovered in a wide range of metazoa, including vertebrates, insects, worms, and molluscs. In early vertebrate embryos, Snail is first expressed in the portion of the ectoderm that is destined to become mesoderm following gastrulation. During embryogenesis,

Table 14.4 Transcription factors orchestrating an EMT

Name	Where first identified	Type of transcription factor	Cancer association
Snail (SNAI1)	mesoderm induction in *Drosophila*; neural crest migration in vertebrates	C2H2-type zinc finger	invasive ductal carcinoma
Slug (SNAI2)	delamination of the neural crest and early mesoderm in chicken	C2H2-type zinc finger	breast cancer cell lines, melanoma
Twist	mesoderm induction in *Drosophila*; emigration from neural crest	bHLH	various carcinomas, high-grade melanoma, neuroblastoma
Goosecoid	gastrulation in frog	paired homeodomain	various carcinomas
FOXC2	mesenchyme formation	winged helix/forkhead	basal-like breast cancer
ZEB1 (δEF1)	postgastrulation mesodermal tissue formation	2-handed zinc finger/homeodomain	wide variety of cancers
ZEB2 (SIP1)	neurogenesis	2-handed zinc finger/homeodomain	ovarian, breast, liver carcinomas
E12/E47 (Tcf3)[a]	associated with E-cadherin promoter	bHLH	gastric cancer
Prrx 1	chick mesoderm formation	paired homeobox	various carcinomas

[a]It remains unclear whether E12/E47 can function on its own to induce an EMT, or whether this bHLH protein functions as a subunit of a heterodimeric TF complex formed with other well-validated EMT-TF proteins such as Twist.

Snail, working together with two other EMT-TFs—Slug, and Twist—convert epithelial cells into the migratory mesenchymal cells that form the mesoderm. Snail and its relative Slug are involved in yet other embryonic steps in which one type of tissue is transformed into another. The truly ancient origins of these TFs are illustrated dramatically by Figure 14.19I, in which the embryo of a very primitive organism—a sea anemone, which lacks bilateral symmetry and three germ layers—expresses the Snail EMT-TF at the site of its future embryonic blastopore.

Some of these TFs are expressed transiently in adult tissues during the tissue remodeling that underlies wound healing. This is illustrated nicely by the behavior of confluent monolayers of normal epithelial cells in culture that are wounded experimentally by scraping away a swath of these cells as was seen in Figure 14.15. Slug expression is induced in the surviving epithelial cells sitting at the edge of the wound, which appeared to enable these cells to acquire motility and migrate into the wound site. This pattern of expression helps to explain how epithelial cells at the edges of wounds undergo a transient EMT in order to reassemble epithelial cell sheets (see also Figure 13.15A). Observations like these broaden our perspective on the normal biological roles of these EMT-TFs: in addition to programming key steps in early embryogenesis, the expression of some of these TFs is resurrected transiently in adults in order to help reconstruct damaged epithelial tissues.

Snail and Slug (sometimes called SNAI1 and SNAI2) are members of the C2H2-type family. These two TFs seem to operate largely as repressors of transcription. Thus, both have been found to be able to repress transcription of the gene encoding E-cadherin, which serves as the keystone of the epithelial cell state. (As we read earlier, the loss of E-cadherin expression can, on its own, cause epithelial cells to assume many of the phenotypic changes associated with an EMT.)

The Snail TF has been found to be expressed in the invasive fronts of chemically induced mouse skin carcinomas, and its expression is associated with the degree of lymph node metastasis of human breast cancers. Moreover, embryonic expression of Snail, its cousin Slug, and Goosecoid is induced by contextual signals, such as TGF-β and Wnts, that are known to be responsible for inducing the EMT conversion of mouse tumor cells. Twist is expressed during the gastrulation of *Drosophila* embryos (see Figure 14.19C) and the out-migration of neuroepithelial cells from the neural crest of chordate embryos. Its expression is also induced by exposure to TGF-β.

Although the micrographs of Figure 14.19 give the impression that these and other EMT-TFs are involved largely in early embryonic morphogenetic steps, in fact they continue to play roles throughout the entire process of embryonic development and postnatal tissue homeostasis (see Supplementary Sidebar 14.1). The multiplicity of these TFs, along with their expression in various combinations in human cancers, helps to further explain the nomenclature cited earlier, namely that the term "EMT" actually refers to a group of cell-biological programs rather than a single, uniformly expressed program.

Some of the EMT-TFs participate in unusually interesting modes of regulation. For example, the EMT was portrayed as generating a succession of phenotypic states arrayed along a spectrum between highly epithelial and highly mesenchymal (see Figure 14.12). This was cited as an example of how and why the EMT does not operate as a simple binary switch that shuttles cells between two alternative phenotypic states, that is, one fully epithelial, the other fully mesenchymal. Nevertheless, certain binary controls appear to contribute to the regulation of EMT expression. One was seen in the behavior of *ras*-transformed cells that have been triggered by TGF-β exposure to enter into a more mesenchymal, EMT-induced state (see Figure 14.13); the resulting cell state causes them to produce their own, endogenously synthesized TGF-β, which operates, in turn, in an autocrine manner, driving a self-sustaining positive-feedback (sometimes termed "feed-forward") signaling loop. Such controls can yield phenotypic states that are stably maintained and expressed over many cell generations and operate in a binary fashion, that is, either they are or are not activated. Another manifestation of binary controls operates in the case of the Zeb1 and Zeb2 EMT-TFs, which are important, if not dominant EMT-TFs in imposing a more mesenchymal state on cancer cells (**Sidebar 14.2**). Somehow, these binary switching mechanisms must be integrated into the well-documented notion that EMT programs produce a spectrum of phenotypic states (rather than just two) arrayed along the E to M axis.

14.5 Signals released by an array of stromal cell types contribute to the induction of invasiveness and intravasation

The diversity of the EMT-inducing signals described above (see Table 14.3 and Figure 14.17) draws attention to the stromal cells that are the sources of these signals. The most prominent cells in the stroma of a high-grade tumor are the carcinoma-associated fibroblasts (CAFs), which form the reactive stroma of the tumor and typically resemble myofibroblasts, releasing among other paracrine signals large amounts of TGF-β. As seen in **Figure 14.21A**, when clusters of murine colorectal carcinoma (CRC) cells, sometimes termed **spheroids**, are introduced into a three-dimensional collagen matrix, small numbers of CRC cells invade the nearby matrix; the same behavior is noted when the CRC cell spheroids are propagated in a collagen matrix containing normal fibroblasts adjacent to an excised tumor. However, when fibroblasts are prepared instead from the tumor-associated stroma (and may be termed CAFs, carcinoma-associated fibroblasts), there is a dramatic increase in invasiveness, ostensibly echoing the behavior of carcinoma cells in a highly progressed primary tumor.

The roles played by macrophages in the tumor-associated stroma were already cited briefly earlier. Their influence on the invasive and metastatic behavior of primary cancer cells can be demonstrated by studying genetically altered mice that lack the ability to make colony-stimulating factor-1 (CSF-1). As was discussed in the last chapter (Section 13.3), mammary carcinomas arising in cancer-prone transgenic mice usually recruit large numbers of tumor-associated macrophages (TAMs). However, when the tumor cells in such mice lack the ability to make CSF-1, TAMs are virtually absent (see Figure 13.25). The absence of CSF-1 and TAMs has no effect on primary tumor growth (Figure 14.21B), but such tumors show a benign, noninvasive behavior, in contrast to tumors that succeed in recruiting TAMs (Figure 14.21C). The influence of these macrophages on metastatic behavior is striking: primary mammary tumors usually succeed in seeding large numbers of metastases in the lungs, but without recruited TAMs in their stroma, they fail entirely to seed detectable lung metastases (Figure 14.21D).

These experiments (Figure 14.21B–D) provide compelling evidence that the invasive and metastatic behavior of mouse breast carcinoma cells can be strongly influenced by the actions of stromal cells, in this case specific heterotypic signals released by macrophages. They fail, however, to reveal the precise nature of these signals. Macrophage-derived TNF-α, as argued earlier, is likely to contribute to induction of the EMT by cancer cells, and therefore to the invasive and metastatic behavior described in **Figure 14.22**.

Another key macrophage-derived signal is conveyed by EGF, as mentioned earlier. Some of the evidence favoring EGF as a key inducer of cancer cell invasiveness comes from studies of mouse breast cancer cells both *in vivo* and *in vitro*. Like most epithelial

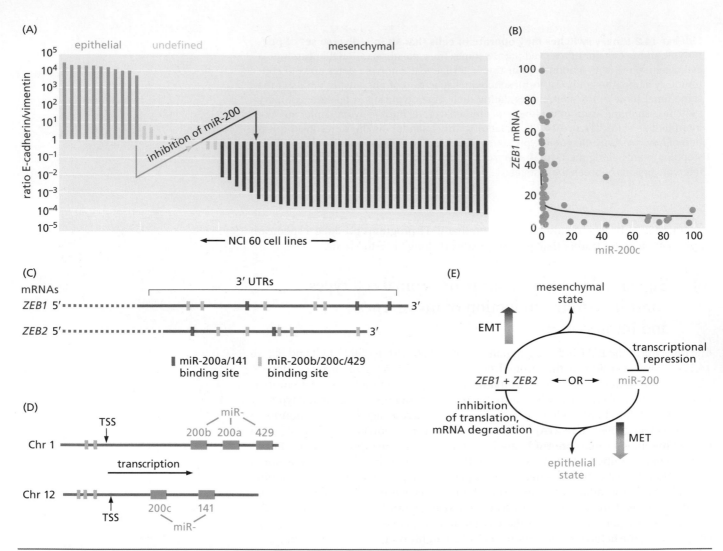

Figure 14.20 Alternation between the epithelial and mesenchymal states Although carcinoma cells growing *in vivo* apparently can switch reversibly between epithelial and mesenchymal states, the cells of many cancer cell lines, when propagated *in vitro*, appear to reside stably in one or the other of these states, being controlled seemingly by a governor that functions as a bistable switch. (A) The levels of E-cadherin and vimentin, representative markers of the epithelial and mesenchymal states, respectively, were measured by immunoblotting of lysates of the 60 distinct cancer cell lines that constitute the collection termed the NCI 60 cell lines. The ratio of E-cadherin vs. vimentin (i.e., epithelial vs. mesenchymal) is indicated on the ordinate. As is apparent, almost all of the NCI 60 cells resided either in the epithelial (*green bars*) or mesenchymal (*red bars*) state when propagated in culture. The arrow indicates the shift of one of these cell lines after it was induced experimentally to undergo an EMT as described in panel C. (B) Examination of a series of microRNAs (see Section 1.9) expressed in either the epithelial or the mesenchymal state revealed that several of these belonging to the miR-200 family of microRNAs were expressed in a fashion that was mutually exclusive with the ZEB1 and ZEB2 EMT-TFs. Shown (in arbitrary units) are the expression levels of ZEB1 mRNA and of miR-200c in ~60 cell lines, with each point representing a single cell line; regulation of ZEB2 mRNA behaved in an essentially identical fashion (*not shown*). (C) Sequencing of the mRNAs encoding ZEB1 and ZEB2 revealed multiple sites in the 3′ UTRs (3′ untranslated regions; *light and dark blue boxes*) that could be targeted by miR-200 family members, resulting in inhibition of translation or degradation of these mRNAs. Alteration of these target sequences by site-directed

mutagenesis rendered these mRNAs resistant to the inhibitory actions of miR-200 microRNAs. Moreover, experimental inhibition in epithelial cells of endogenously expressed miR-200 microRNAs, achieved by transfection of an antisense locked nucleic acid (LNA) construct, allowed accumulation of ZEB1 and ZEB2 and loss of epithelial gene expression. (D) Each of the two genes encoding miR-200 microRNAs specifies multiple miRNAs (*green boxes*) in its primary transcript (*not shown*), which is then cleaved post-transcriptionally to generate several mature microRNAs. The promoters of both genes contain evolutionarily conserved binding sites for ZEB1 and ZEB2 (*orange boxes*), which bind a sequence in these promoters and proceed to repress transcription of these genes. Transcription is from left to right (*horizontal arrow*) and the diagram is not drawn to scale. TSS, transcription start site. (E) The observations in panels C and D, when taken together, indicate the existence of mutually antagonistic interactions that operate as a bistable switch, thereby ensuring the metastable residence of a cell in either the mesenchymal or the epithelial state. Once one group of regulatory molecules gets the upper hand (i.e., either ZEB1/2 or miR-200 microRNAs), it shuts down the other and thereby ensures long-term residence in either the mesenchymal or the epithelial state. In principle, each group of regulators can act on yet other genes and mRNAs; e.g., ZEB1/2 can repress E-cadherin and other epithelial genes. The actions of the miR-200 microRNAs, however, seem to be focused more narrowly on suppressing just *ZEB1/2* expression. (A–C, from S.-M. Park et al., *Genes Dev.* 22:894–907, 2008. With permission from Cold Spring Harbor Laboratory Press. D, from S. Brabletz and T. Brabletz, *EMBO Rep.* 11:670–677, 2010. With permission from John Wiley & Sons.)

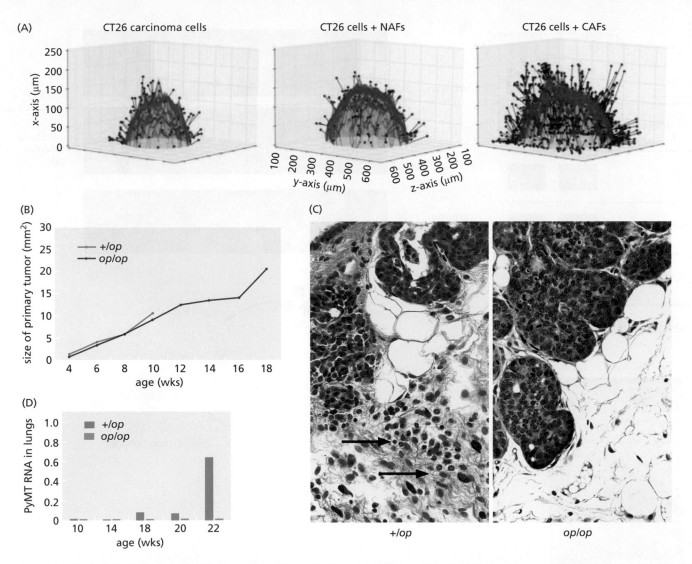

Figure 14.21 Effects of fibroblasts and macrophages on invasion and metastasis Both cancer-associated fibroblasts (CAFs) and macrophages can have strong effects on the biology of nearby carcinoma cells. (A) In an *in vitro* model of invasiveness, clusters of CT26 murine colorectal carcinoma (CRC) cells were embedded in a 3-dimensional collagen I matrix on their own (*purple, left panel*), in the presence of normal fibroblasts (NAFs) adjacent to a CRC tumor excised at surgery from a patient (*middle panel*), or carcinoma-associated fibroblasts (CAFs) prepared from the stroma of the excised human CRC tumor (*right panel*). The fibroblast populations were mixed into the collagen matrix. After 3 days, the nuclei of the invasive cells were visualized (*dark red dots*) by image-processing software. (B) The PyMT transgene (see Figure 14.18B) has been introduced, through breeding, into mice that can (*Csf^{+/op}*) or cannot (*Csf^{op/op}*) make colony-stimulating factor-1 (CSF-1), which is needed to recruit macrophages into tumors (see Figure 13.25). The presence (in *Csf^{+/op}* mice) or absence (in *Csf^{op/op}* mice) of recruited tumor-associated macrophages (TAMs) has no effect on the ability of the primary breast tumors to grow in these transgenic mice. (C) Such mammary tumors arising in *Csf^{+/op}* mice (whose tumors contain many TAMs, *not shown*) develop a highly invasive phenotype, in which individual carcinoma cells invade the nearby stroma in large numbers (*left panel, arrows*). However, if tumors develop in *Csf^{op/op}* mice, in which macrophages cannot be recruited into the tumor-associated stroma (*right panel*), the tumor cells do not break through the basement membrane and the tumor as a whole remains encapsulated and benign, indicating that the macrophages contribute in essential ways to tumor invasiveness. (D) In the *Csf^{+/op}* mice, metastases in the lungs begin to appear at 18 weeks of age and increase progressively thereafter (*blue bars*), as gauged by the amount of polyomavirus middle T RNA (expressed exclusively in tumor cells) present in the lungs (*ordinate*). However, in the *Csf^{op/op}* TAM-negative mice (*orange bars*), lung metastases are virtually absent. (A, from Y. Attieh, A.G. Clark, C. Grass et al., *J. Cell Biol.* 216:3509–3520, 2017. With permission from Elsevier. B–D, from E.Y. Lin et al., *J. Exp. Med.* 193:727–740, 2001. With permission from Rockefeller University Press.)

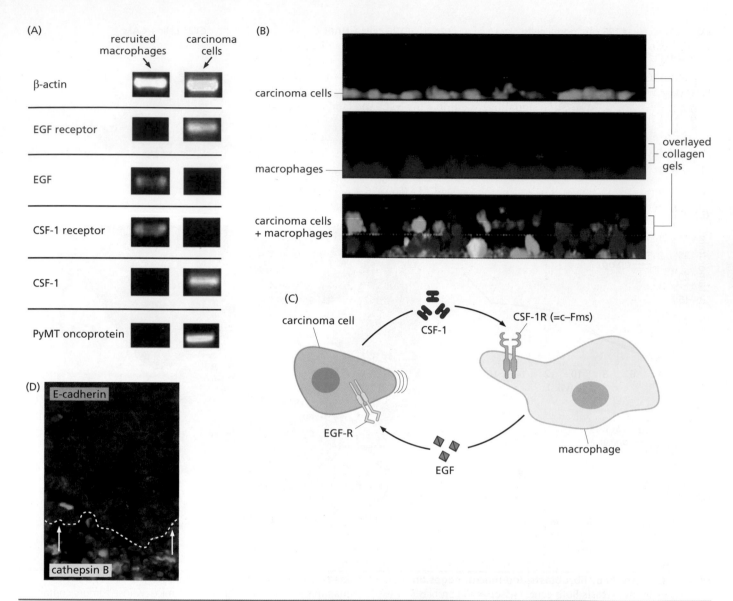

Figure 14.22 Reciprocal stimulation by breast cancer cells and macrophages A variety of experiments indicate that macrophages are the major source of EGF in breast cancers. EGF is known to be able to stimulate epithelial cancer cells to proliferate and invade through extracellular matrix. In addition, EGF exposure causes breast cancer cells to release CSF-1, which allows them to recruit macrophages and stimulates production by the macrophages of more EGF, resulting in a positive-feedback loop between these two cell types. (A) Using PCR analysis, the mRNA levels of these two growth factors and their receptors are found to be reciprocally expressed in mammary carcinoma cells arising in PyMT tumor-prone transgenic mice (see Figure 14.18B) and in recruited stromal macrophages. (B) Breast cancer cells (labeled here with green fluorescent protein, GFP; *green*) were placed at the bottom of a Petri dish (seen here in side view) below a layer of collagen gel (*top image*), where they remained. Similarly, macrophages (*red*) also remained where they were initially placed at the bottom of the Petri dish below a collagen gel (*middle image*). However, when the two populations were co-cultured at this location, the breast cancer cells (*green*) were now induced to move upward and invade the overlying collagen gel (*lower image*).(C) The reciprocal paracrine interactions between carcinoma cells and macrophages operates by the release of EGF by the macrophages, which stimulates the carcinoma cells, which proceed to release CSF-1, which recruits macrophages. (D) The possibility that macrophages can contribute directly to triggering an EMT program in carcinoma cells is indicated by studies of macrophages that have been recruited to the invasive edge (*dashed line, arrows*) of a pancreatic islet cell tumor (*above*) in the Rip-Tag transgenic mouse model (see Figure 13.35). As seen here, the cathepsin B secreted protease (*green*) produced by the recruited macrophages causes the carcinoma cells to lose expression of E-cadherin (*red*). The resulting carcinoma cells (*below dashed line*) are revealed here only by their DAPI-stained nuclei (*blue*). (Inactivation of cathepsin B production by the macrophages permitted the carcinoma cells to regain their E-cadherin expression and caused them to lose invasiveness; *not shown*. Cathepsin B is also responsible for activating a number of distinct matrix metalloproteinases, whose activation leads to amplification of the cathepsin B-mediated proteolysis.) (A, from J. Wyckoff et al., *Cancer Res.* 64:7022–7029, 2004. B, from S. Goswami et al., *Cancer Res.* 65:5278–5283, 2005. Both with permission from American Association for Cancer Research. D, courtesy of V. Gocheva and J.A. Joyce; see also V. Gocheva et al., *Genes Dev.* 20:543–556, 2006.)

cells, these carcinoma cells express the EGF receptor, and activation of this receptor by EGF causes them to acquire both motility and invasiveness (to be discussed later in Supplementary Sidebar 14.9) and to secrete CSF-1, a key attractant and stimulant of macrophages. Macrophages respond reciprocally to CSF-1 by proliferating and releasing EGF, which activates the cancer cells. These effects all proceed through paracrine rather than autocrine signaling, as demonstrated by the fact that the breast cancer cells do not express the CSF-1 receptor and the macrophages do not express the EGF receptor (see Figure 14.22A). Hence, these two cell types collaborate by reciprocally stimulating one another, yielding another type of positive-feedback loop (see Figure 14.22C).

Importantly, the behaviors depicted in Figures 14.7 and 14.22B suggest that macrophages actually play critical roles in two stages of the invasion–metastasis cascade—initial invasiveness and intravasation. Although cancer cell motility and invasiveness are clearly demonstrated in such experiments, the induction of an EMT program in the cancer cells can only be inferred from their acquisition of motile, invasive behavior. More direct evidence comes, however, from the Rip-Tag tumor model (see Figure 13.35), in which the secretion by recruited macrophages of the cathepsin B protease is correlated with the loss—ostensibly through proteolytic degradation—of E-cadherin expression from the cell surface of carcinoma cells; such loss, on its own, can serve as a trigger for initiating an EMT (see Figure 14.22D).

These bidirectional interactions seem to play a critical role in a poorly studied step of the invasion–metastasis cascade, the process of intravasation. As cited earlier, the invasiveness of primary tumor cells leads, with a certain probability, to intravasation, that is, entrance into the lumina of blood and lymphatic vessels. Many had assumed that intravasation is nothing more than an accidental, undirected by-product of invasion. However, recent studies have pointed to more specificity and organization of this process.

These studies indicate that in many breast cancers, and likely in other carcinomas as well, a triad of three distinct cell types—carcinoma cells, macrophages, and endothelial cells—assembles to enable the cancer cells to intravasate (**Figure 14.23**). The macrophages in these multicellular clusters appear to be responsible for creating "doorways" in the endothelial walls of microvessels that allow invasive carcinoma cells to enter the lumina of these vessels. In more detail, the macrophages appear to deploy the VEGF-A angiogenic growth factor to induce retraction of the endothelial cells from one another. These opened doorways therefore represent discrete sites of greatly increased permeability in the vascular endothelial wall and are formed where macrophage-induced endothelial cell retraction creates significant gaps in the otherwise-intact endothelial lining of microvessels (see Figure 14.23D).

Importantly, the density of these microscopically visualized tripartite cell clusters (see Figure 14.23C) represents a strong predictor of eventual systemic metastatic dissemination in breast cancer patients. Thus, in women bearing estrogen receptor-positive, HER2-negative primary breast cancers (which in aggregate constitute ~60% of all breast tumors), the third of the women whose tumors exhibited the highest density of these triads exhibited a 2.7 times higher risk of developing distant metastases than the third of patients whose tumors carried the lowest density of these triads. In fact, this measurement of triad density was found to be a more accurate indicator of future metastatic risk than was the overall histopathological grade of the primary tumor.

To summarize, these diverse lines of evidence indicate that the acquisition of malignant traits by carcinoma cells, involving notably induction of an EMT, is not governed solely by the genomes of these cells. Instead, these profound shifts in cell phenotype are often initiated by a collaboration between specific mutant alleles harbored in cancer cell genomes (for example, mutant *ras* or *p53* genes) and the signals that these cancer cells receive in certain tissue microenvironments, specifically at the boundaries between islands of tumor cells and reactive stroma. In many tumors, these contextual signals are conveyed by certain factors, such as TGF-β, Wnts, PGE$_2$, and TNF-α, that are released by cells in the reactive stroma. Yet other stromal signals, such as those carried by EGF and HGF, may also help to elicit many of the changes that

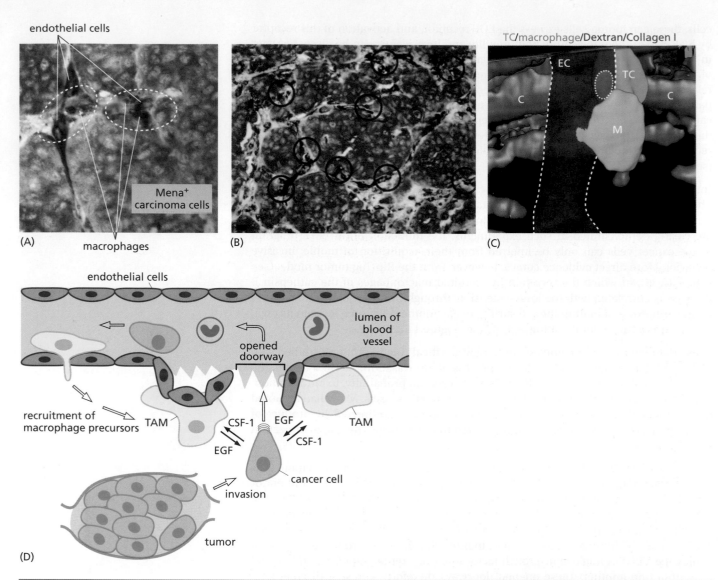

Figure 14.23 Intravasation Like local invasion, the process of intravasation by carcinoma cells often—and perhaps always—requires help from stromal cells, in this case macrophages. (A) Detailed investigation of carcinomas, in both laboratory animal models and human tumors, reveals triads (*ovals*) of carcinoma cells, macrophages, and endothelial cells (sometimes termed TMEM, for tumor microenvironment of metastasis). The macrophages help the carcinoma cells to intravasate through the capillary wall formed by the endothelial cells. A triad is registered only if all three cellular components are directly apposed and if the carcinoma cells express significant levels of the Mena actin-cytoskeleton–regulating protein that is associated with cell motility, seen here as a *red* stain. (B) Although two human carcinomas may have a very similar histopathological appearance, the counting of the density of TMEM triads (*circles*) within sections of these tumors generates a highly useful prognostic marker of metastasis. (C) A 3-dimensional reconstruction of real-time intravital imaging showing collagen I fibers (C, *purple*) behind a doorway in the wall of a microvessel composed of a tripartite cell complex involving a tumor cell (TC, *green*), a macrophage (M, *cyan*), and an endothelial cell (EC). The endothelial cells were not labeled in this image but form a thin 1- to 2-micron thick cell layer (*dashed white lines*) around the intraluminal space of a microvessel, which is defined by fluorescently labeled dextran (*red*) that was intravascularly injected. The carcinoma cell has partially intravasated into the lumen of the micro vessel with the *yellow dotted line* separating its intraluminal portion (*oval dotted line*) from its portion that remains outside of the vascular lumen. (D) The reciprocal interactions between breast cancer cells and macrophages are illustrated schematically here. Because macrophages are often found in close proximity to microvessels, the stimulation by tumor-associated macrophages (TAMs) of breast cancer cell motility and invasiveness may also contribute to cancer cell intravasation, as suggested by the observations of Figure 14.7 and depicted here in more mechanistic detail. (A, B, from B.D. Robinson et al., *Clin. Cancer Res.* 15:2433–2441, 2009. With permission from American Association for Cancer Research. C, courtesy of D. Entenberg, A.S. Harney, P. Guo, and J. Condeelis. D, from W. Wang et al., *Trends Cell Biol.* 15: 138–145, 2005. With permission from Elsevier.)

we associate with cancer cell invasiveness and the EMT. Importantly, these descriptions likely apply to the biology of all carcinomas, which together account for ~80% of life-threatening human tumors; there is also evidence that neuroectodermal tumors of the brain also exhibit EMT-like behavior, possibly testifying to their formation in a tissue derived from an epithelium—the neuroectoderm—present in the early embryo. However, the relevance of our discussions in this chapter to the other main classes of human tumors—mesenchymal tumors (that is, sarcomas) and hematopoietic cancers—remains unclear.

14.6 EMT-inducing transcription factors may enable entrance into the stem cell state

The cellular traits that are induced by the actions of one or another EMT-TF suggest that these master regulators can orchestrate many of the steps of the invasion–metastasis cascade. There is, however, one trait that is critical for metastasis that has not yet been considered here: in order for a disseminated cancer cell to seed a metastasis, it must possess the tumor-initiating ability that we previously ascribed to cancer stem cells (CSCs; see Section 11.7). Recall that these cells are defined operationally through their ability to seed new tumors following experimental implantation into suitable hosts. In principle, the seeding of a metastatic colony, when it occurs as a consequence of primary tumor progression, represents a very similar process: in both cases, small numbers of cells—sometimes just a single cell—are able to spawn the large cell populations that form macroscopic tumors, including the secondary growths that we term metastases.

The trait of self-renewal that is so central to the stem cell state would seem to be far removed mechanistically from the aggressive cell-biological traits programmed by EMT transcription factors. Evidence first reported in 2008, however, indicates otherwise: this work revealed that immortalized human mammary epithelial cells (MECs) that are forced to pass through an EMT acquire many of the attributes of stem cells (**Figure 14.24**). Related studies later extended this work in two directions by showing that mammary carcinoma cells that have undergone an EMT exhibit vastly increased tumor-initiating capacity; similar dynamics apply to normal MECs and their stem cell-like behavior (**Figure 14.25**).

In the analysis provided in Figure 14.24A, a population of experimentally immortalized human MECs could be fractionated, as judged by FACS analysis, into a minority and a majority subpopulation, with cells in the minority subpopulation exhibiting cell surface markers of the stem cell-like state; we encountered the same fractionation procedure previously in Figures 11.19 and 11.20, where it was used to resolve tumor-initiating cells (TICs) from non-TIC subpopulations. Because all the cells in this analysis appeared to share the same genome, this suggested, on its own, that cell populations like this one spontaneously sort themselves into majority and minority subpopulations, doing so via a process that did not require genetic change.

When two EMT-TFs were introduced separately into the cells analyzed in Figure 14.24A, the cells that had previously resided in the majority, non-stem-cell population migrated into the position in this FACS plot of stem-like cells. This behavior suggested that activation of an EMT program facilitated entrance into a stem-like state, but on its own did not prove this because the cellular products of this conversion were not tested biologically for stemness.

Perhaps more persuasive were the observations of Figure 14.24B, in which mRNA levels were quantified in the cells of these two subpopulations, this analysis being measured in the absence of any experimentally forced expression of EMT-TFs. This analysis made it clear that the cells residing in the position (in the FACS plot) of stem-like cells expressed mesenchymal markers at greatly elevated levels relative to cells in the non-stem-like position. Included among these mesenchymal markers were mRNAs specifying a group of EMT-TFs. Here, the affiliation of EMT-inducing TFs with an SC-enriched state reflected regulation by natural, endogenous transcriptional controls rather than regulation imposed by experimentally introduced genes.

Epithelial cells acquire certain properties associated with stem cells (SCs) following activation of EMT programs. (A) When experimentally immortalized human mammary epithelial cells, termed HMLE, are viewed in monolayer culture by phase microscopy (*left, above*), they exhibit the cobblestone morphology that is typical of epithelial cells. These cells can be fractionated using fluorescence-activated cell sorting (FACS) into a minority subpopulation that has a CD44^{hi} CD24^{lo} cell surface antigen profile and a majority subpopulation that has the opposite antigen profile—CD44^{lo} CD24^{hi} (*left, below*). The minority population, although not homogeneous, is greatly enriched for stem cells (SCs), which are absent from the majority population; this appears to be true for normal, immortalized, and neoplastically transformed human mammary epithelial cells. When these cells were then forced to express either the Snail (*middle column*) or the Twist (*right column*) EMT-inducing transcription factor (EMT-TF), their morphology changed in monolayer culture to the dispersed elongated phenotype typical of mesenchymal cells such as fibroblasts (*below*). In addition, virtually all the cells that were previously in the CD44^{lo} CD24^{hi} non-SC state were converted to the CD44^{hi} CD24^{lo} population, which is greatly enriched in mammary SCs. (B) The relative levels of mRNAs expressed in each of the two populations in (A) (*left*) were quantified using RT-PCR analysis; the results are shown in the histogram (*right*). The mRNA encoding E-cadherin, the key epithelial marker, was expressed at ~1/100 the level per cell in the CD44^{hi} CD24^{lo} cells, whereas N-cadherin, vimentin, and fibronectin mRNAs (all encoding mesenchymal proteins) were expressed at levels ranging from ~65- to ~280-fold more abundantly in these cells. Of greater interest, cells that resided in the CD44^{hi} CD24^{lo} state naturally expressed levels of mRNAs specifying four EMT-TFs (FOXC2, ZEB2, Twist, and Snail) that were between 9- and 200-fold higher per cell than the CD44^{lo} CD24^{hi} cells. (A, B, from S.A. Mani et al., *Cell* 133:704–715, 2008. With permission from Elsevier.)

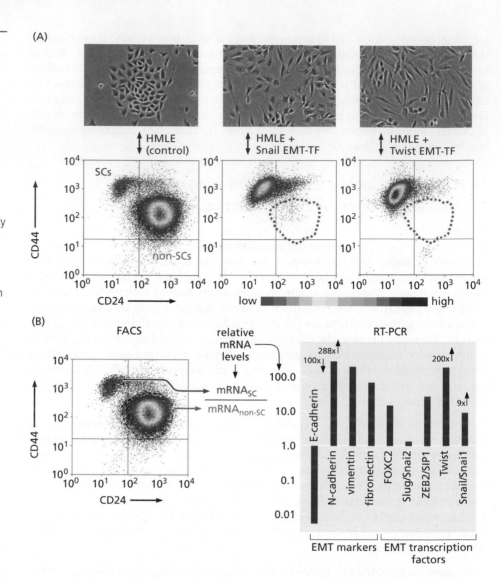

This association could also be extended in two directions with fully neoplastic, transformed mammary carcinoma cells and with non-neoplastic normal mammary stem cells. In one set of experiments, distinct populations of mammary carcinoma cells formed spontaneously in a transgenic model of breast cancer pathogenesis were implanted at different dilutions (diluted in tenfold increments over three orders of magnitude) in the mammary glands of syngeneic mouse hosts. This procedure (see Figure 14.25A) is termed **limiting dilution analysis** (LDA) and can be employed to quantify the number of tumor-initiating cells (TICs), also termed cancer stem cells (CSCs), that are present in a larger population of cells isolated directly from a tumor.

As can be seen in Figure 14.25A, ~1 out of 143 tumor cells that expressed high levels of the Snail EMT-TF and resided in a more mesenchymal state could seed a new tumor upon implantation in the mammary glands of syngeneic mouse hosts; in contrast, the carcinoma cells in a cell population that expressed low levels of the Snail EMT-TF and resided in a more epithelial state (as indicated by the high levels of the EpCAM epithelial marker) harbored a far lower representation of TICs, specifically ~1 in 46,014 cells, that is, a greater than 300-fold difference in the relative concentrations of TICs. When taken together, these experiments showed a *correlation* between EMT-TF expression and concentrations of TICs. Other experiments to be described below relate to the issue of *causation,* that is, by demonstrating directly that an experimentally induced EMT program can actually drive an increase in the formation of CSCs and metastases. (Inevitably, demonstrations of causation are more compelling than those showing only correlation.)

(A)

frequency of TICs in subpopulations of mammary carcinoma cells

number of implanted cells	1×10^5	1×10^4	1×10^3	1×10^2	estimated TIC frequency (95% confidence interval)
Snail-YFP^{hi} EpCAM^{lo}	6/6	6/6	6/6	3/6	1/143 (1/428~1/47)
Snail-YFP^{lo} EpCAM^{hi}	5/6	2/6	0/6	0/6	1/46,014 (1/111,052~1/19,065)

(B)

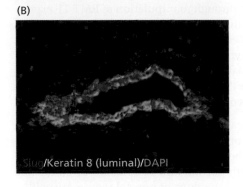

Slug/Keratin 8 (luminal)/DAPI

(C) remove rudimentary ducts including SCs

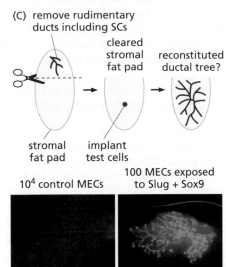

cleared stromal fat pad → reconstituted ductal tree?

stromal fat pad   implant test cells

10^4 control MECs   100 MECs exposed to Slug + Sox9

As mentioned earlier, association of an EMT program with stemness could also be found in non-neoplastic mammary epithelial cells (MECs) (see Figure 14.25B). A cross-section of a normal mouse mammary duct reveals its bilayered structure, much like a number of other epithelial tissues throughout the body. The cells in the **luminal** layer directly confront the lumen of this duct, whereas those of the **basal** layer underlie the luminal cells, being one cell layer removed from the lumen. Figure 14.25B reveals that the basal cells, which are known from other work to contain the normal stem cells of the mammary duct, naturally express the Slug EMT-TF. This observation is striking for the following reason: until now, we have depicted EMT-TFs as master regulators that function in embryonic morphogenesis, in wound healing, and in tumor progression. Now one of them is found to be routinely expressed in an adult tissue that is operating in an unperturbed, fully normal fashion.

Once again this observation, on its own, was also correlative because it did not demonstrate directly that the Slug-expressing cells are functional stem cells (and if so, whether Slug was responsible for their stem-like properties). In this instance, functional tests of stemness could be undertaken using mammary gland reconstitution experiments. This procedure gauges the ability of small numbers of implanted normal mammary cells to generate entire mammary ductal trees after their implantation in mammary stromal fat pads (see Figure 14.25C). Here, the presence of stem cells is gauged by their ability to spawn an entire normal mammary gland—a stringent test of stemness. When MECs prepared from a mouse mammary gland that had been forced to transiently express Slug and Sox9 were implanted into a cleared mammary stromal fat pad, their ability to generate a normal mammary ductal tree was increased >100-fold relative to MECs that had not been exposed transiently to these transcription factors. (Sox9 is a second TF that potentiates in this case the actions of the Slug EMT-TF.)

We are only beginning to learn the identities of the stromal cells that cooperate with one another to create both normal and neoplastic stem cell (SC) niches. These signals clearly overlap significantly with the EMT-inducing signals described in Section 14.5 because activation of an EMT program places MECs in a position to advance into the SC state. Interestingly, the macrophages that assist in the intravasation

Figure 14.25 Contributions of EMT-TFs to mammary carcinoma and normal mammary gland formation (A) Implantation of populations of cancer cells, performed here in syngeneic host mice and at different concentrations, makes it possible, using the algorithm of limiting dilution analysis (LDA), to estimate the representation of tumor-initiating cells (TICs) in the undiluted cancer cell population. As is apparent, when gauging the percentage of TICs/CSCs within these tumors, among the cells expressing high levels of the Snail EMT-TF (Snail-YFP^{hi}EpCAM^{lo}) 1 in 143 can be registered as a TIC, whereas among cells that express much lower levels of Snail (Snail-YFP^{lo}EpCAM^{hi}) only 1 in 46,014 cells was calculated to be a TIC. In these tumors, the latter class of cells expressed significant levels of the Slug EMT-TF. (B) In this cross section of a duct of the normal mouse mammary gland, abluminal (situated away from the lumen) cells in a location known to contain both myoepithelial and stem cells are seen to express the Slug EMT-TF protein at significant levels (*pinkish red*), in contrast to the luminal epithelial cells, which express cytokeratin 8 (*green*). Nuclei are stained *blue* with DAPI. (C) A rigorous test of stemness comes from implanting as few as one mouse mammary epithelial SC into a mammary stromal fat pad that lacks its own endogenous MECs (*above*). Success in mammary gland reconstitution demonstrates the presence of SCs among the implanted cells. In this test, 10,000 control mouse MECs were implanted in a cleared fat pad (*below, left*), resulting in no reconstituted mammary ductal tree. However, If the Slug EMT-TF and a second collaborating TF, termed Sox9, were transiently co-expressed in mouse MECs, as few as 100 of these cells (*below, right*) were able to reconstitute a mammary ductal tree. (A, from X. Ye, W.L. Tam, T. Shibue et al., *Nature* 525:256–260, 2015. With permission from Nature. B and C (bottom), from W. Guo, Z. Keckesova, J.L. Donaher et al., *Cell* 148:1015–1028, 2012. With permission from Elsevier. C (top), adapted from L. Hennighausen and G.W. Robinson, *Nat. Rev. Mol. Cell Biol.* 6:715–725, 2005.)

of primary tumor cells, as described in Figure 14.23, are also reported to induce carcinoma cells to enter into the CSC state during the course of this assisted intravasation.

Evidence from numerous sources now indicates that carcinoma cells residing in a hybrid phenotypic state expressing a mixture of epithelial and mesenchymal attributes (see Figure 14.12), which can be annotated as E/M or termed "quasi-mesenchymal," are placed optimally to function as cancer stem cells. Conversely, transformed cells residing at either end of the E-to-M spectrum of phenotypes, that is, fully epithelial and fully mesenchymal, have far less if any tumor-initiating activity. Moreover, carcinoma cells that lack phenotypic plasticity and have been trapped experimentally in such an E/M phenotypic state (achieved through manipulation of EMT-TF expression) exhibit a high degree of tumor-initiating activity. This indicates that the CSC phenotype exhibited by such E/M cells does not depend on their plasticity, that is, their ability to interconvert to other phenotypic states arrayed along the E-to-M spectrum. Instead, a specific mixture of epithelial and mesenchymal cell-biological traits exhibited by such cells suffices to empower them to function as tumor-initiating cells (TICs) and thus, quite possibly, as founders of metastatic colonies.

These observations hold important implications for the role of EMT programs in epithelial stemness. Although they were only performed in the context of mammary gland biology, it is likely that similar behaviors will be found in other bilayered epithelial tissues throughout the body, both normal and neoplastic. To enumerate these conclusions: (1) The SC machinery that operates in normal mammary epithelial tissue operates in a similar fashion in mammary carcinoma cells; hence, breast cancers do not seem to invent entirely novel SC programs during tumor formation but instead appropriate versions of an SC program that operates in a normal precursor tissue. (2) Contrary to intuition, the SCs in the basal cell layers of the mammary gland express a series of mesenchymal markers. Accordingly, in bilayered epithelia, to the extent that self-renewing stem cells are associated with basal cells, it seems likely that the luminal epithelial cells are the more differentiated progeny of quasi-mesenchymal, less differentiated SCs. It remains unclear whether this lesson also applies to single-layer epithelia. (3) From the perspective of multi-step tumor progression (see Figure 11.7A), because both fully normal and fully neoplastic cells employ EMT-associated stem cell programs, it is plausible if not likely that each of the distinct cell populations formed at intermediate stages of multi-step tumor formation do so as well, that is, they harbor subpopulations of quasi-mesenchymal SCs (see Figures 11.22 and S11.4). Stated differently, it seems that epithelial cell populations, independent of their state of neoplastic transformation, naturally form their own subpopulations of SCs. The latter, in turn, employ versions of EMT programs to help organize themselves, at least in tumors arising in bilayered epithelial tissues.

Finally, in the context of metastasis formation, the tumor-initiating powers associated with the CSC phenotype hold important implications for the seeding of metastases. Intuition would dictate that the same cell-biological functions required to seed tumors upon experimental implantation (as analyzed in Figure 14.25A), are also critical to the founding of metastatic colonies following spontaneous dissemination of cells from primary tumors. Indeed, deletion from carcinoma cells of an important regulator of stemness—the p53-related p63 protein—while compatible with ongoing growth of established primary tumors, deprives these tumors of the ability to seed metastases, which they would otherwise spawn in large numbers. Still, all this leaves us with a major conceptual puzzle that has never been satisfactorily resolved: Why are the EMT program and epithelial stemness so tightly interconnected in certain tissues?

14.7 EMT-inducing transcription factors help drive malignant progression including metastatic dissemination

Once certain EMT-TFs are expressed in cancer cells, they act in various combinations to induce multiple cellular changes associated with invasion and metastasis. They are invariably expressed in combinations in cancer cells, which suggests that no one of them, acting on its own, is able to organize all of the cell-biological changes associated with passage through an EMT program. Indeed, these master regulators have been found to induce the synthesis of one another, often by the direct binding of one EMT-TF to the transcriptional promoter of another. For example, Twist binds directly to the promoters of the genes encoding Twist, Snail, Slug, and Zeb1, and actually depends on Slug induction in order to activate an EMT program. In addition, expression of Slug, which is frequently co-expressed with Twist in human breast cancers, is lost when Twist expression is shut down in some breast cancer cells. Also, some of these TFs can act in a redundant fashion; for example, Snail, Slug, Zeb1, and Zeb2 can all bind to the promoter of the E-cadherin gene (*CDH1*) and thereby repress E-cadherin expression. Accordingly, Twist may delegate the job of shutting down E-cadherin expression to Slug. Although Snail and its cousin Slug function very similarly, their expression is regulated very differently, giving us clues about the dynamics of EMT induction (**Sidebar 14.3**).

Although experiments described above related to the connections between EMT programs and the stem-cell state, they did not directly confront the key question of metastasis: Do the EMT-TFs simply confer traits associated with high-grade malignancy on cultured cells, or are their powers also apparent *in vivo*? More importantly, do they actually *cause* metastatic dissemination or merely accompany it? The most direct route for demonstrating their role in causing various steps of invasion and metastasis involves either overexpressing them or shutting down their expression in tumorigenic cells that have been implanted in suitable host mice.

In fact, some research shows that EMT-TF function is *necessary* for efficient metastasis. Thus, in one set of experiments, expression by carcinoma cells of either the Twist1

Sidebar 14.3 Snail stands out from its colleagues Like many other cell-biological processes, the induction of an EMT program depends on increases in EMT-inducing TFs in the cell nucleus. These increases generally depend on the relatively slow processes of inducing the transcription of the genes that encode these EMT-TFs, the processing of the resulting pre-mRNA transcripts, and the translation of the mRNA products into the proteins that thereafter function as TFs in the nucleus; these yield, in turn, mRNAs that, with the passage of time, enable the accumulation of proteins related to the mesenchymal state. Taken together, these steps dictate that EMT-TF proteins and their downstream gene products proteins only respond to inductive signals over periods of several hours, possibly days.

This account describes the regulation of Slug. However, its close cousin, Snail is an outlier: Snail is usually produced at a high rate and is rapidly degraded thereafter, yielding a short lifetime of 25 minutes; this so-called "futile cycle" resembles the cycles governing the levels of the Myc and p53 proteins (see Sections 8.9 and 9.5). If Snail degradation is blocked, the ongoing high rate of Snail synthesis allows for rapid increases in protein levels that can be measured over the course of minutes and the resulting rapid induction of EMT programs.

Under conditions of cell quiescence, the GSK-3β kinase phosphorylates recently synthesized Snail, thereby causing it to be tagged (by ubiquitylation) for rapid destruction in proteasomes. Conversely, activation of mitogenic signaling via growth factor ligands and their receptors and resulting activation of the Ras–PI3K–Akt/PKB pathway results in the inactivation of GSK-3β by Akt/PKB, permitting in turn survival and thus accumulation of Snail. (We first encountered GSK-3β in its role in driving β-catenin degradation; see Figure 6.19B.)

These dynamics operate in other situations as well. For example, interactions of cancer cells with macrophages, as described earlier in this chapter, result in, among other things, the release of tumor necrosis factor-α (TNF-α) by these myeloid cells, which in turn helps to initiate an EMT in nearby carcinoma cells. (TNF-α, as we read earlier, is an important player in the process of tumor promotion via inflammation; see Figures 11.35 and 11.36). By activating the NF-κB transcription factor in carcinoma cells, TNF-α causes expression of the CSN2 protein, which associates with and blocks the ubiquitylation of Snail, once again allowing the latter to accumulate rapidly in periods of less than an hour. Finally, we should note that certain aspects of the tumor-promoting effects of inflammatory cells, as described in Section 11.13, can be nicely explained by the EMT-inducing effect of TNF-α.

Figure 14.26 Dependence of metastasis in mouse models on EMT-inducing TFs
(A) 4T1 mouse mammary carcinoma cells are usually highly metastatic following implantation in the orthotopic site—the mammary gland (*blue bars*). However, when these cells were infected with a retrovirus vector specifying a potent siRNA directed against the Twist EMT-TF, primary tumors grew as rapidly as before, but metastases to the lungs were reduced by ~85%. The residual metastases expressed significant levels of Twist, indicating that they derived from primary tumor cells in which Twist expression had not been successfully knocked down (*red bars*). (B) Experimentally transformed human melanocytes were found to express high levels of the Slug EMT-inducing TF. When these cells were implanted in a subcutaneous site, they generated large numbers of lung metastases. However, when an siRNA was expressed that reduced Slug mRNA levels by 80%, the number of pulmonary metastases decreased by ~90%, while the growth of the corresponding primary tumors was hardly affected. (C) Rat epithelial cells were transformed by forced constitutive expression of both TrkB (a tyrosine kinase receptor) and its ligand, BDNF, achieved in both cases by retrovirus vectors. The number of mice bearing subcutaneously implanted TrkB-transformed cells was reduced significantly by anti-ZEB1 shRNA relative to those bearing tumors expressing control (anti-EGFP) shRNAs. (A, from J. Yang et al., *Cell* 117:927–939, 2004. With permission from Elsevier. B, from P.B. Gupta et al., *Nat. Genet.* 37: 1047–1054, 2005. C, from M.A. Smit and D.S. Peeper, *Oncogene* 30:3735–3744, 2011. Both with permission from Nature.)

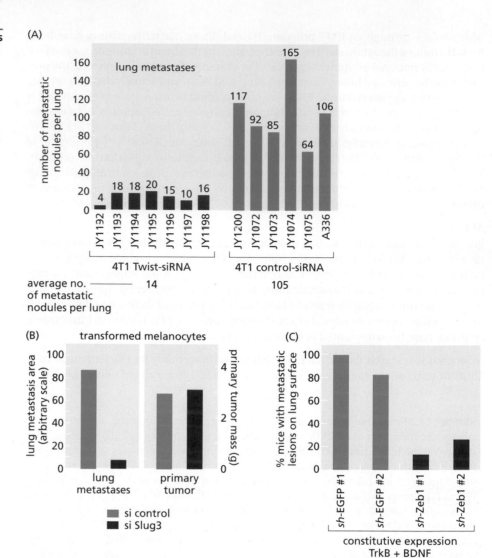

or the Slug EMT-TF was suppressed through the use of siRNA constructs. As seen in **Figure 14.26A and B**, suppression of Twist1 expression in mammary carcinoma cells or of Slug in melanoma cells resulted, in both cases, in significant decreases in the lung metastases formed by primary tumors. In the experiment involving Twist (see Figure 14.26A), the few metastases that formed following siRNA-mediated suppression of Twist1 were found to have escaped complete Twist shutdown by the siRNA. These experiments address the necessity of EMT in metastatic dissemination of pulmonary metastases from primary tumor sites. A similar outcome was seen when the Zeb1 EMT-TF was shut down in epithelial cells transformed by a combination of forced expression of a receptor and its ligand (Figure 14.26C).

Experiments like those of Figure 14.26 do not address whether expression of EMT-TFs is *sufficient* to trigger the latter steps of the invasion–metastasis cascade. One critical rate-limiting step of metastasis formation is extravasation—escape of cancer cells from microvessels into the parenchyma of nearby tissue (see Figure 14.3). The efficiency of this step can be gauged by injecting cancer cells into the tail vein of a mouse, which results in their rapid trapping in the microvessels of the lungs—the procedure termed "experimental metastasis". In this instance, extravasation rates can be determined by microscopic examination of lung tissue, specifically by examining whether injected cancer cells remain trapped intraluminally in the microvessels or have escaped into the lung parenchyma. The experiments shown in **Figure 14.27A** reveal that when the tumors cells used in these experiments were injected without any induced EMT-TF, about 20% succeeded in extravasating; however, when expression of the Twist EMT-TF was induced in the tumor cells for a week or more while these cells were still in culture,

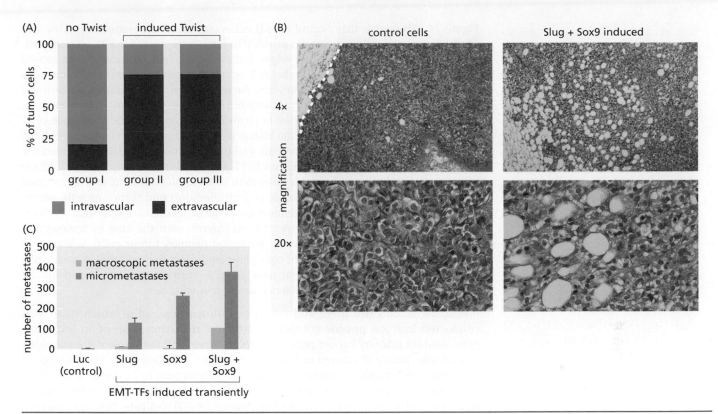

Figure 14.27 Contributions of EMT-inducing TFs to invasion, extravasation, and metastasis (A) The effects of the Twist EMT-TF specifically on the process of extravasation was tested in a mouse model of cutaneous squamous cell carcinoma. Primary tumor cells were propagated in culture and then injected into the tail vein of a mouse with microscopic monitoring 36 hours later in order to determine whether the carcinoma cells had remained trapped in microvasculature of the lungs or had extravasated into the lung parenchyma. In the absence of any induced Twist expression, ~20% of the cells successfully extravasated (*left bar*). However, if Twist was expressed for a week or more in culture prior to tail-vein injection and was either kept on or shut off after tail-vein injection (achieved through use of an inducible Twist expression construct), ~80% of the carcinoma cells were found to have successfully extravasated (*middle, right bars*). (B) Human breast cancer cells of the MCF7-ras cell line are weakly invasive and metastatic. Thus 12 weeks after implantation into an immunocompromised mouse host, these cells formed tumors (*left panels*) that had smooth outer margins (*white line, top left panel*) and the absence

of histopathological traits that were indicative of invasiveness. In contrast, cells in which the combination of the Slug and Sox9 EMT-TFs was induced for 2 weeks after implantation *in vivo* and then shut down (*right panels*), the tumors that formed 10 weeks later showed invasive margins, indicative of invasion into adjacent adipocyte-rich stromal tissue. Hence, these cells continue to be affected by the actions of the EMT-TFs long after expression of these TFs had been de-induced. (C) Lung metastases were counted 12 weeks after orthotopic implantation of MCF7-ras breast cancer cells. Neither micrometastases nor macroscopic metastases were present in significant numbers in the Luc (control) cells. However, these EMT-TFs, induced either singly or in combination, yielded as many as ~400 micrometastases and ~100 macroscopic metastases, demonstrating the powers of the EMT program to drive otherwise-benign primary tumor cells through most of the steps of the invasion–metastasis cascade. (A, from J.H. Tsai, J.L. Donaher, D.A. Murphy et al., *Cancer Cell* 22:725–736, 2012. With permission from Elsevier. B, C, courtesy of Z. Keckesova and J. DeCock.)

~80% of the subsequently injected cells succeeded in extravasating within 36 hours of tail-vein injection, whether or not Twist expression continued after tail-vein injection. Hence, the process of extravasation, which represents one of the last steps of the invasion-metastasis cascade, is greatly facilitated by the actions of an EMT program.

Yet other experiments deal with the larger question of whether expression of EMT-TFs and resulting induction of EMT programs is, on its own, *sufficient* to force otherwise-nonmetastatic human breast cancer cells growing in primary tumors to form metastases and therefore complete most of the steps of the invasion–metastasis cascade. These experiments relied on the reversible induction of the Slug EMT-TF together with a second TF, Sox9; this pair of transcription factors had been previously found to induce stemness of normal mouse mammary epithelial cells, as illustrated in Figure 14.25C. Expression of this pair of TFs was induced in orthotopically implanted tumors (that is, in this case in mammary stromal fat pads) for 2 weeks, after which their expression was shut down and the tumor-bearing mice were observed for another 10 weeks.

Figure 14.27B reveals that control cells (Luciferase), in which expression of the two TFs had not been induced, formed primary tumors at the end of the 12-week experiment that exhibited smooth outer edges, indicative of a lack of invasiveness, as well as harboring cells with more epithelial cell morphologies. In contrast, the tumor cells that had been exposed 10 weeks earlier to the two TFs generated tumors whose outer edges were highly invasive; also, the associated individual carcinoma cells were more mesenchymal in morphology. Hence, once an EMT had been induced in these cells, their descendants "remembered" the effects of the induced EMT-TFs 10 weeks later, even in the absence of ongoing induction of the TFs that had initially triggered the EMT. This result indicated that once an EMT program had been induced, its effects were transmitted in a cell-heritable fashion to its lineal descendants and maintained during the subsequent long period of cell proliferation and tumor growth, ostensibly under the auspices of self-reinforcing positive-feedback loops such as those driven by autocrine signaling. Of even greater interest were the lung metastases that were spawned by the subcutaneously implanted primary tumor cells. As shown in Figure 14.27C, the parental control cells generated essentially no metastases, while the cells that had experienced transient Slug + Sox9 expression formed almost 400 micrometastases and almost 100 macroscopic metastases.

Preclinical studies like these and others not shown here, all of which have been conducted in mice, provide compelling evidence that expression of an EMT program enables primary cancer cells to complete almost all the steps of the invasion–metastasis cascade illustrated in Figure 14.3. Included among these are the steps of localized invasion, the formation of circulating tumors cells (CTCs) that results from intravasation, subsequent extravasation, and the formation of micrometastases. Although the experiments described in Figure 14.27B indicate that macroscopic metastases, which result from the process of colonization, could be accomplished by expression of an EMT program, as discussed further below, execution of this final step of the cascade cannot generally be achieved simply by expression of an EMT program. Nonetheless, these various experiments reveal that, at the very least, EMT programs can physically convey carcinoma cells from the confines of a primary tumor into the parenchyma of a distant tissue; the execution of the last step of the invasion–metastasis cascade (see Figure 14.3), entails unrelated, complex phenotypic shifts that we will encounter later in this chapter.

Reassuringly, the results of the experiments described here have correlates in the observations of human cancer patients occurring in the oncology clinic. As seen in a variety of carcinomas, expression of the Snail, Twist, and Slug EMT-TFs in primary tumors is associated with poor prognoses in at least four types of human carcinomas (**Figure 14.28**). These studies and others like them provide confidence that our discussions in this chapter, focused for historical reasons largely on breast cancer, will eventually prove applicable to a broad range of human carcinomas, possibly all of them. Moreover, the strong effects of EMT, as indicated by the expression of EMT-TFs and EMT phenotype (Figure 14.28D), validate the use of mouse models of carcinoma progression for understanding human cancer pathogenesis.

To summarize, the discovery of the EMT-inducing TFs and the mechanisms that control their expression suggest at least three important ideas about malignant progression. First, many malignant cell phenotypes, notably those displayed by carcinoma cells, may be induced by *nongenetic* changes—heterotypic signals of stromal origin—rather than somatically acquired genetic changes occurring within carcinoma cell genomes; hence the cells within primary tumors may already possess the genetic alterations required for dissemination and therefore only require activation of an EMT program in order to spread to distant sites (Supplementary Sidebar 14.6). Second, because expression of these TFs and the resulting EMT is often dependent on heterotypic signaling from the reactive stroma of primary tumors, carcinoma cells may revert from a more mesenchymal state to an epithelial state once they have left the primary tumor and encounter the normal stromal microenvironments present in sites of metastasis; by definition, such normal stromata do not release the heterotypic signals that can induce activation of EMT programs in nearby carcinoma cells. Third, cancer cells do not need to cobble together all of the phenotypes associated

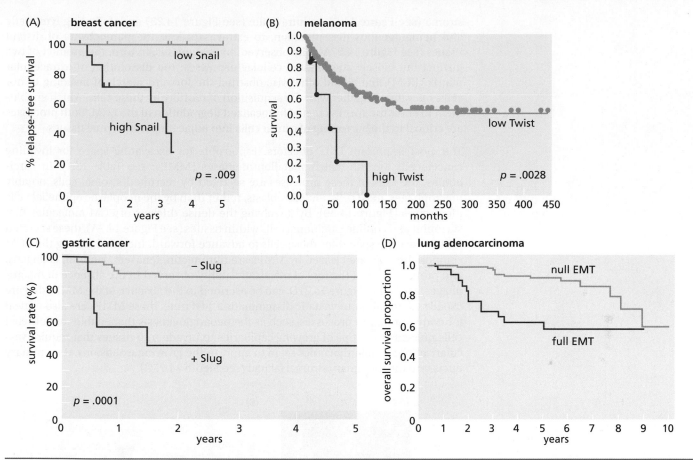

Figure 14.28 Indications of the role of EMT-inducing transcription factors in human tumor progression Increasing evidence correlates the expression of EMT-inducing TFs with the induction of malignant behavior in cancer cells in patients. (A) Among a group of women with locally advanced breast cancers (tumors that appeared histologically to be aggressive without evidence of disseminated disease), those whose tumors expressed either high or low levels of Snail were followed for a period of 5 years after initial surgery. The percentage of women who enjoyed relapse-free survival varied dramatically, depending on whether or not the carcinoma cells in their primary tumors expressed high levels of the Snail EMT-TF protein. (B) Sections of primary melanomas were immunostained for Twist EMT-TF expression and correlated retrospectively with the long-term survival of the patients bearing these tumors. (C) Shown are the postoperative survival rates for stomach cancer patients whose primary tumors retained detectable E-cadherin expression together with or in the absence of Slug EMT-TF expression. (D) The clinical progression of patients suffering from lung adenocarcinomas was tracked following their initial presentation in the oncology clinic. Patients whose tumors were CD133-negative (a stem cell marker) were classified into two groups: null EMT–E-cadherin-positive, vimentin-negative; or Full EMT (E-cadherin-negative, vimentin-positive.) (A, from S.E. Moody et al., *Cancer Cell* 8:197–209, 2005. With permission from Elsevier. B, from K. Hoek et al., *Cancer Res.* 64:5270–5282, 2004. With permission from American Association for Cancer Research. C, from Y. Uchikado et al., *Gastric Cancer* 14:41–49, 2011. With permission from Springer. D, from T. Sowa, T. Menju, M. Sonobe et al., *Cancer Med.* 4:1853–1862, 2015. With permission from John Wiley & Sons.)

with highly malignant cells by acquiring multiple mutant genes or activating multiple cell-biological programs; instead, many of the malignancy-associated traits enabling completion of most of the steps of the invasion–metastasis cascade can be acquired concomitantly through the actions of a single cell-biological program—the EMT.

14.8 The invasiveness of carcinoma cells depends on clearance of obstructing ECM

The discussions of EMT programs and the TFs that induce them have overlooked the individual steps of the invasion–metastasis cascade that yield the final products— fully formed, life-threatening metastases. So, now we backtrack to revisit these steps and revisit how they are manifested in aggressive primary tumors cells. To begin, and as cited on multiple occasions earlier, among the critical consequences of EMT activation are its effects on the invasiveness of carcinoma cells. This invasiveness plays critical roles in enabling islands of primary carcinoma cells to invade the nearby stroma (see Figure 14.5), to intravasate (see Figure 14.23) and, following lymphatic and hematogenous dissemination, to extravasate into the parenchyma of distant tissues (see Figure 14.27A). The observed invasiveness is, in turn, the product of two distinct but closely coordinated cellular processes: the dissolution of extracellular matrix (ECM) that would otherwise obstruct the forward march of invading tumor cells, operating together with the acquisition of motility by these cells, which enables them to enter the spaces created by localized degradation of the ECM. Both processes are critical to the burrowing of cancer cells into adjacent parenchymal tissues.

The most important EMT effectors responsible for excavating space for invading cancer cells are the matrix metalloproteinases (MMPs; see Table 13.1). In carcinomas, the bulk of these proteases are secreted by recruited stromal cells, notably macrophages, mast cells, and fibroblasts, rather than by the neoplastic epithelial cells themselves (**Figure 14.29**). By dissolving the dense thickets of ECM molecules that surround and confine carcinoma cells within tissues (see Figure 14.7A), these secreted MMPs excavate spaces for these cells to advance forward. Included among the ECM components that are cleaved by MMPs are fibronectin, tenascin, laminin, collagens, and proteoglycans. (The loss of metastatic ability observed in the absence of macrophages, as seen in Figure 14.21D, can be ascribed to the absence of the MMPs that are usually essential for metastatic dissemination.) Of note, these MMPs are also critical for nonpathological processes, such as the neoangiogenesis that enables endothelial cells present at the tips of growing capillaries to invade into tissues that require vascularization; similar dynamics seem to apply to the growing extensions of mammary ducts into the mammary stromal fat pad (see Figure 14.25C).

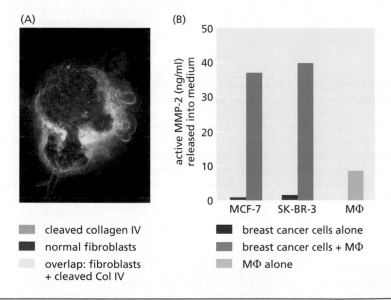

cleaved collagen IV
normal fibroblasts
overlap: fibroblasts + cleaved Col IV

breast cancer cells alone
breast cancer cells + MΦ
MΦ alone

Figure 14.29 Matrix metalloproteinases produced by tumor-associated stromal cells
(A) The ability of tumors to degrade collagen IV, a key component of the basement membrane, can be measured by generating a modified collagen IV substrate that fluoresces *green* upon cleavage. In this experiment, human mammary carcinoma cells (*not seen*) and human mammary fibroblasts (*red*) showed relatively weak ability to degrade the collagen IV substrate when these two cell populations were cultured on their own. However, when these two cell populations were co-cultured, regions of collagen IV cleavage (*green*) were evident, often in areas where fibroblasts were also present (*yellow: overlap of green and red*). This cleavage was essentially eliminated in the presence of MMP inhibitors (*not shown*). (B) An even more important source of MMPs is the macrophages (MΦs) that are recruited into the tumor stroma. In this *in vitro* experiment, the presence of MMP-2 was measured in the culture medium of MΦs that were cultured alone (*green*) or together with either of two human breast cancer cell lines—MCF-7 or SK-BR-3. Neither of these cancer cell types released significant levels of MMP-2 on its own (*red*), but in the presence of MΦs (*blue*), the production and release of MMP-2 increased 4- to 5-fold. (The increase could be traced to the induction of MMP-2 mRNA expression by the MΦs, *not shown*.) The released MMP-2 imparted increased invasiveness to these breast cancer cells (*not shown*). (A, from M. Sameni et al., *Mol. Imaging* 2:159–175, 2003. With permission from SAGE. B, from T. Hagemann et al., *Carcinogenesis* 25:1543–1549, 2004. With permission from Oxford University Press.)

During the course of degrading ECM components, MMPs also mobilize and activate certain growth factors that have been tethered in inactive form to the ECM or to the surfaces of cells. These actions are actually directly intertwined with the control of EMT programs: On the one hand, EMT-induced transcriptomes include mRNAs encoding an array of secreted MMPs, such as MMP-2 and MMP-9. On the other, these MMPs act in the intercellular space to cleave the ectodomains of E-cadherin molecules and to activate previously latent TGF-β; both of these actions stimulate, in turn, the expression of EMT-TFs, creating self-reinforcing, positive-feedback loops that help to sustain expression of EMT programs in cells that have initiated components of these programs.

Complementing these secreted MMPs is a critically important one—MT1-MMP (membrane type-1 MMP)—that remains tethered directly to the plasma membrane surfaces of cancer cells and is wielded by them in order to cleave cell ECM components, cell surface adhesion molecules (for example, cadherins and integrins), as well as growth factor receptors and chemokines. Most importantly, MT1-MMP can also cleave inactive **pro-enzymes**, such as pro-MMP-2, into enzymatically active MMP isoforms. By activating soluble, diffusible MMPs like MMP-2, MT1-MMP can amplify and extend its proteolytic effects to regions in the pericellular space that are not immediately adjacent to the cell surface of cancer cells. (MT1-MMP is one of six MMPs that are membrane-anchored and therefore limited to cleaving substrate proteins in the immediate vicinity of the cells that produce them.) Taken together, these properties explain how ectopic expression of MT1-MMP, acting via its downstream MMP effectors, can induce EMTs in cultured epithelial cells.

As we have read, the initial steps of invasion by a carcinoma cell are obstructed by the basement membrane (BM), which may have pores as small as 40 nm in diameter (see Figure 13.10). Although cells can distort their cytoplasms in order to squeeze through small openings, their nuclei are relatively rigid and thus govern the size of the smallest pores through which cells can pass. Cell nuclei are generally in the size range of 3 to 10 μm, about 100 times larger than the pores in the BM; this explains why dissolution of the BM is critical to carcinoma cell invasion (see also Figure 14.4).

The activities of MT1-MMP, which plays a leading role in BM breakdown, seem to be confined through its concentration at discrete cell surface foci, initially termed **podosomes** but increasingly called **invadopodia** because of the involvement of these structures in cancer cell invasion (**Figure 14.30**). Early in malignant progression, MT1-MMP displayed on the surface of carcinoma cells can cleave collagen type IV, the collagen that imparts rigidity to the basement membrane (BM; see Figures 2.3 and 13.10). The resulting weakening of the BM allows cancer cells to begin invading the underlying stroma (see Figure 14.4). Once in the stroma, an invading carcinoma cell confronts a dense network of cross-linked collagen type I fibers that obstructs further advance (see Figure 14.30B); here, MT1-MMP once again plays a central role. It initiates collagen I degradation and then recruits an inactive pro-enzyme (pro-MMP-2) of stromal origin, which it activates by cleavage as mentioned above (**Figure 14.31**). The resulting active MMP-2 then operates in the peri-cellular space to further reduce collagen I into lower–molecular-weight fragments. Without these proteolytic steps, the dense networks of collagen I fibers that are present in the stromal ECM totally block cancer cell invasion.

Membrane-bound and secreted proteases clearly play important roles in normal cell survival and proliferation. After all, each time a cell within a normal tissue goes through a cycle of growth and division, spaces within the ECM must be carved out for its daughters. Once formed, each daughter cell must, in turn, reassemble a new ECM around itself. Hence, the remodeling of the ECM takes place continuously in mitotically active normal tissues. Therefore, rather than being aberrations of invasive cancer cells, the activities of MMPs and other extracellular proteases are part of the program associated with normal cell proliferation. Of relevance here are clinical trials of certain MMP-inhibiting drugs, which were launched in order to curtail cancer cell invasiveness and thus a key aspect of high-grade malignancy. These trials were terminated due to the effects of these inhibitors on a variety of normal tissues; because these agents suppressed the normal ongoing remodeling of cartilage and other

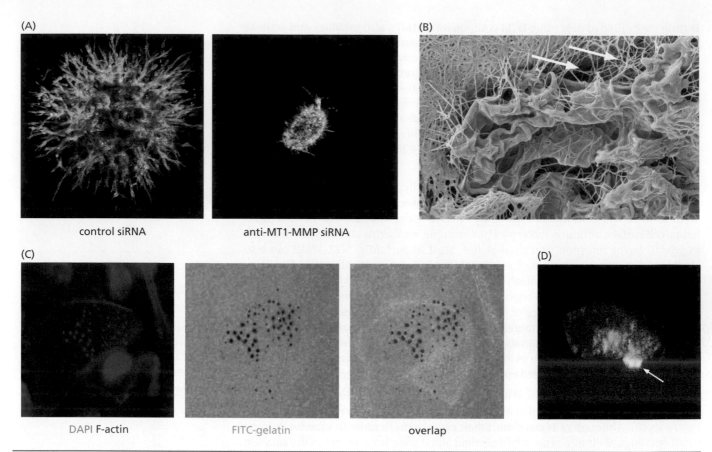

control siRNA   anti-MT1-MMP siRNA

DAPI **F-actin**   FITC-gelatin   overlap

Figure 14.30 MT1-MMP, invadopodia, and the degradation of the extracellular matrix Invadopodia, sometimes termed podosomes, are small, focal protrusions from the cell surface that degrade localized areas of extracellular matrix (ECM) in their immediate vicinity. In the case of tumor development, invadopodia are deployed by invasive cancer cells to drive localized, highly controlled degradation of the ECM near the leading edges of these cells. (A) The critical contribution of MT1-MMP, which is tethered to the surface of invadopodia, in cancer cell invasion can be demonstrated *in vitro* through the behavior of HT1080 human fibrosarcoma cells. When these cells, seen here growing as a colony in a 3D matrix of collagen I, were forced to express a control siRNA (*left panel*), they demonstrated a high degree of invasiveness, as indicated by the numerous protruding columns of cancer cells invading the matrix. However, when MT1-MMP production was knocked down through the actions of a specific siRNA, these cells were confined to a small volume surrounding the site at which the founding cell was initially introduced into the matrix. Knockdown of the two secreted metalloproteinases also made by these cells—MMP-1 and MMP-2—had no effect on their invasiveness (*not shown*). Cells were treated with a phalloidin derivative, which labels filamentous actin (*green*), and di-I (3,3-dioctadecyl indocarbocyanine), which stains membranes (*red*). (B) The effects of the proteases produced by the invasive HT1080 cells are seen in fine detail in this scanning electron micrograph in which the collagen I fibers in the ECM have been pseudo-colored *blue* and the fibrosarcoma cells are colorized *pink*. Note that the collagen matrix is dense beyond the immediate vicinity of the fibrosarcoma cells (*top left*), whereas the collagen fibers are sparse in the immediate vicinity of the cells (*arrows, center, right*) from the actions of the proteases, largely and perhaps exclusively the MT1-MMP made by these cells. (C) MT1-MMP is the major protease displayed on invadopodia. These organelles are defined by dense concentrations of filamentous actin, stained here with phalloidin (*red, left panel*) on the ventral surface (i.e., the surface directly apposed to the underlying substrate) of immortalized human mammary epithelial cells; nuclei are labeled with DAPI (*blue*). These immortalized human mammary epithelial cells were plated above a thin matrix of gelatin labeled with FITC dye (*green, middle panel*). The single cell seen in the middle of the left panel has eroded numerous holes (*black dots, middle panel*) in the gelatin (thereby solubilizing FITC) precisely under the focal areas where invadopodia were seen; the overlap between the F-actin foci and the eroded holes in the gelatin is seen here in red (*right panel*). Formation of these invadopodia was dependent on activation of the Twist EMT-TF as well as expression of the PDGF-Rα that resulted from activation of the EMT program. (D) The invadopodia of a small cluster of cells are visualized from the side (a z-axis view) through confocal microscopy. The staining of filamentous actin (*red*) and cortical actin (*green*; see Figure 14.40) overlap in the invadopodia (*yellow*), which are protruding through a layer of the ECM protein fibronectin (*blue*) through which they have eroded an opening (*white arrow*). (A and B, from F. Sabeh et al., *J. Cell Biol*. 185:11–19, 2009. With permission from Rockefeller University Press. C, from M.A. Eckert et al., *Cancer Cell* 19:372–386, 2011. With permission from Elsevier. D, courtesy of S. Vitale and M. Frame; see also S. Vitale et al., *Eur. J. Cell Biol*. 87:569–579, 2008.)

Figure 14.31 Collaboration of MT1-MMP and MMP-2 in carcinoma cell invasion
Membrane-tethered proteases, notably MT1-MMP, initially enable carcinoma cells to degrade the basement membrane (*green*) that separates them from the adjacent stroma. Once in direct contact with the stroma (*pink*), MT1-MMP can cleave pro-MMP-2 made by the stromal cells, converting it to an active, soluble protease. MT1-MMP also partially cleaves collagen I, which normally forms a densely interwoven network in the ECM of the stroma; the activated MMP-2 then proceeds to degrade the initially formed collagen I cleavage products to lower–molecular-weight fragments. Dissolving the collagen I network creates a channel immediately in front of the carcinoma cell that allows it to invade more deeply into the stroma.

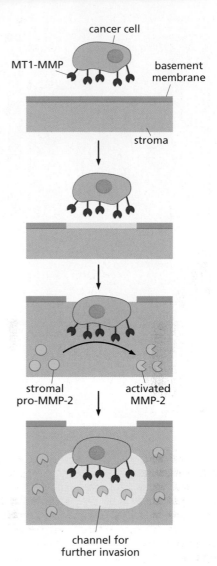

components of the ECM in joints, they created unacceptable levels of joint stiffness and pain.

Each type of MMP usually acts on a well-defined set of substrates (see Table 13.1), doing so in a highly regulated and localized fashion. It is likely that these enzymes continue to show such substrate specificity during the process of cancer cell invasion. However, in the case of invasive cancer cells, such proteolysis seems to proceed continuously rather than in the well-controlled, brief spurts that accompany normal cell growth and division.

As cited earlier, many of the MMPs found in tumors originate in various cellular components recruited to the tumor-associated stroma. For example, the best-studied of the matrix metalloproteinases, MMP-9, is expressed largely by the macrophages (see Section 13.5), neutrophils, and fibroblasts that are recruited to the invasive fronts of tumors. MMP-9 expression at these fronts correlates positively with the metastatic ability of a primary tumor, suggesting that MMPs like this one can act at several stages of the invasion–metastasis cascade, including local invasion of the primary tumor stroma, intravasation, and extravasation, the latter steps described in more detail below. *In vitro* assays indicate that MMP-9 can degrade collagens that are prominent components of the ECM including basement membranes, specifically collagen types IV, V, XI, and XIV. Other targets of MMP-9 include laminin (another key constituent of the basement membrane; see Figure 13.10), chemokines, fibrinogen, and latent TGF-β. In the case of the latter, cleavage by MMPs converts this factor from a latent form into its activated form, as mentioned earlier.

These widely ranging functions of MMPs indicate that their enzymatic activity must be tightly controlled, at least in normal tissues. One critical control mechanism has already been described: Soluble MMPs, such as MMP-2 and MMP-9, are initially synthesized as inactive pro-enzymes that can only function, like the caspases (see Section 9.13), following their activation by other proteases (see, for example, Figure 14.31). Negative regulation is also provided by a class of proteins termed tissue inhibitors of metalloproteinases (TIMPs), which bind MMPs and place them in an inactive configuration.

Although MMPs have been depicted as one of a group of downstream effectors of certain steps in invasion and metastasis, it is clear that the deregulation of MMPs can, on its own, drive the progression of cells through all of the stages of multi-step tumorigenesis including completion of the invasion–metastasis cascade. Thus, when expression of MMP-3 is forced in the mammary gland of transgenic mice, these mice initially develop mammary hyperplasias (**Figure 14.32**). Some of these growths eventually progress to carcinomas that become invasive and metastatic. These mice reveal how critical the regulation of MMP function is and why it must be kept under control in normal tissues.

These brief vignettes of proteases and their contributions to cancer cell invasiveness describe only small parts of what are surely highly complex networks of interacting proteases, protease inhibitors, and substrates. The total number of proteases made by mammalian cells is vast and rivals the number of proteins that form the highly complex intracellular signal-processing circuits described in Chapter 6. To date, the actions of only a tiny proportion of these enzymes have been studied in the context of cancer pathogenesis (see Supplementary Sidebar 14.7).

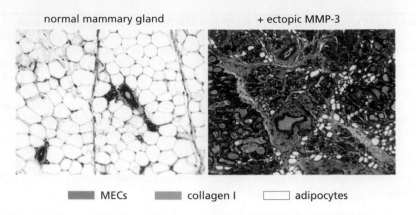

normal mammary gland   + ectopic MMP-3

■ MECs   ■ collagen I   □ adipocytes

Figure 14.32 Ectopic expression of MMP-3 and mammary tumor progression The normal mouse mammary gland (*left panel*) is composed of resting ducts formed by mammary epithelial cells (MECs, *purple*) and abundant adipose tissue (lipid-filled cells, *white*), as well as collagen fibers (*light blue*). However, when the gene encoding MMP-3 (also known as stromelysin-1) is expressed constitutively as a transgene in the mammary epithelium, the mice develop abundant hyperplasias (*right panel*), including extensive islands of hyperplastic epithelial cells (*purple*) that form ducts, as well as a fibrotic, collagen-rich stroma (*light blue*) and abnormal adipocytes (*white oval structures, lower right*). Many of these areas subsequently progress to invasive, metastatic tumors (*not shown*). (From M.D. Sternlicht et al., *Cell* 98:137–146, 1999. With permission from Elsevier.)

14.9 Motility enables cancer cells to move into space excavated by MMPs

The actions of extracellular proteases, notably the MMPs, explain at the biochemical level how paths are cleared for the advance of invasive cancer cells through the extracellular matrix and thus through tissues. They fail, however, to tell us how individual cancer cells take advantage of these cleared paths to move ahead—the trait of cell motility. The motile behavior of cells has been studied extensively with cultured cells, and it is presumed that their crawling on solid substrates *in vitro* reflects the *in vivo* behavior of cancer cells as they invade nearby cell layers and intravasate. Such motility is also presumed to be important for cancer cells' escape from blood vessels or lymph ducts—the process of extravasation.

Motile behavior can be induced in cultured cells by exposing them to a variety of growth factors. (Those GFs able to induce such locomotion are sometimes designated as being **motogenic** in addition to being mitogenic.) In the case of epithelial cells, the most potent inducer of motility is hepatocyte growth factor (HGF); this protein is also called scatter factor (SF) in recognition of its ability to induce multidirectional movement of cells in monolayer culture. Many types of epithelial cells express Met, the receptor for HGF, and such cells have been found to acquire motility in response to HGF treatment. Similarly, EGF is clearly able to induce motility of breast cancer cells.

The cellular machinery that responds to motogenic signals and operates as the engine of motility is extraordinarily complex at the molecular level. Cell motility involves continuous restructuring of the actin cytoskeleton in different parts of a cell, as well as the making and breaking of attachments between the migrating cell and the extracellular matrix (**Figure 14.33**). (In the case of cultured cells, the ECM in question is the network of proteins that has previously been laid down by these cells on the surface of the Petri dish; see Figure 1.11.) The outlines of these complex cellular and molecular mechanisms are described in Supplementary Sidebars 14.7 and 14.8.

We can summarize our understanding of the motility of cancer cells by relating it to the actions of EMT programs. As was apparent from earlier discussions of epithelial cells, they are largely immotile unless induced to move by processes such as wound healing (see, as an example, Figure 14.15).

To present a highly interesting example of motility mechanisms, among the many mRNAs altered by EMT programs, is the one that encodes Mena, a cytoplasmic protein that plays a key role in regulating the extension of actin fibers (**Figure 14.34**).

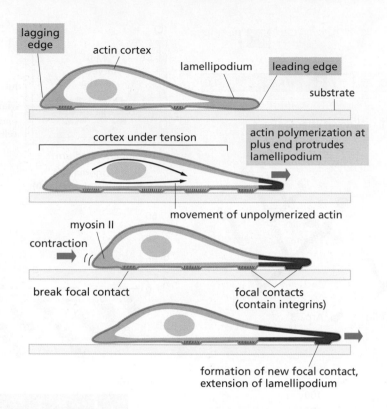

Figure 14.33 Locomotion of cells on solid substrates The locomotion of a cultured cell depends on the coordination of a complex series of changes in the cytoskeleton, as well as the making and breaking of focal contacts with the underlying solid substrate. The cell organizes actin fibers at their plus ends in order to extend lamellipodia at its advancing/leading edge and to establish new focal contacts. At the same time, stress fibers, also consisting of actin, are used to contract the lagging/trailing edge of the cell, where focal contacts are being broken. The making and breaking of the focal contacts depend on localized modulation of the affinities of various integrins for extracellular matrix (ECM) components, represented here by the *yellow* substrate. (From B. Alberts et al., *Molecular Biology of the Cell*, 5th ed. New York: Garland Science, 2008.)

Carcinoma cells that have been driven through an EMT by the actions of the Twist EMT-TF exhibit changes in the splicing patterns of hundreds of pre-mRNAs, these being products of alternative splicing. Among the altered mRNAs is the one encoding Mena. The normally synthesized *Mena* mRNA is replaced by a new version that specifies a form of the Mena protein (see Figure 14.34A) containing an invasion-associated domain (Mena^{INV}). Mena^{INV}, in turn, affects a variety of cell behaviors associated with invasion including responsiveness to EGF-induced cell motility (see Supplementary Sidebar 14.9) and trans-endothelial invasiveness. These changes conspire to generate significant increases in metastatic dissemination by the carcinoma cells within primary tumors (see Figure 14.34B–E). Resulting changes in the actin cytoskeleton, mediated in part by **filopodia** and **lamellipodia**, then enable the motility that underlies the invasive behavior of carcinoma cells (**Figure 14.35**). Lamellipodia are thin, sheet-like protrusions from the plasma membrane (Supplementary Sidebar 14.8) that enable the cytoplasm of a cell to extend forward in the direction of migration while filopodia protrude from lamellipodia and serve to sense guidance cues ahead of migrating cells and to provide attachment to the ECM. The detailed understanding of cell motility proteins garnered from studies like these is increasingly illuminating how an EMT works at the molecular level to power high-grade malignancy.

14.10 Intravasation and the formation of circulating tumor cells: first steps in perilous journeys

The invasiveness that we read about immediately above allows carcinoma cells to take advantage of the "doorways" in microvessels that we encountered in Figure 14.23. Recall that these openings in the walls of microvessels are formed by three-way collaborations between carcinoma cells, macrophages, and endothelial cells. These structures provide an entryway for carcinoma cells, resulting in the process of intravasation. In addition, and not mentioned previously, as individual carcinoma cells enter the lumina of microvessels, they may also acquire stemness, which may qualify them to serve later as founders of metastatic tissues. Still, these intravasation events are only the beginnings of long and difficult journeys. Similar processes, not yet studied, may also permit intravasation into the lumina of lymphatic vessels. Equally unclear are the cellular mechanisms that allow clumps of neoplastic cells to enter into the circulation. Such clumps, more than single intravasated cells, may actually represent the founders of most metastatic colonies.

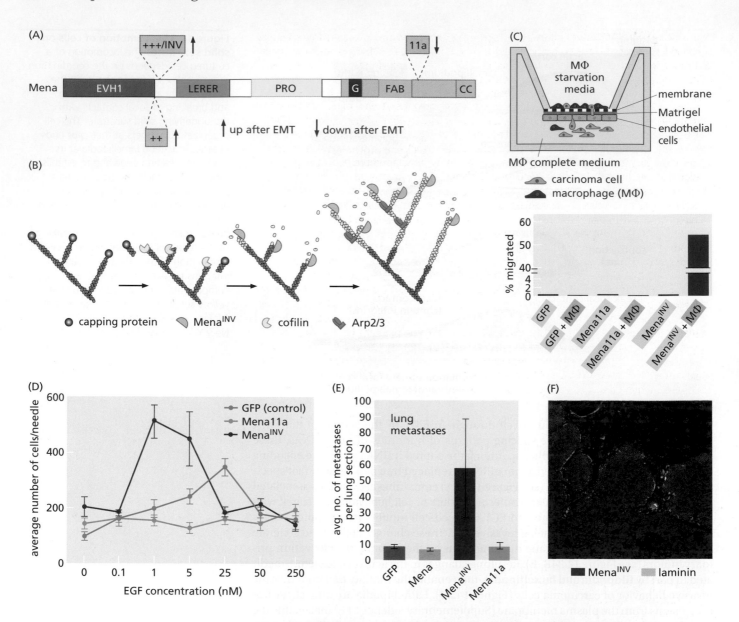

Figure 14.34 How EMT drives alternative splicing, cell motility, and increased invasiveness When carcinoma cells undergo an EMT, the result is a major shift in the expression of certain splicing factors that govern alternative splicing patterns. For example, several splicing factors that favor the epithelial phenotype are depressed, resulting in the synthesis of alternatively spliced mRNAs that encode proteins fostering a mesenchymal cell phenotype. (A) The usually synthesized version of the Mena protein seen both in normal epithelial cells and in localized carcinomas are composed of multiple domains and are specified by mRNAs encoding all the colored domains with the exception of the three light brown domains (termed simply "Mena") or these colored domains plus the 11a segment ("Mena11a"). In contrast, the form of Mena seen in cancer cells that have undergone an EMT and, in addition, are growing *in vivo* within a tumor contains segments encoded by exons INV (also known as +++) and ++ and lacks segment 11a. The Mena domain encoded by the INV exon seems to play an especially important role in carcinoma cell invasiveness. The arrows indicate the changes seen in association with activation of an EMT program. (B) The various forms of Mena promote the elongation of actin fibers that are essential for the growth of both lamellipodia and filopodia; only lamellipodial growth is illustrated here. Normally, a capping protein (*red circle*) blocks the further extension of actin fibers. Cofilin (*yellow*) cleaves the ends of actin fibers, removing the capping protein, and Mena (*green*) then prevents the reattachment of capping protein, permitting further fiber elongation; the INV isoform of Mena is especially effective in promoting this elongation, thereby driving actin fiber elongation and the growth of both lamellipodia and filopodia. Yet another protein—Arp2/3 (*blue*)—enables the branching of actin fibers that is critical for stable lamellipodial extension. (C) As described in Figure 14.22, macrophages (*continued*)

Figure 14.34 How EMT drives alternative splicing, cell motility, and increased invasiveness (*continued*) interact with breast carcinoma cells via bidirectional signaling that promotes carcinoma cell invasiveness. (As shown in Figure 14.23, such invasiveness was manifested by an increased ability to intravasate.) Invasiveness through an endothelial cell layer can be measured in an *in vitro* assay (*above*), in which carcinoma cells (*green*) are introduced into the upper chamber of a two-chamber "transwell" incubation vessel, either together with or in the absence of macrophages (*red*). These cells are separated from the lower chamber by a perforated membrane (*dotted black line*), below which are a layer of ECM formed by Matrigel (a mixture of basement membrane proteins, *solid gray line*) and a monolayer of endothelial cells (*pink*). Conditioned medium in the bottom chamber attracts the carcinoma cells to migrate from the upper chamber, requiring their invasion through the endothelial cell monolayer. Carcinoma cells were genetically modified by vectors expressing either GFP, Mena11a, or Mena^{INV}. As shown *below*, a control expression vector encoding GFP (green fluorescent protein) has no effect on trans-endothelial migration, even in the presence of added macrophages (Mφ). Similarly, no effect was seen when these carcinoma cells were forced to express the Mena11a (epithelial) isoform. Moreover, the presence of both macrophages and the Mena/Mena11a isoforms fails to induce trans-endothelial migration. Only when the carcinoma cells are incubated with macrophages and forced to express the Mena^{INV} isoform does this migration proceed efficiently, resulting in a ~200-fold increase in invasiveness. (D) Expression of the Mena^{INV} isoform also sensitizes carcinoma cells to the presence of EGF, such as that secreted by nearby macrophages (see Figure 14.22). Insertion of a large-bore EGF-filled needle into a carcinoma can be used to assay invasiveness by counting the number of carcinoma cells that are sensitized to motogenic EGF signals, which respond by migrating into the bore of the needle. As seen here, carcinoma cells that express the Mena^{INV} isoform are recruited by a 25-fold lower concentration of EGF than is required to recruit carcinoma cells expressing the control GFP plasmid, whereas expression of the Mena11a isoform actually suppresses the control/basal level of recruitment; this indicates that profound changes in sensitization to EGF are elicited by the different Mena isoforms. (E) These various effects on increased invasiveness and EGF responsiveness collaborate to affect the rate of metastasis to the lungs from primary tumors that express either a GFP control vector, a Mena11a vector, or a Mena^{INV} vector. (F) Mena^{INV} can be seen to be widely expressed among cells at the outer, invasive edge of an island of carcinoma cells (*pink*), where the remnants of the former basement membrane, visualized here by an antibody stain specific for laminin (*green*) are barely apparent, most of it having already been degraded by the invasive carcinoma cells. (A, from S. Goswami et al., *Clin. Exp. Metastasis* 26:153–159, 2009. With permission from Springer. B, from U. Philippar et al., *Dev. Cell* 15:813–828, 2008. With permission from Elsevier. C and D, from E.T. Roussos, M. Balsamo, S.K. Alford et al., *J. Cell Sci.* 124:2120–2131, 2011. With permission from The Company of Biologists. E, adapted from U. Philippar, E.T. Roussos, M. Oser et al., *Dev. Cell* 15:813–828, 2008, and E.T. Roussos et al., 2011. F, Courtesy of F. Gertler.)

The entrance into blood vessels in principle allows the neoplastic cells to spread via the process of **hematogenous dissemination**. Primary tumor cells may also seek out the alternative route of spreading via the lymphatic vessels that drain tumors and carry cells together with tumor-associated debris into the "draining lymph nodes." As will be discussed in Chapter 15, such lymph nodes are the sites where antigen-presenting cells, notably dendritic cells, present tumor-associated antigenic peptides to T lymphocytes as an initial step in the formation of an adaptive immune response. It is unclear whether the lymph nodes represent dead ends for disseminating cancer cells or, alternatively, staging areas for further spread to distant tissues (**Sidebar 14.4**). One study of the genomes of distant metastases found that two-thirds of cancer cells in these metastases arose from subpopulations of primary tumor cells that were genetically distinct from the subpopulations that spawned the lymph node-associated cancer cells. This might suggest two independent, parallel routes of dissemination from primary tumors to either draining lymph nodes or, alternatively, to blood vessels that lead from primary tumors directly into the venous circulation and ultimately to distant anatomical sites throughout the body.

The absence of functional lymphatic vessels within tumor masses must influence the paths used by metastasizing cancer cells to leave the primary tumor. Without ready access to lymph ducts, most motile cancer cells are likely forced to emigrate via the far more numerous functional capillaries, which are threaded throughout the tumor mass. In spite of such limited access, some cancer cells do indeed succeed in entering the lymphatic system. In the specific case of mammary carcinomas, some metastasizing cancer cells enter into the lymphatic vessels that directly drain the mammary gland and collect in the nearby downstream lymph nodes (see Figure 14.36A). These

Figure 14.35 The actin cytoskeleton and the restructuring of the cell surface The formation of both lamellipodia and filopodia depends on assembly of specific configurations of the actin cytoskeleton that lies immediately beneath the plasma membrane. (A) Use of platinum replica electron microscopy reveals the configuration of the actin fibers that form both a lamellipodium and a filopodium. All membranes have been removed to reveal the actin cytoskeleton. The dashed line indicates the imputed location of the plasma membrane, which previously separated the cytoplasm (*below*) from the extracellular matrix (*above*). The lamellipodium (see also Supplementary Sidebar 14.8) extends broadly across the entire micrograph, and a single filopodium protrudes from the leading edge of the lamellipodium. (B) The actin fibers within a lamellipodium, seen here at higher magnification (*above left*), are extended at their growing "barbed" ends by addition of actin monomers (*yellow balls, top in lower diagram*) and are disassembled at their retreating "pointed" ends by liberation of individual monomers (*below*). The same dynamic operates to enable the extension of filopodia (*right*). Ancillary proteins involved in forming the branched fiber networks of the lamellipodium are depicted in *pink*, whereas those involved in forming the tightly bundled, parallel actin fibers of the filopodium are depicted in *green*. The plasma membrane wraps over and around these structures in a fashion that is poorly understood. (A, from T.M. Svitkina, E.A. Bulanova, O.Y. Chaga et al., *J. Cell Biol.* 160:409–421, 2003. With permission from Rockefeller University Press. B, from D. Vignjevic and G. Montagnac, *Semin. Cancer Biol.* 18:12–22, 2008. With permission from Elsevier.)

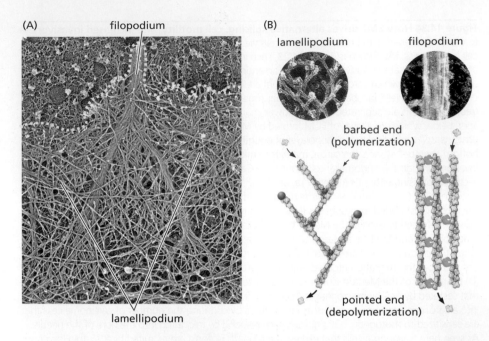

wandering carcinoma cells are readily detected because their appearance differs so strongly from the surrounding lymphoid cells (see Figure 14.36C) and they express epithelial proteins, such as cytokeratins, that are otherwise absent from lymphatic tissues (Figure 14.36D). Histological examination of draining lymph nodes is routinely used to determine whether a primary breast cancer has begun to dispatch metastatic pioneer cells to distant sites in the body.

The fact that lymph nodes operate as collection points for subcellular debris and cells shed by tissues explains why draining nodes are routinely examined to determine whether cancer cells have been released by primary tumors growing in those tissues (for example, the mammary gland; see Figure 14.36A and B). As examples, upon initial diagnosis, about one-third of breast, colorectal, cervical, and oral carcinoma patients have disseminated cancer cells in the lymph nodes near their primary tumors (see Figure 14.36C and D).

Sidebar 14.4 The lymphatic system represents an alternative route to metastasis After invasive, motile cells enter into the vessels of blood or lymphatic systems—the process of intravasation—they disperse and, should they survive the rigors of the voyage, eventually settle in tissue sites that lie at some distance from the primary tumor. The hematogenous route of dissemination is clear and well established. Less obvious is the alternative route via the lymphatic system. Virtually all tissues in the body carry networks of lymphatic vessels that are responsible for continuously draining the interstitial fluid that accumulates in the spaces between cells (see Figure 13.8C). Most of these vessels converge on a major abdominal vessel that empties its lymph into the left subclavian vein near the heart and thence into the general circulation. Consequently, cancer cells present in lymphatic vessels may occasionally enter through this cross connection into the general circulation.

Tumor cells and recruited stromal companions may secrete VEGF-C, which drives lymphangiogenesis—the formation of new lymphatic vessels (see Section 13.6). Moreover, experimental tumors forced to secrete increased levels of VEGF-C will seed larger numbers of metastatic cells in nearby draining lymph nodes—the lymph nodes associated with the lymphatic ducts that drain the tissue in which the tumor lies (**Figure 14.36**). However, detailed histological analyses of spontaneously arising solid tumors indicate that functional lymphatic vessels are rarely found throughout tumor masses. Instead, they largely occupy a zone at the periphery of solid tumors. Those few lymphatic vessels detected in the central regions of tumors are usually collapsed (see Figure 13.34). As discussed in the previous chapter, it seems that the expanding masses of cancer cells within a tumor press on these vessels; because the lymphatic ducts have little internal hydrostatic pressure, they cannot resist these forces and collapse.

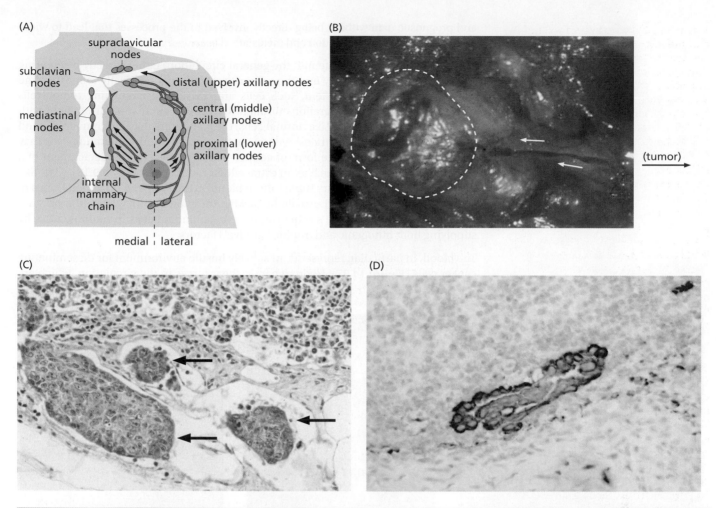

Figure 14.36 Draining lymph nodes of the mammary gland and its tumors (A) The lymphatic ducts (*red*) and the lymph nodes draining the breast (swellings along ducts) are initial sites of metastatic spread, carcinoma cells being carried there by the flow of lymph (*arrows*) leaving various sectors of the breast. The discovery of carcinoma cells in these lymph nodes, which is observed in more than 30% of human breast carcinoma patients at the time of initial diagnosis, suggests the possibility of deposits of metastatic cells in more distant sites in the body, particularly if large numbers of draining nodes are found to carry breast cancer cells. (B) The lymph node that serves as the sentinel node of a tumor can usually be identified among all of the lymph nodes draining the breast by injecting a blue dye into the tumor (outside photographic field to right) and following the trail of the dye via the lymphatic duct (*arrows*) to the draining node (outlined in dashed line, *left*). (C) Hematoxylin–eosin (H&E) staining of a section of an axillary lymph node reveals that three micrometastases (*arrows*) arising from a primary breast tumor have grown in the space between the capsule surrounding this node (not seen, *below*) and the mass of lymphocytes within the node (small cells, dark nuclei, *above*), displacing the latter upward. (D) Immunohistochemistry using an antibody specific for cytokeratins (*brown*) reveals this small micrometastasis in a sentinel node. This procedure is far more sensitive than H&E staining (panel C) in detecting micrometastases, because the mesenchymal cells of the lymph node do not express cytokeratin, which is made by epithelial cells and thus by most carcinoma cells. (A–C, from A.T. Skarin, *Atlas of Diagnostic Oncology*, 4th ed. Philadelphia: Elsevier Science Ltd., 2010. With permission from Elsevier. D, from J.P. Leikola, T.S. Toivonen, L.A. Krogerus et al., *Cancer* 104:14–19, 2005. With permission from John Wiley & Sons.)

These draining lymph nodes serve as proverbial "canaries in the mine," by providing early warning of the presence of metastasizing cells in the body. Among these regional lymph nodes, the single node that directly drains the primary tumor growing in the breast is often termed the "sentinel" node (see Figure 14.36B). Patients with small numbers of affected nodes often have only localized spread of the breast cancer and may never develop metastatic disease, whereas those with many affected nodes are far more likely to harbor other deposits of metastatic cells in distant sites in the body. For example, in one study, 90% of long-term survivors of a variety of carcinomas had one, two, or occasionally three "positive" lymph nodes at diagnosis. Conversely, fewer than 5% of patients with more than five positive lymph nodes when their primary tumors were removed enjoyed long-term, disease-free survival. Accordingly, in multiple types of primary tumors, cancer cell–positive lymph nodes represent "**surrogate markers**" of metastasis by providing useful diagnostic and prognostic data without being directly involved in the processes that lead to widespread cancer cell dissemination and metastatic disease.

Once cancer cells intravasate into the general circulation, either directly by entering the microvessels that empty into the venous circulation or indirectly following time spent in the lymphatic system, such cells are considered circulating tumor cells (CTCs). The long-range migrations through the general circulation are fraught with great danger for the CTCs. Like normal cells, the wanderers may continue to depend on anchorage to solid substrates; without such attachment, the migrating cells may die rapidly from anoikis, the form of apoptosis that is triggered by detachment of a cell from a solid substrate such as an extracellular matrix (see Section 9.13). Also, like their forebears in the primary tumor, these pioneers may depend on various types of stromal support, which will usually be lacking the moment they leave the primary tumor mass. Recall that the stroma can benefit carcinoma cells in multiple ways by supplying both mitogenic and trophic (survival) factors.

The blood, in particular, represents an actively hostile environment for disseminating cancer cells (**Figure 14.37**). Hydrodynamic shear forces in the circulation, which are

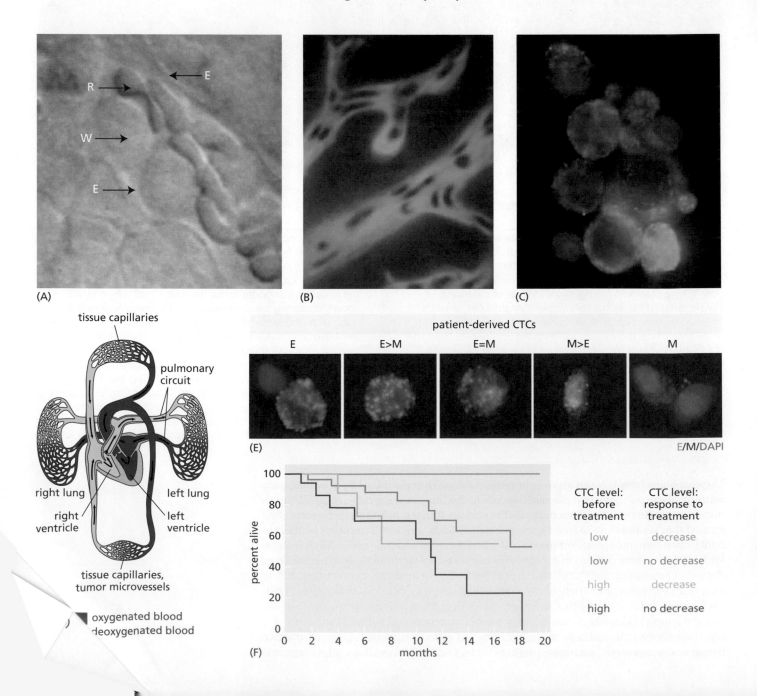

often substantial in smaller vessels, may tear apart the wandering cancer cells. Some experimental models of metastasis in the mouse provide clear indication that the survival of metastasizing cancer cells in the general circulation is greatly enhanced if they can attract an entourage of blood platelets to escort them through the rapids into safe pools within tissues (see Supplementary Sidebar 14.5). These tightly adhered passengers form a protective cloak that reduces the likelihood that the CTCs will be ambushed by natural killer (NK) lymphocytes lying in wait in the circulation (to be described in Chapter 15). In addition, after adhering to carcinoma cells, platelets may release a variety of growth factors that are usually instrumental in supporting wound repair, among them mitogenic and trophic signals. Some of the platelet-secreted factors may activate or sustain EMT programs in the adhered carcinoma cells, thereby rendering them more resistant to various forms of cell death.

The concentrations of CTCs in the blood of patients under treatment often represent powerful prognostic markers (Figure 14.37F). In the analyses illustrated in this figure, individual patients were monitored **longitudinally**, that is, at different times during the course of their treatment and diagnosis. In the analyses shown, the responses of CTC concentrations to patients' therapy were monitored. As is clear, patients who began treatment with low CTC concentrations that decreased following treatment fared well, showing no mortality during the 20 months of clinical observation. Conversely, patients who began treatment with high CTC concentrations that failed to respond to treatment by decreasing showed a high level of mortality, all dying by the end of the observation period. The monitoring of CTC concentrations rather than direct observations of tumor masses is often termed a form of **liquid biopsy**.

Figure 14.37 Passage of tumor cells through the circulation Cancer cells moving through the circulation are found amid densely packed cells of hematopoietic origin. (A) Intra-vital microscopy reveals the two endothelial walls of a capillary (E), erythrocytes (R), and some leukocytes (W). The small size (~7 μm diameter) and deformability of the erythrocytes and leukocytes allow them to pass through capillaries without becoming trapped. (B) The vessels in this intra-vital fluorescence micrograph are slightly larger than capillaries. The plasma has been stained with a *green* dye, and the erythrocytes have been colorized *red*. The high deformability of these red blood cells is apparent. Most cancer cells have more than twice the diameter of erythrocytes and are not deformable, and they are less able to negotiate narrow passages, such as those created by the lumina of capillaries. (C) The difficulties of negotiating passage through the narrow microvessels of various tissues are compounded by the fact that intravasated cancer cells often attract clouds of platelets around them (see Supplementary Sidebar 14.5) and the fact that they are often found in the circulation as multicellular aggregates, as is seen in this cluster of circulating tumor cells (CTCs) isolated from the circulation of a patient with metastatic prostate cancer. Cells were immunostained with an antibody reactive with the prostate-specific membrane antigen (*green*) and another reactive with the CD45 displayed by many types of hematopoietic cells (*red*). DNA was stained *blue* with DAPI. (D) This diagram of the mammalian circulation indicates that venous blood (*blue*) leaving a tissue (and thus cancer cells that have escaped from a primary tumor and intravasated, *below*) must first pass through the right ventricle of the heart and thence through the lungs before it enters the left ventricle and is pumped into the general arterial circulation. Because passage through the pulmonary circulation of the lung requires passage through its capillaries, many metastasizing cells entering into the venous circulation are rapidly trapped in the pulmonary capillary beds. (E) Circulating tumor cells present in the blood of breast cancer patients were analyzed for their expression of epithelial (E, *green*) and mesenchymal (M, *red*) RNAs by *in situ* hybridization. As is apparent, cells expressing various proportions of E and M marker-expressing CTCs were present in the blood of various patients. Not shown is the fact that as tumor progression proceeded in individual patients, the proportions of M-marker-expressing cells increased in the CTCs. (F) While a measurement of the levels of CTCs in a patient relative to other patients often does not provide clinically useful information. Instead, it is useful to monitor individual patients longitudinally, that is at various time points over the course of treatment. As is seen here, the responses of individual patients to therapy can be gauged by measuring levels of CTCs following therapy relative to their initial levels of CTCs. (A and B, from I.C. MacDonald, A.C. Groom and A.F Chambers, *BioEssays* 24:885–893, 2002. With permission from John Wiley & Sons. C, from S.L. Stott, C.H. Hsu, D.I. Tsurkov et al., *Proc. Natl. Acad. Sci. USA* 107:18392–18397, 2010. With permission from National Academy of Sciences, U.S.A. E, from M. Yu, A. Bardia, B.S. Wittner et al., *Science* 339:580–584, 2013. With permission from AAAS. F, from T.T. Kwan, A. Bardia, L.M. Spring et al., *Cancer Discov.* 8:1286–1299, 2018. With permission from American Association for Cancer Research.)

Importantly, longitudinal observations of CTC concentrations in individual patients often provide highly useful prognostic information, as revealed by Figure 14.37F. However, different patients harboring tumors of comparable sizes and degrees of progression may often exhibit very different levels of CTCs, indicating the limitations of using liquid biopsies to assess the sizes of their primary tumors and the numbers of derived metastases.

CTCs may either be present in the blood of cancer patients as single cells or as multicellular clumps, sometimes termed **microemboli** (see Figure 14.37C), their migration being termed **collective metastasis**. As mentioned earlier, accumulating evidence indicates that the CTC clumps are actually more effective than individual CTCs in founding metastatic colonies in distant organ sites. Their multicellularity may confer several benefits. Cells buried within these clumps may be shielded from attacks by NK cells. Because of the larger diameter of these clumps, they are likely to be physically trapped in small-bore microvessels far more quickly than individual CTCs traveling on their own; this short lifetime in the general circulation may, on its own, confer significant benefit, allowing these cells to quickly find refuge away from the raging torrents of the general circulation.

Related molecular evidence indicates that many metastases can be initiated by multicellular aggregates. This insight comes from labeling the individual cancer cells within primary tumors growing in a mouse with a diversity of unique DNA markers, sometimes termed **barcodes**, that have been stably introduced into their genomes. In individual metastases spawned by these primary tumors, one finds multiple barcodes, indicating that the descendants of multiple distinct cells within a primary tumor served as cofounders of each metastatic colony (**Figure 14.38**). Such evidence suggests that these polyclonal metastases were initiated by multicellular, genetically heterogeneous cell clumps. (Additional evidence indicates that these metastases are unlikely to arise

Figure 14.38 Genetic barcodes reveal the multicellular origins of many metastases An important question concerns the origins of metastatic colonies, specifically whether they arise from single, individual founder cells or multiple founders, yielding monoclonal or polyclonal metastases. A powerful strategy to demonstrate the polyclonal origins of metastases derives from the strategy of "confetti labeling," in which one of a series of fluorescent proteins of different colors are expressed in a random fashion in the individual cells of a primary tumor and in the lineal descendants of each of these initially labeled cells. (A) As seen in this map of a vector used to generate expression of one of four randomly activated fluorescent marker proteins, the reading frames of all four proteins are randomly expressed in various alternative configurations. Upon activation of the Cre-Lox recombinase, two successive excisions and reorientations occur, resulting in the expression of only one of the four fluorescent proteins and being expressed within a cell. (B) The confetti labeling procedure was employed to label the cells within a mouse model of primary pancreatic cancer, yielding sectors that express the four alternative fluorescent protein markers. (C) Liver metastases from these pancreatic primary tumors were analyzed for their marker protein expression, revealing the oligoclonality of these metastases. (A, from R. Maddipati and B.Z. Stanger, *Cancer Discov.* 5: 1086–1097, 2015. With permission from American Association for Cancer Research. B and C, courtesy of Ben Stanger and Ravi Maddipati.)

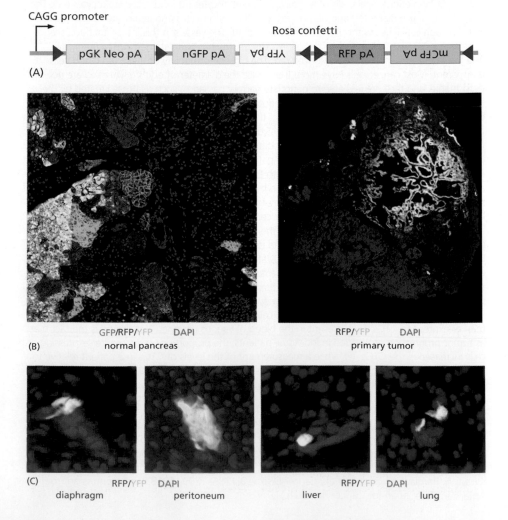

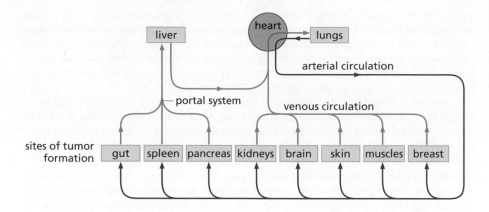

Figure 14.39 The venous circulation and sites of trapping of CTCs
The layout of the general circulation dictates that cancer cells disseminating from most primary tumor sites via the venous circulation (*blue lines*) will soon be trapped, at least transiently, in the microvessels within the lungs. However, the veins draining the gut, pancreas, and spleen are organized differently, in that their venous blood empties directly into the liver via the hepatic portal circulation before being dispatched thereafter to the heart and lungs. Consequently, vast numbers of metastasizing colorectal carcinoma and pancreatic adenocarcinoma cells are trapped in the microvessels of the liver within seconds of leaving the colon and pancreas. (As drawn, this diagram omits mention of the liver as a site of primary tumor formation.) (Adapted from I.C. MacDonald, G.C. Groom and A.F. Chambers, *BioEssays* 24:885–893, 2002.)

from individual disseminated tumor cells (DTCs) that converged independently of one another on the site of dissemination.) Moreover, in these barcode experiments, at least 80% of the metastases gave evidence of oligoclonality, that is, at least two distinct clonal subpopulations in a given metastatic colony. Hence, many metastases represent oligoclonal growths rather than the monoclonal growths that would be expected if they traced their origins to single founder cells.

The CTCs have been the objects of intensive investigation in recent years. They presumably are representative of the cancer cells moving from primary tumors to sites of metastasis formation. But this notion is at best an inference, since only a tiny proportion of CTCs cells ever succeed in founding new metastatic colonies, explaining why individual CTCs have rarely, if ever, been viewed in the act of extravasating and thereafter spawning macroscopic metastases. In one analysis, the CTCs were measured in the portal circulation (**Figure 14.39**) of 18 patients bearing pancreatic ductal adenocarcinomas. These measurements yielded an average of ~15 CTCs/ml. of portal blood, which could yield in aggregate 3×10^7 carcinoma cells entering from a pancreatic tumor into a patient's liver each day. However, the number of pancreatic metastases formed after several months' time (including the extended time prior to diagnosis) could be numbered on the fingers of one hand.

Another calculation revealed that even after arriving in the bone marrow, only one in a thousand disseminated breast cancer cells (DTCs) ever forms a micrometastasis, and only a single macroscopic metastasis arises from a thousand previously formed micrometastases. These calculations, which by necessity are only rough estimates, illustrate the phenomenon of **metastatic inefficiency**, that is, the profound discordance between the numbers of metastasizing cells released by primary tumors and the eventual outgrowth of clinically detectable metastases.

A variety of techniques are being developed to measure the concentrations of CTCs in the circulation of cancer patients (Supplementary Sidebar 14.10). These numbers may prove useful in gauging the efficacy of therapies directed against a primary tumor within a single patient; thus, a decrease in viable cells in the primary tumor following therapy will presumably be accompanied by a corresponding decrease in the release of such cells into the general circulation and a resulting decline in the number of CTCs. Techniques like these may also be used to determine an elusive number—the lifetime of CTCs in the general circulation (**Sidebar 14.5**).

Sidebar 14.5 Lifetimes of CTCs in the population may vary dramatically Multiple factors contribute to the short lifetime of CTCs in the circulation; in a mouse, most of these cells spend less than 2 minutes in the circulation. Unlike red and white blood cells, the cells released by solid tumors are ill-suited to negotiate the passage through microvessels in various tissues (see Figure 14.37A–C). Thus, the internal diameters of most capillaries—3 to 8 μm—are far too small to accommodate the >20 μm diameter of individual cancer cells. Erythrocytes, for example, are only about 7 μm in diameter and are easily deformed, facilitating their passage through capillaries. Most cancer cells, in contrast, are not especially deformable. [Moreover, if cancer cells in the blood are coated with platelets (see Supplementary Sidebar 14.5), their effective physical diameters become even larger, causing them to be trapped in vessels larger than capillaries, such as the small arteries known as **arterioles**.] These factors dictate that within seconds of entering into the venous circulation many cancer cells will encounter the capillary beds of the lungs (see Figure 14.37D), in which they lodge.

Indeed, the lifetimes of CTCs as truly free-floating cells in the circulation may be so short that CTCs cannot respond via anoikis (see Section 9.13) to their loss of anchorage to solid substrates and the absence of stromal support. If so, this particular type of attrition sustained while these cells are swimming through the circulation may not play a major role in limiting their lifetime. Some measurements have been made by examining the attrition rate of carcinoma cells in the mouse by injecting such cells into the venous circulation via the tail vein, that is, using the procedure termed "**experimental metastasis**." Once in the circulation, the great majority of such cells are rapidly trapped in the microvasculature of the lungs (see Figures 14.37D and 14.39). As an example, in one experiment, within 10 minutes of injection into the venous circulation of a mouse more than 98% of the cells were already cleared out of the circulation; the corresponding half-life in humans is longer (~1–2 hours), because the human systemic circulation time is ~1 minute compared with the ~4 seconds within a mouse. Although convenient, such experiments may represent artifacts of experimentation because other work utilizing conjoined mice suggests that CTCs spend several minutes in the circulation before disappearing. These latter experiments would seem to more closely represent the behavior of CTCs released by primary tumors, since they gauge the lifetimes of cell released by **autochthonous** primary tumors, that is, tumors initiated and growing within the hosts in which these measurements are made.

Importantly, the dynamics of physical trapping in the narrow-bore microvessels of many tissues hold important implications for measuring the numbers of the multicellular clumps of cells seen in Figure 14.37C and studied genetically in Figure 14.38. Because of their far larger effective diameters, these microemboli are likely to be trapped far more quickly in the microvasculature of various tissues and accordingly to exhibit far shorter half-lives in the circulation. This, in turn, will proportionately reduce their steady-state concentrations in the general circulation, leading to underestimates of the rates at which these clumps are actually formed and released into the circulation.

Once trapped within the lungs, some metastasizing cancer cells may attempt to found metastases there. However, the metastases spawned by many types of human tumors are often found elsewhere throughout the body, indicating that cancer cells usually succeed in escaping from initial trapping in the lungs and travel further to other sites in the body. How they do so is unclear. In some experiments, cancer cells trapped in capillaries have been observed to pinch off large amounts of cytoplasm, leaving behind cells that, although greatly reduced in size, seem to be viable; once they have undergone this amputation, such slenderized cells may succeed in negotiating passage through the narrow straits of the lung capillaries. A more plausible explanation is that wandering cancer cells may avoid being trapped altogether: in many organs, including the lungs, metastasizing cells can bypass capillaries by traveling through arterial–venous **shunts**, which form large-bore, direct connections between the two halves of the circulatory system. Finally, it is worthwhile noting that multicellular CTC clumps have been observed to work their way through microvessels whose luminal diameter is far smaller than the dimensions of these clumps.

Having snaked their way through the lungs and arrived in the general arterial circulation, roaming cancer cells, either as individual cells or in clumps, can then scatter to all tissues in the body. Some experiments have suggested that cancer cells use specific cell surface receptors to initially adhere to the luminal walls of arterioles and capillaries in certain tissue sites. However, far more extensive evidence indicates that simple physical trapping within small vessels throughout the body, as discussed above in the context of the lungs, provides most wandering cancer cells with their first solid foothold within a tissue.

Once lodged in the blood vessels of various tissues, cancer cells must escape from the lumina of these vessels and penetrate into the surrounding tissue—the step termed

Sidebar 14.6 Cancer cells are clumsy escape artists The complex task of escaping from the circulation into the surrounding tissue parenchyma is accomplished routinely by leukocytes, which must be able to enter into the parenchyma of tissues throughout the body in response to certain inflammatory stimuli, including the presence of infectious agents. Through a sequence of steps known as **diapedesis**, leukocytes are able to induce endothelial cells in postcapillary venules to retract and create portals into the underlying tissue. The entire process from attachment to the endothelial wall to entrance into the tissue parenchyma takes less than a minute and involves an elaborately choreographed program of biochemical and cell-biological changes!

In contrast, the vast majority of metastasizing cancer cells are not endowed with the receptors and biochemical response mechanisms required to execute diapedesis. Accordingly, if neoplastic cells do succeed in penetrating through the wall of a capillary or slightly larger vessel, they seem to do so by brute force, perhaps by degrading patches of endothelium in a process that may require many hours or even a day rather than a minute to complete (see Figure 14.40). Of additional interest is that the thrombin produced during the formation of microthrombi (see Supplementary Sidebar 14.5) is quite effective in cleaving the various proteins used by endothelial cells to attach to the underlying vascular basement membrane; this activity may cause endothelial cells to retract from microemboli, thereby exposing the capillary basement membrane to direct attack by invasive cancer cells and the proteases that they and recruited neutrophils produce.

extravasation (Figure 14.40). The process of extravasation depends on complex interactions between cancer cells and the walls of the vessels in which they have become trapped. Cancer cells can use several alternative strategies to extravasate. They may proceed immediately to elbow their way through the vessel wall. Alternatively, and possibly more typical, their ability to extravasate may depend on their secretion of angiopoietin-like protein-4 (Angptl4), which is released by carcinoma cells that are experiencing TGF-β signaling. This secreted factor stimulates the retraction of endothelial cells from one another, leaving gaps in the vessel walls as the initial step of extravasation (Figure 14.40B). The loosening of endothelial tight junctions may also be facilitated by recruited myeloid cells, notably neutrophils, which can supply proteases that dissolve the junctions between adjacent endothelial cells lining the walls of blood vessels. The gaps in the endothelium created by this loosening expose vascular basement membrane (BM). The ability of the cancer cells to proceed further with extravasation seems to depend on their expression of integrin β1 (see Table 5.4 and Figure 5.22) which, together with associated α integrin subunits enables the tethering of the cancer cells to the laminin in the BM (Figure 14.40C; see also Figure 13.10). The cancer cells seem thereafter to deploy invadosomes (see Figure 14.30) to invade through the BM after which they may confront pericytes (see Figure 13.8) before they finally enter into the tissue parenchyma. Alternatively, trapped CTCs may begin to proliferate while still living within the lumen of a vessel, creating a small tumor that grows and eventually obliterates the adjacent vessel wall (see Figure 14.40D). In doing so, they push aside endothelial cells, pericytes, and smooth muscle cells that previously separated the vessel lumen from the surrounding tissue parenchyma (Sidebar 14.6).

14.11 Colonization represents the most complex and challenging step of the invasion–metastasis cascade

Once they have arrived within the parenchyma of a tissue, metastasizing cancer cells may form micrometastases—small clumps of disseminated cancer cells, some of which are able to expand into clinically detectable masses; this growth of microscopic into macroscopic metastases is often termed **colonization**. It is plausible, if not likely, that CTCs are trapped in diverse tissues throughout the body and can seed micrometastases in most of them. If so, then the observed tendencies of different types of primary tumor cells to preferentially form macroscopic metastases in certain organs but not others may only reflect the efficiency of post-extravasation colonization, which is likely to vary drastically from one site of dissemination to another and reflect how permissive different tissues are for the colonization by various types of disseminated

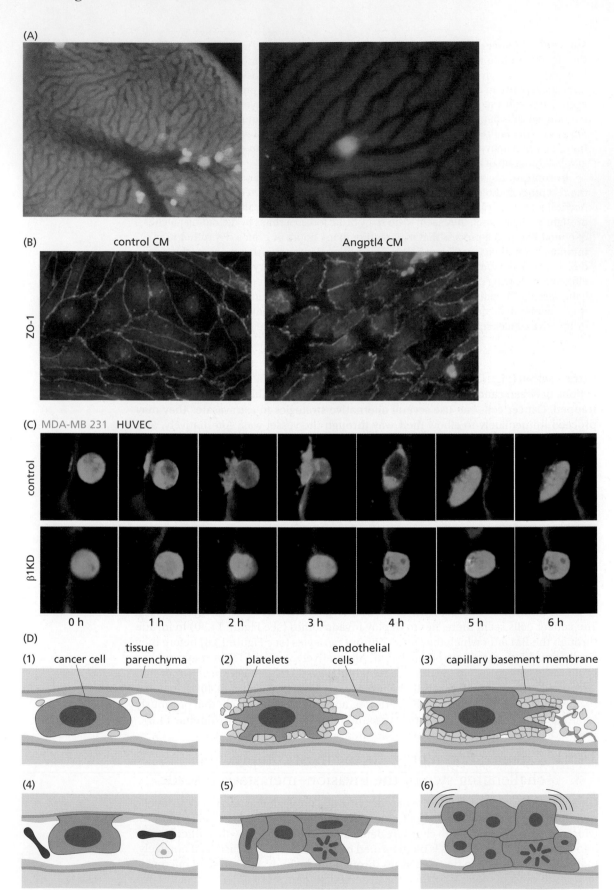

(A)

(B)   control CM   Angptl4 CM

ZO-1

(C) MDA-MB 231   HUVEC

control

β1KD

0 h   1 h   2 h   3 h   4 h   5 h   6 h

(D)

(1) cancer cell   tissue parenchyma   (2) platelets   endothelial cells   (3) capillary basement membrane

(4)   (5)   (6)

Figure 14.40 Lodging and extravasation of circulating tumor cells (A) Tumorigenic, green fluorescent protein (GFP)–labeled Chinese hamster ovary cells were injected via the portal vein into a mouse liver. Large numbers of these cells (*green*) soon became arrested in the microvessels of the liver (*left panel*). However, within 24 hours, only an individual cell (*right panel; higher magnification*) has extravasated out of the microvessels (*dark red*) into the liver parenchyma (*light brown*). (B) Mammary carcinoma cells within a primary tumor experience TGF-β, which is produced in abundance in cells that have undergone an EMT and causes them to produce angiopoietin-like protein 4 (Angptl4). When these cells disseminate and become lodged in microvessels, the Angptl4 that they secrete induces nearby endothelial cells to retract from one another, leaving gaps in the capillary walls that facilitate extravasation. As seen here, cultured human vascular endothelial cells, when exposed to conditioned culture media (CM) produced by Angptl4-secreting cells (*right panel*), retract their associations with one another, as indicated by the dissolution of the ZO-1–containing tight junctions (*green*) between adjacent cells; CM produced by control cells (*left panel*) fails to induce this retraction. (C) The actions of Angptl4, as described in panel B, expose the underlying vascular basement membrane, a specialized BM that differs slightly from that underlying epithelial cell layers. In this *in vitro* microfluidic model of the microvasculature, MDA-MB-231 human breast cancer cells (*green*) have been introduced into an already-formed microvasculature formed by human umbilical vein endothelial cells (HUVECs; *red*). The ability of a carcinoma cell in the lumen of a microvessel to successfully extravasate (*above*) depends on their expression of integrin β1, which allows them to attach to the laminin of the BM. Subsequent invasion through the BM depends on their expression of invadopodia and podosomes. Conversely, in the absence of normal integrin β1 expression by the carcinoma cell (β1KD *below*), achieved by shRNA knockdown, the carcinoma cannot attach to and advance beyond the vascular BM (*not visible*). (D) Electron-microscopic observations of metastasizing lung cancer cells injected into the venous circulation of mice (*not shown*) suggest that the process of extravasation often proceeds via steps involving the intraluminal proliferation of the cancer cells as drawn here. (1) A disseminated cell (*brown*) is trapped physically in a capillary. (2) Within minutes, a large number of platelets (*blue*) become attached to the cancer cell, forming a microthrombus. (3) The cancer cell pushes aside an endothelial cell (*green*) on one wall of the capillary, thereby achieving direct contact with the underlying capillary basement membrane (*orange*). (4) Within a day, the microthrombus is dissolved by the proteases that are responsible for dissolving clots. (5) The cancer cell begins to proliferate in the lumen of the capillary. (6) Within several days, sometimes sooner, the cancer cells break through the capillary basement membrane and invade the surrounding tissue parenchyma (*gray area*). (A, from G.N. Naumov, S.M. Wilson, I.C. MacDonald et al., *J. Cell Sci.* 112:1835–1842, 1999. With permission from The Company of Biologists. B, from D. Padua, X.H. Zhang, Q. Wang et al., *Cell* 133:66–77, 2008. With permission from Elsevier. C, from M.B. Chen, L.M Lamar, R. Li et al., *Cancer Res.* 76:2513–2524, 2016. D, from J.D. Crissman, J.S. Hatfield, D.G. Menter et al., *Cancer Res.* 48:4065–4072, 1988. Both with permission from American Association for Cancer Research.)

cancer cells. To be sure, there are clear exceptions to this thinking, specifically where the layout of the vasculature (as depicted, for example, in Figure 14.39) ensures that tumors arising in the gastrointestinal tract and pancreas dump vast numbers of cells via the portal circulation directly into the liver (as described in Section 14.10). In this situation, even if the liver were an intrinsically inhospitable host tissue for DTCs, the sheer numbers of cancer cells deposited directly into the liver from the portal vein may ensure that even low-probability events, specifically colonization, become possible.

The predilection to form metastases in one or another organ site was noted as early as 1889 by the British pathologist Stephen Paget (**Figure 14.41A**). He proposed the "seed and soil" hypothesis, in which he likened the seeding of cancer cells to the dispersal of the seeds of plants. After studying the clinical course of 735 breast cancer patients, Paget concluded that the patterns of metastasis formation in these patients could not be explained either by random scattering throughout the body or by the patterns of dispersal from the breast through the general circulation. He therefore proposed that the metastasizing cancer cells (the seed) find a compatible home only in certain especially hospitable tissues (the soil). He wrote, "a plant goes to seed, its seeds are carried in all directions; but they can only live and grow if they fall on congenial soil."

These tendencies are often termed **metastatic tropism** or **organotropism**. In some cases, the proclivity of primary tumors to seed certain sites of metastasis formation can be ascribed trivially to the organization of the circulation (see Figure 14.39). In other cases, such as the bone marrow metastases formed by prostate and breast tumors, more

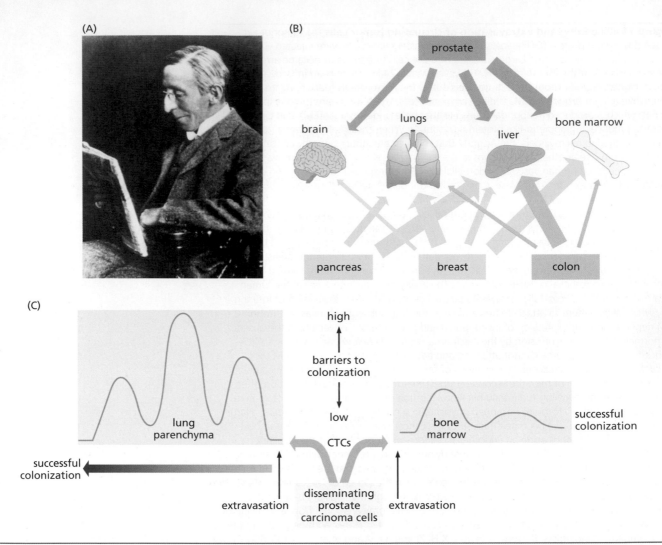

Figure 14.41 Primary tumors and their metastatic tropisms
(A) The British physician Stephen Paget (1855–1926) was the first to enunciate the "seed and soil" hypothesis, which states that the ability of a disseminated cancer cell to successfully found a metastasis depends on whether a distant tissue offers it a hospitable environment to survive and proliferate. (B) In this diagram, the relative width of each arrow indicates the relative proportion of clinically apparent metastases in various tissue sites of metastatic colony outgrowth that are generated by four types of common primary human carcinomas: prostate, breast, pancreas, and colon. In some cases, a tumor's tendency to spawn metastases in one or another tissue reflects the facility of the cancer cells from the primary tumor to adapt to (and thus colonize) the microenvironment of distant tissues, which likely explains the strong tendencies of prostate and breast cancers to generate metastases in the bone marrow. In other cases, the layout of the circulation (see Figure 14.39) may strongly influence the site of metastasis. For example, the high proportions of liver metastases deriving from primary colon cancers and pancreatic adenocarcinomas likely reflect the drainage via the portal vein of blood from the colon and pancreas directly into the liver. (C) One way to conceptualize the biological bases of metastatic colonization portrays the adaptations that disseminating carcinoma cells (circulating tumor cells, CTCs) must undertake in order to surmount multiple barriers to their successful adaption, given their initial differentiation program. Thus, prostate carcinoma cells encounter relatively few low biological barriers in the marrow (*right*) but multiple high barriers in the lung parenchyma (*left*). (A, from I.J. Fidler, *Nat. Rev. Cancer* 3:453–458, 2003.)

subtle biological mechanisms must be invoked, as described below. Moreover, use of the term "metastatic tropism" is actually misleading, because "tropism" usually implies the preferential, directed migration of agents to certain sites but not to others, which clearly does not explain the anatomical sites of most types of metastasis formation.

Metastatic colonization is, without question, the most challenging step of the invasion–metastasis cascade, ostensibly because the foreign tissue environments encountered by recently arrived cancer cells do not provide them with the collections of familiar growth and survival factors as well as extracellular matrix components that allowed their forbears to thrive while still residing within the confines of the primary tumor in which they arose. Without these various types of familiar physiological support, poorly adapted disseminated cancer cells may rapidly die or, at best, survive for extended periods of time as micrometastases that can only be detected microscopically and rarely increase beyond this size.

Of note, the actions of EMT programs, as described in Sections 14.3 and 14.4, would not seem to explain the metastatic tropisms of cancer cells. Instead, it seems that, at least in the case of carcinoma cells, EMT programs are capable of moving these cells physically from primary tumors into the parenchyma of distant tissues. However, once arrived in these distant sites, disseminated tumor cells (DTCs) must devise their own tissue-adaptive programs, an undertaking that seems to be highly challenging for most types of DTCs. In these situations, the DTCs clearly experience enormous selective pressure, with variants succeeding only on rare occasion in contriving adaptive programs that permit survival and subsequent robust proliferation of these cells and their descendants.

In fact, at present, the problem of metastatic colonization represents the major puzzle for cancer researchers intent on understanding the pathogenesis of human neoplastic diseases. As will be apparent below, this is a problem of daunting complexity, given the large number of distinct subtypes of human primary tumors, the number of distinct sites at which metastases may grow out, and the possibility that each successful adaptive program may vary from others that arise independently in other patients and in other anatomical sites (Figure 14.41B).

Note, by the way, that these tissue-specific "adaptations" may be of two sorts. DTCs may acquire the ability to respond to the contextual signals naturally provided by their newly encountered tissue microenvironments; at the same time, they may need to learn how to perturb these microenvironments, forcing the latter to undergo shifts that create congenial metastatic niches.

As depicted here, successful adaptations and outgrowth represent a form of Darwinian evolution in which rare variants successfully adapt to newly encountered environments and associated stresses. Not addressed by this thinking is precisely *how* DTCs succeed in assembling biological programs that ensure their success as the founders of macroscopic metastases. Darwinian evolution is thought to depend largely if not exclusively on heritable changes in the genomes of organisms under selective pressure. However, extensive searches for recurring somatic mutations that function to allow the outgrowth of DTCs in certain destination sites have provided few insights. These failures suggest the alternative mechanism: that effective adaptive programs are largely cobbled together at the epigenetic level, that is, they depend on changes in the transcriptomes of cells without underlying changes in the nucleotide sequences of their genomes.

Such epigenetic programs, once assembled, must be transmitted in a cell-heritable fashion, that is, being perpetuated from generation to generation of cells in cell lineages over extended periods of time. In addition, the invention of such programs by DTCs must depend on their phenotypic **plasticity**, that is, their ability to readily shift from one phenotypic state to another. Such plasticity would seem to echo another type of phenotypic plasticity—the behavior of cells that have activated EMT programs and, as a consequence, shift frequently between multiple phenotypic states lying between the extremes of fully epithelial and fully mesenchymal, as depicted in Figure 14.12.

At the same time, the ability of a DTC to succeed in cobbling together an effective adaptive program (that is, one that adapts them to their new tissue microenvironment) is likely to be governed by this cell's starting point—the differentiation program that it brought along from the primary tumor. This starting point may place the DTC in a position that allows it to adapt with ease to the foreign tissue that now hosts it or, conversely, may require a complex, challenging set of changes in transcriptional programs (and thus shifts in cell state).

Stated in this way, circulating tumor cells (CTCs) originating from certain primary tumors may already be well poised to acquire effective tissue-adaptive programs simply because relatively few adjustments need to be made to their existing transcriptomes in order to thrive in certain sites of eventual dissemination. In contrast, CTCs originating in other types of primary tumors may confront multiple, almost-insurmountable barriers to developing successful adaptation programs in certain destination sites (Figure 14.41C). As an example, metastasizing cutaneous melanoma cells may already be poised to readily acquire adaptive programs that function in a variety of tissues, but distantly related uveal melanomas of the eye may be hard-pressed to devise diverse adaptive programs. As a consequence, cutaneous melanomas may found metastatic colonies in diverse tissue sites, whereas uveal melanomas, because of their limited phenotypic plasticity and unique differentiation program, may be limited to successfully seeding metastases only in the liver.

At present, we understand almost nothing about the mechanisms that enable DTCs within micrometastases to develop novel adaptive programs that power their proliferation thereafter in their newfound homes, doing so ostensibly by shifting from one transcriptomic state (and thus phenotype) to another. Moreover, it is possible in principle that some of the changes in DTCs that optimize colonization efficiency in certain tissues may be achieved by genetic alterations, that is, mutations of their genomes occurring long after they have left the primary tumor; at present, however, there are no anecdotal accounts supporting this notion.

The barriers to post-extravasation proliferation are encountered almost immediately in the tissue parenchyma. To begin, carcinoma cells need to anchor to the surrounding extracellular matrix (ECM) via the integrins displayed on their surface (see Section 5.9). Without this anchorage, they are unable to activate a key source of mitogenic signals transduced by focal adhesion kinase (FAK, see Figure 6.17). This obstacle dooms the extravasated cells to remain as single-cell or several-cell micrometastases. Yet other barriers to successful colonization may also prevent them from thriving in their sites of dissemination. For example, the ability of a growing micrometastatic colony to successfully acquire access to a neovasculature may differ greatly from one site of dissemination to another (see Section 13.6). Moreover, for complex reasons, the stromal microenvironments of certain tissues may be fundamentally incompatible with the needs of DTCs originating in certain types of primary tumors.

Our descriptions of metastatic inefficiency in the previous section can be revisited here but from a slightly different angle. A frequently quoted statistic describes the clinical progression of breast cancer patients: more than 30% of these patients harbor hundreds, likely thousands of micrometastases in their bone marrow at the time of initial clinical presentation, but only half of the women in this group will ever develop metastatic disease. These numbers have implied to many that the bone marrow is the major site of dissemination of breast cancer cells, whereas in fact, it only represents the tissue in which micrometastases involving single DTCs and small clusters of DTCs are readily visualized and quantified (**Figure 14.42**). In reality, micrometastases are likely to be widely disseminated throughout many other tissues of a cancer patient, invariably in tissues where their detection is far more challenging than in the bone marrow, explaining our almost-total ignorance of where most DTCs end up, at least transiently, in the body; some of these occult micrometastases may lead, on occasion, to disastrous outcomes (Supplementary Sidebar 14.11).

Although the probability that a given disseminated cancer cell will succeed in generating a macroscopic metastasis is extremely low, the very fact that a patient's primary tumor is actively dispatching neoplastic cells to sites throughout the body is itself an indication that the tumor carries a bad prognosis (Figure 14.42D). Indeed, at least at present, the number of DTCs in the marrow seems to represent a far more useful prognostic marker than the concentration of circulating tumor cells (CTCs) in the blood. This would seem to reflect the fact that the number of DTCs represents the accumulation of disseminated cancer cells over an extended period of time, whereas the concentration of CTCs may be dictated by complex kinetic processes governing their lifetimes—and thus their **steady-state** concentrations—in the circulation at one point in time.

Support for the existence of dormant micrometastases that persist in a nongrowing but viable state for extended periods of time comes from experiments in which living cancer cells were marked by brief exposure to fluorescent label–containing particles; the latter persist for long time periods within cells but do not affect the viability these cells, explaining why they are often termed **vital dyes**. However, the intracellular concentration—and thus the fluorescence intensity of the dye particles—is diluted

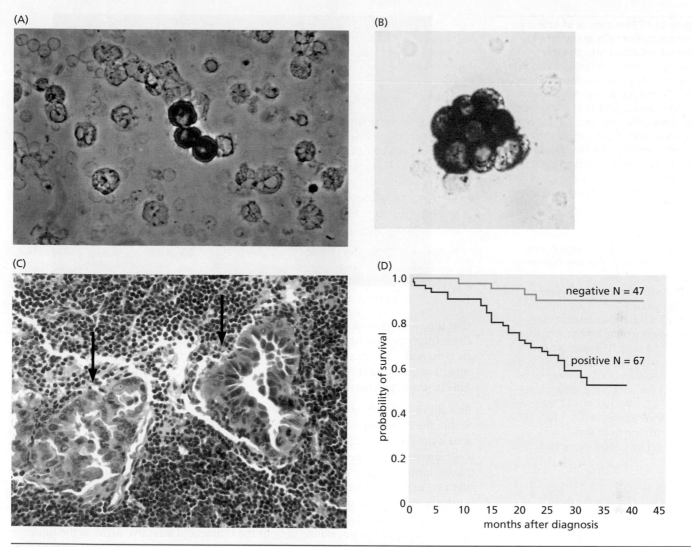

Figure 14.42 Presence of micrometastases in the marrow and lymph node (A) Metastasis of cancer cells to the bone marrow is usually determined by withdrawing marrow from the iliac crest of the pelvis. In this micrograph, the presence of a micrometastasis containing several cancer cells in the bone marrow of a colon cancer patient has been detected by staining with an anti-cytokeratin antibody (*red*). This staining is highly specific because normal marrow cells are of mesenchymal/hematopoietic origin and thus do not express cytokeratins. In contrast, epithelial cells and their neoplastic descendants usually express cytokeratins at significant levels. (B) Clusters of mammary carcinoma cells (*red*), viewed at slightly lower magnification, can often be found in the marrow of breast cancer patients. Although not demonstrated here, these cells are apparently the products of post-extravasation proliferation in the marrow. (C) Micrometastases often form in a draining lymph node—one that is directly connected with the primary tumor via a lymphatic duct that drains the tissue in which this tumor arose (see Figure 14.36). These growths are usually detectable by their strongly contrasting appearance from surrounding lymphocytes. Here two micrometastases of a mouse lung adenocarcinoma (*arrows*) to a lymph node are seen amid a sea of lymphocytes. Note the formation of a ductlike structure by the right micrometastasis, providing clear indication of proliferation after initial dissemination. (D) Cytokeratin-positive cells in the marrows of a cohort of breast cancer patients were counted, allowing the patients to be placed into two groups having either cytokeratin-positive or -negative marrows. As is evident, those with cytokeratin-positive cells in their marrow had a far worse prognosis, with almost 50% of them dying within three years of initial diagnosis. (A, courtesy of I. Funke and G. Riethmüller. B, from S. Braun, K. Pantel, P. Müller et al., *N. Engl. J. Med.* 342: 525–533, 2000. With permission from Massachusetts Medical Society. C, courtesy of K.P. Olive and T. Jacks. D, from J.-Y. Pierga, C. Bonneton, A. Vincent-Salomon et al., *Clin. Cancer Res.* 10: 1392–1400, 2004. With permission from American Association for Cancer Research.)

by a factor of 2 each time a cell divides; hence, the residual fluorescence intensity in cells allows the experimenter to estimate how many times the cells have divided since they were initially marked.

In one influential experiment, such dye-labeled cancer cells were introduced via the portal circulation into mouse livers, in which they formed large numbers of single-cell micrometastases. Eleven weeks later, cancer cells were recovered from these livers, and many of these still possessed full fluorescence intensity (**Figure 14.43**), indicating

Figure 14.43 Persistence of solitary dormant tumor cells many weeks after introduction into the liver Tumorigenic mouse breast cancer cells were labeled by inducing them to take up styrene nanoparticles (48 nm diameter) that had been tagged with a fluorescent dye. These cells were then injected into a mesenteric vein, which carried them via the portal vein into the liver. Eleven weeks later, tumor cells could still be detected in the liver (*white arrows*). Importantly, the fluorescence intensity of many of these cells did not differ significantly from the intensity of cells shortly after labeling, indicating that these cells had not divided even once following labeling and implantation in the liver. Following isolation and *in vitro* culturing, the descendants of many of these cells were tumorigenic when injected into the mammary fat pads of host mice. (Moreover, the dormant cancer cells were found to be fully resistant to a chemotherapy that reduced by 75% the size of metastases that were growing rapidly in the same mice.) (From G.N. Naumov, I.C. MacDonald, P.M. Weinmeister et al., *Cancer Res.* 62:2162–2168, 2002. With permission from American Association for Cancer Research.)

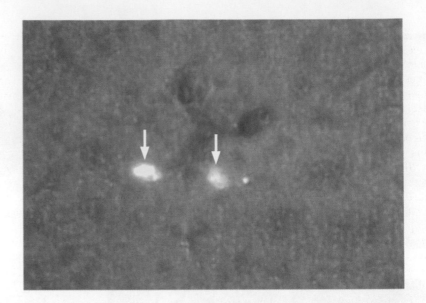

that they represented cells that had not divided even once since their arrival in the liver. Importantly, these recovered cancer cells remained capable of proliferating *in vitro* and were able to generate new tumors when injected subcutaneously into other host mice, that is, into an anatomical site where, for various reasons, they were able to proliferate more readily.

The above experiment shows dramatically that disseminated cancer cells can remain viable for extended periods of time in a nondividing, dormant state within foreign tissue sites. A quite different type of micrometastasis arises when disseminated cancer cells succeed in proliferating and forming colonies of a very small size within a foreign tissue; however, these subclinical micrometastases may never increase in size because the rate of cell proliferation in these clumps is counterbalanced by an equal rate of attrition, possibly as a result of the failure of these cells to execute the angiogenic switch (see Section 13.7). Alternatively, the outgrowth of such micrometastases may be held in check by the actions of both the innate and adaptive arms of the immune system (to be discussed in Chapter 15).

These micrometastatic colonies of various types are essentially undetectable clinically and represent what is often termed "residual disease." The eruption of metastases that renders them clinically apparent may occur months or years after initial seeding; these delayed relapses are most notorious in the case of breast cancer, in which women who have been deemed cured may suddenly develop macroscopic metastases years after their initial, ostensibly successful treatment. As an example, as many as one-third of women treated years earlier by removal of their primary breast cancers exhibit occasional mammary carcinoma DTCs, even though they may remain symptom-free years after their primary tumor surgery. Such women may nonetheless experience metastatic relapses 10–20 years after their initial surgery.

We can imagine that such relapses can be triggered by at least two alternative mechanisms. DTCs may *evolve* on site, solving the problem of colonization by acquiring the ability to thrive in the foreign tissue microenvironments in which they happen to have landed and to resist ongoing attrition by immune-mediated attacks in those sites. Importantly, ongoing cell proliferation over extended periods of time would seem to be essential in order for the cells within disseminated micrometastases to spontaneously evolve to an aggressive state, that is, to generate the genetic or epigenetic heterogeneity that is essential for such diversification (and subsequent selection) via a form of Darwinian evolution (**Sidebar 14.7**).

An entirely different, alternative mechanism of metastatic awakening and colonization is also plausible: changes in the tissue microenvironment surrounding dormant micrometastases, such as localized inflammation, might also trigger their outgrowth, perhaps by re-creating the stromal conditions that operated in the primary tumors

Sidebar 14.7 Genetic analyses suggest that the evolution of metastatic ability can occur outside of the primary tumor In perhaps 30% of breast, prostate, and colon cancer patients whose primary carcinomas have been surgically removed, one can still detect micrometastases in the marrow, lymph nodes, or blood; these patients are considered to have "minimal residual disease." At this stage, analysis of the micrometastases indicates that they are genetically heterogeneous. However, years later, when a patient develops disease relapse and manifests readily detectable metastatic masses, the patient's single-cell micrometastases are now much more similar to one another genetically (see Figure 11.16C).

This sequence of events suggests a stage of tumor progression in which the ability to colonize is acquired long after the acquired ability to disseminate to distant organ sites. Initially, genetically diverse cancer cells are seeded by a primary tumor throughout the body, but none of these succeeds in establishing a macroscopic metastasis simply because none is intrinsically capable of doing so. After a period of time, however, genetic and epigenetic evolution occurring in a micrometastasis somewhere in the body yields a clone of cells with the newly acquired ability to colonize efficiently. As this clone expands, it also begins to seed cancer cells throughout the body, and therefore generates a secondary wave of metastatic dissemination.

The individual cancer cells that are released by this colonizing clone soon constitute the majority of the single-cell micrometastases of a cancer patient, and these micrometastases are genetically very similar to one another because of their shared descent from the same clonal cell population. Importantly, because the cells in these new micrometastases all have inherited the ability to colonize, many may grow rapidly into macroscopic metastases, creating a life-threatening burden of disseminated cancer cells in the patient.

This **cascade model** suggests that the final evolution toward advanced malignancy often occurs at anatomical sites far removed from the primary tumor and that a secondary wave of dissemination (emanating from initially formed metastatic colonies) is responsible for most metastasis-associated death. (This secondary wave of dissemination is sometimes termed a "**metastatic shower**" to indicate the large number of new, robustly growing metastases that seem to appear simultaneously.)

A 1986 study of the autopsies of more than 1500 patients who died from colon carcinoma provided strong indications that metastases initially formed in the livers of these patients and spread from there to the lungs and finally to other sites in the body. Nevertheless, even with such evidence in hand, the cascade model is not yet validated by the genomic sequencing techniques and currently available bioinformatic algorithms.

in which the DTCs originated. Whatever their nature, micrometastases represent an imminent threat because they are often present in vast numbers throughout the body and may erupt years after a cancer has been thought to be cured.

The above discussions of sites of metastasis imply that the microenvironment of each potential target tissue (that is, a tissue in which a metastasis may actually form) represents a relatively stable constellation of biochemical and cellular traits that is typical of that particular tissue and of the niches encountered by recently arrived cancer cells.

The reality is, however, quite different, because certain tissue microenvironments may actually represent dynamically changing surroundings, that is, moving targets. Thus, rapidly accumulating evidence indicates that systemic signals released by primary tumors can actively reshape the microenvironments of distant tissues, often prior to metastatic dissemination, changing the parenchyma of these tissues and making them more hospitable to circulating tumor cells that happen subsequently to land in these tissues. These changes have been described as creating **pre-metastatic niches** in the destination tissues. Alternatively, as DTCs gain a foothold in the distant tissues and begin proliferation, they can then generate progressive changes in the tissue microenvironments around them that re-create the reactive stroma that was present in their corresponding primary tumors prior to the earlier dissemination. Hence, changes in the microenvironment at the site of eventual colony formation may occur both before and after the CTCs arrive.

Moreover, the generation of a reactive, EMT-inducing stroma associated with a given metastatic colony implies that this colony could itself become the source of a new generation of CTCs and a resulting secondary wave of metastatic dissemination, as cited above. If so, the resulting secondarily generated CTCs may actually be more dangerous to the patient than the initially disseminated cells arriving earlier from the primary tumor, simply because these newer CTCs derive directly from precursors that have already undergone evolution in a site of dissemination and have therefore developed tissue-adaptive programs; these programs may allow this second wave of CTCs and derived DTCs to take root almost immediately in various newer sites of dissemination. Such behavior may explain the above-cited clinical observations of metastatic showers, in which individual patients seem to develop multiple metastases synchronously throughout the body (see Sidebar 14.7).

There are diverse ways in which primary tumors could favor the formation of pre-metastatic niches. One frequently observed mechanism involves the creation by primary tumors of **neutrophilia**, the state of elevated neutrophil concentrations present in the circulation. Doing so must involve the release by primary tumors of neutrophil-stimulating factors, such as G-CSF (granulocyte colony-stimulating factor), that impinge on the bone marrow and greatly increase the production of these myeloid cells, their concentrations in the general circulation, and their deposition in tissues throughout the body (**Figure 14.44**).

The resulting neutrophils may actually operate in at least three distinct ways to increase the numbers of successfully launched metastases. Thus, these myeloid cells have been found to protect circulating tumor cells (CTCs) from attacks and thus attrition launched by NK cells in the general circulation. Neutrophils can also facilitate extravasation by CTCs. And finally, neutrophils may create changes in the parenchyma of a destination site that favor rapid outgrowth of extravasated cells.

A variety of noncellular messengers have also been found to serve as agents that are dispatched by primary tumors and contribute to pre-metastatic niche formation. By 2016, more than a dozen signaling molecules had been associated with the formation of these niches, in each case substances that are released by primary tumors, travel systemically to various sites throughout the body, and facilitate the establishment of metastases by subsequently arriving cancer cells. To extend the seed-and-soil metaphor of Paget, it would seem that the agents creating pre-metastatic niches function by tilling the soil, preparing tissues for possible seeding by subsequently arriving cancer cells.

This concept of pre-metastatic niche formation implies a purposive action on the part of a primary tumor that undertakes to pave the way for its metastatic offspring to spread throughout the body. Although attractive in theory, this notion ascribes an almost-intentional behavior to primary tumors, suggesting that preparing destination sites

Figure 14.44 Role of neutrophils in facilitating the formation of pulmonary metastases Neutrophils operate in multiple ways to facilitate metastasis formation by circulating tumor cells, a process best documented in the context of the lungs. A variety of primary tumor types induce neutrophilia through the elaboration directly or indirectly of the G-CSF factor, which stimulates neutrophil expansion in the marrow and then in the systemic circulation. Neutrophils thereafter contribute to the formation of pre-metastatic niches in the lungs. (A) In a mouse model of spontaneous mammary carcinoma metastasis, large numbers of metastatic colonies are readily detected in the lungs. (B) However, if circulating neutrophils are largely depleted using the anti-Ly6G antibody, which is used to specifically deplete neutrophils and reverse neutrophilia, the number of pulmonary metastases is greatly reduced. (From S.B. Coffelt, K. Kersten, C.W. Doornebal et al., *Nature* 522:345–348, 2015. With permission from Nature.)

(A)

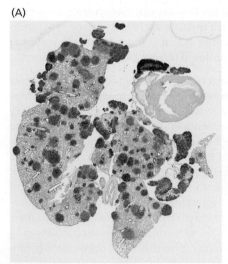

(B)

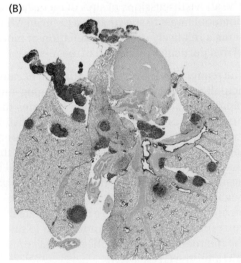

for eventual metastatic outgrowth is advantageous for the growth of cancer cells in the primary tumors. Instead, it is far more plausible that mechanisms that create these niches, like the process of metastasis itself, are unintended side-products of primary tumor growth, representing consequences of the formation of reactive stroma within primary tumors and associated systemic changes in a variety of bone marrow-derived cells that occur in connection with localized wound-healing and inflammation.

14.12 Successful metastatic colonization often involves complex adaptations

Abundant evidence supports the notion that metastatic cancer cells that have colonized a certain target organ must become highly specialized to do so: (1) Of the 75% of young patients with papillary thyroid carcinomas that are accompanied by significant lymph node metastases, only 3% will ever develop distant metastases. Hence, adaptation to the lymph nodes by metastasizing thyroid carcinoma cells does not empower them to colonize other tissues in the body. (2) Similarly, duodenal carcinoid tumors greater than 1 cm in diameter (containing $>10^9$ cells) have a high rate of lymph node metastasis, yet they rarely proceed to metastasize to the liver, which is the common site of metastasis of the tumors that arise in the nearby colon. (3) Cancer cells isolated from human lymph node metastases have been found, after injection into the venous system of mice, to grow preferentially in the lymph nodes of their mouse hosts rather than in other possible sites of colonization, suggesting specialized adaptations to these sites in humans that persist and influence tissue tropism in mice. (4) Mouse melanoma cells can be selected that metastasize preferentially to the lungs and human breast cancer cells that metastasize either to the lungs or to the bone and continue to do so each time they are implanted in mouse hosts. (5) Cells that formed dormant micrometastases in the liver proliferate vigorously when removed from the liver and implanted subcutaneously, as described in Figure 14.43.

These disparate observations reinforce the notion that the ability to colonize a certain organ represents an acquired specialization. Arguably the most dramatic example of the need for such specialization comes of the clinical practice of introducing peritoneovenous shunts to reduce the ascitic fluid accumulated in patients afflicted with various types of inoperable primary tumors. In such patients, billions of cancer cells arising in peritoneal ascites are transferred into the general venous circulation but few if any metastases seem to be seeded by these cells (Supplementary Sidebar 14.12).

To summarize, it seems that the frequency of metastases to an organ site is governed by two parameters—the frequency with which metastasizing cells are physically trapped in an organ, which is dictated in part by the layout of the circulation (see Figure 14.39), and the ease with which they can thereafter develop specialized adaptations that allow them to thrive in the microenvironment of that organ. Aside from metastases to the liver from the gut and pancreas, the layout of the circulation appears to represent a minor factor in governing the sites of metastasis formation.

There are also indications that tissues that are normally not hospitable sites for colonization may become so through specific pathological processes, such as localized wounding (Supplementary Sidebar 14.13). This suggests that areas of chronic inflammation within the body of a cancer patient may occasionally create congenial environments for metastasizing cancer cells, simply because they provide a spectrum of mitogenic and trophic signals, as discussed in Chapter 13. Such essential preconditions for metastatic seeding may explain why the microenvironment of the fully normal **contralateral** breast of a breast cancer patient is rarely the site of metastatic seeding, simply because it is not in a physiological state, such as one of inflammation, that is a key ingredient for effective colonization (**Sidebar 14.8**).

Such metastasis to areas of wounding and inflammation may explain a phenomenon first described in 2009—**tumor self-seeding**. Circulating tumor cells (CTCs) released by an initially formed macroscopic metastasis may find that the most congenial tissue microenvironment to colonize is the stroma of the corresponding primary tumor.

Sidebar 14.8 Contralateral metastases are relatively rare Possibly the greatest embarrassment for the seed-and-soil hypothesis comes from its failure to explain the rarity of contralateral metastases. For example, cancer cells disseminated from a primary tumor in one breast should find that the contralateral breast provides the most hospitable environment for colonization by these cells. In fact, only about 2% of breast cancer cases result in contralateral metastases, comparable to the frequency of tumors in the breast that arise as metastases of primary tumors located elsewhere in the body. Similarly, primary kidney cancers metastasize infrequently to contralateral kidneys. These behaviors are clearly incompatible with the simplest version of the seed-and-soil hypothesis and require explanation. One possibility is that, in addition to landing in an intrinsically compatible tissue, disseminating cancer cells prefer to take root in a tissue that has an activated stroma, such as is seen in sites of chronic inflammation or wound healing (see also Supplementary Sidebar 14.13).

On the one hand, the reactive stroma of the primary tumor (see Section 13.3) would seem to provide the factors that are highly supportive of the survival and proliferation of many types of cancer cells. On the other, these CTCs do not need to undergo the adaptive changes that seem to be required when cancer cells arising in one organ attempt to colonize a different organ; here these traveling cells are returning to familiar territory.

Tumor self-seeding also holds implications for the genetic makeup of the primary tumor: if cancer cells undergo adaptive genetic and epigenetic changes in distant sites of metastasis in order to colonize those sites, secondary metastatic showers of cells released by those distant tumors (see Sidebar 14.7) may carry these genetic and epigenetic changes *back* to a primary tumor, causing its cells to increasingly exhibit the genotypic and phenotypic alterations developed elsewhere in the body!

One interesting and still-unresolved question involves the timing of metastatic dissemination: When during the course of primary tumor progression do cells begin to disperse to distant sites in which they eventually form metastases (**Sidebar 14.9**)?

Sidebar 14.9 Some cancer cells leave home early Embedded in the discussions in this chapter is the notion that multi-step tumor progression occurs in the primary tumor site, and that only after cancer cells in these tumors have evolved to a certain state of aggressiveness do they begin to disseminate and attempt to found metastatic colonies. In fact, clinical observations have demonstrated the presence of disseminated tumor cells in certain organs, such as the bone marrow, long before primary tumor progression has generated aggressive tumor cells. (In the case of carcinomas, these observations may one day be explained by the activation of EMT programs in the cells of relatively benign, early-stage tumors.)

This early dispersal suggests that once relatively early-stage cancer cells have settled in distant tissues, they may evolve within those tissues through the later steps of multi-step tumor progression (see Chapter 11). Accordingly, such **parallel progression** may occur at a great distance from and independently of the multi-step progression occurring in the primary tumor. If validated, this would force a fundamental rethinking of how multi-step tumor progression usually proceeds.

This model is encumbered, however, by certain realities if only because these early-disseminating cells are doubly handicapped: like their late-stage counterparts, they are poorly adapted, at the moment of their arrival, to proliferate in the foreign tissues in which they have landed. In addition, by definition, they lack the suite of mutations (for example, those affecting oncogenes and tumor suppressor genes) needed to proliferate vigorously wherever they land (see Sections 11.8 and 11.9). Because the genetic and epigenetic diversification of tumor cells (which is critical to tissue adaptation via multi-step cellular evolution) would seem to require active cell proliferation, these dual impediments to proliferation may well dictate that these dispersed cells represent dead ends, remaining as dormant micrometastases and undergoing slow, progressive attrition.

14.13 An example of extreme metastatic specialization: metastasis to bone requires the subversion of osteoblasts and osteoclasts

The adaptations that circulating tumor cells must make in order to succeed in colonizing various tissues have been studied in a variety of destination sites. Arguably the best understood of these concerns the formation of metastases in the microenvironment of the bone marrow and surrounding mineralized bone. Indeed, the development of bony metastases represents one instance in which we understand in extraordinary detail the biochemical and biological mechanisms that permit metastasized cancer cells to thrive in a specific tissue microenvironment. This fact, on its own, justifies a detailed discussion of **osteotropic** metastasis. In addition, the current, very detailed understanding of the mechanisms of bony metastasis formation should serve as a model for how comparable processes operating in other sites of cancer metastasis will be elucidated in the future.

Osteotropic metastases also warrant our attention because, as mentioned repeatedly, several of the most common types of cancer occurring in the Western world—carcinomas of the lung, breast, and prostate—show a strong tendency to metastasize to the bone. In fact, patients with advanced breast and prostate cancer almost always develop bone metastases. And in those patients who succumb to these cancers, the bulk of the tumor cells in their bodies at the time of death is often found among the metastases scattered throughout their bones.

We usually think of bone as being a static tissue which, once formed, retains its structure throughout life. The truth is far more interesting. In mammals, about 10% of skeletal bone mass is replaced each year, resulting in an essentially complete replacement over the course of a decade. This continuous remodeling enables the bones to respond to mechanical stresses by compensatory reinforcing of stressed regions. For example, the bones of the legs are continuously being remodeled in response to the weight-bearing signals that different portions of each leg bone receive.

The turnover of bone is the work of **osteoclasts**, which break down mineralized bone, and of **osteoblasts**, which reconstruct it. The osteoclasts function first to **demineralize** the bone (by dissolving its calcium phosphate crystals) and then to degrade the now-exposed extracellular matrix, which previously formed the organic scaffolding for the calcium phosphate crystals (the process together being termed **resorption**; Figure 14.45). Osteoblasts move in soon after the osteoclasts have completed their work in order to carry out reconstruction, which involves both the assembly of new ECM and the deposition of calcium phosphate crystals in the interstices of this matrix. As can be deduced from this description, the two cell types normally work in close coordination.

Most kinds of metastasizing cancer cells are, on their own, incapable of remodeling bone structure. Instead, they manipulate and exploit these two types of cells normally present in the bone. Thus, breast cancer cells preferentially activate the osteoclasts, resulting in **osteolytic** metastases—literally, metastases that dissolve bone. Prostate cancer cells tend, on the other hand, to activate osteoblasts, yielding **osteoblastic** lesions, in which immature mineralized bone (sometimes termed **osteoid**) actually accumulates in the vicinity of the metastases (**Figure 14.46**).

In fact, these two behaviors represent the extremes of a continuum, since both types of cancers activate both osteoblasts and osteoclasts to a greater or lesser extent. For example, while osteolytic metastases predominate in advanced breast cancer patients, as many as one-quarter of these women also have clearly defined osteoblastic lesions in their bones. Similarly, prostate carcinomas also generate occasional osteolytic metastases scattered among the many osteoblastic growths spawned by these tumors. One exception to this rule of a mingling of both types of bone metastases is provided by myeloma cells (tumors of the B-cell, antibody-secreting lineage), which create exclusively osteolytic lesions.

The normally operative close coordination between osteoblasts and osteoclasts is mediated, at least in part, by the exchange of growth factor signals. An important

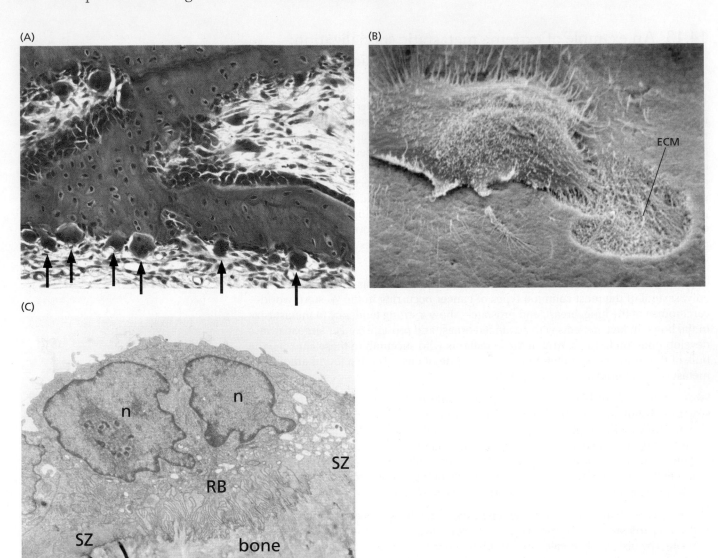

Figure 14.45 Bone degradation by osteoclasts The degradation
of bone, often termed *resorption*, depends on the complex
actions of osteoclasts—large multinucleated cells deriving from
the monocyte lineage that also generates macrophages. (A) This
light micrograph shows osteoclasts (*purple, arrows*) excavating
small pits in the surface of a mouse jawbone (*pink*). (B) At far
higher magnification, this scanning electron micrograph shows
a cat osteoclast that has excavated a shallow pit in the surface
of mineralized bone. The calcium apatite crystals in the bone
have been dissolved away by acid secreted by the osteoclast,
revealing the complex meshwork of collagen-rich extracellular
matrix (ECM) at the bottom of the pit. Associated with this
ECM are mitogens and survival factors that become available to
cancer cells after osteoclasts subsequently break down the ECM.
(C) This transmission electron micrograph reveals at even higher

magnification the details of how osteoclasts resorb bone. This
section through an osteoclast (with multiple nuclei, n, *above*)
and underlying bone (*below*) reveals the osteoclast's complex
ruffled border (RB), which secretes protons to dissolve the mineral
component of the bone and acid proteases to degrade the collagen-
rich extracellular matrix that is exposed following demineralization.
Surrounding this area of contact is a circular sealing zone (SZ)
containing substantial amounts of filamentous actin, which
functions as a gasket to confine these secretions to a small localized
interstitial area between the osteoclast and bone. (A, courtesy of
T.R. Arnett. B, from T.R. Arnett and D.W. Dempster, *Endocrinology*
119:199–124, 1986. With permission from Oxford University Press.
C, from H. Zhao, T. Laitala-Leinonen, V. Parikka et al., *J. Biol. Chem.*
276:39295–39302, 2001. © 2001 ASBMB. This article is available
under a Creative Commons Attribution (CC BY 4.0) license.)

inducer of osteoclast differentiation is RANK (receptor activator of NF-κB) ligand, or
simply RANKL. RANKL is produced by and displayed on the surface of osteoblasts.
When an osteoclast precursor displaying the RANK receptor comes into contact with
an osteoblast and its cell surface RANKL molecules, this results in activation of the
RANK receptors of the osteoclast precursor and its maturation into a functional osteoclast (**Figure 14.47A**). At the same time, osteoblasts produce a soluble **decoy receptor**,
termed osteoprotegerin (OPG), which can bind RANKL and ambush it before it succeeds in activating the RANK receptor on the surface of osteoclast precursors. The
result is a blockage of the RANKL–RANK signaling and the inhibition of osteoclast maturation. Hence, the balance between the RANKL (stimulatory) and OPG (inhibitory)

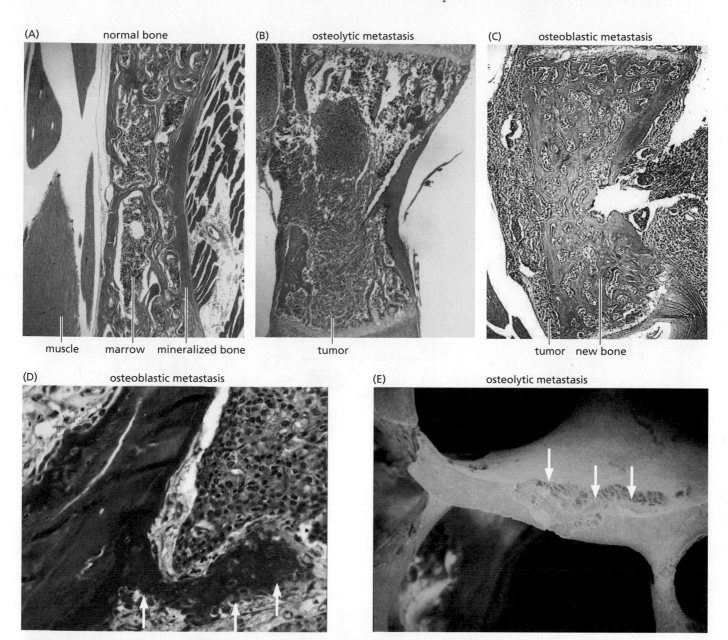

Figure 14.46 Osteolytic and osteoblastic metastases The first three of these micrographs present sections of mouse vertebrae and femurs in which the mineralized bone (*orange*), surrounding muscle (*bright red*), and bone marrow (*dark purple*) are clearly delineated. (A) This vertebra of a control mouse injected only with buffer is seen to be composed of extensive marrow with ribbons of mineralized bone running through the marrow. (B) In a mouse bearing a human breast cancer cell line (MDA-MB-231) that creates osteolytic lesions, much of the mineralized bone is seen to be missing, and the marrow has been displaced by tumor cells (*dark red*). (C) In a mouse bearing a human breast cancer cell line (ZR-75-1) that creates osteoblastic lesions, much of the marrow space is now filled with mineralized bone (*orange*) with tumor masses evident to the *left* and *right*. (D) In the iliac crest of the pelvis of a prostate cancer patient, the native bone (*dark green-blue*) can be readily resolved from the newly synthesized, still-poorly mineralized osteoblastic lesion (*purple*), which is sometimes termed osteoid. The osteoid is protruding into a mass of metastatic prostate cancer cells (*dark purple nuclei*). (E) A scanning electron micrograph reveals how devastating the osteolytic lesions (*arrows*) can be in terms of compromising bone structure in a patient with metastatic osteolytic lesions created by osteoclasts. (A–C, from J.J. Yin, K.S. Mohammad, S.M. Käkonen et al., *Proc. Natl. Acad. Sci. USA* 100:10954–10959, 2003. With permission from National Academy of Sciences, U.S.A. D, courtesy of C. Morrissey and R.L. Vessella. E, courtesy of G.R. Mundy.)

signals displayed by osteoblasts determines the state of activation of osteoclasts (Figure 14.47A and B). This dynamic interaction of osteoblasts and osteoclasts provides the background for the actions of cancer cells that metastasize to bones. Their attraction to the bone derives ultimately from the collagenous extracellular matrix that forms the organic scaffolding in which calcium phosphate crystals are deposited (see Figure 14.45B). As it happens, bone ECM is an unusually rich source of the

(A)

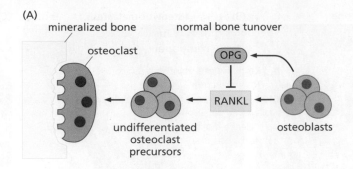

(B)

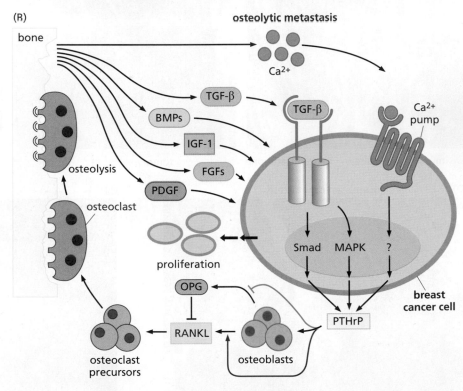

(C)

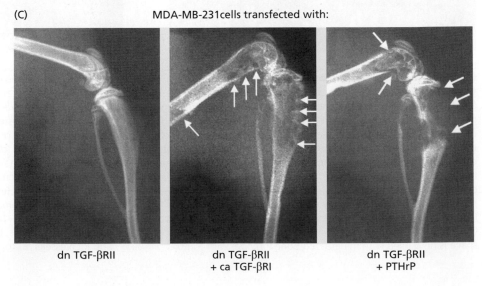

MDA-MB-231cells transfected with:

dn TGF-βRII   dn TGF-βRII   dn TGF-βRII
                     + ca TGF-βRI   + PTHrP

Figure 14.47 Osteoblasts vs. osteoclasts (A) The physiological balance between normal bone formation and resorption is created by signaling between osteoblasts, which assemble bone, and osteoclasts, which dissolve it. In an ongoing cycle, osteoclasts remove mineralized bone by covering and sealing off a section of bone and secreting digestive acid into the bone below them (see Figure 14.45C); this is followed by osteoblastic filling of resulting cavities with new bone. The osteoblasts release RANKL, which acts via the RANK receptor (*not shown*) displayed by osteoclast precursors to induce the latter to mature into functional osteoclasts. The osteoblasts may also secrete osteoprotegerin (OPG), which acts as a *decoy receptor* to ambush RANKL before it has had a chance to activate osteoclast precursors. Hence, the balance between RANKL and OPG determines the net rate of bone growth or loss. (B) Release by a breast cancer cell (*right, gray*) of PTHrP (parathyroid hormone–related peptide, *below*) causes nearby osteoblasts to change the mix of signals that they release: they increase RANKL synthesis (*red arrow*) and decrease OPG synthesis (*blue line*), the result being RANKL-induced maturation of osteoclast precursors into functional osteoclasts. The latter undertake osteolysis, which causes bone demineralization, exposes the extracellular matrix within the bone (see Figure 14.45B), and results in liberation of TGF-β, Ca^{2+}, BMPs, PDGF, FGFs, and IGF-1 (*upper left and middle*). IGF-1 and Ca^{2+} enable cancer cell proliferation and survival, and the additional presence of TGF-β induces the cancer cell to release more PTHrP, resulting in a self-sustaining positive-feedback loop that has been termed the "vicious cycle" of osteolytic metastasis. (C) Evidence supporting the vicious cycle model of osteolytic metastasis comes in part from experiments involving the use of MDA-MB-231 cells, a line of human breast cancer cells that show a high tendency to produce osteolytic metastases (see Figure 14.46B). The ability of these cells to do so is gauged here by X-ray analyses of the hind limbs of mice that have borne MDA-MB-231 tumor xenografts. When a dominant-negative type II TGF-β receptor (dn TGF-βRII) expression construct is introduced into these cancer cells, this mutant protein blocks their ability to respond to TGF-β, specifically the TGF-β that would otherwise be liberated from the extracellular matrix of the osteolytic lesions that they may have created. Without TGF-β stimulation, these cancer cells fail to form osteolytic metastases (*left panel*). However, if this inability to respond to TGF-β is overridden by introducing additionally into these cells an expression construct specifying a constitutively active type I TGF-β receptor (ca TGF-βRI), then the powers of these breast cancer cells to induce osteolytic lesions are restored (*arrows, center panel*). This observation, on its own, does not indicate precisely how the ca TGF-βRI succeeds in restoring the osteolytic activity to these cells. An explanation comes from an experiment in which a vector causing PTHrP expression (instead of ca TGF-βRI expression) is introduced into the cells expressing the dn TGF-βRII, causing them to regain the ability to form osteolytic metastases (*arrows, right panel*). This is compatible with the notion that PTHrP functions downstream of TGF-β signaling and that the latter is no longer important for osteolysis if PTHrP is expressed constitutively, i.e., is no longer under the control of TGF-β. (B, from G.R. Mundy, *Nat. Rev. Cancer* 2:584–593, 2002. With permission from Nature. C, from J.J. Yin, K. Selander, J.M. Chirgwin et al., *J. Clin. Invest.* 103:197–206, 1999. With permission from American Society for Clinical Investigation.)

mitogenic and trophic factors that allow several types of carcinoma cells to thrive. Consequently, by provoking the demineralization of bone, cancer cells gain access to the storehouse of factors sequestered in the bone ECM and use them to support their own proliferation and survival.

Metastasizing cancer cells reach the bone through the vessels feeding the marrow. Once there, they adhere to specialized stromal cells coating the surfaces of the bone facing the marrow. Metastasizing breast cancer cells, upon arrival in bone, revert to a behavior characteristic of their normal precursors (mammary epithelial cells, MECs). During lactation, when producing milk, MECs forming the small sacs (**alveoli**) of the mammary gland release parathyroid hormone–related peptide (PTHrP). PTHrP then travels from the breast through the circulation to the bones, where it triggers a chain of events that encourages the dissolution of bone minerals by osteoclasts. This results in the mobilization of calcium ions, which travel back via the circulation to the mammary gland, where they are incorporated into the milk by the MECs.

This normal calcium-mobilizing mechanism is subverted by metastasizing breast cancer cells that become established in bones (see Figure 14.47B). Having attached to the stromal cells covering the surfaces of mineralized bone, the breast cancer cells, reverting to the habit of normal MECs, release PTHrP. The PTHrP, in turn, impinges directly on its receptors displayed by osteoblasts, causing these cells to release RANKL. RANKL then induces the differentiation of osteoclast precursors into active osteoclasts. The activated osteoclasts degrade nearby mineralized bone, thereby liberating calcium ions together with the rich supply of growth factors attached to the ECM of the bone.

The growth factors liberated from the bone ECM, including PDGF, bone morphogenetic proteins (BMPs), fibroblast growth factors (FGFs), insulin-like growth factor-1 (IGF-1), and TGF-β, fuel the further growth of the breast cancer cells, inducing them to secrete more PTHrP. This PTHrP engenders more osteolysis by the osteoclasts, leading to a self-perpetuating signaling system that has been called a "**vicious cycle**" (see Figure 14.47B) in which TGF-β also plays a key role (**Sidebar 14.10**).

The central role of osteoclasts in this cycle suggests possible points of therapeutic intervention. One highly effective strategy depends on drugs belonging to the class of **bisphosphonates**, which are taken orally and become adsorbed to the apatite crystals that constitute the mineral portion of bone; the drug molecules can persist there for extended periods of time, as long as a decade or more. When bisphosphonate-containing bone is eventually dissolved by osteoclasts, the latter are poisoned by the liberated bisphosphonates, leading to their apoptosis. Hence, bisphosphonates are useful for reducing the burden of osteolytic lesions in patients with various types of metastatic cancer.

When immunocompromised mice carrying human breast cancer cells are treated with bisphosphonates, the number of osteolytic lesions is reduced and, at the same time, the total burden of tumor cells in these animals is decreased. This observation provides additional indication that late in tumor progression, the proliferation of these breast cancer cells depends greatly on osteolysis and the resulting liberation of growth factors from dissolved bone; similar observations have been made in human breast cancer patients treated with bisphosphonates. For example, a clinical study reported in 2011 that premenopausal women who developed breast cancer and were treated with zoledronic acid—a bisphosphonate—experienced a 37% increase in overall survival (relative to those who did not receive this treatment) over a period of seven years. Such clinical responses further reinforce the notion that many breast cancer relapses in patients (1) derive from metastatic deposits in the marrow and (2) depend on the ability of disseminated metastatic cells to generate mitogens by dissolving bone. Moreover, bisphosphonate therapy can provide additional benefits to patients suffering from metastatic breast cancer by reducing the **hypercalcemia**—pathologically high concentration of calcium in the circulation—stemming from the large-scale resorption of mineralized bone. Hypercalcemia usually signals the final stages of malignant disease and causes gastrointestinal, urinary tract, cardiovascular, and neuropsychiatric problems.

Figure 14.47B predicts that the vicious cycle driving osteolytic metastases should be slowed down or even blocked by therapeutic treatment with osteoprotegerin (OPG). In fact, a derivative of OPG has been found to be as effective as a widely used bisphosphonate in slowing down bone resorption in patients with metastatic breast cancer. However, clinical development of this OPG derivative was eventually discontinued, having been displaced by a slightly more effective treatment—a monoclonal antibody (denosumab) that binds and neutralizes RANKL. A **phase III** clinical trial revealed

Sidebar 14.10 TGF-β and PTHrP play pivotal roles in the vicious cycle of breast cancer osteolytic metastases Breast cancer cells that have metastasized to the bone produce far more parathyroid hormone-related peptide (PTHrP) than do others in the same animal that have not because certain growth factors liberated from the bone ECM stimulate PTHrP production by the metastatic cancer cells. The most important of these bone-derived factors is TGF-β, as illustrated by some simple experiments. In one of these, a dominant-negative TGF-β receptor (which blocks a cell's ability to respond to TGF-β) is expressed in the cancer cells; this blocks the ability of these cells to drive osteolytic metastases in the bone; forced expression of PTHrP in these cancer cells reverses this effect (see Figure 14.47C).

In another experiment, breast cancer cells that usually lack the ability to metastasize to bone and fail to secrete TGF-β can be forced (through the use of an expression vector) to secrete TGF-β. The latter then acts in an autocrine fashion to stimulate these cells to produce their own PTHrP, allowing them to form large numbers of bone metastases. Finally, antibodies that bind and neutralize PTHrP are able to block the ability of human breast cancer cells to generate osteolytic lesions in mice. These are some of the disparate observations that have inspired the "vicious cycle" model depicted in Figure 14.47B.

that denosumab was more effective than bisphosphonates in delaying the onset of serious skeletal-related events in patients with metastatic breast disease, while having comparable efficacy in delaying disease progression.

14.14 Occult micrometastases threaten the long-term survival of many cancer patients

Throughout this chapter, we have read repeatedly about the extraordinary inefficiency with which metastases are produced. Some of this metastatic inefficiency is created by the profound difficulties that cancer cells experience as they undertake the initial steps of the invasion–metastasis cascade. Most of those that do manage to reach distant sites and survive in their newfound homes fail to form clinically detectable metastases. The result is the presence of myriad dormant micrometastases seeded throughout the tissues of many cancer patients.

While micrometastases are, with rare exception, unable to expand to form clinically detectable metastases, they do provide clear indication that a primary tumor has seeded cells throughout the body. These micrometastases represent a threat to the long-term survival of cancer patients, if only because some of them may erupt into full-fledged, clinically significant macroscopic metastases years after they become implanted in some distant tissue site. Breast cancers are notorious for yielding relapses one and even two decades after the primary tumor has been removed and the patient has been declared to be free of cancer.

In one study of breast cancer patients, micrometastases were detected by sampling the bone marrow of the iliac crest of the pelvis. About 1% of a population of patients suffering from nonmalignant conditions showed cytokeratin-positive cells (that is, epithelial cells) in their marrow. In contrast, 36% of breast cancer patients carrying tumors of stages I, II, or III had such micrometastases in their marrow. The presence of these micrometastases in the marrow proved to be a highly useful prognostic marker for the risk of relapsing with clinically detectable metastasis (see also Figure 14.42D). Thus, within four years, one-quarter of the marrow-positive patients had died from cancer, whereas only 6% of those lacking cancer cells in their marrow had died from this disease. Another study found a more than tenfold increased risk of death from breast cancer among those whose marrow carried micrometastases composed of single cells or small clumps of cancer cells.

Colon cancer patients who have undergone **resection** (surgical excision) of their primary tumor will often appear in the cancer clinic a year or two later with a small number of metastases in their liver but none elsewhere; these can then be removed surgically, often with significant clinical benefit. Once again, micrometastases in the marrow of the pelvis can be scored. About 90% of those who lack these micrometastases are still alive 15 months later, whereas only 60% of those who carry such micrometastases survive to this point.

A procedure used to treat cancer of the esophagus provides yet another insight into metastatic spread. These tumors are often treated surgically, which necessitates the removal of one or more rib segments, from which marrow can be easily flushed. Two independent studies reported that 79% and 88% of these patients, respectively, harbored carcinoma micrometastases in their rib marrow at the time of their surgery. These numbers, which contrast with the approximately 30% of initially diagnosed breast cancer patients bearing micrometastases, correlated with a far grimmer prognosis for patients suffering this type of cancer, with less than half of them surviving more than three years after diagnosis.

The melanoma literature provides equally dramatic testimony of the long-term dangers posed by **occult**, dormant micrometastases (that is, those that are latent and apparently not growing). In one particularly well-documented case, kidneys were prepared for organ transplantation from the cadaver of a patient who had undergone resection of a small melanoma 16 years earlier. The patient had been followed closely for 15 years after removal of this small primary tumor and had remained symptom-free. However, soon after transplantation, the two recipients of his

Sidebar 14.11 Are all micrometastases truly dormant? The cancer cells borne by many patients with "minimal residual disease" occasionally acquire the ability to colonize a site of dissemination, creating a clinically apparent metastasis and, in turn, a new site from which carcinoma cells can further disseminate. How this occurs depends critically on the state of the disseminated tumor cells (DTCs) in patients with this condition.

The state of metastatic dormancy might be interpreted as one in which the cells are residing in a state of quiescence, that is, out of the active growth-and-division cycle that we encountered in Chapter 8. The alternative is that the cells in such growths are actually in a state of active proliferation, but are unable to drive increases in the size of the micrometastatic deposits, simply because the growth of such deposits is counterbalanced by an equal rate of attrition.

Importantly, a major shift in cell phenotype that is associated with the "awakening" of dormant micrometastases has a low probability of occurring in populations of quiescent cells. Instead, extensive observations made over many years' time indicate that such major changes happen spontaneously only in *proliferating* cell populations. Hence, in many patients with minimal residual disease, some clones of micrometastatic cancer cells are likely to be passing through repeated growth-and-division cycles and occasionally spawning variants that have, through some random accident, acquired colonizing ability.

kidneys (who were immunosuppressed to accommodate the allogeneic kidney grafts) developed aggressive melanomas that were directly traceable to this donor (see Supplementary Sidebar 14.11).

We still understand relatively little about the mechanisms that prevent dormant micrometastases from erupting into clinically threatening growths long after primary tumor removal. In some instances, one can observe micrometastases growing as cuffs around small vessels; this suggests that they lack their own angiogenic capabilities but are nonetheless able to take limited advantage of host capillaries that happen to be nearby. In most micrometastases found in the marrow, the involved cells lack any indication of cell proliferation markers and thus are in a nongrowing, G_0-like state for extended periods of time (see Figure 14.43), perhaps for months and even years (consider, however, **Sidebar 14.11**). (Because such cells are not proliferating, they may be especially resistant to chemotherapeutic treatment designed to eliminate the residual disease that persists following surgical removal of a primary tumor.)

Immune mechanisms may also contribute to suppressing the growth of micrometastases, thereby preventing metastatic disease relapse. This is suggested by the occasionally observed explosive growth of aggressive metastatic tumors in immunosuppressed organ transplant recipients such as the two kidney graft recipients described above. Moreover, the wound-healing response following primary tumor resection may transiently liberate nearby DTCs from immune control, once again enabling their eruption. There is also evidence that localized inflammation in the tissue harboring dormant micrometastases can provoke their awakening and the eruption of rapidly growing metastases. Finally, the possibly critical role of tumor stem cells as founders of macrometastatic colonies may help to explain the inability of most of initially seeded DTCs to generate macroscopic outgrowths (Supplementary Sidebar 14.14). Beyond this, relatively little is known about the details of the mechanisms that preclude most micrometastases from successfully colonizing the tissues in which they have landed.

14.15 Synopsis and prospects

Several major issues have complicated the search for the genetic determinants of aggressive phenotypes of cancer: Are these phenotypes programmed by a small number of pleiotropically acting, master control genes, or do the actions of multiple, independently regulated genes collaborate to create each of these phenotypes? Do these genes undergo mutation during tumor progression, or do they become involved in the late steps of tumor progression through nongenetic, that is, epigenetic mechanisms that control their expression? And how do familiar oncogenes and tumor suppressor genes contribute to invasion and metastasis?

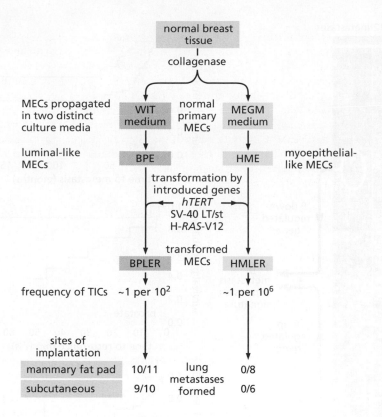

normal breast tissue

collagenase

MECs propagated in two distinct culture media

WIT medium — normal primary MECs — MEGM medium

luminal-like MECs

BPE — HME — myoepithelial-like MECs

transformation by introduced genes
hTERT
SV-40 LT/st
H-*RAS*-V12

transformed MECs

BPLER — HMLER

frequency of TICs   ~1 per 10^2 ~1 per 10^6

sites of implantation

		lung metastases formed	
mammary fat pad	10/11		0/8
subcutaneous	9/10		0/6

Figure 14.48 Influence of cell-of-origin on metastatic propensity
The observations to be presented in Figure 14.49 indicate that a substantial proportion of the cells in a primary tumor share a gene expression signature that was acquired relatively early in the course of tumor progression and is coupled with the propensity of the tumor to eventually metastasize. Such acquisition may have derived from the initial somatic mutations that triggered multi-step tumor progression or, even earlier, in the gene expression signature of the preexisting normal cell-of-origin, the latter reflecting its program of differentiation. In the experiment described here, a heterogeneous population of normal human mammary epithelial cells (MECs) explanted directly from a normal mammary gland were propagated *in vitro* in two alternative tissue culture media, which selected for the outgrowth of MEC populations that expressed either a more luminal (BPE) or a more myoepithelial (HME) gene expression pattern and therefore originated from two distinct differentiation lineages in the mammary gland. These cells were then transformed through the successive introduction of an hTERT gene, the SV40 early region (expressing small and large T antigens), and a RAS oncogene (see Figure 11.28), resulting in BPLER and HMLER tumorigenic cells. Although the introduced transforming genes were expressed at comparable levels in the two cell populations, these transformed cells behaved very differently. The concentration of tumor-initiating cells (TICs), also known as cancer stem cells, in the BPLER cell population was 10^4 times higher than in the HMLER cell population. The BPLER cells formed lung metastases following orthotopic implantation (into mammary stromal fat pads; see Figure 14.25C) or implantation into subcutaneous sites (in 10 of 11 and in 9 of 10 implanted mice, respectively), whereas the HMLER cells, implanted in equal numbers in these sites, failed to do so. Because the two cell populations acquired identical sets of oncogenes that were expressed at very similar levels, the only source of their dramatically different behaviors must be associated with the differentiation programs of the normal MECs in each population prior to experimental transformation. By extension, this implies that the differentiation programs of normal cells-of-origin continue to imprint themselves on the behavior of their neoplastic progeny. (From T.A. Ince, A.L. Richardson, G.W. Bell et al., *Cancer Cell* 12:160–170, 2007. With permission from Elsevier.)

Although many of the genetic elements governing metastasis remain unclear, significant progress is being made in solving another puzzle: Are the cells within a primary human tumor that undertake invasion and metastasis rare variants (among the larger population of tumor cells) that have, through some genetic or epigenetic accident, acquired the ability to execute these steps? Or are all the cancer cells within certain primary tumors equally capable *a priori* of invading and metastasizing (albeit with extraordinarily low efficiency), whereas the great majority of the cancer cells in other tumors lack these abilities?

In fact, such differences in the eventual metastatic behavior of various tumor cell populations may be determined extremely early, even *before* multi-step tumor progression has begun in a tissue: thus, as **Figure 14.48** indicates, the differentiation programs of the normal cells-of-origin can strongly influence the biological behaviors of derived neoplastic descendants and indeed can dominate the actions of any somatically sustained oncogenic alleles. This would suggest that a tendency to eventually metastasize or not is a trait that is shared in common by the neoplastic cells within a primary tumor.

More direct evidence for this scheme (of widespread metastatic competence shared among the cells within a primary tumor) can be found in attempts to predict the metastatic tendencies of primary tumors. Thus, there are multiple examples of cancer types in which the transcriptomes of primary tumor cells can be used as prognostic indicators of eventual tendency to metastasize. By necessity, because the transcriptomes are determined in each case by extracting RNA from large numbers of pooled primary tumor cells, these transcriptomes are reflective of a widely shared state of expression rather than one that is characteristic of a small minority of primary tumor cells. As seen in **Figure 14.49**, an effort to distill the complex transcriptomes of primary tumors and their association with metastases yielded a signature of 17 genes that could be used to predict the likelihood of patients bearing three types of primary tumors to progress to various clinical end points. Observations like these suggest that metastatic cells are drawn from the general population of cells in a primary tumor rather than from small, specialized, genetically unrepresentative subclones of cells.

However, DNA sequencing studies of the genomes of metastases and corresponding primary tumors suggest an apparently opposite conclusion: that primary tumors are

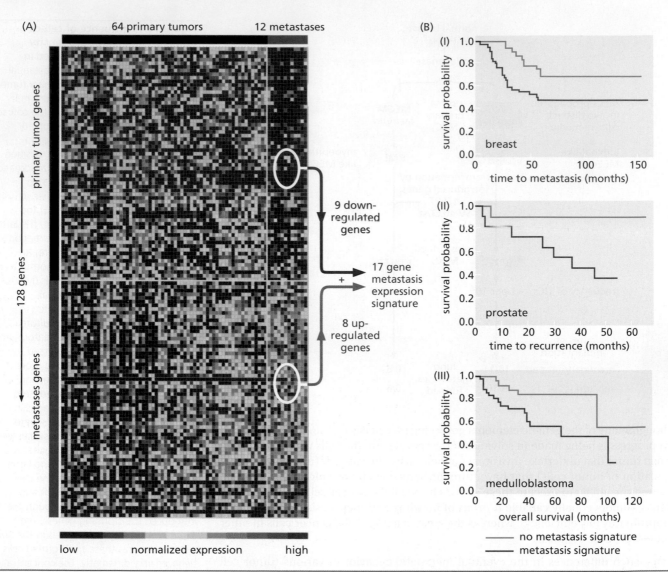

low   normalized expression   high

no metastasis signature
metastasis signature

Figure 14.49 Schedule of acquiring metastatic propensity
The process of metastasis is the endpoint of multi-step tumor progression for many human tumors. This raises the issue of whether the tendency to metastasize is acquired late in tumor progression by a small subpopulation of cells or relatively early. (A) Gene expression microarrays make possible the simultaneous monitoring of the expression of thousands of genes to determine a specific pattern or expression signature that is correlated with a specific phenotype or set of phenotypes. In the expression array analysis seen here, genes that are expressed at high levels are in *red* and those expressed at low levels are in *blue*. RNAs prepared from 64 primary adenocarcinomas (from various tissues) and 12 metastatic nodules of adenocarcinomas (*arrayed across the top*) were analyzed (*horizontal black, red bars*, respectively). Of the thousands of genes analyzed in an initial gene expression array (*not shown*), 128 genes (*arrayed vertically*) were found to be associated—because of over- or underexpression—with metastases and metastatic progression

(*vertical red, black bars, respectively*). Further distillation of the data yielded a set of 17 genes whose expression was as useful as that of the 128-gene set in predicting eventual metastases from primary tumors. (B) When researchers used the metastasis expression signature of panel A to analyze the gene expression patterns of other types of primary tumors, they were able to separate the patients bearing tumors of the breast (I), prostate (II), as well as (III) medulloblastomas into two groups (*blue, red lines*) having markedly different times to clinical progression or relapse following initial surgery. Most cells in certain primary tumors expressed a gene expression signature associated with metastasis, suggesting that this signature was acquired relatively early in primary tumor progression, possibly even before the initiating somatic mutation, and it became widespread in such tumors, being inherited by the great majority of descendants of early founders of the tumor cell population. (A and B, from S. Ramaswamy, K.N. Ross, E.S. Lander et al., *Nat. Genet.* 33:49–54, 2003. With permission from Nature.)

genetically heterogeneous, and that metastases arise from genetically specialized subpopulations within primary tumors. An eventual reconciliation of these two portrayals of metastatic progression may be as follows (**Figure 14.50**): the differentiation program of a normal cell-of-origin is an important determinant of whether a primary tumor will ever have the potential to spawn metastasis-competent subpopulations. In those tumors that happen to inherit a differentiation program that favors eventual

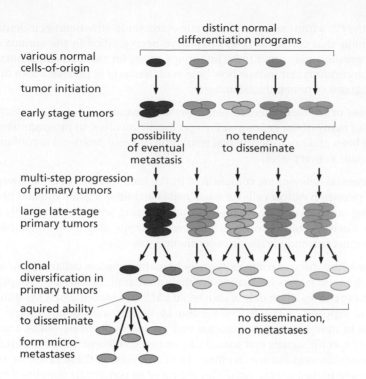

distinct normal
differentiation programs

various normal
cells-of-origin

tumor initiation

early stage tumors

possibility
of eventual
metastasis

no tendency
to disseminate

multi-step progression
of primary tumors

large late-stage
primary tumors

clonal
diversification in
primary tumors

aquired ability
to disseminate

no dissemination,
no metastases

form micro-
metastases

Figure 14.50 Determinants of eventual metastatic dissemination. Observations like those presented in Figure 14.49 suggest that the differentiation programs of the normal cells-of-origin of a tumor continue to strongly influence the fully transformed, tumorigenic descendants cells long after primary tumors are formed. As depicted here, most types of normal cells-of-origin, following neoplastic transformation, will generate primary tumors that have little tendency to disseminate even though the constituent cells are capable of robust proliferation in the context of the primary tumor. A subset of normal cells-of-origin (*left, red*) will, in contrast, yield primary tumors of comparable size and growth rates that contain cells that may well eventually metastasize. As the cells within a primary tumor undergo diversification during primary tumor progression, certain cells within the tumor will form small clonal populations that have acquired the ability, under appropriate circumstances, to disseminate and form micrometastases. Because this last step occurs at random times, most eventually formed metastases will represent the descendants of the clonal populations that were the first to develop the ability to disseminate.

metastasis, diverse, specialized subclones may subsequently arise through somatic mutations or other stochastically acquired cell-heritable changes. Among these subclones will be those that have actually acquired metastatic powers. These subclones will then become the major sources of metastatic dissemination. Some speculations about the evolution of metastatic competence extend and reinforce this model (Supplementary Sidebar 14.15).

As described here, EMT programs offer an attractive mechanistic solution to the problem of how carcinoma cells disseminate from primary tumors to distant tissues. Thus, the multiple distinct phenotypes conferred on neoplastic epithelial cells by this pleiotropically acting program may enable a primary tumor cell to physically translocate from the heart of a primary tumor to the parenchyma of a distant organ. Accordingly, a carcinoma cell that has activated an EMT program may be able to accomplish most of the steps of the invasion–metastasis cascade (see Figure 14.3) except the last one—colonization. If eventually proven, this model would represent a simplifying solution to a problem that has been viewed as one of endless complexity.

Of additional interest, and as illustrated in Figure 14.18, certain observations have demonstrated that once breast cancer cells have passed through an EMT in order to travel from the primary tumor to a foreign tissue compartment and found a new metastatic colony, many of the carcinoma cells in the resulting colony undergo an MET and thereby generate cells of a more epithelial phenotype. Thus, the presence of the resulting epithelial cells in such a metastasis seems to be critical to its robust outgrowth.

Our repeated reference to EMT programs in the context of carcinoma pathogenesis raises the question of whether there are alternative, equally important cell-biological programs that drive malignant progression. At present, it seems gratuitous to search for alternatives. Thus, in light of the detailed surveys of the traits of spontaneously arising, malignant carcinoma cells and the phenotypes of EMT-expressing cultured cells, it seems increasingly likely that EMTs play a central role in the malignant progression of most if not all carcinoma types and that the existence of alternative, equally important cell-biological programs is improbable.

Unanswered by our discussions are the associations of EMT programs with nonepithelial tumors, notably hematopoietic, connective tissue, and central nervous system (CNS) tumors, some of which are able to invade and occasionally metastasize. The origin of the CNS from an early embryonic epithelium—the neuroectoderm—offers the possibility that certain EMT-TFs that have been studied in the context of carcinoma progression may prove to be important actors for this class of tumors as well. Still, this uncertainty reminds us how little is understood at present about the malignancy programs of nonepithelial tumors.

Whether one or another model of metastasis is ultimately validated, it is clear that the identities of many of the genes that are specifically involved in programming metastasis have been elusive. Experimental resolution of these problems is confounded by complications at every level.

1. Experimental analyses are complicated by the inefficiencies of the metastatic process. Even when cancer cells have ostensibly acquired a genotype and phenotype enabling metastasis, they succeed in metastasizing with extraordinarily low efficiency. Such a weak connection between genotype and measurable phenotype derails most currently available experimental strategies.

2. A second dimension of complexity arises from the apparent collaboration of genetic and epigenetic factors in creating the metastatic trait. Recall, for example, that in certain experimental models of cancer, an EMT is achieved when *ras*-transformed cells are exposed to TGF-β (see Section 14.3). This transition, which is likely to operate in many human carcinomas and enable their invasiveness, can be triggered by specific signals that somatically mutated cells encounter in certain tissue microenvironments but not in others. Hence, at least in these cases, invasion and subsequent metastasis can hardly be portrayed as genetically templated traits and, for this reason, cannot be readily studied by many commonly used experimental techniques that focus on cancer cell genetics and cancer genome sequencing.

3. In many tumors, the genes and proteins that participate directly in programming invasion and metastasis may be expressed only at the invasive edges of primary tumors (see Figure 14.14), and the cancer cells in these invasive edges may represent only a minor fraction of the total neoplastic cell populations in these tumors. This organization greatly complicates experiments designed to reveal the biochemical and genetic bases of invasiveness and metastatic ability, which often rely on analyzing bulk populations of cancer cells prepared from large chunks of surgically resected tumors.

4. Carcinomas constitute the most common class of human cancers, and the neoplastic epithelial cells within these tumors may need to undergo an EMT in order to become invasive and metastatic. However, if invading carcinoma cells occasionally advance far through an EMT program and shed almost all epithelial traits, they become the proverbial "wolves in sheep's clothing," because most commonly used histological analyses are unable to distinguish these cells from the non-neoplastic mesenchymal cells of the tumor-associated stroma. (Indeed, this difficulty explains why many tumor pathologists deny the very existence of the EMT as a key process in the development of carcinoma invasiveness.)

5. Metastatic dormancy creates another experimental problem. In breast cancer patients, for example, metastases may suddenly appear as long as 20 years after the initial primary tumor has been removed. Because of this long latency period and the sheer number of micrometastases carried by many patients, it has been difficult to learn how and why only a few of them suddenly acquire the ability to mushroom into macroscopic, life-threatening tumors.

These experimental difficulties have greatly held back the progress of metastasis research, leaving many simple yet fundamental questions unanswered. For example, are there really genes that are specialized to impart an invasive or metastatic phenotype to cancer cells? And in the same vein, are there specialized metastasis suppressor genes that must be inactivated before a population of tumor cells can acquire invasive or metastatic ability? Or do the genes and proteins that affect metastasis operate as components of the regulatory circuits that we have repeatedly encountered throughout this book, namely, the circuits governed by the products of oncogenes and tumor suppressor genes?

The tissue tropisms of metastasizing cancer cells—their tendencies to colonize some but not other organs—represent the major challenge for cancer biologists studying metastasis. Figure 14.41B only begins to suggest the daunting complexity of the colonization problem. There, the tendencies of only four common cancers (those arising in the prostate, pancreas, breast, and colon) are illustrated in terms of their tendencies to form metastases in four organ sites (brain, lungs, liver, and bone marrow). In principle, each type of primary tumor cell depicted in Figure 14.41B must develop a distinct set of adaptations for each organ microenvironment in which it lands, yielding 16 distinct adaptive programs. Of course, there are far more primary tumor types that can metastasize to distant organs, and there are yet other alternative organ sites in which these tumors can found metastatic colonies; in addition, the various subtypes of primary tumors arising in an organ (for example, different subtypes of breast or lung carcinomas) may differ in their requirements for adaptation to different sites of metastatic colonization. In aggregate, these combinatorial interactions suggest the existence of many dozens, likely hundreds of distinct adaptive programs, each composed, in turn, of multiple genetic and epigenetic changes that enable one specific subtype of disseminating primary tumor cell to colonize a certain target organ.

This uncertainty is beginning to change. A powerful strategy for discovering the genes and proteins responsible for specific metastatic tropisms involves the isolation of tumor cells that show preference for colonizing a specific target organ. By retrieving already-metastasized cancer cells from that organ, propagating them *in vitro*, and injecting them into host mice, followed by another round of isolating metastatic cells from that organ, it is possible to select clones of cancer cells that stably express a highly specific tropism for that organ.

Alternatively, single-cell clones (that is, clonal cell populations, each of which is descended from a single isolated cell) can be prepared from a heterogeneous population of cells present in a human cancer cell line. The gene expression profile (see Figure 13.20A) of each clone can then be analyzed, and its tendency to form metastases in one or another target organ can be determined. This analysis can lead to the identification of genes whose expression in a cancer cell is correlated with the metastatic tropism of that cell and may even contribute causally to this behavior (**Figure 14.51**). Indeed, ectopic expression of a group of such genes in otherwise-poorly metastatic clonal cell populations can induce these cells to exhibit potent osteotropic metastasis. Such experiments also indicate that within a heterogeneous tumor cell population, various preexisting gene expression patterns can strongly influence the ability of individual cells to exhibit a variety of metastatic behaviors.

Such research does not address the clinical challenges associated with metastasis, which remain daunting. The existence of micrometastases represents a major challenge for oncologists who would like to prevent disease relapse years after the primary tumor has been eliminated. Micrometastases of less than 0.2 mm diameter may carry several hundred to several thousand cells, and their detection in an organism carrying approximately 5×10^{13} cells represents, at least at present, an almost-impossible undertaking. These micrometastases represent an ongoing threat because some of them may erupt into lethal outgrowths at an unpredictable future time.

This issue leads directly to another: Can the therapies used to treat primary tumors also be used to treat their metastatic derivatives? Or are metastatic cells so different from their progenitors in the primary tumor that they require their own customized therapies? In fact, expression array analyses indicating substantial similarity between the gene expression profiles of primary tumors and their metastatic offshoots provide some hope that metastatic cells may respond to the same therapies that succeed in destroying the primary tumors from which they derive.

To conclude, we go back to the beginning of this chapter: if, as experimental evidence increasingly shows, the epithelial–mesenchymal transition is a critical event in the acquisition of invasiveness and metastatic dissemination, and if cancer cells resurrect embryonic transcription factors to acquire malignant traits, then Lewis Wolpert's statement might require revision, in that gastrulation (and the associated EMT) might well loom as one of the most *dangerous* events in our lives!

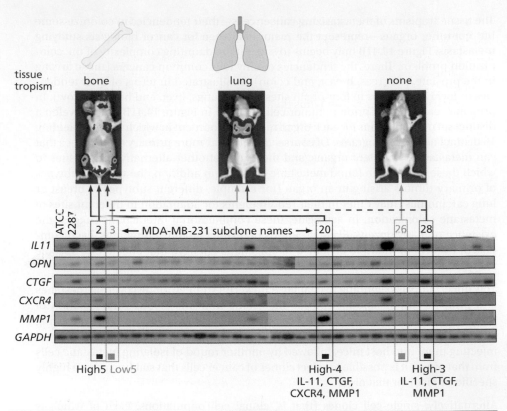

Figure 14.51 Gene expression patterns and metastatic tropism Thirty-three cells from a large population of human MDA-MB-231 breast cancer cells were each expanded into a clonal population in culture. The mRNA expression pattern of each subclone was analyzed (*columns, arrayed left to right*) using probes for the mRNAs of five genes—IL11 (interleukin-11), OPN (osteopontin), CTGF (connective tissue growth factor), CXCR4 (chemokine receptor 4), and MMP1 (matrix metalloproteinase-1)—and, as loading control, a probe for GAPDH (glyceraldehyde-3-phosphate dehydrogenase) mRNA. In addition, the expression patterns of the original tumor cell population (ATCC, *left column*) and a subcloned cancer cell population termed 2287 (which was selected for its ability to generate osteolytic metastases; *second column*) were analyzed. The five experimental genes were chosen because of their overexpression in osteotropic metastatic cells and their known biological properties in promoting osteolytic metastases. Clone 2 cells (*red box*), when injected into the arterial circulation of mice, showed a tendency to produce osteotropic metastases, as indicated by *in vivo* imaging; these cells expressed high levels of all five experimental mRNAs. Clone 3 cells (*yellow box*), in contrast, expressed low levels of all five mRNAs and preferentially formed lung metastases. And clone 26 genes (*yellow box*), which expressed essentially none of these mRNAs, formed no metastases at all. Moreover, when otherwise poorly metastatic cells were forced to express combinations of three of these genes, they acquired the ability to form bone metastases efficiently (*not shown*), pointing to the causal role of these genes in forming these metastases. Metastases were visualized through the presence of a luciferase gene in the tumor cells, which causes cells to release a bioluminescent signal. (Top, from J. Massagué. Bottom, from Y. Kang, P.M. Siegel, W. Shu et al., *Cancer Cell* 3:537–549, 2003. With permission from Elsevier.)

Key concepts

- Invasion and metastasis are responsible for 90% of cancer-associated mortality, and the majority of cancer cells at the time of death may often be found in metastases rather than the primary tumor.

- The invasion–metastasis cascade involves local invasion, intravasation, transport, extravasation, formation of micrometastases, and colonization.

- The sequence of steps in this cascade is completed only infrequently, resulting in metastatic inefficiency. The least efficient of these steps appears to be colonization.

- Many of these steps can be executed by carcinoma cells that activate a cell-biological program called the epithelial–mesenchymal transition (EMT), which is normally used by cells during embryonic morphogenesis and wound healing.

- An EMT can be programmed by pleiotropically acting transcription factors that are normally involved in organizing various steps of early embryogenesis.

- Signals released by the stromal microenvironment of a cancer cell, operating together with genetic and epigenetic alterations of the cancer cell genome, are often responsible for inducing expression of the EMT-inducing transcription factors in the cancer cell and thus the EMT.

- The EMT involves loss of an epithelial cell gene expression program and acquisition of mesenchymal gene expression. The latter enables cells to acquire invasiveness, motility, and a heightened resistance to apoptosis.

- Passage through an EMT places carcinoma cells in a state from which they can progress further to become cancer stem cells, which in turn confers on them powers that are essential for the seeding of new tumor colonies.

- Cell invasiveness is enabled by various matrix metalloproteinases (MMPs) that function to degrade components of the extracellular matrix. These enzymes are often manufactured by inflammatory cells within the tumor-associated stroma.

- Metastatic cancer cells may travel via the lymph ducts to nodes. However, their spread via the blood circulation is responsible for most distant metastases.

- Many cancer cells that are carried through the circulation form microthrombi that lodge in the arterioles and capillaries of various tissues.

- The ability of cancer cells to extravasate may depend on many of the same activities that were used earlier to execute invasiveness and intravasation.

- Although the earlier steps of the invasion–metastasis cascade are likely to be similar in various types of human tumors, the last step—colonization—is likely to depend on complex interactions that are specific to the particular type of metastasizing cells and the microenvironments of the host tissues in which they land.

- The details of colonization mechanisms are best understood in the context of osteotropic metastases, especially the osteolytic metastases initiated by breast cancer cells.

- In some cases, the metastatic tropisms of cancer cells can be explained by the organization of the circulation between the primary tumor site and the target site of metastasis. In many other cases, the reasons that cancer cells metastasize from primary tumors to certain target organs are poorly understood.

- The acquisition of invasive and metastatic powers does not appear to involve major changes in the genotype of cancer cells within the primary tumor.

Thought questions

1. What arguments can be mustered for or against the notion that invasion and metastasis are likely to be orchestrated by specific mutant alleles that are acquired by cancer cells late in tumor progression?

2. What explanations can be offered for the inefficiency of colonization by the cells within micrometastases?

3. What arguments suggest that the ability to metastasize is expressed either by the bulk of cancer cells in a primary tumor or only by a minority of cells that are specialized to do so?

4. What evidence suggests that genetic and phenotypic evolution of cancer cells can occur in sites within the body that are far removed from the primary tumor?

5. What specific types of physiological support might be supplied by tissues that are frequently sites of successful metastasis formation? In what way do these supports affect the ultimate success of the colonization process?

6. How might primary tumors exhibit metastatic powers as soon as they form?

7. Would the ability to prevent metastasis have demonstrable effects on the clinical course of some human tumors but not others?

8. What evidence supports the involvement of an EMT in human tumor pathogenesis, and what evidence argues against it?

9. How might the ability to accurately determine the prognosis of a diagnosed prostate or mammary tumor lead to dramatic changes in the practice of clinical oncology?

10. What mechanisms might be invoked to explain why large primary tumor size is often correlated with a poor prognosis of metastasis?

Additional reading

Alix-Panabières C, Riethdorf S & Pantel K (2008) Circulating tumor cells and bone marrow micrometastasis. *Clin. Cancer Res.* 14, 5013–5021.

Barrallo-Gimeno A & Nieto MA (2005) The Snail genes as inducers of cell movement and survival: implications in development and cancer. *Development* 132, 3151–3161.

Berx G, Raspé E, Christofori G et al. (2007) Pre-EMTing metastasis? Recapitulation of morphogenetic processes in cancer. *Clin. Exp. Metastasis* 24, 587–597.

Blasi F & Carmeliet P (2002) uPAR: a versatile signalling orchestrator. *Nat. Rev. Mol. Cell Biol.* 3, 932–943.

Boccaccio C & Comoglio PM (2006) Invasive growth: a *MET*-driven genetic programme for cancer and stem cells. *Nat. Rev. Cancer* 6, 637–645.

Borovski T, De Sousa e Melo F, Vermeulen L & Medema JP (2011) Cancer stem cell niche: the place to be. *Cancer Res.* 71, 634–639.

Brabletz T, Jung A, Spaderna SK et al. (2005) Opinion: migrating cancer stem cells—an integrated concept of malignant tumour progression. *Nat. Rev. Cancer* 5, 744–749.

Brabletz T, Lyden D, Steeg PS & Werb Z (2013) Roadblocks to translational advances on metastasis research. *Nat. Med.* 19, 1104–1109.

Bracken CP & Goodall GJ (2021) The many regulators of epithelial–mesenchymal transition. *Nat. Rev. Mol. Cell Biol.* December 2021.

Braun S, Pantel K, Muller P et al. (2000) Cytokeratin-positive cells in the bone marrow and survival of patients with Stage I, II or III breast cancer. *N. Engl. J. Med.* 342, 525–533.

Burridge K & Wennerberg K (2004) Rho and Rac take center stage. *Cell* 116, 167–179.

Chambers AF, Groom AC & MacDonald IC (2002) Dissemination and growth of cancer cells in metastatic sites. *Nat. Rev. Cancer* 2, 563–572.

Chiang AC & Massagué J (2008) Molecular basis of metastasis. *N. Engl. J. Med.* 359, 2814–2823.

Christofori G (2006) New signals from the invasive front. *Nature* 441, 444–450.

Coleman RE, Croucher PI, Padhani AR et al. (2020) Bone metastases. *Nat. Rev. Dis. Primers* 6, 82.

Condeelis J & Pollard JW (2006) Macrophages: obligate partners for tumor cell migration, invasion, and metastasis. *Cell* 124, 263–266.

Condeelis J, Singer RH & Segall JE (2005) The great escape: when cancer cells hijack the genes for chemotaxis and motility. *Annu. Rev. Cell Dev. Biol.* 21, 695–718.

Cristofanilli M & Braun S (2010) Circulating tumor cells revisited. *J.A.M.A.* 303, 1092–1093.

De Wever O, Demetter P, Mareel M & Bracke M (2008) Stromal myofibroblasts are drivers of invasive cancer growth. *Int. J. Cancer* 123, 2229–2238.

Drasin DJ, Robin TP & Ford HL (2011) Breast cancer epithelial-to-mesenchymal transition: examining the functional consequences of plasticity. *Breast Cancer Res.* 13, 266.

Dykxhoorn DM (2010) MicroRNAs and metastasis: little RNAs go a long way. *Cancer Res.* 70, 6401–6406.

Fidler IJ (2003) The pathogenesis of cancer metastasis: the "seed and soil" hypothesis revisited. *Nat. Rev. Cancer* 3, 453–458.

Fodde R & Brabletz T (2007) Wnt/β-catenin signaling in cancer stemness and malignant behavior. *Curr. Opin. Cell Biol.* 19, 150–158.

Friedl P & Gilmour D (2009) Collective cell migration in morphogenesis, generation and cancer. *Nat. Rev. Mol. Cell Biol.* 10, 445–457.

Friedl P & Mayor R (2017) Tuning collective cell migration by cell–cell junction regulation. *Cold Spring Harb. Perspect. Biol.* 9, a029199.

Friedl P & Wolf K (2009) Plasticity of cell migration: a multiscale tuning model. *J. Cell Biol.* 188, 11–19.

Gertler F & Condeelis J (2011) Metastasis: tumor cells becoming MENAcing. *Trends Cell Biol.* 21, 81–90.

Gregory PA, Bracken CP, Bert AG & Goodall GJ (2008) MicroRNAs as regulators of epithelial-mesenchymal transition. *Cell Cycle* 7, 3112–3118.

Güc E & Pollard JW (2021) Redefining macrophage and neutrophil biology in the metastatic cascade. *Immunity* 54, 885–902.

Gupta GP & Massagué J (2006) Cancer metastasis: building a framework. *Cell* 127, 679–695.

Haber DA and Velculescu VE (2014) Blood-based analyses of cancer: circulating tumor cells and circulating tumor DNA. *Cancer Discov.* 4, 650–661.

Hay ED (2005) The mesenchymal cell, its role in the embryo, and the remarkable signaling mechanisms that create it. *Dev. Dyn.* 3, 706–720.

Heasman SJ & Ridley AJ (2008) Mammalian Rho GTPases: new insights into their functions from *in vivo* studies. *Nat. Rev. Mol. Cell Biol.* 9, 690–701.

Heldin C-H, Landström M & Moustakas A (2009) Mechanism of TGF-β signaling to growth arrest, apoptosis, and epithelial–mesenchymal transition. *Curr. Opin. Cell Biol.* 21, 1–11.

Huber MA, Kraut N & Beug H (2005) Molecular requirements for epithelial-mesenchymal transition during tumor progression. *Curr. Opin. Cell Biol.* 17, 548–556.

Hugo H, Ackland ML, Blick T et al. (2007) Epithelial-mesenchymal and mesenchymal-epithelial transitions in carcinoma progression. *J. Cell Physiol.* 213, 374–383.

Ilina O & Friedl P (2009) Mechanisms of collective cell migration at a glance. *J. Cell Sci.* 122, 3203–3208.

Jin X, Demere Z, Nair K et al. (2020) A metastasis map of human cancer cell lines. *Nature* 588, 331–336.

Joyce JA & Pollard JW (2009) Microenvironmental regulation of metastasis. *Nat. Rev. Cancer* 9, 239–252.

Kai F, Drain AP & Weaver VM (2019) The extracellular matrix modulates the metastatic journey. *Dev. Cell* 49, 332–346.

Kedrin D, van Rheenen J, Hernandez L et al. (2007) Cell motility and cytoskeletal regulation in invasion and metastasis. *J. Mammary Gland Biol. Neoplasia* 12, 143–152.

Kingsley LA, Fournier PGJ, Chirgwin JM & Guise TA (2007) Molecular biology of bone metastasis. *Mol. Cancer Ther.* 6, 2609–2617.

Klein CA (2020) Cancer progression and the invisible phase of metastatic colonization. *Nature Rev. Cancer* 20, 681–694.

Lamouille S, Xu J & Derynck R (2014) Molecular mechanisms of epithelial-mesenchymal transition. *Nat. Rev. Mol. Cell Biol.* 15, 178–196.

Liu Y & Cao X (2016) Characteristics and significance of the pre-metastatic niche. *Cancer Cell* 30, 668–681.

Lu X & Kang Y (2010) Hypoxia and hypoxia-inducible factors, master regulators of metastasis. *Clin. Cancer Res.* 16, 5928–5935.

Mehra N, Zafeiriou Z, Lorente D et al. (2015) CCR 20th anniversary commentary: circulating tumor cells in prostate cancer. *Clin. Cancer Res.* 21, 4992–4995.

Micalizzi DS, Maheswaran S & Haber DA (2017) A conduit to metastasis: circulating tumor cell biology. *Genes Dev.* 31, 1827–1840.

Mundy GR (2002) Metastasis to bone: causes, consequences, and therapeutic opportunities. *Nat. Rev. Cancer* 2, 584–593.

Munro MJ, Wickremesekera SK, Peng L et al. (2018) Cancer stem cells in colorectal cancer: a review. *J. Clin. Pathol.* 71, 110–116.

Murphy DA & Courtneidge SA (2011) The "ins" and "outs" of podosomes and invadopodia: characteristics, formation and function. *Nat. Rev. Mol. Cell Biol.* 12, 413–426.

Nguyen DX, Bos PD & Massagué J (2009) Metastasis: from dissemination to organ-specific colonization. *Nat. Rev. Cancer* 9, 274–284.

Nieto MA (2002) The Snail superfamily of zinc-finger transcription factors. *Nat. Rev. Mol. Cell Biol.* 3, 155–166.

Nieto MA (2011) The ins and outs of the epithelial to mesenchymal transition in health and disease. *Annu. Rev. Cell Dev. Biol.* 27, 347–376.

Nieto MA (2013) Epithelial plasticity: a common theme in embryonic cells and cancer cells. *Science* 342, 1234850.

Nieto MA, Huang RYJ, Jackson RA & Thiery JP (2016) EMT:2016. *Cell* 166, 21–45.

Odenthal J, Takes R & Friedl P (2016) Plasticity of tumor cell invasion: governance by growth factors and cytokines. *Carcinogenesis* 37, 1117–1128.

Oudin MJ & Weaver VM (2016) Physical and chemical gradients in the tumor microenvironment regulate tumor cell invasion, migration, and metastasis. *Cold Spring Harb. Symp. Quant. Biol.* 81, 189–205.

Page-McCaw A, Ewald AJ & Werb Z (2007) Matrix metalloproteinases and the regulation of tissue remodeling. *Nat. Rev. Mol. Cell Biol.* 8, 221–233.

Paget S (1889) The distribution of secondary growths in cancer of the breast (re-publication of 1889 *Lancet* article). *Cancer Metastasis Rev.* 8, 98–101.

Pandya P, Orgaz JL & Sanz-Moreno V (2017) Actomyosin contractility and collective migration: may the force be with you. *Curr. Opin. Cell Biol.* 48, 87–96.

Pantel K & Alix-Panabieres K (2019) Liquid biopsy and minimal residual disease—latest advances and implications for cure. *Nat. Rev. Clin. Oncol.* 16, 409–424.

Pari AAA, Singhal M & Augustin H (2021) Emerging paradigms in metastasis research. *J. Exp. Med.* 218, e20190218.

Pollard JW (2004) Tumour-educated macrophages promote tumour progression and metastasis. *Nat. Rev. Cancer* 4, 71–78.

Sahai E (2007) Illuminating the metastatic process. *Nat. Rev. Cancer* 7, 737–749.

Savagner P (2014) Epithelial-mesenchymal transitions: from cell plasticity to concept elasticity. *Curr. Topics Devel. Biol.* 112, 273–300.

Sekiguchi R & Yamada KM (2018) Basement membranes in development and disease. *Curr. Top. Dev. Biol.* 130, 143–191.

Shamir ER & Ewald AJ (2014) Three-dimensional organotypic culture: experimental models of mammalian biology and disease. *Nat. Rev. Mol. Cell Biol.* 15, 647–664.

Skrypek N, Goossens S, De Smedt E et al. (2017) Epithelial-to-mesenchymal transition: epigenetic reprogramming driving cellular plasticity. *Trends Genet.* 33, 943–959.

Stacker SA, Achen MG, Jussila L et al. (2002) Lymphangiogenesis and cancer metastasis. *Nat. Rev. Cancer* 2, 573–583.

Svitkina T (2018) The actin cytoskeleton and actin-based motility. *Cold Spring Harb. Perspect. Biol.* 10, a018267.

Thiery JP, Acloque H, Huang RYG & Nieto MA (2009) Epithelial-mesenchymal transitions in development and disease. *Cell* 139, 871–890.

Thiery JP & Sleeman JP (2006) Complex networks orchestrate epithelial-mesenchymal transitions. *Nat. Rev. Mol. Cell Biol* 7, 131–142.

Turajlic S & Swanton C (2016) Metastasis as an evolutionary process. *Science* 352, 169–175.

van de Stolpe A, Pantel K, Sleijfer S et al. (2011) Circulating tumor cell isolation and diagnostics: toward routine clinical use. *Cancer Res.* 71, 5955–5960.

van Roy F (2014) Beyond E-cadherin: roles of other cadherin superfamily members in cancer. *Nat. Rev. Cancer* 14, 121–134.

Vignjevic D & Montagnac G (2008) Reorganisation of the dendritic actin network during cancer cell migration and invasion. *Semin. Cancer Biol.* 18, 12–22.

Wang X & Thiery JP (2021) Harnessing carcinoma cell plasticity mediated by TGF-β signaling. *Cancers* 13, 3397.

Weigelt B, Peterse JL & van't Veer LJ (2005) Breast cancer metastasis: markers and models. *Nat. Rev. Cancer* 5, 591–602.

Weilbaecher K, Guise TA & McCauley LK (2011) Cancer to bone: a fatal attraction. *Nat. Rev. Cancer* 11, 411–425.

Wendt MK, Tian M & Schiemann WP (2012) Deconstructing the mechanisms and consequences of TGF-β-induced EMT during cancer progression. *Cell Tissue Res.* 347, 85–101.

Wirtz D, Konstantopoulos K & Searson PC (2011) The physics of cancer: the role of physical interactions and mechanical forces in metastasis. *Nat. Rev. Cancer* 11, 512–522.

Wren E, Huang Y & Cheung K (2021) Collective metastasis: coordinating the multicellular voyage. *Clin. Exptl. Metastasis* 38, 373–399.

Yang J, Antin P, Berx G et al. (2020) Guidelines and definitions for research on epithelial-mesenchymal transition. *Nat. Rev. Mol. Cell Biol.* 21, 341–352.

Yang J & Weinberg RA (2008) Epithelial-mesenchymal transition: at the crossroads of development and tumor metastasis. *Dev. Cell* 14, 818–829.

Yeung KT & Yang J (2017) Epithelial-mesenchymal transition in tumor metastasis. *Mol. Oncol.* 11, 28–39.

Yilmaz M & Christofori G (2009) EMT, the cytoskeleton, and cancer cell invasion. *Cancer Metastasis Rev.* 28, 15–33.

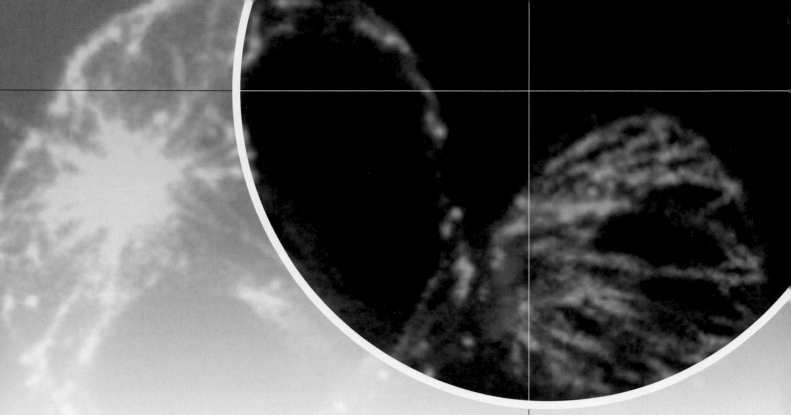

Chapter 15

Crowd Control: Tumor Immunology

> It is by no means inconceivable that small accumulations of tumour cells may develop and, because of their possession of new antigenic potentialities, provoke an effective immunological reaction with regression of the tumour and no clinical hint of its existence.
>
> Macfarlane Burnet, immunologist, 1957

Throughout this text, we have studied various defenses that the body erects against the appearance of cancerous growths. Many of these defenses are inherent in mammalian cells, more specifically in the hard-wired regulatory circuitry operating within individual human cells, and can influence the decision of a cancer cell to commit to division or even undergo programmed cell death. The most obvious of these are the controls imposed on cells by apoptosis or by the mechanisms of other death programs poised to trigger the death of cells that are misbehaving or, alternatively, suffering certain types of damage or physiological stress. Similarly, the pRb and p53 signaling circuits (see Chapters 8 and 9) are configured to impede the outgrowth of incipient cancer cells.

Even the organization of some tissues places constraints on how and where incipient cancer cells can proliferate. For example, normal epithelial cells that lose their tethering to the basement membrane activate the form of apoptosis called anoikis. This mechanism limits the ability of epithelial cells to stray from their normal locations within tissues and grow in ectopic (that is, abnormal) sites. Beyond these cell- and tissue-specific mechanisms, there is another line of defense: the immune system.

The powers of the immune system to detect and eliminate infectious agents have been known for well over a hundred years. In the 1950s, Macfarlane Burnet and Lewis Thomas independently speculated that the immune system can recognize tumor cells as altered forms of the normal cells in our tissues that, in the context of immunology, we hereafter will term **self**; having done so, the immune system would proceed to recognize and eliminate the tumor cells as foreign, unnatural entities that we will term **non-self**. This hypothesized process was codified as **immunosurveillance**, which implies that the immune system continuously monitors tissues for the presence of aberrant cells, including cancer cells, recognizes them, and quickly eliminates them. Subsequent work has confirmed the validity of this fertile but, until recently, speculative hypothesis. Indeed, it is now clear that the immune system can distinguish tumor cells from normal ones and can mount responses to eliminate them. This insight reveals why so many anti-cancer immune therapies have emerged in recent years from efforts to harness the immune system to treat and manage a variety of cancers.

Still, the immune system frequently fails to enforce effective and sustained crowd control over cancer cells, as evidenced by the approximately 19.3 million new cases of cancer worldwide in 2021 and the fact that cancer results in 1 in 5 deaths annually in the United States. Looking beyond these failures, in the next chapter we will see how the body's multi-layered immune system's diverse network of interacting defenses has been marshalled and manipulated to treat cancer, offering multiple opportunities for novel anti-cancer therapies.

Since the immune system has evolved to recognize and eliminate foreign agents from the body while leaving the body's own tissues unmolested, its purported ability to recognize and attack cancer cells long seemed implausible. As we have read repeatedly throughout this book, cancer cells are native to the body and are therefore, in so many respects, indistinguishable from the body's normal cells. In fact, considerable progress has been made toward determining precisely how the immune system distinguishes cancer cells as different from their normal counterparts and, as a consequence, recognizes these cells as appropriate targets of immune-mediated killing. This understanding has provided approaches for harnessing the immune system to effectively target cancer cells and given rise to the burgeoning field of cancer immunotherapy.

15.1 The immune system continuously conducts surveillance of tissues

Early insights in the area of tumor immunology research came from attempts to implant cells from a variety of sources into mouse hosts. Normal human cells were rapidly eliminated, and even normal cells from one strain of inbred mice were equally unable to gain a footing when implanted in the tissues of mice belonging to another inbred strain. Thus, these cells were **rejected** by the host tissues in which they were implanted, ostensibly by the immune systems of the hosts.

These now-ancient findings established a principle that is central to all that follows in this chapter: the immune system can recognize certain molecules displayed on the surface of certain implanted cells that somehow provide a clear and unambiguous signal that can provoke an attack on these cells, often leading to their elimination. As we will see, identification of the precise nature of these cell surface molecules and their recognition by the immune system required decades of research.

The observed rejections of implanted cells in mouse hosts were accompanied by the realization that the immune system usually tolerates the body's own cells, as well as implanted cells prepared from genetically identical mice, that is, cells from the same inbred strain of mice. Such **tolerance** implied an ability to recognize the "familiar," distinguishing it from the "unfamiliar and foreign." (Cells originating in mice of one inbred strain and implanted in mouse hosts of the same strain are said to be **syngeneic** with respect to these hosts.)

Thus, the major problem of anti-cancer immunosurveillance became clear: cancer cells originate from normal cells in the body's own familiar tissues. Given the immune system's ability to tolerate normal tissues (that is, the state of immunological tolerance), its ability to recognize the foreignness of cancer cells became difficult to explain because it was widely assumed that cancer cells were essentially identical to normal cells in the body.

Figure 15.1 Immunization of mice by exposure to killed cancer cells Mice from one strain were initially injected with irradiated, killed cancer cells (*red*) derived from a tumor that had been chemically induced in a mouse of the same strain (i.e., a syngeneic host). When these mice were subsequently injected with live cells prepared from the same tumor, the cells failed to grow (*lower left*). However, when these mice were injected with live cells from a second, independently induced tumor (*blue*), the cells proliferated and formed a tumor mass (*lower right*). The reciprocal experiment (*not shown*) yields the converse results, i.e., implantation of killed blue cancer cells subsequently rendered mice immune to the blue tumor but not to the red tumor. Because these experiments were all performed using tumor cells and mice from the same genetically defined strain, the rejection of these tumors could not be attributed to histoincompatibility, that is, to the rejection of allogeneic cells. Instead, these experiments demonstrated that the chemically induced tumor cells had developed distinctive antigens, apparently acquired during the chemical carcinogenesis (and associated intense mutagenesis) that led to their transformation. These results suggested, in turn, that cancer cells can express mutant proteins that may be recognized as foreign by the immune systems of tumor-bearing hosts.

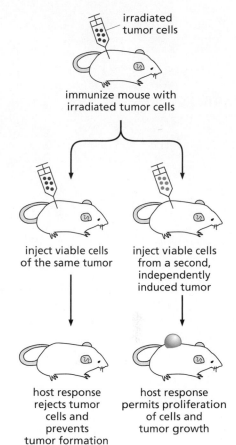

irradiated
tumor cells

immunize mouse with
irradiated tumor cells

inject viable cells
of the same tumor

inject viable cells
from a second,
independently
induced tumor

host response
rejects tumor
cells and
prevents
tumor formation

host response
permits proliferation
of cells and
tumor growth

In fact, a small number of early studies in mice provided evidence that the immune system can indeed play a role in the immunosurveillance of tissues directed specifically toward the detection of aberrant cells including, notably, cancer cells. An important line of investigation relied on observations showing that certain chemically induced tumors (see Sections 2.9 and 11.11) in mice were **immunogenic**, that is, capable of provoking an immune response, and therefore could be recognized as foreign, making them vulnerable to elimination by the immune systems of the host mice bearing these tumors.

In one set of experiments outlined in **Figure 15.1**, cells from a 3-methylcholanthrene (3-MC)–induced tumor that arose in one strain of mice (see Section 11.11) were irradiated prior to implantation in mouse hosts of the same inbred strain. The irradiation served to prevent the proliferation of these cells and the resulting rapid outgrowth of life-threatening tumors in the hosts while, at the same time, permitting these cells to continue expressing certain immunogenic proteins. Because these cancer cells were prepared from tumors that had been induced in the same strain of mice as the recipient hosts, the implanted cancer cells did not elicit an immediate aggressive response.

Subsequently, mice previously implanted with the syngeneic, irradiated tumor cells received a second injection of live tumor cells originating from the same chemically induced tumor or, alternatively, from a different, independently induced 3-MC tumor. As before, all implanted cells were of syngeneic origin. As it turned out, the second set of implanted cancer cells originating from the original chemically induced tumor did not grow out and were ostensibly being attacked by the immune systems of these mouse hosts, that is, they were rejected. In contrast, the cells derived from another, independently induced tumor could indeed grow and form new tumors in the recipients.

These experiments left two lessons. First, under certain conditions, implanted cancer cells could indeed evoke long-term immunity; this response appeared to be similar to the one achieved by vaccinating a host against one or another type of infectious agent, notably some type of virus. Second, somehow, tumor cells originating from one chemically induced tumor could be readily distinguished by the immune system from cells prepared from another independently induced tumor. Thus, the initial exposure to dead cancer cells from one tumor had immunized the mice against live cells originating from the same tumor, while failing to elicit some type of immunity against the second, independently arising tumor.

These observations led, in turn, to the notion that the neoplastic cells in different independently arising tumors exhibit distinct sets of **antigens**, that is, molecules with the potential to provoke an immune response. Exposure to a particular set of such antigens displayed by one type of cancer cell could apparently stimulate the immune system to remember a prior encounter with cells displaying that particular set of antigens, recognize these cells, and kill them. Conversely, such immune-mediated protection did not arise when the prior encounter involved cells from a different, independently formed tumor. As described below, these "antigens" are almost invariably different proteins.

15.2 The human immune system plays a critical role in warding off various types of human cancer

Because the biology of mice and humans differs in so many respects, we need to interpret the results described above with caution when attempting to understand the role of the human immune system in defending us against cancer. In addition, it is possible that the chemical carcinogens used to induce mouse cancers may well create tumors that are far more capable of provoking a strong immune response than are provoked by spontaneously arising human tumors. In fact, there is compelling evidence that the human immune system can indeed recognize cancer cells arising in the body of an individual who has not been exposed to potent carcinogens and that the actions of our immune systems do indeed play an important role in warding off the outgrowth of some tumors. As we will see, this evidence is drawn from a diverse set of observations, including the elevated cancer risk borne by **immunocompromised** individuals, the presence of tumor-infiltrating lymphocytes across many different types of cancers, and the clinical successes achieved by targeting tumors with anti-cancer immune-based therapies that will be described in the next chapter.

To begin, the study of cancer incidence in immunocompromised individuals reveals that dysfunctional or suppressed immune systems are indeed associated with an elevated risk for many types of cancer. Generally, the immunocompromised state responsible for this elevation arises from one of three sources. First, some individuals are born with certain germ-line mutations that render their immune systems dysfunctional. With improvements in hospital care, these immunocompromised individuals now live longer but still experience a reduced life expectancy. Second, in 2019, nearly 38 million people worldwide were infected by human immunodeficiency virus (HIV), which can lead to **acquired immunodeficiency syndrome** (AIDS). Those afflicted with AIDS have increased susceptibility to **opportunistic infections** by various tumor viruses, resulting in higher rates of cancer incidence. Third, organ transplantation, involving kidneys, hearts, and livers, has become common throughout the developed world. Because the transplanted organs are typically derived from donors who are genetically distinct from the transplant recipients (that is, **allogeneic**), the transplanted organs are recognized as foreign and thus subject to rejection by the immune systems of engrafted patients; such donor grafts are termed **histoincompatible** (literally, containing tissues that are incompatible with the tissues of graft recipients). To avoid rejection of the transplant by the recipients' immune systems, these patients are routinely rendered immunocompromised by treatment with several types of potent immunosuppressive drugs and, as an unintended consequence, they experience markedly elevated rates of cancer incidence. This finding suggests that the mechanisms that allow the immune system to recognize implanted allogeneic cells (that is, cells derived from the tissues of genetically non-identical members of the same species) overlap with the mechanisms employed in the body to recognize incipient cancer cells (**Figure 15.2**).

At this stage in our discussion, the processes of anti-cancer immune surveillance must be divided into two classes: tumors that are of viral etiology and those that are not. Virus-induced malignancies (see Chapter 3) present clear biochemical evidence of their viral origin, because tumors that have been triggered by tumor virus infections (such as viruses of the herpesvirus and papillomavirus classes) almost always continue to express virus-encoded oncoproteins during the course of tumor outgrowth. These viral proteins are structurally distinct from normal host proteins and are therefore intrinsically capable of being recognized as foreign by the host immune system; the latter can readily respond by mounting immune responses against them, more specifically against the cancer cells that express them. These facts, on their own, explain how the cells of virus-induced malignancies can attract the attention of the immune system, which may in turn be able to control their outgrowth, and how failure of this control accounts for the 20% of cancer mortality worldwide that derives from this class of human tumors.

(A) cancers not associated with infectious agents

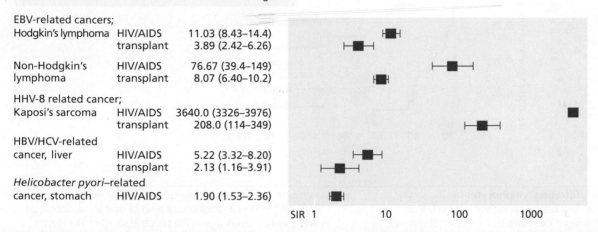

breast	HIV/AIDS	1.03 (0.89–1.20)
	transplant	1.15 (0.98–1.36)
prostate	HIV/AIDS	0.70 (0.55–0.89)
	transplant	0.97 (0.78–1.19)
colon and rectum	HIV/AIDS	0.92 (0.78–1.08)
	transplant	1.69 (1.34–2.13)
ovary	HIV/AIDS	1.63 (0.95–2.80)
	transplant	1.55 (0.99–2.43)
trachea, bronchus,	HIV/AIDS	2.72 (1.91–3.87)
and lung	transplant	2.18 (1.85–2.57)

(B) cancers known to be associated with infectious agents

EBV-related cancers;		
Hodgkin's lymphoma	HIV/AIDS	11.03 (8.43–14.4)
	transplant	3.89 (2.42–6.26)
Non-Hodgkin's	HIV/AIDS	76.67 (39.4–149)
lymphoma	transplant	8.07 (6.40–10.2)
HHV-8 related cancer;		
Kaposi's sarcoma	HIV/AIDS	3640.0 (3326–3976)
	transplant	208.0 (114–349)
HBV/HCV-related		
cancer, liver	HIV/AIDS	5.22 (3.32–8.20)
	transplant	2.13 (1.16–3.91)
Helicobacter pyori–related		
cancer, stomach	HIV/AIDS	1.90 (1.53–2.36)

Figure 15.2 Effects of compromised immune systems on cancer incidence This meta-analysis pooled the conclusions of a group of epidemiologic studies in order to calculate the standardized incidence ratio (SIR)—the number of cancer cases actually observed in populations of HIV/AIDS patients and immunosuppressed individuals divided by the number expected in the age-matched general population. Note that the SIR values are plotted logarithmically on the abscissa. All patients in the transplant category were presumed to be subject to significant immunosuppression in order to prevent rejection of the transplanted organ. The numbers associated with each entry indicate the calculated mean SIR, followed by the confidence interval (which shows, with 95% probability, the interval within which the measured mean value will fall). Similarly, in these "box-and-whisker" plots (*right*), the filled boxes are centered on the median measurements, and the outer points of the whiskers indicate the bounds of the 95% confidence interval. (A) Among a group of five commonly occurring epithelial cancers, the incidence of breast, prostate, and colorectal carcinomas was not elevated in HIV-infected/AIDS patients relative to the general population, whereas carcinomas of the ovary and lungs were modestly elevated and at a statistically significant level. (B) Strikingly different values are seen, however, among a range of tumors that are known to be associated with either chronic viral or bacterial infections, where the lack of a fully functional immune system often leads to dramatic increases in tumor incidence. Included here are EBV-related cancers, non-Hodgkin lymphomas, HHV-8-induced Kaposi's sarcomas, HBV/HCV-related liver cancers, and *Helicobacter pylori*-associated stomach cancers. (EBV, Epstein–Barr virus; HHV-8, human herpesvirus-8; HBV and HCV, hepatitis viruses B and C.) (From A.E. Grulich et al., *Lancet* 370:59–67, 2007. With permission from Elsevier.)

A more challenging task is understanding how the immune system can recognize and eliminate *nonviral* malignancies (that is, those not induced by infecting tumor viruses), and whether failure to do so results in the tumors responsible for nearly 80% of cancer-associated mortality. Here, the major difficulty derives from the phenomenon of tolerance described above—from the fact that the immune system's attentions are directed away from attacking the body's own tissues (that is, self) and thus away from the cancer cells that are, to a large extent, very similar to the cells in our normal tissues.

As we will see in more detail below, many of the actions of the immune system are mediated by a class of immune cells (**immunocytes**) collectively termed **lymphocytes**, which are responsible for organizing attacks against tumor cells. Much of the evidence supporting a role of lymphocyte-mediated immune surveillance in warding

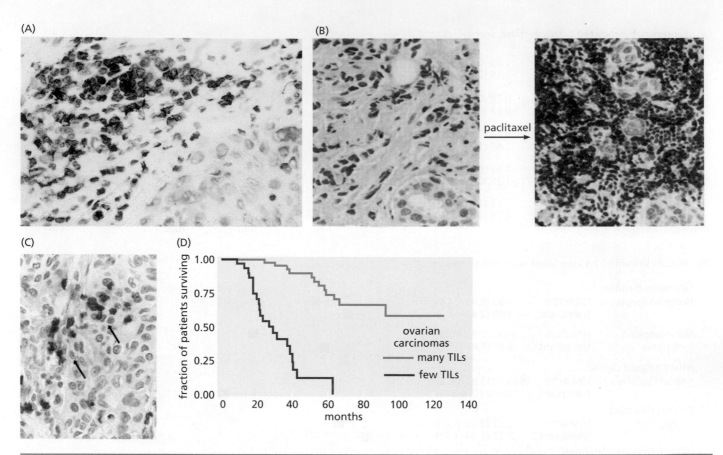

Figure 15.3 Tumor-infiltrating lymphocytes (A) This immunostaining of an oral carcinoma with an antibody that recognizes the CD3 antigen expressed by T lymphocytes reveals an abundance of tumor-infiltrating lymphocytes (TILs; *brown*) in certain areas of the tumor. More detailed characterizations revealed several T-lymphocyte subtypes among these cells (*not shown*). (B) This immunostaining demonstrates that TILs, detected once again with an anti-CD3 antibody (*dark purple*), are relatively rare in an untreated breast tumor (*left*) but become abundant in areas of the tumor following chemotherapy with the drug paclitaxel (*right*). The increased presence of TILs following chemotherapy is assumed to be indicative of an immune response mediated by the immigrant TILs. (C) TILs are also frequently found in invasive non-small-cell lung carcinomas (NSCLCs; *arrows*). The expression here of the CD8 antigen (*dark pink*) indicates that these cells are largely cytotoxic lymphocytes (CTL). (D) The clinical prognosis of a set of ovarian carcinoma patients was strongly correlated with the concentrations of TILs in their tumors. In this Kaplan–Meier plot, the proportion of patients surviving after initial diagnosis (*ordinate*) is plotted versus the months of survival (*abscissa*). Those patients whose tumors had high levels of TILs (*blue line*) fared significantly better than did those whose tumors lacked significant concentrations of TILs (*red line*). (A, from T.E. Reichert et al., *Clin. Cancer Res.* 8:3137–3145, 2002. B, from S. Demaria et al., *Clin. Cancer Res.* 7:3025–3030, 2001. Both with permission from American Association for Cancer Research.C, from A. Trojan et al., *Lung Cancer* 44:143–147, 2004. With permission from Elsevier. D, adapted from L. Zhang et al., *N. Engl. J. Med.* 348:203–213, 2003. With permission from Massachusetts Medical Society.)

off non-virus-induced tumors comes from the histopathological studies of a broad spectrum of human tumors. These neoplasms often contain large numbers of tumor-infiltrating lymphocytes (TILs; **Figure 15.3**), providing a clear indication that immune recognition and associated recruitment of these immune cells into the interstices of tumors has occurred. Importantly, this recognition, on its own, does not guarantee a successful attack on and elimination of tumor cells. Instead, it only suggests that such recognition has taken place, setting the stage for possible attack on the associated tumor masses by the recruited immune cells.

In fact, the presence of TILs in a human tumor usually tells us more: it provides a strong indication that the immune system may be actively launching attacks against neoplastic cells. Such attacks may often succeed in constraining tumor outgrowth, but by definition, they fail to eliminate the tumor, as evidenced by the continued presence of a large, clinically detectable mass in a patient's body. As an example, clinical observations on ovarian carcinoma patients have demonstrated that the presence of large numbers of TILs in their tumors prior to **debulking** of the tumor through surgery and chemotherapy is associated with dramatically better survival outcomes compared to patients whose ovarian tumors lacked TILs (Figure 15.3D). Similar outcomes are

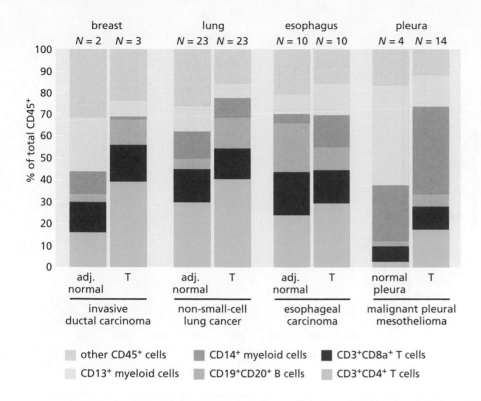

breast   lung   esophagus   pleura
$N = 2$ $N = 3$ $N = 23$ $N = 23$ $N = 10$ $N = 10$ $N = 4$ $N = 14$

% of total CD45$^+$

| | adj. normal | T |
| invasive ductal carcinoma | | |

| | adj. normal | T |
| non-small-cell lung cancer | | |

| | adj. normal | T |
| esophageal carcinoma | | |

| | normal pleura | T |
| malignant pleural mesothelioma | | |

■ other CD45$^+$ cells   ■ CD14$^+$ myeloid cells   ■ CD3$^+$CD8a$^+$ T cells

■ CD13$^+$ myeloid cells   ■ CD19$^+$CD20$^+$ B cells   ■ CD3$^+$CD4$^+$ T cells

Figure 15.4 Tumor-specific infiltration by immunocytes Analysis of immunocyte populations present in different tumors shows dramatic differences among the immune cells recruited by these various tumors. In the analyses presented here, normal and tumor tissue samples excised from patients were enzymatically digested; single cells in the resulting suspensions were then detected with dye-labeled monoclonal antibodies and analyzed by flow cytometry. Only those cells expressing CD45, an antigen present on all leukocytes, were included in these analyses. The number (*N*) of each tumor type and corresponding normal tissue samples analyzed is given above the bars; in each case, normal adjacent tissue (adj. normal) was analyzed in parallel with tumor samples (T) with the aid of fluorescently labeled antibodies specific for each of the indicated displayed cell-surface antigens: CD3, a component of the TCR complex found on all T cells; CD4, helper T cells, and regulatory T cells; CD8, cytotoxic T lymphocytes; CD13, many myeloid cell types, including granulocytes and monocytes; CD14, monocytes and macrophages; CD19, B cells; CD20, B cells. (Courtesy of L.M. Coussens.)

associated with the clinical courses of melanoma and colorectal carcinoma patients whose tumors carry large numbers of TILs; in the case of melanomas, these patients live 1.5 to 3 times longer following diagnosis than do patients lacking comparable populations of TILs in their tumors. Similar correlations have been made between the presence of these infiltrating lymphocytes and the survival of patients bearing carcinomas of the breast, bladder, prostate, and rectum. Still, such observations, although striking, are only correlative, and therefore do not prove that these lymphocytes are actually important agents responsible for launching attacks on tumors, thereby constraining tumor growth and the progression to life-threatening cancers.

The identities of the immune cells that are attracted into a growing tumor mass are actually quite complex, involving multiple distinct types of immunocytes (**Figure 15.4**). The composition of these immunocytes differs greatly when comparing individual tumors with one another. Such heterogeneity complicates attempts at drawing generalizable conclusions about the role of the immune system in controlling all types of cancer. Moreover, each class of immunocytes may be represented by multiple subtypes that can play distinct, even conflicting roles in fostering or suppressing tumor growth. In some cases, the presence of specific subtypes of immune cells is closely associated with the stage of tumor progression, providing strong but still-correlative evidence of a causal role in controlling or failing to control clinical behavior (**Figure 15.5**).

The evidence presented in Figure 15.5 points to apparently important roles of subsets of T lymphocytes in controlling colorectal carcinoma progression; similar evidence exists for other carcinoma types as well. In addition, the phenotypic state of the cancer cells within a tumor may exert strong influence on the spectrum of immunocytes in its stroma. As one example, the carcinoma cells within a tumor will attract quite different groups of immunocytes, depending on whether these individual neoplastic cells reside in the epithelial or mesenchymal phenotypic states that are described in Section 13.3 and, in more detail, in Chapter 14.

As we will see in this chapter, considerable progress has been made in uncovering some of the mechanisms by which the immune system recognizes and responds to, or fails to respond to, cancer. At present, this knowledge is still incomplete and the underlying mechanisms remain imperfectly understood.

(A)

marker genes of
T$_H$1 adaptive immunity

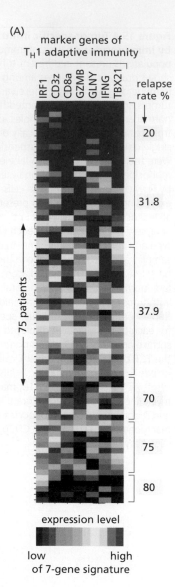

expression level

low   high
of 7-gene signature

(B)

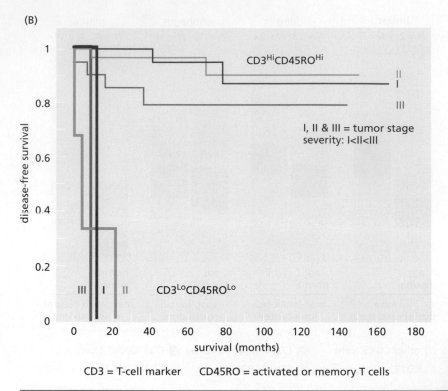

CD3 = T-cell marker   CD45RO = activated or memory T cells

Figure 15.5 T-cell infiltration and the course of colorectal carcinoma progression
(A) The clinical course of colorectal (CRC) tumor progression was monitored in a group of 75 patients over periods up to 15 years and correlated with the presence of specific subtypes of immunocytes in their tumors. Whole tumors, encompassing both their epithelial and stromal compartments, were analyzed for their patterns of RNA expression, using qRT-PCR. An expression signature composed of seven genes (*top*) was generated in order to gauge specifically the representation of immune cells that had infiltrated colorectal carcinomas and also expressed genes of the T$_H$1 subset of helper T cells, specifically memory T cells. These cells persist for many years after exposure to antigen and can be reactivated quickly in response to a subsequent challenge by this antigen. Those patients whose tumors carried many cells having high expression of this set of T$_H$1 markers (*top*) experienced a vastly lower rate (20%, *right side*) of post-surgical relapse (generally liver metastases) than did those whose tumors contained low levels of such T cells (80%, *bottom*). (B) An even more dramatic correlation between T-cell infiltration and clinical progression came from analysis of a group of 406 CRC patients, who were stratified into four subgroups according to the stage of histopathological progression of their primary CRC tumors, with stage I tumors representing the least aggressive and stage IV representing the most aggressive subtype. Each tumor was analyzed for the presence of CD3$^+$ (overall T cells) and CD45RO$^+$ (memory T cells) both at the invasive margins of tumors and at the centers of such tumors. As the Kaplan–Meier curves illustrate, most stage I, II, and III tumor patients showed long-term survival if observation revealed that their tumors exhibited high levels of CD3-positive and CD45RO-positive cells at both the tumor centers and invasive margins, whereas those patients whose tumors lacked such T cells in both sites declined precipitously after initial surgery. The clinical progression of stage IV patients, almost all of whom did poorly, is not shown here. (From J. Galon et al., *Science* 313:1960–1964, 2006. With permission from AAAS.)

15.3 The immune system functions to destroy foreign invaders and abnormal cells in the body's tissues

The structures of our tissues clearly provide physical barriers and thus a measure of protection against infections by various pathogens. Beyond these barriers, immune systems developed early in metazoan evolution in order to defend against infections by viruses, bacteria, and fungal parasites. In vertebrates like ourselves, this protection is implemented by the diverse cellular components of the immune system, including lymphocytes, macrophages, neutrophils, dendritic cells, and natural killer (NK) cells as described below.

Table 15.1 Innate and adaptive immunity

Property	Innate immunity	Adaptive immunity
Phylogenetic distribution	All phyla	Demonstrated only in vertebrates and in bacterial species endowed with CRISPR
Receptor diversity	Many types of germ-line-encoded receptors, but little or no diversity within each receptor type	Antigen receptors including antibodies and T-cell receptors encoded by genes that arise from somatic diversification of germ-line genes to generate an exceedingly large diversity of antigen receptors
Specificity	Molecular patterns	Highly specific. Can discriminate subtle differences between molecular structures. Each member of a species is capable of recognizing $>10^7$ antigen specificities.
Immunological memory	Generally absent, but some cells of the innate immunity system may show increased responses to subsequent encounters with the same stimulus	Capacity for more rapid and more potent immune responses to subsequent encounters with the same antigen
Self/non-self discrimination	Excellent discrimination of microbial molecular patterns from host molecular patterns	Very good but imperfect. Failures can occur and result in autoimmune disease.
Major soluble components	Antimicrobial peptides and proteins, cytokines, chemokines, and other soluble mediators including complement proteins	Antibodies, cytokines, and chemokines
Major cell types	Many different types of phagocytes including macrophages, neutrophils, dendritic cells; NK cells; some epithelial cells	T cells, B cells, and antigen-presenting cells (primarily dendritic cells, but in some cases, activated macrophages and B cells)

Vertebrate immunity is mediated by two distinct but interacting physiologic systems, the **innate immune system** and the **adaptive immune system**. **Table 15.1** summarizes some of the salient features and differences in these two systems. Most fundamentally, the innate immune system recognizes molecular patterns characteristically associated with pathogens and some cancer cells and responds with a variety of measures to prevent, limit, or eliminate infections and to eliminate incipient tumors as well. Such innate immunity operates "instinctively," that is, without any prior exposure of an individual organism to such pathogens, cancer cells, and their associated antigens; accordingly, this type of immunity is not learned during the lifetime of the individual, but instead exists constitutively, being orchestrated entirely by genes inherited through the germ line.

This innate behavior contrasts with that of the adaptive immune system, whose actions are indeed triggered by exposure to a given antigen or the antigens borne by an infectious agent or by aberrant cells, including cancer cells. The learned behavior of the adaptive immune system often enables an individual to respond quickly and strongly to subsequent challenges posed by exposure to the same antigens or infectious agents years or decades later. Such delayed responses are indicative of **immunological memory**, that is, the ability of the adaptive immune system to "remember" over extended periods of time the prior exposure to an antigen or set of antigens.

The innate immune response can be construed as the first line of defense against attacks launched by pathogens, because its components preexist in the body and therefore largely do not require an extended period of development. The second, more powerful line of defense is ultimately afforded by the adaptive immune response, which is highly specific for the antigen-displaying targets that it recognizes and undertakes to attack. Moreover, the adaptive response to these targets can undergo great amplification and is capable of recognizing the essentially unlimited diversity of molecular structures, which are the physical embodiments of antigens.

Strikingly, after an initial encounter and response to a given novel antigen (or antigen-displaying cell), subsequent adaptive immune responses to that particular antigen are typically far stronger and more rapid in onset than the initial response. These secondary responses, sometimes occurring months and even years after the initial exposure to an infecting pathogen or foreign antigen, reflect the phenomenon of immunological memory cited above and are the basis of vaccination. Thus, lymphocytes that participated in a successful immune response to an initially encountered infectious agent (or tumor cell) and associated antigens will leave behind descendants that will persist in the body for years or decades and confer ongoing protection against subsequent exposure to this agent. It is these long-lived lymphocytes that preserve the record of initial encounters with an antigen and therefore provide the immunity conferred by vaccination.

Two classes of lymphocytes, **T lymphocytes (T cells)** and **B lymphocytes (B cells)** and their products, are the cellular mediators of adaptive immunity. B cells produce antibody molecules (also known as **immunoglobulins**) and T cells are key effectors and regulators of immune responses. Distinct classes of antigen-specific receptors displayed on the surface of a T cell—**T-cell receptors** (TCRs)—and on the surface of a B cell—**B-cell receptors** (BCRs)—endow these lymphocytes with the ability to recognize the particular molecular configurations of antigens, doing so with exquisite specificity. In addition, many of the soluble antibody molecules secreted by B cells and present in our blood exhibit high specificity for recognizing and binding tightly to their cognate antigens, that is, the antigens that previously provoked their formation. The B-cell receptors displayed at the cell surface by B lymphocytes are membrane-bound versions of the soluble antibody molecules that these cells secrete.

The functions of these various membrane-bound and secreted antigen-recognizing proteins, or **antibodies**, can be understood from **Figure 15.6**, which reveals that these antigen-recognizing immune molecules contain subdomains domains known as **variable regions** (V regions), which are the sites of their antigen-recognizing and antigen-binding functions. V regions are composed of amino acid sequences that vary according to their antigen specificity, allowing them to recognize and bind an extraordinarily diverse array of antigens. In effect, V regions carry the antigen-binding sites that recognize and bind antigen molecules having complementary structures, much like the fit of individual keys (antigens) in appropriate cognate locks (V regions).

In addition, TCRs and BCRs, like antibody molecules, contain **constant regions** (C regions) that differ little between different molecules of a given class. Thus, the C regions of a diverse array of TCRs contain amino acid sequences that are identical or similar to one another; the same can be said of BCRs and their antibody counterparts. In effect, the constant regions represent the structural scaffolds required for the multiple functions of these three classes of immune molecules—secreted antibody molecules, TCRs, and BCRs. As should be apparent, the C regions of these molecules are not involved in antigen recognition.

Although all three classes of molecules are members of the same immunoglobulin superfamily of proteins, TCRs, BCRs, and antibodies differ markedly from one another in certain critical structural features. As an example, while T-cell receptors are heterodimers, the basic structure of antibodies and BCRs can best be described as dimers of identical heterodimers. Thus, as illustrated in Figure 15.6, each hetero-tetramer forming an antibody molecule is composed of four immunoglobulin chains, two identical larger **heavy chains** (H) and two identical smaller **light chains** (L). Note that when antibody molecules remain anchored to the surfaces of B cells that made them, they function as antigen-recognizing, B-cell receptors (BCRs). When present in soluble, secreted form, the almost-identical immunoglobulin molecules are referred to as antibodies and are found throughout the blood, lymph, and tissue fluids. In contrast to antibodies and BCRs, the antigen receptors found on most T cells recognize a complex formed between an antigen-derived peptide and an MHC protein, a type of cell surface molecule known to be encoded by the

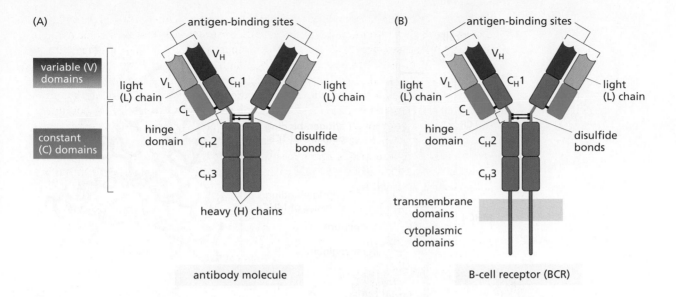

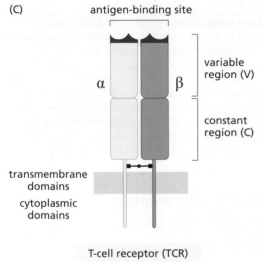

T-cell receptor (TCR)

Figure 15.6 The structure of antigen-binding receptors and antibodies (A) Most human antibody molecules are tetramers composed of four immunoglobulin chains. Each of the larger heavy (H) chains is covalently joined via disulfide linkage to a light chain forming an H–L heterodimer. Two identical H–L heterodimers are joined via disulfide linkage to form the basic four chain (H_2L_2) structure found in all antibody molecules. The amino acid sequences of an immunoglobulin molecule of the same class or type show little variation in the portion of the protein that is termed the constant region and labeled C_H in heavy chains and C_L in light chains. The remainder of each immunoglobulin chain shows great variation within a class of antibodies and is called the variable region (V_H in heavy chains and V_L in light chains). The two antigen-binding regions of each H + L heterodimer are assembled from regions of both the V_H and V_L chains and together are referred to as the antigen-binding site. (B) In contrast to the antibody molecules described in (A), which are soluble, the B-cell receptor (BCR) displayed by B cells is an antibody molecule that also contains a transmembrane segment at the C′ termini of the constant regions, which anchors these molecules to the plasma membrane and enables them to function as transmembrane receptors. (C) The T-cell receptor (TCR) is a heterodimer made up of two polypeptide chains, α and β, both of which are members of the immunoglobulin superfamily of proteins. [Although most T cells bear $\alpha\beta$ T-cell receptors (termed $\alpha\beta$ TCRs), a small but significant fraction of the overall T-cell population display a different heterodimer (*not shown*) made up of γ and δ chains and called $\gamma\delta$ T-cell receptors ($\gamma\delta$TCRs).] Like immunoglobulins, each polypeptide subunit of a TCR has a constant and a variable region, and portions of the variable (V) domains of each $\alpha\beta$ heterodimer collaborate to form antigen-binding sites (*red*).

major histocompatibility complex of genes for which they are named (**Figure 15.7**); they will be discussed in detail in Sections 15.5 and 15.6.

The binding of antigens to the antigen-binding sites of TCRs or BCRs triggers adaptive immune responses. Although antigen binding is necessary to trigger an adaptive immune response, it is almost never sufficient. As a rule, the immune system must receive additional signals in order to assemble a biological program that permits the successful initiation of an actual adaptive immune response. Whether or not an antigen can function as an **immunogen**, that is, a substance capable of provoking an immune response, depends on the context of the encounter with this substance. As we will see, a major problem of anti-cancer immunotherapies derives from the difficulty in rendering cancer cell-associated antigens immunogenic, that is, creating the conditions that make a cancer antigen an immunogen and thus capable of provoking an effective immune attack.

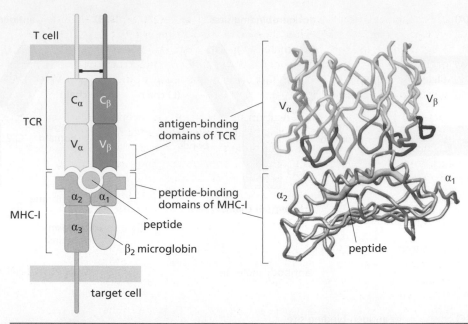

Figure 15.7 T-cell receptors are specific for a complex of peptide and MHC T-cell receptors (TCRs) recognize the structure formed by the noncovalent interaction of a peptide (*yellow*) with the binding pocket of an MHC-I or MHC-II molecule. An MHC-I complex is illustrated here (*left*). Displayed here is the recognition of such a complex (i.e., a peptide in the cleft formed together with the peptide-binding domains of an MHC-I molecule) by the αβ TCR of a T cell. The structures (*right*) of the polypeptide backbones of a TCR antigen-binding domain (*above*) bound to an MHC-I molecule and its associated bound peptide (*below*) complex has been determined by X-ray crystallographic analysis. (Adapted from K. Murphy and C. Weaver (eds.), Janeway's Immunobiology, 9th ed. New York: Garland Science, p.162, Figure 4.24; ribbon diagram from K.C. Garcia, et al. *Science* 274:209-219, 1996. With permission from AAAS.)

15.4 The diversity of B cell and T cell receptors arises from the stochastic diversification of the genes that encode them

During the generation of T lymphocytes and B lymphocytes, a highly unusual process of immunoglobulin diversification (including diversification of soluble antibodies as well as the membrane-tethered TCRs and BCRs) is achieved by the intensive diversification of the genomic nucleotide sequences that encode the V regions of these various proteins. Unlike essentially all other genes in the human genome, the genetic sequences encoding antibodies and thus B-cell receptors are not inherited in the germ line and therefore are not present in the lymphocyte progenitor cells at birth. Instead, a few hundred distinct subgenic segments that are indeed inherited in the genome can combine with one another almost randomly in any one of thousands of different ways as the immune system matures (Supplementary Sidebar 15.1). A similar process of rearrangement and assembly of subgenic segments of T-cell receptor genes generates a diverse repertoire of T-cell receptors. The resulting combinatorially assembled, fused subgenic sequences underlie the staggering diversity of T- and B-cell populations; the potential diversity of T-cell and the B-cell repertoires of an individual human each exceed 10^9 potential antigen specificities. In aggregate, these specificities enable T and B cells to recognize and bind almost any antigen that might be encountered during a lifetime. (Actually, only a tiny subset of these possible antigens is ever encountered by an individual human.)

In fact, the organization of the process of random gene rearrangement ensures that only one of the very large number of possible gene rearrangements is allowed to assemble within a given lymphocyte precursor. Once a new antigen-combining site has been assembled in a lymphocyte precursor (through the process of stochastic combinatorial fusion of encoding DNA segments described above), subsequent recombination events are blocked in that cell and its descendants. Therefore, a lymphocyte precursor cell and its lineal descendants carry genes that encode the ability to recognize only one specific cognate antigen (or set of structurally similar antigens). Accordingly, all antigen receptors or all of the antibody molecules made by a given lymphocyte are identical to one another at the molecular level and all exhibit identical antigen-recognizing specificity.

Importantly, the chance diversification of these antigen receptor genes occurs in the absence of any direct provocation by encounters with specific antigens. Instead, diversification simply allows the immune system to anticipate future challenges arising from an essentially unlimited number of antigens, each composed of a molecular structure that might not have been encountered previously by an individual but might well be encountered during the individual's lifetime. Among these novel antigens are some that have never been encountered during the evolutionary history of the animal species to which the individual belongs.

An encounter with an antigen can provoke the proliferation of a lymphocyte whose antigen receptors (antibody molecules, BCRs, or TCRs) recognize the antigen's detailed molecular conformation. As a consequence, repeated cycles of cell division of an antigen-provoked lymphocyte and its direct descendants produce a large clonal population of cells that all share the same antigen-binding specificity as the parental lymphocyte in which an initial immunoglobulin or T-cell receptor gene rearrangement took place. The proliferation of a clone of B or T lymphocytes in response to an encounter with a specific antigen is referred to as **clonal expansion**. In effect, the antigen that provokes this clonal expansion functions much like a mitogenic growth factor impinging on non-immune cells throughout the body (see Chapter 5). Antigen-provoked clonal expansions are formally similar to the clonal expansions that occur during multi-step tumor progression (see Figure 11.14) as a consequence of the acquisition of advantageous, somatically generated, mutant alleles. The clonal expansions of lymphocytes generate the large populations of these cells that can mediate actual responses to a newly encountered antigen, replacing the initially small number of cells that were incapable of doing so.

An inherent hazard of the system that allows the stochastic generation of antibodies and T-cell receptors is the likelihood that such a process will generate receptors that exhibit a specificity for antigens expressed by the body's own normal cells, that is, "self." The formation of such antigen receptors may open the door to an attack within a host organism by its own immune system, resulting in an **autoimmune** disease. Such an attack is said to be launched by **autoreactive** lymphocytes and antibody molecules.

In fact, most lymphocytes that happen to have developed strong reactivity to the antigens expressed in normal tissues are eliminated soon after their development; this elimination occurs in the tissues termed primary lymphoid organs where they were generated, these being either the thymus in the case of T cells or the bone marrow in the case of B cells. Deletion of autoreactive cells must operate efficiently lest life-threatening autoimmune diseases occur. Moreover, it is this deletion that lies at the heart of immunological tolerance (**Sidebar 15.1**).

Sidebar 15.1 Inappropriate gene expression may yield protein antigens that escape immune tolerance The development of immune tolerance depends on the ability of the immune system to recognize self-antigens and eliminate the lymphocytes that happen to have developed reactivity with these antigens. This process is highly effective but never flawless. As an example, deviations from normal patterns of antigen synthesis and presentation may be undertaken by cancer cells, which may well derive, in turn, from aberrant patterns of gene expression in these cells; such shifts in the transcriptomes of cancer cells may yield, among other aberrations, the inappropriate expression of proteins normally restricted to earlier stages of development and thus no longer expressed in adult tissues. Such aberrantly expressed antigens may not have provoked the development of immune tolerance toward them.

Alternatively, cancer cells may develop ectopic expression of proteins whose synthesis is normally restricted to special sites in the body, such as the testes; such proteins may not have previously elicited the development of immune tolerance, simply because they are expressed in anatomical sites that are usually inaccessible to the components of the adaptive immune system that are responsible for the development of tolerance. These various types of antigens that have failed (for one reason or another) to induce tolerance and are displayed by tumor cells represent excellent candidates for recognition and attack by components of the adaptive immune system.

15.5 MHC molecules play key roles in antigen recognition by T cells

Only those members of the huge, diversified lymphocyte population encountering an antigen that happens to bind their antigen receptor (that is, a BCR or TCR) will be activated to divide and differentiate into cells that perform the **effector** functions (the actual work) of their lymphocyte type. Antibody production is the primary effector function of B cells, whereas the effector functions of T cells are more variable, depending on which of the broad categories of helper, regulatory, or cytotoxic T-cell functions they are destined to perform; these subcategories of T cells are usually denoted as T_H, T_{reg}, and T_C cells, respectively.

An understanding of the molecular mechanisms through which T cells are cued to play their assigned roles in controlling cancer begins with an appreciation of precisely how T cells recognize antigens, specifically antigens expressed by other cells, such as cancer cells, including those proteins that reside within these other cells. This recognition is complicated by the fact that cells of the immune system do not have the capacity to peer inside other cells including cancer cells. Instead, in order to circumvent this problem, the contents of cells must be displayed on the surface of such cells, allowing them to be sensed by the immune system. The ability to surveil the internal contents of cells is critical given the fact that the majority of cancer-associated antigens are intracellular proteins rather than proteins that are naturally displayed on the surface of cancer cells.

This prerequisite for effective immune surveillance provides the rationale for the cellular machinery that has evolved specifically to display samples of the intracellular contents on the cell surface. In the case of protein-associated antigens—essentially all of the antigens discussed here—this surveillance depends, more specifically, on fragments of such proteins being presented at the surface. These fragments are generated by ongoing proteolysis in the cytoplasm of a portion of recently synthesized proteins.

Major histocompatibility complex (MHC) molecules are responsible for displaying fragments of intracellular proteins on the surfaces of all the body's cells. These molecules are tasked with the function of acquiring peptide fragments of recently synthesized proteins and conveying them to the cell surface, where they are subsequently displayed and may, in principle, be recognized by TCRs.

MHCs are transmembrane proteins that are topologically much like the growth factor receptors and integrins described in Chapter 5. However, the functions of these classes of cell surface molecules are totally different: growth factor receptors transduce signals into cells informing the intracellular circuitry of conditions in the outside world, in particular the extracellular space immediately surrounding a cell. In stark contrast, MHC molecules function in precisely the opposite fashion by conveying signals (in the form of peptide:MHC complexes) to the cell surface, where they are displayed to the outside world, more specifically to components of the immune system.

The cells of an individual mammal can muster a diverse array of MHC molecules. These molecules are encoded by highly polymorphic alleles in the genome. (Recall that a genetic polymorphism is a variant allele whose protein products support normal cellular and organismic function.) The polymorphic variants of MHC molecules carry functional pockets that are capable of binding many of the diverse peptides generated by intracellular proteolysis and thereafter conveying the bound peptides to the cell surface for display to the outside world (**Sidebar 15.2**).

Importantly, the polymorphic variability of MHC molecules is not generated by rearrangements of DNA segments inherited through the germ line, as described earlier (see Supplementary Sidebar 15.1), but preexists in the encoding germ-line sequences inherited from parents. Moreover, this diversity of MHC molecules displayed by a given vertebrate cell exists in order to enable a subset of them to bind, with some degree of affinity, a portion of the peptides that are generated by

Sidebar 15.2 MHC polymorphism has functional consequences The MHC complex is a closely linked cluster of genes that is present in the genome of all vertebrate species. Together with bound peptides, proteins encoded by the MHC form the antigenic structures recognized by T cells. In humans, MHC molecules are termed instead HLA molecules. There are two classes of MHC molecules, **MHC class I** (MHC-I) and **MHC class II** (MHC-II). MHC molecules all have a functional pocket, a molecular cleft that binds some peptides but not others in order to form the peptide:MHC complexes that TCRs recognize. MHC molecules are encoded, as mentioned, by highly polymorphic genetic loci, each of which hosts multiple alternative alleles. By now, more than 10,000 distinct MHC alleles have been catalogued in the human gene pool. This diversity is a direct reflection of the need to maximize the chances that an MHC molecule of one variant subtype or another will contain a molecular cleft that selectively binds some members of the diverse population of peptides generated by post-translational intracellular proteolysis of recently synthesized proteins.

Bear in mind that although peptide binding by MHC molecules is somewhat selective, it is highly promiscuous, and any particular MHC molecule is capable of binding a large number of different peptides. This means that in an outbred vertebrate population, such as humans, it is virtually impossible for a pathogen to evolve proteins that will not, upon proteolysis, generate peptides that can be bound by at least some MHC class I and class II molecules displayed by certain individual members of a mammalian species. This combination of MHC polymorphism and peptide-binding promiscuity would seem to guarantee that the species as a whole cannot be driven to extinction by exposure to one or another novel infectious agent. This versatility of peptide binding extends as well to those generated from the proteins expressed by cancer cells: even in a single individual, it is likely that at least some peptides generated from novel antigens arising in cancer cells have a probability of being bound by at least one of the class I or class II MHC molecules encoded in this individual's genome.

Given the large number of polymorphic alleles and the number of loci that carry these alleles, each individual member of a species is likely to carry a combination of MHC molecules that differs significantly from the combination of MHC molecules expressed by almost everyone else. The most obvious exceptions to this are identical twins in humans and members of specific inbred strains of mice in which all individual mice are genetically identical to one another. This polymorphic variability has another immunological consequence: every individual's cells will express at the surface a combination of MHC molecules that is unique (or almost so). Although the assigned task of MHC molecules is to *present* intracellular peptides, enabling the latter to be recognized as antigens, these MHC molecules themselves, independent of their peptide-displaying powers, become antigens in their own right that are readily recognized by components of the immune system. Within an individual's body, these MHC molecules are, quite naturally, the object of immunological tolerance; however, when this individual's cells or tissues are transplanted into the body of a genetically nonidentical person, they will almost inevitably be recognized as foreign because these donor cells and tissues express MHC molecules that the recipient's immune system has not previously experienced.

These dynamics explain historically the origin of the term *major histocompatibility complex,* which derived from the observation that tissues and cells of some organisms can be readily transplanted into other syngeneic organisms; the transplanted donor tissues are therefore "histo" (meaning "tissue") compatible with the recipient hosts. Conversely, transplanted donor tissue that is nonsyngeneic with a potential graft recipient was seen to be rejected by the recipient and was therefore deemed to be *histoincompatible.* (Indeed, the MHC genetic complexes were discovered because of the strong immunogenicity of their encoded proteins, which were found to be major sources of graft recognition and rejection.)

Given the high degree of polymorphism in MHC molecules, differences in MHC repertoires among members of an outbred species, such as our own, are assured. Consequently, transplantation of cells, tissues, or organs between members of a population that are not genetically identical are likely to fail unless powerful immunosuppressive measures are taken to prevent anti-graft rejection responses in tissues of the recipients.

proteolysis of that cell's intracellular contents. In all vertebrates, the loci of genes encoding the two major types of MHC molecules—class I and class II—as well as several other genes encoding many of the proteins essential for antigen presentation, are closely linked to one another in a region of the genome referred to as the MHC complex.

As cited above and described in detail later in this chapter, the recognition of cancer cells by components of the immune system depends on the ability of immune effector cells to recognize MHC-bound peptides displayed on the surfaces of the neoplastic cells. As indicated earlier, these peptides are produced continuously in cells and, in principle, are created by the systematic cleavage of a portion of all the proteins that are synthesized inside a cell; accordingly, they include proteins made by normal cells, proteins made by cancer cells, and proteins made by pathogen-infected cells. One immediate and direct implication is that the failure of a cancer cell to display cancer-associated peptides borne by its MHC molecules may enable that cancer cell to evade detection and subsequent attack by T cells of the immune system.

There are two types of cell surface MHC molecules, MHC class I (class I or MHC-I) and MHC class II (class II or MHC-II), and there is a marked difference in the patterns of their expression among the diverse cell types in the body.

Essentially all nucleated cells display class I MHC molecules, revealing how far-reaching is the ability of the immune system to monitor intracellular proteins—including those made by infectious agents within infected cells—throughout the body. Class II expression, however, is mostly restricted to only a few types of cells with specialized immune-related functions, specifically dendritic cells, B cells, and macrophages, as well as others we need not consider here. As will become apparent, these differences have important functional consequences. The restricted expression of the class II MHC molecules and the specialized function of these class II molecules have caused this small number of immune cell types to be called **antigen-presenting cells** (APCs).

Both MHC-I and MHC-II molecules carry the clefts cited earlier that serve as selective binding sites for the peptides generated by intracellular proteolytic degradation. In effect, these peptide-binding clefts serve as the palms of the hands of these MHC molecules that grab peptides inside the cell and display them outside the cell at its surface. The resulting complex of a peptide and MHC molecule forms a three-dimensional structure that is unique for any particular combination of peptide and MHC and is the structural feature on the surface of a cancer cell that may be recognized by a T-cell receptor (see Figure 15.7).

The relatively small size of the peptide-presenting cleft together with the size of captured peptides, typically 8 to 10 amino acid residues for MHC-I and 13 to 20 or so for MHC-II, create a problem: the interaction domains between the peptide and the MHC cleft that hosts it depend, by necessity, on a small number of noncovalent interactions between the two. Indeed, complex bioinformatic algorithms have been developed to predict whether a given peptide generated by proteolytic cleavage within a cell will find a firm roost in one of a cell's repertoire of MHC molecules and thus have a chance of being carried to a cell surface and displayed there.

These facts suggest that when both a wild-type and corresponding mutant peptide are properly displayed on the cell surface, a subtle structural difference between them may suffice to provoke an immune recognition of the difference between them. Thus, during the development of the immune system, the wild-type peptide should have elicited tolerance and should thereby be overlooked by the multiple arms of the adaptive immune system; in contrast, a single amino acid-substituted peptide version of the wild-type peptide may stand out and be highly successful in attracting the attention of the immunocytes that form the adaptive arm of this system. This explains how a cell bearing a single cancer-associated mutant allele can generate peptide:MHC complexes that have the potential to be recognized by T cells, the agents of adaptive immunity responsible for such recognition.

Importantly, the peptides presented by the two classes of MHC molecules have different origins. MHC-I selectively binds peptides generated from proteins initially synthesized within the *same cell* that later displays this peptide:MHC complex on its surface. Given the almost universal expression of MHC class I molecules throughout all cell types in the body, this implies that a portion of each protein made by a cell anywhere in the body is diverted from its dedicated function and destination site and is, instead, subjected to intracellular proteolysis and subsequent cell surface display via MHC-I molecules.

The MHC-II molecules displayed by a cell operate very differently: they selectively bind peptides derived from proteins initially synthesized in *other cells* in the body that were previously acquired by antigen-presenting cells (APCs), internalized into these cells (by **phagocytosis**, **endocytosis**, or **pinocytosis**), subsequently degraded into peptides, and only then loaded onto an MHC-II molecule and presented at the cell surface. The ability to perform this function, involving antigen acquisition and subsequent cell surface presentation, is not ubiquitous; rather this function is restricted to the APCs enumerated earlier, specifically B cells, macrophages, and most importantly, dendritic cells (DCs).

These unique capacities enable itinerant, "professional" antigen-presenting cells to survey the contents of cells and fluids throughout the body, including infecting microbes and cancer cells, ingest them, digest them, and ultimately present resulting peptides to other components of the immune system, making possible the subsequent development of an effective immune response. To be more specific, this MHC-II–mediated presentation by APCs of acquired peptides is made to T cells, which respond in various ways to create various aspects of the immune response.

Dendritic cells represent the most potent class of APCs, because they possess, by far, the greatest capacity to present certain peptides generated from proteins synthesized in other cells. In addition, because DCs share traits with essentially all other cells in the body, they can also present oligopeptides on their surface via their MHC-I molecules. Hence, DCs employ both MHC-I and MHC-II to display peptides at their surface. In both cases, T cells represent the audience that views and responds to these two types of antigen presentation. The T cells, in turn, proceed to operate as the key organizers, amplifiers, and in many cases, the effectors of adaptive immune responses to microbes and cancer. The pathways leading to presentation of peptides on class I or class II MHC molecules are presented schematically in **Figure 15.8**.

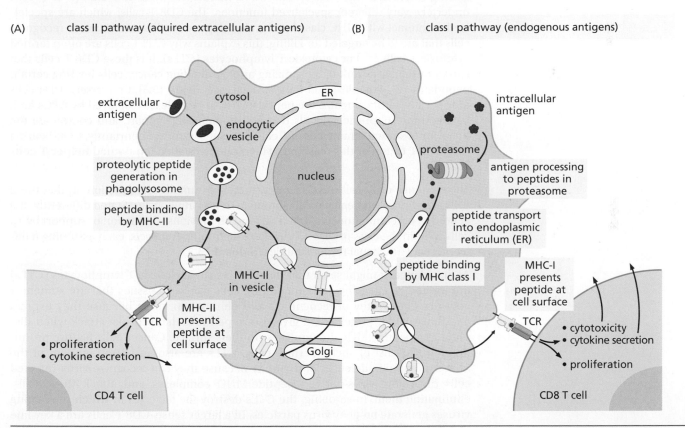

(A)   class II pathway (aquired extracellular antigens)

(B)   class I pathway (endogenous antigens)

Figure 15.8 Different pathways of antigen processing present antigen-derived peptides to CD4 and CD8 T cells
(A) Class II pathway: presentation to CD4 T cells. Extracellular antigens—notably those produced by other cells (*red oval, upper left*) and then internalized by phagocytic cells, especially dendritic cells (via phagocytosis, endocytosis, or pinocytosis)—soon become enclosed in endocytic vesicles, some of which fuse with phagolysosomes in the cytoplasm, where their contents are subjected to proteolysis, resulting in the generation of peptides (*red dots*), some of which are then bound by MHC-II. The resulting peptide:MHC-II complexes are exported to the cell surface by vesicular transport and presented to T-cell receptor (TCR) displayed by an adjacent CD4+ T cell (*light green, lower left*). Recognition of these cell-surface peptide:MHC-II complexes by the TCR of a CD4+ T cell can induce its proliferation and its secretion of certain cytokines. (B) Class I pathway: presentation to CD8 T cells. A portion of the recently synthesized proteins within a cell routinely undergo proteasome-mediated proteolysis, generating peptides that are transported into the endoplasmic reticulum (ER), where some are bound within the lumen of the ER by newly synthesized MHC-I molecules. Subsequent transport first to the Golgi apparatus and then to the cell surface results in the presentation of the peptide:MHC-I complexes at the cell surface. Recognition of peptide:MHC-I complexes by the TCRs of nearby CD8 cells (*light blue, lower right*) can induce these cells to undergo proliferation, secrete cytokines, and differentiate into cytotoxic T lymphocytes (CTLs, *not shown*) that can kill cells bearing peptide–MHC complexes recognized by their TCRs. (Adapted from P. Parham, The Immune System, 5th ed. New York: W.W. Norton & Company, 2021, Figure 5.17.)

15.6 T cells that recognize MHC-I have different roles from those that recognize MHC-II

The T cells that respond to antigen presentation via either MHC-I or MHC-II fall into two distinct subtypes. In both cases, the responses depend on TCRs displayed by the T cells. Whether a TCR recognizes MHC-I–bound or MHC-II–bound peptides is determined by a **co-receptor** protein that is associated closely but noncovalently with this TCR on the cell surface adjacent to this TCR. The co-receptors do not have any antigen-recognizing function of their own, but instead ensure that the TCR with which they are partnered is directed toward the appropriate type of MHC molecule, that is, toward either MHC-I or MHC-II. More specifically, CD4 and CD8 molecules, functioning as co-receptors, direct their TCR partners to bind either MHC-II complexes (CD4) or MHC-I complexes (CD8) displayed by other cells. Indeed, the two dominant classes of T lymphocytes are named CD4$^+$ or CD8$^+$ T cells after their co-receptors (or, more simply, CD4 T cells and CD8 T cells). Thus, a given T lymphocyte will display either a CD4 or CD8 co-receptor and belong to either the CD4$^+$ or CD8$^+$ class of T cells.

Having interacted with appropriate MHC molecules, these two classes of T cells undertake very different specialized functions. The CD8 T cells, which are specialized to interact with MHC class I molecules, depend on this interaction to recognize cells that are to be targeted for killing; this explains why CD8 T cells are often termed cytotoxic T cells (T_C) or cytotoxic T lymphocytes (CTLs). It is these CD8 T cells that carry out the hard work of recognizing and eliminating cancer cells bearing certain peptide:MHC-I complexes on their surface (see Figure 15.8). Conversely, CD4 cells are specialized to recognize the peptide:MHC-II complexes displayed by APCs and, having done so, provide signals that help to activate, support, and coordinate the behavior of several other types of T cells—including, most importantly, CD8-bearing cytotoxic T cells. For this reason, CD4 lymphocytes are often termed **helper T cells** (T_H cells).

The existence of T_H cells reveals a new mode of immunocyte function, in that these cells act primarily to regulate and stimulate other lymphocytes. Stated differently, the actions of other immunocytes often depend on previous stimulation or support by T_H cells. Within the class of CD4 T cells there are multiple subsets, each assuming a different set of functional responsibilities (**Sidebar 15.3**).

At first glance, it might seem that the CD8-bearing cytotoxic T lymphocytes (CTLs) are limited to attacking and eliminating cells from our tissues that are somehow aberrant and worthy of recognition and elimination, often because they express aberrant cellular proteins and give rise to peptides that provoke attack when displayed as peptide:MHC-I complexes on their surface. Cancer cells feature prominently in the list of such aberrant cells. CTLs are also an important arm of the *antiviral* immune defenses of the body because they can recognize virus-infected cells displaying virus-derived peptide:MHC complexes and attack these cells, eliminating them. In so doing, the CTLs destroy the factories in which replicating viruses generate progeny virus particles. In a larger sense, CD8 T cells are a key line of defense against a variety of intracellular pathogens that depend on their cellular hosts to multiply (**Sidebar 15.4**).

15.7 Dendritic cell activation of naive T cells is a key step in the generation of functional helper and cytotoxic T cells

As mentioned earlier, continuous sampling of the body's tissues is a signature capacity of the dendritic cell (DC) lineage. Having acquired fragments of cells, cellular debris, and pathogens in whatever tissue locations they happen to be found, these versatile cells then migrate with their cargo to nearby **draining lymph nodes**, that is, to the lymph nodes that are connected directly via lymph ducts to the tissue sites of antigen acquisition (see Figure 14.36).

Sidebar 15.3 CD4 + T-cells belong to multiple subtypes T lymphocytes of the CD4 category are responsible for a broad range of activities that stimulate, coordinate, or otherwise regulate the responses of multiple types of immune cells. This is accomplished at the time of their activation by the differentiation of naive T cells into one of five major effector CD4⁺ T-cell subsets. Each of these effector subsets orchestrates a distinct function or set of functions. The specific path of differentiation into each of these subsets is determined by the cytokines present during the initial antigen-triggered activation of the naive CD4 cell and involves the expression of a key transcription factor that determines subset identity.

Each T-cell subset is distinguished by the secretion of its own distinct set of cytokines, each providing the means to direct the behavior of specific target cell populations under its control, that is, the particular cell populations that are being helped by the specialized CD4 T-cell subset. **Figure 15.9** provides a summary profile of five widely recognized T-cell subsets. Four of the five CD4 T-cell subsets play a role in activating or intensifying the responses of the immunocytes they target, thereby supporting or augmenting their functional capabilities. In sharp contrast, the fifth subset is composed of regulatory T cells, T_reg's, which *suppress* the activities of the immune cells that they target.

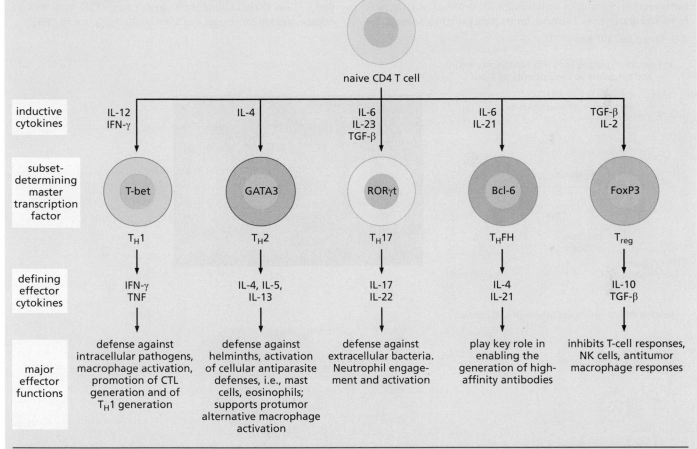

Figure 15.9 T-cell subsets Five of the most studied of the known T-cell subsets derived from naive T cells (*above*) are shown, and the cytokine or cytokines essential for their induced differentiation are identified. As indicated, the expression of a key and distinctive transcription factor or factors is required for differentiation into each subset. Once formed, each subset secretes a characteristic profile of effector cytokines that promotes the indicated effector functions (*bottom*). (Adapted from P. Parham, The Immune System, 5th ed., New York: W.W. Norton & Company, 2021.)

After arriving in these nodes, the DCs encounter **naive** T cells that have traveled to these nodes via a totally independent route—the bloodstream; naive T cells by definition have not experienced contact with an antigen and undergone a resulting activation step.

The two routes of travel—via the lymph ducts and the bloodstream—dictate that in most instances, naive T cells that are destined to eventually launch anti-tumor responses do not initially encounter tumor antigens directly in the tumor itself. Instead, this encounter occurs in lymphoid tissues at sites physically

Sidebar 15.4 CTLs use well-defined mechanisms to kill their targets A cell's display of certain peptide antigens on its surface (via its MHC class I molecules; see Figure 15.7) may subsequently attract the attention of certain cytotoxic T lymphocytes (CTLs or T$_C$), which migrate out of the lymph nodes following their activation by APCs, enter into the circulation, and thereafter enter into various peripheral tissues to search for appropriate targets. When a CTL encounters a cell displaying the specific peptide:MHC complex recognized by its TCR, it will launch an attack on this cell with the goal of killing it. In fact, cytotoxic T cells deploy two distinct mechanisms to kill their targets, relying either on granule exocytosis (**Figure 15.10A**) or on death-inducing ligands (see Figures 9.31 and 9.32). Thus, CTLs can attack their intended victims through the secretion of granules containing toxic proteins, such as perforin and granzymes. Perforin, for its part, punches holes in the

plasma membranes of targeted cells, allowing granzymes to enter the target's cytosol, where they can activate the intrinsic apoptotic pathway shown in Figures 9.31 and 9.32. The alternative killing mechanism involves the Fas death receptor, which is displayed on some cell types. CTLs can present FasL, the ligand of the Fas receptor, to their target cells, activating the Fas death receptors on the surfaces of the targeted cells and triggering the extrinsic apoptotic pathway within these cells (see Section 9.14).

Such CTL-mediated attacks on target antigen-displaying cells (see Figure 15.10A) represent a major component of the cellular arm of the adaptive immune response and play an important role in the elimination of aberrant cells, notably virus-infected cells and cancer cells. Finally, we should note that CTL-mediated killing is efficient; a single CTL may serially engage and kill one target cell after another (Figure 15.10B).

(A) target cell killing by CTL

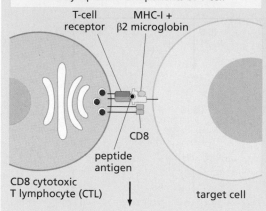

antigen recognition redistributes cytoskeleton and cytoplasmic components of T cell

T-cell receptor   MHC-I + β2 microglobin

CD8

peptide antigen

CD8 cytotoxic T lymphocyte (CTL)

target cell

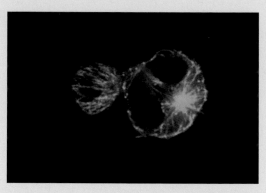

release of lytic granules at site of cell contact

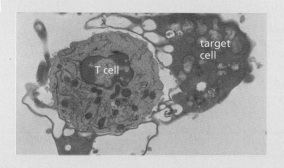

T cell

target cell

(B) serial killing by CTL

| Cytotoxic CD8 T cell recognizes virus-infected cell | CD8 T cell programs first target cell to die | CD8 T cell moves on to second target cell | First target cell dies, second is dying, and the third is being attacked |

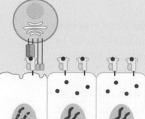
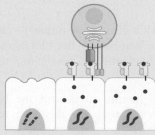

Figure 15.10 Killing of target cells by cytotoxic CD8-expressing T cells (CTLs) (A) Recognition of peptide:MHC-I complexes on the surface of target cells by the T-cell receptor together with the associated CD8 co-receptor of a CTL (*diagram*) causes a reorganization of cytoplasmic elements within the CTL (*micrograph*). As seen in the micrograph, the engagement prompts alignment of the CTL's Golgi apparatus and granules containing cytotoxins, such as granzymes, toward the target cell. The micrograph shows the microtubular networks of both cells (*green fluorescence*) and the lytic granules of the attacking CTL (*red fluorescence*). Subsequent fusion of these lytic cytotoxin-containing granules with the plasma membrane of the CTL results in the release and deposition of their cargo directly onto the membrane of the closely apposed target cell (*middle left*). The accompanying electron micrograph (*middle right*) shows the breakdown of intracellular elements suffered by a target cell sometime after being exposed to a granule-mediated attack by a T cell. (B) Serial killing by a CTL. A cytotoxic CD8 T cell can recognize peptide:MHC-I on one cell, program it for apoptotic death by delivering lytic granules, release it, and serially repeat the process of cell recognition, programming for death, and release. The course of the apoptosis program is illustrated by the changing configuration of the cells (*bottom, left to right*). This capacity allows a single CTL to kill other cells in the vicinity that bear peptide:MHC-I complexes recognized by its TCR. (A, Gillian Griffiths. B, Republished with permission of Springer, Kluwer/Plenum, Mechanisms of Cell-Mediated Cytotoxicity II, Second International Workshop on Mechanisms in Cell-Mediated Cytotoxicity. Annapolis, Maryland, June 10-13, 1984.)

removed from the tumor, where peptides derived from proteolysis of tumor cell debris or components are presented on the surface of DCs as peptide: MHC complexes.

In fact, DCs have the unusual ability to present peptides derived from the proteolysis of proteins on both their MHC-I and MHC-II molecules. Thus, they can present peptides from their own native internal proteins on their MHC class I molecules (like all other cells). DCs can also present peptides derived from previously scavenged (that is, foreign) proteins on their MHC-I molecules. Hence, the MHC-I molecules of DCs are employed to present peptides of both native and foreign origin. This unusual ability of DCs to concomitantly present acquired peptides on both sets of MHC molecules is called **cross-presentation**. This ability is central to the generation of CD4 helper T cells and cytotoxic CD8 T cells (**Figure 15.11**).

15.8 Tumor antigens are targets of the immune response to cancer

The above discussions lead inevitably to the questions surrounding the nature of the antigens recognized by the adaptive immune system. In most human tumor types, somatic mutations of cellular genes produce cell-heritable changes in the genomes of cancer cells, some of which result in structurally aberrant cell proteins. A subset of the novel proteins generated by these mutations, such as those creating *ras* oncogenes, may provoke an immune response, in which case they may be called **neoantigens**, a term implying both their newness and the fact that they were not the objects of tolerance developed during early immune system development.

The presence of neoantigens can be demonstrated experimentally by taking cells that are suspected of expressing them, implanting such cells in an appropriate mouse host, removing or killing these cells, and then determining whether this initial exposure has effectively immunized the host against subsequent implantation with the same strain of tumor cells. This secondary implantation of tumor cells involves live tumorigenic cells and the process of secondary implantation is used to "challenge" the immune system of such mice to determine whether long-term immunity against these cells was actually developed following the initial implantation. (Such an experiment testing for immunity against the cells originating from a particular tumor was described in Section 15.1. We repeat this story in order to emphasize that, at the molecular level, such anti-tumor immunity, if it exists, can be understood only in terms of specific antigens that must have been displayed initially by the immunizing tumor cells as well as by the subsequently introduced, challenging cells.)

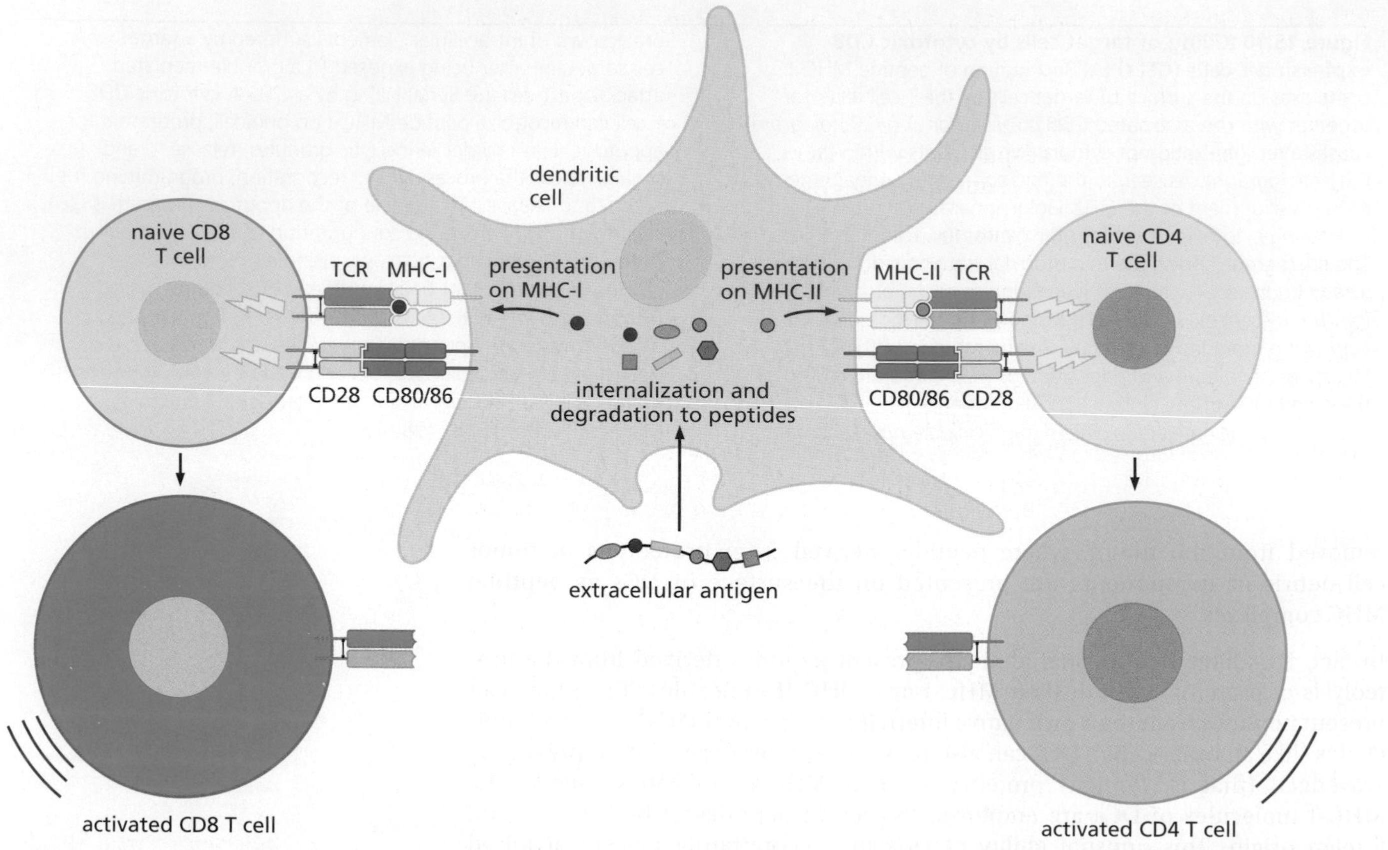

Figure 15.11 Dendritic cells employ cross-presentation to present extracellular antigens to both CD4 and CD8 T cells As described in Figure 15.8, proteins acquired through phagocytosis, endocytosis, or pinocytosis by dendritic cells (DCs) can undergo proteolysis and generate peptides, some of which are presented on the cell surface as peptide:MHC-I complexes to CD8 T cells (*blue, left*), whereas others are presented as peptide:MHC-II complexes to CD4 T cells (*green, right*). This concomitant presentation by two classes of MHC molecules is termed cross-presentation. Although the detailed cell biology of cross-presentation is complex and still unclear, the result is the capacity of DCs to present peptides derived from acquired antigens to both CD4 and CD8 T cells. As indicated in the figure, the amino acid sequences of the peptides presented following binding to MHC-I are different from those presented by MHC-II molecules.

In the event that such anti-tumor immunity has developed, the immune systems of the mice initially exposed to these tumor cells will attack the secondarily implanted cells; that is, they will reject these tumor cells. As controls, other lines of cancer cells that were independently induced previously and unlikely to express the same set of neoantigens can be used to challenge these mice. In principle, exposure to the first population of cells should not confer cross-immunity against the secondarily implanted cells, indicating that the neoantigens displayed by the first populations of cells were *specific* to these cells and not displayed by other independently arising cancer cells or, of course, by normal cells in the body. Such behavior explains the term **tumor-specific transplantation antigens** (TSTAs), which are implied to be expressed by cancer cells in certain tumors but not necessarily by the cells forming other tumors and not by normal cells throughout the body.

Indeed, tumor-specific antigens may be uniquely expressed by one set of cancer cells and not by any others that are examined thereafter. As should be apparent, these altered proteins are likely capable of inducing certain types of adaptive immune responses.

Note that the proteins resulting from either viral infection or somatic mutation of the cellular genome (as described above) will differ from those encoded by the unmutated genome of the host's cells and therefore fall in the class of tumor-specific transplantation antigens (TSTAs). Because certain TSTAs can be created by randomly occurring somatic mutations, each tumor may display its own TSTA or set of

TSTAs; that is, each TSTA will represent a unique immunological marker of the cells in that particular tumor.

Importantly, because TSTAs are not present during the early development of the immune system when immunological tolerance develops, they may well be recognized as foreign entities by the adaptive immune system, and the cancer cells bearing these TSTAs may become objects of immune attack and elimination.

As we have learned (for example, in Chapter 11), most somatically mutated genes formed during the course of tumor progression are unlikely to drive this progression forward, simply because they affect genes whose alteration does not confer advantageous phenotypes on the cells that have acquired such mutant alleles. Instead, as described in Section 11.6, these mutations are categorized as creating mutant *passenger* alleles, in contrast to mutations that affect genes whose alteration is indeed advantageous for cancer cell survival and proliferation and are, in that sense, considered to affect *driver* genes. Thus, the passenger alleles are only "going along for the ride," being present in cells that happen to have acquired advantageous driver mutations and are undergoing clonal expansion during the course of multi-step tumor pathogenesis.

Both mutant driver and mutant passenger genes may encode neoantigens (**Sidebar 15.5**). Therefore, the fact that an allele is either a driver or a passenger is unrelated to its ability to encode a neoantigen. One example of a mutation that is functionally important for tumor growth and, at the same time, able to create a neoantigen, is the mutation of the gene encoding the normal K-Ras protein into a driver oncogene that encodes a K-Ras oncoprotein (see Section 4.3). Once created, the structurally novel oncoproteins have the potential to generate peptides that may evoke T-cell responses when presented on appropriate MHC molecules.

Yet another example derives from the numerous types of chromosomal translocations that lead to fusion proteins (see Section 4.4). Such translocations are frequently found in hematopoietic malignancies, such as chronic myelogenous leukemia (CML). This well-studied translocation, which is created by the fusion of one portion of the *Bcr* gene with a portion of the *Abl* proto-oncogene, generates a novel peptide sequence at the site where the two coding sequences are joined. In fact, the highly immunogenic peptides generated from this novel peptide segment can indeed provoke T-cell responses.

In addition to TSTAs, there are some normal cellular antigens that are associated with certain tumors that may actually attract the attention of the immune system. By implication, expression of these antigens is not limited uniquely to an individual spontaneously arising tumor, and these antigens are therefore not tumor-specific. These antigens are designated as **tumor-associated transplantation antigens** (TATAs) or more simply called **tumor-associated antigens** (TAAs), with the implication that

Sidebar 15.5 Neoantigens encoded by passenger genes have implications for the success of anti-tumor therapies Most neoantigens expressed by tumor cells are encoded by mutant passenger alleles rather than by driver alleles and are, by definition, not critical to continued cancer cell proliferation and survival. The distinction between the two classes of neoantigens may be functionally critical for the outgrowth of a tumor, in particular its responses to anti-tumor immunotherapy, because cancer cells may attempt to evade immune attack by down-regulating expression of certain neoantigens.

This evasive maneuver raises the question of whether or not cancer cells will pay a price for down-regulating the expression of a particular neoantigen. If the neoantigen is encoded by a mutant driver gene allele, a cancer cell cannot afford to lose this antigenic protein, as its loss is likely to compromise further

survival and proliferation of this cell. The contrasting situation pertains to neoantigens encoded by passenger genes: in this instance, a cancer cell may freely shut down expression of the neoantigens without paying a price in terms of its selective fitness. This may, in turn, allow this cell and its progeny to evade further immune attack launched by lymphocytes that previously recognized this neoantigen and, provoked by its display, undertook to eliminate this cell.

As we will learn later, such immunoevasive maneuvers are critically important in determining whether the immune system can sense and eliminate spontaneously arising tumors, thereby preventing their clinical appearance in certain tissues. Similarly, successes of anti-tumor immunotherapy may depend on whether or not targeted tumor cells can down-regulate expression of certain neoantigens, doing so in order to escape therapy-driven killing.

Table 15.2 Examples of tumor antigens

Antigen type	Example	Nature of antigen	Occurs in indicated cancers
Oncogenic virus-encoded nonvirion protein	HPV E6 and E7 proteins	Nonstructural oncogenic proteins	Cervical carcinoma
Product of chromosomal translocation; gene rearrangements, insertions, or deletions	Bcr-Abl protein	Novel oligopeptide at site of fusion of reading frames of two otherwise unlinked genes	Chronic myelogenous leukemia
Neoantigens	Changes in primary structure resulting from point mutations in *Ras*	Oncogene generated by mutation of proto-oncogene counterpart	Many types of cancer
	Changes in primary structure of P53 resulting from any of many dozens of alternative point mutations in *p53*	Alternative amino acid substitutions at sites within or outside of DNA-binding domain	Many types of cancer
	Point mutations that alter primary structures of diverse cellular proteins	Novel protein sequences conferring potential T-cell-recognized epitopes	Many types of tumors with distinctive landscapes of mutationally generated neoepitopes
Differentiation antigens	Tyrosinase	Enzyme in melanin synthesis pathway	Melanoma
	PSMA	Cell surface protein with carboxypeptidase activity expressed by cells of prostatic epithelium	Prostate cancer
Cancer testis antigens	MAGE-1 MAGE-3	Protein normally expressed in testes; aberrantly expressed in several types of cancer	Melanoma, NSCLC, esophageal carcinoma
	NY-ESO	Protein normally expressed in testes; aberrantly expressed in several types of cancer	
Overexpressed antigens	HER-2	Receptor tyrosine kinase expressed on epithelial cell surface	Breast cancer, ovarian cancer
	MUC-1	Glycosylated protein found on epithelial surfaces	Breast cancer Pancreatic cancer
	WT1	Transcription factor	Leukemia
Abnormal post-transcriptional modification	NA17	Abnormal protein resulting from retention of introns during splicing of pre-mRNA	Leukemia

Adapted from K. Murphy, C. Weaver and L. Berg (eds.), Janeway's Immunobiology 10th ed. New York: W. W. Norton & Company, Inc., 2022.

such antigens may also be expressed by other cells in the body rather than being unique, novel antigens created during the formation of a single tumor (**Table 15.2**).

The ability of an immune system to focus on a TATA will depend, by necessity, on whether tolerance has been developed by the immune system of the tumor-bearing host. For example, such tolerance may not have been developed, as discussed earlier, simply because the particular normal antigen was expressed in a tissue, such as the testis, that is shielded from intensive immunosurveillance and therefore did not participate in the generation of immune tolerance early in the development of the immune system. Other TATAs may derive from proteins that are normally expressed only transiently during early embryonic development, long before the functional immune system and associated tolerance have been established. For example, MAGE, a protein expressed in normal cells of the testes, is expressed in certain melanomas and will behave as a TATA (see Table 15.2; **Figure 15.12**). Because certain TATAs are plausibly expressed by a variety of normal cells in the tissues of a tumor-bearing host, an immunotherapy directed against them may often result in unacceptably high levels of therapy-associated toxicities.

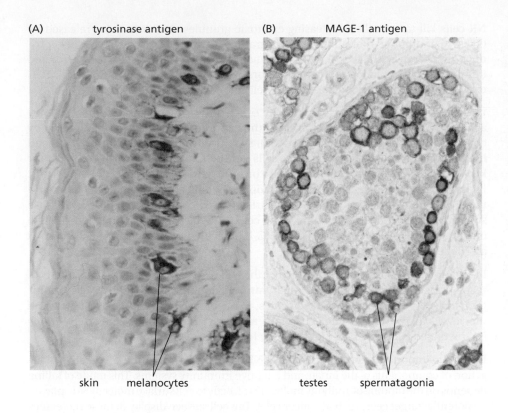

(A) tyrosinase antigen

(B) MAGE-1 antigen

skin   melanocytes

testes   spermatagonia

Figure 15.12 Normal proteins displayed as tumor-associated melanoma antigens (A) A monoclonal antibody has been used here to detect the enzyme tyrosinase, which is involved in pigment production in melanocytes and melanomas. Though present in melanocytes (*dark red*) this enzyme is not detectable in other normal tissues. Its expression by melanoma cells can cause them to become immunogenic and the targets of killing by CTLs. (B) The spermatogonia in the testes have been stained here with a monoclonal antibody against the MAGE-1 antigen (*red*); normally this antigen is only seen in one additional site—the placenta. Its expression has been detected in a variety of human tumor types and has been studied in detail in melanomas because it is often immunogenic when expressed by these tumors. (A, from Y.T. Chen et al., *Proc. Natl. Acad. Sci. USA* 93:5915-5919, 1996. With permission from National Academy of Sciences, U.S.A. B, from J.C. Cheville and P.C. Roche, *Mod. Pathol.* 12:974–978, 1999. With permission from Nature.)

Some of the TATAs may be expressed as domains of cell surface proteins. For instance, normal B cells display a protein called CD20 that is also present as a TATA in many B-cell lymphomas that derive from the normal B-cell compartment of immune cells. In theory, TATAs like CD20 could be used as targets to provoke and guide attacks by the immune system on lymphoma cells that express CD20. However, such attacks rarely happen spontaneously because normal cell proteins like CD20 are shielded by immune tolerance.

We should note that the spontaneous targeting of various normal self-antigens by a host immune system (which could lead to autoimmune conditions) is the rare exception rather than the dependable rule, as might be predicted from the usually efficient actions of the tolerance-inducing mechanisms operating in the immune system. However, in the next chapter we will discuss the deployment of technologies that are indeed capable of enlisting a variety of self-antigens as useful targets for innovative anti-cancer immunotherapies. Table 15.2 lists a variety of TATAs that have been identified in human tumors.

15.9 Natural killer cells contribute to anti-cancer immunity

Cytotoxic T cells are the most formidable weapons that the immune system deploys in the struggle to eliminate cancer cells. Confronted by the strong selective pressure applied by CTLs, it is not surprising that some cancer cells might contrive to shut down display of their class I MHC molecules, thereby depriving CTLs of the peptide:MHC complexes needed by the CTLs to recognize (via their TCRs) these cancer cells and target them for killing. Such down-regulation, like the down-regulation of TATAs described above, may allow cancer cells to elude CTL-mediated death.

Natural killer cells (NK cells) can block this potential avenue of escape from killing. These lymphocytes operate as key components of the innate immune system that are programmed to kill cells throughout the body that fail to display normal levels of MHC-I molecules (independent of the presence or absence of tumor antigens). During the evolution of the immune system, the development of NK cells was clearly selected in order to anticipate immune-evasive maneuvers undertaken by cells infected with viruses and certain other intracellular pathogens.

NK cells kill target cells by releasing cytotoxic granules similar to those released by CTLs. In addition to mounting cytotoxic granule–mediated attack, NK cells also produce interferon γ (IFN-γ), a key inflammatory cytokine that favors cell-mediated immunity by stimulating the up-regulation by potential target cells of class I MHC expression, thereby making the targeted cells more vulnerable to CTL attack.

In fact, potential target cells can display other outward-facing indicators that they are, in one way or another, aberrant and thus appropriate targets for elimination by components of the innate immune system. Specifically, the display of **stress-induced ligands** on the cell surface of cancer cells can be detected by an NK-cell receptor known as NKG2D. These ligands appear to be induced by a variety of cellular stressors, including genetic damage caused by radiation or chemical agents, viral infections, and neoplastic transformation. The NK cell surface receptor NKG2D interacts with these stress-induced ligands on potential target cells, and the resulting interaction, like the one triggered by inadequate levels of MHC-I molecules, results in strong activation of an NK cell's cytotoxic response.

In addition to the surveillance provided by the NKG2D receptor, an additional family of receptors, generically termed KIR (*K*iller *I*mmunoglobulin-like *R*eceptors), also monitors the levels of certain MHC-I molecules; many of these KIRs *inhibit* NK function if the levels of the MHC-I molecules are relatively normal and, conversely, *allow* NK-mediated killing if MHC-I levels slip below a certain threshold.

Finally, an important alternative avenue of NK cell killing involves the cooperative interaction of components of the innate and adaptive immunity systems. This mode of killing depends on the binding of antibody molecules directly to cell surface molecules displayed by potential target cells, such as cancer cells. The cell surface display of these molecules (which are tethered directly to the plasma membrane) reflects their normal localization and does not in any way involve MHC-I molecules as agents of antigen presentation.

NK cells, as agents of innate immunity, bear Fc receptors—a class of receptors that recognize and bind the constant (Fc) region of common antibody molecules. A chance encounter between NK cells and a cell whose surface is studded, even partially, by antigen-bound antibody molecules may, on its own, trigger NK cell–mediated killing of the antibody-coated cell. This process is known as **antibody-dependent cellular cytotoxicity** (ADCC; Sidebar 15.6). Many of the activities of NK cells are summarized in **Figure 15.13**.

15.10 Macrophages make multiple contributions to tumor development

Macrophages represent a category of versatile phagocytic cells that are also involved in tissue repair and often secrete cytokines capable of directing and modifying the behavior of a variety of other cell types. They are essential cellular elements of innate

Sidebar 15.6 Antibody-dependent cellular cytotoxicity (ADCC) is a major pathway of antibody-mediated killing ADCC has attracted great interest over the past two decades, largely because of the development of monoclonal antibodies (mAbs) that bind antigens displayed on the surface of various types of tumor cells. This binding, in turn, can trigger the killing of the antibody-bound cells, often through the actions of NK cells. The killing process reflects a mechanism of antibody-dependent cellular cytotoxicity that is normally operative in the body and depends on naturally arising antibody molecules that bind cell surface proteins expressed by various cells. As an example, many RNA virus particles are cloaked in a lipid bilayer from which virus-encoded glycoprotein spikes protrude. These spikes, which engender an antibody response in infected individuals, are also expressed on the surface of virus-infected cells. As a consequence, such cells may attract antiviral antibody

binding and, in turn, become victims of ADCC. Such an ADCC response also may be triggered by mAbs that are introduced into the body as part of an anticancer immunotherapy.

ADCC, as it operates naturally within the body, represents a convergence of the adaptive immune system, which forms antigen-binding antibodies, and the innate immune system, represented by NK cells that recognize antibody-coated cells. In more detail, many NK cells express on their surface the Fcγ receptors (FcRγ) that bind the constant (C) regions of antibody molecules that are already bound via their V regions to cell surface antigens. One direct consequence is the activation of a cytotoxic attack by the NK cells on the antibody-coated cells, which results in the death of the latter. The killing mechanisms depend, as cited earlier, on the exocytosis by NK cells of cytotoxic granules, including those carrying granzymes that trigger the rapid death of the antibody-bound cell.

immunity and exist in a range of specialized, functionally activated states that span a spectrum of alternative phenotypes. These phenotypes are often described, simplistically, as ranging from the M1 state at one pole of a spectrum of phenotypes to M2 at the other.

To focus on the salient traits of the macrophages: cells in the **M1 state** are cytotoxic and highly phagocytic, and they can produce IL-12, a cytokine that supports the generation of one subset of helper T cells termed T_H1. In the context of cancer pathogenesis, these attributes favor tumor control and elimination. In contrast, macrophages in the **M2 state** (sometimes referred to as **alternatively activated macrophages**) produce IL-10 and TGF-β, both of which are immunosuppressive cytokines and, accordingly, protect cancer cells from immune attack, thereby favoring tumor persistence and progression. The M2 macrophages also favor tissue repair and regeneration, support vascularization by secretion of VEGF (see Section 13.5), and secrete proteases that contribute to tissue remodeling.

Actually, most macrophages do not reside at the extreme poles of the M1/M2 spectrum of phenotypes. Instead, depending on conditions, many adopt phenotypes that are aligned more toward one or the other of these states and thus enter mixed phenotypic states located at various intermediate positions along this spectrum. Accordingly, the extent to which macrophages contribute to tumor control and elimination or, alternatively, to tumor maintenance and progression, depends on where they reside along this continuum of phenotypes. Some of the many roles played by macrophages in normal and tumor physiology are illustrated in **Figure 15.14**.

As an example, T cells are often important regulators of the direction and degree of macrophage polarization. Members of the T_H1 subset of helper T cells (see Sidebar 15.3) and CTLs secrete interferon-γ, a cytokine that favors M1-like behavior, including phagocytic behavior and the generation of oxidizing agents, such as nitric oxide, that macrophages deploy to kill targeted cells. In contrast, other cytokines, specifically the IL-4 and IL-13 secreted by another set of helper T cells termed T_H2 cells, can drive macrophages toward tumor-promoting M2-like phenotypes.

As multi-step tumor progression proceeds, at least during the formation of common, well-studied carcinomas, cancer cells begin to release immunosuppressive cytokines, the most prominent of which is TGF-β. This cytokine, on its own, exerts direct immunosuppressive effects by inducing the formation of M2 macrophages as well as the formation of an immunosuppressive subset of lymphocytes termed **regulatory T cells** (T_{reg}'s; to be described in Section 15.11).

Macrophages can attack and eliminate cancer cells simply by consuming them through the process of phagocytosis (see Figure 15.14C). Some cancer cells, however, having attracted the attentions of innate immune cells such as macrophages, may contrive a strategy for eluding macrophage-dealt phagocytic death; they do so by exaggerating an interesting aspect of normal tissue physiology. In order to protect themselves against inadvertent phagocytosis by roving macrophages, normal cells throughout the body display the CD47 protein—a "don't-eat-me" signal—on their surface. The CD47 receptor is one of several known receptors employed by normal cells throughout the body, including erythrocytes, to avoid inadvertent macrophage attack. CD47 levels are also used to determine the lifetimes of specific cell types in the body. For example, recently formed erythrocytes display high levels of CD47 that decline progressively in such cells until they approach the end of their pre-ordained lifetimes; at this stage CD47 has lost its ability to stave off attack by macrophages and closely allied immune cells, resulting in the removal of the aged red cells via phagocytosis.

Not surprisingly, many types of cancer cells up-regulate display of their CD47 don't-eat-me signal on their surfaces, thereby discouraging would-be macrophage attackers and overriding any tendency of macrophages to kill them via phagocytosis. Indeed, clinical observations have shown that circulating cancer cells, often referred to as circulating tumor cells (CTCs), released from certain human breast carcinomas display high levels of CD47, presumably offering protection against macrophages that are present in the circulation.

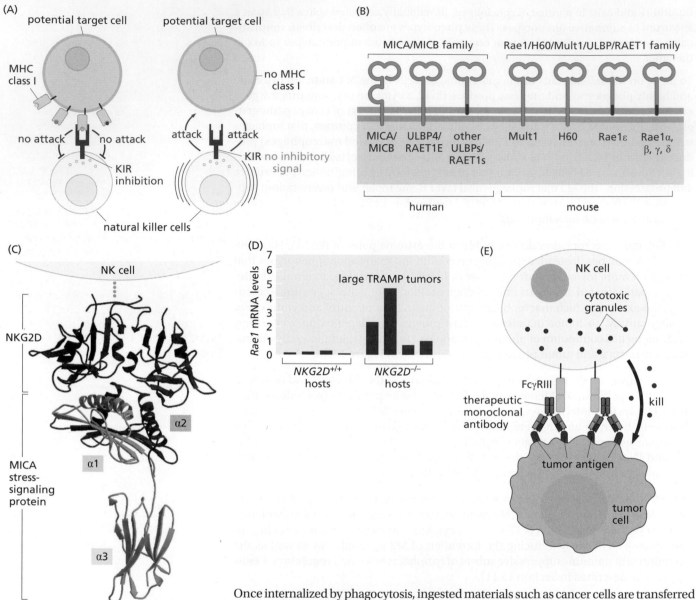

Once internalized by phagocytosis, ingested materials such as cancer cells are transferred to the lysosomes where they are enzymatically degraded. Phagocytosis can be facilitated if a potential target cell is coated by antibodies attached to cell surface molecules displayed by the target cell, much like the ADCC that is executed by NK cells (Section 15.9 and Sidebar 15.6). Phagocytic killing of target cells by this antibody-assisted route is called antibody-dependent cellular phagocytosis (ADCP). Like NK cells that carry out ADCC, macrophages display Fc receptors that, when engaged by antibodies bound to cell surface antigens, enable macrophages to kill the antibody targeted cells.

In fact, antibodies directed against tumor-associated cell surface antigens are employed by a variety of anti-cancer immunotherapies but do not seem to play major roles in the natural control and elimination of cancers. This explains why macrophage-mediated ADCP and NK-mediated ADCC are unlikely to be major factors in determining the clinical progression of most cancers. However, in the context of clinical settings, where monoclonal antibodies (MoAbs) targeting cancer cells are employed as therapeutic agents to kill cancer cells, both ADCP and ADCC can indeed be important contributors to the elimination of cancer cells. Examples of these antibody therapies will be presented in the next chapter.

Following the early stages of tumor development, a period when M1-like macrophages may contribute to cancer cell death, **tumor-associated macrophage** (TAM) populations progressively shift their phenotypes in ways that are more supportive of further tumor outgrowth, that is, increasingly transitioning from an M1 into an

Figure 15.13 Control of killing by natural killer cells is mediated by inhibitory and activating signals (A) Natural killer (NK) cells display *inhibitory* receptors that can recognize MHC class I molecules being displayed on the surface of cells. When MHC-I molecules are present at levels above a threshold, these inhibitory receptors (*red*) provide negative signals that discourage NK-mediated cell killing. One class of inhibitory receptors are those of the KIR (*Killer Immunoglobulin-like Receptor*) family, which recognize some MHC-I molecules but not others. Another type is CD94–NKG2A, a heterodimer (*not shown*). In both cases, recognition by NK cells of MHC-I molecules displayed by a nearby potential target cell can dampen or block NK-mediated attack (*left*); this recognition of MHC-I molecules can be achieved by one or the other and sometimes both types of these inhibitory receptor types displayed by the NK cell. In the absence of such recognition and functional silencing by either of these inhibitory receptors, NK-mediated attack is unimpeded (*right*). (B) In contrast to the actions of these inhibitory receptors, other NK-associated receptors are involved in *activation* of NK cells by proteins; as in (A), this type of response is triggered by cell surface proteins displayed by potential target cells. Thus, when many types of cells are suffering certain types of cell-physiological stress, they may display alarm proteins on their surface. Notable causes of such stresses include genetic damage, viral infection, and neoplastic transformation. Shown here is a series of frequently displayed human and mouse cell surface alarm proteins (*blue*). (C) Seen here are ribbon diagrams of the molecular structures of the components of the complex formed between NKG2D, an NK-associated homodimeric receptor (*light purple, dark purple, above*) and its cognate ligand, the MICA alarm protein displayed by a human cell, such as a cancer cell (*α1 brown, α2 magenta and α3 green, below*), under stress. (The domains that anchor these various proteins to the surfaces of the NK and the target cell are not shown.) (D) Tumor cells arising in the TRAMP transgenic mouse model of prostate adenocarcinoma development express on their surface the Rae1 alarm ligand (see panel B), which serves as an activating ligand for the NKG2D receptor displayed by NK cells. The tumor cells formed in TRAMP host mice that carry wild-type versions of the NKG2D-encoding gene expressed low levels of *Rae1* mRNA. In contrast, in TRAMP host mice in which the *Klrk1* gene encoding NKG2D had been knocked out, the tumor cells expressed far higher levels of the *Rae1* transcript. This suggests that TRAMP tumor cells can down-regulate their Rae1 expression in order to avoid attack and elimination by the NK cells that display NKG2D. (E) The binding of FcγRIII receptors displayed by NK cells to antibodies previously bound to antigens displayed on the surfaces of potential target cells (represented here by a tumor cell), causes the directed release of cytotoxic granules from the NK cell into the target cell, resulting in the death of the latter; this process is termed antibody-dependent cytotoxicity (ADCC). (B, from D. Raulet, *Nat. Rev. Immunol.* 3:781–790, 2003. C, from P. Li et al., *Nat. Immunol.* 2:443–451, 2001. Both with permission from Nature. D, from N. Guerra et al., *Immunity* 28:571–580, 2008. With permission from Elsevier. E, adapted from P. Parham, The Immune System, 5th ed. New York: W.W. Norton & Company, 2021, p. 522.)

M2-like state. This seems to explain why clinical observations have shown that extensive macrophage infiltration of a clinically detectable tumor is generally an indicator of the presence of M2 macrophages and poor clinical prognosis. As an example, one meta-analysis of 55 studies of human cancers showed that dense infiltration of macrophages within a tumor mass correlated with poor overall survival in a wide variety of solid tumors, including breast, bladder, thyroid, and ovarian carcinomas. Images of macrophages presented in Figure 15.14 illustrate some of the many roles played by this class of versatile cells.

15.11 Regulatory T cells are indispensable negative regulators of the immune response that are co-opted by tumors to counteract immune attack

In contrast to the contributions made by most T-cell types to the elimination of cancer cells, one subtype of CD4 T cell, termed regulatory T cells (T_{reg}'s), actively aid tumor maintenance and progression, doing so by inhibiting various kinds of adaptive anti-tumor immune responses. More generally, T_{reg}'s play key immunoregulatory roles that are critical to the maintenance of normal tissue homeostasis. They do so by eliminating or suppressing autoreactive T cells, thereby playing a key role in ensuring immune tolerance. This particular function of T_{reg}'s is illustrated by studies using genetically altered mice that lack T_{reg} cells; such mice are afflicted by a lethal autoimmune disease that develops early in life. Similarly, humans who are unable for one reason or another to form T_{reg}'s develop an aggressive and ultimately fatal autoimmune disease.

Like the negative-feedback loops that modulate proper levels of mitogenic signaling in cells (as described in Section 6.14), the adaptive immune system requires counterbalancing inhibitory control mechanisms to constrain its actions. In the context of cancer development, T_{reg}'s can compromise the ability of the immune system to restrain tumor progression or eliminate tumor cell populations. We now know that they do so by inhibiting the effector functions and the proliferation of helper T cells and of CTLs. In addition, the actions of T_{reg}'s extend beyond their effects on T cells. Thus, they can inhibit effective antigen presentation by dendritic cells (see Section 15.7) as well as suppressing the anti-tumor activities of macrophages (see Section 15.10).

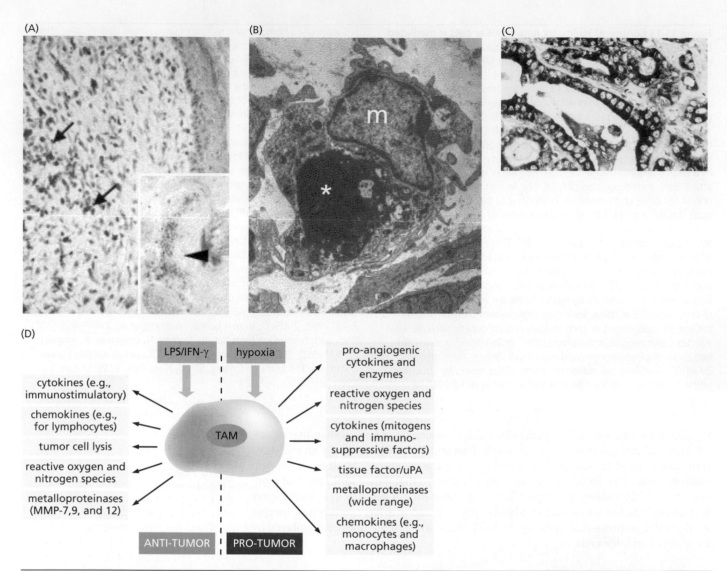

Figure 15.14 Multiple roles of macrophages Macrophages play critical roles in promoting wound healing and tumor progression. They also act in an opposing fashion as agents of the immune system to present tumor antigens and to consume tumor cells. These different functions are carried out by distinct subtypes of macrophages. (A) In this wound-healing site in the skin of a mouse, macrophages (*brown*) are detected in abundance (*arrows*) through use of immunochemical procedures that recognize the F4/80 macrophage-specific antigen and the CSF-1R receptor marker (*arrowhead, inset*). (B) Transmission electron microscopy reveals a macrophage (m) that is engorged with the phagocytosed corpses of apoptotic cells (*asterisk, dark body*) in the middle of a wound-healing site. (C) A macrophage (*pinkish*) is seen as it begins to phagocytose a tumor cell in the lumen of a duct within a papillary thyroid carcinoma. (D) Macrophages can be activated by diverse signals, such as interferon-γ (IFN-γ) and bacterial lipopolysaccharide (LPS), the latter being used as an adjuvant to potentiate the immune response. Once activated, these macrophages, dubbed M1 or M1-like, can function to acquire and present antigens to T_H cells and to effect tumor-cell killing by antibody-dependent cellular phagocytosis (ADCP) and antibody-dependent cellular cytotoxicity (ADCC; *left*). Acting in the opposite direction, hypoxia and a variety of tumor-derived signals direct the activation and differentiation of macrophages into the M2 phenotype, which causes them to foster rather than antagonize tumor formation (*right*); especially important among these M2 functions are those that create immunosuppressive tumor microenvironments. (A and B, from P. Martin et al., *Curr. Biol.* 13:1122–1128, 2003. With permission from Elsevier. C, from A. Fiumara et al., *J. Clin. Endocrinol. Metab.* 82:1615–1620, 1997. With permission from Oxford University Press. D, from L. Bingle, N.J. Brown and C.E. Lewis, *J. Pathol.* 196:254–265, 2002. With permission from John Wiley & Sons.)

The formation of T_{reg}'s is strongly encouraged by TGF-β, which can induce their differentiation from CD4 lymphocyte precursors. Such differentiation can be monitored by the acquired expression of yet another cell surface marker protein, termed CD25. This induction of the T_{reg} state becomes important in the context of tumor pathogenesis because many types of transformed cells, including more mesenchymal carcinoma cells (see Section 13.3), actively secrete TGF-β, which can proceed to encourage the formation of more immunosuppressive, pro-tumor tissue microenvironments.

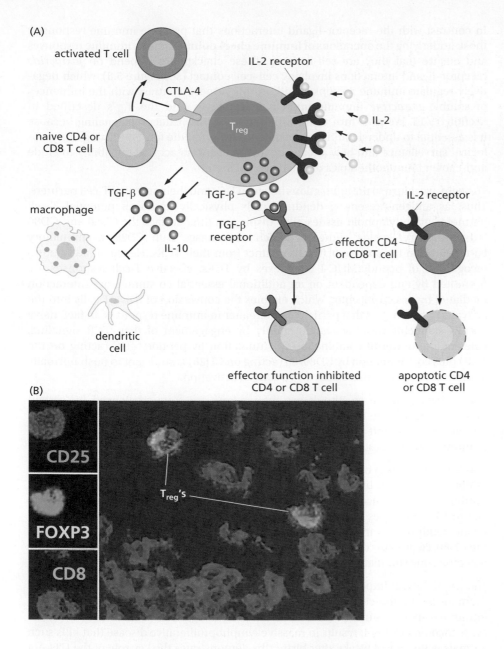

Figure 15.15 Regulatory T cells
(A) Regulatory T (T_{reg}) cells act on other immunocytes in multiple ways. The CTLA-4 receptor has a higher affinity for the co-stimulatory ligands CD80/86 than does CD28 on other T cells. Consequently, CTLA-4 can bind to these co-stimulatory ligands and inhibit the activation of naive T cells to functional effector T cells by preventing them from receiving the co-stimulation on which their priming depends (*above left*). Interleukin-2 (IL-2) is an essential factor for the survival and proliferation of effector T cells. T_{reg}'s display high densities of the IL-2 receptor (IL-2R) on their surface, allowing them to bind and sequester IL-2, thereby denying the essential support by IL-2 to other T cells in their vicinity (*above right*). TGF-β secreted by T_{reg} cells can inhibit the activity of multiple types of effector T cells, including subsets of CD4 helper T cells, as well as CTLs. T_{reg}'s also secrete IL-10, a potent inhibitory cytokine that collaborates with TGF-β in reducing the activity of dendritic cells and macrophages (*lower left*). By inhibiting helper T cells with secreted cytokines like TGF-β, these T_{reg} cells can reduce antibody production by B cells (*middle left*). By depriving CD4 and CD8 T cells of IL-2 (*middle right*), T_{reg} cells can also trigger apoptosis of CD4 and CD8 cells. (B) Immunofluorescence staining reveals the presence of T_{reg}'s through their expression of the CD25 surface antigen (*red*) and the FOXP3 transcription factor (*green*), which is localized to the nuclei of these cells. They are seen here amid CD8+ cytotoxic T cells (*blue*), whose actions they are ostensibly inhibiting; these lymphocytes were present in the ascites of a patient suffering from ovarian cancer. (A, adapted from X. Meng, et al., *Nat. Rev. Cardiol.* 13:167–179, 2016. B, adapted from E.M. Shevach, *Nat. Med.* 10:900–901, 2004. With permission from Nature.)

T_{reg}'s often constitute a significant proportion of the tumor-infiltrating lymphocytes (TILs) in lung, ovarian, breast, and pancreatic carcinomas as well as the immunocytes present in tumor **ascites**—the fluid accumulated in the peritoneal cavities of patients with several types of advanced cancers. Many tumors actively recruit T_{reg} cells to their associated microenvironments by releasing the **chemokine** CCL22. Regulatory T cells display a receptor for CCL22 and can sense CCL22 in their surroundings and advance toward the tumor cells that are generating the CCL22, doing so by following gradients of increasing concentration of this chemokine. Thus, this chemokine serves as a chemoattractant for these cells. These various aspects of T_{reg}-mediated immunosuppression are summarized in **Figure 15.15**.

15.12 Immune checkpoints act to limit immune responses

Once set in motion, immune responses can pose threats to the normal host tissues comparable to those created by infectious pathogens or tumor cells. An elaborate set of controls, including the T_{reg}'s just described, ensures that these responses are limited and eventually shut down once immune responses have completed their work. Moreover, as argued repeatedly above, over extended periods of time, organismic survival depends on mechanisms of immune tolerance to spare normal tissues from immune attack.

In contrast with the receptor–ligand interactions that promote immune responses, those underlying the operations of **immune checkpoints** dampen immune responses and ensure that they are self-limiting. These checkpoints depend on *juxtacrine* receptor–ligand interactions involving cell–cell contact (see Figure 5.8), which negatively regulate immune responses and, in this respect, contrast with the influences of soluble *paracrine* immunosuppressive cytokines and the T_{reg}'s described in Section 15.11. We now turn to an exploration of this juxtacrine signaling because it is essential to understanding how cancers develop in the face of existing immunological surveillance and how already developed cancers acquire the ability to elude anti-cancer immunotherapies directed against them.

To begin, we return to the interactions between DCs and their potential T-cell partners. Thus, as antigen-presenting dendritic cells physically approach potential T-cell partners within lymphoid tissues, they display on their surface the CD80 and CD86 "co-stimulatory ligands" either of which can engage with CD28, a stimulatory co-receptor on the surface of T cells distinct from their TCRs. Hence, in addition to recognition of peptide:MHC-I complexes by TCRs, effective T-cell responses are promoted by and dependent on an additional essential co-stimulatory interaction mediated by this co-receptor, which enables the conversion of naive T cells into the activated lymphocytes that participate thereafter in immune responses. In fact, naive T cells staunchly resist activation simply by engagement of their TCR signaling, whereas the powerful combination of stimulation by peptide:MHC (acting on the TCR) plus co-stimulation by CD80/86 (acting on CD28) is sufficient to push normally activation-resistant naive T cells into a state of activation.

The actions of these co-stimulatory interactions are counterbalanced by certain immune checkpoint mechanisms, which include **co-inhibitory** receptor–ligand interactions that limit the *intensity* of T-cell activation during the initial induction of immune response described above (**Figure 15.16**).

This initial activation of a T-cell in a lymphoid tissue is limited by a counterbalancing inhibition mediated by the CTLA-4 T-cell co-inhibitory receptor, this action representing an initial critical checkpoint control occurring within the lymphoid tissue. As Figure 15.16 indicates, CTLA-4 is activated by its cognate CD80/86 ligand displayed by the dendritic cell and proceeds thereafter to damp down T-cell activation. Hence, the CD80/86 proteins displayed by DCs serve as ligands for two mutually antagonistic receptors, one stimulatory, the other inhibitory (**Sidebar 15.7**).

The physiological importance of the CTLA-4 inhibitory receptor is illustrated most dramatically by the consequences of its deletion from the control circuitry of the immune response. Studies in mice demonstrate that germ-line deletion of *ctla4*, the gene encoding CTLA-4, results in massive lymphoproliferative disease that kills such animals within 3 to 4 weeks after birth; this demonstrates the key role of the CTLA-4 response in constraining T-cell *proliferation* following an initial encounter with activating antigens.

Later, when the now-activated T cell leaves a lymphoid tissue and disperses to tissues throughout the body, the T cell's inhibitory PD-1 co-receptor comes into play. Thus, if and when the TCR of this CTL recognizes cognate antigens displayed by these target cells, this T cell will launch an attack on the latter. Only then does the T cell's PD-1 co-receptor confront its PD-L1/2 ligands displayed by the target cells in one or another peripheral tissue. Thus, the juxtacrine PD-1:PD-L1/2 interaction will eventually intervene to curtail further T-cell activity. This shutdown of further CTL-mediated attack may also occur, for example, upon protracted antigen stimulation or exposure to certain inflammatory conditions.

These two sets of interactions—CTLA-4 with CD80/86 in lymphoid tissues and PD-1 with PD-L1/2—represent the signaling agents controlling the two successive immune checkpoint mechanisms that have been most thoroughly studied to date, and both underlie now well-established cancer immunotherapies as we will read in the next chapter.

In the presence of a sufficiently high level of PD-1 ligands, persistent antigen stimulation in a peripheral tissue may drive a T cell into a state of **T-cell exhaustion**.

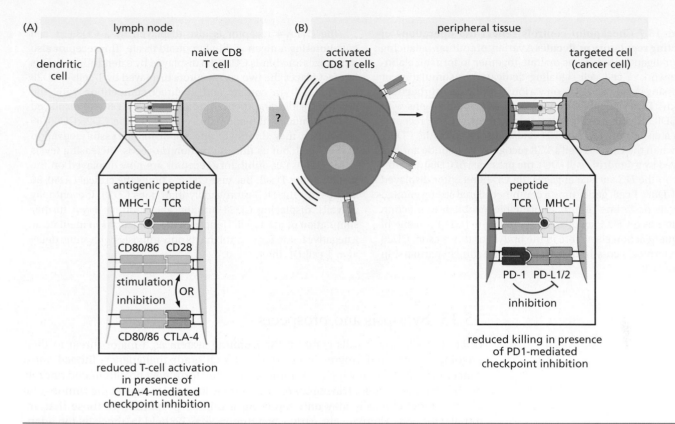

Figure 15.16 Checkpoint inhibition of T-cell activation and effector function Checkpoint controls ensure the limiting or shutdown of the T-cell response and operate in two successive phases of response development. (A) Activation phase. A naive T cell is primed within a lymph node by engagement of its TCRs with cognate peptide:MHC-I complexes presented by a dendritic cell (DC) (*left*). The essential co-stimulation that is necessary to augment signaling from the TCR into the T cell is provided by engagement of the T cell's co-stimulatory receptor, CD28, with stimulatory ligands (CD80/86) displayed on the surface of the dendritic cell (*inset, above*). The combined strength of these signals is sufficient to force activation of the naive T cell, causing it to undergo proliferation and develop effector functions. An alternative counteracting response occurs when the CTLA-4 receptor of a T cell, rather than its CD28 receptor, binds to CD80/86 of a dendritic cell: rather than stimulating the T cell and causing its activation, this CTLA-4:CD80/86 binding generates signals that inhibit T-cell activation (*inset, below*). CTLA-4 binding to the CD80/86 ligands is stronger than is CD28 binding; hence, the inhibitory effects of CTLA-4 dominate if both are expressed. As the activation of

T cells mediated by CD28 proceeds, the levels of CTLA-4 increase progressively, ultimately ensuring the shutdown of further CD28 signaling and resulting T-cell activation. Those T cells that undergo successful activation leave the lymph node via the circulatory system and are carried to peripheral tissues. This negative-feedback control of T-cell activation operates independently of the T_{reg} inhibition described in Figure 15.15. (B) Effector phase. Those CD8 T cells that are successfully activated by antigen exposure (*broad blue arrow*) are exported from the lymph node and disperse into peripheral tissues, in this case a tumor (*right*). There they enter the tumor microenvironment (TME), where recognition of peptide:MHC complexes displayed by cancer cells can provoke a cytotoxic attack by the activated CD8 T cells. However, the presence of PD-1, an inhibitory receptor also displayed on the surface of the cytotoxic CD8 cells, renders them susceptible to inhibition if they happen to encounter the inhibitory ligands PD-L1 or PD-L2 (PD-L1/2) displayed either by cancer cells or by other noncancerous bystander cells present in the TME (*inset*). (Adapted from R.J. Lee, J.S. Abramson and R.A. Goldsby, Case Studies in Cancer. New York: W. W. Norton & Company, Inc., 2018.)

Exhausted T cells typically display high levels of PD-1, together with other inhibitory checkpoint receptors, and at the same time exhibit significantly reduced functional capacity. However, despite the implications of the term "exhausted," such T cells have not lost all functional capacity and may still contribute to T-cell responses, including anti-tumor responses, albeit at greatly reduced levels.

To summarize, the two checkpoint controls, involving the CTLA-4 and PD-1 co-receptors operating one after the other, exert a moderating influence that is necessary to (1) avoid excessive, acute immune responses, (2) discourage the development of auto-immunity and, as a consequence, (3) prevent or limit the tissue damage and systemic pathology that can result from prolonged or chronic T-cell activation and accompanying chronic inflammation.

Sidebar 15.7 Checkpoint controls reflect the operations of conflicting regulatory molecules A variety of counter-balancing receptor–ligand interactions operate together to form an elaborate network of cell–cell signaling dedicated to stimulating or suppressing immune attacks on various normal cells throughout the body. To review, this sequence of interactions begins with naive CD8+ T cells in lymphoid tissue, where the CD28 receptor acts as a co-receptor with the TCR also displayed by the CD8+ T cell; when the CD8+ T cell's TCR recognizes a peptide antigen displayed by a dendritic cell (DC), the presence of CD80/86 molecules on the DC surface engages the CD28 receptor displayed by the CD8+ T cell, generating a combined signal strong enough to activate or "prime" it and, in principle, enable it to function thereafter as a CD8+ cytotoxic T lymphocyte (CTL) capable of subsequent action elsewhere in the body. For this reason, CD28 is often termed a **co-stimulatory** receptor, acting in partnership with the TCR.

The CTLA-4 receptor is also displayed by a CD8+ T cell encountering antigen within lymphoid tissue. This receptor also recognizes and binds CD80/86 displayed by a dendritic cell. This last fact places the two co-receptors displayed by T cells—CD28 and CTLA-4—in conflict and in direct competition with one another for binding their common ligands CD80/86 displayed by DCs. Initially, the stimulatory effects of the CD28:CD80/86 association may dominate and result in successful activation of the T cell, but as this interaction proceeds, increasing levels of the CTLA-4 co-inhibitory receptor are now displayed on the surface of the T cell. Because CTLA-4 binds the critical CD80/86 ligands (on the DC) more avidly than does CD28, it eventually prevails, displacing CD28 binding and shutting down further stimulation of the T cell. Hence, CTLA-4 functions to mediate as a negative feedback control that is only imposed with some delay after T cell:DC interaction.

15.13 Synopsis and prospects

Once formed, cancer cells confront the immune system as a major threat to their ongoing survival and progression to states of high-grade malignancy. Indeed, most cancer cells and the tiny incipient tumors that they form may be eliminated entirely by the immune system, leaving no record of their formation; hence, the tumors that are registered clinically may only represent a minute proportion of those that are initially formed. Alternatively, early-stage tumors may be held in check by the adaptive immune system and limited to forming small indolent growths that may never become clinically manifest. The cells in such tumors may continue to proliferate slowly, with immunogenic subpopulations being continually eliminated by CTLs and by other cytotoxic immune cells. These mechanisms, to the extent that they operate, effectively preclude us from gauging precisely how effective the immune system truly is in protecting us against cancer.

Nonetheless, macroscopic tumors do erupt, underscoring why avoidance of immune destruction is considered a fundamental hallmark of cancer (see Section 11.3). Eventually, an initial delicate balance between survival and immune destruction may be disrupted, tipping away from susceptibility to immune attack and toward expansion of the tumor cell population. Ironically, the powerful selective force exerted by the immune system that is designed to eliminate cancer cells can provide a mechanism that promotes the evolution of cancer cell populations that are resistant to immune attack.

Tumor cell populations are heterogeneous and some cells within individual tumors have antigenic profiles that make them more immunogenic than others. One consequence of the operation of immune selection is likely to be the elimination from populations of heterogeneous, early-stage tumor cells of the more immunogenic variant cells, leaving behind populations that may be more capable of resisting immune attack, in part via some of the mechanisms described in this chapter.

The selective processes that favor the emergence of less immunogenic cancer cells have been termed **immunoediting**. Immunoediting represents a form of Darwinian selection in which fitness is determined by absence of effective antigenicity while the adaptive immune system works as the selective force. One inevitable consequence often follows: over extended periods of time, immunoediting can result in the evolution and eventual outgrowth of tumor cell populations against which the immune system is powerless.

The most obvious strategy mounted by tumor cells to avoid such selection via immunoediting involves shutdown of the expression of immunogenic antigens. In fact, tumors often harbor cell clones that have entirely lost expression of certain tumor antigen-encoding genes as a consequence of DNA methylation of the promoter regions of these genes (see Section 7.7) or outright deletion of the encoding chromosomal DNA sequences. Many of the tumor antigens affected by these types of cancer cell-intrinsic immunoevasive maneuvers derive from proteins whose continued expression is not critical to continued cancer cell viability and proliferation. (Such proteins are often, for example, the products of mutant passenger gene alleles.)

In many cases, however, cancer cells cannot simply resort to down-regulation of the expression of an immunogenic tumor antigen if its continued functions are essential for cellular function or neoplastic proliferation. For example, the overexpressed but otherwise-normal HER2/Neu protein—a growth factor receptor displayed by more than 20% of human breast cancers—may provoke an anti-tumor immune attack (see Figure 4.3). In this case, however, the neoplastic cells in these breast carcinomas *depend* on continued expression and functioning of this receptor protein to drive their proliferation and avoid apoptosis, and for this reason, they simply cannot afford to shed this critical tumor antigen. One response by a cancer cell unable to down-regulate expression of a key tumor antigen might involve the remodeling of this antigen through changes in its primary amino acid sequence that erase an antigenic determinant; this maneuver often does not represent a viable solution, given the potentially disruptive effects of many structural changes on the continued functioning of an essential cancer-forming protein.

Yet another alternative immunoevasive strategy may derive from reducing or halting the cell surface presentation of a targeted tumor peptide antigen by MHC-I proteins. This approach avoids attack by cytotoxic T cells (CTLs) whose TCRs specifically recognize the peptide:MHC-I complexes formed by such an antigen. This strategy can be executed by either a shutdown of MHC class I protein expression or a failure to properly load peptides onto the MHC-I carrier molecules during their maturation in the endoplasmic reticulum. There is evidence that both these evasive maneuvers are employed.

Suppression of MHC class I protein expression is often achieved through transcriptional repression of their encoding genes. In fact, many types of human cancer cells have been found to lack normal levels of the mRNAs encoding these MHC molecules, thereby preventing their synthesis and thus display of antigens at the cell surface (**Figure 15.17**). Alternatively, cancer cells can also use post-translational mechanisms to reduce MHC class I-mediated antigen presentation. For instance, the migration of MHC molecules from the endoplasmic reticulum to the cell surface depends on their association with β_2-microglobulin protein (β_2m). Normally, β_2m escorts the MHC molecules and their peptide cargo to the cell surface (**Figure 15.18A**). In certain high-grade tumors, however, the absence of β_2m synthesis prevents the peptide-loaded MHC class I molecules from reaching the cell surface (Figure 15.18B).

An even earlier step in antigen presentation may also be compromised: some tumors have defective TAP1 or TAP2 (*t*ransporter *a*ssociated with *a*ntigen *p*resentation) proteins (see Figure 15.18A). These proteins are essential for the transport of peptides generated by the proteasomes in the cytosol to the ectodomains of MHC class I molecules present in the lumen of the endoplasmic reticulum. Without help from both TAP proteins, antigen presentation by MHC molecules once again fails (Figure 15.18C).

These various attempts at immunoevasion can be countered, however, through dispatch by the immune system of NK cells, which are programmed to attack cells that lack adequate levels of MHC class I protein expressed on their surface. The tumor, for its part, may respond in turn by launching a counter-counterattack, doing so by secreting TGF-β, which proceeds to shut down NK cells. The back and forth, cat-and-mouse games may continue for extended periods of time with ultimate outcomes that are hardly predictable.

Another defense against attacks launched by the immune system depends on cancer cells modifying the extended tumor microenvironments (TME) around them, using diffusible substances to weaken various types of immune attacks. Interleukin-10 (IL-10) and TGF-β are broadly immunosuppressive agents that are produced by some cancer cells and by certain types of immune cells present in the TME. They suppress T-cell

(A)

(B)

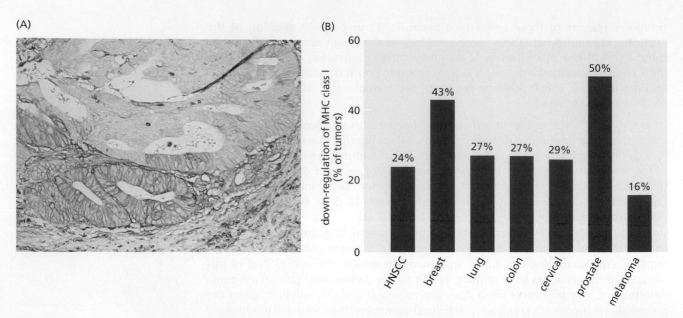

Figure 15.17 Immunoevasion through suppression of MHC class I expression Cancer cells will often down-regulate expression of their MHC class I molecules, ostensibly in order to avoid recognition and attack by cellular components of the adaptive immune response. (A) Shown here are cells of a human colorectal tumor that have been immunostained with an antibody that recognizes a specific subtype of human MHC class I molecule termed HLA-A. The carcinoma cells in the lower part of the tumor strongly express HLA-A (*dark red*), whereas the cells in the upper part of the tumor have partially or totally shut down HLA-A expression. (B) Shown on the ordinate is the percentage of tumor samples in which down-regulation of MHC class I expression occurred in a significant segment of the tumors examined (HNSCC, head-and-neck squamous cell carcinoma.) (A, from A.G. Menon et al., *Lab. Invest.* 82:1725–1733, 2002. With permission from Nature. B, courtesy of S. Ferrone.)

proliferation and can inhibit antigen presentation by dendritic cells. Under appropriate conditions, TGF-β can induce the generation of the strongly immunosuppressive regulatory T cells and inhibit NK function, both being strategies cited earlier.

Carcinoma cells that have activated portions of the epithelial-mesenchymal transition (EMT) cell-biological program described in Chapters 13 and 14 often secrete significant amounts of TGF-β that they depend on, since it operates via an *autocrine* signaling mechanism to help these cells maintain their continued residence in more mesenchymal phenotypic states. However, this secreted TGF-β can also extend its reach further by acting in a *paracrine* manner to suppress a variety of immune cells that have ventured too close to islands of tumor cells. In the case of IL-10, study of the Epstein–Barr virus (EBV) genome (see Section 3.4) has revealed the presence of an IL-10-like viral gene of clear cellular origin that was acquired by the distant ancestors of the virus. In an earlier chapter, we read that this virus is an important etiologic agent of Burkitt's lymphomas, other lymphomas of the B-cell lineage, and nasopharyngeal carcinomas. It appears that EBV forces an infected cell to release an IL-10-like immunosuppressive cytokine, protecting this cell from attack by cytotoxic immune cells, which serves in turn to preserve the infected host cell in a viable state for an extended period of time while the virus exploits its continued viability to ensure viral proliferation and progeny virus formation.

Tumors host a broad variety of immunosuppressive cell populations. The repertoire of suppressive populations may include one or more of the following cell types: regulatory T cells, tumor-associated macrophages (TAMs), tumor-associated neutrophils (TANs), and myeloid-derived suppressor cell (MDSC) populations. Increased numbers of these immunocyte populations provide immunosuppressive environments that are hostile to immune control and elimination of tumor cells. In fact, a subset of B cells, known as regulatory B cells (B_{reg}'s) has recently come to light. Similar to their T_{reg} cousins, B_{reg}'s secrete the immunosuppressive cytokines IL-10 and TGF-β. They have also been found in the tumor microenvironment and are associated with the suppression of immune responses. Further study will provide more indications of their impact on tumor maintenance and progression and of their utility as prognostic clinical indicators.

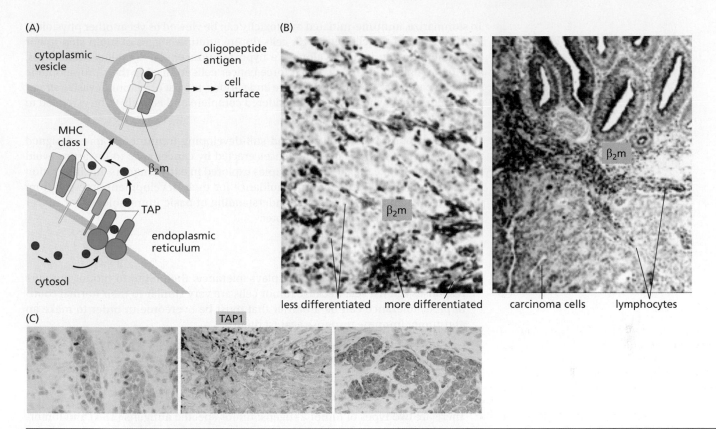

Figure 15.18 Failure of antigen presentation from loss of TAP1 or β₂-microglobulin (A) The TAP complex is a heterodimer of the TAP1 and TAP2 proteins (transporters associated with antigen processing; *dark green*) that is involved in transporting oligopeptides (*red dots*) from the cytosol into the lumen of the endoplasmic reticulum, where the oligopeptides can be loaded onto MHC class I molecules (*yellow*). Independent of this, the β₂-microglobulin (β₂m, *light brown*) molecule associates with MHC class I molecules and helps to shepherd them, together with their captured oligopeptide antigens, to the cell surface. The MHC class I molecules require both a bound oligopeptide antigen and an associated β₂m molecule in order to reach the cell surface via cytoplasmic vesicles (*above*). Hence, defects in either the β₂m or TAP protein expression can preclude MHC-I-mediated presentation of oligopeptide antigens on the cell surface. (B) In one human colorectal carcinoma (*left panel*), β₂m staining (*reddish brown*) is present in the cancer cells forming the more differentiated tubular structures (*below*), while the less differentiated carcinoma cells (*light blue, above*) fail to express any β₂m. In a second colorectal carcinoma (*right panel*), β₂m is present in infiltrating lymphocytes (*dark red*) but is totally absent in many of the carcinoma cells (*light blue, bottom half*). In 5 of 17 microsatellite-unstable (MIN) colorectal tumors (see Section 12.3), the β₂m gene was found to have suffered inactivating mutations and was transcriptionally silenced through unknown mechanisms in a number of other tumors. (C) Defective antigen processing is frequently seen in head-and-neck squamous cell carcinomas (HNSCCs), in which expression of TAP1 or TAP2 is often lost. The immunohistochemical staining shown here reveals the expression of TAP1 (*beige*) in a group of HNSCCs to be variable, being absent in some tumors (*left*), present in patches in others (*middle*), and normally expressed in yet others (*right*). (A, from C.A. Janeway Jr. et al., Immunobiology, 6th ed. New York: Garland Science, 2005. B, from M. Kloor et al., *Cancer Res.* 65:6418–6424, 2005. With permission from American Association for Cancer Research. C, courtesy of Barbara Seliger.)

As an alternative strategy for immunoevasion, some cancer cells may alter their intracellular biochemistry in order to render themselves intrinsically less vulnerable to cytotoxic T cell attacks. For example, some cancer cells become resistant to being killed following an initial encounter with the FasL death receptor ligand wielded by cytotoxic T cells; they do so by altering their intracellular signaling pathways downstream of their Fas death receptor. This pathway dysregulation results in the inactivation of the extrinsic apoptotic pathway (see Figure 9.31), providing cancer cells with yet another way to escape death.

A related immunoevasive strategy reflects strategies used to escape the major mechanism employed by cytotoxic T cells and NK cells to kill their target cells. Thus, CTLs and NK cells introduce proteases—granzymes—into targeted cells to induce apoptosis through cleavage and the resulting activation of a caspase pro-enzyme involved in triggering this cell death program (see Figure 9.31). Cancer cells can escape this killing mechanism by increasing the levels of certain inhibitor-of-apoptosis proteins (IAPs; see Figure 9.27), which sequester and inactivate key pro-apoptotic caspases, thereby blunting any activation of the extrinsic apoptotic pathway.

To summarize, immune-initiated cytotoxicity can be viewed as yet another physiological stressor experienced by cancer cells during multiple stages of multi-step tumor progression, taking its place alongside hypoxia, intracellular signaling imbalances, and loss of anchorage, all of which force cancer cells to disable their pro-apoptotic signaling pathways. Although extensive and various, the list of immunoevasive strategies explored here should not be considered complete and is likely to be extended in various directions by future research.

The next chapter reveals powerful and still-developing immunotherapies designed to defeat or bypass many of the defenses erected by cancer cells to thwart or avoid immune attack. Like the targeted therapies explored in early chapters, the inspiration for these immunotherapies and the guidance for their development emerged from the increasing and still-developing understanding of basic immunology and cancer immunobiology described in this chapter.

Key concepts

- Normally, the immune system displays tolerance, the failure to mount immune responses against self. Because tumor cells are very similar to their normal counterparts, tolerance can be a barrier that must be overcome in order to make an anti-tumor immune response against them.

- Immunocompromised individuals experience elevated levels of various cancers, strongly suggesting that the immune system is continuously monitoring the body's tissues for the presence of tumors and attempting to eliminate them—the process of immunosurveillance.

- There are two types of tumor antigens: tumor-specific antigens (TSA) and tumor-associated antigens (TAAs). Tumor-specific antigens are expressed by some or all of the cells of a given tumor. Tumor-associated antigens are expressed by tumor cells and certain other normal tissues throughout the body; they often reflect the differentiation programs of the tissues in which these tumor cells arose.

- In vertebrates, immunity is provided by the innate immune system and the adaptive immune system, two distinct, but interacting physiological systems. Innate immune responses are triggered by broadly specific pattern-recognition receptors. Adaptive immune responses are triggered by engagement of receptors that are highly specific for a given antigen.

- Two classes of lymphocytes, B lymphocytes (B cells) and T lymphocytes (T cells), are the signature cell types of adaptive immunity. Antibodies, their B-cell receptor counterparts, T-cell receptors and, in complexes with their peptide ligands, MHC-I and MHC-II molecules enable the adaptive immune system to recognize and respond to diverse antigens.

- The extensive diversity of antibodies and T-cell receptors is achieved by the stochastic rearrangement of a few hundred subgenic segments that are inherited in the germ line and can combine almost randomly during the development of B cells and T cells. The resulting molecules, assembled combinatorially, are enormously diverse and generate T cell and B cell repertoires exceeding 10^9 potential antigen specificities in each individual human.

- The T-cell receptors (TCRs) of the body's major T-cell population do not recognize intact protein antigens but instead respond to complexes of peptide fragments derived from intracellular degradation of peptide fragments presented as complexes with MHC molecules.

- Somatic cells throughout the body routinely present peptide fragments of proteins synthesized within their confines as peptide:MHC class I on their surface. A very few cell types—macrophages, B cells, and most notably, dendritic cells—have the ability to present peptides derived from proteins acquired from outside their confines as complexes of peptide:MHC class II molecules.

- Whether T cells recognize MHC-I–bound or MHC-II–bound peptides is determined by a co-receptor protein, either CD4 or CD8, that is associated closely, but noncovalently, with TCRs on their surface. The co-receptors ensure that the TCR with which they are partnered is directed toward the appropriate MHC molecule. CD8 T cells interact with MHC class I molecules and CD4 T cells interact with MHC class II molecules.

- T cells that have not undergone an encounter with antigen that resulted in their activation are referred to as naive T cells. They are unresponsive, except under the special circumstances designated as priming. Conditions for priming are demanding and include antigenic stimulation, an additional signal called co-stimulation, and the presence of appropriate cytokines. During priming, naive T cells almost always receive antigenic stimulation and co-stimulation from dendritic cells.

- Dendritic cells are able to present peptides on both MHC-I and MHC-II molecules, a distinctive feature of their physiology called cross-presentation, which enables presentation to either naive CD4 or CD8 T cells. Their display of the co-stimulatory ligands, CD80/86, also allows dendritic cells to deliver co-stimulatory signals. Given the presence of appropriate cytokines, this combination of TCR and co-stimulatory signaling, effects the transformation of T cells from the naive state to the state of activation.

- Different functions are assumed by CD8 and CD4 T cells. Cytotoxicity is the function of CD8 T cells, and CD4 T cells are primarily involved in the positive or negative regulation of other T cells and some other types of immunocytes. CD4 T cells that deliver signals supporting cell functions are referred to as helper T cells (T_H cells) and those that inhibit other immunocytes are called regulatory T cells (T_{reg}'s).

- Regulatory T cells play a key immunoregulatory role by eliminating or suppressing autoreactive T cells. Despite their indispensable role as negative regulators of excessive or autoreactive immune responses, T_{reg}'s can be co-opted by tumors to diminish or block anti-tumor immune responses.

- In addition to many receptor–ligand interactions that induce or intensify immune responses, there are immune checkpoints that negatively regulate immune responses, dampening their intensity and ensuring that they are self-limiting. Two receptors, CTLA-4 and PD-1, are key receptors for the mediation of checkpoint inhibition. Though essential agents of normal immunophysiology, they can also reduce anti-tumor responses to the point that they become ineffective.

- The CTLA-4 and PD-1 mediated checkpoints exert their effects in different anatomical locations. In lymph nodes, the stimulatory effect of engagement of CD28 on T cells with the co-stimulatory ligands, CD80/86, is counterbalanced by engagement of CD80/86 with the inhibitory receptor CTLA-4, preventing excessive T-cell activation. In peripheral tissues, the intensity and duration of the T-cell response is dampened by engagement of the inhibitory receptor PD-1, borne by responding T cells with its PD-L1/2 ligand displayed by other cells within various tissues.

- Natural killer (NK) cells are cytotoxic cells that kill cells that lack proper levels of cell surface MHC class I molecules. Given the strong selective pressure imposed by CTLs, some cancer cells contrive to shut down their display of MHC-I and escape cytotoxic T-cell attack. NK cells provide a bulwark against this strategy. In addition, NK cells are endowed with receptors that can detect the display of certain stress-induced ligands on the surface of cells and kill them.

- NK cells have receptors for the Fc regions of certain antibody classes, allowing them to bind via their FcRs to antibody-coated cells and kill them via the process of antibody-dependent cellular cytotoxicity (ADCC).

- Macrophages represent a versatile category of phagocytic cells that display a spectrum of phenotypes ranging from the anti-tumor highly phagocytic M1 state to the protumor M2 state that supports vascularization and tissue repair and secretes the immunosuppressive cytokines IL-10 and TGF-β. Macrophages possess Fc receptors and can be induced to kill antibody-targeted cells by antibody-dependent cellular phagocytosis (ADCP) or ADCC.

Thought questions

1. What clinical data and experiments using animal models do you believe provide the most compelling evidence that the immune system plays a significant role in suppressing the appearance of solid tumors?

2. Most of the antigenic differences between normal cells and tumor cells are not displayed on the cell membrane but reside in the cell's interior. Nevertheless, the presence of these antigens, most of which are proteins, are often detected by the immune system and responses are mounted against them. How and on what platforms are specific portions of these interior antigens displayed on the cell surface?

3. Some individuals have congenital defects in antibody production that render them unable to make antibody responses. Despite this deficiency, would you expect many of these individuals to retain the ability to mount many of the most consequential immune responses to cancer? Explain why you answered as you did.

4. The CD4 category of T cells encompasses multiple subtypes of cells. Do any of these make important contributions to the immune response against cancer, and which, if any, would you expect to be a powerful inhibitor of anti-tumor responses? Provide justifications for your responses.

5. Even after tumors come under immune attack, all too often they manage to escape elimination. How would you account for the failures of the immune system to eradicate many of the tumors it recognizes and mounts an initially effective attack against?

6. Adaptive and innate immunity are distinct, but often collaborative, arms of the immune system. Can you describe specific examples that illustrate the harnessing of innate/adaptive collaborative interactions to mount attacks on cancer?

7. Consider the following immune responses: (1) the activation of anti-tumor CD8+ T cells; (2) the activation of anti-tumor CD4 T cells. What effect, if any, would you expect blockade of CTLA-4 by anti-CTLA-4 antibodies to have on either or both of these processes? Justify your response.

8. Tumors often provoke immune responses that result in attacks upon them. However, in many instances these immune attacks fail to eliminate the tumor and become increasingly ineffective at limiting tumor growth. How would you explain the ability of cancers to evade control or elimination by the immune system?

Additional reading

Cassetta L & Pollard JW (2018) Targeting macrophages: therapeutic approaches in cancer. *Nat. Rev. Drug Discov.* 11, 887–903.

Coffelt SB, Wellenstein MD & de Visser KE (2016) Neutrophils in cancer: neutral no more. *Nat. Rev. Cancer* 16, 431–446.

Gonzalez H, Hagerling C & Werb Z (2018) Roles of the immune system in cancer: from tumor initiation to metastatic progression. *Genes Dev.* 32, 1267–1284.

He X & Xu C (2020) Immune checkpoint signaling and cancer immunotherapy. *Cell Res.* 30, 660–669.

Joffre OP, Segura E, Savina A & Amigorena S (2012) Cross-presentation by dendritic cells. *Nat. Rev. Immunol.* 12, 557–569.

Mantovani A, Marchesi F, Malesci A, Laghi L & Allavena P (2017) Tumour-associated macrophages as treatment targets in oncology. *Nat. Rev. Clinical Oncol.* 14, 399–416.

Market E & Papavasiliou FN (2003) V(D)J recombination and the evolution of the adaptive immune system. *PLoS Biol.* 1, e16.

Mauri C & Menon M (2017) Human regulatory B cells in health and disease: therapeutic potential. *J. Clin. Invest.* 127, 772–779.

Miller JS & Lanier LL (2019) Natural killer cells in cancer immunotherapy. *Annu. Rev. Cancer Biol.* 3, 77–103.

Nagarsheth N, Wicha MS & Zou W (2017) Chemokines in the cancer microenvironment and their relevance in cancer immunotherapy. *Nat. Rev. Immunol.* 17, 559–572.

Olingy CE, Dinh HQ & Hedrick CC (2019) Monocyte heterogeneity and functions in cancer. *J. Leukoc. Biol.* 106, 309–322.

Plitas G & Rudensky AY (2020) Regulatory T cells in cancer. *Annu. Rev. Cancer Biol.* 4, 459–477.

Schreiber RD, Old LJ & Smyth MJ (2011) Cancer immunoediting: integrating immunity's roles in cancer suppression and promotion. *Science* 331, 1565–1570.

Sharonov GV, Serebrovskaya EO, Yuzhakova DV, Britanova OV & Chudakov DM (2020) B cells, plasma cells and antibody repertoires in the tumour microenvironment. *Nat. Rev. Immunol.* 20, 294–307.

Topalian SL, Drake CG & Pardoll DM (2015) Immune checkpoint blockade: a common denominator approach to cancer therapy. *Cancer Cell* 27, 450–461.

Veglia F, Perego M & Gabrilovich D (2018) Myeloid-derived suppressor cells coming of age. *Nat. Immunol.* 19, 108–119.

Wan YY & Flavell RA (2007) 'Yin-yang' functions of TGF-β and Tregs in immune regulation. *Immunol. Rev.* 220, 199–213.

Wculek SK, Cueto FJ, Mujal AM et al. (2020) Dendritic cells in cancer immunology and immunotherapy. *Nat. Rev. Immunol.* 20, 7–24.

Wei SC, Duffy CR & Allison JP (2018) Fundamental mechanisms of immune checkpoint blockade therapy. *Cancer Discov.* 8, 1069–1086.

Wing JB, Tanaka A & Sakaguchi S (2019) Human FOXP3+ regulatory T cell heterogeneity and function in autoimmunity and cancer. *Immunity* 50, 302–316.

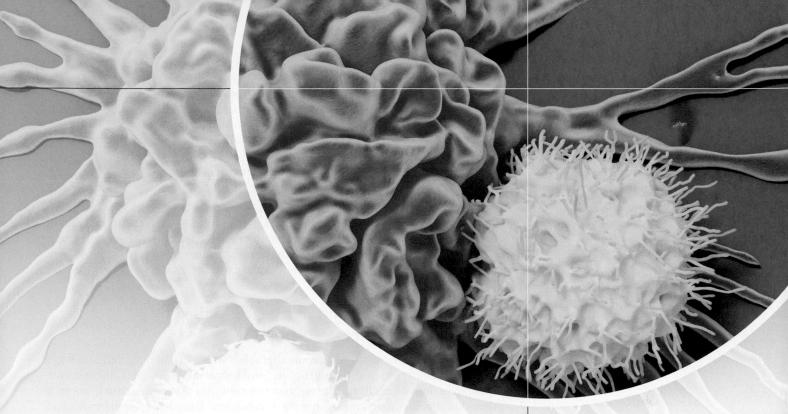

Chapter 16

Cancer Immunotherapy

Chapter opener photo: Juan Gaertner/Science Photo Library/ Getty Images.

> [I]n animals, ... inheritable genetic changes must be common in somatic cells and a proportion of these changes will represent a step toward malignancy. It is an evolutionary necessity that there should be some mechanism for eliminating or inactivating such potentially dangerous mutant cells and it is postulated that this mechanism is of immunological character.
>
> Macfarlane Burnet, immunologist, 1964

As described in the last chapter, research conducted over the past several decades has left us with a rich, detailed accounting of the workings of the immune system. Although much remains to be learned, the existing body of information on immunological functions has provided us with a broad and deep foundation for translation into clinical care, that is, the development of immunotherapies focused on harnessing the powers of the immune system to eliminate clinically diagnosed tumors. At present, the field of cancer immunotherapy is exploding, with hundreds of different immunotherapies being considered or developed, all of which draw directly from the lessons we learned in Chapter 15.

Much of the current enthusiasm for cancer immunotherapies comes from the perception that most of the low-molecular–weight pharmacological treatments of cancer have seemingly reached the limits of what they are capable of providing, and that truly curative therapies will need to be sought from an entirely different therapeutic approach, specifically immunotherapy. This explains the depth and breadth with which cancer immunotherapies are laid out in this chapter.

Drawn in broad strokes, these immunotherapies can be placed into three categories. The most straightforward are those that provoke the immune system to develop reactivity with tumor cells and neutralization of the infectious agents that cause their formation; these protocols proceed via clinical strategies that resemble traditional vaccination by providing antigens that either prevent tumor formation or provoke the development of immunity against already established tumors. A second major strategy can be depicted as **passive immunization**, in which patients under treatment are supplied with antibodies or cells of various sorts that, for one or another reason, the patients' own immune systems have been incapable of developing on their own.

And finally, the third thrust of modern immunotherapy derives from treatments that reveal otherwise-inapparent immunities that patients' own immune systems have successfully developed but that, for various reasons, have been blocked by various immunosuppressive mechanisms. As we shall see, tumor immunology, including immunotherapy, represents a rapidly advancing field; indeed, it is the most dynamic and innovative area of contemporary cancer research. At present, we have only begun to understand its potential for developing truly effective treatments of neoplastic diseases, many of which have proven resistant to existing therapies.

16.1 Vaccination can prevent cancer caused by infectious agents

We begin our exploration of immunotherapies with vaccination, immunology's oldest, safest, most widely used, and to date most successful technology. These immunotherapies derive directly from more than two centuries of experience in generating protection against infectious agents via vaccination. Decades of epidemiological, clinical, and laboratory studies have shown that infectious agents are responsible for about 20% of cancers. As shown in **Table 16.1** at least seven viruses, one bacterium, and three parasites have been implicated as etiological agents in cancer pathogenesis. Some of these are directly involved and others play important but indirect roles in oncogenesis. Given the demonstrated power of immunization to neutralize infectious agents, these agents represent attractive targets for immunization undertaken to *prevent* the appearance of the cancers that they cause.

In fact, all of the infectious agents that are directly involved in driving oncogenesis are viruses. The five known, widely acting oncogenic viruses—HPV, HTLV-1, EBV, MCPyV, and KSHV—all express viral oncoproteins. As argued in Section 8.5, these viruses deploy virus-encoded oncoproteins in order to convert the intracellular environments of infected cells into ones that support efficient viral replication. Accordingly, the evolution of these viruses has been dictated by their need to optimize their mechanisms of proliferation rather than optimizing cell transformation and the formation of tumors. Each of these viruses is considered *directly oncogenic* because one or more of its virus-encoded gene products operates to perturb the intracellular signal transduction machinery, as described previously in Chapters 8 and 9. Central to our thinking here is the notion that the ability of these viruses to transform cells in a fashion that leads to tumorigenesis requires the continued presence of the viral genomes in the descendants of initially infected cells—a mechanism first described in Section 3.3 in the context of Rous sarcoma virus transformation. In the absence of continued expression of virus-encoded oncoproteins, during the earlier stages of tumorigenesis, some virus-transformed cells are likely to revert to a non-neoplastic state.

The other six agents, including two viruses (HBV and HCV), one bacterium (*Helicobacter pylori*), and three flatworm parasites (*Schistosoma haematobium*, *Opisthorchis viverrini*, and *Clonorchis sinensis*), are *indirectly oncogenic*, in that infection by these agents induces localized, persistent inflammation within a tissue, which in turn favors the outgrowth of tumors. The inflammation induced by these agents reflects normal responses by the immune system that are designed to rapidly and effectively eliminate infectious agents; however, because these infections are not successfully cured, chronic localized inflammation ensues, which in turn creates a hospitable tissue microenvironment for the outgrowth of pre-neoplastic and neoplastic growths, as described in Section 11.13.

These persistent inflammatory states usually feature the continuing release of a variety of tumor-promoting cytokines, chemokines, and prostaglandins as well as the generation of mutagenic reactive oxygen species (ROS; see Section 12.4). This explains why prevention of infections by these agents or their timely eradication from the

Table 16.1 Infectious agents implicated in human cancer causation

Classification	Infectious agent	Mode of carcinogenesis (mediator)	Human malignancy	Transmission
herpesvirus	EBV	direct (oncogene)	lymphoma; some nasopharyngeal carcinomas	saliva, transfusion, and sexual contact
delta retrovirus	HTLV-1	direct (oncogene)	adult T cell lymphoma	sexual contact, needle sharing and transfusion
papilloma virus	HPV	direct (oncogene)	cervical cancer, some oropharyngeal, genital and anal cancers	sexual contact
herpesvirus	KSHV (also called HHV8)	direct (oncogene)	Kaposi sarcoma and others	salivary spread in endemic area; sexual spread in nonendemic areas
polyoma virus	MCPyV	direct (oncogene)	skin cancer	skin to skin contact and contact with virus-contaminated surfaces
hepatitis virus	HBV	indirect (chronic inflammation)	liver cancer	transfusion, needle sharing, and sexual contact
hepatitis virus	HCV	indirect (chronic inflammation)	liver cancer	transfusion, needle sharing, and sexual contact
lentivirus	HIV	(immunosuppression)	increased risk of many cancers, including Kaposi sarcoma, lymphoma, and others	sexual contact and transfusion
bacterium	*H. pylori*	(inflammation)	some stomach cancers and gastric tissue associated lymphomas	perhaps by consuming contaminated water and foods
helminth	*Opisthorchis viverrini*	(generation of oxidizing agents?)	cholangiocarcinoma	consumption of raw or undercooked fish from fresh waters of Southeast Asia
helminth	*Shistosoma hematobium*	(generation of oxidizing agents?)	bladder cancer	Contact with fresh water snails infected with S. hematobium
helminth	*Clonorchis sinensis*	(generation of oxidizing agents?)	cholangiocarcinoma	consumption of raw or undercooked fish from fresh waters of many regions of Asia

EBV, Epstein–Barr virus; HTLV-1, human T cell leukemia virus-1; HPV, human papilloma virus; KSHV, Kaposi sarcoma herpes virus, also known as human herpesvirus 8 (HHV8); MCPyV, Merkel cell polyoma virus; HBV, hepatitis B virus; HCV, hepatitis C virus; HIV, human immunodeficiency virus.

body can avoid the cancers that they might otherwise cause. In principle, this might be accomplished in either of two ways: the use of antimicrobial drugs, such as antibiotics being employed to eliminate *H. pylori*, or vaccinations, such as those used to confer immunity to future infections by neutralizing various cancer-inducing agents. Importantly, in the case of the class of tumors induced by indirectly oncogenic infectious agents, the clearing of these agents from the body's tissues will not necessarily affect the continued growth and viability of already-established tumors simply because, once formed, these tumors no longer depend on the inflammatory signals that fostered their development in the first place.

Prophylactic vaccines, which are designed to prevent infections by cancer-inducing agents, represent by far the largest group of vaccines currently used in the clinic. If successful, they prevent the formation of the tumors that are caused by these agents. Therapeutic vaccines, in contrast, are used to treat already-established disease; there are very few examples of this type of vaccine employed clinically, in part for reasons

Sidebar 16.1 Hepatitis B virus infections led to hepatomas in Taiwanese government workers Epidemiological studies usually yield outcomes that make it difficult to discern a strong correlation between occasional exposure to certain environmental factors and a subsequent moderately increased risk of cancer. The incidence of liver cancer, however, varies dramatically throughout the world, making it possible to strongly associate this disease with specific causative factors. In certain parts of the world, including much of Asia—especially China and sub-Saharan Africa—hepatocellular carcinomas (HCCs, also known as hepatomas) are one of the leading causes of death due to cancer. In stark contrast, liver cancer ranks 25th as a cause of cancer-related deaths in the United States.

A stunning correlation can be made between hepatitis B virus (HBV) infections and susceptibility to liver cancer. At the beginning of an epidemiologic study undertaken in 1975, the HBV status of 22,707 men in government service in Taiwan was determined by measuring, among other parameters, the presence of viral antigens in their blood. The causes of deaths in this cohort were then chronicled over the next decade. By 1986, death from hepatocellular carcinoma had claimed 152 of the 3454 men who were initially positive for HBV viral antigen, but only 9 of the 19,253 men who were negative for this antigen in their blood. The relative increased risk of dying from his disease if one carried viral antigen in the circulation (which is indicative of an active HBV infection in the liver) was calculated to be 98.4. This means that an individual afflicted with chronic, active HBV infection experienced an almost 100-fold increased risk of contracting and dying from this cancer compared with someone who lacked viral antigen in his liver and was apparently uninfected. These numbers contrast strongly with most epidemiologic correlations made between specific exposures of patients to suspected carcinogens and disease incidence, where relative risk of disease is often only two to threefold, often hovering on the borderline of statistical significance.

described above. When considering and comparing these two alternative approaches, we must keep in mind that, in terms of reducing morbidity and mortality from neoplastic disease, prevention is always vastly more effective than attempts to treat existing disease, sometimes by orders of magnitude.

Two prophylactic vaccines have proven to be highly effective in lowering the incidence of cancers caused by two infectious agents. One of these prevents infection by hepatitis B virus (HBV), the agent responsible for most hepatocellular carcinomas (**Sidebar 16.1**). The other vaccine prevents infection by oncogenic human papillomaviruses (HPVs), the cause of most (~99%) cases of cervical cancer and cancers at other sites as well. Both of these vaccines have had significant impact in reducing the incidence of the cancers caused by the infectious agents they protect against.

16.2 Vaccination against human papillomaviruses prevents cervical cancer

Because the oncogenic effects of HPVs are direct rather than indirect, we will examine vaccination strategies to prevent HPV infection in some detail. HPVs represent a major human carcinogen, being sexually transmitted and responsible for at least 99% of all cervical cancers and about 4.5% of total cancer incidence worldwide. In addition to cervical carcinoma, other anogenital cancers (88% of carcinomas of the anus, 25% of the vulva, 78% of the vagina, and 50% of the penis) and a significant fraction of head-and-neck squamous cell carcinomas (HNSCC) can be traced back to HPV infections. In the United States, the numbers of HPV-linked oropharyngeal carcinomas (OPCs), a subset of HNSCCs, now exceed the numbers of cervical cancers diagnosed annually. Although more than 150 distinct HPV types have been identified, only 12 of these have been associated with cervical cancer and other HPV-induced cancers, with just two HPV types—HPV16 and HPV18—being responsible for 75% of HPV-induced cancer.

Epidemiological studies demonstrate that in the United States, the prevalence of HPV infection in females increases rapidly in the mid-teens, a consequence of the onset of sexual activity (**Figure 16.1A**). Many years elapse between the times of infection and the progression to cancer by the small proportion of HPV-infected people who actually develop clinically detectable cancers. By now it has become clear that vaccination to prevent infection should occur in boys and girls during childhood or early adolescence, at a time before men (the vectors of the disease) and women become sexually active.

The small genome of HPV (see Section 3.4) limits the number of proteins encoded by HPV to the capsid proteins and several other proteins that foster the replication of its DNA genome. L1, the more abundant of HPV's capsid proteins, can be induced

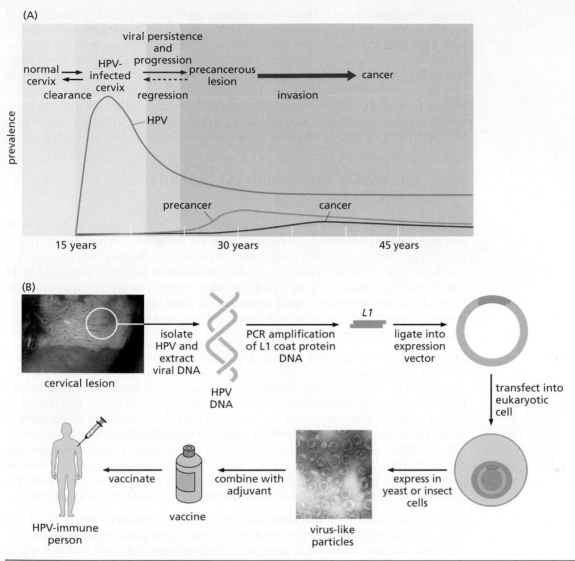

Figure 16.1 Prevention of HPV infections More than 99% of human cervical carcinomas are caused by human papillomavirus (HPV) infections, which has led to the development of multiple anti-HPV vaccines that block viral infections and thereby prevent this cancer. (A) The natural history of HPV and cervical cancer. In women, the prevalence of transient HPV infection (*blue curve*) begins in the early teens and peaks during the 20s, paralleling the initiation of sexual activity. Cervical precancerous conditions (*green curve*) begin to appear approximately 10 years later. The peak prevalence of invasive cancers (*red curve*) is seen between ages 40 and 50. The development of precancerous lesions depends on the persistence of viral infection, because viral oncogenes and encoded oncoproteins are essential for the progression of the precancerous lesions to cancer and in many cases remain important for the growth and maintenance of the resulting cancers. (B) Anti-HPV vaccines were produced by extracting DNA from the virions of various HPV strains, and in each case the segment encoding L1, the major capsid protein, was amplified by PCR using specific primers; the amplified DNA was ligated into an expression vector and introduced into yeast cells for protein production. The resulting L1 protein assembles spontaneously into virus-like particles (VLPs), which are then introduced together with a suitable adjuvant (a substance that enhances the immune response) to formulate a vaccine that is subsequently injected into humans. (A, adapted from M. Schiffman & P.E. Castle, *N. Engl. J. Med.* 353:2101–2104, 2005. With permission from Massachusetts Medical Society. B, adapted from I.H. Frazer, G.R. Leggatt & S.R. Mattarollo, *Annu. Rev. Immunol.* 29:111–138, 2011. With permission from Annual Reviews.)

to assemble into virus-like particles (VLPs) *in vitro* in the absence of viral DNA and other capsid proteins. These particles are immunogenic and, upon injection, induce neutralizing antibodies, which has been exploited for the development of a variety of vaccines against HPV (Figure 16.1B). The VLP-based vaccines are safe because the injected vaccine virions lack the viral DNA needed for viral replication; at the same time, they are capable of inducing a strong protective immune response because the VLPs display the L1 target antigen in a stereochemical configuration that is virtually identical to the way it is displayed by intact virus particles.

The effectiveness of the VLP-based vaccine strategy has been repeatedly demonstrated in large clinical trials, including tests of versions of the vaccine that involve L1 capsid proteins (and resulting VLPs) from an increasing number of HPV types. Although 70% of cervical cancer is caused by just two types, HPV 16 and HPV 18, immunizations against additional virus types have been added with each successive iteration of the vaccine, eventually including VLPs produced by HPV types 31, 33, 45, 52, and 58; these additional HPV types are known to make modest but real contributions to the causation of cervical cancer. In addition, VLPs from HPV types 6 and 11 antigens have been included in these **polyvalent** vaccines because these types cause anal and genital warts. In 2014, the FDA approved Gardisil9™, a nonavalent vaccine containing VLPs made by all nine of these HPV types. This vaccine is 96.8% effective in preventing cervical cancer.

16.3 Therapeutic vaccination is a potential treatment for cancer

An alternative strategy for developing anti-cancer vaccines holds the promise of treating established tumors by developing immune recognition of cancer cells rather than the agents that previously caused them. As we read in Section 15.7, the observation that CD8+ cytotoxic T cells can develop the ability to recognize and lyse cancer cells displaying certain antigens has encouraged attempts to use therapeutic vaccination to induce potent anti-tumor responses. In these cases, vaccination is accomplished by administration of specific tumor antigens under conditions that provoke adaptive immune responses.

In principle, the more general strategy of targeting already-formed cancer cells of both viral and nonviral origin offers the advantage that it enables attacks on both primary tumors and metastatic colonies disseminated throughout the body of a cancer patient. Furthermore, should cancers return after being successfully controlled and caused to regress, memory anti-tumor immune responses offer the potential to recognize and thwart such clinical relapses. Despite the appeal of immunization to treat existing cancers, significant obstacles have blocked the path to its widespread use in clinical practice. The most important of these involve the mechanisms of immune escape and evasion discussed in Section 15.13, some of which derive from the ability of tumors to hide within immunosuppressive microenvironments that they assemble around themselves.

Nonetheless, despite clear and formidable challenges, a number of efforts have been launched to develop effective therapeutic cancer vaccines against established tumors. These fall into one of two major strategies: direct delivery of antigen and delivery of antigen loaded on dendritic cells. Direct delivery can provide a wide variety of putative tumor antigens and has the appeal of simplicity. However, it is totally dependent on *in vivo* host processes to perform the key steps of antigen uptake and presentation by professional antigen-presenting cells (APCs). The alternative involves the loading of dendritic cells or dendritic-like cells *ex vivo* and their subsequent delivery to the patient in order to ensure the *in vivo* presentation of antigens by cells capable of activating T-cell-mediated anti-tumor immune responses.

Direct delivery uses a variety of vaccination strategies to provoke tumor antigen responses. These include injection of (1) the antigenic protein or peptide directly into the tumor or the systemic circulation, (2) the RNA encoding this antigen, or (3) the DNA encoding this RNA. While direct injection of the protein antigen represents the most straightforward approach, the use of RNA or DNA has certain advantages. Transfection with either RNA or DNA makes it possible to include in the vaccine inoculum other molecular templates that encode an adjuvant protein, such as a cytokine, in addition to the target antigen. Although promising RNA-based cancer vaccines have begun to appear (**Sidebar 16.2**), there may also be clinical situations in which more stable DNA vaccines have advantages over the more labile RNA vaccines.

While lacking the simplicity of direct delivery, antigen-loaded antigen-presenting cells were used to develop the first therapeutic cancer vaccine to receive approval from the U.S. Food and Drug Administration (FDA). This vaccine, Sipuleucel-T, was developed for the treatment of prostate cancer. This particular vaccine strategy, schematized in **Figure 16.2**, while delivering only limited clinical benefit, has translated the concept of therapeutic vaccination into actual clinical practice.

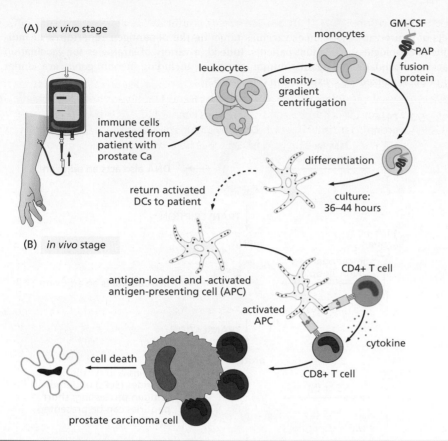

Figure 16.2 Development of therapeutic vaccination strategy for prostate cancer
This therapeutic vaccination involves two stages, the first *ex vivo* and the second *in vivo*.
(A) *Ex vivo* stage: An immunizing antigen is produced by fusing a cDNA encoding granulocyte–macrophage colony stimulating factor (GM-CSF) and a second cDNA encoding prostatic acid phosphatase (PAP). The PAP (*red*)/GM-CSF (*green*) fusion protein generated is introduced into a cancer patient's monocytes. The GM-CSF portion of the fusion protein induces the differentiation of monocytes into antigen-presenting cells (APCs). The PAP portion corresponds to a prostate-specific antigen that is widely expressed in prostate carcinoma cells. PAP undergoes antigen processing in the monocyte-derived APCs, resulting in loading of PAP peptide fragments onto MHC proteins. This population of monocytes and induced APCs is infused into the cancer patient. (B) *In vivo* stage: Following their introduction into the patient, these antigen-loaded and -activated APCs present antigen to CD4 (*blue*) and CD8 (*purple*) T cells, leading in turn to the activation of PAP-specific CD4 and CD8 T cells. Cytotoxic CD8 (*red*) cells are shown engaging and killing a PAP-expressing prostate cancer cell (*brown*). (Adapted from J.A. Garcia, *Ther. Adv. Med. Oncol.* 3:101–108, 2011. With permission from SAGE.)

Sidebar 16.2 RNA vaccination shows promise for the treatment of melanoma A variation on the RNA vaccine technology that has been deployed so effectively to generate highly efficacious coronavirus vaccines shows promise for therapeutic cancer vaccination. Melanoma patients, whose tumors had progressed despite previous treatment and expressed at least one of a set of melanoma antigens, specifically NY-ESO-1, MAGE-A3, TPTE, or tyrosinase, were given a vaccine composed of mRNAs that encoded all four antigens. In an early-stage clinical trial, partial responses (shrinkage of a tumor) were seen in 26% of a cohort of 38 trial enrollees, all of whom were afflicted with unresectable stage III or IV melanomas. Moreover, the vaccine stimulated antigen-specific responses in both CD4+ and CD8+ T-cell populations.

Strikingly, in some enrollees, the levels of antigen-specific CD8+ T-cell clones reactive with one of the vaccine-encoded antigens rose over the course of 8 weeks from undetectable levels prior to immunization to levels approaching 10% of circulating CD8+ T cells. Observations like these have encouraged investigation of the possibility that the addition of an appropriate immune checkpoint inhibitor (described in Section 16.9) may further increase the clinical efficacy of certain mRNA-based therapeutic cancer vaccines.

At this point the development of RNA-based therapeutic cancer vaccines is still in its early stages, and future work will surely extend exploration of this approach to cancer-specific tumor antigens. Efforts will be made to target established tumor antigens, such as the E6 and E7 oncoprotein drivers present in cancers of HPV etiology, notably cervical cancer and some head-and-neck squamous carcinomas (see Section 16.2). In addition, mutations present in tumor cell genomes, such as the point mutations that create Ras oncoproteins and thus potential neoantigens, come to mind as attractive targets of future vaccine development. Finally, the mRNA vaccine technologies developed in the context of anti-COVID vaccine manufacture make them well suited to the fabrication of personalized cancer vaccines targeting the neoantigen repertoires of individual patients. Indeed, a variety of strategies for vaccination against tumor antigens, including neoantigens, are under exploration (**Figure 16.3**).

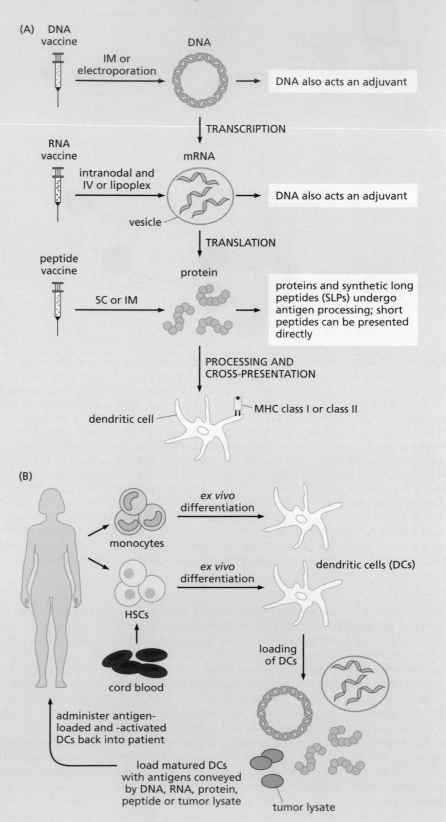

Figure 16.3 Delivery platforms for cancer vaccines Antigen delivery strategies for cancer vaccines fall into two broad categories—direct delivery of antigen or agents encoding it to the patient's tissues and delivery by antigen-loaded dendritic cells. (A) Direct presentation of peptide antigen can be accomplished through three alternative routes. Direct delivery can be accomplished by injection of antigen-encoding DNA or RNA, or by injection of protein itself or a synthetic long peptide (SLP) derived from it. With any of these three alternative delivery routes, the antigen can be processed to peptides in antigen-presenting dendritic cells (DCs). The machinery of cross-presentation by the DCs, which function as APCs, allows presentation of some peptides on MHC-I to CD8 T cells and presentation of some others on MHC-II to CD4 T cells. (B) Antigen-loaded dendritic cells can also be used in lymphoid tissues to deliver peptide antigen fragments in a three-step process. The first step involves induced differentiation *ex vivo* of host monocytes or hematopoietic stem cells (HSCs) into dendritic cells. Second, antigen-encoding construct is delivered to (loaded on) dendritic cells (*lower right*). Third, the resulting antigen-loaded DCs are infused back into the host. Abbreviations: IM = intramuscular, IV = intravenous injection, SC = subcutaneous injection, lipoplex = liposome-based and RNA delivery particles. (Adapted from M. Saxena, S.H. van der Burg, C.J.M. Melief et al., *Nature Rev. Cancer* 21:360–378, 2021. With permission from Nature.)

As much a process as a product, preparing the Sipuleucel-T vaccine involves the harvesting of peripheral blood mononuclear cells from a prostate cancer patient. These cells are then activated in culture by exposing them to recombinant human GM-CSF molecules that have been chemically linked to human prostatic acid phosphatase (PAP), an antigen that is expressed at elevated levels in advanced prostate carcinomas. The GM-CSF portion of the hybrid PAP-GM-CSF protein binds as a ligand to cognate GM-CSF receptors displayed by the patient's monocytes and induces their differentiation into antigen-presenting dendritic cells (DCs). At the same time, some of the receptor-bound hybrid protein–ligand molecules, like many receptor–ligand complexes, are internalized into these DCs, where they undergo proteolysis and processing for subsequent antigen presentation on MHC molecules.

Prostate cancer patients are infused with the resulting *ex vivo* differentiated, antigen-loaded DCs, which should then migrate to the recipient's lymphoid tissues. Once arrived in these tissues, these DCs, using their cross-presentation powers (see Section 15.7), should present resulting peptide antigens—notably those derived from cleavage of PAP—via their MHC class II molecules to CD4+ helper T cells and via their class I molecules to prime the naive CD8+ T cells. Among other consequences, this antigen presentation should result in the generation and proliferation of PAP-specific cytotoxic T lymphocytes (CTLs) in the patient's lymphoid tissues, which thereafter have the potential to attack prostate cancer cells displaying the PAP protein antigen. To be sure, normal prostate cells will be targeted as well, but other cell types will be spared from PAP antigen-specific attack.

Sipuleucel-T is described here largely because of the interesting logic that led to its development and introduction into the clinic. In reality, the clinical benefit that it offers has been modest, with increases in **time to progression** and overall survival of several months. At present it is difficult to know whether similar strategies will prove useful in developing immunotherapies that yield truly striking responses in patients afflicted with a variety of other cancer types.

Several factors will be important in determining the degree to which these and similar vaccines are successful. These include the target antigen's degree of foreignness, the widespread expression of the target antigen within a tumor, and whether its proteolytic cleavage gives rise to peptides that can be effectively presented by the MHC molecules displayed on the surface of a patient's cells.

Also, in the case of tumor antigens that are formed as a consequence of somatic mutations, the identity of the affected gene—whether it is a driver or passenger gene—will be a crucial determinant of success (see Section 11.6). Recall that continued expression of a mutant driver gene is generally critical to ongoing tumor growth and therefore cannot be shut down by cancer cells as an immunoevasive maneuver without incurring the cost of reducing their fitness. In contrast, cancer cells can indeed shut down expression of a mutant passenger gene without paying a fitness price in terms of their own proliferation or survival, thereby evading further immune attack and resulting in the rebound of a tumor's growth.

16.4 Passive immunization with antibodies can be used to treat cancer

The immunotherapeutic strategies described above are focused on energizing the immune system to deploy its humoral and cellular arms to attack targeted tumor cells. An alternative and often more productive strategy is to provide a patient with antibody molecules that have been formed outside the patient's body, that is, to supplement or replace limited endogenous immune responses by providing cellular agents or immune molecules that arise elsewhere. (In fact, a form of passive immunization operates naturally during gestation, when the immune system of the developing fetus is augmented by maternal antibodies acquired via placental transfer; similarly, in the first postnatal years, the immune systems of the very young are strengthened and diversified by antibodies supplied via maternal milk.)

To date, antibody-based passive immunization technologies have yielded the most successful and diverse immunotherapies. The power of antibody-mediated attacks on cancer cells derives from the specificity and high affinity of antigen–antibody reactions and the many weeks-long persistence of introduced antibodies in a patient's circulation.

The advent of technologies for the manufacture of monoclonal antibodies (moAbs) has made the generation of homogeneous populations of antibody molecules routine (Supplementary Sidebar 16.1). Unlike the heterogeneous mixtures of antibodies generated during the natural immune response, solutions of moAbs contain antibodies that are, at the molecular level, identical to one another. Moreover, moAb technology makes it possible to generate antibodies of virtually any desired specificity. Equally important, recombinant DNA technology has enabled the reengineering of antibody molecules, making them substantially more effective and versatile, even creating, as we will see, highly useful antibody-like molecules that do not occur in nature.

In more detail, the use of therapeutic moAbs offers two alternative strategies for attacking cancer cells. Some antibodies exert their anti-tumor activity directly by binding to tumor cells and killing them or, more commonly, impairing their further proliferation. Others act indirectly by modifying the behavior of components of the tumor microenvironment in ways that lead to the impairment or death of tumor cells. Thus, such indirectly acting moAbs may target certain cells or factors that are essential for tumor survival and progression without targeting the tumor cells themselves.

There are actually five distinct ways in which antibodies, such as moAbs, can bind directly to antigens displayed on the surface of targeted cancer cells and thereafter provoke the death of these cells. In all these strategies, the critical antigens are displayed directly on the plasma membrane rather than being presented by the MHC-I/II molecules described in Section 15.6.

Several of these treatments involve recruitment of cellular agents of the innate immune system that can recognize cell-bound antibody molecules, doing so by recognizing the Fc constant region of the already-bound antibody molecules. Such Fc-dependent attacks arise because cellular components of the innate immune system, specifically natural killer (NK) cells, macrophages, and neutrophils, naturally display **Fc receptors** (FcRs), which specifically recognize and bind to the Fc regions of certain subclasses of immunoglobulin gamma (IgG) molecules that have previously bound to the surfaces of cells. Once these cell-bound IgG molecules are recognized and bound by any one of these three classes of innate immune cells (via the various FcRs displayed by these cells), the process termed antibody-dependent cellular toxicity (ADCC) is triggered and the targeted cells, in our case cancer cells, are killed by the cytotoxic powers of the attacking immune cells (**Figure 16.4A**).

A second, related strategy of attack can be mounted by macrophages and neutrophils, both of these being phagocytes whose engagement with the Fc regions of cell-bound IgG molecules results in the engulfment via phagocytosis of antibody-tagged cells (Figure 16.4B).

Yet a third type of lethality is dealt by the activation of the **complement system**, a group of soluble plasma proteins that can be activated by antibodies bound to cells, causing the death of target cells in either of two ways. In one mode of antigen-directed cell killing, the protein components of the complement system recognize and bind cell-bound antibody molecules (once again recognizing the Fc portion of the bound antibody). The resulting multi-subunit protein assemblies create pore-forming complexes, that is, complexes of proteins that insert pores into the adjacent plasma membrane of the cell under attack. Such pores, once formed, deal a death blow to the targeted cell, because they prevent the cell from maintaining its normal osmotic gradients and thus its viability, doing so by allowing various low–molecular-weight substances, notably ions, to flow freely into and out of the cytoplasm. Such a pore-forming assembly of proteins is known as the **membrane attack complex** (MAC) (Figure 16.4C).

A fourth mechanism of antibody-mediated cell killing is also complement dependent. The instigating C3b protein is a by-product of the activation of complement by its interaction with antigen–antibody complexes (see Figure 16.4D), such as those formed between anti-tumor antibodies and cell surface tumor antigens. Complement activation proceeds by proteolytic cleavage, a reaction triggered by antigen–antibody

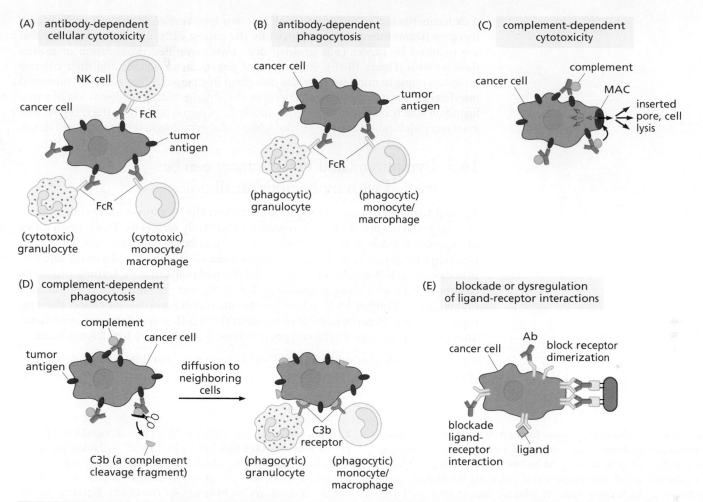

Figure 16.4 Multiple distinct mechanisms of cancer cell killing by antibodies and complement The binding of antibody molecules to cell surface antigens expressed by targeted cancer cells affords multiple alternatives routes of attack by the immune system. (A) Antibody-dependent cellular cytotoxicity (ADCC). Binding of tumor antigen-reactive antibodies to surface-displayed tumor antigens is followed by recognition of the Fc regions of the bound antibodies via Fc receptors displayed by phagocytes and NK cells, triggering their killing of the targeted cell. (B) Antibody-dependent phagocytosis (ADP). Engagement of the Fc regions of antibodies bound to tumor cells with the Fc receptors of phagocytes induces phagocytosis of tumor cells. (C) Complement-dependent cytotoxicity (CDC). Interaction of complement with the Fc region of tumor cell-bound antibodies triggers the assembly of complement-derived components, collectively known as the membrane attack complex (MAC), on the surface of the tumor cell, which punches holes in its plasma membrane (*right*), allowing the free exchange of ions and soluble molecules between the cell interior and the external milieu, leading in turn rapidly to cell death. (D) Complement-dependent phagocytosis (CDP). An alternative response occurs after the activated antibody:complement complex has bound to a cell surface antigen: one subunit of the complement complex may be cleaved, yielding C3b, a polypeptide fragment of a key complement component subunit. Once formed, C3b can be deposited on sites near its site of generation including the surfaces of neighboring cells. Any cells decorated with these C3b fragments can then recognized by C3b receptors on phagocytes and engulfed. (E) Blockade or dysregulation of ligand–receptor interactions. Interference with receipt by the targeted cancer cell of essential growth or survival signals, leading to induction of apoptotic signaling or prevention of receptor dimerization, also resulting in cell death. (A–C, E adapted from G.J. Weiner, *Nat. Rev. Cancer* 15:361–370, 2015.)

complexes, that results in the formation of a number of cleavage products, including C3b. Following their generation, these C3b proteins, which are highly hydrophobic, diffuse to a nearby cell and bind to its plasma membrane in a fashion that is completely nonspecific. Now membrane-bound, C3b attracts the attention of C3b receptor-bearing macrophages and other phagocytes, such as neutrophils, prompting their phagocytosis of cells decorated with this complement-derived protein fragment (Figure 16.4D). Notably, this avenue of complement-mediated cell killing has the potential to kill nearby tumor cells as well, because cytotoxic components of the activated complex can diffuse from their sites of formation and eliminate these bystanders. Even nearby members of the tumor population that may have contrived to lose expression of the tumor antigen targeted by the antibody will be subject to bystander killing.

A fifth mechanism of antibody attack involves interference at the cell surface with receptor–ligand interactions employed by the cancer cells, usually interactions that are required by cancer cells to drive their own continued proliferation or ensure their survival (Figure 16.4E). The actions of growth factor ligands and their cognate receptors come to mind here, as was described in Chapter 5. The antibody-mediated interference may take any one of a number of forms, including sequestration of a vital ligand on which a cancer cell depends, blockade of ligand binding by the receptor, and even receptor binding that triggers the release of aberrant signals that cause cell death.

16.5 Lymphoma and breast cancer can be treated with monoclonal antibodies

The cell surface glycoprotein CD20 is characteristically expressed by cells at multiple stages of normal B-cell differentiation and rarely on T cells. This limited range of expression made it an attractive target for antibody-mediated attack on non-Hodgkin's lymphomas, 85% of which derive from cells of the B-cell lineage and display this antigen. A genetically engineered monoclonal antibody, termed rituximab (**Sidebar 16.3**), with binding specificity for CD20, was generated for the treatment of this cancer (**Sidebar 16.4**). It became the first moAb approved for cancer therapy, having gained FDA approval for clinical use in 1997 and being widely employed thereafter for the treatment of certain types of relapsed or refractory B-cell lymphoma.

Rituximab was first introduced into clinical practice to treat diffuse large B-cell lymphoma (DLBCL), a prevalent form of non-Hodgkin's lymphoma. At the time of its

---

Sidebar 16.3 The conventions for naming monoclonal antibodies are systematic but complex A complex set of naming conventions has been adopted to name monoclonal antibodies, each of which carries one name reflecting use of these conventions and a second name by which it may be marketed. For example, trastuzumab is the technical name of the antibody molecule marketed as Herceptin™.

The "mab" suffix refers to a monoclonal antibody, while the first syllable, for example, the "tras" in trastuzumab, has no particular meaning and is given to confer a unique identity. The "tu" in trastuzumab refers to reactivity with a tumor antigen, while the "zu" syllable refers to a *humanized* monoclonal antibody (see Sidebar 16.4). The "xi" syllable indicates a chimeric antibody in which the constant region sequences are human and most or all of the variable region sequences are nonhuman. As an example, in rituximab, working backward, "mab" indicates monoclonal antibody, "xi" denotes chimeric origin, "tu" indicates targeting of a tumor antigen, and "ri" is the individual unique identifying prefix of this antibody that lacks further meaning. Additional, highly specialized syllables, not listed here, indicate yet other properties of a given moAb.

Sidebar 16.4 Devising useful monoclonal antibodies is a complex process Monoclonal antibodies (mAbs) were initially derived by immunizing mice with antigens of interest, such as cancer cell-associated proteins, deriving moAb-producing cells of mouse origin (see Supplementary Sidebar 16.1), and introducing the secreted mAbs, as a form of passive immunization, into the circulation of patients being treated for various conditions. These early treatments made it apparent that such antibodies can have exquisite antigen specificity, being able to bind to targeted antigens while ignoring the tens of thousands of other proteins made in the body. However, their usefulness was often reduced because the supplied mAb molecules were themselves immunogenic, being of mouse origin and displaying amino acid sequences and conformations that directly induced human anti-mouse antibody (HAMA) responses by the immune systems of the patients under treatment. Once a patient's immune system developed HAMA reactivity, the introduced mouse mAb became useless because HAMA neutralized the response.

The problem of reactivity against intact mouse moAbs was largely addressed by genetic engineering, in which portions of the immunogenic constant regions of mouse-derived mAb molecules were replaced by corresponding human segments. This maneuver produced chimeric antibodies, that is, ones in which the large sections of mouse constant region were replaced by their human equivalents but retained the variable regions of the mouse antibody; such antibody molecules preserved the antigen specificity of the original mouse antibody derived from its variable regions. These "chimeric" antibody molecules typically contain about 30% of murine antigen-binding sequences together with about 70% of human amino acid sequences, while even more extensively altered "humanized" moAbs have eliminated almost all mouse sequences, retaining only those forming the antigen-binding sites of murine origin. However, as explained in Supplementary Sidebar 16.1, technologies now exist for the derivation of fully human antibodies of any desired specificity. Indeed, most of the moAbs currently being developed for clinical use are formed by entirely human amino acid sequences.

introduction, DLBCL was usually treated with the chemotherapy protocol known as CHOP (a combination of the cytotoxic drugs **c**yclophosphamide, **h**ydroxydoxorubicin, **o**ncovin (vincristine sulfate), and **p**rednisone). Among its other limitations, the use of CHOP in elderly patients was constrained because **dose escalation** could not be undertaken due to the severity of side effects experienced in this population.

A clinical trial that compared treatment of elderly DLBCL patients with CHOP alone or CHOP plus rituximab showed that adding rituximab to standard CHOP chemotherapy significantly improved event-free survival and prolonged overall survival compared to treatment with CHOP alone (**Figure 16.5**). More generally, as clinical experience with moAb therapies has grown, it has become apparent that, like their chemotherapeutic forerunners, these agents are usually more effective as components of combination therapies than when they are employed as stand-alone monotherapies.

Monoclonal antibodies have also had a substantial impact on the immunotherapy of a subset of breast cancers. As a prominent example, as many as 15 to 20% of human mammary carcinomas overexpress human epidermal growth factor receptor-2 (HER2), the close cousin and signaling partner of the EGF receptor (see Sections 4.2 and 5.3). When its expression is elevated in breast cancer cells, this receptor molecule serves as a tumor-associated antigen that can be targeted by anti-HER2 antibodies. In fact, it serves as more than a tumor cell-associated antigen target, because various studies have demonstrated that overexpression of HER2 provides critical mitogenic and anti-apoptotic signals to breast cancer cells.

Trastuzumab, an anti-HER2 humanized monoclonal antibody, was the first successful antibody-based therapy for breast cancer and showed efficacy in prolonging progression-free and overall survival in patients with HER2-positive breast cancer. Studies of the mechanisms responsible for the anti-tumor effects of trastuzumab, also termed Herceptin, showed that its use led to attacks on HER2-positive cancer cells via multiple mechanisms, including recruitment of innate immune cells by antibody-dependent cellular cytotoxicity (ADCC), induction of complement-dependent cellular cytotoxicity (CDCC), and interference with growth factor signaling.

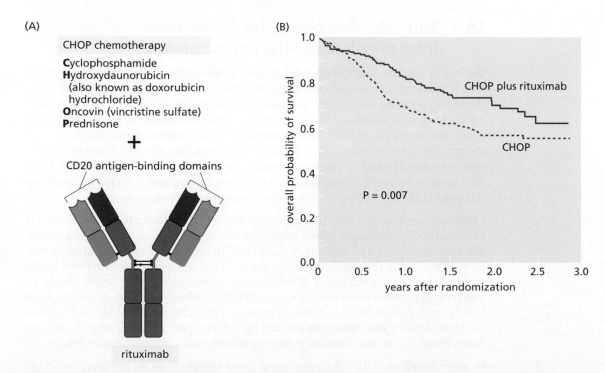

Figure 16.5 Combination therapy with anti-tumor antibody and chemotherapy (A) CHOP therapy directed toward B-cell–derived lymphomas involves a combination of cytotoxic compounds and rituximab, an anti-CD20 monoclonal antibody. (B) Overall survival among elderly patients (median age, 69 years) with diffuse large B-cell lymphoma (DLBCL) treated with CHOP or CHOP plus rituximab. (B, from B. Coiffier, E. Lepage, J. Briere et al., *N. Engl. J. Med.* 346:235–242, 2002. With permission from Massachusetts Medical Society.)

Sidebar 16.5 The peculiarities of HER2 function limit the efficacy of trastuzumab alone A key feature of HER2 signaling is its requirement for dimerization with another member of the family of receptor tyrosine kinases (see Section 5.6). As it turns out, a HER2 receptor molecule can dimerize with another HER2 receptor molecule or, alternatively with either HER1 (=EGFR) or HER3, a closely related member of this receptor family. Trastuzumab interferes with the formation of HER2–HER2 homodimers, but it does not block the formation of HER2–HER3 heterodimers, which represent the active signaling complexes following binding of EGF-like ligands. (Trastuzumab fails to block heterodimer formation because it binds to a site on the HER2 molecule that is distinct from the site involved in interaction with HER3.) This failure to block HER2–HER3 heterodimer formation therefore provides an escape pathway for cancer cells under attack by trastuzumab. The development of pertuzumab—the mAb that blocks the formation of HER2–HER3 heterodimers—closed this escape hatch. The complementary action of trastuzumab and pertuzumab was found to be more effective than either antibody alone, and when the trastuzumab–pertuzumab couple was paired with the powerful anti-mitotic agent, docetaxel, a therapy for HER2-positive metastatic breast cancer emerged that became, for a time, the **standard of care**, especially for HER2-positive metastatic breast cancer. While this combination was surely a milestone in the management of this disease, resistance to this therapy did eventually emerge in a number of patients.

Overall, in a **meta-analysis** that included the clinical histories of 13,864 breast cancer patients over the course of a decade, when applied together with chemotherapy, trastuzumab treatment reduced patient mortality by 33% relative to treatment with chemotherapy alone, a significant advance by any standard.

More recently, alternative versions of anti-HER2 mAbs have been generated and deployed in the cancer clinic. One of these, termed pertuzumab, targets a distinct domain of the HER2 protein and prevents its heterodimerization with HER3 (another EGFR-related receptor protein), thereby blocking HER2 signaling. Its distinct mechanism-of-action explains why it functions synergistically with trastuzumab in treating a variety of breast cancer subtypes (Sidebar 16.5). In early-stage breast cancers, this combination of moAbs often generates **pathological complete responses**, that is, the elimination of all cancerous tissue from the site of treatment, including comparable effects on draining lymph nodes (see Figure 14.36), judged in both cases by histopathological assessment. Such responses are often, but not always, indicative of cures.

16.6 Antibody–drug conjugates deliver toxic drugs to cells displaying tumor antigens

The highly specific targeting afforded by anti-tumor mAbs can be exploited to direct a variety of cytotoxic drugs to the relatively small number of tumor cells in the body, thereby sparing the normal cells and tissues elsewhere in the body. By doing so, such antibody–drug conjugates (ADCs) address a major limitation of existing cytotoxic chemotherapies, that of limited **therapeutic index** (TI), which represents the ratio of the concentration at which a drug generates unacceptable side-effect toxicities to the concentration at which it creates desired therapeutic responses; high TIs are desirable because they reflect the specificity with which various agents act preferentially on intended targets, including tumors, without creating major side effects.

Chemical coupling with toxic molecules can also greatly enhance the cytotoxic effects of the antibody molecules. These responses ensue when the antibody–drug conjugates are internalized by antibody-bound cells. In most instances, the antibody portions of the ADCs should ideally retain many of the functions of unmodified moAb molecules, including the ability of cell-bound antibodies to trigger lethal attacks by antibody-dependent cellular cytotoxicity (ADCC), antibody-dependent phagocytosis (ADP), and complement dependent cellular cytoxicity (CDCC). Similarly, those moAb molecules that disrupt or perturb ligand–receptor signaling should, in most cases, retain this ability following their conjugation with cytotoxic drug molecules.

In more detail, each ADC is formed from three components, an antibody molecule against a tumor antigen, a highly potent cytotoxic drug molecule, and a molecular linker between the two. Stringent requirements are placed on the cytotoxic drug and on the linker. Because relatively small numbers of the cytotoxic drug molecules are actually delivered into individual targeted cells, the intrinsic cytotoxicity of the drug must be very high, explaining why such agents, when used as unmodified soluble molecules, often exert toxicity in the subnanomolar-to-picomolar concentration range.

The linker must be stable in the bloodstream, allowing the antibody to retain its cargo until it binds to and is internalized by a targeted cell. Without this stability, ADCs can release their toxic payload molecules directly into the bloodstream, causing off-target toxicity to normal tissues—precisely the outcome that ADCs were invented to avoid. However, once internalized by a targeted cell, the cytotoxic payload must be released through cleavage from the linker or by the intracellular digestion of the ADC, liberating in either case biologically active drug moieties. As shown in **Figure 16.6**, a great many permutations of the theme of antibody, linker, and cytotoxic drug payload are possible. Most ADCs are built on a human IgG1 backbone, a subclass of antibody molecules that, in unmodified form, enjoy a long serum half-life of 21 days. Their Fc region should be able to trigger complement activation and, because they represent a ligand for binding by the Fc receptors of NK cells, macrophages, and neutrophils, should also enable target cell killing by these innate immune cells.

ADCs often offer additional benefits. Leakage of the introduced toxic molecules from cells directly targeted by an ADC may result in the killing of additional tumor cells present in the nearby tissue microenvironment, acting by the so-called **bystander effect**. This ideally offers the benefits of killing neighboring tumor cells that may have escaped targeting by the ADC, either by chance or because they have adopted evasive strategies including down-regulation of a targeted tumor antigen.

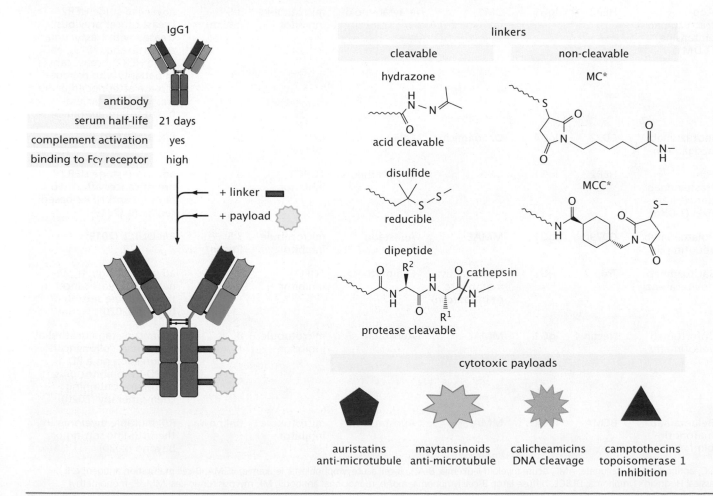

Figure 16.6 Construction of an antibody–drug conjugate
Antibody molecules, usually of the IgG1 subclass, are used because they have a long half-life and their Fc regions are bound by Fc receptors displayed by phagocytes and NK cells. These antibody molecules are modified by the covalent attachment of highly toxic drug molecules referred to as "payloads." A variety of cleavable linkers, five of which are shown here (*right*), are used to conjugate toxic payload molecules to antibody molecules. The figure shows the four major classes of chemical species currently used as cytotoxic payloads (*lower right*). (Adapted from J.Z. Drago, S. Modi and S. Chandarlapaty, *Nat. Rev. Clin. Oncol.* 18:327–344, 2021. With permission from Nature.)

Antibody–drug conjugates constitute a rapidly expanding category of cancer drugs and several have received FDA approval for the treatment of a broad range of cancers (**Table 16.2**).

As an example, trastuzumab emtansine, an ADC derivative of trastuzumab, targets cells that express HER2, doing so via the trastuzumab moiety of the conjugate; after binding, this ADC delivers DM1, a highly cytotoxic microtubule inhibitor to the HER2-expressing cell. In addition to its role in the delivery of a cytotoxic payload, the antibody retains most of the preexisting potent anti-tumor properties associated with the

Table 16.2 Antibody–drug conjugates approved for clinical use

ADC	Target antigen	moAb isotype	Payload	Payload class	Payload action	Drug-to-antibody ratio (DAR)	Disease indication (year of approval)
Gemtuzumab ozogamicin	CD33	IgG4	Ozogamicin	Calicheamicin	DNA cleavage	2–3	CD33[+] R/R AML (2000)[a]
Brentuximab Vedotin	CD30	IgG1	MMAE	Auristatin	microtubule inhibitor	4	R/R sALCL or cHL (2011) R/R pcALCL or CD30[+] MF (2017) cHL, sALCL or CD30[+] PTCL (2018)[b]
Ado-trastuzumab emtansine (T-DM1)	HER2	IgG1	DM1	Maytansinoid	microtubule inhibitor	3.5 (mean)	advanced-stage HER2[+] breast cancer previously treated with trastuzumab and a taxane (2013); early stage HER2[+] breast cancer in patients with residual disease after neoadjuvant trastuzumab–taxane-based treatment (2019)
Inotuzumab ozogamicin	CD22	IgG4	Ozogamicin	Calicheamicin	DNA cleavage	5–7	R/R B-ALL (2017)
Fam-trastuzumab deruxtecan-nxki (T-DXd)	HER2	IgG1	DXd	Camptothecin	TOPO1 inhibitor	8	advanced-stage HER2[+] breast cancer after two or more anti-HER2-based regimens (2019)
Polatuzumab vedotin-piiq	CD79b	IgG1	MMAE	Auristatin	microtubule inhibitor	3.5 (mean)	R/R DLBCL (2019)[c]
Sacituzumab govitecan-hziy	Trop-2	IgG1	SN-38 (active metabolite of irinotecan)	Camptothecin	TOPO1 inhibitor	8	advanced-stage, triple-negative breast cancer in the third-line setting or beyond (2020)
Enfortumab vedotin-ejfv	Nectin 4	IgG1	MMAE	Auristatin	microtubule inhibitor	4	advanced-stage urothelial carcinoma, following progression on a PD-1 or PD-L1 inhibitor plus platinum-containing chemotherapy (2020)
Belantamab mafodotin-blmf	BCMA	IgG1	MMAF	Auristatin	microtubule inhibitor	unknown	R/R multiple myeloma in the fifth-line setting or beyond (2020)

ADC, antibody–drug conjugate; AML, acute myeloid leukemia; B-ALL, B-cell acute lymphoblastic leukemia; BCMA, B-cell maturation antigen; cHL, classical Hodgkin's lymphoma; DLBCL, diffuse large B-cell lymphoma; MF, mycosis fungoides; MMAE, monomethyl auristatin E; MMAF, monomethyl auristatin F; pcALCL, primary cutaneous anaplastic large-cell lymphoma; PTCL, peripheral T-cell lymphoma; R/R, relapsed and/or refractory; sALCL, systemic anaplastic large-cell lymphoma; TOPO1, topoisomerase I; Trop-2, tumor-associated calcium signal transducer 2.

[a]As a single agent or in combination with daunorubicin and cytarabine. Gemtuzumab ozogamicin was withdrawn from the market in 2010 and reapproved in 2017 for newly diagnosed or R/R CD33-positive AML.

[b]In combination with cyclophosphamide, doxorubicin, and prednisone for newly diagnosed sALCL or CD30[+] PTCL and in combination with doxorubicin, vinblastine, and dacarbazine for newly diagnosed cHL.

[c]In combination with bendamustine and rituximab.

Adapted from Z.J. Drago, S. Modi and S. Chandarlapaty, *Nat. Rev. Clin. Oncol.* 18:327–334, 2021. With permission from Nature.

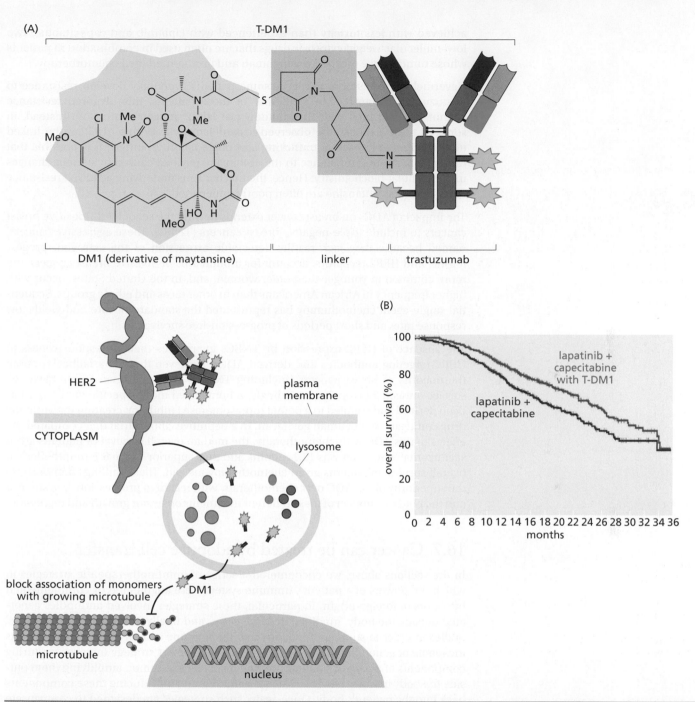

Figure 16.7 Use of an antibody–drug conjugate (ADC) to treat breast cancer (A) The cytotoxic drug DM1, a chemical derivative of the potent microtubule inhibitor, maytansine (*above left*), is chemically coupled via a thioether linker to one of four alternative sites in the constant region of an anti-HER2 moAb directed against this receptor molecule (*above right*), which is frequently overexpressed in breast cancers. After being administered to a patient afflicted with a HER2-overexpressing breast cancer, the ADC can bind to a HER2 molecule (*purple, lower left*) displayed on the surface of HER2+ cancer cells. The ADC–HER2 complex is then internalized via endocytosis and is thereafter delivered to a lysosome within which the linker is cleaved. This liberates the maytansine moiety, which can then escape the lysosome and enter into the cytosol, where it inhibits the growth of microtubules and therefore the assembly of the mitotic apparatus in the M phase of the cell cycle. (B) The Kaplan–Meier graph presents the results of a clinical trial demonstrating that treatment with the ADC, TDM1 (*blue curve*) produced significant gains in short-term survival compared with a standard chemotherapy (lapatinib-capecitabine, *red curve*) employed with this type of breast cancer, although long-term survival was not significantly affected. (A, Adapted from P.M. Lorosso, D. Weiss, E. Guardino et al., *Clin. Cancer Res.* 17: 6437–6447, 2011. With permission from American Association for Cancer Research. B, from S. Verma, D. Miles, L. Gianni et al., *N. Engl. J. Med.* 367:1783–1791, 2012. With permission from Massachusetts Medical Society.)

parental moAb (**Figure 16.7A**). Compared to a standard chemotherapy for patients who had been previously treated with a combination of trastuzumab and a taxane (another class of small-molecular-weight microtubule inhibitors), the ADC provided significant gains in overall survival (Figure 16.7B). Furthermore, these gains were achieved with less toxicity than experienced with lapatinib and capecitabine, two low–molecular-weight cytotoxic drugs that are often used in combination in patients whose tumors have escaped trastuzumab and taxane-mediated chemotherapy.

Nevertheless, the tumors borne by some patients eventually develop resistance to trastuzumab emtansine. In a number of these patients, this acquired resistance cannot be associated with the shutdown of HER2 antigen expression. Instead, in some of these patients, the observed loss of sensitivity to this ADC can be linked to alterations in lysosomal trafficking and degradation. In others, it is possible that tumor cells develop resistance to the emtansine payload caused by certain changes in intracellular biochemistry. Hence, the mechanisms underlying acquired resistance to trastuzumab emtansine are often poorly understood.

The impact of ADCs on breast cancer extends beyond the set of HER2-positive breast cancers to include triple-negative breast cancers (TNBC). These aggressive cancers, named because they lack readily detectable expression of the estrogen, progesterone, and HER2 receptors, account for around 15% of invasive breast cancers, are more common in younger than older women, and, in the United States, occur with higher frequency in African Americans than in other races and ethnic groups. Sequential single-agent chemotherapy has represented the standard of care and yields low response rates and short periods of progression-free survival.

The absence of HER2 expression by TNBCs guarantees their unresponsiveness to HER2-targeting antibodies and derived ADCs. However, there is a fallback option: the majority of breast cancers (including TNBCs) express human trophoblast cell surface antigen 2 (Trop-2). Accordingly, a humanized moAb specific for Trop-2 has been derived and coupled to a potent topoisomerase I inhibitor, creating the antibody drug conjugate sacituzumab govitecan. In a definitive clinical trial that compared the effects of this ADC with chemotherapy, the median overall survival of patients given sacituzumab govitecan was 12.1 months and thus superior to the 6.7-month median overall survival of patients given chemotherapy instead. These findings demonstrate the superiority of the ADC over chemotherapy with regard to progression-free survival and the potential impact of these different treatments on tumor growth and regression.

16.7 Cancer can be treated by adoptive cell transfer

In the sections above, we encountered a series of immunotherapeutic strategies in which the powers of a patient's immune system were augmented or complemented by agents of foreign origin. In particular, these strategies involved antibodies generated outside the body, modified in some cases, and finally introduced into patients' bodies in order to supplant functions that the immune system of these patients was incapable of generating on its own. An entirely different strategy involves capturing components of a patient's native immune response to a tumor, amplifying them outside the body in ways that do not occur naturally, and introducing these components back into the patient's body. Once again, such strategies are designed to compensate for the inability of a patient's native immune system to mount certain types of effective anti-tumor responses on its own.

As we saw in Chapter 15, T cells are the major governors of immune responses to cancer and therefore are key determinants of whether tumors are eliminated, kept under control, or progress. In fact, diverse studies of tumor histology (see Figures 15.3, 15.4, and 15.5) have revealed that many actively growing tumors harbor significant populations of T cells that coexist with large numbers of neoplastic cells but nonetheless have clearly failed to eliminate the cancer cells.

The repeated detection of such tumor-infiltrating T lymphocytes (TILs) has stimulated a major thrust in the development of anti-cancer immunotherapies, which is inspired by the notion that such infiltrating T cells are not present in adequate numbers or in physiological states that would allow them to effectively kill nearby cancer cells. This notion has led to the development of methods for harvesting T cells from a patient, manipulating them *ex vivo*, and then introducing the altered lymphocytes back into the donor or into other histocompatible recipients. Such a clinical strategy, which also represents a form of passive immunization, is termed **adoptive cell transfer** (ACT).

In adoptive cell transfer, immune cells, almost invariably T cells, are harvested directly from a patient's excised tumor, expanded *ex vivo* into much larger cell populations, and then transferred back into the patient via intravenous infusion. Ideally, among the reintroduced cells are anti-tumor cytotoxic T cells whose TCRs direct cytotoxic attacks on cells that display tumor-associated antigens, while sparing cells that do not. By necessity, this target specificity was initially developed in the donor's body during the period when the tumor was growing out and provoking various types of adaptive immune responses. Two major versions of ACT T-cell immunotherapies have been developed: adoptive transfer of T cells bearing conventional TCRs and adoptive transfer of T cells bearing chimeric antigen receptor T cells (CAR T cells).

To provide some background, we note that during the last quarter of the twentieth century, various experiments demonstrated that certain subpopulations of T cells could be harvested from tumor-bearing mice and thereafter propagated in culture under conditions that greatly increased their number without these cells losing their critical biological functions. The expansion of these murine T-cell populations *ex vivo* depended on the inclusion in the *in vitro* growth medium of the cytokine interleukin-2 (IL-2), a potent T-cell growth factor. Later, it became possible to expand corresponding human lymphocytes *ex vivo*. In both cases, the expanded lymphocyte populations could be re-infused back into the bodies of the donors that previously formed them (or, more precisely, formed their initially harvested progenitors). These re-introduced T lymphocytes should not be recognized as foreign and therefore should not provoke an immune response that would blunt or eliminate them in the bodies of engrafted recipients.

In more recent versions of this strategy, genomic sequencing is used to identify tumor-specific somatic mutations in the coding sequences of a patient's cancer cell genome (**Figure 16.8**). Once identified, these mutant sequences can be used to

Figure 16.8 Overview of adoptive T-cell transfer cancer therapy The schematic outline of this treatment strategy shows the following steps: (1) Samples of normal and tumor tissue are taken from the patient and T cells are obtained from the tumor or from blood (*top left*). (2) DNA is extracted from normal and tumor tissue and is subjected to exomic sequencing. (3) Comparison of the tumor vs. normal exome sequences allows identification of structural mutations, that is, those that encode differences in the amino acid sequences of proteins, leading in turn to identification of candidate tumor-specific neoantigens. (4) Minigenes, each specifying a peptide of 25 amino acids in length flanking and including an identified somatic structural mutation, are synthesized, linked in tandem, and transfected into APCs, enabling expression and presentation of encoded neoantigen peptides by the transfected APCs. Five fused minigenes, each represented by a different color are shown here. Alternatively, neoantigen peptides can be chemically synthesized and added directly to *in vitro* cultures of APCs. The presence of mutations in each of these genes is suggested by the small lighter circles within each. (5) Co-culture of patient T cells harvested in Step 1 allow those APCs displaying neoantigen peptides to activate T cells prepared from the patient (*dark blue*) bearing TCRs that recognize complexes of neoantigen peptide:MHC (*lower right*) presented by mutant oligopeptide-displaying dendritic cell (APC). (6) The minority of T cells that display activation markers after interaction with the peptide-displaying APC are identified by flow cytometry, sorted, and expanded to large numbers (>10^{10}) *ex vivo*. (7) The expanded populations of neoantigen-reactive T cells are infused into the donor, who has previously undergone lymphodepletion by treatment with a cytotoxic drug in order to create a niche for the infused cells. (Adapted from S.A. Rosenberg and N.P. Restifo, *Science* 348:62–68, 2015.)

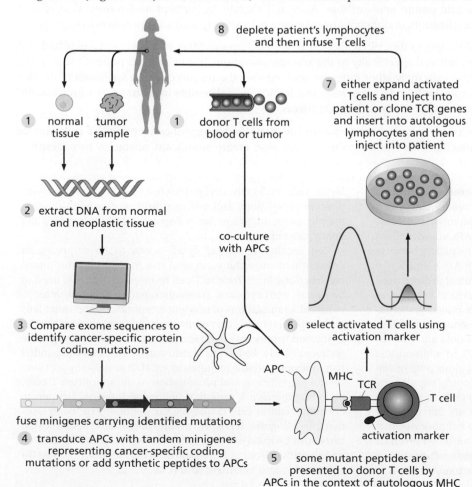

8 deplete patient's lymphocytes and then infuse T cells

1 normal tissue   tumor sample

1 donor T cells from blood or tumor

7 either expand activated T cells and inject into patient or clone TCR genes and insert into autologous lymphocytes and then inject into patient

2 extract DNA from normal and neoplastic tissue

co-culture with APCs

3 Compare exome sequences to identify cancer-specific protein coding mutations

6 select activated T cells using activation marker

APC   MHC   TCR   T cell

activation marker

fuse minigenes carrying identified mutations

4 transduce APCs with tandem minigenes representing cancer-specific coding mutations or add synthetic peptides to APCs

5 some mutant peptides are presented to donor T cells by APCs in the context of autologous MHC

construct recombinant "minigenes" that encode oligopeptides carrying the mutant amino acid sequences (rather than encoding the full-length mutant proteins). Moreover, in a single tumor genome, multiple polypeptide sequences, corresponding to various somatic mutations registered in the patient's tumor genome, can be identified; each of these sequences can be used to construct a recombinant minigene, and the resulting collection of minigenes can be linked to one another to create a tandem string of minigenes. (Such mutant tumor cell sequences can derive from somatically mutated driver genes as well as passenger genes because both are equally capable of specifying tumor-specific antigens.)

These strings of minigenes can then be processed further to encourage the outgrowth *ex vivo* of specific lymphocyte subpopulations that recognize and respond to them. Thus, DNA molecules carrying these minigenes can be transfected into donor-derived antigen-presenting cells (APCs), which should then permit the APCs to present minigene-encoded mutant peptides on their cell surface MHC-I molecules. In principle, co-culture of these transfected APCs with the patient's cultured T cells should trigger activation of a subset of the donor's explanted immunocytes, in particular those lymphocytes bearing T-cell receptors (TCRs) specific for the mutant peptide-antigen:MHC complex that had previously been exhibited *in vivo* by the patient's own tumor cells and therefore experienced by the precursors of the explanted T cells. The resulting *in vitro* activated T cells, ideally expanded to far larger numbers, can then be infused back into the patient's body.

Although creative in exploiting our existing knowledge of T-cell biology, these ACT strategies are complicated by certain realities of immune regulation (**Sidebar 16.6**).

As shown in **Table 16.3**, ACT has been used for the treatment of several types of cancer including melanoma, lymphomas, and carcinomas, such as urothelial and breast tumors. Adoptive T-cell transfer, as outlined here, leverages the capacity of T cells to discriminate between oligopeptide sequences that differ by as little as a single amino acid residue. Although elegant in concept and powerful in clinical application, major challenges remain to its widespread adoption in oncology clinics.

To begin, as described above, this type of immunotherapy is highly personalized, that is, tailored specifically to the somatic mutations borne by each patient's tumor. It is also extremely labor intensive and requires the resources and a spectrum of medical expertise found only in large academic medical centers to manage its administration and respond to possible side effects.

Also, as summarized in the next section, the biology of conventional T cells and the mechanisms leading to their activation create significant obstacles to widespread

---

Sidebar 16.6 Development of effective ACT strategies requires attention to many complicating factors In practice, the design of an adoptive cell transfer protocol requires attention to a number of critical biological details. To begin, the APCs used in these APC experiments should ideally come from the patient because the spectrum of MHC molecules that these APCs employ to present oligopeptide antigens should be identical to those used by the patient's tumor cells to do so. This is critical because, as we learned in the last chapter, the complex molecular structures recognized by TCRs are oligopeptide antigens together with the "hands" of the MHC molecules that hold and present these oligopeptide antigens (see Figure 15.7). In addition, the tumoricidal powers of the lymphocyte populations propagated *ex vivo* are greatly strengthened when the minority populations of these lymphocytes that have been successfully activated by exposure to minigene-encoded peptides are harvested for re-infusion into the donor (discarding those lymphocytes that have not been activated). This enrichment can be achieved by harvesting only those lymphocytes that express on their surface certain antigens associated with T-cell activation. As shown in

Figure 16.8, such cells can be detected readily in cultured lymphocyte populations, isolated, and further expanded *in vitro*, thereby preparing them for subsequent re-infusion into the donor cancer patient.

Yet another biological detail needs to be respected in order to maximize the success of the ACT re-infusion protocol described here: Prior to T-cell re-infusion, patients need to be treated with cytotoxic chemotherapeutic drugs in order to reduce the populations of resident lymphocytes and other leukocytes in their immune systems, thereby "making room" for the re-infused lymphocytes and facilitating their taking root in the recipient's body. Following re-infusion of the *ex vivo* expanded lymphocytes, patients are often given IL-2 to help support successful engraftment and proliferation of the re-infused T cells, including the rare T cells that, following infusion, happen to encounter cancer cells displaying the peptide:MHC complexes that they recognize. It is worth noting that because systemic IL-2 can cause toxic side effects, more recent research has focused on innovative T-cell culture protocols that reduce the need for sustained *in vivo* treatment with IL-2.

Table 16.3 Treatment of cancer by adoptive T-cell transfer

Year	Type of cancer	Molecule targeted	Patients	% objective responses	Comment
1998	melanoma	unknown	15	60	The first clinical trial of TILs; responses were observed but of short duration and TILs did not persist.
2002	melanoma	unknown	13	46%	First use of lymphodepletion before cell transfer, responses were of longer duration and TIL population persisted.
2008	melanoma	NY-ESO-1	9	33%	First demonstration of feasibility of treating cancer with *in vitro* generated antigen-specific TILs.
2011	melanoma	unknown	93	56%	20% of the patients had complete recoveries exceeding 5 years.
2014	cervical cancer	unknown	9	33%	Treatment of carcinoma with TIL therapy.
2014	cholangial carcinoma	mutated ERB2	1	—	Use of *in vitro* selected antigen-specific TILs to effect complete regression >6 years.
2016	metastatic colorectal cancer	mutant KRAS (G12D mutation)	1	—	Use of *in vitro* selected antigen-specific TILs to effect complete regression >5 years.
2018	chemotherapy-resistant metastatic breast cancer	mutant forms of four proteins: SLC3A2 KIAA0368 CADPS2 CTSB	1	—	Use of multiple *in vitro* antigen-specific TIL populations to effect durable regression.

Adapted in part from S.A. Rosenberg and N.P. Restifo, *Science* 348:62–68, 2015.

adoption in oncology clinics. Hence, the ACT therapy described here, as documented in Table 16.3, has been applied to a variety of cancers. At present, it can be regarded as providing a dramatic demonstration of the powers of the adaptive immune system, but nevertheless remains a therapeutic protocol whose logistics and incurred expense keep it from being introduced into the oncology clinic as a routinely employed strategy in the foreseeable future.

16.8 CAR T cells have predetermined specificity and bypass MHC-dependent antigen presentation

An alternative adoptive cell transfer strategy involves the modification of T cells in ways that are simpler than the complex, challenging experimental protocol described above. Recall from Section 15.7 that the conventional T-cell activation triggered by APCs occurs in a lymphoid tissue, and requires both antigen recognition by the TCR complex and co-stimulatory signals generated by the T cell's co-stimulatory receptors. The latter must recognize ligands displayed on the surface of antigen-presenting APCs in order for this activation to proceed to completion.

These two recognition events, mediated by the two sets of receptors, trigger a complex set of signaling events within the naive T cell, with each receptor type acting on a largely distinct intracellular signal-transducing circuitry; the resulting signaling events, operating together, result in activation of the T cell.

Subsequently, following its migration out of lymphoid tissues to peripheral, non-lymphoid sites in the body, the activated T cell may encounter cells displaying peptide:MHC complexes recognized by its TCRs; this T cell will then be induced to respond, specifically by launching a cytotoxic attack on the encountered target cells. This complex process, involving multiple different cell types and several distinct receptor–ligand interactions, is outlined in **Figure 16.9A**.

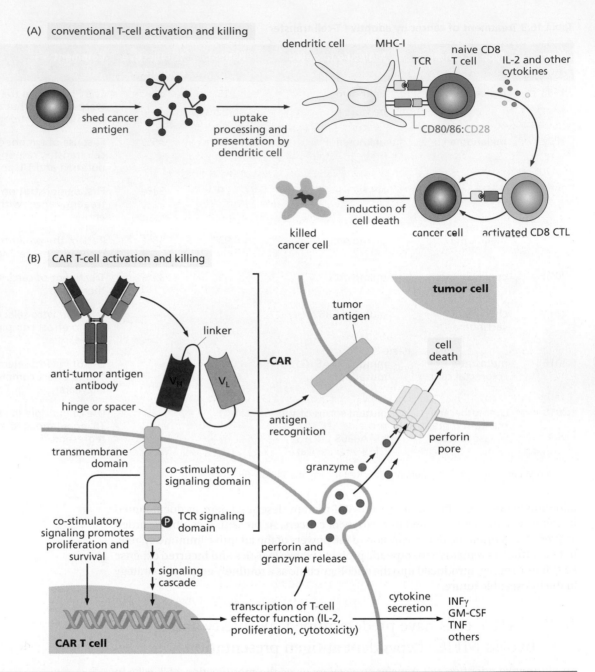

Figure 16.9 Overview of CAR T-cell biology (A) Conventional T-cell activation and killing involves at least three cell types: CD8 T cells (*dark blue*), dendritic cells (*yellow*), and cancer cells (*light red*), the last being the source of tumor antigens for activating T cells and the target of killing. Ligand–receptor interactions between naive T cells and dendritic cells include both TCR recognition of complexes of peptide antigen (*dark red*) formed with MHC-I molecules displayed by dendritic cells and co-stimulation of T cells via C80/86:CD28 interactions. In the presence of IL-2 and other cytokines, these several interactions enable the activation of a naive CD8 T cell (*dark blue*) and its maturation to a functional cytotoxic T lymphocyte (CTL, *light blue*). The resulting activated CD8 T cell can proceed to kill a cancer cell (*light red*) that displays the MHC-I:peptide complex recognized by the T cell's TCR. (B) Chimeric antigen receptor-displaying T-cell (CAR T-cell) responses require only two cell types, a CAR T cell (*lower left*) and its target cell, specifically a tumor antigen-displaying cancer cell (*upper right*). The recognition of the tumor antigen (*green*) on the surface of the target cell is the only type of receptor–ligand interaction required to trigger killing of the targeted cancer cell. Most chimeric antigen receptors include three modules: (1) an extracellular antigen-binding domain comprised of a single-chain variable region fragment (scFv) derived from the V_H and V_L domains of a monoclonal antibody molecule known to bind specifically to a target antigen (*red, pink*), (2) a hinge or spacer segment and a short transmembrane domain derived from any of a variety of cell surface receptor molecules (*light green*), and (3) a cytoplasmic domain assembled from two distinct intracellular signaling domains, one providing co-stimulatory signals usually released by a CD28 co-stimulatory receptor (*light blue*) and a domain containing the TCR signaling domain that is usually activated when a TCR recognizes and binds a cognate antigen on a second cell (*blue, yellow stripes*). Engagement of the CAR with the targeted tumor antigen has three major effects: a perforin–granzyme-mediated cytolytic attack by the CAR T cell on the tumor cell (*upper right*), secretion by the CAR T cell of a variety of cytokines (*lower right*), and proliferation and survival of the CAR T-cell population. (Adapted in part from R.J. Lee, J.S. Abramson and R.A. Goldsby, *Case Studies in Cancer*, New York: W.W. Norton, 2019, p. 275, and in part from R.C. Larson and M.V. Maus, *Nat. Rev. Cancer* 21:145–161, 2021. With permission from Nature.)

The general strategy of **chimeric antigen receptor** (CAR) T-cell therapies bypasses the complexity of this conventional T-cell activation process. These therapies do so by endowing T cells with receptors that enable them to recognize and attack target cells displaying an already-identified cell surface antigen. This is achieved without identifying and assembling the usually required antigen-specific TCR and its associated co-receptor.

To begin, DNA segments encoding the variable (V) antigen-recognizing domains of certain moAbs can be appropriated as starting points. These mAbs have previously been isolated because they recognize known cell surface antigens displayed by cancer cells. Using recombinant DNA, the two V-encoding DNA segments of such a mAb are isolated and strung together, thereby assembling a gene encoding an entirely artificial antigen-binding receptor. This resulting antigen-binding domain can then be incorporated into a larger protein that carries the additional domains associated with a TCR, specifically the transmembrane domains and the cytoplasmic domains of a TCR; finally, the cytoplasmic domain of a co-stimulatory receptor is added. Hence, the antigen-binding ectodomain of the engineered receptor should mimic the functions of a naturally arising TCR.

If constructed properly (**Sidebar 16.7**), the cytoplasmic domains of this newly constructed receptor should enable the release into lymphocytes of the mixture of biochemical signals that normally occurs when a natural TCR recognizes and binds a cognate antigen. Recall that this binding usually elicits functional activation of the lymphocytes and enables them to kill the cells displaying targeted tumor antigens (Figure 16.9B). (The term "chimeric" in chimeric antigen receptor, CAR, comes from the fact that the antigen-recognizing portion of this artificial receptor derives from an already-identified monoclonal antibody, whereas the transmembrane and cytoplasmic domains derive from the signal-emitting domains of a bona fide TCR plus its companion co-stimulatory receptor, such as CD28 or 4-1BB.)

These multiple structural domains of a CAR allow it to incorporate all of the functional features required for its recognition of a target cell and the resulting functional activation of a T cell, all this achieved by a single engineered molecule (see Sidebar 16.7). Once a gene encoding a CAR has been assembled, it can be transfected into a patient's lymphocytes, resulting in the creation of CAR T cells.

An additional attraction of CAR T cells is that the antigen specificity of their CAR can be predetermined to be specific for recognizing and binding a tumor antigen known to be displayed on the surface of targeted cancer cells (see Figure 16.9B). Also, such a protein antigen can be displayed directly on the surface of the cancer cell's plasma membrane without a need for prior antigen processing and association with MHC-I molecules. (Of course, any resulting attack launched by a CAR-expressing T cell will

Sidebar 16.7 T-cell chimeric antigen receptors incorporate functional domains from several different molecules (1)The process of constructing a chimeric antigen receptor begins with the identification of an antigen that is either tumor associated or tumor specific and is reliably displayed by many, if not all, of the neoplastic cells within a tumor. A suitable mAb is then derived against the identified tumor antigen. The DNA sequences that encode the antigen-binding heavy- and light-chain variable regions of the antibody molecule are then subcloned and linked together via a linker-encoding segment into a single reading frame. This yields an expression construct specifying the antigen-binding domain of the parental moAb from which these sequences were derived. Hence, the antigen-recognizing and -binding functions usually achieved collaboratively by the heavy and light chains of a conventional immunoglobulin are now mediated by a single polypeptide, often termed a **single-chain variable region fragment** (scFv). (2) This gene construct is then extended further by adding a segment encoding a membrane-spanning polypeptide, often mirroring the transmembrane sequence of a T-cell protein such as

the CD8 co-receptor. This polypeptide segment serves the dual functions of connecting the scFv ectodomain to the CAR's cytoplasmic signaling domains and, at the same time, operates as a molecular hinge, providing flexibility to the scFv ectodomain module of the CAR. (3) This gene fusion is followed by adding a segment encoding the cytoplasmic domain of the CAR that incorporates the signal-emitting domain of the TCR complex together with the signaling domain of a stimulatory co-receptor molecule, yielding a combined cytoplasmic domain that releases both types of signals into the T-cell cytoplasm following antigen binding.

Ideally, binding of the antibody-derived receptor portion of the CAR by the targeted antigen initiates many of the same T-cell-activating signaling cascades that are triggered within a T cell by engagement of its native TCR. Signals from the co-stimulatory domain contribute to the rate of cell proliferation, the persistence of the CAR T cells, and the durability of the response. As might be apparent, CAR T cells may also kill normal host cells that happen to express the same antigen as the targeted tumor-associated antigen.

fail if and when the cancer cell contrives to shut down expression of the targeted cell surface antigen.)

Like the antibodies from which they derive, antigen-specific CARs can be generated that recognize a variety of chemical classes of antigens, including not only proteins but also carbohydrates and lipids. In addition, the desired CAR T-cell responses involve the participation of only two cell types: the attacking CAR T cell and the targeted antigen-displaying cancer cell. This obviates the need for several different receptor–ligand interactions involving multiple cell types that are normally required for T-cell activation (see Figure 16.9).

Once the chimeric antigen receptor has been generated, the manufacture and administration of CAR T cells is carried out along the general outline illustrated in **Figure 16.10**. Thus, leukocytes are collected via **leukapheresis** of a patient's blood, a procedure in which white blood cells including T cells are separated from blood and the remaining red blood cells and plasma are recycled back to the patient. As illustrated in this figure, T cells are then separated from other leukocytes, transfected with an expression construct encoding the CAR, and stimulated to proliferate by antibody-mediated crosslinking using anti-TCR and anti-CD28 antibodies in the presence of IL-2. Following *ex vivo* expansion, the CAR-displaying T cells are infused back into a patient who, as in the adoptive T-cell transfer therapy described earlier, has been treated with leukocyte-depleting chemotherapy in order to create a niche in the patient's immune system for the re-introduced cells. Patients also are given IL-2 to support longer-term CAR T-cell growth and maintenance. As attractive as CAR T-cell therapy is in concept, it remains encumbered by several significant complications (**Sidebar 16.8**).

A prominent example of the use of CAR T therapy comes from its application to the treatment of B-cell-related lymphomas. Thus, the CD19 antigen is expressed by cells of the B-cell lineage at all stages of their differentiation (including derived lymphoma

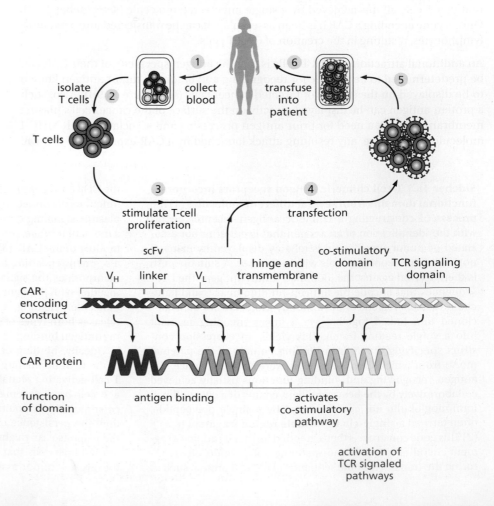

Figure 16.10 Overall strategy of CAR T-cell therapy Production of CAR T cells must be tailored to each individual's tumors and associated tumor antigens. (1) Blood is collected from the patient and leukocytes are prepared by the procedure sometimes termed leukapheresis. (2) T cells are purified from the leukocyte fraction. (3) T cells are stimulated to proliferate by exposure to IL-2 and T-cell activation signals. (4) The proliferating T cells are transfected with a lentiviral or retroviral expression vector construct that encodes the chimeric antigen receptor (CAR, *lower left*). (5) The population of CAR-expressing T cells is harvested. (6) These cells are transfused into the patient who has undergone lymphodepletion prior to the return of these genetically modified T-cells. (Adapted from R.J. Lee, J.S. Abramson and R.A. Goldsby, *Case Studies in Cancer*, New York: W.W. Norton, 2019, pp. 296–297.)

Sidebar 16.8 The utility of ACT therapies is reduced by certain limitations At present, like the adoptive cell transfer (ACT) therapies described earlier, CAR therapies represent personalized therapies designed to respond to each patient's tumor and associated antigens. Moreover, each batch of T cells that is modified to become CAR T cells must be autologous to the patient, that is, derived from a patient's own preexisting T cells. This requirement is critical in order to avoid a **graft-versus-host** (GVH) reaction, which often occurs when transfused donor T cells are derived from an allogeneic donor. Such allogeneic T cells may recognize and react to recipient host antigens displayed widely throughout a patient's body, resulting in a serious and potentially fatal condition.

In the case of CAR T-cell therapy, GVH may occur simply because, in addition to expression of the inserted CAR proteins, the modified T cells continue to express their normal complement of native TCRs. If allogeneic T cells are indeed employed to create CAR T cells (because a patient's lymphocytes are not readily available), these lymphocytes will not be tolerant following infusion of the modified T cells of some of the antigens naturally expressed by the recipient patient's tissues, particularly the protein products of the major histocompatibility complex genes, notably MHC-I and MHC-II. Recognition of these or other host antigens can lead to widespread GVH attacks on the recipient's cells that must be controlled with a variety of immunosuppressive treatments.

Another potential complication is that the CAR expressed by the introduced lymphocytes is itself a novel molecular structure not expressed by the patient's normal cells and therefore represents, on its own, a potentially immunogenic antigen capable of eliciting an immune response from the immune system of this recipient. This response can be minimized, however, by the immunosuppressive preparative lymphodepletion protocol mentioned above.

---

cells) and therefore has represented an attractive candidate for antigen-targeted CAR T therapy. Indeed, anti-CD19–CAR T-cell therapy has become the current standard of care for patients suffering acute lymphoblastic leukemia (ALL) and large B-cell lymphoma. In spite of its demonstrated therapeutic efficacy, this treatment is often accompanied by significant side effects that on their own can be severe and even life threatening. Thus, within the first several weeks of therapy, patients often experience fevers as high as 40°C, with spikes in the production of a variety of inflammatory cytokines, particularly interferon-γ, IL-6, TNFα, GM-CSF, and IL-1. Together, these systemically released cytokines represent the central features of the condition termed **cytokine release syndrome** (CRS), which is manifested as an acute systemic inflammatory syndrome that can lead to the dysfunction if not destruction of multiple tissues, organs, and even neurological disorders. The life-threatening consequences of CRS can usually be avoided or treated by use of a variety of immunosuppressive treatments (**Figure 16.11**).

To date, CAR T therapies targeting the tumor-associated antigen described above and several others listed in **Table 16.4** have made their greatest contribution to the treatment of hematologic cancers. Efforts to deploy CAR T against solid tumors have been largely frustrated by the challenges of identifying target antigens that are strongly expressed within tumors but exhibit little or no expression in normal tissues. Still, several candidate antigens have emerged. These include the B7-H3 antigen, which is expressed at low levels on the cells of many normal tissues, but at high levels by the cells of several pediatric cancers, including medulloblastoma, Ewing sarcoma, and osteosarcoma.

Brain cancers have represented a challenge for mAb-based cancer therapies because of the difficulties imposed by the blood–brain barrier, specifically its ability to block the transfer of large molecules from the circulation into the brain parenchyma. Conveniently, however, a quirk in brain physiology allows intravenously injected T cells to pass through the blood–brain barrier and gain access to the brain parenchyma. This fact has provoked considerable interest in developing CAR therapies for the treatment of glioblastomas.

16.9 Checkpoint inhibition is a distinct type of immunotherapy that modifies the behavior of immune cells

The antibody and T-cell-mediated immunotherapies we have examined so far are designed to directly engage and kill tumor cells. Checkpoint inhibitory therapies operate quite differently. Rather than launching direct cytopathic attacks on these cells, this class of treatments manipulates the immune system in ways that increase the likelihood and potency of *existing* anti-tumor immune responses, particularly

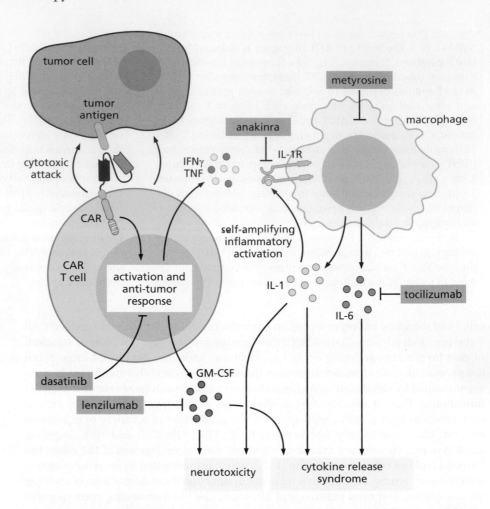

Figure 16.11 Cytokine release syndrome (CRS) and some pharmacological approaches to its treatment Activation of a CAR T cell (*light blue, lower left*; see also Figures 16.9 and 16.10) triggered by its encounter with a targeted tumor cell (*pink, upper left*) causes the CAR T cell to synthesize and secrete a number of powerful cytokines, including tumor necrosis factor (TNF, *dark blue*), interferon-γ (IFNγ, *yellow*), and granulocyte-macrophage colony-stimulating factor (GM-CSF) (*purple*). Macrophages (*light yellow, right*) respond to the highly pro-inflammatory cytokines in a number of ways, including secretion of IL-1 (*light blue*) and IL-6 (*pink*). In addition to its neurotoxic effects, IL-1 (*light blue*) interacts in an autocrine fashion with IL-1 receptors displayed by the macrophages (*light green*) to drive them to higher levels of activation, contributing to a positive feedback loop that intensifies CRS and its impact on the central nervous system (CNS). Anti-CRS therapies include treatment with anakinra, an IL-1 mimic that blocks the IL-1 receptor; lenzilumab, a monoclonal antibody that binds GM-CSF; and tocilizumab, a monoclonal antibody that binds IL-6. In addition, small-molecule therapies deploying dasatinib and metyrosine have been used. Dasatinib, an inhibitor of the Src family of tyrosine kinases as well as Bcr-Abl, lowers the level of activation and cytokine secretion in CAR T cells. Metyrosine acts on macrophages to reduce their production of catecholamines, which function in turn as inducers of IL-6 secretion. (Adapted from R.C. Larson and M.V. Maus, *Nat. Rev. Cancer* 21:145–161, 2021. With permission from Nature.)

responses that may already exist in a patient's body but cannot be launched or effectively sustained because they have been shut down by certain immune regulatory mechanisms.

Checkpoint inhibition was introduced in Chapter 15 as a mechanism that operates to restrain excessive T-cell activation that might otherwise lead to persistent inflammatory states or autoimmune diseases. In effect, immune checkpoints of various types operate to provide negative-feedback controls on normal immune function (see also Section 6.14), thereby ensuring that the occurrence and strength of various immune responses are constrained in order to avoid inadvertent, uncontrolled, highly destructive autoimmune attacks on normal cells and tissues.

Table 16.4 Use of CAR T cells to treat a variety of hematologic cancers

Target antigen	Disease	CAR	Number of patients analyzed	Median age (years)	Response	Patients with CRS (%)	Patients with neurotoxicity (%)
CD19	B-ALL (pediatric)[a]	Tisagenlecleucel; 4-1BB co-stimulation	75	11	6-month relapse-free survival rate of 80%	77	40
CD19	relapsed or refractory DLBCL[a]	Axicabtagene ciloleucel; CD28 co-stimulation	101	58	83% objective response; 58% complete response	93	67
CD19	refractory B-cell lymphomas[a]	Tisagenlecleucel; 4-1BB co-stimulation	28	58.5	64% overall response; 43% complete remission	57	39
CD19	mantle-cell lymphoma[a]	Axicabtagene ciloleucel; CD28 co-stimulation	68	65	93% objective response rate; 67% complete response	91	63
CD19	B-ALL	CD28 co-stimulation	53	44	83% complete remission; median overall survival 12.9 months	85	44
CD22	relapsed or refractory pre-B-ALL	4-1BB co-stimulation	21	19	73% complete remission treated with higher dose	76	unreported
BCMA	relapsed or refractory multiple myeloma	Idecaptagene vicleucel; 4-1BB co-stimulation	33	60	85% objective response rate; 45% complete response rate	76	42
BCMA	multiple myeloma	4-1BB co-stimulation	25		48% overall response rate	88	32

B-ALL, B-cell acute lymphoblastic leukemia; BCMA, B-cell maturation antigen; CAR, chimeric antigen receptor; CRS, cytokine release syndrome; DLBCL, diffuse large B-cell lymphoma; 4-1BB TNF-related ligand protein.
[a]U.S. Food and Drug Administration (FDA) approval.
Adapted from: R.C. Larson and M.V. Maus, *Nat. Rev. Cancer* 21:145–161, 2021. With permission from Nature.

The concept of immune checkpoint regulation and associated anti-tumor immunotherapy emerged from a foundational experiment reported in 1996. As illustrated in **Figure 16.12A**, this work demonstrated the principle of checkpoint immunotherapy by showing that antibody-mediated blockade of CTLA-4 function inhibits tumor growth. This result was most provocative, because it showed that an antibody directed against a T-cell protein (rather than a cancer cell-associated antigen) would impede the growth of a tumor.

While we have discussed the importance of checkpoint controls in limiting the direct attacks by CTLs on potential target cells, these controls also operate to constrain an earlier step in the development of CTLs that operates *centrally* in the lymph nodes rather than *peripherally* at sites of immune attack on tumors. Recall that CTLA-4 acts as a cell surface receptor displayed by T lymphocytes that is activated by binding its cognate ligands displayed by antigen-presenting cells in the lymph nodes (see Sections 15.7 and 15.12). Since its ligands (usually termed CD80 and CD86) are cell surface proteins displayed by other cells, specifically APCs, they have sometimes been termed "counter-receptors." Upon engagement with CD80/86, which requires close physical contact between T cells and APCs displaying their cognate antigens, the CTLA-4 receptor normally acts directly to shut down T-cell receptor signaling and thus T-cell activation in both CD4+ and CD8+ T cells. Hence, introduction of an anti-CTLA-4 mAb severs

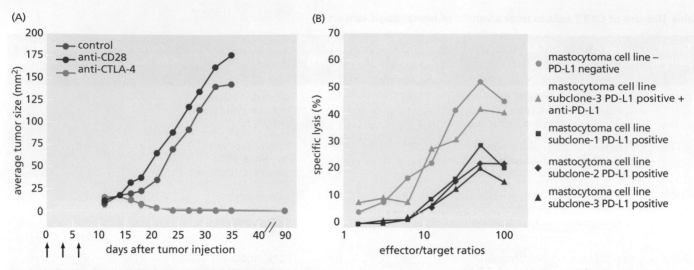

Figure 16.12 Enhancement of anti-tumor immunity by antibody-mediated checkpoint blockade (A) Effect of CTLA-4 blockade. Cells of a colon cancer cell line were implanted into three groups of mice on day zero. The treated groups received either anti-CTLA-4 moAb (*blue*) or anti-CD28 moAb (*red*) at the three times indicated by the arrows, while the control group (*brown*) received neither. Tumor growth was assessed by measuring the size of the tumor nodule. (B) Effect of PD-1 blocking PD-L1 interaction. The targeted cancer cells used were from a cell line that had been derived from a mouse mastocytoma, which arises from mast cells and represent a subtype of granulocytes of the myeloid lineage. These cells did not naturally express PD-L1. Three subclones of the line were transfected with a PD-L1 expression construct to force their expression of PD-L1. Each of these populations was subjected to attack by a cytotoxic T-cell line *in vitro* at increasing ratios of the attacking cytotoxic T-cell line (effectors) to the mastocytoma target cell line (target) and the lysis of targeted mastocytoma cells was measured. The three PD-L1-expressing mastocytoma subclones showed reduced sensitivity to killing (*red lines*) relative to parental mastocytoma cells that were PD-L1 negative (*green filled circles*). Moreover, when the subclone-3 cells, which had been forced to express PD-L1 were treated with anti-PD-L1 moAb (*green filled triangles*), their sensitivity to killing rivaled that of the parental cell line (*green filled circles*) that lacked forced PD-L1 expression. (A, adapted from D.R. Leach, M.F. Krummel and J.P. Allison. *Science* 271:1734–1736, 1996. With permission from AAAS. B, adapted from Y. Iwai, M. Ishida, Y. Tanaka et al. *Proc. Natl. Acad. Sci. USA* 99:12293–12297, 2002. With permission from National Academy of Sciences, U.S.A.)

an important negative-feedback loop that normally operates within lymphoid tissues to constrain T-cell activation. Accordingly, this moAb treatment enables the activation of both CD8+ T cells that are capable of becoming cytotoxic T lymphocytes (CTLs, T_C) and CD4+ helper T-cell (T_H) subsets that make important contributions to anti-tumor immunity.

While the 1996 work did not reveal these mechanistic details, it did show unequivocally that manipulations of immune cells, in this case CTLs, led to major changes in the ability of tumors to persist and continue to grow in host mice. This suggested, in turn, that the immune systems of the tumor-bearing hosts intrinsically possessed the power to kill cancer cells. However, this hypothesized power could not be realized in the absence of experimental intervention, in this case an intervention achieved through the use of an antibody recognizing a cell surface protein (CTLA-4) displayed by T cells.

In fact, the 1996 work could have been interpreted in two ways: either the applied antibody directly *stimulated* the anti-tumor response of the mice treated with it or, alternatively, it *interrupted* a normally operative inhibitory mechanism that blocked their ability to mount an effective anti-tumor response. As is clear from the last chapter and the above text, it was this second mechanism that was found to underlie and explain the mechanism of action of the anti-CTLA-4 moAb.

The observed successes in halting tumor growth suggested a larger principle: checkpoint controls analogous to the CTLA-4:CD80/86 regulation highlighted in these early experiments operate to protect cancer cells from attack and elimination, and conversely, blocking these negative controls can free certain arms of the immune system from suppression by these controls. This liberation from such postulated controls would then give T cells the freedom to launch effective attacks on cancer cells, ideally resulting in elimination of the latter.

Control experiments, undertaken in parallel, involved the CD28 co-receptor, which is also displayed on the surface of T cells. This second set of experiments demonstrated that CD28 acts in a manner that is precisely opposite to CTLA-4, doing so by *enabling* the generation of CTL-mediated immune attack. Thus, if the CD28 stimulatory receptor co-displayed with the TCR on the surface of a T lymphocyte was able to engage CD80/CD86 displayed by APCs, it would provide the co-stimulation necessary to enable the activation of CD8 T cells and their maturation into CTLs capable of mounting cytolytic attacks, notably attacks on targeted tumor cells. (As might be anticipated, blocking this CD28 receptor with an appropriate mAb led to a result opposite to that achieved by anti-CTLA-4 mAbs, namely that tumor growth actually *benefited* by squelching the actions of this activation-promoting CD28 protein.

These functions of CTLA-4 represent a two-edged sword. Normally, CTLA-4 provides critical protection against excessive or inappropriate initiation of immune responses launched by T cells, such as those that can lead to autoimmune diseases. However, at the same time, it can reduce or even prevent the initiation of desirable T-cell responses including those mounted against cancer cells.

In addition, subsequent work has shown that anti-CTLA-4 moAb has another mechanism of action: It contributes to the killing of regulatory T cells (T_{reg}) cells that, when allowed to function, are potently immunosuppressive cells that inhibit both CD4+ and CD8+ T cells. (Regulatory T cells display CTLA-4 on their surface, making them targets for the binding of anti-CTLA-4 antibodies, which leads to their death mediated by either natural killer (NK) cells or phagocytes as described in Section 16.4 and illustrated in Figure 16.4.)

An entirely independent set of studies uncovered PD-1, a second, distinct inhibitory checkpoint receptor displayed by CTLs and other T cells that operates at a later stage of T-cell development, after activated T cells have ceased their interactions with APCs, left lymphoid tissues, and dispersed into tissues throughout the body (**Sidebar 16.9**).

The discovery of the PD-1 receptor was followed by research that revealed and characterized PD-L1 and PD-L2, two ligands of the PD-1 receptor displayed on the surface of potential target cells. These ligands are routinely expressed on the surfaces of a variety of cell types throughout the body that might fall victim to CTL cell attack. Figure 16.12 describes an experiment conducted *in vitro* and employing a PD-1-expressing cytotoxic T-cell line whose TCRs were specific for an antigen expressed by the target P815 tumor cells employed in this particular experiment.

Sidebar 16.9 CTLA-4 and PD-1 act at distinct steps of CTL activation As described here, the two checkpoint controls actually operate at distinct, successive stages of CTL development. Initial activation of naive T cells occurs within lymph nodes, where antigen-presenting cells (APCs) present antigen to developing T cells (see Sections 15.7 and 15.12). This initial confrontation arms the T cells, allowing them to develop into functional CTLs and to disperse to peripheral, non-lymphoid tissues throughout the body. CTLA-4 damps down this initial priming in lymph nodes and other lymphoid tissues, explaining why anti-CTLA-4 can potentiate the priming of T cells that normally occurs within lymphoid tissues.

PD-1 appears on the surface of T cells that have already been primed in lymphoid tissues, as described above, but operates instead at a later phase of T-cell development, doing so in the context of non-lymphoid peripheral tissues. This explains how PD-1 actions interfere directly with the interactions between CTLs and their potential targeted cells in various anatomical sites throughout the body; these PD-1 functions therefore operate only *after* primed CTLs have emigrated from lymphoid organs and dispersed throughout the body.

As mentioned, engagement of PD-1 with its cognate ligands PD-L1 and PD-L2 displayed by target cells only occurs if and when CTLs encounter antigen-expressing target cells. Once this encounter has taken place, the PD-1:PD-L1/2 interaction proceeds to inhibit both T-cell cytotoxicity and the secretion of inflammatory cytokines, such as IFNγ and TNFα, by antigen-activated CTLs. In addition to its inhibition of T-cell cytotoxicity, we now know that the display of PD-1 ligands by potential target cells can be up-regulated by inflammatory cytokines, such as IFNγ and TNFα, that are secreted by CTLs, pointing to the PD-1:PD-L1/2 axis as a negative regulator of inflammation in peripheral, non-lymphoid tissues.

Importantly, because these two checkpoint controls (involving CTLA-4 and PD-1) operate at successive stages of T-cell activation, inhibiting both of them with mAbs directed against the two signaling axes can have additive or even synergistic effects in terms of activating ongoing immune attack. However, while the use of mixtures of moAbs interrupting both checkpoints may have increased potency in eliminating tumors, it can often elicit unacceptably high levels of autoimmune conditions in patients under treatment.

In more detail, the PD-1-expressing CTLs killed their P815 targets perfectly well if the cognate PD-L1 was missing from the surface of the targeted P815 tumor cells. However, forcing PD-L1 expression in the P815 tumor cells, achieved experimentally with an introduced expression vector, rendered these cells less susceptible to killing by the cytotoxic T cells. Hence, the previously observed killing of the target cells proceeded efficiently as long as these P815 target cells failed to express the PD-L1 ligand, but was greatly reduced once these P815 cells were forced to express PD-L1.

Importantly, as shown in Figure 16.12B, in the presence of anti-PD-1 antibody, even PD-L1-displaying cells fall victim to attack by PD-1-expressing CTLs. These experiments demonstrated that blockade of the PD-1/PD-L1 checkpoint by anti-PD-1 relieved inhibition of CTL-mediated immune responses. The PD-1/PD-L1 work established the principle that there are checkpoint regulatory mechanisms beyond the CTLA-4:CD80/86 signaling axis. Indeed, subsequent work has found yet other receptor:ligand pairs that operate to inhibit T-cell responses (Table 16.5). As might have been anticipated, these initial *in vitro* experiments could be extended to demonstrations *in vivo*, which confirmed that attacks on tumors formed by targeted cells could be predicted on the basis of their *in vitro* responses to certain CTLs. To summarize, use of anti-CTLA-4 or anti-PD1 moAbs demonstrated that the survival of mouse hosts bearing various tumors could be extended by antibody-mediated blockade of these two checkpoint controls (see Sidebar 16.9).

Table 16.5 Immune checkpoint inhibitors in clinical use and under development

Immune checkpoint		Immune checkpoint inhibitor (ICI)	FDA approval	Treatment indications
ICIs in clinical use				
Receptor	Ligand			
CTLA-4	CD80/86	anti-CTLA-4 (ipilimumab)	first in 2011, other ICIs since	metastatic melanoma and others
PD-1	PD-L1, PD-L2	anti-PD-1 (pembrolizumab, nivolumab, others)	first in 2014, other ICIs since	metastatic melanoma and subsequently many others
		anti-PD-L1 (durvalumab, atezolizumab)	first in 2016, other ICIs since	metastatic urothelial carcinoma and others
ICIs in advanced clinical trials				
Receptor	Ligand	Nature of IC	Immune process impacted by the immune checkpoint	
LAG-3	MHC-II	moAb[a]	inhibition of T-cell proliferation and cytokine secretion; may promote expansion of T_{reg}'s	
TIM-3	Galectin-9, CEACAM-1	moAb	induces T-cell exhaustion and can affect other immunocytes including NK cells and some dendritic cells	
PVRIG	PVRL2	moAb	engagement of T-cell-borne PVRIG with PVRL2 on dendritic cells inhibits T-cell activation	
VISTA	PSGL-1, VISIG-3	moAb	inhibits T-cell activation and cytokine production; inhibits NK-cell responses; inhibits secretion of TNF and certain other proinflammatory cytokines by myeloid cell types	
TIGIT	CD155	moAb	inhibits T-cell and NK-cell effector functions; can increase T_{reg}-mediated immunosuppression	
CD47	SIRPα	moAb[a]	inhibition of phagocytosis by macrophages	

[a]In addition to monoclonal antibodies (moAb), recombinant fusion proteins joining the binding domains of inhibitory receptors or ligands to constant region domains of IgG have also been employed as immune checkpoint inhibitors (ICIs). Abbreviations: LAG-3, lymphocyte activation gene 3; TIM-3, T-cell immunoglobulin and mucin-domain containing-3; PVIRG poliovirus receptor-related immunoglobulin domain containing; PVRL2, poliovirus receptor related 2; VISTA, V domain Ig suppressor of T-cell activation; PSGL-1, P-selectin glycoprotein ligand 1; VSIG3, V-set and Ig domain-containing 3; TIGIT, T-cell immunoglobulin and ITIM domain; SIRPα, signal regulatory protein alpha.

16.10 Checkpoint immunotherapies based on mouse studies have been applied in the oncology clinic

We turn now to an examination of the translation of these studies of checkpoint immunology to actual clinical applications of these immunotherapies. As shown in Table 16.5, ipilimumab, an anti-CTLA-4 mAb, was the first checkpoint inhibitor to receive FDA approval for the treatment of cancer. It showed survival benefits and improvements in overall survival when used to treat melanoma patients. Soon thereafter, an anti-PD-1 antibody also showed substantial clinical benefit for the treatment of melanoma, and it too received FDA approval (see Table 16.5). In the years since the introduction into clinical practice of these first immune checkpoint blockade inhibitors (ICIs), many clinical trials against a wide variety of cancers have shown a highly variable efficacy of these immunotherapies for the treatment of many types of cancer, with few recapitulating the striking long-term therapeutic benefits of the anti-melanoma therapies (**Sidebar 16.10**).

The data in **Figure 16.13** illustrate the changes in tumor burden, a term referring to the size of the tumor or number of tumor cells carried by each patient in each of the treatment groups. These observations reveal a substantial contrast in the efficacy of the anti-PD-1 versus anti-CTLA-4 blockade therapies. We can also note that a significant percentage of the patients treated with only anti-PD-1 checkpoint immunotherapy experienced durable progression-free survival (PFS) and overall survival (OS). A 5-year follow-up study of the patients enrolled in Checkmate 067 revealed leveling off ("plateauing") in survival beginning around 3 years after initiation of therapy that

Sidebar 16.10 Immune checkpoint blockade can yield unprecedented responses in human cancer patients We can learn much from reviewing a critical clinical trial, aptly titled Checkmate 067, that compared the clinical performance of an anti-CTLA-4 ICI with an anti-PD-1 ICI and with a combination therapy employing both ICIs. Checkmate 067 enrolled into a clinical trial a patient population of 945 individuals who bore histologically confirmed stage III or stage IV metastatic melanomas and had received no prior systemic treatment such as chemotherapy or moAb therapy. Historical controls, that is, the documented clinical responses of patients to various earlier anti-melanoma therapies, indicated that patients under treatment for these conditions would have an overall survival rate of 23% at five years after initial diagnosis.

Each of the participants in this trial was assigned to one of the following groups: anti-PD1 alone, anti-CTLA-4 alone, or anti-CTLA-4 + anti PD-1. As anticipated from previous studies of anti-CTLA-4 and anti-PD-1 ICIs, some patients in each of the three treatment groups showed significant clinical responses. However, the data in **Table 16.6** indicate that responses differed significantly among each of the treatments. Thus, patients enrolled in the anti-CTLA-4 cohort showed inferior responses to therapy relative to those enrolled in the other two in all respects; at the same time, patients whose tumors were better controlled also experienced higher incidence of side effects.

Table 16.6 Responses of melanoma patients to ICI monotherapy or combination therapy

	Nivolumab plus ipilimumab (N = 314)		Nivolumab (N = 316)		Ipilimumab (N = 315)	
	N	%	N	%	N	%
Complete response[a]	69	22	60	19	18	6
Partial response[a]	114	36	81	26	42	13
Total responding	183	58	141	45	60	19
Patients experiencing high-grade adverse events[b]	186	59	73	23	69	28
5-year overall survival (patients alive 5 years after initiation of treatment)	151	52	130	44	67	26
Median treatment-free interval in months[c]	18.1		1.8		1.9	

[a]Responses were evaluated using response evaluation criteria in solid tumors (RECIST). [b]Treatment-related adverse events (rash, diarrhea, fever, hypothyroidism, and others) were evaluated on a scale of 1 (minor) to 4 (severe). [c]A period when no medications or procedures to treat the patient's cancer are applied. Data from J. Larkin, V. Chiarion-Sileni, R. Gonzalez et al., *N. Engl. J. Med.* 381:1535–1546, 2019.

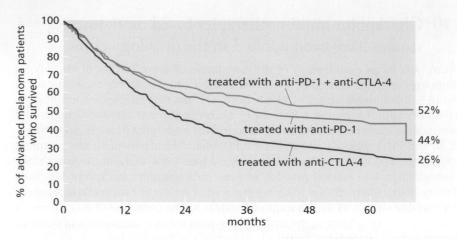

Figure 16.13 Therapeutic efficacy of immune checkpoint inhibitors used singly or in combination The overall survival at 5 years was studied in three groups of melanoma patients treated with different versions of anti-checkpoint blockade moAbs. One group was treated with anti-CTLA-4 moAb (*red curve*), a second with anti-PD-1 moAb (*blue curve*), and a third group received both anti-PD1 and anti-CTLA-4 moAbs (*green curve*). The combination was superior to either anti-CTLA-4 or anti-PD-1 used alone. (Adapted from J. Larkin, V. Chiaron-Seleni, R. Gonzalez et al. *N. Engl. J. Med.* 381:1535–1546, 2019. With permission from Massachusetts Medical Society.)

was maintained for another 5 years. In fact, the absence of additional patient deaths over this period, which yielded the horizontal plateaus in overall survival curves (see Figure 16.13), is often interpreted to represent definitive cures of neoplastic disease, which otherwise are rarely claimed in most types of currently available treatments of high-grade cancers of various types.

The 5-year survival rates for patients presenting in the clinic with metastatic melanomas has, until recently, hovered in the range of 23%, as mentioned above. In contrast, the therapies involving anti-PD-1 alone, anti-CTLA-4 alone, and the combination of anti-CTLA-4 + anti-PD-1, were 44%, 26%, and 52%, respectively (see Figure 16.13). These survival rates of metastatic melanoma patients are unmatched by any non-ICI therapies. So far, only adoptive T-cell therapy, a highly customized and resource-intensive adoptive cell transfer immunotherapy (see Section 16.7), has produced comparable rates of success. At present, ICI therapy is used for many types of cancer and several ICIs have received FDA approval for such therapy (see Table 16.5).

Unfortunately, the impressive success delivered by ICIs in treatment of melanoma have not yet been seen in the treatment of most other types of cancer. Nevertheless, the list of cancers responsive to treatment with ICIs is increasing. More generally, because mutation can generate neoantigens, some of these being subject to targeting by the immune system, cancers that harbor elevated numbers of mutations are likely to provoke immune attack. Therapy of these cancers may benefit from the use of ICIs.

16.11 Resistance to immune checkpoint inhibitors commonly arises

Even among patients being treated for melanoma, a disease for which ICIs have had their greatest successes, the tumors borne by roughly half of these patients are either entirely unresponsive or the responses that are observed are only transient. Such responses are reminiscent of the responses of many types of tumors to a wide variety of anti-cancer therapies, as will be described in Chapter 17. As with chemotherapy, resistance to ICIs may be primary or acquired. Thus, in the cases of **primary resistance**, a significant, measurable response to **immune checkpoint blockade** (ICB) therapy is never observed, whereas patients with **acquired resistance** respond initially and only thereafter become refractory to treatment as therapy continues. Nonetheless, such acquired resistance after months of initial clinical responses may buy patients many months of symptom-free survival that they otherwise would not have enjoyed.

Cases of primary resistance may be ascribed to a variety of mechanisms. These may occur if there has not been any preexisting recognition of tumor cells by the adaptive immune system and therefore no low-level immune attacks that could be amplified by ICI therapy. This might derive from the failure of tumor cells, for whatever reason, to generate or present, via MHC-I proteins, peptide antigens to T cells, leading to the absence of a T-cell response of any kind. Similarly, there may be no T-cell response to enhance if the diverse array of T cells that succeed in entering into the interstices of a tumor lack subpopulations bearing TCRs that can recognize antigen-derived peptide:MHC complexes displayed by the tumor cells.

Even if all these obstacles to ICB inhibition were to be removed in the future, clinical responses to ICIs might still be compromised by changes *within* tumor cells, specifically interruptions of signaling pathways that are essential for effective killing by CTLs. As a prominent example, the disruption in tumor cells of the signaling pathway activated by the interferon-γ (IFNγ) receptor comes to mind. Cancer cells lacking an intact IFNγ signaling pathway would be oblivious to the powerful anti-neoplastic effects of this T-cell-derived cytokine, which normally include a heightened susceptibility to apoptosis, an inhibition of proliferation, and an increased MHC class I-mediated antigen presentation (making them more vulnerable to immune recognition and attack). Such crippling of a cancer cell's capacity to respond to IFNγ denies patients the beneficial effects of increased IFN production, which ideally should occur once inhibitory checkpoints are eliminated by immune checkpoint inhibitors.

All of the mechanisms underlying primary resistance may also be developed by tumor cells during the course of ICI treatment, yielding multiple alternative paths to acquired resistance to immune checkpoint blockade. Finally, in addition to tumor cell-intrinsic pathways to ICI resistance, the tumor microenvironment (TME) can become a major tumor cell-extrinsic contributor to acquired ICI resistance. Populations of immunosuppressive cell types, including regulatory T cells (T$_{reg}$'s), M2-like macrophages, and myeloid-derived suppressor cells (MDSCs), create immunosuppressive TMEs that develop high levels of immunosuppressive cytokines such as TGFβ. This dictates that the evolution of an increasingly immunosuppressive TME during the course of checkpoint immunotherapy will deny a patient the therapeutic benefit that a T-cell-mediated anti-tumor response might otherwise provide. In recent years a variety of markers have been developed that help to predict responses to checkpoint immunotherapy (Supplementary Sidebar 16.2).

16.12 Lethal encounters between T cells and cancer cells can be encouraged by constructing bi-specific antibodies

As discussed earlier in the context of CAR T-cell construction (see Sidebar 16.7), recombinant DNA procedures can be employed to join the reading frames encoding the variable domains of the heavy (H) and light (L) chains of a single antibody, the two being linked via a segment encoding a flexible linker to enable the formation of a functional antigen-combining site. If constructed properly, the fusion protein encoded by these joined engineered DNA segments should have antigen-binding power equivalent to that of the corresponding naturally arising antibody molecule.

This sequence of steps can be repeated with the DNA segments encoding a second, independently arising antibody having a different antigen-binding specificity (**Figure 16.14**). Thereafter, DNA segments encoding the H–L fusion proteins derived from the two distinct antigen-specific antibodies can be fused together, resulting in a DNA construct that now encompasses four reading frames and encodes two sets of H–L fusion proteins. The resulting bi-specific protein should now be able to simultaneously recognize and bind two distinct antigens and, by doing so, cross-link, for example, two cells, each expressing one of the two antigens recognized by this bi-specific protein.

Such cross-linking can be highly useful in the case of cancer immunotherapy because it may be used to physically link a CTL (or other cytotoxic cell) to a targeted cancer cell, thereby circumventing the complex biological steps that are normally required to place the two cells in close juxtaposition. Constructs like these can be loosely termed "bi-specific antibodies," even though they contain only the antigen-binding domains of two parental antibodies.

In one well-studied application of this procedure, a DNA construct encoding a bi-specific antibody was created by joining the reading frames encoding the binding sites of an anti-CD19 antibody with the binding sites of an anti-CD3 antibody. The resulting dual-specificity antibody was denoted anti-CD3xCD19. As mentioned earlier in this chapter (see Section 16.8), many tumors of the B-cell lineage, including B-cell lymphomas, display the CD19 B-cell-specific cell surface antigen; at the same time, virtually all

Figure 16.14 Construction and clinical trial of a bi-specific anti-CD3xCD19 antibody (A) The DNA sequences encoding the variable regions (V$_H$ and V$_L$, *red, pink*) of a monoclonal antibody specific for CD3 (*above left*), a component of the T-cell receptor displayed by all T cells, and the sequences encoding the V$_H$ and V$_L$ sequences (*light green, dark green*) of a monoclonal antibody specific for CD19 (*above right*), a cell surface antigen displayed by B cells, were determined and used to design expression constructs encoding, in each case, single-chain variable region fragments (scFvs) that replicated the antigen specificity of each of the parent antibodies (*middle, left and right*). The DNA sequences encoding these two scFvs were linked to one another via a DNA segment encoding a connecting linker (*light blue*) which then expresses a protein termed a bi-specific T-cell engager (BiTE, *below*) that can bind to a B cell via its anti-CD19 specificity and to a T cell using its anti-CD3 arm. Such molecules can force the physical engagement of a T cell with a CD19-bearing cell. (B) Clinical trial of a BiTE. The efficacy of blinatumomab, an anti-CD3xCD19 BiTE was compared to standard chemotherapy for the treatment of advanced acute lymphoblastic leukemia. As shown in the graph, the bi-specific antibody was superior to chemotherapy by increasing short-term survival but not overall long-term survival. (A, adapted from Slaney et al., *Cancer Discov.* 8:924–934, 2018. With permission from American Association for Cancer Research. B, adapted from H. Kantarjian, A. Stein, N. Gökbuget et al., *N. Engl. J. Med.* 376: 836–847, 2017. With permission from Massachusetts Medical Society.)

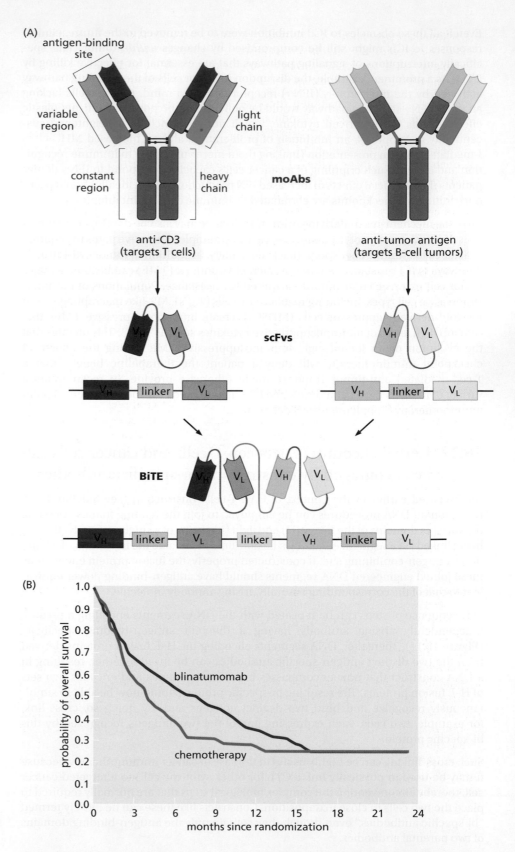

T cells display CD3, which is a component of the multi-subunit TCR complex. Accordingly, a bi-specific anti-CD3xCD19 antibody should be able to simultaneously bind a T cell to a potential target cell of the B-cell lineage (see Figure 16.14A). Such an antibody that is specialized to link a T cell to potential cell targets is sometimes termed a **bi-specific T-cell engager** (BiTE). The T cell in this forced marriage can be a CTL, and this association, once formed, can be used, in the case of B-cell lymphomas, to drive a functional engagement of T cells with lymphoma cells. Some of the cross-linked T cells may even carry a TCR (which continues to be expressed in these modified cells) that recognizes an antigen displayed by the targeted B cells, resulting in an immune-mediated cytotoxic attack and elimination of the B cells and the tumors that they form.

Clinical trials have demonstrated the effectiveness of BiTEs for the treatment of certain cancers, including one landmark study that compared the effectiveness of blinatumomab—an anti-CD3xCD19 BiTE consisting of an anti-CD19 single-chain variable fragment (scFv) joined to an anti-CD3 scFv—applied together with chemotherapy for the treatment of advanced acute lymphoblastic leukemia (ALL). Historically, although chemotherapy produces complete remission in 85–90% of patients newly diagnosed with ALL, most of these patients will relapse sooner or later. **Salvage chemotherapy**, that is, a second round of treatment used with the hope of generating a new round of therapeutic responses, produces remissions in 18–44% of these patients, but these remission periods are short, often between 4 to 5 months. Such periods of remission, while they are occurring, provide an opportunity to implement a second therapy involving bone marrow transplantation (BMT), with the removal by strong chemo-therapy of a patient's entire repertoire of immune cells (including the ALL cells) and the replacement of these cells with those coming from a healthy bone marrow donor.

This explains why a treatment was sought that would induce longer-lasting remission periods than the existing chemotherapy protocol, thereby making more BMTs possible. The results, summarized in Figure 16.14B, found the median 7.3-month period of remission achieved by blinatumomab superior to the median 4.6 months survival afforded by chemotherapy and paved the way for blinatumomab to become the first bi-specific T-cell engager to receive FDA approval. This approval provided validation of an entirely novel type of immunotherapy and prompted the entry of a flock of candidate T-cell engagers into the research and development pipeline directed toward treatment of a variety of hematological and solid tumors.

In the aforementioned clinical trial comparing a BiTE with chemotherapy, 61% of the patients receiving the BiTE experienced a serious adverse event (AE) relative to the 45% in the chemotherapy group to which it was compared. These adverse events included cytokine release syndrome (CRS), which is also encountered during CAR T-cell therapy as described earlier. This BiTE therapy also yields a second type of undesired side effect, specifically the depletion of normal B cells, termed B-cell **lymphopenia**.

Notably, T-cell engager therapies do not require the complex steps that are involved in CAR T-cell therapies, which include the arduous processes of harvesting T-cell populations followed by manipulating and expanding them *ex vivo* prior to re-infusing them into patients. Nonetheless, as seen with almost all anti-cancer therapies including immunotherapies, resistance to BiTE therapy often develops. The primary causes of resistance are predictable. Thus, loss or down-regulation of target antigen display is commonly encountered and, as we read in Section 15.13, is employed by tumor cells in order to evade or escape immune attack. The up-regulation of the PD-1 checkpoint receptor on the surface of the T cells responding to engager therapy can also reduce therapeutic responses. Finally, the phenomenon of T-cell **exhaustion**, discussed in the next section, can also compromise ongoing T-cell engager immunotherapy. Hence, like other immunotherapies, this novel treatment strategy holds the promise for overcoming shortcomings of existing therapies, but, at the same time, carries its own set of risks and side effects.

16.13 T-cell–dependent immunotherapies can be hampered by T-cell exhaustion

The effectiveness of the various types of immunotherapy described here ultimately depends on various regulatory controls that allow T cells, often CD8+ cytotoxic T cells (CTLs), to attack and kill targeted cancer cells. All of these therapies may be ineffective, however, if the CTLs are functionally compromised in one way or another, often by the process that leads to their state of exhaustion. Like the immune checkpoints described above, the T-cell exhaustion response originally evolved as yet another means to protect against the damaging effects on normal tissues generated by chronically activated T-cell populations.

Our current understanding of the controls leading to T-cell exhaustion stems historically from studies in mice of the immune responses following various types of viral infection. Thus, when the infections were rapidly cleared, the populations of CD8 T cells that remained behind were heterogeneous and could be classified into three major subpopulations on the basis of their expression of certain cell surface markers, epigenetic changes, and patterns of gene expression. These three broad groups were effector T cells (T_{eff}), memory T cells (T_{mem}), and exhausted T cells (T_{ex}). Cells belonging to the effector T-cell subpopulation were found to be actively cytotoxic, to secrete IFNγ and other cytokines, to be nonproliferating, and ultimately to undergo cell death. Memory T-cell populations (T_{mem}), in stark contrast, persisted long after the clearance of viral antigens and retained the ability to proliferate in response to a new challenge by the viral antigens that had previously provoked their formation.

The exhausted T cells (T_{ex}) arose under a variety of conditions. Most importantly, initial viral infections that were not successfully cleared could persist as chronic infections. Under such conditions, a T-cell population eventually emerged in which cells exhibited greatly reduced or no effector functions, lost proliferative capacity, and displayed inhibitory receptors such as PD-1. Such T cells were deemed to be "exhausted" and, additionally, were found to have little capacity to self-renew and could not generate T_{mem} populations (**Figure 16.15**).

Further research in this area demonstrated that exhausted T-cell populations could be found in humans suffering from HIV infection and in those suffering from chronic hepatitis B or C viral infections; all of these viral infections can persist for years in infected patients and with unabated viral replication. As we now realize, T-cell exhaustion is also an important element of the immune dysfunction that allows tumors to grow out. This state often develops in populations of T cells that have entered into growing tumors and experience chronic stimulation by various tumor antigens. Under

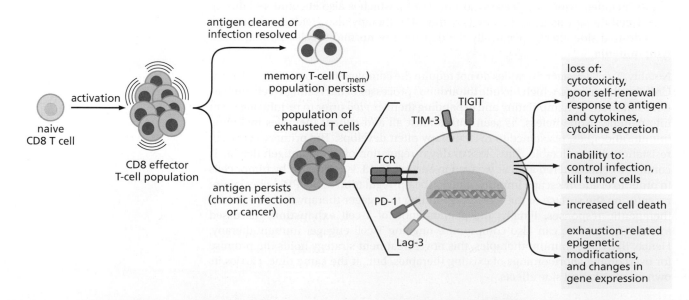

Figure 16.15 Overview of T-cell exhaustion Following activation, T cells may experience different fates depending on the duration of the antigenic stimuli that they receive. When the antigen is presented to the immune system and is cleared, as occurs after immunization via vaccination or after an infection is resolved by the actions of the innate and adaptive immune systems (*above*), some members of the responding T-cell population differentiate into effector T cells, most of which eventually die. Some, however, differentiate into various types of memory T-cell populations, which may survive for protracted periods of time. In contrast, chronic stimulation, which operates in persistent infections and in cancer (*below*), results in protracted stimulation and the eventual differentiation of responding T cells into a state termed "exhaustion." Exhausted T cells (*lower right*) display one or more inhibitory markers, such as PD-1, Lag-3, Tim-3, and TIGIT, and also show alterations in gene expression and loss of several functions. (Adapted in part from K.E. Pauken and E.J. Wherry, *Cell* 163:1038, 2015. With permission from Elsevier.)

Sidebar 16.11 Exhaustion compromises the effectiveness of ACT and CAR T-cell therapies Two immunotherapy strategies described earlier—adoptive T-cell transfer and CAR T-cell therapy—involve the *ex vivo* expansion of small populations of T cells to very large numbers during the course of many weeks. Critical to the success of the expansion protocols is the continuous stimulation of T-cell signaling pathways by engagement of the T-cell receptor in the presence of IL-2, the cytokine essential for T-cell proliferation. Although being indispensable for the generation of the large numbers of cells needed for these therapies, such persistent IL-2-driven expansion *ex vivo* can, on its own, force the proliferating T cells into the early stages of exhaustion. This compromises, in turn, the capacity of any subsequently infused T cells or CAR T cells to maintain robust and durable cytotoxic activity against tumor targets. Moreover, such infused cells have lost the ability to generate persisting memory T-cell (T_{mem}) populations that would be capable of future proliferation and the generation of cytotoxic activity in response to the reappearance of cancers after extended periods of clinical remission.

This significant complication of *ex vivo* T-cell expansion has stimulated the exploration of strategies designed to prevent or at least minimize the development of exhaustion. As one example, some of these strategies involve the use of CRISPR-Cas9 (see Sidebar 12.11) to delete from the genomes of proliferating T cells several of the genes specifying transcription factors that are known to play key roles in orchestrating T-cell exhaustion. Also, chronic rather than periodic signaling induces exhaustion, which has suggested the appealing approach of expanding T-cell populations by periodic cycles of stimulation and stimulus interruption, the latter being termed "rest." In addition, our increasing understanding of the mechanisms underlying T-cell exhaustion may suggest yet other approaches, including pharmacological ones, for the prevention or reversal of this state.

these conditions, such T cells are simply unable to control or eliminate tumors and enter into a state of exhaustion. Furthermore, the development of functional exhaustion can compromise the effectiveness of adoptive T-cell transfer (see Section 16.7) and CAR T-cell immunotherapies (**Sidebar 16.11**; see also Section 16.8).

The PD-1 receptor that operates to enable one type of checkpoint control (see Section 16.9) is also a marker of T-cell exhaustion that appears on responding T cells and on their CAR T-cell counterparts during prolonged antigen stimulation. The development of a fully exhausted state is not inevitable however. Thus, at early stages of T-cell exhaustion, treatment with the anti-PD-1 immune checkpoint inhibitor can halt its development, and previously exhausted T cells can be coaxed back into fully functional states. However, as exhaustion develops further, T cells become increasingly resistant to reversal of exhaustion achieved by interrupting PD-1 signaling, which at this stage has therefore become an ineffective strategy for invigorating various forms of immunotherapy (see Sidebar 16.11).

16.14 Synopsis and Prospects

Cancer immunotherapy provides an impressive demonstration of the translation of basic preclinical research into effective clinical interventions. As it turned out, most of the immunotherapies described in this chapter and now deployed for the treatment of cancer did not emerge from organized programs of drug discovery directed toward the development of novel cancer treatments. Instead, their origins can be traced to efforts designed to understand the biology of immune responses at the most fundamental level with no preconception of how resulting insights might one day impact clinical medicine.

Adoptive cell transfer (ACT) was initially developed as an experimental technique to determine and study the cell–cell collaborations essential for normal immune responses. Only later was it harnessed to capture a sample of a cancer patient's small population of anti-tumor T cells, expand their numbers *ex vivo*, and return the greatly increased cell population to the donor to treat their cancer. Chimeric antigen receptor (CAR) T-cell protocols arose as a marriage of adoptive cell transfer and the capacity, borrowed from monoclonal antibody technology, to generate antibodies that specifically bound tumor antigens.

Antibody–drug conjugates (ADCs) and bi-specific T-cell engagers (BiTEs) represent yet other offspring of monoclonal antibody technology. ADCs enable the targeting of highly cytotoxic drugs to tumors; this technology was designed to avoid the extensive systemic toxicity that often accompanies administration of these agents. BiTEs play a very different role, bringing T cells together with tumor cells that they would not otherwise engage. And finally, we note that fundamental research on how immune function is regulated has led to the truly revolutionary therapies based on immune checkpoint blockade (ICB) that have already yielded two Nobel Prizes!

But all of this, impressive as it is, should be viewed as no more than a beginning, representing very promising early steps toward the goal of effective management of cancers. The challenge is to find ways to improve immunotherapies so that, for many cases of cancer, cure becomes an expected outcome rather than an aspirational goal that is achieved only infrequently. At present, research focused on improving existing immunotherapies is rapidly evolving and future trajectories are already apparent as described below. Also, in light of the rapid progress of immunotherapy over the past generation, it would be surprising if existing treatment strategies were not joined in the near future by some entirely novel and entirely unanticipated ones.

Many of the existing cancer immunotherapies depend on antibodies or antibody-derived binding sites employed to direct immune attacks on tumor cells displaying tumor-associated antigens (TAAs). Unmodified antibodies, antibody–drug conjugates, CAR T cells, and bi-specific T-cell engagers all target antigens that are not exclusively found on tumor cells but instead are shared with certain normal cells. Consequently, collateral attacks on normal cells are inevitable, yielding side-effect toxicities that can prove to be life-threatening. This consideration helps to explain why, for example, antibody-driven therapies to treat hematologic cancers are restricted at present to lymphomas of the B-cell lineage. Although normal B cells are killed together with targeted lymphoma and leukemia cells, the B-cell lymphopenia that develops is manageable, and the affected normal cell population recovers once therapy is completed.

The problems associated with targeting of tumor-associated antigens (TAAs; Section 15.8) persist simply because it is impossible or unfeasible to identify tumor-specific antigens (TSAs) in many tumors. These problems also hobble the ADC, CAR T-cell, and bi-specific antibody treatments employed to treat carcinomas. Although these tumors create ~80% of cancer-associated mortality, immunotherapies targeting their TAAs inevitably affect the cells forming normal epithelia. In many cases, the resulting collateral side effects of attacks on these normal tissues, causing immune-mediated destruction of critical tissues, are intolerable, generating significant morbidity and even life-threatening losses of normal tissue functions. Even if not life-threatening, the necessity of limiting collateral damage to normal tissues is likely to compromise dosage levels and, in many cases, forces withdrawal of the therapy before its treatment objectives have been fully realized.

Nevertheless, despite these inherent liabilities, antibodies and ADCs (antibody–drug conjugates) have proven useful for the treatment of certain subtypes of breast cancers (see Section 16.6). Progress in extending cytotoxic antibody therapies to the treatment of additional breast cancer subtypes will depend on identifying additional TAAs, such as trophoblast cell surface antigen 2 (Trop2); even now, anti-Trop2-mediated attack by the ADC sacituzumab govitecan is used to treat some triple-negative breast cancers. Identification of alternative target antigens displayed by the cells of normal tissues but significantly upregulated by tumor cells will enable extension of these therapies to the treatment of a wider variety of carcinomas and other solid tumors.

The power of antibodies and ADCs to selectively target cells that display their cognate antigen creates a vulnerability that is likely, sooner or later, to render each of these therapies ineffective. Thus, as we have read time and again, antigen-guided attacks on tumor cells select for members of tumor cell populations that contrive to reduce or cease display of the targeted antigens, rendering them unresponsive to the immunotherapy being applied. In principle, application of a bi-specific antibody targeting two distinct TAAs displayed concomitantly on the surface of individual tumor cells should trigger attacks of the type that are not so easily evaded by loss of one of the two targeted TAAs. This, in turn, should avoid or significantly delay the development of therapeutic resistance.

Currently employed clinical practice yields other instructive examples of future therapeutic strategies. For instance, rather than using an antibody as a monotherapy,

Sidebar 16.12 Initial attempts at using CAR T cells to target multiple tumor-associated antigens The observation that many B-cell malignancies that express CD19 also display the CD22 antigen has already inspired the construction of CAR T cells with dual specificity, bearing one CAR capable of recognizing CD19 and a second CAR specific for CD22. Initial clinical findings in patients of lymphoma patients infused with such CD19–22 CAR T cells showed that the dual-specificity CAR T cells induced various degrees of remission in a patient cohort and, in a handful of these patients, PET scans revealed dramatic reductions of tumor masses. Furthermore, some of the treated patients exhibited clear evidence of the persistence of populations of the dual-specificity CAR T cells, suggesting the possibility of launching future attacks on persisting or returning tumor cell populations displaying the targeted antigens. However, though an encouraging beginning, the efficacy of these dual-specificity CAR T cells did not actually exceed that of monospecific anti-CD19–CAR T therapies already in clinical use. Still, even now we can anticipate the development of CAR T-cell therapies involving their further modification from functional monospecificity to multispecificity and resulting substantial increases in therapeutic efficacy.

current clinical practice for the treatment of some B-cell lymphomas combines antibodies that are directly cytotoxic with chemotherapies, which act through mechanistically unrelated ways. The design of such two-pronged attacks on these tumors is motivated by a similar consideration, that is, the low probability that lymphoma cells can concomitantly develop resistance to a chemotherapeutic agent applied together with antibody-driven attacks on one or more B-cell surface antigens.

Antigen loss is also a major cause of the acquired resistance to CAR T-cell therapy. One study described the clinical courses of large B-cell leukemia (LBCL) in 16 patients being treated with anti-CD19–CAR T-cell therapy. Eleven of these patients eventually showed down-regulation or complete loss of CD19 display. In the near future, this problem of immunoevasion will be approached in either of two ways. One may involve the development of a collection of monospecific CAR T cells, each one recognizing a distinct antigen, with infusion of these cell lines either together or sequentially, one after the other. An attractive alternative to this cumbersome approach involves construction of a single CAR T-cell line bearing not one but a repertoire of multiple distinct CARs, each specific for a different tumor antigen (**Sidebar 16.12**).

Nonetheless, even with the adoption of therapeutic strategies that enable CAR T cells to anticipate and overcome eventual evasion arising from antigen loss, the creation of these cells requires an elaborate technology involving harvest of each patient's T cells, their conversion to CAR-expressing cells, and the reinfusion of resulting CAR T cells into the donor. In addition, this immunotherapeutic technology, at least as presently configured, is encumbered by a number of additional significant problems.

Thus, in addition to their forced display of chimeric antigen receptors, CAR T cells retain their complement of preexisting conventional TCRs. The ongoing display of these indigenous TCRs imposes limits on the application of CAR T-cell therapies, simply because these receptors continue to recognize allogeneic MHC complexes as foreign. As a direct consequence, when CAR T cells carrying these TCRs are infused into allogeneic recipients, the cells may attack a broad array of these recipients' tissues, often resulting in life-threatening graft-versus-host (GVH) reactions. For this reason, at least at present, each batch of infused CAR T cells represents, in effect, a personalized drug, offering great possible therapeutic benefit to a syngeneic recipient while, at the same time, triggering potentially lethal responses in allogeneic recipients. In addition, the assembly of CAR T-cell populations requires periods of at least 2–3 weeks between harvest and reinfusion, leading to significant delays in the initiation of a patient's treatment. Finally, quite often a cancer patient's physical condition may make it difficult to harvest sufficient numbers of T cells for CAR T-cell preparation.

A promising alternative to CAR T cells is poised to emerge. Natural killer (NK) cells have been engineered to express chimeric antigen receptors conferring on these innate immune cells a predefined specificity for a particular tumor antigen. As we read previously (Section 15.9), NK cells are capable of launching a variety of potent cytotoxic attacks on the cells they target. Importantly, unlike their CAR T-cell counterparts, the initially unmodified NK cells lack preexisting TCRs and are therefore incapable of mounting inadvertent GVH reactions. For these reasons, CAR-NK cells can be generated from an allogenic donor and deployed in many different allogenic recipients without the danger of life-threatening GVH reactions. Moreover, NK cells, are already known to mount stronger attacks against allogeneic tumors than against syngeneic ones.

In fact, CAR-NK cells can be generated from NK cells retrieved from umbilical cord blood, thereby leveraging the availability of relatively abundant frozen stores of cord blood cells to prepare CAR-NK cells for use in allogeneic recipients. Preliminary studies have already shown that a single unit of umbilical cord blood yields sufficient NK cells to prepare to enough CAR-NK cells for treating several patients. In the longer term, if proven efficacious and safe, previously prepared batches of CAR-NK cells could provide an "off the shelf" therapy, being instantly available, less expensive, and available to a wider range of patients than the conventional CAR T-cell treatments currently available.

Indeed, anti-CD19–specific CAR-NK cells prepared from frozen cord blood-derived NK cells have already been employed in a clinical trial that enrolled patients with relapsed or treatment-refractory CD19-positive non-Hodgkin's lymphoma or chronic lymphocytic leukemia. In this early-stage clinical trial, 7 of 11 members of the patient group showed elimination of detectable tumor cells at 14 months after CAR-NK therapy. Of note, all of these patients were allogenic with respect to the CAR-NK preparation administered. In addition, cytokine release syndrome (CRS), which is often observed as a side effect of existing CAR T-cell therapies, was not observed in this study. If these striking initial clinical studies are sustained and reinforced in larger trials, CAR-NK cells will become a highly useful addition to the repertoire of adoptive cell transfer therapies for the treatment of cancer.

All of the adoptive cell transfer therapies cited here—adoptive T-cell transfer, CAR T-cell, and CAR-NK therapies—as well as the bi-specific T-cell engager (BiTE) therapy, involve continuous functional activation of the introduced cells and their attendant proliferation. This suggests that all of these therapies will eventually be compromised by the functional exhaustion of the introduced immunocytes, indeed an outcome that has already been observed. This suggests an urgent need to develop strategies for preventing, damping, or reversing exhaustion in both T cells and NK cells, and in their modified derivatives.

In the arena of the ICI treatments described here, involving both CTLA-4 and PD-1 strategies (as described in Section 16.9), further improvements are also desperately needed: both treatments are already in widespread clinical use even though their demonstrated utility is actually limited to a relatively small subset of the tumors encountered in the oncology clinic. Moreover, in the years since the development of the CTLA-4 and PD-1 ICIs, a number of other similarly acting immune checkpoints have been discovered (see Table 16.5), several of which appear promising in terms of their potential to generate clinically useful immunotherapies. Nonetheless, it is already apparent that none of these will be sufficiently powerful on its own to be used as a monotherapy. Instead, echoing the theme regularly encountered in cancer immunotherapy and chemotherapy, they will most likely find roles as components of combination therapies.

Beyond combination therapy, as more has been learned about the dynamics of immune checkpoint blockades, appreciation has grown for the complexity of the multiple host and environmental factors that influence responses to ICIs. Some of these factors are discussed in Supplementary Sidebar 16.2.

To summarize, it is clear that much of contemporary cancer research is influenced by the realization that an in-depth understanding of initial cancer pathogenesis (as described in Chapter 11) and the development of improved treatments of existing tumors (as described in this chapter and Chapter 17) will depend on elucidating and exploiting the formidable powers of the immune system. At the same time, development of novel immunotherapies is occurring in the face of an increasing awareness that the development of non-immune-based, truly effective anti-cancer therapeutics has become increasingly challenging. Together, these trends explain the explosive increase in the number of anti-cancer immunotherapies that are currently under development and will surely be introduced into clinical practice in the years to come.

Key concepts

- Most cancer immunotherapies fall into one of three categories: vaccination against tumor cells and oncogenic microbes; passive immunization via administration of anti-tumor cells or antibodies; and release of immunosuppressive mechanisms, notably those imposed by immune checkpoints.

- Prophylactic cancer vaccines prevent infection by microbial agents that directly or indirectly cause cancer, thereby lowering the incidence of the cancers induced by these agents. Vaccination against Hepatitis B virus (HBV) prevents HBV-induced liver cancer and human polyoma virus (HPV) vaccines prevent most cervical cancer.

- Therapeutic cancer vaccines stimulate the immune system to make adaptive immune responses against tumors that have already established themselves in the body, arresting the growth of these tumors and ultimately rejecting them.

- Passive immunization by administration of antibodies against tumor antigens borne by a patient's tumor can kill tumor cells in a variety of ways, many of them involving recruitment of cells of the innate immune system.

- Antibody drug conjugates (ADCs) combine the powers of antibody-mediated passive immunization with antibody-guided delivery of highly potent cytopathic drugs.

- Passive immunization can also involve capturing cells of a patient's immune system, amplifying, and in some cases modifying them, and then infusing them back into the patient. This process is called adoptive cell transfer (ACT).

- CAR T-cell therapy is a form of ACT that is already in widespread use. In CAR T cells, the antigen recognition function of the T-cell receptor is assumed by an engineered chimeric antigen receptor (CAR), a single molecule that incorporates specific antigen recognition, and that has the capacity to trigger T-cell activation and the generation of T-cell proliferation and survival signals.

- Immune checkpoints normally provide essential homeostatic negative feedback controls necessary to prevent excessive inflammation and destructive autoimmune responses. However, these essential homeostatic controls can dampen or prevent strong anti-tumor responses.

- Immune checkpoint inhibition manipulates the behavior of immune cells in ways that increase the likelihood and potency of their anti-tumor responses. Monoclonal antibodies that interfere with either the CTLA4:CD80/86 or the PD-1:PD-L1/2 axis have proven useful for the treatment of melanoma and a variety of other cancers.

- Bi-specific antibodies (and antibody-like molecules) constructed to simultaneously bind to two cells displaying different cell surface antigens, bring cells that would not otherwise closely engage, into sustained physical contact with each other. A bi-specific antibody, known as a bi-specific T- cell engager (BiTE), can link a T cell to another cell, such as a tumor cell. Such forced juxtaposition of a T cell with a second cell can result in the T-cell-mediated death of the second cell, such as a targeted cancer cell.

- T-cell-mediated therapies depend on sustained T-cell activation but can be compromised by T-cell exhaustion. Therefore, it is important to develop a deeper understanding of this phenomenon in order to make activation-dependent T-cell-mediated cancer therapies more enduring and thus more effective.

Thought questions

1. Initial treatment of a patient's cancer with anti-PD-1 was ineffective. Following irradiation, the patient's tumor showed modest shrinkage and then resumed growth. However, a second course of treatment with anti-PD-1 resulted in regression of tumor growth. Offer reasonable hypotheses for: (a) why treatment with anti-PD-1 was initially ineffective and (b) why anti-PD-1 may have been effective after radiation therapy.

2. CAR T cells specific for CD20 were generated from T cells harvested from a lymphoma patient, frozen, and subsequently used in a successful treatment of the patient's cancer. However, several frozen ampoules of this effective CAR T-cell preparation were not used. An enthusiastic oncology resident suggested that these surplus CAR T cells could be used to treat future lymphoma patients whose profile of surface antigens indicated they might benefit from treatment with anti-CD20 CAR T cells. A second oncology resident, who had paid attention during her immunology rotation, explained to her colleague why this was not a good idea. What did she say?

3. Distinguish between prophylactic and therapeutic cancer vaccination. What etiology is shared by cancers that can be prevented by prophylactic immunization? Briefly outline the advantages offered by therapeutic cancer vaccination and suggest the form such vaccines might take.

4. Antibodies have become mainstays of cancer immunotherapy. Some anti-tumor antibodies directly kill tumor cells. Other highly effective antibodies have no direct cytotoxic effect on tumor cells. How does one class of molecules play such different roles in cancer therapy?

5. What advantages, if any, do bi-specific T-cell engagers have over CAR T cells for the treatment of cancer? What feature of CAR T cells might give them an advantage over these bi-specific agents?

6. Adoptive T-cell transfer has shown a capacity to treat a wider variety of cancers than CAR T cells or BiTEs. What is the basis of the ability of this approach to target a particularly broad spectrum of cancers? Given its demonstrated ability to target many different kinds of cancer, why is it not more widely practiced?

7. Antibody–drug conjugates (ADCs) are highly effective anti-tumor agents. Briefly identify the many ways in which an ADC can kill or induce the death of its target cell. Given their many strengths, identify a strategy of evasion used by some tumor cells to escape killing by antibodies and CAR T cells that also allows them to evade killing by ADCs. Suggest how you might create an ADC that would be less susceptible to this commonly adopted strategy of escape from immune-mediated killing.

8. T-cell exhaustion is a well-established immunological phenomenon and can blunt the therapeutic effects of adoptive cell transfer strategies, bi-specific T-cell engagers, and release of checkpoint blockade by ICIs. Although these are very different approaches, why are all of them susceptible to loss of efficacy because of T-cell exhaustion?

Additional reading

Bagchi S, Yuan R, & Engleman EG (2021) Immune checkpoint inhibitors for the treatment of cancer: clinical impact and mechanisms of response and resistance. *Annu. Rev. Pathol.* 16, 223–249.

Bardia A, Hurvitz A, Tolaney S et al. (2021) Sacituzumab Govitecan in metastatic triple-negative breast cancer. *N. Engl. J. Med.* 384, 1529–1541.

Drago ZJ, Modi S & Chandarlapaty S (2021) Unlocking the potential of antibody-drug conjugates for cancer therapy. *Nat. Rev. Clin. Oncol.* 18, 327–344.

Freeman GJ, Long A, Iwai Y et al. (2000) Engagement of the Pd-1 immunoinhibitory receptor by a novel B7 family member leads to negative regulation of lymphocyte activation. *J. Exp. Med.* 192, 1027–1034.

Iwai Y, Ishida M, Tanaka Yet al. (2002) Involvement of PD-L1 on tumor cells in the escape from host immune system and tumor immunotherapy by PD-L1 blockade. *Proc. Natl. Acad. Sci. USA* 19, 12293–12297.

Jenkins RW, Barbie DA & Flaherty KT (2018) Mechanisms of resistance to immune checkpoint inhibitors. *Br. J. Cancer* 118:9–16.

Joura EA, Guilana A, Iverson O et al. (2015) A 9-valent HPV vaccine against infection and intraepithelial neoplasia in women. *N. Engl. J. Med.* 372:711–723.

Kantarjian H, Stein A, Gokbuget N et al. (2017) Blinatumomab versus chemotherapy for advanced acute lymphoblastic leukemia. *N. Engl. J. Med.* 376:836–847.

Labrijn AF, Janmaat ML, Reichert JM & Parren PWHI. (2019) Bi-specific antibodies: a mechanistic review of the pipeline. *Nat. Rev. Drug Discov.* 18, 585–608.

Larkin J, Chairion-Sileni V, Gonzalez R et al. (2019) Five-year survival with combined nivolumab and ipilimumab in advanced melanoma. *N. Engl. J. Med.* 381, 1535–1546.

Larson RC & Maus MV (2021) Recent advances and discoveries in the mechanisms and functions of CAR T cells. *Nat. Rev. Cancer* 21, 145–161.

Leach DR, Krummel MF & Allison JP (1996) Enhancement of antitumor immunity by CTLA-4 blockade. *Science* 271, 1734–1736.

Leko V & Rosenberg SA (2020) Identifying and targeting human tumor antigens for T cell-based immunotherapy of solid tumors. *Cancer Cell* 38, 454–472.

Liu E, Marin D, Banerjee, P et al. (2020) Use of CAR-Transduced natural killer cells in CD19-positive lymphoid tumors. *N. Engl. J. Med.* 382, 545–553.

McLane LM, Abdel-Hakeem MS & Wherry EJ (2019) CD8 T cell exhaustion during chronic viral infection and cancer. *Annu Rev Immunol.* 37, 457–495.

Sahin U, Petra O, Derhovanessian E et al. (2020) An RNA vaccine drives immunity in checkpoint-inhibitor-treated melanoma. *Nature*. 585, 107–112.

Slaney CY, Wang P, Darcy PK & Kershaw MH (2018) CARs versus BiTEs: a comparison between T cell-redirection strategies for cancer treatment. *Cancer Discov*. 8, 924–934.

Saxena M, van der Burg SH, Melief CJM & Bhardwaj N (2021) Therapeutic cancer vaccines. *Nat. Rev. Cancer* 21, 360–378.

Weber EW, Parker KR, Sotillo E et al. (2021) Transient rest restores functionality in exhausted CAR-T cells through epigenetic remodeling. *Science* 372(6537) DOI: 10.1126/science.aba1786.

Wei SC, Anang N, Sharma R et al. (2019) Combination anti-CTLA-4 plus anti-PD-1 checkpoint blockade utilizes cellular mechanisms partially distinct from monotherapies. *Proc. Natl. Acad. Sci. USA* 116, 22699–22709.

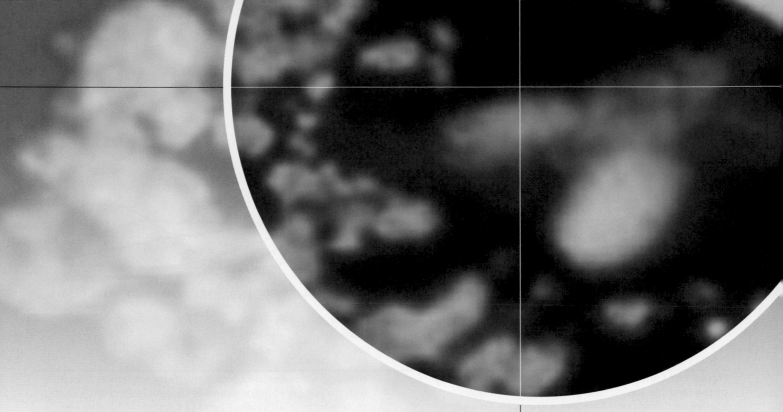

Chapter 17

The Rational Treatment of Cancer

> All substances are poisonous, there is none that is not a poison; the right dose differentiates a poison from a remedy.
>
> Paracelsus (Auroleus Phillipus Theostratus
> Bombastus von Hohenheim), alchemist and physician, 1538

> Doctors are men who prescribe medicines of which they know little, to cure diseases of which they know less, in human beings of whom they know nothing.
>
> Voltaire (François-Marie Arouet), author and philosopher, 1760

The research described throughout this book represents a revolution in our understanding of cancer pathogenesis. In 1975, there were virtually no insights into the molecular alterations within human cells that lead to the appearance of malignancies. Two generations later, we possess this knowledge in abundance. Indeed, the available information and concepts about cancer's origins can truly be said to constitute a science with a logical and coherent conceptual structure.

In spite of these extraordinary leaps forward, more limited progress has been made in exploiting these insights into etiology (that is, the causative mechanisms of disease) in order to prevent the disease and, equally important, to treat it. Most of the anti-cancer treatments in widespread use today were developed prior to 1975, at a time when the development of therapeutics was not yet informed by detailed knowledge of the genetic and biochemical mechanisms of cancer pathogenesis. Consequently, there is a widely felt frustration among molecular oncologists that the potential of their research for contributing to new anti-cancer therapeutics has not yet been realized.

The promise—still unrealized—of the new therapeutics needs to be juxtaposed with the overall progress in treating advanced tumors using the traditional strategies of

Movies in this chapter

17.1 PI3K Inhibitor
17.2 Drug Export by
      the Multi-Drug
      Resistance Pump

781

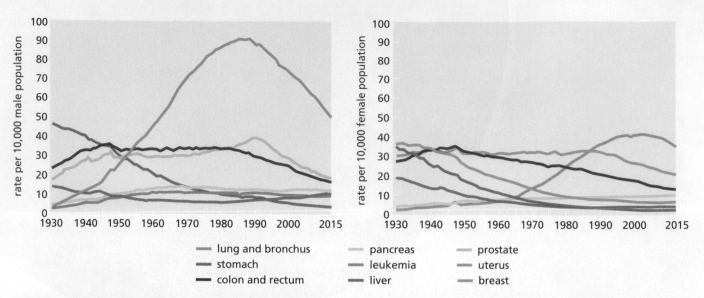

Figure 17.1 Cancer mortality from 1930 to 2015: the lay of the land The statistics compiled in the United States on the age-adjusted death rates for males and females from various types of cancer reveal two different long-term trends. Mortality from several major killers has declined significantly since 1930. This resulted from changes in food storage practices and a possible reduction in *Helicobacter pylori* infection rates, in the case of stomach cancer, and to screening, in the case of colorectal cancer. The decline in mortality from lung cancer is attributable, almost entirely, to decreases in tobacco usage. However, tumors associated with the number of major sources of cancer-related death, including those associated with various forms of lung cancer, have proven resistant to most forms of traditional therapies, especially when these tumors progress to a highly malignant, metastatic stage. (Uterus includes uterine cervix and uterine corpus combined, due to limitations in COD coding prior to 1970. From American Cancer Society. Cancer Facts & Figures, 2018. Atlanta: American Cancer Society; 2018. With permission from American Cancer Society.)

surgery, chemotherapy, and radiotherapy. Significant progress in treating many tumors has been slow. For example, in 1970 in the United States, 7% of the patients diagnosed with lung cancer were still alive five years after their initial diagnosis. Five decades later, this number had risen to only 19%, a relatively minor improvement. And even this degree of therapeutic success may be illusory, because modern diagnostic techniques often detect tumors far earlier in their natural course, creating a greater time span between initial diagnosis and ultimate progression to end-stage disease.

Death rates for colon cancer have begun to fall, because of early detection and surgical removal of growths that have advanced through only the early stages of tumor progression (**Figure 17.1**; see also Figure 11.8B). However, mortality caused by the more advanced colorectal tumors has changed little—a testimonial to the failures of chemotherapy and radiation to eliminate these malignancies once they have invaded and begun to metastasize (Supplementary Sidebar 17.1). Moreover, age-adjusted mortality from other types of tumors has remained constant or declined relatively little. Statistics like these suggest that the potential of the traditional therapies to cure high-grade malignancies has been largely realized and that major progress in the future can only come from the novel therapeutics whose designs are informed by our now-extensive knowledge of the molecular, biochemical, and cellular mechanisms of cancer pathogenesis.

The problems confronted here are manifold. To begin, in the case of certain common cancers, we lack a good estimate of the size of the problem. How many of the cases that are diagnosed each year (yielding age-adjusted incidence) are likely to grow into life-threatening growths? And how often do treatments performed on patients with relatively benign tumors yield more morbidity than clinical benefit in terms of subsequent well-being and gains in life span?

The problem is explained by a critical lesson we learned in Chapter 11: as we grow older, small tumors appear spontaneously in a wide variety of tissues. At the same time, as diagnostic procedures become increasingly sensitive (**Figure 17.2**), we begin to detect more and more cases of cancer without a clear indication of the proportion of these that will grow into life-threatening tumors. This problem is immense, given the limited ability of oncology clinics and funds to respond to an ever-increasing number of cases (**Sidebar 17.1**).

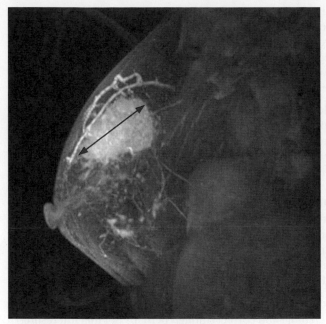

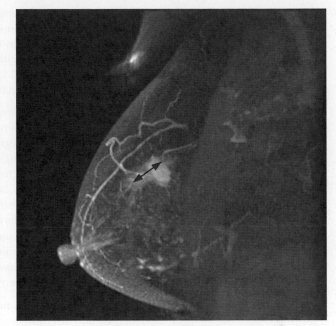

pre-chemotherapy
longest dimension = 47 mm

post-chemotherapy
longest dimension = 16 mm

Figure 17.2 High-resolution noninvasive imaging of human tissues The development of magnetic resonance imaging (MRI) has enabled increasingly higher-resolution noninvasive visualization of living tissues. MRI now allows breast tumors of very small size (several millimeters in diameter) to be detected and, as shown here, makes it possible to view the progress of anti-tumor therapy—in this case chemotherapy with an anthracycline cytotoxic agent—in exquisite detail. Widespread use of such highly sensitive imaging techniques has resulted in increases in the registered incidence rate of breast cancer (resulting from detection of disease that would previously have eluded detection) with unclear effects on the ultimate mortality of the disease. (Courtesy of N.M. Hylton and L.J. Esserman.)

As discussed in the next section, our rapidly evolving understanding of disease pathogenesis may soon allow us to judge more accurately how many cancers are deserving of aggressive treatment and how many can be safely left untreated. At the same time, our improving insights into specific disease processes, such as the development of metastases, have revealed how traditional, widely used forms of therapy have not served many cancer patients well (Supplementary Sidebar 17.2).

In most of this chapter, we explore therapeutic strategies under development or recently introduced into the clinic and examine how their development has been informed by what we have learned in the earlier chapters of this book. The goal here is

Sidebar 17.1 How common are cancers that require clinical intervention? The development and use of diagnostic procedures of ever-increasing sensitivity led to steep increases in the incidence of breast and prostate cancers during the second half of the twentieth century. In the case of the breast, the incidence of carcinomas increased progressively during the late twentieth century and then leveled off (see Figure 17.3). Some of these growths will eventually become life-threatening and many others will not. However, incidence rates are an artifact of technology, and as diagnosis improves (see Figure 17.2), the age-adjusted incidence will increase in lockstep unless the images generated by these new diagnostic techniques are interpreted with caution. Hence, the only truly rigorous and solid measurements are those associated with mortality.

A recent study of breast cancer performed post mortem on women who died of a variety of causes estimated that at the age of 80, at least two-thirds of women carry breast carcinomas (whereas only ~4% will die from breast cancer). An even larger number (~80%) applies to post mortem studies of prostate carcinomas in men of this age. Thus, if the diagnostic procedures performed on living patients were as sensitive as those implemented upon autopsy, the estimated incidence of breast and prostate carcinomas would increase dramatically, because even more non-life-threatening cancers would be registered (see also Supplementary Sidebar 17.2).

not to survey the full range of current research in these areas. That would be unreachable: a 2018 compilation of anti-cancer therapies in pre-clinical development or in clinical testing listed more than 1100 potential therapeutic agents that were being pursued by pharmaceutical companies and biotechnology firms. The agents under development included low–molecular-weight drugs, proteins, antibodies and antibody conjugates, and gene therapy strategies, including viral vectors.

Rather than being encyclopedic, we will concentrate here on a small number of recently developed therapies that illustrate how discoveries described in the previous chapters have inspired novel strategies for treating cancer and how molecular diagnosis will increasingly play a part in the development and clinical introduction of novel therapies. These therapeutic strategies hold great promise, and invariably their true potential is yet to be realized. The anecdotes surrounding the development of each of these agents are interesting and provocative because they teach important lessons about the triumphs and pitfalls of developing novel anti-cancer treatments. Note that the recent development of immunotherapies has profoundly influenced both the current treatment of multiple cancer types and the future prospects for the treatment of malignancies that have traditionally been refractory to therapy. These dramatic developments have been described in some detail in Chapter 16.

Almost all the research findings described throughout this textbook will stand the test of time and be considered credible and correct (though perhaps not that interesting) a generation from now. However, those who love certainty and eternal truths will find the stories that follow below to be unsatisfying for a very simple reason: the work reported is in great flux and many outcomes are uncertain. Many of the newer therapies will seem quaint and anachronistic a decade after this chapter is written. The campaign to convert insights about cancer's molecular causes into new ways of curing disease has just begun.

Note also that we will pass over descriptions of how molecular biology is changing cancer *prevention* strategies. Thus, in this chapter we will not examine the major advances that have been made in developing vaccines that protect against hepatitis B virus (HBV) and human papillomavirus (HPV) infections (for example, see Supplementary Sidebar 15.5); these vaccines should be highly effective in reducing the incidence of hepatomas and cervical carcinomas, which are major sources of cancer-associated mortality in certain parts of the world. (If past public health measures are any guide, the prevention of cancer, that is, the reduction of disease *incidence*, will ultimately yield far greater reductions in overall disease-related *mortality* than will therapies of the sort discussed in this chapter.)

Simple logic would dictate that the newer, "rationally designed" agents, because they attack specific, identifiable molecular targets, are likely to be far more effective in eradicating tumors and less encumbered with toxic side effects than are older agents; the latter were discovered through empirical trial-and-error testing that was undertaken without any foreknowledge of the biochemistry and molecular biology of cancer cells. However, as we will see in Section 17.2, the more traditional ways of treating neoplastic disease have proven to be highly effective in halting disease progression and reducing cancer-associated mortality, explaining why these treatments continue to merit our attention. Such curative outcomes are, ironically, still not the endpoints achieved by most of the recently developed therapies, as we will see.

17.1 The development and clinical use of effective therapies will depend on accurate diagnosis of disease

In previous chapters, we repeatedly categorized cancers in terms of their tissues of origin and their stages of clinical progression. Almost always, these assignments have been dictated by the appearance of normal and malignant tissues under the microscope. On many occasions, to be sure, we have refined these classifications by describing certain molecular markers [for example, expression of HER2 in breast cancers and the implications that they hold for prognosis (see Figure 4.3A)], but in general, histopathology has reigned supreme in our discussions, as it has in the practice of clinical oncology for almost a century.

It has become increasingly clear, however, that the traditional ways of classifying cancers have limited utility. Truly useful diagnoses must inform the clinician about the underlying nature of diseases and, more important, how each disease entity will respond to various types of therapy. As we have learned more about human cancers, we have come to realize that many human cancers that have traditionally been lumped together as examples of a single disease entity should, in fact, be separated into several distinct subcategories of a disease. This tendency to view cancer as a uniform entity helps to explain why many existing anti-cancer therapeutic strategies used over the past three decades have had such low overall success rates, in that they have treated tumors of a certain class as representatives of a homogeneous disease rather than of multiple distinct subtypes of disease, some of which might be responsive to targeted therapeutic attack, whereas others might be quite resistant. These response rates also have important implications for the development of new drugs (Supplementary Sidebar 17.3).

Stated differently, the clinical oncologist really confronts two issues: Should a diagnosed tumor be treated, and if so, what available therapies are appropriate for the subtype of tumor that has been identified? In an ideal world, the decision to proceed with treatment should be a challenging one requiring detailed knowledge of the idiosyncrasies of the diagnosed tumor, but in the real world, it is often simplified: treat almost all tumors (with the exception of skin cancers) aggressively in order to reduce as much as possible the likelihood of eventual life-threatening clinical progression.

Ideally, the complexity of the decision to proceed with treatment should take into account that diagnosed tumors fall into three classes:

1. Indolent tumors that have low invasive and metastatic potential and will remain in such a state during the lifetime of the patient.

2. Highly aggressive tumors with a propensity to metastasize that have, with high probability, disseminated by the time that the primary tumor has been diagnosed.

3. Tumors of intermediate grade that have the potential to disseminate but can be excised or treated with cytotoxic therapies before dissemination occurs and life-threatening metastases are formed.

Tumors of the first class are generally not worthy of treatment, including surgery, and should be left undisturbed. (Indeed, surgery may, under certain conditions, provoke otherwise-indolent growths to become clinically apparent if not aggressive.) One illustrative example of these is pancreatic neuroendocrine tumors (NETs)—that is, tumors of the pancreatic islets (see Figure 13.36)—which constitute about 3% of pancreatic carcinomas and are discovered incidentally during high-resolution imaging, often undertaken for other conditions. Asymptomatic patients with these low-grade "incidentalomas," as the tumors are termed amusingly by oncologists, confront a 90% five-year progression-free survival. Like any type of pancreatic surgery, excision of these tumors is accompanied by a high degree of post-surgical morbidity; such morbidity may eclipse the clinical benefit that these patients receive from surgery, given the low likelihood of clinical progression. [Contrast their prognosis with that of a recently diagnosed pancreatic exocrine adenocarcinoma patient (see Supplementary Sidebar 11.5), whose overall five-year survival is only ~5%.] At present, diagnostic criteria do not distinguish between the small subset of NETs that will eventually progress and the great majority that are unlikely to progress during a patient's lifetime. Papillary thyroid carcinomas represent another class of incidentalomas: in one of the few systematic post mortem studies of this disease, a Finnish research group found these growths in one-third of those autopsied, whereas only 0.07% of annual deaths (in the United States) can be attributed to this disease.

Truly effective treatments for most kinds of metastases are not available at present, raising the question whether tumors of the second class are worth treating, as they will become lethal no matter what therapies are attempted. (The proviso here is that existing treatments may ameliorate symptoms over an extended period of time and may even forestall the inevitable, thereby extending patients' symptom-free life spans significantly.) Cynics argue simplistically that patients carrying these two classes of

Table 17.1 Examples of alkylating agents or other DNA-modifying agents used to treat cancer

Name	Chemical structure	Effects on DNA	Examples of clinical use
bleomycin	nonribosomally synthesized oligopeptide	strand breaks	Hodgkin's lymphoma, squamous cell carcinoma
cyclophosphamide	nitrogen mustard analog	cross-links N^7 of guanine	lymphoma, leukemia
dacarbazine	nitrogen mustard analog	methylation of guanine	Hodgkin's lymphoma, melanoma
melphalan	nitrogen mustard analog	alkylates N^7 of guanine	multiple myeloma
chlorambucil	nitrogen mustard analog	base alkylation, cross-links	chronic lymphocytic leukemia
carboplatin/cisplatin	alkylating-like agent	intra-strand cross-links	breast, ovarian, testicular cancer
temozolomide	alkylating agent	methylation of guanine	glioblastoma multiforme

tumor should be left untreated, because the long-term outcome is predictable no matter what clinical strategy is chosen. Unappreciated by the cynics is the possibility that the sub-optimal responses of the second class of tumors to existing therapies will serve as foundations for the development in the future of vastly more effective therapies that will save those who, at present, confront inevitable disease progression and death.

The third class of tumors—those of intermediate grade—are those that should and will command most of our attention in this chapter—the "in-between" tumors whose treatment will actually prevent metastatic progression and achieve long-term, even curative responses. In reality, in countries like the United States where concerns about neglectful treatment verging on medical malpractice dominate, tumors of all classes are treated, often aggressively. Such unfocused treatments will soon become issues of the past, if only because the costs of unnecessary therapies will no longer be economically sustainable.

At present, such uniformly applied therapies may have other downsides beyond economic costs. Indeed, anti-cancer treatments often cause numerous side effects and some may actually increase the incidence of second-site cancers arising years later. For example, in the early 1980s, breast cancer patients receiving the then-standard dose of cyclophosphamide (a chemotherapeutic drug that is also an alkylating agent; **Table 17.1**) experienced a 5.7-fold increased risk of subsequently developing acute myelogenous leukemia (AML), ostensibly the consequence of the mutagenic actions of this drug. (Many current treatment protocols use lower drug dosages and result in greatly decreased incidence of such second-site cancers.) All this points to the great need for more refined diagnostic tools—ones that can accurately predict responsiveness to various anti-tumor therapies and allow the oncologist to avoid use of therapies when they are not needed because they are highly unlikely to succeed.

We can focus for the moment on female breast cancers in the United States to provide an illustrative example. In 2018, about 266,000 cases of invasive breast cancer were diagnosed together with 64,000 noninvasive (*in situ*) forms of the disease. That year, about 41,000 women were predicted to die from the disease. Most patients with invasive mammary carcinomas were treated with aggressive chemotherapy. The age-adjusted death rate from breast cancer in the United States did not change significantly throughout most of the twentieth century (**Figure 17.3**) during a period when truly effective therapies were not available, suggesting that a comparable annual frequency of life-threatening breast cancers continues to be formed today and that many of the vastly larger numbers of invasive breast carcinomas currently being diagnosed are not likely to cause death, even without therapeutic intervention, much like the prostate cancers that are diagnosed in vast numbers in the West (Supplementary Sidebar 17.4). As screening for breast cancer increases and the power to detect small, previously overlooked tumors improves (see Figure 17.2), this disparity between breast cancer incidence and mortality is likely to grow.

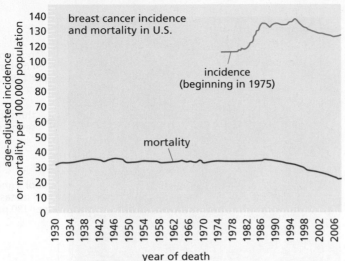

Figure 17.3 **Breast cancer incidence vs. mortality in the United States** The age-adjusted incidence of breast cancer has been increasing steadily over the past several decades, whereas mortality from this disease was quite constant until the end of the twentieth century, when it began to decline significantly. Most of the increase in incidence appears to be attributable to increased screening, but a small proportion of it may have resulted from real changes in the rate at which the disease strikes because of changes in reproductive practices, nutrition, obesity, age of menarche and menopause, and so forth. (From A. Jemal et al., *CA Cancer J. Clin.* 55:10–30, 2005. With permission from John Wiley & Sons.)

Statistics like these underscore the desperate need to develop molecular markers that enable oncologists to distinguish between those tumors that truly require aggressive treatment and those that can be monitored repeatedly for signs of progression (the practice termed "watchful waiting"). In the case of other types of cancer, equally important distinctions must also be made, but of a far grimmer sort—between those cancers that are likely to show some significant response to therapy and those that will not, in which case compassionate care dictates that the disease should be allowed to run its natural course.

Gene expression analysis, of the type first described in Figure 13.20, has provided clinicians with an additional approach to stratify cancers—to classify them into subgroups having distinct biological properties and prognoses. RNAseq, a key analytical tool of the science of functional genomics, allows a researcher to survey the expression levels of a set of genes (often involving virtually all genes in the genome whose expression can be measured in a tissue preparation). Subsequent computerized analyses of these expression data using bioinformatics make it possible to identify patterns of gene expression characteristic of specific tumor types and subtypes. Additionally, these analyses can reveal small subsets of the genes whose expression (at characteristically high or low levels) correlates with a specific biological phenotype, drug responsiveness, or prognosis. For example, the expression of a cohort of several dozen genes by a tumor may suffice to serve, on its own, as a strong predictor of its degree of progression or its association with one or another specific subtype of cancer.

In the case of breast cancers, there has been a crying need to distinguish those primary tumors that are likely to become metastatic from those that will remain indolent and are therefore not likely to spread during the lifetime of the patient. Traditionally, the main prognostic parameters that have been used to predict the course of breast cancer development have been patient age, tumor size, number of involved axillary lymph nodes, histological type of the tumor, pathological grade, cell proliferation rate, and receptor status (that is, the expression of the estrogen, progesterone, and HER2 receptors). Because these factors, when used singly or in combination, do not yield prognoses with a high degree of accuracy, most patients diagnosed with primary invasive breast cancers in the United States have been treated aggressively, even though only ~19% of such patients will ever develop life-threatening disease.

The combined use of gene expression analysis and bioinformatics has made it possible to predict the clinical course of breast cancer progression with more than 90% accuracy (**Figure 17.4**). Such a high prognostic accuracy holds the promise of sparing many women exposure to unnecessary chemotherapy; in these women, gauged statistically, the risks incurred by the therapy may actually exceed the risk of life-threatening cancer development. And in the future, the details of gene expression

Figure 17.4 Stratifying breast cancers using functional genomics (A) Expression arrays were used to analyze the gene expression of 295 primary breast cancers diagnosed in women under 53 years old. The group included patients with metastatic cells in the axillary lymph nodes draining the mammary glands as well as patients whose lymph nodes were free of cancer cells. Bioinformatics analyses of these tumors were then employed to choose a set of 70 "prognosis genes" whose expression could be used to stratify these breast cancer patients (arrayed along vertical axis), whose clinical course had been followed for a mean time of 7 years. The expression levels of these 70 genes (arrayed along horizontal axis, names not given) together with information about the patients' clinical history was then used to set a threshold that separated tumors that had a "good expression signature" from tumors that had a "poor expression signature." (B) This Kaplan–Meier plot reveals the stratification of a group of 151 breast cancer patients whose survival had been followed for 10 years following initial diagnosis. Using the criteria of panel A, they could be separated into two groups with dramatically different clinical courses. Taken together with other factors (such as the efficacy of chemotherapy), calculations indicate that women whose tumors carry a good expression signature derive virtually no benefit from adjuvant chemotherapy. (From M.J. van de Vijver et al., *N. Engl. J. Med.* 347:1999–2009, 2002. With permission from Massachusetts Medical Society.)

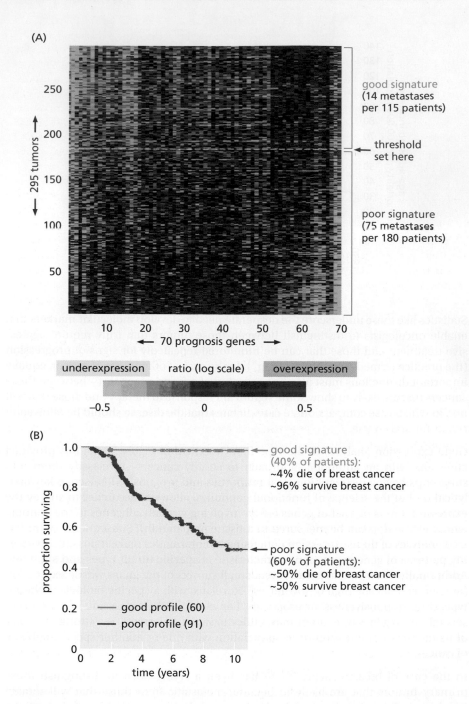

analyses are likely to inform the oncologist about the treatment protocol that is most likely to yield a durable clinical response or even a cure (**Sidebar 17.2**).

Analyses of this sort are only the beginning steps in a large-scale effort to examine a variety of human cancer types by means of gene expression. The resulting information should make it possible to stratify the cancer types into subtypes and, based on the resulting information, to devise therapies tailored to each specific subtype. As an illustrative example, diffuse large B-cell lymphomas (DLBCLs) have presented a quandary to the oncologist because their outcomes are so variable in the clinic, with some patients dying soon after diagnosis and many others being cured, or at least achieving 10-year remissions without any clinical symptoms. At the same time, all these tumors have a very similar appearance under the microscope (**Figure 17.5A**). Through the use of gene expression analysis in combination with additional high-resolution approaches, we now know that this malignancy is actually composed of several distinct molecular disease subtypes, seemingly originating from cells that have passed through different stages of B-cell differentiation

Sidebar 17.2 RNA expression and tumor origins As we read earlier in Chapter 11, the somatic mutations and epigenetic alterations that tumors sustain en route to full-blown malignancy rarely eradicate the influence of the differentiation programs of their normal cells-of-origin (see Sidebar 11.2 and Figures 11.13 and 14.50). In most tumors, these programs are transmitted heritably from one cell generation to another, in spite of the profoundly disruptive effects of various mutations and heritable epigenetic alterations acquired during the course of tumor progression. These differentiation programs work hand in hand with the somatic mutations and epigenetic alterations to dictate cancer cell phenotype and thus responsiveness of a tumor to therapies.

These complex interactions between genetic and nongenetic determinants of cancer cell behavior cannot be discerned by sequencing tumor cell genomes, creating a need to systematically monitor the nongenetic determinants of cancer cell behavior. In principle, cell phenotype could be surveyed in a systematic and quantitative way through proteomic analyses that reveal the complex array of proteins expressed by cancer cells, the levels of these proteins, and their states of post-translational modification. At present this is challenging at the single cell level, indicating the need to use gene expression analyses as the practical alternative for surveying cancer cell phenotype. In the future, complex bioinformatic algorithms will need to be developed to integrate the results of gene expression analyses with those of tumor genome sequencing in order to produce a more complete picture of the molecular and biochemical state of diverse sets of cancer cells within a tumor that should prove even more useful than the results currently achieved by RNAseq and scRNAseq analyses alone.

(Figure 17.5A). These subtypes, which include germinal-center B-cell-like (GCB) lymphomas, activated B-cell-like (ABC) lymphomas, and primary mediastinal B-cell lymphomas (PMBL; Figure 17.5B), show distinct clinical outcomes following standard chemotherapy.

The combination of next-generation DNA sequencing with expression-based classification has further defined a molecular architecture of these lymphoma subtypes. Of these three subtypes, both the activated ABC lymphomas and the PMBLs exhibit genetic and epigenetic alterations that lead to high levels of NF-κB activity (see Figure 6.22A and 17.5C); this transcription factor (see Section 6.12) appears to be driving their proliferation and protecting them from apoptosis. In ABC lymphomas, NF-κB activity can by driven by constitutive B-cell receptor signaling or mutations, like those present in Myd88 or TNFAIP3, which promote or inhibit repression of NF-κB induction. In contrast, GCB tumors depend upon a distinct set of anti-apoptotic signals, including the overexpression or constitutive activation of Bcl2 or Bcl6 (see Section 9.13), or mutations in epigenetic regulators, such as EZH2, that maintain the germinal center cell state, that is, the phenotypic state characteristic of cells in the lymphoid organs in which B cells mature.

The identification of deregulated signaling and driver mutations in lymphoma subtypes has prompted clinical efforts to target these particular pathways in primary tumors and in those resistant to chemotherapy. To summarize, expression analysis of tumor-derived RNA can provide a molecular pathological characterization (rather than a microscope-based histopathological characterization) of a tumor that can then be combined with DNA sequencing analysis to identify molecular targets within these cancer cells that might be targeted by appropriate pharmacological agents. It remains to be seen whether the identification of these vulnerabilities of distinct lymphoma subtypes will result in new first-line treatments for most lymphoma patients.

Interpretation of the gene expression patterns of tumors of complex histology, such as carcinomas composed of epithelial, endothelial, immune, and other stromal cell types, is often confounded because the RNA transcripts being measured represent mixtures deriving from multiple cell types rather than the transcripts produced by the neoplastic cells in these tumors. This problem may be addressed by bioinformatic analyses that are able to deconvolute the contributions of the various distinct cell types in a tumor being analyzed. More precise assessments can be made through use of

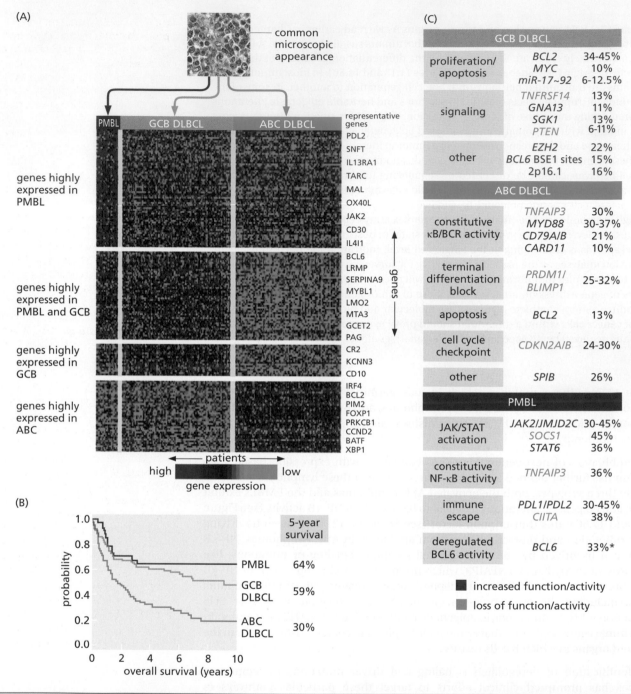

Figure 17.5 Stratification of diffuse large B-cell lymphomas
(A) DLBCLs represent a group of several subtypes of B-cell neoplasia that are essentially indistinguishable in the microscope from one another and from the microscopic appearance of primary mediastinal B-cell lymphomas (*above*). However, use of gene expression analysis (*below*) allows these tumors to be stratified into three subtypes, termed primary mediastinal B-cell lymphomas (PMBLs), germinal-center B-cell (GCB) DLBCLs, and activated B-cell-like (ABC) DLBCLs. In this expression array, a set of genes whose expression levels have been found to be useful in making this stratification is plotted along the vertical axis, and a set of patient tumors (*unlabeled*) is plotted along the horizontal axis. (B) This Kaplan–Meier curve illustrates the greatly differing disease courses that patients with the three subtypes experience. (C) The tumors that are classified through the expression analyses shown in panel A exhibit distinct genetic alterations (shown as present in the indicated percentage of each tumor subtype), including mutations resulting in a loss of gene function (*blue*) and mutations and chromosome translocations resulting in enhanced gene activity or expression (*red*). (A, top, from M. Roschewski, L.M. Stuadt, and W.H. Wilson, *Nat. Rev. Clin. Oncol.* 11:12-13, 2014. With permission from Nature. Bottom, from L.M. Staudt and S. Dave, *Adv. Immunol.* 87:163–208, 2005. With permission from Elsevier. B, from A. Rosenwald et al., *J. Exp. Med.* 198:851–862, 2003. With permission from Rockefeller University Press. C, from L. Pasqualucci and R. Dalla-Favera, *Cancer Cell* 25:132, 2014. With permission from Elsevier.)

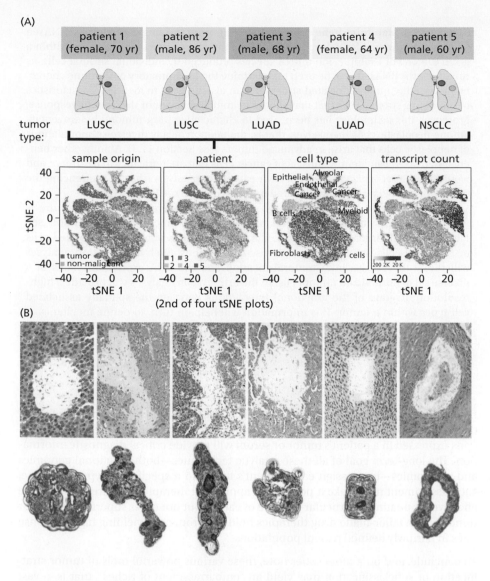

Figure 17.6 scRNAseq and laser-capture microdissection of tumors
High-resolution techniques are increasingly used to examine the cellular pathology of human cancer. (A) Lung tumors excised during surgery are disassociated into single cells and analyzed for mRNA content. The transcriptome of each cell can then be used to demarcate which cells are neoplastic cells vs. those deriving from adjacent nonmalignant tissue, which specific stromal cell types are present within individual tumors, and how the representations of tumor and stromal cell types vary within individual tumors and between patients. The individual cells with similar transcriptomes can be grouped together using a statistical algorithm that displays them in a two-dimensional tSNE plot, examples of which are presented here. The combined expression data provides a new kind of tumor pathological analysis that extends beyond that afforded by more established cell staining and immunohistochemical approaches.
(B) Laser-capture microdissection (LCM) of tumor sections allows for the precise isolation of small sets of tumor cells within a topologically defined area of a tumor. Here each removed patch of tumor cells (*right*) is aligned with the remainder of the tumor section that is left behind following this removal (*left*). This approach can be combined with scRNAseq to determine the gene expression states and the level of heterogeneity that exists between cells in distinct areas of the same tumor. (A, from D. Lambrechts et al., *Nat. Med.* 24:1277–1289, 2018. With permission from Nature. B, Reprinted with permission of Springer Nature: Humana Press, Testicular Germ Cell Tumors, Methods and Protocols by Aditya Bagrodia and James F. Amatruda. 2021; permission conveyed through Copyright Clearance Center, Inc.)

recent advances in RNA sequencing (**Figure 17.6A**). Specifically, the quantification of multiple transcripts produced by individual cells—both normal and neoplastic—can now be performed. This experimental procedure, termed single-cell RNA sequencing or scRNAseq, involves isolating single cells and generating amplified cDNA libraries representing the transcripts of each of these cells, which can then be sequenced using conventional DNA sequencing procedures (**Sidebar 17.3**).

In addition to these various analyses of cell transcripts, technical advancements in DNA sequencing have enabled genome-scale analyses to be performed at an increasingly accessible cost. Thus, next-generation DNA sequencing has become an increasingly common cancer diagnostic tool. Targeted mutational analysis seeks to identify common cancer alterations in specific malignancies (that is, known recurring mutations) that may inform patient prognosis or the use of a drug that targets a specific mutation. Conversely, whole genome or whole exome sequencing seeks to identify rare or previously uncharacterized mutations in the genome of a tumor that may reveal an unexpected vulnerability of this tumor to an existing targeted therapeutic. Whether such an approach will identify targetable alterations for many cancer patients or have a transformative impact in cancer care remains to be seen. For cancer patients whose tumors carry mutations that are targetable by an available therapy, whole-genome sequencing clearly leads to better treatment, outcome, and survival. However, at this point, the number of cancer-causing mutations identified by DNA sequencing vastly exceeds the number of available treatments that specifically target tumors bearing these mutations.

Sidebar 17.3 Examining gene expression in tumors at the single-cell level scRNAseq technology now enables a deep exploration of the range of distinct cell types within a given tumor. For example, the mRNA species produced by individual stromal cells, as revealed by scRNAseq, can be used to determine the inflammatory or immune characteristics of the tumor-associated stroma. It can also be used to identify characteristics of infiltrating stromal cells that may modulate tumor behavior or therapeutic response. Moreover, this technology has been used to characterize intra-tumoral heterogeneity among neoplastic cells themselves, that is, the phenotypically diverse subpopulations of neoplastic cells that arise in advanced tumors (see Section 11.6). Among other benefits, scRNAseq can identify subsets of cancer cells that are refractory to treatment and would not be recognizable by bulk RNA expression analysis (that is, sequencing the transcripts produced by unfractionated mixtures of cells; see Figure 17.6A).

An extension of scRNAseq that will be employed increasingly in the future is its combination with the technique of laser-capture microdissection (LCM; see Figure 17.6B) in order to characterize the transcriptomes of physically isolated cells or groups of cells prepared from microscopic sections. Such combination analyses will reveal the precise intra-tumoral location and the RNA expression states of a diverse set of normal and tumor cells. In the future, such analyses will provide a high-resolution diagram of the functional interactions between the spatially associated cell types within a tumor. This information will help, in turn, to define mechanisms by which the tumor microenvironment promotes phenotypic heterogeneity among tumor cells or supports processes like tumor cell metastasis, immune protection, or therapeutic resistance.

Beyond these gene expression and genomics analyses stands a generation of novel diagnostic tools involving the science of proteomics, in which the spectrum of proteins expressed in a patient's tumor or serum will provide critical diagnostic information. The long-term goal of all these analytic techniques—both functional genomics and proteomics—is to assign each patient's tumor to a specific subtype of disease; such assignment may make it possible to apply drug therapies that are proven to be effective for treating a particular subtype of cancer but not other superficially similar tumors. Such tailor-made drug therapies hold the promise of yielding high response rates in narrowly defined patient populations.

To conclude, and on a more sober note, these various powerful tools of tumor stratification or subclassification may yield an "embarrassment of riches," that is, a vast number of distinct subtypes of human tumors whose diversity and distinctive properties greatly exceed our ability, in the near and distant future, to effectively develop treatments for most of these subtypes. After all, appropriate tailor-made therapies are usually validated in large and expensive clinical trials, as described later in Section 17.11. Moreover, as tumor types are sliced into increasingly thin sections, the number of patients affected by each subtype of tumor will grow correspondingly small, complicating attempts at recruiting enough patients to conduct clinical trials yielding statistically significant readouts. Thus, how will we possibly design and carry out clinical trials of diverse agents on diverse cancer subtypes? The answers to these quandaries are not yet in hand.

17.2 Surgery, radiotherapy, and chemotherapy are the major pillars on which current cancer therapies rest

An exclusive focus in this chapter on the newer, rationally designed agents would not be warranted because of one simple fact: the older treatments—surgery, chemotherapy, and radiotherapy—have been shown over many decades to be highly effective in extending the survival of cancer patients and even eradicating certain types of tumors with curative outcomes, whereas the newer, "rational" agents—whose design and implementation have been informed by the research described in this book—can rarely boast such successes. So, we begin this chapter with these older agents and describe their successes and failures.

The multi-step nature of cancer development, specifically the notion that primary tumors spawn metastases, has been a central concept in anti-cancer therapy since the end of the nineteenth century. Radiotherapy developed only slowly during the course of the twentieth century, and chemotherapy was initiated only after World War II. Through much of the century, then, surgery was the only form of therapy that offered any hope of reversing the disease, perhaps even achieving cures.

The discipline of surgical oncology embraced two notions as fundamental truths: small tumors will inevitably develop into large tumors and the resection of primary tumors represents an effective means of reducing the risk of metastatic relapse— the resurgence of disease long after the primary tumor has been removed. For example, resection of early-stage colorectal carcinomas yields 95% patient survival after five years—essentially a curative outcome—even without **adjuvant** (post-surgical) chemotherapy.

In fact, the accepted truths governing surgical practice were rarely subjected to critical tests until the last decades of the twentieth century, when the specialty of *outcomes research*, often referred to as evidence-based medicine, arose. The evidence in these cases came from clinical trials in which two or more alternative surgical strategies were compared with one another in randomized patient populations. These randomized trials revealed that some surgical procedures produced highly effective clinical outcomes, whereas others were of dubious utility, as described in Supplementary Sidebar 17.2.

Radiation oncology began soon after Wilhelm Röntgen's 1895 discovery of X-rays and the frequent observations soon thereafter that exposure to this form of electromagnetic radiation resulted in burn damage to normal tissues, including even tissue necrosis. The widespread use of radiotherapy in the oncology clinic awaited the development, after World War II, of the technical means of directing these rays in relatively narrow beams of radiation that were focused on diagnosed, clearly delineated tumors, almost invariably primary tumors. Focused adjuvant radiotherapy has become an essential clinical tool in reducing post-surgical relapse in the surgical field (that is, the tissues adjacent to or surrounding resected primary tumors). Indeed, nearly half of all cancer patients will receive radiation or radioactivity as a component of their therapy. A dramatic example of the utility of radiotherapy comes from post-**lumpectomy** (see Supplementary Sidebar 17.2) follow-up of older women with early-stage breast cancer: adjuvant (post-surgical) radiation reduces by 50% the rate of subsequent mastectomy (resection of the entire breast rather than just the area immediately around the tumor) during the decade following initial surgery.

The use of radionuclides—radioactive isotopes administered intravenously as components of a pharmaceutical drug or coupled to monoclonal antibodies— underlies systemic radiotherapy and is often used when all or parts of a tumor cannot be readily resected or when disseminated cancer cells need to be eradicated. Brachytherapy, involving the implantation of radioactive sources carried in small capsules into an organ site or tumor, is also used in a variety of cancers. Because brachytherapy does not generally reach uninvolved normal tissue, its use allows for the delivery of higher localized doses than those that are achievable by radiation with **external beam therapy**.

Chemotherapy has more recent origins: an often-related anecdote describes how the 1943 bombing of an American warship in the harbor of Bari, Italy, led to release of a large cloud of mustard gas of the sort used in World War I chemical warfare. Almost a thousand people died, sooner or later, from the effects of this explosion and the released cloud of gas. The clinical deterioration of some survivors of this catastrophe reawakened interest in the 1919 discovery that exposure to mustard gas leads to depletion of bone marrow cells and thus anemia. In fact, research begun independently at Yale University in 1942, a year before the Bari disaster, had revealed that intravenous doses of mustard gas—named after its characteristic odor—led to temporary regression of a lymphoma. We have encountered chemicals of this class in the form of alkylating agents (see Section 12.5) that are used, among other applications, to treat glioblastomas (see Table 17.1).

[Recall that in addition to their cytotoxicity, these alkylating agents are highly mutagenic, leaving the genomes of exposed cells with hundreds if not thousands of point mutations (see Figure 12.7B); this outcome reveals another side of a number of anti-cancer treatments: in addition to their effects in reducing or eliminating tumors, X-rays and certain cytotoxic agents are also carcinogenic, and their short-term successes in producing clinical remissions may be counterbalanced by the appearance years later of independently arising, second-site tumors that are direct consequences of their mutagenic actions.]

The transient clinical response of a lymphoma observed at Yale was followed by rapidly expanding interest in similar agents in the years that followed. The emphasis here was on cytotoxic agents that, for unknown reasons, killed neoplastic cells preferentially by sparing normal tissues, or at least by inflicting tolerable side effects on the patient. Almost always, discovery of the biochemical and cell-biological mechanisms of these drugs came decades after their introduction into the clinic, and in many cases, as discussed below, we still do not understand the precise mechanisms-of-action of a number of highly effective cytotoxic drugs.

By 1947, the utility of another class of cytotoxics was discovered, initially in the form of an agent for treating pediatric leukemias and lymphomas. In this case, the compound in question—aminopterin—acted as an antagonist of folate metabolism, inhibiting the formation of tetrahydrofolate and thereby blocking some of the critical biosynthetic reactions that are required for the assembly of nucleotides and thus the synthesis of DNA and RNA. A young boy treated that year for lymphoma survived and lived another half century, representing the first documented cure of this disease.

Compounds like aminopterin are often termed **antimetabolites** because they interfere with the normal functioning of specific metabolites or the enzymes that produce them in the cell. Most of these agents closely resemble and are therefore chemical analogs of normal metabolites. For example, a number of highly effective antimetabolites are purine or pyrimidine analogs that operate either by preventing normal biosynthesis or by becoming incorporated into DNA, whose function they then inhibit (**Table 17.2; Figure 17.7**).

Table 17.2 Examples of antimetabolites used to treat cancer

Name	Chemical structure	Targeted reaction	Examples of clinical use
methotrexate	folate analog	formation of tetrahydrofolate	breast cancer, lymphomas
6-mercaptopurine	purine analog	purine biosynthesis	leukemia, NHL
doxorubicin	natural product[a]	intercalating agent, inhibits topoisomerase	wide range
thioguanine	guanine analog	purine biosynthesis	acute granulocytic leukemia
fludarabine	purine analog	ribonucleotide reductase, DNA replication	chronic lymphocytic leukemia, NHL
cladribine	adenosine analog	adenosine deaminase	hairy-cell leukemia
bortezomib	peptide analog	proteasomal degradation	multiple myeloma
paclitaxel	natural product[a]	microtubule destabilization	lung, ovarian, breast cancer
etoposide	natural product[a]	DNA unwinding	lung cancer, sarcomas, glioblastoma
mitoxantrone	topoisomerase inhibitor	DNA unwinding	AML, breast cancer, NHL
irinotecan	topoisomerase inhibitor	DNA unwinding	colorectal carcinoma
vinblastine	natural product[a]	microtubule assembly	Hodgkin's lymphoma
vorinostat	hydroxamic acid	histone deacetylation	cutaneous T-cell lymphoma
azacitidine	pyrimidine analog	DNA methylation	myelodysplastic syndrome

Abbreviations: NHL, non-Hodgkin's lymphoma; AML, acute myelogenous leukemia.
[a]Complex structure.

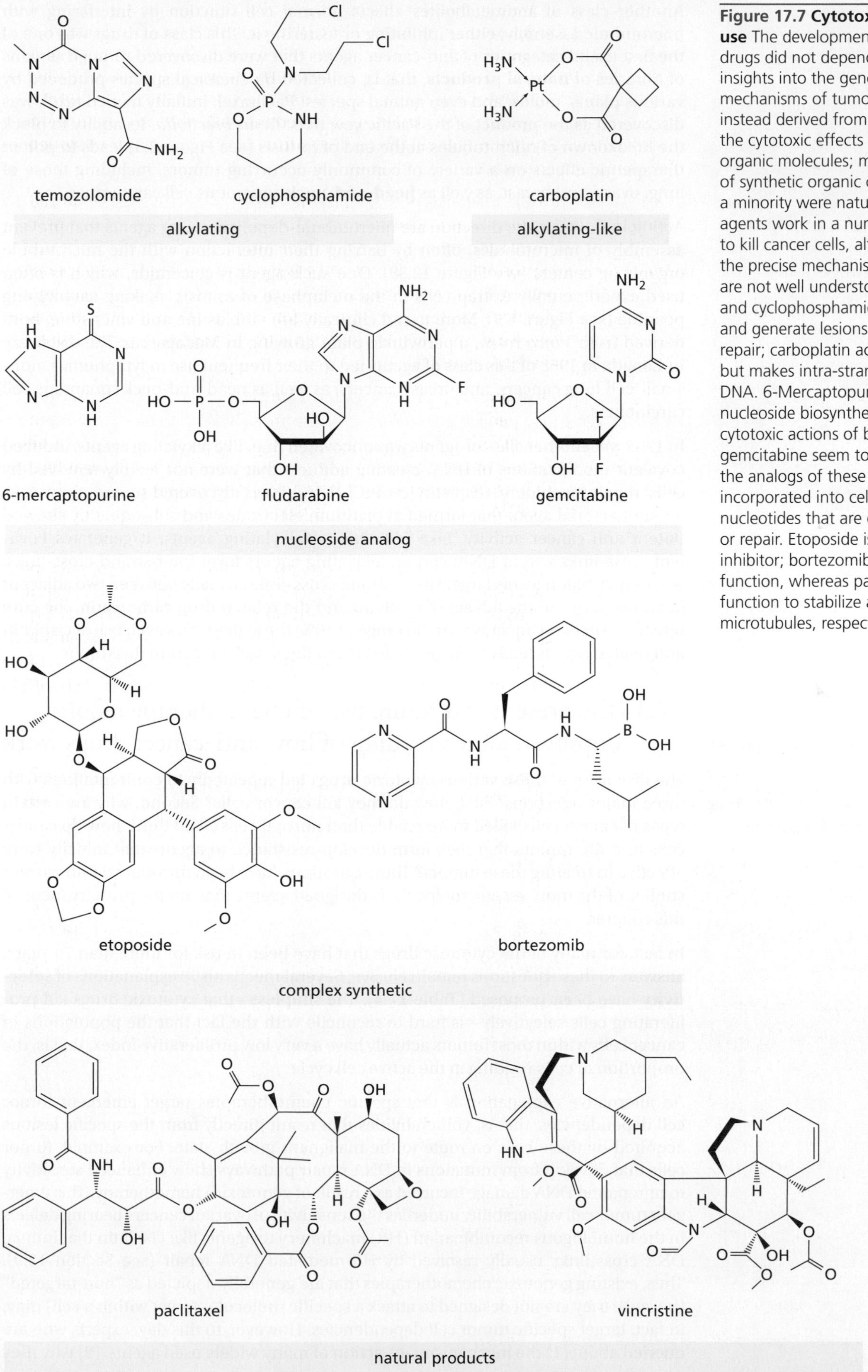

temozolomide   **cyclophosphamide**   **carboplatin**

alkylating   alkylating-like

6-mercaptopurine   **fludarabine**   **gemcitabine**

nucleoside analog

etoposide   **bortezomib**

complex synthetic

paclitaxel   **vincristine**

natural products

Figure 17.7 Cytotoxic drugs in current use The development of most cytotoxic drugs did not depend on detailed insights into the genetic and molecular mechanisms of tumor pathogenesis, but instead derived from investigations of the cytotoxic effects of a diverse array of organic molecules; most were products of synthetic organic chemistry, although a minority were natural products. These agents work in a number of distinct ways to kill cancer cells, although in many cases the precise mechanisms of cytotoxicity are not well understood. Temozolomide and cyclophosphamide alkylate DNA and generate lesions that are difficult to repair; carboplatin acts in a similar fashion but makes intra-strand cross-links within DNA. 6-Mercaptopurine interferes with nucleoside biosynthesis, whereas the cytotoxic actions of both fludarabine and gemcitabine seem to depend largely on the analogs of these nucleosides becoming incorporated into cell DNA, yielding nucleotides that are difficult to replicate or repair. Etoposide is a topoisomerase inhibitor; bortezomib disrupts proteasome function, whereas paclitaxel and vincristine function to stabilize and destabilize microtubules, respectively.

Another class of antimetabolites affects normal cell function by interfering with microtubule assembly, either inhibiting or fostering it. This class of drugs was one of the first useful categories of anti-cancer agents that were discovered through screens of libraries of **natural products**, that is, collected biochemical species produced by various plants, molds, and even animal species. Paclitaxel, initially named taxol, was discovered as the product of the Pacific yew tree, *Taxus brevifolia*. Its ability to block the breakdown of microtubules at the end of mitosis (see Figure 8.3) leads to potent therapeutic effects on a variety of commonly occurring tumors, including those of lung, ovary, and breast, as well as **head-and-neck** squamous cell carcinomas.

Acting in the opposite direction are microtubule-depolymerizing agents that prevent assembly of microtubules, often by barring their interaction with the microtubule organizing centers (see Figure 12.38). One such agent is colcemide, which is often used experimentally to trap cells in the metaphase of mitosis, making karyotyping possible (see Figure 1.9). More useful clinically are vinblastine and vincristine, both derived from *Vinca rosea*, a periwinkle plant growing in Madagascar. The discovery in Canada in 1958 of this class of agents led to their frequent use in lymphomas, non-small-cell lung cancers, and breast cancers, as well as head-and-neck squamous cell carcinomas.

In 1965, yet another class of agents was uncovered that, like alkylating agents, induced covalent modifications of DNA, creating adducts that were not readily removed by cells' repair machinery. Cisplatin [cis-PtCl$_2$(NH$_3$)$_2$] was discovered serendipitously as an antibacterial agent that formed at platinum electrodes and subsequently showed potent anti-cancer activity. Like bifunctional alkylating agents, it generates covalent cross-links within DNA; certain alkylating agents form inter-strand cross-links, whereas cisplatin forms largely intra-strand cross-links, usually between two adjacent guanines. Prior to the advent of cisplatin and the related drug carboplatin, the cure rate for testicular cancer was in the range of 10%; these days, however, use of cisplatin and related agents leads to curing 90 to 95% of those suffering from this tumor.

17.3 The present and future use of chemotherapy requires improved understanding of how anti-cancer drugs work

The discovery of these various cytotoxic drugs led repeatedly to confrontations with three major questions. First, how do they kill cancer cells? Second, why are certain types of cancer cells killed more readily than normal cells? And third, how do cancer cells and the tumors that they form develop resistance to agents that initially were effective in treating these tumors? These questions have been introduced into current studies of the more recent, molecularly designed agents that are the primary focus of this chapter.

In fact, for many of the cytotoxic drugs that have been in use for more than 70 years, answers to these questions remain elusive. Several mechanistic explanations of **selectivity** have been proposed (Table 17.3). The simplest—that cytotoxic drugs kill proliferating cells selectively—is hard to reconcile with the fact that the populations of cancer cells within most tumors actually have a very low proliferative index, that is, the proportion of cells residing in the active cell cycle.

An alternative explanation is that specific chemotherapies target emergent tumor cell dependencies, that is, vulnerabilities that result directly from the specific lesions acquired by these cells en route to the malignant growth state. For example, tumor cells that develop from mutations in DNA repair pathways show enhanced sensitivity to unrepaired DNA damage incurred as a result of genotoxic chemotherapy. This emergent tumor cell vulnerability underlies the sensitivity of ovarian cancers bearing defects in the homologous recombination (HR) machinery to agents like cisplatin that induce DNA cross-links usually resolved by HR-mediated DNA repair (see Section 12.9). Thus, existing genotoxic chemotherapies that are generally depicted as "non-targeted" (because they are not designed to attack a specific molecular target within a cell) may, in fact, target specific tumor cell dependencies. However, to this day, experts who are queried about (1) the mechanisms-of-action of many widely used agents, (2) why they

Table 17.3 Known or suspected mechanisms of anti-tumor selectivity[a]

Name	Mechanism conferring selectivity
cyclophosphamide	detoxified by ALDH in normal bone marrow
cladribine	detoxified by non-hematopoietic cell types
taxol/paclitaxel	high proliferation index,[b] other mechanisms unknown
multiple cytotoxic drugs	high proliferation index,[b] high sensitivity to apoptotic stimuli
cisplatin	intact p53 function in testicular germ-cell tumors[c]
DNA-damaging agents	inability to halt cell cycle advance in response to DNA damage[d]
PARP inhibitors	defective homology-directed repair
DNA-damaging agents	various types of defective DNA repair

[a]Drug selectivity, which leads to therapeutic indices larger than 1, implies that cancer cells are, for various reasons, hypersensitive to a therapeutic agent. Lack of selectivity and the absence of a significant therapeutic index imply that cancer cells are as sensitive to an agent as are cells in normal tissues.
[b]Proliferation-dependent cytotoxicity is not well understood. In certain cases, rapidly dividing cells are presumed to lack the time required to repair therapy-induced DNA damage before DNA replication occurs.
[c]Intact p53 function is known to render these cells highly susceptible to DNA damage–induced apoptosis.
[d]Defective p53 function, as well as nonfunctional cell cycle checkpoints, is presumed to allow cells bearing damaged, still-unrepaired genomes to advance into S phase, M phase, or both, resulting in stalled replication forks or mitotic catastrophe.
Abbreviations: ALDH, aldehyde dehydrogenase; PARP, poly(ADP-ribose) polymerase.

selectively kill cancer cells, and (3) how tumors develop drug resistance readily admit to the minimal progress made in solving these fundamental questions.

Early on, medical oncologists (that is, oncologists who treat cancer with various drugs) recognized that drug resistance developed sooner or later for almost all the drugs and the cancer types under treatment, being manifested as the outgrowth of drug-resistant variant cell populations. This drug resistance was soon understood by applying the lessons learned from bacterial genetics, in which resistance to various types of antibiotic treatment was known to occur via the selective outgrowth of mutant variants, some of which may have already existed in bacterial populations prior to being challenged by drugs. Given the large numbers of cells in clinically detectable human tumors—almost 1 billion in a growth of 1 cm diameter—and the frequency of resistant mutants—perhaps one in a million—the emergence of drug-resistant variants seemed almost inevitable.

The response to this was the development of multi-drug protocols, often using combinations of drugs with distinct and complementary modes of cell killing, such as an antimetabolite, an alkylating agent, and a microtubule antagonist. In theory, the likelihood that variants preexisted in tumor cell populations that were simultaneously resistant to all three agents appeared to be astronomically low—the product of three probabilities, in this case one in 10^{18}. Indeed, the introduction of combination drug regimens resulted in dramatic and unprecedented clinical responses in certain tumors, even cures. For example, the ABVD regimen, which was developed in Italy in the mid-1970s and is still used to treat advanced Hodgkin's lymphomas (HLs), involves the administration of *a*driamycin (doxorubicin), a DNA-intercalating drug; *b*leomycin, a DNA-cleaving agent; *v*inblastine, a microtubule antagonist; and *d*acarbazine, an alkylating agent (**Table 17.4**). Depending on the stage at which the tumor is initially diagnosed, 5-year progression-free survival (PFS) of treated patients ranges from 85 to 98%, with many of these patients essentially cured of the disease. Another multi-drug protocol elicits an almost 90% cure rate in childhood (younger than 15 years of age at diagnosis) acute lymphoblastic leukemia (ALL).

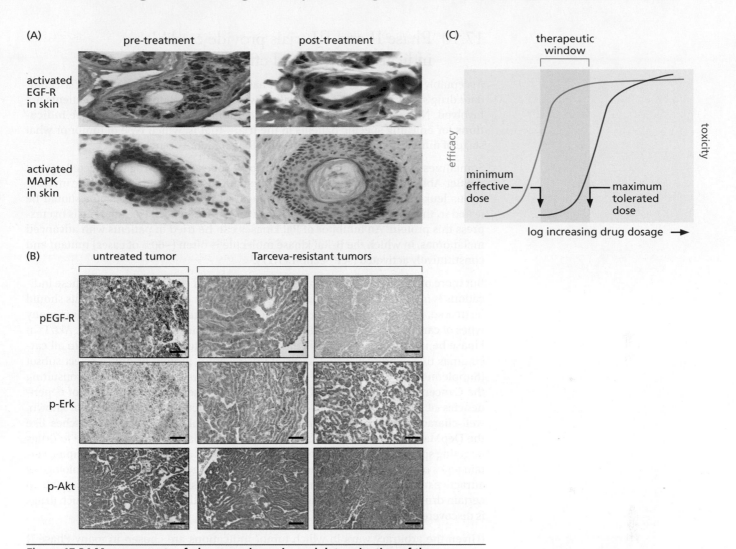

Figure 17.24 Measurements of pharmacodynamics and determination of the therapeutic window The extent of inhibition of the EGF-R in a tumor can, in principle, be gauged by measuring effects of drug treatment on the EGF-R in the skin; the latter is readily assessed through small skin biopsies. In the cases illustrated here, patients were suffering from a variety of tumors, including carcinomas of the ovary, lung, colon, and prostate and head-and-neck cancers. (A) Shown here are the effects of treating a cancer patient with Iressa, a low–molecular-weight EGF-R tyrosine kinase inhibitor (see, as an example, Figure 17.17). The upper panels show immunohistochemistry using an antibody against phospho-EGF-R (*brown*), i.e., the activated form of the receptor. The lower panels used an antibody against phospho-MAPK, the activated form of this downstream kinase (see Figure 6.9A). Both measurements depended on the normally intense mitogenic signaling occurring in hair follicle keratinocytes. (B) The effects of Tarceva, another low–molecular-weight EGF-R tyrosine kinase inhibitor, gauged by immunohistochemical staining of a mouse lung adenocarcinoma. In this case, long-term treatment resulted in a reduction in the overall level of phosphorylated EGF-R (*brown, upper row*). The level of phosphorylated Erk and Akt/PKB (*brown; lower rows*) was, however, elevated in tumor cells that survive Tarceva treatment, suggesting that tumors that escape EGFR inhibition can find alternative mechanisms to activate pathways downstream of active EGFR. (C) Measurements of pharmacodynamics such as these, taken together with studies of pharmacokinetics and toxicity, define the therapeutic window in which a drug should be given—the range of concentrations that are efficacious without creating an unacceptable level of toxic side effects. (A, top from J. Albanell et al., *J. Clin. Oncol.* 20:110-124, 2002. With permission from American Society of Clinical Oncology; bottom from J. Albanell, F., Rojo, and J. Baselga, *Semin. Oncol.* 28:56-66, 2001. With permission from Elsevier. B, Katerina Politi, Pang-Dian Fan, Ronglai Shen, Maureen Zakowski, Harold Varmus; Erlotinib resistance in mouse models of epidermal growth factor receptor-induced lung adenocarcinoma. Dis Model Mech 14 January 2010; 3 (1-2): 111–119. With permission of the Company of Biologists.)

17.10 Phase II and III trials provide credible indications of clinical efficacy

Acceptably low levels of toxicity in a Phase I trial will encourage testing a candidate drug's efficacy in a Phase II trial, in which larger groups of cancer patients are involved. Now, for the first time, critical decisions must be made about the **indications** for enlisting specific patients in the trial—that is, which type of tumor or what stage of tumor progression will justify enrolling patients in such a trial?

Sometimes the clinical indications are obvious. For example, an agent targeted against the Bcr-Abl oncoprotein should be tested in patients diagnosed with chronic myelogenous leukemia (CML). A drug directed against the HER2/neu receptor should be tested in the approximately 30% of breast cancer patients whose tumor cells overexpress this protein. An inhibitor of Raf kinases can be tried in patients with advanced melanomas, in which the B-Raf kinase molecule is often (~60% of cases) mutant and constitutively activated (see Section 17.17).

But more often than not, the choice of indications and clinical response to these indications is neither entirely rational nor optimal. Which class of cancer patients should be treated, for example, with a drug that acts as a general inducer of apoptosis in many types of cancer cells? How should a drug directed against the anti-apoptotic Akt/PKB kinase be used in the clinic? Will an anti-EGF receptor drug prove useful in all carcinomas that express elevated levels of this receptor protein or only a select subset (Supplementary Sidebar 17.9)? Significant help here may come from consulting the Cancer Dependency Map (DepMap), which presents a compendium of dependencies of various tumor types on the ongoing functioning of a variety of known, well-characterized signaling proteins. Information obtained from approaches like the DepMap may demarcate sets of tumors that are particularly susceptible to drugs targeting specific signaling pathways. However, as we will see later in this chapter, certain types of cancer that would never be identified by genetics or molecular biology as attractive targets for drug treatment turn out, on occasion, to be highly susceptible to certain drugs under development. In these cases, the therapeutic utility of such drugs is discovered only by chance.

(Given the arbitrary ways in which tumor indications are chosen in many Phase II trials, we can wonder how many truly useful candidate drugs have been discarded in the past, simply because good luck did not favor them in the design of these trials. Thus, it may well be that a drug has the potential to elicit truly spectacular responses when used to treat gastric carcinomas but this effect is never realized, if it is tested in Phase II trials for its effects on pancreatic or lung carcinomas, where it fails to show any useful effects and is therefore dropped entirely from further development and clinical testing.)

If Phase II trials yield clear signs of efficacy for treating certain types of cancer with a candidate drug, Phase III trials, undertaken in far larger patient populations, will be launched. These trials are very costly but are ultimately critical, if only because they may show, for the first time, whether any clinical responses ascribed to a drug are statistically significant. The results of these trials usually become compelling only if control experiments are performed by treating equally large populations of patients with another therapy in parallel, usually one that is already licensed and in widespread use. Importantly, the licensing of a candidate drug for a specific disease indication (in the United States by the Food and Drug Administration, FDA; in the European Union by the European Medicines Agency, EMA) usually depends on whether it yields a therapeutic benefit that is measurably greater than the existing **standard of care** (Sidebar 17.12).

Patients in Phase III trials of anti-cancer agents usually have gone through previous rounds of chemotherapy with various types of cytotoxic agents, each ending with a relapse and the appearance of tumors that are refractory to the already-applied therapy. Moreover, these tumors are often highly aggressive. This patient profile helps to explain why the bar is not set too high for FDA approval of a new drug or drug combination because the drugs in Phase III trials are dispatched to attack the

Sidebar 17.12 When do we decide that a drug works? It is worth considering how one evaluates success of a given therapy in the context of a clinical trial. Certainly the most relevant and compelling metric is whether patients treated with a new agent show increased overall survival relative to those treated with a standard-of-care therapy. However, in many cases, drug efficacy is assessed as the "objective response" of the tumor, itself, to treatment. Alternatively, one can assess the "time to tumor progression" after treatment. Both "objective response" and "time to tumor progression" seek to use changes in tumor size or number as an endpoint upon which efficacy conclusions can be inferred. However, both endpoints are only useful if they are based on widely agreed upon and commonly applied standards. The RECIST (or Response Evaluation Criteria In Solid Tumors) criteria provide general guidelines for how these endpoints are determined. These criteria, which have been periodically updated based on patient response and outcome data and advancing tumor imaging modalities, define a set of four possible outcomes for tumors of a certain minimum size (at least 10 mm) following treatment. There can be a complete response (CR), in which all tumors shrink below the 10 mm minimum size; a partial response (PR), in which all tumors show at least a 30% decrease in size; progressive disease (PD), in which tumors show at least a 20% increase in size; and stable disease (SD), in which tumors do not show a sufficient volume change to be placed in the PR or PD categories. These criteria have been very useful in comparing the potential efficacy of distinct therapies across broad patient populations. However, these criteria also have significant limitations. Most notably, tumor size changes following treatment correlate but are not necessarily linked with patient outcome. Thus, drugs that elicit a PR may not extend overall survival.

most difficult-to-treat types of cancer. Thus, improvements in patient quality of life or temporary shrinkage of a tumor may suffice even without improvement in long-term survival.

One illustration of this comes from a current treatment for pancreatic cancer. This disease is an extreme example, to be sure, in that the five-year survival rate of this disease (from the time of initial diagnosis) is consistently less than 5%. Gemcitabine (difluorodeoxycytidine; see Figure 17.6), which is widely employed as a therapy for pancreatic carcinoma, received initial FDA approval because in some patients it resulted in an improvement in symptoms, weight gain, and a temporary stabilization in tumor growth, although it offered only a modest increase in survival time (from the time of diagnosis): patients treated with gemcitabine had a median survival time of 5.65 months compared with those given the existing standard of care—5-fluorouracil (5-FU), which afforded them a 4.41-month median survival time (**Figure 17.25**). This and similar anecdotes reveal how desperate is the need for truly effective means of treating solid tumors.

Nonetheless, even with these relatively modest regulatory requirements, the other obstacles in drug development described here keep the current success rate for anti-cancer drug development extremely low. Perhaps one drug in a hundred advances all the way through the drug development "pipeline" from initial *in vitro* testing through a Phase III trial that culminates in some clear improvement in patient outcome and licensing by the FDA. (After licensing has occurred, a Phase IV trial may be conducted to determine the efficacy of drug treatment over an extended period of time, how

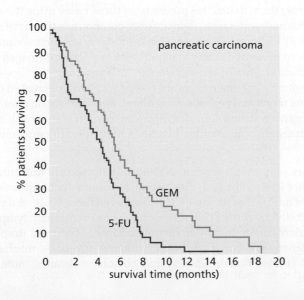

Figure 17.25 Gemcitabine as a treatment for pancreatic cancer This Kaplan–Meier plot illustrates the high mortality exacted by pancreatic cancer. Patients treated with gemcitabine (GEM; see Figure 17.7) lived slightly longer than those treated with 5-fluorouracil (5-FU)— the standard treatment in the 1990s. Both agents are pyrimidine derivatives whose cytotoxicity derives from their ability to inhibit DNA synthesis, in part through misincorporation into the DNA. (5-FU also interferes with pyrimidine biosynthesis.) As is apparent, gemcitabine treatment offered only a modest increase in patient survival in this study reported in 1997, but this nevertheless sufficed to allow its approval by the U.S. Food and Drug Administration. (From H.A. Burris III et al., *J. Clin. Oncol.* 15:2403–2413, 1997. With permission from American Society of Clinical Oncology.)

certain subgroups of patients respond to the drug, and whether concerns about a drug's safety eventually emerge from its use in very large patient populations.)

17.11 Tumors often develop resistance to initially effective therapy

A complication that dogs all anti-cancer drugs is illustrated by the behavior of *HER2/ neu* transgenic mice, in which the mutant, oncogenic transgene has been programmed to induce mammary tumors on a predictable schedule and can be shut down thereafter. Although transgene-induced primary breast tumors and metastases all collapsed when the *HER2/neu* transgene was shut down, new tumors recurred in most of these mice between 1 and 9 months later. These tumors clearly represented variants of the initially observed ones that had developed alternative means of propelling their growth—that is, had become independent of *HER2/neu* oncogene expression.

As we saw in Chapter 12, the unstable, mutable genomes of cancer cells continually generate new alleles and novel genetic configurations (see Section 12.15). Evolving cancer cells can pick and choose among these genetic variations, searching for combinations that improve their ability to survive and proliferate. In this *HER2/neu* example, the relatively small number of cancer cells that survived oncogene shutdown seem to have spent months thereafter trolling for newly arising oncogenes (or other cancer-causing alleles) in their genomes that might enable them to re-launch their program of aggressive proliferation. The rare cells that happened to acquire such advantageous alleles or epigenetic states then began clonal expansions that led to relapsing tumors.

A similar dynamic complicates almost all types of human cancer therapies, where initial clinical successes in reducing tumor cell populations are usually followed by the re-emergence in patients of tumor cell populations that have, through one means or another, developed resistance to the treatment (see Table 17.5). Much of this acquired resistance is attributable to the genetic and therefore phenotypic plasticity of cancer cell populations. Importantly, much of this "acquired" resistance may derive from variants that already existed within a tumor at the time the therapy was initiated.

The acquired mechanisms of drug resistance are quite variable and illustrate the ingenuity of cancer cells (see Table 17.5). Perhaps the most intuitive of all resistance mechanisms are alterations that block drug–target interactions. These alterations include mutations that change the structures of targeted proteins and directly prevent drug binding. They can also confer a decreased ability to import drug molecules through the plasma membrane or an acquired ability to pump out drug molecules through this membrane.

Other resistance mechanisms depend on the acquired ability to metabolize drug molecules, converting them to inactive products, in some cases using the same classes of enzymes that normally operate to detoxify toxic compounds that have entered the cell (see Section 12.5). Cells may also neutralize components of their apoptotic machinery or may acquire an increased ability to repair DNA molecules damaged by chemotherapeutics or radiation. Finally, tumor cells can undergo state changes—that is, transitions from one phenotypic state to another—that result in a decreased dependence on a particular drug target and perhaps, an acquired dependence on another. Sometimes multiple alternative resistance mechanisms can arise in response to a given therapy and multiple resistance-mediating alterations may operate concomitantly within a given tumor.

These behaviors represent a general challenge to all types of anti-tumor therapy, as discussed earlier in Section 17.2. The response to such acquired resistance was the development of multi-drug therapies. However, even these multi-drug therapy strategies are often foiled by cancer cells. For example, as mentioned, lymphomas treated with combination chemotherapy frequently relapse. Moreover, despite 60 years of clinical experience with lymphoma chemotherapy, the basic mechanisms of drug relapse in this context remain unclear and obvious resistance mutations have not been identified.

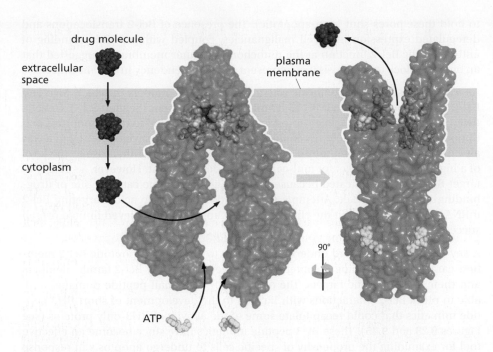

extracellular space

drug molecule

plasma membrane

cytoplasm

90°

ATP

Figure 17.26 Multi-drug resistance and the P-glycoprotein The MDR1 gene encodes P-glycoprotein (P-gp, *green*), a protein that is often present at elevated levels in cancer cells being treated with various types of chemotherapy. A structural model of P-gp is shown here. It is a 170-kD ATP-dependent transmembrane protein that can pump out of cells a wide variety of molecules with molecular weights ranging from 330 up to 4000. Its elevated expression is often associated with and likely causes multi-drug resistance. It is a member of a large family of mammalian ABC (ATP-binding cassette) transporter molecules, of which 49 have been discovered. As shown, a substrate such as a chemotherapeutic drug (*magenta*) enters the cell and is bound by the cytoplasmic domains of P-gp (*left*). In response to this binding and the binding of ATP molecules (*yellow*), the P-gp undergoes a profound conformational shift that collapses its cytoplasmic substrate-binding domain and opens its domain facing the extracellular space (*right*), resulting in the extrusion of the drug molecule from the cell. The drug-binding pocket is shown in *cyan*. The right P-gp molecule has been rotated 90 degrees to better illustrate the stereochemistry of the drug-extrusion step. (From S.G. Aller et al., *Science* 323: 1718–1722, 2009. With permission from AAAS.)

In recent years we have learned much about how multi-drug resistance can be achieved at the molecular level. For example, high-level expression of the *MDR1* gene, which encodes a normal transmembrane drug efflux pump, enables cancer cells to efficiently excrete a variety of chemically unrelated drugs, thereby lowering intracellular drug concentrations to subtoxic levels (**Figure 17.26**). Similarly, inactivation of certain parts of the apoptotic machinery (see Table 17.5) may also confer concomitant resistance to a number of distinct cytotoxic agents. However, whether these processes represent major mediators of multi-drug resistance in cancer patients remains unclear.

The advent of molecularly targeted agents, as discussed in the following sections, has highlighted the tremendous challenge of acquired drug resistance. Tumors almost invariably relapse with resistance-causing alterations after treatment with targeted therapy. This grim clinical reality has led to widespread consensus among drug developers: monotherapies involving either low–molecular-weight drugs or biological molecules are unlikely to cure most types of cancer, and effective multi-agent therapies must be devised if definitive, durable clinical responses are to be achieved in the future.

With these considerations in mind, we will read about a series of illustrative anecdotes in the ensuing sections. Each concerns a type of drug and its targets within cancer cells. The stories are arranged in an order, beginning with a well-established therapy and ending with a speculative one that holds great promise but is still far from clinical validation. In some cases, the specific therapy was inspired by discoveries of malfunctioning proteins within cancer cells; these discoveries allowed drug development to be pursued logically and methodically. In other cases, strokes of good luck or intuitive leaps enabled the development of highly active compounds. Inevitably, these anecdotes represent arbitrary choices and draw from a vast pool of agents currently under investigation or development. They represent the forerunners of many such drugs that will surely be developed and licensed for clinical use in the years to come.

17.12 Targeting Bcl-2 to induce cell death

Arguably the most effective way to treat a tumor is to induce death in its constituent neoplastic cells, specifically to activate the apoptotic cell death program described in detail in Chapter 9. As revealed there, decisions of life versus death in the cell often depend on the functional balances between countervailing pro- and anti-apoptotic proteins, most directly depicted in Figure 9.25C. As described there, these mutually antagonistic proteins are all members of the Bcl-2 family, and they all operate in one or another way to open outer mitochondrial membrane pores (pro-apoptotic) or to hold these pores shut (anti-apoptotic). The presence of Bcl-2 translocations and deregulated expression in B-cell malignancies, coupled with an understanding of anti-apoptotic Bcl-2 function at the mitochondrial outer membrane, suggested that an ongoing block to apoptosis may represent a core dependency in certain cancers.

As was discussed, this hypothesis was tested in mouse models in which Myc and Bcl-2 coordinately promoted B-cell lymphoma development (Section 9.13). Genetic deletion of the Bcl-2 transgene expression in these tumors led to rapid apoptosis and tumor regression. Thus, not only could Bcl-2 promote tumor development, but it was also necessary for ongoing tumor maintenance and, consequently, rose to the level of a highly attractive target for anti-cancer drug development. However, a strategy to target Bcl-2 was complicated because it lacked any targetable catalytic site or drug-binding structural pocket. Attempts to use antisense nucleic acids targeting Bcl-2 mRNA were undertaken in the clinic, but these approaches achieved limited overall success.

A key breakthrough toward understanding how to target anti-apoptotic Bcl-2 members came from examining the interaction between BH3-only Bcl-2 family members and their anti-apoptotic targets. The observation that small peptide domains were able to block Bcl-2 interactions with Bax led to the development of short BH3 peptide mimetics that could recapitulate some of the activity of BH3-only proteins (see Figures 9.23 and 9.25). These BH3 peptide mimetics have since become an effective tool for examining the propensity of specific cells to undergo apoptosis in response to diverse stimuli including anti-cancer agents (Supplementary Sidebar 17.10). Using a chemical screening approach based on **nuclear magnetic resonance**, a small molecule BH3 mimetic termed ABT-737 was developed. This molecule represented one of the first anti-cancer agents that was developed based on structural prediction and simulation. Molecules were identified that could recapitulate two distinct binding interfaces of the BH3-only protein Bim bound to the BH2 and BH4 pockets of Bcl-X_L. These two binding fragments were then cleverly linked together to generate the final ABT-737 molecule (**Figure 17.27A**). ABT-737 bound Bcl-2 and Bcl-X_L with high affinity and showed potent single-agent killing in a range of tumor cell lines and murine tumors (Figure 17.27B).

An orally available version of ABT-737, termed ABT-263 (Figure 17.27C), was subsequently developed and introduced into clinical trials as a **monotherapy** in patients with chronic lymphocytic leukemia (CLL). Here, as occurs in the context of many clinical trials, the agent produced an unanticipated toxicity. Although effective in eliminating leukemia cells, ABT-263 also demonstrated significant platelet toxicity. The resulting **thrombocytopenia** (and associated risk of hemorrhages) restricted the ability to introduce larger doses of ABT-263 and limited the overall efficacy of this drug.

Although this type of setback in a clinical trial represents the end of many drug development programs, a broad understanding of Bcl-2-family protein biology provided a key insight into this toxicity. Specifically, Bcl-X_L was determined to be essential for the survival of platelets and, indeed, ABT-263 was found to target both Bcl-2 (its intended target) and Bcl-X_L. This finding led to further efforts to apply the same approach used previously to formulate ABT-737 in order to generate a Bcl-2 specific inhibitor. The result was the development of ABT-199 (or venetoclax), a selective, orally bioavailable and potent inhibitor of Bcl-2 that lacks significant platelet toxicity (**Figure 17.28A**).

Venetoclax has achieved clinical success in the treatment of CLLs that lack chromosome 17p, an alteration that most notably results in the loss of p53 expression (Figure 17.28B). Notably, 17p loss does not, per se, sensitize these cancers to Bcl-2 inhibition. Instead, CLLs with 17p loss are highly resistant to existing standard-of-care therapies, making it possible to rapidly identify a subset of patients who would not benefit from existing chemotherapy. As is the case for many cancers, it was the absence of effective alternative treatments that provided the opportunity and impetus to try new approaches.

Soon extensive experimental data showed that overexpressed Bcl-2 could promote neoplastic progression in a broad set of hematopoietic cell types, explaining why

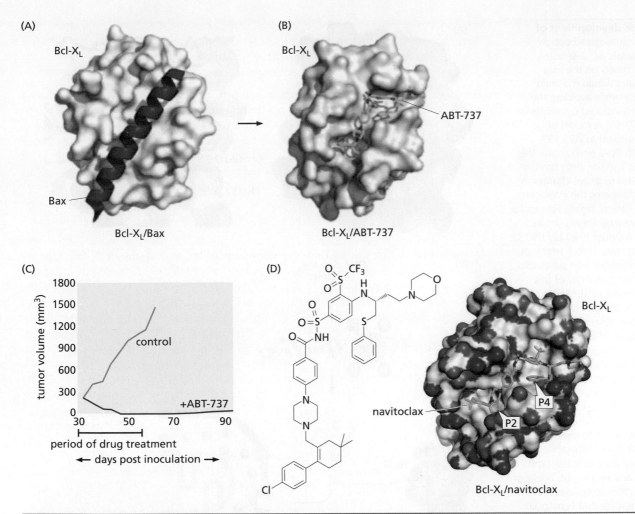

Figure 17.27 The development of anti-apoptotic Bcl-2-family inhibitors (A) The drug termed ABT-737 was developed using nuclear magnetic resonance (NMR)-based screening, parallel synthesis (a form of combinatorial chemistry), and structure-based design, undertaken to visualize the interacting structures of pro-apoptotic Bcl-2 family members with the anti-apoptotic protein Bcl-X_L. The α-helix of the pro-apoptotic Bax protein (*red helix*; see also Figures 9.23 and 9.25) binds to a groove in Bcl-X_L protein (*white/gray surface*) and inhibits its anti-apoptotic function. (B) This groove is also occupied by ABT-737, a small molecule (*green stick figure*)—specifically designed to mimic and thus compete with this protein–protein interaction. (C) The effects of ABT-737 on the growth of a human small-cell lung carcinoma (SCLC) xenograft are shown in the graph. The black bar indicates the time window during which ABT-737 was applied. (D) An orally bioavailable compound navitoclax (also known as ABT-263, *green*), an inhibitor of the Bcl-2, Bcl-X_L and MCL1 anti-apoptotic proteins, was subsequently generated. Shown is an X-ray co-crystal structure of navitoclax (*green stick figure*) bound to Bcl-X_L (*red, white, blue*) with critical drug-binding pockets indicated as P2 and P4. Because navitoclax inhibits Bcl-X_L, a protein important for blood platelet life span, dose-limiting thrombocytopenia has presented a barrier for expanded clinical use of this agent. (A, from N. Chessum et al., *Prog Med Chem*.54:1–63, 2015. With permission from Elsevier. B, from A. Ashkenazi et al., *Nat. Rev. Drug Discov.* 16:273–284, 2017. With permission from Elsevier. C, from T. Oltersdorf et al., *Nature* 435:677–681, 2005. With permission from Nature. D, A. Ashkenazi et al., *Nat. Rev. Drug Discov.* 16:273–285, 2017. With permission from Nature.)

much of the pre-clinical and clinical efforts to explore the clinical utility of venetoclax has focused on blood cancers. These treatment studies included patients with acute myelogenous leukemia (AML), a cancer that is currently treated with conventional combination chemotherapy and has a generally poor prognosis. Venetoclax, in combination with other agents, has shown significant clinical efficacy in AML patients who cannot tolerate the toxicity associated with high-dose chemotherapy. Here again, it was the inability to treat patients with a standard-of-care therapy that provided the opportunity and incentive to test a new agent. Although it remains to be seen if this treatment can supplant standard-of-care regimens in this disease, the introduction of an alternative to toxic combination-chemotherapy represents an important development for the treatment of AML.

Figure 17.28 The development of venetoclax Structure-based design was used to generate Bcl-2-selective small molecules, based on the idea that such molecules would maintain anti-tumor activity while avoiding the thrombocytopenia induced by navitoclax (see Figure 17.27). (A) A model of venetoclax (also known as ABT-199, *yellow*) binding to Bcl-2 (*left*) shows the Bcl-2 electrostatic surface, with positive charges in *blue* and negative charges in *red*. The image compares the interacting residues in Bcl-2 (*green*, *right*), venetoclax (*yellow*), and navitoclax (*blue-purple*). Here, the amino residues Phe112, Thr132, and Glu136 (latter two in *red, white*) of Bcl-2 specifically interact with venetoclax, which underlies the specificity of this agent as a Bcl-2 inhibitor. (B) The diagram shows pathways leading to induction of apoptosis. In the absence of p53-dependent induction of BH3-only proteins (notably Bax), Bcl-2 holds mitochondrial outer membrane channels shut, preventing release of cytochrome c, and cellular stresses fail to induce cell death; this explains why malignancies like p53-deficient chronic lymphocytic leukemia (CLL) are highly resistant to genotoxic agents. Venetoclax blocks the ability of Bcl-2 to inhibit pro-apoptotic Bax (and Bak, *not shown*) and renders p53-deficient tumor cells vulnerable to cellular stress. Results to date indicate that venetoclax functions most effectively to sensitize cancer cells to the killing effects of FDA-approved cytotoxic drugs. (A, from J. Ramos et al., *Int. J. Mol. Sci.* 20:860–879, 2019. B, adapted from M. Suzuki, R.J. Youle and N. Tjandra, *Cell* 103:645–654, 2000. With permission from Elsevier.)

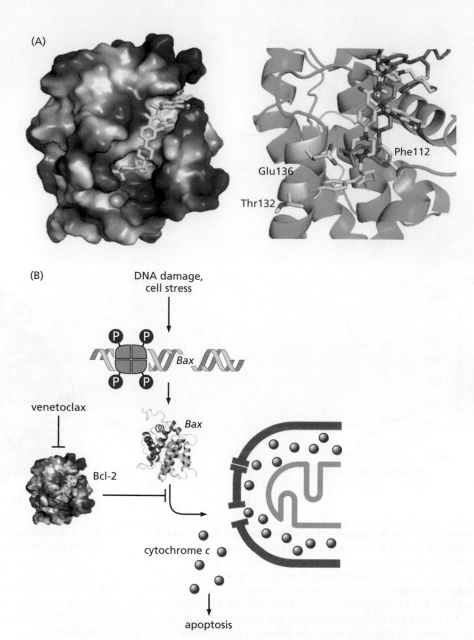

17.13 Gleevec paved the way for the development of many other highly targeted compounds

In the previous sections, we made repeated reference to the Bcr-Abl oncoprotein and to experimental strategies for antagonizing it. Now, we backtrack and review the history of how the Bcr-Abl oncoprotein was discovered and validated as an attractive drug target and finally used as an object of rational drug design. This story is valuable, if only because it illustrates vividly the long course through which drug development passes from initial discovery at the laboratory bench to the oncology clinic.

This particular story begins in 1914, when the German cytologist Theodor Boveri proposed that chromosomal defects might cause a cell to proliferate abnormally, resulting ultimately in the formation of some type of cancer. Almost half a century passed before Boveri's idea received some validation. In 1960, two cytologists working in Philadelphia noted that an abnormal, unusually small Chromosome 22 was characteristically present in the great majority of cells of chronic myelogenous leukemia (CML); since that time, it has been called the Philadelphia chromosome or simply Ph. It took another dozen years before a researcher in Chicago demonstrated that a reciprocal translocation between Chromosomes 9 and 22 was responsible for creating the

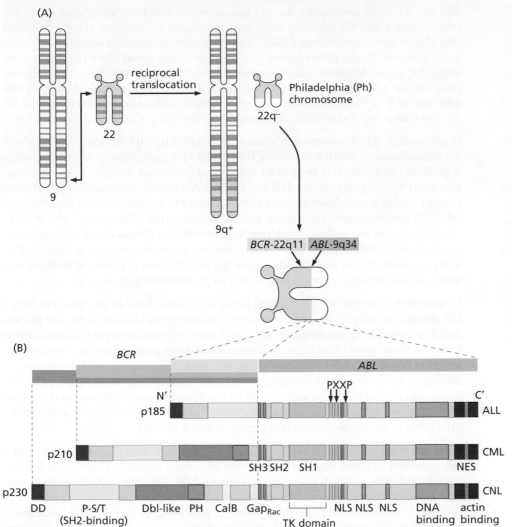

Figure 17.29 Origin and structure of the Bcr-Abl protein
(A) More than 95% of cases of chronic myelogenous leukemia (CML) exhibit the Philadelphia chromosome, which results from a reciprocal translocation between Chromosomes 9 and 22. The q34 region of Chromosome 9 carrying most of the *ABL* gene is transferred to the q11 region of Chromosome 22, replacing a larger segment of Chromosome 22 that is translocated reciprocally to Chromosome 9. The net result is a truncated Chromosome 22 (i.e., 22q⁻), often termed the Philadelphia chromosome (Ph), and a fusion of the 5′ portion of the *ABL* gene with a 3′-proximal portion of the *BCR* gene, which normally resides at 22q11. (B) Depending on the precise location of the breakpoint in *BCR*, three distinct Bcr-Abl fusion proteins may be formed; these are found in ALL (acute lymphoblastic leukemia), CML, and CNL (chronic neutrophilic leukemia). Each of these *BCR-ABL* fusion genes encodes a multidomain (and thus multifunctional) protein. (From A.S. Advani and A.M. Pendergast, *Leuk. Res.* 26:713–720, 2002. With permission from Elsevier.)

Ph chromosome (see Section 4.5). (A larger chunk of Chromosome 22 is donated to the tip of Chromosome 9 than is received from this chromosome, leaving the already-tiny Chromosome 22 even more diminished in size; this remnant of 22 plus the small translocated segment constitutes the Ph chromosome; **Figure 17.29A**). The chromosomal aberration—clearly representing a somatic mutation—was proposed as a potential cause of this malignancy. As mentioned earlier, we now know that this particular translocation is present in more than 95% of cases of CML.

The genes that were fused through this translocation remained unknown until 1982, when molecular biologists discovered that *ABL*, the human homolog of the mouse c-*abl* proto-oncogene (see Tables 4.1 and 4.5), participates directly in these chromosomal translocations, becoming fused with a second, still unknown gene. The breakpoints of this other gene (the chromosomal sites at which it becomes fused to the *ABL* gene) were soon found to be scattered over many kilobases of DNA, yielding the name "**b**reakpoint **c**luster **r**egion" or simply *BCR*. In fact, three distinct fusion proteins arise through the inclusion of variously sized Bcr proteins at the N-termini of the fusion proteins, with almost the entire Abl protein at the C-termini (see Figures 4.11 and 17.29B). As indicated in the figure, the different fusion proteins tend to be associated with distinct types of leukemia.

Within two years of its discovery, the Bcr-Abl protein was found to function as a constitutively activated tyrosine kinase. In this respect, it functions like the Abl oncoprotein of Abelson mouse leukemia virus. Like other acutely transforming retroviruses, the genome of this virus carries an oncogene—in this case *abl*—derived from a corresponding proto-oncogene residing in the normal mouse genome (see Section 3.10).

By 1990, a cDNA encoding the Bcr-Abl fusion protein had been introduced into a retrovirus vector, and the resulting virus was then found to induce in mice a leukemia that closely resembled human CML. Like the human disease, this leukemia involved large numbers of fully differentiated granulocytes in the blood. Under certain conditions, the mouse leukemia, like its human counterpart, progressed to a "blast crisis," involving the accumulation of immature cells of the lymphoid or myeloid lineages (see also Sidebar 8.7). These observations in mice represented the first formal proof that the Bcr-Abl fusion protein operates as the central motive force of leukemogenesis in CML.

Unfortunately, this demonstration of the critical role of Bcr-Abl revealed nothing about the mechanisms by which it functions. The bewildering complexity of Bcr-Abl signaling is indicated by the diverse array of structural and functional domains in the two contributing proteins (see Figures 4.11B and 17.29B). Altogether, the domains of this fusion protein enable it to activate the Ras pathway, the PI3 kinase–Akt/PKB pathway, the Jak–STAT pathway, and transcription factors, including Jun, Myc, and NF-κB. In addition, the Ras-like Rac protein, which regulates activities as diverse as cellular migration, survival, and proliferation, is activated, as are two nonreceptor tyrosine kinases, Hck and Fes. These various associations enable the Bcr-Abl protein to extend its reach into almost all the regulatory circuits governing cell proliferation and survival.

In spite of this complexity, the tyrosine kinase domain of Bcr-Abl was found to be the key element in leukemogenesis. For example, subtle alterations of the Bcr-Abl protein that inactivated its tyrosine kinase catalytic activity led to total loss of its transforming function. In the early 1990s, a research program was begun to develop low–molecular-weight antagonists of the Bcr-Abl tyrosine kinase activity. A drug molecule emerged, termed variously imatinib mesylate, STI-571, Glivec, and Gleevec (see Figure 17.13A), which was able to bind the catalytic cleft of the Bcr-Abl tyrosine kinase. As is the case with all other kinases of this family, the cleft is located between the two major structural lobes of the kinase protein (see Figures 6.6 and 17.13B).

Even though the Abl kinase domain shares roughly 42% amino acid identity with a large number of other tyrosine kinases, the inhibitory effects of Gleevec on Bcr-Abl were found to be relatively specific (see Figure 17.16). Subsequently, four other tyrosine kinases—those belonging to the PDGF (α and β) and Kit receptors as well as the Arg (Abelson-related gene) protein—were also found to be inhibited by Gleevec. Accordingly, this drug, when used at therapeutic concentrations, appears to target only 4 of the 90 or so human tyrosine kinases. Like most other kinase inhibitors, the Gleevec molecule associates directly with the ATP-binding pocket of the Abl kinase domain (see Figure 17.13). Whereas other kinase inhibitors block ATP binding in this cleft, Gleevec works differently: it binds and stabilizes a catalytically inactive conformation of this enzyme (much like the small drug molecule that inhibits the G12C mutant Ras protein; see Section 17.4).

The success with Gleevec encouraged other attempts at creating low–molecular-weight kinase antagonists, which were realized to have certain therapeutic advantages over anti-receptor monoclonal antibodies (Table 17.7). In addition, this advance stimulated efforts to make narrowly targeted tyrosine kinase inhibitors, some of which have indeed demonstrated extraordinary specificity. For example, even though the tyrosine kinase domains of the insulin receptor (IR) and insulin-like growth factor receptor (IGF-1R) are very similar in structure (84% identity in kinase domain residues), a low–molecular-weight inhibitor has been developed that targets preferentially the IGF-1R, doing so at a concentration that is 27-fold lower than that required for it to inhibit the IR.

By 1996, Gleevec had been found able to inhibit the growth of CML cells *in vitro* while having no effect on normal bone marrow cells. More specifically, the proliferation of Bcr-Abl–dependent cells could be inhibited at drug concentrations as low as 40 nM, indicating a high affinity of Gleevec for the catalytic cleft of the tyrosine kinase domain. (Cells that depend on Bcr-Abl for survival can be forced into apoptosis by Gleevec's inhibition of Abl kinase function.) The initial clinical trials, begun in 1998, revealed remissions from disease in all 31 treated CML patients, with only minimal side effects, even when taken daily for many years. Four years later, 6000 patients were participating in Gleevec clinical trials.

Table 17.7 Strengths and weaknesses of anti-receptor antibodies versus low–molecular-weight tyrosine kinase inhibitors as anti-cancer agents

	Small molecule	Antibody
Target	tyrosine kinase domain	receptor ectodomain
Specificity	+++	++++
Binding	most are rapidly reversible	receptor internalized, only slowly regenerated
Dosing	oral daily	intravenous, ≤ weekly
Distribution in tissues	more complete	less complete
Toxicity	rash, diarrhea, pulmonary	rash, allergy
Antibody-dependent cellular cytotoxicity	no	possibly

Courtesy of N.J. Meropol and from N. Damjanov and N. Meropol, *Oncol. (Huntington)* 18:479–488, 2004.

Treatment of early-stage (chronic) CML with Gleevec leads to a hematologic response in 90% of cases: microscopic analysis of blood smears reveals a profound shift in the cellular composition of the blood (**Figure 17.30A**), and PCR analysis reveals an extraordinary decline in the levels of the *Bcr-Abl* mRNA detectable in blood cells (Figure 17.30B). In 50% of these cases, the translocated Philadelphia chromosome

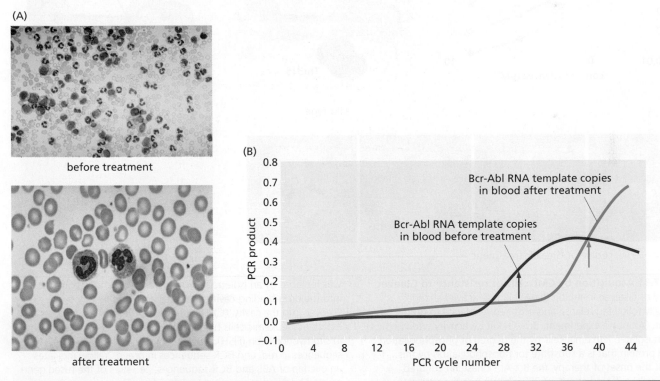

Figure 17.30 Measuring clinical responses to Gleevec treatment (A) The successes of Gleevec in treating chronic myelogenous leukemia (CML) can be gauged from cytological analyses of the patients' blood. As seen here, treatment with Gleevec converted the blood from a state with many leukemia cells (large, dark nuclei, *above*) to one in which only normal granulocytes are visible (*below*) among the red blood cells. (B) A more sensitive and quantitative measure of therapeutic success comes from use of quantitative polymerase chain reaction (qPCR) measurements of the level of Bcr-Abl mRNA (which is reverse-transcribed prior to PCR amplification). In an untreated patient (*red curve*), 50% of maximum (*red arrow*) PCR-mediated gene amplification is observed at about the 29th cycle of gene amplification (in which each cycle results in the doubling of the amplified sequence). Following Gleevec treatment (*blue curve*), a comparable degree of amplification is achieved only at about the 39th cycle (*blue arrow*), indicating that the Bcr-Abl mRNA–expressing cells are present at a level that has been reduced by a factor of about 2^{10}. PCR-based assays can detect as few as one CML cell amid 10^5 to 10^6 normal blood cells. (A and B, courtesy of B.J. Druker.)

is no longer detectable by karyotypic analyses of patients' white cells. About 60% of the patients who have already progressed to blast crisis respond to Gleevec, but they generally relapse after some months. In contrast, a clinical study reported in 2006 that five years after chronic-phase patients began long-term treatment with Gleevec, fewer than 5% of patients had died of CML-related effects. By this fifth year, the annual relapse rate (that is, progression to blast crisis) in this population was ~0.6%.

The molecular mechanisms that allow tumor cells to eventually escape Gleevec inhibition are interesting, in that they shed further light on the Bcr-Abl oncoprotein and its actions and, more generally, they reveal how cancer cells can acquire resistance to highly targeted drugs. Initial analyses of *Bcr-Abl* sequences in the tumors of patients with Gleevec-resistant, relapsed disease revealed that 29 of 32 tumors harbored mutations in the *Bcr-Abl* gene; altogether these yielded substitutions of 13 distinct amino acid residues in the kinase domain. Another dozen have been cataloged in subsequent studies.

Some of these mutations prevent Gleevec from binding to the Abl catalytic cleft, either by directly interfering with its binding or, less directly, by promoting protein conformational changes that affect the dynamics of binding (**Figure 17.31A and B**). In a minority of patients, Gleevec resistance was achieved through amplification of the

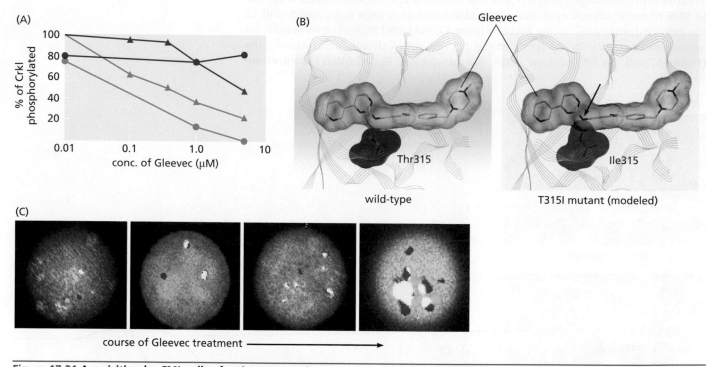

Figure 17.31 Acquisition by CML cells of resistance to Gleevec The ability of Gleevec to inhibit Bcr-Abl kinase activity changes dramatically following relapse and acquired resistance to drug treatment. (A) In this experiment, Bcr-Abl kinase activity was gauged in isolated leukemia cells by the degree of phosphorylation of Crkl, a protein that is a substrate for phosphorylation by this kinase. At the onset of therapy, the Bcr-Abl kinase (in cultured leukemia cells) suffered about a 50% inhibition in the presence of approximately 0.1 μM Gleevec in two patients (*blue triangles, blue circles*). However, after Gleevec resistance developed in these patients, about an 8-μM concentration of the drug was required to inhibit one patient's Bcr-Abl kinase (*red triangles*), while the other patient was totally resistant to the drug (*red circles*). (B) The Gleevec molecule is able to nestle tightly in a molecular cavity created in part by a threonine residue (*dark blue*) in residue position 315 of the wild-type Bcr-Abl oncoprotein (*left*; see also Figure 16.10C). However, in a mutant Bcr-Abl found in the leukemic cells of a Gleevec-resistant patient (*right*), this threonine residue was replaced by an isoleucine (*brown, arrow*), which protrudes into the drug-binding cavity and interferes with insertion of Gleevec into the cavity. (C) The number of *Bcr-Abl* gene copies in a patient's leukemic cells has been gauged here using fluorescence *in situ* hybridization (FISH). Nuclei are visualized here in *blue*, ABL sequences in *red*, and BCR sequences in *green*. Yellow indicates an overlap of ABL and BCR sequences, i.e., sites of the fused gene created by the chromosomal translocation. The copy numbers of the fused gene (*yellow*) at the beginning of therapy (*left*) were quite low, but as treatment proceeded (*rightward*), the copies of the fused gene (and therefore Bcr-Abl fusion protein) increased progressively until the patient's leukemia became resistant to Gleevec treatment. In this particular patient, Gleevec resistance was acquired by the tumor cells because the fusion protein became overexpressed, thereby exceeding the ability of the normally used therapeutic concentration of Gleevec to bind and fully inactivate it. (From M.E. Gorre et al., *Science* 293:876–880, 2001. With permission from AAAS.)

Bcr-Abl gene in their leukemia cells, yielding increased levels of the encoded oncoprotein that apparently could no longer be inhibited by the concentrations of drug used to treat patients (see Figure 17.31C).

The observations that acquired resistance to Gleevec is usually accompanied by either structural alterations or overexpression of the Bcr-Abl protein provide compelling proof that Gleevec's therapeutic responses can be attributed directly to its effects on the Bcr-Abl protein. This insight was further explored by introducing random mutations into a vector encoding the Bcr-Abl protein and then determining which of the resulting mutant forms of this protein were able to resist inhibition by Gleevec (**Figure 17.32**). Such an experimental strategy, which uses cultured cells whose growth and viability are dependent on Bcr-Abl (see Figure 17.21), can in principle reveal the full spectrum of structural alterations of Bcr-Abl that are capable of rendering it resistant to Gleevec inhibition and thereby advance our understanding of the molecular mechanisms underlying acquired drug resistance.

A number of second-generation Bcr-Abl inhibitors have been synthesized that succeed in inhibiting many of the mutant Gleevec-resistant Bcr-Abl proteins that arise in CML patients and trigger clinical relapse; among these are nilotinib and dasatinib, which are actually more potent than Gleevec/imatinib. However, they and other newly developed Abl inhibitors fail to shut down the most formidable of the mutant proteins, termed T315I (**Figure 17.33A and B**). Detailed structural analyses of this mutant protein and its effects on drug binding have, after much effort, yielded novel compounds, including ponatinib, that can also shut down this particular mutant kinase (Figure 17.33C), illustrating once again how a convergence of structural biology and synthetic organic chemistry can often cope with such refractory mutant enzymes.

The ability of Gleevec to also inhibit the platelet-derived growth factor receptors (PDGF-Rα and β) suggested that it might prove useful in treating other types of malignancies as well. For example, translocations of the genes encoding these two receptors that cause constitutive receptor activation have been found in a number of chronic **myeloproliferative** diseases, that is, conditions involving elevated levels in the circulation of one or another cell type arising from the myeloid lineage of hematopoiesis. Indeed, patients suffering from **hypereosinophilic** syndrome that have PDGF-R translocations have shown a **complete response** following Gleevec treatment, with virtual disappearance of their eosinophils.

The growth of many of the far more common glioblastomas is driven by PDGF–PDGF-R autocrine loops (see Figure 5.8C). In this instance, however, the use of Gleevec administered together with cytotoxic drugs has not proven encouraging, possibly because it cannot efficiently penetrate the **blood–brain barrier** and gain access to the tumor parenchyma.

Gleevec's effects on a third tyrosine kinase—the Kit receptor—also make it an attractive agent for attacking gastrointestinal stromal tumors (GISTs), a relatively uncommon sarcomatous tumor for which few therapeutic options have been available. The Kit receptor is mutated in the majority (~85%) of these cancers, while a minority (3–5%) exhibit mutant PDGF-Rα—both targets of Gleevec. The mutant receptors fire constitutively in these tumors and seem to represent the primary mitogenic drivers operating in the tumor cells (see Figure 5.17). In one study, clear regression of the tumor was observed in almost 70% of treated patients (**Figure 17.34A**). By 2005, SU11248—a second inhibitor of Kit tyrosine kinase function—was approved by the FDA for the treatment of GISTs, including those that had developed a resistance to Gleevec. When used as an adjuvant following surgical excision of localized primary tumors, Gleevec treatment offers long-term survival that may result in a cure. With metastatic disease, however, Gleevec can cause regression of GIST lesions, but within a year or two, tumors reappear (see Figure 17.34B).

The clear successes of Gleevec represented the first validation that rational drug design can succeed in producing agents that are highly useful for treating various types of human cancer. It was initially viewed as a disadvantage that Gleevec interferes with multiple tyrosine kinases, because it was feared that this broader activity would lead

Figure 17.32 Screening *in vitro* for Gleevec-resistant mutant forms of Bcr-Abl One strategy for detecting drug-resistant variants of Bcr-Abl involves cultured cells such as BaF3 cells (see Figure 16.15), whose continued survival can be made dependent on the presence of a functionally active Bcr-Abl oncoprotein. When such Bcr-Abl–expressing cells are treated with Gleevec, they are killed because of their dependence on continued Bcr-Abl signaling. (A) A cDNA clone expressing the "wild-type" Bcr-Abl protein (i.e., the direct product of the chromosomal translocation) can be mutagenized by passage through *E. coli* bacteria that are highly error-prone in DNA replication and therefore generate mutant variants of the introduced, plasmid-borne *Bcr-Abl* sequence. The resulting collection of randomly mutated Bcr-Abl–expressing clones is then introduced, via a retrovirus vector, into BaF3 cells, which are then exposed to Gleevec. The rare cells that resist being killed by Gleevec are isolated, and the sequence of the mutant Bcr-Abl protein that conferred Gleevec resistance is determined. (B) When these Gleevec-resistant mutant Bcr-Abl proteins are analyzed, many are found to have single amino acid substitutions of residues located throughout the Abl kinase domain. The "back" and "front" of the Abl kinase domain are shown here, together with the sites of these mutant residues and the identities of the normally present amino acid residues. Surprisingly, the individual mutations, each of which confers Gleevec resistance, alter residues at many sites in the ABL domain, indicating that CML cells have multiple options for developing drug resistance. Many of these mutant residues are found on the back side of Abl opposite the catalytic cleft; some of these residues (*red*) participate in interactions between the kinase (i.e., SH1) domain and the SH2 and SH3 domains of Abl (*not shown*). Other mutant sites (*blue*) produce drug resistance through poorly understood mechanisms. This *in vitro* screen for Gleevec-resistant Bcr-Abl mutants revealed most of those discovered in patients plus a number of previously undocumented ones. (From M. Azam, R.R. Latek and G.Q. Daley, *Cell* 112:831–843, 2003. With permission from Elsevier.)

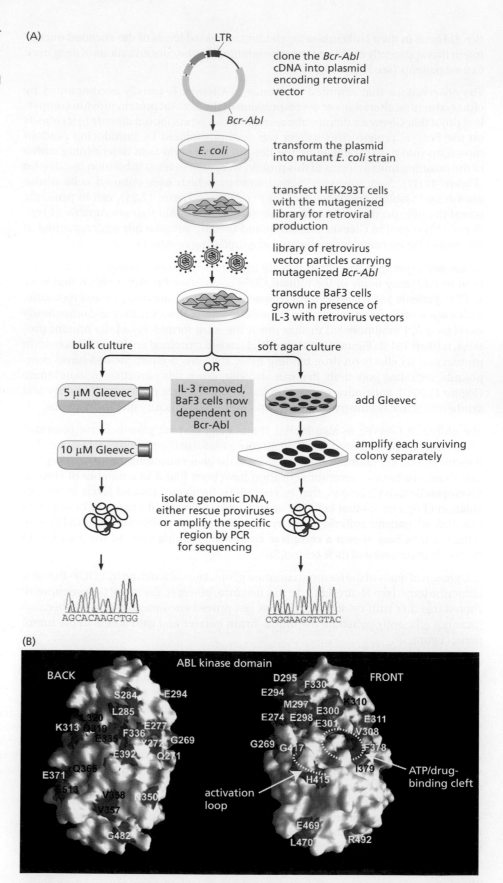

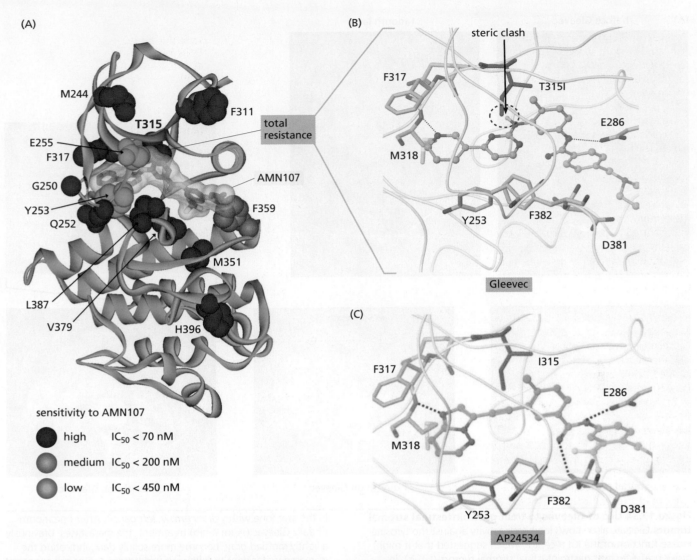

(A)

M244

F311

T315

total
resistance

E255
F317

G250

Y253

Q252

AMN107

F359

M351

L387

V379

H396

sensitivity to AMN107

● high $IC_{50} < 70$ nM

● medium $IC_{50} < 200$ nM

● low $IC_{50} < 450$ nM

(B)

steric clash

F317

T315I

E286

M318

Y253   F382

D381

Gleevec

(C)

F317

I315

E286

M318

Y253   F382   D381

AP24534

Figure 17.33 Backup inhibitors of Bcr-Abl for patients with Gleevec-resistant tumors The fact that patients in the acute (blast crisis) phase of CML often develop resistance to Gleevec (see, for example, Figure 17.31) has stimulated the development of alternative inhibitors of the Abl tyrosine kinase. (A) One of these inhibitors, AMN107, which is ~20 times more potent than Gleevec against unmutated Bcr-Abl, is shown here (*orange stick figure in yellow envelope*) in complex with the tyrosine kinase domain of Bcr-Abl, on which are also indicated the sites of a number of amino acid substitutions found in the mutant forms of Bcr-Abl in tumors of Gleevec-resistant patients. (The number of colored spheres at a site indicates the number of atoms present in the side chain of the substituted amino acid.) The locations of the amino acid substitutions carried by the mutant, Gleevec-resistant Bcr-Abl proteins are indicated by the red, orange, and green spheres and show different levels of sensitivity to inhibition by AMN107

(*below*). One Gleevec-resistant mutant form of Bcr-Abl in which the normally present threonine is replaced by an isoleucine (T315I; *blue spheres*) is also totally resistant to AMN107 and most other second-generation Bcr-Abl inhibitors. (IC_{50}, concentration required for 50% inhibition of the target molecule.) (B) As seen here, the side chain of the isoleucine (*red*) present in the T315I mutant Bcr-Abl protein creates a steric clash with a hydrogen atom of Gleevec in the drug-binding pocket (see Figure 17.31B), effectively precluding binding by Gleevec (*left*). (C) In response to this resistance of the T315I mutant, a pharmacologically active derivative of Gleevec, termed AP24534, avoids the steric clash and therefore can bind the T315I mutant protein. (A, from T. O'Hare et al., *Cancer Res.* 65:4500–4505, 2005. With permission from American Association for Cancer Research. B and C, from T. O'Hare et al., *Cancer Cell* 16:401–412, 2009. With permission from Elsevier.)

to unacceptable side effects. However, with the passage of time, it is increasingly clear that such multi-target effects may actually prove useful in treating certain malignancies. Indeed, second and third generation Bcr-Abl inhibitors engage an even broader range of kinase targets. Dasatinib, for example, targets Abl, Src kinase, and c-Kit at dose ranges of less than 1 nanomolar. Here, as in many examples of kinase inhibitors, higher affinity interaction for a given target is accompanied by increased affinity to many targets (and, consequently, decreased overall drug specificity). This broader

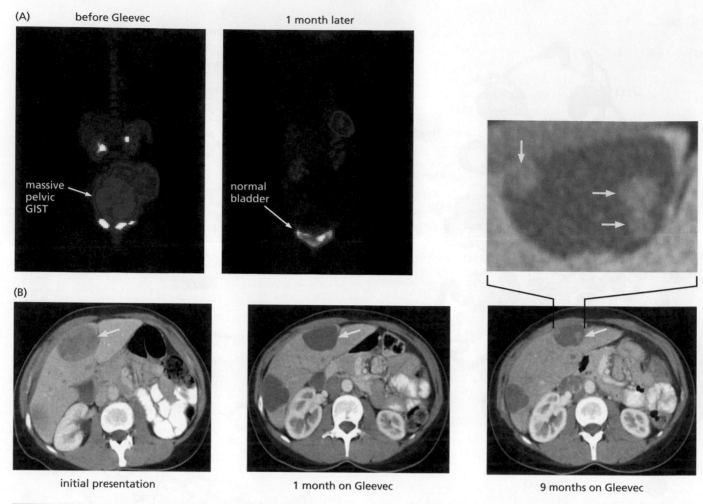

Figure 17.34 Use of Gleevec to treat gastrointestinal stromal tumors Gleevec also shows inhibitory activity against the tyrosine kinase function of the Kit receptor, which suggested that it might prove useful against gastrointestinal stromal tumors (GISTs), in which mutant, constitutively active Kit receptors are commonly found. (A) As seen here, this patient's GIST (red mass, pelvic region, *left image*), which was visualized because of its uptake of a labeled glucose analog, responded dramatically to Gleevec treatment (*right image*). (The residual labeling following treatment reflects the accumulation of the labeled dye in the patient's bladder.) (B) Despite initial anti-tumor activity in nearly 90% of patients, resistance to Gleevec develops in most GIST patients; after two years, about 50% of tumors worsen despite continuing Gleevec treatment. Seen here are computed tomography (CT) scans of a patient with advanced GIST demonstrating several metastatic lesions growing in the liver (one with *yellow arrow, left panel*). After one month of daily Gleevec (= imatinib) treatment, the metastases, previously a lightly mottled *gray*, become more solidly dark, indicating the increased water density following widespread tumor cell apoptosis and degeneration (*middle panel*). However, after nine months of Gleevec (*right panel*), three light spots have appeared within the dark mass that was the original tumor (*yellow arrow*), indicating the eruption of robustly growing subpopulations that signal the onset of clinical resistance to targeted therapy (*yellow arrows, top right*). In this particular patient, each of the three recurring growths carried its own novel secondary point mutation in the *Kit* gene in addition to the initiating *Kit* mutation that launched the tumor in the first place, leading in each case to a doubly mutated *KIT* allele. (Courtesy of G.D. Demetri.)

activity plays a critical role in the treatment of another Bcr-Abl-driven malignancy, a Philadephia chromosome-positive subset of B-cell leukemia, termed Ph+ B-ALL. Unlike CML, Gleevec-treated Ph+ B-cell acute lymphocytic leukemias frequently relapse during treatment and, in roughly one-third of these cancers, relapses occur in the absence of Bcr-Abl mutations. Dasatinib and ponatinib have proven more effective in treating these leukemias, presumably because they are able to inhibit both Bcr-Abl and other pro-survival kinase-driven pathways operating in these cells.

The successes of ponatinib in targeting CMLs bearing T315I mutations and Ph+ B-ALLs led to attempted use of this agent as **first-line therapy** for these cancers. However, clinical trials of this agent were halted after bleeding issues from thrombocytopenia and stenosis in small and large blood vessels appeared. Use of this agent was subsequently limited to the treatment of treatment-resistant CML and Ph+ B-ALL. Thus, the viability and proliferation of the neoplastic cells in many tumors depend on the coordinated actions of multiple tyrosine kinases, and the ability to strike at several of these simultaneously may one day be found to confer great therapeutic advantage. However, broadly acting kinase inhibitors can also yield unexpected toxicity that limits or prevents their use in the oncology clinic.

Unfortunately, in CML, the continued presence of cancer stem cells (CSCs) limits the ability of Gleevec to definitively cure patients (Sidebar 17.13). Recall that research on human cancers, including hematopoietic tumors, breast carcinomas, and brain tumors, has revealed that CSCs often constitute only a small proportion (<<5%) of the neoplastic cells in these tumors and that their existence can be revealed only by the biological test of their tumor-forming ability or by the use of fluorescence-activated cell sorting (FACS; see Section 11.7). As it happens, Gleevec is quite potent in killing actively cycling leukemia cells (that is, the "transit-amplifying" or "progenitor" cells; see Section 11.7). However, most of the cells in the neoplastic stem cell population, which are outside the active cell cycle at any single point in time (see Section 8.1), are not dependent on Bcr-Abl-mediated signaling for their continued survival and have proven to be quite resistant to drug treatment.

The CSCs are usually present in minute numbers in a CML patient undergoing Gleevec therapy, being detectable only by highly sensitive PCR-based assays. However, if treatment is halted, the CSCs usually re-enter into the growth-and-division cycle and regenerate transit-amplifying progeny, soon leading to regrowth of the malignancy and clinical relapse (Figure 17.35A). This seems to explain why Gleevec treatment needs to be chronic and why, in the future, drug development needs to be focused on agents that strike at the cores of tumors by destroying their stem cells (Sidebar 17.13).

As mentioned, Gleevec also fails to effectively treat CML that has progressed to an advanced **blast crisis** phase. In this phase, tumor cells no longer exhibit a myeloid-like appearance and resemble, instead, undifferentiated progenitor cells. Indeed, these cells cytologically resemble tumor cells present in treatment-refractory Ph+ B-ALL. These cells express Bcr-Abl but fail to incur durable responses to Bcr-Abl inhibitors. This explains why patients need to be treated with Gleevec prior to leukemia progression to blast crisis, at which point Gleevec and similarly acting drugs lose much of their efficacy.

Nonetheless, if a drug such as Gleevec succeeds in generating clinical remissions that are durable over many years' time, its inability to kill the stem cells of a CML and the requirement to administer drug early in disease progression is clearly an acceptable shortcoming. In any case, Gleevec represents a major triumph of anti-cancer drug development, because it is vastly superior to all alternative treatments of this otherwise-inexorably progressing disease.

Of note, the striking success of Gleevec in treating CML persuaded many that comparable successes would soon emerge for treating a number of other malignancies driven by deregulated tyrosine kinases. As it turned out, CML is a rather unique entity among adult tumors, in that almost all these leukemias share a common genetic driver (the Bcr-Abl translocation) to which they are all addicted, while carrying few if any additional chromosomal aberrations and oncogenic mutations. In addition, rather than being full-fledged malignant cells, the leukemic cells present in the chronic phase of CML have been likened to *in situ* carcinomas in the breast and early adenomas in the gut (see Section 11.2). These attributes set the chronic-phase cells apart from most adult tumors and help to explain the striking differences between the clinical responses of CML to Gleevec versus the mixed responses of many carcinomas to the other tyrosine kinase inhibitors (TKIs) described below. (In retrospect, the stunning successes of Gleevec inspired unrealistically high expectations for the many TKIs that followed.)

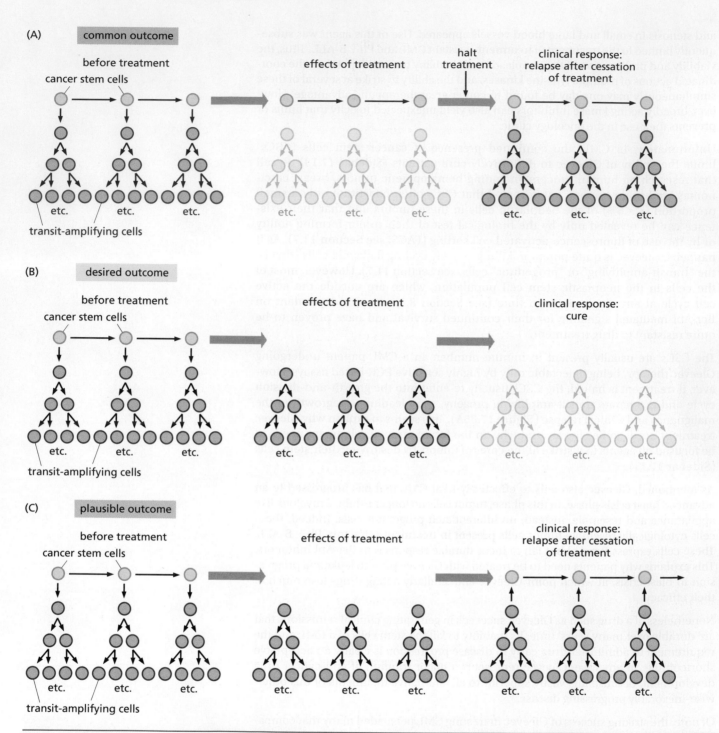

Figure 17.35 Role of tumor stem cells in the response to anti-cancer treatments Although the evidence is still fragmentary, it appears that in many and perhaps all tumors (see Section 11.6), a small proportion of the neoplastic cell population is composed of self-renewing cancer stem cells (*gray*). These spawn the bulk of the cancer cells in tumors (*pink*), which have many of the properties of normal progenitor/transit-amplifying cells, as well as more differentiated progeny. (A) If, as occurs with Gleevec, an anti-cancer therapy results in the depletion of the neoplastic transit-amplifying cells without eliminating the cancer stem cells (CSCs), then the latter can regenerate the tumor soon after that therapy is halted (*right*). Indeed, diverse sources of evidence indicate that in many carcinomas the CSCs are significantly more resistant to therapy than the majority of carcinoma cells. Hence, CSCs are doubly dangerous in that they can survive treatment and are capable of generating entirely new tumors in both primary tumor sites and in distant tissues. (B) In response to the risk posed by CSCs, some researchers have undertaken to discover agents that preferentially eliminate these cells, with the thought that their elimination will lead to the disappearance of entire tumors, because their source of self-renewal has been eliminated (*right*). (C) Elimination of CSCs, by itself, may not lead to a cure because newer evidence has indicated (*right*) that the non-CSCs in certain tumors can spontaneously generate new CSCs. Hence, the targeting of both populations (CSCs and non-CSCs) may be required to develop durable clinical responses following treatment of a variety of tumors.

Sidebar 17.13 Cancer stem cells greatly complicate the evaluation of anti-cancer therapies The existence of cancer stem cells (CSCs) in many solid tumors has profound implications for the evaluation of many types of anti-cancer treatments. These lessons are illustrated graphically in the case of CML and Gleevec. [In the hypothetical case shown in Figure 17.35A, the re-emergence of a tumor is triggered by cessation of treatment, but for most tumors under treatment, such relapses occur because tumor cells have developed resistance to the therapeutic agents.] In carcinomas, for example, the CSCs, which display certain mesenchymal characteristics, seem generally to be more resistant to existing therapeutic regimens than most of the neoplastic cells in these tumors (see Table 17.5). Hence, a therapy may eliminate the bulk of a tumor (leading to a significant clinical response), while leaving behind a residue of clinically undetectable CSCs; the latter may then re-launch tumor growth when the therapy is halted or when drug resistance develops. Clinically, this behavior is sometimes called the "dandelion effect," referring to the rapid re-emergence of weeds in a lawn following mowing, which cuts off their leaves but leaves their roots intact.

This logic suggests that the ability to eliminate CSCs will be the key to generating durable clinical responses for many types of tumors (Figure 17.35B). However, such elimination on its own may not suffice, because there are indications from studies of tumor xenografts that the transit-amplifying/progenitor cells can dedifferentiate and thereby regenerate new CSCs under certain physiological conditions (Figure 17.35C). Together, this would indicate that both the CSCs and the non-CSCs within a tumor need to be eliminated in order to prevent recurrence and clinical relapse. Importantly, identifying agents that specifically eliminate CSCs may be challenging, because such drugs may, on their own, have minimal effects on the overall sizes of tumor masses and may therefore be judged to be unworthy of further development.

In fact, rituximab, the anti-CD20 monoclonal antibody that is used to treat B-cell tumors (see Section 15.19), shows just this behavior: it eliminates the CSCs of multiple myelomas but not the more differentiated, far more abundant antibody-secreting cells that form the bulk of these tumors. (Fortunately, early in its development, rituximab was recognized to have efficacy for treating a wide range of B-lymphocyte–lineage tumors and was therefore approved for further development and introduction into the clinic.)

17.14 EGF receptor antagonists may be useful for treating a wide variety of tumor types

The proposal to develop Gleevec initially met with considerable resistance in the pharmaceutical company where it originated, simply because the market for this drug was judged to be too small to justify the high costs of its development and testing in the clinic. The same could not be said of the class of drugs designed to inhibit the epidermal growth factor receptor (EGF-R). Carcinomas are common tumors, and this receptor is believed to play a key role in the development of as many as one-third of them, being frequently overexpressed.

At least six distinct EGF-related ligands, including EGF itself, have been found to bind and activate the EGF-R. This means that even in those carcinomas in which the EGF-R is not overexpressed, it may nonetheless emit critical oncogenic signals through autocrine or paracrine signaling loops that are driven by the presence of one or more of its ligands in the interstitial space surrounding individual tumor cells. As one example of paracrine signaling, recall the release of EGF by macrophages that confers invasiveness on nearby breast cancer cells (see Figure 14.22).

An alternative role of EGF-R in carcinoma pathogenesis is illustrated by breast carcinomas that overexpress the HER2/neu receptor. The oncogenic actions of this protein depend on its ability to form heterodimers with its cousin, the EGF-R; in such heterodimers, the EGF-R can phosphorylate the C'-terminal tail of HER2/neu, thereby activating signaling by the latter through the formation of multiple phosphotyrosines on its C-terminal tail (see Section 5.6).

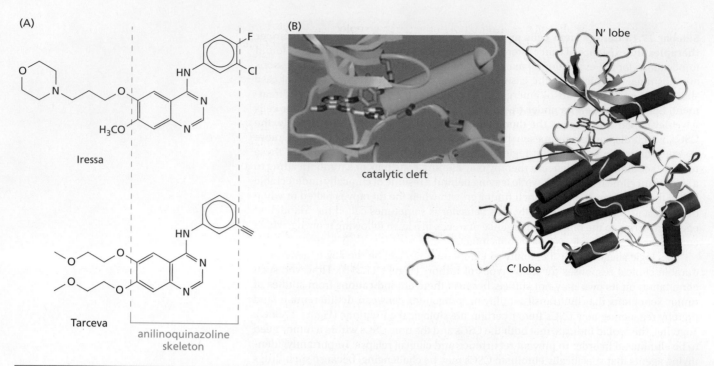

Figure 17.36 Iressa and Tarceva (A) The two epidermal growth factor receptor (EGF-R) antagonists are constructed from a common anilinoquinazoline skeleton, which confers on them an affinity for the ATP-binding site of the receptor tyrosine kinase. The chemical side groups that are attached to this skeleton have biological effects, because the effectiveness of the two drugs differs in treating, for example, non-small-cell lung carcinomas (NSCLCs). (B) Iressa, also termed ZD1839, binds to a very similar region of the EGF-R tyrosine kinase as does Tarceva (see Figure 17.17). A blown-up view (*left*) of the drug-binding site of the EGF-R tyrosine kinase domain (*right*) is shown here, with the drug molecule shown as a colored stick figure. This binding is so strong that 50% inhibition of the tyrosine kinase (TK) enzyme activity is achieved at a concentration of about 0.030 μM. (B, courtesy of A.C. Kay, AstraZeneca.)

The first clinically effective inhibitors developed to target the EGF-R tyrosine kinase were the drugs Iressa, also known as gefitinib/ZD1839, and Tarceva, also called either erlotinib or OSI-774 (**Figure 17.36A**). The two drugs have very similar but not identical properties and act by blocking the ATP-binding site of the receptor-associated kinase (Figure 17.36B; see also Figure 17.17). Once cancer cells are deprived of receptor signaling through inhibition of the EGF-R, they should lose the benefit of the strong mitogenic and anti-apoptotic signals that it generates. For example, in many types of epithelial cells, the continuous firing of the EGF-R sustains expression of Bcl-X_L (the potently anti-apoptotic cousin of Bcl-2) and, acting via MAPK, drives phosphorylation and attendant functional inactivation of the pro-apoptotic Bad protein (see Section 9.13, Figure 9.23B).

Because Iressa and Tarceva target a cell surface receptor, their therapeutic utility must be compared with that of the monoclonal antibodies that also affect this receptor (see Supplementary Sidebar 16.1). In principle, these low–molecular-weight compounds should be able to penetrate into all the interstices of a solid tumor, including those where the far larger antibody molecules may have trouble gaining access (see Table 17.7). Also, it is generally far easier and less expensive to produce low–molecular-weight compounds on an industrial scale than it is to generate large quantities of monoclonal antibodies.

There are yet other possible advantages of low–molecular-weight tyrosine kinase inhibitors. For example, as we have read, in many human carcinomas, truncated forms of the EGF-R are expressed that lack the ectodomain; these mutant EGF-Rs may signal in a ligand-independent, constitutive fashion and therefore can function as potent oncoproteins (see Figure 5.8A). Similarly, about half of high-grade (that is, advanced) gliomas, termed glioblastoma multiforme (GBM), exhibit overexpressed EGF-R, and of these, about 40% display a form of the receptor that lacks the ectodomains specified by exons 2 through 7 of the EGF-R coding sequence. Such decapitated receptors cannot be bound by the monoclonal antibodies (MoAbs) that have been developed to recognize antigenic epitopes present in the ectodomain of the normal receptor protein. However, these aberrations should not derail the low–molecular-weight tyrosine kinase inhibitors, which target the cytoplasmic, signal-emitting domain of the receptor. Weighing against these drugs are their pharmacokinetic properties: as an example, Tarceva has a half-life of ~36 hours, whereas the half-life of cetuximab (an anti-EGF-R monoclonal antibody) is >5 days and of Herceptin ~28 days.

Iressa has approximately a 50-fold more potent activity against the EGF-R–associated tyrosine kinase than against a number of other tyrosine kinases (see Supplementary Sidebar 17.9), and its initial use in the oncology clinic was reasonably encouraging. In the first clinical trials, 10% of patients with non-small-cell lung carcinomas (NSCLCs) showed **partial responses** to the drug, including disease stabilization of tumor growth; these patients tended to be women, nonsmokers, and those with the **bronchioalveolar** subtype of lung cancers. A parallel study in Japan found a far higher rate (27%) of partial responses to Iressa; this difference, which continues to be observed, appears to represent a difference in the genetic constitutions of the Japanese versus Caucasian populations. (More generally, NSCLC tumors have represented difficult diseases to treat, as fewer than 15% of patients survive for five years following initial diagnosis.)

These outcomes were gratifying, if only because they represented clear responses in patients who otherwise had few if any other treatment alternatives. However, the hoped-for synergistic actions of Iressa with standard chemotherapeutic agents did not provide any survival advantage over standard chemotherapy used alone for the treatment of NSCLC tumors, which together constitute almost 80% of lung cancer cases in the United States. When used on its own, Tarceva (but not Iressa) increased by two months the overall survival time of patients whose NSCLCs had become refractory to treatment by standard chemotherapeutic drugs.

Some valuable lessons were learned from these initial trials that may improve the responses in future clinical trials of these and similar drugs: First, the specific contribution of the EGF-R to the growth of the tumors under treatment was not documented. Hence, far greater response rates might have resulted from a stratification of the NSCLC patients and limiting the use of Iressa only to those tumors having specific molecular signatures indicating a key contribution of EGF-R to disease pathogenesis. A second, related consideration derived from the possible contribution of other mutant proteins to the mitogenic and anti-apoptotic signaling operating in tumor cells; here again, these factors were not assessed. There is evidence, for example, that PTEN-negative tumors (which have a hyperactive PI3 kinase pathway; see Section 6.6) do not respond to Iressa, and that inhibitors of Akt/PKB (the downstream beneficiary of PTEN inactivation) can act synergistically with Iressa to halt tumor growth. Third, relatively few pre-clinical studies were undertaken in order to optimize the dosage and the schedule of treatments with this drug.

In 2004, four years after results of the initial clinical trials with Iressa were first reported, two research collaborations in Boston independently provided a molecular explanation for the observed responses to Iressa. Previously, the status of the EGF receptor in NSCLC cells was assessed by determining whether it was overexpressed and whether it was present in truncated, constitutively active form, as is the case in human glioblastomas. In the 2004 studies, however, investigators undertook detailed sequencing of the reading frames of the EGF-R–encoding gene in the NSCLC patients who had been treated with Iressa.

Quite dramatically, they found that almost all of the small group (~10% of the total) of NSCLC patients who had responded well to Iressa treatment (**Figure 17.37A**) bore tumor cells displaying structurally altered EGF-Rs. Such mutant receptors were not found among the tumors that failed to respond to Iressa, including those that expressed elevated levels of this receptor (see, for example, Supplementary Sidebar 17.9). For example, 78% of patients who responded to anti-EGF-R therapy expressed mutant EGF-R protein, whereas among the nonresponders only 6% showed such structurally altered receptor proteins.

The mutations found in the altered EGF-Rs created amino acid substitutions and small deletions in the kinase domain (Figure 17.37B) rather than the major deletions of the receptor ectodomain typically found in glioblastomas. These mutant receptors

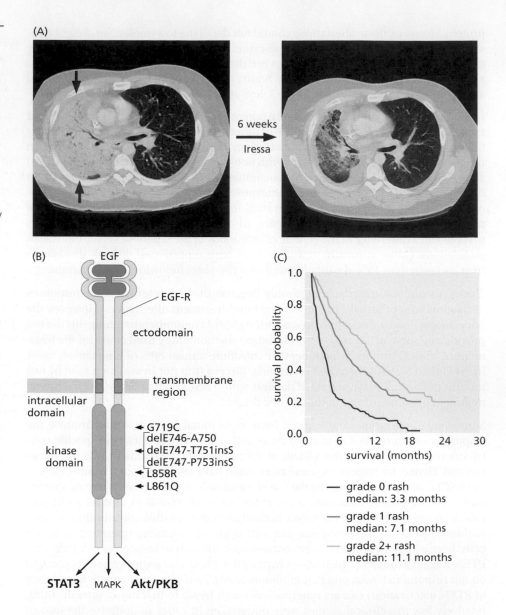

Figure 17.37 Determinants of responsiveness of NSCLCs to Iressa and Tarceva treatment (A) A minority of patients with refractory non-small-cell lung cancers (NSCLCs)—tumors that failed to respond or ceased to respond to standard chemotherapy—show dramatic responses to treatment by Iressa. These computerized tomographic images reveal dramatic regression of a large mass (*left*) in a patient's right lung following six weeks of Iressa treatment (*right*). (B) A substantial proportion of NSCLCs that respond to Iressa have been found to carry deletions ("del") and point mutations in the EGF-R gene that affect the cytoplasmic domain of the receptor. These alterations in EGF-R structure deregulate and activate the tyrosine kinase function of the receptor, thereby stimulating the Akt/PKB and STAT signaling pathways, which protect these tumor cells from apoptosis (see Figures 6.10 and 6.16). (C) Experience with a variety of anti-EGF-R treatments has indicated that, in addition to the genetic state of the tumor cells, the development of an acne-like rash by patients under treatment is a strong positive indicator of the eventual response of tumors to therapy. (This response clearly reflects the role of the EGF-R in skin keratinocyte biology and the precise mechanistic connection between the normal keratinocyte response of patients to therapy and the response of their tumors remains obscure.) In this case, the responses of NSCLC patients to Tarceva therapy are shown in this Kaplan–Meier graph. "Median" indicates median survival. (A and B, from T.J. Lynch et al., *N. Engl. J. Med.* 350:2129–2139, 2004. With permission from Massachusetts Medical Society. C, from B. Wacker et al., *Clin. Cancer Res.* 13:3913–3921, 2007. With permission from American Association for Cancer Research.)

showed distinctive patterns of tyrosine phosphorylation of their C-terminal tails (see Section 6.3) and selectively stimulated the downstream Akt/PKB and STAT5 pathways, leaving the MAPK signaling pathway unaffected.

These observations provided compelling evidence that the EGF-R played a central role in driving the growth of these small groups of tumors. In addition, they demonstrated the value of stratification of tumors using molecular markers when treating patient populations with targeted molecular therapeutics such as Iressa and Tarceva. Still, these experiments did not reveal why Iressa and Tarceva had such strong effects on these particular tumors (**Sidebar 17.14**).

An additional puzzle comes from the frequency of the NSCLCs found to express the structurally altered EGF receptors. They have been found in the tumors of ~10% of Western patients suffering from NSCLC, although in certain Asian populations, as many as 30% of the NSCLCs analyzed express these mutant receptor proteins; these altered EGF receptors would appear to explain the far higher response rates to Iressa in Japan cited above. (In all cases these altered receptors result from somatic mutations rather than inherited germ-line alleles.) The reasons for these inter-ethnic discrepancies and the greater frequency of these tumors in female nonsmokers represent additional puzzles.

One mystery seems to have been solved, however. In certain East Asian populations, as many as half of CML patients show an incomplete response to Gleevec therapy

Sidebar 17.14 Oncogene addiction may explain how Iressa and Tarceva succeed in killing NSCLCs The mutant EGF-Rs that are found in certain NSCLCs cause the cells from these particular tumors to be approximately 100 times more sensitive to Iressa than tumors expressing the wild-type receptors (Figure 17.31A). Importantly, the actual drug concentrations in the plasma of patients being treated fall in the range that allows such selective inhibition to operate. The mechanism of oncogene addiction, introduced in Section 17.4, may explain the selective effects of both Iressa and Tarceva on tumors expressing structurally altered EGF-Rs. Recall that oncogene addiction refers to the strict dependence of certain cancer cells on a certain oncogene or oncoprotein for their growth and survival, whereas other types of cancer cells can lose this gene or protein without suffering significant consequences.

To explain this behavior, we can imagine that some oncogenes are generally deleterious when expressed in wild-type cells but are actually beneficial in cells that previously acquired certain mutant alleles of another oncogene. A good example is provided by the *myc* oncogene, which has pro-apoptotic effects on cells unless they are protected from apoptosis by an acquired anti-apoptotic allele of some sort (for example, a *ras* oncogene); in the presence of the anti-apoptotic mutation, the strongly mitogenic effects of the *myc* oncogene then become apparent. Hence, tumor cells that carry both the *ras* and *myc* oncogenes would behave as if they were "addicted" to the *ras* expression, because they would die quickly by apoptosis if they were deprived of the *ras* oncogene.

Similarly, early in tumor progression, the acquisition of a certain oncogene, such as a mutant EGF-R gene, might create a cellular environment that permits acquisition of other oncogenes (or losses of tumor suppressor genes) that would, on their own, be highly deleterious for tumor cells. If the mutant receptor is now lost, then the deleterious effects of these other oncogenes, such as those favoring apoptosis, would become apparent and result in rapid loss of cell viability.

In the case of non-small-cell lung cancer (NSCLC), those tumors bearing mutant receptors may have come to depend on the firing by their mutant EGF-Rs in order to survive and proliferate; that is, they are "addicted" to the mutant receptors. Conversely, the far more numerous NSCLCs expressing the wild-type EGF-R, often at elevated levels, may have developed alternative means of securing mitogenic and survival signals, as is indeed suggested by the observation of receptor-independent firing of the MAPK and PI3K pathways in some lung cancers.

This scenario is further supported by experiments using siRNAs to inhibit the expression of wild-type or mutant receptors (see Supplementary Sidebar 1.5): NSCLC cells with mutant EGF-R die quickly, whereas those displaying wild-type receptor are only slightly affected (see Figure 17.31B and C). Consequently, the death of cancer cells with mutant EGF-Rs does not occur from some unknown, off-target effect of Iressa or Tarceva, but instead is caused directly by the loss of beneficial signals released by these receptors. Moreover, experiments like these suggest that EGF-R inhibitors may have far greater effects on NSCLCs expressing wild-type receptors if they are applied together with a second drug that inhibits another, functionally redundant signaling pathway, such as the one controlled by PI3K.

compared with only one-fourth of Western patients; a similar therapy-resistant subpopulation exists among NSCLC populations and their responses to EGF-R inhibitors. In both cases, the patients' tumors are resistant from the beginning of treatment to the effects of the tyrosine kinase inhibitors (TKIs) used to treat these two diseases. Among oncologists, this preexisting resistance is often called **primary resistance** to distinguish it from the resistance that develops during the course of therapy, which is termed **secondary resistance**.

In both NSCLC and CML, primary resistance has been traced to an allele of the pro-apoptotic *Bim* gene (see Section 9.13) that encodes a defective version of the Bim protein and is carried by 12.3% of East Asians but is absent in African and European populations. Other work has shown that expression of the wild-type Bim protein is suppressed by oncogenic tyrosine kinases, and its activation following TK inhibition contributes importantly to the apoptotic cancer cell death that drugs like Gleevec, Iressa, and Tarceva succeed in inducing in responsive cell populations. Lacking both copies of the wild-type *Bim* gene and thus normal levels of the Bim protein, the tumors of these East Asian patients show substantially reduced tendency to enter into apoptosis and thus reduced responsiveness to the killing effects of these inhibitors.

In addition to the germ-line and somatically acquired alleles cited above, there are yet other indications that are useful for predicting responsiveness to anti-EGF-R TKIs. Echoing the results from use of anti-receptor MoAbs (see Supplementary Sidebar 17.9), the presence of a mutant, activated K-*ras* or *PI3K* oncogene or an inactivated *PTEN* tumor suppressor gene is strongly correlated with the failure of tumors to respond to these TKIs. These findings can all be rationalized in terms of the signal transduction cascades that we encountered in Chapter 6. Thus, if a downstream cytoplasmic effector of a cell surface receptor becomes activated by mutation, its firing no longer depends on the upstream receptor and can continue unabated, even if the receptor itself is inactivated, in this case by targeted therapies.

There is, however, a striking finding that cannot be rationalized in terms of our current understanding of cellular signaling and therefore represents an abiding mystery: the best surrogate marker of a tumor's responsiveness to various anti-EGF-R therapies is the development by the cancer patient of a readily observable severe skin rash (see Figure 17.37C). Because the skin is clearly not involved in the pathogenesis of lung and colorectal carcinomas, this rash might mean that there exist significant inter-individual differences in pharmacokinetics (PK) that determine therapeutic outcome and, in parallel, affect the skin; in truth, however, there is no correlation between PK and skin rash development. In addition, the positive correlation between skin rash severity and therapeutic response applies to a variety of low–molecular-weight drugs as well as anti-receptor monoclonal antibodies (see Supplementary Sidebar 17.9). These observations provide a strong indication that some subtle, inter-individual variability in the behavior of the basic signal-processing machinery operating in various epithelial cells throughout the body strongly influences responsiveness to anti-EGF-R therapies. This behavior reminds us of how little we understand about the regulators of cell signaling, whose behaviors are more than minor determinants of the success or failure of current, so-called rational therapies.

Unfortunately, even positive responses to these tyrosine kinase inhibitors have been short-lived, and most patients relapse in 6 to 18 months, having developed a secondary resistance to drug treatment. Acquired resistance to erlotinib and gefitinib can occur via a diverse set of mechanisms that highlight both the promise of new therapies as well as the fundamental challenges in achieving durable clinical responses, including cures. The first and most common mechanism of EGF-R inhibitor resistance is mutation of the receptor itself. The T790M amino acid substitution, the product of a so-called **gatekeeper** mutation, alters the ATP binding pocket of EGF-R (**Figure 17.38**). Rather than blocking drug interaction with EGF-R, this amino acid change increases the receptor affinity for ATP such that the effect of inhibitors is

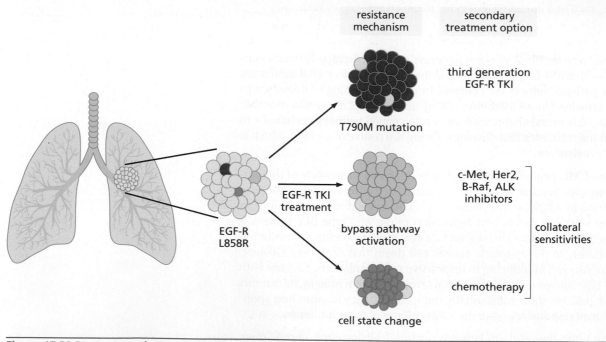

Figure 17.38 Routes to resistance to EGF-R inhibitors Non-small-cell lung cancers (NSCLC) driven by mutation and amplification of EGF-R are sensitive to EGF-R inhibitors including erlotinib and gefitinib. In this illustrated case, an L858R amino acid substitution in a mutant EGF-R drove initial outgrowth of the NSCLC. Resistance to EGF-R-inhibitory drugs could occur via a variety of alternative mechanisms (*middle column*), including second-site mutations in EGF-R (most frequently the T790M "gatekeeper" mutation, *red cells*), activation of parallel bypass pathways or downstream signal transducers (e.g., c-MET, Her2, ALK, or B-Raf activation, *green cells*), or cell state changes (e.g., EMT or transdifferentiation to small-cell lung cancer, *blue cells*). Each resistance state is, in turn, targetable by a subsequent cancer therapy (*right column*). In the case of the activation of bypass pathways and cell state changes, these resistance mechanisms create "collateral sensitivities" or emergent sensitivities that were not present in the initial pre-treatment tumor. The existence of such emergent sensitivities presents opportunities for effective sequential cancer therapy, but also creates challenges in determining how and when such therapies should be applied.

diminished. The common appearance of these mutations following treatment with first generation EGF-R inhibitors led to the development of second and third generation EGF-R inhibitors that can target the T790M mutant. Drugs like osimertinib are now used in the treatment of T790M-mutant NSCLC. Because this agent can also target unmutated EGF-R, it is often used as a first-line therapy that can preempt the eventual emergence of T790M-mutant tumors.

A second, alternative mechanism of acquired EGF-R inhibitor resistance involves the activation of parallel growth signaling pathways. For example, the overexpression—often resulting from gene amplification—of the c-Met TK receptor can substitute for EGF-R signaling in treated tumor cells. Similarly, alterations in kinases downstream of EGF-R (see Section 6.5) can also promote cell survival in the face of EGF-R inhibition. Activating mutations in B-Raf kinase (discussed in Section 17.16 of this chapter), a downstream target of EGF-R signaling, can obviate the requirement for upstream EGF-R activity.

Finally, as mentioned earlier in the context of Bcr-Abl signaling, cell state changes can also promote resistance to EGF-R inhibitors. In one such example, NSCLCs have been shown to undergo an EMT-like switch to a more mesenchymal phenotypic state following treatment (see Sections 13.3 and 14.3). In another, treated NSCLC patients may suffer relapses concomitant with a shift in the phenotype of their tumors and tumor cells, suggesting their entrance into a state closely resembling that of small-cell lung cancer (SCLC), a tumor thought to have a developmental origin distinct from that of the more common adenocarcinomas. Precisely how this kind of developmental transition—essentially a **transdifferentiation**—occurs remains unanswered.

In all these cases, the specific resistance mechanisms that appear in relapsed tumors inform the choice of specific actionable therapies. Thus, T790 mutant tumors are treated with osimertinib; c-Met–amplified tumors are treated with c-Met inhibitors combined with EGF-R inhibitors; B-Raf mutant tumors are treated with B-Raf inhibitors; and tumors that have undergone the cell state change described above are treated with chemotherapy, to which SCLC tumors respond much more effectively than do NSCLCs. This connection between resistance mechanisms and actionable therapeutic strategies designed to target resistance states represents a major advance in cancer therapy, one that has extended and improved the lives of cancer patients. However, none of these treatments is ultimately curative, highlighting the fundamental, ongoing challenge of drug resistance: while we have substantially increased our arsenal of agents to target a variety of tumor-associated alterations, tumor cells invariably contrive more alternative paths to resistance than we have drugs to treat them.

17.15 Proteasome inhibitors yield unexpected therapeutic benefit

Serendipity plays an unusually prominent role in the world of drug discovery. On occasion, the development of an anti-cancer drug is launched as part of a rational drug design program and ultimately yields an agent that turns out to be highly useful, albeit for reasons quite unrelated to those that inspired its development in the first place. This best describes the development of the drug known as Velcade, also called PS-341 and bortezomib (**Figure 17.39A**).

On many occasions throughout this book, we have seen how the levels of key cellular regulatory proteins are determined by the balance between their synthesis and their degradation. Much of this degradation is mediated by the ubiquitin–proteasome system (see Supplementary Sidebar 7.5). Recall that the tagging of a protein by polyubiquitylation results in its transport to proteasomes and its degradation in these intracellular machines.

The phenomenon of cancer-associated **cachexia** initially stimulated interest in inhibitors of proteasome function. Cachexia occurs late in tumor progression and represents a progressive wasting of the cancer patient's tissues through mechanisms that remain poorly understood. It was speculated that a proteasome inhibitor could be useful in retarding the widespread degradation of proteins occurring in the tissues of cachectic patients. At least five distinct classes of proteasome inhibitors have

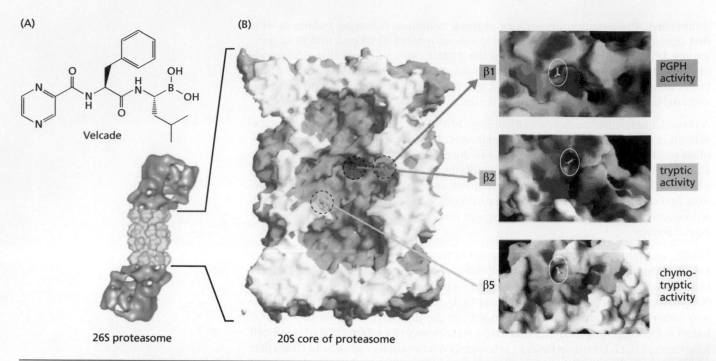

Figure 17.39 Velcade and its effects on proteasomes (A) The chemical structure of Velcade reveals the presence, unusual among drugs, of a boron atom. Peptide boronic acids like Velcade were known to bind to the active site of serine proteases of the chymotrypsin class (which cleave substrate proteins adjacent to phenylalanine and tyrosine residues) by mimicking the normal substrates of these enzymes. This finding suggested to the drug developers that such compounds might inhibit the chymotrypsin-like active site in the 20S core of the proteasome. (B) A cross section (*center*) through the central (20S) core of the yeast proteasome (*left*) reveals the locations of three distinct catalytic sites, involving the β1, β2, and β5 subunits of the proteasome (*far right*), which are responsible for its PGPH (peptidyl-glutamyl-peptide hydrolyzing), tryptic, and chymotryptic proteolytic activities, respectively. Velcade shows a strong preference for inhibiting the β5 chymotryptic activity (*lower right*); a weak interaction with the β2 tryptic activity (*middle right*); and no interaction with the β1 PGPH activity (*top right*). The key nucleophilic threonine residue in each of these catalytic sites is shown as a stick figure (*white inside white ovals*); basic amino acids in the catalytic clefts are in *blue*, acidic amino acids are in *red*, and hydrophobic residues are in *white*. (B, from M. Groll et al., *Chembiochem* 6:222–256, 2005. With permission from John Wiley & Sons.)

been developed, although most have been abandoned as candidates for clinical use because of metabolic instability, lack of specificity, or irreversible binding and inactivation of proteasomes.

Velcade, one of these proteasome inhibitors, is a boronic acid dipeptide that was designed as a specific inhibitor of the **peptidase** (peptide-cleaving) activity that occurs in the 20S core of the proteasome (Figure 17.39B). It has extraordinary potency, being able to inhibit 50% of the proteasome's **chymotryptic** activity at a concentration (that is, its K_i) of only 0.6 nM. Functioning as a competitive inhibitor of this enzyme activity, Velcade slows the flux of substrates through proteasomes, which soon become clogged and dysfunctional.

Proteasome-mediated degradation was subsequently found to play a critical role in regulating a number of key cellular signaling pathways. Because other proteasome inhibitors had been found to be especially potent in killing a variety of cultured cancer cells, Velcade was used in Phase I trials to treat cancer patients who had failed other available therapies. Those with solid tumors showed few striking responses. However, among a group of patients with hematological malignancies was an individual suffering from **multiple myeloma** (MM), a malignancy of the B-cell lineage in which a single clone of antibody-producing plasma cells undergoes clonal expansion and ultimately dominates the bone marrow. The myeloma cells create osteolytic bone lesions that lead to fractures, and they ultimately crowd out the remaining cellular components of the marrow, resulting in severe immune depression and, typically, death from overwhelming infection. Survival after initial diagnosis is usually three to five years. The myeloma carried by this initially treated patient showed a dramatic regression

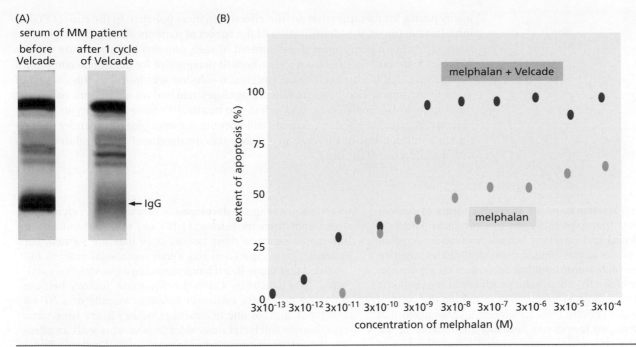

Figure 17.40 Multiple myeloma and the biological effects of Velcade (A) Velcade can have profound effects on the cellular contents of the marrow and thus the composition of antibody molecules in the blood. In one multiple myeloma (MM) patient, after eight doses of Velcade, the neoplastic plasma cells in the marrow declined from 41% of total cells to 1%. At the same time, the level of the single species of immunoglobulin γ (IgG) made by these myeloma cells declined drastically. As seen here, upon gel electrophoresis, the small number of IgG species present before treatment (indicative of the homogeneous antibody population produced by a monoclonal tumor, *left*) resolved into the heterogeneously migrating pattern of IgGs that are present in the circulation of a healthy individual (*right*). (B) Melphalan, an alkylating chemotherapeutic drug used routinely to treat MM, was added at various concentrations to an MM cell line *in vitro*, either on its own (*green*) or in the presence of a noncytotoxic dose of Velcade (*red*). In the presence of Velcade, melphalan was able to induce widespread apoptosis at approximately 3 nM concentration, but when applied on its own, melphalan was unable to induce this degree of apoptosis, even at vastly higher concentrations. (A, from R.Z. Orlowski et al., *J. Clin. Oncol.* 20:4420–4427, 2002. With permission from American Society of Clinical Oncology. B, from M.H. Ma et al., *Clin. Cancer Res.* 9:1136–1144, 2003. With permission from American Association for Cancer Research.)

(**Figure 17.40A**), which soon led to the inclusion of other myeloma patients in this Phase I trial and eventually to large-scale clinical trials.

In a subsequent Phase I clinical trial with a group of multiple myeloma patients suffering from rapidly progressing disease, Velcade showed clear "**objective responses**" in slowing disease progression in 55% of the patients and halted progression in another 25%. In a Phase II clinical trial, half of patients were given Velcade while the other half, who served as controls, were given dexamethasone, a standard treatment for multiple myeloma. Most of these patients had already failed the chemotherapies commonly used for myeloma. This trial was stopped prematurely in 2003 because Velcade demonstrated a clear superiority over existing treatments, with the disease showing a complete response in a small number of patients (that is, myeloma cells disappeared completely from the blood for at least six weeks) and a partial response (at least 50% reduction in myeloma cell–secreted antibody in blood and 90% reduction of this protein in urine over the same time period) in 35% of the Velcade-treated patients. As a consequence, the control patients were then allowed to take the drug as well. In a subsequent trial, the progression of myeloma to a higher stage of disease occurred with a median time of seven months in Velcade-treated patients compared with three months in a control group studied in parallel. Moreover, pre-clinical studies indicate that relatively low doses of Velcade can sensitize myeloma cells to chemotherapeutic drugs, making the latter far more effective (Figure 17.40B and **Sidebar 17.15**).

The lack of a clear mechanism underlying Velcade efficacy has confounded attempts to identify biomarkers for drug response. The absence of biomarkers represents a challenge to clinicians in providing optimal treatment to patients, and it also represents a common challenge in determining how to pay for such treatments. Specifically, given the extraordinarily high cost of some cancer therapies, it can be difficult to justify paying for therapies that are not effective in most patients. In the case of EGF-R inhibitor treatment, the identification of the subset of patients likely to be responsive to therapy not only improved the treatment of such patients but also made the use of these treatments feasible from a cost-benefit perspective for insurance companies and government health agencies. A pragmatic solution was found in the context of Velcade treatment: European healthcare agencies reached an agreement with drug developers to pay for Velcade only when the treatment showed efficacy in specific patients. Whether such arrangements will become commonplace for cancer therapy remains unclear, but in the case of Velcade this arrangement was fundamental to broad clinical use of the drug.

Sidebar 17.15 Myelomas as an attractive target of proteasome inhibitor therapies Inclusion of a myeloma patient in the initial clinical trial involving Velcade was hardly accidental. Multiple myeloma was thought to be an attractive target for treatment by a proteasome inhibitor because of the known elevated activity of the NF-κB signaling pathway in the myeloma cells and its physiological importance in driving the survival and proliferation of these cells. In Section 6.12 we noted that NF-κB transcription factors are normally sequestered in the cytoplasm by a class of inhibitors termed IκBs (inhibitors of N-κB). When these IκBs are phosphorylated by a group of specialized kinases termed IκB kinases, or simply IKKs, the IκBs undergo polyubiquitylation and resulting degradation; this liberates the NF-κBs, allowing them to migrate into the nucleus, where they activate a number of anti-apoptotic genes as well as growth-promoting genes (**Figures 17.41 and 17.42**; see also Figures 6.22A and 11.38B).

Like myriad other polyubiquitylated proteins, the IκBs end up being degraded in proteasomes. Hence, by inhibiting proteasome action, IκBs should be protected from degradation, survive in the cytoplasm, and continue to sequester the NF-κBs, thereby blocking NF-κB nuclear translocation and activation of transcription. Nuclear, functionally active NF-κB was known to be important for inducing the expression of IL-4 and IL-6, two interleukins that operate as important autocrine factors required for the growth and survival of myeloma cells. In addition, as was learned later, NF-κB plays a prominent role in anti-apoptotic signaling in a number of cancer cell types; hence, loss of active NF-κB might well tilt the signaling balance within these cells toward apoptosis. More specifically, once cancer cells lose the potently anti-apoptotic Bcl-2, cIAP-2, and XIAP proteins (all of whose expression is induced by NF-κB), they are in grave danger of slipping into the apoptotic abyss.

All this does not explain, however, why Velcade is far more potent against myelomas than other tumors that rely on NF-κB signaling to protect them from apoptosis. A possible clue comes from observations that the growth and viability of myeloma cells is highly dependent on their ability to synthesize VEGF (vascular endothelial growth factor; see Section 13.1) and adhesion molecules; the latter enable myeloma cells to attach to bone marrow stem cells (BMSCs), with which the myeloma cells establish critically important heterotypic interactions. The genes encoding these various proteins are all under NF-κB control.

NF-κB antagonists are likely to have utility for treating a number of other kinds of cancer as well. Recall that NF-κB plays a key role in the development of a variety of carcinomas (see Supplementary Sidebar 11.12) and may be required for the maintenance of these tumors once they are formed. For example, gene expression and tumor mutational analysis has revealed that many B-cell lymphomas have constitutively activated NF-κB activity. Consequently, these tumors become attractive targets for treatment by either Velcade or a NF-κB treatment, and for one treatment-refractory B-cell lymphoma type, mantle cell lymphoma, Velcade now represents an effective therapy for many patients. However, other B-cell lymphoma types that have constitutive NF-κB signaling fail to respond to Velcade, suggesting that this tumor dependency may not be the sole determinant of Velcade response.

A second, alternative theory of Velcade's mechanism of action, which also has numerous proponents, does not attribute its effects specifically to inactivation of the NF-κB signaling pathway. According to this proposed mechanism, myeloma cells, like normal antibody-secreting plasma cells, continuously synthesize and secrete antibody molecules at a prodigious rate—estimated to be thousands of molecules per second. A certain portion of these molecules are routinely degraded in the proteasomes because of misfolding or other mishaps occurring during their post-translational maturation. As seen in Figure 17.41B, in addition to proteasomal degradation, accumulation of unfolded and misfolded proteins in the endoplasmic reticulum (ER) creates the state of **ER stress**, which in turn activates the **unfolded protein response** (UPR). This complex program depends on signaling by receptor-like UPR transducers that convey information of the presence of significant numbers of misfolded proteins in the lumen of the ER. One component of the UPR, active during acute episodes of ER stress, results in molecular chaperones being dispatched from the cytosol into the lumen of the ER in order to properly fold the aberrant proteins. Another part of the UPR program operates in response to chronic ER stress and triggers apoptosis. Myeloma cells are especially sensitive to inhibitors of protein degradation and become engorged with misfolded protein molecules, specifically subunits of antibodies. In some poorly understood fashion, the failure of proteasomal function in the cytosol triggers activation of the UPR within the ER, and under many circumstances, this results eventually in the activation of apoptosis. The truly amazing potency and therapeutic index of Velcade—killing myeloma cells at concentrations as low as 1 nM—may be explained by the highly specialized secretory phenotype of these cells.

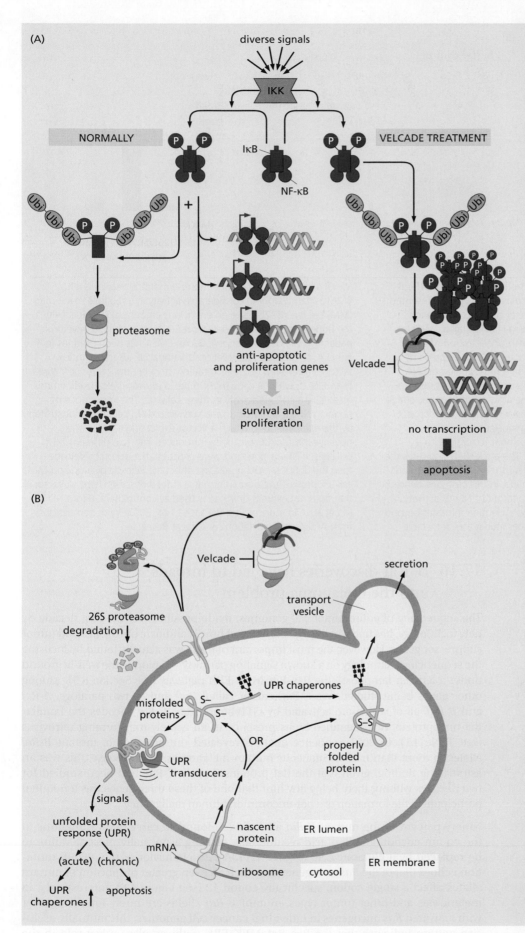

Figure 17.41 Mechanisms of Velcade action Two alternative mechanisms of action of Velcade have been proposed, and both are supported by extensive observations. (A) In normal and neoplastic cells (*left*), a variety of stress, mitogenic, and trophic (survival) signals activate IκB kinase (IKK; *purple, above,* see Figure 6.22A). Once activated, IKK phosphorylates IκB (*red*), the inhibitor of NF-κB. This causes IκB to become ubiquitylated (*left*) and degraded in proteasomes (*lower left*). NF-κB (*blue*) is then free to move into the nucleus, where it activates the expression of numerous proliferation and anti-apoptotic genes. In the presence of Velcade (*right*), the ubiquitylated IκB cannot be degraded in the proteasomes because the latter have become engorged with unprocessed polypeptides. The IκB that accumulates in the cytoplasm binds and sequesters NF-κB (*right*). As a consequence, NF-κB is prevented from moving into the nucleus and activating expression of key anti-apoptotic genes. (B) The striking toxicity of Velcade for MM cells may arise because these cells, like normal plasma cells, are specialized to continuously synthesize and secrete enormous amounts of antibody molecules, in both cases on membrane-bound polyribosomes (*below, middle*). Like other secreted glycoproteins, antibody molecules are processed in the lumen of the endoplasmic reticulum (ER) prior to being secreted via exocytosis (*upper right*). Inevitably, a certain proportion of recently synthesized antibody molecules are misfolded following their insertion into the lumen of the ER. In the absence of proper folding, many of these misfolded proteins are extruded from the ER and broken down, following ubiquitylation, in cytosolic proteasomes (*upper left*). However, when proteasome function is blocked by Velcade (*above*), misfolded proteins accumulate in the lumen of the ER and trigger the unfolded protein response (UPR). This complex program includes the induction of apoptosis when misfolded proteins accumulate to toxic levels (*lower left*). (B, adapted from Y. Ma and L.M. Hendershot, *Nat. Rev. Cancer* 4:966–977, 2004. With permission from Nature.)

Continued

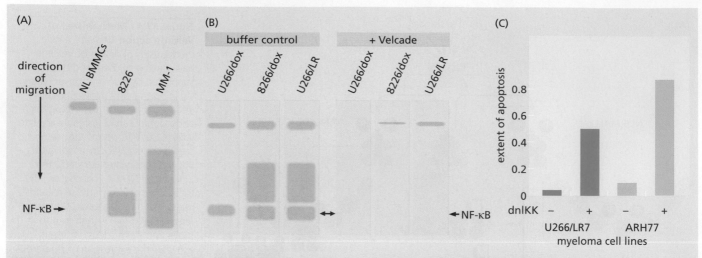

Figure 17.42 Evidence supporting the importance of NF-κB signaling in Velcade-induced apoptosis The scheme presented in Figure 17.41A is supported by a number of lines of evidence. (A) In an electrophoretic mobility shift assay (EMSA; also called a gel retardation assay), the presence and concentration of a functional, DNA-binding transcription factor (TF) are assessed by mixing an extract of nuclear proteins with a radiolabeled dsDNA oligonucleotide that carries a binding site for the TF. The presence of the DNA-binding TF is reflected by the amount of oligonucleotide that has formed a nucleoprotein complex with the TF; the large mass of protein associated with the oligonucleotide retards its migration during electrophoresis, causing it to migrate to a characteristic higher position in the gel. The arrow (*left*) indicates the expected location of a complex containing the NF-κB transcription factor and the radiolabeled oligonucleotide, in this case one derived from the promoter of the TNF-α gene, a target of NF-κB activation. (In the absence of bound DNA, the oligonucleotide migrates beyond the bottom of the gel.) The assay indicates little if any detectable NF-κB activity in normal bone marrow mononuclear cells (NL BMMCs, *left channel*), considerable activity in a multiple myeloma (MM) cell line (8226), and an enormous amount of NF-κB activity in the bone marrow cells prepared directly from a multiple myeloma patient (MM-1, *right channel*). (B) An EMSA has been used, as in panel A, to measure the level of functional NF-κB transcription factor in three MM cell lines treated with a control buffer (*left three channels*) or with Velcade (*right three channels*). Velcade eliminates essentially all NF-κB activity in these cells. (C) The importance of ongoing NF-κB signaling to the survival of MM cells is demonstrated by this experiment, in which a vector expressing a dominant-negative IKK (dnIKK) has been introduced into two different MM cell lines. If NF-κB signaling were critical to the action of Velcade, then the dnIKK should mimic the effects of Velcade by inducing MM cell apoptosis—the outcome that is indeed observed here. A vector that does not express dnIKK was used as control here. (From M.H. Ma et al., *Clin. Cancer Res.* 9:1136–1144, 2003. With permission from American Association for Cancer Research.)

17.16 B-Raf discoveries have led to inroads into the melanoma problem

The sequencing of entire tumor cell genomes, made possible over the past decade by new technology, has uncovered large numbers of novel candidate oncogenes and tumor suppressor genes. However, the most important of these was actually found by focusing the sequencing technology on a known signaling pathway—exploring the well-lit ground under a familiar lamppost—the Ras–Raf–MEK–ERK pathway (see Section 6.5). Among other genes being studied was *C-Raf* (earlier called *Raf*) and its two paralogs, *A-Raf* and *B-Raf*, all of which are activated by GTP-bound Ras; *C-Raf* encodes the familiar Raf oncoprotein, first identified by its presence in an acutely transforming retrovirus (see Table 4.1). In 2002, sequence analysis revealed the presence of mutant *B-Raf* alleles in more than half of cutaneous human melanomas. In one sense, this was an astonishing finding, given that the Raf proteins, largely C-Raf, had been studied for two decades without there being any hint that one of these three genes was a frequent participant in the formation of a not-uncommon human malignancy.

Ninety percent of the mutant *B-Raf* alleles were found to carry point mutations in the codon encoding amino acid residue 600, causing a normally present valine to be replaced by a glutamic acid (**Figure 17.43A**). The behavior of this V600E mutation echoes that of the *Ras* oncogenes, in which an even greater proportion of mutant alleles affect a single codon, specifically codon 12 (see Figure 4.6). Interestingly, in melanomas and other tumor types, mutant *B-Raf* alleles are rarely found together with activated *Ras* oncogenes in individual cancer cell genomes; this mutually exclusive pattern indicates that the Ras–Raf–MEK–ERK pathway plays a key role in the

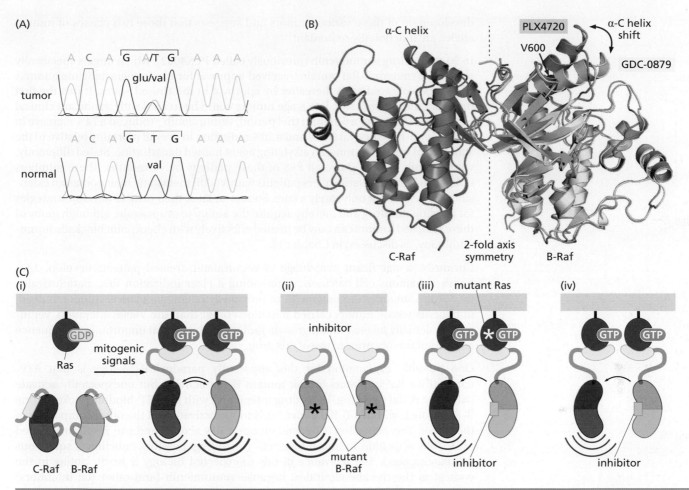

Figure 17.43 The mutant B-Raf and its treatment A search for mutant alleles of the genes controlling the main mitogenic signaling pathway in cells—the Ras–Raf–MEK–MAPK pathway—has revealed the frequent presence of mutant alleles of one of the three genes encoding Raf family kinases, which operate as serine/threonine kinases immediately downstream of the Ras proteins (see Figure 6.9A). (A) The discovery of point-mutated B-Raf alleles is shown here, in which sequencing of the genome of an ovarian tumor (*above*) revealed heterozygosity in codon 600 of the protein, which normally specifies valine (*below*). The mutant allele in the tumor genome was found to specify glutamic acid instead. This V600E mutant allele has since been found to be the predominant mutant B-Raf allele in more than half of all malignant melanomas and a variety of other tumors, including ~10% of colon carcinomas. (B) A molecular model of the B-Raf/C-Raf heterodimer structure is shown here. The two kinase molecules associate with one another in a parallel, side-to-side fashion, with residues from both the N- and C-terminal lobes contributing to the heterodimer interface. The V600 residue of B-Raf (*arrow, red*) is the most common structural alteration encoded by oncogenic mutations (usually V600E) and renders the enzyme constitutively active. Acting via allosteric interactions, the B-Raf protomer is able to activate C-Raf signaling even when B-Raf lacks kinase activity (as a result of mutation or drug action). The orientation of the α-C helices can affect B-Raf/C-Raf dimerization and thus C-Raf kinase activation. As seen here, the binding of two distinct B-Raf inhibitors, PLX4720 and GDC-0879, in the ATP-binding pocket of B-Raf shifts the orientation of this α-helix and thus the interactions between the two protomer subunits. (C) The behavior of Raf dimers greatly complicates the therapeutic effects of the B-Raf V600E inhibitor. (i) In normal cells, mitogen-activated GTP-bound Ras proteins bind and activate B-Raf/C-Raf heterodimers, with much of the downstream signaling being emitted by the more potent B-Raf partner; this results in activation of MEK and MAPK/ERK (see Figure 6.9A). (ii) In many melanomas, a mutant V600E B-Raf protein, acting in a Ras-independent fashion, can signal constitutively as a monomer (*right*), driving strong activation of this signaling cascade. A V600E-specific inhibitor (*green rectangle*) blocks this signaling (*right*), leading to temporary tumor regression of many melanomas. (iii) In cancer cells in which a mutant Ras oncoprotein is present (for example, in many colorectal carcinomas), the V600E-specific inhibitor, while shutting down B-Raf signaling, causes the latter to activate, via allosteric interactions, signaling by C-Raf. (iv) In certain normal cells in which B-Raf is wild type, this inhibitor can stimulate signaling, possibly by activating C-Raf. (A, from H. Davies et al., *Nature* 417:949–954, 2002. B, from G. Hatzivassiliou et al., *Nature* 464:431–435, 2010. Both with permission from Nature. C, courtesy of M.B. Yaffe.)

development of these various tumors and suggests that these two classes of mutant alleles are functionally redundant.

In 2011, the drug vemurafenib (previously called PLX4032), which targets specifically the V600E mutant B-Raf protein, received approval for use in metastatic melanomas in Europe, followed soon thereafter by approval in the United States. It was the first time that these aggressive, late-stage tumors could be treated with significant clinical responses. Thus, over a 6-month trial period, vemurafenib produced a 74% increase in progression-free survival (PFS) and a 20% reduction in overall mortality relative to the standard of care at the time, an alkylating agent named dacarbazine. Stated differently, vemurafenib increased overall PFS of these patients by ~4 months. These numbers indicate that vemurafenib offers patients with V600E metastatic melanomas an extension of life span but only rarely a cure. Sooner or later, their tumors develop strategies for evading this drug and thereby acquire the ability to erupt again, although many of these relapsed tumors can now be treated effectively with checkpoint blockade immunotherapy, as discussed in Chapter 16.

Curiously, a significant percentage of vemurafenib-treated patients develop cutaneous squamous cell carcinomas, providing a clear indication that, paradoxically, this drug *stimulates* the outgrowth of previously latent, subclinical tumors. Indeed, in patients whose tumors carried mutations other than the V600E alteration, vemurafenib actually *favored* tumor growth, indicating the critical importance of sequence analysis of tumors prior to use of this drug.

One possible explanation for this apparently paradoxical biology is that ATP-competitive Raf inhibitors inhibit mutant B-Raf activity, but unexpectedly activate wild-type B-Raf signaling. Here, drug interaction with the ATP-binding pocket of one B-Raf protein within a B-Raf dimer leads to the activation of the other component of the dimer. This activation is dependent upon Ras activity and can result in elevated MEK/ERK activation in non-tumor cells (including, possibly, subclinical squamous cell carcinomas). The relevance of this unexpected biology is worth noting in the context of the therapeutic index. Because vemurafenib (and other Raf inhibitors) activates the Ras/MAPK pathway in normal cells, rather than inhibiting it, it is possible to apply very high doses of drug to treat the tumor. This therapeutic index also allows co-treatment with a MEK inhibitor to increase the potency of vemurafinib in the tumor without the usual toxicity of a MEK inhibitor in normal tissue.

Another explanation for this phenomenon relates to the propensity of deregulated Ras expression to promote the process of cellular senescence, as described in Section 9.11. Studies of transplanted melanomas in mice showed that tumors that acquired the ability to grow in the presence of vemurafenib actually developed a dependence on continued drug treatment. Here, resistance to vemurafenib treatment resulted from hyperactivation of Ras signaling and cessation of vemurafenib treatment, by reactivating B-Raf, led to elevation of already high levels of Ras signaling to a level at which Ras-induced senescence occurred. Conversely, in the context of squamous cell carcinomas, it is possible that vemurafenib treatment decreases senescence induction in pre-neoplastic lesions, allowing skin tumors to progress. The ability to engage a senescence response by halting vemurafenib treatment has promoted the idea that intermittent **drug holidays**, periods in which patients are not administered therapy, may promote better overall tumor control. This result highlights the idea, supported largely by empirical data, that the timing and scheduling of drug treatments can decidedly affect the therapeutic response.

The development of anti-tumor immunotherapies (as described in Chapter 16) has profoundly altered the treatment of melanoma. Still, vemurafenib has the prospect of becoming part of a multi-drug treatment protocol that one day will stop the previously unstoppable metastatic melanomas in their tracks. Indeed, vemurafenib is now used in combination with Mek inhibitors to treat B-Raf mutant melanoma. Additionally, an increasing number of tumors beyond melanoma have been shown to rely on signaling pathways that require B-Raf, and B-Raf mutation and activation can promote tumor cell resistance to targeted therapies, including EGF-R inhibitors in lung cancer. Thus, although not as effective as Gleevec before it, this drug confers significant benefit in melanoma patients and provides support for the goal of treating multiple primary and drug-resistant cancers with rationally developed drugs.

17.17 Synopsis and prospects: challenges and opportunities on the road ahead

"When is cancer going to be cured?" This is the simple and reasonable question posed most often to cancer researchers by those who are not directly involved in this area of biomedical research. In their minds are the histories of other public health measures. Infectious diseases, such as polio and smallpox, can be prevented, and bacterial infections are, almost invariably, cured. Heart disease is, in the eyes of many, well on its way to being prevented. Why should cancer be any different?

The information in this book provides some insights into the answers to these questions. As much as we have invoked unifying concepts to portray cancer as a single disease, the reality—at least in the eyes of clinical oncologists—is far different. Cancer is really a collection of more than 200 diseases, each affecting a distinct cell or tissue type in the body.

Pathological analyses have led us to embrace this number, or one a bit larger. (For example, there are at least eight distinct histopathological categories of breast cancer.) However, even the expanded number, large as it may be, represents an illusion: the current use of molecular diagnostics, specifically gene expression arrays, is leading to an explosion of subcategories, so that several hundred distinct neoplastic disease entities are likely to be recognized, each following its own, reasonably predictable clinical course and exhibiting its own responsiveness to specific forms of therapy. With the passage of time, cancer diagnoses will increasingly be made using bioinformatics rather than the trained eyes of a pathologist.

So the initial response to questions about "the cure" is that there won't be a single major breakthrough that will cure all cancers—a decisive battlefield victory—simply because cancer is not a single disease. Instead, there will be many small skirmishes that will steadily reduce the overall death rates from various types of cancer. And because certain molecular defects and pathological processes (for example, immune evasion) are shared by multiple human cancers, there will be occasions when therapeutic advances on a number of fronts will be made concomitantly.

Before we speculate on the future of cancer therapy, it is worthwhile to step back and assess the scope of the challenge: (1) How large is the problem of cancer and, in the future, how desperate will the need be to cure various types of neoplastic disease? (2) How well are we doing now in curing the major solid tumors?

Epidemiology and demographics provide some answers to the first question. They yield sobering assessments of the road ahead. The statistics illustrated in Figure 11.1 and cited in **Figure 17.44** demonstrate that cancer is largely a disease of the elderly, whose numbers are growing rapidly and will continue to do so, generating progressive increases in the numbers of cases of cancer-related deaths (mortality) over the coming decades.

Equally important, we still have only very imperfect ways of measuring incidence—how often the disease strikes. This greatly complicates assessments of the effectiveness of current therapies and future needs for therapy. As indicated in **Figure 17.45**, perceptions of the incidence of certain types of neoplastic disease are strongly influenced by diagnostic practices.

For many types of cancer, the more one looks, the more one finds. Statistics like those in Figure 17.45 suggest that, in the past, many cancers remained undiagnosed and asymptomatic for the lifetimes of those who carried them and that tumors of this class are contributing the lion's share to the perceived increases in disease incidence, notably of common tumors, such as those arising in the breast and prostate. (The major exceptions here are the cancers related to tobacco use, whose increased incidence in certain populations is real and beyond dispute because the incidence rates are closely paralleled by the rates of mortality.) Such statistics indicate that for certain common

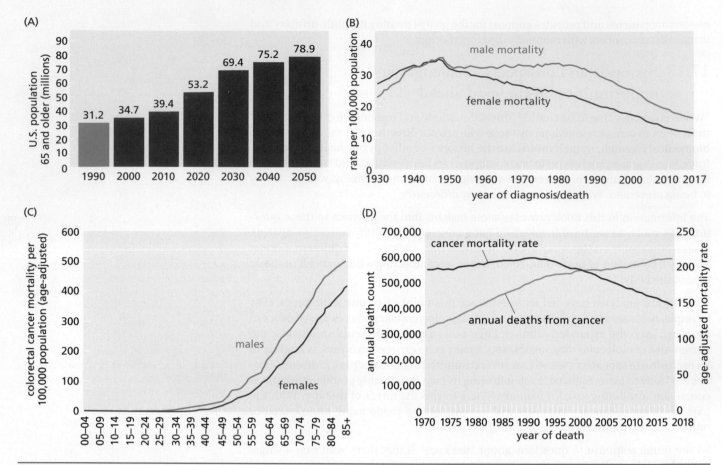

Figure 17.44 The demographics and epidemiology of cancer
(A) Because of dramatic decreases in midlife mortality, populations in industrialized countries are aging rapidly. In the United States, the number of individuals over 65 years of age has increased elevenfold since 1900, while the number of those under 65 has increased by a factor of 3. Comparable increases in the aged population are likely to occur worldwide over the next generation. (B) The age-adjusted death rate from colorectal cancer has changed only slightly over the past several decades. Other major cancers show similar curves. (C) Like many other diseases, cancer is uncommon during early and midlife and then increases rapidly. Shown here is the age-dependent death rate of colorectal cancer. (D) Because (1) the number of elderly individuals has increased steadily over the past century and will continue to do so (see panel A), (2) cancer is a disease of the elderly (panel C), and (3) the age-adjusted death rate of most cancers has

been declining only slowly (as suggested in panel B), the absolute number of annual deaths from cancer has increased dramatically over the past three-quarters of a century in the United States (red line) despite an overall decrease in age-adjusted mortality (blue line, a death rate that controls for the effects of differences in population age distributions). (The number of individuals afflicted with Alzheimer's disease has increased in parallel for the same reasons.) Because these trends are likely to continue, the burden of cancer cases in industrialized societies will continue to climb for many decades, albeit at a slower rate than before 1995. (A, courtesy of D. Singer and R. Hodes, from U.S. Bureau of the Census, Projections of 1996. B, from R. Siegel et al., *CA Cancer J Clin.* 70(3):145–164, 2020. With permission from John Wiley & Sons. C and D, from Natl. Cancer Inst, DCCPS, Surveillance Res. Program, May 2020 and the American Cancer Society.)

tumors—prostate and breast cancer being prominent examples—we have only a poor appreciation of the number of tumors that truly require treatment (see Supplementary Sidebar 17.2). Given these numbers, the current practice in the West of aggressively treating all individuals diagnosed with cancer (whether or not they truly require such treatment) will soon exceed the ability of national economies to support such care.

Data like those in Figure 17.45 also undermine the notion, deeply entrenched in the thinking of many cancer biologists and clinical oncologists, that all benign growths are in danger of becoming, sooner or later, highly malignant ones (see Chapter 11). Cancer epidemiology now makes us confront the alternative possibility: many kinds of early-stage tumors are unlikely to progress to high-grade malignancy during an average human life span, and the long-term side effects of aggressive treatment may often be more serious than the statistical likelihood of dying from these tumors. Unfortunately, we are only beginning to learn how to segregate those tumors that are truly deserving of aggressive treatment from those that are not (see, for example, Figure 17.4).

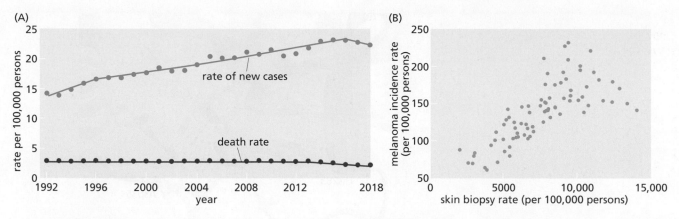

Figure 17.45 Melanoma incidence and mortality Melanoma incidence in the United States has increased by a factor of 6 over the past half century, raising the question whether this trend has created a comparable increase in the number of cases that truly require aggressive clinical treatment. (A) The age-adjusted breakdown of melanoma incidence, collected from 13 locations within the United States, indicates a dramatic increase over the past three decades in the overall incidence rate of disease (*blue curve*). The age-adjusted mortality rate (*red curve*) has been relatively constant during these years. These data raise the question: Has the real incidence of life-threatening disease been relatively constant over this period and has the registered incidence of this disease increased from changes in screening practices, revealing increasing numbers of non-life-threatening early-stage disease? (B) If melanoma incidence rates are plotted against the rates of screening of people aged 65 or over for melanoma in nine areas of the United States, as registered during several time periods between 2002 and 2009, the resulting scatter plot reveals a close correlation between the two. This provides a strong indication that the incidence of the disease is strongly influenced by diagnostic practices. It is therefore possible that (1) the true incidence of life-threatening melanomas (panel A) has not changed significantly over the past two decades; or (2) the true incidence of these tumors has increased, but intensification of screening has held the mortality rate at levels observed two decades ago by allowing the removal of early-stage tumors before they progress to become invasive and metastatic. (A, from https://seer.cancer.gov/statfacts/html/melan.html and H. G. Welch et al., *BMJ.* 331:481–484, 2005; B, from M. A. Weinstock et al, *Br. J. Dermatol.* 176: 949–954, 2016. With permission from John Wiley & Sons.)

As to the second question, which deals with the effectiveness of current cancer therapies, our perceptions have been strongly influenced by the fact that people are living longer with their cancers. This increase in longevity would seem to provide some reassurance that real progress is being made. However, some of these perceived improvements in therapy may, once again, be artifacts of increased screening and more sensitive detection techniques that increasingly uncover tumors relatively early in their development, giving the patient additional years of survival before tumor progression advances through its natural course, whether or not treatments are applied. This logic forces the conclusion that the efficacy of therapies can be accurately gauged only by well-controlled experiments: comparisons of several patient populations that are afflicted by the same malignancy and exposed in parallel to different agents or treatment protocols. Such side-by-side comparisons have, until recently, yielded only incremental gains in treating most solid tumors (see, for example, Figure 17.25), but this is beginning to change as new drugs and immune therapies are introduced into the clinic. Indeed, the age-adjusted mortality from breast cancer in Western countries has decreased by 40% in certain cohorts of women, and most of this decrease is likely attributable to aggressive clinical intervention.

The major hope at present is that agents such as Gleevec, erlotinib, vemurafenib, and rituximab are the forerunners of dozens and eventually hundreds of targeted, highly efficacious drugs. Computer-driven procedures, including high-throughput screening (HTS), automated determinations of the stereochemical structures of target molecules, and computer-aided design of drug molecules (**Figure 17.46**), are proving useful in this endeavor. Without such automated procedures, the current costs of developing novel anti-cancer drugs will soon become economically unsustainable, and the development of agents for less common subtypes of common cancers will never be realized.

Ultimately, the biggest challenge of drug development now and in the future is to demonstrate long-term efficacy: Does a drug being tested have significant effects on extending the life expectancy of cancer patients, doing so with acceptable levels of

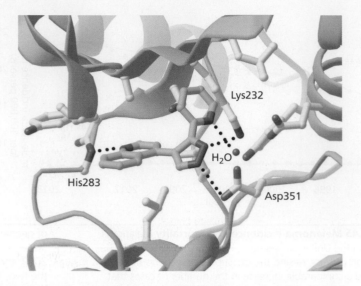

Figure 17.46 Virtual screening—designing drugs using the desktop computer High-throughput drug screening using various cancer cells as targets has yielded a large number of highly attractive drug candidates, many of which have advanced into clinical trials. In the future, however, much of this screening, which is very expensive, may be obviated by the development of powerful algorithms that enable pharmacologists to design drugs using the known stereochemical structures of target proteins. The development of the drug illustrated here began with a compound that showed very weak affinity (IC_{50} of 30 μM) for the ATP-binding site of the TGF-β type I receptor kinase. Researchers then queried a database of 200,000 known compounds for ones that shared some chemical features with the starting compound and conformed to the constraints imposed by the known structure of the ATP-binding site. This yielded the structures of 87 chemical species that satisfied these criteria and were then screened using conventional biochemical techniques. One of these, named HTS466284, is shown here ensconced in the ATP-binding site of the TGF-β type I receptor kinase; it exhibits an IC_{50} of 27 nM and thus functions as a potent inhibitor of TGF-β signaling. Significantly, another research group, working independently, arrived at the identical inhibitor molecule using conventional high-throughput screening. Hydrogen bonds are indicated by dotted lines. (From J. Singh et al., *Bioorg. Med. Chem. Lett.* 13:4355–4359, 2003. With permission from Elsevier.)

side-effect toxicities? And do we dare to hope that it can achieve durable responses, including cures?

As mentioned repeatedly here, rapidly evolving populations of cancer cells develop drug-evading mechanisms that protect them from many types of therapeutic attack. Indeed, recent evidence demonstrates directly that tumors that undergo more genetic diversification and thus evolution are likely to be more resistant to therapy (**Figure 17.47**). Some populations of cancer cells may develop drug evasion mechanisms by increasing their anti-apoptotic defenses (see Table 17.5). Yet others may lose checkpoint controls that previously made them sensitive to certain types of drug treatment, or evade the effects of administered drugs either by altering drug metabolism or enhancing drug efflux. Tumors whose cells were previously being killed through antibody-mediated attack may simply down-regulate expression of the telltale antigen. At present, it remains unclear whether we will, one day, be able to devise treatment strategies that anticipate the plasticity and evasiveness of cancer cells, allowing us to develop definitive cures of malignancies that have long been incurable.

These evasive maneuvers of cancer cells underlie the growing conviction of cancer researchers that most monotherapies are unlikely to yield curative treatments because they are destined to select for the outgrowth of variant cells within tumors that happen to have developed resistance to the single agent being applied. Such logic dictates that truly successful clinical outcomes and durable clinical responses will depend in the future on the development of multi-drug therapies and on the low probability that a clinically detectable tumor will harbor individual cancer cells that exhibit concomitant resistance to all of the cytotoxic agents dispatched to kill them.

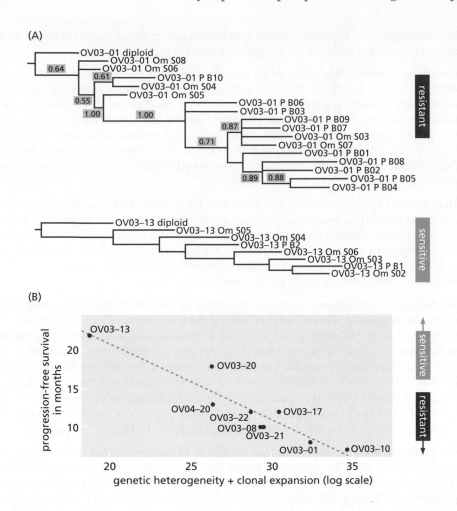

Figure 17.47 Genetic diversification and resistance to therapy A widely embraced notion is that tumors develop resistance to therapy through a process of genetic diversification followed by the selection and outgrowth of more therapy-resistant subpopulations of cells—a model formally akin to Darwinian evolution. This notion has rarely been tested experimentally in a clinical setting. As shown here, the ovarian tumors of individual patients were sampled successively during the course of their treatment, and the interrelatedness and thus genetic diversification of subpopulations within their tumors were gauged using genome sequence analyses and a bioinformatics algorithm. (A) The genetic relatedness of the subclones within two patients is indicated here. As is apparent, the tumor subpopulations within a patient whose ovarian carcinoma became resistant to therapy (*above*) exhibit extensive diversification, with multiple branch points illustrating the relationships among cell subpopulations within her tumor. Conversely, a patient whose tumor was sensitive to treatment (*below*) showed minimal branching and thus minimal genetic diversification of cancer cell populations. (B) The genetic behavior of nine ovarian tumors and the clinical progression of the patients carrying them are plotted here. As is apparent, a metric of intra-tumor genetic heterogeneity within individual tumors (*abscissa*) is correlated inversely with the extent of progression-free survival (*ordinate*). (Courtesy of R. Schwarz, J.D. Brenton, and F. Markowetz.)

This development of therapeutic resistance occurs sooner or later following long-term treatment of a variety of common human cancers. Included among these are high-grade tumors of the lung, breast, prostate, pancreas, ovary, and liver. Such acquired resistance represents the secondary resistance described earlier and indicates the grim scenario confronting medical oncologists at present, because it means that most therapeutic protocols (including those involving multi-drug therapy) will provide only a temporary respite from disease progression. As a consequence, the medical oncologist needs to respond, time after time, to the latest clinical relapse in a patient by devising a new protocol that responds to the most recent evasive maneuver developed by the patient's tumor.

The replacement of one therapeutic protocol that has failed a patient with another requires some understanding of the resistance mechanisms that have provoked clinical relapse and allowed therapy-resistant tumors to grow out. Indeed, in certain cancers, like EGF-R- or ALK-driven lung cancer or Bcr-Abl-driven leukemia, we have a very good understanding of the multiple alternative pathways toward resistance that are sought out by the neoplastic cells under treatment, including treatments involving targeted agents. Moreover, we have myriad new therapies that target the acquired resistant states.

Still, fundamental questions remain regarding how we should treat patients exposed to front-line therapies. For example, can we predict which pathway of resistance will eventually emerge in a given tumor? Can we direct tumors to evolve toward specific resistance states that we are better able to treat? And can we preempt the development of resistance by adding a second-line therapy drug that targets the emergent vulnerabilities of therapy-resistant tumor cells? Additionally, in an ideal world, each of this succession of treatment protocols would provoke tolerable levels of side-effect toxicities, and the sequence of treatments would allow the cancer patient to live in a relatively symptom-free state, surviving in many cases to the end of what would be a normally expected life span. We are still very far from reaching this goal.

Clearly, cell-intrinsic (as opposed to microenvironment-mediated) drug resistance derives from the instability of cancer cell genotypes and phenotypes that allow the emergence of therapy-resistant variant cell clones, as detailed in Chapter 12. The evidence presented in Figure 17.47 provides possibly the first direct demonstration that genetic plasticity plays a central role in generating secondary resistance. In truth, however, we still do not understand how much of this secondary resistance derives from genetic instability and how much can be traced to shifts in the cell-heritable epigenetic programs (see Section 1.8) of these cells.

Here we also confront the fact that there are still no clear and logical strategies for designing multi-drug therapies, beyond the strategy that has been in place for half a century: choose combinations of drugs having cytotoxic mechanisms of action that are distinct from one another, doing so with the hope that they will act synergistically to reduce the burden of cancer cells in a patient's body below the threshold level at which the clinical resurgence of the disease becomes likely.

The lack of clear therapeutic strategies regarding the construction of multi-drug regimens also suggests a related question: What if the real powers of a drug under development are only realizable when it is used in combination with several others? In theory, many anti-cancer agents should fall into this category. If true, it suggests that many truly useful drug candidates have been discarded in the past, having failed tests in monotherapy trials, and that many others will suffer this fate in the future, simply because their true utility as components of multi-drug treatment protocols will never be tested. Here again, one hopes that our increasing understanding of the subcellular signaling circuitry will ameliorate this situation (**Sidebar 17.16**).

The existence of tumor stem cells (see Section 11.6) creates another major challenge for anti-cancer drug development. Recall that the self-renewing cancer stem cells (CSCs) can seed new tumors, whereas their far more numerous transit-amplifying progeny, with limited self-renewal capacity, cannot. Such a hierarchical organization was demonstrated initially in leukemias, breast carcinomas, and brain tumors, but it probably exists in most tumors and holds important implications for the development of anti-cancer drugs. Traditionally, clinical validation of the therapeutic efficacy of these drugs has depended on demonstrations of their ability to halt further tumor growth or decrease tumor size. Many anti-tumor agents in clinical trials may shrink tumor masses by eliminating populations of transit-amplifying and more differentiated neoplastic cells, which together constitute the great bulk of the tumors' cellularity. However, if such drugs leave the minority populations of self-renewing CSCs untouched (see Figure 17.35A), the tumor has a high probability of regrowing, leading sooner or later to clinical relapse. Indeed, as mentioned, there is clear evidence that CSCs are more resistant to a variety of modes of current anti-cancer therapy (see Sidebar 17.13).

Sidebar 17.16 Legal and financial disincentives decrease the likelihood of drugs being tested in combination At present, the biological difficulties of testing candidate drugs in combination with others are compounded by economic forces that often create disincentives for pharmaceutical companies to test their own proprietary drugs in combination with those produced by their competitors. Patent regulations have also discouraged certain uses of patented compounds by firms that are in direct competition with the patent holders.

In the past, the difficulties of organizing early multi-drug clinical trials were further compounded in the United States by the Food and Drug Administration's (FDA's) insistence that the clinical efficacy of drug candidates be demonstrated first as monotherapies. But this practice has begun to change, albeit slowly. For example, Erbitux, the anti-EGF-receptor monoclonal antibody (see Sidebar 15.5), was initially approved for use together with irinotecan because the two used in concert yielded far better responses than did irinotecan on its own. (Irinotecan is a more traditional cytotoxic agent that functions as a topoisomerase I inhibitor; see Table 17.2.)

Such evidence suggests that durable remissions and cures can only derive from therapies that strike both the majority of non-CSCs in tumors and, importantly, those that lie at the heart of the tumor—the self-renewing CSCs. Currently, assays for the presence of these cells in most tumor types are poorly developed. Hence, drug development efforts are hobbled in part because researchers lack key analytical tools that are essential for developing truly efficacious therapies.

But there is another problem that has only recently emerged: the non-CSCs within a tumor, more specifically the transit-amplifying/progenitor cells (see Figure 11.20C), are able in some tumors, and quite possibly in most tumors, to dedifferentiate and thereby form new CSCs (see Figure 17.35C). If validated as a general phenomenon, this tumor cell plasticity adds emphasis to the necessity of eliminating both the existing CSCs within tumors as well as their more differentiated progeny, which may wait in the wings, always poised to regenerate new CSCs.

A major, still-unsolved problem concerns the biological models that are used in pre-clinical drug development. Some xenograft models of human cancer are useful in predicting the behavior of tumors encountered in the oncology clinic, but most are not (see Sidebar 13.13). Similarly, genetically engineered mouse models of human cancer have been limited in their ability to predict clinical responses. However, we now have an expanding array of tumor cell lines, organoid models, primary human patient-derived xenograft (PDX) models, and "humanized" mouse tumors, as well as gene-editing technologies that permit the re-creation of complex tumor genomes. The use of these systems to create predictive pre-clinical models of human cancer will surely reduce the expense of drug development and, quite possibly, even obviate certain early-phase clinical trials.

The non-neoplastic stromal cell types within a tumor may be major determinants of responsiveness to most drug therapies, yet their proportions and respective contributions are not recognized in the design of most pre-clinical models of human cancer. In Chapter 13, we learned, for example, that the most radiosensitive cells in some tumors are likely to be the endothelial cells forming their neovasculature (see Supplementary Sidebar 13.10). By the same token, it seems increasingly likely that many widely used anti-cancer chemotherapeutics have strong effects on the tumor-associated endothelial cells that were never suspected in the past. In fact, some researchers are redesigning chemotherapeutic treatment protocols in order to optimize their toxic effects on the tumor-associated neovasculature. Recognition of these two classes of tumor-associated cells—cancer stem cells and stromal cells—as critically important biological targets of chemotherapy will surely change the entire landscape of the drug development field. Moreover, the concept of "stroma" will surely be expanded to include the microenvironments of tumor cells throughout the body as we learn that certain sites of metastatic dissemination offer particularly hospitable homes for cancer cells, protecting them from the toxic effects of various therapies.

The ultimate test of many candidate anti-cancer drugs comes when they are first tested in substantial numbers of cancer patients. Here, the drug developer is often confronted by the dilemma of not knowing which types of tumors are likely to respond. Should a candidate drug be tested on patients suffering from pancreatic carcinomas or neuroblastomas? The molecular lesions discovered in these and other types of cancer cells would seem to be highly useful indicators for informing this decision. But quite often, tumors respond for reasons that cannot be predicted by the mutant genes and deregulated signaling pathways known to be present in these growths, and so the choices of patients recruited into clinical trials are arbitrary and sub-optimal. Once again, we can only hope that our increasing insights into the molecular etiologies of various cancers will provide truly useful guidelines for oncologists to follow.

Independent of the challenges of cancer drug development and testing are the more transcendent problems created by the complex biology of human cancers: Will different types of drugs need to be developed for different classes of cancers or will a small number of treatments find wide applicability? Will different tumors within a given class (for example, colon carcinomas) require distinct, tailor-made treatments based on their particular genotypes and phenotypes? And will we one day be able to provide "personalized molecular medicine" at an affordable cost, in which the detailed characteristics of each patient's tumor genome, transcriptome, and proteome inform the design of a customized therapy?

Will anti-neoplastic drugs ever be developed that have lethal effects on malignant growths while having minimal side effects on normal tissues? And should drug designers undertake to develop anti-cancer drugs that keep tumors under control rather than attempting to wipe them out? The goal of curing many kinds of tumors, cited above, may well be an unreachable one, and for these tumors, reducing cancer to a chronic but bearable disease may be a more realizable goal. (Such is the thinking of researchers developing new types of anti-HIV treatment protocols, some of which have achieved this goal.)

A recent census of genes mutated in human cancer cell genomes describes 576 distinct genes—more than 2% of the genes present in the human genome. Of these, 533 become involved in cancer largely through somatic mutation, while the remaining 43 are exclusively inherited as germ-line determinants of cancer susceptibility. This list—generated in significant part from the results of cancer genome sequencing—will likely grow somewhat in the future. The list of somatically altered genes will be increased by inclusion of myriad tumor suppressor genes that are shut down by promoter methylation; recall that promoter methylation is as effective as genetic mutation in eliminating genes and proteins from the regulatory circuitry of cells. At the same time, the census of germ-line determinants of cancer will expand as alleles (and thus genes) having less penetrant effects on cancer susceptibility are uncovered.

This census suggests many druggable targets for researchers. At the same time, it represents bewildering complexity. Some of these genes are mutated only in rare cancers, and the costs of developing therapeutic drugs directed against their protein products are unlikely to ever be recouped through drug sales. Most of the encoded mutant proteins contribute in still-obscure ways to the neoplastic growth of various types of human cancer cells. How will we ever ascribe key roles to each of the more than 570 mutant genes that are already known or suspected to be involved in the pathogenesis of many cancers? [Of relevance here is that only a limited portion of the **human kinome** (see Supplementary Sidebar 17.7) has been extensively explored by laboratory researchers and pharmaceutical chemists. A 2009 survey indicated that half of the kinome (a total of ~522 protein kinases) had been described by 20 or fewer research publications, in contrast to 10 kinases (all discussed in some detail in this book) that had been the topics of 5000 or more reports. Additionally, less than 10% of the kinome has been successfully targeted in the form of FDA-approved therapies. Many of the under-studied, potentially druggable enzymes are likely to participate in tumor pathogenesis, even when present in wild-type configuration within cancer cells. Hence, a large array of potentially targetable kinases has yet to be intensively explored.]

Many cancer researchers would like to understand the entirety of a biological system, such as a living cancer cell, rather than its individual functional components. In their eyes, reductionist biology, which focuses on the individual, isolatable components of complex systems, has had its day, and the time has come for the vast amounts of information known about these components to be integrated into complex interacting *systems* whose behavior can be predicted by bioinformatics.

Successes in these efforts, involving the new discipline of "systems biology," will surely benefit cancer research. Imagine a day when the biological responses of various human cells, normal and malignant, can be predicted by mathematical models of these cells and their internal control circuits. Such advances will render many current practices in experimental biology, including many steps of drug development, unnecessary. If this ever becomes possible, drug development will be more a matter of dry bioinformatics than wet biology at the laboratory bench.

But for the moment, most of this remains a pipe dream, still far in the future. For now, at least, we need to wrestle with the grim realities of drug development, the inadequate animal models, our ignorance of the behavior of cellular regulatory circuitry, and the confounding biological complexities of human cancer. And most importantly, we must never give up. If our ancestors had, we would still be living in the Stone Age.

Key concepts

- The science of molecular oncology has revealed hundreds of proteins whose malfunction contributes to the formation and maintenance of tumors. Many of these proteins exhibit molecular properties that make them attractive targets for novel anti-cancer therapeutic agents, such as monoclonal antibodies or low–molecular-weight drugs.

- Proteins that are attractive targets for attack by antibodies are invariably located at the cell surface or in the extracellular space.

- Most proteins that are attractive targets for attack by low–molecular-weight compounds are enzymes that possess druggable catalytic clefts.

- Recent advances have expanded the range of druggable targets to include certain protein–protein interactions and protein conformational states that can also be inhibited by low–molecular-weight drugs.

- The proteins that are most suitable as targets are those whose inactivation is predicted to make tumor cells cease proliferating, to promote cell differentiation, or to lead to their death by apoptosis.

- The most successful targeted anti-cancer drugs developed to date have been those that interfere with the functioning of various growth- and survival-promoting kinases, specifically, receptor-associated tyrosine kinases.

- Successful drugs must have a high therapeutic index, appropriate pharmacokinetics and pharmacodynamics, and minimal side effects on major organ systems.

- Studies of drugs in Phase I, II, and III clinical trials are essential because preclinical studies of drugs' efficacy and tolerability are poorly predictive of the drugs' behavior in humans.

- Clinical use of certain drugs under development may be indicated by the known behavior of targeted proteins in cancer cells (as in the case of Gleevec) or by empirical tests of how various types of human tumors respond to treatment (as with Velcade).

- Stratification of outwardly similar tumors into narrow subclasses greatly helps researchers and clinicians to match drugs to the specific tumor cell types they can most effectively treat.

- The greatest benefit of certain drugs, such as Gleevec, may eventually prove to derive from a broad target specificity that allows them to target multiple pathways in tumor cells and adjacent stromal cells in a wide range of cancers.

- The treatment of tumors with targeted therapies is almost universally accompanied by the development of drug resistance, and so treatment strategies that target the resistant state are critical for durable responses.

Thought questions

1. What are the therapeutic advantages and disadvantages of using a drug that affects a broad range of molecular targets?

2. Given the large and heterogeneous collection of signaling molecules that have been portrayed in this book as playing key roles in the pathogenesis of various cancers, which classes of molecules do you think might become the targets for the development of a new range of anti-cancer therapeutics besides the much-studied kinases?

3. How might tumors that were initiated by the formation of a certain oncogene become independent of this oncogene later in tumor progression?

4. What strategies might you implement to preempt the development of tumor drug resistance or target the drug resistant state?

5. Having concluded that natural products represent a rich resource of potential anti-cancer drugs, what obstacles might limit the search for and testing of such drugs?

6. What obstacles stand in the way of developing drugs for tumors that represent only a very small proportion of the total cancer burden in a population?

7. In the oncology clinic of the future, what types of information might be included in the assembly of anti-cancer therapies that are tailor-made specifically to respond to an individual patient's tumor?

8. Given the new tools available (including gene editing technology) for cancer cell target identification, what criteria should be used to nominate potential drug targets for clinical development?

9. What strategies would you pursue to develop pre-clinical models of human tumors that are highly useful in predicting patient responses to candidate drugs?

10. Given that current combination chemotherapy regimens are highly toxic and provided without a clear sense of which patients will respond and which of the component drugs have efficacy in individual patients, how would you develop strategies to reduce or replace these front-line treatments with more personalized therapies?

Additional reading

Alison MR, Lim SML & Nicholson LJ (2010) Cancer stem cells: problems for therapy? *J. Pathol.* 223, 147–161.

Arbiser J (2007) Why targeted therapy hasn't worked in advanced cancer. *J. Clin. Invest.* 117, 2762–2765.

Barouch-Bentov R & Sauer K (2011) Mechanisms of drug-resistance in kinases. *Expert Opin. Investig. Drugs* 20, 153–208.

Barr S, Thomson S, Buck E et al. (2008) Bypassing cellular EGF receptor dependence through epithelial-to-mesenchymal-like transitions. *Clin. Exp. Metastasis* 25, 685–693.

Black WC & Welch HG (1993) Advances in diagnostic imaging and overestimations of disease prevalence and the benefits of therapy. *N. Engl. J. Med.* 328, 1237–1243.

Cantor JR & Sabatini DM (2012) Cancer cell metabolism: one hallmark, many forms. *Cancer Discov.* 2, 881–898.

Cantwell-Dorris ER, O'Leary JJ & Sheils OM (2011) BRAF^V600E: implications for carcinogenesis and molecular therapy. *Mol. Cancer Ther.* 10, 385–394.

Chabner BA & Roberts TG Jr (2005) Chemotherapy and the war on cancer. *Nat. Rev. Cancer* 5, 65–72.

Cohen P (2002) Protein kinases: the major drug targets of the twenty-first century? *Nat. Rev. Drug Discov.* 1, 309–315.

Creighton CJ, Chang JC & Rosen JM (2010) Epithelial-mesenchymal transition (EMT) in tumor-initiating cells and its clinical implications in breast cancer. *J. Mammary Gland Biol. Neoplasia* 15, 253–260.

Curran T (2018) Reproducibility of academic preclinical translational research: lessons from the development of Hedgehog pathway inhibitors to treat cancer. *Open Biol.* 8:180098.

Deininger MW (2008) Nilotinib. *Clin. Cancer Res.* 14, 4027–4031.

Di Cosimo S & Baselga J (2010) Management of breast cancer with targeted agents: importance of heterogenicity. *Nat. Rev. Clin. Oncol.* 7, 139–147.

Easton JB & Houghton PJ (2006) mTOR and cancer therapy. *Oncogene* 16, 6436–6446.

Felsher DW (2008) Oncogene addiction versus oncogene amnesia: perhaps more than just a bad habit? *Cancer Res.* 68, 3081–3086.

Fojo T & Grady C (2009) How much is life worth: cetuximab, non-small cell lung cancer, and the $440 billion question. *J. Natl. Cancer Inst.* 101, 1044–1048.

Frei E III (1985) Curative cancer chemotherapy. *Cancer Res.* 45, 6523–6537.

Garber K (2009) Trial offers early test case for personalized medicine. *J. Natl. Cancer Inst.* 101, 136–138.

Garraway LA & Jänne PA (2012) Circumventing cancer drug resistance in the era of personalized medicine. *Cancer Discov.* 2, 214–226.

Gerard L, Duvivier L, Gillet JP (2021) Targeting tumor resistance mechanisms. *Fac. Rev.* Jan 26,10:6.

Gilbert LA & Hemann MT (2011) Chemotherapeutic resistance: surviving stressful situations. *Cancer Res.* 71, 5062–5066.

Glunde K, Pathak AP & Bhujwalla ZM (2007) Molecular-functional imaging of cancer: to image and imagine. *Trends Mol. Med.* 13, 287–297.

Hammerman PS, Jänne PA & Johnson BE (2009) Resistance to epidermal growth factor receptor tyrosine kinase inhibitors in non-small cell lung cancer. *Clin. Cancer Res.* 15, 7502–7509.

Hellman S (2005) Evolving paradigms and perceptions of cancer. *Nat. Rev. Clin. Oncol.* 2, 618–624.

Hill C & Wang Y (2020) The importance of epithelial-mesenchymal transition and autophagy in cancer drug resistance. *Cancer Drug Resist.* 3:38–47.

Hirst GL & Balmain A (2004) Forty years of cancer modeling in the mouse. *Eur. J. Cancer* 40, 1974–1980.

Horn L & Sandler A (2009) Epidermal growth factor receptor inhibitors and antiangiogenic agents for the treatment of non-small cell lung cancer. *Clin. Cancer Res.* 15, 5040–5048.

Humphrey RW, Brockway-Lunardi LM, Bonk DT et al. (2011) Opportunities and challenges in the development of experimental drug combinations for cancer. *J. Natl. Cancer Inst.* 103, 1222–1226.

Hunter T (2007) Treatment for chronic myelogenous leukemia: the long road to imatinib. *J. Clin. Invest.* 117, 2036–2043.

Kaelin WG Jr (2005) The concept of synthetic lethality in the context of anticancer therapy. *Nat. Rev. Cancer* 5, 689–698.

Khazak V, Astsaturov I, Serebriiskii IG & Golemis EA (2009) Selective Raf inhibition in cancer therapy. *Expert Opin. Ther. Targets* 11, 1587–1609.

Klein S & Levitzki A (2009) Targeting the EGFR and the PKB pathway in cancer. *Curr. Opin. Cell Biol.* 21, 1–9.

Klein S, McCormick F & Levitzki A (2005) Killing time for cancer cells. *Nat. Rev. Cancer* 5, 573–580.

Konopleva M, & Letai A (2018) BCL-2 inhibition in AML: an unexpected bonus? *Blood* 132:1007–1012.

Lane HA & Breuleux M (2009) Optimal targeting of the mTORC1 kinase in human cancer. *Curr. Opin. Cell Biol.* 21, 219–229.

Letai, A (2017) Apoptosis and Cancer. *Annu. Rev. Cancer Biol.* 1:275–294.

Leverson JD, Sampath D, Souers AJ et al. (2017) Found in translation: how preclinical research is guiding the clinical development of the BCL2-selective inhibitor Venetoclax. *Cancer Discov.* 7:1376–1393.

Lynch TJ, Bell DW, Sordella R et al. (2004) Activating mutations in the epidermal growth factor receptor underlying responsiveness of non-small-cell lung cancer to gefitinib. *N. Engl. J. Med.* 350, 2129–2139.

Massagué J (2007) Sorting out breast-cancer gene signatures. *N. Engl. J. Med.* 358, 294–297.

Milojkovic D & Apperley J (2009) Mechanisms of resistance to imatinib and second-generation tyrosine inhibitors in chronic myeloid leukemia. *Clin. Cancer Res.* 15, 7519–7527.

Moore AR, Rosenberg SC, McCormick F & Malek S (2020) RAS-targeted therapies: is the undruggable drugged? *Nat. Rev. Drug Discov.* 19:533–552.

Orlowski RZ & Kuhn DJ (2008) Proteasome inhibitors in cancer therapy: lessons from the first decade. *Clin. Cancer Res.* 14, 1649–1657.

Palmer AC, Sorger PK (2017) Combination cancer therapy can confer benefit via patient-to-patient variability without drug additivity or synergy. *Cell* 171(7):1678–1691.

Pritchard JR, Bruno PM, Gilbert LA et al. (2013) Defining principles of combination drug mechanisms of action. *Proc. Natl. Acad. Sci.* 110:E170-9.

Rauch J, Volinsky N, Romano D & Kolch W (2011) The secret life of kinases: functions beyond catalysis. *Cell Commun. Signal.* 9, 23, doi: 10.1186/1478-811X-9-23.

Riese DJ 2nd, Gallo RM & Settleman J (2007) Mutational activation of ErbB family receptor tyrosine kinases: insights into mechanisms of signal transduction and tumorigenesis. *Bioessays* 29, 558–565.

Sawyers C (2004) Targeted cancer therapy. *Nature* 432, 204–207.

Scaltriti M & Baselga J (2006) The epidermal growth factor receptor pathway: a model for targeted therapy. *Clin. Cancer Res.* 12, 5268–5272.

Schiller JT & Lowy DR (2006) Prospects for cervical cancer prevention by human papillomavirus vaccination. *Cancer Res.* 66, 10229–10232.

Schlessinger J (2005) SU11248: genesis of a new cancer drug. *The Scientist* 19, 17–18.

Sebolt-Leopold JS & Herrera R (2004) Targeting the mitogen-activated protein kinase cascade to treat cancer. *Nat. Rev. Cancer* 4, 937–947.

Sharpless NE & DePinho RA (2006) The mighty mouse: genetically engineered mouse models in cancer drug development. *Nat. Rev. Drug Discov.* 5, 741–754.

Shibue T & Weinberg RA (2017) EMT, CSCs, and drug resistance: the mechanistic link and clinical implications. *Nat. Rev. Clin. Oncol.* 10:611–629.

Singh A & Settleman A (2010) EMT, cancer stem cells and drug resistance: an emerging axis of evil in the war on cancer. *Oncogene* 29, 4741–4751.

Sotiriou C & Pusztai L (2009) Gene-expression signatures in breast cancer. *N. Engl. J. Med.* 360, 790–800.

Staudt LM (2003) Molecular diagnosis of the hematologic cancers. *N. Engl. J. Med.* 348, 1777–1785.

Stegmeier F, Warmuth M, Sellers WR & Dorsch M (2010) Targeted cancer therapies in the twenty-first century: lessons from imatinib. *Clin. Pharmacol. Ther.* 87, 543–552.

Teodori E, Dei S, Martelli C et al. (2006) The functions and structure of ABC transporters: implications for the design of new inhibitors of Pgp and MRP1 to control multidrug resistance (MDR). *Curr. Drug Targets* 7, 893–909.

Tinker AV, Boussioutas A & Bowtell DDL (2006) The challenges of gene expression microarrays for the study of human cancer. *Cancer Cell* 9, 333–339.

Tsou LK, Cheng Y & Cheng Y-C (2012) Therapeutic development in targeting protein–protein interactions with synthetic topological mimetics. *Curr. Opin. Pharmacol.* 12, 403–407.

van't Veer LJ & Bernards R (2008) Enabling personalized cancer medicine through gene-expression patterns. *Nature* 452, 564–570.

Welch HG & Black WC (2010) Overdiagnosis in cancer. *J. Natl. Cancer Inst.* 102, 605–613.

Welch HG, Woloshin S & Schwartz LM (2008) The sea of uncertainty surrounding ductal carcinoma in situ: the price of screening mammography. *J. Natl. Cancer Inst.* 100, 228–229.

Zhou B-B, Zhang H, Damelin M et al. (2009) Tumour-initiating cells: challenges and opportunities for anticancer drug discovery. *Nat. Rev. Drug Discov.* 8, 806 823.

Zhu J, Chen Z, Lallemand-Breitenbach V & de Thé H (2002) How acute promyelocytic leukaemia revived arsenic. *Nat. Rev. Cancer* 2, 705–714.

Abbreviations

2D	two-dimensional
3D	three-dimensional
3-MC	3-methylcholanthrene
13cRA	13-*cis*-retinoic acid
A	(1) adenine; (2) adenosine
AA	aristolochic acid
ABC	(1) activated B-cell-like subtype of DLBCL; (2) adenoid basal carcinoma; (3) ATP-binding cassette (transporter)
Abl	Abelson leukemia virus oncoprotein
ABVD	adriamycin, bleomycin, vinblastine, dacarbazine combined chemotherapy
ACF	aberrant (intestinal) crypt focus/foci
ACT	adoptive cell transfer
ACTH	adrenocorticotropic hormone
ADC	antibody–drug conjugate
ADCC	antibody-dependent cellular cytotoxicity
ADCP	antibody-dependent cellular phagocytosis
AE	adverse (clinical) event
AEV	avian erythroblastosis virus
AFB1	aflatoxin B1
Ag	antigen
AGMK	African green monkey kidney (cells)
AGT	O^6-alkylguanine-DNA-alkyltransferase; *also called* **MGMT**
AID	activation-induced cytidine deaminase
AIDS	acquired immunodeficiency syndrome
Akt	T-cell lymphoma of Ak mouse strain (serine-threonine kinase); *also* PKB
ALDH1	aldehyde dehydrogenase isoform 1
ALK	anaplastic large-cell lymphoma kinase
ALL	acute lymphocytic (or lymphoblastic) leukemia
α-SMA	α-smooth muscle actin
ALT	alternative lengthening of telomeres
ALV	avian leukosis virus
AML	acute myelogenous leukemia
AMP	adenosine monophosphate
AMPK	adenosine-5′-monophosphate-activated kinase
AMV	avian myelocytomatosis virus
Ang	angiopoietin
ANGPTL	angiopoietin-like (protein)
AP	(1) apurinic; (2) apyrimidinic
AP-1	activator protein-1 transcription factor (Fos + Jun complex)
Apaf-1	apoptotic protease-activating factor-1
APC	(1) antigen-presenting cell; (2) anaphase-promoting complex; (3) adenomatous polyposis coli
APE	apurinic/apyrimidinic endonuclease
APL	acute promyelocytic leukemia
APOBEC	apolipoprotein B mRNA editing enzyme catalytic peptide-like
AR	androgen receptor
ARE	antioxidant response element (DNA sequence)
ARF	alternative reading frame (of protein-encoding sequence)
ASCT	allogeneic stem cell transplantation
asmase	acid sphingomyelinase
AT	ataxia telangiectasia (syndrome)
ATACseq	assay for transposase-accessible chromatin (followed by sequencing)
ATM	mutated in ataxia telangiectasia
ATP	adenosine triphosphate
ATR	(1) ATM-related kinase; (2) ATM and Rad3-related kinase
ATRA	all-*trans*-retinoic acid
a.u.	arbitrary units
AUC	area under the curve
BCC	basal cell carcinoma
BCG	Bacillus Calmette–Guérin (mycobacterial vaccine)
Bcl-2	B-cell lymphoma gene-2
BCNS	basal cell nevus syndrome
BCR, bcr	(1) breakpoint cluster region; (2) B-cell receptor
BDNF	bone-derived neurotrophic factor
BER	base-excision DNA repair
β-gal	β-galactosidase
β2m	β2 microglobulin
BFB	breakage–fusion–bridge (cycle of chromosomes)
BH	Bcl-2–homologous (domain)
BHK	baby hamster kidney (cells)
bHLH	basic helix–loop–helix (transcription factor)
BiTE	bi-specific T cell engager
BL	Burkitt's lymphoma
BM	(1) bone marrow; (2) basement membrane
BMI	body-mass index
BMP	bone morphogenetic protein
BMPR1	BMP receptor-1
BMSC	bone marrow stem cell
BMT	bone marrow transplantation
BP	benzo[*a*]pyrene
bp	base pair
BPDE	benzo[*a*]pyrenediolepoxide
BPV	bovine papillomavirus
BrdU	bromodeoxyuridine
B$_{reg}$	regulatory B cell
BTK	Bruton's TK
bZIP	basic leucine zipper (transcription factor)
C	(1) carboxy (terminus of polypeptide); (2) constant region (of an antibody molecule); (3) cytidine, cytosine
ca	(1) cancer; (2) carcinoma
CAD	caspase-activated DNase
CAF	cancer-associated fibroblast
CagA	cytotoxin-associated gene A
CAK	CDK-activating kinase

CalB	calcium–phospholipid binding domain
cAMP	cyclic adenosine monophosphate
CAR	chimeric antigen receptor
CAR T	T lymphocyte expressing a CAR
CASP	caspase-encoding gene
Cas9	CRISPR-associated protein-9
CAT	computed axial tomography; *see also* **CT**
CBL	Casitas B-lineage lymphoma (protein)
CBP	cyclic AMP response element–binding protein
CCK	cholecystokinin
CCL	chemokine ligand with two adjacent cysteines
CCLE	Cancer Cell Line Encyclopedia
CD	cluster of differentiation (cell surface antigen)
CDC	complement-dependent cytotoxicity
CDH1	E-cadherin
Cdh1	regulatory subunit of the anaphase-promoting complex
CDK	cyclin-dependent kinase
Cdki	CDK inhibitor
CDKN1A	p21^{Cip1} Cdki
CDKN1B	p27^{Kip1} Cdki
CDKN1C	p57^{Kip2} Cdki
cDNA	complementary DNA copy of mRNA
CDR	complementarity-determining region (of an antibody)
CEA	carcinoembryonic antigen
C/EBPβ	CCAAT/enhancer-binding protein β
CEF	chicken embryo fibroblast
CEP	circulating endothelial progenitor (cell)
CFP	cyan fluorescent protein
CGH	comparative genomic hybridization
cGMP	cyclic guanosine monophosphate
CHFR	checkpoint with forkhead and RING finger domains (protein)
ChIP	chromatin immunoprecipitation
Chk2	checkpoint kinase 2
CHK2	gene encoding Chk2 protein
CHOP	cyclophosphamide, doxorubicin, vincristine, and prednisone (chemotherapy cocktail)
CI	confidence interval
CIMP	CpG island methylator phenotype
CIN	(1) cervical intraepithelial neoplasia; (2) chromosomal instability
circRNA	small circular RNA
CIS	carcinoma *in situ*
CK1	casein kinase I
CK2	casein kinase II
CKI	cyclin-dependent kinase inhibitor
CLL	chronic lymphocytic leukemia
CM	(1) conditioned medium; (2) culture medium
CML	chronic myelogenous leukemia
CMV	cytomegalovirus
CNA	(gene) copy number alteration
CNK	connector enhancer of KSR
CNL	chronic neutrophilic leukemia
CNS	central nervous system
CNV	(gene) copy number variation
CoA	coenzyme A
CoO	the (normal) cell-of-origin (of a population of cancer cells)
COSMIC	Catalogue of Somatic Mutations in Cancer
COX-2	cyclooxygenase-2
CPD	cyclobutane pyrimidine dimer
CPE	cytopathic effect
CpG	cytidine phosphate guanosine (DNA sequence)
cpm	counts per minute
CR	complete (clinical) response
CRC	colorectal carcinoma

cre	causes recombination
CRISPR	clustered regularly-interspaced short palindromic repeats
CRISPRa	CRISPR-mediated activation (of transcription)
CRISPRi	CRISPR-mediated inhibition (of transcription)
Crkl	Crk-like; *also* CrkL
CRS	cytokine release syndrome
CRT	calreticulin
CS	Cockayne syndrome
CSC	cancer stem cell
CSF-1	colony-stimulating factor-1
Csk	C-terminal Src kinase
CSR	core serum response (gene expression signature)
CT	(1) cancer-testis (antigen); (2) computed tomography (usually using X-ray scanning); (3) chemotherapy
CTC	circulating tumor cell
CTGF	connective tissue growth factor
CTL	cytotoxic T lymphocyte
CTLA-4	cytotoxic T lymphocyte–associated antigen-4
CTVS	canine transmissible venereal sarcoma
CUP	(metastasized) cancer of unknown primary origin
CXCL	ligand of chemokine receptor
CXCR	chemokine receptor
CYP	cytochrome P450 enzyme
d5′-mC	deoxy-5′-methylcytosine
dA	deoxyadenosine
DAAM	diaphanous-related formin protein
DAG	diacylglycerol
DAPI	4′,6′-diamidino-2-phenylindole (DNA stain)
DBD	DNA-binding domain
DC	(1) dendritic cell; (2) dyskeratosis congenita
DCIS	ductal carcinoma *in situ* (of the breast)
DDR	discoidin domain receptor
del	genetic deletion
DepMap	Cancer Dependency Map
DES	diethylstilbestrol
DFS	disease-free survival
dG	deoxyguanosine
DIABLO	direct IAP-binding protein with low pI
di-I	3,3-dioctadecyl indocarbocyanine (dye)
DISC	death-inducing signaling complex
DKK	Dickkopf (inhibitor of Wnts)
DLBCL	diffuse large B-cell lymphoma
DLC	(1) dynein light chain; (2) deleted in liver cancer
DLL	Delta-like ligand
DM	double-minute chromosome
DM1	mertansine anti-microtubule drug
DMBA	dimethylbenz[*a*]anthracene
dn	*DN* dominant-negative (allele)
DNMT	DNA methyltransferase
DP	differentiation-related transcription factor protein
DPC4	deleted in pancreatic cancer-4
dR	deoxyribose
DRG	dorsal root ganglion
ds	double-stranded (DNA or RNA)
DSB	double-strand (DNA) break
DTC	disseminated tumor cell
dTg	deoxythymidine glycol
DUB	de-ubiquitylating enzyme
dUTP	deoxyuridine triphosphate
Dvl	dishevelled (protein)
E2	17-β-estradiol/estrogen
E2F	transcription factor activating adenovirus E2 gene

4E-BP	eIF4E-binding protein
E	(1) epithelial; (2) gene expressed early in a viral replication cycle
EBV	Epstein–Barr virus
EC	embryonal carcinoma (cell)
ECM	extracellular matrix
EFS	event-free survival
e.g.	for example
EGF	epidermal growth factor
EGF-R	EGF receptor
eIF	eukaryotic (translation) initiation factor
eIF4E	eukaryotic (translation) initiation factor 4E
EM	(1) electron micrograph; (2) electron microscope
EMA	European Medicines Agency
EMSA	electrophoretic mobility shift assay
EMT	epithelial–mesenchymal transition; *also* epithelial-to-mesenchymal transition
EMT-TF	EMT-inducing transcription factor
ENU	*N*-ethyl-*N*-nitrosourea
env	retrovirus envelope glycoprotein
EPC	(1) endothelial precursor cell; (2) endothelial progenitor cell
EpCAM	epithelial cell adhesion molecule
EPO	erythropoietin
εdA	1,N^6-ethenodeoxyadenosine
ER	(1) estrogen receptor; (2) endoplasmic reticulum
erb	(oncogene of) avian erythroblastosis virus
ERG	ETS-related gene
ERK	extracellular signal–regulated kinase
ERV	endogenous retrovirus
ES	embryonic stem (cells)
ESA	epithelial surface antigen
ET-1	endothelin-1
EZH2	enhancer of zeste homolog 2
5-FU	5-fluorouracil
F	filamentous (actin)
FA	Fanconi anemia (syndrome)
FACS	fluorescence-activated cell sorting
FADD	Fas-associated death domain (protein)
FAK	focal adhesion kinase
FANC	defective in Fanconi anemia syndrome (protein)
FAP	familial adenomatous polyposis
FasL	ligand of the Fas death receptor
Fc	fragment crystallizable (region of IgG antibody molecule)
FcR	Fc receptor
FDA	U.S. Food and Drug Administration
FDG	2-deoxy-2-(^{18}F)fluoro-D-glucose
FGF	fibroblast growth factor
FH	fumarate hydratase
Fig	fused in glioblastoma (gene)
FISH	fluorescence *in situ* hybridization
FITC	fluorescein isothiocyanate (dye)
FKBP12	FK506-binding protein of 12 kD
FL	(1) follicular (B-cell) lymphoma; (2) ligand of the Flt-3 receptor
FLICE	FADD-like interleukin-1β–converting enzyme; *also called* caspase 8
FLIP	FLICE-inhibitory protein
Flk-1	murine VEGF-R2
Flt	Fms-like tyrosine kinase
FLT-1	human VEGF-R1 (gene)
FLV	Friend murine leukemia virus
FOX	member of the Forkhead family of transcription factors
FPT	farnesyl protein transferase
FRB	FKBP12 + rapamycin–binding (domain of protein)
FSH	follicle-stimulating hormone
FTI	farnesyltransferase inhibitor
Fzd	frizzled (Wnt receptor)
G	(1) guanine; (2) guanine nucleotide; (3) guanosine; (4) organismic generation
G1	gap 1 phase (of cell cycle)
G2	gap 2 phase (of cell cycle)
G6PD	glucose-6-phosphate dehydrogenase
GAG	glycosaminoglycan
gag	group-specific antigen (retrovirus capsid protein)
GAK	cyclin G–associated kinase
GAP	GTPase-activating protein
GAPDH	glyceraldehyde-3-phosphate dehydrogenase
GBM	glioblastoma multiforme
GCB	germinal-center B-cell lymphoma
GDNF	glial cell–derived neurotrophic growth factor
GDP	guanosine diphosphate
GDSC	Genomics of Drug Sensitivity in Cancer
GEF	guanine nucleotide exchange factor
GEMM	genetically engineered mouse model, *also termed* GEM
GF	growth factor
GFAP	glial fibrillary acidic protein
GFP	green fluorescent protein
GGF	glial growth factor
GGR	global genomic repair
GH	growth hormone
GIST	gastrointestinal stromal tumor
GLUT	glucose transporter
GM-CSF	granulocyte–macrophage colony-stimulating factor
gp	glycoprotein
GPI	glycosylphosphatidylinositol
GPCR	G-protein–coupled receptor
GPX4	glutathione peroxidase 4
Grb2	growth factor receptor-bound-2
GRIP-1	glucocorticoid receptor interacting protein-1
gRNA	guide RNA (in CRISPR-Cas9 procedure)
GRP	gastrin-releasing peptide
GSDMD	gasdermin D
GSEA	gene set enrichment analysis
GSK-3β	glycogen synthase kinase-3β
GSTP1	glutathione *S*-transferase π1
GTP	guanosine triphosphate
GTPase	enzyme that cleaves GTP
GVHD	graft-versus-host disease (immunological reaction)
GVT	graft versus tumor (immunological reaction)
Gy	gray (radiation dose unit)
2-HG	2-hydroxyglutarate
5hmC	5-hydroxymethylcytosine
H	histone
h	prefix referring to the human form of a gene or protein
^{3}H	tritium
H&E	hematoxylin–eosin (tissue stain)
hABH	human AlkB homolog
HAMA	human anti-mouse antibody
HAT	(1) hypoxanthine–aminopterin–thymidine (culture medium); (2) histone acetyltransferase
HB-EGF	heparin-binding EGF
HBV	hepatitis B virus
HCA	heterocyclic amine
HCC	hepatocellular carcinoma
hCG	human chorionic gonadotropin

HCL	hairy-cell leukemia
HCMV	human cytomegalovirus
HCV	hepatitis C virus
HDAC	histone deacetylase
HDM2	human homolog of MDM2
HDR	homology-directed (DNA) repair; *also* **HR**
HEK	human embryonic kidney (cells)
HER	human EGF receptor
HERV	human endogenous retrovirus
HES	hairy and enhancer of split
HGF	hepatocyte growth factor; *also called* scatter factor (SF)
HGFL	hepatocyte growth factor-like
HGPRT	hypoxanthine guanine phosphoribosyltransferase
Hh	Hedgehog
HHV	human herpesvirus
HHV-8	human herpesvirus type 8; *also called* Kaposi's sarcoma herpesvirus (KSHV)
HIF	hypoxia-inducible (transcription) factor
HIP1	huntingtin-interacting protein-1
HIPK2	homeodomain-interacting protein kinase 2
HIV	human immunodeficiency virus
HLA	human leukocyte antigen; equivalent to MHC
HMEC	human mammary epithelial cell
HMT	histone methyltransferase
HNPCC	hereditary non-polyposis colon cancer
hnRNA	heterogeneous nuclear RNA
HNSCC	head-and-neck squamous cell carcinoma
HPE	human prostate epithelial (cell)
HPGD	15-hydroxyprostaglandin dehydrogenase
HPRT	hypoxanthine phosphoribosyltransferase
HPV	human papillomavirus
HR	homology-directed (DNA) repair; *also* **HDR**
HRE	hormone response element (in DNA)
HRG	Heregulin
HRP	horseradish peroxidase
HSA	human serum albumin
HSC	hematopoietic stem cell
HSIL	high-grade squamous intraepithelial lesion
HSR	homogeneously staining region
HSV-1	herpes simplex virus type 1
hTERT	human telomerase reverse transcriptase
HTLV-I	human T-cell lymphotropic virus type-I
hTR	human telomerase-associated RNA
HTS	high-throughput screening
HU	hydroxyurea
IAP	(1) intracisternal A particle; (2) inhibitor of apoptosis
-ib	low–molecular-weight drug (suffix)
IC$_{50}$	concentration required to obtain 50% inhibition
ICAD	inhibitor of caspase-activated DNase
ICB	immune checkpoint blockade; *see also* **ICI**
ICGC	International Cancer Genome Consortium
ICI	immune checkpoint blockade inhibitor; *see also* **ICB**
Id, ID	inhibitor of DNA binding (transcription factor)
IDH	isocitrate dehydrogenase
i.e.	that is
IEG	immediate early gene
IEL	intraepithelial lymphocyte
IF	immunofluorescence
IFN	interferon
IFN-R	interferon receptor
IgH	immunoglobulin heavy chain
IGF-1	Insulin-like growth factor-1
IGF-1R	IGF-1 receptor
IGFBP	Insulin-like growth factor–binding protein
IgL	immunoglobulin light chain
IHC	immunohistochemistry
IκB	inhibitor of NF-κB
IKK	Iκ kinase
IL	interleukin
ILK	Integrin-linked kinase
iMCs	immature myeloid cells
int	gene identified through insertional mutagenesis
IP3	inositol (1,4,5)-triphosphate
IR	Insulin receptor
IRES	internal ribosome entry site
IRK	Insulin receptor kinase
ISH	*in situ* hybridization
iSH2	inter SH2 domain (of PI3K)
Jak	Janus kinase
JCV	JC virus
JM	juxtamembrane
Jmj	Jumonji (enzyme)
kb	kilobase
kbp	kilobase pair
kD	kilodalton
KDR	human VEGF-R2
KEAP1	Kelch-like ECH-associated protein 1
KGF	Keratinocyte growth factor
K$_i$	concentration at which 50% inhibition is achieved
KIP	kinase inhibitor protein
KIR	killer-cell immunoglobulin-like receptor
KO	(1) knockout (of gene); (2) knocked-out (gene)
KSHV	*see* **HHV-8**
KSR	kinase suppressor of Ras
LAK	lymphokine-activated killer (cell)
LANA	latency-associated nuclear antigen (of KSHV)
LATS	large tumor suppressor homolog (kinase)
LBCL	large B-cell lymphoma
LCM	laser capture microdissection
LDA	limiting dilution analysis
LDL	low-density lipoprotein
LEF	lymphoid enhancer-binding factor
LGR5	leucine-rich repeat-containing G protein-coupled receptor 5
LH	luteinizing hormone
LIF	leukemia inhibitory factor
lincRNA	long intergenic noncoding RNA; usually lncRNA
LLC	Lewis lung carcinoma
LNA	locked nucleic acid (RNA)
lncRNA	long noncoding RNA
LOH	loss of heterozygosity
loxP	locus of crossover in bacteriophage P1 genome
LPA	lysophosphatidic acids
LPS	lipopolysaccharide
LRC	label-retaining cell
LRP	LDL receptor–related protein
LSIL	low-grade squamous intraepithelial lesions
LT	large T antigen (of SV40 or polyomavirus)
LTR	long terminal repeat (of a retroviral provirus)
3-MC	3-methylcholanthrene
5mC	5-methylcytosine
μM	micromolar (10^{-6} molar)
M	(1) mitosis; (2) mesenchymal
m	prefix referring to the murine form of a gene or protein
M1	pro-inflammatory macrophage subtype
M2	immunosuppressive macrophage subtype
-mab	therapeutic monoclonal antibody (suffix)
MAC	membrane attack complex
MAF	musculoaponeurotic fibrosarcoma (transcription factor)

MALT	mucosa-associated lymphoid tissue
MAPK	mitogen-activated protein kinase; *also called* MAP-kinase
MAPKK	kinase that phosphorylates a MAPK
MAPKKK	kinase that phosphorylates a MAPKK
mb	megabase; *also* MB
MCP-1	monocyte chemotactic protein-1
MCPyV	Merkel cell (polyoma) virus
mCRP	membrane-bound complement regulatory protein
M-CSF	macrophage colony-stimulating factor
MCT	monocarboxylate transporter
MCV	Merkel cell (polyoma) virus
MDCK	Madin–Darby canine kidney (cells)
MDM2	mouse double-minute chromosome 2
MDR	multi-drug resistance (phenotype)
MDS	myelodysplastic syndrome
MDSC	myeloid-derived suppressor cell
MDV	Marek's disease virus
MEC	mammary epithelial cell
meCpG	methyl-cytosine within a CpG dinucleotide
MEF	mouse embryo fibroblast
MeIQx	2-amino-3,8-dimethylimidazo [4,5-f]-quinoxaline mutagen
MEK	MAPKK/Erk kinase
MEN	multiple endocrine neoplasia
MET	mesenchymal–epithelial transition; *also* mesenchymal-to-epithelial transition
Met	receptor of hepatocyte growth factor
mFISH	multicolor fluorescence in situ hybridization
MGMT	O^6-methylguanine-DNA methyltransferase; *also called* **AGT**
MHC	major histocompatibility complex; *see also* **HLA**
-mib	low–molecular-weight pharmacologic agent (suffix)
min	multiple intestinal neoplasias
MIN	microsatellite instability
miRNA	microRNA
Miz-1	Myc-interacting zinc finger protein-1
MKP	MAP kinase phosphatase
MLKL	mixed lineage kinase-like protein
MLL	(1) mixed lineage leukemia; (2) myeloid/lymphoid leukemia
MLV	murine leukemia virus
MM	multiple myeloma
mM	millimolar (10^{-3} molar)
MMP	matrix metalloproteinase
MMR	mismatch repair (of DNA)
mms	methylmethane sulfonate
MMTV	mouse mammary tumor virus
MNNG	N-methyl-N'-nitro-N-nitrosoguanidine
MNU	N-methylnitrosourea
MoAb	monoclonal antibody
MOCA	4,4′-methylenebis(2-chloroanilinc)
MOMP	mitochondrial outer membrane permeabilization
mp	muscularis propria (of intestinal wall)
mPGES	microsomal prostaglandin E2 synthase
MPNST	malignant peripheral nerve sheath tumor
MRD	minimal residual disease
MRE11	meiotic recombination 11 (protein)
MRI	magnetic resonance imaging
MRN	MRE11/Rad50/Nbs1 protein complex
mRNA	messenger RNA
MSC	mesenchymal stem cell
MSH	melanocyte-stimulating hormone
MSP	methylation-specific polymerase chain reaction
MT	middle T antigen of polyomavirus

MT1-MMP	membrane type 1-MMP
MTD	maximum tolerated dose
mTERT	murine TERT
MTH1	mammalian MutT homolog-1
mTOR	mechanistic target of rapamycin (kinase)
mTR	murine telomerase RNA
MUP	major urinary protein
myc	(oncogene of) avian myelocytomatosis virus
N	ploidy number (where haploid = 1)
N'	(1) N amino terminus of polypeptide; (2) N-terminal (adj.)
NAT	N-acetyltransferase
NBS1	Nijmegen break syndrome 1
NCAM	neural cell adhesion molecule
NCI	National Cancer Institute of the United States
NCI-60	Collection of 60 cancer cell lines prepared by the NCI
NECD	Notch extracellular domain
neo	neomycin
NER	nucleotide-excision DNA repair
NES	nuclear export signal
NET	neuroendocrine tumor
Neu	rat ortholog of HER2 (receptor)
Nf/NF	neurofibromatosis
NF-κB	nuclear factor-κB
NGF	nerve growth factor
NGS	next generation sequencing
NHEJ	nonhomologous end joining
NHL	non-Hodgkin's lymphoma
NICD	Notch intracellular domain
NK	natural killer (lymphocyte)
NKG2D	natural killer group 2D (receptor)
NKT	natural killer lymphocyte expressing T-cell receptor
NLS	nuclear localization (import) sequence
nM	nanomolar (10^{-9} molar)
NMR	nuclear magnetic resonance
NO	nitric oxide
NOD	non-obese diabetic
NOD/SCID	non-obese diabetic/severe combined immunodeficient (mouse strain)
NotchL	Notch ligand
NR	no (clinical) response
NRF2	nuclear factor erythroid 2-related (transcription factor)
NRG	neuregulin
NSAID	nonsteroidal anti-inflammatory drug
NSCLC	non-small-cell lung carcinoma
nt	nucleotide
NURF	nucleosome remodeling factor
4-OHT	4-hydroxytamoxifen
8-oxo-dG	8-oxo-deoxyguanosine
OD	optical density
OIS	oncogene-induced senescence
OPC	oropharyngeal carcinoma
OPG	osteoprotegrin (decoy receptor protein)
OPN	osteopontin (protein)
ORF	open reading frame
OS	overall survival
OSCC	oral squamous cell carcinoma
OSE	ovarian surface epithelium
6-4 PP	pyrimidine (6-4) pyrimidinone (photoproduct)
p	(1) short arm of a chromosome; (2) probability of an event
Pa	Pascal (unit of measured pressure, stress)
PAH	polycyclic aromatic hydrocarbon
PAI	(1) plasminogen activator inhibitor; (2) pathogenicity-associated island

PAM	protospacer adjacent motif (in CRISPR-Cas9 procedure)
PanIN	pancreatic intraepithelial neoplasia
PAP	prostatic acid phosphatase
PAR-1	protease-activated receptor protein-1 (GPCR)
PARP	poly (ADP-ribose) polymerase
PBS	phosphate-buffered saline
PCNA	proliferating cell nuclear antigen
PCP	planar cell polarity
PCR	(1) polymerase chain reaction; (2) pathological complete response
PD	(1) pharmacodynamics; (2) population doubling; (3) progressive disease
PD-1	programmed death-1 (receptor of T cells)
PDAC	pancreatic ductal (exocrine) adenocarcinoma
PDGF	platelet-derived growth factor
PDE	phosphodiesterase
PDGF-R	PDGF receptor
PDH	pyruvate dehydrogenase
PDK1	(1) phosphoinositide-dependent kinase 1; (2) pyruvate dehydrogenase kinase 1
PD-L1	ligand of the PD-1 receptor
PDM	mean population doublings
PDX	patient-derived xenograft
PEG	polyethylene glycol
pEMT	partial EMT
PEP	phosphoenolpyruvate
PET	positron-emission tomography
PFS	progression-free survival
PGDH	15-hydroxyprostaglandin dehydrogenase
PGE$_2$	prostaglandin E$_2$
P-gp	P-glycoprotein
PH	pleckstrin homology (phosphoinositol-binding protein domain)
PHD	proline hydroxylase
PhIP	(1) 2-amino-1-methyl-6-phenylimidazo[4,5-b] pyridine; (2) PH-interacting domain protein
PhK	phosphorylase kinase
pI	isoelectric point
PI	(1) phosphatidylinositol; (2) propidium iodide
PI(3,4,5)P3	*equivalent to* PIP3
PI3K	phosphatidylinositol 3-kinase
PIK3CA	catalytic subunit of PI3K
PIK3R	regulatory subunit of PI3K
PIN	prostatic intraepithelial neoplasia
PINCH	particularly interesting new cysteine-histidine-rich (adaptor protein)
PIP2	phosphatidyl inositol (4,5) diphosphate/ PI(4,5)P$_2$
PIP3	phosphatidyl inositol (3,4,5) triphosphate/ PI(3,4,5)P$_3$
PK	(1) pyruvate kinase; (2) pharmacokinetics
PKA	protein kinase A
PKB	protein kinase B (=Akt)
PKC	protein kinase C
PLC	phospholipase C
PLD	phospholipase D
PMA	phorbol-12-myristate-13-acetate; equivalent to TPA
PMBL	primary mediastinal B-cell lymphoma
PML	(1) promyelocytic leukemia; (2) progressive multifocal leukoencephalopathy
PND	paraneoplastic neurological degeneration
PNI	perineural invasion
pol	(1) polymerase; (2) retrovirus reverse transcriptase
pol II	RNA polymerase II
POT1	protection of telomere 1 (protein)

PP1	protein phosphatase 1
PP2A	protein phosphatase 2A
PR	(1) progesterone receptor; (2) partial (clinical) response.
PRAS	proline-rich Akt substrate
pRb	retinoblastoma protein; *also* **Rb**, RB
PRC1	polycomb repressive complex 1
pre-mRNA	RNA precursor to mRNA
PRL	prolactin
pro-B ALL	pro-B-cell acute lymphocytic leukemia
PS	phosphatidylserine
PSA	prostate-specific antigen
PTB	phosphotyrosine-binding (protein domain)
PTC	(1) papillary thyroid carcinoma; (2) Patched
PTEN	phosphatase and tensin homolog deleted on Chromosome 10
PTH	parathyroid hormone
PTHrP	parathyroid hormone–related peptide
PTM	post-translational modification
PTP	phosphotyrosine phosphatase
PUMA	p53-upregulated modulator of apoptosis
pY	phosphotyrosine
PyMT	polyomavirus middle T oncoprotein
q	long arm of a chromosome
qPCR	quantitative polymerase chain reaction
qRT-PCR	quantitative real-time PCR; *see also* **RT-PCR**
r	resistant to specified agent
R	restriction point (cell cycle landmark)
RA	(1) retinoic acid; (2) Ras-association (domain)
Rad	radiation repair (gene)
RAE	retinoic acid early transcript (ligand)
RAG	recombination-activating gene
RANK	receptor activator of NF-κB
RANKL	ligand of RANK (receptor)
RAR	retinoic acid receptor
ras	(oncogene of) rat sarcoma virus
RAS	protein of *ras* oncogene
Rb	retinoblastoma protein
RBD	Ras-binding domain
RCC	renal cell carcinoma
REF	rat embryo fibroblast
rel	reticuloendotheliosis (oncogene)
RFLP	restriction fragment length polymorphism
RFP	red fluorescent protein
Rheb	Ras homolog enriched in brain
RING	really interesting new gene (protein domain)
RIPK	receptor-interacting protein kinase
RISC	RNA-induced silencing complex
RNAi	RNA interference pathway (involving miRNAs, siRNAs, shRNAs)
RNA-seq	RNA sequencing (via sequencing of cDNAs)
RNS	reactive nitrogen species
ROS	reactive oxygen species
RPA	replication protein A
RR	(1) relative risk; (2) response rate
RSV	Rous sarcoma virus
RT	reverse transcriptase
RTK	receptor tyrosine kinase
RT-PCR	reverse transcription PCR; *see also* **qRT-PCR**
S	(1) DNA synthesis phase (of cell cycle); (2) Svedberg (unit of sedimentation in centrifuge); (3) serine
S1P	sphingosine-1-phosphate
S6	protein 6 of the small ribosomal subunit
S6K1	p70 S6 kinase-1
SA-β-gal	senescence-associated β-galactosidase
SAC	spindle assembly checkpoint
SAHF	senescence-associated heterochromatic foci

SASP	senescence-associated secretory phenotype
SC	stem cell
SCC	squamous cell carcinoma
SCE	sister chromatid exchange
SCF	stem cell factor
scFv	single-chain variable region fragment
SCID	severe combined immunodeficiency (syndrome)
SCLC	small-cell lung carcinoma
scRNAseq	single-cell RNA sequencing
SDF-1	stromal cell–derived factor-1
SDH	succinate dehydrogenase
SEM	scanning electron micrograph/microscope/microscopy
ser	serine
SERM	selective estrogen receptor modulator
SF	(1) scatter factor, *see also* **HGF**; (2) splicing factor
SFK	Src family kinases
SFRP/sFRP	secreted frizzle-related protein (Wnt inhibitor)
SGK1	serum- and glucocorticoid-inducible kinase 1
sgRNA	single-guide RNA (in CRISPR-Cas9 procedure)
SH1	Src homology 1 domain (tyrosine kinase domain)
SH2	Src homology 2 domain (phosphotyrosine binding)
SH3	Src homology 3 domain (proline-rich binding)
SHP	SH2-containing phosphatase
shRNA	small hairpin RNA
SIP1	Smad-interacting protein 1
SIR	standardized incidence ratio
siRNA	small interfering RNA
Skp2	S-phase kinase-associated protein 2
SKY	(multicolor) spectral karyotyping
Sky1	SR-protein–specific kinase of budding yeast
SLE	systemic lupus erythematosus
SMA	smooth muscle actin
Smac	second mitochondria-derived activator of caspase
SMDF	sensory and motor neuron-derived factor
Smo	Smoothened
SNP	single-nucleotide polymorphism
SOD	superoxide dismutase
Sos	Son of Sevenless
SPP1	gene encoding osteopontin
Srf	serum response factor
ss	single-stranded (DNA or RNA)
STAT	signal transducer and activator of transcription
Str-1	stromelysin-1
STS-1	suppressor of T-cell receptor signaling
SUMO	small ubiquitin-like modifier
SV	simian virus
SWI/SNF	mating type switch/sucrose non-fermenting
T	(1) thymine; (2) thymidine; (3) threonine
t	referring to a chromosomal translocation
$T_{1/2}$	half-life
T4SS	type IV secretion system
T_{50}	50% of hosts bearing tumors
T_{mem}	memory T cell
T_{reg}	regulatory T cell
TAA	tumor-associated antigen
Tag	SV40/polyoma T-antigen
T-ALL	T-cell acute lymphocytic leukemia
TAM	tumor-associated macrophage
TAP	transporter associated with antigen processing
TATA	tumor-associated transplantation antigen
TAZ	transcriptional activator with PDZ-binding motif
T_C	cytotoxic T cell
TCF	(1) T-cell factor; (2) T cell-specific transcription factor

TCGA	The Cancer Genome Atlas
TCR	(1) T-cell receptor; (2) transcription-coupled DNA repair
TdT	terminal deoxynucleotidyl transferase
TEAD	transcription-enhanced associate domain (transcription factor)
T_{eff}	Effector T cell
TEM	transmission electron microscopy
TERRA	telomere repeat-containing RNA
TERT	telomerase reverse transcriptase
TET	ten eleven (chromosomal) translocation
T_{ex}	Exhausted T cell
TF	(1) transcription factor; (2) Tissue Factor
T_{FH}	T follicular helper (lymphocyte)
TFSS	Type IV secretion system
TGF-α	transforming growth factor-α
TGF-β	transforming growth factor-β
TGF-βR	TGF-β receptor
T_H	helper T cell
thr	threonine
TI	therapeutic index
TIAM-1	T-cell lymphoma invasion and metastasis-1
TIC	tumor-initiating cell (*see also* **CSC**)
TIF	telomere dysfunction-induced focus
TIL	tumor-infiltrating lymphocyte
TIMP	tissue inhibitor of metalloproteinase
TK	(1) tyrosine kinase; (2) thymidine kinase
TKI	tyrosine kinase inhibitor
TM	transmembrane (domain of a protein)
Tm	melting temperature
TMB	tumor mutational burden
TME	*see* **TMEN**
TMEN	tumor microenvironment
TNBC	triple-negative breast cancer
TNF	tumor necrosis factor
TOR	target of rapamycin
TORC	mTOR-containing complex
*TP*53	gene encoding p53 TSG
TPA	tPA (1) 12-O-tetradecanoylphorbol-13-acetate (*see* PMA); (2) tissue plasminogen activator
TPO	thrombopoietin
TRAIL	TNF-related apoptosis-inducing ligand
TRAMP	transgenic adenocarcinoma of the mouse prostate
TRAP	telomeric repeat amplification protocol
Trop-2	human trophoblast cell surface antigen 2
TRF	telomeric restriction fragment
tRNA	transfer RNA
TrpRS	tryptophanyl-tRNA synthetase
ts	temperature-sensitive
TSA	(1) trichostatin A; (2) tumor-specific antigen
TSC	tuberous sclerosis
TSG	tumor suppressor gene
TSHR	thyroid-stimulating hormone receptor
t-SNE	t-distributed stochastic neighborhood embedding (algorithm)
TSP	tissue-specific (gene) promoter
Tsp-1	thrombospondin-1
TSS	transcription start site
TSTA	tumor-specific transplantation antigen
TUNEL	TdT-mediated dUTP nick end labeling
tx	transformed (cell)
U	(1) uracil; (2) uridine
UC	ulcerative colitis
UDG	uracil-DNA glycosylase
uPA	urokinase (type) plasminogen activator
uPAR	receptor of uPA
UPR	unfolded protein response

UTR	untranslated region (of an mRNA)
UV	ultraviolet
UV-B	ultraviolet-B radiation
V	variable region (of an antibody molecule)
VDAC	voltage-dependent anion channel
VE	vascular endothelial
VEGF	vascular endothelial growth factor (=VPF)
VHL	von Hippel–Lindau (protein)
VPF	vascular permeability factor (=VEGF)
vs.	versus
VSMC	vascular smooth muscle cell
WES	whole exon sequencing

WGS	whole genome sequencing
WM	Waldenström's macroglobulinemia
Wnt	wingless/integration site (growth factor)
wt	wild type
XAF1	XIAP-associated factor 1
XIAP	X-chromosome–linked inhibitor of apoptosis
XP	xeroderma pigmentosum
Y	tyrosine
YAP	Yes-associated protein
YFP	yellow fluorescent protein
zTERT	zebrafish telomerase; *see also* **TERT**
-zumab	suffix indicating chimeric antibody MoAb

Glossary

3′ UTR The region of an mRNA molecule that is downstream of its reading frame.

5-hydroxymethylcytosine The oxidation product of 5-methylcytosine created by TET enzymes.

5′ UTR The region of an mRNA molecule that is upstream of its reading frame.

abasic Referring to a nucleotide that has lost its purine or pyrimidine base. *See also* **apurinic**; **apyrimidinic**.

abdomen (adj., **abdominal**) Referring to the cavity in vertebrates containing the digestive organs.

ablate To eliminate.

abluminal Located away from the lumen of a duct or other hollow structure.

abscissa Horizontal or *x*-axis of a Cartesian graph. *See also* **ordinate**.

abscission Final step of cytokinesis, when the remaining connections, including those created by microtubules, are severed, allowing complete separation of the two daughter cells. *See also* **cytokinesis.**

acellular Lacking or deprived of cells.

acetylation Covalent attachment of an acetyl group to a second molecule such as a protein.

acetylsalicylic acid Chemical name for aspirin.

acetyltransferase An enzyme that covalently links acetate groups to substrates, such as the amino acid side chains of proteins. *See also* **deacetylase**.

acid test The most rigorous test of an experimental hypothesis.

acinus (pl., **-i**) (1) A hollow saclike structure usually emptying into a duct. (2) A hollow spherical structure formed in culture by certain epithelial cells.

acquired immunodeficiency syndrome Disease caused by infection with human immunodeficiency virus (HIV) that causes depletion of CD4+ lymphocytes and resulting collapse of adaptive immunity.

acquired resistance Referring to the resistance to therapy of a tumor that arises with some delay following and in response to implementation of therapy; *also termed* **secondary resistance**. *See also* **intrinsic resistance**.

acromegaly Pathological condition of excessive growth of certain tissues, notably the hands, feet, and face, usually due to the elaboration of excessive growth hormone by a pituitary tumor.

actin Family of proteins that polymerizes to form microfilaments that help assemble the cytoskeleton of a cell and the thin filaments in contractile structures such as those in muscle.

activation-induced deaminase A cytidine deaminase that introduces point mutations in already-rearranged immunoglobulin genes, preferentially in their V regions, thereby diversifying antigen-binding specificities of the encoded immunoglobulins.

activation loop An oligopeptide loop of a protein kinase molecule that is normally positioned to block access by the catalytic cleft of the kinase to its substrates; phosphorylation of this loop, often by another kinase, causes it to swing out of the way, permitting free access of the catalytic cleft to its substrates.

activator A protein that favors the expression of a gene. *See also* **repressor**.

adaptive immune response An immune system response that is acquired or learned following exposure of an organism to an antigen or antigen-bearing agent. *See also* **innate immune response**.

addiction *See* **oncogene addiction**.

adduct Product of covalently linking a chemically reactive molecule to a targeted substrate molecule, such as the addition of a reactive mutagenic molecule to a base in the DNA.

adenocarcinoma Tumor derived from secretory epithelial cells. *See also* **squamous cell carcinoma**.

adenoma (adj., **-omatous**) Any of a series of premalignant, noninvasive growths in various epithelial tissues, many of which have the potential to progress further to carcinomas. *See also* **polyp**.

adherens junction Lateral cell–cell junction between adjacent cells, notably epithelial cells, that is formed by cadherins displayed on the opposing surfaces of participating pairs of cells; these junctions include the cytoplasmic proteins that are physically associated with the cytoplasmic domains of both participating cadherins and serve to link these junctions, in turn, to the actin cytoskeletons of the two cells.

adipocytes Specialized cells of the mesenchymal lineage, closely related to fibroblasts, that create fat and store it in large globules in the cytoplasm; the dominant cell type in fatty tissues.

adipogenic Causing differentiation into adipocytes.

adjuvant   (1) A treatment that is given in concert with or following another treatment. (2) A substance that is given together with an antigen to enhance the immune response. *See also* **neoadjuvant**.

adoptive transfer   Procedure in which immune cells are transferred from a donor to a recipient, undertaken with the goal that the donor's cells will be able to mediate an immune function in the recipient.

adrenal   Referring to the hormone-producing secretory glands that sit above the kidneys.

aerobic glycolysis   The use by cancer cells of glycolysis as a major source of energy production under aerobic conditions that would normally favor use of the far more efficient Krebs/citric acid cycle; *also termed* **Warburg effect**.

afferent   Referring to incoming signals. *See also* **efferent**.

affinity chromatography   Procedure for enriching from a complex mixture of biological molecules the subpopulation of biomolecules that binds specifically to a cognate ligand or molecular binding partner, the latter being immobilized on a solid matrix. Molecules that are retained on the matrix, in contrast to the bulk population of molecules that are not, can subsequently be eluted from the matrix and prepared for various additional enrichment or analytic techniques.

age-adjusted   Referring to an epidemiologic measurement, such as disease incidence or mortality, that has been adjusted to compensate for the age distribution of the population under study, allowing rates to be compared between populations having different age distributions.

agonist   Activating agent; opposite of antagonist.

alkylating   Capable of attaching covalently an alkyl group or similarly structured organic chemical group to a substrate such as a DNA base, usually resulting in the replacement of a proton in the substrate with the organic group, thereby forming a covalent bond.

allele   One alternative among different versions of a gene that may be defined by the phenotype that it creates, by the protein that it specifies, or by its nucleotide sequence.

allelic deletion   Referring to the loss of one of two alternative alleles within a cell and its descendants; such loss can occur via several alternative mechanisms. *See* **allele**, **loss of heterozygosity**.

allogeneic   Referring to a pair of tissues or associated cells that differ genetically and, quite possibly, antigenically.

allogeneic stem cell transplantation   Clinical procedure involving the transplantation of bone marrow cells from a donor that are genetically and thus antigenically nonidentical with those of a recipient.

allograft   Implantation of cells of an animal of one genetic background into a host animal of another genetic background but from the same species.

allograft rejection   Process occurring when a donor tissue graft is rejected by the immune system of a recipient because the donor and recipient, although members of the same species, are genetically distinct. *See also* **rejection**.

allosteric   Referring to a mechanism for regulating the activity of a protein, notably an enzyme, in which its activity or function is modulated by changes in its three-dimensional structure; the change results from binding of a regulatory molecule at a site within the protein that is distant from the functionally important site, such as, in the case of enzymes, from its catalytic cleft.

alternative splicing   Process whereby a pre-mRNA may be spliced in several alternative ways, resulting in mRNAs composed of different combinations of exons.

alternatively activated macrophages   *See* **M2 macrophages**.

***Alu* repeat**   Sequence block of about 300 bp that is found in almost 1 million copies scattered throughout the human genome.

alveolus (pl., alveoli)   A small sac within a tissue that is connected to a duct, such as the sacs seen in the lung and in the mammary gland during pregnancy.

ambient   Referring to conditions in the immediate surroundings.

Ames test   An experimental protocol and versions thereof that enable the quantification of the mutagenic potency of candidate chemicals through their ability to induce mutations in bacteria growing in a Petri dish.

amino   Referring to the end of an oligopeptide or protein chain that is formed first. *See also* **carboxy**.

amoeboid   Exhibiting the shape and motility reminiscent of the behavior of the amoeba protozoan.

amphipathic   Describing a molecule containing distinct hydrophobic and hydrophilic domains.

AMPK   Kinase regulator of autophagy that responds to low intracellular ATP levels by negatively regulating components of the mTOR complex while stimulating certain enzymes that contribute to ATP production and others that decrease ATP consumption.

amplicon   A defined stretch of (chromosomal) DNA that undergoes amplification flanked on both sides by DNA segments that are present in normal copy number.

amplification   Genetic mechanism by which the copy number of a gene is increased above its normal level in the diploid genome.

anabolism (adj., **anabolic**)   Processes by which new biological molecules are assembled via biosynthetic reactions. *See also* **catabolism**.

analogous   Referring to two genes or proteins that have similar functions whether or not they share common evolutionary descent. *See also* **homologous**.

anaphase   Third subphase of mitosis, during which the paired chromatids are segregated to the two opposite poles of the cell.

anaphase bridge   A connection between two clusters of chromosomes that forms during anaphase and results from the simultaneous association of an individual chromatid with both chromosomal clusters and their respective sets of spindles. *See also* **spindle**.

anaplastic   (Referring to a tumor) having a tissue and cellular architecture lacking the differentiated characteristics of an identifiable tissue-of-origin or tumor type.

anastomosis   End-to-end connection formed between two ductal structures, such as a connection formed directly between an artery and a vein.

anchorage dependence   Requirement of cells for tethering to a solid substrate before they will grow.

anchorage independence   Ability of a cell to proliferate without attachment to a solid substrate.

androgen   Steroid hormone that stimulates development of male characteristics.

aneuploid   (1) Describing a karyotype that deviates from diploid because of increases or decreases in the numbers of certain chromosomes. (2) Less commonly, describing a karyotype that carries structurally abnormal chromosomes.

angiogenesis   Process by which new blood vessels are formed by sprouting from existing vessels; *also called* **neoangiogenesis** in certain circumstances. *Compare* **vasculogenesis**.

angiogenic factor   Type of secreted growth factor that is specialized to induce angiogenesis.

angiogenic switch   Shift by a mass of tumor cells from a state in which they are unable to induce neovascularization to one in which they exhibit this ability.

angiosarcoma   Tumor of the cells that are precursors to endothelial cells.

annexin V   A cellular protein that has strong affinity for binding of phosphatidylserine (PS); binding of this protein to the surface of intact cells indicates that the usual segregation of PS to the inner leaflet of the bilayered plasma membrane has been disrupted, as occurs during apoptosis and other forms of cell death.

anoikis   Form of apoptosis that is triggered by the failure of a cell to establish anchorage to a solid substrate, such as the extracellular matrix, or by loss of such anchorage.

anonymous (marker)   Referring to a segment of genomic DNA that is not affiliated with any known gene or biological function.

anoxia   State or environment in which oxygen is essentially absent.

antagonist   Agent acting against another agent or process of interest. *See also* **agonist**.

antibody   A soluble tetrameric protein that is produced by plasma cells of the immune system and is capable of recognizing and binding particular antigens with high specificity. *See also* **immunoglobulin**.

antibody-dependent cellular cytotoxicity   A form of attack by cytotoxic immune cells, such as NK cells, that depends on the presence of antibody molecules bound to antigens displayed on the surface of the targeted cells.

antibody-mediated cellular phagocytosis   Attack on targeted cells by phagocytic cells, such as macrophages, that is triggered by antibody molecules bound to antigens displayed on the surface of the targeted cells.

antigen (adj., **-genic**)   Molecule or portion of a molecule, often an oligopeptide, that can be specifically recognized and bound by an antibody or a T-cell receptor or that provokes the production of an antibody. *See also* **immunogenic**.

antigen-presenting cells   A class of cells—often termed professional antigen-presenting cells, including dendritic cells, macrophages, and B cells—that present oligopeptide antigens via class I and II MHC molecules to other immunocytes, notably helper and cytotoxic T cells.

antimetabolite   Compound, typically used as an anti-tumor cytotoxic drug, that acts through its ability to interfere with the synthesis or function of a normal cellular metabolite.

antisense RNA   RNA molecules that, following introduction into or expression within cells, are capable of inhibiting the functions of targeted RNA molecules in those cells because the introduced RNAs have sequences complementary to those of the targeted RNAs and presumably form dsRNA molecules with them.

antiserum   Serum that is produced by an animal exposed to a specific antigen or agent and contains antibodies that are able to recognize and bind that antigen or agent.

apatite   Mineral component of bone composed of calcium phosphate crystals.

apex (adj., **apical**)   Uppermost point of an object or collection of objects. *See also* **basal**.

apheresis   Procedure involving the drawing of blood from a patient, the fractionation of the blood into plasma and various cellular components, the removal and purification of certain cellular components of the blood such as specific types of leukocytes, and the reintroduction of the remaining blood components into the patient's circulation. *See also* **leukapheresis.**

apical   Referring to the surface of an epithelial cell that is facing an exposed surface of the epithelium such as its luminal surface.

apico–basal polarity   Referring to the asymmetric organization of epithelial cells, often termed polarization, in which many proteins and subcellular structures, notably cytoplasmic and cell surface molecules, are preferentially localized toward either the apical surface (facing a lumen) or the basal surface (facing a basement membrane). *See* **polarity**.

APOBEC   A class of cellular enzymes that mutate genomic sequences by deaminating genomic cytidine residues, thereby converting them into uridine residues, leading to C-to-T mutations.

apoptosis   Complex program of cellular self-destruction, triggered by a variety of stimuli and involving the activation of caspase enzymes, which drive rapid fragmentation of diverse cellular structures and elimination of all breakdown products of a cell and phagocytosis of resulting cell fragments by neighboring cells or macrophages. *See also* **ferroptosis**, **necroptosis**, **pyroptosis**.

apoptosome   Multiprotein complex that consists of cytochrome *c* molecules and Apaf-1 and helps to initiate apoptosis by activating procaspase 9 into caspase 9. *See also* **caspase**.

apposed   Physical entity located directly adjacent to a second one.

apposition   Referring to the close physical positioning of one entity with another, such as one cell with another.

apurinic   Referring to the product of depurination, in which the glycosidic bond linking a deoxyribose or ribose to a purine base is broken, leaving behind only the deoxyribose or ribose in the DNA or RNA, respectively. *See also* **depurination**.

apyrimidinic   Referring to the product of depyrimidination, in which the glycosidic bond linking a deoxyribose or ribose to a pyrimidine base is broken, leaving behind only the deoxyribose or ribose in the DNA or RNA, respectively. *See also* **depyrimidination**.

aromatic   Referring to an organic molecule that contains one or more benzene rings.

arrestin   A protein involved in inhibiting or redirecting downstream signals of ligand-activated GPCRs.

arteriole   A small artery that empties into capillaries.

artifact (adj., **artifactual**)   A misleading observation or finding that results from the conditions or design of an experiment rather than from an intrinsic property of the entity being studied.

ascertainment bias   Tendency of some conclusions or measurements to be strongly influenced by the observational tools that are used to generate them, including the ease or difficulty of applying such tools to a specific biological situation.

ascites (adj., -**itic**)   Fluid that accumulates in the peritoneal cavity of some cancer patients, often containing malignant cells.

astrocytoma   A glioma tumor deriving from astrocytes. *See* **glioma**.

asymmetric division   Process whereby a mother cell, notably a stem cell, generates two daughters that reside in distinct states of differentiation; usually one daughter retains the phenotype of the mother cell while the second daughter acquires novel phenotypes or launches a program of differentiation. *See also* **self-renewal**, **symmetric division**, **stem cell**.

asynchronous   Referring to a population of cells that are dispersed throughout the cell cycle during any point of time and therefore do not execute specific cell-cycle steps in a coordinated or synchronous manner.

ATAC-seq   Procedure for mapping open chromatin by determining stretches of genomic DNA that are accessible for modification by the Tn5 transposase.

ataxia   Loss of muscle coordination, often caused by cerebellar dysfunction.

atelomeric   Lacking a telomere.

atrophy   Shrinkage of a tissue, often caused by loss of viability of its component cells and resulting loss of normal cell numbers.

atypia   Histopathological term for an abnormally appearing cell or collection of such cells.

autochthonous   (1) Of native origin. (2) Referring to a tumor that arises within an organism (rather than from implanted cells or tumor fragments).

autocrine   Referring to the signaling path of a hormone or factor that is released by a cell and proceeds to act upon the same cell (or same cell type) that has released it. *Compare* **endocrine**, **exocrine**, **intracrine**, **juxtacrine**, **paracrine**.

autoimmune   Referring to a process or disease in which the immune system attacks an organism's own normal cells and tissues. *See also* **autoreactive**.

autologous   Referring to biological material, usually cells or tissues, that originates in a patient's own body (and may be reintroduced into that patient following some manipulation *ex vivo*).

autophagic cell death   A form of cell death associated with autophagy in which dying cells sequester large portions of their cytoplasmic contents in autophagosomes. This form of cell death lacks the characteristic chromatin condensation seen in apoptosis. *See* **autophagosome, autophagy**.

autophagosome   The specialized intracellular vesicle that is formed during the course of autophagy and encloses cytoplasmic organelles and other cellular products prior to fusion with lysosomes. *See also* **autophagy**.

autophagy   A catabolic process whereby selected intracellular organelles and molecules are degraded by engulfment in autophagosomes that proceed to fuse with lysosomes in which degradation occurs, yielding metabolites that can be recycled for biosynthetic reactions.

autophosphorylation   Phosphorylation of a protein molecule by its own associated kinase activity. *See also* **transphosphorylation**.

autoradiography   Procedure for detecting radiolabeled molecules by placing them (or the samples carrying them) adjacent to a radiographic emulsion, which responds to radioactive decay by producing silver granules. *Also termed* **radioautography**.

autoreactive   Referring to the ability of components of the adaptive immune system—including antibody molecules and lymphocytes—to recognize and react with antigens displayed by normal tissues, resulting in autoimmunity. *See also* **autoimmune**.

autosome   A chromosome that is not a sex chromosome, i.e., neither an X nor a Y chromosome.

avidity   Referring to the strength with which one molecule binds another, such as the interaction of an antibody molecule binding its cognate antigen(s).

axillary   Referring to the armpit.

axon   Long slender protrusion from the cell body of a neuron through which nerve signals are transmitted, often to other neurons.

B-cell receptor   An immunoglobulin molecule that is displayed on the surface of a B cell that enables this cell to recognize and respond to specific antigens.

B cells   Lymphocytes, also termed B lymphocytes, that develop in the bone marrow and are involved in generating the humoral immune response. *See also* **T cells**.

bacteriophage   Virus that infects bacteria.

barcode   Distinct labels, usually in the form of specific oligonucleotide DNA sequences, that are introduced into the individual cells within a cell population and used subsequently to identify each of these cells and their lineal descendants; the DNA sequences within a barcode library often represent randomly generated oligonucleotide sequences.

Barr body   The condensed, inactive X chromosome found in each of the cells of females of placental mammals.

Barrett's esophagus   Metaplasia in which squamous epithelium of the esophagus is replaced by secretory epithelial cells of a type normally found in the stomach. *See also* **metaplasia**.

basal   (1) Referring to a lower physical location. (2) Referring to cells in an epithelium that are located away from an exposed surface of the epithelium such as its luminal surface. (3) Referring to the surface of an individual epithelial cell that is located away from an exposed surface of the epithelium such as its luminal surface. (4) Referring to a (low) rate of activity or function observed in the absence of any activating stimulus. *See also* **apical**.

basal lamina   *See* **basement membrane**.

basaloid   Having the characteristics of a basal epithelial cell. *See also* **basal**.

base-excision repair   A form of DNA repair that initially involves cleavage by a repair enzyme of the glycosidic bond between a base and a deoxyribose, leaving behind an abasic nucleotide. *Compare* **nucleotide-excision repair**.

basement membrane   The specialized extracellular matrix that forms a sheet separating epithelial from stromal cells or endothelial cells from pericytes. *Also termed* **basal lamina**.

basophilic   Referring to entities in a histopathological section, such as cell nuclei, that are stained preferentially by basic dyes such as hematoxylin.

benign   (1) Describing a growth that is confined to a specific site within a tissue and gives no evidence of invading adjacent tissue. (2) Referring to an epithelial growth that has not penetrated through the basement membrane.

biallelic   Referring to a state in which both copies of a gene are expressed or exert effects on phenotype. *See also* **monoallelic**.

bioavailability   Proportion or amount of a substance that is available for absorption into cells and tissues or is able to exert biological effects on the cells of those tissues.

bioinformatics   Science of using computational methods for analyzing biological information, notably complex sets of biological data.

biomarker   A measurable property or parameter of a cell, tissue, or organism that provides information about the biological state of the entity being analyzed; biomarkers can be used for stratification of disease subtypes and, in the clinic, for disease diagnosis or prognosis.

biopsy   Retrieval of a sample of living tissue performed in order to understand the properties of the tissue using various analytical procedures.

biosynthesis   Process by which new biological molecules are synthesized from precursor substrates via various biochemical reactions.

biotinylation   Process of covalently attaching a biotin molecule to a protein or nucleic acid.

bispecific   Able to specifically recognize or bind two objects simultaneously.

bispecific T-cell engager   A re-engineered antibody-derived molecule that simultaneously binds a T cell displaying a T cell-specific cell surface antigen, often CD3, to a second cell, such as a targeted cancer cell that displays a distinct cell surface antigen; such an antibody is used to drive the physical linking and thus juxtaposition of the T cell with the targeted cell.

bisphosphonates   Class of drugs, characterized by a chemical backbone with the structure P–C–P, that are incorporated into bone apatite and subsequently become available to poison osteoclasts that may later undertake to dissolve the bone, thereby liberating the previously incorporated toxic bisphosphonates.

bistable   Referring to a control device or a system that resides stably in either one of two alternative states.

blast   Term used as prefix or suffix, indicating a relatively undifferentiated or embryonic precursor cell.

blast crisis   Clinical state occurring when a hematopoietic tumor that has been under therapeutic control relapses into a highly aggressive stage of disease progression in which the erupting tumor cells are in a primitive state of differentiation. *See* **blast**.

blastocoel   Inner cavity present in a blastocyst.

blastocyst   An early stage of vertebrate embryogenesis in which the embryo consists largely of an outer layer of cells enclosing an inner cavity that contains a cell mass destined to form the embryo.

bleb   A small, bubble-like herniation through a membrane such as the plasma membrane.

blood–brain barrier   The barrier created by specialized endothelial cells and basement membrane that operates, together with astrocytes, to prevent transport of solutes between the blood and the parenchyma of the brain.

body-mass index   Anatomical metric calculated by dividing the weight of the body (measured in kilograms) divided by the square of height in meters.

bona fide   Genuine, authentic.

Boyden chamber   A cylindrical two-chambered plastic vessel in which the upper and lower chambers are separated by a membrane whose porosity allows molecules or, alternatively, whole cells to pass between them.

brachytherapy   Form of radiotherapy in which a radioactive source carried within a small capsule or vessel is implanted in a patient's tissue in order to generate localized high-intensity irradiation.

breakpoint   Location in a chromosomal region or gene at which it becomes fused through chromosomal translocation with another chromosomal region or gene.

bronchial   Referring to a major airway of the lung.

bronchoalveolar   Subtype of lung adenocarcinoma usually arising in peripheral areas of the lung and referring to tissues derived from small airways termed bronchioles and associated alveoli. *See also* **alveolus**.

bronchoscope   A flexible tube that can be inserted into the major air passages of the lungs and allows visual surveillance of the walls of these passages.

buccal   Referring to the tissues of the oral cavity, specifically the epithelial lining of the cheek.

bypass polymerase   A DNA polymerase that will copy over an unrepaired lesion in the template strand of the DNA, "guessing" (often in an error-prone fashion) the nucleotides that should be incorporated into the nascent complementary DNA strand in order to avoid undesirable stalling of replication forks.

bystander effect   Mechanism through which normal cells or tissues near to those directly targeted are also affected.

bystander mutation   *See* **passenger mutation**.

C-terminus (adj., **-al**)   End of a protein chain that is synthesized last. *See also* **N-terminus**.

cachexia   Physiological state, often seen late in cancer development, in which the patient loses appetite and suffers wasting of tissues throughout the body.

café au lait spots   Coffee-with-milk–colored spots on the skin that are seen characteristically in patients suffering the neurofibromatosis type 1 syndrome.

calreticulin   A cell surface protein whose display by a cell incites the phagocytosis of the cell by macrophages.

canaries in the mine   Referring to the use of a highly sensitive sentinel to detect and report an early warning.

cancer   (1) A clinical condition that is manifested by the presence of one or another type of neoplastic growth. (2) A malignant tumor.

Cancer Dependency Map   The product of an effort focused on systematically cataloging the key roles of genes throughout the entire human genome whose inactivation by CRISPR/Cas9 leads to the death of cells belonging to specific cancer cell lines.

cancer of unknown primary   A tumor, often a metastasis, whose histopathological appearance does not permit a determination of its tissue-of-origin within a patient. *See also* **primary tumor**.

cancer stem cell   A neoplastic cell that resides within a tumor, is able to both self-renew and generate non-stem-cell progeny, and can be defined experimentally by its ability to seed a tumor when implanted in a suitable host such as a mouse. *Also termed* **tumor-initiating cell**.

canonical   Referring to the most widely accepted representation of an entity or process, often the one that is initially discovered and characterized. *See also* **non-canonical**.

capsid   Protein coat of a virus particle that envelops and protects the viral genome.

carboxy   (1) Referring to the end of an oligopeptide or protein chain that is completed last. (2) Referring to the ionizable group of an organic acid. *See also* **amino**.

carcinogen   An agent that contributes to the formation of a tumor.

carcinogenic   Capable of causing or contributing to the causation of cancer.

carcinoma (adj., **-omatous**)   A cancer arising from epithelial cells.

cardia   Region of the stomach adjacent to its junction with the esophagus.

caretaker   A gene that encodes a protein that maintains the integrity of the genome and thereby prevents the accumulation of mutations and, in turn, the formation of neoplastic cells. *See also* **gatekeeper**.

cascade model (of metastasis)   A hypothetical model that posits that initially formed metastatic colonies rather than corresponding primary tumors are the sources of a secondary wave of metastatic dissemination and thus metastatic colonies.

caspase   A protease that contains cysteine in its catalytic site and cleaves substrate proteins after aspartic acid residues. *See also* **apoptosome**.

CAT scan   A clinical procedure in which multiple successive two-dimensional sections of a patient's body or specific tissue are imaged by X-rays and integrated with one another using informatics in order to provide a three-dimensional survey of an extensive portion of a patient's body or tissue.

catabolism (adj., **catabolic**)   Metabolic process involving the breakdown of biological molecules, often used to generate energy or the precursors of new molecules. *See also* **anabolism**.

catalase   An enzyme that catalyzes the reaction by which hydrogen peroxide is decomposed to water and oxygen.

catalytic cleft   Region of a protein, usually an accessible cavity, in which enzymatic catalysis is accomplished by this protein.

CD3   Antigen representing a protein complex associated with the T-cell receptor and used to identify T cells and a subset of NK cells.

CD4   Cell surface protein displayed by helper T cells that enables them to recognize MHC-II proteins on the surface of professional antigen-presenting cells.

CD8   Cell surface protein displayed by cytotoxic T cells that enables them to recognize MHC-I proteins on the surface of cells that they may thereafter target for destruction.

CD20   Antigen that is displayed on the surface of multiple cell types of the B-cell lineage.

CD45   The common leukocyte antigen expressed by all cells of the hematopoietic lineage except platelets and erythrocytes.

CD47   Cell surface protein displayed by various cell types that discourages attack and phagocytosis by macrophages.

CD antigens   Cell surface antigens that were initially characterized in human leukocytes in order to determine their immunophenotypes and later determined for non-leukocyte cell types as well; by 2016, 371 distinct human CD antigens and comparable number of mouse CD antigens had been identified.

celecoxib   A clinically employed nonsteroidal anti-inflammatory drug that strongly inhibits the COX-2 enzyme.

cell-autonomous   Referring to a trait or behavior of a cell that is governed by its own genome and internal physiology and not by its ongoing interactions with other cells and its tissue microenvironment.

cell culture   Procedure for propagating cells in various types of vessels using chemically defined media, usually with admixed serum.

cell cycle   The sequence of changes in a cell from the moment when it is created by cell division, continuing through a period in which its contents including chromosomal DNA are doubled, and ending with the subsequent cell division and formation of two daughter cells.

cell cycle clock   Network of signaling proteins in the nucleus that regulate and orchestrate progression of the cell through its growth-and-division cycle. *See also* **cell cycle**.

cell fate   The entrance of a less-differentiated cell into a specific differentiation lineage and eventual differentiated state, often determined by a cell long before such a cell manifests specific markers of such a state.

cell-heritable   Trait that is transmitted from a mother cell to one or both of its daughters; such heritability may result from genetic mutation or nongenetic mechanisms.

cell line   Strain of cells that has been adapted for indefinite propagation in culture.

cell-of-origin   The fully normal cell that serves as the ancestor of all the neoplastic cells within a tumor, i.e., their last common normal ancestor.

cellular immune response   The arm of the immune system that depends on the ability of specific cell types, such as B cells, cytotoxic T cells, natural killer cells, and macrophages, to recognize and respond to specific entities or agents, notably those expressing certain antigens; such recognition may be followed by the immune cell-mediated destruction of the recognized entities, including abnormal cells and infectious agents. *See also* **humoral immune response**.

cellularity   The proportion of the volume of a tumor that is represented by living cells rather than extracellular matrix and cellular debris.

central tolerance   Process by which immature T-cells recognizing self-antigens are eliminated in the thymus. *Compare* **peripheral tolerance**.

centriole   Component of a centrosome from which spindle fibers radiate during mitosis and meiosis.

centromere   Region of a chromosome that holds the two chromatids together and that binds, via a kinetochore, with mitotic or meiotic spindle fibers.

centrosome   A body in the cytoplasm containing a pair of centrioles, ancillary proteins, and outward radiating microtubules that functions to organize one-half of a mitotic spindle. *See also* **centriole**, **spindle**.

chain-terminating nucleotide   A nucleotide that lacks a 3′-OH group, which is required for the formation of a phosphodiester bond between two nucleotides. The insertion of a chain-terminating nucleotide causes DNA polymerase to cease extension due to the inability to form a new phosphodiester bond with an incoming nucleotide.

chaperone   *See* **molecular chaperone**.

checkpoint   (1) Control mechanism that ensures that the next step in a process does not proceed until a series of prerequisites and conditions have been fulfilled. Such prerequisites may include, as examples, the successful completion of all previous steps during cell-cycle progression or the integrity of the cell genome. (2) Control mechanism that limits or constrains the immune-mediated attack on targeted cells. *See also* **immune checkpoint**.

chemokine   (1) A polypeptide that serves as an attractant for motile cells, notably leukocytes, via chemotaxis. (2) More narrowly, a member of a family of low–molecular-weight (8–10 kD) secreted proteins, most of which contain four cysteine residues in evolutionarily conserved sites. *See also* **chemotaxis**.

chemo-refractory   Resistant or unresponsive to treatment by chemotherapy.

chemotaxis (adj., **-tactic**)   Movement of a cell toward high concentrations of a soluble attractant such as a chemokine or growth factor. *See also* **chemokine**, **haptotaxis**.

chemotherapy   Use of introduced low–molecular-weight drugs to treat a cancer, often because they have cytotoxic effects on proliferating cancer cells.

chimera   Organism or molecule in which different parts or tissues derive from genetically distinct parents or genetically distinct origins.

chimeric antibody   *See* **chimerized**.

chimeric antigen receptor   Cell surface antigen-binding receptor generated by recombinant DNA in which antigen-binding and transmembrane domains have been fused with segments of intracellular signal-emitting domains, resulting in strong potentiation of antigen-provoked signaling within a T cell.

chimerization   Process whereby a donor cell introduced into a host embryo—usually of a different genotype—is able to insert itself into that embryo and participate in the formation of some of the subsequently formed tissues arising from that embryo.

chimerized   (1) Referring to the protein product of a re-engineered gene in which two portions of the protein derive from two distinct sources, such as two distinct species. (2) Describing an antibody molecule whose constant (C) region amino acid sequences have been replaced by the homologous sequences from another species, e.g., one in which a mouse C region is replaced by a human C region.

ChIP-seq   Procedure for determining chromosomal DNA sequences that are bound to specific proteins by immunoprecipitating the proteins of interest together with bound DNA followed by sequencing of co-precipitated DNA.

cholangiocarcinoma   Tumor of the bile ducts in the liver, sometimes termed cholangiosarcoma.

chromatid   A half chromosome that exists after S phase and before M phase; paired chromatids are separated at M phase, whereupon each becomes a chromosome.

chromatin   Complex of DNA, RNA, and protein molecules that together constitute chromosomes.

chromogenic   A chemical that generates a visible pigment upon enzymatic alteration; *sometimes called* **chromagenic**.

chromophore   A chemical entity that serves to emit light, usually colored light visible via microscopy.

chromothripsis   The presence of multiple chromosomal rearrangements usually limited to a single chromosomal region and suggesting the involvement of an initial catastrophic shattering of a chromosomal domain followed by rejoining of the fragmented segments in random order and orientation. *See also* **kataegis**.

chronic myelogenous leukemia   An excess of the cells originating in the bone marrow that are precursors of granulocytes; these cells accumulate in large numbers in the blood of patients. *See also* **granulocyte**.

chymotryptic   Referring to proteolytic reaction similar to that of chymotrypsin that cleaves protein substrates adjacent to hydrophobic amino acids.

cilium (pl., **cilia**)   A cellular organelle that protrudes from the surface of a eukaryotic cell, is wrapped by a lipid bilayer, and contains in its core a unique arrangement of cytoskeletal proteins; a cilium may either be motile or non-motile. *See also* **primary cilium**.

circos plot   Graphic convention in which the chromosomes of a cell, usually of human origin, are plotted in a circular array (often arrayed by increasing chromosome number ranging from Chromosomes 1 to 22 plus X and Y), in which translocations are indicated as arcs connecting participating chromosomal regions, and in which localized changes in copy number, chromosome structure, and mutational status are indicated in concentric outer circles.

circulating tumor cell   A neoplastic cell that has entered into the general circulation following intravasation and dwells in the circulation for a brief period of time before being trapped in microvessels or eliminated by components of the immune system.

cis   (1) Referring to an effect on a genetic segment exerted by the actions of a physically linked genetic segment. (2) Referring to a structural configuration of a molecule, often a complex organic molecule, in which two moieties are oriented in the same direction or on the same side of a plane. *See also* **trans**.

cistrome   The entire set of targeted genes present in a genome that are acted upon by (a trans-acting) transcription factor. *See* **trans-acting**.

citric acid cycle   *See* **tricarboxylic acid cycle**.

class switching   Gene rearrangement occurring in an immunoglobulin gene locus in which an already-formed antigen-binding variable region-encoding sequence is switched from its juxtaposition with one constant region-encoding segment to another constant region-encoding segment; such switching enables one variable region of an immunoglobulin to be expressed in a number of alternative classes of immunoglobulins and related receptors.

clinical presentation   *See* **presentation**.

clonal expansion   Process by which a cell that has acquired a more advantageous phenotype generates a large population of lineal descendant cells, which in aggregate constitute a clone; such clonal populations may be generated during the process of tumor progression or in lymphocytes that are stimulated to proliferate by encounter with an antigen that is cognate with the cell surface receptor that they display.

clonal hematopoiesis   Development in hematopoietic cell lineages of clonal populations of cells in which the constituent cells share a common somatic mutation or set of mutations that appears to favor their preferential clonal expansion in the hematopoietic system; such cells have nonetheless not developed a frankly neoplastic phenotype although they may be poised to do so.

clonal succession   Process by which one clonal population of cells displaces and thus succeeds a preexisting clonal population of cells; by implication, the displacing cell population has acquired a more advantageous phenotype relative to that of the cell population that is being displaced. *See also* **clonal expansion**.

clonality   Condition describing the state of a population of cells, all of which descend from a common progenitor. The cells within this population share a common genetic or epigenetic alteration inherited from this progenitor.

clone (adj., **clonal**)   (1) Copy of a gene that has been isolated by recombinant DNA procedures and amplified into a large number of identical copies. (2) Population of cells, all of which descend from a common progenitor cell. (3) Offspring of a procedure of asexual reproduction in which the genome of a somatic diploid cell of an organism is used to form a cell that functions equivalently to a fertilized egg that may itself develop thereafter into an entire organism.

co-activator   Protein that associates with a DNA-binding transcription factor and causes, directly or indirectly, the modification of nearby chromatin proteins and a resulting stimulatory effect on the transcription of an adjacent gene. *See also* **co-repressor**.

co-carcinogen   An agent or substance that, while not carcinogenic on its own, collaborates with another agent to enable carcinogenesis to proceed.

cognate   Corresponding to or related to.

co-inhibitory   (1) Referring to a molecule or process that operates in concert with a major inhibitory process to ensure completion of the latter or to amplify its effects. (2) Receptor that, when displayed in conjunction with a T-cell receptor, will dampen the duration of subsequent T-cell activation; such a receptor will recognize a ligand presented by a potential target cell. *See also* **co-stimulatory**.

cold tumor   A tumor that lacks evidence of extensive immune cell infiltration and inflammation.

collective invasion   Process whereby a cohesive mass of invasive cancer cells invades in a coordinated fashion into nearby tissue.

collective metastasis   Process whereby a cluster of circulating tumor cells function as the founders of a metastatic colony, yielding a polyclonal growth formed by the lineal descendants of the multiple clustered disseminating tumor cells.

colonization   Proliferation of cells within a micrometastasis that leads to the formation of a macroscopic metastasis.

colonoscopy   Clinical procedure used to visually survey the wall of the colon via insertion of an optical tube through the rectum, employed used to detect previously undiagnosed polyps.

colony   (1) Cluster of cells, usually of clonal origin. (2) Cluster of cells that is able to proliferate in the absence of anchorage to a solid substrate.

colorectal   Referring to the lower gastrointestinal tract including the colon and rectum.

commensal Referring to a long-term interaction between two organisms, such as gastrointestinal bacteria and their mammalian hosts, which reflects symbiosis between the organisms and may accrue to the benefit of both of them.

commitment   Decision made by a relatively undifferentiated cell to enter into one or another differentiation lineage and generate a limited repertoire of descendant differentiated cell types.

committed progenitor   Less differentiated progenitor cell that is committed to spawning a population of descendant cells that are able to differentiate into a limited repertoire of distinct cell types. *See also* **progenitor cell**.

comparative genomic hybridization   Procedure in which the copy numbers of a large array of genomic sequences from cells of interest are compared with the copy numbers of the corresponding sequences in normal reference DNA in order to determine whether the various sequences being analyzed are present in the cells of interest in increased or decreased copy number.

compartment   Physical or virtual functional space that contains all cells of a given type within a tissue, e.g., a stem cell compartment.

competitive inhibition   Mechanism whereby an inhibitory molecule reduces enzyme function by binding to the same site in the enzyme that is bound by the usual substrate of this enzyme.

complement   Group of collaborating plasma proteins that can associate with antibody molecules bound to cell surface antigens, such as antigens displayed by bacterial, yeast, or mammalian cells; once attached via these antibody molecules to a cell surface, complement can kill the cell by introducing pores into the nearby cell membrane, including the plasma membrane of mammalian cells.

complementarity-determining region   The variable domains of the heavy and of the light chains of antibody molecules and of the T-cell receptors that recognize and bind antigens; each of these regions is formed from three oligopeptide loops, meaning that, as an example, an antibody molecule possesses altogether six such loops in each of its two antigen-binding domains.

complementarity   State in which two nucleic acid molecules can anneal to one another via hydrogen bonding because their respective base sequences are complementary with one another.

complementary DNA   DNA strand that is produced on an RNA template by reverse transcriptase.

complementation   Ability of two mutant genotypes, when coexisting in the same cell or organism, to compensate for each other's defects and thereby create a wild-type phenotype, indicating that the two genotypes carry changes in distinct genes and therefore can be classified as belonging to distinct complementation groups. *See also* **complementation groups**.

complementation group   A set of genes or alleles, all of which behave identically when tested for their ability to complement the genetic defects present in a set of mutant organisms or cells. *See also* **complementation**.

complete carcinogen   An agent that can act as both an initiator and a promoter of tumor progression. *See* **initiation, promotion**.

complete response   Elimination of all detectable neoplastic tissue following an anti-cancer therapy. *See also* **partial response**.

computed tomography   Procedure in which imaging generated by successive X-ray scans of sections of a tissue or of the entire body is processed digitally to generate a three-dimensional image of the tissue or of the entire body, sometimes termed computed axial tomography.

concatemer   (1) A molecule resulting from end-to-end joining of multiple copies of a molecular species, such as a DNA sequence or a viral DNA genome. (2) DNA molecules that are topologically intertwined in a fashion that can only be resolved by the actions of a topoisomerase. *See also* **topoisomerase**.

conditional   Referring to a phenotypic state or genetic change that only occurs under certain conditions, often in response to physiological signals or those induced experimentally. *See also* **constitutive**.

conditioned medium   Culture medium that arises after cells have been propagated *in vitro* for a period of time; the secretion of various factors into the medium by these cells then imparts to this medium biological or biochemical properties.

confluence   State reached when cells in monolayer culture proliferate until they fill available space at the bottom surface of a Petri dish.

congenic   Referring to a genotype that differs from another only by the presence of a single or a small number of identified alleles.

congenital   Referring to a condition that is already present at birth and may persist thereafter.

constant region   Portion of an antibody molecule that is present identically by all members of a class of these molecules, e.g., all immunoglobulin gamma molecules. *Compare* **variable region**.

constitutive   Describing a state of activity that occurs at a constant level and is therefore not responsive to modulation by physiologic regulators, or a type of control that yields such a constant output. *See also* **conditional**.

constitutive heterochromatin   Regions of chromosomes that are permanently in a heterochromatic and thus transcriptionally inactive state, in contrast to facultative heterochromatin, which may become euchromatic in response to developmental or physiological signals. *See also* **facultative heterochromatin, euchromatin**.

constitutive proteasome   Term used to describe the proteasome that is present in a wide variety of cell types and is responsible for the degradation of most cellular proteins; this class of proteasomes is distinguished from immunoproteasomes that degrade intracellular proteins in order to generate the oligopeptides that are loaded onto Class I MHC/HLA proteins and presented on the cell surface. *See also* **proteasome, immunoproteasome**.

contact inhibition   Behavior exhibited by cells propagated in monolayer culture, reflecting the halt in cell proliferation when adjacent cells touch one another. *Also termed* **density inhibition**.

contralateral   Referring to the opposite side of the body of a bilaterally symmetrical organism.

convergent evolution   The independent development by multiple species of a common trait during the course of evolution.

copy choice   Genetic recombination mechanism whereby, during the replication of DNA on one chromosome, the replication machinery switches to using the DNA sequences on another chromosome as template.

copy number variation   Alteration involving the number of copies of a gene or chromosomal region present in a genome, usually the genome of a neoplastic cell that has undergone somatic mutation.

co-receptor   Cell surface protein that is displayed in conjunction with a nearby receptor and modulates the function of the latter by modulating its ligand specificity or signaling powers.

co-repressor   Protein that associates with a DNA-binding transcription factor and causes, directly or indirectly, the modification of nearby chromatin proteins and a resulting inhibitory effect on the transcription of an adjacent gene. *See also* **co-activator**.

cortex (adj., **cortical**)   The region of the cytoplasm that immediately underlies the plasma membrane. *See also* **cortical actin**.

cortical actin   Referring to the outer layer of actin filaments that lie just below and in a plane parallel with that of the nearby plasma membrane. *See also* **cortex**.

co-stimulatory   (1) Referring to molecules or processes that operate in concert with a major stimulatory process to ensure completion of the latter or to amplify its effects. (2) Referring to a receptor that, when acting in concert with a T-cell receptor, can act to stimulate the resulting T-cell response; such a receptor will recognize a ligand displayed by a potential target cell. *See also* **co-inhibitory**.

co-translational   Process occurring during the course of and coordinated with translation.

counterstain   Histological procedure in which the microscopic entity of interest is viewed in the context of or is contrasted with other entities by staining the latter with a different dye or other substance.

CpG island   Cluster of CpG dinucleotide sequences located in the vicinity of a gene promoter; the increased methylation of these CpGs may lead to transcriptional repression of the nearby gene. *See also* ***de novo*** **methyltransferase**.

Cre   A bacterial recombinase enzyme that catalyzes the site-specific recombination between two loxP sites, resulting in the excision of the DNA sequences between them. *See* **loxP**.

crisis   (1) Cellular state arising when cells lose telomeres of a length that is required to protect the ends of chromosomal DNA, resulting in the end-to-end fusion of chromosomes, karyotypic chaos, and widespread cell death by apoptosis. (2) Clinical state when patients whose tumors have been under therapeutic control relapse into a highly aggressive phase of disease.

CRISPR/Cas9   A genetic procedure in which a targeted gene within a cell is altered with specificity and accuracy achieved by associated sequence-specific guide RNAs and a specialized nuclease. *See also* **knock-out**, **knock-in**.

cross-immunity   Immunity that has initially arisen against a particular antigen or infectious agent that is subsequently found to confer immunity against another distinct antigen or infectious agent.

cross-presentation   The presentation by antigen-acquiring cells, notably dendritic cells, of peptides acquired from other cells and cellular debris on the MHC-I cell surface molecules of the antigen-acquiring cells; such presentation occurs concomitantly with peptide presentation by these cells via their MHC-II molecules.

cross-reactive   Referring to the ability of an antibody to recognize and/or bind antigens other than the antigen of interest.

cruciferous   Referring to a closely related group of vegetables, including kale, collard greens, cabbage, brussels sprouts, kohlrabi, and cauliflower.

cryoablation   The elimination of a tissue or a group of cells by directed localized freezing.

crypt   A deep cavity in the wall of the small or large intestine in which enterocyte stem cell proliferation and initial differentiation occur.

cutaneous   Referring to the skin.

cyclin   A protein that associates with a cyclin-dependent kinase and serves as a regulatory subunit of this kinase by activating its catalytic activity and directing it to appropriate substrates. *See also* **cyclin-dependent kinase**.

cyclin-dependent kinase   Type of serine/threonine kinase deployed by the cell cycle machinery that depends on an associated cyclin protein for proper functioning. *See also* **cyclin**.

cycloheximide   Drug that prevents the forward movement of ribosomes down mRNA templates, thereby blocking protein synthesis.

cyclooxygenase   An enzyme that converts arachidonic acid to a number of prostaglandins including prostaglandin E_2.

cyclopia   Malformation of the head resulting in embryos with only a single, centrally located eye.

-cyte   Suffix indicating a type of cell.

cytoarchitecture   Physical structure of a cell.

cytochrome P450 enzymes   A family of enzymes that function to oxidize steroids, fatty acids, and xenobiotics, and thereby aid in the clearance of various compounds.

cytocidal   Referring to an effect or influence that causes cell killing. *See also* **cytopathic**, **cytostatic**, **cytotoxic**.

cytogenetics   The study of cell or organismic genetics by examining the microscopic appearance of the chromosomes within individual cells.

cytokeratin   Forms of the intermediate filament protein keratin that constitute part of the cytoskeleton of an epithelial cell; term used to distinguish it from the keratins that constitute hair, nails, and feathers.

cytokine   (1) Secreted protein that operates like growth factors to stimulate or inhibit various cell types, notably those constituting various branches of the hematopoietic and immune systems. (2) Less commonly, secreted proteins that convey signals between cells, including mitogenic growth factors.

cytokine release syndrome   Condition representing acute systemic inflammation associated with fever and often the dysfunction of multiple organs throughout the body, which is often life-threatening.

cytokinesis The late step of mitosis involving the separation of the two daughter cell cytoplasms from one another.

cytology (adj., **-logical**) (1) Analysis of subcellular structure under the microscope. (2) The microscopic appearance of a cell.

cytopathic Causing damage or death to a cell. *See also* **cytocidal, cytostatic, cytotoxic.**

cytoskeleton Network of filamentous proteins in the cell cytoplasm that provides it with structure and rigidity enables it to exhibit motile behavior and is formed from polymerized actin, tubulin, and intermediate filament proteins such as cytokeratins.

cytosol Portion of the cytoplasm that contains soluble material untethered to either cytoskeleton or membranes including those associated with organelles.

cytostasis The halting of cell proliferation. *See also* **cytostatic.**

cytostatic Referring to an influence or a force that inhibits cell proliferation without necessarily having any effect on cell viability. *See also* **cytocidal, cytopathic, cytotoxic.**

cytotoxic Referring to the ability of an agent to kill cells; such an agent might be, for example, a drug or another type of cell. *See also* **cytocidal, cytopathic, cytostatic.**

D-loop The loop of ssDNA created in telomeres through invasion of the telomeric dsDNA by the 5′ ssDNA overhang of the G-rich strand; *also termed* **displacement loop.**

de novo (1) Arising or formed anew. (2) Occurring for the first time.

de novo **methyltransferase** An enzyme that attaches a methyl group to the C of a CpG dinucleotide in the absence of methylation of the complementary CpG.

deacetylase Enzyme that removes acetate groups from substrates, notably those groups that were previously attached by acetyltransferases. *See also* **acetyltransferase.**

deamination Loss or removal of an amine group from a larger molecule such as a DNA base.

death receptor A family of mammalian transmembrane receptors that function at the cell surface to initiate apoptosis through the extrinsic apoptotic program following binding of their cognate ligands. Examples of death receptors are TNFR1, FAS, DR3, DR4, and DR5.

debulk To reduce in overall size with the intent of removing the great majority of a mass such as a solid tumor.

decatenation Untangling and separation of topologically intertwined circular dsDNA molecules from one another, often accomplished by the topoisomerase II enzyme. *See also* **topoisomerase.**

deconvolute Mathematical manipulation that allows the dissection of complex datasets into distinct subsets of data, often yielding an assessment of the contributions of the individual subsets to the initial data set being analyzed.

decoy receptor A nonsignaling protein that binds and sequesters a ligand (e.g., a growth factor), thereby diverting the ligand from binding and activating its functional cognate signaling receptor.

dedifferentiation Reversion of a more differentiated cell to the phenotype of a less differentiated cell, such as its stem cell precursor.

degranulation Discharge of the contents of cytoplasmic granules into the extracellular space in response to a physiological stimulus.

delayed early Referring to genes whose expression depends on prior *de novo* protein synthesis and is induced with some delay following growth factor stimulation of a factor-starved cell. *See also* **immediate early.**

demineralize To dissolve the inorganic apatite (i.e., calcium phosphate) component, notably of bone.

denaturation Process that causes a molecule, such as a macromolecule (DNA, RNA, or protein), to lose its natural three-dimensional structure.

dendritic cell An innate immune cell that phagocytoses fragments of cells or infectious agents and then proceeds within lymphoid tissues to present oligopeptides derived from these phagocytosed particles to several types of T cells.

dendrogram Diagram in which closely related entities, such as genes, cells, or organisms, are placed close to one another on the branches of a multibranched tree on which the lengths of branches and the number of branch points provide an indication of the degree of relatedness. *See also* **hierarchical clustering.**

density inhibition *See* **contact inhibition.**

denuding Removal or stripping away of a tissue or population of cells, such as those forming an epithelium.

dependency Property of a tumor or tumor cell that reflects its continued dependence on an alteration acquired during the course of its formation and, in turn, renders it vulnerable to a specific type of therapeutic intervention.

dependency map The product of an effort to systematically catalog the effects of chemical and genetic perturbations on a large set of human cancer cell lines. The chemical perturbations derive from measurements of large collections of low–molecular-weight drug molecules, whereas the genetic perturbations derive from CRISPR/Cas9-mediated systematic inactivation of a large collection of genes in the human genome.

depurination Breakage of the glycosidic bond that links purine bases to the deoxyribose or ribose of DNA or RNA, respectively.

depyrimidination Breakage of the glycosidic bond that links pyrimidine bases to the deoxyribose or ribose of DNA or RNA, respectively.

derivitization The alteration via synthetic organic chemistry of the molecular structure of an existing drug molecule, usually through the covalent linkage of side groups, and undertaken with the intention of improving its pharmacological properties.

dermis (adj., **dermal**) Thick layer of stromal cells, largely fibroblasts, found beneath the epidermal keratinocyte layer of the skin.

desmoplastic Referring to a hard extracellular matrix that is rich in collagen and myofibroblasts.

detoxify To render a previously toxic substance harmless.

development The processes of growth, differentiation, and morphogenesis that enable a zygote to develop into a fully formed embryo and thereafter a newborn.

diagnostic bias Tendency of an observation or conclusion to be influenced by the specific properties or applications of a diagnostic technique rather than to accurately reflect an intrinsic property of the entity (such as a disease) under examination.

diapedesis Complex sequence of steps enabling leukocytes to extravasate rapidly from the lumen of a blood vessel through the vessel wall into the underlying tissue parenchyma.

dicentric Referring to a chromosome or chromatid that bears two distinct centromeres.

differentiation Process whereby a cell acquires a specialized phenotype, such as a phenotype characteristic of cells in a particular tissue.

differentiation antigen Protein or other product that is expressed only in a specific stage of differentiation of a normal tissue and that, for various reasons, may not induce immune tolerance.

dimer Molecular complex composed of two subunits.

dioxygenase An oxygenase that catalyzes the introduction of both atoms of molecular oxygen into the substrate being oxidized.

diploid Describing a genome in which all chromosomes are present in pairs, one of each pair being inherited from a father and the other from a mother, with the exception of the sex chromosomes, which in placental mammals are paired in either the XX or the XY configuration.

dirty drug A drug species that inhibits molecules within a cell in addition to its intended major target.

discoldin domain receptor A cell surface tyrosine kinase receptor that recognizes and binds extracellular collagen as its activating cognate ligand.

disease-free survival Time period elapsed following initiation of treatment during which there are no apparent symptoms of the disease under treatment. *See also* **overall survival** and **progression-free survival**.

displacement loop *See* **D-loop**.

disseminated Spread or seeded, often widely throughout the body.

DNA chip An array of thousands of sequence-specific DNA segments, constituting a DNA microarray, that is mounted on a solid support, such as a microscope slide. *See also* **microarray**.

dominance *See* **dominant**.

dominant (1) Referring to one of several alternative traits (phenotypes) that can be specified by a genetic locus; when the locus is heterozygous and carries information specifying two distinct traits, the dominant trait will be the one actually exhibited. (2) Describing an allele of a gene that determines phenotype in spite of the concomitant presence of a second allele, considered to be recessive, that specifies a different, alternative phenotype. *See also* **recessive**.

dominant-interfering *See* **dominant-negative**.

dominant-negative Referring to a mutant allele of a gene that, when co-expressed with the wild-type allele of the gene, is able to interfere with the functioning of the latter. *Also termed* **dominant-interfering**.

dorsal (1) Term for the surface of a cell that is exposed upward away from the underlying physical substrate to which the cell is attached, e.g., the bottom surface of a Petri dish. (2) Anatomical term referring to the back side of an organism such as a mammal. *See also* **ventral**.

dose escalation The practice of progressively increasing the dosage of a drug in an investigatory clinical trial in order to determine efficacy of the drug under study at various applied doses and, at the same time, revealing the toxicity of such doses in order define the maximum tolerated dose. *See also* **maximum tolerated dose.**

dose-limiting toxicity Side effect toxicity that dictates the highest dosage of a drug that can be applied without creating unacceptable levels of such toxicity in a patient. *See also* **maximum tolerated dose**.

double-minute A chromosomal segment that becomes separated from the chromosome in which it normally resides and is able to perpetuate itself as an extrachromosomal particle unlinked to a centromere, so termed because it is visualized at metaphase as small spots that are doubled like the chromatids prior to anaphase.

draining lymph node A lymph node that receives cells and debris via a lymph duct that drains lymphatic fluid from a specific tissue or sector of a tissue.

driver gene A gene that is affected by driver mutations and is identified because the mutations affecting this gene occur more frequently in tumors than would be expected by random chance; driver genes are mutated on a recurring basis in the genomes of multiple, independently arising tumors, ostensibly because these mutations confer advantageous cell phenotypes. *See* **driver mutation**.

driver mutation Mutation creating an allele of a gene that is advantageous for the cell that sustains it and results in clonal expansion of the progeny bearing mutant alleles of this gene. *See also* **driver gene**, **passenger gene**.

drug holiday Defined period when drug treatment is halted.

druggable Referring to a molecular species, such as a protein, that has the structural and functional properties suggesting that low–molecular-weight therapeutic drug compounds can be developed that specifically interact with and perturb its functioning; the structural properties include, most prominently, a drug-binding pocket that should permit high-affinity, multivalent associations between a drug molecule and the targeted protein.

dual-address Referring to signaling that involves the translocation of a protein from one intracellular site to another, e.g., from the cytoplasm to the nucleus.

duodenum (adj., **duodenal**) The section of the small intestine beginning immediately after the stomach and extending as far as the jejunum.

dysbiosis Referring to a shift in the representation of microbial species forming a microbiome away from its natural or optimal configuration.

dyskeratosis congenita   An inherited syndrome involving compromise of multiple organs caused by a congenital inability to maintain telomeres in multiple cell types; the disease is often manifest in bone marrow failure and elevated risk of developing certain types of cancer.

dyskerin   Protein that is a component of the shelterin complex whose absence causes, among other defects, a loss of proper telomere maintenance and the syndrome of dyskeratosis congenita. *See* **dyskeratosis congenita, shelterin.**

dysplasia (adj., **-plastic**)   A premalignant tissue composed of abnormally appearing cells forming a tissue architecture that deviates from normal.

ectoderm   Outermost layer of cells in an early chordate embryo that gives rise to the skin and nervous system.

ectodomain   Portion of a cell surface protein that protrudes from the plasma membrane into the extracellular space.

ectopic   (1) Referring to the expression of a gene or protein in a location or physiological situation where it normally would not be expressed. (2) Referring to the presence of cells or tissues in an anatomical site where they would not naturally be found in the body. *See also* **orthotopic.**

effector   An agent (such as a protein) that carries out the actual work of a biological process rather than simply regulating it.

effector function   The functions of an immune cell that enable it to interact with various antigen-displaying entities in the body and, in response, execute a specific physiological operation.

effector loop   Region of a Ras protein that participates in direct physical interaction with one of its effector proteins; such interaction usually leads to the physical relocalization and/or functional activation of an effector protein.

effector T cells   Referring to a group of relatively short-lived T cells types that actively respond to physiological stimuli by executing one or another immunological function; these include CD4+, CD8+, and T_{reg} cells.

efferent   Referring to outgoing signals. *See also* **afferent.**

efficacy   Measure of the ability of a therapeutic agent to elicit a desired clinical response.

electrophilic   Referring to a molecule that seeks out and reacts with electron-rich substrates or molecular partners.

electrophoresis   Procedure used to resolve molecules of differing masses, usually macromolecules, by placing them in an electric field that causes them to migrate at different rates through a matrix, such as an agarose or acrylamide gel; larger molecules are retarded by the pores of the gel, while smaller molecules migrate rapidly through the gel.

electroporation   Procedure used to introduce various types of molecules into cells via the application of an electrical pulse that briefly opens pores in the plasma membranes of cells.

embolization   Process of forming an embolus. *See* **embolus.**

embolus (pl., **emboli**)   A blood clot or other solid body, such as a clump of cells, that can travel from one site in the body via the circulation to a second site in which it may become lodged within a blood vessel. *See also* **microembolus.**

embryonic stem cell   A cell that derives from the inner cell mass of an early embryo and exhibits pluripotent powers, being able to spawn derivatives that can differentiate into all the cell types of an embryo except the placenta. *See also* **pluripotent.**

emergent   Arising out of.

-emia   Suffix denoting an excess of a substance or a cell type in the blood.

encapsidation   Process of packaging a viral genome in a capsid.

end-replication problem   Inability of the main DNA replication machinery to completely copy the end of one of two DNA strands being replicated, specifically the strand bearing a 3′ nucleotide at its end.

endocrine   (1) Referring to a gland that secretes substances into the general circulation and thus to the rest of the body. (2) Referring to the signaling path of a hormone or factor that is made by cells in a tissue, passes through the circulation, and affects the behavior of cells in another tissue at a distant anatomical site. *Compare* **autocrine**, **exocrine, intracrine, juxtacrine, paracrine.**

endocytosis   Uptake by a cell of extracellular material together with associated extracellular fluid that is achieved through the invagination of a patch of plasma membrane, thereby yielding membrane-bound vesicles termed endosomes within the cytoplasm. Pinocytosis and phagocytosis are forms of endocytosis. *See also* **exocytosis, endosome, phagocytosis, pinocytosis.**

endoderm   Innermost layer of cells in an early embryo, which in chordates serves as precursor of the gastrointestinal tract and associated tissues, including the lungs, liver, and pancreas.

endogamy   The practice of marrying within one's own ethnic group, tribe, or clan.

endogenous   Originating from within a cell, tissue, or organism.

endogenous provirus   The provirus of a retrovirus that is integrated into the germ line DNA of an organism, thereby assuming a configuration much like all cellular genes in the organism's genome.

endogenous retroviral genome   *See* **endogenous provirus.**

endometrium   Epithelial lining of the uterus.

endonuclease   An enzyme that cleaves a single- or double-strand nucleic acid at an internal site rather than degrading it from an end. *See also* **exonuclease.**

endoplasmic reticulum   Elaborate network of membranous structures in the cytoplasm in which glycoproteins are assembled by specific proteolytic cleavage and covalent attachment of carbohydrate side chains prior to being transported via membranous vesicles to the cell surface.

endoreduplication   Process whereby a genome or chromosomal segment is replicated without subsequent segregation of the replicated copies during mitosis.

endosome   Membranous vesicle lying beneath the plasma membrane that is formed after a patch of the plasma membrane invaginates into the cytoplasm via the process of endocytosis and pinches off from the plasma membrane into the cytoplasm, carrying with it extracellular fluid and associated molecules. *See also* **endocytosis.**

endothelial cells   (1) Mesenchymal cells that form the walls of capillaries or lymph ducts by assuming shapes that allow the assembly of tubelike structures. (2) Mesenchymal cells lining the luminal walls of larger blood vessels or lymph ducts.

enhancer   A relatively short sequence of nucleotides near or within a gene to which a transcription factor may bind in a sequence-specific manner and in turn influence the transcription of this gene. *See also* **super-enhancer, transcription factor**.

enterocyte   Epithelial cell lining the luminal wall of the gastrointestinal tract or associated structures such as villi.

eosinophil, eosinophile   A motile phagocytic granulocyte that can migrate from blood into tissue spaces, displays cell surface IgE receptors, and plays a role, among other functions, in eliminating parasitic and bacterial infectious organisms.

ependymoma   A tumor deriving from glial cells and resembling the ependymal cells lining the ventricles of the brain.

epiblast   Outer layer of primitive mammalian embryo that gives rise to the three major cell lineages following gastrulation.

epidermis (adj., **-dermal**)   The epithelial layer of the skin, composed largely of keratinocytes at various stages of differentiation.

epigenetic   (1) Referring to a process that affects phenotype without the involvement of heritable changes in the nucleotide sequence of the genome. (2) Referring to a mechanism enabling the transmission of a heritable trait from a cell or organism to its progeny that does not depend directly on changes of specific nucleotide sequences in its genome. (3) Referring to a mechanism that creates a cell-heritable trait, as in (2), and maintains expression of this trait in the absence of ongoing exposure to the stimulus that initially induced the expression of this trait. *See also* **epigenetics, genetic**.

epigenetics   (1) The study of cell-heritable changes in gene expression that are not attributable to alteration in the primary DNA sequence. (2) More broadly, the study of changes in phenotype that reflect nongenetic alterations in the cell or organism; *also termed* **epigenomics**. (3) The study of the covalent modifications of histones forming the chromatin and resulting changes in transcriptional regulation.

epigenome   The entire ensemble of modifications of a genome that do not affect its nucleotide sequences, such as methylation of CpG dinucleotides in the DNA and modifications of chromatin-associated proteins, notably histones. The epigenome is presumed to dictate the transcriptome of a cell, that is, the pattern of expression of a cell's repertoire of genes. *See also* **transcriptome**.

episome (adj., **episomal**)   A genetic element, implicitly formed of dsDNA, that exists in a cell, often over multiple cell generations, but is not physically linked via covalent bonds to the cell's chromosomal DNA.

epistasis (adj., **epistatic**)   The masking of the phenotypic effect of the allele(s) of one gene by the allele(s) associated with another gene.

epithelial–mesenchymal transition   Cell-biological program enabling acquisition by epithelial cells of some of the phenotypes of mesenchymal cells such as fibroblasts. *Compare* **mesenchymal–epithelial transition**.

epithelium (pl., **epithelia**)   A layer of cells that forms the lining of a cavity or duct and the specialized epithelium that forms the skin.

epitope   (1) A specific chemical structure—often a short oligopeptide segment of a larger protein antigen—that is recognized and bound by an antibody molecule or other antigen-specific immune molecules. (2) Less commonly, the portion of a molecule that stimulates a specific immune response.

ER stress   Physiological stress in the lumen of the endoplasmic reticulum often triggered by misfolded or unfolded protein domains. *See also* **unfolded protein response**.

eraser   Enzyme that removes specific covalent modifications from histones within nucleosomes, thereby influencing processes such as transcription, DNA repair, further histone modifications, or DNA methylation. *See also* **writer**, **reader**.

erythroblastosis   Malignancy of the precursors of red blood cells, usually referring to a condition of birds.

erythrocyte   Red blood cell.

erythroleukemia   A leukemia of the nonpigmented precursors of red blood cells.

erythropoiesis   Process by which red blood cells are formed.

erythropoietin   Growth factor that stimulates the production of red blood cells, often in response to inadequate oxygen transport by the blood.

essential genes   The collection of genes that is required for normal organismic development and viability or for normal cellular viability or behavior. *See also* **essentialome**.

essentialome   The collection of genes present in the genome of a cell or organism that is critical to its viability. *See also* **essential genes**.

estradiol   17β-Estradiol, *commonly called* **estrogen**. *See also* **estrogen**.

estrogen   Steroid hormone that controls development of a variety of tissues including those involved in female development and reproductive function. *See also* **estradiol**.

etiology (adj., **-ologic**)   (1) Mechanism or agent that is responsible for causing a specific pathological state. (2) The study of causative mechanisms of pathology.

euchromatin   Chromatin that contains transcriptionally active genes and is therefore relatively expanded (rather than being condensed) and stains lightly. *See also* **heterochromatin**.

eukaryotic   Referring to the large, complex, nucleated cells of metazoa, metaphyta, and many protozoa. *See also* **prokaryotic.**

euploid   (1) Referring to a collection of chromosomes that corresponds precisely in number and structure to the array present in normal, wild-type cells. (2) Describing a karyotype having such a complement of chromosomes. *See also* **karyotype.**

event-free survival Period after initial treatment during which no additional clinical indications or episodes of disease are registered.

ex vivo Occurring outside of a living body or organism. *Compare* **in vitro, in vivo**.

executioner caspases A family of proteases (including caspases 3, 6, and 7) that become activated by initiator caspases through proteolytic cleavage and subsequently cleave a number of cellular target proteins, leading to cell death. *See also* **caspase**.

exhaustion (adj. **exhausted**) Physiological state of a T cell that has lost many of its normally exercised effector functions; provoking signals include chronic antigen stimulation and prolonged contact with antigen-presenting cells. Less commonly, NK cell exhaustion and macrophage exhaustion.

exocrine Referring to a gland that secretes fluids via a duct, often into the gastrointestinal tract or to the surface of the skin, rather than into the bloodstream. *Compare* **autocrine**, **endocrine, intracrine, juxtacrine, paracrine**.

exocyclic Referring to a chemical group that protrudes from the ring moiety of a molecule such as a DNA base.

exocyst Multi-subunit cytoplasmic protein complex that enables the tethering to the inner surface of the plasma membrane of vesicles destined for exocytosis, thereby regulating the cell surface display and secretion of a broad range of signaling molecules. *See* **exocytosis**.

exocytosis (1) Process by which cells secrete soluble products by storing them in the lumina of cytoplasmic exocytic vesicles that are caused to fuse with the plasma membrane, allowing the products carried within these vesicles to be released into the extracellular space. (2) Process by which cells achieve the display of proteins on the cell surface by embedding them in the membranes of exocytic vesicles prior to fusion of the latter with the plasma membrane. *Compare* **endocytosis**. *See also* **exocyst**.

exogenous Originating from outside a cell, tissue, or organism.

exome The collection of all exon segments present in a cellular transcriptome. *See also* **exon, transcriptome**.

exon Portion of a primary RNA transcript that is retained in the RNA product of splicing. *See also* **intron**.

exonuclease An enzyme that degrades a polynucleotide, often by proceeding processively (nucleotide-by-nucleotide) from one end or the other. *See also* **endonuclease**.

exosomes Small (30- to 100-nm) membrane vesicles that are released by many cell types into the extracellular space, following which they may adsorb to and fuse with other cells, thereby conveying molecular cargo into the latter.

experimental metastasis Procedure for introducing cancer cells into the circulation via injection into the tail vein of a mouse; this procedure largely yields cancer cells trapped in the microvessels of the lungs.

expression (1) Transcription of an active gene or synthesis of a protein from its encoded mRNA. (2) Forced release of interstitial fluid from a tissue caused by contraction of cells and/or extracellular matrix in that tissue. *See also* **repression**.

expression array A collection of probes that enables the concomitant measurement of the expression levels of a large group of genes, often encompassing more than 10,000 genes, that are transcribed in a cell type or tissue.

expression program The coordinated expression of a series of genes.

expression signature A constellation of up-regulated and down-regulated genes that can be correlated with a defined biological phenotype.

expressivity The quantitative extent to which a measurable or observable phenotype is expressed by an individual or cell carrying a specific allele or gene. *See also* **penetrance**.

external beam therapy Form of radiotherapy, in which the radiation source is usually a linear accelerator used as the source of radiation that is usually directed in a highly focused fashion to a region of tissue that is being treated.

extracellular matrix Meshwork of secreted proteins, largely glycoproteins and proteoglycans, that surrounds most cells within tissues and creates a structural scaffold in the intercellular space.

extravasation Process of leaving a blood or lymphatic vessel and invading the parenchyma of the adjacent tissue. *See also* **intravasation**.

F-actin The polymerized form of actin that forms filament. *See also* **actin**.

facultative heterochromatin Referring to chromatin that can switch from a heterochromatic to a euchromatic state in response to physiologic or developmental signals. *See also* **constitutive heterochromatin, euchromatin**.

familial Referring to an organismic trait or a syndrome that is heritable and therefore found in clusters in certain families.

familial adenomatous polyposis An inborn condition resulting in the formation of many hundreds of colorectal polyps, some of which are poised to progress to frank colorectal carcinomas. *See also* **hereditary non-polyposis colorectal cancer**.

Fc receptor A cell surface protein displayed by immunocytes, such as NK cells, dendritic cells and macrophages, that recognizes and binds the constant regions of antibody molecules, notably IgG molecules, the latter being bound via their variable regions to antigens on the surfaces of potential target cells such as microbes or cancer cells; such binding enables killing of the antibody-bound cells by the NK cells or macrophages.

Fc region Domain of an antibody molecule, notably that of IgG molecules, that is formed by the H-chain constant region and is recognized, when bound to a cell surface, by Fc receptors. *See also* **Fc receptor**.

feeder layer A population of cells, usually grown as a monolayer, that continuously secretes factors that benefit the growth and survival of a second cell population that is propagated above this monolayer.

feedback A regulatory interaction in which the stimulation of one component of a regulatory circuit enables this component to either stimulate or shut down further signaling by the source of the initial stimulation. *See also* **positive feedback, negative feedback, feedforward**.

feedforward   Alternative term for **positive feedback**.

fenestration   An opening in a surface, such as one found in the wall of a blood vessel.

ferroptosis   A form of programmed cell death that depends on iron and the formation of lipid peroxides and is distinct from other forms of cell death such as apoptosis. *See also* **apoptosis, necroptosis, pyroptosis**.

fibrin   Protein that is formed by the cleavage of plasma fibrinogen and assembles to form the fibers binding the platelets together in clots.

fibrinogen   *See* **fibrin**.

fibrinolysis   Dissolution of fibrin bundles achieved by certain proteases.

fibroblast   Mesenchymal cell type that is common in connective tissues and in the stromal compartment of epithelial tissues and is characterized by its secretion of collagen.

fibrocyte   A relatively undifferentiated circulating cell derived from monocytes in the bone marrow that expresses collagen, adheres tightly to substrates when cultured *in vitro*, exhibits certain features usually associated with macrophages, and serves as the precursor of fibroblasts and/or myofibroblasts in tissues to which it has been recruited.

fibroma   A benign tumor of mesenchymal tissue containing fibroblast-like cells.

fibrosis   Development within a tissue, often following chronic inflammation, of dense fibrous stroma that replaces normally present epithelium, resulting in loss of function of that tissue.

field cancerization   Phenomenon in which an outwardly normal region of an organ or tissue produces multiple, ostensibly independently arising premalignant growths or frank neoplasms that actually share a common clonal origin and share a common set of abnormalities, including somatic mutations. Field cancerization is presumed to reflect the proliferative and survival advantage of somatic mutations (or heritable changes in the epigenome) that are not manifested in the outward behavior of the affected cells, but were already widely dispersed in this region prior to the appearance of these clearly preneoplastic or frankly neoplastic outgrowths. *See also* **clonal hematopoiesis**.

filopodium   Filamentous, spikelike protrusion from the surface of a cell, usually extending from the leading edge of a lamellipodium, that may be used by the cell to explore territory that lies ahead in its path and to forge focal adhesions with the extracellular matrix lying in this path. *See also* **lamellipodium**.

filtrate   Liquid that has passed through a filter.

first-degree relative   A parent, sibling, or offspring of an individual (all of whom share 50% of their genome with the individual).

first-line therapy   The mode of therapy that is initially employed to treat a patient, usually because it is considered most likely to elicit a desired clinical response; failure of this initial therapy often leads to the use of second-line treatments. *See also* **second-line therapy**.

fitness   (1) Term to reflect the likelihood that, during the course of evolution, an organism will leave more descendants within a species as a consequence of its increased adaptation to the environment in which it exists or elevated ability to compete successfully with other similar individual organisms within the species. (2) Similar dynamics applying to the ability of an individual cancer cell or group of cancer cells to outcompete other such cells within a tumor.

fixed   To become stably established, such as established within the genome of a cell or species.

floxed   Referring to a gene or genetic segment that is flanked on both sides by LoxP sites. *See* **loxP**.

fluorescence *in situ* hybridization   Procedure in which a sequence-specific DNA probe linked to a fluorescent chromophore is annealed to the DNA or RNA of cells that have been immobilized on a microscope slide, revealing via resulting fluorescing spots the presence and often the number of copies of homologous DNA/RNA sequences carried by such cells that have annealed to the DNA probe.

fluorophore   A molecule that fluoresces in response to exposure to light of a certain wavelength.

focal   Referring to a change that affects only a small region or area, such as a relatively short segment of a chromosome.

focal adhesion   A multiprotein complex containing many dozens of protein species that is assembled after a cell surface integrin binds macromolecular components of the extracellular matrix (ECM); focal adhesions mediate physical anchoring of a cell to the adjacent ECM, as well as transmitting information about this anchoring to the cell interior.

focus   (pl., **foci**) A cluster of transformed cells growing amid a surrounding monolayer of normal untransformed cells in culture. Individual foci are usually composed of the lineal/clonal descendants of a founding cell that was transformed by one or another transforming agent.

founder effect   Phenomenon in which multiple individuals within a species, often not known to share any genealogical relationship with one another, are found to carry a common germ-line allele that testifies to their actual descent from a common ancestor that bore the allele.

fragile site   A site along a chromosome that is intrinsically unstable and thus especially susceptible to breakage, the existence of which is often revealed when DNA replication forks proceeding through the site are stalled.

frameshift   A mutation that disrupts the reading frame of the coding sequence of a gene, doing so by adding or subtracting nucleotides that are not multiples of 3. *See also* **reading frame**.

Frizzled   Referring to a class of 7-membrane-spanning cell-surface proteins that function as receptors for Wnt ligands.

frontline therapy   *See* **first-line therapy**.

functional genomics   (1) Technology in which cell phenotypes are gauged by measuring the expression level of multiple genes, usually numbering in the thousands, in the cells. (2) Such analysis performed comparatively by examining various cell types or tissues that exist in different physiologic, pathological, or developmental conditions. (3) Study of how specific transcriptional programs contribute to the expression of specific functions or phenotypes.

fusogenic Capable of causing the fusion of membranes, such as the plasma membranes of two closely apposed cells.

G-protein A signaling protein that actively emits signals when it has bound GTP and shuts itself off by hydrolyzing the GTP, yielding bound GDP and transition of the protein to its inactive, non-signaling state. *See also* **guanine nucleotide exchange factor**.

gain-of-function Referring to a mutation that imparts novel activity to a gene and encoded protein. *See also* **loss-of-function**.

gametogenesis Processes that yield gametes, i.e., sperm or eggs.

ganglion (pl., **ganglia**) Cluster of neuron cell bodies formed in various parts of the peripheral nervous system and, to a smaller extent, in the central nervous system.

gasdermin A membrane pore-forming protein that is activated as part of the pyroptosis program. *See also* **pyroptosis**.

gastric Referring to the stomach.

gastrulation An early stage of embryogenesis in which cells of the outer ectodermal layer invaginate into the interior of the embryo, where they serve subsequently as precursors of the endodermal and mesodermal cell layers.

gatekeeper (1) An oligopeptide domain that controls access by small molecules to the catalytic cleft of an enzyme such as a kinase. (2) A gene that operates to regulate cell proliferation or to regulate cell number by controlling cell differentiation or cell death. Loss of a gatekeeper gene removes an impediment to cell proliferation and thus to the appearance of populations of neoplastic cells. *Compare* **caretaker**, **tumor suppressor gene**.

gatekeeper mutation A mutation that restricts the accessibility of the ATP-binding pocket in the catalytic cleft of a protein kinase and thereby reduces the efficacy of a kinase-inhibiting drug designed to be bound in that cleft.

gene amplification *See* **amplification**.

gene conversion Process whereby the genetic information on one chromosome is transferred to the homologous chromosome without genetic recombination or crossing over occurring between the two chromosomes.

gene family Group of genes all of which are descended evolutionarily from a common ancestral gene. The members of a gene family generally encode distinct, structurally related proteins.

gene pool Collection of genes and associated alleles that are carried by the members of a species or by a group of individual organisms.

gene set enrichment analysis A computational algorithm designed to determine whether a set of genes is associated with a specific biological function or phenotype.

gene therapy Therapeutic strategy that involves the introduction of a cloned gene into cells of a diseased tissue, undertaken with the anticipation that the introduced gene will reverse a disease condition.

general transcription factor A protein that associates with sequence-specific transcription factors throughout the genome to enable transcription to occur; general transcription factors usually do not possess sequence-specific DNA-binding domains and are presumed to be responsible for regulating the expression of large numbers of genes, possibly in certain cases all of the actively transcribed genes in a cell. *See also* **transcription factor**.

genetic (1) Involving the action of genes and the information that they carry. (2) Depending directly on the DNA and the nucleotide sequences that it contains. *See also* **epigenetic**.

genetic background The entire array of alleles carried in a cell or organismic genome with the exception of a gene or a small number of genes that are the subject of study.

genetic polymorphism A genetic element that is present in the gene pool of a species in multiple variant forms. Polymorphisms are often implied to have small or no effects on phenotype or evolutionary fitness. *See also* **polymorphism**.

genomic clone The product of a procedure involving recombinant DNA techniques that produces many copies of a specific segment of an organism's genome (e.g., its chromosomal DNA) through experimental isolation and amplification.

genotoxic Referring to an agent or mechanism that is capable of damaging a genome, i.e., is mutagenic.

genotype Genetic constitution of a cell or an organism.

germ cell An egg (ovum) or sperm (spermatocyte) or its immediate precursor within the ovary or testis, respectively.

germ line (1) The collection of genes that is transmitted from one organismic generation to the next. (2) The cells within a multicellular organism that are responsible for carrying and transmitting genes from one organismic generation to its offspring.

germ-line chimera An organism in which the germ cells derive from more than one genetic background.

germinal center Site in spleen or lymph node in which B cells differentiate, assemble, and diversify their antibody genes, and proliferate.

Giemsa A blue stain that is specific for DNA, binding preferentially to segments that are high in adenine and thymine.

glioblastoma A malignant high-grade glioma that often originates from a low-grade astrocytoma. *Also termed* glioblastoma multiforme. *See also* **glioma**.

glioma A group of central nervous system tumors that arise from glial cells and include ependymomas, astrocytomas, and oligodendrogliomas. *See also* **glioblastoma**.

global genomic repair Form of nucleotide excision repair that operates on all parts of the genome.

glutathione-S-transferase A family of enzymes that catalyze the conjugation of the reduced form of glutathione (GSH) to xenobiotic compounds for the purpose of detoxification.

glycolysis The series of enzymatic reactions in the cell cytoplasm that leads to the progressive degradation of a glucose molecule into three- and two-carbon fragments with the generation of ATP and NADH; glycolysis can proceed in the absence of available oxygen.

glycoprotein A protein that has been modified post-translationally through the covalent linkage of polysaccharide trees to the amino-acid side chains of the protein, usually those of asparagine, arginine, or serine. *See also* **glycosylation**.

glycosaminoglycan A charged polysaccharide of the extracellular matrix, such as chondroitin sulfate, hyaluronic acid, or heparin, that is attached covalently to a protein core and is composed of repeating monosaccharides, some of which are amino sugars. *See also* **proteoglycan**.

glycosylase An enzyme that cleaves a glycosidic bond, such as that linking a purine or pyrimidine base to a ribose or deoxyribose sugar.

glycosylation Covalent attachment of a carbohydrate side chain, usually a covalently linked branched network of monosaccharides, to a second molecule, e.g., to the side chain of an asparagine residue of a protein; such attachment occurs in concert with or following initial synthesis of the polypeptide backbone of a protein.

grade Degree or extent to which a tumor has advanced toward a highly aggressive state, as assessed by a pathologist usually on the basis of its histopathological appearance; high-grade tumors are more progressed and generally carry worse prognoses.

graft-versus-host disease A condition, often life-threatening, in which cells of an allogeneic bone marrow graft generate significant immune-mediated destruction of various normal tissues throughout the body of the graft recipient.

graft-versus-tumor An immunological process in which engrafted allogeneic bone marrow cells attack neoplastic cells in the graft recipient, usually the cellular constituents of hematopoietic malignancies; such attack may augment other anti-tumor therapies being applied to the graft recipient and may occur with acceptably low levels of graft-versus-host disease. *See also* **graft-versus-host disease**.

granulocyte Leukocyte that contains cytoplasmic granules, such as a basophil, an eosinophil, or a neutrophil.

granzyme Serine protease that is introduced by cytotoxic lymphocytes and natural killer cells into the cytoplasm of their cellular targets, where it triggers apoptosis by cleaving various cellular proteins, among them caspases.

gray (Gy) Unit of radiation exposure equal to the absorption of 1 joule of energy per kilogram of exposed tissue.

growth (1) Process involving accumulation of cellular constituents and associated increases in cell size such as cell volume. (2) Commonly but inappropriately used term for cell proliferation. *Compare with* **proliferation**.

growth arrest Halting of the progression of a cell through its growth-and-division cycle. *See also* **cytostatic**.

growth factor Secreted protein that is able to stimulate the growth and/or proliferation of a cell by binding to a specific cell-surface receptor displayed by that cell.

GTPase Enzymatic activity that hydrolyzes GTP to GDP.

GTPase-activating protein A protein that interacts with a G-protein and induces the latter to activate its intrinsic GTPase, thereby causing it to hydrolyze its bound GTP and enter into its GDP-bound state. *See also* **G-protein**, **guanine nucleotide exchange factor**.

guanine nucleotide exchange factor Protein that interacts with a G-protein, inducing the latter to release whatever guanine nucleotide it is binding, thereby creating space for another to be bound. Since GTP is present in the cytosol in great molar excess over GDP, this induced change usually results in the replacement of a GDP with a GTP and the associated functional activation of the G protein. *See also* **G-protein**, **GTPase-activating protein**.

hairpin Configuration of a single-stranded RNA or DNA molecule that is able to generate a double-stranded domain through the complementarity of two regions of such a molecule, in effect allowing the two regions to anneal with one another.

half-life (1) Time during which half of a population of metabolically unstable molecules decays or is eliminated or the time required for half of a physiological signal to decrease. (2) Time in which 50% of the atoms of a quantity of a radioactive isotope decay. (3) Time during which half of the entities forming a large population disappear or are removed.

Hallmarks of Cancer A set of cell-biological traits that are shared in common by the neoplastic cells forming a wide range of distinct types of tumors, are required for expression of the neoplastic cell phenotype, and distinguish such cells from the traits of corresponding normal cells.

hamartoma Benign overgrowth of tissue that normally involves mesenchymal cells; for example, gastrointestinal hamartomas are characterized by a benign expansion of stromal cells, often with concomitant hyperplasia of adjacent epithelial cells.

haploid Describing a genome in which all chromosomes are present in a single copy.

haploinsufficiency Genetic condition in which the presence of only a single functional copy of a gene yields a mutant or partially mutant phenotype.

haptotaxis Process in which the direction of migration of a cell is influenced by insoluble molecules in the extracellular space, specifically structural components of the extracellular matrix as well as molecules tethered to this matrix. *Compare* **chemotaxis**.

hard-wired (1) Describing a component of a regulatory circuit that is fixed and not readily changed. (2) Describing a response or behavior that is innate and not readily revised or altered.

hazard ratio The rate at which a given condition or state arises in an experimental population divided by the rate at which this occurs in a matched control population.

head-and-neck cancers A group of carcinomas of the larynx, throat, mouth, nose, and salivary glands.

heat map Graphic representation in the form of a matrix of the results of gene or protein expression analyses, in which a series of gene-specific probes is arrayed in one dimension, a series of biological samples is arrayed in a second dimension, and the intensity of expression is represented by two colors and intermediate shades.

heavy chain Larger of the two polypeptide chains that assemble to form immunoglobulins and related proteins. *Compare* **light chain**.

helicase   An enzyme that functions to unwind helices, usually the double helices formed by DNA.

helper T cell   A subtype of T cell that is identified by its expression of the CD4 co-receptor and functions to stimulate and amplify the functions of B cells and effector T cells, most importantly the CD8+ cytotoxic T cells.

hemangioblastoma   Benign tumor of the precursors of the endothelial cells forming blood vessels within the central nervous system.

hemangioma   Benign tumor of endothelial cell origin often seen in infants.

hematogenous   Depending upon or facilitated by circulating blood.

hematogenous dissemination   Metastatic dissemination occurring via the circulating blood.

hematopoiesis   (adj., -**poietic**) Process that results in the formation of all of the cells in the blood, including its red and white cells, the latter including various cells of the immune system.

hemidesmosome   A cluster of integrin molecules displayed on the basal surface of epithelial cells and used to anchor these cells to the underlying basement membrane.

hemi-methylated   Referring to a DNA molecule in which only one of the two complementary strands of a particular DNA segment is methylated, usually at the deoxycytidine moiety of a CpG dinucleotide sequence.

hemizygosity   Presence of only one copy of an autosomal gene per cell.

heparin   An extracellular matrix glycosaminoglycan.

hepatectomy   Procedure whereby part or all of the liver is surgically excised.

hepatoblast   Embryonic liver cell.

hepatocellular carcinoma   *See* **hepatoma**.

hepatocyte   Epithelial cell type that forms the bulk of the liver and is responsible for virtually all of its metabolic activities.

hepatoma   Tumor of the liver; also known as **hepatocellular carcinoma**.

Herceptin   Chimeric anti-HER2/Neu monoclonal antibody bearing murine antigen-combining (variable) domains and a largely human constant domain. *Also called* **trastuzumab**.

hereditary non-polyposis colon cancer   Inborn condition predisposing to colorectal cancers and distinct from familial adenomatous polyposis, being generally caused by inherited defects in the DNA mismatch repair machinery. *Also termed* **Lynch syndrome**; *see also* **familial adenomatous polyposis**.

heterochromatin   Chromatin that contains transcriptionally inactive genes or no genes at all and is condensed and stains darkly. *See also* **euchromatin**.

heterochromatinization   The process whereby euchromatin is converted to heterochromatin, notably that occurring during the process of X-chromosome inactivation.

heterocyclic amine   Any of various carcinogenic amines formed when creatine or creatinine reacts with free amino acids and sugars in meat cooked at high temperatures.

heterodimer   Molecular complex composed of two distinct subunits. *See also* **homodimer**.

heterogenous nuclear RNA (hnRNA)   The collection of primary transcripts synthesized by pol II and derived processed intermediates arising from the actively transcribed genes of a cell.

heterokaryon   Cell carrying two (or more) genetically distinct nuclei, often arising from the experimental fusion of two cells.

heterotrimer   A molecule that is composed of three distinct subunits such as three distinct protein chains.

heterotypic   Referring to interactions between two or more distinct cell types. *See also* **homotypic**.

heterozygote   Individual cell or organism that carries two distinct alleles of a gene of interest. *See also* **heterozygous**.

heterozygous   Referring to the configuration of a genetic locus in which the two copies of the associated gene carry different versions (alleles) of the gene that represents this locus. *See also* **homozygous**.

hierarchical clustering   Statistical method used to place more similar samples closer together on a tree (*also called* dendrogram) than less similar samples, which are placed on more distant branches of the tree. *See also* **dendrogram**.

high-grade   Referring to a tumor that has progressed through many steps of multi-step tumorigenesis and become highly malignant. *Compare with* **low-grade**.

high-throughput screening   Technology involving robotic screening of large collections of pharmacological compounds to identify individual compounds that exhibit properties of a sought-after agent, such as potent inhibition of an enzyme.

histo-   Prefix referring to tissue.

histochemistry   Use of specific chemicals and chemical procedures to specifically react with and stain subcellular and cellular structures as well as tissues to reveal, when examined under the microscope, the presence of specific types of biological molecules present in these entities.

histocompatible   Ability of tissues or cells to be tolerated by the immune system of the host organism into which they are engrafted. *See also* **histoincompatible**.

histocompatibility antigen   Cell surface protein that determines whether or not an engrafted cell or tissue will be tolerated by the immune system of a host organism. *See also* **major histocompatibility antigen**.

histoincompatible   Failure of tissues or cells to be tolerated by the immune system of the host organism into which they have been engrafted. *See also* **histocompatible**.

histology   Study of tissue structure at the microscopic level. *See also* **histopathology**.

histomorphology   The shapes and forms of cells and assemblies of cells within a tissue as revealed by microscopic examination of sections of that tissue.

histopathology   Study of tissue structure at the microscopic level, often with reference to abnormal tissue. *See also* **pathology**.

Holliday junction   Junction formed between four DNA double helices that can serve as an intermediate in certain forms of homologous recombination.

holoenzyme   A multi-subunit complex containing an enzyme and associated protein subunits that collaborate to mediate and regulate enzymatic activity.

homeostasis (adj. **homeostatic**)   The product of physiological control mechanisms that operate to maintain an optimal, often unchanging level of a certain signaling process, metabolic product, physiological function, or cell type.

homodimer   Molecular complex composed of two identical subunits. *See also* **heterodimer**.

homogeneously staining region   A region of a chromosome consisting of amplified copies of a chromosomal segment that have become fused end to end; because each copy contains a characteristic sequence of DNA, the region as a whole appears homogenous following staining.

homolog (*also* **homologue**)   A gene that is related to another because of evolutionary descent from a common ancestral gene. *See also* **ortholog, paralog**.

homologous   (1) Referring to the relationship between a pair of chromosomes that carry the same set of genes within a diploid cell or organism. (2) Referring to genes or characteristics that are similar in related organisms because of shared descent from a common precursor. (3) Referring to two nucleic acids having similar or closely related nucleotide sequences. *See also* **analogous**.

homology-directed repair   Form of DNA repair in which the reconstruction of damaged DNA, often involving a double-strand break, is instructed by the corresponding homologous nucleotide sequences present in the paired, intact sister chromatid.

homophilic   Describing a molecule that binds preferentially to one or more additional molecules of the same type.

homopolymer   A polymer that is assembled from monomers of a single type, such as an oligonucleotide segment formed entirely from one or another of the four possible deoxyribonucleotides.

homotetramer   An assembly of four identical subunits, usually referring to proteins.

homotypic   (1) Referring to interactions between two cells of the same type; *see also* **heterotypic**. (2) Referring to interactions between two or more molecules of the same type.

homozygous   Referring to the configuration of a genetic locus in which the two copies of the associated gene carry identical alleles of the gene. *See also* **heterozygous**.

horizontal transmission   Transmission of an agent, such as an infectious agent, throughout a population in ways that do not depend on parent-offspring interactions. *See also* **vertical transmission**.

hormone response element   An oligonucleotide sequence in genomic DNA to which hormone-activated nuclear receptors bind.

host   (1) The animal or human that bears a tumor as distinguished from the tumor itself. (2) The animal or human into which cells or molecules are introduced, often by experimentation or infection.

housekeeping gene   A gene that is used universally in all cells throughout the body independent of their differentiated state and is assumed to be essential for their continuing viability.

humanize   To impart human properties to something, e.g., an antibody molecule.

humanized antibody   An antibody molecule of a nonhuman species whose constant (C) region and variable (V) region amino acid sequences outside of the antigen-combining site have been replaced by the homologous sequences of human origin, leaving only the antigen-combining sequences unmodified.

humoral   Referring to a soluble substance or a fluid.

humoral immune response   The arm of the immune system whose functions are mediated by the antibodies that it produces. *See also* **cellular immune response**.

hyaluronic acid   A polymer assembled from disaccharide monomers formed from glucuronic acid plus *N*-acetyl-D-glucosamine.

hybrid   (1) An experimentally formed double helix of two complementary nucleic acid strains. (2) A cell arising as a consequence of the experimental fusion of two parental cells that are distinct from one another. (3) An organism that results from the interbreeding of two genetically distinct parents.

hybridization   Molecular process involving the annealing of two single-strand nucleic acid molecules to one another that depends on their having mutually complementary sequences and yields a double-stranded helical structure, such as DNA:DNA, DNA:RNA, or RNA:RNA helices.

hybridoma   A clonal cell line that derives from the fusion of a plasma cell producing an antibody of interest with a myeloma cell and that is employed for the production of a distinct monoclonal antibody species. *See also* **monoclonal antibody**.

hydrocarbon   An organic molecule composed of hydrogen and carbon atoms.

hydrogel   A three-dimensional network of highly hydrophilic polymer molecules that can absorb 10- or even 100-fold its mass of water and can serve as an experimental model of natural matrices, such as the extracellular matrix.

hydrolysis   Chemical breakdown of a compound through chemical reactions of covalent bonds with water molecules.

hydrophilic   Referring to a chemical moiety or environment that prefers direct association with water. *Compare* **hydrophobic**.

hydrophobic   Referring to a chemical moiety or environment that avoids direct interaction with water. *Compare* **hydrophilic**.

hypercalcemia   Presence of elevated concentrations of calcium ions in the blood.

hypereosinophilic   Referring to a syndrome with elevated levels of eosinophils in the circulation. *See* **eosinophil**.

hyperinsulinemia   Level of circulating insulin that is higher than would be expected or is physiologically anticipated.

hypermethylation   Referring to a state in which the number of CpG sites that are methylated is elevated relative to their number in normal or reference cells. *Compare* **hypomethylation**.

hyperoxia   State of oxygen tension that is elevated above physiological levels.

hyperphosphorylated   Referring to the elevated phosphorylation state (e.g., of a protein).

hyperplasia   (adj., -**plastic**) (1) Accumulation of excessive numbers of normal-appearing cells within an otherwise normal-appearing tissue. (2) An increase in the size of a tissue or organ, usually due to increased cell number rather than increased size of constituent cells.

hyperthyroidism   Condition in which the thyroid gland releases inappropriately high levels of thyroid hormone into the circulation, leading to an elevated metabolic rate and multiple secondary effects on the body.

hypertrophy   An increase in the size of a tissue or organ due to increased sizes of component cells rather than increased numbers of cells.

hypomethylation   Referring to a state in which the number of CpG sites that are methylated is reduced relative to their state in normal or reference cells. See **hypermethylation**.

hypomorphic   Referring to an allele that specifies reduced gene function while not eliminating it entirely. See also **null allele**.

hypopharyngeal   Referring to the region between the pharynx and the entrance to the esophagus.

hypophosphorylated   Referring to relatively low level of phosphorylation (of a protein).

hypoxia (adj., **hypoxic**)   State of lower-than-normal oxygen tension. See also **normoxia**.

idiotype   The structure that is created by the amino acid sequences forming the antigen-combining pocket of an antibody molecule; these sequences can themselves function as antigens that provoke the synthesis of new antibody molecules reactive with the initial antibody molecule.

illegitimate recombination   See **nonhomologous recombination**.

immediate early genes   Referring to the group of genes whose expression is induced within half an hour of growth factor stimulation of previously factor-deprived cells, occurring even when protein synthesis is inhibited. See also **delayed early**.

immortal   See **immortalization**.

immortality   Trait of a cell or population of cells that reflects the ability of these cells and their descendants to proliferate indefinitely, usually observed in cell culture.

immortalization   Process whereby a cell population normally having limited replicative potential acquires the ability to multiply indefinitely.

immune checkpoint   Control mechanism ensuring that immune attacks on cells and tissues are constrained or limited, achieved by modulating downstream signaling responses such as those triggered by receptor–ligand interactions between immunocytes and target cells. See also **checkpoint**.

immune tolerance   Physiological state that reflects the ability of the immune system to ignore or overlook the body's normal complement of antigens and the cells expressing such antigens.

immunoblot   See **Western blot**.

immunocompetent   Referring to an organism whose immune system is fully functional.

immunocompromised   Describing an organism lacking a fully functional immune system. Also termed **immunodeficient**.

immunocyte   A cell associated with one of the multiple types of cells that form the functional arms of the immune system.

immunodeficient   See **immunocompromised**.

immunoediting   Process in which the selective pressure imposed by immune surveillance and attack eliminates strongly immunogenic tumor cells while permitting the survival and outgrowth of those that are weakly immunogenic.

immunoevasion   A biological strategy that enables an abnormal cell or an infectious agent to evade detection and/or elimination by the immune system.

immunofluorescence   Use of antibody molecules linked directly or indirectly to fluorescent dyes used in order to stain tissue sections displaying antigens that are specifically recognized by such antibodies.

immunogen (adj., -**genic**)   A chemical structure that is capable of provoking some type of immune response, e.g., an antigen that can provoke the synthesis of antibody molecules capable of recognizing and binding it.

immunoglobulin   An antibody molecule assembled from two heavy and two light chains.

immunoglobulin superfamily   Proteins belonging to a diverse array of cell surface and secreted soluble molecules, including antibody molecules and cell surface B-cell and T-cell receptors. The structural features shared among these proteins indicate that they have evolved from a common precursor and that their respective encoding genes represent paralogs of one another.

immunohistochemistry   Procedure in which expression of an antigen is localized in a histological section through the use of an antibody that has been coupled to an enzyme (e.g., horseradish peroxidase) capable of generating a product that is visible in the light microscope.

immunological memory   The process by which cells of the adaptive immune system store information of an encounter with an antigen and exploit this stored information to rapidly mount a second immune response against this antigen long after the initial encounter with the antigen has taken place.

immunophenotyping   The process of classifying a cell by determining the spectrum of specific antigens it expresses, usually on its surface.

immunoprecipitation   Process of precipitating a molecule or molecular complex using an antibody that specifically recognizes and binds such a molecule or a component of such a molecular complex.

immunoprivileged   Referring to sites in the body that are shielded by various mechanisms from routine surveillance conducted by components of the immune system.

immunoproteasome   A proteasome that is specialized to generate oligopeptides for eventual presentation by MHC/HLA molecules at the cell surface. See also **proteasome**, **constitutive proteasome**.

immunostaining The use of antigen-specific antibodies linked directly or indirectly to fluorophores or dye-activating enzymes to reveal specific cells or subcellular structures present in histological sections.

immunosuppressive Referring to mechanisms and agents that constrain or inhibit the immune response, often involving its adaptive arms

immunosurveillance Process by which the immune system continuously monitors tissues for the presence of aberrant antigens or cells, including cancer cells, and in response, mounts responses to eliminate the antigens or cells expressing such antigens.

immunotoxin Toxin that is targeted to certain tissues or cells because it has been coupled to an antigen-specific antibody, usually a monoclonal antibody.

imprinting Process by which either the paternally or the maternally derived copy of a gene is transcriptionally inactivated during embryogenesis through promoter methylation, ensuring that only a single copy of the gene is expressed in somatic cells.

in situ (1) Occurring in the site of origin. (2) Localized to a specific site. (3) In the case of carcinomas, confined to the epithelial side of the basement membrane.

in situ **hybridization** Procedure in which a nucleic acid probe is annealed to a nucleic acid (DNA or RNA) while the latter remains localized to its original site within a cell or tissue section; the probe is often linked to a fluorescent dye that can be visualized microscopically or to a radioisotope that can be detected by autoradiography. *See* **autoradiography**.

in utero Occurring in the womb during embryonic or fetal development.

in vitro (1) Referring to biochemical processes, notably reactions, that occur outside of living cells, including in cell lysates or under conditions involving purified reactants. (2) Referring to the propagation of living cells in a vessel (e.g., a Petri dish) rather than in living tissues. *Compare* **in vivo, ex vivo**.

in vivo (1) Occurring in a living organism. (2) Occurring in a living, intact cell. *Compare* **in vitro, ex vivo**.

inborn Inherited, congenital.

inbred Referring to a population generated by inbreeding. *See also* **inbreeding**.

inbreeding (1) Breeding of a strain of organisms, such as a strain of mice or rats, with one another in order to achieve genetic identity among all individuals of the strain (with the exception of male/female genetic differences). (2) Breeding between closely related individual organisms within a genetically heterogeneous species.

incidence Frequency with which a condition or a disease occurs or is diagnosed in a population.

incomplete penetrance Situation in which a dominant allele fails to affect phenotype because of the actions of other genes present in an organism's genome. *See also* **penetrance**.

indel Insertion or deletion mutation.

indication A clinically observed trait or property that represents a valid reason for the making of a certain diagnosis or use of a certain therapy.

indolent Referring to a tumor that is apparently nongrowing or growing at a very slow rate.

inflammasome Multi-subunit complex that assembles in the cytosol in response to detection of pathogens or danger-associated cellular changes and proceeds to trigger cell death via caspase activation. *See also* **caspase**.

inflammation A process in which certain cellular components of the immune system are involved in the remodeling of a tissue in response to wounding, irritation, or infection.

inhibitor of apoptosis Cellular protein that blocks apoptosis by sequestering caspases.

initiation (1) Process of changing a cell, usually in a stable, cell-heritable fashion, so that it and its descendants are able to respond subsequently to the stimulatory actions of a tumor-promoting agent. (2) Such a process, with the implication that the change involves a mutation. (3) The first step in multi-step tumorigenesis. *See also* **promotion**.

initiator Agent that triggers the first step in multi-step tumorigenesis. *See also* **initiation**.

initiator caspases A family of proteases (including caspase 2, 8, 9, and 10) that become activated by upstream activating signals in the apoptotic pathway and functions to activate other downstream members of the caspase family (so-called executioner caspases), which proteolytically cleave a number of cellular target proteins. *See also* **caspase**.

innate Referring to an entity or process that is inborn and therefore does not require specific experience or provocation in order to arise.

innate immune response An immune system response toward an antigen or a cell that occurs in the absence of prior exposure of an organism to this particular antigen or cell. *See also* **adaptive immune response**.

innervated Referring to a tissue that is penetrated by components of the nervous system, specifically neurons and their axons.

inoculum Inoculated material.

inoperable A tumor or disease condition that cannot be treated surgically because of its anatomical location or other property. *See also* **operable**.

inositol A polyalcohol configured as a six-membered ring that has a chemical composition similar to that of hexose and is generated by reduction of the hexose aldehyde group to an alcohol.

insertional mutagenesis Alteration of a gene and its function through the integration of a retroviral provirus or transposon into the gene itself or a closely linked chromosomal site.

integrase An enzyme that is specialized to integrate an episomal DNA, such as a retroviral DNA genome, into host-cell chromosomal DNA. *See also* **integration, episome**.

integration Insertion of a fragment of foreign double-stranded DNA (e.g., the DNA genome of an infecting virus) into chromosomal DNA so that the foreign DNA becomes covalently linked to the chromosomal DNA segments flanking it on both sides and can be replicated together with the flanking DNA during subsequent rounds of cell growth and division.

integrin   A heterodimeric cell surface receptor that binds components of the extracellular matrix and transmits information about this binding to the cell interior; the cytoplasmic domain of an integrin may also be coupled with components of the cytoskeleton, thereby linking the extracellular matrix to the cytoskeleton.

interactome   (1) The collection of all protein–protein interactions that operate within a cell.   (2) The subset of these that can be determined experimentally.

intercalation   Insertion of one molecule between two other molecules; intercalation in DNA involves insertion of a planar hydrophobic molecule between two adjacent base pairs.

interferon   A member of one of three classes of secreted signaling cytokines that participate in activation of various type of immune responses.

intergenic   Referring to genomic sequences located between identified genes. *Compare* **intragenic**.

interleukin   A secreted growth and differentiation factor that stimulates various cellular components of the immune system.

intermediary metabolism   The collection of biochemical reactions within a cell that allow the interconversion of various molecular species into one another, involving both anabolic and catabolic reactions. *See anabolic, catabolic*.

internalization   Process by which proteins and other molecules are imported into a body, usually referring to importation of molecules into cells.

International Cancer Genome Consortium   A multi-national effort to characterize the genetic changes in the genomes of multiple types of human cancer including multiple cases of specific cancer types. *See also* **The Cancer Genome Atlas; Sanger Cancer Genome Project.**

interphase   The portion of the cell cycle outside of M phase.

interstitial (n. pl., **interstices**)   (1) Referring to the space between an array of objects. (2) The space within a tissue that lies between cells.

interstitial deletion   Genetic alteration, often observed through analysis of karyotype, that causes deletion of an internal segment of a chromosomal arm.

interstitial matrix   Portion of the extracellular matrix that resides between cells and is not a component of the basement membrane.

intracrine   A mode of signaling in which a receptor and its bound ligand interact entirely within a cell to release signals, often anticipating or following a similar interaction occurring in the extracellular space. *Compare* **autocrine, endocrine, exocrine, juxtacrine, paracrine.**

intraepithelial   (1) Referring to a change or an attribute confined to the epithelial cell layer. (2) Referring to a neoplasia that remains on the epithelial side of a basement membrane.

intragenic   Process, event, or DNA sequence occurring within the genomic boundaries of a gene. *Compare* **intergenic**.

intravasation   Process of invading into a blood or lymphatic vessel from the surrounding tissue. *See also* **extravasation.**

intravital   Referring to a process that occurs in living tissue, such as imaging a process as it occurs in living tissue.

intrinsic resistance   Referring to the resistance to therapy of a tumor that appears to have existed prior to the application of that therapy, *also termed* **primary resistance**. *Compare* **acquired resistance**.

intron   Portion of a primary RNA transcript that is deleted during the process of splicing. *See also* **exon**.

invadopodium (pl., **-ia**)   An organized structure on the surface of a cell in which plasma membrane–bound proteases degrade adjacent extracellular matrix, apparently to enable subsequent invasion by the cell as a whole. *Sometimes termed* **podosome**.

invagination   Process by which planar structure, such as a membrane, is folded back on itself to generate a cavity or pouch.

invasion   (1) Process by which cancer cells or groups thereof move from a primary tumor into adjacent normal tissue. (2) In the case of carcinomas, a movement that involves breaching of the basement membrane.

invasion–metastasis cascade   A succession of steps that enables primary tumor cells to spawn metastases, doing so via local invasion, intravasation into blood or lymphatic vessels, dissemination to distant tissues, extravasation into the parenchyma of those tissues, formation of micrometastases, and ultimately colonization leading to the formation of macroscopic metastases.

invasive   (1) Referring to the increased aggressiveness of a tumor or its associated cells. (2) Referring to a procedure that involves the insertion of medical instruments into the body of a patient. *See also* **invasion**.

involution   Regression or disappearance of a tissue, notably, the regression of the mammary epithelium upon weaning.

ionizing radiation   A type of high-energy radiation that has enough energy to remove an electron from an atom or molecule, causing it to become ionized.

ischemia   State within a tissue caused by inadequate access to circulating blood, resulting in hypoxia, decreased pH, and inadequate supply of nutrients.

isochromosome   A chromosome in which one of the two chromosomal arms has been lost and replaced by a duplicated copy of the surviving arm.

isoelectric point   The pH value at which a molecule, such as a protein, exhibits no net charge.

isoform   Protein that is functionally and structurally similar but not identical to another protein. Multiple distinct isoforms of a protein are usually encoded either by a single gene (via alternative splicing of a pre-mRNA) or by closely related paralogous genes. *See* **paralog**.

isopycnic   A form of centrifugation in which a medium or solution composed of components of different densities is established within a centrifuge tube in a continuous gradient of densities from most dense solution at the bottom to least dense at the top; during centrifugation, analyte molecules, multi-subunit complexes, or organelles of a specific density will, at equilibrium, settle into a zone within the centrifuge tube that is equivalent to their own density, enabling the resolution of analytes of different densities.

junk DNA   Genomic DNA that cannot be associated with any biological function.

juxtacrine   Type of cell–cell signaling in which the signal-emitting cell must be directly apposed to the signal-receiving cell in order for the signal to be properly transmitted. *Compare* **autocrine**, **endocrine**, **exocrine**, **intracrine**, **paracrine**.

juxtamembrane   Located immediately adjacent to a membrane.

Kaplan–Meier plot   A convention for graphing various clinical observations in which the percentage of surviving patients (or another clinical parameter such as disease-free or progression-free survival) is plotted on the ordinate (y-axis), while the time course after initial diagnosis or treatment is plotted (usually in increments of months or years) on the abscissa (x-axis).

kataegis   Localized areas of point mutations arising near sites of chromothripsis and affecting preferentially cytidine and guanine nucleotides. *See* **chromothripsis**.

karyotype   (1) The array of chromosomes carried by a cell, as determined by detailed microscopic examination of these chromosomes usually performed with condensed chromosomes at metaphase. (2) Images of the metaphase chromosomes of a cell arrayed systematically by homologous pairs from the largest to the smallest pair.

karyotypic   Referring to the array of chromosomes of a cell that can be visualized by microscopy.

keratin   (1) Protein component of the intermediate filaments forming the cytoskeleton of epithelial cells; *also termed* **cytokeratin**. (2) Extracellular protein forming hair, nails, and feathers left behind and deposited after keratin-forming cells die.

keratinocyte   Epithelial cell type found in a variety of epithelial tissues, notably the skin. These cells are named after the cytokeratins that they express, which form key components of the cytoskeleton of these cells. *See also* **cytokeratin.**

keratosis   A benign lesion of the keratinocyte lineage in the skin usually caused by exposure to ultraviolet solar radiation.

Ki-67   A nuclear protein that serves as a marker of actively proliferating cells and is expressed in all phases of the active cell cycle.

kinase   Enzyme that removes the γ phosphate from ATP and covalently attaches the phosphate moiety to substrate molecules, often but not exclusively proteins.

kinetochore   Nucleoprotein complex that is associated with the centromeric DNA of a chromosome and is responsible during mitosis or meiosis for forming physical connections between the chromosome and the microtubules of the spindle fibers.

kinome   The complete repertoire of kinases encoded by a genome, such as the human genome.

knock-down   Procedure whereby the expression of a gene is reduced through the introduction into a cell of an inhibitory molecule such as an shRNA or siRNA that interferes with expression of its cognate mRNA.

knock-in   A targeted mutation achieved through the homologous recombination of an experimentally introduced, cloned DNA fragment containing, in part, sequences of the targeted gene; the resulting integration of the introduced fragment into the genome of a cell usually results in the addition of novel sequences that affect the function of the targeted gene. *See also* **knock-out**.

knock-out   (1) A targeted mutation that is achieved by a knock-in genetic strategy that results in insertion of a genetically defective copy of a gene or portion of a gene into the homologous gene sequence residing in the genome of a cell or organism. (2) A targeted mutation that is achieved by introduction of a CRISPR-Cas9 construct into a cell. *See also* **knock-in, CRISPR-Cas9**.

Krebs cycle   *See* **tricarboxylic acid cycle**.

labile   Highly susceptible or vulnerable to being changed, including suffering inactivation or destruction.

lamellipodium (pl., **-ia**)   A broad, sheetlike ruffle extending from the plasma membrane into the extracellular space that is typically found at the leading edge of a motile cell.

lamin   A fibrillar protein component of the nuclear membrane.

laminin   A fibrillar protein component of the basement membrane.

laser capture microdissection   Procedure in which a laser beam is used to delineate a patch of cells away from other cells present in a tissue section mounted on a microscope slide; the cells forming the patch are subsequently selectively removed for further analysis.

latency   The time period that defines the delay before a process reaches completion and becomes manifest or apparent.

latent   An entity or process that is not readily detectable or apparent.

lead-time bias   A bias (e.g., in measurement of patient survival time) created when a disease condition is diagnosed at an earlier stage of its natural multi-step progression schedule because newly developed and more sensitive or effective diagnostic techniques have been deployed; resulting lead-time biases can lead to the impression that the temporal course of disease has been increased, whereas disease progression is only being monitored starting at an earlier step of the naturally occurring process.

leiomyoma   Benign tumor of the mesenchymal cells forming the wall of the uterus.

leukapheresis   Procedure whereby leukocytes are prepared *ex vivo* from drawn blood prior to returning the remaining portions of the blood to the donor or to another recipient.

leukemia   Malignancy of any of a variety of hematopoietic cell types, including the lineages leading to lymphocytes and granulocytes, in which the tumor cells are nonpigmented and dispersed throughout the circulation. *Compare* **lymphoma**.

leukemogenesis (adj., **-genic**)   Process that creates a leukemia.

leukocyte   A nonpigmented white blood cell such as a lymphocyte, monocyte, macrophage, neutrophil, or mast cell.

leukosis   A leukemia-like disease of chickens.

library   (1) A collection of DNA clones derived from a cell's or an organism's genome or from cDNAs prepared from its expressed mRNAs, in which each component clone ideally derives from a distinct gene in this genome. (2) A collection of chemical compounds such as those synthesized because of their possible therapeutic utility.

ligand   Molecule that binds specifically to a receptor and thereby activates its signaling powers. *See also* **receptor**.

ligase   An enzyme that covalently joins two molecules together; in the context of DNA, ligases join the 3′ end of one ssDNA to the 5′ end of another other via a phosphodiester linkage.

light chain   Smaller of the two polypeptide chains that assemble to form immunoglobulins and related proteins. *Compare* **heavy chain**.

limiting dilution analysis   Procedure in which the representation or percentage of a given cell type of interest within a larger, often heterogeneous cell population is gauged by diluting the sample of interest and extrapolating from subsequent biological assays the concentration of the cell type of interest in the initial sample being analyzed.

lineage   A continuous, linear succession of cells extending from an ancestral cell via multiple intervening generations to its most recent descendants.

liquid biopsy   Sampling of the contents of the blood rather than the contents of a solid tissue such as a tumor.

lobe   (1) A cluster of alveoli formed in the mammary gland. (2) One of seven major sectors of the lungs or four major sectors of the liver. *See* **alveoli**.

lobular   Referring to a lobe.

locked nucleic acid   A chemically synthesized oligonucleotide in which the ribose residues contain methylene bridges between their 2′ oxygens and 4′ carbons, which greatly increase the ability of the oligonucleotide as a whole to anneal to complementary sequences.

locus   (1) Chromosome site that can be studied genetically and is presumed to be associated with a specific gene. (2) A genetic element that can be mapped by genetic analysis.

long terminal repeat   A genetic segment that is present in two identical copies at the ends of a retroviral genome or transposon and carries a strong transcriptional promoter and sequences important for chromosomal integration.

longitudinal   Referring to the repeated, successive monitoring of a single patient or a group of patients over an extended period of time.

loss-of-function   Referring to a mutation that deprives a gene and/or its protein product of an existing function. *See also* **gain-of-function**.

loss of heterozygosity   A genetic event in which one of two alleles at a heterozygous locus is lost; the lost allele may simply be discarded or be replaced with a duplicated copy of the surviving allele. *See also* **allelic deletion**.

low-grade   Referring to a tumor that has progressed minimally and is still relatively benign. *Compare with* **high-grade**.

loxP   A bacteriophage DNA sequence that, when present in pairs within a genome, can be acted upon by a Cre recombinase, resulting in a recombination event between the two loxP sequences and loss of the intervening DNA segment located between the two loxP sites. *See also* **Cre**.

luciferase   An ATP-dependent light-emitting protein.

lumen (pl., **lumina**; adj., **luminal**)   (1) The bore of a hollow, tubelike structure, such as the gut, a bronchiole in the lung, a blood vessel, or a duct in a secretory organ. (2) The enclosed cavity within a spherical structure, such as a membranous vesicle.

luminal   Referring to a cavity in a tissue or the cells that line and face a lumen. See also **lumen**.

lumpectomy   Surgical procedure in which a tumor is removed together with immediately surrounding normal tissue while leaving the bulk of the affected organ intact; usually used in the context of breast cancer surgery.

lymph   Interstitial fluid between cells that is drained via the lymph nodes to larger lymphatic vessels that eventually empty into the venous circulation.

lymph node   Small kidney-shaped structure through which lymph passes and where antigens scavenged by antigen-presenting cells are presented to B and T lymphocytes.

lymphangiogenesis   The process of forming new lymphatic vessels.

lymphatic   Referring to vessels and nodes that drain interstitial fluid from tissues and eventually deposit this fluid in the venous circulation.

lymphedema   Swelling of a tissue due to accumulation of interstitial fluid, which in turn is caused by failure of proper drainage via lymphatic ducts.

lymphocytes   Class of leukocytes that mediate humoral or cellular immunity, encompassing B cells, T cells, and NK cells and derivatives thereof.

lymphoid   (1) Referring to the lymphatic system and its associated organelles. (2) Referring to the lineage of hematopoietic cells that yields B and T lymphocytes as well as natural killer cells. *Compare* **myeloid**.

lymphokine   A cytokine or growth factor specialized to attract and/or activate lymphocytes.

lymphoma   Solid tumor of lymphoid cells. *Compare* **leukemia**.

lymphopenia   Deficiency of lymphocytes in the circulation. *Also termed* **lymphocytopenia**.

lymphotropic   Capable of infecting lymphocytes.

Lynch syndrome   A form of familial colorectal cancer caused by inherited mutations in one of several mismatch repair genes; also known as hereditary non-polyposis colorectal cancer (HNPCC). Beyond colorectal cancer, individuals with Lynch syndrome are also at increased risk of developing brain, stomach, and endometrial cancer as well as other cancer types. *See also* **hereditary non-polyposis colorectal cancer**.

lysate   Product of experimentally dissolving the structure of a tissue or a population of cells, usually generated in order to liberate the internal contents of the tissue or cells.

lysosome   A cytoplasmic vesicle that contains in its lumen degradative enzymes in a solution of low pH, allowing it to degrade various molecules that are introduced into it; in addition, cells may sense nutrient availability by gauging the concentrations of certain amino acids in the lumina of their lysosomes.

lytic   Dissolving a cell or tissue; often associated with the potent cytopathic effects of certain viruses on specific host cells.

lytic cycle   Cycle of viral infection and replication that results ultimately in the death of the infected host cell.

M1 macrophage   A subtype of macrophages that is associated with highly cytotoxic and phagocytic powers and can function to assist the immune attack on cancer cells. *Compare* **M2 macrophage**.

M2 macrophage   A subtype of macrophages that releases immunosuppressive cytokines, notably IL-10 and TGF-β, and also favors tissue repair and regeneration. *Compare* **M1 macrophage**.

macrometastasis   A metastatic tumor of a size that allows it to be detected by a variety of clinically used imaging procedures and thus does not require microscopy for its detection. *See also* **micrometastasis**.

macronucleus   Larger of the two nuclei in many ciliate cells, which carries multiple copies of each gene and is used for the production of mRNAs.

macrophage   A phagocytic cell that either resides over extended periods of time within a tissue or arises within the tissue via the differentiation of recently arrived bone marrow–derived monocytes and proceeds to engulf and digest various infectious agents and cellular debris, often in order to present resulting oligopeptides to the adaptive immune system; additional normal functions include the release of angiogenic factors, mitogens, and proteases as part of a wound-healing program. *See also* **phagocytic**.

MAF transcription factors   A family of bZip-containing transcription factors involved in regulating tissue-specific transcription programs.

maintenance methyltransferase   An enzyme that attaches a methyl group to an unmethylated CpG that is complementary to an existing methylated CpG in a DNA double helix.

major histocompatibility antigen   One of a group of cell-surface proteins that are responsible for the presentation of oligopeptide antigens derived from proteolysis of intracellular proteins to responding cells of the immune system. *See also* **histocompatibility antigen**.

malignant   Describing a growth that shows evidence of being locally invasive and possibly even metastatic.

malignant peripheral nerve sheath tumor   A tumor, usually displaying mesenchymal traits, that often arises from a neurofibroma and derives from cells that form the cellular sheaths protecting peripheral nerves.

malpractice   Clinical treatment that is not appropriate or optimal given existing standards of therapy and implying incompetence or willful neglect.

mammary   Referring to the breast and its milk-producing glands.

mammogram   The image produced by mammography. *See* **mammography**.

mammography   A clinical procedure using X-rays to detect incipient breast cancers.

mass action   Referring to a situation in which the rate of a chemical or biochemical reaction is governed by the concentration of one or more of its reactants.

mass cytometry   Procedure whereby antigen-specific antibody molecules are coupled chemically (conjugated) to metal atoms of various isotopes followed by detection of resulting antigen-metal-antibody conjugates by mass spectrometry, enabling the simultaneous detection of several dozen antigens. *See also* **mass spectrometry**.

mass spectrometry   Analytical procedure by which the mass-to-charge ratio of ions is measured. The presence of ions, which are generated by bombarding molecules with electrons, reveals the nature of the molecules being analyzed or fragments of such molecules, often leading to the identification of their chemical structure.

Masson's trichome stain   A mixture of staining chemicals that, in one alternative version, stains collagen blue, muscles red, and nuclei dark brown.

mast cell   Granulocyte of bone marrow origin that displays Fc receptors for binding IgE antibody molecules and undergoes IgE-mediated degranulation following encounter with antigens bound by the associated IgE molecules.

mastocytosis   A benign excess of mast cells, which are normally involved in allergic responses, wound healing, and defense against pathogens. *See also* **mast cell**.

Matrigel   A preparation of ECM components, largely of basement membrane origin, that is secreted by mouse sarcoma cells and is often used as a gelatinous matrix in which to suspend cells in three-dimensional culture.

matrisome   The collection of molecular species that together form the extracellular matrix of a tissue or of cultured cells.

matrix metalloproteinases   Secreted proteases that employ zinc ions in their catalytic sites and degrade extracellular proteins, largely components of the extracellular matrix.

maximum tolerated dose   The dosage of an administered drug above which a patient suffers unacceptable levels of toxic side effects.

mechanotransduction   Process by which a cell senses mechanical stimuli, such as tension in the adjacent extracellular matrix, and converts this physical information into intracellular biochemical signals.

median   The numerical value of a measurement that resides halfway through a list of such measurements, the latter arrayed in an order of increasing or decreasing value.

medical   Referring to the use of drugs, rather than surgery, to treat disease.

medical oncology   The practice of treating cancer with drugs. *See also* **surgical oncology**, **radiation oncology**.

medium   The fluid in which cells are propagated *in vitro*. *See also* **tissue culture**.

medulloblastoma   Tumor of the primitive precursors of neurons in the cerebellum.

megakaryocyte   Large cell of the hematopoietic system that produces platelets by pinching off fragments of its cytoplasm.

melanin   A brown-black or red-brown pigment that is synthesized by melanocytes and, in the skin, transferred to basal keratinocytes, thereby creating pigmentation of the skin; alternatively, pigment is transferred to cells within the iris of the eye.

melanocyte   Cell of neural crest origin that creates pigmentation of the skin and iris.

melanoma   Tumor arising from the melanocytes present as the pigmented cells of the skin, iris, and retinal pigmented epithelium.

melanosome   Melanin-containing body in the cytoplasm of a melanocyte that is transferred to a keratinocyte or cells of the iris in order to impart pigmentation to the latter.

membrane attack complex   The complex of multiple proteins including cell surface–bound antibody molecules and complement proteins that can introduce pores in the nearby plasma membrane, thereby killing an involved cell.

memory   *See* **immunological memory**.

menarche   Timing in life of the first menstrual cycle. *See also* **menopause**.

menopause   Time in life when menstrual cycles halt. *See also* **menarche**.

merotely   Condition established during the prophase of mitosis when a single kinetochore becomes linked simultaneously to two opposing spindle fibers.

mesenchymal   (1) Referring to tissue composed of cells of mesodermal origin, including fibroblasts, smooth muscle cells, endothelial cells, and adipocytes. (2) Referring to an individual cell type belonging to this class of cells. (3) Referring to a phenotypic state associated with mesenchymal markers and created by the epithelial–mesenchymal transition cell-biological program. *See* **epithelial–mesenchymal transition**.

mesenchymal–epithelial transition   The reversal of the epithelial–mesenchymal transition program that converts phenotypically more-mesenchymal cells into more-epithelial derivatives. *Compare* **epithelial–mesenchymal transition**.

mesenchymal stem cell   An oligopotent mesenchymal cell, often of bone marrow origin, that can be recruited to various tissues in which it can differentiate into chondrocytes, osteoblasts, adipocytes, and possibly other mesenchymal cell types.

mesoderm   Middle layer of cells in an early chordate embryo lying between the ectoderm and endoderm, which is the precursor of mesenchymal tissues, including connective tissues and the hematopoietic system. *See also* mesenchymal.

mesothelioma   Tumor of the epithelial membranes lining the lungs, peritoneal cavity, and heart.

messenger RNA (mRNA)   The transcript of a gene that has been processed by splicing and exported to the cytoplasm where it serves as the template for protein synthesis conducted by ribosomes.

meta-analysis   Statistical procedure for pooling and integrating the results of a number of independent experimental, clinical, or epidemiologic studies in order to generate a larger data set and greater statistical significance than is afforded by the data of only a single study.

metabolite   Chemical species resulting from the metabolic conversion by enzymes of a precursor chemical species.

metalloproteinase   A protease that contains a metal atom, usually zinc, in its catalytic site.

metaphase   Second major phase of mitosis, during which chromosomes complete condensation and attach to the mitotic spindle after the nuclear membrane disappears; chromosomes are now readily observed in the light microscope following appropriate staining.

metaphase plate   A plane or region that is approximately equidistant from the two poles of a dividing cell at which chromosomes align prior to the anaphase portion of the cell cycle.

metaphyta   Plants composed of many cells.

metaplasia   Replacement within a tissue of cells of one differentiation lineage by cells belonging to another lineage.

metastable   State of a cell or process that appears over a period of time to be stable and unchanging but can in fact be changed by various signals or perturbants.

metastasis (pl., -es; adj., **metastatic**; v. **metastasize**)   A neoplastic growth forming at one site in the body, the cells of which derive from a malignancy located elsewhere in the body, such as the site of primary tumor formation. (2) The process leading to the formation of metastases. *See also* **primary tumor**.

metastatic colonization   The phenomenon in which disseminated tumor cells forming micrometastases succeed in spawning macroscopic metastases.

metastatic inefficiency   Phenomenon in which only a minute portion of disseminated tumor cells succeed in spawning macroscopic metastases.

metastatic shower   Clinical observation of an apparent synchronous formation of multiple new metastases at multiple sites of metastatic dissemination.

metastatic tropism   The observed tendency of cancer cells from one specific primary tumor type to colonize preferentially one target organ or another.

metazoan   Referring to animals composed of many cells. *Compare* **protozoan**.

methylator phenotype   Tendency of the DNA methylation apparatus in a cell to drive excessive CpG methylation, leading to the hypermethylated state of its genome.

methyltransferase   An enzyme that attaches methyl groups covalently to substrates, such as the cytosine bases of DNA or lysine residues of histones.

MHC class I   A class of cell surface proteins that is specialized to acquire (during its maturation within the endoplasmic reticulum) oligopeptide fragments of its own intracellular proteins and thereafter present these fragments on the cell surface where they may be recognized by components of the adaptive immune system such as CD8+ lymphocytes. *Compare* **MHC class II**.

MHC class II   A class of cell surface proteins that is specialized to acquire (during its maturation within the endoplasmic reticulum) oligopeptide fragments of acquired phagocytosed cells or particles and thereafter present these fragments on the cell surface where they may be recognized by components of the adaptive immune system such as CD4+ lymphocytes. *Compare* **MHC class I.**

microarray A collection of sequence-specific DNA probes that are attached at specific sites to a solid substrate, such as a glass microscope slide; these probes may derive from specific genes scattered throughout a cell genome or from a particular subset of chosen genes. *See also* **DNA chip.**

microbiome The collection of microbes, largely bacteria, that exist naturally as commensal organisms in normal organs, such as the gastrointestinal tract and the skin; each of these and other tissues carries its own spectrum of microbes and thus its own microbiome.

microembolus (pl., **microemboli**) Small body formed from cells or products of the clotting machinery that travels through the circulation and may be trapped in small vessels, often blocking further circulation through these vessels. *See also* **embolus.**

microenvironment (1) The biological environment within a tissue that is created by nearby cells that influence the behavior of a cell of interest. (2) In a tumor, the environment that is generated by the recruited stromal cells that are closely apposed to the neoplastic cells such as carcinoma cells.

micrometastasis A single cell or small clump of disseminated cancer cells that can only be visualized using one or another form of microscopy. *See also* **macrometastasis.**

micronucleus (1) A small fragment of a nucleus that has its own nuclear membrane and results from certain aberrations in cell division or from damage inflicted on a cell; micronuclei often carry only a single chromosome and arise when this chromosome fails to be incorporated into the previously formed mitotic spindle, thereby escaping inclusion in the normal karyotypic array of a daughter cell. (2) Smaller of the two nuclei in many ciliate cells, which is used to carry and transmit the ciliate genome to progeny cells.

microRNA Endogenously synthesized RNA transcribed by RNA polymerase II and processed into 21- to 23-nt-long ssRNA that interferes with translation of an mRNA or causes its degradation, depending on the degree of complementarity with the mRNA.

microsatellite A DNA relatively short oligonucleotide segment of simple sequence that resides within a cell genome and is repeated hundreds, even thousands of times in this genome, often at sites scattered though this genome. *See also* **satellite.**

microsatellite instability A condition caused by mutations in genes involved in DNA mismatch repair that is manifested when microsatellite sequences undergo frequent expansion or shrinkage in length. *See also* **microsatellite, mismatch repair.**

microthrombus *See* **thrombus, embolus.**

mimetic An agent, such as a protein or chemical, that imitates or functionally mimics another agent.

minimal residual disease Following treatment of diagnosed tumors, the persistence of derived cancer cells that are undetectable by clinical or radiological measurements but nevertheless persist as potential sources of eventual clinical relapses.

mismatch repair Class of DNA repair processes that depend on proofreading a recently synthesized segment of DNA and removing any misincorporated bases that may have escaped previous correction by DNA polymerase-associated proofreading functions. Defects in mismatch repair lead to microsatellite instability. *See also* **microsatellite instability, proofreading.**

mis-segregation The failure to properly segregate chromosomes during mitosis or meiosis.

missense codon A triplet codon in the genetic code, often created by a point mutation, that specifies an amino acid residue different from that specified by the codon that it replaces. *See also* **nonsense codon.**

missense mutation Point mutation causing an amino acid substitution. *See* **missense codon.**

mitogen (adj., -**genic**) An agent that provokes cell proliferation.

mitosis (1) Nuclear division, composed of the four steps of prophase, metaphase, anaphase, and telophase, the last including cytokinesis. (2) Process by which a single cell separates its complement of chromosomes into two equal sets in preparation for the division of the cell into two daughter cells achieved by cytokinesis.

mitotic catastrophe Process by which unrepaired damage to DNA or chromosomes or mis-segregation of chromosomes causes cells to enter into M phase and die there through non-apoptotic mechanisms because of an inability to properly complete the critical final steps of mitosis.

mitotic recombination Recombination between homologous chromosomal arms occurring during somatic cell proliferation, often during the G2 phase of the cell cycle (rather than during mitosis).

mitotic spindle The complex array of microtubules that is responsible for separating the two sets of chromatids during the anaphase of mitosis.

molecular chaperone A protein that functions to facilitate the proper folding of unfolded or misfolded proteins, doing so without making or breaking covalent bonds.

monoallelic Referring to a state in which only one of the two copies of a gene is expressed or exerts effects on phenotype. *See also* **biallelic.**

monoclonal Describing a population of cells, all of which derive by direct descent from a single common ancestral cell. *See also* **oligoclonal, polyclonal.**

monoclonal antibody An antibody that is made via secretion by a population of cells that has been produced through the immortalization of a single antibody-producing plasma cell and subsequent clonal expansion of its descendants, so that all antibody molecules in a preparation are biochemically identical to one another and show identical antigen specificity. *See also* **hybridoma.**

monocyte Phagocytic leukocyte of bone marrow origin that circulates briefly in the blood before migrating into tissues, where it can differentiate into macrophages, osteoclasts, dendritic cells, Langerhans cells, and possibly other phagocytic cell types.

monogenic   Referring to a phenotype or disease condition that can be traced to the actions of a single gene and its alleles. *See also* **polygenic**.

monolayer   A population of cells growing *in vitro* as a layer one cell thick.

monomer   Molecule composed of a single unit or a single subunit of a chemical entity that is capable of forming higher–molecular-weight polymeric structures by association with or covalent linkage to other identical or similar units.

monotherapy   A therapy in which only a single drug or treatment is used at one time.

monozygotic   Referring to two or more organisms derived from a single fertilized egg, thereby referring to identical twins, triplets, or multiple births involving higher numbers of genetically identical newborns.

morbidity   The existence of a medical condition or disease.

morphogen   Low–molecular-weight substance that induces cells to assemble and participate in the construction of a tissue of a certain shape and form.

morphogenesis   Process whereby shape is created, usually referring to the creation of shape of various tissue and organ structures during embryonic development.

morphology   Shape and form of a cell, tissue, or organism.

mortal   Referring to a cell population having limited proliferative capacity. *Compare* **immortalization**.

mortality   The rate or frequency of death from a particular disease condition.

motility (adj., **motile**)   Tendency for movement, such as that of individual cells, from one location to another.

motogenic   Referring to an agent or signal that stimulates cell movement or motility.

mTOR   Serine/threonine kinase that forms the catalytic core of the mTORC1 and mTORC2 complexes; by responding to nutrient availability, mTORC1 acts as the central regulator of cell growth, biosynthesis, and autophagy.

mucin   Member of a family of heavily glycosylated glycoproteins that are secreted by luminal cells in a variety of epithelial tissues and form gel-like aggregates in order to lubricate or erect barriers to protect the luminal and underlying epithelial cell layers.

mucosa   Epithelial cell layer that secretes a mucus-like substance that forms a protective layer above the secreting cells.

multi-drug resistance   Development by cancer cells under treatment of concomitant resistance to multiple chemically distinct species of therapeutic drugs.

multi-drug transporter   Molecular pump embedded in the plasma membrane of a cell that is capable of exporting multiple distinct molecular species including drug molecules from the cell. *See also* **multi-drug resistance**.

multimeric.   Referring to a complex of multiple subunits. *See also* **monomer, dimer, trimer, tetramer**.

multiparous   Referring to a woman who has given birth multiple times.

multiple myeloma   *See* **myeloma**.

multiplicity of infection   In virology, the ratio of number of infectious particles to the number of infectable cells.

multipotent   Referring to the ability of a stem cell to differentiate into multiple distinct cell types. *See also* **totipotent, pluripotent**.

mural   Referring to the outer cell layers of blood vessels, which are composed of pericytes and smooth muscle cells and surround the luminal endothelial cells.

murine   Referring to the subgroup of rodents that includes mice and rats.

muscularis propria   Layer of muscles immediately underlying the submucosal layer in the gastrointestinal tract.

mutability   The susceptibility of a cell and its genome to alteration via mutation.

mutable   Referring to an entity that is subject to change, notably a genome subject to genetic mutation.

mutagen (adj., -**genic**)   An agent that creates a mutation. *See also* **genotoxic**.

mutagenicity   The property of being mutagenic.

mutation   Change in the genotype of a virus or cell that involves an alteration in the nucleotide sequence of its genome, in the arrangement of a segment within a chromosome, in the number of copies of a segment, in the physical structure of a chromosome, or in the number of copies of a structurally normal chromosome.

mutational burden   *See* **tumor mutational burden**.

mutator phenotype   Cell state in which defects in DNA repair or mutagen detoxification lead to elevated numbers of mutations sustained per cell generation.

myelin   Substance composed of lipid and protein that is applied around axons by Schwann cells and oligodendrocytes in order provide electrical insulation, thereby facilitating rapid signal conduction. *See also* **Schwann cells**.

myelo-   Prefix referring to bone marrow or an entity of bone marrow origin.

myeloablation   Clinical procedure involving the elimination of the hematopoietic system, including its cellular components residing in the bone marrow, usually achieved through the application of a toxic drug or whole-body radiation.

myelocytomatosis   A malignancy of avian bone marrow cells.

myelodysplasia   Condition in which incompletely differentiated cells of one or another myeloid lineage accumulate in the blood, often accompanied by a reduction in the concentration within the circulation of properly differentiated cells of bone marrow origin.

myelodysplastic syndrome   Hyperproliferative condition of cells of the myeloid lineage in the bone marrow which often progresses to acute myelogenous leukemia.

myelogenous   Originating in the bone marrow; *also termed* **myelocytic**. *See also* **myeloid**.

myeloid (1) Referring to the lineages of hematopoietic cells that yield cells that are not lymphocytes and includes erythrocytes, platelets, granulocytes, monocytes, mast cells, and their derivatives. (2) Pertaining to or resembling bone marrow; often used as a synonym for myelogenous. *Compare* **lymphoid**.

myeloid derived suppressor cell Relatively undifferentiated cell of myeloid origin, deriving from cells of the granulocytic or monocytic lineages, that exerts potent immunosuppressive functions on other immune cells of both the innate and adaptive arms of the immune system.

myeloma A malignancy of the antibody-producing cells of the bone marrow, often called multiple myeloma to indicate the large number of osteolytic lesions that are encountered in patients with advanced disease.

myeloproliferative Referring to excessive proliferation and resulting elevated levels of one of the several cell types generated by the myeloid branch of the hematopoietic system.

myoblasts The undifferentiated precursors of the myocytes present in differentiated muscle. *See also* **myocytes**.

myocytes The individual contractile cells that assemble to form functional muscles. *See also* **myoblasts**.

myofibroblast A type of fibroblast that is normally involved in wound healing and inflammation and is often defined by its expression of α-smooth muscle actin which, together with myosin, imparts to it contractile powers.

N-terminus (adj., -**al**) End of a protein chain at which synthesis is initiated. *See also* **C-terminus**.

naive Referring to immune lymphocytes that have not experienced contact with cognate antigens and have not, as a consequence, undergone the biological activation associated with such contact.

nascent Referring to an entity, such as a molecule, that is in the course of being formed or synthesized.

natal Referring to birth. *See also* **prenatal**.

natural killer cells A type of lymphocyte whose ability to attack target cells is not dependent on any form of adaptive immunity. *See also* **innate immune response**.

natural product A chemical species that is a product of the natural metabolism of an organism, such as a bacterium, fungus, or plant.

NCI-60 cell lines (often **NCI-60**) The collection of 60 human cancer cell lines prepared from nine commonly occurring tumor types and provided by the U.S. National Cancer Institute that is often used to gauge the responsiveness of different types of cancer cells to therapeutic agents under development.

necroptosis A programmed form of necrosis that leads to cell death and the release of cellular debris that may persist and trigger local tissue inflammation. *See also* **apoptosis**, **ferroptosis**, **pyroptosis**.

necrosis A non-programmed form of cell death triggered by the collapse of cellular energy production, involving the breakdown of a cell's constituents through steps that are distinct from those in the neocroptosis and apoptotic death programs. *See* **necroptosis**, **apoptosis**.

necrotic Referring to a region of a normal or neoplastic tissue in which the cells have died.

negative feedback A regulatory process whereby a stimulatory signal upstream in a signaling pathway activates a downstream target that proceeds to inhibit the source of the initial upstream signal, thereby reducing further overall signaling. *See also* **positive feedback**.

neoadjuvant Referring to a treatment that is undertaken before the main therapy is applied, e.g., to reduce tumor size before surgery. *See also* **adjuvant**.

neoangiogenesis *See* **angiogenesis**.

neoantigen An antigen that is newly formed by tumor cells, usually during the course of tumor development, and is created by somatic mutations affecting the reading frames of cellular mRNAs.

neomorphic Referring to a mutation that confers a novel function on a gene and encoded protein.

neoplasia (adj., -**plastic**) (1) The state of cancerous growth. (2) Benign or malignant tumor composed of cells having an abnormal appearance and abnormal proliferation pattern.

neoplasm A tumor. *See also* **neoplasia**.

neovasculature Vasculature that is newly developed, often by a growing tumor. *See also* **vasculature**.

neural crest Dorsal region of the early chordate embryo that contains cells serving as precursors of various specialized tissue and cell types, including certain cells of the peripheral nervous system, bones of the face, melanocytes, and several types of neurosecretory cells.

neurite A thin provisional protrusion from the cell body of a neuron that has the potential to develop into a fully formed, functional axon. *See also* **axon**, **neuron**.

neuroblastoma Tumor of primitive neuronal precursor cells of the peripheral nervous system and adrenal medulla.

neuroectodermal Referring to the components of the nervous system, which derive from the embryonic ectoderm.

neuroendocrine Referring to a cell type that releases certain hormones in response to neuronal signals.

neurofibroma Benign tumor of cells forming the sheath around nerve axons.

neurofibrosarcoma Malignant tumor of cells forming the sheath around nerve axons.

neuron A nucleated cell in nervous tissue from which neurites and axons extend. *See* **axon**, **neurite**.

neuropeptide A small protein or oligopeptide that is produced by cleavage of a higher–molecular-weight precursor and is released by certain cells of the central nervous system and neuroendocrinal cells and that impinges upon and thereby signals to other cells inside and outside of the central nervous system.

neurosecretory Referring to a cell type that will secrete a substance, such as a hormone, in response to neuronal signals.

neurotrophic Referring to secreted factors that support the growth, survival, and differentiation of nervous tissue.

neutral mutation Change in DNA sequence that has no effect on phenotype, including mutations that have no effect on protein structure.

neutralization Inactivation of a biological activity, such as inactivation of viral infectivity by antibody molecules.

neutropenia Deficiency of neutrophils in the marrow and circulation.

neutrophil An abundant granulocyte in the circulation that expresses Fc receptors and is responsible for multiple functions, including recognizing and engulfing various types of infectious agents and participating in various types of inflammation.

neutrophilia A condition in which the concentration of neutrophils in the circulation is elevated above its normal value.

nevus (pl., **nevi**) A benign growth of melanocytes in the skin yielding moles or collections of pigmented cells that may, at low probability, progress into a melanoma.

next-generation sequencing A set of high-throughput methods used to determine a portion of the nucleotide sequence of DNA. These techniques utilize DNA sequencing technologies that are capable of determining multiple DNA sequences in parallel. *Also called* massively parallel sequencing.

niche A functional locale that supports the survival and proliferation of specialized cells. Specific niches support the proliferation and survival of stem cells in normal tissues and, quite possibly, cancer stem cells in tumors.

NKG2D Receptor expressed by natural killer cells and other immunocytes that recognizes MHC-like molecules displayed by other cells undergoing various types of physiological stress and may trigger a cytolytic attack on such cells in response.

non-canonical Referring to a representation of an entity or process that differs from the most widely accepted, often initially established representation. *See also* **canonical**.

nondisjunction (1) Failure of two chromatids to separate from one another during mitosis or two homologous chromosomes to separate from one another during meiosis. (2) State created by this failure of separation.

nongenotoxic Incapable of damaging a genome. *See also* **genotoxic**.

nonhomologous end joining Type of DNA repair consisting of fusion of two dsDNA ends, in which the precise joining of the two ends is not informed or directed by sequences present in a normal sister chromatid or homologous chromosome.

nonhomologous recombination Process of recombination between two DNA molecules in which the two participating molecules do not share significant sequence identity. *Also called* **illegitimate recombination**.

noninvasive Referring to procedures that allow diagnosis or treatment without the need to enter the body with diagnostic instruments or surgery.

nonpermissive (1) Describing a physiological state that does not permit the survival and/or proliferation of a cell or infectious agent such as a temperature-sensitive virus. (2) Describing a type of host or host cell that does not permit cells or infectious agents to proliferate. *See* **permissive**.

non-self Constituent or agent in the body that is of foreign, non-native origin. *See also* **self**.

nonsense codon A triplet codon in the genetic code that specifies termination of the growing polypeptide chain; nonsense codons generated by point mutations within reading frames cause premature termination of translation and resulting formation of truncated proteins. *See also* **missense codon**.

nonsense mutation Mutation causing premature termination of a growing polypeptide chain. *See also* **nonsense codon**.

non-small-cell lung carcinoma Any of several types of lung cancers with the exception of small-cell lung carcinoma.

nonsynonymous A nucleotide mutation that alters the amino acid sequence of an encoded protein. Nonsynonymous mutations differ from synonymous mutations, which do not alter amino acid sequences of encoded proteins. *See also* **synonymous**.

normoxia A level of oxygen that corresponds to that normally experienced by cells in a specific tissue environment. *See also* **hypoxia**.

Northern blot Adaptation of the Southern blotting procedure in which RNA (rather than DNA) is resolved by gel electrophoresis and transferred to a filter that is subsequently incubated with a sequence-specific, radiolabeled DNA probe. *See also* **Southern blot**, **Western blot**.

nuclear magnetic resonance Analytical procedure for determining molecular structures and interactions in which molecules are placed under strong magnetic fields and the influence of radio frequency waves; the frequencies at which the atoms within a molecule resonate yields information on the molecular interactions of individual atoms and, in turn, the overall structure of a complex molecule.

nuclear receptor One of a family of nuclear DNA-binding proteins that bind lipid-soluble ligands and, in response, undergo a conformational shift that enables them to directly control the transcription of genes to which they are bound.

nucleolus The organelle within the cell nucleus that is the site of ribosomal subunit biosynthesis.

nucleophilic Referring to a molecule that seeks out and reacts with electron-poor substrates. *Compare* **electrophilic**.

nucleoplasm The unstructured portion of the nucleus that does not include the nucleoli and chromosomes.

nucleosome Protein octamer, composed of two each of histones H2A, H2B, H3, and H4, around which DNA is wrapped in chromatin.

nucleotide-excision repair Type of DNA repair in which the initial step involves the excision of nucleotides (rather than bases). *Compare* **base-excision repair**.

null allele An allele of a gene that lacks all normal functions of the gene. *See also* **hypomorphic**.

nulliparous Referring to a female who has never given birth.

objective response Demonstration of clinical response to therapy as gauged by changes in progression-free survival, overall survival, time to progression, size of tumor, marked improvement in patient symptoms, or other measurable response to therapy.

obviate   To render unnecessary.

occlude   To block access to.

occult   Hidden, inapparent.

ocular   Referring to the eye.

off-target effects   Effects of a therapeutic agent on molecules or cell types other than the intended targeted molecules or cell types.

Okazaki fragment   A short DNA fragment of 150–200 nucleotides that enables lagging strand polymerization and is joined soon after its formation to adjacent Okazaki fragments by a DNA ligase.

oligoclonal   Referring to a cell population that contains the descendants of a small number of founding cells. *See also* **polyclonal**, **monoclonal**.

oligomer   A polymer of several monomeric subunits.

oligonucleotide   A polymer composed of a relatively small number of nucleotides.

oligopeptide   A short protein chain consisting of a relatively small number of amino acids.

oligopotent   Referring to the ability of a less differentiated cell, such as a stem cell, to differentiate into two more distinct progeny.

-oma   Denoting a benign or malignant growth.

-ome   Suffix referring to the entire spectrum of a type of entity being expressed in a given cell or tissue, e.g., genome, proteome, (DNA) methylome, transcriptome, orfeome, matrisome or secretome.

ommatidium (pl., **-ia)**   The individual compartment within the multi-compartment arthropod eye in which light is sensed.

oncofetal   Referring to an antigen that is normally expressed transiently during embryonic development and is also expressed by some tumor cells.

oncogene (1) A cancer-inducing gene. (2) A gene that can transform cells from a normal to a neoplastic state.

oncogene addiction   (1) The physiological state of a cancer cell in which it continues to be dependent for its proliferation or survival on the ongoing function of a certain oncogene, usually a mutant oncogene, often long after the oncogene has been acquired during the course of multi-step tumorigenesis. (2) More generally, ongoing dependence on any type of previously acquired genetic or epigenetic alteration.

oncogenic   Referring to the ability to induce or cause cancer.

oncologist   A physician who treats cancer, usually through medical rather than surgical means.

oncometabolite   A chemical species that accumulates within cancer cells and contributes to their neoplastic transformation, e.g., 2-hydroxyglutarate.

oncomiR   A microRNA associated with or causing cell transformation or contributing to tumor progression.

oncoprotein   A protein specified by an oncogene that mediates its effects on cell physiology, often including neoplastic transformation.

ontogeny   The developmental history of an individual organism.

open reading frame   A sequence that encodes a continuous polypeptide segment, usually encompassing the entirety of the coding information of a gene.

operable   Capable of being treated successfully by surgery. *See also* **inoperable**.

opportunistic   A process, such as an infectious event, that exploits an environment that normally would be inhospitable to that process. An opportunistic infection by a virus occurs in host organisms that have compromised antiviral defense mechanisms.

opsonization   Process of coating a cell, including a bacterial cell, with antibody molecules that recognize and bind cell surface antigens expressed by the cell; such binding often leads to the phagocytosis of the antibody-coated cell by phagocytes, such as macrophages, that display Fc receptors on their surface.

ordinate   The vertical or *y*-axis of a Cartesian graph. *See also* **abscissa**.

orfeome   The collection of many or all of the open reading frames expressed within a cell or cell type.

organelle   A subcellular structure, such as a ribosome, mitochondrion, or Golgi apparatus.

organoid   An experimentally created model of a normal adult tissue, usually generated through incubation of organ-specific stemlike cells suspended in a three-dimensional gelatinous matrix containing a cocktail of factors designed to induce a tissue-specific differentiation program in the organoid.

organotropism   The tendency of cancer cells in a specific type of primary tumor to spawn metastases in a specific subset of normal tissues in the body.

organotypic   Tending to reflect or recapitulate a state or condition mimicking the one operating within a specific organ or tissue.

origin of replication   DNA segment that determines the site of initiation of replication in a chromosome, viral genome, or episome, often resulting in two diverging replication forks.

oropharynx (adj., **oropharyngeal**)   The cavity extending from the back of the mouth down the respiratory tract to the larynx (voice box).

ortholog (*also* **orthologue**)   A gene in one species that is the closest relative of a gene in another species, with the two genes arising from a common ancestral gene in a common ancestral species; usually orthologs represent direct counterparts of one another in the genomes of two species, often retaining very similar functions. *See also* **homolog**, **paralog**.

orthotopic   Referring to an anatomically proper or native site. *See also* **ectopic**.

osteoblast   Mesenchymal cell type related to fibroblasts that constructs mineralized bone through the deposition of a collagenous matrix and apatite crystals.

osteoblastic   Referring to a class of bone lesions that involve localized increases in the amount of mineralized bone. *See also* **osteolytic**.

osteoclast   Cell type of monocyte origin that functions to degrade and demineralize already-assembled bone.

osteoid   The extracellular matrix that forms during bone formation prior to mineralization.

osteolytic   Referring to a bone lesion that involves localized dissolution of mineralized bone usually driven by the localized actions of osteoclasts. *See also* **osteoblastic**.

osteotropism   The trait of cells, notably metastasizing cancer cells, to migrate to and colonize the bone marrow.

outcomes research   The practice of evaluating treatment protocols by comparing clinical outcomes achieved by several alternative protocols, usually in controlled clinical trials.

overall survival   Proportion of patients who are still alive at a certain time following initiation of treatment, often measured after a certain interval, such as five years, after such treatment initiation. *See also* **progression-free survival**, **disease-free survival**.

overexpression   Expression of an RNA or a protein at higher-than-normal levels.

oxidative phosphorylation   The process of generating ATP from ADP driven by electron gradients in the mitochondrion.

p value   Statistical parameter describing the probability that an event or quantity occurred simply by random chance.

Paneth cell   Secretory cells that reside in the bottom of intestinal crypts and also help to create a stem-cell niche for adjacent intestinal stem cells. *See also* **niche**.

papilloma   Benign, adenomatous proliferation of epithelial cells; term often used to describe benign lesions of the skin.

papovaviruses   Class of viruses that includes SV40, polyomavirus, and papillomaviruses.

paracrine   Referring to the signaling path of a hormone or factor that is released by one cell and acts on a nearby cell, ostensibly after transport through the interstitial fluid of a tissue. *Compare* **autocrine**, **endocrine**, **exocrine**, **intracrine**, **juxtacrine**.

parallel progression   Process whereby cancer cells advance through multi-step tumor progression at sites distant from and independently of the primary tumor. *See also* **primary tumor**.

paralog (*also* **paralogue**)   A gene or protein that is related to another gene or protein through evolution from a common ancestral precursor; paralogs usually arise via gene duplication followed by sequence divergence of the two resulting genes, often with the development of two distinct functions, with both resulting genes continuing to be present in the genome of a descendant organism. *See also* **homolog, ortholog**.

paraneoplastic   Referring to a biological effect evoked in the body by a tumor at a site in the body that is located some distance from the tumor itself and is apparently not directly involved in the pathogenesis of the tumor.

parenchyma   The portion of a tissue that lies outside the circulatory system and often is responsible for carrying out the specialized functions of the tissue.

parenteral   Referring to a route of substance administration (usually injection) or infectious-agent transmission not involving the mouth or gastrointestinal tract.

parity   (1) The condition of having given birth. (2) The number of times that a female has given birth.

parous   Referring to a female who has given birth at least once.

partial response   Halting of further tumor growth or a 50% or greater reduction in tumor mass following anti-cancer therapy. *See also* **complete response**.

passaging   (1) Practice of transferring cells from one culture vessel to another, often performed because the cell population has filled up the first vessel. (2) Similarly, the transfer of tumor cells or infectious agents, notably viruses, from one host cell or organism to another. *See also* **serial passaging**.

passenger gene   Gene altered by passenger mutation. *See* **passenger mutation**.

passenger mutation   Mutation that confers no selective advantage on a cell carrying it but whose presence in a population of cells is increased because it resides in the same genome as other mutant alleles that do indeed confer advantage; sometimes termed bystander mutation. *Compare* **driver mutation**.

passive immunization   Procedure in which the immune responses of an organism are supplemented or strengthened through the introduction of immunological agents, usually antibodies or immunocytes, of foreign origin.

pathogen Agent that causes disease, often as an infectious agent.

pathogenesis   Process that leads to the creation of a disease state or diseased tissue.

pathological   (1) Diseased or associated with a disease. (2) Referring to the study of a disease process, often at the level of light microscopy.

pathological complete response   The response of a tumor to a therapy that involves the complete elimination of microscopically detectable remnants of a tumor under treatment. *See* **complete response**.

pathologist   Physician who examines tissues microscopically to study and classify disease.

pathology   (1) A disease or disease condition. (2) The science of analyzing a tissue by studying its microscopic structure. *See also* **histopathology**.

patient-derived xenograft   A population of cancer cells that is derived by implanting fragments of patients' tumors directly into immunocompromised mouse hosts, thereby avoiding initial *ex vivo* propagation of tumor cells in tissue culture.

pausing   *See* **polymerase pausing**.

pediatric   Referring to an attribute or condition of children.

penetrance   Extent to which an allele of a gene can or cannot influence phenotype, e.g., the likelihood that a germ-line allele will or will not induce a clinical phenotype in a carrier of this allele. *See also* **expressivity, incomplete penetrance**.

-penia   Suffix denoting a deficiency or abnormally low levels of a substance or entity, usually detected in the circulation.

peptidase   Peptide-cleaving enzyme.

perforin   A protein made by cytotoxic immune cells and inserted by them into the plasma membrane of a targeted cell; once inserted, perforin molecules assemble to create a channel through the membrane that causes the death of the cell, often by allowing the introduction of pro-apoptotic proteins such as granzymes into the cell.

peri-   Prefix implying surrounding or around.

pericytes   Cells closely related to smooth muscle cells that surround capillaries and provide the capillary walls formed by endothelial cells with tensile strength and contractility; pericytes also provide trophic signals to nearby endothelial cells. *See also* **trophic**.

perineural invasion   Invasion of cancer cells along tracts established by neurons and axons.

perineurial   Referring to the fibroblast-like cells that wrap around and protect nerve axons.

perinuclear   Surrounding or near to the cell nucleus.

peripheral neuropathy   Damage to peripheral nerves, usually in the limbs, that affects muscle coordination or sensation and may even create sensed pain.

peripheral tolerance   Process by which T cells recognizing self-antigens are eliminated or functionally inactivated in lymph nodes throughout the body rather than centrally in the thymus. *Compare* **central tolerance**.

peritoneal   Referring to the cavity in the abdomen that is limited by an enclosing membrane and includes the lower gastrointestinal tract and associated organs, including pancreas and liver.

peritoneovenous shunt   A tube that is inserted into the peritoneal space that allows the collection of cancer cells forming the ascites fluid in patients suffering high-grade cancers; the cancer cells, which enter through holes in the tube, are eventually emptied directly by the tube into the venous circulation, often via the vena cava.

permissive   (1) Describing a physiological state that allows the survival and/or proliferation of a cell or infectious agent, such as a temperature-sensitive virus. (2) Describing a type of host or host cell that allows cells or infectious agents to proliferate. *See also* **nonpermissive**.

peroxidase   An enzyme, often prepared from horseradish, that reduces peroxides and is often used in immunohistochemistry, when coupled to an antibody, to generate via oxidation products that exhibit a characteristic color.

peroxisome   A cytoplasmic vesicular organelle that is involved in the oxidation of various substrates, notably lipids.

Peto's paradox   The discrepancy, first noted by Richard Peto, that large bodied, long-lived organisms do not confront greater cancer incidence than corresponding smaller, short-lived animals in spite of vast differences in cumulative cell divisions experienced during an average lifetime.

phage   *See* **bacteriophage**.

phagocyte, (adj. **phagocytic**)   Cell of the immune system—e.g., a macrophage or dendritic cell—that is specialized to engulf and destroy other cells, cellular fragments, and other debris.

phagocytosis   Endocytotic process by which a cell, usually a component of the immune system, engulfs a particle (which may be another cell), internalizes this particle, and usually proceeds to degrade it by introducing it into lysosomes. *See also* **endocytosis**.

phagolysosomes   Vesicular structure in the cytoplasm formed by the fusion of a phagosome with a lysosome. *See also* **phagosome**.

phagosome   A vacuole or vesicle formed in the cytoplasm directly by the process of phagocytosis.

pharmacodynamics   Time course of biological responses within a tissue or its cells that are induced by a drug, usually gauged at different times after initial administration.

pharmacokinetics   Results of measurements gauging the rise and fall in the concentration of an administered drug in the body, usually measured in the plasma and gauged at different times after initial administration.

pharmacological   Referring to the study and use of drugs, usually those of low molecular weight.

phase I   A clinical trial with a relatively small group of patients that gauges the safety/tolerability of a new experimental drug as well as its pharmacokinetics and pharmacodynamics. *Compare* **phase II**, **phase III**.

phase II   A clinical trial in a large group (e.g., hundreds) of patients in which the safety of a new drug is further examined after completion of a phase I trial as well as its potential therapeutic efficacy, e.g., its ability to affect the growth of certain tumors. *Compare* **phase I**, **phase III**.

phase III   A clinical trial in which very large numbers (e.g., hundreds to thousands) of patients are exposed to a new drug, given in a randomized fashion with placebo controls, and tested for its efficacy relative to existing approved therapeutic modalities, with the intent of demonstrating statistically significant clinical benefit, usually one greater than existing approved therapies. *Compare* **phase I**, **phase II**.

phasing   The process controlling the precise positioning of DNA-bound nucleosomes relative to specific nucleotide sequence landmarks along the DNA.

phenocopy   (1) A biological entity—cell, tissue, or organism—that replicates the phenotype of another entity even though the two entities may have distinct underlying genotypes or arise through distinct mechanisms. (2) To create a phenocopy. *See also* **phenotype**.

phenomenon (pl., **phenomena**)   An observable process or entity.

phenotype   (1) A measurable or observable trait of an organism. (2) The sum of all such traits of an organism.

pheochromocytoma   Tumor of the neural crest-derived cells of the adrenal glands that secrete elevated levels of certain hormones through which they may instigate a number of abnormal clinical symptoms.

phosphatase   An enzyme that removes phosphate groups from phosphorylated substrates, such as the phosphoamino acid residues in a protein or the phosphorylated inositol of a phospholipid.

phosphoprotein   A protein to which one or more phosphate groups have been covalently attached, usually to threonine, serine, or tyrosine residues.

phosphoproteome   A census of phosphoproteins, such as all those that can be shown to be present within a given cell.

phosphorylation   Covalent attachment of a phosphate group to a substrate, often the side chain of an amino acid residue of a protein. *See also* **phosphoprotein**.

phylogeny   The evolutionary history and development of a species.

physiological   (1) Referring to the function of a biological system, such as a cell or organ. (2) Referring to the normal state of function of a biological system or process. *Also termed* **physiologic**.

physiology   (1) Biological functioning of cells, tissues, organs, and organisms. (2) The study thereof.

phytochemical   A chemical species produced by a plant as part of its normal metabolism.

pimonidazole   1-[(2-hydroxy-3-piperidinyl)propyl]-2-nitroimidazole hydrochloride, a chemical used to detect regions of hypoxia within a tissue.

pinocytosis   Process similar to endocytosis in which a small volume of extracellular fluid together with its dissolved contents is internalized by the invagination of patches of plasma membrane, yielding cytoplasmic vesicles. *See also* **endocytosis**.

placebo   A therapy or agent that is thought or known to be ineffective but is intended to deceive patients into believing that they are being treated with a therapy or agent that is or may be truly effective.

planar cell polarity   Process by which the individual cells forming an epithelial cell sheet acquire a polarity, specifically along axes parallel to the plane of the sheet, and is coordinated with other cells in the plane of the sheet, resulting in an ordered, organized epithelium.

-plasia   Suffix denoting a growth.

plasma   The fluid component of blood that is left behind when its cellular components are removed, usually by centrifugation.

plasma cells   Cells of the B-cell lineage that secrete antibodies into the blood plasma.

plasma membrane   Lipid bilayer membrane that surrounds a eukaryotic cell and separates the aqueous environment of the cytoplasm from that of the extracellular space.

plasminogen   The inactive pro-enzyme in the plasma that is converted into the active plasmin protease through proteolytic cleavage as part of the process of clotting.

plasticity   Describing an ability to readily transit between alternative states, such as distinct phenotypic states.

pleiomorphic   Referring to a population of cells in which individual cells exhibit diverse morphologies seen upon microscopic examination.

pleiotropy   Ability of a gene or protein to concomitantly evoke multiple distinct downstream responses within a cell or organism.

pleura   The membrane that covers the surface of the lungs and lines the inner wall of the chest. *See also* **pleural**.

pleural   Referring to the spaces between the membrane covering the lungs and the wall of the chest and the underlying tissues covered by this membrane.

pleural effusion   Accumulation of cancer cells and fluid in the space between the lungs and the surrounding pleural membrane.

plexiform   Forming an interwoven network.

ploidy   The number of haploid genome sets of chromosomes in a cell or organism, e.g., diploid, triploid, tetraploid.

pluripotent   Referring to the ability of a stem cell to generate progeny that can participate in the formation of all of the tissues of an embryo except the extraembryonic membranes. *See also* **multipotent, totipotent**.

pneumocyte   Epithelial cell lining the alveoli of the lungs that is responsible for gas exchange with the blood and the secretion of surfactant protein.

pocket protein   Term referring to pRb and two related proteins, p107 and p130.

podosome   *See* **invadopodium**.

-poiesis   Suffix referring to the formation of a cell or tissue, often used to refer to the formation of various types of cells in the blood. *See* **hematopoiesis**.

point mutation   Substitution of a single base for another in a DNA sequence.

polar   Referring to a chemical compound that contains charged groups and is therefore hydrophilic.

polarity   Phenomenon by which a cell can organize its cytoplasm, and therefore its general axis of orientation, into distinct regions, such as the behavior of epithelial cells forming a cell sheet and acquiring apical versus basal polarity or planar cell polarity. *See also* **apico–basal polarity, planar cell polarity**.

polyclonal   Describing a population of cells that trace their origins to multiple founding ancestral cells. *See also* **monoclonal, oligoclonal**.

polycomb   Group of multi-subunit protein complexes that remodel chromatin and place it in a transcriptionally repressed state. *See also* **polycomb domain**.

polycomb domain   A region of chromatin that has been placed in a transcriptionally repressed state through the actions of polycomb complexes. *See also* **polycomb**.

polycyclic   Referring to molecules with a structure that contains multiple covalently closed rings.

polycyclic aromatic hydrocarbon   A hydrocarbon molecule that carries multiple benzene rings.

polycythemia   Condition involving higher-than-normal levels of circulating red blood cells.

polygenic   Referring to a phenotype or disease condition that is thought to be caused by or can be traced to the collaborative actions of multiple genes and their alleles. *See also* **monogenic**.

polykaryon   Cell carrying multiple nuclei in a single cytoplasm.

polymer   A molecule composed of a large number of monomeric subunits, usually joined covalently end-to-end in a linear array.

polymerase   An enzyme that can join end-to-end multiple nucleotides, thereby assembling a DNA or an RNA molecule.

polymerase chain reaction   Procedure by which large copy numbers of a double-strand DNA segment can be generated *in vitro* by repeated cycles of denaturation, followed by DNA replication that is enabled by the presence of short primers of the appropriate sequence at its ends; the DNA segment being copied may be a naturally existing DNA molecule, one generated by reverse transcription, or one arising from a chemically synthesized DNA molecule.

polymerase pausing   The process in which an RNA polymerase, usually Pol II, initiates elongation of an RNA transcript and halts after producing 25–50 nucleotide-long transcripts until it is released from the pause site by various anti-pausing regulators and permitted to complete transcription of a gene.

polymorphism   A variant genetic allele or phenotype that does not appear to be associated with any pathology and, by implication, is a reflection of normal intraspecies variability and thus normal function.

polyp   A tumor, usually presumed to be premalignant, protruding into the lumen of an organ, such as gut or bladder, often equated with an adenoma. *See also* **adenoma, polyposis**.

polypectomy   Surgical removal of a (colonic) polyp.

polyploid   Referring to a genome in which the entire haploid array of chromosomes is represented in copy numbers larger than two. *See also* **haploid, diploid**.

polyploidy   The state of a karyotype that contains an array of chromosomes that exceeds the normal number.

polyposis   Tendency to develop multiple preneoplastic polyps in the gut. *See also* **polyp**.

polyprotein   Initial product of translation that is cleaved subsequently into multiple smaller proteins, each of which may then exhibit its own distinct function.

polyubiquitin   Chain of covalently joined multiple ubiquitin subunits. *See also* **ubiquitin**.

polyvalent   Having multiple distinct reactivities, such as multiple reactive chemical groups within a single molecule or multiple antigens administered together in a single vaccine.

pool   A population or collection of similar entities, e.g., a gene pool or a pool of stem cells.

porcupine   An *O*-acyltransferase enzyme that attaches lipid tails to Wnt proteins post-translationally, preparing them for secretion and signaling by binding cognate Frizzled receptors.

portal   Referring to the properties of the vena porta and the veins that it drains, which convey venous blood directly from the pancreas, spleen, and intestine to the liver.

positive feedback   A regulatory process whereby an upstream stimulatory signal provokes a downstream target that proceeds to sustain and/or amplify signaling by the initial stimulatory signal. *See also* **feedforward, negative feedback**.

post mortem   Referring to a biological process or medical analysis that occurs after death.

post-mitotic   Describing a nonproliferating cell that has given up the option of ever again re-entering into an active growth-and-division cycle.

post-translational modification   Covalent alteration of a protein occurring concomitantly with or after the initial polymerization of the polypeptide backbone of the protein, including cleavage of the initially synthesized polypeptide and covalent modification of its amino-acid side chains.

pre-clinical   Referring to all of the steps in the research and development of an agent or therapeutic protocol leading up to but not including initiation of clinical testing.

pre-metastatic niche   A site within a tissue that is altered in some fashion to facilitate the survival and/or proliferation of subsequently arriving disseminated cancer cells.

pre-mRNA   The class of nuclear RNA molecules that are destined to become mRNA following their processing via splicing and export to the cytoplasm.

prenatal   Occurring before birth. *See also* **natal**.

pre-neoplastic   Referring to a tissue that has acquired certain features of a frank tumor but has itself not yet developed into a neoplasia. *See also* **neoplastic**.

presentation   (1) The constellation of traits and symptoms exhibited by a patient upon examination in the clinic, often upon first encounter with a physician. (2) The set of symptoms and diagnostic parameters accompanying a disease when initially diagnosed.

primary   Initial or initially formed.

primary cells   (1) Literally, cells that have been recently explanted from living tissue into culture vessels and have not been propagated thereafter in culture. (2) More commonly, cells that have been explanted from living tissue into culture dishes and have been subjected to a relatively small number of successive *in vitro* passages thereafter.

primary cilium   Single nonmotile cilium displayed by many types of mammalian cells which bears a specific set of receptors that confer responsiveness to various cytokines. *See also* **cilium**.

primary lymphoid organs   The tissues in which immunoglobulin rearrangements initially occur, these being the thymus in the case of T lymphocytes and the bone marrow in the case of B lymphocytes.

primary resistance   Term denoting the behavior of a tumor that is refractory to treatment from the onset of that treatment; *also termed* **intrinsic resistance**. *See also* **secondary resistance**.

primary tumor   Tumor growing at the anatomical site where tumor formation began and proceeded to yield this mass. *See also* **metastasis**.

primase   An enzyme that initiates DNA synthesis by laying down a short RNA segment on the template strand; the 3′-hydroxyl end of this RNA primer then serves as the site for attachment of the initial deoxyribonucleotide by a DNA polymerase. *See also* **primer**.

primer   A DNA or RNA molecule whose 3′ end serves as the initiation point of DNA synthesis by a DNA polymerase.

priming The process that allows a specific set of converging signals to convert a naive T cell into a functionally activated derivative.

primitive (1) Referring to the relatively undifferentiated phenotype of a cell. (2) Referring to an embryonic cell.

private mutation A mutation that is present in a distinct subpopulation of cells or organisms and is not shared with the majority of cells or organisms in the population as a whole.

probe (1) An RNA or DNA, often radiolabeled, that anneals specifically with a complementary nucleic acid being analyzed, enabling the detection of the targeted nucleic acid sequence. (2) An immunological reagent used to detect an antigen or antibody.

procarcinogen A chemical compound that is relatively nonreactive chemically, but can be converted into a highly reactive carcinogen, usually through metabolic processes. *See also* **ultimate carcinogen**.

procaspase Protein that is converted to an active caspase enzyme by proteolytic cleavage. *See also* **caspase**.

processive Referring to a function of an enzyme in which it undertakes a succession of alterations of a substrate, doing so without dissociating from this substrate—e.g., a DNA polymerase extending a polynucleotide chain by hundreds of nucleotides without dissociating from the growing chain.

pro-drug An inactive precursor of a biologically active drug.

pro-enzyme Catalytically inactive precursor form of an enzyme that requires some type of alteration (often proteolytic cleavage) in order to become catalytically active.

professional antigen-presenting cell An immune cell (usually a dendritic cell or macrophage) that has, as one of its salient functions, the presentation within lymphoid tissues of antigenic peptides to T cells in lymphoid tissues. *See* **antigen-presenting cells**.

progenitor cell A cell derived from a stem cell whose descendants undergo a limited number of proliferative divisions before undergoing end-stage differentiation. *Also termed* **transit-amplifying cell**.

progeria Syndrome in which an individual undergoes premature or accelerated aging.

prognosis A prediction about the future clinical course of a disease, often influenced by detailed analyses of its existing attributes such as histopathology and biochemical markers.

programmed cell death One of several cell alternative death programs, such as apoptosis, that coordinate the process of cell death.

progression (1) The process whereby cells evolve progressively from a normal state to a state that enables them to form a high-grade malignancy. (2) The evolution of benign cancer cells to malignant derivatives. (3) The evolution of an initiated cell that is dependent on continued exposure to a tumor-promoting agent for its further proliferation into one that continues to proliferate in the absence of further exposure to the promoter. *See also* **initiation**, **promotion**.

progression-free survival Time elapsed following initiation of treatment during which a clinical condition does not worsen. *See also* **overall survival**, **disease-free survival**.

prokaryotic Referring to the relatively small, nonnucleated cells of bacteria and archaea. *See also* **eukaryotic**.

prolactin A polypeptide hormone made largely by the pituitary gland that stimulates mammary gland growth and milk production among other functions.

proliferation Process leading to increase in cell number and requiring passage through cells' growth-and-division cycle, often termed the **cell cycle**. *Compare* **growth**; *see also* **cell cycle**.

proliferation index Proportion of cells in a population that are proceeding through the active growth-and-division cycle.

prometaphase The phase of mitosis in the early part of metaphase when the nuclear membrane breaks down and the chromosomes form kinetochores. *See* **metaphase, kinetochore**.

promoter (1) An agent that furthers the progression of multi-step tumorigenesis by nongenetic mechanisms, notably those involving inflammation and/or mitogenesis, sometimes termed promoting agent. Tumor promoters are presumed to work by accelerating the rate of clonal expansion of already-mutated cells. (2) The DNA sequence associated with a gene that controls its transcription. *See also* **promotion**.

promotion Process that stimulates or accelerates tumor progression, usually presumed to do so without directly damaging the genomes of cells. *See also* **promoter**, **initiation**.

promyelocyte An immature cell of the myeloid lineage that functions as precursor of various differentiated granulocyte cell types. *See* **granulocyte**.

proofreading Process by which a DNA polymerase scans the deoxyribonucleotide segment that it has just synthesized in order to ensure that the sequence of this segment was properly synthesized and thus is precisely complementary to that of the template strand. *See also* **mismatch repair**.

prophase First phase of mitosis in which chromosomes begin to condense and centrosomes begin to assemble.

prophylactic Preventative.

protease An enzyme that cleaves protein substrates.

proteasome A specialized intracellular multi-subunit machine that internalizes polyubiquitylated proteins and degrades them using specialized proteases, generating oligopeptides. *See also* **immunoproteasome, constitutive proteasome**.

proteoform A specific biochemical form of a protein that is defined by its primary amino acid residues together with a specific set of post-translational modifications. A single initially synthesized protein may be associated with multiple alternative proteoforms, each defined by its own combination of post-translational modifications.

proteoglycan Molecule with one or more glycosaminoglycan chains attached to a protein core. *See also* **glycosaminoglycan**.

proteolysis (adj., **-lytic**) Process, usually mediated by protease enzymes, of cleaving a polypeptide to lower–molecular-weight fragments including individual amino acids.

proteome   The collection of all the protein species expressed by a cell, tissue, or organism.

proteomics   (1) Technologies by which systematic analyses are made of large numbers of distinct protein species in a biological sample, such as a cell lysate or a biological fluid. (2) Enumeration of the set of proteins synthesized by a cell or tissue.

protomer   A subunit—usually a single polypeptide chain—of a multi-subunit (oligomeric) protein.

proto-oncogene   A normal cellular gene that, upon alteration by DNA-damaging agents or viral genomes, can acquire the ability to function as an oncogene. *See* **oncogene**.

protozoan   Referring to single-cell eukaryotic animals that feed on other microorganisms or microbe-derived debris. *Compare* **metazoan**.

provirus   The dsDNA copy of a retroviral genome that is the product of reverse transcription; it can exist transiently, as an episomal (nonchromosomal) plasmid, or stably, following its integration into the chromosomal DNA of an infected host cell.

pseudogene   A gene present in the genome that shares extensive sequence similarity with known functional genes but encodes a defective function, such as a catalytically inactive enzyme; such pseudogenes invariably derive evolutionarily from functional genes that have sustained inactivating mutations in the distant past but have nonetheless persisted in an organism's genome.

pseudokinase   A protein domain that exhibits many of the structural features of a bona fide kinase and is a paralog of bona fide kinases but that lacks the catalytic activity of the latter because of amino acid replacements in the catalytic cleft.

pseudoligand   A compound or substance that, although quite different from the actual normal ligand, can nevertheless bind to the ligand-binding site of a receptor, often doing so without activating signaling by the receptor.

pseudopregnant   Referring to a female that has been placed in a physiological state that closely resembles that of pregnancy, achieved through exposure to certain hormones.

pulmonary   Referring to the lungs.

purifying selection   Process occurring during Darwinian evolution in which negative selection eliminates less-fit variants, leaving behind a population of cells or organisms that is most fit.

pulse-chase   An experimental protocol in which a period of labeling, such as that involving incorporation of radioisotopes, is followed by a period in which further labeling is discontinued and may be actively blocked by agents that prevent further labeling.

pyknosis   (also **pycnosis**) Chromatin condensation and shrinkage of nuclei creating small, densely staining structures observed as products of apoptosis and necrosis.

pyroptosis   A cell death program that is often provoked by infection by intracellular pathogens and results in cell death and the release of pro-inflammatory cytokines. Pyroptosis can also be induced following exposure to certain chemotherapeutic agents.

qRT-PCR   Procedure for quantifying the concentration of an RNA molecule by determining the number of PCR amplification cycles of its reverse transcript that are required to generate a certain threshold level of PCR-generated DNA products. *See also* **polymerase chain reaction**.

quartile   Any one of four equal groups into which a large sample of individuals or test objects has been subdivided, being guided through the use of measurements of the properties of the individuals or test objects.

quasi-mesenchymal   Referring to a phenotypic state between the fully epithelial and the fully mesenchymal states and involving the expression of a mixture of genes characteristic of both states.

R point   *See* **restriction point**.

rad   Unit of radiation corresponding to 0.01 joule of absorbed radiation per kilogram of exposed tissue or 0.01 gray.

radiation oncology   The practice of treating tumors with various types of radiation. *See also* **medical oncology**, **surgical oncology**.

radioautography   *See* **autoradiography**.

radionuclide   An atomic species that undergoes radioactive decay, thereby emitting radioactivity. *Also termed* **radioisotope**.

radiosensitive   Describing cells or tissues that are particularly sensitive to killing by radiation including that inflicted by radiotherapies.

radiotherapy   Treatment of a disease, notably cancer, through irradiation, usually in the form of X-rays that, in the case of cancer treatment, are focused on a targeted tumor or tumor-bearing tissue.

rapalog   A chemical analog of rapamycin. *See* **rapamycin**.

rapamycin   A naturally occurring, low–molecular-weight inhibitor of mTOR. *See also* **mTOR**.

rate-limiting   Referring to a step in a multi-step process that governs the overall rate at which the process as a whole reaches completion because this particular step is kinetically slower than all other steps.

reactive oxygen species   A chemically reactive, oxidizing molecule that results from the incomplete reduction of oxygen, including notably superoxide, hydrogen peroxide, and hydroxyl radical; these reactive oxygen species damage a wide range of molecules in the cell, requiring compensatory anti-oxidant defenses to prevent or reverse this damage.

reactive stroma   The stroma of a highly progressed tumor that has recruited a number of cell types that are typically associated with inflammation and wound healing, including prominently activated fibroblasts and myofibroblasts. *See also* **stroma**.

reader   A protein that recognizes existing post-translational covalent modifications of histones within nucleosomes and responds by influencing the structure and/or function of the associated chromatin, thereby regulating processes such as transcription, DNA repair, and further histone modification. *See also* **writer**, **eraser**.

reading frame   (1) The nucleotide sequence within a gene or mRNA that encodes the amino acid sequence of a protein. (2) The registration of triplet codons within this sequence that enables the proper translation of this protein sequence.

receptor   Protein found on the plasma membrane or in the nucleus within a cell that is capable of specifically binding a signaling molecule (its cognate ligand). Most types of receptors emit signals, such as those inducing cell proliferation, in response to such binding. *See also* **ligand**.

recessive   (1) Referring to one of several alternative traits that can be specified by a genetic locus; when the locus is heterozygous and carries information specifying two distinct traits, the dominant trait will be exhibited phenotypically by the organism and the recessive will not. (2) Referring to an allele of a gene that is unable to dictate phenotype when expressed in the presence of a second allele that acts dominantly. *See also* **dominant**.

reciprocal translocation   Exchange via a process resembling recombination of chromosomal segments between two chromosomes from different (i.e., usually nonhomologous) chromosome pairs, resulting in the conservation of all participating chromosomal segments.

recombinant   Referring to a protein that has been produced through the procedures of recombinant DNA.

recombinant DNA   Technology that allows the isolation of genetic elements from organisms, often via the procedures of gene cloning, the experimental amplification of such segments, and their subsequent manipulation via various techniques of DNA alteration, with the intent of understanding the function of the isolated elements, altering them to generate useful gene products, or sequencing them in order to understand disease processes, genetic predilections, and relationships between organisms.

recombinase   An enzyme that executes one or more steps of genetic recombination.

recurrent mutation   Somatic mutations of a particular gene or nucleotide sequence that is observed in the genomes of a number of independently arising tumors, providing strong indication that mutations of this gene or sequence confer selective advantage to the cells that sustain them. *See also* **driver mutation**.

reductionism   A scientific philosophy and research strategy that involves the study of individual, relatively simple components or subsystems of complex systems rather than the systems as a whole.

refractory   Unresponsive or resistant to some type of signal or therapeutic agent.

regulatory B cells   A subset of B cells that releases immunosuppressive signals. *Also termed* B_{reg}'s.

regulatory T cells   A subset of CD4 lymphocytes that releases various immunosuppressive signals. *Also termed* T_{reg}'s.

rejection   Process whereby the immune system of a host organism prevents the outgrowth of implanted donor cells or tissues, usually by the killing donor cells. *See also* **allograft rejection**.

relapse   (1) (n.) Reoccurrence of a disease state, such as the reappearance of a tumor, after treatment with an initial, apparently successful therapy. (2) (v.) To sustain such a reoccurrence.

remission   Retreat or disappearance of a disease state with the implied possibility of its eventual reappearance or worsening.

renal   Referring to the kidney.

replication   (1) Process involving the copying of a nucleic acid into more nucleic acid. Thus, DNA can be copied into progeny DNA and RNA can be copied into progeny RNA. (2) Proliferation of a virus in an infected cell.

replication cycle   The program of molecular and biochemical changes that enables an infecting virus particle to multiply within a cell, generate progeny virus particles, and release these particles so that they may infect other cells.

replication fork   The structure formed when a helicase enzyme locally unwinds a DNA double-helix, enabling the resulting single-strand DNAs to serve as templates for the synthesis of progeny double-helices. *See also* **stalled replication fork**, **replication stress**.

replication stress   A physiological cell state that occurs when the normal coordination of DNA synthesis is perturbed, often in response to stalled replication forks, resulting in, among other consequences, the collapse of replication forks arising because of breakage of the persisting ssDNA and the resulting formation of dsDNA breaks. *See also* **replication fork**, **stalled replication fork**.

replicative immortality   *See* **immortality**.

replicative senescence   The cell state entered when chromosomal telomeres have undergone erosion or, more commonly, when cells have experienced cumulative physiological stress over multiple cell generations in culture due to suboptimal conditions of propagation. *See* **senescence**.

replicon   A defined segment of (chromosomal) DNA that is replicated by the firing of a single origin of replication, in principle occurring once during each normal cell cycle. *See also* **origin of replication**.

repressed   Referring to a gene that is not being expressed and therefore is not being transcribed.

repression   Regulatory mechanism that causes shutdown of the expression (transcription) of a gene. *See also* **expression**.

repressor   A protein involved in inhibiting gene transcription. *Compare* **activator**.

rescue   To restore a phenotype, usually by specific genetic or epigenetic manipulation.

resection   Removal by surgical excision.

resorption   The osteoclast-mediated process of dissolving mineralized bone with attendant mobilization of calcium into the circulation.

restriction fragment length polymorphism   Variation in DNA sequence that can be detected through its effect of allowing or preventing cleavage of a chromosomal DNA segment by a restriction enzyme.

restriction point   Decision-making point in the late G_1 phase of the cell cycle at which a cell commits itself to either completing the remaining phases of the cell cycle, remaining in G_1, or exiting the active cell cycle and entering into G_0.

reticuloendotheliosis   Tumor of monocyte/macrophage lineage common in chickens and turkeys.

retinoblastoma   Pediatric tumor of the oligopotential stem cells of the retina.

retrotransposon A transposon that propagates throughout a cellular genome by specifying an RNA transcript that is reverse transcribed by the transposon-encoded reverse transcriptase followed by chromosomal integration of the resulting reverse transcript. *See* **reverse transcriptase, transposon**.

retrovirus A class of viruses that uses a reverse transcriptase enzyme to copy its genomic single-strand RNA into double-strand DNA that, after integration into a host cell chromosome, serves as template for transcription of viral mRNA and progeny viral genomes.

reverse transcriptase Enzyme capable of making a DNA complementary copy of an RNA molecule using the RNA molecule as template.

reverse transcription Enzymatic reaction whereby an enzyme, such as reverse transcriptase, copies an RNA template into a complementary DNA copy.

rheostat An electrical controller that can vary resistance continuously over a certain defined range, thereby moderating current flow over a continuous range.

ribosome The ribonucleoprotein complex that operates in the cytoplasm to translate the coding information of an mRNA to which it is bound into an amino acid sequence that yields as an end product a complete protein molecule.

risk factor A lifestyle practice or experience or inherited genetic determinant that is judged to affect the likelihood of the development of one or another pathological state. The importance of such a factor is gauged quantitatively and contrasted with the quantitative risk confronted by control populations, such as the general population, in developing such a condition.

Rituxan The chimeric anti-CD20 monoclonal antibody bearing a murine antigen-combining variable domain and a human constant domain. *Also called* **rituximab**.

RNAi The molecular pathway within cells that results in the post-transcriptional silencing of gene expression effected by microRNAs, siRNAs, and shRNAs. *See also* **microRNA, siRNA, shRNA**.

RNAseq A procedure used to survey the RNAs expressed within a cell population by reverse transcribing these RNAs followed by sequencing the resulting cDNA molecules. *See also* **scRNAseq**.

rosette A set of repeating objects, such as cells, arrayed in a circle, reminiscent of the shape of a rose.

RT-PCR Procedure of reverse transcription of mRNAs followed by PCR to amplify resulting reverse transcripts; this procedure employs the powers of DNA sequencing to gauge the representation of various mRNAs within the transcriptome of a cell or tissue. *See also* **polymerase chain reaction, qRT-PCR**.

sagittal Referring to a geometric plane through an organism that divides the organism into a right and a left half.

salvage chemotherapy A round of chemotherapeutic treatment given to a patient who initially responded well to an initial round of chemotherapy but then relapsed.

Sanger Cancer Genome Project The program undertaken by the Wellcome Trust Sanger Institute in the United Kingdom to identify sequence variants that are critical to the development of human cancers, including the assembly of the Catalogue of Somatic Mutations in Cancer.

sarcoma Tumor derived from mesenchymal cells, usually those constituting various connective tissue cell types, including fibroblasts, osteoblasts, endothelial cell precursors, and chondrocytes.

satellite Highly repetitive sequences present in genomic DNA. These sequences often have a different composition of bases compared to that of the bulk of the genomic DNA and thus form separate "satellite" bands that can be resolved upon isopycnic density centrifugation. *See also* **isopycnic, microsatellite**.

scaffold The structural backbone of a complex organic molecule that can be covalently modified (derivatized), often through the attachment of side groups, in order to create variants with differing pharmacological properties.

Schwann cells A cell that wraps itself around axons and secretes myelin, forming the myelin sheath that electrically insulates axons and facilitates transmission of electrical signals by the axons. *See also* **myelin**.

schwannoma Tumor of the Schwann cells.

scRNAseq *See* **single-cell RNA sequencing, RNAseq**.

second-line therapy The mode of therapy that is employed after a patient has been previously treated with a first-line therapy that has failed. *See also* **first-line therapy**.

second messenger A low–molecular-weight molecule that functions as an intracellular hormone, conveying signals from one part of a cell to another.

secondary resistance Term denoting the behavior of a tumor that was initially responsive to a treatment but subsequently becomes refractory to that treatment; *also termed* **acquired resistance**. *See also* **primary resistance**.

secretome The collection of proteins that are released by a cell into the extracellular space under specific physiological conditions or states of differentiation, often focused on signaling proteins.

section A slice, e.g., through a tissue.

seed-and-soil Hypothesis enunciated by Stephen Paget that describes the tendency of disseminated cancer cells from a specific type of primary tumor to preferentially colonize one or another distant tissue based on the biological compatibility between the disseminated cancer cells and the tissue in which they have landed.

segregation (1) Separation of chromosomes at the end of mitosis or meiosis. (2) Separation of alleles during meiosis.

selectin A cell surface receptor that enables a cell to bind to carbohydrate moieties expressed by other cells, such as those forming the luminal walls of blood vessels.

selective estrogen receptor modulator A pharmacological agent that binds as a pseudo-ligand to the estrogen receptor and perturbs its behavior in one way or another. *See* **pseudoligand**.

selectivity Relative ability of a therapy to affect targeted cells or tissue compared with its (side) effects on non-targeted normal cells or tissue. *See also* **therapeutic index**.

self A constituent of the body that is native to the body's normal tissues and associated cells. *See also* **non-self**.

self-antigens Antigens that are expressed in the body's normal tissues.

self-reactive Referring to the ability of certain components of the immune system of an organism to recognize and react with the normal tissue and normal cellular antigens of that organism, potentially resulting in the development of an autoimmune condition.

self-renewal Trait that enables a cell, usually a stem cell, to generate, upon cell division, at least one daughter cell that retains all of its traits. *See also* **stem cell**.

self-seeding *See* **tumor self-seeding**.

seminoma A tumor of the epithelial cells forming in the seminiferous tubules of the testes.

senescence A nongrowing state of cells in which they exhibit distinctive cell phenotypes and remain viable for extended periods of time but are generally unable to proliferate again; senescence often arises after extended passaging of cells *in vitro*, notably under suboptimal conditions of culture. *See* **replicative senescence**.

senescence-associated heterochromatin foci Domains of heterochromatin that form in senescent cells and contribute to the silencing of proliferation-promoting genes in such cells.

senescence-associated secretory phenotype A condition observed in some cells undergoing senescence that is associated with the secretion of certain pro-inflammatory cytokines, immune modulators, proteases, and growth factors.

sequence motif (1) Short oligonucleotide sequence in DNA that is characteristically associated with one or another biological function including recognition and binding by sequence-specific transcription factors. (2) Amino acid sequence that is characteristically associated with a structural or functional aspect of a protein.

sequencing depth The degree of redundant sequencing of a DNA segment or a DNA genome.

serial passaging Practice of transferring a cell population or viral population from one culture vessel or host cell to another, taking the products of this initial passage, and transferring a portion of them to yet another host cell or culture vessel to allow a new cycle of proliferation, doing so serially and repeatedly. *See* **passaging**.

serpentine Referring to the class of membrane proteins, such as G protein-coupled receptors, that wend their way back and forth (in a snakelike pattern) through a membrane multiple times.

serum (pl., **sera**) The fluid left behind when blood clots.

set point A state to which a complex regulatory system naturally reverts in the absence of strong external perturbants; set points are often maintained through homeostatic regulatory mechanisms. *See* **homeostasis.**

shelterin A complex of at least six proteins that binds to the ssDNA and dsDNA regions of telomeres and functions together with the telomeric DNA to protect the ends of chromosomal DNA from end-to-end fusions.

shRNA A sequence of ~60 nucleotides that is cloned into an expression vector that produces a hairpin transcript. This transcript is cleaved by cellular enzymes into a 21- to 25-nt-long dsRNA that is partially complementary to the mRNA target that it is designed to inhibit. *See* **hairpin**.

shunt A passageway or connection that allows fluid to pass directly from one place to another, such as the direct connection between an artery and a vein that occurs without blood passing through intermediate microvessels, i.e., capillaries.

side effect A biological or biochemical effect of an applied treatment that elicits a response other than the intended one.

single-cell RNA sequencing A procedure in which the RNA transcripts of individual cells are reverse transcribed; the resulting DNA molecules are then amplified and subjected to next-generation DNA sequencing, revealing the transcriptomes of each of such cells. *See also* **next-generation sequencing**.

single-chain variable region fragment An antigen-binding polypeptide that is formed by connecting the variable domains of the heavy and light chains of an immunoglobulin via an intermediate linker; this fusion protein can function like the parental, naturally formed antibody to recognize and bind a specific antigen.

single-nucleotide polymorphism A polymorphism within the gene pool of a species that differs by a single nucleotide from homologous sequences that are present in the genomes of other individuals of the species. Such polymorphisms are generally assumed to be functionally and thus phenotypically silent. *See* **polymorphism**.

sinusoid A capillary-like channel located between liver hepatocytes that is lined with endothelial cells but lacks mural cells as well as a capillary basement membrane. *See also* **mural**.

siRNA A 21- to 25-nt-long dsRNA molecule that is synthetically produced and introduced into cells in order to interfere with the expression of a targeted mRNA with which it typically shares partial complementarity. *See also* **RNAi**.

sister chromatids The two chromatids that are formed from the two double helices synthesized following the most recent round of DNA replication and that remain joined via a pair of kinetochores associated with their respective centromeres until they are eventually separated from one another during mitosis.

small-cell lung carcinoma Lung cancer of specialized pulmonary cells having neurosecretory properties.

smoldering Referring to slow combustion releasing smoke but no flames.

soft-tissue sarcoma A sarcoma arising in a tissue other than bone.

soma All the tissues in the body outside of the germ cells (sperm and egg) and the immediate precursors of the germ cells.

somatic hypermutation Enzyme-mediated process, notably that effected by activation-induced deaminase, that creates point mutations in the already-rearranged immunoglobulin genes, largely focused on their V region-encoding segments and generating greatly increased antigen-binding diversity. *See also* **activation-induced deaminase**.

somatic mutation Mutation that strikes the genome of a cell outside of the germ line; such a mutation cannot, by definition, be transmitted to the next organismic generation. *See also* **germ line**.

Southern blot Procedure in which DNA molecules, usually produced by restriction enzyme cleavage, are resolved by gel electrophoresis and transferred to a filter to which they adsorb; the filter is subsequently incubated with a sequence-specific radiolabeled DNA probe to reveal, upon subsequent autoradiography, the sizes of the DNA fragments recognized by the probe. *See also* **Northern blot**, **Western blot**, **autoradiography**.

speciation The formation during the course of evolution of a new species from a preexisting one or the development of two new species from a preexisting one, the latter often enabled by the geographical isolation of two subpopulations of the original species from one another.

spectral karyotyping A procedure that allows the "painting" of individual chromosomes in a metaphase spread using chromosome-specific DNA probes coupled to five or more specific fluorochromes, achieved via fluorescent *in situ* hybridization and informatics to process and generate the colors exhibited in the finished karyotypic image.

spheroid Small, sphere-shaped cluster of cells typically observed when cells are propagated *in vitro* in three-dimensional matrices.

spindle The array of microtubule spindle fibers that originate at and radiate from the two oppositely located centrosomes and extend to the kinetochores of chromosomes, being involved in segregating chomosomes during mitosis and meiosis. *See also* **centrosome**, **kinetochore**, **spindle fiber**.

spindle assembly checkpoint The complex mechanism that monitors proper attachment of spindle fibers to kinetochores and blocks entrance into the anaphase of mitosis and meiosis if these attachments are not properly formed.

spindle fiber A microtubule fiber that extends from a centrosome to the kinetochore of a chromosome and is formed during meiosis and mitosis. *See also* **kinetochore**, **centrosome**, **spindle**.

splenocyte A leukocyte isolated from the spleen, which may belong to B or T cell populations as well as myeloid-derived populations, such as macrophages and dendritic cells.

splice site A sequence on a nuclear RNA molecule that determines the precise location of one of the two sequences that are fused to one another during the process of splicing.

splicing (1) Process that causes the deletion of a defined internal segment of a primary RNA transcript and the subsequent fusion of the two RNA segments flanking the deleted RNA segment. (2) Process occurring in the nucleus whereby a pre-mRNA precursor is converted into an mRNA through the deletion of introns and the fusion of remaining exons.

splicing factor Protein involved in regulating the process of splicing including the recognition of appropriate splice sites in the RNA molecule serving as the substrate of the splicing. *See also* **splice site**, **splicing**.

sporadic Describing a disease or condition that appears to occur randomly in a large population without any apparent inborn predisposition.

squamous Referring to epithelial cells that line a duct, hollow organ, or the skin and lack secretory function.

squamous cell carcinoma Tumor arising from squamous cells. *See also* **adenocarcinoma**.

staging A determination of the extent to which a tumor has progressed to a specific stage of multi-step tumor progression, involving consideration of an array of gross/macroscopic, histopathological, and karyotypic parameters.

stalled replication fork A DNA replication fork that, because of various impediments, cannot proceed further to replicate a DNA double helix; such stalled forks represent a potential physiological stress on a cell because of the persistence of single-strand DNA at the fork and the attendant danger of its breakage, resulting in turn in double-strand DNA breaks in the DNA being replicated. *See* **replication fork**.

standard of care The therapeutic agent or protocol that is widely accepted in the treatment of disease and often serves as the reference against which candidate novel treatments are measured.

standardized incidence ratio Ratio of the observed incidence of an event in a subpopulation of interest, such as a disease diagnosis, with the incidence of such an event that would be observed in a comparable control population, such as the general population.

start site The site in a gene at which an RNA polymerase initiates elongation of a nascent RNA transcript, also termed transcription start site.

steady state The condition reached when a series of dynamic, often countervailing processes have placed a complex system in a balanced, relatively constant, unchanging state.

stem cell Cell type within a tissue that is capable of self-renewal and is also capable of generating daughter cells that develop new phenotypes, notably those that are more differentiated than those displayed by the stem cell. *See also* **asymmetric division**.

stereochemistry Description of the three-dimensional structure of a molecule (such as a protein or drug) and the influence that this structure exerts on the chemical behavior or biochemical function of the molecule.

stochastic Referring to an event that occurs randomly with a certain probability rather than in a precisely predetermined fashion.

stock A solution of viruses that is used experimentally to infect cells or organisms.

stoichiometric Referring to a relationship or reaction between two or more molecular species in which the relative molarities of the participating species are precisely specified.

stratify (n., **stratification**) To classify superficially similar entities (e.g., a group of tumors) into several distinct categories or subclasses based on the shared properties of the entities within a given subclass.

stress-induced ligand   A cell surface protein that signals that the cell displaying this protein is suffering from some type of cell physiological stress.

stressor   An agent that imposes some type of physiological stress.

stroma (pl., **stromata**; adj., **stromal**)   (1) The mesenchymal components of normal epithelial and hematopoietic tissues, which may include fibroblasts, adipocytes, endothelial cells, and various immunocytes as well as associated extracellular matrix. (2) A spectrum of similar cell types that constitute the stromal cells within a tumor and are recruited into the tumor from host tissues and lack the somatic mutations born by the tumor cells. *See also* **reactive stroma**.

stromalization   Referring to the process by which stroma is generated in a normal or neoplastic tissue including the formation of stroma-associated extracellular matrix.

structure-based drug design   Use of the three-dimensional structure of a drug target, such as a protein, often determined by X-ray crystallography or nuclear magnetic resonance spectroscopy, in order to predict the chemical structures of compounds that will bind the target specifically and with high affinity, usually inhibiting its function.

subclinical   Referring to a state or process that does not elicit specific symptoms and/or eludes detection by available diagnostic tools.

subcutaneous   Beneath the skin.

submicroscopic   Too small to be seen through the light microscope.

substrate   (1) Molecule that is acted upon by an enzyme, usually resulting in the covalent modification of the substrate. (2) A physical structure upon which entities such as cells reside.

subtelomeric   Referring to a segment of DNA that contains imperfect copies of the telomeric hexanucleotide sequence, does not function to protect the ends of chromosomal DNA, and lies between the functional telomeric DNA and the bulk of chromosomal DNA.

SUMO   One of a group of four ubiquitin-like polypeptides that, like ubiquitin, are attached covalently to substrate proteins, thereafter affecting intracellular localization and functions of the modified proteins but not their degradation.

sumoylation   The process of modifying a protein by covalently attaching one or more SUMO molecules to it. *See also* **SUMO**.

super-enhancer   A large cohort of transcription factors, each bound to its cognate enhancer sequence, that assemble in a physical complex and drive the high-level expression of a nearby gene. *See also* **enhancer**.

supernatant   Fluid that is formed after solids are removed via centrifugation or solutes by precipitation or crystallization.   *Also termed* **supernate**.

supernumerary   Referring to a greater-than-normal number of some object.

superoxide   Reactive ion arising from addition of an electron (reduction) to diatomic oxygen.

superoxide dismutase   An enzyme that catalyzes the conversion of 2 superoxide molecules into one diatomic oxygen molecule (O_2) and one hydrogen peroxide (H_2O_2) molecule. *See also* **superoxide**.

supranormal   Higher than normal physiological levels.

surgical oncology   The practice of treating tumors surgically, usually by excision. *See also* **medical oncology**, **radiation oncology**.

surrogate marker   A measurable parameter, often a diagnostic parameter, that serves to indicate the behavior of another process or agent whose behavior it parallels and reflects.

survivin   Member of the IAP family of proteins that functions by inhibiting certain pro-apoptotic caspases.

symmetric division   The process whereby a mother cell yields, following cell division, two identical daughter cells. *See also* **asymmetric division**.

synapse   Physical connection formed between two interacting immune cells or between a cytotoxic lymphocyte and a targeted cell that facilitates exchange of signals between them and, in the case of cytotoxic cells, the transfer of cytotoxic granules from the cytotoxic cell to the targeted cell.

synchronous   (1) Occurring at the same time; i.e., in a temporally coordinated fashion. (2) Referring to a population of cells that enter a specific phase of the cell cycle at the same time. *See also* **asynchronous**.

syncytium   Cell formed when the plasma membranes of two or more cells fuse, initially resulting in a multi-nucleated cell.

syndrome   Collection of symptoms and pathologies that are associated with and together define a specific disease condition.

syngeneic   (1) Referring to two organisms that share the identical genetic background, such as two members of an inbred strain of mice. (2) Describing the relationship between two sets of cells or tissues, or between a set of cells and an organism, deriving from identical genetic backgrounds.

synonymous   Referring to a mutation in a reading frame that does not change the identity of an encoded amino acid. *See also* **nonsynonymous**.

synthetic lethal   (1) Describing phenotypes that result from the combined effects of two alleles, each of which is nonlethal on its own, but which, when present in combination, result in lethality. (2) Referring to a state in which an existing genetic or epigenetic condition renders a cell vulnerable to killing by one or another applied agent, such as a therapeutic agent.

synthetic lethal screen   A screen of a large number of genes, often those present in an entire genome, in which the cells introduced into the screen bear one or another mutation; the subsequent screen is designed to detect, in an unbiased fashion, other genes that, when disrupted, lead to lethality in these cells but not in wild-type cells lacking the mutation.

T antigen   Tumor-associated antigen, such as that encoded by the genome of a transforming tumor virus such as SV40 and the genomes of the cells that it has transformed. *See* **tumor-associated antigen**.

T cell   Class of lymphocytes that develops largely in the thymus and includes T_H cells, T_C cells, and T_{reg}'s.

T-cell exhaustion   State of a T cell that has been induced by prolonged, possibly excessive antigen exposure as well as by exposure to certain cytokines, and is associated with loss of its various effector functions.

T-cell receptor   The immunoglobulin-like molecule that is displayed on the surface of T cells and used by them to recognize antigens displayed by HLA/MHC class I or class II proteins on the surfaces of other cells, including professional antigen-presenting cells and potential target cells.

T-loop   *See* **activation loop**.

t-loop   Lasso-like structure at the end of a telomere that serves to protect the termini of a chromosomal DNA molecule from end-to-end fusions and degradation by exonucleases.

T lymphocyte   *See* **T cell**.

tamoxifen   A synthetic pharmacological analog of estrogen which can bind the estrogen receptor and inhibit a subset of the normal signaling functions of this estrogen-activated receptor, thereby antagonizing the actions of estrogen.

telomerase   An enzyme specialized to extend telomeric DNA; those telomerases characterized to date carry an RNA subunit and a reverse transcriptase-like catalytic subunit. *See also* **telomere**.

telomere   Nucleoprotein structure at the end of a eukaryotic chromosome that protects this end from degradation and from fusion with the ends of other chromosomes.

telophase   Fourth subphase of mitosis, during which chromosomes de-condense and the nuclear membranes reassemble in anticipation of separation of the two daughter cells.

temozolomide   A chemotherapeutic drug used primarily in the treatment of certain brain tumors. It is a type of alkylating agent that causes direct DNA damage.

temperate   A type of behavior (e.g., of an infectious agent, such as a virus) that creates minimal damage or pathology in an infected host cell or organism. *See also* **virulent**.

temperature-sensitive   Describing a phenotype that is apparent when cells or viruses grow at one temperature but not at another.

tetramer   Complex composed of four subunits.

teratogen   An agent that causes malformations by perturbing embryonic morphogenesis.

teratoma   Benign tumor formed by embryonic stem cells in which a wide variety of differentiated cell types are formed.

term   The period of time required to complete a normal pregnancy.

ternary   (1) Referring to a complex of three components. (2) Referring to the third step in a multi-step process.

TET   A class of enzymes that operate to oxidize the methyl moiety of 5-methyl cytosines in the DNA, ultimately yielding demethylation of this base.

tetraploid   Describing a karyotype having precisely four haploid complements (or two diploid complements) of chromosomes.

tetraspanin   A relatively small protein with four transmembrane domains that operates in various membranes to modulate signaling by a wide variety of other transmembrane proteins such as receptors.

The Cancer Genome Atlas   A project to identify the complete set of DNA changes in many different types of cancer, led by the U.S. National Cancer Institute and the U.S. National Human Genome Research Institute.

therapeutic index (pl., **indices**)   The ratio of the concentration of a drug that elicits an unacceptable level of toxicity compared with the concentration of the drug that elicits a desirable therapeutic response.

therapeutic range   *See* **therapeutic window**.

therapeutic window   Range of concentrations of a drug that are higher than that needed to elicit a therapeutic effect and lower than the maximum tolerated dose. *Also termed* **therapeutic range**.

thrombin   A plasma protease that is activated following wounding and triggers blood coagulation by activating platelets and cleaving fibrinogen to fibrin.

thrombocyte   A platelet.

thrombocytopenia   A deficiency of platelets in the circulation.

thromboembolus   *See* **embolus**.

thrombopoiesis   Process leading to the formation of blood platelets from megakaryocytes.

thrombopoietin   A growth factor that stimulates the production of megakaryocytes and thus of derived blood platelets.

thrombus (pl., **thrombi**)   A blood clot.

thymocyte   A leukocyte residing in the thymus.

thymus   The immune organ lying in the upper chest and front of throat in which the maturation of thymocytes occurs, including the elimination of autoreactive thymocytes

time-lapse microscopy   Procedure in which the same microscope field is photographed repeatedly, usually at regular intervals.

time to progression   The period between inception of a treatment and the subsequent progression of a tumor under treatment to a stage of worse clinical presentation or higher stage of malignancy.

tissue culture   Procedure of propagating cells outside of living tissues in various types of vessels including flasks and Petri dishes.

Tissue Factor   A cell-surface glycoprotein expressed by many cell types in the body that interacts with clotting factors in the plasma, thereby triggering the coagulation cascade.

tissue-specific gene   Gene that is expressed only in cells of certain individual tissue types.

titer   Concentration of a substance in solution, usually referring to the concentration of viruses or antibodies as gauged by certain biological measurements involving dilution.

tolerance   *See* **immune tolerance**.

tomography   A computerized image-processing technique that integrates images of successive sections obtained by X-rays, ultrasound, or by other imaging procedures in order to generate a cross-sectional image or a three-dimensional image of an object, such as the human body.

topoinhibition   *See* **contact inhibition**.

topoisomerase   (1) An enzyme that relieves torsional tension in a DNA double helix by cleaving one or both strands, permitting winding or unwinding of the helix to relieve tension; thereafter this enzyme, ligates the DNA strand(s), thereby restoring the covalent integrity of the helix. (2) An enzyme that disentangles two intertwined, concatenated dsDNA molecules.

totipotent   Referring to the ability of a stem cell to generate all the differentiated cell lineages existing in the embryo as well as the extraembryonic membranes. *See also* **multipotent**, **pluripotent**.

toxicity   The undesired side effect(s) of a drug on normal tissues and normal metabolism.

trafficking   The process of translocating biological entities, usually vesicles, from one intracellular locale to another.

trans   (1) Term to indicate the fact that a gene is acted upon by the product of a second, unlinked gene. (2) Term used in structural organic chemistry to indicate that two moieties are oriented in opposite directions or located on opposite sides of an imagined plane. *See also* **cis**.

trans-acting   Referring to a protein, often a transcription factor, that is encoded by one gene and acts upon or regulates a second gene.

transactivation   Process by which one molecule is activated by another that arrives from elsewhere in the cell; e.g., expression of a gene that is induced by a diffusible transcription factor made by another gene. *See also* **cis**, **trans**.

transactivation domain   Domain of a transcription factor that serves to activate the transcription of genes, usually by attracting other transcription-regulating proteins.

transcription   Copying of DNA sequences into RNA molecules mediated by RNA polymerases.

transcription-coupled repair   A form of nucleotide excision repair (NER) in which transcribed regions of the genome are preferentially repaired. RNA polymerase stalled at DNA lesions mediates the recruitment of NER enzymes to the damage site, initiating transcription-coupled repair.

transcription factor   Protein that is involved in regulating the transcription of a gene, often by associating with specific sequences in the promoter region of the gene. *See also* **general transcription factor**.

transcriptional pausing   The process causing an RNA polymerase II molecule that has initiated transcription of a gene to halt at a downstream site relatively close to the transcriptional start site; formation of a full-length transcript depends on release of the polymerase from the site of pausing.

transcriptome   (1) The assembly of all of the transcripts that are produced by one cell or cell type. (2) More commonly, the assembly of all of the mRNAs produced by one cell or cell type.

transdifferentiation   Transition by a cell from one differentiation lineage into the phenotypic state characteristic of cells from another, distinct differentiation lineage.

transduction   (1) Process whereby a signaling element, such as a protein, receives a signal and, in response, processes the signal and emits another signal. (2) Process by which a gene is introduced into a cell, usually by a vector such as a viral vector.

transfectant   A recipient cell that has taken up and incorporated into its genome transfected donor DNA.

transfection   A procedure for introducing DNA or RNA molecules into cells which may thereafter be expressed transiently in such cells or, in the case of DNA, stably in such cells.

transferase   An enzyme that attaches a complex molecule, such as glutathione, to its substrate.

transformant   A neoplastic cell that is formed by the experimental transformation of a non-neoplastic cell. *See* **transformation**.

transformation   (1) Process of converting a normal mammalian cell into a cell having some or many of the attributes of a cancer cell. (2) Alteration of a cell through the introduction of a genetic element, usually referring to a process occurring in prokaryotic cells.

transgene   (1) A cloned gene that has been inserted experimentally into the germ line of an animal, often that of a laboratory mouse. (2) Less commonly, any experimentally altered gene in the germ line.

transgenic   (1) Referring to an animal or breed of animal whose germ line has been experimentally altered, usually through the insertion of a cloned gene. (2) Less commonly, referring to an animal or breed of animal whose germ line has been altered through any of a variety of genetic manipulations, including the addition of a cloned gene or the alteration of a resident gene through homologous recombination or other procedure.

transit-amplifying cell   A relatively undifferentiated cell that is initially generated by division of a stem cell and is capable of exponential proliferation for a limited number of successive cell generations before spawning highly differentiated progeny, which in many tissues become post-mitotic. *See also* **progenitor**.

transition   Point mutation in which one purine base replaces another, or in which one pyrimidine base replaces another. *See also* **transversion**.

translation   Synthesis of proteins by ribosomes in which the amino acid sequences of the synthesized proteins are dictated by the nucleotide sequences of the ribosome-associated mRNA molecules. *See also* **ribosome**.

translocation   (1) Rearrangement of chromosomes that results in the fusion of two chromosomal segments that are not normally attached to one another, often resulting in a microscopically visible alteration of karyotype. (2) Movement of a physical entity from one part of the cell to another. (3) Movement of a ribosome down an mRNA being translated.

transmembrane   Referring to the domain of a protein, usually composed of a stretch of 12 to 35 hydrophobic amino acids, that is threaded through a lipid bilayer membrane and therefore resides largely or entirely within the hydrophobic environment of the lipid bilayer.

transphosphorylation   Phosphorylation of one protein molecule by another, such as the phosphorylation of one receptor subunit by the kinase carried by another. *See also* **autophosphorylation**.

transposable element   DNA segment residing in a cellular genome that is capable of either jumping from one integration site to another or dispatching new copies of itself to novel sites of integration, doing so via one of several alternative molecular mechanisms.

transposase   Enzyme that binds transposon DNA and causes its copying and insertion into novel integration sites in a cellular genome. *See also* **transposon**, **transposable element**.

transposon   Genetic element or DNA segment that is able to move from one chromosomal integration site to another within a cell. *See also* **retrotransposon**.

transversion   Point mutation in which a purine base replaces a pyrimidine or vice versa. *See also* **transition**.

trastuzumab   *See* **Herceptin**.

tricarboxylic acid cycle   The succession of biochemical reactions occurring within the mitochondria that results in the oxidation of a glucose molecule, yielding the equivalent of ~38 ATP molecules. *Also termed* **Krebs cycle**, **citric acid cycle**.

trimeric   Complex composed of three subunits.

triploid   Describing a karyotype or a genetic state having precisely three haploid complements of a chromosome, chromosomal array, or copies of a gene.

tritium   Radioactive isotope of hydrogen.

trophic   Aiding in or supporting cell growth, proliferation, or survival.

-tropic   Referring to a tendency of a cell or an organism to move toward or turn toward some object or source or to direct its actions toward that source.

tropism   (1) Tendency of a cell to face or move toward a specific location or signaling source. (2) Tendency of a cell to migrate in a specific direction or, in the case of metastatic cancer, to appear to home to a specific tissue site in the body.

truncal   Referring to a genetic alteration that is present in the genomes of all of the otherwise-heterogeneous neoplastic cells within a tumor and is indicative of their shared descent from a common ancestral cell; such mutations are associated with the trunk of a branched tree drawn to depict graphically the evolutionary history of these cells from their common ancestor.

t-SNE   A graphical convention in which the data present in high-dimensional datasets can be reduced via a statistical method to a two- or three-dimensional depiction; t-SNE plots are often used to depict results of single-cell RNA sequencing, such that cells with similar transcriptomes are clustered together with one another. *See also* **scRNAseq**.

tumor-associated antigen   An antigen that is associated with the cells of a tumor and may also be displayed by normal cells or tissues throughout the body; *also termed* **tumor-associated transplantation antigen**. *Compare* **tumor-specific antigen**.

tumor-associated macrophage   A macrophage that has been recruited into a tumor.

tumor-associated transplantation antigen   An antigen encoded by a normal cellular allele that is expressed by a certain class of tumors and may have, for various reasons, not elicited immune tolerance. *See also* **tumor-specific antigen**.

tumor-initiating cell   *See* **cancer stem cell**.

tumor microenvironment   The areas within a tumor that are populated by recruited non-neoplastic cells, including various types of immunocytes and inflammatory cells, often synonymous with stroma. *See also* **stroma**.

tumor mutational burden   The total number of mutations found in the genomic DNA of tumor sample, often presented as number per haploid genome or per megabase of genomic DNA.

tumor progression   (1) Process of multi-step evolution of a normal cell into a fully neoplastic, malignant tumor cell. (2) Evolution of a benign into a malignant cancer cell or growth. (3) Evolution of a premalignant cell from a promoter-dependent to a promoter-independent state.

tumor promoter   *See* **promoter**.

tumor rejection   Process by which an organism prevents the formation of a tumor (including tumor formation by engrafted cells), usually achieved through the action of its immune system.

tumor self-seeding   Process whereby circulating tumor cells originating in a primary tumor or in its derived metastases return to the primary tumor and seed populations of cells within that tumor.

tumor-specific   *See* **tumor-specific antigen**.

tumor-specific antigen   An antigen that is associated with the cells of a tumor that is unique to this tumor and is not shared by other cells or tissues throughout the body; *also termed* **tumor-specific transplantation antigen**. *See also* **tumor-associated antigen**.

tumor-specific transplantation antigen   An antigen that can, through transplantation analyses, be associated with one particular tumor cell population but not with other similar tumor cell populations. *See also* **tumor-specific antigen**.

tumor suppressor gene   (1) A gene whose partial or complete inactivation, occurring in either the germ line or the genome of a somatic cell, leads to an increased likelihood of cancer development. (2) Such a gene that is responsible for constraining cell proliferation. *See also* **gatekeeper**.

tumoricidal   Able to kill cancer cells and/or destroy a tumor.

tumorigenesis   The process of forming a tumor, often involving a succession of steps.

tumorigenic   (1) Referring to the ability of cells to form tumors when introduced into appropriate animal hosts. (2) Less commonly, pertaining to an agent such as a tumor virus that imparts this ability to cells.

tumorigenicity   The ability of cells to serve as founders of new tumors, often gauged experimentally by transplanting cells into appropriate hosts.

tumorsphere A colony of cells that is formed in semi-solid culture medium by cells that are deprived of contact with solid substrate; the ability to form such colonies is sometimes used to predict tumor-forming ability *in vivo*.

turnover number The rate with which an enzyme processes its substrate, usually presented as the number of enzymatic reactions catalyzed per second by a single enzyme molecule.

type IV secretion system A collection of structures formed by bacteria and archaea that create a channel through which bacterial and archaean proteins are delivered through the outer membranes of other cells.

ubiquitin A polypeptide composed of 76 amino acid residues that undergoes covalent attachment to a substrate protein, often signaling that the resulting ubiquitylated protein is destined for degradation and usually present as a linear or branched polyubiquitylated modification of the substrate protein.

ubiquitylation Process by which one or more ubiquitin molecules are attached covalently to a protein substrate molecule, which often results in the degradation of the tagged protein in a proteasome, *alternatively termed* **ubiquitination**.

ultimate carcinogen A chemical compound that is able to directly contribute to the induction of cancer without further chemical modification, usually by direct chemical interaction with DNA, thereby altering the chemical structure of the latter. *See also* **procarcinogen**.

undruggable Referring to a molecular species within a cell, usually a protein, whose function cannot be readily inhibited by existing drug molecules or, ostensibly by alternative drug molecules that might be developed in the future; such a property is often judged by the absence of potential drug-binding pockets in the known three-dimensional structure of the species.

unequal crossing over A mechanism of genetic recombination in which a DNA sequence recombines with a nonidentical but similar sequence on either the sister chromatid or a homologous chromosome; because the recombination occurs between two nonidentical DNA sequences, one of the resulting recombinant chromosomes will gain sequence (and physical length), whereas the other will lose sequence (and physical length).

unfolded protein response A cell-biological program, which is the main source of ER stress, that is activated by the accumulation of unfolded or misfolded proteins in the lumen of the endoplasmic reticulum; this program halts further translation, induces synthesis of molecular chaperones, and, if unsuccessful in reducing the level of unfolded and misfolded proteins, induces apoptosis. *See* **ER stress**.

urothelium The specialized epithelial cell lining of the urinary bladder.

uveal melanoma A tumor arising from the melanocytes of the uvea, more specifically located in the iris, the ciliary body, or the choroids of the eye.

vacuoles Small, fluid-filled, bubble-like structures, often seen in the cytoplasm of cells that are under physiological stress and in cells infected by certain viruses.

variable region Portion of an antibody molecule or T-cell receptor that contains the antigen-binding domain and, within an individual organism, is present in millions of alternative versions (and thus alternative amino acid sequences) in the various antibody molecules that together comprise a given class of antibody molecules such as gamma immunoglobulin. *Compare* **constant region**.

vascular ZIP code The display by the luminal surfaces of endothelial cells of specific proteins that reflect or are specific to the tissue in which the endothelial cells and the vessels they form reside; one theory of metastasis proposes that circulating cancer cells adhere to the vessel walls by recognizing the specific homing address created by these displayed proteins.

vascularized Referring to the presence of blood vessels throughout a tissue such as a tumor.

vasculature Network of blood vessels.

vasculogenesis Process by which vessels are assembled *de novo* through the differentiation program of primitive mesenchymal proteins termed angioblasts. *Compare* **angiogenesis**.

vasculogenic mimicry The process that enables the neoplastic cells within a tumor to assemble into blood-carrying microvessels without these cells becoming *bona fide* endothelial cells.

vasoactive Referring to a regulator of vascular function, such as a regulator of vascular permeability or constriction.

vector (1) Agent, often a virus, that is able to carry (transduce) a gene from one cell to another. (2) An infected organism that serves to transmit and distribute an infectious agent to other organisms.

vehicle The solvent that is used to deliver a drug.

venereal Involving or resulting from sexual intercourse.

Venn diagram Diagram of two or more circles, each of which encompasses a group of items such as genes; the overlap between two circles indicate the items that are shared in common between the groups represented by each of these circles.

ventral (1) Referring to the underside of a cell that is apposed to a solid substrate such as the surface of a culture dish. (2) Referring to the side of an organism opposite to its dorsal side and, in chordates, referring to the abdomen. *See also* **dorsal**.

venule A small vein that conducts blood from capillaries to larger veins.

versus In contrast to.

vertical transmission The transmission of an agent, such as an infectious agent, from parent to offspring. *See also* **horizontal transmission**.

vesicle (adj., **vesicular**) A structure, usually within the cell cytoplasm, that is composed of a lipid bilayer that surrounds and encloses an interior lumen and that is used to transport various biomolecules to various destinations inside and outside a cell. As examples, cytoplasmic vesicles may be generated by endocytosis and soluble proteins may be exported by a cell through the exocytosis of cytoplasmic vesicles.

viability   Referring to the state of a cell or organism that enables its continued survival.

vicious cycle   Term used to describe the self-perpetuating, complex interactions between disseminated cancer cells, osteoclasts, and osteoblasts that has been used to explain the mechanisms underlying osteolytic metastases.

villus (pl., **villi**)   Fingerlike structure that is covered by epithelial enterocytes and protrudes from the wall of the small intestine into its lumen.

vimentin   Intermediate filament protein of the cytoskeleton of mesenchymal cells such as fibroblasts.

viremia   Presence of high concentrations of virus in the bloodstream.

virion   Virus particle including a capsid (coat) and the viral genome.

virulent   Referring to an infectious agent, such as a virus, that exhibits toxicity and creates damage in an infected host cell or organism. *See also* **temperate**.

virus stock   A solution of virus particles used experimentally to infect cells or organisms.

vital dye   Dye that can be used to stain living cells or tissues and is retained for extended periods of time in these objects without compromising viability.

vitiligo   A skin disorder, often of autoimmune origin, that leads to loss of patches of melanocytes from the epidermis and resulting loss of pigmentation.

volcano plot   Convention of graphing in which the individual data points of a large data set are plotted in two dimensions; individual data points are located by the statistical significance (p-value) of the measured value data points indicated on the ordinate (y-axis) versus extent of measured change from a reference data set on the abscissa (x-axis). Both axes are usually presented as the logarithm of the values indicated and both positive and negative values of extent of change are indicated to the right or left of the graph, respectively. *See also* **p value**.

vulnerability   Property of a tumor or tumor cell that reflects its responsiveness to a specific type of cytostatic or cytotoxic therapy.

Warburg effect   *See* **aerobic glycolysis**.

waterfall plot   Convention of graphing the responses of individual patients to a treatment protocol by depicting each patient's response as a bar that extends either above or below the abscissa indicating ongoing growth or shrinkage in the measured size of a tumor, respectively; by convention, lower to higher patient responses to a therapy are plotted left to right on these graphs.

Western blot   Procedure whereby proteins are resolved by gel electrophoresis and transferred to a filter, whereupon they are detected by incubation with appropriate monoclonal antibodies. *Also termed* **immunoblot**. *See also* **Southern blot**, **Northern blot**.

whole-exome sequencing   Method used to determine the nucleotide sequence primarily of the exonic (including protein-encoding) regions of genomic DNA sample.

wild type   The allele of a gene that is commonly present in the great majority of individuals in a species and is usually presumed to be fully functional.

writer   An enzyme that adds covalent modifications to histones within nucleosomes, thereby regulating processes such as transcription, DNA repair, and further histone modification. *See also* **eraser**, **reader**.

xenobiotic (n., adj.)   A chemical species that originates outside of the body of an organism and is foreign to its normal metabolism.

xenograft   A normal or neoplastic tissue derived from one species that has been implanted into a host animal from another species.

xenotropic   Referring to a class of retroviruses from one species that can infect and replicate in cells of another species.

xeroderma pigmentosum   Syndrome resulting from an inherited inability to repair UV-induced DNA lesions in the skin, resulting in severe burning of the skin following exposure to sunlight and the development of skin cancers at a high rate.

zinc finger   A protein structural domain that binds and is stabilized by one or more zinc ions. Zinc fingers are employed by certain transcription factors to recognize and bind specific oligonucleotide sequences in DNA.

zygote   The cell created by the fusion of sperm and egg.

zymogen   An inactive precursor form of an active enzyme.

zymogram   Analytic technique in which the migration rates of various proteins upon gel electrophoresis are gauged by their localized enzymatic activity following such electrophoresis.

Index